Eckert

ANIMAL PHYSIOLOGY

Eckert

ANIMAL PHYSIOLOGY

Mechanisms and Adaptations

David Randall
City University of Hong Kong

Warren Burggren
University of North Texas

Kathleen French
University of California, San Diego

W. H. Freeman and Company
New York

Acquisitions editor: JASON NOE

Development editor: MORGAN RYAN

Marketing director: JOHN BRITCH

Project editor: JANE O'NEILL

Cover and text designer: BLAKE LOGAN

Layout artist: MARSHA COHEN

Illustration coordinator: BILL PAGE

Illustrations: FINE LINE ILLUSTRATIONS

Photo editor: MEG KUHTA

Production coordinator: PAUL ROHLOFF

Supplements coordinator: REBECCA PEARCE

Composition: PROGRESSIVE INFORMATION TECHNOLOGIES

Front cover: Joe McDonald/DRK Photo
Back cover: Patricio Robles Gil/Bruce Coleman
Part opening and chapter opening photo credits:

Part 1 and all chapters within Part 1
Pronghorn doe and twins running in sage low light
J. Cooney/Animals Animals

Part 2 and all chapters within Part 2
Pronghorn Antelope herd, Oregon
Donna Ikenberry/Animals Animals

Part 3 and all chapters within Part 3
Running Pronghorn herd, Montana
Alan and Sandy Carey

Silhouettes of various animals from these images appear on the half-title page, brief contents page, and after part titles in the table of contents.

Library of Congress Cataloging-in-Publication Data

Randall, David J., 1938-
Eckert animal physiology : mechanisms and adaptations / David Randall, Warren Burggren, Kathleen French.
p. cm.
Includes bibliographical references and index.
ISBN 0-7167-3863-5 (cloth)
1. Physiology. I. Burggren, Warren W. II. French, Kathleen. III. Eckert, Roger. IV. Title.

QP31.2 .R36 2001
571.1--dc21 2001040270

Printed in China
Second printing 2002

For our students: past, present and future

About the authors

DAVID RANDALL is a prominent fish physiologist and a leading expert in respiratory and circulatory physiology. His research interests concentrate on oxygen, carbon dioxide, and ammonia transfer in fish. He is also interested in aquatic toxicology, especially in relation to ammonia and hypoxia. Randall collaborated with the late Roger Eckert on earlier editions of *Animal Physiology*. Currently Professor and Chair of Biology and Chemistry at the City University of Hong Kong, he received his Ph.D. from the University of Southampton, U.K., in 1963 and then joined the faculty of the University of British Columbia, where he was appointed Professor in 1973. Randall has studied and taught around the world, working at institutions in Brazil, France, Italy, Germany, Kenya, Russia, Singapore, the United States, Canada, and the United Kingdom. He was a member of the Evaluation Committee for Biology for the Swedish Natural Science Research Council (1994), was elected president of the Canadian Society of Zoologists in 1985, and served as president of the Western Canadian Universities Marine Biological Society in 1998–2000. He has been a Guggenheim Fellow (1968) and a Killam Fellow (1981) and was elected a fellow of the Royal Society of Canada in 1981. He received the prestigious Fry Medal from the Canadian Society of Zoologists in 1993 and an Award of Excellence from the American Fisheries Society in 1994. Dr. Randall has authored about 200 original papers and has edited and contributed to many books, including the noted series *Fish Physiology* (Academic Press, 19+ volumes). He has been a member of several editorial boards and many peer review committees.

WARREN BURGGREN has helped students learn physiology for nearly three decades. He has been a professor of biological sciences at the University of North Texas since 1998 and was previously a professor at the University of Nevada, Las Vegas (1991–1998) and before that at the University of Massachusetts, Amherst (1978–1991). He has taught human anatomy and physiology, bioenergetics, introductory zoology, and comparative physiology and has contributed to textbooks and student study guides on these subjects. Burggren's research interests include physiology from developmental, comparative, ecological, and evolutionary perspectives. In particular, his research focuses on the ontogeny of respiratory and cardiovascular processes in animals and how the systems that regulate them change over the course of development and in the face of environmental challenge to the growing animal. Burggren has been involved in symposia, seminars, and formal extramural research and training activities in numerous countries. His collaborations with David Randall date back to the late 1970s, and in 1981 Burggren and Randall co-authored *The Evolution of Air Breathing in Vertebrates* (also with Tony Farrell and Steve Haswell). In the intervening period, Burggren has co-authored several scholarly works ranging from *New Directions in Ecological Physiology* (Cambridge University Press, 1987), *Environmental Physiology of the Amphibia* (University of Chicago Press, 1992), and *Development of Cardiovascular Systems: Molecules to Organisms* (Cambridge University Press, 1997), as well as published more than 130 research papers, reviews, and book chapters in animal physiology.

KATHLEEN FRENCH has been a neurobiologist at the University of California, San Diego, since 1985 and has taught upper-division courses in embryology, comparative physiology, cellular neurobiology, and mammalian physiology for preprofessional students for 15 years. In addition, French teaches an intensive neurobiology laboratory course for neuroscience graduate students at UCSD and serves on the faculty of the Neural Systems and Behavior course at the Marine Biological Laboratory in Woods Hole, Massachusetts, an intensive summer course designed primarily for graduate students and postdoctoral fellows. French has been a co-author of *Animal Physiology* since the fourth edition. As an Associate Project Scientist at UCSD, French investigates the development of neuronal circuitry and the development of behavior, topics that she has studied in various invertebrate species. Her current research focuses on the cellular events that control differentiation of identified neurons in the medicinal leech, emphasizing the cellular physiology of embryonic neurons, the effects of cell-cell contacts, and the development of neuronal circuitry. She has been author and co-author of numerous published research and review articles in many journals, including *Developmental Biology*, *Journal of Comparative Physiology*, *Journal of Neuroscience*, and *Journal of Neurophysiology*.

Brief Contents

Contents

Preface

The animal kingdom comprises several million species, adapted to every possible niche on the planet's surface and oceans. Naturalists under sail, in jungles and caves, in balloons and diving bells, have spent hundreds of years discovering and cataloging the animal species of the Earth—the job is nowhere near finished—and have brought home descriptions and specimens of amazing diversity. Fish that stunned with electricity, fish that could thrive in both salt and fresh water, animals that seemed never to drink, creatures that hunted without vision, mammal-like monotremes that, surprisingly, laid eggs . . . the task of understanding animal physiology was enormous. Yet from careful study, principles have emerged, and the more that was learned, the clearer it became that evolution is conservative as well as inventive. The laws of physics and chemistry govern animal evolution as they do the rest of the universe, and animal adaptations can be understood by referring to a manageable number of concepts and principles that govern how biological organisms process materials and energy, interact with their environments, grow, and reproduce.

The basic principles and mechanisms of animal physiology form the central theme of this book, originally the product of the fertile mind of Roger Eckert, who died on June 13, 1986, while revising the third edition. Although only one of us knew and worked with Roger Eckert, we have attempted to keep the spirit of his book alive through the fourth, and now fifth, editions.

A beginning course in physiology is a challenge for both teacher and student because of the interdisciplinary nature of the subject. In preparing this edition, we found ourselves once again between the devil and the deep blue sea. On the one hand, students need to have mastered a very wide range of chemical, physical, and biological subject matter to understand the breadth of physiology. On the other hand, most students are eager to plunge directly into the exciting topics of animal specializations, lifestyles, and behaviors. For this reason, *Eckert Animal Physiology* presents the essential background material in early chapters, making it possible for students to review this material on their own if need be and move on quickly to the substance of animal function and to an understanding of how scientists develop knowledge by experimentation.

Eckert Animal Physiology stresses principles and mechanisms over vast inventories of information. The reader will see patterns in the functional strategies that animals have evolved within the bounds of chemical and physical possibility. Math is used when it is essential, but priority is given to the development of a qualitative and intuitive understanding. Common threads are developed to explain and compare the interactions between regulated physiological systems and coordinated responses to environmental change in a wide variety of animal groups. Cellular and molecular topics are integrated early in the book and then emphasized throughout, reflecting the fact that many biologists at the moment are applying themselves to the task of integrating physiological knowledge with the explosion of information about the genome and protein chemistry.

ORGANIZATION

The chapters are organized into three parts, with the goal of promoting an understanding of animals as integrated systems at every level of organization. Each part is introduced by an opening statement that gives students an overview of the material to follow. The four chapters in Part 1 are concerned with the central principles of physiology. Part 2 (Chapters 5–11) deals with physiological processes, and Part 3 (Chapters 12–17) discusses how these basic processes are integrated in animals living in a variety of environments. All 17 chapters have been extensively reworked and reorganized for this new edition to stay abreast of new scientific developments.

OLD AND NEW STRENGTHS

- The *experimental basis of physiology* continues to be a main unifying theme in *Eckert Animal Physiology*, leading students to an understanding of physiology and the scientific process by emphasizing how scientific knowledge is acquired.
- The *zoological context of physiological mechansisms* is brought forward in this new edition, as students are led to appreciate how physiological mechanisms have evolved in response to the selective pressures encountered by animals in their natural environments.
- The *comparative approach* of previous editions is extended, teaching students how basic principles can be uncovered by observing patterns of physiological function across different species.
- Above all, this new edition continues the deep *integrative approach* that has served users of this book over the years. Physiological systems are examined at all levels of organization, from *molecular* to *cellular* to *organismal* and further to the interactions of animals within their environment.

PEDAGOGY

- The ideas developed in the text are illuminated and augmented by liberal use of illustrations and figure legends. All figure legends begin with declarative sentences that convey the central point of the illustration.
- Full color is introduced with this edition. Care has been taken to use color consistently and with sound pedagogical purpose. The reader will find that an arrow indicating the flow of calcium is always green; a given cellular substructure is the same color throughout the book. We hope that you will find the illustrations lush and beautiful, but our goal was always to make them clear, uncluttered, and instructive.
- Spotlight boxes provide in-depth information about experiments and individuals associated with important advances in the subject matter, the derivations of some equations, and historical background. Icons designate the theme of the box:

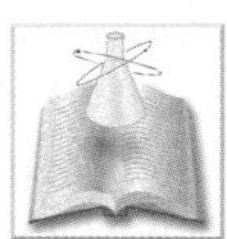
Technical Foundations

A Deeper Look

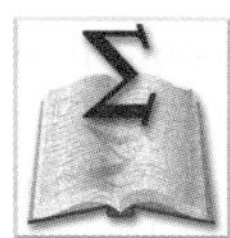
Mathematical Foundations

Historical Background

- Thought questions within chapter text, designated with the icon below, encourage problem-based learning and stimulate discussion on various aspects of the material presented.

- The foundation of animal physiology as a scientific discipline is a treasure trove of classic experiments, some venerable and some recent, that have elegantly revealed physiological mechanisms and opened new ways of thinking about how animals function. Many classic experiments are described in figures, and they are designated with an icon in the legend.

- To cultivate in students an awareness of the rich heritage upon which our topic is founded, classic literature is also cited with an icon in the Suggested Readings at the end of each chapter. References to the literature within the body of the text and in figure legends are made unobtrusively, but with enough frequency so that students can become aware of the role of scientists and their works as a subject is developed.
- Key terms appear in boldfaced type at their first mention in the text and are formally defined in a useful, comprehensive glossary at the back of the book.
- End-of-chapter materials include a bulleted Summary, which provides the student with a quick review of important points covered in the chapter, Review Questions, and Suggested Readings. Students will find appendixes, references cited, and an index at the back of the text.

SUPPLEMENTS

Instructor Supplements

- All figures in the text are available to adopters on a CD. The CD will also include Presentation Manager Pro software so instructors can quickly prepare play lists for display during lectures. Instructors can acquire the CD from their sales representative.
- Instructors will be able to access all text images from the password-protected instructor's side of the Web site located at http://www.whfreeman.com/animalphys5 Instructors can obtain passwords by contacting their local sales representative.

Student Supplements

- Students who go to http://www.whfreeman.com/animalphys5 will find links to World Wide Web sites related to animal physiology. Animal physiologists have been among the most prolific creators of fascinating topical Web sites. We encourage students to explore!

More information may be obtained by contacting your W. H. Freeman and Company sales representative or by accessing our Web site located at http://www.whfreeman.com

ACKNOWLEDGMENTS

Eckert Animal Physiology has benefited greatly from the contributions of several people whom we wish to acknowledge. Further, we are grateful for the informal comments of associates around the world and for the formal reviews of chapters provided by the following colleagues:

Bill Allen, *Humboldt State University*
Philip Brownell, *Oregon State University*
David G. Butler, *University of Toronto*
John Cameron, *Wellesley College*
Joseph F. Crivello, *University of Connecticut*
Lynn Davis, *University of Virginia*
Larry Didow, *University of Winnipeg*
Douglas Eagles, *Georgetown University*
Dale Erskine, *Lebanon Valley College*
Carl S. Hoegler, *Marymount College*
Todd C. Holmes, *New York University*
Dennis C. Haney, *Furman University*
Angela B. Lange, *University of Toronto*
Louise Milligan, *University of Western Ontario*
Richard H. Moore, *University of Texas at Austin*
David L. Nelson, *University of Wisconsin-Madison*
Eric Peters, *Chicago State University*
Lorrie Rea, *University of Central Florida*
Gerald Robinson, *Towson State University*
C. Nelson Sinback, *East Carolina University*
J.E. Steele, *University of West Ontario*
Richard Stephenson, *University of Toronto*
Murray Wiegand, *University of Winnipeg*

Finally, the good sense, kind words, and attention to detail of Morgan Ryan, Jason Noe, and Jane O'Neill, our editors, and the artistic skill of Blake Logan, our designer, have much improved the book and ensured its publication. We heartily thank them.

Our goal has been to produce a balanced, up-to-date treatment of animal function that is clear, complete, and exciting. We hope that readers will find *Eckert Animal Physiology* valuable, and we welcome constructive criticism and suggestions.

David Randall
Warren Burggren
Kathleen French
September 2001

Principles of Physiology

Animal physiology is the study of how animals function. The cheetah racing after a gazelle, the rattlesnake striking at a desert rat, and the mosquito quietly sucking blood from the ear of a caribou all coordinate specialized anatomic features and physiological processes to find their prey and, in turn, to evade predators and prolong their own lives. An arctic fox possesses a luxurious coat of fur, as well as finely tuned physiological mechanisms, to protect it from the bitter cold of its environment. Even animals living in apparently ideal environments, with benign temperatures year-round, ample food sources, and regular day/night cycles, must meet numerous challenges.

Meeting the demands of survival has resulted in numerous evolutionary variations on the basic theme of life. The environments in which life expresses itself are equally varied (there is even tantalizing evidence of microbial life existing outside of our own planet). As a result, animal physiologists have a vast array of species and environments available for their investigations of how animals work. Even so, the broad range of philosophical and technological approaches to the study of animal physiology rest on a relatively small number of fundamental concepts, which are presented in Part 1 of this book.

Chapter 1 explores the central themes in animal physiology, including the close relationship between structure and function, the processes of adaptation and acclimation, and the concepts of homeostasis and its maintenance by feedback-control systems.

Scientific knowledge comes from experiments; consequently, in **Chapter 2** we discuss the nature of physiological experimentation and the various approaches animal physiologists use to design and test hypotheses. We briefly describe many of the major experimental methods that are currently used by physiologists, including new and rapidly evolving molecular techniques. Physiology is grounded in physics and chemistry, and **Chapter 3** reviews basic physical and chemical principles that underlie the physiological mechanisms discussed in the rest of the book. This chapter focuses particularly on the processes of metabolism, the biochemical reactions that are the basis for all physiological processes.

The membranes that surround cells and their internal organelles provide an important example of how physical and chemical mechanisms combine in living cells to produce biological processes. In **Chapter 4** we investigate the nature of cell membranes. We pay special attention in this chapter to how the outer membrane of a cell helps to stabilize its internal environment. Active transport of materials across cell membranes is discussed in detail because this process is crucial for numerous physiological processes as diverse as conduction of nerve impulses, regulation of body fluids, and uptake of nutrients, all discussed in later sections. How fundamental biochemical, molecular, and cellular processes are combined to produce the integrated regulation of physiological systems throughout an animal's body is discussed in Parts 2 and 3 of the book.

Studying Animal Physiology

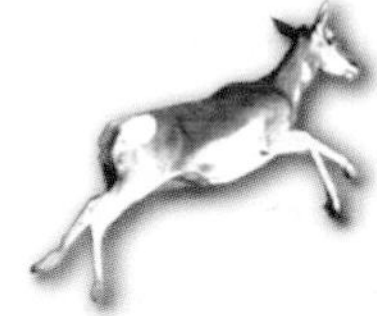

Animal physiology focuses on the functions of tissues, organs, and organ systems in multicellular animals. The animal physiologist investigates the mechanisms that operate in living organisms at all levels, ranging from molecules to the whole organism. Understanding how animals work requires detailed knowledge of the molecular interactions that underlie cellular processes. Armed with this knowledge, animal physiologists design experiments to learn about the control and regulation of processes within groups of cells and how the combined activities of these cell groups affect the function of the animal. The activities of cells in specialized organs are coordinated. It is this coordination that provides the basis for the behavioral capabilities and physiological processes that distinguish cell colonies from single cells and animals from plants, including movement, relative independence from environmental conditions, use of sensory information about the environment, and often complex social interactions, to name but a few.

Animal physiology is, above all, an *integrative science.* It attempts to bring together everything known about an animal's function to create an integrated picture of how that animal operates in its environment. We now know, for example, that the seemingly straightforward process in a mammal of maintaining a stable body temperature requires the temperature-control system of the brain to manage information about a multitude of contributing factors: the heat load or heat sink presented by the external environment, the rate of heat production by metabolically active tissues, the transport of heat by blood flow between the body core and its periphery, the contribution of evaporative cooling, the insulating efficiency of the fur, and many other physiological and anatomic variables. An integrated understanding of how all these factors combine to maintain body temperature has been hard-won through decades of experimentation by physiologists typically concentrating on a single facet of the problem, but ultimately seeking an integrated understanding of the phenomenon as a whole. It is the theme of integration that sets physiology apart from many other biological sciences, adding to its complexity—and its fascination.

Physiology is firmly rooted in the laws of chemistry and physics. As Table 1-1 on the next page reveals, physiological processes can be understood in terms of the physical constraints or chemical properties that are exploited or overcome by a physiological system. The relationship between physiology and the physical sciences is indivisible. Life's processes obey the laws of nature, and we investigate and describe the mechanisms of life using concepts and language from chemistry and physics, many of which are already familiar to you.

The central concepts of evolution—natural selection and speciation—underlie animal physiology, just as they do any other study of life on earth. Natural selection has led to the production in mammals and birds of enzymes that can tolerate high body temperatures, to the modified gills land crabs use for breathing air, and to the tolerance of migrating salmon to both freshwater and seawater. The history of animal physiology is rich in studies of adaptations by individual species to particular environmental constraints and demands; such studies have produced a great "knowledge matrix" of environments and adaptations. More than a million animal species have been described, inhabiting nearly every distinct environment on earth. In recent years, animal physiologists have sought to identify patterns in this vast array of physiological data by incorporating powerful new tools from evolutionary biology and molecular biology into their studies. One of the goals of this book is to describe the general patterns in animal physiology even as we use specific examples to illustrate physiological principles.

Table 1-1 **Common physiological processes and the chemical and physical principles that relate to them**

Process	Related chemical/physical principle
Nerve conduction and ionic current; cell membrane capacitance	Ohm's law
Gas exchange in lungs and gills	Boyle's law; ideal gas law
Animal locomotion; blood flow	Gravity, inertia, momentum, velocity, drag
Muscle contraction and limb movement	Kinetic and potential energy
Anabolic and catabolic metabolism	Laws of thermodynamics

THE SUBDISCIPLINES OF ANIMAL PHYSIOLOGY

Animal physiology has several important subdisciplines. In **comparative physiology**, species are compared in order to discern physiological and evolutionary patterns. The comparative approach can be extraordinarily powerful, but it involves more than simply using a variety of animals. Comparative physiologists use analytic and statistical techniques to make carefully structured multi-species comparisons (e.g., comparing a salmon to a rainbow trout and a brook trout to understand the adaptations of teleost fishes to salt water). Thus, the findings of a physiologist working on kidney function in armadillos or lizards are quite likely to add depth to the knowledge of a medical physiologist working with more typical laboratory animals such as mice or rabbits.

Environmental physiology examines animals in the context of the environments they inhabit. Environmental physiology focuses on evolutionary adaptations—the thick fur of Arctic animals, the high blood volume of diving seals, the waterproof cuticle of cockroaches—to understand how they allow animals to function in environments that can range from benign to supremely hostile.

Evolutionary physiology uses the techniques of evolutionary biology and systematics (e.g., the construction of taxonomic family trees, or **cladograms**) to understand the evolution of animals from a physiological viewpoint, focusing on physiological markers (e.g., maintenance of constant body temperature) rather than anatomic markers (e.g., feathers).

Developmental physiology concentrates on how physiological processes unfold during the course of animal development from embryo through larva or fetus to adulthood. Much of this work has focused on the cardiovascular and respiratory systems, in part because the dramatic and often sudden changes that mark birth or hatching in mammals and birds can reveal much about how these physiological systems operate in the adult. Increasingly, however, studies in developmental physiology are expanding to other physiological processes in a wide variety of animals to gain insights into both physiological and evolutionary processes.

Finally, **cell physiology**, though not strictly a subdiscipline of animal physiology, provides vital information on the physiology of the cells themselves, which can be woven into an integrated understanding of the physiological responses of tissues, organs, and organ systems.

Befitting an integrative science, fundamental questions in animal physiology are often most thoroughly answered through the integration of two approaches—for example, by employing the approaches of both developmental and evolutionary physiology. Thus, an apparently perplexing change in physiology during development might be more easily understood by simultaneously considering the evolutionary history of the species.

THE HISTORY OF ANIMAL PHYSIOLOGY

The study of animal physiology can be traced to the earliest writings of learned persons. Socrates (470–399 B.C.), Aristotle (384–322 B.C.), and, much later, Galileo (1564–1642) all documented the heart rate in a developing chick embryo. Marcello Malpighi (1628–1694), the first serious biological student of the microscope, described the structural underpinnings of many tissue functions. The metabolism of animals and plants was studied by European chemists such as Antoine Lavoisier (1743–1794) to understand oxygen-consuming and oxygen-producing reactions. The era of modern medical science, heralded by the open discussion and examination of the human body, and perhaps best typified by the British physician William Harvey (1578–1657), led to the rapid development of animal physiology as a discipline of the natural sciences. This time-honored and time-tested central position of animal physiology continues to this day, particularly as animal physiologists expand their scientific arsenal to include the exciting new tools offered by molecular biology and computer technology. Moreover, the central role of animal physiology gives every indication of continuing unabated into the future. Some of the very first scientific experiments in space involved measuring the effects of microgravity on physiological processes. Physiological experiments are a regular fixture of U.S. space shuttle missions, and they will have a prominent place as laboratory work commences on the International Space Station.

Animal physiologists are typically very conscious of the rich history of their discipline. While new physiological knowledge unfolds daily, there are nevertheless some truly classic studies that are cited repeatedly because of their groundbreaking approach and elegant simplicity. An excellent example is the study whose results are illustrated in Figure 17-16, which described the basic principles of the thermal physiology of tunas nearly four decades ago. Although many investigators have since corroborated and expanded on this study, current ongoing work in tuna physiology still traces back to F. G. Carey and his colleagues. We have chosen to recognize these classic studies and the pioneering data they have produced by using this icon *(right)* to identify them in the chapter figures and suggested readings that conclude each chapter.

Classic Work

WHY STUDY ANIMAL PHYSIOLOGY?

What is the allure of animal physiology, felt by Aristotle and countless individuals since his time? Several factors account for our fascination with animal physiology over the millennia.

Scientific Curiosity

Underlying all studies of animal physiology—even those designed for practical, applied purposes—is curiosity about how animals work.

> How can a hummingbird sustain a heart rate of 20 beats per second during hovering flight?
>
> How do insects see in the ultraviolet spectrum?
>
> How do kangaroo rats survive in the desert with no access to drinking water?

There is no limit to the curiosity of animal physiologists, for a common opinion among us is that the more we learn, the more we realize how little we still know about physiological systems of animals.

Commercial and Agricultural Applications

Animal physiology studies have led to many commercial and agricultural advances during the last few decades. Veterinarians, for instance, now offer veterinary care that rivals the medical treatment available to humans. Farmers have been able to improve the yield and quality of the milk, eggs, and meat they produce. Improved breeding techniques now include widespread use of artificial insemination. And at the frontier of current knowledge, genetic engineering can now be used to instill desirable physiological traits in domesticated animals; the implementation of genetic engineering is limited at present more by political and ethical considerations than by technological ones.

Insights into Human Physiology

Finally, animal physiology can teach us much about physiological processes in humans. This is hardly surprising, for the human species shares with all other animal species

- the same fundamental biological processes that, in total, are called "life"
- the same constraints established by the laws of physics and chemistry
- the same principles and mechanisms of Mendelian and molecular genetics
- a common origin in the branching tree of evolutionary history

The beating of your heart results from physiological mechanisms fundamentally no different from those that underlie heart function in fishes, frogs, and birds. Likewise, the molecular events that produce an electrical nerve impulse in your brain are fundamentally the same as those that produce an impulse in the nerve of a squid, rat, or the free-swimming larva of a sea squirt. For these reasons, animal physiology has made innumerable contributions to our understanding of human physiology. In fact, most of what we have learned about the function of human cells, tissues, and organs was known first (or is still known only) through the study of various species of vertebrate and invertebrate animals.

Animal physiology, especially as it applies to the human body, is the cornerstone of scientific medical practice. Our understanding of how living tissue functions—and malfunctions—is the foundation for effective, scientifically sound treatment for human disease. The contributions of animal physiology to medicine have been greatly expanded by new techniques for generating unique animal models of specific human diseases, such as diabetic mice, congenitally fat rats, and zebrafish embryos with specific heart defects. The insights provided by experimentation with such animal models build upon a fundamental understanding of underlying physiological processes. A physician or medical researcher who understands animal physiology—both its potential contributions and its limitations—is better equipped to make intelligent and perceptive use of information from such model systems.

CENTRAL THEMES IN ANIMAL PHYSIOLOGY

Our goals for this book are to explore physiological processes that are basic to all animal groups and to show how they have been shaped by natural selection.

Comparing the ways in which different organisms have become adapted to survive similar environmental challenges provides useful insights about patterns of physiological evolution and the adaptive value of physiological innovations. As you study animal physiology and physiological adaptation, you will see several basic themes emerge. We briefly discuss a few of them here. Others will be highlighted as you progress through the following chapters.

Structure/Function Relationships

Function flows from structure. We can illustrate this central principle of animal physiology with a familiar example. A frog leaps for a passing fly by contracting powerful skeletal muscles attached to its leg bones. Once the fly has been eaten, the smooth muscle in the frog's stomach slowly massages and mixes the stomach's contents. The nutrients derived from the fly are absorbed into the blood, which is propelled throughout the body by the regular beating of the cardiac muscle of the heart. Throughout this daily occurrence in a frog's life, three structurally distinct forms of muscle reliably carry out three distinct functions. Such relationships between tissue structure and function are found not only in muscle, but in every tissue of an animal's body.

A strong relationship between function and structure occurs at all levels of biological organization. As shown in Figure 1-1, structure/function relationships are clearly evident at the molecular level in muscle tissue. Indeed, the contractile machinery of skeletal muscle represents one of the most intensively studied examples of the dependence of function on structure at the molecular and biochemical levels. As you will learn in Chapter 10, the movement of a frog's leg is the end result of a chain of biochemical events that depend critically on interactions between thousands of rodlike structures composed of the contractile proteins actin and myosin within each muscle cell. The molecular structures of each of these proteins are fashioned to interact in a way that moves one protein along the length of the other. These protein movements lead to contraction (shortening) of individual muscle cells. Compounded over the thousands of muscle cells that form each leg muscle, muscle cell contraction causes overall shortening of the leg muscle. Because of the structural relationship between the powerful contracting muscle and the long bones of the frog's leg, muscle shortening moves the leg. This produces the jumping movement, which moves the frog's entire body.

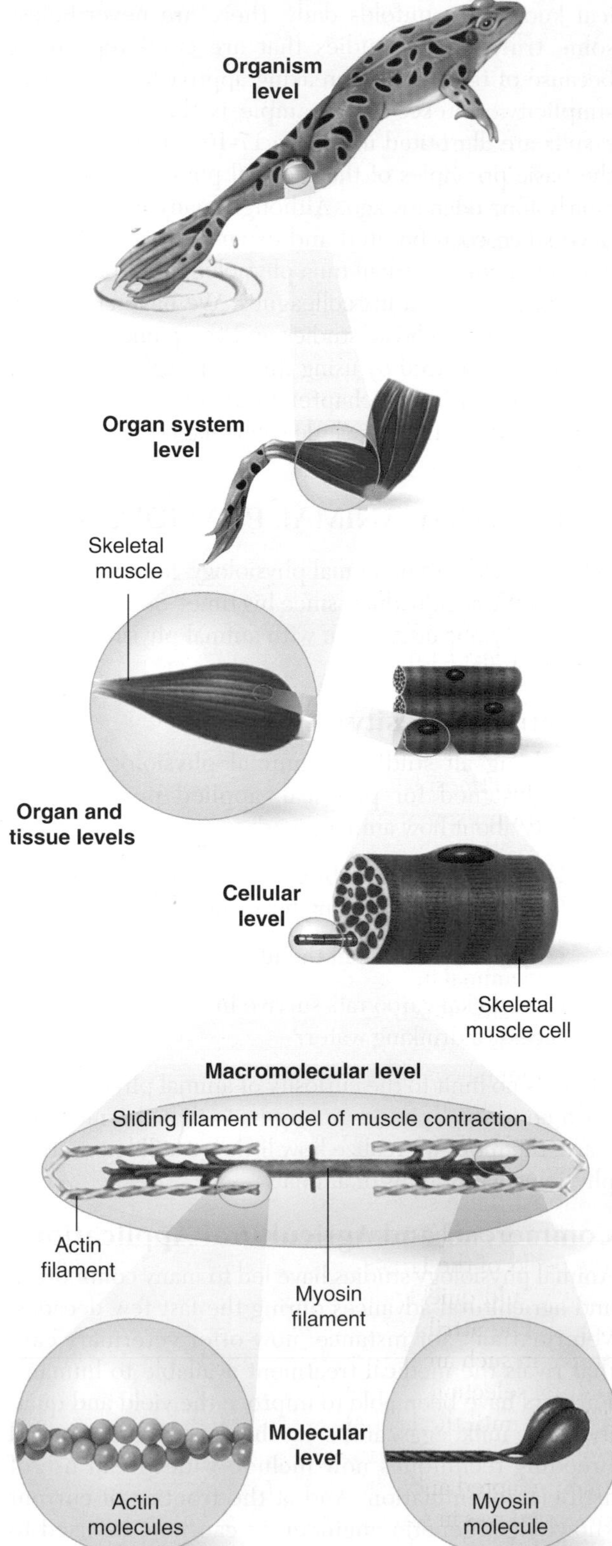

Figure 1-1 Biological function at each level of organization depends on the structure of that level and the levels below. Beginning with the whole animal, this principle can be traced from complex physiological systems through cells down to macromolecular assemblages. The dozens of muscle systems present in the adult frog allow it to move its eyes and limbs, to swallow, and to carry out all of the numerous activities of frogs. In this example, groups of skeletal muscles form a system for moving the frog's leg. Skeletal muscles themselves are composed of skeletal muscle cells, which in turn are formed from thousands of macromolecular assemblages formed from a pair of contractile proteins—actin and myosin. These macromolecular assemblages form the basic unit of muscle contraction.

The principle that function depends on structure holds true across the whole range of physiological processes. In fact, you will see that a consideration of structure/function relationships is virtually unavoidable for each of the physiological processes explored in this book.

Adaptation, Acclimatization, and Acclimation

The physiology of an animal is usually very well matched to the environment that the animal occupies, thereby contributing to its survival. Evolution by natural selection is the accepted explanation for this condition, called **adaptation.** Adaptation typically occurs in a gradual manner over many generations; it is generally not reversible. Be aware that adaptation is frequently confused with two other processes, acclimatization and acclimation. **Acclimatization** is a physiological, biochemical, or anatomic change within an individual animal during its life that results from that animal's chronic exposure in its native habitat to new, naturally occurring environmental conditions. **Acclimation** refers to the same process as acclimatization when the changes are induced experimentally in the laboratory or in the wild by an investigator.

Generally, both acclimation and acclimatization are reversible. If a bear migrates from a low-lying valley near sea level up several thousand feet to the high slopes of a tall mountain (a voluntary change in a natural environment), its lung ventilation rate initially increases so that it can acquire adequate oxygen in the rarified mountain atmosphere. Within a few days, however, lung ventilation drops back toward the rate near sea level because other physiological mechanisms that facilitate gas exchange at high altitudes begin to operate. The animal is then said to be acclimatized to the high-altitude conditions. If an animal physiologist places an unacclimatized animal in a hypobaric chamber, thus simulating the high-altitude conditions of mountain slopes, a similar result is seen: the animal becomes acclimated to the experimental conditions within a few days. In either situation, a return to low-altitude oxygen levels typically causes a reversal of the physiological changes originally induced through acclimatization or acclimation. Contrast these short-term responses with the physiology of the bar-headed goose, which is able to fly above the peaks of the Himalayas (while honking!). Other animals would quickly die where the goose so athletically soars. The attributes that allow this goose species to thrive in such an extreme environment are the result of natural selection.

Up until the last few decades, animal physiologists operated under the assumption that animals were optimally adapted and that every physiological process they observed was in some way maximized to ensure the survival of the animal. More recently, armed with theories and observations developed by evolutionary and developmental biologists, animal physiologists now realize that even though evolution by natural selection leads to change in physiological processes, the result is often not "optimal," but simply good enough to contribute to the survival of the animal. Mammals, for example, typically maintain a relatively constant body temperature within a range of 1–2°C. Given the precision of some known physiological control systems, it is conceivable that a more precise temperature-control system could exist, but such a system has not been fixed by selection. That is, a 1–2°C temperature range is good enough for survival.

Clearly, adaptation is a central concept in animal physiology, yet it can be difficult to establish whether some characteristic of an animal actually has adaptive value. A physiological process is **adaptive** if it is present at high frequency in the population because it results in a higher probability of survival and reproduction than alternative processes. Proof of the adaptive value of some physiological processes can be difficult to obtain, but the comparative approach can generate evidence that supports the case. In this approach, the researcher examines a physiological process in distantly related species living in identical environments or in closely related species living in strikingly different environments. The presence of a similar physiological process supported by a similar anatomic structure in several distantly related animal species occupying a single environment would suggest that that process/structure combination is adaptive. A classic example of the power of the comparative approach involves the llama and its close relative, the camel. Originally, researchers were convinced that the unusually high affinity of llama blood for oxygen was an adaptation to the rarified air at the high altitudes at which llamas live. To their surprise, animal physiologists then discovered that the blood of camels, which live at low altitudes, also has high affinity for oxygen. Thus, the high oxygen affinity of llama blood is not a specific adaptation to high altitude. That is, the blood characteristics of llamas and camels have little to do with the altitude at which they live, and much to do with their both being in the family Camelidae. Such indirect criteria are typically accepted in judging adaptiveness, particularly when set in the framework of a carefully designed comparative study.

Physiological and anatomic adaptations are genetically based, passed on from generation to generation, and constantly shaped and maintained by natural selection. Animals inherit genetic information from their parents in the form of **deoxyribonucleic acid (DNA)** molecules. Spontaneous alterations **(mutations)** can occur in the nucleotide sequence of DNA, potentially causing changes in the properties of the encoded ribonucleic acids (RNAs) or proteins. Mutations in the germ-line DNA that enhance the animal's survival, and thus its chances of reproducing, are retained by selection, and the frequency with which they occur in the population of organisms increases over time. Conversely, those mutations that render animals less well adapted to their environment will lessen their chances to reproduce. If deleterious enough, such mutations generally are eliminated over time.

Genetic material in the form of DNA is passed on from multicellular parents to their offspring. This DNA is contained in a line of germ cells that, in each generation, are derived directly from parent germ cells, creating an uninterrupted lineage. The blind, nondirected process of evolution is centered on the survival of the germ-line DNA, since the information it encodes defines a species. Failure by a species to reproduce that genetic information leads to the species' immediate, irreversible extinction. From the biological viewpoint, then, the major goal in an animal's life is to reproduce and propagate its DNA, and all behaviors, physiological processes, and anatomic structures are ultimately subservient to the survival of the germ line. Adaptation to the constraints and demands of the environment is best appreciated and understood in this context of an animal's struggle to maintain and reproduce its DNA.

Homeostasis

Many animals seem to live comfortably in their environments, but most habitats are actually quite hostile to animal cells. The water surrounding many aquatic animals, for example, is more dilute (freshwater) or more salty (seawater) than their own body fluids. Both terrestrial and aquatic animals may live in environments where the temperature is too hot or cold to sustain normal physiological processes. Moreover, with only a few exceptions (such as the deep abyss of the oceans), most environments are characterized by fluctuations in their physical and chemical properties (especially temperature). The environmental changes raging about an animal's exterior would disrupt cellular, tissue, and organ function were it not for physiological control systems that maintain relatively stable conditions within an animal's body tissues. This tendency of organisms to regulate and maintain relative internal stability is called **homeostasis.**

Claude Bernard, the nineteenth-century French pioneer of modern physiology, first recognized the importance to animal function of maintaining stability in the *milieu intérieur,* or internal environment. Bernard noted the ability of mammals to regulate the condition of their internal environment within rather narrow limits. This ability is familiar to most of us from measurements of our own body temperature, which in healthy humans is maintained within a degree of 37°C. Cells in the body experience a relatively constant environment with respect to not just temperature, but also glucose concentration, pH, osmotic pressure, oxygen level, ion concentrations, and so forth. Bernard (1872) concluded, "Constancy of the internal environment is the condition of free life," arguing that the ability of animals to survive in often stressful and varying environments directly reflects their ability to maintain a stable internal environment. In the early 1900s Walter Cannon extended Bernard's notion of internal consistency to the organization and function of cells, tissues, and organs. It was Cannon (1929), in fact, who coined the term *homeostasis* to describe the tendency toward internal stability.

Homeostasis, one of the most influential concepts in the history of biology, provides a conceptual framework within which to interpret a wide range of physiological data. This phenomenon is nearly universal in living systems (Figure 1-2). The evolution of homeostasis and the physiological systems that maintain it were essential factors in allowing animals to venture from relatively "physiologically friendly" environments and invade habitats more hostile to life processes. One fascination of physiology is discovering the different adaptations evolved among species to maintain homeostasis in the face of environmental challenges.

Complex, multi-organ physiological mechanisms are often involved in maintaining homeostasis, but homeostasis also pervades physiology at the cellular level. In fact, some degree of homeostasis is found even in the simplest unicellular organisms. Protozoans, for instance, have been able to invade freshwater and other osmotically stressful environments because their intracellular concentrations of salts, sugars, amino acids, and

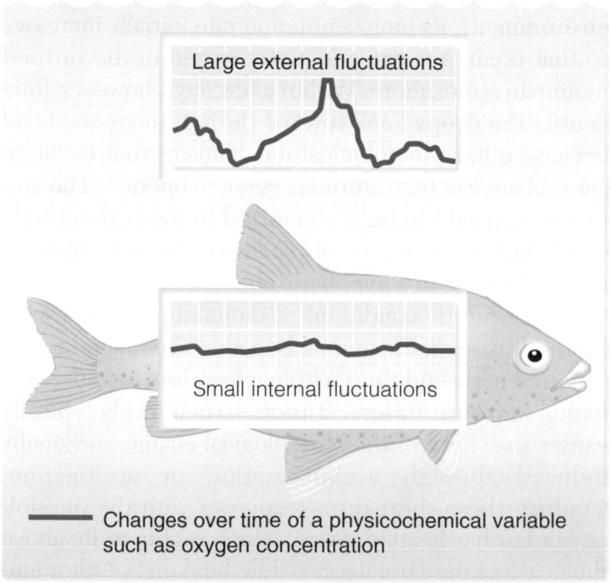

Figure 1-2 Physiological regulatory systems maintain internal conditions within a relatively small range. Large variations in the external environment induce equally large responses of the control system to offset the disturbance. The net effect is that internal fluctuations of a variable in an animal are usually far smaller than the environmental fluctuations in that variable. As will be frequently evident in subsequent chapters, homeostasis of almost all critical internal variables is maintained in this fashion.

other solutes are regulated by selective cell membrane permeability, active transport, and other mechanisms. In all animal cells, including the single cells of protozoans, these processes maintain intracellular conditions—typically quite different from those of the extracellular environment—within limits favorable to the metabolic requirements of the cell.

Feedback-Control Systems

The regulatory processes that maintain homeostasis in cells and multicellular organisms depend on **feedback**, which occurs when sensory information about a particular variable (e.g., temperature, salinity, pH) is used to control processes in cells, tissues, and organs that influence the internal level of that variable. Homeostatic regulation requires continuous sampling of the controlled variable coupled with immediate corrective action, a process termed **negative feedback**. To understand this concept, imagine a motorist at the beginning of an absolutely straight 10-kilometer stretch of highway. The driver is forbidden to touch the steering wheel. Clearly it is impossible for the driver to simply step on the gas and travel the 10-kilometer distance without deviating from his lane. The slightest asymmetry in the steering mechanism of the car—not to mention wind or unevenness of the road surface—causes the car to go off course. An ordinary driver uses visual information to stay in the lane. A gradual drift in either direction is corrected by a compensatory movement applied to the steering wheel. The visual system of the driver acts as the sensor in this case, and the driver's neuromuscular system causes a correctional movement in the direction opposite to the perceived error and toward the **set point** (in this case, the center of the lane). The motion of the car is thus governed by negative feedback.

Another example of regulation by negative feedback can be demonstrated with a thermostatic device that maintains the temperature of a hot-water bath at or near a set point (Figure 1-3). When the water temperature is below the set point, the sensor maintains the heater switch in the closed, "on" position. When the set-point temperature is achieved, the heater switch opens, and further heating ceases until the temperature again drops below the set point. This example suggests that regulation of body temperature requires a "thermostat" whose information must be provided to a temperature-control system, which either heats or cools the body depending on the temperature signal. Physiological investigations have discovered a great deal about temperature regulation, including the location of the thermostat, as we'll see in Chapter 17.

The characteristics of negative and positive feedback are summarized in Spotlight 1-1 on page 11. You will find examples of feedback-control systems appearing throughout this text, especially in our discussions of intermediary metabolism (Chapter 3), endocrine control (Chapter 9), neural control of muscle (Chapter 10), circulatory and respiratory control (Chapter 13), and regulation of ionic balance (Chapter 14). Indeed, you will discover that feedback is a central concept throughout the study of physiological systems.

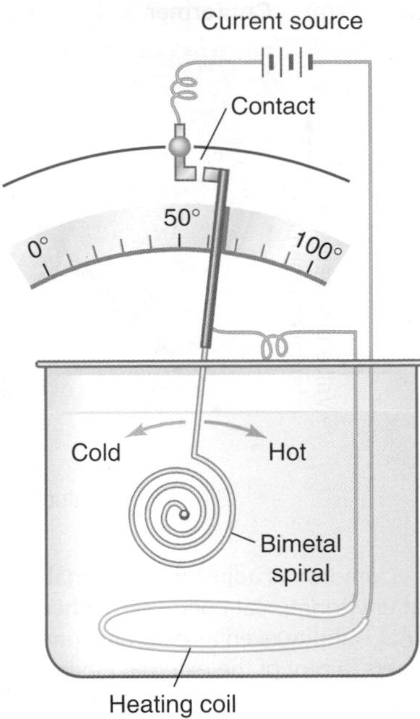

Figure 1-3 The temperature of a mechanically thermoregulated hot-water bath is maintained by a negative-feedback-control system. The bimetal spiral, whose center is attached to the water bath wall, winds slightly as the temperature of the water bath drops below the desired (set-point) temperature. This winding action causes electrical contacts to touch, completing an electric circuit and allowing electric current to flow through a heating coil in the bath. As the water warms, the coil unwinds slightly and the contacts separate. The water temperature, at or slightly above the set point, now stops rising. When the bath begins to cool off again, the cycle is repeated. Many physiological control systems operate on the same principles.

Conformity and Regulation

When an animal is confronted with changes in its environment (for example, changes in oxygen availability or salinity), it can respond in one of two ways: conformity or regulation. In some species, environmental challenges induce internal body changes that simply parallel the external conditions (Figure 1-4 on the next page). Such animals, called **conformers**, are unable to maintain homeostasis for internal conditions such as body fluid salinity or tissue oxygenation. Echinoderms such as the starfish *Asterias*, for example, are osmoconformers whose internal body fluids quickly come to equilibrium with the seawater that surrounds them. They experience an increase in body fluid salinity when placed in high-salt water and a decrease in body fluid salinity when placed in low-salt water. Similarly, the oxygen consumption of oxyconformers such as annelid worms rises and

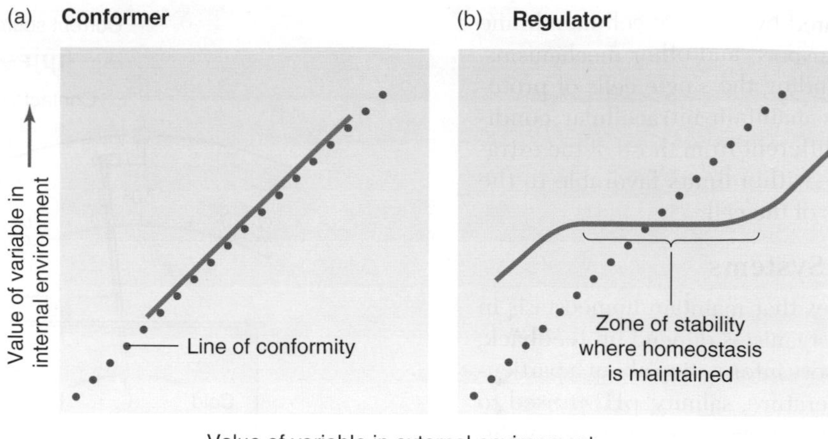

Figure 1-4 Conformers adjust their internal conditions to reflect external environmental conditions, whereas regulators maintain internal stability even as external conditions change. **(a)** For conformers, a plot of the external environmental value of a variable (e.g., salinity, oxygen availability) versus its internal value (green line) is typically a straight line with a slope of 1. When an animal is unable to mount the physiological or other responses necessary to counteract external changes in a variable, its internal value varies directly with its external value, mimicking the "line of conformity" (blue dotted line). **(b)** For regulators, a plot of the external value versus the internal value of a variable shows that they are able to maintain internal stability over a wide range of external change. The line of conformity is shown for comparison. At environmental extremes, however, regulators are unable to regulate internal conditions and are forced to become conformers. The breadth of the zone of stability of a regulator depends on the particular species and environmental variable.

falls as oxygen availability waxes and wanes. The degree to which conformers can survive in changing environments depends largely upon the tolerance of their body tissues to externally imposed internal changes.

Regulators, as their name implies, use biochemical, physiological, behavioral, and other mechanisms to regulate their internal environment over a broad range of external environmental changes—that is, they maintain homeostasis. Thus, an osmoregulator maintains the ion concentrations of its body fluids above environmental levels when placed in dilute water and below environmental levels when placed in a concentrated solution. Oxyregulators, which include crayfishes, most mollusks, and almost all vertebrates, maintain their oxygen consumption at nearly steady levels as environmental oxygen availability falls. Eventually, however, oxygen may become so limited that oxygen consumption cannot be maintained, and the animal reverts to oxygen conformity. Thus, physiological systems must acclimatize to lessen the requirement for oxygen, to the extent tolerable by the organism.

It is tempting to make broad generalizations about taxonomy based on whether animals are conformers or regulators. Although most invertebrates are conformers and almost all vertebrates are regulators, there are many exceptions. For instance, decapod crustaceans (e.g., crabs, crayfishes, shrimps, lobsters) tend to be accomplished regulators, as are many mollusks and most insects. Moreover, among regulators, the acceptable ranges for physiological variables may be very broad or very narrow depending on the species.

THE LITERATURE OF THE PHYSIOLOGICAL SCIENCES

All the information in this book is based on the experimental results of scientists who published their work in reports and articles, commonly referred to as scientific papers. Such papers, which include descriptions of the experimental methods used and the results, followed by a discussion, are published in scientific journals, many of which focus on particular disciplines or specialized research areas.

After preparing the manuscript of a scientific paper, the author(s) send it to the editorial office of the journal they feel has the most appropriate readership for their paper. This is the start of a long, sometimes arduous process that may lead to publication. The journal editor sends each submitted manuscript to two or more scientists expert in the topic of the paper for their review and critical comments. The reviewers recommend acceptance or rejection on grounds of scientific quality and often make numerous suggestions for the improvement of acceptable papers. This process, called *peer review*, helps to ensure that papers accepted for publication are based on accepted research methods and that their conclusions are valid. Once a paper is

Spotlight 1-1 THE CONCEPT OF FEEDBACK

Any effective control system, whether the human brain, a computer, or a household thermostat, is vitally dependent on **feedback,** which is the return of information to a **controller** that regulates a controlled variable. Feedback can be either positive or negative, with each producing profoundly different effects. Feedback is widely employed in both biological and engineering control systems to maintain a preselected level of the controlled variable.

Negative Feedback

Consider the model system shown in part a of the accompanying figure. Assume for the moment that the **controlled system** experiences a new **disturbance** (e.g., a change in length, temperature, voltage, or concentration). The output of this system is detected by a **sensor,** which sends a signal to an amplifier. Now, imagine an amplifier that inverts the signal it receives (an "inverting amplifier") so that the sign or direction of its output is opposite to that of its input (e.g., plus is changed to minus or vice versa). Such **signal inversion** provides the basis for negative feedback, which can be used to regulate the **controlled variable** (e.g., length, temperature, voltage, or concentration) within a limited range.

When the sensor detects a change in the state (e.g., length, temperature, voltage, concentration) of the controlled system, it produces an **error signal** proportional to the difference between the **set point** at which the system is to be held and the actual state of the system. The error signal is then both amplified and inverted (i.e., changed in sign). The inverted output of the amplifier, fed back to the system, counteracts the disturbance and reduces the error signal. Consequently, the system tends to stabilize near the set point. The inversion of sign is the most fundamental feature of negative-feedback control.

A hypothetical negative-feedback loop with infinite amplification would hold the system precisely at the set point, because the slightest error signal would result in a massive output from the amplifier to counteract the disturbance. Since no amplifier, electronic or biological, produces infinite amplification, negative-feedback control only approximates the set point during a disturbance. The less amplification the system has, the less accurate the control will be.

Positive Feedback

In the model system shown in part b of the figure, an applied disturbance acts on the controlled system just as in part a. In this case, however, the signal is amplified, but its sign (plus or minus) remains unchanged. Therefore, the output of the amplifier, when fed back to the controlled system, has the same effect as the original disturbance, reinforcing the disturbance to the controlled system. This **positive-feedback** system is highly unstable because the output becomes progressively stronger as it is fed back and reamplified. You are probably familiar with positive feedback in public address systems: when the output of the loudspeaker is picked up and reamplified by the microphone, a loud squeal is generated. Thus, a tiny disturbance at the input can cause a much larger effect at the output. The output of the system is usually limited in some way; for example, in the public address system, the intensity of the output is limited by the power of the audio amplifier and speakers or by saturation of the microphone signal. In biological systems, the response may be limited by the amount of energy or substrate available.

Positive feedback is most often encountered in pathological conditions that affect normal negative-feedback control. Congestive heart failure provides a classic example. In this disorder, the inability of the heart to pump blood causes blood to accumulate in the ventricles, which further impairs their ability to pump blood, which causes more blood to accumulate in the ventricles, which further impairs their pumping ability, and so on. Unless such vicious cycles of positive feedback are interrupted, they will quickly lead to complete failure of the controlled system.

In normal, healthy animals, positive feedback generally functions only to produce a regenerative, explosive, or autocatalytic effect. This type of control is often used to generate the rising phase of a cyclic event, such as the upstroke of the nerve impulse or the explosive growth of a blood clot to prevent blood loss. Another example of positive feedback translating an initial signal into an amplified response is seen in the rapid emptying of body cavities, as in expulsion of the fetus from the uterus, vomiting, and swallowing.

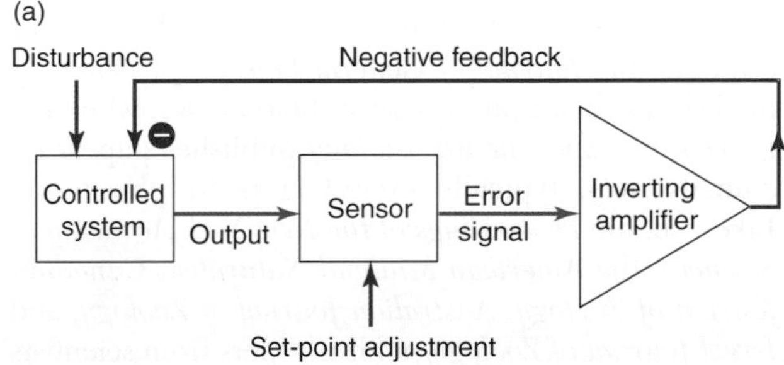

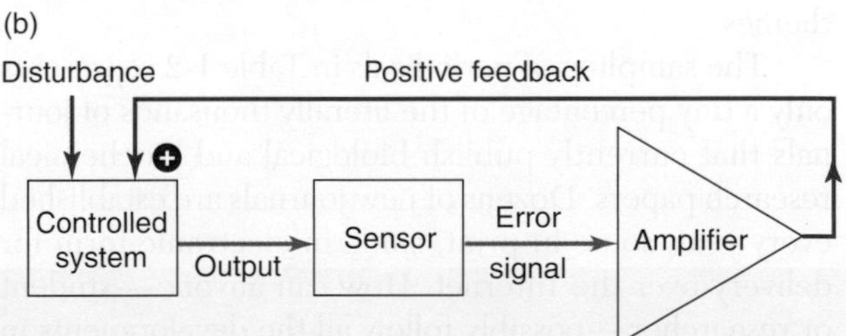

Biological control systems depend on negative or positive feedback. The unique features of negative-feedback systems ***(a)*** *are generation of an error signal and sign inversion by the amplifier. Positive-feedback systems* ***(b),*** *which amplify the signal without sign inversion, lead to a rapid departure from the set point of the controlled variable. The basic elements illustrated here occur in a number of variations in biological systems. In some cases, sensor and amplifier functions are performed by a single element (e.g., response of a single cell to environmental disturbances).*

published, members of the scientific community are free to test its conclusions by repeating key experiments and to accept or reject individually the conclusions stated in the paper. Healthy skepticism and attempts to improve on the work of other scientists are vital for maintaining the self-correcting nature of an experimental science such as physiology.

Numerous scientific journals publish papers on research in animal physiology. Many of the most widely read journals are listed in Table 1-2. Some journals accept papers dealing with a wide range of topics, whereas numerous specialty journals provide in-depth coverage of more limited areas of interest. Many journals publish *reviews,* articles that summarize and evaluate recent papers from a variety of journals on some particular topic; some journals publish only reviews. Another category of journals deals with organisms primarily from a taxonomic perspective. These journals publish physiological and other research papers dealing with specific animal groups. Finally, weekly scientific news journals publish preliminary reports on physiological research that the editors believe will capture the interest of the general scientific community.

As you become familiar with the physiological research literature, be aware that journals (printed or virtual), like animals, have undergone evolution from their original forms. This is especially apparent when considering the names of journals, which in a few instances seem to reflect their contents only loosely. For example, the *Journal of General Physiology* primarily publishes cellular physiology and biophysics, while the *Journal of Experimental Biology* publishes papers on animals only, typically excluding plant physiology. Likewise, the *Proceedings of the New York Academy of Sciences,* the *American Midland Naturalist, Canadian Journal of Zoology, Australian Journal of Zoology,* and *Israel Journal of Zoology* publish papers from scientists around the world even while retaining their regional themes.

The sampling of periodicals in Table 1-2 represents only a tiny percentage of the literally thousands of journals that currently publish biological and biochemical research papers. Dozens of new journals are established every year, some in print, some in electronic form for delivery over the Internet. How can anyone—student or researcher—possibly follow all the developments in a specific area of biology? Fortunately, along with the explosion of information over the last few decades, technology has been developed that allows us to sit in front of a computer terminal and, with a few keystrokes, search millions of documents, roaming freely through the libraries of thousands of universities and research institutes worldwide. Additionally, the World Wide Web is leading to an accelerating shift in the way scientific information is disseminated. Many journals now distribute their issues in both conventional printed format and electronically to World Wide Web subscribers, either as abstracts or entire electronic versions. Journals that publish *only* a digital version are already becoming important to the scientific community. Couple these changes with World Wide Web pages that distribute a laboratory's latest research to all who care to read it, and we see a coming revolution in how information is accessed and processed. (See the Web page accompanying this book for a list of current links to the leading journals and other relevant information from the Web.)

What has not been replaced by technology is the need for the student to read descriptions of original experiments in order to understand the process that generates physiological data. Consequently, at the end of each chapter of this text you will find a Suggested Readings section, listing a few key articles that offer greater detail on certain topics. In addition, the original sources for much of the material presented in the text, figures, and tables are listed in the References Cited in the back of the book. We encourage you to visit the periodical section of your library and search online for some of these articles.

ANIMAL EXPERIMENTATION IN PHYSIOLOGY

Much of the information presented in this book comes from experiments in which animals were used to answer specific questions about how physiological processes work. Because basic physiological properties are similar among different animal species, the results from animal experiments are broadly applicable to other species, including humans. Indeed, almost all of the important medical treatments in use today are based directly on the results of animal experimentation. Even though data from experiments on humans would be the most directly relevant, in most cases such research would be ethically unacceptable. Consequently, animal experimentation is absolutely vital to many types of medical research, as well as in its own right. Through animal experimentation, enormous gains have been made in human health, including an increased life span, a reduced death rate from stroke, the development of prosthetic devices (e.g., artificial heart valves), and more effective treatments for formerly debilitating diseases such as diabetes.

Early editions of this book assumed that the benefits of animal experimentation were understood by all. This assumption can no longer be made. Despite the general benefits animal research has produced for all humans, controversy exists about the ethics of using animals in experiments. In describing this controversy, it is important to begin by making the distinction between animal welfare and animal rights. **Animal welfare** refers to the humane treatment of animals with respect for their comfort and well-being. **Animal rights** refers to the idea that animals have intrinsic and unassailable "rights," much like the inalienable human rights to life and liberty propounded in the founding

Table 1-2 A sampling of scientific journals that publish physiological research papers

Name	Abbreviation*	Topics covered
General journals		
American Journal of Physiology	*Am. J. Physiol.*	
Pflügers Archiv für Physiologie (now *European Journal of Physiology*)	*Pflugers Arch. Physiol.* (*Eur. J. Physiol.*)	Broad areas of physiology from the cell to organ systems
Journal of Physiology	*J. Physiol.*	
Journal of General Physiology	*J. Gen. Physiol.*	Physiological and biophysical studies at the cellular and subcellular level
Comparative Physiology and Biochemistry	*Comp. Physiol. Biochem.*	
Journal of Comparative Physiology	*J. Comp. Physiol.*	Many different areas, with emphasis on lower vertebrates and invertebrates
Journal of Experimental Biology	*J. Exp. Biol.*	
Physiological and Biochemical Zoology	*Physiol. Biochem. Zool.*	
Specialty journals		
Brain, Behavior, and Evolution	*Brain Behav. Evol.*	
Cell		
Circulation Research	*Circ. Res.*	
Evolution and Development	*Evol. Dev.*	
Endocrinology		
Gastroenterology		
Journal of Cell Physiology	*J. Cell Physiol.*	Research related to specific areas or processes indicated by journal's name
Journal of Membrane Biology	*J. Membr. Biol.*	
Journal of Neurophysiology	*J. Neurophysiol.*	
Journal of Neuroscience	*J. Neurosci.*	
Molecular Endocrinology	*Mol. Endocrinol.*	
Nephron		
Respiration Physiology	*Respir. Physiol.*	
Annual reviews		
Annual Review of Neuroscience	*Annu. Rev. Neurosci.*	
Annual Review of Physiology	*Annu. Rev. Physiol.*	Summaries and evaluations of original papers on particular topics published in other journals
Federation Proceedings	*Fed. Proc.*	
Physiological Reviews	*Physiol. Rev.*	
Taxonomy-oriented journals		
Auk		
Condor		Physiology and other topics related to birds
Emu		
Crustaceana		Physiology and other topics related to crustaceans
Copeia		
Herpetologica		Amphibian and reptilian physiology
Journal of Herpetology	*J. Herpetol.*	
Journal of Mammalogy	*J. Mammal.*	Physiology and other topics dealing with mammals
Weekly journals		
Nature		Preliminary reports about topics of general interest to the scientific community
Science		

*Single-word journal names are not abbreviated.

documents of progressive, free countries. The concept of animal rights, in its most extreme interpretation, would prohibit domestication of animals for food production and their being kept as pets.

The scientific community strongly supports the concept of animal welfare. Scientists recognize their professional obligations to safeguard and improve the welfare of laboratory animals. Indeed, scientific researchers concerned about the care and treatment of laboratory animals were the first to set voluntary care standards at the turn of the century, long before federal regulations were instituted in the United States. Now, government regulations strictly control animal research, requiring all animal care facilities to meet strict standards of cleanliness, veterinary care, and humaneness. Professional societies as well as scientific journals have stringent requirements regarding animal experimentation, and publication of research results requires evidence that these requirements have been met.

All research institutions that receive federal government funding are required to have Laboratory Animal Care Committees, which evaluate proposed experiments to ensure that they are performed with the minimum pain and discomfort to animal subjects and that they use the minimum number of animals necessary to achieve conclusive results. Such committees include scientists, veterinarians, and community representatives. These individuals are empowered to limit or prohibit experiments that do not adequately address the issue of animal pain, and ill-conceived proposals are quickly rejected.

Despite these internal and external safeguards, the issue of animal rights appeals to some who oppose animal research in any form. The presence of sanctioned animal care facilities and of strictly enforced regulations does not satisfy those advocating animal rights, nor are they swayed by the obvious benefits of animal research. The ongoing debate over animal welfare versus animal rights is a healthy one, if the available data are evaluated objectively and the intentions of the participants are clear. Through such debate and evaluation we can ensure that our need for animal experimentation is balanced with broadly shared concerns for the well-being of animals.

SUMMARY

- Animal physiology deals with the functions of tissues, organs, and organ systems, particularly how these functions are controlled and regulated.

- Although this book concentrates on presenting the functional basis of animal physiology, it also looks at the shaping of physiological processes by environmental constraints through natural selection.

- The subdisciplines of animal physiology include comparative physiology, environmental physiology, evolutionary physiology, developmental physiology, and cell physiology.

Why study animal physiology?

- Biologists study animal physiology because they are curious about how animals work and also because they can learn much about human physiology by observing other animals.

- Animal physiology is a cornerstone of scientific medical practice as well as veterinary practice.

Central themes in animal physiology

- Function depends on structure at all levels, from molecules to organisms. Specialized structures often produce specialized functions.

- Natural selection has led to physiological adaptation, promoting the emergence of processes suited to helping animals survive in often challenging environments.

- The adaptive cell, tissue, and organ functions that have arisen during evolution are genetically determined and encoded in DNA.

- Many animals exhibit homeostasis, the tendency toward relative stability in the internal environment of an organism.

- Feedback-control systems are critical to maintaining homeostasis.

- Animals respond to changes in external environmental conditions in two general ways. The internal environment of conformers adjusts to reflect external conditions; that is, these animals do not maintain homeostasis. Regulators keep their internal environment within narrow limits as environmental conditions change; that is, they maintain homeostasis.

Animal experimentation in physiology

- Almost all of our knowledge about animal physiology—and most of what we know about human physiology—is derived from experiments on animals.

- Numerous regulations, strictly enforced by local, state, and federal agencies, have been adopted to assure that researchers follow accepted standards for animal experimentation.

REVIEW QUESTIONS

1. Give an example of a simple structure/function relationship in animal physiology and describe how structure relates to function in this system.
2. What evolutionary advantage is conferred on an animal that can successfully maintain homeostasis?
3. Compare and contrast negative and positive feedback, giving original examples of each. Explain why negative rather than positive feedback is required for maintenance of homeostasis.

4. Go to your library and use its electronic data base to search for the word *homeostasis* among the catalogued books and articles. Do the same with the World Wide Web and compare their effectiveness in finding this type of information.
5. Distinguish between the concepts of animal welfare and animal rights. Ask your professor about the composition and function of the Laboratory Animal Care Committee at your college or university.

SUGGESTED READINGS

Adolph, E. F. 1968. *Origins of Physiological Regulations.* New York: Academic Press. (The many innovative ideas in this classic work form a bridge between the original work of Cannon and that of modern physiologists and engineers working in control systems theory and practice.) Classic Work

Benison, S. A., A. C. Barger, and E. L. Wolfe. 1987. *Walter B. Cannon: The Life and Times of a Young Scientist.* Cambridge: Harvard University Press. (An intriguing, insightful biography about the distinguished scientist who introduced the concept of homeostasis in 1929.)

Dworkin, B. R. 1993. *Learning and Physiological Regulation.* Chicago: University of Chicago Press. (A very thorough treatment of the theory and mechanisms underlying physiological regulation and behaviors.)

Futuyma, D. J. 1997. *Evolutionary Biology.* 3d ed. Sunderland, Mass.: Sinauer Associates. (One of several comprehensive undergraduate textbooks that introduce the basic concepts of evolutionary biology as they apply to physiological process.)

Garland, T., Jr., and P. A. Carter. 1994. Evolutionary physiology. *Annu. Rev. Physiol.* 56:579–621 (A review that lays out many of the fundamental concepts in this emerging physiological discipline).

Knobil, E. 1999. The wisdom of the body revisited. *News Physiol. Sci.* 14:1–11. (Reviews the concept of homeostasis, using endocrinological examples to describe how sharply pulsatile physiological processes are regulated.)

Contemporary Experimental Methods for Exploring Physiology

Our knowledge of animal physiology is based on information derived from experimentation. We gain knowledge by gathering information about key variables (e.g., metabolic rate, blood flow, urine production, muscle contraction) in an animal, or in its cells or tissues, while the animal is in a known state such as resting, exercising, digesting, or sleeping. Experiments on living animals are challenging and have inspired the invention of techniques and measuring devices that have passed the test of time to become fundamental tools of the working physiologist. These tools include pressure transducers to measure blood or other pressures, catheter implantation to draw blood or inject samples, respirometers for determining metabolic rates based on the rate at which O_2 is consumed or CO_2 is produced, and numerous others. Details of many of these now classic techniques will be introduced later in this book, where the description of a technique will often yield insights into the nature and use of the information that is acquired.

Historically, techniques for exploring physiological problems at the level of the whole animal were developed first. Techniques for experimenting at the cellular and molecular levels came later. Conceptually, however, we generally operate in the reverse order, starting at the molecular level, then moving successively upward to the whole-animal level, as outlined in Figure 1-1. In this chapter, we focus on a few of the many molecular and cellular techniques that form part of the physiologist's toolbox, briefly describing them and illustrating their use in physiological research. Much of the information presented in other chapters of this book is based on experimental results obtained with these techniques.

FORMULATING AND TESTING HYPOTHESES

Scientists use experimental data to create general laws of physiology. Some of these laws are centuries old; others are still emerging. There are formal laws bearing the discoverer's name, and other laws that are known and depended upon but not formally recognized as laws of nature or physiology. An example of such an informal "law" supported by much existing data is that water-breathing animals regulate their acid-base balance by adjusting the amount of bicarbonate ion (HCO_3^-) excreted in exhaled water, while air-breathing animals regulate their acid-base balance by modifying the elimination of CO_2 gas in exhaled air. Such general laws serve as the basis for formulating new hypotheses, which are specific predictions that can be tested by performing further experiments. The following statement is one of many testable hypotheses that could be derived from this general law: *CO_2 is eliminated as HCO_3^- in water-breathing tadpoles but as gaseous CO_2 in air-breathing adult frogs.* Hypotheses are framed as statements rather than as questions. The goal of experimentation is to test the validity of the statement, thus answering the implied question of whether the hypothesis is supported.

Physiological experiments should begin with a well-formed, specific hypothesis that focuses on a particular level of analysis and is amenable to experimental verification. Although a hypothesis such as *Killer whales have a very high cardiac output while in pursuit of seals* may be interesting and in fact true, it is merely an intellectual exercise to suggest this hypothesis unless a feasible experiment can be designed to support or disprove it. The search for ways to test novel hypotheses has been an important stimulus for development of new experimental techniques and measuring instruments. For example, telemetry devices have been devised for gathering data on blood flow in small to medium-sized animals such as ducks, fish, and seals, and are now being modified for use on animals as large as killer whales.

Animal Physiology and the August Krogh Principle

August Krogh was a Danish animal physiologist with extremely broad interests in comparative physiology. Dozens of key research articles bearing his name have served as the basis for whole areas of further experimentation in the area of respiration and gas exchange. Krogh's work in the late 1800s and early 1900s eventually earned him the Nobel Prize in physiology. One reason for Krogh's extraordinary success as a physiologist was his uncanny ability to choose just the right experimental animal with which to test his hypotheses. His view was that for every well-defined physiological problem, there was an animal optimally suited to yield an answer. This idea has come to be known as the **August Krogh principle** (Krebs, 1975), and it remains viable to this day (Burggren, 2000).

In the early 1990s, for example, a group of American and Brazilian animal physiologists interested in the physiological changes that take place during metamorphosis in amphibians were studying relatively small North American frog larvae (tadpoles), but the small size of these animals made extracting their physiological secrets a frustrating challenge. Evoking the August Krogh principle, these physiologists converged in São Paulo, a state in Brazil that is home to the "paradoxical frog" *(Pseudis paradoxus)*. Relatively modest in size as an adult frog, its name derives from the fact that it has the world's largest tadpole stage, which looks like a green golf ball with a tail! The monstrous size of these tadpoles allowed numerous experiments yielding important new data that could not have been acquired using any other frog species.

As another example, animal physiologists interested in cardiac performance in fishes often have a difficult time measuring pressure and flow in and sampling blood from the heart because of its typically inaccessible location in bony fishes (teleosts). Yet the sea raven *(Hemitripterus americanus)*, a deep-water (benthic) marine teleost that is quite unremarkable in most respects (although it is downright ugly!), has an unusually large heart that is much easier to access than in other fishes. By following the August Krogh principle and using the sea raven as the basis for their experiments, comparative cardiovascular physiologists discovered more about heart function in fishes than they would have if they had continued to struggle with the relatively unforgiving anatomy of the trout, salmon, or catfish.

The August Krogh principle applies equally well at the cellular and molecular levels. Studies on nerve function blossomed with the discovery of the conveniently oversized squid giant axon. The single giant chromosome of the maggot recommended this organism for eukaryotic genetic studies. Illustrations of the August Krogh principle abound in this book and throughout modern animal physiology.

Experimental Design and Levels of Biological Organization

When designing an experiment, the first and most important decision a physiologist must make is at which level of biological organization—molecules, cells, organs, or organisms—to analyze the physiological problem. The choice of level determines the methodology (and choice of animal) appropriate for measuring the experimental variables of interest.

No level of analysis is intrinsically more valuable or important than any other. Indeed, the best understanding of animal physiology comes from integrating knowledge about physiological activities at all levels, from molecules to organs. Having said this, we recognize the strong trend in animal physiology (and in the rest of biology) toward **reductionism**, the reduction of complex systems to their constituent parts, followed by detailed analysis of the components rather than the integrated whole. The most revealing experiments at the reductionist level of inquiry are those that allow insights about processes at adjacent organizational levels.

Researchers and students are often fascinated by new and frequently expensive methodologies. We must remember, however, that incisive results can also be obtained with well-designed experiments using relatively simple instruments and techniques. A well-conceived experimental design at any level of organization is a more potent scientific tool than a poorly conceived experiment using cutting-edge equipment.

POPULAR MOLECULAR TECHNIQUES IN PHYSIOLOGY

The past few decades have seen an explosion in the availability of sophisticated techniques for probing molecular events. New methods and refinements emerge constantly. In this section we describe a few of the powerful molecular techniques that are being used to answer questions in animal physiology.

Tracing Molecules with Radioisotopes

Physiological processes can often be unraveled by tracing the movements of molecules within and between cells. For example, we can more easily understand the role of a particular neurohormone in regulating physiological processes if its movements can be traced from its site of synthesis to its site of release and on to its site of action. Many types of experiments that follow the movement of physiologically important molecules employ **radioisotopes**, the relatively unstable radioactive isotopes of the chemical elements. The natural disintegration of radioisotopes is accompanied by release of high-energy particles that can be detected and quantified by appropriate instruments.

A radioisotope of an element can be incorporated *in vitro* or *in vivo*, either directly into a molecule of interest or into a precursor molecule that will be converted by metabolic reactions into the molecule of interest. The resulting radiolabeled molecule has the same biochemical properties as the unlabeled molecule. Particles emitted from the radiolabeled molecule can then be used to detect its presence even at very low concentrations.

In one type of tracing experiment, a radiolabeled metabolite is administered to an animal, an isolated organ, or a population of cells growing in culture, and samples are removed periodically for measurement of particle emission. Two types of instruments are used to detect emitted particles. A **Geiger counter** detects ionization produced in a gas by the emitted energy. A **scintillation counter** detects and counts tiny flashes of light that the energetic particles create as they pass through a specialized "scintillation fluid." The amount of radiation detected by either instrument is related directly to the amount of the radiolabeled molecule present in the sample.

In another type of experiment, the location of radiolabeled molecules within a tissue section is pinpointed by **autoradiography.** In this technique, a thin slice of tissue containing a radioisotope is laid on a photographic emulsion. Over the course of days, particles emitted from the radioisotope expose the photographic emulsion, producing black grains that correspond to the locations of the labeled molecules in the tissue (Figure 2-1). This qualitative record can be quantified by measuring the amount of exposure of the emulsion in a **densitometer** and comparing it with exposures caused by samples of known concentrations. In this way, the actual concentration of a radiolabeled molecule in the tissue can be determined.

While radioactive tracers are still favored when minimal invasiveness and great sensitivity is necessary, obvious hazards and disposal problems attach to the use of radioactive materials. These problems have led to the increasing use of spectroscopic probes, nonradioactive substances that exploit chemoluminescence or fluorescence to reveal the presence of molecules of interest. Using a variety of techniques, spectroscopic probes can be selectively attached to specific molecules or regions of the cell, effectively staining targets of interest with a high degree of specificity.

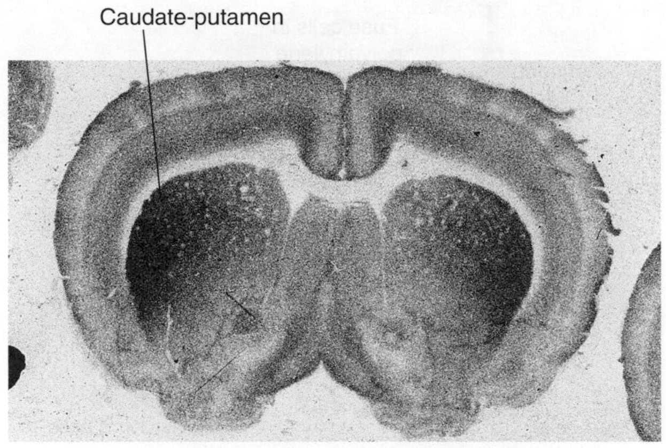

Figure 2-1 Autoradiographs reveal biochemical and structural details that cannot be seen with traditional techniques for tissue fixation and staining. This autoradiograph shows a frontal section through the rat brain after cannabinoid receptors have been bound by a radiolabeled synthetic cannabinoid (closely resembling the active agent in marijuana). The most radioactive areas (that is, the areas with the most cannabinoid receptors) have most heavily exposed the photographic film on which the brain slice was laid, and show up primarily as dark areas in the striatum (caudate-putamen), which mediates motor functions. [Courtesy of Miles Herkenham, NIMH.]

Visualization Techniques Using Antibodies

Examination of a biological structure in a fixed tissue slice on a microscope slide can be daunting, even with selective staining of subcellular structures using dyes. Improved visualization of the structural details of cells is made possible by antibody staining. This remarkable technique permits molecules to be located precisely even when present in concentrations so low that they are difficult to study by any other technique.

Antibody staining generally involves covalently linking a fluorescent dye to an antibody that recognizes a specific determinant on an antigen molecule. (We often think of antigens as disease-causing microbes or invading foreign materials like pollen. However, normal biologically active molecules, such as neurotransmitters and cell growth regulators, can also act as antigens and induce production of specific antibodies when injected into other animals.) The crucial advance that made antibody staining feasible was the development of a method for producing large amounts of antibody selective for a specific target. Antibodies are produced by B lymphocytes (B cells), which proliferate when activated by foreign molecules. Each B cell produces an antibody that is specific for a single antigen, so all the cells in the new population, or **clone,** derived from a single parent B cell synthesize the same antibody, which is thus termed a **monoclonal antibody.**

Isolation and purification of specific antibodies from antiserum is not practical because the antibodies

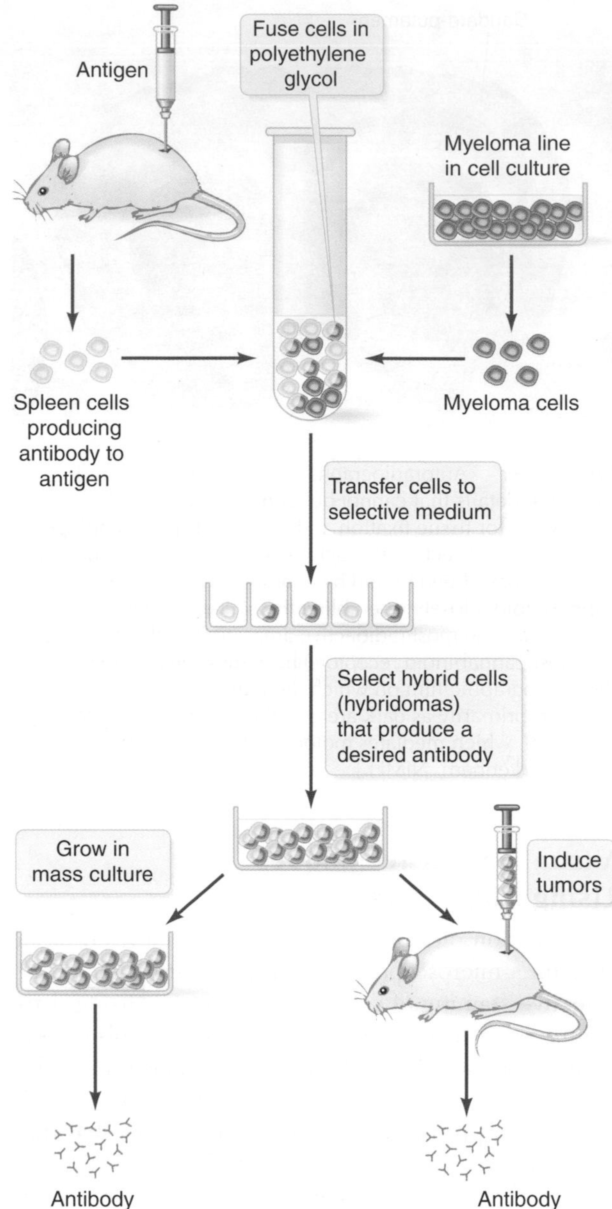

Figure 2-2 Hybridoma cell lines secrete monoclonal antibodies. To prepare monoclonal antibodies, antibody-producing spleen cells are fused with myeloma cells originally derived from B lymphocytes. The hybrid cells, or hybridomas, that secrete antibody specific for the protein of interest are separated out. They can be maintained in culture, where they secrete large quantities of the specific antibody, or they can be injected into a host animal, where they induce the production of the antibody.

are present in only tiny amounts. Another possibility for manufacturing antibodies, the culture of activated B cells, is impractical because cultured B cells normally die within a few days. In the mid-1970s, G. Köhler and C. Milstein discovered that normal B cells raised against a specific antigen could be fused with cancerous lymphocytes, called *myeloma cells,* which grow indefinitely in culture (i.e., they form an "immortal" cell line). The resulting hybrid cells, termed *hybridomas,* are spread out on a solid growth medium in a culture dish, where they proliferate to form clones of identical cells that each secrete a monoclonal antibody. The clones are then screened to identify those that secrete the desired antibody; these self-perpetuating cell lines can be maintained in culture and used to obtain large quantities of homogeneous monoclonal antibody (Figure 2-2). Alternatively, a mixture of antibodies that recognize different sites on the same target molecule can be screened for, resulting in **polyclonal antibody.** Although individual investigators can make and maintain their own hybridoma cell lines, many now choose to have specific monoclonal antibodies prepared by companies specializing in their production. (The next time you are in your university or college library, look at the classified ads in the back of *Science* or *Nature* to see how vigorous the market in custom monoclonal antibodies has become.) The development of monoclonal antibody technology by Köhler and Milstein so revolutionized molecular studies that they received the Nobel Prize for their research.

Once antibodies that recognize discrete sites on a molecule of interest have been produced, they can be linked to a fluorescent dye and injected into the cells or tissues under study, where they bind their target molecules and act as a particularly selective stain. Researchers routinely use a combination of monoclonal and polyclonal antibodies for antibody staining, particularly in immunofluorescent microscopy (Figure 2-3).

As an alternative to fluorescence techniques, radiolabeled monoclonal antibodies can be used in a technique called **radioimmunoassay (RIA).** In this particularly sensitive technique, antibodies are first labeled

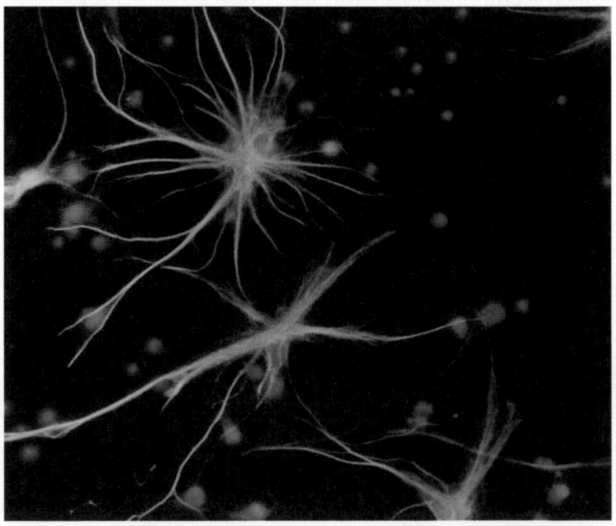

Figure 2-3 Both monoclonal and polyclonal antibodies are frequently used in antibody staining. In this immunofluorescent micrograph of rat spinal cord cultured 10 days, a mouse monoclonal antibody (green) and a rabbit polyclonal antibody (red) that is specific for a single protein, along with a blue fluorescent dye that binds DNA directly, are used. Here we see neurons (red), astrocytes (green), and DNA (blue). [Courtesy of Nancy L. Kedersha/ImmunoGen.]

(a)

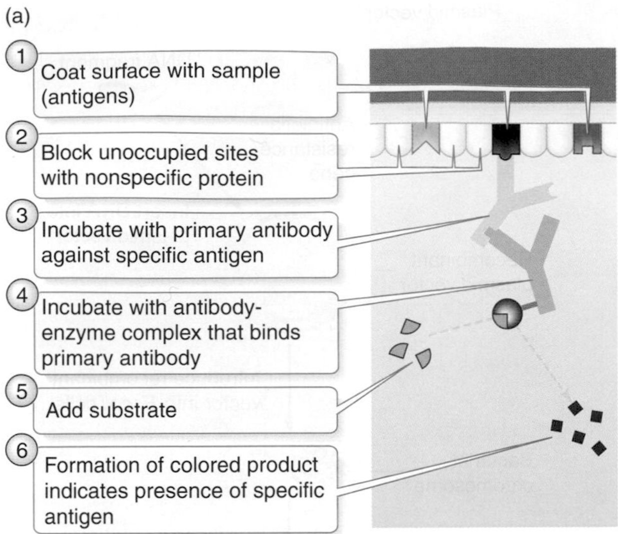

(b)

ELISA assay

Figure 2-4 ELISA (enzyme-linked immunosorbent assay) can be used to identify and quantify the presence of proteins. **(a)** ELISA comprises 6 steps, beginning with adsorption of the protein of interest onto a polystyrene surface. After blocking unoccupied sites with nonspecific proteins, the protein of interest is bound to a primary antibody. A second antibody-enzyme complex is then linked to the primary antibody. The enzyme catalyzes a colorimetric reaction when exposed to the proper substrate, with the intensity of the color produced proportional to the initial concentration of the protein of interest. **(b)** Typically, the surface on which proteins are adsorbed consists of a polystyrene plate with up to 96 embedded wells, each of which can hold a specific sample to be tested. In this ELISA test for herpes simplex virus, the intensity of the yellow staining is indicative of the blood concentration of antibody to the virus. [From Nelson and Cox, 2000.]

with an appropriate radioisotope and then injected into an animal, eventually finding their way to the tissue containing the suspected antigen. The location and concentration of the resulting antibody-antigen complexes can then be revealed by autoradiography. This approach has been used to localize the hormones epinephrine and norepinephrine within certain cells of the adrenal medulla, as described in Chapter 9. Antibodies can be used not only to track down specific molecules within cells, but also to purify them. A later section of this chapter describes how the binding properties of antibodies can be used to separate molecules from mixtures with extraordinary selectivity.

Identification and quantification of antigens can be achieved using an **enzyme-linked immunosorbent assay (ELISA).** A sample thought to contain a protein of interest is adsorbed onto an inert surface, typically a polystyrene plate covered with small wells into which different samples are inserted (Figure 2-4). The plate is then treated with a nonspecific protein to block further protein binding to its surface. Next it is washed with a solution containing a primary antibody against the antigen of interest, which will bind to the antigen if it is present. The surface of the plate is then incubated with an antibody-enzyme complex that will bind to the primary antibody. The enzyme used to create this antibody-enzyme complex is one that catalyzes a colorimetric reaction in the presence of the proper substrate. In the final step, the plate is exposed to the substrate to develop the colorimetric reaction. The intensity of the color that results is proportional to the concentration of the initial protein absorbed onto the polystyrene plate. ELISAs, which are relatively inexpensive, are typically used for rapid mass screening of samples. For example, by initially coating the polystyrene surface with herpes simplex virus (HSV) antigen, then washing the plate with a sample of human blood, antibodies in the blood will bind with the antigen, and the presence of HSV in the blood sample can thus be detected. ELISAs are also used in pregnancy testing to detect human chorionic gonadotropin, whose presence signifies pregnancy.

Immunoblot assays operate similarly to ELISAs, but the proteins of interest are first separated into distinct bands by gel electrophoresis (a technique described in detail in a later section of this chapter). The separated proteins are then transferred to a nitrocellulose membrane. The protein bands on the membrane are then subjected in sequence to primary antibody, secondary antibody linked to an enzyme, and then substrate for the colorimetric reaction that the enzyme catalyzes, just as in ELISA. The formation of colored precipitate signals the presence and approximate molecular weight of specific proteins.

Genetic Engineering

Genetic engineering encompasses various techniques for manipulating the genetic material of an organism. This approach is increasingly used in both agriculture and medicine, and it offers considerable promise for investigators in animal physiology. The techniques of genetic engineering make it possible to produce large quantities of biologically important molecules (e.g., hormones) that are normally present at very low concentrations and to create animals with mutations that affect specific physiological processes.

Genetic engineering begins with identification of the structural gene that codes for a specific protein within the DNA isolated from an organism of interest. The gene that encodes human insulin, for example, can be identified in DNA isolated from human cells. The section of DNA containing the insulin gene can be snipped from the long human DNA strands and inserted into a cloning vector (either a retrovirus or a DNA element). The vector is used to deliver the insulin gene into a host cell, where it can be reproduced using the host cell's genetic machinery. Insertion of a fragment of foreign DNA (e.g., the human insulin gene) into a cloning vector yields **recombinant DNA,** a DNA molecule containing DNA from two or more different sources.

Bacterial plasmids, a common type of cloning vector, are extrachromosomal circular DNA molecules that replicate themselves within bacterial cells. Under certain conditions, a recombinant plasmid containing a gene of interest is taken up by a host cell, typically *E. coli,* in a process called transformation (Figure 2-5). Within a transformed cell, the incorporated plasmid can replicate. As the cell divides, a group of identical cells, or clone, develops. Each cell in the clone contains at least one plasmid with the gene of interest. This general genetic engineering procedure, called DNA or gene cloning, can be used to obtain a DNA "library" consisting of multiple bacterial clones, each of which contains a specific gene from humans or other species. Several variations of DNA cloning are used depending on the size and number of genes in the organism being studied. Increasingly, bacterial artificial chromosomes (BACs) and yeast artificial chromosomes (YACs), as well as retroviral vectors, are being used for cloning in animal cells.

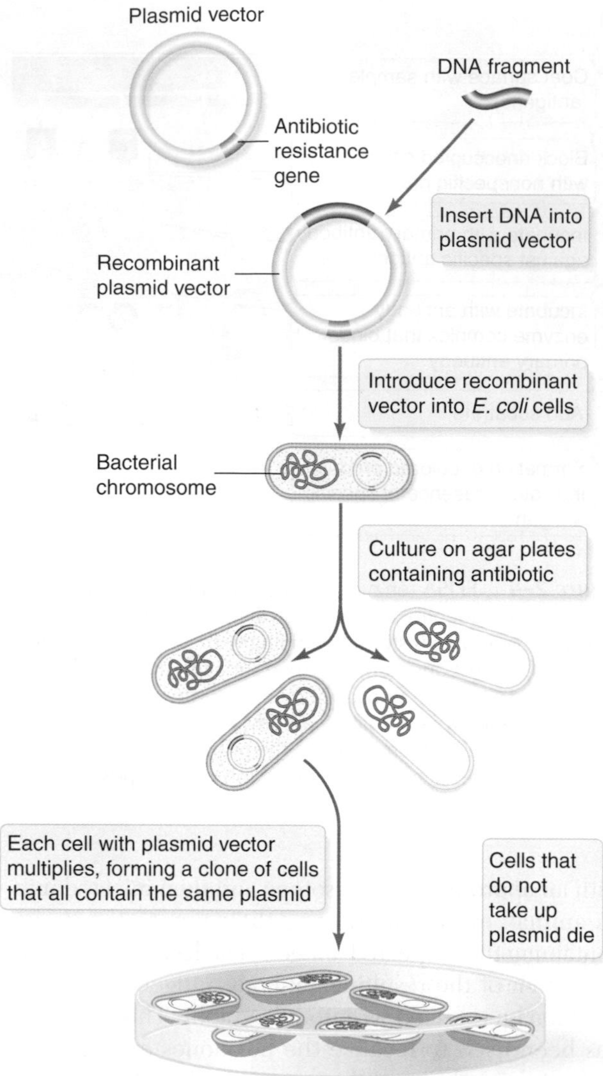

Figure 2-5 DNA cloning can be used to isolate and maintain individual genes. In the cloning procedure illustrated, the specific DNA fragment to be cloned is inserted into a plasmid vector, which also contains a gene conferring resistance to the antibiotic ampicillin. *E. coli* cells are then induced to take up a plasmid, which can replicate within the cells. If the cells are placed in a medium containing ampicillin, only those that have taken up the vector will grow. As each selected cell multiplies, it eventually forms a clone, a colony of cells all possessing the same genetic endowment, including the recombinant plasmid.

Clonal populations for medicine and research

Recombinant DNA has been used in a variety of settings to produce substances—usually proteins—used in treating human diseases. In the past, hormones needed for treating humans with endocrine disorders were extracted from the tissues of other mammals, such as cows and pigs. Because hormones are present in such low concentrations, this was a time-consuming and expensive process. Moreover, hormones isolated from other mammalian species often induce an immune response in humans. Producing these hormones with genetically engineered bacteria has proved to be a safer and less expensive alternative. Under appropriate environmental conditions, the recombinant DNA in an "engineered" bacterial clone is transcribed into messenger RNA, which is used to direct the synthesis of the encoded protein. Some biotech companies now ferment huge vats of *E. coli* cells carrying recombinant DNA containing the gene for human insulin or other hormones; after the bacterial cells are harvested, large quantities of the human hormone can be isolated relatively easily. Recombinant DNA technology has also been used to engineer sheep and cattle that produce human proteins that are subsequently purified quite easily from their milk.

Recombinant DNA technology is also a powerful tool in basic research on human genetic disorders. By isolating and studying genes associated with hereditary diseases, scientists can determine the molecular basis of these diseases. This strategy will certainly lead to better methods of controlling or even curing them. During the past decade, publicly and privately funded laboratories worldwide have engaged in a massive project that has mapped the locations of all human genes on the long strands of DNA in our chromosomes and determined their nucleotide sequences. This Human Genome Project is now providing invaluable data for researchers studying genetic diseases.

DNA cloning and recombinant DNA technology also form the basis for gene therapy. In this approach to treating genetic disorders, the normal form of the gene that is missing or defective is introduced into patients. For example, persons with cystic fibrosis have a defective *CFTR* gene and thus cannot produce the normal protein encoded by this gene. One result of this defect is production of very thick mucus in the airways of the lungs, which leads to potentially lethal breathing problems. Molecular biologists have engineered common cold viruses containing the normal *CFTR* gene. When patients with cystic fibrosis were infected with an engineered cold virus, the viral particles carried the normal human gene into the patients' lung cells, where it became established. Subsequent synthesis of the normal gene product helped alleviate most of the symptoms of cystic fibrosis in the treated patients. While the therapeutic effects were not permanent, these experiments verified the feasibility of the technique, which can now be refined. Gene therapy has also been used to treat immune diseases (e.g., "bubble boy disease"), and many new genetic therapeutic tools are just around the corner.

As genetic engineering becomes more commonplace, animal physiologists will increasingly use genetically engineered animals to probe basic physiological process. It will be feasible, for example, to generate genetically engineered freshwater trout carrying modified genes that code for the branchial ion-pumping proteins of a salmon. This technique would help tease apart the role of branchial versus renal mechanisms for maintaining electrolytic balance in fish migrating between freshwater and salt water.

Considerable controversy surrounds the use of genetic engineering in producing biochemical products, primarily because of fears that genetically modified microorganisms might escape into the environment and produce unexpected effects, such as human or domestic crop diseases. However, many genetically engineered microorganisms are modified to make them unable to live outside of the chemical factory for which they were designed. Assuming you could modify anything about a bacterium's physiology or biochemistry (e.g., the temperature range it tolerates or the chemicals it uses for metabolic substrates), how would you go about ensuring that a genetically engineered bacterium could not flourish in the natural environment if it escaped?

"Made-to-order" mutants

Mutations are permanent changes in the nucleotide sequence of DNA. Mutations can occur spontaneously or be induced experimentally. At the time of cell division, mutations are duplicated and passed on to daughter cells. Mutations in germ cells (sperm and eggs) are passed on to all cells in the resulting offspring, and may be propagated indefinitely in future generations. Many mutations produce abnormal effects only in the homozygous state (i.e., when an individual receives a mutated form of a gene from both parents). Even when a mutation causes a lethal condition, it can be preserved in parents who are heterozygous for the mutation, carrying one normal and one mutated form of the gene. When a pair of these parents reproduces, some of their offspring will be homozygous for the mutation and show its abnormal effects. Thus, heterozygous parents are a "living gene library" of these mutations.

The specific disruption in a physiological process resulting from a single mutant gene can pinpoint the functions controlled by that gene, information that may not be revealed by conventional physiological techniques. For example, cardiovascular physiologists are producing and analyzing the effects of mutations in zebrafish to understand heart development. In research described by J.-N. Chen and M. Fishman (1997), dozens of specific cardiovascular mutations have been produced and examined. The process starts when adult zebrafish are exposed to powerful **mutagens**—compounds that produce permanent mutations in the germ cell line. Subsequent matings of these adults and then of their offspring yield zebrafish with a large variety of mutations, many of which prove lethal. Very rarely, an embryo will appear with just one specific mutation in a structure or process of interest to the investigator. For instance, zebrafish embryonic mutants have been developed in which the heart has abnormally thin ventricular walls. Other mutants have a constriction of the arterial outflow tract of the heart. Both of these conditions mimic human disease states and serve as models for studying the physiological underpinnings of cardiovascular disease.

Transgenic animals

Transgenic animals are another type of genetically engineered organism with the potential to make great contributions to physiology. A **transgenic** animal is one whose genetic constitution has been experimentally altered by the addition or substitution of genes from other animals of the same or other species. Transgenic animals (especially mice) are at the forefront of the menagerie of animal models that are helping researchers understand basic physiological processes and the disease states that results from their dysfunction.

Numerous techniques have been employed to produce transgenic animals. In one method, foreign DNA containing a gene of interest, called a *transgene,* is injected into the pronucleus of a fertilized egg, which is then implanted into a pseudopregnant female. The success rate is low, but in some cases the transgene is incorporated into the chromosomal DNA of the developing embryo, leading to offspring that carry the transgene in all their germ-line cells and somatic cells (Figure 2-6). Animals expressing the transgene are then mated to produce a transgenic line. This approach is used to add functional genes, either extra copies of a gene already present in the animal (leading to overexpression of the gene product) or a gene not normally present (leading to expression of a new product in that animal). Subsequent analysis of the morphology and physiology of the transgenic animals can provide considerable insight into the link between genetic endowment and physiology. Transgenic manipulations by direct, "surgical" intervention into gene structure is being made possible by the continuing development of laser-based "tweezers" and "scissors." Using finely focused laser beams, chromosomes can be held in place while genes are removed and inserted.

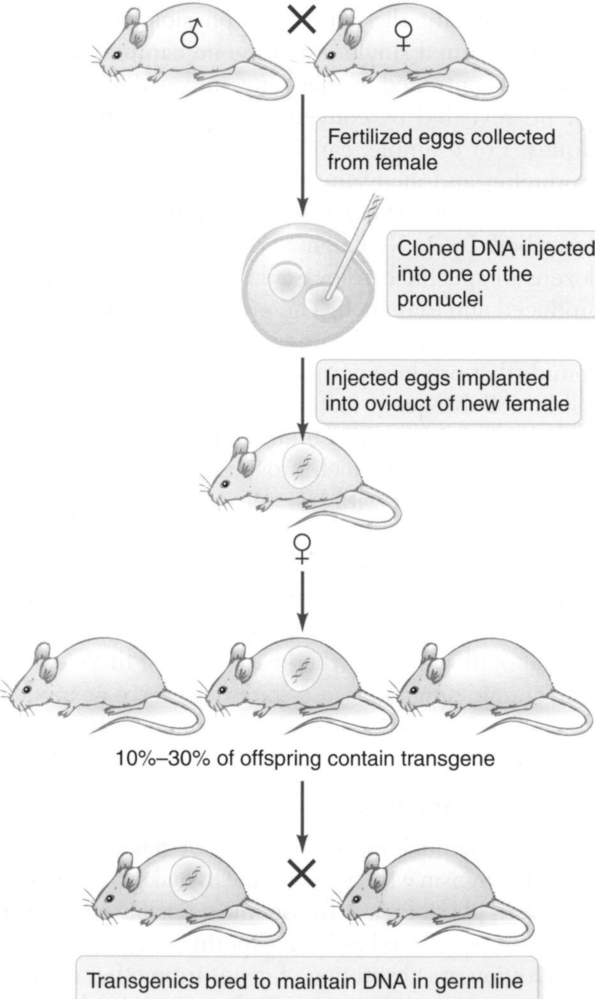

Figure 2-6 A transgenic animal is produced by adding or substituting genes from another organism. To introduce a transgene into mice, cloned foreign DNA is injected into the pronucleus of a fertilized egg, which is then implanted in a female. Offspring from this procedure will retain the transgene, which can be maintained in the germ line by selective breeding.

Transgenic animals characterized by underexpression or complete lack of expression of a particular gene can be equally informative. Functional genes can be replaced by defective ones, thereby producing so-called knockout animals (typically mice, although other knockouts are being produced). These animals cannot express the protein originally encoded by the replaced gene and thus lack the functions mediated by the missing protein. The molecular and genetic basis of physiological processes can be determined by examining the effects of such functional ablation of genes. Knockout mice are used extensively to unravel human physiological processes because human and mouse genes are more than 98% identical. To cite a couple of examples, researchers are investigating the normal genes that regulate early heart development in the embryo and the oncogenes responsible for some types of cancer in studies with knockout mice.

CELLULAR TECHNIQUES

Understanding cells and cellular behavior is a goal of many experiments in physiology. With a knowledge of cellular behavior and communication, we can begin to understand how communities of cells function as tissues and how groups of tissues function as organs. Physiological analysis at the cellular level has been pursued most vigorously using several standard techniques. In this section we discuss three very common and productive cellular techniques: recording with microelectrodes, microscopy, and cell culture.

Uses of Microelectrodes and Micropipettes

Many experiments in cellular physiology make use of micropipettes or various types of microelectrodes. These tiny glass "needles" can be inserted into tissues or even individual cells, where they can serve as sensors or injectors. The technology used to make them is decades old. A region in the middle of a glass capillary tube is heated to the point of melting, and the ends of the tube are then carefully pulled apart, which draws the soft spot in the glass down to an invisibly small diameter before it breaks and separates. The result is two micropipettes, each with a drawn-down tip that may be smaller than a micron in diameter. When a micropipette is filled with a KCl solution or some other appropriate electrolyte, it can function as a **microelectrode.** Typically, a micropipette or microelectrode is mounted in a micromanipulator, a mechanical device that holds the pipette steady and allows its tip to be moved by tiny increments in three different planes.

Measuring electrical properties

Since nerve cells communicate via electrical signals, microelectrodes can be used to "eavesdrop" on their communication by measuring changes in the electric charges across the cell membrane under different conditions. The microelectrodes used to measure the electric potential (voltage) across the cell membrane cause virtually no flow of current from the cell into the electrode. Thus, little or no disruption of the nerve cell occurs even as its communication with neighboring cells is being detected.

A microelectrode for recording electrical signals from neurons or muscle cells is made by filling a micropipette with an ionic conducting solution (typically KCl) and connecting it to an appropriate amplifier. The tip of the electrode is then pushed through the cell membrane into the cytoplasm. A second electrode connected to the amplifier is placed in the fluid surrounding the cell, completing an electric circuit whose properties (voltage, current flow) can be measured.

Microelectrode recording techniques were introduced in the 1930s and 1940s, and refinements have been made ever since. A revolutionary advance in microelectrode recording methodology was the development of **patch clamping** (Figure 2-7). With this technique, the behavior of a single protein molecule constituting an ion channel can be recorded *in situ* (Latin for "in its normal place"). Patch clamping lies at the heart of the recent explosion of knowledge about membranes, including their channels and how they regulate the movement of materials (see Chapters 4–7).

Measuring ion and gas concentrations

Specially constructed microelectrodes can be used to measure intracellular concentrations of common inorganic ions, including H^+, Na^+, K^+, Cl^-, Ca^{2+}, and Mg^{2+}. Cells use movements of ions across their membranes to communicate and perform work, and the magnitude, direction, and time course of these ion movements provide important information about cellular processes.

The tip of a microelectrode for measuring the concentration of a particular ion (e.g., Na^+) is plugged with an ion-exchange resin that is permeable only to that ion. The remainder of the electrode (the "barrel") is filled with a known concentration of the same ion. The electric potential measured by the microelectrode when no current flows reflects the ratio of the ion concentrations on the two sides of the ion-exchange barrier in the tip. Proton-selective microelectrodes are particularly useful for measuring the pH of blood and other body fluids. Microelectrodes also exist that can measure the partial pressure of gases (e.g., O_2 and CO_2) dissolved in very small volumes of fluid.

Measuring pressure

Microelectrodes can be used to measure mechanical fluid pressures within individual cells and microscopic

(a)

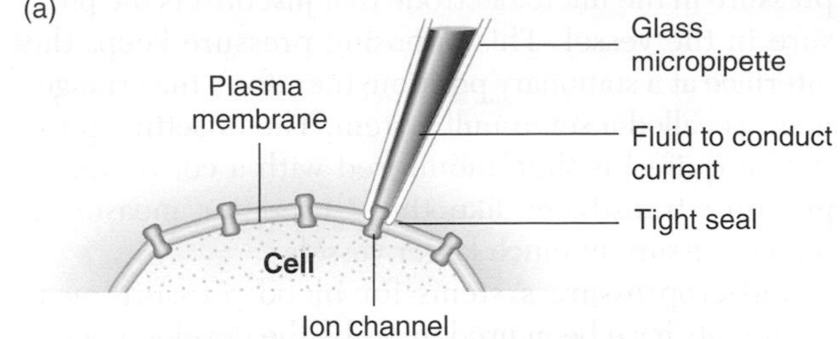

(b)

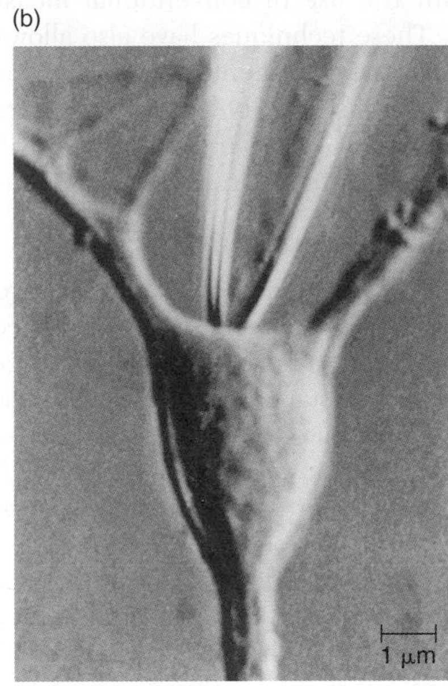

Figure 2-7 Patch-clamp recording permits determination of ion movement across a small patch of membrane containing transmembrane ion channels. **(a)** Diagram of a patch clamp in place. When a fire-polished microelectrode is placed against the cell surface, a high-resistance seal forms between the electrode tip and the membrane. This seal allows direct measurement of the membrane features beneath the tip. Typically, only a few transmembrane ion channels lie beneath the tip, allowing the current flow through them to be measured directly. **(b)** Photomicrograph showing the tip of a patch micropipette applied to the cell body of a nerve cell. The tip has a diameter of about 0.5 μm. [Part b from Sakmann, 1992.]

blood vessels—indeed, in any fluid-filled space into which the tip of a microelectrode can be inserted. To understand the principle of such micropressure systems, let's consider a small blood vessel. A microelectrode filled with a 0.5 M NaCl solution and mounted in a micromanipulator is inserted into the vessel. The higher pressure inside the vessel causes the interface between the blood plasma and the solution filling the electrode to move into the electrode. This results in increased resistance across the electrode tip, because the resistance of plasma is higher than that of the NaCl solution. The change in resistance, which is easily measured, is proportional to the associated change in pressure. Extending this technique, a motor-driven pump added to the micropressure system can produce a pressure in the microelectrode that just offsets the pressure in the vessel. This opposing pressure keeps the interface at a stationary position; therefore, the arrangement is called a servo-null system. The offsetting pressure required is then monitored with a conventional pressure transducer like those used for measuring blood pressure in much larger vessels.

Micropressure systems for blood pressure measurements have been used to probe the development of cardiovascular function in embryos and larvae, animals too small for the use of conventional measurement techniques. These techniques have also allowed direct cardiovascular measurements in adults of very small animals such as insects. Hydrostatic pressures in the tubules of kidneys have also been measured with micropressure systems, as have intracellular pressures.

Microinjecting materials into cells

In addition to serving as microelectrodes, micropipettes can also be used to inject into individual cells substances that produce a measurable change in cell or tissue function. For instance, drugs that influence blood pressure and heart rate can be injected into very small blood vessels (e.g., those lining the shell of a bird egg) or into the beating heart of a moth larva. Microinjection can also be used to introduce macromolecules directly into cells. One might use this technique, for example, to induce osmotic shock, overcoming the natural ability of cells to osmoregulate in the face of changes in the extracellular environment.

Alternatively, the injected substance may be a dye used to mark specific cells, helping to reveal cell shape to trace cells as they divide. A classic variation of this technique involves horseradish peroxidase, an enzyme derived from the horseradish plant that forms a colored product from certain colorless substrates. When this enzyme is injected via a micropipette into the extensions (especially axons) of nerve cells, it is taken up and transported back to the cell body. Subsequent injection of the enzyme's substrate generates a colored "trail" between the injection site and the cell body. By this technique, peripheral nerves can be traced back to their origin in the central nervous system, a task that would defy even the most skilled neuroanatomist using more traditional techniques.

Microinjection can also be used to introduce chemoluminescent or fluorescent dyes into cells. In this way, one can trace the changes in intracellular Ca^{2+} or transmembrane voltages during the contraction of individual muscle cells, or the changes in intracellular pH during acidification of the extracellular environment.

Microscopic Techniques

Cellular function is dependent on cellular structure, reaffirming the central theme discussed in Chapter 1. Physiologists commonly use structural analyses at the cellular level to complement physiological measurements in order to discover how animals function. Such analyses depend on various types of microscopy because animal cells are typically about 10–30 μm in diameter, which is well below the smallest particle visible to the human eye.

Light microscopy

Light microscopy, as its name implies, uses the photons of visible or near-visible light to illuminate specially prepared cells. Under optimal conditions, the **resolution**, or resolving power, of a light microscope is a few microns; two objects that are located closer to each other than a microscope's resolution will appear as one. The resolution of light microscopes has reached a practical limit, primarily because it is capped by the wavelength of visible light and the optical properties of glass lenses. Impressive advances continue to be made, however, in how we view, digitally enhance, store, and analyze light microscopic images. Consequently, light microscopy continues to contribute to our growing understanding of the structure of cells and their components.

Cells are typically nearly colorless and translucent, and must be specially treated to reveal their internal structure. Cells removed from a living animal rapidly die, so tissue must be prepared quickly to prevent degradation of cellular constituents. **Fixation** is the addition of a specialized chemical, such as formalin, that kills the cells and immobilizes their constituents, typically by cross-linking amino groups of proteins with covalent bonds. The fixed cells are then treated with dyes or other reagents that stain particular cellular features, allowing visualization of the cells' internal structures.

Fixation and staining of large blocks of tissue is impractical and does not allow visualization of individual cells. Typically, small blocks of tissue are cut into sections, or slices, just 1–10 μm thick using a special slicing device called a microtome. Because most tissue is fragile even when fixed, it is embedded in a stabilizing medium such as wax, plastic, or gelatin for support while it is sectioned. Such media surround and infiltrate the tissue and then harden to make sectioning possible. The tissue sections are then placed on glass slides for staining and subsequent viewing under a microscope (Figure 2-8a,b). In some instances, tissue embedding compromises the structure of the cell or its contents such that they can no longer be stained or labeled with special compounds prior to viewing. This problem can be circumvented by freezing the tissue rather than embedding it; the frozen tissue is rigid enough to be sectioned. Once prepared, the tissue is viewed with a high-power light microscope (Figure 2-8c).

Improvements in staining techniques have kept pace with improvements in analysis of light microscopic images. Many organic dyes originally developed for use in the textile industry have been found to stain only particular cellular constituents. Some of these dyes stain according to electric charge, such as hematoxylin, which marks negatively charged molecules such as DNA,

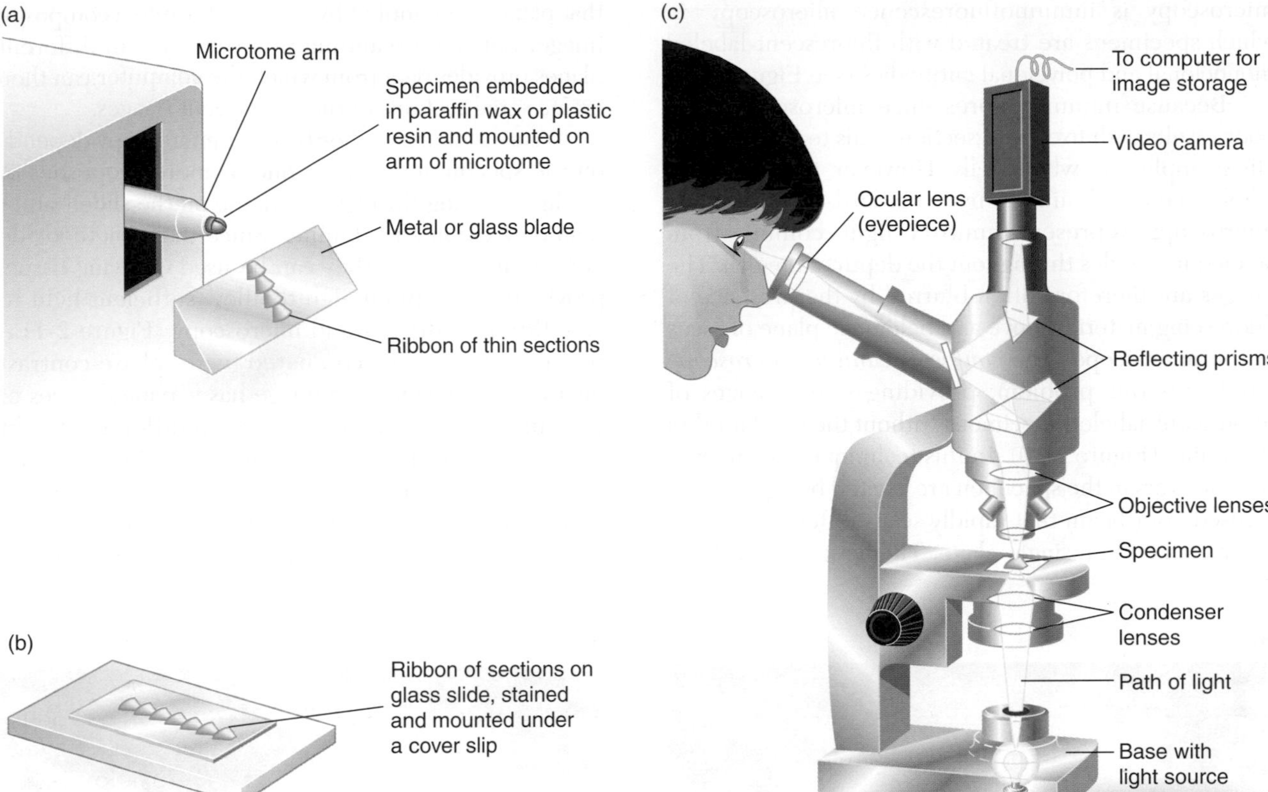

Figure 2-8 Specimens are prepared for light microscopy by cutting them into thin sections and staining. **(a)** Cells and tissue removed from a living organism are first fixed to preserve their structure and then cut into thin sections using a metal or glass knife. **(b)** These sections are mounted on a glass slide, where they can then be stained. **(c)** The compound optical microscope transmits light vertically up through a condenser lens, the specimen on the slide, an objective lens, and finally the ocular lens in the eyepiece, from which the specimen is viewed. [Adapted from Lodish et al., 1995.]

RNA, and acidic proteins. However, the basis of the specificity of many dyes is not known.

Staining with fluorescent-labeled reagents rather than traditional dyes increases the sensitivity of visualization. Fluorescent molecular labels absorb light of one wavelength and emit it at another, longer wavelength. When a specimen treated with a fluorescent reagent is viewed through a fluorescence microscope, only those cells or cellular constituents to which the label is bound are visualized (Figure 2-9). Probably the most common and useful type of fluorescence

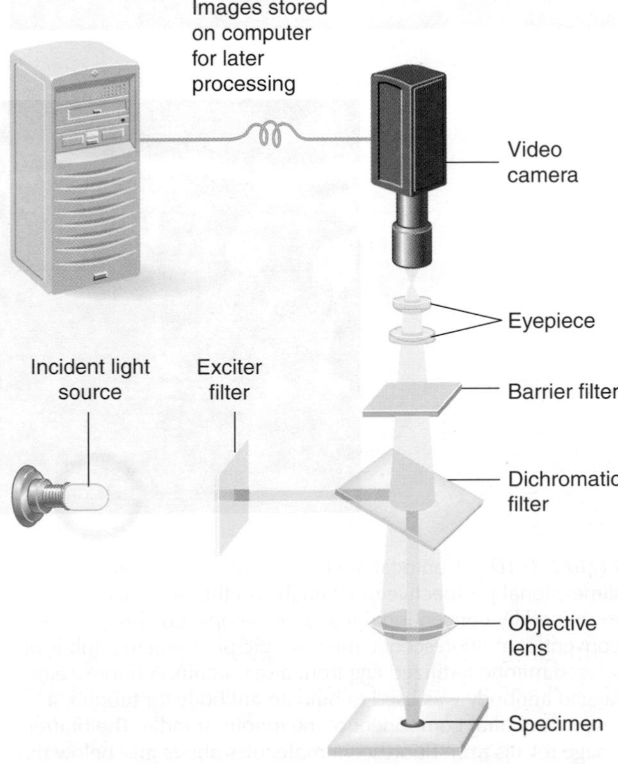

Figure 2-9 A specimen stained with a fluorescent label is viewed through a fluorescence microscope, which produces an image only of structures that bind the label. The incident light source is passed through an exciter filter that passes light within a specific range of wavelengths to provide optimal illumination for the specimen. The incident light is directed toward the specimen by a beam-splitting mirror that reflects light below 510 nm downward but transmits light above 510 nm upward. The fluorescent signals emitted from the labeled specimen pass upward through a barrier filter that removes unwanted fluorescent signals not corresponding to the wavelengths emitted by the label used to stain the specimen. The images are captured with a video or still camera and manipulated on a computer.

microscopy is immunofluorescence microscopy, in which specimens are treated with fluorescent-labeled monoclonal and polyclonal antibodies (see Figure 2-3).

Because immunofluorescence microscopy gives poor results with fixed thin sections, this technique usually is applied to whole cells. However, the images of whole cells obtained by standard fluorescence microscopy represent emitted light coming from labeled molecules throughout the depth of the cell. The images are therefore often blurred by the presence of fluorescing material above and below the plane of focus of the microscope. The *confocal scanning microscope* eliminates this problem, providing sharp images of fluorescent-labeled specimens without the need for thin sectioning (Figure 2-10). In this technique, the fluorescent markers in the specimen are excited by light from a focused laser beam that rapidly scans different areas of the specimen in a single plane. The light emitted from that plane is assembled by a computer into a composite image. Repeated scanning of a specimen in different planes provides data from which the computer can then create serial sections of the fluorescent images.

Visualization by other types of microscopy depends on the specimen changing one or more properties of the light passing through the tissue on the slide, rather than on fixation and staining. Since these methods do not require staining, they can be used on living tissue, provided it is thin enough to allow sufficient light to pass through. Bright-field microscopy (Figure 2-11a) reveals few details compared with phase-contrast microscopy, in which the image has varying degrees of both brightness and contrast due to differential light refraction by different components of the specimen (Figure 2-11b). In Nomarski microscopy, also called differential-interference contrast microscopy, an illuminating beam of plane-polarized light is split into parallel

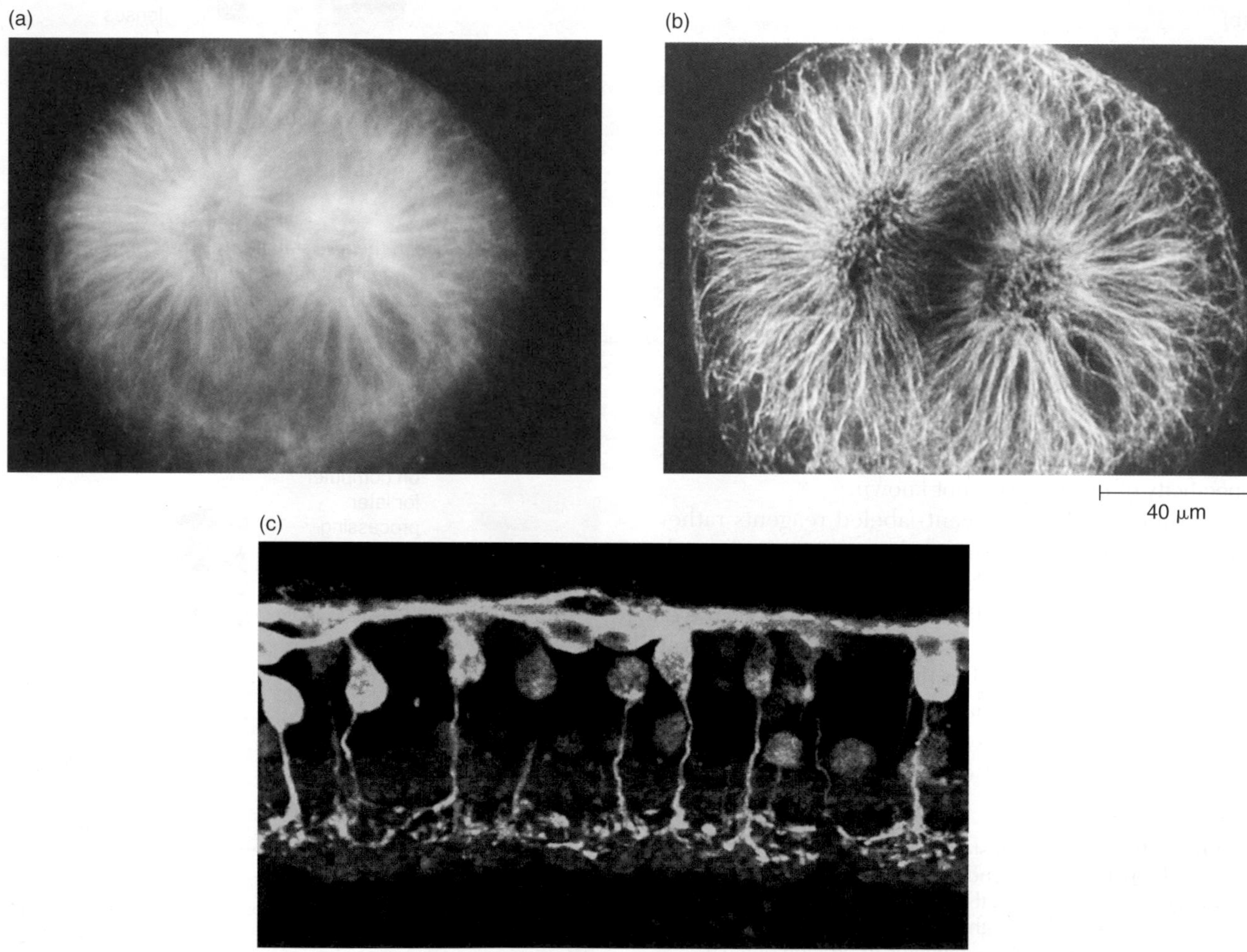

Figure 2-10 Confocal microscopy provides a three-dimensional perspective unattainable in thin sections examined by conventional light microscopy. **(a)** This conventional fluorescence microscopic photomicrograph is of a lysed mitotic fertilized egg from a sea urchin. A fluorescein-tagged antibody was used to bind an antibody for tubulin, a major structural component of the mitotic spindle. The blurred image results from fluorescein molecules above and below the plane of focus. **(b)** Confocal microscopy, which detects fluorescence only from within the plane of focus, produces a much sharper image of the same sea urchin egg. [From White et al., 1987.] **(c)** Computer-assisted generation of confocal images produced this image of immunolabeled bipolar cells in the retina of a ferret. [From Shields, Tran, Wong, and Lukasiewicz, 2000.]

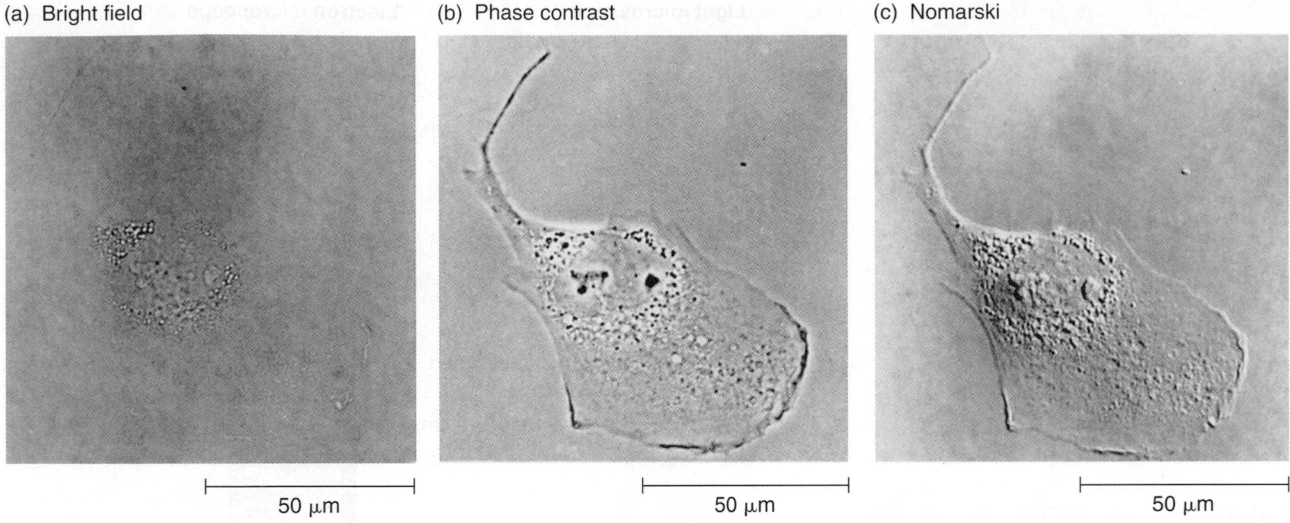

Figure 2-11 Different light microscopic techniques give strikingly different images of the same cell. **(a)** A bright-field image, typical of that obtained with an unstained cell viewed through a compound light microscope, exhibits little contrast and few details. **(b)** A phase-contrast image heightens the visual contrast between different regions of the cell. **(c)** Nomarski (differential-interference contrast) microscopy provides the illusion of depth. [Courtesy of Matthew J. Footer.]

beams before it passes through the tissue specimen and the exiting beams are reassembled into a single image. Slight differences in the refractive index or thickness of adjacent parts of the specimen are converted into degrees of brightness, dark if the light beams are out of phase when they recombine, bright if they are in phase. The final image gives an illusion of depth to the specimen (Figure 2-11c). In dark-field microscopy, light is directed toward the specimen from the side so that the observer sees only light scattered from cellular constituents. The image therefore appears as if the specimen has numerous sources of light within it.

Microscopic images are now routinely stored and manipulated electronically after collection by digital or video cameras. With a digital camera, a color image is collected in its entirety on a two-dimensional array of photosensitive elements. Digital color cameras can capture optical information with a high degree of fidelity, but they often require an intense light source. Gray scale images can be acquired with black-and-white cameras that typically are more sensitive to light. Alternatively, video cameras with lower light requirements can be used to sample the image according to a preset scanning pattern. Cameras that are more sensitive to light permit viewing of cells for longer periods without associated damage caused by photons. Such image intensification is particularly important for viewing live cells that contain fluorescent labels, which can be toxic to cells at high light intensities.

Electron microscopy

The limit of resolution in microscopy is directly related to the wavelength of the illuminating light. That is, the shorter the wavelength of the illumination, the shorter the minimal distance between two distinguishable objects (i.e., the greater the resolution). In electron microscopy, a high-velocity electron beam, rather than visible light, is used for illumination. Because the wavelength of electron beams is much shorter than that of visible light, electron microscopes have much better resolution. Indeed, modern transmission electron microscopes typically have a resolution of 0.5 nm (5 angstroms, Å), whereas light microscopes have a resolution of no less than about 0.2 μm). Because the effective wavelength of an electron beam decreases as its velocity increases, the resolution of an electron microscope depends on the voltage available to accelerate the illuminating electrons.

The transmission electron microscope (TEM) forms images by sending electrons through a specimen and focusing the resulting image on an electron-sensitive fluorescent screen or photographic film (Figure 2-12 on the next page). The electron beam is aligned and focused by magnets, just as light is focused by the condenser lens in a compound light microscope. As the electrons pass through the specimen, some are scattered, and the remainder are focused by the objective lenses onto the viewing screen. Because an electron beam passes almost unimpeded through an unstained sample, little differentiation of cellular components is possible without staining. The most common stains for electron microscopy are salts of heavy metals (e.g., osmium, lead, or uranium), which increase electron scattering. In photographs of an electron microscopic image, components stained with such electron-dense materials appear dark. Since air molecules would deflect the focused electron beam aimed at the sample, the specimen must be held in a vacuum during imaging.

Specimens must be carefully fixed to preserve their biological structure during electron bombardment in

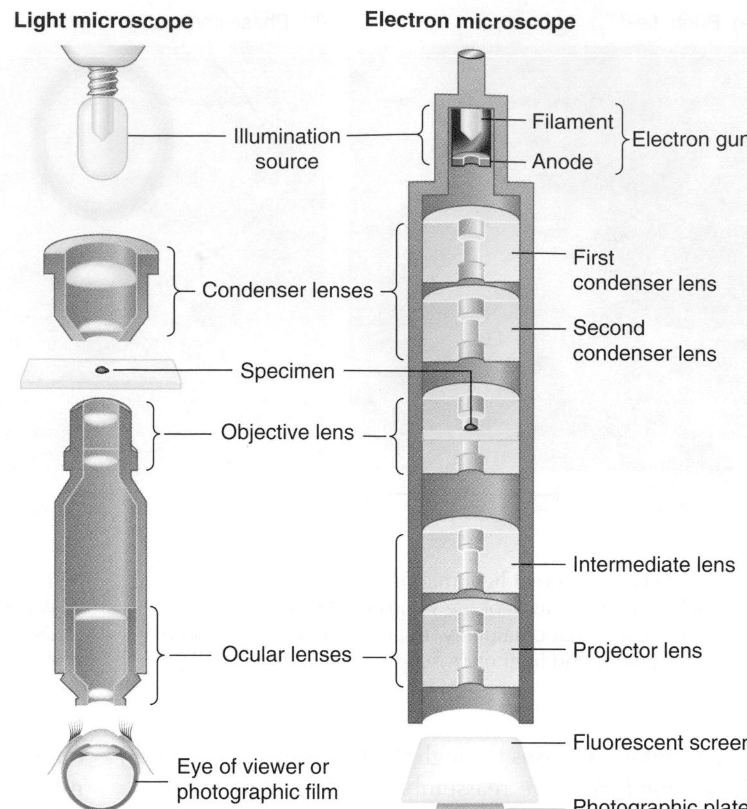

Figure 2-12 Electron microscopes share features such as lenses with compound optical microscopes, but use an electron beam rather than a light beam to illuminate the specimen. In a transmission electron microscope, shown on the far right, an image is formed by blasting electrons through a specimen and then focusing and projecting them onto a fluorescent screen. In a scanning electron microscope, electrons reflected from the surface of a specimen coated with a reflective metal film are collected by lenses and viewed on a cathode ray tube.

the electron microscope. Glutaraldehyde is used to covalently cross-link proteins and osmium tetroxide to stabilize lipid bilayers. After fixing, the specimens are infiltrated with a plastic resin. Specimens must be sectioned into extremely thin slices (50–100 nm thick) to allow penetration by the electron beam. Only diamond or glass knives can be constructed fine enough to cut tissue sections into such thin slices. Glass knives are formed by breaking on the diagonal a 2.5 cm glass square that is about 5 mm thick. The tip of the diagonal becomes the cutting surface. Because glass is actually a slow-moving liquid, the edge formed is only sharp enough to cut tissue for a few hours before molecular flow of glass dulls the edge. Although extremely expensive, diamond knives do not suffer from this problem and thus are the preferred tool for cutting ultrathin sections. Once cut from the resin block, the sections are stained and finally placed on a metal grid in the transmission electron microscope.

The exquisite detail in TEM images can provide important insights into the structure of biological tissue (Figure 2-13a). Unfortunately, the need for ultrathin sectioning means that only very small specimens can be examined. In the past, it was difficult to develop an understanding of the three-dimensional structure of a sample without the truly tedious procedure of manually reconstructing an image from a series of individual sections. Computer technology now facilitates the reconstruction of serial images into three-dimensional structures.

The scanning electron microscope (SEM) collects electrons scattered from the surface of a specially prepared specimen, rather than transmitted directly through it as in the TEM. Before examination in a scanning electron microscope, the specimen is coated with an extraordinarily thin film of a heavy metal such as platinum. The tissue is then dissolved away with acid, leaving a metal replica of the tissue's surface, which is viewed under the microscope. Scanning electron microscopy provides excellent three-dimensional images of the surfaces of cells and tissues, but it cannot reveal features beneath the surface (Figure 2-13b). Resolution of SEM images has a limit of about 10 nm, not as high as the resolution of TEM images.

Cell Culture

The rearing of cells *in vitro* (from the Latin, "in glass") is known as cell culture. This technique has revolutionized our ability to study cells and the physiological processes they support at the tissue and organ levels. Historically, explants (small pieces of tissue from a donor animal) were kept alive and grown in a flask filled with an appropriate mixture of nutrients and other chemicals. Today, the most common procedure is to break up (dissociate) tissues into individual cells and then suspend the cells in a nourishing chemical broth in which they grow and divide as separate entities. For tissues with a multiplicity of cell types, automatic cell sorters can be used to separate cell types before they are separately cultured.

(a)

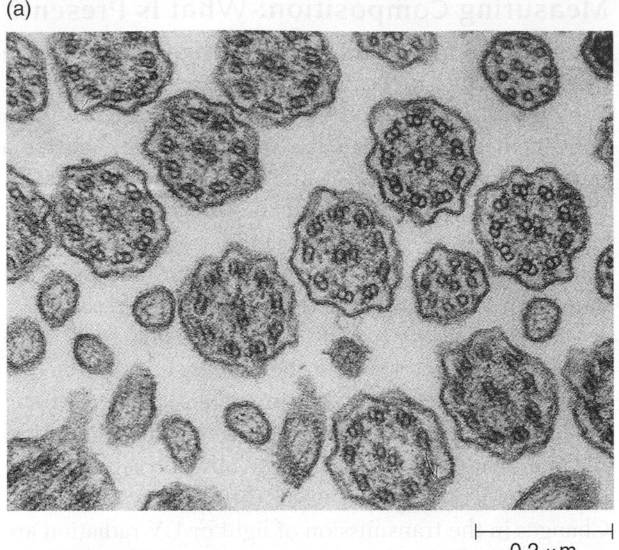

(b)

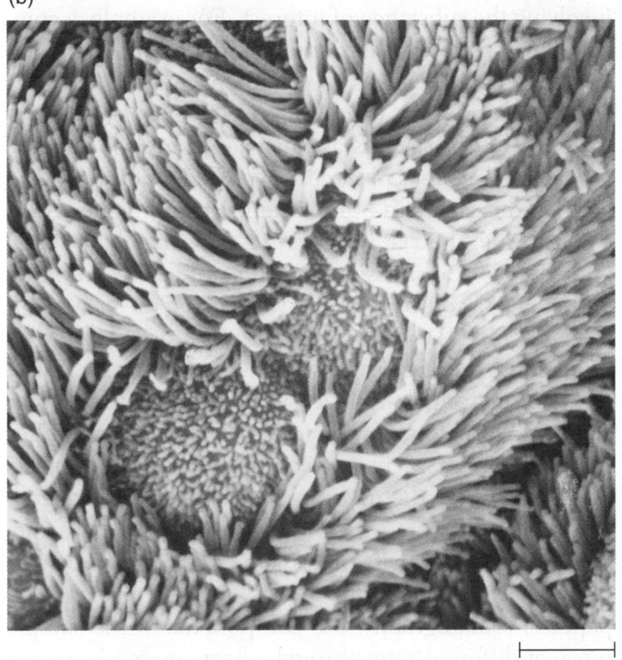

Figure 2-13 Transmission electron microscopy provides a two-dimensional view of the interior of biological tissue, while scanning electron microscopy emphasizes surface features in three dimensions. **(a)** Transmission electron micrograph of cilia in the mouse oviduct. **(b)** Scanning electron micrograph of cilia in the mouse oviduct. [Courtesy of E. R. Dirksen.]

Growing cells successfully *in vitro* requires the right **culture medium**, the liquid in which the cells are suspended. Up until the early 1970s, cells from all animals were routinely grown in liquid media consisting largely of either serum (a clear component of blood plasma) from horses or fetal calves or an unrefined chemical extract made from homogenized chick embryos. These media, however, were poorly defined chemically, containing numerous unidentified compounds. Moreover, luck played too great a part in determining whether selected cells would grow in one of these media and how its composition should be adjusted if the first attempt was unsuccessful. Growing cells *in vitro* was largely a matter of trial and error. Today, defined culture media manufactured according to precise chemical recipes are available for research. However, the successful culture of many cell types still requires the addition of a small amount (less than 5%) of horse or bovine serum to the media, suggesting that some growth factor in blood is necessary for the growth and division of animal cells *in vitro* (Figure 2-14).

Even with the availability of defined culture media, growing animal cells *in vitro* is challenging. Normal animal cells generally grow for only a few days *in vitro,* then stop multiplying and eventually die out. A relatively homogeneous population of such cells is referred to as a cell strain. Cultured cell strains are useful for many kinds of experiments, but their usefulness is limited by their life span. In addition, many types of animal cells have not yet been successfully cultured. However, the list of cells that can be cultured is constantly growing as a result of refinements in media and culture techniques. Cell strains that can now be grown in culture include bone and connective tissue; skeletal, cardiac, and smooth muscle; epithelial tissue from liver, lung, breast, skin, bladder, and kidney; some neural tissue; and some endocrine glands (e.g., adrenal, pituitary, islets of Langerhans in pancreas). The culturing of stem cells derived from fetuses is becoming particularly important because stem cells have the ability to divide and differentiate into more specialized cells and tissues. Cell culture techniques may soon lend themselves to the production of tissues and even whole organs that can be transplanted into human patients.

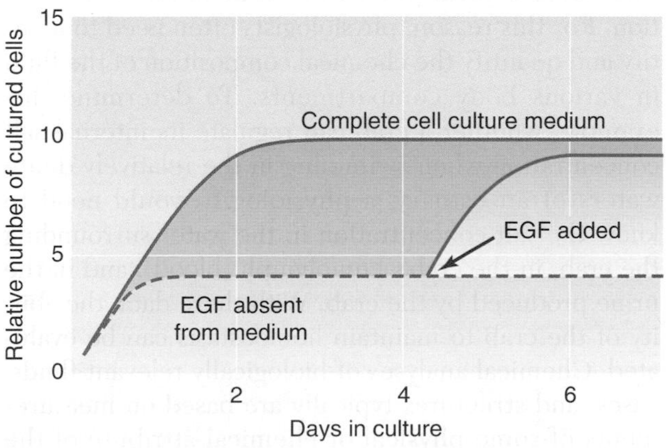

Figure 2-14 Cells grown in culture often require specific factors to stimulate maximal rates of division and growth. In this cell culture, maximal cell numbers are reached only in the presence of epidermal growth factor (EGF). Addition of EGF (arrow) to a culture lacking this substance results in immediate further growth of the cell colony (red line). [Adapted from Lodish et al., 1995.]

In contrast to normal animal cells, cancer cells commonly exhibit rapid, uncontrolled growth in the body and are capable of indefinite growth in culture. Treatment of normal cultured cells with certain agents may cause transformation, a process that makes them behave like cancer cells that have been isolated from tumors. Cells transformed in this way can be cultured indefinitely. Homogeneous populations of such "immortal" cells are termed cell lines. Although normal cells differ from cancer cells and transformed cells in many ways, the culture of immortal cell lines has permitted many types of studies that are not feasible with normal cells.

Cell culture has many uses in animal physiology. New devices, such as silicon wafer sensors that measure acidity and other variables, have been combined with cell culture techniques to provide important insights into cellular and organismal physiology. For example, the hormonal regulation of H^+ secretion from a variety of cell types can be studied by treating cultured cells with stimulating or inhibiting factors and measuring changes in the rate of acidification of the medium. This approach has also been used to study tissues and organs with unusual rates or properties of H^+ secretion, such as the swimbladder tissues of fishes.

Smooth, skeletal, and cardiac muscle cells can all be cultured *in vitro*. Suppose you are observing an individual muscle cell under a microscope as it lies in a culture dish. How could you tell if the cell was still capable of contracting?

BIOCHEMICAL ANALYSIS

Most biochemical processes occur in aqueous solution. For this reason, physiologists often need to identify and quantify the chemical composition of the fluid in various body compartments. To determine, for example, whether a crab can regulate its internal salt concentration when swimming in the relatively dilute waters of an estuary, a physiologist would need to know the salt concentration in the water surrounding the crab, in the crab's hemolymph (blood), and in the urine produced by the crab. With these data, the ability of the crab to maintain homeostasis can be evaluated. Chemical analyses of biologically relevant fluids, gases, and structures typically are based on measurements of some physical or chemical attribute of the materials of interest (e.g., Na^+ in the crab's urine). The substantial increase in the sensitivity and accuracy of such measurements in the recent past has allowed physiologists to probe deeply the nature of subtle physiological functions that previously could not be measured at all.

Measuring Composition: What Is Present?

Numerous time-honored and emerging methods are available for measuring chemical composition. Sometimes, animal physiologists are interested only in knowing whether a particular substance (e.g., ammonia or hemoglobin) is present in a sample. At other times, they may want to identify all the different proteins or carbohydrates or other molecular species present. In other words, the nature of the problem being studied determines which compositional data are relevant.

A wide variety of **colorimetric assays** have been developed for determining the presence or absence of specific substances in a solution. These assays depend on subjecting the substance of interest to a chemical reaction that changes its ability to absorb visible light or ultraviolet (UV) radiation at different wavelengths. Changes in the transmission of light or UV radiation are then detected by a spectrophotometer. Many biochemical assays employ an enzyme that catalyzes a reaction involving the substance of interest. For example, a common assay for lactate (a product of the anaerobic metabolism of glucose) makes use of an enzyme that converts lactate into products with different UV absorption properties. To perform this assay, a solution suspected of containing lactate is placed in a small reaction vial along with the enzyme and other reaction components. After a short time, the vial is placed in a spectrophotometer, and the UV transmission of the solution is compared with that of a control reaction vial lacking the enzyme. A difference in UV transmission between the two vials indicates that lactate is present in the sample.

Chromatography is a widely used technique for separating proteins, nucleic acids, sugars, and other molecules present in a mixture. Often physiologists wish to correlate the presence, if not the actual concentration, of a substance in physiological fluids (urine, blood) with the activity of a tissue or organ; chromatography is a vital tool in this process. In its simplest form, paper chromatography, the components of the sample move at different rates through chromatography paper, depending on their relative solubility in a solvent (Figure 2-15a). In order to visualize the separated components, the chromatogram commonly is sprayed with a colorimetric reagent that stains the components of interest. More complex mixtures can be separated by column chromatography, in which the sample solution is passed through a column packed with a porous matrix of beads (Figure 2-15b). The different components of the sample pass through the column at varying rates, and the resulting fractions are collected in a series of tubes. Assays are then used to determine the presence of specific components in the fractions collected.

Many different kinds of matrices are employed in column chromatography. Matrices are available that sort components according to their charge, size, insolubility in water (hydrophobicity), or binding affinity for the matrix. The last type of matrix is used in affinity

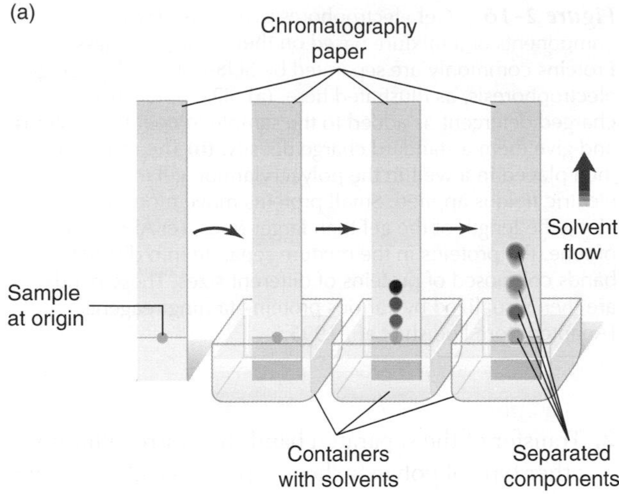

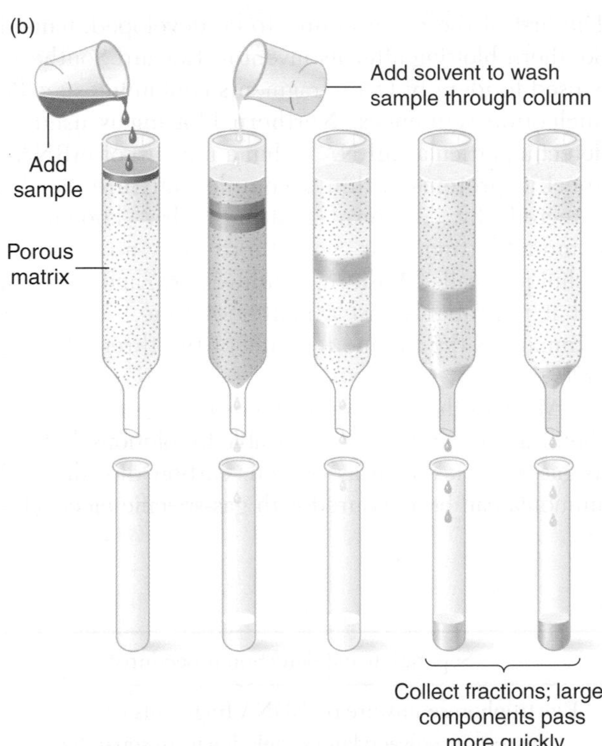

Figure 2-15 Chromatography is a powerful technique for separating the components of a mixture in solution. **(a)** In paper chromatography, the sample is applied to one end of a piece of chromatographic paper and dried. The paper is then placed into a solution containing two or more solvents, which flow upward through the paper via capillary action. Different components of the sample move at different rates in the paper because they have different relative solubilities in the solvent mixture. After several hours, the paper is dried and stained to determine the locations and relative amounts of the separated components. **(b)** In column chromatography, the sample is applied to the top of a column that contains a permeable matrix of beads. A solvent is then pumped slowly through the column and is collected in separate tubes (called fractions) as it emerges from the bottom. Components of the sample travel at different rates through the column and are thus sorted into different fractions.

chromatography, in which the matrix beads are coated with molecules (e.g., antibodies or receptors) that bind to the component of interest. When a sample mixture is applied to the column, all the components pass through except the one recognized and bound by the affinity matrix. Affinity chromatography is a powerful technique for purifying proteins and other biological molecules that are present at very low concentrations.

While the principles of chromatography have been evident to humans since the first hominid observed a mashed berry leaching its color into the underlying sand, refinements continue to be made in both the matrix materials and the columns in which they are housed. **High-performance liquid chromatography (HPLC)** employs high-pressure pumps operating at 1000–4500 psi (70–306 atmospheres) to force a solution through a column of suitable matrix, speeding the movement of the sample down through the column. HPLC yields higher-resolution results because the matrix provides a huge surface area for absorption and separation, and also because the rapid passage of the sample molecules through the column allows less time for them to diffuse from their bands of separation. HPLC can be used to examine not only proteins, but also lipids, fragments of DNA and RNA, and other macromolecules.

Electrophoresis is a general technique for separating molecules in a mixture based on their rate of movement through a porous matrix to which an electric field is applied. The net charge of a molecule, as well as its size and shape, determines its rate of migration during electrophoresis. This technique works well on small molecules such as amino acids and nucleotides, but by far the most common use of electrophoresis is to separate mixtures of proteins or nucleic acids. In this case, the sample is placed at one end of an agarose (sugar) or polyacrylamide gel (polymer), an inert matrix with fixed-diameter pores that impedes the migration of different molecules to varying degrees when an electric field is applied. Protein mixtures usually are exposed before and during electrophoresis to sodium dodecyl sulfate (SDS), a negatively charged detergent that gives each protein the same density of electric charge. This causes the proteins to separate purely on the basis of their size (molecular weight). The rate of migration of SDS-coated proteins through the gel is proportional to their molecular weight; the lower the molecular weight of a protein, the faster it moves through the gel (Figure 2-16 on the next page). When a protein-binding stain is applied to the gel, the separated proteins are visualized as distinct bands.

Three related procedures employing gel electrophoresis are used to separate and detect specific DNA fragments, messenger RNAs (mRNAs), or proteins. Each of these procedures involves three steps:

1. Separation of the sample mixture by gel electrophoresis

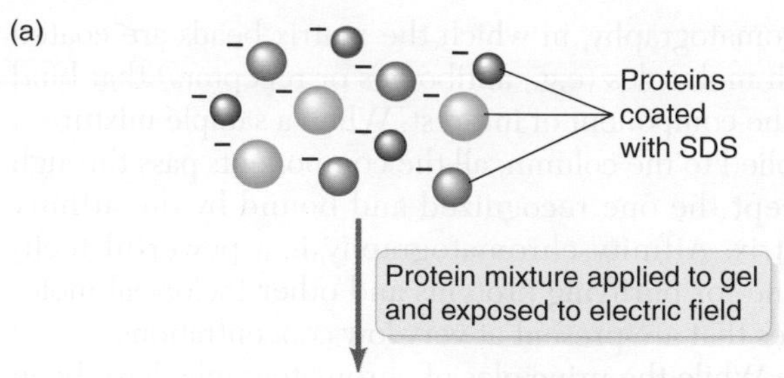

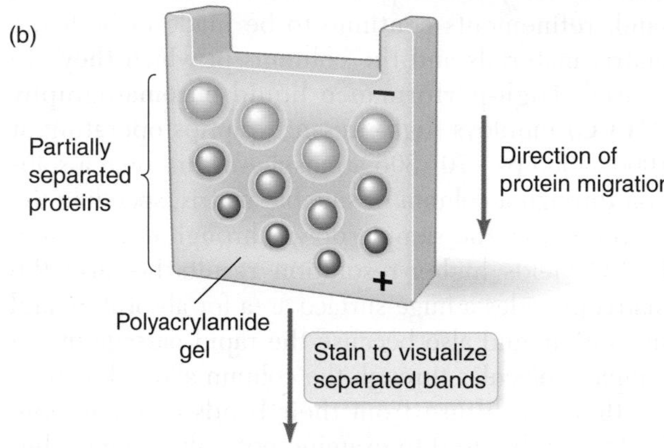

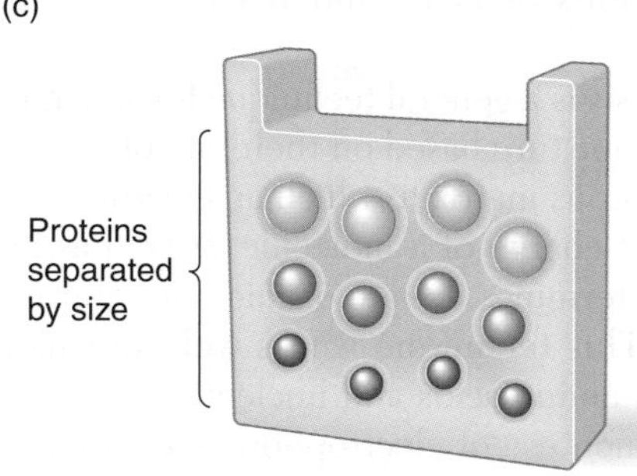

Figure 2-16 Gel electrophoresis separates the components of a mixture based on their charge or mass. Proteins commonly are separated by SDS-polyacrylamide gel electrophoresis, as illustrated here. **(a)** SDS, a negatively charged detergent, is added to the sample to coat the proteins and give them a standard charge density. **(b)** The sample is then placed in a well in the polyacrylamide gel and an electric field is applied. Small proteins move more quickly along the length of the gel than larger ones. **(c)** After a period of time, the proteins in the mixture separate into distinct bands composed of proteins of different sizes. These bands are then visualized by various protein-staining reagents. [Adapted from Lodish et al., 1995.]

2. Transfer of the separated bands to a nitrocellulose or other type of polymer sheet, a process called blotting
3. Treatment of the sheet (or blot) with a probe that reacts specifically with the component of interest

The first of these procedures to be developed, named Southern blotting after its inventor, Edward Southern, is used to identify DNA fragments containing specific nucleotide sequences. Northern blotting is used to detect a particular mRNA within a mixture of mRNAs. Specific proteins within a complex mixture can be detected by Western blotting, also known as immunoblotting. (As yet, there are no blotting techniques called Eastern, Southwestern, and so forth, but it is probably just a matter of time.) Table 2-1 summarizes the unique features of the three blotting procedures.

Many of the common methods of determining biochemical composition are applicable to solutions, but not to gases. Gases such as oxygen, carbon dioxide, and ammonia can be measured with gas-specific electrodes

Table 2-1 **Electrophoretic blotting procedures**

	Molecules detected	Separation and detection procedure*
Southern blotting	DNA fragments produced by cleavage of DNA with restriction enzymes	Electrophorese mixture of dsDNA fragments on agarose or polyacrylamide gel; denature separated fragments into ssDNA and transfer bands to polymer sheet; use radiolabeled ssDNA or RNA to label fragment of interest; detect labeled band with autoradiography.
Northern blotting	Messenger RNAs	Denature sample mixture; electrophorese on polyacrylamide gel and transfer separated bands to polymer sheet; use radiolabeled ssDNA to label mRNA of interest; detect labeled band with autoradiography.
Western blotting	Proteins	Electrophorese sample mixture on SDS-polyacrylamide gel and transfer separated bands to polymer sheet; use radiolabeled monoclonal antibody to label protein of interest; detect labeled band with autoradiography.†

*dsDNA = double-stranded DNA; ssDNA = single-stranded DNA.
†If radiolabeled monoclonal antibody is not available, then the band containing the antibody-protein complexes can be detected by adding a secondary antibody that binds to any monoclonal antibody. This secondary antibody is covalently linked to an enzyme, such as alkaline phosphatase, that catalyzes a colorimetric reaction. When substrate is added, a colored product forms over the band with the protein of interest, generating a visible colored stain in this region of the blot.

that rely upon current-producing reactions at the face of the electrode exposed to the sample gas. Another technique, **mass spectrometry**, can distinguish between the different gases composing a gaseous mixture by determining their mass and charge. Animal physiologists most often use this technique to determine the composition of respiratory gases while an animal is resting or exercising in an experimental setting. Figure 2-17 illustrates the basic design of a mass spectrometer. The gas sample is first ionized by intense heating and passage through an electron beam. The charged ions are then focused and accelerated by an electric field into an analyzer, where the beam of ions is deflected either by an applied magnetic field or by passage through tuned rods emitting specific radio frequencies that deflect ions. The amount of deflection depends on the mass of the ion and its charge. The readings at an array of detectors reveal the presence and quantity of the gases in the original sample.

An increasingly important technique for the measurement of molecules in living cells is nuclear magnetic resonance spectroscopy (NMR), which capitalizes on the quantum mechanical properties of atomic nuclei, in particular ^{1}H, ^{13}C, ^{15}N, ^{19}F, and ^{31}P. The pattern of absorption (absorption spectrum) of a very brief pulse of electromagnetic energy by these atoms in cells, tissues, or even living, intact animals can reveal their atomic concentration. NMR has been used for many purposes, such as describing the levels of inorganic phosphates and the pH within resting and active skeletal muscle. A new twist on this technique, magnetic resonance phase-velocity mapping, uses NMR to produce striking images of the movement of fluids, such as blood flowing through the heart, in real time in intact animals.

Measuring Concentration: How Much Is Present?

Most instruments and analytic techniques used to determine the composition of a fluid or gas mixture also provide data about the concentrations of the components present. For example, the degree of color change produced in a colorimetric assay depends on how much of the substance being measured is present in the sample. Likewise, the output signal from a mass spectrometer depends not only on the types of gases present in a mixture, but on how much of each is present.

Typically, the analytic technique being used to determine the concentration of a particular substance is

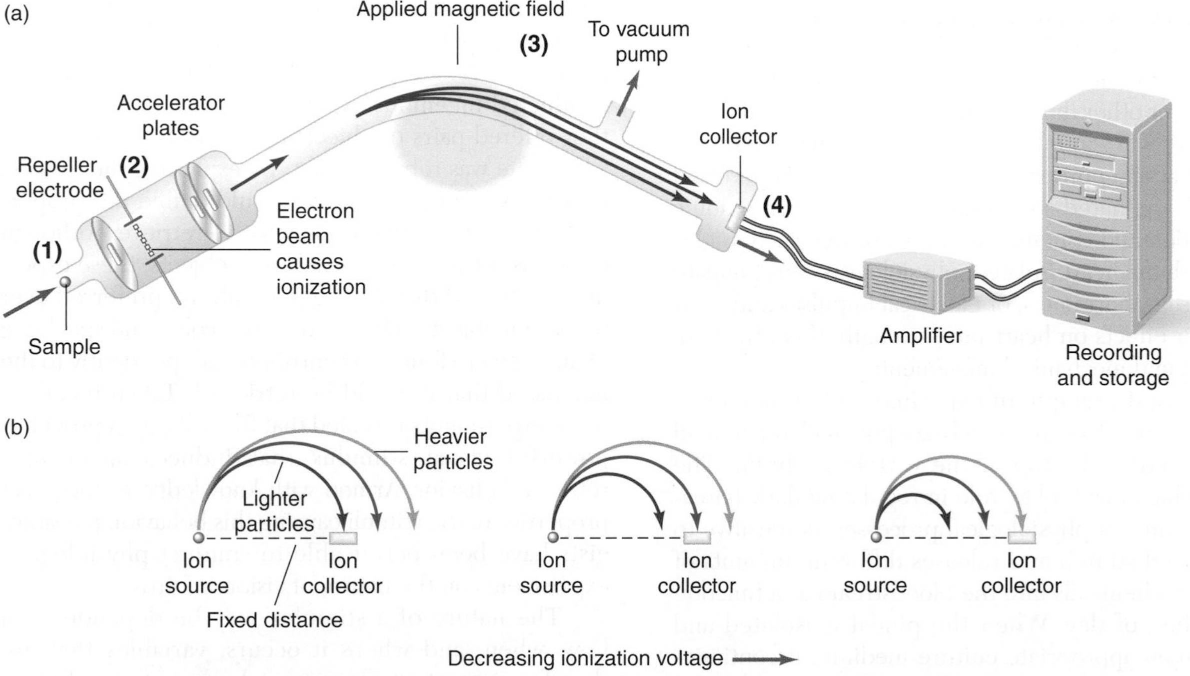

Figure 2-17 The identities and concentrations of gases in a mixture can be determined by gas mass spectrometry. **(a)** The fixed-collector mass spectrometer detects how much of an ionized sample is deflected by an imposed magnetic field. This device has four essential parts. First is a carefully constructed inlet device (1) through which the sample is steadily aspirated into the system. Second is an ionization chamber (2), kept under high vacuum and at high temperature (about 190°C), where the sample passes through an electron beam and is accelerated via application of an electric field. The gas molecules leave this chamber as negatively charged ions. Third is an analyzer tube (3) in which the accelerated beam of ions is subjected to a magnetic field that causes the ions to flow in a curved path. Finally, the ion beam is detected using an ion collector situated at the end of the analyzer tube (4). The extent to which the path of an ion is bent by the applied magnetic field depends on the strength of the field and the ion's mass, charge, and velocity. Only those ionic species whose paths parallel the sides of the analyzer tube will reach the ion collectors and be detected. **(b)** With the magnetic field held constant, particles can be distinguished by changing the strength of the ionization voltage. Heavier particles will collide with the ion collector as the ionizing voltage is lowered. [Adapted from Fessenden and Fessenden, 1982.]

carried out on several samples of the substance at different known concentrations; the measured outputs are then plotted against the known concentrations, yielding a standard calibration curve. The actual concentration of an experimental sample is then determined by comparison with this standard curve.

EXPERIMENTS WITH ISOLATED ORGANS AND ORGAN SYSTEMS

All animals have several different major organ systems that must be coordinated and controlled to maintain homeostasis. As we'll see in later chapters, these organ systems are regulated primarily by neuronal and hormonal inputs. To understand physiological control mechanisms, these key regulatory signals and their sources must be characterized. In many cases, intact organs cannot be studied *in situ*. Instead, experiments are conducted on isolated organs removed from an animal by surgery and maintained in an artificial environment. In this way, important variables such as temperature, oxygen availability, and nutrient levels can be controlled, mimicking homeostasis, or varied to test particular hypotheses. Two examples illustrate the power of this experimental approach.

When the heart of almost all vertebrates, including mammals, is surgically isolated and placed in a bath of saline, it continues to beat and perform work by pumping saline or other fluid supplied to it. An isolated vertebrate heart will continue to beat for hours or even days if it is kept at an appropriate temperature and is perfused with an oxygenated solution that has the correct ionic composition and contains an energy source such as glucose. With the heart isolated, physiologists can stimulate it with drugs, hormones, or electrical impulses and measure their effects on heart rate, strength of contraction, flow rate, and mechanical movements.

A second example of experimentation on an isolated organ involves the vertebrate pineal gland, a small organ found at the top of the vertebrate brain. The pineal, which plays a key role in regulating daily (circadian) rhythms of physiological processes, is sensitive to light-related stimuli and releases different amounts of regulatory chemicals into the bloodstream as a function of the time of day. When the pineal is isolated and placed in an appropriate culture medium, it continues to exhibit a circadian rhythm. Direct experimentation with this *in vitro* preparation has provided answers to specific questions concerning pineal gland regulation of physiological systems.

OBSERVING AND MEASURING ANIMAL BEHAVIOR

Scientists studying animal physiology may complement their experiments with observations of animal behavior. Care must be taken to ensure that the physiological state of the animal (e.g., breeding condition, rearing young, digesting a meal) is known precisely. Further, the observations should focus on natural behaviors that are stereotypic and readily observed in the animal. Within these strictures, experiments in which behavior is controlled or stimulated can provide important insights into physiological processes that are not always amenable to direct physiological investigation. Analysis of the total time spent performing each behavior and the temporal sequence of behaviors, in conjunction with information about the behavior of other animals and key environmental variables, may reveal how closely behavior is related to the internal physiological state of the animal. The prerequisite for such experimentation is a thorough knowledge of the natural behavior of the animal in its habitat.

The Power of Behavioral Experiments

Research in the 1950s and 1960s on the retrieval behavior of ground-nesting birds illustrates how behavioral studies can contribute to physiological knowledge. K. Z. Lorenz and N. Tinbergen discovered that geese not only recognize their eggs and recover them if they lie outside the nest, but will also retrieve a wide variety of objects (grapefruits, light bulbs, baseballs) lying near their nests. Tinbergen and his students subsequently conducted ingenious experiments with gulls in which they offered pairs of objects to the birds and recorded which one was retrieved first. By exploiting the process of pairwise comparison, they could define the properties that gulls use to choose what to retrieve. Although the birds retrieved many different objects, these experiments showed that real eggs are always preferred over unnatural objects. The relative size, color, and speckling of an egg were found to contribute independently to the likelihood that it would be retrieved. Taken together, these experiments revealed that for gulls, eggs provide a powerful visual stimulus that induces specialized retrieval behavior. Armed with knowledge of the exact properties of the stimuli causing this behavior, physiologists have been better able to conduct physiological experiments on the nature of vision in birds.

The nature of a stimulus may be dependent on how, when, and where it occurs, variables that are therefore important experimental parameters in behavioral experiments. In stickleback fish, for example, the display of a red belly by a male signals to other male sticklebacks that he is defending a nest and to females that he is interested in spawning. The meaning of the signal depends on the sex of the receiver. The red belly arises from physiological processes triggered by the onset of the breeding season. The coordination between behavior and physiology in this species was investigated using behavioral analysis to guide physiological investigation.

Methods in Behavioral Research

A variety of instruments are used to record and analyze the physiological basis of specific behavioral acts. High-speed video cameras may be used in conjunction with electrophysiological detectors of neural or muscular activity to capture simultaneously both fleeting behaviors and their physiological underpinnings. X-ray cameras may reveal the interaction of skeletal components during specific behaviors (e.g., feeding, running on a treadmill). As in so many other aspects of physiology, the availability of inexpensive, fast computers with ever-increasing data storage capabilities has revolutionized the acquisition and analysis of this kind of data.

Figure 2-18 illustrates how a selection of techniques commonly used to study animal behavior and its underlying physiological processes can be brought to bear on a specific behavior—in this case, the feeding strike of a venomous snake and the underlying roles of the musculature of its trunk and jaws. The rapid strike is recorded in two views, dorsal and lateral, using a video camera viewing the animal directly and via a mirror set at 45° above the snake. Quantification of the animal's position is assisted by a grid image in the background that is included in the video images. The snake is placed on a platform that records the force exerted along three orthogonal axes. By measuring this set of external parameters, the investigator can record the forces associated with the snake's movement across the surface. The force exerted by the snake's jaws is recorded by a strain gauge mounted on its head, and its muscle activity is measured by electrodes inserted into the four lateral jaw muscles. All of the data are recorded on videotape and in a computer using data-acquisition hardware and software. The values of the measured variables are typically displayed as a function of time and related to the behavioral analysis recorded on videotape.

Data from such an experiment reveal how many different skeletal muscles in the trunk and head act together in a highly orchestrated fashion to position the fangs to strike the prey and to close the jaws around the prey. These experimental measurements can be used to test hypotheses about which structures and muscles are involved in a strike and how their temporal relationships change during the behavior. This experiment also suggests how many physiological systems contribute to the production of a complex behavioral act. A more complete functional analysis of this behavior and a greater understanding of the performance of the animal is possible if other variables are measured in repeat experiments performed under identical conditions. This experimental setup could be used, for example, to measure differences in the strike behavior as a function of the size and type of prey species. Such measurements can also suggest hypotheses about the neural control of muscle activity, visual feedback guiding the behavior, and a host of other interesting topics.

Animal physiologists can even make use of the ability to train animals to repeat certain natural behaviors. For example, physiologists interested in the physiological changes that occur during diving in seals and water birds have trained animals fitted with instruments for recording heart rate, depth of dive, and even blood gas levels to dive to specific depths or for specific times, allowing repeated measurements of the physiological changes associated with diving.

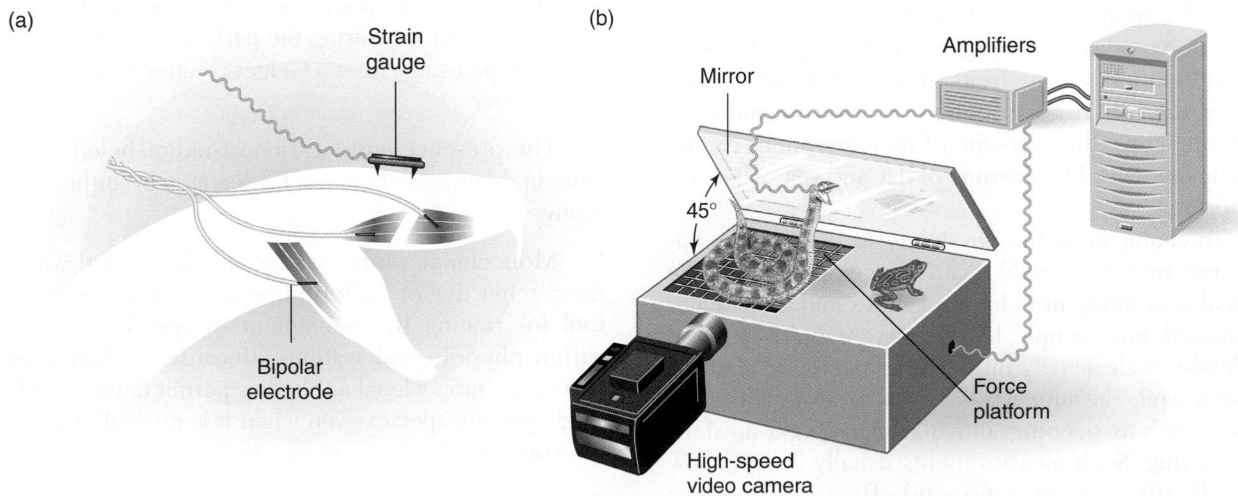

Figure 2-18 The feeding strike of a venomous snake can be analyzed to determine the muscles used and the pattern in which they contract. **(a)** To record electric potentials from the jaw musculature, fine bipolar wire electrodes are placed surgically into the four lateral jaw muscles in a procedure that is performed under anesthesia. A strain gauge is also attached to the top of the snake's head to measure the motion of the underlying skull bones. **(b)** The snake is placed on a force-recording platform and videotaped as it strikes its prey. The leads from the electrodes and strain gauge are connected to electronic amplifiers, which boost their low-voltage signals. The amplified signals are displayed and recorded on a computer.

Humans can be instructed to behave in certain ways during physiological experiments (e.g., to breathe deeply, to run, or to flex muscles). Some animals can be trained to perform as needed for a particular experiment (e.g., to run on a treadmill or to dive), whereas others will only engage in the behavior of interest at irregular intervals while the investigator waits, hoping to capture data at the right moment. What are the relative strengths and weaknesses of data from experiments in which the subject is instructed to perform, is trained to perform, or is simply allowed to behave spontaneously?

THE IMPORTANCE OF PHYSIOLOGICAL STATE IN RESEARCH

Research studies at all physiological levels, from molecular to behavioral, must take into account the animal's physiological state at the time of the experiment. Some physiological states are obvious to the investigator, as when an animal is diving (breath holding) or sprinting at top speed on a treadmill. Other physiological states, such as digesting a meal or coming into reproductive condition, may be far more subtle but have just as great an influence on physiological processes. Of course, the obvious or subtle nature of a physiological state depends on the animal. For instance, a mouse that is curled up with its eyes closed and is showing relatively regular breathing with no locomotor activity can probably be assumed to be asleep. But what about a relatively sluggish species of fish that is motionless? Is it asleep, or merely not moving? Physiological state may be greatly influenced by environmental variables such as the season and time of day. To illustrate, stimulation of the vagus nerve causes a much greater slowing of the heart rate in temperate-zone frogs examined at night in the spring than in frogs examined in the afternoon in autumn. Thus, the outcome of an experiment can be greatly influenced by the time of day and year when it is performed.

To characterize the physiological states of an animal, one or more variables can be measured while the animal is in different behavioral states and their values compared. For example, blood pressure, pulse rate, and skeletal-muscle activity might all be measured simultaneously while the animal is observed in several different states, such as sleeping, moving, digesting a meal, or hibernating. Such measurements usually do not allow the identification of cause-and-effect relationships among the measured variables, but inferences can be drawn and testable hypotheses about those relationships can be formulated based on such data.

A typical experiment might measure key physiological variables during intermittent bouts of hibernation in a ground squirrel. Comparison of behavior with body temperatures and metabolic rates recorded over time reveals that the increased activity in the waking state is correlated with increased body temperature and metabolic rate. These findings suggest that as the animal becomes active, key physiological systems become active at the same time. Although it makes intuitive sense that the animal will need more blood circulating when it is physically active, it is not clear from these data how that increase in blood flow is achieved or how it is regulated. Does the physiological change precede the behavioral activity or result from it? A new experiment suggests itself.

SUMMARY

Formulating and testing hypotheses

- Physiological research begins with a well-formed, specific hypothesis related to a particular level of analysis and capable of being tested experimentally.

- Testing of hypotheses can be greatly facilitated by choosing an animal optimally suited for carrying out those experiments needed to answer particular questions (i.e., by employing the August Krogh principle).

- A key issue in designing physiological experiments is the level at which each physiological problem studied will be analyzed. The choice of level determines the methodology and experimental animal appropriate for measuring the physiological variables of interest.

Molecular techniques in physiology

- Radioisotopes can be incorporated into physiologically important molecules or their precursors. After a radiolabeled molecule is injected into an animal, its movements can be determined by subsequently sampling tissue and measuring the particles emitted by the radioisotope using either a Geiger counter or a scintillation counter.

- The presence and location of radiolabeled molecules in thin tissue slices can be detected by autoradiography.

- Monoclonal antibodies covalently labeled with a fluorescent dye or radioisotope are another powerful tool for tracing the movement of specific proteins within physiological systems. Because of their great specificity, monoclonal antibodies permit detection of a single protein species even when it is present at a very low concentration.

Genetic engineering

- Genes cloned in easily grown bacterial cells can be used to produce large quantities of the gene products (e.g., human insulin and other hormones).

- Transgenic animals (commonly mice) contain a new gene or additional copies of a gene of interest.

■ In knockout animals, a normal gene is eliminated, so that the animals cannot produce a functional protein. Analysis of the effects of either deletion or addition of specific genes can provide insights into the mechanism and regulation of a physiological process.

Microelectrodes and micropipettes

■ The concentration of ions and some gases and the fluid pressure within cells or blood vessels can be determined with specially constructed microelectrodes. A common use of microelectrodes is in recording electrical signals from neurons or muscle cells.

■ Micropipettes are used to inject materials (e.g., dyes, radiolabeled compounds) into individual cells or fluid-filled tissue spaces.

Microscopic techniques

■ Structural analyses of cells, and the physiological processes that derive from these cells, depends heavily on microscopy.

■ Light microscopy uses photons of visible or near-visible light to illuminate specially prepared tissue samples. Specimens to be viewed under a light microscope are fixed (preserved), embedded in plastic or wax, and then cut into extremely thin slices (sections) with a microtome. Finally, the sections are treated with organic dyes or fluorescent-labeled antibodies that differentially bind to and stain various cell components.

■ Electron microscopes use electrons to form images, greatly increasing the resolution of microscopic analysis and permitting visualization of intracellular structural details not apparent with light microscopes.

■ In transmission electron microscopes, a beam of electrons is directed straight through ultrathin tissue slices stained with electron-dense heavy metals.

■ In scanning electron microscopes, electrons are reflected from the surface of the specimen, producing a three-dimensional image of the surface features of cells and other structures.

Cell culture

■ Cell culture, the rearing of cells *in vitro*, allows the propagation of relatively short-lived cell strains and "immortal" cell lines, which can grow indefinitely.

■ Cell cultures, which usually are quite homogeneous, are useful in experiments designed to examine the functions, secretions, responses, and other properties of particular cell types.

Biochemical analysis

■ Many experiments depend on biochemical analysis to determine the composition of sample mixtures derived from cells as well as the concentrations of the constituents present.

■ Among the most commonly used techniques in biochemical analyses are colorimetric assays, transmission spectrophotometry, paper and column chromatography, electrophoresis, and mass spectrometry.

Experiments at higher levels

■ Maintenance of isolated organs or entire organ systems *in vitro* allows the function of intact tissues to be examined in an artificial, controlled environment. Important variables such as temperature, oxygen availability, and nutrient levels can be controlled, mimicking homeostasis, or varied to test particular hypotheses.

■ Animal physiologists frequently supplement their experiments with observations of animal behavior, which can provide important insights into physiological processes.

The importance of physiological state in research

■ In all experimental approaches at any level, the animal's physiological state at the time of experimentation (or tissue sampling) is an important consideration. Physiological state can depend upon internally regulated factors (e.g., sleep, hibernation, activity) or environmental influences.

■ To characterize the physiological states of an animal, one or more variables can be measured and the values of these key variables correlated with different behavioral states.

REVIEW QUESTIONS

1. What is the difference between a scientific question, a hypothesis, and a law?
2. An investigator carries out experiments on crickets, bullfrogs, and rattlesnakes, but is testing a single hypothesis related to a single physiological process. Explain how this investigator could be embracing the August Krogh principle.
3. Discuss the relative merits of radioisotopes and antibody-linked fluorescent markers as cellular probes.
4. What is a clone, how is it produced, and how could it be useful to an animal physiologist?
5. A mutation involving a physiological system proves interesting and useful in revealing physiological process, yet is ultimately lethal before an animal reaches the reproductive stage of its life cycle. How can the mutation be perpetuated in the laboratory to allow long-term study and repeated experiments?
6. Why would an air bubble in a microelectrode used in recording nerve signals disrupt the recording?
7. What are the major differences between light and electron microscopy? What are the advantages and disadvantages of each?

8. Describe the difference between experiments done *in vivo, in vitro,* and *in situ.* What are the advantages and disadvantages of each experimental approach?
9. Imagine and describe an experiment in which you must determine only the presence, and not the actual concentration, of an element or compound. Describe an experiment in which both must be determined.
10. How would you determine whether the resting heart rate of an animal was influenced by daily rhythms?

SUGGESTED READINGS

Burggren, W. W. 1987. Invasive and noninvasive methodologies in physiological ecology: A plea for integration. In M. E. Feder, A. F. Bennett, W. W. Burggren, and R. Huey, eds., *New Directions in Physiological Ecology,* 251–272. New York: Cambridge University Press. (A description of two major approaches to animal experimentation along with their advantages and disadvantages.)

Burggren, W. W. 2000. Developmental physiology, animal models, and the August Krogh principle. *Zoology* 102:148–156. (An essay on the use of the Krogh principle and its use in selecting popular animal models for physiological research.)

Cameron, J. N. 1986. *Principles of Physiological Measurement.* New York: Academic Press. (A short but detailed introduction to several important methods of physiological measurement; the descriptions of the underlying principles of physiological measurement are timeless, making this book worth reading.)

Lodish, H., A. Burk, et al. 2000. *Molecular Cell Biology.* 4th ed. New York: W. H. Freeman. (A well-written, very comprehensive text that describes many techniques used for molecular analyses of the cell.)

Lorenz, K. Z. 1970. *Studies in Animal and Human Behavior.* Vol. 1. Cambridge, Mass.: Harvard University Press. (Classic collection of research papers, translated from the original German, describing the early research of Lorenz, who won the Nobel Prize in physiology in 1973.)

Classic Work

Nelson, D. L., and M. M. Cox. 2000. *Lehninger Principles of Biochemistry.* 3d ed. New York: Worth. (An extensive, contemporary treatment of biochemistry describing many of the biochemical techniques used by animal physiologists.)

Ream, W., and K. G. Field. 1998. *Molecular Biology Techniques.* New York: Academic Press. (Outlines in great detail the various tools in molecular biological research.)

Wilmut, Ian. 1998. Cloning for medicine. *Scientific American,* December. (A possible future involving cloning technology and its use in medicine is outlined in this review article.)

Molecules, Energy, and Biosynthesis

The living organisms found on our planet form a vast and varied array, ranging from viruses, bacteria, and protozoans to flowering plants, invertebrates, and vertebrates. In spite of this immense diversity, all forms of life as we know it have in common the same chemical elements and the same categories of organic (carbon-containing) molecules. Moreover, all life processes take place in a milieu of water and depend on the physicochemical properties of this abundant and unique solvent. The common biochemistry of all living organisms is powerful evidence in support of evolutionary kinship, one of the common threads that runs through all areas of biological study. As we will now see, the role of water in biology is central to the biochemical continuity of living organisms.

WATER: THE UNIQUE SOLVENT

Earth has been called the "water planet," a label that proved to be particularly appropriate given the spectacular views of our planet from space. Because water is so common, it is often regarded with indifference, as some sort of inert filler that simply occupies space in living systems. In truth, water is a vitally important solvent that is directly and intimately involved in all details of animal physiology. Water is a highly reactive substance, quite different both physically and chemically from most other liquids. Indeed, life as we know it would be impossible if water did not have the properties it does. The first living systems presumably arose in the aqueous environment of shallow seas. It is therefore not surprising that the living organisms of the present are intimately adapted at the molecular level to the special properties of water. Today, even terrestrial animals are composed of more than 75% water. Much of their energy expenditure and physiological effort is devoted to the conservation of body water and the regulation of the chemical composition of the internal aqueous environment, as you will learn in Chapter 14.

The Water Molecule

The special properties of water that are so important to life stem directly from its molecular structure. Water molecules are held together by polar covalent bonds between one oxygen atom and two hydrogen atoms. The polarity (i.e., uneven charge distribution) of the covalent bonds results from the strong tendency of the O atom to acquire electrons from other atoms, such as H. This high **electronegativity** causes the electrons of the two H atoms to tend to occupy positions closer to the O atom than to the parent H atoms. The O—H bond is therefore about 40% ionic in character, and the water molecule has the following partial charge distributions (δ represents the local partial charge of each atom):

$$\overset{\delta^+}{\mathrm{H}} \diagdown \underset{2\delta^-}{\mathrm{O}} \diagup \overset{\delta^+}{\mathrm{H}}$$

The angle between the two O—H bonds in the water molecule is 104.5° (Figure 3-1 on the next page). This configuration can be ascribed to crowding of the hydrogen atoms by the two nonbonding electron orbitals of oxygen. Because of the semipolar nature of O—H bonds, H_2O differs greatly, both chemically and physically, from related hydrides such as H_2S. Why? The uneven distribution of electrons in the water molecule causes it to act like a **dipole**. That is, it behaves somewhat like a bar magnet, but instead of having two opposite magnetic poles, north and south, it has two opposite electric poles, positive and negative (see Figure 3-1). This accounts for one of the most important chemical features of water: its ability to form networks of **hydrogen bonds** between the nearly electron-bare, positively charged protons (H atoms) of one water molecule and the electron-rich, negatively charged oxygen atom of neighboring water molecules (Figure 3-2 on the next page). In each water molecule, four of the eight electrons in the outer shell of the oxygen atom are covalently bonded with the two hydrogen atoms. This leaves two

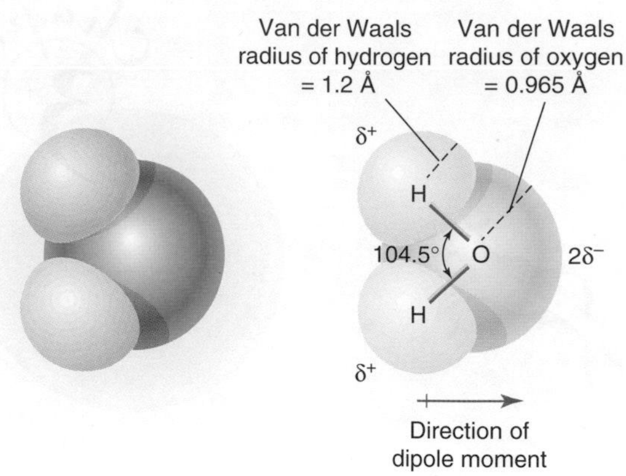

Figure 3-1 In the water molecule, the electron density is greater around the oxygen atom than around the hydrogen atoms, giving the O—H bond a semipolar character. The symbols δ^+ and δ^- indicate a partial positive and negative charge, respectively.

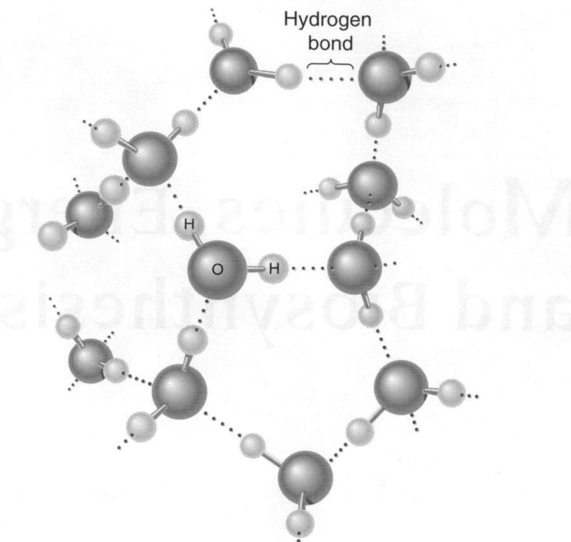

Figure 3-2 Because of the semipolar nature of the O—H bonds in water, adjacent water molecules form hydrogen bonds. These noncovalent bonds (indicated by black dots) represent the electrostatic interaction between the partially positively charged hydrogens on one molecule and the electronegative oxygens in neighboring molecules.

pairs of electrons free to interact electrostatically (i.e., to form hydrogen bonds) with the electron-poor H atoms of neighboring water molecules. Since the angle between the two covalent bonds of water is about 105°, groups of hydrogen-bonded water molecules form tetrahedral arrangements. This arrangement is the basis for the crystalline structure of the most common form of ice.

Properties of Water

The hydrogen-bonded structure of water is highly labile and transient; the lifetime of a hydrogen bond in liquid water is only about 10^{-10} to 10^{-11} seconds. This transience is due to the relatively weak nature of the hydrogen bond. It takes only about 4.5×10^3 calories (4.5 kcal) of energy to break a mole of hydrogen bonds, whereas 110 $\text{kcal} \cdot \text{mol}^{-1}$ are required to break the covalent O—H bonds within a water molecule. As a result of the weakness of hydrogen bonds, no specific groups of H_2O molecules remain hydrogen-bonded for more than a brief instant. Statistically speaking, however, a constant fraction of the molecular population is joined together by hydrogen bonding at any given time, with the fraction dependent on the temperature.

Despite the modest strength of the hydrogen bond, it increases the total energy (i.e., heat) required to separate individual molecules from the rest of the population. For this reason, the melting point, boiling point, and heat of vaporization of water are much higher than those of other common hydrides of elements related to O (e.g., NH_3, HF, H_2S). Of the common hydrides, only water has a boiling point (100°C) far above temperatures commonly found on Earth's surface. The loose bonding between water molecules also endows water with unusually high surface tension and cohesiveness, which has major implications for both biochemical and biological events that occur at, or depend upon, air/water interfaces.

Water as a Solvent

Medieval alchemists, looking for the universal solvent, were never able to find a more effective and "universal" solvent than water. The solvent characteristics of water are due largely to its high dielectric constant, a manifestation of its electrostatic polarity. The dielectric constant is a measure of the effect that a polar dielectric substance, such as water, has in diminishing the electrostatic force between two charges separated by the dielectric substance.* This effect is illustrated especially well by the behavior of ionic compounds, or **electrolytes**, which dissociate (ionize) when placed in water, thereby increasing the conductivity of the solution. Common electrolytes include salts, acids, and bases. In contrast, solutes that undergo no dissociation, and therefore do not increase the conductivity of a solution, are called nonelectrolytes. Common examples of nonelectrolytes are the sugars, alcohols, and oils.

Figure 3-3a illustrates the arrangement of the ions Na^+ and Cl^- in a sodium chloride crystal. This highly structured array is held together firmly by the electrostatic attraction between the positively charged sodium ions and the negatively charged chloride ions. A nonpolar liquid, such as hexane, cannot dissolve the crystal, because no source of energy exists in the nonpolar solvent to break an ion away from the rest of the crystal. Water, however, can dissolve the NaCl crystal, just as it can dissolve most other ionic compounds. The dissolv-

* The electrostatic force between two charges separated by water or another dielectric medium is given by Coulomb's law:

$$f = q_1 q_2 / \epsilon d^2$$

where f is the force (in dynes) between the two electrostatic charges q_1 and q_2, d is the distance (in centimeters) between the charges, and ϵ is the dielectric constant.

Figure 3-3 Water disrupts the crystalline structure of salts by interacting electrostatically with the ions composing the salt. **(a)** Diagram of the highly organized crystalline structure of sodium chloride, showing the relative ionic sizes of Na^+ and Cl^-. **(b)** Hydration of sodium chloride. The oxygen atoms of the water molecules are attracted to the cations, and the hydrogen atoms are attracted to the anions.

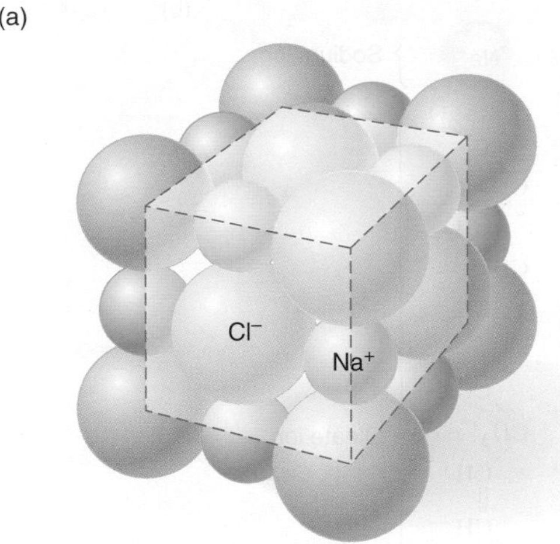

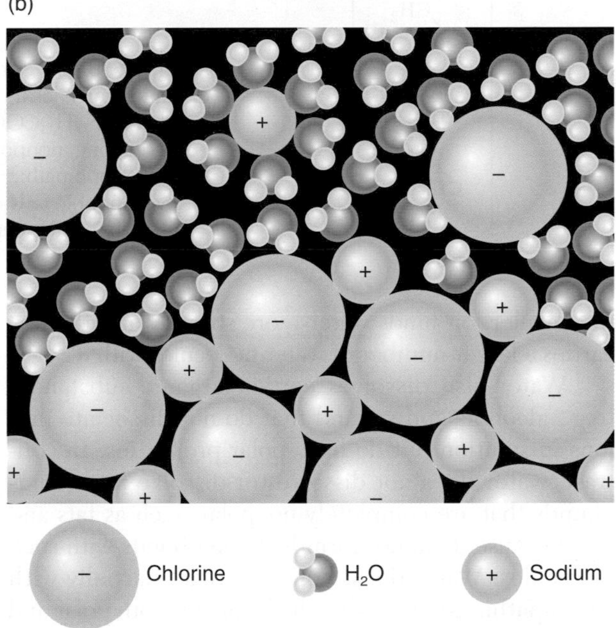

ing power of water arises because the dipolar water molecules can overcome the electrostatic interactions between individual ions (Figure 3-3b). Weak electrostatic binding occurs between the partial negative charge of the oxygen atoms and positively charged cations (Na^+ in this case). Such binding also occurs between the partial positive charge of the hydrogen atoms and negatively charged anions (Cl^- in this case). The clustering of water molecules about individual ions and polar molecules is called **solvation, or hydration.**

As water molecules surround ions, they orient themselves so that their positive poles face anions and their negative poles face cations. This orientation further reduces the electrostatic attraction between the dissolved cations and anions of an ionic compound—in other words, the H_2O molecules act as "insulators." The first shell of water molecules surrounding an ion attracts a second shell of less tightly bound, oppositely oriented water molecules. The second shell may attract even more water in a third shell. Thus, the ion may carry a considerable quantity of water of hydration. The effective diameter of hydrated ions of a given charge varies inversely with their diameter. For example, the ionic radii of Na^+ and K^- are 0.095 and 0.133 nm, respectively, whereas their effective hydrated radii are 0.24 and 0.17 nm. The reason for this inverse relationship is that the electrostatic force between the nucleus of the ion and the dipolar water molecule decreases markedly as the distance between them increases (Figure 3-4). Thus,

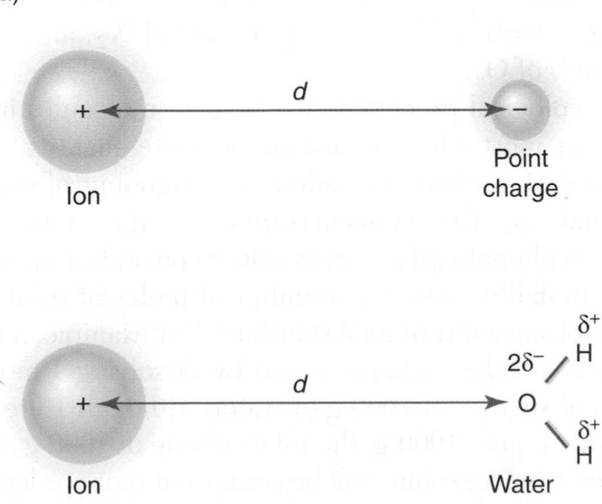

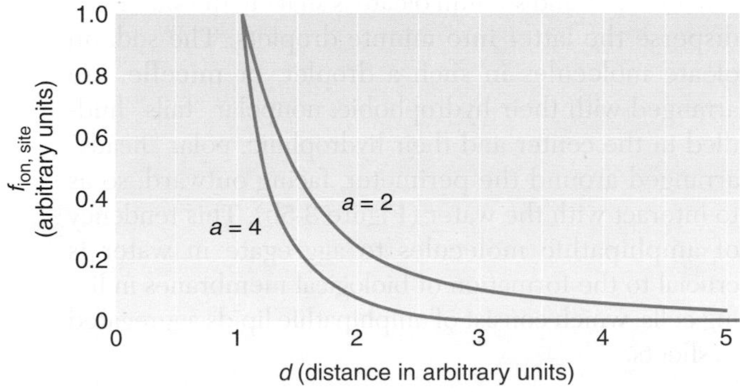

Figure 3-4 Interactions between ions and charged sites are influenced by the distance separating them. The electrostatic force, *f*, between an ion and a site of opposite charge varies inversely with the distance, *d*, raised to some power, *a*, between them: $f \propto 1/d^a$. **(a)** For a point charge, the exponent *a* equals 2.0, so that the force drops inversely with the square of the distance. For a dipole such as the water molecule, the value of *a* can be as high as 4.0. **(b)** The drop in electrostatic force as a function of distance is illustrated for these two values of *a*. In the case of water and a positive point charge, the actual value of *a* is closer to 3.0.

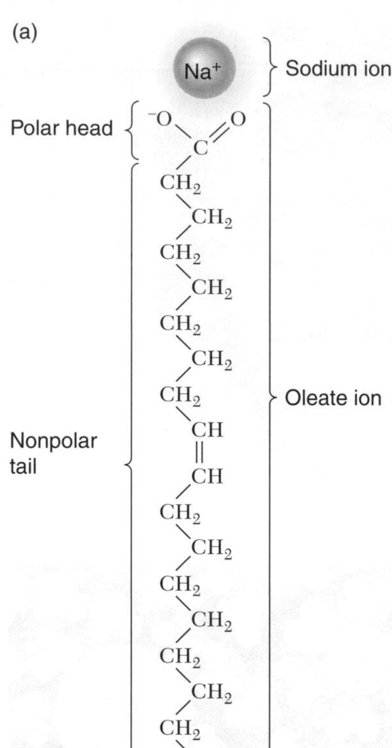

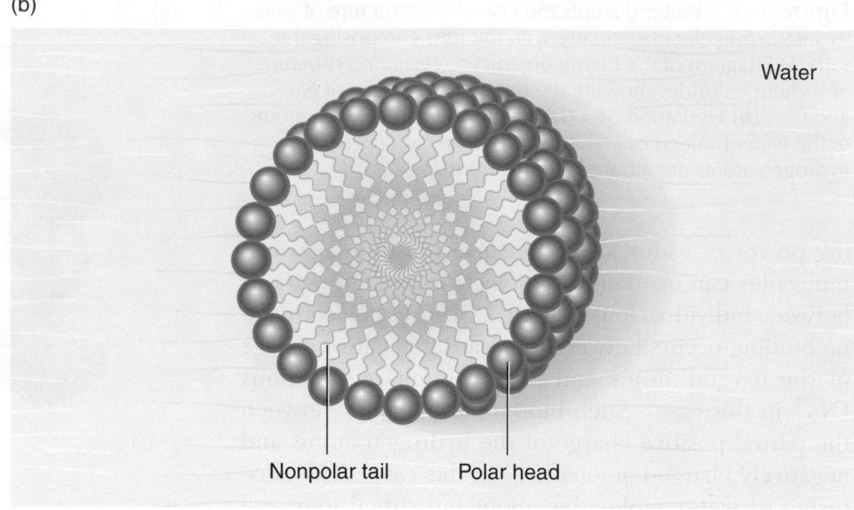

Figure 3-5 Sodium oleate is an amphipathic lipid that forms spherical structures called micelles in a polar solvent, such as water. **(a)** The chemical structure of sodium oleate, $C_{17}H_{33}COONa$. The oleate ion has a polar (hydrophilic) head (shown in red) and a long, nonpolar (hydrophobic) tail (shown in black). **(b)** Diagram of a micelle with amphipathic lipid molecules represented by conventional symbols. The hydrophobic tails of the molecules tend to avoid contact with the polar solvent by grouping at the center of the micelle.

the smaller ion binds water more strongly and thereby carries a larger number of water molecules with it.

Water also dissolves certain organic substances (e.g., alcohols and sugars) that do not dissociate into ions in solution but do have polar properties. In contrast, water does not dissolve (or dissolve into) compounds that are completely nonpolar, such as fats and oils, because it cannot form hydrogen bonds with such molecules. Water does, however, react partially with **amphipathic** compounds, which contain both polar and nonpolar groups. A good example is sodium oleate, a common constituent of soap, which has a **hydrophilic** (water-attracting) polar head and a **hydrophobic** (water-repelling) nonpolar tail (Figure 3-5a). If a mixture of water and sodium oleate is shaken, the water will disperse the latter into minute droplets. The sodium oleate molecules in such a droplet, or **micelle,** are arranged with their hydrophobic, nonpolar "tails" huddled in the center and their hydrophilic, polar "heads" arranged around the perimeter, facing outward, so as to interact with the water (Figure 3-5b). This tendency of amphipathic molecules to aggregate in water is crucial to the formation of biological membranes in living cells, which consist of amphipathic lipids aggregated in sheets.

PROPERTIES OF SOLUTIONS

As noted already, water plays a critical role in living systems. Indeed, many of the physical and chemical processes of cells occur in aqueous solution. The fluids within the cells and tissues of animals, as well as the environments in which aquatic animals live, are critically influenced by the solutes—particularly the electrolytes—that they contain.

Concentration, Colligative Properties, and Activity

Conventionally, the quantity of a pure substance is expressed in **moles** (abbreviated mol). A mole is Avogadro's number of molecules (6.022×10^{23}) of an element or compound; it is equivalent to the molecular weight expressed in grams. Thus, 1 mol of ^{12}C consists of 12.00 g of the pure nuclide ^{12}C, or 6.022×10^{23} carbon atoms. Likewise, there are 6.022×10^{23} molecules in 2.00 g (1 mol) of H_2, in 28 g (1 mol) of N_2, and in 32 g (1 mol) of O_2.

For biological processes that involve molecules in solution, the most relevant measure of solute quantity is the amount of a solute in relation to the amount of solvent—that is, the *concentration* of the solute. Sometimes physiologists express solute concentration in terms of **molality** (m)—the number of moles of solute in 1000 g of solvent (not total solution). For example, a 1 molal sucrose solution is produced by dissolving 1 mol (342.3 g) of sucrose in 1000 g of water. Although 1 liter (L) of water equals 1000 g, the total volume of 1000 g of water plus 1 mol of solute will be somewhat more or less than 1 liter, with the difference depending on the nature of the solute. Molality, therefore, is generally an inconvenient way of stating concentration. A more useful measure of concentration in physiology is **molarity** (M). A 1 molar solution is one in which 1 mol of solute is dissolved in a total final volume of 1 liter; this is written as 1 mol/L, 1 mol · L^{-1}, or 1 M. In the laboratory, a 1 M solution is made by simply adding enough water to 1 mol of the solute to bring the volume of the final solution up to 1 liter. A millimolar (mM) solution contains 1/1000 mol · L^{-1}, and a micromolar (μM) solution contains 10^{-6} mol · L^{-1}. If a solution contains equimolar concentrations of two solutes, then the number of molecules of one solute equals the number of molecules of the other solute per unit volume of solution.

The **colligative properties** of a solution are those that depend on the total number of solute particles in a given volume, irrespective of their chemical nature. These properties include osmotic pressure, depression of the freezing point, elevation of the boiling point, and depression of the water vapor pressure. All of a solution's colligative properties are intimately related to one another and are quantitatively related to the number of solute particles dissolved in a given volume of solvent. Thus, 1 mol of an "ideal" solute—that is, one in which the particles neither dissociate nor associate—dissolved in 1000 g of water at standard pressure (760 mm Hg) depresses the freezing point by 1.86°C, elevates the boiling point by 0.54°C, and exhibits an osmotic pressure of 22.4 atm at standard temperature (0°C) when measured in an ideal apparatus. Measurement of any of these colligative properties can be used to determine the sum of the concentrations of solutes in a solution. Concentrations determined in this way are expressed in osmoles per liter, a measure called **osmolarity**. In theory, osmolarity and molarity are equivalent for solutions of ideal nondissociating solutes exhibiting the same colligative properties.

The theoretical equivalence of osmolarity and molarity, however, does not hold for electrolyte solutions because of ionic dissociation. This is true because a dissociating electrolyte solution will contain more individual particles than a nonelectrolyte solution of the same molarity. For example, a 10 mM NaCl solution contains nearly twice as many particles as the same volume of a 10 mM glucose solution, because NaCl is a strongly dissociating electrolyte. Thus, the colligative properties, and hence the osmolarity, of a 10 mM NaCl solution will be nearly equivalent to those of a 20 mM glucose solution.

Because of electrostatic interaction between the cations and anions of a dissolved electrolyte, there is a statistical probability that at any instant some cations will be associated with anions. For this reason, an electrolyte in solution behaves as if it were not 100% dissociated. The effective free concentration of an electrolyte, as indicated by its colligative properties, is referred to as **activity**. The **activity coefficient**, γ, of an electrolyte is defined as the ratio of its activity, a, to its molal (not molar) concentration, m—that is, $\gamma = a/m$. As we saw earlier, however, the electrostatic force between ions decreases with the distance between them. Thus, as an electrolyte solution becomes more dilute, the extent of dissociation increases. In other words, an electrolyte's activity and activity coefficient depend on both its tendency to dissociate in solution and on its total concentration. The lower the concentration, the higher the activity coefficient. Table 3-1 lists the activity coefficients of some common electrolytes. Those electrolytes that dissociate to a large extent (i.e., have a large activity coefficient) are called *strong electrolytes* (e.g., KCl, NaCl, HCl); those that dissociate only slightly are called *weak electrolytes* (e.g., $MgSO_4$). It should be noted that although the activity coefficient is useful as an index of a solute's tendency to dissociate, and thus of its ability to impart colligative properties to a solution, the activity coefficient is not directly related to the osmotic pressure or other colligative properties of that solute. This value is given by the osmotic coefficient, which must be determined empirically for each solution.

Table 3-1 **Activity coefficients of representative electrolytes at various molal concentrations***

	Molal concentration				
Electrolyte	0.01	0.05	0.10	1.00	2.00
KCl	0.899	0.815	0.764	0.597	0.569
NaCl	0.903	0.821	0.778	0.656	0.670
HCl	0.904	0.829	0.796	0.810	1.019
$CaCl_2$	0.732	0.582	0.528	0.725	1.555
H_2SO_4	0.617	0.397	0.313	1.150	0.147
$MgSO_4$	0.150	0.068	0.049	—	—

* Activity coefficients are given at various molal concentrations rather than molar concentrations. At low concentrations, however, molality and molarity are nearly identical.
Source: West, 1964.

Ionization of Water

Along with the ever-changing nature of the hydrogen bonds between adjacent water molecules, there is a finite probability that a hydrogen atom from one water molecule will become covalently bonded to the oxygen atom of another, forming a **hydronium ion,** H_3O^+. The water molecule that loses a hydrogen atom is converted to a **hydroxyl ion,** OH^- (Figure 3-6a on the next page). The probability of H_3O^+ and OH^- ions forming is actually quite small. At any given time, a liter of pure water at 25°C contains only 1.0×10^{-7} mol of H_3O^+ and an equal number of OH^- ions. The positive charges on the hydrogen atoms of the hydronium ion form hydrogen bonds with the electronegative (oxygen) ends of surrounding nondissociated water molecules, yielding a hydrated hydronium ion (Figure 3-6b).

The dissociation of water is conventionally written as

$$H_2O \rightleftharpoons H^+ + OH^-$$

Bear in mind that the proton (H^+) is not, in fact, free in solution, but becomes part of a hydronium ion. A

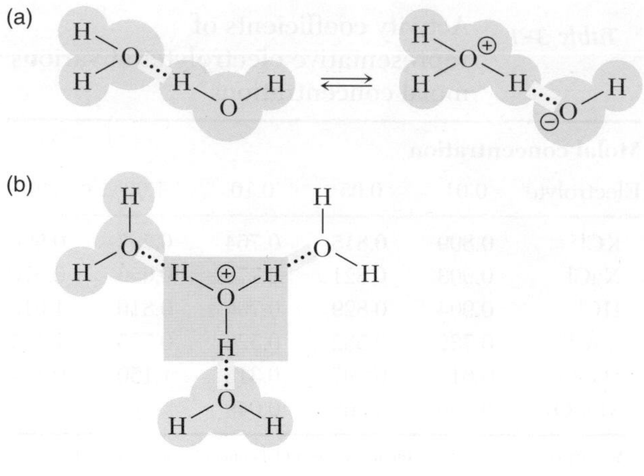

Figure 3-6 The bonding between adjacent water molecules is highly dynamic. **(a)** A hydrogen atom from one water molecule may bond with the oxygen atom of another, producing hydronium ions, H_3O^+, and hydroxyl ions, OH^-. **(b)** In solution, the hydronium ion (shaded in pink) may become associated by hydrogen bonds (dotted lines) with three water molecules.

proton can, however, hop among surrounding H_2O molecules, converting each in turn briefly to a H_3O^+ ion. A sequence of such migrations and displacements can, in the fashion of falling dominoes, conduct a proton equivalent over relatively long distances, with any one proton traveling only a short distance. There is evidence that such proton conduction plays an important role in some biochemical processes, such as metabolic chain reactions in photosynthesis and respiration.

Acids and Bases

Any substance that can donate a proton is called an **acid**, and any substance that combines with a proton is called a **base.** An acid-base reaction always involves such a conjugate acid-base pair—a proton donor and a proton acceptor (H_3O^+ and OH^- in the case of water). Water is said to be amphoteric, meaning that it can act as either an acid or a base. Amino acids also have amphoteric properties. Some common acids are hydrochloric acid, carbonic acid, ammonium, and water:

hydrochloric acid $HCl \rightleftharpoons H^+ + Cl^-$

carbonic acid $H_2CO_3 \rightleftharpoons H^+ + HCO_3^-$

ammonium $NH_4^+ \rightleftharpoons H^+ + NH_3$

water $H_2O \rightleftharpoons H^+ + OH^-$

Common bases include ammonia, sodium hydroxide, phosphate, and water:

ammonia $NH_3 + H^+ \rightleftharpoons NH_4^+$

sodium hydroxide $NaOH + H^+ \rightleftharpoons Na^+ + H_2O$

phosphate $HPO_4^{2-} + H^+ \rightleftharpoons H_2PO_4^-$

water $H_2O + H^+ \rightleftharpoons H_3O^+$

The dissociation of water into H^+ and OH^- ions is an equilibrium process that can be described by the law of mass action, which expresses the relationship between the concentrations of reactants and products at equilibrium for any reaction. For example, the equilibrium constant for the reaction

$$H_2O \rightleftharpoons H^+ + OH^-$$

is given by

$$K_{eq} = \frac{[H^+][OH^-]}{[H_2O]} \tag{3-1}$$

(Brackets denote the concentrations of the reactants enclosed.) The concentration of water remains virtually unaltered by its partial dissociation into H^+ and OH^- because the concentration of each of the dissociated products is only 10^{-7} M (10^{-7} mol·L^{-1}), whereas the molar concentration of water in a liter of pure water (equal to 1000 g) is 1000 g·L^{-1} divided by the gram molecular weight of water (18 g·mol^{-1}), or 55.5 M (55.5 mol·L^{-1}). Equation 3-1 can thus be simplified to

$$55.5\,K_{eq} = [H^+][OH^-]$$

Recall that a consequence of the law of mass action is the reciprocal relation between the concentrations of two compounds in an equilibrium system. This reciprocity is apparent in the constant $[H^+][OH^-]$, which may be lumped with the molarity of water (55.5) into a constant that will be termed the ion product of water, K_w. At 25°C, this has a value of 1×10^{-14}:

$$K_w = [H^+][OH^-] = 10^{-14}$$

This equation follows from the fact, noted above, that $[H^+]$ and $[OH^-]$ each equal 10^{-7} mol·L^{-1}. If $[H^+]$ increases for some reason, as when an acid substance is dissolved in water, $[OH^-]$ will decrease so as to keep $K_w = 10^{-14}$. This reaction is the basis for the **pH scale,** the standard for acidity and basicity, measured as the concentration of H^+ (actually H_3O^+) and defined as

$$pH = -\log_{10}[H^+]$$

Note that the pH scale is logarithmic and typically ranges from 1.0 M H^+ to 10^{-14} M H^+ (Table 3-2). Thus, a 10^{-3} M solution of a strong acid such as HCl, which dissociates completely in water, has a pH of 3.0. A solution in which $[H^+] = [OH^-] = 10^{-7}$ M has a pH of 7.0, and so forth. A solution with a pH of 7 is said to be neutral—that is, neither acidic nor basic. However, the ratio of $[H^+]$ to $[OH^-]$ depends on temperature, so the true "neutral pH" (called the pN) at which $[H^+] = [OH^-]$ actually rises above 7.0 at temperatures below 25°C and falls below 7.0 at temperatures above 25°C. The pH of a solution can be conveniently measured as the voltage produced by H^+ diffusing through the proton-selective glass envelope of an electrode immersed in the solution.

Table 3-2 The pH scale

	pH	$[H^+]$ $(mol \cdot L^{-1})$	$[OH^-]$ $(mol \cdot L^{-1})$	Examples
	0	10^0	10^{-14}	
	1	10^{-1}	10^{-13}	Human gastric fluids
↑ Increasing acidity	2	10^{-2}	10^{-12}	
	3	10^{-3}	10^{-11}	Household vinegar
	4	10^{-4}	10^{-10}	
	5	10^{-5}	10^{-9}	Interior of lysosomes
	6	10^{-6}	10^{-8}	Cytoplasm of muscle
Neutral	**7**	$\mathbf{10^{-7}}$	$\mathbf{10^{-7}}$	**Pure water at 25°C**
	8	10^{-8}	10^{-6}	Seawater
	9	10^{-9}	10^{-5}	
↓ Increasing alkalinity	10	10^{-10}	10^{-4}	Alkaline lakes
	11	10^{-11}	10^{-3}	Household ammonia
	12	10^{-12}	10^{-2}	Saturated lime solution
	13	10^{-13}	10^{-1}	
	14	10^{-14}	10^0	

Question ?

Many animals that normally experience temperature fluctuations in their body fluids have homeostatic mechanisms to maintain pH at a constant fraction of a pH unit above neutral pH, rather than at a set pH per se. This is because the pN (pH at neutrality) rises as solution temperature falls. A problem thus arises with some types of open heart surgery in humans that require cooling the body temperature by several degrees, which changes the neutral pH of water-based fluids such as blood. Should an anesthesiologist maintain a patient's blood pH at 7.4, the normal level for humans, or allow blood pH to rise as body temperature falls?

The Biological Importance of pH

The concentrations of H^+ and OH^- ions are important in biological systems because protons freely move from H_3O^+ to associate with, and thereby neutralize, negatively charged groups, and OH^- ions are available to neutralize positively charged groups. This ability to neutralize ionized groups is especially important in amino acids and proteins, which are amphoteric molecules containing both carboxyl (i.e., $-COOH$) and amino (i.e., $-NH_2$) groups.

In solution, amino acids normally exist in a dipolar configuration called a zwitterion:

$$\underset{\text{Undissociated}}{R{-}\overset{\displaystyle NH_2}{\underset{\displaystyle H}{C_\alpha}}{-}COOH} \qquad \underset{\text{Zwitterion}}{R{-}\overset{\displaystyle NH_3^+}{\underset{\displaystyle H}{C_\alpha}}{-}COO^-}$$

Amino acids and other amphoteric molecules have a characteristic **isoelectric point**, which is the pH at which the net charge of the molecule is zero. If the pH of an amino acid solution is decreased, the H^+ concentration of the solution increases. As a result, the probability of a proton neutralizing a carboxyl group will be greater than the probability of a hydroxyl ion removing the extra proton from the amino group. A large proportion of the amino acid molecules will then bear a net positive charge:

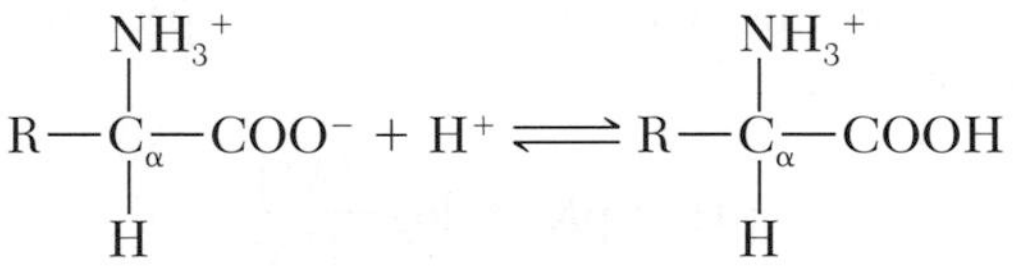

$$R{-}\overset{\displaystyle NH_3^+}{\underset{\displaystyle H}{C_\alpha}}{-}COO^- + H^+ \rightleftharpoons R{-}\overset{\displaystyle NH_3^+}{\underset{\displaystyle H}{C_\alpha}}{-}COOH$$

Raising the pH will, of course, have the opposite effect, with many of the amino acid molecules bearing a net negative charge.

The only amphoteric groups in some amino acids are the $-COOH$ and $-NH_3$ attached to the alpha-carbon atom (C_α); these groups enter into peptide bonds. Other amino acids, however, have additional ionizable groups in their side chains that can accept or donate protons. Dissociable side groups in a macromolecule will determine to a large extent the electrical properties of the molecule and will additionally render it sensitive to the pH of its environment. This sensitivity is most dramatically evident in the influence of pH on the properties of an enzyme, particularly on the chemical state of its active site, where interaction occurs with substrates. Since the binding of a substrate to the active site of an enzyme generally includes electrostatic interactions, the formation of the enzyme-substrate complex is highly pH-dependent. Functional groups within the enzyme active site may also become ionized by changes in pH. **Optimal pH** is the pH at which there is the highest probability of catalysis.

The Henderson-Hasselbalch Equation

Strong acids such as hydrochloric acid dissociate completely, whereas weak acids such as acetic acid dissociate only partially. The generalized chemical equation for the dissociation of an acid can be written as

$$HA \rightleftharpoons H^+ + A^-$$

in which A^- is the anion of the acid HA. Accordingly, the dissociation constant derived from the law of mass action is given by

$$K' = \frac{[H^+][A^-]}{[HA]} \tag{3-2}$$

It is convenient to use the logarithmic transformation of K', namely, pK', which is analogous to pH:

$$pK' = -\log_{10} K'$$

Hence, if pK' = 11, then $K' = 10^{-11}$. A low pK' indicates a strong acid; a high pK' indicates a weak acid.

Acid-base problems can be simplified by rearranging Equation 3-2. Taking the log of both sides, we obtain

$$\log K' = \log[H^+] + \log\frac{[A^-]}{[HA]} \tag{3-3}$$

Rearranging gives us

$$-\log[H^+] = -\log K' + \log\frac{[A^-]}{[HA]} \tag{3-4}$$

Substituting pH for $-\log[H^+]$ and pK' for $-\log K'$, we obtain

$$pH = pK' + \log\frac{[A^-]}{[HA]} \tag{3-5}$$

In other words,

$$pH = pK' + \log\frac{[\text{proton acceptor}]}{[\text{proton donor}]}$$

Equation 3-5 is the **Henderson-Hasselbalch equation,** which permits the calculation of the pH of a conjugate acid-base pair, given the pK' and the molar ratio of the pair. Conversely, it permits the calculation of the pK', given the pH of a solution of known molar ratio.

Buffer Systems

Changes in pH affect the ionization of basic and acidic groups in enzymes and other biological molecules. Consequently, the pH of intracellular and extracellular fluids must be held within the narrow limits within which enzyme systems have evolved if these enzymes are to carry out their normal functions. Deviations of one pH unit or more generally disrupt the biochemistry of organisms, in part because the reaction rates of different enzyme systems become mismatched and uncoordinated. Particularly large pH changes can even inactivate critical enzymes, slowing or halting metabolic pathways. Maintaining the pH of blood is a major goal of the body's homeostatic mechanisms because large changes in blood pH can be rapidly transmitted to other body fluids, including intracellular fluids.

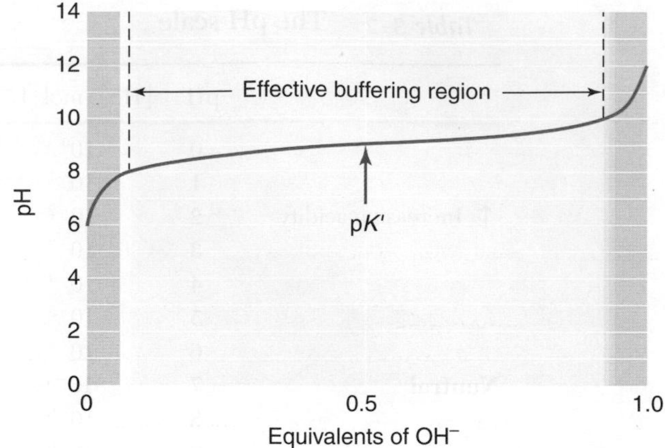

Figure 3-7 The greatest buffering capacity of a conjugate acid-base system is obtained when pH = pK'. On the graph, this point corresponds to the part of the curve with the shallowest slope (small pH changes with large amounts of OH^- added).

The pH of body fluids is maintained within normal ranges with the help of natural pH **buffers.** A buffered system is one that tolerates the addition of relatively large amounts of an acid or base with little change in pH over an intermediate pH range. A buffer must contain an acid (HA) to neutralize added bases and a base (A^-) to neutralize added acids. The properties of buffered systems can be determined by adding small amounts of an acid or base and recording the pH after each addition. A plot of pH versus the amount of added acid or base is called a titration curve. The greatest buffering capacity of a conjugate acid-base pair occurs when the concentrations of HA and A^- are both large and equal. Referring to Equation 3-5, we see that this situation exists when pH = pK' (since $\log_{10} 1 = 0$). This point corresponds to the portion of a titration curve along which there is the smallest change in pH (Figure 3-7).

The most effective buffer systems are combinations of weak acids and their salts. Weak acids dissociate only slightly, thus ensuring a large reservoir of HA. The salts of weak acids dissociate completely, providing a large reservoir of A^-. Added H^+ therefore combines with A^- to form HA, and added OH^- combines with H^+ to form H_2O. As H^+ is thereby removed, it is replaced by dissociation of HA. The most important inorganic buffer systems in the body fluids are the bicarbonates and phosphates. Amino acids, peptides, and proteins, because of their weak-acid functional groups, form an important class of organic buffers in the cytoplasm and extracellular plasma.

Electric Current in Aqueous Solutions

The importance of electrical phenomena in animal physiology will become abundantly clear in later chapters, especially those dealing with the nervous system. Familiarity with the basic concepts of electricity is also useful for an appreciation of laboratory instruments used in recording biological signals and other data.

Water conducts electric current, which is why we are frequently cautioned against using electrical appliances in wet conditions. Water's **conductivity**, the rate of charge transfer caused by the migration of ions under a given charge difference, is far greater than that of oils or other nonpolar liquids. The conductivity of water depends entirely on the presence of charged atoms or molecules (ions) in solution. Electrons, which carry electric current in metals and semiconductors, play no direct role in the flow of electric current in aqueous solutions.

Because the concentrations of H^+ and OH^-, the ions present in pure water, are quite low (10^{-7} M at 25°C), the electrical conductivity of pure water is relatively low, though far higher than that of nonpolar liquids. The conductivity of water is greatly enhanced by the addition of electrolytes, which dissociate into *cations* (positive ions) and *anions* (negative ions) in water (see Figure 3-3b). Thus, seawater conducts electric current far more readily than freshwater. Spotlight 3-1 reviews some common

Spotlight 3-1 ELECTRICAL TERMINOLOGY AND CONVENTIONS

The main electrical properties and electrical units that you will encounter in this and later chapters are defined here. Common symbols used to diagram electric circuits are shown in the accompanying figure.

- **Electric charge,** *q*, is measured in **coulombs** (C). To convert 1 g equivalent weight of a monovalent ion to its elemental form (or vice versa) requires a charge of 96,500 C (1 **faraday,** 1 F). Thus, in loose terms, a coulomb is equivalent to 1/96,500 g equivalent of electrons. The charge on one electron is -1.6×10^{-19} C. If this value is multiplied by Avogadro's number, the total charge is one faraday (i.e., $-96{,}487\ \mathrm{C \cdot mol^{-1}}$).
- **Current,** *I*, is the flow of charge, which is measured in **amperes** (A). A current of 1 $\mathrm{C \cdot s^{-}}$ equals 1 ampere. By convention, the direction of current flow is the direction in which a positive charge moves (i.e., from the anode to the cathode).
- **Voltage,** *V* or *E*, is the electromotive force (emf) or electric potential expressed in **volts.** When the work required to move 1 C of charge from one point to a point of higher potential is 1 joule (J), or about 0.24 calories (cal), the potential difference between these points is said to be 1 volt (V).
- **Resistance,** *R*, the property that hinders the flow of current, is measured in **ohms** (Ω). A resistance of 1 Ω allows exactly 1 A of current to flow when a potential drop of 1 V exists across the resistance. An ohm is equivalent to the resistance of a column of mercury 1 mm^2 in cross-sectional area and 106.3 cm long. R = resistivity × length/cross-sectional area.
- **Resistivity,** ρ, is the resistance of a conductor 1 cm in length and 1 cm^2 in cross-sectional area.
- **Conductance,** *g*, is the reciprocal of resistance: $g = 1/R$. The unit of conductance is the **siemen** (S) (formerly the mho).
- **Conductivity** is the reciprocal of resistivity.

Ohm's law states that current is proportional to voltage and inversely proportional to resistance:

$$I = V/R \text{ or } V = I \times R$$

Thus, a potential of 1 V across a resistance of 1 Ω will result in a current of 1 A. Conversely, a current of 1 A flowing through a resistance of 1 Ω produces a potential difference across that resistance of 1 V.

Capacitance, *C*, is a measure of the ability of a nonconductor to store electric charge. A capacitor consists of two plates separated by an insulator. If a battery is connected in parallel with the two plates, charges will move up to one plate and away from the other until the potential difference between the plates is equal to the emf of the battery, or until the insulation breaks down. No charges move "bodily" across the insulation between the plates in an ideal capacitor, but charges of one sign accumulating on one plate electrostatically repel similar charges on the opposite plate. The unit of capacitance is the **farad** (F). If a potential of 1 V is applied across a capacitor and 1 C of positive charge is thereby accumulated by one plate and lost by the other plate, the capacitor is said to have a capacity of 1 F:

$$C = \frac{q}{V} = \frac{1 \text{ coulomb (C)}}{1 \text{ volt (V)}} = 1 \text{ farad (F)}$$

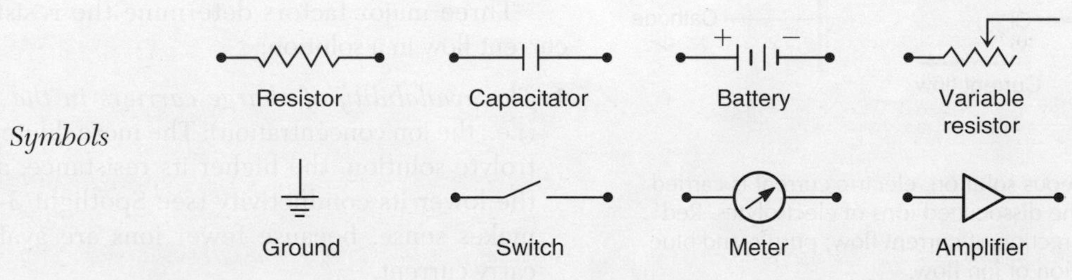

terms, units, and conventions that apply to electrical properties.

The role of ions in conducting electric current in solution is illustrated in Figure 3-8. In this example, two electrodes are immersed in a solution of KCl and connected by wires to a voltage source, the two terminals marked + and −. The voltage generates an *electromotive force*, or *emf*, which can be thought of as a kind of electron pressure that causes a current (i.e., a unidirectional displacement of positive electric charge) to flow through the electrolyte solution from one electrode to the other. What constitutes this electric current? In wires, it consists of the displacement of electrons from the outer shell of one metal atom to another, then to another, and so on. In the KCl solution, electric charge is carried primarily by K^+ and Cl^-; because the concentrations of OH^-, H_3O^+, and H^+ are so low, their contribution to the current can be ignored. When a potential difference (voltage) is applied to an electrolyte solution, the cations migrate toward the **cathode** (the electrode with the negative potential) and the anions migrate toward the **anode** (the electrode with positive potential).

The rate at which each ion species migrates in solution is termed its **electrical mobility.** It is determined by the ion's hydrated mass and by the amount of charge that it bears. The mobility of H^+ is considerably higher than the mobilities of other common ions. The movement of ions that constitutes an ionic current is roughly analogous to a wave of falling dominoes, in which each domino (ion) is displaced just enough to cause a displacement of the next domino. Instead of interacting mechanically, like falling dominoes, however, ions influence each other through electrostatic interactions, with like charges repelling each other.

By convention, the current in a solution is considered to flow in the direction of cation migration. Anions flow in the opposite direction. The rate at which charge is displaced past a given point in the solution determines the intensity of the electric current, measured in amperes. Thus electric current is analogous to the volume of water that flows in 1 second past a point in a pipe (Figure 3-9).

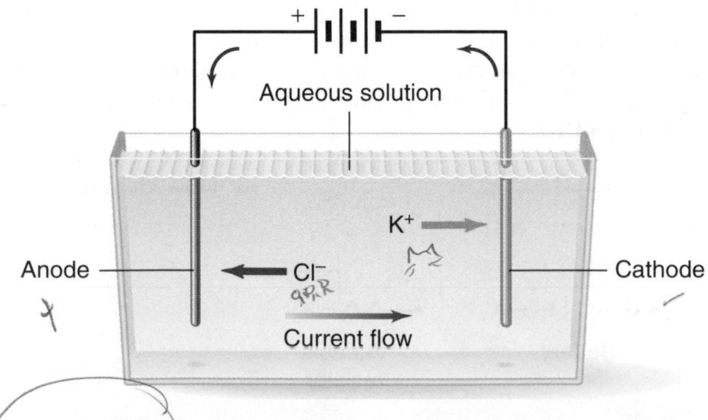

Figure 3-8 In aqueous solution, electric current is carried by the movement of the dissociated ions of electrolytes. Red arrows indicate the direction of current flow; purple and blue arrows indicate direction of ion flow.

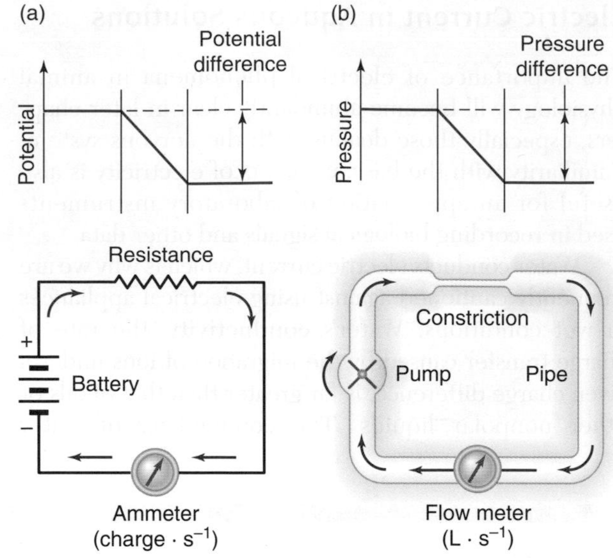

Figure 3-9 The flow of electrons in a wire **(a)** can be compared to the flow of water in a pipe **(b)**. An electric current always meets some resistance, analogous to friction acting on water in a pipe. The resistance of the entire circuit is symbolized in part a by a resistor symbol and in part b by a constriction in the pipe.

An electric current always meets some electrical **resistance** to its flow, just as water flowing through a pipe meets mechanical resistance owing to such factors as friction with the pipe's walls and other water molecules. In order for the charges to flow through an electrical resistance, there must be an electrostatic force acting on the charges. This force (analogous to hydrostatic pressure driving water through a pipe) is the difference in electric pressure, or potential, V, between the two ends of the resistive pathway (see Figure 3-9a). A difference in potential, or **voltage,** exists between separated negative and positive charges. This potential difference is related to the current, I, and resistance, R, as described by Ohm's law (see Spotlight 3-1). To force a given current through a pathway of twice the resistance requires twice the voltage (Figure 3-10a). Similarly, the current will be reduced to half its value if the resistance it encounters is doubled while the voltage is kept constant (Figure 3-10b).

Three major factors determine the resistance to current flow in a solution:

1. *The availability of charge carriers in the solution* (i.e., the ion concentration): The more dilute an electrolyte solution, the higher its resistance, and thus the lower its conductivity (see Spotlight 3-1). This makes sense, because fewer ions are available to carry current.

(a)

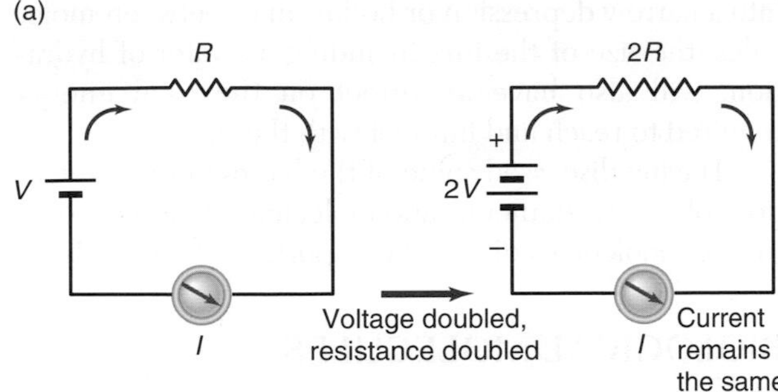

(b)

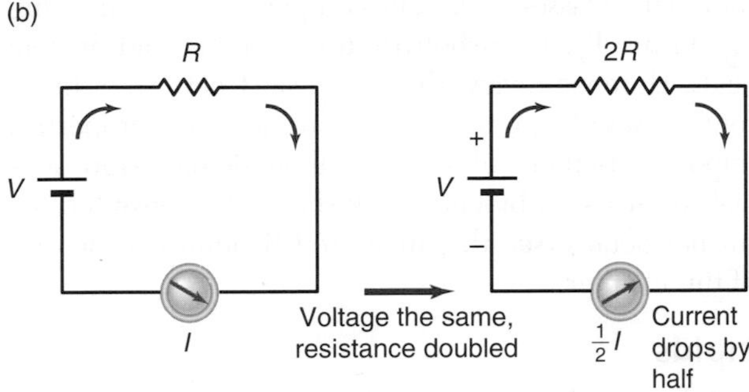

Figure 3-10 Ohm's law describes the relationship between electric current, *I* (number of charges moving past a point per unit time), potential difference, *V*, and resistance, *R*. **(a)** The current intensity, indicated by readings on the ammeter, remains unchanged if both voltage and resistance are doubled. **(b)** Current drops by half if resistance alone is doubled.

2. *The cross-sectional area of the solution* in a plane perpendicular to the direction of current flow: The smaller this cross-sectional area, the higher the resistance encountered by the current. This, again, is analogous to the effect of the cross-sectional area of a pipe carrying water.

3. *The distance traversed in solution by the current:* The total resistance encountered by a current passing through an electrolyte solution is directly proportional to the distance the current traverses.

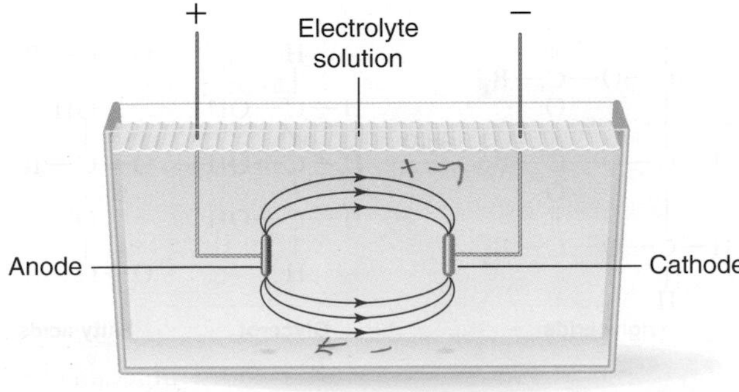

Figure 3-11 Flowing current spreads through a volume of electrolyte solution, involving more ions and lowering the effective resistance.

The ions carrying current are distributed evenly throughout a solution, and current flowing between two electrodes arches out in curved paths rather than simply flowing in a direct pathway (Figure 3-11). This behavior brings far more ions into play than are present in a direct path between the electrodes, thus providing a lower effective resistance to the flow of electric current (point 1 on the list of factors), even though the curved pathway is longer.

Binding of Ions to Macromolecules

Ions free in solution inside or outside living cells interact electrostatically with one another and with a variety of ionized or partially ionized portions of molecules, especially proteins. The ion-binding sites of biological macromolecules carry electric charges, and their interactions with free inorganic ions are based on the same principles that determine ion exchange at sites on such nonbiological materials as soil particles, glass, and certain plastics. Interactions between fixed ion-binding sites on macromolecules and various ions are essential to certain physiological mechanisms, such as enzyme activation and the selectivity of membrane channels and carriers for particular ions.

The energetic basis for interaction between an ion and an ion-binding site is the electrostatic attraction between the two. This interaction is identical in principle to the interactions that occur between anions and cations in free solution. Thus, a site with a full or partial negative charge (recall the partial charge on the oxygen atom of the water molecule) attracts cations; a site with a positive charge attracts anions. Two or more species of cations in solution will compete with each other to bind electrostatically to an anionic (i.e., negatively charged) site. The site will show an order of binding preference among cation species, ranging from those that bind most strongly to those that bind least strongly. This order of preference is called the **affinity sequence**, or **selectivity sequence**, of the site.

Cation-binding sites on organic molecules are generally oxygen atoms in functional groups such as silicates ($—SiO^-$), carbonyls ($R—C{=}O$), carboxylates ($R—COO^-$), and ethers ($R_1—O—R_2$). As noted earlier, the oxygen atom is electron-hungry and draws electrons from surrounding atoms in the molecule. The oxygen atoms in such neutral groups as the carbonyls or ethers can be treated as having a partial negative charge due to the statistically higher number of electrons around them (Figure 3-12 on the next page). Since the group itself is neutral, there must also, of course, be partial positive charges on the other atoms. When silicate and carboxylate groups are ionized, their oxygen atoms carry a full negative charge.

The energetics of electrostatic interaction of a site with an ion are expressed in terms of potential energy—namely, the energy, U, of bringing together

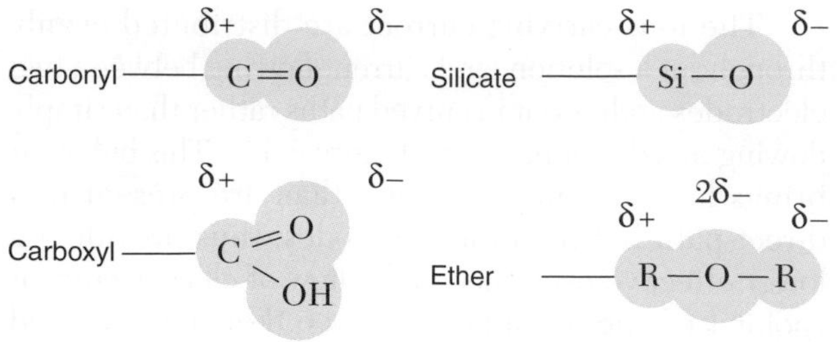

Figure 3-12 Many biological molecules contain groups that exhibit a partial charge separation. Most common are oxygen-containing groups in which the highly electronegative oxygen atom draws electrons from neighboring atoms. The electron cloud distributions of several molecular side groups are indicated by blue shading. Silicate, although not present in animals, is a major component of the skeleton of diatoms.

two charges, q^+ and q^-, in a vacuum from a separation of infinity to the new distance of separation d:*

$$U = \frac{(q^+q^-)}{d^a} \quad (3\text{-}6)$$

The exponent a equals 1 in the case of two single charges (i.e., a monovalent anion and a monovalent cation). For a dipolar molecule such as water, in which there are centers of both negative and positive charge (but no net charge), the energy of interaction falls off more rapidly with distance (i.e., a in Equation 3-6 is greater than 1). This phenomenon plays an important role in the electrostatic tug-of-war experienced by an ion dissolved in water attracted to a site of opposite charge.

In an aqueous environment (i.e., in a solution as opposed to a vacuum), the relation between the atomic radius of a cation and its affinity for a given fixed electronegative site (Equation 3-6) is modified by the electrostatic interaction of the cation with dipolar water molecules. The cation is attracted to the positive charge of both the binding site and water—molecules of water and the site engage in competition for binding of the cation. The more successfully the site competes with water for a given ionic species, the greater the "selectivity" of the site for that ionic species. The selectivity sequence of a site for a group of different ions will be determined by the field strength and the distribution of electrons near the site. In addition, small monovalent cations will interact more strongly with a particular electronegative site than will large monovalent cations because they carry the same unit charge but can approach more closely.

In addition to the principles of electrostatic interaction briefly described here, there are steric constraints on the binding of ions with some sites. If, for example, a site is situated so that an interacting ion must squeeze into a narrow depression or hollow in or between molecules, the size of the ion, including its water of hydration, will also have an effect on the total energy required to reach and interact with the site.

Having discussed some of the basics of the interactions of atoms, elements, and molecules, let us now turn to those molecules of specific importance to animals.

* Equation 3-6 should not be confused with Coulomb's law, which is given in the footnote on page 42.

BIOLOGICAL MOLECULES

Carbon atoms form the "backbone" structure for the four major classes of organic compounds found in living organisms: lipids, carbohydrates, proteins, and nucleic acids. Here we review the chemical structures of these four classes of substances and consider some properties important to their roles in animal physiology. More specialized texts on biochemistry should be consulted for further details (see the Suggested Readings at the end of this chapter).

Lipids

Lipids are a diverse group of water-insoluble biological molecules with relatively simple chemical structures. The various lipids have a variety of functions. Fats, for example, serve as energy stores, while phospholipids and sterols are major components of membranes (see Chapter 4).

Fats are composed of **triglyceride** molecules, each of which consists of a glycerol molecule to which three fatty acid chains are connected through ester linkages. When triglycerides are hydrolyzed (cleaved) by the insertion of H^+ and OH^- into the ester bonds, they break down into glycerol and three fatty acid molecules (Figure 3-13). The three fatty acids in a triglyceride may or may not be the same, but they all contain an even number of carbon atoms. If all the carbon atoms in a fatty acid chain are linked by single bonds (i.e., each carbon except the carboxyl carbon bears two hydrogens), the fatty acid is said to be **saturated.** If the fatty acid chain contains one or more double bonds between carbon atoms, the fatty acid is said to be **unsaturated.**

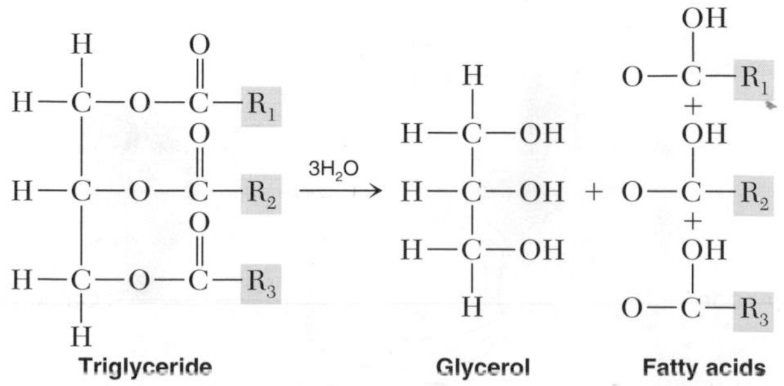

Figure 3-13 Fats are composed of triglyceride molecules, which can be hydrolyzed to glycerol and fatty acids. This reaction is catalyzed by the enzyme lipase. R represents a fatty acid. The fatty acid groups in a particular triglyceride may be the same or different.

The degree of saturation and the length of the fatty acids (i.e., number of carbon atoms) determine the physical properties of the molecule.

Fats containing unsaturated fatty acids generally have low melting points and form oils or soft fats at room temperature, whereas fats containing saturated fatty acids form solids at room temperature. That is why the process of hydrogenation (saturating the fatty acid chains with hydrogens and thereby breaking the double bonds) converts vegetable oil into shortening. In addition, if the number of double bonds is constant, the shorter the chain length of a fatty acid, the lower its melting point. Saturated fatty acids are more readily converted by metabolic processes into **sterols** such as cholesterol. Because excess cholesterol appears to be a risk factor for cardiovascular disease in humans, many dietary guidelines recommend limiting consumption of saturated fats. Some intake of cholesterol, however, is necessary, as it is a vital component of biological membranes and is also the precursor for synthesis of the steroid hormones (see Figure 9-23).

Triglycerides typically accumulate in the fat vacuoles of specialized adipose cells in vertebrates. Because of their low solubility in water, these energy-rich molecules can be stored in high concentrations in the body without requiring large quantities of water as a solvent. Triglyceride energy stores are also rendered highly compact by the relatively high proportions of hydrogen and carbon and low proportions of oxygen in the molecule. Thus, 1 g of triglyceride will yield about twice the energy upon oxidation as 1 g of carbohydrate (Table 3-3).

In **phospholipids**, one of the outer fatty acid chains of a triglyceride is replaced with a phosphate-containing group (see Figure 4-3). Thus the phospholipids are amphipathic molecules, with a hydrophilic portion (the phosphate-containing group) and a hydrophobic portion (the fatty acid chains) that is soluble in lipids (or **lipophilic**). This property allows phospholipid molecules in biological membranes to form a transition layer between an aqueous phase and a lipid phase. As we will see in the next chapter, biological membranes consist largely of two phospholipid layers, with the nonpolar "tails" of each layer oriented inward toward each other and the polar "heads" oriented toward the aqueous phases (see Figure 4-6).

Other types of lipids found in membranes are glycolipids, which contain one or more sugar groups, and sphingolipids, which contain a long-chain amino alcohol called sphingosine. Sphingolipids are present in particularly high concentrations in brain and nerve tissue. Another group of lipids, the *waxes*, form a waterproofing layer in certain insects (see Chapter 14).

Table 3-3 **The energy content of the three major categories of foodstuffs**

Substrate	Energy content ($kcal \cdot g^{-1}$)
Carbohydrates	4.0
Proteins	4.5
Fats	9.5

Carbohydrates

Carbohydrates are polyhydroxyl aldehydes and ketones with the general chemical formula of $(CH_2O)_n$. The simplest carbohydrates are the **monosaccharide sugars**, the most common of which contain six carbons (**hexoses**) or five carbons (**pentoses**). Monosaccharides typically exist as ring structures containing four or five carbon atoms and one oxygen, with the remaining carbon(s) outside the ring (Figure 3-14a). Green plants manufacture the hexose **glucose** from H_2O and CO_2 by the process of photosynthesis. All the energy trapped by photosynthesis and transmitted as chemical energy to the living world is channeled through six-carbon sugars such as glucose. As noted later in this chapter, the complete or partial degradation of glucose to H_2O and CO_2 during cellular respiration releases the chemical energy that

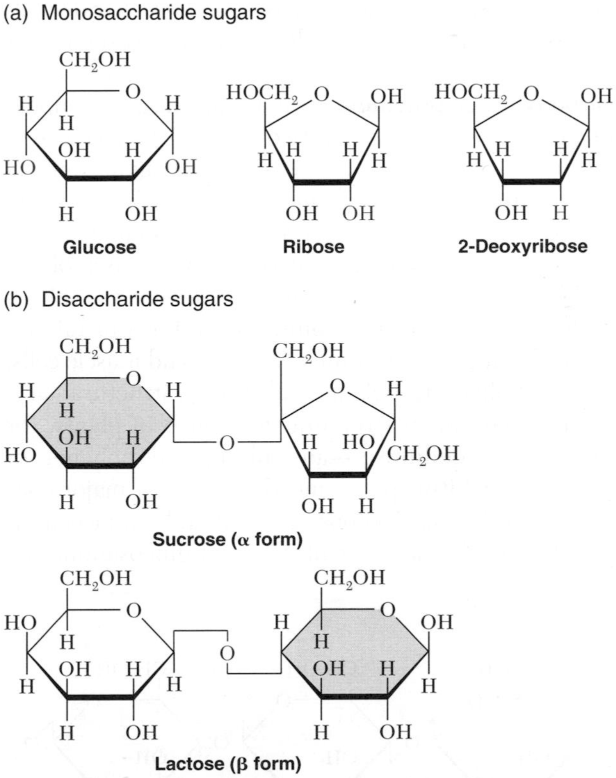

(a) Monosaccharide sugars

Glucose Ribose 2-Deoxyribose

(b) Disaccharide sugars

Sucrose (α form)

Lactose (β form)

Figure 3-14 The simple sugars are the monosaccharides and disaccharides. **(a)** Glucose, the most prevalent hexose in cells, is degraded to provide energy. The hydroxyl groups shown in red can form a covalent bond with another sugar molecule, forming a disaccharide. Two pentoses, ribose and 2-deoxyribose, are constituents of nucleic acids. **(b)** Disaccharides are formed by condensation of two monosaccharide units. Sucrose and lactose both contain one glucose unit (shaded in pink) plus a second monosaccharide. The glycosidic bond (red) linking two monosaccharide units can have two different orientations, designated α and β.

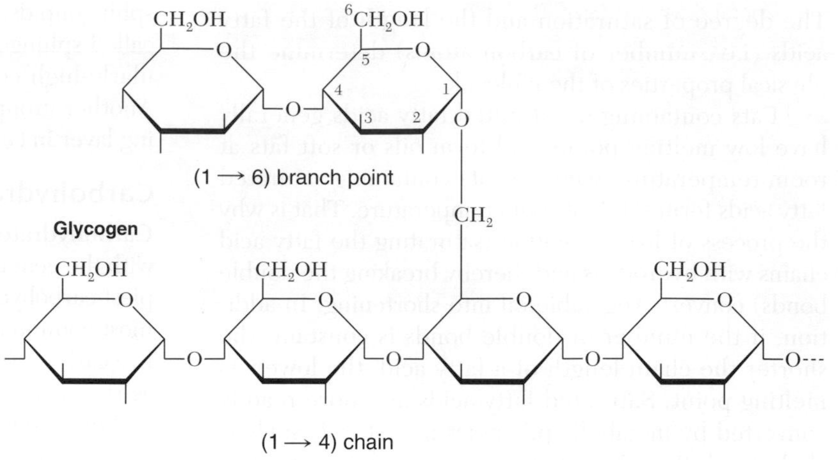

Figure 3-15 Glycogen is the primary carbohydrate storage form in animal cells. A glycogen molecule is a long chain of glucose residues in which carbons 1 and 4 in adjacent molecules are linked (1 → 4), with branches extending from carbon 6 every eight to ten glucose residues. Only a small portion of a glycogen molecule is depicted.

was stored in the molecular structure of glucose during photosynthesis. The two most important pentose sugars are **ribose** and **2-deoxyribose.** These pentoses constitute a repeating unit in the backbones of all nucleic acid molecules (DNA and RNA) and are constituents of a number of central metabolic cofactors.

Cells contain enzymes that can convert glucose to other monosaccharides or link two monosaccharide molecules to form a **disaccharide sugar** such as sucrose or lactose (Figure 3-14b). Cells also can synthesize various carbohydrate **polymers** containing large numbers of monosaccharide units. Two branched polymers of D-glucose—**starch** in plant cells and **glycogen** in animal cells—are the primary forms for carbohydrate storage (Figure 3-15). Like fats, these high-molecular-weight polymers require a minimal amount of water as a solvent and serve as a concentrated energy reserve in the cell. In vertebrates, glycogen is found in the form of minute intracellular granules, primarily in liver and muscle cells.

Carbohydrate polymers also form structural substances. The main structural substance in plants, for example, is **cellulose**—an unbranched polymer of D-glucose. **Chitin** (pronounced "ky-tin"), a major constituent of the exoskeletons of insects and crustaceans, is a cellulose-like polymer of *N*-acetyl glucosamine, an amino derivative of D-glucose (Figure 3-16). Both cellulose and chitin are flexible, elastic, and insoluble in water.

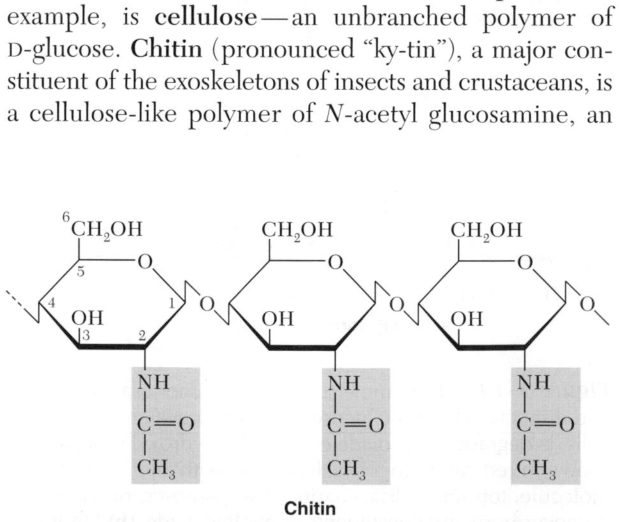

Figure 3-16 Chitin, a structural carbohydrate polymer, consists of *N*-acetyl glucosamine units joined by 1 → 4 glycosidic bonds. In *N*-acetyl glucosamine, an acetamide group (in pink) replaces the hydroxyl group on carbon 2 of glucose.

Proteins

Proteins are the most complex and most abundant organic molecules in the living cell, making up more than half the dry mass of the cell. Although the basic structure of all proteins is similar, a vast array of different proteins with diverse functions is found in biological systems.

Primary structure

Proteins are composed of linear chains of amino acids, which are amphoteric molecules containing at least one carboxyl group and one amino group. The 20 common amino acids that make up proteins are all alpha-amino acids, in which the amino group is bonded to the alpha-carbon (C_α) atom—that is, the carbon atom adjacent to the carboxyl group. Amino acids differ from one another in the structure of their side groups, generically referred to as R groups (Figure 3-17a). The protein-synthesizing machinery of cells joins amino acid molecules together via covalent **peptide bonds,** forming long **polypeptide chains.** Adjacent C_α atoms in a polypeptide chain are separated by a **peptide group** (Figure 3-17b). The specific linear sequence of amino acids in a polypeptide is termed its **primary structure.** Since the amino acid components of a polypeptide chain differ only in their side groups, these groups are like letters in the protein alphabet, defining the primary structure of a protein. A protein molecule may consist of one, two, or several polypeptide chains, either covalently or noncovalently linked.

The amino acid sequence of a polypeptide (i.e., its primary structure) is encoded in an organism's genetic material. The amino acid sequence laid down during protein synthesis is the expression of this information and is the primary determinant of the properties of any protein molecule. Since there are about 20 different amino acid building blocks, an impressive variety of different amino acid sequences is possible. Suppose, for example, that we were to construct a polypeptide

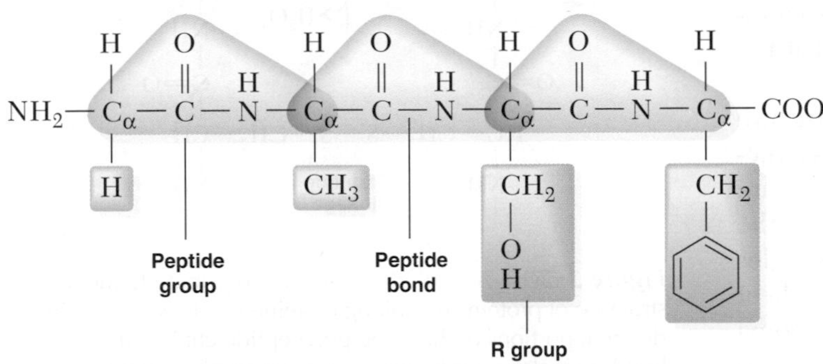

Figure 3-17 The primary structure of a protein is a linear sequence of α-amino acids linked by peptide bonds. **(a)** All the amino acids found in proteins have a common fundamental structure. Each has a characteristic side group commonly indicated by R. **(b)** The peptide bonds (red lines) linking the amino acid residues in a polypeptide have a partial double-bond character. As a result, the peptide group (blue shading) is planar. All proteins have in common a polypeptide backbone; they differ in the sequence of side groups. This sequence, the primary structure, is the defining property of each protein.

molecule consisting of one of each of those 20 building blocks. How many different linear arrangements could we make without ever repeating the same sequence of amino acids? This is determined by multiplying $20 \times 19 \times 18 \times 17 \times 16 \times \ . \ . \ . \ \times 2 \times 1$, which gives 10^{18}. But this enormous number, which describes a relatively small protein with a molecular weight of only about 2400, pales in comparison to the possibilities for a more typical protein with a molecular weight of 35,000. For a protein of this size, containing just 12 kinds of amino acids, the number of possible sequences exceeds 10^{300}!

Higher levels of structure

The primary structure of a polypeptide chain determines the three-dimensional conformation, or shape, that the protein assumes in a given environment. This conformation depends on the nature and position of the side groups that project from the peptide backbone. In addition to the primary structure, proteins exhibit additional levels of structure, designated secondary, tertiary, and quaternary. **Secondary structure** refers to the local organization of parts of the polypeptide chain, which can assume several different arrangements. **Tertiary structure** refers to the foldings of the chain to produce globular or rodlike molecules. **Quaternary structure** refers to the joining of two or more polypeptide chains to form dimers, trimers, and larger aggregates.

Because the C—N peptide bond has a partial double-bond character, it is not free to rotate; hence, the atoms of the peptide group (see Figure 3-17b) are confined to a single plane. The remaining bonds of the peptide backbone, however, are free to rotate. Linus Pauling and Robert Corey, using precisely constructed atomic models, found that the simplest stable secondary structure of a polypeptide chain is a helical arrangement called the **alpha (α) helix** (Figure 3-18). In this

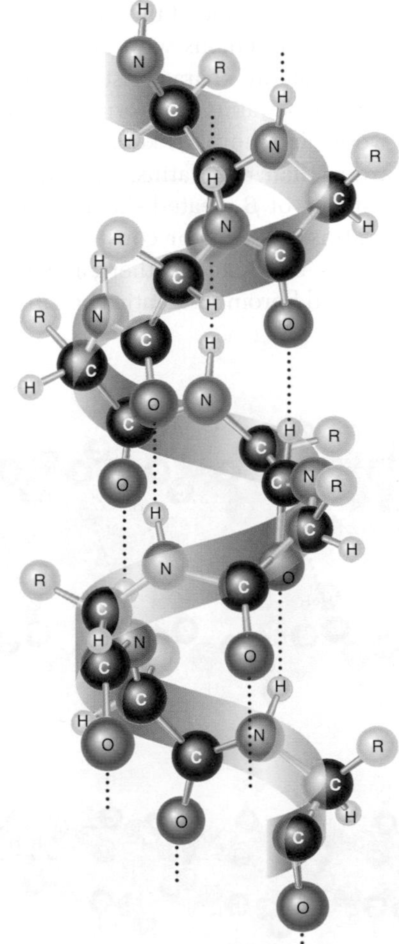

Figure 3-18 The α helix is a common and very stable type of secondary structure in proteins. This helical arrangement, containing 3.6 amino acids per turn, is stabilized by hydrogen bonds (black dots) between the oxygen atom of a carbonyl group and the hydrogen atom attached to the amide nitrogen four residues away in the backbone. The side groups (R) extend outward from the axis of the backbone.

structure, the plane of each amide group is parallel to the major axis of the helix, and there are 3.6 amino acid residues per turn. The side group of each amino acid residue extends outward from the helical backbone, free to interact with other side groups or other molecules. The stability of the α helix is enhanced substantially by hydrogen bonding between the oxygen atoms of a carbonyl group and the hydrogen atoms attached to the amide nitrogen four residues ahead. Because of the stability of the α helix, a polypeptide chain spontaneously assumes this conformation, provided that the side groups do not interfere.

Another major type of protein secondary structure is the **beta (β) pleated sheet** (Figure 3-19). This structure consists of laterally associated β strands, which are fairly short, nearly fully extended stretches of the polypeptide chain. The "pleat" arises from the angles between adjacent planar peptide groups, and the sheet arises when hydrogen bonds form between carbonyl oxygen atoms and amide hydrogen atoms in adjacent β strands. The side groups of the amino acid residues project above or below the plane of the sheet.

Long polypeptide chains with an uninterrupted α-helix conformation are characteristic of fibrous proteins, such as the α-keratins that form hair, fingernails and claws, wool, and horn. β-keratins, which form materials harder than α-keratins, have a secondary structure consisting of β pleated sheets rather than α helices. β-keratins are a major constituent of reptile scales, turtle shells, claws, and feathers, and in a distinct configuration called fibroin, β-keratin contributes to the immense strength of spider and caterpillar silk. Nonstructural intracellular proteins are typically constructed of domains containing short segments in the α-helix or β-pleated-sheet conformation, connected by random coils.

Two types of relatively weak intramolecular interactions help to stabilize tertiary structure: coulombic (electrostatic) interactions between the charged side groups and **van der Waals forces,** weak attractions between nonpolar groups. Some proteins are further stabilized by covalent bonds between the sulfhydryl (—SH) side groups of the amino acid cysteine. The side chains of two cysteine residues react to form a **disulfide linkage** (—S—S—), which covalently joins the residues (Figure 3-20). A disulfide linkage can covalently cross-link different portions of a polypeptide chain, thereby stabilizing its folded tertiary structure, or connect two separate chains.

Some, but not all, proteins join with other protein subunits to form a quaternary structure. The association of protein subunits involves noncovalent interactions between complementary regions on their surfaces. For example, negatively charged groups of one subunit may fit against positively charged groups of another subunit; hydrophobic, nonpolar side groups on the subunits may meet to the mutual exclusion of water molecules; or residues in each subunit may be oriented so that they can form hydrogen bonds. The subunits of multimeric proteins assemble spontaneously, as demonstrated when they are added separately to an aqueous solution and mixed.

Heating a protein disrupts the noncovalent interactions (electrostatic interactions, hydrogen bonds, and van der Waals forces) that stabilize protein structure and reduces the protein conformation to a disordered state, a process called **denaturation.** Hair-curling irons

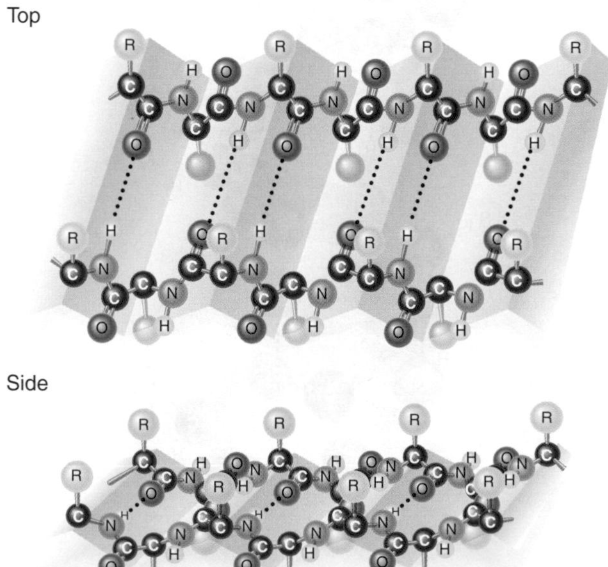

Figure 3-19 The β pleated sheet is a type of secondary structure found in silk fibers and some other fibrous proteins. Beta sheets are formed by the lateral association of two or more β strands stabilized by hydrogen bonds (black dots). The side groups (R) extend above and below the plane of the sheet.

Cysteine residues within peptide chain(s)

$HC—CH_2—HS$ $SH—CH_2—CH$

$\downarrow O_2 \rightarrow H_2O_2$

$HC—CH_2—S—S—CH_2—CH$

Disulfide linkage

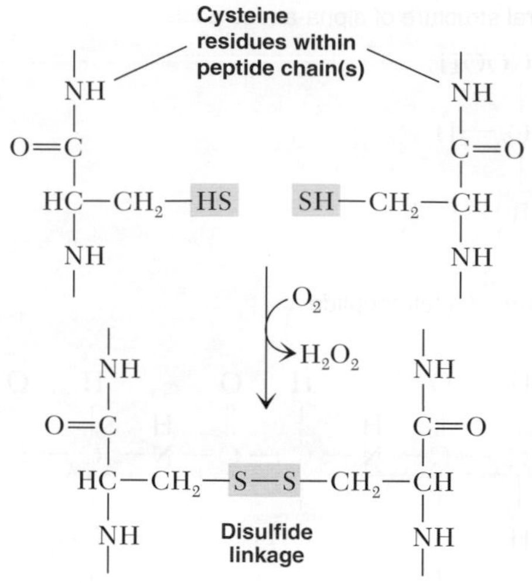

Figure 3-20 A disulfide bond can contribute to the tertiary structure of proteins by linking cysteine residues present in different portions of the same polypeptide chain. Disulfide bonds can also form between cysteine residues in different polypeptide chains.

work in this way, temporarily denaturing the proteins in the hair shaft by heating them and then letting them cool in slightly new configurations that alter the shaft's orientation. In this same way, high temperatures can disrupt the structures of enzymes, rendering them inactive and killing the cells in which they reside.

Molecular chaperones and stress proteins

Although protein folding occurs quite slowly and inefficiently *in vitro*, it is carried out efficiently in the cell. This enigma was recently solved with the discovery of **molecular chaperones**, a family of proteins that features prominently in the folding of other proteins and the preservation of their complex folded states. Molecular chaperones, as well as genes that encode and express them, have been discovered in every organism in which they have been sought, and so are among the most ancient of biological molecules. Chaperones have several modes of action:

- They assist in the folding of newly synthesized proteins.
- They bind to and stabilize proteins that are partially unfolded or improperly folded, protecting them from degradation by protein-digesting enzymes and redirecting them along the correct folding pathway.
- They have an important role in rescuing the cell after an environmental insult such as heat shock, which can cause cytosolic proteins to denature.

The protective effect of molecular chaperones comes at a price—the binding of molecular chaperones to their proteins is energy-dependent, requiring the hydrolysis of ATP. As we have seen, the native structures of proteins are stabilized by weak chemical bonds. A number of stressors, such as changes in temperature, changes in solute concentrations, and toxic metabolic end products, can disrupt these weak bonds, altering the tertiary structure of the protein in ways that can completely destroy biological function. Among other functions, molecular chaperones recognize and bind proteins in a non-native state (typically at exposed hydrophobic residues), preventing further denaturation and allowing a return to the native folded state. Alternatively, some molecular chaperones mark denatured proteins specifically for degradation. By contributing to the elimination of these dysfunctional proteins, chaperones help to preserve the overall integrity of the remaining intracellular protein pool.

Molecular chaperones were first discovered through their specific action on proteins in danger of being denatured by the stresses of high temperature. Consequently, they are commonly called **heat-shock proteins (HSPs)**. While molecular chaperones certainly are expressed in large numbers when cells are under heat stress, they are expressed in quantity in response to a variety of additional intracellular stressors; thus, the broader term **stress proteins** is now also in common usage.

Cellular and molecular biologists have categorized the biochemical and genetic underpinnings of many molecular chaperones, and their role in the ultimate fitness of organisms is becoming increasingly clear. Comparative physiologists working at the organismal level are teaming up with ecologists to determine the ultimate adaptive significance of the numerous families of molecular chaperones.

The proteins of most animals begin to denature at temperatures above 43–45°C. Yet some species of fish, insects, algae, and bacteria inhabit hot-water springs that are about 48°C. A few species of bacteria live at temperatures of up to 100°C! What structural specializations do you think could account for the continuing function at high temperatures of the proteins of these heat-tolerant species?

Nucleic Acids

Deoxyribonucleic acid (DNA) was first isolated from white blood cells and fish sperm in 1869 by Friedrich Miescher. During the next century the chemical composition of DNA was gradually worked out, and evidence slowly accumulated that implicated it in the mechanisms of heredity. We now know that DNA carries coded information, arranged into **genes**, that is passed from each cell to its daughter cells and from one generation of organisms to the next. A second group of nucleic acids, **ribonucleic acid (RNA)**, was subsequently discovered. RNA is now known to be instrumental in translating the coded message of DNA into sequences of amino acids during synthesis of protein molecules.

The **nucleic acids** are polymers of **nucleotides**, each of which consists of a **pyrimidine** or **purine** base, a pentose sugar, and a phosphate residue (Figure 3-21). The nucleotides composing DNA contain deoxyribose, whereas those composing RNA contain ribose (see

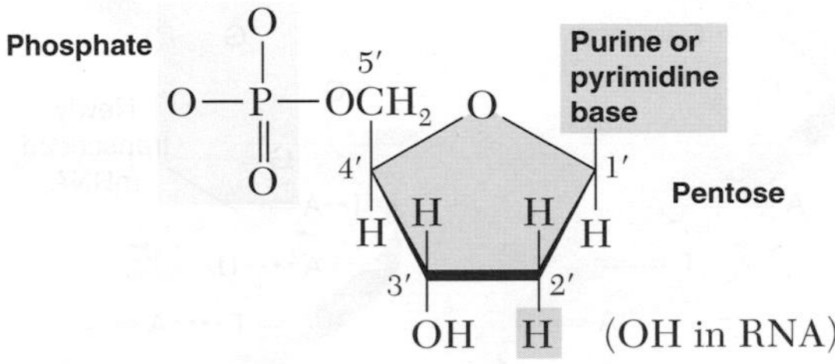

Figure 3-21 The nucleotides composing the nucleic acids have a common structure consisting of a purine or pyrimidine base, a pentose sugar, and a phosphate residue. In DNA, the pentose is 2-deoxyribose, which has two hydrogen atoms attached to the C-2′ atom; in RNA, one of these hydrogens, indicated by the pink-shaded square, is replaced by a hydroxyl group.

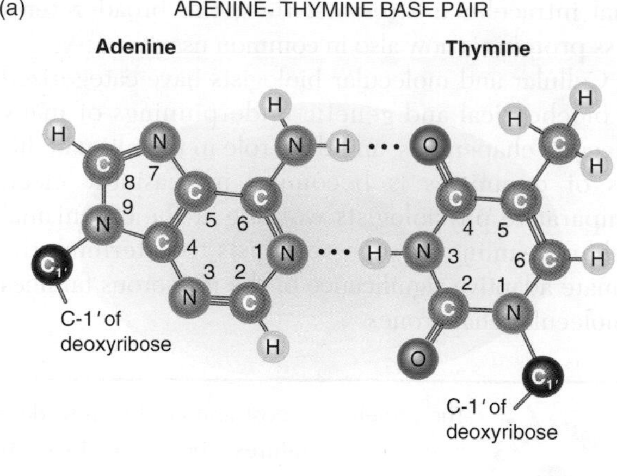

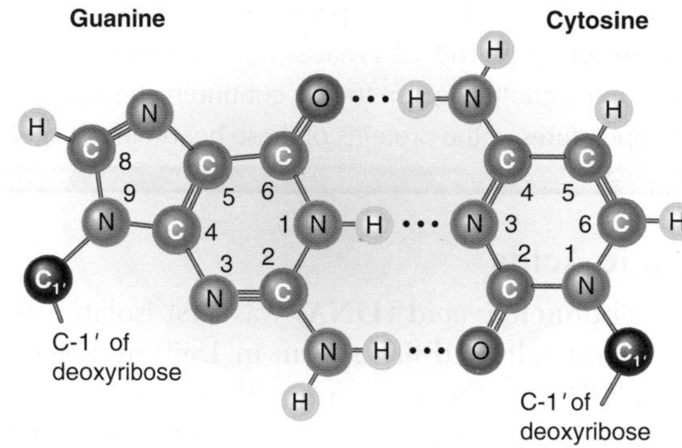

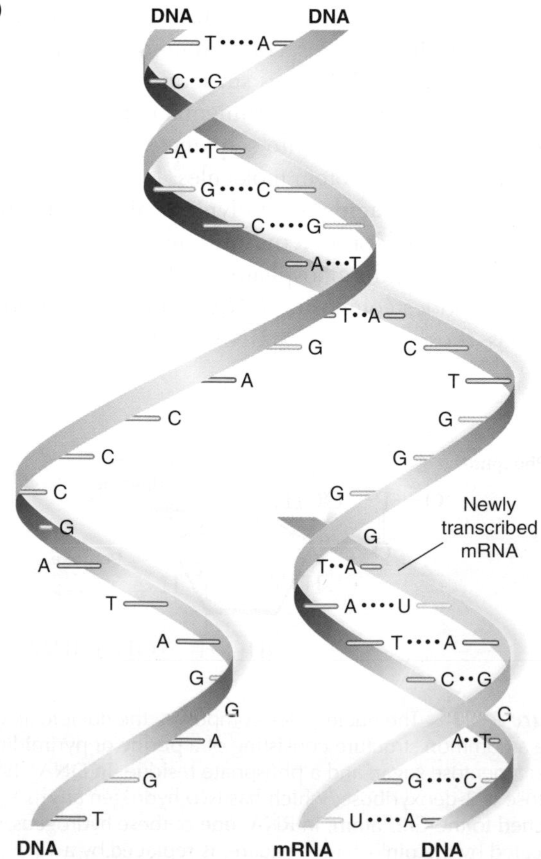

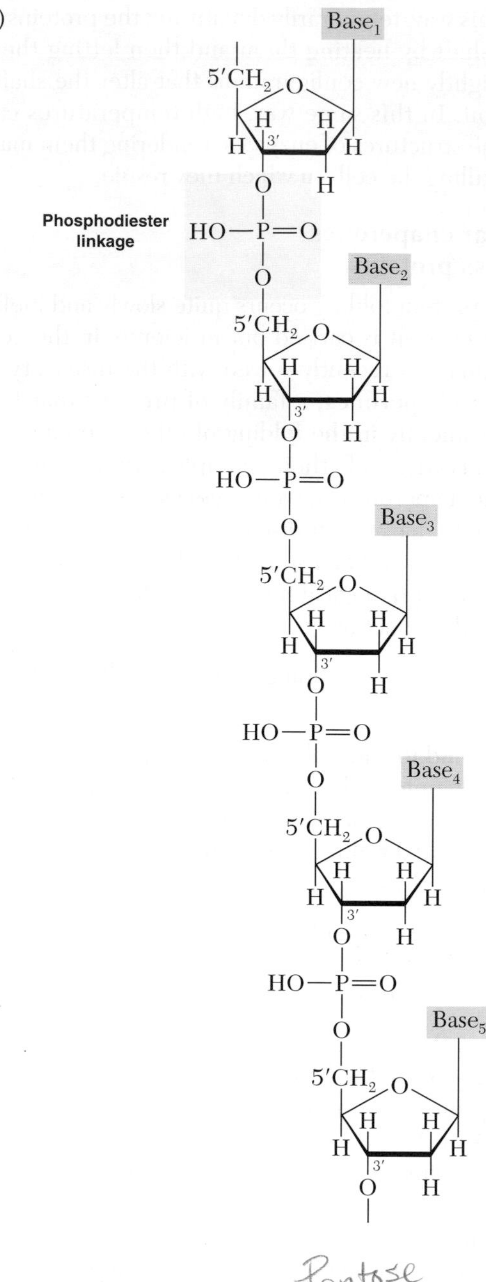

Figure 3-22 **(a)** In nucleic acids, hydrogen bonding (black dots) between purine and pyrimidine bases forms the stable base pairs G-C, A-T (in DNA), and A-U (in RNA). The structure of uracil (U) is the same as that of thymine except that the methyl (—CH_3) group on C-5′ is replaced with a hydrogen. **(b)** The backbone structure of a polynucleotide chain consists of pentose residues linked by phosphodiester bonds. The bases extend away from the backbone. This diagram shows a small portion of a single strand of DNA. **(c)** Native DNA contains two strands wound around each other in a double helix. Hydrogen bonding (black dots) between complementary bases stabilizes the structure. The molecule unwinds during transcription, and one of the strands acts as a template for synthesis of mRNA complementary to the DNA.

Figure 3-14a). The major nucleotides found in nucleic acids contain the following bases: **adenine, thymine, guanine, cytosine,** and **uracil.** Thymine occurs only in DNA, and uracil only in RNA; the other three bases are found in both nucleic acids. Stable base pairs, linked by hydrogen bonds, can form between adenine and thymine (A-T), guanine and cytosine (G-C), and adenine and uracil (A-U), as depicted in Figure 3-22a. In a polynucleotide chain, **phosphodiester linkages** join the 3′ carbon of one pentose ring and the 5′ carbon of the next (Figure 3-22b). The purine and pyrimidine bases extend outward from the repetitive, nonvarying backbone.

Native DNA consists of two polynucleotide chains (or strands) in which the sequence of bases is complementary (e.g., an adenine in one strand is matched by a thymine in the other). Each complementary strand is coiled into a helical staircase, and the two strands are intertwined, forming the familiar DNA double helix, with the hydrogen-bonded base pairs on the inside of the molecule (Figure 3-22c). During replication of DNA, the strands separate from each other, and each of the strands acts as a template for the formation of its complementary strand, thereby yielding two identical molecules of double-stranded DNA.

The genetic information of an organism is encoded in the sequence of bases in its DNA. In a process called **transcription,** a DNA strand acts as a template for the synthesis of **messenger RNA (mRNA)** in the cell nucleus (see Figure 3-22c, bottom). The mRNA strand, which contains the informational sequence present in its DNA template, leaves the nucleus and enters the cytoplasm, where it is decoded by a ribosome into the amino acid sequence of a polypeptide chain. In this process, called **translation,** certain sequences of three bases in the DNA code for certain amino acids. For example, GGU, GGC, CGA, and GGC all code for the amino acid glycine; GCU, GCC, GCA, and GCT code for the amino acid alanine. Thus, the genetic code consists of a four-letter alphabet, A, G, C, and T (in DNA) or U (in RNA), combined into three-letter "words." Once a polypeptide chain is synthesized, it curls and folds, assuming the characteristic secondary and tertiary structure of a protein molecule, and in some cases associates with other chains to form a multi-subunit protein.

More detailed accounts of the major steps relating protein synthesis to nucleic acids may be found in the references listed under Suggested Readings at the end of this chapter.

ENERGETICS OF LIVING CELLS

Animals are fueled by the intake of organic food molecules, which are degraded by digestive and metabolic processes. The chemical energy inherent in the molecular structures of foods is thereby released and made available for the energetic needs of the organism. Animals require energy for such obvious activities as muscle contraction, ciliary movement, and the active transport of molecules by membranes. However, chemical energy is also required for the synthesis of complex biological molecules from simple chemical building blocks and for the subsequent organization of those molecules into organelles, cells, tissues, organ systems, and complete organisms. A living organism must take in fuel frequently and expend energy continuously to maintain its function and structure at all levels of organization. If energy intake drops below the amount required for its maintenance, the organism consumes its own energy stores. When these stores are exhausted, the organism dies, because it cannot perform energy-requiring functions or, ultimately, stave off the tendency of all matter to become disorganized.

The material and energy transactions that take place in an organism constitute its **metabolism.** At the intracellular level, these transactions take place via intricate reaction sequences called **metabolic pathways,** which in a single cell can involve thousands of different kinds of reactions. These reactions do not occur randomly, but in orderly sequences, regulated by a variety of genetic and chemical control mechanisms.

The processes of cell metabolism in animals are of two kinds:

- extraction of chemical energy from food molecules and the channeling of that energy into useful functions
- chemical alteration and rearrangement of food-derived molecules into precursors of other kinds of biological molecules

An example of extraction is the acquisition of amino acids during the digestion of food proteins and their subsequent oxidation within cells, which releases their chemical energy. An example of alteration and rearrangement is the incorporation of amino acids from food into newly synthesized protein molecules in accord with the specifications of the genetic information of the cell. In this section we are concerned less with the biochemical details of cell metabolism than with the thermodynamic and chemical principles that underlie the transfer and utilization of chemical energy within the cell. Thus, we consider the mechanisms by which chemical energy is extracted from food molecules and the manner in which it is made available for the energy-requiring processes discussed in subsequent chapters.

Energy: Concepts and Definitions

Energy may be defined as the capacity to do work. **Work,** in turn, may be defined as the product of force times distance ($W = F \times d$). As an example, when a force lifts a 1 kg mass a height of 1 m, the force is 1 kg, and the mechanical work done is 1 m · kg. The energy expended to do this work (i.e., the useful energy, not

Table 3-4 Forms of energy

Type of energy	Everyday example	Example in animal physiology
Mechanical potential	Stretched spring, lifted weight	Stretched tendon
Chemical potential	Gasoline, natural gas	Glucose, ATP
Mechanical kinetic	Rebounding spring, falling weight	Rebounding tendon
Thermal energy	Actual kinetic energy at molecular level	Heat produced by metabolizing tissue
Electrical energy	Chemical battery	Polarized cell membrane
Radiant energy	Sunlight, heat from a radiator	Heat lost from the skin's surface

including that expended in overcoming friction or expended as heat) is also 1 m·kg. Once the mass is raised to the height of 1 m, it possesses, by virtue of its position, a **potential energy** of 1 m·kg. This potential energy can be converted to **kinetic energy** (energy of movement) if the mass is allowed to drop. Table 3-4 lists the various forms of energy, with examples.

The various forms of energy can power different types of work, as summarized in Table 3-5. We are concerned in this chapter primarily with **chemical energy**, the potential energy stored in the structures of molecules. We will return to mechanical energy in later chapters, especially when we deal with animal locomotion. Before delving into the energy relationships involved in the biochemical reactions of cellular metabolism, it will be useful to review the first and second laws of thermodynamics and the concept of free energy.

Thermodynamic laws

The **first law of thermodynamics** states that energy is neither created nor lost in the universe. Thus, if we burn wood or coal to fuel a steam engine, this does not create new energy but merely converts one form of energy to another—in this example, chemical energy is converted to thermal energy, thermal energy to mechanical energy, and mechanical energy to work.

The **second law of thermodynamics** states that all the energy of the universe will inevitably be degraded to heat and that the organization of matter will become totally randomized. Put more simply, your car will spontaneously break down (become less organized) but will never spontaneously repair itself (become more organized)! In more formal terms, the second law states that the **entropy**—a measure of randomness—of a closed system will progressively increase and that the amount of energy within the system capable of performing useful work will diminish. A system that is ordered (nonrandom) contains potential energy in the form of its orderliness, because in becoming disordered (i.e., as a result of an increase in entropy), it can perform work.

Orderliness increases as an organism develops from a fertilized egg into an adult. In this sense, living systems seemingly defy the second law. It should be recalled, however, that the second law refers to a closed system, and animals (and plants) in their environments are not closed systems. Living organisms maintain relatively low entropy at the expense of energy obtained from their environment. The highly ordered carbohydrate, protein, and fat molecules in the grass are converted in the animal to CO_2, H_2O, and low-molecular-weight nitrogen compounds, releasing energy trapped in the organization of the larger molecules. The carbon, hydrogen, and oxygen atoms in cellulose are in a much more highly ordered state than they are in CO_2 and H_2O; thus the metabolic breakdown of cellulose in the grass by the rhinoceros represents an increase in entropy. At the same time, the cells of the rhinoceros utilize for their own energy requirements a portion of the chemical energy originally stored in the molecular organization of the food molecules. This situation is not in conflict with the second law because the decrease in entropy that results from the animal's synthesis of complex molecules occurs at the expense of an increase in the entropy of molecules produced by plants using the energy of the sun. Ultimately, of course, the rhinoceros dies, and the entropy of its body greatly increases as it undergoes microbial decay or is consumed by other animals. The energy lost from the rhinoceros tissue is then incorporated into a new cycle of order in the cells of the microbes or scavengers.

Free energy

Living systems must function within a relatively narrow range of temperatures and pressures. For this reason, biological systems can utilize only that component of

Table 3-5 Forms of work

Type of work	Driving force	Displacement variable
Expansion work	P (pressure)	Volume
Mechanical work	F (force)	Length
Electrical work	E (electric potential)	Electric charge
Surface work	Γ (surface tension)	Surface area
Chemical work	μ (chemical potential)	Mole numbers

the total available chemical energy capable of doing work under isothermal conditions. This component is called the **free energy**, symbolized by the letter G. Changes in free energy are related to changes in heat and entropy by the equation

$$\Delta G = \Delta H - T\Delta S \qquad (3\text{-}7)$$

in which ΔH is the heat (or enthalpy) produced or taken up by the reaction, T is the absolute temperature, and ΔS is the change in entropy (in units of $cal \cdot mol^{-1} \cdot K^{-1}$). From this equation it is evident that in a chemical reaction that produces no change in temperature ($\Delta H = 0$), there will be a decline in free energy (i.e., ΔG is negative) if there is a rise in entropy (i.e., ΔS is positive), and vice versa. Since the direction of energy flow is toward increased entropy (second law), chemical reactions proceed spontaneously if they produce an increase in entropy (and thus a decrease in free energy). In other words, the reduction of free energy is the driving force in chemical reactions.

The inevitable trend toward increased entropy, with the inevitable degradation of useful chemical energy into useless thermal energy (useless in that it cannot be used to perform work), requires that living systems trap or capture new energy from time to time in order to maintain their structural and functional status quo. In fact, the ability to extract useful energy from their environment is one of the remarkable features that distinguish living systems from inanimate matter, and is the subject of Chapter 15.

All life on Earth ultimately depends on radiant energy from the sun, with the exceptions of chemoautotrophic bacteria and algae and some hydrothermal vent organisms, which obtain energy by the oxidation of inorganic compounds, and those animals that obtain their nourishment from these organisms. Electromagnetic energy from the sun (including visible light) has its origin in nuclear fusion, a process in which the energy of atomic structure is converted to radiant energy. In this process, four hydrogen nuclei are fused to form one helium nucleus, with the release of an enormous amount of radiant energy. A very small fraction of this radiant energy reaches Earth, and a small portion of that is absorbed by chlorophyll molecules in green plants and algae. The energy trapped by photically activated chlorophyll molecules eventually is used to synthesize glucose from H_2O and CO_2. The chemical energy stored in the structure of glucose is available to the plant for controlled release during the processes of cellular respiration.

Later in this chapter we will consider the metabolic pathways by which animal cells release energy through the oxidation of food molecules. First, however, it will be useful to examine some general principles of energy transfer in biochemical reactions and also some features of enzymes, the cellular proteins that allow biochemical reactions to proceed rapidly at biological temperatures.

Transfer of Chemical Energy by Coupled Reactions

There are several categories of biochemical reactions, but the features of reaction rates and kinetics can be illustrated by a simple combination reaction in which two reactant molecules, A and B, react to form two new product molecules, C and D:

$$\underset{\textbf{(reactants)}}{A + B} \rightleftharpoons \underset{\textbf{(products)}}{C + D} \qquad (3\text{-}8)$$

As the double arrow indicates, this reaction is reversible. In theory, any chemical reaction can proceed in either direction provided that the products are not removed. Sometimes, however, the tendency for a reaction to go forward (reactants $\longrightarrow$ products) is so much greater than the tendency for it to go in reverse that for practical purposes the reaction may be considered irreversible.

A reaction tends to go forward if it shows a free-energy change, ΔG, that is negative—in other words, if the total free energy of the reactants exceeds that of the products. Such reactions are said to be **exergonic** and typically liberate heat. The oxidation of hydrogen to water is a simple exergonic reaction:

$$2\ H_2 + O_2 \longrightarrow 2\ H_2O + \text{heat}$$

The energy-requiring reverse reaction occurs during photosynthesis, with the energy supplied by chlorophyll-trapped light quanta:

$$2\ H_2O \xrightarrow{\text{light quantum}} 2\ H_2 + O_2$$

This reaction, which requires the input of energy, is an example of an **endergonic** reaction. Exergonic and endergonic reactions are sometimes referred to as "downhill" and "uphill" reactions, respectively.

The amount of energy liberated or taken up by a reaction is related to the equilibrium constant, K'_{eq}, of the reaction. This is a constant of proportionality relating the concentrations of the products to the concentrations of the reactants when the reaction has reached equilibrium—that is, when the rate of the forward reaction is equal to the rate of the reverse reaction, and the concentration of reactants and products has stabilized:

$$K'_{eq} = \frac{[C][D]}{[A][B]} \qquad (3\text{-}9)$$

Here [A], [B], [C], and [D] are the equilibrium molar concentrations of the reactants and products in Equation 3-8. It is evident that the greater the tendency for the reaction in Equation 3-8 to go to the right, the higher the value of its K'_{eq}. As noted already, this tendency depends on the difference in free energy, ΔG, between the products C and D and the reactants A and B. The greater the drop in free energy, the more completely the reaction proceeds to the right, and the higher its K'_{eq}. The equilibrium constant is related to $\Delta G°$, the change in the free energy of the system under

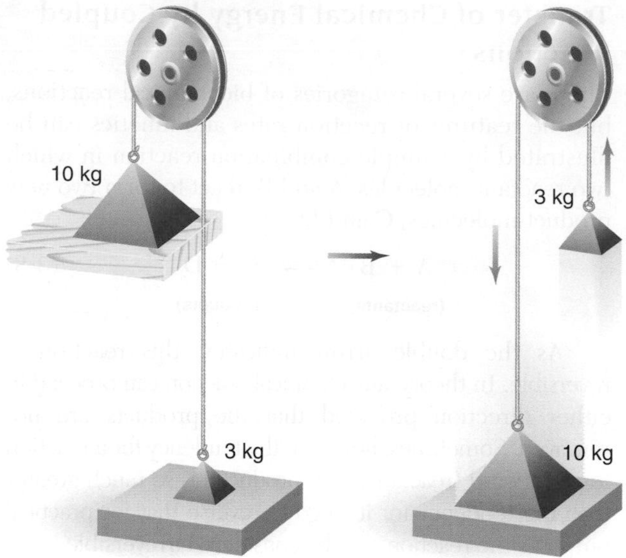

Figure 3-23 In this mechanical analogy to a coupled reaction, the fall of the 10 kg weight easily provides the energy required to lift the 3 kg weight. The pulley and rope constitute the mechanism for connecting the two weights, just as a coupled reaction connects substrates and products.

standardized conditions (25°C, standard pH 7.0, and all reactants present at a concentration of 1 M), by the equation

$$\Delta G^\circ = -RT \ln K'_{eq} \qquad (3\text{-}10)$$

where R is the gas constant and T is temperature in degrees kelvin. It is evident from this equation that if K'_{eq} is greater than 1.0, ΔG° will be negative; and if K'_{eq} is less than 1.0, ΔG° will be positive. Exergonic reactions have a negative ΔG° and therefore occur spontaneously without the need of external energy to "drive" them. Endergonic reactions have a positive ΔG°; that is, they require the input of energy from a source other than the reactants.

Some biochemical processes in living cells are exergonic and others are endergonic. Exergonic processes can proceed on their own under the appropriate conditions. Endergonic processes, however, must be "driven." This requirement is generally fulfilled in the cell by means of coupled reactions, in which common intermediates transfer chemical energy from a molecule with a relatively high energy content to a reactant with a lower energy content. As a result, the reactant is converted into a molecule with a higher energy content and can undergo the required reaction by releasing some of this energy.

A mechanical analogy to a coupled reaction is seen in Figure 3-23. The 10 kg weight on the left can lose its potential energy (10 m · kg) by dropping a distance of 1 m, in which case it will lift the 3 kg weight on the right the same distance. Because the two weights are connected with a rope over a pulley, the fall of the 10 kg weight is coupled to the rise of the 3 kg weight, which initially had no potential energy of its own. It is evident that the falling weight can raise the other one only if it weighs more. Likewise, an exergonic reaction can "drive" an endergonic reaction only if the former liberates more free energy than the latter requires. As a consequence, some energy is necessarily lost, so the efficiency of a coupled reaction is always less than 100%.

ATP: Energy Source for the Cell

The most widely used energy-rich common intermediate in cellular metabolism is the nucleotide **adenosine triphosphate (ATP)**, which can donate its terminal energy-rich phosphate group to any of a large number of acceptor molecules (e.g., sugars, amino acids, other nucleotides, or water). Coupled to an endergonic reaction, this phosphate transfer supplies enough energy to the reacting system to allow the endergonic reaction to proceed.

The ATP molecule consists of an adenosine group, made up of the pyrimidine base adenine and the five-carbon sugar residue ribose, and three linked phosphate groups (Figure 3-24a). Much of the free energy of the molecule resides in the mutual electrostatic

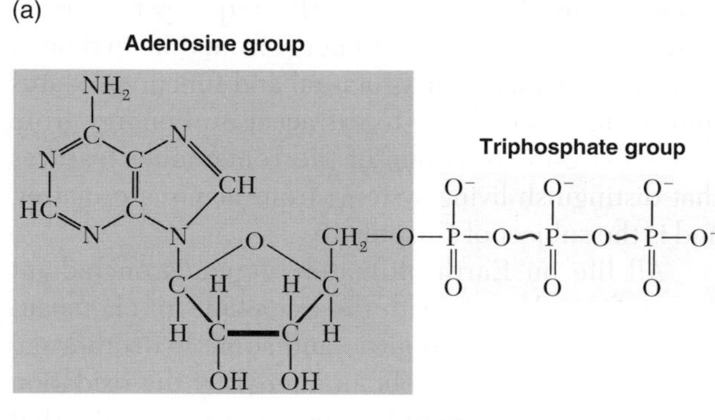

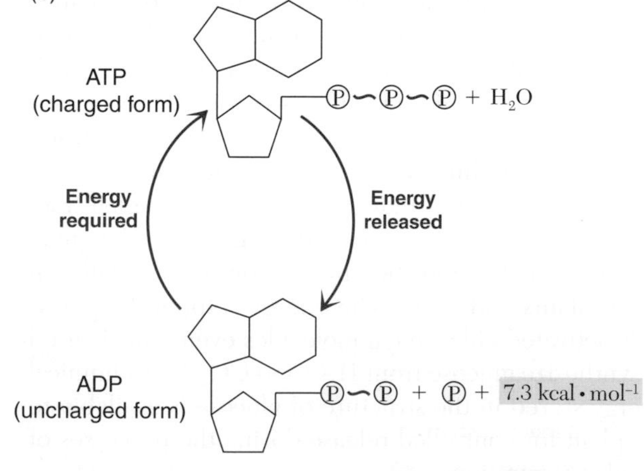

Figure 3-24 ATP is the most common energy carrier in cells. **(a)** Structural formula of ATP, showing the adenosine and triphosphate groups and two high-energy phosphate bonds (red). **(b)** Schematic depiction of interconversion of charged and uncharged forms of ATP. Hydrolysis of ATP to ADP and inorganic phosphate, P_i, releases about 7.3 kcal of free energy per mole of ATP. This reaction can be conveniently monitored by measuring the concentration of inorganic phosphate.

repulsion of the three phosphate units, with their positively charged phosphorus atoms and negatively charged oxygen atoms. The mutual repulsion of these phosphate units is analogous to the repulsion of bar magnets, with their north and south poles aligned, held together by a sticky wax. If the wax, which is analogous to the high-energy bonds in ATP (often symbolized by O~P), is softened by warming, the energy stored by virtue of the proximity of the mutually repelling magnets is released as the magnets spring apart. Likewise, the breaking of the bonds between the phosphate units of ATP results in the release of free energy (Figure 3-24b). Once the terminal phosphate group of ATP is removed by hydrolysis, the mutual repulsion of the two products, adenosine diphosphate (ADP) and inorganic phosphate (P_i), is such that the probability of their recombining is very low. That is, their recombination is highly endergonic. The standard free-energy change, $\Delta G°$, for the hydrolysis of ATP under standard conditions is -7.3 kcal·mol^{-1}.

The role of ATP in driving otherwise endergonic reactions by means of coupled reactions is illustrated by the endergonic formation of X—phosphate from X + phosphate:

$$\text{X} + \text{P}_i \longrightarrow \text{X—phosphate}$$

$$\Delta G° = +4.0 \text{ kcal}\cdot\text{mol}^{-1}$$

This reaction can be coupled with the hydrolysis of ATP to ADP + P_i, which is strongly exergonic:

$$\text{ATP} + \text{H}_2\text{O} \longrightarrow \text{ADP} + \text{P}_i$$

$$\Delta G° = -7.3 \text{ kcal}\cdot\text{mol}^{-1}$$

The total free energy liberated in the coupled reaction will be equal to the sum of the free-energy changes of the two parent reactions. The combined reaction is written

$$\text{X} + \text{P}_i \longrightarrow \text{X—phosphate}$$
$$\text{ATP} + \text{H}_2\text{O} \longrightarrow \text{ADP} + \text{P}_i$$
$$\text{ATP} + \text{X} \longrightarrow \text{ADP} + \text{X—phosphate}$$

The free energy released is

$$\Delta G° = +4.0 \text{ kcal}\cdot\text{mol}^{-1} + (-7.3 \text{ kcal}\cdot\text{mol}^{-1}) = -3.3 \text{ kcal}\cdot\text{mol}^{-1}$$

Note that the $\Delta G°$ for the condensation of X and P_i has a positive value (+4.0 kcal·mol^{-1}), so normally this reaction would not proceed. However, because the $\Delta G°$ for the hydrolysis of ATP is larger and negative (-7.3 kcal·mol^{-1}), the net $\Delta G°$ of the coupled reaction is negative, allowing it to proceed.

Although ATP and other nucleotide triphosphates are responsible for the transfer of energy in many coupled reactions, it should be stressed that the mechanism of a common intermediate is widely employed in biochemical reaction sequences. Portions of molecules—and even atoms, such as hydrogen—are transferred, along with chemical energy, from one molecule to another by common intermediates in many such sequences of reactions. The high-energy nucleotides are special only in that they act as a general energy currency in a large number of energy-requiring reactions. In this role, ADP is the "discharged" form, and ATP is the "charged" form (see Figure 3-24b). Numerous other high-energy phosphorylated compounds occur in the cell, some with higher free energies of hydrolysis than ATP (Figure 3-25). The cell can use these compounds in the formation of ATP. As we will see, the cell also has other biochemical mechanisms for channeling chemical energy into the formation of ATP.

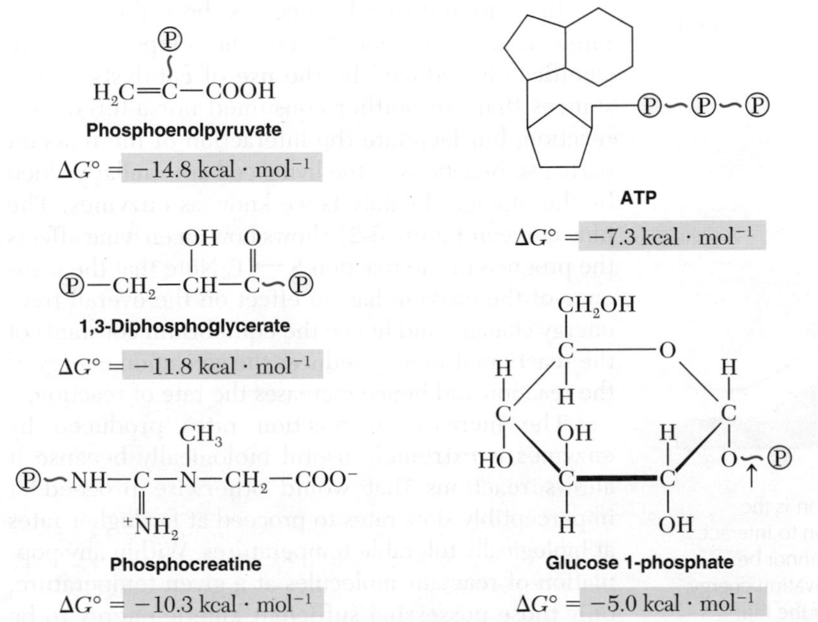

Figure 3-25 Hydrolysis of compounds containing high-energy phosphate bonds (red) provides cells with energy for energy-requiring reactions and processes. Although ATP is the most common energy currency in biological systems, several other phosphorylated compounds have higher free energies of hydrolysis. These compounds can be used by cells to synthesize ATP from ADP and inorganic phosphate. The $\Delta G°$ values are the standard free energies at pH 7 for hydrolysis of the bonds indicated by the red squiggle bonds.

Arginine phosphate and creatine phosphate are special reservoirs of chemical energy for the rapid phosphorylation of ADP to reconstitute ATP during vigorous muscle contraction. These compounds are called **phosphagens.** In vertebrate muscle, which contains only creatine phosphate, the following transphosphorylation reaction occurs:

$$\text{creatine phosphate} + \text{ADP} \xrightleftharpoons[\text{enzymes}]{\text{Transphosphorylase}} \text{creatine} + \text{ATP}$$

$$\Delta G^\circ = -3.0\ \text{kcal} \cdot \text{mol}^{-1}$$

In invertebrate muscle, either or both phosphagens may be found.

Temperature and Reaction Rates

The rate at which a chemical reaction proceeds depends on temperature. This dependency is not surprising, because temperature is an expression of molecular motion. As temperature increases, so does the average molecular velocity. This greater velocity increases the number of molecular collisions per unit time and thereby increases the probability of successful interaction of the reactant molecules. Furthermore, as their velocities increase, the molecules possess higher kinetic energies and thus are more likely to react upon collision. The kinetic energy required to cause two colliding molecules to react is called the free energy of activation, or **activation energy** (ΔG^*). It is measured as the number of calories required to bring all the molecules in a mole of reactant at a given temperature to a reactive (or activated) state.

The requirement for activation energy applies to exergonic as well as endergonic reactions. Although a reaction may have the potential for liberating free energy, it will not proceed unless the reactant molecules possess the necessary kinetic energy. This situation can be compared to one in which it is necessary to push an object over a low ridge before it is free to roll downhill (Figure 3-26).

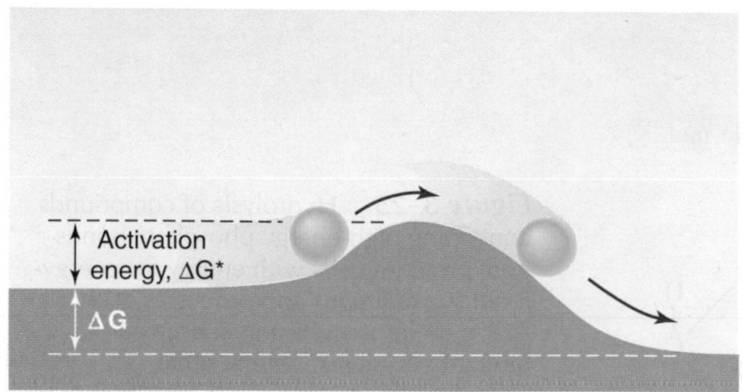

Figure 3-26 The activation energy of a reaction is the energy required to bring the reactants into position to interact. In this analogy, the potential energy of the rock cannot be liberated until some energy, referred to as the activation energy, is expended to bring it into position at the crest of the hill.

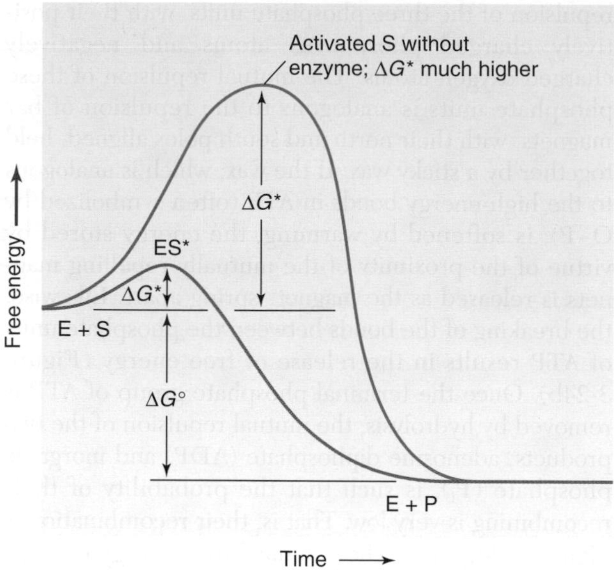

Figure 3-27 Enzymes lower the activation energy, ΔG^*, of a reaction. An enzyme (E) binds with a substrate (S) to form an activated enzyme-substrate complex (ES*). In the ES* state, the path to the product (P) has a lower activation energy than the nonenzymatic reaction. Note that the overall free-energy change, ΔG°, is the same in the nonenzymatic reaction (red curve) and in the enzymatic reaction (blue curve).

The red curve in Figure 3-27 shows the relationship between free energy and the progress of a reaction in which a reactant, or substrate (S), is converted to a product (P). The substrate must first be raised to an energy state sufficient to activate it, allowing it to react. Since the reaction yields free energy, the energy state of the product is lower than that of the substrate. Note that the overall free-energy change of the reaction is independent of the activation energy required to produce the reaction.

In many industrial processes, both the reaction rate and activation energy (i.e., the temperature) are significantly reduced by the use of **catalysts**—substances that are neither consumed nor altered by a reaction, but facilitate the interaction of the reactant particles. Reactions in the living cell are similarly aided by the biological catalysts we know as **enzymes.** The blue curve in Figure 3-27 shows how an enzyme affects the progress of the reaction S → P. Note that the presence of the enzyme has no effect on the overall free-energy change (and hence the equilibrium constant) of the reaction; it merely reduces the activation energy of the reaction and hence increases the rate of reaction.

The increase in reaction rates produced by enzymes is extremely useful biologically because it allows reactions that would otherwise proceed at imperceptibly slow rates to proceed at far higher rates at biologically tolerable temperatures. Within any population of reactant molecules at a given temperature, only those possessing sufficient kinetic energy to be activated will react. If an enzyme that reduces the energy required for activation is added, a far larger number of molecules can react in a given time at the same temperature. The rates of various enzyme-catalyzed reactions range from 10^3 to 10^{17} times the rate of the corresponding uncatalyzed reactions, representing an enormous acceleration.

An extremely important advantage of catalyzed reactions is the possibility of regulating the rate of reaction by varying the activity of the catalyst. In the next two sections, we first discuss how enzymes operate and then see how cells regulate their metabolic reactions by controlling the synthesis and catalytic activity of enzymes.

ENZYMES: GENERAL PROPERTIES

Enzymes were first discovered more than a century ago when substances that increased the rate of alcoholic fermentation were isolated from yeasts. These substances were found to be inactivated by heating, whereas the reactants in these fermentation reactions were unaffected by heating. This finding was the first indication that enzymes are protein molecules. It was subsequently discovered that enzymes are proteins of very specific amino acid composition and sequence. All of these proteins, or at least their enzymatically active portions, have a globular conformation. Each cell in an organism contains literally thousands of types of enzyme molecules, which catalyze all the synthetic and metabolic reactions of the cell.

Enzyme Specificity and Active Sites

Each enzyme is, to some degree, specific for a certain **substrate** (reactant molecule). Some enzymes act at certain types of bonds and therefore act on many different substrates that have such bonds. Trypsin, which is a **proteolytic** (protein-hydrolyzing) enzyme found in the digestive tract, acts on any peptide bond in which the carbonyl group is part of an arginine or lysine residue, regardless of the position of those bonds in the polypeptide chain of a protein. Another intestinal proteolytic enzyme, chymotrypsin, specifically catalyzes the hydrolysis of peptide bonds in which the carbonyl group belongs to a phenylalanine, tyrosine, or tryptophan residue (Figure 3-28).

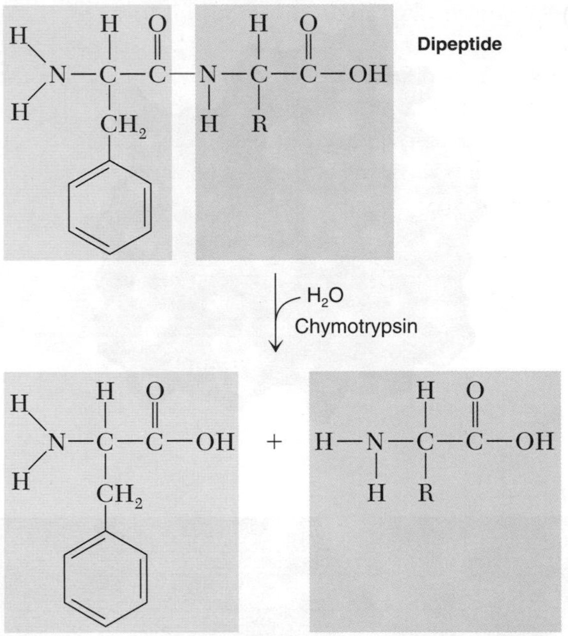

Most enzymes, however, exhibit far more substrate specificity than do proteolytic enzymes. The enzyme sucrase, for example, catalyzes the hydrolysis of the disaccharide sucrose into glucose and fructose, but it cannot attack any other disaccharides, such as lactose or maltose. These substrates are hydrolyzed by enzymes that are specific for them (lactase and maltase, respectively). Many enzymes are specific for a single stereoisomer of a substrate molecule. Stereoisomers are molecules that are chemically identical but whose functional groups are attached in different configurations around central carbon atoms. Two molecules that have identical

Dipeptide

H_2O
Chymotrypsin

Figure 3-28 Chymotrypsin hydrolyzes any peptide bond in which the carbonyl group belongs to a phenylalanine, tyrosine, or tryptophan residue. Shown here is the action of chymotrypsin on a dipeptide containing phenylalanine (pink).

functional groups but different configurations may be chemically identical, but it is unlikely that they will both bind effectively to the same enzyme active site, just as the left and right hand have the same number of digits but don't fit in the same glove.

The highly specific nature of most enzymes arises from the close and complementary fit between enzyme and substrate in a special portion of the enzyme surface called the **active site** (Figure 3-29 on the next page). The enzyme molecule is made up of one or more peptide chains folded to form a more or less globular protein of a specific conformation. The active site consists of the side groups of certain amino acid residues that are brought into proximity by this tertiary structure, even though they may be widely separated in the amino acid sequence of the enzyme. The complementarity between substrates and enzyme active sites has been well established by experiments with substrate analogs (i.e., molecules structurally related to the substrate but slightly different). The ability of an enzyme's active site to interact with analogs decreases as the interatomic distances, number and position of charged groups, and bond angles of the analog molecules depart from those of the normal substrate.

Mechanism of Catalysis by Enzymes

Enzyme activity, the catalytic potency of an enzyme, can be expressed as the **turnover number**, which is the number of reactions catalyzed per second by the enzyme. In an enzymatic reaction, the substrate first interacts with the active site of the enzyme, forming an

(a)

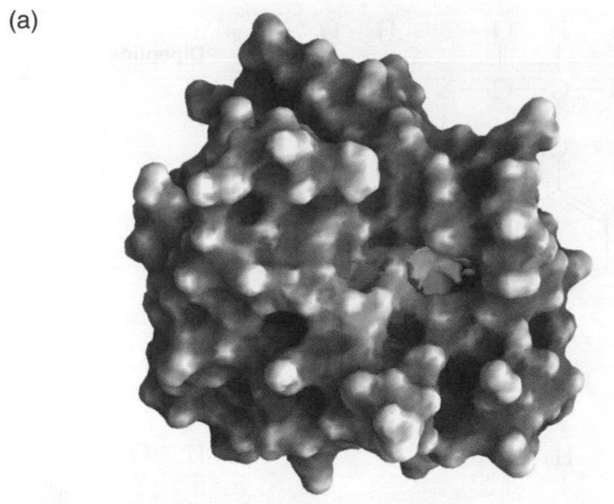

(b)

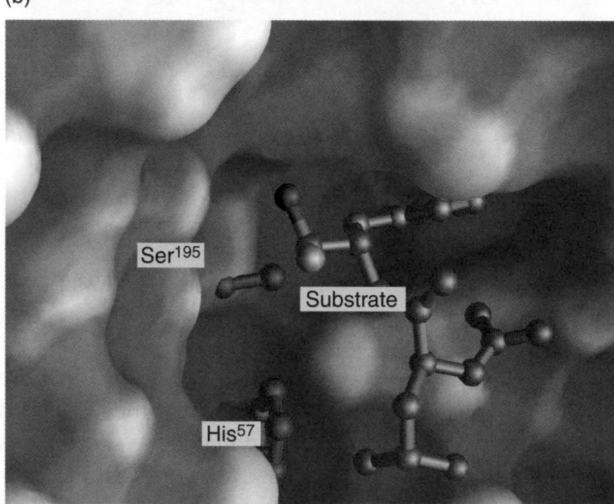

Figure 3-29 Computer-generated models of the enzyme chymotrypsin. **(a)** A surface model of the enzyme; the active site is red. **(b)** A close-up of the active site with bound substrate (mostly green). The two amino acid residues that are most important for substrate binding (residues 195 and 57, shown in red) are situated far apart in the primary structure of the protein, as indicated by their numbers. [From Nelson and Cox, 2000.]

enzyme-substrate complex (ES). As noted earlier, this interaction reduces the activation energy of the reaction, thereby increasing the rate of the reaction (see Figure 3-27).

Enzymes use several catalytic mechanisms to accelerate reaction rates. An enzyme may

- hold the substrate molecules in a particular orientation in which the reacting groups are sufficiently close to one another to enhance the probability of reaction
- react with the substrate molecule to form an unstable intermediate that then readily undergoes a second reaction, forming the final product
- have side groups within the active site that act as proton donors or acceptors in acid-base reactions

Whatever the catalytic mechanism for a particular reaction, once the substrate molecules have reacted, the product separates from the enzyme, freeing the enzyme to form an ES complex with a new substrate molecule. Essentially all of the enzyme molecules present can become tied up as ES if the substrate concentration is high enough relative to the enzyme concentration, a condition called **saturation.**

Effects of Temperature and pH on Enzymatic Reactions

Any factor that influences the structural conformation of an enzyme, and hence the arrangement of amino acid side groups in the active site, will alter the activity of the enzyme. Temperature and pH are two common factors within cells that influence the rates of enzymatic reactions through enzyme conformational changes.

As we saw previously, an increase in temperature increases the probability of protein denaturation, which disrupts the conformation of polypeptide chains. In the case of enzymes, denaturation destroys catalytic activity. For this reason, enzyme-catalyzed reactions exhibit a characteristic curve of reaction rate versus temperature (Figure 3-30a). As temperature increases, the reaction rate initially increases due to the increased kinetic energy of the substrate molecules. As temperature increases further, however, the onset of denaturation causes the reaction rate to decrease. At the optimal temperature, the reaction rate is maximal. The temperature sensitivity of enzymes and other protein molecules contributes to the lethal effects of excessive temperatures.

Electrostatic bonds often participate in the formation of an ES complex. Since H^+ and OH^- can act as counterions for electrostatic sites, a drop in pH exposes more positive sites on an enzyme for interaction with negative groups on a substrate molecule. Conversely, a rise in pH facilitates the binding of positive groups on a substrate to negative sites on the enzyme. Thus, it is not surprising that the activity of enzymes typically is exquisitely sensitive to the pH of the environment and that each enzyme has an optimal pH range (Figure 3-30b).

Cofactors

Some enzymes require the participation of small molecules called **cofactors** in order to perform their catalytic function. One class of cofactors consists of small organic molecules called **coenzymes,** which participate along with substrates in enzymatic reactions. The enzyme glutamate dehydrogenase, for example, requires the coenzyme **nicotinamide adenine dinucleotide (NAD^+)** to catalyze the oxidative deamination of the amino acid glutamate:

$$\underset{\textbf{(oxidized)}}{\text{Glutamate} + NAD^+} \rightleftharpoons \alpha\text{-ketoglutarate} + \underset{\textbf{(reduced)}}{NADH + NH_4^+}$$

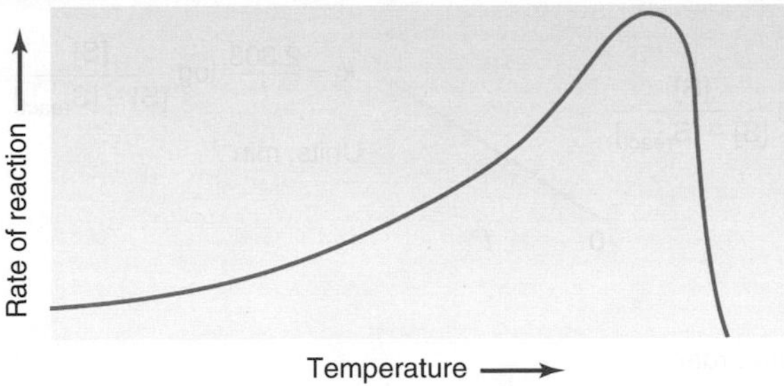

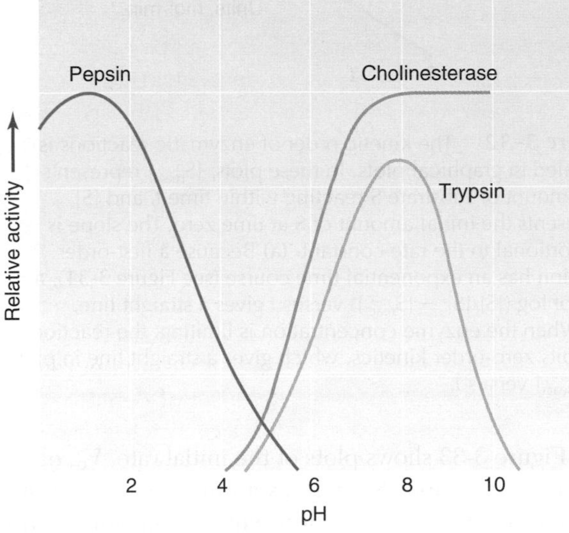

Figure 3-30 Both temperature and pH influence the activity of enzymes. **(a)** Temperature affects the reaction rates of most enzymes in the same way. The sharp decline at higher temperatures is the result of denaturation of the enzyme. **(b)** The effect of pH on catalytic activity varies among enzymes, but nearly all have a distinct optimal pH.

A number of enzymes contain covalently attached cofactors as part of the enzyme molecule. Many of these cofactors are vitamins, as we will see in Chapter 15. An enzyme minus its cofactor is called an **apoenzyme.** Since an apoenzyme cannot function without its cofactor, or coenzyme, it is not surprising that vitamin deficiencies can have profound pathological effects.

Other enzymes require metal ions as cofactors. The most important of the ions that function as cofactors, along with examples of enzymes that require them, are listed in Table 3-6. Especially interesting is Ca^{2+}, whose intracellular concentration ($<10^{-6}$ M) is much lower than that of most other common physiologically important ions. Unlike Mg^{2+}, Na^+, K^+, and other cofactor ions, the calcium ion is present in concentrations that are limiting for certain enzymes. As discussed in Chapter 9, the Ca^{2+} concentration of the **cytosol** (the unstructured fluid phase of the cytoplasm in which many metabolic reactions occur) is regulated by the plasma membrane of the cell and by internal organelles, such as the endoplasmic reticulum. The activity of calcium-activated enzymes can be regulated by controlling calcium transport at these points. Cellular processes regulated by Ca^{2+} concentration include muscle contraction, secretion of neurotransmitters and hormones, ciliary activity, assembly of microtubules, and amoeboid movement.

Table 3-6 **Metal ions functioning as cofactors**

Metal ion	Some enzymes requiring this cofactor
Ca^{2+}	Phosphodiesterase
	Protein kinase C
Cu^{2+} (Cu^{+})	Cytochrome oxidase
	Tyrosinase
Fe^{2+} or Fe^{3+}	Catalase
	Cytochromes
	Ferredoxin
	Peroxidase
K^+	Pyruvate phosphokinase (also requires Mg^{2+})
Mg^{2+}	Phosphohydrolases
	Phosphotransferases
Mn^{2+}	Arginase
	Phosphotransferases
Na^+	Plasma membrane ATPase (also requires K^+ and Mg^{2+})
Zn^{2+}	Alcohol dehydrogenase
	Carbonic anhydrase
	Carboxypeptidase

Source: Adapted from Nelson and Cox, 2000.

Enzyme Kinetics

The rate at which an enzymatic reaction proceeds depends on the concentrations of substrate, product, and active enzyme. For purposes of simplicity, we will assume that the product is removed as fast as it is formed. In that case, the rate of reaction will be limited by the concentration of either the enzyme or the substrate. If we further assume that the enzyme is present in excess, then the rate at which a single substrate species, S, is converted to the product, P, is determined by the concentration of S:

$$S \xrightarrow{k} P$$

where k is the **rate constant** of the reaction. The rate of conversion of S to P can be expressed mathematically as

$$\frac{-d[S]}{dt} = k[S] \tag{3-11}$$

in which [S] is the instantaneous concentration of the substrate, k is the rate constant of the reaction, and $d[S]/dt$ is the rate at which S is converted to P with respect to time.

The disappearance of S and the appearance of P are plotted as functions of time in Figure 3-31. Note that the concentration of S decreases exponentially as the concentration of P increases exponentially. An exponential time function is always generated when the rate of change of a quantity ($d[S]/dt$ in this example) is proportional to the instantaneous value of that quantity ([S] in this example).

The relationship expressed by Equation 3-11 is more usefully presented as

$$\log \frac{[S]}{[S] - [S_{react}]} = \frac{k_1 t}{2.303} \qquad (3\text{-}11a)$$

where [S] is the initial concentration of substrate and $[S_{react}]$ is the amount of substrate that has reacted within time t. A plot of the left side of Equation 3-11a versus time yields a straight line whose slope is proportional to the rate constant, k_1 (Figure 3-32a). A reaction that exhibits this behavior has **first-order kinetics.** The rate constant of a first-order reaction has the dimension of reciprocal time—that is, "per second," or s^{-1}. The rate constant can be inverted to yield the **time constant,** which has the dimension of time. Thus, a first-order reaction with a rate constant of $10 \cdot s^{-1}$ has a time constant of 0.1 seconds.

In a reaction with two substrates, A and B, in which excess enzyme is present and the product, P, does not accumulate, the rate of disappearance of A will be proportional to the product [A][B].

$$A + B \xrightarrow{k} P$$

This reaction will proceed with **second-order kinetics.** It is noteworthy that the order of the reaction is not determined by the number of substrate species participating as reactants, but instead by the number of species present in rate-limiting concentrations. Thus, if B were present in great excess over A, the reaction A + B → P would become first-order, since its rate would be limited by only one substrate concentration.

The rate of an enzymatic reaction is independent of substrate concentrations when the enzyme is saturated—that is, the enzyme is present in limiting concentrations and all the enzyme molecules are complexed with substrate. Such reactions proceed with **zero-order kinetics** (Figure 3-32b).

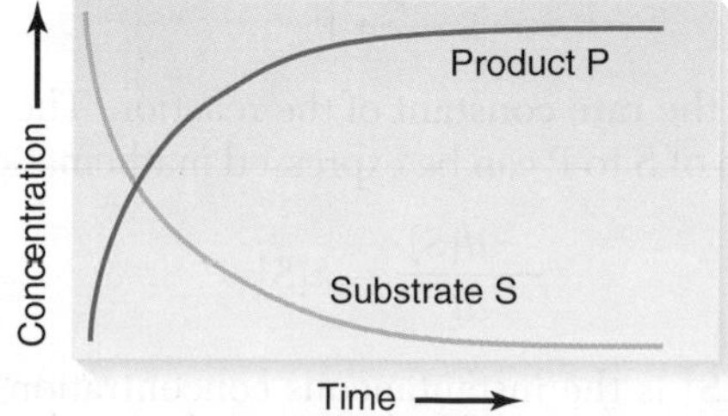

Figure 3-31 The concentrations of substrate S and product P change in a nonlinear fashion during the reaction S → P.

(a) First order

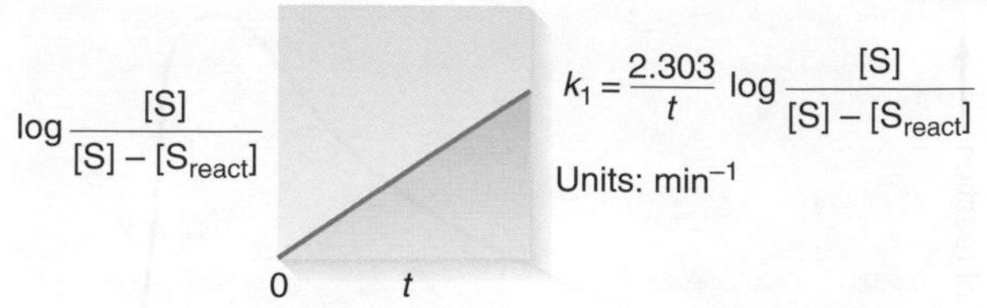

(b) Zero order

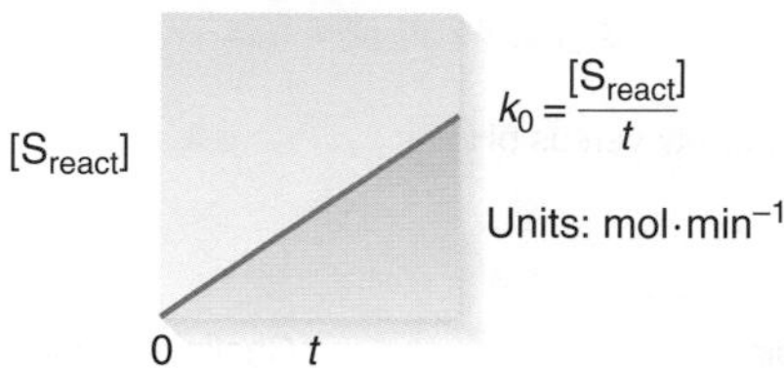

Figure 3-32 The kinetic order of enzymatic reactions is revealed in graphical plots. In these plots, $[S_{react}]$ represents the amount of substrate S reacting within time t, and [S] represents the initial amount of S at time zero. The slope is proportional to the rate constant. **(a)** Because a first-order reaction has an exponential time course (see Figure 3-31), a plot of log ([S]/[S] − $[S_{react}]$) versus t gives a straight line. **(b)** When the enzyme concentration is limiting, the reaction exhibits zero-order kinetics, which gives a straight line in plots of $[S_{react}]$ versus t.

Figure 3-33 shows plots of the initial rate, V_0, of an enzymatic reaction (S → P) as a function of substrate concentration, [S], at two different enzyme concentrations. At both enzyme levels, the reaction is first-order (i.e., V_0 is proportional to [S]) at low substrate concentrations. At higher substrate concentrations, however, the reaction becomes zero-order, because all the enzyme molecules are complexed with substrate. In this situation, the concentration of enzyme, not substrate, limits V_0. In the living cell, all orders of reaction, as well as mixed-order reactions, occur.

The maximum rate of any enzymatic reaction, V_{max}, occurs when all the enzyme molecules catalyzing that reaction are bound to substrate molecules—that is, when the substrate is present in excess and the enzyme concentration is rate-limiting (see Figure 3-33). For each enzymatic reaction, there is a characteristic relationship between V_{max} and enzyme concentration. Although all enzymes can become saturated, they show great variation in the concentration of a given substrate that will produce saturation. The reason for this is that enzymes differ in affinity for their substrates. The greater the tendency for an enzyme and its substrate to form an ES complex, the higher the percentage of total enzyme, E_t, tied up as ES at any given concentration of substrate. Thus, the higher this affinity, the lower the substrate concentration required to saturate the enzyme.

The relationship between the kinetics of an enzyme-catalyzed reaction and the affinity of the

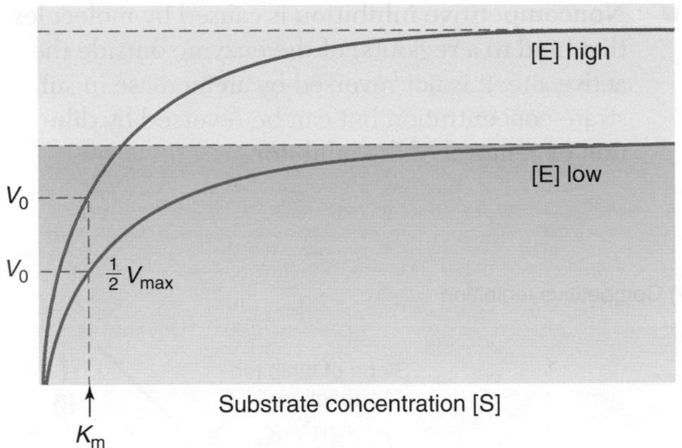

Figure 3-33 At a given enzyme concentration, the initial rate, V_0, of the reaction S → P rises linearly with increasing substrate concentration. Eventually all of the enzyme becomes saturated, at which time the enzyme (E) becomes rate-limiting (rate = V_{max}). The Michaelis-Menten constant, K_m, is equal to the substrate concentration at which the reaction rate is one-half the maximum. The blue and red curves are for different enzyme concentrations. Note that K_m is independent of the enzyme concentration, [E], whereas V_{max} depends directly on [E].

enzyme for substrate was worked out early in the twentieth century. The general theory of enzyme action and kinetics was proposed by Leonor Michaelis and Maud L. Menten in 1913, and later extended by George E. Briggs and John B. S. Haldane. The rate for a reaction catalyzed by a single enzyme is given by the **Michaelis-Menten equation:**

$$v_0 = \frac{V_{max}[S]}{K_m + [S]} \tag{3-12}$$

where v_0 is the initial reaction rate at substrate concentration [S], V_{max} is the reaction rate with excess substrate, and K_m is the **Michaelis-Menten constant.**

Consider the special case when $V_0 = \frac{1}{2}V_{max}$. Substituting for v_0, we get

$$\frac{V_{max}}{2} = \frac{V_{max}[S]}{K_m + [S]}$$

Dividing by V_{max} gives

$$\frac{1}{2} = \frac{[S]}{K_m + [S]}$$

On rearranging, we obtain

$$K_m + [S] = 2\,[S] \tag{3-13}$$

or

$$K_m = [S] \tag{3-14}$$

Therefore, K_m equals the substrate concentration at which the initial reaction rate is half what it would be if the substrate were present to saturation.

Thus, the Michaelis-Menten constant, K_m (in units of moles per liter), depends on the affinity of the enzyme for a substrate. For a given enzyme and substrate, K_m is equal to the substrate concentration at which the initial reaction rate is $\frac{1}{2}V_{max}$. By inference, then, K_m represents the concentration of substrate at which half the total enzyme present is combined with substrate in ES; that is, $[E_t]/[ES] = 2$. The value of K_m can be determined from a plot of V_0 versus [S], as illustrated in Figure 3-33. The greater the affinity between an enzyme and its substrate, the lower the K_m of the enzyme-substrate interaction. Stated inversely, $1/K_m$ is a measure of the affinity of the enzyme for its substrate. As illustrated by the plots for two enzyme concentrations in Figure 3-33, K_m does not depend on the enzyme concentration, whereas V_{max} does.

The relationship between V_0 and substrate concentration described by the Michaelis-Menten equation (Equation 3-12) is a hyperbolic function. For this reason, numerous data points are required to accurately plot V_0 against [S] as in Figure 3-33. The Michaelis-Menten equation can be algebraically transformed, however, into a more useful linear form called the **Lineweaver-Burk equation:**

$$\frac{1}{V_0} = \frac{K_m}{V_{max}[S]} + \frac{1}{V_{max}} \tag{3-15}$$

We can conclude from this equation that a plot of $1/V_0$ versus $1/[S]$ will give a straight line with a slope of K_m/V_{max} and with intercepts of $1/V_{max}$ on the vertical axis and $-1/K_m$ on the horizontal axis (Figure 3-34). Because of the linear nature of this curve, only two experimental data points (i.e., V_0 at two values of [S]) are needed to construct a Lineweaver-Burk plot. V_{max} and K_m can then be determined from the two intercepts.

Note that the Michaelis-Menten analysis is not limited to enzyme-substrate interactions but can be (and often is) applied to any system that displays hyperbolic saturation kinetics as illustrated in Figure 3-33.

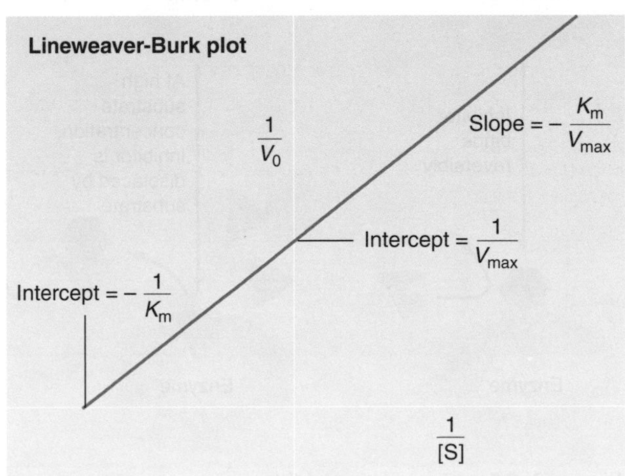

Figure 3-34 In a Lineweaver-Burk plot, the reciprocal of the reaction rate, $1/V_0$, is plotted against the reciprocal of the substrate concentration, $1/[S]$. For an enzymatic reaction with first-order kinetics, this plot intercepts the horizontal axis at $-1/K_m$ and the vertical axis at $1/V_{max}$.

Enzyme Inhibition

The activity of most enzymes can be inhibited by certain molecules. As we'll see in the following section, enzyme inhibition is used in the living cell as a means of controlling enzymatic reactions. By studying enzyme inhibitors, both physiological and nonphysiological, biochemists have discovered important features of the active sites of enzymes and of the mechanisms of enzyme action. The therapeutic effects of many drugs depend on their ability to inhibit specific enzymes, thereby blocking metabolic or physiological processes involved in disease. The antidepressant drug tranylcypromine (Parnate), for example, inhibits the enzyme monoamine oxidase (MAO). High levels of serotonin in the brain are associated with positive mood, and MAO breaks down serotonin. Inhibiting MAO thus increases the level of serotonin in the brain and improves mood.

Enzymes can be irreversibly inhibited by agents that form highly stable covalent bonds with groups inside the active site, thereby blocking formation of the ES complex. While some types of animal toxins fall into this category, two types of reversible inhibition are more relevant to normal cell function:

- **Competitive inhibition** is caused by molecules that appear to react directly with the active site of the enzyme; it can be reversed by an increase in substrate concentration.
- **Noncompetitive inhibition** is caused by molecules that bind to a region(s) of the enzyme outside the active site; it is not reversed by an increase in substrate concentration but can be reversed by dilution or removal of the inhibitor.

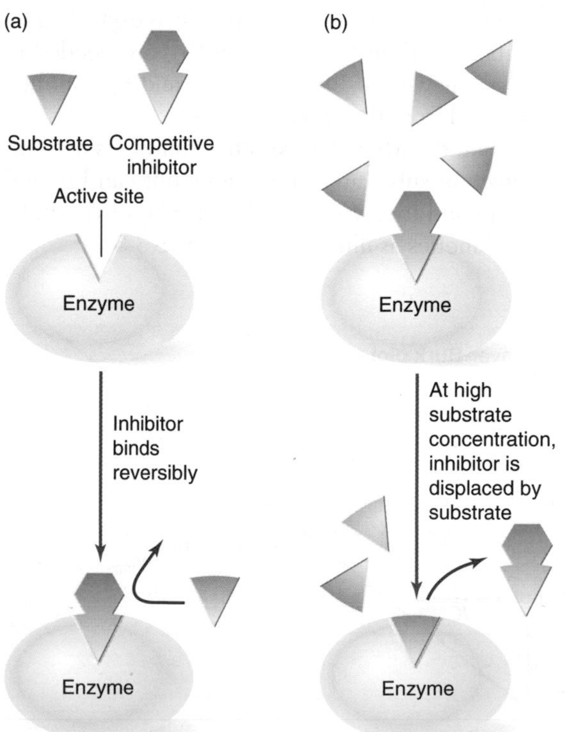

Figure 3-35 Competitive inhibitors compete with substrate for binding sites. **(a)** Binding of a competitive inhibitor to the active site of an enzyme interferes with the binding of the substrate. **(b)** If the substrate concentration is increased, substrate molecules can displace bound inhibitor molecules.

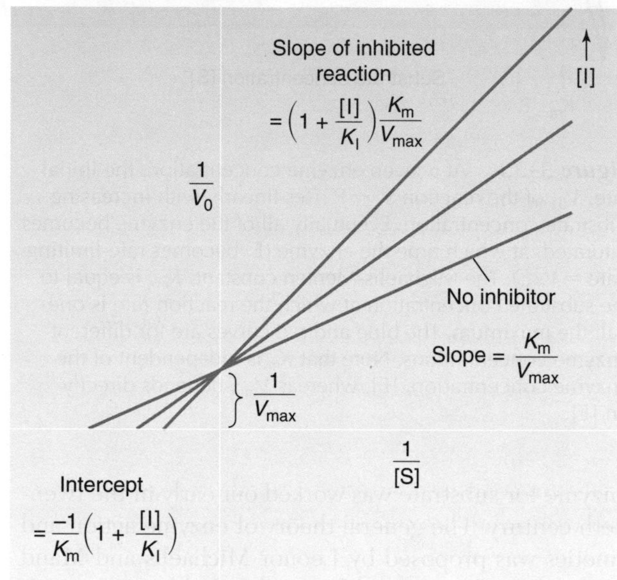

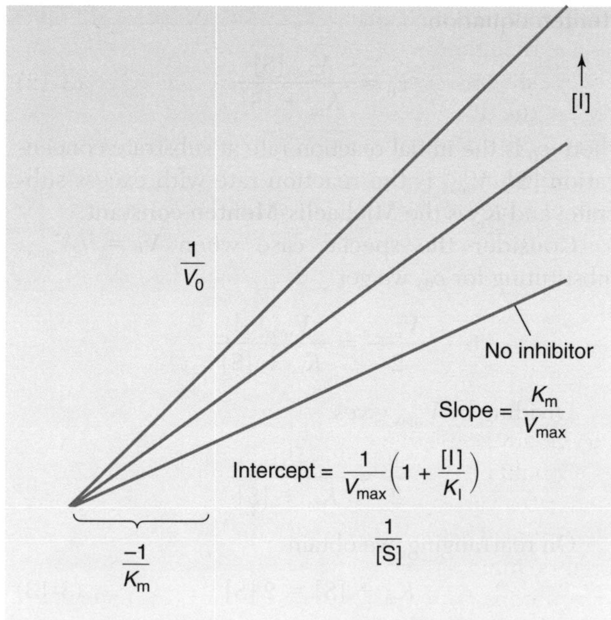

Figure 3-36 Competitive and noncompetitive inhibitors produce different effects on Lineweaver-Burk plots. **(a)** A competitive inhibitor increases K_m but does not affect V_{max}. **(b)** Conversely, a noncompetitive inhibitor produces no change in K_m but decreases V_{max}. It is kinetically similar to a reduction in enzyme concentration. [I], inhibitor concentration; [S], substrate concentration; K_I, dissociation constant of inhibitor-enzyme complex.

Most competitive inhibitors are substrate analogs and compete with substrate molecules for the active site (Figure 3-35). Thus, increasing the concentration of the substrate reduces the probability of the inhibitor binding—more substrate reduces the probability that the next molecule to bind to the active site will be an inhibitor. In contrast, a noncompetitive inhibitor binds the enzyme at a site other than the active site, and its chemical structure typically differs from that of the substrate. Noncompetitive inhibition is not reversed by an increase in substrate concentration because inhibitor and substrate are not competing for the same binding site. As illustrated in Figure 3-36, competitive and noncompetitive inhibitors produce readily distinguishable alterations in Lineweaver-Burk plots. Both types of inhibitors increase the slope of a Lineweaver-Burk plot—the hallmark of inhibition—but competitive inhibitors change the x-intercept and noncompetitive inhibitors change the y-intercept.

A competitive inhibitor does not change the V_{max} of an enzymatic reaction; that is, when the substrate concentration is extrapolated so that 1/[S] approaches 0, the substrate will displace all of the inhibitor molecules from the enzyme (see Figure 3-36a). However, it takes a higher concentration of substrate in the presence of the competitive inhibitor to keep half of the enzyme molecules at any instant complexed with the substrate. Essentially, a competitive inhibitor increases the "apparent" K_m—in other words, the intercept on the 1/[S] axis is shifted toward a higher substrate concentration in the presence of a competitive inhibitor. Because a noncompetitive inhibitor does not directly interfere with binding of substrate to an enzyme, it has no effect on the intercept on the 1/[S] axis in a Lineweaver-Burk plot (see Figure 3-36b); that is, the K_m of an enzyme is not affected by a noncompetitive inhibitor. However, the kinetic effect of a noncompetitive inhibitor is to reduce the catalytic potency, or turnover number, of an enzyme; that is, it reduces the effective concentration of the enzyme.

REGULATION OF METABOLIC REACTIONS

Without any regulation of reaction rates, cellular metabolism would be uncoordinated and undirected, proceeding at rates varying somewhere between explosive and nonexistent. Growth, differentiation, and maintenance would be impossible, to say nothing of subtle homeostatic compensatory responses of cells to externally imposed stresses. Fortunately, metabolism is regulated by several different mechanisms, including the control of enzyme synthesis and the control of enzyme activity. Let us consider these two mechanisms in turn.

Control of Enzyme Synthesis

The quantity of an enzyme in a cell depends on the balance between its rate of destruction and its rate of synthesis. As we saw earlier, proteins are denatured at high temperatures and are broken down by the action of proteolytic enzymes. The rate of synthesis of an enzyme can be limited under certain conditions that reduce protein synthesis generally (e.g., inadequate diet or the unavailability of necessary amino acid precursors). Normally, however, the rate of synthesis of a particular enzyme is regulated at the molecular level by modulation of the rate of transcription of the gene encoding it.

Some cells synthesize certain enzymes (e.g., those involved in metabolizing lactose) only after they are exposed to the initial substrate in the reaction pathway (or molecules related to it). In these cases, the substrate interacts with a protein that interacts directly with the chromosomal DNA molecule, changing the rate at which the DNA is transcribed into RNA, and therefore the rate at which RNA is translated into protein. This process is an example of metabolic economy, since enzymes are synthesized only when they are needed (i.e., when their substrate is present). Similarly, metabolic conditions can result in the activation of repressor proteins that have the opposite effect on DNA transcription.

Cells possess additional mechanisms for regulating the production of enzymes, such as controlling access to specific genes by controlling how the DNA is packaged in the chromosome and controlling the rate at which RNA is translated into protein. All of these mechanisms are important in the differentiation of cells during development. Each somatic cell in an organism contains the same information in its DNA, yet cells in different tissues contain widely divergent amounts of the different enzymes encoded by the genetic material. It is evident that in any given tissue some genes are turned on and others are turned off. The inventory of proteins in cells is monitored and maintained on a moment-to-moment and day-to-day basis to ensure metabolic and structural continuity and to maintain intracellular homeostasis.

Control of Enzyme Activity

The activity of some enzymes can be regulated by modulator (regulator) molecules, which interact with a part of the enzyme molecule distinct from the active site. This part of the enzyme, called an allosteric site, can bind a modulator molecule, altering the tertiary structure of the enzyme and thus changing the conformation of the active site (Figure 3-37 on the next page). As a result, the affinity of the enzyme for its substrate decreases or increases. Allosterically regulated enzymes operate at key points in metabolic pathways, and modulation of their activities plays an important role in the regulation of these pathways. Let's take a closer look at allosteric mechanisms for controlling enzyme activity.

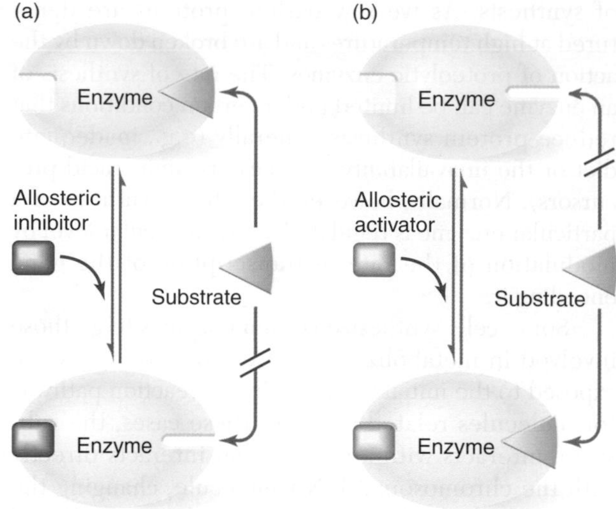

Figure 3-37 Allosteric interactions can result in either activation or inhibition of enzyme activity. **(a)** Binding of an allosteric inhibitor molecule to an allosteric site can indirectly alter the configuration of the active site of an enzyme, thereby rendering the enzyme inactive. Noncompetitive inhibitors act by this mechanism. **(b)** Conversely, binding of an allosteric activator can alter the active site so that the enzyme becomes catalytically active.

End-product (feedback) inhibition

In most metabolic pathways, a regulatory enzyme catalyzes a reaction that is virtually irreversible under cellular conditions; for this reason, accumulation of the end product does not slow the rate of the reaction. However, some important metabolic pathways have built-in mechanisms for regulating the reaction rate by end-product accumulation, independently of enzyme quantity. In these pathways, it is usually the first enzyme of the sequence that acts as a regulatory enzyme. Most commonly, the end product of the pathway feeds back to inhibit the activity of this first enzyme in the pathway (Figure 3-38). Such end-product inhibition limits the rate of accumulation of the end product by slowing the entire sequence from the beginning. The interaction of the end product in a biosynthetic pathway with a regulatory enzyme occurs at its allosteric site, making the end product an allosteric inhibitor (see Figure 3-37a). An example of end-product inhibition occurs in the biosynthesis of the catecholamine norepinephrine, which functions as both a neurotransmitter and a hormone. High concentrations of norepinephrine inhibit the enzyme tyrosine hydroxylase, an essential enzyme in the sequence of reactions leading to the production of norepinephrine.

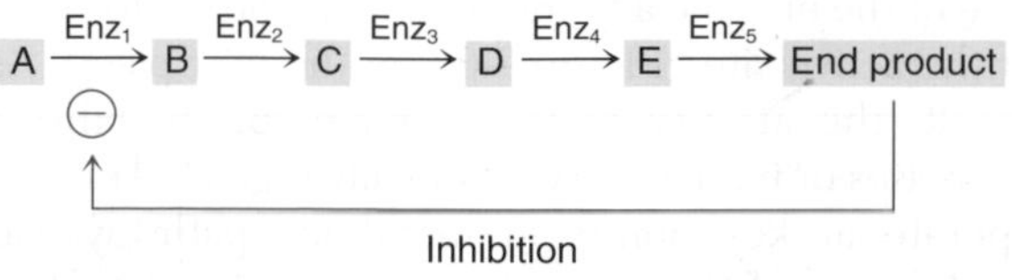

Figure 3-38 Products of a reaction sequence can allosterically inhibit a rate-limiting enzyme in the pathway. The metabolic end product in this case directly reduces the activity of the initial enzyme.

Enzyme activation

The requirement for cofactors exhibited by some enzymes provides another means of regulating enzyme activity and hence the rate of biochemical reactions. As noted earlier, Ca^{2+} and several other cations act as cofactors for various enzymes (see Table 3-6). These cation cofactors appear to act as allosteric activators (see Figure 3-37b) for some enzymes.

The intracellular free concentration of certain ions depends on diffusion and active transport across membranes separating the cell exterior from intracellular ion storage sites. By regulating the levels of cofactor ions in the cytosol, the cell can modulate the activity of certain enzymes. Ca^{2+} is an important and common regulatory cofactor. This cation is present at much lower concentrations than other cofactor cations within the cytosol. Because extracellular concentrations of Ca^{2+} are typically 1000 times higher than its concentration in the cytosol (commonly far less than 10^{-6} M), extremely small changes in the net flux of Ca^{2+} across the cell membrane (or the membranes of cytoplasmic organelles) can produce substantial changes in the intracellular free Ca^{2+} concentration (see Figure 9-16). The special role of Ca^{2+} as an intracellular regulatory molecule is discussed in Chapters 6, 9, and 10.

Now that we have discussed the principles underlying cellular energetics and the characteristics of enzyme-catalyzed reactions, let's consider how cells produce ATP, the key molecule in the energy transactions of cells.

METABOLIC PRODUCTION OF ATP

If we compare the energy utilization of an animal with that of an automobile, we note that both "machines" require the intermittent intake of chemical fuel to energize their activities. Their use of fuel differs, however, in at least one important aspect. In the automobile engine, the organic fuel molecules in gasoline are oxidized (ideally) to CO_2 and H_2O in one explosive step. The extreme heat generated by this rapid oxidation produces a great increase in the pressure of gases in the engine's cylinders. In this way the chemical energy of the fuel is converted to mechanical movement (kinetic energy).

Since living systems are capable of sustaining only small temperature and pressure gradients, heat provided by the simple one-step combustion of fuel would be essentially useless for powering their activities. For this reason, cells have evolved metabolic pathways that carry out the conversion of chemical energy in small increments. The energy of food molecules is recovered for useful work through the formation of intermediate compounds of progressively lower energy

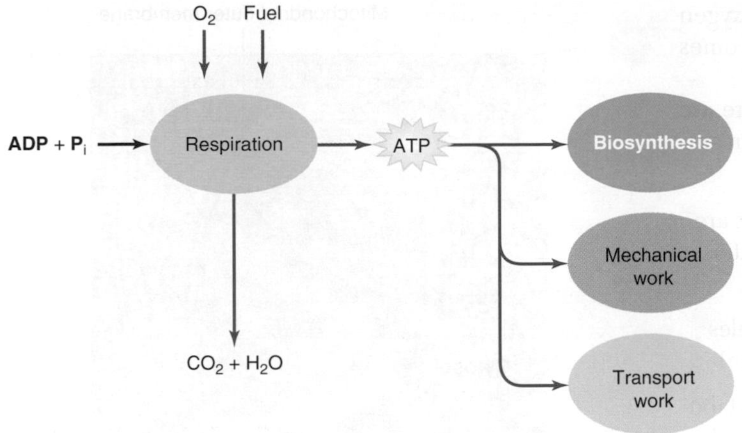

Figure 3-39 The hydrolysis of ATP powers numerous energy-requiring processes in biological systems, including biosynthesis, mechanical work, and transport work. The ADP produced by hydrolysis is recycled to ATP by rephosphorylation energized by the oxidation of food molecules to CO_2 and H_2O.

content. At each exergonic step, some of the chemical energy is liberated as heat, while the rest is transferred as free energy to the reaction product. Chemical energy conserved and stored in the structure of these intermediate products is then transferred to the common high-energy intermediate ATP and to other similar high-energy molecules. The chemical energy of these molecules is readily available for a wide variety of cellular processes (Figure 3-39).

As mentioned earlier, carbohydrates, lipids, and proteins ingested in foods are the primary fuels for animals. After digestion, these molecules generally enter a distribution system (usually the circulatory system in complex metazoans) as five- or six-carbon sugars, fatty acids, and amino acids, respectively (Figure 3-40). These small molecules then enter the tissues and cells of the animal, where they may be (1) immediately broken down into smaller molecules for the extraction of chemical energy or for rearrangement and recombination into other types of molecules, or (2) built up into larger molecules, such as polysaccharides (e.g., glycogen), fats, or proteins. With few exceptions, these molecules, too, will eventually be broken down and eliminated as CO_2, H_2O, and urea (or other nitrogenous waste products). Nearly all molecular constituents of a cell are in dynamic equilibrium, constantly being replaced by components newly synthesized from simpler organic molecules.

Some simple organisms, including certain bacteria and yeasts as well as a few invertebrate animal species, can live indefinitely under totally **anaerobic** (i.e., essentially oxygen-free) conditions. The anaerobes fall into two groups: obligatory anaerobes, which cannot grow in the presence of oxygen (e.g., *Clostridium botulinum,* the botulism bacterium, and some vertebrate intestinal gut parasites), and facultative anaerobes, which survive and reproduce equally well in the presence or absence of oxygen (e.g., yeasts). All vertebrates and most invertebrates, however, are **aerobic,** meaning that they require a supply of molecular oxygen for cellular respiration. Even these aerobic animals generally possess tissues such as locomotor muscles that can metabolize for some

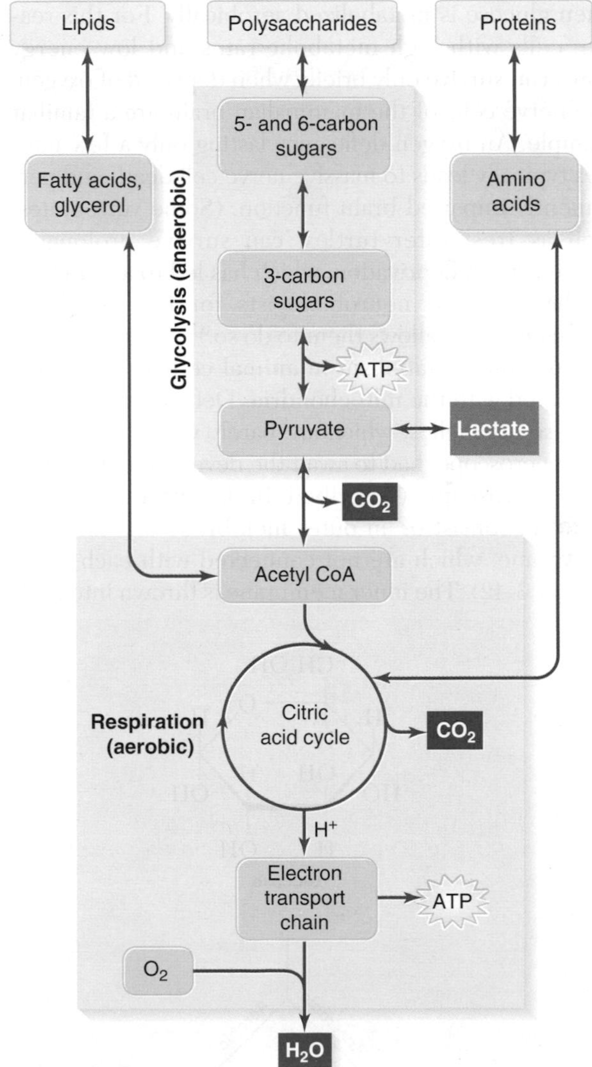

Figure 3-40 Carbohydrates, lipids, and proteins can all be degraded by cells to yield usable energy in the form of ATP. All three classes of food molecules feed intermediate products into the citric acid cycle, which is linked to the electron-transport chain. During aerobic metabolism, these products are oxidized completely to CO_2 and H_2O. During anaerobic metabolism, however, cellular respiration is impossible, and lactate accumulates.

period of time anaerobically, building up an "oxygen debt" that is repaid when sufficient oxygen becomes available, as we'll see at the end of this chapter.

As these metabolic observations suggest, there are two kinds of energy-yielding metabolic pathways in animal tissues (Figure 3-41):

- **aerobic metabolism,** in which food molecules are completely oxidized to carbon dioxide and water by molecular oxygen
- **anaerobic metabolism,** in which food molecules are oxidized incompletely to lactic acid (lactate)

The energy yield per molecule of glucose metabolized anaerobically by the pathway of glycolysis is only a tiny fraction of the energy yield in aerobic metabolism: 2 mol of ATP per mol of glucose versus 38 mol of ATP when glucose is metabolized aerobically. For this reason, cells with high metabolic rates and low energy stores can survive only briefly when deprived of oxygen. The nerve cells of the mammalian brain are a familiar example. An oxygen deficiency lasting only a few minutes typically leads to massive nerve cell death and permanently impaired brain function. (Some vertebrates, such as freshwater turtles, can survive prolonged, severe oxygen deprivation, which has led to a search by biochemists and neurobiologists for the metabolic mechanism that allows them to do so.)

Aerobic metabolism in animal cells is intimately associated with the **mitochondria.** Detailed description of these organelles, which are barely visible under the light microscope, had to await the development of electron microscopy (described in Chapter 2). Mitochondria consist of an outer membrane and an inner membrane, which are not connected with each other (Figure 3-42). The inner membrane is thrown into folds

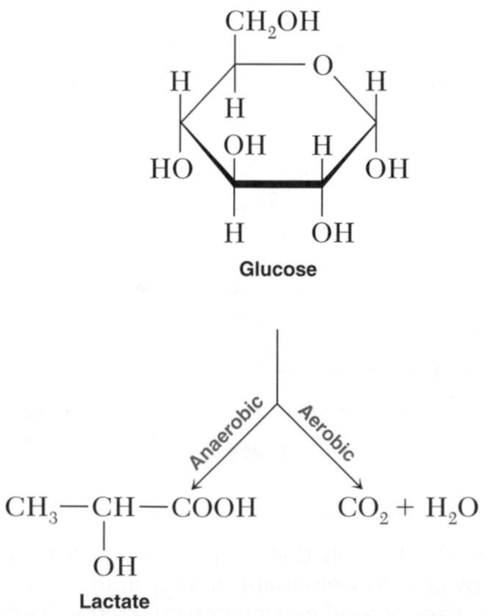

Glucose ($C_6H_{12}O_6$ ring: CH_2OH, O, H, H, OH, H, HO, OH, H, OH)

Anaerobic → $CH_3—CH(OH)—COOH$ (Lactate)

Aerobic → $CO_2 + H_2O$

Figure 3-41 Glucose catabolism can occur by either an anaerobic or an aerobic pathway. The energy yield from aerobic metabolism is far greater than that from anaerobic metabolism.

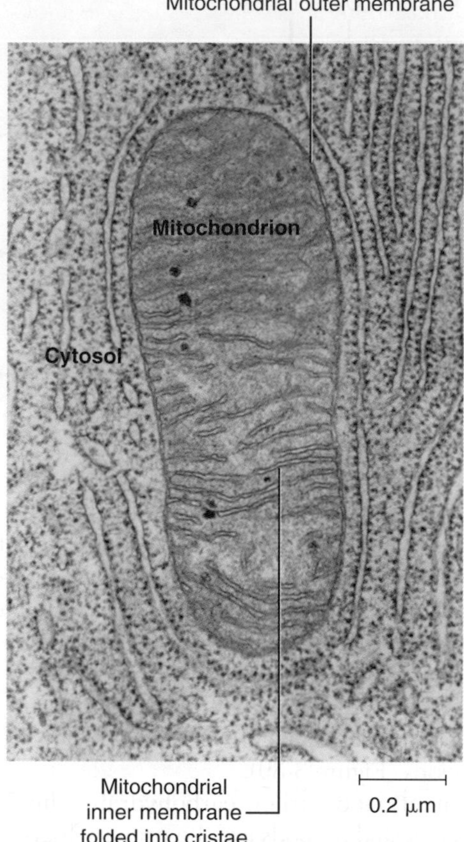

Figure 3-42 Cellular respiration occurs in mitochondria. This electron micrograph of a mitochondrion in a bat pancreas cell reveals the outer and inner membranes. The folding of the inner membrane to form cristae creates a large surface area. [Courtesy of K. R. Porter.]

called **cristae,** which make the surface area of the inner membrane much greater than that of the outer membrane. The interior space bounded by the inner membrane is called the **mitochondrial matrix,** and the space between the two membranes is the intermembrane space. As we'll see below, the inner membrane and the matrix contain enzymes that catalyze the final oxidation of foods and the production of ATP during cellular respiration. The matrix contains ribosomes, dense granules (consisting primarily of salts of calcium), and mitochondrial DNA, which is involved in replication of the mitochondria. Mitochondria are quite numerous in most cells, ranging from 800 to 5000 per mammalian liver cell. They also tend to congregate most densely in those portions of a cell that are most active in utilizing ATP.

Oxidation, Phosphorylation, and Energy Transfer

How is the chemical energy liberated during metabolism conserved and channeled into high-energy intermediates? Recall that when a complex organic molecule is taken apart, free energy is liberated. This increases the entropy (degree of randomness) of the constituent matter. Such a release of energy occurs when glucose is oxidized to carbon dioxide and water by combustion in the overall reaction

$$C_6H_{12}O_6 + 6\ O_2 \longrightarrow 6\ CO_2 + 6\ H_2O$$
$$\Delta G° = -686\ \text{kcal} \cdot \text{mol}^{-1}$$

The 686 kcal liberated by the oxidation of 1 mol of glucose is the difference between the free energy incorporated into the structure of the glucose molecule during photosynthesis and the total free energy contained in the carbon dioxide and water produced. If 1 mol of glucose is oxidized to carbon dioxide and water in a single step by combustion (i.e., if it is burned), the free energy is released as 686 kcal of heat. If the same 1 mol of glucose is oxidized by cellular respiration, however, a portion of this energy is conserved as useful chemical energy rather than appearing as a flash of heat, and the conserved energy is channeled into ATP through the phosphorylation of ADP. The overall reaction for the metabolic oxidation of glucose by the cell, including the coupled conversion of ADP to ATP, can be written as

$$C_6H_{12}O_6 + 38\ P_i + 38\ ADP + 6\ O_2 \longrightarrow 6\ CO_2 + 6\ H_2O + 38\ ATP$$

$$\Delta G° = -420\ \text{kcal (as heat)}$$

Subtracting the energy lost as heat (420 kcal) from the total initial energy (686 kcal) leaves the energy stored in ATP, about 266 kcal, or about 38 mol of ATP (266 kcal divided by the amount of energy stored per mole of ATP, about 7 kcal · mol^{-1}).

How is the all-important free energy of the glucose molecule transferred to ATP during respiration? To understand this, we must first recall that the **oxidation** of a molecule is most broadly defined as the transfer of electrons from that molecule to another molecule. Conversely, the **reduction** of a molecule is the acceptance of electrons from another molecule. In an oxidation-reduction reaction, the **reductant** (electron donor) is oxidized by the **oxidant** (electron acceptor). Together they form a **redox pair:**

$$\text{electron donor} \rightleftharpoons e^- + \text{electron acceptor}$$

or

$$\text{reductant} \rightleftharpoons e^- + \text{oxidant}$$

Whenever electrons are accepted from a reductant by an oxidant, energy is liberated, because the electrons move into a more stable (higher-entropy) situation when transferred to the oxidant. This transfer is akin to water dropping from a higher to a lower level. It is the difference between the two levels that determines the amount of energy liberated. A molecule with a greater tendency to accept electrons than the molecule with which it undergoes a redox reaction is said to have a greater (more positive) **reduction potential** and will act as an oxidant. If it has a more negative reduction potential (sometimes described as a higher **electron pressure**, or tendency to pass on electrons), it will act as a reductant. The free-energy change in each reaction is proportional to the difference between the electron pressures of the two molecules of the redox pair.

In aerobic cell metabolism, electrons move sequentially from compounds of higher electron pressure to compounds of lower electron pressure. The final electron acceptor in the chain is molecular oxygen. Since oxygen acts merely as an electron acceptor, it is possible in theory to support aerobic metabolism without oxygen, provided that a suitable electron acceptor is supplied in place of oxygen.

In being transferred from glucose to oxygen, electrons undergo an enormous drop in both reduction potential and free energy. One of the functions of cell metabolism is to transport electrons from glucose to oxygen in a series of small steps instead of one large drop. This transport is carried out by two mechanisms found in all aerobic cells. First, as we noted earlier, the chemical conversion of food molecules such as glucose to the fully oxidized end products (e.g., CO_2 and H_2O) occurs in many steps and involves many intermediate compounds. Second, electrons removed from these compounds are passed to oxygen via a series of electron acceptors and donors of progressively lower electron pressure. These mechanisms allow energy to be channeled into the synthesis of ATP in "packets" of appropriate size.

Efficiency of Energy Metabolism

If water is boiled by the heat of burning glucose to produce steam pressure for a steam engine, the mechanical output of the engine divided by the free-energy drop of 686 kcal · mol^{-1} represents the efficiency of the conversion of chemical to mechanical energy. Steam engines have attained efficiencies of approximately 30%. How efficiently do living cells transfer chemical energy from glucose to ATP? Let's see.

Under standard conditions, it takes about 7 kcal to phosphorylate 1 mol of ADP to form ATP. If the free energy of glucose were conserved with an efficiency of 100%, each mole of glucose could energize the synthesis of 98 mol of ATP from ADP and inorganic phosphate (686/7 = 98). In fact, during the metabolic oxidation of 1 mol of glucose, only 38 mol of ATP are synthesized, giving an overall efficiency of about 42% or more.* The remaining free energy is liberated as metabolic heat, which accounts for a part of the heat that warms and thereby increases the metabolic rate of the tissue. According to the laws of thermodynamics

* The 42% calculated here is for standard conditions. The efficiency of energy conservation by the cell may in fact be as high as 60%, because the free energy of hydrolysis of ATP under intracellular conditions has been estimated to be greater than that under standard conditions. The energetic efficiency of ATP production is therefore substantially better than that of a steam engine—in fact, better than that of any method yet devised by humans for converting chemical energy to mechanical energy.

discussed earlier, all the energy incorporated into ATP and transferred to other molecules is eventually degraded to heat. The oxidation of fossil fuels represents a long-delayed return of stored chemical energy to the original low-energy, high-entropy state of CO_2 and water.

It is interesting to compare the efficiencies of anaerobic and aerobic glucose metabolism. As noted earlier, only 2 mol of ATP are produced per mol of glucose during anaerobic glycolysis. Thus, whereas aerobic respiration conserves a minimum of about 42% of the free energy of the glucose molecule, glycolysis conserves only about 2%. Stated differently, the energy conservation of glucose metabolism via aerobic glycolysis and the citric acid cycle is about 20 times more efficient than via anaerobic glycolysis. It is not surprising, then, that most animals carry out aerobic metabolism and require molecular oxygen for survival.

SUMMARY

Water and the concept of pH

■ The polarity of the water molecule is responsible for hydrogen bonding.

■ Hydrogen bonding between water molecules confers on water many special properties that have profoundly shaped the evolution of animals.

■ Water dissociates spontaneously into H^+ and OH^-; in 1 liter of pure water, there is 10^{-7} mol of each ion.

■ Many substances in solution contribute to an imbalance in the concentrations of H^+ and OH^-, giving rise to acid-base behavior (i.e., donation and acceptance of protons).

■ H^+ and OH^- concentrations are measured by the pH system, defined as $pH = -\log_{10}[H^+]$.

■ The pH of biological fluids influences the charges carried by the side groups of amino acids and hence the conformation and activity of proteins.

■ Physiological buffering systems maintain intracellular and extracellular pH within a narrow range.

Biological molecules

■ Animal cells are composed of four major groups of organic molecules: lipids, carbohydrates, proteins, and nucleic acids.

■ Lipids include triglycerides (fats), fatty acids, phospholipids (the primary structural components of membranes), waxes, and sterols. Lipids serve as membrane components, energy stores, and signal molecules.

■ Carbohydrates include sugars, storage carbohydrates (glycogen and starch), and structural polymers such as chitin and cellulose. Sugars, glycogen, and starch are major substrates of energy metabolism by cells.

■ Proteins are made up of linearly arranged amino acid residues.

■ The nucleic acids DNA and RNA encode the genetic information necessary for the synthesis of all the proteins in the cell.

Energetics of living cells

■ Biological systems maintain a state of low entropy—that is, they are highly and improbably organized.

■ Animals derive energy from food molecules by the processes of energy metabolism.

■ In the living cell, metabolism occurs as orderly, regulated sequences of chemical reactions catalyzed by enzymes organized in metabolic pathways.

■ Chemical reactions tend to proceed spontaneously down an energy gradient, decreasing free energy and increasing entropy.

■ Living systems appear to defy entropy, but they do not; they exist at the expense of chemical energy obtained from their environment, and at the expense of an increase in the entropy of the environment.

ATP: Energy source for the cell

■ Energy-requiring biological reactions utilize ATP, a nucleotide triphosphate that serves as a common reaction intermediate capable of contributing chemical energy stored in its terminal phosphate bond. This energy transfer is accomplished by means of coupled reactions in which an endergonic (energy-requiring) reaction is driven by an exergonic (energy-releasing) reaction, the hydrolysis of ATP to ADP and inorganic phosphate.

■ ATP is reconstituted from ADP by the oxidation of food molecules, which largely have their origin in the radiant solar energy trapped during the process of photosynthesis in green plants. Thus, animals depend on energy ultimately derived from the sun.

Enzymes: General properties

■ Enzymes act as biological catalysts. They work by lowering the energy of activation for a reaction, thereby increasing the rate of the reaction at a given temperature. With the aid of enzymes, cell metabolism can proceed at biological temperatures.

■ The catalytic action of an enzyme arises from its ability to bind specific reactant molecules at the active site; the close steric fit required for this interaction is largely responsible for enzyme specificity. This binding produces favorable spatial relations between the reacting molecules.

■ Regulation of metabolic activity occurs at the level of enzyme synthesis (controlled at the level of transcription and translation) and enzyme activity.

■ The activity of some enzymes is controlled by the binding of regulatory molecules or ions to the enzyme molecule at an allosteric site that is distinct from the active site. This binding results in a conformational change that affects the properties of the active site.

Metabolic energy production

■ Liberation of the free energy stored in foods during metabolism occurs by transfer of electrons from an electron donor (reductant) to an electron acceptor (oxidant).

■ The release of free energy in the cell is budgeted in small steps compatible with the amounts of free energy required to phosphorylate ADP to ATP.

■ Electrons from the reduced coenzymes NADH and $FADH_2$ are transported in increments down a chain of electron acceptors and electron donors, yielding enough energy for synthesis of ATP at three points in the chain.

■ Because an electron-pressure gradient exists along the electron-transport chain of cytochromes, electrons flow to the ultimate electron acceptor, molecular oxygen.

REVIEW QUESTIONS

1. What important physical and chemical characteristics of H_2O can be directly related to the dipole nature of the water molecule?
2. What is the pH of a 1 M solution of an acid that is 10% dissociated?
3. Why is a weak acid rather than a strong one required for a pH buffer system?
4. What is the difference between molality and molarity?
5. How many grams does a mole of CO_2 weigh?
6. Approximately how many particles are in a 1 M solution of NaCl?
7. What is the approximate boiling point of a 1 molal solution of NaCl?
8. Why do some liquids conduct electricity whereas others do not?
9. How many ions flow past a point (equivalents per second) at a current of 1 mA?
10. What are the primary factors that govern the binding of two cations, a and b, to an electronegative binding site? Write the expression that integrates these factors into a meaningful quantity.
11. Does the force of attraction fall off most rapidly with distance between a monovalent cation and (a) a monopolar binding site or (b) a multipolar site? Give the expression relating force and distance for each site.
12. What factors define each level of protein structure—primary, secondary, tertiary, and quaternary? What factors contribute to stability at each level?
13. Why do proteins become denatured (structurally disorganized) at elevated temperatures?
14. At a particular temperature, will a reaction with $\Delta S > \Delta H$ be endergonic or exergonic?
15. Under what conditions will an endergonic reaction proceed?
16. What is ΔG for a system at equilibrium?
17. How does ATP "donate" stored chemical energy to an endergonic reaction?
18. What is meant by the term *coupled reaction*?
19. How does increased temperature increase the rate of a chemical reaction?
20. What factors can influence the optimal temperature for an enzymatic reaction?
21. How does a catalyst increase the rate of a reaction?
22. Why is catalysis necessary in living organisms?
23. How do enzymes exhibit substrate or bond specificity?
24. How does pH affect the activity of an enzyme?
25. What evidence supports the theory that active-site specificity is due to complementary fit between enzyme and substrate?
26. What factors can influence the rate of enzyme-catalyzed reactions?
27. The Michaelis-Menten constant, K_m, is equal to the substrate concentration at which a particular reaction proceeds at half its maximum velocity, V_{max}. Does a high K_m indicate a greater or a lesser enzyme-substrate affinity?
28. Why does a high substrate concentration reverse the effects of a competitive inhibitor and yet have no effect on a noncompetitive inhibitor?
29. How does each type of inhibition affect the Michaelis-Menten constant, K_m? Explain why.
30. Why does aerobic metabolism yield much more energy per glucose molecule than anaerobic metabolism?

SUGGESTED READINGS

Atkins, P. W. 1998. *Physical Chemistry.* 6th ed. New York: W. H. Freeman. (A complete treatment at the undergraduate level of many of the basic physicochemical and biochemical concepts introduced in this chapter.)

Cooper, G. 1997. *The Cell: A Molecular Approach.* Sunderland, Mass.: Sinauer Associates. (Molecular biology is the unifying theme for this treatment of the cell.)

Lodish, H. D., et al. 2000. *Molecular Cell Biology.* 4th ed. New York: W. H. Freeman. (A comprehensive textbook describing many of the basic biochemical processes that occur in the cell.)

Nelson, D. L., and M. M. Cox. 2000. *Lehninger Principles of Biochemistry.* 3d ed. New York: Worth. (Presents the principles of biochemistry.)

Stryer, L. 2001. *Biochemistry.* 5th ed. New York: W. H. Freeman. (A highly readable reference providing information about biochemical structures and mechanisms.)

Membranes, Channels, and Transport

The complex chemical reactions responsible for animal life can proceed in a regulated fashion only under stable, restricted conditions. The required stability is maintained within cells largely through the action of biological membranes, which form a protective barrier that allows only certain materials to pass into or out of the cell. Animal tissue contains an astounding amount of biological membrane. The chimpanzee brain, for example, is estimated to have about 100,000 m^2 of cell membrane, an area equal to three full-sized soccer fields.

Though we now know that cell membranes are a major constituent of all living matter and are essential to all life process, their existence was questioned up until the 1930s. There was little or no direct anatomic evidence for biological membranes at the time, and few techniques were available to study something so miniscule, so their existence could only be inferred from physiological studies. The first important observations on the diffusion-limiting properties of the cell surface were made in the mid-nineteenth century by Karl Wilhelm von Nägeli, who noticed that the cell surface acted as a barrier to free diffusion of dyes into the cell from the extracellular fluid. From these experiments, he deduced the presence of a "plasma membrane." He also discovered the osmotic behavior of cells, noting that they swell when placed in dilute solutions and shrink in concentrated solutions. Eventually, the advent of the electron microscope (see Chapter 2) yielded the first structural evidence for the existence of a distinct cell membrane.

Understanding membrane structure and the functions that derive from it is critical to the study of animal physiology. In this chapter we discuss membrane structural features and their role in maintaining cell integrity and controlling cell activities. In Chapter 5, we consider how the electrical behavior of cell membranes is responsible for cell-to-cell signaling, which, in turn, coordinates action in animals.

MEMBRANE STRUCTURE AND ORGANIZATION

At their surfaces, all animal cells are surrounded by a **plasma membrane**, an extraordinarily thin (6–23 nm), lipid-based structure that encloses the cytoplasm (consisting of the cytosol and all cell organelles) and the cell nucleus. (The internal organelles, such as the ATP-producing mitochondria that we discussed in Chapter 3, have their own surface membranes.) The plasma membrane may be compared to the sheen of a drop of gasoline on water—both consist of an ultrathin layer of molecules, organized by the interaction of the molecules with water at the interface. Despite its incredible thinness, the plasma membrane regulates molecular traffic between the orderly interior of the cell and the more disorderly, potentially disruptive external environment. Plasma membranes sustain different concentrations of certain ions on their two sides, leading to **concentration gradients** of several ionic species across membranes. Protein structures in the membranes actively participate in the transport of substances between compartments and ultimately regulate the cytoplasmic concentration of dissolved ions and other molecules quite precisely. This regulation allows the maintenance of the stable intracellular milieu required for the finely balanced catabolic and synthetic chemical reactions of the cell.

Membrane Composition

All biological membranes, including the internal membranes of organelles of eukaryotic cells, have essentially the same structure: lipid and protein molecules kept

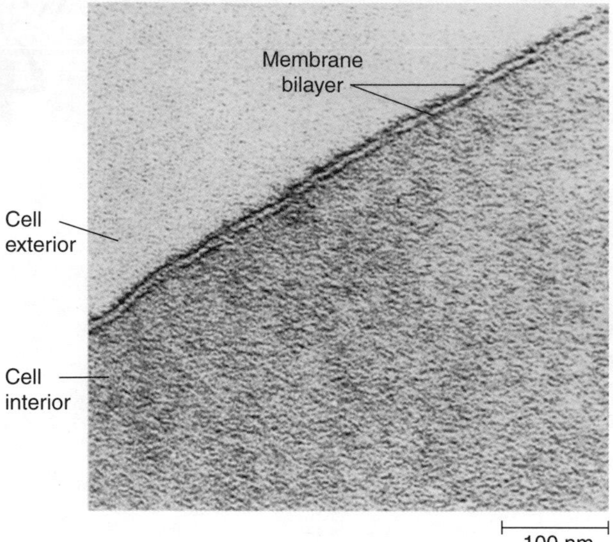

Figure 4-1 The plasma membrane creates a barrier between the interior and exterior of the cell, as revealed in this electron micrograph. The cell interior is separated from the cell exterior by the membrane bilayer, which is seen in cross-section as a dark-light profile about 10 nm thick. The dark-light-dark sandwichlike appearance is due to the differential staining of various components of the "unit membrane" by an electron-opaque substance during preparation of the tissue. [Courtesy of J. D. Robertson.]

together by noncovalent interactions (Figure 4-1). As explained in Chapter 3, the lipid molecules are arranged in a continuous double layer, called the **lipid bilayer,** which is relatively impermeable to the passage of most water-soluble molecules. In 1925, using a simple but elegant experiment, Gorter and Grendel provided the first evidence that cell membranes are lipid bilayers. First, they dissolved the lipids from red blood cell ghosts (the empty membrane sacs left when red blood cells have been induced to burst open). They then allowed the extracted lipids to spread out on the surface of water in a trough. Because of their asymmetry, the lipid molecules became oriented so that their polar head groups formed hydrogen bonds with the water and their hydrophobic hydrocarbon tails stuck up into the air. When the dispersed film of lipid molecules was gently compressed into a continuous monomolecular film, it occupied an area about twice the surface area of the original red blood cells. Since the only membrane in mammalian red blood cells is the plasma membrane, it was concluded that the lipid molecules in the membrane must be a continuous bilayer. The bilayer has since been visualized in cross-section by the electron microscope, as well as with freeze fracture methods, in which the membrane is split through the center of the bilayer (see Chapter 2).

Membranes are remarkably fluid structures. Most of their lipid and protein molecules "float" in the plane of the bilayer (Figure 4-2). The relative proportion of lipids and proteins present in a membrane depends on the kind of cell or organelle the membrane encloses. The lipid molecules, which are smaller and simpler than the proteins, provide the fundamental structure of the membrane. The membrane is spanned by **integral proteins** that serve as passive-transport pores and channels, active-transport pumps and carriers, membrane-linked enzymes, and chemical signal receptors and transducers. **Peripheral proteins** are associated with the surface of the membrane via electrostatic interactions.

Lipid molecules are insoluble in water but can be dissolved in organic solvents. They constitute about half the mass of plasma membranes in animal cells, most of the rest being protein. Each square micron of membrane has about 10^6 lipid molecules, meaning that typical small cells have about 10^9 lipid molecules. The three primary types of lipids in cell membranes are

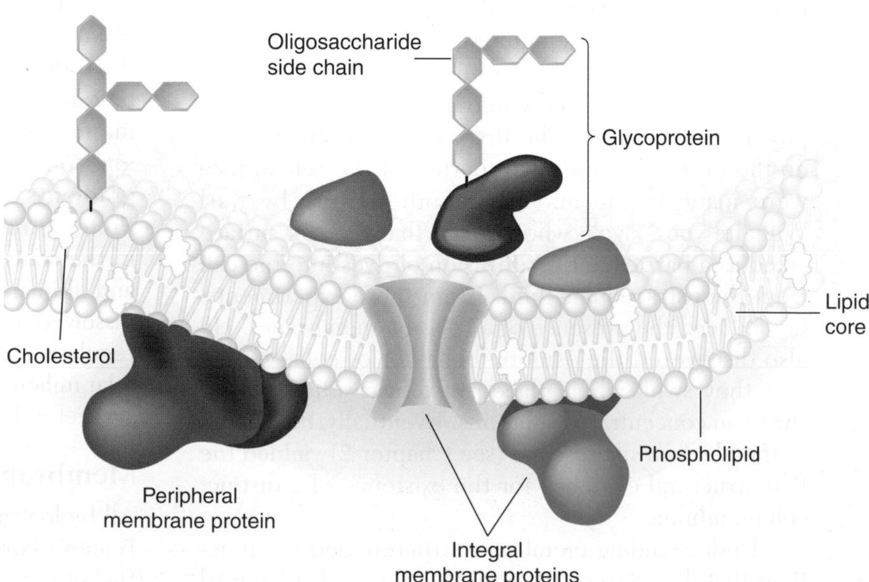

Figure 4-2 A biological membrane has multiple components based on a lipid bilayer. Globular integral proteins embedded in the lipid bilayer provide a mechanism for transmembrane transport. Most plasma membranes have considerably more proteins than indicated in this figure. The glycoproteins bear oligosaccharide side chains and are vital for cell recognition and communication. Cholesterol molecules lie close to the heads of the phospholipid molecules, where they reduce membrane flexibility. The inner ends of the phospholipid tails are highly mobile, giving the membrane fluidity.

- **Phosphoglycerides,** characterized by a glycerol backbone
- **Sphingolipids,** which have backbones made of sphingosine bases
- **Sterols,** such as cholesterol, which are nonpolar and only slightly soluble in water

The first two lipid types are amphipathic, meaning that one end of the molecule is hydrophilic (water-soluble) and the other end is hydrophobic (water-insoluble; Figure 4-3). The dual nature of these amphipathic membrane lipids is crucial to the organization of biological membranes. The polar heads of membrane lipids seek water, and the nonpolar hydrocarbon tails aggregate together as they are excluded from water by water-water interactions, just as, on a larger scale, oil is excluded from water. Bunched together in the interior of the lipid bilayer, the hydrophobic tails are further stabilized by weak interchain van der Waals interactions. The same forces that cause lipid bilayers to form also allow them to reseal themselves when they are torn—they are self-repairing. Differences in the lengths of the two fatty acid tails and in their composition influence lipid packing and hence *fluidity* (the ability of membrane components to move relative to one another), causing subtle differences in lipid bilayer characteristics. The hydrophobic properties of the hydrocarbon tails are responsible for the low permeability of membranes to polar substances (e.g., inorganic ions and polar nonelectrolytes such as sucrose and insulin) and for their correspondingly greater permeability to nonpolar substances (e.g., steroid hormones).

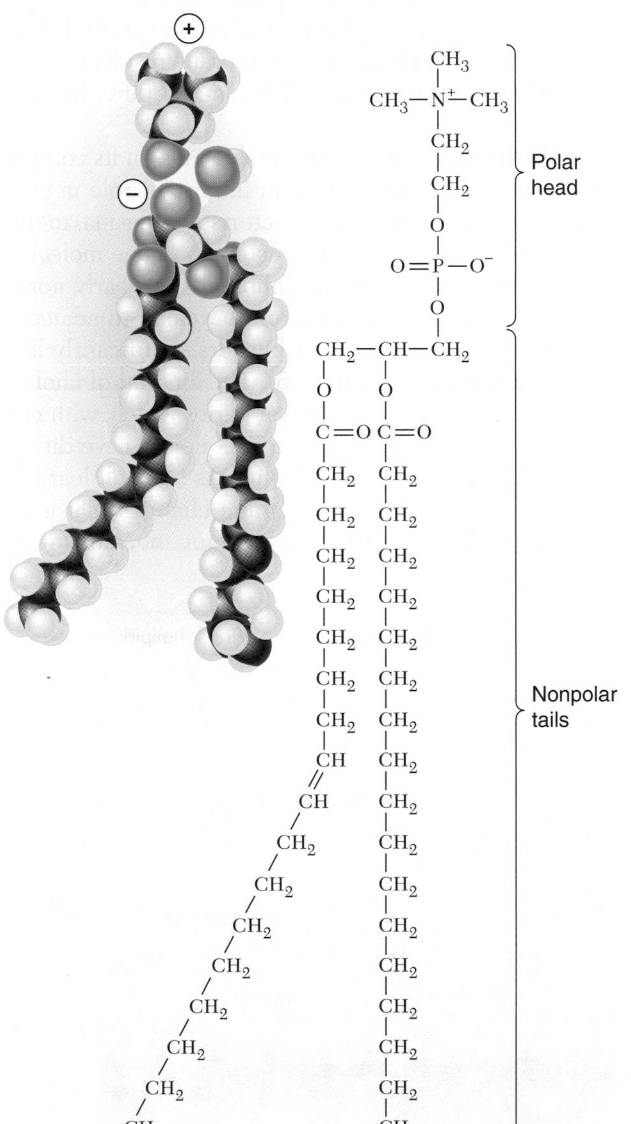

Figure 4-3 Phosphatidylcholine, a phosphoglyceride, has charges that give the head group its polar character. Note that unsaturated fatty acid chains, such as the left hydrocarbon chain in this figure, are conventionally drawn with a distinct bend in the chain where double bonds occur. Only the double bond is rigid in the unsaturated fatty acid; the rest of the chain has considerable rotational mobility.

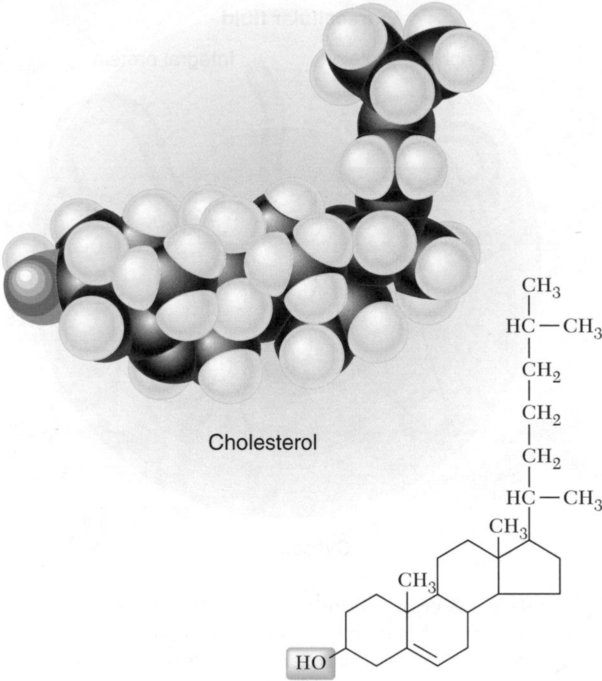

Figure 4-4 Cholesterol, a sterol, is an important stabilizing component of the lipid membrane.

The third class of membrane lipids, the sterols, are largely nonpolar and only slightly soluble in water (Figure 4-4). In aqueous solution they form complexes with proteins that are far more water-soluble than sterols alone. Once in the membrane, a sterol molecule fits snugly between the hydrocarbon tails of the phospholipids and glycolipids, increasing the viscosity of the hydrocarbon core of the membrane.

The Fluid Mosaic Model of Membranes

The concept of a lipid bilayer membrane enclosing most cells gained wide acceptance midway through the

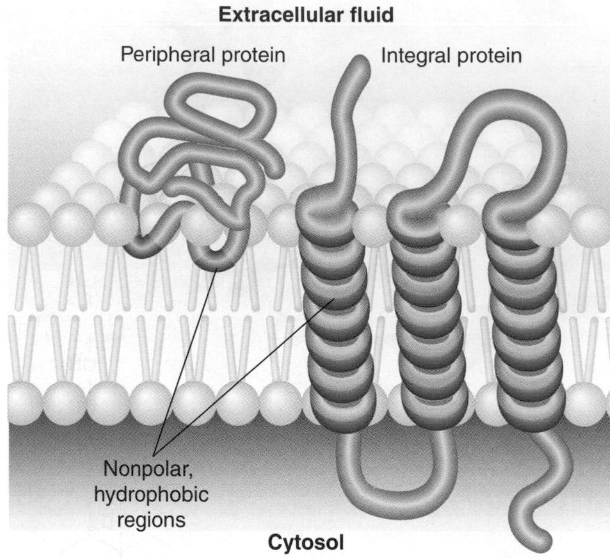

Figure 4-5 A cross-sectional view of the Singer-Nicolson fluid mosaic model of membranes reveals proteins with polar and nonpolar segments. The polar segments (those containing charged hydrophilic amino acid side groups) project into the aqueous phase (the water-based intra- and extracellular fluid). The hydrophobic nonpolar domains are buried in the lipid phase of the bilayer.

twentieth century. Chemical fractionation of membranes and immunochemical studies confirmed that proteins, as well as lipids, are important components of membranes. Moreover, the enzymatic properties of membranes, such as active transport and other metabolic functions, require the participation of proteins. An example is the protein complexes responsible for electron transport and oxidative phosphorylation, described in Chapter 3.

Despite this early progress in characterizing biological membranes, it took additional decades for researchers to recognize just how fluid and heterogeneous membranes really are. It was discovered that some integral protein molecules are free to diffuse laterally along the membrane, presumably because of the fluidity of the lipid matrix. In addition, radioactive labeling studies demonstrated that protein molecules or parts of molecules facing one side of the membrane differ from those facing the other side, and that they normally do not "flip-flop" across the membrane as previously suspected. Additionally, in many membranes, the distribution of lipid species differs in the two lipid layers.

On the basis of evidence that had emerged during the 1950s and especially the 1960s, Singer and Nicolson (1972) proposed the **fluid mosaic model** of the membrane, in which globular proteins are integrated with the lipid bilayer, with some protein molecules penetrating the bilayer completely and others penetrating it only partially (Figure 4-5). These membrane-associated proteins have polar and nonpolar portions. The nonpolar portions of transmembrane proteins generally consist of α-helical segments of approximately seven turns that are rich in amino acids with hydrophobic side chains. These segments are buried in the hydrocarbon core of the bilayer while the polar, hydrophilic portions protrude from the bilayer and interact with water in the aqueous phase. The hydrophobic nature of the embedded portions is important in keeping the integral proteins from leaving the lipid bilayer.

Membrane fluidity

A variety of techniques have been used to demonstrate that lipid molecules in a membrane very rarely move from one surface of the membrane to the other (about once a month), but exchange places with adjacent molecules in a monolayer about 10^7 times per second. This brisk lipid exchange within a membrane results in rapid migration along the plane of the membrane, but not across it.

The fluidity of a membrane depends on its composition, and **cholesterol** plays an important role in governing this membrane characteristic. Plasma membranes in animals can contain as much as one molecule of cholesterol for every phospholipid, or nearly none. Cholesterol, when present, binds weakly to adjacent phospholipids, making lipid bilayers significantly less fluid, but stronger (Figure 4-6). The amount of cholesterol present in the lipid bilayer varies widely with cell type, and different cell types consequently have differing degrees of membrane fluidity. As you will learn in Chapter 17, structural membrane changes that alter membrane fluidity are an important mechanism by

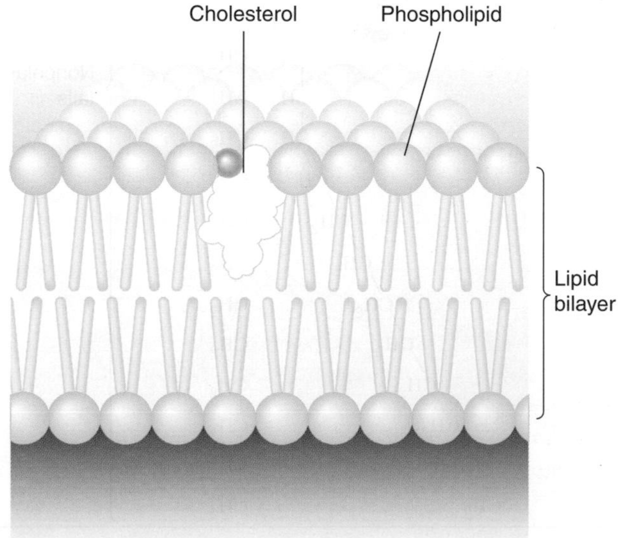

Figure 4-6 Cholesterol interacts weakly with adjacent phospholipids in the membrane, partially immobilizing their fatty acid chains. As a result, the membrane is less fluid, but mechanically stronger. (The structural formula of cholesterol is shown in Figure 4-4.)

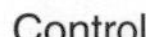

which animals adjust to long-term temperature changes in their environments. On the pathological side, cholesterol accumulation is the mechanism underlying "hardening of the arteries," a major cause of human cardiovascular disease, in which the membranes of the endothelial cells lining the arteries become abnormally rigid (with additional cholesterol plaques stored within the vessel walls as well).

The chemical nature of the head groups in lipids, the lengths of the nonpolar tails, and the amount of cholesterol in the membrane all affect the structural properties of the membrane and its interactions with proteins. Indeed, some integral proteins function only in the presence of a specific ratio of lipid types. Hence, cells must regulate the distribution of lipid species in their membranes during their development and rearrange lipid concentrations according to specific cellular functions.

Heterogeneity of the integral membrane proteins

The integral proteins found in the plasma membrane take many functional forms and may act as ion channels, carriers and pumps, receptor molecules, and recognition molecules. The number of integral proteins in a membrane varies; in some membranes the protein content is so high that the membrane is essentially a sheet of proteins with lipids tucked between them.

Morphological evidence for the heterogeneous mosaic arrangement of globular proteins in a lipid bilayer can be seen in freeze-etch electron micrographs of the surface of a membrane (Figure 4-7). When subjected to digestion by proteolytic (protein-digesting) enzymes, the globular units seen in the membrane are progressively removed, demonstrating that they are indeed proteins.

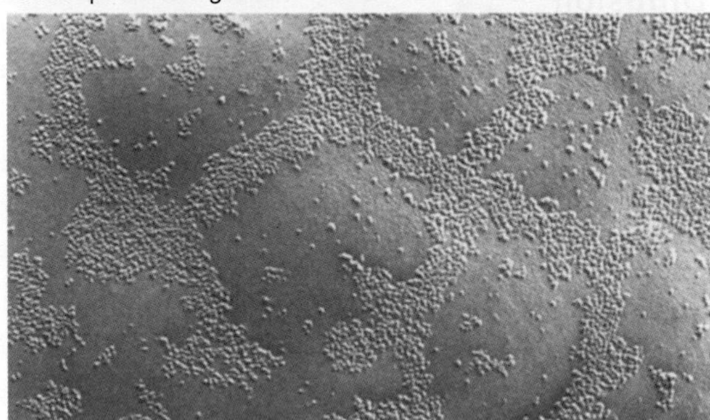

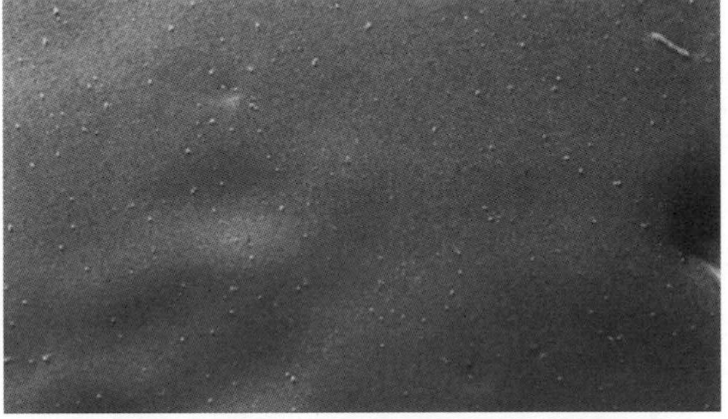

Figure 4-7 Freeze-etch electron micrographs yield morphological evidence for the fluid mosaic membrane model. This plasma membrane has been split along the middle of the bilayer, exposing membrane-embedded particles with diameters of 5 to 8 nm. Digestion with a proteolytic enzyme produced progressive loss of these particles, indicating that they are globular proteins inserted into the lipid phase of the membrane. [Courtesy of L. H. Engstrom and D. Branton.]

Variation in Membrane Form

Membrane composition varies greatly among cell types. At one extreme is the metabolically inert myelin sheath surrounding the axons of many nerve cells, in which the lipid bilayer is largely uninterrupted by proteins. At the other extreme are cells with membranes that have structures of repeating nonlipid macromolecular units that nearly obliterate the lipid bilayers. Such membranes evolved for highly specialized purposes, such as cell-to-cell signaling or enzymatic activity. In visual receptor cells, for example, the repeating macromolecular units are molecules of the visual pigment rhodopsin. Mitochondrial membranes, specialized for enzymatic activity, are composed almost entirely of repeating subunits of ordered enzymatic aggregates. Between these two extremes are the plasma membrane and most intracellular membranes, in which the bilayer is interrupted frequently by integral protein molecules. Thus the basic structure of the lipid bilayer with integral proteins has been highly modified as required for functional specialization.

CROSSING THE MEMBRANE: AN OVERVIEW

The structure of biological membranes makes them quite selective about which molecules can pass through them and which are blocked. The hydrophobic interior of the lipid bilayer makes membranes highly impermeable to most polar molecules. This impermeability prevents water-soluble components of the cell from easily entering or escaping. However, the movement of such components is sometimes necessary, so mechanisms for transferring them across membranes have evolved in all cells. Macromolecules such as proteins must also be transported across plasma membranes by specialized mechanisms.

To understand membrane transport in animal cells, let's first review the physical principles of solute and solvent movement in solution and across artificial semipermeable membranes. Such membranes closely resemble those found in animal cells, and the principles explained here apply in many physiological situations.

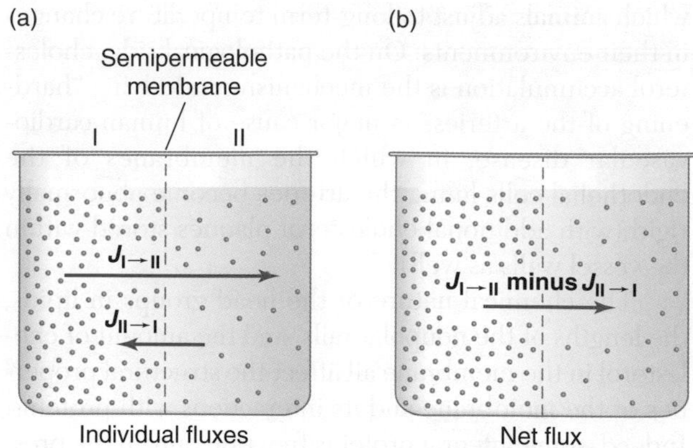

Figure 4-8 Solutes may move through a permeable membrane in either direction, depending on prevailing physical and chemical conditions. **(a)** The arrows represent the actual fluxes of a substance between compartments I and II. **(b)** The single arrow indicates the resulting net flux, from compartment I to II.

Diffusion

Random thermal motion of suspended or dissolved molecules causes their dispersion from regions of higher concentration to regions of lower concentration, a process called **diffusion.** Diffusion is extremely slow when viewed on a tissue, rather than a cellular, scale. A crystal of blue-tinted copper sulfate, for example, dissolves in unstirred water so slowly that it may take a whole day to tint a liter of water uniformly. Diffusion is rapid enough, however, to distribute material effectively on the microscopic scale of cells.

The rate of diffusion of a solute s can be defined by the **Fick diffusion equation:**

$$\frac{dQ_s}{dt} = D_s A \frac{dC_s}{dx} \tag{4-1}$$

in which dQ_s/dt is the rate of diffusion (i.e., the quantity of s diffusing per unit time), D_s is the diffusion coefficient of s, A is the cross-sectional area through which s is diffusing, and dC_s/dx is the concentration gradient of s (i.e., the change in concentration with distance). The concentration gradient is clearly very important because it determines the rate at which s diffuses down the gradient. D_s varies with the nature and molecular weight of s and of the solvent, which is water in most physiological situations.

Membrane Flux

If a solute occurs on both sides of a membrane through which it can diffuse, it exhibits a unidirectional flux in each direction (Figure 4-8a). The flux, or rate of diffusion, J, is the amount of the solute that passes through a unit area of membrane every second in one direction, so that

$$J = \frac{dQ_s}{dt} \tag{4-2}$$

where J would typically have units of moles per square centimeter per second ($M \cdot cm^{-2} \cdot s^{-1}$). The flux in one direction (say, from cell exterior to cell interior) is considered independent of the flux in the opposite direction. Thus, if the **influx** and **efflux** are equal, the **net flux** is zero. If the flux is greater in one direction than in the other, there is a net flux, which is the difference between the two unidirectional fluxes (Figure 4-8b).

The **permeability** of a membrane to a substance is the rate at which that substance passively penetrates the membrane under a specified set of conditions. A greater permeability is accompanied by a greater flux if other factors remain equal. If we assume that the membrane is a homogeneous barrier and that a continuous concentration gradient exists for a nonelectrolyte substance between the side of higher concentration (I) and the side of lower concentration (II), then

$$\frac{dQ_s}{dt} = P(C_I - C_{II}) \tag{4-3}$$

in which dQ_s/dt is, again, the amount of substance s crossing a unit area of membrane per unit time (e.g., moles per square centimeter per second), C_I and C_{II} are the concentrations of the substance on the two sides of the membrane, and P is the **permeability constant** of the substance, with the dimension of velocity ($cm \cdot s^{-1}$).

Note that Equation 4-3 applies only to molecules that are not being actively transported or influenced by any forces other than simple diffusion. This excludes electrolytes, since they are electrically charged when dissociated and, consequently, their flux depends not only on the concentration gradient but also on the electrical gradient (i.e., the electric potential difference across the membrane). As is evident from Equation 4-3, the flux of a nonelectrolyte should be a linear function of the concentration gradient ($C_I - C_{II}$). This linear relationship is characteristic of simple diffusion, and it can be used to distinguish between passive diffusion and active transport of a substance.

The permeability constant incorporates the properties inherent in the membrane and substance in question. These factors determine the probability that a molecule of a particular substance will cross a particular membrane. This relationship can be expressed formally as

$$P = \frac{D_m K}{x} \qquad (4\text{-}4)$$

In this equation, D_m is the **diffusion coefficient**, an expression of the rate of diffusion of the substance through the membrane. The more viscous the membrane or the larger the molecule, the lower the value of D_m. K is the **partition coefficient**, an indication of the ratio between dissolved and undissolved substance in the membrane, and x is the thickness of the membrane.

Permeability constants for different substances vary greatly. The permeability of red blood cells to different solutes, for example, ranges from 10^{-12} cm·s^{-1} to 10^{-2} cm·s^{-1}. The permeability of some membranes to certain substances *in vivo* can be altered greatly by hormones and other molecules that react with receptor sites on the membrane and thereby influence channel size or carrier mechanisms. Antidiuretic hormone, for example, can increase the water permeability of the renal collecting duct in mammals by as much as 10 times. Similarly, neurotransmitters, acting on specialized integral membrane proteins in nerve and muscle cells, induce large increases in permeability to ions such as Na^+, K^+, Ca^{+2}, or Cl^-.

Osmosis

In 1748 Abbé Jean Antoine Nollet noted that if pure water is placed on one side of a membranous tissue (e.g., a bladder wall) and a solution of water containing electrolytes or other molecules is placed on the other side, the water passes through the membrane into the solution. This movement of water down its concentration gradient was called **osmosis** (from the Greek *osmos*, "to push"). We don't usually think of "water concentration," but in fact, water acts just like any other substance by diffusing down its concentration gradient.

Osmosis is a colligative property of great importance to living systems. The movement of water by osmosis can produce a **hydrostatic pressure** (i.e., a fluid mechanical pressure), resulting in a pressure gradient across a semipermeable membrane. As can be seen in Figure 4-9, a net imbalance in hydrostatic pressures on either side of the membrane, as would have been the case in Nollet's experiment, causes a rise in the level of the solution as the water diffuses through the semipermeable membrane into it. This rise continues until the hydrostatic pressure of the solution in compartment II is sufficient to force water molecules back through the membrane to compartment I at the same rate that osmosis causes them to diffuse from I to II. At that point, the net rate of water movement is zero. The hydrostatic back-pressure produced in compartment II at equilibrium is a measure of the **osmotic pressure** of the fluid in compartment I. Osmotic pressures are of great importance to the integrity of animal cells, and as

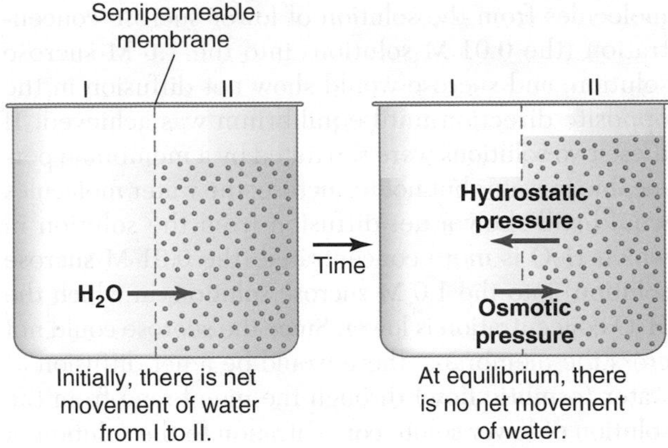

Figure 4-9 Water flow produced by osmosis generates hydrostatic pressure. Compartment I contains pure water; compartment II, water with a solute that cannot penetrate the membrane. Osmotic pressure forces water to enter compartment II from compartment I, raising the height of the fluid in compartment II until the hydrostatic pressure difference equals the opposing osmotic pressure difference. When the pressures equalize, the net flux of water falls to zero.

we will see in Chapter 14, they are crucial for the uptake and elimination of water and salts.

In 1877, Wilhelm Pfeffer made the first quantitative studies of osmotic pressure. He deposited a "membrane" of copper ferrocyanide on the surface of porous clay cups, producing membranes that would allow water molecules to diffuse through them far more freely than sucrose molecules could. These artificial membranes were also strong enough to withstand relatively high pressures without rupturing because of the clay substratum. Using these membranes, Pfeffer was able to make the first direct measurements of osmotic pressure. Some of his results are shown in Table 4-1. Note from the table that the osmotic pressure of a solution is proportional to its solute concentration.

Osmosis is responsible for the net movement of water across cell membranes and epithelia. To understand how this occurs, consider a 1.0 M aqueous solution of sucrose in a beaker, onto which has been poured, carefully and slowly, a 0.01 M aqueous solution of sucrose. There would be net diffusion of water

Table 4-1 **Osmotic pressure of sucrose solutions of various concentrations***

Sucrose (%)	Osmotic pressure (atm)	Ratio of osmotic pressure to percentage of sucrose
1	0.70	0.70
2	1.34	0.67
4	2.74	0.68
6	4.10	0.68

* Results were obtained by Pfeffer (1877) in experimental measurements.

molecules from the solution of lower sucrose concentration (the 0.01 M solution) into the 1.0 M sucrose solution, and sucrose would show net diffusion in the opposite direction until equilibrium was achieved. If these two solutions were separated by a membrane permeable to water but not to sucrose, the water molecules would still show a net diffusion from the solution in which H_2O is more concentrated (the 0.01 M sucrose solution) into the 1.0 M sucrose solution, in which the H_2O concentration is lower. Since the sucrose could not cross the membrane, there would be a net diffusion of water (**osmotic flow**) through the membrane from the solution of lower solute concentration to the solution of higher solute concentration.

Osmotic pressure, π, is proportional not only to the concentration of the solute, C (moles of solute particles per liter of solvent = osmolarity), but also to its absolute temperature, T:

$$\pi = K_1C \tag{4-5}$$

and

$$\pi = K_2T \tag{4-6}$$

where K_1 and K_2 are constants of proportionality. Jacobus van't Hoff related these observations to the gas laws and showed that molecules in solution behave thermodynamically like gas molecules. Thus,

$$\pi = RTC$$

or

$$\pi = \frac{nRT}{V} \tag{4-7}$$

where n is the number of mole equivalents of solute, R is the molar gas constant ($0.082 \text{L} \cdot \text{atm} \cdot \text{K}^{-1} \cdot \text{mol}^{-1}$),* and V is the volume in liters. Like the gas laws, however, this expression for osmotic pressure holds true only for dilute solutions and for completely dissociated electrolytes.

Large concentration gradients across plasma membranes can generate surprisingly high osmotic pressures—on the order of several atmospheres. Such pressures, if allowed to develop, would be large enough to explode a cell. Consequently, mechanisms for regulating osmotic balance have evolved that minimize osmotic pressure gradients across plasma membranes and through tissues (see Chapter 14).

Osmolarity and Tonicity

Two aqueous solutions that exert the same osmotic pressure through a membrane permeable only to water are said to be **isosmotic** to each other. If one solution exerts less osmotic pressure than the other, it is **hypoosmotic** with respect to the other solution; if it exerts greater osmotic pressure, it is **hyperosmotic. Osmolarity** is thus defined on the basis of an ideal osmometer in which the osmotic membrane allows water to pass but completely prevents the solute from passing. All solutions with the same number of dissolved particles per unit volume have the same osmolarity and are thus defined as isosmotic.

The **tonicity** of an aqueous solution, in contrast to its osmolarity, is defined by the response of cells or tissues immersed in the solution. A solution is considered to be **isotonic** to a given cell or tissue if the cell or tissue neither shrinks nor swells when immersed in it, meaning that there is no osmotic pressure difference between the cell interior and the extracellular solution, and thus no net water gain or loss across the plasma membrane. If the tissue swells because it absorbs water, the solution is said to be **hypotonic** to the tissue. If it shrinks because it loses water, the solution is said to be **hypertonic** to the tissue.

If cells actually behaved as ideal osmometers, tonicity and osmolarity would be equivalent, but this is not generally true. For example, sea urchin eggs maintain a constant volume in a solution of NaCl that is isosmotic to seawater, but they swell if immersed in a solution of $CaCl_2$ that is isosmotic to seawater. The NaCl solution therefore behaves isotonically relative to the sea urchin egg, whereas the $CaCl_2$ solution behaves hypotonically. The tonicity of a solution depends on the rate of intracellular accumulation of the solute in the tissue in question as well as on the concentration of the solution. The more readily the solute accumulates in cells of the tissue, the lower the tonicity of a solution of a given concentration or osmolarity. As the cell gradually loads up with the solute, water follows according to osmotic principles, causing the cell to swell. Thus, unlike *osmolarity*, the terms *isotonic, hypertonic,* and *hypotonic* are meaningful only in reference to actual experimental determinations on living cells or tissues.

Electrical Influences on Ion Distribution

Membrane permeability to charged particles depends both on the membrane permeability constant and on the electric potential across the membrane. Understanding the interaction of charged particles with membranes is extremely important for understanding how electrically excitable cells function. Neurons are the most highly specialized of this class of cells. Since neurons will be discussed in the next few chapters, only a few important observations will be summarized here.

Two forces can act on charged atoms or molecules (such as Na^+, K^+, Cl^-, Ca^{2+}, and amino acids) to produce net passive diffusion across a membrane:

1. The chemical gradient arising from the difference in the concentrations of the substance on the two sides of the membrane

2. The electric field, or difference in electric potential, across the membrane

* R is the constant of proportionality in the gas equation $PV/T = R$ when referring to 1 mol of a perfect gas, and it has the value of 1.985 $\text{cal} \cdot \text{mol}^{-1} \cdot \text{K}^{-1}$; P is in atmospheres and V is in liters.

An ion, like any other solute, will move away from regions of higher concentration. If that ion is positively charged, it will also move toward regions of more negative potential. The sum of the combined forces of concentration gradient and electrical gradient determines the net **electrochemical gradient** acting on that ion.

When an ion is at equilibrium across a membrane (that is, when there is no net transmembrane flux of that ionic species), there will exist a potential difference just sufficient to balance and counteract the chemical gradient acting on the ion. The potential at which an ion is in electrochemical equilibrium is called the **equilibrium potential,** measured in volts (or, more realistically, millivolts). Several factors influence the value of the equilibrium potential, but the most prominent is the ratio of the ion concentrations on opposite sides of the membrane. For a monovalent ion such as Na^+ or K^+ at 18°C, the equilibrium potential (in volts) is equal to 0.058 × $\log_{10}$ of the ratio of the extracellular to intracellular concentrations of the ion. Thus, a potential difference of 58 mV across the membrane has the same effect on the net diffusion of that ion as a transmembrane concentration ratio of 10 : 1.

An apparently paradoxical situation therefore arises in which an ionic species can passively diffuse against its chemical concentration gradient (that is, move "uphill" to an area of higher concentration) if the electrical gradient (i.e., the potential difference) across the membrane is in the opposite direction to and exceeds the concentration gradient. For example, if the interior of a cell has a negative charge greater than the equilibrium potential for K^+, potassium ions will diffuse into the cell even though the intracellular concentration of K^+ is much higher than the extracellular concentration. The distribution of ions across membranes and the attendant equilibrium potential is described by the Nernst equation, which is discussed in detail in the next chapter.

Electrical forces cannot act directly on uncharged molecules such as sugars. The movements of these substances are influenced primarily by their concentration gradient.

Donnan Equilibrium

If diffusible solutes are separated by a membrane that is freely permeable to water and electrolytes but totally impermeable to one ionic species, the diffusible solutes become unequally distributed between the two compartments. This phenomenon, called a **Donnan equilibrium,** was discovered in 1911 by Frederick Donnan.

To understand the Donnan equilibrium, imagine starting with pure water in two compartments and adding some KCl to one of them (Figure 4-10). The dissolved salt (K^+ and Cl^-) will diffuse through the membrane until the system is in equilibrium—that is, until the concentrations of K^+ and Cl^- become equal on both sides of the membrane (see Figure 4-10a). Now imagine adding the potassium salt of an impermeant anion (a macromolecule A^- having multiple negative charges) to

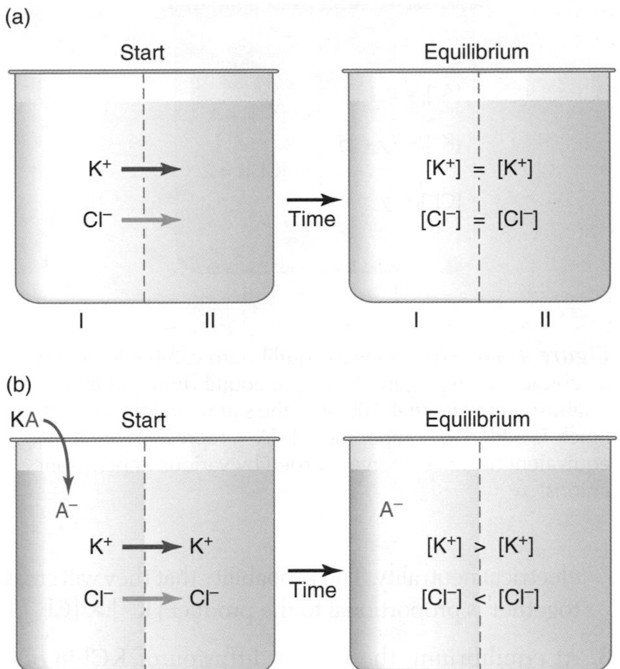

Figure 4-10 The Donnan equilibrium describes ion distribution across a semipermeable membrane. **(a)** When KCl is added to compartment I of a container divided by a permeable membrane, K^+ and Cl^- diffuse across the membrane until their concentrations are equal on either side. **(b)** If the potassium salt of an impermeant anion is added to compartment I, some K^+ and Cl^- diffuse into compartment II until electrochemical equilibrium is reestablished. Note that these compartments, unlike the living cell, are not distensible.

the KCl solution in compartment I. The K^+ and Cl^- quickly become redistributed until a new equilibrium is established by movement of some K^+ and some Cl^- from compartment I to compartment II (see Figure 4-10b). Donnan equilibrium is characterized by a reciprocal distribution of the anions and cations so that

$$\frac{[K^+]_I}{[K^+]_{II}} = \frac{[Cl^-]_{II}}{[Cl^-]_I}$$

At equilibrium, the permeant cation K^+ is more concentrated in the compartment containing the impermeant anion A^-, whereas the permeant anion Cl^- becomes less concentrated in that compartment than in the other.

We can better understand this situation by considering the consequences of the following physical principles:

1. There must be electroneutrality within both compartments; that is, in each compartment, the total number of positive charges must equal the total number of negative charges. Thus, in the example shown in Fig 4-10a, $[K^+] = [Cl^-]$ in compartment II.
2. Considered statistically, the permeant ions K^+ and Cl^- cross the membrane in pairs to maintain

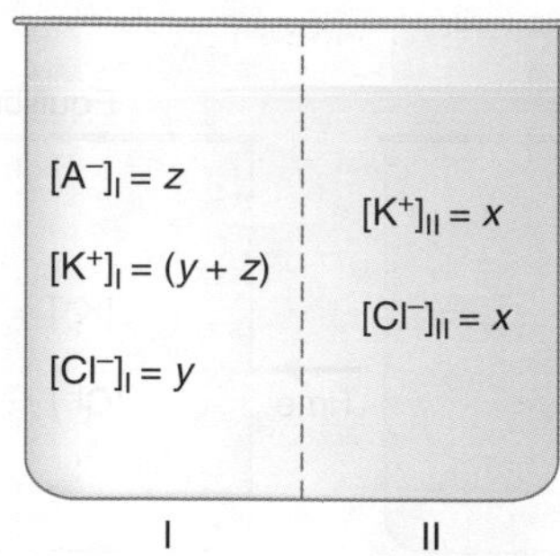

Figure 4-11 The Donnan equilibrium can be described algebraically. This figure shows the equilibrium condition established in Figure 4-10b after the salt of an impermeant anion is added to compartment I. $[A^-]_I$ represents the molar equivalent negative charges carried by various impermeant anions.

electrical neutrality. The probability that they will cross together is proportional to the product $[K^+] \times [Cl^-]$.

3. At equilibrium, the rate of diffusion of KCl in one direction through the membrane must equal the rate of KCl diffusion in the opposite direction. It follows, then, that at equilibrium, the product $[K^+] \times [Cl^-]$ in one compartment must equal the product in the other compartment. Letting x, y, and z represent the concentrations of the ions in compartments I and II, as shown in Figure 4-11, we can express the equilibrium condition (i.e., equality of the product $[K^+] \times [Cl^-]$ in the two compartments) algebraically:

$$x^2 = y(y + z) \tag{4-8}$$

This equation also holds, of course, if A^- is not present. In that case, K^+ and Cl^- are equally distributed, $z = 0$, and $x = y$. By rearranging Equation 4-8, we can see that, at equilibrium, the distributions of the permeant ions in the two compartments are reciprocal:

$$\frac{y + z}{x} = \frac{x}{y} \quad \text{or} \quad \frac{[K^+]_I}{[K^+]_{II}} = \frac{[Cl^-]_{II}}{[Cl^-]_I} \tag{4-9}$$

From this relation, it is clear that as the concentration of the impermeant anion, z, is increased, the concentrations of the permeant ions (x and y) will become increasingly divergent. This unequal distribution of permeant ions is the hallmark of Donnan equilibrium.

At Donnan equilibrium, the osmotically unequal distribution of solute particles makes water move in the direction of the compartment of higher osmolarity (compartment I in Figure 4-10). This osmotic pressure difference plus any resultant increase in hydrostatic pressure of that compartment is called the **oncotic pressure.** This concept is important in understanding the balance of hydrostatic and osmotic pressures across certain biological barriers such as capillary walls.

This explanation of a Donnan equilibrium depends on the occurrence of an ideal—and simple—set of conditions. The living cell and its plasma membrane are, of course, far more complex. For example, the plasma membrane is best regarded as "semipermeable" to a variety of ions and molecules, and there will almost never be a single "impermeant anion," which here represents various anionic side groups of proteins and other large molecules. Although the physical and mathematical principles recognized by Donnan play a role in regulating the distribution of electrolytes in living cells, clearly nonequilibrium mechanisms must modify the distribution of many substances across the plasma membrane. In particular, the permeability of the plasma membrane to particular ions changes over time, changing the conditions dramatically. Thus, cells cannot be considered passive "osmometers," and the distribution of substances across biological membranes generally cannot be predicted entirely by Donnan equilibrium principles.

OSMOTIC PROPERTIES OF CELLS

We can now use the physical principles outlined above to analyze the properties of the plasma membrane that maintain different concentrations of ions inside and outside the cell (Figure 4-12). Plasma membranes must closely regulate cell volume and thus intracellular osmotic pressure.

Ionic Steady State

Every cell maintains concentrations of inorganic solutes inside the cell that are quite different from those outside the cell (Table 4-2). The most concentrated inorganic ion in the cytosol is K^+, which is typically 10–30 times as concentrated in the cytosol as in the extracellular fluid.

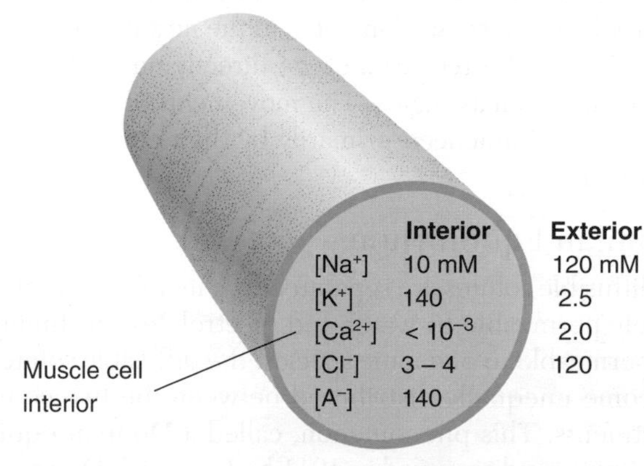

Figure 4-12 Concentrations of common ions, shown here in millimoles per liter, are very different inside and outside a vertebrate skeletal muscle cell. The concentration given for intracellular Ca^{2+} is for the free, unbound, and unsequestered ion in the muscle cell cytoplasm. Because the list of ions is incomplete, the totals do not balance out perfectly.

Table 4-2 **Internal and external concentrations of some electrolytes in specific nerve and muscle tissues**

	Internal concentrations (mM)			External concentrations (mM)			Ratios, inside/outside		
Tissue	Na^+	K^+	Cl^-	Na^+	K^+	Cl^-	Na^+	K^+	Cl^-
Squid nerve	49	410	40–100	440	22	560	1/9	19/1	1/14–1/6
Crab leg nerve	52	410	26	510	12	540	1/10	34/1	1/21
Frog sartorius muscle	10	140	4	120	2.5	120	1/12	56/1	1/30

Conversely, the intracellular concentrations of free Na^+ and Cl^- are typically less (approximately one-tenth or lower) than the extracellular concentrations. Another important generalization is that the Ca^{2+} concentration in the cytosol is maintained several orders of magnitude below the extracellular concentration. This difference is due in part to active transport of Ca^{2+} out of the cell across the plasma membrane and in part to the sequestering of Ca^{2+} within organelles such as the mitochondria and endoplasmic reticulum. As a result, the Ca^{2+} concentration in the cytosol is generally well below 10^{-6} M.

Plasma membranes typically are about 30 times more permeable to K^+ than to Na^+. Membrane permeability to Cl^- varies, being similar to that of K^+ in some cells while lower in others. The permeability of the plasma membrane to Na^+ is low, but not low enough to prevent Na^+ from leaking steadily into the cell.

Certain features of the plasma membrane, particularly its differential permeability to different ionic species, suggest that under some conditions the Donnan equilibrium might apply. To understand when the Donnan equilibrium is useful in determining membrane characteristics in living cells, three related factors are important:

1. Inside the cell, carboxyl groups and other anionic sites found on impermeant peptide and protein molecules contribute most of the net negative charge. These charges must be balanced by positively charged counterions such as Na^+, K^+, Mg^{2+}, and Ca^{2+}.

2. The anionic sites trapped inside the cell make it similar to the artificial case presented above (see Figure 4-10), in which Donnan equilibrium applies. If K^+ and Cl^- were the only permeant ions, an equilibrium situation similar to that shown in Figure 4-10b would indeed develop in the cell. However, the plasma membrane is slightly permeable to Na^+ and other inorganic ions, and with time, the cell would load up with these ions if they were simply allowed to accumulate. This, in turn, would cause osmotic movement of water into the cell, causing it to swell and eventually burst.

3. Such osmotic disasters are avoided because the cell pumps out Na^+, Ca^{2+}, and some other ions at the same rate as they leak in, keeping the intracellular Na^+ concentration about an order of magnitude lower than the extracellular concentration. This active pumping, which will be discussed later, has the same effect as would membrane impermeability to Na^+ and Ca^{2+}. As a result, the concentrations of Na^+ and Ca^{2+} are not allowed to come into equilibrium, and the cell in fact behaves very much as if it were in a state of Donnan equilibrium. In precise terms, however, the unequal distribution of ions represents a steady state, not a true equilibrium, because the pumping of ions requires a continual expenditure of energy.

Since K^+ and Cl^- are by far the most concentrated and most permeant ions in animal tissue, they distribute themselves in a way similar to Donnan equilibrium. That is, the KCl product $[K^+] \times [Cl^-]$ of the cell interior approximately equals the KCl product of the extracellular solution (Figure 4-13), provided the membrane permeabilities of K^+ and Cl^- are both high relative to those of other ions present.

Cell Volume

Plant and bacterial cells have rigid walls lying outside the plasma membrane, which place an upper limit on the size of the cell and allow the osmotic buildup of

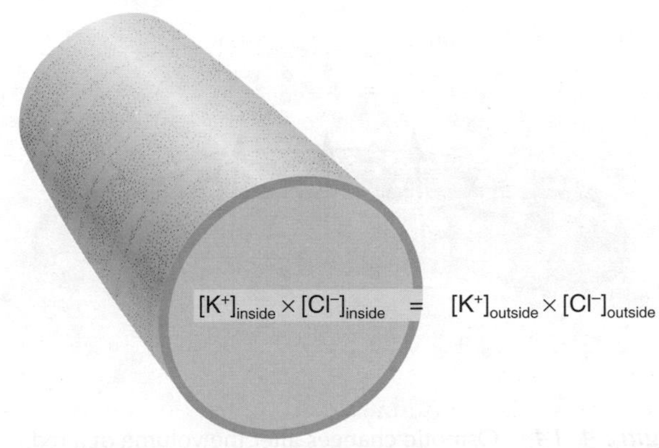

Figure 4-13 The KCl product is governed by the Donnan equilibrium. The distribution of K^+ and Cl^- inside and outside this skeletal muscle cell follows Donnan equilibrium principles, since the membrane is permeable to both K^+ and Cl^-.

turgor pressure in the cell. (The loss of turgor pressure is evident in the wilting of a cut flower.) In contrast, animal cells do not have rigid walls and therefore cannot resist any buildup of intracellular pressure. As a result, animal cells will change size when placed in solutions containing different concentrations of impermeable substances. Their shrinkage or swelling under these circumstances is due to osmotic movement of water (Figure 4-14). The plasma membrane might prevent osmotic swelling if it could pump water back out of the cell as fast as it leaked in. However, there is no evidence of any animal cell with an active water transport mechanism (although a similar effect is achieved by the contractile vacuole of certain protozoans). The only mechanism for regulating cell volume, then, is to pump out solutes that leak into the cell (Figure 4-15). Thus, at steady state, Na^+, the major osmotic constituent outside the cell, is expelled from the cell by active transport as rapidly as it leaks in. In effect, there is no net Na^+ entry, and the situation is osmotically equivalent to complete sodium impermeability, with a relatively fixed concentration of Na^+ trapped in the cell. Because Na^+ is not allowed to further accumulate in the cell, there is no compensatory osmotic influx of water.

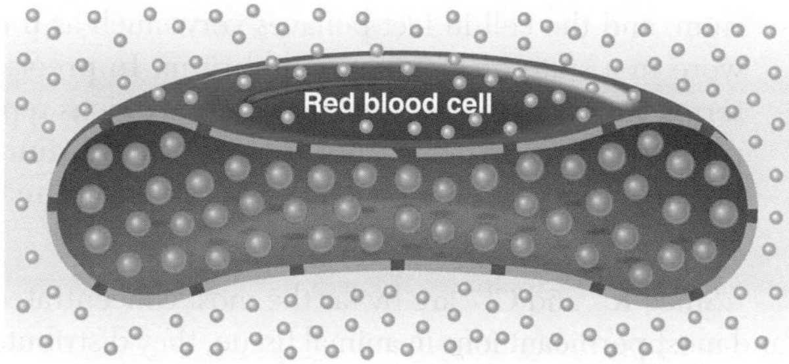

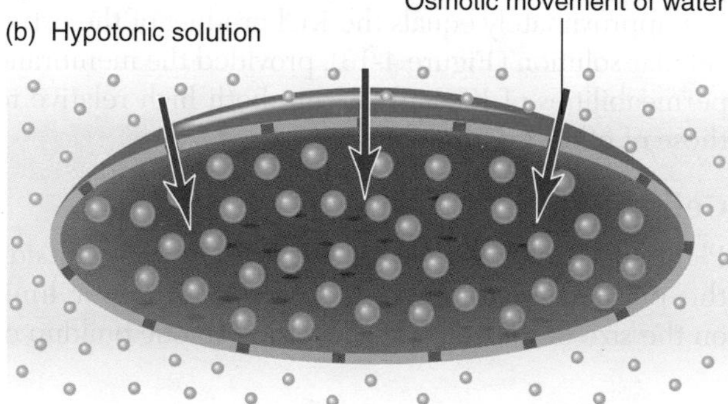

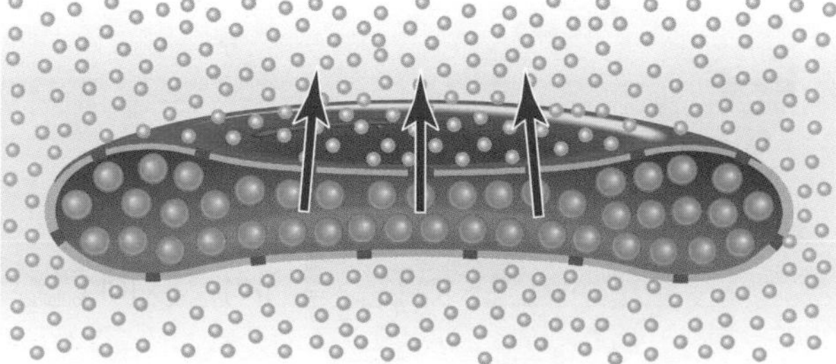

Figure 4-14 Osmotic changes alter the volume of a red blood cell. **(a)** In an isotonic solution, the cell volume remains unchanged. **(b)** In a hypotonic solution, water (arrows) enters the cell because of the higher osmotic concentration of the cytoplasm with respect to the solution, producing swelling. **(c)** In a hypertonic solution, water leaves the cell, causing it to shrink.

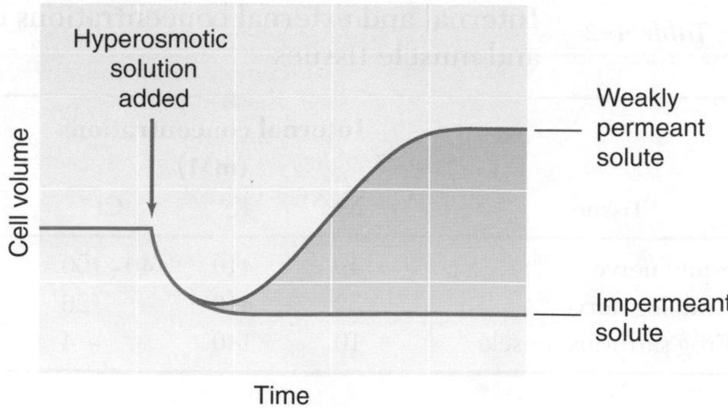

Figure 4-15 Hyperosmotic solutions containing an impermeant solute (unable to enter cell) and a weakly impermeant solute both cause initial cell shrinkage because water is drawn out of the cell by osmosis. A completely impermeant solute maintains cell shrinkage because the solution remains basically hypertonic in this situation. If the solute is only weakly impermeant (that is, it can slowly enter the cell), then the external solution becomes hypotonic over time as water flows into the cell by osmosis. This process eventually produces swelling, even though the solution is hyperosmotic to the cell.

The low sodium concentration in animal cells (relative to the extracellular concentration) is important in balancing the other osmotically active solutes in the cytoplasm. The importance of active transport in maintaining the sodium gradient, and thereby the osmolarity of the cell and its volume, is seen when the energy metabolism of the cell is interrupted by metabolic poisons (Figure 4-16). Without ATP to energize the "uphill" extrusion of Na^+, both Na^+ and its Cl^- counterion leak into the cell, and water follows osmotically, causing the cell to swell.

MECHANISMS FOR TRANSMEMBRANE MOVEMENTS

The plasma membrane evolved to provide separation between the external environment and the highly specialized internal environment of the cell, where metabolic reactions occur. At the same time, nutrients must be able to enter the cell, and products and wastes resulting from metabolic reactions must be able to escape. Materials can cross the plasma membrane by three general mechanisms:

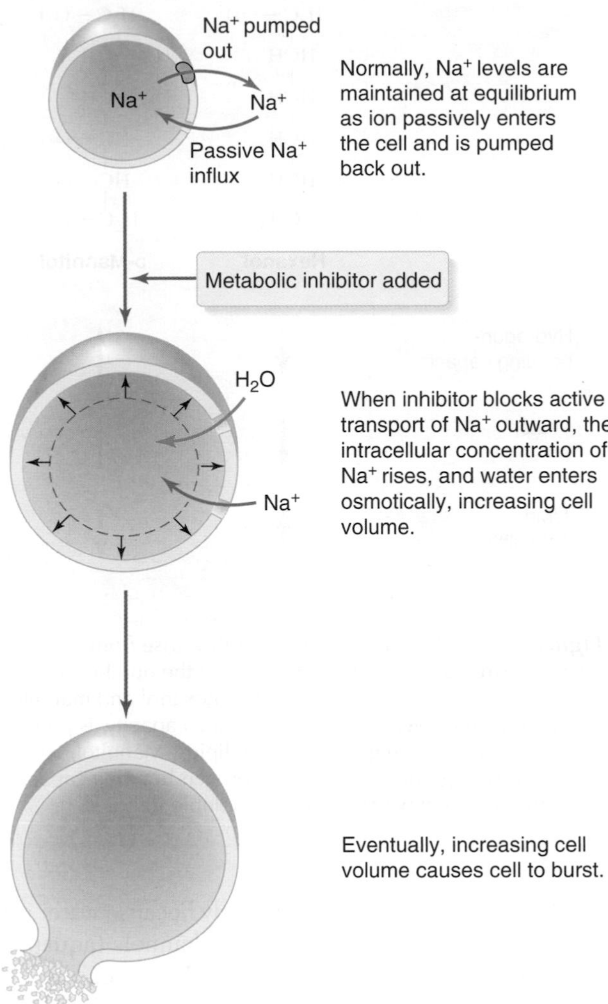

Figure 4-16 The maintenance of cell volume is disturbed by a metabolic inhibitor that interferes with Na^+ pumping.

- **passive diffusion** (also called simple diffusion)
- **passive transport** (also called facilitated diffusion)
- **active transport**

Passive diffusion and passive transport do not require a direct input of metabolic energy; instead, they are powered by a concentration gradient or electrical gradient across a cell membrane. However, one should understand that the gradient is itself created and maintained by an expenditure of metabolic energy (Figure 4-17). The potential energy stored in the gradient is ultimately responsible for the translocation of molecules across the membrane. Active transport, on the other hand, involves the direct use of ATP as an energy source. Let's consider each of these mechanisms in turn.

Passive (Simple) Diffusion

If a solute molecule comes into contact with the lipid portion of the plasma membrane and its thermal energy is high enough, it may enter and cross the lipid phase and finally emerge into the aqueous phase on the other side of the membrane. To leave the aqueous phase and enter the lipid phase, a solute must first break all of its hydrogen bonds with water. Breaking a hydrogen bond requires about 5 kcal of kinetic energy per mole of solute. Moreover, a solute molecule crossing the membrane must dissolve in the lipid phase, so the lipid solubility of a substance also plays a major role in determining whether or not it will cross the membrane. Consequently, those molecules having a minimum of hydrogen bonding with water will most readily enter the lipid bilayer, whereas polar molecules such as inorganic ions will almost never dissolve in the bilayer.

A number of factors, such as molecular weight and molecular shape, influence the mobility of nonelectrolytes within the membrane, but the empirically measured partition coefficient is the primary predictor of the diffusion of a nonelectrolyte across a pure lipid bilayer. To measure this property, a test substance is

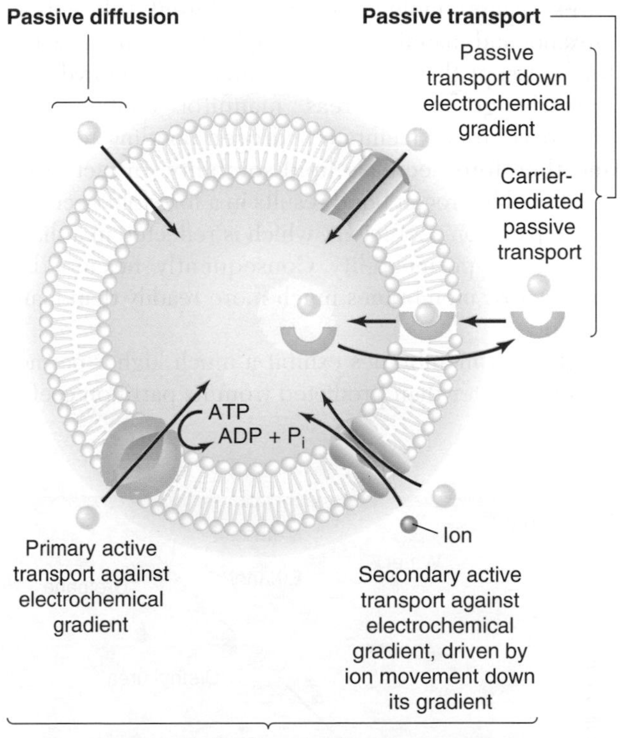

Figure 4-17 Substances cross the plasma membrane by three mechanisms: passive diffusion, passive transport, and active transport. Passive diffusion is the simple diffusion of a substance across the membrane. Passive transport, also called facilitated diffusion, occurs when a substance moves through a channel down an electrochemical gradient. Substances may be assisted in passive transport by ionophores, which are carrier proteins specific for ions. Primary active transport uses ATP to move a substance (often an ion) against an electrochemical gradient. Secondary active transport does not consume energy by itself, but uses an electrochemical gradient created earlier by primary active transport to move a substance against its concentration gradient.

shaken in a closed tube containing equal amounts of water and olive oil, and the partition coefficient *K* is determined from its relative solubilities in water and oil at equilibrium, using the equation

$$K = \frac{\text{solute concentration in lipid}}{\text{solute concentration in water}} \quad (4\text{-}10)$$

Collander (1937) systematically tested the idea that nonelectrolyte membrane permeability is related to the partition coefficient of the solute. Using the giant algal cell *Chara,* he plotted the permeability coefficient (Equation 4-4) against the partition coefficient (Equation 4-10) for a number of solutes. He found that lipid solubility is almost linearly related to the permeability of a substance, independent of molecule size (Figure 4-18).

Nonelectrolytes exhibit a wide range of partition coefficients. The value for urethane, for example, is 1000 times that for glycerol (see Figure 4-18). These differences depend on particular features of the molecular structure, as illustrated in Figure 4-19, which compares related molecules with different solubilities. Hexanol and mannitol, for example, have similar structures, except that hexanol contains only one hydroxyl (—OH) group, whereas mannitol contains six. Hydroxyl groups facilitate hydrogen bonding to water and therefore decrease lipid solubility. In fact, each additional hydrogen bond results in a fortyfold decrease in the partition coefficient, which is reflected in a sharp decrease in permeability. Consequently, hexanol diffuses across membranes much more readily than mannitol does.

Plasma membranes exhibit a much higher permeability to water than predicted from its partition coefficient (see Figure 4-18). This is partly because water can pass through selective permanent channels (aquaporins) that penetrate the lipid bilayer. However, even in channel-free, artificial lipid bilayers, water permeability is still several times higher than predicted from the solubility of water in long-chain hydrocarbons. A possible explanation is that the small, uncharged water molecules can pass through transient gaps between lipid molecules. Other small, uncharged polar molecules, such as CO_2, NO, and CO, also have relatively high permeabilities across artificial and natural membranes.

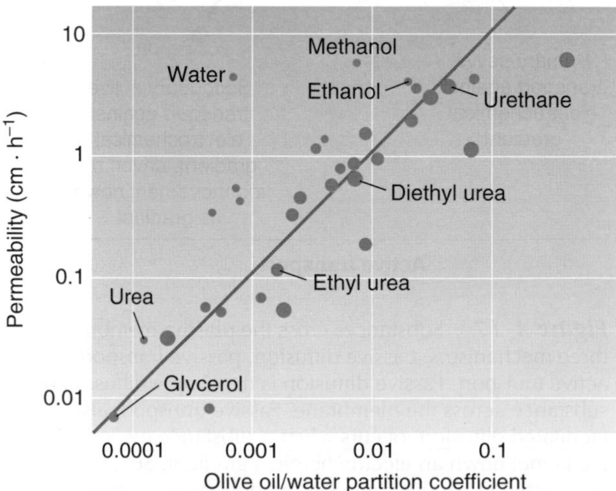

Figure 4-18 The membrane permeability of a nonelectrolyte is highly dependent on its oil/water partition coefficient—a measure of its ability to dissolve in and diffuse through the lipid bilayer. Note that the permeability of nonelectrolytes is independent of molecular size (roughly indicated by symbol size).

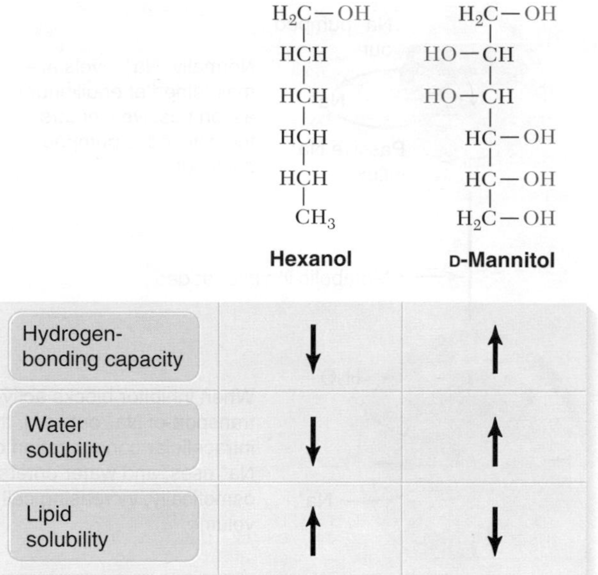

Figure 4-19 Water and lipid solubility arise from chemical structure. Note the difference in the number of hydroxyl (—OH) groups (red) between hexanol and mannitol. Hexanol, with its weak hydrogen-bonding capacity, is poorly soluble in water and highly soluble in lipids. Mannitol, with many hydroxyl groups and strong hydrogen-bonding capacity, is highly soluble in water and poorly soluble in lipids.

In passive diffusion, the rate of influx increases in proportion to the concentration of the solute in the extracellular fluid (Figure 4-20a). This pattern is observed because the net rate of influx is determined only by the difference in the number of solute molecules on the two sides of the plasma membrane. This proportionality between external concentration and rate of influx over a large range of concentrations distinguishes passive diffusion from channel- or carrier-mediated transport mechanisms, which show **saturation kinetics** (Figure 4-20b,c).

Passive Transport (Facilitated Diffusion)

Passive transport, unlike passive diffusion, involves specialized protein molecules that assist in the transmembrane movements of solute. A passive transport mechanism can be as simple as the creation of pores or channels providing a specialized route across the plasma membrane. In a more complex form of passive

(a) Passive diffusion through membrane

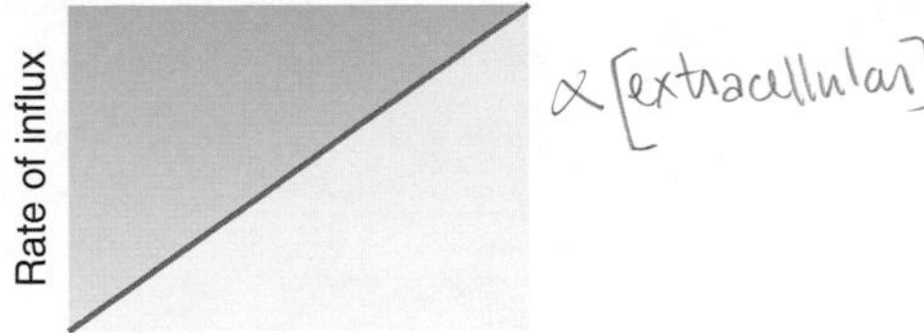

(b) Passive transport through channels

(c) Carrier-mediated transport (passive or active)

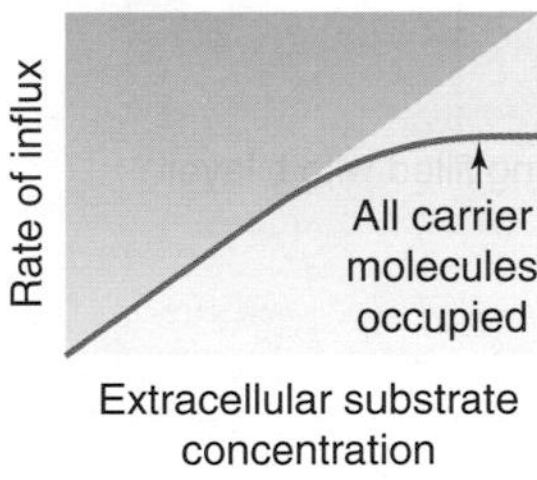

Figure 4-20 The kinetics of the influx of a substance crossing a plasma membrane and entering a cell depend on the mechanism for that substance's movement. **(a)** For passive diffusion through the lipid phase of a membrane, the rate of influx is proportional to the extracellular concentration of the substance (within physiological concentrations). **(b)** The movement of a substance through specialized channels is proportional to concentration at low extracellular concentrations, but at higher concentrations the number of channels begins to limit influx. **(c)** Carrier-mediated transport shows clear saturation kinetics, in which the rate of influx cannot rise above a certain rate, reflecting the occupation of all carrier molecules involved in the transport of that substance.

transport, the solute molecule combines with a **carrier** (transporter) protein dissolved in the membrane. This carrier "mediates" or "facilitates" the movement of the solute molecule across the membrane. Such carriers can "mask" the electrical properties of even a polar solute and, because of their lipid solubility, allow the solute to diffuse rapidly across the membrane down its concentration or electrochemical gradient. This mechanism is called **carrier-mediated** (or facilitated) **passive transport.**

All channel proteins and most carrier proteins allow solutes to cross the membrane passively at no energetic cost (other than the original cost of generating potential energy in the form of different solute concentrations on opposite sides of the membrane, as mentioned earlier). The electrochemical gradient determines the direction of passive transport. As passive transport proceeds, the solute concentrations in the two compartments approach equilibrium, at which point no further net diffusion will occur. While passive transport does not directly require energy (hence, "passive"), energy has been consumed at some time to create the electrochemical gradient that provides for solute movement.

Charged molecules can cross membranes by diffusing through specific water-filled channels (see Figure 4-17). Since inorganic ions such as Na^+, K^+, Ca^{2+}, and Cl^- cannot diffuse through lipid bilayers, highly specialized protein molecules have evolved that extend across the plasma membrane and act as channels or pores. When these channels are open, they allow specific solutes to pass through them (Figure 4-21a). The functioning of ion channels can be demonstrated directly in artificial lipid bilayer membranes that are by themselves highly impermeable to even the smallest charged molecules (Spotlight 4-1 on the next page). A dramatic increase in ion permeability occurs upon addition of small amounts of channel proteins extracted from cellular membranes. This increase can be measured as discrete pulses of current carried by ions from one side of the membrane to the other, just like those measured in biological membranes. These **unitary currents** are due to the sudden opening of individual channels that allow thousands of ions per second to stream down their gradients and across the membrane.

Membrane channels can be created artificially to study their effects on membranes. For example, rod-shaped molecules of the antibiotic nystatin applied to both sides of an artificial or natural membrane will

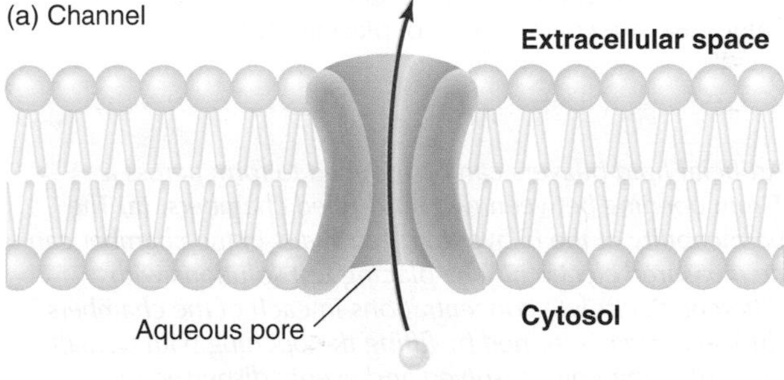

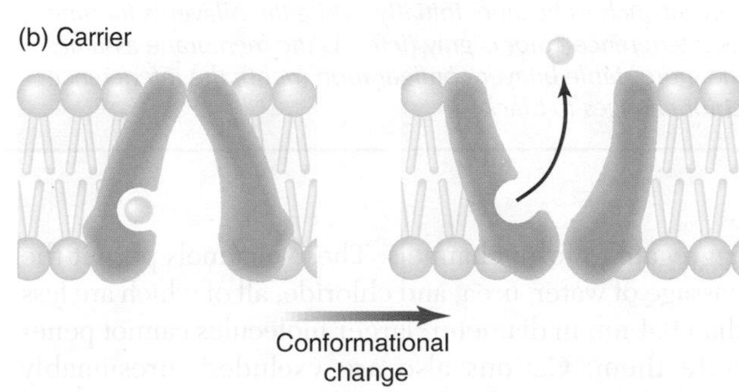

Figure 4-21 Materials can be transported across membranes via channels or carrier proteins. **(a)** A channel protein forms a water-filled pore across the bilayer through which specific ions can diffuse. **(b)** A carrier protein alternates between two extremes in conformation, so that the solute binding site is sequentially accessible on one side of the bilayer and then on the other.

Spotlight 4-1 ARTIFICIAL BILAYERS

Many of our ideas about how molecules and ions pass across membranes have grown out of experiments and observations using artificial bilayers that are similar to the phospholipid bilayer that forms the basis of biological membranes. Artificial bilayers are extremely useful in studies of permeation mechanisms because they can be made from chemically defined mixtures of lipids. Selected substances can be added to test their effects on permeability. Channel-forming substances, such as antibiotics that act as ionophores (molecules that facilitate the diffusion of ions across membranes), thereby disrupting normal ionic gradients in microbes, and membrane channel components of excitable tissues have been incorporated into artificial bilayers, allowing their properties to be studied in isolation under highly controlled conditions.

The accompanying figure shows the principle of artificial bilayer formation. The most stable configuration attained consists of two layers of lipid molecules whose hydrophobic, lipophilic hydrocarbon tails are loosely associated to form a liquid-lipid phase sandwiched between the hydrophilic polar heads of the molecules, which are directed outward toward the aqueous medium. The thickness of the lipid film is easily determined from the interference color of light reflected from the two surfaces of the film. Membranes with thicknesses of approximately 7 nm (for which the interference color is black) are most commonly used. These membranes have electrical conductances (ion permeabilities) and capacitances consistent with their thickness and lipid composition. Although their permeability to ions is much lower than that of natural plasma membranes, the addition of certain ionophores increases it to values that are characteristic of plasma membranes.

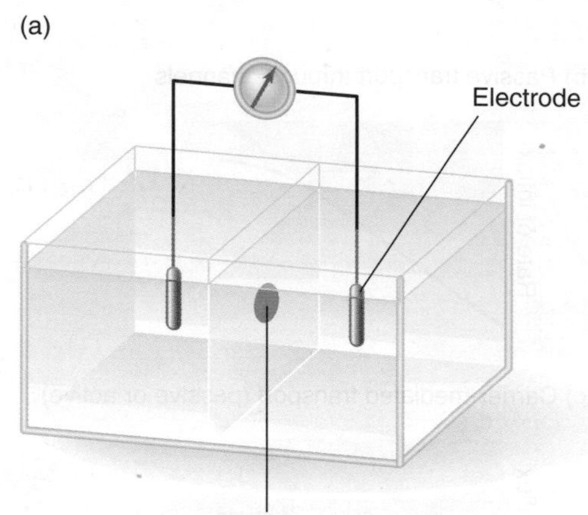

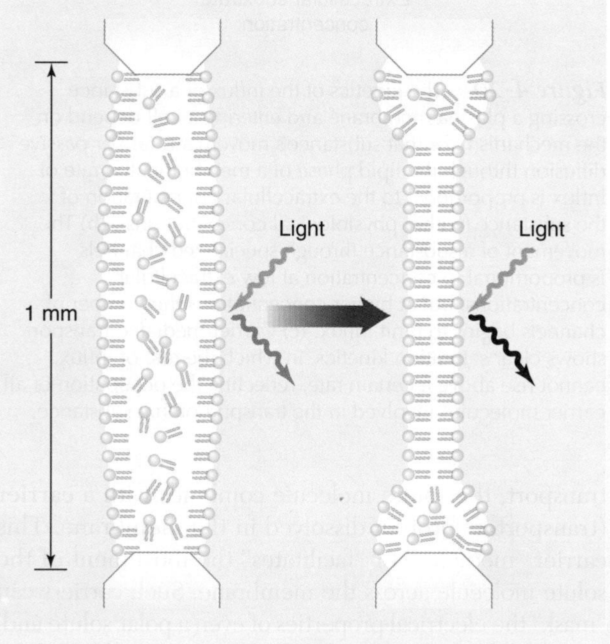

Artificial lipid bilayers can be induced to form across a 1 mm opening between two fluid-filled chambers. ***(a)*** *The permeability of the bilayer to electrolytes in the chamber can be measured electrically by placing test solutions with different electrolyte concentrations in each of the chambers.* ***(b)*** *The bilayer is formed by filling the opening with a small amount of the lipid dissolved and evenly dispersed in a solvent such as hexane. Initially, while the bilayer is forming, its interference color is gray (left). As the membrane assumes the more stable bilayer configuration (right), the interference color changes to black.*

aggregate to form channels. These channels permit the passage of water, urea, and chloride, all of which are less than 0.4 nm in diameter; larger molecules cannot penetrate them. Cations also are excluded, presumably because there are fixed positive sites along the channel walls. Incorporation of nystatin into an artificial membrane produces a negligible increase in the membrane area occupied by fixed channels (0.001%–0.01%), but it produces a 100,000-fold increase in membrane permeability to chloride ions. We learn from this observation that very little membrane area need be devoted to channels to account for the significant ion permeabilities of natural membranes. This conclusion is supported by the observation that the electrical capacitance of the plasma membrane (which depends on membrane area) remains relatively unchanged during the large changes in permeability that follow the excitation of some membranes. (This phenomenon is discussed further in Chapter 5.)

Cloning and genetic expression studies during the last decade or so have identified a family of membrane channels that specifically permit the passive diffusion of water but exclude ions and other substances. These **aquaporins** typically appear as an hourglass-shaped arrangement of proteins that spans the entire thickness of the membrane. Interestingly, aquaporins do not appear to be permanent residents in the cell membrane, but rather membrane vesicles containing aquaporins can be inserted and removed under active hormonal regulation. In the mammalian kidney, for example, the peptide hormone arginine vasopressin (AVP) greatly increases water permeability in the epithelial cells of the inner medullary collecting duct by changing the number of aquaporins in the membrane.

Biological membranes must be permeable to various polar molecules such as sugars, amino acids, nucleotides, and certain cell metabolites that would cross lipid bilayers by diffusion only very slowly. The movement of these molecules through membranes is facilitated by carrier proteins (Figure 4-21b). **Ionophores** are small organic compounds that specifically transport ions across the plasma membrane. Carrier proteins, which exist in many forms in all types of membranes, are exquisitely selective about which species of molecules they transport. Carrier proteins that transport a single solute from one side of the membrane to the other are called **uniporters,** while those that transfer one solute and simultaneously or sequentially transfer a second solute are called **coupled transporters.** Coupled transporters that transfer two solutes in the same direction are called **symporters,** while those that transfer solutes in opposite directions are called **antiporters** (Figure 4-22). These terms can also be applied to the active transport systems discussed below.

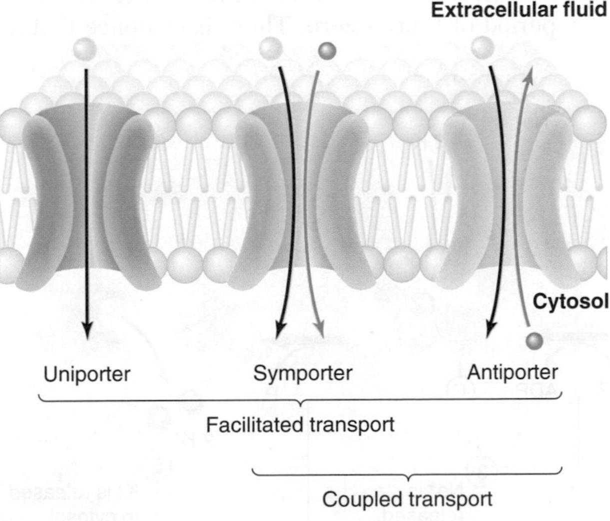

Figure 4-22 Carrier proteins can be configured as uniporters, symporters, or antiporters. Uniporters transport a single type of solute in one direction across the lipid bilayer membrane. Symporters simultaneously transport two different solutes in the same direction. Antiporters transport two solutes in opposite directions across the membrane. Symporters and antiporters both carry out coupled transport, in which both solutes must move in concert.

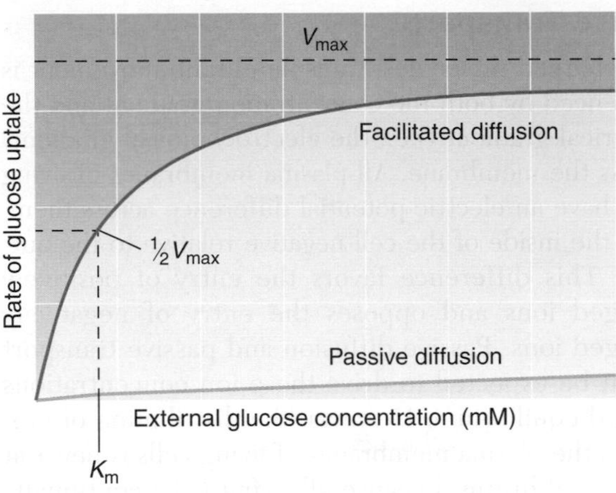

Figure 4-23 The kinetics of simple diffusion differ from those of carrier-mediated (facilitated) diffusion. In this example of glucose movement, the rate of passive diffusion is always proportional to the glucose concentration. However, the rate of carrier-mediated glucose diffusion reaches a maximum (V_{max}) when the glucose transporter is saturated. The Michaelis constant (K_m) of the glucose transporter, a measure of binding affinity that is analogous to the K_m of an enzyme for its substrate, is measured when transport is at half its maximal value. [Adapted from Lodish et al., 2000.]

The existence of carriers was initially inferred from kinetic studies of molecule transfer across membranes (Figure 4-23). For some solutes, the measured rate of influx reaches a plateau beyond which an increase in solute concentration produces no further increase in flux. This finding reveals that a rate-limiting step must be occurring. Experiments elucidating the kinetics of such permeation patterns led to the conclusion that transport occurs through the formation of a carrier-substrate complex similar in concept to an enzyme-substrate complex. Each carrier protein has a characteristic binding constant for its solute equal to the concentration of solute when the transport rate is half its maximum value. The carrier and solute molecules temporarily form a complex based on bonding, steric specificity, or both. Like enzyme-substrate binding, the binding of solute to carrier can be blocked by specific inhibitors.

The specificity of transporters was first established in studies in which single-gene mutations abolished the ability of bacteria to transport specific sugars across their cell membranes. Similar mutations have now been found in many other circumstances, including human inherited diseases that affect the transport of specific solutes across the kidneys, intestine, or lungs. In cystic fibrosis, for example, a defect in the chloride transport channel protein (CFTR) appears to be responsible for fluid imbalance in the lungs. Cystic fibrosis is a prime candidate for treatment by gene therapy, in which genes that correctly express CFTR are introduced into patients to create populations of cells with normal chloride transport (see Chapter 2).

Active Transport

For charged molecules, transmembrane movement is influenced by both the concentration gradient and the electrical gradient (i.e., the electrochemical gradient) across the membrane. All plasma membranes of living cells have an electric potential difference across them, with the inside of the cell negative relative to the outside. This difference favors the entry of positively charged ions and opposes the entry of negatively charged ions. Passive diffusion and passive transport might be expected to drive these ion concentrations toward equilibrium. However, the distribution of ions across the plasma membranes of living cells is never at true equilibrium, because all living cells continually expend chemical energy to maintain a stable differential of transmembrane ion concentrations. The energy required for this process of active transport is typically supplied in the form of ATP.

Active transport can be classified as either primary active transport or secondary active transport. In **primary active transport**, ATP-dependent membrane pumps are used to transport substances against a gradient—primary active transport requires the direct expenditure of metabolic energy. When the energy source for the pumps is cut off, active transport ceases and passive diffusion governs the distribution of substances. The concentrations of the substances thus gradually return to equilibrium. Many of the physiological changes that occur in the hours or minutes following an animal's death (e.g., rigor mortis) can be attributed directly to the development of equilibrium across cell membranes.

Secondary active transport is the movement of a substance against an electrochemical gradient because the substance is moving down its own concentration gradient. While ATP is not directly consumed in secondary active transport, the conditions necessary for this type of transmembrane movement are initially set up by primary active transport at some other point in the membrane.

The Na^+/K^+ Pump as a Model of Primary Active Transport

Many of the features of primary active transport are readily evident in the mechanism by which animal cells maintain their characteristically steep transmembrane concentration gradients for Na^+ and K^+. The concentration of K^+ is about 10–20 times higher inside cells than outside, while the opposite is true for Na^+ (see Figure 4-12). These concentration differences are sustained by a **Na^+/K^+ pump** found in the plasma membrane of virtually all animal cells. This pump is an enzyme—specifically, an ATPase—with binding sites for Na^+ and ATP on its cytoplasmic surface and binding sites for K^+ on its external surface (Figure 4-24). The operation of the pump is thought to depend on a series of conformational changes that allow the cotransport of K^+ and Na^+ in opposite directions across the membrane.

The Na^+ concentration gradient created by the Na^+/K^+ pump drives the transmembrane movement of other substances, as we will see below. Two opposing factors determine the size of this concentration gradient: the rate of active transport of Na^+ and the rate at which Na^+ can leak (i.e., diffuse passively) back into the cell. In the steady state, the number of Na^+ ions pumped, or transported, out of the cell is equal to the number of Na^+ ions that leak in. Thus, even though there is a continual turnover of Na^+ (and other ionic species) across the membrane, the net Na^+ flux over any period of time is zero. There is evidence that an

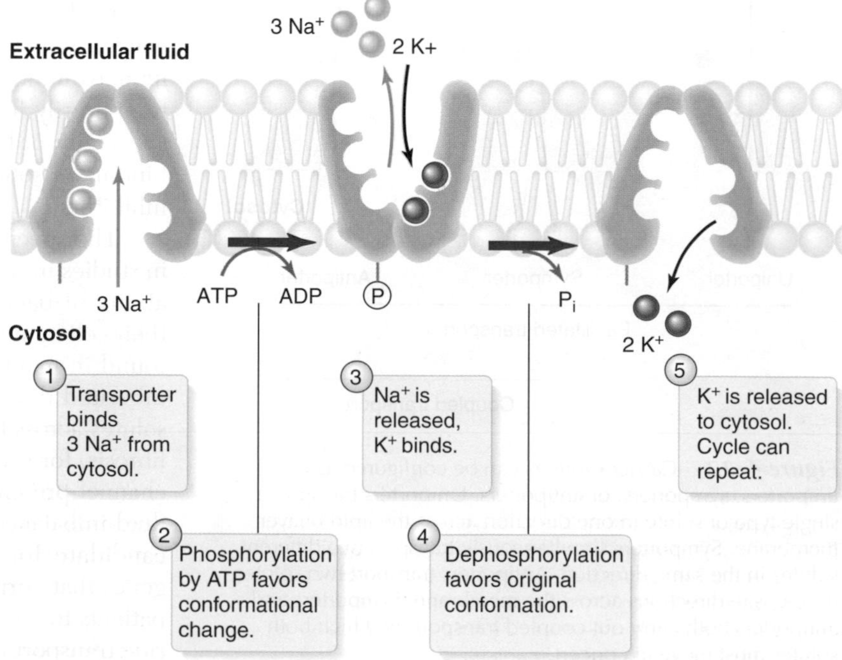

Figure 4-24 The Na^+/K^+ pump actively transports Na^+ out of the cell and K^+ into the cell against their respective electrochemical gradients. For every molecule of ATP consumed, three Na^+ ions are pumped out and two K^+ ions are pumped in. The specific pump inhibitor ouabain and K^+ compete for the same sites on the extracellular side of the pump. [Adapted from Nelson and Cox, 2000.]

increase in the intracellular concentration of Na^+ leads to an increase in the rate of Na^+ expulsion by the membrane pump (which may merely be a mass action effect due to the increased availability of intracellular Na^+ to the pump).

Cotransport

The movement of some substances up a concentration gradient is driven by the movement of another substance down its concentration gradient, as we have just seen for the active movements of Na^+ and K^+ in opposite directions across the membrane. However, the ubiquitous Na^+ gradient is also used to carry certain sugars and amino acids into the cell by a symport mechanism and to drive Ca^{2+} out of the cell by an antiport mechanism. Let's consider these coupled transport mechanisms in detail.

Symporters

Symporters are carrier proteins that run on energy stored in electrical gradients. An example is the transport of the amino acid alanine, which is coupled to the transport of Na^+ (Figure 4-25). In the presence of Na^+, the amino acid is taken up by the cell until the intracellular concentration is 7–10 times that of the extracellular concentration. In the absence of Na^+, the intracellular concentration of alanine merely approaches the extracellular concentration. In both cases, the rate of influx of alanine shows saturation kinetics, indicating a carrier mechanism. The effect of extracellular Na^+ is to enhance the activity of the alanine carrier. Increasing the intracellular Na^+ concentration by blocking the Na^+/K^+ pump with ouabain, a specific inhibitor of this pump, has the same effect as decreasing the extracellular Na^+ concentration. Thus, it appears to be the Na^+ gradient that is important for inward alanine transport, and not merely the presence of Na^+ in the extracellular fluid.

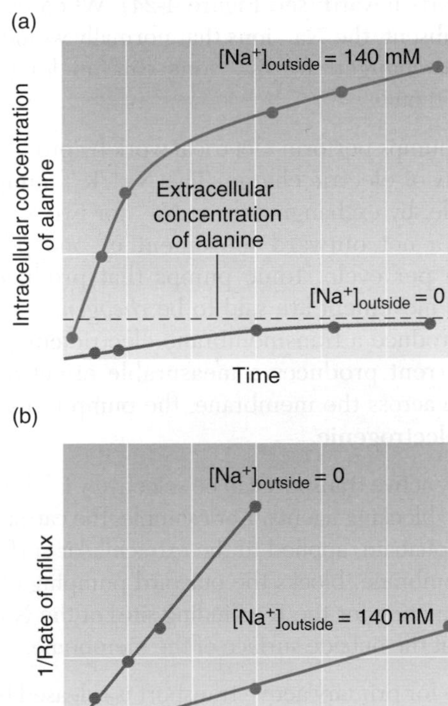

Figure 4-25 The cellular uptake of the amino acid alanine depends on the Na^+ concentration gradient. **(a)** Intracellular concentrations of alanine as a function of time with and without extracellular Na^+ present. The dashed line represents the extracellular concentration of alanine. **(b)** Lineweaver-Burk plots of alanine influx with and without extracellular Na^+ present. The abscissa is the reciprocal of the extracellular alanine concentration. The common intercept indicates that at infinite concentration of alanine, the rate of transport is independent of $[Na^+]$. [From Schultz and Curran, 1969.]

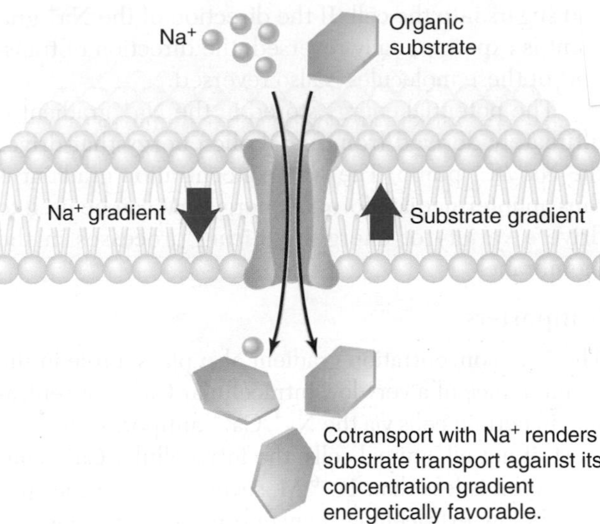

Figure 4-26 Sugar and amino acid transport can occur by sodium-mediated cotransport. The carrier must bind both the Na^+ and the organic substrate before it will transport either. The substrate shown here is moving against its concentration gradient, driven by the reservoir of energy stored in the Na^+ gradient.

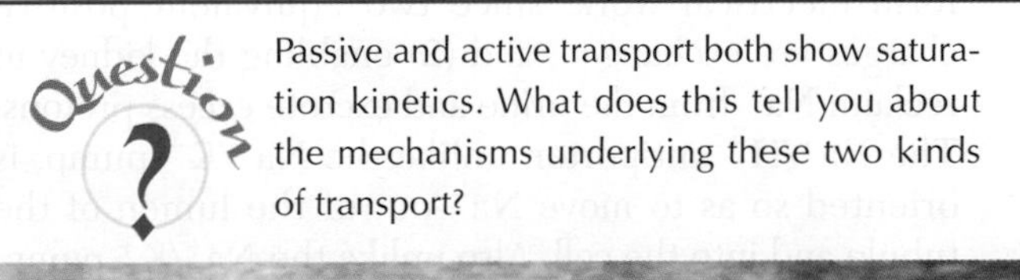

Passive and active transport both show saturation kinetics. What does this tell you about the mechanisms underlying these two kinds of transport?

The transport of amino acids and sugars is often coupled to inward flow of Na^+ by means of a common carrier. The carrier must bind both Na^+ and the organic substrate molecule before it can transport either (Figure 4-26). The tendency for Na^+ to diffuse down its concentration gradient drives this transport system. Anything that reduces the concentration gradient of Na^+ (low extracellular Na^+ or increased intracellular Na^+) reduces the inwardly directed driving force and thereby reduces the coupled transport of amino acids and sugars into the cell. If the direction of the Na^+ gradient is experimentally reversed, the direction of transport of these molecules is also reversed.

The potential energy stored in the Na^+ gradient is ultimately derived from metabolic energy that drives the Na^+/K^+ pump. The Na^+ concentration gradient is thus an intermediate form of energy that can be used to drive a variety of energy-requiring processes in the membrane.

Antiporters

The Na^+ concentration gradient also plays a role in the maintenance of a very low intracellular Ca^{2+} concentration in certain cells via the Na^+/Ca^{2+} antiport system. In most, if not all, animal cells, the intracellular Ca^{2+} concentration is less than 10^{-6} M, several orders of magnitude below extracellular concentrations. Yet many cell functions are regulated by changes in the intracellular Ca^{2+} concentration. Efflux of Ca^{2+} from cells is reduced when extracellular Na^+ is reduced, because Ca^{2+} is expelled from the cell in exchange for Na^+ leaking in. The opposing movements of these two ions are coupled to each other by an antiporter. One view is that Ca^{2+} and Na^+ both compete for this carrier, but that Ca^{2+} competes more successfully inside the cell than on the outer cell surface, so that there is a net efflux of Ca^{2+}. Here, again, the immediate source of energy is the Na^+ gradient, which ultimately depends on the ATP-energized active transport of Na^+. Calcium is also transported independently of the Na^+ gradient by an ATP-energized Ca^{2+} pump, which is the major source of Ca^{2+} extrusion under normal conditions.

The Na^+/H^+ antiporter in the proximal tubule of the mammalian kidney provides another example of cotransport in opposite directions (see Chapter 14). Here the extrusion of H^+ from inside the cells lining the renal tubule into the urine contained within the tubule is coupled to Na^+ uptake into the cell in a 1:1 stoichiometry. That is, for each H^+ expelled, one Na^+ is taken up into the cell. This mechanism has the advantages of (1) avoiding the expenditure of energy to perform electrical work, since two equivalent positive charges are exchanged, and (2) enabling the kidney to reclaim Na^+ from the urine and excrete excess protons. The Na^+/H^+ antiporter, unlike the Na^+/K^+ pump, is oriented so as to move Na^+ out of the lumen of the tubule and into the cell. Also unlike the Na^+/K^+ pump, this mechanism is not an example of primary active transport, in which ATP is the immediate source of energy. Instead, the Na^+/H^+ antiporter is an example of secondary active transport, in which the source of energy is the electrochemical gradient of one or both exchanged ions. In this case, the energy driving the exchange arises from the Na^+ concentration gradient, directed from the lumen into the cell. This gradient is maintained by the removal of Na^+ from the cell by the Na^+/K^+ pump located in the membrane on the other side of the cell, which faces the plasma and blood.

Summary of Primary Active Transport

Primary active transport is characterized by several important features:

1. ATP is required as a source of chemical energy for primary active transport. Metabolic poisons that stop the production of ATP bring active transport to a halt.

2. Primary active transport can take place against substantial concentration gradients. The Na^+/K^+ pump, for example, transports Na^+ from the cell interior to the external fluid against a 10:1 Na^+ concentration gradient.

3. Primary active transport systems generally exhibit a high degree of selectivity. The Na^+/K^+ pump, for example, fails to transport lithium ions, which have ionic properties very similar to those of sodium ions.

4. Certain membrane pumps exchange one kind of molecule or ion from one side of the membrane for another kind of molecule or ion from the other side. The Na^+/K^+ pump is an antiporter that transports three Na^+ ions outward for every two K^+ ions it transports inward (see Figure 4-24). When external K^+ is absent, the Na^+ ions that normally would have been exchanged for K^+ ions can no longer be pumped out.

5. Some pumps perform electrical work by producing a net flux of electric charge. The Na^+/K^+ pump, for example, by exchanging three Na^+ for two K^+, produces a net outward movement of one positive charge per cycle. Ionic pumps that produce net charge movement are said to be *rheogenic* because they produce a transmembrane electric current. If the current produces a measurable effect on the voltage across the membrane, the pump is also said to be **electrogenic.**

6. Primary active transport can be selectively inhibited by specific blocking agents. For example, the cardiac glycoside ouabain, applied to the extracellular surface of the membrane, blocks the outward pumping of Na^+ by competing for the K^+ binding sites of the Na^+/K^+ pump at the outside surface of the membrane.

7. Energy for primary active transport is released by the hydrolysis of ATP by enzymes (ATPases) present in the membrane. Primary active transport exhibits Michaelis-Menten kinetics and competitive inhibition by analog molecules. Both behaviors are characteristic of enzymatic reactions (see Chapter 3). Associated with the Na^+/K^+ pump are Na^+- and K^+-activated ATPases that have been isolated from red blood cell membranes and other tissues. These enzymes catalyze the hydrolysis of ATP into ADP and inorganic phosphate only in the presence of Na^+ and K^+, and they bind the specific Na^+/K^+ pump inhibitor ouabain. The fact that ouabain blocks the

Na^+/K^+ pump is evidence that these ATPases are involved in primary active transport of Na^+ and K^+.

The actual process of active transport takes place across the cell membrane, with molecules pumped either into or out of the cell. At a higher level of organization, the arrangement of cells in an epithelial sheet makes possible the active transport of substances from one side of a body surface to the other because the plasma membranes on each side are asymmetric in their transport properties. This characteristic enables tissues to move salts and other substances across the epithelia of amphibian skin and bladder, fish gills, the vertebrate cornea, kidney tubules, the intestine, and many other tissues. Consequently, primary active transport can be viewed as one of the major molecular transport mechanisms that underpin almost all of the physiological functions to be described in subsequent chapters.

ION GRADIENTS AS A SOURCE OF ENERGY

Electrochemical gradients across biological membranes are vital to the energy economy of living cells. In humans, the activity of the Na^+/K^+ pump alone accounts for as much as 25% of total energy expenditures and nearly half of the energy expended by the kidneys! The energy of ion gradients can be used to drive passive transport or secondary active transport and is also used to conduct information along the surface of plasma membranes (see Chapter 5). The amount of free energy stored in an electrochemical gradient depends on the ratio of the ion concentrations—or, more accurately, the ratio of the chemical activities of an ionic species—on each side of the membrane. Energy release occurs when the ions are allowed to flow down their gradient across the membrane. Three important cellular processes utilize the free energy of these gradients: production of electrical signals, chemiosmotic energy transduction, and uphill transport of other molecules.

Production of Electrical Signals

Electrochemical energy is stored primarily as Na^+ and Ca^{2+} gradients across the plasma membrane. The release of this electrical energy is under the control of "gated" ion channels. These channels are normally closed, but in response to certain chemical or electrical signals, they switch to an open state in which they exhibit selective permeability to specific ions. These ions then flow passively across the membrane down their electrochemical gradients. Because ionic species carry charge, the movement of ions across a membrane produces an electric current that changes the potential difference existing across the membrane. This electrical activity is the functional basis of the nervous system (the subject of Chapter 5).

Chemiosmotic Energy Transduction

The energy recovered from the metabolism of food culminates in the passage of electrons along the respiratory chain in mitochondria. The energy released during this transport of electrons is stored as a gradient of proton concentration formed across the mitochondrial inner membrane (see Chapter 3). The nature of this energy storage mechanism eluded biologists for many years because they were seeking more conventional high-energy chemical intermediates. The mechanism finally became clear when Peter Mitchell proposed the chemiosmotic coupling hypothesis. The term *chemiosmotic* refers to the direct link between chemical ("chemi") and transport ("osmotic") processes. Two ideas are central to this concept:

- Redox enzymes are oriented within the inner membrane of the mitochondrion so that the electron-transport system of the respiratory chain pumps hydrogen ions from inside the mitochondrial matrix across the inner membrane into the intermembrane space (Figure 4-27). The inner mitochondrial membrane has a low intrinsic permeability to H^+, so this active pumping produces an H^+ gradient across the inner membrane, with a high pH (low $[H^+]$) in the

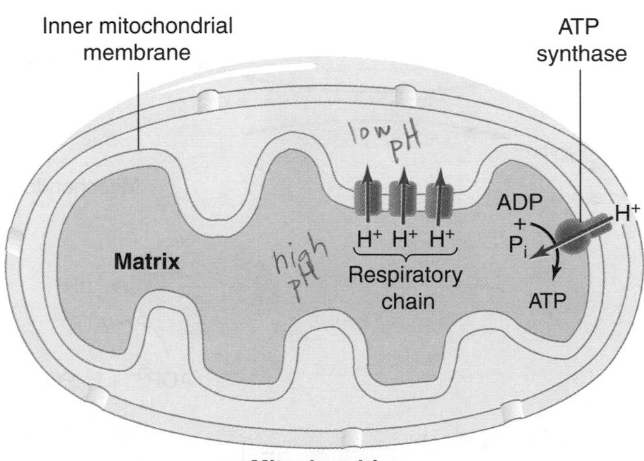

Figure 4-27 The chemiosmotic coupling hypothesis describes energy transduction in the mitochondrion. The membrane-embedded protein complexes of the respiratory chain use energy derived from the transfer of electrons to pump H^+ against its concentration gradient out of the mitochondrial matrix. H^+ reenters the mitochondrial matrix through ATP synthase embedded in the inner mitochondrial membrane. Energy released by the movement of H^+ back down its concentration gradient is employed by ATP synthase in the catalytic formation of ATP from ADP and inorganic phosphate. Note that the outer mitochondrial membrane contains large pores that allow free movement of small molecules between the intermembrane space and the cytosol. The inner membrane, however, is impermeable to H^+, which is an essential feature of the hypothesis—the energy stored in the H^+ concentration gradient (low concentration inside the matrix, high outside) is preserved until H^+ is channeled through the ATP synthase.

matrix relative to the intermembrane space (and cell cytosol, with which the intermembrane space is continuous via large, nonselective pores).

- The energy-rich H^+ gradient across the inner membrane provides the free energy required for the synthesis of ATP:

$$ADP + P_i \longrightarrow ATP + H_2O$$

$$\Delta G^\circ = +7.3\ \text{kcal} \cdot \text{mol}^{-1}$$

This reaction requires that an ATPase complex be oriented on the inner mitochondrial membrane so as to take advantage of the H^+ gradient across the membrane, since the movement of H^+ drives newly formed ATP off of the ATPase:

$$ADP + P_i \longrightarrow ATP$$

Chemiosmosis also involves a phosphate transporter that transports $H_2PO_4^{2-}$ into the matrix from the intermembrane space, as well as an ADP/ATP exchanger, both of which are involved in maintaining the stoichiometry of charge balance and ATP substituents in the two mitochondrial compartments (Figure 4-28).

Chemiosmotic energy transduction similar to that proposed for oxidative phosphorylation in mitochondria is now known to be the mechanism for energy transduction during photosynthesis in chloroplasts and photosynthetic bacteria. In addition, there is evidence that the Na^+/K^+ pump, which normally utilizes ATP to produce the Na^+ gradient, can in special circumstances run in reverse, so that the movement of Na^+ down its gradient causes the pump to synthesize ATP from ADP and P_i.

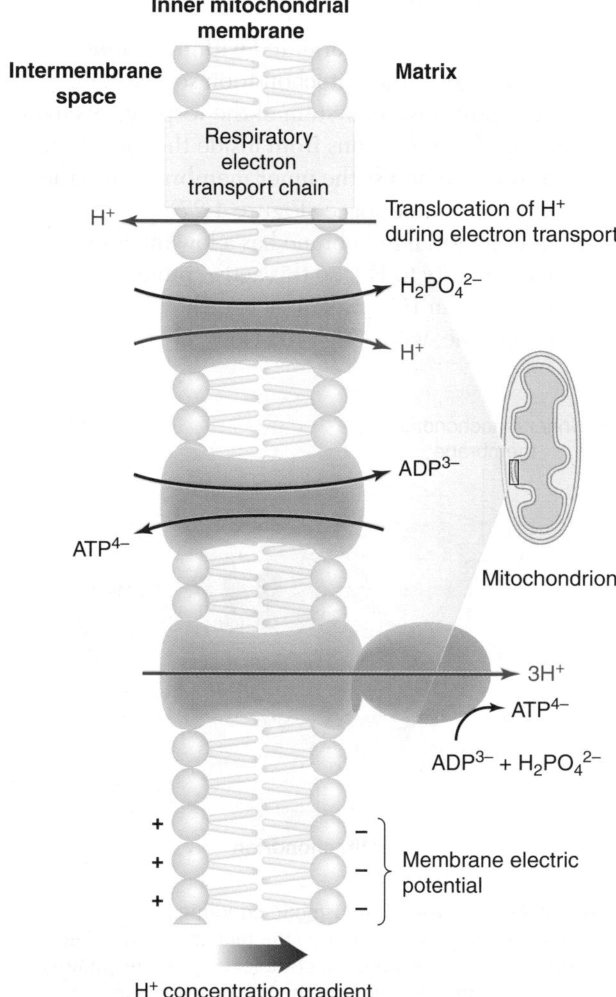

Figure 4-28 The phosphate and ATP-ADP transport system that generates ATP is located in the inner mitochondrial membrane. The phosphate transporter couples the uptake of one $H_2PO_4^{2-}$ (inorganic phosphate) with the simultaneous uptake of H^+ translocated during respiration and now moving back down its concentration gradient. At the same time, the ATP-ADP antiporter exchanges one incoming ADP^{3-} for one ATP^{4-} exported from the matrix. As a result, there is a net uptake of one ADP^{3-} and one $H_2PO_4^{2-}$ in exchange for one ATP^{4-}. For every four H^+ translocated outward, three are used to synthesize one ATP molecule and one is used to export ATP in exchange for ADP and P_i. [Adapted from Lodish et al., 2000.]

MEMBRANE SELECTIVITY

The utility of biological membranes lies in their selectivity—their ability to allow the passage of only specific types of molecules. Selectivity is a property of each specific transport system and ensures that only correct ionic species in appropriate amounts enter the cell. Some transport systems are less specific than others. For example, when the Na^+ in a physiological saline solution bathing a nerve cell is replaced with lithium ions, the Li^+ readily passes through the Na^+ channels, which open during electrical excitation of the nerve cell membrane. The other alkali metal cations, K^+, Rb^+, and Cs^+, are blocked by these channels. On the other hand, the ATPase of the Na^+/K^+ pump in the same membrane is highly specific for intracellular Na^+ and will not pump Li^+ out of the cell. Lithium ions that enter the cell through the Na^+ channels will therefore gradually accumulate in the cell until it comes into electrochemical equilibrium. This situation is an example of electrolyte selectivity by the transport system, but not by the membrane channels. Let's see how this selectivity for both electrolytes and nonelectrolytes is achieved.

Selectivity for Electrolytes

How do transmembrane channels discriminate between different ions, even those of essentially identical size and shape? Na^+ and K^+ have almost the same shape and size (K^+ is a little larger), yet the resting nerve cell membrane is about 30 times more permeable to K^+ than to Na^+. At first glance, we might conclude that these ions are distinguished on the basis of their hydrated size, with K^+ passing freely through channels that are too small for Na^+. Size can explain how the K^+ channel excludes Cs^+ or Rb^+ (Table 4-3) but not Na^+, particularly in light of the fact that permeability to Na^+ can change dramatically. During the excitation of nerve or muscle membrane, the Na^+ permeability of the membrane increases about 300-fold to a value about 10 times greater than its K^+ permeability at rest. If, during excitation, the membrane were suddenly to develop channels that pass the Na^+ ion on the basis of size alone, there should be a simultaneous increase in permeability to K^+ through the same channels, given their comparable sizes. Since this increase does not occur, the membrane's selectivity must rest on properties other than size. Indeed, estimated pore sizes for different membrane channels illustrate that size alone cannot be the agent of membrane selectivity.

Two interesting features other than size appear to be important in governing membrane channel selectivity: ease of dehydration and interaction with charges within the channel pore. For an ion to enter a pore, it must dissociate from water molecules. Ease of dehydration appears to be an important factor in governing selectivity, particularly if the charges within the pore are weak. Since large ions dehydrate more easily than small ones (see Table 4-3), a pore with weak polar sites along it will admit large ions preferentially over small ones.

In channels with strongly charged protein subunits, the interaction of the dehydrated ion with these subunits may be more important for conferring specificity than ease of dehydration. Thus, a transmembrane channel lined with predominantly positively charged amino acid residues will selectively repel positively charged ions but permit negatively charged ions to pass through (Figure 4-29). In such cases, smaller ions can approach the polar sites more closely, and hence interact with them more strongly, than large ions, exaggerating the effect. The configuration of K^+ channels is particularly well understood. K^+ channels from the bacterium *Streptomyces* consist of four identical protein subunits, arranged to form a funnel that opens out into the extracellular space. K^+ ions pass through the channel in single file, aided by negatively charged carboxyl termini that obtrude into the channel pore. Na^+ ions are excluded from this channel, in part because they are too

Table 4-3 Ionic radii and hydration energies of the alkali metal cations

Cation	Ionic radius (Å)	Free energy of hydration ($kcal \cdot mol^{-1}$)
Li^+	0.60	2122
Na^+	0.95	298
K^+	1.33	280
Rb^+	1.48	275
Cs^+	1.69	267

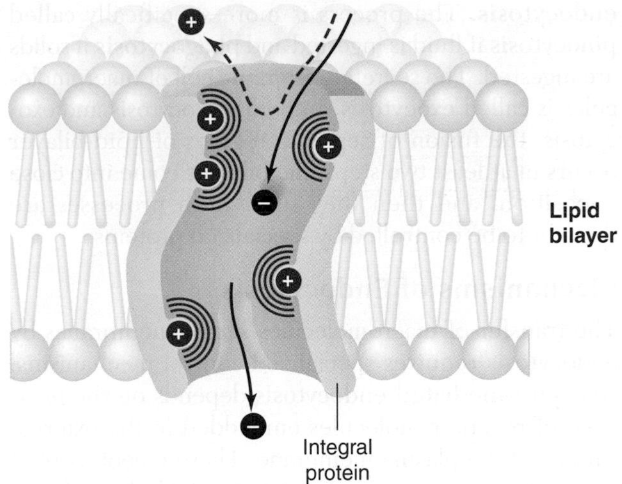

Figure 4-29 Positive charges lining a channel pore allow anions to pass but retard the diffusion of cations through the channel, as shown in this hypothetical cross-section through a highly simplified membrane channel.

small to interact like K^+ with the protein subunits forming the channel.

Selectivity for Nonelectrolytes

Many nonelectrolytes cross the membrane by dissolving in the lipid bilayer and simply diffusing across it. Since the relationship between permeability and the partition coefficient K is essentially linear (see Figure 4-18), selectivity is completely determined by the molecular properties responsible for the partition coefficient. Those few nonelectrolytes that deviate from the linear relation between partition coefficient and permeability all have greater than predicted permeability. Some of these substances cross the membrane by carrier-mediated transport. Alternatively, small molecules such as ethyl alcohol, methyl alcohol, and urea can cross via both the lipid layer and water-filled channels. All of these deviant molecules are small and water-soluble, regardless of their relative solubilities in water versus lipid (i.e., their partition coefficients). It is important to note that mechanisms for restricting nonelectrolyte access through membranes have not evolved, making cells vulnerable to penetration by these molecules. Drugs applied to the human skin, such as anti-nauseants or nicotine delivered by skin patches, can use this route to enter the body.

ENDOCYTOSIS AND EXOCYTOSIS

The transport processes described above for small polar molecules cannot transport macromolecules such as proteins, polynucleotides, or polysaccharides across membranes. Yet cells do manage to ingest and secrete such macromolecules via the sequential formation and fusion of membrane-bounded vesicles. The intake of material into the cell in this manner is generally called

endocytosis. The process is more specifically called **pinocytosis** if fluid is ingested and **phagocytosis** if solids are ingested. The secretion from the cell of macromolecules is called **exocytosis.** In both endocytosis and exocytosis, the fusion of separate regions of lipid bilayer occurs in at least two steps: the bilayers come into close apposition, and then they fuse. Both processes are thought to be controlled by specialized proteins.

Mechanisms of Endocytosis

The transfer of macromolecules across membranes by endocytosis requires specialized control mechanisms. **Receptor-mediated endocytosis** depends on the presence of receptor molecules embedded in the external surface of the plasma membrane. These receptors bind certain ligand molecules or particles, including plasma proteins, hormones, viruses, toxins, immunoglobulins, and certain other substances that cannot pass through membrane channels. The receptors are free to diffuse laterally in the plane of the membrane, but upon binding of ligand, the receptor-ligand complex tends to accumulate within depressions in the membrane called **coated pits.** The coated pit invaginates and pinches off, forming a **coated vesicle**, internalizing the ligands (Figure 4-30). Coating these structures is the protein **clathrin**, which covers the cytoplasmic surface of the vesicle membrane. The clathrin is organized into pentagonal or hexagonal latticelike arrays on the membrane surface. It appears to have several functions, including binding the ligand-occupied receptor molecules and directing the subsequent budding off of the vesicle from the plasma membrane. Once the coated vesicle buds off into the cytoplasm, it delivers its contents by fusing with other organelles, such as lysosomes. The clathrin and receptors are then recycled to the plasma membrane.

Mechanisms of Exocytosis

The release of chemicals from cells through exocytosis plays a crucial role in the endocrine and nervous systems. The presynaptic terminals of nerve cells, for

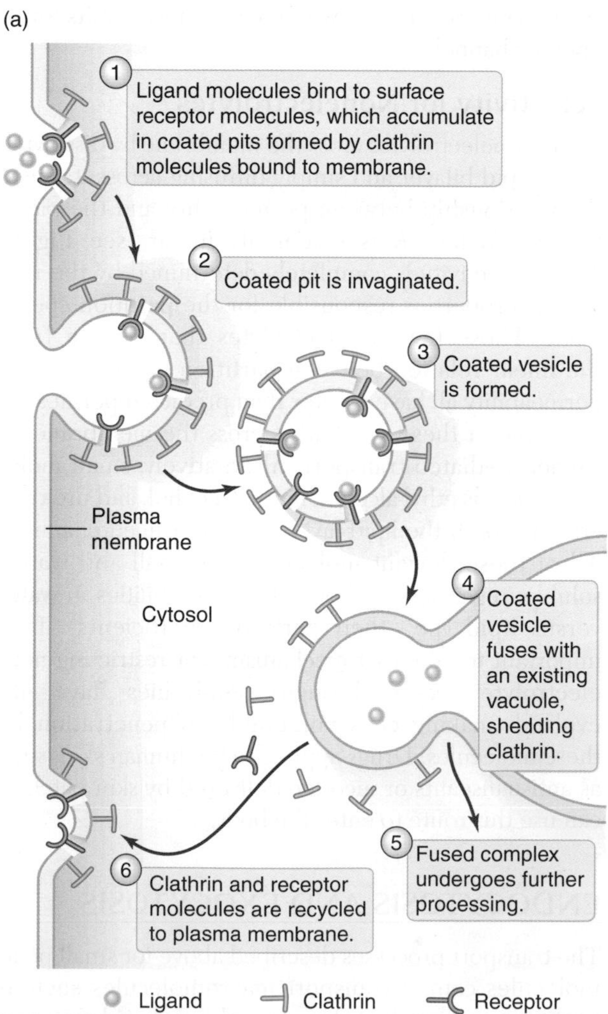

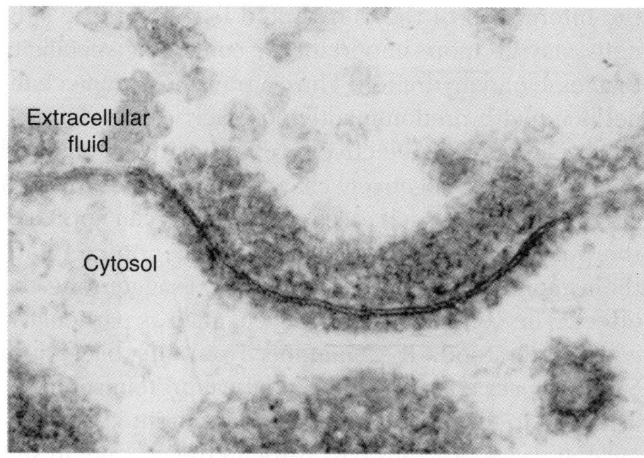

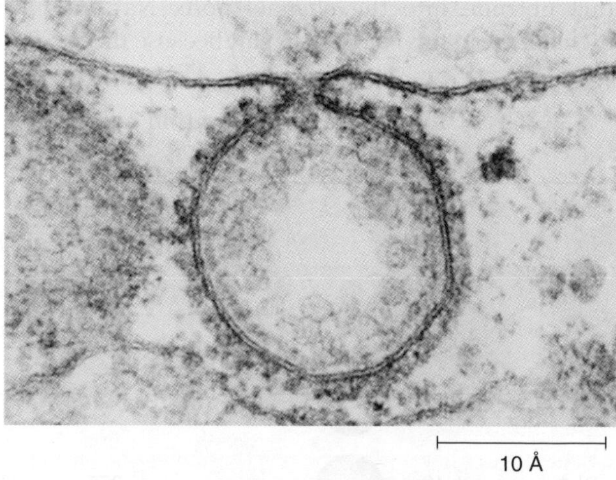

Figure 4-30 Coated vesicles form during receptor-mediated endocytosis. **(a)** The six major steps in the process are diagrammed. **(b,c)** Electron micrographs of a coated pit **(b)** and a coated vesicle **(c)** in a chicken oocyte. Note the dense clathrin coat on the cytoplasmic surface of the membrane. [(a) From Pearse, 1980; (b) and (c) from Bretscher, 1985.]

example, contain many membranous internal vesicles about 50 nm in diameter, which contain neurotransmitter substance. These vesicles fuse with the plasma membrane of the nerve terminal and release their contents to the cell exterior, the typical method of exocytosis. This activity occurs with greatly enhanced probability when the terminal is invaded by a nerve impulse, and serves to release the neurotransmitter, which interacts with the postsynaptic membrane. Similar mechanisms are involved in the secretion of hormones.

An important feature of both endocytosis and exocytosis is that the secreted or ingested macromolecules are sequestered in vesicles and hence do not mix with macromolecules or organelles in the cell. Since the vesicles can fuse only with specific membranes, they assure the directed transfer of their contents. In exocytosis, once the membrane of the vesicle is incorporated into the plasma membrane, the freed contents—hormones, neurotransmitters, or accessory molecules—diffuse away into the interstitial space.

Exocytosis requires recovery of the relatively large amounts of secretory vesicle membrane that initially surround the macromolecules being expelled. In the absence of retrieval of this newly incorporated membrane, the surface area of the plasma membrane would continually grow. Conveniently, endocytosis recovers excess membrane for the formation of new secretory vesicles. Evidence for such **membrane recycling** through endocytosis comes from experiments in which protein molecules, such as horseradish peroxidase, are introduced into the extracellular fluid and their movement into the cell is traced with electron-microscopic methods. In these experiments, horseradish peroxidase shows up inside the cell, but only within vesicles. Since the large size of the horseradish peroxidase molecule prevents its penetration by direct passage across biological membranes, it must have been taken up in bulk during the formation of endocytotic vesicles budding off from the plasma membrane into the cytoplasm.

Calcium ion triggers the exocytotic secretion of neurotransmitters from nerve cells and of hormones from endocrine cells, as we will see in Chapter 6. An elevation of intracellular Ca^{2+} enhances the probability of exocytotic activity by permitting the fusion of vesicles with the inner surface of the plasma membrane. The membrane regulates exocytotic activity by regulating the intracellular accumulation of Ca^{2+}. As enhanced calcium influx allows Ca^{2+} levels to rise, the rate of exocytotic secretion increases.

Before fusion of the two membranes can take place, a secretory granule must contact the plasma membrane. Release of secretory products from glandular secretory cells can be blocked by colchicine, an antimitotic agent that leads to the disassembly of microtubules, or by cytochalasin, an agent that disrupts microfilaments. This pharmacological evidence suggests that microtubules or microfilaments participate in the movement of secretory granules toward sites of exocytotic release on the inner side of the plasma membrane. Motor proteins attached to the surfaces of secretory granules direct their movement along microtubles, ensuring efficient delivery of the vesicle contents to specific sites at the cell surface.

JUNCTIONS BETWEEN CELLS

Cells in animals are organized into cooperative assemblies called **tissues.** In certain tissues, including epithelium, smooth muscle, cardiac muscle, central nervous system tissues, and many embryonic tissues, neighboring cells are connected by special structural adaptations of their abutting surfaces. These specialized connections are divided into two major groups: gap junctions and tight junctions. Gap junctions enhance cell-cell communication through minute water-filled channels that connect adjacent cells, while tight junctions "sew" into sheets those cells involved in transepithelial transport.

Gap Junctions

Gap junctions provide a means of communication between cells by allowing inorganic ions and small water-soluble molecules to pass directly from the cytosol of one cell to the cytosol of another. These junctions couple cells both electrically and metabolically, with important functional consequences for the tissue. The distance between the two plasma membranes of a gap junction is only 2 nm (Figure 4-31). The two

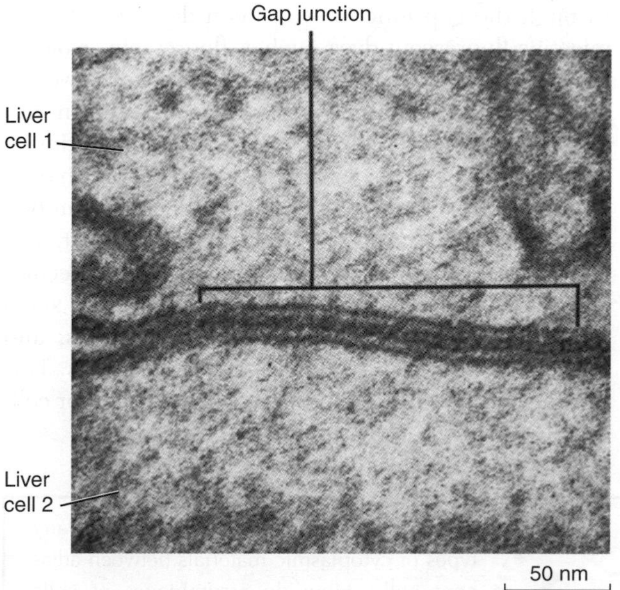

Figure 4-31 The gap between neighboring cells connected by gap junctions is about 2 nm, near the lower limit for electron microscopic resolution. This electron micrograph reveals a gap junction between the membranes of two neighboring mouse liver cells. [Courtesy of D. Goodenough.]

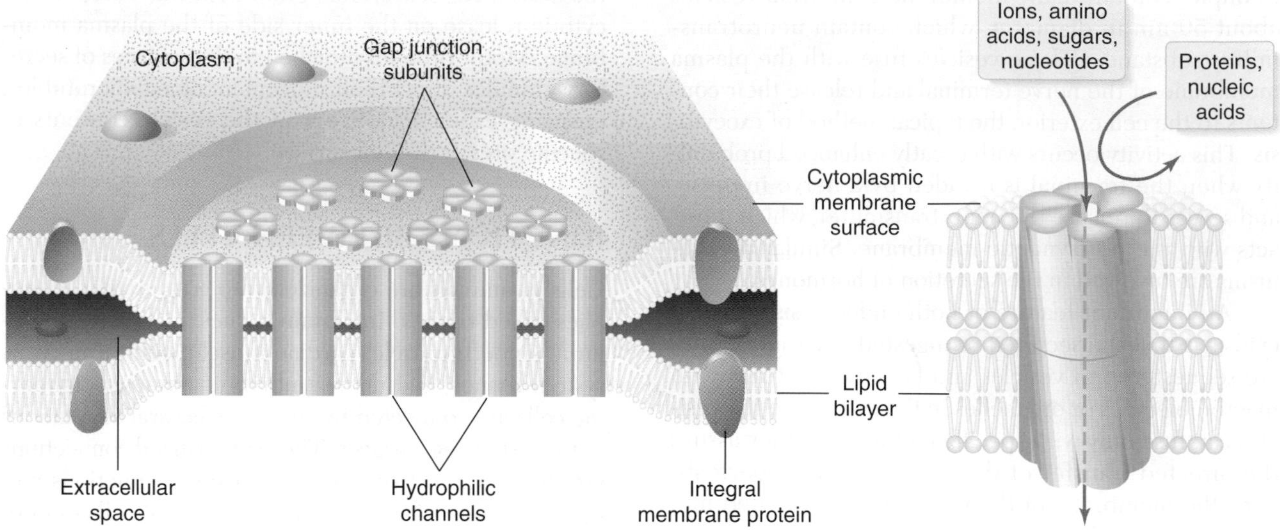

Figure 4-32 Gap junctions permit the passage of molecules between neighboring cells. The two membranes belonging to the two coupled cells each contain an array of hexagonal subunits, each of which connects with a matching subunit in the apposed membrane. A central pore or channel penetrates both subunits, providing a path of communication between the connected cells. Details of a channel complex are shown at right. Molecules smaller than about 2 nm can pass between coupled cells through the channels. Molecules larger than 2 nm, such as proteins and nucleic acids, are too large to enter the channels.

adjoining membranes both contain clusters of hexagonal channels, each composed of six subunits, that span this narrow space (Figure 4-32). The channels are about 5 nm in diameter and resemble miniature doughnuts whose holes form pores between the interiors of the neighboring cells.

The continuity of these cell-cell passageways through the gap junction has been demonstrated by injecting fluorescent dyes, such as fluorescein (molecular weight 332) and procion yellow (molecular weight 500), into one cell and following their diffusion into neighboring cells (Figure 4-33). Electrical continuity between the cells has been confirmed by the finding that electric current readily passes directly from one cell into another when gap junctions are present. The intercellular channels in these junctions pass molecules with a molecular weight of as much as 500, so small molecules, such as ions, amino acids, sugars, and nucleotides, are easily exchanged between cells. This exchange of small molecules is the mechanism for cell-cell communication via gap junctions.

Question

Gap junctions allow the exchange of many types of cytoplasmic materials between adjacent cells. How do assemblages of cells linked by gap junctions—freely exchanging ions, amino acids, sugars, and nucleotides—challenge the concept of cells? What is the functional difference between a single cell, a series of cells linked by gap junctions, and a tissue?

Gap junctions are not permanently open; they close rapidly (within seconds) in response to any treatment that increases intracellular Ca^{2+} or H^+ concentrations. A functional uncoupling of cells from their neighbors can be produced by injecting Ca^{2+} or H^+ into a coupled cell, by lowering the temperature, or by using poisons that inhibit energy metabolism. The subsequent loss of electrical transmission between cells confirms the

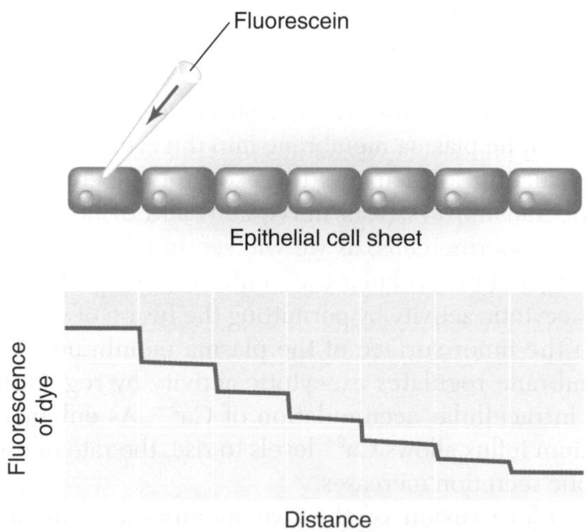

Figure 4-33 Gap junctions between coupled cells can be demonstrated by following the flow of fluorescent dye (fluorescein) injected into one cell in a group of coupled epithelial cells. Subsequent diffusion of the dye into neighboring cells without loss into the extracellular space indicates direct pathways from the cytosol of one cell to the cytosol of the adjacent cell.

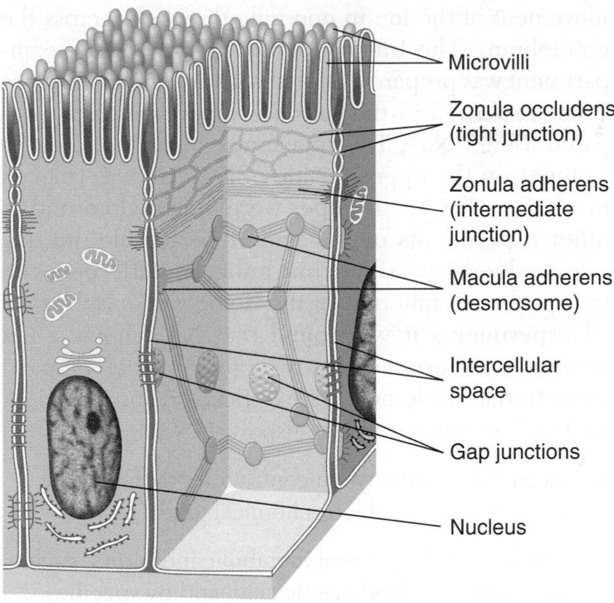

Figure 4-34 Adjacent epithelial cells such as those that line the mammalian small intestine are connected by tight junctions, as well as by other types of intercellular junctions. The membranes and associated structures are drawn disproportionately large in this diagram.

uncoupling. Thus, gap junctions are maintained intact only if the metabolic activity of the plasma membrane maintains sufficiently low concentrations of intracellular free Ca^{2+} and H^+. The closing mechanism of the gap junction channel is not clearly understood, but the channel appears to be open or closed depending on the relative positions of its six subunits.

Tight Junctions

Tight junctions seal cells together into an epithelial sheet but do not provide a channel; even small molecules cannot get from one side of the sheet to the other. The two apposing plasma membranes make intimate contact, fully occluding the extracellular space in between. In epithelial tissues, tight junctions are found most commonly as a **zonula occludens,** a thin band of protein molecules that encircles a cell like a gasket. The zonula occludens is in tight contact with the zonulae of the surrounding cells, forming an impermeable seal that prevents the passage of substances from one side of the epithelium to the other via leakage into gaps between cells (Figure 4-34). Considered en masse, the zonulae are analogous to a continuous rubber sheet, penetrated only by the ends of the epithelial cells. Substances can pass through the ends of the cells (the **transcellular pathway**), but not around them (the **paracellular pathway**). In some epithelial tissues, such as those of the mammalian small intestine, gallbladder, and proximal tubule of the nephron, these zonulae are not fully continuous and thus are not really very "tight." These tissues are so leaky that they do not produce a transepithelial potential difference, even though their cells contain ion pumps capable of generating transepithelial ion fluxes.

Two other types of cell junctions are shown in Figure 4-34: the **zonula adherens** and the **desmosome,** which serve primarily to stabilize the structural bonding of neighboring cells.

EPITHELIAL TRANSPORT

Epithelial cell sheets line the cavities of the body and its hollow organs, as well as the free surfaces of animal bodies, where they form barriers affecting the movement of water, solutes, and cells from one body compartment to another. Some of these sheets serve only as passive barriers between compartments and do not preferentially transport solutes or water. Others are involved in active transport, performing regulatory functions. The osmoregulatory activities of animals, for example, are carried out by actively transporting epithelia in a variety of specialized tissues and organs (see Chapter 14).

All epithelia have several features in common. First, they occur at surfaces that separate the internal space of the organism from the environment. Included are the surfaces lining deep invaginations such as the lumen of the intestine, which, despite being in the body's interior, nevertheless constitutes an interface with the external environment. Second, the cells forming the outermost layer of the epithelium are generally sealed together by tight junctions, which, to varying degrees in different epithelia, eliminate paracellular pathways between the **serosal** (internal) and **mucosal** (external) sides of the epithelium (Figure 4-35). In the epithelia of capillary walls, for example, leaky

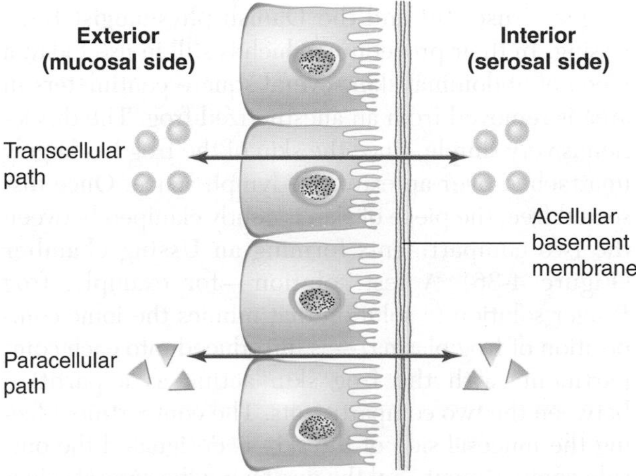

Figure 4-35 Substances cross epithelial layers by either paracellular or transcellular pathways. Active transport takes place only across plasma membranes, suggesting that all actively transported molecules follow the transcellular pathway.

junctions permit water and solute molecules to cross the epithelial layer by diffusion within the passages that exist between the cells. This diffusion through paracellular pathways is not coupled to any metabolically energized transport mechanism, so these pathways allow only passive movement of water and ions. Substances that are actively transported across an epithelium must follow transcellular pathways, in which the plasma membrane participates. Such substances must cross the cell membrane first on one side of the cell and then on the other. As we will see in the following section, the functional properties of the serosal and mucosal plasma membranes of an epithelial cell differ in important aspects that contribute to epithelial active transport.

Active Salt Transport across an Epithelium

Active transport of ions from one side of an epithelium to the other has been demonstrated in a number of epithelial tissues, including the amphibian skin and urinary bladder, the gills of fishes and aquatic invertebrates, insect and vertebrate intestines, and the vertebrate kidney tubule and gallbladder. Much of the initial work on epithelial active transport was done on frog skin, which acts as a major osmoregulatory organ. Salts are actively transported from the mucosal side (i.e., the side facing the pond water) to the serosal side of the skin to compensate for the salts that leak out of the skin into the freshwater surrounding the frog. Similar salt uptake occurs in the gut. Water that enters the skin because of the osmotic gradient between the hypotonic pond water and the more concentrated internal fluid is eliminated in the form of a copious dilute urine that is hypotonic relative to the body fluids (see Chapter 14).

Frog skin was first used in the study of epithelial transport in the 1930s and 1940s by the German physiologist Ernst Huf and the Danish physiologist Hans Ussing. In their procedure, which is still in use today, a piece of abdominal skin several square centimeters in area is removed from an anesthetized frog. The dissection is very simple, since the skin of the frog lies largely unattached over an extensive lymph space. Once dissected free, the piece of skin is gently clamped between the two compartments forming an **Ussing chamber** (Figure 4-36). A test solution—for example, frog Ringer solution (a solution that mimics the ionic composition of frog plasma)—is introduced into each compartment, with the frog skin acting as a partition between the two compartments. The compartment facing the mucosal side of the skin is designated the outside compartment, and the one facing the serosal side is the inside compartment. Air is bubbled through the two solutions to keep them well oxygenated.

In 1947 Ussing reported the first experiments in which two different isotopes of the same ion were used to measure bidirectional fluxes (i.e., the simultaneous movement of the ion in opposing directions across the epithelium). The Ringer solution in the outside compartment was prepared using the isotope $^{22}Na^+$, and the Ringer solution in the inside compartment was prepared using $^{24}Na^+$. The appearance of each of the two isotopes on the opposite side of the skin was tracked over time. The two isotopes were switched around in other experiments of the same type to rule out any effects due to possible (but unlikely) differences in transport rates inherent in the isotopes themselves. In all experiments it was found that Na^+ shows a net movement across the skin from the outside compartment to the inside one. Active transport must account for this Na^+ flux because transport

- occurs without any concentration gradient, and even against an electrochemical gradient
- is inhibited by general metabolic inhibitors, such as cyanide and iodoacetic acid, and by specific transport inhibitors, such as ouabain
- displays a strong temperature dependence
- exhibits saturation kinetics
- shows chemical specificity, such that Na^+ is transported while the very similar lithium ion is not

How can ions be actively transported across a layer of cells contained in an epithelium? The adjacent cells of the epithelium are intimately tied together with tight junctions. Assume for the sake of simplicity that this

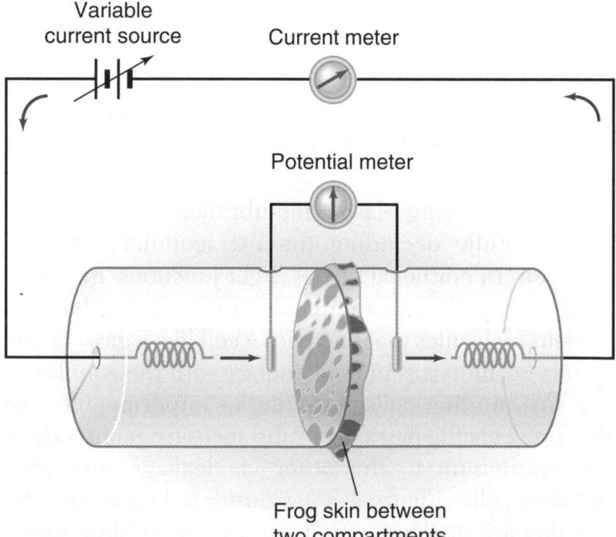

Figure 4-36 A piece of frog skin separates the two compartments of this Ussing chamber. Each compartment is filled with a physiological saline or other test solution. An electric current source is adjusted until the potential difference across the skin is zero. Under these conditions, the current flowing through the circuit (and thus through the skin) is equivalent to the rate of charge transfer by the active movement of sodium ions across the skin.

closeness eliminates all paracellular passageways for the diffusion of ions between the two sides of the epithelium. In that case, all substances that cross the epithelium would be forced to traverse the epithelial plasma membrane twice, first entering the cell by crossing the membrane on one side of the cell and then leaving the cell through the membrane on the other side. Active transport by this route requires differentiation of the plasma membrane of each epithelial cell, so that the portion of the membrane facing the serosal side of the epithelium has different functional properties from the portion facing the mucosal side. Experiments on frog skin in Ussing chambers have provided several lines of evidence to support the hypothesis of such a differentiated plasma membrane:

- Ouabain, which blocks the Na^+/K^+ pump, inhibits transepithelial Na^+ transport only when applied to the inner (serosal) side of the epithelium. This blocker is ineffective on the outer (mucosal) side. Conversely, the drug amiloride, a powerful inhibitor of carrier-mediated passive transport, blocks Na^+ movements across the skin only when applied to the mucosal side of the skin.
- For active Na^+ transport to take place, K^+ must be present in the solution on the inner side of the chamber, but is not required on the outer side.
- Transport of Na^+ exhibits saturation kinetics as a function of Na^+ concentration in the outer solution, but it is unaffected by Na^+ concentration in the inner solution.

Such evidence led to the model of epithelial Na^+ transport shown in Figure 4-37. According to this model, a Na^+/K^+ pump is located in the plasma membrane of the serosal side of the epithelial cell (together with Na^+/H^+ and Na^+/NH_4^+ exchange pumps in the intact animal). This membrane behaves in the manner typical of many plasma membranes, pumping Na^+ out of the cell in exchange for K^+, thus maintaining a high intracellular K^+ concentration and a low intracellular Na^+ concentration. The outward diffusion of K^+ across the membrane on this side of the cell produces an inside-negative resting potential.

The situation on the mucosal side must be different. The plasma membrane on this side of the cell is relatively impermeable to K^+. Moreover, a net inward diffusion of Na^+ across this membrane (apparently facilitated by carriers or channels in the membrane) replaces the Na^+ pumped out of the cell on the serosal side. This model explains why inhibitors of Na^+/K^+ pumps exert an effect only from the serosal side of the epithelium and why changes in the concentration of K^+ on only that side influence the rate of Na^+ transport.

Thus, net flow of Na^+ across the frog skin from the mucosal side to the serosal side stems from the functional asymmetries of the plasma membranes on the two sides. The driving force is none other than the active transport of Na^+ that is common to the plasma membranes of all tissues.

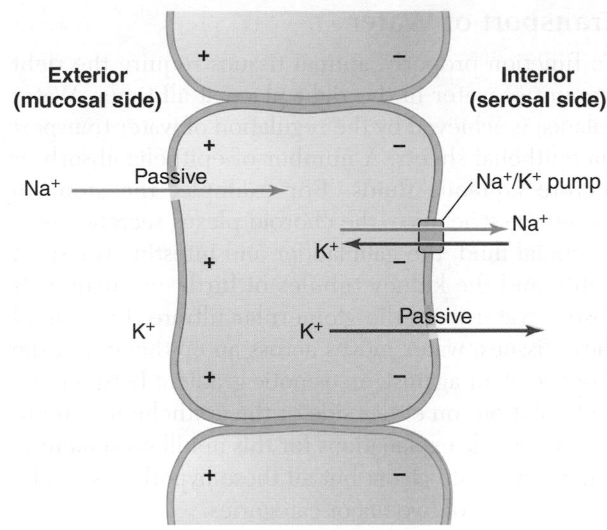

Figure 4-37 Transepithelial Na^+ transport depends on a combination of diffusion and active transport. In this model of an isolated frog skin bathed in Ringer solution, Na^+ diffuses passively down its concentration gradient into the cell from the mucosal solution. K^+ diffuses out of the cell into the serosal space as it is displaced by Na^+ influx. In the face of these leaks, a Na^+/K^+ exchange pump in the serosal membrane of the cell transports Na^+ out of and K^+ into the cell, maintaining the high internal K^+ and low internal Na^+ concentrations.

The frog skin has served for many decades as a model system for studying the general problem of epithelial salt transport. Although details may differ from one type of epithelial tissue to another, the major features, listed below, are probably common to all transport epithelia.

1. To varying degrees, tight junctions obliterate paracellular pathways. As a result, transport through transcellular pathways assumes major importance in epithelial transport.
2. Mucosal and serosal portions of the plasma membrane exhibit functional differences, being asymmetric in both pumping activity and membrane permeabilities.
3. The active transport of cations across an epithelium is typically accompanied by transport (passive or active) of anions in the same direction or by exchange for another species of cation, minimizing the buildup of electric potentials. The converse applies to actively transported anions.
4. Epithelial transport is not limited to the pumping of Na^+ ions. Various epithelia are known to transport Cl^-, H^+, HCO_3^-, K^+, and other ions.

Transport of Water

To function properly, animal tissues require the right amount of water in the right place at all times. Water balance is achieved by the regulation of water transport via epithelial sheets. A number of epithelia absorb or secrete aqueous fluids. For example, the stomach secretes gastric juice, the choroid plexus secretes cerebrospinal fluid, the gallbladder and intestine transport water, and the kidney tubules of birds and mammals absorb water from the glomerular filtrate. In some of these tissues, water moves across an epithelium in the absence of, or against, an osmotic gradient between the bulk solutions on either side of the epithelium. A number of possible explanations for this uphill movement of water have been given, but all these hypotheses can be placed in one of two major categories:

- Water is transported by a specific carrier mechanism driven by metabolic energy.
- Water is transported by osmosis as a consequence of solute transport.

So far, there has been no convincing evidence to indicate that water is actively transported by a primary water carrier or pump. This leaves us with the osmotic hypothesis, in which water undergoes net diffusion in one direction owing to concentration gradients built up by solute transport.

The osmotic hypothesis of water transport was validated by Peter Curran's work in 1965, in which he showed that an osmotic gradient produced by active salt transport from one side of an epithelium to the other could, in theory, result in a net flow of water across the epithelium (Figure 4-38). Biological correlates of Curran's model were subsequently found in the epithelium of the mammalian gallbladder. This finding led to the **standing-gradient hypothesis** of solute-coupled water transport, presented by Diamond and Bossert (1967).

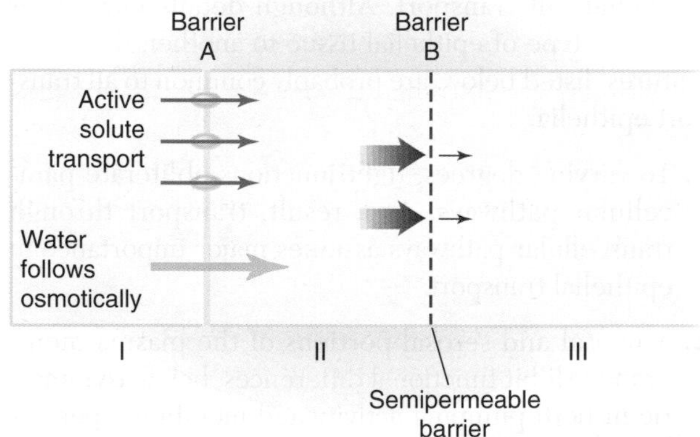

Figure 4-38 Curran's model of solute-linked water transport, first introduced in 1965, depends on active transport of a solute across a water-permeable membrane, labeled barrier A. A solute (e.g., Na^+) is pumped through barrier A from compartment I to compartment II. Semibarrier B slows diffusion of the solute into compartment III and thereby keeps the osmolarity high in compartment II. The rise in osmolarity in compartment II causes water to be drawn from I to II. In the steady state, both water and solutes diffuse into compartment III at the same rate at which they appear in II. Compartment III is much larger than II.

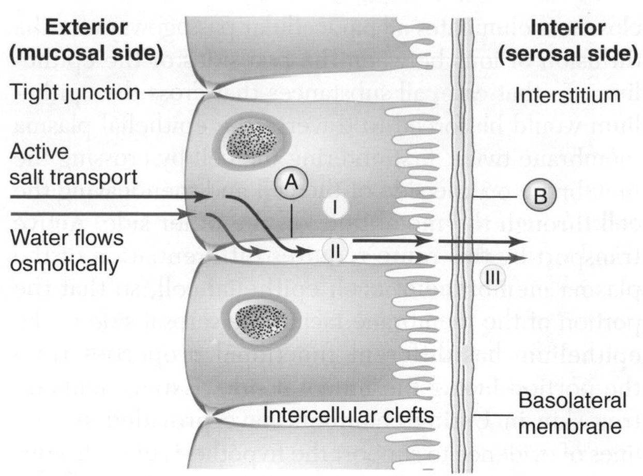

Figure 4-39 The biological counterpart of Curran's model for solute-coupled water transport. The compartments corresponding to those in Figure 4-38 are numbered I, II, and III. Salts transported actively into the intercellular clefts produce a high osmolarity within the clefts. Water flows osmotically into the clefts across the cell, and the bulk solution flows through the freely permeable basolateral membrane and into the bulk fluid of the interstitium. The barriers A and B are analogous to barriers A and B in Figure 4-38. [Adapted from Diamond and Tormey, 1966.]

A simplified schematic version of the standing-gradient hypothesis is shown in Figure 4-39. Two anatomic features are of major importance. First, the tight junctions near the luminal (mucosal) surface block paracellular pathways through the epithelium. Second, the lateral intercellular spaces, or **intercellular clefts,** between adjacent cells are restricted at the luminal ends by the tight junctions and are freely open at the basal (serosal) ends.

The basis for the standing-gradient hypothesis is the active transport of salts across the portions of the epithelial cell membranes facing the intercellular clefts. The membranes bordering the clefts have been shown to be especially active in pumping Na^+ out of the cell. It is suggested that as salts are transported out of the cell into these long, narrow clefts, the salt concentration in the clefts sets up an osmotic gradient between the extracellular spaces on either side of the tight junctions that join the epithelial cells. There may also be an osmotic gradient within the clefts, with the salt concentration highest near their closed ends and diminishing toward their open ends, where it comes into equilibrium with the bulk phase on the serosal side. As a consequence of the high extracellular osmolarity in the clefts, water is osmotically drawn into the clefts across the "not so tight" tight junction, or possibly from within the cell across the plasma membrane. The water leaving the cell is replaced by water drawn osmotically into the cell at the mucosal surface. The water that enters the clefts gradually moves, together with solute, out into the bulk phase. In this way, the steady, active extrusion of salts by one plasma membrane of the cell produces an elevated salt concentration in the narrow intercellular clefts. This, in turn, results in a steady osmotic flow of water from one side of the epithelium to the other.

The general applicability of the standing-gradient mechanism of solute-coupled water transport is supported by ultrastructural studies showing that the necessary cellular geometry—namely, narrow intercellular clefts closed off at the luminal end by tight junctions—is present in all the water-transporting epithelia that have been examined. Also important in this regard are the deep basolateral clefts and infoldings typical of transporting epithelial cells (see Figure 4-39). These spaces are dilated in epithelia fixed during conditions that produce water transport. In epithelia fixed in the absence of water transport, the intercellular clefts are largely obliterated.

SUMMARY

Membrane structure and organization

■ Biological membranes are fundamental features of living cells. Their functions include formation of cellular and subcellular compartments; maintenance of a stable intracellular milieu using selective permeability and transport mechanisms; regulation of intracellular metabolism by maintaining intracellular concentrations of enzyme cofactors and substrates; metabolic activities carried out by enzyme molecules in ordered arrays in or on the membrane; sensing and transduction of extracellular chemical signals via surface receptor molecules and regulatory molecules located in the membrane; propagation of electrical signals that conduct messages and/or regulate the transport of substances across the membrane; and endo- and exocytosis of bulk material.

■ The basic structure of membranes is a lipid bilayer in which the hydrophilic "heads" of phospholipid molecules face outward and their lipophilic "tails" face inward. A mosaic of globular proteins, including enzymes, penetrates the bilayer.

Crossing the membrane

■ An unequal distribution of solutes between cell interior and cell exterior can cause water to enter the cell, following its tendency to flow from a region of lower to a region of higher osmotic pressure. Osmotic pressure is equal to the hydrostatic pressure balancing osmotic flow (water movement across a semipermeable membrane) down a concentration gradient at equilibrium.

■ Osmolarity describes the number of dissolved particles per volume of solvent, as well as the behavior of a solution in an ideal osmometer. Tonicity describes the osmotic effects that a solution has on a given tissue.

■ Permeability is a measure of the ease with which a substance traverses a membrane.

Mechanisms for transmembrane movements

■ Nonpolar molecules can move readily through the lipid phase of a membrane by passive diffusion. Water and some small polar molecules passively diffuse through transient aqueous channels created by thermal motion.

■ Some substances move across a membrane through channel proteins that are more or less specific for certain ions and molecules. This process is called passive transport (facilitated diffusion).

■ Passive transport also occurs via carrier proteins that bind to a substance and shuttle it through the lipid phase of the membrane.

■ Primary active transport is responsible for the movement of substances across a membrane against a concentration gradient. The expenditure of energy is required for primary active transport.

■ The Na^+/K^+ pump is the most familiar primary active transport system. It maintains the intracellular Na^+ concentration below that of the cell exterior. The energy stored in the form of this Na^+ concentration gradient drives the uphill movement of calcium ions, amino acids, sugars, and other substances by means of exchange diffusion and coupled transport.

■ Primary active transport compensates for the tendency of substances such as Na^+ to leak into cells and thereby cause uncontrolled increases in osmotic pressure and subsequent swelling of the cell. Continual removal of Na^+ by the Na^+/K^+ pump is therefore a major factor in controlling cell volume.

■ Na^+ and K^+ gradients are also important for the production of electrical signals, such as nerve impulses.

Epithelial transport

■ Transport across an epithelial cell layer depends on an asymmetry in the permeability and pumping activities of the mucosal and serosal regions of epithelial plasma membranes. On the serosal side of the cell, ions are actively transported across the membrane against an electrochemical gradient; on the mucosal side, ions cross the membrane by passive transport. Diffusion of ions back through the epithelial layer is slow because the spaces between cells are restricted by tight junctions.

■ Water is transported across some epithelia by being drawn osmotically down a standing salt concentration gradient built up by active salt transport into long, narrow intercellular clefts. There is no true active transport of water.

REVIEW QUESTIONS

1. What are some of the physiological functions of membranes?
2. What is the evidence for the existence of membranes as real physical barriers?
3. What is the evidence for a mosaic of globular proteins set into the lipid bilayer of the membrane?
4. Explain the meanings of isotonic and isosmotic. How can a solution be isosmotic but not isotonic to another solution?
5. What factors determine the permeability of a membrane to a given electrolyte? To a given nonelectrolyte?
6. Describe the probable mechanisms by which water and other small (less than 1 nm in diameter) polar molecules pass through a membrane.
7. Why do nonpolar substances diffuse through a membrane more easily than polar substances?
8. There is no convincing evidence for direct active transport of water. Explain one way in which water is moved by epithelia against a concentration gradient—that is, from a concentrated salt solution to a more dilute salt solution.
9. How does passive transport differ from simple diffusion?
10. What factors influence the rate of passive transport of ions across a membrane?
11. How does active transport differ from passive transport?
12. Why can the sodium ion concentration gradient be considered a common cellular energy currency?
13. What are some parameters by which a membrane discriminates between ions of the same charge?
14. Explain the osmotic consequences of poisoning the metabolism of a cell.
15. How does a cell maintain a higher concentration of K^+ inside the cell than in the extracellular fluid?
16. Describe the morphological and functional distinctions between gap junctions and tight junctions.
17. A given cell is 40 times as permeable to K^+ and Cl^- as to any other ions present. If the inside-to-outside ratio of K^+ is 25, what would the approximate inside-to-outside ratio of Cl^- be?
18. Given that cell membranes can transport substances only into or out of a cell, explain how substances are transported through cells.
19. Describe the experiments that first demonstrated active transport of Na^+ across an epithelium.
20. What is some of the evidence that active transport of Na^+ and K^+ occurs only across the serosal membranes of epithelial cells?

SUGGESTED READINGS

Goodsell, D. S. 1991. Inside a living cell. *Trends Biochem. Sci.* 16:203–206. (This article takes the reader on a tour through the amazing structures and processes found in the living cell.)

Heymann, J. B., and A. Engel. 1999. Aquaporins: Phylogeny, structure and physiology of water channels. *News Physiol. Sci.* 14:187–193. (A brief review of the latest research on aquasporins.)

Lodish, H., A. Burk, et. al. 2000. *Molecular Cell Biology.* 4th ed. New York: W. H. Freeman. (This comprehensive textbook describes many of the basic biochemical processes involving biological membranes and their role in transcellular transport.)

Singer, S. J., and G. L. Nicolson. 1972. The fluid mosaic model of the structure of cell membranes. *Science* 175:720–731. (The original description of the fluid mosaic model for the structure for cell membranes.)

Classic Work

Steel, A., and M. Hediger. 1998. The molecular physiology of sodium- and proton-coupled solute transporters. *News Physiol. Sci.* 13:123–130. (A summary of current knowledge.)

Physiological Processes

If an animal is to survive and prosper, it must respond appropriately and effectively to its environment and to its own internal states. Effective responses often require that different parts of the body, which may be quite far apart, act together in a coordinated fashion. The nervous and endocrine systems both initiate and regulate coordinated responses. They control activity in muscles and glands, which generate an animal's behavioral responses. In Part 2, we focus first on the sensing and signaling systems (nervous and endocrine) and then on the effector systems (glands and muscles). These tissues are composed of highly specialized cells that work together in groups to acquire and integrate information and to generate responses that are suitable to the perceived situation.

The tasks of collecting information from outside and inside the body and of integrating that information belong largely to the cells of the nervous system. In Chapter 5 we discuss the properties of nerve cells that allow them to gather, transduce, and transmit information. Neurons in all species that have been studied have a large number of features in common, from the nature of the molecules underlying their function to the physical principles that determine how they work. This basic conservation of cellular properties has made it possible to study neurons that are particularly convenient for experimental manipulation with the conviction that the knowledge gained will be broadly applicable.

All nervous systems consist of large numbers of cells that must receive and transmit information in order to function effectively. Chapter 6 considers the processes that allow signals to travel along the plasma membrane of a single neuron and the processes by which signals are passed between neurons. Within a single neuron, a signal is encoded electrically, and in some cases transmission between neurons also is accomplished electrically. In most cases, however, the electrical signal in one neuron is transformed into a chemical signal to transmit information from one cell to another. Understanding the mechanisms that allow neurons to communicate with one another and with other cells provides a basis for understanding the power and the limitations of the nervous system.

At the interface between an animal and its environment are many cells that are specially tuned to receive information; other specialized cells monitor conditions within the body. These sensory cells, and the properties that make them particularly suited for gathering information, are discussed in Chapter 7.

All nervous systems consist of many neurons, and the anatomic arrangement of these neurons is correlated with their function. Thus, knowledge of the anatomic organization of nervous systems provides a basis for understanding the functional organization of the system. Principles of anatomic and functional organization of nervous systems in several groups of animals are considered in Chapter 8.

The second major system that contributes to coordination within an animal's body is the endocrine system. The cells of this system are collected into organs called endocrine glands. (Other glands, called exocrine glands, produce chemicals that are secreted by way of ducts into particular regions of the body.) Endocrine glands send signals by way of molecules that are released into the bloodstream. The signaling molecules of endocrine glands, called hormones, can influence widely separated parts of the body simultaneously because they are carried throughout the body in the circulatory system. Hormones act on their target cells through specific receptor molecules, and the effect of a hormone on its targets depends on both the nature of the receptor molecules and the mechanisms by which activated receptors influence internal processes in the target cell. Several different types of glands (representing both endocrine and exocrine glands), the mechanisms that control the release of hormones, and the mechanisms by which hormones modulate the physiological state of their targets are discussed in Chapter 9.

The outwardly visible behavior of an animal, as well as much of the activity that goes on inside the body, depends on contractions of muscle cells. In Chapter 10 we discuss the cellular properties of muscles that allow them to move the body or to change the shape of internal organs. We then turn to the properties of groups of muscle fibers and consider how muscular movements are coordinated to produce effective behavior.

Finally, in Chapter 11, we consider some examples of how specific behaviors are produced. Intensive experimental investigation has elucidated details of how particular behaviors are initiated and shaped. A modern description of a behavior includes information concerning sensory input, its processing by the nervous and endocrine systems, and the generation and control of movements that allow an animal to accomplish goals such as finding food or a mate or fleeing a potential predator.

One emphasis in Part 2 is on the properties of single cells that allow them to perform their particular tasks and to work together effectively and harmoniously. Another emphasis is on the mechanisms that coordinate cellular function into higher levels of organization to enhance an animal's overall fitness.

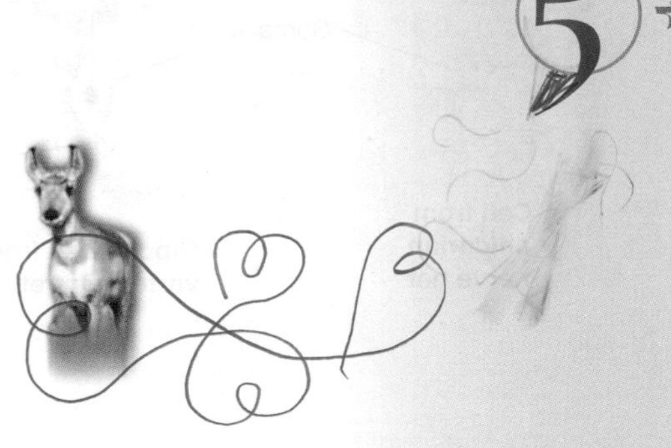

The Physical Basis of Neuronal Function

In all multicellular animals, activity depends on the precise coordination of many individual cells. The most important system for producing rapid coordination in the body is the nervous system. The major players in the nervous system are **neurons,** or nerve cells, which communicate information using a combination of electrical and chemical signals. Most neurons are electrically excitable; that is, electrical signals can be generated across the plasma membrane of these cells and transmitted along the length of the cells without loss of signal strength, as a result of the movement of ions. Neurons convey electrical signals rapidly and accurately to organize and direct physiological responses throughout an animal's body. All of the neurons in an organism's body, along with supporting cells called glial cells, make up the nervous system, which collects and processes information, analyzes it, and generates output to control the animal's responses from the simplest to the most complex.

Although the patterns of neuronal activity underlying behavior are understood completely only for some small and relatively simple neuronal circuits (see Chapter 11), individual neurons have been among the most thoroughly studied of all cell types, for several reasons. First, their electrical properties can be conveniently investigated with tools from the physical sciences. Second, findings from such studies have revealed that neurons function similarly in all types of animals; results obtained from studying the neurons of one species are therefore readily applicable to the neurons of all other species. Finally, neurons process information in a highly sophisticated manner, but in doing so they rely on a surprisingly small number of basic physical and chemical processes. Neurons provide the pathways connecting stimulus and response for the physiological systems upon which the life of an animal depends. In this chapter, we introduce the physical and molecular mechanisms that allow neurons to function so effectively in acquiring and transmitting information.

OVERVIEW OF NEURONAL STRUCTURE, FUNCTION, AND ORGANIZATION

Neurons have evolved specialized properties that allow them to receive information, process it, and transmit it to other cells. These functions are performed by identifiable and anatomically distinct regions of the cell. The different regions of a neuron are characterized by specializations within the plasma membrane and in the subcellular architecture. Although neurons vary greatly in shape and size, each neuron typically has a **soma,** or cell body, which is responsible for the metabolic maintenance of the cell. Several thin fibers, or **nerve processes,** emanate from the soma (Figure 5-1 on the next page). There are two main types of processes: dendrites and axons. Most neurons possess multiple dendrites and a single axon.

Dendrites generally extend from the soma, are branched, and serve as receivers that gather signals from other neurons and carry them toward the soma. Neurons with an extensive and complex dendritic tree typically receive input from many other neurons.

Axons (also called nerve fibers) are specialized extensions of the neuron that conduct signals away from the soma. Some axons extend remarkably long distances. In a whale, for example, the axon of a single motor neuron, which carries information from the nervous system to muscle fibers, may extend many meters from the base of the spine to the muscles that control movement of the tail. Axons have evolved mechanisms that allow them to carry information over these distances with high fidelity and without loss of signal strength. At its termination, an axon typically divides into numerous branches, called **axon terminals,** allowing its signals to be sent simultaneously to many other neurons, to glands, or to muscle fibers (see Figure 5-1).

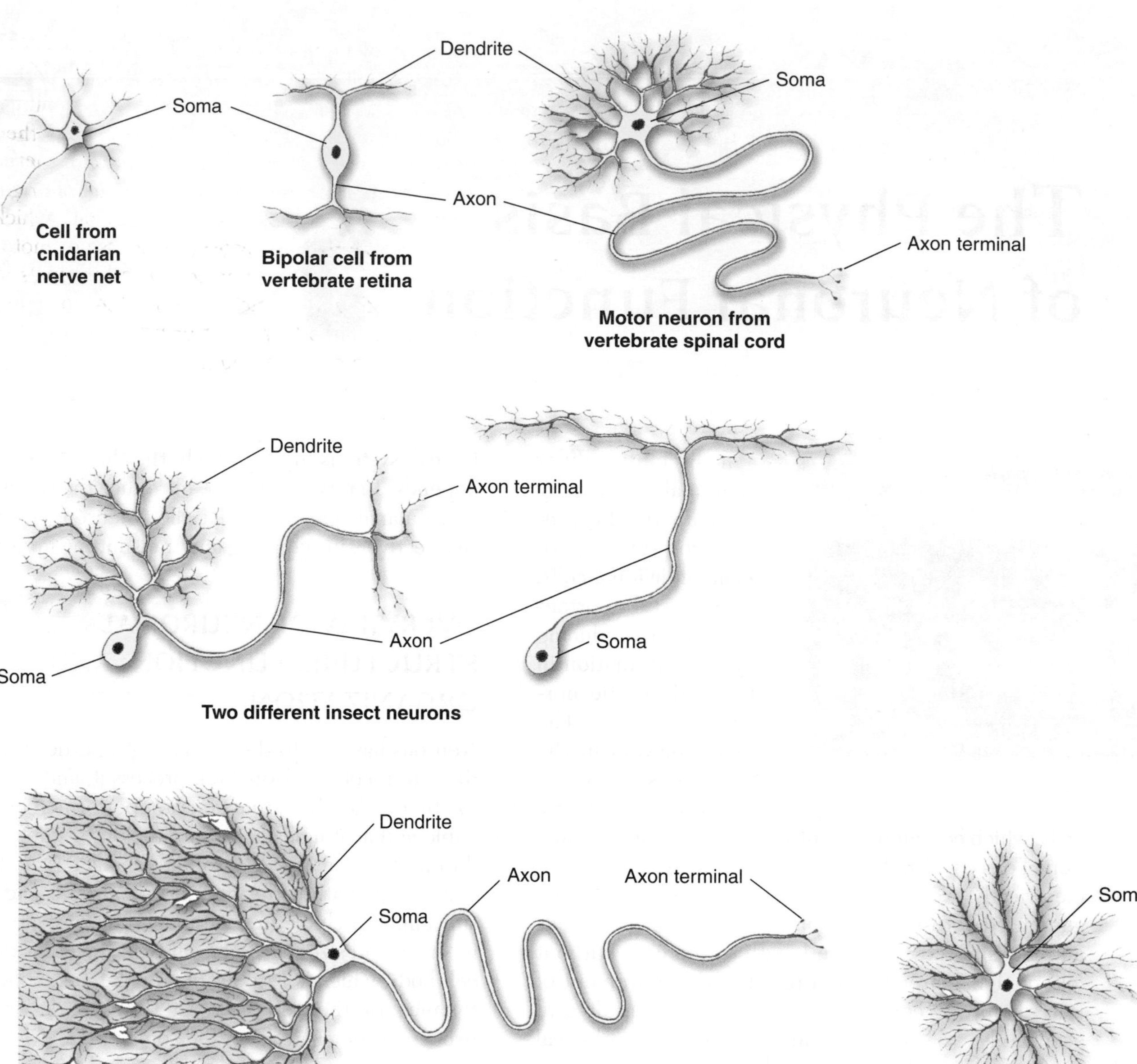

Figure 5-1 The morphology of neurons varies from simple to very complex, but most neurons have a soma, dendrites, and an axon. Notice that there is little correlation between phylogeny and the complexity of neuronal structure. Although simple animals have simple neurons (e.g., the coelenterate neuron), some neurons in more complex animals are also simple in structure (e.g., the vertebrate retinal bipolar cell). Complex animals have neurons with a very complex structure (e.g., the Purkinje cell from the mammalian cerebellum), but so do insects and other invertebrates. In some neurons (e.g., cerebellar Purkinje cells and vertebrate motor neurons), the dendrites and the axon are easily distinguished. In other neurons (e.g., retinal bipolar cells and mammalian association cells), no morphologic features readily distinguish the axon from the dendrites.

Many vertebrate axons are surrounded along their length by supporting cells that provide an insulating layer called the myelin sheath, which we will examine in Chapter 6.

During the embryonic development of each neuron, the dendrites and the axon grow outward from the soma. Throughout the life of the neuron, the maintenance of these fibers depends on a steady flow of proteins and other constituents that are synthesized in the soma and transported down the processes. If an axon in an adult animal is severed or even just damaged, it typically degenerates back to the soma within a few days or weeks. In mammals, regeneration or regrowth of axons is generally limited to nerves in the periphery of the body, whereas in cold-blooded vertebrates, some regeneration may take place within the central nervous system (i.e., in the brain and spinal cord). In contrast, damaged neurons in many invertebrates readily regenerate and reestablish connections with their original targets.

Transmission of Signals in a Single Neuron

A typical vertebrate spinal motor neuron, which has its soma in the spinal cord and carries signals to skeletal muscle fibers, illustrates the structural and functional features that characterize most neurons (Figure 5-2). The plasma membrane of the motor-neuron dendrites and of the soma receives signals from, or is innervated by, the terminals of other neurons. Typically, a region of membrane called the **spike-initiating zone** integrates signals from many input neurons to determine whether the neuron will initiate its own signal, called an **action potential** or **AP**. In an action potential, the voltage across the plasma membrane rapidly rises and then falls, so APs are sometimes called *spikes* or *nerve impulses*. The axon carries an AP from its point of origin in the spike-initiating zone to the axon terminals, which transmit the signal to other cells; in the case of motor neurons, the signal travels from the axon terminals to skeletal muscle fibers. In many types of neurons, the spike-initiating zone is located at or near the junction between the axon and the soma, a region called the **axon hillock**, although it may lie elsewhere.

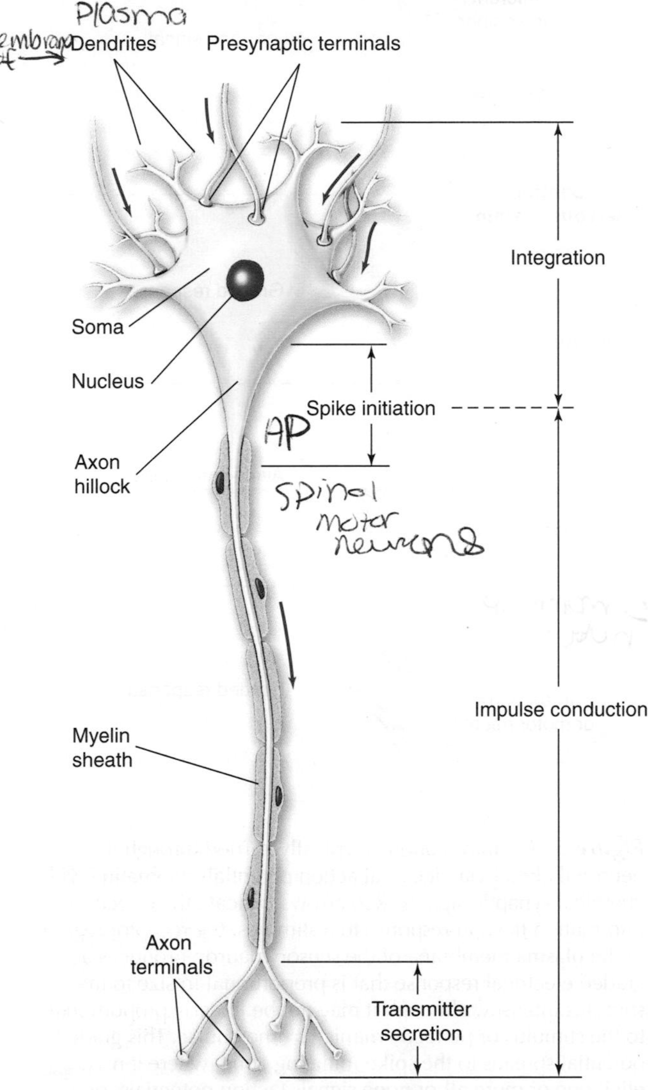

Figure 5-2 A vertebrate spinal motor neuron exemplifies the functionally specialized regions of a typical neuron. The flow of information is indicated by small red arrows. Information is received and integrated by the plasma membrane of the dendrites. In some neurons, like this one, the soma also receives information and contributes to signal integration. In spinal motor neurons, action potentials are initiated at the spike-initiating zone, located in or near the axon hillock, and then travel along the axon to the axon terminals, where a chemical neurotransmitter is released to carry the signal on to another cell. In some other types of neurons, the spike-initiating zone is located in a different part of the cell. The axon and surrounding myelin sheath cells are shown in longitudinal section in this diagram.

What are the potential advantages and disadvantages of a neuronal structure in which signal inputs are located in one region (the dendrites and soma), the output channel is located in another region (the axon), and the spike-initiating zone occupies only a small fraction of the cell surface?

Much of the physiological behavior of a neuron depends on **passive electrical properties**, such as capacitance and resistance, that the cell has in common with wires and other electrical conductors. These properties will be discussed later in this chapter. Unlike wires, neurons also possess **active electrical properties** that allow them to conduct electrical signals without decrement (i.e., with no loss of signal strength). The active electrical transmission of signals by neurons and other excitable cells depends on the presence of specific proteins, called voltage-gated ion channels, in the plasma membrane. These channels allow ions to move across the plasma membrane in a regulated fashion.

Neurons possess various types of ion channels, each permitting passage of a particular ionic species. The ion channels found in neurons are not distributed uniformly over the surface of the cell, but are instead localized to the regions that have specialized signaling functions. The axonal membrane, for example, is specialized for the conduction of APs by virtue of fast-acting, voltage-gated ion channels that selectively allow Na^+ and K^+ to cross the membrane. In addition, the plasma membrane of the axon terminals contains voltage-gated Ca^{2+} channels and other specializations that allow neurons to transmit signals to other cells when APs invade the terminals.

Transmission of Signals Between Neurons

Information processing by any nervous system begins when sensory neurons collect information from the outside world or from the interior of the animal and send it to other neurons. The axon of a sensory neuron is called an **afferent** fiber—the term means that it conducts a signal "inward" toward higher processing centers in the brain. Sensory neurons pass information on to other neurons, and the signal is then transferred from neuron to neuron in the animal's nervous system. Interneurons—the most numerous type of neuron—lie entirely within the central nervous system and carry information between other neurons. Information is passed between neurons, or between neurons and other target cells, at specialized locations called **synapses.** If an animal is to respond to the sensory information it has acquired, a sensory signal must be followed by the activation of neurons that control effector organs, such as muscles or glands. Neurons that carry information from the processing regions of the central nervous system outward to effectors are called **efferent** neurons. Together the afferent neurons and efferent neurons, along with any interneurons that participate in processing the information, make up a **neuronal circuit** (Figure 5-3).

A cell that passes information to a particular neuron is said to be **presynaptic** to that neuron; a cell that receives information transmitted across a synapse is said to be **postsynaptic.** Most synaptic transmission is carried by chemical neurotransmitters, which are specific molecules released from the axon terminals of the presynaptic neuron in response to APs in its axon. The plasma mem-

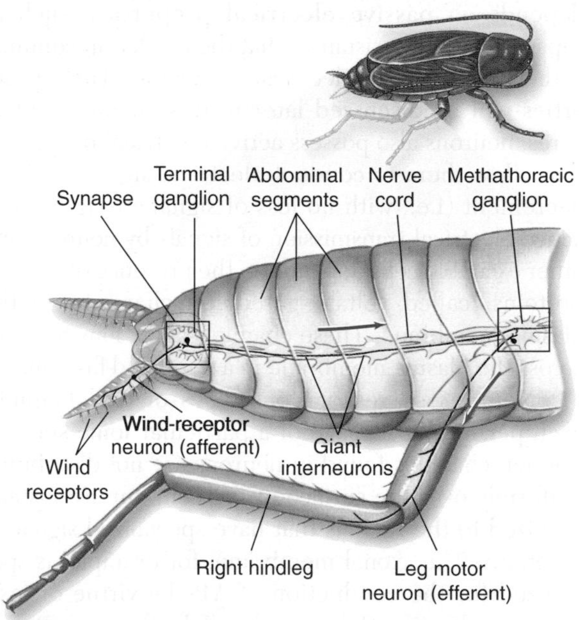

Figure 5-3 In a simple neuronal circuit, an afferent neuron carries sensory information to interneurons in the central nervous system, and an efferent neuron carries the processed information to effector organs. This figure showing the posterior of a cockroach illustrates a neuronal circuit consisting of wind-receptor (afferent) neurons in the tail, giant interneurons in the central nervous system, and motor (efferent) neurons controlling the muscles of the legs. The wind-receptor neurons contact the giant interneurons at synapses in the terminal ganglion of the nervous system, and the giant interneurons contact the leg motor neurons at synapses in the thoracic ganglia (for example, the metathoracic ganglion shown here). Stimulation of the wind-receptor neurons sets up activity in this circuit that causes the cockroach to run away from the stimulus.

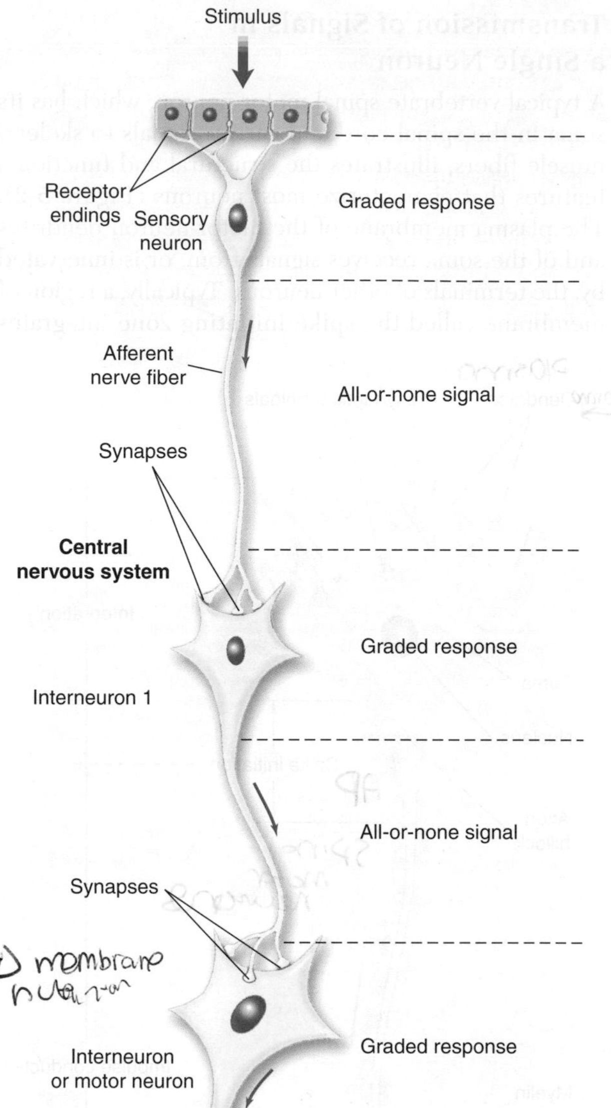

Figure 5-4 Information is typically carried through a neuronal circuit via electrical action potentials alternating with chemical synaptic signals. Red arrows indicate the direction of information flow. In response to a stimulus, the receptor region in the plasma membrane of the sensory neuron produces a graded electrical response that is proportional in size to the stimulus intensity, although it may not be linearly proportional to the stimulus or perfectly mimic its time course. This graded potential spreads to the spike-initiating zone, where it may elicit one or more all-or-none signals (action potentials, or APs), which are propagated along the axon. When an AP arrives at the axon terminals, it causes the release of chemical neurotransmitter molecules from the presynaptic cell. The neurotransmitter produces a graded potential in the next (postsynaptic) neuron. If the change in membrane potential in the postsynaptic neuron is large enough, one or more all-or-none APs will be generated in the postsynaptic neuron. Thus, graded and all-or-none electrical signals alternate in the pathway. A single pathway through the central nervous system may include only a few or many neurons connected by synapses. In this diagram, the sensory neuron is presynaptic to interneuron 1. Interneuron 1 is presynaptic to other neurons, which could be other interneurons or motor neurons. Thus, interneuron 1 is both presynaptic and postsynaptic to other neurons.

brane of the postsynaptic neuron's dendrites and soma contains ligand-gated ion channels that bind neurotransmitters and cause the postsynaptic cell to respond to the presence of the chemical signal. This mechanism will be discussed in much more detail in Chapter 6. The numerous synaptic inputs arriving at a neuron in this way are integrated along the plasma membrane of its dendrites and soma to produce a change in membrane potential at the spike-initiating zone. As indicated in Figure 5-4, information is carried in neuronal circuits via alternating chemical and electrical signals. Along the chain of neurons, the electrical signals alternate between signals that are invariant in amplitude (i.e., APs) and signals whose amplitude depends on stimulus strength or some other variable. Signals whose amplitude is invariant are called **all-or-none signals;** signals whose amplitude varies are called **graded signals.** These two kinds of signals will also be discussed in more detail in Chapter 6.

Organization of the Nervous System

The nervous system is composed of two basic cell types: neurons and supporting glial cells. As we have seen, neurons can be classified functionally into three types:

- **sensory neurons,** which transmit information collected from external stimuli (e.g., sound, light, pressure, or chemical signals) or which respond to stimuli inside the body (e.g., blood oxygen level, the position of a joint, or the orientation of the head)
- **interneurons,** which link other neurons within the central nervous system
- **motor neurons** (or motoneurons), which carry signals to effector organs, causing contraction of muscles or secretion by gland cells

Sensory neurons transform the energy of a stimulus into electrical signals used by the nervous system. Networks of interneurons exchange information and perform most of the complex computations that produce thought and behavior. Motor neurons constitute the output portion of a neuronal circuit, carrying specific instructions to the muscles or other effector organs that they innervate. (Neurons that innervate gland cells and other effector targets are called motor neurons even though they do not control bodily movement.)

Neurons are grouped into clusters in almost all phyla. Typically, the cell bodies (somata) of many—even most—neurons are contained within the **central nervous system (CNS).** In most animals, the CNS consists of a **brain,** located in the head, and a nerve cord that extends posteriorly along the midline of the animal. Many invertebrates have a brain located in the head and collections of neuronal somata, called **ganglia,** distributed along the nerve cord; neurons in the ganglia typically control local regions of the animal's body (Figure 5-5). Vertebrates also have ganglia, which are groups of neuronal somata that are located outside the CNS. In vertebrates, the nerve cord, called the **spinal cord,** is

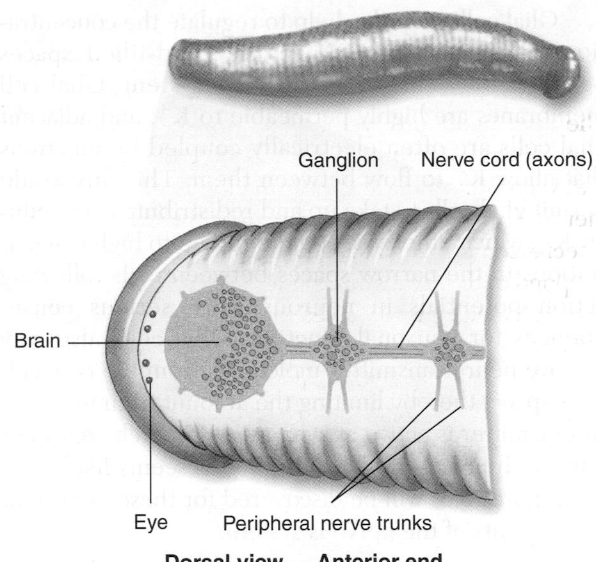

Figure 5-5 The central nervous system, typically consisting of a brain and nerve cord, is the site of most information processing and usually contains most of an animal's neuronal somata, as well as many axons. The brain, usually in the head of the animal, contains a large number of neurons and their interconnections. In many invertebrates, such as the leech shown in this figure, the somata of neurons outside the brain are grouped within the nerve cord into structures called ganglia. In a segmented animal like the leech, each segment commonly contains a ganglion, which controls many behavioral responses that are confined to that segment. Nerves composed of axons connect structures within the CNS and connect the CNS to peripheral structures.

located along the dorsal midline, whereas in many invertebrates (e.g., insects, crustaceans, and annelids), the major nerve cord lies along the ventral midline. Many neurons whose somata lie in the CNS send processes into the rest of the body (the periphery) to collect sensory information or to deliver motor signals that control the activity of muscles or glands.

The other main cell type in the nervous system, the **glial cells** (or neuroglia), fill the space between neurons, except for a very thin fluid-filled space (about 20 nm wide) that separates the glial and neuronal membranes. In general, more complex animals have a larger number of glial cells relative to neurons than do simpler ones. The vertebrate central nervous system contains 10–50 times more glial cells than neurons, and these cells occupy about half the volume of the nervous system.

Although some glial cells have voltage-gated ion channels in their membranes, glial cells generally do not produce APs, and their role in the nervous system has long been a puzzle. At the least, it is clear that glial cells provide intimate structural, and probably metabolic, support for neurons. In vertebrates, for example, **oligodendrocytes** in the CNS and **Schwann cells** in the periphery are glial cells that wrap axons in an insulating myelin sheath, which contributes to reliable and rapid transmission of APs (see Figure 5-2 and Chapter 6).

Glial cells may also help to regulate the concentration of K^+ and the pH in the fluid-filled spaces between the cells of the nervous system. Glial cell membranes are highly permeable to K^+, and adjacent glial cells are often electrically coupled by junctions that allow K^+ to flow between them. This flux would permit glial cells to take up and redistribute extracellular K^+, which otherwise could build up to high concentrations in the narrow spaces between cells following action potentials in neurons, with serious consequences for neuronal function. Glial cells also may remove neurotransmitter molecules from the extracellular space, thereby limiting the amount of time a neurotransmitter is active at synapses. Research continues into the function of glial cells, and it seems likely that even more roles will be discovered for these important components of the nervous system.

MEMBRANE EXCITATION

Although a stable **voltage** (or electric potential difference) exists across the plasma membrane of all animal cells, only the membranes of **electrically excitable** cells (e.g., neurons and muscle fibers) can respond to changes in their transmembrane potential difference by generating APs. The study of electrically excitable tissues has a long history, the start of which is reviewed in Spotlight 5-1. To understand both the basis of excitability and its consequences for neuron function, we first need to see how changes in the electric potential difference across the plasma membrane are measured.

Measuring Membrane Potentials

Whenever there is a net flux of charged particles across a plasma membrane, it can be measured as an electric current. These currents can be detected directly by using two electrodes to measure the change in electric potential that is caused by current flow across the membrane. One sensing electrode is placed in electrical contact with the cytosol, and the other is placed in contact with the extracellular medium, so that the two electrodes will measure any potential difference that exists across the plasma membrane. The potential difference measured in this manner (the **membrane potential**, V_m, in volts) can then be electronically amplified and displayed on a recording instrument, such as an oscilloscope or computer monitor (Figure 5-6).

Much of what we know about how APs are generated rests on experiments carried out by A. L. Hodgkin and A. F. Huxley in the 1940s and 1950s. They recorded membrane potentials from squid giant axons, which are large enough that a thin silver wire can be inserted longitudinally along the inside of the cylindrical process (Hodgkin and Huxley, 1952). We will consider their remarkable work later in this chapter.

Electrical activity of other types of neurons, most of which are much smaller than the squid giant axon, has been studied using glass capillary microelectrodes,

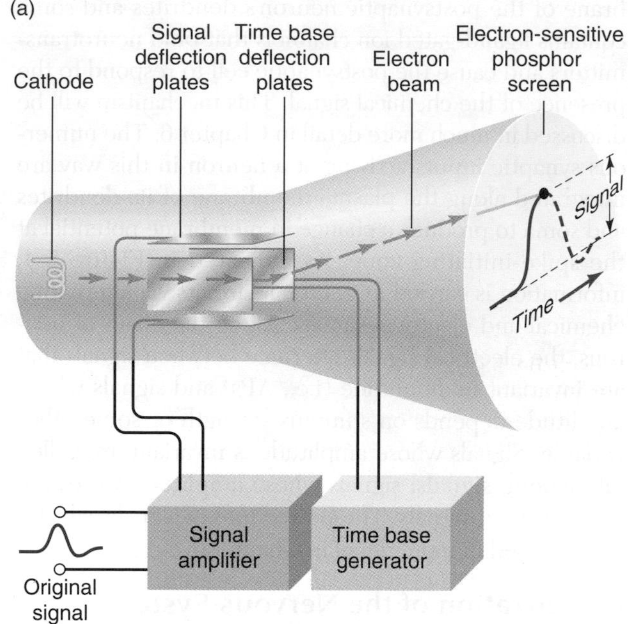

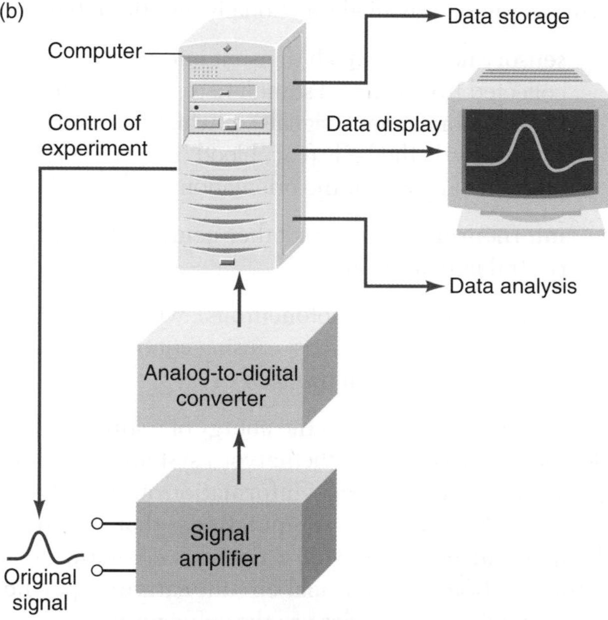

Figure 5-6 Potential differences measured across plasma membranes can be amplified and displayed using standard electronic instruments available in any physics laboratory. **(a)** An oscilloscope amplifies and plots the amplitude of an electrical signal in the vertical displacement of a beam of electrons emitted by a cathode ray tube. The passage of time is indicated as the beam of electrons is driven from left to right by a time base generator, and its position is visualized as it "writes" on the phosphor screen. Thus, the size of a signal fed into the oscilloscope is plotted on the screen as a function of time. **(b)** Today a computer frequently supplies the display functions of an oscilloscope. In addition, data can be stored directly as they are recorded or even analyzed in real time as the experiment progresses. An analog signal sensed by electrodes is converted to digital information by sampling the signal at discrete time intervals that are sufficiently short to capture all the necessary information. Computers can also control aspects of an experiment, such as the delivery of a specific pattern of stimulation to cells.

THE DISCOVERY OF "ANIMAL ELECTRICITY"

Electrical excitability is a fundamental property of neurons and muscles. Today we understand this phenomenon in great detail, but electrically excitable animal tissues have been studied for centuries. Both the study of "animal electricity" and the origin of electrochemical theory can be traced to observations made late in the eighteenth century by Luigi Galvani, a professor of anatomy at Bologna, Italy. Working with a muscle and its attached nerve dissected out of a frog leg, Galvani noticed that the muscle contracted if the nerve and muscle were touched by metal rods in a particular way. The two rods had to be made of different metals (e.g., one of copper and the other of zinc). In Galvani's experimental setup, one rod contacted the muscle and the other contacted the nerve to that muscle; when the two rods were brought together, the muscle contracted, as illustrated in part a of the accompanying figure.

Galvani and his nephew Giovanni Aldini, a physicist, ascribed this response to a discharge of "animal electricity" that was stored in the muscle. They hypothesized that an "electrical fluid" passed from the muscle through the metal and back into the nerve, and that the discharge of electricity from the muscle triggered the contraction. We now understand that this creative interpretation, which was published in 1791, is largely incorrect. Nevertheless, this work stimulated many inquisitive amateur and professional scientists to investigate two new and important areas of science: the physiology of excitation in nerve and muscle and the chemical origin of electricity.

Alessandro Volta, a physicist at Pavia, Italy, took up Galvani's experiments. In 1792 he proposed an alternative explanation for Galvani's results. He suggested that the electrical stimulus causing the muscle to contract in Galvani's experiments was generated outside the tissue by the contact between the dissimilar metals and the saline fluids of the tissue. It took several years for Volta to demonstrate unequivocally the electrolytic origin of electric current from dissimilar metals because no physical instrument was available that was sensitive enough to detect these weak currents. Indeed, the nerve-muscle preparation from the frog leg was probably the most sensitive indicator of electric current available at that time and for many years afterward.

In his search for a method to produce stronger electric currents, Volta found that he could increase the amount of electricity produced electrolytically by placing metal-and-saline cells in series. The fruit of his labor was the so-called voltaic pile, a stack of alternating silver and zinc plates separated by saline-soaked papers. This first "wet-cell" battery produced higher voltages than can be produced by a single silver-zinc cell, and its design is still commonly used in today's batteries.

Although Galvani's original experiments did not really prove the existence of "animal electricity," they did demonstrate that some living tissues can respond to minute electric currents. In 1840, Carlo Matteucci advanced the study of electricity in living biological tissues by using the electrical activity of one contracting muscle to stimulate a second nerve-muscle preparation, as illustrated in part b of the figure. His experiment was the first recorded demonstration that excitable tissues actually produce electric current. We now know that the production of signals in the nervous system and in other excitable tissues depends on the electrical properties of plasma membranes.

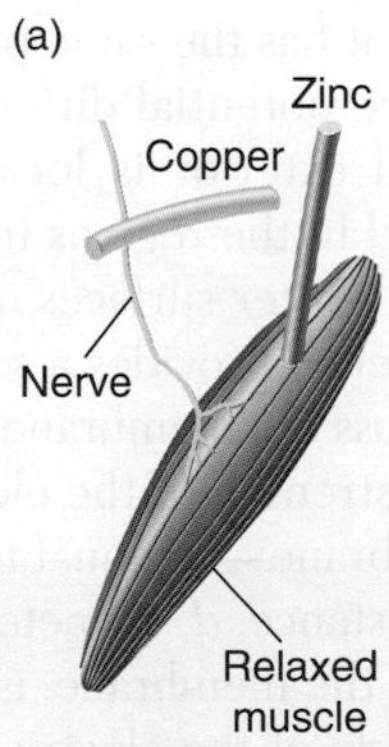

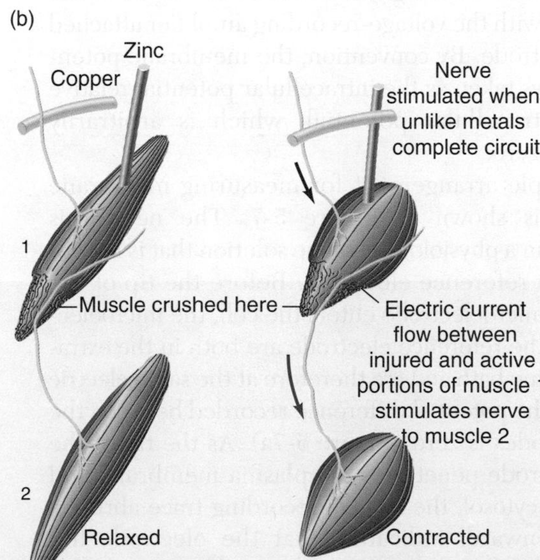

*Early electrical experiments were carried out on nerves and muscles. **(a)** In Galvani's experiments, which were carried out in a saline solution, a muscle and its nerve were contacted by rods of two different metals, such as copper and zinc. When the two rods were touched together, the muscle contracted. **(b)** Matteucci elaborated on Galvani's experiment by connecting one nerve-muscle preparation (1) to a second one (2) via a nerve. When muscle 1 was stimulated by making the dissimilar metal rods contact each other, the electrical activity of muscle 1 stimulated the nerve to muscle 2, causing muscle 2 to contract. In this experiment, muscle 1 had to be injured at the point of contact with the nerve from muscle 2 in order for ionic currents flowing in the fibers of muscle 1 to stimulate the nerve.*

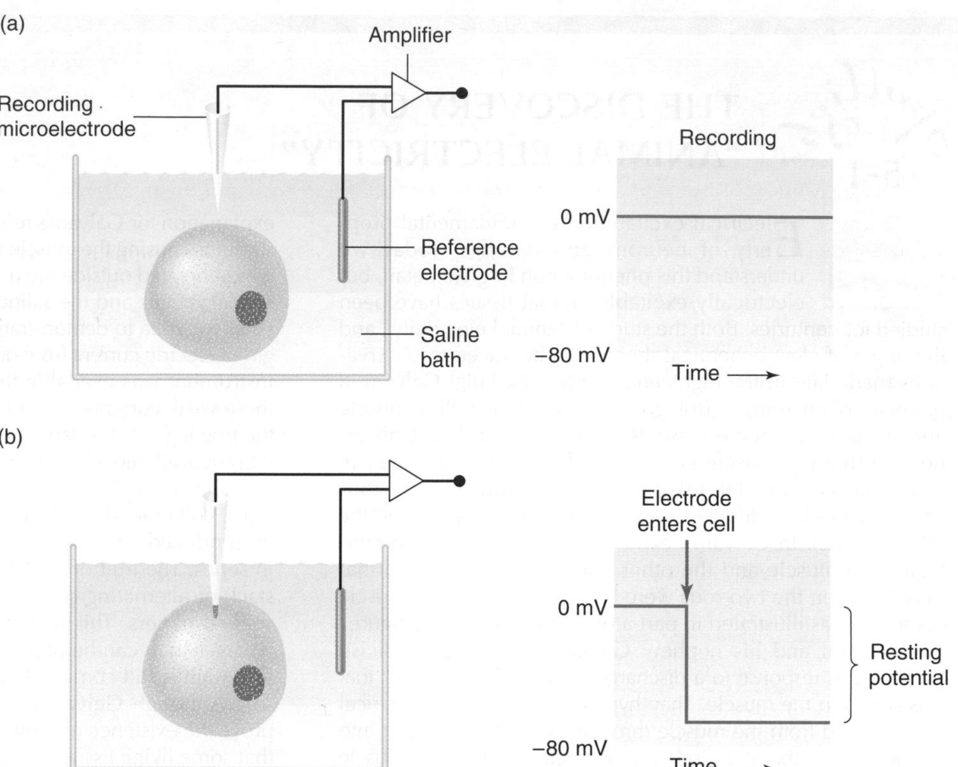

Figure 5-7 When a microelectrode penetrates a neuron's plasma membrane, there is a shift in the recorded voltage. **(a)** No potential difference is recorded between the reference electrode and the recording microelectrode when both are in the saline solution bathing the neuron. **(b)** As soon as the microelectrode tip penetrates the neuron's plasma membrane, the electrode records an abrupt shift in the negative direction, which is shown as a downward deflection on the oscilloscope screen or computer monitor. This deflection corresponds to the resting potential across the membrane.

which are described in Chapter 2. A cell sustains minimal damage when it is impaled by a microelectrode because the lipid bilayer of the plasma membrane seals itself around the electrode tip after penetration. Inserting the electrode tip through the plasma membrane into the cell brings the cell interior into electrical continuity with the voltage-recording amplifier attached to the electrode. By convention, the membrane potential is always taken as the intracellular potential relative to the extracellular potential, which is arbitrarily defined as zero.

A simple arrangement for measuring membrane potential is shown in Figure 5-7. The neuron is immersed in a physiological saline solution that is in contact with a reference electrode. Before the tip of the recording microelectrode enters the cell, the microelectrode and the reference electrode are both in the extracellular saline bath and are therefore at the same electric potential; the potential difference recorded between the two electrodes is zero (Figure 5-7a). As the tip of the microelectrode penetrates the plasma membrane and enters the cytosol, the voltage recording trace abruptly shifts downward, indicating that the electrode has entered the cell and is now measuring the potential difference across the plasma membrane (Figure 5-7b). By electrophysiological convention, an inside-negative potential is shown as a downward displacement of the recording trace on an oscilloscope or computer monitor.

The steady inside-negative potential recorded when no action potentials or postsynaptic events are occurring is the **resting potential,** V_{rest}, of the plasma membrane. It is most conveniently expressed in millivolts (mV, or thousandths of a volt). Virtually all cell types that have been investigated have an inside-negative resting potential with a value between -20 mV and -100 mV. The potential sensed by the intracellular electrode does not change as the tip is advanced farther into the cell because, in the resting state, the cytoplasm of the cell is *isopotential;* that is, it has the same potential everywhere. Thus, the entire potential difference between the cell's interior and exterior is localized across the plasma membrane and in the regions immediately adjacent to the inner and outer surfaces of the membrane. This potential difference provides a source of energy that can move ions across the membrane. The amount of energy—that is, the strength of the electric field, E, across the plasma membrane—is equal to the voltage in volts divided by the distance, d, in meters ($E = V/d$). Since d, the thickness of the membrane, is only 5 nm (5×10^{-9} m), the magnitude of the electric field across the plasma membrane is on the order of 60 mV/10^{-9} m, or 60 million volts $\cdot$ m^{-1}—a very large field!

What would keep a cell's cytosol isopotential? What would happen if local potential differences arose in the cytosol?

Electrical Properties of Membranes

As noted earlier, the membranes of neurons and other excitable cells display both passive and active electrical properties. Both must be considered if we are to understand the basis of electrical integration and signaling in neurons.

The passive electrical properties of the plasma membrane can be measured by delivering a pulse of current across the membrane to produce a slight perturbation of the membrane potential. Two microelectrodes are inserted into one cell, as illustrated in Figure 5-8. Either inward (bath-to-cytosol) or outward (cytosol-to-bath) electric current is delivered from a current generator to the *current electrode.* A *recording electrode* records the effect this current has on the membrane potential. Note that all the current carried in solution and across the membrane is in the form of migrating ions. (Current along the connecting wires is, of course, carried by electrons.) By convention, the flow of current is from a region of relative positivity to one of relative negativity and corresponds to the direction of cation migration. Thus, if the current electrode is made positive, this current will, by definition, flow from the current electrode into the cytosol and out of the cell across its membrane. Conversely, if the current electrode is made negative, it will draw positive charge out of the cytosol and cause current to flow into the cell across the membrane; this situation is depicted in Figure 5-8a.

When a current pulse causes positive charge to exit the cell through the current electrode (i.e., when the current electrode is made more negative), the negative interior of the cell becomes even more negative; this increase in the magnitude of the potential difference across the plasma membrane is called **hyperpolarization.** For example, the membrane potential might change from a resting potential of −60 mV to a hyperpolarized potential of −70 mV. Neuronal membranes generally respond passively to hyperpolarization, producing no response other than the change in potential caused by the applied current. Small negative pulses to the current electrode produce a small amount of hyperpolarization; larger negative pulses produce greater hyperpolarization (Figure 5-9, traces 1 and 2 on the next page).

If the current electrode adds positive charge to the interior of the cell, this additional positive charge will diminish the potential difference across the membrane, causing **depolarization** of the membrane (Figure 5-9, traces 3 and 4). That is, the membrane potential becomes less negative (e.g., it may shift from −60 mV to −20 mV). As the amount of applied current is increased, the degree of depolarization will also increase. Depolarization causes some of the membrane's voltage-gated channels that are selectively permeable to sodium ions (Na^+) to open. When electrically excitable cells become sufficiently depolarized to open a critical number of Na^+-selective channels, an action

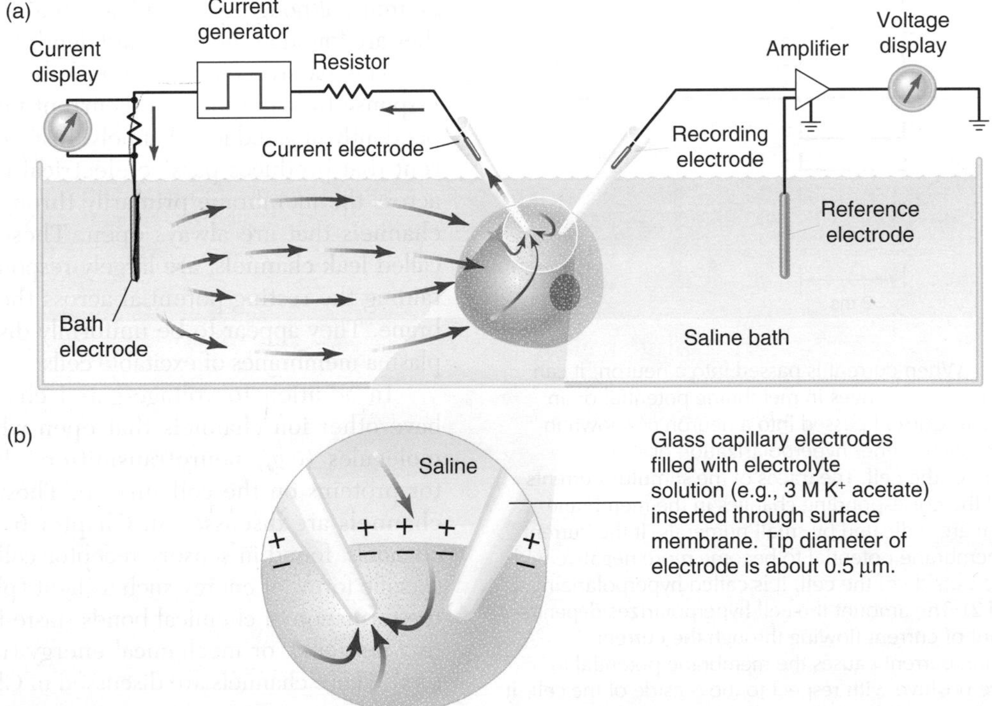

Figure 5-8 Impaling a cell with two microelectrodes allows the passive electrical properties of the plasma membrane to be measured. **(a)** Diagram of an experimental setup to measure passive membrane properties. Current flows in a circuit from the current generator through the "bath electrode," the saline bath, the plasma membrane, the current electrode, and the resistor. In this case, the current electrode has been made negative, so positive charges move out of the cytosol and into the electrode. The recording amplifier has a very high input resistance, preventing any appreciable current from leaving the cell through the recording electrode. **(b)** Magnified view of glass capillary microelectrodes inserted through the membrane of a cell. The electrode at the left passes current into, or out of, the cell. As it crosses the plasma membrane, this current changes the membrane potential, V_m, which is recorded by the second electrode.

potential—a dramatic and rapid change in the membrane potential—is triggered (Figure 5-9, trace 4).

The value of the membrane potential at which an AP is triggered 50 percent of the time is called the **threshold potential.** As we will see below, the value of the threshold potential can vary depending on the precise state of the cell at a given moment. The opening of voltage-gated Na^+ channels in response to depolarization, which allows Na^+ ions to enter the cell from outside, is one example of membrane excitation. The mechanisms responsible for the AP and other examples of membrane excitation are discussed in more detail later in this chapter.

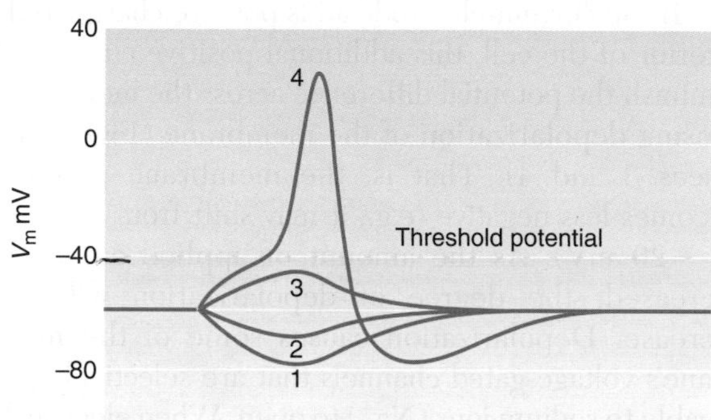

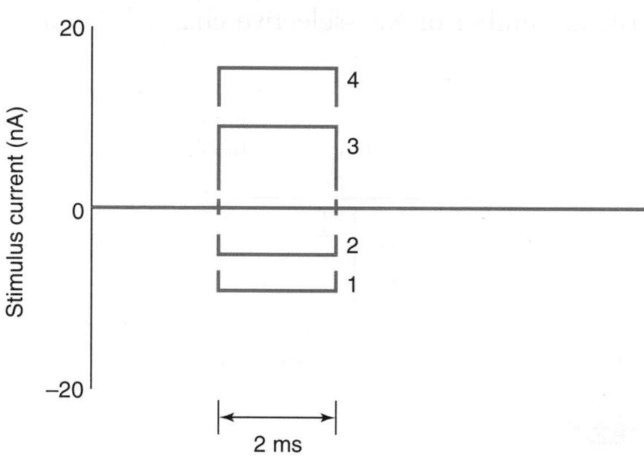

Figure 5-9 When current is passed into a neuron, it can evoke either passive changes in membrane potential or an action potential. Current passed into a neuron as shown in Figure 5-8 produces either hyperpolarization or depolarization of the cell. The traces of the stimulus currents (bottom) and the corresponding changes in the membrane potential (top) are indicated by small numerals. If the current causes the membrane potential to become more negative with respect to the outside of the cell, it is called hyperpolarizing (traces 1 and 2). The amount the cell hyperpolarizes depends on the amount of current flowing through the current electrode. If the current causes the membrane potential to become more positive with respect to the outside of the cell, it is called depolarizing (traces 3 and 4). Hyperpolarizing currents and small depolarizing currents produce passive shifts in V_m (traces 1–3). If a depolarizing current is sufficiently large, it triggers an action potential in the cell. Note that the size of the passive responses is more or less linearly proportional to the stimulus current, whereas in the active response, the change in membrane potential is much larger than would be expected for a linear response. The threshold potential is defined as the membrane potential at which an AP is produced 50% of the time.

The Role of Ion Channels

Table 5-1 summarizes the properties of some ion channels that contribute to the passive and active electrical properties of neurons. Most membrane channels allow only one or a few ionic species to cross the membrane; that is, they exhibit **ion selectivity.** Ions are driven through these channels by an electrochemical gradient. Traffic through some channels is continuous; other channels act like gates, opening in response to a change in conditions. For example, the ion channels that produce action potentials open when the plasma membrane becomes depolarized, so these channels are called **voltage-gated channels.**

The ion channels responsible for action potentials are named for the ionic species that normally moves through them. Na^+ is the ion that normally moves through voltage-gated Na^+ channels, but some other ions (e.g., lithium ions) can also pass through them. The dramatic depolarization that takes place during an AP depends on the nearly simultaneous opening of many voltage-gated Na^+ channels when initial depolarization of the membrane reaches the threshold potential. Ion flow through these channels provides the current across the plasma membrane during the first phase of an AP. Voltage-gated ion channels may be localized to particular areas in a cell. For example, voltage-gated Na^+ channels are typically confined to the axonal membrane in neurons, although we now know that in some neurons they are found in the soma and dendrites, too.

The passive change in membrane potential in response to hyperpolarizing current takes place independently of gated ion channels. Instead, the ionic current that produces passive electrical responses flows across the membrane primarily through K^+-selective channels that are always open. These K^+ channels, called **leak channels**, are largely responsible for maintaining the resting potential across the plasma membrane. They appear to be uniformly distributed in the plasma membranes of excitable cells.

In addition to voltage-gated channels, neurons have other ion channels that open when messenger molecules (e.g., neurotransmitters) bind to receptor proteins on the cell surface. These **ligand-gated channels** are discussed in Chapter 6. Still other ion channels, found in sensory receptor cells, are gated by specific forms of energy such as light (photoreceptors), the formation of chemical bonds (taste buds and olfactory neurons), or mechanical energy (mechanoreceptors). These channels are discussed in Chapter 7.

PASSIVE ELECTRICAL PROPERTIES OF MEMBRANES

The capacitance and conductance of plasma membranes—in other words, their passive electrical properties—correspond to particular membrane structural elements (Figure 5-10). The lipid bilayer of the mem-

Table 5-1 Examples of ion channels found in axons

Channel	Current through channel	Characteristics	Selected blockers	Function
Leak channel (open in resting axon)	$I_{K\,(leak)}$	Produces relatively high P_K of resting cell	Partially blocked by tetraethylammonium (TEA)	Largely responsible for V_{rest}
Voltage-gated Na^+ channel	I_{Na}	Rapidly activated by depolarization; becomes inactivated even if V_m remains depolarized	Tetrodotoxin (TTX)	Produces rising phase of AP
Voltage-gated Ca^{2+} channel	I_{Ca}	Activated by depolarization but more slowly than Na^+ channel; inactivated as function of cytoplasmic $[Ca^{2+}]$ or V_m	Verapamil, D600, Co^{2+}, Cd^{2+}, Mn^{2+}, Ni^{2+}, La^{3+}	Produces slow depolarization; allows Ca^{2+} to enter cell, where it can act as second messenger
Voltage-gated K^+ channel ("delayed rectifier")	$I_{K(V)}$	Activated by depolarization but more slowly than Na^+ channel; inactivated slowly and not completely if V_m remains depolarized	Intra- and extracellular TEA, amino pyridines	Carries current that rapidly repolarizes the membrane to terminate an AP
Ca^{2+}-dependent K^+ channel	$I_{K(Ca)}$	Activated by depolarization plus elevated cytoplasmic $[Ca^{2+}]$; remains open as long as cytoplasmic $[Ca^{2+}]$ is higher than normal	Extracellular TEA	Carries current that repolarizes the cell following APs based on either Na^+ or Ca^{2+} and that balances I_{Ca}, thus limiting depolarization by I_{Ca}

brane is impermeable to ions and is therefore an insulator. Two conductors separated by an insulator form an electrical capacitor, a device that can store energy in the form of separated electric charges. The amount of charge stored on the two sides of a capacitor (positive charges on one side and an equivalent number of negative charges on the other) is measured in units of coulombs per volt, or farads (F). Because the lipid bilayer of the plasma membrane acts as an insulator between two saline solutions—the cytosol on the inside and the extracellular fluid on the outside—it has the properties of a capacitor and can store charges on its two faces. Channel proteins that allow ions to pass across the membrane give the membrane its electrical conductance. These two electrical properties of a membrane, capacitance and conductance, can be

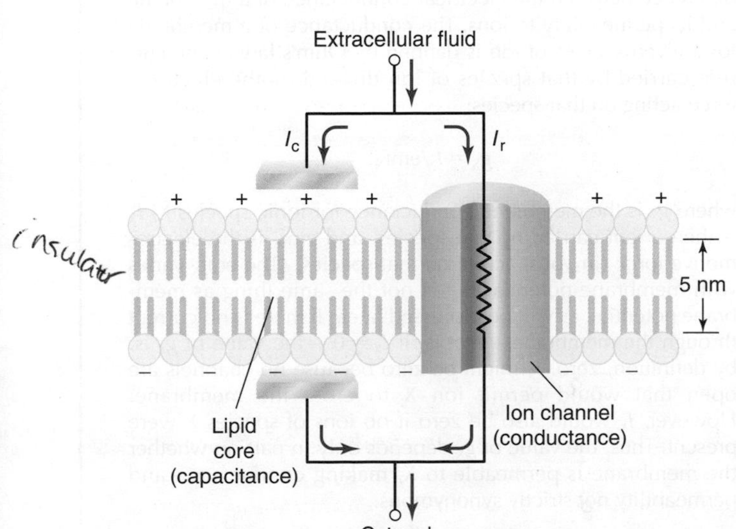

Figure 5-10 The passive electrical properties of a plasma membrane can be represented by a simple electric circuit. The membrane has a capacitance because the lipid bilayer is an insulator, bathed on its two surfaces by conducting fluids. The membrane's ionic conductance depends on the presence of open ion channels. Arrows indicate flow of capacitative current, I_c, and resistive current, I_r, when a current pulse is applied across the membrane in an experiment such as the one shown in Figure 5-8.

represented by an equivalent circuit in which an electrical capacitor is connected in parallel with a resistor (see Figure 5-10). The resistor represents the conductance through the ion channels, and the capacitor represents the capacitance of the lipid bilayer. Such an equivalent circuit is useful for making an explicit model of current flow through a membrane, allowing us to determine likely values for membrane conductance and capacitance.

Membrane Resistance and Conductance

Electrical resistance (R) and electrical conductance (g) both characterize how readily charges can move in a circuit (see Chapter 3). The resistance of a membrane—a measure of its impermeability to ions—is equal to the inverse of the conductance—which reflects its permeability ($R = 1/g$). For a given transmembrane voltage, the lower the resistance of the membrane (i.e., the greater its conductance), the more ionic charges will cross the membrane through open ion channels per unit time. The relationship between current, resistance, and steady-state voltage across a membrane is described by Ohm's law, which states that the voltage drop produced across a membrane by a current that passes through it is directly proportional to the current multiplied by the resistance of the membrane:

$$\Delta V_m = \Delta I \times R \quad (5\text{-}1)$$

where ΔV_m (in volts) is the voltage drop across the membrane, ΔI (in amperes) is the current across the membrane, and R (in ohms, Ω) is the electrical resistance of the membrane.

The total resistance encountered by current flowing into or out of the cell is called the cell's **input resistance.** Input resistance depends in part on the surface area of a cell's membrane, A, because the membrane of a larger cell typically contains more ion channels than does the membrane of a smaller cell. The input resistance also depends on the density of channels per unit area. We can compare the membrane properties of different cells by removing the effect of size from the comparison, as described in Spotlight 5-2.

Membrane Capacitance

The maximum rate at which ions can cross a pure lipid bilayer is less than 10^{-8} times their average rate of diffusion through water. As a result, ion fluxes through the lipid bilayer of a plasma membrane are negligible. Ions,

COMPARING MEMBRANE PROPERTIES OF DIFFERENT CELLS

The surface area of a cell affects its input resistance and its capacitance, so in order to compare intrinsic membrane properties of different cells, we need to remove the effect of size from the comparison. To do this, we define the specific resistivity, R_m, of the membrane as

$$R_m = R \times A$$

where A is the membrane area, R is the input resistance, and R_m is the resistance of a unit area of membrane. Rearranging Ohm's law gives

$$R = \Delta V_m/\Delta I$$

Substituting yields

$$R_m = (\Delta V_m/\Delta I) \times A$$

where $\Delta V_m/\Delta I$ is in ohms and A is in square centimeters; thus R_m is in units of ohms·cm^2. Note that membrane area and input resistance, R, are reciprocally related. The specific resistivity, R_m, of the membrane is a property of the population of ion channels carrying ionic current across the membrane. Specific resistivities of various plasma membranes range from hundreds to tens of thousands of ohms·cm^2.

The reciprocal of resistance, R, is conductance, g (in units of siemens, S):

$$g = 1/R$$

Substituting this expression into Ohm's law (Equation 5-1) gives

$$\Delta V_m = \Delta I/g_{input}$$

The reciprocal of the specific resistivity of a membrane is the specific conductance, g_m (in units of siemens·cm^{-2}). The electrical conductance of a membrane is closely related to its ionic permeability because the current across the membrane is in the form of ions moving through channels. However, there is a difference between the electrical conductance of a membrane and its permeability to ions. The conductance of a membrane for a given species of ion is defined by Ohm's law as the current carried by that species of ion divided by the electrical force acting on that species:

$$g_X = I_X/\text{emf}_X$$

where g_X is the membrane conductance for ionic species X, I_X is the current carried by that species, and emf$_X$ is the electromotive force (in volts) acting on that species. (The emf$_X$ varies with membrane potential but is not the same thing as membrane potential, as will be discussed later.) If there is no current through the membrane—that is, if $I_X = 0$—the value of g_X is, by definition, zero. I_X might be zero because no channels are open that would permit ion X to cross the membrane. However, I_X would also be zero if no ions of species X were present. Thus, the value of g_X depends only in part on whether the membrane is permeable to X, making conductance and permeability not strictly synonymous.

however, can interact across the membrane to produce a **capacitative current** (also called a displacement current). How is such a current generated? When a voltage is applied across a membrane, ions pile up on its two surfaces. Excess cations move onto the positive side of the membrane (i.e., the side connected to the cathode of the current source) and interact electrostatically with ions on the other side of the membrane. Cations on the other side are repelled, and an excess of anions accumulates on that side. The movement of ions up to one side of the membrane and away from the other side is an ionic current. The membrane thus stores ions in the same way that charges are stored by a capacitor in an electric circuit (Figure 5-11). The interactions between oppositely charged ions that have accumulated on the two sides of the membrane are quite strong because the membrane is so thin.

The amount of charge that can be separated by a layer of insulating material depends on its thickness and on its **dielectric constant**, a property that reflects the inherent ability of a particular insulating material to store charge. It is possible to calculate an expected value for the capacitance of neuronal plasma membranes if the membrane thickness and the dielectric constant of the membrane lipids are known. Based on a lipid-layer thickness of about 5 nm and a dielectric constant of 3, which is about that of an 18-carbon fatty acid, the membrane capacitance has been calculated to be about 1 microfarad ($1\ \mu F = 10^{-6}$ F) per square centimeter. Indeed, measured values of the capacitance of biological membranes have generally proved to be close to $1\ \mu F \cdot cm^{-2}$.

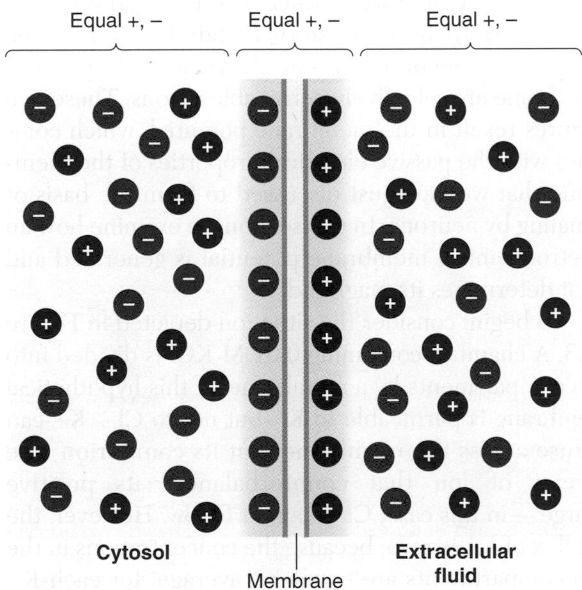

Figure 5-11 A plasma membrane acts as a capacitor. It can maintain a separation of charges, with cations in a thin layer on one side of the membrane and an equal number of anions forming a thin layer on the opposite side. Electrostatic interaction between these charges holds them in a narrow region immediately adjacent to the two surfaces of the membrane. The bulk solutions on either side of the membrane remain neutral (that is, they conform to the physical principle of electroneutrality). The excess charges on the two sides of the membrane are located very close to its two surfaces, and the total charge near the membrane (the region shaded in deeper blue) is also zero because the opposite charges gathered near the membrane cancel each other.

If the surface area of a neuron increased, would it affect the cell's input resistance? If so, how? Would it affect the cell's capacitance? If so, how?

The capacitance of a membrane affects how it responds to a change in applied voltage or current. Remember that as positive ions pile up on one side of the membrane and repel positive ions from the other side, there is a transient flow of capacitative current. Since the movement of charged particles near the membrane takes time, capacitance effectively limits how fast the voltage across a membrane can change under specified conditions. This effect can be illustrated by an equivalent circuit representing the neuronal plasma membrane, as shown in Figure 5-12a on the next page. In this example, a 1-ampere current, I_m, is applied suddenly to the membrane. According to Ohm's law (see Equation 5-1), such an applied current will produce a change in voltage across the membrane, as shown in Figure 5-12b (trace V_m). This passive potential change is produced when a current flows across the plasma membrane.

As shown in Figure 5-12a, the current passing across the membrane must be distributed between the resistive and the capacitative pathways, which are arranged in parallel across the membrane. When this sudden pulse of current, I_m (called a *square pulse* because a record of the current has square corners), is forced across the membrane, the relative proportion of the current that passes through the membrane capacitance and through the resistance changes with time. At first, charges move relatively easily onto the capacitor component of the circuit, so most of the current is carried by the capacitance as it charges, and little current flows through the membrane resistance. This buildup of charge on the capacitor is a form of electric current, although no charges physically move across the capacitor. As time passes and the capacitance charges, a voltage appears across the capacitance, which opposes the accumulation of more charge on the capacitor and slows the rate at which further charging occurs. This reduction in the rate at which charges can move onto the capacitor forces an increasingly large fraction of the total current to flow through the membrane resistance. Hence, the capacitative current, I_c, falls with an exponential time course, while the resistive current, I_r, which passes through the membrane conductance (i.e., through ion channels), increases exponentially, because the sum of I_r and I_c necessarily equals the total current applied. V_m rises with the same time course as I_r (see Figure 5-12b).

(a)

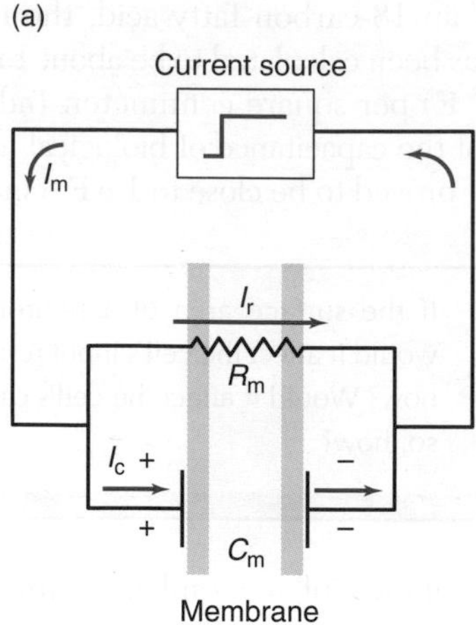

(b)

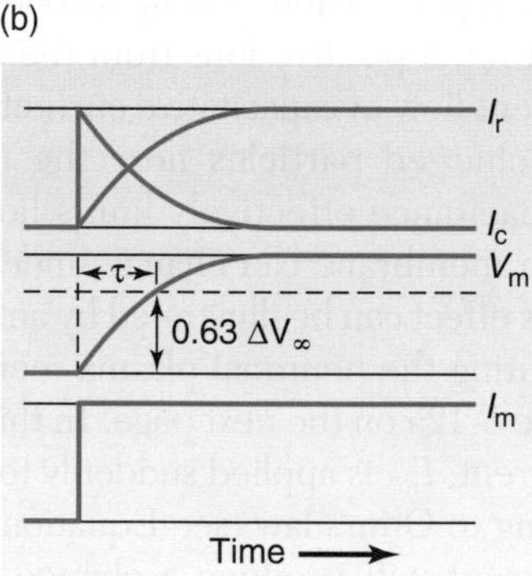

Figure 5-12 The capacitance and resistance of the plasma membrane shape the passive change in potential in response to applied current. **(a)** Equivalent circuit for a plasma membrane showing the current flow (red arrows) when an abrupt, sustained pulse of constant current is passed across the membrane. **(b)** Time courses for the currents and voltage produced by the sustained stimulus shown in part a. The top record shows how the total current is divided between the resistive current, I_r, and the capacitative current, I_c. The middle record shows the membrane potential, V_m (i.e., the potential across both the membrane resistance and the membrane capacitance). The bottom record shows the total current across the membrane, $I_m = I_c + I_r$. The time required for I_r and V_m to reach 63% of their asymptotic values is proportional to the product of the resistance and the capacitance of the membrane. This product is the time constant, τ, of the membrane.

The relationship between potential and time during the charging of the capacitance is given by the equation

$$V_t = V_\infty(1 - e^{-t/RC}) \tag{5-2}$$

where V_∞ is the potential across the capacitor at time $t = \infty$ when a constant current is applied to the network, t is the time in seconds after the onset of the current pulse, R is the resistance of the circuit in ohms, C is the capacitance of the circuit in farads, and V_t is the potential across the capacitor at any time, t. Ohms multiplied by coulombs yields units of seconds. When t is equal to the product RC, then $V_t = V_\infty(1 - 1/e) = 0.63\ V_\infty$. The value of t that equals RC is termed the **time constant** (τ) of the process. It is the time required for the voltage across a charging capacitor to reach 63% of its asymptotic value, V_∞ (see Figure 5-12b, trace V_m). Note that τ is independent of both V_∞ and current strength; it depends only on the resistance and capacitance arranged in parallel in this circuit.

In summary, the passive electrical properties of the plasma membrane are its equivalent resistance and capacitance, which are connected in parallel. Together they give the membrane a time-dependent response to changes in voltage; the time course of the response depends on the sizes of the resistance and the capacitance. To understand how these parameters influence changes in membrane potential, we first need to examine the origin of potential differences across the membrane.

ELECTROCHEMICAL POTENTIALS

All electrical phenomena in neurons and other cells depend fundamentally on the membrane potential, V_m. This voltage difference across the plasma membrane is an **electrochemical potential** that depends on two features found in all eukaryotic cells. First, the concentrations of several ionic species inside the cell are different from the concentrations of those same ions in the fluids outside the cell; these concentration gradients are maintained at the expense of metabolic energy (see Chapter 4). Second, the ion channels that span the membrane are selectively permeable to ions. These two features result in the membrane potential, which combines with the passive electrical properties of the membrane that we have just discussed to form the basis of signaling by neurons. In this section we examine how an electrochemical membrane potential is generated and what determines its magnitude.

To begin, consider the situation depicted in Figure 5-13. A chamber containing 0.01 M KCl is divided into two compartments by a membrane. If this hypothetical membrane is permeable to K^+ but not to Cl^-, K^+ can diffuse across the membrane, but its counterion (the species of ion that counterbalances its positive charge—in this case, Cl^-) cannot follow. However, the net flux of K^+ is zero, because the concentrations in the two compartments are equal; on average, for each K^+ ion that passes in one direction across the membrane, another passes in the opposite direction, there is no net flux of K^+ ions, and the membrane potential remains zero (Figure 5-13a).

If we now increase the concentration of KCl in compartment I tenfold to 0.1 M (Figure 5-13b), more K^+ will diffuse from compartment I to compartment II than in the opposite direction, producing an increase in positive charge in compartment II. A positive potential will thus quickly develop in compartment II, and the voltmeter will indicate a potential difference between

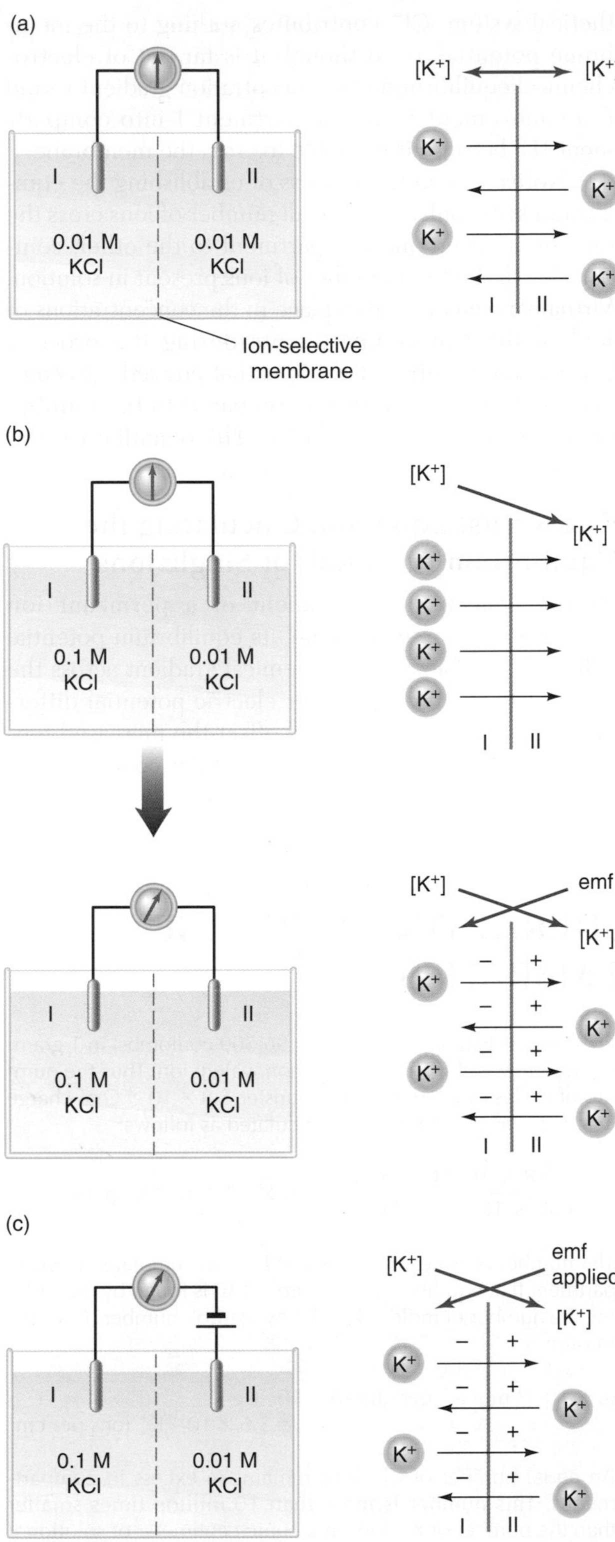

Figure 5-13 At electrochemical equilibrium, an electric potential difference exactly balances the chemical concentration gradient. **(a)** A membrane that is permeable only to K^+ separates compartments I and II, each of which contains 0.01 M KCl. Although K^+ moves in both directions across the membrane, there is no net flux. The red arrows in the schematic illustrations on the right side indicate the orientation of the driving forces acting on the K^+ ions. Black arrows indicate actual flow of ions. **(b)** When the KCl concentration in compartment I is increased to 0.1 M, there is a small net movement of K^+ into compartment II (see Spotlight 5-3). The electromotive force (emf) produced by the resulting charge separation impedes further movement of K^+. When the emf just balances the concentration gradient, the net movement of K^+ again becomes zero. **(c)** An emf generated by an outside source, such as a battery, can just counterbalance the concentration gradient, so that despite the concentration difference between compartments I and II, there is no net flux across the membrane. Because energy is expended by the battery to counterbalance the concentration-driven flux, this system is said to be in a steady state, rather than at equilibrium.

the two compartments, an equilibrium state that will be maintained indefinitely, provided there is no leakage of Cl^- across the membrane. Note, too, that the concentrations in the two compartments are now unequal. In this new equilibrium, two different forces act on K^+: the difference in concentration favors movement of K^+ out of compartment I, but the electric potential difference across the membrane pushes K^+ back into compartment I. Let's look at this situation in more detail.

After we increase the KCl concentration in compartment I, for every K^+ that is available for diffusion through K^+ channels from compartment II to compartment I, 10 K^+ ions are available in compartment I to pass across the membrane into compartment II. The difference in K^+ concentrations—that is, the chemical potential difference—causes an initial net diffusion across the membrane from I to II. However, each K^+ that diffuses from compartment I to compartment II adds a net positive charge to compartment II, because Cl^- cannot accompany K^+ across the membrane. As K^+ ions accumulate in compartment II, the potential difference across the membrane quickly rises, because the membrane then separates an excess of positive charges on the compartment II side from an excess of negative charges on the compartment I side (see Figure 5-11). As K^+ enters compartment II and produces a buildup of positive charge in that compartment, further movement of K^+ becomes less likely because the extra positive charges in compartment II repel additional positively charged ions, while the electrostatic attraction of the excess negative charges in compartment I holds K^+ back. Every K^+ entering the membrane via a K^+ channel therefore has two forces acting on it: (1) a chemical gradient favoring a net flux of K^+ from I to II, and (2) an electric potential difference favoring a net flux of K^+ from II to I. After some time, the opposing forces come into equilibrium and remain balanced: the electromotive force (emf) due to the electric potential difference across the membrane precisely offsets the tendency for K^+ to diffuse down its concentration gradient (bottom tank in Figure 5-13b). At this point, K^+ is in electrochemical equilibrium. A potential difference across the membrane that is established in this way is called the

equilibrium potential for the ion in question—in this case, the potassium equilibrium potential, E_K.

Alternatively, an outside battery could be used to establish a voltage difference between the two compartments (Figure 5-13c). If the voltage difference produced across the membrane by the battery is equal to the equilibrium potential for the ion, it can balance the tendency for ions to diffuse across the membrane, just as the diffusion-based electric potential does. Notice, however, that in this situation, the battery continuously uses energy to generate an emf to counteract the diffusion of K^+. Because maintenance of this condition requires expenditure of energy, it is not an equilibrium state. On the other hand, as long as the battery continues to operate, the system will remain invariant. Physiologists call this condition a **steady state.** As we will see, the membrane resting potential is a steady-state potential.

Once an ion is in electrochemical equilibrium, there will be no further net flux of the ion across the membrane, even if the membrane is freely permeable to that ion. Conversely, if an ion present in the system cannot diffuse across the membrane, its presence does not influence the equilibrium state. Thus, in our hypothetical system, Cl^- contributes nothing to the membrane potential, even though it is far out of electrochemical equilibrium (its concentration gradient would favor movement from compartment I into compartment II), because it is unable to cross the membrane.

Notice that in the process of establishing the equilibrium state, only a very small number of ions cross the membrane from one compartment to the other, compared with the total number of ions present in solution. Virtually no change takes place in the concentrations of KCl in the two compartments during the process, because the number of K^+ ions that crossed into compartment II is insignificant compared to the number originally present in the solution. This quantitative relation is discussed further in Spotlight 5-3.

The Nernst Equation: Calculating the Equilibrium Potential for Single Ions

As the concentration gradient of a permeant ion increases across a membrane, its equilibrium potential will also increase: a larger chemical gradient across the membrane requires a greater electric potential difference across the membrane to offset the increased tendency for the ions to diffuse down their concentration

A QUANTITATIVE CONSIDERATION OF CHARGE SEPARATION ACROSS MEMBRANES

It takes only a small number of ions diffusing across 1 cm^2 of a membrane, as in Figure 5-13b, to bring the membrane potential to E_K. In fact, the actual number of excess ions that cross the membrane can be calculated for a system that contains only one permeant ion. Using the system shown in Figure 5-13, the number of excess K^+ ions that accumulate in compartment II (and hence the number of excess Cl^- ions left behind in compartment I) depends on two factors: E_K and the capacitance, C, of the membrane. The charge, Q, that accumulates across a capacitor equals the capacitance times the voltage, V

$$Q = C \times V$$

where Q is in coulombs (C), C is in farads (F), and V is in volts (V).

Biological membranes typically have a capacitance of about 1 $\mu F \cdot cm^{-2}$. According to the Nernst equation (Equation 5-3), the voltage at equilibrium, E_K, when the membrane separates a tenfold difference in the concentrations of a monovalent cation such as K^+, is 58 mV. By substituting these values into the above equation, we can calculate the coulombs of charge that diffuse across 1 cm^2 of a biological membrane when the membrane separates a tenfold difference in the concentrations of K^+:

$$Q = (10^{-6}\ F \cdot cm^{-2})(5.8 \times 10^{-2}\ V)$$

$$5.8 \times 10^{-8}\ C \cdot cm^{-2}$$

There is one Faraday of charge (= 96,500 coulombs) in 1 gram-equivalent weight (1 mole) of a monovalent ion. Thus the number of moles of K^+ required to transfer 5.8×10^{-8} C of charge across 1 cm^2 of membrane is calculated as follows:

$$\frac{5.8 \times 10^{-8}\ C \cdot cm^{-2}}{9.65 \times 10^4\ C \cdot (mol\ K^+)^{-1}} = 6 \times 10^{-13}\ mol\ K^+ \text{ per } cm^2$$

The number of excess K^+ ions that have accumulated in compartment II at equilibrium in Figure 5-13b is found by multiplying the number of moles of K^+ by Avogadro's number (6×10^{23} ions/mol):

$$(6 \times 10^{-13}\ mol\ K^+ \text{ per } cm^2)(6 \times 10^{23}) = 3.6 \times 10^{11}\ K^+ \text{ ions per } cm^2$$

An equal number of Cl^- ions remains in excess in compartment I. This number is more than 10 million times smaller than the number of K^+ ions in a cubic centimeter of solution II (6×10^{18} K^+ ions per cm^3, calculated from Avogadro's number). So the concentrations in compartments I and II are essentially unchanged as a result of the charge separation across the membrane. Even though there is a slight separation of anions from cations across the membrane, the segregation exists only on a microscopic scale, separated by the thickness of the membrane. Electroneutrality (i.e., an equal number of + and − charges) is maintained on the macroscopic scale.

gradient. In fact, the equilibrium potential is proportional to the logarithm of the ratio of the concentrations in the two compartments (for a brief review of logarithms, see Appendix 2). The relation between the chemical gradient and the electric potential difference across a membrane at equilibrium was derived by Walther Nernst from the ideal gas laws in the nineteenth century. The **Nernst equation** states that the equilibrium potential depends on the absolute temperature, the charge on the permeant ion, and the ratio of the ionic concentrations on the two sides of the membrane:

$$E_X = \frac{RT}{zF} \ln \frac{[X]_I}{[X]_{II}}$$

where R is the gas constant, T is the absolute temperature in kelvins (formerly degrees Kelvin); F is the Faraday constant (96,500 coulombs/gram-equivalent charge); z is the charge on each ion X; $[X]_I$ and $[X]_{II}$ are the concentrations (or more accurately, the chemical activities) of ion X on sides I and II of the membrane; and E_X is the equilibrium potential for ion X (i.e., the potential on side II minus the potential on side I). At a temperature of 18°C, standard room temperature and hence the body temperature of poikilotherms, and converting from natural logarithms, ln, to common logarithms, log, the Nernst equation reduces to

$$E_X = \frac{0.058}{z} \log \frac{[X]_I}{[X]_{II}} \qquad (5\text{-}3)$$

where E_X is expressed in volts. (At 38°C, which is the approximate body temperature of many mammals, the multiplication factor is $0.061/z$.) Note that E_X will be positive if X is a cation and the ratio of $[X]_I$ to $[X]_{II}$ is greater than unity. The sign will become negative if the ratio is less than 1. Likewise, the sign will be reversed if X is an anion, rather than a cation, because z will then be negative.

In summary, diffusion of a single ionic species down its concentration gradient, unaccompanied by its counterion, can set up an electric potential difference between the two sides of a semipermeable membrane. At equilibrium, the emf developed across the membrane will be equal to the equilibrium potential as determined by the Nernst equation. Similarly, if an emf equal to the equilibrium potential is established between the two compartments by an outside current source (see Figure 5-13c), the emf can precisely compensate for the concentration gradient, causing the net flux of ions between the two compartments to be zero.

By convention, the membrane potential, V_m, is given as $V_{in} - V_{out}$, with the potential of the cell exterior arbitrarily defined as zero. For this reason, when determining the equilibrium potential across the plasma membrane, we place the extracellular concentration of the ion in the numerator and the intracellular concentration in the denominator of the concentration ratio. Applying the Nernst equation (Equation 5-3), we can calculate the potassium equilibrium potential, E_K, in a hypothetical cell in which $[K^+]_{out} = 0.01$ M and $[K^+]_{in} = 0.1$ M:

$$\begin{aligned} E_K &= \frac{0.058}{z} \log \frac{[K^+]_{out}}{[K^+]_{in}} \\ &= \frac{0.058}{1} \log \frac{0.01}{0.1} \\ &= 0.058 \times (-1) \\ &= -0.058 \text{ V or } -58 \text{ mV} \end{aligned}$$

Note that the calculated E_K has a negative sign. The inside of the cell will become negative when a minute amount of K^+ leaks out of the cell, driven by the concentration gradient of K^+. The Nernst equation predicts a rise in the equilibrium potential of 58 mV when the concentration ratio of the permeant ion is increased by a factor of 10. When E_K is plotted as a function of log $[K^+]_{out}/[K^+]_{in}$, the relation has a slope of 58 mV per tenfold increase in the concentration ratio (Figure 5-14a on the next page). Equation 5-3 also implies that if the ion in question is a divalent cation such as Ca^{2+} (i.e., $z = +2$), the slope of the relation will be 29 mV per tenfold increase in the concentration ratio.

The Goldman Equation: Calculating the Steady-State Potential for Multiple Ions

If there is a concentration gradient across the membrane for a particular ion, the Nernst equation gives the equilibrium potential *for that ion.* However, the Nernst equation is valid only for one ionic species at a time. In reality, plasma membranes are permeable to several ionic species, all of which may be distributed asymmetrically across the membrane. All of the different permeant ions contribute to setting the potential difference across the membrane. The Nernst equation cannot be applied to this more complicated situation.

In addition, the concentrations of at least some ionic species are very different inside and outside of cells; these differences are maintained by active transport of ions at the expense of metabolic energy (see Chapter 4). These concentration gradients are quite stable over time, but because energy is required to maintain them, they are said to be in a steady state, and the Nernst equation, which applies only to equilibrium conditions, is not strictly applicable. In biological systems, in which exchanges of matter and energy occur continuously across the boundaries of the system, the stability of the system is referred to as a **dynamic steady state.** It is possible to predict the steady-state membrane potential of a cell in this more complex situation using an equation that is formally similar to the Nernst equation.

In 1943, D. E. Goldman derived a quantitative representation of the membrane potential when more than

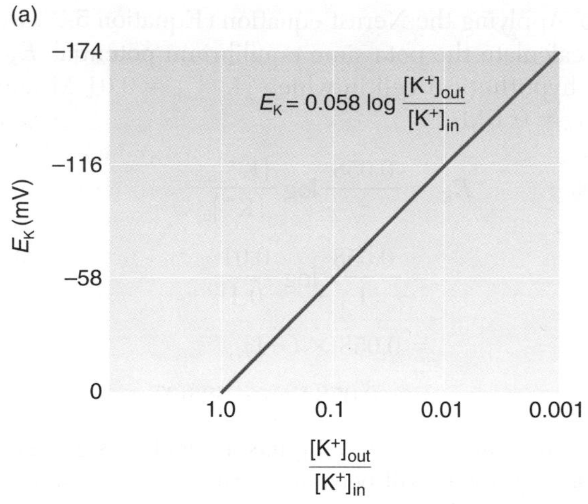

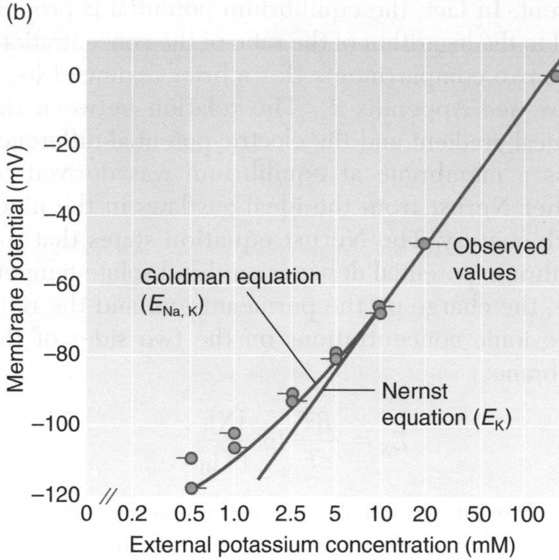

Figure 5-14 The concentration ratio of K^+ across the plasma membrane affects both the calculated membrane potential and V_{rest} measured experimentally. **(a)** Semilog plot of the relationship between the equilibrium potential for K^+, E_K, and the ratio of K^+ concentrations on the two sides of a membrane, $[K^+]_{out}/[K^+]_{in}$, calculated from the Nernst equation. At 18°C, the slope of the line is 58 mV for every tenfold increase in the concentration ratio.

Classic Work **(b)** Semilog plot showing how the measured V_{rest} of frog muscle changes with the external K^+ concentration. The measured values are presented as blue circles. These values are compared with two calculated values (solid blue lines). The curved line shows $E_{Na,K}$ calculated using the Goldman equation, assuming P_{Na} is 1% of P_K and $[K^+]_{in}$ is 140 mM, and ignoring any contribution made by Cl^-. The straight line represents the predicted change in E_K, calculated using the Nernst equation, as $[K^+]_{out}$ changes. Note that both the relationship predicted by the Goldman equation and the measured V_{rest} values deviate from the straight line at low $[K^+]_{out}$. At higher $[K^+]_{out}$, the calculated E_K and $E_{Na,K}$ are nearly identical, and both predict V_{rest} very well. [Part b adapted from Hodgkin and Horowicz, 1959.]

one ionic species can cross the membrane. The **Goldman equation** (sometimes called the Goldman-Hodgkin-Katz equation) is an approximate generalization of the Nernst equation, extended to include the relative permeability of each species of ion:

$$E_{Na,K,Cl} = \frac{RT}{F} \log \frac{P_K[K^+]_{out} + P_{Na}[Na^+]_{out} + P_{Cl}[Cl^-]_{in}}{P_K[K^+]_{in} + P_{Na}[Na^+]_{in} + P_{Cl}[Cl^-]_{out}} \tag{5-4}$$

P_K, P_{Na}, and P_{Cl} are the permeability constants for the major ionic species in the intracellular and extracellular compartments. That is, these values indicate the relative permeabilities of the membrane to each of these three ions. $[K^+]_{out}$ and $[K^+]_{in}$ indicate the concentrations outside and inside the cell, respectively. Notice that although we designate the calculated potential as E, it is really a steady-state potential, rather than a true equilibrium potential. Notice also that only monovalent ions are included in the equation, so the multiplier z has an absolute value of 1 and has not been included explicitly in the expression. The concentration for Cl^- inside the cell, $[Cl^-]_{in}$, is in the numerator, rather than in the denominator as it is for $[Na^+]_{in}$ and $[K^+]_{in}$, because the charge on Cl^- is negative, so for Cl^- the value of z is -1. Recall that $\log a/b = -\log b/a$; thus the chloride concentrations inside and outside the cell must be "upside down" in the Goldman equation.

In this equation, the probability that one ionic species will cross the membrane is taken to be proportional to the product of its concentration (more accurately, its thermodynamic activity) on that side and the permeability of the membrane to that ionic species. This assumption suggests a useful simplification. In frog muscle cells, for example, the permeability constant for sodium has been measured to be about 1/100 that of potassium, and the membrane is nearly impermeable to chloride. As a result, the Goldman equation can be simplified to:

$$E_{Na,K} = \frac{RT}{F} \log \frac{1[K^+]_{out} + 0.01[Na^+]_{out}}{1[K^+]_{in} + 0.01[Na^+]_{in}}$$

If we then substitute into this equation the concentrations of K^+ and Na^+ inside and outside frog muscle cells (expressed in millimoles) at 18°C, we get

$$E_{Na,K} = 0.058 \log \frac{2.5 + (0.01 \times 120)}{140 + (0.01 \times 10)}$$

$$= -0.092 \text{ V}$$

$$= -92 \text{ mV}$$

If the membrane potential of frog muscle cells depends most heavily on the diffusion of Na^+ and K^+, then the Goldman equation predicts, based on measurements of concentration and permeability, that the value of V_m should be near -92 mV. Indeed, measured values for

V_m in frog muscles do lie close to this value, but it is possible to test further the accuracy of the hypothesis that V_m depends on the diffusion of specific ionic species by changing external ion concentrations and observing the effects on V_m, as shown in Figure 5-14b.

THE RESTING POTENTIAL

Every cell that is in a nonexcited or "resting" state has a potential difference, V_{rest}, across its membrane. Typically, V_{rest} lies between -20 and -100 mV, depending on the kind of cell and on the ionic environment. Two factors govern this resting potential: first, the presence of open ion channels in the membrane that are permeable to some—but not all—of the ionic species present, and second, the unequal distribution of inorganic ions between the cell interior and cell exterior, maintained by active transport across the membrane and by a Donnan equilibrium (see Chapter 4). The unequal distribution of ions drives diffusion of ions and thus allows a membrane potential to be established.

The Role of Ion Gradients and Channels

As predicted by the Goldman equation, ions influence the potential across a membrane roughly in proportion to the permeability of the membrane to each of the ionic species present. Notice that any ions that cannot cross the membrane—for example, the many large organic anions in the cytosol—have no effect on the membrane potential because they cannot carry charge from one side of the membrane to the other. Conversely, if a membrane is permeable to only one species of ion, the distribution of that ion will dictate the membrane potential, which can be predicted by using the Nernst equation for that ionic species. If a membrane is permeable to more than one ion—as most biological membranes are—and the permeabilities are known, it is possible to predict the value of V_{rest} using the Goldman equation. In the previous section, we calculated a predicted $E_{Na,K}$ of -92 mV using measured values for the permeabilities and ion concentrations in frog skeletal muscle fibers. Measurements of V_{rest} in frog skeletal muscle fibers range from -90 to -100 mV, supporting the hypothesis that the resting potential depends in large part on the diffusion of Na^+ and K^+ ions.

The contribution that a particular ionic species makes to the membrane potential diminishes as its concentration gradient is reduced. This point is illustrated in Figure 5-14b, in which values of the membrane potential calculated from the Goldman equation (assuming $P_{Na} = 0.01\ P_K$) are plotted against the external K^+ concentration. At high external K^+ concentrations, the slope of the plot is about 58 mV per tenfold increase in $[K^+]_{out}$, a value that is accurately predicted by the Nernst equation for K^+. At low external K^+ concentrations, however, the curve deviates from this slope because the product $P_{Na}[Na^+]_{out}$ approaches the value of the product $P_K[K^+]_{out}$, allowing Na^+ to make a more important contribution to the potential in spite of the low permeability of the membrane to Na^+. The measured values for V_{rest} in living frog muscle cells closely parallel the values calculated from the Goldman equation. It is interesting to note that these relationships apply equally well to neurons and muscle cells, suggesting conservation of important functional mechanisms among excitable cells during the course of evolution.

Resting potentials of muscle, nerve, and most other cells have been found to be far more sensitive to changes in $[K^+]_{out}$ than they are to changes in the concentrations of other cations. This experimental result is consistent with the relatively high permeability of plasma membranes to K^+ as compared with other cations. This high permeability is thought to depend on the K^+-selective leak channels mentioned earlier, which remain open in the resting membrane. Large changes in $[Na^+]_{out}$ have little effect on the resting potential because the resting membrane is relatively impermeable to Na^+.

Why don't ions to which the cell membrane is relatively impermeable affect the value of the resting potential?

The Role of Active Transport

The second factor contributing to V_{rest} is the asymmetric distribution of inorganic ions across plasma membranes, which depends on the active transport of particular ions across the membrane. Because biological membranes are more or less leaky to solutes, cells must spend energy to maintain this asymmetry. Key ions are actively transported against their concentration gradients.

Consider Na^+ in frog muscle. The concentrations of extracellular and intracellular Na^+ are about 120 mM and 10 mM, respectively. From these values we can calculate the sodium equilibrium potential, E_{Na}, using the Nernst equation (Equation 5-3) as follows:

$$\begin{aligned} E_{Na} &= \frac{0.058}{1} \log \frac{120}{10} \\ &= 0.063 \text{ V} \\ &= 63 \text{ mV} \end{aligned}$$

The force acting on an ion depends on how different the membrane potential is from the equilibrium potential for that ion:

$$\text{emf}_X = V_m - E_X \qquad (5\text{-}5)$$

Because V_{rest} in frog muscle ranges from -90 to -100 mV, the Na^+ ions are more than -150 mV out of equilibrium ($V_m - E_{Na} = -100$ mV $- 63$ mV, or -163 mV). In other words, there are strong electrical and chemical forces driving Na^+ into the cell. Even with only a small permeability to Na^+, there will be a steady influx of Na^+, driven by the large electric driving force acting on that ion. If Na^+ were not removed from the cell's interior at the same rate at which it leaks in, it would gradually accumulate in the cell. Such a rise in the intracellular Na^+ concentration would depolarize the cell; the resulting reduction in internal negativity would make it less able to hold K^+ inside, and internal K^+ would leak out, moving down its concentration gradient. In fact, the high intracellular concentration of K^+ and low intracellular concentration of Na^+ are maintained by the action of a specific membrane protein, the Na^+/K^+ pump (Figure 5-15; see also Chapter 4). This protein is an ATPase that transports Na^+ out of the cell and K^+ in, using energy supplied by the hydrolysis of ATP. This active transport is not stoichiometrically balanced: for each molecule of ATP hydrolyzed, three Na^+ ions are transported out and two K^+ ions transported in.

The unequal stoichiometry of the Na^+/K^+ pump can have important consequences for V_{rest}. Because the pump produces a net transport of charge across the membrane, it is considered electrogenic and can contribute to the membrane potential. The net effect of the pump should be to cause V_{rest} to be more negative than the equilibrium potential calculated, using the Goldman equation, for the highly permeant K^+ ions and less permeant Cl^- ions. It has been found, however, that the action of the Na^+/K^+ pump rarely contributes more than a few millivolts to the value of V_{rest} because some positive charge leaks back into the cell (or negative charge leaks out of the cell), partially offsetting the effect of the pump.

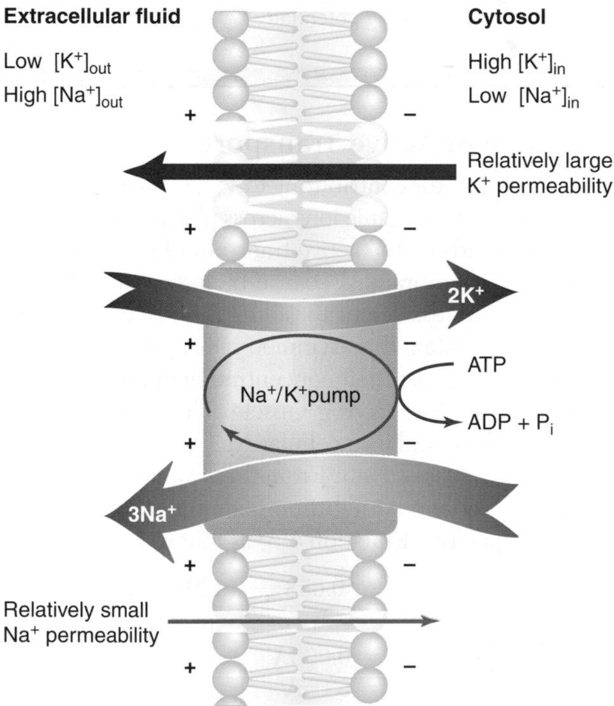

Figure 5-15 The Na^+/K^+ pump contributes to V_{rest} in two ways. The major source of V_{rest} is diffusion of K^+ through the membrane from inside the cell to outside, driven by the unequal concentrations of K^+ inside and outside the cell. The pump indirectly contributes to the resting potential by maintaining the high internal K^+ concentration. Because the Na^+/K^+ pump transports Na^+ and K^+ in a ratio of 3Na : 2K, it may also contribute directly to the resting potential by removing a small net amount of positive charge from the cell's interior.

When metabolically driven sodium transport is eliminated by an inhibitor of oxidative metabolism (e.g., cyanide or azide) or by a specific inhibitor of sodium transport (e.g., ouabain), a net influx of Na^+ occurs and internal K^+ is gradually displaced. As a consequence, the resting potential decays as the ratio of $[K^+]_{in}$ to $[K^+]_{out}$ gradually decreases. Thus, over the long term, it is the metabolically energized transport of Na^+ and K^+ that keeps the Na^+ and K^+ concentration gradients from running downhill to a final equilibrium in which $V_m = 0$. By continually maintaining the K^+ concentration gradient, the Na^+/K^+ pump plays an important *indirect* role in determining the resting potential.

To summarize the process, the major portion of the inside-negative V_{rest} across a plasma membrane arises directly from the high internal K^+ concentration relative to the extracellular K^+ concentration, combined with the membrane's relatively high P_K. As a result, a small amount of K^+ leaks out of the cell through numerous K^+-selective channels that are open at rest, and a net negative charge is left behind. Because the resting membrane has relatively few open Na^+-selective channels, Na^+ makes a very small contribution to the resting potential. In some cells, both the P_{Cl} and the electrochemical gradient for Cl^- are small, so Cl^- makes little contribution to V_{rest}. In other cells, the membrane is quite permeable to Cl^-, and the flux of Cl^- across the membrane contributes to stabilizing V_{rest}. The ultimate, although indirect, basis for the resting potential is the metabolically energized active transport of Na^+ out of the cell in exchange for K^+. By maintaining a low intracellular Na^+ concentration, the Na^+/K^+ pump allows K^+ to be the dominant intracellular cation. In addition, a small fraction of the resting potential may arise directly from the action of the pump as it moves a net amount of positive charge (in the form of Na^+) out of the cell.

ACTION POTENTIALS

Most neurons use just one type of signal, the action potential (AP), to send information along the axon of the cell. APs in the nervous system are the basis for every sensation, every memory, every thought, every impulse to act in the environment. An action potential is a large, brief change in V_m that is propagated along an axon, sometimes over a long distance, without decrement. Once initiated, the AP travels along the plasma membrane, producing the same rise and fall in V_m at every point. There are no intermediate-sized APs—that is, they are said to be all-or-none events.

The production of an AP depends on three key elements:

- Active transport of ions by specific proteins in the plasma membrane generates asymmetric concentrations of ionic species across the membrane.
- The unequal distribution of ions generates an electrochemical gradient across the membrane that provides a reservoir of potential energy.
- This electrochemical gradient drives ions across the membrane when ion-selective channels open, making the membrane permeable to certain ions, and this ionic current dramatically changes V_m.

As we shall see, two types of voltage-gated ion channels, Na^+ channels and K^+ channels, are responsible for essentially all features of the AP.

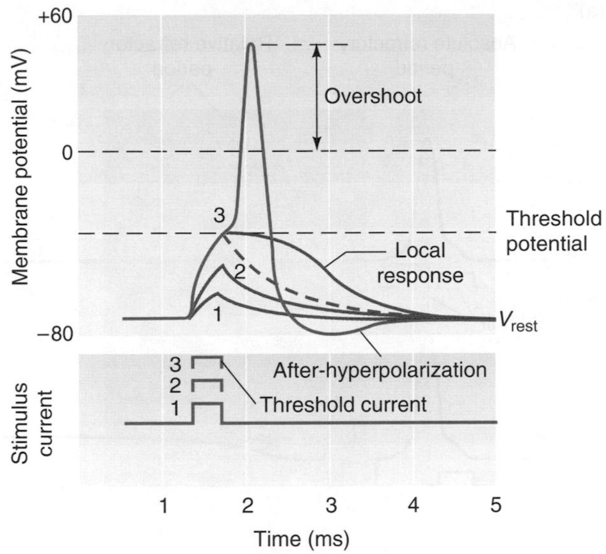

Figure 5-16 A neuron produces an action potential when V_m reaches or exceeds the threshold potential. The magnitudes of three depolarizing current pulses are shown at the bottom of the figure, and the corresponding responses in the stimulated neuron are shown above. The numbers indicate which stimulus produced which response. Only stimulus 3 depolarized the membrane sufficiently to trigger an AP. Smaller changes in V_m evoked much smaller responses in the neuron (traces 1 and 2). The dashed curve illustrates the falling phase of the change in V_m that would be recorded if the neuron produced only a passive response to stimulus 3, rather than an AP. Sometimes a stimulus that is just at the threshold value evokes an abortive, nonpropagated excitation, called a local response, rather than an all-or-none AP.

General Properties of Action Potentials

Action potentials are generated by the membranes of neurons and muscle cells, as well as by some receptor cells, secretory cells, and single-celled animals. Depolarization of an excitable membrane past a threshold value triggers a rapid and continuously increasing depolarization until the membrane potential briefly becomes inside-positive and then rapidly repolarizes to a potential near V_{rest}. In many types of cells, the repolarization process continues until the cell is transiently hyperpolarized, then V_m returns more slowly to its original resting value.

To illustrate the general features of an AP, we might pass short pulses of depolarizing current across the membrane of a neuron (Figure 5-16). These pulses produce only passive depolarization until the current delivered is strong enough to depolarize the membrane to its threshold potential, whereupon an AP is triggered. If the depolarization brings the membrane almost, but not quite, to threshold, there may be an abortive, nonpropagated excitation called a *local response,* which is the beginning of an AP that died out before it was irreversibly under way.

The **threshold current** is the intensity of stimulating current that is just sufficient to bring the membrane to the threshold potential and elicit an AP. Although most neurons have threshold potentials between −30 mV and −50 mV, no absolutely consistent value can be assigned either to the threshold current or to the threshold potential, because the threshold depends on electrical events in the immediate past, which can modify the state of the membrane. We will come back to this topic shortly when we describe refractory periods, during which the cell is less than usually responsive to depolarization.

Once the threshold potential is reached, the AP becomes *regenerative;* that is, the event becomes self-perpetuating, and V_m continues to become more inside-positive without any further stimulus. As the cell interior rapidly gains positive ions, the membrane potential actually reverses polarity, and the interior continues to become increasingly positive, typically until it reaches a peak of +10 mV to +50 mV. The very brief period when V_m is inside-positive is called the **overshoot** (see Figure 5-16). In mammalian neurons, neuronal APs typically last only a millisecond or so, although in many invertebrate species, APs can last 10—or even 100—milliseconds. In other types of excitable cells in vertebrate animals (e.g., heart muscle cells), each AP can last as long as half a second (see Figure 12-7).

After the peak, V_m plummets to its original inside-negative state and even beyond. The transient period when V_m is more negative than V_{rest} is called the **after-hyperpolarization** or **undershoot.** For a short time after an AP has occurred, it is difficult or impossible to trigger another AP. During this period, the cell is said to be refractory. If a second stimulus is delivered during an AP or immediately afterward, no AP is triggered, and the cell is said to be in the **absolute refractory period.** If stimulation is delivered slightly later, an AP may be triggered as long as the stimulus is more intense than usual, but the amplitude of the AP may be smaller than

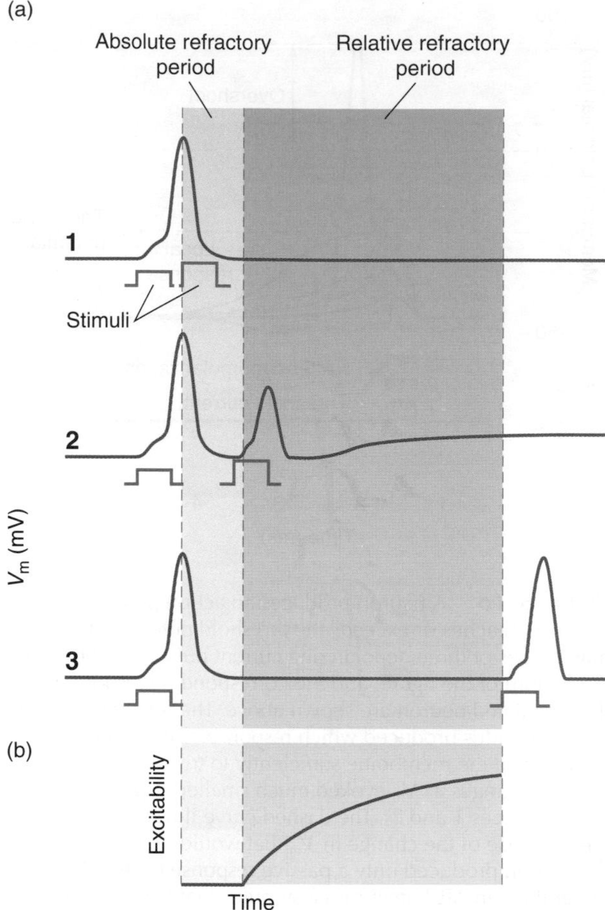

Figure 5-17 During and after one AP, the neuron is refractory to producing another AP. **(a)** Recorded changes in V_m in response to three pairs of stimuli delivered to a neuron. The timing and magnitude of each stimulus are shown under the V_m traces. In trace 1, the second stimulus was intense but produced no AP, indicating that it was delivered during the absolute refractory period. In trace 2, a stronger-than-normal second stimulus was required to trigger an AP (i.e., the threshold was still elevated at this time), and the second AP was smaller, indicating that the second stimulus was delivered during the relative refractory period. When the two stimuli were separated by a sufficiently long interval, both normal-sized stimuli produced normal-sized APs (trace 3). **(b)** Time course of the change in membrane excitability during the refractory period. During the absolute refractory period (blue), the neuron cannot be excited to produce another AP, regardless of how strong the stimulus is. During the relative refractory period (orange shaded area), excitability is reduced (i.e., the threshold is elevated), so stronger stimuli are required to reach the threshold. Over time, membrane excitability returns to normal.

usual. During this period, the cell is said to be in the **relative refractory period.** The relative refractory period is one of the very few instances when the all-or-none property of APs does not apply.

The triggering of an AP is somewhat like the flushing of a toilet. Once the flush is initiated, it continues to completion, independent of how much pressure was applied to the triggering lever. The flush is thus an all-or-none phenomenon, like an AP. On the other hand, if after one flush, the lever is pressed before the tank has filled completely, the second flush that is produced will be smaller than normal. If the lever is pressed during or immediately after one flush, a second flush may not take place at all.

In an electrically excitable cell, no stimulus, however large, is sufficient to evoke an AP during the absolute refractory period (Figure 5-17a, trace 1). Following this period, the membrane enters the relative refractory period, during which the threshold potential is elevated above normal, but a strong stimulus may evoke an AP (Figure 5-17a, trace 2). An AP initiated during this phase sometimes has a reduced amplitude (i.e., the overshoot may be smaller). Excitability progressively increases—that is, the threshold potential decreases—during the relative refractory period until the membrane returns to its normal resting state (Figure 5-17b). Refractoriness during and immediately after the AP permits the propagation of closely spaced yet discrete impulses.

If a neuron is stimulated by a series of subthreshold depolarizations, excitability progressively decreases. For example, if the membrane is depolarized gradually with a current of steadily increasing intensity, a greater depolarization is required to elicit an AP than when the stimulus has an abrupt onset (Figure 5-18). The slower the rate of increase in the intensity of the stimulating current, the greater the increase in threshold potential. This characteristic of excitable membranes, called **accommodation,** is caused by time-dependent changes in the way membrane channels respond to depolarization.

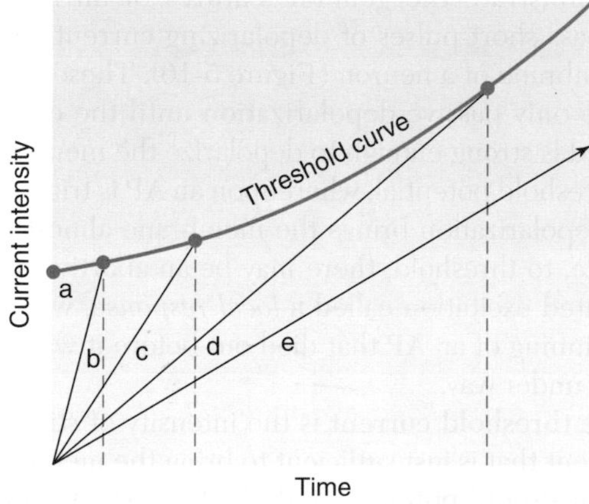

Figure 5-18 When a neuron is subjected to subthreshold stimuli of slowly increasing intensity, the threshold potential increases. This phenomenon is called accommodation. In this experiment, a ramplike stimulating current that was gradually increased in intensity was passed into a neuron, with different rates of rise on different trials (a–e). For the most rapid rate of rise, the threshold was closest to the threshold potential in response to a square pulse (a), in which the current's rate of rise is infinite. When the intensity of the stimulating current increased more slowly, the intensity necessary to reach threshold became larger (lines b through d). If the intensity of the stimulating current rose sufficiently slowly, threshold was never reached (line e).

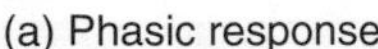

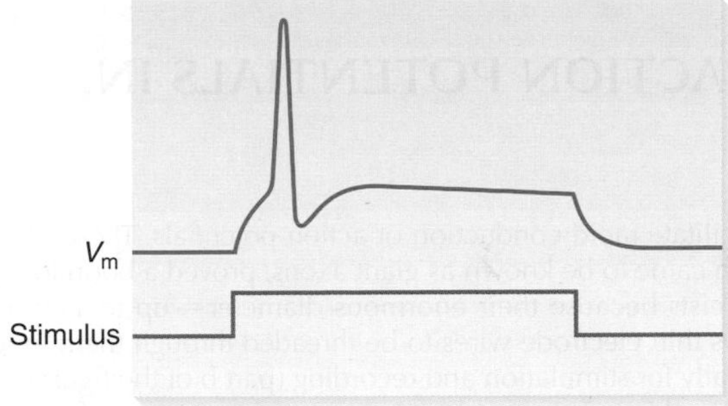

(b) Tonic response

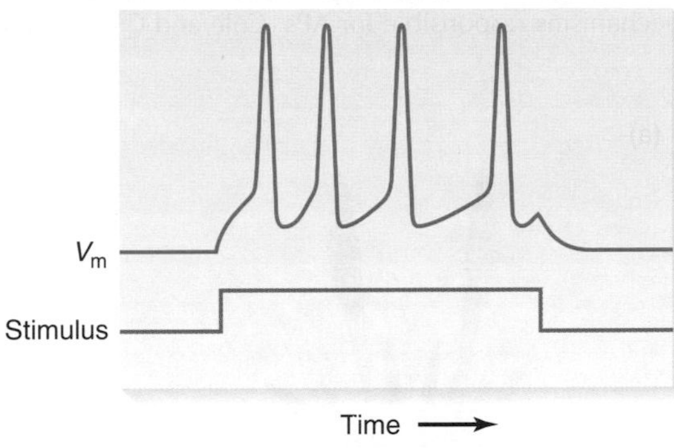

Figure 5-19 Many, but not all, neurons accommodate to sustained stimulation. This figure shows intracellular recordings of V_m while a sustained current is passed into a neuron. **(a)** Some neurons exhibit strong accommodation to a prolonged, constant stimulus, producing only one or two impulses at the beginning of the stimulus. This pattern is called a phasic response. **(b)** Other neurons accommodate very little, except for a progressive lengthening of the interspike interval; this pattern is called a tonic response.

Accommodation determines how individual neurons respond to input—whether they are continuously active or produce only bursts of APs. When they are stimulated continuously by a current of constant intensity, some neurons accommodate rapidly and generate only one or two APs at the beginning of the stimulus period (Figure 5-19a). These neurons are said to have a **phasic response.** Other neurons accommodate more slowly and fire repeatedly, although with gradually decreasing frequency, in response to a prolonged constant stimulus (Figure 5-19b). These neurons are said to have a **tonic response.** This difference among neurons plays a key role in how sensory neurons transmit information (see Chapter 7). The reduction in the frequency of APs that is typically seen in a neuron that responds tonically during a sustained stimulus, visible as the increasing distance between APs in Figure 5-19b, is termed **adaptation.**

Ionic Basis of the Action Potential

The electrical signature of an AP is a rapid depolarization of the plasma membrane. We now know that this change in V_m depends on an inward Na^+ current caused by a sudden large increase in Na^+ conductance (g_{Na}) (Figure 5-20). Although both g_{Na} and g_K (K^+ conductance) are increased by depolarization, during the rising phase of the AP g_{Na} grows much faster than g_K. As a result, early in an AP, almost all of the ionic current crossing the membrane is inward Na^+ current, which moves the membrane potential toward the equilibrium potential for Na^+, E_{Na}. As g_{Na} reaches its peak, V_m peaks, too. Notice, however, that at this time g_K is still increasing. Now the outward K^+ current greatly exceeds the inward Na^+ current, causing the membrane potential to repolarize back toward E_K. These dramatic changes in ionic conductances are caused by structural changes that first open and then close voltage-gated ion channels.

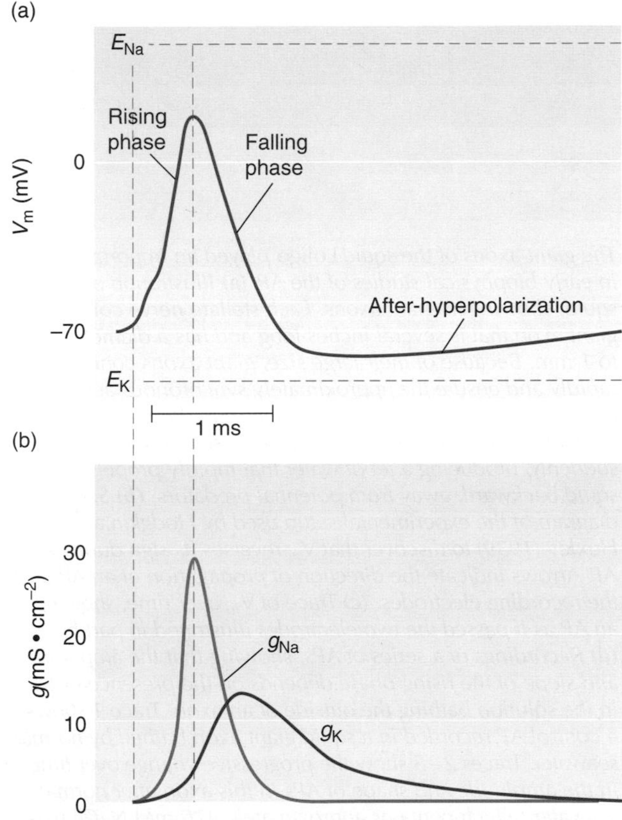

Figure 5-20 An action potential is caused by transient changes in ionic conductances across the membrane. **(a)** shows V_m, and **(b)** shows the conductances as they change over time. The dashed lines show the temporal relationships between V_m and the conductances at key points. The AP in this figure (red trace in part a, recorded from a squid giant axon) consists of three phases: a rising phase, which begins at the vertical dashed line on the left and depends on an increase in g_{Na}; a falling phase, which begins at the vertical dashed line on the right and depends on both a drop in g_{Na} and a rise in g_K; and an after-hyperpolarization, which occurs because g_K remains elevated for some time. The decrease in g_{Na} is due to inactivation of Na^+ channels, whereas the decrease in g_K is caused by repolarization. When g_{Na} is high, V_m approaches (but may not reach) E_{Na}; when g_K is high, V_m approaches E_K. Notice that initially g_{Na} increases rapidly; then, with some delay, g_K also increases, but with a lower slope.

Spotlight 5-4 PIONEERING STUDIES OF ACTION POTENTIALS IN THE SQUID GIANT AXON

Our knowledge about how changes in the state of ion channels generate APs is due in large part to the efforts of several pioneering physiologists. In 1936, the distinguished English zoologist J. Z. Young, working at the marine station in Naples, Italy, first reported that some longitudinal structures in squids and cuttlefishes were not blood vessels, as had previously been thought, but were instead very large axons (part a of the accompanying figure). The large size of these axons, which control the swift escape response of the animal, is thought to have evolved to facilitate rapid conduction of action potentials. These fibers, which came to be known as giant axons, proved a boon to biophysicists because their enormous diameter—up to 1 mm—allows thin electrode wires to be threaded through them longitudinally for stimulation and recording (part b of the figure).

Working with squid giant axons, two groups of experimenters—K. S. Cole and H. J. Curtis in Woods Hole, Massachusetts, and Alan Hodgkin and Andrew Huxley in Plymouth, England—made major discoveries in 1939 about the mechanisms responsible for APs. Cole and Curtis demon-

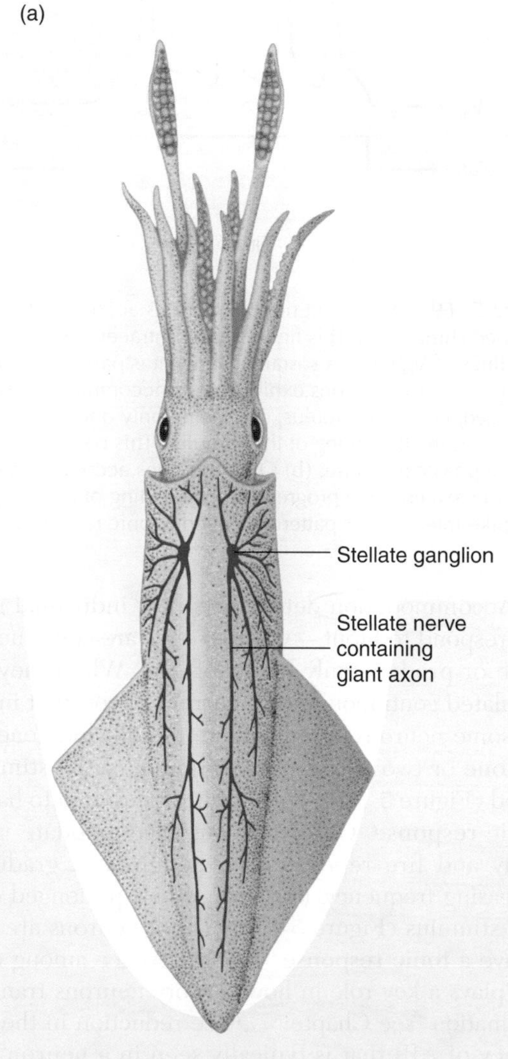

The giant axons of the squid Loligo *played an important role in early biophysical studies of the AP.* ***(a)*** *Illustration of a squid, showing its giant axons. Each stellate nerve contains a giant axon that is several inches long and has a diameter of up to 1 mm. Because of their large size, giant axons conduct APs rapidly and ensure the approximately synchronous activation of all muscles in the mantle. When the squid is startled and the giant axons are activated, mantle muscles contract suddenly, producing a jet of water that rapidly propels the squid backward, away from potential predators.* ***(b)*** *Schematic diagram of the experimental setup used by Hodgkin and Huxley (1939) to discover that V_m reverses its sign during an AP. Arrows indicate the direction of propagation of an AP past the recording electrodes.* ***(c)*** *Trace of V_m over time, showing an AP as it passed the two electrodes illustrated in part b.* ***(d)*** *Recordings of a series of APs, showing that the amplitude and slope of the rising phase depends on the presence of Na^+ in the solution bathing the outside of an axon. Trace 1 shows a control AP recorded in a squid giant axon bathed by normal seawater. Traces 2–5 show the progressive change over time in the amplitude and shape of APs in this axon after normal seawater (which contains approximately 470 mM NaCl) was replaced by artificial seawater containing only choline chloride instead of NaCl. The giant axon is enclosed by a layer of cells, and as time progressed, the Na^+ concentration inside the coating and near the membrane gradually decreased. Trace 6 was made after the axon was returned to normal seawater at the end of the experiment. [Part a adapted from Keynes, 1958; part c adapted from Hodgkin and Huxley, 1939; part d adapted from Hodgkin and Katz, 1949.]*

To put it another way, at rest the membrane is most permeable to K^+, but in the early phase of an AP, it becomes very much more permeable to Na^+ than it is to K^+ at rest. As a result of this greatly increased permeability and the strong force pushing Na^+ into the cell, a spurt of positive charge enters the cell, making it less inside-negative. When voltage-gated Na^+ channels close and permeability to Na^+ drops, the membrane is much more permeable to K^+ than it is at rest, because voltage-gated K^+ channels are still open. Later, the

strated that during an AP the membrane resistance decreases, but the membrane capacitance remains constant. This observation implied that a change in conductance must be entirely responsible for the change in ionic current. Hodgkin and Huxley found that V_m does not simply go to zero during an AP, but instead actually reverses sign during the impulse (part c of the figure). Although this observation might now seem to be a mere detail, it was at odds with the prevailing belief that the increased ionic current measured during excitation was nonspecific, allowing all ions present to move according to their emf. Before Hodgkin and Huxley observed that V_m overshoots zero during an AP, it was believed that a nerve impulse consisted of a simple collapse of V_m to zero. In fact, the overshoot of the AP approaches the E_{Na} calculated from the Nernst equation (see Equation 5-3) using a 10 : 1 ratio of external-to-internal Na^+ concentrations, which has been measured for the squid giant axon:

$$E_{Na} = \frac{0.058}{1} \log 10 = 0.058 \text{ V} = 58 \text{ mV}$$

Further confirmation of the role of Na^+ in the AP was obtained through experiments by Hodgkin and Bernard Katz (1949). In their experiments, a squid giant axon was bathed in artificial seawater in which choline chloride replaced NaCl. Choline, a large organic cation, cannot cross the plasma membrane, and its presence, coupled with the low $[Na^+]_{out}$, decreased the magnitude of the AP, exactly as expected if Na^+ were the major cation responsible for carrying ionic current across the membrane (part d of the figure).

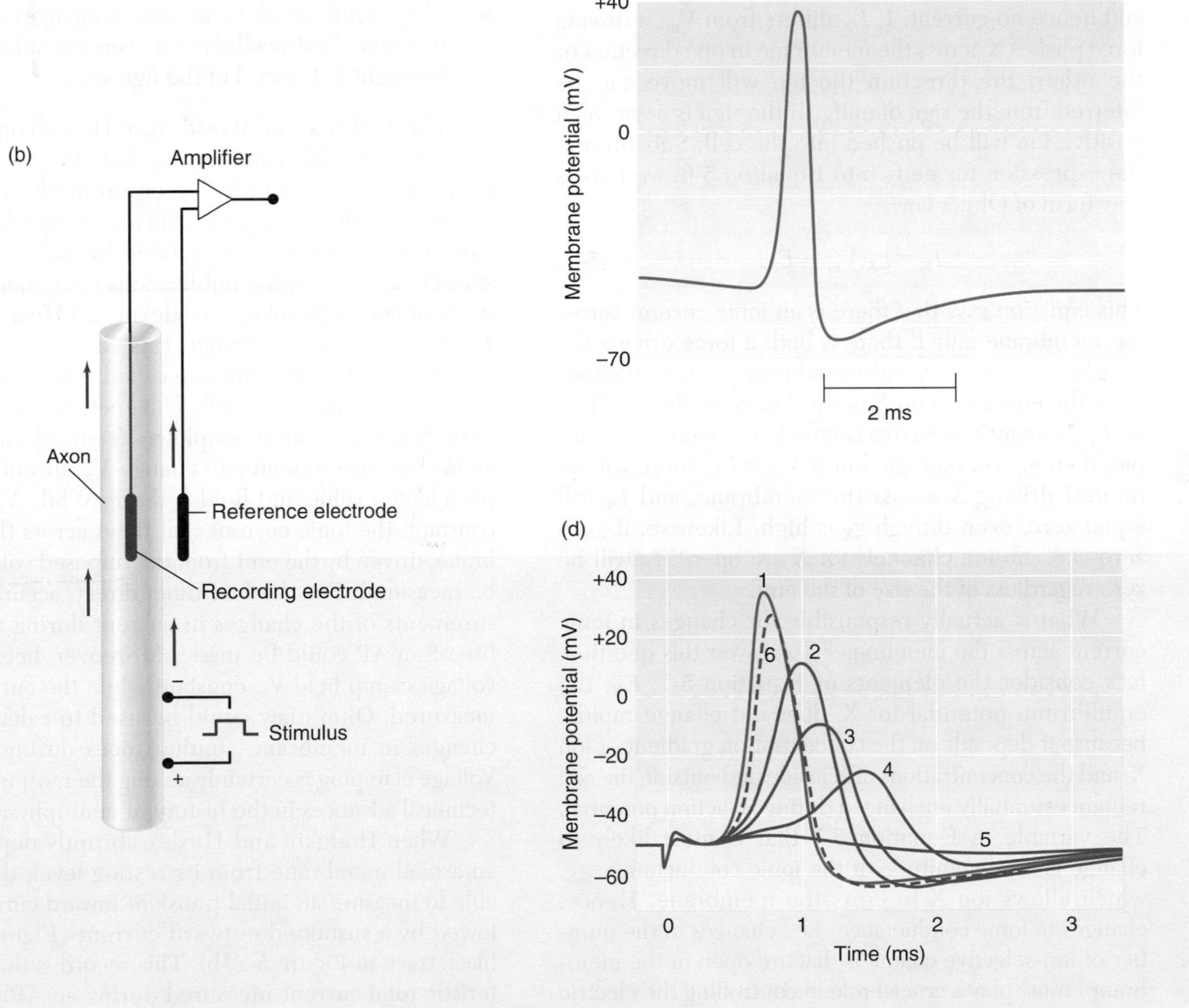

permeability to K^+ drops back to its resting value, with only the passive leak channels remaining open. The permeability of the membrane to Cl^- does not change during an AP. This brief overview of the mechanisms responsible for APs summarizes several decades of painstaking work and creative insight by many biophysicists.

To understand how changing membrane characteristics might account for the AP, it is useful to return to the model of a membrane as a capacitor arranged in parallel with a conductor (see Figure 5-12a). However, because we now know that Na^+ and K^+ cross the membrane through separate types of ion-selective channels, we will modify the model to include a separate, parallel conductance for each species of ion. Remember, ionic currents through ion channels obey Ohm's law (Equation 5-1). Thus, the current carried by ionic species X, I_X, is given as

$$I_X = g_X \times \text{emf}_X \tag{5-6}$$

where g_X is the membrane conductance for X and is proportional to the number of open channels selective for X. The emf_X is the electromotive force acting on X, which is the difference between the membrane potential, V_m, and the equilibrium potential of X, E_X (Equation 5-5).

If V_m is equal to E_X, there is no driving force on X, and hence no current. If E_X differs from V_m, a driving force pushes X across the membrane in one direction or the other; the direction the ion will move can be inferred from the sign of emf_X. If the sign is negative, a positive ion will be pushed into the cell. Substituting the expression for emf_X into Equation 5-6, we have a new form of Ohm's law:

$$I_X = g_X(V_m - E_X) \tag{5-7}$$

This equation says that there is an ionic current across the membrane only if there is both a force driving the translocation of ion X and a conductance for X. If either g_X or the emf acting on X is equal to zero, there will be no I_X. If many X-selective channels are open, for example, then g_X will be high, but if $V_m = E_X$, there will be no emf driving X across the membrane, and I_X will equal zero, even though g_X is high. Likewise, if g_X is zero (i.e., no ion channels for X are open), I_X will be zero regardless of the size of the emf.

What is actually responsible for changes in ionic current across the membrane? To answer this question, let's consider the elements of Equation 5-7. E_X, the equilibrium potential for X, does not change rapidly because it depends on the concentration gradient of ion X, and the concentrations of X inside and outside the cell remain essentially unchanged during an action potential. The variable in Equation 5-7 that is most likely to change is the magnitude of the ionic conductance, g_X, which allows ion X to cross the membrane. Hence, changes in ionic conductance (i.e., changes in the number of ion-selective channels that are open in the membrane) must play a crucial role in controlling the electric currents carried across biological membranes. Notice that changes in V_m affect the ionic current carried by X because the driving force on X is equal to $V_m - E_X$. Thus, as V_m changes during the AP, the driving force on Na^+ and K^+ changes, too, but not because E_X changes.

The results from many biophysical experiments with squid giant axons performed by Hodgkin, Huxley, and others in the 1930s and 1940s (Spotlight 5-4 on pages 136–137) provided four pieces of evidence that Na^+ is the major ionic species responsible for the AP:

- Because $[Na^+]_{out}$ exceeds $[Na^+]_{in}$ by a factor of about 10, the E_{Na} calculated using the Nernst equation is about +55 mV to +60 mV. Thus the emf acting on Na^+ ($V_m - E_{Na}$) will be large (about 100 mV) and will drive Na^+ into the cell across the membrane.
- Entry of positively charged Na^+ into the cell could produce the positive shift in V_m that was reported by Alan Hodgkin and Andrew Huxley in 1939 (see Spotlight 5-4, part c of the figure).
- The observed overshoot of the AP approaches the calculated value of E_{Na}.
- The magnitude of the overshoot changes as a function of extracellular Na^+ concentration (see Spotlight 5-4, part d of the figure).

The outbreak of World War II interrupted the research of Hodgkin and Huxley, but after the war they continued their electrical measurements of action potentials in the squid giant axon and were able to measure directly the currents carried by individual ionic species. Their resulting publications revolutionized the study of neurophysiology (Hodgkin and Huxley, 1952a, 1952b). They accomplished this feat using a newly invented electronic technique called **voltage clamping** (Spotlight 5-5 on page 140). This method, first applied to the squid giant axon, employs a feedback circuit that allows the experimenter to change V_m abruptly to any preselected value and hold it there. While V_m is held constant, the ionic current that flows across the membrane, driven by the emf from the imposed voltage, can be measured. For the first time, direct, accurate measurements of the changes in current during the brief life of an AP could be made. Moreover, because the voltage clamp held V_m constant while the current was measured, Ohm's law could be used to calculate the changes in membrane conductances during the AP. Voltage clamping is certainly among the most important technical advances in the history of neurophysiology.

When Hodgkin and Huxley abruptly depolarized an axonal membrane from its resting level, they were able to measure an initial transient inward current, followed by a sustained outward current (Figure 5-21a; black trace in Figure 5-21b). This record is the characteristic total current measured during an AP. In a key experiment, Hodgkin and Huxley demonstrated that the transient initial inward current is due to the passage of Na^+ ions across the membrane. They clamped the transmembrane voltage at E_{Na}, a value at which there is no driving force on Na^+ because ($V_m - E_{Na}$) equals zero. In addition, they lowered the external Na^+ concentration by substituting choline for Na^+ in the sea-

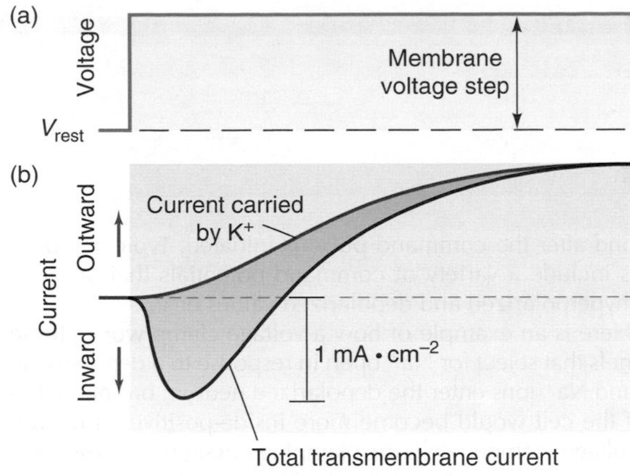

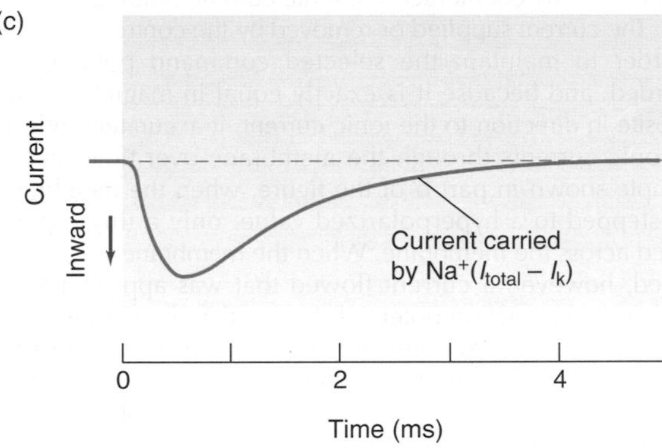

Figure 5-21 Voltage-clamping experiments allow the time course of ionic currents to be determined during an action potential. **(a)** In this experiment, the membrane of a squid giant axon was clamped at + 60 mV for at least 5 ms. **(b)** The black trace shows the total transmembrane current during the voltage-clamp pulse shown in part a; this current is carried by both Na^+ and K^+. The purple trace shows the current carried by K^+ alone, recorded in low-Na^+ seawater with V_m held at E_{Na} by the voltage clamp. With this protocol, the sodium current, I_{Na}, equals zero, because there is no emf acting on Na^+. **(c)** Subtracting the K^+ contribution in part b from the total transmembrane current reveals the time course of I_{Na}. [Adapted from Hodgkin and Huxley, 1952a.]

water bathing the axon, making less Na^+ available. In this state, there was no measured inward current, but a delayed outward current remained (Figure 5-21b, purple trace). Hodgkin and Huxley found that this outward current was influenced, but not eliminated, by stepping the membrane potential to E_{Cl}. Putting all of these data together, they proposed that the outward current is carried by K^+. When the solution bathing the axon was changed back to normal Na^+-containing seawater, the inward current returned, indicating that the inward current is produced by a transient influx of Na^+ across the membrane. Later experiments, described below, showed that the delayed outward current was indeed carried by K^+. When the current measured in the absence of Na^+ flux was subtracted from the total current obtained in normal seawater, the difference between the two currents (shaded area, Figure 5-21b) shows the time course of the inward current carried by Na^+, which is plotted in Figure 5-21c.

These experiments led to the hypothesis that a sudden depolarization causes a large number of Na^+-selective channels to open transiently, producing an increase in the Na^+ conductance across the membrane and allowing Na^+ to flow into the axon. In the normal extracellular environment, the electrochemical gradient acting on Na^+ ($V_m - E_{Na}$) drives Na^+ into the cell. Thus, according to Ohm's law, when g_{Na} rises, I_{Na} should rise as well:

$$I_{Na} = g_{Na}(V_m - E_{Na}) \tag{5-8}$$

What does the time course of the Na^+ current tell us about the behavior of the membrane? Based on Equation 5-7, the time course of I_{Na} depends both on changes in the conductance of the membrane to Na^+, or g_{Na}, and on changes in the emf acting on Na^+, or ($V_m - E_{Na}$). Clamping V_m at a constant value holds the emf on Na^+ constant; Na^+ concentrations inside and outside the cell do not change, so ($V_m - E_{Na}$) remains constant. As a result, the time course of I_{Na} must directly reflect how g_{Na} changes over time in response to depolarization. An important feature of I_{Na}, illustrated in Figure 5-21c, is that even when V_m is held constant at a depolarized value, I_{Na} reaches a maximum within 1 millisecond and then rapidly returns to its low prestimulus value. Thus I_{Na}, and hence g_{Na}, must consist of two separate processes: *activation*, the time-dependent increase in g_{Na} caused by depolarization, and *inactivation*, the time-dependent return of g_{Na} to its baseline level.

Hodgkin and Huxley used the voltage-clamping technique to measure the individual ionic currents that contribute to the total current as they change over time during an AP. They concluded that each ionic current reflects a separate conductance through the membrane and that each conductance is selective for the particular ion carrying that current. Using these data, they formulated equations to describe each conductance as a function of V_m and time. These equations predicted the electrical behavior of a neuronal membrane under many different conditions. It has subsequently been established that the properties of many excitable membranes can be accounted for largely or entirely by these time-dependent changes in Na^+ and K^+ conductances across the membrane.

THE NATURE OF ION CHANNELS

Hodgkin and Huxley's detailed electrical measurements in squid giant axons revealed that g_{Na} and g_K change during an AP. They hypothesized that these changes in conductance permit ions to move across the plasma membrane and that the resulting ionic currents produce the change in V_m that takes place during an AP. Their elegant experiments clearly defined the aggregate, macroscopic properties that characterize these

Spotlight 5-5 THE VOLTAGE-CLAMPING METHOD

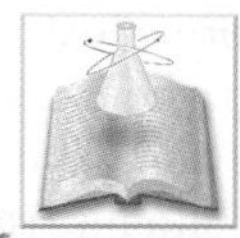

In voltage clamping, an electronic feedback circuit holds the voltage difference across a plasma membrane at a constant value. This method, first described in 1949 by Kenneth Cole, allows the experimental measurement of ionic currents that flow across a membrane when ion channels are activated. The results are interpreted based on Ohm's law, $I = V \times g$. When the voltage is held constant, any changes in the current, I, must reflect changes in conductance. We now know that the changes in conductance seen during an AP depend on the opening and closing of particular ion channels. Stepping the membrane to a series of different voltages and holding it at each voltage for some time reveals how channel conductances depend on voltage and time. By performing voltage-clamping experiments on neurons that have been exposed to solutions of different ionic composition, or to agents that specifically block particular ion channels, researchers have directly determined which ionic currents produce neuronal activity.

In a voltage-clamping experiment, an electrode is inserted into the neuron, and a control amplifier compares the potential that is recorded across the membrane with an electronically generated "command" potential, chosen by the experimenter from a broad range of values (part a of the accompanying figure). If V_m differs from the command signal, a control amplifier produces a current that passes across the membrane in the direction that will make V_m equal to the command signal. The adjustment of V_m occurs rapidly—within a fraction of a millisecond after the command pulse is initiated. Typical experiments include a variety of command potentials that produce both hyperpolarized and depolarized values of V_m.

Here is an example of how a voltage clamp works. If the channels that select for Na^+ open in response to a depolarizing step and Na^+ ions enter the depolarized neuron, ordinarily the V_m of the cell would become more inside-positive. However, in a voltage-clamped cell, the clamping circuit produces a current that exactly counteracts the ionic current, holding V_m constant. The current supplied or removed by the control amplifier in order to maintain the selected command potential is recorded, and because it is exactly equal in magnitude and opposite in direction to the ionic current, it accurately reflects the ionic currents through the membrane over time. In the example shown in part b of the figure, when the membrane was stepped to a hyperpolarized value, only a tiny current flowed across the membrane. When the membrane was depolarized, however, a current flowed that was approximately equivalent to an action potential. Notice that the current trace first moved downward from zero and then moved to a position above zero. Comparing this record with Figure 5-16, notice that the early current corresponds to the rising phase of the action potential—that is, to the period during which Na^+ enters the cell. The later current corresponds to the falling phase of an action potential—that is, to the period when K^+ leaves the cell. By convention, in voltage-clamp records, inward positive currents are plotted in a downward direction.

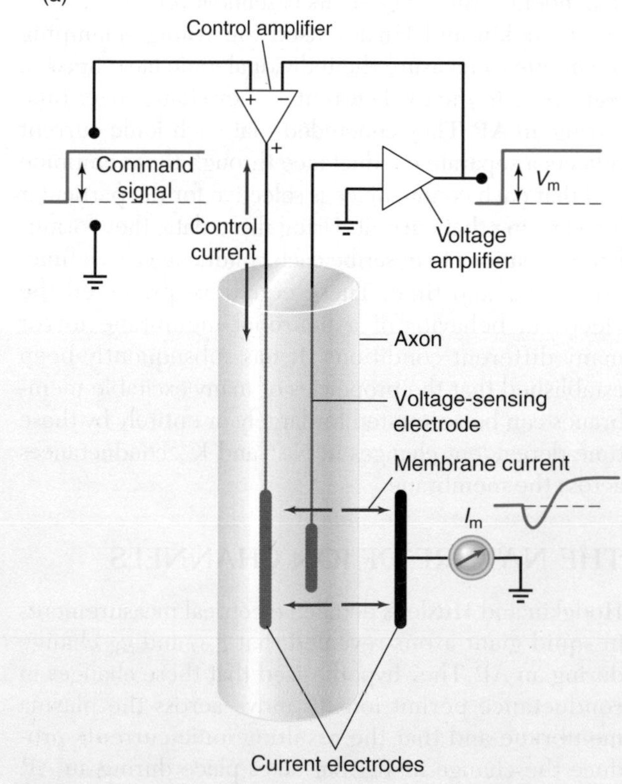

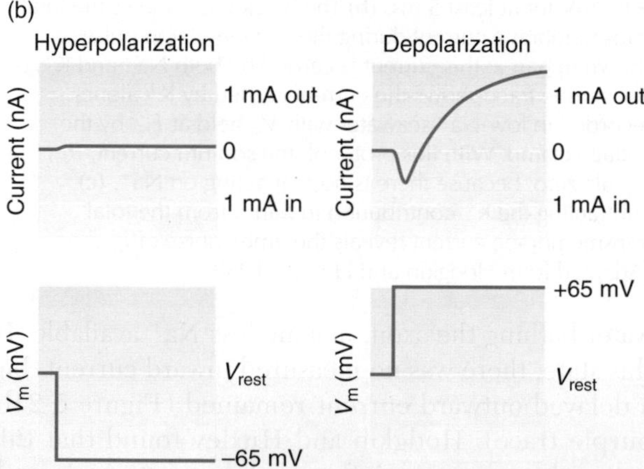

*In voltage clamping, an electronic feedback circuit allows the potential across the membrane, V_m, to be held constant. **(a)** The control amplifier compares V_m with a command signal. If V_m is different from the preset command potential, an electric current is quickly passed through the clamping circuit to make V_m equal to the command potential. If the membrane's permeability to ions changes, more or less current will be required to maintain V_m at a constant level. **(b)** Results of a voltage-clamp experiment illustrate the different effects of hyperpolarizing and depolarizing voltage steps. When the membrane is stepped to a hyperpolarized value, only a tiny current flows across the membrane. However, when the membrane is depolarized, a current flows that is approximately equivalent to an action potential. [Part b adapted from Hodgkin, Huxley, and Katz, 1952.]*

conductances in neurons, but they had no evidence concerning the physical basis for the conductances, their selectivity, or how they are activated. They suggested that the physical basis of these conductances could be pores through the membrane, but they could not rule out some kind of molecules in the membrane that rapidly transported ions across. Observing the properties of current flow through a single membrane channel was not possible at that time, because ordinary voltage-clamping techniques were not sufficiently sensitive. A conventional voltage-clamping device necessarily collects current from thousands of channels residing in a large area of membrane, and the precision of recording is limited by substantial electrical background noise that arises from various sources. These constraints were lifted in the late 1970s when E. Neher and B. Sakmann developed a technique called **patch clamping**, which allowed currents passing through individual ion channels to be recorded (Spotlight 5-6 on pages 144–145). In addition, the techniques of protein chemistry and molecular biology have made it possible to investigate the structures of the channel proteins themselves.

To relate the molecular properties of ion channels to their role in the generation of APs, we need to understand four key features of ion channels:

- the distribution of ion channels in neuronal membranes
- the nature of current flow through a single channel
- the mechanism by which depolarization of the membrane opens a voltage-gated channel
- the physical basis for channel selectivity

Let's discuss each of these features. While we focus here on voltage-gated Na^+ and K^+ channels, keep in mind that other voltage-gated ion channels have similar properties.

Voltage-Gated Na^+ Channels

Localization and characterization of voltage-gated channels have been facilitated by the use of several naturally occurring neurotoxins that bind to specific channels. One particularly potent and useful toxin is **tetrodotoxin (TTX)**, obtained from the viscera of the Japanese puffer fish *Sphoeroides rubripes* and related species. TTX selectively blocks fast-acting, voltage-gated Na^+ channels. When radioactively labeled TTX molecules are added to the extracellular fluid bathing a neuron, they bind to Na^+ channels. Examination of neurons labeled by this technique has allowed the density of bound TTX molecules, and hence of Na^+ channels, to be estimated. More recently, antibodies to channel proteins have been developed, allowing the proteins to be labeled and viewed directly (see Chapter 2). In nonmyelinated axons from several different types of neurons, the density of Na^+ channels has been measured at about 100 to 500 channels per μm^2; Na^+ channels therefore occupy only a small fraction of the total surface area. However, each channel can pass 10^7 Na^+ ions or more per second, providing enough I_{Na} to account for the macroscopic currents that have been measured in various neurons.

Results from patch-clamp recordings of voltage-gated Na^+ channels have given us a detailed picture of how these channels operate. From Ohm's law, Faraday's constant, and Avogadro's number, it can be calculated that one activated Na^+ channel carries Na^+ ions at a rate of about 10,000 ions per millisecond at an emf ($V_m - E_{Na}$) of -100 mV, which is approximately the driving force as an AP gets under way. The summed activity (i.e., openings and closings) of thousands of Na^+ channels, each allowing a minute *unitary current* (i.e., a current through a single channel) to cross the membrane, gives rise to the macroscopic I_{Na} that produces the rising phase of the AP (see Spotlight 5-6, part c of the figure). The number of Na^+ channels open at any instant depends on time, because neither channel activation nor inactivation is instantaneous, and both also depend on V_m.

How does depolarization of the membrane influence the opening of voltage-gated ion channels? Hodgkin and Huxley originally suggested that changes in V_m might regulate g_{Na} and g_K by causing a conformational change in a gating molecule. According to their proposal, this gating molecule would bear a net charge at physiological pH values, so that any change in membrane potential would produce an emf on the charge, causing it to move in space and changing the shape of the molecule. Consider a typical resting neuron with a potential difference of about -75 mV across its plasma membrane. A depolarization of 50 mV (to $V_m = -25$ mV) generally activates a large fraction of the Na^+ channels present in such a membrane. These channels consist of integral protein molecules in the lipid bilayer of the membrane, which is about 5 nm thick. It can be calculated that the portions of the channel proteins within the 5 nm bilayer of the membrane sense a voltage change of 10^{-3} V per 10^{-8} cm, or 100,000 $V \cdot cm^{-1}$, during the 50 mV depolarization. Driven by this huge transmembrane electric field, charged groups on the channel proteins certainly might move, producing conformational changes in the proteins.

Hodgkin and Huxley proposed that membrane depolarization would lead to a movement of a gating charge from the inside toward the outside of the membrane. They further suggested that this movement of charge should correspond to a small, but measurable, gating current (I_g) that would be associated in time with the opening and closing of Na^+ channels. This hypothetical I_g was detected in the early 1970s, when sensitive new techniques finally allowed it to be measured. I_g can be observed only when the much larger ionic current through the Na^+ channels, I_{Na}, is pharmacologically blocked by tetrodotoxin or a similar agent. Since that time, gating currents also have been detected for voltage-gated K^+ channels and for voltage-gated Ca^{2+} channels in several tissues.

Figure 5-22 presents a model of the opening and closing of a voltage-gated Na^+ channel when a depolarizing step of voltage is applied to the membrane. Detailed analysis of Na^+ channel gating currents has revealed that (1) gating of the channel takes place in several distinct steps, each of which is associated with movement of charges, and (2) activation and inactivation of the channel are coupled processes.

Once a Na^+ channel is open, only certain ions can pass through it. A channel's selectivity can be characterized experimentally by determining its relative permeability for various ionic species. For instance, if the permeability for Na^+ of the voltage-gated Na^+ channel is set at 1.00, then its permeability for Li^+ is found to be 0.93, whereas for K^+ it is only 0.09. The channel acts as a filter that selects at least partly on the basis of size, but as we saw in Chapter 4, size cannot be the whole story (Figure 5-23). Current hypotheses explaining how a channel selects among ions are based partly on ionic size and partly on other properties of the permeant species, such as net charge and, importantly, ease of dehydration. The region of the channel that determines its selectivity is called the **selectivity filter.** Spherical cations larger than 4 Å in diameter are too large to pass through the pore of either Na^+ or K^+ channels, although some asymmetric ions with one dimension larger than that can move through the Na^+ channel (Figure 5-23b). Cations smaller than 4 Å may pass through the pore of the Na^+ channel, but only once they have lost the shell of water molecules (the *water of hydration*) that normally surrounds charged ionic species in free aqueous solution (Figure 5-23b). The ease with which polar oxygen or charged functional groups that line the pore of the channel can substitute for the water of hydration helps to determine how readily ions pass through the channel. In addition, negative charges located at the opening of a cation-selective channel attract cations and repel anions.

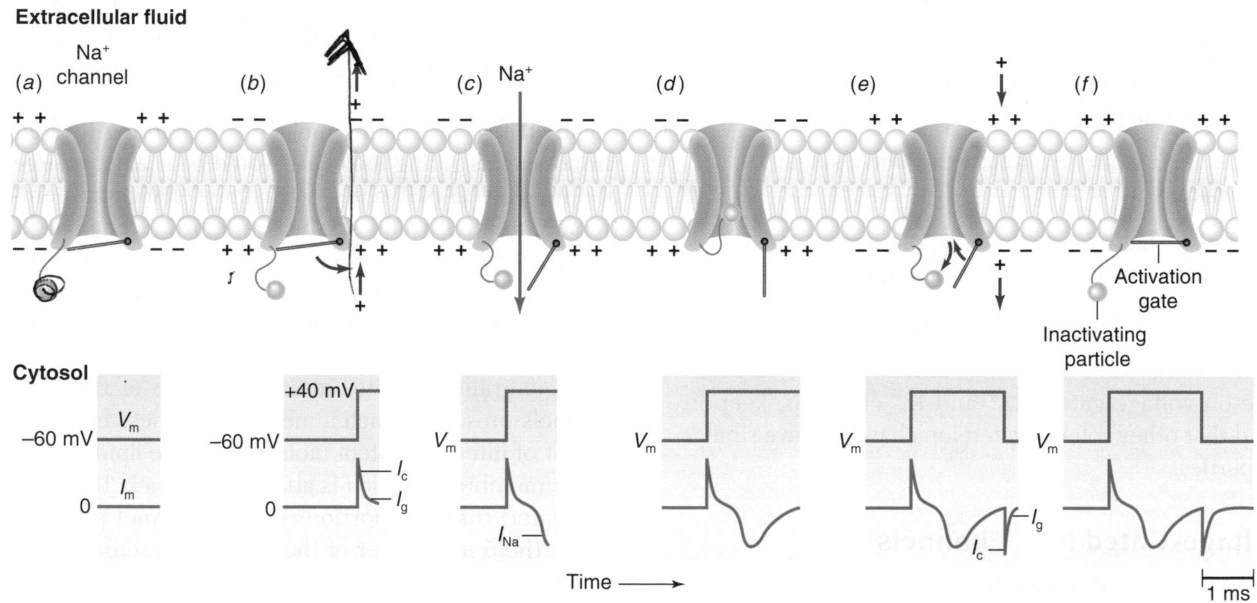

Figure 5-22 The structural conformation of the voltage-gated Na^+ channel changes during membrane excitation. A schematic model representing the opening and closing of the channel is shown above, with cumulative current records corresponding to each step below. V_m across the voltage-clamped membrane is shown in blue, and the current recorded by the voltage-clamping device is shown in red. Although only a single Na^+ channel is shown in the diagram, the currents show the summed effects of activity in thousands of channels. In the resting state (*a*), the activation gate is closed and the inactivating particle is located away from the pore. When V_m is voltage-clamped to a depolarized level (*b*), the activation gate moves to the open configuration (curved arrow) in response to the new electric field across the membrane. With sufficiently sensitive equipment, movement of the charged gate, like the movement of any charged particle, can be detected as a current. Notice that the first current to be recorded when the transmembrane voltage is stepped to +40 mV is a capacitative current, I_c, and that I_g is superimposed on I_c. (Although I_c is a part of every voltage-clamp record, it has been subtracted out of other records shown in this chapter.) (*c*) When the majority of the Na^+ channels have opened, the inward Na^+ current, I_{Na}, is maximal. As depolarization continues (*d*), movement of the inactivation particles blocks the open channels. After the membrane is repolarized (*e*), the gating charges of the Na^+ channel reorient, giving rise to another gating current superimposed on a capacitative current. When V_m is once again inside-negative, the inactivation particle moves out of the channel, and the activation gate closes the pore, returning the channel to its resting state (*f*). Removal of inactivation takes place only after the membrane has returned to the inside-negative condition.

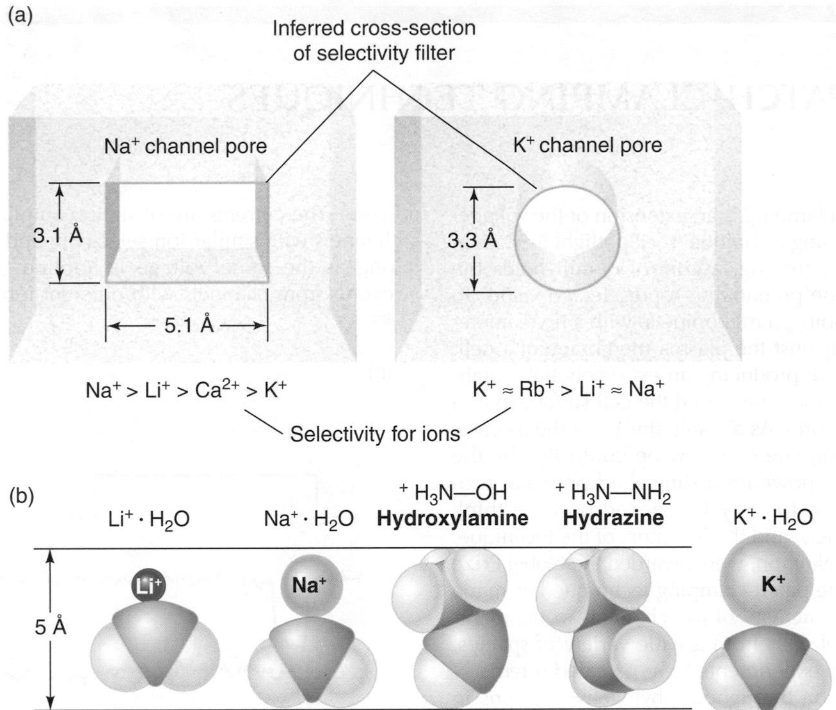

Figure 5-23 Ion channels are selective, allowing some ions to pass through the membrane and holding others back. Selectivity depends on the properties of a critical region of the channel pore, called the selectivity filter, combined with the ease of dehydrating the permeant ions. **(a)** Schematic diagram of a cross-section through the selectivity filters of Na^+ and K^+ channels inferred from the relative permeabilities of various ions. Based on this evidence, the K^+ channel is thought to have a round selectivity filter that is smaller than at least one dimension of the selectivity filter in the Na^+ channel. The sizes and shapes of the organic ions that can pass through the selectivity filter of the Na^+ channel suggest that the filter is rectangular, rather than round. The selectivity filter of an ion channel is thought to contact permeant ions directly, indicating that the channel strips the water of hydration from the ions. **(b)** Schematic diagrams of several partially hydrated inorganic and organic ions. All of these ions can pass through the Na^+ channel, but among these ions only K^+ can pass through the K^+ channel. The channel sizes and ions are drawn to the same scale. [Part b adapted from "Ion channels in the nerve cell membrane" by Richard D. Keynes. Copyright © 1979 by Scientific American, Inc. All rights reserved.]

During an AP, Na^+ channels respond to an initial depolarization by opening, thus allowing Na^+ to enter the cell, which further depolarizes the membrane. This depolarization causes more channels to open, allowing still more Na^+ to enter the cell and triggering an explosive, regenerative event. Once an AP is started, it needs no additional stimulus to continue. The relationship between membrane potential and sodium conductance, which has been called the Hodgkin cycle, is an example of positive feedback (Figure 5-24). Positive feedback systems are rarely found in biological tissues because such systems are inherently unstable. The effects of positive feedback in an AP are limited, however, in two ways. First, as the membrane potential approaches E_{Na}, the driving force on Na^+ is reduced, so less Na^+ is driven into the cell. More importantly, open Na^+ channels become inactivated, independently of V_m, after a short time (see Figure 5-22), even if the depolarization is maintained (see Figure 5-21c). The spontaneous termination of the Na^+ current by the intrinsic inactivation of Na^+ channels would be sufficient to end an AP. In fact, current through voltage-gated K^+ channels also contributes to the recovery of the membrane potential. We now turn to those channels.

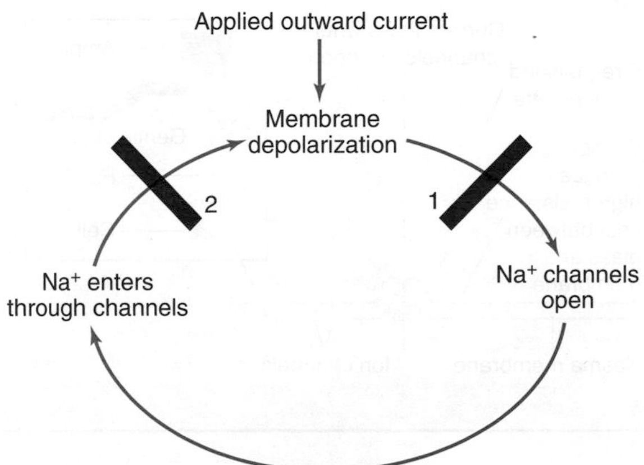

Figure 5-24 The Hodgkin cycle is the positive-feedback loop connecting membrane depolarization and sodium conductance; it is responsible for the rising phase of an AP. The cycle is normally initiated by a depolarization of the membrane that is triggered from outside the neuron and that is independent of the voltage-gated Na^+ channels. Under normal conditions, the positive feedback is interrupted by intrinsic inactivation of Na^+ channels, which terminates the rising phase of the AP (black line 1). Alternatively, the cycle can be experimentally interrupted by voltage-clamping the membrane, preventing I_{Na} from changing V_m (black line 2).

PATCH-CLAMPING TECHNIQUES

Patch clamping is an extension of the voltage-clamping technique (see Spotlight 5-5). As in voltage clamping, a control circuit holds the membrane potential at a preselected value. To accomplish patch clamping, a micropipette with a tip diameter of 1–2 μm is placed against the plasma membrane of a cell. Gentle suction is applied, producing an extremely tight, high-resistance seal between the pipette and the cell surface (part a of the accompanying figure). As a result, the V_m of the patch of membrane under the pipette can now be controlled by the electronic circuitry. This powerful technical advance has produced data that have profoundly influenced how we think about membrane channels, and the inventors of the technique, Erwin Neher and Bert Sakmann, were awarded the Nobel Prize for this contribution. The patch-clamping technique has made it possible to record the activity of ion channels in the membranes of many types of cells from a wide variety of species, and the results of these experiments have revealed a remarkable level of conservation throughout phylogeny. Neurons in animals as widely separated as jellyfishes and mice use essentially the same ionic mechanisms to produce APs.

Using one type of patch-clamping technique, researchers can record ionic currents passing through a single membrane channel while V_m in that small region is clamped to a chosen value. When a depolarizing voltage step larger than the threshold is applied, the recorded events are all-or-none currents with a square shape that indicates abrupt opening and closing of the channel whose activity is being recorded (part b of the figure). The currents are of similar amplitude for all individual channels with similar ion selectivity and kinetic properties, as long as the same voltage is applied. The general form of records from channels with different ionic selectivities is also

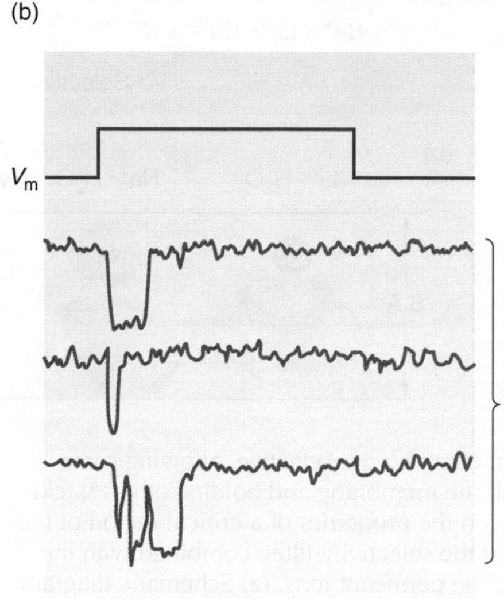

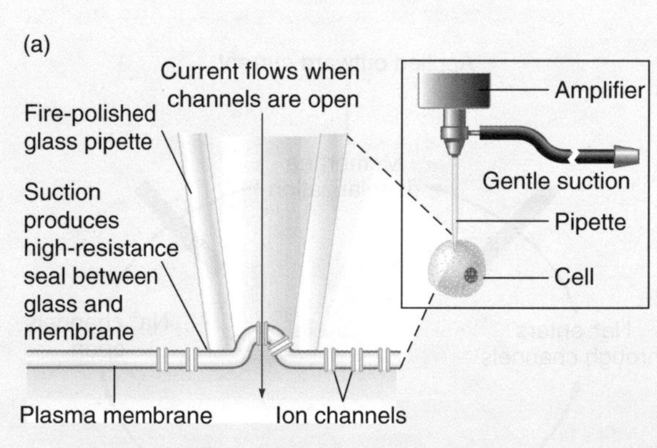

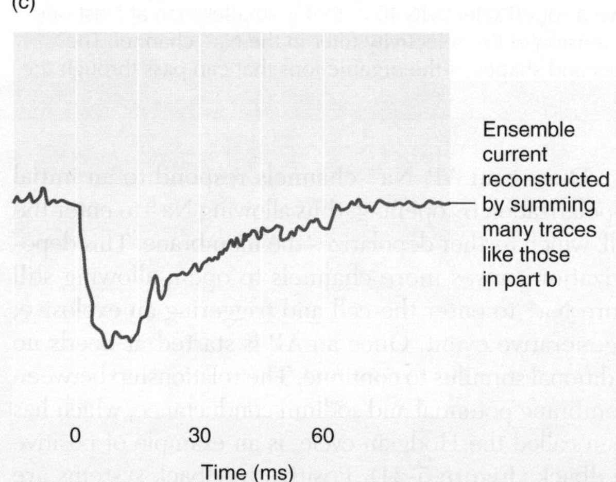

Voltage-Gated K^+ Channels

Voltage-gated K^+ channels are another major component of neuronal membranes. These channels respond more slowly to voltage changes than do Na^+ channels (see Figure 5-20b). The membrane g_K increases very little until the AP is near its peak, and g_K remains elevated during the falling phase. The net outward current through the K^+ channels brings V_m back toward its resting value, and in neurons that have an after-hyperpolarization, the membrane potential moves even closer to E_K than it is at rest (see Figure 5-20).

The properties of voltage-gated K^+ channels vary much more widely than those of voltage-gated Na^+ channels. Some types of voltage-gated K^+ channels inactivate spontaneously, as do voltage-gated Na^+ channels. Others do not; instead, their conductance to ions is tightly linked to V_m. The I_K through open voltage-gated K^+ channels speeds repolarization in the falling phase

very similar, suggesting that all voltage-gated ion channels function in the same general way. All channels, for example, are either open or closed; channels typically do not open partway. The time that individual channels remain open varies randomly over a wide range. The conductance of a single Na^+ channel does not depend significantly on V_m; its value ranges from 5 to 25 picosiemens, pS (10 pS = 10×10^{-12} S of conductance, or 10^{11} Ω of resistance) for different Na^+ channels.

When a large number of single-channel recordings from a patch, such as the traces shown in part b of the figure, are added together, the result is very similar to the macroscopic current that would be recorded with more conventional voltage-clamping methods (part c of the figure). This result is consistent with the hypothesis that macroscopic currents recorded using voltage clamping are the sum of many tiny currents flowing through individual channels.

Since the invention of patch clamping, many variations have evolved that are useful for different types of research. Part d of the figure illustrates five variations that have proved useful and that will be referred to in succeeding chapters.

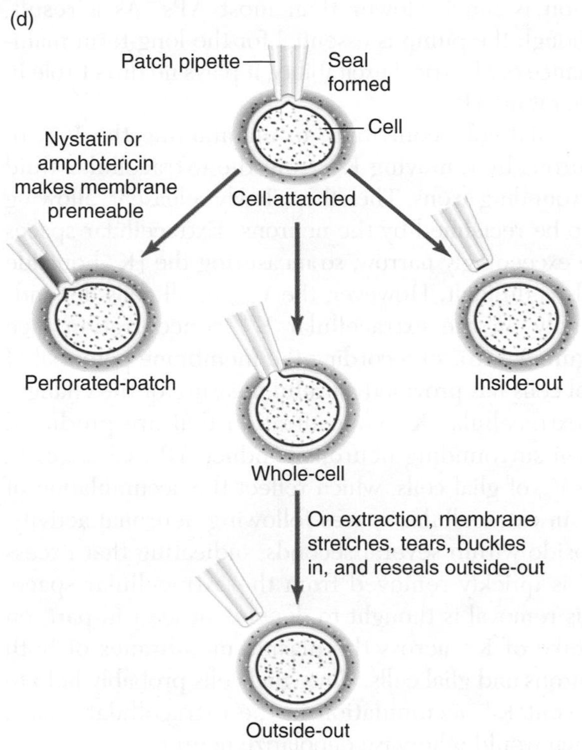

The patch-clamping technique is a variation on voltage clamping that allows currents through a single membrane channel to be recorded. ***(a)*** *A fire-polished micropipette, with a tip diameter of about 2 μm and containing the same solution as the one bathing the rest of the neuron, is sealed with gentle suction against the plasma membrane of a cell until a very high resistance seal is achieved. The high resistance of the seal prevents the loss of current from the pipette to the outside saline bath and reduces the background noise. Current flow through an open channel is detected by a sensitive electronic circuit. The voltage across the patch of membrane that is surrounded by the tip of the pipette is held constant using an electronic feedback circuit similar in principle to that used in the original voltage-clamping experiments described in Spotlight 5-5.* ***(b)*** *Depolarization (red trace) of a patch of rat muscle fiber membrane caused a single* Na^+ *channel to open transiently several times, producing* Na^+ *currents that varied in duration and latency, but not in amplitude (the three blue traces).* ***(c)*** *Summation of 144 such records from one patch produced an "ensemble current" whose time course reflects the temporal distribution of individual openings of that one channel. This summed current resembles the* Na^+ *current measured with more conventional voltage clamping, suggesting that the conventionally measured currents depend on the activity of many channels opening randomly in response to a single depolarizing step (see Figure 5-21c).* ***(d)*** *Variations of the patch-clamping technique permit single-channel properties—and even currents across the entire plasma membrane—to be recorded under a wide variety of conditions. In some cases, a patch of membrane is pulled from the cell and its channel currents are recorded in isolation. In* perforated-patch *recordings, an agent such as amphotericin, which increases the membrane's permeability but leaves its structure intact, is added to the inside of the pipette. In the* cell-attached *method, the membrane remains intact, single-channel currents are recorded, and cytoplasmic contents are unchanged. Similarly, currents through individual channels can be recorded in the* outside-out *configuration, but in this case the cytoplasmic face of the channels can be efficiently exposed to drugs by including them in the solution filling the pipette. In contrast, the* inside-out *configuration and the* perforated-patch *configuration allow the solution bathing the cytoplasmic face of the whole membrane to be changed. The* whole-cell *configuration allows ionic currents across the entire plasma membrane to be measured, making this method similar to impaling the soma with a sharp microelectrode, but with less damage to the cell. [Parts b and c adapted from Patlack and Horn, 1982; part d adapted from Cahalan and Neher, 1992.]*

of the AP. Thus, as I_K causes V_m to return toward its resting value, g_K decreases. As noted above, K^+ channels are not needed by the membrane to generate APs, and indeed, some myelinated mammalian neurons appear to lack them entirely. However, the rapid membrane repolarization produced by current through K^+ channels does shorten the AP, allowing neurons to generate APs at a higher frequency than they otherwise could.

Absolute and Relative Refractory Periods

As we saw earlier, immediately following an AP, the neuronal plasma membrane is first unresponsive to stimulation and then less responsive than it is at rest (see Figure 5-17). These absolute and relative refractory periods result from the same mechanisms responsible for the falling phase of the AP and the afterhyperpolarization: inactivation of Na^+ channels and continued activation of K^+ channels. In the falling phase of an AP and immediately afterward, the majority of Na^+ channels remain inactivated and thus cannot be opened by depolarization. Return to an inside-negative membrane potential removes this inactivation (see Figure 5-22), so at a slightly later time, some, but not all, of the Na^+ channels are no longer inactivated. During this period, depolarization can open the activatable Na^+ channels. However, because this smaller number of channels carries much less inward current, an above-normal depolarization is required to generate enough current to initiate an AP. Moreover, outward current through the K^+ channels that are still open from the previous AP opposes the inward Na^+ current. Together, these two ionic mechanisms produce the macroscopic effects measured as refractory periods.

Intracellular Ions and the Na^+/K^+ Pump in Action Potentials

Although the membrane potential may change by more than 100 mV during an AP, the actual number of ions that move across the membrane is much too small to appreciably change the concentration of either Na^+ or K^+ inside the neuron. Calculations similar to the one in Spotlight 5-3 have shown that a single AP in a squid giant axon changes the internal concentration of Na^+ and K^+ by only about 0.001%. For example, 10^{-12} mol of Na^+ crossing 1 cm^2 of membrane with a capacitance of 1 $\mu F \cdot cm^{-2}$ is sufficient to produce an AP of 100 mV amplitude. That is, only 160 Na^+ ions per μm^2 of membrane are needed. Since a single Na^+ channel can pass 10^7 Na^+ ions per second, the I_{Na} required for an AP could be supplied by a remarkably small number of channel openings. Our calculation of the number of Na^+ ions that cross the membrane during one AP is probably an underestimate, because some K^+ ions move out of the cell, partially canceling the electrical effect of the influx of Na^+. The actual number of ions is likely to be closer to 500 Na^+ ions per μm^2 per impulse. If this value is correct, then during a single AP in a squid giant axon with a diameter of 1 mm, the intracellular Na^+ concentration would still change by only 1 part in more than 100,000. Indeed, it has been shown that a squid giant axon can still generate thousands of impulses after its Na^+ pumps have been incapacitated by a metabolic poison. Of course, the concentrations, and hence the equilibrium potentials, of Na^+ and K^+ will eventually change, and the axon will no longer be able to produce normal action potentials.

The situation is somewhat different in very small mammalian axons, such as mammalian C fibers, which have a diameter of only about 1 μm. Their small size gives these axons a much larger surface-to-volume ratio, and the concentrations of Na^+ and K^+ in such small axons may change by as much as 1% following a single AP. This shift in ionic concentrations can affect the axons' physiological properties until the Na^+/K^+ pump restores the ion concentrations to their normal resting values. Indeed, in all excitable cells, the Na^+/K^+ pump is responsible for maintaining the driving force behind the Na^+ and K^+ currents that flow during an AP, but its action is much slower than most APs. As a result, although the pump is essential for the long-term maintenance of electrical excitability, it plays no direct role in generating APs.

Glial cells contribute to maintaining the V_{rest} of neurons by removing K^+ from the extracellular fluid surrounding axons. They then slowly release it, allowing it to be reclaimed by the neurons. Extracellular spaces are exceedingly narrow, so measuring the $[K^+]$ outside cells is difficult. However, the V_{rest} of all cells depends heavily on the extracellular K^+ concentration (see Figure 5-14b), so recording the membrane potential of glial cells has provided a useful measure of the changes in extracellular K^+ concentration that are produced when surrounding neurons conduct APs. Changes in the V_m of glial cells, which reflect the accumulation of K^+ in extracellular spaces following neuronal activity, subside within several seconds, indicating that excess K^+ is quickly removed from the extracellular space. This removal is thought to depend, at least in part, on uptake of K^+ across the plasma membranes of both neurons and glial cells. Thus glial cells probably help to prevent K^+ accumulation in the extracellular space, which would otherwise depolarize neurons.

Molecular Structure of Voltage-Gated Ion Channels

A combination of protein chemistry and molecular biology has yielded detailed information about the channels that mediate action potentials. Spotlight 5-7 on page 148 discusses some methods that have been particularly useful in studying channels at this level. Voltage-gated Na^+ and K^+ channels have now been identified, cloned, and sequenced in a variety of species, from the nematode *C. elegans* and the fruit fly *Drosophila* to mice and humans. The homology among channels from different species is striking. In addition, there is considerable homology among the various types of voltage-gated cation channels.

In 1998, in a long-awaited breakthrough, the structure of a K^+ channel from the bacterium *Streptomyces lividans* was resolved in atomic detail. This channel serves as a model for other K^+ channels, including those in animal neurons, although there are some obvious structural differences. Each K^+ channel is an oligomeric complex of four monomeric α-subunits that are arranged symmetrically around a central pore (Figure 5-25a). Four β-subunits associated with each channel modulate channel function. In eukaryotes, each monomeric α-subunit of most K^+ channels has a molecular mass of about 70 kD and has six probable transmembrane helices (Figure 5-25b). These six transmembrane regions are called S1 (toward the amino-terminal end of the molecule) through S6 (toward the carboxyl end). In contrast, in the bacterial K^+ channel whose structure has been worked out in such great

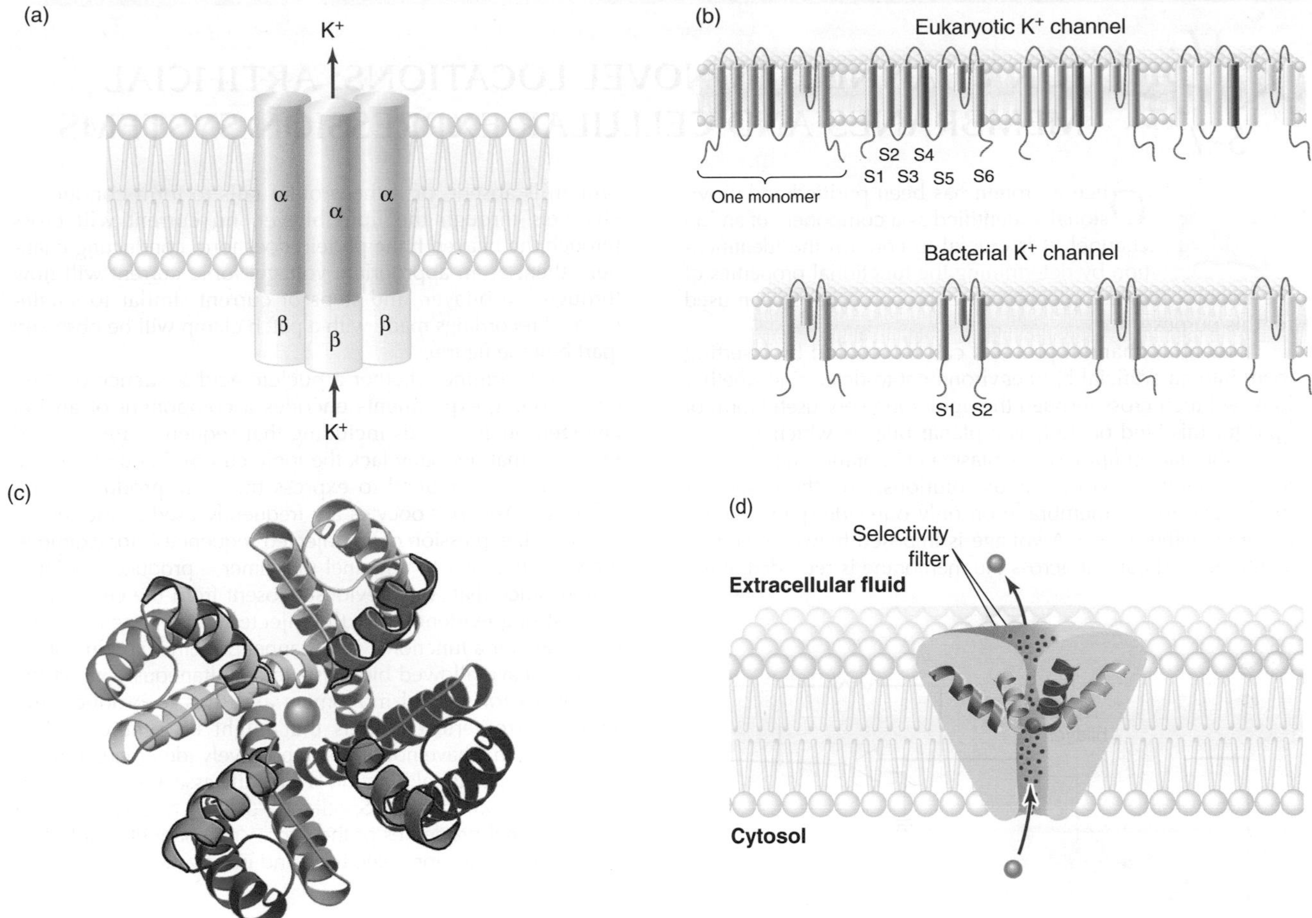

Figure 5-25 Each K^+ channel is a complex of four monomeric α-subunits plus associated β-subunits. **(a)** The four α-subunits each span the membrane and are arranged symmetrically around a central pore. Each monomer is associated with an auxiliary β-subunit on the cytoplasmic face of the membrane, which affects the functional properties of the channel. **(b)** In most K^+ channels of eukaryotes (top), each monomer includes six membrane-spanning helices as well as intracellular and extracellular segments that link the membrane-spanning regions. The membrane-spanning helix S4 (red) is thought to be important for sensing the transmembrane voltage. A pH-gated bacterial K^+ channel (bottom) has been shown to consist of four monomers, each of which includes only two membrane-spanning regions (called S1 and S2) that are homologous to segments S5 and S6 of K^+ channels in eukaryotes. **(c)** X-ray analysis of the bacterial K^+ channel reveals that the transmembrane pore is lined by the transmembrane segments homologous to S6. Each of the four monomers is shown in a separate color. The selectivity filter consists of the sequences that link the two membrane-spanning helices of each monomer, and it is so narrow at one point that it accommodates a K^+ ion (central purple sphere) only if the water of hydration has been removed. **(d)** The entire bacterial K^+ channel is cone-shaped, with the selectivity filter located closer to the extracellular face of the membrane. [Part c adapted from Doyle et al., 1998; part d adapted from Nelson and Cox, 2000.]

detail, each α-subunit has only two transmembrane domains, called S1 and S2. These two membrane-spanning regions are most homologous to S5 and S6 of the eukaryotic channels. Other data suggest that the region of the eukaryotic monomers that includes S5 and S6 contributes to the conducting pore and selectivity filter of these channels, a conclusion supported by the finding that the bacterial channel contains only the homologs of these two regions.

Evidence from X-ray crystallography indicates that the conducting pore of the bacterial K^+ channel is lined by the helical transmembrane segment S2 from each of the four monomers, arranged symmetrically around a pore (Figure 5-25c). In addition, part of the sequence that links the two transmembrane segments lines the pore and contributes to the selectivity filter. Measurements suggest that the diameter of the pore varies along its length, becoming so narrow at the selectivity filter that it can just accommodate a dehydrated K^+ ion (Figure 5-25d).

One interesting aspect of the family of voltage-gated K^+ channels is its heterogeneity. At least 18 different genes for voltage-gated K^+ channels are known to be expressed in the mammalian nervous system alone, and even in *Drosophila* several different classes of K^+ channel genes have been discovered. Additional

ION CHANNELS IN NOVEL LOCATIONS: ARTIFICIAL MEMBRANES AND CELLULAR EXPRESSION SYSTEMS

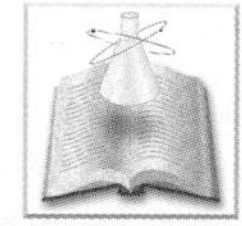

Once a protein has been purified and provisionally identified as a component of an ion channel, it is crucial to confirm the identification by determining the functional properties of the protein. Two complementary approaches have been used for this purpose.

Potential channel proteins can be studied by inserting them into an artificial lipid environment to determine whether ions will then cross through the lipid. One very useful form of lipid for this kind of study is a planar bilayer, which approximates the state of lipids in the plasma membrane. The bilayer is formed between two aqueous solutions, and the protein is introduced to the membrane on only one side (part a of the accompanying figure). A voltage is imposed between the two solutions, and current across the membrane is recorded. If the protein is absent, or if the protein fails to form conducting channels through the lipid bilayer, no current will cross through the bilayer. If the protein does form conducting channels, then, at an appropriate voltage, ionic current will flow through the bilayer, and steps of current similar to single-channel recordings made with a patch clamp will be observed (part b of the figure).

To determine whether a nucleic acid sequence derived from cloning experiments encodes a component of an ion channel, nucleic acids including that sequence are injected into cells that normally lack the ionic current in question. The cells are then induced to express the gene product of the sequence. *Xenopus* oocytes are frequently used as the recipient cells. If expression of the injected sequence—for example, the sequence of a K^+ channel monomer—produces an ionic conductance that was previously absent from the cell, it provides strong evidence that the injected nucleic acid can by itself code for a functional ion channel protein. Experiments of this type have allowed biologists to simultaneously inject the sequences for several monomers and thus to produce and study heteromeric channels that might be synthesized in nature, but that have not yet been positively identified. In these experiments, channel function is typically assayed using patch-clamping techniques (see Spotlight 5-6), which reveal in detail the functional properties of the expressed channels, including voltage sensitivity, ion selectivity, and inactivation.

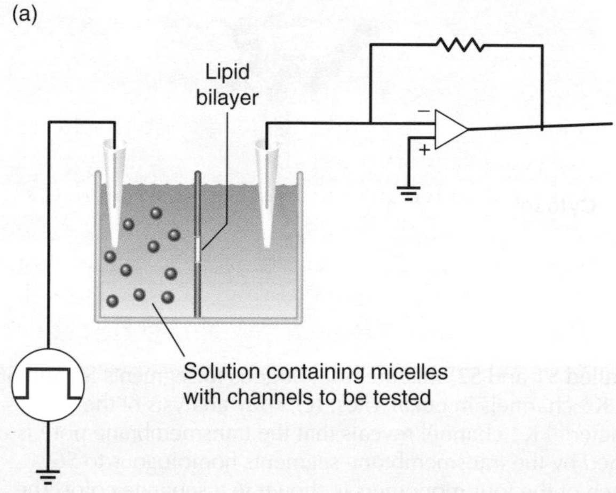

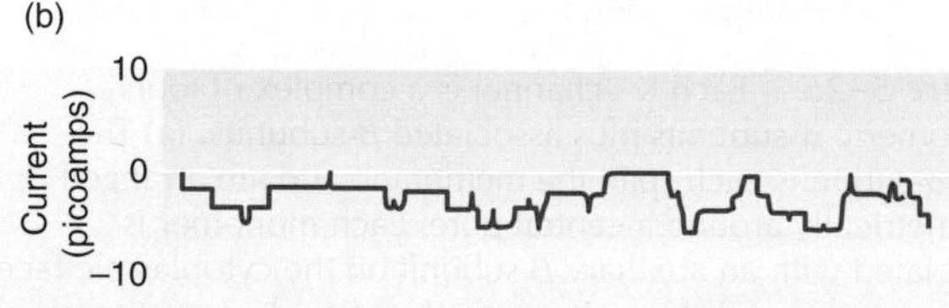

Inserting proteins in planar lipid bilayers tests their channel-forming properties. ***(a)*** *A lipid bilayer is formed across a hole in a partition separating two solutions, and the material to be tested is added to one of the solutions. After several minutes, the current that can pass between the two compartments is measured using a voltage clamp. A lipid bilayer has a very high resistance, so with no addition of channels, little or no current passes through.* ***(b)*** *If the substance being tested forms channels that open periodically, the current passing through the bilayer varies over time in steps. In this record, a maximum of three channels are open at any one time, and sometimes only one or two are open. Compare this record with those obtained from patch clamping (see Spotlight 5-6). The steplike appearance of the record implies that each channel is either open or closed; there are no partially open states. [Adapted from Miller, 1983.]*

heterogeneity may result when different kinds of monomers are assembled into a single oligomeric channel, and there is evidence that different forms of heteromeric channels may be localized in different regions of a single neuron.

In contrast to the tetrameric voltage-gated K^+ channels, each voltage-gated Na^+ channel has been found to consist of one large α-protein (about 260 kD) that contains four internally homologous transmembrane domains (Figure 5-26). Although the α-protein normally is associated with several smaller β-proteins, it can, by itself, function as a voltage-gated channel when it is expressed in *Xenopus* oocytes (see Spotlight 5-7). The transmembrane domains of the α-protein are thought to be arranged around a central conductance pore that is similar to the pore of the K^+ channel. Experiments have identified particular amino acids that are likely to confer the properties of ion selectivity, inactivation, and voltage gating on these Na^+ channels. For example, using the technique of site-directed mutagenesis to replace charged amino acid residues along the S4 helices with neutral residues changes the way in which

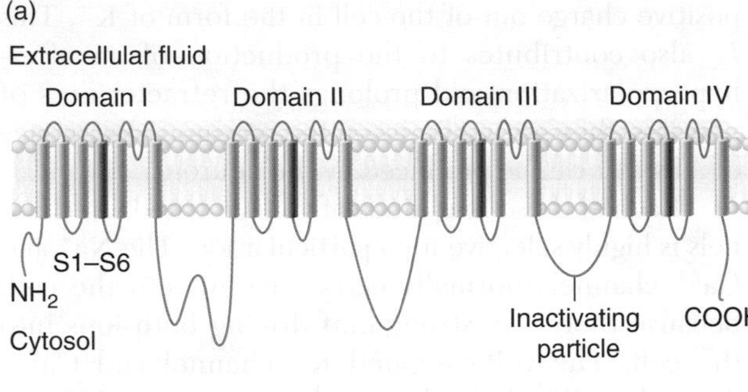

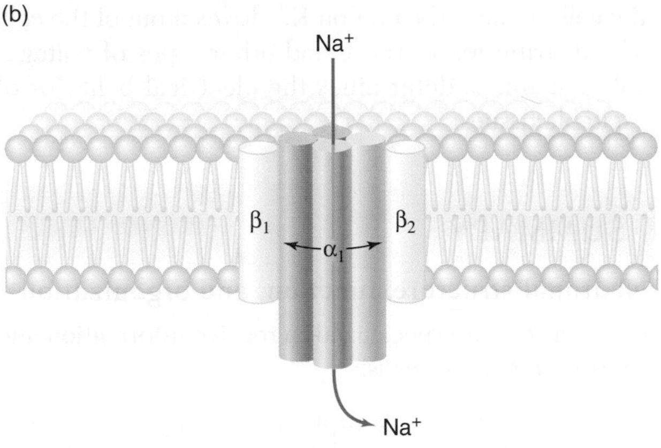

Figure 5-26 The molecular structure of a voltage-gated Na^+ channel resembles four linked voltage-gated K^+ channels. **(a)** Each Na^+ channel contains a large protein subunit (α or α_1) that can form functional channels when expressed by itself in *Xenopus* oocytes. The α-protein in each channel includes four homologous repeats (I–IV) of helical regions (numbered S1–S6). Each set of six transmembrane helices is homologous to a single voltage-gated K^+ channel subunit. The voltage sensor is thought to be located on transmembrane domain S4 (red). Just as four monomers associate to form the K^+-conducting pore, the four homologous repeats in a single Na^+ channel sequence are thought to associate with one another to form the ion-conducting pore of the channel. **(b)** Each long cylinder in this diagram corresponds to one repeat in the α-protein. In addition, one or more smaller proteins (designated β) normally associate with the α-protein to form the complete channel complex. These small subunits, which differ among channel types, contribute to the physiological properties of the channels.

channels respond to transmembrane voltage; therefore the charged residues are thought to be important in voltage sensing and activation. Similarly, site-directed mutagenesis and antibody studies have indicated that inactivation (see Figure 5-22) depends on the intracellular segment that links domain III and domain IV (see Figure 5-26). Continuing studies are expanding our understanding of the details of ion selectivity, voltage gating, inactivation, and other properties of these voltage-gated channels.

Other Electrically Excitable Channels

Transmembrane ion channels are found in virtually all cell types. Although voltage-gated Na^+ and K^+ channels collaborate in the production of typical APs, Ca^{2+} channels may be of more widespread importance in cell function (see Table 5-1). Ca^{2+}-selective channels carry at least part of the depolarizing current during APs in crustacean muscle fibers; in vertebrate smooth muscle and cardiac muscle cells; in the somata, dendrites, and terminals of many neurons; in embryonic neurons during the process of differentiation; and in ciliates such as *Paramecium*, to list just a few examples. In many of these membranes, Ca^{2+} carries inward current along with Na^+, but in a few cell types it carries all of the inward current. An I_{Ca} typically is not strong enough to produce an all-or-none AP without help from an I_{Na}, and in most membranes that contain voltage-gated Ca^{2+} channels, the rising phase of an all-or-none AP is generated by a strong I_{Na} that rapidly depolarizes the membrane. The Ca^{2+} channels, which open more slowly and conduct less current, are activated by this depolarization. Ca^{2+} ions that enter the cell through the Ca^{2+} channels often have two functions: propagating an electrical signal and acting as an intracellular messenger that triggers subsequent intracellular events. For example, an increase in intracellular Ca^{2+} is responsible for the release of neurotransmitter substances from presynaptic axon terminals and also causes contraction of muscles (see Chapters 6 and 10).

The molecular structure of voltage-gated Ca^{2+} channels is strikingly similar to that of voltage-gated Na^+ channels. Like Na^+ channels, Ca^{2+} channels consist of a large protein (α_1) that includes four homologous transmembrane regions, which are thought to associate and form the ion-conducting pore. The Ca^{2+} channel α_1-protein typically associates with several smaller proteins (Figure 5-27 on the next page). Many differentiating neurons express for a time both voltage-gated Na^+ and voltage-gated Ca^{2+} channels. Usually the Ca^{2+} channels appear first, and the Na^+ channels become functional at later stages. The remarkably homologous molecular structure of Na^+ and Ca^{2+} channels, the frequency with which the two kinds of channels are expressed in the same cells, and the greater prevalence of Ca^{2+} channels in simpler organisms such as protozoans all suggest that Na^+ channels may be a more recent evolutionary specialization for impulse conduction. In addition, Ca^{2+} acts as an intracellular messenger in most types of cells, providing further evidence that Ca^{2+} channels may have had an earlier evolutionary origin than the channels that pass monovalent cations.

Some neurons have APs based on voltage-gated Ca^{2+} channels, but most do not. What kinds of effects might be seen in neurons with Ca^{2+}-based APs? Could these effects produce selective pressure to change the ionic basis of APs to currents carried by monovalent cations?

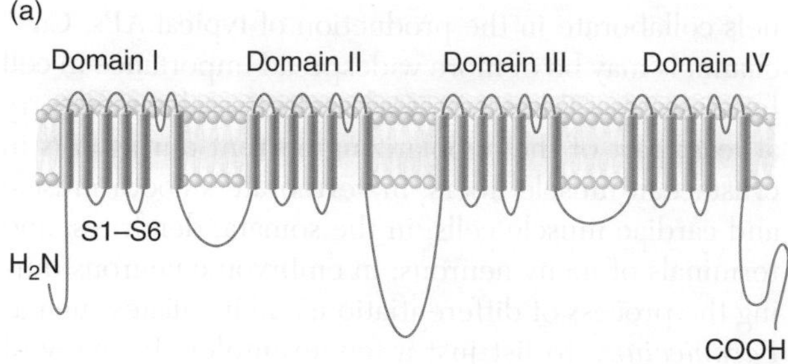

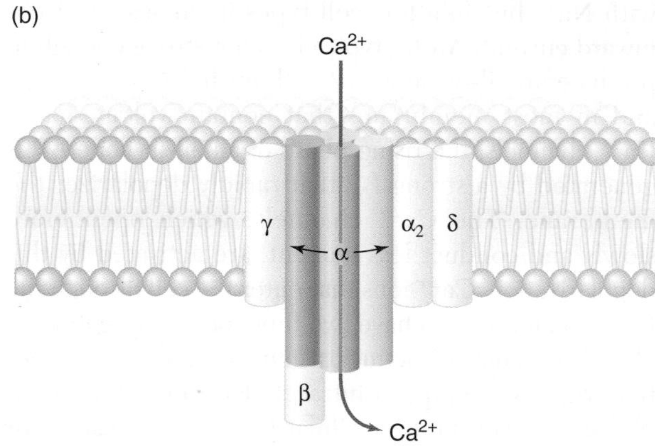

Figure 5-27 The structure of voltage-gated Ca^{2+} channels is very similar to that of voltage-gated Na^+ channels. **(a)** The α_1-protein of voltage-gated Ca^{2+} channels includes four sequential sets of six transmembrane helices that are homologous to the transmembrane domains of voltage-gated Na^+ channels. **(b)** Like other ion channels, voltage-gated Ca^{2+} channels typically associate with a set of smaller proteins (designated by various Greek letters), which differ among the different types of channels.

Unlike voltage-gated Na^+ channels, many types of Ca^{2+} channels fail to inactivate fully under maintained depolarization alone. Instead, in at least one type of Ca^{2+} channel, the probability of inactivation increases as the intracellular concentration of free Ca^{2+} rises. These Ca^{2+} channels become inactivated during maintained depolarization because I_{Ca} through the channels increases the intracellular Ca^{2+} concentration near the membrane, closing the channels and providing negative feedback control of intracellular free $[Ca^{2+}]$.

The intracellular concentration of free Ca^{2+} also regulates function in a class of voltage-gated K^+ channels that differs from the voltage-gated K^+ channels previously described. These K^+ channels, called Ca^{2+}-dependent K^+ channels, are found in many different tissues. They are activated by membrane depolarization, but only if the concentration of intracellular free Ca^{2+} is higher than normal (see Table 5-1). In these cells, Ca^{2+} enters through Ca^{2+} channels and accumulates near the inner surface of the membrane; it then causes the Ca^{2+}-dependent voltage-gated K^+ channels to open if the membrane is depolarized. Where these channels are present, the entry of Ca^{2+} fosters repolarization as the result of the enhanced I_K, which carries positive charge out of the cell in the form of K^+. This I_K also contributes to the production of an afterhyperpolarization and prolongs the refractoriness of neurons, thus setting limits on the maximum frequency of APs that can be produced by the neurons.

Each of these four kinds of voltage-gated ion channels is highly selective for a particular ion. The Na^+ and Ca^{2+} channels normally carry current into the cell, because there is a strong emf driving both ions into the cell. The voltage-gated K^+ channel and Ca^{2+}-dependent K^+ channel generally carry current out of the cell, because the emf on K^+ drives it out of the cell. The distribution of these and other types of voltage-gated channels determines the electrical behavior of excitable tissues.

SUMMARY

Neuronal structure, function, and organization

- Neurons are specialized to receive information and transmit it to other cells.

- Most neurons are morphologically asymmetric cells consisting of a soma (or cell body), dendrites, and a single axon. Information is typically received by the dendrites and in some cells by the soma, and it is transmitted along the axon.

- Information is conveyed along the axon as an action potential. Action potentials arise at a specialized region of the axon called the spike-initiation zone.

- Information is passed from a neuron to another cell across an intercellular gap called a synapse.

- In almost all animals except the very simplest, neurons are grouped into a central nervous system, which includes the somata of most of the neurons in the animal's body. Axons connect these central neurons with the rest of the body.

- Neurons can be classified into three categories: sensory neurons, which gather information about the internal and external environments; interneurons, which are entirely contained within the central nervous system and which process information; and motor neurons, which carry information from processing centers to the muscles that produce behavior and visceral movement.

Membrane excitation

- A steady potential difference, called the resting potential, is maintained across the plasma membrane of all living cells, with the cytoplasmic side of the membrane being negative with respect to the extracellular side. Standard electrical equipment, such as that used in the average physics laboratory, can be used to record the membrane potential.

- When the membrane potential is more inside-negative than the resting potential, the cell is said to be hyperpolarized; when it is more inside-positive than the resting potential, the cell is said to be depolarized.

■ Selective ion channels allow ions to travel across the plasma membrane. Voltage-gated ion channels change their conducting properties when the membrane potential changes. Ligand-gated ion channels change their conducting properties when specific molecules bind to receptor sites on the protein.

Passive electrical properties of membranes

■ Plasma membranes have an electrical capacitance, provided by the insulating lipid bilayer of the membrane, and an electrical conductance (or resistance), provided by the ion channels that are embedded in the membrane.

■ Ionic currents across a plasma membrane obey Ohm's law, $I = V/R$.

■ The specific resistivity of a membrane characterizes the resistance of a unit area of membrane to ionic flow. It depends on the number of open ion channels in the membrane.

■ The time constant, τ, of a membrane determines how rapidly the membrane potential can change in response to a change in current across the membrane. The time constant is equal to the product of the resistance and the capacitance ($\tau = RC$), so a change in either the resistance or the capacitance of the membrane will change the temporal response of a cell to ionic current.

Electrochemical potentials

■ When ions diffuse across a semipermeable membrane, they carry a net charge and generate an electric potential difference across the membrane, called an electrochemical potential.

■ Several ions are asymmetrically distributed across the membrane of a resting cell. These ions tend to diffuse from the region of higher concentration to the region of lower concentration, and this tendency is equivalent to a force driving the ions.

■ The Nernst equation describes the equilibrium condition for electrochemical potentials—the potential across a semipermeable membrane at which no net movement of ions will take place. At the equilibrium potential, the tendency of an ion to move under the influence of a concentration gradient is exactly balanced by its tendency to move under the influence of an equal, but opposite, electrical gradient.

■ Cells at rest are permeable to K^+, and the resting membrane potential approaches, but typically does not equal, the Nernst potential for K^+.

■ The resting potential for most cells can be more accurately predicted using the Goldman equation for the steady-state potential, calculated for a combination of K^+, Na^+, and Cl^- diffusion. This equation provides a better estimate of resting potential because the resting conductances for Na^+ and Cl^- are low, but not zero, causing the resting potential to be somewhat more positive than the equilibrium potential for K^+.

The resting potential

■ The action of the Na^+/K^+ pump keeps the concentration of Na^+ low in the cytosol and the concentration of K^+ high. Energy is required to maintain this steady state because concentration gradients continually drive some Na^+ into the cell and some K^+ out. The Na^+/K^+ pump must do work to move ions against these gradients.

Action potentials

■ During an action potential, the membrane potential briefly becomes inside-positive, then returns to its usual inside-negative condition.

■ An action potential is initiated only after the membrane potential has been depolarized by some minimum amount, called the threshold. In a particular neuron, all action potentials typically are the same amplitude, which is called the all-or-none property.

■ The depolarizing phase (or rising phase) of an action potential is caused by an inward Na^+ current moving through voltage-gated Na^+ channels, which open transiently and close spontaneously.

■ The repolarizing phase (or falling phase) of an action potential is typically caused by an outward K^+ current moving through voltage-gated K^+ channels, which do not inactivate spontaneously, but instead are closed when the K^+ current returns the membrane potential to its resting value.

■ Following an action potential, an axon is refractory—it is less able to generate a second action potential. For a brief period after an action potential—the absolute refractory period—it is impossible to generate a second action potential; then for a short time thereafter—the relative refractory period—a larger than normal stimulus is required.

■ During the absolute refractory period, Na^+ channels are inactivated and voltage-gated K^+ channels are open. During the relative refractory period, voltage-gated K^+ channels remain open, making it more difficult for the membrane potential to reach threshold.

■ The Na^+/K^+ pump slowly extrudes the Na^+ that entered the axon during an action potential and replaces it with K^+ to restore the resting concentrations of both ions. The pump thus maintains the conditions required to generate action potentials, but it plays no direct role in their conduction.

The nature of ion channels

■ The voltage-gated ion channels that produce action potentials have been cloned and sequenced. All consist of sets of homologous sequences comprising six transmembrane domains. The three-dimensional structure of a pH-gated K^+-selective bacterial channel, determined by X-ray crystallography, reveals the mechanism of ion selectivity. The structures of other voltage-gated channels are expected to be similar.

■ Perhaps the most ancient voltage-gated ion channels are those that selectively allow Ca^{2+} to cross the membrane. These channels are found throughout the animal kingdom and in most types of cells.

REVIEW QUESTIONS

1. What are the major anatomic regions of a neuron, and what is the principal physiological function of each?
2. Does a nerve signal that reaches the brain accurately encode all features of the original stimulus? Why or why not?
3. The plasma membrane separates electric charge so that one side has an excess of positive charge and the other side has an excess of negative charge. The net effect is to produce an electric potential difference across the membrane. Does this arrangement violate the general physical principle of electroneutrality? Why or why not?
4. What is the structural basis for capacitance in biological membranes? For conductance?
5. What is the relation between the time course of potential changes across the plasma membrane and the resistance and capacitance of the membrane?
6. Cells are typically inside-negative. Explain how this condition arises from the diffusion of ions.
7. You have on your laboratory bench an artificial system of two aqueous saline solutions separated by a semipermeable membrane. The solutions contain the same ions, but in different concentrations, and the membrane is permeable to only one of the ions in solution. Will there be a stable electric potential difference across the membrane? Why or why not? Does the sign of the permeant ion affect the outcome of this experiment?
8. Living cells are typically quite permeable to K^+ and at least slightly permeable to other cations as well. What maintains the high K^+ concentration inside the cell by preventing the K^+ inside from gradually being displaced by other cations?
9. What are the equilibrium potentials for each of the following ions at the given concentrations? (a) $[K^+]_{out} = 3$ mM, $[K^+]_{in} = 150$ mM; (b) $[Na^+]_{out} = 100$ mM, $[Na^+]_{in} = 10$ mM; (c) $[Ca^{2+}]_{out} = 10$ mM, $[Ca^{2+}]_{in} = 10^{-2}$ mM.
10. For a typical cell that is 100 times more permeable to K^+ than to any other ion, use the Goldman equation to determine the potential change that would be produced by a doubling of the extracellular K^+ concentration. (Use appropriate concentrations from question 9.)
11. In 1939, Cole and Curtis reported that membrane conductance increases during an AP, but capacitance remains essentially unchanged. Relate these findings to membrane structure and to the changes in that structure that are now thought to take place during excitation.
12. Describe two observations suggesting that Na^+ carries the inward current responsible for the rising phase of the action potential.
13. Does the Na^+/K^+ pump play a direct role in any part of the AP? Explain. Is the pump important for the production of an AP?
14. What limits the flux of Na^+ across the membrane during an AP? What limits the flux of K^+ across the membrane during an AP?
15. Calculate the approximate number of Na^+ ions entering through each square centimeter of axon surface during an AP with an amplitude of 100 mV. (Recall that 96,500 C is equivalent to 1 mole-equivalent of charge; that membranes have a typical capacitance of 1 $\mu F \cdot cm^{-2}$; and that Avogadro's number is 6.022×10^{23} atoms $\cdot$ mol^{-1}.)
16. The following properties of action potentials were discovered decades before physiologists knew about ion channels and their role in action potentials. Explain each of the properties in terms of the behavior of ion channels: (a) threshold potential; (b) all-or-none overshoot; (c) refractoriness; (d) accommodation.
17. Why is it that an axon of large diameter undergoes essentially no change in intracellular ion concentrations after many APs, whereas the very thinnest axons may undergo significant changes after only a few impulses? How would these changes affect the function of a small axon, and what mechanisms might help to prevent them?
18. The rising phase of an AP is an example of positive feedback in a biological system. How does positive feedback take place in this situation? Positive feedback is inherently unstable, so how can you account for the limited amplitude of the rising phase?
19. How do the properties of ionic currents through a single channel compare with the properties of macroscopic ionic currents measured across the plasma membranes of real cells?
20. Name four kinds of voltage-gated ion channels that contribute to neuronal function. How do these channels select for particular ionic species?
21. Compare and contrast the molecular properties of voltage-gated Na^+ channels with the properties of voltage-gated K^+ channels and of voltage-gated Ca^{2+} channels.

SUGGESTED READINGS

Aidley, D. J. 1998. *The Physiology of Excitable Cells.* 4th ed. New York: Cambridge University Press. (A thorough examination of the physiology of nerves, muscles, and unusual features such as the electric organs of some fishes.)

Aidley, D. J., and P. R. Stanfield. 1996. *Ion Channels: Molecules in Action.* New York: Cambridge University Press. (A very readable book, similar to a textbook, that considers the structure and function of these integral membrane proteins.)

Doyle, D. A., J. M. Cabral, et al. 1998. The structure of the potassium channel: Molecular basis of K^+ conduction and selectivity. *Science* 280:69–77. (The original description of K^+ channel structure; certain to become a classic.)

Hille, B. 2001. *Ionic Channels of Excitable Membranes.* 3d ed. Sunderland, Mass.: Sinauer Associates. (An excellent compendium of information about the biophysics of ion channels, written for professionals in the field or advanced students.)

Hodgkin, A. L. 1964. *The Conduction of the Nervous Impulse.* Springfield, Ill.: Thomas. (A classic description of the electrophysiology of signal conduction in neurons.)

Classic Work

Hodgkin, A. L. 1994. *Chance and Design: Reminiscences of Science in Peace and War.* New York: Cambridge University Press. (An extremely readable and personal view of biophysical history. Highly recommended!)

Jan, L. Y., and Y. N. Jan. 1997. Cloned potassium channels from eukaryotes and prokaryotes. *Annu. Rev. Neurosci.* 20:91–123. (A review of the molecular and physiological properties of this extremely large family of ion channels.)

Kandel, E. R., J. H. Schwartz, and T. M. Jessell. 2000. *Principles of Neural Science.* 4th ed. New York: Elsevier. (An enormous and authoritative compendium of information about the function of the nervous system, from the biophysics of membrane channels to the physiological basis of memory and learning.)

Marban, E., T. Yamagishi, and G. F. Tomaselli. 1998. Structure and function of voltage-gated sodium channels. *J. Physiol.* 508:647–657. (A concise discussion of a huge amount of information on voltage-gated channels, including consideration of their phylogeny.)

Yellen, G. 1999. The bacterial K^+ channel structure and its implication for neuronal channels. *Curr. Opin. Neurobiol.* 9:267–273. (A discussion of the broader implications of the Doyle, Cabral, et al. paper.)

Communication Along and Between Neurons

The survival of all animals depends on their ability to recognize and respond to challenges from other animals and from the environment. Nervous systems evolved to permit organisms to absorb information, process it, and respond rapidly. They are found in essentially all animals, from the simplest coelenterates to the most complex mammals.

The complexity of nervous systems is amply illustrated by the human nervous system, which contains more than 10^{12} neurons plus an even larger number of supportive cells (glial cells, or neuroglia). The functional units that allow animals to respond effectively to the environment are sets of neurons connected so that information can be passed among them. These arrays are called **neuronal circuits,** and their interconnections are in many ways analogous to the connections among elements in an electric circuit. All the complex capacities of the nervous system—perception, control of movement, learning, memory, thought, and consciousness—arise from the physical and chemical processes considered in Chapter 5 and examined further in this chapter. Understanding how neuronal activity results in behavior, consciousness, and creative thought is undoubtedly one of the greatest of all challenges to biologists.

In spite of the enormous complexity of most nervous systems, a great deal has already been learned about the physiology and biophysics of single neurons, as described in Chapter 5. We know that all neurons carry information by means of electrical signals that are based on the movement of particular ions across the plasma membrane. These ionic currents encode signals that travel along axons by mechanisms that constitute the first topic of this chapter. Although we consider events in a single neuron when we examine the transmission of information by action potentials (APs), behavior always depends on the activity of many neurons working together. If the signal carried by any one neuron is to generate effective and adaptive behavior, it must be transmitted to other neurons or cells. Information is passed to other cells at structures called synapses, and synaptic transmission—the mechanisms that allow information to be passed from one neuron to another cell at a synapse—constitutes the second major topic of this chapter.

TRANSMISSION OF SIGNALS IN THE NERVOUS SYSTEM: AN OVERVIEW

Signals move from point to point along the plasma membrane of a single neuron in either of two ways: via graded, electrotonically conducted potentials—that is, electrical signals that are conducted just as they would be along a wire—or via action potentials (all-or-none impulses). These two basic methods of transmission are used alternately as information passes along one neuron and is then transferred to another neuron. The pattern is illustrated schematically in Figure 6-1 on the next page, which shows the path traveled by information received at the interface between an animal and its surroundings (for example, at the skin). Energy from a physical stimulus (such as pressure on the skin) is received and changes the membrane potential, V_m, at a specialized receptor region of the plasma membrane in a sensory neuron. The amplitude of this change in V_m, which is called a **receptor potential,** is graded (i.e., varies in a continuous fashion) in proportion to the strength of the stimulus; in other words, a more intense stimulus produces a larger change in V_m. The receptor potential generally lasts for the duration of the stimulus, although its amplitude may decrease with time. For example, a pressure stimulus that persists for a long time typically produces a long, slightly attenuated depolarization of the receptor membrane. This signal spreads along the plasma membrane away from the location of the stimulus very much as an electrical signal spreads along a wire. Membranes that are specialized to receive sensory stimuli lack the voltage-gated ion

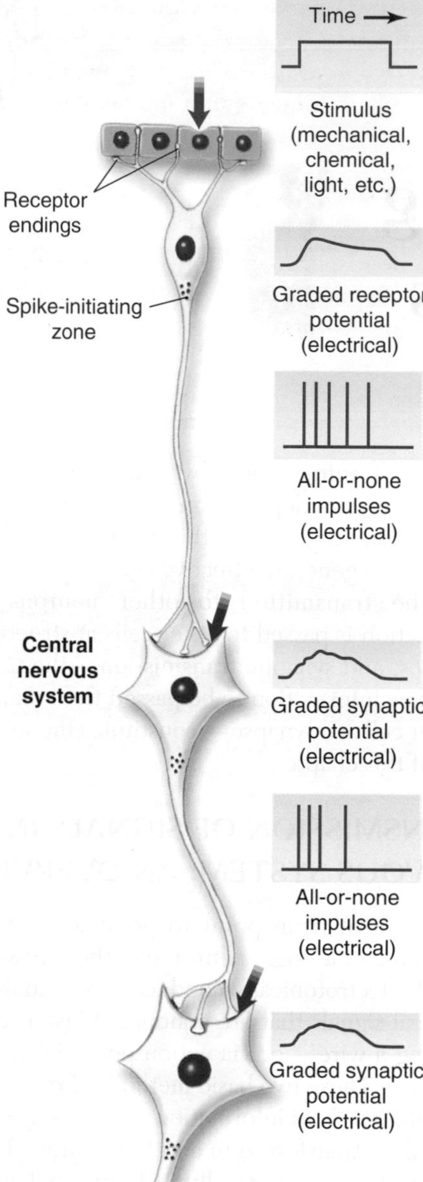

Figure 6-1 Graded and all-or-none electrical signals alternate with one another as a signal is carried along a neuronal circuit. A stimulus (represented by the red arrow) applied to the receptor endings of a sensory neuron produces a graded change in membrane potential that reflects the amplitude and duration of the stimulus. This change in V_m spreads passively through the first part of the sensory neuron, and if it is sufficiently large when it reaches the spike-initiating zone, it elicits all-or-none propagated APs in the axon. When the APs arrive at the terminals of the sensory neuron, they cause the release of a chemical neurotransmitter that induces a graded change in V_m in the next neuron. If the membrane potential in the second neuron reaches threshold, an AP or a train of APs is produced. Thus, graded and all-or-none potentials alternate along the pathway. Within each neuron, the signal is carried electrically; between neurons, the signal is carried by neurotransmitter molecules. Blue arrows indicate the sites of chemical signal transmission.

channels that produce all-or-none APs, so signals cannot be propagated regeneratively in this part of a sensory neuron; receptor potentials progressively decay with distance from the site of origin. This kind of signal transmission is called **passive electrotonic**, or **decremental, transmission.** The term *electrotonic* refers to conduction in which the amplitude of the signal is diminished with distance from the source; note that *electrotonic* is not synonymous with *electronic.* In contrast, the amplitude of a signal conducted *regeneratively* along an axon, such as an AP, remains constant as it travels because, as we will soon see, the signal is generated anew at each point along the axon.

Because they decay, passively transmitted signals cannot travel between widely separated parts of the body. For long-distance transmission, the sensory signals must be transformed into APs, which can conduct information without decrement for long distances. The membrane at the spike-initiating zone of a sensory neuron contains a high density of the voltage-gated ion channels that permit APs to start. If the passively propagated depolarization of a receptor potential is large enough at the spike-initiating zone, the signal will be transformed from passive into active propagation. Once that occurs, the signal will be carried by APs without decrement along the sensory neuron's axon, which may span many meters.

The next transformation of the signal takes place at the axon terminals of the sensory neuron, where it must be passed across synapses to other neurons. Transfer of information between neurons is typically (but not always) accomplished through chemical signals, carried by molecules called **neurotransmitters.** If the transfer of information across a synapse is to be effective, the neurotransmitter must cause a change in the V_m of the next neuron in line (called the **postsynaptic neuron**). The amount of transmitter released, and thus the amplitude of the response in the next neuron along the pathway, depends on the number and frequency of APs arriving in the terminals of the signaling neuron (called the **presynaptic neuron**). Within limits, more APs and higher frequencies of APs in the presynaptic neuron cause more neurotransmitter molecules to be released, producing a greater change in V_m of the postsynaptic neuron. The change in V_m of the postsynaptic neuron, called the **postsynaptic potential (psp)**, is a graded signal that reflects at least some properties of the original stimulus, although it may be distorted. If the psp is sufficiently large, it can bring the spike-initiating zone of the postsynaptic neuron to threshold, triggering one or more all-or-none APs in the postsynaptic neuron.

Thus, as a signal is received from the environment and transmitted by neurons, it is coded alternately in graded potentials and all-or-none APs, changing from one to the other repeatedly. Graded potentials are produced at sensory and postsynaptic membranes, and all-or-none APs are largely confined to structures that are specialized for long-distance conduction, such as axons. With minor exceptions, both types of signals are generated when the ionic conductance through specific types of membrane channels changes.

Between neurons, the signal is carried by neurotransmitter molecules. Transmission along an axon provides high fidelity, but signals transmitted across synapses are subject to many influences that can change their features This plasticity of synaptic transmission allows the nervous system to be flexible and to change over time. For example, modifications in synaptic connections are thought to underlie such complicated behavioral changes as learning. Synaptic plasticity, neuronal modification, and their meaning for higher brain functions such as learning and memory will be our theme at the end of this chapter.

TRANSMISSION OF INFORMATION WITHIN A SINGLE NEURON

In this section we consider the two types of signal conduction within a single neuron, passive electrotonic conduction and active or regenerative conduction. Electrotonic conduction, like conduction of electricity along a wire, depends only on the physical properties of a cell. Conduction along a wire depends on the movement of electrons; conduction along a neuron depends on the movement of ions along the two faces of the plasma membrane. In contrast, regenerative conduction depends on the presence and activity of biological molecules such as voltage-gated ion channels, which allow ions to move across the membrane.

Passive Spread of Electrical Signals

The passive spread of changes in V_m occurs in all neurons and depends primarily on the resistance and capacitance of the plasma membrane. In a hypothetical spherical cell, potentials would spread uniformly through the cytoplasm with minimal decay, because the electrical resistance of the saline cytosol would be much lower than the electrical resistance of the membrane. Real neurons, however, have more complicated shapes, so the spread of a change in V_m is more complex. Many neurons have long, thin processes extending from the soma. If current is injected at a point on a long, thin, cylindrical structure (e.g., a wire, or an axon, dendrite, or muscle fiber), the electrical signal spreads away from that point in a manner that depends on the **cable properties** of the structure—those electrical properties that affect conduction of a signal over distance (Figure 6-2). Any current flowing longitudinally along an axon

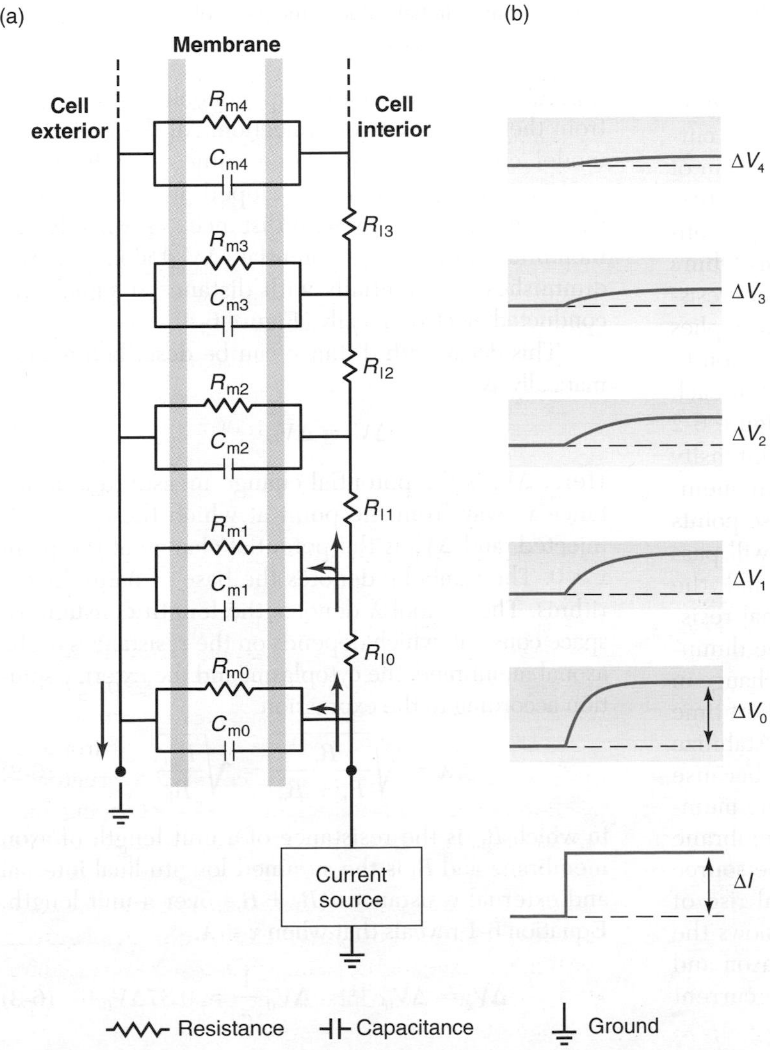

Figure 6-2 The cable properties of a long, thin cylinder determine how current spreads along it. **(a)** Equivalent circuit of an axon. Resistance and capacitance elements, representing properties of the membrane, are connected by the longitudinal resistances of the interior (cytosol) and exterior (extracellular) fluids. In this figure, the membrane resistance, R_m, the longitudinal resistance, R_l, and the membrane capacitance, C_m, have been arbitrarily divided into discrete circuit elements, labeled 0, 1, 2, 3, 4. The red arrows show the direction of current flow. Notice that some current flows across the membrane through each of the circuit elements. **(b)** The voltage across each of the circuit elements in response to a square pulse of current produced by the current source. The amplitudes of the potential changes, ΔV_0 to ΔV_4, decrease exponentially with distance from the source; in addition, the signals rise more slowly farther from the source.

(or other long, narrow fiber) decays with distance because (1) the cytoplasm has some resistance to the flow of electrical signals, and (2) the resistance of the plasma membrane to the passage of electrical signals is high, but not infinite. Along a perfectly insulated electrical wire, a longitudinal current moves uniformly without decrement. In contrast, a longitudinal current moving passively along a nerve fiber becomes smaller as it travels because some charges leak out of the cell across the plasma membrane. Charges that leak out no longer contribute to current flow in the cytoplasm; instead, they contribute to current returning along extracellular pathways to complete the electric circuit. Understanding the implications of the cable properties of neurons is important for understanding how current spreads through cells and how impulses are conducted along axons.

The current along axons can be modeled using circuit elements with properties that match those of ionic currents. Current entering an axon distributes itself along the axon according to the passive electrical properties depicted in the equivalent circuit shown in Figure 6-2. The components R_m and C_m are the same as those illustrated in Figure 5-10 and represent the uniformly distributed resistance and capacitance of the resting membrane. (In Figure 6-2, the elements are depicted as discrete entities for convenience.)

In an electric circuit, as in any physical system, energy must be conserved. Conservation of energy in an electric circuit requires that the sum of all the currents leaving a point within a circuit equals the sum of all the currents entering that point (Kirchhoff's first law). In addition to meeting the requirement for conservation of energy, the current flow must satisfy Ohm's law, which states that voltage equals current times resistance in a circuit (see Equation 5-1). Ohm's law implies that current distributes itself in inverse proportion to the resistances of the various routes open to it at each branch point. When the current source in Figure 6-2 generates a square pulse, the new constant-intensity current, ΔI, will flow across the "equivalent membrane," dividing at each branch point. At these points (labeled 0, 1, 2, 3, 4), some of the current will pass through the membrane resistance (R_m), and the remainder will travel through the longitudinal resistance (R_l). The current along the axon will be diminished by the loss through the membrane. The change in V_m that results from the current flow takes a short time to build up. The time necessary for V_m to stabilize depends on the membrane capacitance, because charges must accumulate on either side of the membrane to produce a given V_m. Because of membrane capacitance, the square pulse delivered at the source appears a few millimeters away as a gradual rise of potential. Thus, the membrane capacitance slows the passive transmission of the signal along the axon and distorts the signal, while the transmembrane current

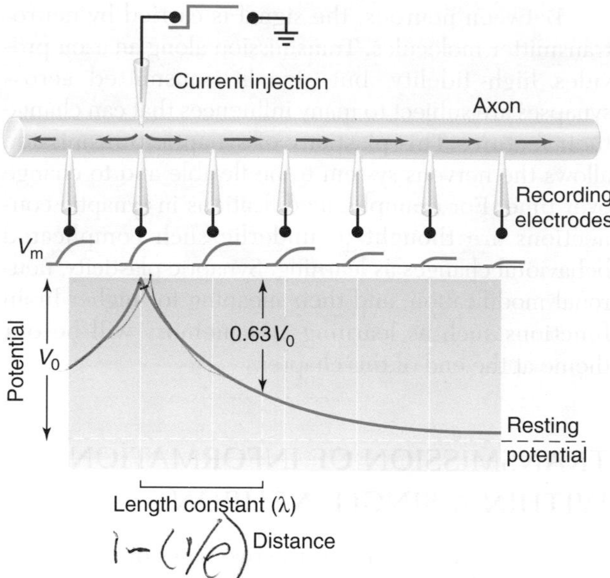

Figure 6-3 When current is transmitted electrotonically, the change in the membrane steady-state potential decays exponentially with distance from the source. The steady-state ΔV_m that is caused by a square pulse of current injected at one point diminishes along a nerve or muscle fiber with increased distance from the point of current injection. The length constant (or space constant), λ, is defined as the distance over which the potential falls by $1-(1/e)$, a reduction of 63% from its initial value at the point of injection (V_0).

through the R_ms decreases exponentially with distance from the point of current injection. All the R_ms in our model circuit have the same value, so Ohm's law requires the potential developed across them to decrease exponentially with distance. As a result, the membrane steady-state potential (ΔV_m) along an axon diminishes exponentially with distance if signals are conducted electrotonically (Figure 6-3).

This decay with distance can be described mathematically as

$$\Delta V_x = \Delta V_0 e^{-x/\lambda} \tag{6-1}$$

Here, ΔV_x is the potential change measured at a distance x away from the point at which the current is injected, and ΔV_0 is the potential change at the point $x = 0$. The symbol e denotes the base of natural logarithms. The symbol λ denotes the **length constant**, or space constant, which depends on the resistances of the axonal membrane, the cytoplasm, and the external solution according to the expression

$$\lambda = \sqrt{\frac{R_m}{R_i + R_o}} = \sqrt{\frac{R_m}{R_l}} \tag{6-2}$$

in which R_m is the resistance of a unit length of axon membrane and R_l is the summed longitudinal internal and external resistances ($R_i + R_o$) over a unit length. Equation 6-1 reveals that when $x = \lambda$,

$$\Delta V_x = \Delta V_0 e^{-1} = \Delta V_0 \frac{1}{e} = 0.37\Delta V_0 \tag{6-3}$$

As a result, λ is defined as the distance over which a steady-state potential shows a 63% drop in amplitude (see Figure 6-3).

Length constants for real axons depend critically on R_m and range from less than 0.1 mm for a small axon with a low-resistance membrane to 5 mm for a large axon with a high-resistance (nonleaky) membrane. Note that the value of λ is directly proportional to the square roots of both R_m and $1/R_l$ (Equation 6-2), so the spread of electric current along the interior of an axon is enhanced by a high membrane resistance or by a low longitudinal resistance.

The cable properties of neurons affect many aspects of neuronal function. For example, we will see shortly that the velocity at which APs are conducted along an axon is closely related to how effectively current can spread along the interior of the axon. The cable properties of neurons also shape the ways in which sensory information is processed in the nervous system, as we will see in Chapter 7.

Myelination of an axon increases the resistance of the membrane, R_m, and decreases the effective membrane capacitance. Both of these effects speed up transmission along the axon. How? (Hint: See Figures 6-2 and 6-3.)

Propagation of Action Potentials

Some neurons are so small that electrotonic conduction is sufficient for their needs. In fact, many such cells are incapable of producing APs and are therefore referred to as **nonspiking neurons.** Their graded signals are conducted electrotonically to the axon terminals without the aid of all-or-none impulses. In these nonspiking (or local-circuit) neurons, the amplitudes of the signals are attenuated as they spread through the cell, but the signals are still large enough at the terminals to modulate the release of a neurotransmitter. Nonspiking neurons are widely distributed throughout the animal kingdom; they are found in locations such as the vertebrate retina and other parts of the vertebrate central nervous system, the barnacle eye, the insect central nervous system, and the crustacean stomatogastric ganglion. Nonspiking neurons are seldom more than a very few millimeters in overall length, and they are generally characterized by a high specific membrane resistance, which contributes to a large length constant and hence to the relatively efficient and undecremented electrotonic spread of signals over these distances.

In most cases, however, communication between different parts of the nervous system depends critically on the propagation of APs along the axons of neurons, because most axons are too long for the electrotonic spread of signals to be effective. In Chapter 5, we considered the events that produce an AP at a single point in an excitable cell. In order for APs to carry information, these events must be regenerated—that is, they must take place over and over again along an axon, producing a signal that moves along the axon without decrement (Figure 6-4a on the next page). Thus, as an AP travels along an axon, each activated patch of membrane excites neighboring patches.

As we saw in Chapter 5, an action potential is typically produced by two classes of ion channels, one selective for Na^+ and the other selective for K^+. At the initiation of an AP, voltage-gated Na^+ channels open, increasing the permeability of the plasma membrane to Na^+. When the Na^+ channels are open, Na^+ ions carry a large but transient current into the axon. This inward current spreads longitudinally along the axon and then leaks out across the membrane to complete the circuit of current flow. The longitudinal spread of current depends on the cable properties discussed above. This electrotonic spread of current away from active patches of membrane is crucial for the propagation of an AP.

Let's take a closer look at what happens. When positively charged ions enter the neuron through open Na^+ channels, the potential difference across the membrane becomes less inside-negative. This change causes charges grouped near the neighboring region of the membrane to move. Positive charges in the cytosol are pushed away from the open channels, and positive charges in the extracellular fluid are pulled toward the outside mouths of the open channels. As a result, the patch of axonal membrane that is immediately ahead of the AP partially depolarizes (Figure 6-4b). Electric current can flow in any circuit only if the circuit is complete; this physical law must be obeyed in axons as well. The current that flows longitudinally within an axon must flow out of the axon across unexcited parts of the membrane, away from where Na^+ is entering, and then back into the active region of the axon to complete the circuit. (Because the conductance of the resting membrane depends primarily on open K^+ channels, the outward current is carried primarily by K^+.)

As the membrane just ahead of the impulse becomes depolarized by these local currents, voltage-gated Na^+ channels in that portion of the membrane open, bringing that region of the membrane to threshold and initiating an AP in this new patch of membrane. The newly excited region then generates local currents that depolarize, and thereby excite, the region of the axon just ahead of them. This mechanism was initially recognized by Alan Hodgkin while he was still an undergraduate (Spotlight 6-1 on page 161).

Note how different this method of signal transmission is from the decremental electrotonic conduction seen in passive transmission. The amount of depolarization required to bring a patch of inactive membrane to threshold is about 20 mV, whereas the total depolarization during an AP is typically about 100 mV. The

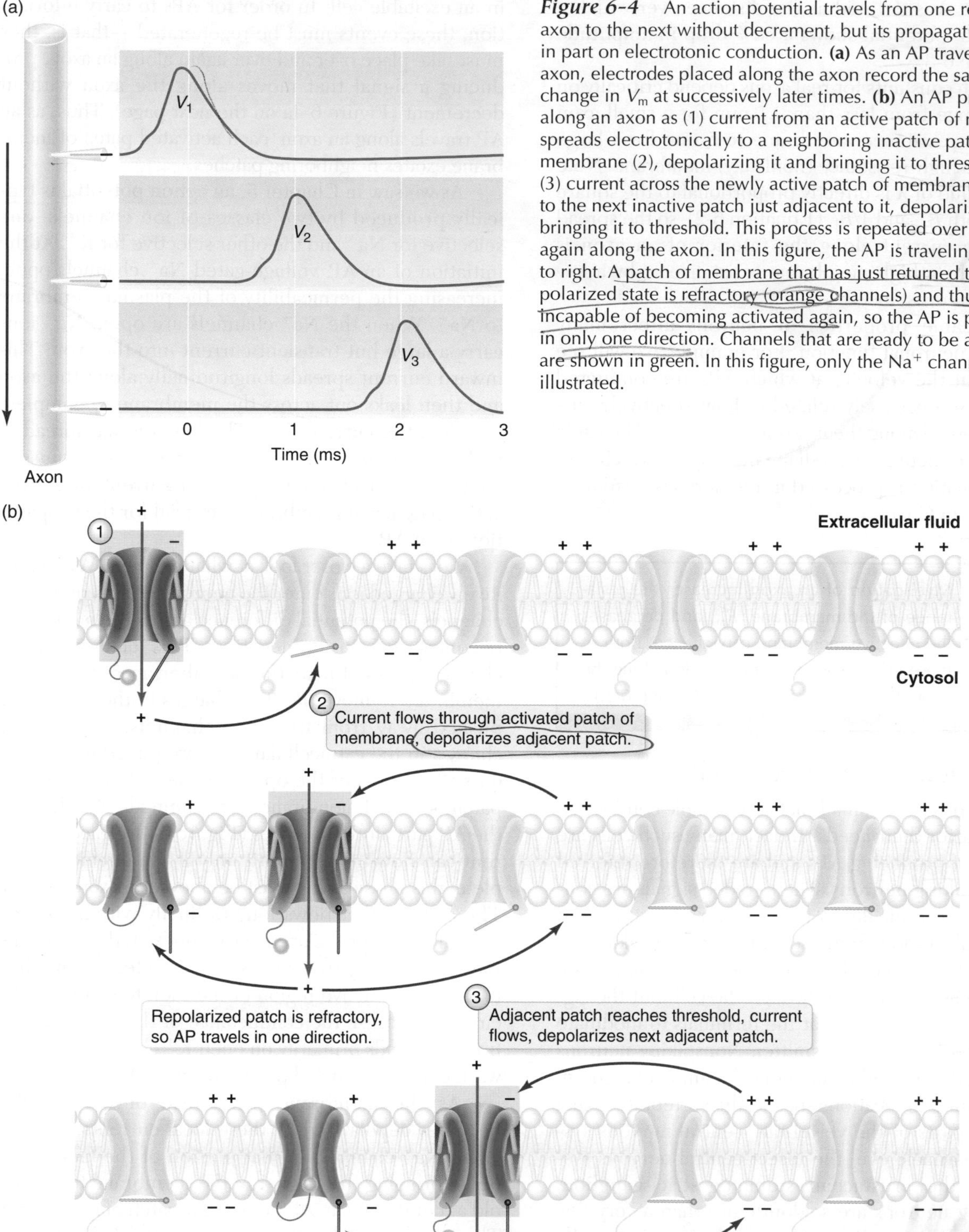

Figure 6-4 An action potential travels from one region of an axon to the next without decrement, but its propagation depends in part on electrotonic conduction. **(a)** As an AP travels along an axon, electrodes placed along the axon record the same-sized change in V_m at successively later times. **(b)** An AP propagates along an axon as (1) current from an active patch of membrane spreads electrotonically to a neighboring inactive patch of membrane (2), depolarizing it and bringing it to threshold. Then (3) current across the newly active patch of membrane spreads to the next inactive patch just adjacent to it, depolarizing it and bringing it to threshold. This process is repeated over and over again along the axon. In this figure, the AP is traveling from left to right. A patch of membrane that has just returned to its polarized state is refractory (orange channels) and thus incapable of becoming activated again, so the AP is propagated in only one direction. Channels that are ready to be activated are shown in green. In this figure, only the Na^+ channels are illustrated.

membrane all along the axon undergoes this same depolarization. Thus, an AP produces an approximately fivefold "boost" of the electrotonic signal along the axon. The energy for this amplification is derived from the unequal concentrations of Na^+ inside and outside the membrane.

Nothing prevents the current that enters an axon at the excited region from spreading backward along the axon. Indeed, if an axon has been stimulated somewhere in the middle of its length, the AP will propagate in both directions, away from the point of

6-1 CONFIRMATION THAT LOCAL CIRCUITS CONTRIBUTE TO ACTION POTENTIALS

As an undergraduate in physiology at the University of Cambridge in 1937, Alan Hodgkin confirmed the hypothesis that the inactive membrane ahead of an AP becomes depolarized by electrotonically conducted local current. His experiment is illustrated in the accompanying figure. Hodgkin blocked the conduction of APs by cooling a small part of a frog nerve; (cooling prevents the conduction of APs) He then recorded the potentials at points located successively farther from the stimulus. Any potential changes recorded beyond the blocked region must have been conducted electrotonically. Under these conditions, the depolarization of the membrane decreased exponentially with distance from the region of the block, as would be expected of electrotonically conducted current.

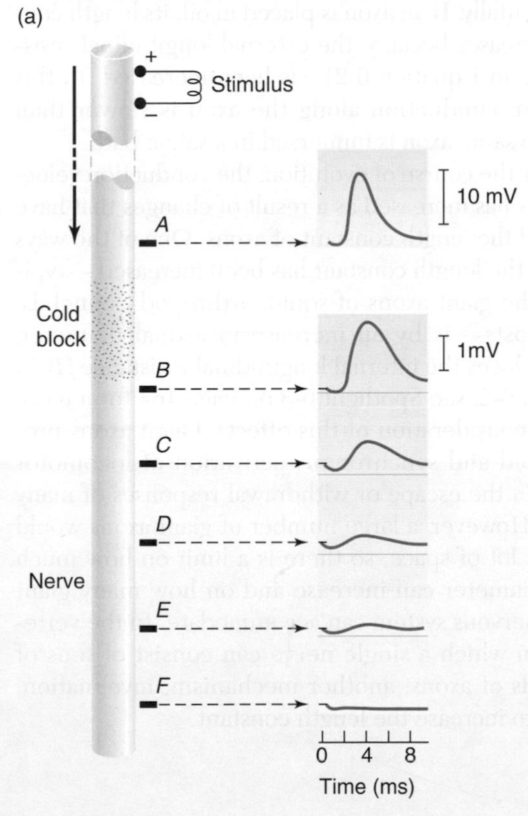

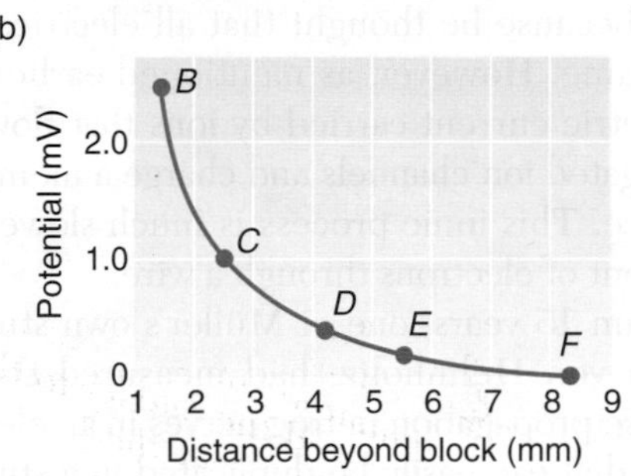

Alan Hodgkin discovered that currents are conducted electrotonically ahead of a propagated AP. ***(a)*** *Hodgkin stimulated an isolated frog nerve at one end and recorded changes in potential at several locations (points A-F) farther along the nerve. He cooled the axon at a point between the stimulating and recording electrodes (blue stippled area), blocking conduction of APs past that point. Under these conditions, he recorded changes in potential with amplitudes that dropped with distance from the cooled region.* ***(b)*** *The amplitude of the change in potential dropped exponentially with distance, suggesting that electrotonically conducted currents produced these changes in voltage. [Adapted from Hodgkin, 1937.]*

stimulation. However, in a nervous system, backward-moving current normally cannot produce a backward-traveling AP, because the membrane just behind a region of advancing excitation is in a refractory state (see Chapter 5). The Na^+ channels in the refractory region are inactivated, and the K^+ channels are still open, so the local current is simply carried out of the cell as an efflux of K^+, preventing depolarization in that region. The opening of K^+ channels is important because it allows the membrane to return quickly to a condition in which it can again be activated, priming it for future APs.

To summarize, propagation of a nerve impulse depends primarily on two factors:

1. The passive cable properties of an axon. These properties permit the electrotonic spread of local currents from a region of Na^+ influx to neighboring regions of inactive membrane.
2. The electrical excitability of Na^+ channels in the axon's membrane. The Na^+ channels carry the current that produces regenerative amplification of the passive depolarization caused by local currents.

You might ask why the extracellular currents from an axon that is conducting APs do not excite other, nearby axons, creating "cross talk" between the axons. The answer, in short, is that the resistance of inactive membranes is so much greater than the resistance of the extracellular current path that only a tiny fraction of the total current produced by the active membrane of one axon can flow into a neighboring inactive axon. This tiny current is not sufficient to bring the neighboring axon to threshold. However, extracellular currents generated by an AP can be detected by extracellular electrodes (Spotlight 6-2), which provide physiologists with a convenient way to monitor activity in the nervous system.

Speed of Propagation

Johannes Müller, a leading nineteenth-century physiologist, declared in the 1830s that the velocity of the AP would never be measured. He reasoned that the AP, being an electrical impulse, must travel at a speed approaching that of light (3×10^{10} cm $\cdot$ s^{-1}), too fast to resolve over biological distances, even with the best instruments available at that time. His reasoning is understandable because he thought that all electrical signals are the same. However, as mentioned earlier, the AP is an electric current carried by ions that flow through voltage-gated ion channels and charge a membrane capacitance. This ionic process is much slower than the movement of electrons through a wire.

Indeed, within 15 years, one of Müller's own students, Hermann von Helmholtz, had measured the velocity of impulse propagation in frog nerves in an elegant experiment that can easily be duplicated in a student laboratory (Figure 6-5). Using a frog nerve-muscle preparation, Helmholtz stimulated the nerve at two locations 3 cm apart, and measured the latency to the peak of the muscle twitch. Suppose that the latency increases by 1 ms when the stimulating electrode is moved from location 1 to location 2 in the figure. The velocity of propagation, v_p, can then be calculated as

$$v_p = \frac{\Delta d}{\Delta t} = \frac{3 \text{ cm}}{1 \text{ ms}}$$
$$= 3 \times 10^3 \text{ cm} \cdot \text{s}^{-1} = 30 \text{ m} \cdot \text{s}^{-1}$$

This value is 10 million times slower than the speed at which current flows through a copper wire or an electrolyte solution. From such experiments, Helmholtz correctly concluded that the nerve impulse is more complex than a simple longitudinal flow of current along the nerve fiber.

The conduction velocity of an AP depends primarily on how fast the membrane ahead of the active region is brought to threshold by the local currents. The greater the length constant, the farther the local currents can flow before they become too weak to bring neighboring membrane regions to threshold, and hence the more rapidly the membrane ahead of the excited region depolarizes. The importance of the length constant for conduction velocity can be demonstrated experimentally. If an axon is placed in oil, its length constant decreases because the external longitudinal resistance (R_o in Equation 6-2) has been increased. In this condition, conduction along the axon is slower than when the same axon is immersed in a saline bath.

Over the course of evolution, the conduction velocity of APs has increased as a result of changes that have increased the length constant of axons. One of the ways in which the length constant has been increased—typified by the giant axons of squid, arthropods, annelids, and teleosts—is by an increase in axonal diameter, which reduces the internal longitudinal resistance (R_i in Equation 6-2; see Spotlight 6-3 on page 164 for a more detailed consideration of this effect). Giant axons produce rapid and synchronous activation of locomotor reflexes in the escape or withdrawal responses of many species. However, a large number of giant axons would occupy a lot of space, so there is a limit on how much axonal diameter can increase and on how many giant axons a nervous system can accommodate. In the vertebrates, in which a single nerve can consist of tens of thousands of axons, another mechanism, myelination, evolved to increase the length constant.

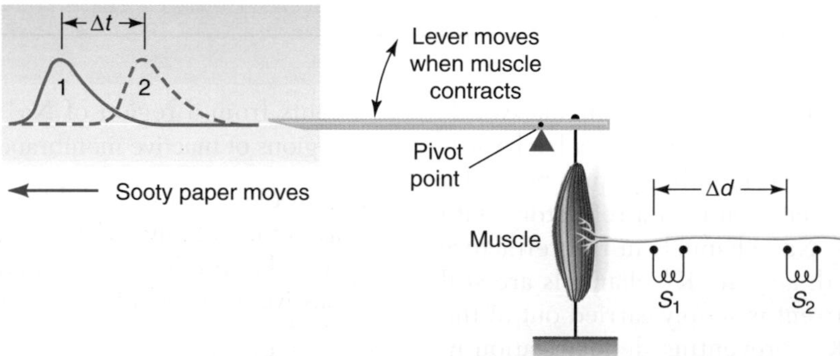

Figure 6-5 Hermann von Helmholtz measured the time from stimulus to contraction in a frog nerve-muscle preparation to determine conduction velocity along a nerve. Electrodes to stimulate the nerve were first placed at position S_1. Contraction of the muscle moved a lever that scratched a record on a sheet of sooty paper moving at a calibrated speed. (Notice that time is plotted as the paper moves.) Then the electrodes were moved to position S_2, the paper having been aligned so that the time of the second stimulus exactly coincided with the time of the first stimulus. The tracings made by the lever showed the change in latency, Δt, when the location of the electrodes was changed. The conduction velocity along the nerve was calculated using the difference in latency of the muscle twitches evoked by stimulating the nerve at the two different locations.

Spotlight 6-2 EXTRACELLULAR SIGNS OF IMPULSE CONDUCTION

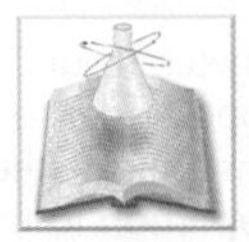

Nerve impulses can be recorded with a pair of extracellular electrodes. The recording electronics are typically arranged so that a negative potential recorded by electrode 1 causes an oscilloscope beam to go up, whereas a similar potential recorded by electrode 2 makes the beam go down; positive potentials do the opposite. An AP passing along an axon is seen from the outside as a wave of negative potential, because the cell exterior becomes more negative than its surroundings when Na^+ ions flow into the cell during the rising phase of the AP (see Figure 6-4b). As a result, when an AP passes by first one and then the other of the electrodes, it produces a biphasic waveform on the oscilloscope (part a of the accompanying figure).

The recording is simplified if the AP can be prevented from invading the part of the axon that is in contact with electrode 2, which can be accomplished by anesthetizing, cooling, or crushing that part of the axon (part b of the figure). A similar effect can be obtained by placing electrode 2 in the bath at some distance from the axon (part c of the figure).

Extracellular recordings are frequently made from nerve bundles that contain many axons (part d of the figure). In this situation, the summed activity of many axons gives a compound recording, the characteristics of which depend on the *number* of axons conducting, on the *relative timing* of their impulses, and on their *current strength*. Larger axons generate larger extracellular currents because the amount of current flowing across a membrane increases in direct proportion to the area of the membrane. The size of an AP recorded with extracellular electrodes is proportional to the amount of current flowing through the extracellular fluid, so APs from axons with a large diameter appear larger when recorded extracellularly, even though the value of ΔV_m measured with an intracellular electrode would be no greater than that of ΔV_m in smaller nerve fibers. Frequently these amplitude differences in extracellularly recorded APs allow signals carried by particular axons to be distinguished by their size.

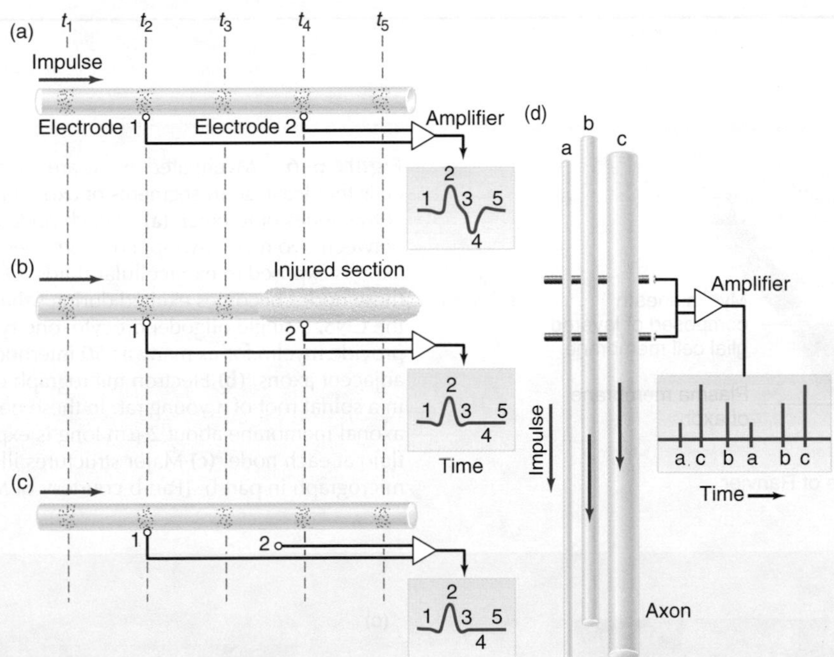

The activity of axons in a nerve bundle can be recorded with extracellular electrodes. The red stippled areas in each diagram show the location of the depolarized membrane at five different times, from t_1 to t_5. The records show what an observer would see on an oscilloscope for each method of recording illustrated. The numbers on the records correspond to the five different times. In all cases, the oscilloscope signal is produced by electronic subtraction of the potential measured by electrode 2 from the potential measured by electrode 1. ***(a)*** *A biphasic recording compares the electric potential at two locations along the nerve bundle. As an AP passes along an axon in the nerve, first one and then the other electrode records the AP.* ***(b)*** *Recording with electrode 2 placed against a crushed part of the nerve produces a monophasic record.* ***(c)*** *Recording with electrode 2 placed at a distant point in the bath also produces a monophasic record.* ***(d)*** *Extracellular recordings can pick up signals from many axons in a nerve bundle. Sometimes the sizes of the extracellular signals enable an experimenter to distinguish between fibers of different diameters. For example, the signal from axon c is always large in the record shown here, whereas the signal from axon a is always small. A larger current flowing along a larger axon produces a larger voltage difference between the recording electrodes. (Notice, though, that the distance between the electrode and the axon also can affect the amplitude of the signal.)*

AXON DIAMETER AND CONDUCTION VELOCITY

The velocity at which an AP is propagated depends in part on how far the current arising from the Na^+ influx can spread at any instant. This distance depends on the relation between the longitudinal resistance (within the axon) and the transverse resistance (across the axon membrane) encountered by currents flowing in a unit length of axon (see Equation 6-2):

$$\lambda = \sqrt{\frac{R_m}{R_i + R_o}}$$

The transverse resistance, R_m, of a unit length, l, of axon membrane is inversely proportional to the radius, *r*, of the axon, because the surface area, A_s, of a cylinder of unit length l is equal to $2\pi rl$. The longitudinal resistance of a unit length of axonal cytoplasm, R_i, is inversely proportional to the cross-sectional area, A_x, of the axon. Because $A_x = \pi r^2$, the resistance R_i is inversely proportional to the square of the radius. It follows, then, that for any increase in radius, the drop in R_i will be greater than the drop in R_m. Thus an increase in axon diameter produces an increase in λ. Typically, $R_i >> R_o$, so λ is proportional to *k* times the square root of *r*, where *k* is simply a constant; in other words, the length constant increases in direct proportion to the radius of the axon. As the radius increases, λ increases.

Because the velocity of propagation depends on the rate of depolarization at each point ahead of the AP, membrane capacitance cannot be ignored. Note that the time constant ($R_m \times C_m$) of a unit length of axon membrane remains constant as axon diameter changes, because capacitance (C_m) increases in direct proportion to increased surface area, whereas resistance (R_m) decreases in proportion to increased surface area. Thus, increased axonal diameter increases λ while leaving the membrane time constant unchanged. An increase in diameter produces a greater outward membrane current at distance *x* without an increase in the membrane time constant, and the resulting increased rate of depolarization brings the membrane to threshold sooner at every distance and increases the conduction velocity.

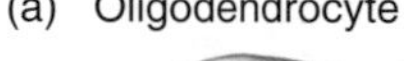

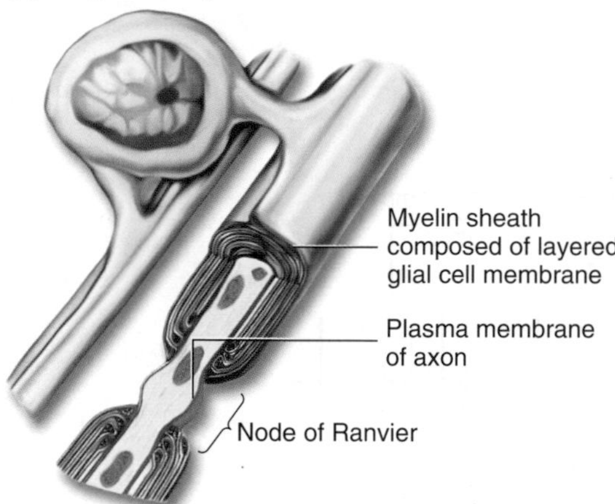

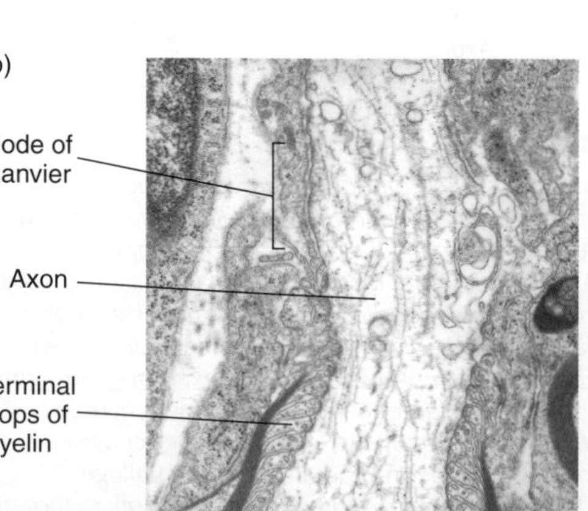

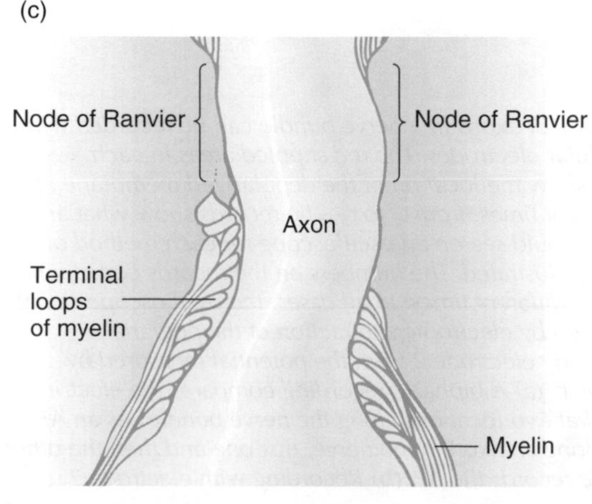

Figure 6-6 Myelinated axons are wrapped by supporting cells that leave short segments of axonal membrane exposed at the nodes of Ranvier. **(a)** At each node of Ranvier, located between two myelin-wrapped internodes, a short segment of axon is exposed to extracellular fluids. Only the membrane at these nodes becomes excited during saltatory conduction. In the CNS, a single oligodendrocyte (one type of glial cell) can provide myelin for as many as 50 internodes on several adjacent axons. **(b)** Electron micrograph of a node of Ranvier in a spinal root of a young rat. In these nerves, a segment of axonal membrane about 2 μm long is exposed to extracellular fluid at each node. **(c)** Major structures illustrated by the micrograph in part b. [Part b courtesy of Mark Ellisman.]

Rapid, Saltatory Conduction in Myelinated Axons

Some types of glial cells are wrapped around segments of axons to produce layers of fatty membranes collectively known as **myelin** (Figure 6-6a). These fatty layers have two effects on the cable properties of neurons: they increase the effective transmembrane resistance, and they decrease the effective membrane capacitance. The resistance between the cytoplasm and the extracellular fluids increases with the number of membrane layers wrapped around the axon, which may be as many as 200. The capacitance decreases because the myelin layer is very thick. This reduction in C_m means that less capacitative current is required to change V_m, so more charge can flow down the axon to depolarize the next segment. These changes in resistance and capacitance greatly increase the length constant, λ, of the axonal membrane that is covered by myelin, thus enhancing the efficiency with which longitudinal current spreads.

Myelin, however, would not improve conduction if it completely covered the axon, because the electrotonically conducted current would still eventually decrease to zero as a function of distance. Instead, the myelin sheath is segmented, with the length of each wrapped segment typically about 100 times the external diameter of the axon, ranging from 200 μm to 2 mm. The myelinated regions are separated by short, unmyelinated gaps called **nodes of Ranvier**, at which about 10 μm of axon is exposed to the extracellular fluid (Figure 6-6b,c). The regions of axon that lie under the myelin wrapping are called *internodes.*

In the course of development, myelin is laid down around the axons of vertebrates by two kinds of glial cells: Schwann cells in peripheral nerves and oligodendrocytes in the central nervous system. Between nodes of Ranvier, the myelin sheath is so snug that it nearly eliminates the extracellular space surrounding the axonal membrane. Moreover, the internodal axonal membrane has been found to lack voltage-gated Na^+ channels and K^+ channels. Thus, when a local current flows in advance of the AP, it exits the axon almost exclusively through the nodes of Ranvier. As noted earlier, very little current is expended in discharging membrane capacitance along the internodes because the capacitance of the thick myelin sheath is low. An AP that is initiated at one node electrotonically depolarizes the membrane at the next node; thus, in myelinated axons, APs do not propagate continuously along the axonal membrane, as they do in unmyelinated nerve fibers. Instead, APs are produced only in the small areas of the membrane exposed at the nodes of Ranvier. The result is **saltatory conduction,** a series of discontinuous and regenerative depolarizations that take place only at the nodes of Ranvier, as illustrated in Figure 6-7. The velocity of signal transmission is greatly enhanced because the electrotonic local currents spread rapidly along the internodal segments.

The nervous systems of all vertebrates include both myelinated and unmyelinated neurons. Typically, a single nerve contains many myelinated axons of different diameters as well as a large number of unmyelinated fibers. As a result, signals travel along a single nerve at many different velocities (Figure 6-8). However, notice

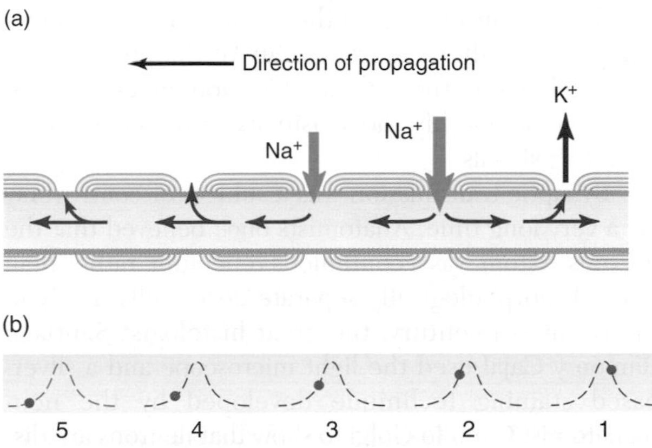

Figure 6-7 In saltatory conduction, an action potential jumps from node to node. **(a)** Current spreads longitudinally between nodes by electrotonic conduction. The orange arrows indicate Na^+ influx through open Na^+ channels at the nodes. The thinner purple arrow represents the later efflux of K^+ through activated K^+ channels. **(b)** The dots indicate the value of V_m at each node shown in part a, all at the same instant. At site 1, the membrane is in the falling phase of an AP; at site 2, the membrane is in the rising phase. At sites 3, 4, and 5, the membrane is in successively earlier phases of an approaching AP.

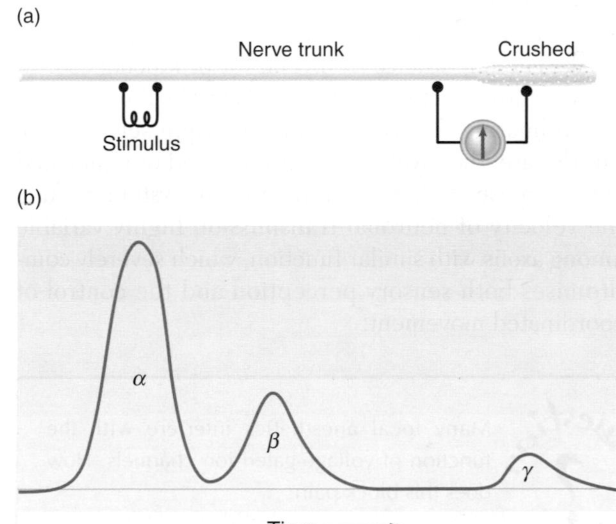

Figure 6-8 A single frog nerve contains axons with many different conduction velocities. **(a)** Experimental setup for stimulating and recording from a nerve. One end of the nerve is crushed to act as a neutral reference site (see Spotlight 6-2). Classic Work **(b)** Externally recorded "compound APs" (i.e., the summed signals from all active fibers in the nerve). The α-fibers have the largest diameter and the fastest conduction velocity. The γ-fibers have the smallest diameter and the lowest conduction velocity of those shown (but see Table 6-1). The nerve was stimulated before the beginning of the record shown here. [Adapted from Erlanger and Gasser, 1937.]

Table 6-1 The diameter of frog axons and the presence or absence of myelination control the conduction velocity.

Fiber type	Average axon diameter (μm)	Conduction velocity ($m \cdot s^{-1}$)
Myelinated fibers		
Aα	18.5	42
Aβ	14.0	25
Aγ	11.0	17
B	Approximately 3.0	4.2
Unmyelinated fibers		
 C	2.5	0.4–0.5

Source: Erlanger and Gasser, 1937.

that the conduction velocities illustrated in Figure 6-8 fall into categories, rather than varying continuously. Axons typically can be sorted into distinct groups, and the axons within each group carry particular classes of information. Interestingly, C fibers, which carry signals, are among the smallest and slowest. The conduction velocity of myelinated fibers varies from a few meters per second to more than 120 $m \cdot s^{-1}$; in contrast, unmyelinated fibers of similar diameter conduct at a fraction of a meter per second (Table 6-1).

The evolution of saltatory conduction and the resulting rapid transmission of APs was probably crucial for permitting large muscles in distant parts of the vertebrate body to be effectively coordinated. In addition, myelin allows even slender axons to conduct impulses rapidly, so that axons can be thin enough for a large number of them to be packed into a compact nerve trunk. The importance of myelin for producing coordinated neuronal activity is demonstrated by the effects of demyelinating diseases such as multiple sclerosis. In this disease, the myelin sheath is reduced or eliminated along some axons in the central nervous system, making the velocity of neuronal transmission highly variable among axons with similar function, which severely compromises both sensory perception and the control of coordinated movement.

Many local anesthetics interfere with the function of voltage-gated ion channels. How does this block pain?

TRANSMISSION OF INFORMATION BETWEEN NEURONS

All information processing done by neurons depends on the transmission of signals from one neuron to another, which is accomplished at structures called **synapses.** At **electrical synapses,** the presynaptic neuron is electrically connected to the postsynaptic neuron by proteins within the membranes. Transmission across electrical synapses proceeds very much like signal transmission along a single axon. However, electrical synapses are relatively rare. Most signaling between neurons takes place at **chemical synapses.** At a chemical synapse, APs in the presynaptic neuron cause the release of neurotransmitter molecules that diffuse across a narrow space (about 20 nm wide), called the **synaptic cleft,** that separates the membranes of the pre- and postsynaptic neurons. As recently as the 1970s, only a handful of chemicals were known to be synaptic neurotransmitters, and all synaptic transmission was thought to resemble transmission at **neuromuscular junctions,** synapses that connect motor neurons and the skeletal muscle fibers that they control. Today, more than 50 neurotransmitters have been identified in a wide range of animals, more are being discovered all the time, and we now know that their modes of action vary greatly. Whereas it was once thought that neurotransmitters simply caused a postsynaptic cell to depolarize or hyperpolarize, it is now known that they can increase or decrease the number of ion channels inserted into the membrane of the postsynaptic cell, alter the excitability of the postsynaptic cell by changing the rate at which ion channels open and close, or modify the sensitivity of the channels to activating signals.

Synaptic transmission was a subject of controversy for a very long time. Anatomists once believed that the nervous system was a continuous reticulum, rather than a set of morphologically separate nerve cells. Early in the twentieth century, the great histologist Santiago Ramón y Cajal used the light microscope and a silver-based staining technique developed by the neuroanatomist Camillo Golgi to show that neurons are discrete units, but this evidence was not immediately accepted. It was not until electron microscopy was developed in the 1940s that unequivocal evidence was obtained that neurons are indeed separate from one another and that particular regions of neurons are specialized for communication between cells.

In 1897, however, long before the ultrastructural basis of neuron-neuron interactions was determined, the functional junction between two neurons was given the name *synapse* (from the Greek, meaning "to clasp") by Sir Charles Sherrington, who is widely regarded as the founder of modern neurophysiology. It was his conclusion that "the neurone itself is visibly a continuum from end to end, but continuity fails to be demonstrable where neurone meets neurone—at the synapse. There a different kind of transmission may occur" (Sherrington, 1906). Although Sherrington had no direct information about the microstructure or microphysiology of these specialized regions of contact between excitable cells, he had extraordinary insight, the sources of which were his cleverly designed experiments exploring functions of the nervous system in whole animals.

In this section, we begin by considering synaptic transmission across electrical synapses. We then turn to chemical synapses, first investigating transmission at the neuromuscular junction, then at other, more recently studied types of chemical synapses.

Synaptic Structure and Function: Electrical Synapses

Electrical synapses transfer information between cells by direct ionic coupling. At an electrical synapse, the plasma membranes of the pre- and postsynaptic cells are in close apposition, and communication between cells takes place by way of protein channels called **gap junctions** (Figure 6-9a on the next page; see also Chapter 4). Ions can flow directly from one cell into the other through gap junctions, so an electrical signal in the presynaptic cell produces a similar, although somewhat attenuated, signal in the postsynaptic cell by simple electrotonic conduction through the junction (Figure 6-9b). A key feature of electrical synaptic transmission is its rapidity. As we will soon see, signal transmission across chemical synapses is always slower than purely electrical signal transmission.

Electrical transmission can be illustrated experimentally by injecting current into one cell and measuring the effect in a connected cell (see Figure 6-9b). A subthreshold current pulse injected into cell A elicits a transient change in the membrane potential of that cell. If enough of the current injected into cell A spreads through gap junctions into cell B, it will cause a detectable change in the V_m of cell B as well. There is a potential drop as the current crosses the gap junctions, so the subthreshold potential change recorded across the membrane of cell B will always be less than that recorded in cell A. Current generally flows through gap junctions equally well in either direction. At some electrical synapses, however, ionic current flows more readily in one direction than the other (Figure 6-9c). Such junctions are said to be *rectifying*.

The transmission of an AP through an electrical synapse is basically no different from propagation along a single axon, because both phenomena depend on the passive spread of local current ahead of the AP to depolarize and excite a neighboring region, although at an electrical synapse, the synaptic ΔV_m in the postsynaptic cell is smaller than the ΔV_m in the presynaptic cell. Notice that electrical synapses provide little flexibility in synaptic transmission, which may be one evolutionary reason why they are less common than chemical synapses. However, in places where rapid and faithful signal transmission is important, electrical synapses offer definite advantages over chemical synapses, which are much slower.

Electrical transmission between excitable cells was first discovered in 1959 by Edwin J. Furshpan and David D. Potter, who were studying the nervous system of the crayfish. Since their early work, electrical transmission has been discovered in many locations; for example, in the vertebrate retina and in other locations within the vertebrate central nervous system, between smooth-muscle fibers, between cardiac-muscle fibers, and between sensory receptor cells. The rapidity with which current crosses electrical synapses makes this means of information transfer very effective in synchronizing electrical activity within a group of cells. These synapses are also effective for rapidly transmitting information across a series of cell-cell junctions. They are found, for example, in the giant nerve fibers of the earthworm, which are composed of many segmental axons (one axon in each body segment) connected end-to-end by electrical synapses, and in the walls of the vertebrate heart, in which signals are passed between muscle cells by way of gap junctions (see Chapter 12).

At some synapses, transmission is both electrical and chemical. Combined synapses were first identified in cells of the avian ciliary ganglion. They have also been found in a circuit controlling the fish escape response, in synapses made by some neurons onto spinal interneurons of the lamprey (a primitive fish), and among the synapses onto spinal motor neurons in the frog. Combined synapses appear to offer the animal the best of both kinds of transmission. Such a connection between two neurons can be activated very rapidly, but it also offers the flexibility that we will discover when we consider chemical transmission.

Synaptic Structure and Function: Chemical Synapses

One common mode of synaptic transmission, known as **fast**, or **direct**, **chemical synaptic transmission**, is found at the neuromuscular junction and at many synapses in the central nervous system. (Although this transmission is called "fast," it is in fact considerably slower than transmission across electrical synapses.) The sequence of events at fast chemical synapses is summarized in Figure 6-10 on page 169. Briefly, when an AP travels down an axon and spreads into the axon terminals, neurotransmitter molecules that are stored in membrane-bounded spheres, called **synaptic vesicles**, are released by exocytosis into the synaptic cleft. The liberated neurotransmitter molecules diffuse across the cleft and bind to specific protein receptor molecules in the postsynaptic membrane, opening

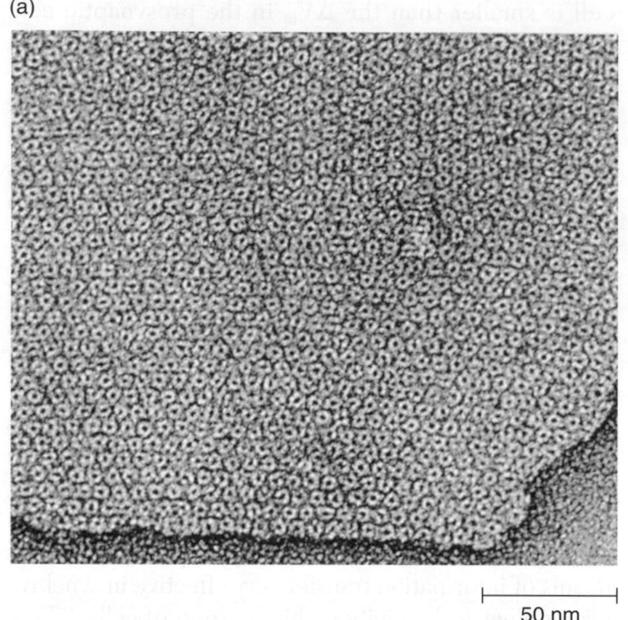

Figure 6-9 At electrical synapses, pre- and postsynaptic cells are electrically connected, permitting rapid signal transmission between the cells. **(a)** Electron micrograph of densely packed gap junctions in a plasma membrane. Each "doughnut" in this micrograph is a complex of protein subunits, called connexins in vertebrates, that form a pore through the membrane, allowing ions and small molecules to move across the membrane. In invertebrates, the subunits are called innexins and appear to be different in molecular structure. **(b)** In electrically coupled cells, the injection of current into one cell elicits a potential change in both cells. Usually, the coupling at electrical synapses is equal in both directions. Injecting current into either cell A or cell B produces a change in V_m in both cells, and the amplitude of ΔV_m is independent of the direction in which current was passed. The change in V_m is larger in the cell into which current was injected than it is in the coupled cell. Classic Work **(c)** There are, however, asymmetric electrical synapses, as illustrated by the giant electrical synapse in the crayfish. (Top) An AP in the presynaptic axon (A) is transmitted across the electrical synapse, bringing the postsynaptic cell (B) to threshold and eliciting an AP with only a small delay. This recording is a typical example of signal transmission across an electrical synapse. (Bottom) At this asymmetric electrical synapse, an AP in axon B fails to produce a significant potential change in cell A. This type of electrical synapse is said to be rectifying. [Part a courtesy of N. Gilula; part c adapted from Furshpan and Potter, 1959.]

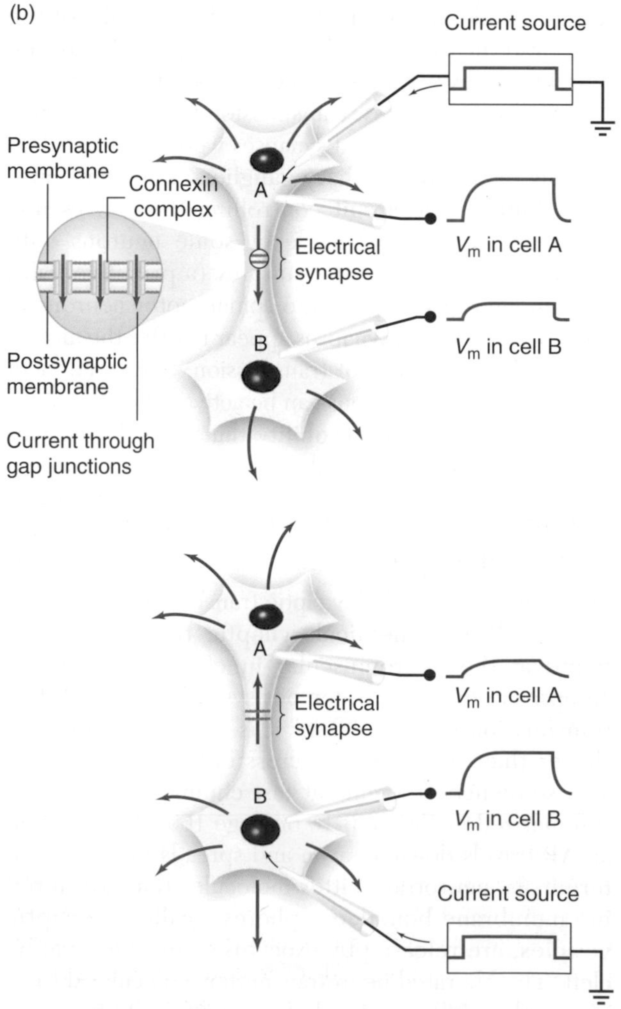

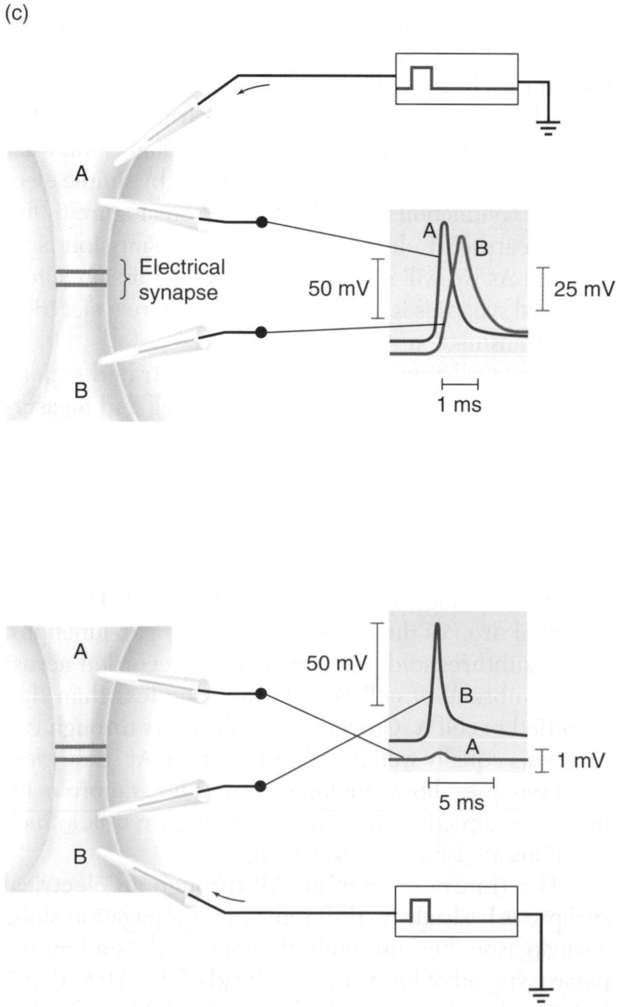

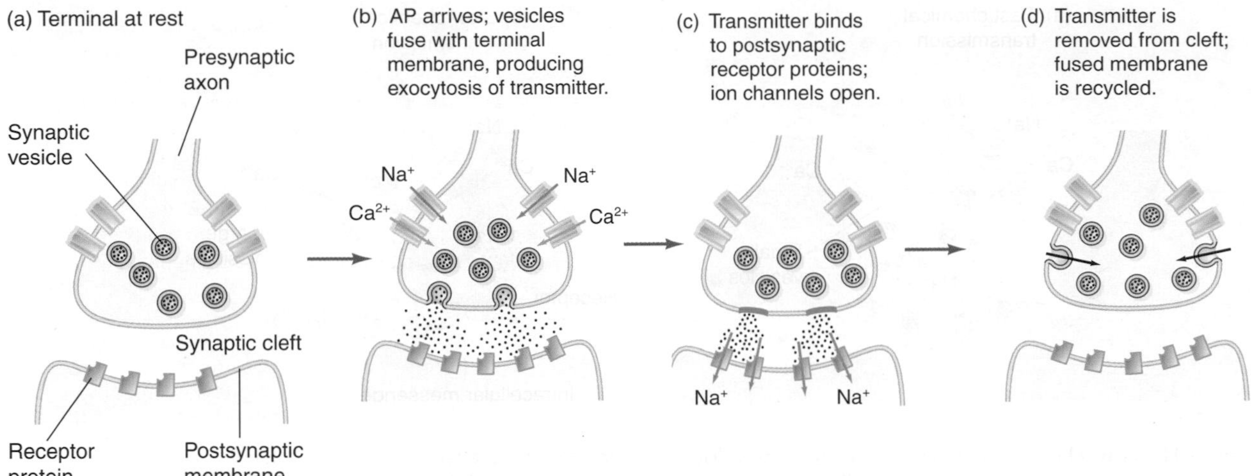

Figure 6-10 In fast chemical synaptic transmission, signals in the pre- and postsynaptic cells are linked by chemical neurotransmitters. **(a)** At rest, transmitter molecules are packaged in membrane-bounded synaptic vesicles contained in the axon terminals. **(b)** When an AP enters the presynaptic terminal, it causes voltage-gated Ca^{2+} channels in the membrane to open, allowing Ca^{2+} ions to flow into the terminal. The increase in intracellular free Ca^{2+} causes synaptic vesicles to fuse with the presynaptic membrane, releasing neurotransmitter into the synaptic cleft by exocytosis. **(c)** Neurotransmitter molecules diffuse across the synaptic cleft, driven by their concentration gradient, and some bind to receptor proteins in the postsynaptic membrane, opening ligand-gated ion channels. In this illustration, Na^+ flows through the open channels into the postsynaptic cell. The vesicle membrane (shown in orange) remains fused with the membrane of the terminal, but it moves to the sides of the terminal. **(d)** Transmitter molecules are removed from the cleft, the postsynaptic ion channels close, and the membrane that was added to the presynaptic terminal when synaptic vesicles fused is eventually recycled into the terminal (small arrows) and may be reused in new vesicles.

ligand-gated ion channels. The open channels allow a brief ionic current to flow through the membrane of the postsynaptic cell. This mechanism is the basis for fast chemical synaptic transmission in all animals.

The existence of chemical transmission and transmitter substances was the subject of intense scientific debate in the first six decades of the twentieth century. The earliest direct evidence for a chemical transmitter substance was obtained by Otto Loewi in 1921. In his experiments, he isolated a frog heart with the vagus nerve attached. When he electrically stimulated the vagus nerve, the heart rate slowed down. Taking a closer look, he found that in this process a substance was released into the surrounding saline solution that could cause a second frog heart to beat more slowly, too. Loewi's finding led to the discovery that acetylcholine is the transmitter substance released by postganglionic neurons of the parasympathetic nervous system in response to stimulation of the vagus nerve, as well as by motor neurons innervating skeletal muscle in vertebrates.

For much of the twentieth century, all synaptic transmission was thought to operate by mechanisms that were very similar to those at the neuromuscular junction. That view has changed. It is now known that, in addition to fast chemical synapses, most species also have synapses that produce **slow, or indirect, chemical synaptic transmission.** Slow transmitters affect the postsynaptic cell by activating receptors that alter the levels of signal molecules within the postsynaptic cell, which eventually modify ion channels, rather than by directly changing the conductance through a ligand-gated ion channel. The multiple steps required by this method of signaling make it slower. Physiologists believed for decades that each synaptic terminal could synthesize and release only a single neurotransmitter, another misconception recently overturned. Both physiological and anatomic evidence indicates that a single presynaptic neuron may participate in both kinds of chemical neurotransmission. In such neurons, one transmitter substance typically produces fast transmission and another produces slow transmission.

In both fast and slow chemical synaptic transmission, the transmitter molecules are packed into vesicles in the presynaptic terminal and are released by exocytosis. However, there are many differences between these two transmission mechanisms (Figure 6-11 on the next page). The transmitters used in fast transmission are small molecules, which are typically synthesized and packaged within the axon terminals. The neurotransmitters used in slow transmission are larger molecules, typically synthesized from one or more amino acids. They are called *biogenic amines* if they contain a single amino acid or *neuropeptides* if they consist of several amino acid residues. They are usually synthesized in the soma, packaged into large vesicles, and transported to the axon terminals. As the name implies, the onset of the postsynaptic response is slower (the latency of the response may be hundreds of milliseconds), and it can last much longer (from seconds to hours).

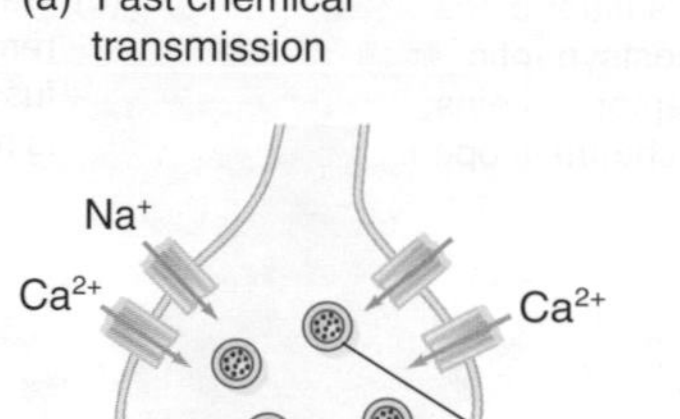

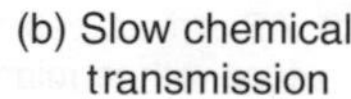

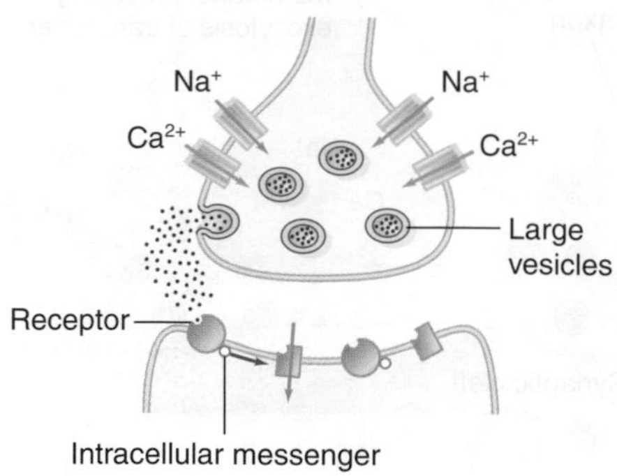

Figure 6-11 Fast chemical synaptic transmission and slow chemical synaptic transmission act through different postsynaptic mechanisms. **(a)** In fast transmission, neurotransmitters are synthesized in the axon terminals and stored in small vesicles that are clear when viewed with an electron microscope. These transmitters are typically small organic molecules. The vesicles are located near the plasma membrane, and transmitters are released by exocytosis into the synaptic cleft through specialized sites (active zones) on the terminal membrane. When they are released, these neurotransmitters act on ligand-gated ion channels in the postsynaptic membrane. **(b)** In slow transmission, the transmitters are typically larger molecules—for example, peptides containing several amino acids. These transmitters are stored in large, distinctive vesicles. They are released from sites lacking morphological specialization and located away from the active zones. In the postsynaptic cell, these neurotransmitters typically act indirectly by binding to receptors that activate intracellular signaling molecules, eventually modifying ion channels and other intracellular processes. Single neurons may produce both kinds of transmission, and a single neurotransmitter may interact with both ligand-gated channels and receptors linked to intracellular signaling pathways.

The release of neurotransmitter into the synaptic cleft is controlled by mechanisms that are common to both fast and slow chemical transmission. When an AP arrives at the axon terminal, it activates voltage-gated Ca^{2+} channels in the plasma membrane of the terminal, allowing Ca^{2+} to enter the terminal (see Figure 6-10b). The increased concentration of Ca^{2+} initiates exocytosis of the vesicles containing the transmitter, dumping transmitter molecules into the synaptic cleft, where they diffuse away from the presynaptic terminal. In fast synaptic transmission, neurotransmitter-containing synaptic vesicles fuse with the plasma membrane and release their contents at specialized sites called **active zones.** The vesicles that mediate slow synaptic transmission may release their transmitter molecules at many sites in the presynaptic terminal.

Neurotransmitters affect postsynaptic cells by modifying ionic currents traversing the postsynaptic membrane, hence producing a change in the membrane potential of the postsynaptic cell. These currents may either increase or decrease the probability that APs will occur in that cell; that is, chemical synaptic signals can be either *excitatory* or *inhibitory.* What makes a synaptic signal one or the other is examined later in this chapter.

Fast Chemical Synapses

The most extensive studies of synaptic transmission have been done on fast chemical transmission at neuromuscular junctions (also called *motor terminals* or *motor endplates*), which link motor neurons and the skeletal-muscle fibers that they control. We will use the well-studied neuromuscular junction as our primary example because fast chemical synaptic transmission between neurons within the central nervous system closely resembles transmission at the neuromuscular junction, although in many cases the transmitters are different.

Structural features

The frog motor endplate (Figure 6-12) includes structural specializations of the presynaptic terminal, postsynaptic membrane, and associated Schwann cells. At the terminal, the axon of the presynaptic motor neuron forms a small number of branches (each approximately 2 μm in diameter), each of which lies in a longitudinal depression along the surface of the muscle fiber. The muscle membrane lining the depression is thrown into transverse folds, called *junctional folds,* at intervals of 1 to 2 μm. Directly above each fold in the postsynaptic membrane is an active zone, in which are clustered many synaptic vesicles. The vesicles are released along the active zones by the process of exocytosis (Figure 6-13). There are thousands of vesicles, each about 50 nm in diameter, in a presynaptic terminal. The branches of the nerve terminal innervating a single frog muscle fiber, for example, typically contain about 10^5 synaptic vesicles. When vesicles fuse with the plasma membrane and release transmitter molecules into the synaptic cleft, the transmitter molecules reach the postsynaptic membrane by diffusing down their concentration gradient. The cleft itself is filled with a mucopolysaccharide that "glues" together the pre- and postsynaptic membranes, both of which usually show some degree of thickening at the synapse when viewed with an electron microscope (see Figure 6-12c).

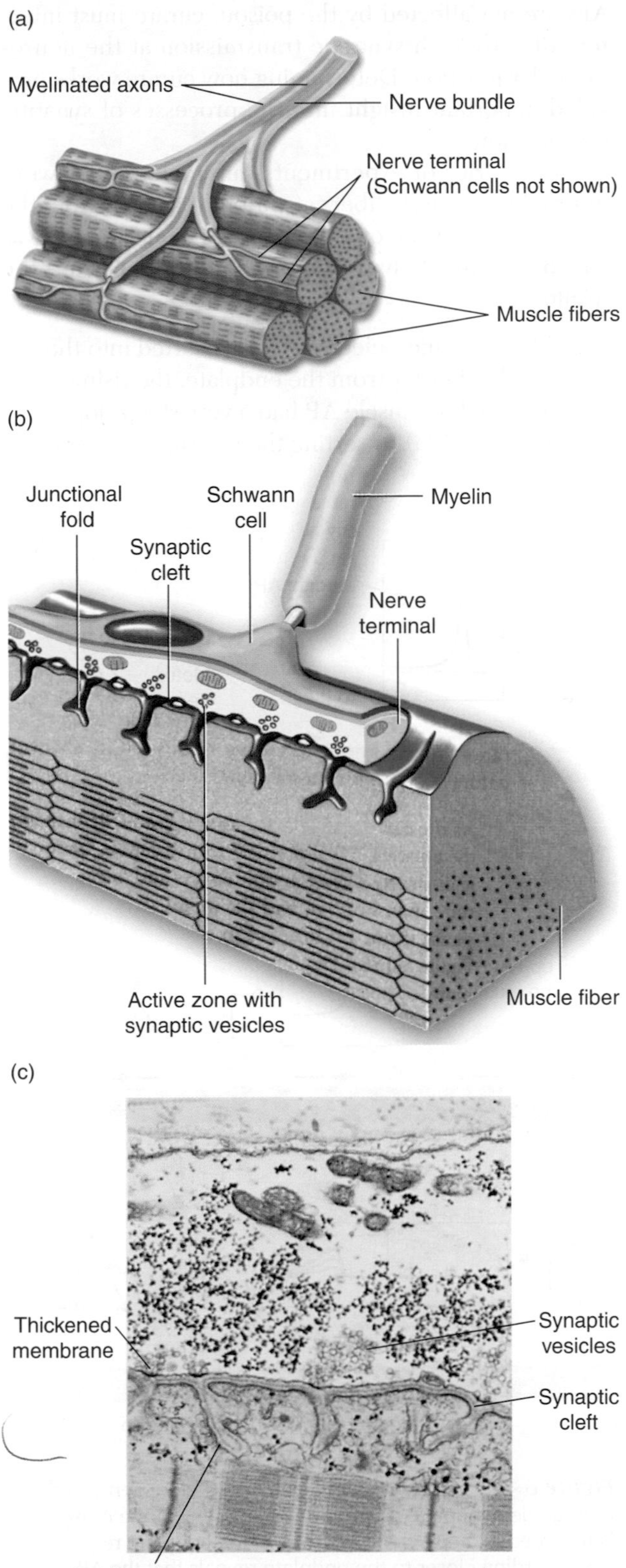

Figure 6-12 Structural specializations are found in the pre- and postsynaptic cells of the frog neuromuscular junction. **(a)** Diagram illustrating the pattern of innervation of frog muscle. Each neuron innervates several muscle fibers, but a muscle fiber is innervated by only a single neuron. The terminal Schwann cells have been removed to reveal the structure of the axon terminals. **(b)** Diagram of the neuromuscular junction. The nerve terminal lies within a longitudinal depression in the surface of the muscle fiber. The depression contains transverse junctional folds in the membrane of the muscle fiber. In the presynaptic membrane, an active zone, which is rich in synaptic vesicles, is located over each junctional fold in the muscle fiber membrane. A Schwann cell covers the terminal. Classic Work **(c)** Electron micrograph of the neuromuscular junction (false color; compare with part b). The muscle cell is shown at the bottom of the micrograph and contains striated myofibrils (see Chapter 10). The membrane of the muscle fiber is thrown into numerous junctional folds. The axon terminal is seen in longitudinal section above and contains pale synaptic vesicles grouped in bunches. Notice that below each group of vesicles the presynaptic membrane is thickened, a signature of the active zone. Denser granules and mitochondria lie above the active zones. The synaptic cleft is filled with an amorphous mucopolysaccharide. [Part c from McMahan et al., 1972.]

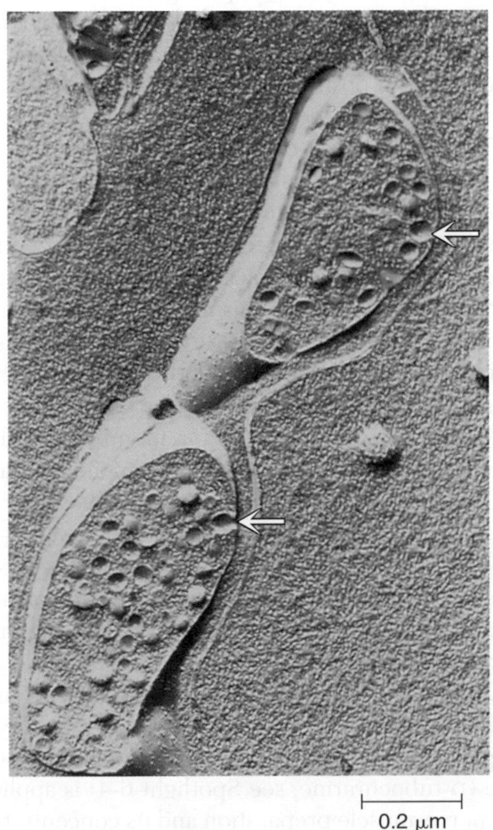

Figure 6-13 The presynaptic terminal at a neuromuscular junction contains thousands of vesicles. In this transverse section of a freeze-etched specimen from the electric organ of the ray *Torpedo,* synaptic vesicles can be seen in the terminal. Two vesicles (white arrows) had fused with the presynaptic membrane when the tissue was fixed, illustrating the process of exocytosis. [From Nickel and Potter, 1970.]

Vesicular membrane that has fused with the plasma membrane of the presynaptic terminal is taken up into the terminal and recycled (see Figure 6-10d).

Acetylcholine (ACh) is the neurotransmitter released at the neuromuscular junction. When it diffuses across the cleft, it binds to ACh-specific receptor molecules in the postsynaptic membrane, causing ion channels that are selective for Na^+ and K^+ to open briefly. At the same time, the ACh is hydrolyzed by the enzyme acetylcholinesterase (AChE) in the synaptic cleft. In essence, there is a race between diffusion of the ACh molecules across the cleft and breakdown of ACh by acetylcholinesterase. Breakdown of neurotransmitter molecules in the synaptic cleft limits the time during which the transmitter is active. In contrast to ACh, some neurotransmitter molecules are inactivated by being taken back up into presynaptic terminals, a process mediated by specialized transporter molecules.

Synaptic potentials

In 1942, Stephen W. Kuffler used extracellular electrodes to record electric potentials from single fibers of frog muscle. He discovered depolarizations intimately associated with the motor endplate that took place in response to motor neuron APs and that preceded the AP generated in the muscle cell. The potential changes were greatest in amplitude at the endplate and gradually became smaller with distance from the endplate. Given their apparent site of origin, Kuffler named them **endplate potentials (epps)**, a term still in use to refer specifically to the postsynaptic potentials in muscle fibers. Kuffler correctly concluded that the arrival of an AP in the presynaptic terminal could cause local depolarization of the postsynaptic membrane and thus initiate an AP along the membrane of the muscle fiber.

The development of the glass capillary microelectrode in the late 1940s made it possible to record potentials produced within a much smaller tissue volume and, hence, to identify more precisely the source of endplate potentials. Numerous intracellular studies of synaptic transmission at the frog neuromuscular junction, performed largely in the laboratory of Bernard Katz in England, have provided a detailed picture of electrical events at this synapse. Like a neuron, a muscle fiber has a resting potential across its plasma membrane. When a fiber is impaled by a microelectrode at a point several millimeters away from the motor endplate, the microelectrode records this resting potential and any all-or-none muscle APs that pass through. The muscle AP occurs with a delay of several milliseconds after an AP arrives in the terminal of the innervating motor axon, and the muscle fiber responds to each muscle AP with a twitch.

Katz and others used pharmacological agents to explore the nature of the nerve-muscle synapse. For example, if the South American blow-dart poison curare (D-tubocurarine; see Spotlight 6-4) is applied to a frog nerve-muscle preparation and its concentration is increased incrementally, at some particular concentration there is a sudden, all-or-none failure of the muscle APs, and the muscle fails to contract. The APs in the motor axon, however, remain unaffected. Furthermore, the muscle fiber retains the ability to generate an AP and contract if an electrical stimulus is applied directly to the fiber. Because the presynaptic and postsynaptic APs are not affected by the poison, curare must interfere directly with synaptic transmission at the neuromuscular junction. Determining how curare works provided important insight into the processes of synaptic transmission.

In a series of experiments, microelectrodes were inserted into muscle fibers at various distances from the motor endplate region (Figure 6-14), and curare was added incrementally to the preparation. Here are the results:

- When the microelectrode was inserted into the muscle fiber far from the endplate, the rising phase of the muscle AP had a very steep slope (Figure 6-14a). Inserting the electrode closer to

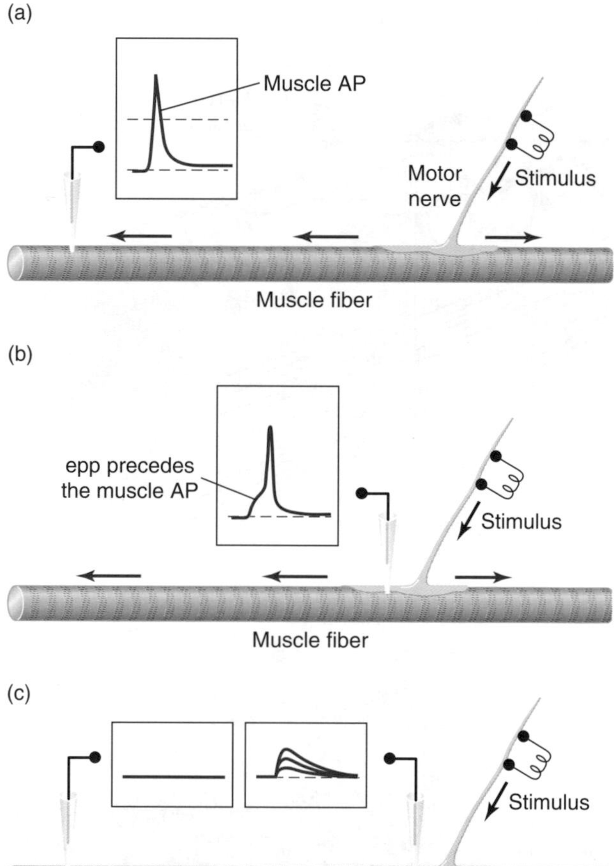

Figure 6-14 Action potentials in muscle are generated from graded endplate potentials. **(a)** An all-or-none muscle AP is recorded in a muscle fiber far from the endplate region. **(b)** Recording closer to the endplate reveals that the AP arises out of an endplate potential. **(c)** Endplate potentials can be recorded without superimposed APs if the epps can be made too small to bring the muscle fiber to threshold. Curare, a drug that blocks receptor channels in the postsynaptic membrane, provides one way to reduce the amplitude of endplate potentials. When a preparation is bathed in saline containing curare, the membrane far away from the endplate (left record) remains at its resting potential when the motor neuron fires, while at the same time graded endplate potentials are recorded near the endplate (right record).

PHARMACOLOGICAL AGENTS USEFUL IN NEUROPHYSIOLOGICAL STUDIES

Studies of signal transmission in the nervous system have been greatly aided by the discovery and application of natural toxins—from animals, plants, or fungi—that either selectively interfere with or mimic certain steps in the process of transmission. Toxins have been found that interact with ion channels, with receptors, and with enzymes important for nervous system function. Some commonly used agents that have been valuable in studies of axonal transmission of APs and synaptic transmission are described here.

Channel Toxins

Several toxins interact specifically with particular types of voltage-gated ion channels. Tetrodotoxin (TTX), derived from puffer fish (*Sphoeroides* sp.), binds to a site on voltage-gated Na^+ channels and blocks Na^+ current across the membrane. Similarly, saxitoxin (STX), derived from dinoflagellates, blocks voltage-gated Na^+ channels, although by a slightly different mechanism. Finally, μ-conotoxins, from the venom of piscivorous (fish-eating) snails of the genus *Conus,* block voltage-gated Na^+ channels from the extracellular side. Voltage-gated K^+ channels can be blocked by several agents. For example, tetraethylammonium (TEA), a synthetic organic compound, blocks most types of K^+ channels from either inside or outside the membrane, and 4-amino pyridine blocks several types of K^+ channels. Voltage-gated Ca^{2+} channels are blocked by any of several ω-conotoxins derived from cone snails (the venom of *Conus geographus* is particularly well studied). The various subtypes of this toxin block different classes of Ca^{2+} channels.

Several agents are known to act on postsynaptic ion channels. Toxins that act on ionotropic glutamate receptors have proved invaluable in distinguishing among the variety of iGluR channel types. Kainate, from a red alga *(Digenea simplex),* is an effective agonist for one subtype of glutamate receptors. AMPA is a second potent agonist that is selective for another receptor subtype. A particularly useful category of antagonists has been the conatokins, from cone snails, which are noncompetitive antagonists of a third class of iGluR receptors, called NMDA receptors for *N*-methyl-D-aspartate, a chemical that activates them.

Presynaptic Toxins

Several toxins act on presynaptic terminals to inhibit transmitter release. β-Bungarotoxin (β-BuTX), derived from the venom of the krait (a relative of the cobras), inhibits transmitter release by permeabilizing the axon terminal. Notexin from the tiger snake also inhibits transmitter release, causing lethal paralysis. Two toxins from bacteria of the genus *Clostridium* prevent the release of neurotransmitters by affecting the SNAREs in the presynaptic terminals (see Figure 6-29). Botulinum toxin blocks the release of ACh at the neuromuscular junction, producing flaccid paralysis. Tetanus toxin, from another species of the genus, blocks the release of glycine from inhibitory neurons that synapse onto spinal motor neurons, producing rigid paralysis. Evolution has made these toxins highly effective for incapacitating prey (i.e., only very small amounts are needed), and they must be handled with great caution in the laboratory.

Postsynaptic Receptor Toxins

Agonists and antagonists for different receptor subtypes have contributed importantly to defining the roles of these receptors in neuronal communication. γ-Aminobutyric acid (GABA), which usually acts as an inhibitory neurotransmitter, has been studied extensively with a pair of chemicals, one an agonist of GABA and the other an antagonist. The agonist, muscimol, is derived from the mushroom *Amanita muscaria.* It specifically activates $GABA_A$ Cl^- channels. Bicuculline, produced from the plant *Dicentra cucullaria,* is a competitive antagonist of the same channel.

A huge collection of reagents exists for ACh receptors. Muscarine and pilocarpine activate muscarinic ACh receptors. In vertebrates, muscarinic ACh receptors are most prevalent in the visceral tissues that are innervated by the cholinergic axons of the parasympathetic nervous system. Atropine (belladonna) is a plant-derived alkaloid that blocks muscarinic synaptic transmission. The name *belladonna* is derived directly from the Italian expression "beautiful woman." Due to its effects on muscarinic synapses in the iris of the eye, atropine produces widely dilated pupils, which was thought to be a mark of great beauty in Renaissance Italy.

Nicotine, another plant alkaloid, and carbachol act as agonists of nicotinic ACh receptors. D-Tubocurarine is the active principle of curare, the South American blow-dart poison, made from the plant *Chondodendron tomentosum.* This molecule blocks transmission postsynaptically by competing with ACh for the ACh-binding site of nicotinic receptors. It binds competitively to this site without opening the channel, thereby interfering with the generation of a postsynaptic current. Another nicotinic receptor blocker, α-Bungarotoxin (α-BuTX) is isolated from the venom of the krait. This protein molecule binds irreversibly and with very high specificity to nicotinic ACh receptors. With the use of radioactively labeled α-BuTX, it has been possible to determine the number of ACh receptors present in a membrane, as well as to isolate and purify the receptor protein.

Eserine (physostigmine) is an anticholinesterase; that is, it blocks the action of acetylcholinesterase. Use of this alkaloid has enabled physiologists to measure the amount of ACh released at a synapse because the drug prevents the rapid enzymatic destruction of the transmitter molecules. Partial doses accentuate the postsynaptic potential at cholinergic synapses.

the endplate revealed that the AP arises from a slower change in potential: the endplate potential (Figure 6-14b).

- Adding curare to the saline bathing the muscle fiber modified the endplate potential. At a low concentration of curare, some of the postsynaptic receptors were blocked and the synapse was thus weakened, causing the endplate potential to rise more slowly.
- When the concentration of curare was high enough to depress the amplitude of the endplate potential below the threshold potential for an AP

in the muscle, there was an abrupt failure of muscle APs. However, endplate potentials were still recorded (Figure 6-14c). Further increases in the concentration of curare reduced the initial slope and amplitude of the epp even more.

These results suggest that curare interfered with synaptic transmission in proportion to its concentration and thus reduced the size of the epp more and more as its concentration increased. If the concentration of curare was sufficient to reduce the size of the epp to just below threshold, the AP was eliminated, and the epp was revealed. If the recording electrode was now reinserted into the muscle fiber at progressively greater distances from the motor endplate, the amplitude of the measured epp dropped approximately exponentially with distance from the endplate (Figure 6-15). In contrast to the AP, which propagates without attenuation because it is regenerative, the epp spreads electrotonically and thus decays with distance. Normally the epp is masked by the much greater potential change of the AP, but by exploiting the effects of curare, physiologists were able to distinguish between these two elements of the synaptic response, launching new exploration into nerve-muscle communication.

(a)

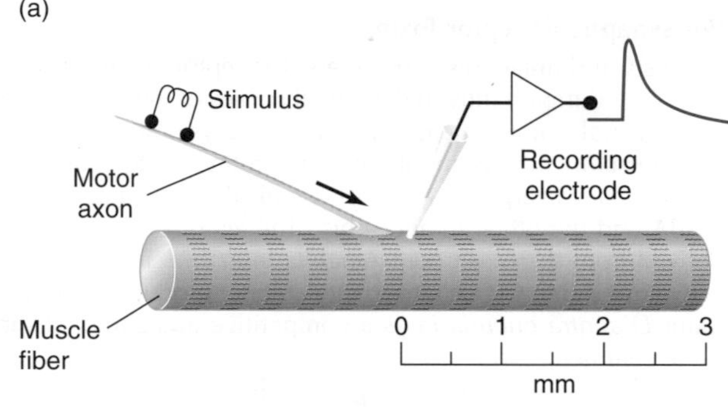

(b)

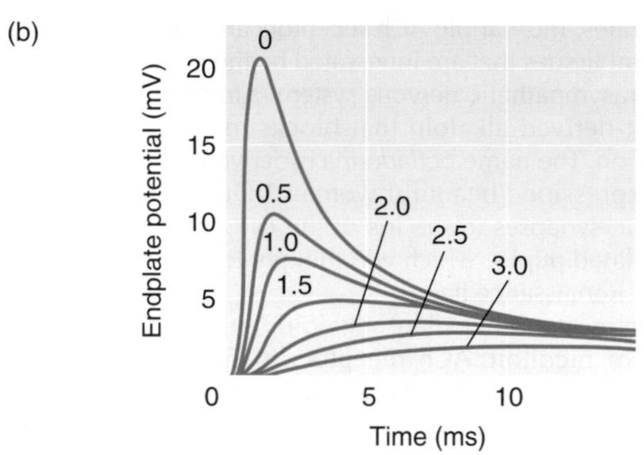

(c)

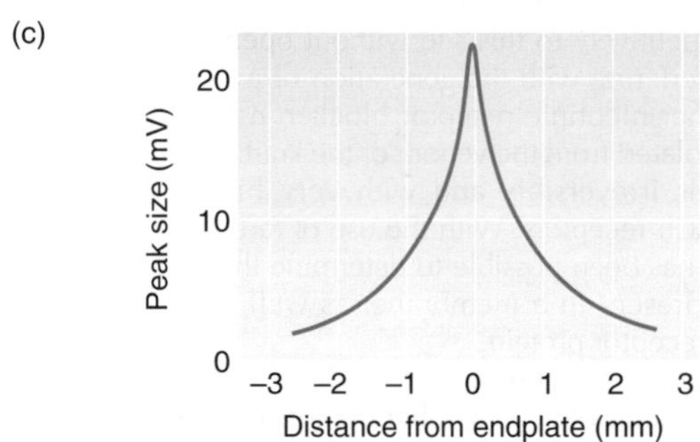

Figure 6-15 The amplitude of an endplate potential decays exponentially with distance from the motor endplate. **(a)** The endplate potential was recorded with a microelectrode that was sequentially inserted at 0, 0.5, 1.0, 1.5, 2.0, 2.5, and 3.0 mm from the endplate in a partly curarized frog muscle fiber.

Classic Work **(b)** Recordings of epps at each location. The distance away from the endplate (in millimeters) is indicated for each recording. Notice that the amplitude and initial slope of the record decrease with distance, similar to the signal in Figure 6-2.

Classic Work **(c)** A plot of the peak potential of each recording shows that the amplitude of the epp decreases approximately exponentially with distance from the endplate. [Adapted from Fatt and Katz, 1951.]

Synaptic currents

As described in Chapter 5, a change in membrane permeability to one or more ionic species (i.e., the opening or closing of a population of membrane channels that selectively passes those ions) typically shifts the membrane potential toward a new level. The change in the flow of ions changes the transfer of charge from one side of the membrane to the other, which in turn causes the transmembrane voltage, V_m, to change. The change in the rate of ion flow across the postsynaptic membrane is called the **postsynaptic current** (or **psc**). In chemical synaptic transmission, ion-specific channels in the postsynaptic membrane open (or close) when neurotransmitter molecules bind to receptor proteins, changing the amount of ionic current crossing the membrane. The direction and intensity of the postsynaptic current, which are controlled by the size of the conductance through the open channels and by the electrochemical driving force and charge on the permeant ions, determine the polarity and the amplitude of the postsynaptic potential.

The ionic currents that underlie postsynaptic potentials can be recorded by voltage-clamping the postsynaptic membrane, thus holding the postsynaptic potential constant (see Spotlight 5-5). In a nerve-muscle preparation, this procedure must be carried out close to the motor endplate (Figure 6-16a). The motor nerve (the presynaptic element) is stimulated while V_m of the postsynaptic membrane is voltage-clamped at some predetermined value. The release of transmitter by the presynaptic terminal is quickly followed by a postsynaptic current of ions moving down their electrochemical gradients through open channels in the postsynaptic membrane (Figure 6-16b).

The ions responsible for carrying the postsynaptic current at particular synapses have been identified through experiments in which the extracellular concentrations of each ionic species were changed one at a time and the resulting effect on the postsynaptic current was measured. Such measurements have demonstrated that the depolarizing postsynaptic current at the vertebrate neuromuscular junction consists of an influx of Na^+ that is partly canceled under normal conditions by a simulta-

(a)

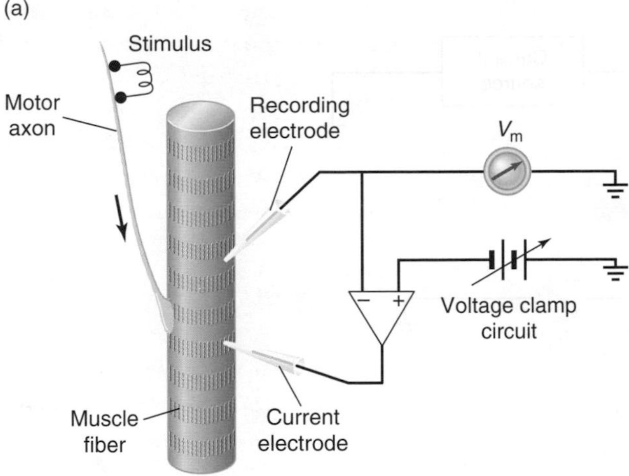

(b)

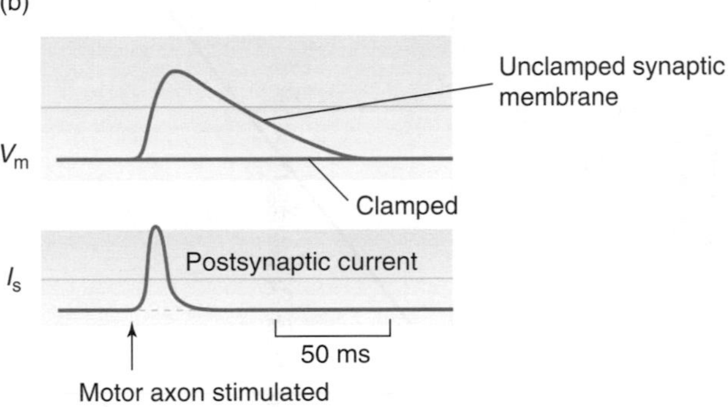

Figure 6-16 Voltage-clamping the postsynaptic membrane allows the postsynaptic current to be measured. **(a)** Setup for voltage-clamping the muscle membrane so that the postsynaptic potential can be held constant while ionic current flowing across the postsynaptic membrane is recorded (see Spotlight 5-5). **(b)** The upper record shows an endplate potential (or epp) when the nerve is stimulated and the muscle is not voltage-clamped. It also shows that under voltage clamping, the transmembrane voltage remains constant following stimulation of the neuron. The lower record shows a postsynaptic current (or psc) when the muscle fiber is voltage-clamped under the same conditions. The postsynaptic current rises and decays much faster than the endplate potential.

neous smaller efflux of K^+. At this synapse, both Na^+ and K^+ ions pass through the very same postsynaptic ACh-activated channels, indicating that these channels have a broader ion selectivity than do the highly selective, voltage-gated Na^+ and K^+ channels that underlie APs.

Postsynaptic currents are shorter-lived than postsynaptic potentials (see Figure 6-16b). Acetylcholine-activated channels open only momentarily because the transmitter at the neuromuscular junction is rapidly removed from the cleft by enzymatic degradation, after which the channels close and the postsynaptic current ceases to flow. A postsynaptic potential lasts longer than a postsynaptic current because its time course depends not only on the duration of the postsynaptic current, but also on the time constant of the membrane (see Chapter 5). Because the membrane consists of both resistive and capacitative elements, time is required to charge and discharge the membrane, slowing down changes in V_m.

Reversal potential

At every fast chemical synapse, one or more ionic species carries current across the postsynaptic membrane, and the ΔV_m caused by this current determines whether the synapse is excitatory or inhibitory—that is, whether the postsynaptic potentials will make APs more or less likely in the postsynaptic cell. Measuring the properties of the postsynaptic current provides an experimenter with clues to which ions are in motion. These measurements can be made by injecting current into the postsynaptic cell to set the membrane potential at different values and then observing the sign and amplitude of the postsynaptic potential produced by synaptic inputs (Figure 6-17a,b on the next page). The amplitude and sign of the postsynaptic potential depend on the transmembrane voltage prior to the psp and on the species of ion or ions carrying the current. Remember: activating membrane channels that select for a given ionic species, X, causes V_m to move closer to the equilibrium potential, E_X, for that ion (see Chapter 5).

Consider the experiment illustrated in Figure 6-17 for a synapse at which only one ionic species, X, carries the postsynaptic current. As the membrane potential, V_m, is shifted toward the equilibrium potential, E_X, the driving force on X (that is, $V_m - E_X$) decreases. When $V_m = E_X$, no current flows across the membrane, even though the channels are open, because there is no driving force on the ions. If V_m is then set on the other side of E_X, current will once again flow, because $V_m - E_X$ will again be nonzero, but X will flow through the open channels in the opposite direction, and the sign of the postsynaptic potential will be reversed (Figure 6-17b,c). Because the direction of the ionic current and the sign of the postsynaptic potential reverse as V_m passes through E_X, E_X is called the **reversal potential,** E_{rev}, of that current. When postsynaptic channels open, the postsynaptic current causes V_m to shift toward the E_{rev} of the current, no matter where V_m was set experimentally before the synapse was activated (as long as the cell is not being voltage-clamped). The reversal potential has proved to be a useful property because it provides a hint about which ions carry the postsynaptic current. In fact, before membrane patch-clamp recording was introduced (see Spotlight 5-6), measuring the reversal potential of a current was the primary method of distinguishing the ionic species that produced a particular postsynaptic potential, although it was, by itself, not conclusive.

If a single ionic species carries the postsynaptic current, the reversal potential, E_{rev}, can be calculated by using the Nernst equation (Equation 5-3) for that ionic species. However, many postsynaptic channels are permeable to more than one ionic species, as is the acetylcholine channel. In that case, E_{rev} depends on the concentrations and relative permeabilities of all of the

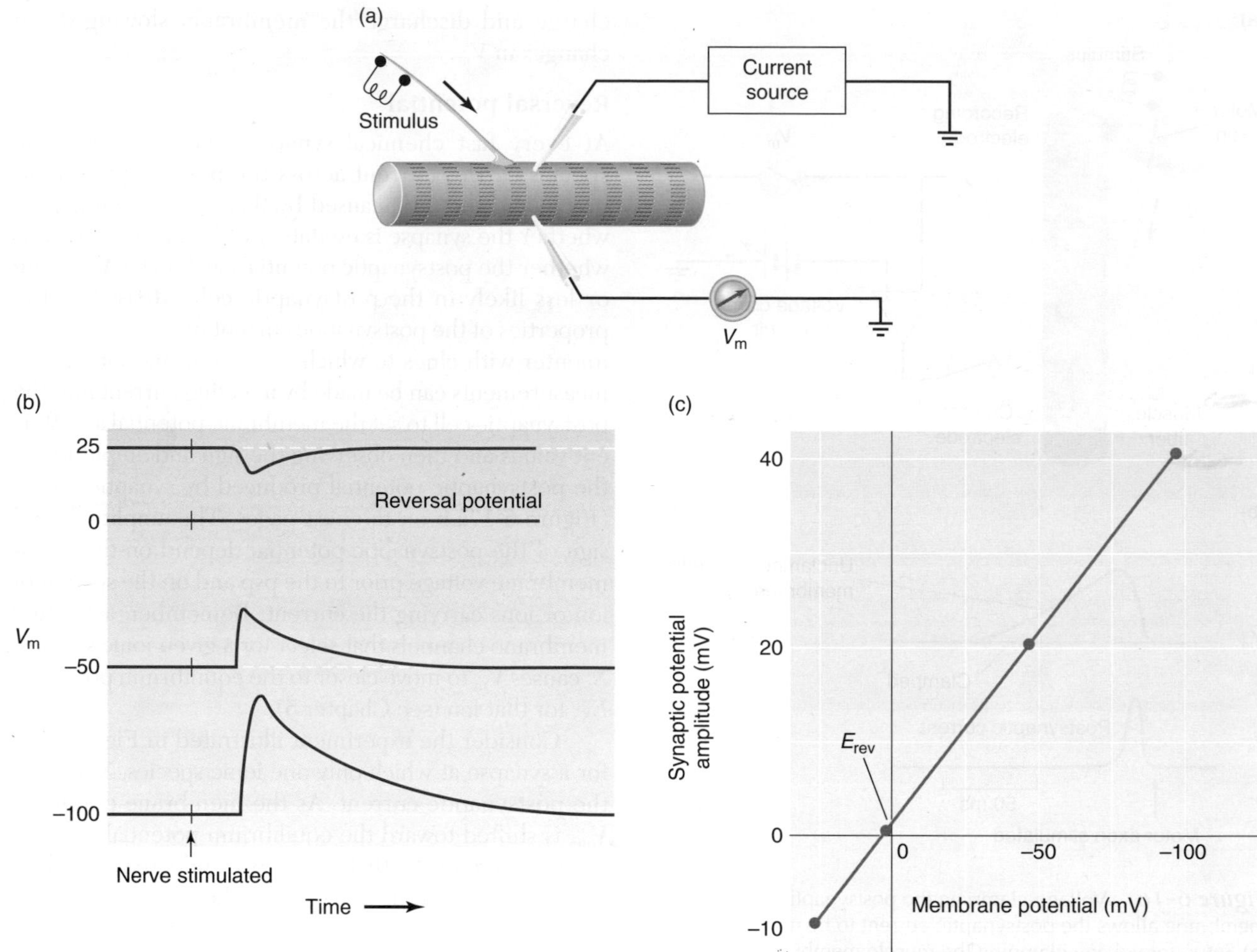

Figure 6-17 The reversal potential is measured by changing the membrane potential and recording the postsynaptic potential. **(a)** Method for determining the reversal potential (E_{rev}) at a synapse. Steady current is injected into the postsynaptic cell (here, a muscle fiber) with an electrode to set the initial value of V_m. (Notice that V_m is *not* voltage-clamped; it can vary if transmembrane currents change.) At each value of V_m, an endplate potential is produced by stimulating the presynaptic nerve. The resulting endplate potentials are recorded with a second electrode inserted into the postsynaptic muscle fiber. **(b)** At first (bottom), V_m is set at values that are more negative than the equilibrium potential, E_X, for the ions carrying the synaptic current. As V_m is made less negative, the amplitude of the endplate potential becomes smaller, because the driving force on the ions traversing the membrane through postsynaptic channels is lower. When V_m is set equal to E_X, no postsynaptic current flows, and the amplitude of the postsynaptic potential is zero, even though the postsynaptic channels are open. When V_m is set at values more positive than E_X, the driving force on the ions carrying the postsynaptic current is opposite to its direction when V_m is more negative than E_X. As a result, ion flow changes direction, and the sign of the endplate potential reverses (in this case, it becomes hyperpolarizing). **(c)** The results of this type of experiment can be plotted to show the amplitude of the endplate potentials as a function of V_m. The line fitting the experimental points crosses the abscissa at E_{rev}. In this case, E_{rev} is about 0 mV.

participating ions. If the concentrations of the various ionic species and their permeabilities through the postsynaptic channels are known, E_{rev} can be predicted using the Goldman equation (Equation 5-4) rather than the Nernst equation. Alternatively, if the current is carried by only two ionic species, E_{rev} can be calculated from Ohm's law (Spotlight 6-5).

The ACh-activated channels at vertebrate neuromuscular junctions provide an example of this calculation. When these channels open, they become permeable to both Na^+ and K^+, and the reversal potential of the current, E_{rev}, will lie between the equilibrium potentials of the two permeant ions (Figure 6-18). In the experiment shown in Figure 6-18, V_m was electronically held at each of several different values, as shown in Figure 6-17b, and the synapse was activated at each value. When V_m was held at E_{Na} (trace 1), the driving force on Na^+ was zero ($V_m - E_{Na} = 0$), but there was a large driving force on K^+ ($V_m - E_K >> 0$). In this condition, the postsynaptic current is carried entirely by an outward flux of K^+, which drives V_m to a more negative value. In contrast, when V_m is held at E_K (trace 5), there is no driving force on K^+, but there is a large driving force on Na^+. In this case, all of the current through the

CALCULATION OF REVERSAL POTENTIAL

The value of the reversal potential of an ionic current elicited by a stimulus or a neurotransmitter depends on the relative conductances of the ions carrying the current as well as on their equilibrium potentials. If it is likely that only Na^+ and K^+ carry current through a particular set of postsynaptic ion channels, the reversal potential of that current can be related to the conductances for these ions by using Equation 5-7, with the values g_K and g_{Na} representing transient changes in the conductances for the two ions.

$$I_K = g_K(V_m - E_K) \tag{1}$$

$$I_{Na} = g_{Na}(V_m - E_{Na}) \tag{2}$$

At the reversal potential, I_K and I_{Na} must be equal and opposite regardless of their relative conductances, because the net current must be zero. Thus, when V_m *is at the reversal potential,* E_{rev},

$$-I_K = I_{Na} \tag{3}$$

Substituting from Equations 1 and 2, at the reversal potential we have

$$-g_K(V_m - E_K) = g_{Na}(V_m - E_{Na}) \tag{4}$$

From this equation, if g_K is greater than g_{Na}, then V_m must be closer *to* E_K than to E_{Na}, and vice versa. Solving Equation 4 for $V_m = E_{rev}$ gives

$$E_{rev} = \frac{g_K}{g_{Na} + g_K} E_K + \frac{g_{Na}}{g_{Na} + g_K} E_{Na} \tag{5}$$

From Equation 5, it is apparent that E_{rev} will not be simply the algebraic sum of E_{Na} and E_K, but will lie somewhere between the two, depending on the ratio g_{Na}/g_K. Thus, if g_{Na} and g_K become equal to each other (as they might when endplate channels are activated by ACh in frog muscle), the membrane potential will shift toward a reversal potential that lies exactly halfway between E_{Na} and E_K:

$$E_{rev} = \left(\frac{1}{2}\right) E_K + \left(\frac{1}{2}\right) E_{Na} = \left(\frac{1}{2}\right)(E_K + E_{Na})$$

For frog muscle, E_K is about -100 mV, and E_{Na} is about $+60$ mV. Hence, we would predict that during synaptic activation of a frog muscle, $E_{rev} = \frac{1}{2}(-100 + 60) = -20$ mV. The measured reversal potential of the postsynaptic current at the frog neuromuscular synapse, -10 mV, is somewhat more positive than this value, possibly because g_{Na} is actually somewhat greater than g_K.

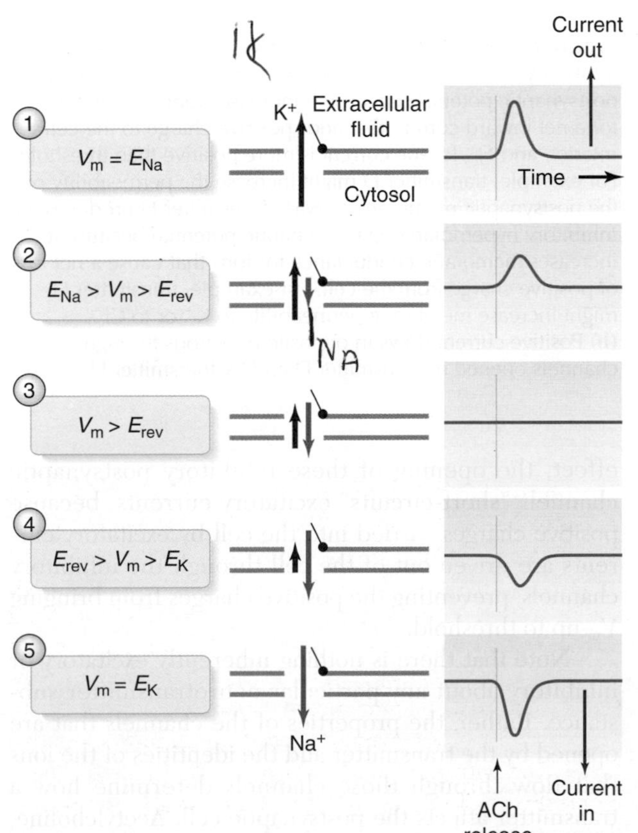

Figure 6-18 The postsynaptic current at the vertebrate neuromuscular junction is carried by both Na^+ and K^+ ions. Sodium and potassium currents through activated acetylcholine (ACh) channels are measured at different membrane potentials, beginning at E_{Na} (trace 1). The ACh-activated channels are approximately equally permeable to Na^+ and K^+, so the magnitudes of I_{Na} and I_K depend on the driving force on each ion. The relative magnitudes of the Na^+ and K^+ currents are represented by the lengths of the arrows, purple for I_K and orange for I_{Na}. The amplitude and time course of the net current through the channels are shown on the right. At E_{rev} for the combined currents, the net current through the channels is zero.

ACh-activated channel is carried by an influx of Na^+, and V_m becomes more positive. Somewhere between E_{Na} and E_K, there must be a value of V_m at which the Na^+ and K^+ currents through this channel will be equal and opposite to each other, so that, although both ions flow through the channel, there will be no net current (trace 3). This value of V_m is the reversal potential for the ACh-activated current. For frog endplate channels, the conductances for the two permeant ions, Na^+ and K^+, are approximately equal. Notice that the postsynaptic current cannot drive V_m past E_{rev}, regardless of how many channels become activated. When V_m reaches E_{rev}, the net driving force on the permeant ions drops to zero, and V_m cannot change further. As a result, E_{rev} sets the maximum change in V_m that can be produced by activation of the channels.

The reversal potential has a special functional significance at synapses, because the relation between E_{rev} and the threshold for excitation in the postsynaptic cell determines whether a synaptic potential will be excitatory or inhibitory. We will discuss this topic next.

Postsynaptic excitation and inhibition

Any change in V_m at a postsynaptic membrane that increases the probability that an AP will be initiated in the postsynaptic cell is called an **excitatory postsynaptic potential (epsp)**. Conversely, any change in V_m that reduces the probability of an AP in the postsynaptic cell is an **inhibitory postsynaptic potential (ipsp)**. If the reversal potential, E_{rev}, of a postsynaptic current is more positive than the threshold of the postsynaptic cell, that synapse is excitatory (Figures 6-19a and 6-20a). If E_{rev} is more negative than threshold, the synapse is inhibitory, even if E_{rev} is more positive than V_{rest}. We will consider this statement further below.

At fast chemical synapses, excitatory currents are typically carried through channels that conduct Na^+ or Ca^{2+}. These channels may be permeable to K^+ as well—the ACh channel of the vertebrate neuromuscular junction is one example—and at such synapses any K^+ that moves through the channels carries positive charge out of the cell, reducing the depolarization caused by Na^+ or Ca^{2+} (see Figure 6-18). Inhibitory postsynaptic currents are typically carried through channels that are permeable either to K^+ or to Cl^-. The reversal potential, E_{rev}, for K^+ or Cl^- typically lies near V_{rest}, so it is more negative than the threshold. If E_{rev} for inhibitory channels is more negative than V_{rest} in the postsynaptic cell, the postsynaptic current will cause V_m to become more negative than V_{rest}, hyperpolarizing the cell toward E_{rev} (see Figure 6-19a).

Although all excitatory synapses generate depolarizing postsynaptic currents, the situation may be more complicated for inhibitory synapses. For example, if E_{rev} for a postsynaptic current happens to be identical with V_{rest} ($V_m - E_{rev} = 0$), no net postsynaptic current will flow even when postsynaptic channels are open. The net current will be zero, as in Figure 6-18, trace 3. In this case, when the postsynaptic channels open, V_m will not change. Such a synapse is inhibitory because the open channels hold the membrane potential at its resting value, preventing it from changing toward threshold. In some cases, E_{rev} is more positive than V_{rest}, but more negative than threshold (Figure 6-20b). In this situation, the postsynaptic potential is depolarizing, but it is, nonetheless, inhibitory because it increases the difficulty of bringing V_m up to threshold. In each of these two special cases, the synapses have an inhibitory action, because activation of the inhibitory postsynaptic channels can counteract a simultaneous activation of excitatory channels (Figure 6-20c). In effect, the opening of these inhibitory postsynaptic channels "short-circuits" excitatory currents, because positive charges carried into the cell by excitatory currents are driven out of the cell through the inhibitory channels, preventing the positive charges from bringing V_m up to threshold.

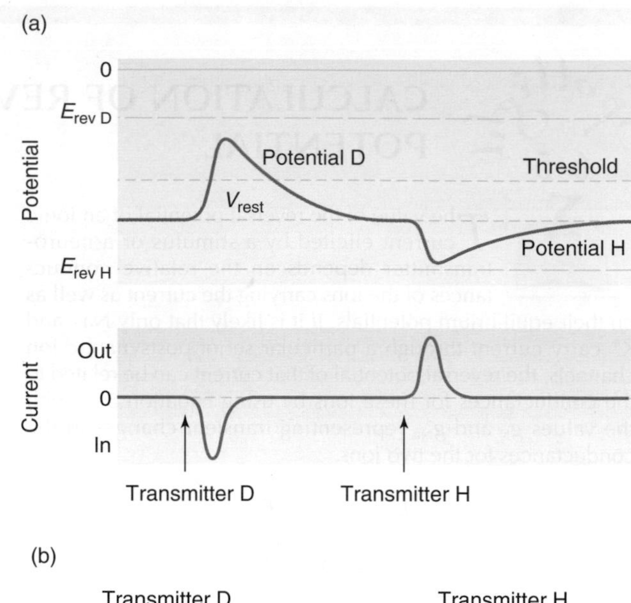

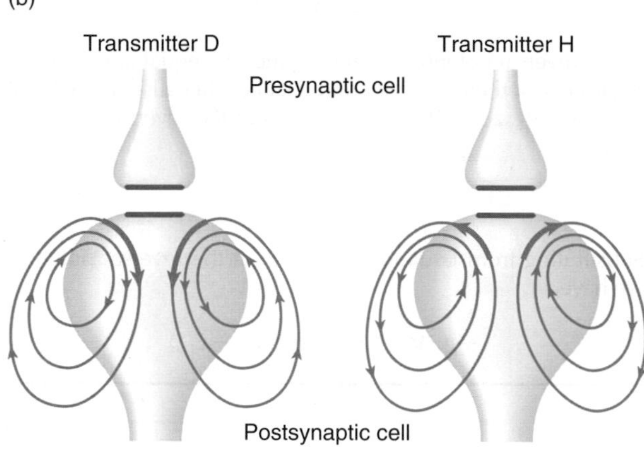

Figure 6-19 Postsynaptic currents can be excitatory or inhibitory. **(a)** Transmitter D evokes an excitatory depolarizing postsynaptic potential because it increases ionic conductance for a net inward current that adds positive charge to the cell's interior, and E_{rev} for the current is more positive than threshold. For example, transmitter D might increase the permeability of the postsynaptic membrane to Na^+. Transmitter H produces an inhibitory hyperpolarizing postsynaptic potential because it increases membrane conductance for ions that cause a net loss of positive charge from the cell. For example, transmitter H might increase membrane permeability to K^+ or to Cl^-. **(b)** Positive current flows in opposite directions through channels opened by transmitter D and by transmitter H.

Note that there is nothing inherently excitatory or inhibitory about any particular neurotransmitter substance. Rather, the properties of the channels that are opened by the transmitter and the identities of the ions that flow through those channels determine how a transmitter affects the postsynaptic cell. Acetylcholine, for example, is an excitatory transmitter at the verte-

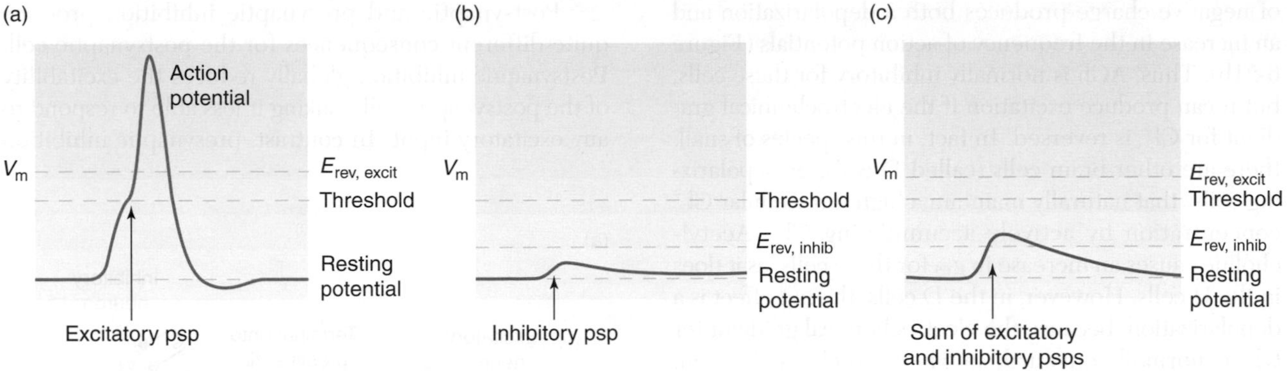

Figure 6-20 Excitatory and inhibitory postsynaptic signals are integrated in the postsynaptic cell. **(a)** An action potential arises out of an excitatory postsynaptic potential if that postsynaptic potential brings the membrane potential, V_m, above the threshold for impulse generation. **(b)** A postsynaptic potential is inhibitory, even if it depolarizes V_m, if its E_{rev} is more negative than the threshold. **(c)** An inhibitory postsynaptic potential, as in part b, may reduce the depolarization produced by an excitatory postsynaptic potential, as in part a, sufficiently to keep the postsynaptic potential from reaching threshold, thus preventing an AP. Notice that a hyperpolarizing postsynaptic potential would similarly reduce the effect of the excitatory postsynaptic potential.

brate neuromuscular junction, where it opens channels that allow Na^+ and K^+ to cross the postsynaptic membrane with an E_{rev} that is near 0 mV. On the other hand, ACh is inhibitory at the terminals of parasympathetic neurons innervating the vertebrate heart, where it causes K^+-selective channels to remain open longer, thus reducing the frequency of the spontaneous depolarizations that drive the heartbeat.

This discussion suggests that an inhibitory transmitter could be made excitatory if the ionic gradients across the postsynaptic membrane were changed. Such an experimental manipulation has been accomplished for neurons of the mammalian spinal cord and for neurons in the brain of a snail (Figure 6-21). In some of the neurons in this snail, ACh increases the conductance for Cl^-, g_{Cl}, of the postsynaptic membrane. In one group of these cells (called H cells, or hyperpolarizing cells), the intracellular Cl^- concentration is relatively low, making E_{Cl} more negative than V_{rest}. When ACh is applied to an H cell, it opens Cl^- channels, allowing Cl^- to flow into the cell down its electrochemical gradient. This current shifts V_m toward E_{Cl}, hyperpolarizing the cell (Figure 6-21a). If all extracellular Cl^- is replaced by SO_4^{2-}, which cannot pass through the Cl^- channels, application of ACh leads to an *efflux* of Cl^-, because Cl^- now has an outwardly directed electrochemical gradient. This efflux

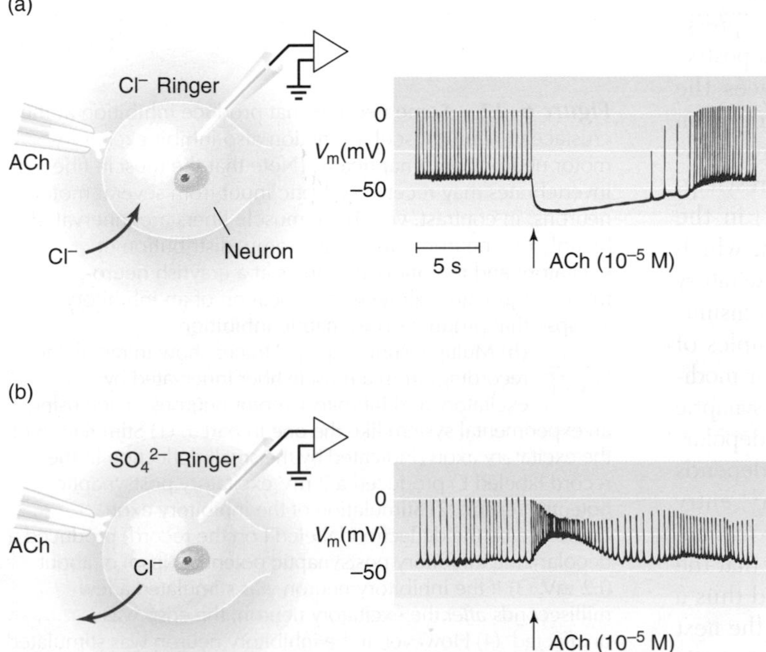

Classic Work *Figure 6-21* Experimentally changing ionic gradients across the membrane of a postsynaptic cell can change the sign of a synapse. **(a)** Acetylcholine (ACh) applied to H-type cells in the snail brain activates Cl^- channels, producing a hyperpolarization because Cl^- brings negative charges into the cell as it moves down its electrochemical gradient. **(b)** When the extracellular Cl^- ions are entirely replaced by SO_4^{2-}, leaving Cl^- inside the cell, the electrochemical gradient for Cl^- is reversed, causing the direction of the postsynaptic current to reverse. As a result, the postsynaptic potential becomes depolarizing, and the synapse becomes excitatory. Electrical activity of the cell before, during, and after synaptic activation is shown at the right. [Adapted from Kerkut and Thomas, 1964.]

of negative charge produces both a depolarization and an increase in the frequency of action potentials (Figure 6-21b). Thus, ACh is normally inhibitory for these cells, but it can produce excitation if the electrochemical gradient for Cl^- is reversed. In fact, in this species of snail, there are other brain cells (called D cells, or depolarizing cells) that naturally maintain a high intracellular Cl^- concentration by actively accumulating Cl^-. Acetylcholine causes an increase in g_{Cl} for these cells, as it does in the H cells. However, in the D cells, the net effect is a depolarization, because the electrochemical gradient for Cl^- is normally outward, so when the channels open, negative charges leave the cell, and V_m becomes more inside-positive. Hence, in this example, excitation and inhibition depend critically on the nature of the local ionic gradients and not on the identity of the signaling molecule.

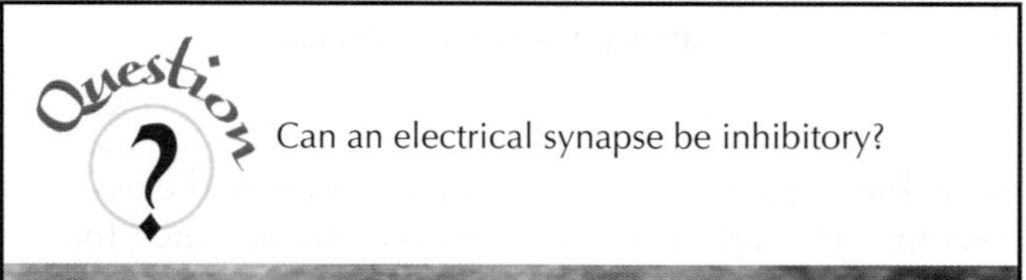

Presynaptic inhibition

Experiments performed in the 1960s on neurons of the mammalian spinal cord and on the crustacean neuromuscular junction revealed an additional inhibitory mechanism at some synapses. This type of inhibition is probably most important in the central nervous system, although it was discovered at a neuromuscular junction in a crayfish. In this type of inhibition, called **presynaptic inhibition,** an inhibitory transmitter is released from a terminal that ends on the presynaptic terminal of an excitatory axon (Figure 6-22a). In this case, the presynaptic terminal of the excitatory axon is itself a postsynaptic element. Presynaptic inhibition reduces the amount of transmitter released from the excitatory axon terminal, which in turn reduces excitation in the postsynaptic cell (Figure 6-22b). In some cases, the presynaptic inhibitory transmitter increases g_K or g_{Cl} in the presynaptic terminals of the excitatory axon, which reduces the amplitude of any AP invading the excitatory terminal and hence diminishes the amount of transmitter released from the terminal. In other examples of presynaptic inhibition, the inhibitory transmitter modifies some property of Ca^{2+} channels in the presynaptic membrane, rendering them less responsive to depolarization. The release of transmitter molecules depends on Ca^{2+} entry into the terminal, so reducing Ca^{2+} entry reduces transmitter release. Regardless of the mechanism, the net effect of presynaptic inhibition is that the postsynaptic cell receives less transmitter, and thus a smaller postsynaptic potential is generated. In the next section we will consider how events in the presynaptic terminal control the release of neurotransmitter.

Postsynaptic and presynaptic inhibition produce quite different consequences for the postsynaptic cell. Postsynaptic inhibition globally reduces the excitability of the postsynaptic cell, making it less able to respond to any excitatory input. In contrast, presynaptic inhibition

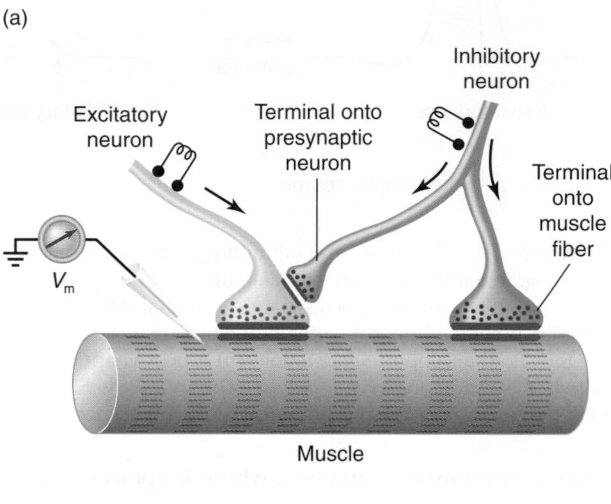

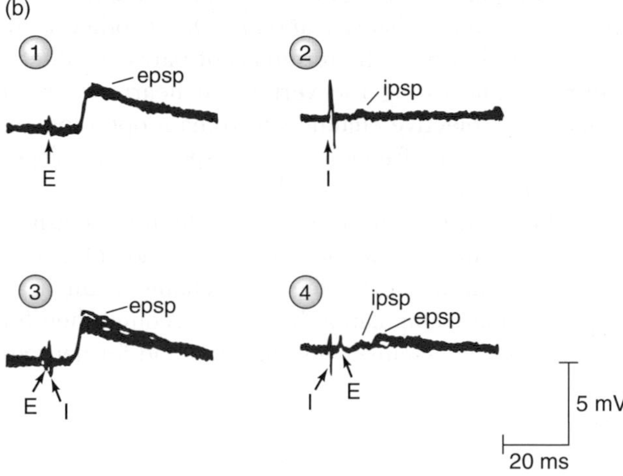

Figure 6-22 Some neurons that produce inhibition at the crustacean neuromuscular junction also inhibit excitatory motor neurons presynaptically. (Note that the muscle fibers of invertebrates may receive synaptic input from several motor neurons; in contrast, vertebrate muscle fibers are innervated by only one neuron.) **(a)** The anatomic distribution of excitatory and inhibitory terminals at a crayfish neuromuscular junction, showing the location of an inhibitory synapse that produces presynaptic inhibition.

Classic Work **(b)** Multiple superimposed traces show intracellular recordings from a muscle fiber innervated by excitatory and inhibitory motor neurons, made using an experimental system like the one in part a. (1) Stimulation of the excitatory axon (indicated by the small deflection in the record labeled E) produced a 2 mV excitatory postsynaptic potential (epsp). (2) Stimulation of the inhibitory axon (indicated by the deflection labeled I on the record) produced a depolarizing inhibitory postsynaptic potential (ipsp) of about 0.2 mV. (3) If the inhibitory neuron was stimulated a few milliseconds *after* the excitatory neuron, the epsp was unaffected. (4) However, if the inhibitory neuron was stimulated a few milliseconds *before* the excitatory neuron, the epsp was almost abolished. [Adapted from Dudel and Kuffler, 1961.]

acts only on specific inputs to the postsynaptic cell, allowing the cell to remain normally responsive to all other inputs. Thus presynaptic inhibition provides a mechanism for narrowly targeted and subtle control of **synaptic efficacy** (the effectiveness of a presynaptic impulse in producing a postsynaptic potential change) among the many synaptic connections onto a particular neuron.

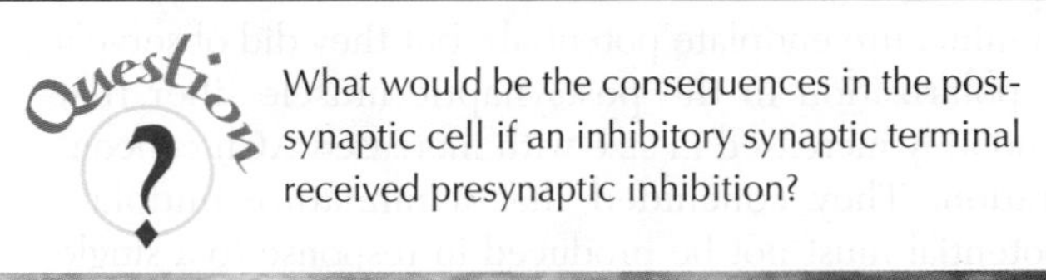

What would be the consequences in the postsynaptic cell if an inhibitory synaptic terminal received presynaptic inhibition?

PRESYNAPTIC RELEASE OF NEUROTRANSMITTERS

Cellular mechanisms that control the release of neurotransmitters from presynaptic terminals play a major role in determining the effectiveness of synaptic transmission because the number of transmitter molecules released affects the size of the postsynaptic response. Understanding transmitter release is thus of central importance for understanding chemical synaptic transmission and its role in neuronal communication. In addition to its importance for physiology, the history of experimentation on transmitter release provides classic examples of the scientific method and effective experimental strategies, so we will consider it in some detail.

A particularly striking example of such a classic experiment was the demonstration in the 1950s and 1960s, by Sir Bernard Katz and his coworkers, that neurotransmitters are generally released in tiny packets called quanta. More recent experiments have shown that synaptic release is closely related to other forms of exocytosis used by cells to release a variety of chemicals, such as the hormones secreted by glandular cells (see Chapter 9). The conservation of this mechanism has contributed to elucidating the details of exocytosis in all cells.

Quantal Release of Neurotransmitters

While investigating neuromuscular transmission, Paul Fatt and Bernard Katz (1952) discovered that spontaneous "miniature" depolarizations (about 0.1 mV in amplitude) could be recorded near the postsynaptic membrane at the motor endplate of a frog muscle (Figure 6-23). These spontaneous signals became progressively smaller when the intracellular recording electrode was inserted farther from the endplate. Because these potentials have a shape, time course, and drug sensitivity similar to those of endplate potentials, they were called **miniature endplate potentials (mepps).** Katz and his collaborators asked themselves, ingeniously, whether miniature endplate potentials might represent a "unit" of transmitter release—and whether endplate potentials evoked by stimulating the motor neuron were the result of many such units being released simultaneously when an AP invades the presynaptic terminal. It had been found earlier that progressively increasing the concentration of extracellular Mg^{2+} or decreasing that of extracellular Ca^{2+}, or both, caused the evoked endplate potential to become smaller in amplitude. Katz and his coworkers used this observation to find concentrations of these cations at which the evoked endplate potential became as small as a single spontaneously occurring miniature endplate potential. They recorded postsynaptic responses evoked by presynaptic motor neuron impulses in a high-Mg^{2+}, low-Ca^{2+} solution and obtained the following results:

- Some motor neuron impulses produced no response at all. These trials were labeled "failures."
- Some impulses produced endplate potentials that had approximately the same amplitude as single spontaneous miniature endplate potentials.
- Some impulses produced endplate potentials with amplitudes that were integer multiples (e.g., two, three, four, etc.) of the mean amplitude of single spontaneous miniature endplate potentials (Figure 6-24 on the next page).

These findings supported the hypothesis that a normal endplate potential is produced when a large number of discrete transmitter units are released simultaneously. Calculations showed that in frog muscle, approximately 100–300 such units could account for the amplitude of normally evoked excitatory postsynaptic potentials.

Because synaptic release appeared to occur in the form of discrete units—known as packets, or *quanta,* of

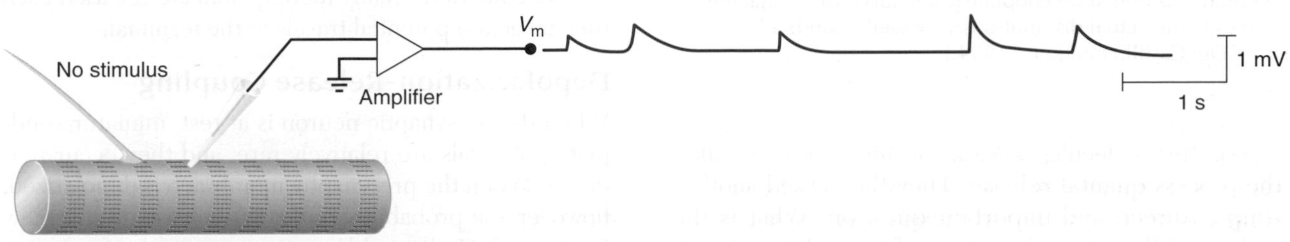

Classic Work **Figure 6-23** Spontaneous miniature endplate potentials (mepps) can be recorded from the motor endplate region of a skeletal-muscle fiber. Note that the amplitudes of these miniature endplate potentials are both small and variable.

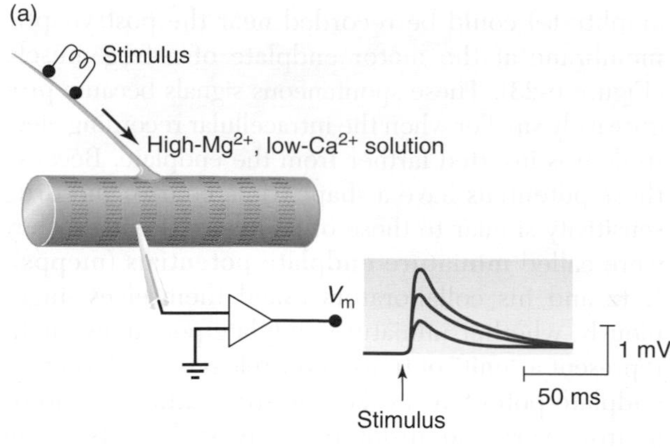

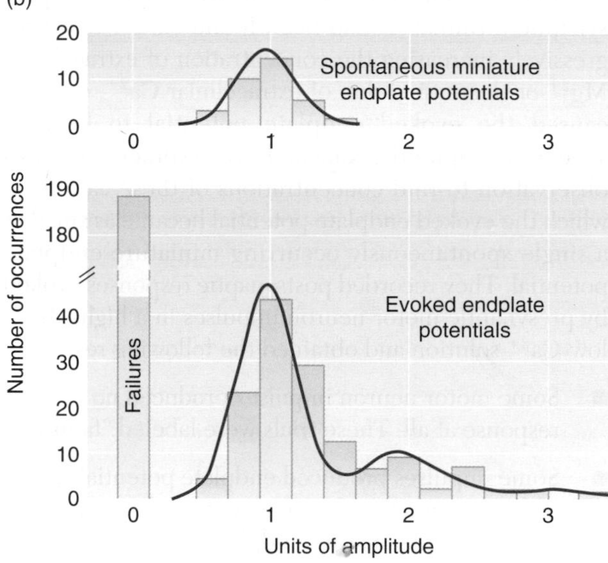

Figure 6-24 Under appropriate conditions, action potentials in a motor neuron produce small endplate potentials, similar to miniature endplate potentials, in the postsynaptic muscle fiber. **(a)** A nerve-muscle preparation is placed in a high-Mg^{2+}, low-Ca^{2+} solution, which reduces the amount of transmitter released at the neuromuscular junction when the neuron is stimulated. The evoked endplate potentials then have small and variable amplitudes.

Classic Work **(b)** Amplitude distribution of spontaneous miniature endplate potentials (top) and of endplate potentials that were evoked by motor neuron stimulation (bottom) in high-Mg^{2+}, low-Ca^{2+} saline. Note the similar shapes of the two distributions, as well as the many failures in the distribution of evoked potentials. The continuous curves in the upper and lower histograms were calculated from a theoretical Poisson distribution, assuming that the evoked endplate potentials are made up of units corresponding to the spontaneous miniature endplate potentials. Notice that the curves fit the actual distributions very well. [Part b adapted from Del Castillo and Katz, 1954.]

transmitter molecules—Katz and his associates called the process **quantal release.** They then asked another simple, direct, and important question: What is the makeup of the unit, or quantum, of transmitter release? Is it a single molecule of ACh? If not, how many molecules are in a quantum? They reasoned that if the cause of a spontaneous miniature endplate potential was a single ACh molecule that leaked out of the presynaptic terminal, then adding a very small amount of ACh to the saline bathing the muscle should greatly increase the number of miniature endplate potentials, because a mepp would be seen every time a molecule of ACh bound to the postsynaptic membrane. Beginning with very low concentrations of ACh and working their way up to higher concentrations, they never saw an increase in miniature endplate potentials, but they did observe a depolarization in the postsynaptic muscle fiber that smoothly increased in size with increased ACh concentration. They concluded that a miniature endplate potential must not be produced in response to a single ACh molecule. Instead, they calculated that each miniature endplate potential is produced by the release of a packet of about 10,000 molecules of ACh, and that these molecules activate about 2000 postsynaptic channels.

At about this time, electron-microscopic studies revealed the presence of membrane-encased packets, or vesicles, in presynaptic axon terminals (see Figure 6-12c) corresponding to the quanta of transmitter that had been inferred physiologically by Katz and his group. They concluded that a miniature endplate potential is caused by the release of a single vesicle, and that full endplate potentials are generated when many vesicles are released simultaneously. An additional piece of evidence that supports this view is the finding that membrane capacitance at the presynaptic terminal increases measurably during exocytosis. When vesicles fuse with the plasma membrane, they increase its surface area, which explains the increase in capacitance.

At any given time, only a fraction of the vesicles inside an axon terminal are available for immediate release; this group of vesicles is called the *readily releasable pool.* In fast transmission, the readily releasable pool seems to consist of some of the vesicles located at the active zones. Whether or not a vesicle in the readily releasable pool will fuse with the plasma membrane depends heavily on the intracellular concentration of Ca^{2+}, which in turn can depend on external conditions. The high-Mg^{2+}, low-Ca^{2+} solution employed by Fatt and Katz modified transmitter release by changing the probability of vesicle fusion. If the probability is sufficiently low, as in Figure 6-24b, presynaptic stimulation leads to many failures (i.e., no vesicles are released) or to the release of only a few vesicles. Under normal conditions, many more quanta are released each time an action potential travels to the terminal.

Depolarization-Release Coupling

When the presynaptic neuron is at rest, miniature endplate potentials are relatively rare, and they occur randomly. When the presynaptic membrane is depolarized, however, the probability of quantal release dramatically increases, as indicated by an increase in the frequency of miniature endplate potentials (Figure 6-25). This relation between presynaptic membrane potential and transmitter release was examined by Bernard Katz and

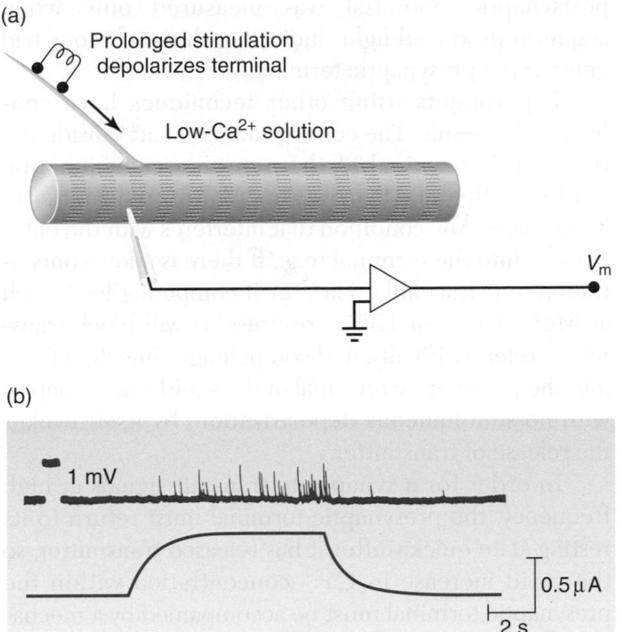

Figure 6-25 When the presynaptic terminal is depolarized, miniature endplate potentials in the postsynaptic muscle fiber become more frequent. **(a)** Experimental setup. **(b)** Prolonged depolarization of the presynaptic terminal by application of current (lower trace) increases the probability of transmitter release, which is shown by an increase in the frequency of miniature endplate potentials recorded in the muscle fiber (upper trace). [Adapted from Katz and Miledi, 1967.]

Ricardo Miledi at an unusually large synapse, called the *squid giant synapse* (Figure 6-26a). The relatively enormous size of the presynaptic axon terminal at this synapse, in which the postsynaptic cell is the squid giant axon, allows microelectrodes for passing current and for recording membrane potential to be inserted into the terminal very close to the synaptic region. This arrangement would be technically impossible in most other synapses, because most axon terminals are tiny. In this experiment, Na^+ channels were blocked by tetrodotoxin (TTX), and K^+ channels by tetraethylammonium (TEA), so that V_m of the presynaptic terminal could be controlled without interference from all-or-none APs. The postsynaptic potential, recorded with a third microelectrode inserted into the giant axon near the synapse, provided a sensitive bioassay of how much transmitter was released from the presynaptic cell. The following results are shown in Figure 6-26b,c:

- When the presynaptic membrane was depolarized, transmitter was released (detected as depolarization in the postsynaptic cell) despite the absence of an AP.
- As the presynaptic membrane was increasingly depolarized, the amplitude of the postsynaptic potential increased, implying that the amount of transmitter released varied directly with depolarization of the presynaptic terminal: more depolarization caused more transmitter to be released.
- For each value of presynaptic depolarization, the postsynaptic response was smaller when the concentration of Ca^{2+} in the extracellular fluid was lower.

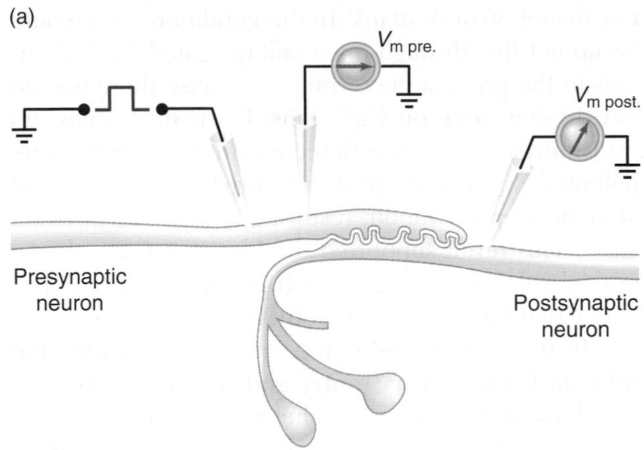

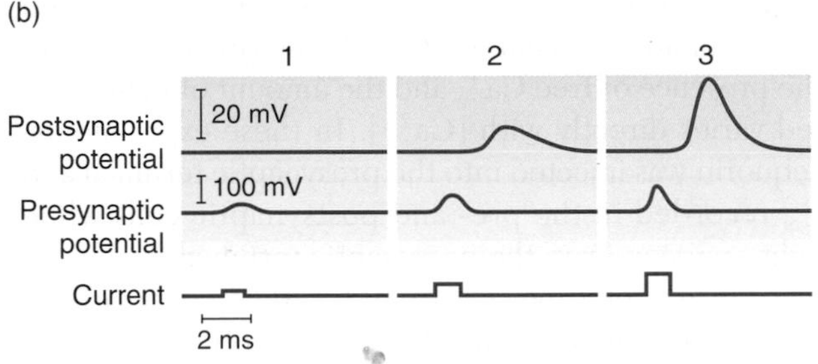

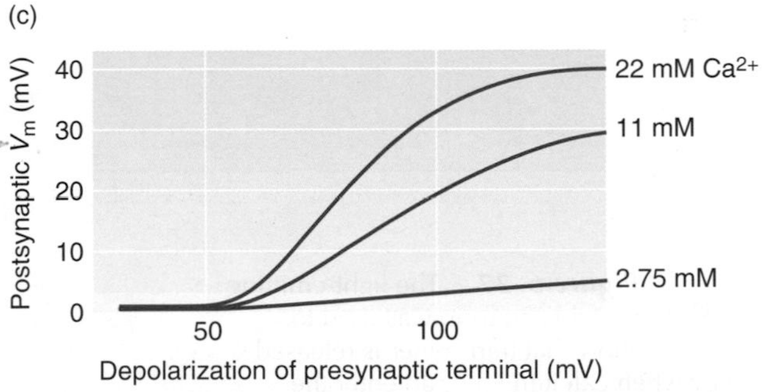

Figure 6-26 The relation between presynaptic depolarization and transmitter release can be studied directly at the squid giant synapse. **(a)** The unusually large size of this synapse allows the presynaptic membrane to be depolarized with current from an intracellular microelectrode while V_m of both the presynaptic and the postsynaptic regions is recorded with two more microelectrodes. **(b)** Depolarizing current passed into the presynaptic terminal was increased in steps from 1 to 2 to 3, producing correspondingly larger postsynaptic potentials in the postsynaptic neuron. **(c)** Depolarizing the presynaptic terminal was increasingly effective as the extracellular $[Ca^{2+}]$ was increased. For a constant value of $[Ca^{2+}]_{out}$, a larger depolarization produced more transmitter release and hence larger postsynaptic potentials. As $[Ca^{2+}]_{out}$ was decreased in the extracellular solution, the amplitude of postsynaptic potentials dropped. [Adapted from Katz and Miledi, 1966, 1970.]

These three results supported the hypotheses that transmitter release depends on a depolarization of the presynaptic membrane, not on the chemical identity of the ions that cause the depolarization, and that Ca^{2+} ions play a role in neurotransmitter release. The role of Ca^{2+} ions was further emphasized when the presynaptic terminal was depolarized all the way to the Ca^{2+} equilibrium potential, E_{Ca}, which is typically more positive than +30 or +40 mV. In this condition, there could be no net flux through open voltage-gated Ca^{2+} channels in the presynaptic terminal, because there was no net driving force on Ca^{2+} ions. Correspondingly, no transmitter release was detected until the membrane potential, V_m, was allowed to return to its resting level after the depolarization, restoring the driving force on Ca^{2+}. This result supported the idea that Ca^{2+} plays a crucial role in causing neurotransmitter release, but notice that the evidence was still indirect.

In further work using the squid giant synapse, the relation between Ca^{2+} entry and transmitter release was demonstrated directly. The calcium ion concentration in the presynaptic terminal was measured using aequorin, which is a Ca^{2+}-sensitive protein extracted from a bioluminescent jellyfish. Aequorin emits light in the presence of free Ca^{+}, and the amount of light emitted varies directly with $[Ca^{2+}]$. In these experiments, aequorin was injected into the presynaptic terminal and V_m recorded in the pre- and postsynaptic cells while light emission from the presynaptic terminal was monitored by a phototube (Figure 6-27). When depolarizing current was injected into the presynaptic terminal, a postsynaptic potential was measured only when aequorin produced light, indicating that Ca^{2+} ions had entered the presynaptic terminal.

Experiments using other techniques have confirmed this result. The concentration of Ca^{2+} inside the presynaptic terminal of the neuromuscular junction must rise after an AP arrives in order for transmitter to be released. Any condition that interferes with the entry of Ca^{2+} into the terminal (e.g., if there is a low concentration of extracellular Ca^{2+} or if competing ions, such as Mg^{2+}, Co^{2+}, or La^{3+}, are present) will block transmitter release. Finally, and compellingly, injecting Ca^{2+} into the presynaptic terminal of the squid giant synapse, with no simultaneous depolarization, by itself evokes the release of transmitter.

In order for a synapse to transmit signals at high frequency, the presynaptic terminal must return to its resting state quickly after it has released transmitter, so the rapid increase in Ca^{2+} concentration within the presynaptic terminal must be accompanied by a mechanism for its rapid removal. Although ultimately Ca^{2+} ions that entered the terminal must be pumped back into the extracellular space by an exchanger in the plasma membrane, this process is too slow to account for the rapid disappearance of Ca^{2+} ions from the cytosol of the terminal. It seems likely that the extra Ca^{2+} ions are stored in internal compartments until they can be extruded from the terminal.

Exocytosis of synaptic vesicles can occasionally be seen in electron micrographs (Figure 6-28), and recent work has revealed more details of the synaptic release

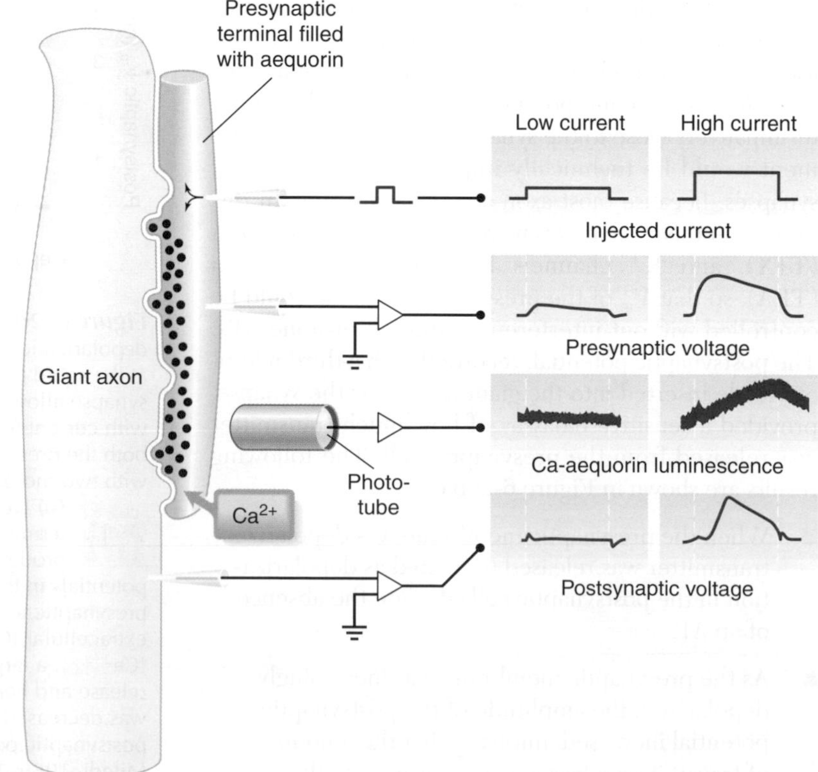

Classic Work **Figure 6-27** The light-emitting protein aequorin was used to show that transmitter is released only when calcium ions can enter the presynaptic terminal. In this experiment, Na^{+} and K^{+} currents at the squid giant synapse were blocked with TTX and TEA. Aequorin was injected into the presynaptic terminal (darker yellow), and the terminal was stimulated by injecting depolarizing current through a microelectrode. Pre- and postsynaptic potentials were recorded using intracellular microelectrodes. Light emitted by aequorin was monitored by a photoelectric device. The responses to weak and strong presynaptic stimulating currents are shown at the right. A postsynaptic potential was recorded only when aequorin produced light, indicating that Ca^{2+} had entered the presynaptic terminal. Synaptic vesicles in the terminal are shown as dots. [Adapted from Llinás and Nicholson, 1975.]

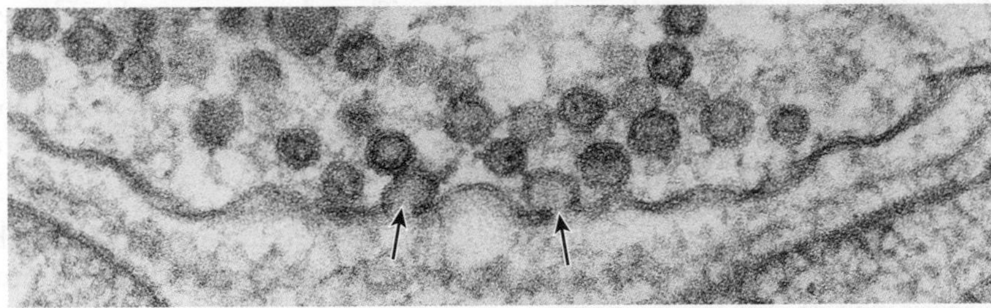

Figure 6-28 Exocytosis at the synaptic terminal can be visualized with an electron microscope. This electron micrograph shows exocytosis at a nerve terminal of a frog neuromuscular junction. The arrows point to vesicles that have fused with the membrane and are emptying into the synaptic cleft. Notice that the lipid bilayer of each fused vesicle has become continuous with the plasma membrane of the terminal. [Courtesy of J. Heuser.]

process. Dieter Bruns and Reinhard Jahn (1995) reported the first real-time measurement of transmitter release from single synaptic vesicles in cultured leech neurons that contain two kinds of vesicles. Small, clear vesicles discharge about 4700 transmitter molecules with a time constant of about 260 μs. Other larger vesicles, similar to the ones illustrated in Figure 6-11b, release about 80,000 molecules with a time constant of about 1.3 ms. The time constant of these processes is a measure of the speed at which fusion takes place. These measures indicate that the fusion phase of synaptic transmission is very fast, contributing very little to the total time required for transmission, which is called the *synaptic delay*.

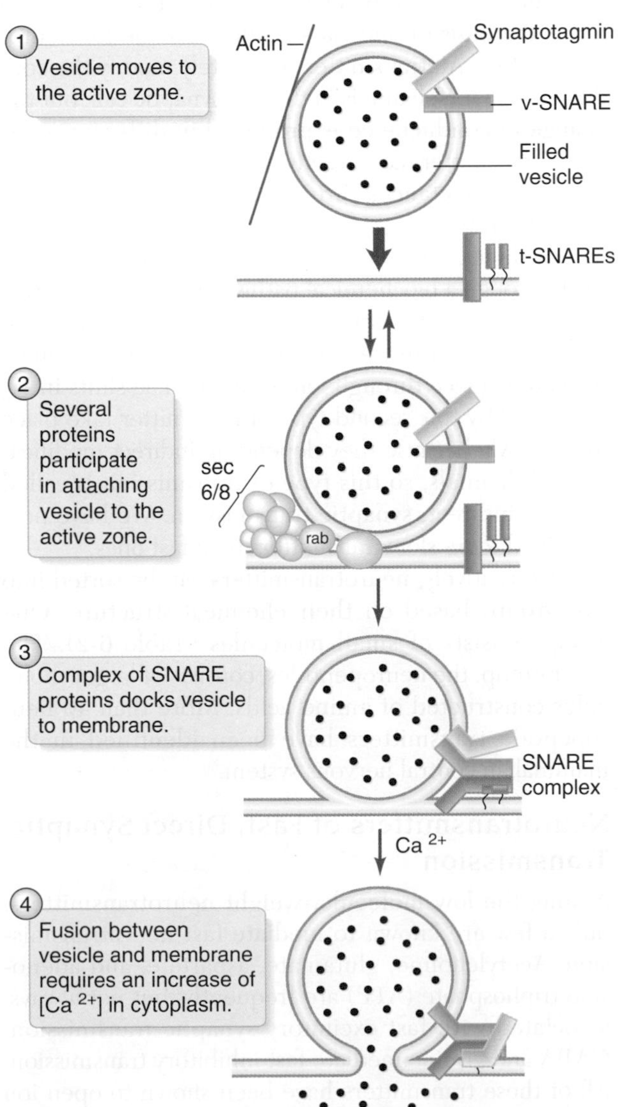

We now know that transmitter release proceeds through identifiable steps. A filled synaptic vesicle must first travel to an active zone in the plasma membrane of the terminal (Figure 6-29, step 1). This process depends on an interaction between the proteins actin and myosin, molecules that belong to the same families as the molecules that generate force in muscle contraction. Next, the vesicle must become attached to the membrane. Filled synaptic vesicles attach preferentially at active zones, which are marked by particular protein complexes, among them a large one called sec 6/8 and another called rab. At first the attachment is reversible (Figure 6-29, step 2), but once the vesicle is attached to the membrane, it may dock irreversibly. This process is mediated by proteins called SNAREs, which are located in the vesicular membrane and in the plasma membrane at the active site. The SNARE molecules in a vesicle (called v-SNAREs) form a complex with related molecules in the active-site plasma membrane (called t-SNARES for "target membrane SNAREs"),

Figure 6-29 Several identified molecules play a role in the process of synaptic release. The release of neurotransmitter depends on a series of events that take place in a predictable sequence. Mature vesicles move up to active zones through a process that depends on the cytoskeletal protein actin (step 1). Vesicles are targeted to active zones by the presence of the large protein complex sec 6/8 and reversibly tethered there by another protein called rab (step 2). Tethered vesicles may become irreversibly docked to the membrane if SNARE proteins in the vesicle and in the plasma membrane form a complex (step 3). A Ca^{2+}-binding vesicular protein, synaptotagmin, interacts with the SNARE complex to produce rapid fusion between the vesicular and terminal membranes, releasing neurotransmitter into the synaptic cleft (step 4). [Adapted from Bajjalieh, 1999.]

and formation of this complex docks the vesicle to the plasma membrane (Figure 6-29, step 3). Another protein associated with the membrane of mature vesicles, synaptotagmin, interacts with the proteins of the SNARE docking complex to permit fast Ca^{2+}-dependent membrane fusion (Figure 6-29, step 4). Recent experiments have revealed that this presynaptic molecular machinery is remarkably similar to the molecules that control exocytosis in yeast. Apparently exocytosis arose as a cellular mechanism early in evolution and has been maintained with only minor modifications since.

Many synapses include several separate active sites (see Figure 6-12), and many vesicles cluster around each active site. Recently acquired evidence indicates that only one vesicle per active site is actually docked and ready to be released, so only one of the many vesicles clustered at an active zone fuses with each AP. In addition, each AP is likely to cause fusion in only some of the active sites of a terminal. Notice that the best assay for whether neurotransmitter was released, and in what amount, remains the postsynaptic potential, so some of these conclusions await direct confirmation.

Nonspiking Release

As mentioned earlier, some neurons, called nonspiking neurons, never carry APs; they transfer information by means of electrotonically conducted voltage signals alone. They transmit signals to other cells by releasing neurotransmitter from their terminals in the absence of APs. This mechanism is called **nonspiking release.** The amount of transmitter a nonspiking cell releases into the synaptic cleft depends on the membrane potential, V_m, which controls the activity of voltage-gated Ca^{2+} channels. When the cell is more strongly depolarized, it releases more transmitter. Because the amount of transmitter released is a direct function of the presynaptic membrane potential, the amplitude of the postsynaptic potential can indicate how strongly the presynaptic neuron was stimulated.

THE CHEMICAL NATURE OF NEUROTRANSMITTERS

Once it became clear that most synaptic transmission is carried by chemical signals, the race was on to discover the molecular identities of these transmitter substances. By the mid-1960s, only three compounds had been unequivocally identified as neurotransmitters: acetylcholine, norepinephrine, and γ-aminobutyric acid (GABA, pronounced "gab-uh"). In the process of identifying and characterizing these compounds, three criteria were established to distinguish neurotransmitters from other candidate molecules:

- When the candidate substance is applied to the membrane of a postsynaptic cell, it must elicit in this cell precisely the same physiological effects produced by presynaptic stimulation.
- The substance must be released when the presynaptic neuron is active.
- The action of the substance must be blocked by the same agents that block natural transmission at that synapse.

Using these criteria, physiologists have identified many other neurotransmitters, but enormous effort has been required. Identifying the transmitters in the vertebrate central nervous system has been very difficult because very little transmitter is released at most synapses (only about 10^4 molecules per synapse per AP). Moreover, neuronal tissue is a nonhomogeneous collection of tightly packed and diverse cell types, which complicates the collection of transmitter molecules. However, enough transmitter substances have been identified that patterns have begun to emerge. A striking discovery that comes out of this painstaking research is that the same neurotransmitters are used throughout the animal kingdom, providing an impressive example of evolutionary conservation of molecular identity.

Neurotransmitters exert their effects by either of two very different mechanisms, and the difference forms the basis of one classification scheme for neurotransmitters. All transmitters ultimately modify the conductance of ion channels in the postsynaptic cell, but the change in conductance is produced in different ways. Some transmitters act directly on ion channel proteins to change conductances through the postsynaptic membrane, thereby changing V_m; this type of transmission is fast, or direct, synaptic transmission. Other transmitters work through a biochemical pathway in the postsynaptic cell, changing the state of membrane-associated or cytosolic second messengers that subsequently change the conductance through ion channels. The shifts in V_m generated by this second type of transmitter take place more slowly because they depend on indirect modification of channels, so this type of transmission is called slow, or indirect, synaptic transmission. We have now identified more slow transmitters than fast ones.

Alternatively, neurotransmitters can be sorted into two groups based on their chemical structure. One group consists of small molecules (Table 6-2). The other group, the neuropeptides, consists of larger molecules constructed of amino acids. More than 40 neuropeptide transmitters have been identified in the mammalian central nervous system.

Neurotransmitters of Fast, Direct Synaptic Transmission

Among the low-molecular-weight neurotransmitters, only a few are known to mediate fast neurotransmission. Acetylcholine, glutamate, aspartate, and adenosine triphosphate (ATP) are frequently, but not always, associated with fast excitatory synaptic transmission. GABA and glycine mediate fast inhibitory transmission. All of these transmitters have been shown to open ion channels in the membrane of the postsynaptic cell.

Table 6-2 Typical small neurotransmitters, their structures, and functions

Neurotransmitter	Typical effects*	Structure
Acetylcholine (ACh)	Fast excitation; slow inhibition	$H_3C{-}C({=}O){-}OCH_2CH_2{-}N^+(CH_3)_3$
Glycine (Gly)	Fast inhibition	$^+H_3N{-}CH(COO^-){-}H$
γ-Aminobutyric acid (GABA)	Fast inhibition; slow inhibition	$^+H_3N{-}CH_2{-}CH_2{-}CH_2{-}COO^-$
Glutamate (Glu)	Fast excitation; slow change in postsynaptic metabolism	$^+H_3N{-}CH(COO^-){-}CH_2{-}CH_2{-}COO^-$
Norepinephrine (Nor-epi)	Slow excitation; slow inhibition	3,4-(HO)$_2$-ring$-CH(OH)CH_2NH_2$
Dopamine	Differs with location but causes slow postsynaptic effects	3,4-(HO)$_2$-ring$-CH_2CH_2NH_2$
Serotonin (5-HT = 5-hydroxytryptamine)	Slow excitation or slow inhibition	HO-indole (N–H) with $-CH_2-CH_2-NH_2$
Nitrogen oxide (NO)	Synaptic modulation	$N{=}O$
Adenosine triphosphate (ATP)	Both fast and slow synaptic transmission	$^-O{-}P(O)(O^-){-}O{-}P(O)(O^-){-}O{-}P(O)(O^-){-}O{-}CH_2$-ribose (HO, OH)-adenine (NH_2)
Histamine	Slow modulation	$HC{=}C{-}CH_2{-}CH_2{-}NH_3^+$ (imidazole ring: N, NH, C–H)

*Notice that the effect of a neurotransmitter depends on the properties of the postsynaptic cell. For most neurotransmitters, however, it is possible to identify their most probable effect.

Acetylcholine (see Table 6-2) is the most familiar of the established transmitter substances. Neurons that release ACh, which are said to be **cholinergic**, are widely distributed throughout the animal kingdom. To list just a few of the known examples, ACh is the neurotransmitter used by vertebrate motor neurons, the preganglionic neurons of the vertebrate autonomic nervous system, the postganglionic neurons of the parasympathetic division of the autonomic nervous system, and many neurons of the vertebrate central nervous system. Acetylcholine is also the transmitter in a number of invertebrate neurons, including some cells in the

$H_3C-\overset{O}{\overset{\|}{C}}-OCH_2CH_2-\overset{CH_3}{\overset{|}{N}}{}^{\oplus}-CH_3$ (CH_3)

Acetylcholine (ACh)

$H_2N-\overset{O}{\overset{\|}{C}}-OCH_2CH_2-\overset{CH_3}{\overset{|}{N}}{}^{\oplus}-CH_3$ (CH_3)

Carbachol, an ACh agonist

D-Tubocurarine, an ACh antagonist

Figure 6-30 Many different molecular species that are structurally similar to acetylcholine can interact with postsynaptic receptors at fast acetylcholine synapses. Acetylcholine (ACh) is the natural ligand. The ACh analog carbachol is an agonist; it mimics the action of ACh at fast ACh synapses. D-Tubocurarine, an antagonist of ACh action, blocks the activation of receptors at these synapses. Red type highlights identical structures in ACh and carbachol; blue type indicates homologous region in all three molecules.

molluscan central nervous system, motor neurons of annelid worms, and sensory neurons of arthropods.

Molecules that have crucial structural features in common with ACh, called structural analogs, can also act at cholinergic synapses. The structural analog carbachol, for example, can activate cholinergic synapses. Molecules that mimic the action of a neurotransmitter in this manner are said to be **agonists** at the synapse (Figure 6-30). Alternatively, structural analogs can block transmission by binding to receptor sites but not causing activation. One example is D-tubocurarine, the active agent in curare, which competes with ACh at receptor binding sites. Molecules that block the action of a neurotransmitter in this manner are called **antagonists.**

Just as the Ca^{2+} concentration in the terminals must rise only transiently in order to permit high-frequency synaptic signaling, neurotransmitter molecules must occupy the synaptic cleft only briefly. Transmission is terminated at cholinergic synapses when ACh is hydrolyzed to choline and acetate by the enzyme acetylcholinesterase (AChE), which is present in abundance in the synaptic cleft near the surface of the postsynaptic membrane (Figure 6-31a). Some of the acetate and choline diffuse away from the cleft, but choline that remains in the cleft is actively reabsorbed by the presynaptic membrane and recycled by condensation with acetyl coenzyme A (acetyl CoA) to form new molecules of ACh. Blocking the activity of AChE produces dramatic and dangerous effects, as illustrated by the action of *anticholinesterases,* which include some nerve gases used as weapons and many insecticides. In

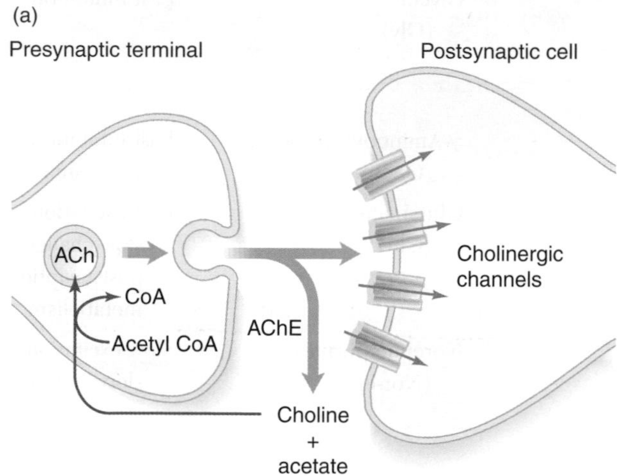

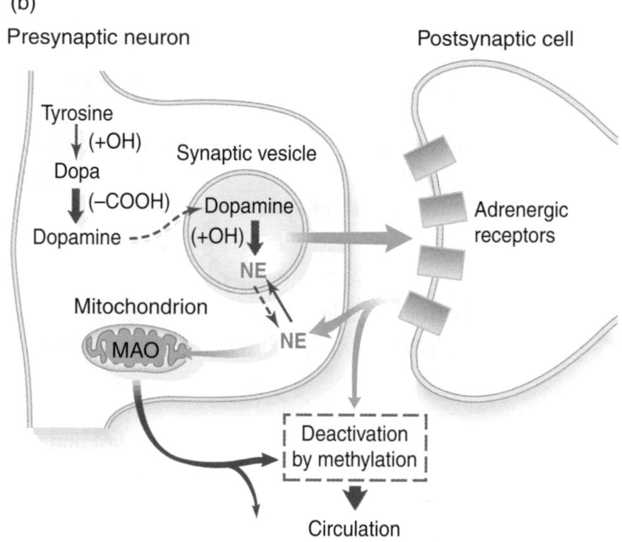

Figure 6-31 Synaptic transmission is terminated by a variety of mechanisms. **(a)** Acetylcholine (ACh) is synthesized in synaptic terminals from choline and acetyl CoA. Once released into the synaptic cleft, ACh is hydrolyzed to acetate and choline by the enzyme acetylcholinesterase (AChE), which is plentiful in the cleft. The liberated choline can be taken up by the presynaptic terminal and reacetylated to form new ACh molecules. **(b)** Norepinephrine (NE) is synthesized in neurons through a series of steps beginning with the amino acid tyrosine. The final stage of synthesis takes place within the synaptic vesicles. After it is released into the synaptic cleft, some norepinephrine is taken back up into the presynaptic terminal, and some is deactivated by methylation and carried away in the blood. Cytoplasmic norepinephrine is either repackaged into synaptic vesicles or degraded by monoamine oxidase (MAO) located within mitochondria in the presynaptic neurons. [Adapted from Mountcastle and Baldessarini, 1968.]

the presence of these agents, ACh lingers in the synaptic cleft, and its concentration builds up. At some synapses, the postsynaptic cell cannot repolarize and thus remains activated. At many other synapses, the ACh receptor molecules become inactivated in a process called **synaptic desensitization.** At desensitized synapses, the postsynaptic membrane fails to respond to ACh, even if it is present at a high concentration. In either case, the function of the nervous and muscular systems is disrupted, and death can follow. In vertebrates, death from this kind of poison is typically caused by paralysis of the respiratory muscles.

Several amino acids have been found to act as fast neurotransmitters (see Table 6-2). Glutamate (glutamic acid) is the most common transmitter at excitatory synapses in the vertebrate central nervous system, and it is the transmitter at fast excitatory neuromuscular junctions in insects and crustaceans. GABA is the transmitter at inhibitory motor synapses onto crustacean and annelid muscle and is an important inhibitory transmitter in the vertebrate central nervous system. Glycine is the most common inhibitory transmitter in the vertebrate spinal cord.

It has been found that each neuron synthesizes only one fast neurotransmitter and releases that transmitter at all of its synapses. However, in addition to a fast transmitter, many neurons synthesize other synaptically active agents, such as the molecules considered next.

Neurotransmitters of Slow, Indirect Synaptic Transmission

The biogenic amines constitute an important class of neurotransmitters (see Table 6-2) that act through second messengers to produce slow synaptic transmission. They include norepinephrine, dopamine, serotonin, and histamine, which act as neurotransmitters for some invertebrate neurons and in the central and autonomic nervous systems of vertebrates.

Norepinephrine (also known as noradrenaline) is the primary transmitter released by postganglionic cells of the vertebrate sympathetic nervous system. Neurons that use norepinephrine (or epinephrine) as transmitters are called **adrenergic neurons.** At some synapses, norepinephrine is excitatory; at others, it is inhibitory. Its effect depends on the properties of the postsynaptic cell. Norepinephrine is synthesized in the synaptic terminals from the amino acid tyrosine, and it is inactivated by methylation within the synaptic cleft or by reuptake into the synaptic terminal, where some of it is repackaged into synaptic vesicles for rerelease and some of it is inactivated by the enzyme monoamine oxidase (Figure 6-31b).

In addition to these relatively small, "classic" transmitter molecules, there is a growing list of peptide molecules that are produced and released in the vertebrate central nervous system. Many of these molecules, or very similar analogs, have also been found in the nervous systems of invertebrates. Some of these peptides act as transmitters; others act as modulators that influence synaptic transmission at the synapse from which they were released or at neighboring synapses. Interestingly, a number of these neuropeptides are produced in many tissues, not just in neurons. Thus, a single molecular species may be released from intestinal endocrine cells, from autonomic neurons, from various sensory neurons, and in several parts of the central nervous system. In fact, some neuropeptides were initially discovered in visceral tissues and were only later found in neurons. The gastrointestinal hormones glucagon, gastrin, and cholecystokinin are prime examples.

It is not yet clear how many peptide neurotransmitters there are. We know that some neuropeptides act in a neurosecretory fashion; that is, they are liberated into the circulation and are carried by the blood to their targets, rather than being released into the confined space of a synaptic cleft. The pituitary hormone–releasing factors released by neurons in the hypothalamus operate in this manner (see Chapter 9). There is evidence that a single neuropeptide species may be released as a transmitter by some neurons, as a neurosecretory substance by other neurons, and as a hormone by nonneuronal tissue. This multiplicity of function is not really all that novel. It has long been known that norepinephrine (as well as its close relative, epinephrine) acts as a hormone when released by the adrenal medulla and as a transmitter when released at synapses. However, it has become clear—much to the surprise of neurophysiologists—that a neuropeptide can be released as a cotransmitter by axon terminals that also release a more familiar transmitter such as ACh, serotonin, or norepinephrine. Several combinations of a classic transmitter and a paired cotransmitter have been identified in the mammalian brain.

The first neuropeptide was discovered in 1931 by U. S. von Euler and John H. Gaddum while they were assaying for ACh in extracts of rabbit brain and intestine. The extracts stimulated contraction of the isolated intestine, much as ACh does, but the resulting contractions were not blocked by ACh antagonists. Von Euler and Gaddum discovered that the contractions were produced in response to a polypeptide, which the researchers named *substance P.* Since then, substance P and a growing list of other neuropeptides have been found in various parts of the central, peripheral, and autonomic nervous systems of vertebrates and in many invertebrate nervous systems. Investigators have typically used immunological labeling with fluorescent antibodies that recognize specific neuropeptides to determine where these molecules are found. The labels can be detected in histological sections using a fluorescence microscope, revealing the distribution of specific peptides in the nervous system. Some well-known neuropeptides are antidiuretic hormone (see Chapter 14), the hypothalamic releasing hormones (see Chapter 9), and various gastric hormones (see Chapter 15).

Unlike small neurotransmitters, which may be synthesized in the synaptic terminals, neuropeptides are made in the soma and are transported along the axons to the terminals. Neuropeptides are typically synthesized as parts of larger polypeptides, called *propeptides*, each of which may contain the sequences for many biologically active peptides and proteins. Specific enzymes cleave the propeptide into individual peptide molecules. This method of production can limit the amount of peptide neurotransmitter available at a synapse compared with that of a small neurotransmitter synthesized on site. Peptides are, however, more potent than small neurotransmitters for three reasons. First, they bind to receptors with greater affinity than do other neurotransmitters. Their dissociation constants typically are about 10^{-9} M, versus 10^{-5} M for typical neurotransmitters, so very small amounts of neuropeptide can be effective. Second, they act through intracellular pathways that typically include the activation of enzymes that catalyze many subsequent reactions, producing significant amplification of the original signal. Thus, even a small amount of peptide transmitter can produce a large effect. Third, the mechanisms that terminate the actions of neuropeptides are slower than those for other neurotransmitters, so they remain available to their receptors for a longer time.

Although particular transmitter molecules can in some cases be associated with fast or slow synaptic transmission exclusively, some transmitters participate in both types. The most important distinction between fast and slow transmission lies not in the identity of the transmitter, but in the target cells and how they respond to the transmitter. Let's consider how neurotransmitter molecules exert their effects on postsynaptic cells.

POSTSYNAPTIC MECHANISMS

Neurotransmitter molecules act on postsynaptic target cells by way of specific receptor proteins in the plasma membranes of the postsynaptic cells. The properties of these postsynaptic receptors form a crucial link in the chain of events that begins when an AP arrives at the terminal of a presynaptic neuron and ends when the response of the postsynaptic neuron is complete. In this section, we will consider the properties of the receptor molecules that mediate the two major classes—fast and slow—of chemical synaptic transmission and the events that take place after a neurotransmitter molecule binds to these receptors.

Receptors and Channels in Fast, Direct Neurotransmission

As we have seen, the neurotransmitters at fast, direct synapses act by directly changing the permeability of the postsynaptic membrane to certain ions; they do so by binding to and opening ion channels that are included in the structure of the receptor protein molecules. When a postsynaptic channel opens, a minute ionic current passes through it. Many such single-channel currents sum to form the macroscopic postsynaptic current that produces postsynaptic potentials. Much of what we know about these events has been revealed in studies of the ACh-activated channels at the vertebrate neuromuscular junction.

The fast, directly activated acetylcholine receptor channel

The number of postsynaptic channel protein molecules is very small relative to the numbers of other proteins in a membrane; as a result, the isolation, identification, and characterization of these important proteins has been difficult. In early studies, physiologists used a variety of agents to create a pharmacological taxonomy of receptor types, in which various ion channels were named for substances that could modify their activity. Acetylcholine receptors, for example, have been classified into two types. Nicotine, an alkaloid produced by tobacco and some other plants, mimics the action of ACh on the channels found at the vertebrate neuromuscular junction and on postganglionic cells of the vertebrate autonomic nervous system, so these ACh receptors (AChRs) are called **nicotinic AChRs.** Muscarine, a toxin isolated from some kinds of mushrooms, activates the other type of AChR, which is found in the target cells of parasympathetic neurons in the vertebrate autonomic nervous system. These AChRs are called **muscarinic AChRs.** We now know that only the nicotinic AChRs participate in fast, direct chemical synaptic transmission; muscarinic AChRs work through indirect mechanisms.

Our understanding of nicotinic AChRs was given a huge boost when it was discovered that extremely high densities of these receptors are present in the electroplax organs of particular elasmobranch and teleost fishes. These organs generate high-intensity electrical discharges that the fishes use to stun their prey. The receptors are found on one side of the electroplax organ, which consists of many flattened cells that originate during development from embryonic muscle tissue. The unusually high density of nicotinic AChRs in the electroplax tissue allowed the nicotinic AChR to be the first ligand-gated channel to be purified chemically. More recently, its molecular structure has been resolved; we even have images of how the receptor channel looks as it opens.

A second important aid to the study of the nicotinic AChR has been its sensitivity to α-bungarotoxin (α-BuTX; see Spotlight 6-4), a component of snake venom that binds irreversibly and with high specificity to nicotinic AChRs. α-Bungarotoxin can be isotopically labeled and used to tag AChR molecules, facilitating their chemical isolation and purification. Physiological and biochemical studies have shown that the receptor site to which ACh molecules bind at nicotinic synapses is an integral part of the channel protein complex.

Each AChR at the neuromuscular junction consists of five homologous subunits that associate to form a channel at the center of the complex (Figure 6-32). There are two identical α-subunits plus one each of three different subunits termed β, γ, and δ. (AChR complexes in the central nervous system are also pentameric, but they typically consist of only two or in some cases three different kinds of subunits.) Each channel protrudes from both sides of the membrane, with a funnel-shaped opening bulging outward from the cell surface.

Acetylcholine binds to the AChR where the receptor complex extends into the extracellular space. This location was first deduced because ACh injected into a muscle fiber near the endplate failed to modify the postsynaptic potentials. Since then, experiments have revealed that there are ligand-binding sites on the extracellular part of each of the two α-subunits. When both

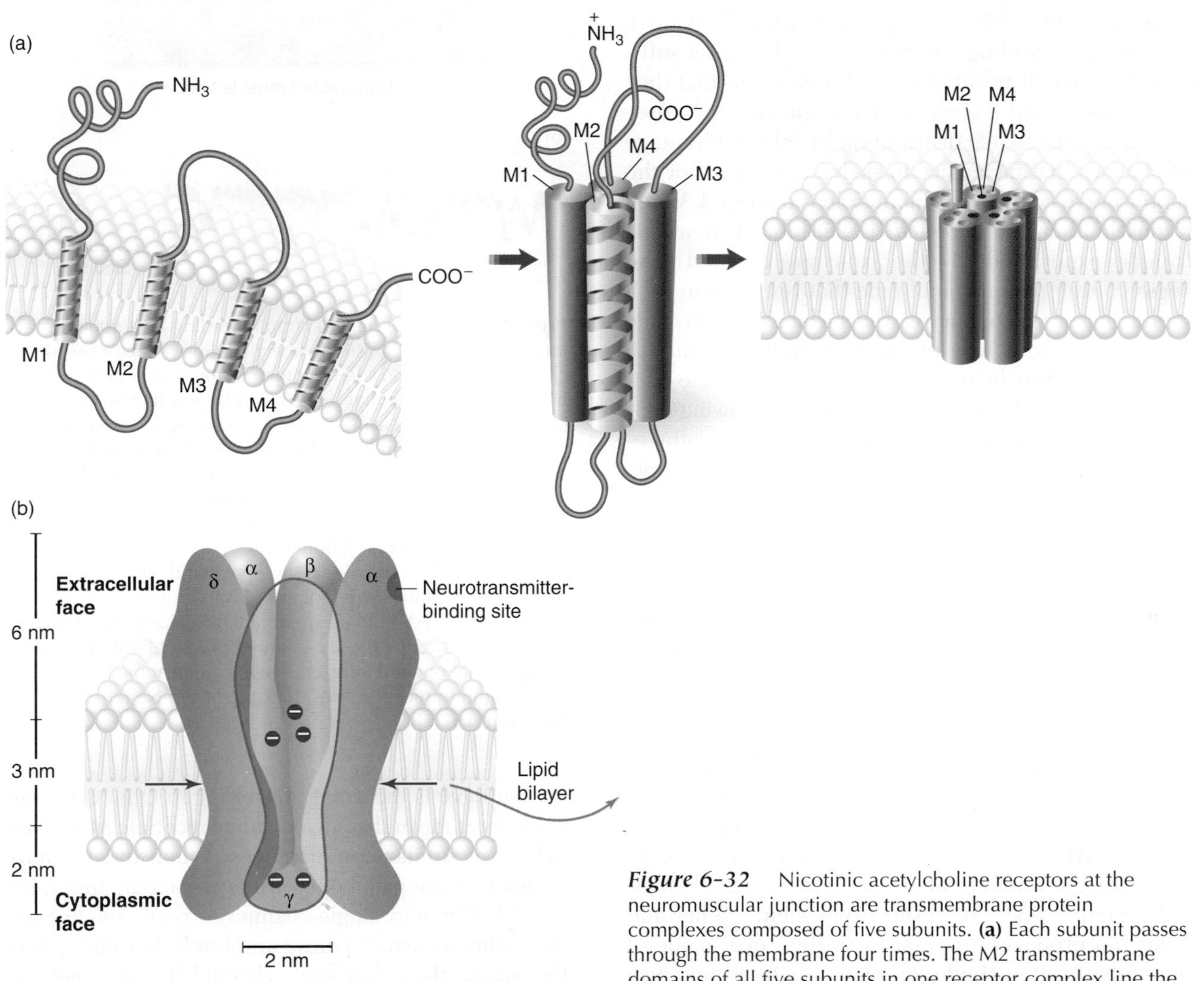

Figure 6-32 Nicotinic acetylcholine receptors at the neuromuscular junction are transmembrane protein complexes composed of five subunits. **(a)** Each subunit passes through the membrane four times. The M2 transmembrane domains of all five subunits in one receptor complex line the pore of the channel and contribute significantly to the selectivity of the channel. **(b)** Acetylcholine binds to the two α-subunits, one molecule per subunit. The entry to the channel from outside the cell is a broad funnel that becomes narrower and bears a net negative charge toward the cytoplasm, forming the selectivity filter, indicated by red arrows. **(c)** When both ACh-binding sites are occupied, the channel opens. Channel opening depends on a change in the spatial relations among the five subunits, as shown by this diagram of the channel at the level of the selectivity filter. The blue and green circles mark identifiable locations on each subunit to illustrate how the subunits change their position. [Parts a and c from Nelson and Cox, 2000; part b adapted from Unwin, 1993.]

sites are occupied by ligand molecules, the channel opens. The nature of this gating process has been studied most extensively at the neuromuscular junction of frog skeletal muscle.

Normally, AChRs are confined to the postsynaptic membrane in the endplate region. The density of ACh-activated channels in the postsynaptic membrane of the frog endplate is about 10^4 per square micrometer, and this dense packing initially impeded work to explore the behavior of individual channels. Analysis of single channels was made possible by the invention of the patch-clamping technique by Erwin Neher and Bert Sakmann (1976; see Spotlight 5-6). Their work depended on finding a region of muscle with a sufficiently sparse distribution of AChR channels that they could isolate and record from a single channel. They generated this sparse distribution by taking advantage of changes that take place in skeletal muscle after the motor nerve controlling the muscle is damaged. When a muscle fiber is *denervated* (i.e., it loses its neuronal input—experimentally, the axons are crushed), the region of the plasma membrane that responds to ACh gradually spreads out across the surface. Initially, only the membrane at the endplate region responds to ACh, but eventually most or all of the membrane contains AChRs and can respond to ACh. Following denervation of a muscle, Neher and Sakmann voltage-clamped an area of muscle membrane with sparsely distributed ACh channels at a hyperpolarized potential to increase the driving force for inward current. They then filled a large-bore micropipette (tip diameter 10 μm) with Ringer solution containing a low concentration of an ACh agonist, then brought the polished tip of the pipette to the surface of the muscle fiber, exposing any AChRs under the pipette tip to the action of the agonist. The extracellular pipette was connected to a highly sensitive, low-noise amplifier with which they could record currents flowing in it. When it was applied snugly to the surface of the denervated muscle fiber, the pipette detected minute inward currents (less than 5×10^{-12} A) produced by the transient opening of single ACh-activated channels (Figure 6-33). With this experiment, Neher and Sakmann produced the first recordings ever made of currents through single ion channels in a biological membrane. Indeed, this work produced the first direct evidence that ionic currents cross the membrane through discrete, gated channels, rather than by some other means, such as carrier molecules.

Single-channel currents such as those recorded by Neher and Sakmann are more or less rectangular in shape; they turn on and off abruptly, and they are all-or-none. This observation strongly suggests that the channels can exist only in one of two states, completely closed or completely open. Moreover, the unitary currents recorded from all nicotinic ACh-activated channels are about the same size when the electrochemical

(a)

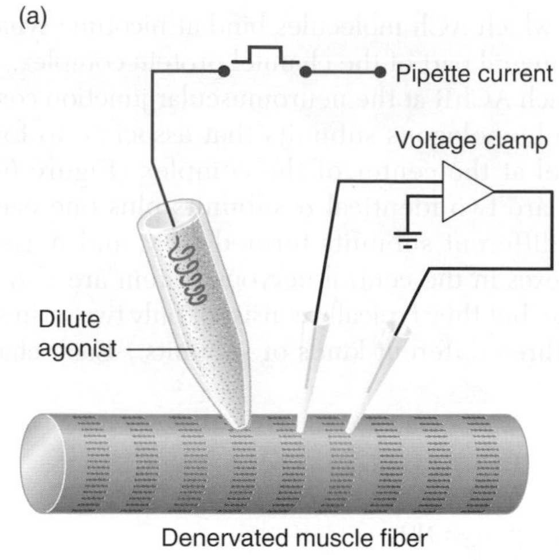

(b)

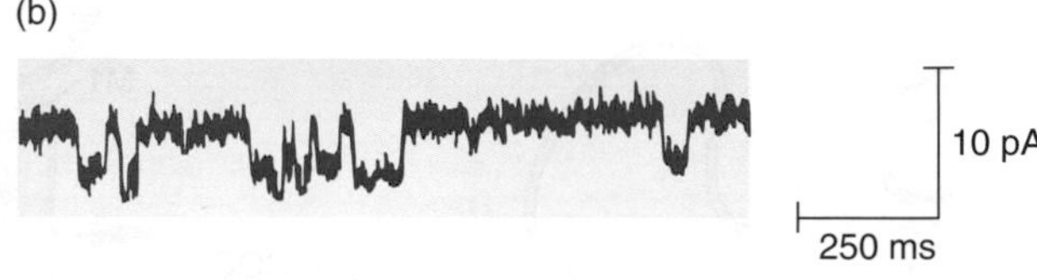

Figure 6-33 The patch-clamp recording technique reveals ionic currents through single acetylcholine receptor (AChR) channels. **(a)** The muscle fiber membrane is held at a hyperpolarized potential (−120 mV) by a voltage-clamp circuit, greatly increasing the inward driving force on Na^+ ions, and exerting an inward force on K^+ ions as well, because −120 mV is more negative than E_{rev} for K^+. The surface of the muscle is then explored with a patch pipette filled with Ringer solution containing an ACh agonist (in this case 2×10^{-7} M suberylcholine).
Classic Work **(b)** When the pipette tip is sealed tightly against the membrane, brief, minute inward currents are recorded. In this experiment, the pipette recorded current flow through the ion channel of a single AChR protein complex that opened transiently when agonist molecules were bound to receptor sites. [Adapted from Neher and Sakmann, 1976.]

driving force is the same. According to Ohm's law, this result *must* mean that all individual nicotinic ACh channels have similar conductances. When two or more channels in the patch open with overlapping times, the individual, unitary single-channel currents sum linearly, producing a current whose amplitude is a multiple of the current through a single channel. These changes in current are not seen unless the external pipette contains ACh or an ACh agonist, and the frequency of their occurrence depends on the concentration of the transmitter or agonist in the pipette. From Ohm's law, the conductance of a single open nicotinic AChR channel was calculated to be about 2×10^{-11} S, which is usually expressed as 20 picosiemens (20×10^{-12} S; put another way, each channel has a resistance of 5×10^{10} Ω).

Many ligand-gated postsynaptic ion channels have now been studied intensively by this method of recording single-channel currents. Statistical analyses of these unitary currents indicate that the channels can fluctuate

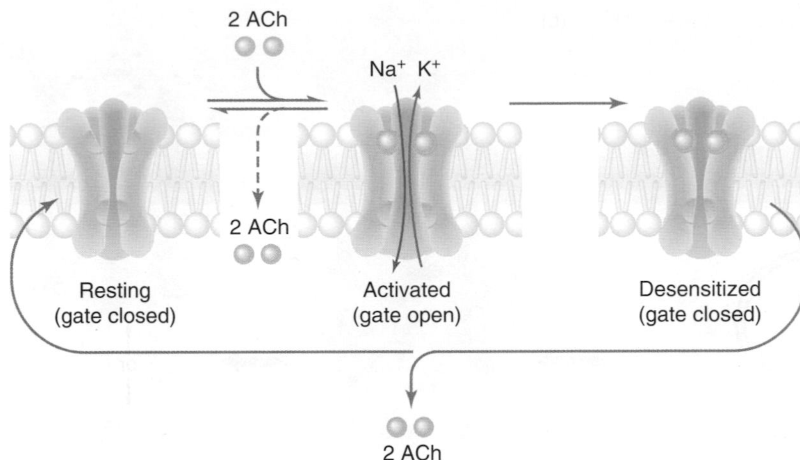

Figure 6-34 The nicotinic acetylcholine receptor channel exists in at least three distinct functional states. The channel opens when two acetylcholine (ACh) or agonist molecules bind to the channel protein complex. The channel can "flicker' between closed and open states while ACh molecules remain bound. When ACh molecules unbind (dashed arrow), the channel closes, and it remains in the closed state until two more ACh molecules bind. After a channel has been open for about 1 ms, it may close, even if ACh molecules are still bound. The channel is then said to be desensitized, and it cannot be reopened until the ACh molecules unbind and new ones bind.

between an open state and several closed states. For example, when two agonist molecules bind to the receptor sites of a closed ACh channel, the probability greatly increases that the channel will change to an open state (Figure 6-34) and briefly allow ions to flow through the channel. The channel remains open for only about 1 ms and then closes, even if ACh is still bound to the receptor sites. After a short time, the agonist molecules leave the binding sites, and the channel remains closed until two more molecules of ACh bind. The macroscopic currents and postsynaptic potentials recorded by conventional methods from a postsynaptic cell represent the sum of many such single-channel events in the postsynaptic membrane.

Other ligand-gated channels

Several types of ligand-gated channels have now been isolated from neurons and characterized, including ACh, glutamate, glycine, and $GABA_A$ receptors, all of which mediate fast, direct postsynaptic responses. These receptors are all multimeric protein complexes, and each receptor typically is composed of two to four different kinds of subunits. As in the muscle ACh channel, only one type of subunit binds the ligand. The DNA sequences of ACh, GABA, and glycine receptor subunits are closely related, suggesting that all ligand-gated ion channels may have a common ancestral origin. Different combinations of subunit types produce receptors with slightly different properties, and in the mammalian brain different regions express receptors of characteristic composition. The large number of permutations that are possible, even among receptors that respond to a single neurotransmitter, suggests the subtlety of the mechanisms that allow the brain to achieve its highly differentiated functional states.

DNA sequence analysis has revealed that the ligand-gated glutamate receptors (called **ionotropic glutamate receptors** or **iGluRs**) belong to a family bearing only slight sequence homology to the nicotinic receptors. Currently, there is intense interest in this receptor family, both because glutamate is the most ubiquitous excitatory neurotransmitter in the mammalian central nervous system and because glutamate receptors participate in modifications of synaptic strength that may underlie learning and memory. At present, three types of fast-acting glutamate receptors have been identified; they are named for their sensitivity to specific agonists that activate each type selectively. The agonists that typify the three receptor classes are kainate, AMPA (α-amino-3-hydroxy-5-methylisoxazole-4-propionic acid), and NMDA (*N*-methyl-D-aspartate). Like other ligand-gated receptors, the glutamate receptors are multimeric protein complexes consisting of either four or five monomers per complex, and several varieties of each type of monomer are known. Each class of glutamate receptors has a unique set of monomers. The monomers are somewhat different in structure from those of other ligand-gated ion channels, and two of the extracellular domains of a monomer act together to bind ligand and open the channel (Figure 6-35 on the next page). In the central nervous systems of mammals, AMPA and NMDA receptors are closely associated with each other in the postsynaptic membrane and appear to work together.

The NMDA receptors are unusual in several ways. In mammals, both glycine and glutamate must bind to a receptor before it opens. Even with both ligands bound, the channel still will not allow ions to cross the membrane through the pore as long as the membrane potential is near V_{rest}. At values of V_m near V_{rest}, the ion channel is blocked by a Mg^{2+} ion that prevents cations from entering the channel. When the membrane is depolarized by postsynaptic current through AMPA channels, the Mg^{2+} ion is dislodged, and ions can flow through the NMDA channel. AMPA channels are quite selective for Na^+ ions; NMDA channels are larger and allow both Ca^{2+} and Na^+ to cross the membrane. Thus, activation of NMDA channels increases the intracellular $[Ca^{2+}]$, and this free Ca^{2+} can serve as an intracellular messenger.

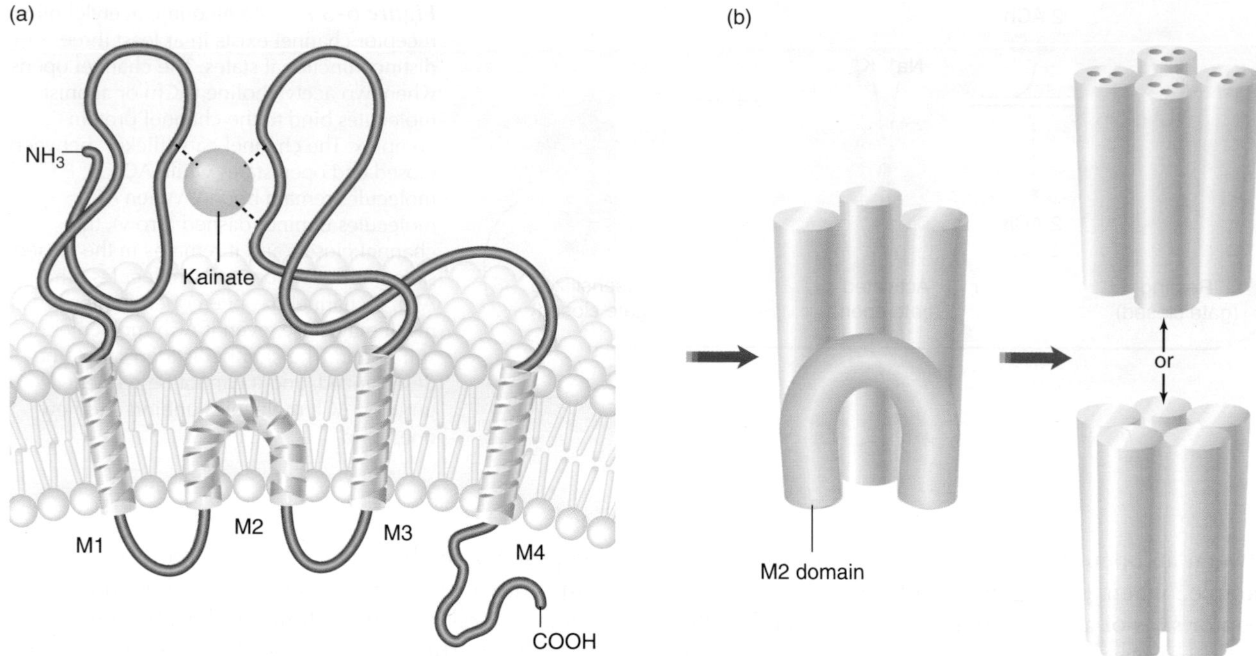

Figure 6-35 The monomers of ionotropic glutamate receptors have a structure distinct from those of other ionotropic neurotransmitter receptors. **(a)** Structure of an iGluR monomer. The M2 domain, rather than extending through the membrane, doubles back to the intracellular face of the membrane. It plays an important role in the ion selectivity of the channel. **(b)** Four or five associated monomers, typically of at least two different subtypes, compose an active receptor. [Part a adapted from Bigge, 1999.]

Receptors in Slow, Indirect Neurotransmission

A large family of receptors participates in slow synaptic transmission. These receptors are linked to **G proteins,** a class of membrane-linked molecules that play important roles in signal transduction. G proteins are covered in more detail in Chapter 9.

At least three separate proteins contribute to G protein–mediated synaptic transmission (Figure 6-36). The neurotransmitter receptor protein, which spans the membrane, binds the neurotransmitter on its extracellular face and activates the G protein on its cytoplasmic face. The activated G protein then regulates the activity of effector proteins, which can be ion channels, enzymes that control the concentrations of intracellular second messengers, or both. We now know of more than a hundred receptors that act through G proteins, and these signaling molecules respond to a wide variety of external stimuli, ranging from peptides to light and odors. The G proteins themselves constitute a family of at least 20 different proteins. The combinatorial richness of this system provides yet another mechanism for producing subtle control within the nervous system.

A well-studied example of indirect neurotransmission that regulates conductance through ion channels is found in heart atrial cells, the system Otto Loewi used more than 75 years ago to first demonstrate that neurons can transfer information by way of chemical sig-

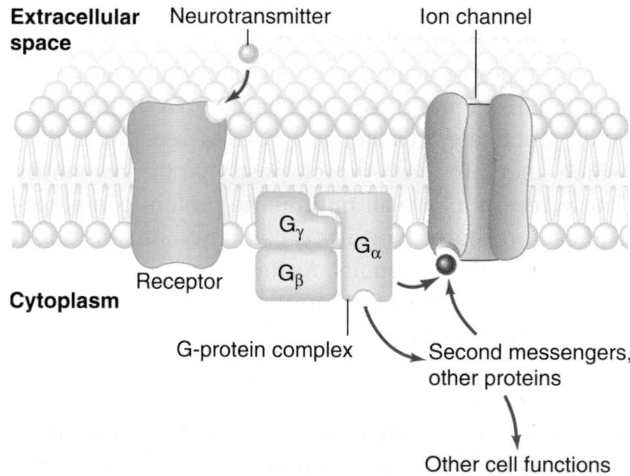

Figure 6-36 Intracellular second messengers modify channel conductances at slow, indirect chemical synapses. G proteins participate in signal transduction at these synapses. The receptor protein spans the plasma membrane. Neurotransmitter molecules bind to the extracellular domain of the receptor, which activates a G protein that resides on the cytoplasmic side of the membrane. The activated G protein regulates the activity of other intracellular proteins, which directly or indirectly change the conductance through ion channels in the membrane. Activated G proteins can also have other effects on cellular functions, such as changing metabolic pathways, modifying transcription, or altering the structure of the cytoskeleton.

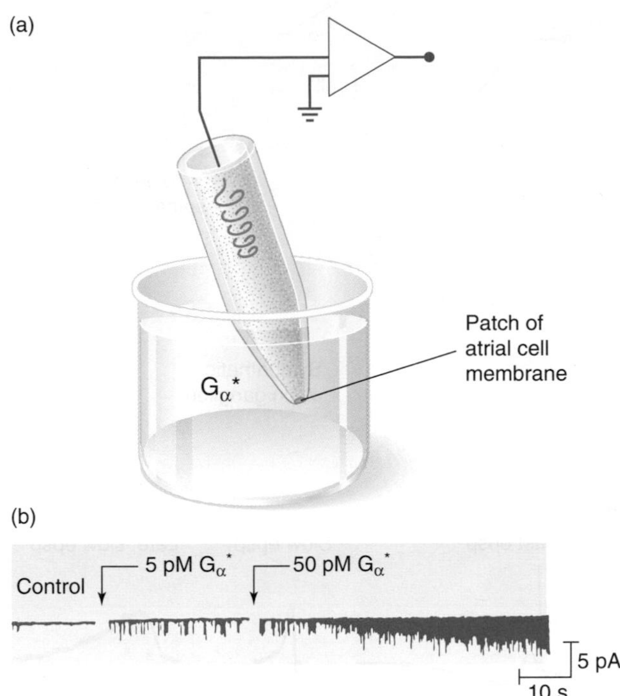

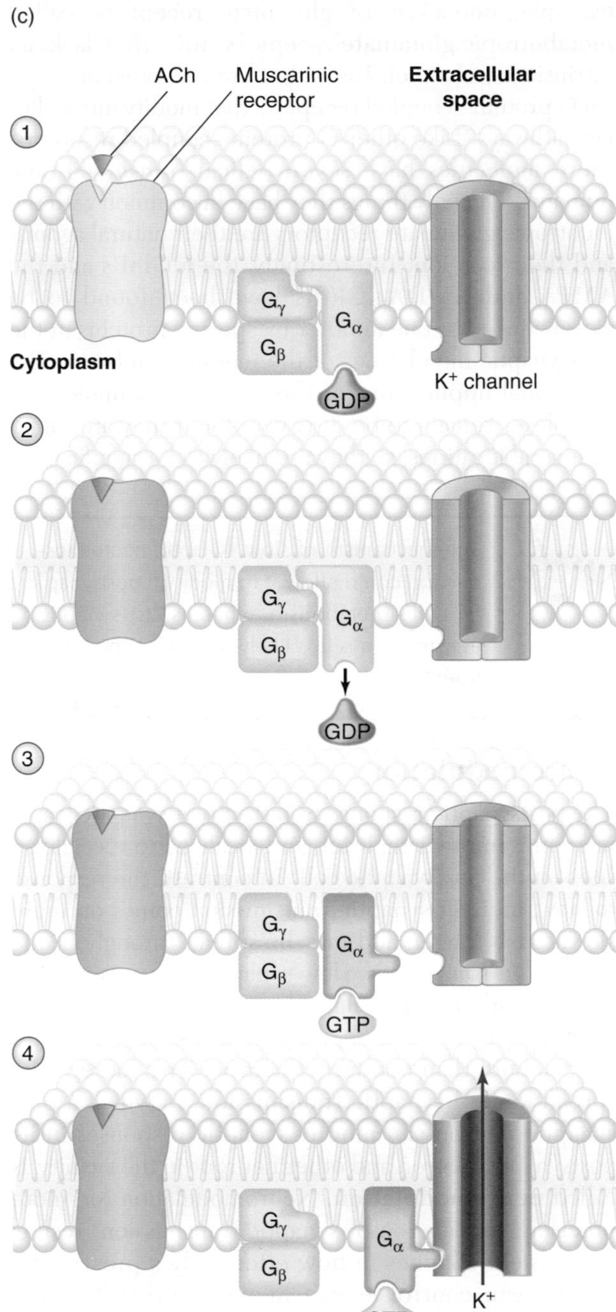

Figure 6-37 Muscarinic acetylcholine receptors in cardiac cells indirectly cause K^+ channels in the plasma membrane to open. **(a)** Experimental setup for measuring the effect of slow synaptic activation on guinea pig heart atrial cells. A nonhydrolyzable analog of GTP, GTPγS, was bound to α-subunits of the G protein to activate them, and the activated α-subunits (the activated state is indicated by an asterisk) were applied to the intracellular surface of an inside-out patch of atrial cell plasma membrane. This treatment mimics receptor-mediated activation of the endogenous G protein. **(b)** Typical recordings from an experiment like the one shown in part a. A low concentration of activated α-subunits (5 pM, or 5×10^{-9} molar) produced periodic channel openings. When the concentration of activated α-subunits was increased tenfold, the K^+ channels opened more frequently, producing more frequent current steps in the single-channel records. **(c)** Schematic representation of events at a muscarinic synapse in an intact cell. When ACh binds to a muscarinic receptor (step 1), a G protein associated with the receptor releases bound GDP (step 2) and binds GTP (step 3). This activates the subunits, and the activated α-subunits then bind to K^+ channels (step 4), causing them to open. [Data adapted from Codina et al., 1987.]

nals. Acetylcholine acts on muscarinic ACh receptors in the heart to hold K^+-selective channels open longer than usual, reducing excitability. Several different kinds of experiments contributed to establishing that this action of ACh depends on a G protein. For example, acetylcholine has been found to act on heart atrial cells only if there is guanosine triphosphate (GTP) inside the cells. G proteins can function only if GTP is available to activate them. Furthermore, the effect produced when ACh binds to muscarinic ACh receptors in the membranes of these cells is blocked by pertussis toxin, which inactivates many G proteins. In a direct test of the hypothesis that ACh acts on these cells through a G protein, Codina and colleagues (1987) prepared G-protein α-subunits that had been activated by a nonhydrolyzable analog of GTP, GTPγS, and applied them to the inside of membrane patches from cardiac-muscle cells (Figure 6-37a). The result mimicked the stable activation of G proteins in the cell. As the amount of activated α-subunit was increased in the solution bathing the cells, the number of open K^+ channels increased, as indicated by the increased number of single-channel currents (Figure 6-37b).

Similar experiments have identified a large variety of postsynaptic channels whose activity is regulated by receptor-activated α-subunits of G proteins. For example, one class of glutamate receptors, called **metabotropic glutamate receptors (mGluRs)**, lacks an intrinsic ion channel. Instead, these receptors are typical G protein–coupled receptors that modify intracellular pathways. Like other G protein–coupled receptors, these molecules have seven helical transmembrane domains. The only things they have in common with the ionotropic glutamate receptors are their natural agonist and their location. Interestingly, the mGluRs and the AMPA and NMDA iGluRs have been found to be closely linked to one another by specific proteins in the postsynaptic membranes of the mammalian brain. The functional implications of these large multimolecular complexes remain to be discovered, but they are found in several locations, so they are likely to be significant.

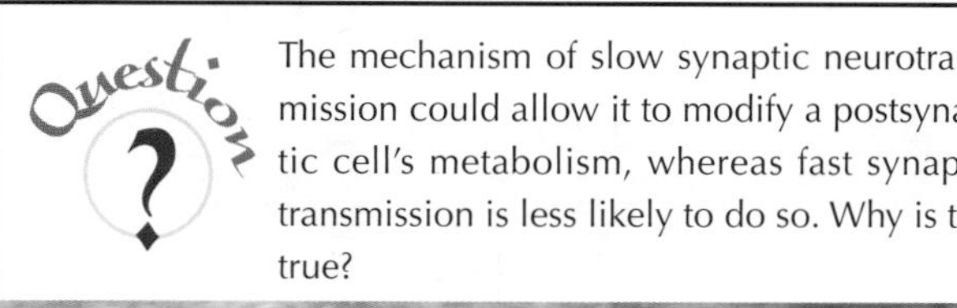

The mechanism of slow synaptic neurotransmission could allow it to modify a postsynaptic cell's metabolism, whereas fast synaptic transmission is less likely to do so. Why is this true?

Neuromodulation

The postsynaptic response to fast, direct synaptic transmission is immediate, brief, and localized to specialized sites on the postsynaptic cell. In contrast, the response to slow, indirect synaptic transmission comes on more slowly, lasts longer, and may be spread throughout the neuron. In some cases, slow synaptic transmission can interact with and modulate the effects of fast synaptic transmission. In addition, slow transmission sometimes affects more than one target neuron. This more widespread kind of synaptic transmission, in which a presynaptic neuron can modify synaptic responses in its postsynaptic neuron and other neurons in the vicinity, is called **neuromodulation.** Neuromodulation (or, more precisely, modulation of synaptic transmission) results in transient changes in how effectively a presynaptic neuron can control events in the neurons that are postsynaptic to it (i.e., its synaptic efficacy). Neuromodulatory changes last from seconds to minutes. Changes in synaptic efficacy that are longer lasting or even permanent are typically called **synaptic plasticity,** described later in this chapter.

One of the best-understood examples of neuromodulation and its role in normal synaptic excitation is found in the neurons of sympathetic ganglia of frogs. This system is complex because these neurons receive three classes of synaptic inputs that are mediated by two different neurotransmitters acting on three distinct types of receptors. Three distinguishable excitatory postsynaptic potentials are produced: a fast epsp, a slow epsp, and a late slow epsp. A typical experimental setup for measuring these potentials is shown in Figure 6-38a. Both the fast and the slow epsps (Figure 6-38b) are pro-

(a)

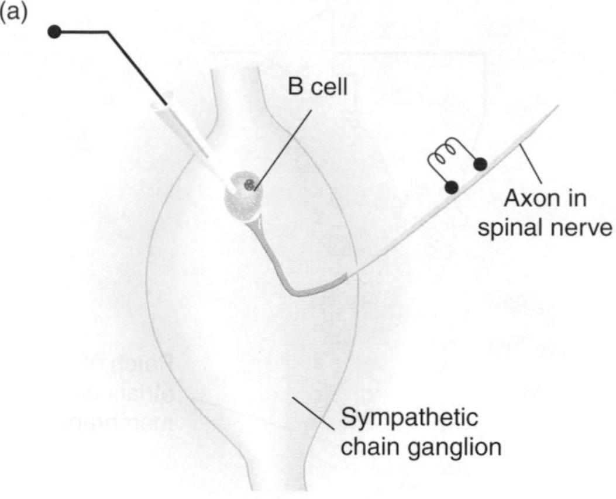

(b)

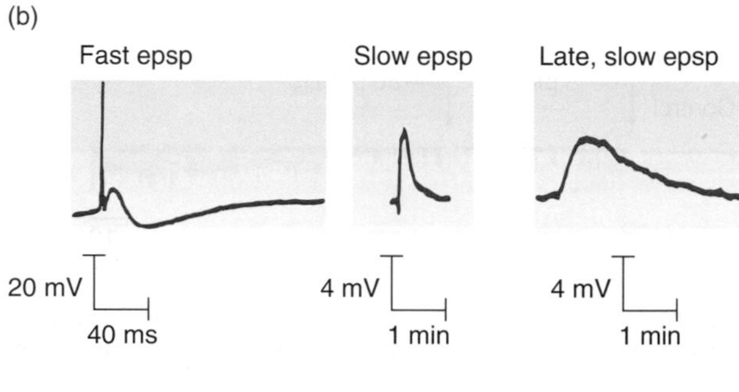

(c)

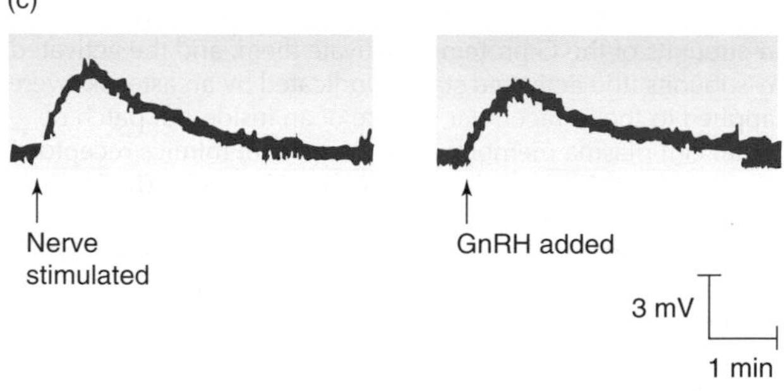

Figure 6-38 Three kinds of postsynaptic potentials with very different time courses can be recorded in neurons of the bullfrog sympathetic chain ganglion following presynaptic stimulation. **(a)** The sympathetic chain ganglia are located on either side of the spinal cord (see Chapter 8). The responses of the large B cells (one class of neurons in the ganglia) can be recorded while nerves that innervate the ganglia are stimulated. Anterior is up in this diagram. **(b)** Three different kinds of synaptic responses can be recorded in B cells following nerve stimulation: (1) a fast excitatory postsynaptic potential (with a latency of 30–50 ms) when ACh activates nicotinic receptors in the postsynaptic membrane; (2) a slow, longer-lasting epsp (with a latency of 30–60 ms) when ACh binds to muscarinic receptors in the postsynaptic membrane; and (3) a late, slow epsp (with a latency of more than 100 ms) caused by a decapeptide molecule—found in the brains of cold-blooded vertebrates—that is closely related to the hypothalamic releasing hormone GnRH. When this GnRH-like peptide binds to postsynaptic receptors, it produces a depolarization in the B cells that lasts many minutes. (Notice the calibration bars below the records.) **(c)** When exogenous GnRH is applied to B cells, the effect is identical in onset, magnitude, and duration with the late, slow epsp in part b. [Adapted from Jan and Jan, 1983.]

duced by ACh released from presynaptic nerve terminals. The postsynaptic cells have both nicotinic receptors (fast response) and muscarinic receptors (slow response) in their plasma membranes. In contrast, the late, slow epsp is produced by a neuropeptide very similar to mammalian gonadotropin-releasing hormone (GnRH). This neuropeptide is released from the presynaptic neurons, but not directly onto the postsynaptic neurons. The three postsynaptic potentials depolarize the postsynaptic cell by different amounts and at different times after stimulation, and they act through different but not entirely independent mechanisms.

When ACh binds to a nicotinic receptor, the ion channel in the receptor complex opens, allowing Na^+ and K^+ to pass through and producing the fast epsp (Figure 6-38b). This excitatory postsynaptic potential can be elicited by a single stimulus that lasts only a few tens of milliseconds. The slow epsp is produced when ACh binds to muscarinic receptors, and it can be elicited only after several trains of APs have arrived at the presynaptic terminal and caused the release of ACh. The muscarinic receptors act through a G protein to cause a special type of K^+ channel, called an M channel, to close (Figure 6-39a). When M channels close, the low steady-state influx of Na^+ is no longer balanced by an efflux of K^+ and, as a result, the cell depolarizes. The depolarization is small (only about 10 mV; see Figure 6-38b), because it depends on the small steady-state Na^+ current. By itself, it cannot produce an AP in the postsynaptic cell, but it can significantly change the response of the cell to subsequent fast synaptic signals, particularly when it acts in concert with the late, slow epsp.

The late, slow epsp results from the release of the GnRH-like peptide, which acts through a transmembrane receptor to close the same M channels that are affected by the muscarinic receptors. Adding GnRH to the postsynaptic neurons produces the same kind of response (Figure 6-38c). The time course of the response to the GnRH-like peptide is even slower than the muscarinic response; it begins 100 ms after the stimulus and can last for as long as 40 minutes.

The similarities and differences between the two slow responses are important for understanding how neuromodulation operates in animals. The late, slow epsps modulate the other synaptic responses. To explore the role of the modulatory late, slow epsps in sympathetic ganglion cells, the efficacy of an injection of depolarizing current into the presynaptic cell was

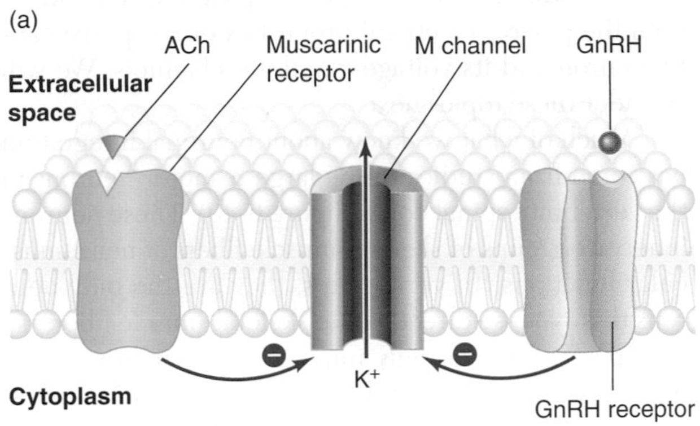

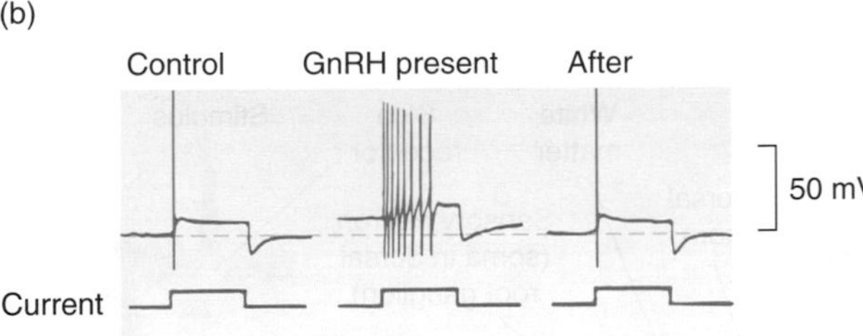

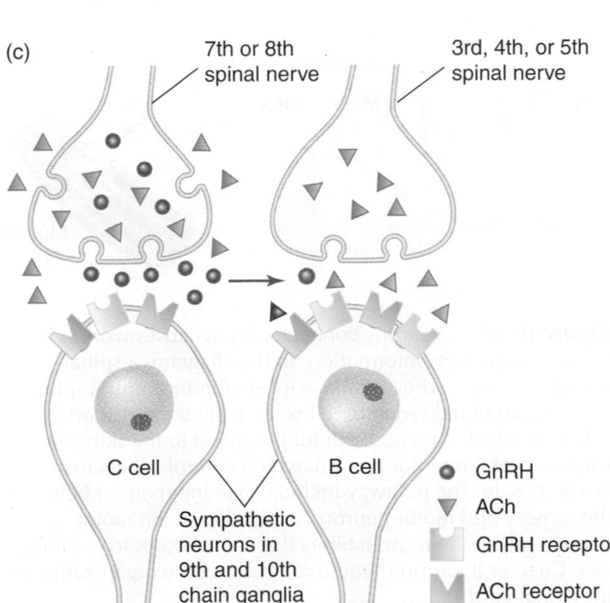

Figure 6-39 Both muscarinic acetylcholine receptors and GnRH receptors depolarize a postsynaptic cell by closing M-type K^+ channels. **(a)** When acetylcholine (ACh) binds to muscarinic receptors or when the GnRH-like neuropeptide binds to its receptor, M channels close, reducing the K^+ current across the membrane and allowing residual Na^+ current to depolarize the neuron. **(b)** The effect of a fast excitatory postsynaptic potential (epsp) in a postsynaptic B cell before, during, and after a late, slow epsp. During the late, slow epsp, the decrease in K^+ current through M channels increases the excitability of the B cell, producing a train of action potentials in response to the fast epsp. **(c)** The GnRH-like peptide is likely to reach B cells from a distant site, rather than by direct synaptic contact. Cholinergic neurons from the seventh and eighth spinal nerves innervate C cells of the ninth and tenth sympathetic ganglia of the bullfrog, whereas neurons from the third, fourth, and fifth spinal nerves innervate only B cells in those ganglia. Only C cells receive axon terminals that are immunoreactive for the GnRH-like peptide, but stimulating the seventh and eighth spinal nerves produces a late, slow epsp in both B and C cells, suggesting that the GnRH-like peptide diffuses from its site of release at the surface of C cells and activates receptors on B cells. [Part b adapted from James and Adams, 1987; part c adapted from Jan and Jan, 1983.]

evaluated before, during, and after a late, slow epsp (Figure 6-39b). Before the modulatory epsp, a presynaptic stimulus caused a single postsynaptic AP; during the modulatory epsp, the same stimulus elicited a burst of APs. After the modulatory epsp had subsided, only a single AP was elicited. Clearly, the modulatory epsp increased the synaptic efficacy of the fast psps.

Normally, the K^+ current through M channels is activated by membrane depolarization and acts to repolarize the cell by shunting any depolarizing currents that enter through the open channels, thus reducing the effectiveness of excitatory postsynaptic potentials. When the M channels are closed by ACh muscarinic receptors, repolarization of the membrane by the K^+ current is prevented, and incoming excitation is enhanced. The late, slow epsp acts similarly, but with a longer latency and for a longer time, and it shares the M channel as a final common pathway. However, there is an additional twist, because the GnRH-like peptide diffuses to nearby neurons, which it influences if the appropriate receptors are present (Figure 6-39c). Only some of the presynaptic neurons in the sympathetic ganglia release the GnRH-like peptide, but most postsynaptic cells seem to have GnRH receptors, which strongly suggests that neuromodulation is a normal part of this neuronal circuitry. Taken together, these mechanisms can generate a variety of postsynaptic effects following presynaptic transmitter release. A brief burst of activity in the presynaptic cells typically causes only the fast excitatory postsynaptic response. More prolonged stimulation might also activate the slow pathway, which would amplify somewhat the response of a postsynaptic cell to its fast excitatory postsynaptic potentials. With still greater stimulation, the late, slow pathway would even further increase the effectiveness of fast excitatory postsynaptic potentials. It might also enhance responses in neighboring neurons (see Figure 6-39c), increasing the efficacy of neurotransmission in cells that are not directly postsynaptic to the GnRH-releasing neurons. Moreover, this modulation could be relatively long lived, given the long duration of the late, slow response.

Within the past few years, studies in the crustacean stomatogastric ganglion have demonstrated the extreme power of neuromodulatory mechanisms. This ganglion contains only about 30 identified neurons, whose interconnections have been characterized in detail and whose output patterns are well known. When certain neuromodulatory transmitters, such as proctolin or cholecystokinin, are added to the saline bathing the stomatogastric ganglion, the properties of at least some of the membrane channels change dramatically, effectively rewiring the entire ganglion and generating circuits and outputs that are never seen in the absence of the modulator. Interestingly, at least some of the neuromodulators in this system work to change conductances through identified specific subtypes of K^+ channels in postsynaptic cells. Thus, neuromodulators afford a means of remodeling neuronal circuitry, allowing a set of neurons to interact in several distinctly different ways, even though their physical synaptic contacts remain unchanged.

INTEGRATION AT SYNAPSES

Up to this point, we have been discussing events at single synapses, but even the simplest behavior typically requires that many neurons—even many thousands of them—act in a coordinated fashion. Each neuron integrates all the various excitatory and inhibitory synaptic signals that impinge on it and then either produces an AP or does not. Even neurons that receive thousands of individual synaptic inputs respond in this simple way. However, the processes that link postsynaptic potentials to the production of APs are complex and depend on both the passive electrical properties of the postsynaptic neuron and its voltage-gated ion channels. We will consider these topics next.

Much of what we know about neuronal integration has come from studies of the large α-motor neurons in the vertebrate spinal cord (Figure 6-40). These neurons innervate groups of skeletal-muscle fibers at neuromuscular junctions. In vertebrates, they are the only neurons that synapse directly onto skeletal-muscle fibers, so they play an exceedingly important role in generating overt behavior. Thousands of inhibitory and excitatory synaptic terminals contact the dendrites and soma of each α-motor neuron. The net effect of all this synaptic

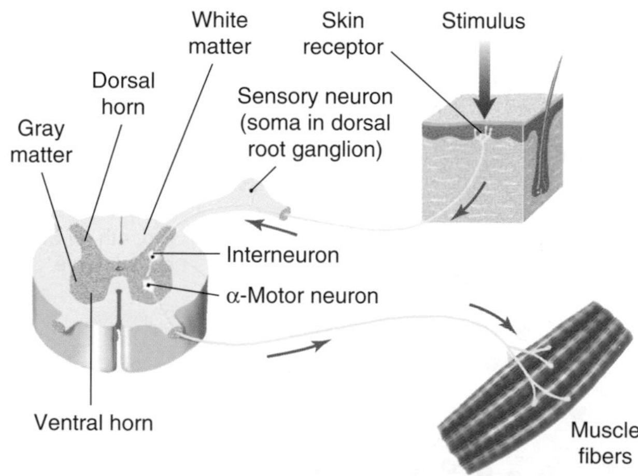

Figure 6-40 Neurons connected by synapses work together to process information. In this diagram, a spinal α-motor neuron, whose soma is located in the ventral spinal cord, is part of a disynaptic reflex arc (called the *flexion reflex*) in which a noxious stimulus applied to the skin causes excitation of the motor neuron, which controls fibers in a flexor muscle. The pathway includes one interneuron between the sensory and motor neurons. Activation of the motor neuron causes the muscle fibers that it innervates to contract. (See Chapter 8 for more information about the organization of the spinal cord.)

activity is to control the frequency with which APs are generated in the α-motor neuron. Its frequency of firing (typically expressed as APs per second) determines the strength of contraction in the set of muscle fibers it innervates.

All of the integrative activity in a neuron is centered on producing APs (excitation) or suppressing them (inhibition). Because APs are the only events that can carry information over distances greater than a few millimeters, only synaptic inputs that cause APs in α-motor neurons can generate behavior. Any excitatory input that fails to bring the motor neuron to threshold, either by itself or acting in concert with other inputs, is lost, because no AP is produced in the postsynaptic cell and the signal dies out.

In an α-motor neuron, APs are generated at the spike-initiating zone, which in these neurons is located just beyond the axon hillock (see Figure 5-2). This region is more sensitive to depolarization than are the soma and dendrites and thus has a lower threshold for producing APs; in some neurons it has been determined that the membrane here has a higher density of voltage-gated Na^+ channels. To generate an AP, a postsynaptic current must bring the membrane of the spike-initiating zone to threshold. Each post-synaptic potential is conducted decrementally from the synapse to the spike-initiating zone, so bringing the neuron to threshold is by no means assured.

How do the many thousands of individual synaptic inputs onto a motor neuron influence its activity? Postsynaptic currents spread electrotonically from synapses on dendrites and the soma (Figure 6-41). How much the current decays over distance is determined by the length constant of the postsynaptic neuron, but in all cases, postsynaptic potentials become smaller as they spread away from their sites of origin and toward the spike-initiating zone. A postsynaptic current set up at the end of a long, slender dendrite will decay more than a current closer to the spike-initiating zone, so synapses distant from the spike-initiating zone exert a relatively smaller influence on the activity of the postsynaptic neuron. As a result, the location of synapses, as well as the initial size of synaptic currents, can influence how much control they exert. Surprising new evidence suggests that in the mammalian brain there are voltage-gated Na^+ channels in dendritic membranes of at least some neurons. These channels can boost synaptic currents, preventing them from decaying as rapidly as they would if they were conducted only electrotonically. These channels thus help to counteract the relatively long distances that must be traveled by synaptic inputs onto the dendrites of these large neurons. The location of synapses is important in another context as well. The density of inhibitory synapses contacting many neurons is highest near the axon hillock, making these synapses powerfully effective in preventing excitatory postsynaptic currents from depolarizing the spike-initiating zone to threshold. In effect, these inhibitory synapses have veto power over input from the more distant excitatory synapses.

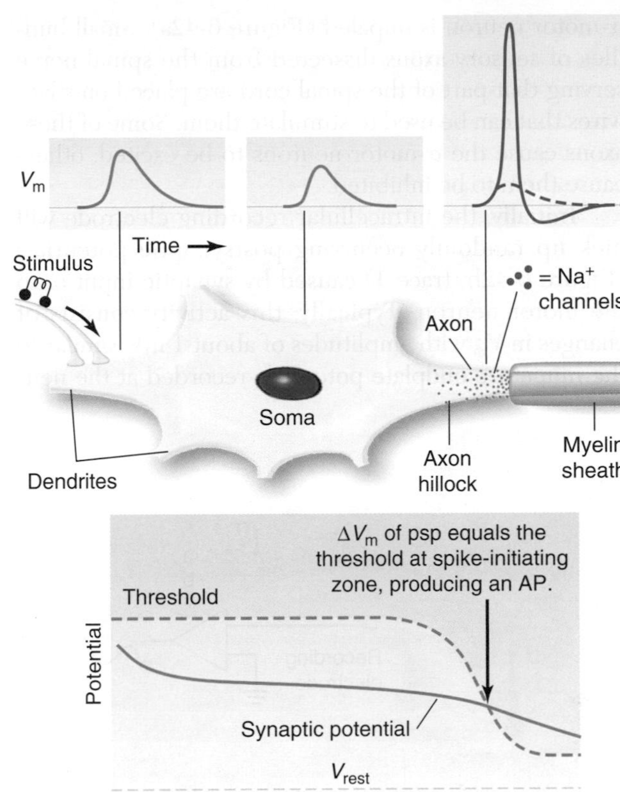

Figure 6-41 Each synaptic input decays with distance as it travels to the spike-initiating zone. An excitatory postsynaptic potential originating in a dendrite spreads electrotonically, decreasing in amplitude with distance traveled (top). The amplitude of the signal is shown at successive times as it travels from the synapse to the spike-initiating zone. No AP is generated until the signal reaches the densely packed voltage-gated Na^+ channels in the spike-initiating zone, where the firing threshold is lowest. The dashed red line in the top right record shows how the amplitude of the signal would continue to decline in the absence of these Na^+ channels. In all parts of the neuron, the density of Na^+ channels in the membrane (shown as red dots) determines the threshold (dashed orange trace in lower graph). The graph at the bottom of the figure shows the relation between the threshold potential and the amplitude of a postsynaptic potential as it moves along the membrane between the synapse and the spike-initiating zone. The solid line in this graph shows how the amplitude of the excitatory postsynaptic potential would decay as it traveled farther down the axon if the AP were blocked.

Much of what we know about synaptic summation has emerged from experiments in frogs of the genus *Rana*. In a typical experiment, the spinal cord of an anesthetized frog is exposed by opening the vertebral column. A microelectrode is then lowered into the ventral horn of the gray matter (which contains the cell bodies of motor neurons), and the soma of a single α-motor neuron is impaled (Figure 6-42a). Small bundles of sensory axons dissected from the spinal nerve serving that part of the spinal cord are placed on silver wires that can be used to stimulate them. Some of these axons cause the α-motor neurons to be excited: others cause them to be inhibited.

Initially, the intracellular recording electrode will pick up randomly occurring postsynaptic potentials (Figure 6-42b, trace 1) caused by synaptic input onto the motor neuron. Typically, this activity consists of changes in V_m with amplitudes of about 1 mV, similar to the miniature endplate potentials recorded at the neuromuscular junction (see Figure 6-23). An AP in a single neuron that is presynaptic to these α-motor neurons causes only one or a few quanta of transmitter to be released.

(a)

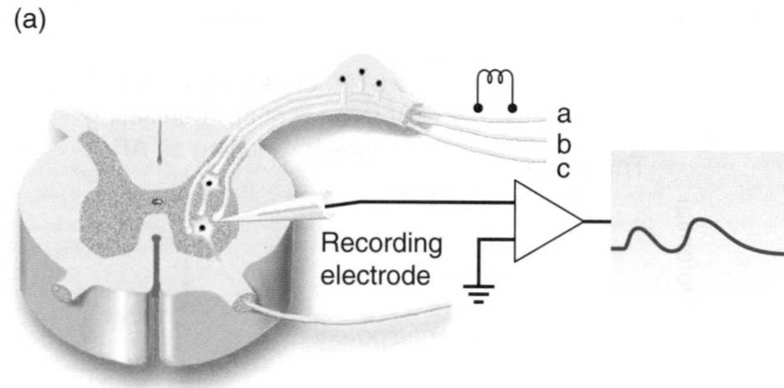

(b)

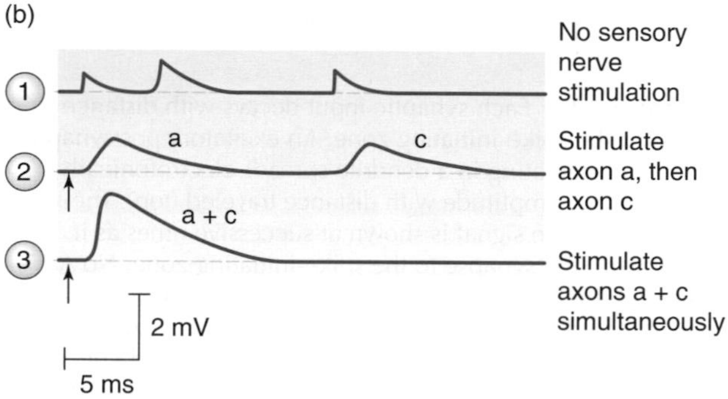

Figure 6-42 Individual excitatory postsynaptic potentials sum together to produce larger postsynaptic potential changes. **(a)** Diagram of a typical synaptic experiment in a frog spinal cord. The spinal cord is exposed, along with the sensory nerve supplying the exposed neurons. Wire electrodes on the sensory nerve are used to stimulate sensory axons. A spinal α-motor neuron is impaled with a microelectrode, and electrical activity in the neuron is recorded. **(b)** Intracellular recordings of synaptic potentials in a motor neuron. Arrows indicate the time at which sensory axons were stimulated. If no stimulation is delivered to the sensory nerve (trace 1), ongoing activity similar to miniature endplate potentials is recorded in the motor neuron, caused by synaptic input from unidentified presynaptic neurons. If either axon a or axon c is stimulated by the wire electrodes (trace 2), an excitatory postsynaptic potential is recorded in the motor neuron. Neither of these epsps brings the neuron to threshold. If axons a and c are both stimulated simultaneously (trace 3), their epsps sum, and the resulting epsp is larger than either epsp in trace 2. Spatial summation of postsynaptic potentials from many synapses is typically required to produce a postsynaptic potential that exceeds the threshold of a motor neuron. If too few excitatory inputs are active simultaneously, V_m at the spike-initiating zone fails to reach threshold, and no APs are produced.

In this respect, excitatory synapses in the central nervous system are very different from the neuromuscular junction. At the neuromuscular junction, a single AP in the presynaptic motor neuron causes the release of 100 to 300 quanta and can produce an excitatory postsynaptic potential of 60 mV or more. As a result, the vertebrate neuromuscular junction is an unusually powerful synapse and transmits in a one-to-one manner. In other words, there will be one AP in the muscle fiber for each AP in the motor neuron. In contrast, activation of a single axon terminal that synapses onto an α-motor neuron releases few quanta and depolarizes the motor neuron by only about 1 mV, far less than the amount required to bring the membrane potential to threshold. Bringing the spike initiation zone of a spinal α-motor neuron to threshold requires more or less simultaneous input from many excitatory presynaptic terminals. Each small postsynaptic current is ineffective by itself, but the activity at each presynaptic terminal contributes to controlling activity in the postsynaptic neuron. This rather democratic behavior prevents activation of motor neurons by trivial input. In addition, it provides a means of integrating inputs from a variety of sources, both excitatory and inhibitory, to determine when the neuron will produce APs and how many there will be. The power of the vertebrate neuromuscular junction makes this filtering important because each AP in the motor neuron will cause contraction in the muscle fibers it controls.

In the experiment shown in Figure 6-42, as the strength of the stimulating current is increased, more and more excitatory axons become active; that is, they are recruited by the increasing intensity of the stimulus. When these axons fire in unison, the total amount of transmitter released onto the motor neuron rises, producing more individual postsynaptic currents that add together to cause a larger excitatory postsynaptic potential (compare Figure 6-42b, traces 2 and 3).

When several postsynaptic potentials add together to change V_m in the postsynaptic neuron, the process is called **synaptic summation.** If all of the summed synaptic inputs are excitatory, they produce a large depolarization. If inhibitory transmitter is released onto the postsynaptic cell at some synapses while excitatory transmitter is released at others, the excitatory and inhibitory synaptic currents sum together (Figure 6-43), but in that case the summed psp is smaller than it would be in the absence of summation. Open inhibitory synaptic channels can short-circuit the depolarizing current passing through excitatory channels. Activation of inhibitory synapses thus reduces the depolarization at the spike-initiating zone and decreases the probability that an AP will be produced.

Summation at synapses is traditionally divided into two categories: **spatial summation** and **temporal**

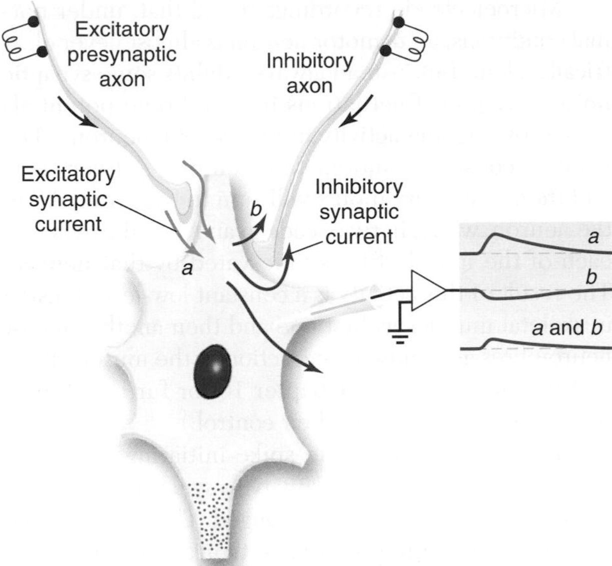

Figure 6-43 Excitatory and inhibitory synaptic currents are summed. Stimulation of separate axons that synapse with a target neuron gives rise to (*a*) excitatory and (*b*) inhibitory synaptic currents. The traces at the right show synaptic potentials recorded when either *a* or *b* is stimulated individually and when they are stimulated together. Simultaneous stimulation of the two axons causes their postsynaptic effects to sum.

summation. If the summed synaptic inputs are contributed by two or more different neurons, summation is typically considered spatial. If the summed synaptic inputs arise from a series of high-frequency APs arriving at a single synaptic terminal, the summation is typically considered temporal. We have already seen an example of spatial summation (see Figure 6-42); let's briefly consider the basis of temporal summation. If a second postsynaptic potential arrives within a very short time after a psp has been elicited, it can add to, or "ride piggyback" on, the first, even when the two synaptic events are caused by APs in the same presynaptic neuron (Figure 6-44). The shorter the interval between two successive postsynaptic potentials, the higher the second response rides upon the first, and the larger the postsynaptic potential can become. If additional stimuli arrive in rapid succession, the third postsynaptic potential rides on the second, and so forth. The maximum amplitude of the psp is determined by E_{rev} of the combined synaptic currents, as we discussed earlier. Notice that the categories of spatial and temporal summation are useful in describing the source of synaptic inputs, but these categories are unrelated to the biophysics of the postsynaptic cell. The changes in V_m produced by postsynaptic currents combine in the same way for both kinds of summation.

Both spatial and temporal summation of postsynaptic potentials depend on the passive electrical properties of neurons. Spatial summation takes place because postsynaptic potentials that originate simultaneously, but at different synapses, spread electrotonically away from the synapse, so these changes in V_m add together at the spike-initiating zone. Temporal summation takes place even though the individual currents do not overlap in time (see Figure 6-44c) because the time constant of the membrane is long relative to the time course of postsynaptic currents (see Figure 6-16b). The first postsynaptic current brings positive charge into the cell, partially discharging the capacitance of the plasma membrane. The positive charge carried into the neuron by the postsynaptic current then slowly leaks out through the resistance—that is, in the form of K^+ flux through I_K leak channels—and the capacitance of the membrane. The membrane potential gradually returns to its resting value some time after the postsynaptic current has ceased; typically, a postsynaptic potential outlasts the postsynaptic current by several milliseconds. If a second postsynaptic current flows before the first postsynaptic potential has subsided, it causes a second

(a)

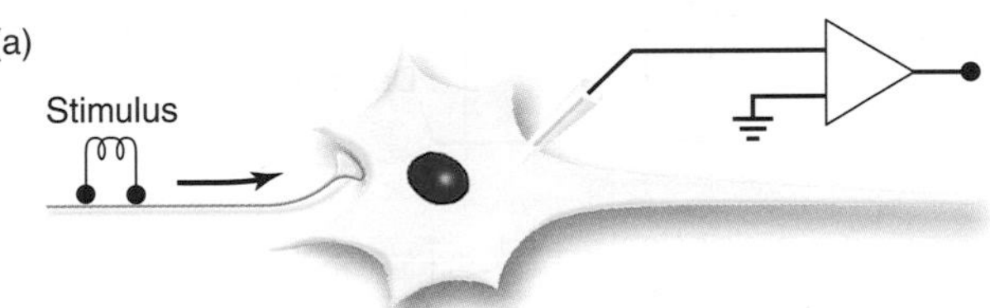

(b)

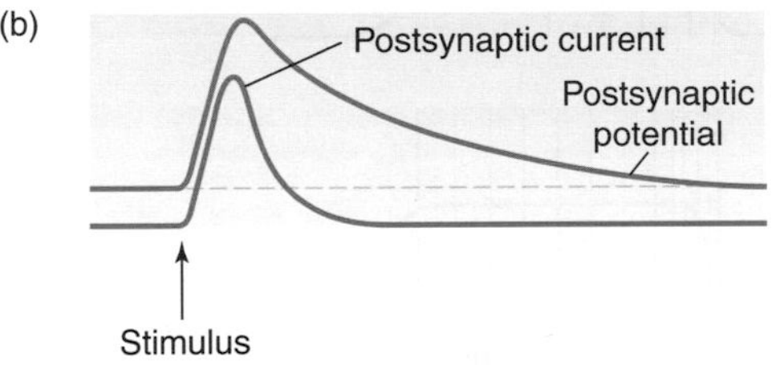

(c)

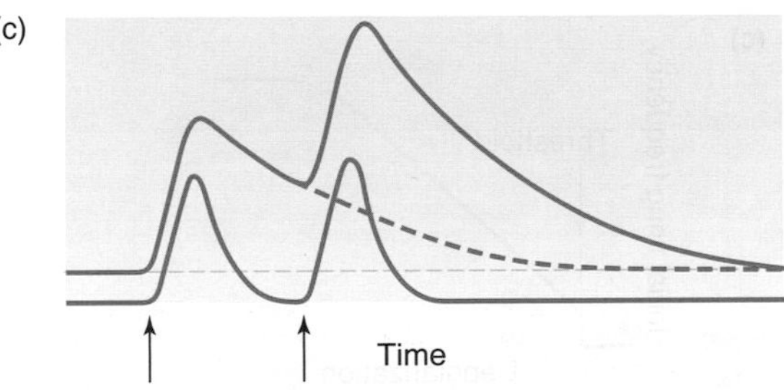

Figure 6-44 Temporal summation occurs when multiple presynaptic signals arrive at a synapse in rapid succession. **(a)** Experimental setup for recording postsynaptic events. **(b)** A single stimulus evokes a postsynaptic current (lower signal) and a more slowly decaying postsynaptic potential. **(c)** Two stimuli are delivered in rapid succession, and the psps sum, although the postsynaptic currents do not overlap in time. Summation of postsynaptic currents is not required for summation of postsynaptic potentials, because the postsynaptic potential outlasts the postsynaptic current. Recall that the duration of the postsynaptic potential depends on the time constant, τ, of the plasma membrane. Arrows indicate the times at which presynaptic impulses arrived at the synapse.

depolarization that adds to the falling phase of the first, even though the two postsynaptic currents do not overlap in time. Thus, the membrane's charge-storing capacity allows the voltage effect of postsynaptic currents to sum over time. The longer the time constant, τ, of the membrane, the slower the decay of postsynaptic potentials will be, and the more effective the temporal summation of asynchronous synaptic inputs can be. The membrane time constant of vertebrate motor neurons is typically about 10 ms, but it can range from 1 ms to 100 ms in other neurons.

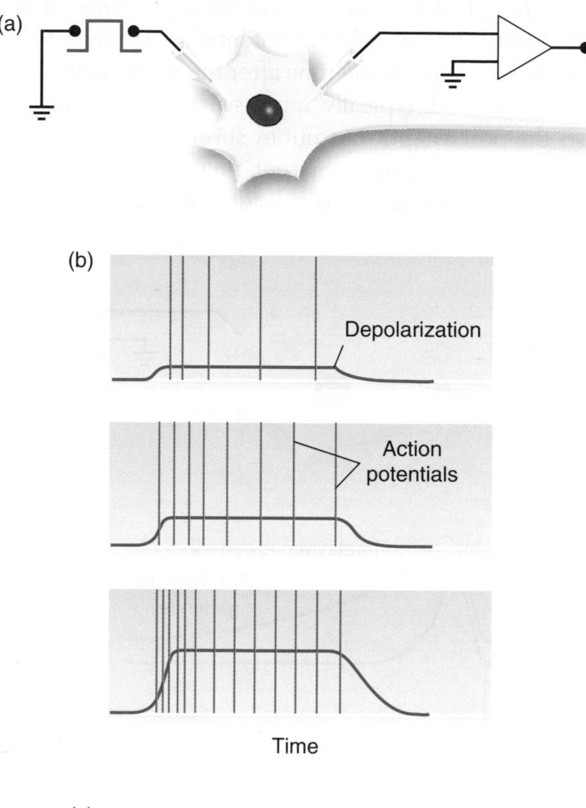

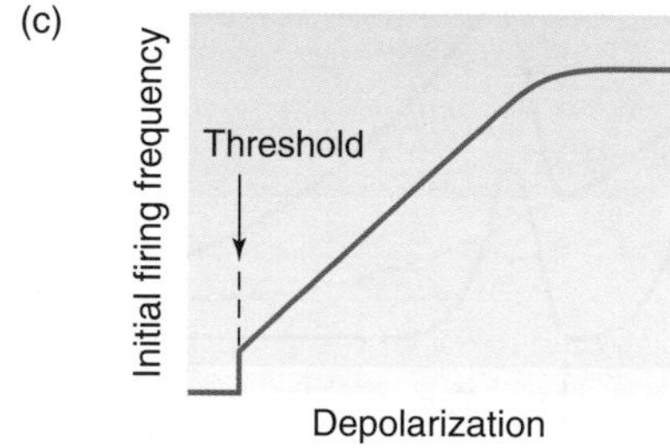

Figure 6-45 The initial frequency of impulses generated in a motor neuron is approximately proportional to the amplitude of the membrane depolarization. **(a)** Two electrodes, one for injecting depolarizing current and one for recording membrane potential, are inserted into a spinal α-motor neuron. **(b)** Three idealized traces show that increasing the current delivered through the current electrode, and thus increasing the amplitude of depolarization (top to bottom), causes an increased rate of firing. **(c)** Initial firing frequency plotted against depolarization. As the depolarization is increased, the frequency of APs increases, up to some maximum value. The maximum frequency of APs is set by the refractory period of the motor neuron.

Microelectrode recordings reveal that, under normal conditions, an α-motor neuron is almost never electrically silent, but instead always exhibits some synaptic noise (irregular fluctuations in membrane potential) caused by ongoing activity in presynaptic neurons. The result is constant, random variation in V_m. Every now and then, excitatory inputs will sum to trigger an AP in the neuron, which in turn leads to an AP and a twitch in each of the muscle fibers innervated by that neuron. The result of this activity is a constant low-level tension in skeletal muscles as first one and then another motor neuron fires and causes contraction in the muscle fibers that it innervates. (See Chapter 10 for further discussion of muscle fibers and their control.)

The membrane at the spike-initiating zone in a motor neuron loses its sensitivity to synaptic input only briefly, even if the input is prolonged. Therefore, a long, strong postsynaptic potential—typically the result of many summed individual inputs—causes the motor neuron to fire a sustained train of APs. The frequency of impulses in a train depends on how depolarized the spike-initiating zone becomes (Figure 6-45) and on the refractory period of the neuron. As a result, both the number and the frequency of APs produced in a neuron carry information about the input to the neuron.

To summarize, APs are the only way most neurons can carry information. They are generated in a neuron when the low-threshold spike-initiating zone (often at the axon hillock) is depolarized to threshold or beyond. As the spike-initiating zone is depolarized more, the frequency of APs in the neuron rises, up to some maximum firing frequency set by the length of the refractory period of the neuron, as discussed in Chapter 5. The amount of depolarization at the spike-initiating zone depends on summation of synaptic inputs, which in turn is affected by the relative timing and magnitudes of excitatory and inhibitory synaptic currents and where those currents originate.

SYNAPTIC PLASTICITY

Neuronal plasticity is the modification of neuronal function as a result of experience. It is of premier importance for the survival of any organism. Neuronal plasticity lies behind much of human intelligence, as well as the ability of all higher animals to respond adaptively to stimuli in ways that allow them to go beyond fixed reflexes programmed into their developing nervous systems by genetic mechanisms. Virtually all animals demonstrate some degree of behavioral plasticity, much of which is thought to depend on changes in synapses, called **synaptic plasticity.** The mechanisms that underlie synaptic plasticity are currently the subject of many experiments. Several mechanisms that produce synaptic plasticity are found in widely separated phylogenetic groups, suggesting that they have been strongly conserved in evolution. Synaptic plasticity may also take place as the result of developmental events over the course of a lifetime. Synaptic connections that are established in embryos are later refined into adult patterns, and throughout life, changes in synaptic strengths at mature synapses are thought to be important mechanisms for learning and memory.

In fully developed adult organisms, neuronal plasticity is based largely on changes in synaptic efficacy. Notice that either an increase or a decrease in synaptic efficacy can modify behavior. In fact, we know of both kinds of changes as a result of experience. A change in synaptic efficacy is not necessarily the only way in which neuronal function might be modified, but at present it is the one for which there is the most experimental support. D. O. Hebb suggested in 1949 that the effectiveness of an excitatory synapse should increase if activity at the synapse is consistently and positively correlated with activity in the postsynaptic neuron. Such changes in synaptic efficacy could depend on modification of either the presynaptic terminals or the postsynaptic neuron. A presynaptic change might alter the amount of transmitter released in response to a presynaptic AP. A change in the postsynaptic neuron might alter the amplitude of the postsynaptic potential produced in response to a particular amount of transmitter. Since Hebb first stated his concept of learning, one challenge has been to identify the cellular mechanisms that could underlie such changes at synapses. We will first consider presynaptic mechanisms of neuronal plasticity.

Changes on the presynaptic side can be classified into two categories. In **homosynaptic modulation**, activity in the terminal itself causes a change in its release of transmitter. In **heterosynaptic modulation**, changes in presynaptic function are induced by the action of a modulator substance released from another, closely apposed axon terminal. Heterosynaptic modulation has typically been found to last longer than homosynaptic modulation.

Short-Term Homosynaptic Modulation: Facilitation, Depression, and Potentiation

Several different activity-dependent changes in synaptic efficacy can be seen in a partly curarized motor endplate region of a frog skeletal-muscle fiber when two stimuli are applied to the motor axon in rapid succession. If two APs arrive at the motor endplate in rapid succession, and the second postsynaptic potential begins before the first has subsided, the two postsynaptic potentials will always sum. However, in some frog muscle fibers, the amplitude of the second response is greater than can be accounted for by summation alone (Figure 6-46a). This phenomenon is called **synaptic facilitation.** Even if the second postsynaptic potential begins soon after the first has completely subsided, precluding temporal summation, the second postsynaptic potential may still reach a higher amplitude than the

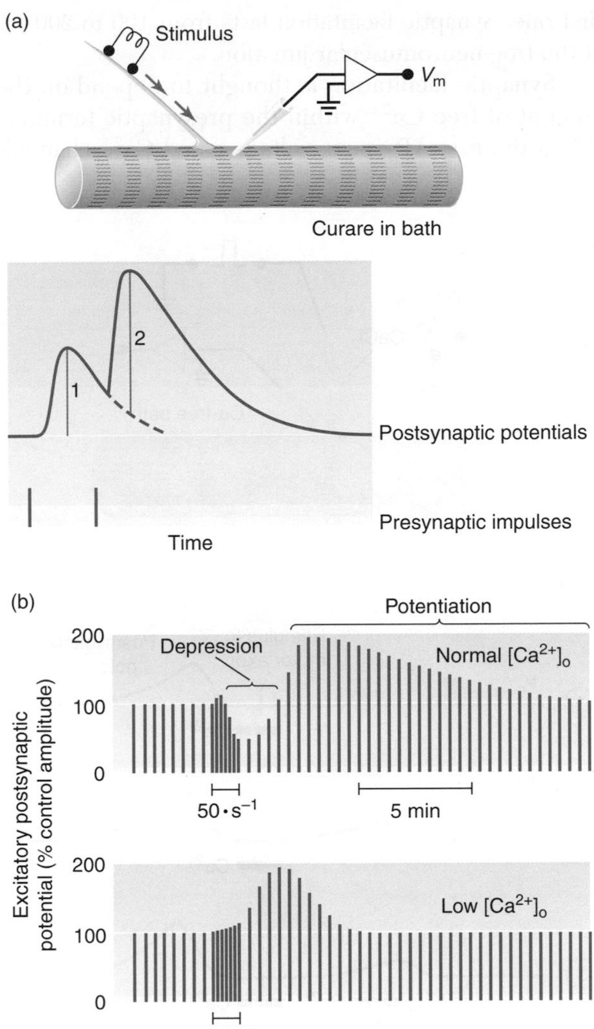

Figure 6-46 Action potentials that arrive in rapid succession produce a variety of effects at the frog neuromuscular junction, depending on their timing and pattern. In these experiments, curare in the bathing saline blocked some ACh receptors, reducing the amplitude of excitatory postsynaptic potentials to below the firing threshold. **(a)** In this experiment, two stimuli were delivered to the motor axon in rapid succession. Notice that the second postsynaptic potential summed with the falling phase of the first, producing a larger postsynaptic potential. However, the amplitude of the second response (indicated by the line labeled 2) was greater than could be accounted for by summation alone. That is, the change in V_m indicated by line 2 is larger than the change in V_m indicated by line 1. This increase in synaptic efficacy is called synaptic facilitation. Classic Work **(b)** After a series of test pulses, the motor axon was stimulated tetanically (that is, at high pressure), and then test pulses were presented every 30 seconds. Immediately after tetanic stimulation, the responses to the test pulses were depressed. However, after about a minute, the responses increased until they reached a larger amplitude than before tetanic stimulation. In the experiment shown in the top trace, the nerve and muscle were bathed in normal frog Ringer solution, which has a Ca^{2+} concentration of about 2 mM. However, when the concentration of extracellular Ca^{2+} was reduced to 0.225 mM (bottom trace), only potentiation, and no depression, followed tetanic stimulation. [Part b adapted from Rosenthal, 1969.]

first one. Synaptic facilitation lasts from 100 to 200 ms at the frog neuromuscular junction.

Synaptic facilitation is thought to depend on the amount of free Ca^{2+} within the presynaptic terminal. When the first AP opens voltage-gated Ca^{2+} channels in the presynaptic terminal, the intracellular concentration of free Ca^{2+} ions briefly rises. If the Ca^{2+} concentration is still somewhat elevated in the terminal when the second impulse arrives, Ca^{2+} ions that enter following the second AP are added to the residual Ca^{2+} ions, and an even higher Ca^{2+} concentration is produced in the terminal. Because the release of transmitter depends so strongly on the Ca^{2+} concentration near the presynaptic release sites, a small increase in Ca^{2+} concentration inside the terminal can produce a large increase in the amount of transmitter released in response to the second impulse. Experimental evidence for this hypothesis was obtained by Bernard Katz and Ricardo Miledi (1968). They used a carefully positioned micropipette to supply pulses of Ca^{2+} ions to the external solution near a motor endplate in a frog muscle that was immersed in Ca^{2+}-free Ringer solution (Figure 6-47a). They found that synaptic facilitation was greatest when a pulse of extracellular Ca^{2+} ions coincided with the arrival of the first AP (Figure 6-47b). The first pulse of Ca^{2+} did not result in facilitation if it was given after the first AP had arrived at the terminals. Synaptic facilitation was seen only if Ca^{2+} could enter the presynaptic terminal when the first AP invaded the terminal.

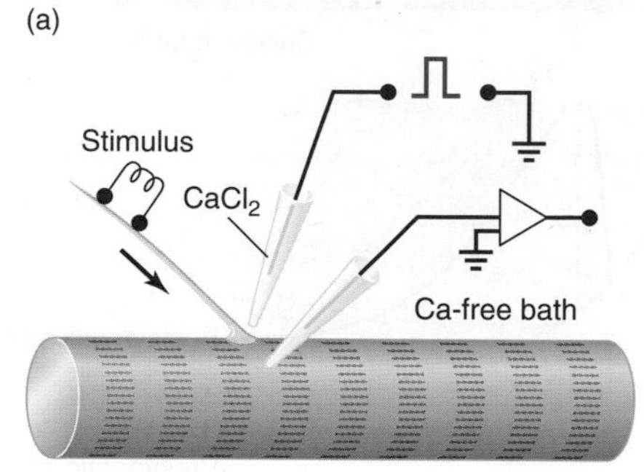

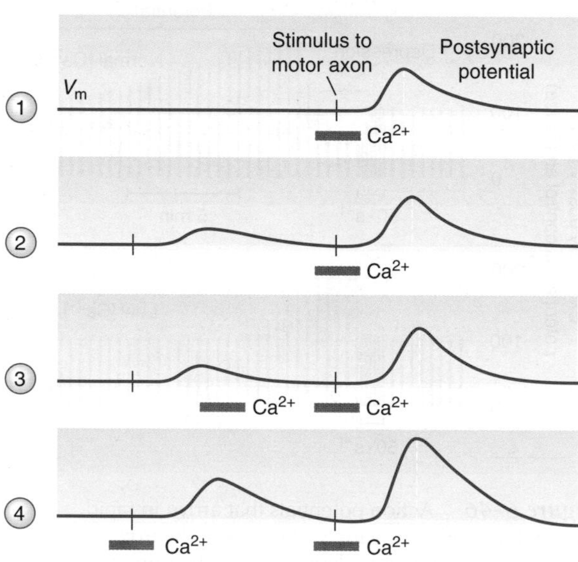

Figure 6-47 Synaptic facilitation depends on the presence of calcium ions in the extracellular fluid. **(a)** A muscle and nerve are bathed in calcium-free Ringer solution, so the only available Ca^{2+} is provided in small pulses extruded from a $CaCl_2$-containing pipette positioned just above the motor endplate region. The motor neuron innervating the muscle fiber is stimulated, and the resulting postsynaptic potential in the muscle fiber is recorded. In this experiment, the relative timing between stimuli to the motor neuron and the delivery of $CaCl_2$ is varied. Classic Work **(b)** Diagrammed postsynaptic potentials in the muscle fiber. Horizontal green bars show the timing of Ca^{2+} pulses; thin black vertical lines indicate stimuli delivered to the presynaptic neuron. Trace 1 shows the amplitude of a postsynaptic potential in response to a single AP in the motor neuron. In the three other traces, two stimuli were delivered to the motor neuron, and the temporal relation between the first AP and the pulse of Ca^{2+} was varied. Notice that in the absence of a Ca^{2+} pulse delivered to the endplate, only a small postsynaptic potential was elicited (trace 2). The amplitude of the second postsynaptic potential varied depending on the timing of the Ca^{2+} pulse. In all cases, Ca^{2+} ions were available at the time of the second AP. Facilitation was seen only when Ca^{2+} ions were present at the endplate when both APs reached the endplate. [Adapted from Katz and Miledi, 1968.]

If a frog motor axon is stimulated *tetanically* (i.e., at a high frequency for a relatively long time), synaptic efficacy at the neuromuscular junction is reduced immediately after the tetanic stimulation. This phenomenon is called **synaptic depression.** It is thought to reflect a depletion of readily releasable synaptic vesicles at the presynaptic active zones. Recovery from synaptic depression is thought to depend on the arrival of new vesicles at the active zone to replace the ones lost during the tetanic stimulation. After recovery from synaptic depression, responses to test pulses are typically found to be larger than normal. This increase in the amplitude of the response lasts for as long as several minutes. This **posttetanic potentiation** is another example of an activity-dependent change in presynaptic efficacy, and it is found in one form or another at many types of synapses.

Figure 6-46b illustrates the results of one such experiment. Initially, excitatory postsynaptic potentials (epsps) were evoked at a frog neuromuscular junction by stimulating the motor nerve at a low control rate (one stimulus every 30 seconds). The rate of stimulation was then increased to 50 per second for a period of 20 seconds, after which a series of test stimuli were administered at the control rate. In Ringer solution that contained a normal concentration of Ca^{2+} (Figure 6-46b, top), posttetanic depression of the evoked epsps was seen immediately after the tetanic stimulation. However, within 1 minute, the amplitude of the epsps increased; in other words, posttetanic potentiation had occurred. The amplitude of the epsps returned to the control level after about 10 minutes. In Ringer solution containing a lower than normal concentration of Ca^{2+}

(Figure 6-46b, bottom), there was no depression, and the posttetanic potentiation subsided more rapidly.

What is the mechanism of posttetanic potentiation? According to one hypothesis, during tetanic stimulation, the large number of Ca^{2+} ions that have entered the presynaptic terminals load up and even saturate all available Ca^{2+}-binding sites that ordinarily buffer the intracellular concentration of Ca^{2+}. As a result, the load of Ca^{2+} ions lingers within the terminals until they are gradually pumped out by active transport across the plasma membrane. Many neurobiologists believe that posttetanic potentiation and its slow decay reflect this increase and subsequent decrease in the concentration of Ca^{2+} inside the terminals.

There are two different results in the low-Ca^{2+} experiment that must be explained. First, why is no synaptic depression elicited? In low-Ca^{2+} Ringer solution, fewer Ca^{2+} ions are available to enter the terminals, so fewer synaptic vesicles fuse with the membrane and release transmitter. As a result, there is less depletion of available synaptic vesicles, and thus no posttetanic depression. Second, what causes posttetanic potentiation to be just as pronounced immediately after the stimulation, but to decay more rapidly? In the low-Ca^{2+} solution, tetanic stimulation does bring extra Ca^{2+} ions into the terminals, so potentiation is seen initially following the tetanic stimulation. The more rapid decay of potentiation may occur because the concentration of Ca^{2+} inside the terminals is less elevated, or because the presynaptic terminal is able to pump the extra Ca^{2+} out more rapidly because less has accumulated.

Heterosynaptic Modulation

The release of transmitter from axon terminals can be influenced at some synapses by neuromodulators. Such modulatory agents include serotonin in mollusks and in vertebrates, octopamine in insects, and norepinephrine and GABA in vertebrates. All of these agents also act as neurotransmitters (see Table 6-2). In addition, endogenous opioids have been shown to act as modulatory agents in vertebrate neurons (see Spotlight 6-6 on pages 208–209). When these modulatory agents are released into the circulation or liberated by nerve endings near a synapse, they appear to modify the release of transmitter from presynaptic terminals. When they are liberated near, but not at, a presynaptic terminal, they are said to act on that terminal heterosynaptically, because transmission at that synapse is altered not by the presynaptic neuron, but by the neuron that released the modulator. One class of heterosynaptic action that has already been discussed in this chapter is presynaptic inhibition; another, in which the amount of transmitter released is increased by the presence of the modulator, is called **heterosynaptic facilitation.**

In at least some heterosynaptic modulation, the modulator is thought to alter the number of Ca^{2+} ions that enter the presynaptic terminals following an AP. Synaptic modulators usually do not directly open (or close) ion channels. Instead, they increase or decrease the ionic currents carried through channels that are activated by a presynaptic AP. This action by modulators is typically mediated by one or more intracellular second messengers that act on the ion channels.

The most extensively studied example of heterosynaptic modulation at a synapse is found in the sea hare *(Aplysia californica)*, a sluglike gastropod mollusk that has been widely used in studies of neuronal plasticity. Eric Kandel (who won a Nobel Prize in 2000 for his work) and his associates have found that excitatory transmission between specific identified neurons in the central nervous system of *Aplysia* is enhanced during behavioral sensitization. In behavioral sensitization, an animal responds to a particular stimulus more strongly or at a lower threshold following some other event. For example, you might respond more strongly to a tap on your shoulder if it followed soon after a very loud explosion. The sound of the explosion would have sensitized you to further stimulation.

In *Aplysia,* behavioral sensitization occurs through heterosynaptic facilitation of transmitter release. The facilitation is triggered by the release of serotonin near the synapse (Figure 6-48 on the next page). In this case, serotonin binding to the presynaptic terminal appears to elevate the concentration of the intracellular second messenger 3′,5′-cyclic adenosine monophosphate (cAMP), which influences the opening of a specific type of K^+ channel known as the S channel. Specifically, when cAMP is elevated in the presynaptic neuron, S channels are more likely to be closed at any given V_m. The efflux of K^+ through S channels contributes to repolarization after an AP, so the closing of S channels prolongs the presynaptic AP and allows more Ca^{2+} ions to enter the terminal through voltage-gated Ca^{2+} channels. An increase in the influx of Ca^{2+} ions causes more transmitter to be released and increases the amplitude and duration of the postsynaptic potential.

Long-Term Potentiation and Depression

The examples of synaptic modification that we have discussed so far last up to several minutes, but if your synapses could not be modified for longer than that, you would never manage to do well in your next physiology examination. Clearly, very long lasting modifications in synaptic efficacy must underlie much of learning and memory. In the search for physiological mechanisms that could generate very long lasting changes in the strength of synaptic connections, intense interest has been focused on synaptic connections within the mammalian hippocampus (Figure 6-49a on page 225). Clinical and experimental evidence indicates that the hippocampus plays a crucial role in generating at least some types of memory. Repetitive stimulation delivered to particular regions of the hippocampus has been shown to produce long-term increases or decreases in synaptic efficacy between identifiable classes of neurons. These effects are called **long-term potentiation**

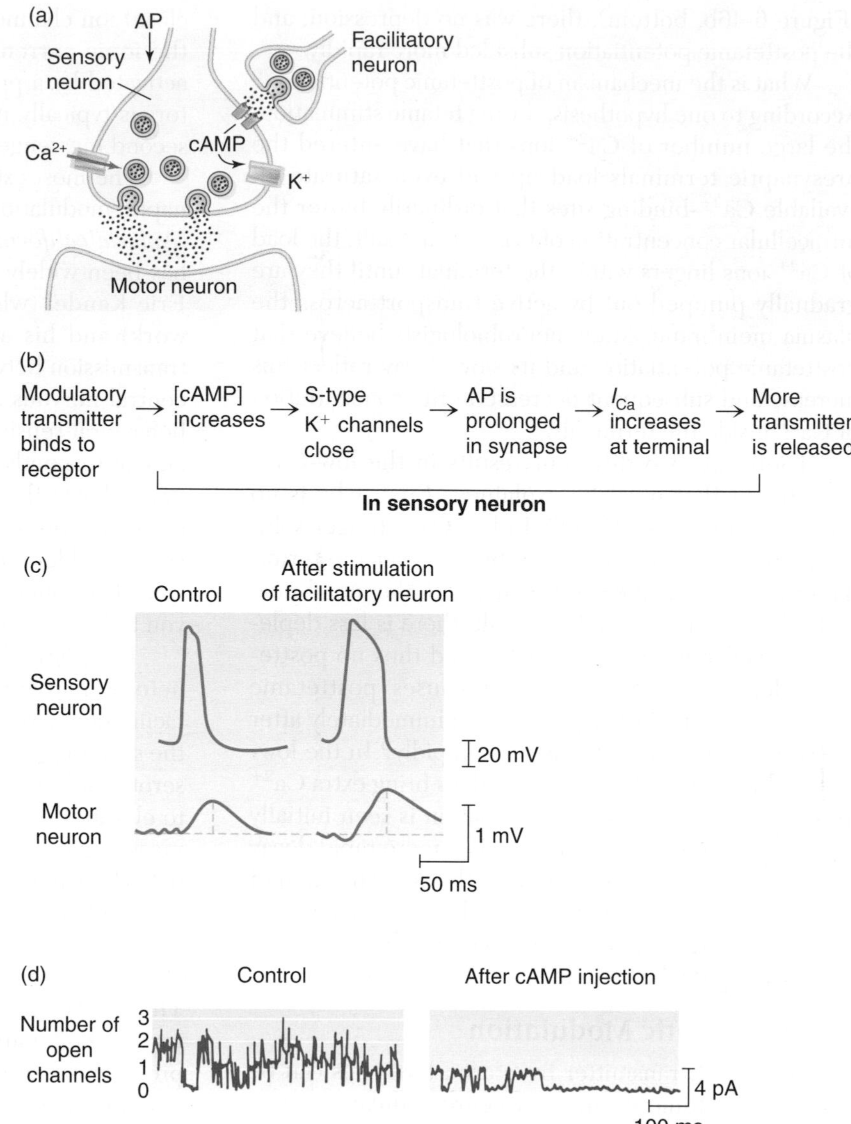

Figure 6-48 In heterosynaptic facilitation at a synapse in *Aplysia*, more K^+ channels are closed in the presynaptic terminals, allowing more calcium ions to enter and causing more transmitter to be released. **(a)** Diagram illustrating the three neurons that participate in this facilitation. If the facilitatory neuron is active when APs arrive at the terminal of the sensory neuron, the amount of transmitter released by the sensory neuron increases. Serotonin from the facilitatory neuron binds to a receptor that causes an increase in the level of cAMP in the terminal. Elevated cAMP causes S-type K^+ channels in the terminal to close, prolonging the depolarization of the terminal and holding voltage-gated Ca^{2+} channels open. **(b)** Summary of events at the sensory (presynaptic) neuron terminal. **(c)** An AP in the sensory neuron (upper traces) produces an excitatory postsynaptic potential in the motor neuron (lower traces). Stimulating the facilitatory neuron prolongs the AP in the sensory neuron and produces a concomitant facilitation of the synaptic response in the motor neuron. **(d)** Currents through single channels in the presynaptic membrane of the sensory neuron were recorded by the patch-clamping technique before and after cAMP was injected into the neuron. The activity of S-type K^+ channels was reduced after cAMP was injected, as indicated by fewer and smaller single-channel currents. [Part c adapted from Kandel et al., 1983; part d adapted from Siegelbaum et al., 1982.]

(LTP) and **long-term depression (LTD)**. Studies of other parts of the mammalian cortex and in other animal groups have found that LTP and LTD are very broadly distributed in the animal kingdom.

Long-term potentiation in the mammalian hippocampus is induced when neurons that innervate the hippocampus are stimulated at *high frequency* (e.g., 50 stimuli per second, or 50 Hz) for many minutes, in a procedure similar to the one that produces posttetanic potentiation (see Figure 6-46b). This procedure causes an increase in the amplitude of excitatory postsynaptic potentials elicited in the postsynaptic neurons after the period of intense stimulation ends (Figure 6-49b). In an intact animal, the increased amplitude appears to last for hours—even days or weeks—after the potentiating stimulation. At different sites in the brain, long-term potentiation may require different patterns of stimulation, may decay at different rates, and may depend on different underlying mechanisms. The outcome is quite different when repeated *low-frequency* stimulation (e.g., at 2 stimuli per second, or 2 Hz) is applied for many minutes. Then the result is long-term depression (Figure 6-49b).

Most experiments on LTP and LTD are done in cells or tissues that have been removed from an animal, rather than in an intact organism. One preparation that has proved very powerful is called a "slice," which is well named, because to make this preparation the living brain is removed from an anesthetized animal and then cut into slabs, each of which may be 200 μm or more thick. The slice technique makes individual cells much more accessible for electrical recording, but at the same time the tissue is thick enough that connections among neurons within the tissue are preserved. Stimulating an appropriate input pathway reliably induces LTP or LTD in a slice, just as it does in an intact animal. As long as the experimenter provides the right conditions, a slice can remain alive for several hours, and LTP or LTD typically lasts the lifetime of the tissue.

By studying hippocampal slices, great progress has been made in elucidating the underlying cellular mechanism of LTP. The principal excitatory transmitter at

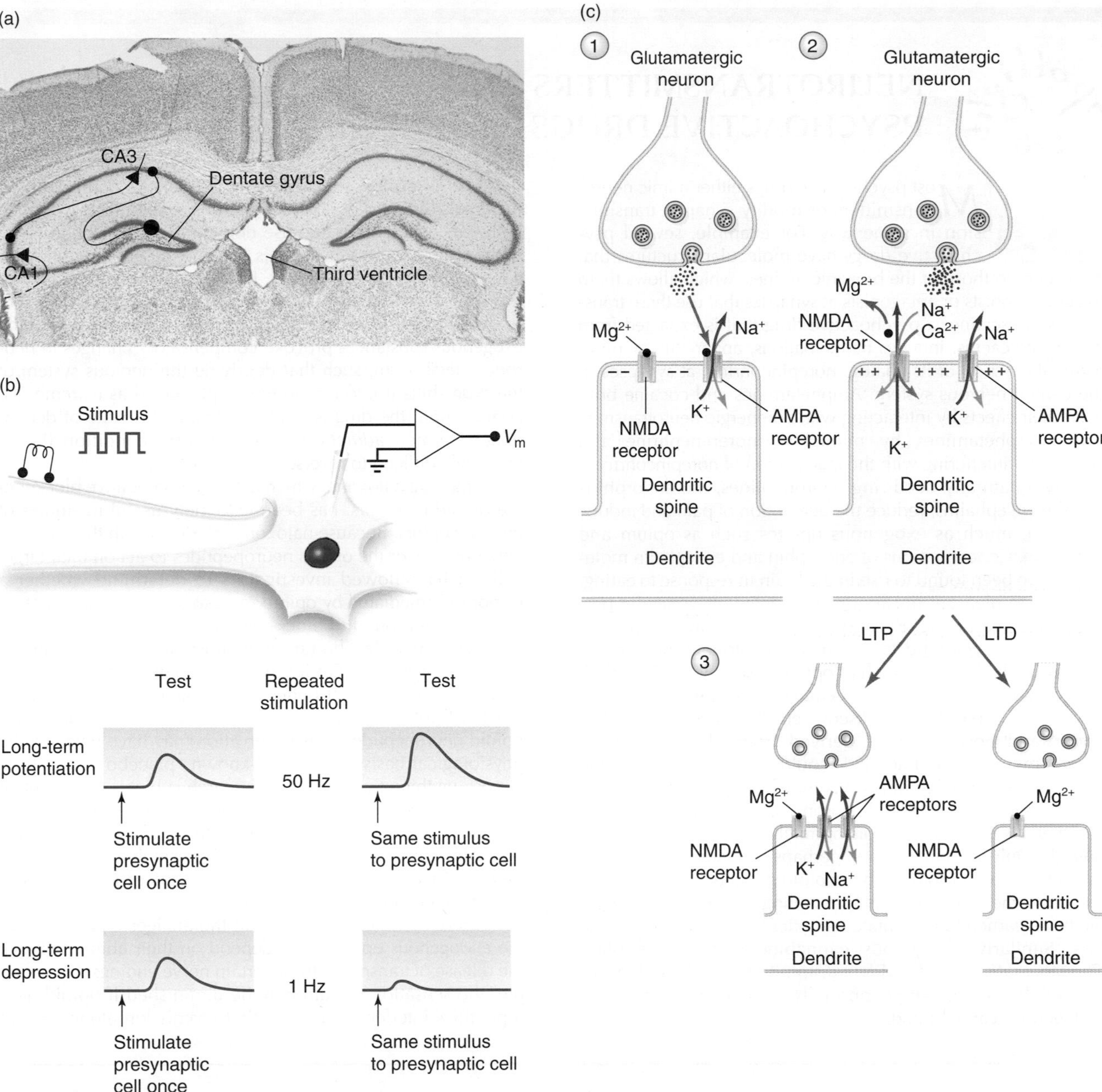

Figure 6-49 Long-term potentiation and long-term depression at glutamatergic synapses may underlie learning and memory. **(a)** Photomicrograph of a section through a hamster's hippocampus. The structure of this part of the brain, which is important in memory storage, allows discrete stimulation of only a few neurons. Synaptic connections between neurons in the CA3 and CA1 regions of the hippocampus show LTP and LTD following prolonged stimulation of axons coming from the dentate gyrus. **(b)** In long-term potentiation, tetanic stimulation lasting several minutes produces a long-lasting increase in the amplitude of the postsynaptic potential in response to synaptic input. In long-term depression, many minutes of low-frequency stimulation reduces the synaptic efficacy onto the target neuron. **(c)** Possible explanation for LTP and LTD. If the postsynaptic membrane is at rest, glutamate opens AMPA channels, but does not open NMDA channels (1). However, if the postsynaptic membrane is depolarized, dislodging Mg^{2+} from NMDA channels, Ca^{2+} can enter the postsynaptic cell (2) through open NMDA channels. Ca^{2+} produces either an increase (LTP) or a decrease (LTD) in the number of AMPA receptors in the postsynaptic membrane by as yet unknown mechanisms (3). The net effect is a larger (LTP) or a smaller (LTD) synaptic response to subsequent synaptic inputs. [Part a courtesy of David Rapaport; part c adapted from Malenka and Nicoll, 1999.]

hippocampal synapses displaying LTP has been found to be glutamate. Of the three pharmacologically distinct ionotropic glutamate receptors, both AMPA and NMDA receptors are thought to contribute to LTP. At many synapses in the hippocampus, activation of NMDA receptors is required for the induction of long-term potentiation, although it is not required for normal neurotransmission.

The mechanism underlying LTP continues to be the focus of intense research effort, and several

NEUROTRANSMITTERS AND PSYCHOACTIVE DRUGS

Most psychoactive drugs either mimic neurotransmitters or modify synaptic transmission in some way. For example, several psychoactive drugs have molecular structures that are similar to those of the biogenic amines, which allows them to act as agonists or antagonists at synapses that use these transmitters. Mescaline, a psychoactive drug that is extracted from the peyote cactus, induces hallucinations, apparently by mimicking the action of its analog, norepinephrine, at synapses in the central nervous system. Amphetamines and cocaine both exert their effects by interacting with adrenergic neurotransmission—amphetamines by mimicking norepinephrine and cocaine by interfering with the inactivation of norepinephrine.

Two naturally occurring neuropeptides, the endorphins and the enkephalins, reduce the perception of pain and induce euphoria, much as exogenous opiates such as opium and heroin do. Concentrations of endorphin and enkephalin molecules have been found to rise in the brain in response to eating, listening to pleasant music, or other activities generally perceived as pleasurable. Because of their functional properties, and because they bind to the same receptors in the nervous system to which opiates such as opium and its derivatives bind, these neuropeptides are called **endogenous opioids.** Until the endogenous opioids were discovered, it was very difficult to understand how alkaloids derived from plants—such as opium, morphine, and heroin—could so powerfully affect the nervous systems of animals. We now know that the plasma membranes of many central neurons contain opioid receptors, and that this class of receptors normally binds the endogenous opioids. Only secondarily, and perhaps coincidentally, do they bind exogenous opioids. When opioid molecules bind to these receptors, they elicit such intense feelings of pleasure that people have learned to use opiate narcotics to stimulate the receptors. Similarly, exogenous cannabinoids from the plant *Cannabis sativa* bind to widely distributed cannabinoid receptors in the brain, which normally bind molecules called endogenous cannabinoids.

Use of a pleasure-producing exogenous compound by mammals (humans or laboratory animals) stimulates identified pathways in the brain that use dopamine as their neurotransmitter, whether the substance is cocaine, alcohol, nicotine, or a large number of other drugs. It has been shown that laboratory animals will work very hard to maintain repeated delivery of these drugs. Repeated doses of many pleasure-producing exogenous substances provoke compensatory changes in neuronal metabolism, such that depriving the nervous system of the drug shifts it into a state that is perceived as extreme discomfort until the drug is readministered. This state of dependence, termed *addiction,* is a long-term condition that is extremely difficult to reverse.

The drug naloxone, which acts as a competitive blocker of the opioid receptors, has been extremely useful in studies of these receptors. Because naloxone interferes with the ability of either opiates or the opioid neuropeptides to act on their target cells, it has allowed investigators to determine whether a response is mediated by opioid receptors. For example, naloxone has been found to block the analgesic effect that can be produced by a placebo (an inert substance given to patients with the suggestion that it will relieve pain). Apparently, the very fact that a subject believes a medication or other treatment will relieve pain can induce the release of endogenous opioid neuropeptides. This observation may have revealed the physiological basis for the well-known "placebo effect" (i.e., almost anything that you do to research subjects will produce whatever effect you promise in at least some of the subjects). Similarly, naloxone renders acupuncture ineffective in relieving pain, which has led to the hypothesis that acupuncture treatment causes the release of natural opioid neuropeptides within the central nervous system.

There is some indication that the analgesic properties of the endogenous opioids may depend on their ability to block the release of transmitter from certain nerve endings. For example, the sensation of pain may be diminished if opioid neuropeptides interfere with synaptic transmission along afferent

hypotheses have been generated to explain the phenomenon. The hypotheses fall into two classes, emphasizing either the postsynaptic cell or the presynaptic cell. Both classes have strong proponents. However, a recent explanation may incorporate all of the observations that must be accounted for by the final correct description. This hypothesis is illustrated in Figure 6-49c.

As described earlier in this chapter, when glutamate binds to receptors on the postsynaptic membrane, AMPA receptors rapidly open, allowing cations (primarily Na^+) to cross the membrane, depolarizing the postsynaptic neuron. At least initially, the NMDA receptor channels do not open because Mg^{2+} blocks them until the membrane is quite depolarized. If both glutamate and glycine are present in the synaptic cleft *and* the postsynaptic cell is depolarized, as it might be due to current through AMPA receptors, the Mg^{2+} is dislodged, allowing Na^+ and Ca^{2+} to enter the postsynaptic cell through the NMDA receptor channels. The core of the explanation rests on the suggestion that Ca^{2+} entering the postsynaptic cell through NMDA channels acts either directly or indirectly to activate previously "silent" AMPA receptors, or even causes more AMPA receptors to be inserted into the postsynaptic membrane. The net result would be an increase in the size of the postsynaptic current and the postsynaptic potential in response to future presynaptic input—precisely the effect seen in LTP. As long as this higher number of active AMPA receptors was maintained, the epsp would remain elevated. Interestingly, long-term depression (LTD) also depends on NMDA receptors, whose behavior is modified differently by the low-frequency stimulation that evokes LTD. The current hypothesis proposes that LTD works by exactly the

pathways that carry information about noxious stimuli. Indeed, enkephalins and endorphins have been found in the dorsal horn of the vertebrate spinal cord, part of the pathway carrying sensory input within the spinal cord.

Other psychoactive drugs modify the inactivation of a neurotransmitter. One clinically important class of such drugs modulates transmission at serotonergic synapses. Serotonin is inactivated in a manner similar to norepinephrine, both by reuptake into the presynaptic terminal and by inactivation catalyzed by monoamine oxidase (see Figure 6-31b). Low activity at serotonergic synapses has been implicated in major clinical depression. One common drug currently used to treat this condition is fluoxetine hydrochloride (Prozac), which reduces the rate of reuptake of serotonin, thus prolonging the residence time of serotonin in the synaptic cleft. Another class of antidepressant drugs inhibits the enzyme monoamine oxidase, prolonging the life of serotonin molecules.

Psychoactive drugs—both those that have been consciously designed by pharmacologists and those that have been discovered during the course of human existence on earth—exert their effects by acting on synaptic transmission. Their powerful behavioral consequences are a dramatic demonstration of the profound importance of synaptic transmission in the function of nervous systems.

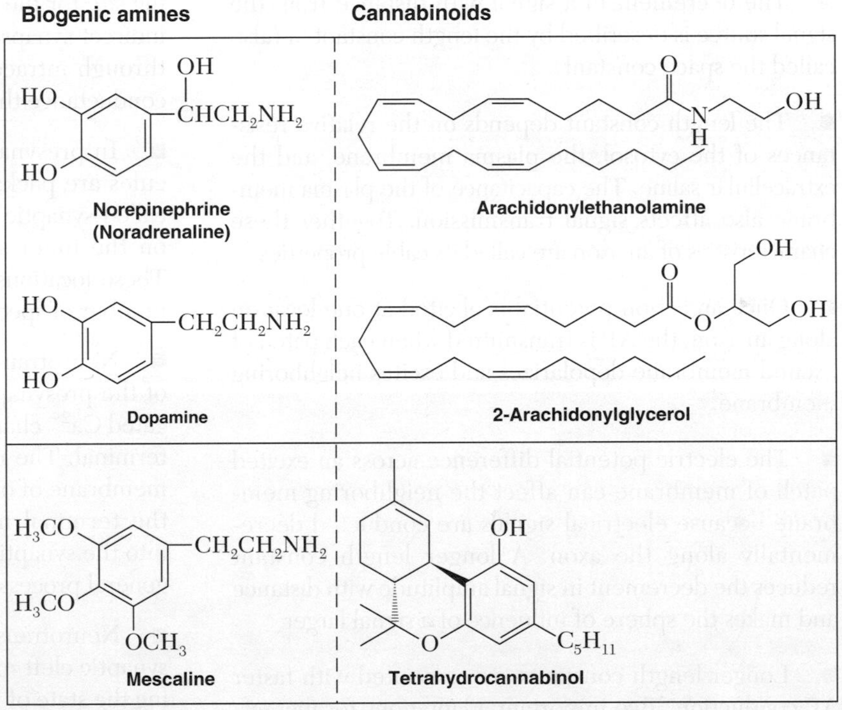

	Biogenic amines	Cannabinoids
Endogenous ligands	Norepinephrine (Noradrenaline)	Arachidonylethanolamine
	Dopamine	2-Arachidonylglycerol
Exogenous ligands	Mescaline	Tetrahydrocannabinol

Many psychoactive drugs act by mimicking endogenous neurotransmitters. This figure shows the structures of some endogenous neurotransmitters mimicked by exogenous chemicals that some humans use for their psychoactive effects.

reverse process of LTP; that is, in LTD, the number of active AMPA receptors in the postsynaptic membrane is reduced.

Whether these mechanisms can account for the modifications in synaptic transmission that underlie learning remains to be seen, but recent progress in identifying the molecular substrates of memory has been encouraging, and the topic continues to be the target of an enormous research effort.

Can you devise an experiment to determine whether LTP and LTD are responsible for learning?

SUMMARY

Transmission of signals in the nervous system: an overview

- Signals are carried electrically along each individual neuron and chemically between neurons.
- Signals are transmitted from neurons to other cells at locations called synapses.
- Signals are conducted passively (or electrotonically) along dendrites and over the membrane of the soma. They are conducted actively (or regeneratively) along the axon of a neuron.
- Action potentials travel along axons without decrement, so they are called "all-or-none" signals.

■ Synaptic signals normally can vary in amplitude, so they are called “graded” signals.

Transmission of information within a single neuron

■ Voltage signals travel passively along the plasma membrane, as they would in a wire, although transmission along a neuron is based on the movement of ions through the cytosol, across the membrane, and in the extracellular fluids.

■ The amplitude of a signal traveling electrotonically, or passively, in a neuron is diminished with distance as charge leaks out across the plasma membrane.

■ The decrement in a signal with distance from the signal source is described by the length constant, λ (also called the space constant).

■ The length constant depends on the relative resistances of the cytosol, the plasma membrane, and the extracellular saline. The capacitance of the plasma membrane also affects signal transmission. Together these characteristics of an axon are called its cable properties.

■ Once an action potential is elicited at one location along an axon, the AP is transmitted when each patch of excited membrane depolarizes and excites neighboring membrane.

■ The electric potential difference across an excited patch of membrane can affect the neighboring membrane because electrical signals are conducted decrementally along the axon. A longer length constant reduces the decrement in signal amplitude with distance and makes the sphere of influence of a signal larger.

■ Longer length constants are associated with faster AP conduction. Two important adaptations for increasing conduction velocities are large axonal diameter and myelination of axons.

Electrical synapses

■ At electrical synapses, the presynaptic cell and postsynaptic cell are connected by protein complexes called gap junctions, which are made up of subunits called connexins. There is continuity between the cytosolic compartments of the two cells through the gap junctions, so ions and small molecules pass easily from one cell to the other.

■ Electrical synapses may be symmetric (that is, signals pass equally readily in either direction across the junction), or they may be rectifying (that is, signals pass more easily in one direction than the other).

■ Electrical synapses provide rapid and faithful signal transmission between cells, but a less flexible response than chemical synapses.

Chemical synapses

■ Chemical synapses are characterized by structural and physiological specialization in the presynaptic axon terminal and in the membrane of the postsynaptic cell.

■ Signals are carried across the synaptic cleft between the presynaptic and postsynaptic cells by the diffusion of neurotransmitter molecules, which are released from the presynaptic terminals and bind to receptor proteins in the postsynaptic membrane.

■ Chemical synapses can be categorized as fast, direct synapses or slow, indirect synapses, depending on events in the postsynaptic cell. At fast, direct synapses, the transmitter receptor proteins include both the binding site for the transmitter and an ion channel. At slow, indirect synapses, the transmitter receptor proteins act through intracellular messenger systems to affect the conductance through ion channels.

■ In presynaptic terminals, neurotransmitter molecules are packaged into membrane-bounded spheres called synaptic vesicles. Vesicles cluster at active zones on the inner surface of the presynaptic membrane. These locations are characterized by the presence of a number of specific proteins.

■ Neurotransmitter is released when depolarization of the presynaptic terminal membrane causes voltage-gated Ca^{2+} channels to open, allowing Ca^{2+} to enter the terminal. The resulting increase in $[Ca^{2+}]_{in}$ causes the membrane of one or more synaptic vesicles to fuse with the terminal membrane, releasing neurotransmitter into the synaptic cleft. This event is one example of the general process of exocytosis.

■ Neurotransmitter molecules that diffuse across the synaptic cleft may bind to postsynaptic receptors, altering the state of ion channels and changing the ionic current across the postsynaptic membrane. The result is a change in the potential across the postsynaptic membrane, called a postsynaptic potential, or psp. The change in ionic current across the postsynaptic membrane is called a postsynaptic current, or psc.

■ Because the amount of transmitter that is released or that binds to postsynaptic receptors can vary, psps and pscs are graded. That is, unlike action potentials, their amplitude can vary from signal to signal.

■ The equilibrium potential of a postsynaptic current is called its reversal potential and is calculated using the Nernst equation or the Goldman equation, depending on how many ionic species can pass through the postsynaptic ion channels.

■ A postsynaptic potential is excitatory if it increases the probability that an AP will be elicited in the postsynaptic cell. Such signals are called epsps, and they depolarize the postsynaptic cell.

■ A postsynaptic potential is inhibitory if it decreases the probability that an AP will be elicited in the postsynaptic cell. Such signals are called ipsps, and they typically hyperpolarize the postsynaptic cell, although in some cases they depolarize the cell.

■ The effect of a psp depends on the relation between the reversal potential of the synaptic current and the threshold potential of the postsynaptic cell.

■ A psp produced by exocytosis of a single synaptic vesicle is called a miniature postsynaptic potential. At the neuromuscular junction, a normal psp is caused by exocytosis of many synaptic vesicles.

■ Some presynaptic neurons release neurotransmitter even in the absence of APs. Graded depolarization of the presynaptic membrane is sufficient to cause this release of transmitter, which is called nonspiking release.

The chemical nature of neurotransmitters

■ A large variety of neurotransmitters have been identified. They range from small molecules, such as acetylcholine, and unaltered amino acids, such as glutamate and glycine, through large polypeptides.

■ Acetylcholine (ACh) is the neurotransmitter at the vertebrate neuromuscular junction; glutamate is the excitatory neurotransmitter at the crustacean and insect neuromuscular junction. Glutamate is the most ubiquitous excitatory neurotransmitter in the vertebrate central nervous system. At the neuromuscular junction and in many parts of the central nervous system, these transmitters act through receptors that include an ion channel.

■ GABA (γ-aminobutyric acid) is typically an inhibitory neurotransmitter; it is found in the vertebrate central nervous system and at inhibitory neuromuscular junctions of insects and crustaceans. Glycine is the most prevalent inhibitory neurotransmitter in the vertebrate spinal cord. At many synapses these transmitters act through receptors that include an ion channel.

■ Neurotransmitters that are modified amino acids, such as epinephrine, norepinephrine, and serotonin, as well as most peptide and protein transmitters, typically act by way of slow, indirect mechanisms.

■ In neuromodulation, transmitters affect target neurons away from the site of the synapse. Neuromodulation allows a presynaptic neuron to affect the behavior of many of its neighbors with which it does not form conventional synapses.

Integration at synapses

■ The neuromuscular junction is a very strong synapse, producing an AP in the target muscle fiber for every AP in the motor neuron. Most neuron-neuron synapses are not this powerful. Instead, it takes many simultaneous inputs from several presynaptic neurons to cause a postsynaptic neuron to fire an AP.

■ Most neurons receive inputs from a large number of other neurons. Production of an AP depends on integration of all of the inputs—excitatory and inhibitory—impinging on the cell, a process called summation.

■ In spatial summation, the membrane at the spike-initiating zone of the postsynaptic cell integrates the psps from many different presynaptic neurons.

■ In temporal summation, the membrane at the spike-initiating zone of the postsynaptic cell integrates psps that arrive within a very small time window, which could have been generated by a train of APs along a single presynaptic axon.

Synaptic plasticity

■ Variation in the strength of chemical synaptic connections provides flexibility and plasticity in the nervous system, from relatively simple examples such as sensitization to very complex examples such as learning physiology.

■ Synaptic facilitation, synaptic depression, and post-tetanic potentiation are examples of short-term homosynaptic modulation.

■ Long-term potentiation and long-term depression are found in many locations of the vertebrate brain and may provide insight into the processes of learning and memory.

REVIEW QUESTIONS

1. Compare and contrast the two basic kinds of signal transmission found in the nervous system.
2. Action potentials are carried along neurons by electric currents. Why are they so much slower than electricity traveling along a wire?
3. How can an AP travel over long distances without decrement, whereas synaptic potentials cannot?
4. Explain why, all else being equal, an axon of large diameter will conduct impulses at higher velocity than will an axon of small diameter.
5. Calculate the relative conduction velocities for unmyelinated axons that are 10 μm and 25 μm in diameter, with all other parameters being equal in the two kinds of axons.
6. Explain why, all else being equal, a myelinated axon will conduct impulses at a higher velocity than will an unmyelinated axon.
7. Explain how loss of myelination, which happens in the demyelinating disease multiple sclerosis, disrupts signal transmission in the nervous system.

8. Design an experiment to test whether a synapse between two neurons is electrical or chemical.
9. What factors determine whether a transmitter depolarizes or hyperpolarizes the postsynaptic membrane?
10. What determines whether a neurotransmitter is excitatory or inhibitory? Explain depolarizing inhibitory synaptic transmission.
11. Marine invertebrates typically have much higher concentrations of inorganic ions in their body fluids than do freshwater invertebrates. For example, the table below gives the intracellular and extracellular K^+ concentrations for two mollusks: *Limnaea,* a freshwater snail, and *Sepia,* a marine cuttlefish.

	Limnaea	*Sepia*
Intracellular	14.8 mM	188.3 mM
Extracellular	1.8 mM	21.9 mM

If these two species had in common a neurotransmitter that opened K^+ channels in the postsynaptic membrane, how would the values for E_{rev} at those synapses compare in the two species? If the postsynaptic neurons had a resting potential of -70 mV and had a threshold of -55 mV, would the transmitter be excitatory or inhibitory in each of these species?
12. What is the evidence that an endplate potential is composed of smaller units called miniature endplate potentials?
13. What places an absolute limit on the amplitude of a postsynaptic potential? What places an absolute upper limit on the frequency of APs conducted along an axon?
14. What prevents ACh released from the presynaptic terminal from persisting and interfering with subsequent synaptic transmission? What happens if ACh remains in the synaptic cleft?
15. Compare the synaptic effects of ACh at the neuromuscular junction (where it acts on nicotinic ACh receptors) and on heart atrial cells (where it acts on muscarinic ACh receptors).
16. The amplitude of postsynaptic potentials decays with distance, so where on a neuron is the most effective site for a synapse to be located?
17. Compare and contrast fast and slow chemical neurotransmission.
18. What is meant by neuromodulation?
19. Discuss the role of Ca^{2+} in each of the following events: depolarization-release coupling, facilitation, posttetanic potentiation, heterosynaptic modulation of transmitter release, and long-term potentiation.
20. How is signal intensity encoded in graded signals, such as receptor potentials and synaptic potentials? How is signal intensity encoded in APs?

SUGGESTED READINGS

Aidley, D. J. 1998. *The Physiology of Excitable Cells.* Cambridge: Cambridge University Press. (A recent edition of a classic text in the field.)

Bajjalieh, S. M. 1999. Synaptic vesicle docking and fusion. *Curr. Opin. Neurobiol.* 9:321–328. (Briefly summarizes our current understanding of many major players in synaptic exocytosis.)

Cooper, J. R., F. E. Bloom, and R. H. Roth. 1996. *The Biochemical Basis of Neuropharmacology.* 7th ed. New York: Oxford University Press. (A frequently updated book describing the chemistry of neurotransmission and neuromodulation.)

Cowan, W. M., T. C. Südhoff, and C. F. Stevens. 2001. *Synapses.* Baltimore: Johns Hopkins University Press. (An excellent collection of reviews covering current work on synaptic transmission, written for professionals and advanced students by many major players in this field.)

Di Marzo, V., D. Melck, T. Bisogno, and L. De Petrocellis. 1998. Endocannabinoids: Endogenous cannabinoid receptor ligands with neuromodulatory action. *Trends Neurosci.* 12:521–528. (A brief review that explains how marijuana affects the mammalian nervous system.)

Fossier, P., L. Tauc, and G. Baux. 1999. Calcium transients and neurotransmitter release at an identified synapse. *Trends Neurosci.* 22:161–166. (A detailed description of the role of Ca^{2+} in neurotransmitter release, studied at a particularly tractable synapse in the mollusk *Aplysia.*)

Hille, B. 2001. *Ionic Channels of Excitable Membranes.* 3d ed. Sunderland, Mass.: Sinauer Associates. (An up-to-the-minute, authoritative resource for the biophysics of signal transmission.)

Hodgkin, A. L. 1964. *The Conduction of the Nervous Impulse.* Springfield, Ill.: Thomas. (A classic summary of the original work on how APs are initiated and conducted, written by one of the primary contributors to the field.) Classic Work

Hodgkin, A. L. 1994. *Chance and Design, Reminiscences of Science in Peace and War.* New York: Cambridge University Press. (A personal account of work that had a profound impact on our understanding of action potentials. A great read.)

Johnston, D., and S. M.-S. Wu. 1995. *Foundations of Cellular Neurophysiology.* Cambridge, Mass.: MIT Press. (A thorough textbook for the advanced student.)

Katz, B. 1966. *Nerve, Muscle, and Synapse.* New York: McGraw Hill. (A clearly written little book describing the work in Katz's lab. Hard to find, but well worth reading for the light it sheds on this example of first-rate physiological research.) Classic Work

Katz, P. S. 1999. *Beyond Neurotransmission: Neuromodulation and Its Importance for Information Proces-*

sing. New York: Oxford University Press. (A recent collection of papers considering neuromodulation from several points of view and in many phyla.)

Koob, G. F., P. P. Sanna, and F. E. Bloom. 1998. Neuroscience of addiction. *Neuron* 21:467–476. (A review of the biological basis of addiction, written by major contributors to our understanding of this complex and socially important problem.)

Levitan, I. B., and L. K. Kaczmarek. 1997. *The Neuron: Cell and Molecular Biology.* 2d ed. New York: Oxford University Press. (Particularly good chapters on synaptic transmission and neuromodulation, written by researchers who have contributed greatly to this field.)

Malenka, R. C., and R. A. Nicoll. 1999. Long-term potentiation—A decade of progress? *Science* 285:1870–1874. (A description of recent thought in this field.)

Ozawa, S., H. Kamiya, and K. Tsuzuki. 1998. Glutamate receptors in the mammalian central nervous system. *Prog. Neurobiol.* 54:581–618. (A detailed description of the structure, physiological properties, and pharmacology of this important class of neurotransmitter receptors.)

Scales, S. J., J. B. Bock, and R. H. Scheller. 2000. The specifics of membrane fusion. *Nature* 407: 144–146. (A discussion of aspects of membrane fusion found in a wide variety of cell types.)

Südhof, T. C. 1995. The synaptic vesicle cycle: A cascade of protein-protein interactions. *Nature* 375:645–653. (A useful review of vesicular neurotransmitter release.)

Unwin, N. 1995. Acetylcholine receptor channel imaged in the open state. *Nature* 373:37–43. (High-resolution images of the nicotinic acetylcholine receptor channel as it opens.)

Walmsley, B., F. J. Alvarez, and R. E. W. Fyffe. 1998. Diversity of structure and function at mammalian central synapses. *Trends Neurosci.* 21:81–88. (A brief introduction to the variety of synaptic functions in the central nervous system.)

Sensing the Environment

Everything an animal does depends on receiving and correctly interpreting information from its external and internal environments. A bird listening for calls from its competitors, a gazelle sniffing the air as a lion passes upwind, or a hawk hovering over a meadow and peering with one eye and then the other at the brush below—all need accurate information about their surroundings in order to choose what to do next. The decision can be appropriate only if data gathered from the environment are faithfully encoded in signals that can be received and processed by neurons in the brain.

Sensory organs provide the only channels of communication from the external world to the nervous system. Sensory input, which is gathered constantly from the environment and from within the body, is processed by the nervous system, whose properties depend on a combination of the genetic inheritance of the individual animal, events during its embryonic development, and its experience over a lifetime. This concept was recognized two millennia ago by Aristotle when he said, "Nothing is in the mind that does not pass through the senses." An understanding of how environmental information is converted into neuronal signals, and of how those signals are then processed, is therefore of deep philosophical interest to humans, as well as an important focus of physiological research.

Sensory reception begins in organs containing sensory receptor cells that are specialized to respond to particular kinds of stimuli. Sensory receptors are positioned at many locations on the surface and in the interior of the body, and they constitute the first step in gathering sensory information. In contrast with this initial physiological coding step, *sensations* are part of our subjective experience. They arise when signals that are initiated in sensory receptor cells are transmitted through the nervous system to particular parts of the brain, producing signals in the brain that we experience as subjective phenomena closely associated with the stimulus.

Human subjects generally agree on the kind of sensation that is produced by a particular stimulus, even though such subjective sensations are not really inherent in the stimuli themselves. For example, when sugar is placed on the tongues of many subjects, all are likely to report that it is "sweet." Similarly, light with a wavelength of 650–700 nm is described by most subjects as "red." In both cases, these perceptions are not inherent in the stimuli themselves. Instead, they depend on the subjects' neuronal processing of the stimulus. Thus, a description of sensory physiology must include both the properties of sensory receptor cells that allow them to receive information from the environment and a consideration of how the nervous system processes information from the receptor cells to produce recognizable sensations. This chapter presents the general principles of how sensory receptors encode and transmit information and compares events in the receptors of several major sensory systems. The structure and organization of the parts of the nervous system that process sensory information are considered in Chapter 8, and the ways in which sensory information is used to generate and shape behavior are considered in Chapter 11.

In the course of evolution, sensory systems have developed from single, independent receptors into specialized sense organs in which receptor cells are arranged in well-organized spatial arrays and are associated with accessory structures. The cellular organization of sensory organs allows stimuli to be sampled more accurately than can be accomplished by isolated receptor cells. One very complex example is the vertebrate eye, which includes several structural adaptations (considered later in this chapter) that improve both our sensitivity to light and our ability to perceive images. Most animals have eyes, but only some eyes allow the organism to perceive images. The importance of precise sensory information—for example, seeing well—is suggested by the apparent contribution of good vision to evolutionary success. About 85% of all living animal species have image-forming eyes.

Until very recently, the extraordinary diversity of stimuli and the corresponding receptor cell types were considered a tribute to the wide variety of solutions that could be generated by natural selection, because no unifying principles were apparent among these receptors. However, recent research has revealed a remarkable consistency among the molecules and mechanisms that mediate sensory reception.

GENERAL PROPERTIES OF SENSORY RECEPTION

In this chapter, we consider chemoreceptors, mechanoreceptors, electroreceptors, thermoreceptors, and photoreceptors. Although the different sensory modalities may seem quite distinct from one another, several general features have been found to characterize sensory processing in all of them. We begin this chapter with a discussion of several of these common features to provide background for the consideration of specific sensory modalities that follows.

Properties of Receptor Cells

We are able to distinguish among different types of sensory stimuli because they selectively activate different specialized types of receptor cells. The mechanical stimulation that produces the sensation of touch, for example, is a different form of energy than the light that evokes a visual response, and touch receptors are much more sensitive to mechanical stimulation than are visual receptors. Each category of sensation—such as touch, light, or sound—is called a sensory *modality.* In addition, stimuli of a particular type may differ in some features. Light can be red or blue; sounds can be high or low in pitch. The features that characterize stimuli within a particular modality are called *qualities.*

A listing of sensory modalities typically includes vision, hearing, touch, taste, and smell, but this list leaves out important internal sensory systems, as well as sensory modalities possessed by nonhuman animals. For example, many **interoceptive** (or internal) **receptors** respond to signals from within the body and communicate this information to the brain by pathways that typically are not brought into consciousness. **Proprioceptors,** for example, monitor the positions of muscles and joints, and other interoceptive receptors monitor the chemical and thermal state of the body and its orientation. Normally we are not consciously aware of these signals. Imagine how complicated walking would be if we had to pay conscious attention to the position of every muscle and joint taking part in the process.

Sensation begins in sensory receptor cells—or more exactly, in regions of these cells in which the

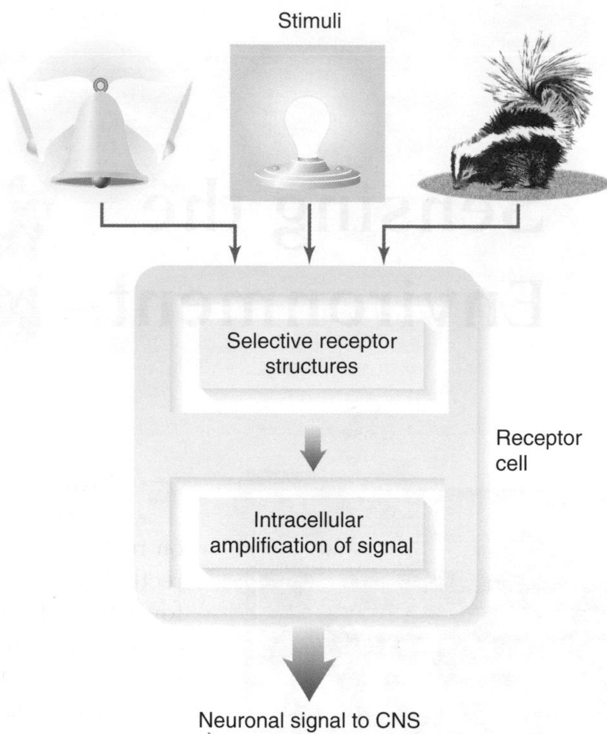

Figure 7-1 Sensory receptors are specialized to respond only to certain stimuli (represented here by a ringing bell, a light flash, and a skunk). A photoreceptor, for example, will respond to light energy, but not to sound or olfactory stimuli, because the properties of a photoreceptor cell make it most sensitive to light energy. In some situations an alternative stimulus will activate receptor cells, but only if the stimulus is unusually intense. A sharp blow to the eye, for example, will elicit the sensation of light because the strong mechanical stimulation activates photoreceptor cells. Similarly, a blow to the ear causes "ringing of the ears." In many receptors, a sensory signal is chemically amplified within the receptor cell. In order for it to be effective, the intracellular chemical signal must cause membrane channels to open (or, in some cases, to close), producing a neuronal signal that is carried to the central nervous system (CNS).

membrane is specialized for reception (Figure 7-1). Notice that each class of receptors is named for the form of energy to which the cells are most sensitive: chemical, mechanical, electrical, thermal, and light. Although the energetic nature of light and sound may be clear—light is electromagnetic energy and sound is mechanical energy—the nature of the energy transaction when a chemical signal binds to a receptor molecule may be less obvious. The important feature of any stimulus is its ability to modify the conformation of a receptor molecule, which then triggers cellular events leading to a neuronal signal. When a chemical stimulus binds to a receptor molecule in the membrane of a chemoreceptor cell, the conformation of the receptor changes. Conformational changes require energy to drive them, so the interaction between a chemical stimulus and its receptor molecule must include

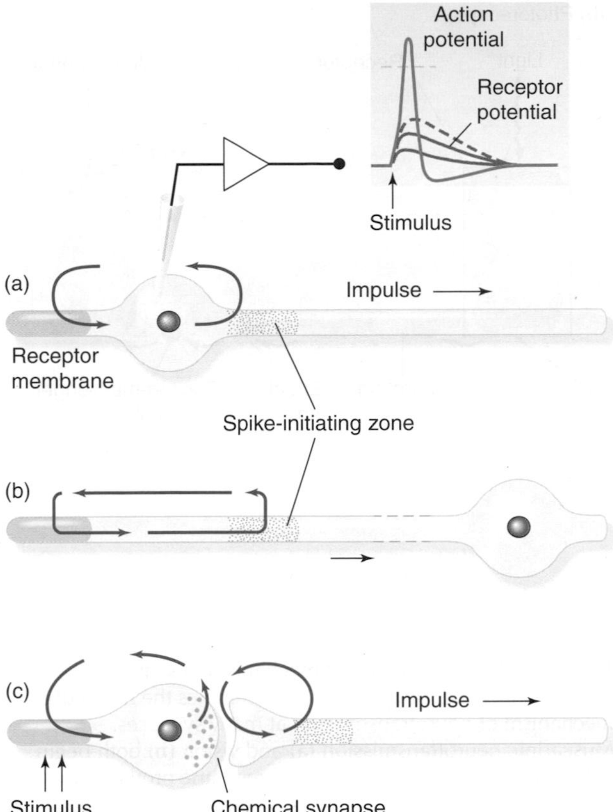

Figure 7-2 There are three basic patterns for transmission of sensory information to the central nervous system by action potentials. **(a,b)** In two types of receptor cells, the receptor current arises at a specialized zone of the receptor membrane and spreads electrotonically to depolarize the spike-initiating zone. In both cell types, the axon of the receptor cell extends into the central nervous system. The difference between the cells in parts a and b is that the soma of the cell in part a is located relatively far away from the central nervous system, whereas that in part b is located near the central nervous system. Many invertebrate sensory neurons are like the cell in part a; vertebrate touch receptors are like the cell in part b. **(c)** In this arrangement, the receptor cell does not produce APs itself, but instead releases neurotransmitter at a synapse that modulates AP production in an afferent neuron. In many examples of this pattern (such as taste receptors), the receptor cell is not itself a neuron; instead, it synapses onto a primary afferent neuron. Red arrows show the pattern of current flow.

a release of energy—the characteristic energy of chemoreception.

Some sensory receptor cells are neurons (for example, vertebrate photoreceptors and touch receptors); others are epithelial cells (for example, taste receptors and the hair cells in the vertebrate inner ear). In all cases, the signal that arises in the receptor cell is carried to the central nervous system by one of several pathways (Figure 7-2). In some receptors, a depolarizing receptor potential spreads electrotonically from its site of origin in the specialized receptor membrane to the spike-initiating zone in the axonal membrane, which then generates APs. In other sensory systems, the receptor cell forms a chemical synapse with a neuron. In this case, a depolarizing or hyperpolarizing receptor potential spreads electrotonically from the sensory region of the receptor cell membrane to the presynaptic part of the cell and modulates the release of a neurotransmitter. The neuron whose axon carries the signal to the central nervous system is called the *primary afferent neuron,* regardless of whether it contains the specialized sensory membrane.

Two general features are found in sensory receptor cells. First, each kind of receptor cell is highly selective for a specific kind of energy. Second, many receptors are exquisitely sensitive to their selected modality because they are specialized to amplify that type of signal. For example, energy from the external environment, such as light, may strike any external surface of a mammal, but only the eyes and the pineal gland (a small gland located in the brain) contain sensory cells that can change photons of light into neuronal energy. The process by which receptor cells change stimulus energy into the energy of a nerve impulse is called **transduction.**

Photoreceptor cells, for example, contain a visual pigment called rhodopsin, which consists of a protein called opsin coupled to a light-absorbing molecule called retinal (Figure 7-3 on the next page). Molecules of rhodopsin capture energy from photons of light, producing a transient structural change in the rhodopsin molecule. This change leads to a cascade of associated enzymatic activity, ultimately changing the state of ion channels in the receptor cell membrane. Similarly, the plasma membranes of mechanoreceptors contain protein molecules that respond to slight distortions in the membrane by changing the conductance through ion channels. Evidence from the molecular biological study of these receptor molecules indicates that many sensory systems probably have a common ancestry. The basic molecular mechanisms have been evolutionarily conserved; the specific functions are wonderfully diverse.

Sensory receptor cells typically amplify the sensory signal. In some cases, activation of receptor molecules in the plasma membrane initiates a cascade of chemical reactions in the cell that effectively amplifies the signal by many orders of magnitude. The final step in all sensory receptor cells is the opening (or closing) of ion channels, which changes the amount of ionic current crossing the plasma membrane and directly or indirectly modifies the number of APs sent to the central nervous system. In summary, each sensory receptor cell transduces a particular form of stimulus into a membrane current that produces a change in the membrane potential, V_m, of the receptor cell, whether the receptor cell is a neuron or an epithelial cell. In this way, a receptor cell is analogous to a common electrical device such as a microphone, which transduces the mechanical energy of sound into modulated electrical signals, which can then be amplified.

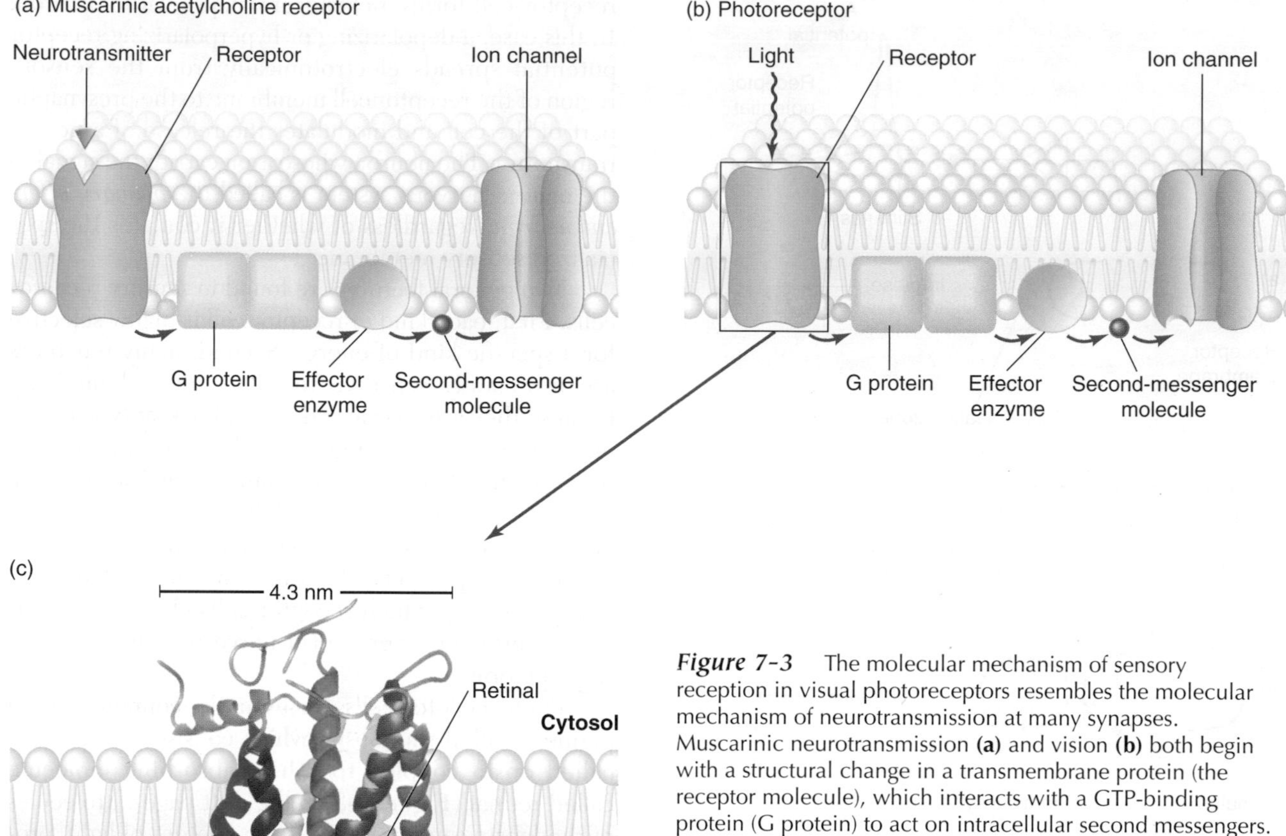

Figure 7-3 The molecular mechanism of sensory reception in visual photoreceptors resembles the molecular mechanism of neurotransmission at many synapses. Muscarinic neurotransmission **(a)** and vision **(b)** both begin with a structural change in a transmembrane protein (the receptor molecule), which interacts with a GTP-binding protein (G protein) to act on intracellular second messengers. The second messengers modify conductance through ion channels, either directly or indirectly, and can thus modify the pattern of APs in afferent neurons. **(c)** The detailed molecular structure of opsin and its relationship to retinal have recently been determined. A single opsin molecule contains seven helical domains that span the membrane. This motif of seven sequential transmembrane helices is common in sensory receptor proteins, as well as in many receptor molecules that respond to hormones or to neurotransmitters, including muscarinic receptors. [Parts a and b adapted from Bear et al., 1996; part c adapted from Bourne and Meng, 2000.]

Common Mechanisms and Molecules of Sensory Transduction

All sensory transduction systems perform the same basic operations, and it is now clear that many types of sensory receptors operate through similar cellular mechanisms and contain similar molecules. Table 7-1 summarizes typical events in sensory transduction as it is carried out by many different kinds of receptors. Some of the processes occur within single receptor cells; others depend on interactions among many cells. The basic events in a receptor cell are detection of a sensory stimulus, amplification of the stimulus, and encoding of the stimulus into electrical signals.

The initial event in all sensory transduction is detection. The weakest stimulus that will produce a response in a receptor 50% of the time is called the **threshold of detection.** Significant technical advances have enabled physiologists to measure transduction events at extremely low stimulus intensities, providing accurate estimates of the threshold of detection and of the time course of the response. Many sensory receptor cells are capable of detecting inputs that are very near the theoretical limits of the stimulus energy: photoreceptors can be activated by single photons, mechanoreceptor hair cells by displacements equal to the diameter of a hydrogen atom, and odor receptors by binding only a few odorant molecules. The response time of a sensory receptor cell is important because in order to convey accurate information about rapidly changing stimuli, receptors must be able to respond quickly and repeatedly. Alternatively, receptor cells must be interconnected in a way that allows the population of receptors to extract information about very rapid events on the basis of their collective activity. Interestingly, the response latencies of the various known receptor cells vary over five orders of magnitude. Hair cells in the auditory system respond within several microseconds; olfactory receptors respond only after several tenths of a second.

Consider the nature of sound and olfactory stimuli along with the roles played in the life of an animal by hearing and smell. What kinds of evolutionary pressures might have produced the very different response times of these two important sensory modalities?

Recent evidence indicates that receptor cells for vision in vertebrates and flies, for olfaction in vertebrates and flies, and for sweet and bitter tastes in mammals all contain receptor proteins with a common structural motif: a sequence of seven transmembrane helices (see Figure 7-3c). Furthermore, transduction in all of these sensory modalities requires G proteins as intermediaries. This pattern is also found in the receptor molecules for several neurotransmitters, including muscarinic acetylcholine receptors (see Chapter 6) and hormone receptors (see Chapter 9). The close evolutionary relationship among the receptor mechanisms in many sensory modalities was underscored when the DNA sequence that codes for opsin was used to identify putative olfactory receptor molecules (Chess et al., 1992).

In all of these sensory systems, an enzymatic cascade amplifies sensory signals. Once again, vertebrate photoreceptor cells provide a convenient example. One photon of red light contains about 3×10^{-19} joules (J) of radiant energy, but the capture of a single photon by a receptor cell has been found to produce a current across its plasma membrane that is equivalent to about 5×10^{-14} J of electrical energy—the cell amplifies the signal by a factor of 1.7×10^5. The exquisite sensitivity of human photoreceptor cells allows a completely dark-adapted human to detect a flash of light containing as few as 10 photons delivered simultaneously over a small region of the retina, a feat that is equivalent to being able to see the light from a candle flame that is 19 miles away.

On the other hand, not all sensory modalities are transduced through G protein–mediated mechanisms. Detection of salt and sour tastes depends on much simpler mechanisms. Sour reception is mediated by a ubiquitous pH-sensitive K^+ channel, and the response to salt is caused by passive movement of Na^+ ions across the plasma membrane through ion channels, depolarizing the taste cell directly. In both of these cases, the stimuli are themselves abundant ions, and no intermediate amplification is required.

Table 7-1 **General features and processes common to many types of sensory receptors**

Transduction operations*	Found within single cells	Found in cell populations
Detection ↓	Mechanisms that select stimulus modality: filters, carriers, tuning, inactivation	Mechanisms that select stimulus modality: filters, carriers, tuning, inactivation
Amplification ↓	Positive feedback among chemical reactions or membrane channels Signal-to-noise enhancement Active processes in membranes	Positive feedback among cells Signal-to-noise enhancement
Encoding and discrimination ↓	Intensity coding Temporal differentiation Quality coding	Different dynamic ranges among cells Independent coding of quality and intensity Center-surround antagonisms Opponent mechanisms
Adaptation and termination ↓	Desensitization Negative feedback Temporal discrimination Repetitive responses	Temporal discrimination
Gating of ion channels ↓	Channels open or close	
Electrical response of membrane ↓	Depolarization or hyperpolarization	
Transmission to brain	Electrotonic spread Number and frequency of APs Synaptic transmission	Spatial patterns: maps and image formation Temporal patterns; directional selectivity, etc.

*Arrows indicate that these operations are a series of steps.

Can there be signal amplification in sensory receptor cells in which the stimulus directly opens ion channels without any intermediary intracellular messengers? What could provide a source of energy to generate amplification?

The encoding of sensory information into a neuronal signal to be transmitted to the brain depends on changes in the conductance through ion channels in the plasma membrane of the receptor cell. When channel conductances change, the result is a shift in the probability that the receptor cell—whether it is a neuron or an epithelial cell—will release neurotransmitter at its synapses with the neuron that is next in line.

A single sensory receptor neuron can encode information about the intensity of a stimulus in how much its V_m changes, but it cannot directly report the quality of the stimulus. A single photoreceptor, for example, cannot report whether a stimulating light is red or blue. Instead, patterns of activity in many receptor cells encode the quality of the stimulus. The wavelength of light is conveyed by activity patterns among groups of receptor cells. Certain photoreceptors respond maximally to red light, whereas others respond maximally to blue light. The same is true for the detection of frequency in sound: some auditory receptors respond most strongly to high-frequency sound, others to low-frequency sound. Typically, sensory organs contain a variety of receptor cells that respond differentially to stimuli with different qualities. As a result, sensory organs can provide significantly more information about a stimulus than a single receptor could, including its absolute intensity, its spatial distribution, and other characteristics such as its quality.

Each sensory system must be able to detect stimuli that persist in time, while at the same time retaining the ability to respond to further changes. The process of adaptation, described in Chapter 5 for single neurons, is also found in the responses of many receptor cells. From the perspective of an organism, sensory adaptation allows detection of new sensory stimuli in the presence of ongoing stimulation; it thus makes a sensory system much more useful. Wearing clothing, for example, stimulates touch receptors at all points where our garments touch the skin, and we typically adapt to this touch input. Yet we can easily detect any new touch stimuli that impinge on our skin, even at locations covered by our clothing. Many mechanisms underlying sensory adaptation take place within individual receptor cells, and several of them appear to depend on changes in intracellular $[Ca^{2+}]$ (e.g., in vision, olfaction, and mechanoreception). In addition, some sensory adaptation depends on feedback pathways from higher brain centers.

From Transduction to Neuronal Output

Electrical measurements have provided crucial insight into the steps that lie between sensory transduction and the generation of neuronal responses. One of the first such experiments was done on receptor cells called *stretch receptors* that sense muscle length in the abdomen of crayfishes and lobsters (Figure 7-4). Because each stretch receptor is a relatively large cell, its soma can be impaled with glass microelectrodes. It is also possible to record APs extracellularly along the axon of the cell (see Spotlight 6-2). The dendrites of each stretch receptor are embedded in a bundle of muscle fibers. If the muscle is stretched, a steady train of impulses can be recorded from the stretch receptor axon. The frequency of the APs varies directly with the amount of stretch applied. To understand the source of the APs, the intracellular potential can be recorded by a microelectrode inserted into the soma. A small stretch applied to the relaxed muscle leads to a small depolarization, called a **receptor potential** (Figure 7-4b). Greater stretch produces a larger depolarizing receptor potential. This change in V_m indicates that the stretch caused an ionic current to flow across the membrane, and that this receptor current must have carried positive charge into the cell to produce the depolarization. If receptor potentials are sufficiently large, they trigger one or more APs in the cell.

What is the relation between the stimulus, receptor current, receptor potential, and APs? Action potentials can be eliminated by blocking voltage-gated Na^+ channels with tetrodotoxin (TTX; see Figure 7-4b). When APs are blocked, the receptor potential remains, and its amplitude varies directly with the strength of the stimulus (i.e., receptor potentials are graded). In these respects, receptor potentials resemble excitatory postsynaptic potentials at the postsynaptic membrane of neurons and muscle cells and are quite distinct from APs. Both observations indicate that receptor potentials must be produced by a different mechanism from the one that generates the all-or-none depolarizing phase of an AP. Finally, the size of the maintained depolarization in a stretch receptor determines the number and frequency of APs it produces, whether the depolarization was produced by stretch or by electrical depolarization (Figure 7-4c). Thus, the size of the stimulus determines the size of the receptor current, which affects the size of the receptor potential, and the size of the receptor potential determines the number and frequency of APs carried along the axons of the stretch receptor.

Sensory receptors differ in how faithfully they reflect the timing of a stimulus. A **phasic receptor** (such as the one shown in Figure 7-4c) produces APs during only part of the stimulation—typically only at the onset or at the offset of the stimulus, or in some cases both—and thus cannot by itself convey information about the duration of the stimulus. **Tonic receptors** continue to

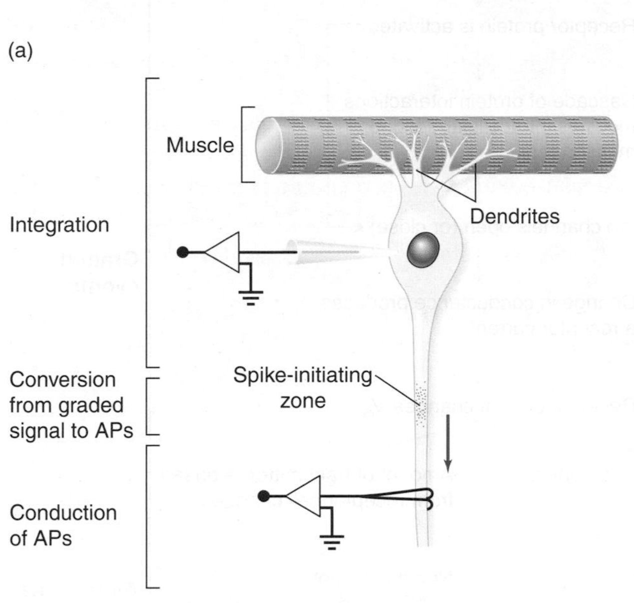

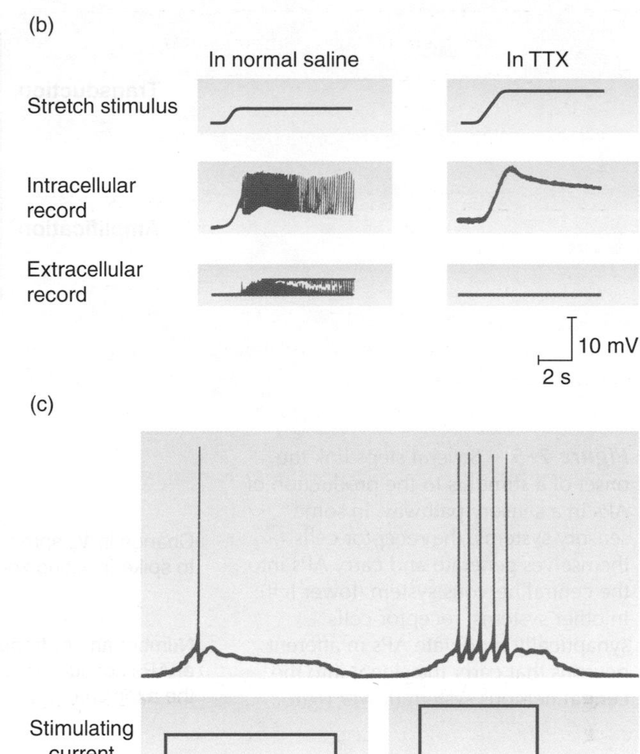

Figure 7-4 Stretch receptors in the abdomen of a crayfish transmit information about the amount that the abdominal muscles are stretched. Each receptor consists of a sensory neuron that has stretch-sensitive dendrites embedded in a bundle of special muscle fibers located on the dorsal surface of the abdominal muscles. When the abdomen of the crayfish bends, the muscle is stretched, and the receptor is activated. **(a)** Recordings can be made either intracellularly from the soma or extracellularly along the axon. The parts of the neuron are functionally differentiated, as labeled at the left. Graded receptor potentials from the stretch-sensitive membrane of the dendrites are converted into all-or-none APs at the spike-initiating zone. The red arrow indicates the direction of AP propagation. **(b)** When the muscle is stretched, a receptor potential is recorded by the intracellular electrode, and APs are recorded by both intracellular and extracellular electrodes. If the preparation is bathed in tetrodotoxin (TTX), the APs are blocked, but the receptor potential remains. In this panel, greater stretch produces a larger receptor potential. **(c)** A small sustained depolarization of a stretch receptor (produced by passing current into the soma) leads to only one AP, whereas a larger sustained depolarization causes multiple APs. In both cases, the stimulus outlasts the production of APs. [Part b adapted from Loewenstein, 1971; part c adapted from Eyzaguirre and Kuffler, 1955.]

fire APs throughout the duration of stimulation and can thus directly convey information about how long the stimulus lasts (see Figure 7-4b).

In summary, a general sequence of steps leading from a stimulus to a train of impulses in a sensory neuron can be formulated based in part on the results obtained with the crayfish stretch receptor. The sequence is summarized in Figure 7-5 on the next page. The stimulus produces an alteration in a receptor protein, generally located in the plasma membrane of the sensory receptor cell. The receptor protein may be part of an ion channel, or it may modulate the activity of transmembrane channels indirectly through an enzyme cascade, which amplifies the signal. In either case, activation of a receptor molecule eventually causes a population of ion channels to open or to close. This change in membrane permeability produces a shift in V_m in accord with the principles presented in Chapter 5. As the intensity of the stimulus increases, more channels respond, producing an increased (or decreased) receptor current and hence a larger receptor potential. Thus, all the steps leading up to, and including, the receptor potential are graded in amplitude.

Unlike the current carried by Na^+ during an AP, the receptor current is not regenerative, even though it typically is carried by Na^+, and it therefore must spread through the receptor cell electrotonically—that is, decrementally. If sensory information is to be propagated over long distances into the central nervous system, the information contained in the receptor potential must be converted into APs. In some sensory systems, the receptor zone is part of the same neuron that carries APs to the central nervous system (see Figure 7-2a,b). When the receptor potential spreads to the spike-initiating zone and directly modulates the generation of APs, it is called a **generator potential.** In

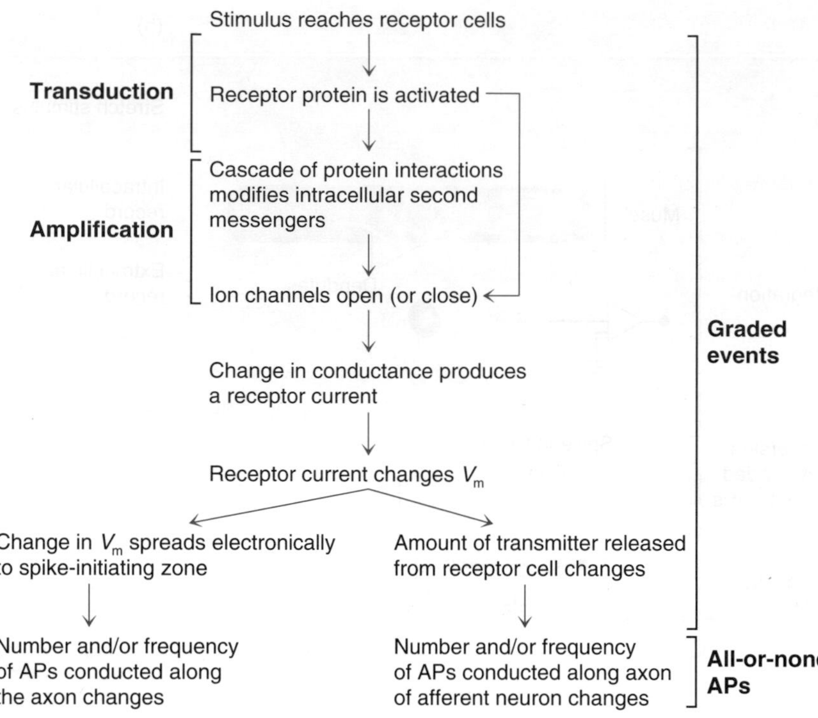

Figure 7-5 Several steps link the onset of a stimulus to the production of APs in a sensory pathway. In some sensory systems, the receptor cells themselves generate and carry APs into the central nervous system (lower left). In other systems, receptor cells synaptically modulate APs in afferent neurons that carry the signal into the central nervous system (lower right).

sensory modalities in which the receptor cell lacks an axon and instead synapses onto an afferent neuron, the receptor potential modulates neurotransmitter release onto the primary afferent neuron (see Figure 7-2c). The transmitter produces a postsynaptic potential, modulating the frequency of APs in the postsynaptic neuron.

Encoding Stimulus Intensities

Individual APs originating in different sense organs are indistinguishable from one another, as first noted in the 1830s by Johannes Müller, who called this observation the "law of specific nerve energies." Müller hypothesized that the modality of a stimulus is not encoded by any characteristics inherent in the individual APs, but instead depends on the anatomic region in the brain to which the information is sent. Thus, stimulation of photoreceptors in the eye produces the sensation of light, whether the photoreceptors are stimulated in the normal way by incident light or abnormally by an intense blow to the eye.

Because APs are all-or-none phenomena, the only way in which information carried along a single nerve fiber can be encoded, other than by specificity of anatomic connections, is in the number and the timing of the impulses. A high frequency of impulses typically represents a strong stimulus, and a reduced frequency of ongoing impulses signals a reduction in the strength of the stimulus. Because the relations between a stimulus and the sensory response differ in different kinds of receptors, there is no simple quantitative rule describing sensory coding. For instance, some receptors for a single modality are tonic; others are phasic. Nonetheless, we can say in general that as the intensity of a stimulus is increased, the receptor current

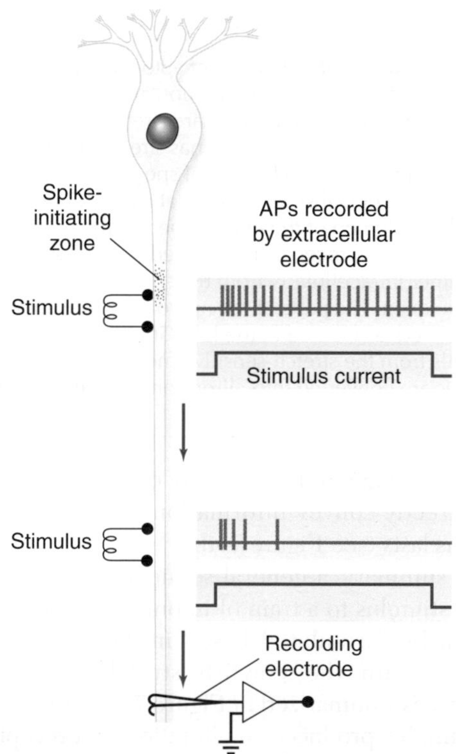

Classic Work *Figure 7-6* Sustained stimulation of a crayfish stretch receptor produces a long train of APs when the stimulus depolarizes the spike-initiating zone, but not when the electrode contacts other parts of the axon. In this experiment, different regions of a stretch-receptor neuron were stimulated electrically using a moveable extracellular electrode, and the resulting APs were recorded by a second extracellular electrode. Stimulation of the spike-initiating zone produced APs for as long as the stimulation lasted. Other areas of the cell adapted rapidly to steady stimulation. [Adapted from Nakajima and Onodera, 1969.]

increases, and a greater receptor potential is produced. In tonic receptors, the spike-initiating zone continues to produce a steady train of impulses as long as it is depolarized. Experimentally depolarizing other regions of the same axon typically elicits a rapidly adapting response (Figure 7-6).

Input-output relations

An ideal sensory system would be able to translate stimuli of all intensities into useful signals. However, biological sensory systems actually encode stimuli over only a limited range of intensity, called the **dynamic range** of the receptor cell (or sense organ). Below this range, the receptors fail to respond; above this range, they saturate—that is, they cannot increase their response when the stimulus becomes more intense. Three major factors combine to set the maximum response that a receptor cell can produce to strong stimulation:

- There is an upper limit on the receptor *current* that can flow in response to a strong stimulus because there are a finite number of ion channels in the specialized receptive membrane of the receptor cell.
- There is an upper limit on the *amplitude* of the receptor potential because it cannot exceed the reversal potential of the receptor current (see Chapter 6).
- There is an upper limit on the *frequency* of APs carried along each axon because refractoriness (see Chapter 5) determines the minimum time separating APs propagating along the axon. Typically, the maximum frequency of APs is several hundred per second or less.

Within the dynamic range, the amplitude of the receptor potential in most receptor cells is approximately proportional to the logarithm of the stimulus intensity (Figure 7-7a). In addition, the frequency of APs in an afferent neuron varies approximately linearly with the amplitude of the receptor potential (Figure 7-7b), up to the limit set by the length of its refractory period. As a consequence of these two independent functional relations, the frequency of APs in a slowly adapting receptor is typically linearly related to the *logarithm* of stimulus intensity, within the limits set by the reversal potential of the receptor current and by the refractory period of the axon (Figure 7-7c). This relationship is generally maintained once the signal reaches the central nervous system. When APs reach the synapses made in the central nervous system by the afferent neuron, they generate postsynaptic potentials in the next cell in line, and these potentials sum and undergo synaptic facilitation in ways that depend on the frequency of presynaptic impulses. Thus, the postsynaptic potentials produced in central sensory neurons are graded as a function of the initial stimulus intensity, and as the signal travels through the central nervous system

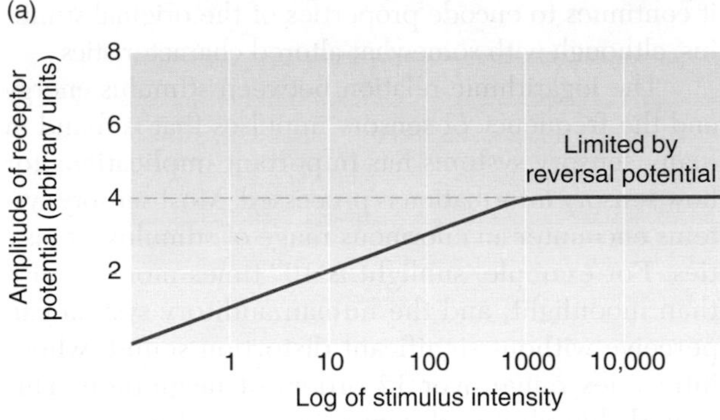

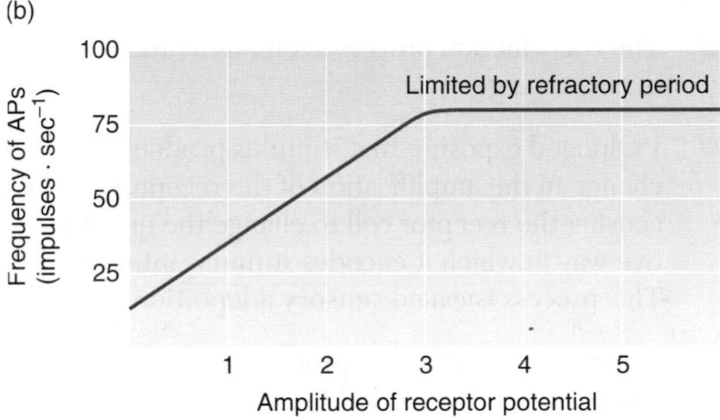

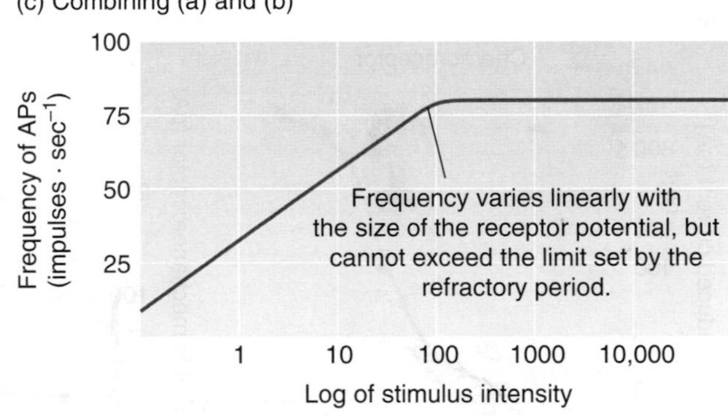

Figure 7-7 The response in most sensory receptors is proportional to the logarithm of stimulus intensity. **(a)** In many receptors, the amplitude of the receptor potential is linearly related to the logarithm of stimulus intensity over a large—but finite—range. The amplitude of the receptor potential cannot increase infinitely because it is limited by the reversal potential of the receptor current and by other biophysical properties of the receptor cell. **(b)** In addition, the frequency of APs in an afferent neuron depends linearly on the amplitude of the receptor potential, but the refractory period of the neuron determines the upper limit of AP frequency. **(c)** Based on the independent biophysical properties illustrated in parts a and b, the frequency of APs in many sensory axons should vary linearly with the log of the stimulus intensity. However, notice that in this hypothetical receptor cell, the frequency of APs reaches the limit set by the cell's refractory period at a stimulus intensity for which the amplitude of the receptor potential could still increase linearly with the intensity. Thus, the receptor potential determines the frequency of APs, but only up to the maximum value set by the refractory period.

it continues to encode properties of the original stimulus, although with somewhat altered characteristics.

The logarithmic relation between stimulus energy and the frequency of sensory impulses that is found in many sensory systems has important implications for how sensory information is processed. Most sensory systems encounter an enormous range of stimulus intensities. For example, sunlight is 10^9 times more intense than moonlight, and the human auditory system can perceive without significant distortion sounds whose intensities range over 12 orders of magnitude. This remarkable ability of sense organs to function over broad ranges of stimulus intensity is based on several physiological mechanisms:

- The transduction process itself has a broad dynamic range.
- Prolonged exposure to a stimulus produces a change in the amplification of the receptor events, causing the receptor cell to change the quantitative way in which it encodes stimulus intensity. This process is called **sensory adaptation.**
- Neuronal networks that process sensory signals have features that extend the dynamic range of the system beyond the capabilities of individual receptor cells.

At low stimulus intensities, the receptor potential in a nonadapted receptor cell represents a very large amplification of energy. The amplification factor is, however, progressively reduced as the intensity of the stimulus increases. The logarithmic relation between intensity of a stimulus and the amplitude of the receptor potential is explained, at least in part, by the Goldman equation (see Equation 5-4), which predicts that V_m should vary with the *log* of the membrane's permeability, P_X, to the ion (or ions) underlying the receptor potential. In many receptor cells, Na^+-selective ion channels are opened by the stimulus. Following stimulation of these cells, the change in V_m should thus be proportional to the log of the change in the permeability to sodium (P_{Na}) that was produced by the stimulus. Normal stimulus intensities typically lie within the logarithmic part of the input-output curve (see Figure 7-7a), although not all receptors follow this general rule. In some, there is, instead, a power-function relation: the *log* of the receptor potential's amplitude is proportional to the log of the stimulus intensity. For practical purposes, either a logarithmic function or a power function describes the relation between stimulus intensity and receptor response very well in the range of stimulus intensities commonly encountered by animals. The differences between the two functions become apparent only at extreme values of stimulus intensity.

(a)

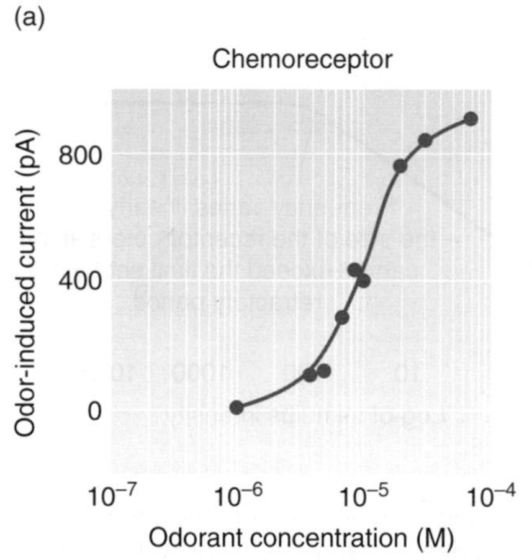

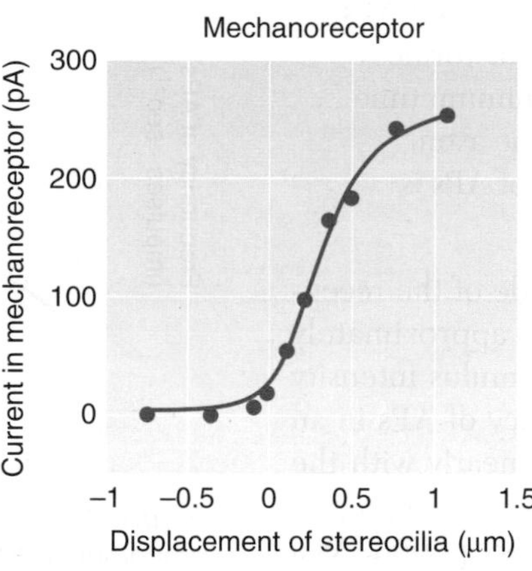

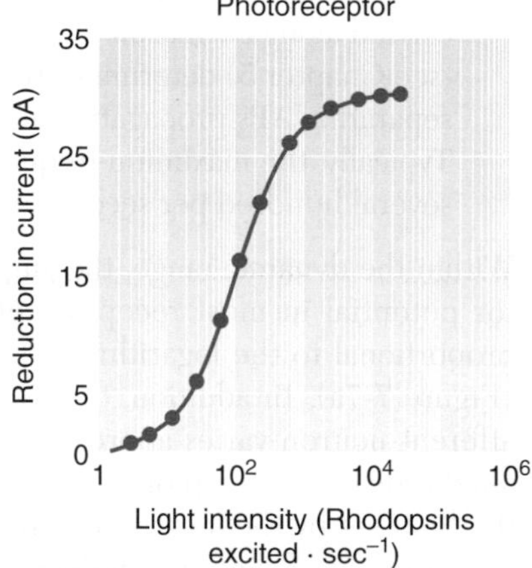

(b)

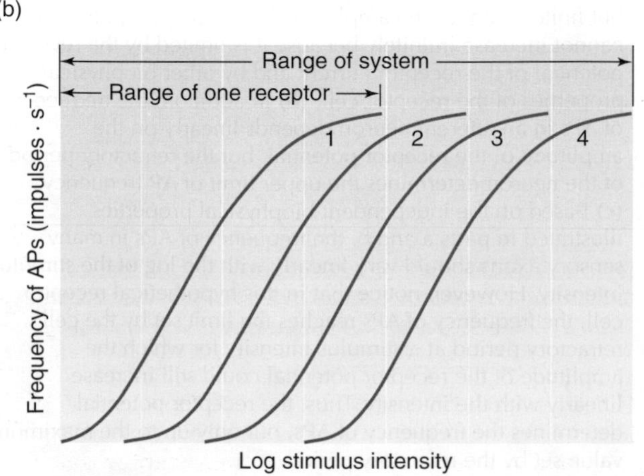

Figure 7-8 Sensory adaptation and range fractionation increase the dynamic range of a sensory system. **(a)** Sensory adaptation compresses the response of a receptor cell, making the cell more sensitive to changes in a weak stimulus than to the same absolute size of change in a strong stimulus. These records show the behavior of three individual patch-clamped receptor cells: a salamander olfactory chemoreceptor, a turtle mechanoreceptor, and a newt rod cell (a photoreceptor). **(b)** Idealized records from a collection of sensory receptors of a single type. Each curve in this graph represents the discharge frequency of an individual sensory afferent neuron, plotted as a function of stimulus intensity. In this hypothetical example, each of the four sensory fibers, labeled 1 through 4, has a dynamic range of about three to four log units of stimulus intensity, whereas the overall dynamic range of the four neurons taken together covers seven log units of intensity. [Part a adapted from Torre et al., 1995.]

As a consequence of the logarithmic relation between the intensity of a stimulus and the amplitude of the receptor potential, any given percentage of change in stimulus intensity evokes the same increment of change in the receptor potential over a large range of intensities. In other words, a doubling of the stimulus intensity at the low end of the intensity range will evoke the same increment in the amplitude of the receptor potential as a doubling of the intensity near the high end, up to the limit where the receptor potential cannot increase. So we have

$$\frac{\Delta I}{I} = K \quad (7\text{-}1)$$

where I is the stimulus intensity and K is a constant. The logarithmic relation between the intensity of the stimulus and the frequency of APs (see Figure 7-7c) thus "compresses" the high-intensity end of the scale, which greatly extends the range of discrimination. The net effect is that the relation between stimulus intensity and the response of individual receptor cells is typically sigmoidal in a variety of sensory modalities (Figure 7-8a).

The sigmoidal relation between intensity and response in receptor cells is similar to the *Weber-Fechner law* of psychophysics, which states that equal increments in the log of the intensity of a stimulus produce equal increments in the subjectively perceived change in stimulus intensity. This feature of sensory systems confers great advantages. For example, it allows us to recognize the objects in a particular scene even when we see the scene under very different lighting conditions. If we observe a scene in bright sunlight, each object is distinguished by its relative brightness. If we observe the same scene by moonlight, the absolute brightness of each object is very different from its brightness in sunlight—in fact, the difference in an object's brightness under the two lighting conditions may be far greater than the difference in brightness of various objects within the scene when illuminated by bright sunlight. However, we are able to recognize the objects in the scene on the basis of their *relative* intensities, independent of the absolute level of illumination. Detecting relative intensities and changes in intensity within a given scene is far more informative to a viewer than is the absolute energy content of each stimulus. Thus, a deer is acutely attuned to movements (changes in the *pattern* of visual stimuli) whether in a field or a forest—that is, independent of changes in the overall lighting conditions.

Range fractionation

The dynamic range of a multireceptor sensory system is typically much broader than the range of any single receptor or afferent sensory neuron. This extension of the dynamic range is possible because individual receptors respond in different parts of the full spectrum of sensitivity. The most sensitive receptors may produce a maximal response at a stimulus intensity below the threshold of other, less sensitive receptors in the system. Above that intensity, the most sensitive receptors become saturated, but the stimulus is then large enough to activate the less sensitive receptors. Thus, at the very lowest stimulus energies, a few especially sensitive sensory fibers will respond weakly. If the stimulus energy is increased a little, the frequencies of their APs will increase, and new, less sensitive fibers may join in weakly, and so on up the range of intensity. Receptors that are less and less sensitive become active in turn with increasing stimulus intensity—a phenomenon called *recruitment*—until the least sensitive sensory fibers are finally recruited, and all receptors are responding maximally. At that point, the system is saturated and is therefore unable to detect further increases in intensity. This **range fractionation**, in which individual receptors or sensory afferents cover only a fraction of the total dynamic range of the sensory system (Figure 7-8b), enables the sensory processing centers of the central nervous system to discriminate stimulus intensities over a broad range. The photoreceptors of the vertebrate eye provide a clear example of range fractionation. Rod photoreceptors are more sensitive to light and respond to dimmer stimuli; cone photoreceptors respond to bright light that saturates the rods.

Control of Sensory Sensitivity

How accurate are our sensations? That is, how do sense organs compare with physical transducers such as thermometers, light meters, and strain gauges? From our own experience, we know that biological sensory systems may not be trustworthy indicators of absolute energy levels. Moreover, many sensations change over time. For example, when a person dives into an unheated pool for a swim, the water initially feels colder than it does several seconds later. A pleasantly sunny day may seem painfully bright for a few minutes after a person emerges from a dimly lit house, but soon the light level is comfortable; for this reason, even an experienced photographer requires a light meter to make accurate judgments of camera exposure settings. These changes of *perceived* intensity, even when the *physical* intensity of the stimulus has not itself changed, are lumped under the general term *sensory adaptation.*

Mechanisms of sensory adaptation

Different classes of receptors exhibit different degrees of sensory adaptation. Tonic receptors exhibit little adaptation. They continue to fire steadily in response to a constant stimulus, as illustrated in Figure 7-9a on the next page, which shows a record from a tonic mechanoreceptor that responds to the displacement of a hair. This receptor produces APs at an almost constant frequency when the hair is displaced and held in a new position. In contrast, phasic receptors adapt quickly. In some phasic receptors, APs are generated only when the intensity of the stimulus is changing. Figure 7-9b shows activity in a phasic mechanoreceptor that fires

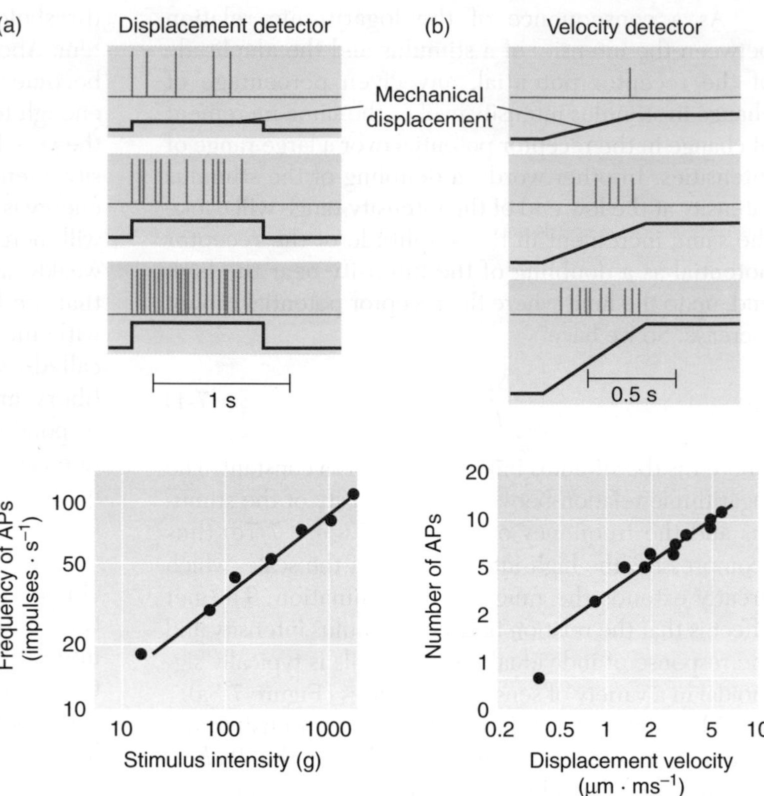

Classic Work ***Figure 7-9*** A displacement detector fires tonically, whereas a velocity detector fires phasically. **(a)** Behavior of a tonic displacement detector. This mechanoreceptor responded to the steady displacement of a hair by generating APs at a relatively constant frequency for as long as the hair was held in the new position. (Top) The figure shows extracellularly recorded APs elicited by three different amounts of displacement, which increased from top to bottom. The amount of displacement is indicated in the red trace below each record. (Bottom) Steady-state frequency of APs plotted against the number of grams of force applied to the hair. **(b)** Behavior of a phasic velocity detector. This rapidly adapting mechanoreceptor responded to the *rate* at which the position of a hair changed. (Top) Action potentials elicited by three different rates of change (red traces). Higher velocities produced more APs. (Bottom) The number of impulses produced during a 0.5-second stimulus was proportional to the log of the velocity of displacement. [Adapted from Schmidt, 1971.]

only when the displacement of a hair changes. The frequency of APs in this receptor depends on the rate of change, rather than on the absolute amount of displacement, making it a *velocity* detector rather than a *displacement* detector.

Where does sensory adaptation take place? There is no simple answer to this question; adaptation can take place at any of a number of stages in the transmission of a sensory signal to the central nervous system.

1. The mechanical properties of the receptor cell or its accessory structures may act as a filter that preferentially transmits information from transient, rather than sustained, stimuli. This mechanism is common among mechanoreceptors.
2. The receptor molecules themselves may "run down" during a constant stimulus. For example, a significant percentage of visual pigment molecules become bleached when they are exposed to continuous light and must be regenerated metabolically before they can again respond to illumination.
3. The enzyme cascade activated by a receptor molecule may be inhibited by accumulation of a product or an intermediate substance.
4. The electrical properties of the receptor cell may change in the course of sustained stimulation. In some receptors, activation of receptor channels diminishes because the concentration of free Ca^{2+} in the cell increases during sustained stimulation. Accumulation of intracellular free Ca^{2+} can also activate calcium-dependent K^+ channels, producing a shift in V_m back toward the resting potential.
5. The membrane of the spike-initiating zone may become less excitable during sustained stimulation.
6. Sensory adaptation may take place in higher-order cells in the central nervous system.

A good example of mechanical filtering by accessory structures (mechanism 1 in the list above) is the rapid adaptation of a Pacinian corpuscle, a pressure and vibration receptor found in the skin, muscles, mesentery, tendons, and joints of mammals (Figure 7-10a). Each Pacinian corpuscle includes a region of receptor membrane that is sensitive to mechanical stimuli, surrounded by concentric layers of connective tissue resembling the layers of an onion. When something presses on a corpuscle, deforming it, the disturbance is transmitted mechanically through the layers to the sensitive receptor membrane. The receptor membrane responds with a brief, transient depolarization at both the onset and offset of the deformation (Figure 7-10b). However, if the layers of the corpuscle are peeled away and a mechanical stimulus is applied directly to the naked membrane, the receptor potential obtained is sustained much longer, producing a more accurate representation of the stimulus (Figure 7-10c). Although

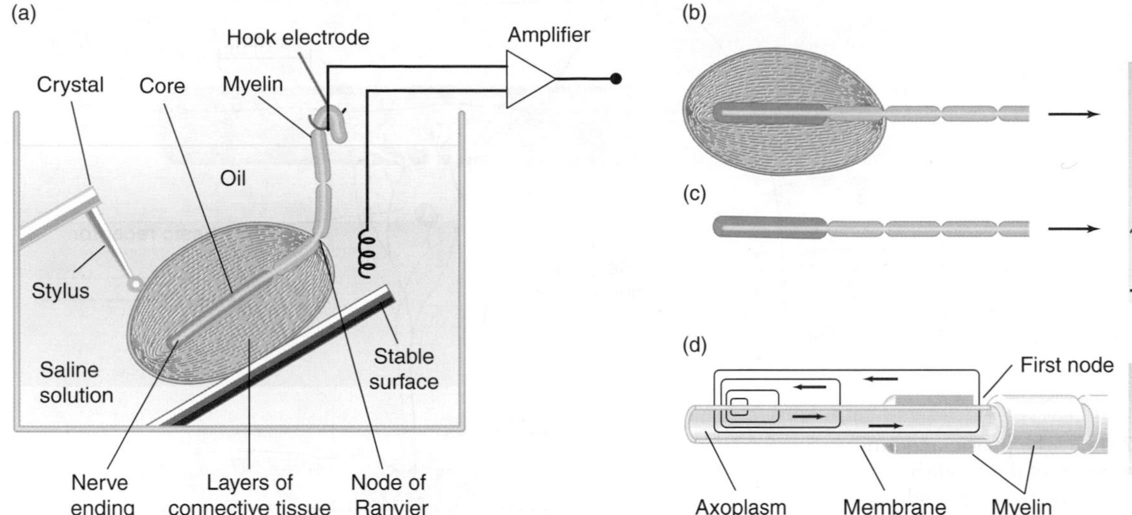

Classic Work **Figure 7-10** Adaptation in a Pacinian corpuscle depends primarily on the mechanical properties of accessory structures. **(a)** Experimental arrangement for tapping a Pacinian corpuscle receptor with a piezoelectric crystal-driven stylus. An electrical recording is made between the hook electrode on the axon and the oil-water interface. **(b)** Electrical response of an intact corpuscle. The neuron depolarized transiently at the onset and offset of the stimulus (dashed black lines). **(c)** In contrast, after the accessory layers were removed, the neuron remained depolarized during most of the stimulus. **(d)** Receptor current flow in response to deformation at the sensory zone of the axon. As in Figure 7-2b, the receptor potential is conducted electrotonically to the spike-initiating zone, which is located at the first node of Ranvier. If the receptor potential is sufficiently large, it brings the spike-initiating zone to threshold, producing APs in the axon. [Adapted from Loewenstein, 1960.]

the receptor potential still shows some degree of adaptation (there is sag in the record shown in Figure 7-10c), there is no distinctive response at the offset of the stimulus. The layers of the intact corpuscle, which preferentially pass on rapid changes in pressure while filtering out prolonged steady pressure, confer on this kind of receptor neuron its normally phasic response. This behavior explains, in part, why we quickly lose awareness of moderate, sustained pressures, such as the stimuli that wearing clothing produces on our skin.

A cellular mechanism of adaptation, the fifth in the list above, is illustrated by the abdominal stretch receptors of crayfishes and lobsters. These receptors are present in pairs in the abdominal musculature, each pair consisting of one phasic receptor and one tonic receptor. Stretching the muscle fiber to a constant length produces a transient response in the phasic receptor and a sustained response in the tonic receptor (Figure 7-11 on the next page). When these receptors are stimulated by direct injection of a depolarizing current into the soma through a microelectrode, rather than by stretching the muscle fiber, each cell retains some of its characteristic properties. That is, when the stimulating current is prolonged, the tonic receptor produces a longer train of APs than does the phasic receptor, suggesting that the electrical properties of each cell contribute to its adaptation.

Regardless of its site or mechanism of origin, sensory adaptation plays a major role in extending the dynamic range of sensory reception. Together with the logarithmic nature of the primary transduction process, sensory adaptation allows an animal to detect changes in stimulus energy against background stimulation that can range over many orders of magnitude.

Mechanisms that enhance sensitivity

Several mechanisms enhance the sensitivity of sensory receptors to sustained stimulation. One such mechanism modifies a receptor's ongoing spontaneous activity. Many receptor cells produce APs—or release neurotransmitter independent of APs—spontaneously in the absence of stimuli. (The amount of transmitter released from nonspiking receptors varies directly with the membrane potential, V_m.) When these spontaneously active receptors are stimulated, the frequency of their APs, or the rate at which they release transmitter in the absence of APs, is increased or decreased away from the baseline level. Such a change has two important consequences. First, any small increase in stimulus energy will produce an increase in the rate of firing of the receptor cell, or the afferent neuron with which it synapses, above the baseline level. Small receptor currents in response to weak stimuli modulate the impulse frequency by shortening the intervals between APs (Figure 7-12a on the next page). This modulation of impulse frequency allows the receptors to be much more sensitive to changes in stimuli than they could be if the receptor current had to bring a completely quiescent spike-initiating zone to threshold. The input-output relation of such a sensory fiber is described by a sigmoidal curve, as shown in Figure 7-12b. Notice that

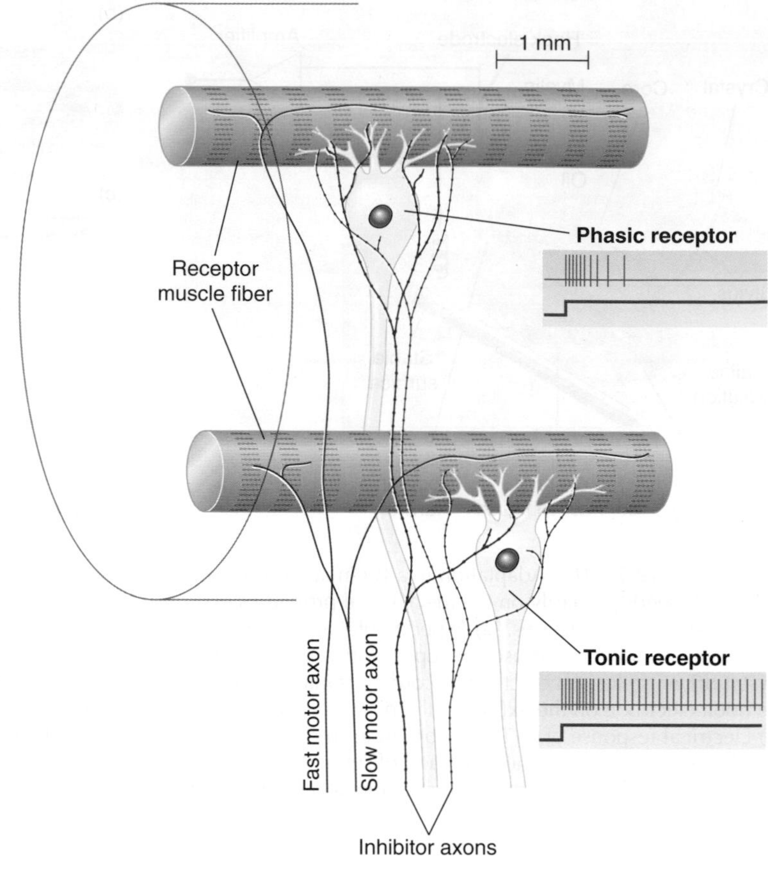

Figure 7-11 The phasic and tonic stretch receptors of the crayfish respond differently to sustained electrical stimulation. This diagram shows stretch receptors (tonic and phasic) and motor neurons (fast, slow, and inhibitory) that innervate crayfish abdominal muscles. The phasic receptor adapts quickly to a constant stretch or depolarizing current (red trace), producing only a short train of impulses. The tonic receptor fires steadily during maintained stretch or depolarization. In this receptor, as in many tonic receptors, the frequency of APs is highest at the beginning of the stimulation and drops off during sustained stimulation. [Adapted from Horridge, 1968.]

in the absence of stimulation, the firing frequency is already on the steep part of the curve, so that even a small input will significantly increase it.

Second, in some spontaneously active sensory neurons, stimuli can either increase or decrease impulse frequency, permitting the receptor to convey information about the polarity or direction of a stimulus (see Figure 7-12b). For example, in some mechanoreceptors, such as hair cells, movement in one direction increases the rate of firing in the sensory fiber, whereas movement in the other direction decreases the rate of firing. If these receptors were silent when they were not stimulated, it would be impossible to encode information about movement in the second direction.

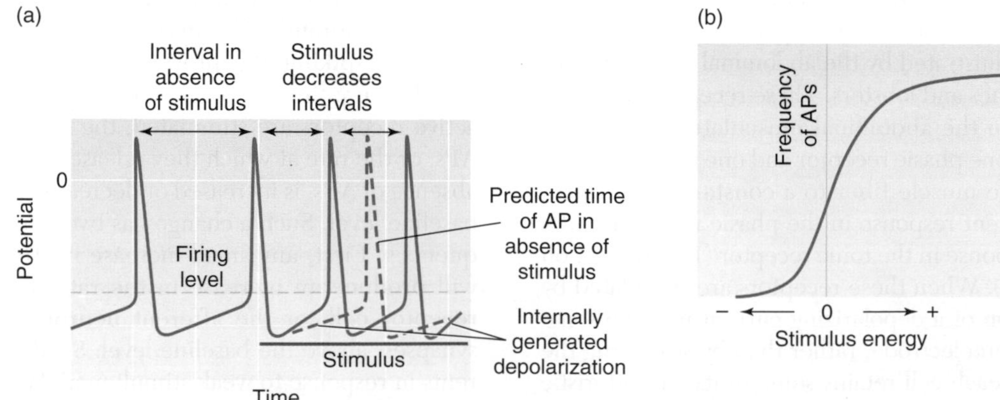

Figure 7-12 The frequency of spontaneous APs in some sensory neurons is modulated by stimulation. **(a)** In a spontaneously firing receptor cell, the interval between APs depends on stimulus conditions. The interval can be decreased by extremely small stimuli because a stimulus increases the rate at which the internally generated depolarization rises. (Compare the slopes of the solid and dashed lines between APs.) In cases in which APs are generated in a primary afferent neuron, rather than in the receptor cell itself, synaptic potentials are similarly enhanced. **(b)** The input-output relation of a spontaneously active receptor cell (or primary afferent neuron) is sigmoidal. In the absence of any input (zero on the abscissa), the cell fires spontaneously. This spontaneous output lies on the steep part of the curve relating stimulus intensity to the frequency of APs, so even a very small stimulus will increase the rate of firing. In some receptors, stimuli can also decrease the rate of firing, indicated by the negative part of the curve.

The existence of numerous parallel sensory pathways provides a third mechanism for enhancing the distinction between a signal and ongoing background noise. In this situation, signals from many receptor cells can be summed by the central nervous system. All of the signals produced by a stimulus will arrive at central neurons nearly simultaneously, whereas noise will be random and will thus tend to be canceled out at central synapses. By reducing noise, this arrangement allows small signals to be detected. For example, a human observer will not reliably perceive a single photon absorbed by a single photoreceptor cell, but if several photoreceptor cells each simultaneously absorb a single photon, the observer experiences the sensation of light.

Efferent control of receptor sensitivity

The responsiveness of some sense organs is influenced by the central nervous system by way of efferent axons that innervate the sense organ itself. In lobsters and crayfishes, for example, stretch receptors end in only a few fibers of the abdominal extensor muscle. When the whole extensor muscle shortens, the muscles containing the mechanosensory endings of the stretch receptors shorten, too. This shortening is driven by efferent neurons. If there were no such mechanism, the stretch receptors would go slack when the abdominal extensor muscle shortened and would thus be unable to detect any further change in the length of the extensor muscle. Instead, contraction of the muscles to which the stretch receptors are attached, driven by efferent input, maintains a fairly constant tension on the sensory parts of the receptors, and the receptors retain their sensitivity to flexion of the abdomen, regardless of the abdomen's position in space. In addition to this mechanism, the abdominal stretch receptors are innervated by efferent neurons that form inhibitory synapses directly on the stretch receptor cells (see Figure 7-11). When the inhibitory efferent neuron is active, the size of the receptor potential in the stretch receptor is diminished, reducing or even abolishing APs in the axon. The interplay of these two mechanisms—one that enhances the responsiveness of the system and another that reduces it—allows activity in the central nervous system either to increase or to decrease the sensitivity of these stretch receptors.

In a similar way, in the skeletal muscles of vertebrates, efferent fibers innervate tiny muscles called intrafusal fibers to which muscle stretch receptors are attached. The stretch receptors and the attached intrafusal fibers run parallel to the rest of the contractile fibers in the muscle. The efferent fibers control contraction of the intrafusal fibers, modifying the length of the stretch receptors themselves and thereby setting the sensitivity of the stretch receptors. Thus, whether a muscle is contracted or relaxed, its stretch receptors remain sensitive to stretching of the muscle.

Feedback inhibition of receptors

The sensitivity of sensory receptors is also controlled through feedback inhibition. In this mechanism, activity in the receptors produces signals that are sent more or less directly back to the receptors, inhibiting them. The crustacean abdominal stretch receptors provide an example of this mechanism (Figure 7-13). Activation of the sensory neuron by stretching produces a reflex output—by way of the central nervous system—that travels through efferent inhibitory nerves leading to the stimulated sensory neuron (**autoinhibition**) and to its anterior and posterior neighbors (**lateral inhibition**).

(a)

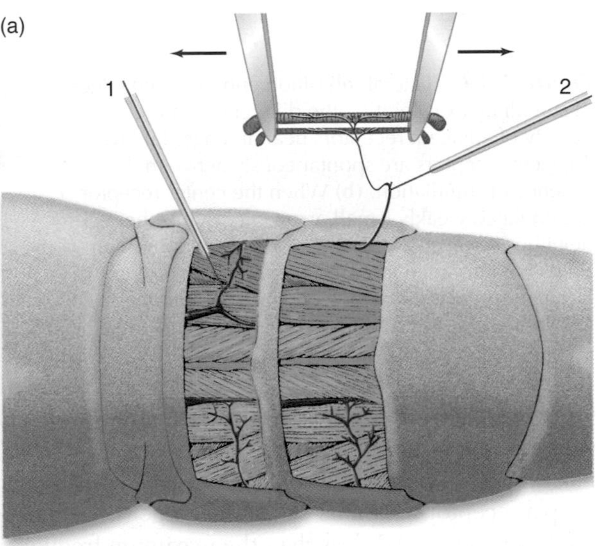

(b)

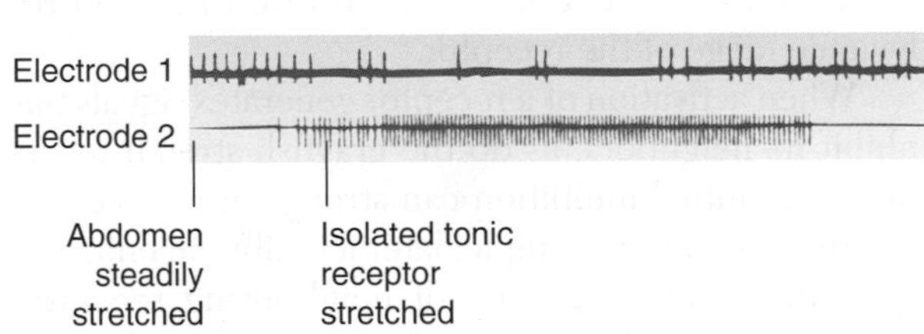

Classic Work **Figure 7-13** Steady stretching of crayfish muscles produces reflex inhibition of the tonic stretch receptors. **(a)** The muscle stretch receptors of one abdominal segment can be removed with their innervation intact, and extracellular recordings are made as shown. Electrode 1 records activity in a stretch receptor *in situ;* electrode 2 records activity in a receptor that has been removed from the abdomen along with its associated muscle. **(b)** At the beginning of the record, in response to a steady stretch of the whole abdomen, electrode 1 records a steady train of APs from the tonic receptor *in situ*. Then the isolated tonic receptor is stretched, an action that has no direct effect on the receptor recorded by electrode 1 because the stretch of the abdomen remains constant. Electrode 2 then records a train of sensory impulses, and there is a simultaneous drop in the frequency of APs in the receptor recorded by electrode 1. This effect arises from activity in a reflex pathway. Activity in the stretch receptor monitored by electrode 2 increases activity in an inhibitory neuron, which reduces the steady output of the neuron that is monitored by electrode 1. [Adapted from Eckert, 1961.]

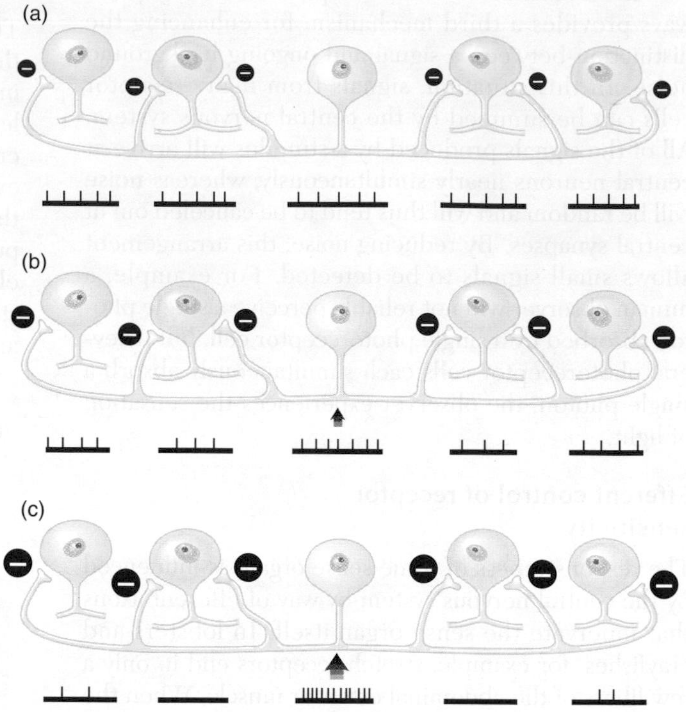

(a)

(b)

(c)

Figure 7-14 Lateral inhibition enhances the edges of stimuli by exaggerating the differences in the activity of adjacent receptors near an edge. **(a)** Five adjacent receptors are spontaneously active in the absence of stimulation. **(b)** When the center receptor is stimulated weakly (small arrow), it inhibits the neighboring receptors. The strength of the inhibition drops with distance from the activated receptor. **(c)** Stronger stimulation of the center receptor (large arrow) increases its inhibition of its neighbors.

Only relatively strong sensory signals evoke reflex activation of the inhibitory neurons—the stronger the signal, the stronger the inhibition. This mechanism acts to keep a receptor within its operating range (i.e., it keeps the frequency of APs less than the maximum frequency set by the refractory period of the cell). The net effect of the inhibitory feedback is therefore to extend the dynamic range of the receptor.

When activation of a receptor generates signals that inhibit its neighbors, as do the crayfish stretch receptors, this mutual inhibition can strongly influence sensory reception. For example, lateral inhibition enhances any differences in activity of neighboring receptors (Figure 7-14). Although this phenomenon was first discovered in visual systems, it is found in a number of other sensory systems as well. The net effect of the interaction between neighboring cells is to exaggerate the differences between activity levels in weakly and strongly stimulated receptors, producing an increase in the perceived contrast between regions of weak and strong stimulation. Lateral inhibition, for example, emphasizes edges in a visual scene.

THE CHEMICAL SENSES: TASTE AND SMELL

Although unicellular organisms were on Earth 3.6 billion years ago, the first multicellular organisms did not arise until 2.5 billion years later. This enormous time lag may indicate, at least in part, how long it took to evolve mechanisms for cell-cell signaling, which are required to coordinate the development and activity of the different parts of a multicellular body. Most cells respond to a variety of signaling molecules; for example, chemical signals such as hormones evoke changes in the metabolic patterns of many cell types. Organisms use some of these same mechanisms to respond to chemical stimuli in the external environment. Even simple organisms such as bacteria detect and respond to specific substances in the environment; some of them move toward or away from particular chemicals, a behavior called *chemotaxis*. In multicellular animals, cells called **chemoreceptors** are specialized for acquiring information about the chemical environment and transmitting it to neurons.

According to one traditional classification scheme, chemoreceptors can be divided into two categories: **gustatory** (taste) **receptors**, which respond to dissolved molecules, and **olfactory** (smell) **receptors**, which respond to airborne molecules. This dichotomy, however, rapidly breaks down. By these definitions, aquatic organisms, such as fishes, could have no smell receptors; all of their chemical sensation would be taste. Furthermore, even in terrestrial organisms, airborne molecules must first pass through a layer of aqueous solution before they reach the olfactory receptors. If taste and smell are legitimately different senses, there must be a more useful distinction between them. Indeed, as we will see, the receptors of taste and smell are distinctly different from each other at the cellular and molecular levels. In addition, alternative schemes can distinguish between taste and smell at a global level. For example, "smell" can be considered to be chemoreception of signals from distant sources, and "taste" to be chemoreception of signals from material that directly contacts the receptive structure (e.g., masticated food in the mouth or material at the bottom of a pond on which a catfish is resting).

Some chemosensory systems are extraordinarily sensitive. The antennal chemoreceptors of the male silkworm moth (*Bombyx mori*) provide a spectacular example. In the laboratory, a male moth responds behaviorally to the pheromone bombykol (the female sex attractant) at concentrations as low as one molecule per 10^{17} molecules of air. The receptors that detect bombykol are extremely specific, responding only to bombykol and a few of its chemical analogs. This highly evolved odorant receptor system allows a male *Bombyx* moth to locate a single female at night from several miles downwind, an ability that confers obvious reproductive advantage in a widely dispersed species.

To investigate the sensitivity of the bombykol receptors, their electrical responses have been recorded. When only about 90 bombykol molecules per second impinge on a single receptor cell, the rate at which the cell fires APs increases significantly. However, the ensemble of receptors is even more sensitive: a male moth reacts behaviorally (e.g., flaps his wings excitedly) when only about 40 receptor cells (out of a total of 20,000 per antenna) each intercept one molecule per second. The moth's central nervous system apparently senses very slight increases in the average frequency of impulses arriving along numerous parallel chemosensory channels and generates behavioral changes based on the averaged input.

Mechanisms of Taste Reception

Electrophysiological studies of the contact chemoreceptors (or taste receptors) of insects have revealed useful information about how chemoreception works in general. These receptor cells send fine dendrites to the tips of hollow hairlike projections of the cuticle, called *sensilla* (plural; singular, *sensillum*). Each sensillum has a minute pore that allows stimulant molecules to reach the sensory cells (Figure 7-15). In the proboscis or on the feet of an ordinary housefly (*Musca domestica*), every sensillum contains dendrites from several receptor cells, each of which is sensitive to a different chemical stimulus (e.g., water, cations, anions, or sugars). The electrical activity of these chemoreceptors can be recorded through a crack made in the wall of the sensillum; such records have revealed both a receptor potential and APs. The receptor potential arises at the ends of the dendrites that extend to the tip of the sensillum, whereas the APs originate near the cell body.

Appropriate chemical stimulation of even a single sensillum evokes a behavioral response in a fly. For example, a small drop of sugar solution applied to a single sensillum on the foot causes the fly to lower its proboscis as if to feed. The effectiveness of various compounds in evoking this stereotypic behavior varies. All compounds that release the feeding reflex also evoke electrical activity in sugar receptor cells. This type of receptor cell in the fly is known to respond only to certain carbohydrates, and those compounds that do not trigger feeding behavior, such as D-ribose, also fail to stimulate the sugar receptors.

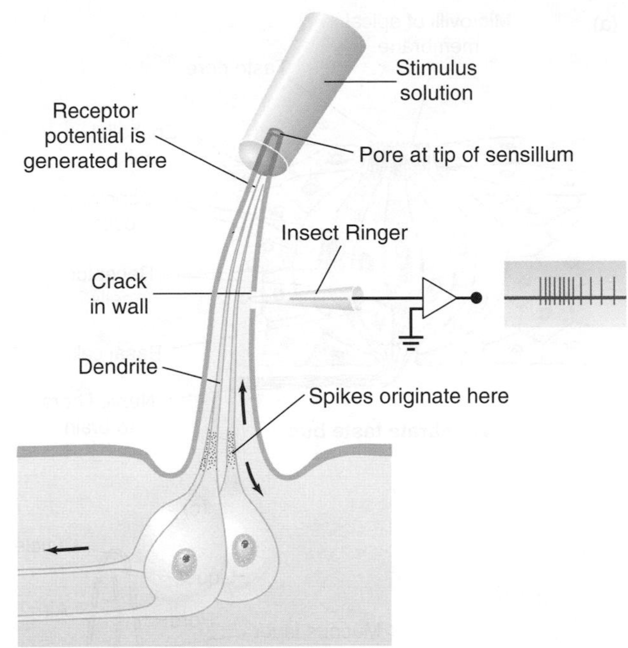

Figure 7-15 The response of a housefly's contact chemoreceptors can be recorded extracellularly. The dendrites of several neurons share a single sensillum. The dendrite of each individual neuron within the sensillum is sensitive to a particular class of substances (e.g., sugars, cations, anions, or water). In this experiment, stimuli are presented through a fine glass tube slipped over the tip of the sensillum, and electrical responses (in blue at right) are recorded through a crack made in the cuticle covering the sensillum.

Like insects, many vertebrates have taste receptors on the surface of the body. The sea robin, a bottom-dwelling fish, has modified pectoral (anterior) fins with taste receptors at the tips of the fin rays, which the fish uses to probe the muddy bottom for food. In terrestrial vertebrates, taste receptors are found internally on the tongue and the epiglottis, in the back of the mouth, and in the pharynx and upper esophagus.

Taste receptor cells in vertebrates are grouped into taste buds, which have some organizational features in common with olfactory organs (Figure 7-16 on the next page). The taste receptors are surrounded by support cells and by *basal cells*, which are thought to be progenitor cells that give rise to new taste receptors. Each taste receptor cell lives for only about 10 days, and the basal cells, which are derived from epithelial cells, regularly generate new taste receptor cells. This turnover of sensory receptor cells is also found in vertebrate olfactory organs and in specialized parts of photoreceptor cells. In all three locations, the cells or cell parts that are regularly renewed interact directly with physical stimuli

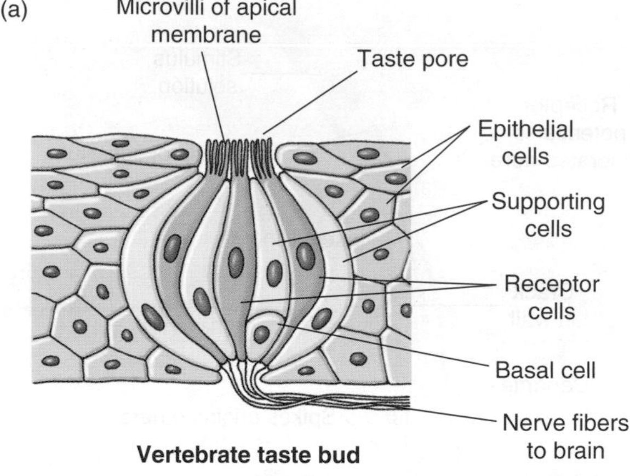

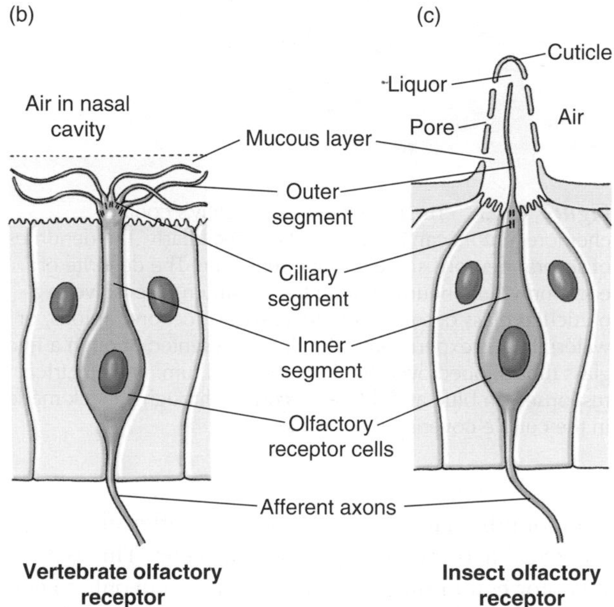

Figure 7-16 Chemosensory organs typically consist of receptor cells surrounded by supporting structures. **(a)** In vertebrate taste buds, the receptor cells are surrounded by basal cells, which generate new receptor cells, and by supporting cells. Transduction takes place across the apical membrane. The receptor cells do not themselves send axons to the central nervous system, although they can produce APs. Instead, they synaptically excite primary afferent neurons that carry the information to the CNS. In contrast, the olfactory receptors of vertebrates **(b)** and insects **(c)** themselves send primary afferent axons to the CNS. Structures that are analogous between vertebrates and insects are drawn similarly in parts b and c. All three types of receptors extend fine processes into a mucous layer that covers the epithelium. [Part a adapted from Murray and Murray, 1970; part c adapted from Steinbrecht, 1969.]

from outside the organism. The turnover of sensory cells poses a problem for the maintenance of sensory specificity in an organism because, unless the new cells are precisely integrated into the existing neuronal network, specificity will be lost. Just how the integrity of taste and smell sensations is maintained remains an unsolved but actively studied mystery.

Although our subjective experience would suggest that there is a wonderfully large spectrum of possible tastes, taste sensations have been grouped by physiologists into just four distinct qualities: sweet, salt, sour, and bitter. Recently, gustatory physiologists have recognized a fifth distinct taste in vertebrates, called *umami*, taken from the Japanese word for "delicious." Umami is described as savory or "meaty." It is evoked by monosodium glutamate and some peptides, and the mechanism behind its reception is still under study, although it appears to be distinct from the other four tastes. In evolutionary terms, these categories may be related to some basic properties of foods. Sweet foods are likely to be rich in calories and thus useful; salt is essential for maintaining water balance (see Chapter 14). A sour or bitter taste can signal danger—many sour or bitter substances are toxic, and many toxic substances are bitter. The discovery that vertebrates respond to no more than four or five fundamental categories of tastes suggests that all perceived tastes must depend on various combinations of these fundamental qualities. In addition, it generated the hypothesis that there is a separate, identifiable sensory pathway associated with each of these tastes.

How do molecules interact with membranes to produce distinct tastes? In the past few years, the mechanisms responsible for each taste modality have been identified by the use of patch-clamp recording. At the tip (or apex) of each receptor cell, a segment of membrane that is especially sensitive to chemical stimuli is exposed to the environment. In vertebrates, the apical membrane of each individual taste receptor cell reacts to particular stimuli, and each class of taste stimuli activates a distinctive cellular pathway in the receptors that respond to it (Figure 7-17). Salty *tastants*, such as NaCl, readily dissociate in water, and the Na^+ ions enter receptors through a particular type of Na^+ channel that is found in the apical membrane. The net effect is to depolarize the receptor cell. These Na^+ channels are distinctive because they are constitutively open and can be blocked by the drug amiloride, unlike the voltage-gated Na^+ channels that mediate most APs. Similar amiloride-sensitive Na^+ channels are found in the mucosal membranes of renal and intestinal epithelial cells. Sour stimuli, which are characterized by excess H^+ ions, act either through this same amiloride-sensitive Na^+ channel (observed in the hamster) or by blocking a K^+ channel (observed in the salamander *Necturus*). In either case, the apical membrane is depolarized when the tastant encounters the receptor.

The transduction of sweet and bitter tastes is generally more complex. Some sweet compounds, including the amino acid alanine (Ala), bind to receptors that act through a G protein to initiate an intracellular cascade that closes K^+ channels in the plasma membrane along the sides and base of the cell (the basolateral

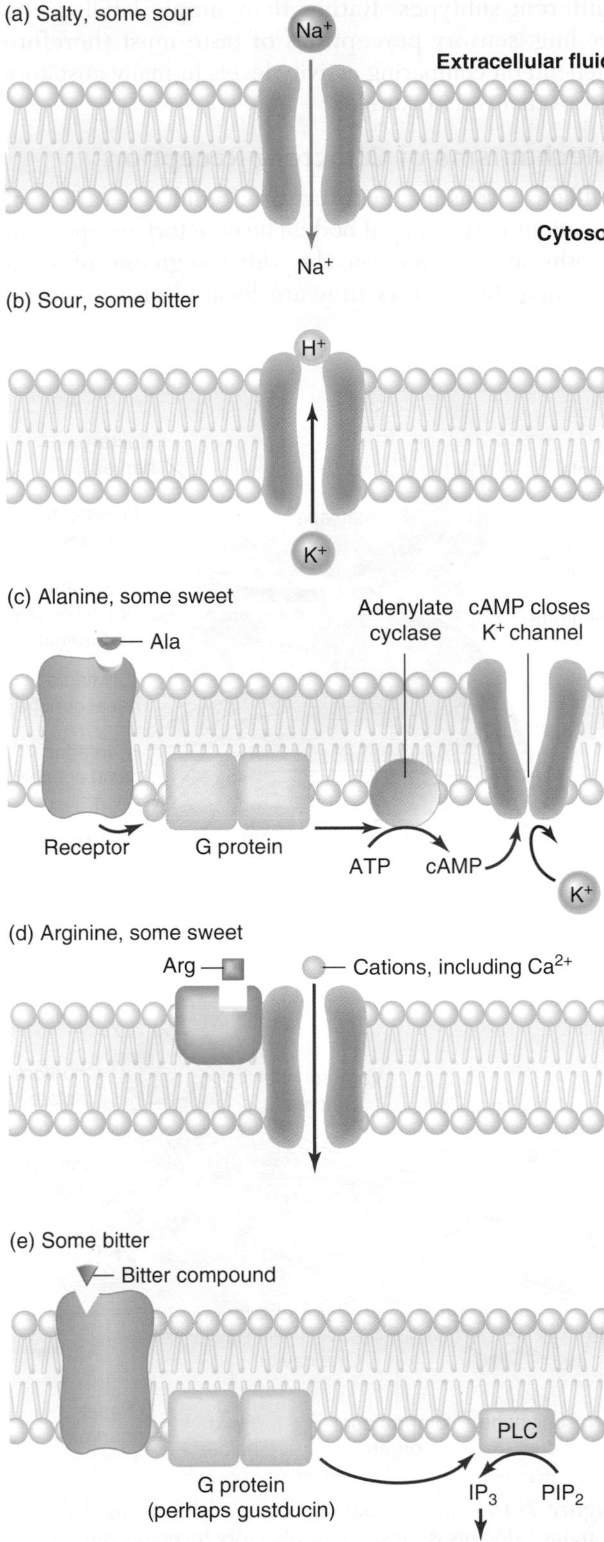

Figure 7-17 Each quality of taste is transduced by a distinctive mechanism. **(a)** In the transduction of salty and some sour tastes, Na^+ (or H^+) ions pass through amiloride-sensitive Na^+ channels in the apical membrane of the taste receptor cell, directly depolarizing the receptor cell. **(b)** In the transduction of other sour tastes and some bitter tastes, protons (sour) or certain bitter compounds block K^+ channels, allowing the slow resting leakage of Na^+ into the cell to depolarize the receptor. **(c)** L-Alanine (Ala) and some other sweet compounds bind to receptors that activate a G protein. The activated G protein then activates adenylate cyclase, and the resulting increase in cAMP closes K^+ channels in the basolateral membrane, allowing the small resting influx of Na^+ to depolarize the cell. **(d)** L-Arginine (Arg) and some sweet compounds bind to and open a ligand-gated, nonselective cation channel. **(e)** Some bitter compounds bind to a receptor and activate a G protein (which might be the recently identified G protein gustducin) that is thought to increase the activity of phospholipase C (PLC), producing an increase in the synthesis of intracellular inositol trisphosphate (IP_3) from phosphoinositol 4,5-bisphosphate (PIP_2). An increase in IP_3 releases Ca^{2+} from intracellular stores, and the increase in intracellular [Ca^{2+}] increases the release of neurotransmitter from the receptor cell. [Adapted from Bear et al., 1996, and from Herness and Gilbertson, 1999.]

membrane), depolarizing the receptor. In catfishes, other sweet substances, including the amino acids arginine and proline, activate nonspecific cation-selective channels in taste receptors, causing the cells to depolarize and directly increasing the intracellular [Ca^{2+}] as well. Transduction of bitter tastes is complex, and the mechanisms remain controversial. Some bitter tastants, such as quinine, directly block K^+ channels in the apical membrane, allowing the small resting Na^+ current to depolarize the cell. Transduction of other bitter substances is less well understood but appears to rely on intracellular second-messenger systems to excite the cell. Taste receptors that respond to quinine have been shown to contain a specific G protein, *gustducin*, that may be coupled to a class of bitter receptors. The G protein appears to activate the enzyme phospholipase C, which increases the intracellular level of inositol trisphosphate (IP_3), triggering the release of Ca^{2+} from intracellular stores. In all cases, the initial contact of a tastant with transmembrane molecules in the receptor cell eventually causes an increase in the concentration of intracellular Ca^{2+} and thus increases the release of neurotransmitter onto the primary afferent neurons.

Vertebrate taste receptors generate APs, but they have no axons, so they cannot themselves carry information to the central nervous system. Instead, they synapse onto, and modulate activity in, neurons whose axons run in the facial, glossopharyngeal, and vagus nerves (the VIIth, IXth, and Xth cranial nerves; see Chapter 8). The existence of four or five kinds of taste sensations and the specificity of membrane transduction mechanisms for each kind of taste suggested that each receptor subtype might be connected to a particular set of axons. In such an arrangement, information about "sweetness," for example, would be carried by one specific subset of axons that project to particular regions of the brain. Such a pattern is called **labeled line coding.** Recordings have revealed, however, that

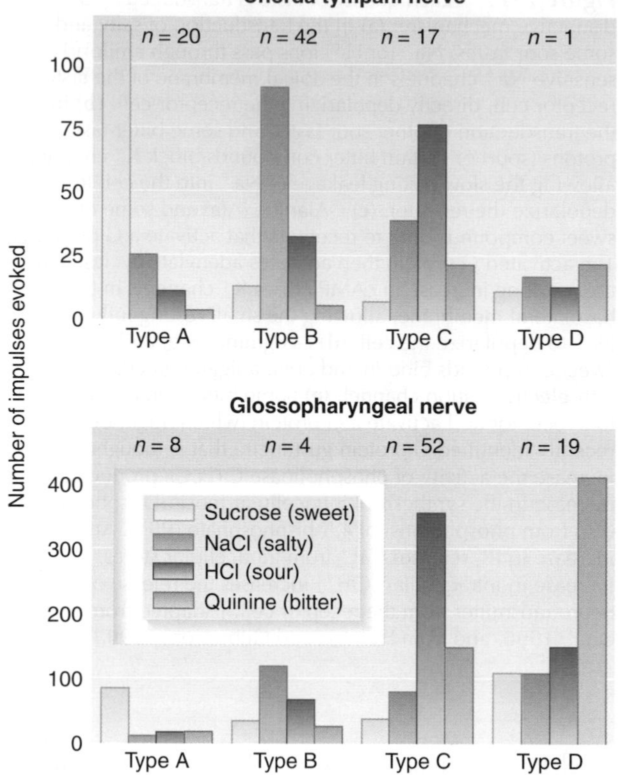

Figure 7-18 Each afferent taste neuron is most effectively stimulated by one type of taste stimulus, but it also responds to other taste stimuli. The responses to four different tastants were recorded from single taste afferent axons in two different nerves in hamsters. Each neuron responded maximally to one of the four taste stimuli; different neurons responded maximally to different stimuli. However, all of the axons responded at least weakly to all four stimuli, indicating that any one taste afferent neuron carries information about more than just one taste. In this graph, neurons were grouped according to the relative effects of the four stimuli, and each group has arbitrarily been given a name. Notice that the majority of neurons recorded in the chorda tympani nerve in this study fell into type B, whereas in the glossopharyngeal nerve, most neurons were of type C, suggesting that different populations of afferent axons run in these nerves. The two nerves innervate different regions of the tongue. In humans, the chorda tympani nerve (a branch of the facial nerve) carries taste information from the outer two-thirds of the tongue, whereas the glossopharyngeal nerve carries information from the inner third of the tongue. The number of axons from which recordings were made is indicated for each group. [Adapted from Hanamori et al., 1988.]

taste information is not nearly this neatly organized. Recordings from single afferent taste neurons show that a single neuron typically responds optimally to a particular type of taste stimulus (Figure 7-18), but that many afferent neurons also respond suboptimally to stimuli in other classes. The data thus suggest that a single neuron innervating receptors in a taste bud receives input from multiple receptors belonging to different subtypes. Rather than simple labeled line coding, sensory perceptions of taste must therefore depend on comparing activity levels in many gustatory axons that run in parallel.

Mechanisms of Olfactory Reception

Olfactory sensory receptors are found at a variety of locations in the animal body. The olfactory receptors of moths are located on the third segment of each antenna. In lobsters they are located on the anten-

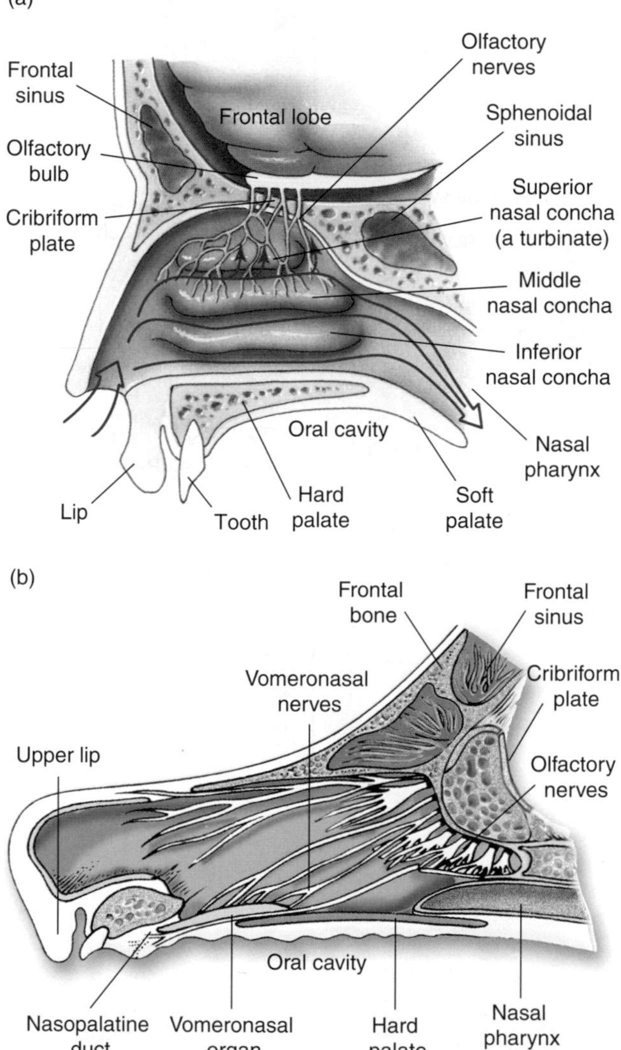

Figure 7-19 In vertebrate olfactory organs, air (or water) carrying odorants is moved past olfactory receptors during ventilation. **(a)** Midline sagittal section through a human head to illustrate the location of the olfactory epithelium, which covers part of the surface of air passages in the nose. The arrows indicate the route followed by inhaled air as it passes over the olfactory epithelium that lines the rostral recesses of the nasal cavity. Olfactory receptor cells are located in the olfactory epithelium. **(b)** Lateral sagittal section through the head of a dog illustrating the location of a vomeronasal organ. Notice that odorants reach the organ by way of the nasopalatine duct opening into the oral cavity, rather than through the nostrils.

nules. In vertebrates, the majority of olfactory receptors are located inside the nasal cavity, arranged so that a stream of air or water flows over them during ventilation (Figure 7-19a). Animals that are particularly dependent on olfactory cues have complex nasal cavities, called *turbinates,* that are lined by sheets of receptors.

In addition, many terrestrial vertebrates (but perhaps not humans) possess a second kind of olfactory chamber, the **vomeronasal organ,** which also is lined by an olfactory epithelium (Figure 7-19b). The vomeronasal organs are bilaterally paired blind-ended cavities that open into either the oral cavity or the nasal passages. A major function of the vomeronasal organ is to mediate chemical communication among animals of the same species. Signals to conspecific animals regarding territory and reproductive condition are carried by pheromones, chemical signaling molecules to which vomeronasal receptor cells respond selectively. Interestingly, the olfactory receptors in the nasal cavity and those in the vomeronasal organ are physiologically and molecularly distinct from each other, even though they have a common embryonic origin.

Each receptor neuron in the main olfactory epithelium has a long, thin dendrite that terminates in a small knob at the surface (Figure 7-20a). Emanating from the knob are several thin cilia, about 0.1 μm in diameter and about 200 μm long, which are covered by a protein solution called *mucus.* Molecules that enter the nasal cavity are absorbed into the mucous layer and diffuse to the cilia. Receptor neurons in the vomeronasal organ typically lack cilia; instead, they have microvilli on the cell surface that contacts odorants.

Two lines of evidence suggest that the cilia of receptors in the nasal cavity are the location of olfactory transduction. First, only ciliated olfactory neurons respond to odors. The second piece of evidence comes from experiments in which olfactory neurons were grown in culture and exposed to odorants while the receptor current was recorded by an intracellular electrode in the soma (Figure 7-20b,c). When a solution of odorant molecules was applied to only the cilia of a receptor cell, the cell responded strongly, whereas when the same solution was applied to only the soma, there was no more than a small response. In contrast, applying a solution of KCl (which would depolarize the membrane of the receptor, opening voltage-gated ion channels) onto only the cilia produced a small response, whereas applying KCl onto only the soma produced a large response. These data imply that only the cilia responded to the odorant.

The nasal olfactory transduction cascade includes an enzyme, adenylate cyclase, that when activated by a G protein catalyzes an increase in intracellular cAMP. Increased [cAMP] opens cation-selective ion channels, depolarizing the receptor cell (Figure 7-21a on the next page). Action potentials along receptor axons carry the signal to the olfactory bulb of the brain (see Figure 7-19a).

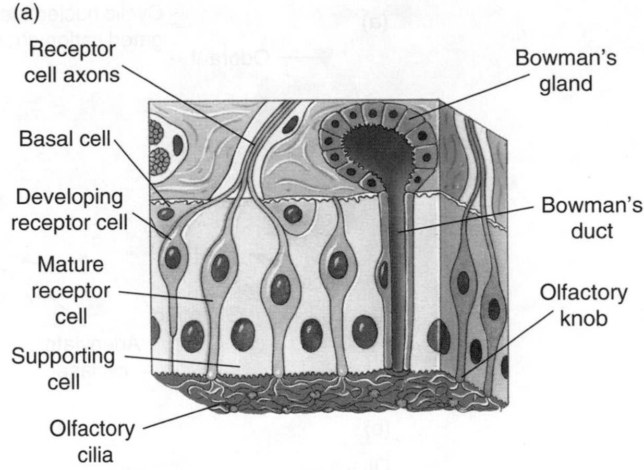

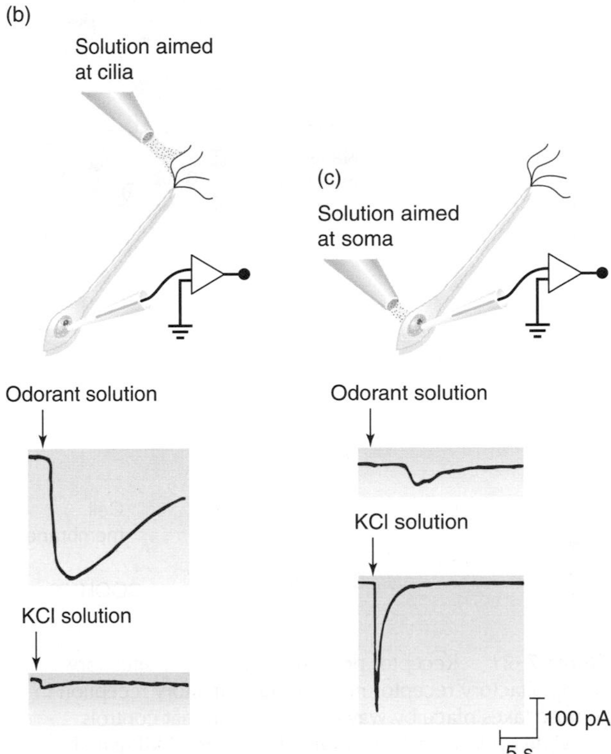

Figure 7-20 Receptors of the vertebrate nasal olfactory epithelium depolarize in response to odorant molecules. **(a)** The organization of cells within the mammalian olfactory epithelium. **(b)** Response of an isolated cultured salamander olfactory receptor neuron to small, directed pulses of an odorant. When the stimulating chemical pulse was directed at the receptive membrane of the cilia, it produced a large current (top record). When a solution containing a high concentration of KCl was delivered at the same location, the response was small (bottom record). Exposing a cell to a high concentration of KCl transiently depolarizes the membrane and opens voltage-gated ion channels. This tiny response to KCl suggests there are very few voltage-gated channels in the ciliary membrane. **(c)** In contrast, when the pulse of odorant was directed at the soma, rather than at the cilia, the response was small (top record), suggesting that there are very few receptor molecules in the somatic membrane. However, when the KCl solution was directed at the soma, it produced a large response (bottom record), indicating the presence of many voltage-gated channels in the membrane of the soma. [Part a adapted from Shepherd, 1994; part b adapted from Firestein et al., 1990.]

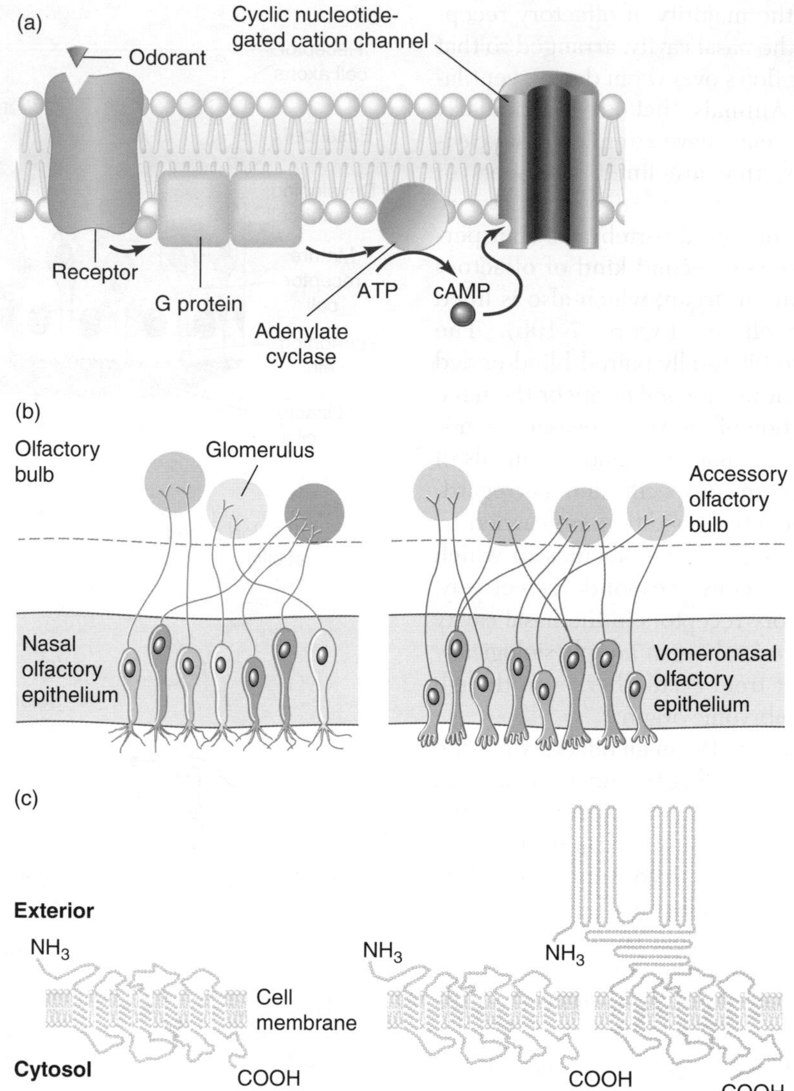

Figure 7-21 Receptor proteins and synaptic sites vary among olfactory receptor neurons. **(a)** Olfactory reception in the nose takes place by way of a G protein that controls adenylate cyclase and thus indirectly opens cAMP-gated cation channels. The identity of the odorant is revealed by the particular class of receptors that are stimulated. Activation of the receptor leads to inward cationic current that depolarizes the neuron, generating APs in the axon. **(b)** Axons of receptor neurons synapse onto cells in the brain that are grouped into clusters called glomeruli. Nasal olfactory receptors (left) synapse in the olfactory bulb, and receptors that express a particular receptor protein (indicated by color in this diagram) all synapse onto one or a few glomeruli. Each shade of orange represents a member of this molecular family that uniquely recognizes a specific ligand. Vomeronasal receptors (right) synapse onto glomeruli in the accessory olfactory bulb. These receptors fall into two distinct groups whose nuclei are located at two different depths in the vomeronasal olfactory epithelium. Within each category, each cell expresses only one or a few of a large family of proteins. As with nasal receptors, cells expressing a particular receptor protein converge onto corresponding glomeruli, but each vomeronasal receptor typically projects to a larger number of glomeruli than do nasal receptors. **(c)** All olfactory receptor proteins have seven transmembrane helices, a motif typical of receptors coupled to G proteins. In mammals there are three distinct families of these proteins—one expressed in the nasal olfactory cells and two in the vomeronasal organ—with little or no homology between the families except for sequences associated with the seven transmembrane helices. The color of each sequence in this diagram corresponds to the colors of the cells in part b, indicating the sequences associated with each type of receptor cell. Olfactory receptor molecules identified from the nematode *C. elegans* and the fruit fly *Drosophila* also have this general configuration. [Part c adapted from Matsunami and Buck, 1997, and Tirindelli et al., 1998.]

A very large family of proteins—as many as 1000 independent sequences are known in the rat, mouse, and human—has been identified that are expressed only in nasal olfactory receptor cells. These proteins are thought to be olfactory receptor molecules. The family of genes encoding these proteins represents about 1% of the entire rodent genome. Approximately 70 members of this same gene family have recently been identified in *Drosophila,* and members of the family are found in the nematode *C. elegans,* indicating that olfactory receptor proteins are both ancient in evolution and broadly distributed across the animal kingdom. Each individual receptor cell expresses only one or a few of these proteins, conferring great stimulus specificity on the cell. In the olfactory bulb, axons from receptors expressing a particular receptor protein synapse onto one or at most a few spherical clusters of neurons, called *glomeruli* (Figure 7-21b). The olfactory bulbs of many insects and some other invertebrates, as well as vertebrates, are organized into glomeruli, suggesting that this structural arrangement is particularly well suited for analyzing complex sensory input.

The structure of each olfactory receptor protein includes the seven transmembrane helices that are common among membrane signaling proteins (Figure 7-21c). This structure, along with additional features, indicates that these molecules are related to the proteins that mediate other transduction processes. Although the regions of these proteins that reside within the membrane are strongly conserved, the sequences that stick out into the extracellular fluid are highly variable, which is not surprising given the very large number of odorants that can be distinguished and the specificity with which each receptor protein appears to bind odorants. The huge number of distinct olfactory receptor proteins and the glomerular arrangement of postsynaptic cells within the olfactory bulb suggest that there are a multitude of individual receptor subtypes, each of which is tuned to receive distinct odors, in contrast with the small number of receptor types that code for taste.

Interestingly, the receptor cells of the vomeronasal olfactory epithelium operate somewhat differently from the nasal olfactory receptors at both the cellular and molecular levels. Proteins from two different families of seven-transmembrane-helix receptor molecules are expressed in the receptor membrane of these vomeronasal cells, and neither of these families is expressed in nasal olfactory receptor cells. The vomeronasal receptor cells that express these two different protein families are separated spatially. Although the total number of proteins in the two families is smaller than the number of proteins found in nasal olfactory receptor cells, the two vomeronasal families together include more than 100 different proteins in the mouse. Just as in the nasal olfactory epithelium, it appears that each vomeronasal receptor cell expresses only one or a few receptor proteins. The axons of the vomeronasal receptors project to a separate location in the brain, called the accessory olfactory bulbs, where they, too, end in glomeruli (see Figure 7-21b).

In the nematode *C. elegans,* about 10% of all neurons in the body participate in olfaction, and 500–1000 putative olfactory receptor proteins have been identified (about 5% of the entire genome). In both *C. elegans* and the fruit fly *Drosophila,* the receptor proteins include seven transmembrane helices, just as they do in vertebrates, so the general protein structure is similar. However, there is little or no homology in the details of the protein sequences when they are compared with vertebrate sequences, or between the nematode and fly proteins. One receptor-ligand pair has been firmly determined: a gene of *C. elegans* (named *odr10*) has been shown conclusively to encode receptor function for the odorant diacetyl. Much remains to be done in order to understand the relationship between receptor protein sequence and ligand identity, and many surprises are sure to lie ahead.

The manner in which vertebrates encode olfactory information into patterns of neuronal activity has been studied electrophysiologically in the olfactory epithelium of the frog (Figure 7-22a on the next page). In these experiments, activity in single receptor axons was recorded by one electrode while at the same time the summed potential from large numbers of olfactory receptors in the epithelium (the *electro-olfactogram,* or EOG) was recorded by another electrode (Figure 7-22b). Impulses from individual receptors were then superimposed electronically on the electro-olfactogram. This technique permits the activity in a single receptor to be compared with the total response of the complete population of receptors when a single odorant, or a combination of odorants, is presented.

The results indicate that the encoding of information in the vertebrate nose is far more complex than in the contact chemoreceptors of houseflies. Individual receptors respond differently to the same odorant. In some olfactory axons, a particular odorant increased the impulse frequency (Figure 7-22c). Some odorants that smell alike to humans have similar effects on some frog olfactory receptor cells, suggesting that they might smell alike to frogs, too. However, these same odorants have different effects on other receptor cells (see Figure 7-22c, cell a versus cell b), suggesting that they would not smell alike to the frog if these receptors were the sole basis for the judgment. Neurons in the olfactory bulb may respond to an odorant with decreased activity or with increased activity (Figure 7-22d). Surprisingly, it has proved impossible to establish a one-to-one relation between classes of odorants and types of olfactory receptor cells by recording in this way in the frog. Instead, each olfactory receptor cell appears to respond to a small number of different odorant molecules, suggesting that each receptor cell expresses more than one, but only a few, receptor proteins. Further work may reveal that the ability of many organisms to distinguish among a wide variety of odors resides at least in part in the ability of higher olfactory centers in the brain to decode a combinatorial signal from a large number of olfactory receptor cells, each of which responds to a small number of odorants.

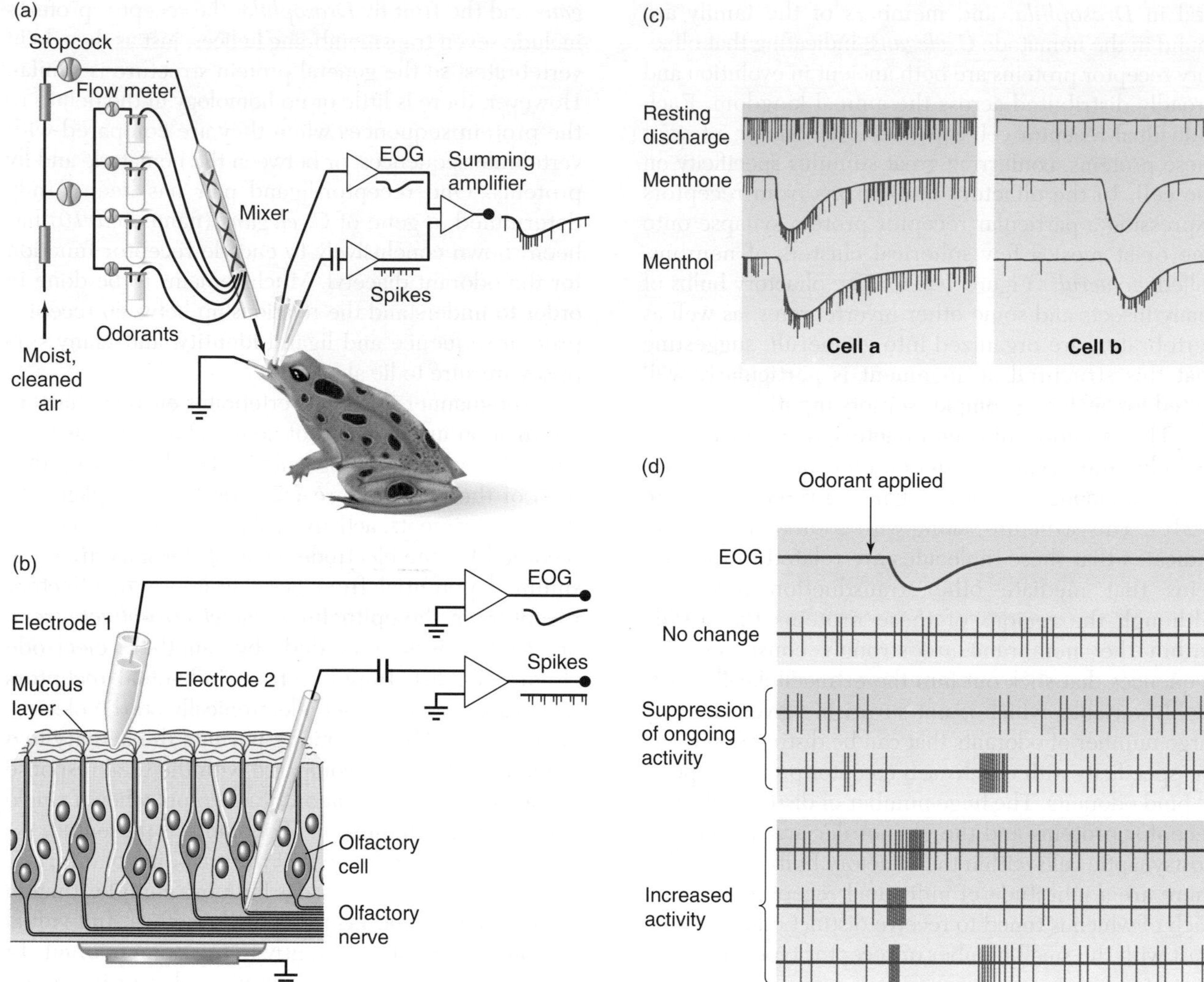

Classic Work **Figure 7-22** Responses to individual odorants can be studied at the cellular and organ levels simultaneously in the olfactory epithelium of the frog. **(a)** Various odorants are applied to the nasal epithelium while both the summed electro-olfactogram (EOG) and spikes from individual receptor axons are monitored using extracellular recording techniques. The two kinds of records can then be electronically summed to give a composite recording (shown at the right). **(b)** Detail of tissue and electrodes. Electrode 1 records the overall EOG potential, because it is far from any one axon, whereas electrode 2 records the activity of the single axon to which it is closest. **(c)** Recordings from two frog olfactory receptors, labeled cell a and cell b. Ongoing activity in cell a was slightly suppressed by both menthone and menthol, indicating that cell a could not distinguish between the two substances. In contrast, cell b responded differentially to the two substances. Compared with its resting activity, it produced fewer APs in response to menthone and more in response to menthol. Notice that the EOG has been summed with the individual record for each cell, as in part a. **(d)** Recordings from six second-order olfactory neurons in the olfactory bulb of another amphibian, a tiger salamander, showing responses to a single odorant. Notice that although the reception of an odorant causes excitation of the primary afferent neurons, it may either reduce or increase ongoing activity in these second-order neurons. [Parts a and b adapted from Gesteland, 1966; part c adapted from Gesteland, 1966; part d adapted from Kauer, 1987.]

MECHANORECEPTION

All animals can sense physical contact with the surface of their bodies. Such stimuli are detected by **mechanoreceptors**, the simplest of which consist of morphologically undifferentiated nerve endings found in the connective tissue of skin. More complex mechanoreceptors include accessory structures that transfer mechanical energy to the part of the receptor cell's membrane that is specialized to transduce mechanical energy into a change in V_m. Typically these accessory structures also filter the mechanical energy in some way, as described earlier for the mammalian Pacinian corpuscle (see Figure 7-10). Other mechanoreceptors include the muscle stretch receptors of various kinds found in arthropods and vertebrates (see Figure 7-11), and the hairlike sensory bristles that extend from the exoskeletons of arthropods (Figure 7-23a). The most elaborate accessory structures associated with mechanoreceptive cells are found in the auditory and vestibular structures of the vertebrate ear, both of which are considered later in this chapter.

(a)

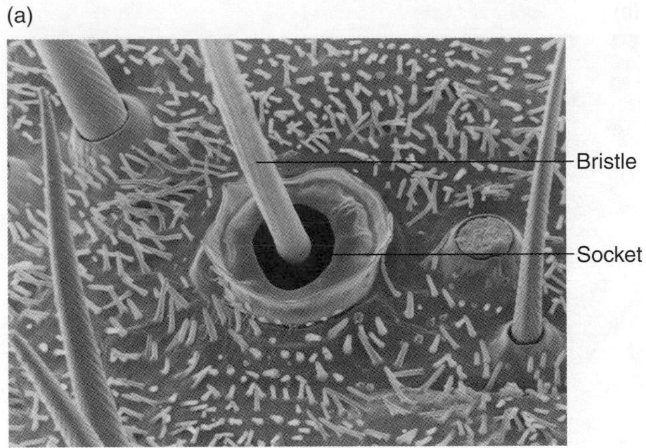

(b)

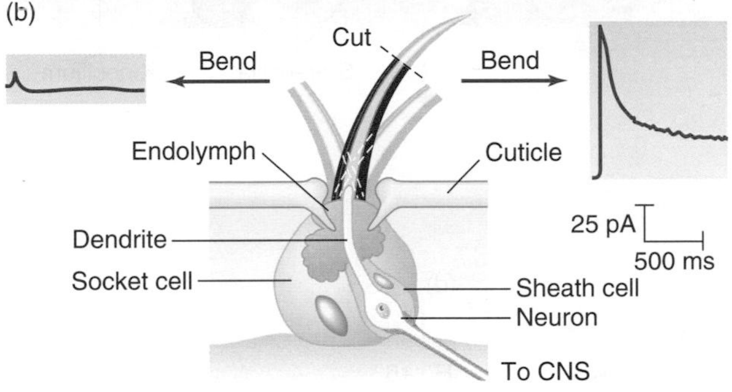

(c)

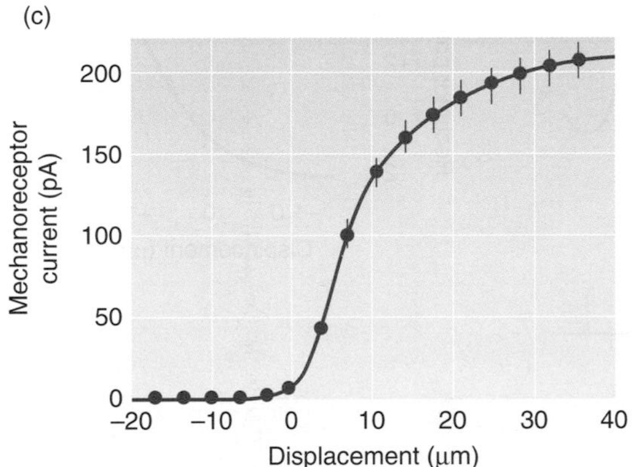

Figure 7-23 Mechanosensory bristles are found at several locations on the bodies of insects. **(a)** A scanning electron micrograph of several types of sensory hairs on the body of a cricket *(Gryllus)*. The sensory bristle shown at the center of the figure is of the same type shown in part b. **(b)** Longitudinal section through a bristle mechanoreceptor of the fruit fly *Drosophila*. Cutting the shaft allows electrical access to the sensory neuron of the receptor. Bending the bristle of this sensory hair toward the surface of the body produces a large sensory current; bending the bristle an equal amount in the opposite direction produces only a small response. **(c)** The amount of receptor current elicited by bending the shaft toward the body varies with the distance the shaft is moved. Movements as small as 10 μm elicit a receptor current. [Part a courtesy of Thomas A. Keil; parts b and c adapted from Walker et al., 2000.]

The stimulus that activates a mechanoreceptor cell is a stretch or distortion of its plasma membrane. Ion channels that are activated by stretch are probably found in all types of organisms. Patch-clamp data indicate that conductance through these channels changes when there is a change in tension within the plane of the membrane; the channels may be either activated or inactivated by stretch. The transduction of mechanical stress probably depends on some combination of the cytoskeleton, signal cascades, and the ion channels themselves; the combination is likely to differ among various categories of mechanoreceptor cells. Recently a stretch-gated channel protein was described in *Drosophila,* and a close homolog was identified in the nematode *C. elegans.* Future work will explore how widespread these receptor proteins are in the animal kingdom; we will discuss them in more detail later in this chapter.

Mechanoreceptors can be exquisitely sensitive; some of them respond to mechanical displacements of as little as 0.1 nm. This motion is equivalent to displacing the Eiffel Tower by the width of a human thumb. It is a continuing challenge to understand how such small movements can produce changes in ion permeability through the membrane. In addition, responses in these receptors take place within microseconds of stimulation, a remarkably short latency. This temporal feature sets severe limits on how transduction can be accomplished because very few known sensory mechanisms can generate such rapid responses.

Insect Mechanoreceptors

The surface of an insect body displays a variety of sensory bristles and other mechanoreceptors. A typical sensory bristle consists of a stiff hair that protrudes from the insect's body and is associated with a single sensory neuron (see Figure 7-23a). The neuron is bipolar; that is, a dendrite extends from the soma into the base of the bristle, and its axon projects into the central nervous system. The dendrite is bathed in an unusual body fluid, called **endolymph**, that contains a much higher concentration of K^+ than is typical for extracellular fluids. As a result, the value of E_K across the plasma membrane is significantly more inside-positive for this cell type than for most other cells in the body. When the shaft of the bristle is bent, the dendrite is deformed, which opens stretch-activated cation-selective channels. The unusual value of E_K produces an inward driving force on K^+, rather than the more usual outward force, so when cation-selective channels open in the membrane of the dendrite, K^+ enters the cell, carrying positive charge (Figure 7-23b,c). Recent work has identified a channel protein, localized in the base of the sensory dendrite, that has the six transmembrane helices typical of voltage-gated ion channels. The discovery is exciting because this protein is the first mechanosensory channel to be found in animals despite years of searching (bacterial stretch channels have been known for a few years). Future research will reveal how widely homologs of this protein are spread through the animal kingdom.

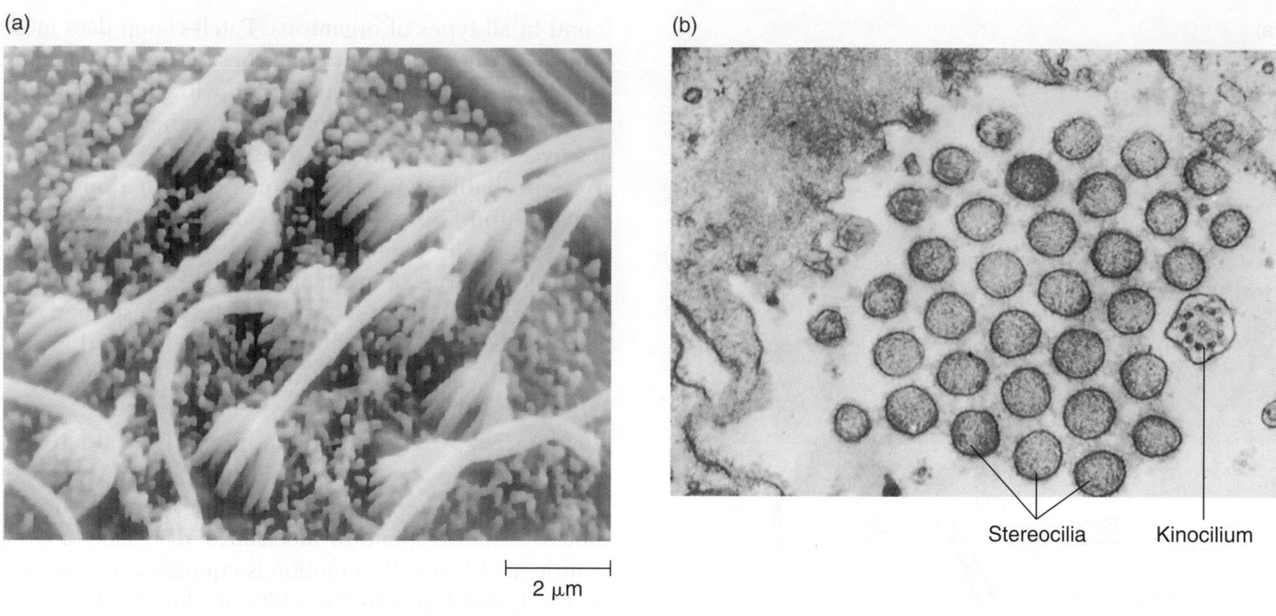

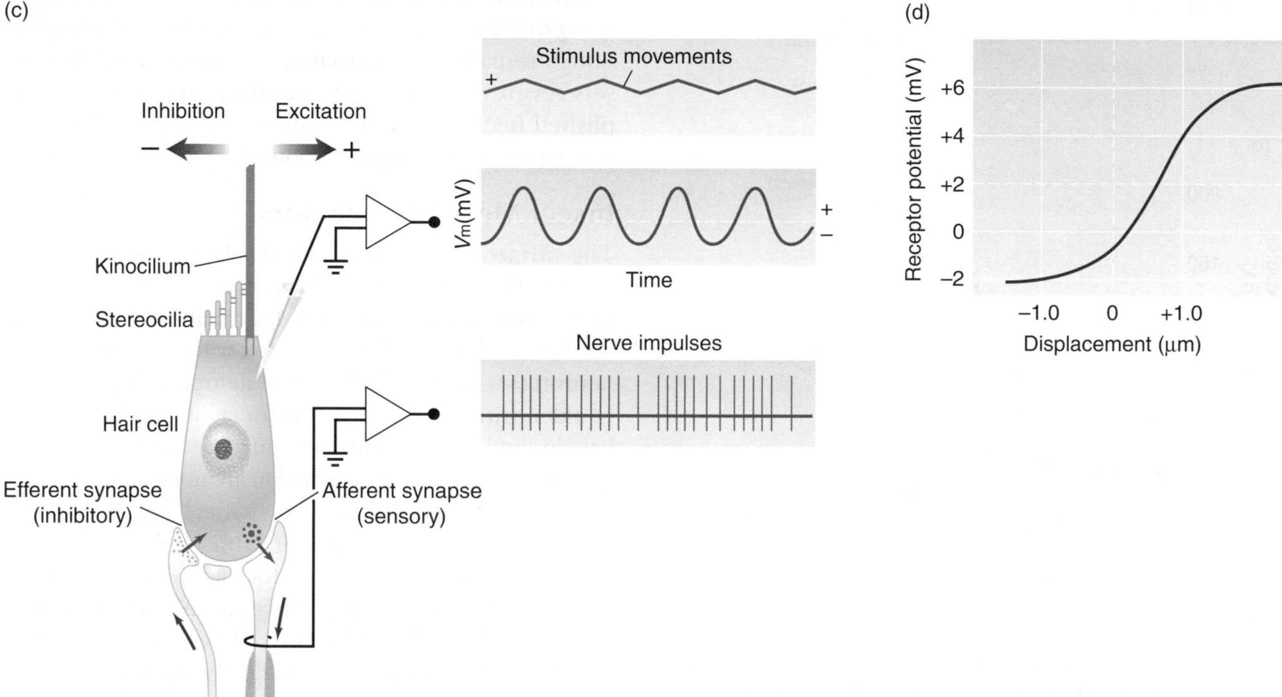

Figure 7-24 The membrane potential of a hair cell changes when the cilia are moved away from their resting position. **(a)** Scanning electron micrograph of several hair cells from a water-wave detector of a giant zebrafish *(Danio)*. **(b)** Electron micrograph of a cross-section through the cilia of a hair cell. The large cilium at the right, containing a typical 9 + 2 microtubule arrangement, is the kinocilium; the others are stereocilia. **(c)** Diagram of a typical hair cell showing the anatomic relations of the stereocilia and the kinocilium. The hair cell releases transmitter onto a primary afferent neuron, which carries the sensory signal to the central nervous system. It also receives synaptic signals from efferent neurons. Depending on the direction in which the cilia are bent, the hair cell can either increase or decrease the frequency of APs in the continuously active afferent fiber. A linear back-and-forth motion applied to the cilia rhythmically changes the transmembrane potential in the hair cell, which can be recorded with an intracellular microelectrode. Extracellular recording from the afferent axon shows the change in the frequency of APs associated with changes in V_m in the hair cell. **(d)** Input-output relation for a hair cell. Note that the depolarization produced by movement toward the kinocilium and tallest stereocilia (positive values of displacement) is larger than the hyperpolarization in response to movement away from the kinocilium. [Part a courtesy of Christopher Braun; part b from Flock, 1967; part c adapted from Harris and Flock, 1967; part d adapted from Russell, 1980.]

Hair Cells

The **hair cells** of vertebrates are extraordinarily sensitive mechanoreceptors that transduce mechanical stimuli into electrical signals. Hair cells are named for the many cilia that project from the apical end of each cell (Figure 7-24a,b). These cilia fall into two classes: each hair cell typically has a single *kinocilium* and 20–300 nonmotile *stereocilia.* The kinocilium has a "9 + 2" arrangement of internal microtubules (Figure 7-24b, middle right) similar to that of other motile cilia found in a variety of cells. The stereocilia contain many fine longitudinal actin filaments, and they appear to be developmentally distinct from the kinocilium. The stereocilia of a hair cell are arranged in order of increasing length from one side of the cell to the other (Figure 7-24c), giving a hair cell a beveled appearance like the tip of a hypodermic needle. In most mechanosensory organs, the hair cells are coupled by their kinocilia to some kind of accessory structure. Stimuli that move the accessory structure are transmitted to the bundles of stereocilia through bonds that connect the accessory structure and kinocilium to the stereocilia. If the tip of the bundle of stereocilia is touched with a fine probe, the bundle moves as a unit, regardless of the direction of stimulation, suggesting that the stereocilia are mechanically linked with one another.

The exact process by which pressure or force from the outside world moves bundles of stereocilia depends on the specific arrangement of the hair cells and their accessory structures within each mechanosensory organ, but ultimately it is the movement of the stereocilia that produces an electrical signal. When the stereocilia bend toward the kinocilium, the hair cell depolarizes; when they bend in the opposite direction, the cell hyperpolarizes (Figure 7-24d). (If the stereocilia bend to either side, V_m remains the same.)

Hair cells are found in several locations. Fishes and amphibians have a set of external receptors called the *lateral-line system,* which is based on hair cells and detects motion in the surrounding water (Figure 7-25). In each of these receptors, the kinocilia of several hair cells are embedded in an accessory structure called a *cupula,* which bends the kinocilia when it is displaced. The vertebrate organs of hearing and the organs that report the position and motion of the body with respect to gravity (called the *vestibular organs* or the organs of equilibrium) also are based on hair cells. Although most hair cells in vertebrates have a kinocilium and stereocilia, some hair cells in the adult mammalian ear lack a kinocilium, suggesting that the kinocilium may not be necessary for transduction. This hypothesis has been tested by the technically remarkable feat of microsurgically removing the kinocilia from hair cells. The procedure does not block transduction, so although kinocilia are a normal component of hair cells, they are not required for mechanotransduction.

At rest, about 15% of the channels in a hair cell are open, producing a resting potential of about −60 mV.

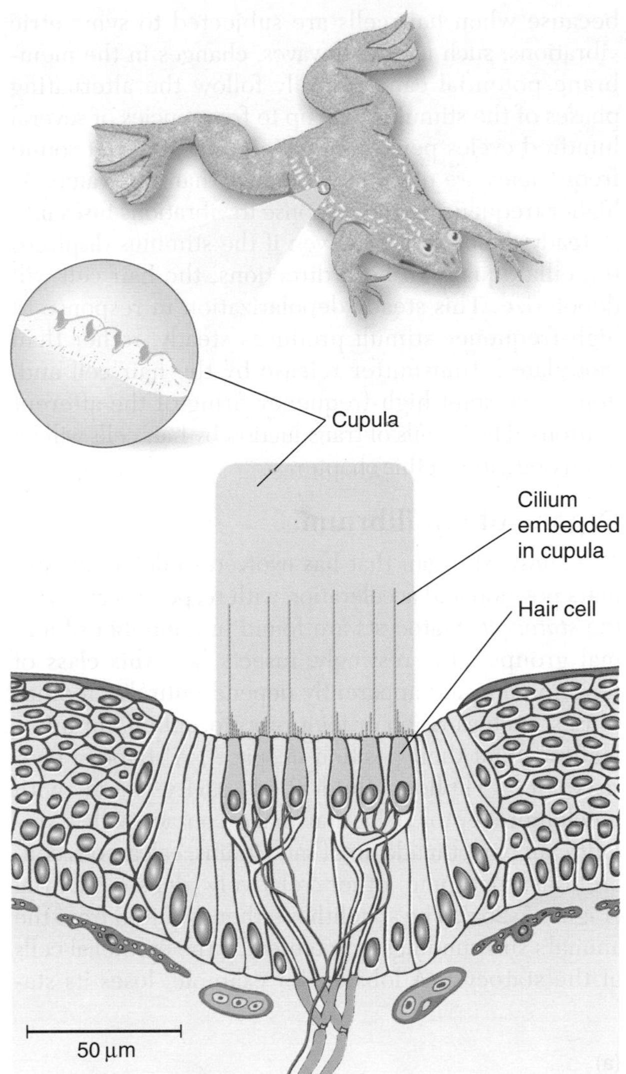

Figure 7-25 The lateral-line sensory system of fishes and amphibians, which detects motions in the surrounding water, is based on hair cells. The drawing shows the location of these mechanosensory organs along the body of an African clawed frog *(Xenopus).* The inset shows four units along the lateral line. The lower diagram shows a cross-section through part of the lateral line, illustrating the cupula, an accessory structure that bends the cilia of hair cells when it is displaced. Compare the structure of this organ with the hair cells shown in Figure 7-24.

Hair cells do not produce APs. Instead, they release neurotransmitter in a graded fashion onto primary afferent neurons. The amount of neurotransmitter released depends on V_m in the hair cell and determines the frequency of APs in the afferent neuron (see Figure 7-24c) Notice that the input-output relation for hair cells is markedly asymmetric (see Figure 7-24d), just as it is for insect mechanosensory bristles. That is, the depolarization produced by a given amount of displacement toward the kinocilium is larger than the hyperpolarization produced by a similar displacement in the opposite direction. This asymmetry is important because when hair cells are subjected to symmetric vibrations, such as sound waves, changes in the membrane potential can faithfully follow the alternating phases of the stimulus only up to frequencies of several hundred cycles per second (Hertz, or Hz), but sound frequencies are often much higher than this value. At higher frequencies, the response to vibrations fuses into a steady depolarization. Even if the stimulus displaces the cilia equally in both directions, the hair cell will depolarize. This steady depolarization in response to high-frequency stimuli produces steady, rather than modulated, transmitter release by the hair cell and, hence, constant high-frequency firing of the afferent neurons. The details of transduction by hair cells will be discussed later in this chapter.

Organs of Equilibrium

The simplest organ that has evolved to detect an animal's position and acceleration with respect to gravity is the *statocyst.* Statocysts are found in a number of animal groups. (Interestingly, insects lack this class of sense organs and apparently depend entirely on other senses, such as vision or joint proprioception, for information about their position in space.) A statocyst consists of a hollow fluid-filled cavity lined with mechanoreceptor cells that make contact with a *statolith,* an object made up of sand grains, calcareous concretions, or some other relatively dense material (Figure 7-26a). The statolith is either acquired from the animal's surroundings or secreted by the epithelial cells of the statocyst. A lobster, for example, loses its statoliths at every molt and replaces them with new grains of sand. In any case, the statolith must have a higher specific gravity than the surrounding fluid.

As the position of the animal changes, the statolith rests on different regions of the statocyst epithelium. When a lobster is tilted to the right about its longitudinal axis, the statolith rests on receptor cells on the right side of the statocyst, stimulating them and causing a tonic discharge in their axons (Figure 7-26b). Recordings from many different fibers of a statocyst reveal that each cell fires maximally in response to a certain orientation of the lobster (Figure 7-26c). Information from these receptors travels to the central nervous system and can set up reflex movements of the appendages. This pattern of information processing was confirmed in a clever experiment in which molting lobsters were presented with iron filings rather than sand. When the lobsters molted, they replaced their statoliths with iron filings, allowing the position of the iron statoliths to be manipulated by a magnet. As the magnet was moved through space, pulling on the iron statolith, the lobster—whose position with respect to gravity had not changed—produced a series of compensatory postural responses. This experiment showed that sensory input from the statocysts was sufficient to elicit postural reflexes; no other input was required.

The Vertebrate Ear

The ears of vertebrates are evolutionary specializations at the anterior end of the lateral line system (see Figure 7-25). They perform two separate sensory functions,

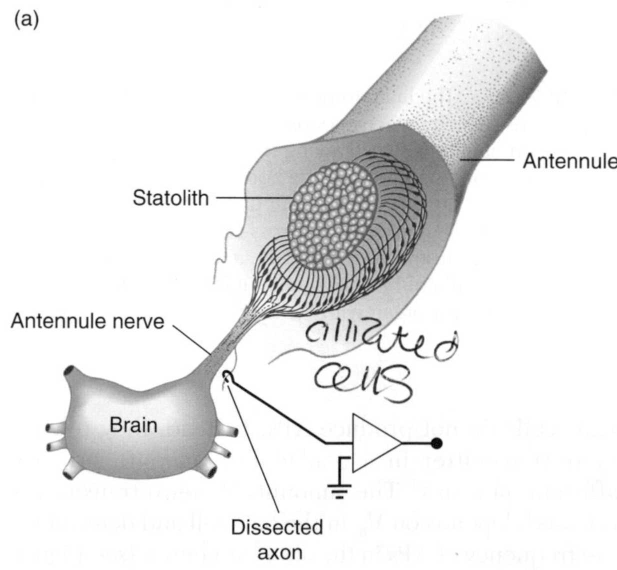

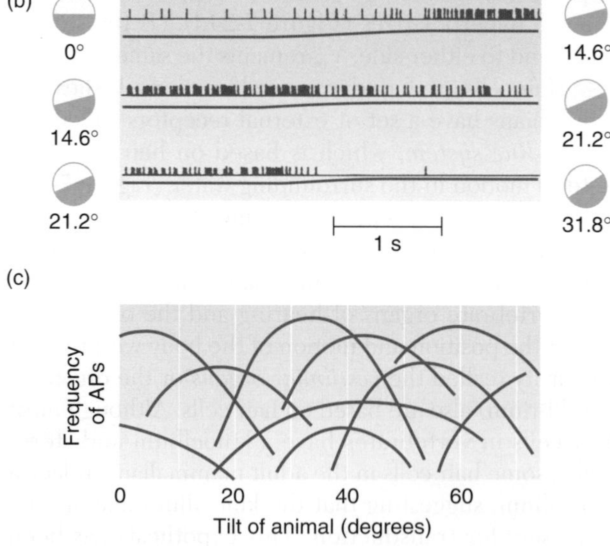

Classic Work **Figure 7-26** Statocysts sense position and acceleration with respect to gravity. **(a)** Structure of a statocyst in a lobster. A statolith rests on an array of ciliated cells that, unlike vertebrate hair cells, themselves send axons to the brain. **(b)** Action potentials recorded from individually dissected nerve fibers while a lobster was being tilted. All three recordings show the response of a single receptor, and the trace below the recording indicates the initial and final angle and the time course of the tilt. Notice that this receptor is tonically active any time the animal is tilted between 10 and 25 degrees. **(c)** Frequencies of APs recorded from different fibers plotted as a function of the position of the animal. Each cell responded with a maximum rate of discharge when the animal's body was held at a particular position, and the orientation that produced a maximal response varied among fibers. [Adapted from Horridge, 1968.]

each of which is based on the activity of hair cells. Some structures of the ear, the organs of equilibrium, perform like the statocysts of invertebrates, reporting on the animal's position with respect to gravity and its acceleration through space. Other structures, the organs of hearing, provide information about vibrational stimuli in the environment—stimuli that are called *sound* if they fall within the frequency range perceived by humans.

Vertebrate organs of equilibrium

In vertebrates, the organs of equilibrium reside in a membranous labyrinth consisting of two chambers, the *sacculus* and the *utriculus*, that are surrounded by bone and filled with endolymph similar to that in insect mechanoreceptors. The utriculus gives rise to the three semicircular canals of the inner ear, which are oriented in three mutually perpendicular planes (Figure 7-27a,b). Hair cells in these three orthogonal semicircular canals detect acceleration of the head. As the head is accelerated in the plane of one canal, the inertia of the endolymph in that canal moves the gelatinous cupula (Figure 7-27c). When a cupula moves, it bends the cilia of the hair cells at its base, and V_m of the hair cells changes. All the hair cells in a canal have their kinocilia on the same side, so that all of them are excited when the fluid moves in one direction and inhibited when it

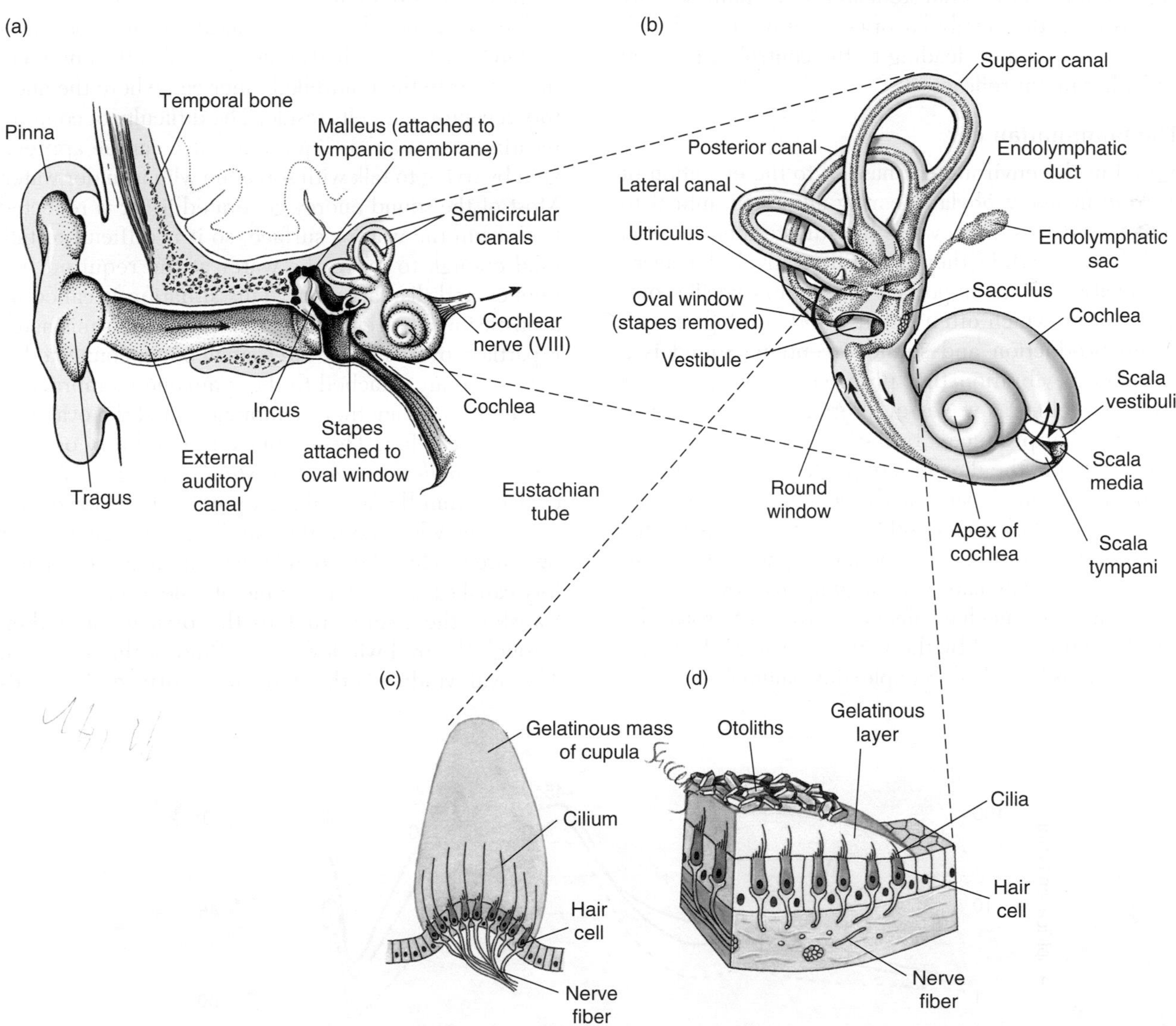

Figure 7-27 The human organs of audition and equilibrium are located in the ear. **(a)** The major parts of the ear. **(b)** The semicircular canals and cochlea. The stapes has been removed to reveal the oval window of the cochlea. At the far right, a section has been removed from the cochlea to reveal the inner structure. Figure 7-29 shows this structure in more detail. The pathway taken by auditory signals is shown by black arrows in parts a and b. **(c,d)** Detailed structure of two parts of the organs of equilibrium. **(c)** The cilia of hair cells in the semicircular canals are embedded in a gelatinous cupula. When fluid moves in the canal, the cupula bends the cilia. **(d)** Particles called otoliths rest on the cilia of hair cells in the sacculus (one part of the vestibular apparatus). Changes in the position of the animal's head cause the otoliths to shift position, changing how much the cilia are bent. [Parts a and b adapted from Beck, 1971; parts c and d adapted from Williams et al., 1995.]

moves in the opposite direction (see Figure 7-24). The orthogonal arrangement of the three semicircular canals allows them to detect any movement of the head in three-dimensional space because any movement will have at least one vector component that coincides with the plane of one or more of the canals.

Below the semicircular canals, larger bony chambers, the utriculus and sacculus, contain three more patches of hair cells called *maculae.* Mineralized concretions termed *otoliths,* similar to the statoliths associated with statocysts (see Figure 7-26), are associated with the maculae. The otoliths signal position relative to the gravitational field; in the lower vertebrates, they also detect vibrations, such as sound waves, in the surrounding medium. In the brain stem and cerebellum, sensory signals from the vestibular organs are integrated with other sensory input, leading to the control of postural and other motor reflexes.

The mammalian ear

Sound in the environment has led to the evolution of hearing in many phyla. Hearing allows an animal to detect predators or prey and to estimate their location and distance while they are still relatively far away. Sound also plays an important role in intraspecific communication, which often entails subtle tuning of both sound production and sound reception. Sound is a mechanical vibration that propagates through air or water, traveling as waves of alternating high and low pressure that produce a back-and-forth movement of the medium in the direction of propagation. The physical nature of sound, particularly the differences in how it is conducted through air and through water, has set distinct constraints on its detection. Comparing the physiology of hearing in many animal groups has revealed that many different mechanisms have evolved to solve the problems presented by the nature of sound. Here we examine a well-studied example: the mammalian ear.

External ear, auditory canal, and middle ear The structure of the external ear acts as a funnel that collects sound waves in the air from a large area and concentrates the oscillating air pressure onto a specialized surface, the eardrum or tympanic membrane. The external structures of the ear—the *pinna* and *tragus*—facilitate the collection of sound waves (see Figure 7-27a). These shell-like outer structures, which are moveable in some species, can modify the directional sensitivity of the auditory system. In many mammals, including humans, the acoustic properties of the external ear amplify sound in particular frequency ranges. In addition, the human ear emphasizes the spatial distribution of sound stimuli by preferentially amplifying sounds from certain directions (Figure 7-28).

To be detected, airborne vibrations must be transmitted from the air-filled auditory canal of the outer and middle ear to the fluid-filled inner ear, where the auditory receptor hair cells reside. The difficulty of communicating across an air-liquid interface can be appreciated by trying to talk with someone who is under water. Most of the sound energy generated in air is reflected back from the water's surface, so it is difficult to talk loud enough to move the water at the required frequency and displacement. This situation is called an acoustic impedance mismatch. In the ear, this mismatch is partially overcome by three small bones connected in series that are attached to the tympanic membrane at one end and to another membrane called the oval window at the other. These bones, the auditory ossicles (labeled *malleus, incus,* and *stapes* in Figure 7-27a), evolved from the articulation points of the lower jaw and are now located in the middle ear. Changes in air pressure produced by sound waves in the external auditory canal cause the typanic membrane to move, which transfers the energy first to the ossicles, and then through the oval window to the fluid of the inner ear. The oval window is the outermost surface of a fluid-

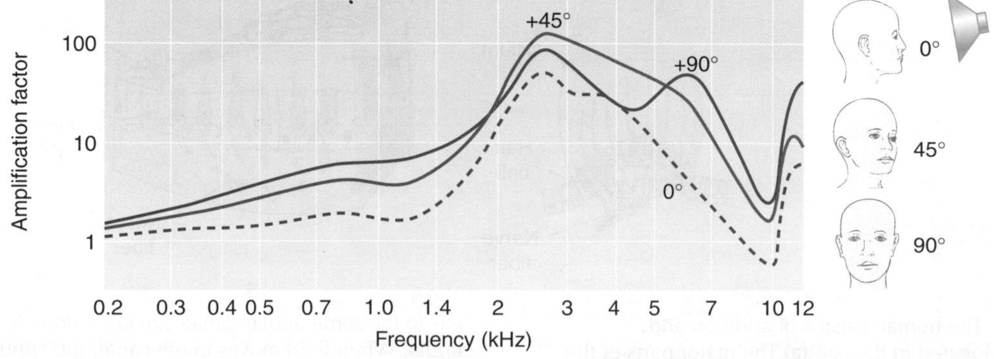

Classic Work *Figure 7-28* The structure of the human pinna and tragus selectively amplifies specific frequencies of sound. This graph shows the gain in pressure at the tympanic membrane over what the pressure would be if sound impinged on the auditory canal with all external ear structures removed. If there were no amplification, the graph would be a horizontal line intersecting with the ordinate at an amplification factor of 1. Values above 1 indicate amplification; values below 1 indicate suppression. The gain varies as a function of frequency, and sounds emanating from different directions are amplified differentially. Zero degrees is straight in front of the face. [Adapted from Shaw, 1974.]

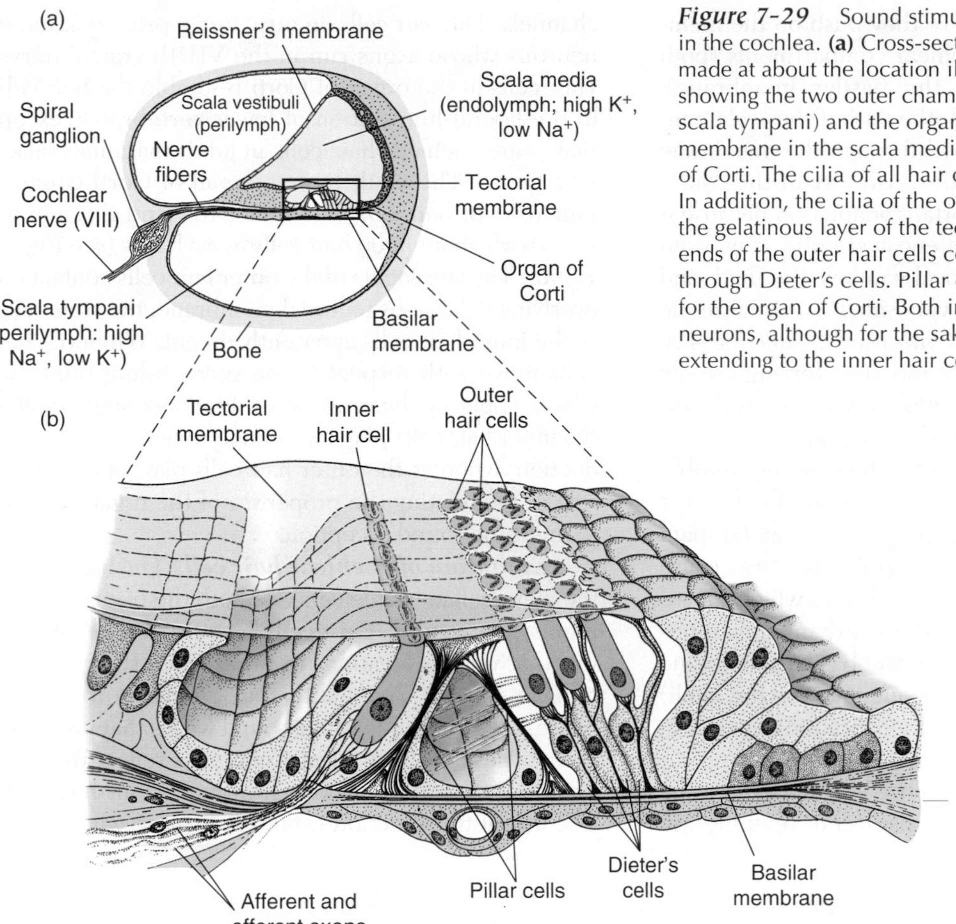

Figure 7-29 Sound stimuli are transduced by hair cells in the cochlea. **(a)** Cross-section through cochlear canal, made at about the location illustrated in Figure 7-27b, showing the two outer chambers (the scala vestibuli and the scala tympani) and the organ of Corti attached to the basilar membrane in the scala media. **(b)** Enlargement of the organ of Corti. The cilia of all hair cells are bathed in endolymph. In addition, the cilia of the outer hair cells are embedded in the gelatinous layer of the tectorial membrane. The basal ends of the outer hair cells connect to the basilar membrane through Dieter's cells. Pillar cells provide structural support for the organ of Corti. Both inner and outer hair cells contact neurons, although for the sake of clarity only the fibers extending to the inner hair cells are shown in this diagram.

filled chamber called the **cochlea**, which contains the receptor hair cells. At the other end of this fluid-filled compartment is another membrane, the round window.

The structures of the middle ear cause the pressure of the sound signal to be greatly amplified because the eardrum has an area of about 0.6 cm^2, whereas the oval window is considerably smaller, about 0.032 cm^2. This ratio of about 19:1 between the areas of the two membranes means that the sound pressure that impinges on the tympanic membrane is concentrated onto the smaller area of the oval window, producing a much greater pressure. This increase in pressure helps to overcome the inertia of the cochlear fluid on the other side of the oval window, which is greater than the inertia of air.

Structure and function of the cochlea The mechanically amplified sound input is transduced into electrical signals by the hair cells located in the cochlea. The cochlea, a tapered tube encased in the mastoid bone, is coiled somewhat like the shell of a snail (see Figure 7-27a,b). It is divided internally into three longitudinal compartments (Figure 7-29a). The two outer compartments (the *scala tympani* and *scala vestibuli*) are actually one long continuous tube folded back on itself at the middle. The bend in the tube, where the scala vestibuli becomes the scala tympani, is located at the apical end of the cochlea and is called the *helicotrema.* The scala tympani and scala vestibuli are filled with an aqueous fluid called *perilymph,* which resembles other extracellular fluids in having a relatively high concentration of Na^+ (about 140 mM in humans) and a low concentration of K^+ (about 7 mM in humans). Between these compartments, and bounded by the basilar membrane and Reissner's membrane, is another compartment, the *scala media,* which is filled with endolymph, the same type of fluid that surrounds the cilia of hair cells in vertebrate organs of equilibrium and the dendrites of mechanoreceptor neurons in the sensory bristles of insects. Recall that endolymph differs from most extracellular fluids because it is high in K^+ (about 150 mM in humans) and low in Na^+ (about 1 mM in humans). The **organ of Corti,** which bears the cochlear hair cells, lies bathed in endolymph on the basilar membrane of the scala media. As we will see, the unusual ionic composition of endolymph contributes importantly to the process of auditory transduction.

A coiled cochlea is found only in the ears of mammals, although birds and crocodilians have a similarly structured straight cochlear duct that includes the basilar membrane and the organ of Corti. Other vertebrates have no cochlea, and many detect sound waves by way of hair cells in the organs of equilibrium.

Sound vibrations are transferred by the ossicles of the middle ear to the oval window and from there into the cochlear perilymph, where they push on the membranes that separate the cochlear compartments (both Reissner's membrane and the basilar membrane). When the sound waves reach the end of the scala tympani, their energy is dissipated through the membrane of the round window. The distensibility of the round and oval windows is an important adaptation because if the fluid-filled cochlea were encased entirely by solid bone, displacements of the oval window, the fluid, and the internal tissues would be very small. To visualize the path taken by the sound vibrations, imagine the eardrum moving in and out and transferring energy through the ossicles of the middle ear to the oval window. As the oval window is pushed in and out, it causes pressure waves to travel through the incompressible perilymph along the scala vestibuli, around the bend at the helicotrema, and back through the scala tympani toward the round window. These pressure waves cause part of the basilar membrane to vibrate, which moves the hair cells with respect to the overlying gelatinous *tectorial membrane.* The part of the basilar membrane that vibrates depends on the frequency (i.e., the pitch) of the sound; the amplitude of the vibration depends in part on the intensity of the sound (Spotlight 7-1).

The movement of fluid in the cochlea causes the stereocilia of hair cells to be displaced, opening ion channels. The hair cells, in turn, excite primary afferent neurons whose axons run in the VIIIth cranial nerve. Hair cells in the organ of Corti resemble the hair cells in the lateral-line system of lower vertebrates, except that some cochlear hair cells in adult mammals lack a kinocilium. The adult human organ of Corti contains four rows of hair cells, one inner row and three outer rows, with about 4000 hair cells in each row (see Figure 7-29b). The stereocilia of the outer hair cells contact the overlying gelatinous tectorial membrane; the stereocilia of the inner hair cells apparently do not. When the hair cells move with respect to the surrounding fluid, the cilia are bent by shearing forces. Evidence suggests that the inner hair cells play the primary role in sound transduction, whereas the outer hair cells play an important role in modulating the properties of the organ of Corti to fine-tune sound reception.

Excitation of cochlear hair cells The hair cells of the mammalian cochlea encode both the frequency and intensity of sound. Extracellular electrical recordings made at various locations in the cochlea show fluctuations in electric potential that are similar in frequency, phase, and amplitude to the sound waves that produced them. These cochlear electrical signals reflect the summation of receptor currents from numerous hair cells. The actual transduction event takes place when a per-

MECHANICS OF THE BASILAR MEMBRANE

Hermann von Helmholtz noted in 1867 that the basilar membrane of the cochlea consists of many transverse bands that increase gradually in length from the proximal end (that is, at the oval window) to the apical end. The bands are about 100 μm long at the oval window and about 500 μm long at the apex, or helicotrema. This structure reminded him of the strings of a piano and inspired him to suggest a resonance theory of audition. He proposed that various locations along the basilar membrane vibrate in resonance with a specific tonal frequency while other locations remain stationary, just as the appropriate string of a piano resonates in response to a tone from a tuning fork.

This theory was later challenged by Georg von Békésy (1960), who developed a technique for studying the exposed cochlea directly. He found that the movements of the basilar membrane are not standing waves, as Helmholtz suggested, but consist instead of traveling waves that move from the narrow base of the basilar membrane toward the wider apical end (see part a of the accompanying figure). These waves have the same frequency as the sound entering the ear, but they move more slowly than sound moves through air ($343\ \mathrm{m \cdot s^{-1}}$).

A familiar example of a traveling wave can be seen by shaking the free end of a rope that is secured at the other end. Sinuous waves move along the rope from the end you are shaking to the secured end. Unlike a rope, the basilar membrane has mechanical properties that change along its length. The compliance of the membrane (the amount that the membrane will stretch in response to a given amount of force) increases from its narrow end to its broad end, which causes the amplitude of each traveling wave to change as it passes along the length of the membrane. For each sound frequency in the range heard by an organism, some location along the basilar membrane will be displaced maximally (part b of the figure). When the stimulating sound has a high frequency, the traveling waves produce maximum displacement near the oval window (called the basal end of the cochlea), whereas the lowest detectable frequencies maximally displace the basilar membrane closest to the helicotrema. Frequencies between these two extremes maximally displace the membrane somewhere in the middle. More intense sounds cause the basilar membrane to be displaced farther, more strongly stimulating the hair cells. The distance by which the basilar membrane is displaced determines how strongly hair cells in the organ of Corti are stimulated, so it also determines the rate of discharge in the primary afferent neurons carrying information from different locations along the basilar membrane. Even at maximum amplitude, the movements of the basilar membrane are very tiny: the loudest sounds displace it only about 1 μm. The movement of the hair cell cilia is much smaller, and the threshold of stimulus detection by hair cells is at the limit imposed by random movements of atoms and molecules.

turbation of the basilar membrane forces the tips of stereocilia to bend laterally (Figure 7-30a on the next page). This mechanical deflection directly causes ion channels in the tips of the stereocilia to open.

The perceptual threshold of cochlear hair cells corresponds to a deflection of 0.1–1.0 nm, which produces a change in receptor current of only about 1 pA through ion channels in the hair cell membrane. These channels have been shown in the laboratory to be permeable to many small, monovalent cations (e.g., Li^+, Na^+, K^+, Rb^+, and Cs^+), but when they open *in vivo*, K^+ ions, and probably some Ca^{2+} ions as well, enter the cell from the endolymph (Figure 7-30b). The high concentration of K^+ in the endolymph produces a net inward driving force on K^+, in contrast to the usual situation in which $V_m - E_K$ is an outward force. This inward K^+ current depolarizes the hair cells.

Based on measured current flow, it has been estimated that there are 30–300 channels per bundle of stereocilia, which implies that as few as one to five channels per stereocilium may be responsible for transduction. The channels are thought to be opened directly by a mechanical stimulus because when isolated bundles of stereocilia are abruptly deflected in experiments, the transduction current increases with an extremely short latency (about 40 μs). A latency this short makes it unlikely that any enzymatic or biochemical step is included in the process. This interpretation is reinforced by the results of patch-clamping experiments that indicate that the channels open faster when the deflection is larger, again suggesting a direct mechanical influence on the state of the channel.

The physical basis of this direct mechanical transduction is thought to be fine strands, called *tip links*, that connect each stereocilium on a hair cell with its next taller neighbor (Figure 7-30c). When the cilia move toward the longest cilium, each tip link pulls on the membrane of the adjacent cilium, stretching ion channels embedded in the membrane. It remains unknown whether the stretch-gated channels on the cilia of cochlear hair cells will prove to be homologous in sequence to the recently cloned stretch-gated channels of *Drosophila* mechanoreceptor cells, but the possibility is exciting because it would provide another example of how evolution has conserved molecular structure while generating a wide variety of applications for particular proteins.

The sensitivity of cochlear hair cells is affected by several factors that differ among animal groups. Each hair cell in the cochlea may be tuned by either mechanical or channel properties to respond maximally to a particular band of sound frequency. Because of their

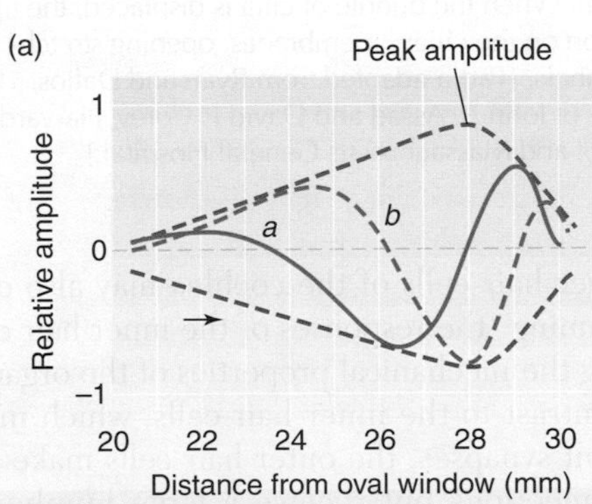

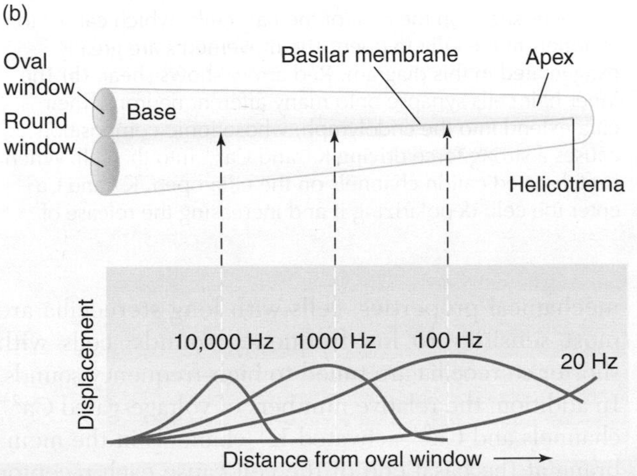

Classic Work

*Sound sets up traveling waves along the basilar membrane. **(a)** Waves caused by a continuous pure tone move in the direction shown by the arrow. Lines a and b indicate the shape of the membrane caused by the same sound at two different times. The blue dashed lines indicate the outline swept out by the movements of the membrane, which for this frequency have the largest amplitude near the apical end. (The amplitudes of the waves are greatly exaggerated in this figure.) **(b)** The cochlea drawn as if it were uncoiled. The pattern of sensitivity to different frequencies is shown in the graph below the diagram. [Part a adapted from von Békésy, 1960; part b adapted from Moffett et al., 1993.]*

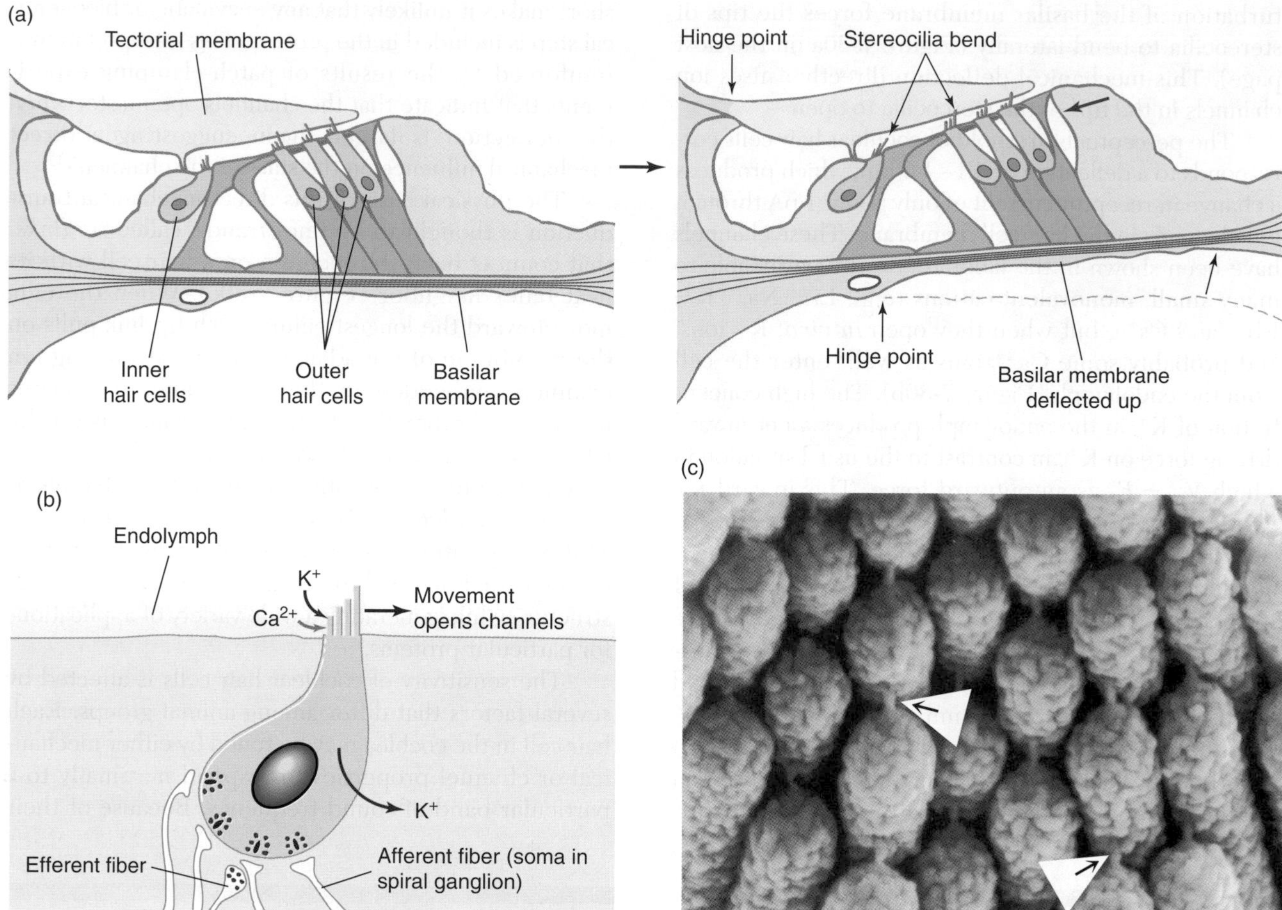

Figure 7-30 Movement of the basilar membrane produces shear on the stereocilia of cochlear hair cells. **(a)** The tectorial membrane and the basilar membrane pivot about different points when they are displaced by pressure waves traveling through the perilymph. Movement of the basilar membrane produces shear on the cilia of the hair cells, which causes ion channels in the cilia to open. The movements are greatly exaggerated in this diagram. Red arrow shows shear. **(b)** The inner hair cells synapse onto many afferent neurons. Their cilia extend into the endolymph, whose ionic composition causes a strong force driving K^+ and Ca^{2+} into the cell. When stretch-gated cation channels on the cilia open, K^+ and Ca^{2+} enter the cell, depolarizing it and increasing the release of neurotransmitter onto the afferent neurons. Each afferent neuron synapses with only one hair cell; each hair cell synapses with many afferent neurons. Efferent axons to the inner hair cells synapse with the afferent axons, not with the hair cells themselves. In contrast, outer hair cells receive synaptic input from many efferent axons. **(c)** Scanning electron micrograph of several stereocilia. Arrows point out fine strands called tip links that connect each stereocilium to the next taller adjacent cilium. When the bundle of cilia is displaced, the tip links put tension on the ciliary membranes, opening stretch-gated ion channels. [Part a adapted from Ryan and Dallos, 1996; part c courtesy of John A. Assad and David P. Corey, Harvard Medical School and Massachusetts General Hospital.]

mechanical properties, cells with long stereocilia are most sensitive to low-frequency sounds; cells with shorter stereocilia are tuned to high-frequency sounds. In addition, the relative numbers of voltage-gated Ca^{2+} channels and Ca^{2+}-activated K^+ channels in the membrane at the basal end of the cell cause each receptor cell to respond most strongly to a particular frequency of oscillatory depolarization. That is, the electrical properties of a receptor cell, as well as its mechanical characteristics, affect its sensitivity to different frequencies of stimulation. The electrical tuning of a cell can be determined experimentally if it is stimulated by oscillating electrical signals of various frequencies. The frequency to which a cell responds most strongly is called its *electrical resonance frequency.*

The outer hair cells of the cochlea may also contribute to "tuning" the responses of the inner hair cells by modifying the mechanical properties of the organ of Corti. In contrast to the inner hair cells, which make many afferent synapses, the outer hair cells make few afferent connections but receive a large number of efferent synapses. When isolated outer hair cells are stimulated electrically during experiments, they shorten when depolarized and elongate when hyperpolarized. Thus, the outer hair cells could modify mechanical stimulation of inner hair cells by actively changing the shape of the basilar membrane. Current research is exploring whether this motility of outer hair cells observed *in vitro* is important in modulating cochlear transduction.

Hair cells adapt to changes in the position of their stereocilia, a process that has been particularly well studied in the bullfrog sacculus. When the cilia of frog hair cells are deflected by a probe and held at the new position, the operating range of the cell adapts within milliseconds to this new tonic position, causing the hair cell to then respond to small changes in position away from this new set point. Calcium ions have been shown to play a pivotal role in this process, evidently by modifying how much a tip link is stretched in any particular steady position. Initially, a displacement will stretch the tip link, but adaptation decreases the stretch, resetting the system. In addition, efferent input onto a hair cell can decrease the cell's response to sound and broaden its frequency selectivity by opening inhibitory K^+ channels, which short-circuit the cell's electrical resonance properties.

Taken together, these attributes of the cochlear hair cells reveal their exquisite tuning. All of the adaptations that make hair cells so extremely sensitive to sound, however, also make them highly vulnerable to overstimulation, which can cause rupture at the base of the stereocilia. Acoustic trauma can produce permanent hearing loss that is worst at the frequencies of sound that damaged the hair cells. Although most vertebrates can recover from such trauma, the loss appears to be permanent in mammals.

The receptor currents of hair cells faithfully transduce the movements of the basilar membrane over the whole spectrum of audible sound frequencies. The hair cells transmit their excitation through chemical synapses onto sensory axons of auditory neurons that have cell bodies located in the cochlear ganglion (also called the spiral ganglion because the cell bodies are distributed along the coiled cochlea). Release of neurotransmitter by the hair cells modulates the firing rate of these neurons, whose axons travel in the VIIIth cranial nerve and synapse onto neurons in the cochlear nucleus of the brain stem. The inner hair cells receive about 90% of the contacts made by neurons of the cochlear ganglion, which suggests that they are largely responsible for detecting sound. For sounds up to about 1 kHz, APs in the auditory sensory axons appear to follow the fundamental frequency of the sound. Above 1 kHz, the time constant of the hair cells and the electrical properties of the axons in the auditory nerve prevent a one-to-one correspondence between sound waves and electrical signals. In this higher frequency range, some other mechanism must inform the central nervous system of the sound frequency.

We have discussed a variety of sensory functions carried out by hair cells in vertebrates, but the range of information provided by these cells is broader still. In the evolution of particular species of fishes, hair cells have lost their cilia entirely and acquired an even more specialized function, that of detecting very small electric fields in the environment (Spotlight 7-2 on the next page). Although the stimulus that excites these cells is unusual, the physiology of the cells and their relationship to the VIIIth cranial nerve closely resemble the properties of more conventional hair cells.

An Insect Ear

Many organisms have ears that operate differently from the mammalian ear, and it is instructive to consider at least one of them to illustrate the variations that have arisen in the evolution of audition. Crickets find their mates through auditory communication: male crickets produce a song whose pattern is specific to their species, and female crickets are attracted by the song of their species. Cricket ears are located on the first thoracic legs, and they are associated with the respiratory passages, called tracheae. Such tympanal organs of hearing (that is, those that are based on the movement of a tympanal membrane) are found at an amazing number of locations when a variety of insects are considered (Figure 7-31). Each ear includes a tympanum, analogous in function to the tympanic membrane of the mammalian ear. The changes in air pressure that constitute sound cause the tympanum to vibrate, just as they cause the tympanic membrane of the mammalian ear to vibrate. In an insect, however, the tympanum can be exposed to changes in air pressure both directly from outside the animal and indirectly from inside the body by way of the tracheae. If a sound arises on the right side of the cricket, it will directly cause the tympanum on the right side of the insect to vibrate. In addition, the sound can be carried through the tracheal system to the left tympanum,

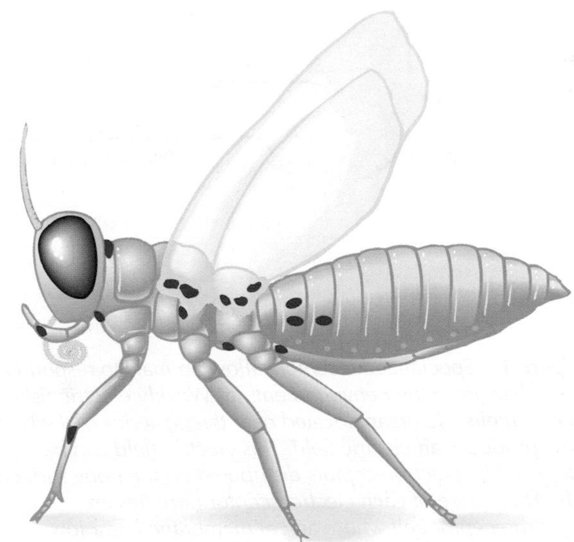

Figure 7-31 Insect tympanal ears are found at a variety of different locations. This diagram shows a typical insect body. The many locations at which ears are found in a variety of insect species are pointed out by red spots. (Typically, a particular insect species will have ears in only one of these locations.) The tympanum is located on the body surface and overlies an expanded air-filled tracheal tube. A hair-cell-based chordotonal organ is associated with the tympanum. Vibrations of the tympanum stimulate the hair cells, sending signals to the central nervous system. [Adapted from Hoy and Robert, 1996.]

Spotlight 7-2 UNUSUALLY SENSITIVE SENSORY RECEPTORS

Although most sensory receptors are quite sensitive to the signals they are specialized to receive, some types of receptors have evolved the ability to detect remarkably small changes in stimulus energy. Here we consider two examples of such highly sensitive receptors, which are found in a relatively small number of vertebrate species.

Electroreception

Hair cells located in the skin of some bony and cartilaginous fishes have lost their cilia in the course of evolution and become modified for the detection of electric currents in the surrounding water. The sources of these currents are either the fishes themselves or currents that originate in the electrically active tissues of other animals in the vicinity. These weakly

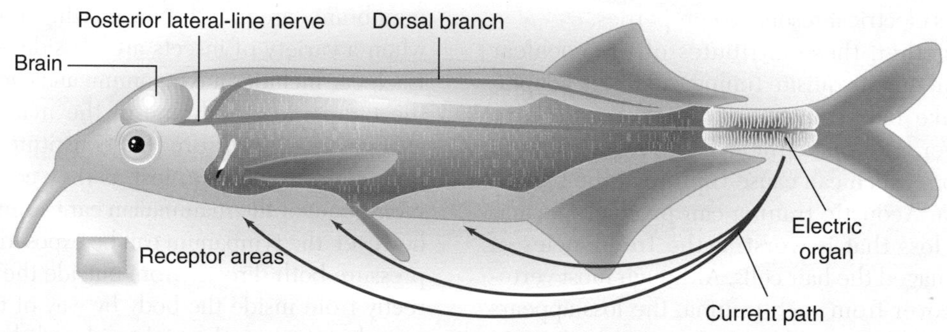

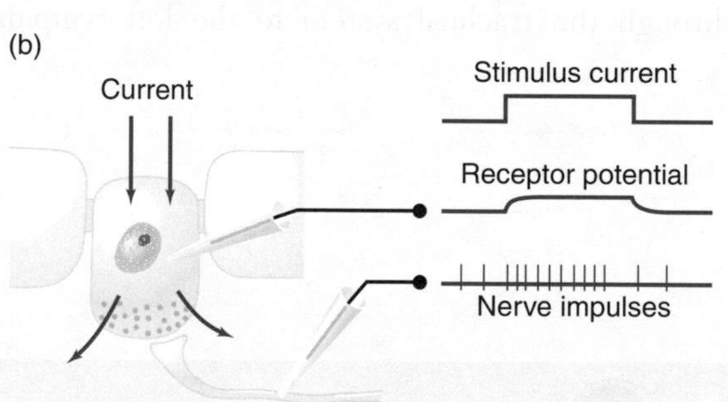

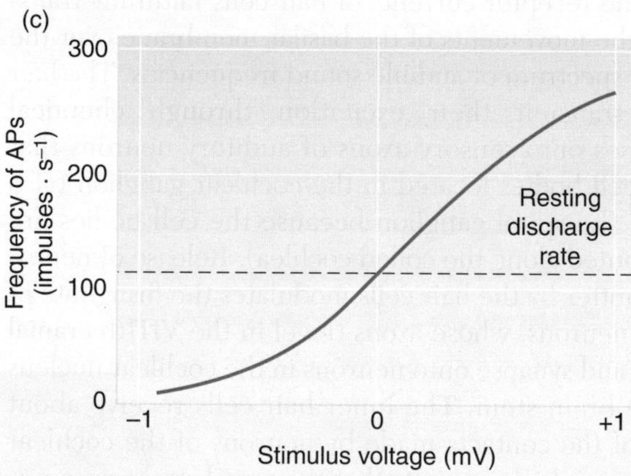

Figure 1 *Specialized receptors allow animals to respond to tiny changes in their environment.* ***(a)*** *Weakly electric fishes have an electric organ, located near the posterior of the body, that produces an electric field. This electric field can be detected by electroreceptors distributed on the body surface.* ***(b)*** *At the base of each electroreceptor pore lies an electroreceptor cell whose apical membrane has a low electrical resistance compared with that of its basal membrane. These receptor cells release transmitter molecules spontaneously. Current entering the cell depolarizes it, increasing the rate of transmitter release and hence the frequency of APs in the primary afferent neuron that contacts the cell. Current leaving the cell decreases the rate of transmitter release. The amount of transmitter released by the receptor cell changes when V_m is altered by only a few microvolts.*

Classic Work ***(c)*** *Relation between stimulating voltage of an electroreceptor cell and the frequency of APs in the primary afferent axon of a neuron that receives input from an electroreceptor. [Parts b and c adapted from Bennett, 1968.]*

causing the left tympanum to vibrate as well. The tympanum is associated with a modified chordotonal organ, a hair-cell-based mechanoreceptive organ that more typically delivers information about the position of a joint. Thus, excitation within the tympanal insect ear may be similar to excitation in the mammalian ear, even though the two kinds of ears have very different anatomies. The insect ear incorporates some of the same features as the

electric fishes possess specialized electric organs, made up of modified muscle or nerve tissue, that generate fields that can be sensed by these receptors. They use the fields for communicating with one another and for navigating in turbid water. In fact, all electrically active tissues, such as neurons and muscles, generate electric fields, and some sharks are especially adept at locating their prey by sensing the electric currents emanating from the active muscles of the animal. The electroreceptors of weakly electric fishes are distributed widely over the head and body (Figure 1a).

In weakly electric fishes (which are distinct from strongly electric fishes, such as the electric eel), electric pulses produced by the electric organ at one end of the body reenter the fish through epithelial pores in the lateral-line system. At the base of each pore, this inward-flowing current encounters an electroreceptor cell that makes synaptic contact with axons running in the VIIIth cranial nerve, which innervates the lateral-line system (Figure 1b). The apical membrane of the receptor cell (the part facing the exterior) has a lower electrical resistance than does the basal membrane, so most of the change in voltage caused by electric current moving across the cell is localized across the basal membrane, depolarizing it. Depolarization of the basal membrane activates voltage-gated Ca^{2+} channels in the membrane, and the resulting influx of Ca^{2+} increases the release of transmitter by the receptor cell. This transmitter increases the frequency of APs in the primary afferent neuron that contacts the receptor. Conversely, a current flowing outward from the pore hyperpolarizes the basal membrane of the receptor cell and decreases the release of transmitter below the spontaneous rate. Thus, the firing frequency in the primary afferent neuron goes up or down, depending on the polarization of the cell caused by the stimulating current (Figure 1c). The sensitivity of these receptors and their sensory fibers, like that of the hair cells of the vertebrate ear, is truly remarkable. As the graph shows, microvolt changes in V_m across the membrane of the receptor cell affect the frequency of APs in sensory axons of the VIIIth cranial nerve.

The train of current pulses flows through the water from the posterior to the anterior end of the fish (see Figure 1a). Any object whose conductivity differs from that of water will distort the lines of current flow, and the distortion is picked up by the electroreceptors. This sensory information is processed in the greatly enlarged cerebellum of the fish, enabling it to detect and locate objects in its immediate environment.

Extraordinary Thermoreception

Thermoreceptors can also be remarkably sensitive, as the infrared (radiant heat) detectors in the facial pits of rattlesnakes demonstrate (Figure 2a). These receptors consist of branched endings of sensory neurons with no obvious structural specializations. They appear to detect changes in tissue temperature, rather than to respond directly to the radiant energy of heat. The mechanisms by which temperature changes alter receptor output are not known. Sensory axons from the facial pit increase their firing rate transiently when the temperature inside the pit increases as little as 0.002°C, and this change in receptor firing rate can modify behavior. For example, a rattlesnake can detect the radiant heat emitted by a mouse that is 40 cm away if the body temperature of the mouse is at least 10°C above the ambient temperature. The temperature receptors lie deep within the facial pits, and this arrangement allows the snake to detect the direction of a source of radiant heat (Figure 2b). Thus the pits provide a great advantage to a snake that is hunting in poor light.

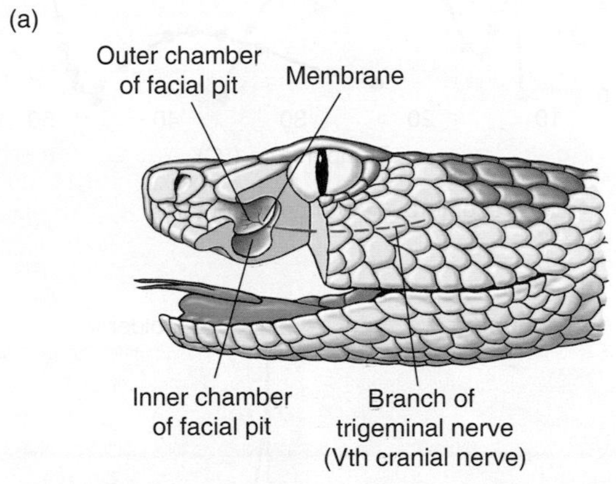

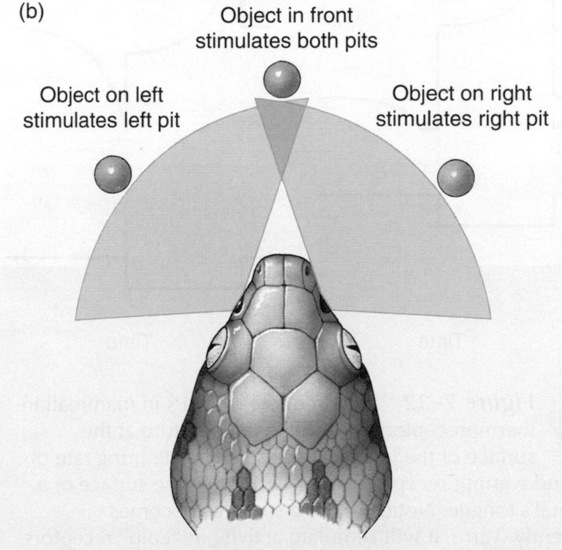

Figure 2 *The facial pits of rattlesnakes contain extraordinarily sensitive thermoreceptors.* ***(a)*** *Structure of a facial pit in the rattlesnake* Crotalus viridis. ***(b)*** *The position of the facial pits makes thermoreception directionally sensitive. [Adapted from Bullock and Diecke, 1956.]*

mammalian ear: sound pressure acts on a movable surface, which vibrates in response to the sound waves. When the tympanum vibrates, it excites receptors either directly or indirectly, sending signals to the central nervous system. In insects, however, sound travels only through air. There is no water-filled channel as there is in the vertebrate ear, which avoids the problem of an air-water acoustic impedance mismatch.

THERMORECEPTION

Temperature is an important environmental variable, and many organisms acquire sensory information regarding temperature from specialized nerve endings, or **thermoreceptors**, in the skin. Higher-order neurons in the central nervous system receive input from these cutaneous thermoreceptors and contribute to the mechanisms that regulate the temperature of the body. In addition, some neurons of the hypothalamus in the vertebrate brain detect changes in body temperature.

In mammals, both the external skin and the upper surface of the tongue contain two kinds of thermoreceptors: those that increase their firing when the skin is warmed ("warmth" receptors) and those that increase their firing when the skin is cooled ("cold" receptors). These receptors are quite sensitive to temperature changes, although they do not have the exquisite sensitivity of the thermoreceptors in the facial pits of snakes (see Spotlight 7-2). Humans, for example, can perceive a change in average skin temperature of as little as 0.01°C. The two categories of thermoreceptors are distinguished from each other by how they respond to temperature changes near the normal temperature of the human body (about 37°C). Both warmth and cold receptors increase their firing rate as the temperature becomes increasingly different from 30–35°C (Figure 7-32a): warmth receptors fire faster as the temperature gets warmer; cold receptors fire faster as the temperature gets colder. However, when the temperature becomes sufficiently different from 30–35°C, the frequency of APs drops in both types of receptors. These responses exemplify a common behavior among sensory receptors: A change in temperature in the appropriate direction elicits a large transient change in the firing rate of the receptor, followed by a longer-lasting steady-state phase. The transient phase emphasizes any *change* in temperature (Figure 7-32b), whereas the steady-state phase encodes the actual temperature. This response pattern is standard: the vast majority of sensory receptors are much more sensitive to changes in stimuli than they are to the magnitude of a sustained stimulus.

(a)

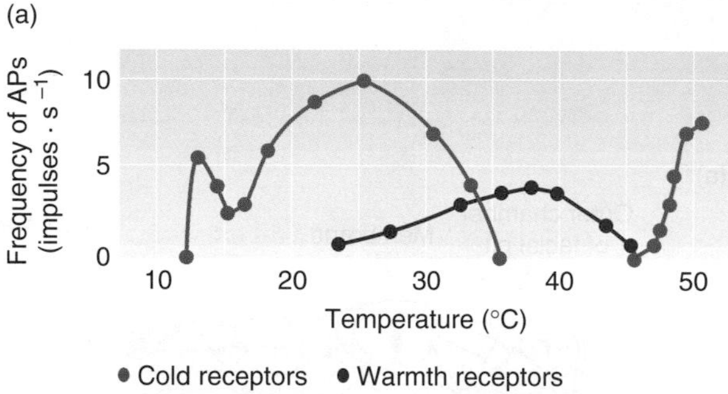

(b)

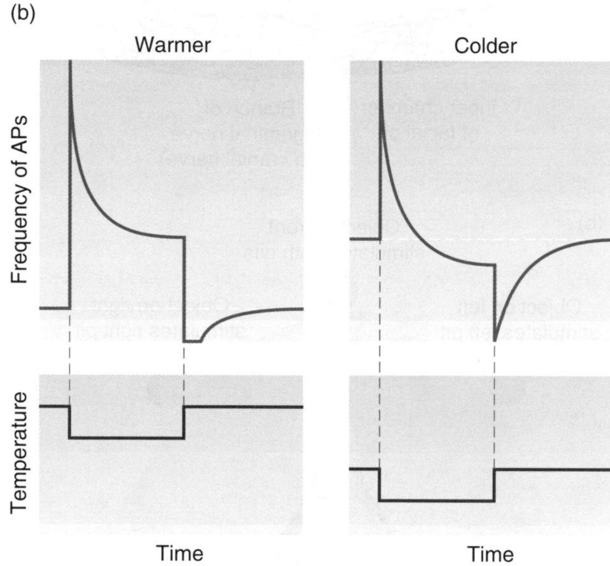

Classic Work **Figure 7-32** The frequency of APs in mammalian thermoreceptors varies with temperature at the surface of the body. **(a)** The steady-state firing rate of cold and warmth receptors that are found in the surface of a mammal's tongue. Notice that if a stimulus becomes sufficiently warm, it will stimulate activity in "cold" receptors. This observation may explain the phenomenon of "paradoxical cold"—that is, the sensation of cold that sometimes immediately follows climbing into a very hot bath. **(b)** Time course of a cold receptor's response when the tongue was first cooled and then rewarmed starting from two different temperatures, as shown by the lower red traces. This figure illustrates the receptor's response to the same amount of change in temperature beginning at two different parts of the temperature range. Notice that the large phasic response to the change in temperature has about the same magnitude independent of the actual temperature value, whereas the steady-state part of the response varies with temperature—a higher frequency of APs is associated with colder temperatures. [Adapted from Zotterman, 1959.]

VISION

Since Earth formed more than 5 billion years ago, sunlight has been an extremely potent selective force in the evolution of living organisms. Most organisms respond to light in some way. **Photoreceptors** transduce photons of light into electrical signals that can be interpreted by the nervous system. Photoreceptive organs—typically called eyes—have evolved in many shapes and sizes and with many distinct designs. Interestingly, although the physical structure of eyes varies greatly among species, visual transduction is based on highly conserved protein molecules that capture photons reaching the photoreceptors.

This conservation of visual molecules suggests that once suitable biochemical means had evolved to solve the problem of capturing light energy, the sequences were conserved, even though they became packaged into organs with highly diverse structures. The **opsins**, discussed earlier in this chapter, are a component of visual pigment molecules. Each opsin molecule includes seven transmembrane helices (see Figure

7-3c), placing this protein in the same family as many neurotransmitter and hormonal receptors. Opsin molecules are coupled to a particular light-absorbing organic molecule, **retinal**, producing visual pigments called **rhodopsins**. Rhodopsins are found widely throughout the animal kingdom, even in photoreceptive structures that are too simple to be recognized by most people as eyes. In many organisms, the structure of the eye has evolved to collect and focus incident light rays before they arrive at the site of transduction. In many eyes, light is bent, or *refracted*, by structures called *lenses*. These refractive structures also have an interesting evolutionary history because, in contrast to the extreme evolutionary conservation of visual pigments, a large number of different optical solutions were generated in the course of evolution. Let's first consider how eyes collect and focus light.

Optic Mechanisms: Evolution and Function

The physics of light tightly constrains the structure of an organ evolved to produce a usable image. Most of the possible designs known to physics have been "discovered" in the course of evolution, giving rise to similar structures in phylogenetically unrelated animals. One of the best-known examples of convergent evolution is the close similarity of eyes in the phylogenetically unrelated squids (which are cephalopod mollusks) and fishes (which are vertebrates). This similarity strongly suggests that optical laws have dictated convergent solutions to the problem of seeing under water. In contrast, the eyes of humans and fishes are similar because they share common evolutionary descent, although they differ to some extent because the two species live in different optical media.

The evolution of eyes has proceeded in two stages. Virtually all major animal groups have evolved simple eyespots consisting of a few receptors in an open cup of screening pigment cells (Figure 7-33a). Biologists estimate that such photon detectors have evolved independently between 40 and 65 times. Eyespots provide information about the surrounding distribution of light and dark, but they do not provide enough information to allow an animal to distinguish either predators or prey. For pattern recognition or for controlling locomotion, animals need eyes with an optical system that can restrict the light acceptance angle of individual receptors and form some kind of image. This stage of optical evolution has happened less frequently. Image-forming eyes are found in only 6 of the 33 metazoan phyla (Cnidaria, Mollusca, Annelida, Onychophora, Arthropoda, and Chordata). However, these phyla account for about 96% of all extant species, so it is tempting to speculate that the possession of image-forming eyes confers significant selective benefits.

Ten optically distinct designs for image-forming eyes have been described to date. They include nearly all the possibilities known from physical optics except Fresnel lenses and zoom lenses. Simple eyespots (see Figure 7-33a) are typically less than 100 μm in diameter and contain between 1 and 100 receptors. Even the simplest eyespots allow some visually guided behavior.

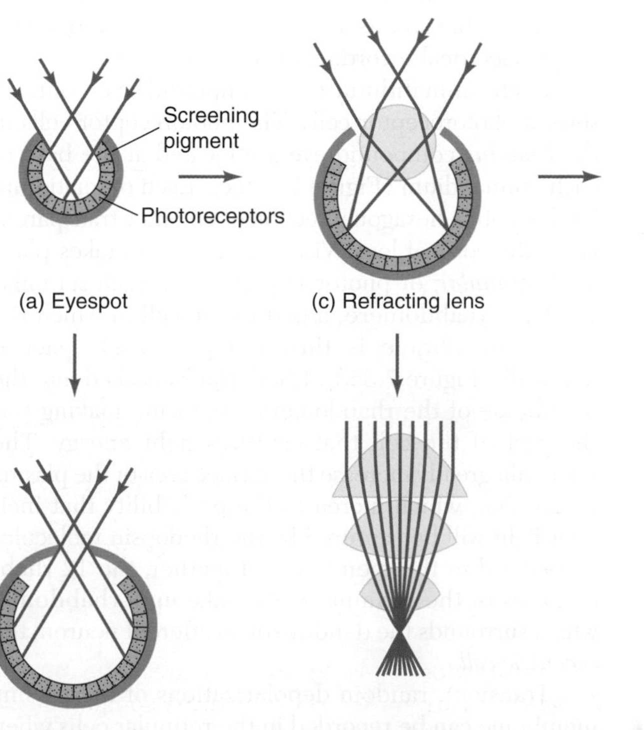

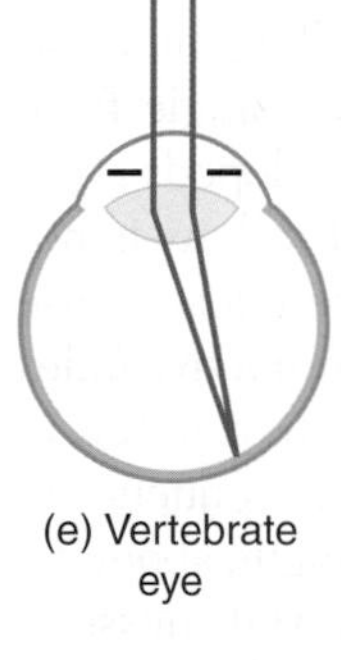

Figure 7-33 The structures of eyes incorporate many different optical principles. **(a)** The simplest eyespot consists of a shallow open pit lined with photoreceptor cells that are backed by screening pigment. **(b)** In slightly more complicated eyes, the aperture that admits light is small in proportion to the size of the eye. This type of eye operates like a pinhole camera, forming an image on the layer of photoreceptor cells. **(c)** An alternative means of improving image quality is the addition of a refracting element (or lens) between the aperture and the layer of photoreceptors. **(d)** Three lenses arranged in a series improve the optical properties of the eye in *Pontella*, a copepod. **(e)** The vertebrate eye is an elaboration of a simple eye to which a small and variable aperture *and* a lens have been added. Arrows indicate possible evolutionary relations among types of eyes. [Adapted from Land and Fernald, 1992.]

Protozoans and flatworms can detect the direction of a light source with the help of a screening pigment that casts a shadow on the photoreceptors. Some flagellates, for example, have near the base of the flagellum a light-sensitive organelle that is shielded on one side by a spot of pigment. This shielded organelle provides a crude but effective indication of directionality. As the flagellate swims along, it rotates about its longitudinal axis roughly once per second. If it enters a beam of light shining from one side and perpendicular to its path of locomotion, the eyespot is shaded each time the shielding pigment passes between the source of the light and the photosensitive part of the eyespot. Whenever the eyespot is shaded, the flagellum changes its motion just enough to turn the flagellate slightly toward the side bearing the shading pigment. The net effect is to turn the flagellate toward the source of the light, a behavior called positive phototaxis.

The simplest eyes are improvements on eyespots that have been achieved by reducing the size of the aperture that admits light (Figure 7-33b) or by adding a refracting structure, or lens (Figure 7-33c). The evolutionarily ancient cephalopod mollusk *Nautilis* has an image-forming pinhole eye that is quite advanced, except that it lacks a lens. It is nearly 1 cm in diameter, and the size of the aperture can be varied, expanding from 0.4 to 2.8 mm. In addition, extraocular muscles compensate for the rocking motion produced when the animal swims, stabilizing the eye and hence the image on the retina.

Most aquatic animals have a single-chambered eye with a spherical lens (see Figure 7-33c). This type of lens provides the high refractive power needed to focus images under water, but it introduces the problem of spherical aberration: parallel light rays falling on a uniform spherical lens fail to converge on a single focal point. The lenses found in fishes and cephalopods avoid this difficulty because the material of the lens is not homogeneous. Instead, it is progressively less dense from the center outward and has a lower refractive index toward the periphery. This pattern was first noted in 1877 by Matthiessen, who showed that a consequence of the lens density gradient is a short focal length, about 2.5 times the radius of the lens (known as Matthiessen's ratio). Such a gradient in the density of a spherical lens has evolved eight different times among aquatic animals, suggesting that it is a good solution to the problems of vision in water, and perhaps the simplest. Other aquatic species, however, have evolved eyes with multiple lenses that get around the same problems in a different way. The eye of the copepod *Pontella* (Figure 7-33d), for example, contains three lenses in a series that together provide the necessary refraction.

The terrestrial vertebrate eye combines a relatively small aperture with a refractive lens (Figure 7-33e). These two features together provide a high-quality image that is focused on the layer of photoreceptors in the retina, located at the back of the eye.

Compound Eyes

The compound eyes of arthropods are image-forming eyes composed of many units, each of which has the features of the eye shown in Figure 7-33c. Each optic unit, called an **ommatidium**, is aimed at a different part of the visual field (Figure 7-34a), and each samples an angular cone-shaped portion of the environment, taking in about 2–3° of the visual field. In contrast, in the vertebrate eye, each receptor may sample as little as 0.02° of the visual field (Figure 7-34b). In addition, the vertebrate lens inverts the image that falls on the retina (illustrated by the image of the arrow in the figure). Because the receptive field of each ommatidium is relatively large, compound eyes have lower visual acuity than do vertebrate eyes. However, although the mosaic image formed by a compound eye is coarser than the image produced by a vertebrate eye (Figure 7-34, right side), it is certainly recognizable.

The eyes of *Limulus*

The eyes of the horseshoe crab *Limulus polyphemus* have taught physiologists an enormous amount about how vision takes place. *Limulus* has paired lateral compound eyes as well as five simple eyes: medial and lateral pairs on the dorsal surface (Figure 7-35a on page 274) and a single unpaired simple eye on the ventral surface. The lateral compound eyes of *Limulus* are typical compound eyes, whereas the simple eyes are similar in structure to the eyespot shown in Figure 7-33a. Most of the early electrical recordings of activity in single visual units were made in the lateral compound eye of *Limulus* because the eye was experimentally accessible and its activity could be monitored with simple electrical recording techniques.

Each ommatidium of a compound eye contains several photoreceptor cells. The photoreceptor cells of the *Limulus* compound eye are located at the base of each ommatidium (Figure 7-35b,c). Each ommatidium lies beneath a hexagonal section of an outer transparent layer, the corneal lens. Visual transduction takes place in 12 *retinular*, or photoreceptor, cells. Each retinular cell has a **rhabdomere**, a part of the cell in which the plasma membrane is thrown into densely packed microvilli (Figure 7-35d). Rhodopsin is packed into the membrane of the rhabdomeric microvilli, making this the part of the cell that captures light energy. The microvilli greatly increase the surface area of the plasma membrane, which increases the probability that incident light will be captured by the rhodopsin molecules embedded in the membrane. Together, the 12 rhabdomeres of the retinular cells make up a **rhabdome**, which surrounds the dendrite of an afferent neuron, the *eccentric cell.*

Transient, random depolarizations of the plasma membrane can be recorded in the retinular cells when the eye is exposed to very dim, steady illumination. These "quantum bumps" in the recording increase in

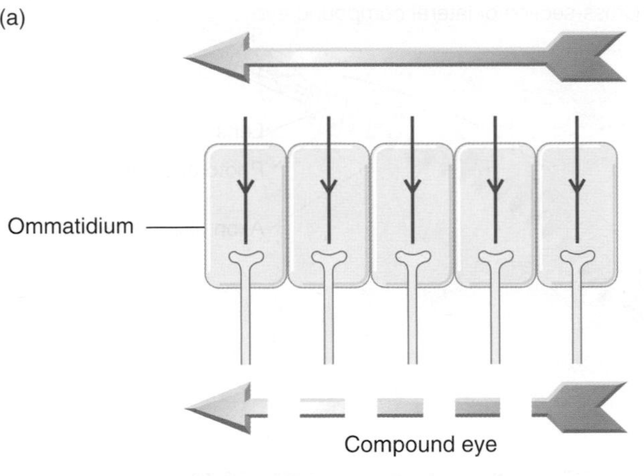

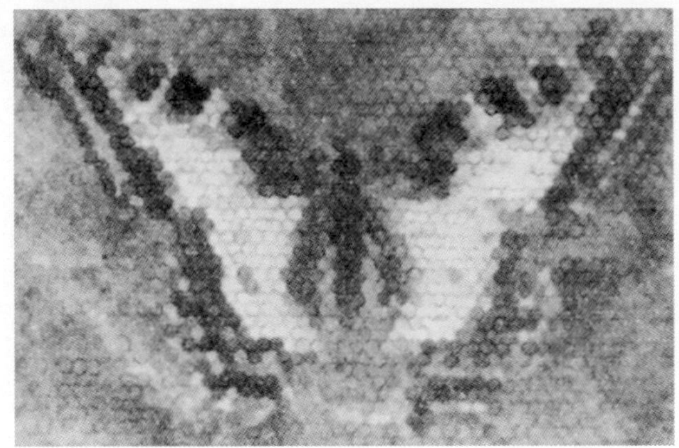

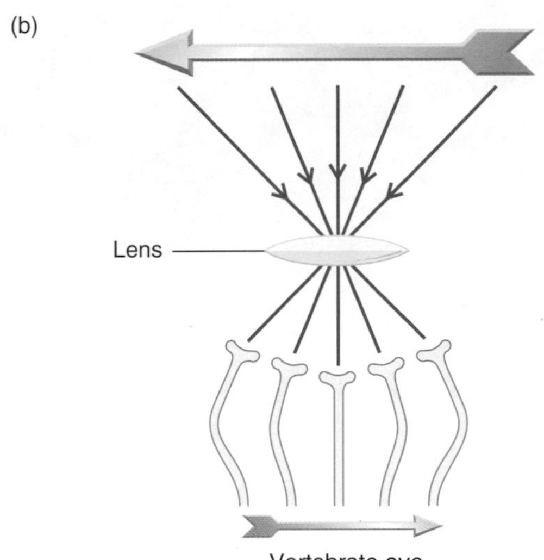

Figure 7-34 Compound eyes produce mosaic images, whereas the optics of simple vertebrate eyes produce higher spatial resolution. **(a)** (Left) In a compound eye, each ommatidium samples a different part of the visual field through its own separate lens. (Right) The image of a butterfly as it might be perceived by a dragonfly through its compound eyes at a distance of 10 cm. **(b)** (Left) In a vertebrate eye, each receptor cell samples a small part of the visual field through a lens that is shared by all receptor cells. (Right) The same butterfly as it might be perceived by the vertebrate eye. Arrows show that the optics of a vertebrate eye invert the image on the retina, whereas the optics of a compound eye do not. [Adapted from Kirschfeld, 1971, and Mazokhin-Porshnyakov, 1969.]

frequency as the light intensity is gradually increased (i.e., as more photons impinge on the receptors). The transient depolarizations are electrical signals generated as the result of the absorption of individual quanta of light by individual visual pigment molecules. In the eyes of *Limulus,* a single photon captured by a single visual pigment molecule produces a receptor current of 10^{-9} A (one nanoampere). This transduction event amplifies the energy of the absorbed photon between 10^5 and 10^6 times. How can capture of a single photon lead to the rapid release of so much energy? In this case, the amplification occurs through a cascade of chemical reactions inside the cell that includes G-protein activation. The net effect is to open ion channels, allowing cations to enter the cell. In *Limulus,* the receptor current through the light-activated channels is carried by Na^+, K^+, and some Ca^{2+}. This current causes a depolarizing receptor potential by a mechanism similar to the one that generates a depolarizing postsynaptic potential when acetylcholine activates the motor endplate channels in muscle (see Chapter 6). When the light goes off, these channels close again, and the membrane repolarizes. The sensitivity of individual photoreceptors drops with exposure to light. This light adaptation is thought to be mediated by Ca^{2+} ions, which enter the cells when light causes ion channels to open and which by some mechanism then reduce the current through light-activated channels.

Although retinular cells have axons, they apparently do not produce APs. Instead, the receptor current arising in the retinular cells spreads through low-resistance gap junctions into the dendrite of the eccentric cell, and from there the depolarization spreads to

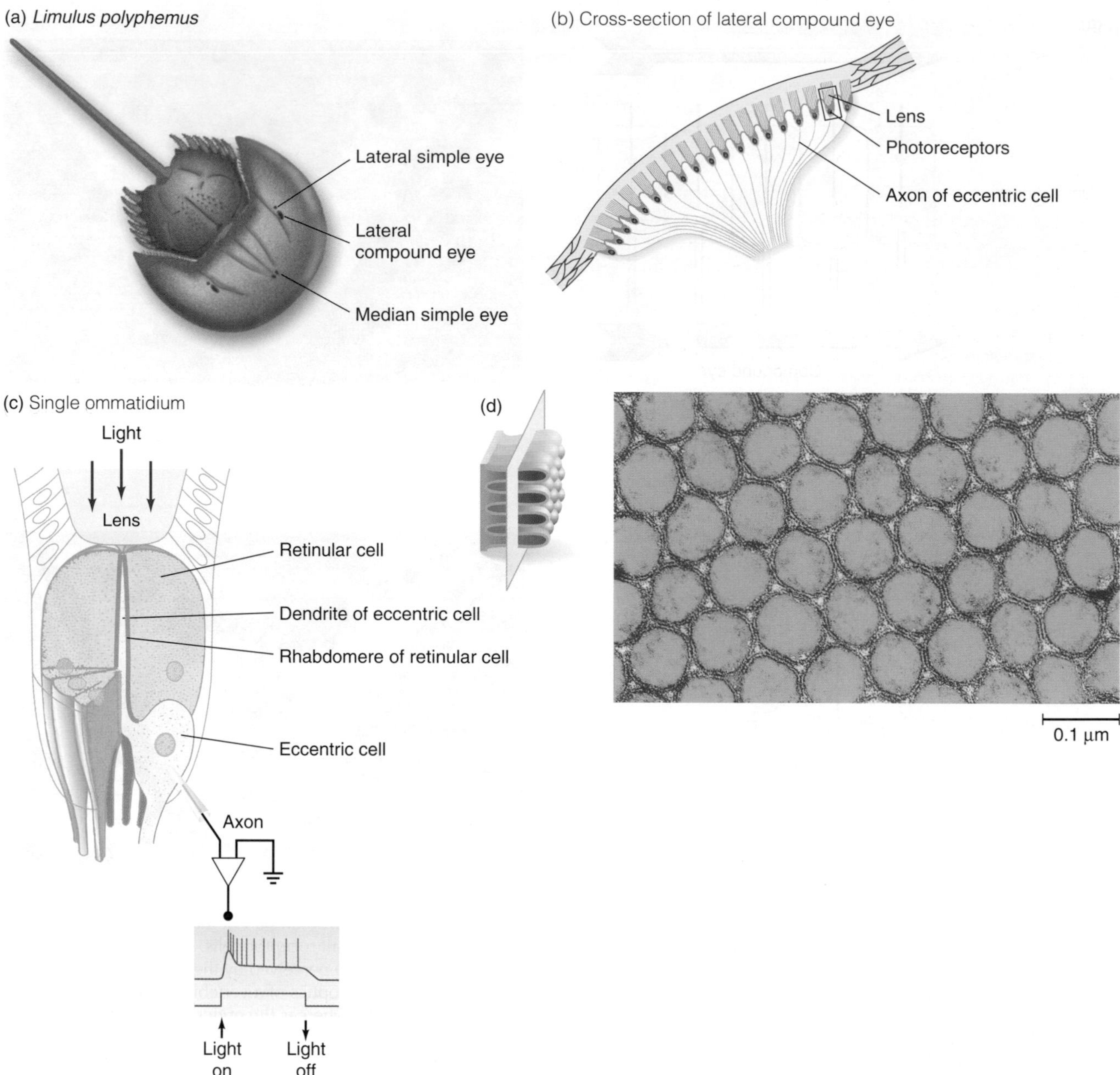

Figure 7-35 Early studies of the compound eyes of the horseshoe crab *Limulus polyphemus* provided insights into visual transduction. **(a)** The lateral compound eyes of *Limulus* are located on the dorsal carapace. Simple eyes are located posterior to the lateral compound eyes, anteriorly near the dorsal midline, and on the ventral midline. **(b)** A cross-section through a lateral compound eye, which is made up of many ommatidia. Individual ommatidia are red; the lens of each ommatidium is blue. **(c)** The structure of a single ommatidium. Light enters through the lens and may be absorbed by visual pigment molecules in the rhabdomeric membrane of the retinular cells. The cells are arranged like the segments of an orange around the dendrite of a single neuron, the eccentric cell. Capture of light by rhodopsin molecules in the rhabdomeres causes the receptor cell and then the eccentric cell to depolarize, and the eccentric cell may then generate APs. **(d)** An electron micrograph of a cross-section through the microvilli of a rhabdomere shows how tightly packed these rhodopsin-containing structures are. The illustration on the left shows the plane of section. [Part c from "How Cells Receive Stimuli," by W. H. Miller, F. Ratliff, and H. K. Hartline. Copyright © 1961 by Scientific American, Inc. All rights reserved. Part d courtesy of A. Lasansky.]

the spike-initiating zone of the eccentric cell axon, where it generates APs. The APs are conducted along the optic nerve to the central nervous system. Although the organization of the *Limulus* eye is simple in comparison with that of vertebrate eyes, the *Limulus* visual system is capable of generating electrical activity that parallels some of the more sophisticated features of human visual perception (Spotlight 7-3).

SUBJECTIVE CORRELATES OF PRIMARY PHOTORESPONSES

Studies on the *Limulus* eye carried out by H. Keffer Hartline and his associates in the 1930s revealed correlations between activity in the photoreceptors and the properties of the stimuli that caused the receptors to respond (called effective stimuli). Although the *Limulus* photoreceptors may appear to be quite different from human photoreceptors, they are similar in fundamental ways, such as the chemical identity of the visual pigment and some electrical properties of the cells. One of the most interesting results of Hartline's work was the finding that many features of human visual perception, as measured in psychophysical experiments, parallel the electrical behavior of single *Limulus* ommatidia. This relation suggests that some properties of visual perception, which might seem to depend on higher centers in the brain, instead originate in the behavior of photoreceptor cells themselves and remain relatively unmodified as information undergoes further processing by the nervous system. Several findings support this conclusion:

1. The frequency of APs recorded from the axons of single ommatidia is proportional to the logarithm of the intensity of the stimulating light. This logarithmic relation is typical of judgments made by a human subject who is asked to compare the intensities of different light stimuli.

2. A receptor's response to flashes of light that are less than 1 second in duration is proportional to the total number of photons in the flash, regardless of the actual duration (part a of the accompanying figure). That is, the number of APs generated remains constant as long as the product of intensity and duration is kept constant. This result might be expected, because the response should be determined—within some limits—by the number of photopigment molecules that absorb light. For such short flashes, a human observer cannot detect a difference between flashes when the intensity and duration of the flashes are changed reciprocally.

3. If a photoreceptor is stimulated with a flickering light, the changes in V_m follow the frequency of the flashes up to nearly 10 Hz (part b of the figure). Beyond this frequency, the receptor potential can no longer follow the flashes; instead, the ripples in V_m fuse into a steady level of depolarization. Action potentials in sensory fibers no longer follow the patterning of the flashes, but instead are generated at a steady rate. When the patterning of APs no longer conforms to the frequency of flicker, the message sent to the central nervous system indicates that the light is constant, even though the actual stimulus is not. In fact, humans cannot tell the difference between a steady light and one that flickers at a rate higher than the frequency that can no longer be encoded by receptors. The lowest frequency at which flickering lights produce constant stimulation of visual sensory fibers is called the *critical fusion frequency*. A standard incandescent light bulb flickers at 60 Hz, for example, but to us it appears as a constant light source. This characteristic of photoreceptors is very important to the film and television industries.

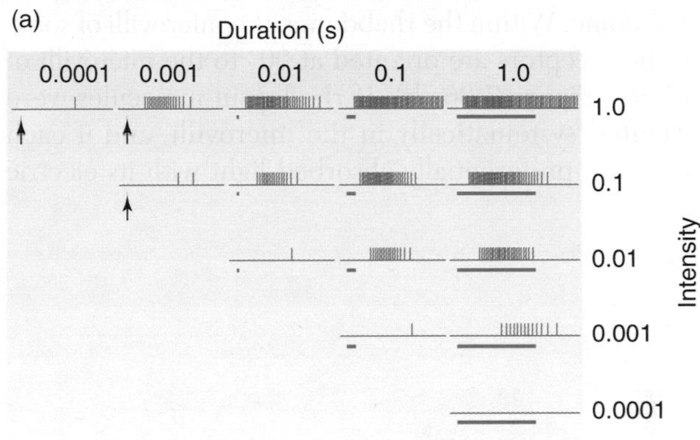

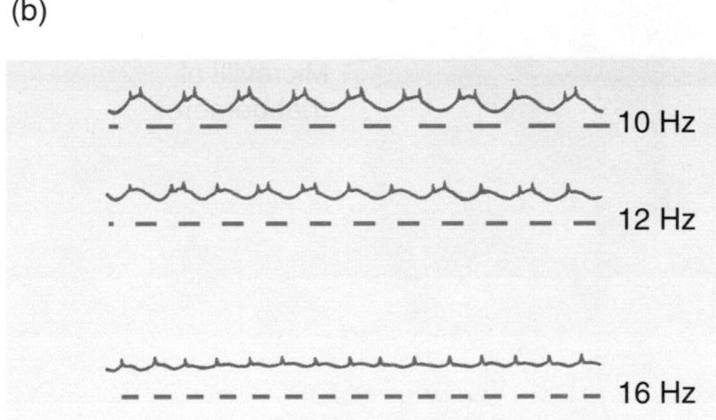

Classic Work *The responses of* Limulus *photoreceptors are similar to psychophysical properties of human vision.* ***(a)*** *When light flashes are shorter than 1 second, the product of intensity and duration determines the number of APs produced by a* Limulus *ommatidium. Short, bright flashes produce a response that is indistinguishable from the response to a dimmer, but longer, stimulus.* ***(b)*** *Lights flickering above a certain frequency cannot be distinguished from constant illumination. The on-off pattern of the stimulus is shown under the response recorded from a* Limulus *photoreceptor. At 10 Hz the photoreceptor follows the flicker faithfully, at 12 Hz the photoreceptor becomes less accurate in reporting the flicker, and at 16 Hz the response of the photoreceptor is continuous. [Part a adapted from Hartline, 1934; part b from "How Cells Receive Stimuli," by W. H. Miller, F. Ratliff, and H. K. Hartline. Copyright © 1961 by Scientific American, Inc. All rights reserved.]*

Perceiving the plane of polarized light

The arrangement of cells within the ommatidia of compound eyes confers special abilities on some arthropods. Some insects and crustaceans, for example, can orient behaviorally with respect to the position of the sun even when the sun itself is blocked from their view. This ability depends on the polarization of sunlight, which is different in different parts of the sky. Many arthropods can detect the plane of the electric vector of polarized light entering the eye, and some use this information for orientation and navigation. Measurements of the *birefringence* (the ability of a substance to absorb light polarized in various planes) of crayfish retinular cells show that the absorption of polarized light is maximal when the plane of the electric vector of light is parallel to the longitudinal axis of the microvilli of the rhabdomeres. Each crayfish ommatidium contains seven retinular cells, and the rhabdomeres of those seven retinular cells interdigitate, forming the rhabdome. Within the rhabdome, the microvilli of some of the receptors are oriented at 90° to the microvilli of others (Figure 7-36a,b). If rhodopsin molecules were oriented systematically in the microvilli, and if each molecule preferentially absorbed light with its electric vector parallel to the microvilli, the anatomic arrangement within the rhabdome could provide a physical basis for arthropods' demonstrated ability to detect the plane of polarized light. Consistent with this hypothesis, electrical recordings from single retinular cells in crayfishes showed that the response to a given intensity of light did indeed vary with the plane of polarization of the stimulating light (Figure 7-36c). Thus, experimental evidence confirms that crayfishes not only could, but do, use the precisely arranged architecture of the rhabdomere to acquire additional information about their environment.

The Vertebrate Eye

The eyes of some vertebrate animals (see Figure 7-33e) have structural features similar to those of a

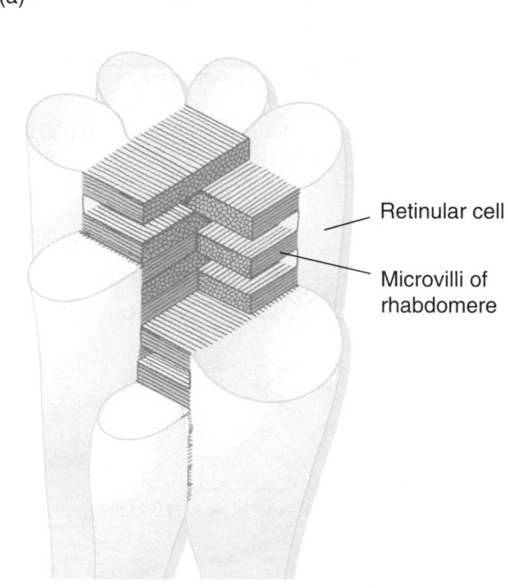

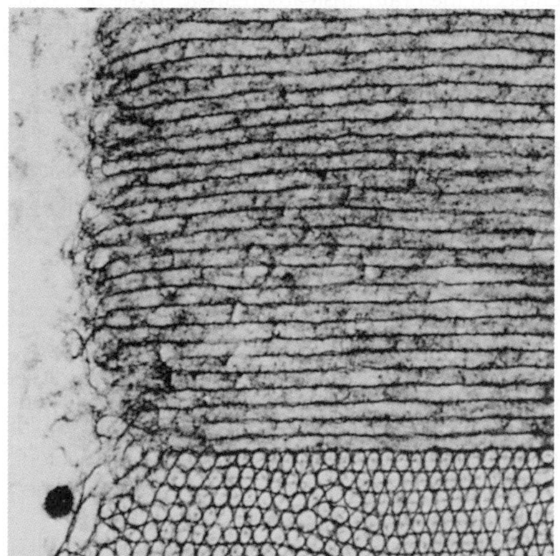

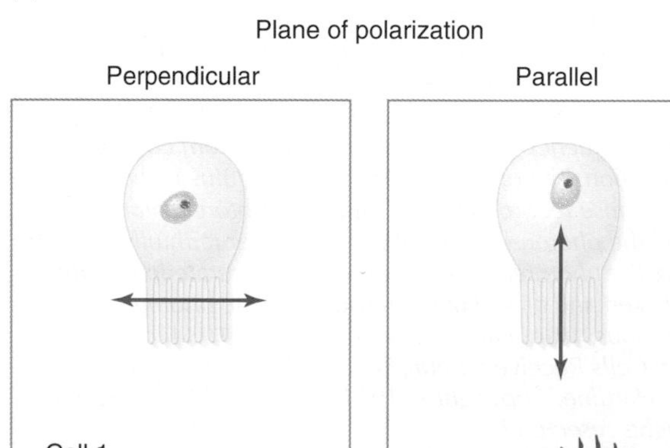

Figure 7-36 The structure of ommatidia allows some arthropods to perceive the plane of polarized light. **(a)** The interdigitating rhabdomeres of separate retinular cells produce two sets of mutually perpendicular microvilli. **(b)** Electron micrograph of a section through the rhabdome. The microvilli from one cell (in the upper part of the micrograph) were sectioned parallel to their longitudinal axis, and those from another cell (in the lower part of the image) were sectioned perpendicular to their longitudinal axis.

Classic Work **(c)** The response of crayfish photoreceptors to polarized light varies with the plane of polarization. Two cells were presented with a series of equal-energy flashes of polarized light at various wavelengths. The color of the light in each flash (identified by its wavelength in nanometers) is indicated along the lower axis. Cell 1 responded maximally to light with a wavelength of about 600 nm, cell 2 to light of 450 nm. When the plane of polarization (shown by a red arrow) was perpendicular to the microvilli (records on the left), the responses in both cells were small. Responses of both cells were enhanced when the plane of polarization was rotated to lie parallel to the microvilli (records on the right). [Part a adapted from Horridge, 1968; part b courtesy of Waterman et al., 1969; part c adapted from Waterman and Fernandez, 1970.]

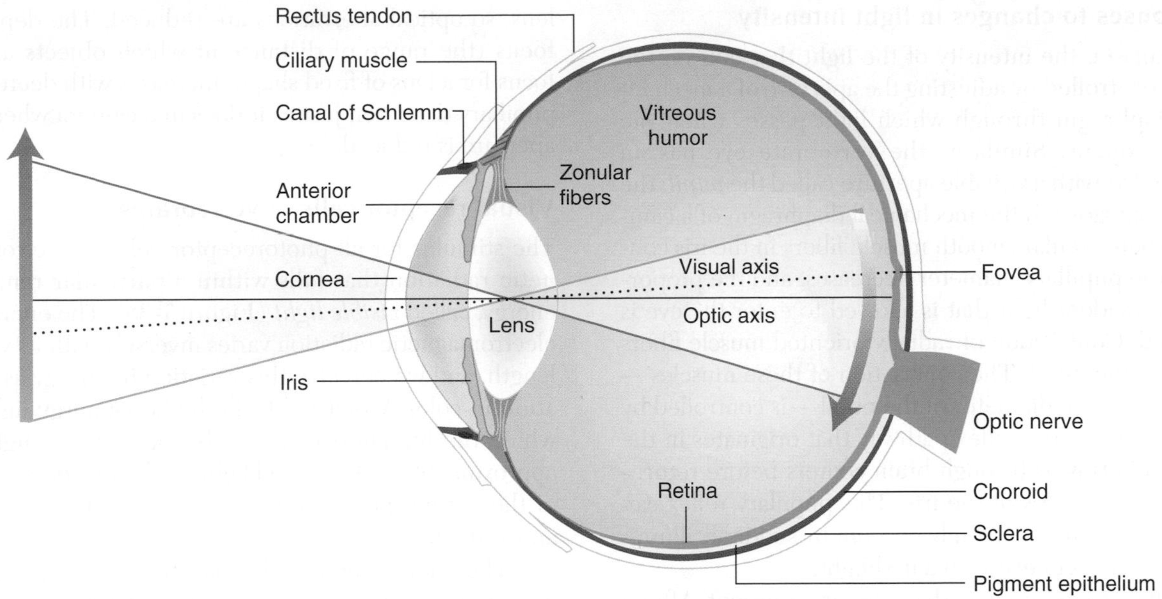

Figure 7-37 In the mammalian eye, incident light is refracted by the cornea and the lens and is focused on the photosensitive retina. In this diagram, the refraction of light has been simplified: refraction at the air-cornea interface is omitted, even though this boundary in reality provides most of the refraction. The image focused on the retina is inverted by the lens. The lens is held in place by the zonular fibers. When ciliary muscle fibers contract, tension on the zonular fibers is reduced, and the elastic properties of the lens cause it to become more rounded, shortening the focal length.

camera. In a camera, an image is focused on the film by moving the lens forward or backward along the optic axis. To bring objects that are close to the camera into focus, the lens must be positioned relatively far away from the film; to focus on distant objects, the lens must be moved closer to the film. In the terrestrial vertebrate eye, incident light is focused in two stages. In the initial stage, incident light rays are bent, or refracted, as they pass through the clear outer surface of the eye, called the **cornea** (Figure 7-37). They are further refracted as they pass through a second structure, the lens, and finally form an inverted image on the rear internal surface of the eye, the **retina.** Most of the refraction that occurs in the eye (about 85% of the total) takes place at the air-cornea interface, and the rest depends on the effect of the lens. The eyes of some bony fishes focus images on the retina like a camera by moving the lens with respect to the retina. This principle of changing the distance between the lens and the light-receptive surface has also been adapted by some invertebrates. In the eyes of jumping spiders, for example, the position of the lens is fixed, and focusing depends on moving the retina.

In contrast, neither the lens nor the retina can be moved in the eyes of higher vertebrates. Rather, images of objects at different distances from the eye are focused by changing the curvature and thickness of the lens, which changes its *focal length* (the distance at which an image passed through the lens comes into focus). The shape of the lens is changed by modifying the tension exerted on the perimeter of the lens. The lens is held in place within the eye by the radially oriented *zonular fibers* (see Figure 7-37), which exert outwardly directed tension around the perimeter of the lens. Radially arranged ciliary muscles adjust the amount of tension the zonular fibers exert on the lens. When the ciliary muscles relax, the lens is flattened by the zonular fibers, which pull the perimeter of the lens outward. In this state, objects far from the eye are focused on the retina, but near objects appear fuzzy. Objects close to the eye are brought into focus on the retina when the ciliary muscles contract, relieving some of the tension on the lens and allowing it to become more rounded. This process is called *accommodation* to close objects. The ability to accommodate decreases with age in humans as the lens becomes less elastic, producing a type of "far-sightedness" called *presbyopia.*

Perhaps the most remarkable thing about accommodation is not the mechanical mechanisms for altering the focal length of the lens, but rather the neuronal mechanisms by which a "selected" image—out of all the complexity in the visual environment—is sharply focused on the retina as a result of nerve impulses to the ciliary muscles. A related neuronal mechanism produces *binocular convergence,* in which the left and right eyes are positioned by the ocular muscles so that the images received by the two eyes fall on analogous parts of the two retinas, regardless of the distance between the object and the two eyes. When an object is close, each of the two eyes must rotate toward the middle of the nose; when an object is far away, the two eyes rotate outward from the midline.

Responses to changes in light intensity

In a camera, the intensity of the light that falls on the film is controlled by adjusting the aperture of a mechanical diaphragm through which light passes when the shutter opens. Similarly, the vertebrate eye has an opaque *iris* with a variable aperture called the *pupil;* the iris is analogous to the mechanical diaphragm of a camera. When circular smooth muscle fibers in the iris contract, the pupillary diameter decreases, and the proportion of incident light that is allowed to enter the eye is reduced. Contraction of radially oriented muscle fibers enlarges the pupil. The contraction of these muscles—and, hence, the diameter of the pupil—is controlled by a central neuronal reflex pathway that originates in the retina and travels through brain centers before returning to the muscles of the iris. This pupillary reflex can be demonstrated in a dimly lit room by suddenly illuminating a subject's eye with a flashlight.

Changes in pupillary diameter are transient. After a response to a sudden change in illumination, the pupil gradually returns after several minutes to its average size. Moreover, the area of the pupil can change only about fivefold, making it no match for the changes in light intensity normally encountered by the eye, which equal six or more orders of magnitude. Thus, although the pupil can produce rapid adjustments to moderate changes in light intensity, other mechanisms must be available. The eye copes with extremes of illumination by changes in the state of visual pigments, by the processes of sensory adaptation, and by range fractionation among receptors. Pupillary constriction provides one more advantage: the quality of the image on the retina improves. The edges of the lens are optically less perfect than the center, and when the pupil is constricted, light cannot pass through the perimeter of the lens, so optical aberrations are reduced. The depth of focus (the range of distance at which objects are in focus for a lens of fixed shape) increases with decreased pupillary diameter, just as it does in a camera when the aperture is reduced.

Visual receptor cells of vertebrates

The stimulus for all photoreceptor cells is electromagnetic radiation that falls within a particular range of energy, called *visible light* (Figure 7-38). The energy of electromagnetic radiation varies inversely with its wavelength, and we perceive this variation in energy as variation in color. Violet light, the highest-energy light to which the human eye responds, has a wavelength of approximately 400 nm. Red light, at the low-energy end of the visible spectrum, has wavelengths between 650 and 700 nm.

The photoreceptor cells that capture the energy of light and transduce it into neuronal signals are located in the retina lining the vertebrate eye. The retina in these animals contains several different types of cells that are interconnected in a network. Vertebrate photoreceptor cells fall into two classes, *rods* and *cones,* which are named for the shapes of the cells as they appear under a microscope (Figure 7-39). Rods and cones initiate the response to light, and the other types of neurons within the retina contribute to vision by processing the information acquired by the receptors and transmitting it to the CNS. In addition, the layer of epithelial cells located behind the photoreceptors plays an important role in keeping the photoreceptors responsive to light.

The physiological properties of rods and cones differ. Cones function best in bright light and provide high resolution, whereas rods function best in dim light but

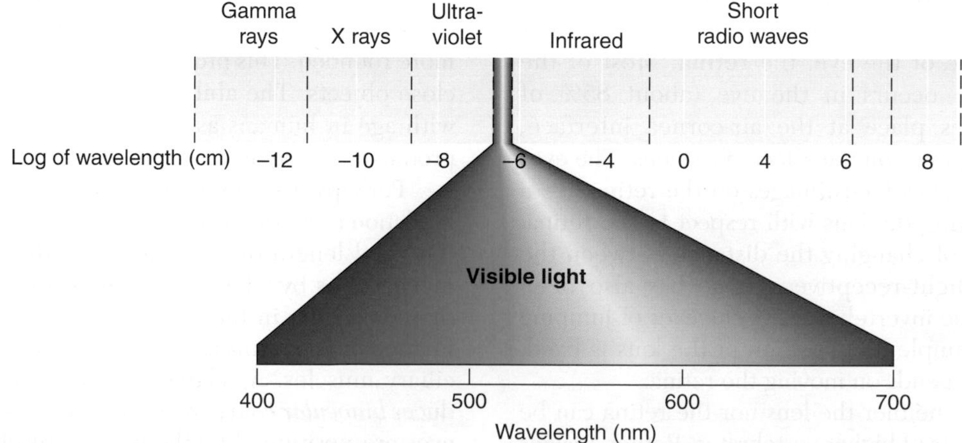

Figure 7-38 The spectrum of electromagnetic radiation encompasses a broad range of energy that is detected by various sensory modalities. Most photoreceptors detect energy in the "visible" range, shown in this diagram, but some can detect light in the ultraviolet range as well. The pit organs of some snakes can detect infrared radiation, although they appear to measure the increased temperature of surrounding tissue that has absorbed the infrared energy, rather than responding directly to the infrared radiation (see Spotlight 7-2).

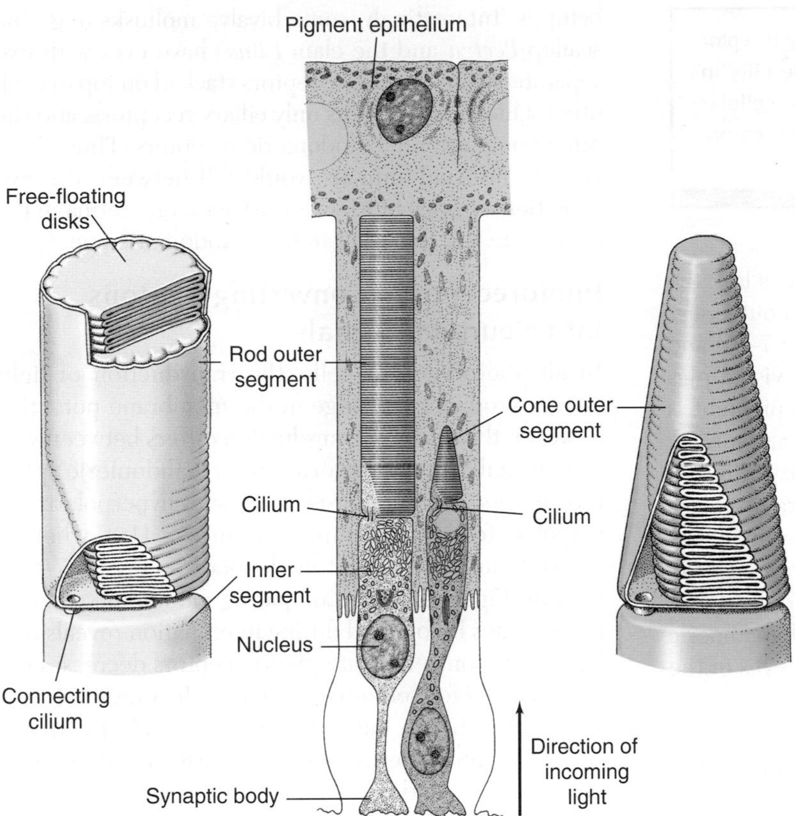

Figure 7-39 Vertebrate photoreceptors are classified as rods or cones on the basis of their morphological and physiological properties. In the retina, the photoreceptors are aligned parallel to the light pathway, with the outer segments of rods and cones, where light is captured, facing away from the source of light. The photopigment is contained in membrane lamellae, and the ends of the outer segments lie against the pigment epithelium.

provide much less resolution. These different properties are used by animals to expand their visual capabilities. Animals that live in flat, open environments (e.g., cheetahs and rabbits) usually have horizontal regions within the retina that contain a high density of cones. This concentration of cones is called a *visual streak*. Such a region corresponds to the horizon in the visual world and is thought to confer maximal resolution in this part of the scene, allowing the animal to interpret shapes on the horizon with great precision. The visual streak also contains a dense population of retinal ganglion cells—the cells that transmit visual information to the brain. In contrast, the retinas of arboreal species (and humans) typically have a radially symmetric density gradient of photoreceptors. An important feature of this kind of retina is the **fovea**, or *area centralis.* This small (about 1 mm^2) central part of many mammalian retinas provides very detailed information about a small part of the visual world, a characteristic called *high visual acuity*. In humans and some other mammals, the fovea contains only cones, whereas the remainder of the retina contains a mixture of rods and cones, with the rods significantly outnumbering the cones. In mammals, cones mediate color vision, and rods, which are more light-sensitive, mediate only achromatic vision. This distinction between rods and cones does not, however, pertain to all vertebrates. In fact, in some species, the retina contains only rod-shaped receptors, but may nonetheless be capable of color vision. Instances of different color classes among rods have been found; some frogs, for example, have two kinds of rods in addition to cones, which could provide the basis for color vision. Color vision has been demonstrated in at least some members of all classes of vertebrates.

Rods and cones are structurally and functionally rather similar to each other when they are compared with the wide variety of photoreceptors found in the invertebrates. Each vertebrate photoreceptor cell contains a segment with an internal structure similar to that of a cilium. This rudimentary cilium connects the outer segment, which contains the photoreceptive membranes, to the inner segment, which includes the nucleus, mitochondria, synaptic contacts, and so forth (see Figure 7-39). The receptor membranes of vertebrate photoreceptors consist of flattened lamellae derived from the plasma membrane near the origin of the outer segment. In the cones of mammals and some other vertebrates, the lumen of each lamella is open to the cell exterior. In rods, the lamellae pinch off completely from the plasma membrane of the outer segment and form flattened sacs, or disks, that are stacked like pieces of pita bread within the rod outer segment. The stack of disks is completely surrounded by the plasma membrane of the outer segment. Visual pigment molecules are tightly packed in the membranes of the disks. Because the visual pigment lies in the disk membranes of the rod outer segment, and not in the plasma membrane, the primary step in photochemical transduction must take place in the disk membranes, rather than in the plasma membrane.

How many different types of sensory receptor cells can you think of that include cilia in their structures? Why might this cellular arrangement be so ubiquitous among sensory receptors?

The photoreceptors of many invertebrates lack the ciliary structure that connects the inner and outer segments of vertebrate rods and cones (Figure 7-40) and the lamellae or stacks of disks containing visual pigment. Instead, as we have seen, the visual pigment is located in microvilli formed by the plasma membrane, and these pigment-containing microvilli are organized into rhabdomeres. Because many invertebrate species have simple eyes in which the photoreceptors are of the rhabdomeric type, it might be tempting to conclude that rhabdomeric photoreceptors are found only in simple eyes. However, several mollusks have rhabdomeric photoreceptors, including the very complex eyes of the octopus. Interestingly, some bivalve mollusks (e.g., the scallop *Pecten* and the clam *Lima*) have eyes with two separate layers of photoreceptors stacked on top of each other. One layer contains only ciliary receptors, and the other contains only rhabdomeric receptors. Thus, these visually unusual animals would fall between the two evolutionary lines in Figure 7-40, with one retina in the ciliary line and the other in the rhabdomeric line.

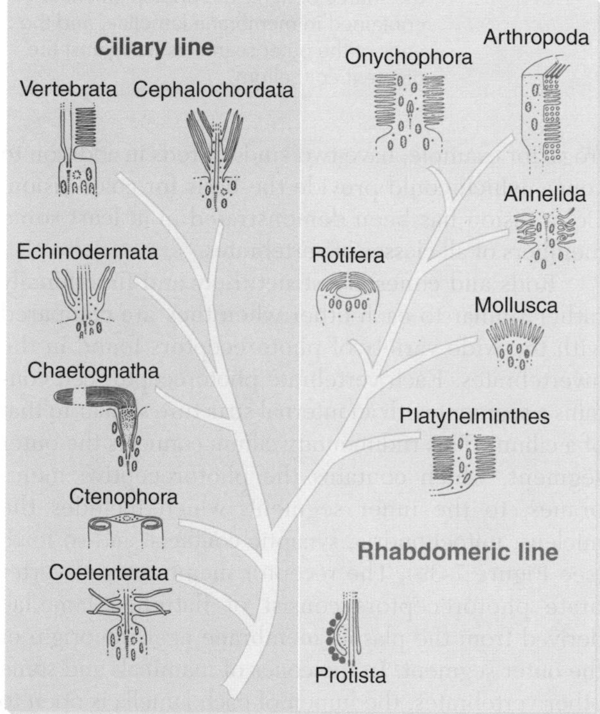

Figure 7-40 Vertebrate photoreceptors contain a cilium that connects the inner and outer segments, whereas many invertebrate photoreceptors lack this ciliary structure and instead contain many microvilli. This diagram illustrates the phylogenetic distribution of ciliary and rhabdomeric eyes. There are, however, exceptions. The scallop *Pecten* and the surf clam *Lima* (both mollusks) have complex eyes with two separate layers of photoreceptors, one containing ciliary photoreceptors, the other rhabdomeric receptors. Notice that all photoreceptors have a region in which the surface area has been greatly increased, either by the presence of microvilli or by elaborate folding of the membrane. [Adapted from Eakin, 1965.]

Photoreception: Converting Photons into Neuronal Signals

In all photoreceptor cells, the transduction of light energy produces a change in the membrane potential; however, the effect of transduction differs between vertebrate (ciliary) and invertebrate (rhabdomeric) photoreceptors. Vertebrate rods and cones hyperpolarize in response to a light stimulus (Figure 7-41a), whereas invertebrate photoreceptors depolarize (Figure 7-41b; see also Figure 7-35c). Comparing membrane conductance values before and during illumination reveals that light falling on vertebrate photoreceptors decreases the conductance for sodium, g_{Na}, across the outer segment membrane. In the dark, the plasma membrane of the vertebrate rod outer segment is nearly equally permeable to Na^+ and K^+, and V_{rest} lies about halfway between E_K and E_{Na}. (In other words, the membrane is much less inside-negative than typical neurons.) In this state, Na^+ ions enter the outer segment through channels that are steadily open (Figure 7-42a). The Na^+ ions that carry this inward current, which is called the *dark current* because it is maximal in the dark, are kept from accumulating in the cell by the steady action of the

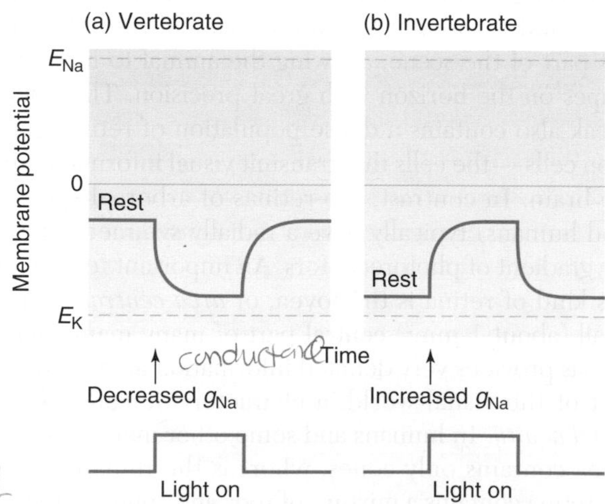

Figure 7-41 Vertebrate photoreceptors hyperpolarize in response to a stimulus, whereas most invertebrate photoreceptors depolarize. **(a)** Vertebrate photoreceptors respond to light with a decrease in g_{Na} across the plasma membrane; the residual g_K remains, and V_m shifts toward E_K. As a result, the cell hyperpolarizes. **(b)** Transduction of light energy in most invertebrate photoreceptors causes an increase in the permeability of the plasma membrane to Na^+ and K^+, depolarizing the cell.

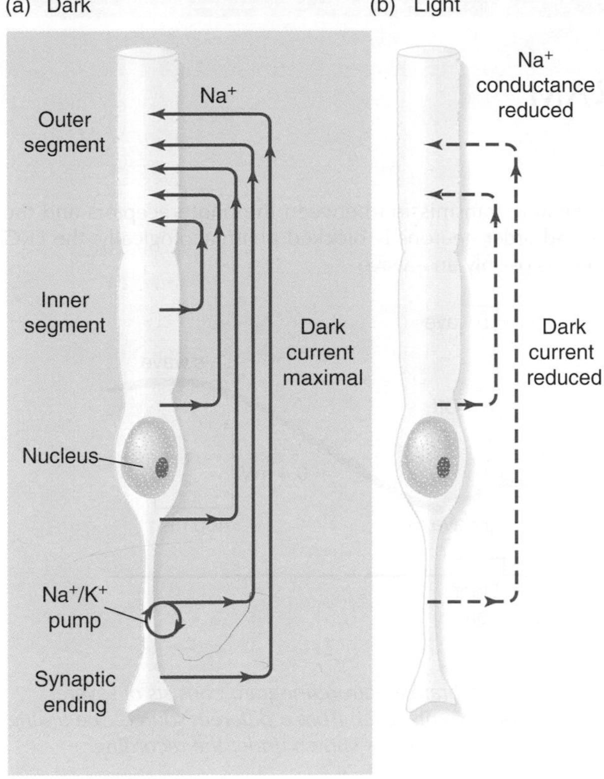

(c)

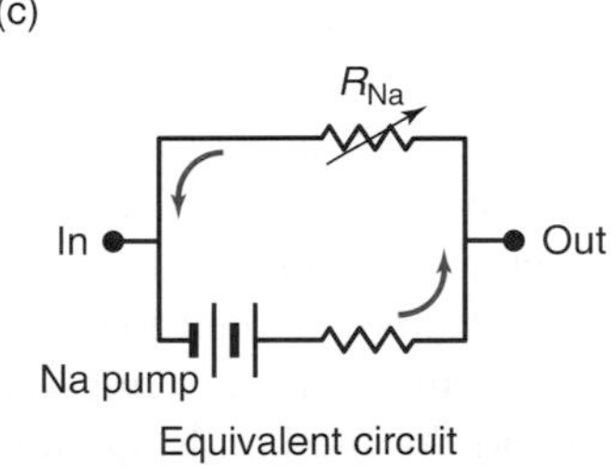

Figure 7-42 Illumination reduces the dark current in vertebrate rods and cones. This diagram illustrates the current in rods. The g_{Na} across the membrane of a rod outer segment is high in the dark **(a)** and lower in the light **(b)**. For this reason, the dark current, which is carried by Na^+ ions entering the outer segment, drops during illumination. By analogy with an equivalent circuit **(c)**, the "battery" providing the driving force for the dark current is the asymmetry of ionic concentrations maintained across the plasma membrane by the Na^+/K^+ pump. The light-inactivated variable resistor (R_{Na}) represents the g_{Na} of the outer segment. [Adapted from Hagins, 1972.]

metabolically driven Na^+/K^+ pump. A dark current is found in vertebrate photoreceptor cells, but not in typical invertebrate photoreceptors.

Like vertebrate auditory receptors, vertebrate photoreceptors lack axons. In the dark, rods and cones steadily release neurotransmitter from presynaptic sites in the basal part of the cell onto the next neurons in line. A light-induced decrease in the dark current causes V_m in the outer segment to hyperpolarize toward E_K (see Figures 7-41a and 7-42b). This change in V_m spreads electrotonically to the inner segment, where changes in V_m modulate the release of neurotransmitter. Hyperpolarization of the photoreceptors decreases the amount of transmitter released, modifying the activity of the second-order neurons. When the light stimulus stops, g_{Na} of the outer segment membrane returns to its high resting level, V_m becomes more positive, and release of transmitter increases again. It is striking that a hyperpolarization, rather than a depolarization, is produced when a vertebrate photoreceptor absorbs light, because in rhabdomeric photoreceptors and in most other sensory receptor cells, reception of a stimulus depolarizes the receptor cell.

The change in membrane potential that is produced in a cluster of photoreceptors when they are illuminated can be recorded by extracellular electrodes, as can action potentials traveling down the axons of a nerve (see Spotlight 6-2). Many photoreceptors are especially tiny cells, making intracellular recording difficult, so this method of extracellular recording—called an electroretinogram—has been extremely useful in the study of vision (Spotlight 7-4).

The cellular processes of visual transduction have received an enormous amount of research attention. Studies of photoreception have been carried out in many different species spanning several phyla, both vertebrate and invertebrate, and have revealed phylogenetic similarities and differences. As we have seen, the capture of a single photon by a photoreceptor in *Limulus* produces a peak receptor current of about 1 nA (10^{-9} amperes). In contrast, the capture of a single photon by a vertebrate rod photoreceptor changes the transmembrane current by about 1 pA (10^{-12} amperes), three orders of magnitude less. Moreover, an individual invertebrate photoreceptor may respond to light intensities spanning seven orders of magnitude, whereas a vertebrate rod responds to an intensity range of only four orders of magnitude. In addition, as we will soon see, the intracellular pathways that lead from photon capture to receptor current differ between vertebrates and invertebrates. Despite these differences in detail, all types of photoreceptors have been shaped by evolution to convert the energy of photons into neuronal energy, and studies of all types of eyes have contributed to our understanding of this process.

Visual pigments

The spectrum of electromagnetic radiation extends from gamma rays, with wavelengths as short as 10^{-12} cm, to radio waves, with wavelengths greater than 10^6 cm (see Figure 7-38). The segment of the electromagnetic spectrum with wavelengths between 10^{-8} cm and 10^{-2} cm is called "light," but only a small part of this region of the spectrum—ranging from about 400 nm to about 740 nm—is visible to humans. Below this range is the ultraviolet (UV) part of the spectrum, and above it the infrared (IR), neither of which is visible to humans or other mammals.

Spotlight 7-4

THE ELECTRORETINOGRAM

It is frequently useful to record the summed electrical activity in an eye, a procedure that is technically much less complicated than recording from single cells with microelectrodes. The recording electrode (which can be a thread or a wick that is saturated with saline) is placed on the cornea, and the reference electrode is attached to another part of the body. When a light is flashed on the eye, a complex waveform is recorded by the electrode (as shown in the accompanying figure). This recording is called an electroretinogram (ERG). It reflects the simultaneous activity of many photoreceptor cells plus other neurons in the retina.

It took several years to sort out the source of each of the components of the ERG, but we now believe that the *a* wave is due to the receptor current produced in the photoreceptor cells. The *b* wave that follows is produced by electrical activity in the second-order neurons that receive input from the receptor cells. The *c* wave is found only in vertebrates and appears to be produced by the pigment epithelial cells, against which the outer segments of the visual cells abut. In the developing eyes of tadpoles, the ERG consists only of an *a* wave before synaptic contacts are established. Similarly, in the eyes of an adult frog, if synaptic transmission between the photoreceptors and the second-order neurons is blocked pharmacologically, the ERG consists of only an *a* wave.

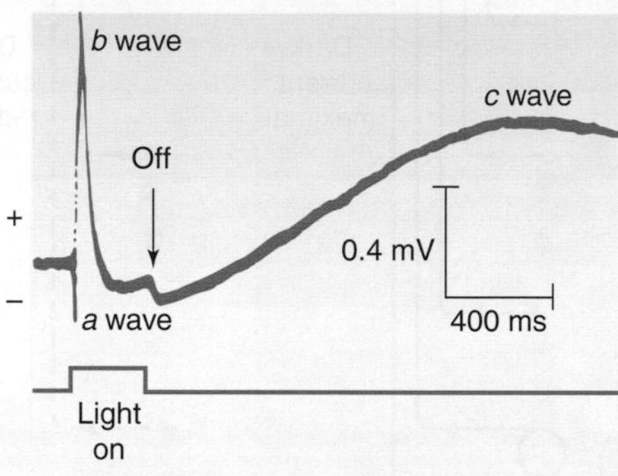

Classic Work A vertebrate electroretinogram consists of several components, each from a different source. The timing of the stimulus is shown under the recording. [Adapted from Brown, 1974.]

There is nothing qualitatively special about the parts of the spectrum that we can see. Rather, what we see depends on which wavelengths are absorbed by our visual pigments. Ordinarily, humans are unable to see light in the UV range, but that can change. Normally, the lens of the eye absorbs UV light, filtering it out before it reaches the photoreceptors. In the condition called a *cataract*, however, the lens becomes opaque. A cataract is treated by surgically removing the lens and in many cases replacing it with a synthetic lens. After this surgery, many patients can see light into the UV range. The range of the spectrum to which vertebrate visual pigments—including those of terrestrial mammals such as humans—are sensitive matches rather closely the spectrum of light that is admitted through water, which heavily filters electromagnetic radiation. It is tempting to speculate that the visual pigments of vertebrates absorb in only this limited range of sunlight's electromagnetic spectrum because vertebrate life evolved in water. In contrast to the eyes of mammals, the compound eyes of many insects can detect light into the UV range, causing some rather plain flowers containing UV-reflecting pigments to look much more spectacular to insects than they do to mammals.

Pigments are molecules that selectively absorb particular wavelengths of light while reflecting others. Light has been found to be divided into packets of energy called quanta, or photons, and the energy of a photon is directly proportional to its frequency and inversely proportional to its wavelength. Because each pigment selectively absorbs only some wavelengths, we see the wavelengths it reflects, which makes it appear colored. All known organic pigments owe their selective absorption to the presence of a carbon chain or ring that contains alternating single and double bonds. Quanta with wavelengths less than 1 nm contain so much energy that when they are absorbed they break chemical bonds or even disrupt atomic nuclei; quanta with wavelengths greater than 1000 nm lack sufficient energy to affect molecular structure. The visual pigments absorb maximally between these two limits. When a quantum of radiation is absorbed by a visual pigment molecule, it raises the energy state of the molecule by moving electrons associated with one or more conjugated double bonds into higher energy orbitals. This same process is the first step in the photosynthetic conversion of radiant energy into chemical energy by plants.

Photochemistry of visual pigments The energy content of visible light is just low enough that it can be absorbed by molecules without breaking them up. The idea that a pigment is essential for the process of absorbing light and transducing its electromagnetic energy into chemical energy originated with John W. Draper, who concluded in 1872 that to be detected, light must be absorbed by molecules in the visual system. R. Boll found soon thereafter that the characteristic reddish-purple color of the frog retina fades (we say it "bleaches") when the retina is exposed to light. The light-sensitive substance that is responsible for the reddish-purple color, rhodopsin, was extracted in 1878 by W. Kühne. Kühne also discovered that once the pigment has been bleached by light, its color can be restored by keeping the retina in the dark, provided that the receptor cells contact the pigment epithelium.

Since then, much has been learned about the chemical nature and physiological effects of rhodopsin. It absorbs light maximally at wavelengths of about 500 nm. It is found in the outer segments of rods in many vertebrate species and in the photoreceptors of many invertebrates. Rhodopsin molecules are packed at high density into receptor membranes; there may be as many as 5×10^{12} molecules per square centimeter, which is equivalent to an intermolecular spacing of about 5 nm.

All known visual pigments consist of two major components: a lipoprotein (opsin) and a light-absorbing molecule. In all instances, the light-absorbing molecule is either retinal or 3-dehydroretinal (Figure 7-43). Retinal is the aldehyde form of the carotenoid vitamin A_1, an alcohol also called retinol; 3-dehydroretinal is the aldehyde of vitamin A_2, also called 3-dehydroretinol. All visual pigments for which retinal is the light-absorbing molecule are called rhodopsins; all human visual pigments, and most other visual pigments, are rhodopsins. Visual pigments in which 3-dehydroretinal is the light-absorbing molecule are called porphyropsins. Rhodopsin contains, in addition to its major components, a six-sugar polysaccharide chain and a variable number of phospholipid molecules (as many as 30 or more). The lipoprotein opsin, which binds the phospholipids and the polysaccharide chain, is an integral membrane protein. During bleaching and regeneration of the visual pigment, the carotenoid molecules move back and forth between the photoreceptor membrane and the pigment epithelium at the back of the retina. (Incidentally, the pigment for which the pigment epithelium is named is photochemically inactive and is unrelated to the visual pigment. Instead, it keeps light from scattering and reflecting diffusely back toward the retina.)

The retinal molecule exists in two sterically distinct states in the retina. In the absence of light, the opsin and the retinal are linked covalently by a Schiff's base bond, and the retinal is in the 11-*cis* configuration. When the 11-*cis*-retinal captures a photon, it isomerizes into the all-*trans* configuration (see Figure 7-43). This *cis-trans* isomerization is light's only direct effect on the visual pigment. The conversion from 11-*cis*-retinal to all-*trans*-retinal initiates a series of changes in the relation between the retinal and the opsin protein, culminating in a change in the conformation of the opsin.

When light strikes rhodopsin, we say that the molecule has been activated. Activated rhodopsin (called metarhodopsin) activates a G protein, called *transducin* in recognition of its key role in the transduction of light (Figure 7-44a on the next page). In vertebrate photoreceptors, an activated subunit of transducin diffuses in the plane of the membrane, encountering and

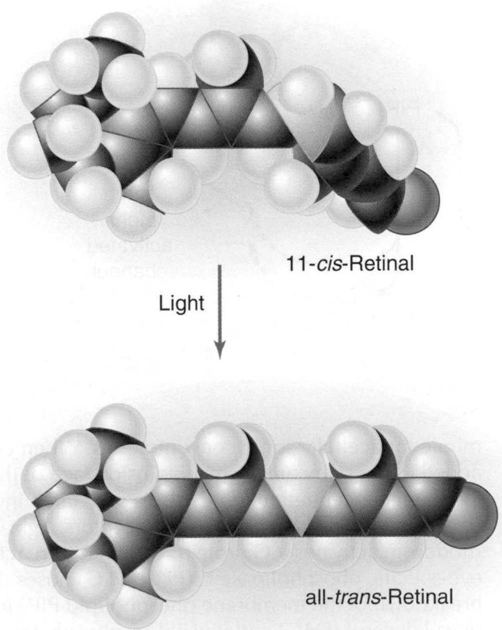

Figure 7-43 Retinal changes its steric conformation when it absorbs a photon. In the dark, the bonds of carbon 11 are arranged in the *cis* configuration. When a photon is captured, these bonds are converted into the straight, all-*trans* configuration. Both space-filling and line diagrams of the molecular structure are shown.

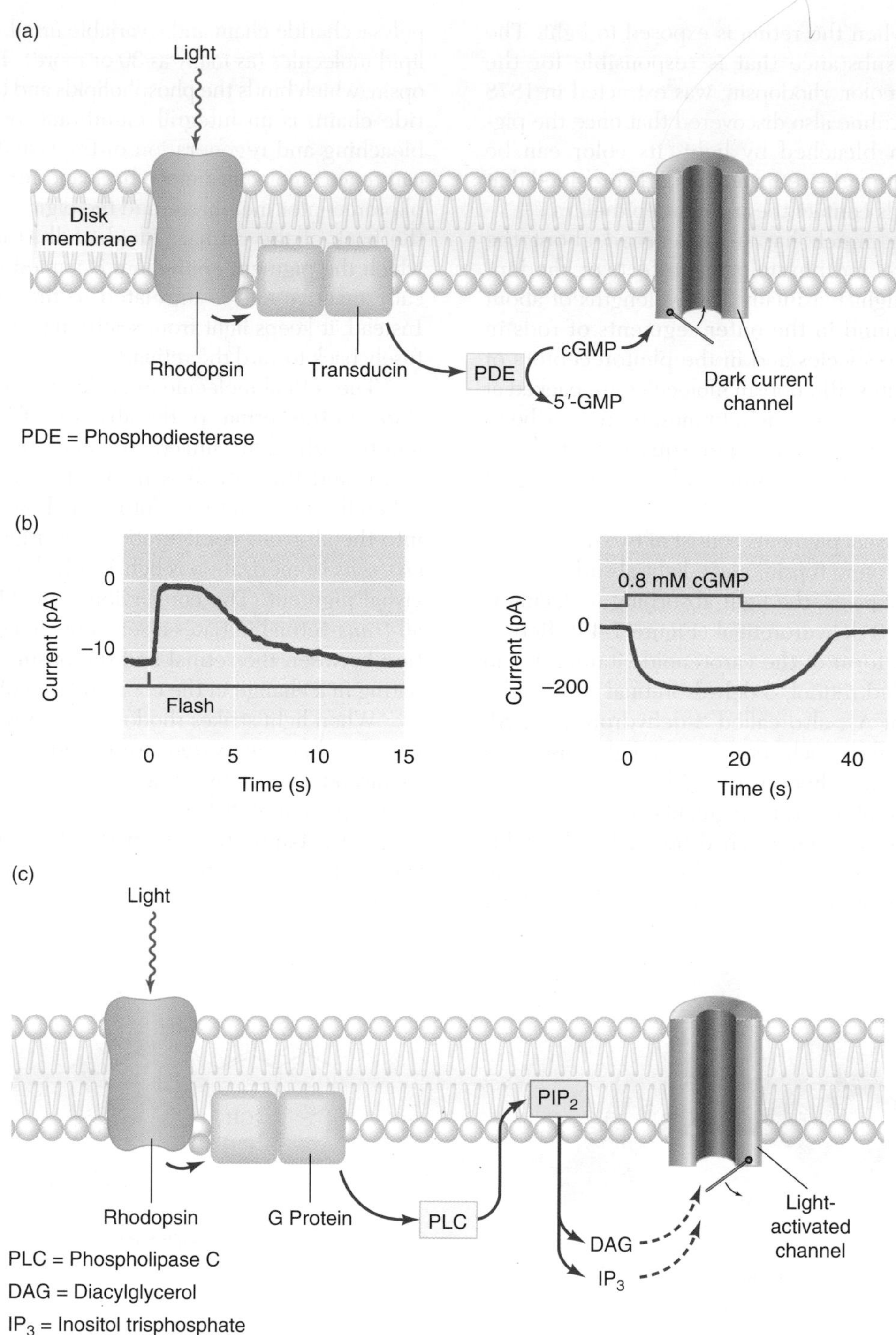

Figure 7-44 When light is absorbed by retinal, a series of reactions produces a change in photoreceptor ion channels. **(a)** In vertebrates, activated rhodopsin increases the activity of transducin, a G protein. Activated transducin then activates many phosphodiesterase (PDE) molecules, reducing the intracellular concentration of cyclic guanosine monophosphate (cGMP), which leads to the closing of the channels that carry the dark current. The receptor cell then hyperpolarizes. **(b)** Recordings of current in isolated rod photoreceptors from a toad retina. (Left) A brief flash of light causes the inward dark current of 10 pA to drop to zero. (Right) The outer segment membrane of the photoreceptor has been opened up, and the external saline has been changed to mimic intracellular ionic concentrations. When cGMP is added to the saline (exposing the inside face of the membrane to a high concentration of cGMP), a very large inward current develops. **(c)** In *Drosophila* photoreceptors, light-activated rhodopsin activates a G protein, which in turn activates an eye-specific phospholipase C (PLC). PLC causes the breakdown of the membrane phospholipid PIP_2 into diacylglycerol (DAG) and inositol trisphosphate (IP_3). Through a mechanism that is not yet understood, one or both of these molecules lead cation-selective channels at the base of the rhabdomere to open, producing a depolarization. [Part a adapted from Baylor, 1996; part b adapted from Yau and Nakatani, 1985; part c adapted from Zuker, 1996.]

activating many molecules of an enzyme called phosphodiesterase (PDE). This enzyme acts on cyclic guanosine monophosphate (cGMP), an intracellular signaling molecule analogous to cAMP, which was discussed in Chapter 6. PDE catalyzes the hydrolysis of cGMP to 5′-guanosine monophosphate (5′-GMP), which has no intracellular signaling properties. In vertebrate rods, the channels that carry the dark current are said to be "cyclic nucleotide–gated" because they are open only in the presence of the cyclic nucleotide cGMP. These channels are permeable to three cations: Na^+, Mg^{2+} and Ca^{2+}. When the level of cGMP drops, the conductance through these channels drops (Figure 7-44b). Most importantly, the inward I_{Na} drops, and residual K^+ current through other channels causes the cell to hyperpolarize. When the light stimulus ends, cGMP is regenerated by the action of another enzyme, guanylate cyclase. As the level of cGMP rises, the dark-current channels reopen, and the dark current returns to its full value. Activated transducin collides with and activates phosphodiesterase molecules at a rate of about 10^6 molecules per second, allowing the capture of a single photon to affect the conductance through an enormous number of ion channels. This numerical relation generates an impressive amplification between the capture of a single photon and the effect that event produces on V_m. The capture of a photon of light in an invertebrate photoreceptor produces a different chain of events, as we will soon see.

After the *cis-trans* isomerization of retinal, further changes in the molecule appear to be irrelevant to the excitation of visual receptor cells, but some of the intermediates are probably related to light adaptation, and the subsequent reactions (Figure 7-45) are necessary for regenerating rhodopsin. Activated rhodopsin is hydrolyzed spontaneously to retinal and opsin, which are both reused repeatedly. Free retinal is isomerized back into the 11-*cis* form and reassembled with an opsin to form rhodopsin. Any retinal that is lost or chemically degraded in the process is replenished from vitamin A_1 (retinol) stored in the cells of the pigment epithelium, which actively take up the vitamin from the blood. A nutritional deficiency of vitamin A_1 decreases the amount of retinal that can be synthesized and hence decreases the amount of available rhodopsin. The result can be reduced photosensitivity of the eyes, which can lead to the condition commonly known as night blindness.

A rod photoreceptor can respond to a single photon hitting the outer segment, partly because rhodopsin is so densely packed in the disk membranes. There are about 20,000 rhodopsin molecules per square micrometer in the rod outer segment, which is much closer packing than, for example, the density of acetylcholine receptors at the neuromuscular junction. By recording from a single rod, Denis Baylor has measured the response to the capture of a single photon (Figure 7-46 on the next page). In these experiments, rods are teased apart and one of them is drawn into a recording pipette, where it is stimulated by a very narrow beam of light. When the stimulating light is very dim, it is possible to record small current fluctuations, each of which corresponds to the photoisomerization of a single rhodopsin molecule by a single photon. Because photoreceptors can respond to a single photon, or quantum, of energy, the sensitivity of photoreception is limited by the quantal nature of light; there is no smaller amount of light than one photon.

The process of visual transduction by invertebrate photoreceptors has been studied in detail in *Drosophila* by combining genetic and molecular techniques with electrophysiology. Once it is activated, *Drosophila* rhodopsin activates a G protein associated with the membrane, but this G protein then activates a phospholipase C, which catalyzes the breakdown of phosphatidyl 4,5-bisphosphate, forming inositol trisphosphate and diacylglycerol (see Figure 7-44c). These two molecules cause Na^+ channels to open, but how they do so remains unresolved, though there are hints that cGMP may play a role in the process.

Interestingly, the dark-current channels in rod and cone outer segments and the transduction channels in

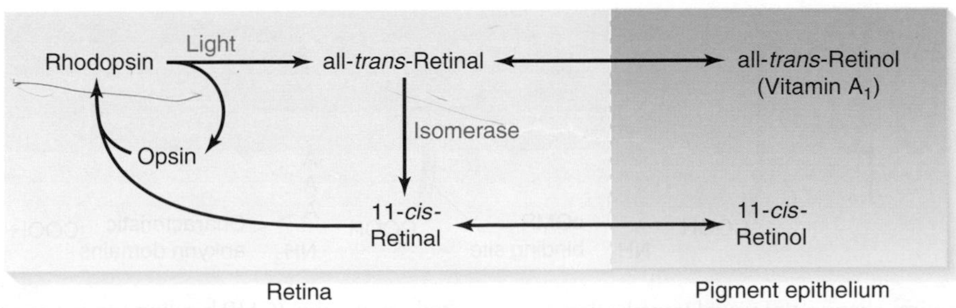

Figure 7-45 When rhodopsin is activated, the all-*trans*-retinal separates from the protein opsin. Rhodopsin is reconstituted after an isomerase returns the retinal to the 11-*cis* configuration. Retinol (vitamin A_1) is stored in the pigment epithelium and can be delivered to the photoreceptors for generating new rhodopsin molecules. The pathway is similar for porphyropsin.

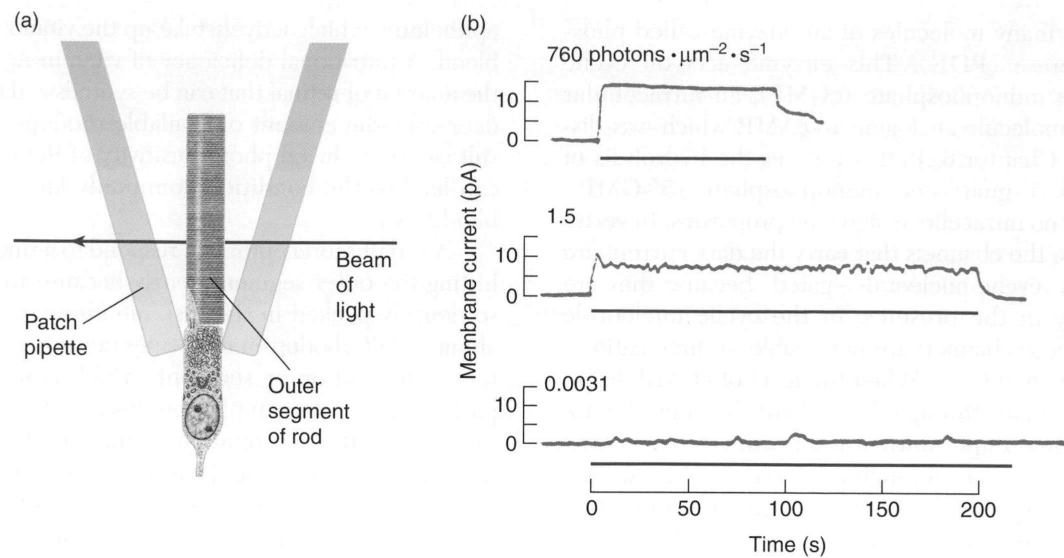

Figure 7-46 Rods can respond electrically to the capture of a single photon. **(a)** A single rod outer segment is sucked into a smooth glass pipette electrode and is illuminated by a narrow beam of light while the ionic current across the membrane is recorded by the pipette. **(b)** The recorded membrane current changes in response to illumination. In very dim light (bottom record), small individual changes in the current accompany the capture of single photons. As the light intensity is increased (intensity is indicated above each recording), the responses become larger and smoother. The duration of the illumination is indicated by the red bar under each recording. [Adapted from Baylor et al., 1979.]

the rhabdome of insects belong to the same molecular family (Figure 7-47), one that includes a ubiquitous class of voltage-gated K^+ channels found in many types of neurons. The first of the visual transduction channels to be cloned was the *trp* (for "transient receptor potential") channel of *Drosophila,* but subsequent work has revealed that the cyclic nucleotide–gated channels in the membranes of vertebrate photoreceptors are members of the same family. Elucidating the process of visual transduction has demonstrated the power of a comparative approach. Although vertebrate and invertebrate photoreceptors seem quite different from each other at the electrophysiological level, many similarities between them have emerged from studies at the molecular level. In addition to the similarities already described, light adaptation in both types of photoreceptors appears to depend on a change in the concentration of intracellular free Ca^{2+} as a result of the changing current through light-dependent channels. The ease with which molecular genetics experiments can be conducted on *Drosophila* and the physiological accessibility of the vertebrate retina have been combined to provide an array of enormously powerful experimental approaches to the question of how visual information is acquired and processed by photoreceptors. If the molecular identities of the players had not been so strongly conserved through phylogeny, extracting the details of visual transduction would probably have taken much longer.

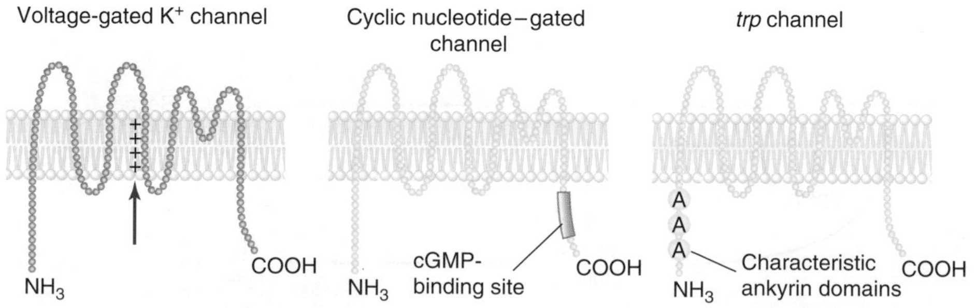

Figure 7-47 The cation channels of visual transduction belong to the same family as some voltage-gated K^+ channels that are found in many types of neurons. The structures of these channel proteins include six transmembrane helices. Differences in sequence provide individual characteristics such as charged amino acids in the voltage-gated K^+ channel (red arrow), the cGMP binding site in vertebrate phototransduction channels (green), and the repeated ankyrin domains in the *trp* channel. (Ankyrin sites typically bind to cytoskeletal elements and suggest that *trp* channels may interact with the cytoskeleton.) [Adapted from Harteneck et al., 2000.]

Cones and rods

The ability to distinguish color is correlated with possession of multiple visual pigments, each of which absorbs maximally at a different wavelength. Each photoreceptor cell expresses only one visual pigment. The magnitude of a photoreceptor's response to different wavelengths constitutes its *action spectrum*. Within a receptor's action spectrum, the electrical response is maximal at a particular wavelength of incident light and falls off when the wavelength is either increased or decreased. In many species for which action spectra have been recorded, three classes of photoreceptors have been found, although species are known in which there are more visual pigments than this. Some birds, for example, have four or even five different classes of photoreceptors, implying that these animals synthesize four or five different visual pigments.

The action spectra for the retinas of some species have been compared with the absorption spectra of individual photoreceptors using a process called microspectrophotometry, in which a tiny beam of light is focused on one photoreceptor at a time and the light absorption properties of that cell are determined. Photoreceptors studied in this way fall into distinct classes for each species; there are no intermediate absorption spectra, which implies that each photoreceptor synthesizes a single visual pigment (Figure 7-48). Both action spectra and absorption spectra of photoreceptors have been determined in many species, and the two kinds of spectra match each other closely, confirming that the action spectrum of a photoreceptor depends on the absorption properties of its visual pigment. In addition, light that contains different wavelengths generates photochemical reactions in a particular photoreceptor cell in proportion to the amount of each wavelength absorbed. In other words, a photoreceptor cell is excited by different wavelengths in proportion to the efficiency with which its pigment absorbs each wavelength. Any photon that is not absorbed has no effect on the visual pigment molecule; any photon that is absorbed transfers its energy to the molecule. Thus, it is possible to restate Young's trichromacy theory (described in Spotlight 7-5 on the next page) in relation to cone photoreceptors and their photopigments: there are in the human retina three classes of cones, each of which contains one visual pigment that is maximally sensitive to blue, green, or orange light. Similarly, in the goldfish retina, there are three classes of cones with absorption maxima in the blue, green, and red regions

(a)

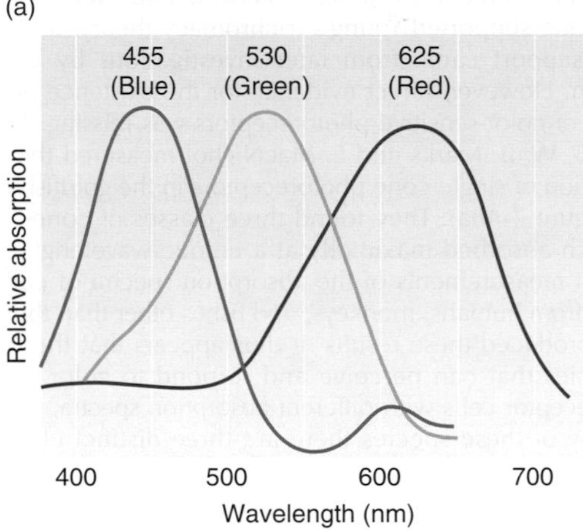

(b)

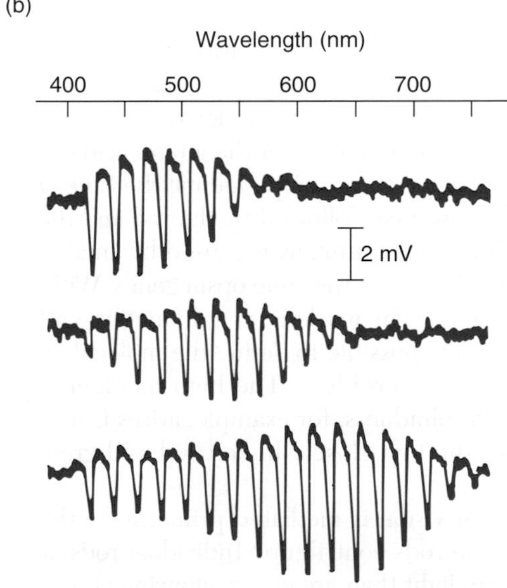

(c)

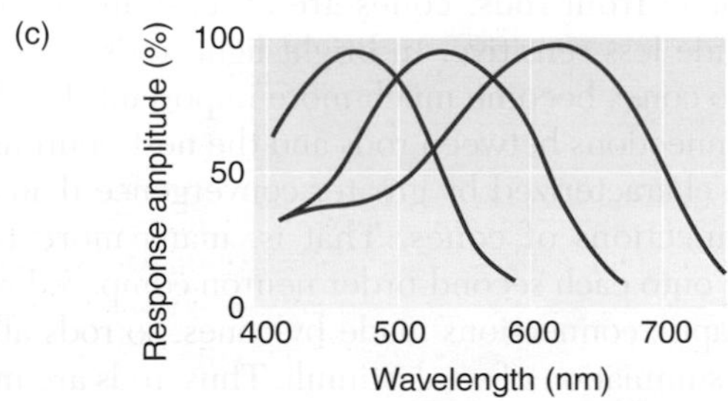

Classic Work

Figure 7-48 The goldfish retina includes three classes of cones, each with a distinctive action spectrum. **(a)** Absorption spectra of individual cones in the retina of a goldfish indicate that there are three separate visual pigments, each with a distinctive absorbance peak. These measurements were made by microspectrophotometry, which allows the absorption spectrum of a single photoreceptor to be measured. There is variation among species: in humans, for example, the class of cones that is equivalent to the red-absorbing cone in the goldfish absorbs maximally closer to 560 nm, which is in the yellow-orange part of the spectrum. **(b)** Electrical responses of three single cones in the retina of a carp to flashes of different wavelengths, as shown by the scale at the top. The wavelength that produces a maximal response is different for each of the three cones. **(c)** Activity in each cell shown in part b was plotted as a function of wavelength. The result reveals three classes of cones, each with an action spectrum approximating one of the absorption spectra measured in part a. [Part a adapted from Marks, 1965; parts b and c adapted from Tomita et al., 1967.]

Spotlight 7-5

LIGHT, PAINT, AND COLOR VISION

In 1666, Sir Isaac Newton demonstrated that white light is separated into an infinite number of colors when it passes through a prism. Each color in the resulting spectrum is monochromatic—that is, it cannot be separated into yet more colors. At that time, however, it was already known that a painter could match any spectral color (e.g., red-orange) by mixing two pure pigments (e.g., red and yellow to make red-orange), each of which reflects a wavelength different from that of the color produced. Thus, there seemed to be a paradox between Newton's demonstration that there are an infinite number of colors in light and the growing awareness among Renaissance painters that all colors could be produced by combinations of only three primary pigments—red, yellow, and blue.

This paradox appeared to be resolved by Thomas Young's suggestion in 1802 that some structures in the eye are selective for the three primary colors: red, yellow, or blue. Young reconciled the infinite variety of spectral colors and the small number of painter's pigments required to produce all colors by proposing that each class of these structures is excited maximally by a specific wavelength of light. We now recognize that Young's hypothetical structures are photoreceptors. According to his *trichromacy theory,* a "red" structure would be stimulated maximally by monochromatic "red" wavelengths, and a "yellow" structure by monochromatic "yellow" wavelengths. Both types of structures would be stimulated simultaneously, but to a lesser degree, by monochromatic orange light. Although he did not know about photoreceptors, Young thus proposed that the sensation for "orange" results from the simultaneous excitation of "red" structures and "yellow" structures. Because he had no notion of the physiology of photoreceptors, Young's insight was truly remarkable.

Extensive psychophysical investigations carried out in the nineteenth century by James Maxwell and Hermann von Helmholtz supported Young's trichromacy theory, and additional support came from later investigations by William Rushton. However, direct evidence for the existence of three classes of color-sensitive photoreceptors was missing. Finally, in 1965, W. B. Marks and E. MacNichol measured the color absorption of single cone photoreceptors in the goldfish retina (see Figure 7-48a). They found three classes of cones, each of which absorbed maximally at a unique wavelength. Subsequent measurements of the absorption spectra of cones in retinas from humans, monkeys, and fishes other than the goldfish reproduced these results. It thus appears that the retinas of species that can perceive and respond to color contain photoreceptor cells with different absorption spectra, and that in many of these species there are three distinct classes of receptors.

of the spectrum. (For consistency, all cones that absorb longer wavelengths maximally are called "red" cones, even if they absorb in the orange part of the spectrum, as in humans.) The electrical output of each class of receptor cells depends on the number of quanta that are absorbed by the pigment. The sensation of color arises when higher-order neurons integrate signals received from the three classes of cones.

Knowledge about the molecular basis of color reception has grown enormously since 1984, when Jeremy Nathans described the molecular structure of human opsins and thus provided an explanation for hereditary color blindness. It appears that 11-*cis*-retinal (or 11-*cis*-3-dehydroretinal) is the light-absorbing molecule in all visual pigments. This prosthetic group is combined with different opsin proteins to produce visual pigments with different absorption maxima; the spectral sensitivity of the photopigment depends on the opsin, not on retinal. Nathans and his coworkers discovered three genes that encode the opsins in human cones. The gene encoding the opsin for the blue-absorbing pigment is located on an autosomal chromosome, whereas the two genes for the red-absorbing and green-absorbing opsins are closely linked on the X chromosome. The red and green opsins differ at only 15 of 348 amino acids, and each shares about half its amino acid sequence with the rhodopsin in rods (Figure 7-49). On the basis of sequence similarity, we can surmise that the genes for these pigments probably arose from a common ancestral gene that underwent duplication and divergence. A comparison of the amino acid sequences suggests that, of the cone pigments, the blue-sensitive pigment arose first, followed by the red and the green. Color blindness in humans is caused by an absence of, or a defect in, one of the cone opsin genes. With the use of these molecular markers in conjunction with visual tests, it is now possible to define the molecular basis of this perceptual problem. The high incidence of red-green color blindness, for example, arises from a defect in one of the two closely linked red and green opsin genes.

If color vision is mediated primarily by the cones, what do the rods contribute? Individual rods are more sensitive to light than are cones, allowing vision even in very dim light. The records in Figure 7-46, for example, were made from rods; cones are at least an order of magnitude less sensitive. In bright light, rods are saturated, so cones become much more important. In addition, connections between rods and the next neurons in line are characterized by greater convergence than are the connections of cones. That is, many more rods synapse onto each second-order neuron compared with the synaptic connections made by cones, so rods allow greater summation of weak stimuli. Thus, rods are most

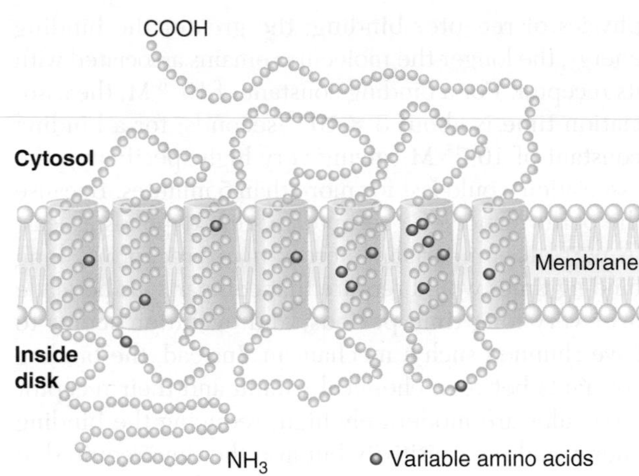

Figure 7-49 The opsin proteins from the red-absorbing and green-absorbing human cones differ by only 15 amino acids. In this diagram illustrating the structure of the human "red" and "green" opsins, these variable amino acids are colored red. Notice that opsin, like other G protein–coupled receptor proteins, has seven transmembrane helical domains. Most of the amino acids that differ between the two opsins appear to be in membrane-spanning helices. [Adapted from Nathans et al., 1986]

effective in dim light. Because dim light stimulates only our rods, in a dark environment we see in shades of black and gray, rather than in color.

In daylight, most images are preferentially focused onto the fovea of the human eye, which contains only tightly packed cones. Rods are found outside the fovea, with cones interspersed between them. This differential distribution of rods and cones causes us to be most sensitive to dim light when an image is focused outside the fovea, where the rod population is higher. A dim star will appear brighter if you adjust your gaze to make its image fall outside the fovea. If you shift your gaze to bring the image onto the fovea (i.e., if you look directly at the star), it will fade or even disappear. The high sensitivity of rods comes at a price: the greater number of connections among rods reduces the acuity of rod-based vision. Our visual *sensitivity* is greatest when an image is focused onto the rods outside the fovea; our visual *acuity* is greatest when an image is focused onto the cones of the fovea.

When visual pigments are explored in many phyla, interesting patterns emerge. The distribution of the retinal-containing rhodopsins and 3-dehydroretinal-containing porphyropsins among species is correlated with environment. All visual pigments of terrestrial vertebrates are rhodopsins, and rhodopsins are also found among many invertebrates, including *Limulus,* insects, and crustaceans. In contrast, porphyropsins are found in the retinas of freshwater fishes, euryhaline fishes, and some amphibians. This distribution suggests that some feature of porphyropsins makes them particularly well adapted to conditions in freshwater. In fact, anadromous fishes, which migrate between freshwater and saltwater during their life cycle, synthesize porphyropsin during their stay in freshwater and rhodopsin while they are in saltwater. The absorption maxima of the porphyropsins are shifted toward the longer-wavelength, red end of the visual spectrum, whereas rhodopsins typically absorb maximally at shorter wavelengths. Perhaps the freshwater environment makes sensitivity to the red end of the spectrum important.

We have traced the transduction of visual information from the absorption of a photon to the production of neuronal signals, but we have not answered the question of how all this information about incident radiation is molded into a coherent view of the world. In Chapters 8 and 11 we will explore how the collected information is passed on to higher neuronal centers, where it is integrated and used to shape behavior.

LIMITATIONS ON SENSORY RECEPTION

You might think that a perfect sensory receptor would be very sensitive to stimulation from the environment and would encode information with perfect accuracy. In fact, because of the physical properties of both stimuli and receptors, no receptor meets these requirements; all receptors represent compromises in how sensory information is received and encoded. The laws of physics necessarily limit the fidelity with which sensory information is received and transmitted by cells. In some sensory modalities, the accuracy of sensory reception is limited primarily by the relative magnitudes of the signals and the background noise. This signal-to-noise ratio limits the performance of all systems that receive and transmit information, whether or not they are living. In other modalities, the performance and sensitivity of the sensory system are limited by the form of energy to which the receptors are tuned. Light, for example, is by its nature quantized into photons. No receptor can respond to less than one quantum of light, because light does not exist in fractions of quanta.

A major source of background noise arises from a corollary of the third law of thermodynamics. That is, at all temperatures above 0°K, molecules have kinetic energy and are in motion. Thermal energy is given by

$$E_{\text{therm}} = kT \quad (7\text{-}2)$$

where k is Boltzmann's constant (1.3805310^{-16} erg K^{-1}) and T is the absolute temperature. This equation gives the energy that is associated with the movement of molecules (i.e., Brownian motion) at an animal's body temperature. It sets a lower limit on the sensitivity of receptors in detecting signals because thermal energy provides a constant noise level against which stimulation occurs. To detect an external signal, a receptor must be able to distinguish the signal from this baseline thermal noise. How easily can receptor cells accomplish this task?

Photoreceptors provide an example. At a body temperature of 25°C, thermal energy is about 0.58 $kcal \cdot mol^{-1}$, or 4×10^{-14} erg. We must compare this energy to the energy of a typical sensory stimulus. The stimulus for a vertebrate photoreceptor is light in the visible part of the electromagnetic spectrum (see Figure 7-38). The energy of a single photon of light is given by the Einstein relation:

$$E = h\nu = \frac{hc}{\lambda} \tag{7-3}$$

where h is Planck's constant and ν, c, and λ are the frequency, speed, and wavelength of light, respectively. Substituting the values for a photon of blue light ($\lambda =$ 500 nm), the energy is calculated to be approximately 57 $kcal \cdot mol^{-1}$—almost a hundredfold greater than the thermal energy. Thus, in vision, the sensitivity of detection is not limited by thermal energy within the detector. Instead, it is limited by the quantal nature of light itself.

By analogy with light, sound energy can be described by the Einstein relation for single "phonons," which are quantum units of sound energy analogous to photons of light—the energy of a phonon equals the frequency of a sound wave times Planck's constant, h. If we consider audition in many phyla, we discover that animals respond to sounds across a remarkably broad range of frequencies, from 10 to 10^5 Hz. The energy of phonons at these frequencies ranges from 7×10^{-26} to 7×10^{-23} erg. In the middle of this range, the energy of a single phonon is 10 orders of magnitude (10^{10}) below the limit of detection that would be set by thermal energy. This result indicates that the detection of acoustic stimuli is fundamentally limited by thermal noise, and that there must be special mechanisms, such as methods for filtering out thermal noise, that permit auditory sensory reception. Some advantage is gained, for example, by tuning the detectors to a limited range of frequencies, which can filter out some of the noise. Numerous mechanisms have evolved to combat the limitations of thermal noise, but direct measurements have also shown that individual sensory cells in auditory systems do faithfully reproduce the thermal noise at their inputs.

As discussed earlier in this chapter, most chemical stimuli (olfaction, taste, chemotaxis) excite receptor cells by binding to specific receptor molecules. Thus, in most chemical transduction, the binding energy between an odorant or tastant and a receptor molecule compared with the background thermal energy determines the limits of detection. The binding energies that have been measured in chemical sensory systems are typically about 1–10 $kcal \cdot mol^{-1}$. This energy is sufficiently greater than thermal energy that chemoreceptors could theoretically count single molecules. There is, however, an important constraint dictated by the physics of receptor binding: the greater the binding energy, the longer the molecule remains associated with its receptor. For a binding constant of 10^{-6} M, the association time is about 3×10^{-3} seconds; for a binding constant of 10^{-11} M (giving very high specificity), the association would last for more than 5 minutes. Because the performance of a receptor system depends at least in part on comparisons across many receptors, long binding times would require that comparisons be made over very long time periods, and evolution seems to have shunned such a mechanism. Instead, the binding constants between chemical stimuli and their receptor molecules are moderately high, reducing the binding energy and the sensitivity but also the time required to transduce and interpret chemical signals.

Signal-to-noise properties can be predicted for stimuli that activate electroreceptors and thermoreceptors. Electroreception is relatively widespread in aquatic organisms, in which it is used for navigation, communication, and predation. The energy of electric fields, carried through water at the frequencies used, is about 10 orders of magnitude below the thermal energy. Thus, the process of electroreception, like that of auditory reception, must be dominated by thermal noise in the detector, and special filters and coincidence detectors are required to separate the signal from the noise. Thermoreception depends on the detection of electromagnetic radiation in the infrared region of the spectrum (which has lower frequencies and longer wavelengths than those in the visible spectrum), and by definition it is limited by the temperature difference between the measured object and the measuring organ. Some animals have apparent adaptations that keep their detectors cooler than the rest of the body, decreasing the background thermal noise at the thermoreceptors.

As scientists have explored the limits of animal senses, it has become clear that many modalities operate at or near the theoretical limits imposed by physical laws. To accomplish this prodigious task, the evolution of many types of receptors has converged on similar molecular mechanisms, and the mechanisms that are available for each sensory modality depend, at least to some extent, on whether sensory reception is limited by thermal energy or by the quantal nature of the stimulus.

SUMMARY

General properties of sensory reception

■ Sensory organs and afferent neurons provide an organism with all of its information about events and conditions in the external world and in its internal environment.

■ Acquisition of this information begins in receptor cells, whose physiological properties are specialized to respond to and encode specific forms of energy, such as light, sound, or heat. Receptor cells change the energy of the stimulus into the energy of an electrical signal, a process that is called sensory transduction.

Common mechanisms and molecules of sensory transduction

- Sensory reception comprises three major events: transduction, amplification, and transmission.

- All sensory receptors have a threshold of detection—that is, the lowest amount of stimulus energy that will produce a response in the receptor 50% of the time.

- Many sensory receptors cells (e.g., photoreceptors and some chemoreceptors) receive energy by way of transmembrane receptor molecules that belong to the large family of seven-transmembrane-helix proteins.

- An activated receptor molecule activates a G protein, which in turn either directly affects ion channels or activates intracellular second messengers that modify the conductance through ion channels.

- Regardless of the specific transduction mechanism, sensory signals are transduced into neuronal signals when the ionic conductance through membrane ion channels is increased or decreased, causing V_m to change. The resulting change in V_m is called a receptor potential. Stronger stimuli cause larger receptor potentials. In other words, the receptor potential is graded.

- Phasic receptors produce APs only at the onset or the offset of a stimulus (or both). Tonic receptors produce APs throughout the duration of a stimulus, although the frequency of APs may drop with time.

Encoding stimulus intensities

- As a sensory signal travels to the CNS, the strength of stimulation is encoded in the number and frequency of APs elicited in a sensory neuron.

- The range of stimulus energy that can be encoded by a sensory cell or sensory organ is called its dynamic range. At energy levels below the dynamic range, receptors fail to respond. Above the dynamic range, the receptors are saturated, so all energy levels are encoded identically. The sensitivity of most sensory receptors drops as exposure to the stimulus continues, a process called sensory adaptation.

- Typically, the relation between the intensity of stimulation and the sensory response is logarithmic. This relation is seen at both the physiological level and the psychophysical level of study.

- In many sensory systems, individual receptors have different dynamic ranges; together they cover a broader range than that of any individual receptor.

Control of sensory sensitivity

- Sensory adaptation resets the sensitivity of individual receptors, allowing them to respond to a broad range of stimulus intensities.

- Mechanisms of sensory adaptation include mechanical properties of the receptors, changes in the transduction molecules, modifications of intracellular signaling pathways, changes in ion channels, changes in the electrical properties of the receptor cell, or shifts in the properties of sensory neurons in the CNS.

- Efferent signals from the CNS modify the sensitivity of sensory reception, allowing the CNS to fine-tune the sensory information it receives. In addition, signals from sensory receptors can modify the sensitivity of neighboring receptors, which in the case of lateral inhibition enhances contrast in the activity of adjacent receptors.

The chemical senses: taste and smell

- The receptor molecules of taste (gustation) and smell (olfaction) and the neuronal pathways carrying taste and smell information are independent of and very different from each other.

- Four classes of taste stimuli are known for vertebrates: sweet, sour, salty, and bitter. A fifth, *umami* (in response to glutamate and some peptides), has recently been recognized. Other animals may recognize other tastes, but in each species there are only a few distinguishable classes of taste stimuli.

- Each category of taste recognized by vertebrates excites receptor cells by a separate mechanism, either by the direct flow of ions across the membrane (e.g., salty and sour) or by intracellular second-messenger cascades (e.g., bitter and sweet).

- Animals can distinguish among a very large number of different olfactory stimuli, making olfaction quite distinct from taste.

- Each olfactory receptor cell expresses one or a few members of a large family of seven-transmembrane-helix receptor proteins. The mammalian genome includes genes for about 1000 different olfactory receptor proteins; the *Drosophila* genome includes somewhat fewer than 100 olfactory receptor genes.

- Binding of an odorant by an olfactory receptor protein activates a G protein, which, through a series of steps, opens ion channels.

Mechanoreception

- Mechanoreceptors are activated when mechanical distortion of the plasma membrane causes ion channels to open. The mechanism of mechanoreception underlies the senses of touch and pressure, as well as the organs of audition and equilibrium.

- Insect mechanoreceptors are frequently associated with accessory structures, such as sensory bristles, that receive, transmit, and even amplify mechanical stimulation. Stretch-activated Na^+ channels have been identified in *Drosophila,* and current research is exploring how widespread these channels are among mechanoreceptor cells in other species.

■ Hair cells, which are characterized by cilia distributed over part of the cell surface, form the basis of vertebrate organs of equilibrium and audition. They typically are associated with a fluid-filled canal (in the case of the inner ear), or directly contact the water in which an animal lives (in the case of the lateral line of fishes and frogs). Ciliated sensory cells are also found in some invertebrates.

■ When the cilia of a hair cell are bent in one direction, the plasma membrane depolarizes. In some cases, bending the cilia in the opposite direction produces hyperpolarization.

■ Many hair cells lack axons; instead, they synapse onto primary afferent neurons. Release of neurotransmitter by hair cells varies with the membrane potential of the cell and determines the rate of APs in the afferent neurons.

Organs of equilibrium

■ Organs of equilibrium provide information about an organism's position with respect to gravity and its motion through space.

■ Statocysts are organs of equilibrium found in many invertebrates. A statocyst is a small fluid-filled chamber lined with ciliated receptor cells that contains a small hard object (a statolith). The statolith rests at the lowest point of the statocyst, contacting the receptor cells at that point. If the animal's position with respect to gravity changes, the statolith moves, exciting other receptor cells.

■ Vertebrate organs of equilibrium are typically located in the inner ear. In most animals, they include the semicircular canals and one or more other fluid-filled structures. The epithelial walls of these structures include hair cells. The fluid within the canals—or in some cases a gelatinous accessory structure in which the cilia of hair cells are embedded—moves when the animal changes its position. This movement bends the cilia, changing ion fluxes across the hair cell membrane.

Organs of audition

■ The hair cells of the cochlea are the receptor cells of audition in many vertebrates. Structures of the outer and middle ear amplify and filter sound signals that are delivered to the cochlea.

■ The cochlea consists of two long perilymph-filled chambers, the scala tympani and the scala vestibuli, separated by a third, endolymph-filled chamber, the scala media. Hair cells are located in the organ of Corti, which rests on the basilar membrane of the scala media.

■ Sound waves are transmitted through the outer ear to the middle ear by the tympanic membrane, through the middle ear by way of a set of three small bones, and thence to the oval window, which separates the air-filled middle ear from the fluid-filled cochlea.

■ Movements of the oval window set up traveling waves in the perilymph of the cochlea. Waves traveling through the perilymph move the basilar membrane and thus the hair cells, causing their cilia to bend and their ion channels to open.

■ Insect ears may be located in any of a variety of places on the body.

■ The tympanic ear of insects is related to the chordotonal organs, hair-cell-based mechanoreceptive organs that more commonly deliver information regarding the position of a joint.

Thermoreception

■ Many animals have receptors that are specialized to receive information about temperature. These thermoreceptors may be widely distributed on the surface of the body or may be localized in particular parts of the nervous system.

■ Some thermoreceptors respond to increased temperature (warmth receptors), whereas others respond to a drop in temperature (cold receptors). All thermoreceptors are responsive in only a limited temperature range.

Vision

■ Photoreceptors are excited by light, which is conducted to the receptor cells through accessory structures that focus light rays and filter out some wavelengths (or energies).

■ Light that reaches a photoreceptor can be absorbed by a visual pigment molecule. Although the structure of eyes varies greatly across the animal kingdom, the visual pigments have been highly conserved.

■ A compound eye is composed of individual units, called ommatidia, each of which has the optical capability to form a focused image on the photoreceptor cells. Light is focused onto a structure called a rhabdome that is made up of rhodopsin-packed microvilli from a number of photoreceptor cells.

■ Inhibitory connections between cells that lie close to one another enhance contrast between stimuli. This pattern, called lateral inhibition, is common to a variety of sensory systems.

The vertebrate eye

■ The cornea and the lens of the vertebrate eye refract and filter incident light, focusing it onto photoreceptors in the back layer of the eyeball, which is called the retina.

■ The retina consists of several types of cells. The photoreceptors carry out visual transduction, and the rest of the cells process the information acquired by the receptors and transmit it to the CNS.

■ Vertebrate photoreceptors fall into two categories, rods and cones. In cones, the membrane containing the visual pigment is continuous with the plasma membrane. In rods, the visual pigment is located in membrane disks that are arranged in a stack inside the rod and are separate from the plasma membrane.

Photoreception: Converting photons into neuronal signals

■ Once a molecule of visual pigment absorbs a photon of light, a series of events is triggered that leads to a change in the ionic conductance of the photoreceptor's plasma membrane. The resulting current changes V_m.

■ The photoreceptors of vertebrates hyperpolarize in response to light because a steady "dark current" of Na^+ ions entering the cell is reduced.

■ Vertebrate photoreceptors lack axons. Instead, their hyperpolarization reduces their steady release of a transmitter, modifying activity in the next neurons in line.

■ All known visual pigments consist of a protein, called an opsin, and one of two prosthetic groups, retinal in most animals and 3-dehydroretinal in some freshwater vertebrates.

■ Absorbing a photon of light causes retinal to isomerize from the 11-*cis* form into the all-*trans* form. The *cis-trans* isomerization causes the opsin to activate a G protein called transducin. In vertebrates, transducin causes cGMP-gated Na^+ channels to close, hyperpolarizing the cell. In invertebrates, activated transducin causes Na^+-selective channels to open, depolarizing the cell.

■ In many vertebrate retinas, rods provide the basis for vision in dim light but become saturated in bright light. Cones provide the basis for vision in bright light and for color vision.

■ Each cone synthesizes only one of a small family of visual pigments, each of which maximally absorbs light of a particular wavelength. Thus there are the same number of cone types as there are visual pigments synthesized by cones. In humans and many other animals with color vision, there are three cone pigments (and thus three types of cones) plus a separate rhodopsin that is synthesized by rods.

Limitations on sensory reception

■ In some sensory modalities, such as vision, the energy of the stimulus arrives in quanta. At least some photoreceptor cells can respond to absorption of a single photon of light. Light cannot be divided into smaller quantities than a photon, so the maximum sensitivity of photoreceptors is determined by the quantized nature of light.

■ In other sensory modalities, such as olfaction, the sensitivity of receptors depends on a complex relationship between the background thermal energy ("noise") and the energy released when a stimulating molecule binds to the receptor molecule (the signal). The sensitivity of these modalities is limited by thermal noise unless specialized features of detectors or neuronal networks act to filter and sharpen the signal.

REVIEW QUESTIONS

1. Visual receptor cells can be stimulated by pressure, heat, and electricity as well as by light if the intensity of these other stimuli is sufficiently great. How can this fact be reconciled with the concept of receptor specificity?
2. For chemoreception, mechanoreception, and photoreception, choose one type of stimulus and outline the steps from energy absorption by a receptor cell to the initiation of action potentials (APs) that will travel to the central nervous system.
3. Why must receptor potentials be converted into APs to be effective?
4. All sensory information enters the central nervous system in the form of APs having similar properties. How, then, do we differentiate among various stimulus modalities?
5. For chemoreception, mechanoreception, and photoreception, choose one type of stimulus and describe how transduction and amplification are related in that modality.
6. Discuss the relation between the intensity of a stimulus and the magnitude of the signal sent to the central nervous system. How is stimulus intensity encoded? What mechanisms allow a sensory system to respond to stimuli whose intensity varies over many orders of magnitude?
7. Describe three mechanisms that contribute to sensory adaptation.
8. Discuss one example in which efferent activity can regulate the sensitivity of receptor cells.
9. What major classes of membrane molecules contribute to the reception and transduction of sensory stimuli? Compare and contrast these molecules in taste, olfaction, mechanoreception, and vision.
10. How does spontaneous firing enhance the sensitivity of certain receptor systems—for example, lateral-line hair cells?
11. How is the presence of an object perceived by electroreceptors of the weakly electric fishes swimming in very turbid water?
12. Describe the pathway that converts movements of the basilar membrane in the cochlea into APs in the auditory nerve.

13. Discuss the functions of inner and outer hair cells in the cochlea.
14. What is the major difference between vertebrate and invertebrate photoreceptor cells in their electrical responses to illumination?
15. What allows some arthropods to respond to the orientation of polarized light? Humans cannot do this—why not?
16. Compare the mechanisms that allow the auditory system to distinguish the frequency of incident sound and the visual system to distinguish the frequency of incident light.
17. Compare the ways in which mammalian and bony fish lenses focus images.
18. Outline the steps in the transduction of light energy in vertebrate visual receptors. Compare transduction of light energy in invertebrate visual receptors.
19. How does our current understanding of the physiology of color vision corroborate Young's trichromacy theory?
20. Compare and contrast the morphological and functional properties of vertebrate rods and cones.

SUGGESTED READINGS

Baylor, D. 1996. How photons start vision. *Proc. Natl. Acad. Sci. USA* 93:560–565. (A paper, presented at a colloquium on vision, giving a concise summary of early events in the visual pathway.)

Corey, D. P., and S. D. Roper. 1992. *Sensory Transduction.* 45th Annual Symposium of the Society of General Physiologists. New York: Rockefeller University Press. (A series of papers that discuss transduction in many different sensory modalities.)

Døving, K. B., and D. Trotier. 1998. Structure and function of the vomeronasal organ. *J. Exp. Biol.* 201:2913–2925. (A description of this heretofore somewhat obscure sensory organ, ranging from anatomy to molecules.)

Dowling, J. 1987. *The Retina.* Cambridge, Mass.: Belknap Press of Harvard University Press. (A very readable compendium of information on the vertebrate retina, written by a major contributor to our understanding of this sensory organ.)

Herness, M. S., and T. A. Gilberteson. 1999. Cellular mechanisms of taste transduction. *Annu. Rev. Physiol.* 61:873–900. (A detailed review of taste from anatomy to receptor molecules, focusing on mammals.)

Hudspeth, A. J. 1989. How the ear's works work. *Nature* 341:397–404. (A beautifully written account of auditory transduction by hair cells, written by a man who has played a major role in exploring the subject.)

Land, M., and R. Fernald. 1992. The evolution of eyes. *Annu. Rev. Neurosci.* 15:1–29. (A consideration of the physical and optical properties of visual organs across all of animal phylogeny.)

Mombaerts, P. 1999. Seven-transmembrane proteins as odorant and chemosensory receptors. *Science* 2886:707–711. (A broad and informative review of our knowledge regarding the chemoreceptor proteins of olfaction.)

Nef, P. 1998. How we smell: The molecular and cellular bases of olfaction. *News Physiol. Sci.* 13:1–5. (A brief, but readable and informative, summary of information on olfactory mechanisms.)

Nobili, R., F. Mammano, and J. Ashmore. 1998. How well do we understand the cochlea? *Trends Neurosci.* 21:159–167. (A review of cochlear physiology, focusing on the physical interactions among and behavior of cochlear components.)

Pichaud, F., A. Briscoe, and C. Desplan. 1999. Evolution of color vision. *Curr. Opin. Neurobiol.* 9:622–627. (A discussion of color vision in both vertebrates and invertebrates, what might have driven its evolution, and how it works.)

Purves, D., G. J. Augustine, D. Fitzpatrick, L. C. Katz, A.-S. LaMantia, J. O. McNamara, and S. M. Williams. 2001. *Neuroscience.* Sunderland, Mass: Sinauer Associates. (A general textbook on neurobiology, with a strong series of chapters on sensory reception, particularly in the vertebrates.)

Shepherd, G. M. 1994. *Neurobiology.* 3d ed. New York: Oxford University Press. (A concise text that considers several sensory modalities in both vertebrates and invertebrates.)

Strausfeld, N. J., and J. G. Hildebrand. 1999. Olfactory systems: Common design, uncommon origins? *Curr. Opin. Neurobiol.* 9:634–640. (A comparison of olfaction in the vertebrates and invertebrates, ranging from anatomy to molecules.)

Torre, V., J. F. Ashmore, T. D. Lamb, and A. Menine. 1995 Transduction and adaptation in sensory receptor cells. *J. Neurosci.* 15:7757–7768. (A highly informative comparison of transduction in vision, olfaction, and audition. Challenging, but highly recommended.)

Zuker, C. S. 1996. The biology of vision in *Drosophila. Proc. Natl. Acad. Sci. USA* 93:571–576. (A review of visual transduction mechanisms by an author who has made major contributions by taking a molecular approach to this topic.)

The Structural and Functional Organization of the Nervous System

The nervous systems of most animals are made up of an enormous number of cells. Even in very simple animals, such as the free-living cnidarian *Hydra* or the tiny nematode *Caenorhabditis elegans,* the nervous system consists of hundreds of separate neurons, and these neurons are not just connected in random order. Rather, particular neurons synapse onto and receive synapses from other neurons in a predictable pattern that varies very little among the individuals of a species. Furthermore, in all members of a species, each neuron with a particular function is found in a predictable location. This regularity of structure and connectivity has permitted neurobiologists to formulate general principles of organization that apply to the wide variety of nervous systems that have been studied. In this chapter we will consider some of these organizational principles and examine specific examples of each.

The main functions of neurons can be divided into three categories: sensory reception, central processing, and motor output. In many simple animals, sensory receptors and neurons controlling motor output are distributed throughout the organism, allowing each region of the body to respond to the environment independently without necessarily activating other regions. This kind of diffusely distributed nervous system is found in modern jellyfishes and in *Hydra.* Remnants of this local control remain a feature of even the most complex nervous systems, as we will see when we consider the spinal cord in vertebrates. However, as more complex animals evolved, neurons became more numerous and neuronal connections became more complex. Concomitantly, the neurons became largely compacted into a central nervous system (CNS). Within the CNS, the cell bodies of neurons are located close to one another, increasing the possibilities for interconnection. With many—even most—neurons located in the central nervous system, receptors and effectors in the rest of the body (called the *periphery*) must connect to the central nervous system by way of long axons.

THE FLOW OF INFORMATION IN THE NERVOUS SYSTEM

Information typically flows into any nervous system from sensory receptor neurons, through a more or less complicated central processing network, and out by way of motor neurons, which produce responses in effectors such as muscles and glands (Figure 8-1 on the next page). The simplest neuronal network that carries information in this way is called a **reflex arc.** It is possible that the primordial reflex arc may have consisted of a single receptor cell that itself directly innervated an effector cell (Figure 8-2a on the next page). In the pharynx of the nematode *Caenorhabditis elegans,* for example, single cells have been identified that probably serve both as sensory receptors and motor neurons. Similarly, several types of neurons that carry out "housekeeping functions" in higher organisms can operate quite autonomously, acquiring sensory information and releasing signaling chemicals that control effectors (Spotlight 8-1 on page 279).

More commonly, a reflex arc consists of at least two, and often many, neurons connected in series. A simple type of reflex arc seen in many species is the **monosynaptic reflex arc** (Figure 8-2b), in which a sensory neuron (the receptor) synapses in the central nervous system onto a motor neuron that innervates an effector, such as a muscle. This type of reflex arc comprises three elements: a sensory neuron, a motor neuron, and the effector. When the sensory neuron becomes sufficiently activated, it excites the motor

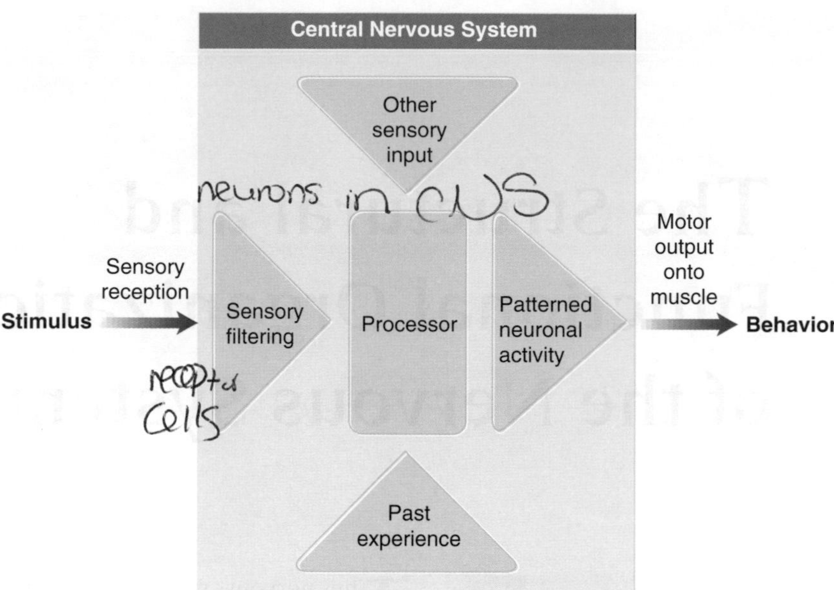

Figure 8-1 Information processing in the nervous system can be divided into functional compartments. Sensory information is acquired and encoded by receptor cells and is filtered and processed by neurons in the central nervous system. This information is integrated with information arriving from other sensory modalities and with memories of past experience, eventually activating motor neurons that control effectors to produce appropriate behavior. The connections among neurons are established during embryonic development based strongly on the genetic program for each species. The connections can be modified during the animal's lifetime depending on experience. Thus a wide variety of influences contribute to linking sensory input to behavioral output.

neuron and, hence, the effector, as shown in Figure 8-2b. All through animal phylogeny, these elementary components of the reflex arc—sensory input pathways and motor neurons that synapse onto muscles—have common features that have been conserved from the most primitive jellyfishes to the most complex vertebrates.

Most reflex arcs include more than one synapse in the pathway and are therefore said to be **polysynaptic.** Such a pathway contains at least one interneuron intervening between the sensory and motor neurons (Figure 8-2c). We will discuss reflex arcs and the role they play in behavior in Chapter 11. In the course of evolution, the number of interneurons increased enormously as animals became more complex, a development that permitted great behavioral complexity in higher animals. Accumulated evidence strongly suggests that if a species has a large number of neurons interposed between input and output neurons, this condition—by itself—confers greater potential for behavioral flexibility and learning.

Our understanding of the nervous system has been aided by a general principle that has emerged from studies of a huge number of species: neurons with similar functions are typically found grouped together. Thus, for example, the somata (cell bodies) of the sensory neurons in a reflex arc are located close together; their axons run in the same nerve and synapse onto interneurons that are clustered together in the central

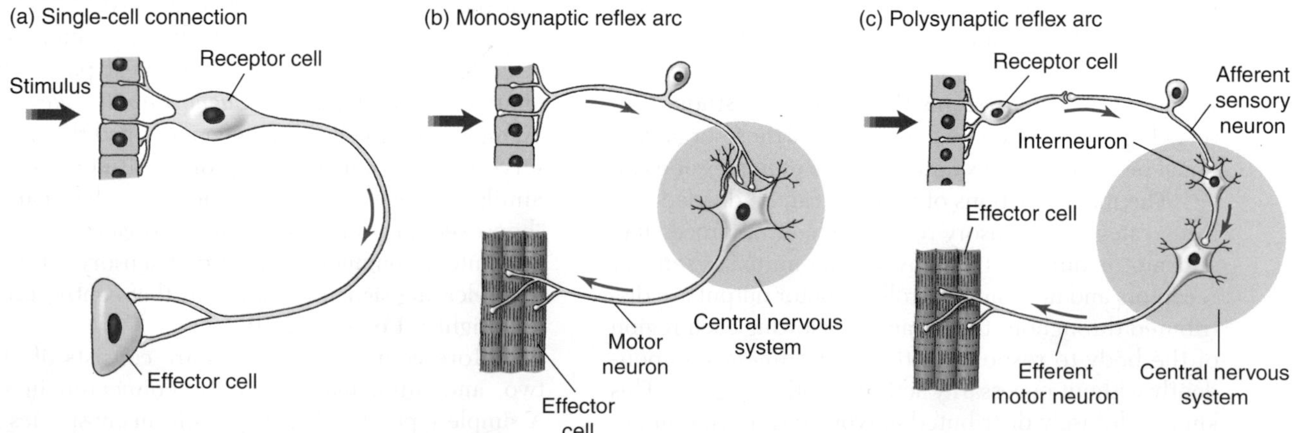

Figure 8-2 In simple reflex arcs, sensory receptors activate effector cells through a small number of synapses. **(a)** In this primitive reflex arc, the same neuron both receives sensory information and controls an effector cell. Some chemoreceptor cells in the pharynx of the nematode *C. elegans* probably act in this way. **(b)** A monosynaptic reflex arc consists of a receptor neuron that synapses onto a motor neuron, which in turn activates muscle fibers. This kind of reflex is called monosynaptic because it includes only a single synapse within the central nervous system. **(c)** This more complicated polysynaptic reflex arc relies on a series of several synapses. In parts b and c, the gray circle indicates the parts of the reflex arc that lie within the central nervous system.

SENSING EFFECTORS AS HOUSEKEEPERS

Even in very complex organisms, some individual cells serve both sensory reception and motor control functions. These multifunctional cells typically regulate crucial physiological variables, such as the level of particular chemical components in the blood. Some examples include vertebrate neurons that release antidiuretic hormone (ADH) to control the osmolarity of blood plasma, leech neurons that modulate excretory function to maintain the [Cl^-] of the blood, and neurons located in the pineal gland of the brain that release melatonin. The accompanying figure summarizes the function of each of these cell types. Interestingly, even though they function autonomously, cells of each type are generally grouped together into a small region, rather than broadly distributed throughout the body, so that they sense conditions in only a small neighborhood. Each of these cell types is narrowly specialized, sensing the level of just one variable and acting to modulate that variable.

Because sensing effectors can act on their own, they can provide rapid and appropriate responses to changes in their local neighborhood. At least some of them, however, receive synaptic input from other neurons, which allows them to integrate information from other parts of the nervous system and respond to changes in more distant parts of the body. The neurons that release ADH, for example, receive synaptic input from other neurons that transmit information about blood pressure, and they modulate their release of ADH based on that input (see Chapter 9 for more information).

The sensing mechanisms employed by sensing effectors are varied, but they are all related to mechanisms found in typical sensory neurons. ADH-releasing neurons, for example, have stretch-inactivated cation-selective channels in their plasma membranes. If the osmolarity of the blood plasma and extracellular fluid increases, water moves out of the cell, the cell shrinks, and the stretch on these cation-selective channels is reduced. As a result, conductance through the channels increases, and Na^+ and Ca^{2+} are driven into the cell, depolarizing it. This depolarization raises the number and frequency of APs in the cell, increasing its release of ADH from its axon terminals, just as increased activity in a typical neuron increases its release of neurotransmitter. Sensory effectors in the pineal of both chickens and trout contain a form of rhodopsin, called pinopsin, that allows them to sense light energy in much the same way the photoreceptors of the retina do. The capture of photons causes these cells to hyperpolarize, which reduces their release of melatonin. Thus melatonin secretion is high in the dark, when the cells are relatively depolarized, and low in the light.

The broad phylogenetic distribution of sensing effectors and their importance for holding critical physiological variables homeostatic over long periods of time suggest that such autonomously acting cells provide a very successful evolutionary solution.

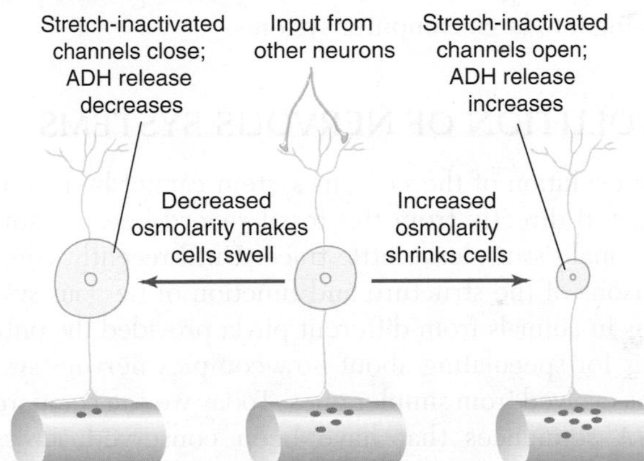

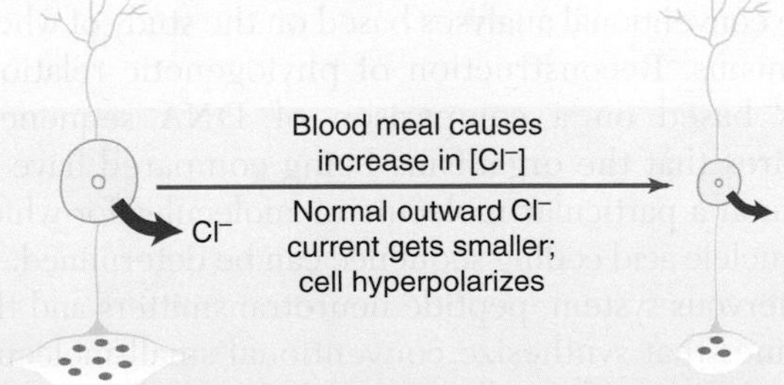

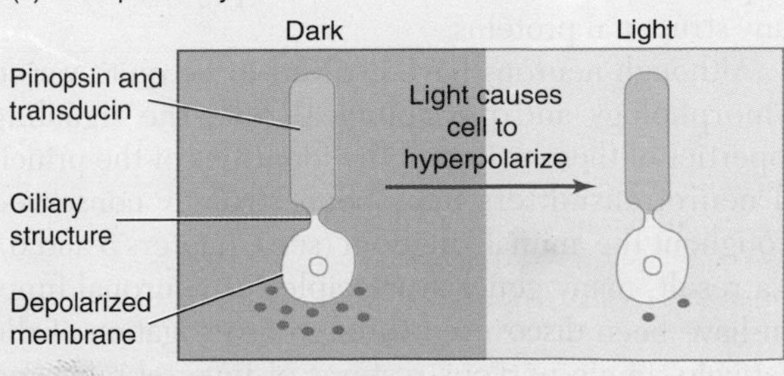

*Sensing effectors use typical sensory mechanisms to acquire information and control the release of chemical messengers to modulate effector activity. **(a)** Magnocellular neurons in the mammalian hypothalamus sense osmolarity by means of stretch-inactivated cation channels. When the osmolarity of the blood plasma and extracellular fluids bathing a magnocellular neuron increases, it causes the neuron to shrink, increasing conductance through the channels. The resulting depolarization of the membrane increases the neuron's release of antidiuretic hormone (see Chapter 9). The neuron releases ADH from its axon terminals, which contact the walls of blood vessels, so that the ADH enters the bloodstream. **(b)** The Cl^--sensitive nephridial neurons of a leech, which are associated with the excretory organs (nephridia) of each body segment, release a peptide that reduces urinary excretion of NaCl. A blood meal typically has a much higher NaCl concentration than leech body fluids. The nephridial neurons sense the increase in [Cl^-] in the plasma that follows a blood meal and reduce their release of the peptide, increasing excretion of NaCl and bringing the [Cl^-] of the plasma back to normal within a day or two. **(c)** In the dark, pinealocytes of the trout brain release the hormone melatonin into the bloodstream. When light shines on these cells, it activates the rhodopsin analog pinopsin and causes the cells to hyperpolarize, reducing the amount of melatonin released. [Adapted from Wenning, 1999.]*

nervous system. The information carried by interneurons eventually arrives at a cluster of motor neurons, all of which innervate the same muscle. These, too, are found in a predictable location.

Information within the nervous system both converges and diverges as it is carried along axons in the system. Convergence of information onto clusters of neurons allows the comparison and integration of information from various sensory modalities, from multiple processing areas of the CNS, from multiple motor centers, or between different levels of processing. Within the CNS, information also diverges. A signal may be propagated along separate pathways to different centers within the CNS, each of which performs a different type of analysis on the information. This ability of the nervous system to carry out different operations on one set of information simultaneously is called **parallel processing**, and it permits the system to analyze information more rapidly than if the different kinds of analysis were performed sequentially. Once the analysis is completed, the information converges again. The resulting increase in efficiency is so great that electronics design engineers now routinely build parallel processing into large computer systems.

EVOLUTION OF NERVOUS SYSTEMS

The evolution of the nervous system cannot be reconstructed directly from the fossil record because soft neuronal tissues leave little trace. Until recently, comparisons of the structure and function of nervous systems in animals from different phyla provided the only basis for speculating about how complex nervous systems evolved from simpler ones. Today we can compare DNA sequences that have been conserved across species, inferring evolutionary relationships based on similarities in the sequences. This method of *molecular phylogeny* provides information that complements more conventional analyses based on the study of whole organisms. Reconstruction of phylogenetic relationships based on a comparison of DNA sequences requires that the organisms being compared have in common a particular molecule or molecules for which the nucleic acid coding sequence can be determined. In the nervous system, peptide neurotransmitters and the enzymes that synthesize conventional small-molecule neurotransmitters (such as acetylcholine and serotonin) can provide the basis for molecular phylogeny, as do many structural proteins.

Although neurons have evolved to be quite varied in morphology and physiological role, the signaling properties of these cells and the identities of the principal neurotransmitters have been strongly conserved throughout the animal kingdom (see Chapters 5 and 6). As a result, many general principles of neuronal function have been discovered through investigation of the relatively simple nervous systems of invertebrates and simple vertebrates, which are more amenable to experimentation than the complex nervous systems of higher vertebrates. In particular, neurons in many invertebrate species are large, accessible, and readily recognizable from animal to animal, making their activity relatively easy to record and analyze. Recently, it has even become possible to do biochemical and molecular analyses of specific single neurons dissected from invertebrate nervous systems. The results from such experiments have been found to apply very well to neurons in more complex systems.

The anatomically simplest nervous systems consist of very fine axons that are distributed in a diffuse network (Figure 8-3). These **nerve nets** are commonly found among the cnidarians (previously called coelenterates, and including hydras, jellyfishes, anemones, and corals). The axons make synaptic contacts at points where they intersect; a stimulus applied to one part of a jellyfish, for example, produces a response that spreads in all directions away from the point of stimulation. If the stimulus is repeated at brief intervals, conduction is facilitated, and the signal spreads farther. Very little is known about synaptic mechanisms in diffuse nerve nets, because the axons are so extremely fine that intra-

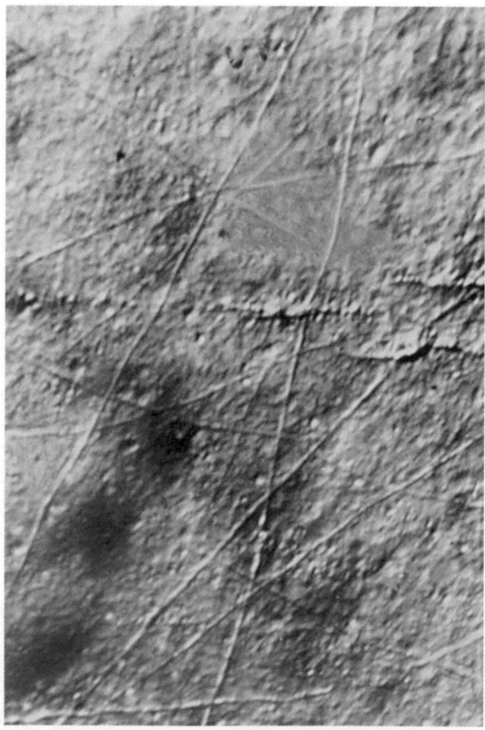

Figure 8-3 Axons in the nerve net of the jellyfish *Aurelia* can be seen on the subumbrellar surface when the organism is viewed in oblique light. The axons are arranged in a diffuse network that carries signals in all directions. The neurons innervate muscle fibers that cause the umbrella to contract. Because the neurons are interconnected, their activity is coordinated, and they can synchronize muscle contraction, permitting more effective locomotion than random muscle contraction would produce. [Courtesy of A. Horridge.]

cellular recording from them is difficult. However, even in the very simple nerve nets of cnidarians and ctenophores (comb jellies), there is evidence that the neurons are organized into reflex arcs.

A major early advance in the evolution of nervous systems was the organization of neurons into **ganglia**, or clusters of neuronal somata. Even animals as simple as cnidarians have some ganglia, and they are common throughout most of the animal kingdom. Many ganglia consist of neuronal somata that are organized around a mass of nerve processes (axons and dendrites) called a **neuropil** (Figure 8-4). This mode of organization permits extensive interconnections to be made among the neurons, while at the same time allowing each

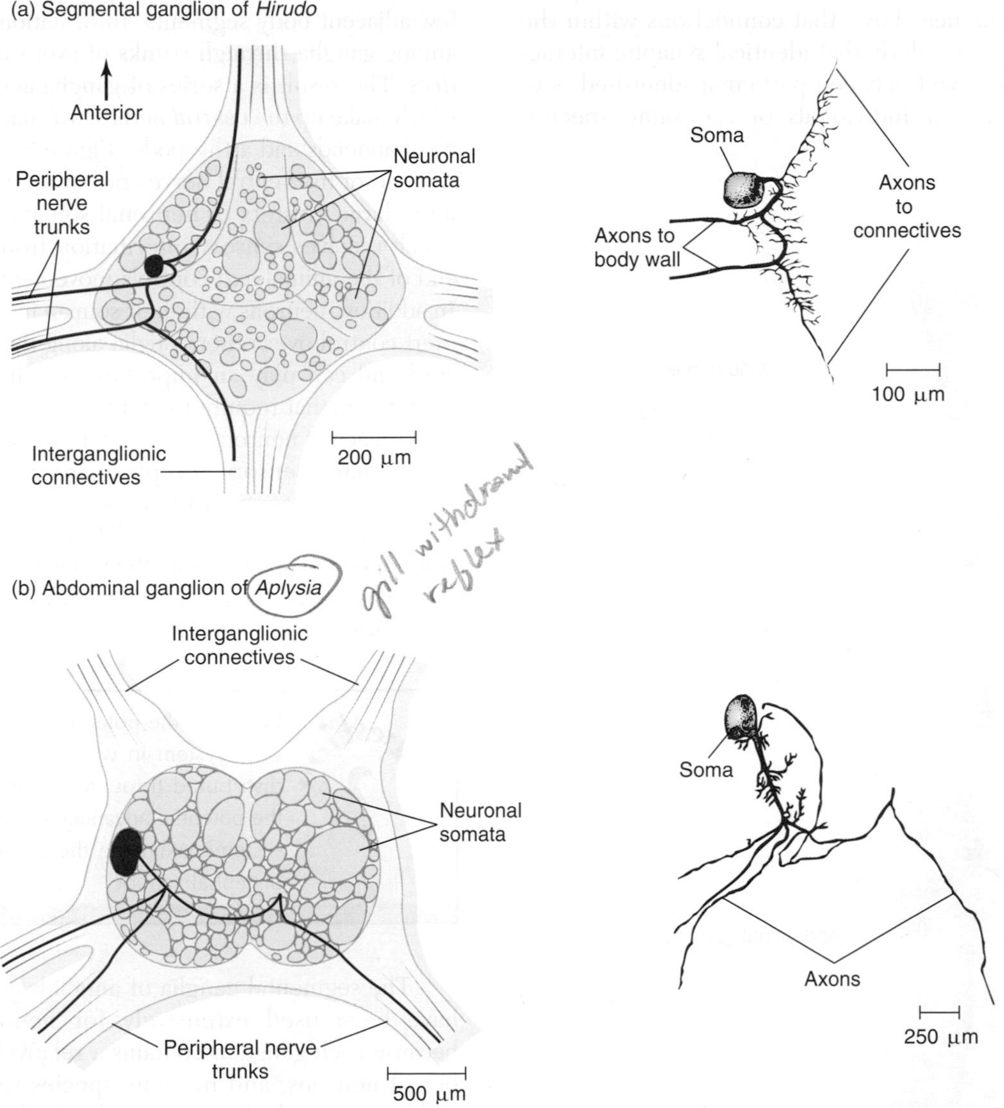

Figure 8-4 In the ganglia of many invertebrates, individual neurons can readily be identified from one individual to another. This figure illustrates ganglia from an annelid and a mollusk. In both parts, anterior is at the top of the figure. **(a)** A segmental ganglion from the leech *Hirudo,* showing the positions of individual neuronal somata. The paired interganglionic connectives (trunks of axons) at the top and bottom of the image contain axons that connect neurons in the ganglion with neurons in the ganglia of other body segments. The peripheral nerve trunks emerging laterally carry motor and sensory axons to the viscera, muscles, and skin. The neuronal somata are all located on the surface of the ganglion, and the processes are contained in the neuropil located inside the ganglion. At right is a mechanosensory receptor neuron (shown in maroon at the left) that has been labeled by injecting it with an intracellular marker that diffuses into all of its branches. Note the numerous small branches that can form synaptic contacts with similar branches of other neurons in the ganglion. The two large axons of this cell enter the peripheral nerve trunks at left, and the two smaller axons enter the interganglionic connectives. **(b)** Abdominal ganglion of the sea hare *Aplysia californica.* Many of the neuronal somata in this ganglion are located on the surface, whereas the interior consists of a neuropil. At right, we see the morphology of a labeled neuron from the ganglion (also shown in maroon at left). This neuron sends axonal branches to all of the peripheral nerves shown at the posterior of the ganglion. [Part a, right, adapted from Muller, 1979; part b, right, adapted from Winlow and Kandel, 1976.]

neuron to produce a minimum number of collateral processes—that is, side branches arising from the axon or axons. The many fine processes in a neuropil seem at first sight to be arranged randomly, but injection of markers or dyes into individual neurons (see Figure 8-4, right) reveals that the major structural features of each particular type of neuron, such as the arrangement of its axons and the location of branches, are similar from one individual animal to another. Moreover, physiological evidence shows that connections within the neuropil are so orderly that identical synaptic interactions are observed between particular identified neurons in different individuals of the same species.

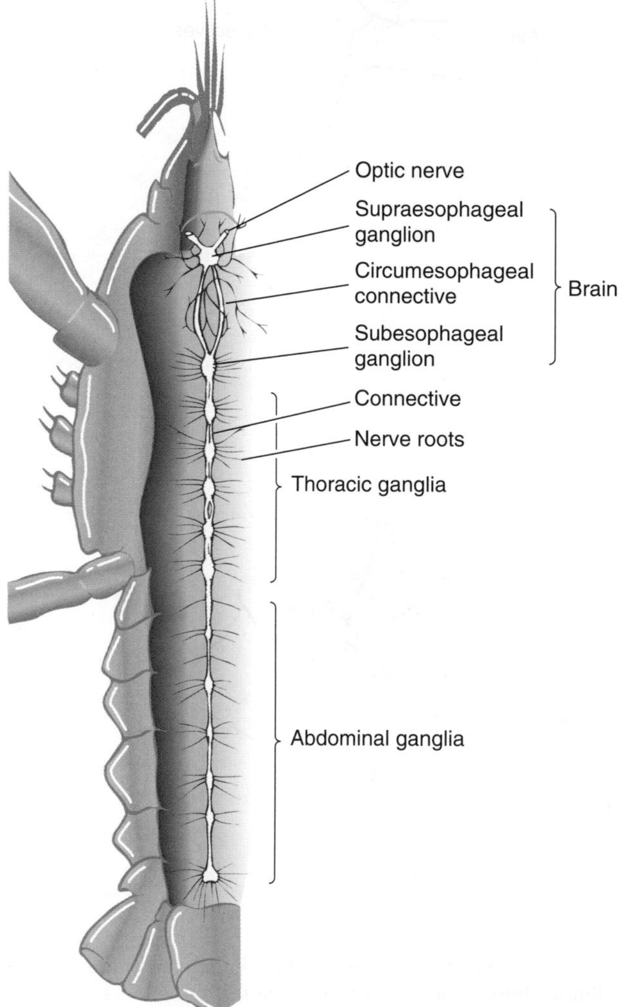

Figure 8-5 The ventral nerve cord of the crayfish *Astacus* illustrates the segmented arrangement of the nervous system in many invertebrates. Each segment of the body houses one ganglion. Connectives contain axons that carry information between ganglia, and the nerve roots that link the ganglia with structures in the periphery contain sensory and motor axons. The bilateral symmetry that is typical of nervous systems in most phyla is apparent in this diagram. (Notice that the abdominal ganglion of *Aplysia* shown in Figure 8-4, like many molluscan ganglia, is asymmetric and is thus an exception to this general rule.)

Relatively early in animal evolution, many of the ganglia became distributed along the line marking the plane of bilateral symmetry of an animal's body. This set of ganglia is called the central nervous system.

Many segmented invertebrates have central nervous systems that are somewhat distributed, consisting of a ganglion within each body segment. Each segmental ganglion usually serves the reflex functions of the segment in which it is located, plus perhaps those of one or a few adjacent body segments. Information is exchanged among ganglia through trunks of axons called *connectives.* The result is a series of ganglia and connectives, which make up the *ventral nerve cord* that is characteristic of annelids and arthropods (Figure 8-5). At the anterior end of the ventral nerve cord in these phyla, one or more large clusters of neuronal somata form a **brain,** which receives sensory information from the anterior end of the animal and controls movements of the head. In addition, neurons with their somata in the brain often exert control over other ganglia along the ventral nerve cord and can play an important role in coordinating movements that require the body to work as a unit. This coalescence of neurons at the anterior end of the animal, where many sensory receptors are typically concentrated, is called **cephalization** and is a common feature of central nervous systems. (Not all brains are located at the anterior end of an organism; the leech, for example, has a tail brain at the posterior end of its ventral nerve cord that is even larger than the head brain.)

What are the potential advantages of a nervous system in which neurons are broadly distributed throughout the body? What are the potential advantages of a central nervous system? What are the potential advantages of cephalization?

The segmental ganglia of annelids and arthropods have been used extensively for neuronal analysis because each ganglion contains a relatively small number of neurons, and in many species several ganglia within an individual contain similar or identical complements of neurons. Thus, in principle, an analysis of interactions among neurons in one segment of the ventral nerve cord can be generalized to many—even to all—other segments. This approach has been useful in studying the nervous system of the leech (see Figure 5-5), in which the ganglia in all segments are very similar to one another. Despite the simplicity of its nervous system, however, a leech can perform such complex tasks as swimming to seek food or crawling to escape danger, which makes it particularly well suited for studies of the neuronal basis of behavior.

The structure of the nervous system varies among the other invertebrate phyla. In contrast to annelids and arthropods, whose body plans are segmentally structured and bilaterally symmetric, an echinoderm typically has a nerve ring and radiating nerves arranged about its axis of symmetry. Perhaps as a result of their radial symmetry, echinoderms lack any brainlike ganglia. Mollusks have a nonsegmental and asymmetric nervous system consisting of several dissimilar ganglia that are connected by long nerve trunks.

Neurons in some molluscan species have contributed greatly to our understanding of neuronal interactions. The ganglia of opisthobranch mollusks (e.g., the sea hare *Aplysia*) and nudibranchs (e.g., *Tritonia*) contain a number of neurons with extraordinarily large somata; in *Aplysia,* the somata of some neurons are more than 1 mm in diameter. Individual giant neurons can be recognized visually from individual to individual, and they lend themselves well to long-term electrical recording, injection of experimental agents, and isolation for microchemical analysis. As in the nervous systems of annelids and arthropods, individual neurons in mollusks can be reliably identified on the basis of the location and size of their somata, enabling an experimenter to determine the properties that characterize a particular type of cell and to measure the amount of variation that can occur from individual to individual.

The most complex nervous system known among the invertebrates belongs to the octopus. The brain alone is estimated to contain 10^8 neurons—compare this number with the 10^3 neurons in a leech's brain and 10^5 neurons in the entire leech body. The neurons in the octopus brain are arranged in a series of highly specialized lobes and tracts that evidently evolved from the more dispersed ganglia of the simpler mollusks. If number of neurons in any way correlates with intelligence, the octopus should be fairly smart, and behavioral studies have shown that, at least by invertebrate standards, it is indeed quite intelligent.

In general, the nervous systems of invertebrate animals, excluding the octopus, contain significantly fewer neurons than those of the vertebrates; for this reason, invertebrate nervous systems are often referred to as "simple." However, superficial appearances can be deceiving, and the functional sophistication of even relatively simple nervous systems becomes apparent on closer examination. In Chapter 11, we will consider examples of invertebrate behavior and the neuronal circuitry underlying that behavior.

From observation of a wide variety of species from throughout the animal kingdom, several principles of the evolution of nervous systems have emerged:

- *Nervous systems in all organisms are based on one type of cell, the neuron.* Although neurons are sculpted into myriad shapes in the course of development, the mechanisms of electrical signaling within the cell and the nature of the chemical signals that allow information to be transmitted between cells have been highly conserved across phylogeny.

- *The organization of the nervous system evolved through the elaboration of one fundamental pattern: the reflex arc.* Just as the neuron is the basic structural unit of the nervous system, the reflex arc is its basic operating unit. In its simplest form, a reflex produces a stereotypic response to a particular sensory stimulus.

- *There has been a trend in evolution toward the gathering of neurons into a central nervous system,* which is connected to peripheral sensory receptors and muscles by long axons. The organization of these networks favors one-way conduction through neurons, from dendrites to the axon and thence to the axon terminals, although the biophysical properties of axons would enable them to conduct signals either toward or away from the soma.

- *Complex organisms have more neurons than simpler organisms.* In complex species, the neurons are concentrated in a brain, usually located in the head.

- *As nervous systems became more complex through evolution, new structures were added on to older structures, rather than replacing them.* As a result, structures that serve functions that have arisen recently in evolution are often seen to be literally layered onto older, more primitive structures. One example is the cerebral cortex of the mammalian brain, which will be discussed in detail later in this chapter.

- *The relative size of each region in the brain of a species is usually related to how important sensory input into that region, or motor control out of that region, is for the survival of that species.* For example, in animals that depend primarily on vision, the regions of the brain that process visual information are typically larger than all other sensory areas. In nocturnal animals, other brain regions—for example, those that process hearing or olfaction, which are independent of light—are largest.

- *Particularly in the vertebrates, many regions of the brain are organized into topological maps* in which neurons processing information related to adjacent regions of the body are located close to one another. Neurons that receive information from the left index finger, for example, are likely to be close to neurons that receive information from the left thumb and farther from neurons that receive information from the left elbow.

ORGANIZATION OF THE VERTEBRATE NERVOUS SYSTEM

The vertebrate nervous system is organized into identifiable structural and functional regions (Figure 8-6), although neurons in these separate regions may work together to process incoming information and generate appropriate behavior. The vertebrate nervous system can be divided into the central nervous system (CNS) and the peripheral nervous system (PNS). The central nervous system contains most of the neuronal somata. It encompasses the entire structure of all interneurons and contains the somata of most neurons that innervate muscles and other effectors. Within the central nervous system, collections of somata of neurons with similar function are called **nuclei**, and bundles of axons extending from those somata are called **tracts.**

The peripheral nervous system includes **nerves,** which are bundles of axons from sensory and motor neurons; ganglia that contain the somata of some autonomic neurons; and ganglia that contain the somata of most sensory neurons. (A notable exception is the retina, which is part of the central nervous system.) Nerves are considered *afferent* if they carry information toward the CNS and *efferent* if they carry information away from it. Many vertebrate nerves are *mixed nerves,* including both afferent and efferent axons.

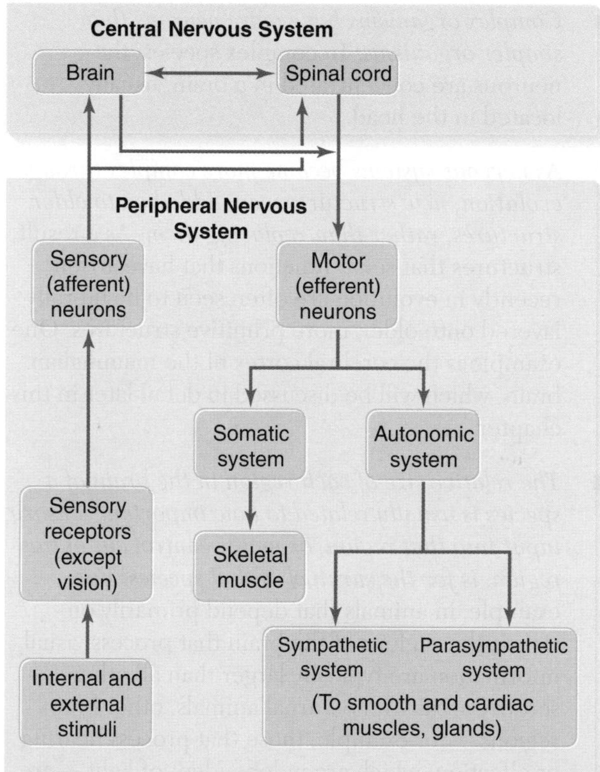

Figure 8-6 The vertebrate nervous system is organized into several identifiable regions that are interconnected. The central nervous system consists of the brain and spinal cord. Information about the environment is acquired by sensory receptors and brought to the central nervous system by sensory (or afferent) neurons. The animal's response is initiated and shaped by motor (or efferent) neurons. Motor neurons of the somatic, or voluntary, nervous system control the contraction of skeletal muscles. Efferent neurons of the autonomic nervous system, which is divided into sympathetic and parasympathetic branches, control the activity of smooth muscles, cardiac muscles, and some glands.

Efferent output from the central nervous system can be divided into two main pathways. The **somatic nervous system** is also referred to as the *voluntary system,* because motor neurons of this pathway control skeletal muscles to produce movements that are under the animal's voluntary control. The **autonomic nervous system** includes efferent neurons that modulate the contraction of smooth and cardiac muscle and the secretory activity of glands. The autonomic nervous system thus controls the animal's "housekeeping" functions, such as heartbeat, digestion, and temperature regulation. The term *autonomic*, which means "self-managed," was introduced when the relation between the autonomic nervous system and the more voluntary parts of the central nervous system was poorly understood. We now know that the apparently automatic responses regulated through the autonomic nervous system are integrated and controlled within the central nervous system, just like the responses produced consciously through more voluntary channels, and that connections between the autonomic and somatic nervous systems allow each to influence the other. As we will see below, the neurons of the autonomic system are themselves divided in two pathways, the sympathetic and the parasympathetic, which differ from each other both anatomically and functionally.

A striking characteristic of vertebrate nervous systems is their enormous redundancy: there are a large number of individual neurons of every identifiable type. In an arthropod nervous system, a single motor neuron may control all of the fibers in a particular muscle. In some cases, a single neuron may even, by itself, control more than one muscle of a limb. In contrast, each skeletal muscle in the vertebrates is typically innervated by a pool of several hundred motor neurons or more, each controlling a group of individual muscle fibers. The motor neurons of each pool have common physiological properties, so physiological information obtained from one motor neuron generally characterizes the whole pool. The fundamental repetitive orderliness of the system is thought to have arisen as natural selection conserved many neuronal properties through time while permitting duplication of individual units.

What are the potential advantages of a large number of neurons and thus a large brain? What are the potential disadvantages? For mammals, what might limit brain size?

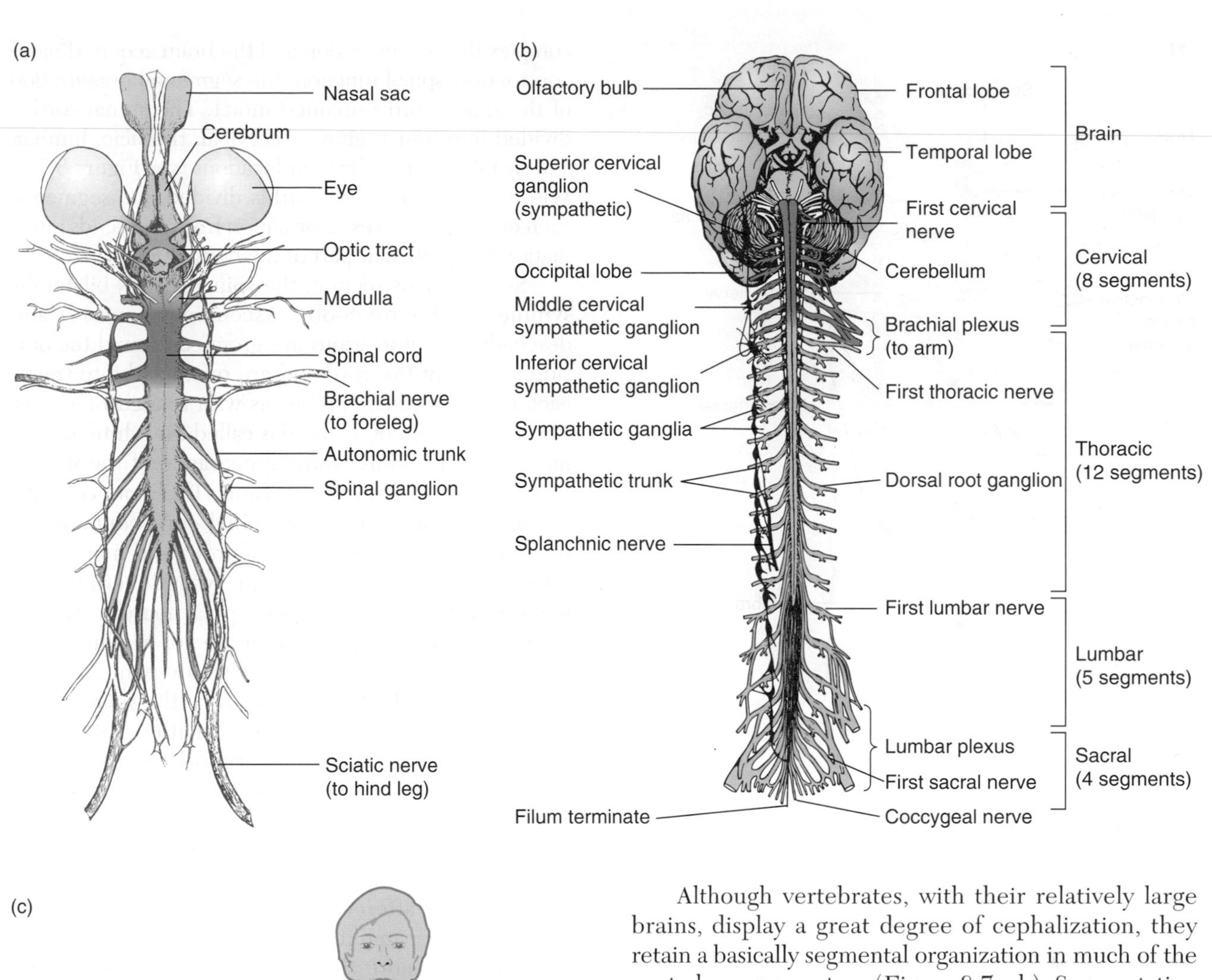

Although vertebrates, with their relatively large brains, display a great degree of cephalization, they retain a basically segmental organization in much of the central nervous system (Figure 8-7a, b). Segmentation is particularly apparent in the organization of the spinal cord (the vertebrate analog of the invertebrate ventral nerve cord) and in the pattern in which the spinal segments send nerves out into the body.

The Spinal Cord

In the simpler vertebrates, the spinal cord, enclosed and protected by the vertebral column (Figure 8-8a on the next page), is a site of reflex action that can act independently of the brain but that also receives input from higher centers in the brain. As vertebrates became more

Figure 8-7 The structure of the vertebrate central nervous system is very elaborate in the head, but a segmental organization is maintained in both the brain and the spinal cord. Diagrams show the central nervous system of **(a)** a frog and **(b)** a human seen from the ventral side. Although segmentation is not obvious in many structures of the brain, it is the basis of organization within the spinal cord. Spinal segments are divided into four regions: cervical (in humans, C1 through C8), thoracic (T1 through T12), lumbar (L1 through L5), and sacral (S1 through S4). **(c)** Each spinal segment receives information from and controls movement in a defined region of the body. Thus both the spinal cord and the body itself are segmented. [Adapted from Wiedersheim, 1907, and Romer, 1955.]

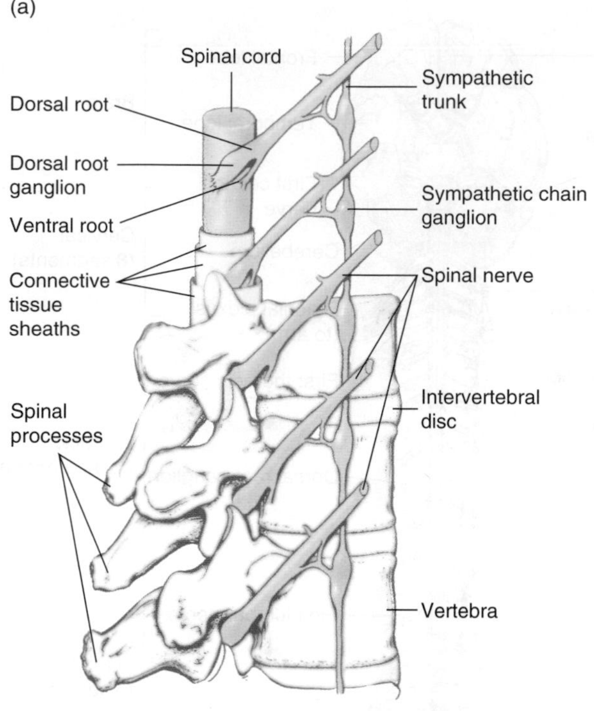

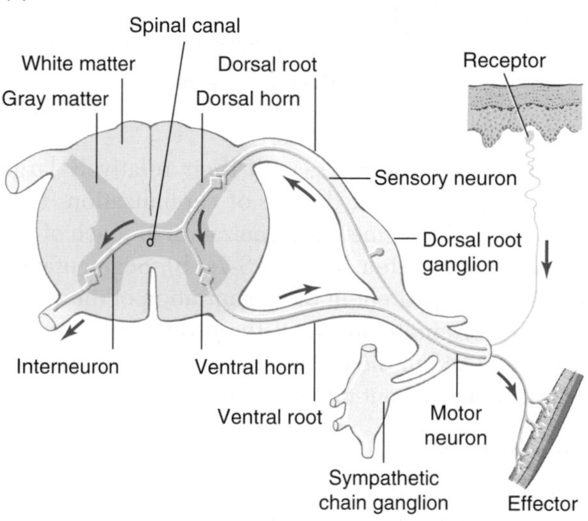

Figure 8-8 Each segment of the spinal cord is connected to structures in the periphery by paired dorsal roots, through which afferent signals enter, and by paired ventral roots, through which efferent signals leave. **(a)** Relation between the vertebral column, the spinal cord, and the spinal nerves, which emerge between the vertebrae. Vertebrae are separated from one another by cushions of connective tissue, the intervertebral discs. Parts of the sympathetic nervous system located in the periphery are included in this diagram. Nervous tissue is shown in yellow. **(b)** Diagram of a cross-section through a single spinal segment. Neurons in a polysynaptic reflex arc that connects input from skin receptor endings with output to skeletal muscles are shown in deep yellow. The direction of signal transmission is indicated by red arrows. The pathway includes an interneuron in addition to the sensory receptor and the motor neuron. Note that the soma of the sensory afferent neuron resides in a dorsal root ganglion outside the spinal cord. The spinal cord is bilaterally symmetric, although for clarity structures from only one side are illustrated in this diagram. [Adapted from Montagna, 1959.]

complex through evolution and the brain acquired more control over spinal function, the segmental organization of the spinal cord remained intact. The spinal cord is divided into four regions—cervical, thoracic, lumbar, and sacral—named for their locations (see Figure 8-7b). Within each region, the cord is divided into segments, each of which receives information from and sends information to a particular part of the body (Figure 8-7c).

Seen in cross-section, the spinal cord is bilaterally symmetric (Figure 8-8b). Ascending (sensory) and descending (motor) axons are grouped around the outside surface of the cord and are organized into tracts, each of which consists of axons with related functions. This outer region of the cord is called the **white matter,** named for the shiny white appearance of the myelin that ensheathes many of the axons. The more centrally located **gray matter** of the spinal cord contains the somata and dendrites of interneurons and motor neurons, as well as the axons and presynaptic terminals of neurons that synapse onto these spinal neurons. Most of the structures in the gray matter are not myelinated, so this central region lacks the shiny whiteness of the tracts. A fluid-filled central cavity, the *spinal canal,* is continuous with fluid-filled cavities within the brain, which are called the *cerebral ventricles.* The **cerebrospinal fluid** in these cavities has a composition similar to that of blood plasma.

The regular organization of the spinal cord has facilitated its study by neurophysiologists. To a very large extent, afferent and efferent pathways are anatomically separated from each other (a pattern that was observed more than a century ago and that has been referred to as the "Bell-Magendie rule"). In each body segment, afferent axons enter the central nervous system through the *dorsal roots* on each side of the spinal cord (see Figure 8-8b); efferent nerve fibers carry information out of the central nervous system through the *ventral roots.* (There are minor exceptions to this rule. In the cat, for example, fine unmyelinated sensory afferent neurons enter the spinal cord through at least some of the ventral roots.) The somata of spinal motor neurons are located in the ventral gray matter, called the *ventral horn,* whereas the somata of interneurons that receive and transmit sensory information are located in the dorsal gray matter, called the *dorsal horn.* Afferent axons that synapse onto sensory interneurons within the cord arise from sensory receptor neurons whose somata are located in *dorsal root ganglia* outside the central nervous system. (There is one dorsal root ganglion on either side of each spinal segment.) The segregation of sensory and motor axons into the dorsal and ventral roots makes it possible to selectively stimulate afferent or efferent neurons in a single spinal segment. Alternatively, input to a segment or output from it can be selectively eliminated by transecting one of the spinal roots.

Many neuronal connections that produce reflex behaviors are located in the spinal cord. (In Chapter 11, we will consider one of these spinal reflexes, the stretch reflex.) In addition, neuronal connections that produce patterned movements of locomotion—for example, walking, running, and hopping—are made in the spinal cord. Although signals from the brain can activate, suppress, or modulate these behavioral patterns, it appears that connections among spinal neurons are sufficient by themselves to generate these complex and coordinated locomotory behaviors.

What advantages might lead the neuronal circuitry underlying locomotory behaviors to have evolved entirely within the spinal cord? Why not have the brain direct the entire behavior? What advantages could arise from having input from the brain modulate the activity in locomotory circuits?

The Brain

In all vertebrates, the brain is organized into groups of neurons with specialized functions, such as reception and processing of information from the eyes or initiation of movements that require coordination of the whole body. In the higher vertebrates, the brain contains many more neurons than does the spinal cord, and it exerts very strong control over the rest of the nervous system.

Structure of the vertebrate brain

Although segmentation is much less obvious in the brain than in the spinal cord, the brain retains a vestige of segmental organization in the bilaterally paired cranial nerves, which connect centers of the brain with structures in the head and the rest of the body (Figure 8-9). The basic structure of the vertebrate brain is similar in all vertebrate classes (Figures 8-10 and 8-11 on the next two pages). Some structures, such as the cerebellum and the amygdala, are found in the brains of all vertebrates; others, such as the cerebral cortex, are found only in higher vertebrates (Figure 8-11). Figures 8-10 and 8-11 are labeled to indicate only a few parts of the brains shown, but the labeled regions illustrate both brain areas that have been preserved throughout evolution and new ones that arose in the course of the evolution. We will consider brain structures from the most *caudal* (toward the tail) to the most *rostral* (toward the forehead).

At the most caudal part of the brain, where the brain joins the spinal cord, the cord enlarges to form the **medulla oblongata** (also called simply the medulla). The medulla contains centers that control respiration and autonomic function. It also contains groups of neurons that receive and relay sensory information from several modalities (e.g., the organs of equilibrium and of hearing) and other neuron clusters that receive and relay information from motor centers.

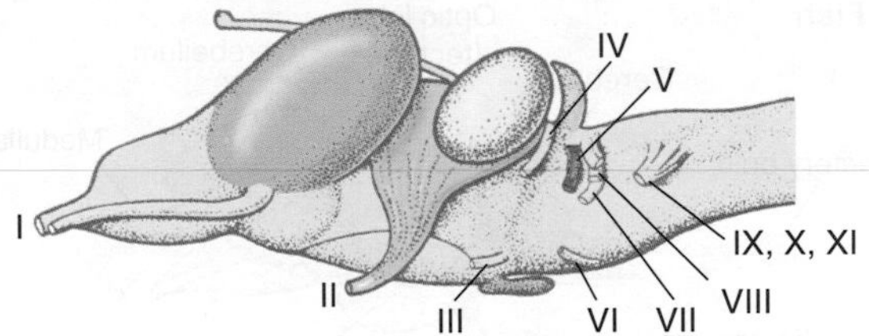

Figure 8-9 Although segmentation is obscured in much of the brain, it is apparent in the organization of the bilaterally paired cranial nerves, seen here in the brain of a frog. This brain is viewed from the left side, so that only the left nerve of each pair is drawn. All vertebrate brains are connected with the periphery by twelve pairs of cranial nerves, each of which emerges from a predictable region of the brain. The order and naming of these nerves is invariant among species. For example, the olfactory nerve is the first cranial nerve in all vertebrates; the optic nerve emerges somewhat more caudally and is the second cranial nerve. Some cranial nerves are purely sensory in function (such as the olfactory and optic nerves); some have purely motor functions (such as the spinal accessory and hypoglossal nerves). Several, such as the trigeminal and vagus nerves, carry both sensory and motor information.

The **cerebellum**, which is dorsal to the medulla, consists of a pair of hemispheres that have a smooth surface in lower vertebrates and a very convoluted surface in higher vertebrates. Neurons with somata in the surface layer are important in cerebellar function, and the convolutions increase the surface area, providing space for many more neurons. The cerebellum has long been known to contribute to the coordination of motor output. It compares and integrates information arriving from the semicircular canals and the neurons that provide information about muscle stretch and the positions of joints (together called *proprioceptors*) and from the visual and auditory systems. Nerve signals leaving the cerebellum help to coordinate the motor output responsible for maintaining posture, orienting an animal in space, and producing accurate limb movements. The relative size of the cerebellum differs greatly among species and exemplifies the principle that the size of a brain region is correlated with its relative importance in the behavior of each species. For example, a bird's cerebellum is significantly larger relative to total body size than the cerebellum of most mammals. This difference in size is thought to reflect the greater complexity of motor control and orientation in space required by an animal that gets around by flying in three dimensions, rather than walking on the two-dimensional surface of the earth.

The cerebellum lacks any direct connection to the spinal cord, so it cannot control movement directly. Instead, it sends signals to regions of the brain that do

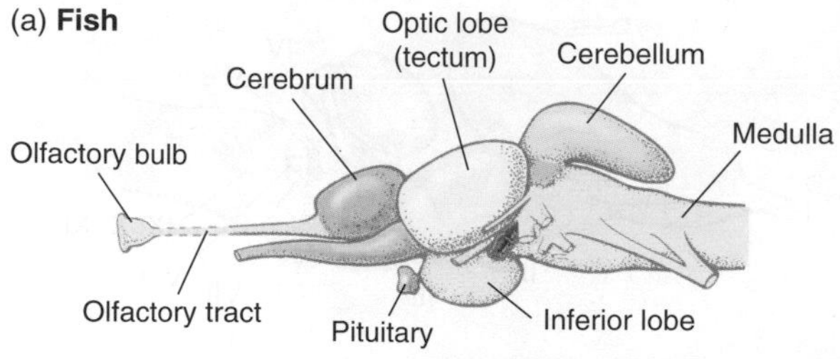

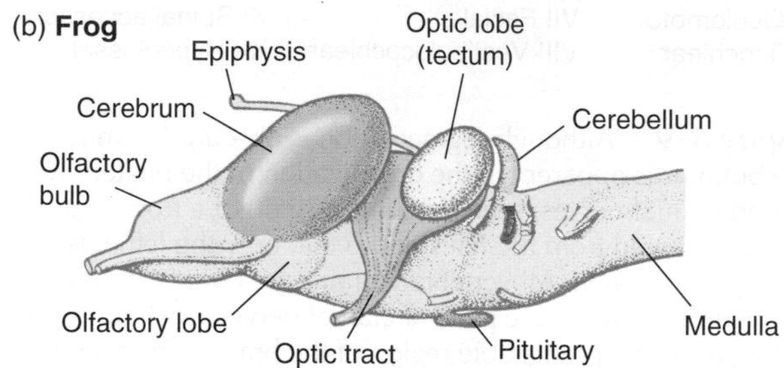

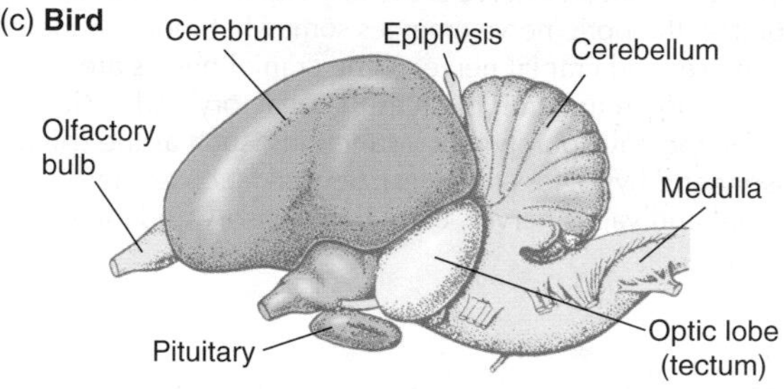

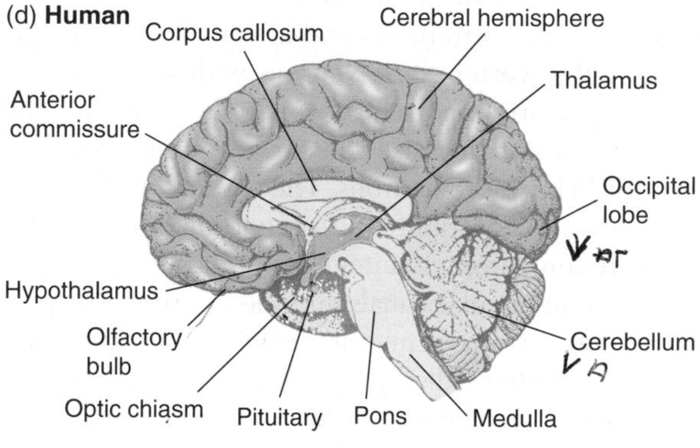

Figure 8-10 The brains of all vertebrate classes have many structures in common, although the relative sizes of particular regions vary among species. These diagrams show lateral views of the brain in **(a)** a fish, **(b)** a frog, **(c)** a bird, and **(d)** a human. The human brain has been sectioned down the midline along the plane of bilateral symmetry. Note that the cerebrum occupies a larger fraction of the brain in higher vertebrates than in lower vertebrates. In contrast, the tectum maintained its relative size during evolution. (In mammals, the tectum is located in the pons and is called the superior colliculus. It is not visible in the human brain because it consists of a bilateral pair of nuclei that lie lateral to the midline.) The cerebellum, which has an important function in the coordination of movement, is highly developed in birds and mammals. [Adapted from Neal and Rand, 1936; Romer, 1955.]

directly control movement. In addition, the cerebellum has been found to participate in the learning of motor skills, and recent observations suggest that abnormalities in cerebellar neurons may contribute to the problems faced by autistic people. As we learn more about cerebellar function, it is possible that this part of the brain will be found to have an even broader role in the regulation of behavior than has been suspected up to now.

Some of the structures of the lower brain act as relay stations, receiving information and transmitting it to other centers. In mammals, the *pons* (whose name means "bridge") consists of fiber tracts that interconnect many different regions of the brain. Tracts in the pons, for example, connect the cerebellum and the medulla with higher centers. The **tectum** (also called the optic lobe, or in mammals the *superior culliculus*), located in the pons, receives and integrates visual, tactile, and auditory inputs. Information from each of the sensory modalities that comes into the tectum is organized into a map that represents some feature of the environment. In the projection of visual input, for example, locations in the environment that are close to one another are represented as being close to one another in the visual map. Maps from different sensory modalities are located in different layers of the tectum, but the maps are congruent with one another. For example, the tectum of the barn owl brain receives both auditory and visual inputs. In the map of auditory inputs, the locations of sound sources are related to one another with the same geometry as objects in the visual world are organized in the visual map. In fishes and amphibians, the tectum is a major source of control over body movements. Indeed, surgical removal of the cerebral hemispheres from a frog's brain does not substantially reduce the frog's behavioral capacity, but removal of the tectum incapacitates the animal. The tectum also plays a major role in reptiles and birds, whereas in mammals it has a more limited function.

In higher vertebrates, the **cerebral cortex**—a multilayered collection of cells on the outer surface of the cerebrum—has taken over many tasks performed by the tectum in lower vertebrates. This region of the brain is most enlarged and elaborated in humans, but it is typical of all mammals. The cerebral cortex is subdivided into functional regions that will be considered later in this chapter.

The **thalamus** is a major coordinating center for sensory and motor signaling. It serves as a relay station for sensory input, providing some information processing as well. In mammals, sensory information is sent by neurons of the thalamus to sensory regions of the cerebral cortex, and motor information is received from the motor regions of the cortex by the thalamus and relayed to other motor centers. The cortex not only receives information from the thalamus; it can also modify thalamic function to change the nature and amount of information the thalamus relays. Such feedback relations

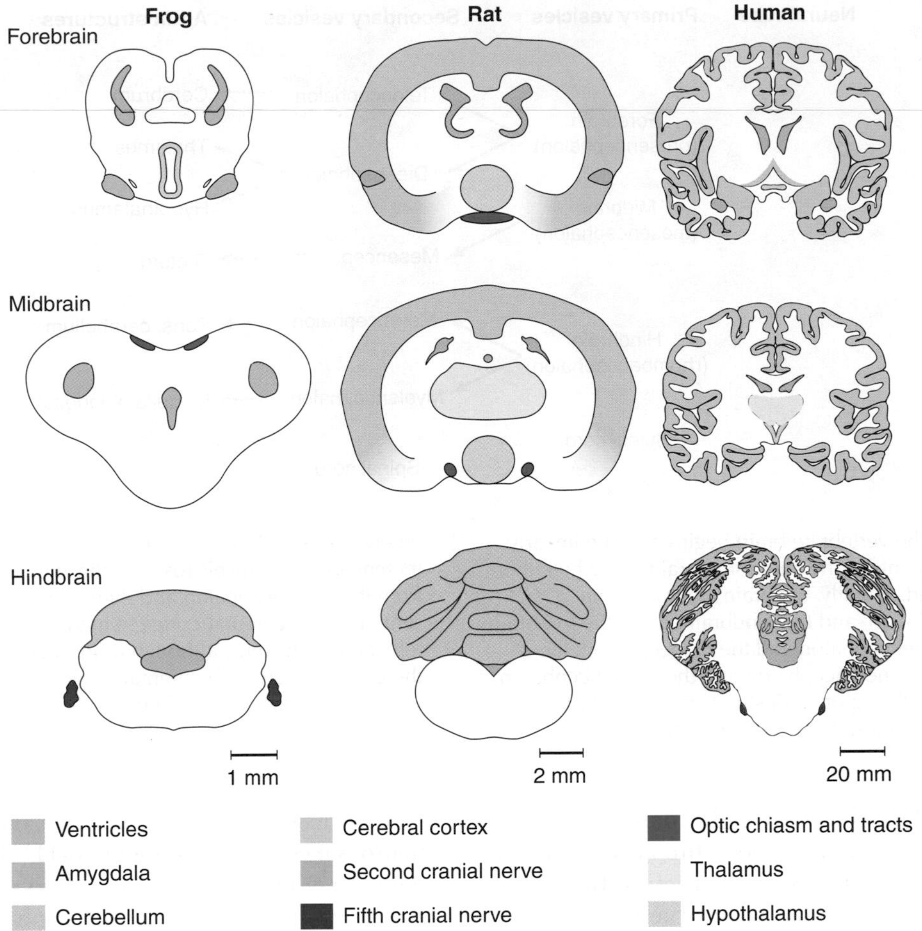

Figure 8-11 Some nerve centers in the brain are found in all vertebrates, whereas others are typical only of the higher vertebrates. This diagram shows transverse sections made at three locations in the brain of a frog, a rat, and a human. The locations of the sections are approximately homologous in the three species, but differences in relative sizes and organization of the brains makes strict homology impossible. Notice that the brain of the frog lacks a cerebral cortex, which is found only in higher vertebrates. The cerebral cortex in the rat is smooth and simple in structure, whereas the cerebral cortex is enormously elaborated in the human brain. The many folds greatly increase the surface area, providing space for a huge number of neurons. It has been estimated that the primate cerebral cortex contains approximately 10^{12} neurons. In contrast, the amygdala and the hypothalamus—centers for regulating visceral function and processing emotional responses such as fear—are present in all three species and in approximately the same location. The cerebellum is present in all vertebrates, but it is more elaborate in the human than in either of the other two species shown. [Adapted from König and Klippel, 1967; Kemali and Braitenberg, 1969; Hanaway et al., 1998.]

between parts of the brain are common, and they can powerfully modify brain function.

The **amygdala** processes information and organizes output that is related to the emotions. In this task, the amygdala is linked to the **hypothalamus**. The hypothalamus includes a number of centers that control functions related to survival of the individual and the species, such as body temperature regulation, eating, drinking, and sexual appetite. Hypothalamic centers also participate in the expression of emotional reactions such as excitement, pleasure, and rage. Neuroendocrine cells in the hypothalamus control water and electrolyte balance and the secretory activity of the pituitary gland (Chapter 9).

The most anterior part of the brain contains structures related to its oldest and newest features. The olfactory system forms the largest part of the anterior brain in many primitive vertebrates, suggesting that detecting food odors and interpreting chemical communication must have been powerful selective agents in the evolution of early vertebrates. In lower vertebrates, the primitive cerebrum integrates olfactory signals and organizes motor responses to them. Perhaps as a manifestation of its ancient importance, olfaction is the only sensory modality that is not processed through the thalamus in mammals, but instead travels directly to the cerebrum. The huge cerebral hemispheres that dominate the human brain evolved from the small cerebrum of fishes and amphibians.

Development of the vertebrate brain

The segmental organization of the vertebrate central nervous system appears to be based on developmental

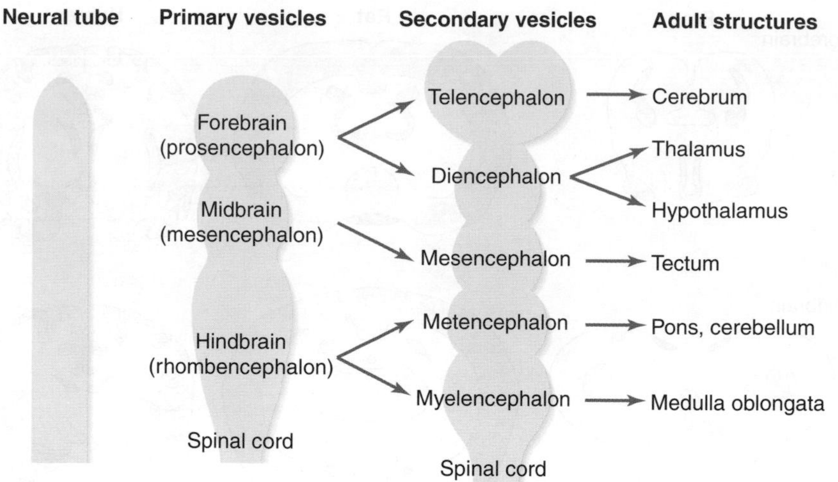

Figure 8-12 The vertebrate brain begins as three linearly arranged enlargements of the anterior neural tube, which then become elaborated. Initially, the brain consists of the forebrain, the midbrain, and the hindbrain. Later, the forebrain divides into the telencephalon and the diencephalon; the hindbrain forms the metencephalon and the myelencephalon. At right is a partial listing of adult structures that arise from these secondary vesicles. A lot of sensory information is received in structures derived from the myelencephalon and metencephalon, travels rostrally through structures derived from the mesencephalon and diencephalon, and is then sent on to the cerebral cortex, which develops from the telencephalon. Thus, although morphogenetic changes in the brain obscure its segmental origins, the pattern of information processing echoes its early segmentation.

processes that have been strongly conserved during evolution. The progenitor of the entire nervous system is the *neural tube,* a fluid-filled structure that arises from part of the outermost layer of a gastrula-stage embryo. The first step in the development of the brain is the formation of three expanded vesicles in the most anterior part of the neural tube; these vesicles are precursors to (anterior to posterior) the *forebrain,* the *midbrain,* and the *hindbrain* (Figure 8-12). The center of the tube is a fluid-filled cavity, which is the precursor of the cerebral ventricles and spinal canal (see Figure 8-11). In later stages, these three regions generate a total of five subdivisions. These five segments grow by cell division, particularly along the surface of the fluid-filled cavity, the *ventricular zone,* followed by migration of cells away from this zone. Embryonic neurons usually migrate within their segment of origin, although some have been found to cross boundaries between segments. The segmental and linear organization that appears early in the development of the brain is eventually obscured by distortions produced through unequal growth within the embryonic subdivisions. However, segmentation remains in the distribution of the cranial nerves, and the linear organization of the primary vesicles is preserved to some extent in the organization of the pathways traversed in the adult brain by incoming and outgoing information.

Organization of the mammalian cerebral cortex

In higher mammals, the cerebral cortex—the layer of cells that covers the two hemispheres of the cerebrum—is thrown into prominent folds that greatly increase its surface area and, hence, the total number of neurons it contains (see Figure 8-11). This surface coating of gray matter is organized into sublayers that run parallel to the surface, each having a recognizable pattern of input and output. In addition, the cortex is organized into functional regions (Figure 8-13). Some areas of the cerebral cortex contain neurons that are purely sensory in function; that is, they receive sensory information, process it, and pass it on. Other areas have purely motor functions. In relatively primitive mammals, such as rats, the sensory and motor areas account for almost all of the cortex. In contrast, the cerebral cortex of humans and other primates contains large regions that are neither clearly sensory nor clearly motor in function. These regions are referred to as **association cortex,** and they are responsible for such complex functions as intersensory associations, memory, planning future behavior, thought, and communication.

The cortical regions that are purely sensory in function include the primary auditory, somatosensory, and visual cortical areas. (The first location in the cortex to which information from a particular sensory modality is transmitted is called the *primary projection cortex* for that modality.) The amount of brain space allotted to each of the sensory modalities is related to the principal habits of an animal species. For example, the primary somatosensory cortex—which receives information from neurons that sense stimuli such as touch, the positions of body hairs, temperature, and pain—is relatively much larger in rats and in tarsiers (a primitive primate) than it is in chimpanzees and humans (see Figure 8-13). The larger somatosensory cortex in rats and tar-

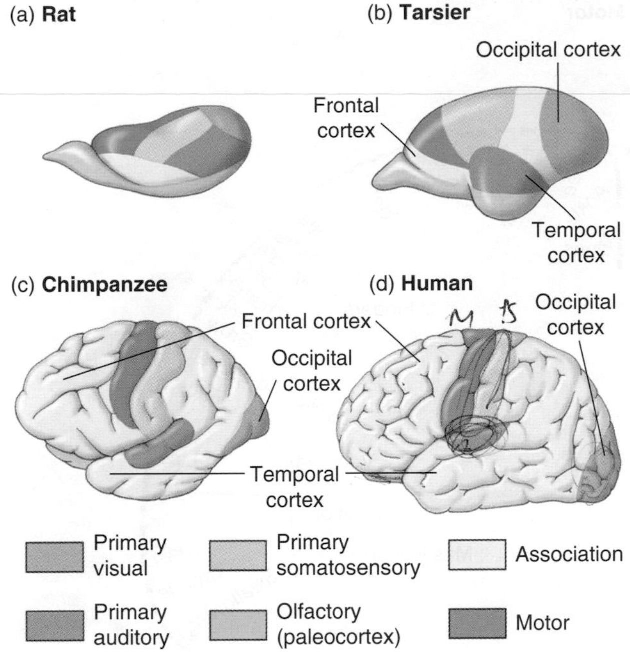

Figure 8-13 The mammalian cerebral cortex is divided into regions that serve specific functions. **(a–d)** Side views of the cerebral cortices of four different mammals, showing functional divisions within the cortex. In all drawings, the posterior pole of the brain is located at the right. Regions that have purely sensory or purely motor function are brightly colored. Regions colored beige serve "association" functions. Notice that the relative amount of association cortex increases from rat to human. The frontal, temporal, and occipital regions are labeled on the brains of the three primates (tarsier, chimpanzee, and human). Notice that the cortex of the tarsier is smooth, whereas the cortices of the chimpanzee and the human are folded, greatly increasing the surface area.

siers correlates with their dependence on collecting and discriminating tactile information. On the other hand, all primates—including humans, chimpanzees, and tarsiers—have much larger primary visual cortices than do rodents such as rats.

The functional organization of cortical regions has been determined in electrophysiological experiments in which activity recorded from neurons within each area is correlated with the application of specific stimuli. Perhaps the most dramatic of these experiments have been performed on conscious humans in the course of therapeutic surgery. Many neurosurgical procedures are performed while patients are awake and relatively alert. (The central nervous system itself has no pain receptors, so local anesthesia in the scalp and skull allows the patient to remain comfortable and awake during the surgery.) In these procedures, stimulating neurons within a particular region of sensory cortex evokes sensations in the patient, who is able to tell the surgeon where in the periphery the sensation seems to arise and which modality is being stimulated. These experiments have supported the hypothesis that all sensory perception occurs in the central nervous system, primarily in the sensory and association areas of the cerebral cortex.

The primary somatosensory cortex exemplifies the mapping of sensory input within sensory cortical areas. Each separate location within the somatosensory cortex receives inputs from a specific area of the body, and sensory information from adjacent areas of the body is transmitted to adjacent cortical regions (Figure 8-14a on the next page). More neurons are dedicated to peripheral areas from which "important" information is received than to other parts of the body. In humans, for example, approximately half of the somatosensory cortex receives input from the face and hands, while the other half is responsible for the entire remainder of the body surface. This dedication of resources to tactile information coming from the face and the hands correlates with the importance of these areas in our daily lives. For example, the human central nervous system requires detailed sensory information from the hands to allow us to carry out the fine manipulations that we perform when we use complicated equipment or identify objects by touch. The receptive field—the area from which a neuron collects information—is extremely small for sensory neurons that innervate the hands, increasing the resolution available. However, if each neuron collects information from only a tiny area, then an enormous number of neurons are necessary to acquire information from the surface of a hand. This very rich sensory innervation of the skin on each hand is translated into a large region of cortical surface area receiving somatosensory information from the hands.

The map of sensory projection onto the somatosensory cortex, also called the *sensory homunculus,* differs in detail among species, reflecting the relative importance of somatosensory input from particular parts of the body (Figure 8-15 on the next page). Notice that in all four species illustrated in this figure, the head occupies more somatosensory cortex than one would expect from its relative anatomic size. However, the index finger of the human is also disproportionately large, whereas projection from the teeth of rabbits and cats is surprisingly large.

The general principle that space in the brain is organized in a strict somatotopic map, and that space is allocated according to the relative importance of incoming sensory information, is particularly well illustrated by the somatosensory cortex of the star-nosed mole (Figure 8-16 on page 293). In this animal, eleven fleshy rays protrude from each side of the nose (Figure 8-16a,b). Each ray is covered with small swellings of skin, under which lie arrays of sensory receptors and nerve endings. The animal probes the ground with these fleshy rays when it hunts for insects. The projections from this remarkable organ have been traced anatomically to the somatosensory cortex. Sensory input from the rays crosses the midline of the mole, and in the

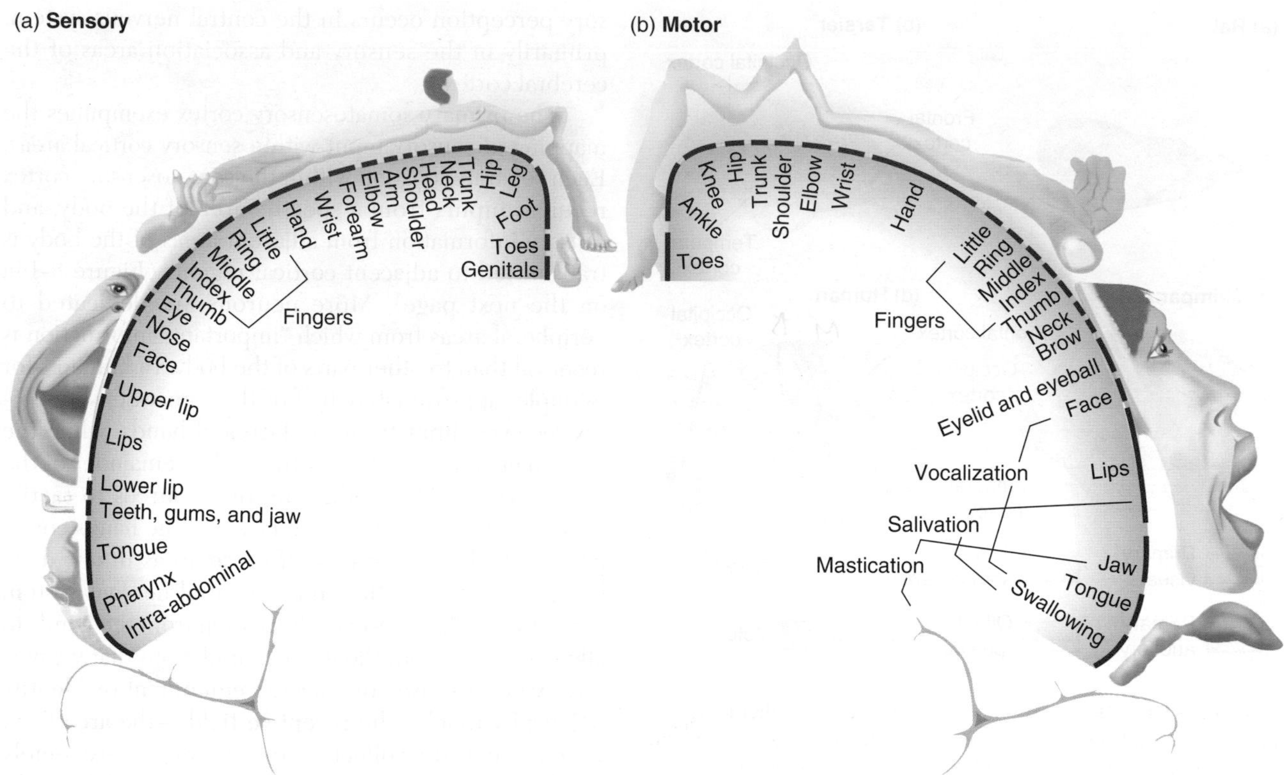

Figure 8-14 Both sensory and motor cortical regions are organized into topological maps of the body. **(a)** Map of a transverse section through the human primary somatosensory cortex, depicting the locations of neurons and their corresponding peripheral projections—that is, the places in the periphery from which stimuli are received and are subjectively "felt." Information coming from the face and hands occupies about half of the entire human somatosensory cortex. **(b)** Map of a transverse section through the human primary motor cortex, showing the distribution of cortical neurons that project to motor neurons in the brain and spinal cord, which in turn control activity in specific skeletal muscles. The sensory and motor maps are similar, with the head occupying a disproportionately large fraction of the total area, but notice the relative sizes of the hands in the two maps.

contralateral somatosensory cortex, there are eleven wedge-shaped cellular regions, each of which receives input from one ray (Figure 8-16c). In a similar fashion, the nervous systems of many species organize sensory input by sending information from specific locations to strictly mapped, distinct structures in the brain. Spatial sorting of incoming somatosensory information appears to be common, at least among the vertebrates, although it is less readily observable in species that lack somatosensory organs as spectacular as those of the star-nosed mole.

The auditory cortex of the temporal lobe and the visual cortex of the occipital lobe (see Figure 8-13) are both purely sensory in function. Direct electrical stimulation of these areas during neurosurgery evokes in the

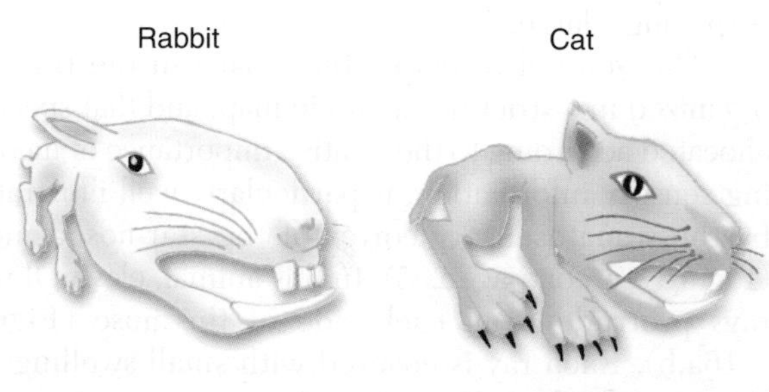

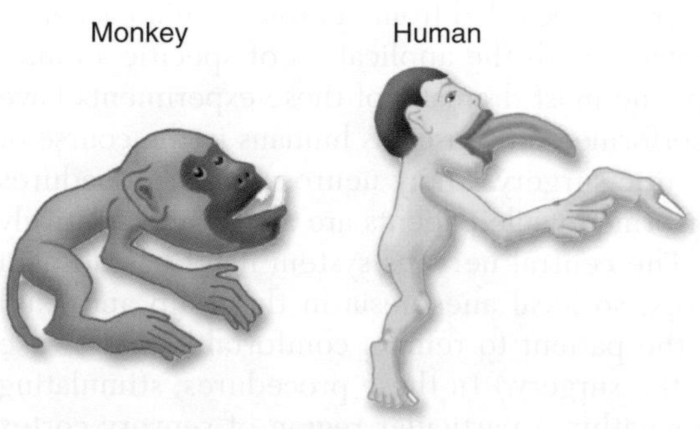

Figure 8-15 The somatosensory projection map varies among species, depending on receptive field sizes throughout the body. This diagram shows four species distorted to reflect the richness of somatosensory innervation in different parts of the body. Small receptive fields and dense innervation are associated with disproportionately large projection areas in the primary somatosensory cortex.

Figure 8-16 Sensory receptors in the elaborate nose of the star-nosed mole project to specific regions of the somatosensory cortex, forming a strict map of the surface receptors. **(a)** The nose of a star-nosed mole, seen from the front. **(b)** A scanning electron micrograph of the fleshy rays surrounding the right nostril. Each of the eleven rays has been numbered to permit comparison with regions of somatosensory cortex. **(c)** A section through the contralateral somatosensory cortex, cut parallel to the surface of the brain, that has been processed for cytochrome oxidase to show cellular organization. To generate this image, afferent neurons from the nose were stimulated, and the activity in these neurons caused the cytochrome oxidase activity of cortical neurons receiving this information to increase. The pattern of raylike stripes of darkly stained tissue corresponds to the pattern of rays on the nose, demonstrating that nearest-neighbor relations of receptors in the periphery are maintained in their projection to the somatosensory cortex. Medial, top; posterior, right. [Courtesy of Ken Catania.]

patient rudimentary auditory and visual sensations. Again, the relative sizes of these areas correlate with their importance to the animal. For example, the visual cortex occupies nearly a third of the entire cortical surface in the tarsier, whereas it takes up only a small fraction of the cortex in the rat (see Figure 8-13). As in the somatosensory cortex, the sensory world is arranged in a highly ordered fashion. In the primary visual cortex, for example, the two-dimensional image of the world that stimulates the retina is mapped directly onto the two-dimensional surface of the cortex; that is to say, points lying near one another in the field of view excite cells that lie near one another in both the retina and the cortex. This *retinotopic* mapping was one of the first features of brain organization to be discovered, and it has guided our exploration of other sensory systems. However, understanding how other sensory modalities are processed has presented challenges. The visual scene and the area of the cortical surface can be directly represented in two-dimensional maps, but mapping of the other senses, such as audition, requires more complex computations by the nervous system, as we will see in Chapter 11.

In the brain, the motor cortex is located adjacent to the somatosensory cortex, and it, too, is organized into a map corresponding to the rest of the body (see Figure 8-14b). As in the somatosensory cortex, the spatial distribution of neurons in the motor cortex correlates with the locations of the muscles that are controlled by those neurons. Neurons that are near neighbors in the motor cortex exert control over muscle fibers that are neighbors in the body. The relative proportions of the motor map reflect another principle of motor control: it takes more neurons to control muscles that make precise movements than it does to control muscles that make only large, imprecise movements. The muscles that move the human fingers, for example, execute very detailed and finely tuned motions, and the amount of motor cortex dedicated to controlling the fingers is very large. In contrast, the movements of the toes are relatively simple and coarsely controlled, and the amount of motor cortex dedicated to the toes is correspondingly quite small. Interestingly, the fine details of motor maps are somewhat plastic; they can change to some extent depending on muscle use.

The control of movement arises from activity in the motor cortex, which is based on input from other areas of the cortex and the brain. A motor control signal travels to muscles in the body by several parallel pathways, including the *corticospinal* tract. Neurons with axons in this tract have somata in the motor cortex and make synapses in the spinal cord. (In the vertebrates, tracts are commonly named based on where the somata and the synapses are located, with the location of the somata mentioned first and the location of the synapses mentioned second.) The number of interneurons that separate neurons in the motor cortex from spinal motor neurons varies among species, with more interneurons in the motor pathways of lower vertebrates and fewer in those of the higher vertebrates, such as primates. In rabbits, for example, a signal from the motor cortex must be transmitted through several serially arranged interneurons in the spinal cord before it reaches the motor neurons. In cats, few interneurons separate the motor cortex from the spinal segment containing the target motor neurons; however, when the signal reaches the appropriate spinal segment, it must traverse several interneurons before it reaches the spinal motor neurons. In primates, some cortical motor neurons synapse directly onto spinal motor neurons, an arrangement that is thought to confer special motor capacities on these animals. However, neurons that directly connect the motor cortex with spinal motor neurons constitute only about 3% of the motor control neurons. Thus, even in humans, most motor control is achieved through less direct pathways.

The neurons controlling motor activity maintain a steady background of low-level synaptic input to the motor neurons. An increase in their activity synaptically activates motor neurons, which can produce forceful movements of the limbs. This behavior occurs naturally when an animal consciously generates a strong contraction of a muscle, which is why the somatic pathway of the nervous system is called the "voluntary" system. Movements can also be evoked experimentally in anesthetized animals when clusters of neurons in the motor cortex are directly stimulated with weak electric current.

What kinds of influences might have led so much of the brain to be organized into topological maps? Would you expect areas of association cortex to be organized into a map?

The Autonomic Nervous System

Visceral function in vertebrates is regulated largely without conscious control, primarily by the autonomic nervous system (see Figure 8-6). The autonomic nervous system is partitioned into two distinct divisions: the **sympathetic** and **parasympathetic** pathways. In general, these two pathways act continuously and in opposition to each other, and the balance between them determines the state of the animal. When an animal is in a relaxed state or is sleeping without impinging stimuli, the parasympathetic pathway dominates, lowering the heart rate and diverting metabolic energy to housekeeping tasks such as digestion. When an animal is very active or frightened, increased activity in the sympathetic pathway inhibits the housekeeping functions and enhances functions that can support physical exertion. The heart rate becomes elevated, the concentration of glucose in the blood rises, and blood flow to skeletal muscles increases. These two states—deep sleep and intense physical activity—are at opposite ends of a continuum. Most of the time, the two divisions are nearly in balance; as a result, physiological variables such as heart rate are held at an intermediate value. A short list comparing some sympathetic and parasympathetic actions is given in Table 8-1.

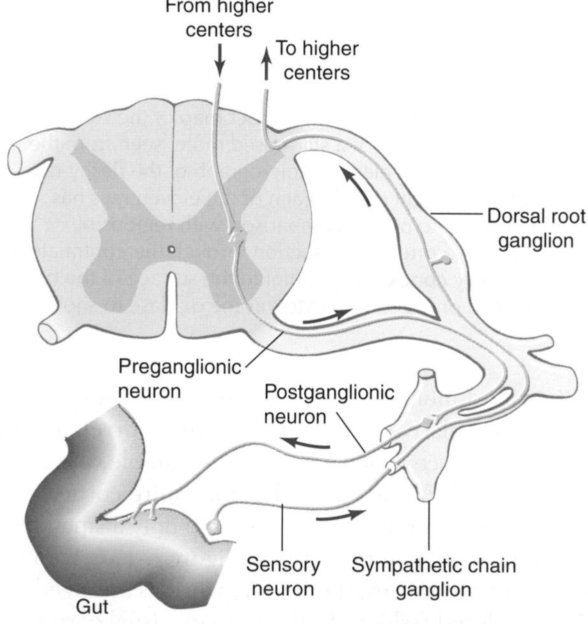

Figure 8-17 The functional unit of the autonomic nervous system is the autonomic reflex arc. This diagram illustrates a reflex arc of the sympathetic nervous system that could modulate the motility of the gut. Sensory input passes directly through the sympathetic chain ganglion, synapsing somewhere in the central nervous system, typically in the brain. Motor output from these higher centers synapses onto a preganglionic neuron whose soma is located in the central nervous system, which in turn synapses onto postganglionic neurons in a sympathetic chain ganglion, located outside the central nervous system. Postganglionic neurons synapse onto the target organ. [Adapted from Montagna, 1959.]

Table 8-1 Opposing effects on target tissues of the sympathetic and parasympathetic divisions of the autonomic nervous system

Target tissue	Sympathetic division	Parasympathetic division
Glands		
Lacrimal (tear) glands	No effect	Stimulates production of tears
Salivary glands	Stimulates production of a small amount of viscous saliva ("dry mouth")	Stimulates production of a large amount of dilute saliva
Adrenal medulla	Stimulates secretion	No effect
Eye		
Radial muscles of iris	Pupillary dilation	No effect
Iris sphincter muscles	No effect	Pupillary constriction
Ciliary muscle (controls thickness of lens)	Relaxation (focuses on distant objects)	Contraction (focuses on close objects)
Heart		
Pacemaker cells	Increases rate of heartbeat	Decreases rate of heartbeat
Ventricular contractile fibers	Increases force of contraction	Little or no effect
Lungs		
Smooth muscles in walls of bronchioles	Dilates bronchioles	Constricts bronchioles
Mucous glands	No effect	Stimulates secretion of mucus
Gastrointestinal tract		
Sphincter muscles	Contraction	Relaxation
Smooth muscles in walls of tract	Reduces tone and motility	Increases tone and motility
Exocrine glands	Inhibits secretion	Stimulates secretion
Gallbladder	Inhibits contraction	Stimulates contraction
Liver	Increases glycogenolysis and therefore blood sugar	No effect
Other tissues		
Urinary bladder	No effect	Stimulates muscle contraction
Arterioles	Vasoconstriction in vessels supplying skin and gut; vasodilation in some vessels supplying skeletal muscle	No effect

The functional unit in both the sympathetic and the parasympathetic divisions is the autonomic reflex arc. Figure 8-17 shows a sympathetic reflex arc. The afferent side of an autonomic reflex arc is largely indistinguishable from that of a somatic reflex arc, although the sensory neurons are likely to respond to different stimuli—for example, the concentration of glucose in the blood, blood pressure, or oxygen content of the tissues. Unlike many somatic muscle reflexes, for which all necessary neurons are located in the spinal cord, autonomic reflexes are typically processed through the brain.

The efferent side of an autonomic reflex arc is quite different from that of a somatic reflex arc. In both divisions of the autonomic nervous system, the motor output is carried by a chain of two neurons (Figure 8-18a on the next page). In the sympathetic nervous system, the soma of the first neuron—the **preganglionic neuron**—is located in the central nervous system, and the soma of the second neuron—the **postganglionic neuron**—typically lies in a *sympathetic chain ganglion* (also called a paravertebral ganglion). The postganglionic neurons lie entirely outside of the central nervous system, and they synapse onto the target cell of the autonomic reflex.

Structure of the sympathetic and parasympathetic divisions

The locations and properties of postganglionic neurons depend on the autonomic division to which they belong (Figure 8-18b). The somata of sympathetic preganglionic neurons are located in the thoracic and lumbar

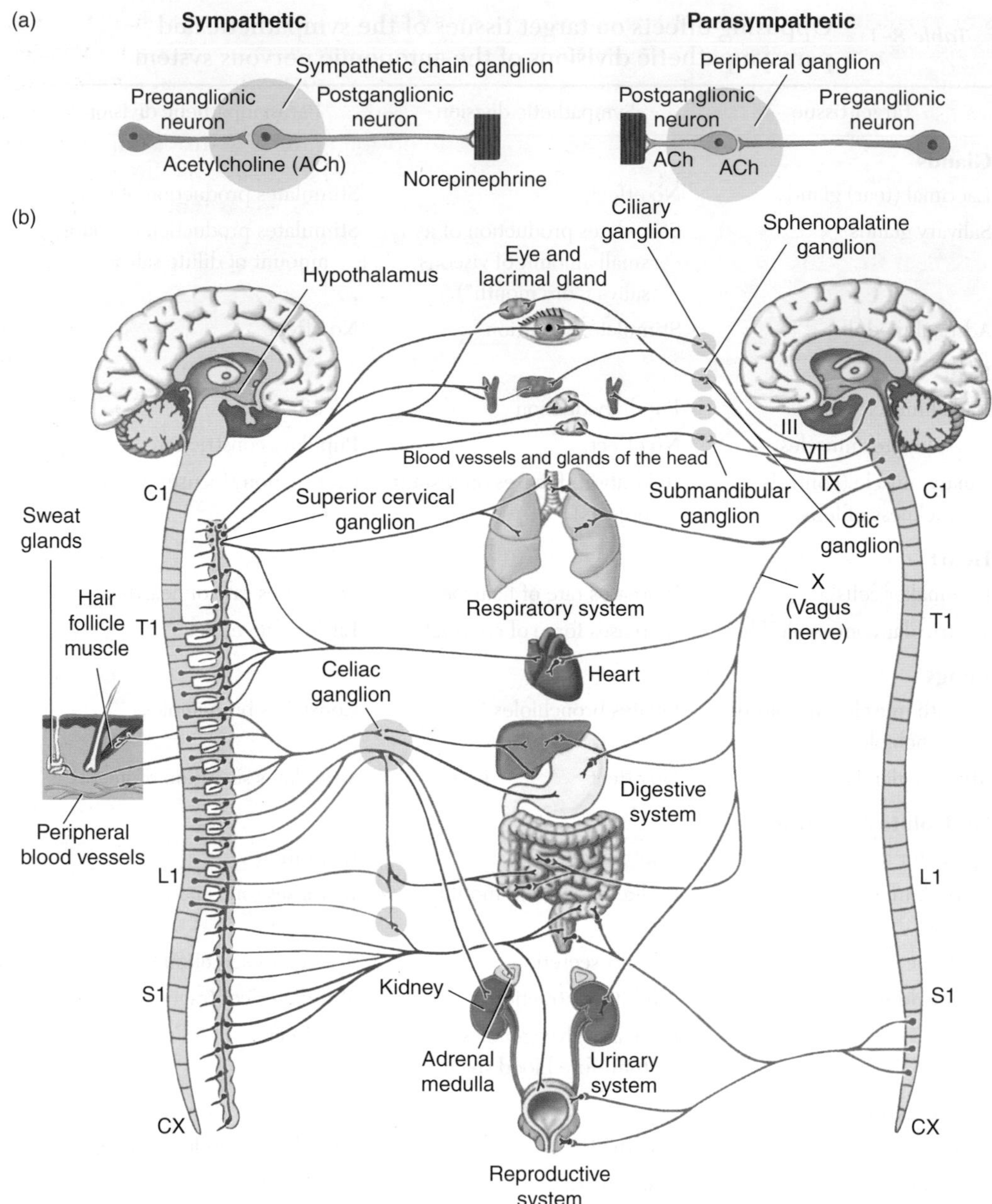

Figure 8-18 The two divisions of the autonomic nervous system have many targets in common. **(a)** Both the sympathetic and the parasympathetic divisions innervate their targets through a two-neuron chain. The soma of each preganglionic neuron lies within the central nervous system. The preganglionic neuron synapses in a peripheral ganglion onto a postganglionic neuron, which contacts the target organ. The preganglionic neurons of both divisions are cholinergic, but the neurotransmitters of the postganglionic neurons differ. The postganglionic neurons of the parasympathetic division are cholinergic, but in the sympathetic division, the postganglionic neurons are adrenergic, primarily using norepinephrine as their transmitter. **(b)** Locations and targets of pre- and postganglionic neurons in the sympathetic (left) and parasympathetic (right) branches of the autonomic nervous system. Preganglionic neurons are shown in violet; postganglionic neurons are shown in orange. This diagram illustrates the human autonomic nervous system, but the system is similar in most vertebrates. Abbreviations: C1, first cervical segment of the spinal cord; T1, first thoracic segment; L1, first lumbar segment; S1, first sacral segment; CX, coxal segment. In the parasympathetic nervous system, several pathways run in cranial nerves, which are indicated by roman numerals (see Figure 8-9).

segments of the spinal cord. Many sympathetic preganglionic neurons synapse onto postganglionic neurons in the sympathetic chain ganglia, which contain the somata of postganglionic neurons. The axons of the postganglionic neurons then extend to the target organs, which may lie quite far from the ganglion. (One exception to this pattern is the celiac ganglion, which contains the somata of sympathetic postganglionic neurons innervating the stomach, liver, spleen, pancreas, kidney, and adrenal gland. It is located in the abdominal cavity.)

Preganglionic neurons of the parasympathetic nervous system, on the other hand, synapse onto postganglionic neurons in ganglia that lie near—or even in—the walls of the target organs. Hence, in the parasympathetic division, axons of preganglionic neurons may be very long, and axons of postganglionic neurons are typically short. The somata of parasympathetic preganglionic neurons are located in the brain and in the sacral spinal cord.

Neurotransmitters of the autonomic nervous system

The two divisions of the autonomic nervous system also differ chemically (Table 8-2). All preganglionic neurons are cholinergic; that is, their neurotransmitter is acetylcholine. The neurotransmitter of a postganglionic neuron depends on the division to which the neuron belongs. The neurotransmitter of parasympathetic postganglionic neurons is acetylcholine; the neurotransmitter of sympathetic postganglionic neurons is norepinephrine. (Sympathetic postganglionic neurons also release a little epinephrine, and a few of them are cholinergic.)

Postganglionic neurons of the parasympathetic and sympathetic divisions typically innervate the same target organs (see Figure 8-18b), but exert opposite effects on their shared targets. For example, pacemaker activity in the heart is slowed down by acetylcholine released by parasympathetic postganglionic neurons, whereas it is accelerated by norepinephrine released by sympathetic postganglionic neurons. These actions of the parasympathetic and sympathetic divisions are reversed in the digestive tract, where acetylcholine from parasympathetic neurons stimulates intestinal motility and secretion of digestive enzymes, whereas norepinephrine from sympathetic neurons inhibits these functions.

Within the autonomic nervous system, the neurotransmitters acetylcholine and norepinephrine each bind two specific kinds of receptors. These receptors can be distinguished by drugs that act as agonists (i.e., mimic the natural transmitter) or as blockers. The two kinds of cholinergic receptors, the nicotinic and muscarinic receptors, were first distinguished from each other on the basis of physiological reactions to two substances—nicotine and muscarine—that are never produced in animals. Nicotine, a plant alkaloid, acts as an agonist at some cholinergic synapses, including the synapses between pre- and postganglionic neurons in both divisions of the autonomic nervous system (see Spotlight 6-4). Muscarine, which is extracted from certain toxic mushrooms, acts as an agonist at other cholinergic synapses, including synapses made by parasympathetic postganglionic neurons onto their target cells. These receptors can also be distinguished by another pair of drugs that act as blockers. Curare (D-tubocurarine) blocks the action of acetylcholine on nicotinic receptors (including those at the endplates of skeletal muscle), whereas atropine blocks muscarinic receptors. Although they both respond to acetylcholine in the body, nicotinic and muscarinic receptors have completely different molecular structures and response mechanisms. Nicotinic receptors consist of protein complexes that bind acetylcholine and contain ion-selective channels within their structure. Muscarinic receptors are also proteins, but they affect ion channels indirectly through the mediation of intracellular second messengers (see Chapter 6).

Norepinephrine released by the adrenergic postganglionic neurons of the sympathetic division binds to two kinds of receptors on the target organs, designated α- and β-adrenergic receptors. Like the cholinergic receptors, the adrenergic receptors are distinguished from each other by their pharmacology. For example, α-adrenergic receptors are more sensitive to norepinephrine than to the drug isoproterenol, and they are selectively blocked by phenoxybenzamine. β-Adrenergic receptors are more sensitive to isoproterenol than to epinephrine, and they are selectively blocked by propranolol. The two types of adrenergic receptors activate separate but parallel intracellular second messenger pathways.

Tobacco smoke introduces nicotine into the body. Predict what effect(s) this chemical would have on the function of a smoker's autonomic nervous system.

Although all vertebrates have autonomic nervous systems, the two divisions are not well defined in all groups. In teleost fishes, for example, the sympathetic and parasympathetic divisions are distinguishable, but

Table 8-2 Pharmacology of neurotransmission in the autonomic nervous system

	Transmitter of preganglionic neuron	Receptors on postganglionic neuron	Transmitter of postganglionic neuron	Receptors on target tissue
Sympathetic division	Acetylcholine (ACh)	Nicotinic ACh receptors	Norepinephrine	α- or β-adrenergic receptors
Parasympathetic division	Acetylcholine	Nicotinic ACh receptors	Acetylcholine	Muscarinic ACh receptors

in cyclostome fishes (such as lampreys), the autonomic nervous system is not partitioned into two divisions. The autonomic nervous system of amphibians appears to be essentially identical to that of mammals, but in reptiles the distinction between the sympathetic and parasympathetic divisions may be less clearly defined. Comparative studies of the autonomic nervous system have been relatively rare, and further research may reveal some phylogenetic surprises.

SUMMARY

■ Sensory neurons acquire information about the internal and external environments and carry it to the central nervous system. Interneurons carry out most information processing. Motor neurons control the activity of effectors, causing the animal to respond appropriately to changes in either the internal or external environment.

Evolution of the nervous system

■ In the simplest nervous systems, such as those of cnidarians, neurons are distributed throughout the body in a diffuse network.

■ As animals and nervous systems became more complex, neuronal somata were gathered into a central nervous system or into peripheral ganglia. Neurons in the central nervous system are connected with the rest of the body by long axons.

■ In addition, many neurons are located at the anterior end of the central nervous system, perhaps at least partly because a number of sensory organs are also located there.

■ Many animals have segmented bodies, and the central and peripheral nervous systems reflect this segmentation. Typically, each segment of the nervous system serves a segment of the body.

Organization of the vertebrate nervous system

■ The vertebrate nervous system is divided into the central nervous system, consisting of the brain and spinal cord, and the peripheral nervous system, consisting of the peripheral ganglia and nerves.

■ Groups of cell bodies in the central nervous system are called nuclei; groups of axons in the central nervous system are called tracts. Groups of cell bodies in the peripheral nervous system are called ganglia; groups of axons in the peripheral nervous system are called nerves. In mammals, most nerves are mixed nerves; that is, they contain both sensory and motor axons.

■ The efferent pathways of the nervous system can be divided into two main systems. The somatic, or voluntary, nervous system controls skeletal muscles to produce movements. The autonomic nervous system modulates the contraction of smooth and cardiac muscle and the secretory activity of glands, thus controlling the body's "housekeeping" functions.

The spinal cord

■ The spinal cord is segmented, and the spinal nerves connect each spinal segment with a defined segmental region of the body.

■ Each portion of the spinal cord is divided into functional regions. The dorsal spinal cord is largely sensory in function, the ventral spinal cord largely motor. The interior of the spinal cord, called gray matter, consists largely of neuronal somata. The exterior of the cord consists of tracts of axons. This external region is called the white matter, due to the shiny white appearance of myelinated axons.

■ Sensory information enters the spinal cord through the dorsal roots. The cell bodies of most sensory neurons are located in the dorsal root ganglia outside the spinal cord. The dorsal region of the spinal gray matter, which receives and processes sensory information, is called the dorsal horn.

■ Motor information leaves the spinal cord through the ventral root. The ventral region of the spinal gray matter, which houses the somata of spinal motor neurons, is called the ventral horn.

The brain

■ Early in its development, the brain is clearly segmented, and the twelve paired cranial nerves reflect this segmentation.

■ The brain is divided into anatomic regions containing neurons that have a common function. Some of these regions process sensory information; some generate signals for controlling motor activity.

■ The size of brain structures is typically correlated with the importance of their particular function in the life of an animal.

■ Both sensory and motor regions of the brain are arranged in maps that reflect the topology of either the body or the outside world.

■ The cerebral cortex, a layer of cells that covers the cerebrum in higher mammals, evolved from an olfactory center in the lower vertebrates. The structure of the cerebral cortex is least complex in lower mammals and most complex in primates and humans.

■ The cerebral cortex is functionally divided into sensory regions, motor regions, and areas that are neither clearly sensory nor clearly motor. These in-between regions are known as association cortex, and the percentage of the cortex that falls into this category is larger in higher mammals than in lower mammals.

The Autonomic nervous system

■ The autonomic nervous system is divided into the sympathetic division, whose activity is associated with a state of physical activity or response to an emergency, and the parasympathetic division, whose activity is associated with "housekeeping" functions such as digestion.

■ The sympathetic and parasympathetic divisions both innervate the same targets, and both are typically tonically active. The physiological state of the animal depends on the balance of activity between the two divisions.

■ Activity in the autonomic nervous system is based on autonomic reflex arcs.

■ The output side of the autonomic nervous system consists of at least two neurons: a preganglionic neuron and a postganglionic neuron. Preganglionic neurons typically have short axons in the sympathetic branch, but long axons in the parasympathetic branch; postganglionic neurons typically have long axons in the sympathetic branch, and short axons in the parasympathetic branch.

■ All preganglionic neurons in the autonomic nervous system are cholinergic, and all postganglionic neurons have nicotinic acetylcholine receptors.

■ Postganglionic neurons of the sympathetic branch are adrenergic; postganglionic neurons of the parasympathetic branch are cholinergic.

■ Postsynaptic target cells of parasympathetic postganglionic neurons have muscarinic acetylcholine receptors. Postsynaptic target cells of sympathetic postganglionic neurons have either α- or β-adrenergic receptors.

REVIEW QUESTIONS

1. Describe the general organization of information processing by a nervous system.
2. List and discuss several trends in the evolution of nervous systems.
3. Given that action potentials in all neurons are fundamentally alike, how is the modality of input from the various sense organs recognized by the central nervous system of an organism?
4. Describe the general organization of the nervous system of a segmented invertebrate such as an annelid or a crustacean.
5. Describe the general organization of the vertebrate brain and spinal cord.
6. What is a somatotopic map? What is a retinotopic map? What general principle of brain organization do these maps illustrate?
7. It has been said that all sensation takes place in the brain. Explain what is meant by this statement.
8. What are the major functions of the autonomic nervous system?
9. Compare and contrast the sympathetic and the parasympathetic divisions of the autonomic nervous system. How are they similar and how do they differ anatomically? How are they similar and how do they differ functionally and biochemically?

SUGGESTED READINGS

Allman, J. 2000. *Evolving Brains.* New York: Scientific American Library. (A highly readable discussion of the evolution of vertebrate brains, with an emphasis on mammals.)

Goodhill, G. J., and L. J. Richards. 1999. Retinotectal maps: Molecules, models and misplaced data. *Trends Neurosci.* 22:529–534. (A discussion of mechanisms by which one well-studied topological map in the brain may be set up during development.)

Holland, L. Z., and N. D. Holland. 1999. Chordate origins of the vertebrate central nervous system. *Curr. Opin. Neurobiol.* 9:596–602. (Detailed anatomic and molecular study of the nervous system in the simple chordates *Amphioxus* and tunicates; sheds light on the evolution of the vertebrate nervous system.)

Kandel, E., J. Schwartz, and T. Jessell. 2000. *Principles of Neural Science.* 4th ed. New York: McGraw-Hill. (A giant compendium of information about the nervous system, with some emphasis on vertebrate—particularly mammalian—species.)

Nauta, W. J. H., and M. Feirtag. 1986. *Fundamental Neuroanatomy.* New York: W. H. Freeman. (A comprehensive description of mammalian neuroanatomy.)

Wenning, A. 1999. Sensing effectors make sense. *Trends Neurosci.* 22:550–555. (An interesting review of a variety of systems in which single cells act both as sensors and effectors to maintain homeostasis in a variety of physiological variables.)

Zigmond, M. J., F. E. Bloom, S. C. Landis, J. L. Roberts, and L. R. Squire. 1999. *Fundamental Neuroscience.* San Diego: Academic Press. (An enormous compendium of information regarding—among other things—the functional anatomy of the mammalian brain with a focus toward clinical applications.)

Glands and Hormones

All cells secrete material into the surrounding environment, either to form a protective barrier around themselves or to communicate with and recognize other cells. Cells that secrete similar substances (for example, a specific hormone) are often collected together to form **glands.** The specialized cells composing a gland act as a unit. The glands possessed by an individual vary not only with the species, but also during various developmental stages. The venom glands of snakes, human sweat glands, the wax glands of insects, the thyroid gland, and the pituitary are just a few of the wide variety of glands found in the animal world. Glandular secretions are synthesized by **secretory cells** and released from the gland in response to an appropriate stimulus. Many of the substances secreted by different species have similar functions and identical or related chemical structures (Spotlight 9-1 on page 303).

Secretions from glands constitute an important response by animals to a variety of situations. Feeding, for example, results in the massive activation of a large range of digestive glands (see Chapter 15). In vertebrates, the secretion of hydrochloric acid into the lumen of the stomach after eating can be large enough to cause a marked increase in blood pH—the so-called *postprandial* (after-dinner) *alkaline tide*. In a voracious feeder like the crocodile, capable of eating a whole gazelle in one meal, this increase in blood pH can be very large indeed. The webs spun by spiders to catch prey are another example of glandular secretions. The nature of the secretion and the pattern of the web vary with the species and the environmental conditions. Mucus nets secreted by deep-sea fishes may play a role similar to that of spiders' webs in the interactions between predator and prey in their unique environment. Secretions from glands often play important roles in mating behavior, as well as reproduction in general.

We saw in Chapter 6 that neurotransmitters released from presynaptic neurons act over a short distance to activate receptors on the postsynaptic cell. Some glandular secretions also mediate communication between cells, but the target cells may be in distant parts of the body. The signaling molecules bind selectively only to target cells bearing specific receptors, initiating a response in the target cells.

In this chapter, we first describe the nature and mechanisms of cellular secretion. We then move to the organ level, discussing the nature of endocrine glands, which secrete substances into the blood, and exocrine glands, which secrete substances onto the body surface (including internal surfaces, such as the intestinal lining). Endocrine glands are described in detail, along with their cellular mechanisms of action. A few examples of exocrine glands are also presented in this chapter; more will be encountered later in this book, as exocrine secretions play an important role in many different physiological mechanisms. Finally, we briefly review the metabolic costs of glandular secretions.

Many abbreviations are used when referring to various hormones and their actions. A list of abbreviations used can be found at the end of this chapter on page 357.

CELLULAR SECRETIONS

All cells secrete a surface coat. The cells of an epithelium also secrete mucus onto its external surface, giving rise to the term *mucosal,* referring to the external surface of the epithelium. The surface coat allows cells to recognize one another, and the coat and mucus together form a protective barrier around epithelial cells, creating a controlled microenvironment between these cells and the surrounding extracellular space.

Cells also secrete signaling substances that are used for intercellular communication. These secretions can be categorized by the distance at which they have an effect (Figure 9-1):

- **Autocrine secretions** affect the secreting cell itself. An example is the autoinhibition of norepinephrine release from adrenergic neurons by norepinephrine itself, as we will see below.
- **Paracrine secretions** affect neighboring cells. An example is the inflammatory response, in which localized vasodilation is induced by histamine released from mast cells in the area of tissue damage; see Chapter 12.
- **Endocrine secretions** are released into the bloodstream and act on distant target tissues.
- **Exocrine secretions** are released onto the surface of the body, including the surface of the gut and other internalized structures.

Some exocrine secretions, called **pheromones,** are produced by one animal to communicate with another and can initiate a range of physiological responses. Among insects, pheromones can identify the members of a colony. They also play an important role in reproduction for many species; bombykol, the powerful sex attractant released by female silkworm moths, is an example (see Chapter 7). It is not uncommon for hormones or their breakdown products to act as pheromones. A steroid hormone that induces molting in crabs also serves as the female sex attractant, producing behavioral responses in males at concentrations in the seawater as low as 10^{-13} M. In certain marine invertebrates, such as clams and starfishes, spawning of eggs and sperm is triggered by pheromones that are liberated into the environment along with the gametes. The spawning of one individual thus triggers spawning by others of both sexes. The adaptive value of such epidemic spawning is that it enhances the probability that sperm and egg will meet and that fertilization will occur.

Pheromones are also used to repel predators. A common example is the foul-smelling musk that makes skunks unpalatable to their enemies. A pair of scent glands near the anus produces this yellowish, oily, malodorous secretion. The encapsulating muscles close to the anal opening are capable of propelling the musk over a meter. Its scent has been detected by humans at sea as far as 30 kilometers from the nearest land.

Some animals produce secretions that act both locally and at a distance—they combine autocrine, paracrine, and endocrine effects. Calcitonin, for instance, is produced in the gills of Pacific salmon, binds in the gills, and modulates calcium flux across the gills. Calcitonin is therefore both an autocrine and a paracrine secretion. Pacific salmon also produce calcitonin in the ultimobranchial gland; in this case, the calcitonin is released into the blood and subsequently acts on the gills as an endocrine secretion.

Exocrine secretions have numerous functions in addition to communication. Saliva produced in the mouth helps food slide down the esophagus. Pancreatic secretions aid digestion. Snails secrete mucus with special elastic properties that enable them to slide and stick as they move over the ground.

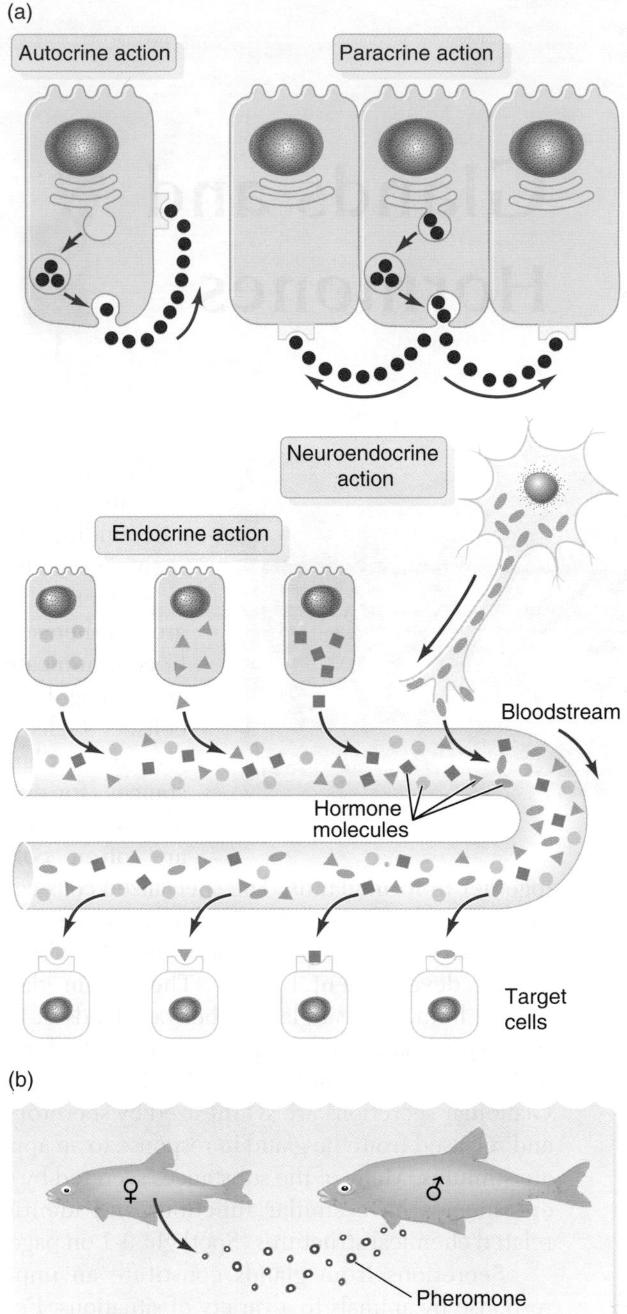

Figure 9-1 Cellular communication by means of cellular secretions can be categorized by the distance between origin and action. **(a)** In autocrine signaling, a cell responds to its own signal; in paracrine signaling, near neighbors respond; in endocrine signaling, hormones are transported over distances via the bloodstream. The same categories apply to neuroendocrine action. **(b)** The release of secretions into the environment by one animal to communicate with other animals is an example of exocrine action. The female fish has liberated a pheromone into the water that is detected by, and alters the behavior of, the male fish.

SUBSTANCES WITH SIMILAR STRUCTURES AND FUNCTIONS SECRETED BY DIFFERENT ORGANISMS

Just as organisms have many biochemical pathways in common, there are many similarities in the substances secreted by different animals. Substances with similar functions that are secreted by quite different organisms often have similar structures. For example, the α-factor secreted by yeast and the gonadotropin-releasing hormone (GnRH) produced in the pituitary gland of mammals are both small peptide hormones and have similar amino acid sequences (see part a of the accompanying figure). Not only are their structures homologous, but both secretions function in reproductive processes. The α-factor acts as a mating pheromone in yeast, whereas GnRH induces the release of luteinizing hormone (LH), which causes ovulation in mammals. When injected into mice, yeast α-factor causes release of LH, but it has a lower affinity for GnRH receptors than does native GnRH (i.e., GnRH from mice). In most, but not all, cases, a native hormone has a higher affinity for its own receptors and is more effective than a similar but foreign substance.

Occasionally, a foreign substance has a greater effect than a native hormone. Calcitonin is produced in humans by the C (clear) cells of the thyroid gland. It acts on bones, the major body store of calcium, to lower extracellular calcium levels under conditions when the calcium levels are raised. Calcitonin accomplishes this by inhibiting osteoclastic (bone-resorbing) activity without affecting osteoblastic (bone-building) activity. Salmon and eels produce a structurally similar calcitonin hormone in their ultimobranchial glands. This foreign hormone is many times more effective than native human calcitonin in preventing bone resorption and thus lowering blood calcium in humans.

(a)

(b)

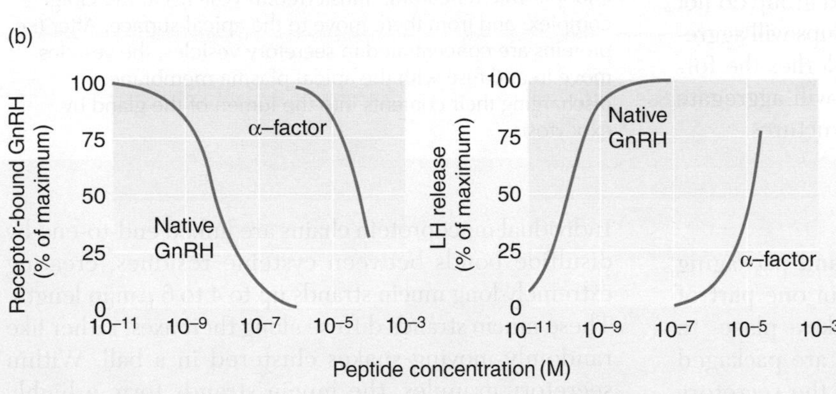

Many species produce secretions with similar structures and analogous functions. ***(a)*** *The amino acid sequences of yeast α-factor and mammalian gonadotropin-releasing hormone (GnRH). These short peptide hormones contain several identical residues (yellow).* ***(b)*** *Binding and activity curves for yeast α-factor and GnRH. The α-factor can bind to mammalian receptors for GnRH (left graph), and when injected into mice it induces release of luteinizing hormone (LH), the normal effect of GnRH (right graph). Compared with GnRH, however, a much higher concentration of α-factor is required for binding and LH release. [Adapted from Loumaye, Thorner, and Catt, 1982.]*

Surface Secretions: The Cell Coat, Collagen, and Mucus

Glandular secretion is a specialization of a general cell function. All animal cells secrete a cell coat, or **glycocalyx**, that is continuously renewed. Composed of glycoproteins and polysaccharides with negatively charged sialic acid termini, the glycocalyx can be visualized using appropriate dyes, revealing a microenvironment around the cell that modulates processes of filtration and diffusion between the cell and its environment.

The glycocalyx around some cells contains mucopolysaccharides that can associate with proteins to form mucoproteins. Mucoproteins have a much longer polysaccharide component than glycoproteins, are amorphous, and form gels able to hold large amounts of water. The jelly coats of frog eggs are a common example of such gels. **Mucus** is a thick, slimy fluid composed of water, inorganic salts, and *mucin* (any mucoprotein that raises the viscosity of the secreted fluid) and often contains leukocytes and cells sloughed off from the secreting epithelium. Although distinct from the glycocalyx, mucus also contains mucoproteins with a large number of sialic acid termini and creates a protective environment around the cell. **Goblet cells**, found in most epithelia, secrete enough mucus to cover the surfaces of many surrounding cells.

The mammalian lung provides a clear example of the protective role of mucus. The airways to the lungs continually secrete a sticky mucus layer that traps dirt particles and bacteria. The mucus is propelled toward the mouth by the action of millions of tiny cilia, to be either swallowed or expectorated. Tobacco smoke inhibits ciliary activity in the airways but enhances the production of mucus. The "smoker's cough" is an attempt to remove the accumulated mucus.

Collagen is a fibrous protein found in the extracellular matrix and is the most abundant animal protein. There are many forms of collagen, but the basic structure consists of three-stranded helical segments that are assembled into either fibrils or lattice networks. Collagen fibrils in the extracellular space give tissues form and rigidity and create surfaces on which cells can slide. The basal lamina of an epithelium has a chicken wire–like network of collagen that gives it mechanical support. To produce collagen, a cell assembles small procollagen molecules in the rough endoplasmic reticulum and Golgi complex and then transports them in vesicles to the cell surface for export via exocytosis. Terminal segments of the procollagen molecules prevent large collagen fibrils from forming inside the cell. Once outside the cell, cleavage of the procollagen molecules allows them to polymerize into normal fibrils in the extracellular space.

The chemical composition of the glycocalyx permits certain cells to recognize and adhere to one another. Red blood cells, for example, have specific cell-surface antigens that are distinguished by their terminal carbohydrates and form the basis of the ABO blood groups. Red blood cells in the same blood group do not aggregate, whereas those in different groups will aggregate when mixed. Cell-cell adhesion underlies the formation of organs; even cells in culture will aggregate with other like cells to form organized structures.

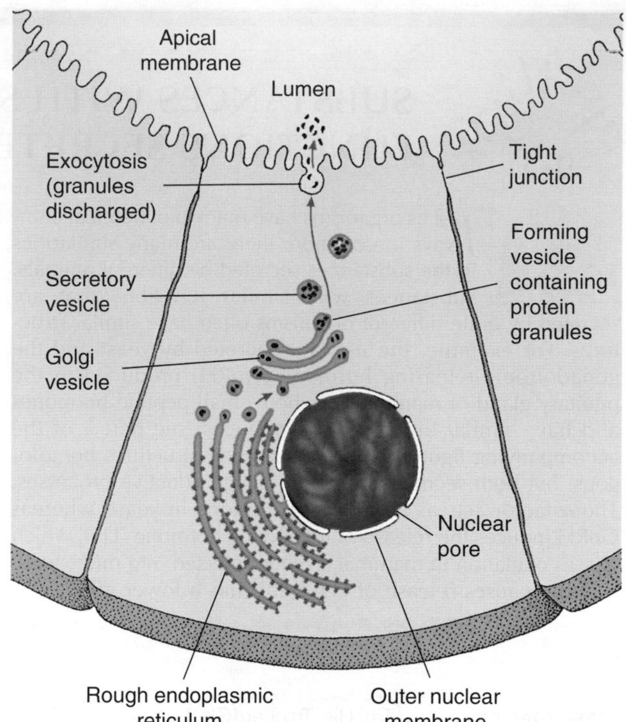

Figure 9-2 Secretory proteins are synthesized in the rough endoplasmic reticulum, transferred in vesicles to the Golgi complex, and from there move to the apical surface. After the proteins are concentrated in secretory vesicles, the vesicles move to and fuse with the apical plasma membrane, discharging their contents into the lumen of the gland by exocytosis.

Packaging and Transport of Secreted Material

In most secretory cells, the synthesis and packaging of the secreted substance takes place in one part of the secretory cell and secretion takes place in another (Figure 9-2). Most substances are packaged in membrane-bounded vesicles within the secretory cell and later liberated into the extracellular space, although some secretions—the steroid hormones, for example—are secreted as free molecules. Electron microscopy of most secretory tissues reveals membrane-bounded secretory granules (secretory vesicles), 100 to 400 nm in diameter, containing the substance to be secreted. The terms **secretory granule** and **secretory vesicle** are used interchangeably, depending on whether the emphasis is on the contents (granule) or the limiting membrane (vesicle). Secretory vesicles are similar in many respects to synaptic vesicles, which are somewhat smaller (50 nm in diameter).

Polymer gels

Mucus, a polymer gel, is packaged and stored along with various other chemicals as compact (condensed) granules in secretory vesicles within goblet cells. Mucus consists of extremely long mucoprotein strands, with a large number of sulfate and sialic acid termini, which are negatively charged at neutral pH (Figure 9-3). Individual mucoprotein chains are linked end-to-end by disulfide bonds between cysteine residues, creating extremely long mucin strands up to 4 to 6 μm in length. These mucin strands diffuse along their axes, rather like randomly moving snakes clustered in a ball. Within secretory granules, the mucin strands form a highly condensed polymer network arranged as a tangled web. When released into water, however, the mucin network rapidly expands in volume by as much as several hundred times. A disturbed hagfish can produce enormous volumes of protective mucus; once released from the body, the mucus expands very rapidly in water and can completely fill a bucket containing the hagfish within minutes. This slimy coat presumably protects the hagfish from attack.

High concentrations of Ca^{2+} ions in mucus can neutralize the negative charges on the sialic acid termini of mucin strands, thus favoring the condensed phase. The typically low pH of mucin vesicles may also play a role in the condensation process, and lipids within the vesicles may help to ensure that the mucin network remains condensed while stored in the vesicle. Ca^{2+} is not the only shielding cation that keeps polymers in the condensed phase in vesicles. In the chromaffin cells of the adrenal medulla, for example, chromogranin is the polymer and catecholamines act as the

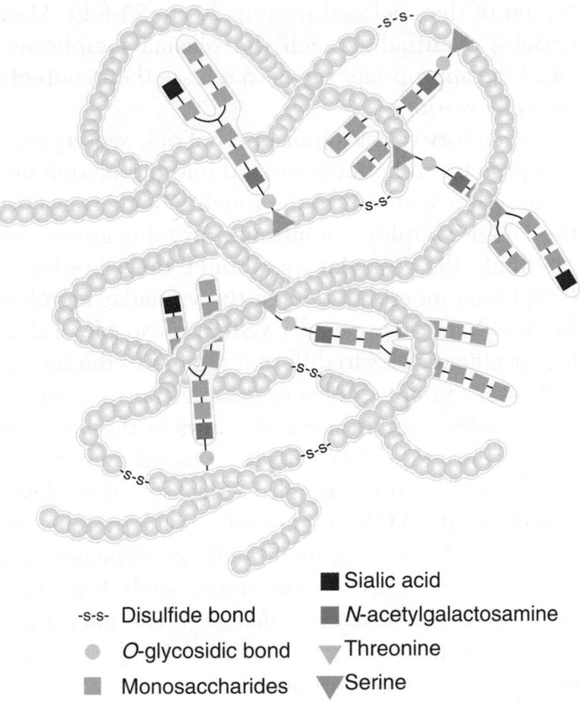

Figure 9-3 Mucus is a polymer gel consisting of mucoprotein chains joined end-to-end by disulfide bonds. The disulfide bonds do not form cross-bridges, which would restrict movement of the polymer chains. Notice the oligosaccharide side chains, many of which have a negatively charged sialic acid terminus. In the condensed phase, the mucin strands form a highly tangled network. [Adapted from Verdugo, 1990.]

shielding cation, and in mast cells, heparin is the polymer and histamine is the cation.

It appears that all exocrine and endocrine secretions that are stored in vesicles are released by exocytosis. In the exocytosis of mucus, a vesicle moves to the cell surface, and a pore forms as the vesicle fuses with the plasma membrane. The increase in conductance through the pore has been shown to be independent of, and therefore not caused by, expansion of the polymer gel. Ion exchange occurs between the contents of the vesicle and the extracellular space, causing the Ca^{2+} level to fall and perhaps the pH to rise within the vesicle. As a result, the release of mucus by exocytosis is explosive, with the mucus popping out of the vesicle like a jack-in-the-box (Figure 9-4). The mucin network in the giant granule of the slug can expand as much as 600-fold in 20 to 30 ms. This extremely rapid rate of swelling is driven by repulsive forces between the negative charges uncovered by the loss of Ca^{2+} and perhaps the rise in pH, rather than by the diffusion of water into the mucus, a process too slow to explain the rapid expansion that occurs. It appears that mucus is present in many different types of secretory vesicles and probably assists in the release of stored material under most conditions.

The shorter the mucin strands in the tangled web of condensed mucus, the faster its expansion. The most important factor determining the rate of expansion, however, is the nature of the environment into which the mucin network is released. The ionic composition, pH, and quantity of the extracellular fluid have a marked effect on the final state of hydration of released mucus. Hyperosmotic solutions, for example, can inhibit swelling of the mucin granule. The state of hydration has a marked effect on the fluidity of mucus. For example, the properties of the mucus secreted onto the surfaces of the airways are strongly influenced by the nature of the fluid that lines them. The abnormally viscous mucus found in humans suffering from cystic fibrosis has its origins in defective ion transport processes across the epithelium of the airways, which change the ionic composition of the extracellular fluid.

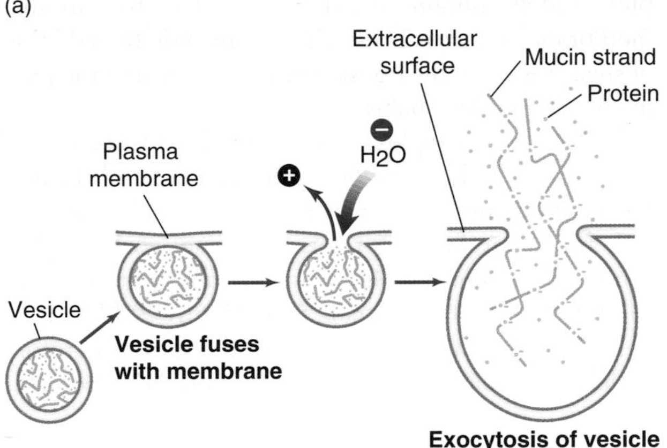

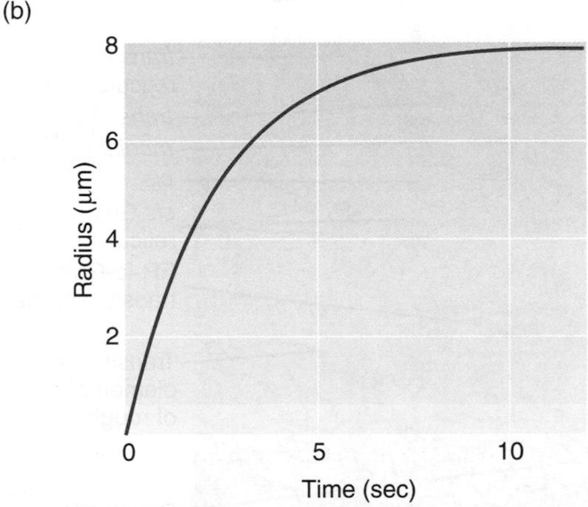

Figure 9-4 Mucus bursts from a vesicle in an explosive manner. **(a)** Mucus release by exocytosis. Following fusion of the vesicle with the plasma membrane, the shielding cations are released from the mucus inside the vesicle, or extracellular anions flow inward. The net result is that the negative charges in the condensed polyionic mucin become unshielded, driving a fast volumetric expansion and release of the vesicle's contents into the extracellular space. Water enters and enlarges the vesicle as the mucus swells. **(b)** Time course of the swelling of mucus in the extracellular space immediately after its release from a goblet cell *in vitro*. The change in radius as a function of time follows first-order kinetics. [Adapted from Verdugo et al., 1987.]

Formation and delivery of vesicles

The intracellular movement of secretory proteins has been studied by pulse-chase radiography, a tracer technique in which radioactively labeled amino acids are incorporated for a short time into newly synthesized proteins. Such studies reveal that secretory proteins are synthesized on the rough endoplasmic reticulum (ER) and accumulate within the ER lumen (interior). The proteins then pass into the smooth ER, regions of which bud off, encapsulating the secretory products in transfer vesicles (Figure 9-5a). The newly formed vesicles migrate to the Golgi complex, which consists of a stacked set of slightly concave, nearly flat membrane saccules, called *cisternae,* with closely associated free vesicles and vacuoles (Figure 9-5b). Microscopic studies indicate that the membranes of the transfer vesicles fuse with the Golgi cisternae. Proteins can undergo alterations within the Golgi complex, which contains enzymes bound to the luminal membrane surfaces. These changes include the addition of sugar residues, the excision of fragments, and the joining of polypeptide chains.

The Golgi complex consists of at least three sets of cisternae, called *cis, medial,* and *trans.* The *cis* face of the Golgi complex receives transfer vesicles from the ER; the *trans* face, or *trans*-Golgi network (TGN), produces secretory vesicles, which subsequently move to the surface of the cell (see Figure 9-5b). It is believed that water is drawn out of the future secretory vesicle as it matures, increasing the effective concentration of the enclosed protein 20- to 25-fold. Mature vesicles eventually reach the plasma membrane to await the appropriate signal to release their contents to the cell exterior.

Secretory vesicles transport not only secretions, but also proteins to be incorporated into the plasma membrane. After synthesis in the rough ER and transport to the Golgi complex, membrane proteins are incorporated into the vesicular membrane, then inserted into the plasma membrane when the vesicular membrane fuses with it. This vesicular system is capable of directing specific vesicles to different regions of the secretory cell—for example, delivering some transport proteins to the apical membrane and others to the basolateral membrane. Newly synthesized membrane proteins transported from the rough ER are sorted by destination within the TGN (Figure 9-6). Some proteins delivered to the basolateral membrane are ultimately transferred to the apical membrane; such transport is referred to as *transcytotic delivery.* The microtubular system appears to play a central role in the movement of vesicles to the cell surface.

Can you propose possible mechanisms within the cell for sorting vesicles to either the basal or apical membranes? How might such a system be influenced by hormonal action?

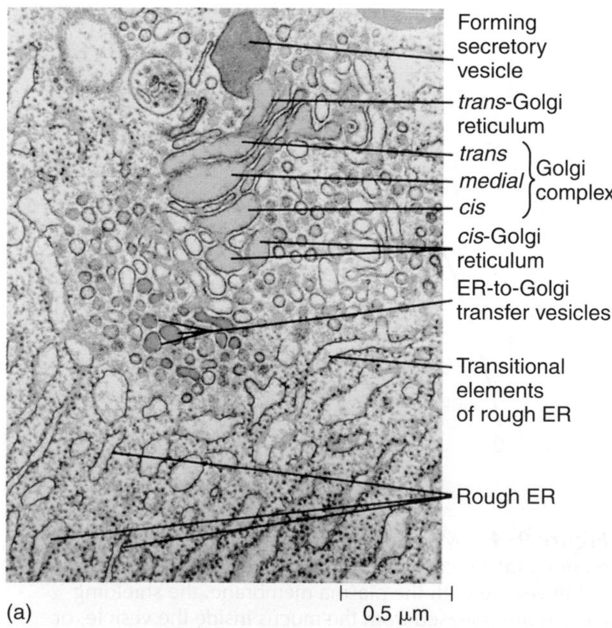

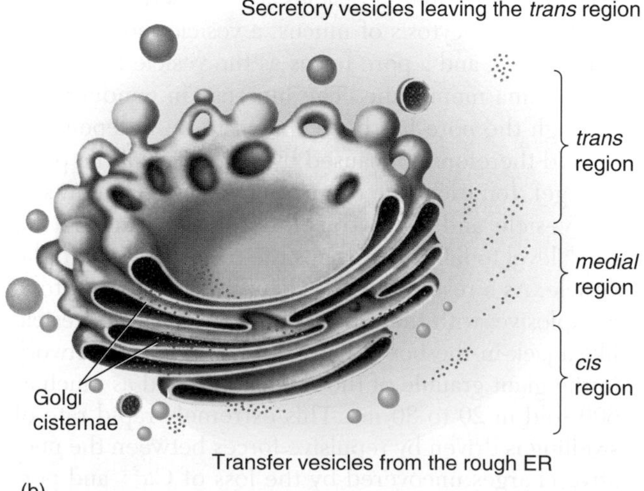

Figure 9-5 Intracellular vesicles transport secretory and membrane proteins. **(a)** Electron micrograph of the Golgi complex and rough ER in an exocrine pancreatic cell. Notice the stacked layers of the Golgi complex and the forming secretory vesicle, as well as the transfer vesicles that shuttle secretory and membrane proteins from the rough ER to the Golgi complex. **(b)** Three-dimensional model of the Golgi complex and intracellular vesicles. Transfer vesicles that have budded off from the rough ER fuse with the *cis* membranes of the Golgi complex. The secretory vesicles that bud off from sacs on the *trans* membranes store secretory and membrane proteins in concentrated form. [Part a courtesy of G. Palade; part b from Lodish et al., 1995, after a model by J. Kephardt.]

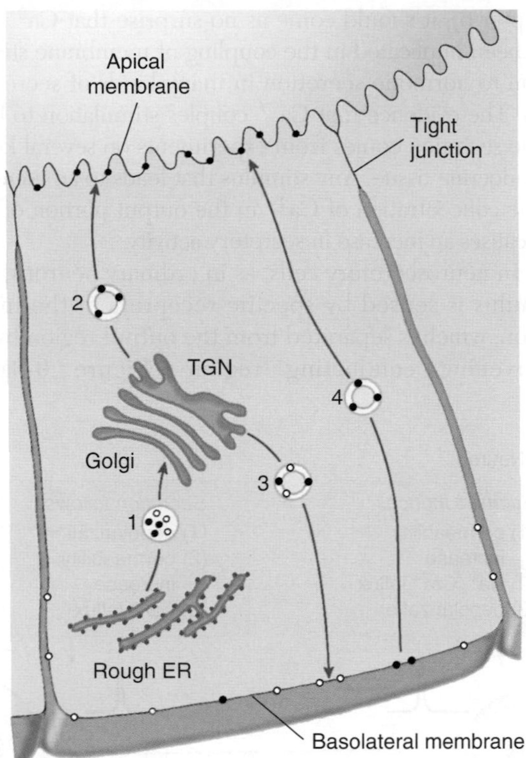

Figure 9-6 The *trans*-Golgi network (TGN) sorts newly synthesized membrane proteins into vesicles destined for the apical or basolateral membrane. After their synthesis in the rough ER, both kinds of membrane proteins move in common transfer vesicles to the Golgi complex (1), where they occupy the same compartments. In the TGN, the apical and basolateral proteins are sorted and incorporated into the membranes of vesicles, which move to the apical membrane (2) or the basolateral membrane (3). Some apically destined proteins are initially delivered to the basolateral membrane but then are retrieved and transported to the apical surface, a process termed transcytotic delivery (4). The proteins are inserted into the plasma membrane when the vesicle fuses with it.

Storage of Secreted Substances

Retention of a secretory substance within vesicles until a releasing signal is received is accomplished by a variety of means. Large protein hormones are retained simply because of their size, which renders them incapable of crossing the vesicular membrane. Some small hormone molecules are bound to larger accessory molecules, usually proteins. There is evidence that secretory vesicles containing catecholamines (norepinephrine and epinephrine) are maintained by continual active uptake into the vesicles from the cytosol. The tranquilizing drug reserpine interferes with this uptake, allowing the catecholamines to leak out of their secretory vesicles and out of the secretory cells.

The duration of storage of hormones within glands varies widely. The steroid hormones, which are not packaged in vesicles and are lipid-soluble, appear to diffuse out of secretory cells across the plasma membrane in a matter of minutes after their synthesis. Most hormones, however, are stored in vesicles until their release is stimulated by the mechanisms discussed in the next section. The thyroid hormones are secreted into the extracellular spaces of spherical clusters of cells, termed *follicles*, and are stored there for up to several months. Even after being secreted into the circulation, a hormone is in effect stored in the bloodstream until it is taken up by cells or degraded. The steroid and thyroid hormones, which are hydrophobic, are carried in the blood attached to binding proteins; these hormones remain inactive until they dissociate from the protein.

SECRETORY MECHANISMS

There are several mechanisms by which substances stored within a cell might find their way to the cell exterior. For most substances stored in secretory vesicles, the entire contents of a vesicle are delivered to the cell exterior by exocytosis. In other cases, portions of the cell, or even whole cells, break open and release material into the extracellular space.

- In **apocrine secretion**, the apical portion of the cell, which contains the secretory material, is sloughed off, and the cell then reseals at its apex. This process occurs in some molluscan exocrine glands and in certain sweat glands in hairy regions of the human body.
- In **merocrine secretion**, the apical portion of the cell pinches off, and this portion, containing the secretory products, breaks open in the lumen of the gland. This process is characteristic of many digestive glands in mammals and the exocrine glands of arthropods and annelids.
- In **holocrine secretion**, the entire cell is cast off and breaks up to release its contents. This process occurs in some insect and molluscan exocrine tissues and is characteristic of mammalian sebaceous (oil) glands in the skin.

Secretion occurs in response to stimulation of the cell. The stimulus may be a hormone or a neurotransmitter arriving at the membrane of the secreting cell; for example, acetylcholine released from sympathetic neurons stimulates the chromaffin tissue of the adrenal medulla to secrete catecholamines. Secretion may also result from direct stimulation; an increase in plasma osmolarity, for example, stimulates certain neurons that have endocrine functions. For other such neurosecretory cells, the stimulus is electrical.

Stimulation of neurosecretory cells evokes action potentials (APs) that travel to the axon terminals and there elicit the release of neurohormones. This effect can be demonstrated experimentally by stimulating such cells electrically at a considerable distance up the axon and monitoring the release of neurohormone from the terminals. The rate of secretion increases with the

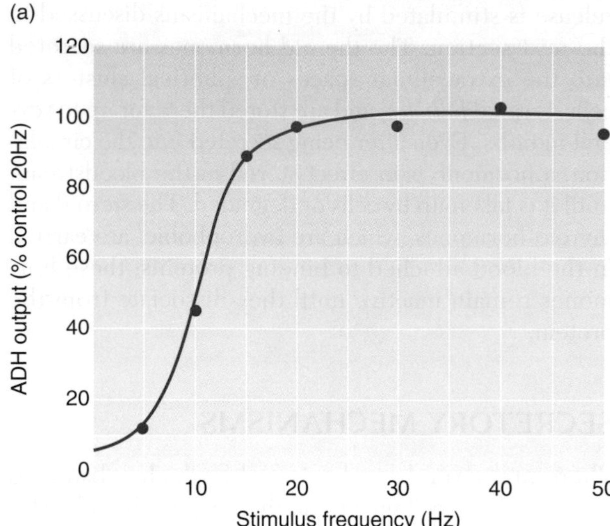

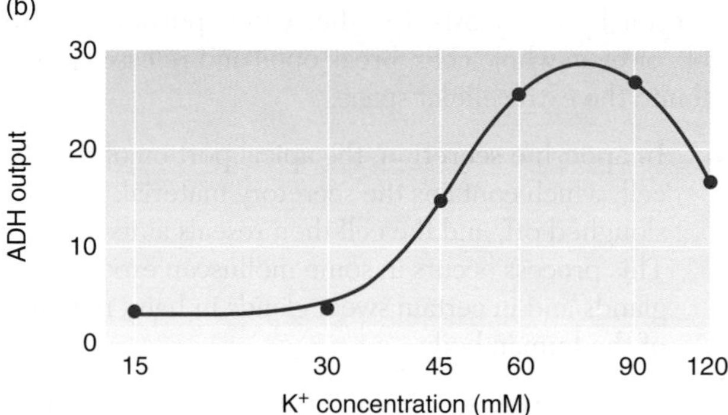

Classic Work **Figure 9-7** Both electrical stimulation (action potentials) and increased extracellular K^+ can induce the release of antidiuretic hormone (ADH) from neurosecretory cells. **(a)** Release of ADH from the rat neurohypophysis as a function of the frequency of electrical stimulation. The stimulus pulses at each frequency were continued for 5 minutes. **(b)** Release of ADH (arbitrary units) as a function of extracellular K^+ concentration. Freshly dissected neurohypophyses were placed in incubation media of different K^+ concentrations (to produce varying degrees of depolarization) for 10 minutes, after which the amount of ADH released into the medium was assayed. [Part a adapted from Mikiten, 1967; part b adapted from Douglas, 1974.]

frequency of the impulses (Figure 9-7a). Neurohormone secretion can also be stimulated in the absence of action potentials if the plasma membrane is depolarized experimentally by increasing the extracellular K^+ concentration. Secretion rises to a maximum with increasing K^+ concentration and hence with increasing depolarization (Figure 9-7b). The effect of such artificial depolarization suggests that depolarization accompanying the action potential is normally the trigger for secretion of this type.

At still higher K^+ concentrations, membrane depolarization exceeds the value for maximal Ca^{2+} entry into the neurosecretory cell, and secretion decreases (see Figure 9-7b, far right). In view of the well-known role of Ca^{2+} in the regulation of neurotransmitter release (described in Chapter 6), it should come as no surprise that Ca^{2+} has also been implicated in the coupling of membrane stimulation to hormone secretion in many kinds of secretory cells. The evidence that Ca^{2+} couples stimulation to hormone secretion comes from experiments on several kinds of endocrine tissue. Any stimulus that leads to an increase in the concentration of Ca^{2+} in the output portion of the cell causes an increase in secretory activity.

In neurosecretory cells, as in ordinary neurons, the stimulus is sensed by specific receptors in the input region, which is separated from the output region by an intervening conducting region (Figure 9-8a,b).

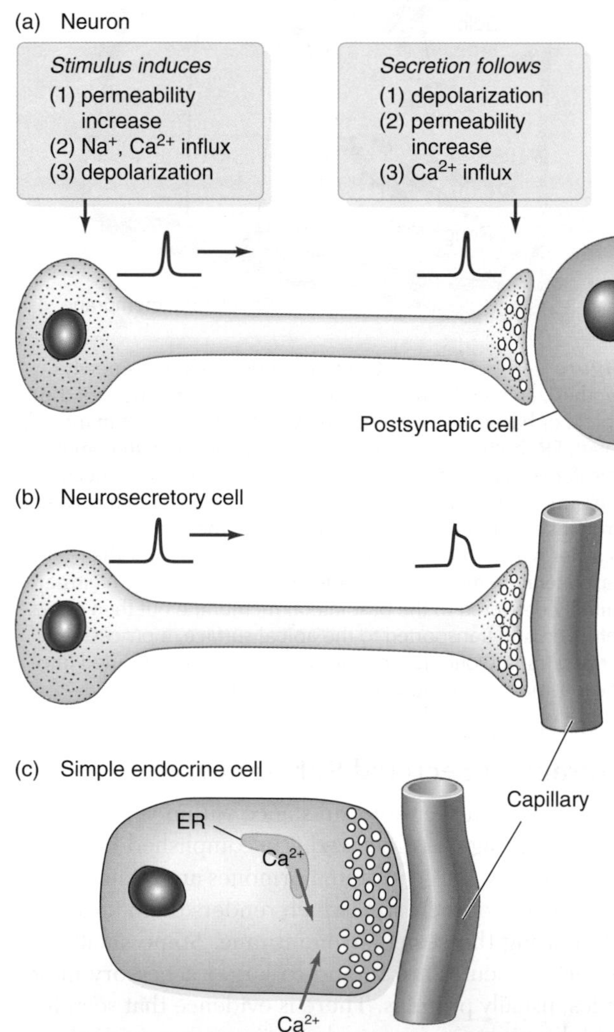

Figure 9-8 The elevation of Ca^{2+} in the output region of a secretory cell triggers exocytosis. **(a)** In ordinary neurons, depolarization is initiated in the input region and spreads to the output region (the axon terminals) by means of action potentials. **(b)** This process also occurs in neurosecretory cells; note the prolonged action potential characteristic of some neurosecretory terminals. In both cases, depolarization results in an influx of Ca^{2+}, which stimulates exocytosis **(c)** Although some simple endocrine cells (those that are not neurons) produce action potentials, many are activated to secrete without membrane depolarization. In these cells, the stimulus causes a release of Ca^{2+} stored in the ER, and the increased cytosolic Ca^{2+} level results in exocytosis.

Incoming stimuli (synaptic input or physical or chemical changes in the plasma membrane) increase the frequency of APs in the axon. By invading and depolarizing the terminal membranes, the action potentials cause Ca^{2+} channels to open. The resulting influx of Ca^{2+} then triggers exocytosis.

Ca^{2+} also triggers exocytosis in some simple endocrine and exocrine cells. Stimulation of these cells initiates a chemical cascade leading to the release of Ca^{2+} sequestered in the ER or to an influx of Ca^{2+} into the cell from the extracellular medium. The resulting rise in free cytosolic Ca^{2+} induces hormone secretion (Figure 9-8c). We will see how this happens when we discuss intracellular signaling pathways below.

Glandular Secretions

Some secretions result from the combined activity of a number of secretory cells that form a **gland.** Secretion from a gland often occurs at a low "resting" level, which can be modulated up or down by signals acting on the gland. Some glands, however, exhibit no secretory activity until they are stimulated into action. Only when a seabird drinks seawater, for example, are its nasal glands activated to excrete salt.

Various kinds of signals regulate glandular activity. Prominent among them are neurotransmitters released from neurons innervating the glandular tissue and hormones released from other endocrine tissues. In addition, some glandular tissues respond directly or indirectly to conditions in the extracellular environment. For example, osmoregulatory neurons in the vertebrate hypothalamus respond to the osmolarity of the extracellular fluid surrounding them, which is about the same as the osmolarity of the blood plasma. The salivary glands are under direct neuronal control, and are influenced by both conditioned and unconditioned reflexes. The sight and smell of food—even the thought of food—can cause a marked increase in salivation, especially if the animal is hungry.

Types and General Properties of Glands

Glands are classified as either endocrine or exocrine glands (Figure 9-9). **Endocrine glands** are organs that secrete hormones directly into the circulatory system and modulate body processes; the thyroid gland, for example, produces thyroid hormone, which modulates growth. Endocrine glands are sometimes referred to as ductless glands; their secretions are called **hormones. Exocrine glands,** on the other hand, secrete fluids through a duct onto the epithelial surface of the body; for example, sweat glands produce sweat for evaporative cooling, and the gallbladder stores bile salts produced in the liver and excretes them into the gut via the bile duct.

Although endocrine and exocrine glands can usually be distinguished by the presence (exocrine) or absence (endocrine) of a duct, there is no such thing as a typical gland. Endocrine tissues are structurally and

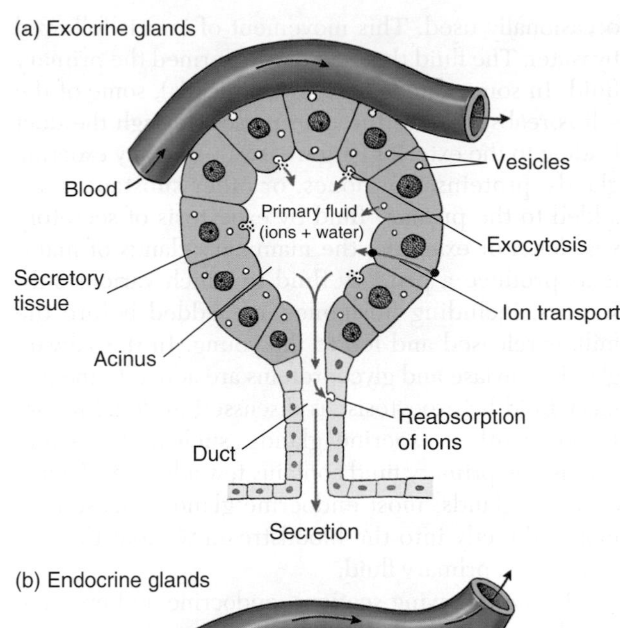

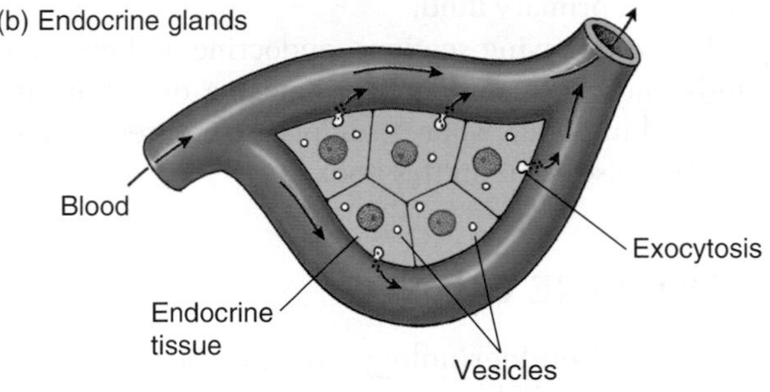

Figure 9-9 Glands can be divided into two broad structural types. **(a)** Exocrine glands release secretions via a duct onto an epithelial surface. The primary fluid is formed by ion transport, with water following osmotically. A variety of compounds may be added to the primary fluid by exocytosis. The resulting primary secretion may be modified by reabsorption of material as the fluid passes down the duct. **(b)** Endocrine glands are ductless and release secretions directly into the bloodstream. Water-soluble secretions are released by exocytosis of secretory vesicles, whereas lipid-soluble secretions may leave secretory cells by diffusion.

chemically diverse, and some contain more than one kind of secretory cell, each secreting a different hormone. Endocrine glands exhibit no distinctive morphologic features (other than rich vascularization). For this reason, and because many hormones are secreted in vanishingly small amounts, it has often been extremely difficult to establish unequivocally that a particular tissue has a specific endocrine function.

Exocrine glands are more easily identified because they possess a duct leading to the body surface. In these glands, secretory cells are grouped around blind cavities, called acini, that are connected by ducts to the exterior. In many exocrine glands (e.g., the rectal gland of sharks, the nasal gland of birds, and the sweat glands of mammals), an asymmetric distribution of ion pumps on the apical and basolateral surfaces of the secretory cells makes it possible for the cells to pump ions from one side to the other. Sodium chloride is usually the secreted salt, although potassium chloride is occasionally used. This movement of ions is followed by water. The fluid thus secreted is termed the **primary fluid.** In some glands (e.g., sweat glands), some of the salt is reabsorbed as the fluid passes through the duct leading to the exterior (Figure 9-9a). In many exocrine glands, proteins, hormones, or other substances are added to the primary fluid by exocytosis of secretory vesicles. For example, the mammary glands of mammals produce a primary fluid to which various substances, including hormones, are added before the milk is released and fed to the young. In the salivary glands, amylase and glycoproteins are added to the primary fluid by exocytosis, as discussed in detail below. In some other exocrine glands, such as the sweat glands, the primary fluid contains few additives. Unlike exocrine glands, most endocrine glands release hormones directly into the bloodstream without the formation of a primary fluid.

In the following sections, endocrine and exocrine glands and their cellular mechanisms of action are described in detail. Many glandular structures are also described elsewhere in this book.

ENDOCRINE GLANDS

The study of **endocrinology** probably began in 1849 when A. A. Berthold reported his classic experiments. When he castrated cockerels (young roosters), they grew small combs and wattles, demonstrated little interest in hens or fighting, and had a weak crow (Figure 9-10). If he replaced the testes into the abdominal cavity, the comb and wattle approached normal size, and the cock had a normal crow and showed normal male behavior. Berthold speculated that the testes secreted something that conditioned the blood, and that the blood then acted on the cockerel to produce male characteristics.

The experiments carried out by Berthold paved the way for many similar experiments in which the effects of removing and then replacing an organ were observed in order to demonstrate an endocrine function for that organ. William Bayliss and Ernest H. Starling described the first hormone to be discovered: secretin, a substance liberated from the mucosa of the small intestine that causes increased flow of digestive juices from the pancreas. In 1905, Starling introduced the term *hormone,* derived from the Greek for "I arouse," and described three characteristic properties that define hormones:

- Hormones are synthesized by specific tissues or glands.
- Hormones are secreted into the bloodstream, which carries them to their site(s) of action.
- Hormones change the activities of target tissues or organs.

	1	2
Experimental groups	Normal cock	Castrated cock
Treatment	Both testes removed	One testis replaced
Results	Comb and wattles small	Comb and wattles normal
	No interest in hens	Interest in hens
	Weak crow	Normal crow
	Listless fight behavior	Aggressive fight behavior
		Testis larger than in controls

Classic Work **Figure 9-10** A. A. Berthold carried out some of the earliest experiments demonstrating endocrine action. When he removed the testes from cockerels, they lost many of the characteristics of cocks (group 1). When one testis was replaced into the abdominal cavity, the male characteristics were retained (group 2). Since Berthold's experiments, the functions of many endocrine glands have been identified by similar removal-and-replacement experiments. [Adapted from Hadley, 1992.]

The effects of castration on cattle and humans were known long before Berthold's experiments. What was the important contribution made by Berthold?

The absence of discrete morphologic markers complicates the identification of endocrine glands. The following criteria have been used to establish whether a tissue has an endocrine function:

- *Ablation (removal) of the suspected tissue should produce deficiency symptoms in the subject.* Experimentally, this effect may be difficult to demonstrate if the tissue is part of an organ that has more than one function, as does atrial tissue of the heart.

- *Replacement (reimplantation) of the ablated tissue elsewhere in the body should prevent or reverse the deficiency symptoms.* Misleading results, however, may be obtained when ablation-and-reimplantation experiments are done with tissues closely associated with the nervous system because of the interruptions of neuronal connections.

- *The deficiency symptoms should be relieved when the suspected hormone is injected.* Successful replacement is the most important criterion for identification of a suspected endocrine tissue and its hormone. It is also the basis of replacement therapy for patients with a dysfunctional endocrine gland.

Following purification of the suspected hormone, the chemical structure of the active substance can be determined. The molecule can then be synthesized and tested for biological potency. Immunohistochemical techniques can be used to determine the cellular location of the hormone in different tissues once it has been isolated. New biochemical techniques have greatly accelerated this process, making important contributions to the development of endocrinology over the last two decades. For example, radioimmunoassays (RIAs) permit detection of specific hormones in minute concentrations with a high degree of accuracy (see Chapter 2). In this technique, antibodies are raised against the hormone in question, usually in a rabbit. A standard curve is then constructed to describe the binding of the hormone to the antibody, using a radiolabeled hormone and a known amount of antibody. Subsequently, the quantity of hormone in a sample can be determined by the extent to which it competes with and reduces the binding of labeled hormone to the antibody. The use of monoclonal antibodies, which recognize only one antigen, has further improved the accuracy of detecting and quantifying hormones and their receptors by RIA.

Endocrinologists also use recombinant DNA techniques in various ways. For instance, genetic material can be inserted into bacteria to produce strains capable of synthesizing human hormones. Foreign genes have also been introduced into mammalian embryos; for example, when the structural gene for rat growth hormone is introduced into mouse embryos, the resulting mice grow to be much larger than normal.

Hormone molecules come into contact with all tissues in the body, but they affect only cells that contain receptors specific for that hormone. Hormones are generally produced in small amounts and greatly diluted in the blood and interstitial fluid. They must therefore be effective at very low concentrations (typically between 10^{-8} and 10^{-12} M). By comparison, if human taste buds could detect sugar at 10^{-12} M, we would be able to taste a pinch of sugar dissolved in a swimming pool full of coffee or tea. The high sensitivity of hormonal signaling is due to the high affinity of target cell receptors for hormones. As we will see below, the binding of a hormone molecule to its receptor leads to a cascade of two or more intracellular signal molecules, or **second messengers,** that greatly amplify the effect of hormone binding; just a few hormone molecules can influence thousands or millions of molecular reactions within a cell.

Although there are many hormones, most of those found in metazoans belong to one of four structural categories (Figure 9-11):

- **Amines,** which include the catecholamines epinephrine and norepinephrine as well as the

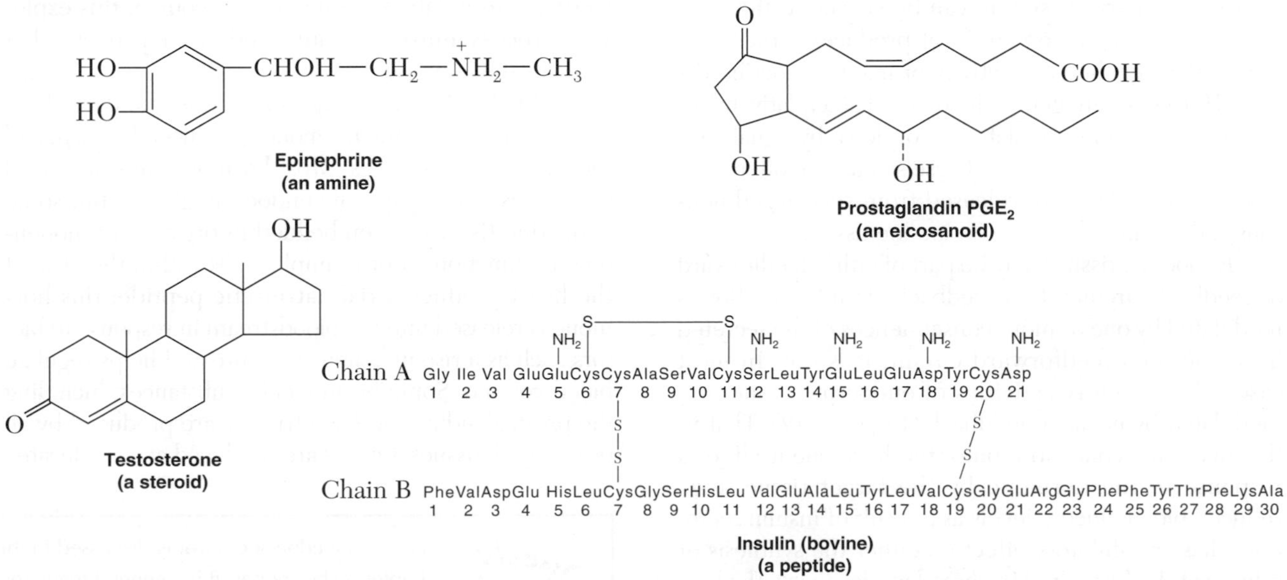

HO, HO—(ring)—CHOH—CH_2—$\overset{+}{NH_2}$—CH_3

Epinephrine (an amine)

Prostaglandin PGE_2 (an eicosanoid)

Testosterone (a steroid)

Chain A Gly Ile Val GluGluCysCysAlaSerValCysSerLeuTyrGluLeuGluAspTyrCysAsp

Chain B PheValAspGlu HisLeuCysGlySerHisLeu ValGluAlaLeuTyrLeuValCysGlyGluArgGlyPhePheTyrThrPreLysAla

Insulin (bovine) (a peptide)

Figure 9-11 Most hormones belong to one of four structural categories. Amine hormones (with the exception of thyroid hormones) and peptide hormones are lipid-insoluble, whereas steroid hormones and eicosanoids such as prostaglandins are lipid-soluble.

thyroid hormones, are small molecules derived from amino acids.

- **Eicosenoids** (prostaglandins, leukotrienes, thromboxanes, and lipoxins) are produced by the metabolism of arachidonic acid.
- **Steroid hormones** (e.g., testosterone and estrogen) are cyclic hydrocarbon derivatives synthesized from cholesterol.
- **Peptide and protein hormones** (e.g., insulin) constitute the largest number of hormones and are the most complex.

The term **autacoid hormone** is used to describe various active endogenous substances, such as histamine and serotonin, that are not included in any of the above categories.

In contrast to neurotransmitters, which mediate rapid signaling over short distances, hormones communicate over longer distances and on a longer time scale. Thus endocrine systems are well suited for regulatory functions that are sustained for minutes, hours, or days. These functions include the maintenance of blood osmolarity (antidiuretic hormone) and blood sugar (insulin and glucagon), regulation of metabolic rates (growth hormone and thyroxine), control of sexual activity and reproductive cycles (sex hormones), and modification of behavior (various hormones). In fact, the rapid activity of the nervous system and the slower, more sustained activity of the endocrine system work together in the integration of physiological and metabolic functions in the body, to the point that a given molecule may serve as a neurotransmitter in some circumstances and a hormone in others. There is a close and overlapping relationship between the nervous and endocrine systems. In many respects the nervous system can be viewed as the most important endocrine organ, for it produces certain hormones that regulate the activity of many endocrine tissues. Hormones are generally secreted at a relatively low resting level that is modulated up or down by signals acting on the endocrine tissue. These signals are often neurohormones, which are released from specialized neurons and act directly on the endocrine tissue.

Endocrine tissues may be part of either feedforward or feedback circuits. In a feedback circuit, secretion is modulated by one or more consequences of the secreted hormone; in a feedforward circuit, it is not. In most cases, the secretory activities of endocrine tissues are modulated by negative feedback (Figure 9-12). That is, the increasing concentration of the hormone itself, or a response to the hormone by the target tissue (e.g., reduced blood glucose levels as a result of insulin secretion), has an inhibitory effect on either the synthesis or release of the hormone. In *short-loop feedback*, the hormone itself, or a direct effect of the hormone, acts on the endocrine tissue to reduce secretion, thereby keeping hormone secretion in check. *Long-loop feedback* is similar in principle, but it includes more elements in series.

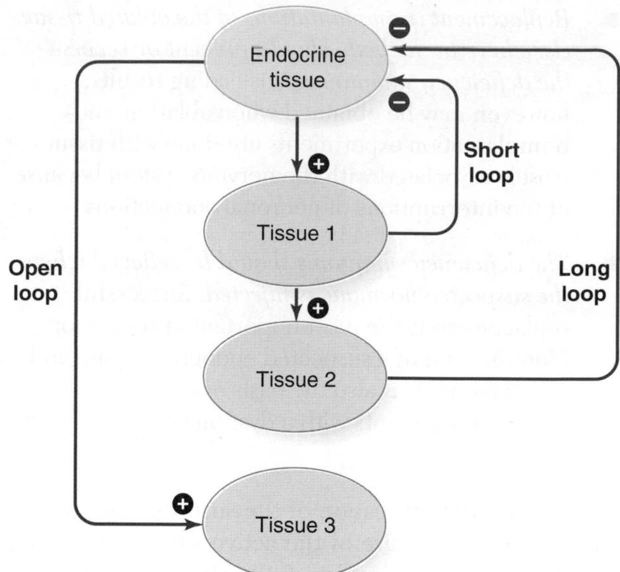

Figure 9-12 Most endocrine tissues are subject to negative-feedback control. In short-loop feedback, the response of the primary target tissue (tissue 1) feeds back onto the endocrine gland. In long-loop feedback, signals from secondary targets (tissue 2) control secretory activity. In an open loop, there is no feedback.

In physiological situations in which an extremely rapid hormonal response is called for, the endocrine tissue may be subject to transient positive feedback; that is, a rise in the level of a hormone stimulates further secretion. Positive feedback is most common in the early phases of a response. For example, early in the reproductive cycle of some vertebrates, the level of luteinizing hormone peaks rapidly, assisted by positive-feedback stimulation. Ultimately, of course, this explosive process must be countered by a process that reverses the increase.

Table 9-1 lists the major endocrine glands and tissues in vertebrates, the hormones produced by each, and their physiological roles. Most hormones are produced and released from specific endocrine glands, but some endocrine tissues are embedded in organs with nonendocrine functions. For example, cells within the atria of the heart produce atrial natriuretic peptide; this hormone is released into the bloodstream in response to factors such as a rise in venous pressure and helps regulate blood volume. Some hormonelike substances, including the prostaglandins and leukotrienes, are produced by all or nearly all tissues. Others are produced at a single site.

Although feedback control is discussed in this chapter on hormones, it is a general regulatory mechanism in physiological systems. What are some examples of feedforward and feedback control in other physiological systems?

Table 9-1 Vertebrate endocrine glands and tissues

Gland/source	Hormone	Major physiological role*
Adrenal gland		
Steroidogenic tissue (cortex)	Aldosterone	↑ Sodium retention
	Cortisol and corticosterone	↑ Carbohydrate metabolism and sympathetic function
Chromaffin tissue (medulla)	Epinephrine and norepinephrine	Multiple ↑ and ↓ effects on nerves, muscles, cellular secretions, and metabolism
Gastrointestinal tract	Cholecystokinin	↑ Secretion of enzymes by pancreatic acinar cell; ↑ gall-bladder contraction
	Chymodenin	↑ Secretion of chymotrypsinogen from the exocrine pancreas
	Gastric inhibitory peptide	↓ Gastric acid (HCI) secretion
	Gastrin	↑ Gastric acid (HCI) secretion
	Gastrin-releasing peptide	↑ Gastrin secretion; ↓ gastric acid (HCI) secretion
	Motilin	↑ Gastric acid secretion and motility of intestinal villi
	Neurotensin	Enteric neurotransmitter
	Secretin	↑ Bicarbonate secretion by pancreatic acinar cells
	Substance P	Enteric neurotransmitter
	Vasoactive intestinal peptide	↑ Intestinal secretion of electrolytes
Heart (atrium)	Atrial natriuretic peptide (ANP)	↑ Salt and water excretion by kidney
Kidney	Calcitriol†	↑ Blood Ca^{2+}, bone formation, and intestinal absorption of Ca^{2+} and PO_4^{23}
	Erythropoietin (erythrocyte-stimulating factor)	↑ Production of red blood cells (erythropoiesis)
	Renin	↑ Conversion of angiotensinogen to angiotensin II
Ovary		
Preluteal follicle	Estradiol	↑ Female sexual development and behavior
	Estrogen	↑ Estrus and female secondary sexual characteristics; prepares reproductive system for fertilization and ovum implantation
Corpus luteum	Progesterone	↑ Growth of uterine lining and mammary glands, and maternal behavior
	Relaxin	↑ Relaxation of pubic symphysis and dilation of uterine cervix
Pancreas (islets of Langerhans)	Glucagon	↑ Blood glucose, gluconeogenesis, and glycogenolysis
	Insulin	↓ Blood glucose; ↑ protein, glycogen, and fat synthesis
	Pancreatic polypeptide	↑ ↓ Secretion of other pancreatic islet hormones
	Somatostatin	↓ Secretion of other pancreatic islets hormones
Parathyroid glands	Parathormone	↑ Blood Ca^{2+}; ↓ blood PO_4^{-3}
Pineal (epiphysis)	Melatonin	↓ Gonadal development (antigonadotropic action)
Pituitary gland	See Table 9-2, 9-3	
Placenta	Chorionic gonadotropin (CG, choriogonadotropin)	↑ Progesterone synthesis by corpus luteum
	Placental lactogen	↑ Fetal growth and development (possibly); ↑ Mammary gland development in the mother
Plasma angiotensinogen‡	Angiotensin II	↑ Vasoconstriction and aldosterone secretion; ↑ Thirst and fluid ingestion (dipsogenic behavior)
Testes		
Leydig cells	Testosterone	↑ Male sexual development and behavior
Sertoli cells	Inhibin	↓ Pituitary FSH secretion
	Müllerian regression factor	↑ Müllerian duct regression (atrophy)

(continued on the next page)

Gland/source	Hormone	Major physiological role*
Thymus gland	Thymic hormones	↑ Proliferation and differentiation of lymphocytes
Thyroid gland		
Follicular cells	Thyroxine and triiodothyronine	↑ Growth and differentiation; ↑ metabolic rate and oxygen consumption (calorigenesis)
Parafollicular cells (or ultimobranchial glands)	Calcitonin	↓ Blood Ca^{2+}
Most or all tissues	Leukotrienes	↑ ↓ Cyclic nucleotide formation
	Prostacyclins	↑ Cyclic nucleotide (cAMP) formation
	Prostaglandins	↑ Cyclic nucleotide (cAMP) formation
	Thromboxanes	↑ Cyclic nucleotide (cGMP?) formation
Selected tissues	Endorphins	Opiate-like activity
	Epidermal growth factor	↑ Epidermal cell proliferation
	Fibroblast growth factor	↑ Fibroblast proliferation
	Nerve growth factor	↑ Neurite development
	Somatomedins	↑ Cellular growth and proliferation

* ↑ means hormone stimulates or increases indicated effect; ↓ means hormone inhibits or decreases indicated effect.
†The final steps in synthesis of calcitriol from vitamin D_3 occur in the kidney, but the skin and liver also play a role in its synthesis.
‡Angiotensinogen is produced in the liver and circulates in the bloodstream, where it is cleaved by renin to form the active hormone angiotensin II.
Source: Adapted from Hadley, 2000.

NEUROENDOCRINE SYSTEMS

As mentioned earlier, the secretion of hormones from some endocrine tissues is regulated by **neurohormones,** which are produced by specialized neurons called **neurosecretory cells.** Most neurosecretory cells are similar to ordinary neurons, but there are some notable differences. First, the secretory vesicles containing neurohormones are typically 100–400 nm in diameter, whereas the presynaptic vesicles containing neurotransmitters in ordinary neurons are much smaller, 30–60 nm in diameter (Figure 9-13). Second, although ordinary neurons use both slow and fast axonal transport systems, neurosecretory cells appear to use only fast axonal transport, moving neurohormones at rates of up to 2800 mm a day. Third, ordinary neurons form synapses with other cells at their terminals, whereas neurosecretory axons generally terminate in clusters within a bed of capillaries, forming a discrete **neurohemal organ** (Figure 9-14). The neurohormones released into the interstitial space diffuse into the capillaries and are carried in the bloodstream to target tissues.

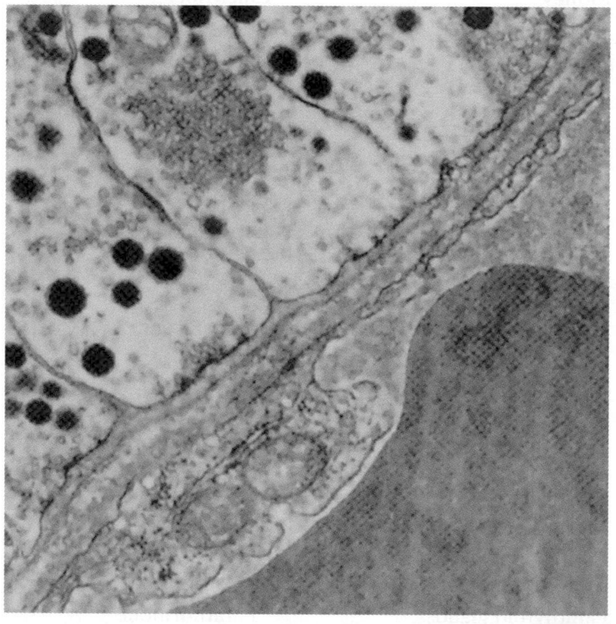

Figure 9-13 The secretory vesicles in neurosecretory axons are much larger than the neurotransmitter vesicles in ordinary presynaptic neurons. In this electron micrograph of the hamster posterior pituitary gland, the large dark bodies are the secretory vesicles (or granules). The axon terminals end on an endothelial basement membrane that separates them from a fenestrated capillary wall. The large dark object at the lower right corner is a red blood cell in the capillary. [Adapted from Douglas et al., 1971.]

Neurosecretory cells in the hypothalamus control hormonal secretion in the two very different parts of the pituitary gland (hypophysis), a small appendage lying just below the hypothalamus (Figure 9-15). Because it secretes at least nine hormones, many which control other glands, the pituitary gland has been called the "master gland." Neurohormones secreted from neurosecretory cells in the hypothalamus regulate the secretion of various glandular hormones from the nonneuronal anterior pituitary (or *adenohypophysis*). Other neurosecretory cells with their somata located in the hypothalamus extend their axons to, and release their neurohormones into, the posterior pituitary (or *neurohypophysis*). The neurosecretory cells in the hypothalamus respond to sensory input from various parts of the body.

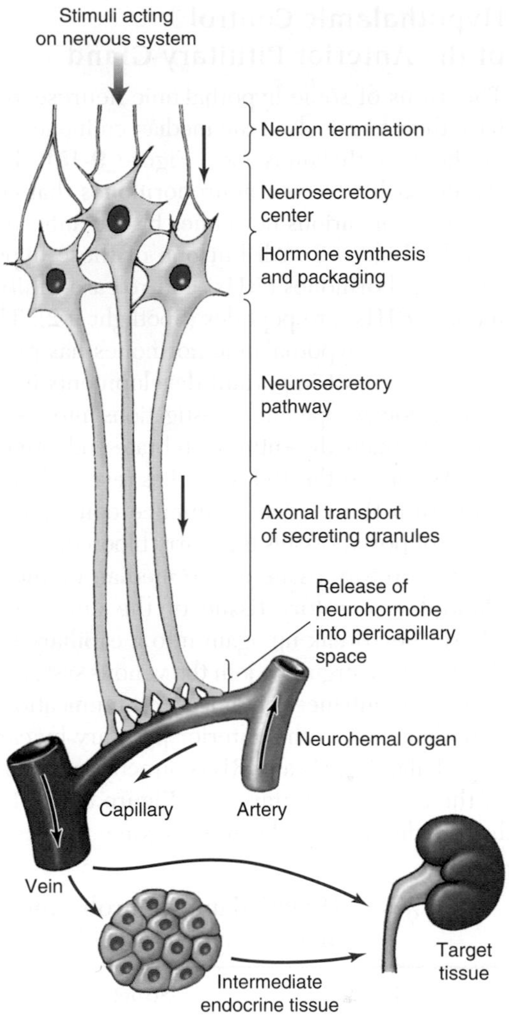

Figure 9-14 Neurohormones are released from the terminals of neurosecretory cells into a bed of capillaries forming a neurohemal organ. After entering the bloodstream, some neurohormones (e.g., oxytocin) act directly on a somatic target tissue, but most activate an intermediate endocrine gland, stimulating secretion of another hormone that acts on the target tissue.

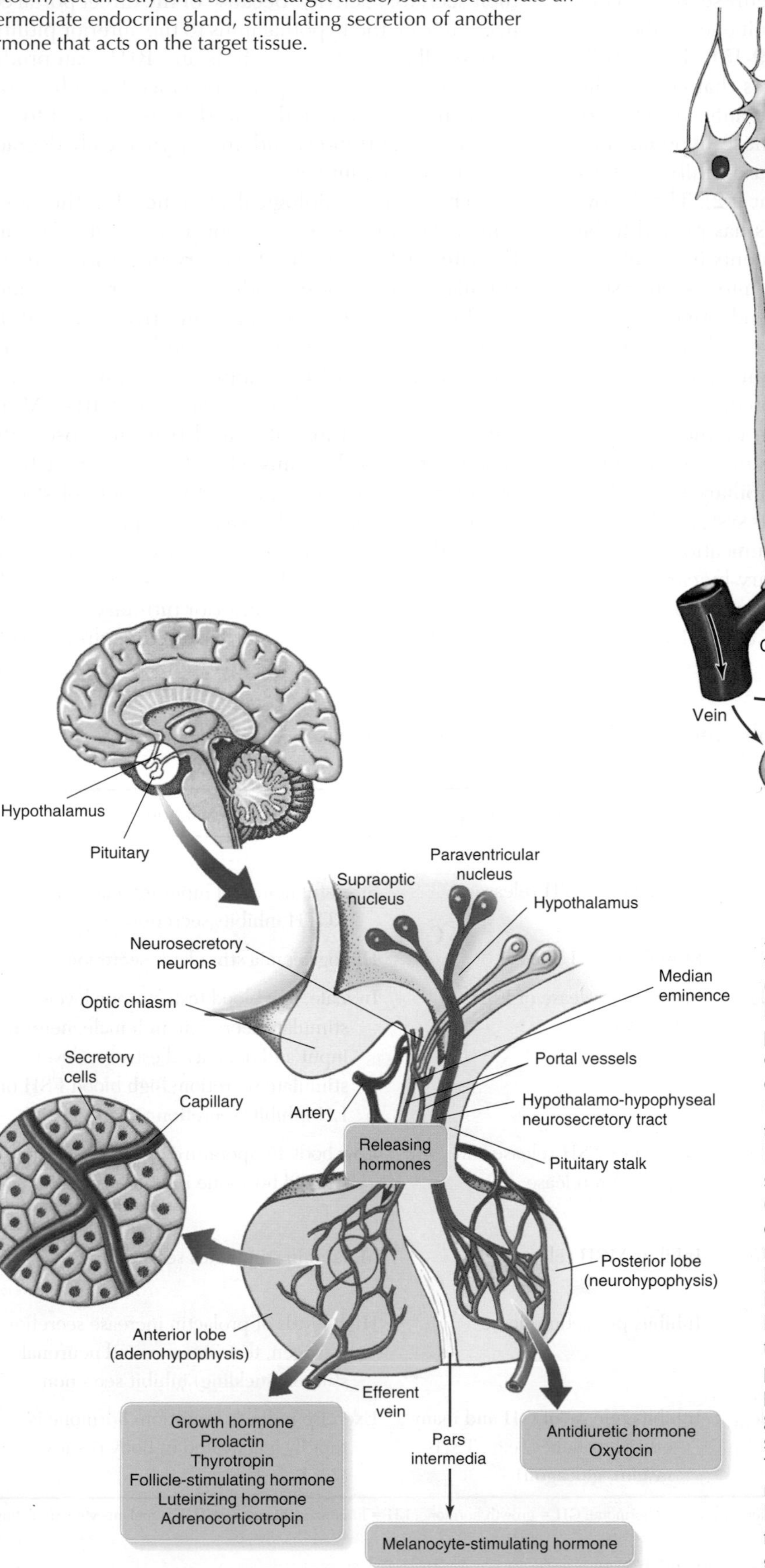

Figure 9-15 Hormonal secretion from the primate pituitary gland (hypophysis) is controlled by the hypothalamus. The anterior lobe of the pituitary gland (adenohypophysis), which consists of nonneuronal glandular tissue, comprises the pars distalis, pars intermedia, and pars tuberalis. (The pars tuberalis, not shown, consists of a thin layer of cells surrounding the pituitary stalk.) The posterior lobe (neurohypophysis or pars nervosa), an extension of the brain, consists of neuronal tissue. Hypothalamic releasing hormones (RHs) or release-inhibiting hormones (RIHs) secreted by hypothalamic neurosecretory terminals in the median eminence are carried via the portal vessels (hypothalamo-hypophyseal portal system) to the anterior pituitary gland, where they stimulate (or inhibit) secretion of several glandular hormones. Two neurohormones produced by neurosecretory cells whose somata are located in the hypothalamus are released from terminals in the posterior pituitary gland.

Hypothalamic Control of the Anterior Pituitary Gland

The axons of some hypothalamic neurosecretory cells have their terminals in the median eminence at the floor of the hypothalamus (see Figure 9-15). These cells secrete at least seven neurohormones that control the secretion of various hormones by the anterior pituitary gland (Table 9-2). All but one of these hypothalamic releasing hormones (RHs) and release-inhibiting hormones (RIHs) are peptides (Spotlight 9-2). The discovery of these hypothalamic hormones has proved to be one of the most important developments in vertebrate endocrinology, opening investigations into the orchestration of virtually the entire vertebrate endocrine system.

As early as the 1930s, studies revealed that capillaries within the median eminence converge to form a series of portal vessels that carry blood directly from the neurosecretory tissue of the median eminence to the glandular secretory tissue of the anterior pituitary. There they break up again into a capillary bed before finally reconverging to join the venous system. This portal system enhances chemical communication from the hypothalamus to the anterior pituitary by carrying the hypothalamic RHs and RIHs directly to the interstitium of the anterior pituitary (see Figure 9-15). Here these hypothalamic neurohormones come into contact with the glandular endocrine cells that secrete adenohypophyseal hormones, either stimulating or inhibiting their secretory activity. Because of the direct portal connection from the hypothalamus to the anterior pituitary, very small amounts of the RHs and RIHs can produce effects on the anterior pituitary. Once these hormones enter the general circulation, they are diluted to ineffective concentrations and are enzymatically degraded within several minutes.

The first physiological evidence for the neurohomonal control of the anterior pituitary gland came in the late 1950s with the discovery of a substance that stimulates the release of adrenocorticotropic hormone (ACTH, or adrenocorticotropin) from the anterior pituitary. This substance, obtained by extraction from the hypothalami of thousands of pigs, was given the name corticotropin-releasing hormone (CRH). Minute amounts of CRH are liberated from neurosecretory cells in the hypothalamus when the cells are activated by neuronal input in response to a variety of stressful stimuli (e.g., cold, fright, sustained pain). The CRH travels through the portal vessels to the anterior pituitary, where it stimulates the release of ATCH. The ACTH released from the anterior pituitary circulates in the bloodstream to its target tissue, the adrenal cortex, where it stimulates release of adrenocortical hormones.

Table 9-2 **Hypothalamic neurohormones that stimulate or inhibit release of adenohypophyseal hormones**

Hormone	Structure	Primary action in mammals	Regulation*
Stimulatory			
Corticotropin-releasing hormone (CRH)	Peptide	Stimulates ACTH release	Stressful neuronal input increases secretion; ACTH inhibits secretion
GH-releasing hormone (GRH)	Peptide	Stimulates GH release	Hypoglycemia stimulates secretion
Gonadotropin-releasing hormone (GnRH)	Peptide	Stimulates release of FSH and LH	In male, low blood testosterone levels stimulate secretion; in female, neuronal input and decreased estrogen levels stimulate secretion; high blood FSH or LH inhibits secretion
TSH-releasing hormone (TRH)	Peptide	Stimulates TSH release and prolactin release	Low body temperatures induce secretion; thyroid hormone inhibits secretion
Inhibitory			
MSH-inhibiting hormone (MIH)	Peptide	Inhibits MSH release	Melatonin stimulates secretion
Prolactin-inhibiting hormone (PIH)	Amine	Inhibits prolactin release	High levels of prolactin increase secretion; estrogen, testosterone, and neuronal stimuli (suckling) inhibit secretion
Somatostatin (GH-inhibiting hormone, GIH)	Peptide	Inhibits release of GH and many other hormones (e.g., TSH, insulin, glucagon)	Exercise induces secretion; hormone is rapidly inactivated in body tissues

*ACTH = adrenocorticotropic hormone; FSH = follicle-stimulating hormone; GH = growth hormone; LH = luteinizing hormone; MSH = melanocyte-stimulating hormone; TSH = thyroid-stimulating hormone.

PEPTIDE HORMONES

The opportunism of biochemical evolution is evident in the distribution and structure of a group of hormones and neurotransmitters consisting of small polypeptide chains. These chains can range from as few as three or four amino acid residues to as many as two or three dozen. Collectively called peptide hormones, most of these substances are widely distributed in the human body and throughout the animal kingdom. Thus, we may find a particular peptide hormone in visceral tissues, such as the digestive tract (see Chapter 15), and in the central nervous system (see Chapter 6). For example, insulin and somatostatin, both of which were originally discovered in the pancreas, are now known to be present in hypothalamic neurons. TSH-releasing hormone (TRH), the hypothalamic hormone originally found to cause the release of thyroid-stimulating hormone (TSH) from the anterior pituitary gland, has recently been found in lampreys (which produce no TSH) and in snails (which have no thyroid or pituitary glands), as well as in many other invertebrates.

The discovery in the 1970s that peptide hormones originally thought to be confined to tissues of the mammalian gut also occurred in various parts of the CNS initially was surprising. Now the concept of "brain-gut" hormones is no longer unusual, and we have grown used to the idea that the gene coding for a regulatory molecule in one type of body tissue may also be utilized by another tissue to make the same hormone, but for a different function. Recall that the action of a hormone depends on the presence of a receptor and the nature of the signaling cascade linked to it, as well as the effector molecules expressed in a particular tissue.

An interesting feature of the peptide hormones is that some of them are produced in variant forms both within an individual and among different taxonomic groups. This phenomenon is well illustrated by the vasopressin-oxytocin family (see Table 9-4). Another example is cholecystokinin: variants of this hormone with 33, 39, or 58 amino acid residues are present in the mammalian digestive tract, but small 4- or 8-residue fragments, cleaved from the carboxyl end of the larger cholecystokinin variants, are found in the brain.

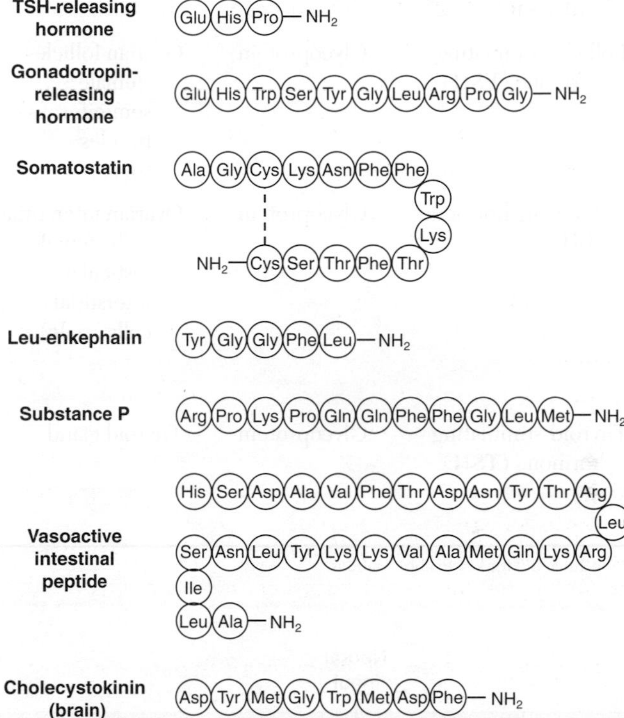

Peptide hormones range in length from as few as three amino acid residues to several dozen residues. Of the representative peptide hormones shown here, the upper three are releasing or release-inhibiting hormones produced by hypothalamic neurons and the lower four are brain-gut hormones.

Glandular Hormones Released from the Anterior Pituitary Gland

The anterior lobe of the pituitary gland consists of the pars distalis, pars tuberalis, and pars intermedia. In mammals, six hormones are released from the pars distalis and one from the pars intermedia (see Figure 9-15). Although the glandular secretory cells of the anterior pituitary are all generally similar in appearance, they can be classified into two histochemically distinct types:

- *Acidophils,* which stain orange or red with acid dyes, secrete growth hormone (GH; also termed somatotropin) and prolactin.
- *Basophils,* which stain blue with basic dyes, secrete ACTH, thyroid-stimulating hormone (TSH), melanocyte-stimulating hormone (MSH), luteinizing hormone (LH), and follicle-stimulating hormone (FSH).

Like ACTH, the hormones TSH, LH, and FSH are primarily **tropic** in their actions. That is, they act on other endocrine tissues (e.g., thyroid, gonads, and adrenal cortex), regulating the secretory activity of these target glands (Table 9-3 on the next page). LH and FSH, acting on the gonads, are often referred to as *gonadotropins.* Thus, the effect of these tropic hormones on nonendocrine somatic tissues is indirect, operating through the hormones released from their target glands. The remaining adenohypophyseal hormones—growth hormone, prolactin, and MSH—act directly on somatic target tissues without the intervention of other hormones.

Table 9-3 Tropic hormones of the anterior pituitary gland

Hormone	Structure	Target tissue	Primary action in mammals	Regulation*
Adrenocorticotropic hormone (ACTH)	Peptide	Adrenal cortex	Increases synthesis and secretion of steroid hormones by adrenal cortex	Cortical-releasing hormone (CRH) stimulates release; ACTH slows release of CRH
Follicle-stimulating hormone (FSH)	Glycoprotein	Ovarian follicles (female); seminiferous tubules (male)	In female, stimulates maturation of ovarian follicles; in male, increases sperm production	GnRH stimulates release; inhibin and steroid sex hormones inhibit release
Luteinizing hormone (LH)	Glycoprotein	Ovarian interstitial cells (female); testicular interstitial cells (male)	In female, induces final maturation of ovarian follicles, estrogen secretion, ovulation, corpus luteum formation, and progesterone secretion; in male, increases synthesis and secretion of androgens	GnRH stimulates release; inhibin and steroid sex hormones inhibit release
Thyroid-stimulating hormone (TSH)	Glycoprotein	Thyroid gland	Increases synthesis and secretion of thyroid hormones	TRH induces secretion; thyroid hormones and somatostatin slow release

*See Table 9-2 for key to abbreviations.

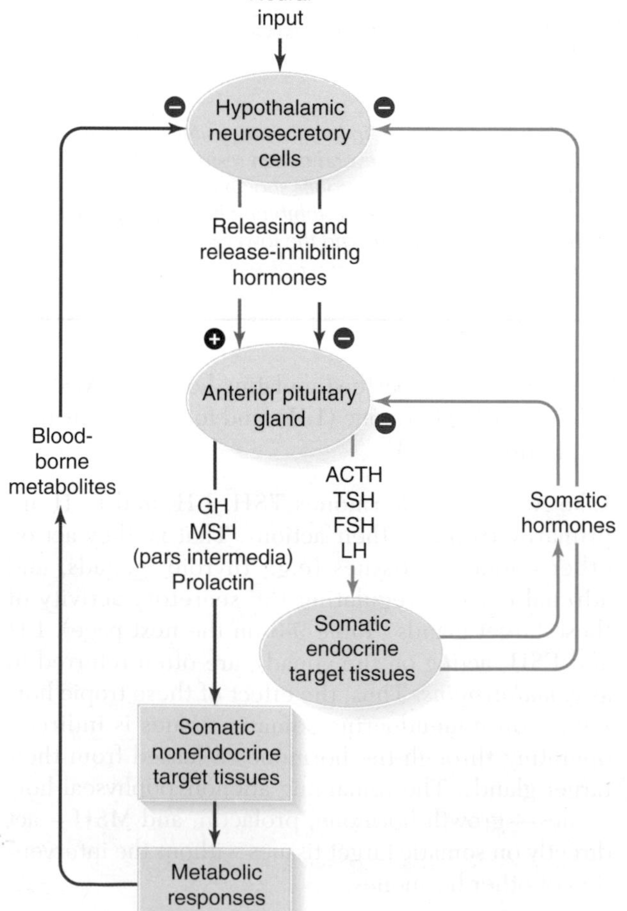

Figure 9-16 Secretion of adenohypophyseal hormones is regulated by hypothalamic releasing and release-inhibiting hormones and is further modulated by feedback loops. Growth hormone (GH), melanocyte-stimulating hormone (MSH), and prolactin act directly on nonendocrine somatic (nonneuronal) tissues. The tropic hormones—adrenocorticotropic hormone (ACTH), thyroid-stimulating hormone (TSH), follicle-stimulating hormone (FSH), and luteinizing hormone (LH)—all stimulate the secretory activity of somatic endocrine tissues. Once released, the corresponding somatic hormones themselves exert negative feedback on the hypothalamic neurosecretory cells and, in some cases, on the corresponding adenohypophyseal cells. The circulating products of some somatic metabolic responses (e.g., blood glucose) also act on the hypothalamic centers, providing additional negative feedback.

The relations between the hypothalamus and the anterior pituitary gland are summarized in Figure 9-16. The three release-inhibiting hormones from the hypothalamus suppress the release from the anterior pituitary gland of MSH, prolactin, and growth hormone. Growth hormone is also under the control of a releasing hormone. Note the short and long feedback loops involving ACTH, TSH, FSH, and LH, which control the hypothalamo-hypophyseal system, and the long feedback loop involving growth hormone, prolactin, and MSH, which controls the hypothalamus.

Neurohormones Released from the Posterior Pituitary Gland

The posterior lobe of the pituitary gland (also called the neurohypophysis or pars nervosa) stores and releases the two neurohormones antidiuretic hormone and oxytocin. These neurohypophyseal hormones are synthesized and packaged in the somata of two groups of neurosecretory cells that constitute the supraoptic and paraventricular nuclei in the anterior portion of the hypothalamus (see Figure 9-15). After their synthesis, the hormones are transported down the axons of the hypothalamo-hypophyseal tract to terminals in the posterior pituitary, where they are released into a capillary bed. This system was the first neurosecretory system discovered in vertebrates.

Both antidiuretic hormone (ADH, also known as vasopressin) and oxytocin are peptides containing nine amino acid residues. Both are mildly effective in fostering contractions of the smooth-muscle tissue in arterioles (Figure 9-17). In mammals, however, oxytocin is best known for stimulating uterine contractions during parturition and for stimulating the release of milk from the mammary gland; in birds, it stimulates motility of the oviduct. The foremost function of ADH is to promote water retention by the kidney, as we will see below.

The amino acid sequences of mammalian oxytocin and ADH differ only at positions 3 and 8 in the peptide chain. Likewise, the sequences of the neurohypophyseal hormones from different vertebrate groups exhibit variations only at positions 3, 4, and 8 (Table 9-4). The sequence of amino acid residues in each pituitary nonapeptide is, of course, genetically determined. Substitution of amino acid residues at positions 3, 4, and/or 8 during evolution has resulted in several forms of these peptide hormones. The residues that are highly conserved (never undergo substitution) are presumably necessary for function; those that are not conserved (positions 3, 4, and 8) seem to be functionally neutral and probably serve only to place the essential residues in the positions appropriate for the biological activity of these neuropeptides.

Within their respective neurosecretory cells, the neurohypophyseal neurohormones are covalently linked in a 1:1 ratio to cysteine-rich protein molecules termed

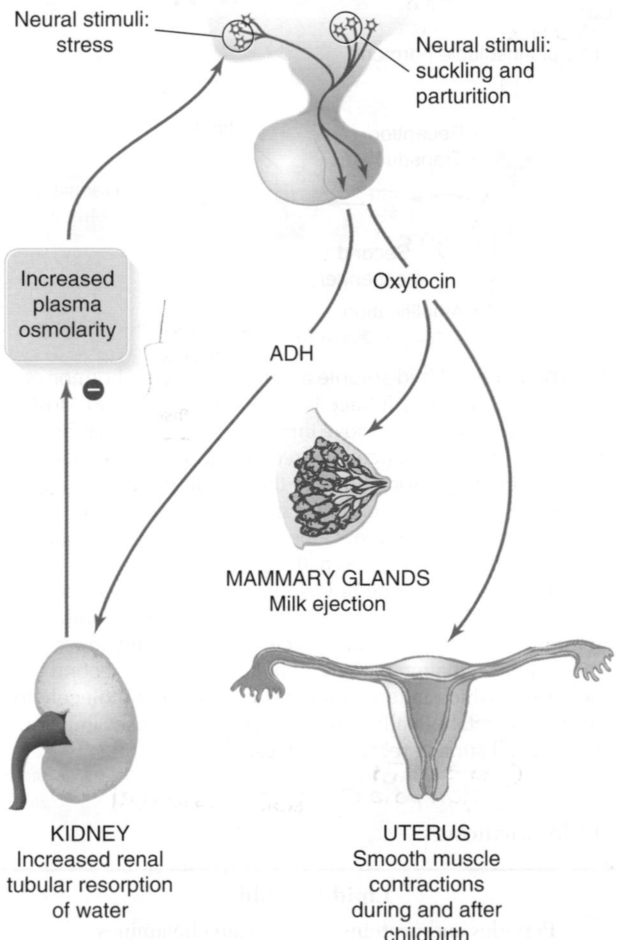

Figure 9-17 The two neurohormones released from the mammalian posterior pituitary gland function primarily in reproduction (oxytocin) and regulation of water balance (ADH). Osmoreceptors in the hypothalamus, pressure receptors in the aorta, and other sources of sensory input influence the neurosecretion of antidiuretic hormone (ADH). High plasma solute concentration and low blood pressure resulting from low plasma volume stimulate ADH output. Oxytocin is released during labor and nursing.

Table 9-4 **Variant forms of neurohypophyseal nonapeptide hormones**

Peptide	Positions of amino acid residues* (1 2 3 4 5 6 7 8 9)	Animal group
Lysine vasopressin	Cys—Tyr—Phe—Gln—Asn—Cys—Pro—Lys—Gly—(NH_2)	Pigs and relatives
Arginine vasopressin	Cys—Tyr—Phe—Gln—Asn—Cys—Pro—Arg—Gly—(NH_2)	Mammals
Oxytocin	Cys—Tyr—Ile—Gln—Asn—Cys—Pro—Leu—Gly—(NH_2)	Mammals
Arginine vasotocin	Cys—Tyr—Ile—Gln—Asn—Cys—Pro—Arg—Gly—(NH_2)	Reptiles, fishes, and birds
Isotocin	Cys—Tyr—Ile—Ser—Asn—Cys—Pro—Ile—Gly—(NH_2)	Some teleosts
Mesotocin	Cys—Tyr—Ile—Gln—Asn—Cys—Pro—Ile—Gly—(NH_2)	Reptiles, amphibians, and lungfishes
Glumitocin	Cys—Tyr—Ile—Ser—Asn—Cys—Pro—Gln—Gly—(NH_2)	Some elasmobranchs

*The cysteine residues in positions 1 and 6 of each peptide are bridged by a disulfide bond.
Source: Frieden and Lipner, 1971.

neurophysins, which exist in two major types, neurophysin I and neurophysin II. Oxytocin is associated with type I and vasopressin with type II. Neurophysins have no hormonal activity, although they are secreted along with the neurohypophyseal hormones. It is conjectured that the hormone-neurophysin molecules are enzymatically cleaved upon release into the blood, yielding the neurohypophyseal hormone and the neurophysin moiety. Thus neurophysins appear to act as storage proteins, serving to retain the hormones in the secretory granules until release.

CELLULAR MECHANISMS OF HORMONE ACTION

As noted already, hormones produce their effects on their target tissues via specialized receptor proteins located either within or on the surface of the target cell (Figure 9-18).

- *Lipid-soluble (hydrophobic) hormones,* such as the steroid and thyroid hormones, bind to cytoplasmic receptors and are transported to the nucleus. Hormone-receptor complexes in the nucleus and the cytoplasm act directly on the DNA of the cell to effect long-term changes lasting hours or days.
- *Lipid-insoluble hormones* cannot penetrate the plasma membrane and therefore bind to cell-surface receptors. This binding often leads to the production of one or more second messengers, which amplify the signal and mediate rapid, short-lived responses via various effector proteins.

The prostaglandins, although lipid-soluble, bind to cell-surface receptors and have a rapid, short-lasting

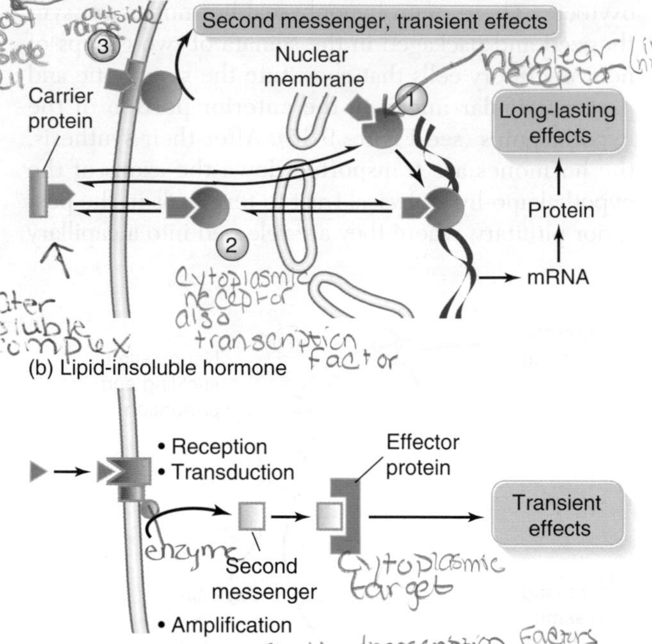

Figure 9-18 Lipid-soluble and lipid-insoluble hormones differ in their primary intracellular mode of action. **(a)** Most lipid-soluble hormones move through the plasma membrane and combine with intracellular receptor proteins, usually in the nucleus (1) but sometimes in the cytoplasm (2), forming active complexes that act on the genetic machinery to modulate gene expression. Generally their effects are long-lasting. An exception to this pattern is the prostaglandins, which, although they are lipid-soluble, bind to cell-surface receptors and have transient effects (3). **(b)** Lipid-insoluble hormones bind to cell-surface receptors, triggering an intracellular signaling pathway that may involve a second messenger, which in turn combines with another molecule to produce a metabolically active complex. Responses initiated through cell-surface receptors are usually transient.

Table 9-5 Comparison of lipid-soluble and lipid-insoluble hormones

	Lipid-soluble		Lipid-insoluble	
Property	Steroids	Thyroid hormones	Peptides and proteins	Catecholamines
Feedback regulation of synthesis	Yes	Yes	Yes	Yes
Binding to carrier proteins	Yes	Yes	Rarely	No
Lifetime in blood plasma	Hours	Days	Minutes	Seconds
Time course of action	Hours to days	Days	Minutes to hours	Seconds or less
Receptor location	Cytosolic or nuclear	Nuclear	Plasma membrane	Plasma membrane
Mechanism of action	Receptor-hormone complex stimulates or inhibits gene expression		Hormone binding triggers second-messenger or activates intrinsic catalytic activity	Hormone binding causes change in membrane potential or triggers second-messenger pathway

Source: Adapted from Smith et al., 1983, p. 358. Used with permission of McGraw-Hill.

effect similar to that of lipid-insoluble hormones. Table 9-5 summarizes some properties characteristic of the major types of lipid-soluble and lipid-insoluble hormones.

Lipid-Soluble Hormones and Their Receptors

The lipid-soluble steroid and thyroid hormones move through the bloodstream complexed with carrier proteins. Without these carriers, only small amounts of these hormones could dissolve in the blood, and they would be taken up completely by the first tissues they encountered. The binding constants of the various carrier proteins differ, ensuring adequate rates of hormone delivery to all target tissues.

Once steroid and thyroid hormones dissociate from their carrier proteins, they can readily enter and leave cells by diffusing across the plasma membrane. Within the cell, these hormones bind specific receptor proteins in the nucleus or the cytoplasm. The resulting hormone-receptor complexes bind to regulatory elements in the DNA, thereby stimulating (and in a few cases inhibiting) transcription of specific genes. Because the lipid-soluble hormones stimulate or inhibit production of particular proteins, their effects persist for hours to days. In contrast, the lipid-insoluble hormones affect the synthesis of shorter-lived metabolites, and their effects usually last only minutes to hours. Some steroid hormones, including aldosterone and estrogen, also act on receptors located on the plasma membrane to initiate rapid responses in addition to their slower responses.

Lipid-Insoluble Hormones and Intracellular Signaling

As noted already, the binding of some hormones to cell-surface receptors triggers the production of second messengers inside the cell, which mediate the cell's response to the extracellular hormone. Many hormones are known to stimulate second-messenger formation, but the most important second messengers fall into just three distinct groups (Figure 9-19):

- *Cyclic nucleotide monophosphates* (cNMPs), such as adenosine 3′,5′-cyclic monophosphate (cAMP) and the closely related guanosine 3′,5′-cyclic monophosphate (cGMP)
- *Inositol phospholipids,* including inositol 1,4,5-trisphosphate (IP_3) and 1,2-diacylglycerol (DAG)
- *Ca^{2+} ions*

We will first examine intracellular signaling systems employing each of these second messengers, as well as membrane-bound enzyme signaling systems that don't involve second messengers. Then we'll see how multiple systems can interact to produce complex tissue responses.

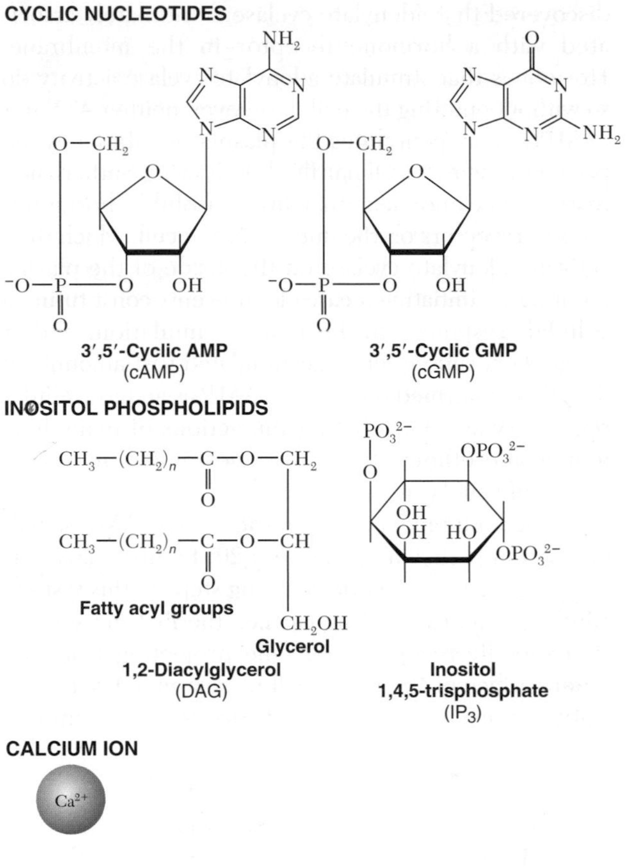

CYCLIC NUCLEOTIDES

3′,5′-Cyclic AMP (cAMP)

3′,5′-Cyclic GMP (cGMP)

INOSITOL PHOSPHOLIPIDS

Fatty acyl groups

Glycerol

1,2-Diacylglycerol (DAG)

Inositol 1,4,5-trisphosphate (IP_3)

CALCIUM ION

Ca^{2+}

Figure 9-19 The three classes of second messengers have very different structures. The cyclic nucleotides are synthesized from ATP and GTP. DAG and IP_3 are produced by hydrolysis of a common precursor.

Cyclic nucleotide signaling systems

The advance of science generally depends on two forms of progress. One is the everyday growth of scientific knowledge by the slow but steady accumulation of data in thousands of laboratories. Such small-scale incremental progress represents by far the greater part of the effort expended by the community of scientists. This type of progress, however, generally builds upon infrequent and often unanticipated breakthroughs that provide revolutionary new insights or points of departure. Such breakthroughs open new paths of inquiry, which are then explored in detail by the small step-by-step mode of progress.

An example of such a giant leap occurred in the mid-1950s when the late Earl W. Sutherland and associates discovered the role of cyclic AMP (cAMP) as an intracellular regulatory agent. In his initial studies on cAMP, Sutherland noted that the activity of adenylate cyclase, which catalyzes the conversion of ATP to cAMP, was enhanced when certain hormones were added to liver homogenate or preparations of intact liver cells. He then separated the cell-free homogenate into fractions and found that the adenylate cyclase activity disappeared if the plasma membrane fragments of the homogenate were removed. It was subsequently discovered that adenylate cyclase is intimately associated with a hormone receptor in the membrane. Hormones that stimulate adenylate cyclase activity do so without entering the cell; moreover, neither ATP nor cAMP readily penetrates the plasma membrane when placed in the extracellular fluid. Following Sutherland's work, it became clear that lipid-insoluble hormones bind to receptors on the outside of the cell, which then activate adenylate cyclase on the inside of the plasma membrane, initiating a cascade of events constituting a cellular response to hormone stimulation. Other researchers subsequently accumulated vast amounts of data that confirmed the role of cAMP as an intracellular regulatory agent mediating the actions of many hormones and other extracellular messengers in a wide variety of cellular responses.

A general model of the events in the cAMP signaling system is shown in Figure 9-20. Figure 9-20a outlines the series of generic coupling steps in this system. Binding of an external signal (i.e., the first messenger) to a specific receptor molecule projecting from the outer surface of the target cell plasma membrane activates a transducer protein that carries signals through the membrane. The transducer protein then activates an amplifier that catalyzes the formation of a second messenger. The second messenger binds to an internal regulator that controls various effectors, leading to the cellular response(s). The pathway employing cAMP as the second messenger is shown in Figure 9-20b. The pathway in the diagram has a stimulatory receptor (R_s) and an inhibitory receptor (R_i), which both communicate with the amplifier adenylate cyclase by way of transducer G proteins: stimulatory G protein (G_s) and inhibitory G protein (G_i). Thus, the message is carried through the membrane by the interactions of three membrane-bound proteins: the receptors, the receptor-linked G proteins, and adenylate cyclase.

Hormone binding to a receptor promotes the binding of guanosine triphosphate (GTP, a close relative of ATP) to the G proteins (hence their name). As Figure 9-21 shows, the G proteins remain activated as long as GTP remains bound; they are inactivated when the GTP is hydrolyzed to guanosine diphosphate (GDP). The activated G proteins activate or inhibit adenylate cyclase, which catalyzes the conversion of ATP to cAMP. As cAMP is produced, it binds to an inhibitory

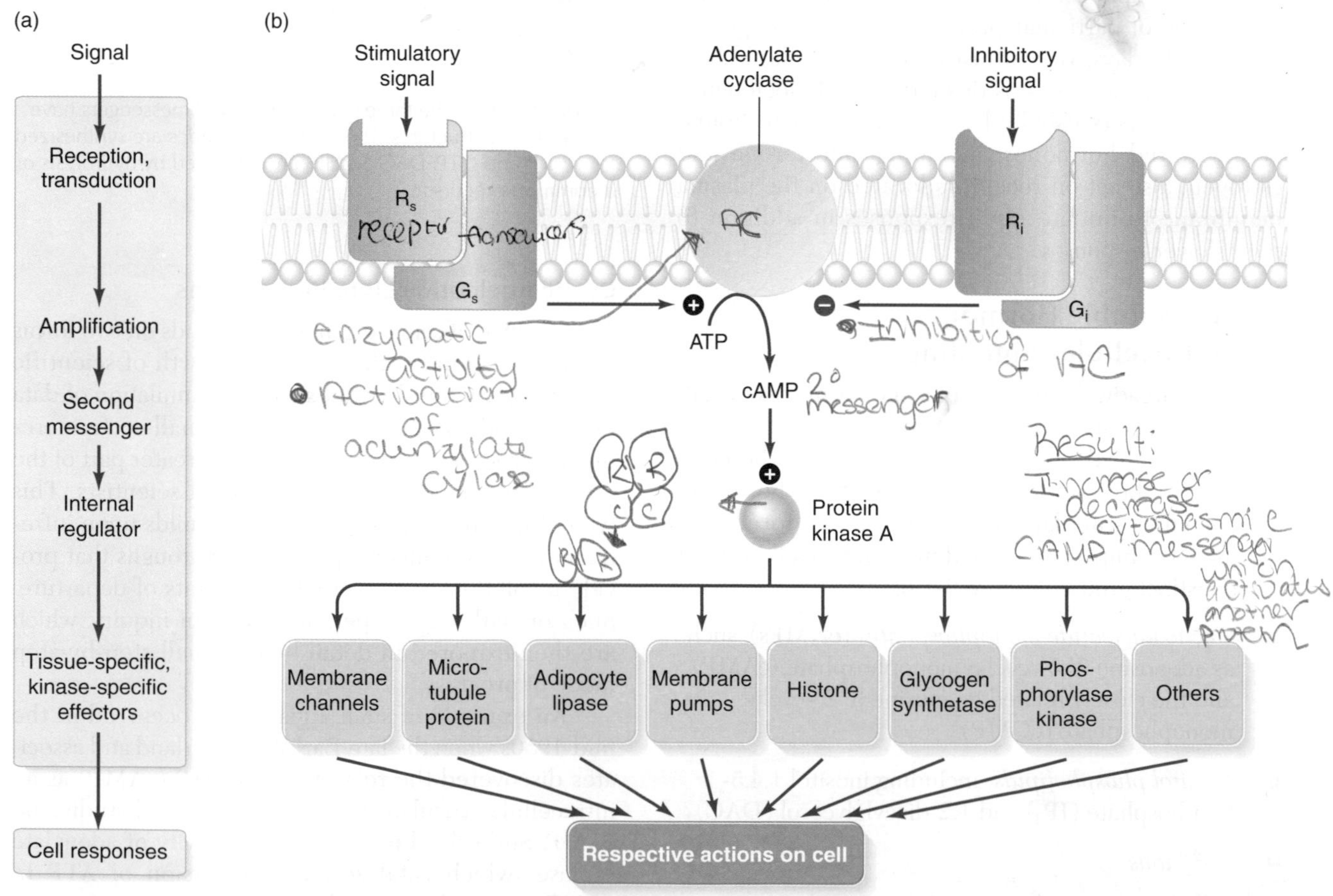

Figure 9-20 The binding of an extracellular signal to a G protein–coupled receptor stimulates or decreases production of the second messenger cAMP, which transduces the signal into cellular responses. **(a)** Generic steps leading from hormone binding by the surface receptor to the cellular response(s). **(b)** Condensed outline of the cAMP second-messenger system. Stimulating and inhibiting receptors are denoted by R_s and R_i, respectively; transducer proteins by G_s and G_i.

regulatory subunit of protein kinase A, causing that subunit to dissociate. This event leaves the catalytic subunit of protein kinase A free to phosphorylate effector proteins, using ATP as the source of high-energy phosphate groups. Phosphorylation of these proteins may either increase or inhibit their activity, thereby inducing the cellular response(s). Some effector proteins are enzymes that catalyze further chemical reactions; others are nonenzymatic proteins such as membrane channels, structural proteins, or regulatory proteins.

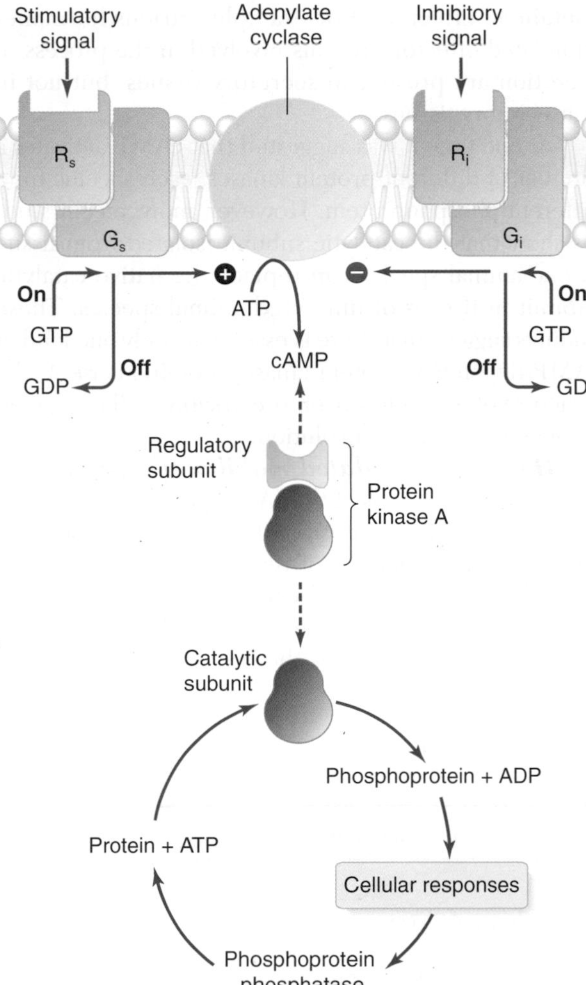

Figure 9-21 Hormone-stimulated regulation of adenylate cyclase within the plasma membrane leads to an increase or decrease in the cytosolic cAMP level. Binding of hormones or other ligands to their stimulatory or inhibitory receptors (R_s and R_i, respectively) induces binding of GTP to the respective transducer proteins, G_s and G_i. The GTP-activated G proteins then activate or inhibit the catalytic activity of adenylate cyclase until the GTP is dephosphorylated enzymatically to guanosine diphosphate (GDP) and the effect on the cyclase ceases. Activated adenylate cyclase catalyzes the conversion of ATP to cAMP, which binds to and removes the regulatory subunit of protein kinase A. The catalytic subunit, once free of the inhibitory regulatory subunit, can phosphorylate various intracellular effector proteins, yielding activated phosphoproteins that mediate cellular responses. In time, cAMP is degraded to AMP by a phosphodiesterase (PDE), and the phosphorylated effector proteins are eventually dephosphorylated to their inactive forms. Both of these mechanisms reduce or terminate the effects of the external signal. [Adapted from Berridge, 1985.]

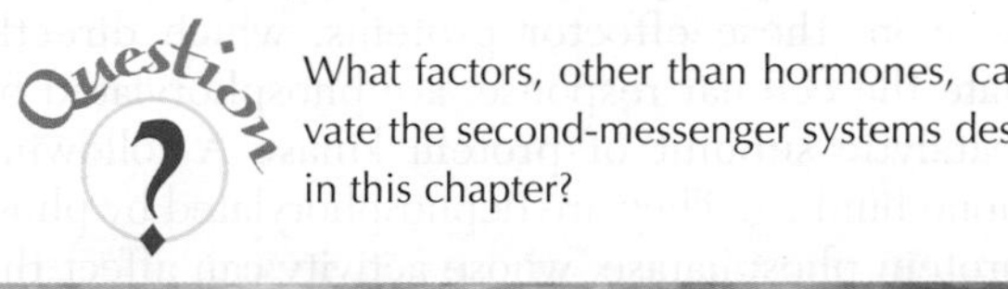

What factors, other than hormones, can activate the second-messenger systems described in this chapter?

Signal amplification in the cAMP pathway An important feature of intracellular signaling pathways is amplification of the signal, in which the binding of hormone to receptor influences the function of many molecules within the cell. Amplification occurs at several stages in the cAMP pathway. First, a single activated receptor protein can activate many G-protein molecules, which then activate many molecules of adenylate cyclase. In some cases, the hormone may remain bound to its receptor for less than 1 second, but the G protein remains active as long as GTP remains bound (e.g., 10–15 seconds), providing sufficient time for considerable amplification to occur. Second, each molecule of activated adenylate cyclase catalyzes the conversion of many ATP molecules into cAMP. Thus, one hormone molecule bound to a receptor for a short period (i.e., 1 second or less) can elicit the generation of a large number of cAMP molecules. Each molecule of cAMP binds to the regulatory subunit of a protein kinase A molecule, liberating a catalytic subunit that in turn catalyzes activation of many effector molecules. Finally, many effectors are themselves enzymes, and thus a fourth amplification occurs as they act on many substrate molecules.

Control of cellular responses Another feature of second-messenger systems is mechanisms by which responses to signals are modulated or terminated so that they have the appropriate magnitude and duration. Three such control mechanisms operate in the cAMP pathway. As we've seen, there are two kinds of receptors, R_s and R_i, which bind stimulatory and inhibitory hormones, respectively. Of course, the two types of transducer proteins, G_s and G_i, are linked to R_s and R_i, respectively. Thus, the activity of adenylate cyclase can be increased by a stimulatory signal (R_s through G_s) or reduced by an inhibitory signal (R_i through G_i). Both stimulation and inhibition of adenylate cyclase can occur in the same cell, with the end result depending on the net intensity of the signal.

A second control mechanism depends on the intracellular level of cAMP, which is determined by the relative rates of cAMP synthesis and degradation:

$$\text{ATP} \longrightarrow \text{cAMP} \longrightarrow \text{AMP}$$

In many tissues, cAMP synthesis is controlled by extracellular signals that modulate the activity of adenylate cyclase. Its degradation is catalyzed by phosphodiesterase (PDE), which is activated by Ca^{2+}.

Finally, the cellular response to extracellular signals transduced by cAMP can be controlled by dephosphorylation of the phosphorylated effector proteins. As we've seen, these effector proteins, which directly mediate the cellular response, are phosphorylated by the catalytic subunit of protein kinase A following hormone binding. They are dephosphorylated by phosphoprotein phosphatase, whose activity can affect the magnitude and duration of the cellular response to hormone stimulation. The activity of phosphoprotein phosphatase is itself indirectly dependent on the cAMP level, decreasing as the cAMP level increases and increasing as cAMP is cleared.

Diversity of cAMP-mediated responses Since Sutherland's discovery that cAMP acts as a second messenger linking hormone action to glucose mobilization in liver cells, cAMP has been shown to function as a second messenger for numerous hormones. To confirm that cAMP is the intracellular second messenger of a hormone, researchers have often used dibutyl cAMP, a lipid-soluble analog that, unlike cAMP, can penetrate the plasma membrane. If the application of dibutyl cAMP to tissues mimics the effects normally induced by a particular hormone, then cAMP is indicated as a second messenger of the hormone. A second approach is to treat tissues with methyl xanthines, which block phosphodiesterase, thereby elevating cAMP levels. The finding that such treatment increases the response to a particular hormone also provides evidence that it operates via the cAMP pathway.

The various hormones linked to cAMP induce multiple, sometimes competing, physiological effects (Table 9-6, see Figure 9-20). How can one second messenger mediate incompatible biochemical and physiological responses? The key lies in the tissue distribution of effector proteins that can be phosphorylated by cAMP-dependent protein kinase A. Not all tissues contain all effectors. For example, various hormone-stimulated effector proteins involved in the process of secretion are present in secretory tissues, but not in nonsecretory tissues.

At one time it was suggested that cAMP activates a number of different protein kinases, each specific for a different phosphoprotein. However, more recent studies show that the catalytic subunit isolated from tissue in one animal species can replace the native catalytic subunit in tissues of unrelated animal species. These findings suggest that there is essentially only one kind of cAMP-dependent protein kinase, protein kinase A, the structure of which has been remarkably well conserved through the course of evolution.

Hormone-stimulated mobilization of glucose Let's take a closer look at the cAMP pathway involved in the hormone-stimulated mobilization of glucose from glycogen. The sequence of reactions in this system—the one originally studied by Sutherland and his associates—has been worked out in great detail. The hormone glucagon stimulates the breakdown of glycogen

Table 9-6 Some hormone-induced responses mediated by the cAMP pathway

Signal	Tissue	Cellular response
Stimulatory		
Epinephrine (β-adrenoreceptors)	Skeletal muscle	Breakdown of glycogen
	Fat cells	Increased breakdown of lipids
	Heart	Increased heart rate and force of contraction
	Intestine	Fluid secretion
	Smooth muscle	Relaxation
Thyroid-stimulating hormone (TSH)	Thyroid gland	Thyroxine secretion
ADH (vasopressin)	Kidney	Reabsorption of water
Glucagon	Liver	Breakdown of glycogen
Serotonin	Salivary gland (blowfly)	Fluid secretion
Prostaglandin I_2	Blood platelets	Inhibition of aggregation and secretion
Inhibitory		
Epinephrine (α_2-adrenoreceptors)	Blood platelets	Stimulation of aggregation and secretion
	Fat cells	Decreased lipid breakdown
Adenosine	Fat cells	Decreased lipid breakdown

Source: Berridge, 1985.

to glucose 6-phosphate (glycogenolysis) in the liver, and epinephrine does the same in skeletal and cardiac muscle; these hormones also inhibit the synthesis of glycogen from glucose (glycogenesis) and stimulate formation of glucose from lactate and amino acids (gluconeogenesis). Thus the net effect of hormone stimulation is a rise in blood glucose.

Figure 9-22 outlines the steps between binding of glucagon (in the liver) and epinephrine (in skeletal and cardiac muscle) and the resulting increase in glucose. Binding of epinephrine to the membrane-bound β-adrenoreceptor activates adenylate cyclase, resulting in an increased rate of cAMP synthesis from ATP (steps 1 and 2). The immediate action of cAMP is activation of protein kinase A (step 3). These three steps appear to be common to all cAMP-regulated systems. Once activated, protein kinase A can catalyze phosphorylation of another enzyme, phosphorylase kinase (step 4), thereby activating it. Phosphorylase kinase-PO_4 in turn catalyzes phosphorylation of glycogen phosphorylase *b* to yield the active form, called glycogen phosphorylase *a* (step 5). It is this enzyme that cleaves glycogen to form glucose 1-phosphate (step 6). In cells, glucose 1-phosphate is readily converted to glucose 6-phosphate, which enters the glycolytic pathway or is dephosphorylated to glucose, which is transported across the plasma membrane into the bloodstream. Binding of glucagon to its receptor produces similar results.

The cAMP-dependent protein kinase A that stimulates the formation of glycogen phosphorylase *a* also acts in an indirect way to inhibit glycogen synthetase, the enzyme that catalyzes the polymerization of glucose into glycogen. Thus, a hormone-stimulated increase in intracellular cAMP stimulates glycogen breakdown and inhibits glycogen synthesis. This synergistic effect is important, for it keeps the rise in glucose from driving by mass action the resynthesis of glycogen from glucose. Conversely, a decrease in cAMP inhibits glycogen breakdown and stimulates glycogen synthesis. This example illustrates that multiple cAMP-mediated effects can occur simultaneously within a single cell.

Cyclic GMP as a second messenger In addition to cAMP, most animal cells also use cyclic GMP (cGMP) as a second messenger (see Figure 9-19). The intracellular concentration of cGMP is about one-tenth that of cAMP. The formation of cGMP from the ATP analog GTP is catalyzed by guanylate cyclase, of which there are two separate species, one bound to the plasma membrane and one free in the cytosol. Guanylate cyclase and adenylate cyclase respond inversely to Ca^{2+}. Studies on isolated guanylate cyclase indicate that the enzyme becomes progressively more active as the Ca^{2+} concentration is increased. Adenylate cyclase, on the other hand, is most active at low concentrations of Ca^{2+}. In principle, then, the relative concentrations of cAMP and cGMP can be influenced by the intracellular concentration of free Ca^{2+}. Like cAMP, cGMP activates a specific protein kinase (protein kinase G), which then phosphorylates effector proteins in the cell.

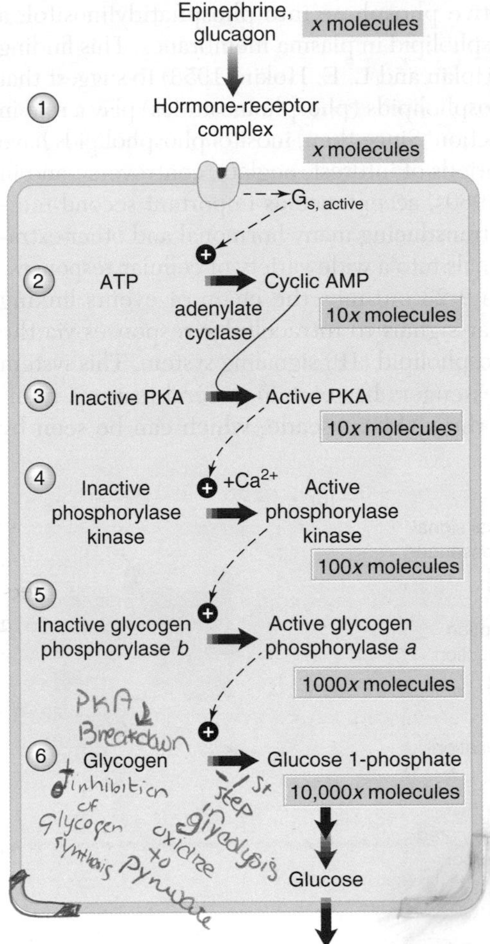

Figure 9-22 Epinephrine and glucagon stimulate breakdown of glycogen to glucose (glycogenolysis) in muscle and liver, respectively. When epinephrine binds to β-adrenoreceptors it triggers a sequence of reactions in which several enzymes are converted from an inactive to an active form. As a result of this enzyme cascade, the original signal is greatly amplified. [From Nelson and Cox, 2000]

Hormonal stimulation of the same type of receptor can simultaneously induce changes in cAMP and cGMP levels. For example, stimulation of the β-adrenoreceptors of the brain, lymphocytes, cardiac muscle, and smooth muscle simultaneously produces a rise in the level of cAMP and a drop in the level of cGMP. Conversely, stimulation of the muscarinic acetylcholine receptors in these tissues results in a drop in the level of cAMP but a rise in the level of cGMP. In some tissues, cAMP and cGMP exert opposing physiological actions. For example, the rate and strength of each heartbeat are increased by an epinephrine-induced rise in cAMP, but decreased by an acetylcholine-induced rise in cGMP.

Inositol phospholipid signaling systems

In the early 1950s it was found that some extracellular signaling molecules could stimulate the incorporation of radioactive phosphate into phosphatidylinositol, a minor phospholipid in plasma membranes. This finding led M. R. Hokin and L. E. Hokin (1953) to suggest that inositol phospholipids (phosphoinositides) play a role in hormone action. Since then, inositol phospholipids have enjoyed periods of interest, neglect, controversy, and in the early 1980s, acceptance as important second messengers in transducing many hormonal and other extracellular signals into a wide variety of cellular responses.

Figure 9-23 outlines the chain of events linking extracellular signals to intracellular responses via the inositol phospholipid (IP) signaling system. This system of lipid messengers has certain general features reminiscent of the cAMP cascade, which can be seen by comparing Figure 9-23a and Figure 9-20a. In both cases, the membrane contains a receptor, a transducer G protein, and an amplifier enzyme, which catalyzes the formation of second-messenger molecules from phosphorylated precursors. These second messengers in turn activate internal regulators, primarily protein kinases, which then activate various tissue-specific, kinase-specific effector molecules.

A closer look at Figure 9-23 reveals the distinguishing features of the IP pathway. Unlike the cAMP system, which has both stimulatory and inhibitory G proteins, the IP pathway has only a stimulatory G protein. Stimulation of this protein, tentatively called G_p, induces activation of phosphoinositide-specific phos-

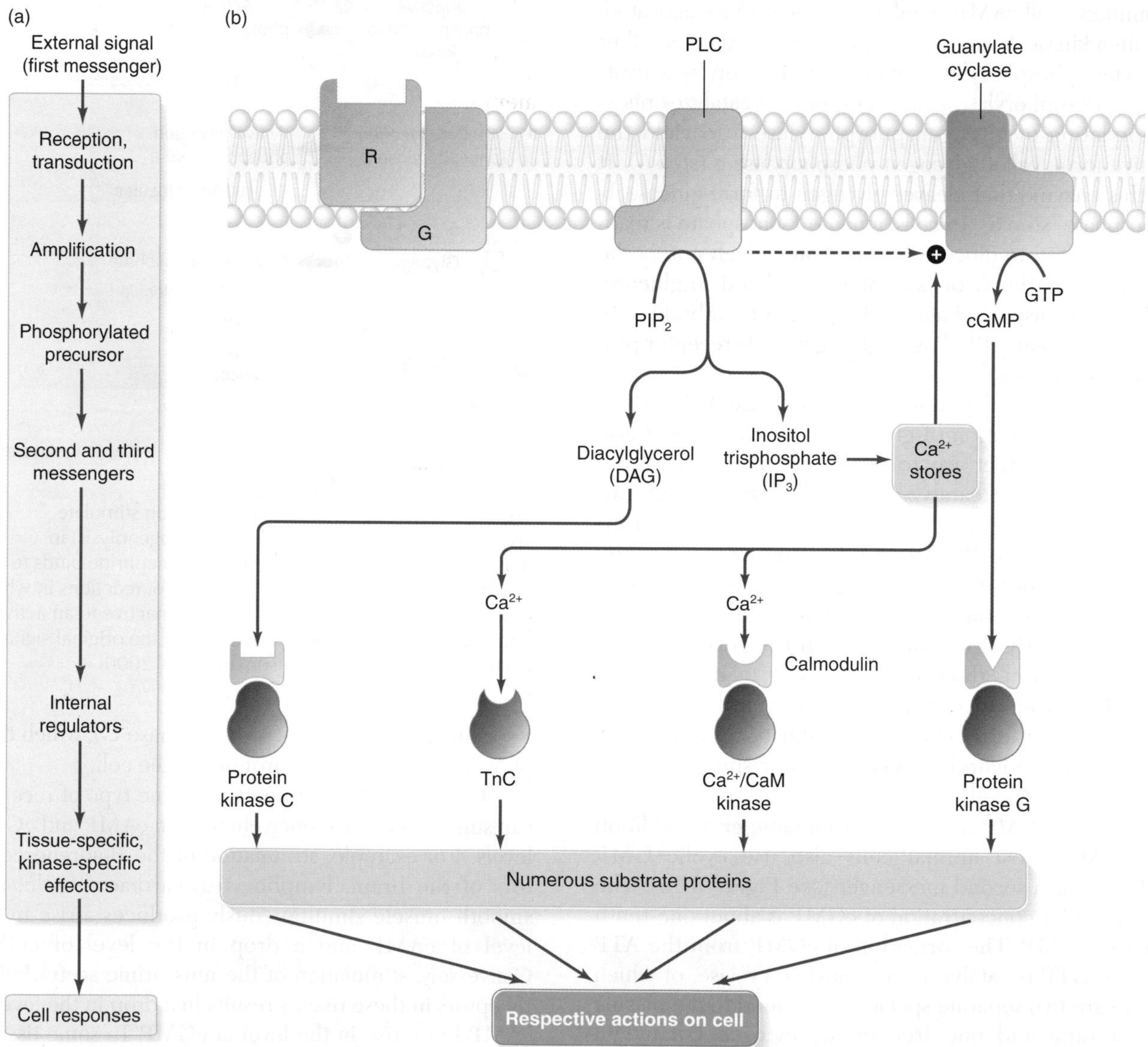

Figure 9-23 Binding of hormones by some G protein–linked receptors induces formation of the phospholipid-derived second messengers diacylglycerol (DAG) and inositol trisphosphate (IP_3). **(a)** Generic steps of the inositol phospholipid pathway, which are similar to those of the cAMP pathway (see Figure 9-20a). **(b)** Condensed outline of the inositol phospholipid second-messenger system. The amplifier enzyme in this pathway is phosphoinositide-specific phospholipase C (PLC). The direct activation of guanylate cyclase by PLC (dashed pathway) is not fully established. Note that Ca^{2+} mobilized from intracellular stores may activate troponin C (TnC), form a complex with calmodulin (CaM) that activates Ca^{2+}/calmodulin-dependent kinase (Ca^{2+}/CaM kinase), promote activation of protein kinase C, or increase cGMP production by stimulating membrane-bound guanylate cyclase.

pholipase C (PLC), the amplifier enzyme in the IP pathway. (G_p is similar, but not identical, to G_s, which activates adenylate cyclase in the cAMP pathway.) PLC hydrolyzes phosphatidyl inositol bisphosphate (PIP_2) into two major second messengers, inositol trisphosphate (IP_3) and diacylglycerol (DAG). A remarkable feature of the IP system is that PIP_2, the precursor of the second messengers, is itself a constituent of the membrane. PIP_2 is located primarily in the cytoplasmic half of the lipid bilayer, where it can come in contact with membrane-bound phospholipase C (Figure 9-24). Once formed, the water-soluble IP_3 diffuses away from the membrane into the cytosol; the other second messenger, DAG, remains in the cytoplasmic half of the plasma membrane. These two second messengers subsequently follow their own pathways, but the two branches of the IP system sometimes collaborate in producing a cellular response. IP_3 and DAG are rapidly metabolized, and their degradation products are used to replenish PIP_2.

IP_3 acts on intracellular calcium stores such as those in the endoplasmic reticulum (ER). Some IP_3 is phosphorylated to form inositol 1,3,4,5-tetraphosphate (IP_4), which enhances the entry of Ca^{2+} from the cell exterior through Ca^{2+} channels in the plasma membrane. The Ca^{2+} released from the ER by IP_3 acts as another messenger, and thus can be considered a third messenger in this system. For example, Ca^{2+} binds to and activates troponin C (TnC) and calmodulin, as well as a number of other regulator and effector molecules (see Figure 9-23). As we'll see in Chapter 10, Ca^{2+}/TnC stimulates muscle contraction directly. Ca^{2+}/calmodulin may act as an effector protein, or it may bind to and activate a number of enzymes and other effector proteins, of which the most studied is Ca^{2+}/calmodulin kinase.

The actions of DAG, the other second messenger in the IP system, occur in the plasma membrane, in which DAG molecules can move laterally by diffusion. DAG has two signaling roles. First, it can be cleaved to release arachidonic acid, a precursor in synthesis of the prostaglandins and other biologically active eicosanoids. Second, and even more important, DAG activates membrane-bound protein kinase C by a mechanism analogous to the activation of protein kinase A by cAMP. Although protein kinase C is located both in the cytosol and on the cytosolic face of the plasma membrane, it can be activated only when associated with the membrane. The activation of protein kinase C by DAG depends on

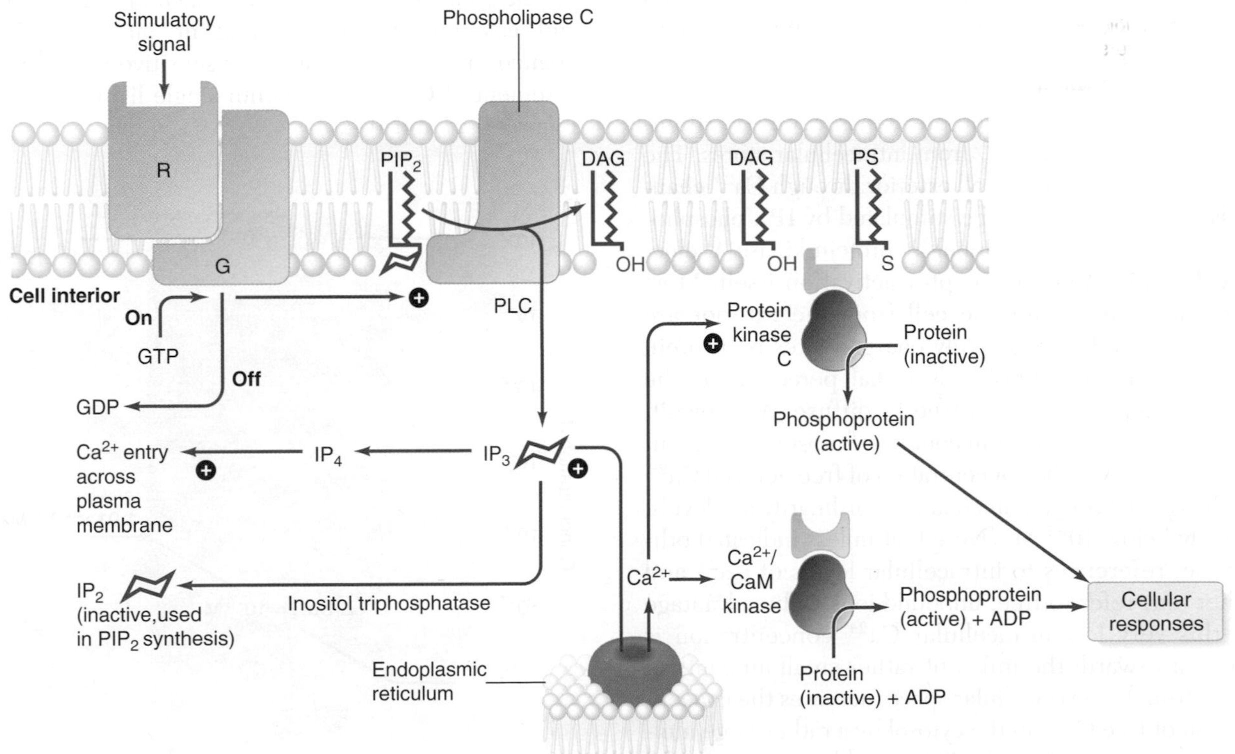

Figure 9-24 In the inositol phospholipid pathway, one second messenger acts in the plasma membrane and the other in the cytosol. Hormone binding induces activation of the G protein, which activates phosphoinositide-specific phospholipase C (PLC). PLC then catalyzes the hydrolysis of membrane phosphatidyl inositol bisphosphate (PIP_2) into diacylglycerol (DAG), which remains in the membrane, and inositol trisphosphate (IP_3), which diffuses into the cytosol. DAG promotes activation of the membrane-bound protein kinase C; this activation also requires Ca^{2+} and phosphatidylserine (PS), another membrane phospholipid. IP_3 promotes the liberation of Ca^{2+} from intracellular storage depots such as the endoplasmic reticulum. The free Ca^{2+} has numerous regulatory functions, including stimulation of Ca^{2+}/calmodulin kinase (Ca^{2+}/CaM kinase). [Adapted from Berridge, 1985.]

the presence of Ca^{2+} and phosphatidylserine (PS), another phospholipid constituent of the membrane. Binding of DAG and PS to protein kinase C in the membrane increases the affinity of the enzyme for Ca^{2+}; as a result, protein kinase C can be activated at the usual low concentrations of Ca^{2+} present in the cytosol. Thus, the activation of protein kinase C requires two intracellular messengers, DAG and Ca^{2+}, both of which can be induced by the same extracellular signal.

Ca^{2+} signaling systems

In recent decades it has become clear that Ca^{2+} plays an important and ubiquitous role as both an intracellular regulatory agent and a messenger linking external signals to cellular responses. Two important characteristics of cells permit Ca^{2+} to function effectively in cellular regulation and signaling: (1) the ability of cells to increase and decrease the intracellular Ca^{2+} concentration over a wide range and (2) the presence within cells of numerous proteins whose activity is modulated by the binding of Ca^{2+}. We first discuss these two aspects of calcium's role in the cell, and then look at how Ca^{2+} functions as a second messenger.

Modulation of intracellular Ca^{2+} concentration The concentration of free Ca^{2+} in the cytosol can be increased in two ways: (1) by release of Ca^{2+} from intracellular calcium stores, such as those in the endoplasmic reticulum, and (2) by influx of Ca^{2+} from the cell exterior through Ca^{2+} channels in the plasma membrane. As described in the previous section, IP_3 stimulates the release of Ca^{2+} from intracellular stores. The entry of Ca^{2+} from the cell exterior through Ca^{2+} channels has been shown to be stimulated by IP_4, phosphorylation of the Ca^{2+} channel by protein kinase A, electrical stimulation, or receptor activation itself. Most Ca^{2+} ions that enter the cell from the exterior are rapidly bound to negatively charged sites on protein molecules in the cytosol; only a small percentage of the ions remain ionized and free to diffuse. As a result, although the total calcium content of most cells is about 1 mM (10^{-3} M), the concentration of free, ionized Ca^{2+} in the cytosol is maintained at extraordinarily low levels, usually below 10^{-7} M. (Note that unless indicated otherwise, references to intracellular levels of Ca^{2+} and other ions refer to free, unbound ions.) The advantage of this very low intracellular Ca^{2+} concentration is straightforward: the influx of rather small amounts of Ca^{2+} from the extracellular space increases the concentration of free Ca^{2+} in the cytosol by a rather large multiple. This concept can be illustrated by comparing the relative changes in the intracellular concentration of Ca^{2+} and Na^+ that result from the entry of equal quantities of these two ion species in response to a transient increase in the permeability of the plasma membrane to both ions (Figure 9-25). Likewise, the release of small amounts of Ca^{2+} from intracellular calcium stores causes a relatively large increase, as much as tenfold, in the concentration of free Ca^{2+} in the cytosol.

Since the extracellular Ca^{2+} concentration is typically about 10^{-3} M, 10^4 times higher than the intracellular concentration, the electrochemical gradient strongly favors entry of Ca^{2+} into cells. The cell has two primary mechanisms for removing excess Ca^{2+} from the cytosol to keep the free Ca^{2+} level low: primary and secondary active transport of Ca^{2+} across the plasma membrane to the exterior (see Chapter 4) and movement of Ca^{2+} ions into the endoplasmic reticulum via a calcium pump in the ER membrane. Two additional mechanisms help keep the intracellular Ca^{2+} level from transiently becoming too high. First, various cytosolic proteins bind Ca^{2+} when the Ca^{2+} level increases and release it when the level decreases. In effect, these proteins "buffer" the Ca^{2+} concentration, limiting perturbations in free Ca^{2+} levels, just as pH buffers limit perturbations in free H^+ levels. Second, when the cytosolic Ca^{2+} level becomes abnormally high, the mitochondria may import Ca^{2+} in exchange for H^+.

Certain technical advances have been essential in studying the physiological effects of changes in intracellular Ca^{2+} concentrations. One was the discovery in 1963 of the jellyfish protein aequorin, which emits light when it complexes with Ca^{2+}. Injection of aequorin into cells provides a means of detecting minute changes in the intracellular free Ca^{2+} level. Calcium-sensitive dyes and calcium-sensitive fluorescent molecules have opened up new possibilities for sensitive optical measurement of Ca^{2+} levels within single living cells (see Figures 6-27 and 10-23).

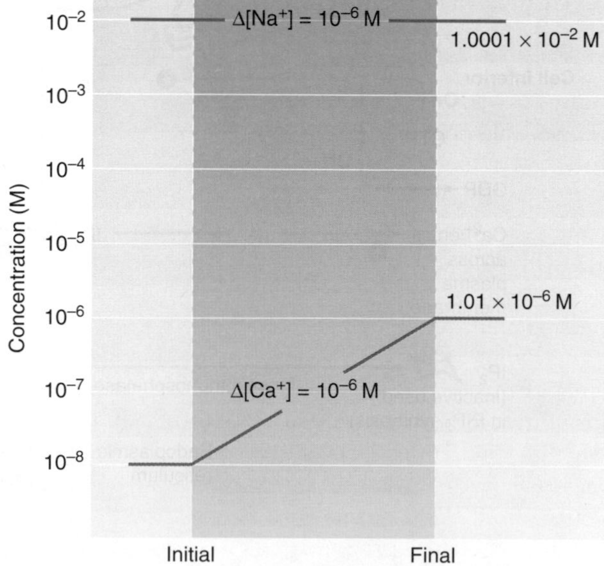

Figure 9-25 The intracellular concentration of free Ca^{2+} is elevated manyfold by the influx of even small amounts of Ca^{2+}. In this comparison, the low initial intracellular concentration of Ca^{2+} of 10^{-8} M is raised a hundredfold by a transient influx, $\Delta[Ca^{2+}]$, equivalent to a 10^{-6} M increment. The initial intracellular concentration of Na^+ is already 10^{-2} M, so a 10^{-6} M increment, $\Delta[Na^+]$, produces virtually no change in the intracellular Na^+ concentration.

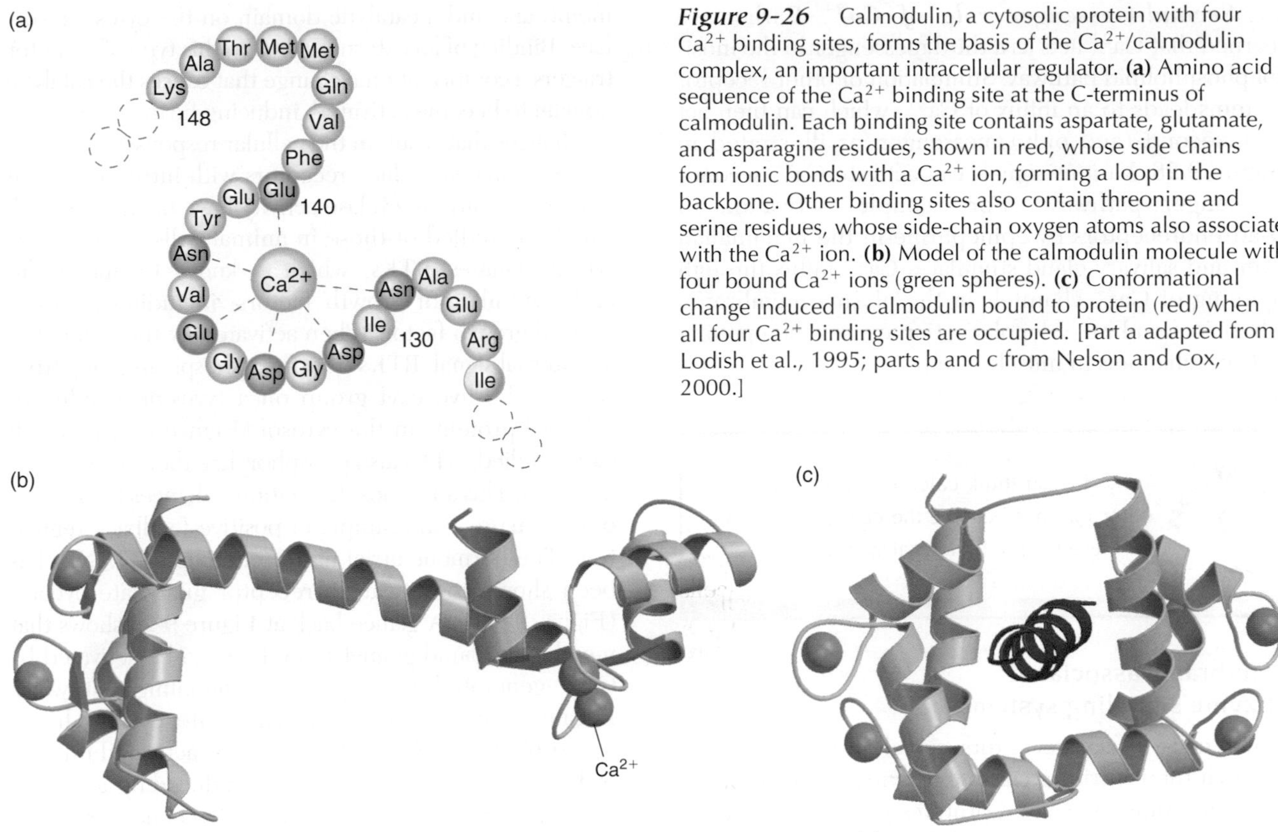

Figure 9-26 Calmodulin, a cytosolic protein with four Ca^{2+} binding sites, forms the basis of the Ca^{2+}/calmodulin complex, an important intracellular regulator. **(a)** Amino acid sequence of the Ca^{2+} binding site at the C-terminus of calmodulin. Each binding site contains aspartate, glutamate, and asparagine residues, shown in red, whose side chains form ionic bonds with a Ca^{2+} ion, forming a loop in the backbone. Other binding sites also contain threonine and serine residues, whose side-chain oxygen atoms also associate with the Ca^{2+} ion. **(b)** Model of the calmodulin molecule with four bound Ca^{2+} ions (green spheres). **(c)** Conformational change induced in calmodulin bound to protein (red) when all four Ca^{2+} binding sites are occupied. [Part a adapted from Lodish et al., 1995; parts b and c from Nelson and Cox, 2000.]

Ca^{2+}-binding proteins The other important feature typical of Ca^{2+}-mediated intracellular regulation and signaling is the presence of multiple Ca^{2+} binding sites in certain enzymes and regulatory proteins. These specialized binding sites have a very high affinity for Ca^{2+}, allowing tight binding of the cation at very low concentrations of free Ca^{2+}. The Ca^{2+} binding sites in all these proteins consist of acidic amino acid residues, which are negatively charged and rich in oxygen atoms. The oxygen atoms, carrying full or partial negative charges, are located in a loop of the peptide chain, so that six to eight oxygen atoms form a cavity of just the right size to harbor the positively charged calcium ion (Figure 9-26a). In fact, about 70% of the entire amino acid sequences of the various Ca^{2+}-binding regulatory proteins are homologous.

Binding of Ca^{2+} to these proteins generally leads to a conformational change in the molecule that alters its properties. For example, binding of Ca^{2+} to troponin C, which is found only in striated muscle, causes a conformational change that initiates a series of steps leading to contraction. We'll discuss troponin C, the first Ca^{2+}-binding regulatory protein to be discovered, in detail in Chapter 10.

Calmodulin, a Ca^{2+}-binding protein closely related to troponin C, is present in relatively large amounts in every eukaryotic tissue examined thus far. It functions as a multipurpose intracellular regulatory protein, mediating most Ca^{2+}-regulated processes. The single polypeptide chain of calmodulin, consisting of 148 amino acid residues, contains four Ca^{2+} binding sites (Figure 9-26b).

Binding of Ca^{2+} to all four sites produces a Ca^{2+}/calmodulin complex that can bind to and activate numerous enzymes and effector proteins (Figure 9-26c). For example, Ca^{2+}/calmodulin binds to the regulatory subunit of Ca^{2+}/calmodulin kinase. Once freed of its regulatory subunit, the catalytic subunit of Ca^{2+}/calmodulin kinase can phosphorylate serine and threonine residues on various effector proteins, which induce cellular responses (see Figure 9-23). Other enzymes and cellular processes regulated by Ca^{2+}/calmodulin are shown in Figure 9-27.

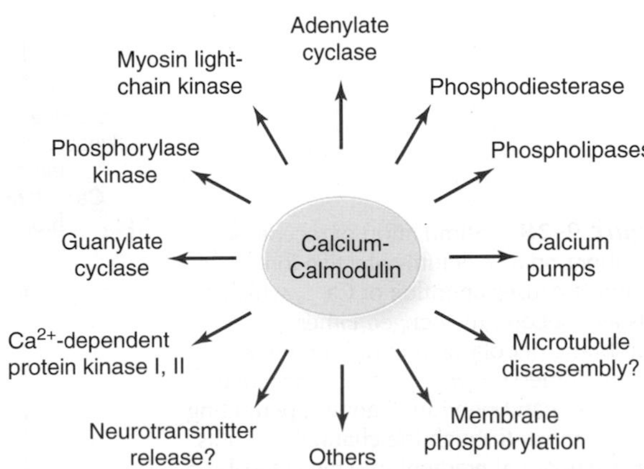

Figure 9-27 Ca^{2+}/calmodulin regulates many processes and enzymes within cells. Among these are adenylate cyclase and guanylate cyclase, which catalyze formation of the cyclic nucleotide second messengers. [Adapted from Cheung, 1979.]

Second-messenger role of Ca^{2+} Earlier we learned that Ca^{2+} acts as a third messenger in the inositol phospholipid pathway. Stimulation of other receptor systems leads to an influx of Ca^{2+}, which can then act as a second (and only) messenger, as illustrated in Figure 9-28. Various signals can activate Ca^{2+} second-messenger pathways. For example, activation of α-adrenoreceptors by epinephrine in the mammalian liver and salivary gland stimulates Ca^{2+} influx through opening of Ca^{2+} channels in the plasma membrane, whereas membrane depolarization causes the opening of Ca^{2+} channels in muscle.

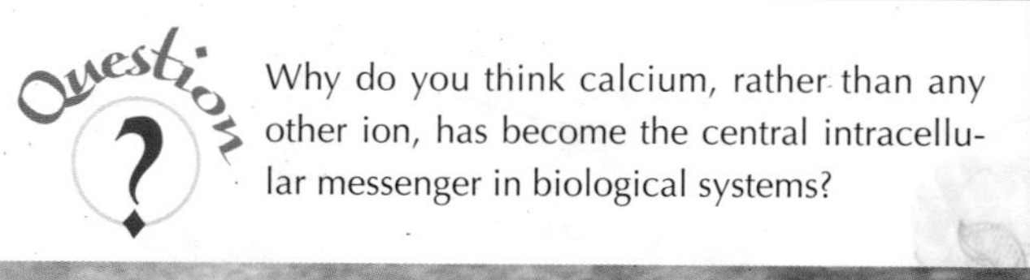

Why do you think calcium, rather than any other ion, has become the central intracellular messenger in biological systems?

Membrane-associated enzyme signaling systems

Some cell-surface receptors signal the cell directly through their intrinsic enzyme activity. Such receptors are membrane-associated proteins that have a ligand-binding domain on the extracellular surface of the plasma membrane and a catalytic domain on the cytosolic surface. Binding of an external signal to this type of receptor triggers a conformational change that causes the catalytic domain to become activated, inducing further intracellular changes that result in the cellular responses.

Several cell-surface receptors with intrinsic protein kinase or guanylate cyclase activity have been identified. The best studied of these in animal cells are receptor tyrosine kinases (RTKs), which are known to bind insulin and a number of growth factors, including platelet-derived growth factor. When activated by the binding of an external signal, RTKs transfer a phosphate group from ATP to the hydroxyl group on a tyrosine residue of selected proteins in the cytosol (Figure 9-29a). In all cases studied, RTKs also phosphorylate themselves when activated. This autophosphorylation enhances the activity of the kinase—an example of positive-feedback regulation. The hormone atrial natriuretic peptide (ANP) has been shown to activate a receptor guanylate cyclase (Figure 9-29b). A glance back at Figure 9-23 shows that membrane-bound guanylate cyclase can be activated by Ca^{2+} generated through other signaling pathways. Because the ANP receptor guanylate cyclase has a ligand-binding domain, it can also be activated directly by hormone binding. The cGMP produced by activation of the receptor can then function as a second messenger to mediate cellular responses, as in other pathways.

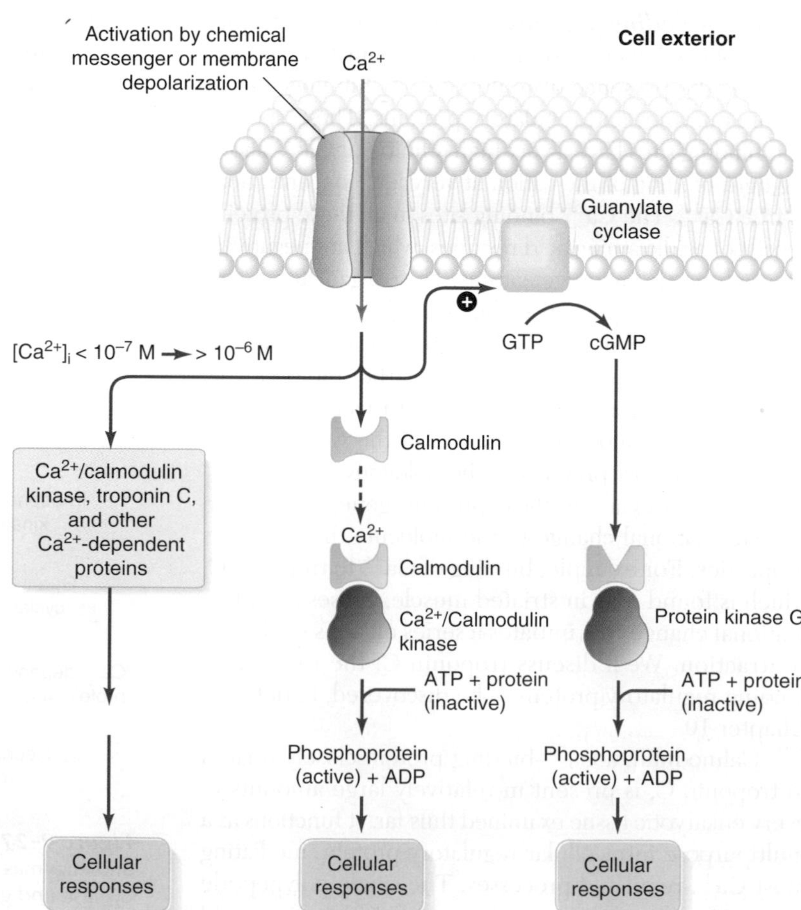

Figure 9-28 Stimulation of receptors that function as calcium-selective ion channels causes an influx of Ca^{2+}, which acts as a second messenger. Either membrane depolarization or binding of a chemical messenger (e.g., an extracellular hormone) can open ion channels, permitting Ca^{2+} to move through the channel down its electrochemical gradient into the cytosol. The resulting local increase in cytosolic free Ca^{2+} from a resting level of 10^{-7} M to 10^{-6} M can activate several different intracellular signaling pathways, leading to various cellular responses.

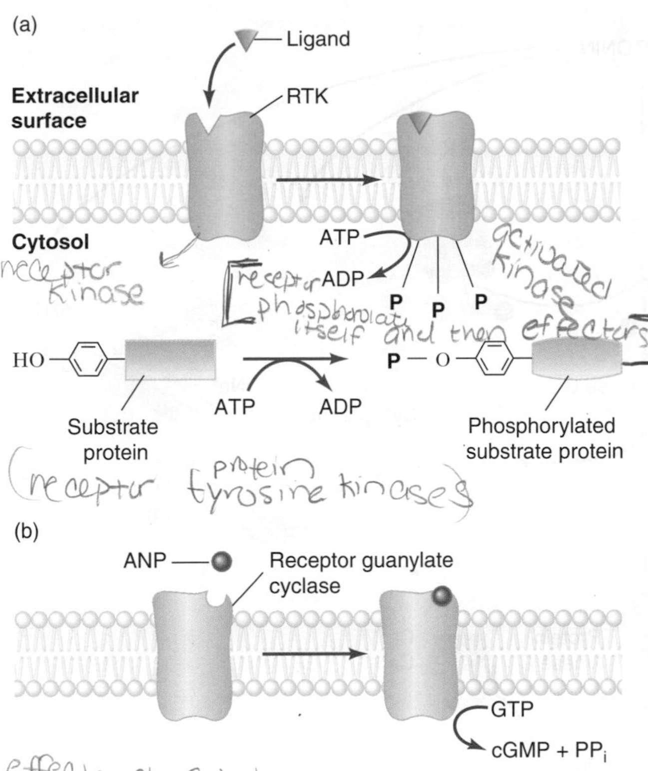

Figure 9-29 Some hormone receptors have intrinsic catalytic activity, which is stimulated by hormone binding. **(a)** Binding of a ligand (e.g., insulin) to a receptor tyrosine kinase (RTK) activates catalytic activity in the cytosolic domain of the receptor. In some cases, the activated receptor may directly phosphorylate certain substrate proteins; in other cases, it binds a transducer protein that initiates a rather complicated signaling pathway. **(b)** The receptor for atrial natriuretic peptide (ANP) has guanylate cyclase activity. Hormone binding leads to production of the second messenger cGMP. [Adapted from Lodish et al., 1995.]

Second-messenger networks

It is important to note that a single hormone can trigger several second-messenger systems by activating different types of receptors, even in the same cell. Binding of epinephrine to α- and β-adrenoreceptors in the mammalian salivary gland is an example of a divergent pathway in which the two second messengers—namely, intracellular free Ca^{2+} and cAMP, respectively—mediate different cellular responses (Figure 9-30a). In the mammalian liver, however, binding of epinephrine to α- and β-adrenoreceptors leads through two second messengers to the same cellular response, an example of a convergent pathway (Figure 9-30b). In this case, both second messengers, Ca^{2+} and cAMP, activate phosphorylase kinase, which in turn stimulates glycogenolysis, as discussed earlier.

A more complicated example of second-messenger networks involves serotonin (5-hydroxytryptamine, 5-HT), a lipid-insoluble amine that functions as both a neurotransmitter and an endocrine hormone regulating gastric secretion and smooth muscle contraction in blood vessels. As shown in Figure 9-31 on the next page, serotonin binds to several receptor subtypes, which are linked to various second-messenger pathways or ion channels; some of these converge, others diverge. Like other lipid-insoluble hormones, serotonin binds to cell-surface receptors; however, its mode of action, which ultimately affects gene transcription, differs from that of most lipid-insoluble hormones (see Figure 9-18b).

As we saw earlier, a particular signaling system may be activated by one class of receptors and inhibited by a

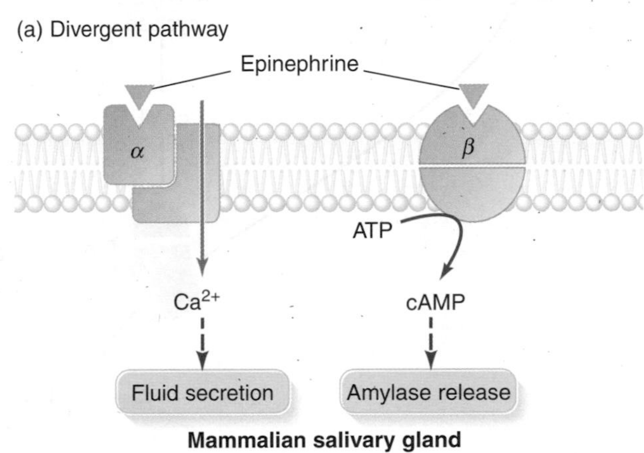

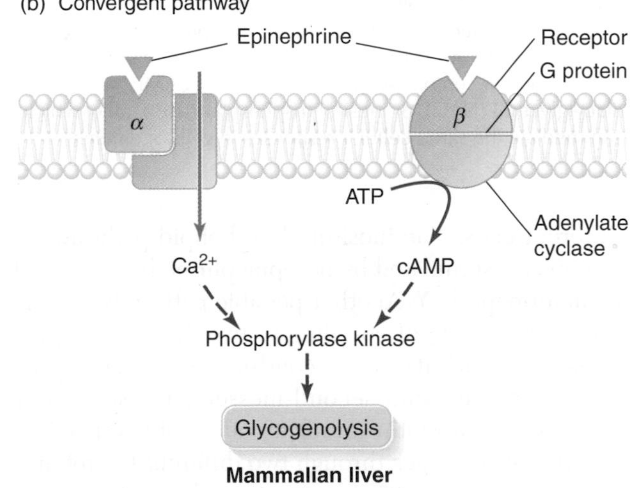

Figure 9-30 A single hormone may bind to different receptors, initiating convergent and/or divergent signaling pathways. Binding of epinephrine to α- and β-adrenoreceptors leads to increases in intracellular Ca^{2+} and cAMP, respectively. **(a)** In the mammalian salivary gland, these two second messengers mediate divergent pathways, leading to different, independent end effects—fluid secretion and amylase secretion by secretory cells in the gland. **(b)** In the mammalian liver, these two second messengers both induce activation of phosphorylase kinase, which catalyzes the breakdown of glycogen to glucose (glycogenolysis; see Figure 9-22). Thus, binding of the same hormone to different receptors triggers convergent pathways leading to the same end response. There is growing evidence that epinephrine is not unique in having multiple receptors in the same animal, or even in the same cell.

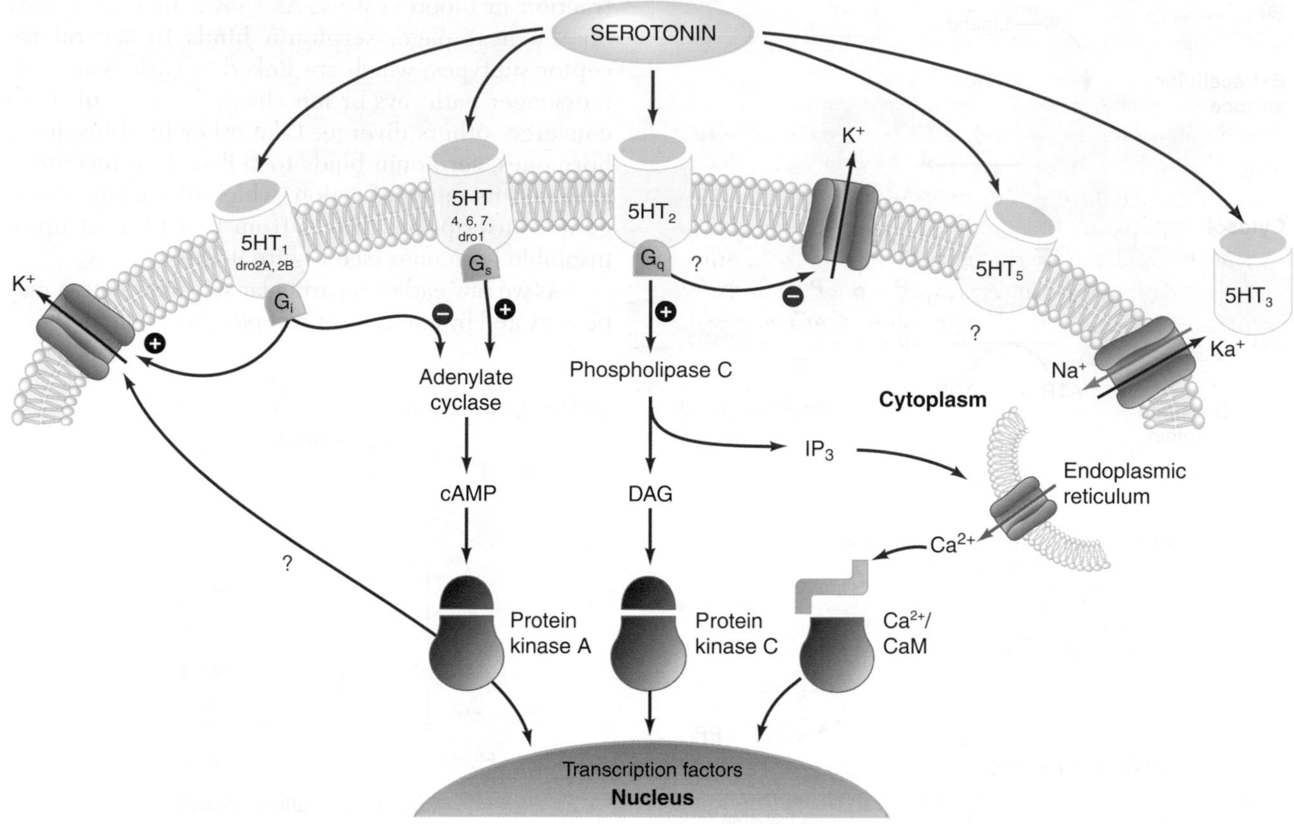

Figure 9-31 Serotonin binds to multiple receptors, which are linked to convergent and divergent second-messenger pathways. Binding of serotonin, also known as 5-hydroxytryptamine (5-HT), to some receptors leads to production of cAMP, diacylglycerol (DAG), or inositol trisphosphate (IP_3), all of which can mediate the same cellular responses in cells of different tissues, or even in the same cells. The various receptors illustrated represent subclasses of the serotonin receptor family (dro = *Drosophila*). G_i = inhibitory G proteins; G_s = stimulatory G proteins; G_q = pertussis toxin–insensitive G proteins; Ca^{2+}/CaM kinase = Ca^{2+}/calmodulin-dependent protein kinase. [Adapted from Saudou and Hen, 1994.]

different class. The inositol phospholipid pathway, for example, is stimulated by norepinephrine but inhibited by neuropeptide Y. Another possible pattern is a single receptor capable of coupling to two different G proteins, each with its own second-messenger system or both with the same second-messenger system. For example, somatostatin stimulates adenylate cyclase in a number of cell types through two different G proteins, one sensitive to pertussis toxin, the other not. Another example of a single receptor linked to several second-messenger pathways is the octopamine/tyramine receptor in the fruit fly *Drosophila.*

NH_2 NH_2 HO OH OH

Tyramine **Octopamine**

Activation of this receptor inhibits adenylate cyclase via one G protein and activates phospholipase C via a different G protein, leading to an elevation in intracellular Ca^{2+}. Interestingly, tyramine has a more potent effect on the adenylate cyclase pathway than octopamine, whereas octopamine has the greater effect on the phospholipase C pathway. Thus two agonists, differing in structure by a single hydroxyl group, can differentially couple this receptor to two second-messenger pathways.

Clearly, although intracellular signaling systems are often described as separate pathways, the boundaries between them fade in the cell. Because extensive interactions occur between many elements of the various signaling pathways, we cannot completely understand their physiological roles if we view them in isolation.

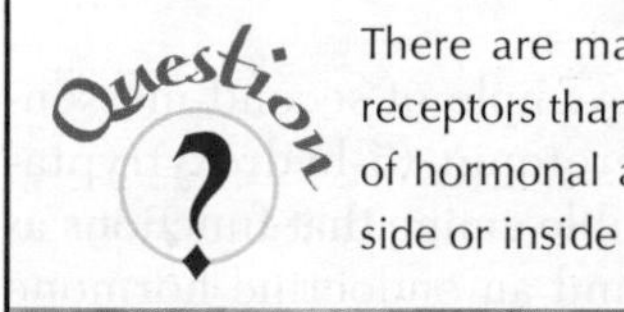

There are many more types of cell-surface receptors than G proteins. Why? Do pathways of hormonal action intersect more often outside or inside cells?

PHYSIOLOGICAL EFFECTS OF HORMONES

As noted earlier, most hormones produce tissue-specific physiological effects. That is, a given hormone generally induces responses only in selected tissues, and may induce different responses in different tissues. This specificity in hormonal action depends partly on the distribution of the components of hormone-triggered signaling pathways (especially receptors) and partly on the preferential expression of effector proteins in different tissues. In the following sections, we'll examine the physiological effects of the various categories of hormones.

Metabolic and Developmental Hormones

Several different hormones regulate metabolism and various developmental processes. These hormones have diverse structures (e.g., steroids, catecholamines, peptides). Table 9-7 on the next page summarizes the properties of the major metabolic and developmental hormones.

Catecholamines and glucocorticoids

The mammalian adrenal glands are paired, with one attached to the rostral end of each kidney (Figure 9-32). Each adrenal gland is in fact two glands: an outer layer, the adrenal cortex, surrounds an inner portion, the adrenal medulla (Figure 9-33). The two portions of the mammalian adrenals are of different developmental origin: the cells of the cortex are derived from mesodermal tissue, whereas those of the medulla are derived from ectodermal tissue. The adrenal cortex produces steroid hormones involved in blood ion and glucose regulation, as we will see below, as well as in anti-inflammatory reactions. The cells of the adrenal medulla, on the other hand, produce the catecholamines epinephrine and norepinephrine. Cells of the adrenal medulla are referred to as **chromaffin cells** because they have an affinity for chromium salts, which can be used as a stain. The chromaffin cells that produce norepinephrine have dark-staining irregular granules, whereas those that produce epinephrine have light-staining, spherical granules. Chromaffin cells are modified postganglionic sympathetic neurons, and under some conditions, they will grow into typical postganglionic sympathetic neurons. In a healthy gland they are prevented from doing so by the presence of high concentrations of glucocorticoid hormones released from the surrounding adrenal cortex into the blood flowing from the cortex to the medulla (see Figure 9-32).

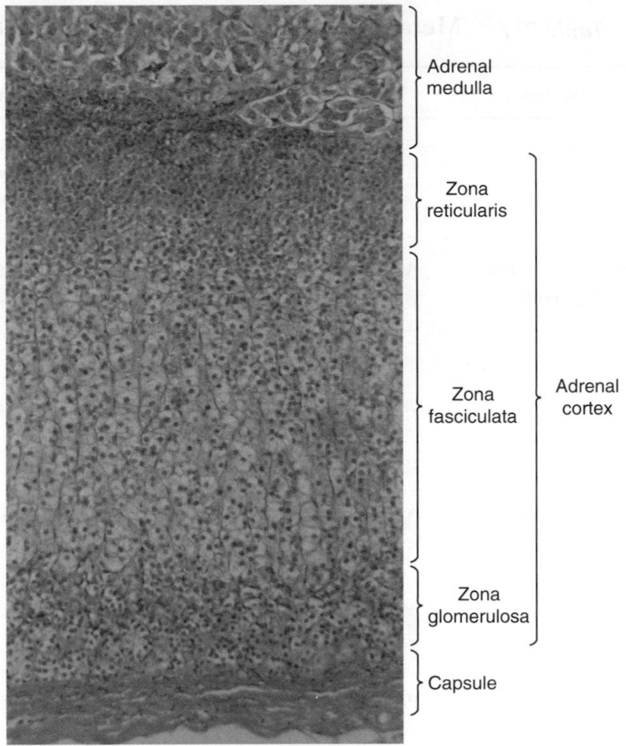

Figure 9-33 Mammalian adrenal glands have a recognizable cortex and medulla, which produce different hormones. This light micrograph reveals the outer capsule, the three concentric layers of the cortex, and the underlying medulla. The zona glomerulosa, the outermost cortical layer, secretes mineralocorticoids; the zona fasciculata and the zona reticularis secrete glucocorticoids. The adrenal medulla secretes two catecholamines, epinephrine and norepinephrine. [Courtesy of Frederic H. Martini.]

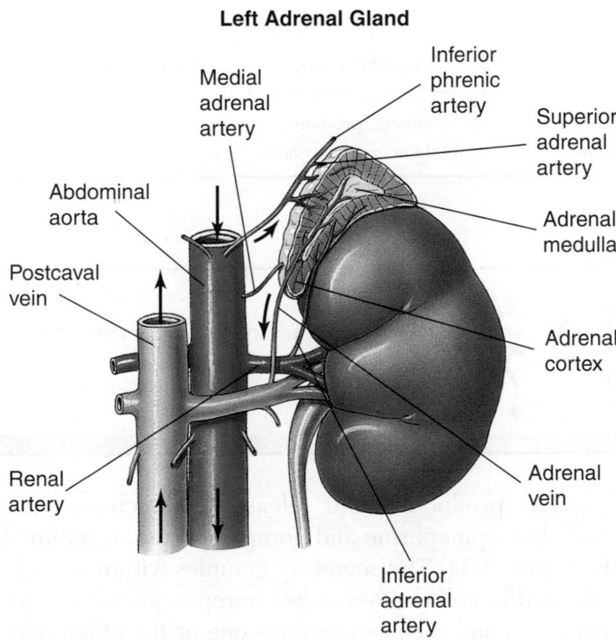

Figure 9-32 The adrenal glands in mammals are attached to the rostral ends of the kidneys. Arteries enter the adrenal cortex through the outer capsule and branch into smaller vessels, which pass into the centrally located medulla. Thus hormones produced in the cortex and released into the blood are carried into the medulla, which is drained by the adrenal vein.

Table 9-7 Metabolic and developmental hormones

Hormone	Tissue of origin	Structure	Target tissue	Primary action	Regulation
Glucagon	Pancreas (alpha cells)	Peptide	Liver, adipose tissue	Stimulates glycogenolysis and release of glucose from liver; promotes lipolysis	Low serum glucose increases secretion; somatostatin inhibits release
Glucocorticoids (e.g., cortisol)	Adrenal cortex	Steroid	Liver, adipose tissue	Stimulate mobilization of amino acids from muscle and gluconeogenesis in liver to raise blood glucose; increase transfer of fatty acids from adipose tissue to liver; exhibit anti-inflammatory action	Physiological stress increases secretion; biological clock via CRH and ACTH controls diurnal changes in secretion
Growth hormone (GH)	Anterior pituitary	Peptide	All tissues	Stimulates RNA synthesis, protein synthesis, and tissue growth; increases transport of glucose and amino acids into cells; increase lipolysis and antibody formation	Reduced plasma glucose and increased plasma amino acid levels stimulate release via GRH; somatostatin inhibits release
Insulin	Pancreas (beta cells)	Peptide	All tissues except most neuronal tissue	Increases glucose and amino acid uptake by cells	High plasma glucose and amino acid levels and presence of glucagon increase secretion; somatostatin inhibits secretion
Norepinephrine and epinephrine	Adrenal medulla (chromaffin cells)	Catecholamine	Most tissues	Increase cardiac activity; induce vasoconstriction; increase glycolysis, hyperglycemia, and lipolysis	Sympathetic stimulation via splanchnic nerves increases secretion
Thyroxine	Thyroid	Tyrosine derivative	Most cells, but especially those of muscle, heart, liver, and kidney	Increases metabolic rate, thermogenesis, growth, and development; promotes amphibian metamorphosis	TSH induces release

Synthesis and release of catecholamines Catecholamines are both neurotransmitters and hormones. When the catecholamine norepinephrine is released from adrenergic neurons, it acts as a neurotransmitter, whereas epinephrine released into the blood by the adrenal medulla acts as a hormone. The catecholamines affect the contraction of smooth muscle and stimulate glycolysis and lipolysis. They also increase heart rate and force of contraction (see Chapter 12). Their numerous cardiovascular and metabolic effects constitute the stress response, or *fight-or-flight reaction,* in which various tissues are activated and the body is mobilized to either attack or flee from a perceived threat. For example, plasma epinephrine levels can be elevated in a cat when it hears a dog bark. Catecholamines are also released under a wide variety of other physiological conditions; for example, during heavy exercise or even when moving from a sitting to a standing position.

What are the advantages and disadvantages of evoking a stress response?

The production and release of catecholamines, including epinephrine and norepinephrine, is outlined in Figure 9-34. The secretory granules within a single chromaffin cell contain either norepinephrine or epinephrine, and each cell secretes one or the other catecholamine. The granules also contain enkephalin, ATP, and several acidic proteins called chromogranins. The catecholamines within the granule are probably bound to the chromogranins, which are polymers maintained in the condensed state by the shielding action of the catecholamines within the granule. Once a pore is opened in the vesicle, the catecholamines begin to diffuse out and

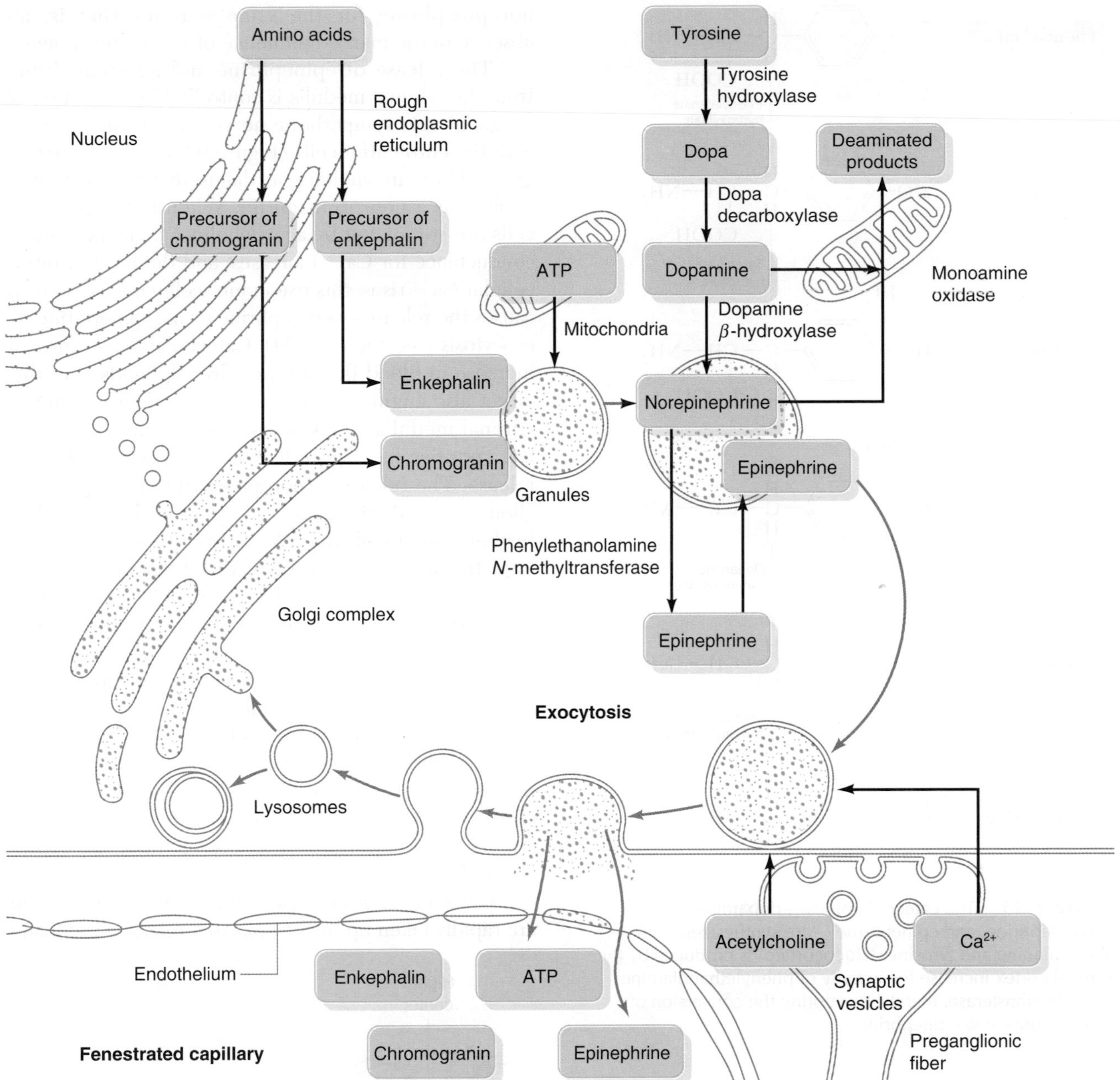

Figure 9-34 Secretory vesicles in the chromaffin cells of the adrenal medulla contain catecholamines, enkephalin, ATP, and chromogranin, all of which are synthesized in different cellular compartments. In epinephrine-producing cells (shown here), norepinephrine leaves the secretory vesicles, is converted to epinephrine, and then is reincorporated into the vesicles. Stimulation of chromaffin cells by acetylcholine, which is liberated from the terminals of preganglionic sympathetic nerve fibers, triggers release of the granule contents by exocytosis. The neuronal stimulus increases the membrane permeability for Ca^{2+}, leading to the increased intracellular Ca^{2+} required for exocytosis. [Adapted from Matsumoto and Ischii, 1992.]

the chromogranin polymer rapidly expands, propelling the contents of the vesicle into the extracellular space.

Norepinephrine is synthesized from tyrosine, with dopa and dopamine as intermediate compounds (Figure 9-35 on the next page). Conversion of tyrosine to dopamine occurs in the cytosol, catalyzed by tyrosine hydroxylase and dopa decarboxylase, which are cytosolic enzymes. Dopamine is then incorporated into the granules and converted to norepinephrine in a reaction catalyzed by dopamine β-hydroxylase contained in the secretory granules. Norepinephrine is methylated to form epinephrine, a reaction catalyzed by phenylethanolamine *N*-methyltransferase, which is found in the cytosol. Thus, norepinephrine must leave the secretory granules to be converted to epinephrine, which then reenters the granules (see Figure 9-34).

Although chromaffin cells and steroidogenic tissues are found together in the adrenal glands in mammals, this is not the case in all vertebrates. In fishes, for instance, chromaffin tissue is separate from

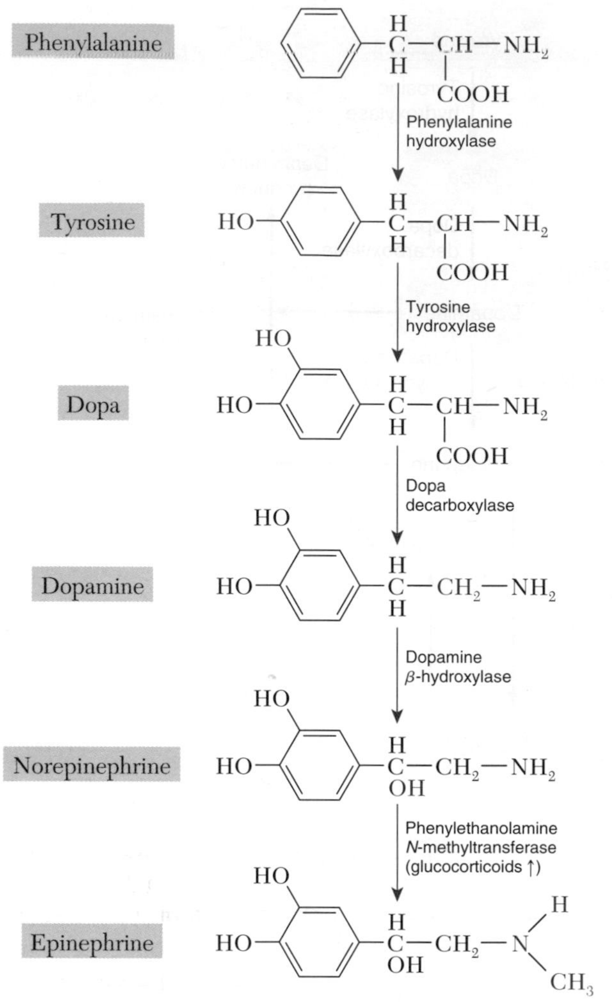

Phenylalanine

↓ Phenylalanine hydroxylase

Tyrosine

↓ Tyrosine hydroxylase

Dopa

↓ Dopa decarboxylase

Dopamine

↓ Dopamine β-hydroxylase

Norepinephrine

↓ Phenylethanolamine N-methyltransferase (glucocorticoids ↑)

Epinephrine

Figure 9-35 The catecholamines—dopamine, norepinephrine, and epinephrine—are synthesized from phenylalanine and tyrosine. Glucocorticoids produced by the adrenal cortex increase the activity of phenylethanolamine *N*-methyltransferase, thereby promoting the conversion of norepinephrine to epinephrine.

steroidogenic cells, but both are still in the general region of the kidney; the chromaffin tissue is associated with blood vessels, whereas the steroid-producing cells are embedded in the kidney. The close association of the adrenal cortex and medulla in mammals is of functional significance. As noted already, blood entering the medulla has already passed through the cortex and thus carries high levels of glucocorticoid hormones (see Figure 9-32). In the medulla, these glucocorticoids promote the synthesis of phenyl-ethanolamine *N*-methyltransferase, the enzyme that catalyzes conversion of norepinephrine to epinephrine. When chromaffin tissue is isolated from the influence of steroidogenic tissue, as in the dogfish, it produces more norepinephrine than epinephrine. The human fetus contains some isolated chromaffin tissue, which synthesizes norepinephrine rather than epinephrine, presumably because of the absence of steroidogenic tissue. Postganglionic sympathetic neurons also produce norepinephrine for the same reason—that is, an absence of the marked influence of steroid hormones.

The release of epinephrine and norepinephrine from the adrenal medulla is controlled by the action of preganglionic sympathetic nerves that form synapses with the chromaffin cells (Figure 9-36). These preganglionic fibers are cholinergic; that is, they release acetylcholine as a neurotransmitter. When the chromaffin cells are stimulated by acetylcholine, their membrane conductance for Ca^{2+} increases, and the level of intracellular Ca^{2+} rises; this rise in intracellular Ca^{2+} in turn causes the release of epinephrine or norepinephrine by exocytosis (see Figure 9-34). Catecholamines cause an increase in blood flow to the adrenal glands, and this effect also augments catecholamine release from the adrenal medulla. Thus the release of catecholamines has a positive-feedback effect on further catecholamine release. The release of norepinephrine from postganglionic sympathetic nerves, on the other hand, inhibits further norepinephrine release from those nerve endings. In this case, negative feedback operates. ATP is stored in the granules of chromaffin cells and released along with catecholamines. ATP and its breakdown product, adenosine, inhibit further release of catecholamines by reducing calcium influx, thereby providing negative-feedback control on catecholamine release from the adrenal medulla. Hypoxia also stimulates catecholamine release from chromaffin cells. In tissues of some species, such as the hagfish heart, the chromaffin cells are not innervated, and hypoxia is among the important alternate stimuli for catecholamine release.

Catecholamines released into the extracellular fluid are rapidly taken up by the cells that released them and

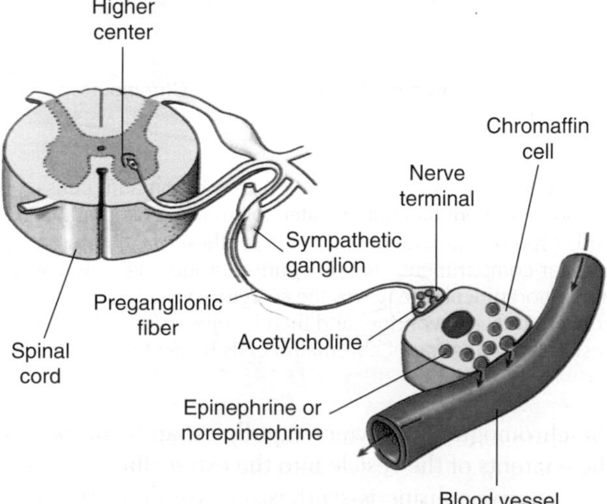

Figure 9-36 Catecholamine secretion by the adrenal medulla is regulated by neurons. Sympathetic nerve axons originating in the spinal cord pass through the sympathetic ganglia without forming synapses, but then synapse on the catecholamine-producing cells. Acetylcholine liberated from these preganglionic nerve terminals stimulates the secretion of medullary hormones.

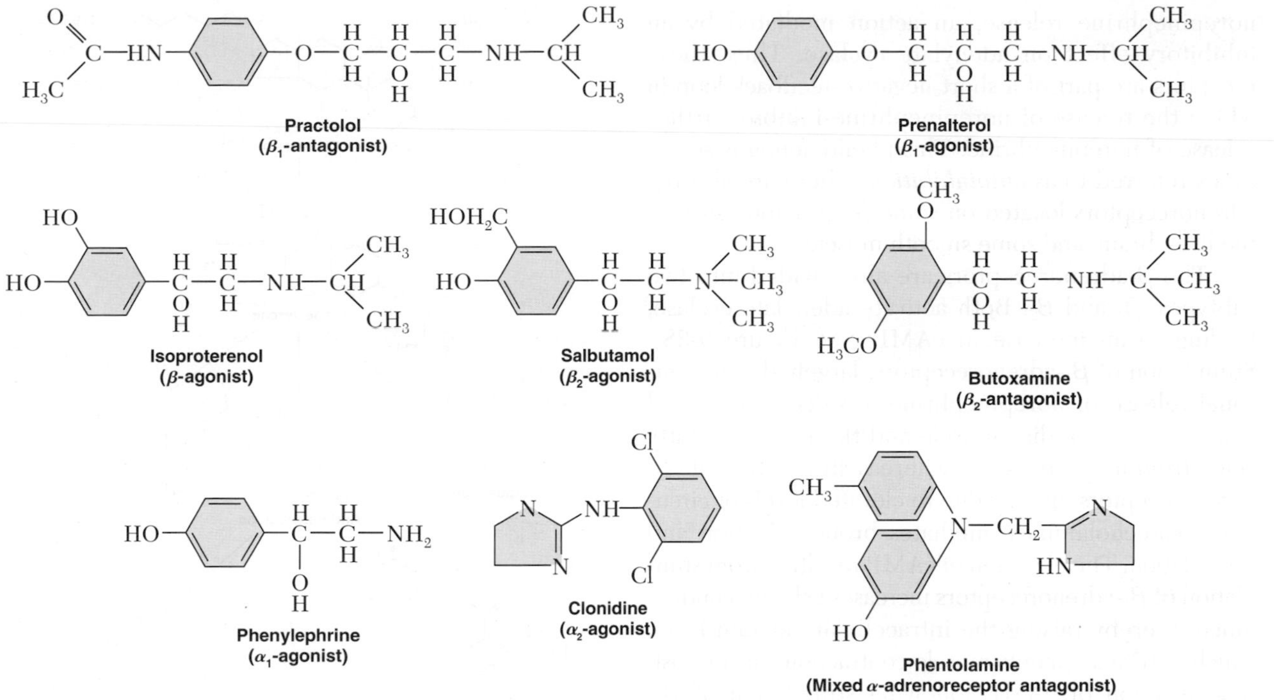

Practolol (β_1-antagonist)

Prenalterol (β_1-agonist)

Isoproterenol (β-agonist)

Salbutamol (β_2-agonist)

Butoxamine (β_2-antagonist)

Phenylephrine (α_1-agonist)

Clonidine (α_2-agonist)

Phentolamine (Mixed α-adrenoreceptor antagonist)

Figure 9-37 A variety of drugs can activate (agonists) or block (antagonists) adrenoreceptors. These drugs have been used to identify adrenoreceptor subtypes and to determine the effects of catecholamines on different tissues.

either stored in secretory vesicles or destroyed by monoamine oxidase located on the outer membrane of mitochondria (see Figure 9-34). Catecholamines in the extracellular space are catabolized by catecholamine *O*-methyltransferase, especially in the liver and kidney, and the breakdown products are excreted. The actual level of catecholamines in the blood thus depends on the balance between their release, uptake, and catabolism. Although the concentration of catecholamines in the blood is determined primarily by the rate of release from the adrenal medulla, release from postganglionic sympathetic nerves also contributes significantly. Adrenergic neurons release norepinephrine, whereas the medulla releases mainly epinephrine, so the relative activity of the nerves and the medulla also influences the relative levels of epinephrine and norepinephrine in the blood. Catecholamine levels in the blood remain elevated for only a few minutes in humans, but they can remain high for several hours following exhaustive exercise in fishes.

Effects and regulation of catecholamines Epinephrine and norepinephrine bind with adrenergic receptors, also termed *adrenoreceptors,* in the plasma membranes of target cells. This binding activates one of a number of intracellular second messengers, leading to a particular tissue response. In a paper published in 1948, R. P. Ahlquist concluded that there are two types of adrenoreceptors—α and β—that differ in their sensitivity to catecholamines. More recent studies have demonstrated that there are several subtypes of both α- and β-adrenoreceptors distinguished by the ability of various drugs to either activate or block their activity (Figure 9-37).

The α_1-adrenoreceptors mediate smooth muscle contraction in many tissues. Stimulation of these receptors results in activation of the inositol trisphosphate (IP_3) pathway, leading to elevation of intracellular IP_3 (Figure 9-38). Elevated IP_3 causes release of calcium from stores within the cell; the resulting rise in cytosolic calcium causes muscle contraction. There appear to be subtypes of α_1-adrenoreceptors in different tissues. The α_2-adrenoreceptors, which are located in presynaptic cells at noradrenergic synapses, cause inhibition of

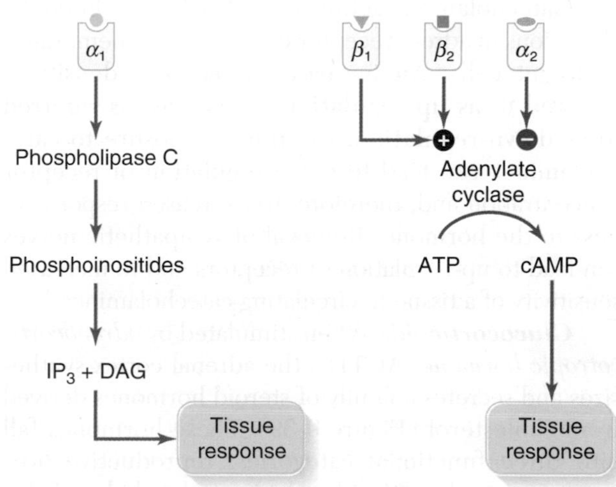

Figure 9-38 Binding of catecholamines to α_1-, α_2-, β_1-, or β_2-adrenoreceptors either activates or inhibits a second-messenger pathway. Adrenoreceptor signal transduction occurs via the adenylate cyclase or the inositol phospholipid pathway. [Adapted from Hadley, 1992.]

norepinephrine release, an action mediated by an inhibitory effect on adenylate cyclase. Thus, these receptors are part of a short negative-feedback loop in which the release of norepinephrine inhibits further release of norepinephrine. This phenomenon is sometimes referred to as *autoinhibition.* There are also α_2-adrenoreceptors located on some postsynaptic sites in the liver, brain, and some smooth muscle.

The β-adrenoreceptors are also divided into two subtypes, β_1 and β_2. Both activate adenylate cyclase, leading to an increase in cAMP (see Figure 9-38). Stimulation of β_1-adrenoreceptors, largely due to neuronal release of norepinephrine, results in increased contraction of cardiac muscle and the release of fatty acids from adipose tissue, whereas stimulation of β_2-adrenoreceptors, usually due to elevated levels of circulating catecholamines, mediates bronchodilation and vasodilation. The elevation of cAMP resulting from stimulation of β_1-adrenoreceptors increases calcium conductance, thereby raising the intracellular calcium level, which in turn augments muscle contraction. In contrast, the elevation of cAMP following stimulation of β_2-adrenoreceptors causes activation of the calcium pump, and calcium is both sequestered within and extruded from the cell, so that intracellular calcium levels fall, promoting muscle relaxation.

The physiological action of catecholamines is quite variable and is influenced by other factors. For example, neuropeptide Y, sometimes co-released with norepinephrine from adrenergic neurons, modulates the action of catecholamines on the IP_3 second-messenger pathway, augmenting the action of catecholamines in some tissues, reducing it in others. Many other factors can modulate both the release and action of catecholamines. Adenosine, for example, has been shown to inhibit catecholamine release from the bovine adrenal medulla by reducing calcium flux.

Catecholamine action can also be modified by alterations in adrenoreceptor density in the membranes of target cells. An increase in receptor density is referred to as **up-regulation;** a decrease is referred to as **down-regulation.** Continual exposure to catecholamines can lead to down-regulation of receptor concentration and, therefore, to decreased responsiveness to the hormone. Removal of sympathetic nerves can lead to up-regulation of receptors and to increased sensitivity of a tissue to circulating catecholamines.

Glucocorticoids When stimulated by *adrenocorticotropic hormone* (ACTH), the adrenal cortex synthesizes and secretes a family of steroid hormones derived from cholesterol (Figure 9-39). These hormones fall into three functional categories: reproductive hormones; mineralocorticoids, which regulate kidney function; and glucocorticoids, which have widespread actions, including the mobilization of amino acids and glucose and anti-inflammatory actions. Here we discuss the glucocorticoids; reproductive hormones and mineralocorticoids will be described in later sections.

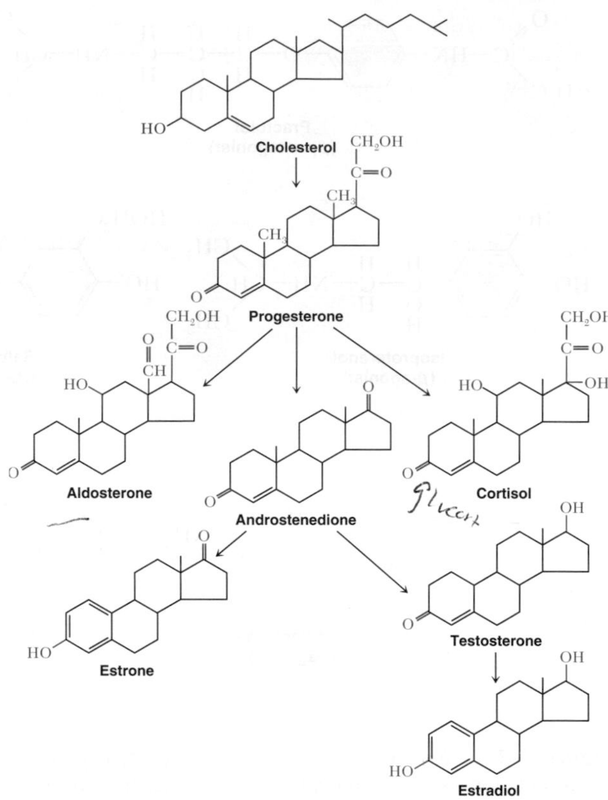

Figure 9-39 Cholesterol is the precursor for three major classes of steroid hormones: mineralocorticoids, glucocorticoids, and reproductive hormones. Modifications to the cholesterol structure, shown in red, yield a large number of related steroid hormones and intermediates. (Some intermediates have been omitted in the synthetic pathway shown here.) Several steroid hormones have mineralocorticoid or glucocorticoid activity; primary among these in mammals are aldosterone and cortisol, respectively. The adrenal cortex is the primary site for secretion of these hormones. The reproductive hormones (progesterone, testosterone, estrone, estradiol) are secreted in the greatest quantities from the gonads, although they are also secreted by the adrenal cortex.

Several adrenocortical hormones have glucocorticoid activity, including *cortisol, cortisone,* and *corticosterone.* Of these, cortisol is the most important in humans. The basal level of secretion of glucocorticoids is regulated via negative feedback by the hormones themselves on the neurosecretory cells of the hypothalamus that secrete corticotropin-releasing hormone and the cells of the anterior pituitary gland that secrete ACTH in response to CRH (Figure 9-40). The basal level of glucocorticoid secretion also undergoes a diurnal rhythm resulting from cyclic variation in CRH secretion, which appears to be influenced by an endogenous biological clock. Basal glucocorticoid levels in humans are maximal during the early hours of the morning prior to waking. This cycle is adaptively useful because of the energy-mobilizing actions of these hormones. In addition to such endogenous regulation of secretion, the adrenal cortex is stimulated to secrete glucocorticoids in response to stress of various types

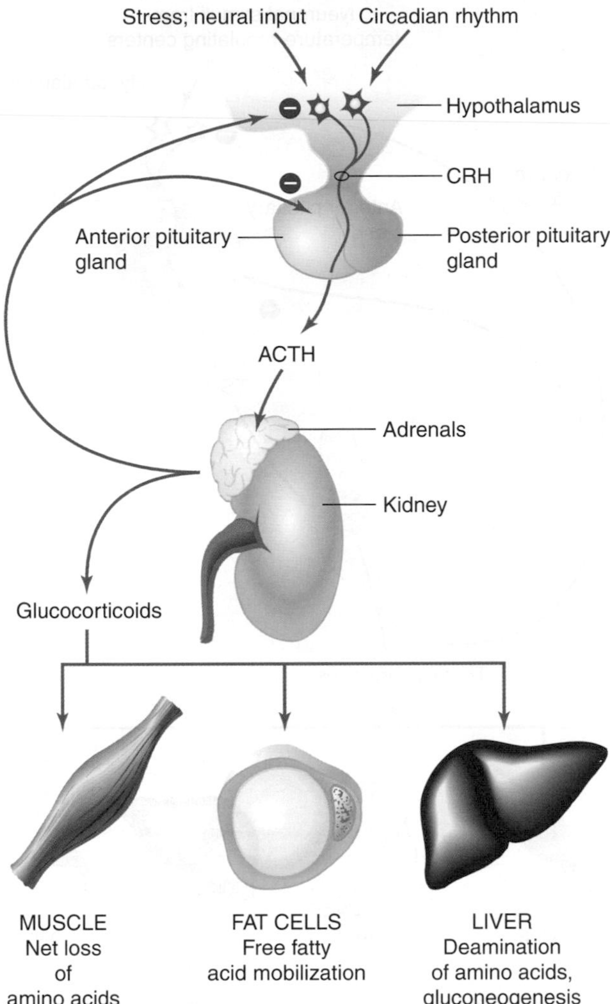

Figure 9-40 The secretion of glucocorticoids, and hence their effects on target tissues, is regulated by neuronal stimuli and negative feedback. Neuronal stimuli induce the release of corticotropin-releasing hormone (CRH) from hypothalamic neurosecretory cells. The resulting release of adreno-corticotropic hormone (ACTH) from the anterior pituitary gland stimulates the secretion of glucocorticoids by the adrenal cortex. These steroids produce an increase in blood glucose and liver glycogen by stimulating conversion of amino acids and fats to glucose. Negative feedback by the glucocorticoids to both the pituitary and the hypothalamus may limit ACTH release.

(including starvation). Stress, acting through the nervous system, causes an elevation in ACTH and hence stimulation of the adrenal cortex.

The glucocorticoids act on the liver, increasing the synthesis of enzymes that promote gluconeogenesis (synthesis of glucose from substances other than carbohydrates, such as amino acids and fats). Some of the newly synthesized glucose is stored as glycogen in the liver and muscle, but most is released into the circulation, causing a rise in blood glucose levels. However, the usage of this glucose is restricted, as the glucocorticoids also act to reduce the uptake of glucose into peripheral tissues such as muscle. The uptake of amino acids by muscle tissue is also decreased by glucocorticoids, and amino acids are released from muscle cells into the circulation. This release increases the quantity of amino acids available for conversion into glucose in the liver under glucocorticoid stimulation, a pathway that is especially important during starvation. The end result is the degradation of tissue proteins to maintain adequate blood glucose to sustain energy production in critical tissues such as the brain. The glucocorticoids also stimulate the mobilization of fatty acids from stores of fat in adipose tissue. All these actions increase the availability of quick energy to muscle and nervous tissue. The glucocorticoids have numerous other actions, including stimulation of gastric secretion and inhibition of immune responses.

As discussed earlier, the glucocorticoids, like other lipid-soluble steroid hormones, bind to specific receptors in the cytosol, forming hormone-receptor complexes that enter the nucleus and regulate the transcription of specific genes (see Figures 9-18 and 9-19).

Thyroid hormones

The follicles of the thyroid gland are stimulated by *thyroid-stimulating hormone* (TSH) to synthesize and release two major thyroid hormones. These hormones—3,5,3′-triiodothyronine (T_3) and thyroxine (T_4)—are formed from two iodinated tyrosine precursors (Figure 9-41 on the next page). Iodine is actively taken up from the blood by the thyroid tissue.

The secretion of thyroid hormones is regulated by negative feedback of these hormones on the hypothalamic neurons that secrete TSH-releasing hormone (TRH) and on the TSH-secreting cells of the anterior pituitary (Figure 9-42 on the next page). Superimposed on this regulation is stimulation of the hypothalamus by stress; low skin temperature, for example, stimulates the release of TRH from the hypothalamus.

The thyroid hormones act on the liver, kidneys, heart, nervous system, and skeletal muscle, sensitizing these tissues to epinephrine and stimulating cellular respiration, oxygen consumption, and metabolic rate. The acceleration of metabolism stimulated by thyroid hormones generates heat, which is of major importance in the thermoregulation of many vertebrates (see Chapter 16).

The thyroid hormones also significantly affect the development and maturation of various mammalian vertebrate groups. Thyroid hormones only affect development in the presence of growth hormone, and vice versa. The synergistic actions of the thyroid hormones and growth hormone promote protein synthesis during development. Hypothyroidism resulting from a lack of dietary iodine during early stages of development in fishes, birds, and mammals results in a deficiency disease (called *cretinism* in humans) in which somatic, neuronal, and sexual development are severely retarded, the metabolic rate is reduced to as little as half the normal rate, and resistance to infection is reduced. Inadequate production of thyroid hormones

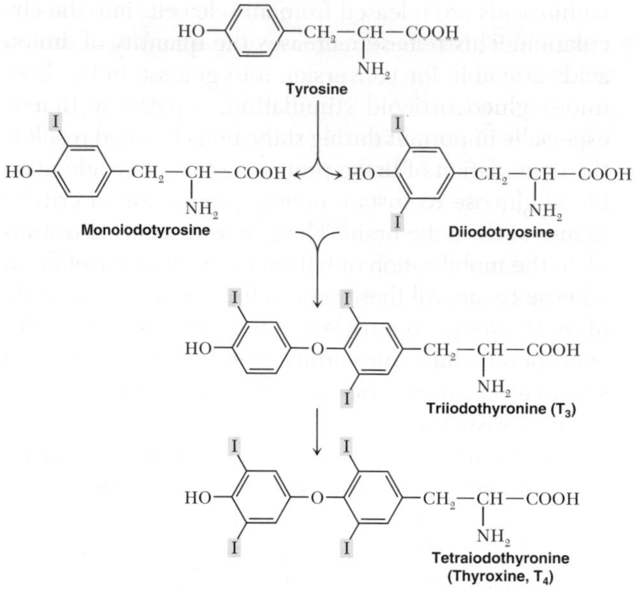

Figure 9-41 Thyroid hormones are produced from iodinated derivatives of the amino acid tyrosine. Condensation of the tyrosine derivatives yields 3,5,3′-triiodothyronine (T_3) and thyroxine (T_4). T_3 is also produced by removal of one iodide from thyroxine.

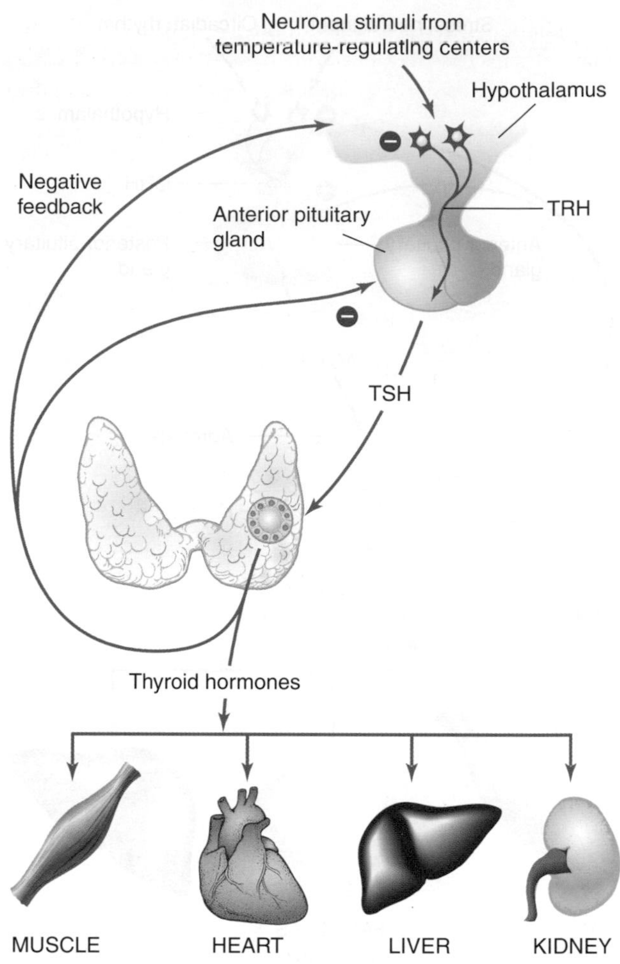

Figure 9-42 Thyroid hormones, which regulate metabolism in various tissues, are themselves regulated by neuronal stimuli and negative feedback. Low skin temperature and stress stimulate the release of TSH-releasing hormone (TRH) from hypothalamic neurosecretory cells. TRH then stimulates secretion of thyroid-stimulating hormone (TSH) from the anterior pituitary gland. The thyroid responds by secreting the thyroid hormones, which cause increased metabolism in skeletal and cardiac muscle, liver, and kidney, leading to the metabolic generation of heat. Feedback inhibition by thyroid hormones apparently occurs at the levels of both the anterior pituitary and the hypothalamus. The follicle shown superimposed on the thyroid gland is drawn at a disproportionately large scale.

leads to excessive production of TSH due to decreased negative feedback to the hypothalamus and the anterior pituitary. The resulting overstimulation of the thyroid gland by TSH causes hypertrophy of the gland (goiter). Raising the level of iodine in the diet increases thyroid hormone production, thereby establishing normal feedback control on TSH production. The incidences of both cretinism and goiter have been greatly reduced by the widespread iodization of table salt, which replaces the requirement for natural trace amounts of iodine in food.

The thyroid hormones, like the steroid hormones, are lipid-soluble and bind to specific receptors within the cytosol. Both types of hormones exert their effects by regulating the transcription of specific genes and, ultimately, the production of the proteins encoded by these genes. For this reason, the effects of these hormones develop slowly. For example, it may take up to 48 hours after a rise in blood levels of thyroid hormones before their effects are seen.

Insulin and glucagon

Insulin signals the presence of high blood glucose levels, whereas *glucagon* signals the presence of low blood glucose levels. These two hormones thus have opposing effects. Insulin is secreted by the β cells of the pancreatic islets of Langerhans, small patches of endocrine tissue that are scattered throughout the exocrine tissue of the pancreas. High blood glucose levels act as the major stimulus to the pancreatic β cells to secrete insulin (Figure 9-43). The release of insulin is also stimulated by glucagon, growth hormone, gastric inhibitory peptide (GIP, also known as glucose-dependent insulin-releasing peptide), epinephrine, and elevated levels of amino acids.

Insulin has important effects on carbohydrate, fat, and protein metabolism. With regard to carbohydrate metabolism, it has two major actions: it increases the rate of uptake of glucose into liver, muscle, and fat cells, and it stimulates glycogenesis (the polymerization of glucose to form glycogen). In lipid metabolism, insulin stimulates the release of fatty acids from the liver and adipose tissue. In protein metabolism, insulin stimulates the uptake of amino acids into the liver and muscles and the incorporation of amino acids into proteins.

Diabetes mellitus in humans, which occurs in two major forms, is characterized by defects in the insulin signaling system. *Type I diabetes mellitus* is associated with a loss of pancreatic β-cell mass, which leads to diminished or decreased insulin production and secretion. *Type II diabetes mellitus* is associated with defective signal reception in the insulin pathway. Whatever their cause, breakdowns in the insulin signaling pathway lead to hyperglycemia (high levels of blood glucose), glycosuria (spillover of excess glucose into the urine, which occurs when the blood glucose levels exceed the renal threshold for glucose: see Spotlight 14-1), and reduced stores of lipids and proteins, which are broken down to supply energy because the cells are deficient in glucose. In addition, mobilized fat particles that cannot be rapidly metabolized accumulate in the blood as ketone bodies. These are excreted in the urine but can also interfere with liver function. These disturbances in carbohydrate, lipid, and protein metabolism also produce a large number of complications in various organs (e.g., cataracts and cardiovascular diseases).

Although the insulin receptor exhibits tyrosine kinase activity, the intracellular signaling pathway triggered by the binding of insulin differs from that associated with other receptors of this type. Phosphorylation of various effector and regulatory proteins by the activated insulin receptor presumably mediates the various short-term and long-term effects of insulin. Insulin binding also induces the formation of peptide insulin mediators, which can inhibit adenylate cyclase and activate cAMP phosphodiesterase. This dual action has the effect of lowering intracellular cAMP levels.

High blood glucose; gastrointestinal hormone secretion (GIP)
Low blood glucose
Islets of Langerhans
Pancreas
Alpha cell (glucagon)
Glucagon
Insulin
Beta cell (insulin)
ALL TISSUES Insulin increases transport of glucose and amino acids into cells
MUSCLE Insulin induces glycogen synthesis
KIDNEY TUBULES Insulin increases absorption of glucose from filtrate
LIVER
Glucagon: Induces glycogenolysis and release of glucose into blood
Insulin: Suppresses glycogenolysis

Figure 9-43 The pancreatic hormones insulin and glucagon play a major role in regulating blood glucose levels. High levels of blood glucose and glucagon and/or gastrointestinal hormones signaling food ingestion (e.g., gastrointestinal inhibitory peptide, GIP) stimulate the pancreatic β cells to secrete insulin, which stimulates glucose uptake in all tissues. Glucagon, secreted by pancreatic α cells, exerts an action that is antagonistic to that of insulin in the liver, where it stimulates glycogenolysis and glucose release. Insulin has several other effects as well.

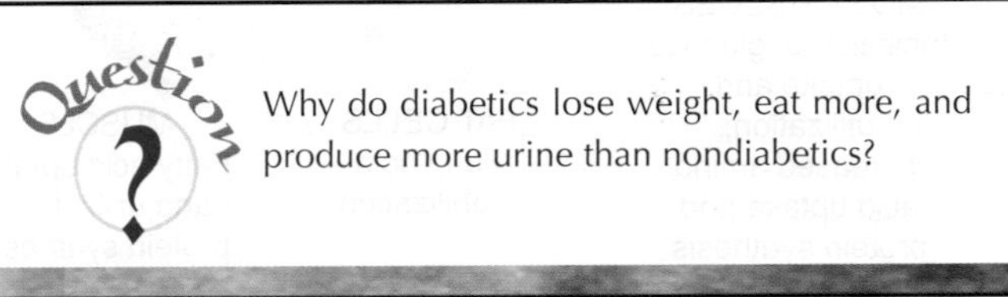

Why do diabetics lose weight, eat more, and produce more urine than nondiabetics?

Glucagon is secreted by the α cells of the pancreatic islets in response to hypoglycemia (low levels of blood glucose). This hormone has the opposite effects of insulin, stimulating the breakdown of glycogen in the liver and the release of glucose into the blood; it also stimulates lipolysis. The antagonistic actions of insulin and glucagon are important in maintaining an appropriate blood glucose level, so that adequate glucose is available for all tissues. Like epinephrine, which also promotes glycogenolysis, glucagon binds to receptors linked to the cAMP second-messenger pathway.

Growth hormone

The production and release of *growth hormone* (GH) by the anterior pituitary gland is under the direct control of two neurohormones released by the neurosecretory cells of the hypothalamus: *GH-releasing hormone* (GRH) and *GH-inhibiting hormone* (GIH), otherwise known as somatostatin (see Table 9-2). The levels of GRH and GIH are regulated in turn by factors such as blood glucose levels (Figure 9-44 on the next page). Reduced glucose levels, for example, indirectly stimulate the release of growth hormone by increasing the secretion of GRH.

Growth hormone exerts both metabolic and developmental effects. Many of its diverse metabolic effects are opposite those of insulin. Whereas insulin causes a decrease in blood glucose levels, growth hormone, like glucagon, causes an elevation of blood glucose. Therefore, growth hormone counteracts hypoglycemia, whereas insulin counteracts hyperglycemia.

Growth hormone elevates blood glucose by three mechanisms: it stimulates gluconeogenesis, blocks glucose uptake by tissues other than the nervous system,

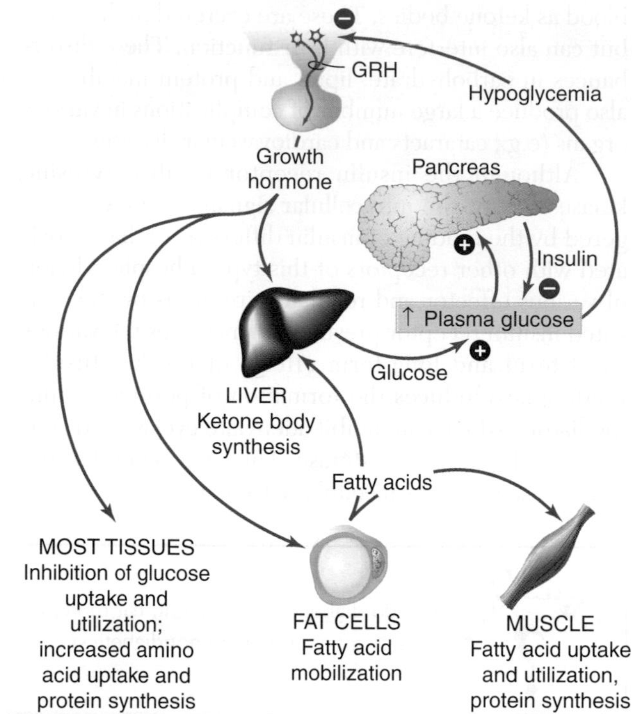

Figure 9-44 Many of the actions of growth hormone are antagonistic to those of insulin and similar to those of glucagon. Output of insulin from pancreatic beta cells occurs in response to high blood glucose, as after a meal. Growth hormone (GH) is released, usually several hours after a meal or after prolonged exercise, in response to insulin-induced hypoglycemia. Growth hormone causes lipolysis and the uptake of fatty acids by muscle tissue for energy and by the liver for ketone body synthesis. The GH-induced general depression of glucose uptake (except in the central nervous system) leads to a rise in plasma glucose, which then stimulates insulin secretion. The insulin stimulates glucose uptake into cells and thus counteracts GH-induced hyperglycemia.

and promotes the utilization of fatty acids as an energy source in place of glucose. In response to growth hormone, adipose tissue releases fatty acids into the bloodstream. These fatty acids are converted in the liver to ketone bodies for release into the circulation. Growth hormone also stimulates fatty acid uptake in muscle. By increasing the utilization of fatty acids, it helps conserve muscle glycogen stores (see Figure 9-44). Growth hormone reaches its peak plasma level several hours after a meal, when immediate energy supplies (e.g., blood glucose, amino acids, and fatty acids) have begun to decrease. Furthermore, growth hormone stimulates insulin secretion both directly, through its action on the pancreatic beta cells, and indirectly, through its effect in elevating blood glucose levels.

Growth hormone stimulates RNA and protein synthesis, which may account for its developmental effects, and it also promotes the growth of tissues—in particular, cartilage and subsequently bone. GH-stimulated tissue growth occurs by an increase in cell number (i.e., cell proliferation) rather than an increase in cell size. As noted above, thyroid hormones and growth hormone work synergistically to promote tissue growth during development. The growth-enhancing effects of growth hormone depend very much on the stage of development of the animal: the neonatal mammal is relatively insensitive to growth hormone but becomes more sensitive as it grows. Growth hormone not only stimulates proliferation of cells directly; it also stimulates the liver to produce other growth-promoting factors, called *insulin-like growth factors,* that also act directly on cells to promote growth. Disturbances in the secretion of growth hormone lead to several patterns of abnormal growth and development in humans: *gigantism,* characterized by excessive size and stature caused by hypersecretion of growth hormone during childhood (before puberty); *acromegaly,* characterized by enlargement of the bones of the head and extremities caused by hypersecretion of growth hormone beginning after maturity; and *dwarfism,* the abnormal underdevelopment of the body caused by insufficient secretion of growth hormone during childhood and adolescence.

Hormones that Regulate Water and Electrolyte Balance

The major organs involved in the regulation of water and electrolyte balance in vertebrates are the kidneys, intestine, and bones; in fishes, the gills also contribute. Most of the hormones that regulate water and electrolyte balance act on the epithelial tissues of these organs, since they are responsible for the uptake or excretion of water and electrolytes. The processes for maintaining water and electrolyte balance are described in more detail in Chapter 14. Here we consider the hormones that play a major role in regulating these processes (Table 9-8).

Antidiuretic hormone (ADH), also called *vasopressin,* regulates water turnover in the mammalian kidney. Secretion of this neurohormone from the posterior pituitary gland is stimulated by high blood osmolarity acting on osmoreceptors in the anterior hypothalamus (see Figure 9-17). By increasing the water permeability of the kidney collecting ducts, ADH stimulates reabsorption of water from the forming urine; the end result is a reduction in urine volume and increased water retention by the body. Increases in venous blood pressure, which often reflect increases in blood volume, stimulate atrial stretch receptors in the heart; these receptors then send an inhibitory signal to the hypothalamus that decreases ADH release and, therefore, enhances urine production, leading to a reduction in blood volume. ADH also enhances the release of ACTH and TSH from the anterior pituitary gland.

Mammals produce ADH, but other vertebrates produce slightly different nonapeptides with similar actions, as noted earlier. Reptiles, fishes, and birds produce a related peptide, called *vasotocin,* which exerts effects similar to those of vasopressin and oxytocin (see Table 9-4). Like vasopressin, vasotocin promotes water

Table 9-8 **Mammalian hormones involved in regulating water and electrolyte balance**

Hormone	Tissue of origin	Structure	Target tissue	Primary action	Regulation
Antidiuretic hormone (ADH, vasopressin)	Posterior pituitary	Nonapeptide	Kidneys	Increases water reabsorption	Increased plasma osmotic pressure or decreased blood volume stimulates release
Atrial natriuretic peptide (ANP)	Heart (atrium)	Peptide	Kidneys	Reduces Na^+ and water reabsorption	Increased venous pressure stimulates release
Calcitonin	Thyroid (parafollicular cells)	Peptide	Bones, kidneys	Decreases release of Ca^{2+} from bone; increases renal Ca^{2+} and PO_4^{3-} excretion	Increased plasma Ca^{2+} stimulates secretion
Mineralocorticoids (e.g., aldosterone)	Adrenal cortex	Steroid	Distal kidney tubules	Promotes reabsorption of Na^+ from urinary filtrate	Angiotensin II stimulates secretion
Parathyroid hormone (PTH)	Parathyroid gland	Peptide	Bones, kidneys, intestine	Increases release of Ca^{2+} from bone; with calcitriol increases intestinal Ca^{2+} absorption; decreases renal Ca^{2+} excretion	Decreased plasma Ca^{2+} stimulates secretion

reabsorption by the animal. Vasotocin may also play a role in sexual behavior and is associated with expulsion of eggs from the oviduct in turtles (somewhat analogous to the action of oxytocin). Both vasopressin and vasotocin are known to cause smooth-muscle contraction. ADH and its analogs exert their effects through the cAMP pathway.

The mineralocorticoids, in particular aldosterone, enhance reabsorption of sodium (and, indirectly, chloride) by the distal tubules and the collecting ducts of the kidney, thereby increasing the osmolarity of the blood. *Aldosterone* is one of the steroid hormones secreted by the adrenal cortex when it is stimulated by ACTH (adrenocorticotropic hormone). Secretion of aldosterone is also stimulated by angiotensin II (see Chapter 14) and high blood [K^+] and is subject to negative feedback by the action of aldosterone on the CRH-secreting neurons of the hypothalamus and on the ACTH-secreting cells of the anterior pituitary (see Figure 9-16). The mineralocorticoids, like other steroid hormones, exert their effects by binding to intracellular receptors and modifying gene expression.

Atrial natriuretic peptide (ANP) acts on the kidney to reduce the reabsorption of sodium, and therefore water, from the ultrafiltrate, leading to an increase in urine production and sodium excretion by the kidney. Thus the effects of this hormone counteract those of aldosterone and ADH. ANP is released by the atrium of the heart into the blood in response to an increase in venous pressure.

As we saw earlier, Ca^{2+} plays a key role as a second messenger and regulatory agent in cells. Thus, careful regulation of the concentration of Ca^{2+} in the blood and the extracellular fluid is critical. In the vertebrates, this ion is actively absorbed through the intestinal wall into the plasma and is deposited in bone, which serves as the major depot for storage of Ca^{2+}. Elimination of Ca^{2+} from the body occurs through the kidney. The balance between these processes, which determines the blood Ca^{2+} concentration, is influenced by three hormones: parathyroid hormone, calcitriol, and calcitonin.

Parathyroid hormone (PTH), also known as parathormone, is secreted from the paired parathyroid glands in response to a drop in plasma Ca^{2+} levels. It acts to increase plasma Ca^{2+} by promoting Ca^{2+} mobilization from bone, increasing Ca^{2+} uptake from the urine forming in kidney tubules, and enhancing intestinal Ca^{2+} absorption (Figure 9-45 on the next page). PTH works in conjunction with *calcitriol,* a steroidlike compound produced from vitamin D, ingested with some foods, and from vitamin D_3, which can be synthesized from cholesterol in the skin. Conversion of these precursors into calcitriol involves reactions in the liver and kidneys. The actions of calcitriol are similar to those of parathyroid hormone.

Calcitonin is secreted from the parafollicular, or C, cells in the thyroid gland in response to high plasma Ca^{2+} levels. It rapidly suppresses Ca^{2+} loss from bone, quickly countering the effects of PTH. Although calcitonin and PTH have opposing actions on bone metabolism, there is no feedback interaction between them. Each hormone, however, exerts negative feedback on its own secretion. The dominance of calcitonin prevents hypercalcemia and extensive dissolution of the skeleton. Essentially, the skeleton acts as a large reservoir and buffer for Ca^{2+} as well as for PO_4^{3-}. The plasma Ca^{2+} and PO_4^{3-} levels are held within narrow limits by the opposing actions of PTH and calcitonin, which regulate the flux of these ions between plasma and bone.

PTH and calcitonin are both peptide hormones and bind to cell-surface receptors. Calcitriol is lipid-soluble and presumably binds to an intracellular receptor.

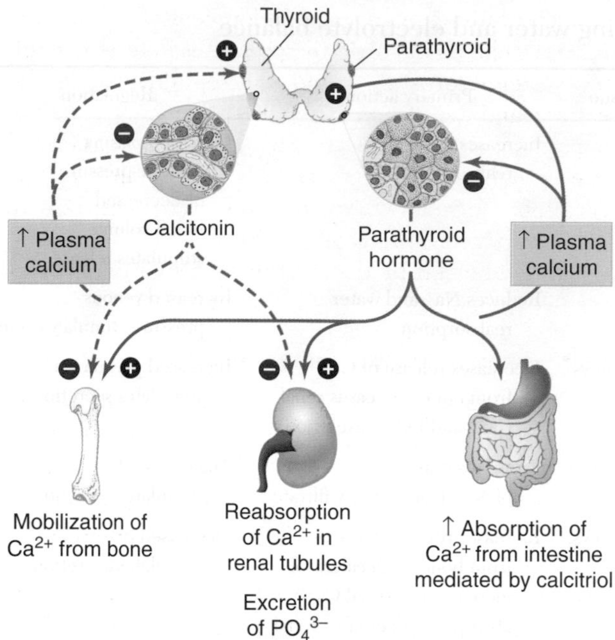

Figure 9-45 Calcitonin and parathyroid hormone (PTH) have opposite effects on plasma Ca^{2+} levels in mammals. Low levels of plasma Ca^{2+} stimulate the cells of the parathyroid glands to release PTH, which has several actions, all tending to increase plasma Ca^{2+}. High concentrations of Ca^{2+} in the blood stimulate parafollicular cells in the thyroid gland to release calcitonin, which acts to increase plasma Ca^{2+}. Calcitriol, the active hormonal form of vitamin D, also increases intestinal absorption of Ca^{2+}.

Reproductive Hormones

In vertebrates, several steroid hormones that affect reproduction are produced from cholesterol in the gonads (testes or ovaries) and adrenal cortex of both sexes (see Figure 9-39). Cholesterol is first converted to *progesterone,* which is then transformed into the *androgens* (androstenedione and testosterone). These hormones are then converted into the *estrogens,* of which estradiol-17β is the most potent. The steroid sex hormones, like other steroid hormones, bind to intracellular receptors and modify the expression of specific genes. In addition to the steroid sex hormones, two peptide hormones produced in the pituitary gland function in parturition and lactation. Table 9-9 summarizes the properties of the steroid and peptide reproductive hormones.

The estrogens and androgens are important in both sexes in various aspects of growth, development, and morphologic differentiation, as well as in the development and regulation of sexual and reproductive behaviors and cycles. However, androgens predominate in the male, estrogens in the female. The production and secretion of the steroid sex hormones in both sexes are promoted by *follicle-stimulating hormone* (FSH) and *luteinizing hormone* (LH), which are synthesized in the anterior pituitary gland (see Table 9-3). These tropic hormones are released from the anterior pituitary in response to *gonadotropin-releasing hormone* (GnRH) from the hypothalamus. The steroid sex hormones exert negative feedback on the GnRH-secreting neurons of

Table 9-9 **Important mammalian reproductive hormones**

Hormone	Tissue of origin	Structure	Target tissue	Primary action	Regulation
Primary sex hormones					
Estradiol-17β (estrogens)	Ovarian follicle, corpus luteum, adrenal cortex	Steroid	Most tissues	Promotes development and maintenance of female characteristics and behavior, oocyte maturation, and uterine proliferation	Increased FSH and LH levels stimulate secretion
Progesterone	Corpus luteum, adrenal cortex	Steroid	Uterus, mammary glands	Maintains uterine secretion; stimulates mammary duct formation	Increased LH and prolactin levels stimulate secretion
Testosterone (androgens)	Testes (Leydig cells), adrenal cortex	Steroid	Most tissues	Promotes development and maintenance of male characteristics and behavior and spermatogenesis	Increased LH level stimulates secretion
Other Hormones					
Oxytocin	Posterior pituitary	Nonapeptide	Uterus, mammary glands	Promotes smooth muscle contraction and milk ejection	Cervical distention and suckling stimulate release; high progesterone inhibits release
Prolactin (PL)	Anterior pituitary	Peptide	Mammary glands (alveolar cells)	Increases synthesis of milk proteins and growth of mammary glands; elicits maternal behavior	Continuous secretion of PL-inhibiting hormone (PIH) normally blocks release; increased estrogen and decreased PIH secretion permit release

the hypothalamus and on the anterior pituitary endocrine cells that produce FSH and LH.

Steroid sex hormones in males

The seminiferous tubules of the mammalian testes are lined with germ cells and Sertoli cells. Binding of FSH to receptors on Sertoli cells stimulates spermatogenesis in the germ cells after sexual maturity, either continuously or seasonally, depending on the species (Figure 9-46). The Sertoli cells support development of the sperm and are responsible for synthesis of androgen-binding protein and inhibin. Lying between the seminiferous tubules are interstitial cells, called Leydig cells, that produce and secrete sex hormones, particularly testosterone. Both testosterone itself and inhibin provide inhibitory feedback to the hypothalamic centers controlling GnRH production and hence diminish release of the gonadotropins FSH and LH from the anterior pituitary gland.

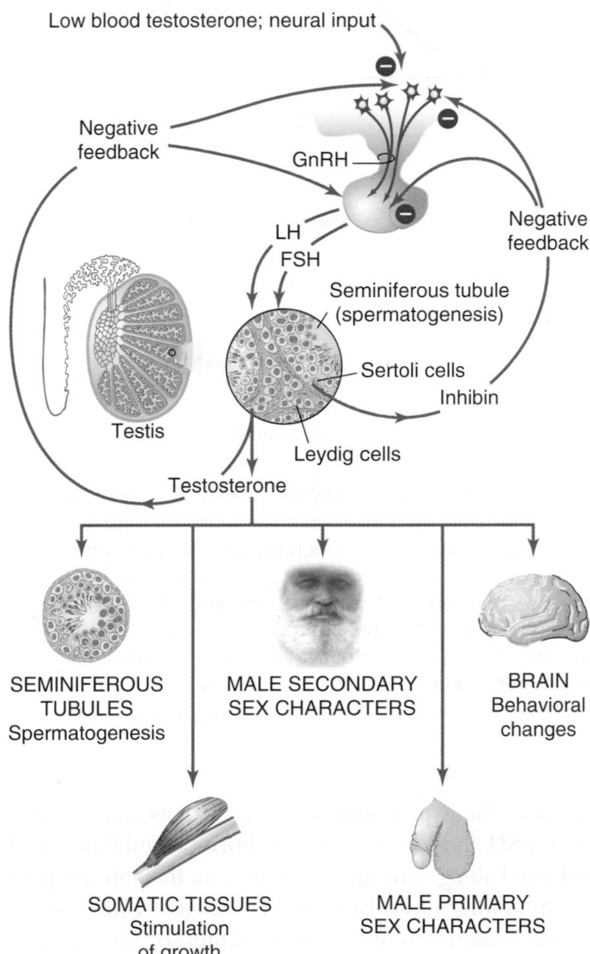

Figure 9-46 Testosterone, the primary sex hormone in males, has numerous actions and is regulated by neuronal stimuli and feedback control. A decrease in blood levels of testosterone stimulates the secretion of gonadotropin-releasing hormone (GnRH), which promotes the release of follicle-stimulating hormone (FSH) and luteinizing hormone (LH). Some of the actions of testosterone are indicated at the bottom of the figure. High testosterone levels and inhibin, also secreted by the testes, inhibit FSH secretion both directly and indirectly.

The androgens trigger development of the primary male sex characters (e.g., the penis, vas deferens, seminal vesicles, prostate gland, and epididymis) in the embryo and the secondary male sex characters (e.g., the lion's mane, the rooster's comb and plumage, and facial hair in men) at the time of puberty. The androgens also contribute to general growth and protein synthesis—in particular, the synthesis of myofibrillar proteins in muscle, as evidenced by the greater muscularity of males relative to females in many vertebrate species.

Steroid sex hormones in females

Androgens in males stimulate prenatal differentiation of the embryonic genital tract; the estrogens play no such role in early differentiation of the female reproductive tract. However, estrogens stimulate later development of primary sexual characteristics such as the uterus, ovaries, and vagina. The estrogens are also responsible for development of the female secondary sex characters, such as the breasts, and for regulation of reproductive cycles (Figure 9-47 on the next page).

Simultaneous reproduction has obvious survival value for a population. The gathering of large numbers of individuals of both sexes for mating, bearing young, and parental care during this period of high vulnerability can be timed to coincide with favorable weather and an adequate food supply. Moreover, the sudden appearance of large numbers of defenseless young individuals of a species can overwhelm even the most voracious of predators, permitting the survival of enough individuals of the new generation to ensure survival of the species. In general, reproductive cycles arise from within the animal under the control of the neuroendocrine system, but these inner cycles are constrained by environmental signals such as the changes in day length that accompany the changing seasons.

Female mammals and birds are born with a full complement of oocytes, each of which becomes embedded in a follicle within the ovary and is capable of developing into one ovum. Most of the follicles and their oocytes degenerate early, but even before puberty some develop just short of yolk formation or maturation. In humans, about 400 ova are available for release between menarche (onset of menstruation) and menopause. In lower vertebrates, oogenesis occurs throughout life.

In mammalian females, the reproductive cycle is composed of the follicular phase and luteal phase (Figure 9-48, left, on page 347). The *follicular phase* begins with the release of FSH, which stimulates the development of 15–20 ovarian follicles. These follicles are fluid-filled cavities containing an ovum and enclosed by a membranous sac of several cell layers, including the theca interna and ovarian granulosa. LH then stimulates the theca interna to synthesize and secrete androgens. FSH stimulates the production of an enzyme that then converts the androgens to estrogens in the ovarian granulosa, leading to a substantial increase in estrogen levels. At the high levels characteristic of the time just

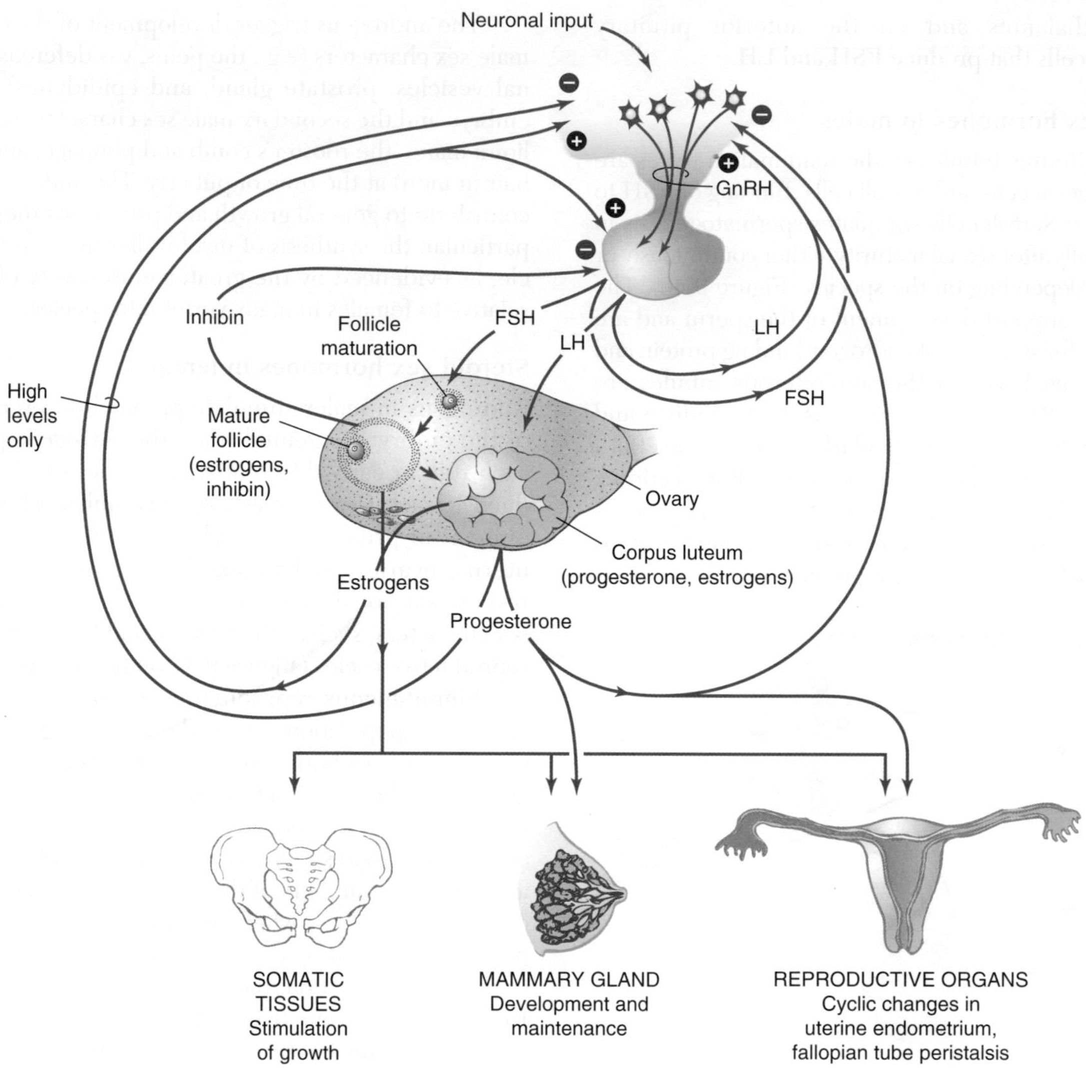

Figure 9-47 Estrogens and progesterone, the primary steroid sex hormones in females, mediate reproductive cycles and other effects under complex regulation. In mammals, a decrease in progesterone and estrogen levels, as well as neuronal inputs, stimulates release of gonadotropin-releasing hormone (GnRH) from the hypothalamus. This hormone acts on the anterior pituitary gland, stimulating secretion of follicle-stimulating hormone (FSH), which promotes maturation of the primordial follicles in the ovary. Estrogens secreted by the follicles and by the interstitial cells of the ovary eventually reach levels that stimulate the release of luteinizing hormone (LH), which triggers ovulation and the subsequent development of the corpus luteum. The corpus luteum secretes primarily progesterone and estrogens, which are needed to maintain pregnancy. The high levels of estrogen and progesterone inhibit the activity of the hypothalamic neurosecretory cells, leading to a decrease in gonadotropin secretion and preventing ovulation during pregnancy.

prior to ovulation, estrogens activate the hypothalamus and anterior pituitary gland, producing a surge in FSH and LH release, an example of positive feedback. The FSH accelerates maturation of the developing follicles. Only one follicle completes its maturation and, under the influence of LH, ruptures at the surface of the ovary, releasing the ovum. The increase in estrogens during the follicular phase also stimulates proliferation of the endometrium, the tissue that lines the uterus.

During the *luteal phase,* which begins with ovulation, estrogen secretion declines and LH transforms the ruptured follicle into a temporary endocrine tissue, the corpus luteum. The corpus luteum secretes estrogens and progesterone, which exert negative feedback on GnRH release by the hypothalamus, leading to decreased secretion of FSH and LH. The ovarian hormone inhibin, which is released along with the ovum, acts on the anterior pituitary, inhibiting FSH (but not LH) release. Progesterone stimulates secretion of endometrial fluid by the endometrial tissue, preparing it for implantation of a fertilized ovum. In the absence of fertilization and implantation of an ovum, the corpus luteum degenerates (in humans, after about 14 days), and secretion of estrogens and progesterone subsides. In humans and some other primates, this precipitates the menses, or shedding of the endometrial lining. With the reduction in estrogen, progesterone, and inhibin levels, FSH and LH secretion by the pituitary increases again, initiating a new cycle.

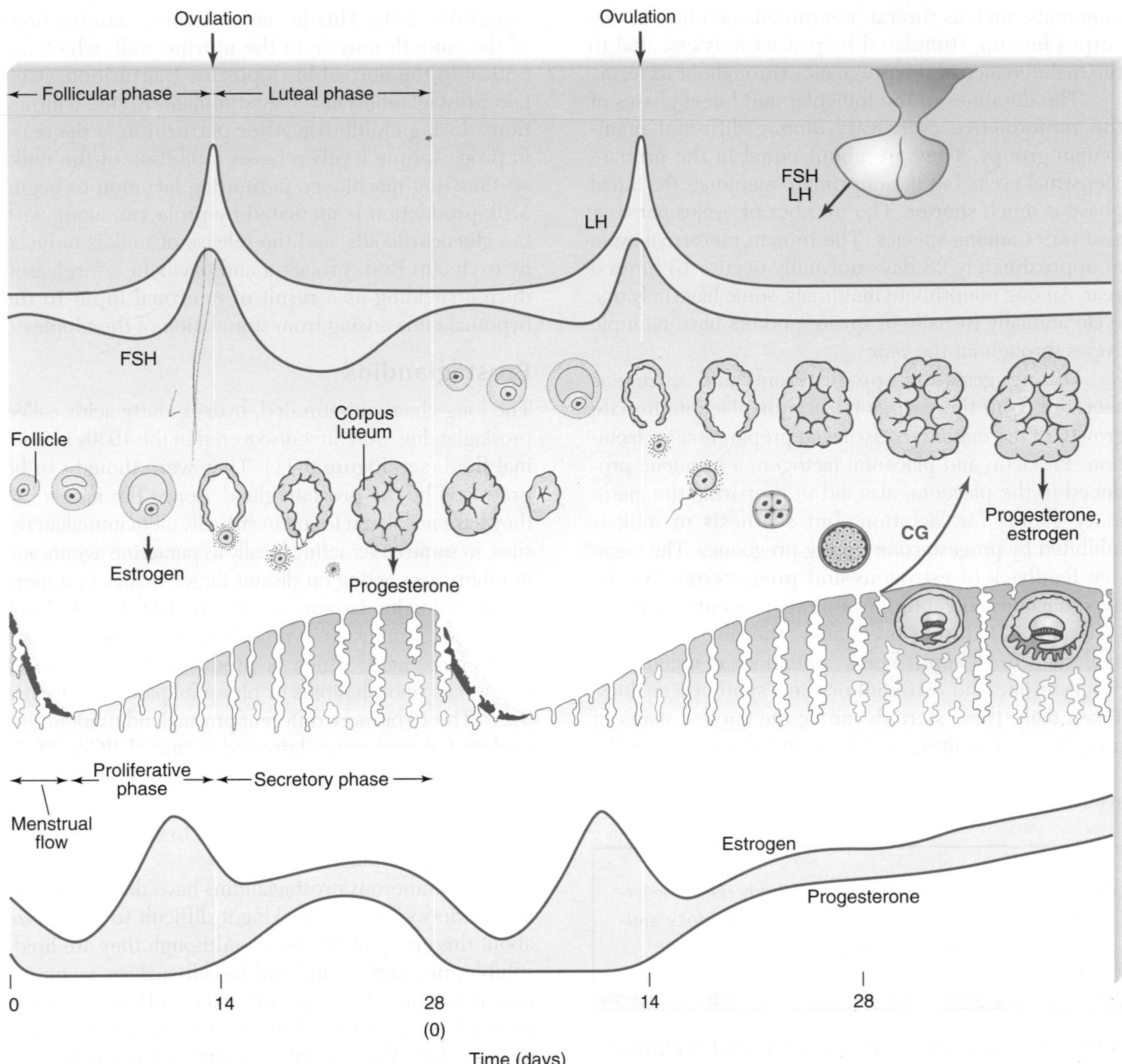

Figure 9-48 The primate menstrual cycle is regulated by periodic changes in the levels of the gonadotropins, estrogens, and progesterone. Before ovulation, follicle-stimulating hormone (FSH) promotes maturation of ovarian follicles, which secrete estrogen. High estrogen levels cause a surge of luteinizing hormone (LH), which triggers ovulation from one follicle. LH promotes development of the corpus luteum and induces it to secrete progesterone and some estrogen. In the absence of implantation (left), the progesterone and estrogen levels peak and then fall, initiating menstruation. The subsequent decrease in estrogen, progesterone, and inhibin levels allows pituitary secretion of FSH and LH to increase again, thus initiating a new cycle. If implantation and pregnancy occur (right), secretion of chorionic gonadotropin (CG) by the placenta "rescues" the corpus luteum, which maintains secretion of estrogen and progesterone for the first two to three months of pregnancy in humans. Thereafter, the placenta itself secretes estrogens and progesterone. [Adapted from McNaught and Callander, 1975.]

If the released ovum is fertilized as it travels down the ciliated fallopian tube, and the fertilized ovum becomes implanted in the endometrium, the developing placenta begins to produce *chorionic gonadotropin* (CG) (see Figure 9-48, red arrow). This hormone, whose action is similar to that of LH, induces further growth of the active corpus luteum, so that estrogen and progesterone secretion continues. The placenta begins secreting CG within about a day of implantation of the ovum and effectively takes over the gonadotropic function of the anterior pituitary during early pregnancy by maintaining the corpus luteum. Pituitary FSH and LH are not secreted again until after parturition (birth of the fetus). In many mammals, including humans, the corpus luteum continues to grow and to secrete estrogen and progesterone until the placenta fully takes over the production of these hormones, at which time the corpus luteum degenerates. In other mammals, such as the rat, continued secretion by the corpus luteum, stimulated by prolactin, is essential to the maintenance of the pregnancy throughout its term.

The durations of the follicular and luteal phases of the reproductive cycle vary among different mammalian groups. They are about equal in the primate menstrual cycle, but in nonprimate mammals the luteal phase is much shorter. The number of cycles per year also varies among species. The human menstrual cycle of approximately 28 days normally occurs 13 times a year. Among nonprimate mammals, some have only one cycle annually (usually in spring); others have multiple cycles throughout the year.

During gestation, progesterone and estrogens secreted from the corpus luteum or placenta initiate growth of the mammary tissues in preparation for lactation. Prolactin and placental lactogen, a hormone produced in the placenta, also aid in preparing the mammary glands for lactation, but synthesis of milk is inhibited by progesterone during pregnancy. The negative feedback of estrogens and progesterone on the hypothalamus and anterior pituitary prevents release of FSH and LH during pregnancy, thereby preventing ovulation. Birth control pills contain small amounts of progesterone and estradiol or their synthetic analogs. Taken daily, these steroids mimic the earliest stages of pregnancy, preventing ovulation and also acting on the endometrium to provide a highly effective means of avoiding conception.

Why doesn't the mother's body reject the fertilized egg? Does the mother produce antibodies against the developing fetus?

Hormones involved in parturition and lactation

As pregnancy nears term, cervical distention stimulates release of oxytocin from the posterior pituitary gland (see Table 9-9). This hormone induces contractions of the smooth muscle in the uterine wall, which are critical to the normal birth process (parturition). Certain prostaglandins also may stimulate uterine contractions during childbirth. After parturition, a decrease in progesterone levels relieves inhibition of the milk-synthesizing machinery, permitting lactation to begin. Milk production is mediated by prolactin, along with the glucocorticoids, and the release of milk is induced by oxytocin. Both prolactin and oxytocin are released during suckling as a result of neuronal input to the hypothalamus arising from stimulation of the nipples.

Prostaglandins

The long-chain, unsaturated, hydroxy fatty acids called prostaglandins were first discovered in the 1930s in seminal fluid (see Figure 9-11). They were thought to be produced by the prostate gland, hence the name, but they have now been found in virtually all mammalian tissues, in some cases acting locally as paracrine agents and in other cases acting on distant target tissues in a more classic endocrine fashion (see Figure 9-1). As noted earlier, prostaglandins are synthesized in membranes from arachidonic acid, which is produced by cleavage of membrane phospholipids by phospholipases (see Figure 9-24). The 16 or more different prostaglandins identified to date fall into nine classes (designated PGA, PGB, PGC, . . . PGI). Some of them are converted to other biologically active prostaglandins. The prostaglandins undergo rapid oxidative degradation to inactive products in the liver and lungs.

The numerous prostaglandins have diverse actions on a variety of tissues, making it difficult to generalize about this group of hormones. Although they are lipid-soluble, prostaglandins bind to cell-surface receptors linked to the cAMP pathway. The effects of some prostaglandins produced in selected tissues are shown in Table 9-10. For example, prostaglandins produced in the kidney act on the smooth muscle of blood vessels to regulate vasodilation and vasoconstriction.

Table 9-10 Selected prostaglandins

Tissue of origin	Target tissue	Primary action	Regulation
Seminal vesicles, uterus, ovaries	Uterus, ovaries, fallopian tubes	Potentiates smooth muscle contraction and possibly luteolysis; may mediate LH stimulation of estrogen and progesterone synthesis	Introduced during coitus with semen
Kidney	Blood vessels, especially in kidneys	Regulates vasodilation or vasoconstriction	Increased angiotensin II and epinephrine stimulate secretion; inactivated in lungs and liver
Neuronal tissue	Adrenergic terminals	Blocks norepinephrine-sensitive adenylate cyclase	Neuronal activity increases release

Prostaglandins also modulate inflammatory responses and the function of blood cells, such as platelets. Aspirin acts as an anti-inflammatory agent by inhibiting prostaglandin synthesis.

HORMONAL ACTION IN INVERTEBRATES

Endocrine cells—in particular, neurosecretory cells—have been identified in all invertebrate groups, including even the primitive hydroid coelenterates. In *Hydra*, for example, neurons secrete what is believed to be a growth-promoting hormone during budding, regeneration, and growth. Hormone actions have been studied in a limited number of invertebrate species, typically those with particularly accessible endocrine systems. Hormonal regulation of development in insects has been widely studied and will serve to illustrate the general principles of hormone action in invertebrates.

Most insects fall into two groups based on their pattern of development: holometabolous insects exhibit complete metamorphosis, and hemimetabolous insects exhibit incomplete metamorphosis. The life cycle of hemimetabolous insects—including the Hemiptera (bugs), Orthoptera (locusts, crickets), and Dictyoptera (roaches, mantids)—begins with the development of the egg into an immature nymphal stage. The nymph eats and grows and undergoes several *ecdyses*, or molts, replacing its old exoskeleton with a soft new one that expands to a larger size before hardening. The stages between molts are termed *instars*. The final nymphal instar gives rise to the adult stage.

The development of holometabolous insects—including the Diptera (flies), Lepidoptera (butterflies, moths), and Coleoptera (beetles)—is more complex. The egg develops into a larva (e.g., maggot, "worm," caterpillar), which grows through several instars. The larva is specialized for eating and is therefore the insect stage that causes the most damage to agricultural crops. The last larval instar molts to become a *pupa*, an outwardly dormant stage in which extensive internal reorganization takes place to give rise to the adult form. The adult, which shows little morphologic resemblance to the pupa or larva, is the reproductive stage, and in some species is not even equipped to feed.

The first experiments demonstrating probable endocrine control of insect development were done between 1917 and 1922 by S. Kopec, who ligated moth larvae at various times during the last instar. He found that when the ligature was tied before a certain critical period, the larva would pupate anteriorly to the ligature but remain larval posteriorly. Cutting the nerve cord had no effect, so he concluded that pupation must be induced by a circulating substance that had its origin in a tissue located in the anterior portion of the larva. By testing various tissues, Kopec found that removal of the brain prevents pupation and that reimplantation of the

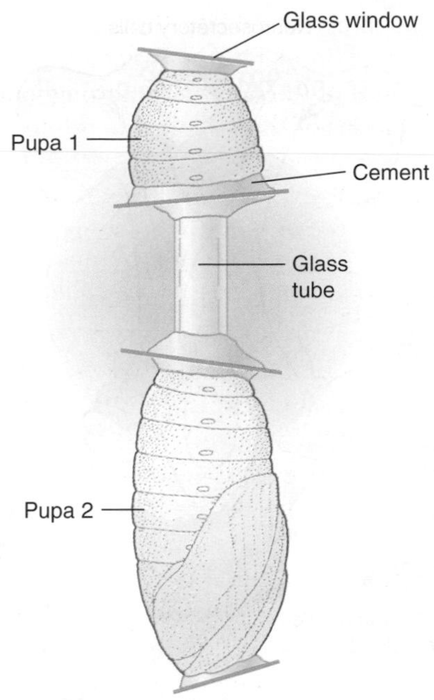

Figure 9-49 Parabiosis, the joining of body parts from different individuals, is a useful experimental method in insect endocrinology. Insect tissues readily survive such radical surgery as transection and decapitation. In this example, the abdomen of one pupa is joined to another pupa by a glass tube. Glass windows at either end permit visual inspection of the developing tissues.

brain allows it to proceed again. It was subsequently found that a neurohormone secreted by cells in the brain stimulates the prothoracic glands, which produce the molt-inducing hormone. Thus, ligating posterior to the prothoracic glands after their activation by the brain-derived hormone prevents pupation of the abdomen. Pupation can be initiated by implanting activated prothoracic glands into the isolated abdomen.

The hardiness of insects makes them ideal subjects for experiments that demonstrate the hormonal control of molting and metamorphosis. It is possible, for instance, to carry out extended parabiosis experiments, in which two insects or two parts of one insect are joined so that they share a common circulation and exchange body fluids (Figure 9-49). Thin glass windows incorporated into the insect's body make it possible to observe developmental changes in the tissues of the separated parts.

Five major hormones, three of them produced by neurosecretory cells, are now known to control development in insects (Figure 9-50 and Table 9-11 on the next page):

- *Prothoracicotropic hormone* (PTTH) is a neurohormone produced by neurosecretory cells that have their cell bodies in the pars intercerebralis of the brain. PTTH appears to be a small protein with a molecular weight of about 5000.
- *Juvenile hormone* is synthesized and released from the corpora allata, which are nonneuronal

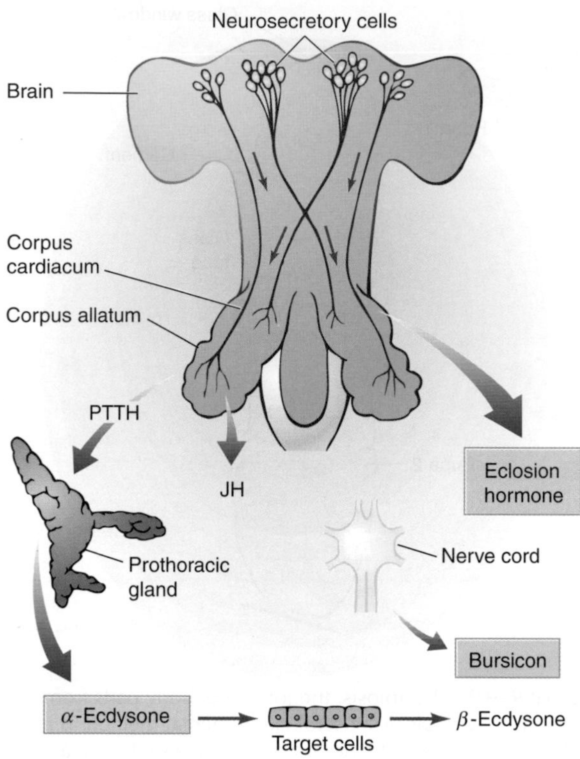

Figure 9-50 Of the five major insect developmental hormones, three are produced by neurosecretory cells and two by endocrine tissues. Neurosecretory cells in the brain synthesize prothoracicotropic hormone (PTTH) and eclosion hormone, which are stored in nerve terminals until their release into blood sinus spaces in the corpus cardiaca and corpus allata. A third neurohormone, bursicon, is released primarily from nerve terminals in the nerve cord. The corpus allatum also contains endocrine cells that release juvenile hormone (JH). Under the stimulus of PTTH, the prothoracic gland produces and secretes α-ecdysone, which is converted in target tissues to the active molting hormone β-ecdysone. [Adapted from Riddiford and Truman, 1978.]

paired glands somewhat analogous to the anterior pituitary gland. Several juvenile hormone homologs occur naturally in insects; they all have a modified fatty acid structure (Figure 9-51a).

- *Ecdysone,* produced by the prothoracic glands, is synthesized from cholesterol. It is structurally similar to vertebrate steroid hormones but contains more hydroxyl groups (Figure 9-51b).
- *Eclosion hormone,* a peptide neurohormone, is released from neurosecretory cells whose terminals are in the corpora cardiaca, which are paired neurohemal organs immediately posterior to the brain.

Table 9-11 **Insect developmental hormones**

Hormone	Tissue of origin	Structure	Target tissue	Primary action	Regulation
Bursicon	Neurosecretory cells in brain and nerve cord	Protein (MW ~40,000)	Epidermis	Promotes cuticle development; induces tanning of cuticle of newly molted adults	Stimuli associated with molting stimulate secretion
Ecdysone (molting hormone)	Prothoracic glands, ovarian follicle	Steroid	Epidermis, fat body, imaginal disks	Increases synthesis of RNA, protein, mitochondria, and endoplasmic reticulum; promotes secretion of new cuticle	PTTH stimulates secretion
Eclosion hormone	Neurosecretory cells in brain	Peptide	Nervous system	Induces emergence of adult from puparium	Endogenous "clock"
Juvenile hormone (JH)	Corpus allatum	Fatty acid derivative	Epidermis, ovarian follicles, sex accessory glands, fat body	In larva, promotes synthesis of larval structures and inhibits metamorphosis; in adult, stimulates synthesis of yolk protein; activates ovarian follicles and sex accessory glands	Inhibitory and stimulatory factors from the brain control secretion
Prothoracicotropin (PTTH)	Neurosecretory cells in brain	Small protein (MW ~5000)	Prothoracic gland	Stimulates ecdysone release	Various environmental and internal cues (e.g., photoperiod, temperature, crowding, abdominal stretch) stimulate release; JH inhibits release in some species

- *Bursicon,* also a neurohormone, is produced by other neurosecretory cells in the brain and nerve cord. It is a protein with a molecular weight of about 40,000.

PTTH is transported along the axons of the neurosecretory cells to neurohemal organs formed by the terminals of the axons in the corpora allata (see Figure 9-50). Upon its release, PTTH activates the prothoracic gland to synthesize and secrete the molt-inducing factor α-ecdysone. Insects require cholesterol in their diets to synthesize this steroid hormone. It is now thought that α-ecdysone is a prohormone that is converted to the physiologically active form, 20-hydroxyecdysone (β-ecdysone), in several peripheral target tissues (see Figure 9-51b).

Juvenile hormone, acting in association with β-ecdysone, promotes the retention of the immature ("juvenile") characteristics of the larva, thereby postponing metamorphosis until larval development is completed. The presence of juvenile hormone in the early nymphal instar was demonstrated in the mid-1930s in experiments by V. B. Wigglesworth, in which parabiotic coupling of a larva in an early instar to a larva in the final instar prevented the latter from becoming an adult. The circulating concentration of juvenile hormone is highest early in larval life, dropping to a minimum at the end of the pupal period (Figure 9-52). Metamorphosis to the adult stage occurs

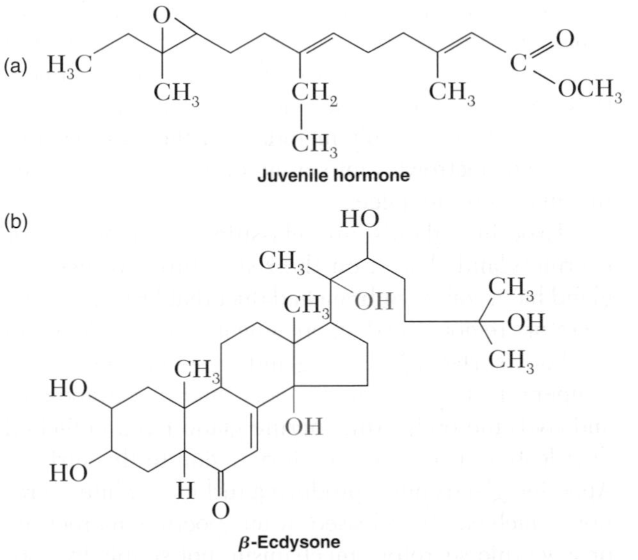

(a) Juvenile hormone (H_3C, O, CH_3, CH_2, CH_3, CH_3, C, O, OCH_3)

(b) β-Ecdysone (HO, CH_3, CH_3, OH, CH_3, OH, CH_3, CH_3, HO, HO, OH, H, O)

Figure 9-51 Juvenile hormone and β-ecdysone play key roles in regulating insect development. **(a)** The structure of juvenile hormone from the cecropia moth *(Hyalophora cecropia).* This hormone promotes the retention of juvenile characteristics in larvae and induces reproductive maturation in adults. Several homologs of juvenile hormone occur naturally in insects. **(b)** The structure of β-ecdysone, the physiologically active molt-inducing hormone. The prohormone α-ecdysone, which lacks the hydroxyl group on C-20 (red), is synthesized from cholesterol in the prothoracic gland. After α-ecdysone is released, it is converted in certain target tissues into the active hormone β-ecdysone.

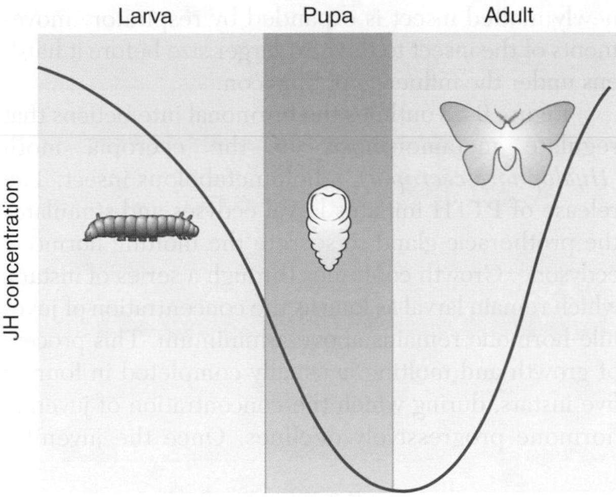

Figure 9-52 Normal progression through the insect life cycle depends on changes in the level of juvenile hormone. Metamorphosis of the juvenile larval form to the pupa occurs when the concentration of juvenile hormone falls below a certain threshold level. After the adult insect emerges and feeds, secretion of juvenile hormone begins again, regulating ovarian activity and stimulating development of male accessory organs. [Adapted from Spratt, 1971.]

when juvenile hormone disappears from the circulation. The concentration then rises again in the reproductively active adult. In the males of some insect species, juvenile hormone promotes development of the accessory sexual organs; in many female insects, it induces yolk synthesis and promotes maturation of the eggs. Thus, the normal development of an insect depends on precisely adjusted concentrations of juvenile hormone at each stage.

During the growth and development of insects, the epidermis undergoes conspicuous changes, including the production of a cuticle, a chitinous, horny outer covering. Considerable attention has been given to the production of the new cuticle, its tanning (hardening), and the shedding of the old cuticle during molting. PTTH, juvenile hormone, and β-ecdysone are all involved in the initiation of ecdysis. Ecdysone, secreted by the prothoracic glands in response to stimulation by PTTH, acts on the epidermis to initiate production of the new cuticle, which begins with apolysis, the detachment of the old cuticle from the underlying epidermal cells. The epithelial cells then begin synthesizing the materials for the new cuticle, while the old cuticle is partially digested from beneath by enzymes in the molting fluid secreted by the epidermis. At high concentrations of juvenile hormone, a new larval cuticle is formed, whereas at low concentrations, an adult cuticle is produced and other events of metamorphosis ensue.

Two additional hormones, eclosion hormone and bursicon, are responsible for promoting the terminal phase of the molting process. Shedding of the cuticle of the pupa is triggered by eclosion hormone in at least some holometabolous species. The pale, soft cuticle of a newly molted insect is expanded by respiratory movements of the insect to the next larger size before it hardens under the influence of bursicon.

Figure 9-53 outlines the hormonal interactions that regulate metamorphosis of the cecropia moth (*Hyalophora cecropia*), a holometabolous insect. The release of PTTH initiates larval ecdyses and stimulates the prothoracic gland to secrete the molting hormone ecdysone. Growth continues through a series of instars, which remain larval as long as the concentration of juvenile hormone remains above a minimum. This process of growth and molting is usually completed in four or five instars, during which the concentration of juvenile hormone progressively declines. Once the juvenile-inducing effects of juvenile hormone are removed, the larva molts to the pupal stage. The pupal stage is the overwintering (diapause) stage in many insects. Prolonged exposure to cold stimulates release of PTTH in the pupa, inducing the release of ecdysone; in the absence of juvenile hormone, ecdysone induces pupal development into the adult moth.

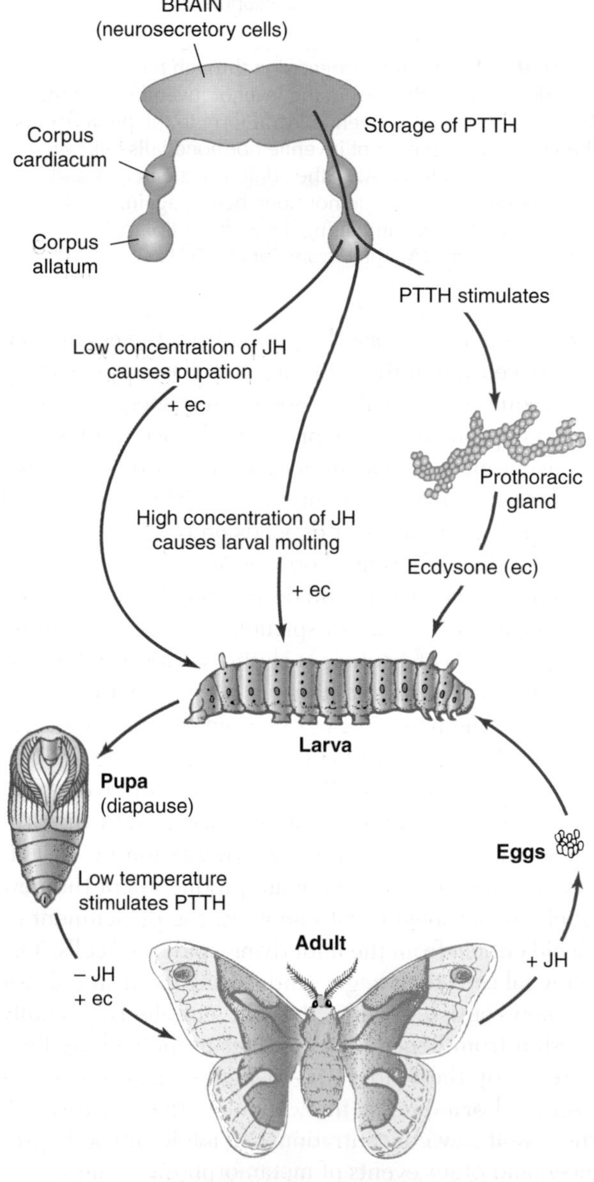

Classic Work **Figure 9-53** Interactions of juvenile hormone and ecdysone regulate metamorphosis in holometabolous insects. This example shows the developmental sequence of the cecropia moth (*Hyalophora cecropia*). [Adapted from Spratt, 1971.]

EXOCRINE GLANDS

Unlike endocrine secretions, the output of an exocrine gland does not ooze into the circulation, but generally flows through a duct into a body cavity (e.g., the mouth, gut, nasal passage, or urinary tract) that is continuous with the exterior. As noted earlier, exocrine secretions are usually aqueous mixtures consisting of a water-based primary fluid and added components, rather than a single substance. In the alimentary canal, these mixtures typically consist of water, ions, enzymes, and mucus. Exocrine tissues of the alimentary canal include the salivary glands, secretory cells in the stomach and intestinal epithelium, and secretory cells of the liver and pancreas.

An exocrine gland typically consists of an invaginated epithelium of closely packed secretory cells lining a blind cavity called the acinus (see Figure 9-9a). Several acini connect to a small duct that, in turn, connects to a larger duct from which secretions are released. The basal surfaces of the epithelial cells are usually in close contact with the circulation. Once the primary secretory products are free in the acinar lumen, they generally undergo secondary modifications in the secretory duct, possibly including further transport of water and electrolytes into or out of the duct to produce the final secretory juice.

Exocrine glands are classified as apocrine or eccrine glands based on their structure. An **eccrine gland** has a coiled, unbranched duct that leads from the secretory region; the duct opening lies perpendicular to the body surface. Eccrine glands respond to elevated temperatures by secreting a clear fluid that evaporates and cools the body. An **apocrine gland** has a branched duct leading from the secretory region to the surface. Apocrine glands often produce a turbid or white secretion, which can be released by an apocrine, merocrine, or holocrine secretory mechanism, not simply by apocrine secretion, as the name of these glands might suggest. This extensive, and often confusing, terminology is a reflection of the vast array of glands and glandular functions found in animals.

The Vertebrate Salivary Gland

The **saliva** present in the human mouth is a complex mixture consisting of secretions from the salivary glands, bacteria normally resident in the mouth, epithelial cells, and the remains of food and drink and whatever else has been in the mouth. This complex liquid is referred to as *whole saliva* to distinguish it from the

Table 9-12 **Inorganic constituents of whole saliva (mg · 100 ml^{-1})**

Constituent	Range	Mean
Sodium	0–80	15 resting 60 stimulated
Potassium	60–100	80
Calcium	2–11	6
Phosphorus (inorganic)	6–71	17 resting 12 stimulated
Chloride	50–100	—
Thiocyanate	—	9 (smokers) 2 (nonsmokers)
Fluoride (parts/10^6)	0.01–0.04	0.03 resting 0.01 stimulated
Bicarbonate	0–40	6 resting 36 stimulated
pH	5.0–8.0	—

Source: Edgar, 1992.

duct saliva released from an individual gland. Whole saliva is about 99.5% water, has a pH of 5.0–8.0, and contains a variety of ions (Table 9-12).

Functions and flow of saliva

Saliva has many functions. First, it lubricates the mouth and surrounding regions, thereby facilitating speech, eating, and swallowing. By dissolving and diluting food and drink, saliva assists in swallowing and allows food to be tasted. Second, saliva controls the bacterial flora in the mouth by inhibiting the growth of some bacteria and promoting the growth of others. The antibacterial action of saliva depends on three components: lysozyme, which causes bacterial lysis; lactoferrin, which removes from saliva the free iron needed for the growth of some bacteria; and sialoperoxidase, which oxidizes thiocyanate to hypothiocyanate, a potent antibacterial agent. A third function of saliva is the initial digestion of starch, which is catalyzed by the enzyme salivary amylase. The high pH of saliva promotes this digestive action, and saliva also acts as an effective buffer that protects the tissues of the oral cavity. Although saliva is not essential for digestion of food, impairment of saliva production makes chewing and swallowing difficult, and leads to tooth decay. In addition, as most of us know, it is difficult to talk or sing with a dry mouth.

Humans produce about a liter of saliva each day, most of which is swallowed and reabsorbed. Flow varies according to a circadian rhythm, probably driven by aldosterone levels in the plasma, with minimum flow during the night, especially during sleep. A slow resting flow of saliva keeps the mouth moist. Dehydration and stress activate the sympathetic nervous system and reduce flow, causing the dry mouth that typically accompanies fear or anxiety. Flow increases with the anticipation or the sight and smell of food, especially if the individual is hungry. The activation of muscle and tendon stretch receptors associated with jaw movements during eating also causes an increase in saliva production. The magnitude of the increase in flow depends on taste sensations. In general, sour tastes cause the greatest increase in salivation, followed by sweet, salt, and bitter in decreasing order of effectiveness. Increased salivation also is common before vomiting; this increased flow presumably protects the oral membranes by diluting and buffering the highly acidic vomit.

Formation of saliva

Saliva is formed as a primary fluid in the acini of the salivary glands and is then modified during passage through the ducts. NaCl is secreted into the acinus with water, which moves down an osmotic gradient. Amylase, mucous glycoproteins, and proline-rich glycoproteins are added to this fluid by exocytosis. Variation in the production of saliva, unlike other digestive exocrine processes, is mainly under neuronal control. Changing levels of circulating antidiuretic hormone and aldosterone, however, modify the ionic composition of saliva. Salivary glands are innervated by sympathetic and parasympathetic nerves. The sympathetic nerves release norepinephrine, which increases the production of amylase and other proteins but causes vasoconstriction and a decrease in saliva production. Parasympathetic stimulation, which is mediated by acetylcholine, the peptide hormone *substance P,* and *vasoactive intestinal peptide* (VIP, a polypeptide hormone of the gut), causes vasodilation and increases salivation. Chewing tobacco, which releases nicotine, mimics the effects of parasympathetic stimulation, causing a large increase in the flow of saliva.

The binding of acetylcholine, substance P, or norepinephrine to the appropriate receptors on the basal membrane of an acinar cell leads to the activation of the inositol phospholipid signaling pathway, in which phospholipase C catalyzes formation of DAG and IP_3 (Figure 9-54 on the next page). IP_3 stimulates the release of Ca^{2+} from the endoplasmic reticulum, which in turn causes the opening of K^+ channels in the plasma membrane and K^+ efflux from the cell. A rise in the external K^+ level activates a $Na^+/2Cl^-/K^+$ cotransporter, and K^+, Na^+, and Cl^- enter the cell. The movements of Na^+ and K^+ are counteracted by a Na^+/K^+ pump, which maintains Na^+ and K^+ levels in the cell. Thus, Na^+ and K^+ are cycled through the membrane, and the only net transfer is the outward movement of Cl^- ions, which leave via the apical (lumenal) membrane of the cell. In other words, there is a net movement of Cl^- across the acinar cell from the blood to the acinar lumen. This ionic movement generates a transepithelial potential that is positive on the blood side and creates the driving force for diffusion of Na^+ through paracellular channels from the blood to the lumen. The movement of NaCl into the lumen establishes the osmotic gradient that generates a flow of water into the lumen.

Binding of norepinephrine to *β*-adrenoreceptors or of VIP to peptidergic receptors activates adenylate cyclase (see Figure 9-54), resulting in the formation of cAMP, which in turn activates a protein kinase that stimulates exocytosis. DAG also promotes exocytosis of amylase, mucous glycoproteins, and proline-rich glycoproteins into the lumen.

The primary fluid, which consists of water, sodium chloride, amino acids, proteins, and glycoproteins, is forced into the salivary duct by the formation of more fluid. As it passes down the duct, potassium bicarbonate is added to the fluid and some sodium is reabsorbed. Because less sodium is reabsorbed at high flow rates, the final product leaving the duct approaches the composition of the primary secretion during heavy salivation. In contrast, the bicarbonate level in the final

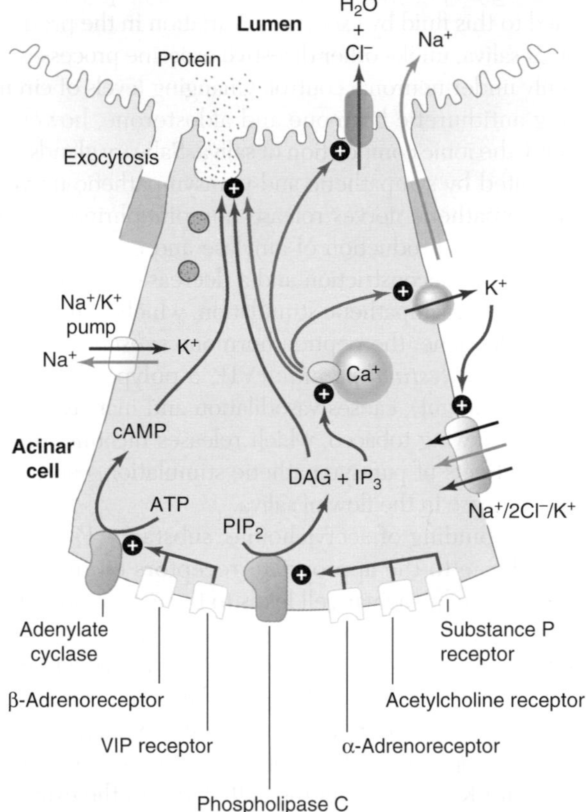

Figure 9-54 Production and release of primary fluid by salivary gland acinar cells is under neuronal control. Ligand binding to *α*-adrenoreceptors, acetylcholine receptors, or substance P receptors activates phospholipase C. This enzyme splits PIP_2 into DAG and IP_3, leading to the release of stored Ca^{2+} and the opening of K^+ channels. As a result of various ion movements, NaCl and water enter the acinar lumen. Exocytosis of amylase and glycoproteins stored in secretory granules is promoted by activation of the cAMP signaling pathway following ligand binding to receptors for vasoactive intestinal peptide (VIP) and *β*-adrenoreceptors. DAG and increased cytosolic Ca^{2+} also promote exocytosis. The primary fluid released into the acinar lumen is modified as it passes through the duct of the salivary gland. [Adapted from Edgar, 1992.]

secretion does not fall, but in fact rises, with increased flow. The addition of bicarbonate to the fluid must somehow be coupled to the flow rate, such that an increase in flow promotes bicarbonate addition to the duct fluid.

Invertebrate Silk Glands

The number and variety of glands in invertebrates is probably larger than that in vertebrates. The silk gland is described here not so much because it is representative of a large number of invertebrate exocrine glands, but because so much is known about it. Many insects and spiders produce silk threads from silk glands to make webs and spin cocoons. The silkworm *(Bombyx mori)* is raised commercially for its larvae, which spin a protective cocoon. Each cocoon produced by a pupating larva consists of about 275 m of silk thread. Commercial silk thread is made by spinning together the threads from many cocoons.

Spider silk and webs

It has been estimated that an acre of meadow in England may contain over 2 million spiders, and there are 30,000 species of spiders worldwide. One contributor to the evolutionary success of this group is silk. Spider silk is made by *spinneret glands* on the underside of the abdomen, which exude a liquid that hardens into silk threads once it leaves the gland. The threads are used for draglines, webs, egg cases, and silk-lined tunnels. The main function of webs is to capture prey, such as insects and other small animals. The spider can detect the position and nature of an animal in the web by the pattern of vibration of the web. Different behavioral responses are evoked by different patterns of vibration. A male, wishing to be recognized as a potential mate rather than food, vibrates the web of a female in a species-specific pattern to evoke the appropriate response from her.

Spider webs undoubtedly evolved from the dragline, with the simplest webs being sticky threads hanging in the breeze to catch insects. More complex webs are constructed in either two or three dimensions and have intricate designs for snaring prey. In many webs, some of the threads break as the insect hurtles into the web; the more the insect struggles, the more entangled it becomes. Other webs have sticky regions that trap prey.

Obviously the characteristics of a web depend on the tensile properties and patterns of the threads. There are many web designs, and spider species are often named for the type of web they construct: ladder-web spiders, funnel spiders, filmy dome spiders, and mesh-web spiders, to name just a few (Figure 9-55).

Silk production by spiders

The abdominal silk glands of spiders are large and open via modified appendages—the spinnerets—each of which has several spigots. The spinnerets are shaped like conical gun towers and are very mobile. The threads, produced in the abdominal glands and extruded through the spigots, consist of alpha-keratin

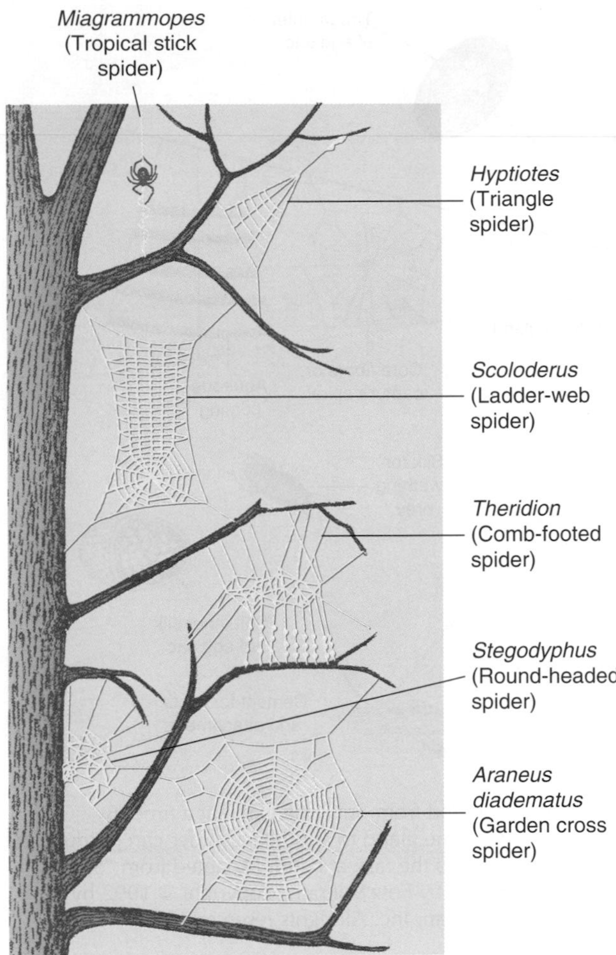

Figure 9-55 Different species of spiders build webs with a characteristic design. The common names of spiders are often based on the appearance of their webs. [Adapted from "Spider Webs and Silks," by Fritz Vollrath. Copyright © 1992 by Scientific American, Inc. All rights reserved.]

crystals embedded in a rubberlike matrix of amino acid chains, which are not cross-linked to the crystalline alpha-keratin structures (Figure 9-56). The silk threads are stiff and brittle when dry, but become pliable when wet. In many cases the extruded thread is dry and remains so because of an oily lipid covering. These dry threads are elastic and have great strength, but they can be extended to only about 25% of their length before breaking.

Dry threads in three-dimensional webs break when an insect collides with the web, thereby snaring the prey by tangling it in a network of threads. In two-dimensional webs, like that of the garden cross spider, dry threads form the spokes of the web, and wet threads form a sticky continuous spiral that defines the shape of the web. The spiral thread has glue droplets surrounding glycoprotein "doughnuts" at intervals along its length; these make the thread adhesive so that prey stick to the web. The cribellate spiders, on the other hand, produce an adhesive dry thread by covering it with a loose network of entangling amino acid chains, rather like Velcro.

Each spider has several different silk glands, each of which produces a unique silk characterized by the

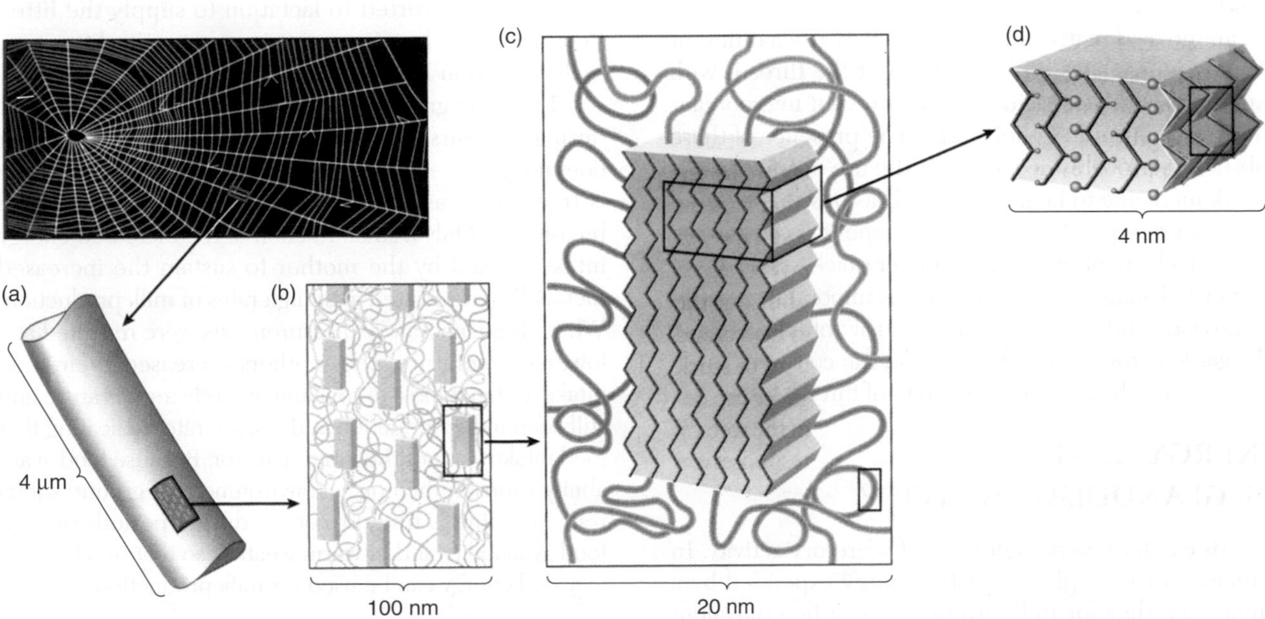

Figure 9-56 Spider silk thread **(a)** is a composed of alpha-keratin crystals embedded in a disordered matrix of amino acid chains **(b and c)**. Each alpha-keratin crystal is composed of several amino acid chains that are pressed into an accordion-like structure called a β pleated sheet **(d)**. The contracted disarray of the matrix provides silk with its elasticity. (Most of what is known about the molecular structure of silk comes from studies of silkworm silk. In this illustration, it is assumed that spider silk resembles that of the silkworm.) [Adapted from "Spider Webs and Silks," by Fritz Vollrath. Copyright © 1992 by Scientific American, Inc. All rights reserved.]

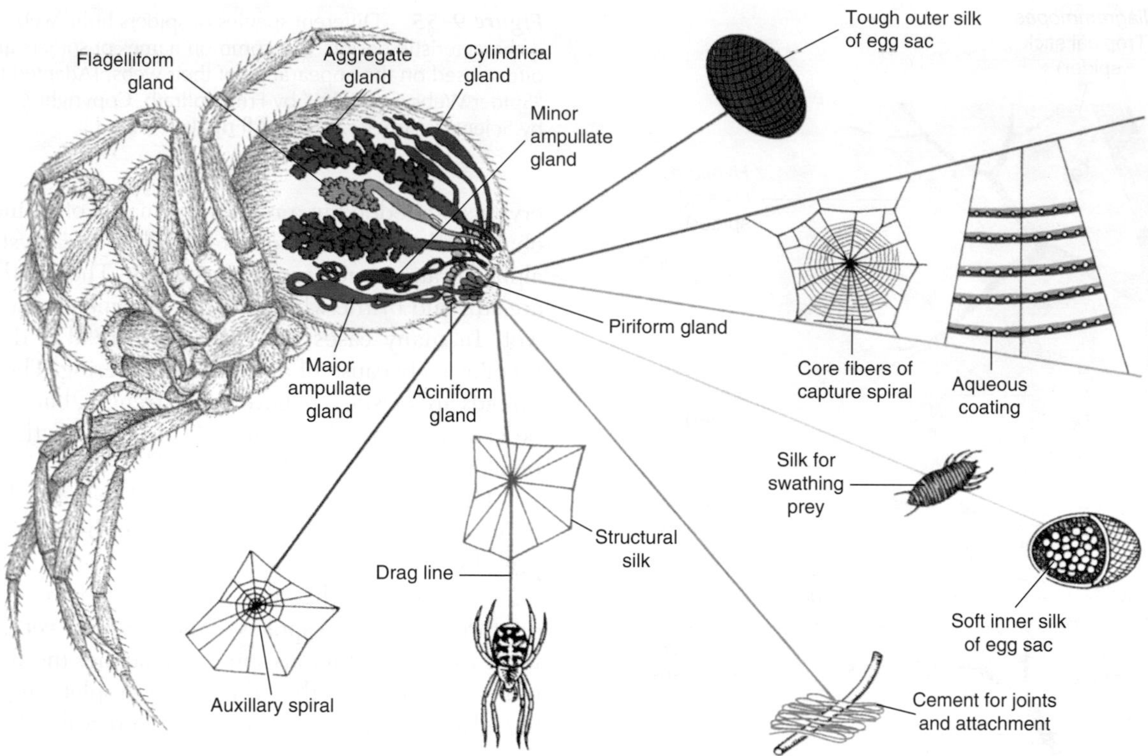

Figure 9-57 Different silks with different functions can be produced by the same spider. The garden cross spider *(A. diadematus)* has seven different abdominal silk glands, each of which produces silk with a characteristic amino acid matrix composition. The various glands open into common spigots, but silk is extruded from only one gland at a time. By switching from one gland to another, a spider can produce the silk appropriate to the task at hand. [Adapted from "Spider Webs and Silks," by Fritz Vollrath. Copyright © 1992 by Scientific American, Inc. All rights reserved.]

composition of its amino acid matrix. Spiders probably can adjust the valving mechanism at the exit of the gland to create thicker threads. If threads with a different amino acid matrix are required, they use a different gland (Figure 9-57). Spiders also coat the threads with fungicides and bactericides, which prevent microorganisms from consuming their webs. The presence of these substances probably accounts for the use of spider webs in folk medicine to heal cuts and abrasions to the skin.

Spiders expend a great deal of energy constructing webs, which are often damaged rather quickly. Spiders eat their own damaged webs, which are an important source of amino acids, and construct new webs daily, often overnight. The garden cross spider, for example, can construct a web in less than an hour using about 20 m of thread.

ENERGY COST OF GLANDULAR ACTIVITY

Glands can have very high rates of secretory activity. In humans, for example, the extra energy expended by a nursing mother for milk production can be equivalent to the energy expenditure of a long-distance runner. In litter-bearing animals, the energy cost of lactation is even greater. Breast milk is the sole source of nourishment for newborn mice until they are almost half the size of the mother mouse. For a litter of eight, the total weight of the babies at weaning is four times that of the mother. Thus the mother must eat enough food to supply all the nutrients for four times her body weight, 75% of which are diverted to lactation to supply the litter. Food intake by lactating mice increases with litter size, as illustrated in Figure 9-58.

Lactating ground squirrels double their food intake during the nursing period (Figure 9-59). The mother does not gain weight because most of the ingested energy is transferred as potential energy to the young in the breast milk. Only a small fraction of the increased energy intake is used by the mother to sustain the increased metabolism associated with high rates of milk production. When these laboratory measurements were repeated at a lower temperature, the mothers increased their food intake and metabolism to maintain body temperature and still keep milk production at the same rate, indicating that food intake was not the limiting factor. Because food availability varies in the normal environment of ground squirrels, breeding is timed to occur during periods of high food availability and in warm weather so that much of the ingested energy can be used for milk production.

Why do squirrels produce young only in the spring, whereas humans produce young throughout the year?

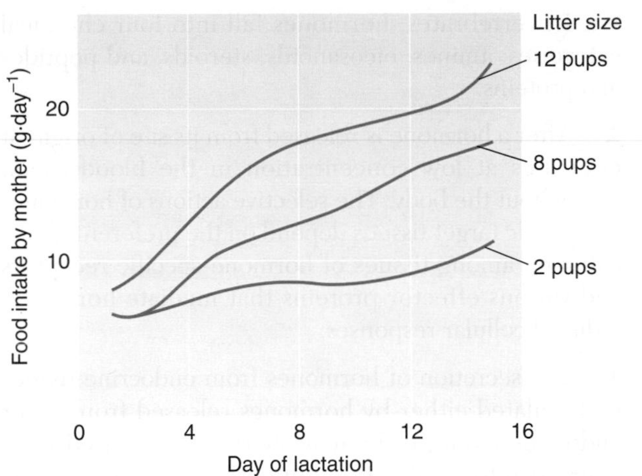

Figure 9-58 Food intake by lactating mice increases with litter size and the number of days of lactation since giving birth. Maternal food intake parallels the demands that the growing pups make on maternal milk production. Milk output reaches its peak on day 15, after which the pups begin to meet their food requirements by nibbling solid food. [Adapted from Diamond and Hammond, 1992.]

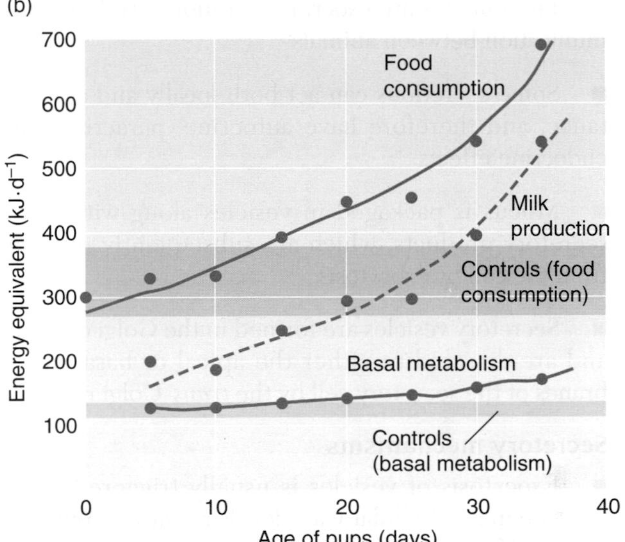

Figure 9-59 Most of the increased energy intake by lactating ground squirrels is stored as potential energy in the milk they produce and thus is transferred to their pups. Data are mean values for females producing an average litter of four pups. **(a)** Maternal body mass as a function of the age of the pups. Note that mothers did not gain weight despite their increased food consumption during the nursing period. **(b)** Comparison of maternal food consumption, milk production, and basal metabolism during the nursing period, expressed as daily energy equivalents in kilojoules (kJ/day). The food consumption and basal metabolism of nonlactating control squirrels are indicated by shaded bars. Note that milk production uses about 75% of the energy intake by mothers. [Adapted from Kenagy, Stevenson, and Masman, 1989.]

Abbreviations used in this chapter

ACTH	Adrenocorticotropic hormone	G_s	Stimulatory G protein	MSH	Melanocyte-stimulating hormone
ADH	Antidiuretic hormone	G_i	Inhibitory G protein	NPY	Neuropeptide Y
ANP	Atrial natriuretic peptide	G_q	Pertussis toxin–insensitive G-protein	PIP_2	Phosphatidyl inositol bisphosphate
AMP	Adenosine 5′-phosphate	GH	Growth hormone	PLC	Phospholipase C
ATP	Adenosine triphosphate	GIH	GH-inhibiting hormone	PTH	Parathyroid hormone
cAMP	Adenosine 3′,5′-cyclic monophosphate	GIP	Gastric inhibitory peptide	PTTH	Prothoracicotropic hormone
cGMP	Guanosine 3′,5′-cyclic monophosphate	GnRH	Gonadotropin-releasing hormone	R_s	Stimulatory receptor
cNMP	Cyclic nucleotide mono-phosphate	GRH	GH-releasing hormone	R_i	Inhibitory receptor
CG	Chorionic gonadotropin	GTP	Guanosine triphosphate	RIA	Radioimmunoassay
CRH	Corticotropin-releasing hormone	5-HT	5-Hydroxytryptamine, serotonin	RTK	Receptor tyrosine kinase
DAG	Diacylglycerol	IP_3	Inositol trisphosphate	TGN	Trans-Golgi network
ER	Endoplasmic reticulum	IP	Inositol phospholipid	TnC	Troponin C
FSH	Follicle-stimulating hormone	JH	Juvenile hormone	TRH	TSH-releasing hormone
		LH	Luteinizing hormone	TSH	Thyroid-stimulating hormone

SUMMARY

■ Glands are organs composed of specialized cells that produce secretions. They vary in function not only among species, but also during different stages of development.

Cellular secretions

■ Many kinds of cellular secretions function in communication between cells. Such secretions are classified into four types based on the distance at which they exert an effect. Autocrine secretions affect the secreting cell itself; paracrine secretions affect neighboring cells. Endocrine secretions are released into the bloodstream and act on distant target tissues. Exocrine secretions are released to the outside of the body.

■ Pheromones are exocrine secretions used for communication between animals.

■ Some secretions can act both locally and at a distance, and therefore have autocrine, paracrine, and endocrine effects.

■ Mucus is packaged in vesicles along with other secretory products, which are subsequently released from the cell by exocytosis.

■ Secretory vesicles are formed in the Golgi complex and are directed to either the apical or basal membranes of the secretory cell by the *trans*-Golgi network.

Secretory mechanisms

■ Exocytosis of vesicles is usually triggered by an increase in intracellular Ca^{2+} levels resulting from neuronal or hormonal stimulation of the secreting cell.

■ The nature and extent of secretions and the type of stimulus that causes exocytosis vary greatly among glands.

■ Glands can be characterized as either endocrine or exocrine glands. Exocrine glands are easier to recognize than endocrine glands because they all possess ducts and secrete material onto the body surface.

■ The contents of secretory vesicles are often released into the acinus of an exocrine gland, where they combine with a primary fluid formed from water and actively transported ions.

■ Because endocrine glands lack any characteristic morphologic markers, a variety of techniques have been used to identify them. The development of RIA and recombinant DNA techniques has led to rapid developments in endocrinology over the past several decades.

Endocrine glands

■ Hormones are blood-borne messenger molecules that are released from endocrine secretory tissues.

■ In vertebrates, hormones fall into four chemical categories: amines, eicosanoids, steroids, and peptides and proteins.

■ After a hormone is released from its site of origin, it circulates at low concentration in the bloodstream throughout the body. The selective actions of hormones on specific target tissues depend on the preferential distribution among tissues of hormone-specific receptors and various effector proteins that mediate hormone-induced cellular responses.

■ The secretion of hormones from endocrine tissues is stimulated either by hormones released from other endocrine tissues or by neurohormones released from neurosecretory cells. In addition, some endocrine tissues respond directly to conditions in the extracellular environment.

■ The secretory activities of most endocrine tissues are modulated by negative feedback. In these cases, the increasing concentration of the hormone itself, or a response to the hormone by the target tissue, has an inhibitory effect on the synthesis or release of the hormone. Positive feedback occurs in some systems; occasionally hormone secretion is feedforward.

Neuroendocrine systems

■ The secretions of some hormones are regulated by neurohormones produced by secretory neurons. The axons of these neurosecretory cells terminate in blood capillaries, forming a neurohemal organ.

■ Hypothalamic neurosecretory cells regulate the production of hormones in the anterior pituitary. These hypothalamic neurosecretions can be divided into releasing hormones and release-inhibiting hormones, all but one of which are peptides.

■ The mammalian posterior pituitary stores and releases two hypothalamic neurohormones, namely antidiuretic hormone and oxytocin.

Cellular mechanisms of hormone action

■ To exert their effects, all hormones must bind to specific receptors. Hormone binding initiates intracellular mechanisms leading to a cellular response.

■ Steroid and thyroid hormones are lipid-soluble. They enter cells freely and bind to receptor proteins in the nucleus or cytoplasm, where the resulting hormone-receptor complexes bind to regulatory elements in the DNA, thereby stimulating (and in a few cases inhibiting) transcription of specific genes.

■ All other hormones bind to receptors located in the plasma membrane of target cells. This binding triggers one or more intracellular signaling pathways leading to the cell's responses.

■ In the cAMP signaling pathway, hormone binding to a receptor activates a G protein, which then stimulates adenylate cyclase to convert ATP into the second messenger cAMP. In a similar pathway, binding of some hormones stimulates guanylate cyclase to produce the related second messenger cGMP. Once formed, the cyclic nucleotide activates a specific protein kinase, which then phosphorylates various effector proteins that mediate cellular responses.

- In the inositol phospholipid signaling pathway, hormone binding to a G protein–linked receptor activates phosphoinositide-specific phospholipase C, which then hydrolyzes PIP_2 into two major second messengers, IP_3 and DAG. IP_3 induces the release of Ca^{2+} from intracellular stores. In addition, IP_3 is converted to IP_4, which promotes entry of Ca^{2+} from the cell exterior into the cell. The resulting increase in free cytosolic Ca^{2+} regulates the activity of a variety of cellular proteins. DAG, on the other hand, remains in the plasma membrane and activates membrane-bound protein kinase C. This kinase then phosphorylates various effector proteins that mediate cellular responses.

- In the Ca^{2+} signaling system, hormone stimulation of the receptor directly activates Ca^{2+} channels in the plasma membrane, thereby stimulating Ca^{2+} influx. Hormone-induced changes in intracellular Ca^{2+} levels regulate diverse cellular processes.

- In membrane-associated enzyme signaling systems, hormone binding activates intrinsic enzyme activity in the cytosolic domain of the receptor. The activated enzymes, in turn, induce intracellular responses.

- Even in the same cell, a single hormone may bind to different cell-surface receptors linked to different second messengers, thereby inducing the same cellular response (convergent pathways) or different responses (divergent pathways). A particular receptor may be coupled to two different G proteins, each linked to its own second-messenger pathway or both linked to the same second-messenger pathway. Other variations in signaling systems, some involving third messengers, are also possible.

Physiological effects of hormones

- Although most hormones have multiple actions, they can usefully be grouped into several functional classes, with the very diverse eicosanoids constituting an additional class.

- Hormones with major roles in regulating metabolism and developmental processes include catecholamines and glucocorticoids, which are produced in the adrenal glands and affect energy metabolism; thyroid hormones, which regulate metabolic rate; insulin and glucagon, which are produced in the pancreas and have opposite effects on blood glucose levels; and growth hormone, which is produced in the anterior pituitary and works synergistically with the thyroid hormones to promote growth and development.

- The hormones most responsible for regulating water and electrolyte balance are antidiuretic hormone (ADH), which increases water reabsorption in the kidney; mineralocorticoids, which promote Na^+ reabsorption in the kidney; atrial natriuretic peptide (ANP), which reduces Na^+ and water reabsorption in the kidney; parathyroid hormone and calcitriol (derived from vitamin D or cholesterol), which act to increase the plasma Ca^{2+} concentration; and calcitonin, which has the opposite action, decreasing the blood Ca^{2+} concentration.

- The reproductive hormones include the androgens (in males) and estrogens (in females), which promote development of sex characters and gametes (sperm or oocytes). In females, progesterone acts to prepare the endometrium for the implantation of a fertilized ovum and helps prepare breast tissue for lactation; oxytocin stimulates uterine contractions during birth and milk ejection after birth; and prolactin promotes formation of milk and maternal behavior.

Hormonal action in invertebrates

- Five major hormones are known to control development in insects.

- Juvenile hormone, which is synthesized and released from the corpora allata, inhibits metamorphosis and promotes growth of larval structures. Ecdysone, produced by the prothoracic glands, promotes the secretion of new cuticle.

- Three neurohormones—prothoracicotropic hormone, eclosion hormone, and bursicon—also influence insect development.

Exocrine glands

- The products of exocrine glands flow through ducts to the body surface. That surface may be enclosed, as in the case of the mouth or gut.

- Salivary glands are exocrine glands that secrete saliva into the mouth. Saliva is about 99.5% water and contains a number of ions. Saliva serves as a lubricant, assisting in eating, swallowing, and talking. It also has an antibacterial action, which helps reduce tooth decay.

- Spiders have abdominal exocrine glands that produce silk threads, which are often used to form webs. Silk threads are extruded from the glands through spigots and consist of alpha-keratin crystals embedded in a rubberlike matrix of amino acid chains. Different silk glands in a spider produce silk with different amino acid compositions.

Energy cost of glandular activity

- Glands can have very high rates of secretory activity and can constitute a large proportion of the energy expended by an animal.

REVIEW QUESTIONS

1. What role does mucus play in exocytosis? Explain the rapid expulsion of mucus from vesicles.
2. What is the role of the *trans*-Golgi network in determining cell polarity?
3. Discuss the differences between autocrine, paracrine, endocrine, and exocrine secretion. What are pheromones?
4. What criteria must be met before a tissue can be unequivocally identified as having an endocrine function?
5. What is the significance of having the adrenal medulla and cortex collected together in a single organ? How does the circulatory pattern in the adrenal glands affect the relative secretion of epinephrine and norepinephrine?
6. Explain how catecholamines can have so many different actions.
7. Give examples of short-loop and long-loop negative feedback in the control of hormone secretion.
8. Discuss examples that show the intimate functional association of the nervous and endocrine systems.
9. Explain how differential activation of sympathetic and parasympathetic nerves can influence the composition of saliva.
10. How does a spider alter the composition of silk produced for making a web and other structures?
11. Describe the differences between dry and wet silk threads.
12. How can a single second messenger (e.g., cAMP or IP_3), induced by binding of different hormones, mediate different cellular responses in different tissues?
13. Explain how a small number of hormone molecules can elicit cellular responses involving millions of times as many molecules.
14. What is the significance of protein phosphorylation in intracellular signaling systems?
15. How can working muscle mobilize glycogen stores in the absence of epinephrine-induced glycogenolysis?
16. Describe two ways in which the concentration of free cytosolic Ca^{2+} becomes elevated. Discuss the role of Ca^{2+} as a second and third messenger.
17. Describe the similarities and differences that characterize four intracellular signaling systems described in this chapter.
18. Describe the interactions between cAMP and the inositol phospholipid pathway.
19. Describe the interrelations between Ca^{2+} and cAMP in the mammalian salivary gland and liver to illustrate convergent and divergent second-messenger pathways activated in response to a single first messenger such as epinephrine.
20. How do the glucocorticoids (as well as growth hormone and glucagon) combat hypoglycemia?
21. How does insulin produce its hypoglycemic effects?
22. What factors influence the secretion of growth hormone?
23. Discuss the endocrine control of the menstrual cycle.
24. Explain how birth control pills prevent conception.
25. Discuss the role of juvenile hormone in the development and metamorphosis of an insect.
26. How much of the total energy budget of an animal is spent on secretion? How could this percentage be estimated?
27. Discuss from an energetics point of view why many mammals give birth in the spring.

SUGGESTED READINGS

Edgar, W. M. 1992. Saliva: Its secretions, compositions and functions. *Br. Entomol. J.* 172:305–312. (A concise description of the functional organization of the mammalian salivary gland.)

Hadley, M. E. 2000. *Endocrinology*. 5th ed. Upper Saddle River, N.J.: Prentice Hall. (A useful general text on endocrinology.)

Lodish, H., A. Berk, et al. 2000. *Molecular Cell Biology*. 4th ed. New York: W. H. Freeman.

Matsumoto, A., and S. Ischii, eds. 1992. *Atlas of Endocrine Organs*. Heidelberg: Springer-Verlag. (Beautiful diagrams of the structure of vertebrate endocrine organs.)

Pimplakar, S. W., and K. Simons. 1993. Role of heterotrimeric G proteins in polarized membrane transport. *J. Cell Sci.* 17(Suppl):27–32. (Plenty of information, but not for the beginner.)

Raymond, J. R. 1995. Multiple mechanisms of receptor-G protein signaling specificity. *Am. J. Physiol.* 269 (*Renal Fluid Electrolyte Physiol.* 38):F141–F158.

Robb, S., T. R. Cheek, et al. 1994. Agonist-specific coupling of a cloned *Drosophila* octopamine/tyramine receptor to multiple second messenger systems. *EMBO J.* 13:1325–1330.

Truman, J. W. 1992. The eclosion hormone system of insects. *Prog. Brain. Res.* 92:361–374.

Verdugo, P. 1994. Molecular biophysics of mucin secretion. In T. Takishima and S. Shimura, eds., *Airway Secretion: Physiological Basis for the Control of Mucous Hypersecretion*, 101–121. New York: Marcel Dekker. (A good review of the subject.)

Vollrath, F. 1992. Spider webs and silks. *Sci. Am.* 266(3):70–76. (All you ever wanted to know about spider webs.)

Muscles and Animal Movement

All animal behavior—locomotion, prey capture, eating, copulation, the production of sound—is generated by three fundamentally different mechanisms: amoeboid movement, ciliary and flagellar bending, and muscle contraction. In this chapter, we focus on muscle contraction because it is responsible for most observable animal behavior and much visceral function as well. The importance of muscle contraction has been recognized from ancient times. In the second century A.D., Galen hypothesized that "animal spirits" flow from nerves into muscles, inflating them and increasing their diameter at the expense of their length. As recently as the 1950s, it was suggested that muscles contract because linear molecules of "contractile proteins" within the muscles become shorter. These molecules were thought to be helical in shape; changes in the pitch of the helix were thought to change the length of the molecules. This hypothesis was based on what was then the recent discovery of helical proteins, but it was quickly supplanted. Through evidence from electron microscopy, biochemistry, and biophysics, we have discovered in minutest detail the structure of the contractile mechanism in muscles and how it produces force and shortening. It has also become clear how contraction is initiated by electrical activity in the membrane of muscle fibers. The mechanisms of muscle contraction and relaxation are major topics in this chapter.

Muscles are classified, on both morphologic and functional grounds, into two major types, **striated muscle** and **smooth muscle.** Vertebrate striated muscle is the best understood, and in this chapter we will consider it in detail. Striated muscle can be subdivided into **skeletal muscle,** the type that underlies movement of the body, and **cardiac muscle,** the type found in the heart. Smooth muscle is found primarily in the walls of hollow organs such as the intestines and the blood vessels. Each of the three main muscle types has unique morphologic and physiological properties, but the mechanism by which all muscles contract is nearly identical. The major differences between the three types are found in how their cells are organized and how contraction is initiated. Recent research on comparative aspects of muscle function has uncovered unexpected diversity among skeletal muscles and reveals an elegant match between evolutionary "design" and function. Several examples of such varied "designs" are discussed in this chapter.

ESSENTIALS OF SKELETAL MUSCLE CONTRACTION

The general organization of skeletal muscle tissue is depicted in Figure 10-1 on the next page. Typically, a muscle is anchored at each end by a tough strap of connective tissue called a *tendon.* Each muscle consists of long, cylindrical, multinucleate cells called **muscle fibers,** which are arranged in parallel with one another. This arrangement allows all of the fibers in a muscle to pull together in the same direction. Striated muscle fibers range from 5 to 100 μm in diameter and may be many centimeters long. This extraordinary size reflects the unusual developmental pathway of muscle. Each fiber arises from many individual embryonic muscle cells, called *myoblasts,* which fuse during embryonic development to form a *myotube.* Each myotube contains many nuclei within a single plasma membrane and differentiates into an adult muscle fiber (sometimes called a *myofiber*). Each muscle fiber, in turn, is composed of numerous parallel subunits called **myofibrils,** which consist of longitudinally repeated units called **sarcomeres.** (Notice that the prefixes *myo-* and *sarco-* both signal that a word is related to muscle; for example, the cytoplasm of a muscle fiber is interchangeably called *myoplasm* or *sarcoplasm.*) The sarcomere is the functional unit of striated muscle. The myofibrils of a striated muscle fiber are lined up with the sarcomeres in register, so the fiber looks striated, or striped, when it is observed with a light microscope.

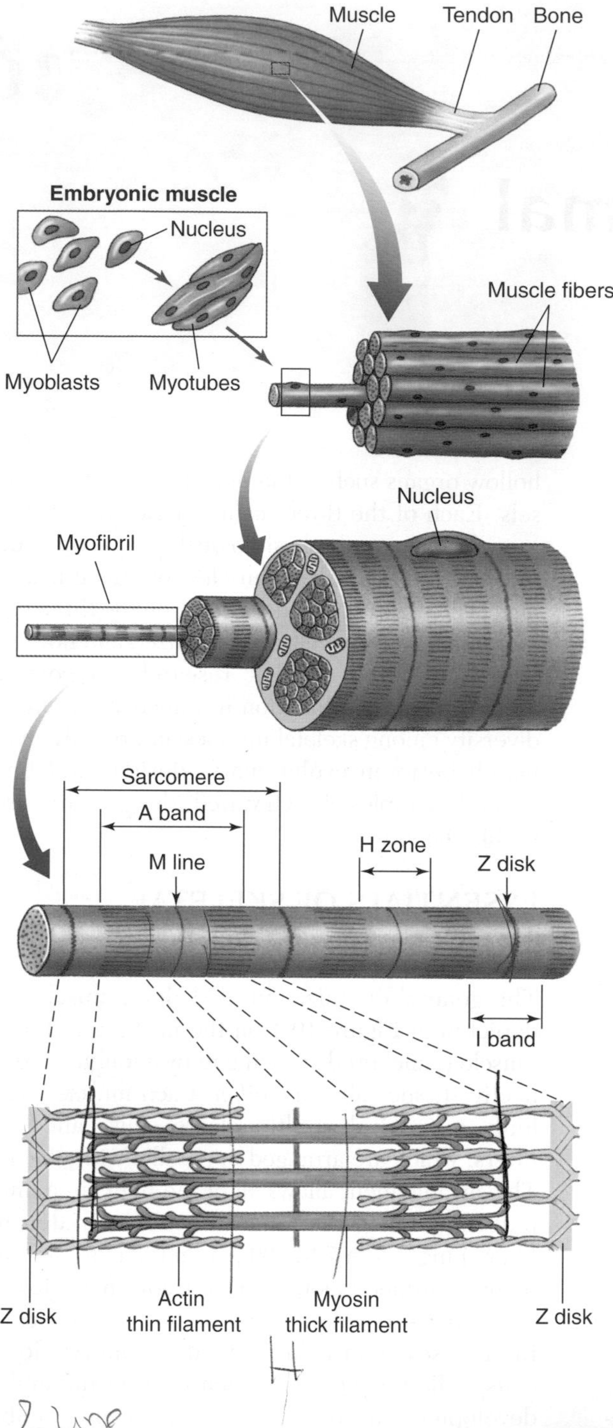

Figure 10-1 Vertebrate skeletal muscles are organized in a stereotyped hierarchy. Each muscle is made up of parallel multinucleate muscle fibers, each of which contains many myofibrils. Muscles are attached to bones or other anchor points by bands of tough connective tissue called tendons. Each muscle fiber is derived embryonically from a group of myoblasts that fuse to form a myotube. After fusion, each myotube synthesizes the proteins characteristic of muscle fibers and differentiates into its adult form. Each myofibril is made up of sarcomeres arranged end-to-end. Each sarcomere contains two kinds of protein filaments: thin filaments consisting primarily of actin, and thick filaments consisting primarily of myosin. The filaments interdigitate in a precise geometric relationship (see Figure 10-3). Thin filaments are anchored in regions called Z disks. [Adapted from Lodish et al., 1995.]

Each sarcomere contains two kinds of long, thin proteins called **myofilaments arranged** in a precise geometric pattern (Figure 10-2). Each sarcomere is bounded at either end by a **Z disk** (or *Z line*), which contains *α-actinin,* one of the proteins found in many motile cells. Extending in both directions from the Z disk are numerous **thin filaments** consisting largely of the protein **actin.** The α-actinin in the Z disk binds actin and anchors the thin filaments firmly to the disk. The thin filaments interdigitate with **thick filaments** made up primarily of the protein **myosin.** Interdigitated thick and thin filaments make up the densest portion of the sarcomere, the *A band.* The lighter portion in the center of the A band, called the *H zone,* contains only thick filaments. In the middle of the H zone is the *M line,* which has been shown to contain enzymes that are important in the energy metabolism of muscle fibers. The portion of the sarcomere between two A bands is called the *I band.*

If cross-sections are made through the various bands of a single sarcomere, the precisely arranged geometric relationship between thick and thin filaments is revealed (Figure 10-3). Only thin filaments are seen in a section through an I band, and only thick filaments are seen in a section through an H zone. At the ends of an A band in vertebrate skeletal muscle, where thick and thin

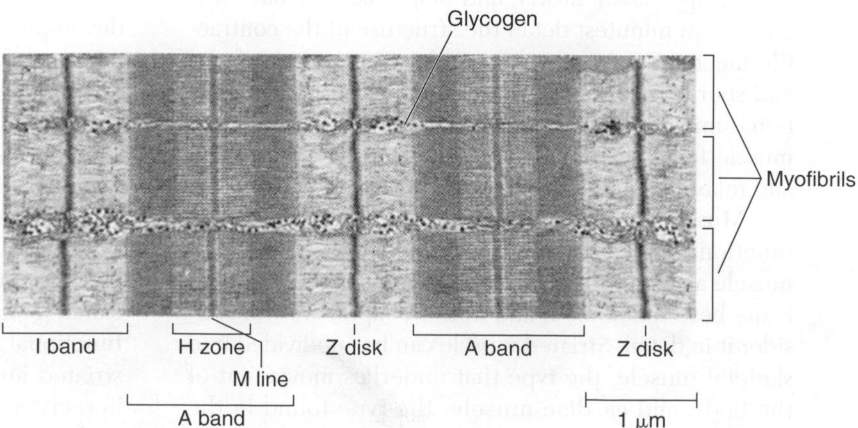

Figure 10-2 The sarcomeres within neighboring myofibrils are aligned in register, producing the characteristic banding of skeletal muscle. This electron micrograph of a longitudinal section through a frog skeletal muscle includes two complete sarcomeres of three myofibrils. The various bands and the Z disks (which appear as lines in longitudinal section, and thus have also been called Z lines) are labeled. Dark granules between fibrils are glycogen deposits. [Courtesy of L. D. Peachey.]

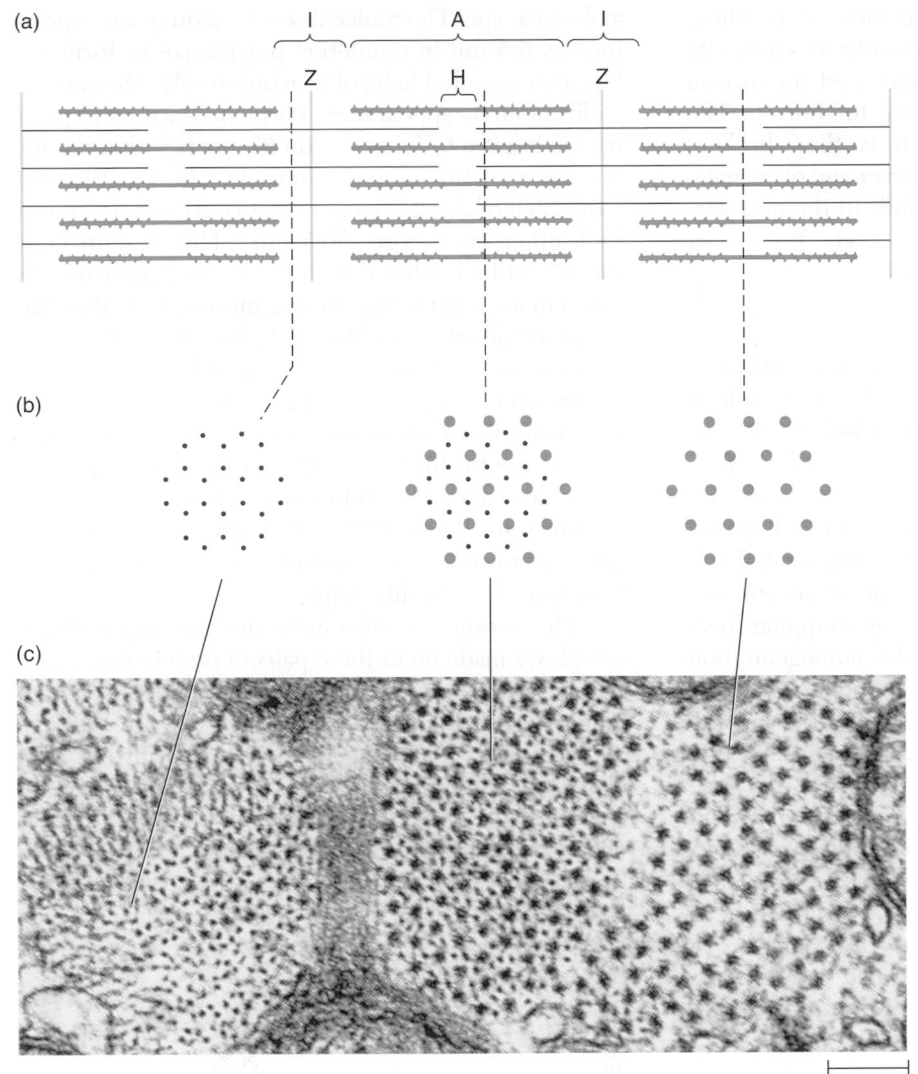

Figure 10-3 Within a myofibril, thin filaments extending from the Z disks overlap with thick filaments in a precise geometric pattern. **(a)** Diagram of three sarcomeres, showing thick and thin myofilaments forming I and A bands, H zones, and Z disks. **(b)** Diagram of the geometric relation between thick and thin filaments in cross-sections made at different locations in a sarcomere. **(c)** Electron micrograph of a cross-section through myofibrils of a spider monkey extraocular muscle, in which the sarcomeres of adjacent myofibrils are out of register so they can be matched with the profiles shown in part b. [Part c courtesy of L. D. Peachey.]

filaments overlap, each thick filament is surrounded by six thin filaments, and it shares these thin filaments with surrounding thick filaments. Each thin filament is surrounded by three thick filaments.

When a section through a sarcomere is examined at high magnification with an electron microscope, small structures, called **cross-bridges**, are visible. These projections extend outward from the thick myosin filaments and contact actin in the thin filaments during muscle contraction (Figure 10-4). Muscles contract when the cross-bridges on myosin molecules bind transiently to sites on actin molecules, causing the myosin

Figure 10-4 Cross-bridges extend from myosin thick filaments toward actin thin filaments. The muscle fiber shown in this figure was flash-frozen in liquid helium, etched to reveal proteins in the tissue, and stained with a heavy metal. Many cross-bridges are seen as small, horizontal projections extending from the thick filaments. [Courtesy of Professor John Heuser, M. D.]

molecules to change their physical conformation. These changes in the shape of the myosin molecules generate force, which drags the actin filaments past the myosin filaments and causes the sarcomere to shorten. The bond between actin and myosin is then broken. The cross-bridges bind and unbind over and over again, generating force each time they bind. In the next section we will consider in detail this cyclic process of cross-bridge binding and unbinding.

Myofilament Substructure

In the mid-nineteenth century, Wilhelm Kühne showed that multiple factors can be extracted from a skeletal muscle when it is minced and then soaked in solutions containing different concentrations of salts. Nonstructural soluble proteins, such as myoglobin, are extracted by distilled water, whereas actin and myosin filaments are solubilized by highly concentrated salt solutions. Fragments of myofibrils that are several sarcomeres in length can be prepared by chopping fresh muscle in a laboratory blender. If this homogenization is carried out in a "relaxing solution" containing magnesium, ATP, and a calcium-chelating agent, the myofibrils fall apart into their constituent thick and thin filaments. This result suggests that ATP and Ca^{2+} are important in regulating chemical bonds between myosin and actin, which we will soon confirm.

An actin filament resembles two strings of beads twisted around each other into a two-stranded helix (Figure 10-5a). Each "bead" in the string is a monomeric molecule of G-actin, so called because of its globular shape. The molecules of G-actin (each approximately 5.5 nm in diameter) polymerize to form the long two-stranded helix of F-actin, so called because of its filamentous appearance. Purified G-actin will polymerize *in vitro* to form F-actin filaments with the same physical structure found in muscle. The F-actin helix has a pitch of about 73 nm, so its two strands cross over each other once every 36.5 nm. (This F-actin helix should not be confused with the much tighter α helix found in many proteins.) In frog muscle, actin thin filaments are about 1 μm long and about 8 nm thick, and they are joined at one end to α-actinin in the Z disk. Positioned in the grooves of the actin helix are filamentous molecules of the protein **tropomyosin.** At intervals along the actin filament, a cluster of globular protein molecules called the **troponin complex** is attached to tropomyosin (Figure 10-5b). Troponin and tropomyosin play a major role in controlling muscle contraction, as we will see later in this chapter.

The myosin molecules in the thick myofilaments are complexes made up of three pairs of protein molecules. Two of the proteins in the complex are identical large molecules called *myosin heavy chains* (Figure 10-6). In addition, each complex contains a pair of myosin light chains called *essential light chains* and a pair of light chains called *regulatory light chains.* Each of the heavy chains is long and thin with a globular "head" region (Figure 10-6a). The long, slender portion of the complex is formed by α-helical regions of the heavy chains, which are twisted around each other to form the neck and tail of the complex.

(a)

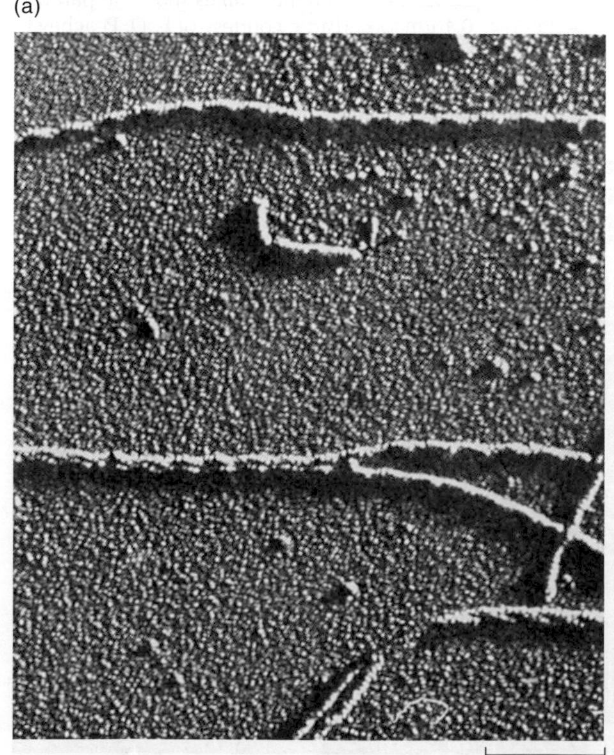

(b)

Troponin complex

Actin

Tropomyosin

20 nm

Figure 10-5 Actin myofilaments are made up of globular actin monomers and other associated proteins. **(a)** Electron micrograph of isolated F-actin filaments. Note the two-stranded helical arrangement of the globular monomers. The specimen was prepared for microscopy by shadowing the actin filaments with a thin film of metal. **(b)** Diagram showing G-actin monomers in the two-stranded helix of F-actin. Intact thin myofilaments contain two other proteins, tropomyosin and troponin; the latter is a complex of three subunits. This structure has been deduced from electron micrographs such as the one in part a and from X-ray diffraction studies. [Part a courtesy of R. B. Rice; part b adapted from Gordon et al., 2000.]

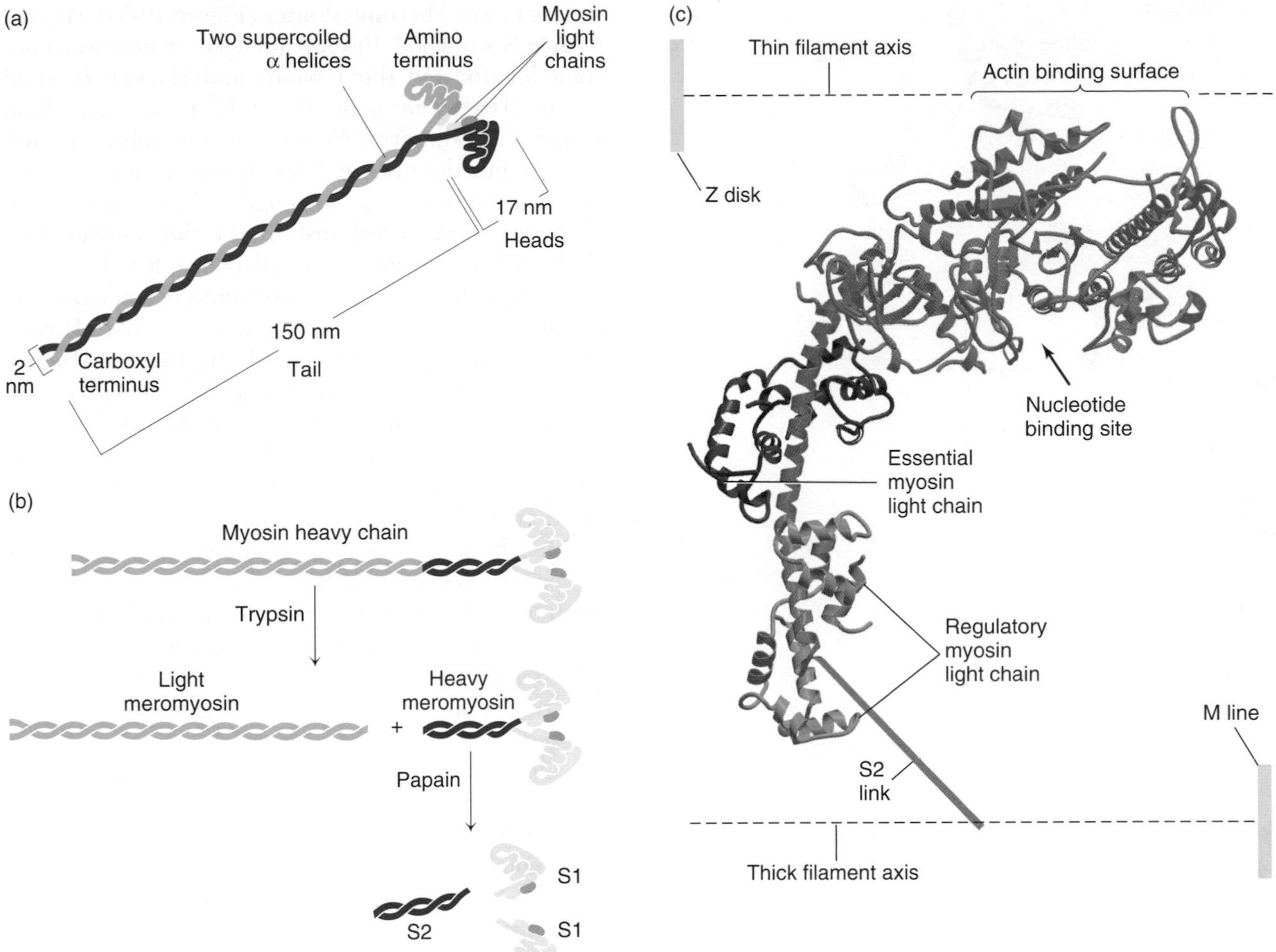

Figure 10-6 In thick myofilaments, myosin is organized into complexes consisting of six protein subunits. **(a)** Each complex has a long, thin tail made of two supercoiled α-helical segments of two myosin heavy chains. At one end are two heads, each belonging to a myosin heavy chain. In addition, two essential myosin light chains and two regulatory myosin light chains associate with the heads. **(b)** The proteolytic enzymes trypsin and papain reliably cleave the myosin complex at particular locations. Trypsin cleaves the complex into two pieces: light meromyosin, which includes most of the tail, and heavy meromyosin, which includes the head and the myosin light chains. Papain cleaves heavy meromyosin into three pieces, two called S1 and one called S2. **(c)** The three-dimensional structure of the S1 fragment has been determined, including the structure of the associated myosin light chains. In a sarcomere, the S1 fragment is connected by way of the S2 fragment to the tail, which is located in a thick myofilament. [From Nelson and Cox, 2000, with adaptations.]

When myosin complexes are treated with the proteolytic enzyme trypsin, they separate into two parts, called *light meromyosin* and *heavy meromyosin* (Figure 10-6b). Light meromyosin constitutes the major part of the tail region, and heavy meromyosin includes the globular head and the neck. Treating heavy meromyosin fragments with another proteolytic enzyme, papain, breaks them into two kinds of fragments, called S1 and S2. The three-dimensional structure of the S1 fragment has been determined, along with the structure of the myosin light chains that bind to it (Figure 10-6c). The light chains are calcium-binding proteins. Several molecular variants are known, which differ among muscle types and influence some functional properties of muscles, such as the speed of contraction.

Isolated myosin complexes in salt solution will aggregate spontaneously *in vitro* to form reconstituted thick filaments if the ionic strength of the solution is reduced. Initially, several myosin complexes aggregate with their tails overlapping and their heads pointing outward from the region of overlap and in opposite directions (Figure 10-7 on the next page). The result is a short filament in which the central region is devoid of heads. This bare zone in the middle of thick filaments has implications for muscle contraction, as we will soon see. As molecules of myosin are added to each end of the filament, all maintain this orientation, with their tails pointing toward the center of the filament and overlapping with the tails of previously added molecules. The head of each newly added myosin molecule projects laterally from the filament. Myosin molecules are added symmetrically to the two ends so that the heads on half the filament are oriented opposite to those of the other half. This reconstitution experiment

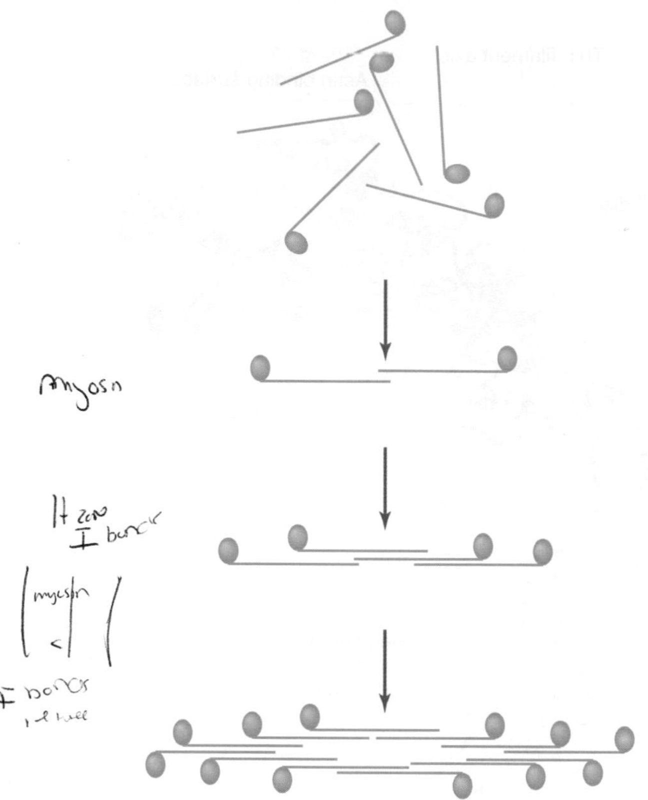

Figure 10-7 Myosin complexes will polymerize spontaneously *in vitro* to reconstitute thick filaments with an organization identical to that found in muscle. Individual myosin complexes join the nascent thick filament with their heads oriented toward either end of the filament and their tails parallel to its long axis. About equal numbers of myosin complexes make up the two ends of the filament at any given time, keeping the filament roughly symmetric in structure.

suggests that the structure of thick filaments depends directly on the physical and chemical properties of myosin molecules, and no other assembly instructions are required. Aggregation continues until, for vertebrate myosin, the filament is about 1.6 μm long and about 12 nm thick. It is as yet unclear why filaments stop growing at this particular length.

Contraction of Sarcomeres: The Sliding-Filament Theory

The striations of sarcomeres were first observed with the light microscope well over a century ago. Around the same time, it was also noted that sarcomeres change in length when a muscle contracts or is stretched, and that these changes in the sarcomeres correspond to the change in muscle length. In 1954, using a specially built light microscope that permitted accurate measurement of the sarcomeres, Andrew F. Huxley and R. Niedergerke confirmed earlier reports that the A bands, which correspond to the length of the thick filaments, maintain a constant length when a muscle shortens. In contrast, the I bands and the H zone (zones where actin and myosin filaments do not overlap in the resting muscle) become shorter (Figure 10-8a). When a muscle is stretched, the A band again maintains a constant length, and the I bands and H zone become longer. That same year, Hugh E. Huxley and Jean Hanson, working with electron micrographs, reported that neither the myosin thick filaments nor the actin thin filaments change in length when a sarcomere shortens or is stretched. Instead, it is the *extent of overlap* between actin and myosin filaments that changes.

Largely based on the observations described in the previous paragraph, H. E. Huxley and A. F. Huxley independently proposed the **sliding-filament theory** of muscle contraction. They suggested that sarcomeres shorten during muscle contraction as the thin filaments actively slide along the thick filaments. The thin filaments are pulled closer to the center of the sarcomere, and because they are firmly anchored in the Z disks, the sarcomeres become shorter (Spotlight 10-1 on pages 368–369). When a muscle relaxes or is stretched, the overlap between thin and thick filaments is reduced, and the sarcomeres elongate. We will consider how evidence was accumulated to support this theory as an example of elegant physiological methodology.

One of the strongest pieces of evidence supporting the sliding-filament theory is the **length-tension relation** for a sarcomere. A length-tension curve relates the amount of overlap between actin and myosin filaments to the tension developed by an active sarcomere under that condition. According to the sliding-filament theory, each myosin head provides a cross-bridge that interacts with an actin filament. Each cross-bridge generates force independently of all other cross-bridges and provides an increment of tension. Thus, the total tension produced by a sarcomere should be proportional to the total number of cross-bridges that can interact with actin filaments, and this number should in turn be proportional to the amount of overlap between thick and thin filaments. The sliding-filament theory thus predicts that no active tension (i.e., no tension beyond the amount due to the elasticity of the muscle fiber) will develop if a sarcomere is stretched so far that actin and myosin filaments no longer overlap.

To test the predicted relationships between filament overlap and tension generated, single frog muscle fibers were stimulated to contract at different fixed sarcomere lengths. First, the sarcomere length was adjusted with the aid of an electromechanical system. Changing the length of the sarcomeres changes the amount of overlap between the actin and myosin filaments (see Figure 10-8a). Then the fiber was electrically stimulated to contract, and the tension generated was measured and plotted as a function of sarcomere length. When the fiber was stretched so that there was no overlap at all between thick and thin filaments, stimulation produced no tension beyond what was required to stretch the fiber (Figure 10-8b). When the fiber was held so that the actin filaments overlapped completely with the parts of

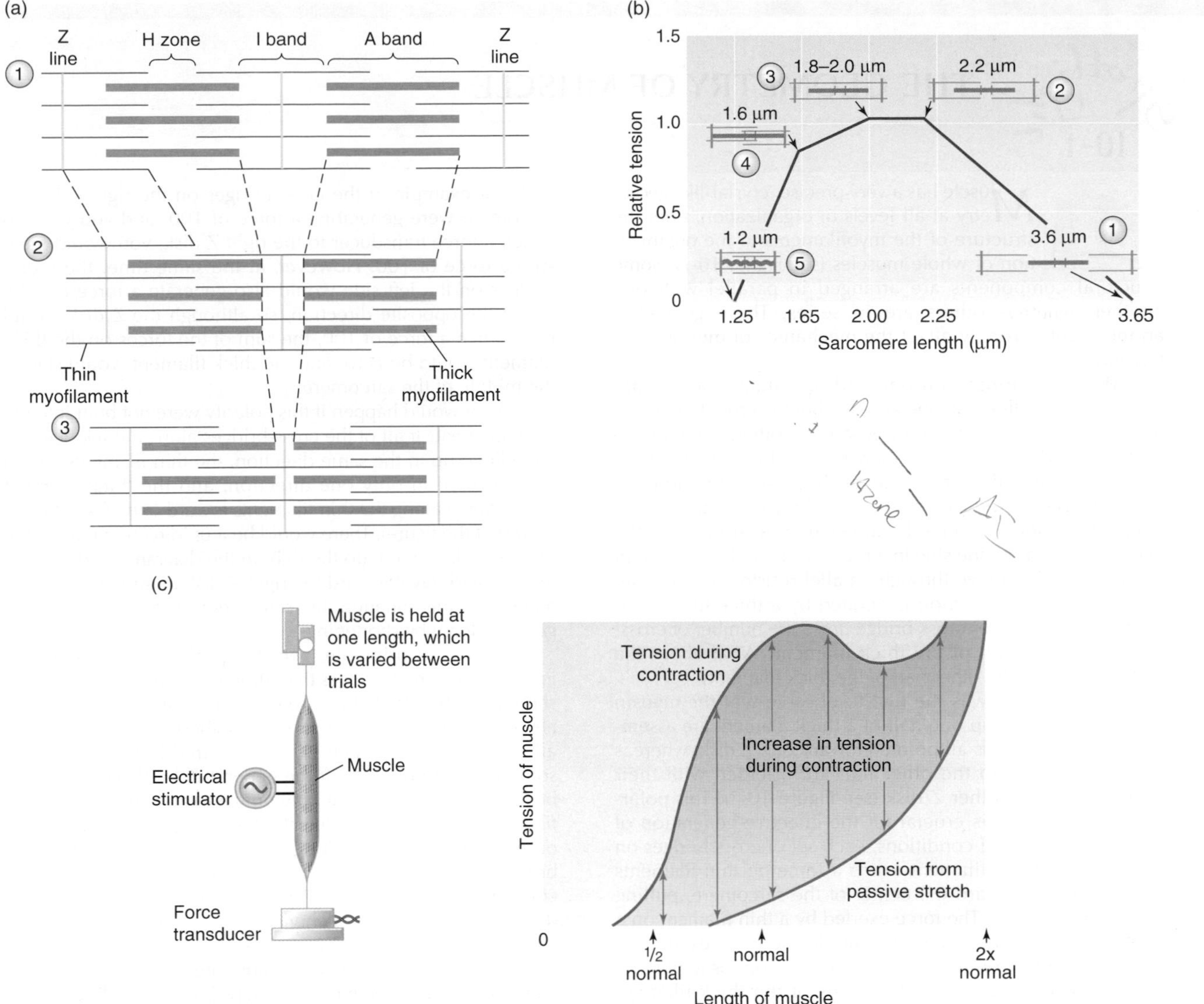

Figure 10-8 The sliding-filament theory states that sarcomeres shorten when their actin and myosin filaments move past each other. **(a)** Relations of the myofilaments when two sarcomeres shorten. Note that the lengths of the thick and thin filaments remain constant; only the amount by which they overlap changes.

Classic Work **(b)** Length-tension curve for a typical vertebrate sarcomere. The length and configuration of the sarcomere are depicted schematically near the curve at critical points. (Lengths exclude the thickness of the Z disks. Adding the width of one-half of a Z disk to each end of the sarcomere would increase each length by 0.05 μm.) The tension produced by the muscle is maximal when the overlap between thick and thin filaments allows the largest number of myosin cross-bridges to bind to actin. Tension drops off with increased length, because the thick and thin filaments overlap less and fewer cross-bridges can bind. It also drops off with decreased length, because thin filaments at the two ends of the sarcomere begin to collide with each other, preventing further shortening. Skeletal muscle rarely operates over such a broad range of sarcomere lengths because the structure of the skeleton and the joints limits the range of movement. Sarcomere length normally remains within the plateau region of this curve. **(c)** The length-tension relation for whole muscle. A muscle is removed from the body and placed in an apparatus that allows the length of the muscle to be set and the tension generated by contraction to be measured (left). The red curve on the right shows the reading of the force transducer produced by manipulation of the muscle's length by the experimenter. Due to the physical properties of the muscle, the experimenter must exert more and more force to stretch the muscle to longer and longer lengths. The blue curve shows the reading of the force transducer when the muscle is stimulated to contract at each preset length. The tension produced by the muscle (called the active tension) is the difference between the red and blue curves. Like the tension produced by a single sarcomere, the active tension rises as length increases and then, at even greater lengths, it decreases.

the myosin filaments that bear the cross-bridges, the tension generated was maximal. When the fiber was so short that the actin filaments in the two halves of the sarcomere collided, the tension decreased. Tension decreased still further if the fiber was so short that the myosin filaments crumpled up against the Z disks.

Before these experiments were actually carried out, some of the properties of the length-tension curve

Spotlight 10-1 THE GEOMETRY OF MUSCLE

Muscle has a very precise, crystal-like geometry at all levels of organization, from the structure of the myofilaments to the organization of whole muscles (see Figure 10-1). Some structural components are arranged in parallel with one another, whereas others are in series. These geometric arrangements strongly affect the mechanics of muscle contraction.

The cross-bridges on one end of a thick filament are arranged in parallel with one another, but the cross-bridges on the two ends of a filament oppose each other. Every cross-bridge extends from a thick filament to a thin filament independently of all other cross-bridges. Because of this arrangement, the forces produced by the cross-bridges along one thin filament are additive, like the forces produced by all of the people lined up on one side in a tug-of-war, or like the current components that move through parallel resistors in a circuit. The force in one direction generated by a thick filament is equal to the force per cross-bridge times the number of cross-bridges on one-half of the thick filament. What about the cross-bridges on the other half of the thick filament?

Hugh Huxley was the first to observe that the myosin molecules making up one-half of a thick filament are assembled with their heads all pointed toward one Z disk, whereas those that make up the other half are oriented with their heads toward the other Z disk (see Figure 10-7). This polarized configuration is crucial for the effective generation of force. Under normal conditions, each set of cross-bridges on one-half of a thick filament exerts a force on thin filaments that is directed toward the center of the sarcomere, pulling the Z disks together. The force exerted by a thin filament on a thick filament is equal and opposite to the force exerted by the thick filament on the thin filament. The opposite polarity of the cross-bridges at the two ends of the thick filament means that a thin filament on one end of a sarcomere exerts a force onto the thick filament that is on average just balanced by the force of a matching thin filament on the other end. Hence the net force exerted on a thick filament by the surrounding thin filaments is zero, and the thick filament stays in the center of the sarcomere (part a of the accompanying figure). For example, if the cross-bridges on the right side of a sarcomere were generating a force of 100, and you were to attach a force transducer to the right Z disk, you would measure a force of 100. However, at the same time, the cross-bridges on the left side would also generate a force of 100, but in the opposite direction, so although the Z disk would experience a force of 100, the sum of the forces on the thick filament would be zero, and the thick filament would stay in the middle of the sarcomere.

What would happen if this polarity were not built into the thick filament? If all of the cross-bridges along a thick filament were lined up in the same direction, the thin filaments would exert a force in only one direction, and the thick filament would travel along the thin filaments toward one of the Z disks (part b of the figure). There would be a unidirectional net force on the thick filament (to the right in the diagram), and the filament would travel toward the right Z disk in an unpredictable manner. In this situation, the sarcomere would not be able to generate force by shortening.

At least in principle, very long thick filaments should include more cross-bridges than short ones; more cross-bridges should thus be able to bind, so long thick filaments should create greater force. We can test this hypothesis with some unusually long myosin filaments that are found in the muscles of some invertebrates. The force generated by these fibers has been found to depend not only on the total length of the thick filaments, but also on the length of the bare zone in the middle of each filament. If the long filaments include more cross-bridges, they can generate a larger maximum force per sarcomere than in vertebrates. This observation confirms the prediction that muscles generate more force when more cross-bridges bind to actin.

Sarcomeres in a single myofibril are arranged in series—that is, end to end (or, more precisely, Z disk to Z disk), just as resistors can be placed in series in a circuit. When resistors are placed in series, the current through each resistor is the same as the current through every other resistor in the series. Similarly, the force generated by a series of sarcomeres is the same all along the chain of sarcomeres. Thus, although a myofibril might contain a huge number of cross-bridges, the

---

shown in Figure 10-8b could be predicted based on aspects of the sliding-filament theory. First, the theory states that the force generated by a sarcomere is proportional to the number of cross-bridges binding thick filaments to thin filaments. Second, it states that cross-bridges are evenly distributed along each thick filament, except in the central bare zone where no cross-bridges are present. (This second statement had been experimentally verified.) From these assumptions and the dimensions of the filaments, which are given in Figure 10-9a on page 370, it is possible to predict or explain the properties of the sarcomere length-tension curve. Let's consider some critical points on the length-tension curve illustrated in Figure 10-8b.

At what sarcomere length would we expect the filaments to be pulled beyond overlap and hence to generate no force? To determine the sarcomere length (excluding the Z disks) for a given amount of overlap between filaments of known lengths, imagine the path of a very tiny ant trying to crawl along the filaments from the face of one Z disk to the face of the next Z disk. In a sarcomere that has been stretched to the point where the filaments just fail to overlap, the ant would crawl along one thin filament (1.0 μm), step up to a thick filament and traverse it (1.6 μm), and step down to a thin filament on the other side of the sarcomere and traverse it (1.0 μm). The total distance traveled would be 3.6 μm (Figure 10-9b, condition 1).

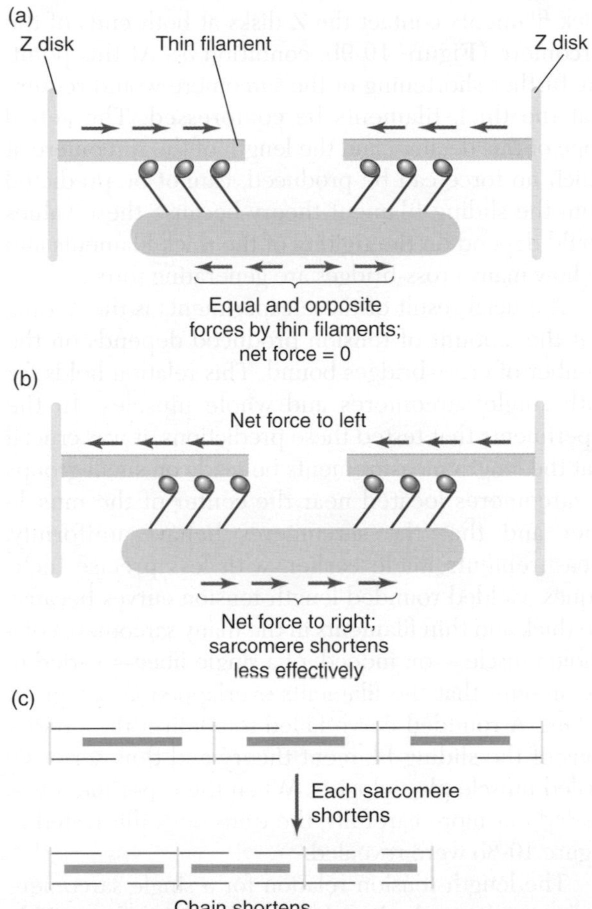

The geometry of myofilaments in a sarcomere strongly affects the contractile properties of muscle. ***(a)*** *Equal numbers of myosin cross-bridges act in opposition to each other, causing Z disks to move toward the center of the sarcomere while the thick filaments remain stationary.* ***(b)*** *If the cross-bridges all worked synergistically, the thick filament would move with respect to the thin filaments, interfering with the production of force.* ***(c)*** *The amount a myofibril shortens is equal to the sum of shortening in all of its sarcomeres.*

force generated by the entire chain of sarcomeres is the force generated by any one of the sarcomeres; that force in turn is determined by the number of cross-bridges working in parallel in one half of the sarcomere. However, because the sarcomeres are arranged in series, changes in length and contraction velocities are additive (part c of the figure). For instance, assume that there are 1000 sarcomeres in series along a myofibril, each 2 μm long. If each sarcomere shortens by 0.1 μm, the whole series shortens by 1000×0.1 μm = 100 μm. Similarly, if thin filaments move past a thick filament in each sarcomere at a rate of 10 $\mu m \cdot s^{-1}$, then the chain of sarcomeres will shorten at $2 \times 1000 \times 10$ $\mu m \cdot s^{-1}$, or 20 $mm \cdot s^{-1}$. (Notice that because each Z disk in a sarcomere is moving toward the center of the sarcomere at 10 $\mu m \cdot s^{-1}$, the overall shortening velocity of each sarcomere is twice that of each half-sarcomere.) The large amplification factor for length change implies that to obtain very rapid shortening, it is necessary to have as many sarcomeres in series as possible.

In a muscle fiber, the myofibrils are arranged in parallel, and in a muscle, the fibers are arranged in parallel. Each muscle fiber typically extends from one tendon to another and generates force between the tendons independently of surrounding muscle fibers. Because the muscle fibers are arranged in parallel, the force generated by each fiber adds to the force generated by the rest. One way to increase the force that can be generated by a muscle is simply to put more fibers together in parallel. This mechanism is used transiently when the nervous system recruits different numbers of muscle fibers to perform different activities.

The precision in the geometry of muscle makes it possible to calculate the force generated by a single cross-bridge if you know the amount of force generated by a whole muscle (or even by a whole animal). Consider a frog muscle with a 1 cm^2 cross-section that can generate 30 newtons of force. There are about 5×10^{10} thick filaments per square centimeter of muscle cross-section, and we know that thick filaments are arranged in parallel. So each thick filament must generate 6×10^{-10} newtons, or 600 piconewtons (pN) of force. There are about 150 cross-bridges at each end of a thick filament, so each cross-bridge must generate approximately 4 pN of force. Experimental measurements of force using microscopic techniques yield measurements of 1.5 to 3.5 pN of force per cross-bridge—very good agreement with the calculated number.

Why is there a plateau in maximal force between 2.0 and 2.2 μm? When the sarcomere is 2.2 μm long, the ends of the thin filaments line up with the beginning of the bare zone on the thick filaments (where there are no cross-bridges). Thus all the cross-bridges on the thick filaments are optimally aligned to interact with binding sites on the thin filaments (Figure 10-9b, condition 2). As the sarcomere shortens further, no more cross-bridges are added to the number that can interact with the thin filaments, so the force generated remains the same. The end of the plateau occurs when the thin filaments meet in the center of the sarcomere (Figure 10-9b, condition 3).

Why does the force fall as the sarcomere continues to shorten? The sliding-filament theory makes no quantitative predictions about muscle force beyond the point of maximal overlap, so it has been necessary to answer this question experimentally. From one perspective, the force might remain constant because all the cross-bridges still overlap with binding sites on the thin filaments and can, at least in principle, generate force. However, two effects could reduce the force generated. First, when the thin filaments overlap at the middle of the sarcomere, binding between myosin cross-bridges and thin filaments could be sterically hindered (Figure 10-9b, condition 4). Second, some cross-bridges might bind with an inappropriate thin filament (one projecting from the

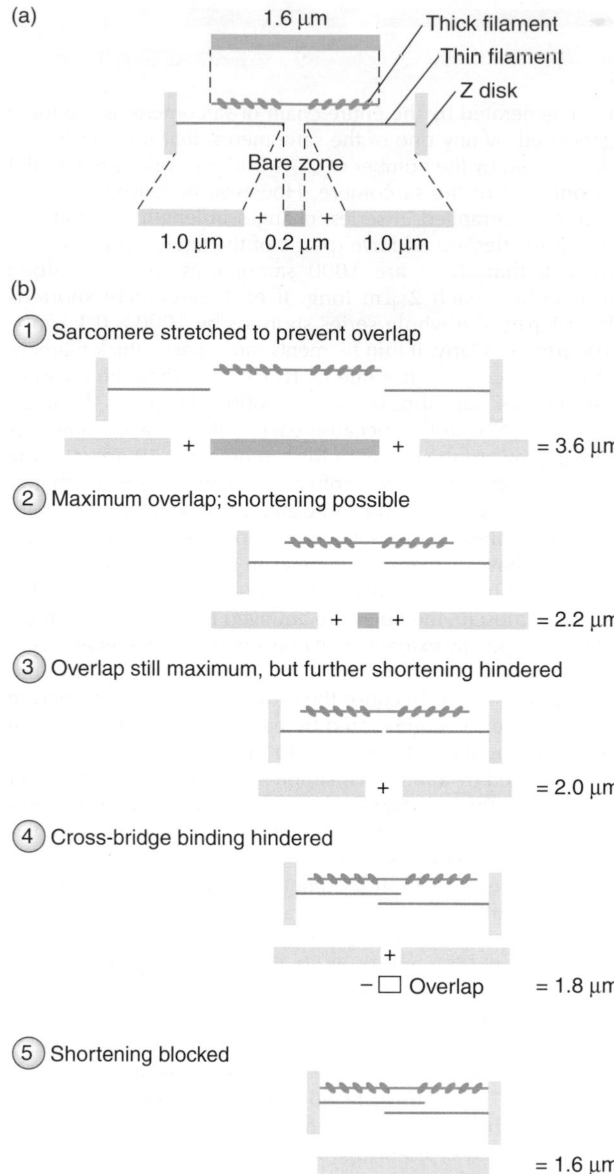

Classic Work *Figure 10-9* If the lengths of thick and thin filaments are known, the sarcomere length can be predicted for various amounts of overlap. Carefully controlled experiments have confirmed these predictions. The measured sarcomere lengths can be correlated with the known length-tension curve, providing experimental support for the sliding-filament theory. **(a)** Filament lengths measured from high-resolution electron micrographs of frog muscle fibers. The lengths of the individual components are represented by colored bars: red for thick filaments and blue for thin filaments. **(b)** Amounts of overlap between thick and thin fibers at different points in the sarcomere length-tension curve. Each condition in this figure is matched to a point on the curve in Figure 10-8b. The graphic equation accompanying each drawing calculates the length of a single sarcomere (excluding the thickness of the Z disks) for that condition of overlap. [Part b adapted from Gordon et al., 1966.]

Z disk at the other end of the same sarcomere) and thus exert a force that pushes the Z disks apart, rather than pulling them together. Such a force would be considered negative and would need to be subtracted from the force generated by the cross-bridges operating normally.

Why does force decline steeply at 1.6 μm and fall to zero at about 1.2 μm? The force generated declines steeply when the sarcomere is so short that the thick filaments contact the Z disks at both ends of the sarcomere (Figure 10-9b, condition 5). At this point, any further shortening of the sarcomere would require that the thick filaments be compressed. The actual slope of this decline, and the length of the sarcomere at which no force can be produced, cannot be predicted from the sliding-filament theory because these values would depend on the rigidity of the thick filaments and on how many cross-bridges are generating force.

A crucial result of these experiments is the finding that the amount of tension produced depends on the number of cross-bridges bound. This relation holds for both single sarcomeres and whole muscles. In the experiments that tested these predictions, it was crucial that the length measurements be made on small groups of sarcomeres located near the center of the muscle fiber and that the sarcomeres behave uniformly. Measurements made earlier, with less precise techniques, yielded rounded length-tension curves because the thick and thin filaments in the many sarcomeres of a whole muscle—or, indeed, of a single fiber—varied in the amount that the filaments overlapped at any given instant. A rounded curve failed to confirm the predictions of the sliding-filament theory and thus seriously misled muscle physiologists. When the experiment was carried out more carefully, the transitions illustrated in Figure 10-8b were revealed.

The length-tension relation for a single sarcomere is very similar to the length-tension relation for a whole muscle made up of millions of sarcomeres. In a typical length-tension experiment, a muscle is removed from the body and mounted in an apparatus that holds it at a constant length. The length of the muscle is varied by the experimenter. Once the length is set, the muscle is stimulated to contract, and the amount of tension produced is measured by a force transducer. Notice that this procedure is analogous to setting the overlap between thick and thin filaments in a single sarcomere. Figure 10-8c shows results obtained in this type of experiment. The red curve shows the reading of the force transducer produced entirely as a result of setting the length of the muscle; it is unrelated to the production of tension by the muscle. As long as the muscle is allowed to be short, the red curve has a value of zero. Once force is required to stretch the muscle to a given length, the force transducer reads out that force, but the value reflects only the activity of the experimenter to set the length of the muscle. When the muscle is stimulated to contract, the force transducer reports a higher value, shown by the blue curve. The amount of tension generated by the activity of the muscle at each length is represented by the difference between the red and blue curves at that length. Notice that subtracting the red curve from the blue produces a curve very like the one in Figure 10-8b, but without sharp transitions.

The striking similarity between the length-tension curves for whole muscle and for a single sarcomere suggests that this relation, which has been known for many decades for whole muscle, is based entirely on the behavior of the sarcomeres within the muscle.

The muscle of the heart is similar to striated skeletal muscle in many ways, including its length-tension curve. If a heart attack causes the heart to stop pumping, blood returning from the circulation can cause the volume of the heart to increase beyond normal limits, stretching its walls. What effect would you expect this to have on the strength of contraction, and why?

Cross-Bridges and the Production of Force

Elucidating precisely how myosin cross-bridges and actin filaments work together to produce force continues to be one of the great challenges facing researchers who study muscle function. According to current versions of the sliding-filament theory, the force driving muscle contraction arises when several different sites on the myosin head bind sequentially to sites on the actin filament, which produces relative motion between the actin filament and the myosin filament. The bond between the head and the actin filament is then broken, freeing the head for another cycle of sequential binding at a site farther along the actin filament. We will now consider these processes in detail.

Cross-bridge chemistry

Myosin cross-bridges must attach to binding sites on actin filaments in order to generate force, but the attachment must be reversible. If the cross-bridges never detached from actin, the filaments could slide no more than a micrometer past each other. In order to shorten the sarcomere and do work, the cross-bridges must attach to the actin filament, pull on it, and then detach in a cyclic fashion.

The first hints about the chemistry of the interaction between myosin cross-bridges and actin filaments came from studies begun several decades ago on crude and purified extracts of muscle. When actin and myosin are mixed together in the absence of ATP, they form a stable complex called **actomyosin.** If ATP is added to the solution, however, it causes rapid dissociation of the complex into actin and myosin-ATP:

$$\text{Actomyosin} + \text{ATP} \rightleftharpoons \text{Actin} + \text{Myosin-ATP}$$

The observation that ATP is required for the *dissociation* of actomyosin explains a phenomenon well known to readers of detective novels. Following death, the body of a human being or other animal gradually becomes stiff and will hold the same position for hours or even days. This condition, called *rigor mortis,* differs from muscle contraction, because in rigor mortis the muscles do not shorten. Instead, they simply remain locked at the same length. This rigidity occurs in part because when all of the ATP is used up after a cell dies, myosin binds irreversibly to actin, rigidly locking the muscles in place.

When ATP binds to isolated myosin, the myosin acts as an ATPase, rapidly hydrolyzing it to ADP and P_i, but these breakdown products unbind from the myosin only slowly. As a result, the rate of this reaction is very slow, and the rate-limiting step is the release of ADP and P_i from myosin. When myosin is bound to actin, however, the release of ADP and P_i proceeds much faster due to an allosteric change in the conformation of myosin. This actin-induced effect greatly increases the rate at which myosin can hydrolyze ATP:

$$\text{Myosin-ATP} \longrightarrow \text{Myosin-ADP-P}_i \xrightarrow{\text{very slow}} \text{Myosin} + \text{ADP} + P_i$$

$$\text{Myosin-ADP-P}_i + \text{Actin} \xrightarrow{\text{fast}} \text{Actomyosin} + \text{ADP} + P_i$$

Because energy is released following the binding of actin to the myosin-ADP-P_i complex, the formation of actomyosin is kinetically favored. When the actomyosin complex forms, P_i and ADP are released and replaced by ATP, which then breaks the actomyosin complex. These reactions proceed in a cycle of binding and unbinding between myosin and actin (Figure 10-10). The net effect of one turn of the cycle is to split one molecule of ATP into ADP + P_i, liberating energy, some of which can be captured to do work.

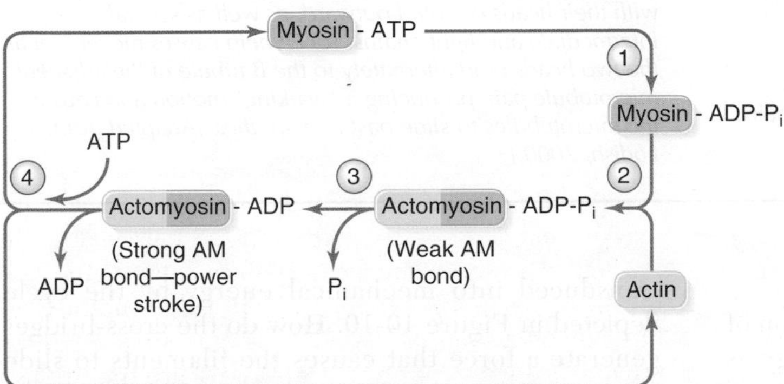

Figure 10-10 In the presence of ATP, myosin and actin filaments associate and dissociate cyclically. Myosin acts as an ATPase, hydrolyzing ATP (1), but the release of the products ADP and P_i is slow unless actin binds to the myosin (2), increasing the rate of release. Initially the actin-myosin bond is weak, but when P_i is released (3), it becomes stronger. The departure of P_i is accompanied by a release of energy, which can be used to generate force. ADP is then released and replaced by ATP (4), which breaks the actomyosin complex into actin and myosin-ATP. The cycle can then start again as long as myosin binding sites are available on actin.

Spotlight 10-2 MOTOR MOLECULES

Myosin is one of three "motor molecules" that drive movements in living organisms. Amazingly, these same three molecules drive movement at all levels of scale, from the movement of particles within a cell through the locomotion of very large animals. Myosin drives the locomotion of almost all multicellular animals, and it also participates in moving cells around, as well as in translocating organelles within a single cell. The other two motor molecules—kinesin and dynein—play important roles in the movement of organelles, and dynein also powers ciliary and flagellar movement.

The key feature of all three molecules is the ability to change chemical energy into mechanical work, a process called *chemomechanical transduction*. To accomplish this task, each molecule has ATPase activity, and each interacts with a filamentous protein to produce mechanical motion. Thus, each molecule must be able to bind ATP and catalyze the release of energy from a high-energy phosphate bond. In addition, each molecule must be able to bind to the appropriate filaments and generate mechanical force that ends in movement.

Initially, it appeared that although the three motor molecules had all these features in common, they were fundamentally unrelated to one another. However, recently acquired evidence suggests that myosin and kinesin are very distantly related molecules, whereas dynein still stands apart in evolution. The function of myosin is discussed at length in the text. Here we will briefly consider kinesin and dynein.

Dynein was initially discovered in the cilia and flagella of eukaryotes. In these organelles, it binds to microtubules that are longitudinally organized along the cilium or flagellum in a conventional 9 + 2 arrangement (Figure 1a). Dynein, itself a large and complex molecule, forms one part of this active molecular complex (Figure 1b). Dynein causes flagella and cilia to bend by moving the microtubule pairs with respect to one another. It is anchored to the A tubule of one tubule pair, and its two heads transiently bind to the B tubule of the adjacent pair. The two heads "walk" along and push the B tubule by alternately binding to it, first one and then the other (Figure 1c).

Dynein also participates in intracellular traffic. In neurons, for example, it carries organelles from axon terminals

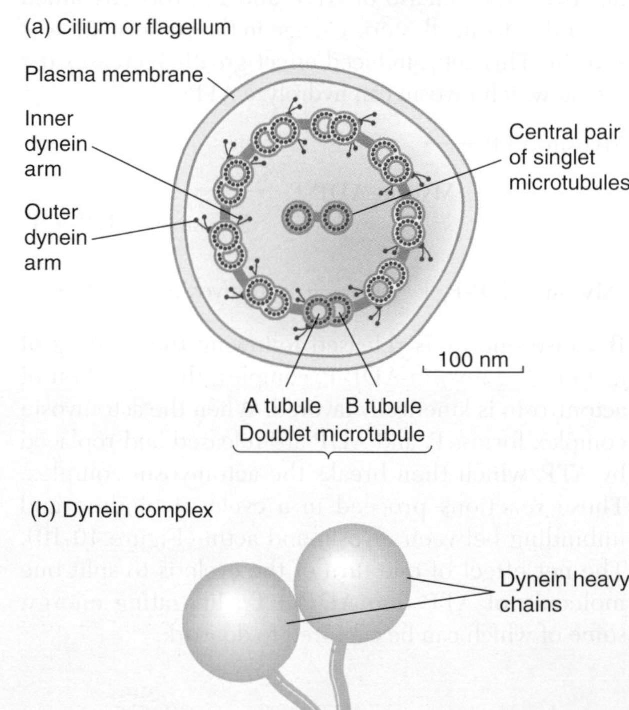

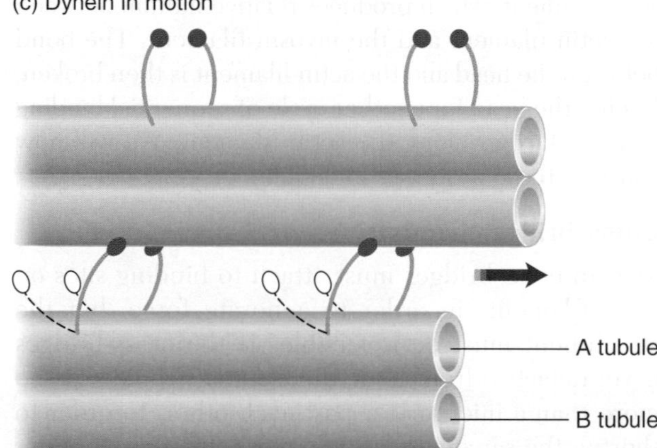

Figure 1 *Dynein drives the movement of cilia and flagella.* ***(a)*** *Typical structure of a cilium or flagellum, showing the "9 + 2" arrangement of microtubule pairs and the location of dynein complexes along the tubules. Each microtubule pair consists of an "A" tubule and a "B" tubule, which are not identical. Dynein complexes are anchored along the A tubules.* ***(b)*** *The dynein complex includes two heavy chains with their heads oriented outward, as well as several intermediate and light chains.* ***(c)*** *Dynein causes movement as the two heads bind alternately to the B tubule of the adjacent microtubule pair, producing a "walking" motion and causing the microtubules to slide past one another. [Adapted from Lodish, 2000.]*

Energy transduction by cross-bridges

One of the major questions regarding the function of myosin cross-bridges is how chemical energy is transduced into mechanical energy by the cycle depicted in Figure 10-10. How do the cross-bridges generate a force that causes the filaments to slide

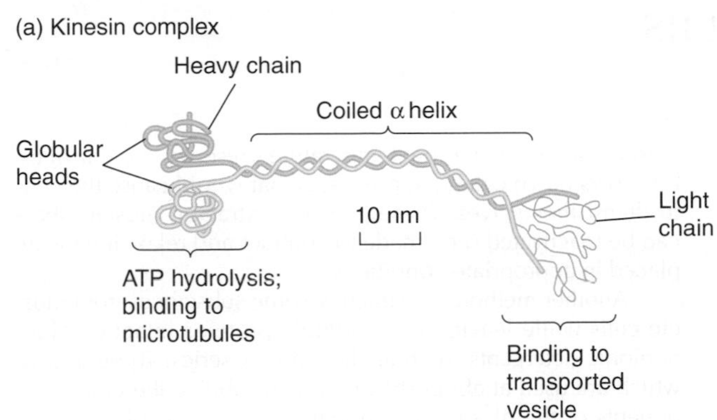

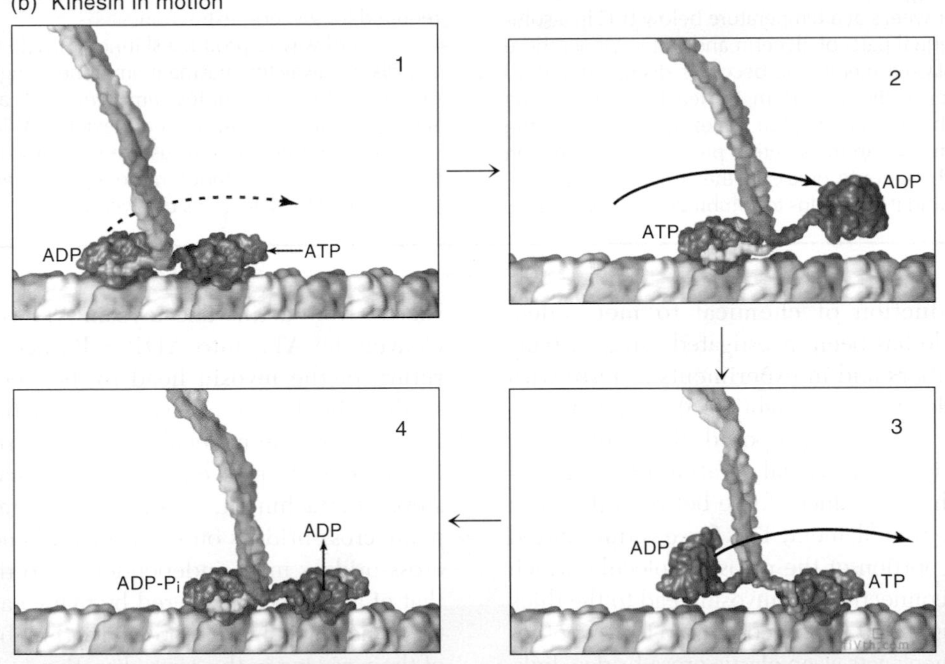

Figure 2 *Kinesin is similar to myosin in structure, but acts more like dynein.* ***(a)*** *A kinesin complex consists of two heavy chains, each of which includes a globular head and a long rod, and two light chains. The head binds to microtubules and to ATP; the opposite end of the rod binds the "cargo" to be carried.* ***(b)*** *The two heads of a kinesin complex bind alternately to a microtubule, "walking" down the structure and pulling their cargo along with them. The red linker represents the active "head." [Part b from Vale and Milligan, 2000.]*

back to the soma, a process called *retrograde axonal transport.* To perform this task, the dynein binds to microtubules arrayed along the axon and to whatever "cargo" it is transporting. In all motions based on dynein, ATP provides the energy to generate mechanical work.

Kinesin, the most recently discovered motor molecule, was identified initially for its role in *orthograde axonal transport* in neurons—that is, the movement of materials and vesicles from the soma out to the axon terminals. The structure of the kinesin complex is reminiscent of the structure of myosin—that is, it consists of paired heavy chains, each of which has a head at one end and a long rod at the other (Figure 2a). In addition, light chains associate with the end of the rod. The head end of kinesin binds to an axonal microtubule and the tail end to whatever it is transporting. The kinesin then "walks" down the microtubule from the soma toward the axon terminal by binding and then moving first one of its two heads and then the other, in a manner similar to dynein (Figure 2b).

Both dynein and kinesin produce movement by walking along microtubules, but they walk in opposite directions with respect to the orientation of the microtubules. In axons, kinesin ferries materials away from the soma; dynein, apparently moving along the same microtubules, carries materials toward the soma.

past each other? The transduction of chemical to mechanical energy in animals is accomplished by three different "molecular motors" that are ubiquitous in the animal kingdom: myosin, dynein, and kinesin (Spotlight 10-2). In muscles, it is myosin that performs this function.

Spotlight 10-3

SKINNED MUSCLE FIBERS

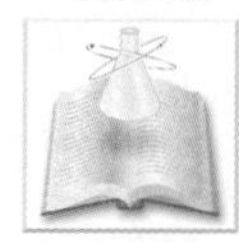

An early advance that contributed significantly to the study of muscle-fiber physiology was the discovery by Albert Szent-Gyorgyi of a procedure for isolating muscle fibers in which the intracellular structure remains intact, but the membrane no longer prevents free exchange of materials between the cytoplasm and the extracellular solution. This kind of preparation is called a "skinned" muscle fiber because the outer membrane of the fiber has been entirely removed or rendered so leaky that it is functionally absent. In skinned fibers, an investigator can control the composition of the intracellular fluids without any interference from regulatory mechanisms that are normally present in an intact muscle fiber.

In Szent-Gyorgyi's procedure, muscle fibers are soaked for several days or weeks at a temperature below 0°C in a solution made up of equal parts of glycerin and water. Under these conditions, the plasma membrane becomes disrupted and all soluble substances in the myoplasm are leached out, leaving intact the insoluble molecules that make up the contractile machinery. The glycerin in the solution prevents the formation of ice crystals, which could break up the structural organization of the fibers, and it also helps to solubilize the membranes. Storing the tissue at a low temperature preserves the enzymes, but slows down catabolic processes that would cause the cells to digest themselves. These glycerin-extracted muscle fibers can be reactivated (i.e., made to contract and relax) if they are placed in appropriate conditions.

Another method of extracting some substances from muscle cells while leaving the insoluble proteins intact employs nonionic detergents, such as the Triton X series. These agents, which are used at about 0°C, rapidly solubilize the lipid components of the plasma membrane, allowing soluble metabolites to diffuse out of the cell and substances in the extracellular medium to diffuse rapidly into the cell. Fibers treated in this way are called "chemically skinned muscle fibers." This process requires only minutes, rather than the days or weeks required for glycerin extraction.

A third way to produce skinned muscle fibers is to manually dissect away the plasma membrane using fine forceps. This process, which resembles removing the casing from a link sausage, requires great manual dexterity. With practice, however, a skilled person can use this method to prepare structurally intact fibers without introducing any artifacts that could be produced by detergents or glycerine.

The transduction of chemical to mechanical energy in muscle has been investigated using partially intact muscle fibers and in experiments *in vitro* with "skinned" muscle fibers (Spotlight 10-3). Although various hypotheses have been proposed, the most widely accepted view is that a partial rotation of the actin-bound myosin head produces force between the thick filament and the thin filament. This force is transmitted through the S2 portion of the myosin molecule, which forms a "neck" connecting the myosin head to the thick filament (see Figure 10-6c). According to this hypothesis, the myosin neck acts as an elastic cross-bridge, linking the myosin head and the thick filament and transmitting to the thick filament the force produced as the head rotates on the actin filament.

Evidence from a huge number of studies, including X-ray diffraction studies to determine the conformation of the myosin head under different conditions, confirms this view. The myosin head, bound to both ADP and P_i, binds to actin (Figure 10-11, step 1). Initially, this bond is relatively weak, but once it is formed, P_i is released accompanied by the release of energy, the strength of the bond between actin and myosin increases, and a major rotational movement of the myosin head with respect to the connecting link occurs (Figure 10-11, step 2). The rotation produces force on the link, which is then translated through the link to the light meromyosin rod in the thick filament. The release of ADP from the myosin head follows, and ATP replaces it in the nucleotide binding site of the head (Figure 10-11, step 3). The binding of ATP causes myosin to dissociate from actin. The myosin ATPase activity then cleaves the ATP into ADP + P_i, accompanied by a return of the myosin head to its "cocked" position, ready to bind again to a site on the actin that is a little farther along the molecule. This cycle repeats, and the filaments slide past each other in small incremental steps of attachment, rotation, and detachment of the many cross-bridges on each thick filament. Individual cross-bridges move independently, so the net effect is that of a long boat powered by many oarsmen, each of them moving in his or her own rhythm, but at least most of them rowing in the same direction.

The elasticity of the cross-bridge allows the rotational movement to occur without an abrupt change in tension. Once it is stretched, the link transmits its tension smoothly to the thick filament, generating force to push the thick filament past the firmly anchored thin filament. A major piece of evidence supporting the hypothesis that the elasticity of cross-bridge links plays a major role in mechanically coupling the myosin heads with the thick filament is the observation that the longitudinal elasticity of a muscle fiber is proportional to the amount of overlap between the thick and thin filaments. Hence it is proportional to the number of attached cross-bridges. In addition, sudden small decreases in fiber length are accompanied by very rapid recovery of tension, which presumably results from rotation of the cross-bridge heads into more stable positions of interaction with actin sites. In other words, the relative positions of the myosin head and actin adjust so that the sites that bind most strongly are favored.

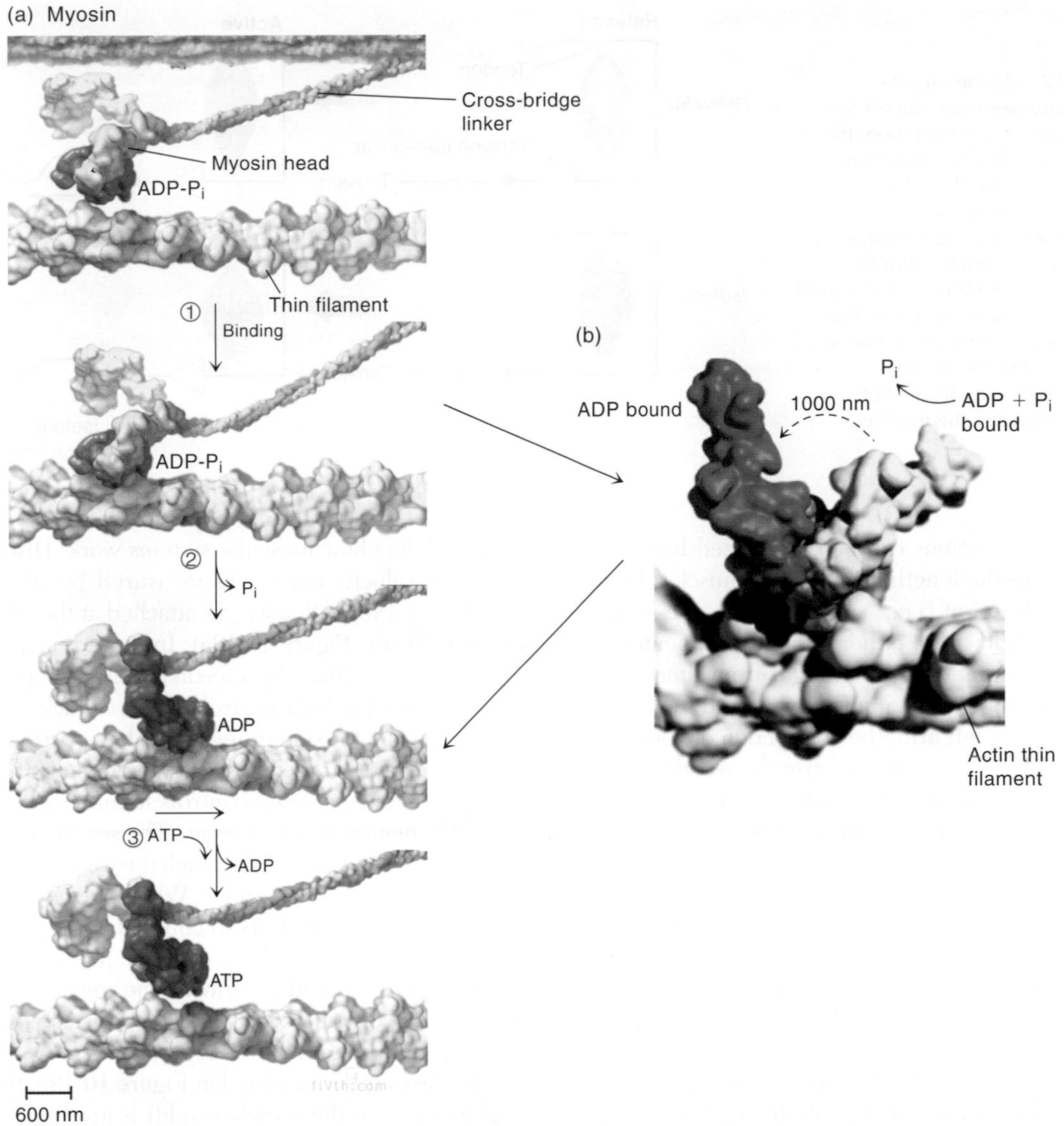

Figure 10-11 The sliding of thick and thin filaments past each other is driven by changes in the geometry of bonds between myosin cross-bridges and actin. **(a)** Space-filling models of myosin heads illustrate the sequence of events in the attachment of myosin cross-bridges to actin filaments. Although each myosin complex includes two heads, only one is active at any given time. In the relaxed state, ADP + P_i occupies the nucleotide binding site on the head (dark blue), and the myosin head is not bound to actin. One myosin head then attaches to actin (step 1) at specific binding sites (green). As a result, P_i is released (step 2), energy is liberated, and the head (now shown in red) rotates with respect to the actin filament, creating a force between the thick and thin filaments and causing them to slide past each other. The rotation is thought to accompany the sequential formation of multiple bonds between the myosin head and the actin binding site; some evidence suggests that four bonds are formed in sequence, each stronger than the last. ADP is then released from the nucleotide binding site and replaced by ATP (step 3), which weakens the bond between the myosin and actin, permitting relaxation. Hydrolysis of the ATP returns the head to its "cocked" position, ready to bind again to a site on actin. **(b)** The action of a myosin cross-bridge is similar to that of an oar pushing a boat through the water. Each swing of the myosin head "rows" the thick filament about 1000 nm past the neighboring thin filament. [Adapted from Vale and Milligan, 2000.]

Working out at a gym causes individual skeletal muscle fibers to become larger in cross-section; no new muscle fibers are formed. How does working out make you stronger? Similarly, what allows an elephant to lift larger weights than you can?

MECHANICS OF MUSCLE CONTRACTION

Many of the mechanical properties of contracting muscle were elucidated before 1950, when the mechanism of contraction was not yet understood. It is useful to consider these classic findings and attempt to explain them in terms of our current understanding of cross-bridge behavior.

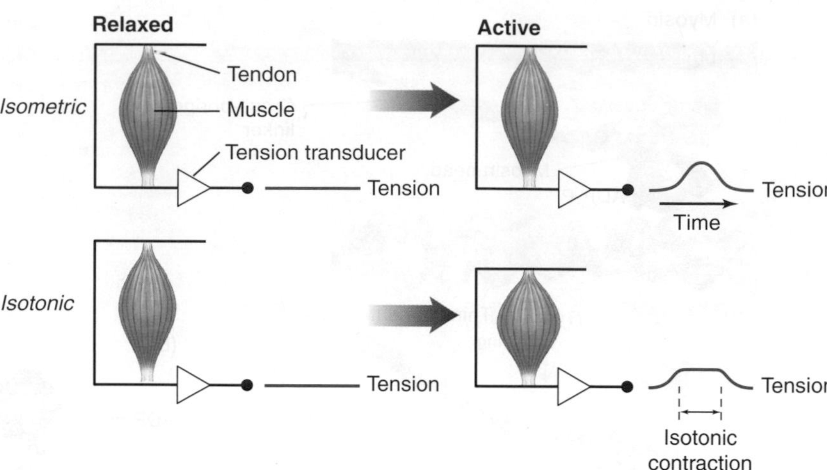

Figure 10-12 Some muscle contractions are isometric; others are isotonic. In isometric contraction, the length of the muscle does not change. Tension increases and then decreases. In isotonic contraction, the muscle is allowed to shorten during the time when tension is being generated. Strictly speaking, after a brief transitional period when the muscle develops force, the tension during isotonic contraction is constant. At the end of isotonic contraction, there is another brief dynamic period when the muscle returns to rest.

Muscle contractions can be categorized based on what happens to the length of the active muscles. In an **isometric contraction** (i.e., "same length"), the length of a muscle is held fixed, either by an experimenter or by the physical situation, preventing it from shortening (Figure 10-12, top). For example, if you tried to pick up your car with your left arm, the contraction of your arm muscles would be isometric, because the weight of the car would prevent them from shortening. The previous discussion of the length-tension relationship for a sarcomere was based exclusively on isometric contractions. Note that although no external shortening is permitted during an isometric contraction, there can be a very small amount of internal shortening (about 1%), which occurs when intracellular and extracellular elastic components—such as cross-bridge links and the connective tissue that is found in series with the muscle fibers—are stretched. Even though the muscle is not shortening, cross-bridge bonds are repeatedly made and broken, and a great deal of energy can be consumed (imagine again trying to pick up your car with one arm). In an **isotonic contraction** (i.e., "same tension"), the muscle shortens as force is generated (Figure 10-12, bottom). Although, strictly speaking, the tension generated during an isotonic contraction would be expected to remain constant, some physiologists use the term to indicate any condition in which the active muscle is allowed to shorten, whether or not the tension is indeed constant. Most natural movements are produced by muscle contractions that are neither isometric nor strictly isotonic, but many studies of muscle function are carried out in the simpler conditions of isometric or isotonic contraction, making this distinction useful.

Relation Between Force and Shortening Velocity

For animals to move, muscles must shorten and pull on parts of the skeleton. Therefore, the relation between the production of force and the rate at which a muscle shortens (the so-called *force-velocity curve*) is crucial for understanding how muscular systems work. Historically, the force-velocity curve was measured by attaching a muscle to a lever with a weight attached at the other side of the fulcrum (Figure 10-13a). In more recent experiments, a motor driven by a feedback circuit replaces the weight, providing finer control. The feedback control of such a servomotor system regulates the apparent weight against which the muscle is shortening in much the same way that a voltage clamp controls membrane potential across the membrane of a neuron. The system is arranged so that there is a limit on how much the weight, or the servomotor, can stretch the muscle. When the muscle is electrically stimulated, it starts to contract. When the force generated by the muscle becomes equal to the force exerted on the weight by gravity, the muscle begins to shorten at a constant velocity (i.e., the muscle no longer contracts isometrically), and the velocity is measured.

In the example depicted in Figure 10-13b, the maximal weight that the muscle can lift is just under 100 g; that is, if the muscle contracts against a load of 100 g or more, it cannot shorten. If the load is less than 100 g, the muscle shortens, but the rate of shortening (ΔL per unit time) depends on the size of the weight. If a 50 g weight is attached, the muscle shortens slowly; if lighter weights are attached, the muscle shortens faster. When it shortens against no weight at all, the velocity of contraction is maximal; this velocity is represented by V_{max}. This characteristic of muscles, V_{max}, will play an important role later in this chapter when we discuss the evolutionary adaptation of skeletal muscles.

Plotting the force generated by a muscle against its shortening velocity generates a hyperbolic curve whose equation was determined empirically in the 1930s by Archibald V. Hill, an important pioneer in muscle physiology:

$$V = \frac{b(P_0 - P)}{P + a} \tag{10-1}$$

where V is the velocity of shortening; P, the force (or load); P_0, the maximal isometric tension of that muscle;

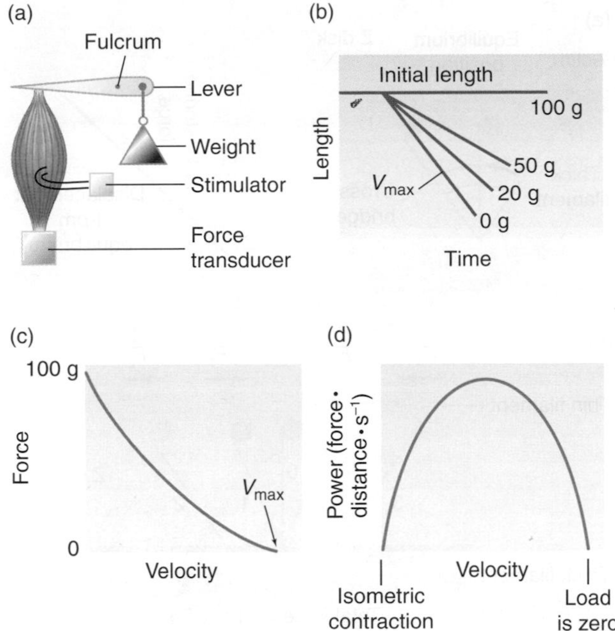

Figure 10-13 The force against which a muscle works and the velocity at which it shortens are reciprocally related. **(a)** Typical setup for measuring the relation between force and velocity for a muscle. The muscle works against a weight that is hung on the other side of the fulcrum of a lever. When the stimulated muscle generates a larger force than the weight, it shortens and pulls down on the lever. Alternatively, a servomotor system can be used to provide finer control of the initial muscle length and the load. At the start of the experiment, the length of the muscle would be set to optimize the overlap between thick and thin filaments in the sarcomeres. **(b)** Contraction of a muscle as it works against four different loads: 100 g, 50 g, 20 g, and 0 g. When the load is small, the muscle contracts more rapidly; that is, ΔL per unit time is larger. The maximal weight this muscle can lift is just under 100 g, so when it works against the 100 g weight, it cannot shorten; contraction against 100 g or more is isometric. **(c)** Force-velocity curve based on the data shown in part b. The data that produce this curve match Equation 10-1 well. At the maximal force of 100 g, the velocity of shortening is zero; that is, contraction is isometric. Shortening of the muscle is fastest when the muscle is completely unloaded. **(d)** Power-velocity curve calculated by multiplying the force and the velocity for each data point in part c. The power is zero if either the force or the velocity equals zero.

b, a constant with dimensions of velocity; and a, a constant with dimensions of force (Figure 10-13c).

Equation 10-1 implies that as the load increases, the shortening velocity decreases. You are probably familiar with this principle from personal experience; you can lift a feather from a table much more rapidly than you can lift a heavy book. Notice that the decline in force with increased velocity does not reflect a change in myofilament overlap. On the contrary, these experiments are purposely performed at the plateau of the length-tension relationship, so that the number of cross-bridges that can interact with actin remains high and constant during shortening.

The relation between power and velocity is as important to an animal's behavior as the relation between force and velocity. For a fish to swim or a frog to jump, its muscles must generate mechanical power. The mechanical work performed by a muscle is the product of force times the change in length (ΔL), and mechanical power is given by

$$\text{power} = \frac{\text{work}}{\text{time}} = \frac{(\text{force}) \times (\Delta L)}{\text{time}}$$
$$= (\text{force}) \times (\text{shortening velocity})$$

Hence, multiplying the force generated by the muscle times the velocity at which it shortens yields the power produced under each condition. As shown in Figure 10-13d, the power generated is maximal at intermediate shortening velocities. Power falls to zero if either the velocity of shortening (i.e., in isometric contraction) or the force generated is zero.

As we will see later in this chapter, it is frequently useful to describe force production or power production in terms of V/V_{max}, where V is the velocity of shortening under a particular condition and V_{max} is the maximal velocity of shortening for a particular muscle. Power production by the muscle shown in Figure 10-13 is maximal at a V/V_{max} of about 0.4; this relation has been found to hold for all muscles, no matter how fast their V_{max} is.

Effect of Cross-Bridges on the Force-Velocity Relation

From the force-velocity curve described in the previous section, we know that the force generated by a muscle drops as its shortening velocity increases. Recall that this relation is not caused by a change in the amount of overlap between thin and thick filaments; rather, it is observed at the maximal overlap. From our earlier discussion of the role played by cross-bridges in isometric contraction, we might suppose that this drop in force with increased velocity could result if fewer cross-bridges were actually attached during rapid shortening, or if each of the cross-bridges that was attached generated a smaller force, or both. This model of cross-bridge kinetics was proposed by Andrew Huxley in 1957. Although it has been superseded in some fine details, it still provides the basic principles for understanding the overall mechanics and energetics of muscle contraction.

In Huxley's model, cross-bridges are considered to be elastic structures that generate zero force when they are at equilibrium. Their behavior is similar to that of a piece of spring steel projecting out from a surface. Deforming the piece of steel by bending it creates a restoring force that can return it to its original position. Similarly, when a cross-bridge is bent toward or away from the Z disk, a restoring force is created that tends to bring it back to its original position; the magnitude of this force is proportional to the displacement of the cross-bridge from its equilibrium position (Figure 10-14a). If a cross-bridge bent toward the Z disk were attached to a thin filament, the restoring force would pull the Z disk toward the center of the sarcomere; this force is considered to be in the "positive" direction. By contrast, if a cross-bridge bent away from the Z disk were attached to a thin filament, the restoring force would push the Z disk away from the center of the sarcomere; this force is viewed as a "negative" force.

Figure 10-14b illustrates how the forces generated by passive cross-bridge displacement cause movement of a thin filament. When the cross-bridge is at subscript the equilibrium position (0), the force generated, F_0, is zero; when the cross-bridge is bent toward the Z disk, the force is positive (F_1 and F_2); and when the cross-bridge is displaced away from the Z disk, the force is negative ($F_{1'}$ and F_3). The force generated by one thick filament is equal to $\Sigma n_i F_i$, the sum of the product of the number of attached cross-bridges at each displacement, n_i, and the force produced per cross-bridge at that displacement, F_i. As the velocity of shortening increases, the number of cross-bridges that are attached drops, and the displacement of the cross-bridges that are attached becomes smaller (Figure 10-14c). In addition, during rapid shortening, some cross-bridges become attached when they are in a position that generates negative force. As a result of all of these changes, the net force produced during rapid shortening is lower than the force produced during slow shortening.

According to Huxley's theory, unattached cross-bridges are moved away from their neutral position by random thermal motion. If cross-bridges attached randomly to thin filaments, no force would be generated as the result of this thermal motion, because the number of cross-bridges generating negative force would equal the number generating positive force. However, cross-bridges can initially attach to thin filaments only when they are in a position that would generate positive force. Thus, when a muscle is loaded maximally and is contracting isometrically, there will be an even distribution of cross-bridges that generate a positive force, and because all attached cross-bridges are generating positive force, the average force per cross-bridge will be positive and large.

If cross-bridges can attach to thin filaments only when they are displaced to a position that produces positive force, how can cross-bridges generate negative force? During shortening, the thin filaments move toward the center of the sarcomere, so any cross-bridges that are attached to them at a sharp angle toward the Z disk (e.g., cross-bridge 2 in Figure 10-14b) will be shifted closer to the equilibrium position. The amount of force they produce will thus be reduced by the movement of the thin filament. A cross-bridge that is attached at a very shallow angle (e.g., cross-bridge 1 in Figure 10-14b) can be dragged over to a position (1′) that causes it to generate a negative force ($F_{1'}$). Of course, this effect could not go on indefinitely because

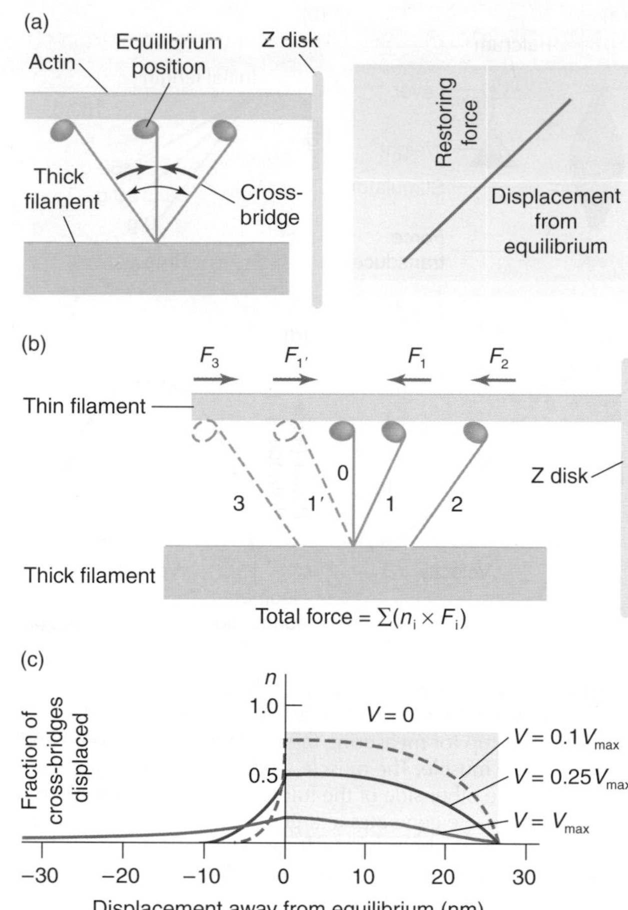

Figure 10-14 Cross-bridges generate a restoring force when they are moved away from their equilibrium position. **(a)** Relations between the position of a cross-bridge and the magnitude and direction of the force generated. Cross-bridges are assumed to behave like a strip of spring steel, fixed at one end. At the equilibrium position of the cross-bridge, no force is generated. Displacement of the cross-bridge away from equilibrium in either direction generates a restoring force that tends to bring the cross-bridge back to equilibrium. **(b)** Cross-bridges attached to a thin filament at different positions. When displaced toward the Z disk (solid lines), they generate a positive force (e.g., F_2 for cross-bridge 2); when displaced away from the Z disk (dashed lines), they generate a negative force (e.g., F_3 for cross-bridge 3). The total force at any time is the sum of the forces generated by all cross-bridges. Movement of the thin filament can change the displacement of some bound cross-bridges (e.g., position 1 to 1′ for cross-bridge 1), causing them to exert negative ($F_{1'}$) rather than positive (F_1) force. **(c)** The fraction of the total number of cross-bridges that are attached and displaced at different shortening velocities. As the velocity at which thick and thin filaments slide past each other increases, fewer cross-bridges are attached, and the average position (and hence force production) of the cross-bridges becomes more negative. At V_{max}, the net force generated by the cross-bridges equals zero, because the positive force generated by some cross-bridges equals the negative force generated by others. Conversely, when the muscle contracts isometrically ($V = 0$), the production of force is maximal because many cross-bridges are attached, and all of the attached cross-bridges are in a position that produces positive force.

such cross-bridges would produce more and more negative force, preventing further sliding of the thin filament. Each cross-bridge must detach, and the time required for a cross-bridge to detach is the key to what limits the maximal velocity of shortening.

Assuming that it takes a fixed amount of time for cross-bridges to detach, then, as the velocity at which the filaments slide past each other increases, more cross-bridges will be dragged to a position from which they can generate a negative force before they can detach. There should then be a velocity at which the negative force generated by cross-bridges that have been dragged to the negative side of the equilibrium position will just balance the positive force generated by the attached cross-bridges on the positive side. At this point, the net force generated by all attached cross-bridges is zero. Because the muscle cannot shorten any faster than this rate, this constitutes the maximal velocity of shortening, V_{max}. Thus, at V_{max}, some cross-bridges are attached, but the net force—or average force per cross-bridge—is zero. It follows that a muscle can have a fast V_{max} if its cross-bridges detach rapidly, breaking their bonds with the thin filaments before they can generate large negative forces.

Two aspects of this model explain the observed decrease in force as the velocity of shortening increases. First, the average force generated by the cross-bridges drops with increased velocity. Second, the total number of cross-bridges attached at any one time drops with increased velocity. The argument supporting this second aspect of the model is based on chemical kinetics: As cross-bridges are dragged to positions in which they generate negative force, they detach faster, which causes fewer cross-bridges to be attached at higher velocities. It is thought that at V_{max}, as few as 20% of the cross-bridges are attached; in high isometric tension, approximately 30% of cross-bridges are attached.

REGULATION OF MUSCLE CONTRACTION

Up to this point, we have considered only how cross-bridges on myosin thick filaments in an activated muscle fiber bind to and unbind from actin thin filaments, thereby generating force. Of course, if our muscles were "on," or activated, all the time, we would be in a constant state of rigidity, unable to move, talk, or breathe. Thus, to perform useful work, muscles must turn on and off at the appropriate times. The mechanisms by which contraction is regulated—that is, turned on and off—are discussed in the following sections.

The Role of Calcium in Cross-Bridge Attachment

The earliest evidence that Ca^{2+} plays a physiological role in muscle contraction came from the work of Sidney Ringer and Dudley W. Buxton in the late nineteenth century. They found that an isolated frog heart stops contracting if Ca^{2+} is omitted from the saline bath in which it is immersed. (This observation marked the origin of Ringer solution and other physiological salines.) The possibility that Ca^{2+} participates in the regulation of muscle contraction was first tested in the 1940s, when several researchers experimentally introduced a variety of cations into the interior of skinned muscle fibers. Of all the ions tested, only Ca^{2+} was found to produce contraction when it was present at concentrations similar to those normally found in living tissue. It was subsequently discovered that skeletal muscle fails to contract in response to stimulation if its internal calcium stores are depleted, although the presence of calcium in the *extracellular* fluids is not required.

The concentration of Ca^{2+} ions is normally very low in the cytosol of relaxed muscle fibers—10^{-6} M or lower. Initial attempts to study the events of contraction in solution were foiled because it was impossible to maintain the Ca^{2+} concentration of experimental solutions as low as it is in the cytosol. Even double-distilled water contains more than 10^{-6} M Ca^{2+}. The discovery of calcium-chelating agents, such as EDTA (ethylenediaminetetraacetic acid) and EGTA (ethylene-*bis*[oxyethylenenitrilo] tetraacetic acid), overcame this obstacle. The development of methods for preparing skinned muscle fibers, which lack a functional outer membrane, also facilitated research on the role of Ca^{2+} in contraction (see Spotlight 10-3).

The quantitative relation between the concentration of free cytosolic Ca^{2+} in muscle fibers and contraction has been determined by exposing skinned myofibrils to solutions containing different Ca^{2+} concentrations. Skinned myofibrils contract only if the solutions contain both Ca^{2+} and ATP, because ATP is required for muscle contraction (see Figure 10-10). If the Ca^{2+} is removed, the myofibrils relax, even if ATP is plentiful (Figure 10-15a on the next page). The amount of tension generated rises sigmoidally from zero at a Ca^{2+} concentration of about 10^{-8} M to a maximum at about 10^{-6} M (Figure 10-15b).

As we've seen already, force can be developed only when myosin cross-bridges bind to actin thin filaments, so anything that facilitates or inhibits this binding will affect contraction. The key to how Ca^{2+} induces muscle contraction lies in two proteins—tropomyosin and troponin—that are associated with actin in thin filaments (see Figure 10-5b). Tropomyosin is a filamentous protein that runs parallel to actin filaments. When a myofibril is relaxed, tropomyosin occupies a position that sterically blocks the myosin binding sites on the actin filament. Troponin is a complex of three protein subunits: troponin C, which binds Ca^{2+}; troponin T, which binds tropomyosin; and troponin I, which binds both actin

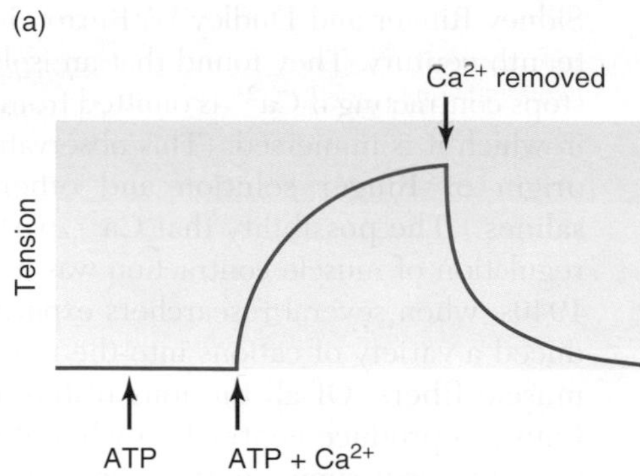

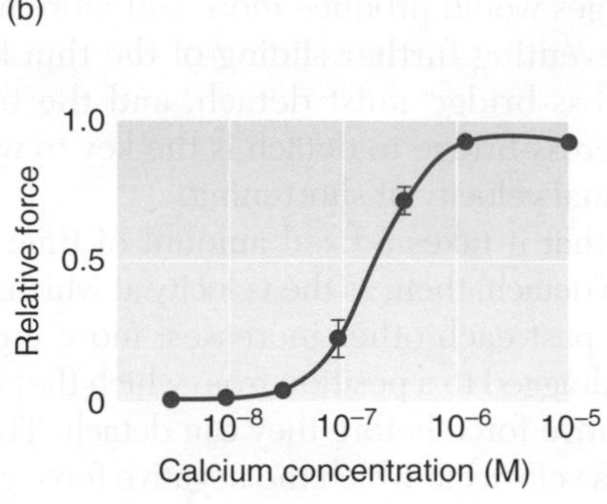

Figure 10-15 Free calcium ions in the cytosol regulate the state of muscle contraction. **(a)** Glycerin-extracted muscle fibers generate tension when they are exposed to Ca^{2+} and ATP. They relax when Ca^{2+} is removed, even if ATP is still present.

Classic Work **(b)** The force generated by a skinned muscle fiber varies with the concentration of Ca^{2+} in the surrounding medium. Force increases with increasing Ca^{2+} concentration, up to about 10^{-6} M. [Part b adapted from Hellam and Podolsky, 1967.]

and troponin C (Figure 10-16a). The troponin complex binds to tropomyosin about every 40 nm along the actin filament (see Figure 10-5b). Its association with troponin causes tropomyosin to change its shape and thus to change its association with the actin filaments. Troponin is the only protein in either the thin or the thick filaments of vertebrate striated muscle that has a high binding affinity for Ca^{2+}. (Myosin light chains bind Ca^{2+}, but less avidly.) When Ca^{2+} binds to troponin, the troponin molecule undergoes a change in conformation that modifies binding strengths in the complex (Figure 10-16b). The net result is that the tropomyosin moves with respect to the actin filaments, allowing myosin heads access to myosin binding sites on the actin (Figure 10-16c). Thus, when Ca^{2+} binds to troponin, it removes an otherwise constant inhibition of attachment between myosin cross-bridges and thin filaments. It is inferred from experimental results like those shown in Figure 10-15b that cross-bridges can bind to actin when the concentration of free Ca^{2+} in the cytosol reaches about 10^{-7} M.

As discussed earlier, myosin hydrolyzes ATP much more rapidly when the heads bind to actin. Ca^{2+} increases the binding of myosin heads, so adding Ca^{2+} to naked myofibrils would be expected to increase the ATPase activity of the myosin. Experiments like the one shown in Figure 10-17a on page 382 have demonstrated that this is exactly what happens. Cross-bridges cycle normally and produce tension only when both ATP and Ca^{2+} are present in the cytosol surrounding the myofibrils (Figure 10-17b). When a glycerin-extracted muscle fiber was initially exposed to Ca^{2+} in the absence of ATP, no tension was generated. When ATP was added, tension developed, and that tension was then maintained even when ATP was removed—that is, when rigor mortis set in. Once the muscle was in rigor, removing Ca^{2+} had no effect because the lack of ATP caused all the attached cross-bridges to be frozen in place. When ATP was added back to the muscle that was in rigor, but Ca^{2+} was absent, the muscle relaxed. Thus, both ATP and Ca^{2+} must be present if the thick and thin filaments are to interact effectively to produce tension, but only ATP is required for relaxation.

If this picture of how Ca^{2+} regulates contraction is correct, Ca^{2+}, troponin, and tropomyosin together would be expected to modify the ATPase activity of myosin. In the experiment shown in Figure 16-17c, S1 fragments of myosin, which are the location of myosin's ATPase activity, were mixed with ATP and other components. If only ATP and actin were present in solution with the S1 fragments, there was some ATPase activity, particularly at higher concentrations of the S1 fragments. Adding troponin and tropomyosin to the mixture reduced the ATPase activity. Troponin and tropomyosin spontaneously bind to actin, occluding potential myosin binding sites. However, adding troponin and tropomyosin along with an appropriate concentration of Ca^{2+} greatly increased the ATPase activity of the S1 fragments, indicating that just as in skinned muscle fibers, Ca^{2+} modulated the inhibitory effect of troponin and tropomyosin on cross-bridge binding.

These results showed that calcium regulates the actin-myosin interaction via troponin and tropomyosin in vertebrate striated muscles. It regulates contraction in other types of muscle as well, but by at least two other mechanisms. In most invertebrate striated muscles, calcium initiates contraction by binding to the myosin light chains of the cross-bridge heads. Contraction of vertebrate smooth muscle and of nonmuscle actomyosin depends on calcium-dependent phosphorylation of the myosin head, as described in the last section of this chapter.

Excitation-Contraction Coupling

In Chapter 6, we stated that an action potential arriving at the synapse between a motor neuron and a skeletal muscle fiber causes an action potential in the fiber, followed by a twitch. This process is summarized in Figure 10-18 on page 382. The depolarization of the muscle

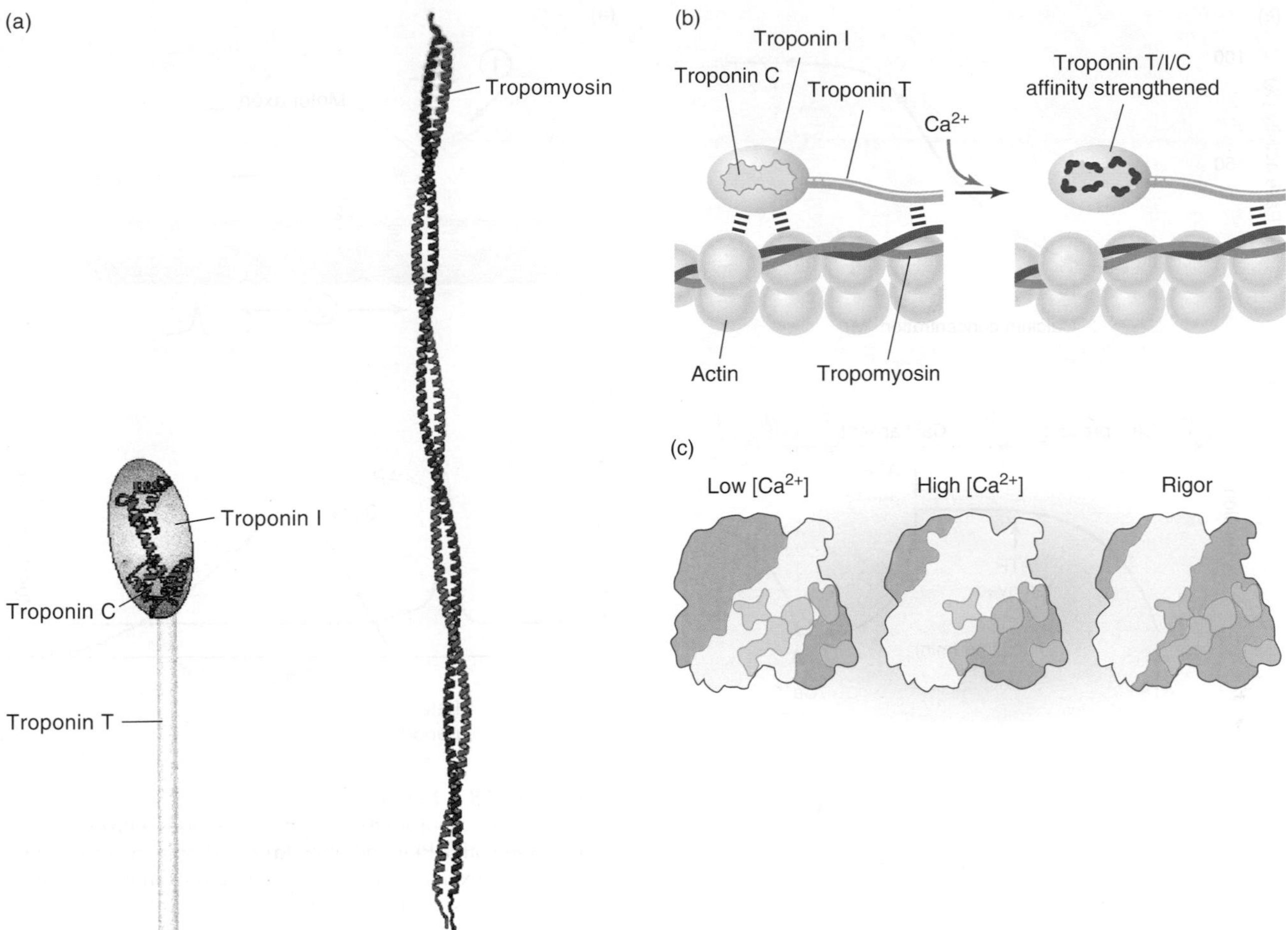

Figure 10-16 Troponin and tropomyosin regulate binding between myosin cross-bridges and actin thin filaments. **(a)** Three-dimensional models depicting the molecular structures of troponin and tropomyosin. Troponin consists of three subunits. The molecular structure of troponin C, which binds Ca^{2+}, has been worked out in detail. Troponin T, which binds to actin and blocks myosin binding sites on actin, is a long chain. The structure of troponin I, which binds to actin and to troponin C, has not yet been determined in detail, but in the complex it is located in the position shown. **(b)** Diagram illustrating bonds along thin filaments. The thickness of the red lines represents the strength of the bonds. In the absence of Ca^{2+}, troponin T binds to tropomyosin, and troponin I binds to actin. Both troponin I and troponin T bind to troponin C. When troponin C binds Ca^{2+}, however, its bond with troponin I becomes stronger, and the bond between troponin I and actin is broken. In addition, the bond between troponin C and troponin T is strengthened. In a way that is not yet understood, this shift in bond strengths exposes myosin binding sites on the actin. **(c)** Structure of a single G-actin subunit (shown in gray), including potential myosin binding sites. The part of the actin subunit covered by tropomyosin under three conditions is shown in blue. Exposed myosin binding sites are red; sites blocked by tropomyosin are purple. When Ca^{2+} is absent, tropomyosin covers most of the potential myosin binding sites. With Ca^{2+} in the cytosol, the position of tropomyosin shifts, exposing some of the myosin binding sites. In rigor, all of the binding sites become available, permitting tight binding between thick and thin filaments. When binding sites are transiently exposed in normal contraction, cross-bridges can bind cyclically until Ca^{2+} is removed from the troponin complex. [Part a adapted from Squire and Morris, 1998; parts b and c adapted from Gordon et al., 2000.]

fiber is typically large enough to elicit an action potential, which travels down its length. Following a latent period of several milliseconds, the muscle fiber then generates tension (Figure 10-18b) Given what is known about cross-bridge attachment, sliding filaments, and the crucial role played by Ca^{2+}, it seems likely that the regulation of muscle contraction includes some mechanism by which an action potential in the muscle fiber changes the concentration of free Ca^{2+} in the cytosol. This process is called **excitation-contraction coupling.** Its net effect is to link an AP in the plasma membrane of the muscle fiber to the concentration of free Ca^{2+} in the cytosol. We examine the details of this critical process in the following sections.

Membrane potential and contraction

As we saw in Figure 5-14, if some of the Na^+ ions in the normal saline bathing excitable cells are replaced by K^+ ions, the membrane potential, V_m, of the cells is depolarized. When muscle fibers are suddenly depolarized

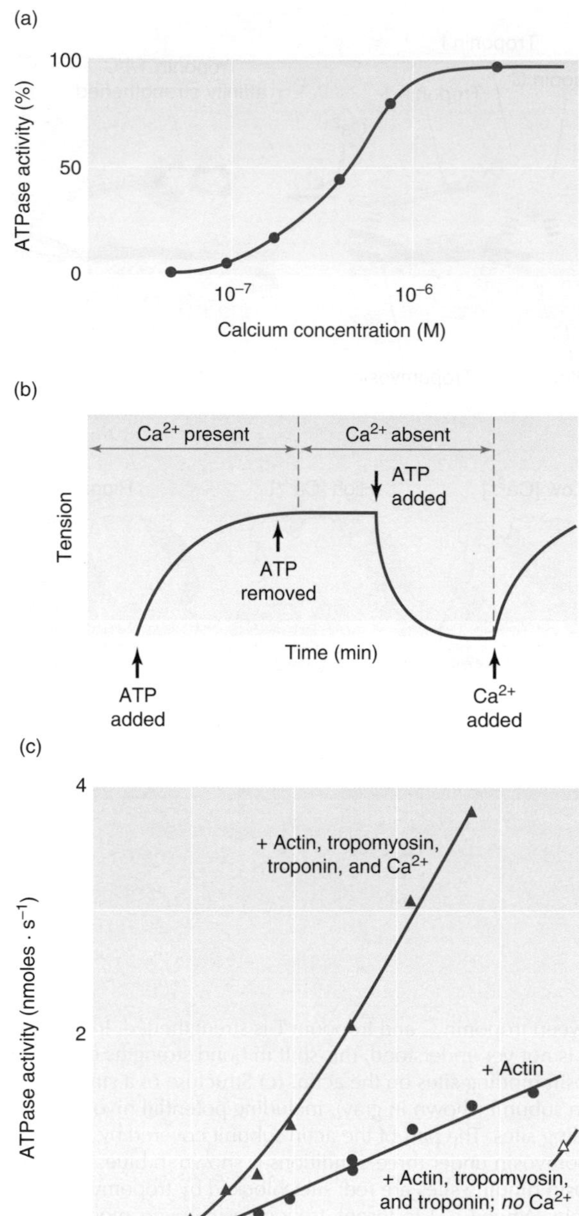

Figure 10-17 Free calcium modulates the activity of both glycerin-extracted muscle fibers and isolated myofibrils. Classic Work **(a)** The ATPase activity of myofibrils increases sigmoidally with the Ca^{2+} concentration of the surrounding solution, with a threshold somewhat lower than 10^{-7} M. **(b)** Both Ca^{2+} and ATP are required for muscles to contract, but relaxation occurs only in the presence of ATP and the absence of Ca^{2+}. If ATP is removed once tension has developed, the fiber enters rigor mortis (flat part of the curve). Rigor is relieved only by removal of Ca^{2+} *and* addition of ATP. **(c)** When actin, S1 fragments of myosin, and ATP are mixed in a solution, ATPase activity of the myosin can be detected. Adding troponin and tropomyosin to the mixture reduces the ATPase activity as troponin and tropomyosin bind to actin, blocking myosin binding sites. If Ca^{2+} is then added, the ATPase activity increases greatly due to the interactions illustrated in Figure 10-16. [Part a adapted from Bendall, 1969; part c adapted from Lehrer and Morris, 1982.]

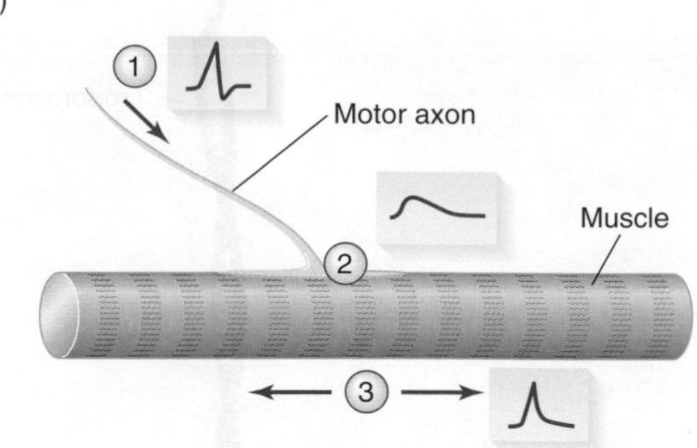

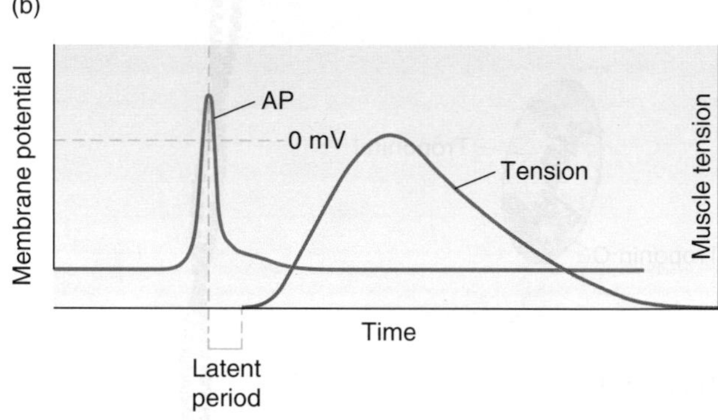

Figure 10-18 Muscle fibers contract when a postsynaptic potential at the neuromuscular junction causes a propagated action potential (AP) in the fiber. **(a)** An AP in a motor neuron (1) causes an excitatory postsynaptic potential in the muscle fiber (2), which gives rise to a propagated muscle AP (3). **(b)** The AP in the muscle fiber (red trace) is followed, after a latent period, by a transient, all-or-none contraction (blue trace) called a twitch.

in this way, they produce a transient contraction, which is called a *contracture* to differentiate it from a normal contraction. In the experiment depicted in Figure 10-19, a single frog muscle fiber was exposed to various concentrations of extracellular K^+ while the membrane potential and muscle tension were monitored. When the membrane was depolarized to about −60 mV, tension began to develop; with further depolarization, tension increased sigmoidally, reaching a maximum at about −25 mV.

This experiment demonstrates that the contractile system can produce graded contraction when the membrane is depolarized to different values. However, a single twitch in response to a single AP is typically an all-or-none event. How can these two observations be reconciled? During an AP in a muscle fiber, the membrane potential swings from a resting value of about −90 mV to an overshoot of about +50 mV. At the peak of the AP, the membrane potential is as much as 75 mV more positive than the potential required to give a maximal contracture. As a result, during an AP, the membrane potential of the muscle fiber exceeds the value at which contraction is fully activated. The twitch is all-or-none because the AP is both large and all-or-none.

(a)

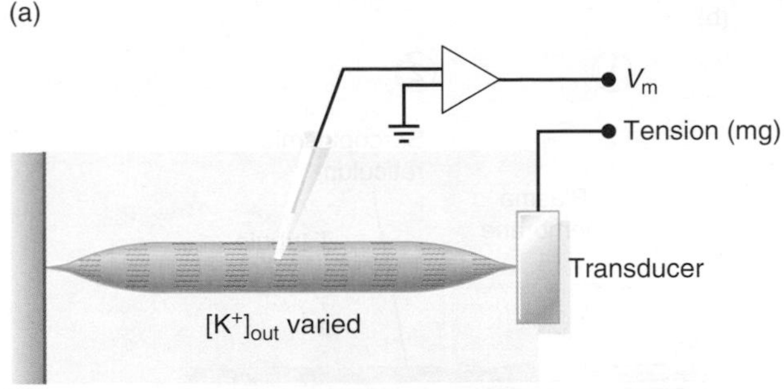

(b)

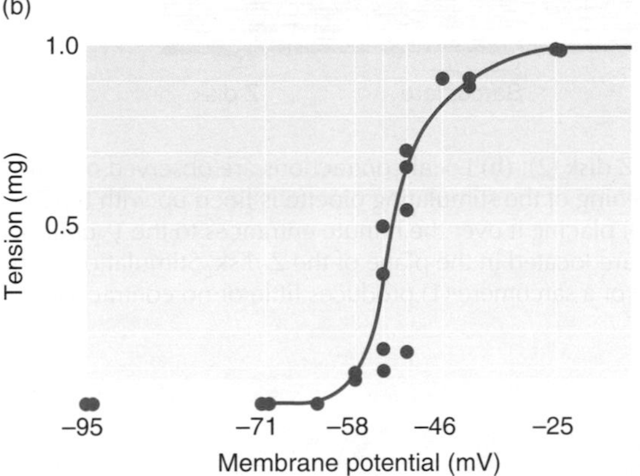

Figure 10-19 The tension developed by a muscle fiber varies with the membrane potential, V_m. **(a)** Setup for measuring the membrane potential of and tension produced by an isolated muscle fiber as the concentration of K^+ is varied in the extracellular solution.

Classic Work **(b)** The tension produced by the muscle fiber as a function of membrane potential. Data points are plotted, and the red curve shows the sigmoidal function that best fits the points. The threshold potential for contraction is about −60 mV. [Adapted from Hodgkin and Horowicz, 1960.]

A potential difference across the plasma membrane of a muscle fiber directly affects an intracellular region that extends at most only a fraction of a micrometer from the inner surface of the membrane. As a result, a potential change across the plasma membrane cannot directly exert any influence on the great bulk of the myofibrils in a typical skeletal muscle fiber, which is 50–100 μm in diameter. There must be something that couples depolarization of the plasma membrane to the activity of myofibrils deep within each muscle fiber. Electrotonic spread of local currents produced by a propagated AP was experimentally ruled out: when currents of physiological magnitude were passed between two microelectrodes inserted into a muscle fiber, they produced no contraction.

The hypothesis that Ca^{2+} might play a role in linking membrane potential and contraction was suggested relatively early. During the 1930s and 1940s, Lewis V. Heilbrunn argued for the importance of calcium in many cellular processes, including muscle contraction. We now know that his hypothesis that the contraction of muscle is controlled by intracellular changes in calcium concentration is essentially correct, although it was widely rejected at first because of a fundamental misunderstanding about the nature of excitation-contraction coupling. It was assumed that calcium would have to enter the cytosol of the muscle fiber (also called the myoplasm) through the plasma membrane to initiate contraction. As A. V. Hill pointed out, the rate of diffusion of an ion or a molecule from the plasma membrane to the center of a muscle fiber that is 25–50 μm in radius is several orders of magnitude too slow to account for the short observed latent period (about 2 ms) between an AP at the plasma membrane and activation of the entire cross-section of the muscle fiber. Using this logic, Hill correctly concluded that a process, rather than a substance, must couple the surface signal to myofibrils that lie deep within the muscle fiber. As we will see, it is the AP itself that is conducted deep into the cell interior, where it causes the release of intracellular Ca^{2+} from internal storage depots that surround the myofibrils. Elevation of the concentration of free Ca^{2+} in the myoplasm permits myosin cross-bridges to attach to the actin thin filaments and generate force.

T tubules

Anatomic and physiological evidence suggesting a link between the plasma membrane and the internal myofibrils came to light about 10 years after Hill's calculation. In 1958, Andrew F. Huxley and Robert E. Taylor studied the details of excitation-contraction coupling by stimulating the outside surface of single frog muscle fibers with tubular glass microelectrodes (Figure 10-20a on the next page). Their most significant findings were the following:

- Pulses of current that were too small to initiate a propagated AP, but sufficient to depolarize the membrane under the pipette opening, led to small local contractions, but only when the tip of the pipette was positioned directly over a Z disk (Figure 10-20b).
- These contractions occurred only around the perimeter of the fiber and very close to the Z disk.
- Contractions spread farther into the fiber as the intensity of the stimulating current was increased.
- Contractions were limited to the two half-sarcomeres immediately on either side of the Z disk over which the electrode was positioned. In other words, contraction occurred only in the sarcomeres lying directly under the stimulated membrane; other sarcomeres attached to the ends of the stimulated sarcomeres remained relaxed.

Electron microscopic studies of amphibian skeletal muscle performed at about the same time provided an anatomic correlate of these physiological findings. Running around the perimeter of each myofibril at the level of the Z disk is a hollow membranous tube called a

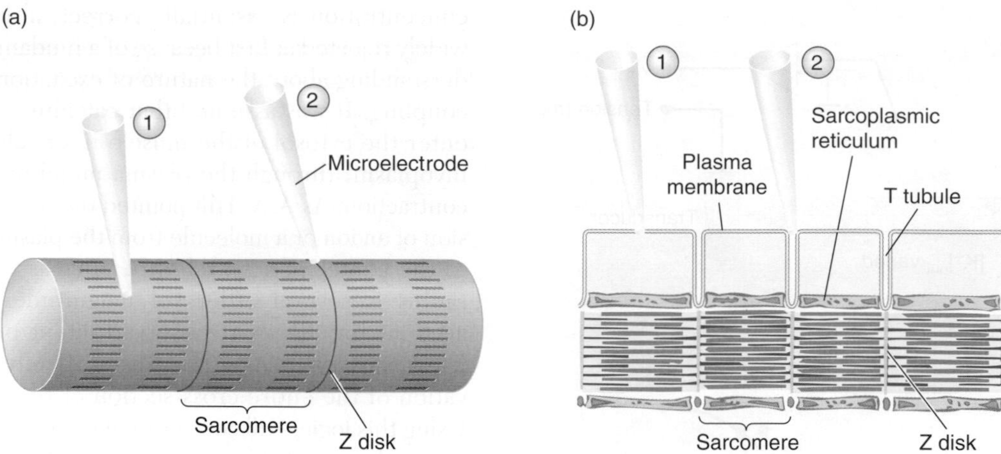

Figure 10-20 When frog muscle fibers are stimulated by an extracellular microelectrode, they can contract, but only when the pipette stimulates the fiber near a Z disk. **(a)** Experimental setup, showing the stimulating pipette positioned either in the center of a sarcomere (1) or directly over a Z disk (2). **(b)** Local contractions are observed only if the opening of the stimulating pipette is lined up with the Z disk (2), placing it over the minute entrances to the T tubules, which are located in the plane of the Z disk. Stimulation at the middle of a sarcomere (1) produces little or no contraction.

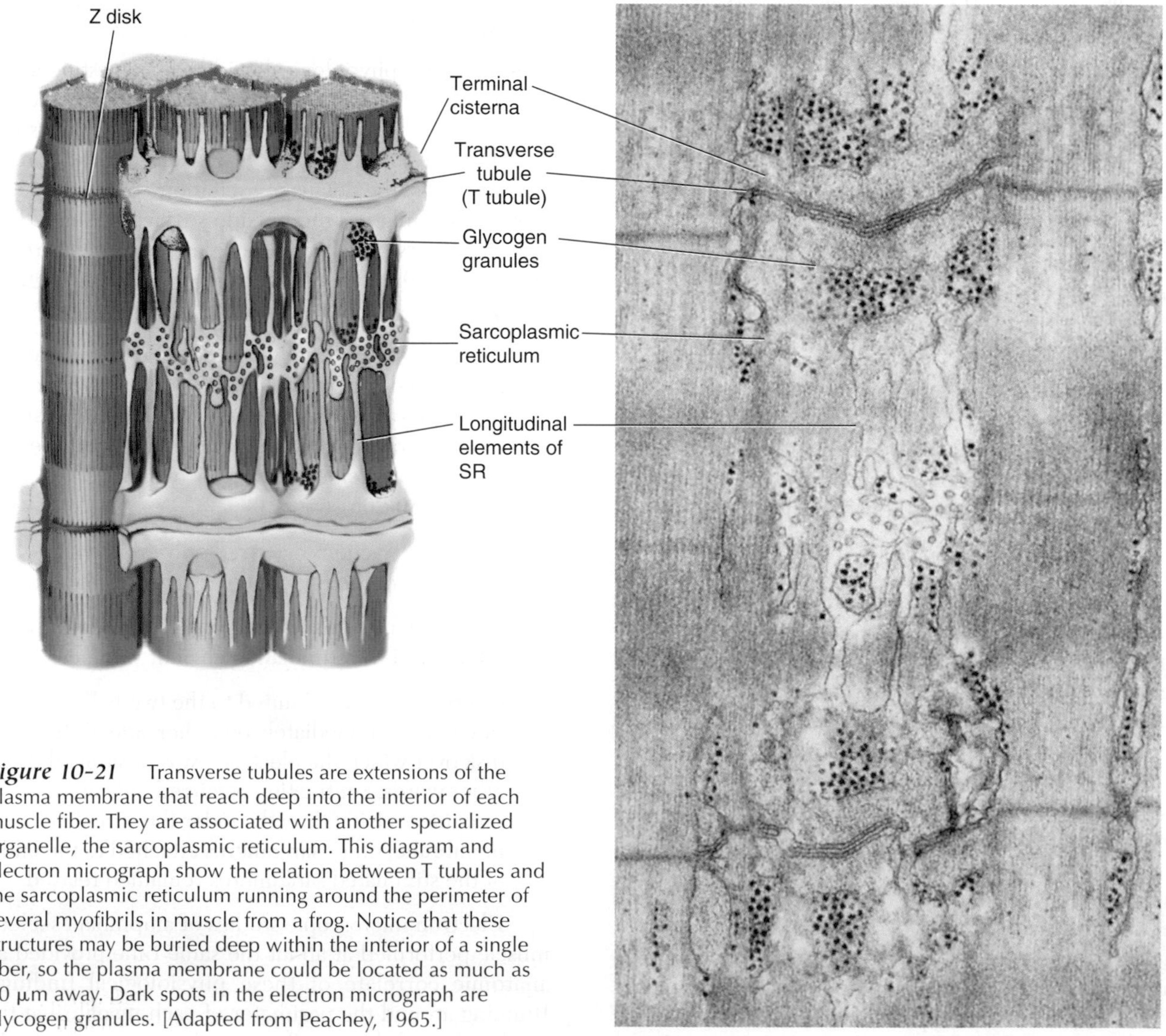

Figure 10-21 Transverse tubules are extensions of the plasma membrane that reach deep into the interior of each muscle fiber. They are associated with another specialized organelle, the sarcoplasmic reticulum. This diagram and electron micrograph show the relation between T tubules and the sarcoplasmic reticulum running around the perimeter of several myofibrils in muscle from a frog. Notice that these structures may be buried deep within the interior of a single fiber, so the plasma membrane could be located as much as 50 μm away. Dark spots in the electron micrograph are glycogen granules. [Adapted from Peachey, 1965.]

transverse tubule (or **T tubule**). T tubules are less than 0.1 μm in diameter and form a network surrounding neighboring myofibrils (Figure 10-21). The membrane of this network of tubules is connected directly to the plasma membrane of the muscle fiber, and the lumen of the T-tubule system is continuous with the solution on the outside of the fiber. This continuity was confirmed by the demonstration that ferritin or horseradish peroxidase—protein molecules that are much too large to cross plasma membranes—nevertheless appear in the lumen of the T tubules if a muscle fiber is soaked in a solution of these molecules before the tissue is fixed and then examined with an electron microscope.

The T-tubule system provides the anatomic link between the plasma membrane and the myofibrils deep inside the muscle fiber. When Huxley and Taylor placed their stimulating pipette at a Z disk, over the entrance to a T tubule (see Figure 10-20), depolarizing current spread along the membrane of the tubule and initiated contraction deep within the muscle fiber. If they produced hyperpolarizing current from their pipette instead, no contraction occurred. Comparative studies have further strengthened the conclusion that T tubules carry excitation into muscle fibers. In crabs and some lizards, the T tubules are located at the ends of the A bands, rather than at the Z disks (Figure 10-22). In these species, contraction is produced when a stimulating pipette is placed at the edge of an A band, rather than over a Z disk. From such results it has been concluded that T tubules, rather than the Z disks or any other part of the sarcomere, are most likely to transmit excitation into muscle fibers.

Further confirmation that T tubules play an important role in excitation-contraction coupling was obtained by osmotically shocking muscle fibers with a 50% glycerol solution, which disconnects the T tubules from the plasma membrane. After this treatment, membrane depolarization no longer evokes a contraction; that is, physically uncoupling the T-tubule system from the plasma membrane functionally uncouples the contractile system from action potentials propagating along the plasma membrane.

The spread of excitation to the center of a muscle fiber is reduced if tetrodotoxin is added to the bath or if the concentration of Na^+ in the extracellular fluid is reduced. Either treatment reduces or eliminates sodium-based APs, suggesting that the APs that are characteristic of the plasma membrane are actively carried deep into the muscle fiber by the membranes of the T tubules.

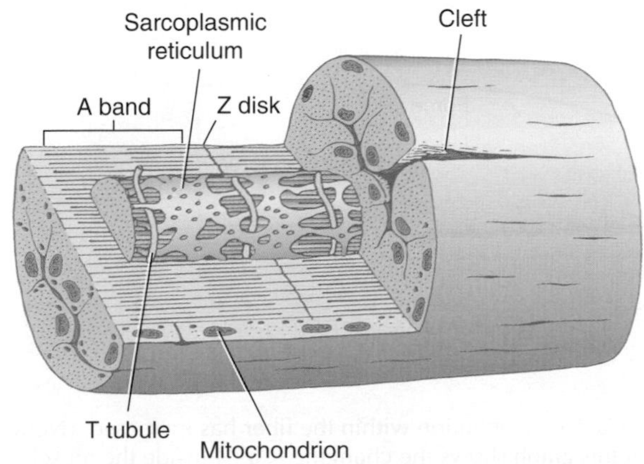

Figure 10-22 In crab muscle fibers, the T tubules are located at the A bands, rather than at the Z disks as in frog fibers. Stimulation with an extracellular micropipette produces a local contraction in crab fibers only when the tip of the pipette is placed over an A band. Compared with frog muscle fibers, crab fibers have a larger diameter and contain deep clefts. [Adapted from Ashley, 1971.]

Sarcoplasmic reticulum

In addition to the T-tubule system, striated muscle fibers contain a second intracellular membrane system, the **sarcoplasmic reticulum** (or **SR**). In frog muscle, the SR forms a hollow collar, called the terminal cisterna, around each myofibril on either side of a Z disk, and extends from one Z disk to the next as well (see Figure 10-21). The T tubules are sandwiched between the terminal cisternae of adjacent sarcomeres, but do not contact them directly.

Several different experimental approaches have demonstrated that the sarcoplasmic reticulum actively transports Ca^{2+} from the surrounding medium and concentrates it. When an AP is conducted along the membrane of a T tubule, it causes the neighboring SR to release stored Ca^{2+} ions into the cytoplasm by a mechanism that will be discussed shortly. The released Ca^{2+} ions are then available to bind to troponin, beginning the contraction process that was described in the previous section.

The calcium-sequestering activity of the sarcoplasmic reticulum is sufficiently powerful to keep the concentration of free Ca^{2+} in the myoplasm of resting muscle fibers below 10^{-7} M—low enough to remove essentially all Ca^{2+} from troponin in the cytoplasm. In other words, the SR is capable of driving the concentration of intracellular free Ca^{2+} so low that contraction is prevented. This ability of the SR to remove Ca^{2+} from the myoplasm depends on the activity of Ca^{2+}/Mg^{2+} ATPases, or calcium pumps, proteins in the SR membrane that bind and transport Ca^{2+} ions. In freeze-fracture electron micrographs, many densely packed inclusions can be seen in the membrane of the longitudinal elements of the SR; these inclusions have been associated with calcium pump molecules. The calcium pump, like other active transport molecule, requires ATP as its energy source—another key role for ATP in muscle contraction.

Under resting conditions in the sarcoplasmic reticulum, Ca^{2+} is bound to a protein called **calsequestrin**.

As a result, even though the total amount of Ca^{2+} within the SR is high, the concentration of *free* Ca^{2+} remains relatively low, reducing the gradient against which the pumps must work.

Combining observations of Ca^{2+} uptake and release by the sarcoplasmic reticulum with the role Ca^{2+} was known to play in the interaction between thin and thick filaments strongly suggested to physiologists that muscle contraction is initiated when the SR releases Ca^{2+} into the myoplasm. The first direct evidence that the concentration of free Ca^{2+} in muscle fibers rises in response to electrical stimulation came from a photometric method using the calcium-sensitive bioluminescent protein aequorin, isolated from a species of jellyfish (see Figure 6-27). The light-emitting reaction of aequorin is complex, however, and its response to changes in the concentration of free Ca^{2+} is rather slow. In more recent experiments, dyes have been used that respond to changes in Ca^{2+} concentration with rapid changes in their fluorescent properties. The dye furaptra, for example, fluoresces in the absence of Ca^{2+}; that is, it emits light of a particular wavelength when it is illuminated with exciting light of a different wavelength. The intensity of furaptra's fluorescence decreases as the Ca^{2+} concentration increases, so furaptra can be used to monitor changes in Ca^{2+} concentration within muscle fibers. When a furaptra-loaded muscle fiber is stimulated with a brief electric shock, the fluorescence of the dye first declines and then returns to its initial value (Figure 10-23). This result indicates that when the muscle is electrically stimulated, the amount of free Ca^{2+} in the myoplasm transiently increases. A very small fraction of the newly released Ca^{2+} binds to the furaptra, and the fluorescence of the dye declines. Most of the Ca^{2+} released is free to bind to troponin. As the released Ca^{2+} is resequestered, Ca^{2+} dissociates from the dye, and the dye's fluorescence rises again.

All of this evidence indicates that contraction is activated when Ca^{2+} ions are released from the sarcoplasmic reticulum, and that this release is linked to APs that are initiated at the plasma membrane and transmitted into the depths of the muscle fiber along the T tubules. Notice that in vertebrate skeletal muscle, essentially

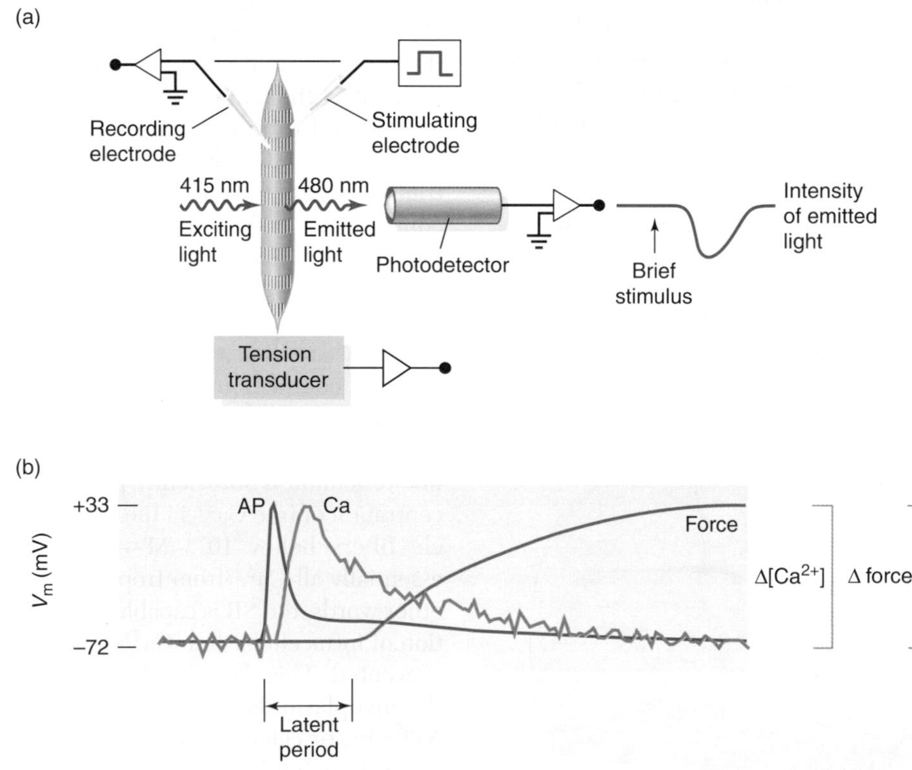

Figure 10-23 The concentration of free Ca^{2+} in a muscle fiber can be measured using a calcium-sensitive fluorescent dye such as furaptra. **(a)** In this experimental setup, a muscle fiber injected with furaptra is electrically stimulated and the subsequent changes in fluorescence, the membrane potential, and the production of tension by the fiber are recorded. **(b)** When the muscle fiber is stimulated, an AP propagates along the plasma membrane and is recorded by the recording microelectrode. A short time later, the fluorescence signal from the calcium-sensitive dye inside the fiber indicates that the Ca^{2+} concentration within the fiber has increased. (Notice that this graph shows the changing $[Ca^{2+}]$ inside the muscle fiber, rather than the fluorescence of the furaptra, although the concentration was inferred from the dye's fluorescence. The fluorescence of furaptra varies inversely with $[Ca^{2+}]$.) Even later, the tension transducer measures the production of tension by the fiber. Notice that the tension begins to rise only after the AP is over and the intracellular Ca^{2+} concentration has already begun to decline. [Part b courtesy of S. M. Baylor.]

none of the regulatory Ca^{2+} enters the cell across the plasma membrane; *the only source of the regulatory calcium is the SR.* The anatomy of the T tubules and SR suggests how this coupling happens. As noted above, each T tubule is located in close apposition to the terminal cisternae of the SR (see Figure 10-21). In fact, histologists have for decades called this portion of a muscle fiber the *triad* because sections through this region consistently revealed three associated tubes or sacks. Two of the sacks are always relatively large, and they are located on either side of a much smaller central tube or sack. We now know that the two large sacks are terminal cisternae of the SR and the smaller central sack is a T tubule. As we will see, the triads play a crucial role in linking a muscle AP to a change in myoplasmic free Ca^{2+}.

Receptor molecules in triads

In 1970, electron microscopy experiments by Clara Franzini-Armstrong revealed electron-dense particles in the part of the SR membrane that lies adjacent to the T tubule (Figure 10-24a on the next page). She called these structures "feet." More recently, these feet have been identified as the cytoplasmic portion of membrane-associated protein complexes, called **ryanodine receptors** because they bind the drug ryanodine. Ryanodine receptors are tetrameric proteins that span the SR membrane (Figure 10-24b). Half of them are lined up with proteins in the T-tubule membrane, called **dihydropyridine receptors** for their ability to bind dihydropyridine drugs (Figure 10-24c). The dihydropyridine receptors have been more precisely identified as a type of voltage-gated calcium channel, called an L-type channel, which explains how they can respond to APs traveling along the membrane of T tubules. It is noteworthy that in skeletal muscle, little or no Ca^{2+} passes from the T-tubule lumen to the myoplasm through these channels, but in vertebrate contractile cardiac muscle, Ca^{2+} entering activated fibers through dihydropyridine receptor channels plays an important role, as we will see. In skeletal muscle it is a *mechanical* interaction between activated dihydropyridine receptors and ryanodine receptors that permits Ca^{2+} to move from the lumen of the sarcoplasmic reticulum into the myoplasm. Figure 10-24d summarizes the important molecular players in the Ca^{2+} economy of the SR.

Shortly after Franzini-Armstrong described the feet located between the T tubule and SR membranes, Knox Chandler and his colleagues suggested that these proteins are Ca^{2+} channels and incorporated them into a "plunger model" for release of Ca^{2+} from the sarcoplasmic reticulum (Figure 10-25 on page 389). They suggested that depolarization of the T tubule causes a plug to be removed from Ca^{2+} channels in the SR membrane, allowing Ca^{2+} to escape into the myoplasm, driven down its steep electrochemical gradient. When the T-tubule membrane repolarizes, the plug is replaced, preventing further Ca^{2+} release. The current version of the plunger model proposes that when the T tubule depolarizes, a change in the conformation of a voltage-sensitive dihydropyridine receptor forces the calcium channel of an associated ryanodine receptor to open, allowing Ca^{2+} to rush into the myoplasm from the lumen of the SR.

Interestingly, only about half of the ryanodine receptors in the sarcoplasmic reticulum are associated directly with dihydropyridine receptors in the T-tubule membrane. The other, independent ryanodine receptors are activated by the increase in myoplasmic free Ca^{2+} following opening of the mechanically linked channels, a process called *calcium-induced calcium release.* Activation of these unlinked ryanodine receptors in turn opens more Ca^{2+} channels in the membrane of the sarcoplasmic reticulum. Calcium-induced calcium release has been found in many cell types, and it plays an important role in excitation-contraction coupling in cardiac muscle.

Time course of calcium release and reuptake

The ability to measure rapid changes in the myoplasmic Ca^{2+} concentration, combined with knowledge of calcium binding by troponin and the kinetics of the calcium pumps in the SR membrane, has permitted the Ca^{2+} fluxes during muscle contraction and relaxation to be modeled. The results suggest that when T tubules become depolarized, Ca^{2+} flows out of the sarcoplasmic reticulum for several milliseconds, then the Ca^{2+} channels close. Most of the Ca^{2+} that leaves the sarcoplasmic reticulum binds very quickly to troponin. The concentration of troponin in muscle fibers is about 240 μM, which represents a large buffer for Ca^{2+} ions. Thus, only a very small amount of the released Ca^{2+} remains free in the myoplasm. During and after the release of Ca^{2+} from the SR, the free Ca^{2+} in the myoplasm is pumped back into the SR lumen, lowering the myoplasmic level of free Ca^{2+}. As the concentration of free Ca^{2+} in the myoplasm becomes very low, Ca^{2+} bound to troponin is released back into the myoplasm and is subsequently pumped back into the SR, where it binds to calsequestrin.

The Contraction-Relaxation Cycle

Starting with a relaxed skeletal muscle fiber, let's summarize the sequence of events that leads to contraction and then relaxation.

1. The plasma membrane of the fiber is depolarized by an AP or, in some muscles, by electrotonically conducted postsynaptic potentials. In the body of an animal, APs in skeletal muscle fibers are generated by postsynaptic potentials, so neuronal input is required for initiating contraction in skeletal muscle.

2. The AP is conducted deep into the muscle fiber along the T tubules.

3. In response to depolarization of the T-tubule membrane, dihydropyridine receptors in the membrane

(a)

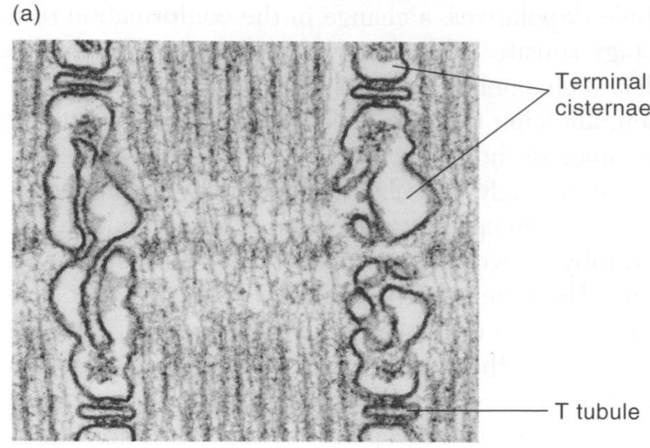

(c)

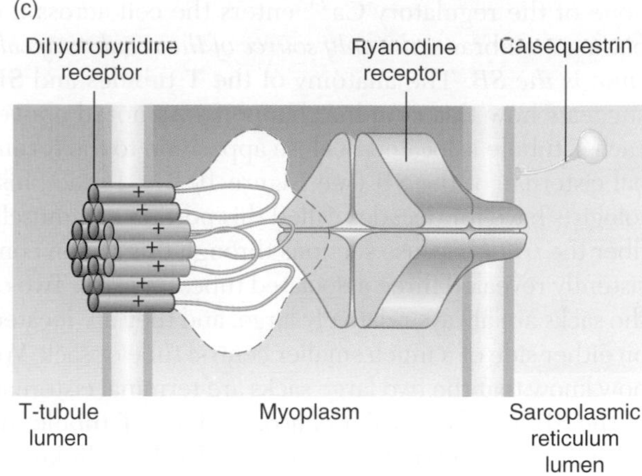

(b) Ryanodine receptor

Viewed from outside SR

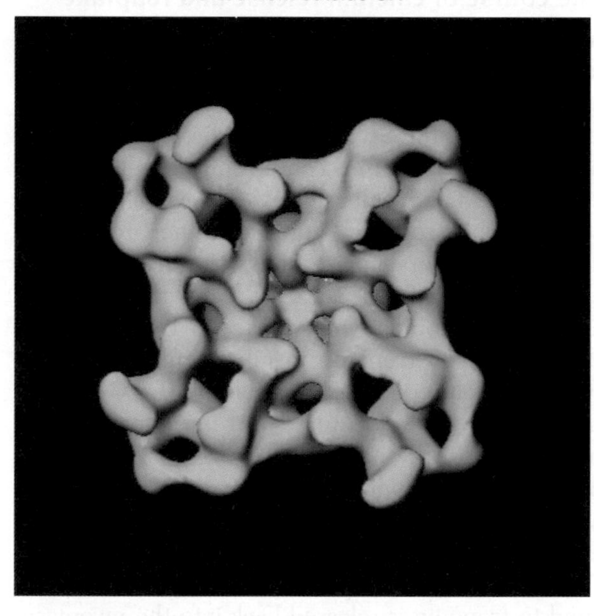

(d)

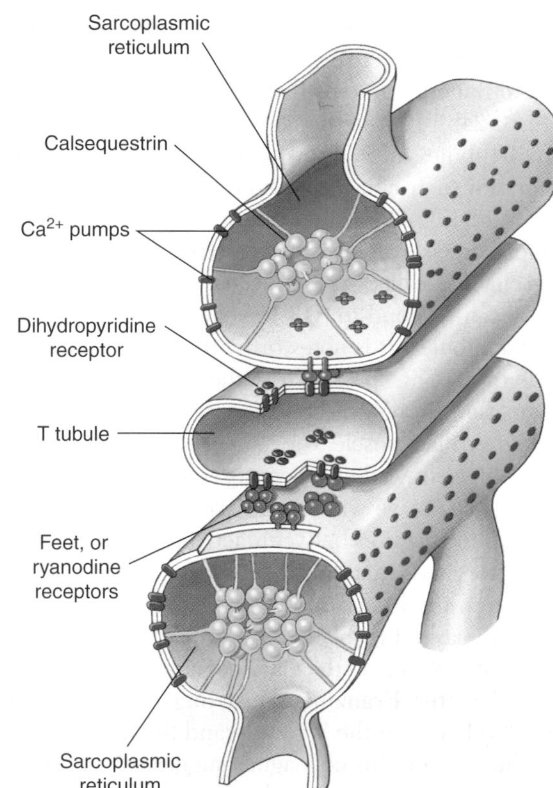

Viewed from side

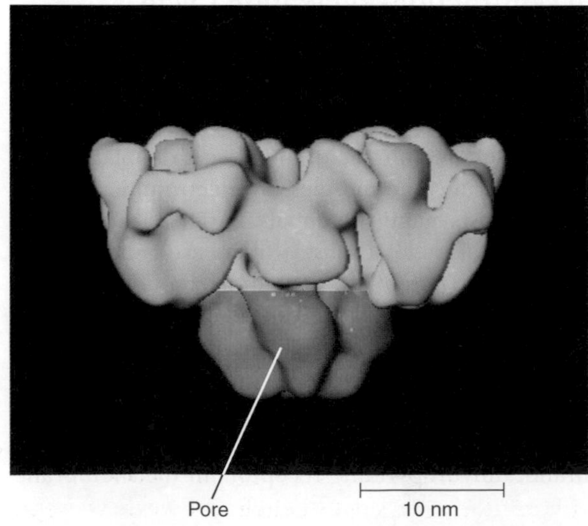

undergo a conformational change that—through direct mechanical linkage to ryanodine receptors in the SR membrane—causes the opening of Ca^{2+} channels in the SR membrane (see Figure 10-25, step 2).

4. As Ca^{2+} flows from the lumen of the sarcoplasmic reticulum into the myoplasm, the free Ca^{2+} concentration of the myoplasm increases within a few milliseconds from a resting value of below 10^{-7} M to an active level of about 10^{-6} M or higher. The Ca^{2+} channels in the SR membrane then close because V_m of the T tubules has returned to V_{rest}.

5. Most of the Ca^{2+} that enters the myoplasm binds rapidly to troponin, inducing a conformational change in the troponin molecules. This conformational change causes a change in the position of the associated tropomyosin molecule, allowing myosin cross-bridges to bind to actin thin filaments (see Figure 10-16c).

6. Myosin cross-bridges attach to the actin filaments and go through a series of binding steps that cause the myosin head to rotate against the actin filaments, pulling on the cross-bridge link (see Figure 10-11). This pulling produces force on, and in some cases sliding of, the thin filaments that is directed toward the center of the sarcomere, causing the sarcomere to shorten by a small amount (see Figure 10-8a).

7. ATP binds to the ATPase site on the myosin head, causing the myosin head to detach from the thin filament. ATP is then hydrolyzed, and the energy of hydrolysis is stored as a "recocking" of the myosin head, which then attaches to the next site along the actin filament as long as binding sites are still available, and the cycle of binding and unbinding is repeated (see Figure 10-10). During a single contraction, each cross-bridge attaches, pulls, and detaches many times as it "rows" itself along the actin filament toward the Z disk.

8. Finally, calcium pumps in the SR membrane actively transport Ca^{2+} from the myoplasm back into the SR lumen (see Figure 10-25, step 3). As the concentration of free Ca^{2+} in the myoplasm drops, Ca^{2+} bound to troponin is released, allowing tropomyosin again to prevent cross-bridge attachment, and as long as ATP is present, the muscle relaxes. The muscle remains relaxed until the next depolarization.

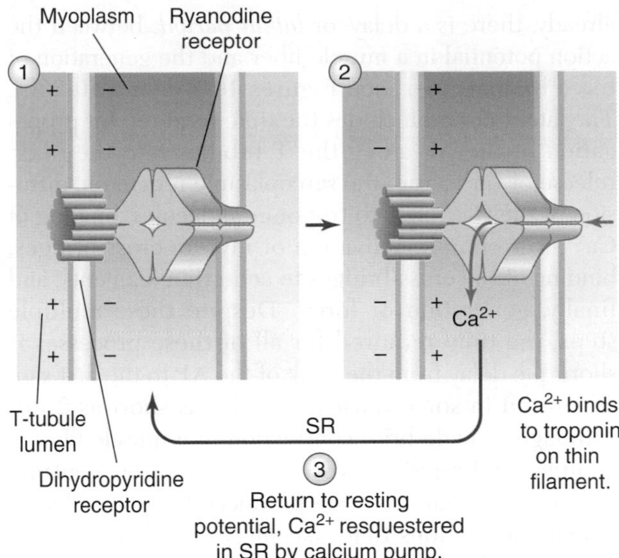

Figure 10-25 Depolarization of the T-tubule membrane indirectly causes calcium channels in the sarcoplasmic reticulum to open. When the membrane of the T tubule is at rest (1), the calcium-conducting channels of ryanodine receptors are closed. When the T-tubule membrane depolarizes (2), dihydropyridine receptors convey the signal to the ryanodine receptors, and their calcium channels open, allowing Ca^{2+} to flow out of the SR lumen into the myoplasm. The free Ca^{2+} binds to troponin, exposing cross-bridge binding sites on actin molecules. When the membrane potential returns to its resting value (3), the ryanodine receptor calcium channels close. Calcium pumps in the SR resequester Ca^{2+}, shifting the equilibrium of Ca^{2+} binding to troponin and causing the cross-bridge binding sites on actin to be concealed. [Adapted from Berridge, 1993.]

Figure 10-24 (*on facing page*) Dihydropyridine receptors in the T-tubule membrane and ryanodine receptors in the SR membrane interact at triads. **(a)** Near the Z disks of mammalian and amphibian skeletal muscle, T tubules and the terminal cisternae of two sarcomeres form triads. Darkly stained "feet" connect the terminal cisternae with the T tubule in each triad. **(b)** The molecular structure of a ryanodine receptor reveals fourfold symmetry, suggesting that it is a tetramer. The boundaries of the monomers have not yet been resolved. In this diagram, the feet would be some part of the green region of the complex, and the Ca^{2+}-conducting pore is colored pink. **(c)** Dihydropyridine receptors span the T-tubule membrane and are organized in clusters of four. Each dihydropyridine receptor associates closely with the extracellular "foot" of a ryanodine receptor, which extends from the SR membrane. The molecular structure of each unit of the dihydropyridine receptor is similar to that of the voltage-gated Ca^{2+} channel illustrated in Figure 5-27. Calsequestrin inside the SR binds Ca^{2+}. **(d)** In an intact triad, several molecules contribute to the control of the myoplasmic Ca^{2+} concentration. The voltage-sensitive dihydropyridine receptors and the ryanodine receptors work together, linking depolarization of the T tubule to the opening of calcium channels in the SR membrane. A calcium ATPase (calcium pump) in the SR membrane resequesters Ca^{2+} from the myoplasm, and calsequestrin inside the SR binds Ca^{2+}, reducing the concentration of ionic free Ca^{2+} inside the SR. [Part a courtesy of Clara Franzini-Armstrong; part b adapted from Samsó and Wagenknecht, 1998; part c adapted from Lamb, 2000; part d adapted from Block et al., 1988.]

THE TRANSIENT PRODUCTION OF FORCE

Up to now, we have considered the mechanics of muscle fibers during a single twitch. As we have seen already, there is a delay, or *latent period,* between the action potential in a muscle fiber and the generation of force by that fiber (see Figures 10-18b and 10-23b). The latent period includes the time required for propagation of the AP along the T tubules into the fiber, release of Ca^{2+} from the sarcoplasmic reticulum, diffusion of the Ca^{2+} ions to troponin molecules, binding of Ca^{2+} to troponin, activation of myosin cross-bridges, binding of the cross-bridges to actin thin filaments, and finally, generation of force. Despite these multiple steps, the time required for all of these processes is short: the delay from the peak of the AP to the first sign of tension in some muscles can be as short as 2 ms. Although a single brief contraction in a muscle fiber is complex and rapid, almost all useful muscle contractions in an animal's body are produced by repeated individual contractions that take place simultaneously in many muscle fibers. We now will consider how the activities of individual fibers and individual twitches are coordinated to do useful work.

Series Elastic Components of Muscle

A muscle can be modeled as a contractile element that is arranged in parallel with one elastic component and in series with another elastic component, as depicted in Figure 10-26a. The *parallel elastic component* in this model represents the properties of the plasma membrane of the muscle fibers and the connective tissues that run in parallel with them. The *series elastic component,* also called the *series elastic elements,* represents tendons, connective tissues that link muscle fibers to the tendons, and perhaps the Z disks of the sarcomeres. An additional important constituent of the series elastic elements appears to be the myosin cross-bridge links themselves, which undergo some stretch when tension is generated (see Figure 10-11). Representing all of the elastic components of a muscle with only these two components greatly simplifies the model, making it easier to manipulate mathematically while maintaining sufficient accuracy to increase our understanding of the mechanics of muscle contraction.

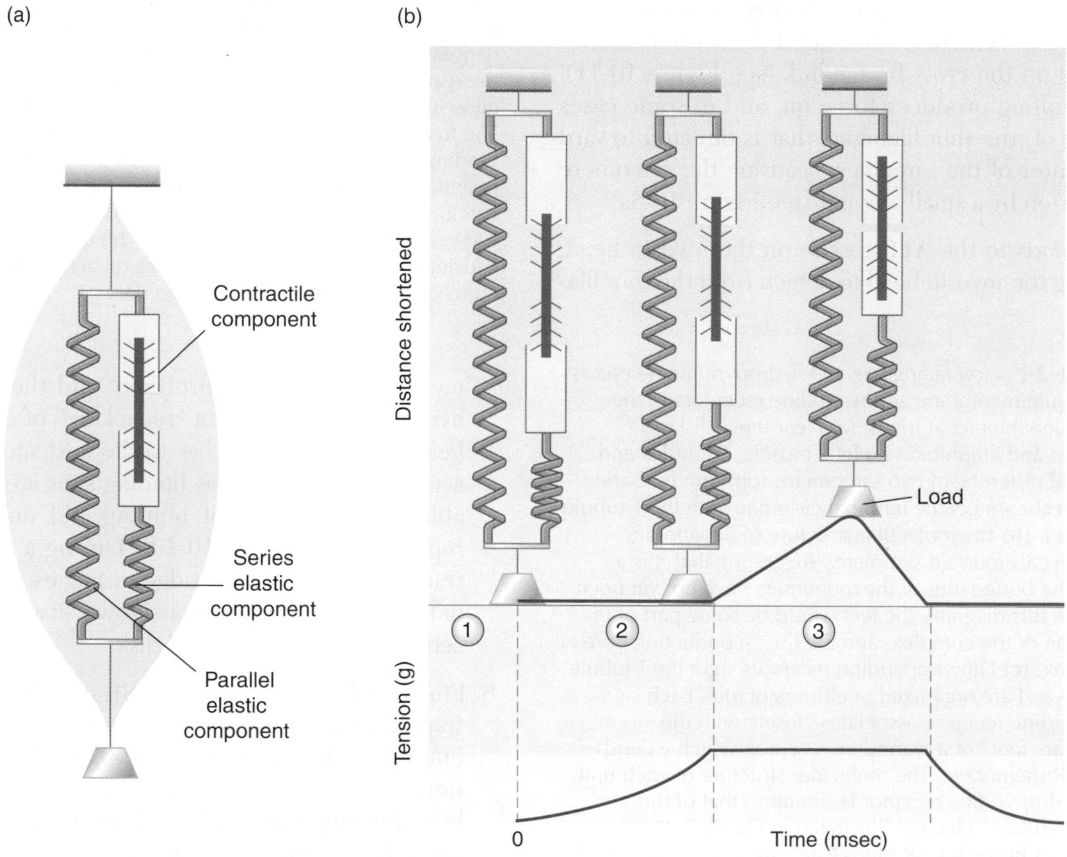

Figure 10-26 A muscle fiber, or an entire muscle, can be represented by a mechanical model that includes a contractile component and elastic components. **(a)** Mechanical model of a muscle consisting of a contractile component (the sarcomeres), in series with one elastic component (e.g., tendons) and in parallel with another elastic component (e.g., connective tissue layers within the muscle). **(b)** Effect of series elastic components on muscle contraction. At the beginning of contraction in this model muscle, the load rests on a surface (1). As the thick and thin filaments begin to slide past each other and tension increases, the series elastic components are stretched (1 ⟶ 2), but the length of the muscle has not yet changed; contraction up to this point (2) is isometric. Once the muscle generates tension that is equal to or greater than the weight of the load, the load is lifted and the contraction becomes isotonic (3). Note that as contraction progresses, the thick and thin filaments overlap increasingly and more cross-bridges can bind. [Adapted from Vander et al., 1975.]

As a muscle becomes activated and the contractile component begins to shorten, the series elastic component must be stretched before full tension can be transmitted to the external load (steps 1 and 2 in Figure 10-26b). When the tension developed equals the weight of the load, the muscle begins to lift the load off of the surface (step 3). In steps 1 and 2, the contraction is essentially isometric, whereas in going from step 2 to 3 it becomes more isotonic as the load is finally lifted. If the load were sufficiently heavy that the muscle never produced tension equal to its weight, the contraction would remain isometric throughout. (At maximal tension during an "isometric" contraction, a small amount of shortening in the contractile component stretches the series elastic component by an amount equivalent to about 1% of the muscle length, even though the external length of the muscle does not change.)

Time is required for the thin and thick filaments to slide past each other and stretch the series elastic component as tension builds up. Thus the series elastic component slows the development of tension in the muscle and smooths out abrupt changes in tension.

The Active State

During contraction, external shortening of the muscle fiber and production of tension reach a maximum within 10 to 500 milliseconds, depending on the kind of muscle, the temperature, and the load. At first glance, this statement might suggest that the contractile mechanism is activated with a similar slowly rising time course. It is important, however, not to confuse the time it takes a muscle to develop tension with the time course of cross-bridge activity. Cross-bridges become activated and attach to thin filaments before the filaments begin to slide past each other. Then once the filaments begin to slide, they must take up the slack in the series elastic component before tension can be fully developed.

The state of the cross-bridges after activation, but before the muscle has had a chance to develop full tension, can be determined by the application of quick stretches with a special apparatus. These stretches can be applied at various times after stimulation, both before and during contraction. The rationale for quick-stretch experiments is that the series elastic elements are stretched by the apparatus when it applies force, eliminating the time that is normally required for the contractile mechanism to take up this slack. This maneuver thus improves the time resolution when measuring the state of cross-bridge activity. The "internal" tension recorded by the sensing device during a quick stretch represents the tensile strength of the bonds between the thick and thin filaments, which depends on the holding strength of the cross-bridges at the instant of the stretch. If the stretch applied is stronger than the holding strength of the cross-bridge bonds, the cross-bridges will slip, and the filaments will slide past each other in a negative direction. Thus, loading during a quick stretch that is just sufficient to make the thick and thin filaments slide apart approximates the load-carrying capacity of the muscle at the time of stretch. This tension should be proportional to the average number of active cross-bridges per sarcomere.

In the relaxed state, a muscle has very little resistance to stretch aside from the contribution made by connective tissue, the plasma membranes, and other elastic components. Quick-stretch experiments revealed that a muscle's resistance to stretch rises steeply very soon after stimulation and reaches a maximum at about the time when external shortening or tension in an unstretched muscle is just getting under way. After a brief plateau, the resistance to stretch returns to the low level characteristic of the relaxed muscle.

The term **active state** is used to describe this state of increased stretch resistance, or internal tension, in a muscle following a brief stimulation (Figure 10-27a on the next page). The active state corresponds to the formation of bonds between myosin cross-bridges and actin thin filaments and to the subsequent slight internal shortening generated by the cross-bridges. Because cross-bridge activity is controlled by the concentration of free Ca^{2+} in the myoplasm, the time course of the active state is believed to approximately parallel changes in myoplasmic Ca^{2+} concentration following stimulation. The brief increase in tension due to cross-bridge activity is called a **twitch.**

If stimulation of a muscle is prolonged, the active state persists. A prolonged active state produced by a barrage of high-frequency APs is called **tetanus.** In this state, the measurable external isometric tension can increase until it reaches the value of internal tension during the active state as measured by quick-stretch experiments (Figure 10-27b).

Contractile States: Twitches and Tetanus

The graphs in Figure 10-27 raise a question: Why is the maximal external isometric tension produced by the muscle during a twitch so much lower than the internal tension associated with the active state, or than the tension produced during tetanus? In other words, during a brief contraction, why does the muscle produce so much less tension than it is actually capable of producing?

During a single twitch, the active state is rapidly terminated by the calcium-sequestering activity of the sarcoplasmic reticulum, which efficiently removes Ca^{2+} from the myoplasm soon after it is released. Thus, the active state begins to decay even before the filaments have had time to slide far enough to stretch the series elastic component to a fully developed tension. For this reason, the tension of which the contractile system is capable cannot be realized in a single twitch.

Before the peak of the twitch tension, the contractile elements store potential energy in the series elastic component by progressively stretching it. If a second

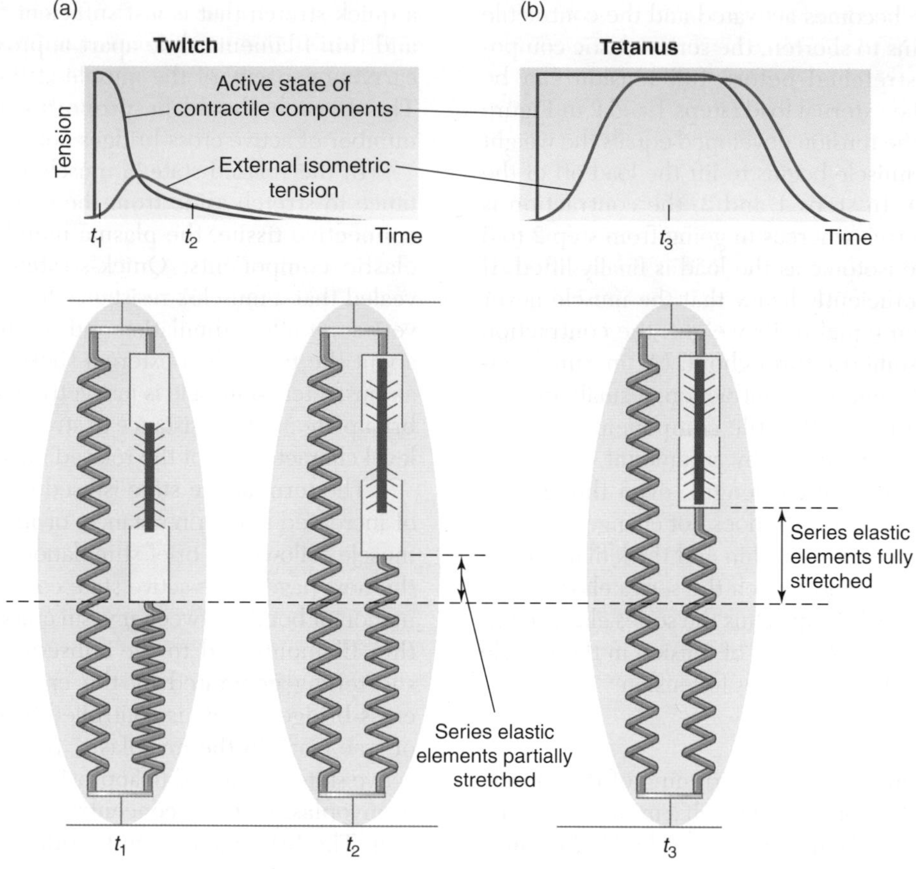

Figure 10-27 The time course of the active state differs from the time course of tension production. **(a)** The active state—as measured in quick-stretch experiments—develops rapidly in response to a short stimulus. This brief response is called a twitch. Measurable external isometric tension develops considerably more slowly, and it fails to reach the same tension that can be measured during a quick stretch because without the quick stretch, the muscle must first do work to take up slack in the series elastic elements. **(b)** In response to a prolonged stimulus, a tetanic contraction develops. In this case, the external isometric tension has time to reach the same value as the internal tension that is measured during quick-stretch experiments. The time period shown in part a is much shorter than the time period shown in part b. [Adapted from Vander et al., 1975.]

AP follows the first before the sarcoplasmic reticulum can entirely remove the previously released Ca^{2+} from the myoplasm, the concentration of Ca^{2+} remains high in the myoplasm, and the active state is prolonged. If the active state continues long enough, the isometric tension increases over time until the tension produced by the internal shortening of the contractile components and the stretching of the series elastic component is just sufficient to cause cross-bridges to slip and prevent further shortening of the contractile components. The muscle has then reached full tetanus.

Depending on the repetition rate of muscle APs, the amount by which individual twitches fuse can vary, reaching a maximal value in tetanus (Figure 10-28). The addition of tension due to repeated rapid stimulation is called **summation of contraction.** Notice that summation depends partly on the release of free Ca^{2+} into the cytoplasm faster than the pumps in the SR membrane can remove it and partly on the inability of the series elastic component to relax back to its slack condition between stimuli.

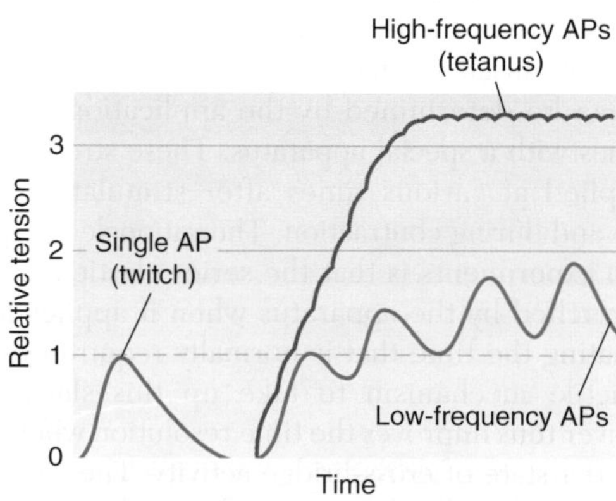

Figure 10-28 Twitches fuse when stimulating APs arrive in rapid succession, generating summed tension. A single AP produces a single twitch. If a set of relatively low-frequency APs is conducted along the muscle fiber, each successive twitch begins before the muscle has time to return to its fully relaxed condition following the previous twitch, so tension builds up moderately. At a maximum frequency of APs, the twitches fuse with one another, producing a long, strong contraction called tetanus.

ENERGETICS OF MUSCLE CONTRACTION

In muscle contraction, two major processes require the expenditure of energy. The most obvious is the hydrolysis of ATP by myosin cross-bridges as they cyclically attach to and detach from actin thin filaments (see Figure 10-10). The other process that consumes ATP is the pumping of Ca^{2+} back into the sarcoplasmic reticulum against a Ca^{2+} concentration gradient (see Figure 10-25, step 3).

Biochemical studies have determined that two molecules of ATP are required to pump each Ca^{2+} ion into the sarcoplasmic reticulum. During a twitch, some amount of Ca^{2+} is released during the few milliseconds following the AP, and exactly that amount of Ca^{2+} must eventually be pumped back into the sarcoplasmic reticulum if the muscle fiber is to relax. During summation (just as in a twitch), the calcium pumps immediately start to resequester the Ca^{2+} released during the first AP, but they do not have time to remove all of it from the myoplasm before the next AP occurs, and each successive AP causes more Ca^{2+} to be released. The buildup of Ca^{2+} in the myoplasm keeps troponin saturated with Ca^{2+} until the APs cease; at that point, the calcium pumps eventually can return all of the released Ca^{2+} to the SR. To maintain the condition of tetanus, ATP is steadily hydrolyzed by both the myosin ATPase and the calcium pumps. Then further ATP is used by the pumps to return the muscle to its relaxed state.

ATP Consumption by the Myosin ATPase and Calcium Pumps

The relative consumption of ATP by myosin ATPase and the calcium pumps was determined in experiments with tetanized frog muscle. In these experiments, muscles were stretched to different degrees, producing different amounts of overlap between thick and thin filaments. As the muscle is progressively stretched, fewer cross-bridges can interact with actin, reducing both the amount of force that can be produced and the amount of ATP hydrolyzed by the myosin ATPase. (Remember that myosin can hydrolyze ATP on its own, but when myosin is not bound to actin, the hydrolysis products ADP and P_i are released very slowly. Therefore, if myosin cross-bridges have few available sites where they can bind to actin, their ATPase activity will be low.) By contrast, stretching the muscle should have little or no effect on the rate at which Ca^{2+} is released from and resequestered by the sarcoplasmic reticulum, because these processes are mediated by membrane proteins whose activity is unrelated to the amount of myofilament overlap. Thus, as the muscle is increasingly stretched, the total ATPase activity declines. At the length where the myofilaments no longer overlap, any ATPase activity can be attributed entirely to the calcium pumps.

Using this experimental approach, researchers determined that the calcium pumps accounted for about 25%–30% of total ATPase activity during muscle contraction. It is generally assumed that this percentage is the same for all muscles; that is, we believe that muscles with a higher maximal contraction velocity, and hence higher myosin ATPase activity, also have faster calcium pumps in their sarcoplasmic reticulum. It is possible, however, that in the very fast sound-producing muscles discussed later in this chapter, pumping of Ca^{2+} may account for a larger fraction of overall energy usage.

Regeneration of ATP During Muscle Activity

As the previous discussion indicates, muscles use exclusively ATP to power their contraction. Yet early measurements of overall ATP usage during muscle contraction produced a surprising result: the ATP concentrations in stimulated and unstimulated muscles (matched as closely as possible for other variables) were nearly identical. For many years, this finding caused some muscle physiologists to hypothesize that muscles use some molecule other than ATP to power their contractions. However, an alternative explanation turned out to be correct. In addition to ATP, muscle fibers contain a second high-energy molecule: creatine phosphate, also known as phosphocreatine (see Figure 3-36). Within muscle fibers, the enzyme creatine phosphokinase transfers a high-energy phosphate from creatine phosphate to ADP, regenerating ATP so quickly that the ATP concentration remains constant, even when the muscle is using energy at a high rate. Because of this reaction, accurately measuring the amount of ATP hydrolyzed by the muscle is best done by measuring either the drop in the concentration of creatine phosphate or the rise in the concentration of P_i.

Beyond the technical issue of accurately measuring the rate of ATP hydrolysis, the creatine phosphokinase reaction is extremely important for effective muscle function. If a muscle runs out of ATP, it goes into rigor (see Figure 10-17b). It is therefore essential that the concentration of ATP in muscles be buffered. Under most circumstances, oxidative or anaerobic metabolism can generate ATP fast enough to power muscle contraction. However, during high-intensity, short-duration activity (e.g., when an animal sprints to run down prey or to avoid becoming prey), ATP may be used up too fast to be replenished by these mechanisms. Under these circumstances, continuous rephosphorylation of ADP by the creatine phosphokinase reaction can keep the muscle supplied with ATP for a short time (Figure 10-29 on the next page). The concentration of creatine phosphate in muscle fibers (20–40 mM) is several times greater than the reserve of ATP (about 5 mM in muscle fibers), so an animal can use this reserve of high-energy phosphate to power muscle contraction under sudden load conditions until anaerobic and oxidative metabolism can catch up to the increased need. An animal's life may depend on this short-lived extra source of energy.

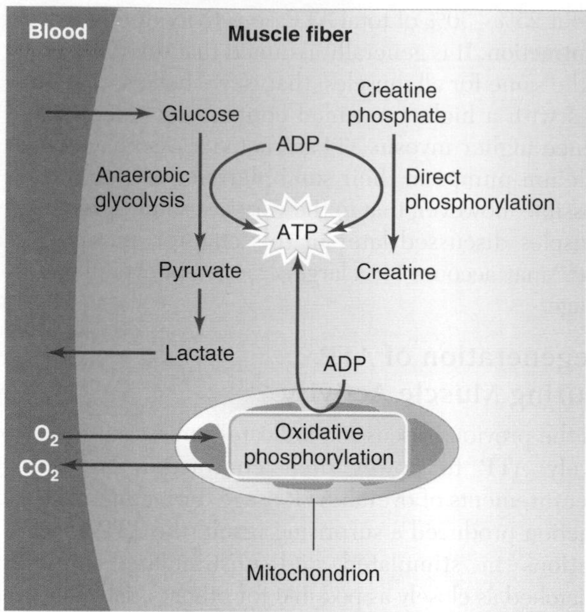

Figure 10-29 The ATP that provides energy for muscle contraction comes from several different sources. In direct phosphorylation, high-energy phosphates are transferred from creatine phosphate to ADP, regenerating ATP. The concentration of creatine phosphate in muscle fibers is higher than the concentration of ATP, so creatine phosphate effectively buffers the ATP concentration during short periods of intense demand. Anaerobic glycolysis metabolizes glucose, rephosphorylating ADP in the process. Lactate accumulates as a by-product and leaks into the blood. Oxidative phosphorylation of ADP regenerates ATP, but produces it more slowly than the other two processes and requires O_2 to proceed (see Chapter 3).

Moreover, the ATP concentration is stabilized because the creatine phosphokinase reaction greatly favors phosphorylation of ADP by creatine phosphate. Under most conditions, only the concentration of creatine phosphate falls in a working muscle, while the concentration of ATP remains nearly constant.

Recently a dietary supplement called creatine monohydrate has been advertised to improve athletic training and performance. Can you propose a physiological mechanism that might support this claim? Would this supplement be equally effective at improving performance in all kinds of activities? Why or why not?

FIBER TYPES IN VERTEBRATE SKELETAL MUSCLE

The muscular systems of animals perform a great variety of motor tasks, ranging from the high-speed movements of sound production, which occur at a frequency of several hundred contractions per second, to sustained locomotion during long-distance migrations, in which an individual may cover thousands of miles. Even a casual observer notices the diversity in the external attributes of muscular systems, such as lobster claws and human legs. There is an equally impressive diversity in the characteristics of the muscles themselves. To produce such a broad range of activities, different muscles must be organized to perform very differently. Recent experiments have shown that the properties of a muscle are in many cases well matched to the other components in a system, optimizing the system for its biological function. To appreciate how well muscles are adapted to their biological roles, we now examine the properties of a variety of muscles in light of their primary functions.

Classification of Fiber Types

The components of an individual muscle fiber confer on it several important characteristics that differ among fibers. The skeletal muscles of vertebrates typically contain muscle fibers of more than one type. Among the biochemical, metabolic, and histochemical properties that distinguish the various fiber types are the following:

- The electrical properties of the plasma membrane determine whether a fiber will respond with an all-or-none twitch or with a graded contraction. If the membrane produces APs, the fiber will contract with all-or-none twitches.
- The rate at which cross-bridges detach from actin thin filaments determines the maximal rate of contraction, V_{max}. The chemical nature of the myosin heavy chains determines the rate of cross-bridge detachment.
- The density of calcium pump molecules in the SR membrane determines how long the myoplasmic free Ca^{2+} remains elevated following an AP.
- The number of mitochondria and the density of a fiber's blood supply determine its maximum rate of sustained oxidative ATP production and hence its resistance to fatigue.

Based on these and other properties, four major groups of vertebrate skeletal muscle are recognized—**tonic fibers** and three types of **twitch** (or *phasic*) **fibers** (Table 10-1).

Tonic muscle fibers contract very slowly and do not produce twitches. They are found in the postural muscles of amphibians, reptiles, and birds, as well as in the muscle spindles that house muscle stretch receptors and in the extraocular muscles (the muscles that move the eyeball in its socket) of mammals. Tonic fibers normally produce no APs, and APs are not required to spread excitation because the innervating motor neuron runs the length of the muscle fiber, making repeated synapses all along it. In tonic muscle fibers, the myosin cross-bridges attach and detach very slowly, accounting for the

Table 10-1 **Properties of twitch (phasic) fibers in mammalian skeletal muscles**

Property	Slow oxidative (type I)	Fast oxidative (type IIa)	Fast glycolytic (type IIb)
Fiber diameter	↓	↔	↑
Force per cross-sectional area	↓	↔	↑
Rate of contraction (V_{max})	↓	↑	↑
Myosin ATPase activity	↓	↑	↑
Resistance to fatigue	↑	↔	↓
Number of mitochondria	↑	↑	↓
Capacity for oxidative phosphorylation	↑	↑	↓
Enzymes for anaerobic glycolysis	↓	↔	↑

Source: Adapted from Sherwood, 2001. Key = ↓ Low ↔ Intermediate ↑ High

fibers' extremely slow shortening velocity and their ability to generate isometric tension very efficiently.

Slow-twitch (or *type I*) *fibers* contract slowly and fatigue slowly; they are found, for example, in mammalian postural muscles. They are characterized by a slow to moderate V_{max} and slow Ca^{2+} kinetics. They generate all-or-none APs, so they respond to motor-neuron input with all-or-none twitches. Like other twitch fibers, type I fibers typically have one or at most a few motor endplates; in most vertebrates, all synapses on a single fiber are made by a single motor neuron. Slow-twitch fibers are used both for maintaining posture and for moderately fast repetitive movements. They fatigue very slowly for two reasons. First, they contain a large number of mitochondria and have a rich blood supply bringing plenty of oxygen, which supports sustained oxidative phosphorylation. Second, they use ATP at a relatively slow rate. They are characterized by a reddish color (examples are the dark-colored meat of fish and fowl) conferred by a high concentration of the oxygen-storage protein myoglobin (see Chapter 13). Muscles that contain a high proportion of type I fibers are often called red muscle.

Fast-twitch oxidative (or *type IIa*) *fibers* have a high V_{max} and activate quickly. They are specialized for rapid repetitive movements, such as sustained, strenuous locomotion—the flight muscles of migratory birds are a striking example. With their many mitochondria, they produce enough ATP by oxidative phosphorylation to support work over long periods. They are thus relatively resistant to fatigue, although they are not as tireless as type I fibers.

Fast-twitch glycolytic (or *type IIb*) *fibers* contract very rapidly and fatigue quickly. They have a high V_{max}, and they activate and relax quickly because of their rapid Ca^{2+} kinetics. These fibers contain few mitochondria, depending instead on anaerobic glycolysis to generate ATP. A familiar example of this type of fiber is found in the white breast muscles of domestic fowl, which are never used for flying and cannot produce sustained activity. (The breast muscles of migratory birds feature type IIa fibers, consistent with their locomotory function.) Ectothermic vertebrates, such as amphibians and reptiles, also make extensive use of glycolytic muscle fibers.

These categories are somewhat arbitrary because some muscle fibers combine properties of different types. In addition, the numerical values for many of the parameters vary among species. The slow-twitch fibers of a mouse, for instance, have a faster V_{max} than the fast-twitch oxidative fibers of a horse. Within a given muscle, however, the fiber types can be distinguished by their histological properties. For example, histochemical staining reveals differences in the properties of the myosin ATPase in different fiber types (Figure 10-30). Another useful histochemical method for distinguishing fiber types is based on the abundance of the protein complexes that carry out oxidative phosphorylation.

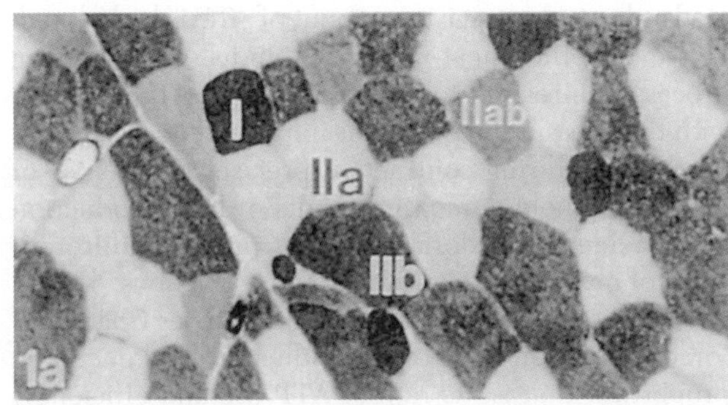

Figure 10-30 Histochemical staining for myosin ATPase activity reveals different types of fibers within a single muscle. This section through a muscle from a horse contains slow oxidative (type I), fast oxidative (type IIa), and fast glycolytic (type IIb) fibers. The fiber labeled IIab has intermediate properties. [Courtesy of L. Rome.]

All mammalian muscles contain a variety of fiber types, but the percentages of the different types vary from muscle to muscle and from animal to animal. Can you predict which type of muscle fibers would predominate in the large leg muscles of the antelope on the cover of this book? Compare this pattern with what you would expect to find in a sprinter, such as a cheetah.

Functional Rationale for Different Fiber Types

What do animals gain by having different types of muscle fibers? Fast-twitch fibers obviously are necessary if an animal is to move its limbs or fins very rapidly, but why, then, have slow-twitch fibers? A basic principle in muscle physiology is that there is always a trade-off between speed and energetic cost. Very fast muscles require a large amount of ATP. Slow muscles perform less rapidly, but they also use relatively little energy. To better understand this trade-off, it is useful to compare the energetic costs and mechanical abilities of fiber types with different values of V_{max}.

The technique that has the best time resolution for measuring energy utilization by muscle and the one on which many conclusions about muscle energetics have been based is measurement of heat production. The hydrolysis of ATP by muscles is exothermic—heat is released by the reaction. This heat production can be used by homeotherms to warm the entire body—for example, in shivering (see Chapter 16). During a typical single contraction, the temperature of a muscle increases by a very tiny amount, about 0.001–0.01°C. Very fast and very sensitive thermometers called *thermopiles* can be used to measure the heat generated. In theory, the amount of ATP hydrolyzed by a muscle could be calculated by measuring the work done during contraction and dividing that value by the energy liberated as heat when a given amount of ATP is hydrolyzed. However, other heat-absorbing and heat-producing processes unrelated to the hydrolysis of ATP interfere with this measurement, so it is impossible to relate heat production during contraction precisely to the use of ATP. Nonetheless, measurements of heat production have yielded considerable insight into how different types of muscle use energy during contractions.

A muscle fiber's mechanical properties (that is, its force generation and power production) and energetic properties (that is, its rate of ATP use and efficiency) depend on both its velocity of shortening, V, and V/V_{max}. For a given rate of shortening, V, the force and mechanical power produced per cross-sectional area can be considerably higher in a fiber with a high V_{max} than in a slower fiber (Figure 10-31a,b). Furthermore, the generation of power is maximal at

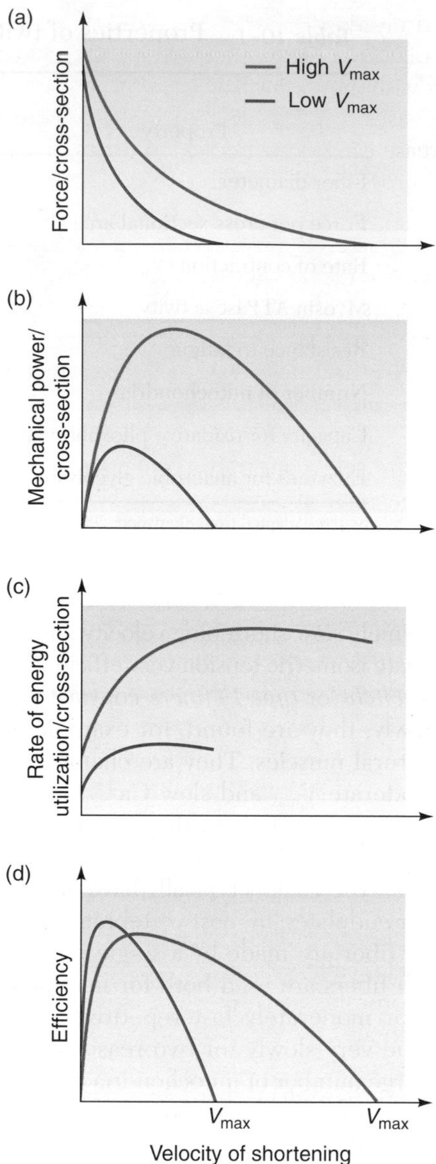

Figure 10-31 Force, power, rate of energy utilization, and efficiency vary as a function of shortening velocity. Fibers with a high V_{max} can generate more force **(a)** and mechanical power **(b)** than those with a low V_{max}, but they also use more energy at all shortening velocities **(c)**. The efficiency of contraction is calculated as the power output divided by the energy used **(d)**. Note that low-V_{max} fibers are more efficient at low shortening velocities, whereas high-V_{max} fibers are more efficient at higher rates of shortening. These curves were derived from heat-production, oxygen-usage, and mechanical measurements of frog muscle contraction. [Adapted from Hill, 1964; Hill, 1938; and Rome and Kushmerick, 1983.]

intermediate values of V/V_{max}. It therefore takes fewer high-V_{max} fibers than low-V_{max} fibers to generate a given amount of power.

It might, then, seem advantageous to have only muscle fibers with high values of V_{max}. There is, however, an energetic price to be paid for a high V_{max}. Measurements of heat liberated and high-energy phosphate hydrolyzed show that use of ATP is also a direct function of V/V_{max}. The rate at which ATP is hydrolyzed increases with increasing V/V_{max} up to a maximum and then decreases as V/V_{max} approaches 1 (Figure 10-31c). This increase can be understood in terms of the Huxley model of cross-bridge function (see Figure 10-10). In muscles with faster V_{max}, cross-bridges detach faster and hence consume ATP molecules faster. Notice in Figure 10-31c that the rate at which ATP is used is considerably higher in fibers with a high V_{max} than it is in fibers with a low V_{max} at all rates of shortening.

Thus, we find an adaptive balance between the mechanics and energetics of contraction. From the combination of mechanical and energetic data, the efficiency of muscle contraction (defined as the ratio between mechanical power output and energy utilization) can be calculated. Efficiency also turns out to be a function of V/V_{max} (Figure 10-31d). Fibers with a low V_{max} are more efficient than fibers with a high V_{max} at low rates of shortening, but less efficient at higher rates of shortening. As a consequence, if an animal is to produce both slow and fast movements efficiently, it must have both kinds of fibers and must use them appropriately to produce the two kinds of movements.

ADAPTATION OF MUSCLES FOR DIVERSE ACTIVITIES

The principles that determine the mechanical properties of muscles can be illustrated by three very different kinds of motor activity: frogs jumping, fish swimming, and toadfish and rattlesnakes producing sound. Here we consider each of these activities and the muscles that are used to produce them. Our discussion focuses on three features of a working muscle:

- the amount of overlap between thick and thin filaments (that is, where on its length-tension curve the muscle is working)
- the relative velocity of shortening, V/V_{max}, during the activity, which determines the power and efficiency of the muscle
- the timing and duration of the muscle's active state

In this section we will draw heavily from the work of Lawrence Rome and his colleagues, who have contributed much to our understanding of comparative muscle physiology.

Adaptation for Power: Jumping Frogs

When a frog jumps, it moves from a crouched to an extended position in just 50 to 100 ms. The work performed per unit time is high, so the muscles that produce the jump must generate considerable power. The distance covered in a single jump depends directly on how much power the muscles produce. From our earlier discussion, we would expect a muscle that generates high power to exhibit three properties: (1) it should operate within the plateau of the sarcomere length-tension curve, where maximal force is generated (see Figure 10-8b); (2) it should shorten at a rate at which maximal power is generated (see Figure 10-13d); and (3) it should become maximally activated (that is, every fiber should be in the active state) before shortening begins. To determine whether a frog's jumping muscles actually have these properties, G. Lutz and L. C. Rome, using *Rana pipiens* as their model system, observed both jumping frogs and frog muscles in isolation, integrating the results of the two kinds of experiments.

Length-tension relation

To examine the length-tension relation of muscles during a jump, Lutz and Rome measured the length and changes in length of the semimembranosus muscle, a hip extensor. They took measurements from videotapes of frogs jumping and from isolated limbs whose positions were manipulated to match the shape of a jumping frog's legs (Figure 10-32 on the next page). By plotting the changes in muscle length against the pelvic angle, they determined the *moment arm* of the muscle. (The moment arm is the distance separating a fixed fulcrum from a point at which a force is exerted that will tend to rotate a mass around the fixed point, as illustrated in the inset in Figure 10-32.) The length of the moment arm is crucial because it determines both the leverage of the muscle (that is, the angular acceleration around the fulcrum point that will be produced by contraction of the muscle) and how much the muscle must shorten to produce any given change in the angle of the hip joint.

The length of sarcomeres in the hip extensor was determined when the hip was in the crouched position and when it was in the extended, jumping position. During a jump, the sarcomere length changes from 2.34 μm in the crouched position to 1.82 μm at the point of take-off. To determine where these lengths fall in the sarcomere length-tension relation, they were compared with a length-tension curve measured in a closely related species of frog *(Rana temporaria)*, as shown in Figure 10-33a on the next page. The measured lengths of the hip extensor sarcomeres fell along the plateau of the sarcomere length-tension curve; thus, as expected for a power muscle, the fibers of the hip extensor operate very near their optimum during a jump. It has been calculated that this muscle generates at least 90% of its maximal tension throughout the jump. In fact, if the initial sarcomere length were either longer or shorter than the measured length, the muscle would produce less power.

A number of factors must be matched to produce this optimal behavior. The lengths of the myofilaments and the number of sarcomeres per muscle fiber must

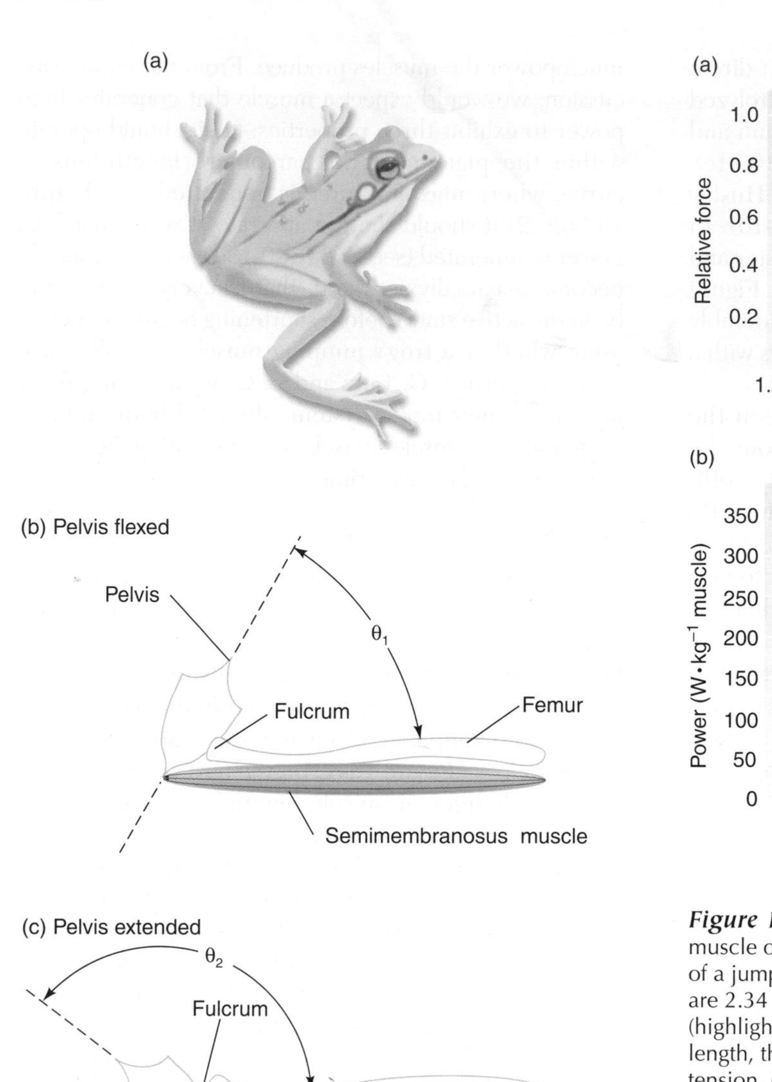

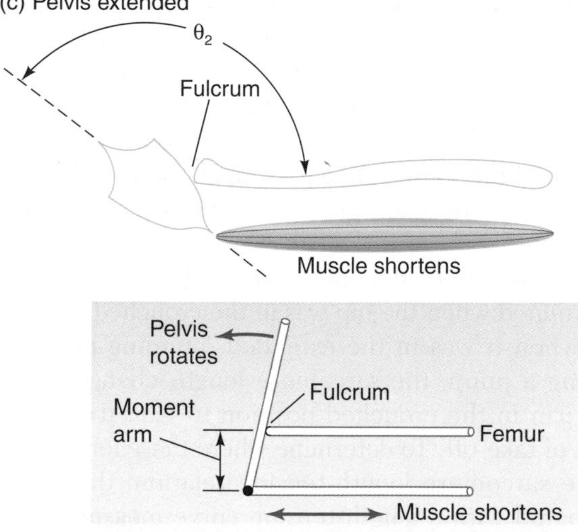

Figure 10-32 When the hip extensor muscle of a frog contracts, the hip joint rotates around the point at which the femur attaches to the pelvis, pushing the frog off the substrate. **(a)** In a sitting frog, the thigh is held at a small angle with respect to the longitudinal axis of the body, as diagrammed in part b. **(b)** When the frog is in a crouched position, the angle of the hip joint (θ_1) is small, and the hip extensor muscle (the semimembranosus muscle) is relaxed. **(c)** When the semimembranosus muscle contracts, the angle of the hip joint increases (θ_2) as contraction of the muscle acts through a moment arm to pull on the pelvis. As a result, the frog's legs push against the substrate, propelling the body upward and forward. Inset: Schematic diagram illustrating geometric relations among the mechanical components contributing to jumping. [Part a adapted from Pough et al., 1996; part b adapted from Lutz and Rome, 1996a, b.]

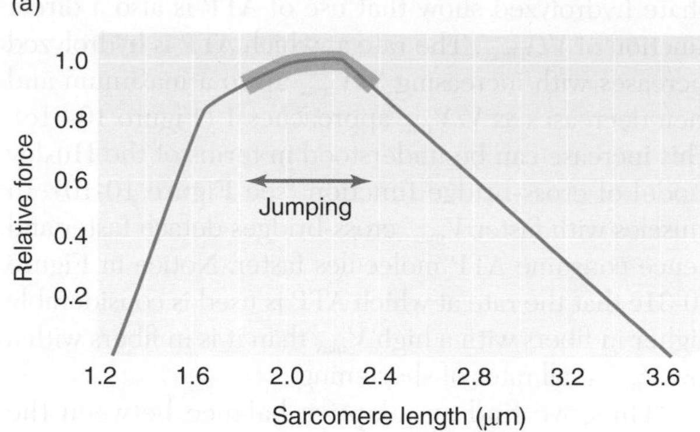

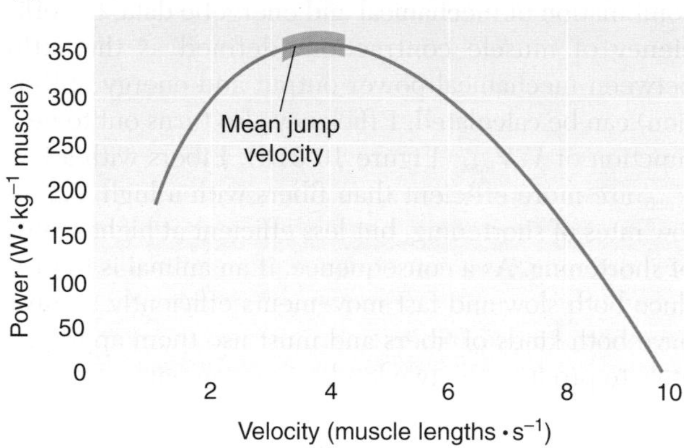

Figure 10-33 The mechanics of the frog's hip extensor muscle operate optimally during a jump. **(a)** At the beginning of a jump, the sarcomeres in the semimembranosus muscle are 2.34 μm long; they shorten to 1.82 μm during the jump (highlighted portion of curve). Even at the shortest sarcomere length, the muscle still generates over 90% of its maximal tension. **(b)** At the shortening velocity used during jumping, the muscle operates in the highlighted portion of the power curve in which at least 99% of maximal power is generated. The velocity of shortening is expressed in terms of muscle lengths per second to take into account the difference in length of muscles taken from different-sized frogs. [Adapted from Lutz and Rome, 1994.]

combine to produce optimal overlap of thick and thin filaments when the frog is in the crouched position. In addition, given the change in the angle of the hip joint observed during jumping, the moment arm of the hip joint must allow the muscle and its sarcomeres to undergo appropriate changes in length while maintaining optimum overlap.

Value of V/V_{max}

The V_{max} of the hip extensor muscle is about 10 muscle lengths per second, and it generates maximal power at 3.44 muscle lengths per second (Figure 10-33b). The mean rate at which the muscle shortens during a jump is 3.43 muscle lengths per second; that is, at a V/V_{max} of 0.33, almost exactly the rate at which the muscle produces maximal power. Thus, the frog's muscles, joint configuration, and mass are all matched to allow the hip extensor muscle to shorten at a V/V_{max} appropriate for maximal generation of power.

State of activation

Even if the hip extensor muscle begins to contract at the optimal sarcomere length and shortens at an optimal rate, it also must be maximally activated if it is to generate maximal power. If the muscle started to shorten before it became fully activated, it would generate a force lower than the maximum possible at that velocity (i.e., the actual force would lie below the force-velocity curve), and its power would also be lower than the maximum. As discussed earlier, the time required for activation to occur depends on the rate at which Ca^{2+} is released from the sarcoplasmic reticulum and binds to troponin and on the rate of cross-bridge attachment. If the frog hip extensor is to become maximally activated before shortening begins, activation must occur rapidly, and movement of the hip joint must be delayed until activation is complete. The optimal lag time will depend on the mass of the frog.

One way to determine whether the extensor muscle shortens only after it becomes maximally activated would be to do the equivalent of a quick-stretch experiment on individual muscle fibers in a frog as it jumped. However, such a measurement would require that force transducers be implanted into the frog to measure the behavior of a single muscle fiber, a strategy that is not yet technically practical. In an alternative approach, the length of the hip extensor muscle and the electrical activity of fibers within the muscle were measured as carefully as possible in an intact frog, and these values were then reproduced in an isolated muscle.

This second approach was used in the experiment depicted in Figure 10-34. Electrical activity of the muscle in an intact frog was measured by tiny electrodes implanted in the muscle; such electrodes record APs in muscle fibers, just as extracellular electrodes record APs in nerve bundles (see Spotlight 6-2). The record obtained from these electrodes is called an *electromyogram* (EMG). The APs in different fibers within a muscle are not synchronous, and the amplitude of the signal from any particular fiber depends on how close the fiber is to the electrode, so an EMG recording can appear very complex. However, the pattern of APs in the biggest units recorded by the EMG electrode can be abstracted from the record, and an isolated muscle can then be stimulated electrically using that temporal pattern (Figure 10-34a). In addition, the temporal pattern of length change in the hip extensor muscle was measured from video images of a jumping frog, and this pattern of length change was mechanically imposed on the isolated muscle while it was simultaneously being stimulated electrically (Figure 10-34b). When treated in this way, the isolated hip extensor muscle generated the

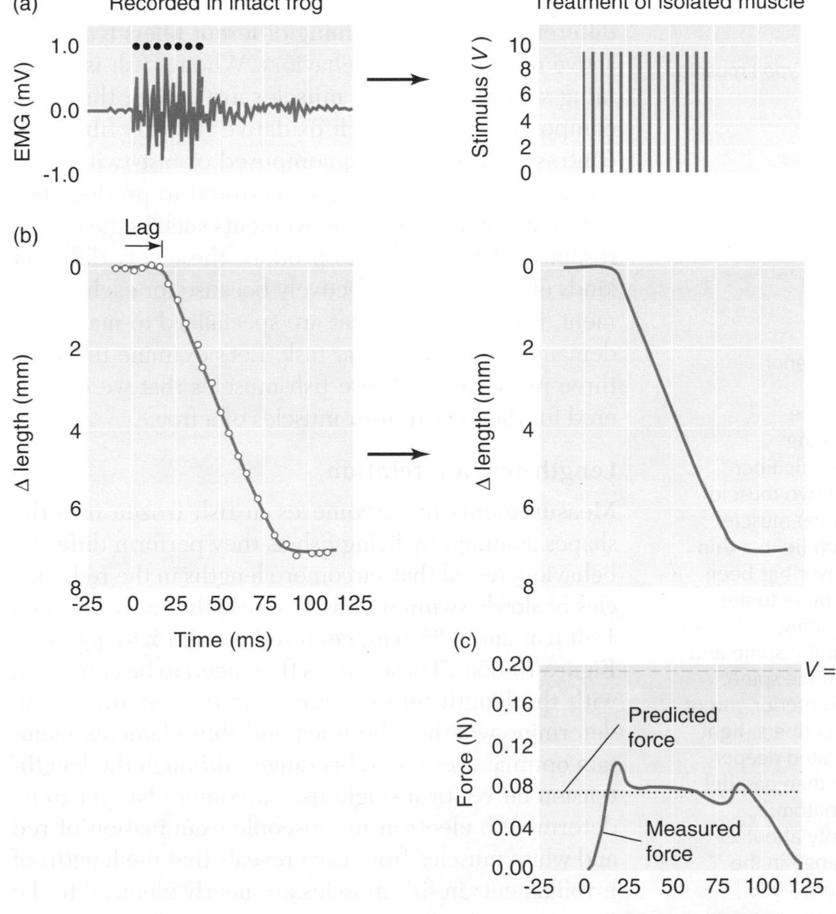

Figure 10-34 The electrical activity patterns and length changes recorded from muscles in an intact frog can be imposed on isolated muscles to study activation and the force generated. **(a)** Electromyogram recorded from the semimembranosus muscle of a frog during a jump (left), and the abstracted pattern of stimulation that was imposed on an isolated muscle (right). A black dot appears over each action potential that was used to generate the abstracted pattern, which persisted longer than the endogenous pattern. **(b)** The rate at which the muscle shortened in an intact frog during jumping and the change in length imposed on the isolated muscle while it was being electrically stimulated as shown in part a. **(c)** Force produced by the isolated muscle (red line) during the experimental manipulation. The dashed line shows the isometric force produced by this muscle when $V = 0$, and the dotted line shows the force that would be expected in the same muscle contracting at the imposed shortening velocity during a force-velocity experiment similar to the one shown in Figure 10-31a. The time axis is identical in all panels. Time equals zero at the start of stimulation. Notice that there is a lag between the start of stimulation and the time when muscle length begins to change in part b. [Adapted from Lutz and Rome, 1994.]

maximal force expected at the imposed shortening velocity (Figure 10-34c), strongly suggesting that it is maximally activated during jumping. The implication of this result is that the molecular components of activation in this muscle are strikingly matched to the biomechanics of jumping.

Adaptation for Contrasting Functions: Swimming Fish

The study of muscles in fish has for two reasons been particularly useful in elucidating how muscular systems are organized. First, fish make many different kinds of movements that can be elicited readily and analyzed quantitatively. Second, different movements are powered by different muscle fiber types, which in fish are anatomically separated, permitting the activity of individual fiber types to be monitored by electromyogram electrodes (Figure 10-35). (This arrangement contrasts with that of the muscles in most other vertebrates, which contain more than one type of fiber, making electrical monitoring of activity in one particular fiber type difficult or impossible.)

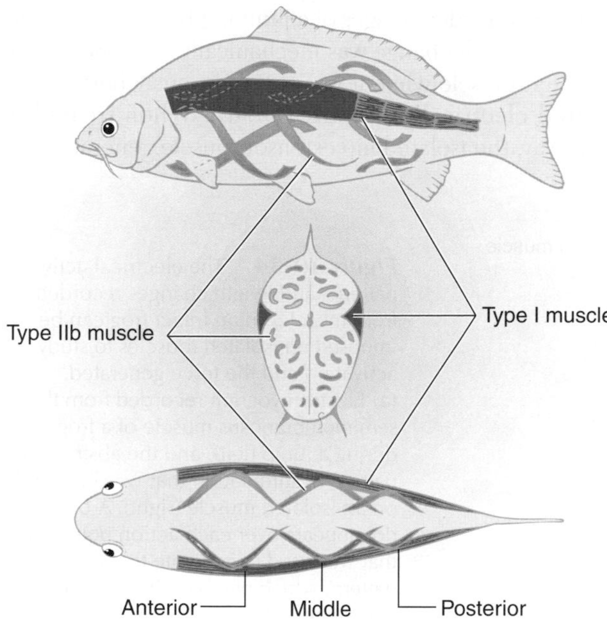

Figure 10-35 In fish, the muscle fiber types are anatomically separate from one another, facilitating electromyographic monitoring of activity in specific fiber types. These diagrams show the arrangement of two muscle fiber types in a carp. Type I (slow-twitch oxidative) muscle fibers (dark red) are found in red muscles, which lie in a thin layer just under the skin; the thickness of this layer has been enlarged in this drawing. These muscles run parallel to the body axis, so the change in sarcomere length during shortening is directly related to the curvature of the spine and to the distance separating the muscle layer from the spine. Notice that details have been omitted from the anterior end of the muscle band. Type IIb (fast-twitch glycolytic) fibers (light red) compose the white muscles, which are located deeper in the body. These muscles run helically, rather than parallel to the long axis of the body. Because of their anatomic arrangement, white muscles need to shorten only about 25% as much as red muscles to produce a given change in the curvature of the body. [Adapted from Rome et al., 1988.]

During the many movements of which fish are capable, the change in sarcomere length is roughly proportional to the curvature of the spine. When a carp is swimming steadily at a velocity of 25 $cm \cdot s^{-1}$, the curvature changes very little along most of its spine (Figure 10-36a), indicating that the lengths of sarcomeres along its body change very little. In contrast, when the fish is startled—for example, by a loud sound—and produces an escape response, its spine curves markedly, indicating that sarcomeres have shortened on one side of the body and lengthened on the other (Figure 10-36b). Notice the difference in time scale between steady swimming and the escape response. During steady swimming, one tailbeat takes about 400 ms, whereas in the escape response, the body of the fish changes from straight to highly curved in only 25 ms.

The muscles of a fish must, then, be able to generate both slow, low-amplitude movements and fast, high-amplitude movements. Earlier in this chapter, it was argued that muscles must be finely tuned to a particular activity in order to perform optimally, but these two behaviors seem to require very different properties. Can muscle fibers perform such different tasks while still operating optimally? If so, how do they do it?

Electromyograms recorded from fish swimming normally and responding to a loud sound revealed that different muscles, containing different fibers types, are active during the two behaviors. When a fish is swimming steadily, only red muscles are active; these are composed of slow-twitch oxidative (type I) fibers. In contrast, white muscles, composed of fast-twitch glycolytic (type IIb) fibers, are recruited to produce fast swimming or large, rapid movements such as the escape response. A fish is able to produce these very different kinds of movements effectively because for each movement, it uses muscles that are specialized to match the demands of the particular task. Let's examine the same three properties of these fish muscles that we considered for the hip extensor muscles of a frog.

Length-tension relation

Measurements of sarcomeres in fish frozen into the shapes assumed by living fish as they perform different behaviors reveal that sarcomere lengths in the red muscles of slowly swimming fish vary rhythmically between 1.89 μm and 2.25 μm, centered around 2.07 μm (see Figure 10-36a). These values then need to be compared with the length-tension curve for fish sarcomeres to determine whether the thick and thin filaments maintain optimal overlap in this range. Although the length-tension curve for a single fish sarcomere has yet to be determined, electron microscopic examination of red and white muscles from carp reveals that the lengths of myofilaments in fish muscles are nearly identical to the

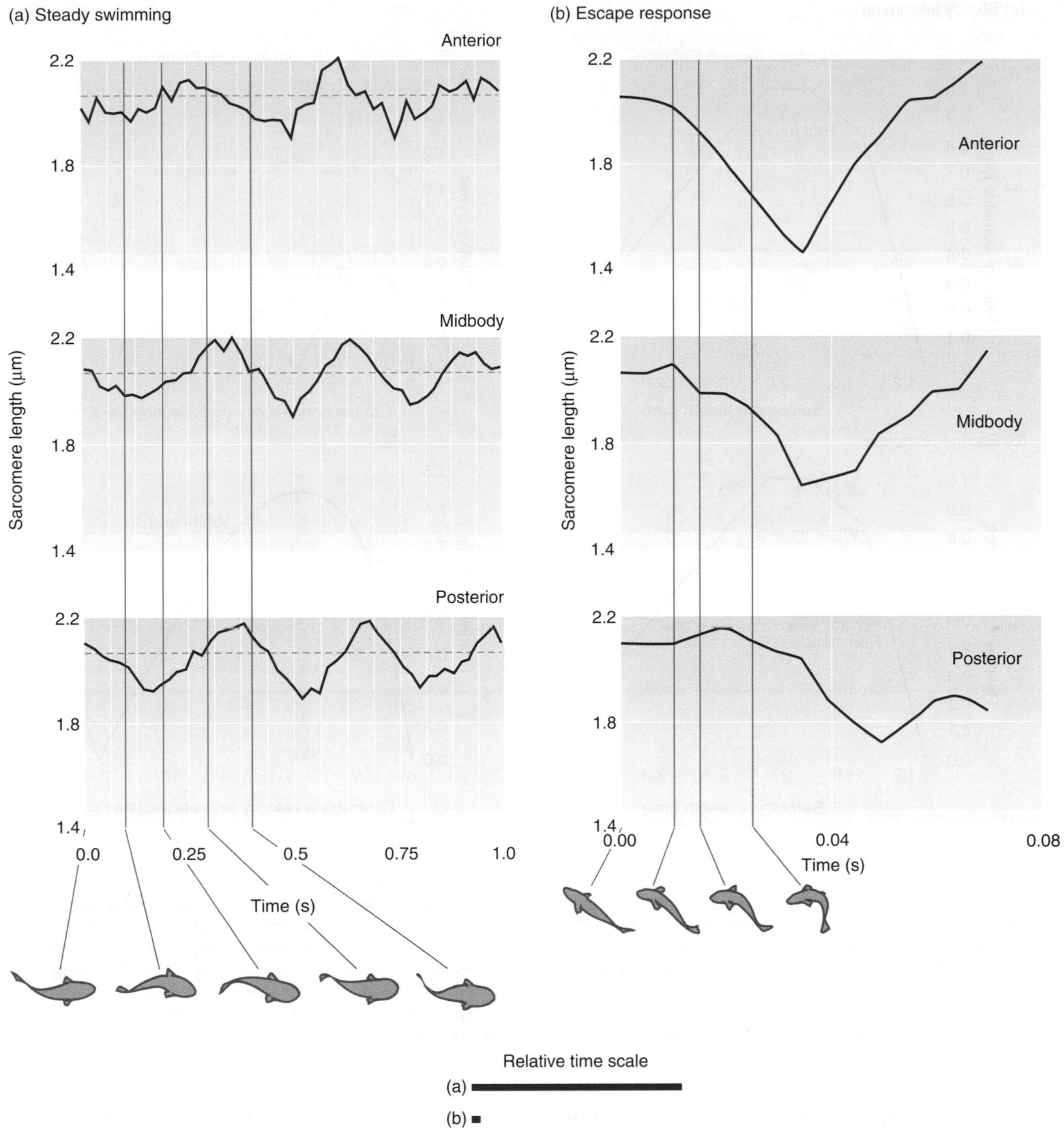

Figure 10-36 Steady swimming movements and the escape response in fish differ greatly in magnitude and time course. This figure shows inferred changes in sarcomere length within the red muscles located on one side of the anterior, midbody, and posterior of a carp engaged in two activities: **(a)** swimming steadily at 25 $cm \cdot s^{-1}$ or **(b)** making an abrupt escape response. The changes in sarcomere length were calculated from the shape of the fish at each time point in a high-speed movie of the behavior. The shape of the body at selected time points is indicated in figures below each graph. Type I (slow-twitch oxidative) muscles are active during steady swimming. In contrast, type IIb (fast-twitch glycolytic) muscles produce the escape response (see Figure 10-37). [Part a adapted from Rome et al., 1990; part b adapted from Rome et al., 1988.]

lengths of these filaments in frog muscles. Therefore, the sarcomere length-tension curve for frog muscle is likely to provide a good approximation of the same relation in carp. Comparison of the sarcomere lengths measured in swimming carp with the frog length-tension curve shows that in swimming fish, the red muscles at peak tension generate at least 96% of their maximal force (Figure 10-37a on the next page).

In the escape response, the fish moves rapidly, and its body curves dramatically. As Figure 10-35 shows, the

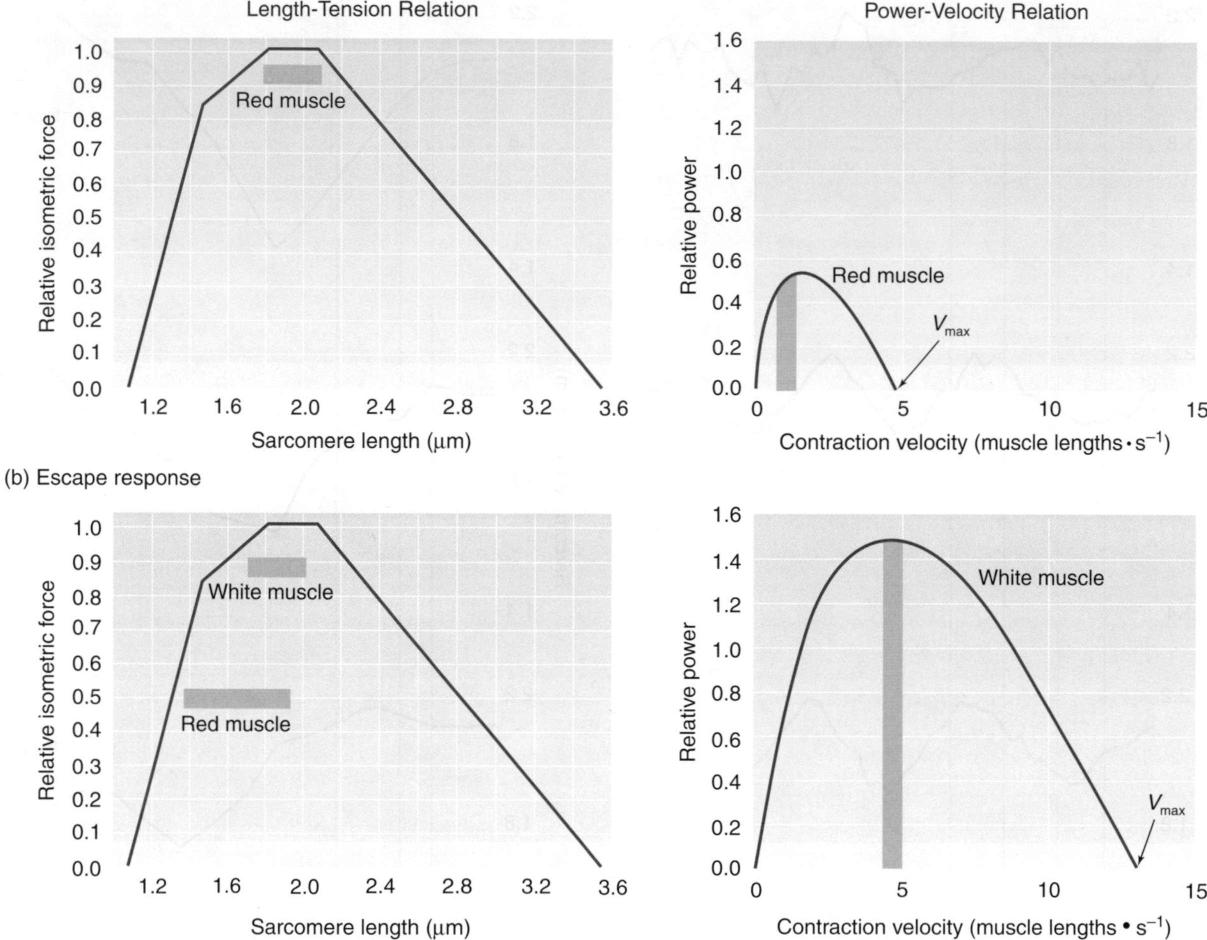

Figure 10-37 The properties of the red and white muscles of fish make the two kinds of muscle optimally suited for different kinds of activities. **(a)** The changes in the sarcomere length of red muscles during steady swimming coincide with the plateau of the sarcomere length-tension curve (left). The orange bar indicates the lengths of red muscle sarcomeres during slow, steady swimming. In addition, the contraction velocities of red muscles during swimming correspond to values of V/V_{max} between 0.17 and 0.36, near the value at which red fibers produce maximal power (right; orange region under curve). **(b)** Because of their anatomic arrangement, white fibers can produce the escape response at a more favorable region of the sarcomere length-tension curve than can red fibers (left). In addition, the high V_{max} of white fibers allows them to generate power when they are shortening very rapidly (right). Indeed, during the escape response, the V/V_{max} for white muscles (orange bar) is 0.38, which is the peak of their power curve. [Adapted from Rome and Sosnicki, 1991.]

red muscles in carp run parallel to the long axis of the fish, whereas the white muscles run helically. To produce the escape response, the sarcomeres of the red muscle would have to shorten to 1.4 μm, a length at which their force production would be low (Figure 10-37). In contrast, the sarcomeres in the helically arranged white muscles need shorten only to about 1.75 μm during this behavior. In other words, the mechanical advantage conferred by the anatomic arrangement of the white muscles allows them to produce any given change in the curvature of the spine with much less sarcomere shortening than would be required by the red muscles. Thus, white muscles are much better suited to produce the escape response, and they generate about 85% of their maximal force during this behavior (see Figure 10-37b). When white muscle is used in less extreme movements (e.g., when a fish is swimming rapidly) and the curvature of the spine is not nearly as extreme, the sarcomeres shorten less, and the muscles generate nearly maximal force.

Because the fish uses different muscles to produce different movements, the myofilament overlap (sarcomere length) is never far from its optimal level, even in the most extreme movements. The lengths of the thick and thin filaments and the anatomic arrangement of the muscle fiber types combine to allow this optimization.

Value of V/V_{max}

In addition to their different anatomic arrangements, the red and white muscles of a carp have different values of V_{max}. The V_{max} of carp red muscle is 4.65 muscle lengths per second, whereas the V_{max} of carp white muscle is 12.8 muscle lengths per second, about 2.5 times higher. During steady swimming, the red muscle shortens at a V/V_{max} of 0.17–0.36, which is near the value at which maximal power is generated (see Figure 10-37a, right). At higher swimming speeds, a fish needs to generate greater mechanical power, but at these higher values of V/V_{max}, the mechanical power output of the red muscle actually declines. In order to swim faster, a fish must activate white muscles as well.

In contrast to steady swimming, the escape response depends entirely on activity in the white muscles. To power the escape response, the red muscles would have to shorten at 20 muscle lengths per second—four times faster than their V_{max}. White muscle in the anatomic orientation of the red muscles would also be unable to power the escape response, because the V_{max} of these muscles is only about 13 muscle lengths per second. However, the helical arrangement of the white muscles allows them to produce the escape response when they shorten at only about five muscle lengths per second, which corresponds to a V/V_{max} of about 0.38, the value at which the carp's white muscles produce the most power (see Figure 10-37b, right).

Perhaps, you might suggest, a fish would be better off with only white muscles. The white muscles could certainly power slow swimming. However, the high V_{max} of white muscles would mean that the V/V_{max} during slow swimming would be so low (0.01–0.03) that they would be extremely inefficient. Red muscles can produce adequate power to generate slow swimming, and they do it much more efficiently than white muscles could. Thus, the anatomic arrangement and the V_{max} of the two kinds of muscles suit each of them to the particular behavior during which they are active. Fish need both kinds of muscles if they are to perform both slow swimming and fast escape responses optimally.

Kinetics of activation and relaxation

In considering jumping frogs, our main concern was to determine whether the muscle becomes maximally activated during the early phase of shortening. The kinetics of muscle relaxation were essentially irrelevant because a frog doesn't jump repeatedly and rhythmically. A fundamentally different problem is faced by animals during cyclical locomotion, such as swimming by fish. Swimming should be most efficient if muscles do not have to work against one another. When muscles on one side of a fish shorten, for example, they should be most efficient in changing the shape of the fish—allowing it to push against the water—if the muscles on the other side of the body are already relaxed.

To better understand how the kinetics of activation and relaxation affect the generation of power during cyclical muscle contractions, Robert Josephson introduced the "workloop" technique to the study of muscles. In this approach, muscles are driven by a servomotor system through the cyclical changes in length that are observed during locomotion, and the investigator delivers a stimulus to the muscle at a particular time in the cycle. In this kind of experiment, the timing and duration of the stimulus, the intrinsic activation and relaxation rate of the muscle, and the value of V_{max} for the muscle interact to determine how much power the muscle generates.

A useful way to quantify these potentially complex interactions is to measure the amount of net work (force × change in length) a muscle generates during one cycle of shortening and lengthening (Figure 10-38a).

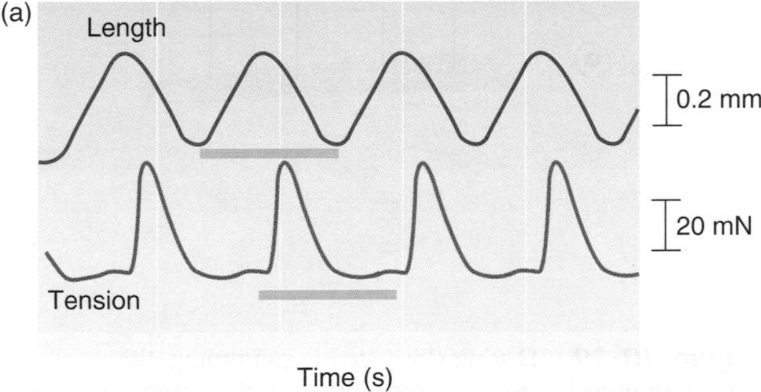

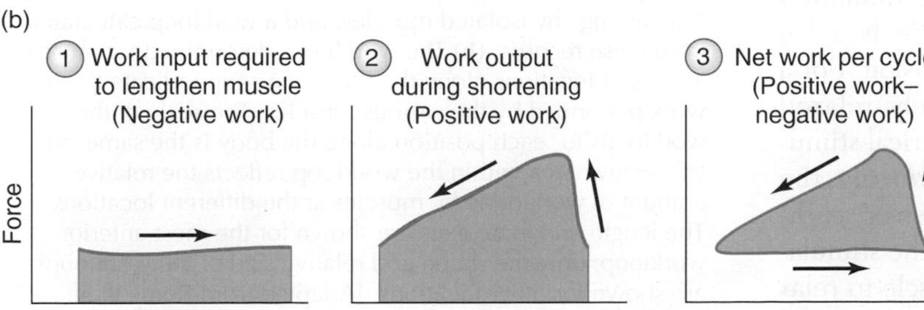

Figure 10-38 Workloops graphically depict the net work done during cyclical muscle contractions. **(a)** The length (red upper record) and tension (blue lower record) of a katydid flight muscle was recorded as it was being driven to shorten and lengthen cyclically. The orange bar indicates the duration of a single cycle. **(b)** The force-length relation during one complete cycle. In (1) the muscle is becoming longer because it is being stretched by an outside force; the shaded area under the curve represents the "negative work" done during this phase. In (2) the muscle shortens; the shaded area represents the positive work done during this phase. The net work (3) is the difference between the negative work and the positive work and equals the area encompassed by the force-length curve or workloop. [Adapted from Josephson, 1985.]

Net work is graphically equivalent to the area contained within a force-length loop (Figure 10-38b). A muscle does positive work only when it is shortening; thus positive work is equal to the area under the force-length curve during the shortening phase of a cycle. A muscle generates "negative work" when it is forcibly lengthened by an antagonist muscle (or a servomotor system); thus negative work is equal to the area under the force-length curve during the lengthening phase of each cycle. The net work—the difference between the positive and negative work done during one cycle—is equal to the area between the positive and negative legs of the force-length curve for one cycle. In other words, the net work is equal to the area inside the force-length loop, or workloop. For a muscle to generate net positive work, it must generate a greater force during shortening than was required to stretch it to its initial length. The net power generated by a cyclical contraction is expressed by the equation

$$\text{net power} = (\text{positive work} - \text{negative work})_{\text{cycle}} \times \text{frequency of cycles}$$

It would seem that muscles might operate optimally if their fibers were fully activated during shortening (as in the frog) and could fully relax before they were forced to elongate by the activity of other muscles. If a muscle could be fully activated instantaneously and then relax instantaneously, the generation of force during shortening would be given by the force-velocity curve. There is, however, a problem. A muscle that was maximally activated throughout shortening and that then relaxed instantly at the end of shortening would be very energetically expensive, for two reasons. First, such a muscle would have to pump Ca^{2+} back into its sarcoplasmic reticulum very rapidly, requiring a huge number of calcium pumps to be continuously active—an unrealistically large expenditure of ATP. Second, instantaneous relaxation would require that cross-bridges detach very rapidly, but rapidly cycling cross-bridges use ATP much faster than cross-bridges that cycle more slowly. A muscle with more modest rates of calcium pumping and cross-bridge cycling will be energetically less expensive, allowing it to work more efficiently. Efficiency of operation is important in muscles that are used almost continuously, such as the swim muscles of an active fish.

If a muscle has a slow relaxation time, allowing it to be metabolically efficient, the timing of stimulation becomes important. Remember that there is a lag between the muscle AP and the onset of tension. For a slowly relaxing muscle to be appropriately relaxed before contralateral muscles pull on it, electrical stimulation activating the muscle must start *during* the lengthening phase and continue into only the very earliest part of the shortening phase. Otherwise the stimulation will continue too long to allow the muscle to relax soon enough. However, this pattern of stimulation reduces the amount of work the muscle can do. Once again, there is a trade-off between two desirable features—in this case, the ability of the muscle to do work versus its metabolic efficiency.

Workloop experiments have been performed on swimming fish to determine whether swim muscles emphasize rapid relaxation, which is metabolically costly, or lower work output, which is less metabolically expensive. The basic experimental approach used in these studies was similar to that described for the frog hip extensor muscle. The electrical activity of muscles

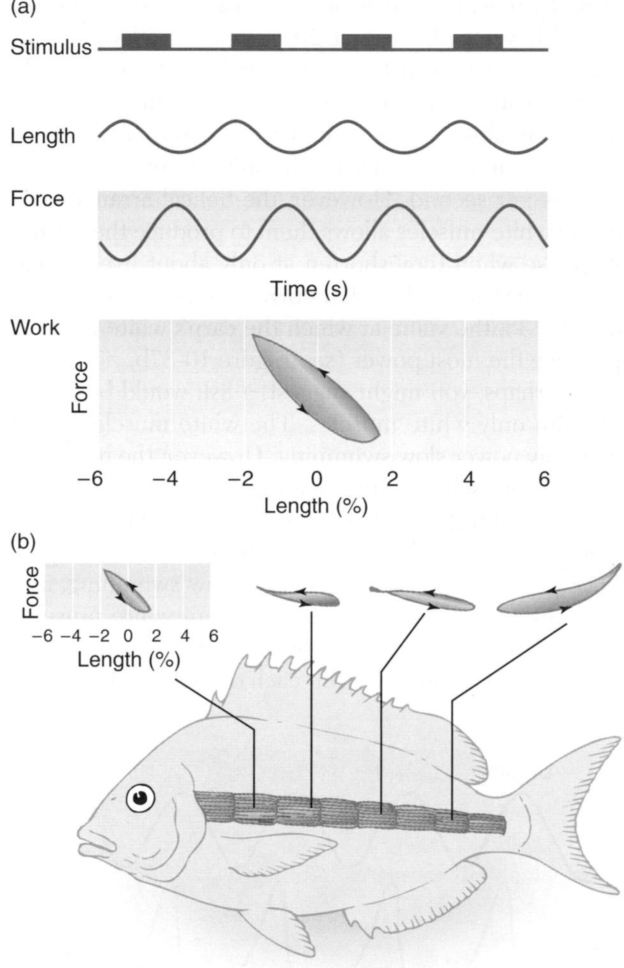

Figure 10-39 During slow, steady swimming, the posterior muscles do more net work than the anterior muscles. **(a)** Typical stimulus, length, and force records during "swimming" by isolated muscles, and a workloop calculated from these records. **(b)** The workloops determined for muscles at several locations along the body of a scup indicate the net work performed by the various muscles. The scale of the workloops for each position along the body is the same, so the relative area within the workloop reflects the relative amount of work done by muscles at the different locations. The length and force axes are shown for the most anterior workloop; only the shape and relative size of the workloops are shown for other locations. [Adapted from Rome et al., 1993.]

and the changes in muscle length were determined in swimming fish. Then, using the type of setup illustrated in Figure 10-34, isolated muscles were stimulated in a temporal pattern identical to that of the electromyogram, and the length of the muscles was controlled to match the changes measured during swimming. The force and power generated by the muscles under these conditions were determined, and the net work done by the muscles was calculated by plotting the workloop. These experiments revealed that during slow, steady swimming, the posterior muscles do more net work than the anterior muscles (Figure 10-39).

During cyclic swimming, muscles at different locations along the axis of the body receive different stimulus patterns and change in length by different amounts, affecting both the force generated and the power produced. The *stimulus duty cycle* (the percentage of one cycle during which the muscle is stimulated) is about 50% in the anterior part of the fish and falls to only about 25% in the posterior part of the fish. In addition, posterior muscles change in length much more than anterior muscles during swimming. The combination of large changes in length and a short duty cycle causes the posterior muscles to generate a great deal of mechanical power. When isolated anterior muscles were exposed to the same set of conditions (i.e., stimulation pattern and length changes), they generated the same amount of power as the posterior muscles. It is the *pattern* of muscle contraction at the back of the fish that generates more power, not an intrinsic property of the posterior muscle fibers themselves.

Examination of the workloops for red muscles during swimming indicates that they activate slowly and relax slowly. Stimulation to the posterior muscles, which generate the most power, begins during lengthening and ends just after the beginning of shortening, as predicted for this kind of muscle earlier in this section. As a result, the muscle must be in relaxing mode through most of its power stroke, reducing the mechanical power it generates, but presumably also decreasing its energetic cost and perhaps thereby increasing its efficiency.

Adaptation for Speed: Sound Production

Some animals produce sound through mechanisms that are not directly coupled to muscle contraction—for example, the movement of a column of air past a vibrating membrane or past the vocal cords. In other animals, however, muscles directly cause structures, such as the swimbladder of a toadfish or the rattle on the tail of a rattlesnake, to vibrate, generating sound. In these animals, the sound-producing (or sonic) muscles must undergo contraction-relaxation cycles at the frequency at which the sound is made, which can be 10 to 100 times faster than most locomotory muscles operate.

In the last section, we saw that the swim muscles of fish have relatively slow rates of relaxation, allowing them to avoid the high energetic cost of excessive calcium pumping. When these swim muscles are experimentally stimulated at the high frequencies needed for sound production, they are unable to relax between stimuli, and they go into tetanus (see Figure 10-28). If sound muscles became tetanic in the same way, the animal would be rendered silent. Sonic muscles must, therefore, have unique properties that allow them to operate at the high frequencies associated with sound production.

Toadfish swimbladder

The male toadfish *(Opsanus tau)* produces a "boatwhistle" mating call ten to twelve times per minute for hours on end to attract females to its nest. This call is generated by rapid oscillatory contractions of the muscles encircling the fish's gas-filled swimbladder. The sonic muscles of the swimbladder must contract and relax at frequencies of several hundred hertz. In contrast, the steady swimming movements of toadfish are made at about 1–2 Hz, and the escape response of these fish operates at 5–10 Hz. To understand the differences among muscles that allow them to function with such different kinetics, the properties of three kinds of muscles in the toadfish—swim, escape, and sonic—have been studied.

The time course of many biological events is characterized by their **half-width**, which is the width of the event on the temporal axis when the measured variable is equal to half of its peak value (Figure 10-40a on the next page). The half-width of a single muscle twitch in the toadfish is 500 ms in red (swim) muscle, 200 ms in white (escape) muscle, and only 10 ms in swimbladder (sonic) muscle (Figure 10-40b).

If a muscle is to activate and relax rapidly, two conditions must be met. First, Ca^{2+}, the trigger for muscle contraction, must enter the myoplasm rapidly and be removed rapidly. Second, myosin cross-bridges must attach to actin and generate force soon after the myoplasmic Ca^{2+} concentration rises, then detach and stop generating force soon after the Ca^{2+} concentration falls. In the red and white muscles of the toadfish, myoplasmic free Ca^{2+} rises and falls with typical kinetics, but in the sonic muscles, these Ca^{2+} transients are the fastest ever measured for any fiber type from any animal. Similarly, force measurements indicate that the sonic muscles both contract and relax about 50 times faster than the red muscles (see Figure 10-40b).

The effect of these very fast Ca^{2+} transients is most obvious during repeated stimulation. When red muscle is stimulated at 3.5 stimuli per second (3.5 Hz), Ca^{2+} lingers in the myoplasm so long that the concentration of myoplasmic free Ca^{2+} cannot return to its resting value between stimuli. Indeed, between stimuli delivered at this rate, the myoplasmic Ca^{2+} concentration remains constantly above the threshold level required for contraction, so a partially fused tetanus is

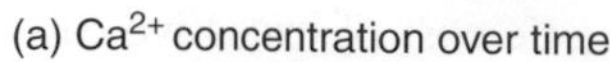

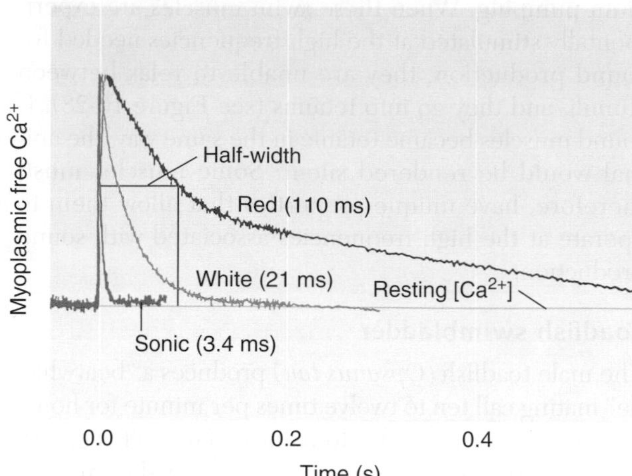

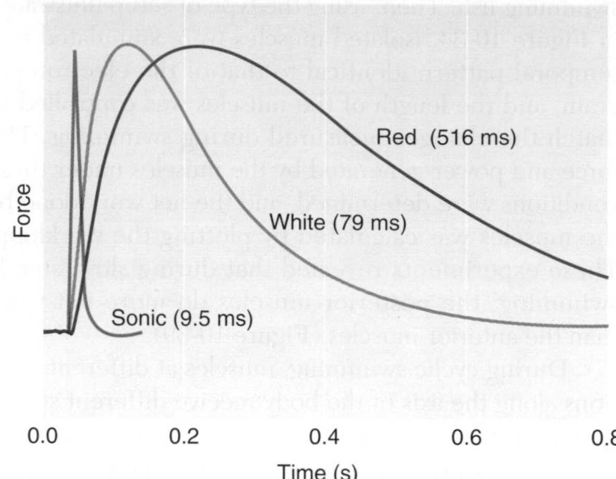

Figure 10-40 Because the sonic muscles of the toadfish activate and relax much more rapidly than its red (swim) and white (escape) locomotory muscles, they can operate at higher stimulation frequencies without becoming tetanic. **(a)** Time course of myoplasmic Ca^{2+} concentration following stimulation at 16°C in three types of muscle fibers isolated from toadfish. The time course of many biological events is characterized by the half-width of the event, illustrated for the red muscle. The half-width is the duration separating the rising and falling phases of the event when the value of the variable is one-half of its peak value. The vertical black line intersects the record at one-half of its peak value; the horizontal black line shows the time between the rising and falling phases. The average half-width of the Ca^{2+} transients in these muscles ranges from 3.4 to 110 ms, as indicated on the graph. **(b)** Time course of twitch tension in the three fiber types measured under the same conditions as in part a. Sonic muscles both contract and relax much faster than do the red or white fibers. The average half-width of twitch tension ranges from 9.5 to 516 ms. Thus sonic muscle operates more than 50 times faster than red muscle. The Ca^{2+} concentration and tension records are normalized to their maximal values for all fiber types. [Adapted from Rome et al., 1996.]

produced (Figure 10-41a). By contrast, the swimbladder sonic muscle has such fast Ca^{2+} transients that even at 67 Hz, the myoplasmic Ca^{2+} concentration returns to its baseline value between stimuli. Because the myoplasmic free Ca^{2+} concentration is below the threshold for the generation of force for much of the time between stimuli, only the first two twitches in the series are fused (Figure 10-41b). The production of an individual twitch in response to each stimulus is required for generating the oscillation of the swimbladder that produces sound.

The ability of a muscle to relax rapidly requires not only a very short Ca^{2+} transient, but also the rapid release of bound Ca^{2+} from troponin. Comparing the time course of force generation and of myoplasmic Ca^{2+} transients in frog white muscle fibers and in toadfish sonic muscles suggests that Ca^{2+} must be released from troponin in the sonic muscles three times faster than in the frog fibers.

Finally, for force to drop quickly following the dissociation of Ca^{2+} from troponin, myosin cross-bridges must detach rapidly from actin filaments. The Huxley model discussed earlier in this chapter suggests that the maximal velocity of shortening of a muscle, V_{max}, must be proportional to the rate at which cross-bridges detach from actin. Indeed, the V_{max} of toadfish sonic muscle (about 12 muscle lengths per second) is exceptionally fast—five times higher than that of toadfish red muscle and two and a half times higher than that of toadfish white muscle. (Notice that toadfish white muscle is slower than the white muscles of carp discussed earlier.)

Ultrastructural and biochemical studies of toadfish sonic fibers have revealed adaptations that allow these muscles to contract at such high frequencies. It appears that the short Ca^{2+} transient depends on an unusually high density of Ca^{2+} channels and calcium pumps in the SR membrane, an increased concentration of calcium-binding proteins (e.g., troponin), and a fiber morphology in which the distance between the SR membrane and the myofilaments is particularly short, reducing the time required for diffusion. The rapid release of Ca^{2+} from troponin probably reflects an unusually low affinity of the troponin for Ca^{2+}. Finally, the rapid detachment of cross-bridges implies that the myosin in sonic fibers also has special molecular properties, although this aspect of the fibers remains to be tested.

To emit continuous sound, sonic muscle must do work to overcome frictional energy losses in the sound-producing system and to produce sound energy. Workloop experiments, similar to those described previously for the swim muscles of fish, show that swimbladder fibers can perform positive work at frequencies above 200 Hz at 25°C, the highest frequency for work production ever recorded in vertebrate muscle. By comparison, the highest frequency known for vertebrate locomotory muscles is 25–30 Hz, measured in mouse and lizard fast-twitch muscles at 35°C.

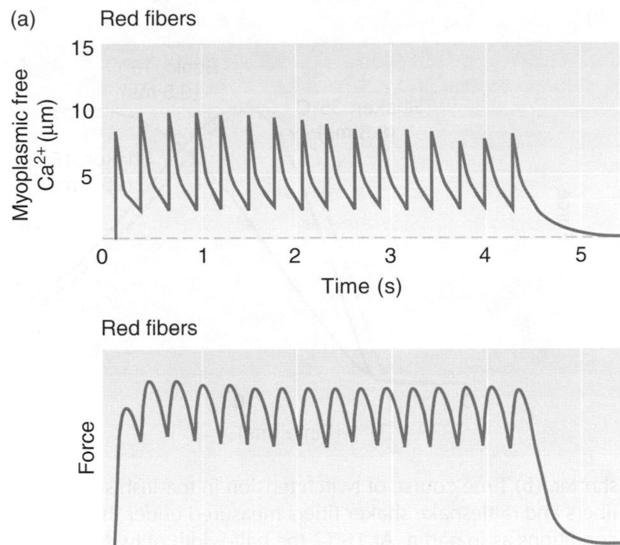

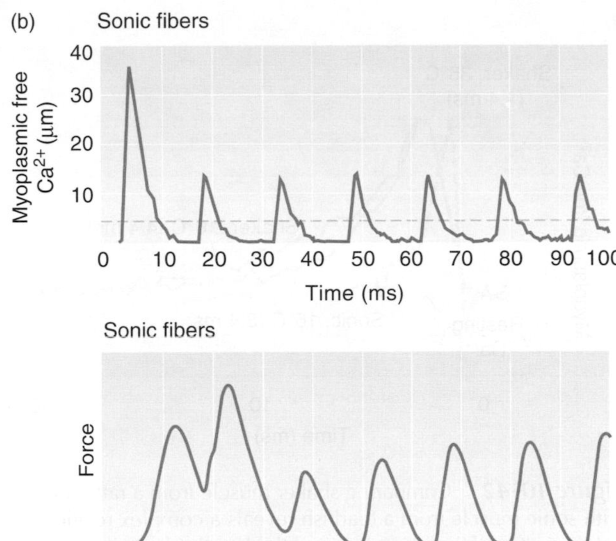

Figure 10-41 Red muscle from a toadfish contracts tetanically in response to relatively low-frequency stimulation, whereas sonic muscle produces individual twitches, even when stimulated at a much higher frequency. **(a)** Myoplasmic free Ca^{2+} in and force generated by a red fiber from a toadfish stimulated at 3.5 Hz. The threshold concentration of myoplasmic free Ca^{2+} necessary for the generation of force is shown by the dashed line on the Ca^{2+} trace. **(b)** Myoplasmic free Ca^{2+} in and force generated by a sonic fiber from a toadfish swimbladder stimulated at 67 Hz. The threshold Ca^{2+} concentration for force production (dashed line) is much higher in the swimbladder sonic fiber than in the red fiber, and the Ca^{2+} transient is fast enough that the concentration falls below the threshold value between each pair of stimuli. Notice the difference in time scale between parts a and b. [Adapted from Rome et al., 1996.]

Rattlesnakes

Rattlesnakes in the genus *Crotalus* also use special noise-making muscles, but as a warning to members of other species rather than to attract conspecifics for mating. Rattling is a loud and effective warning that renders these snakes, like many venomous animals, very conspicuous; the sound is produced when muscle fibers ("shaker fibers") rapidly shake the rattle on the snake's tail. Unlike the periodic toadfish boatwhistle, rattling can go on continuously for as long as 3 hours.

Comparing shaker muscles and sonic muscles illustrates how important individual aspects of fiber function can be. However, because muscle contraction, like most biochemical reactions, is very sensitive to temperature, experimental conditions must be carefully controlled. This comparison is complicated because, whereas toadfish normally live in cool seawater, rattlesnakes are typically desert dwellers. Hence, to match temperature conditions, one muscle type is forced to operate away from its usual temperature. At a typical temperature for a toadfish, 16°C, rattlesnake shaker fibers have a very rapid calcium transient, with a half-width of 4–5 ms, only 1–2 ms slower than that of swimbladder sonic fibers at the same temperature (Figure 10-42a on the next page). In contrast, at 16°C, the half-width of the shaker muscle *twitch* is considerably longer than that of swimbladder muscle (Figure 10-42b). It is likely that the shaker muscle twitch is slower than the sonic muscle twitch, even though the calcium transient is almost the same length, because its cross-bridges detach more slowly. This hypothesis is based on measurements showing that the V_{max} of the shaker muscle at 16°C is about 7 muscle lengths per second, only about half the V_{max} of the swimbladder muscle. In addition, Ca^{2+} may detach from troponin more slowly, although this feature has not been conclusively demonstrated. The properties of rattlesnake shaker muscle demonstrate that a rapid calcium transient is not sufficient for the production of very rapid contractions. The release of Ca^{2+} from troponin and the detachment of cross-bridges from actin filaments must be unusually fast as well. For example, at 16°C, shaker fibers can be stimulated up to only about 20 Hz before summation of contraction begins, with tetanic fusion occurring at about 50 Hz. In contrast, sonic fibers produce individual twitches at 67 Hz at 16°C.

However, we must consider the function of shaker fibers in their usual environment. Many rattlesnakes are active at temperatures above 30°C, and at 35°C *Crotalus* snakes rattle at 90 Hz. At this temperature, the calcium transient and the twitch speed are even faster in the shaker muscle than the frequencies measured in the swimbladder sonic muscle at a toadfish's typical ambient temperature (16°C) (see Figure 10-42). Most likely, both the V_{max} of shaker fibers and the rate at which Ca^{2+} is released from troponin are higher at 35°C than at 16°C. At 35°C, shaker fibers can be stimulated at 100 Hz without going into complete tetanus, and they can perform work at 90 Hz. The similarities in the properties of toadfish swimbladder sonic muscles

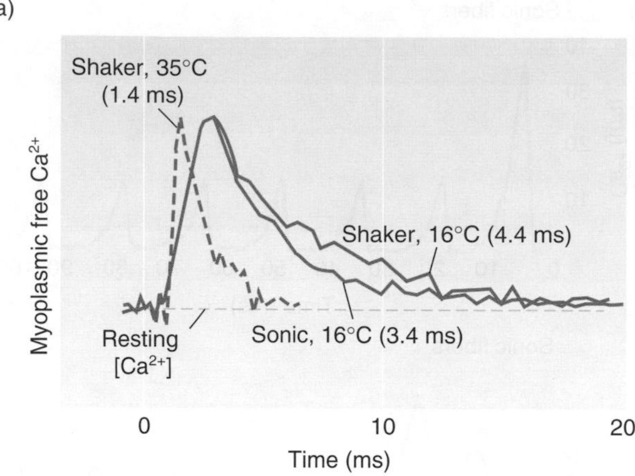

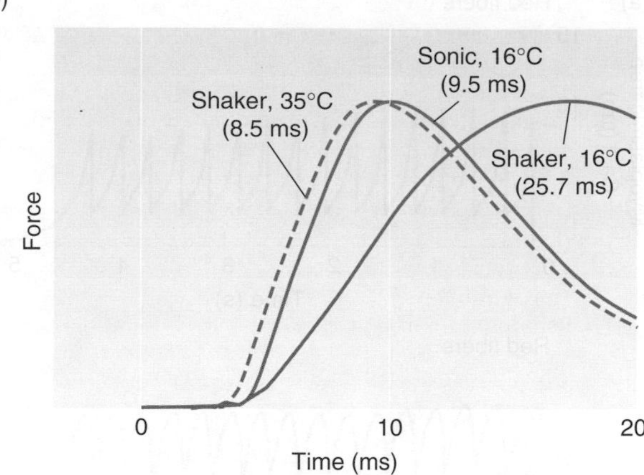

Figure 10-42 Comparing shaker muscle from a rattlesnake with sonic muscle from a toadfish reveals a complex relation between the calcium transient and the length of a twitch. **(a)** Myoplasmic free Ca^{2+} following stimulation of toadfish sonic fibers and rattlesnake shaker fibers at the indicated temperatures. The half-widths of the calcium transient (indicated in parentheses) are quite similar in sonic and shaker fibers at 16°C. At 35°C, a typical ambient temperature for rattlesnakes, the calcium transient in shaker fibers becomes shorter. **(b)** Time course of twitch tension in toadfish sonic fibers and rattlesnake shaker fibers measured under the same conditions as in part a. At 16°C, the half-width of twitch tension is nearly three times longer in shaker fibers than in sonic fibers, even though the calcium transients have about the same half-width. At 35°C, the shaker fibers contract and relax considerably faster than at the lower temperature. The Ca^{2+} concentration and force records are normalized to their maximal values. [Adapted from Rome et al., 1996.]

and rattlesnake shaker muscles suggest that in these species, convergent evolution has arrived at similar solutions to the challenges posed by high-frequency oscillatory contraction.

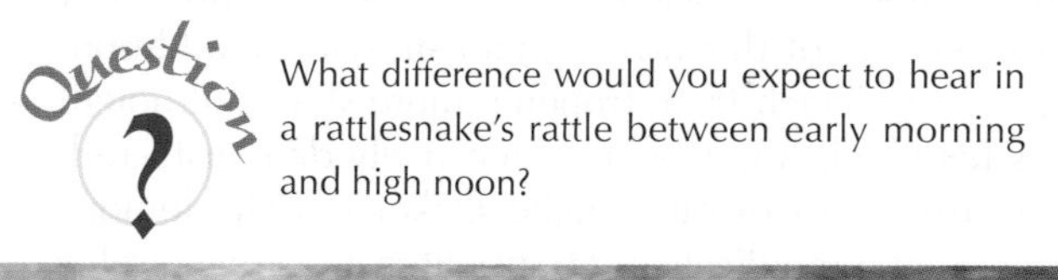

What difference would you expect to hear in a rattlesnake's rattle between early morning and high noon?

High Power and High Frequency: Insect Asynchronous Flight Muscles

In muscle fibers that contract and relax very rapidly, extremely large calcium fluxes must be supported by many SR Ca^{2+} channels and by large numbers of calcium pumps, which must be powered by large amounts of ATP. To meet the requirement for rapid Ca^{2+} flux across the membrane of the sarcoplasmic reticulum, we would expect the surface area and volume of the sarcoplasmic reticulum to be relatively high. Toadfish sonic fibers provide one such example. In these fibers, Ca^{2+} is removed from the myoplasm 50 times faster than in the red fibers, and about 30% of the entire volume of each fiber is occupied by SR. Furthermore, in fast muscles operating continuously, such as the rattlesnake shaker muscle, only aerobic metabolism can generate enough ATP to fuel the high rate of calcium pumping; thus each fiber might be expected to contain many mitochondria. An extensive SR and numerous mitochondria both take up space within the cell at the expense of space for myofilaments—the structures that generate force. In other words, once again we find a trade-off. If too much space is occupied by the calcium regulatory machinery, there is not enough space left for the contractile apparatus. The energetics of such a fiber might be impressive, but there might be too few myofilaments to generate enough power to do the required work.

Fortunately, producing sounds does not require a lot of force, and in many species, the effort lasts for only short periods of time. The situation is quite different for flight in some insects, however, which depends on muscles that can operate at high frequency and produce considerable power at the same time. We have just made the argument that a muscle *cannot* produce both high power and high frequency. Then how do these insects fly? The answer lies in the cellular properties and anatomic arrangement of their highly specialized flight muscles.

Many species among the Hymenoptera (bees and wasps), Diptera (flies), Coleoptera (beetles), and Hemiptera (true bugs) have special flight muscles that are a notable exception to the rule that only one contraction is evoked by a single AP. Although contraction of the fibers in these muscles is **neurogenic**—that is, it is initiated and maintained by activity in motor neurons—unlike more typical striated fibers, they produce many contractions for each arriving nerve impulse. Consequently, these flight muscles are called **asynchronous muscles** (or sometimes *fibrillar muscles*) to distinguish them from other skeletal muscles that contract in synchrony with APs from their motor neuron.

The first clues that asynchronous flight muscles are special came from observing insect flight. In many small insects, the wingbeat frequency (and hence, you might

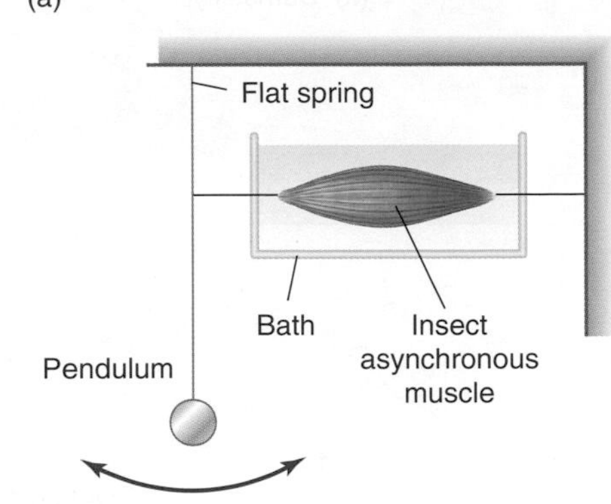

Figure 10-43 A glycerin-extracted asynchronous muscle will contract and relax rapidly and repeatedly if sufficient Ca^{2+} is present *and* it receives mechanical stretch at an appropriate frequency. **(a)** In this experimental setup, the muscle is surrounded by a saline solution and mounted between a pendulum and a fixed surface. A tension transducer monitors the force generated by the muscle. The first time the muscle is stimulated to contract, it pulls on the pendulum, which in turn pulls on the muscle as it returns to its rest position. The system creates a mechanical resonance between the muscle and the pendulum. First, the muscle moves the pendulum by contracting. As it shortens, its activity wanes. Then the pendulum stretches the muscle, reactivating it. If the resonance frequency of the pendulum (that is, the rate at which the pendulum naturally swings back and forth) matches the requirements of the muscle, the muscle will continue to contract and relax rhythmically as long as the concentration of free Ca^{2+} in the saline is sufficiently high. Classic Work **(b)** Contraction of the asynchronous muscle in the setup shown in part a depends on sufficient Ca^{2+} in the solution bathing the muscle. [Adapted from Jewell and Ruegg, 1966.]

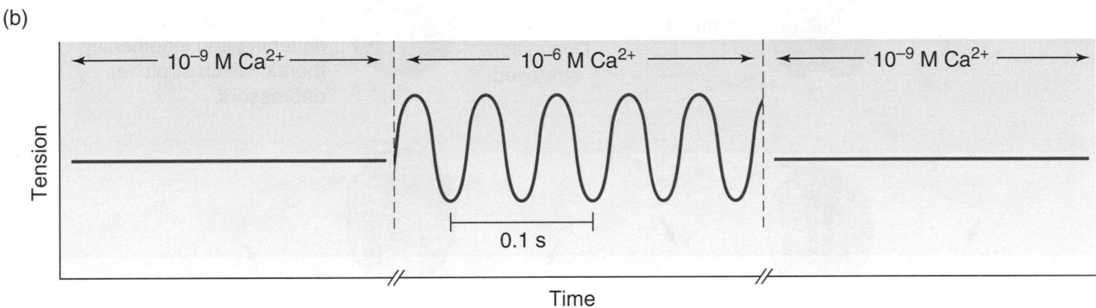

conclude, the frequency of wing muscle contractions) far exceeds the maximal maintained discharge rates of which axons are capable. Wingbeat frequency has been found to vary inversely with wing size. A tiny midge or mosquito, for example, beats its wings at a frequency of 1000 Hz or more, producing the annoying high-pitched sound that causes campers and picnickers to reach for their insect repellent.

Three different aspects of the insect's body combine to produce these rapid oscillations of the wings: first, the physiology of asynchronous flight muscles; second, the anatomic arrangement of the muscles in the thorax; and third, the mechanical properties of the thorax and wing joints. In some respects, asynchronous muscles are very similar to the synchronous flight muscles found in most flying insects. As in all other muscles, the level of myoplasmic free Ca^{2+} must rise above some threshold before the muscle is activated. In addition, neuronal input is required to initiate contraction in both synchronous and asynchronous muscle fibers, and neuronal input regulates the myoplasmic Ca^{2+} concentration in both. However, even if myoplasmic free Ca^{2+} is elevated, the active state is not initiated in asynchronous muscles until the muscle is given a quick stretch. The active state is terminated if tension on the muscle is released. In the presence of a constant concentration of Ca^{2+} higher than 10^{-7} M, a skinned asynchronous muscle actively develops tension following a quick tug, and it will oscillate repeatedly between contraction and relaxation if it is coupled to a mechanical system that oscillates at an appropriate frequency (Figure 10-43). Each asynchronous muscle has a contraction frequency at which it generates the most power, although what tunes the muscle is not yet understood.

In addition to these cellular differences, the mechanics of flight differ considerably between insects with asynchronous flight muscles and most other insects, which fly with conventional skeletal muscles. In insects with synchronous flight muscles (e.g., the damselfly), the wings are elevated and depressed by simple lever mechanics (Figure 10-44a on the next page). One end of the flight muscle is attached to the wing and the other end is attached to the floor of the thorax. Contraction of the elevator muscles causes the wings to move up; contraction of the depressors causes them to move down. The wingbeats of these insects are limited largely to frequencies below 100–200 Hz. (For comparison, the frequency of a hummingbird's wingbeat, which is also driven by synchronous muscles, is in the range of 70 Hz.)

Insects with asynchronous flight muscles have a more complex arrangement. In these insects, contractions of two sets of flight muscles arranged perpendicularly to one another change the shape of the thorax to elevate and depress the wings (Figure 10-44b). Neither set of muscles directly contacts the wings or the wing hinges. Instead, changes in the shape of the thorax act through a complex hinge structure to move each wing. The elevator muscles are attached dorsal to ventral; the depressor muscles run from the anterior to the posterior. Contraction of the elevator muscles flattens the thorax dorsoventrally and lengthens it, and this change in the shape of the thorax acts through the hinges to move the wings up. At the same time, the lengthening

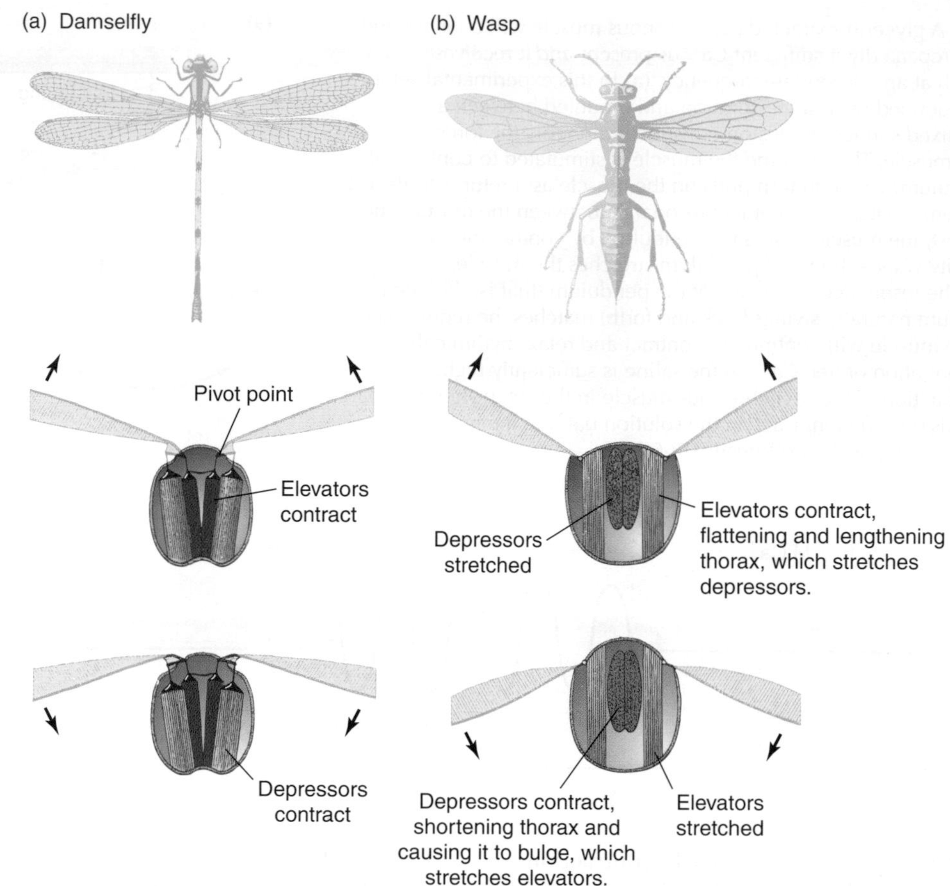

Figure 10-44 The mechanics of insect flight powered by synchronous flight muscles differ from the mechanics of flight powered by asynchronous muscles. **(a)** The wings of a damselfly act like a simple lever, pivoting about a fixed point. The synchronous flight muscles of the wings are arranged so that elevator muscles (contracted in the middle image) and depressor muscles (contracted in the bottom image) work vertically to raise and lower the wings. Both elevator and depressor muscle fibers run dorsal to ventral. **(b)** In contrast, the asynchronous flight muscles of a wasp are arranged roughly perpendicular to each other. In this transverse section through the thorax, elevators have been cut in longitudinal section and depressors cut in cross-section. The elevators run dorsal to ventral, and contraction of these muscles flattens and lengthens the thorax. This change in the shape of the thorax works through a complex hinge to raise the wings. Contraction of the depressors (bottom), which are attached at the anterior and posterior ends of the thorax, causes the roof of the thorax to bulge upward. This change in the shape of the thorax works through the hinge to lower the wings. These alternating changes in the shape of the thorax alternately stretch, and thus activate, the asynchronous elevators and depressors.

of the thorax stretches and activates the depressor muscles, while the reduced length of the elevator muscles reduces the tension on the elevators, causing them to relax. As the depressor muscles contract, the thorax is pulled shorter and becomes thicker from dorsal to ventral; this change in shape acts through the hinges to move the wings downward. In the shortened thorax the elevator muscles are stretched, while the stretch on the depressors is reduced. This alternation between contraction in the elevator and depressor muscles continues as long as conditions are appropriate for the thick and thin filaments of the fibers to slide past each other and generate force. For decades it was believed that the thorax had only two stable conformations, wing-up and wing-down, and that the instability of any state between these two contributed to the mechanical wingbeat. More recent evidence has called this "click" hypothesis into question, and the controversy remains unresolved, but there is agreement that both stretch-based activation of the muscle fibers and the mechanical properties of the thorax contribute to moving the wings.

When electrical stimulation of an asynchronous flight muscle ceases, the muscle membrane repolarizes, the myoplasmic free Ca^{2+} concentration drops, and myosin cross-bridges can no longer bind to actin filaments. If stretch is applied when myoplasmic Ca^{2+} is too low, the muscle fibers cannot be activated, and flight movements cease. Thus the motor neurons innervating asynchronous muscles act largely as an on-off switch, rather than regulating the frequency of contraction. The frequency of the wingbeat depends instead on the mechanical properties of the muscle and the mechanical resonance of the flight apparatus (thorax, muscles, and wings). If a fly's wings are clipped short, for example, its wingbeat frequency increases, even though the frequency of incoming APs remains unchanged. Under more normal conditions, flies adjust their wingbeat frequency by changing tension on the flight muscles through contraction of several small conventional skeletal muscles that attach on or near the wing hinge. These muscles modify the mechanical properties of the thoracic exoskeleton and the hinge, which changes the resonance frequency.

The force-velocity curves of both types of insect flight muscles are similar in shape to those of the striated muscles in vertebrates. Indeed, the workloop method of studying the mechanics of muscle contraction was initially developed for studying asynchronous flight muscles, and the data illustrated in Figure 10-38 were recorded for the flight muscle of a katydid, an insect with synchronous flight muscles.

With their novel mechanical arrangement, the asynchronous flight muscles of insects avoid many of the constraints that limit contractile frequency in most muscle fibers. They are able to produce extraordinarily high-frequency contractions even though the concentration of Ca^{2+} in the myoplasm changes slowly. As a result, these muscles need not have a large sarcoplasmic reticulum to pump Ca^{2+}, as do sound-producing muscles in vertebrates, nor do they require vast amounts of ATP to power a large population of calcium pumps. Calcium regulation is accomplished by a relatively modest amount of SR, which releases Ca^{2+} as APs arrive along the motor neurons and resequesters Ca^{2+} when APs cease. Asynchronous fibers instead devote much of their volume to force-generating myofilaments, containing only enough mitochondria to supply ATP for the myosin ATPase and the modest amount of SR.

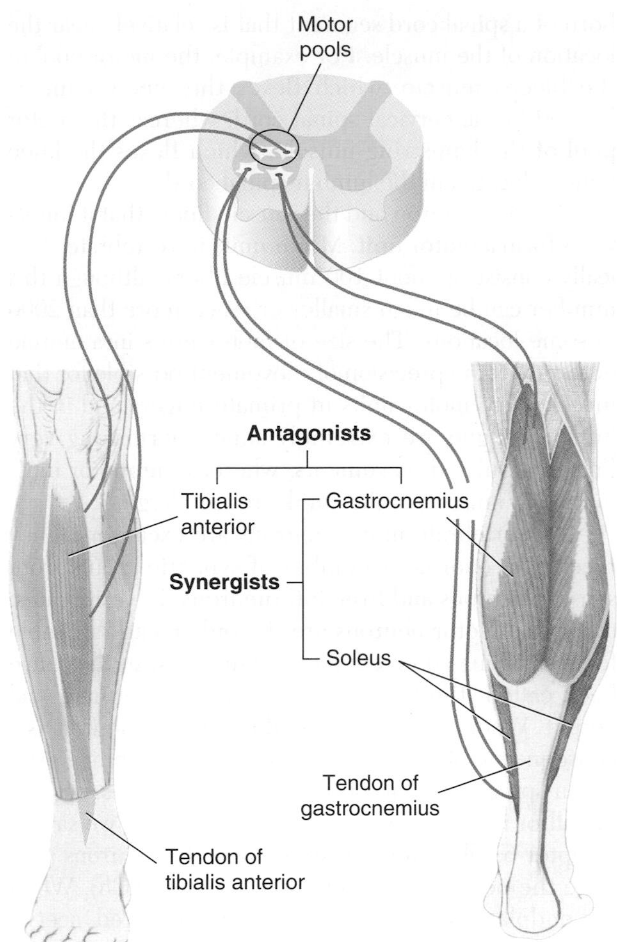

Figure 10-45 Vertebrate muscles are typically arranged in antagonistic pairs; each muscle is innervated by a separate group of motor neurons, called its motor pool. Here the pattern is illustrated for the human calf and foot. The tibialis anterior muscle flexes (closes) the ankle joint when it contracts. Its action is opposed by two other muscles, the gastrocnemius and soleus, which extend (open) the ankle joint. These two muscles are antagonists to the tibialis anterior, and they are synergists to each other because they both produce the same effect on the ankle joint. The motor pools of synergistic muscles tend to be active at the same time. In contrast, if the motor pool of a flexor is active, the motor pool of the antagonistic extensor is likely to be less active.

NEURONAL CONTROL OF MUSCLE CONTRACTION

Effective animal movement requires that the contractions of many fibers within a muscle—and of many muscles within the body—be correctly timed with respect to one another. This coordination is generated within the nervous system, as most muscles contract only when APs arrive at the neuromuscular junction. In addition to controlling the timing of contractions, the nervous system regulates their strength by selecting among different fiber types and by determining how many fibers will be active simultaneously. Several means for achieving fine control of muscle contraction have arisen during the course of evolution. We will use the neuromuscular mechanisms of vertebrates and arthropods as particularly well studied examples of these contrasting mechanisms.

Motor Control in Vertebrates

Vertebrate muscles are typically arranged in antagonistic pairs (Figure 10-45). That is, if a muscle pulls on a joint and causes that joint to close (which would make the muscle a **flexor**), its action is opposed by that of a muscle that causes the joint to open (an **extensor**). Each vertebrate skeletal muscle is innervated by motor neurons whose somata are located in the ventral horn of the gray matter of the spinal cord or in particular parts of the brain. The axon of a spinal motor neuron leaves the spinal cord through a ventral root, continues to the muscle by way of a peripheral nerve, and finally branches repeatedly to innervate skeletal muscle fibers. A single motor neuron may innervate only a few fibers or a thousand or more. However, in vertebrates, each muscle fiber receives input from only one motor neuron.

The collection of motor neurons that innervate a particular muscle is called its **motor pool.** The somata of each motor pool are clustered together in the ventral horn of a spinal cord segment that is relatively near the location of the muscle. For example, the motor pool of the biceps muscle, which flexes the elbow joint, is located in the cervical spinal cord, whereas the motor pool of the hamstring muscle, which flexes the knee joint, is located in the lumbar spinal cord.

A motor neuron and the muscle fibers that it innervates form a **motor unit.** Motor units in vertebrates typically consist of about 100 muscle fibers, although this number can be much smaller or reach more than 2000 in some locations. The size of motor units in a muscle determines the precision of movement possible for that muscle. The motor units in primate fingers and in the human tongue are extremely small, permitting very finely modulated movements, whereas the motor units in the big muscles of the trunk are very large.

All vertebrate motor neurons are excitatory. They receive an enormous number of synaptic inputs from sensory neurons and from interneurons. In vertebrates, the spinal motor neurons are the only means available for controlling contraction of the muscles, so they have been called "the final common pathway" of neuronal output. When an AP is initiated in a motor neuron as a consequence of its synaptic inputs, the membrane excitation spreads into all of its terminal branches, activating all of its endplates (see Figure 6-12). As we saw in Chapter 6, all vertebrate spinal α-motor neurons produce the neurotransmitter acetylcholine (ACh). When the endplates of a motor neuron are activated, acetylcholine is released onto all of the fibers in the neuron's motor unit. In twitch muscle fibers, the postsynaptic membrane of the neuromuscular junction is typically sufficiently depolarized by a single incoming AP to bring the fiber above threshold (see Figure 10-18), and produce an AP and a twitch in all of the muscle fibers of its motor unit (Figure 10-46a). Whether the resulting contractions are single twitches or sustained contractions depends on the frequency of the APs in the motor neuron. Higher-frequency APs in the muscle fiber produce summation of contraction and hence stronger contractions.

Because APs in a twitch muscle fiber are so tightly correlated with the APs in its motor neuron, there is no gradation in a motor unit between total inactivity and a twitch. If many APs are carried in succession along the motor axon, the result is either partial or full tetanus (see Figure 10-28). In the vertebrates, the problem of how to increase muscle tension in a graded fashion is solved by activating increasing numbers of motor units, as well as by varying the average frequency at which neurons in the motor pool fire. For example, if a small number of motor units in a muscle are maximally active, the muscle will contract with a small fraction of its total maximal tension. On the other hand, if all the motor neurons innervating the muscle are recruited to fire at a high rate, all the motor units are brought into a state of full tetanus, producing the maximal contraction of which the muscle is capable. In addition, many vertebrate muscles contain different types of fibers (see Figure 10-30), which vary in their ability to produce

(a) Vertebrate

All-or-none muscle action potential

All-or-none muscle twitch

Single nerve impulse

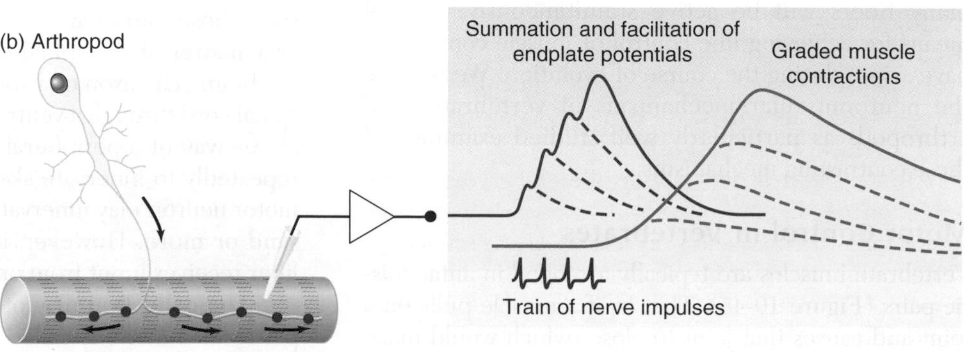

Figure 10-46 Most vertebrate muscles are made up of twitch fibers, whereas fibers in many invertebrate muscles produce graded contractions. **(a)** Vertebrate twitch muscle fibers produce an all-or-none twitch in response to each all-or-none AP traveling along the membrane of the fiber. **(b)** Many arthropod muscle fibers, as well as vertebrate tonic muscle fibers, produce graded contractions in response to the overlapping postsynaptic potentials at multiple motor synapses distributed along the fiber length.

force, so the nervous system can modulate which fibers are active, as well as how many are active. By differentially activating motor neurons, the central nervous system determines the strength and duration of muscle contractions.

The pattern of muscle contraction around a joint depends on the patterns of activity in the motor pools of different muscles. If the motor pool of a flexor muscle at a joint is active, neurons in the motor pool controlling the antagonistic extensor receive inhibitory input. If both motor pools are active simultaneously, the position of the joint will be locked as both muscles pull on the joint. Thus, the orchestration of body movements is rooted in the activity patterns of motor pools within the central nervous system.

The tonic muscle fibers of vertebrates (found primarily in amphibians and lizards) are unusual in that they receive *multiterminal innervation*—that is, a motor neuron makes many synapses along the length of each fiber. (Notice that they do not receive *multineuronal* input.) In these fibers, which lack all-or-none APs, the synaptic potentials produced by the broadly distributed neuromuscular junctions are sufficient to generate graded contractions (Figure 10-46b). The tension produced by these muscles depends strongly on the frequency of incoming motor neuron APs. As we have seen, these tonic muscle fibers generally are found where slow, sustained contractions are required.

As noted already, most vertebrate skeletal muscles contain several types of twitch fibers. Typically, all the fibers within a single motor unit are of the same type. In addition, the properties of the innervating motor neuron are often matched to the properties of the muscle fibers. For example, motor neurons innervating slow-twitch oxidative (type I) muscle fibers typically carry APs at a lower frequency than do motor neurons innervating fast-twitch glycolytic (type IIb) fibers. This matching of motor neurons to muscle fibers is also observed in arthropods, but the mechanisms that generate the match are not well understood in any species.

Motor Control in Arthropods

Arthropod nervous systems consist of a relatively small number of neurons compared with those of vertebrates, so a small number of motor units must generate the full range of muscle contractions, from weak to strong, without relying on extensive recruitment of new motor units. Moreover, many types of arthropod muscles never produce APs, or do so only under limited conditions. In these muscles, as in the tonic muscle fibers of vertebrates, contraction is controlled by graded depolarization of the muscle fiber's plasma membrane, rather than by summation of muscle APs. The pattern of neuronal control that has evolved under these constraints is quite different from the pattern of motor control in vertebrates.

Each vertebrate twitch muscle fiber is typically innervated at only one endplate, at which APs are initiated. The APs then propagate along the muscle fiber. In contrast, crustacean skeletal muscle fibers, like vertebrate tonic fibers, receive many synaptic terminals located along the entire length of the muscle fiber, so no propagated AP is required to spread the signal in the muscle fiber (see Figure 10-46b). Postsynaptic potentials arising along the distributed neuromuscular junctions are summed; the closer together excitatory postsynaptic potentials fall in time and space, the greater the depolarization of the muscle membrane. In vertebrate twitch muscles, Ca^{2+} is released from the sarcoplasmic reticulum in an all-or-none fashion in response to all-or-none APs, whereas in tonic muscles, Ca^{2+} is released from the SR in a graded fashion, because the electrical signals conducted along the membrane are graded rather than all-or-none. Because the coupling between membrane potential and tension is graded, each muscle fiber can produce tension within a wide range, instead of being limited to either all-or-none twitches or tetanus, with few possibilities in between. For this reason, even with very few motor units, arthropod muscles function well over a large range of tensions. The variation in the tension produced by single fibers replaces the recruitment of multiple fibers that is seen in most vertebrate muscles. In some arthropod muscles, one motor neuron may innervate all—or at least most—of the fibers in a muscle, so the entire muscle is one motor unit, yet contraction of the muscle can be finely graded.

In many invertebrates, the flexibility of motor control is further enhanced by multineuronal innervation of muscle fibers. Each muscle fiber receives synapses from several motor neurons, including one or two inhibitory neurons. The effects of inhibitory and excitatory synapses directly sum at the level of the muscle fiber plasma membrane. In these systems, there is typically one excitatory neuron that produces exceptionally large excitatory synaptic potentials in the muscle fiber. This fast exciter axon can generate a strong contraction with little facilitation and summation, whereas the slow exciter axon or axons must fire repeatedly at high frequency to produce similar levels of depolarization and, hence, contraction in the muscle fiber.

The variety and complexity of peripheral motor organization is increased still further by the presence in many arthropod muscles of several types of muscle fibers exhibiting different electrical, contractile, and morphologic properties. At one end of the spectrum are fibers with rapid all-or-none contractions, which resemble vertebrate twitch fibers. When a series of intracellular current pulses are delivered to these fibers experimentally, they produce a series of subthreshold depolarizations until the firing level is exceeded (Figure 10-47a on the next page). Once the plasma membrane is depolarized past threshold, it responds with an all-or-none AP, which then elicits an all-or-none fast twitch. At

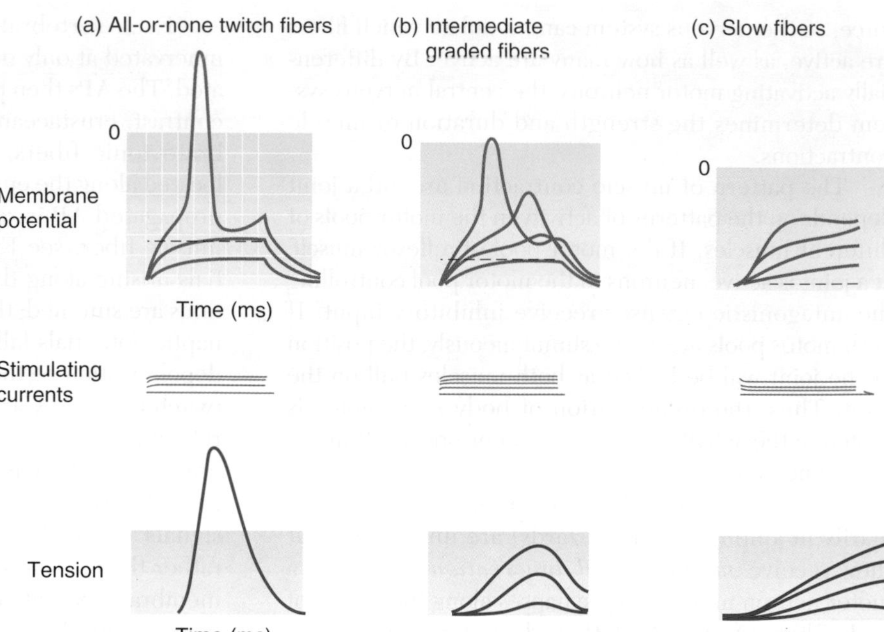

Figure 10-47 Many crustacean muscles contain fibers with highly diverse properties. Shown here are the membrane potentials (top) and tension generated (bottom) in three types of crustacean muscle fibers following intracellular stimulation (middle). **(a)** All-or-none twitch fibers produce APs and fast twitches. **(b)** Intermediate graded fibers produce nonpropagating graded potentials and graded contractions. **(c)** Slow fibers produce only small and slow depolarizations, and they contract very slowly. [Adapted from G. Hoyle, 1967, in *Invertebrate Nervous Systems,* C. A. G. Wiersma, ed. © 1967 by University of Chicago Press.]

the opposite end of the spectrum in crustacean muscle are fibers in which the electrical responses show little sign of regenerative depolarization, and the contractions are fully graded with the amount of depolarization (Figure 10-47c). Between these two extremes is a continuum of intermediate muscle fiber types (Figure 10-47b). The differences in the contractile behavior of these fiber types are correlated with morphologic differences. The slowly contracting fibers have relatively fewer T tubules and less sarcoplasmic reticulum than their more rapidly contracting counterparts. Even more than in the vertebrates, the properties of the motor neurons innervating each type of muscle fiber are at least approximately matched to the properties of the fibers themselves.

Question

What are some potential advantages and disadvantages of the vertebrate pattern of motor control? What are some potential advantages and disadvantages of the arthropod pattern of motor control? Most vertebrates are larger than most arthropods. Might this difference be related to the two patterns?

CARDIAC MUSCLE

The vertebrate heart consists primarily of muscle fibers. Unlike the large and multinucleate skeletal muscle fibers that we have been describing, each cardiac muscle fiber is a small and elongated cell, tapered at both ends, containing a single nucleus. Individual fibers are electrically connected to neighboring fibers by gap junctions, particularly at structures called intercalated disks. In addition, the fibers are tightly bound together by anchoring structures called *desmosomes* (see Figure 4-34). Histologists recognized intercalated disks as a unique structural feature of cardiac muscle long before the role of these junctions was determined. Just like other gap junctions, those of the intercalated disks allow electric current to pass unimpeded between cardiac muscle fibers, an important feature of cardiac physiology.

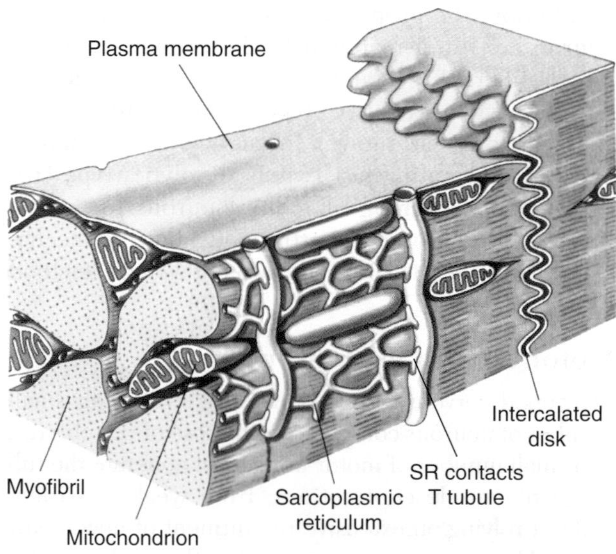

Figure 10-48 Adult mammalian contractile cardiac muscle has an extensive sarcoplasmic reticulum. As in skeletal muscle, T tubules are associated with Z disks and come into close apposition with the SR. Unlike skeletal muscle fibers, cardiac contractile fibers are connected to one another electrically through intercalated disks, at which the extensive membranes of neighboring cells are coupled by numerous gap junctions and held tightly together by desmosomes. [Adapted from Fawcett, 1986.]

Table 10-2 Characteristics of the major types of muscle fibers in vertebrates

	Striated muscle		Smooth (nonstriated) muscle	
Property/component	Skeletal	Cardiac	Multi-unit	Single-unit
Visible banding pattern	Yes	Yes	No	No
Myosin thick filaments and actin thin filaments	Yes	Yes	Yes	Yes
Tropomyosin and troponin	Yes	Yes	No	No
Transverse tubules	Yes	Yes	No	No
Sarcoplasmic reticulum	Well developed	Well developed	Very little	Very little
Mechanism of contraction	Sliding of thick and thin filaments past each other	Sliding of thick and thin filaments past each other	Sliding of thick and thin filaments past each other	Sliding of thick and thin filaments past each other
Innervation	Somatic nerves	Autonomic nerves	Autonomic nerves	Autonomic nerves
Initiation of contraction*	Neurogenic	Myogenic	Neurogenic	Myogenic
Source of Ca^{2+} for activation†	SR	ECF and SR	ECF and SR	ECF and SR
Gap junctions between fibers?	No	Yes	No	Yes
Speed of contraction	Fast or slow depending on fiber type	Slow	Very slow	Very slow
Clear-cut relationship between length and tension	Yes	Yes	No	No

*Neurogenic muscles contract only when stimulated by synaptic input from a neuron. Myogenic muscles endogenously produce depolarizing membrane potentials, allowing them to contract independently of any neuronal input.
†SR, sarcoplasmic reticulum; ECF, extracellular fluid.
Source: Adapted from Sherwood, 2001.

Two very different types of muscle fibers are found in the vertebrate heart. *Contractile fibers* are striated and similar to skeletal muscle fibers in many ways (Table 10-2). Each fiber contains many myofibrils made up of sarcomeres plus an elaborate sarcoplasmic reticulum and T tubules that are associated with Z disks (Figure 10-48). In contrast, *conducting fibers,* which include **pacemaker fibers,** bear little resemblance to most muscle fibers and do not contract. They arise embryonically from myoblasts, as do other muscle fibers, but they lack contractile proteins and the structural features of striated muscle. Instead, they function as a signal-transmission system, rapidly spreading electrical signals through the heart by way of the gap junctions they form among themselves and with contractile fibers.

Heart function will be considered in detail in Chapter 12, but let's look briefly here at the properties of the cardiac muscles. The contraction of heart muscle is **myogenic;** that is, it is initiated in the muscle fibers themselves. An electrical signal arises endogenously in the pacemaker fibers and spreads as APs through the heart by way of gap junctions. Unlike all of the types of skeletal muscle we have been discussing, cardiac muscle does not depend on neuronal input to initiate or sustain contraction. Cardiac muscle fibers do, however, receive input from neurons of the sympathetic and parasympathetic divisions of the autonomic system (see Chapter 8 for more discussion of the autonomic nervous system). This innervation of the contractile fibers produces no discrete postsynaptic potentials; instead, it serves a modulatory role. The strength and rate of cardiac muscle contractions are increased by input from sympathetic neurons and decreased by parasympathetic input.

Although the contractile mechanism of vertebrate cardiac muscle resembles that of skeletal twitch muscle, their membrane APs differ. In contrast to the very brief AP in skeletal muscles, the AP in cardiac contractile muscle fibers has a plateau phase that is hundreds of milliseconds long following the upstroke (Figure 10-49a on the next page; also see Figure 12-7). The long duration of the cardiac-muscle AP, combined with a long refractory period of several hundred milliseconds, prevents tetanic contraction and permits the muscle to relax. As a result of regularly paced, prolonged APs, the

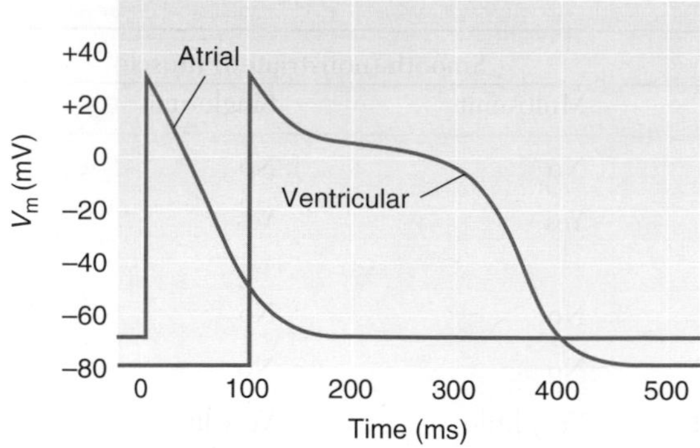

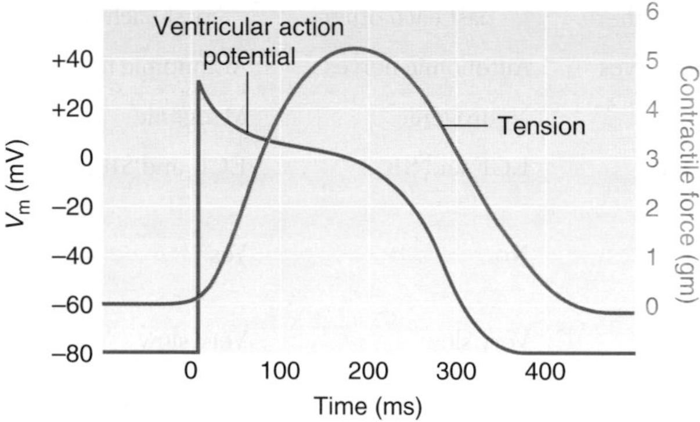

Figure 10-49 Mammalian cardiac action potentials last much longer than skeletal muscle action potentials. **(a)** This graph shows action potentials from muscle fibers in the atria (top two chambers) and ventricles (bottom two chambers) of the heart. Atrial action potentials are shorter than ventricular action potentials and precede them in time by about 100 ms, but the duration of both types of APs greatly exceeds the duration of skeletal muscle APs. **(b)** Action potential and force generation in a ventricular fiber. Notice that the force curve overlaps the AP significantly, which is quite different from skeletal muscle (compare with Figure 10-18). [Adapted from Rhoades and Pflanzer, 1996.]

heart contracts and relaxes at a rate suitable for its function as a pump.

As in skeletal twitch muscle, contraction of cardiac muscle is activated when the cytosolic Ca^{2+} concentration rises. In cardiac fibers, however, the rise in cytosolic Ca^{2+} depends on an influx of Ca^{2+} across the plasma membrane as well as on its release from the sarcoplasmic reticulum. Mammalian cardiac muscle fibers possess an elaborate SR and system of T tubules (see Figure 10-48). Membrane depolarization activates dihydropyridine receptors in the T tubules, allowing an inward flow of Ca^{2+} from the extracellular space. In skeletal muscle, the flux of Ca^{2+} through dihydropyridine receptors is negligible, but in cardiac muscle, the influx of Ca^{2+} triggers the release of a much larger amount of Ca^{2+} from the SR via ryanodine receptors, leading to contraction. Interestingly, the dihydropyridine receptors expressed in cardiac fibers appear to be unable to mechanically affect the function of ryanodine receptors. Instead, release of Ca^{2+} from the SR depends on this calcium-induced calcium release. Calcium is removed rapidly from the cytosol by calcium pumps in the SR membrane and by Na^{+}/Ca^{2+} exchange proteins in the plasma membrane.

In cardiac contractile fibers, the long plateau of the AP depends on the influx of Ca^{2+} through voltage-gated Ca^{2+} channels. Just as in other types of muscles, the amount of tension that can be developed by a cardiac muscle depends on the amount of Ca^{2+} in the myoplasm, and contraction persists in these fibers as long as Ca^{2+} continues to cross the plasma membrane and be released from the sarcoplasmic reticulum, making the timing of contraction quite unlike that in striated skeletal muscle fibers (Figure 10-49b; compare with Figure 10-18b).

The relative importance of the sarcoplasmic reticulum and the plasma membrane for Ca^{2+} regulation of cardiac muscle varies among species. In frogs, cardiac muscle fibers have only a rudimentary SR and T-tubule system. Most of the Ca^{2+} that regulates contraction in amphibian heart cells enters across the plasma membrane when depolarization opens voltage-gated ion channels. The fibers of the frog heart are much smaller than adult mammalian cardiac muscle fibers, and their resulting large surface-to-volume ratio appears to reduce the need for an elaborate intracellular SR to store, release, and resequester Ca^{2+}. When muscle cells from a frog heart are depolarized experimentally, Ca^{2+} flows into the cell through open Ca^{2+} channels in the depolarized membrane. Because the influx of Ca^{2+} is voltage-dependent, the amount of tension developed depends on the amount of depolarization, with greater induced depolarization producing greater tension (Figure 10-50a). Reducing the extracellular Ca^{2+} concentration causes a weaker contraction for a given amount of depolarization, because the reduced pool of available Ca^{2+} reduces the driving force on Ca^{2+}, and fewer ions enter the cell (Figure 10-50b). In contrast, reducing the amount of extracellular Ca^{2+} has little or no effect on the contractile force of a mammalian heart, because most of the rise in intracellular Ca^{2+} in these fibers is caused by release from the SR.

The intracellular Ca^{2+} concentration in cardiac muscle is determined not only by depolarization but also by a number of other factors, including the action of catecholamines on the heart. The catecholamines epinephrine and norepinephrine, which circulate in the blood or are released from neuron terminals, activate α- and β-adrenoreceptors on the surface of cardiac muscle cells and augment cardiac contractile force. Stimulation of α-adrenoreceptors activates the inositol phospholipid second-messenger system (see Figures 9-23 and 9-24), resulting in increased Ca^{2+} release from the sarcoplasmic reticulum. β-adrenoreceptor stimulation activates the adenylate cyclase second-messenger system (see

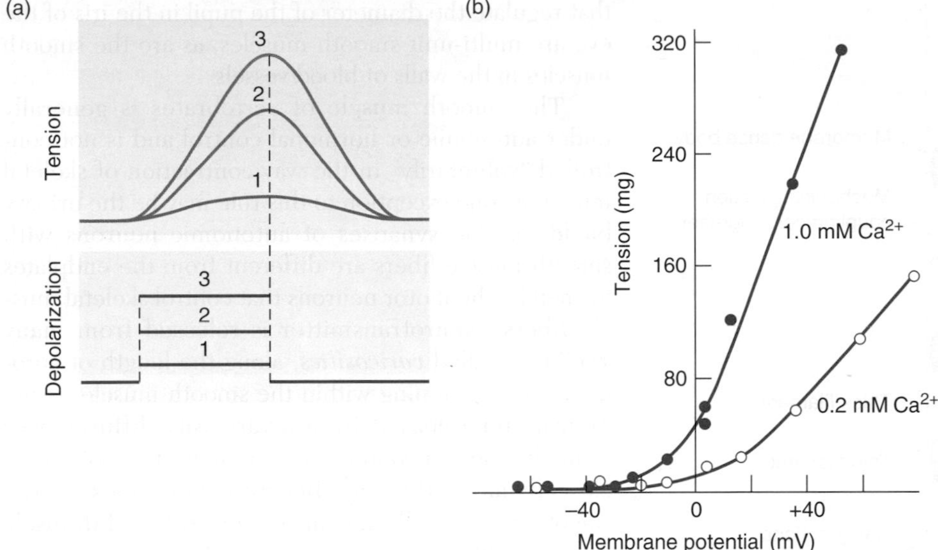

Figure 10-50 Tension developed by isolated frog contractile cardiac muscle fibers depends on the amount of depolarization *and* the concentration of extracellular Ca^{2+}. **(a)** Tension (upper traces) developed at three voltage steps (lower traces). **(b)** Effect of the amount of depolarization and extracellular Ca^{2+} concentration on tension developed. The tension was recorded at the end of each voltage step (indicated by a dashed line in part a), and is plotted against the membrane potential in millivolts. [Adapted from Morad and Orkand, 1971.]

Figures 9-20 and 9-21), resulting in increased calcium flux across the plasma membrane.

SMOOTH MUSCLE

A large class of muscles are called *smooth muscles* because they lack sarcomeres and do not appear striated when viewed with a microscope. Many smooth muscles are found in the walls of hollow organs—for example, in the alimentary canal, the urinary bladder, the uterus, and many blood vessels. Smooth muscle primarily supports visceral functions, rather than locomotion and other behavior. It has some properties in common with skeletal striated muscle, others in common with cardiac striated muscle, and some features that are unique to smooth muscle. A major impediment to a brief description of smooth muscles is their heterogeneity. Unlike striated muscles, all of which function pretty much alike, smooth muscles can be divided into subclasses that are quite different from one another. Smooth muscle is less well understood than striated muscle, but some generalizations can be made. Many smooth muscles function more independently of the nervous system than striated skeletal muscles do, and all are innervated by neurons of the autonomic nervous system, rather than the voluntary nervous system. Many smooth muscles can produce more force per cross-sectional area than striated muscles, and some can generate prolonged contractions that require much less energy per unit time than would be required by striated muscles.

Like striated muscle, smooth muscle contracts when actin and myosin filaments slide past each other, pulled by myosin cross-bridges. As in cardiac muscle, each fiber is an individual small cell containing a single nucleus. Unlike either class of striated muscle fibers, many smooth muscle fibers have little or no sarcoplasmic reticulum, and they lack T tubules. Instead of being organized into sarcomeres, the myofilaments of smooth muscle are gathered into bundles of thick and thin filaments that are anchored in structures called *dense bodies*, or they are connected to the inside surface of the plasma membrane at sites called *attachment plaques* (Figure 10-51 on the next page). Attachment plaques contain high concentrations of α-actinin, which is also found in the Z disks of skeletal muscle, and the protein vinculin, which is not. In smooth muscle fibers, vinculin anchors actin filaments to the plasma membrane by binding α-actinin.

The myofilaments of many smooth muscles are arranged helically within the fibers. Compare this pattern with the musculature of a fish to consider whether this geometric arrangement produces a mechanical advantage for these fibers.

Vertebrate Single-Unit and Multi-Unit Smooth Muscles

Vertebrate smooth muscles can be divided into two categories, single-unit and multi-unit, depending on how contraction arises and is controlled (see Table 10-2). In **single-unit smooth muscles**, the individual muscle cells, like cardiac fibers, are typically small, elongated, and tapered at both ends. They are coupled with one another through electrically conducting gap junctions. If one or a few of the cells spontaneously depolarize, the rest of the cells depolarize soon afterward as excitation is passed from fiber to fiber through the gap

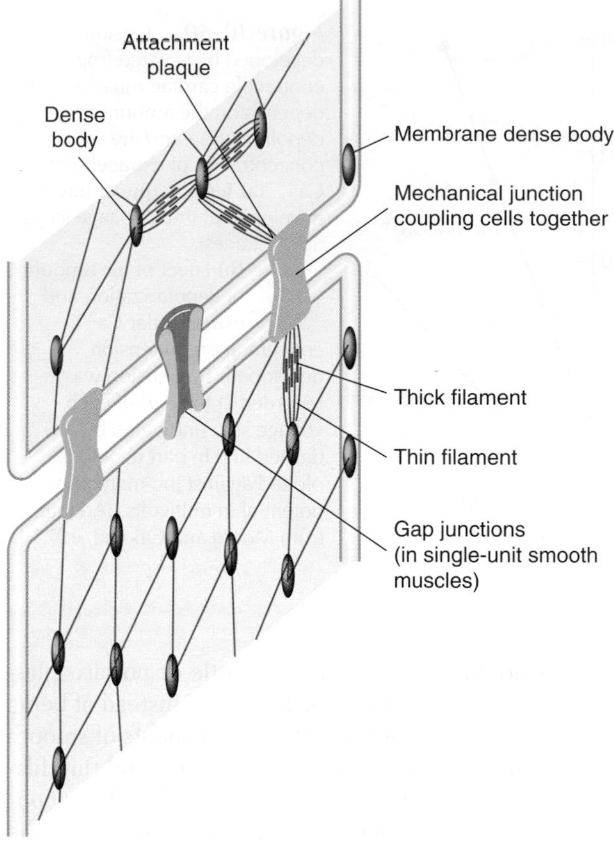

Figure 10-51 In smooth muscle, thin and thick filaments interdigitate, but they are not organized into sarcomeres. Parts of two fibers are shown in this diagram. Thin filaments are anchored to dense bodies and to dense areas abutting the membrane (attachment plaques); thick filaments lie between thin filaments. The two fibers illustrated are coupled by gap junctions, indicating that these fibers are electrically coupled. In addition, fibers are held together by strong mechanical junctions. [Adapted from Berne and Levy, 1998.]

junctions. As in cardiac muscle, the contraction of these muscles is myogenic, and the activation of a few fibers can generate contraction that moves throughout the entire organ in a wave. One example is the wavelike peristaltic contraction of the intestinal tract that propels food along the intestine. These muscles are called "single-unit" because the entire set of fibers behaves as a unit, rather than as a set of independently controlled fibers. Autonomic neurons synapse onto single-unit muscle fibers and can modulate the rate, strength, and frequency of contraction, but neuronal input is not required to initiate contraction. Single-unit smooth muscle forms the walls of vertebrate visceral organs (e.g., the alimentary canal, urinary bladder, ureters, and uterus).

In contrast, the cells in **multi-unit smooth muscles** act independently and contract only when stimulated by neurons, or in some cases hormones; in other words, their contraction is neurogenic. These fibers are not coupled to one another by gap junctions. The muscles that regulate the diameter of the pupil in the iris of the eye are multi-unit smooth muscles, as are the smooth muscles in the walls of blood vessels.

The smooth muscle of vertebrates is generally under autonomic or hormonal control and is not controlled "voluntarily" in the way contraction of skeletal muscle is (one exception to this rule may be the urinary bladder). The synapses of autonomic neurons with smooth muscle fibers are different from the endplates formed by the motor neurons that control skeletal muscle fibers. Neurotransmitter is released from many swellings, called *varicosities,* along the length of autonomic axons running within the smooth muscle tissue. Transmitter released from a varicosity diffuses over some distance, encountering a number of smooth muscle cells along the way. Receptor molecules on the smooth muscle cells appear to be distributed diffusely over the cell surface. Notice that this pattern is somewhat similar to that of neuromodulatory synapses, described in Chapter 6.

Regulation of Smooth-Muscle Contraction

As in striated muscle fibers, the cyclic binding and unbinding of myosin and actin myofilaments in smooth muscle depends on the presence of free Ca^{2+} in the cytoplasm. Most smooth muscle cells contract and relax far more slowly than striated muscle fibers and are capable of more sustained contraction. This difference in contraction kinetics is reflected in the duration and amplitude of the cytosolic Ca^{2+} pulse. Excitation-contraction coupling operates more slowly, and by different mechanisms, in smooth muscle than it does in striated muscle. The slow release and uptake of Ca^{2+} in smooth-muscle cells is associated with a relatively undeveloped sarcoplasmic reticulum, which is composed only of smooth, flat vesicles located close to the inner surface of the plasma membrane. Because the cells are small and slender, no point in the cytoplasm is more than a few micrometers away from the plasma membrane. As a result, diffusion of Ca^{2+} between the membrane and the myofilaments seems sufficient for regulating the slow contraction of smooth muscles, and the plasma membrane of smooth muscle cells performs calcium-regulating functions similar to those of SR membranes in striated muscle.

In smooth muscle cells—as in all cells—Ca^{2+} is actively and continuously pumped outward across the plasma membrane, keeping the cytosolic Ca^{2+} concentration very low. When the membrane is depolarized, voltage-gated Ca^{2+} channels open, permitting an influx of Ca^{2+}, which activates contraction. Relaxation occurs when the Ca^{2+} permeability returns to its low resting level and pumps in the plasma membrane push Ca^{2+} back out of the cell. In some smooth muscles, although not all, depolarization of the plasma membrane generates APs in which Ca^{2+} carries the inward current, causing the cytosolic Ca^{2+} to rise sharply. As in other muscle

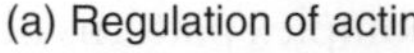

types, the tension generated is proportional to the intracellular level of Ca^{2+}.

The regulation of contraction in smooth muscle has been found to depend on a variety of mechanisms, unlike that in striated muscle, in which troponin and tropomyosin control access to myosin binding sites. Smooth muscles in general lack troponin. Instead, another filamentous protein, **caldesmon**, binds to thin filaments in smooth muscle, preventing binding between myosin and actin. Caldesmon is removed from the thin filaments by either of two events. One depends on yet another protein, **calmodulin**, which is a calcium-binding protein that has been found in a large number of cell types and is implicated in the control of a number of metabolic pathways and other cellular functions regulated by calcium. When Ca^{2+} binds to calmodulin and the Ca^{2+}/calmodulin complex binds to caldesmon, myosin cross-bridges are permitted to bind to the thin filaments (Figure 10-52a). In addition, caldesmon may be phosphorylated by an enzyme called *protein kinase C*. Phosphorylated caldesmon cannot bind to thin filaments and therefore does not inhibit myosin-actin interactions.

Three other mechanisms for regulating smooth-muscle contraction depend on the regulatory light chains of myosin (Figure 10-52b). In vertebrate smooth muscle and in some invertebrate muscles, binding of Ca^{2+} directly to the myosin regulatory light chains induces a conformational change in the myosin head that allows it to bind to actin, and the muscle contracts. Phosphorylation of myosin light chains by *myosin light-chain (LC) kinase* is another pathway to contraction in vertebrate smooth muscle. Myosin LC kinase is activated by Ca^{2+}/calmodulin, making the rate at which the myosin light chains are phosphorylated calcium-dependent. In a third mechanism, phosphorylation of another site on the myosin regulatory light chain by protein kinase C induces a conformational change that prevents actin-myosin interactions, resulting in relaxation (see Figure 10-52b, bottom). The slow actions of the protein kinases, along with the slow changes in cytosolic Ca^{2+} levels, contribute to the slow rate of contraction seen in many smooth muscles.

Smooth-muscle contraction is modulated by a wide variety of factors, both neuronal and humoral, which can inhibit or activate contraction. All of these factors operate to influence the cytosolic Ca^{2+} concentration or the activity of protein kinase C, myosin LC kinase, and myosin phosphatases. The diverse mechanisms that

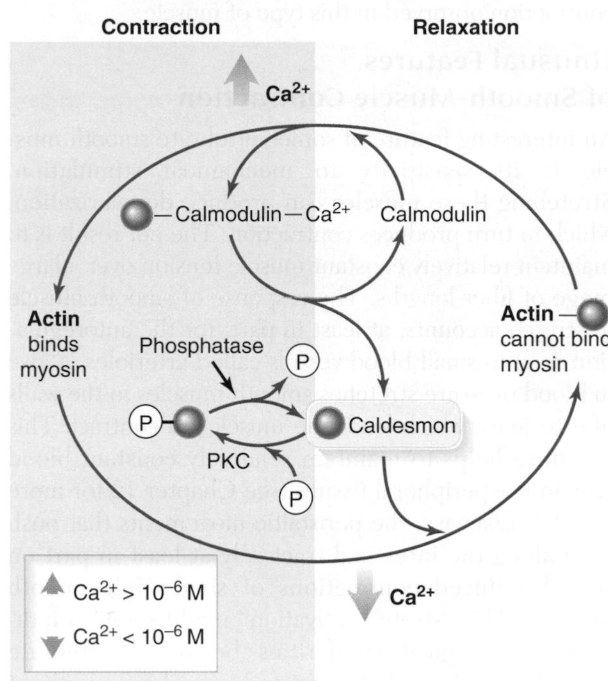

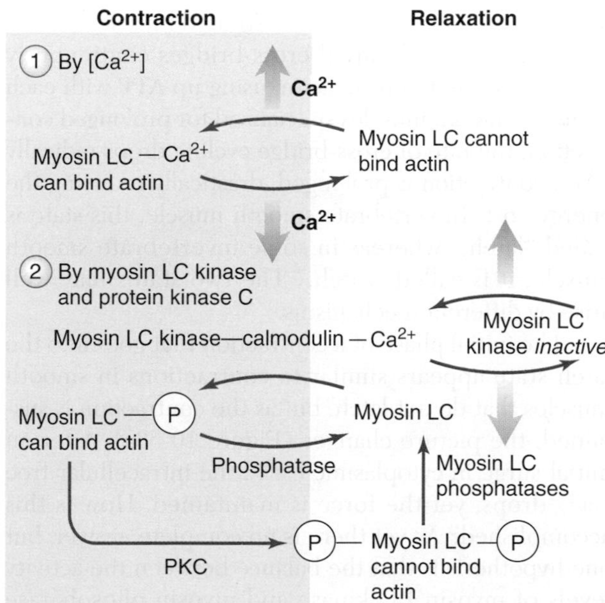

Figure 10-52 Both actin-dependent and myosin-dependent mechanisms control smooth-muscle contraction and relaxation. **(a)** Binding of caldesmon to the actin of thin filaments prevents contraction. At cytosolic Ca^{2+} levels above 10^{-6} M, Ca^{2+}/calmodulin complexes form. When this complex binds to caldesmon, myosin binding sites are revealed on actin, allowing the muscle to contract. Phosphorylation of caldesmon by protein kinase C (PKC) also prevents it from binding to thin filaments and promotes contraction. **(b)** Binding of Ca^{2+} to myosin regulatory light chains (1) allows actin-myosin binding and promotes contraction. Phosphorylation of myosin regulatory light chains (2) by myosin LC kinase, which is activated by Ca^{2+}/calmodulin, also promotes contraction. Phosphorylation of myosin regulatory light chains by PKC at a different site inhibits myosin-actin interactions and causes smooth-muscle relaxation. [Adapted from Lodish et al., 1995.]

control smooth muscle result in the complex patterns of contraction observed in this type of muscle.

Unusual Features of Smooth-Muscle Contraction

An interesting feature of some vertebrate smooth muscle is its sensitivity to mechanical stimulation. Stretching these muscles can produce depolarization, which in turn produces contraction. The net result is to maintain relatively constant muscle tension over a large range of fiber lengths. The response of smooth muscle to stretch accounts, at least in part, for the autoregulation seen in small blood vessels called arterioles. A rise in blood pressure stretches smooth muscles in the walls of arterioles, which leads the muscles to contract. This response helps to maintain relatively constant blood flow in the peripheral tissues (see Chapter 12 for more details). Likewise, the peristaltic movements that push food along the intestinal tract rely at least in part on stretch-induced contractions of single-unit smooth muscles. This "stretch activation" might seem to indicate physiological similarities between vertebrate smooth muscle and the striated asynchronous flight muscles of insects, but we do not yet understand enough about either kind of contraction to know whether the similarity is more than superficial.

In both vertebrates and invertebrates, some smooth muscles are specialized to maintain the contracted state for long periods of time while expending a minimum amount of energy. In striated muscle, tension can be maintained only if cross-bridges continuously bind to and unbind from actin, using up ATP with each cycle. In smooth muscles specialized for prolonged contraction, the rate of cross-bridge cycling drops radically when contraction is prolonged, drastically reducing the energy cost. In vertebrate smooth muscle, this state is called "latch," whereas in some invertebrate smooth muscle, it is called "catch." The two states may well arise by different mechanisms.

The initial phase of a contraction that goes into the latch state appears similar to contractions in smooth muscles that do not latch, but as the contraction is sustained, the picture changes (Figure 10-53a). After an initial surge in cytoplasmic Ca^{2+}, the intracellular free Ca^{2+} drops, yet the force is maintained. How is this accomplished? As yet there is no complete answer, but one hypothesis is that the balance between the activity levels of myosin LC kinase and myosin phosphatase determines the rate at which cross-bridges cycle. According to this hypothesis, as a fiber enters into latch, the activation of both enzymes drops, causing cross-bridges to cycle more slowly. Most cross-bridges enter the bound state and stay there. In this state, the number of bound cross-bridges remains high, so the static force produced by the muscle is large, but the amount of ATP hydrolyzed goes down because few cross-bridges detach. However, evidence is accumulating that components of the thin filaments in smooth muscle—particularly caldesmon and calmodulin—play a role in establishing latch, so a full explanation awaits further experimentation. Whatever the explanation, latch allows very thrifty generation of tension. In this state, a smooth muscle may use only 0.3% of the

(a) Vertebrate smooth muscle

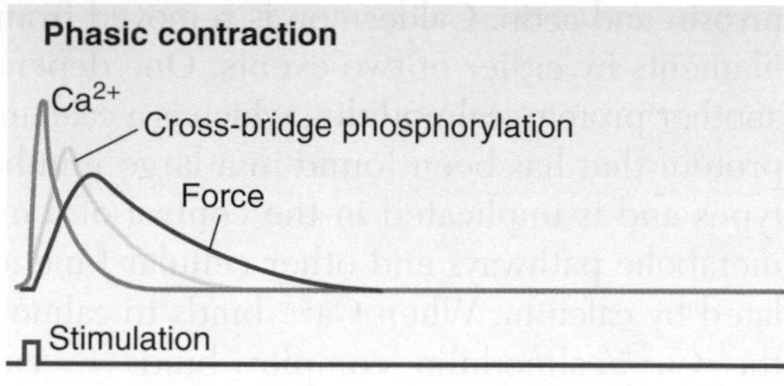

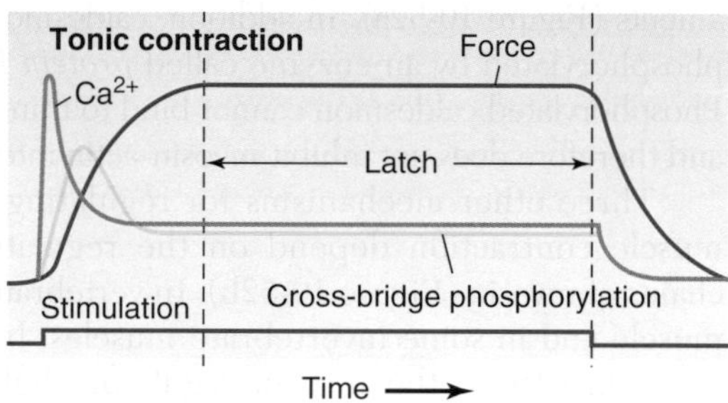

(b) Mollusk catch muscle

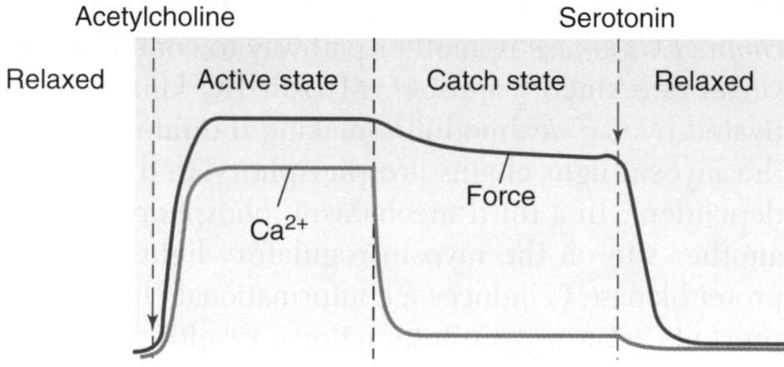

Figure 10-53 Some smooth muscles are able to generate force very economically over long periods of time. **(a)** Events in a vertebrate smooth muscle during a typical short-term (phasic) contraction and during a long-term (tonic), or "latch," contraction. In both cases, contraction begins with a pulse of intracellular Ca^{2+}. In the phasic contraction, cross-bridges are phosphorylated for a short time, and force is generated, but both phosphorylation and force dissipate quickly. When the muscle goes into the latch state, although intracellular Ca^{2+} drops to a level, cross-bridges remain phosphorylated and the generation of force continues for much longer.
(b) Control of the contractile state in a mollusk catch muscle. Contraction is initiated by cholinergic synaptic input, which causes an increase in intracellular Ca^{2+}, permitting cross-bridges to cycle. When intracellular Ca^{2+} drops, the muscle enters the catch state, in which cross-bridges remain bound but cycle very slowly. Input from a serotonergic motor neuron releases the muscle from catch through a process that depends on an increase in intracellular cAMP. [Part a adapted from Berne and Levy, 1998; part b adapted from Withers, 1992.]

energy that a striated muscle would require to do the same job.

The classic catch muscles are the smooth muscle components of the adductor muscles that hold together the two shells of bivalve mollusks. Typically, the adductor muscle also includes a section of striated skeletal muscle. The smooth-muscle in the adductor closes the shell slowly and holds it closed for long periods; the striated muscle can close the shell rapidly. It is thought that acetylcholine, acting as a neurotransmitter, and serotonin, acting as a neuromodulator, determine whether the adductor muscle enters the catch state (Figure 10-53b). Acetylcholine released onto the muscle fibers by motor neurons depolarizes the fibers, allowing Ca^{2+} to enter through voltage-gated Ca^{2+} channels. The result is contraction in which the cross-bridges cycle rapidly, generating force and using ATP. As membrane pumps drive the intracellular concentration of Ca^{2+} down, the muscle moves into the catch state as long as there is no serotonin in the body fluids. Catch is thought to depend on the state of *paramyosin,* a protein that is found along with myosin in the thick filaments of these invertebrates. The details of the mechanism are not yet clear, but a low rate of cross-bridge cycling is part of the picture, as it is in the latch state of vertebrate smooth muscle. When serotonergic motor neurons synapsing onto these fibers are stimulated, the level of the intracellular second messenger cAMP rises, which activates a protein kinase that somehow causes bound cross-bridges to release. More experiments will be needed before we understand either latch or catch well, but the similarities between the two states are intriguing.

SUMMARY

■ Muscles can be classified into striated muscles, which appear striped when viewed with a light microscope, and smooth muscles, which lack this striated appearance.

■ Muscles produce movement of the body or within the body by generating force and in many cases becoming shorter, a process that is based universally on two proteins: actin and myosin. This process is called contraction.

Essentials of skeletal muscle contraction

■ Each skeletal muscle fiber is a large multinucleated cell containing many subunits called myofibrils. Each myofibril is made up of sarcomeres arranged end-to-end.

■ The sarcomere is the functional unit of vertebrate skeletal muscle. Within each sarcomere, thick filaments composed principally of myosin interdigitate with thin filaments composed primarily of actin. Thin filaments are anchored at the ends of each sarcomere in Z disks.

■ Muscle fibers generate force when the sarcomeres shorten. Sarcomeres shorten as thin and thick filaments slide past each other, pulling the Z disks closer together.

■ Thick and thin filaments slide past each other, generating force, as the "heads" of myosin molecules repeatedly make and break bonds with actin in the thin filaments; in effect, the thick filaments "row" themselves along the thin filaments by means of these cross-bridges.

■ The amount of force a muscle generates depends directly on the number of bound cross-bridges.

■ The amount by which thick and thin filaments overlap, which determines how many cross-bridges can bind, determines the maximum amount of tension a muscle can generate. This geometry produces the length-tension relation of both individual sarcomeres and whole muscles.

■ The energy to do the work of muscle contraction is derived from the hydrolysis of ATP by myosin.

Mechanics of muscle contraction

■ In isometric contraction, a muscle generates force but does not shorten. In isotonic contraction, a muscle generates constant force and becomes shorter.

■ A muscle shortens fastest when it is unloaded. As the load increases, the velocity of shortening decreases, until at high loads the muscle cannot shorten at all (isometric contraction).

■ The force-velocity relation of muscle can be explained in terms of cross-bridge binding properties.

Regulation of muscle contraction

■ Myosin cross-bridges can bind to actin only when binding sites are available. In resting muscle, the myosin binding sites on the actin thin filaments are covered by the protein tropomyosin.

■ Tropomyosin moves away from the myosin binding sites when Ca^{2+} binds to troponin, another protein associated with thin filaments.

■ A striated muscle fiber contracts when an AP travels along its plasma membrane. Through the mechanism of excitation-contraction coupling, the AP is linked to a rise in intracellular free Ca^{2+}.

■ During muscle activation, the concentration of intracellular free Ca^{2+} increases as much as a hundredfold, Ca^{2+} binds to troponin, tropomyosin shifts on the thin filaments, myosin binding sites are revealed on actin, and the cyclic binding of myosin cross-bridges pulls the thick and thin filaments past each other.

■ In vertebrate striated muscle, the rise in intracellular free Ca^{2+} depends on release of Ca^{2+} from intracellular stores in the sarcoplasmic reticulum.

■ Calcium is released from the SR when an AP travels along the plasma membrane and into the muscle fiber through a network of T tubules.

■ An AP activates dihydropyridine receptors in the T-tubule membrane. Dihydropyridine receptors are themselves voltage-gated Ca^{2+} channels, but in vertebrate skeletal muscle fibers, little or no Ca^{2+} enters the cell through these channels. Instead, the activated dihydropyridine receptors mechanically open ryanodine receptor Ca^{2+} channels in the SR membrane, allowing Ca^{2+} to escape. In contrast, in cardiac muscle fibers, Ca^{2+} enters through the dihydropyridine receptors and contributes to calcium-induced calcium release from the SR.

■ Following an AP, myoplasmic free Ca^{2+} returns to its resting level when ryanodine receptor channels close and calcium pumps in the SR membrane resequester Ca^{2+} in the SR.

Energetics of muscle contraction

■ ATP is required for two crucial processes in muscle contraction: breaking the myosin-actin bond and powering the calcium pumps that resequester Ca^{2+} in the sarcoplasmic reticulum.

■ ATP in muscle fibers is generated by way of oxidative metabolism, glycolysis, and the direct phosphorylation of ADP by creatine phosphate.

Fiber types in vertebrate skeletal muscle

■ Vertebrate skeletal muscle fibers can be classified based on how fast they contract and how fast they fatigue.

■ Type I slow-twitch fibers contract slowly, fatigue very slowly, and depend largely on oxidative metabolism for ATP production. Type IIb fast-twitch fibers contract rapidly, fatigue rapidly, and depend largely on glycolysis for ATP production. Type IIa fast-twitch fibers contract relatively rapidly, fatigue relatively slowly, and can usually meet their energy needs through oxidative metabolism.

■ There typically are trade-offs between strength of contraction, speed of contraction, and efficiency in using ATP. No one muscle type is best for all tasks. Evolution has produced elegant matching of task and fiber type in many animals.

Neuronal control of muscle contraction

■ Each vertebrate skeletal muscle fiber is controlled by a single motor neuron, although one motor neuron can innervate a few or a large number of muscle fibers.

■ All vertebrate motor neurons are excitatory. Vertebrate skeletal muscles contract only when they receive synaptic input from their innervating motor neuron, making the motor neurons the final common pathway of behavioral control.

■ The pattern of muscle contraction—its strength, duration, and speed—is thus determined by the pattern of activity in motor neurons controlling contraction of the many fibers in a muscle.

■ In contrast, invertebrate muscle fibers are typically innervated by multiple motor neurons; some of these neurons are excitatory and some inhibitory. The pattern of muscle contraction is determined by summation of the many synaptic inputs to the muscle fiber.

Cardiac muscle

■ Cardiac contractile muscle fibers are striated and have many features in common with vertebrate skeletal muscle. However, they have some major differences: they are small, contain a single nucleus, and are coupled to other fibers by way of gap junctions.

■ Cardiac conducting fibers lack contractile machinery. Instead, they serve as a conducting system, spreading depolarization throughout the heart by way of gap junctions that link the conducting fibers to one another and to contractile fibers.

■ The signal initiating contraction in the heart is myogenic; it arises in pacemaker fibers and spreads to the rest of the heart through gap junctions. Autonomic neurons modulate contraction of cardiac muscle, but are not required to initiate it.

■ The APs of cardiac contractile fibers are prolonged and include a significant plateau phase based on a Ca^{2+} current through the plasma membrane. The direct role played by this Ca^{2+} in regulating contraction varies from relatively minor in the mammalian heart to enormously significant in the frog heart.

Smooth muscle

■ Smooth muscles form a large and varied group. They are typically found in the walls of hollow organs.

■ Smooth muscles lack sarcomeres, although their contraction is based on myosin thick and actin thin filaments that slide past each other.

■ Contraction of some smooth muscles (called single-unit muscles) is myogenic and consists of fibers that are electrically coupled to one another. Contraction of other smooth muscles (called multi-unit muscles) is neurogenic; fibers of these muscles are not coupled electrically.

■ Contraction of smooth muscles depends on increased intracellular free Ca^{2+}, but the means by which Ca^{2+} regulates binding between thick and thin filaments are varied and complex.

■ Some smooth muscles are specialized to hold tension for very long periods of time while using much less ATP than a skeletal or cardiac muscle would.

REVIEW QUESTIONS

1. Describe the organization and components of each of these structures: myofilaments, myofibrils, muscle fibers, and muscles.
2. What kinds of evidence led A. F. Huxley and H. E. Huxley to propose the sliding-filament hypothesis?
3. Draw a sarcomere and label its components. Briefly state the function of each component.
4. Discuss the contributions of myosin, actin, troponin, and tropomyosin to contraction of striated muscle.
5. Predict the shape of the sarcomere length-tension graph of a muscle with the following filament dimensions: thick filament, 1.6 μm; bare zone, 0.4 μm; thin filament, 1.1 μm.
6. Why do muscles become rigid several hours after an animal dies (rigor mortis)?
7. How do myosin cross-bridges produce the force that causes the thick and thin filaments to slide past each other?
8. When a muscle is shortening at V_{max}, what is the net force generated by its cross-bridges? What is the power produced? (Hint: See Figure 10-14.)
9. Why does the velocity of shortening decrease as heavier loads are placed on a muscle?
10. Explain how intracellular free Ca^{2+} is regulated in striated muscle fibers and how it controls their contraction.
11. List the steps of muscle activation and relaxation in striated muscle. List the same steps for smooth muscle.
12. How does depolarization of the plasma membrane of a striated muscle fiber cause the release of Ca^{2+} from the sarcoplasmic reticulum? (There are two mechanisms.) What molecules play a role in each mechanism?
13. What are the major processes in muscle function that require ATP?
14. What limits the tension that can be produced by a myofibril? By a muscle fiber? By a muscle?
15. What allows a muscle fiber to produce greater tension during tetanic contraction than during a single twitch?
16. Define *mechanical power*. Define *efficiency*. Why are mechanical power and efficiency equal to zero during isometric contractions and when a muscle shortens at V_{max}?
17. During locomotion, what is the disadvantage of using a muscle that is too slow (V_{max} is too low) to power a movement requiring a given shortening velocity? What is the disadvantage of using a muscle that is too fast (V_{max} is too high)? What is an optimal value of V_{max}?
18. Describe the features of the fish muscular system that enable a fish to produce both relatively slow movements with little backbone curvature and very fast movements with large backbone curvature.
19. Why must the muscles that produce sound relax very rapidly? What are some adaptations that permit a muscle to relax quickly?
20. Why is a large energetic cost associated with rapidly relaxing muscles? How do insect asynchronous muscles avoid some of this cost?
21. What factors determine frequency of contraction in insect asynchronous muscle?
22. Compare and contrast the neuronal control of vertebrate twitch muscle fibers and arthropod muscle fibers.
23. Compare and contrast skeletal, smooth, and the two types of cardiac muscle fibers with respect to structure, excitation-contraction coupling, speed of contraction, relaxation, and energetics.
24. What are the advantages of latch and catch in smooth muscle?

SUGGESTED READINGS

Alexander, R. M., and G. Goldspink. 1977. *Mechanics and Energetics of Animal Locomotion*. London: Chapman and Hall. (A discussion of how a very wide variety of animals move around and what it costs them to do so.)

Huxley, H. E. 1969. The mechanism of muscular contraction. *Science* 164:1356–1365. (One of the landmark papers in the history of muscle physiology.) Classic Work

Josephson, R. E. 1993. Contraction dynamics and power output of skeletal muscle. *Annu. Rev. Physiol.* 55:527–546. (A thorough review of muscle function by the physiologist who first applied work-loop methods to contracting muscles.)

Lamb, G. D. 2000. Excitation-contraction coupling in skeletal muscle: Comparisons with cardiac muscle. *Clin. Exp. Pharmacol. Physiol.* 27:216–224. (A concise review of this complex process in two important types of striated muscle.)

Lutz, G., and L. C. Rome. 1994. Built for jumping: The design of frog muscular system. *Science* 263:370–372. (A brief introduction to studies described in this chapter.)

Marden, J. H. 2000. Variability in the size, composition, and function of insect flight muscles. *Annu. Rev. Physiol.* 62:157–178. (A review of the ways in which insects have solved the puzzle of flight.)

Maughm, D. W., and J. O. Vigoreaux. 1999. An integrated view of insect flight muscle: Genes, motor molecules, and motion. *News Physiol. Sci.* 14: 87–92. (A molecular approach to insect flight.)

Reggiani, C., R. Bottinelli, and G. H. M. Stienen. 2000. Sarcomeric myosin isoforms: Fine tuning of a molecular motor. *News Physiol. Sci.* 15:26–33. (A molecular approach to understanding how different types of striated muscle can behave differently.)

Rome, L. C., R. P. Funke, et al. 1988. Why animals have different muscle fibre types. *Nature* 355:824–827. (More detail on one theme of this chapter.)

Squire, J. M., and E. P. Morris. 1998. A new look at thin filament regulation in vertebrate skeletal muscle. *FASEB J.* 12:761–771. (A summary of interactions among troponin, tropomyosin, and actin, including some great three-dimensional figures presented as stereo pairs.)

Taylor, C. R., E. Weibel, and L. Bolis, eds. 1985. *Design and Performance of Muscular Systems. J. Exp. Biol.*, Vol. 115. Cambridge: The Company of Biologists, Ltd. (A collection of papers on muscle physiology and motor systems.)

Vale, R. D., and R. A. Milligan. 2000. The way things move: Looking under the hood of molecular motor proteins. *Science* 288:88–95. (An enlightening exposition comparing myosin-based and kinesin-based motion.)

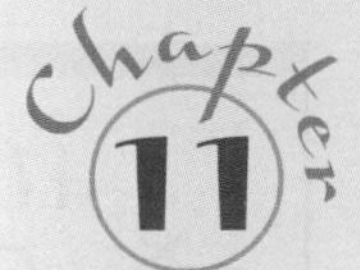

Behavior: Initiation, Patterns, and Control

Human beings have been studying—and making predictions about—animal behavior for as long as the species *Homo sapiens* has existed. Although our current interest in the initiation and control of animal behavior depends in part on curiosity about the origins and workings of our own behavior, our ancestors may have been more interested in optimizing hunting strategies and minimizing the chance that they would themselves become prey. Producing behavior is complex: consider all of the processes that contribute. How do the various sensory organs collect the information that will eventually elicit a behavioral response? Where in the nervous system are decisions made about what to do next, and where is coordinated action organized? How is activity in the nervous system translated into effective behavior? In the simplest of animals, the one-celled Protozoa, all sensory and motor functions are carried out by components of a single cell (Spotlight 11-1 on the next page), but understanding the fundamental processes underlying behavior in complex animals requires that the physiology of the nervous, endocrine, and muscular systems be considered.

The behavior of an animal may change continuously in response to changes in stimulation from the animal's internal and external environments. Some behavior consists of simple and predictable reflexes. Other kinds of behavior depend strongly on information stored from past experience and are therefore less predictable to an observer who has no access to the animal's memories. All behaviors ultimately depend on activity in motor neurons, which form the "final common pathway" for controlling skeletal muscles.

Although many properties of motor neurons have been conserved through evolutionary time, the networks of neurons that process sensory information have offered fertile ground for evolutionary invention. Some elements of sensory transduction are common to many senses (see Chapter 7), but the properties of the central neurons that process sensory signals have been exquisitely tuned by the circumstances of each species. For example, both birds and bats fly—and thus encounter the same kind of environment—but these animals encode information about their environment into sensory signals quite differently. Although bats can see perfectly well, they typically are active from dusk to dawn. As they fly in the dark, bats glean information about their surroundings by emitting sounds and listening for echoes reflected by surfaces. In contrast, most birds are active during the day and rely heavily on vision. If a bird and a bat were to fly in the same area, the sensory systems of the two animals would represent their surroundings quite differently. Hence birds and bats use different regions of the brain and different mechanisms of information processing to interpret sensory signals from the same environment. The auditory regions of the bat's brain are relatively large and complex; in birds, the visual regions are large and elaborate.

Despite such differences across sensory modalities, there are general principles that apply to many sensory processing systems. For example, independent parameters of the stimulus—such as the color, size, and direction of motion of a visual stimulus—are processed in separate, parallel pathways. As information from each sensory modality is transmitted through the brain, the properties of the stimulus that are represented in each brain region become increasingly specific. In addition, most stimuli are organized systematically within each region of the brain, generating a map in which parts of the body, parts of the environment, or some other attribute of the stimulus (e.g., the frequency of sounds) are arranged topologically in an orderly fashion with respect to one another. The original view that these maps are static has been replaced by the understanding that to some extent their topology can be modified dynamically by use-dependent mechanisms.

The production of complex behavior patterns must depend on the enormous amount of neuronal circuitry that lies between the relatively simple afferent sensory

BEHAVIOR IN ANIMALS WITHOUT NERVOUS SYSTEMS

The behavior of multicellular animals depends on activity in the nervous system, but protozoans produce several interesting behaviors, too, although these single-celled creatures lack any neurons or muscles. Instead, events within the single cell serve the same functions as do sensory receptor cells, interneurons, motor neurons, and muscles. The mechanisms that allow these simple organisms to produce surprisingly complex behaviors provide insight into the very conservative nature of evolution.

The ciliate *Paramecium* produces an avoidance response if it bumps into an object as it is swimming or if it is touched. Touching the posterior end of a *Paramecium* causes it to swim forward more rapidly; touching its anterior end causes it to reverse its direction (see part a of the accompanying figure). The direction in which a *Paramecium* swims depends on the direction in which its cilia are beating; it reverses its direction of progress by reversing its ciliary beat. What mechanism could account for this reversal of ciliary action in response to mechanical stimulation?

Experiments in the laboratory of Roger Eckert revealed that the membrane of *Paramecium* includes stretch-activated ion channels that are differentially distributed: the channels in the anterior membrane are Ca^{2+}-selective, whereas the channels in the posterior membrane are K^+-selective. As a result, touching the front end of the organism produces an increase in Ca^{2+} flux into the cell, depolarizing it (part b of the figure), whereas touching the back end of the organism produces an increased K^+ flux out of the cell and hyperpolarizes it (part c).

Thus, in single-celled organisms, behavioral changes are associated with changes in membrane potential produced by specific ion fluxes across membranes. Perhaps the properties of neurons and muscles can be regarded as an extension of some of the capabilities of these multifunctional "simple" organisms.

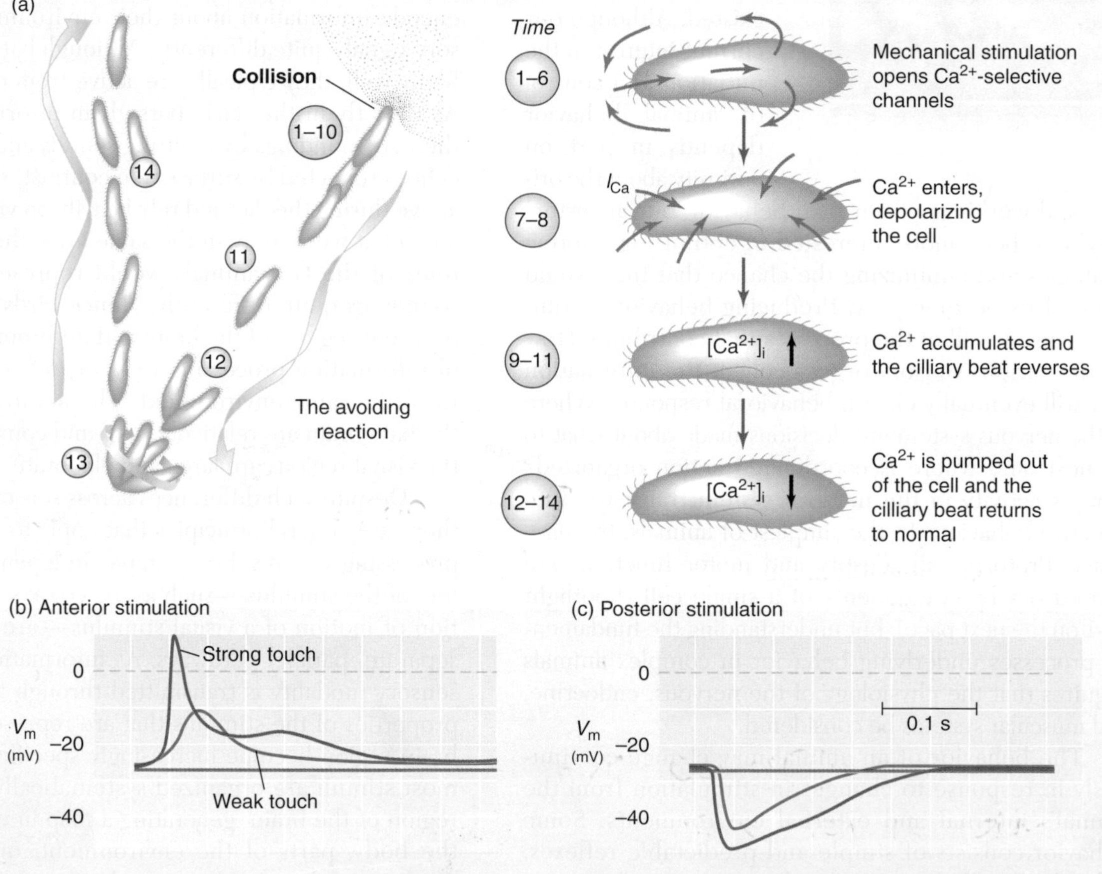

Classic Work *A Paramecium avoids objects with which it has collided by changing its direction and rate of swimming.* ***(a)*** *After it collides with an object, the Paramecium backs up, changes direction, and swims away in a new direction (left). The passage of time is indicated by the sequential numbers. Mechanical stimulation of the anterior end of the organism opens Ca^{2+}-selective channels (right), allowing Ca^{2+} to enter the cell, which in turn reverses the ciliary beat.* ***(b)*** *When anterior Ca^{2+} channels are opened by mechanical stimulation to the anterior end, the membrane depolarizes and produces a graded, but weakly regenerative, change in the membrane potential that is associated with reversal of the ciliary beat.* ***(c)*** *Mechanical stimulation of the posterior end opens K^+-selective channels, producing a graded hyperpolarization of the membrane that accelerates the ciliary beat in the same direction by an unknown mechanism. [Part a adapted from Grell, 1973; parts b and c adapted from Eckert, 1972.]*

pathways and the efferent motor pathways. In higher organisms, most neurons are central interneurons. In Figure 8-1, this complex interface between sensory input and motor output is represented as a single multicomponent box separating input and output neurons. Indeed, although this part of the nervous system is becoming better understood, to a frustrating extent it remains a "black box" that is the focus of intense research effort. In this chapter, we examine some of the systems that are helping us to understand what the nervous system does when it processes sensory information and generates patterned responses.

THE NEURONAL BASIS OF A SIMPLE BEHAVIOR

Behavior can be initiated either in response to sensory stimulation or by activity that arises in the central nervous system. To understand how this can happen, it is necessary to know the properties of neurons and the ways in which information is transmitted from one neuron to the next. The "hardware" underlying all behavior is composed of **neuronal networks**, or interconnecting circuits of neurons. We have already seen how individual neurons are linked by synapses into such circuits (see Chapter 8). Now we need to consider how these neurons work together to acquire information and produce behavior. Unlike electric circuits, which are wired in a single way, neuronal networks are not "hard-wired." Instead, they exhibit *plasticity*, which means they are modified functionally, and even anatomically, in response to experience.

We begin with a discussion of the simplest neuronal circuits: reflex arcs that depend on activity in a small number of neurons. It is useful to begin with these simple circuits because they illustrate the fundamental components of behavioral control. The complicated circuits that produce more complex behaviors are thought to differ from reflex arcs primarily in the number of neurons involved, not in the underlying mechanisms.

Simple Reflexes

The simplest complete circuit that produces behavior is a **reflex arc**, which consists of a sensory receptor and an output motor neuron that synapse directly with each other, or with only a few interneurons in between. Reflex arcs are found in both the somatic (voluntary) nervous system and the autonomic nervous system. In both locations they generate relatively *stereotypic* behavior—that is, behavior that is both predictable and apparently automatic.

Vertebrate skeletal muscle stretch reflex

The classic *patellar tendon reflex*, or knee-jerk reflex, is probably the most familiar example of a reflex: a moderately intense tap on the appropriate part of the kneecap elicits extension of the knee joint. This **myotatic reflex**, or **muscle stretch reflex**, regulates the amount of tonic tension in the muscles that hold an animal in its normal posture. If a joint starts to bend—which could cause the animal to fall to the ground—it is automatically straightened by this reflex. Thus the reflex provides a negative-feedback system that maintains posture.

How does the reflex work? The tap on the knee stretches a tendon of the quadriceps muscle, which is an extensor of the knee, and thus stretches the muscle. Sensory neurons in the muscle respond to the stretch, and the reflex arc is activated. The net effect is contraction of the quadriceps muscle, and the knee joint extends. It takes only two kinds of neurons—muscle stretch receptors (also called *1a-afferent neurons*) and spinal α-motor neurons—connected together to produce the reflex (Figure 11-1a,b on the next page). Because the basic form of this reflex requires only the synapse between afferent and efferent neurons, with no interposed interneurons, it is a **monosynaptic reflex.**

Sensory endings of 1a-afferent neurons are located within each muscle, associated with sensory structures called *muscle spindle organs.* Each spindle organ contains a small bundle of specialized muscle fibers, called **intrafusal fibers** to distinguish them from the majority of contractile fibers, which are called **extrafusal fibers.** Extrafusal fibers are the skeletal muscle fibers discussed in Chapter 10, and they are innervated by α-motor neurons. The intrafusal fibers are small in mass and number, and they make no contribution to the production of tension by the muscle. Instead, they participate in a feedback loop that regulates how sensitive the spindle organs are to stretch.

Muscle spindles lie parallel to the extrafusal fibers; if something happens to stretch the muscle (for example, if a weight is added to an isolated muscle, or a joint bends), the muscle spindles are stretched too. Stretching the central region of a muscle spindle increases the frequency of action potentials in its 1a-afferent axon. In the spinal cord, these axons make excitatory synapses directly on the α-motor neurons that control the muscle containing their spindle organs. When activity in the 1a-afferent axons increases, the motor neurons to that same muscle become more active, causing reflexive contraction in the stretched muscle (Figure 11-1c). The reflex provides negative-feedback control of muscle length because stretch of the muscle initiates neuronal activity that causes the muscle to contract, opposing the stretch. The reflex can be shown to rely on a complete circuit by cutting the dorsal root leading into the appropriate segment of the spinal cord (see Figure 11-1b). Severing the dorsal root leaves all of the motor innervation intact but removes sensory input to the spinal segment. Following the cut, the muscles innervated by that spinal segment go limp, even though their motor input is intact.

Notice that tension is removed from the muscle spindles when a muscle shortens under the influence of the stretch reflex. If nothing else happened, the

Figure 11-1 Only two kinds of neurons are required to produce the muscle stretch reflex. **(a)** The steady state of a muscle that is holding up a light weight. Extrafusal fibers are shown in red; an α-motor neuron is shown in blue. An intrafusal fiber and a 1a-afferent axon are shown in green. **(b)** If a heavier weight is added to the muscle, it stretches the muscle, activating 1a-afferent neurons in the muscle spindle. Those neurons synapse in the spinal cord onto α-motor neurons controlling the same muscle and cause the stretched muscle to contract more forcefully, returning it to its original length. If the 1a-afferent axons were cut where they enter the spinal cord (black dashed line), there would be no feedback onto the motor neurons, and the weight would cause the muscle to elongate. **(c)** Sequence of events (1 to 4) in the muscle stretch reflex.

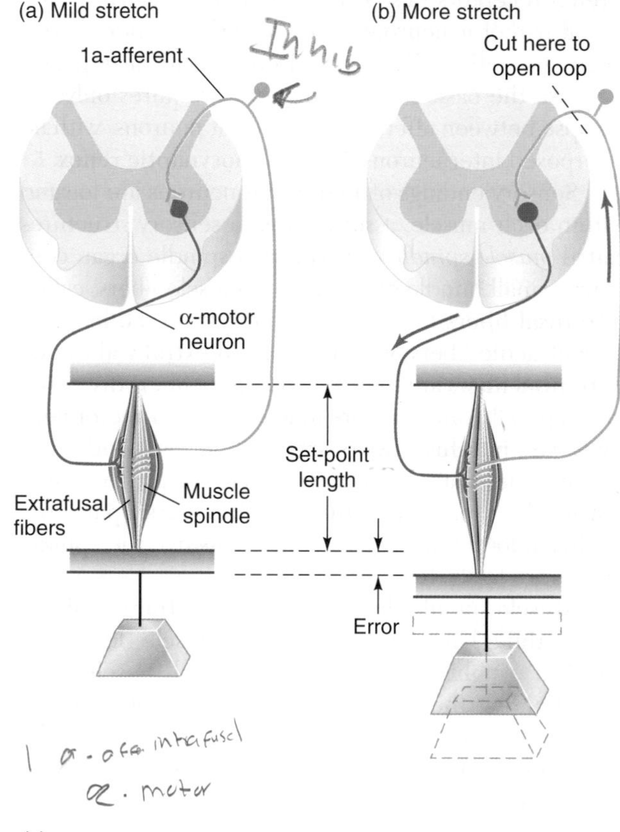

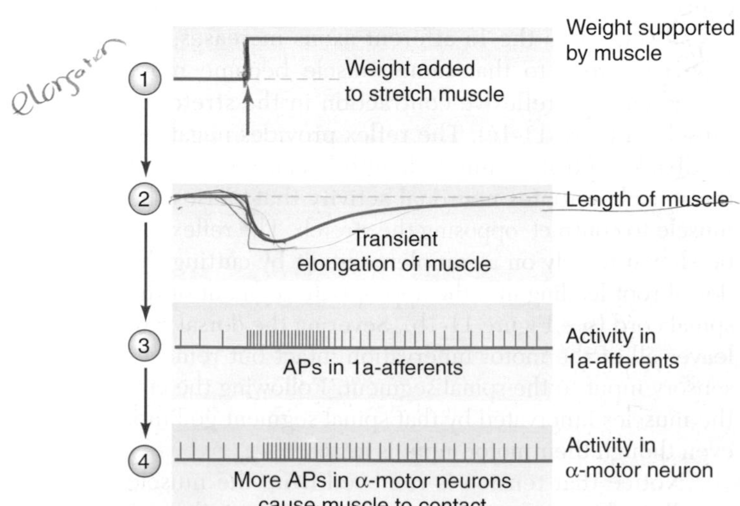

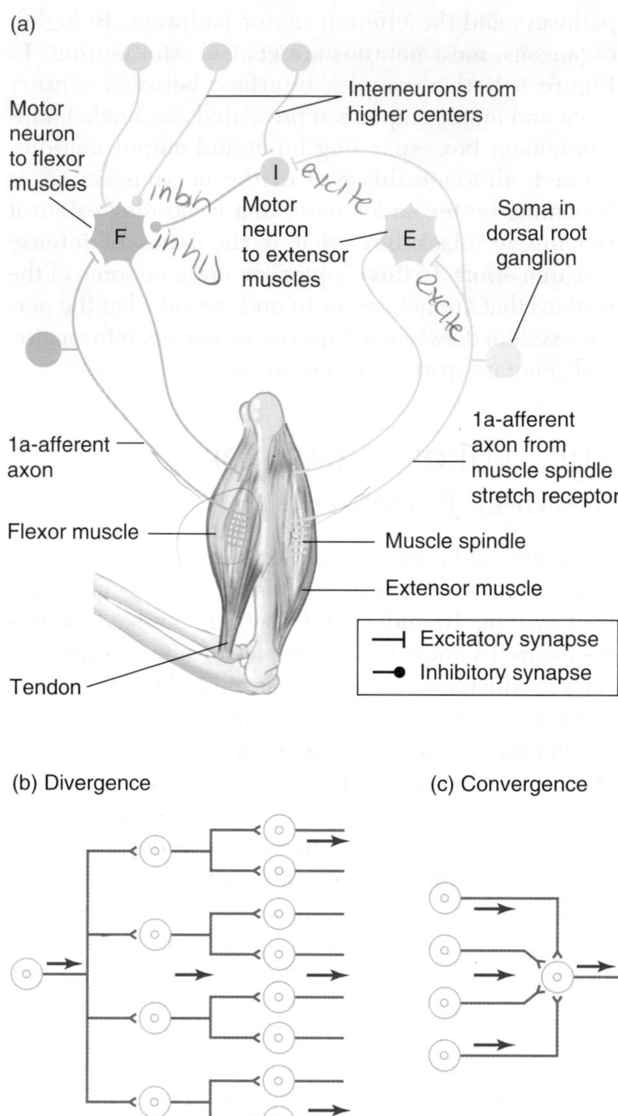

Figure 11-2 Convergent and divergent information modifies the activity of neurons in the muscle stretch reflex. This diagram focuses on connections relevant to stretch of the extensor muscle at a joint, but convergence and divergence are ubiquitous in all nervous systems. **(a)** The two neurons that are required to produce the stretch reflex are 1a-afferents from the muscle spindle stretch receptors and spinal α-motor neurons to the same muscle (shown in blue). The 1a-afferents also terminate on inhibitory neurons in the spinal cord, which in turn synapse onto and inhibit motor neurons controlling the antagonist (flexor) muscle at this joint (neurons shown in purple). Finally, neurons carrying information from the brain and other regions of the spinal cord synapse onto the motor neurons and spinal interneurons with the potential for modifying activity in the spinal segment (shown in yellow). **(b)** Divergence is the branching of one neuron, allowing it to synapse with and modify the activity in several others. In the muscle stretch reflex, the 1a-afferent axon from the extensor muscle spindle diverges to synapse onto both the α-motor neuron and the inhibitory interneuron. **(c)** Convergence refers to the innervation of a single neuron by many presynaptic neurons. In the muscle stretch reflex, many excitatory and inhibitory neurons converge on the flexor motor neuron shown in part a. [Part a adapted from Clarac et al., 2000.]

1a-afferents would then become silent. If the muscle were then stretched a little from its new length, the muscle spindles would fail to respond unless their own length had been restored to its original value. Under the control of another set of motor neurons called the *γ-efferents,* the intrafusal fibers regulate the length of the muscle spindle organs. When a muscle shortens, driven by its α-motor neurons, increased activity in the γ-efferents drives the intrafusal fibers to shorten as well, maintaining a constant tension on the spindle. In this way, the γ-efferents maintain sensitivity in the muscle spindles through a wide range of muscle length.

Although only two kinds of neurons are required to produce the muscle stretch reflex, these neurons are embedded in more complex circuitry within the nervous system, and other neurons can affect the reflex. For example, when the patellar tendon is tapped, the knee joint can extend only if flexor muscles relax at the same time the extensor contracts. If extensors and flexors contract simultaneously, the knee joint is typically locked in one position. The knee flexor (commonly called the *hamstring muscle* in humans) relaxes during the patellar tendon reflex because the motor neuron pool to the flexor receives inhibitory synaptic input as the result of activity in the 1a-afferents of its antagonist (Figure 11-2a). In addition, although you may be able to stand erect for hours, supported by muscles in which tension is maintained by your stretch reflexes, you can also choose to do a deep knee bend. That behavior requires signals coming from your brain to override the stretch reflex from your quadriceps muscle, allowing the knee joint to flex. Summation of all inputs onto the motor pool controlling your quadriceps muscle determines the actual behavior of the muscle.

Notice that even in this small circuit, information is sent along divergent pathways; for example, the 1a-afferent axon from the extensor diverges to synapse onto both the extensor α-motor neurons and the inhibitory interneurons. In addition, synapses from many different neurons converge onto the α-motor neuron innervating the flexor muscle. *Convergence* and *divergence* are widespread in the nervous system and enrich the possibilities for processing information (Figure 11-2b,c). In addition, numerous small feedback loops and other small synaptic circuits contribute to the net activity of a neuron. For example, many α-motor neurons send *recurrent collateral axons* to contact inhibitory interneurons called **Renshaw cells,** which in turn contact the α-motor neurons (Figure 11-3). These small feedback circuits effectively establish limits to how excited the motor neurons can become. Strychnine blocks the glycine-mediated synapses made by Renshaw cells onto the motor neurons—and probably other glycine-mediated synapses as well—producing convulsions, spastic paralysis, and death from paralysis of the respiratory muscles. These gruesome consequences of blocking inhibitory synapses demonstrate the importance of synaptic inhibition in the normal function of the nervous system.

Feedback loops are found throughout the central nervous system. In some small feedback loops, the net effect is to generate positive rather than negative

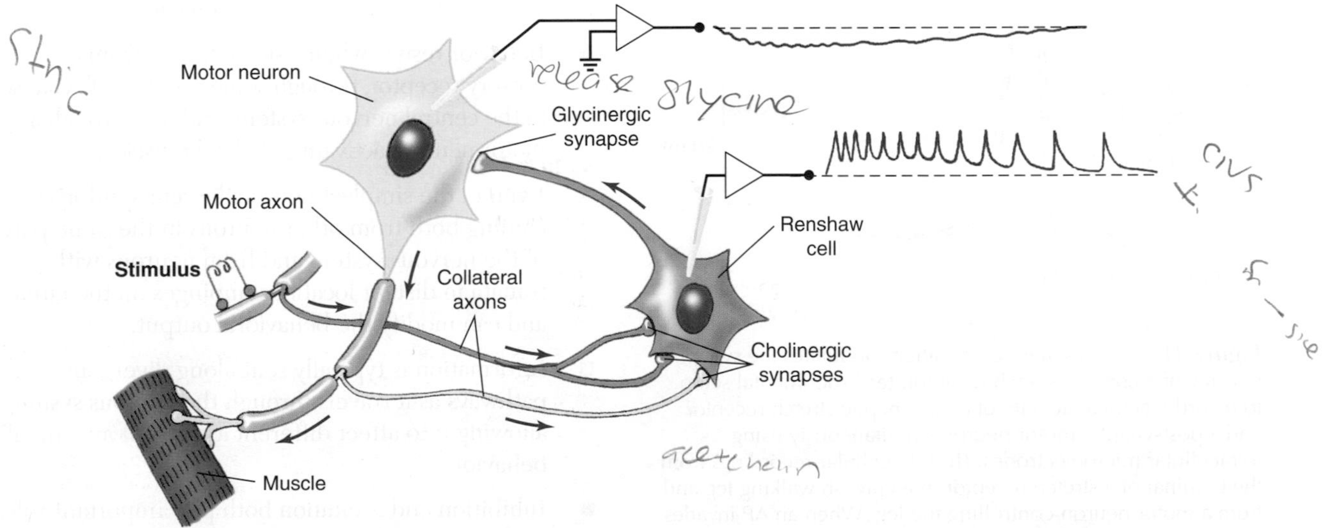

Figure 11-3 The effects of recurrent collateral axons provide an example of the many small convergent and divergent synaptic loops linking neurons in the central nervous system. In this experiment, activity in a motor neuron is recorded intracellularly. The motor neuron innervates one or more skeletal muscle fibers in the periphery. Its axon diverges to make an excitatory synapse onto a Renshaw cell, one class of inhibitory spinal neurons. Notice that the motor neuron releases acetylcholine at both its spinal synapse and the motor endplate. This same Renshaw cell receives input from another motor neuron (axon shown in pink). When the second motor neuron is stimulated electrically, the resultant activity in the Renshaw cell causes the release of glycine onto the first motor neuron, hyperpolarizing it. Notice that activity in the first motor neuron itself would excite the Renshaw cell, so this loop places limits on how excited the motor neuron can become.

feedback so that activity in a neuron excites other neurons whose synapses uphold and reinforce the activity.

Arthropod stretch reflexes

The activity of stretch receptors in the crayfish abdomen was illustrated in Figure 7-13. Much about the muscle stretch reflex is similar in arthropods and vertebrates. In arthropods, the stretch receptors make monosynaptic connections onto motor neurons that control the muscle containing the stretch-receptor endings, and the net effect of the reflex is the same as in vertebrates—the reflex provides negative-feedback control to hold a joint in position. In arthropods, these reflexes are typically called "the resistance reflex," a name that would be appropriate in vertebrates, too. However, each of the components of a stretch reflex circuit in arthropods is quite different from the equivalent components of the vertebrate circuit. For example, the stretch receptors are arranged differently, and many of the motor neurons act through graded membrane potentials, rather than APs.

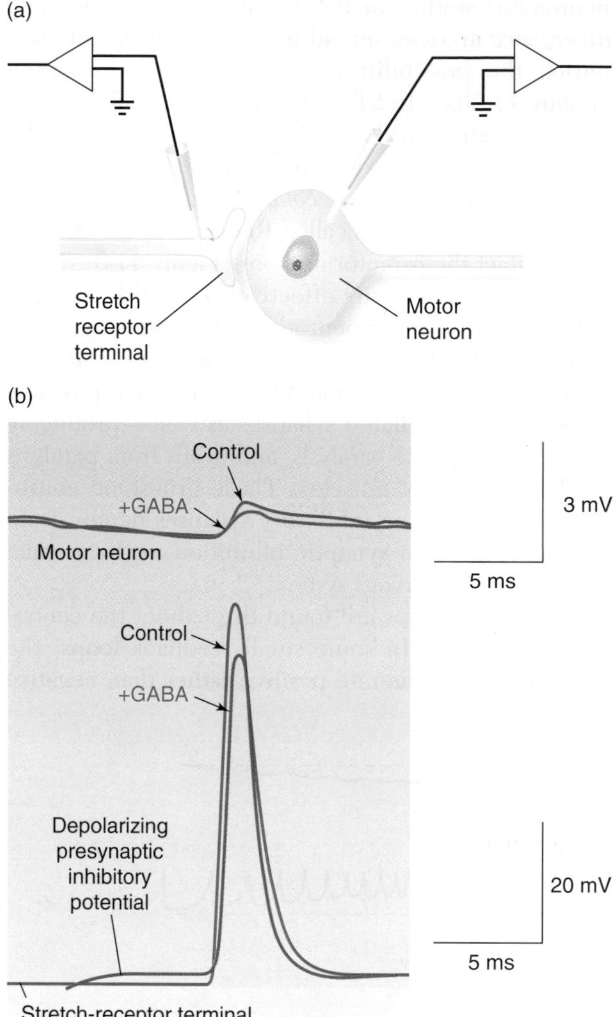

Figure 11-4 Presynaptic inhibition modulates the synaptic efficacy of a crayfish stretch receptor. **(a)** Experimental setup to record electrical activity of a presynaptic stretch receptor and a postsynaptic motor neuron simultaneously using intracellular microelectrodes. **(b)** Intracellular recordings from the terminal of a stretch receptor in a crayfish walking leg and from a motor neuron controlling the leg. When an AP invades the stretch-receptor terminal, a depolarization is recorded in the motor neuron. If the AP in the stretch receptor is preceded by a depolarizing presynaptic inhibitory potential produced by the application of GABA, the amplitude of the impulse is reduced, and the subsequent depolarization in the motor neuron is smaller. These motor neurons control the skeletal muscles they innervate by way of electrotonically conducted graded membrane potentials, so a change in the magnitude of the postsynaptic potential can produce a noticeable effect on contraction of the muscles. [Part b adapted from Cattaert et al., 1992.]

Despite these differences, some of the cellular mechanisms that operate stretch reflexes are common to both animal groups. For example, the efficacy of synapses made by stretch receptors onto motor neurons is modulated by GABA-based presynaptic inhibition in both groups (Figure 11-4). In vertebrates and invertebrates, synaptic release of GABA onto the stretch-receptor terminals causes a small depolarization of the membrane at the presynaptic terminal, which reduces the amplitude of the AP arriving at the terminal, thus reducing the amount of transmitter released. The neurons providing the GABA have not yet been identified in most animals, but pharmacological experiments confirm the identity of the transmitter. Homology at this level of detail suggests that the mechanism is ancient.

Principles of Neuronal Control

From our consideration of the simple muscle stretch reflex, we can deduce several principles that govern how the nervous system controls behavior.

- Behavior results when a signal travels from a sensory receptor, through some number of synapses in the central nervous system, and out through a motor neuron, activating skeletal muscles.
- Even in the simplest neuronal circuits, information coming both from other neurons in the same part of the nervous system and from neurons with somata in distant locations impinges on the circuit and can modify the behavioral output.
- Information is typically sent along divergent pathways as it travels through the nervous system, allowing it to affect different aspects of an animal's behavior.
- Inhibition and excitation both play important roles in shaping behavior.

Although a simple muscle stretch reflex is scarcely an exciting behavior, the concepts exemplified in its description provide a very useful framework for more complex behaviors. In each case, we will find that specialized sensory receptors collect information about the environment, incoming sensory signals are processed by interneurons in the central nervous system, and the output of processing centers travels to motor centers that generate patterned output onto motor neurons, which control contraction of the muscles.

To understand these more complicated circuits, we must consider several questions. First, what exactly about behavior do we want to explain? Second, can we construct this explanation from known circuits and their interactions? Third, do general principles describing behavior emerge, or is every behavior a "special case"? To suggest how clearly and completely these questions can currently be answered, we first consider some complex behaviors and then turn to some properties of networks known to underlie behavior.

THE STUDY OF BEHAVIOR

Basic Behavioral Concepts

A basic problem facing any animal is what to do and what to avoid doing in any given situation. Understanding the mechanisms that allow animals to make these choices is a major challenge. Behavioral scientists have taken two complementary approaches to meeting this challenge. **Neuroethologists** bring an animal into the laboratory and observe its behavior under a drastically simplified set of well-defined circumstances. In the laboratory, there are no predators; the number and sex of conspecifics are controlled by the experimenter; there are unusual lights, smells, and sounds; and the animal is often limited to a relatively small and confined space. In some cases, the nervous system of the animal is exposed surgically to allow the experimenter to record from neurons while behavior is going on. Studies under these simplified and controlled conditions are useful, even necessary, for obtaining answers to circumscribed, well-defined questions about behavior, but the results of such experiments can be difficult to translate into an understanding of how animals deal with the complex challenges they face in their everyday lives. In contrast, **ethologists** go into the field and observe the animal as it goes about its business in its normal environment. Observing animals in their normal state is an ancient practice, and much can be learned from this approach. However, these natural conditions raise challenging problems for the physiologist. While observing an animal in its natural setting, it is difficult to determine the relative importance of all the different things that the animal does, and it is essentially impossible to record activity from the animal's nervous system. However, ethologists have made valuable contributions by recognizing and describing repeating patterns in natural behaviors.

Fixed action patterns

On the basis of observations made in the field and of many ingenious experiments, ethologists have compiled detailed inventories of what animals do and how and where individual animals spend their time. These reports offer a wealth of information as well as some organizing principles. By considering such observations, ethologists Konrad Lorenz and Niko Tinbergen developed an interpretive framework that ultimately crystallized these data into a useful system. They recognized that the behavioral repertoire of an animal seemed to be constructed from elemental motor and sensory "units." They called the elemental motor patterns **fixed action patterns** (also called *modal action patterns*), and they called the corresponding elemental sensory "units" **key stimuli** (also called *sign stimuli* or *releasers*). They were awarded a Nobel Prize in 1973 in recognition of the importance of their work.

Fixed action patterns have six properties:

1. *They are relatively complex motor acts,* each consisting of a specific temporal sequence of components. They are not simple reflexes.
2. *They are typically elicited by specific key stimuli* rather than by general stimuli or by a variety of stimuli.
3. *Once elicited, they are normally enacted in their entirety.* The environmental stimulus that normally elicits a fixed action pattern is frequently stable and prolonged, but if an experimenter removes the stimulus after the animal's movements have begun, the behavior will usually continue to completion. This all-or-none property of fixed action patterns distinguishes them from reflexes, which typically require continued stimulation to maintain movement. It is as though the key stimulus is required to turn the pattern on, but once it has begun, it plays out independently of further stimulation.
4. *The stimulus threshold for fixed action patterns varies with the state of the animal,* and the variation can be quite large. For example, immediately after an animal has copulated, it cannot be induced to copulate again without intense stimulation. The threshold for eliciting copulation then drops as time passes.
5. *They are performed nearly identically by all members of a species* (or sometimes by those that are of a particular age or sex). Some fixed action patterns are common to many species of a genus. The properties of fixed action patterns are so reliable within some taxonomic groups that comparison of variations within these patterns has been used to deduce taxonomic relationships.
6. *They are typically carried out in a recognizable form even by animals that have had no prior experience with the key stimulus.* In other words, these patterns are inherited genetically, not learned, although it is clear that in many species the patterns can change somewhat with experience.

The stereotypic nature of muscle activity during fixed action patterns has made these behaviors extremely useful to physiologists interested in the control of motor activity. The ability to reliably elicit precisely the same muscle contractions over and over again by applying the appropriate key stimulus has provided a

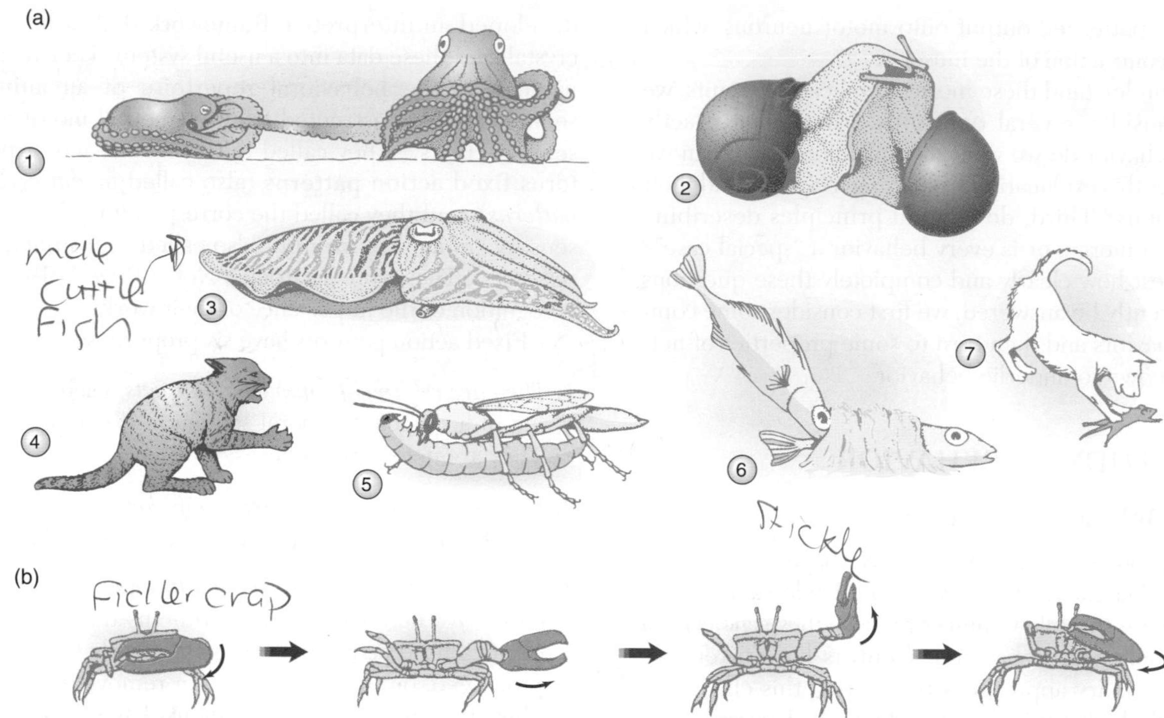

Figure 11-5 Many fixed action patterns contribute to reproductive behavior or prey capture. **(a)** Mating behavior in (1) an octopus and (2) a snail *(Helix pomatia)*. (3) Sexual display of a male cuttlefish *(Sepia officinalis)*. The cuttlefish has pigment-containing cells that are controlled by the nervous system, and when a male is trying to attract a mate, fantastic colored patterns flash and move across his dorsal surface. (4) European wildcat striking with its paw. (5) Digger wasp capturing its prey. (6) Male three-spined stickleback quivering to stimulate a female to spawn. (7) The "mouse jump" carried out by a domestic dog. **(b)** The courtship behavior of a male fiddler crab. The crab attempts to attract a female by waving its large claw forward, laterally, upward, then forward again. [Part a, diagrams 2, 3, 5, and 6 adapted from Tinbergen, 1951, 1 adapted from Buddenbrock, 1956, 4 adapted from Lindemann, 1955, 7 adapted from Lorenz, 1954; part b adapted from Lorenz and Tinbergen, 1938.]

window into the workings of the nervous system, particularly in simple animals. Some typical behaviors are illustrated in Figure 11-5. Notice that many of these behaviors are related to life-or-death circumstances such as capturing prey, attracting a mate, or mating.

Key stimuli

Key stimuli, which specifically stimulate an animal to perform a fixed action pattern, have also been called *releasers,* because they appear to "release" a prepatterned behavioral response that is all set to go within an animal. To sort out the features of a key stimulus that are essential to its effectiveness, ethologists have performed experiments with models that are shaped and colored to mimic the suspected signal. The features are varied to determine which of them is most effective at eliciting the particular fixed action pattern. A classic example of a fixed action pattern that has been studied in this way is the aggressive response of a courting male three-spined stickleback fish. When a courting male sees another male stickleback, it displays aggressively and may even attack. Presenting courting male fish with models that differed in shape and color revealed that the key stimulus for releasing this behavior is the red underside characteristic of male sticklebacks (Figure 11-6a). The shape of the model male fish was relatively unimportant, but the red underside was effective as a sign stimulus only if the model was orientated horizontally (Figure 11-6b). Thus, it is not simply the color red, but the red belly in the proper visual context, that releases the aggressive behavior of the male fish.

The discovery of key stimuli suggested further experimental work, just as the discovery of fixed action patterns had. The highly specific nature of the releasing stimulus for aggressive behavior in the male stickleback fish, for example, suggested that there might be adaptations within the stickleback's visual pathways that were specially tuned to important features, acting perhaps as sensory filters to admit only the appropriate parts of the stimulus. Indeed, the theme of specialized sensory detectors has proved to be an important and powerful organizing concept in sensory physiology.

Fixed action patterns and key stimuli do not occur in isolation, but are instead woven seamlessly into behavioral interchanges between animals. In courtship, caring for offspring, or aggressive encounters, animals use complex sets of movements to communicate their behavioral intent. In courtship, for example, a movement by one animal sends a signal to its partner. The partner may, in turn, move in a way that sends a signal back.

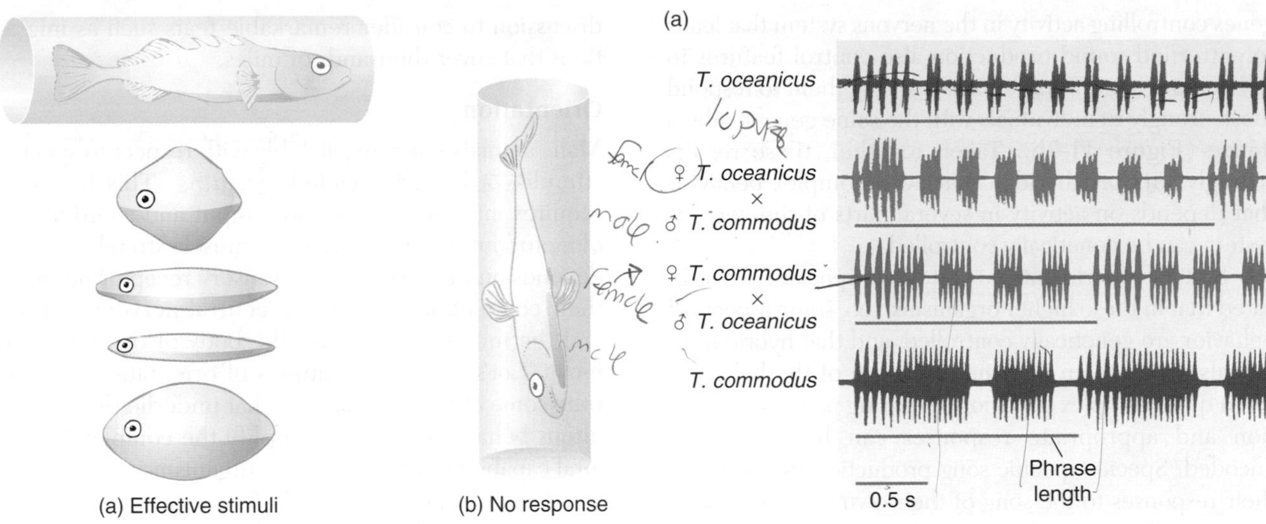

Figure 11-6 Studies of the aggressive behavior of a courting male stickleback fish have been facilitated by the use of models. A variety of models that simulated a rival male fish were used to discover which ones most effectively elicited this fixed action pattern. **(a)** All of the models in this panel elicited aggression from a courting male stickleback, suggesting that the feature releasing aggressive behavior is the horizontal red underside of the intruding male. Notice that the shape of the model fish was unimportant. **(b)** This model mimics a male stickleback very faithfully, but when oriented vertically, it elicited no response. [Adapted from Tinbergen, 1951.]

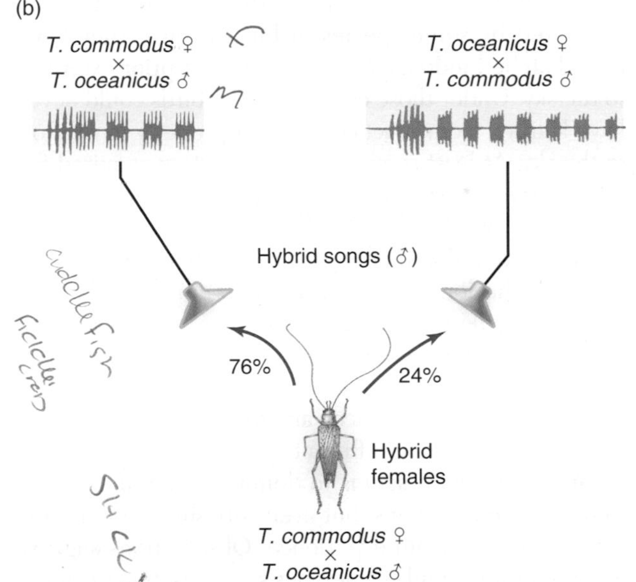

Figure 11-7 The songs of male crickets and the auditory responses of female crickets are genetically controlled. **(a)** Sonograms illustrating the features of male cricket songs from two species, *Teleogryllus commodus* and *Teleogryllus oceanicus,* and from males produced by crosses between the two species. When the two species were crossed in the laboratory, the male F_1 hybrid offspring sang a song with characteristics that combined the properties of the songs of both parental species. The details of the song depended on which parent was male and which was female. For example, the phrase length (indicated by a green bar below each record) was more like that of the mother's species. It was longer in F_1 males whose maternal parent was *T. oceanicus* and shorter in F_1 males whose maternal parent was *T. commodus.* **(b)** Songs of F_1 hybrid male crickets were recorded and played in the laboratory to F_1 hybrid female crickets, and the direction in which the females moved in response was observed. Females were more likely to move toward the speaker playing the song from males of the same genetic cross. For example, females from crosses in which the female parent was *T. commodus* moved toward the speaker playing the recorded song of F_1 male crickets with *T. commodus* mothers in 76% of trials. [Part a adapted from Bentley and Hoy, 1972; part b adapted from Hoy et al.,

Genetic control of behavior

Genetic components of fixed action patterns and key stimuli have been used to ask how species-specific behavior is organized. An example of this approach has been the study of the wing movements that produce the chirps and trills in the courting song of a male cricket. Patterns of cricket songs are species-specific and are largely independent of environmental factors other than temperature. The sound pattern produced by a species is directly related to the pattern of APs in the motor neurons that control the sound-producing muscles. In one set of studies, sound production by males of two closely related species was characterized. One species produces a short trill (consisting of about two sound pulses), the other species a long trill (about ten sound pulses). When a male of one species was crossed with a female of the other, the trills of the resulting F_1 hybrid males (that is, the male progeny produced by crossing individuals of the two species) consisted of an intermediate number of pulses (about four). Backcrosses to produce various genetic combinations demonstrated that the neuronal network producing the song pattern is under rigid genetic control, sufficiently precise to specify differences as small as the exact number of APs going to the chirp-producing muscles. The songs produced by hybrid males differed depending on which species their mothers and fathers belonged to (Figure 11-7a). In addition, female F_1 offspring responded most strongly to the song produced by F_1 males from the same cross. It thus appears that the genes controlling activity in the nervous system that leads to patterned sound production also control features in the nervous systems of females, causing them to respond most strongly to individuals with the same genetic inheritance (Figure 11-7b). Taken together, these results strongly support the idea that even complex behavior that depends on activity in several parts of the nervous system can be genetically controlled.

Similar experiments on vertebrate species have indicated that in these higher organisms, too, some aspects of behavior are genetically controlled, and that hybrid individuals generate an intermediate form of the behavior. Even quite complex behaviors requiring pattern recognition and appropriate responses can be genetically encoded. Species-specific song production by birds and their responses to the song of their own species provide one example. In another example, the ability to orient with respect to stellar patterns, complete with time compensation to account for Earth's rotation, is fully expressed by some species of birds even if they have been hatched indoors and shielded from prior exposure to the sky. Under those conditions, the birds could never have learned from practice to orient properly in relation to the sky. At least some of the information required for proper orientation resides in the genetic material.

This property of heritability has sparked a long-lived debate about the dichotomy between inherited and learned behaviors (i.e., "nature versus nurture") that still flares up regularly. It is reasonable to say that behavior in higher animals generally depends on both genetic and learned components. The relative contributions made by heredity and by experience vary greatly among different behaviors and among different species. Genetically programmed behavior seems to dominate in animals with simple nervous systems, but even very simple organisms appear to learn from experience. Observations suggest that the more complex a nervous system is, the greater is its potential for learning. This potential enables organisms not only to depart from, but also to enlarge upon, their repertoire of inherited, fixed patterns of behavior.

What evolutionary advantages and disadvantages might be conferred by fixed action patterns? What advantages and disadvantages go along with behavioral plasticity?

Examples of Behavior

A brief introduction to the fundamental units of behavior cannot begin to capture the broad range of behavioral capacities known in animals. Various animals have evolved specialized sensory and motor capabilities that allow them to produce many interesting behaviors, and the annals of animal behavior are filled with examples that amaze us. Here we consider only a few behaviors in which locomotion is directed by external stimuli. We begin with relatively simple behaviors, then expand the discussion to consider remarkable feats such as migrations that cover thousands of miles.

Orientation

Many animals move predictably with respect to specific stimuli—a behavior called *orienting*. This behavior requires integration of sensory input and coordination of motor output, so like a simple muscle stretch reflex, it depends on the properties of sensory receptor neurons, their connections within the central nervous system, and the muscles that cause the body of the animal to move. Let's consider examples of orientation to illustrate some of the mechanisms that underlie these ubiquitous behaviors and to exemplify the complex behavioral capabilities of even simple organisms.

A common nocturnal sight for many city dwellers is the scurrying of cockroaches as they abandon their snacks and disappear when the kitchen light is switched on. This behavior is an example of a **taxis**—a movement that is directed with respect to a stimulus. Scurrying cockroaches that move away from a light are displaying negative **phototaxis.** The female crickets moving toward a source of sound in Figure 11-7b were displaying positive **phonotaxis.** Jacques Loeb (1918) suggested early in the twentieth century that simple taxes are caused by asymmetric sensory input that drives asymmetric motor activation, much like rowing a boat with only one oar. According to this view, negative phototaxis occurs when a light impinging on one eye produces a strong *ipsilateral* motor output (i.e., on the same side as the eye), causing the animal to veer away from the source of the light. Positive phototaxis occurs if the light stimulus to one eye stimulates *contralateral* locomotor output (i.e., on the side away from the light), causing the animal to turn toward the source of light. Positively phototactic animals that are experimentally blinded in one eye will orient so that the intact eye points away from the light, consistent with this hypothesis.

Sensory information modulates behavior in other ways as well. It can be used to correct for structural or functional asymmetries that arise in the central nervous system or in the structures (e.g., wings, legs, fins, etc.) that produce locomotion. For example, if one of the four wings of a locust is partly or entirely removed, the locust continues to fly in a straight line, as long as it can use its eyes for orientation. In the dark, even an intact flying locust will roll about its longitudinal axis. (Locusts do not typically fly in the dark, but in the laboratory a locust can be tethered to allow its behavior to be observed and then induced with strong stimulation to fly.) The roll is due to slight asymmetries in the structure of the wings and in the centrally generated motor output. A tethered locust ceases to roll if it is provided with visual cues in the form of an artificial horizon; visual input allows the motor output to the wings to be corrected, stabilizing the insect's position. The visual input provides information to the central neurons that control flight, regulating the relative outputs to the left and right sets of wing muscles so as to hold the animal on an even keel with respect to the perceived horizon.

The importance of sensory feedback for orientation and locomotion is confirmed in our daily experience. For example, as we saw in Chapter 1, the driver of a car continually makes minor steering adjustments: her eyes are the sensors in a feedback system in which her neuromuscular system, coupled to the car's steering mechanism, corrects any deviations that the car makes from the center of the lane. Those deviations may be due to asymmetries in the driver's nervous and muscular systems or to irregularities or imperfections in the car's steering system and in the road. When a person is deprived of this visual feedback, behavior changes. Blindfolded human subjects, either walking or driving a car in an open flat field, typically adopt an approximately circular course, with the size and direction (clockwise or counterclockwise) of the average circle differing among subjects. Similar turning biases are seen in animals at all phylogenic levels. Visual and other forms of sensory feedback compensate for these inherent locomotor biases, some of which may be due to congenital asymmetries in the function of the muscles and nervous system.

Many animals locate their prey using vibrations that are set up in the substrate by the prey. A spider is alerted to prey in its web, as we saw in Chapter 9, by vibrations of the silk threads—vibrations that are detected by mechanoreceptor organs located in the spider's legs. Another group of arachnids, the nocturnal desert scorpions, use sand-borne vibrations produced by movements of their prey to locate and orient toward potential victims that are as far away as 0.5 m. At 15 cm or less, the scorpion can determine the distance, as well as the direction, of the source of vibration. Besides typical mechanoreceptor hairs, a scorpion possesses a set of especially sensitive receptors on each of its eight legs. Although the sensitivity of these vibration receptors is considerably less than that calculated for cochlear hair cells, it is nonetheless impressive. With the use of a finely tuned mechanical displacement stimulator, it has been shown that these receptors respond when the distal segment of the leg is displaced by a nanometer or less, which allows the receptors to detect the direction in which sand-borne vibrations are traveling.

To orient accurately toward the source of vibration, a scorpion keeps all of its eight legs in contact with the substrate, precisely spaced in a circle (Figure 11-8a). With this arrangement, the leg closest to the source will be the first to sense any vibrational waves traveling along the sandy substrate. Central neurons that receive sensory inputs from the vibration receptors appear to compare the timing of impulses that are elicited in the receptors of all the legs. Those legs facing the stimulus source intercept the waves first, and those on the opposite side receive the stimulus about 1 ms later (vibrational waves travel 40–50 $m \cdot s^{-1}$ in the sand). By integrating the timing of APs from the different legs, neurons in the central nervous system calculate the direction of the stimulus source, and the scorpion then produces the appropriate motor output allowing it to orient and respond in the direction of the source (Figure 11-8b).

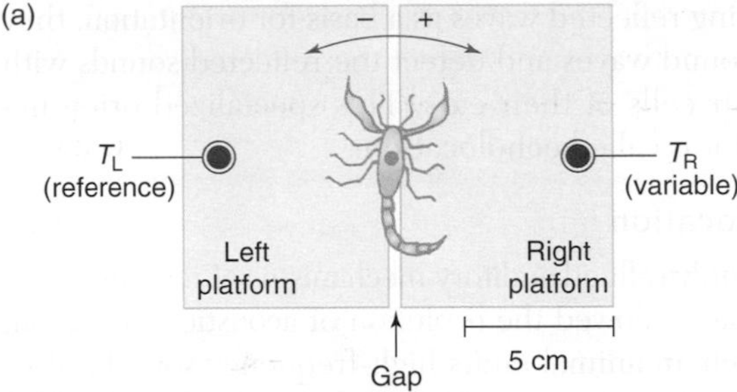

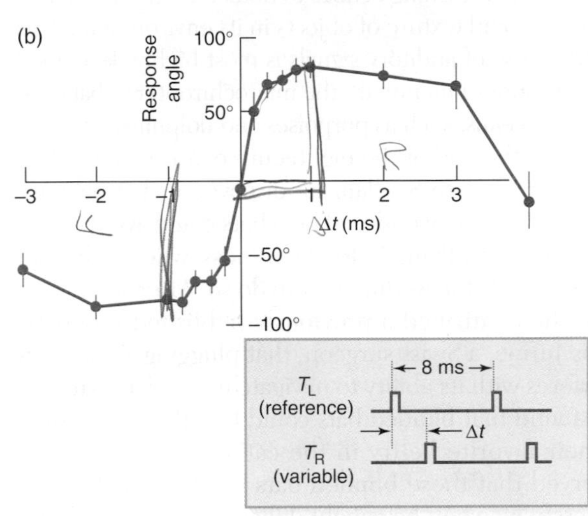

Figure 11-8 Sensory receptors on the legs of desert scorpions allow them to receive vibrations that are produced by potential prey and carried through the sand. Central neurons respond to the disparity in the times at which information arrives from the eight legs and produce motor output that causes the scorpion to orient toward the source of vibration. **(a)** Experimental setup to manipulate the timing of sensory input to the mechanoreceptors (T_L and T_R, left and right transducers). The timing between vibration of the left and right tables can be varied. If the left table is vibrated first, the system mimics an animal moving to the left of the scorpion; if the right table vibrates first, it mimics an animal moving to the right of the scorpion. Even very small time differences separating vibration of the two tables produce behavioral responses by the scorpion.
Classic Work **(b)** Relation between the times at which sensory input from the legs arrives in the central nervous system and the angle of orientation assumed by the scorpion. The scorpion turns toward the legs from which the sensory input was first received. [Adapted from Brownell and Farley, 1979a,b.]

Many aquatic organisms also orient in relation to vibrational stimuli. Some surface-swimming insects and many fishes and amphibians sense reflections of waves, initiated by their own swimming movements, that have bounced off nearby objects. Fishes and amphibians intercept these reflected waves by using the hair cells of their lateral-line system (see Figure 7-25). Some mammals and a few birds have refined the mechanism of detecting reflected waves as a basis for orientation: they emit sound waves and detect the reflected sounds with the hair cells of their ears. This specialized orienting behavior is called **echolocation.**

Echolocation

The highly refined auditory mechanisms of mammals and birds have allowed the evolution of acoustic orientation, in which an animal emits high-frequency sound pulses and uses the returning echoes to detect the direction, distance, size, and texture of objects in its environment. This sonarlike use of auditory signals is most highly developed in two groups of mammals: the microchiropteran bats and some cetaceans, such as porpoises and dolphins.

Near the end of the eighteenth century, the Italian naturalist Lazzaro Spallanzani discovered that bats use echolocation. He wondered how bats could avoid obstacles even when flying in total darkness, whereas his pet owl required at least dim light to do so. After some false leads, he confirmed a previously published report by Louis Jurine, a Swiss surgeon, that plugging a bat's ears interferes with its ability to navigate in the dark. He further found that blinded bats could find their way home to their favorite belfry in the cathedral at Pavia. He observed that these blinded bats caught insects quite successfully: when he caught, killed, and dissected several, he found their stomachs stuffed with insects that they had caught on the wing during their trip back to the belfry. At the time, little was known about the physics of sound, and Spallanzani failed to recognize that the bats themselves emitted vocal sounds that were inaudible to human beings. Instead, he came to the creative but erroneous conclusion that bats navigate by detecting the echoes of sounds from their wingbeats and that they locate their prey by homing in on the buzzing of insect wings.

It was not until 1938 that Donald Griffin and Robert Galambos, both students at Harvard University, used newly developed acoustic equipment to determine that bats emit ultrasonic cries and use the echoes of these sounds to "see in the dark" (Figure 11-9a). Further studies by Griffin and his colleagues revealed the precision of the echolocating capabilities of insectivorous bats. High-speed photography showed that a bat using echolocation can capture two separate mosquitoes or fruit flies in about half a second. The fish-eating bats of Trinidad even use echolocation to find and capture their underwater prey by detecting the ripples that are produced on the water's surface when a fish swims just underneath.

There are three phases to the process of echolocation used by typical insectivorous bats when capturing an insect (Figure 11-9b). The "cruising" phase, which occurs during straight flight, consists of pulsed sounds separated by silent periods of at least 50 ms. Each pulse of sound is frequency-modulated, sweeping downward through a frequency spectrum somewhere between 100 and 20 kHz. (Because humans cannot hear sounds above about 20 kHz, the bat's cries are *ultrasonic.*) In the second phase, which begins when the bat detects its prey, pulses are produced at shorter intervals. The third, and final, phase, in which the bat homes in on its prey, consists of a buzzlike emission in which the intervals between pulses become even shorter, the duration of each pulse drops to about 0.5 ms, and the frequency of the sounds is reduced. Finally, the bat scoops up the insect with its wings or with the webbing between its hind legs, guiding the insect to its mouth.

The sounds produced by these bats are highly energetic, reaching intensities above 200 dynes $\cdot$ cm^{-2} close

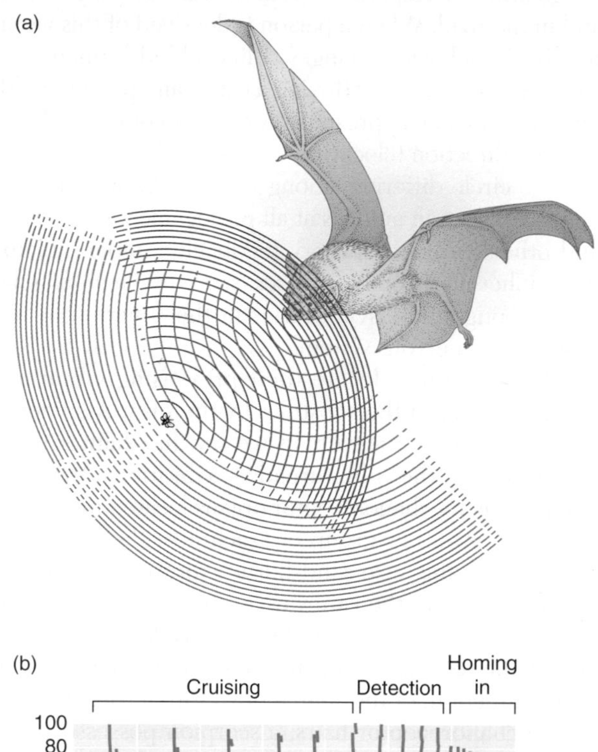

Figure 11-9 Bats find and capture their prey through echolocation. **(a)** The little brown bat *(Myotis lucifugus)* has specialized structures of the snout and large ear pinnae. The bat emits a frequency-modulated sound that is reflected from objects in its environment, including the insects that are its prey. The spacing of the curves in this drawing indicates the changing frequency of the emitted and reflected sound over the duration of a sound pulse. Only a small fraction of the emitted sound (blue curves) is reflected back by objects in the environment (red curves), and only a small fraction of the reflected sound energy is intercepted by the bat. **(b)** Three phases of echolocation in the pursuit of an insect by a different bat, *Eptesicus.* In the initial phase, the bat's cries are separated in time, and they sweep downward through frequencies between 100 kHz and about 40 kHz. When the bat detects an insect, the intervals between cries become shorter, though the cries are still in about the same frequency band. At close range, the cries turn into a buzz, with a very short period between each cry and a smaller range of frequencies scanned. [Part b adapted from Simmons et al., 1979.]

to the mouth of the bat. These sounds are as intense as the noise made when a jet plane takes off and passes only 100 m overhead; they are 20 times as intense as the sound of a pneumatic jackhammer only several yards away. Nonetheless, the sound energy that returns as an echo reflected from a small object is very weak, because sound intensity, like other forms of radiating energy, drops off as the square of distance. The intensities of both the bat's cries and the very much less intense echoes are reduced with distance traveled, so the bat's nervous system faces a formidable task in differentiating the very weak and complex echoes from its own, far more powerful, emitted sounds.

In bats that use echolocation, a number of morphologic and neuronal modifications assist in detecting the echoes. The snout is covered by complex folds, and the nostrils are spaced to produce a megaphone effect. The pinnae of the ears are very large to help capture any echoes. The eardrum and ear ossicles are especially small and light, transmitting sound pressure with high fidelity even at high frequencies. During the emission of sounds, muscles controlling the auditory ossicles contract briefly, reducing the sensitivity of the ear. (This mechanism is common in the ears of mammals.) Blood sinuses (that is, blood-filled cavities), connective tissue, and fatty tissue mechanically isolate the inner ear from the skull, reducing direct transmission of sound from the mouth to the inner ear. Finally, the auditory centers occupy a very large fraction of the bat's relatively small brain. Many regions in the bat's brain receive and interpret auditory signals and, through processes of neuronal computation, construct from these auditory cues a spatial representation of the external world. Similar mechanisms for processing auditory information have been studied in the brain of the barn owl and will be considered later in this chapter.

Navigation

Navigation over long distances is a striking extension of simple orienting behavior. Many kinds of animals—from monarch butterflies to golden plovers and gray whales—migrate over long distances through unfamiliar territory. These navigational abilities have long been surrounded by an aura of mystery because the animals seem to rely on cues that humans cannot detect. In addition, an individual often uses several different sensory cues as navigational guides, and this redundancy has made the experimental study of navigation difficult and confusing. Often, one sensory modality predominates when conditions for others are unfavorable, but if conditions change, the animal can shift its strategy. For example, it is now evident that birds can use particular landmarks, other visual cues such as the plane of polarized light or the position of the sun and stars, odors, sounds, and even Earth's magnetic field to find their way. Experimenters—who typically vary only one parameter at a time—have found themselves mystified by an animal's continued ability to find its way when one of its sensory modalities has been disabled.

Animals from many phyla can navigate. Bees use the sun's position and the pattern of polarized light in the sky to keep track of the direction from their hive to a food source, and their hivemates receive and interpret this information when a foraging bee does a "waggle dance" on the wall of the hive (Figure 11-10).

Several species of birds can navigate over vast stretches of ocean that appear to humans to be devoid of landmarks. As long as the sky is clear, the pattern of stars could serve as a guide. Do these birds perceive the pattern of stars? It would seem so. When they are exposed to a replica of the night sky in a planetarium, some species of nocturnally migrating birds—for example, warblers—orient their bodies with respect to the projected patterns of stars.

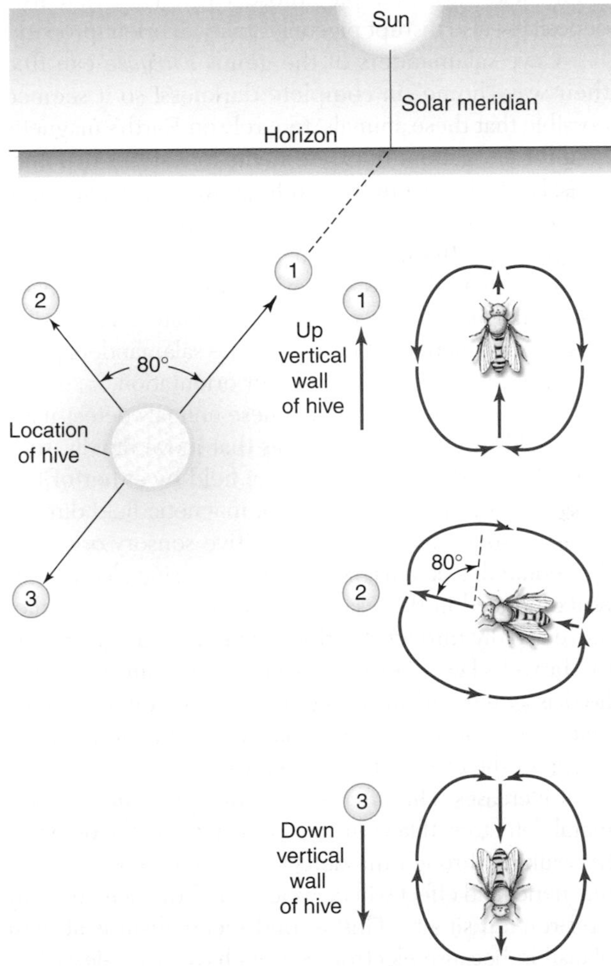

Figure 11-10 Bees can encode information about the location of a food source with respect to the hive by indicating the direction of the source in relation to the position of the sun. In this experiment, three potential food sources (numbered 1 to 3) were placed in different locations with respect to the hive. Scout bees that found a food source indicated its direction with respect to the hive by means of a display called a "waggle dance." The scout moves its abdomen back and forth as it walks in a straight line along the vertical wall of the hive. The direction the bee takes with respect to the vertical during its dance shows the direction of the food source relative to the sun. [Adapted from Camhi, 1984.]

Magnetic fields The best explanation for several apparently mysterious navigational abilities is that many animals orient using cues undetectable by humans. For example, although it has long been suspected that some animals use Earth's magnetic field for orientation and navigation, only recently has this hypothesis been supported by experimental evidence. Homing pigeons deprived of familiar landmarks can still find their way home, even on an overcast day when it seems impossible to detect the position of the sun. Under these conditions, the birds first fly in circles for a few minutes, then head in the correct direction toward home. However, if small magnets are attached to their heads, or if they are transported to the release point in containers that make it impossible for them to detect Earth's magnetic field en route, they become disoriented. Local magnetic anomalies—for example, those caused by rich iron deposits—also disrupt pigeons' ability to orient properly.

Cave salamanders of the genus *Eurycea* can find their way "home" in complete darkness, so it seemed possible that these animals, too, rely on Earth's magnetic field for navigation. In experiments to test this hypothesis, salamanders were trained to go to a particular place in a maze that was always oriented in the same way with respect to Earth's magnetic field. They were then tested in cross-shaped passageways whose orientation to Earth's magnetic field could be manipulated. The results of the experiment confirmed that the salamanders probably use Earth's magnetic field for orientation.

The next question is how these animals detect magnetic fields. In principle, animals that move rapidly, such as birds, could detect a magnetic field by either of two mechanisms: they could detect a magnetic field directly by using some magnetically sensitive sensory organ, or they could detect a magnetic field by sensing electric currents induced in their saline body fluids as their bodies move rapidly through Earth's magnetic field. (According to Maxwell's laws of electricity and magnetism, an electric field is generated in a conductor—including a saline fluid—as it moves through a magnetic field, and the magnitude of the electric field increases as the rate of movement increases. This effect is the principle behind airport metal detectors. If keys in your pocket set off the detector, try walking through the detector again very slowly. The magnetic field effect will be weaker, and the detector may well remain silent.) This second mechanism is at least plausible because electroreceptors have been described in many species. However, the finding that cave salamanders can use Earth's magnetic field as a navigational cue has significant implications. Salamanders move very slowly through their environment compared with birds, so any electric field set up in a salamander's body fluids would be exceedingly small and unlikely to provide a basis for indirectly detecting magnetic fields. It therefore seems most likely that at least some animals are able to detect magnetic fields directly.

What sort of sensory mechanism could underlie a magnetic sense? This question cannot yet be answered with certainty. However, studying species of bacteria that orient with respect to Earth's magnetic field has provided a clue. Individuals of these species in the Northern Hemisphere orient toward the North Pole, whereas those indigenous to the Southern Hemisphere orient toward the South Pole. The difference depends on the orientation of particles of magnetite (Fe_3O_4) in the bacteria. Magnetite has been shown to be crucial to the orienting response of these microorganisms, enabling them to swim downward into deeper mud, following the angle at which the magnetic field enters the earth at each location. If the bacteria are placed in a drop of water within an artificial magnetic field, they collect at the appropriate side of the water drop; if the magnetic field is reversed, they swim to the opposite side of the drop.

Deposits of magnetite have been found in or near the brains of several animals that respond behaviorally to magnetic fields—a finding that is both intriguing and suggestive. For example, magnetite has been found in the cerebral cortex of pelagic whales that are thought to navigate using Earth's magnetic field. Catastrophic strandings of these whales in unfamiliar inshore waters have been shown to correlate significantly with the occurrence of geomagnetic disturbances in the areas in which they went astray. In bobolinks, magnetite has been detected in the upper beak in a region innervated by the ophthalmic nerve, a branch of the trigeminal (Vth cranial) nerve. Anesthetizing the ophthalmic nerve blocks the birds' response to pulsed magnetic fields, and specific neurons in the ophthalmic nerve have been found to change their activity when the magnetic field is changed by as little as 0.5% of the intensity of Earth's magnetic field (Figure 11-11). Although the sensory receptor cells responsible for this activity have yet to be found, these results reinforce the hypothesis that many animals can directly sense even small changes in magnetic fields, and behavioral evidence strongly suggests that they use this information to guide long-distance migration.

Electric fields The American eel *(Anguilla rostrata)* probably uses another navigational method based on Earth's magnetic field that is available only to marine organisms. Larval eels migrate from spawning grounds in the Sargasso Sea to the Atlantic coast of North America, a distance of about 1000 km. When it was initially suggested that these eels employ Earth's magnetic field for guidance, the statement was met with skepticism because the density of the magnetic field is so low. However, the lateral-line system of eels contains extremely sensitive electroreceptors that could allow the eels to respond to electric fields generated by the movement of ocean currents through Earth's magnetic field. Ocean water functions as a conductor moving through Earth's magnetic field, so Maxwell's laws indicate that here, too, an electric field is generated. The geoelectric fields set up by ocean currents such as the Gulf Stream reach intensities of about 0.5 $mV \cdot cm^{-1}$, or a 1.0 V drop over 20 km. The minute electric currents produced by these small voltage gradients can apparently be detected by the eel's electroreceptors. This ability has been demonstrated by training eels to reduce their heart rates when the surrounding electric field changes. The heart rate of a trained eel drops when the surrounding electric field changes by as little as 0.002 $mV \cdot cm^{-1}$. The fields generated in the ocean are two or three orders of magnitude greater than this value, so it is entirely possible that eels can use information gathered by

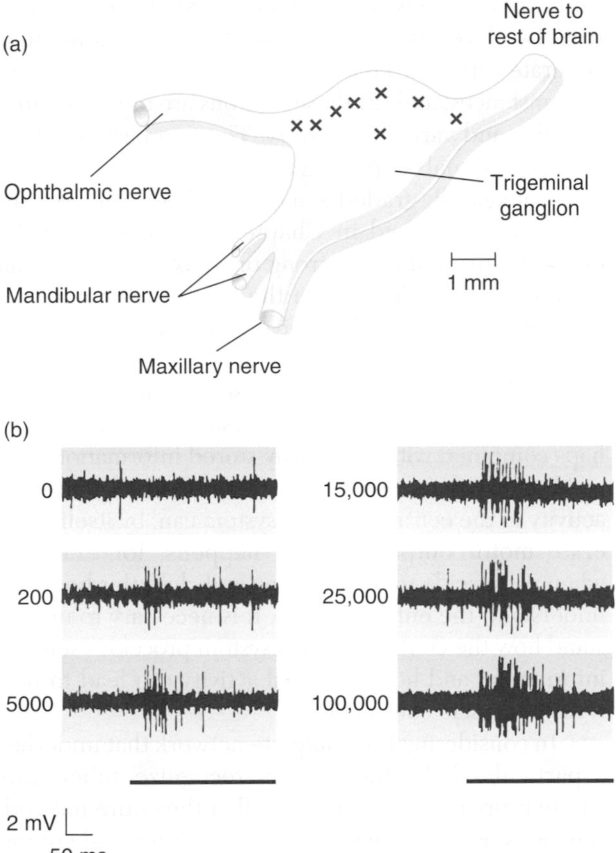

Figure 11-11 Neurons projecting to the brain from a region near the beak of a bobolink change their activity in response to changes in the magnetic field. **(a)** Neurons that innervate a magnetite-containing region above the beak have axons in the ophthalmic nerve (a branch of the trigeminal nerve). Extracellular recordings made in the trigeminal ganglion near where the ophthalmic nerve arises show activity in neurons that become excited when incident magnetic fields increase. Each X marks a location where activity was detected in response to a change in the magnetic field. **(b)** Increases in an imposed magnetic field excited one of these neurons, whose activity was being recorded extracellularly. The changes in magnetic field intensity imposed on the bobolink are shown to the left of each record, expressed in units of nanoteslas, or 10^{-9} T. (A tesla is defined as 1 newton per ampere-meter. The magnetic field at the surface of Earth is approximately 5×10^{-5} teslas.) The smallest change to which this neuron responded was thus about 0.5% of Earth's magnetic field, and the largest change was about twice Earth's magnetic field. The bars under the recordings indicate the duration of each change in magnetic field. [Adapted from Lohmann and Johnsen, 2000.]

their electroreceptors to orient with respect to the geoelectric field.

This brief survey of animal orientation and navigation reveals some of the complexity of the behavior that we see in animals. In view of these capabilities, it may seem daunting to begin a study of the neurons responsible for producing all of this behavior, and as we will see, descriptions of the neuronal basis of behavior commonly lag behind descriptions of the behavior itself. Nonetheless, activity in the nervous system lies behind all behavior, and great progress is being made in neuroethology.

PROPERTIES OF NEURONAL CIRCUITS

Despite the extreme complexity of nervous systems, several generalizations can be made about their organization and function. First, neuronal circuits consist of specific connections between neurons, and the pattern of connectivity is essentially the same in all normal individuals of a species. These connections, which are typically established during embryonic development, are maintained and may be modified by use throughout an organism's life. If a particular circuit remains unused for a long period, connections in that circuit become weaker, and there can be significant loss of function. Recent evidence indicates that the opposite is also true: within certain limits, repeated activity can increase the strength of existing connections.

The crucial importance of appropriate synapses for producing behavior is illustrated by an experiment on a simple reflex connection in a frog (Figure 11-12 on the next page). In this experiment, the consequences of altering neuronal connections were tested by surgically disconnecting the sensory fibers entering the spinal cord from one side of the body and suturing them to the dorsal root of the opposite side. The severed sensory neurons reconnected to the appropriate *kind* of central neurons, but contacted the wrong side of the spinal cord. For example, sensory neurons that innervated the skin of the right foot now synapsed onto neurons in the left side of the spinal cord. In a normal frog, a noxious stimulus to a leg causes the reflexive withdrawal of that leg. After neuronal connections had regenerated in this new configuration, a noxious stimulus to the leg on the surgically manipulated side evoked an inappropriate withdrawal of the other leg, while the stimulated leg remained immobile. In other words, the abnormal connections produced inappropriate behavior.

Connections within the nervous system also play an enormously powerful role in sensory perception. In the nineteenth century, Johannes Müller formulated the *law of nerve-specific energy,* stating that the modality of a sensation is determined not by the nature of the stimulus, but rather by the central connections of the nerve fibers activated by the stimulus. This notion is now universally accepted, and we know that the various senses, as well as the topographic distributions of sensory

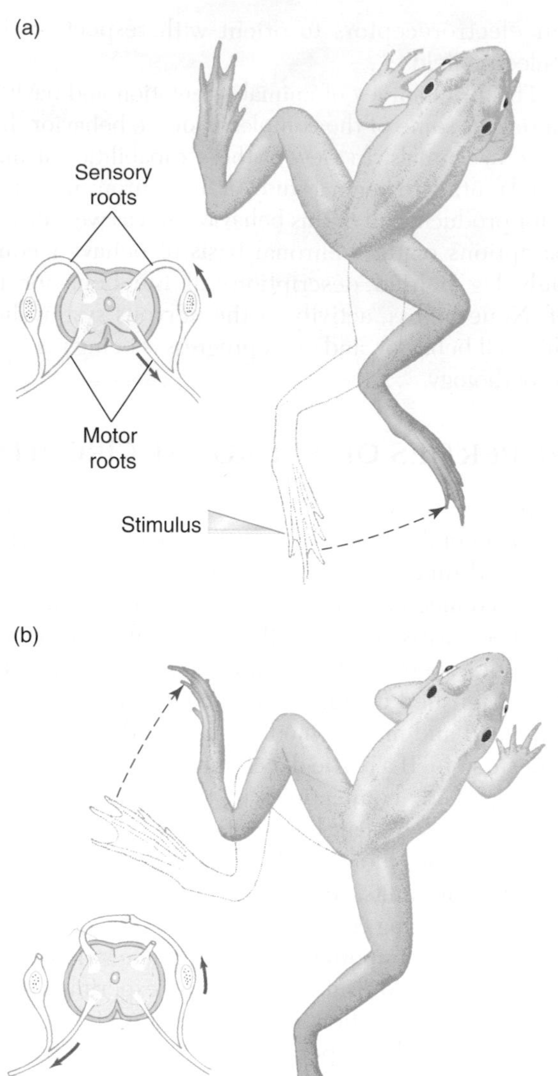

Classic Work **Figure 11-12** The "wiring diagram" of the nervous system is crucial for producing appropriate behavior. **(a)** In a normal frog, when a noxious stimulus is applied to a leg, that leg is reflexively withdrawn. **(b)** If the dorsal roots are cut and are caused to regrow into the contralateral spinal cord, the neurons establish synaptic contacts with neurons in the new location. Following the surgery, the frog responds to a noxious stimulus on the surgically manipulated side by withdrawing the opposite, unstimulated leg. The sensory input from the right leg is now connected to motor neurons controlling the left leg. [From "The Growth of Nerve Circuits" by R. W. Sperry. Copyright © 1959 by Scientific American, Inc. All rights reserved.]

receptors, are segregated into particular regions of the cerebral cortex (see Figures 8-13 and 8-14). In clinical studies of conscious humans, direct electrical stimulation of local areas in the somatosensory cortex is reported by the subjects to evoke sensations that are more or less similar to those produced by stimulation of the corresponding sense organ. Conversely, in many animals, peripheral stimulation has been found to produce electrical signals at points in the somatosensory cortex that correspond to the specific region of the skin that was stimulated.

A second generalization about the nervous system is that the metabolic state of a neuron, its electrical properties, and the summed total of all synaptic inputs impinging on it determine how that neuron will respond at any given moment. Each active neuron, by virtue of its connections, in turn influences activity in other neurons.

A third generalization is that the complexity and variety of functions carried out by nervous systems arise from two levels of organization: (1) individual neurons generate different kinds of signals depending on their circumstances, and (2) those neurons are organized into complex and variable circuitry. The two basic kinds of neuronal signals—propagated, all-or-none APs and nonpropagated, graded synaptic and receptor potentials—are discussed in Chapter 6. Synapses can be excitatory or inhibitory, strong or weak. In the end, all neuronal signals depend on the flow of ionic currents through ion channels, driven by electrochemical gradients.

Behavior is produced when sensory information is received by the central nervous system, processed, perhaps combined with previously stored information, and used to generate motor output. Alternatively, ongoing activity in the central nervous system can, by itself, generate motor output—which happens, for example, when you decide to stop reading and close this book. To understand the entire process, it is necessary to understand how the central nervous system processes sensory information and how neuronal activity can lead to patterns of muscle contraction.

In considering the complete network that underlies a particular behavior, we can recognize subcircuits whose properties affect the way that the entire network functions. Filtering networks on the sensory side of the nervous system and pattern-generating networks on the motor side act in concert to produce the complete behavior. Some simple behaviors, such as the knee-jerk reflex discussed earlier in this chapter, are produced without either sensory filters or motor pattern generators, but most behaviors rely on both. We will consider in detail how some of these subcircuits work.

Sensory Networks

The first step in generating appropriate behavioral responses is the sorting and refining of incoming information by sensory neuronal networks. Individual sensory neurons respond to only a limited range of stimulus energy, which can be described by a *tuning curve* (Spotlight 11-2). This property of receptors combines with other properties of sensory networks to filter incoming sensory information. In addition, sensory networks can amplify, add to, subtract from, and even completely reconfigure the original pattern of sensory input. The visual system has been particularly well studied, so we

SPOTLIGHT 11-2 SENSORY TUNING CURVES

Recordings of activity made from individual neurons in sensory areas of the cerebral cortex indicate that each neuron responds to a range of stimuli, but each responds optimally to very specific features of a stimulus. A plot showing how the response strength of a sensory neuron varies with a feature of the stimulus is called a *tuning curve.* Some neurons, such as neuron *c* in the accompanying figure, are very broadly tuned; that is, they respond to stimuli over a wide range of values for the relevant feature (sound frequency in this case). Others, such as neuron *d* in the figure, are very narrowly tuned. The tuning curve of a central neuron depends on its pattern of synaptic input.

The way in which information flows through a neuronal circuit depends heavily on the tuning curves of the neurons at each level of processing. For example, very narrowly tuned neurons act as filters, permitting only signals with particular properties to pass to the next level.

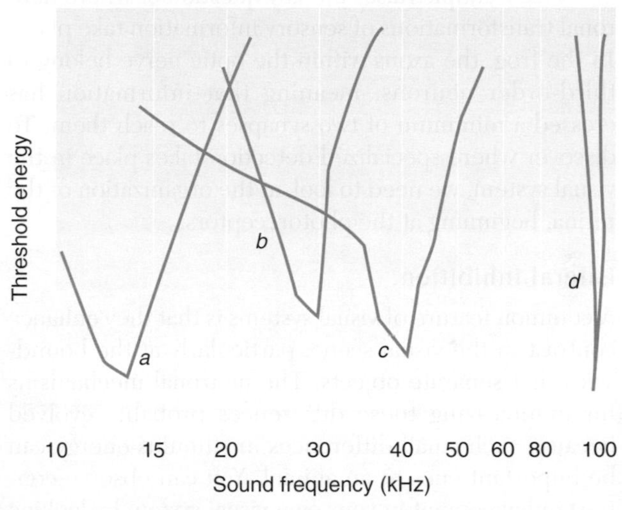

Tuning curves indicate the relation between the activity of a neuron and features of the stimuli that excite that neuron. This set of tuning curves illustrates the ranges of sound frequency to which four primary auditory neurons (labeled a, b, c, *and* d*) in the ear of the bat* Rhinolophus *respond. Each neuron is most sensitive to a particular sound frequency (that is, the threshold energy required to stimulate the neuron is lowest at that frequency), but it can also be stimulated by other frequencies within some range. Receptor* d *is very narrowly tuned and could act as a filter, whereas receptor* c *is broadly tuned. Sounds outside the tuning curve of a neuron fail to activate it at normal energy levels. [Adapted from Camhi, 1984.]*

examine it in some detail to illustrate the general organizational principles that apply to many sensory systems.

In its simplest form, sensory processing can be seen as the abstracting of features from the mass of information about the environment that is continuously available to an animal. A classic example of this process is illustrated in Figure 11-13. Recordings from axons in the optic nerve of a frog show that some of its neurons are remarkably specifically tuned, responding only to certain features of the visual field. For example, one kind of axon carries APs only when the photoreceptors to which it is connected are activated by a small object, such as a fly, that is moving against a light background

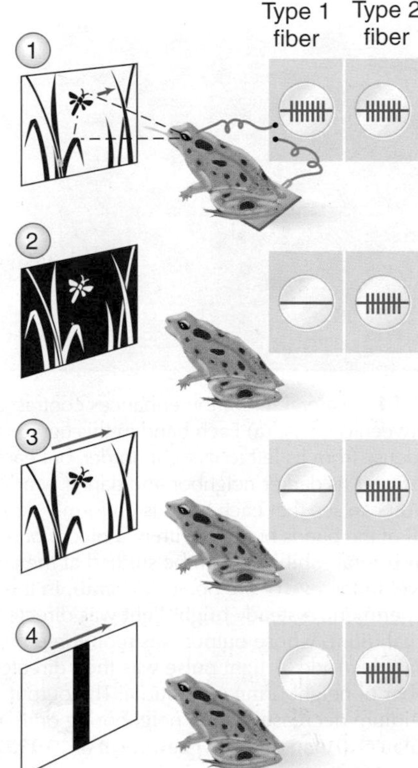

Classic Work **Figure 11-13** Some neurons in the frog retina respond specifically to stimuli that resemble a moving fly. A frog will stick out its tongue in response to a small, dark object moving in its visual field (1). It fails to respond to movement of a light object against a dark background (2), to movement of the background behind a stationary small object (3), or to other nonspecific moving visual stimuli (4). Some axons in the frog's optic nerve (labeled Type 1 in the drawing) become active only in response to a small, dark, sharply outlined object moving across a lighter background. Other fibers (Type 2) are activated by a wide variety of visual stimuli, such as a moving background or a large moving bar. [Adapted from Bullock and Horridge, 1965.]

containing stationary objects. These neurons are quiet if the entire scene moves, or if the background light is simply turned on or off. Because frogs catch and eat moving insects, this class of neurons may tell the frog's brain that supper is in the vicinity—information that is likely to be more significant to the frog than are most of the other details in the visual scene. Anyone who tries to maintain frogs in captivity soon learns that a frog does not recognize a dead insect as food. In other words, a dead fly fails to activate the circuitry that triggers the frog's feeding response, apparently because it is not a small object moving against a stationary background—the salient distinguishing feature of a live fly.

This example raises the key question of where neuronal transformations of sensory information take place. In the frog, the axons within the optic nerve belong to third-order neurons, meaning that information has crossed a minimum of two synapses to reach them. To discover where specialized detection takes place in the visual system, we need to look at the organization of the retina, beginning at the photoreceptors.

Lateral inhibition

A common feature of visual systems is that they enhance contrast in the visual scene, particularly at the boundaries that separate objects. The neuronal mechanisms for emphasizing these differences probably evolved because such small differences in stimulus energy can be important cues to an animal. You can observe contrast enhancement in your own visual system by looking at Figure 11-14a. Each band in the figure appears to be lightest at its border with its darker neighbor and darkest at its border with its lighter neighbor. In fact, the actual luminosity of each band is uniform across its entire width, and the apparent difference in brightness within a band is an illusion.

What causes this illusion? It is a consequence of a phenomenon called **lateral inhibition** that takes place at the level of the photoreceptors. This phenomenon was first discovered in a series of experiments carried out in the laboratory of H. K. Hartline at Rockefeller University in the mid-1950s; the importance of his work was recognized by a Nobel Prize in 1967. The experiments used the compound eye of the horseshoe crab *(Limulus polyphemus)*. The activity of a single ommatidium (one unit of the compound eye; see Chapter 7) was recorded first in a dark room with a bright light focused on only that ommatidium. You might suppose that adding more light to the whole eye would increase the number of APs produced by that ommatidium. Instead, adding more light by switching on the room lights *decreased* the firing rate of the test ommatidium. The diffuse light in the room stimulated the surrounding ommatidia, and connections from them inhibited the test ommatidium. This phenomenon has since been observed in many other visual systems, as well as in other sensory systems.

In another experiment, the source of lateral inhibition in the compound eye of the horseshoe crab was shown conclusively to be neighboring photoreceptors (Figure 11-14b). While the response of a single ommatidium to a steady light was recorded, a light pulse was presented to nearby ommatidia. The onset of light stimulation to adja-

(a)

(b)

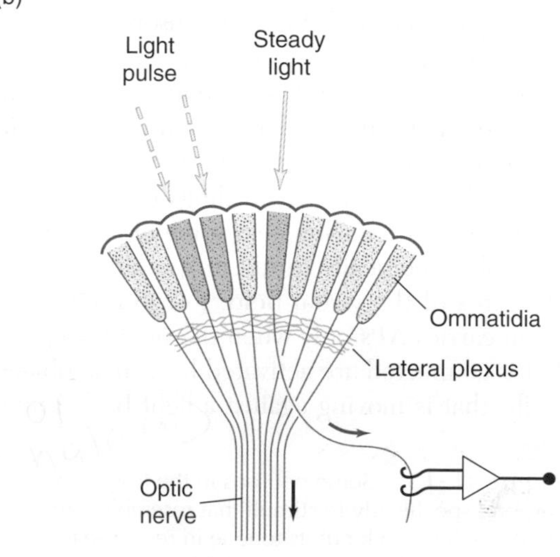

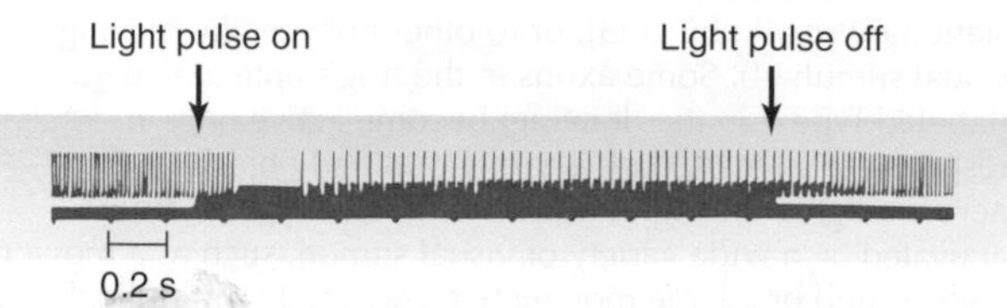

Figure 11-14 Lateral inhibition enhances contrast at borders between objects. **(a)** Each band in this figure is actually uniformly dense from its left to its right border, but each appears to be lighter near its darker neighbor and darker near its lighter neighbor. You can see that each band is uniformly dense by covering all of the bands but one with two pieces of paper. Classic Work **(b)** Lateral inhibition can be studied at the cellular level in the eye of the horseshoe crab. In this experiment, a steady bright light was directed at a single ommatidium, whose output was monitored by an extracellular electrode. A light pulse was then directed at a small number of neighboring ommatidia. The output of the test ommatidium decreased when neighboring ommatidia were stimulated. [Adapted from Hartline et al., 1956.]

cent photoreceptor cells reduced the response of the ommatidium whose activity was being recorded. This inhibition between interacting units is completely reciprocal, and the amount of inhibition decreases with distance; inhibition is strongest between nearest neighbors.

Lateral inhibition in the *Limulus* eye is mediated by the lateral plexus, a set of collateral branches from the axons of eccentric cells (the second-order neurons that receive input from the photoreceptors; see Chapter 7) that form inhibitory synapses on one another. Action potentials in the collateral branches of these eccentric cell cause the release of inhibitory transmitter onto neighboring eccentric cell axons. Because the inhibition exerted by a unit on its neighbors increases as the unit's activity increases, a strongly stimulated ommatidium strongly inhibits neighboring, less strongly stimulated units. At the same time, the strongly stimulated unit receives weak inhibition from its neighbors. This interaction enhances the contrast in activity level between neighboring units that are exposed to different intensities of light (Figure 11-15). Contrast enhancement is greatest for units that lie at the boundary separating a bright and a dim region because lateral inhibition is strongest over short distances. Lateral inhibition thus sharpens the visual detection of edges by increasing contrast at the borders between areas of different luminosities, and such edges are defining features of objects in the visual world. Visual processing thus begins in the very first neurons of the network. Processing of visual information by neurons farther along the chain continues to abstract and accentuate properties of borders and other features of visual stimuli.

Information processing in the vertebrate retina

The image of the world that falls on the vertebrate retina is a relatively accurate representation of the visual field viewed by that eye. Its accuracy is limited only by the quality of the optics in the path that light follows through the eye. The way in which the visual system transforms this raw material into a perceived image has been the subject of intense study at several levels of organization, ranging from visual transduction (see Chapter 7) to brain neurons that might recognize entire perceived objects, such as faces.

The principles that have emerged from the study of vision apply to other sensory systems as well, which suggests that evolution may have settled on certain universal solutions to some problems commonly faced by many or all neuronal networks. In this section we examine some of what is known about the neuronal processes underlying visual perception. However, it is worthwhile to consider at the start the complexity of the task. Figure 11-16 on the next page shows a diagram of some of the centers of a monkey's brain that receive visual information and the known connections among them. It is humbling to remember that vision is only one of the sensory modalities, and that sensory processing is only part of the brain's function.

The visual pathway of vertebrates begins in the retina and continues to the tectum in lower vertebrates (Figure 11-17a on the next page), or to the lateral geniculate nucleus of the thalamus and from there to the visual cortex in birds and mammals (Figure 11-17b). The visual system can be viewed as a series of connected layers of cells (Figure 11-17c). The cells within

(a)

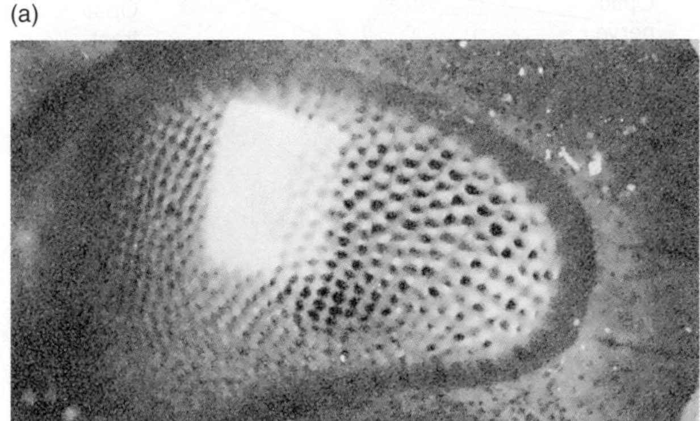

(b)

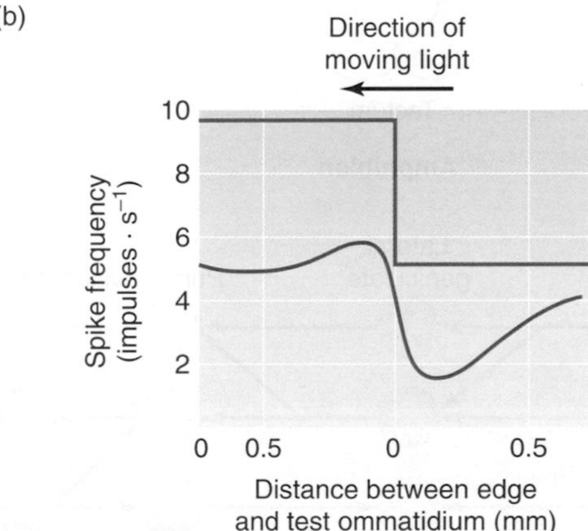

Figure 11-15 Lateral inhibition enhances contrast most strongly at the boundary between dim and bright regions. **(a)** In this experiment, the activity of a single ommatidium was recorded as a small, very bright rectangular field was moved across the moderately lit compound eye.
Classic Work **(b)** Activity in the test ommatidium is plotted against the position of the approaching bright edge. If all other ommatidia are masked, the output of the test ommatidium changes in an abrupt, steplike manner as the edge of the bright rectangle reaches it (blue plot). However, when the mask is removed and the edge of the rectangle is allowed to pass over all the ommatidia, the output of the test ommatidium changes more gradually and in a more complex fashion (shown by the red plot). Ongoing activity in the test ommatidium is first inhibited by input from neighboring ommatidia that are excited by the approaching light; then it, too, is excited by the light. Its activity reaches a peak when the trailing edge of the rectangle continues to stimulate it, but the neighboring ommatidia on one side are again in darkness, reducing the total amount of inhibition received. Notice that the activity shown by the red plot is never as high as the activity shown by the blue plot. [From "How Cells Receive Stimuli" by W. H. Miller, F. Ratliff, and H. K. Hartline.]

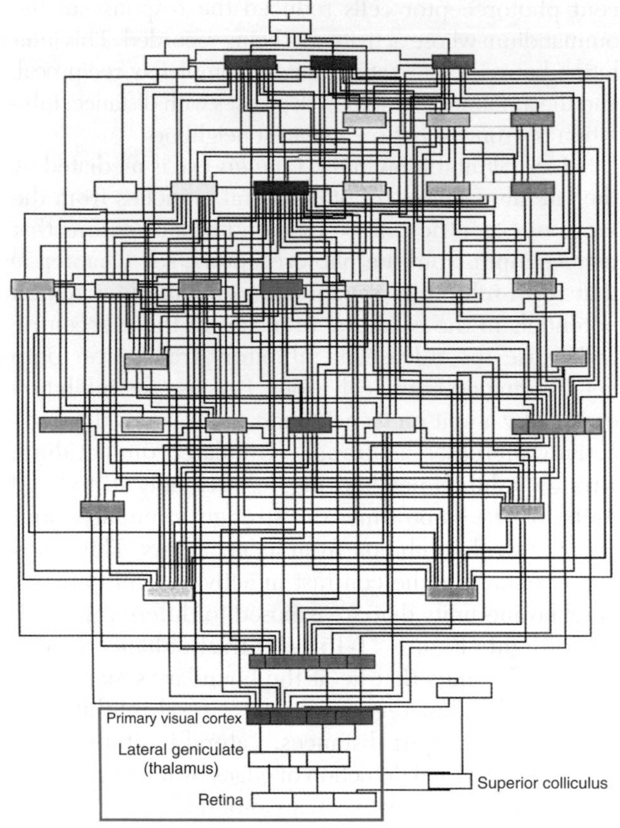

Figure 11-16 A diagram of some of the known interconnections among regions of the primate cortex known to receive and process visual information reveals the complexity of this system. The points to notice in this figure are the large number of separate, identifiable, and nonredundant centers in the brain that participate in processing input from the two eyes and the enormous number of connections that link these areas. Each brain area that receives visual information is represented by a rectangle. The areas have been arranged hierarchically; areas near the top of the figure are activated by very complex combinations of visual features, whereas areas near the bottom respond to simpler, less highly processed information. Notice the large number of parallel vertical lines in the diagram. These represent different streams of information arising from a single input at the retina and traveling through the information processing channels in the brain. It is worthwhile considering the enormous amount of effort hundreds of neurobiologists have expended to produce this description of visual processing. The structures in the box outlined in red are diagrammed in Figure 11-17c. LGN = lateral geniculate nucleus (part of the thalamus); SC = superior colliculus (tectum). [Adapted from Van Essen et al., 1992.]

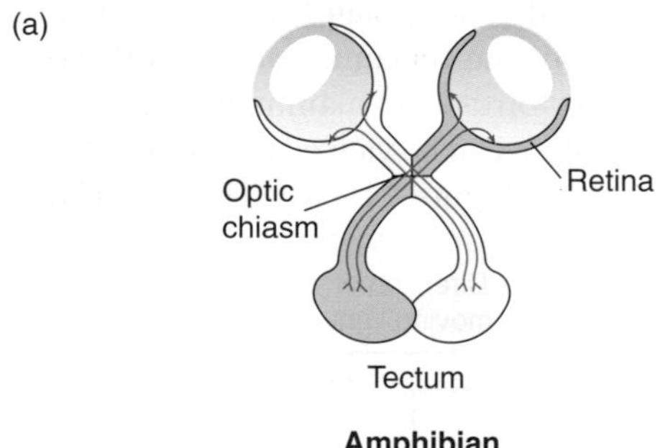

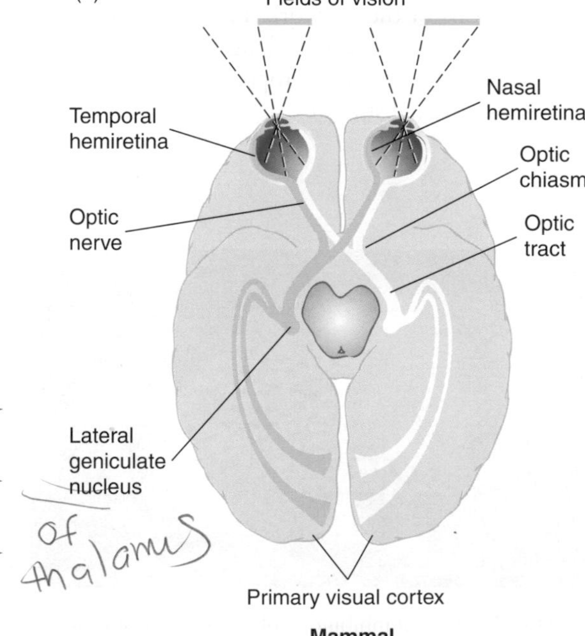

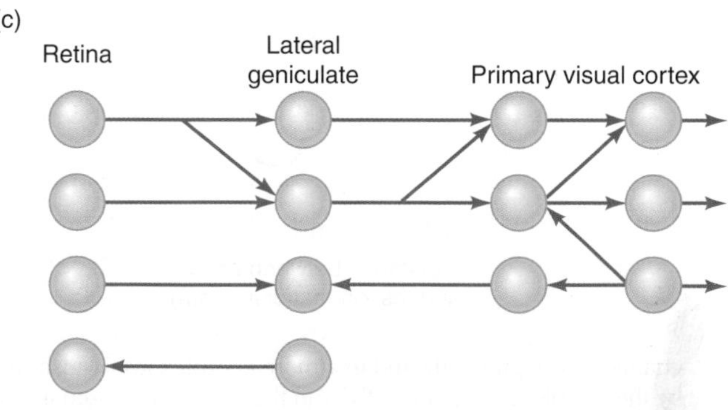

Figure 11-17 In vertebrates, visual information is transmitted from the retina to the brain through connected layers of cells. **(a)** In an amphibian, the left and right sides of the tectum receive projections from the entire field viewed by the contralateral eye. **(b)** In a mammal, each side of the visual field is projected to the opposite side of the visual cortex. For example, the temporal half (the side toward the ear) of the left retina and the nasal half (the side toward the nose) of the right retina project to the left visual cortex. Both of these parts of the retina view the right visual field, represented in the figure by a purple bar. **(c)** The neurons that initially process visual information are organized in layers. The retina contains the first three layers, and the remainder are located in the brain. In birds and mammals, visual signals are sent first to the lateral geniculate nucleus of the thalamus and from there to the cortex. Information converges and diverges between the layers and flows in both directions between them. [Part a from "Retinal Processing of Visual Images" by C. R. Michael. Copyright © 1969 by Scientific American, Inc. All rights reserved. Part b adapted from Noback and Demarest, 1972.]

each layer have properties in common. In projecting from one layer to the next, information both converges and diverges. Signals from the primary visual cortex diverge to several regions of the brain, as shown in Figure 11-16.

Cell types. Visual image processing begins before the signal leaves the retina. In the direct pathway that is followed by visual information traversing the retina, photoreceptors connect with **bipolar cells,** which, in turn, contact **retinal ganglion cells,** whose axons make up the optic nerve (Figure 11-18). The receptors are first-order cells, the bipolar cells are second-order cells, and the ganglion cells are third-order cells in the afferent pathway. This nomenclature is oversimplified, however, because there are two other types of retinal neurons—the **horizontal cells** and the **amacrine cells**—that mediate lateral interactions within the retina. The horizontal cells receive input from neighboring and moderately distant photoreceptors, and they synapse onto bipolar cells. The amacrine cells interconnect bipolar cells and ganglion cells. The pattern of these synaptic connections allows retinal neurons to extract features of sensory stimuli and to carry out other processing of sensory information. Horizontal cells and amacrine cells increase the complexity of synaptic connectivity within the retina, and more intricate synaptic connectivity permits more complex information processing. However, neither cell type is necessary for simply transmitting a visual signal from photoreceptors to ganglion cells and from there to the brain.

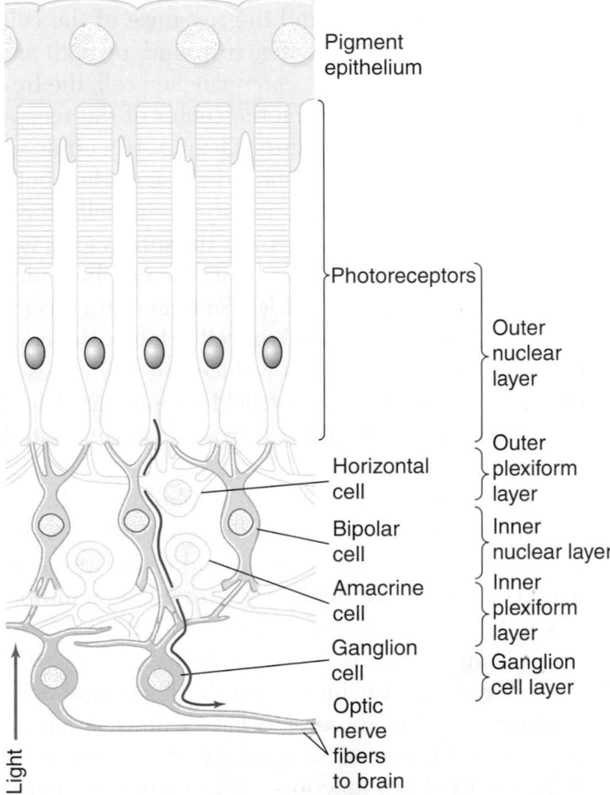

Figure 11-18 The function of the vertebrate retina is based on five types of neurons. Photoreceptors receive light stimuli and transduce them into neuronal signals. Bipolar cells carry signals from the photoreceptors to the ganglion cells, which send their axons to the central nervous system through the optic nerves. Horizontal and amacrine cells, which are located in the outer and inner plexiform layers, respectively, carry signals laterally. An example of the direct pathway is traced by the red arrow. [From "Visual Cells," by R. W. Young. Copyright © 1970 by Scientific American, Inc. All rights reserved.]

Studies combining intracellular recording techniques with the injection of fluorescent marker dyes have revealed the electrical activity typical of each retinal cell type (Figure 11-19 on the next page). Vertebrate photoreceptor cells hyperpolarize when they are illuminated (see Chapter 7). In the dark, they release synaptic transmitter continuously; transmitter release is reduced when they hyperpolarize in response to illumination. Like bipolar cells, horizontal cells receive input from photoreceptors, and like photoreceptors, they produce only graded hyperpolarizations in response to light. Some bipolar cells produce graded depolarizations in response to light; others, such as the one shown in Figure 11-19, produce graded hyperpolarizations. A ganglion cell responds with the same polarity as the bipolar cells that innervate it. It becomes depolarized and fires APs when the bipolar cells synapsing onto it depolarize, and it becomes hyperpolarized and ceases spontaneous firing when the bipolar cells synapsing onto it hyperpolarize. Many amacrine cells, such as the one shown in Figure 11-19, respond transiently at the onset and offset of a light stimulus in response to input from bipolar cells. These individual neuronal responses combine to begin the process of analyzing visual information.

Each bipolar cell typically connects more than one photoreceptor to each ganglion cell, and a single bipolar cell may also connect one photoreceptor cell to several ganglion cells. Thus, bipolar cells provide convergence and divergence at an early stage, between the first- and third-order cells of the visual system. The amount of each depends on retinal location. In primates, both convergence and divergence are minimal in the fovea (the area in the center of the retina on which visual images are most sharply focused). This pattern of connections produces very high visual acuity based on one-to-one-to-one connections between cone photoreceptors, bipolar cells, and ganglion cells. (Cones constitute the majority of photoreceptors in the fovea.) Outside the fovea, each ganglion cell receives input from many photoreceptors—primarily rods—conferring on these ganglion cells a greater sensitivity to dim illumination but a lower degree of visual acuity.

Retinal ganglion cells. Structurally, the output of the retina is carried in the optic nerve by the axons of ganglion cells, but how is this output organized? Understanding the information exported by ganglion cells hinges on the concept of a receptive field, an idea that was first proposed by Sir Charles Sherrington early in the twentieth century and that was applied to visual

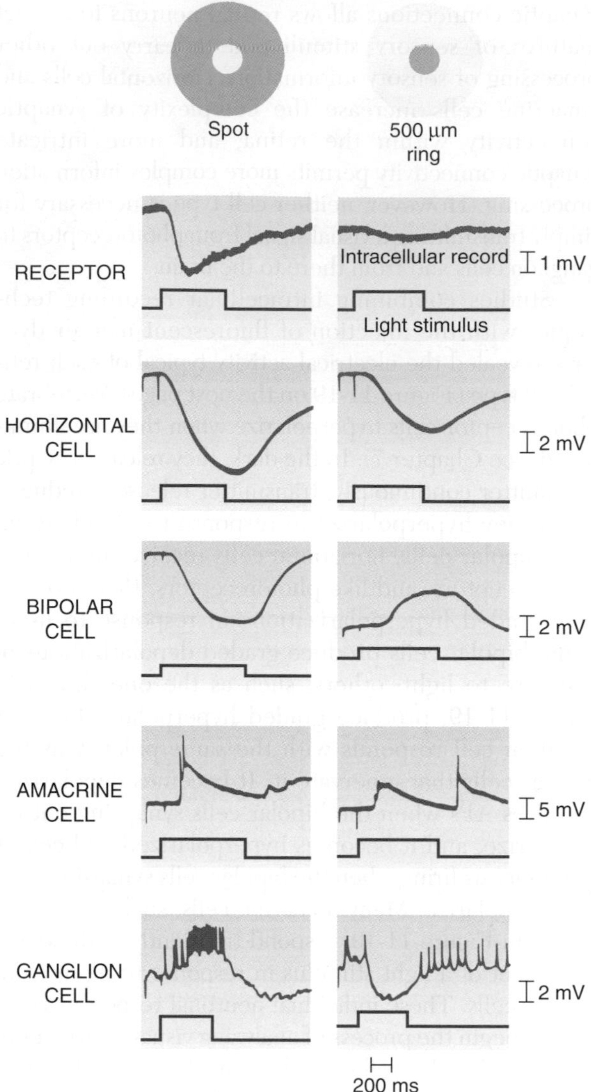

Classic Work **Figure 11-19** Each type of retinal neuron has a distinctive electrical response to light. Activity was recorded intracellularly in an example of each cell type in response to a spot of light focused directly on the photoreceptors in the region of the retina in which the cell's soma was located (left) and in response to a ring of light surrounding those photoreceptors (right). The duration of the light stimulus is indicated by the red trace in each record. In this example, the ganglion cell is activated by the spot and inhibited by the ring. Note that the bipolar cells and ganglion cells produce responses of opposite polarities to the spot and the ring. This effect is believed to depend on lateral inhibition similar to that in the horseshoe crab (see Figures 11-14 and 11-15). Notice also that the bipolar cell and the ganglion cell shown in this figure would not be synaptically connected, because the bipolar cell shown here hyperpolarizes in response to the spot of light, whereas the ganglion cell depolarizes. (See Figure 11-21 for a detailed description of how ganglion cell responses relate to signals from bipolar cells.) [Adapted from Werblin and Dowling, 1969.]

processing by H. K. Hartline in the 1940s. A visual cell's **receptive field** is the area of the retina that, when stimulated by light, affects that cell's activity. Notice that an image of the environment is formed on the retina, so a visual cell's receptive field also corresponds to a particular location in the outside world. The receptive field of a ganglion cell is roughly centered on that cell and varies in size, depending on the degree to which photoreceptor and bipolar cells converge in the pathway to that cell. At the center of the fovea, a ganglion cell's receptive field extends over only one or a few photoreceptors; that is, over only a very small part of the visual scene. At the periphery of the retina, where convergence is great, the receptive field of a ganglion cell can be as large as 2 mm in diameter, which covers a broad angle in the outside world.

Each ganglion cell is spontaneously active in the dark, and its level of activity changes when a spot of light falls within its receptive field. The response of the ganglion cell depends on the type of cell it is and on which photoreceptors in its receptive field are illuminated. The frequency of APs in the ganglion cell may increase if a small spot of light enters the cell's receptive field—an *on response.* Alternatively, the frequency of APs may drop in response to a light stimulus and increase again when the light is turned off—an *off response.* The receptive field of a ganglion cell is typically divided into a *center* and a *surround,* and the response of the cell depends on whether the center, surround, or both are being illuminated. In an *on-center* ganglion cell, the frequency of APs increases when the center of the receptive field is illuminated (Figure 11-20a). If a circle of light shines on the entire receptive field, including both the center and the surround, activity in the cell drops. When a ring of light illuminates only the surround, activity ceases; at the offset of the light, the cell becomes active—an *off response*—but less so than during its *on* response. An *off-center* ganglion cell exhibits the converse behavior: its activity is reduced or even ceases when the center of its receptive field is illuminated, and its firing increases when the surround is illuminated (Figure 11-20b).

The local, direct pathway from photoreceptor to bipolar cell to ganglion cell produces a ganglion cell's center response. The center-surround organization of ganglion cell receptive fields arises from lateral inhibition similar to that found in the compound eye of *Limulus.* Lateral inhibitory connections in the vertebrate retina depend primarily on the horizontal cells (see Figure 11-18). These cells have extensive lateral processes and connect with neighboring horizontal cells by means of gap junctions. They receive chemical synaptic inputs from photoreceptor cells and make chemical synapses onto bipolar cells. Input from a photoreceptor to a horizontal cell causes hyperpolarization that spreads electrotonically in all directions away from the photoreceptor, so every bipolar cell receives input from at least a few surrounding photoreceptors through horizontal cells. However, the strength of this input falls off with distance because the graded potentials in horizontal cells decay as they spread electrotonically, so the photoreceptor cells

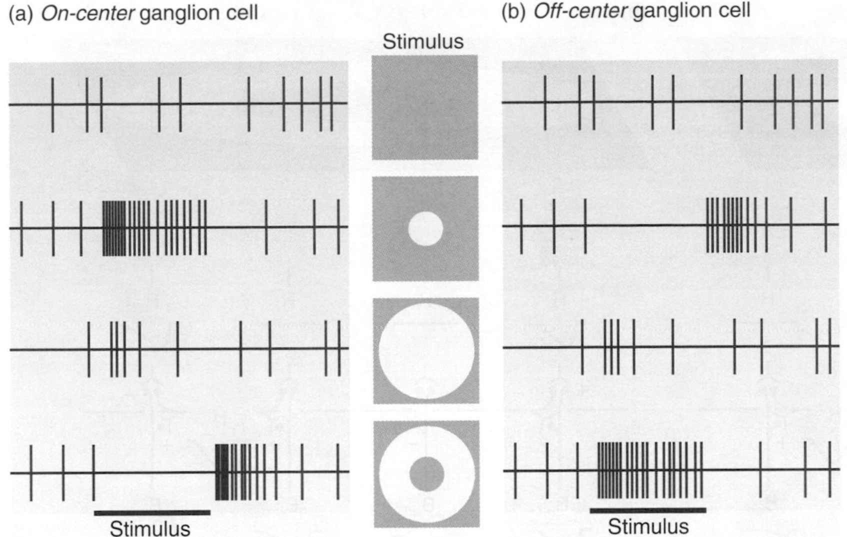

Figure 11-20 Retinal ganglion cells have either *on-center* responses or *off-center* responses to light stimuli. **(a)** Four recordings from a typical *on-center* retinal ganglion cell. Each record shows activity in the ganglion cell during a 2.5-second interval. The configuration of each light stimulus is shown in the middle of the figure. In the dark, the frequency of APs in the cell is low, and they occur more or less at random. The lower three records show responses to a small spot of light centered on the cell, to a large spot of light that includes the center of the cell's receptive field plus its surround, and to a ring of light that is centered on the cell but covers only the surround. **(b)** Responses of an *off-center* retinal ganglion cell to the same set of stimuli. [Adapted from Hubel, 1995.]

closest to a bipolar cell have the strongest effect on it through the horizontal cell network. The indirect pathway from photoreceptors through horizontal cells to bipolar cells and, thence, to ganglion cells mediates the ganglion cell's response to the surround part of its receptive field. The structure of a ganglion cell's direct center pathway and indirect surround pathway shows how particular features of a stimulus can be extracted by even this relatively simple network composed of only a few neurons.

The distinctive responses of *on-center* and *off-center* ganglion cells arise from their connections with two classes of bipolar cells: *on* bipolar cells and *off* bipolar cells. These two types of bipolar cells respond oppositely to synaptic input, both from receptors and from horizontal cells (Figure 11-21 on the next page). *Off* bipolar cells, such as the bipolar cell shown in Figure 11-19, become hyperpolarized by illumination of the photoreceptors in the center of their receptive field, whereas *on* bipolar cells become depolarized. In both types of bipolar cells, a light flashed onto the surround produces a response, mediated by horizontal cells, that is of the opposite electrical sign to the cell's response to illumination of the center. Each bipolar cell causes potential changes in its target ganglion cell or cells that are of the same sign as the potential change in the bipolar cell. Thus, ganglion cells innervated by *on* bipolars have *on-center* receptive fields, whereas those innervated by *off* bipolars have *off-center* fields. An *on-center* ganglion cell is excited by light in the center of its receptive field because it receives direct synaptic input from *on* bipolar cells. It is inhibited by light in the surround of its receptive field because horizontal cells that receive input from surrounding photoreceptors inhibit the *on* bipolar cells lying in the direct pathway from photoreceptors to the ganglion cell.

The responses of *on* and *off* bipolar cells depend on how they respond to the neurotransmitter released by photoreceptor cells and to the different neurotransmitter released by horizontal cells. In the dark, *on* bipolar cells are steadily hyperpolarized by the transmitter that is steadily released by the partially depolarized receptor cells. When a light shines on the photoreceptors and causes them to hyperpolarize, their release of transmitter drops, and *on* bipolar cells are allowed to depolarize. This depolarization causes the *on* bipolar cells to release an excitatory transmitter that depolarizes ganglion cells, increasing the frequency of APs in the ganglion cells. In contrast, the *off* bipolars have a different class of postsynaptic channels with different ionic selectivity, and in the dark, they are steadily depolarized by the neurotransmitter released from photoreceptors. When a light stimulus hyperpolarizes the photoreceptors, reduction in their release of neurotransmitter causes the *off* bipolar cells to hyperpolarize. This hyperpolarization is accompanied by a drop in transmitter release by the *off* bipolar cells, producing a hyperpolarization of postsynaptic ganglion cells.

In summary, the receptive field organization of the vertebrate retina depends on three basic features:

1. Two types of ganglion cells receive input from two corresponding types of bipolar cells. These connections produce *on-center* and *off-center* ganglion cell responses.
2. Photoreceptors in the surround of a ganglion cell's receptive field exert their effects through a network of electrically interconnected horizontal cells that synapse onto the two types of bipolar cells.
3. Direct input to the bipolar cells from overlying photoreceptors and indirect input to the bipolar cells through the horizontal cell network oppose each other and thereby produce the contrasting center-surround organization seen in both *on-center* and *off-center* ganglion cells.

The organization of the retina reveals several general principles that apply to other parts of the central

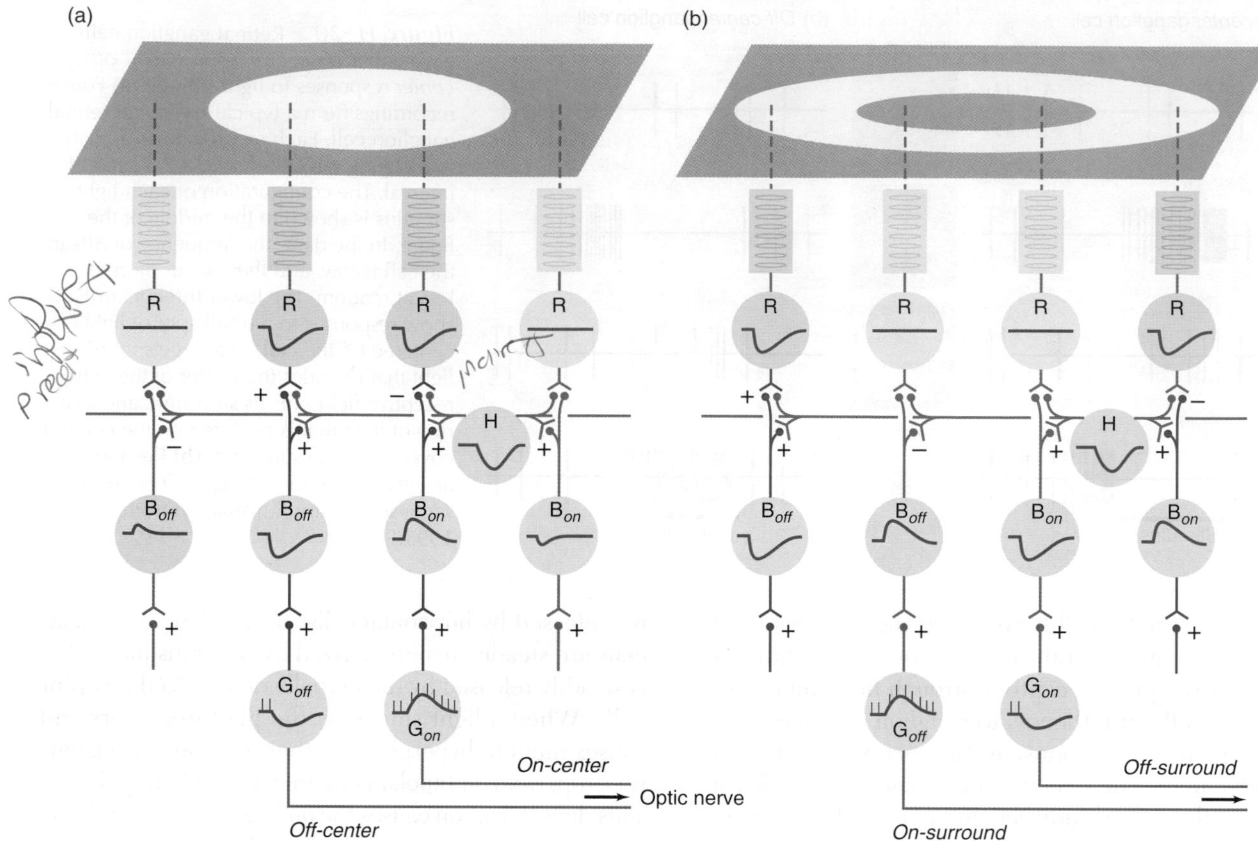

Figure 11-21 Connections within the retina produce the responses characteristic of *on-center* and *off-center* ganglion cells. Two kinds of bipolar cells, B_{on}, and B_{off}, respond oppositely to direct input from photoreceptors (R) and to indirect input carried laterally by horizontal cells (H). The *on* bipolar cells become depolarized when light shines on overlying photoreceptor cells and are weakly hyperpolarized by lateral input from horizontal cells. The *off* bipolars behave in the opposite manner. **(a)** Responses of bipolar cells and ganglion cells to a spot of light that directly stimulates the two central photoreceptors in the diagram. **(b)** Responses of bipolar cells and ganglion cells to a light that stimulates only the two lateral photoreceptors, and not the central cells in the diagram. Amacrine cells have been omitted from the diagram for simplicity. The direct pathway from photoreceptors to ganglion cells (G) is shown in red. The indirect, lateral pathway through horizontal cells is shown in blue. The plus and minus signs indicate synaptic transfer that conserves (+) or inverts (−) the polarity of the signal.

nervous system. First, neurons can effectively send signals to each other electrotonically without APs if the distances between them are small. In fact, it has been found that nonspiking neurons can convey more information more accurately than can all-or-none signals. However, electrotonic signals are attenuated with distance, which limits the spatial range of effects such as retinal lateral inhibition. Second, reception of stimuli is not necessarily synonymous with depolarization. In some neurons (e.g., photoreceptors and some horizontal cells), hyperpolarization is the normal response to stimulation; it modulates synaptic transmission by causing a drop in the steady release of transmitter. Third, the response of a postsynaptic neuron cannot be predicted from the sign of the potential change in the presynaptic neuron. A postsynaptic cell can be either depolarized or hyperpolarized when the presynaptic neuron hyperpolarizes. The postsynaptic response depends on the ionic currents produced in the postsynaptic cell as a result of modulated release of transmitter by the presynaptic neuron.

Information processing in the visual cortex

What happens to a retinal image after it has been transformed into an array of receptive-field responses within the retinal ganglion cells? Physically, the information is carried by the axons of ganglion cells to visual processing areas within the brain. The details of this pathway vary among species. In mammals and birds, the axons of retinal ganglion cells are routed either to the ipsilateral or to the contralateral side of the brain at the optic chiasm, the site where some axons cross the midline (see Figure 11-17b). Typically, in vertebrates other than birds and mammals, all optic fibers are routed to the contralateral side (see Figure 11-17a). To some extent, the degree of crossing at the optic chiasm depends on how much overlap there is between the visual fields of the two eyes. In animals in which the visual field of one eye is entirely different from the visual field of the other, all retinal ganglion cell axons cross the midline.

The lateral geniculate nucleus. In mammals, ganglion cell axons synapse with fourth-order neurons in the **lateral geniculate nucleus**, one of the many nuclei in the thalamus. (A *nucleus* in the brain is a group of neuronal somata that have some function in common.) Lateral geniculate neurons send axons to synapse onto fifth-order cortical neurons (see Figure 8-13) in an area of the occipital cortex called Brodmann's Area 17 (also called the *primary visual cortex* because it is the first region of the cortex in the visual pathway to receive visual information).

The pattern of synaptic relations within the lateral geniculate nucleus is based on both the source and the nature of the information carried by the retinal ganglion cells, and it provides another step in the processing of visual input. Each of the bilaterally paired lateral geniculate nuclei is composed of six layers of cells, stacked like a club sandwich that has been folded (Figure 11-22). The outer four layers contain neurons with small somata, which are called *parvocellular* neurons, and the inner two layers contain neurons with large somata, called *magnocellular* neurons. Input to these neurons is tightly organized. Each lateral geniculate nucleus receives information from only one-half of the visual field (i.e., either the purple or the yellow portion of the visual field as illustrated in Figure 11-17b). Each lateral geniculate neuron receives information from only one retina. The neurons in a given layer receive information from the same eye, and the layers alternate from one eye to the other, with the pattern of alternation changing between the fourth and fifth layers. Across all layers, the topology of the corresponding retinal surface is preserved exactly, and the topology is kept in register among the layers. If we passed an electrode along the path indicated by the dashed line in Figure 11-22, we would encounter cells that respond to a spot of light in precisely the same point within the visual field, but the eye of origin would switch from left to right as our electrode moved from one layer to the next.

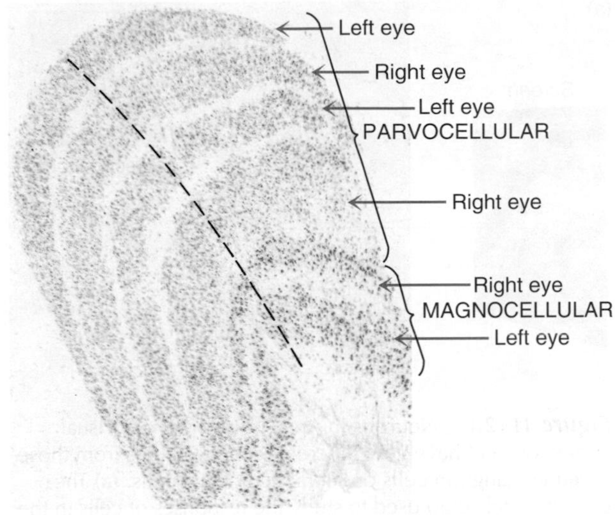

Figure 11-22 Cells of the mammalian lateral geniculate nucleus are organized into layers, each of which receives information from only one eye. To make this histological section of the left lateral geniculate nucleus of a macaque monkey, the section was cut parallel to the monkey's face. The outer four layers, called parvocellular layers, are made up of cells with small somata. The deeper layers, called magnocellular layers, are made up of cells with larger somata. In the left lateral geniculate nucleus, all cells receive information about the right half of the visual field. The outermost layer of the left lateral geniculate nucleus receives input from only the left eye, whereas cells in the next deeper layer receive input from only the right eye, and so forth. The order of alternation is reversed in the magnocellular layers. A recording electrode passed from one layer to the next along the path indicated by the dashed line would reveal that all cells along this path respond to precisely the same location in visual space, but that the eye that receives the information alternates. [Adapted from Hubel, 1995.]

Are there functional differences among the layers that receive information from each eye? Yes, the cells in each layer respond to particular properties of a stimulus, and the response varies from layer to layer. In the monkey, for example, cells within the parvocellular layers respond to the color of a stimulus, whereas cells within the magnocellular layers do not. In contrast, cells in the magnocellular layers respond to movement, whereas those in the parvocellular layers do not. This spatial sorting of outputs from ganglion cells illustrates another principle of brain organization: information about a single stimulus is divided among parallel pathways. This pattern, called **parallel processing,** is a major finding of research into higher brain function. The properties of receptive fields for neurons in the lateral geniculate nucleus are similar to those for retinal ganglion cells. That is, the receptive fields have a concentric center-surround organization of either the *off-center* or *on-center* type.

Visual cortex. The difficult question of how the visual world is organized in the next visual processing area, Area 17 of the cortex, was extensively and insightfully analyzed in the 1960s by David Hubel and Torsten Wiesel, who received the Nobel Prize in 1981. Hubel and Wiesel recorded the activity of individual neurons in the brain of an anesthetized cat while a simple visual stimulus—such as a dot, circle, bar, or border—was projected onto a screen positioned to cover the entire visual field of the cat (Figure 11-23a on the next page). The responses that they recorded from cortical neurons were correlated with the position, shape, and movement of the projected images. In retrospect, Hubel, Wiesel, and their collaborators made two important decisions in designing their experiments that allowed them to uncover order and regularity amid the enormous complexity of the visual parts of the brain. First, they chose to use more complex stimuli than just simple spots, and they asked which of these stimuli was most effective in eliciting a response in each neuron. Second, they recorded from many cells with each electrode penetration, which allowed them to learn what neighboring cells had in common and how cells were grouped in the brain. These strategies allowed them to discover several different kinds of order among the interconnections within the visual cortex, and their

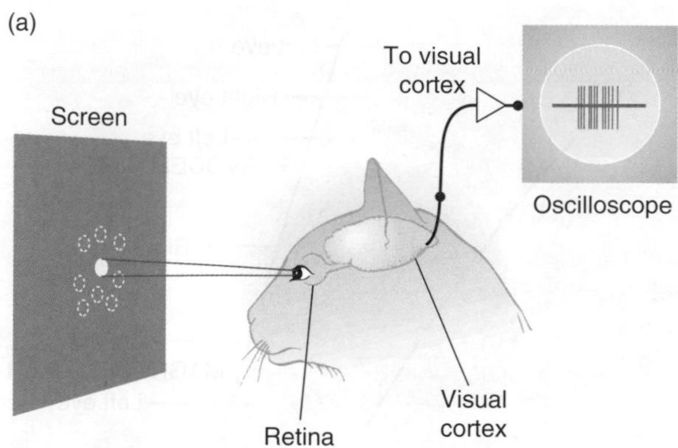

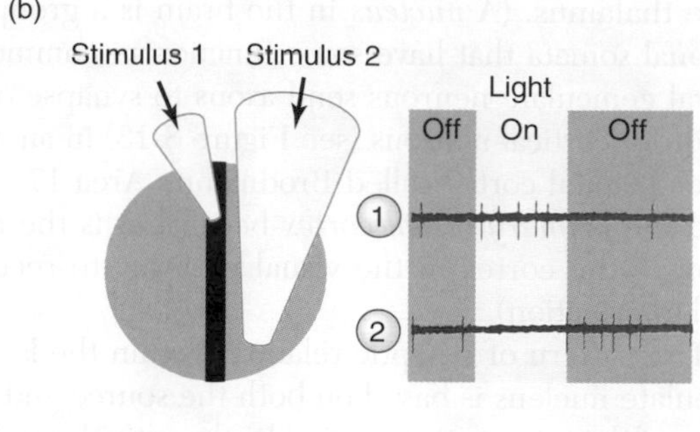

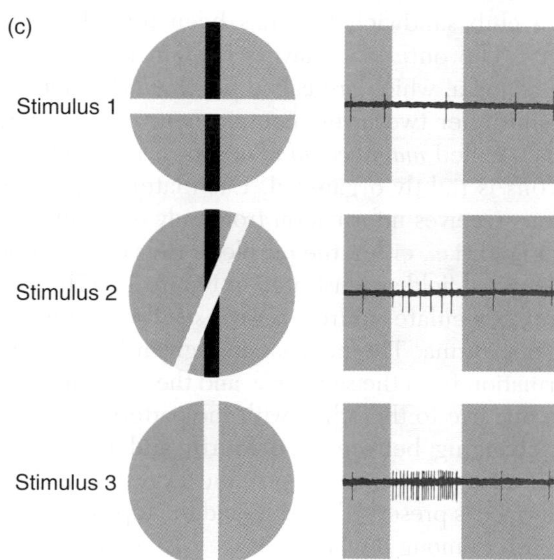

Figure 11-23 Neurons in Area 17 (the primary visual cortex) of a cat have very different receptive fields from those of retinal ganglion cells or lateral geniculate cells. **(a)** The experimental setup used to study the responses of cells in the visual cortex. An electrode is advanced through the cortex while light stimuli are projected onto a screen.

Classic Work **(b)** The receptive field of cortical simple cells is bar-shaped. A spot of light anywhere along the *on* part of this receptive field (stimulus 1) produces a small excitation of the simple cell. A spot of light adjacent to the bar-shaped *on* region (stimulus 2) causes inhibition of APs in these tonically active cells.

Classic Work **(c)** Rotating a bar of light (yellow bar) across the receptive field of a simple cell (black bar) produces maximum activity in the cell when the stimulating bar coincides completely with the *on* region of the cell's receptive field (stimulus 3) and partial excitation at other orientations (e.g., stimulus 2). [Part a from "Cellular Communication" by G. S. Stent. Copyright © 1972 by Scientific American, Inc. All rights reserved. Parts b and c from "The Visual Cortex of the Brain" by D. H. Hubel. Copyright © 1963 by Scientific American, Inc. All rights reserved.]

discoveries have provided a model for examining other sensory systems.

Hubel and Wiesel's major discovery was that cells of the visual cortex respond to entirely different properties of stimuli than do retinal ganglion cells or cells of the lateral geniculate nucleus. They called the two major classes of cells that they found in this region *simple cells* and *complex cells,* based on the nature of their optimal stimulus. They found that cells of each type were arranged systematically in space according to their optimal stimuli.

Simple cells. The receptive fields of simple cells are long and bar-shaped, and the *on* region of the field has a straight border separating it from the *off* region (Figure 11-23b), rather than the circular border found for cells in the retina and in the lateral geniculate nucleus. As in retinal ganglion cells and lateral geniculate cells, the receptive field of a simple cell lies in a fixed position on the retina and, hence, represents a particular part of the total visual field. There is some variation in the receptive fields of simple cells. Some have a bar-shaped *on* region surrounded by an *off* region; for others, the receptive field consists of an *off* bar surrounded by an *on* region. For still others, the receptive field is divided in half by a straight boundary, with an *off* region on one side of the boundary and an *on* region on the other.

A bar of light elicits maximal activity in a simple cell when it overlaps completely with the cell's *on* receptive field (Figure 11-23c). When the bar is rotated so it no longer aligns perfectly with the orientation of the receptive field, it has less effect on the spontaneous activity of the simple cell, or it may even inhibit the cell's activity. If the bar of light is displaced so that it falls just outside the *on* region, the cell is maximally inhibited. The optimal orientation and the *on-off* boundaries differ from one simple cell to another; thus, when a bar of light moves horizontally or vertically across the retina, it activates one simple cell after another as it enters one receptive field after another.

What makes simple cells respond specifically to straight bars or to borders with precise orientations and locations? Hubel and Wiesel suggested—and recent experiments have confirmed—that each simple cell receives excitatory connections from lateral geniculate cells whose *on* centers are arranged linearly on the retina (Figure 11-24a). Simple cells that respond to borders, rather than to bars, are thought to receive inputs as shown in Figure 11-24b. A simple cell would receive maximal input when light fell on all of the pho-

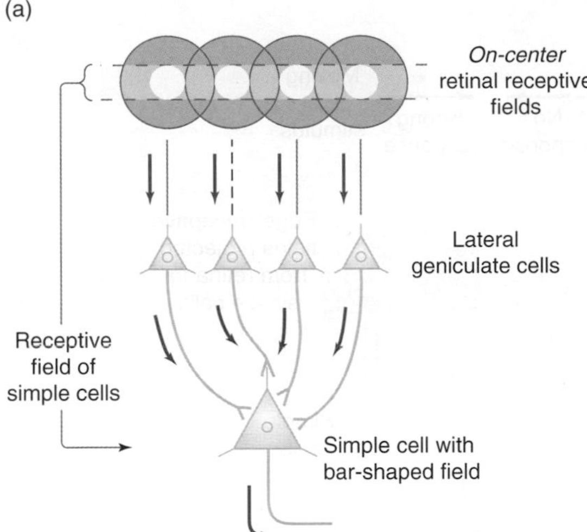

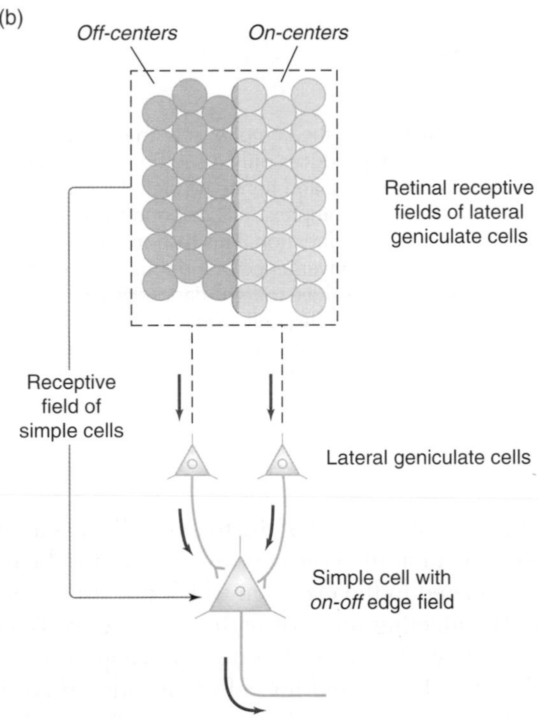

Figure 11-24 The responses of simple cells in the visual cortex arise from the pattern of their synaptic inputs. **(a)** The fixed, bar-shaped receptive field of a simple cell arises from the convergence of outputs from ganglion and lateral geniculate cells whose circular *on-center* receptive fields are linearly aligned. **(b)** An *on-off* straight-edge receptive field results from the convergence of *off-center* and *on-center* geniculate cells onto the simple cell.

toreceptors that activated the *on-center* fields of the ganglion cells and the lateral geniculate cells connected to it. Any additional illumination would fall on the inhibitory surround of the ganglion cells and could only reduce the response of the cortical cell.

Complex cells. Complex cells constitute the next level of abstraction in the processing of visual information. They are believed to be innervated by simple cells, which would make complex cells sixth-order neurons in the hierarchy of visual information processing. Like simple cells, complex cells respond best to straight bars or borders at specific angular orientations on the retina. Unlike the simple cells, however, complex cells do not have topographically fixed receptive fields. Appropriate stimuli presented anywhere within a relatively large region of the retina are equally effective at activating complex cells; as for simple cells, light shining continuously on the whole retina is not an effective stimulus. Some complex cells respond to bars of light of specific orientations (Figure 11-25a on the next page). Others give an *on* response to a straight border when the light is on one side of a boundary and an *off* response when the light is on the other side. Still other complex cells respond optimally to a moving border that progresses in only one direction (Figure 11-25b). For these cells, movement in the opposite direction evokes either a weak response or no response at all.

The receptive fields of complex cells can be explained by a convergence of synaptic inputs from simple cells. As a light-dark border moves through the receptive fields of the simple cells that synapse onto a complex cell, each simple cell excites the complex cell in turn as the border passes through its receptive field. This arrangement could also produce directional sensitivity to movement (see Figure 11-25b). If the border moved so that sequential simple cells were illuminated, one simple cell after another would be excited, exciting the complex cell. When each simple cell became inhibited by the dark side of the moving border, the next one in line would be excited. In contrast, if the border moved so that simple cells were sequentially exposed first to inhibition and only later to stimulation, one simple cell after another would inhibit the complex cell, counteracting any tendency for excitation caused by the bright side of the border.

Notice two things as we move deeper into the nervous system. First, at each progressively deeper level, there are more variations in what excites individual neurons. For example, there are more types of complex cells than there are types of simple cells. Second, cells of deeper levels respond to more complicated combinations of features. Lateral geniculate cells, for example, have circular center-surround receptive fields, whereas cortical simple cells respond to bars of particular orientations, and cortical complex cells may respond only to moving bars. Cells transmit signals only if they are excited, and at each level the cells filter the signal with greater selectivity.

Columns in the cortex. The properties of individual cortical cells suggest that they abstract features of the visual scene, such as edges, as a first step toward analysis and recognition. The spatial relations among visual cortical cells are correlated in an orderly fashion with their functional properties. In their experiments, Hubel and Wiesel discovered that cells adjacent to one another responded to similar features of a stimulus. When they penetrated the visual cortex with an

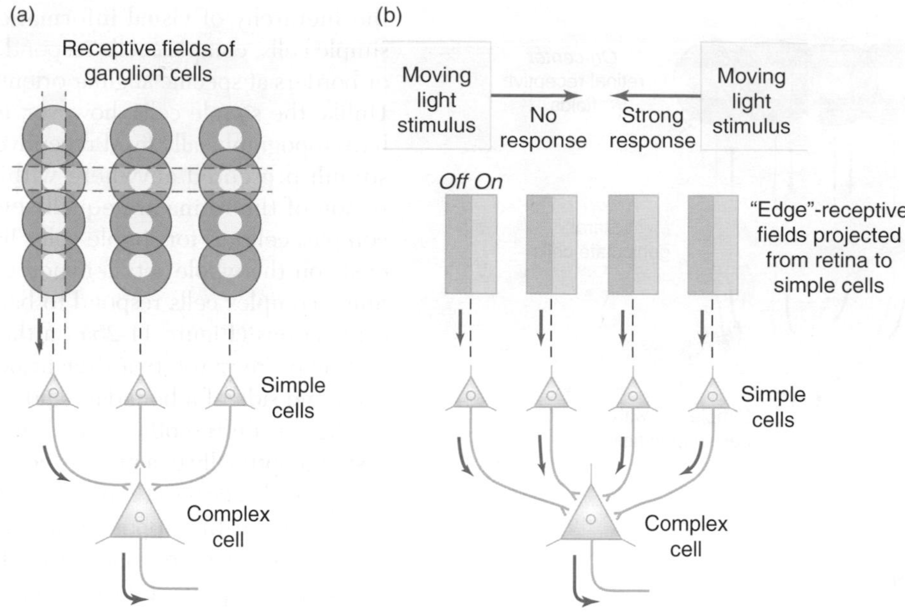

Figure 11-25 Responses in complex cells could be based on their pattern of input from simple cells. **(a)** Some complex cells respond to bars of light that have a specific angular orientation, but the location of the bar can be anywhere within a large visual field. This pattern of response could be evoked by the convergence of many simple cells, each having similarly oriented bar-shaped receptive fields. In this example, the vertical bar of light stimulates one simple cell to fire because it falls on a row of ganglion cells whose receptive fields produce the bar-shaped receptive field of the simple cell. If the bar were moved to the right, it would excite another simple cell that synapses onto the same complex cell, producing activity in the complex cell. In contrast, a horizontal bar of light produces only a subthreshold response in each of the simple cells, and hence no signal is sent to the complex cell. **(b)** Some complex cells respond to edges of light moving in only one direction. This response pattern could be produced by convergence of a population of simple cells, all of which are sensitive to light-dark edges with the same orientation. The complex cell would be excited if the edge were moved in such a way that it illuminated the *on* side of the simple cell receptive fields before it illuminated the *off* side. Movement in the opposite direction would produce only inhibition.

electrode that was perpendicular to the cortical surface and recorded responses from the cells encountered as they advanced the electrode along that pathway, they discovered that the cells along each pathway responded to bars of light having the same orientation. When they moved the electrode laterally and made another penetration, they found a column of cells that responded optimally to a stimulus having a different orientation from that of the optimal stimulus for the neighboring column of cells. Each such set of cells is called a *cortical column.* In contrast, recording from cells along a track parallel to the surface of the cortex revealed an amazingly regular change in the optimal stimulus orientation, with the preferred orientation shifting about 10° each time the electrode advanced 50 μm. This result implies that the cells of the visual cortex are organized in columns according to features of their optimal stimuli and that these features change in an orderly fashion across the cortex (Figure 11-26a).

The columnar organization of cells with similar response properties had been seen earlier in the somatosensory cortex, in which adjacent columns contained cells responding to touch on a particular part of the body or to the bending of a particular joint (see Figure 8-16c). However, the orderly array of orientation columns was only the first of the functionally based subdivisions found in the visual cortex. The next to be discovered related to the eye from which the visual signal came. By injecting one eye with a radioactive tracer molecule that is transported across synapses to the visual cortex, Hubel and his colleagues identified the projection pattern of each eye onto the cortical surface. These experiments revealed a second columnar system in which alternating columns receive information from only one or the other eye. Three-dimensional reconstructions of these columns, called *ocular dominance columns,* show their distribution across the cortical surface (Figure 11-26b).

The spatial map. These experiments revealed that the visual cortex is subdivided into small functional units that separate a stimulus into its constituent features before passing it on to higher levels for further analysis. This modular organization is superimposed on a fundamental topologic map of space, which persists through the layers of the visual cells. To understand the nature of the spatial map at the cortical level, experiments have been done to plot the visual field onto the cortex directly, using a radioactive labeling technique (Figure 11-27). Radioactive 2-deoxyglucose was injected into an anesthetized monkey, and a complex,

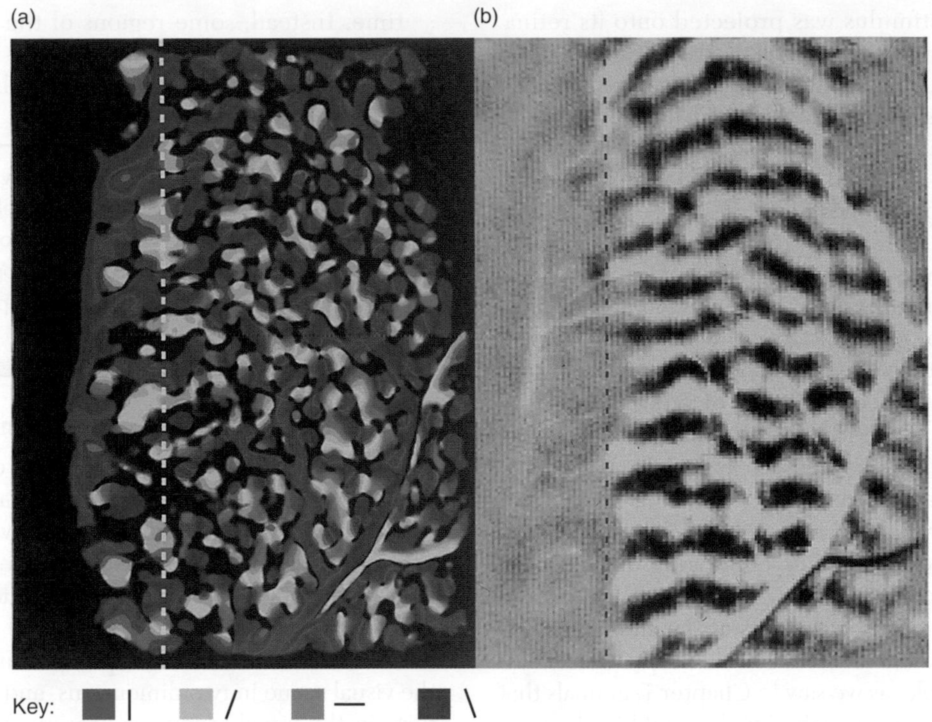

Figure 11-26 The visual cortex is organized into columns of cells with similar functions. These images were made from optical recordings of neuronal activity in the primary visual cortex of a cat. The images show cells at the surface of the cortex, but the cells that lie directly below the surface have functional properties similar to the cells in the top layer. Parts a and b both show the same region of cortex, and the dashed vertical line indicates precisely the same location. Notice that cortical cells are simultaneously organized into orientation and ocular dominance columns. **(a)** The organization of orientation columns of simple cells. Each optimal stimulus orientation is represented by a color (see Key). Notice that as you move from one column to another across the cortex, the orientation of the optimally effective stimulus rotates in an orderly fashion. **(b)** Ocular dominance columns in the same cortical area. Cells driven by one eye are indicated by black; cells driven by the other eye are indicated by white. [Courtesy of Anirudah Das.]

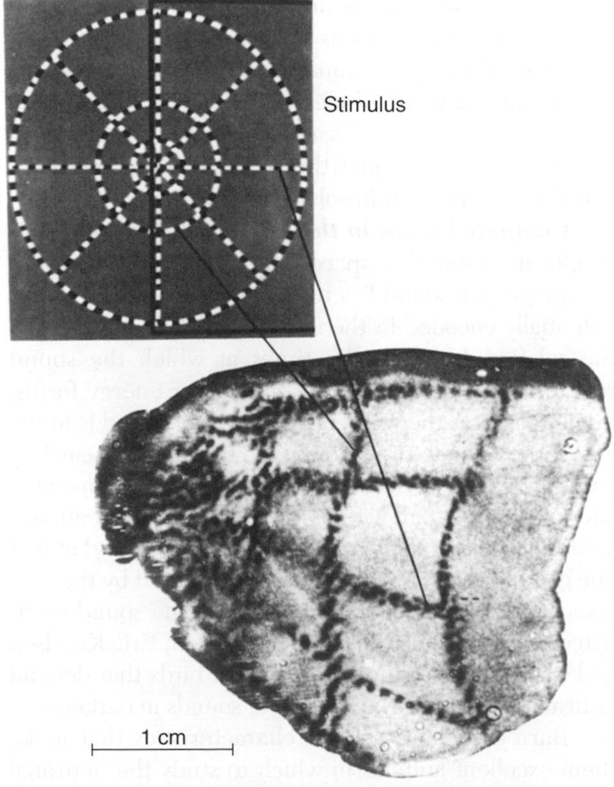

Figure 11-27 The topography of visual space is represented on the surface of the visual cortex, but in a distorted form. This target-shaped stimulus with radial lines, shown at the top of the figure, was centered on the visual fields of an anesthetized macaque monkey for 45 minutes after radioactive 2-deoxyglucose had been injected into the monkey's bloodstream. One of the monkey's eyes was kept closed. The cortex was then removed, flattened, frozen, and sectioned. The lower image shows a section cut parallel to the cortical surface. The roughly vertical lines of label represent the curved lines of the stimulus; the horizontal lines of label represent the radial lines in the right visual field. The lines are broken because only one eye was stimulated. This dotted pattern displays the ocular dominance columns. Notice that the center of the target, which was seen by foveal photoreceptor cells, projects to proportionally more of the cortical area than does the edge of the target. [Adapted from Tootell et al., 1982.]

target-shaped stimulus was projected onto its retina. Active neurons take up more 2-deoxyglucose than do resting neurons, so cortical neurons that were activated by the stimulus would be expected to contain more radioactivity than their inactive neighbors. The pattern of radioactivity observed in the visual cortex revealed that, although the two-dimensional surface of the retina was completely represented on the cortical surface, the cortical pattern was a distorted replica of the spatial features of the retinal stimulus. Retinal regions representing the center of gaze (the fovea) were greatly magnified relative to those representing the peripheral view. There is much less convergence of information from photoreceptors in the fovea, so the pathways from the fovea are much less condensed. The cortical area that receives information from the fovea is thus proportionally much larger than the area that receives information from the edge of the retina, conferring on the fovea maximum visual acuity. This distortion of the map in accord with the needs and habits of the animal is characteristic of all animals with well-developed visual systems. For example, as we saw in Chapter 7, animals that evolved on large, open plains, such as rabbits, have an elongated, horizontal retinal region of specialization, called a *visual streak*, that provides the greatest number of photoreceptors and the least convergence for receiving stimuli along the visual horizon.

All of these various levels of cortical organization must be combined to provide the next set of cortical cells with a complete picture of visual stimuli. The manner in which this synthesis is accomplished is still the subject of intense research. It now appears possible, for example, that some higher-order visual neurons may be active only if a specific object (e.g., a face) enters their receptive field. This synthesis is part of the task performed by the complex interconnections illustrated in Figure 11-16.

The visual cortex has taught physiologists several important principles about the organization of sensory networks:

1. The visual system is organized hierarchically. At each level, cells require more complex stimuli to excite them optimally. This complexity arises from synaptic convergence of cells with simpler receptive fields onto cells with more complex receptive fields.

2. Although convergence is apparent as we follow a stimulus into the system, the parallel analysis of distinct features of a stimulus requires divergence of information as well. The simultaneous analysis of different features of a stimulus, which takes place along parallel pathways, appears to be an important principle of functional organization.

3. The activity of cortical neurons in Area 17 results in abstraction of some features of the visual stimuli.

4. The visual cortex does not receive a simple one-to-one projection from the retina in either space or time. Instead, some regions of the visual field are expanded dramatically in their cortical representation, whereas others are compressed.

The organization of the visual system poses major challenges for developmental neurobiologists. During embryonic development, what kinds of cellular properties and mechanisms could generate the precise connections observed within the cerebral cortex?

Processing of auditory information

The retinotopic and somatotopic maps described above are found in many levels of the brain as sensory information is transmitted through the nervous system. We can recognize these maps because, even in a distorted form, they mimic the spatial organization of objects in the environment. The two-dimensional array of cells on the retinal surface, for example, produces a replica of the visual scene in two dimensions, and the spatial relations in the visual scene are preserved as the image is projected onto the cells of the lateral geniculate nucleus and the cortex. For other sensory systems, the nature of possible maps in the nervous system is not so obvious. In the auditory system, for example, the topologic arrangement of hair cells along the cochlea is correlated with their sensitivity to particular frequencies of sound (see Chapter 7). If the spatial order of these hair cells were preserved in the projection of their axons to the brain, a brain map of sound frequency—a *tonotopic* map—would be the result. Indeed, tonotopic maps have been found in some auditory regions of the brain. However, it is not obvious how sorting sounds by frequency would help an animal acquire information about the arrangement of objects in its environment. We know that humans can locate the source of a sound in space, but knowing just the frequency of the sound would not help much in solving this problem.

Computed maps in the brain How does an animal locate a sound in space? Information about where the source of a sound lies relative to a listener is at least potentially encoded in the intensity of the sound and in the relation between the times at which the sound reaches the two ears. Sounds, like other energy forms, lose intensity as they travel. If a source of sound is to the left of the animal, the sound will reach the left ear first and will be loudest there; the sound will arrive somewhat later at the right ear, with its intensity somewhat attenuated. The time separating the arrival of the sound at first one ear and then the other can be computed by the nervous system as an indication of where the sound originated. To understand how this is achieved, Eric Knudsen and Mark Konishi studied barn owls, birds that depend critically on locating the sources of sounds in darkness.

Barn owls have several characteristics that make them excellent animals in which to study the neuronal mechanisms that underlie sound localization. If light is available, owls use both vision and hearing to guide their hunting, but they can capture a mouse in total darkness, finding their prey just by listening to the sounds it makes (Figure 11-28). In addition, an owl cannot move its eyes within their orbits; instead, it must move its whole head, whether it is orienting to a sound or to a visible object. This orienting response is quite accurate. Owls can point their heads toward the source of a sound with an accuracy of 1–2° in both *azimuth* (lateral distance away from a point straight in front of the owl's head) and *elevation* (vertical distance away from a point straight in front of the owl's head). This angle locates an object to within 6 inches at a distance of 30 feet.

To test its orienting ability, the researchers placed an owl on a perch and generated sounds from a speaker whose location could be varied in space while remaining at a fixed distance from the bird (Figure 11-29a). The position of the owl's head was monitored as it oriented toward the sounds produced by the speaker, and the orientation of the head in response to each sound was expressed in degrees of elevation and azimuth (Figure 11-29b). Careful behavioral observations indicated that the owl was using two kinds of cues in its orienting response: the intensity of sound was used to determine the elevation of the source, and the relative times of arrival at the two ears were used to determine the azimuth of the source.

Figure 11-28 Barn owls can capture mice in total darkness. These images are from a film that was made using only infrared illumination that the owl cannot see. The owl successfully captured the mouse (lower right). [Courtesy of M. Konishi.]

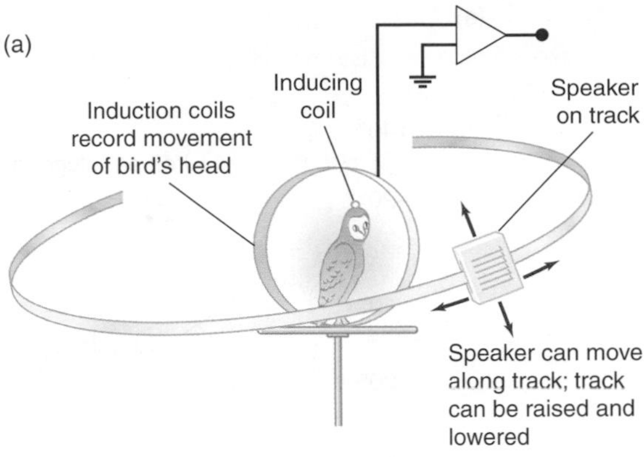

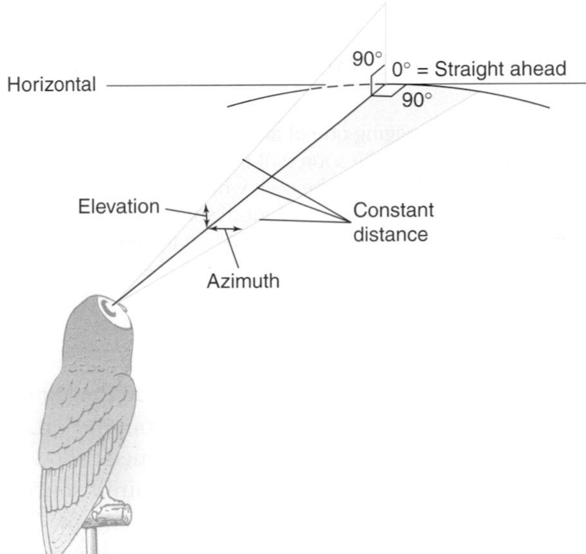

Figure 11-29 Owls move their heads to orient toward a sound, a behavior that is readily observable. **(a)** Experimental setup for studying the ability of an owl to locate the source of a sound. The target speaker can be moved to any location in a hemisphere surrounding the front of the owl. The distance between the owl and the source is kept constant because the speaker travels on a curved track. **(b)** Coordinate system for determining the location of a sound. Elevation indicates the angle along a vertical axis; azimuth indicates the angle in the horizontal plane. [Adapted from "The hearing of the barn owl" by E. I. Knudsen. Copyright © 1981 by Scientific American, Inc. All rights reserved.]

To examine the role played by intensity cues, either the right or the left ear of an owl was plugged, using plugs that either weakly or strongly reduced the sound intensity. The results of this experiment revealed that the owl consistently misdirected its gaze when one of its ears was plugged (Figure 11-30a on the next page). With the right ear plugged, the owl oriented below the actual source and slightly to the left. With the left ear plugged, it oriented above the source and slightly to the right. In other words, when the sound was louder in the right ear, it seemed to the owl to be coming from above and from the

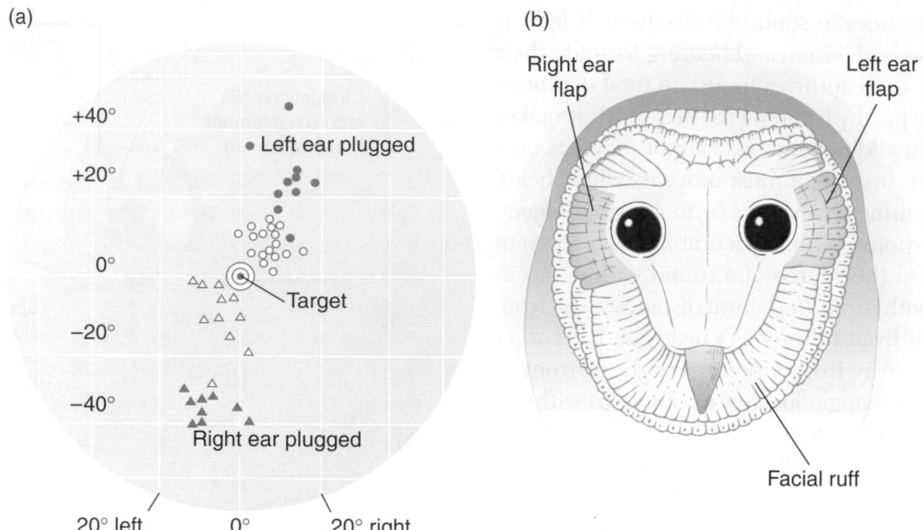

Figure 11-30 Plugging one of an owl's ears caused it to make errors in locating the source of a sound. **(a)** A sound was presented directly in front of the owl, which had either a hard plug in one ear (high attenuation; solid symbols) or a soft plug (less attenuation; open symbols). When the left ear was plugged (circles), the owl judged the sound to be above its real location. When the right ear was plugged (triangles), the owl made mistakes in the opposite direction. **(b)** An owl with its facial ruff partly removed, showing the asymmetry in the auditory openings. Notice that the left ear is higher on the head than the right ear. Although it is not visible in this diagram, the right ear canal points slightly upward, whereas the left ear canal points slightly downward. This small difference is amplified by the position of the feathers in the facial ruff. [Adapted from "The hearing of the barn owl" by E. I. Knudsen. Copyright © 1981 by Scientific American, Inc. All rights reserved.]

right; whereas, when the sound was louder in the left ear, it seemed to be coming from below and from the left.

How can interaural intensity differences allow an owl to discriminate the elevation of a sound source? The answer lies—at least in part—in anatomy. The region around the openings of an owl's ears has a ring of stiff feathers, called the facial ruff, which form a surface that very effectively directs sounds into the ear canals, much like the fleshy pinnae of mammalian ears (Figure 11-30b). Notice that the location of the two ears is slightly asymmetric, with the left ear higher than the right (see Figure 11-30b). In addition, the opening of the right ear is directed slightly upward, whereas the opening of the left ear is directed slightly downward. This arrangement could provide a basis for discriminating elevation from intensity cues. The importance of the facial ruff was revealed by removing it. If an owl lacked its facial ruff, it was no longer able to identify the elevation of sound sources, although its estimates along the azimuthal axis remained as accurate without the facial ruff as with it. Thus, the facial ruff must amplify the directional asymmetry of the ears, a function that is essential for discriminating differences in elevation among sound sources.

How does an owl locate sounds along the horizontal or azimuthal meridian? From behavioral experiments, it was clear that disparity in the times at which sounds arrived at each of the ears was important for this discrimination. However, the relevant cue could have been either disparity in the onset (or offset) of the sound or ongoing disparity that occurred during the duration of sound (Figure 11-31). Disparity of onset refers to the difference in the times at which a given signal first reaches each ear; the ear nearest the source receives the signal first. Disparity can also occur between the signals that are received by the two ears as a sound continues; just as the onset of a sound reaches the two ears at different times, identifiable features of the sound reach first one ear and then the other. These two types of disparity were independently varied by implanting small speakers in an owl's ears. In response to disparity of onset, the owls failed to make "correct" head movements, whereas, in response to ongoing disparities ranging from 10 to 80 μs, they made rapid head movements toward the azimuthal angle correctly corresponding to the time difference (Figure 11-31b). These experiments confirm an owl's ability to orient to sounds in space with remarkable accuracy. The elevation of a sound source is judged from differences in the intensity of sounds arriving at each ear, and the azimuth is judged from the ongoing temporal disparity between sounds arriving at each ear.

How is information about a sound's location in space represented in the owl's nervous system? The ears cannot directly provide the brain with a representation of external space. Instead, as we have seen, cells in the owl's brain must compute the difference in intensity between sound signals sensed by its two ears to determine the elevation of a sound. Brain cells must generate an ongoing evaluation of disparity between the sound signals reaching the two ears to determine the position of the sound in the azimuthal plane. How and where these comparisons are made, and how the output is represented in the brain, were discovered by Eric Knudsen and Marc Konishi in the late 1970s.

(a)

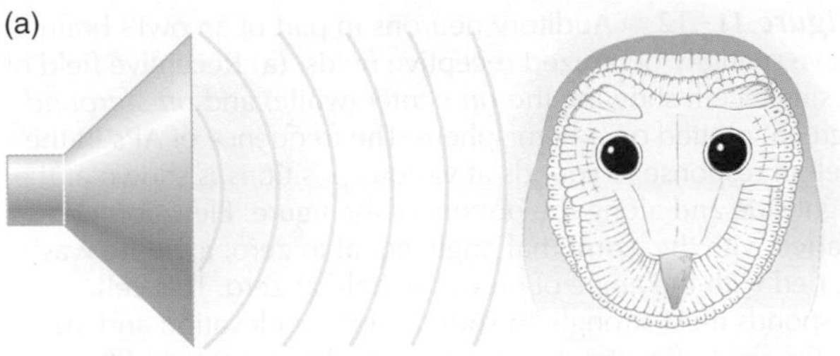

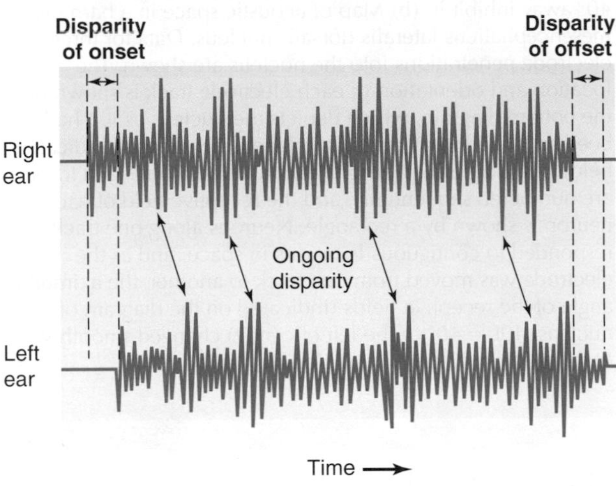

(b)

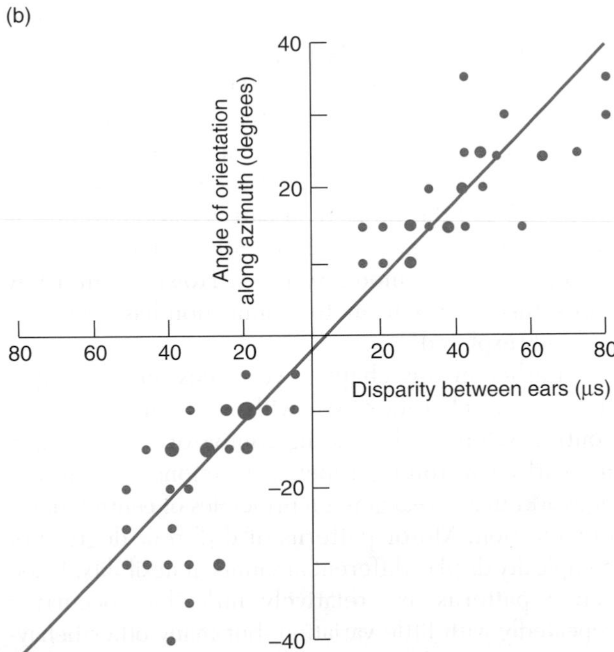

Figure 11-31 Owls judge azimuthal location from the disparity between ongoing sounds that arrive at the ears. **(a)** Onset disparity occurs when a sound reaches one ear before the other. Ongoing disparity refers to the continued difference in sound waves as perceived by both ears. **(b)** Owls use the ongoing disparity between the sounds reaching the two ears to localize a signal accurately in space. The linear relation between azimuth and ongoing disparity between signals at the two ears confirms that this type of disparity is the relevant cue for judging azimuthal angle. [Adapted from "The hearing of the barn owl" by E. I. Knudsen.]

Location of the spatial map Knudsen and Konishi identified a collection of spatial location–specific neurons in a nucleus of the owl's midbrain. Each of these cells responds best to sound signals whose sources are located at a particular point in space, and each cell has a receptive field with an *on-center, off-surround* organization similar to that found in retinal ganglion cells (Figure 11-32a on the next page). Sounds whose sources are located within the center region of the cell's receptive field (which has a mean diameter of about 25°) excite the cell, whereas sounds in the surround of the receptive field inhibit its response. These auditory neurons are arrayed in the nucleus so as to form a spatial map (Figure 11-32b) analogous to the retinotopic map derived from the retina and the somatotopic map derived from the body surface. Cells at each point on the surface of the nucleus fire APs in response to sounds from a particular point in space. Adjacent points in the nucleus respond to sounds that arrive from adjacent locations in space.

Another feature common to this map and other brain maps is that the receptive fields of cells that receive information from directly in front of the animal are smaller than those of cells that receive information from the sides of the animal. The area directly in front of the animal is represented in a larger part of the nucleus and is therefore magnified compared with the area to the sides. This projection pattern is reminiscent of the exaggerated representation of the retinal fovea in the visual cortex and of the large representation of the hands and face on the human somatosensory cortex.

In the barn owl, the nucleus where these spatial receptive fields are recorded is the mesencephalicus lateralis dorsalis (MLD), which is the avian homolog of the mammalian inferior colliculus, a major auditory center that lies just beneath the superior colliculus (which is itself the mammalian homolog of the tectum). The MLD nucleus passes a map of sound location in space to higher brain centers. Ongoing disparity between signals is sensed by neurons in nuclei that lie ventral to the MLD in the midbrain. These neurons, which are called coincidence detectors, receive input from both ears, and their activity changes depending on whether signals from the two ears arrive simultaneously or sequentially. The mechanisms by which differences in sound intensity are computed by the owl's brain are still being investigated.

The barn owl's map of acoustic space was the first example of a topologic map in the brain that is generated *de novo* from the response properties of neurons. Since then, similar computed maps have been found in the brains of bats, which, like owls, hunt by using auditory information. The spatial representation of sound in an owl's brain ultimately projects to the tectum, where it meets—and is congruent with—a map of space generated by the visual system. Adjacent layers of the tectum are topographically correlated, with cells in one

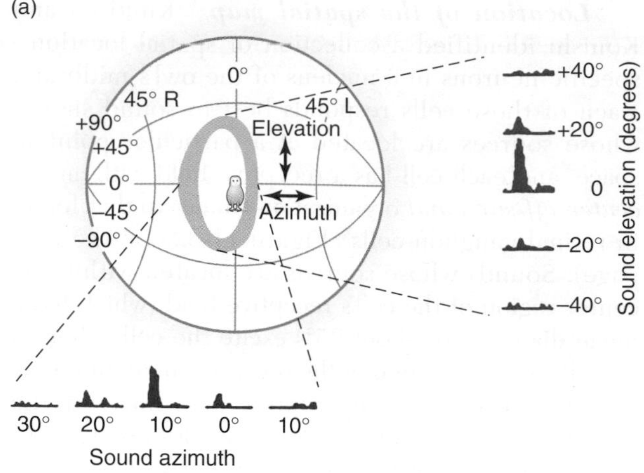

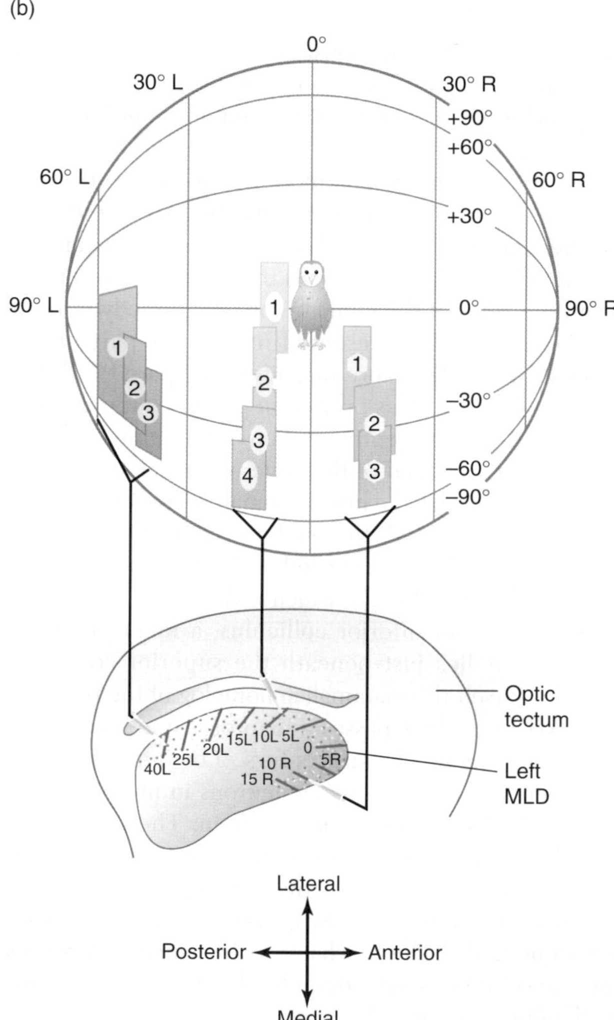

Figure 11-32 Auditory neurons in part of an owl's brain have spatially organized receptive fields. **(a)** Receptive field of a single cell, showing the *on-center* (white) and *off-surround* (green) plotted on a hemisphere. The frequency of APs in the cell in response to sounds at various positions is shown at the right side and along the bottom of the figure. Elevation was varied with the azimuthal angle equal to zero; azimuth was varied with the angle of elevation held at zero. This cell responds most strongly to sounds at 0° of elevation and 10° to the right of center. Sounds that are 20° away from 0° of elevation stimulate the cell only weakly, and sounds that are 40° away inhibit it. **(b)** Map of acoustic space in a barn owl's mesencephalicus lateralis dorsalis nucleus. Data for three electrode penetrations into the nucleus are shown. The location and orientation of each electrode track is shown on the bottom diagram, which depicts the nucleus as if it had been sectioned in a horizontal plane (orientation is indicated below the diagram). Neurons encountered along each track are numbered sequentially, and the receptive field of each neuron is shown by a rectangle. Neurons along one track responded to contiguous locations in space, and as the electrode was moved from one track to another, the azimuthal angle of the receptive fields (indicated on the diagram of the nucleus: 40L = 40° to the left of center) changed smoothly. [Adapted from Knudsen and Konishi, 1978.]

layer processing information about sounds, and cells in the other layer processing information about visual input. This arrangement suggests that behavior can be organized more effectively if all of the sensory information about an object in space is first assembled at one location. The next problem in understanding the production of behavior is a consideration of where and how sensory information leads to the decision to act.

Motor Networks

The sensory side of the nervous system acquires and analyzes information about the outside world, which is essential for producing behavior that is matched to an animal's current circumstances. This information must then be passed on to neurons that are responsible for generating coordinated movement. Relatively little is known about the interface between the sensory and motor sides of this process, in part because most investigators have worked independently at understanding either sensory or motor systems. However, in a few cases, this sensory-to-motor connection has been successfully explored.

Earlier in this chapter we discussed the simple muscle stretch reflex. We will now consider motor control systems of increasing complexity, from simple networks controlling repetitive actions to complex networks that reveal general principles of central motor organization. Motor patterns of different degrees of complexity display different amounts of flexibility. Fixed action patterns are relatively inflexible, occurring repeatedly with little variation, but many other behaviors are extremely plastic. An animal can shape these flexible behaviors to fit each new set of circumstances. One of the challenges in studying motor control is to understand the neuronal activity that allows an organism to produce behavior that changes from moment to moment, as the situation changes.

Levels of motor control

Most studies of how neurons control muscle activity have been focused either on animals with simple nervous systems or on repetitive actions by more complex animals. The neuronal control of fixed action patterns has been a major topic of this work because the all-or-none property of these behaviors suggests that a single neuronal decision might generate them. The use of the term "decision" in this sense does not imply a conscious process, but rather that the activation of a neuronal "switch" in the central nervous system is sufficient to initiate the behavioral pattern. Conceptually, this idea can be formalized as a hierarchical motor control system in which sensory input is used to select some specific motor output. The lowest level of control is the motor neuron that connects to a muscle; activity in motor neurons is determined by the summed synaptic input onto each individual neuron (Figure 11-33).

Initially, some physiologists believed that a short feedback loop between stretch receptors in leg muscles and the spinal motor neurons controlling those muscles might account for the motions of walking in vertebrates. It has become clear, however, that repetitive motor output—such as walking, swimming, or flying—depends on activity in a central neuronal network that generates essential features of the motor pattern. A neuronal network that produces repetitive output is called a **central pattern generator.** The pattern of walking, swimming, or flying can be modified in response to sensory feedback, however, and varies with features of the terrain or with water or wind currents in the environment. Finally, control is exerted by centers higher in the nervous system, whose decisions or commands are also influenced by sensory input. Notice that there may be no strict chain of command in this control hierarchy. A wide variety of distinct environmental inputs can lead to related kinds of motor output, and feedback control operates at all levels of the system to modulate outputs.

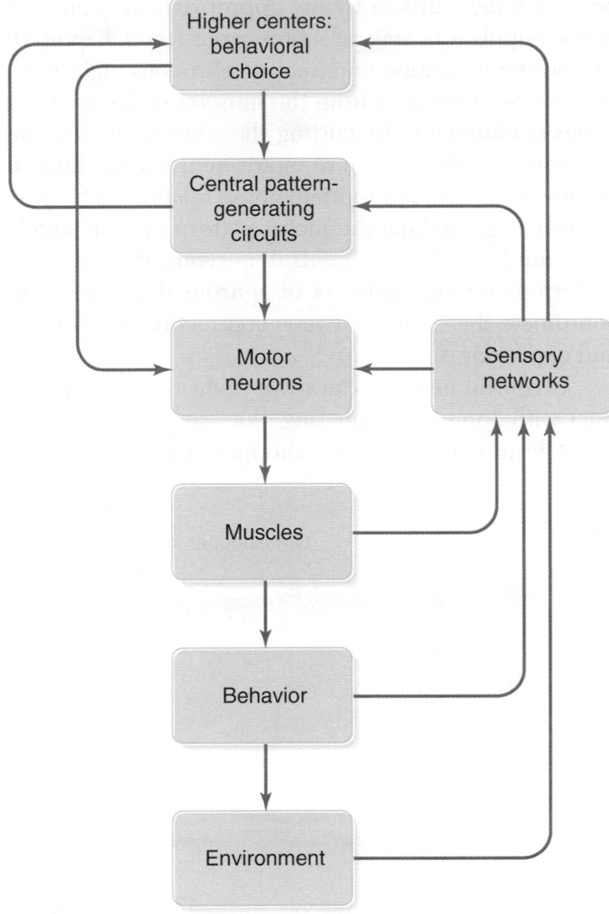

Figure 11-33 Motor control systems are arranged hierarchically. Neurons in the brain can exert control over the entire motor side of the nervous system, generating decisions concerning motor output. These decisions modulate activity within sets of interconnected neurons, called central pattern generators, that activate motor neurons according to more or less preset patterns. Motor neurons provide the only pathway between the nervous system and the muscles. Contraction of the muscles causes behavior, which can change the environment. Sensory input from the environment and from proprioceptors in the muscles and joints can modulate behavior, acting at each level in the hierarchy. Finally, feedback is found at all levels of the hierarchy, potentially shaping the output of each level and of the system as a whole.

Centrally generated motor rhythms

Locomotion and respiration typically consist of rhythmic movements produced by repetitive patterns of muscle contraction. Each phase of such cyclic behavior is both preceded and followed by characteristic activity in motor neurons. The pattern of APs is consistently repeated over and over as the animal produces the behavior. Logically, it would seem that these repetitive acts could depend on moment-to-moment sensory input to the nervous system or, alternatively, on activity in a central pattern generator that operates entirely independently of sensory input. Some combination of these two mechanisms could also be possible. Examination of the regulation of repetitive motor output in many animal systems has revealed that both mechanisms contribute to shaping behavior (see Figure 11-33).

Experiments on repetitive behaviors are frequently carried out using *semi-intact preparations*—that is, animals that can produce recognizable behaviors even though the nervous system has been exposed for electrical recording. In some cases, even an isolated nerve cord can produce all the features of a motor output pattern, suggesting that sensory feedback is not necessary to produce the behavior. The "behavior" produced by an isolated nerve cord is sometimes referred to as "fictive" to distinguish it from the real behavior of the intact animal.

Grasshopper flight: The role of feedback

Centrally generated motor patterns have been most clearly demonstrated in the nervous systems of invertebrates—for example, in the production and control of rhythmic locomotory movements. Grasshopper flight has been studied in intact insects and in semi-intact preparations. It is carried out by synchronous muscles that cause alternate up and down movements of the two pairs of wings, driven by the appropriate sequence of nerve impulses in several motor axons (see Chapter 10 for more information on insect synchronous flight muscles). If sensory input from the muscles or joints of the wings is eliminated by cutting the sensory nerves, the patterns of activity in these motor neurons continue to be produced with appropriate phase relations. This persistence suggests that the motor pattern may be largely generated within the central nervous system by a pattern-generating network of neurons that interact to coordinate the timing between contractions of elevator and depressor muscles.

If central neurons can drive flight muscles to contract with appropriate phasing, does sensory input play any role in controlling grasshopper flight? In normal flight, stretch receptors at the base of each wing are stimulated by the movement of the wings, and they have been found to provide sensory feedback that modifies the motor output, increasing the frequency, intensity, and precision of the rhythm. If these receptors are destroyed, the motor output to the wing muscles slows down and the wingbeat drops to about half its normal frequency, although the phase relations among impulses in the different motor neurons are retained (Figure 11-34a). This phenomenon can be studied in detail in a semi-intact grasshopper (Figure 11-34b). In this situation, the original frequency of the flight rhythm drops when the wing-joint stretch-receptor axons are cut, but it is restored if the nerve roots containing the axons of the receptors are stimulated electrically (Figure 11-34c).

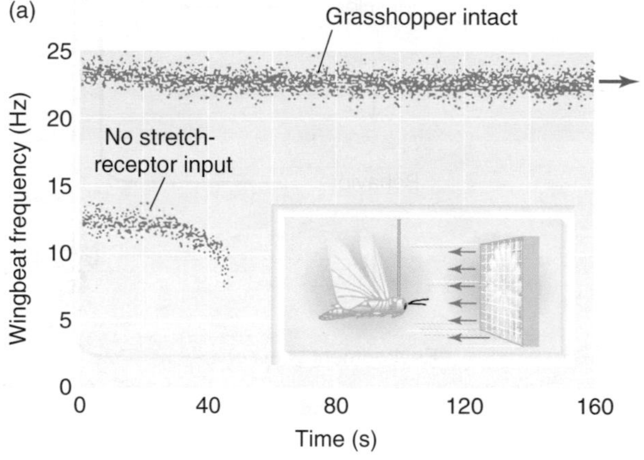

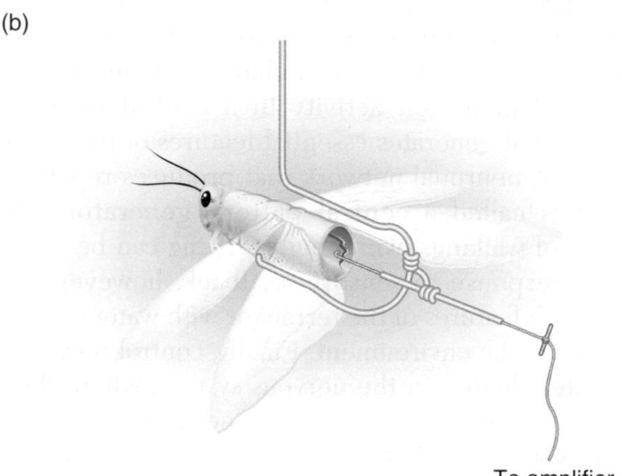

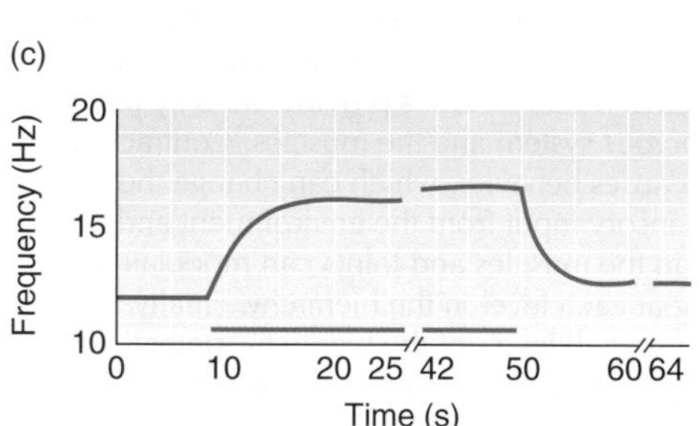

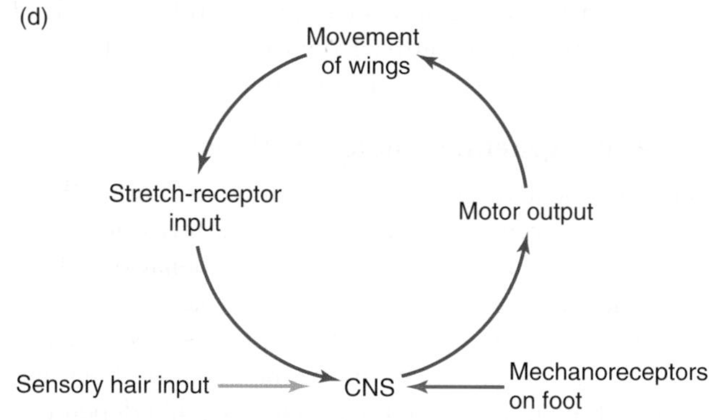

Figure 11-34 Both a central pattern generator and sensory feedback contribute to the production of grasshopper flight. **(a)** In this experiment, an intact grasshopper was mounted in front of a source of wind, and the frequency of the wingbeat was measured before and after sensory afferent axons were cut. When sensory receptor neurons at the base of the wings were destroyed, the central pattern generator produced a low-frequency pattern and the period of flight was shorter, even though air flow over the head was maintained. **(b)** Flight behavior can be studied in a semi-intact grasshopper that is mounted so that it can flap its wings. Flight is initiated by blowing air on facial receptor hairs. Electrodes for recording motor output and for stimulating sensory nerves are fixed in place. Enough of the insect's body remains that flight is easy to observe, but the arrangement allows motor neuronal output and sensory input to be manipulated to measure their effects on the flight pattern.

Classic Work **(c)** Electrically stimulating the stretch-receptor axons in a semi-intact grasshopper increased the frequency of output from the central pattern generator, which would correlate with a faster wingbeat. The time during which the receptor nerve was being stimulated is indicated by the blue line. After the stimulation ceased, the rhythm returned to a low frequency. **(d)** Organization of flight motor output. External sensory input (e.g., a puff of air on the facial receptor hairs) stimulates the flight central pattern generator. Wing movements activate stretch receptors, providing input that stimulates the flight motor. Activity of the muscles produces positive feedback that maintains activity in the flight central pattern generator. Mechanical stimulation of the feet (as when the grasshopper lands) shuts down the flight central pattern generator. [Part a adapted from Pearson and Ramirez, 1997; parts b, c, and d adapted from Wilson, 1964, 1971.]

Interestingly, although the motor rhythm increases in frequency when it is provided with this sensory input, the temporal relation between the rhythm of impulses in the sensory nerves and of the motor output is weak. Even random stimulation of the wing-joint stretch-receptor axons speeds up the motor output, but the sensory input is most effective if it is timed to coincide with a particular *phase* of the wingbeat cycle. Thus, although proprioceptive feedback (feedback from the positions of joints or muscles) is not required for proper phasing of motor impulses to the flight muscles, once the flight central pattern generator becomes activated, sensory feedback reinforces its output.

What turns the flight motor on and off? We get a clue by observing the behavior of an intact grasshopper. When a grasshopper jumps off the substrate to begin flight, sensitive hair receptors on its head are stimulated by its movement through the air. Experiments with semi-intact grasshoppers have shown that this specific sensory input can initiate activity of the flight motor. When an intact grasshopper alights, the flight central pattern generator is turned off by signals from mechanoreceptors on the foot. Thus various forms of sensory information turn on the flight central pattern generator, reinforce its activity, and turn it off, but the essential patterning of flight is controlled by a set of central neurons (Figure 11-34d).

Such endogenous pattern-generating networks have been shown to exist in a number of invertebrate nervous systems. For example, the cyclic motor output that controls the abdominal swimmerets of the crayfish persists not only in an isolated nerve cord, but even in single, isolated abdominal ganglia. In an intact crayfish, this rhythm is initiated and maintained by the activity of "command" interneurons, whose somata are located in the supraesophageal ganglion of the brain. Neurons in each central pattern generator fire APs in periodic groups, called *bursts.* Although the bursting pattern in each abdominal ganglion requires maintained activity in one, or perhaps several, interneurons that are not integral parts of the central pattern generator, there is no simple one-to-one relation between the firing pattern of these interneurons and the pattern of motor output to the swimmerets. The input from the interneurons appears to provide a general level of excitation, which keeps the neurons of the central pattern generators active.

Tritonia swimming: A small network

One very well studied central pattern generator controls escape swimming in the nudibranch mollusk *Tritonia* (Figure 11-35a). This sea slug swims away from noxious

(a)

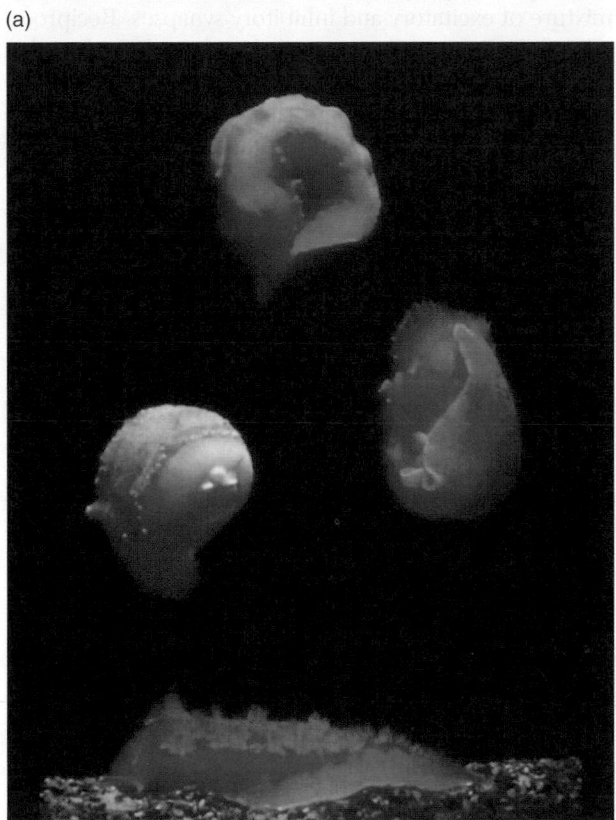

(b)

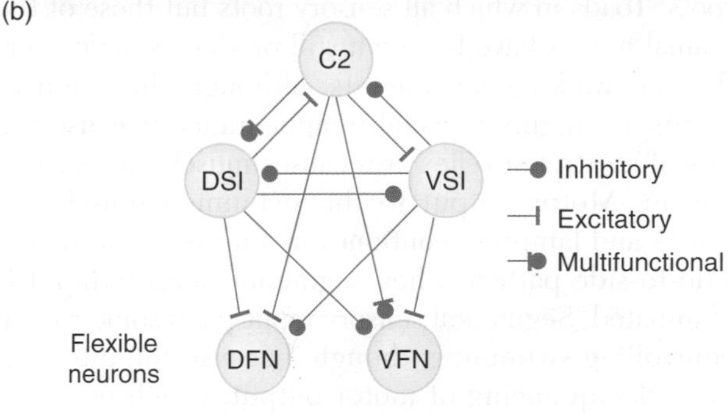

(c)

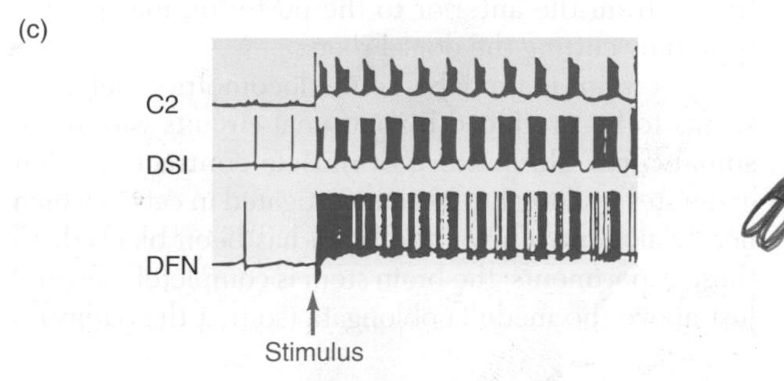

Figure 11-35 Swimming in the nudibranch mollusk *Tritonia* is controlled by a central pattern generator consisting of only three types of neurons. **(a)** If a *Tritonia* is threatened (for example, by a nudibranch-eating starfish), it rises off the substrate and swims by rhythmically contracting dorsal and ventral flexor muscles. **(b)** Three interconnected types of neurons act together to generate the swimming motor pattern. Membrane properties and synaptic interactions determine the swimming motor pattern, which changes if these parameters change. **(c)** Recordings of activity in the swimming central pattern–generating neurons in an isolated brain after electrical stimulation of the pedal nerve, which would carry sensory information from the foot in an intact *Tritonia.* Each type of neuron produces APs in bursts, and the timing of the bursts controls the pattern of muscle contractions in swimming. Abbreviations: C2, cerebral neuron; DSI, dorsal swim interneurons; VSI, ventral swim interneurons; DFN, dorsal flexion neurons; VFN, ventral flexion neurons. [Part a courtesy of P. Katz; parts b and c adapted from Katz et al., 1994.]

stimuli by means of alternating dorsal and ventral flexions of its body, which are produced by alternating contractions of dorsal and ventral flexor muscles. The swimming motor pattern is generated by interconnections among three types of central neurons—a cerebral neuron, the dorsal swim interneurons, and the ventral swim interneurons—all of which synapse onto the dorsal and ventral flexion neurons, which are motor neurons (Figure 11-35b). The cerebral neuron, the dorsal swim interneurons, and the ventral swim interneurons are connected by reciprocal synaptic links, many of which consist of a mixture of excitatory and inhibitory synapses. Reciprocal inhibitory synapses between neurons have been found in many central pattern generators that produce rhythmic outputs. If, for example, the reciprocal inhibitory synapses in the central pattern generator for swimming in *Tritonia* are blocked, the slug is unable to swim. After the initial stimulus, the dorsal and ventral swim interneurons produce alternating bursts of neuronal activity, which alternately activate the dorsal and ventral flexion motor neurons. Intracellular recordings of activity in all five neuron types indicate that the swimming rhythm depends on both the membrane properties of individual neurons and their synaptic connections. Thus, the rhythm is neurogenic, produced by interactions among neurons. Recently, it has been demonstrated that synaptic strengths among the neurons of this network can be modified by neuromodulators during a single episode of swimming to change the properties of the network, even while it is actively producing the swimming output.

Vertebrate central pattern generators

Central pattern generators play important roles in vertebrates, too. Respiratory movements, which are driven by neurons in the brain stem, persist in mammals when sensory feedback from the thoracic muscles is eliminated by cutting the appropriate sensory nerve roots. Toads in which all sensory roots but those of the cranial nerves have been cut still produce simple coordinated walking movements, although these movements are highly unusual in appearance because the loss of all stretch reflex arcs causes muscles to become flaccid. Motor output to the swimming muscles of sharks and lampreys continues in a normal, alternating side-to-side pattern when segmental sensory input is eliminated. Segmental sensory input plays some role in controlling swimming, though, because the intersegmental sequencing of motor output, which normally travels from the anterior to the posterior, may be disrupted by cutting the dorsal roots.

Even in mammals, some locomotory behavior seems to be produced by neuronal circuits within the spinal cord. The pattern of muscle contractions that generates walking has been investigated in cats in which nearly all control from the brain has been blocked. In these experiments, the brain stem is completely severed just above the medulla oblongata (so that the pathways controlling respiration are still intact) and the cat is supported on a treadmill, which moves its legs passively and stimulates muscle stretch and joint position receptor neurons. Such studies reveal that the mechanical sequence of leg movements during walking can be performed without any input from higher brain centers. Moreover, a rudimentary walking rhythm has been seen to continue even after the dorsal roots innervating the limbs have been cut, eliminating sensory feedback. Thus, even in the vertebrates, some aspects of rhythmic movements are programmed into intrinsic connections among neurons within the spinal cord and hindbrain, and patterned motor output continues when sensory feedback and other sensory inputs are disrupted.

Central command systems

Stimulating appropriate individual neurons in the central nervous system can elicit coordinated movements of various degrees of complexity. Electrical stimulation of a region in the ventral nerve cord of a crayfish, for example, causes the animal to assume the stereotypic defense posture, with open claws held high and body arched upward on extended forelegs. Sensory input excites this system through a single identified interneuron, and the output of this interneuron diverges broadly, producing excitation in some motor neurons and inhibition in others. A single neuron that can elicit a behavior by itself is called a **command neuron**, and a small set of neurons that act together to produce a behavior is called a **command system.** Command systems in arthropods characteristically coordinate activity in many muscles simultaneously. That is, if activity in a command neuron or command system causes a particular muscle to contract, muscles that are antagonists to that muscle are simultaneously inhibited and its synergists are excited. Perhaps not surprisingly, the command interneurons that are most effective in eliciting a coordinated motor response are generally least easily activated by simple moderate sensory input.

A crayfish command system

Let's take a closer look at command systems. The crayfish *Procambarus clarkii* produces two different types of escape responses, called *tail-flip responses,* depending on the location of stimulation (Figure 11-36a). In each behavior, at least one giant axon is part of the control circuitry. Giant axons, with their rapid signal conduction, are typical components in the neuronal control of escape responses in many animals. In the crayfish, the *medial* giant interneuron controls flexion to propel the animal backward, and it is active in the response to a touch at the anterior. The *lateral* giant interneuron plays a key role in propelling the animal up and forward, and it is excited by a touch at the animal's posterior. The basic circuitry surrounding the lateral giant interneuron is shown in Figure 11-36b. The neuronal network surrounding the medial giant interneuron is quite differ-

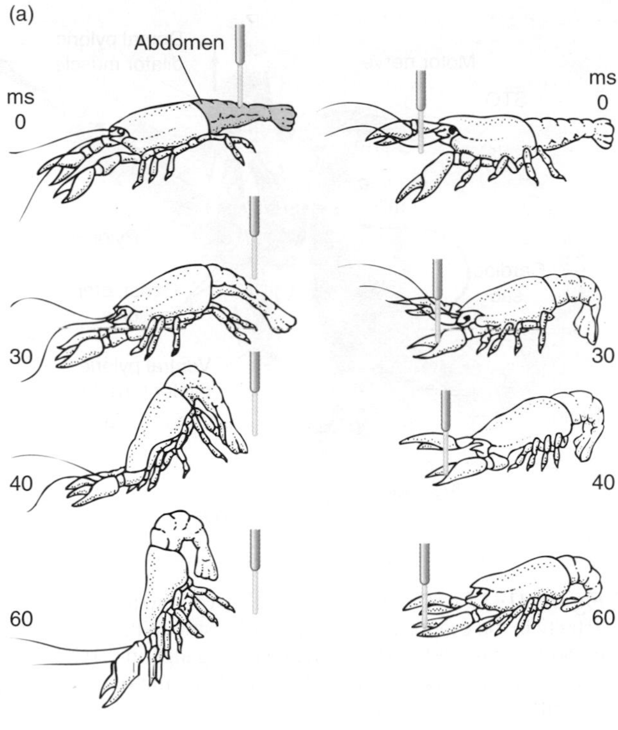

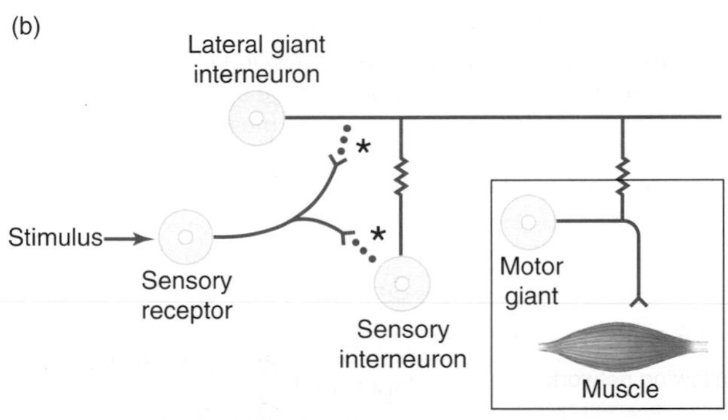

Figure 11-36 Tactile stimulation that produces activity in certain giant interneurons causes a tail-flip escape response in the crayfish. **(a)** Stimulation of the abdomen (upper left) evokes an abdominal flexion that moves the crayfish upward and forward. This behavior is mediated by the *lateral* giant interneuron. Stimulation of the antennae (upper right) evokes an abdominal flexion of a different form that propels the animal backward. This response is mediated by the *medial* giant interneuron. Notice that in both cases, the behavior moves the animal away from the stimulus. Each diagram shows the animal's body shape and location at several times following the stimulation; the time after stimulation is indicated in milliseconds, and time proceeds from top to bottom. **(b)** Simplified diagram of the circuit that allows a crayfish to escape from a touch on the abdomen. Sensory input is carried by chemical (dots) and electrical (resistor symbols) synapses to the lateral giant interneuron, which makes a rapid, electrical synapse onto the motor giant neuron. The motor giant neuron synapses onto the abdominal flexor muscles. The large size of the giant axons produces a high conduction velocity, and the electrical synapses provide rapid communication between neurons. Asterisks indicate the synapses that underlie habituation of this escape response. [Adapted from Wine and Krasne, 1972, 1982.]

ent—on both the input and the output sides of the network—which explains why the behavior is so different when the crayfish is touched on its antenna and its abdomen. The important point about this circuitry is that it produces, by itself, the entire behavioral response.

The discovery of these command interneurons in the crayfish initially caused physiologists to hypothesize that much of an animal's behavior might be controlled by a small population of command neurons, each of which was responsible for producing and shaping a particular behavior. In this case, "choosing" among behaviors would depend on which command neuron was most active. However, further study of the neuronal basis of behavior has suggested that most command functions arise within *networks* of neurons, in which all of the contributing neurons play an important role. To determine experimentally that a neuron fills a command function, it is necessary to show that the activity of that neuron is both *necessary* and *sufficient* for causing the particular motor output. That is, removing that neuron from the network must block, or noticeably modify, the behavior (a test for necessity), and activating only that neuron must produce the behavior (a test for sufficiency).

When the necessity and sufficiency tests are carried out for many neurons that have been found to shape behaviors, three observations are made again and again.

1. *Many neurons are multifunctional,* acting quite differently under different conditions. Some retinal bipolar cells, for example, have been found to carry signals from rod photoreceptors in dim light and from cone photoreceptors in bright light. Their functional synaptic connectivity must shift as the level of ambient light changes.

2. *A single neuron may belong to different levels of a hierarchical control system.* One particular neuron in the *Tritonia* swimming control network, for example, acts both in the central pattern generator for swimming and in the command system for the escape response.

3. *There must be mechanisms that can modify neuronal connectivity,* because networks can change their output depending on circumstances. Anatomic connections may constrain the range of possible outputs for a set of neurons, but functional connections define their output at any given time.

Some sources of behavioral plasticity

One of the best-understood mechanisms for shifting neuronal networks among possible functional configurations is neuromodulation (see Chapter 6). Neuromodulators can cause changes in synaptic efficacy that reconfigure a collection of neurons into an entirely different functional unit without changing any of the anatomic connections. Recognition that "anatomy is *not* destiny" in the nervous system has changed the way in which these systems are analyzed. Here we consider three examples that have been analyzed in sufficient detail to illustrate further organizational principles in command systems.

First, behavioral responses become less probable if stimuli are presented repeatedly; this phenomenon is called **habituation.** If a crayfish is touched repeatedly, at first it responds to every stimulus with escape behavior, but after several minutes of stimulation it fails to respond even if the stimuli continue. Although habituation can depend on functional changes at many different points in a network, it has been found that this particular escape behavior habituates because less neurotransmitter is released from the terminals of the sensory afferent neurons when stimuli are repeated for long periods (indicated by asterisks in Figure 11-36b).

Second, the overall control of the crayfish tail includes a second neuronal pathway, parallel to the one shown in Figure 11-36b, that can also elicit the tail-flip response. The "fast flexor" motor neurons of the second pathway, which are not giant neurons, produce more precise control of the tail flip, although the response is neither as rapid nor as vigorous as the tail flip produced by the motor giant neurons. The slower pathway can operate by itself when the giant neuron pathway is quiet, but when the tail flip is initiated by the motor giant neurons, the fast flexor pathway is activated at the same time. Thus we discover parallel pathways in motor control, just as we saw in sensory processing.

Third, a change in the level of the neuromodulator serotonin in the body fluids of the crayfish dramatically

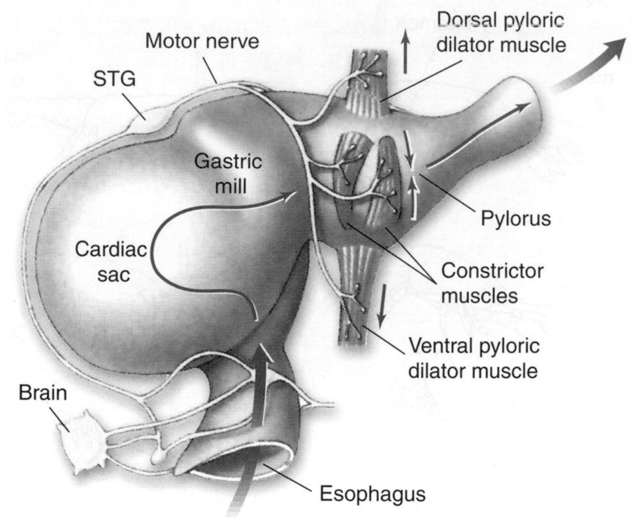

Figure 11-37 The stomatogastric nervous system controls activity in the esophagus, cardiac sac, gastric mill, and pylorus of the lobster. The stomatogastric ganglion (STG) contains only 30 neurons, most of which are motor neurons and all of which have been identified and characterized. The output of these neurons controls the contraction of muscles that cause food to be swallowed, chewed, and moved to the rest of the digestive system. (The muscles that control the pylorus are shown here. Constrictor muscles close the pylorus, preventing food from moving out. Dilator muscles open the pylorus, allowing food to move into the next segment of the digestive system. These muscles receive input from STG neurons.) The blue arrows show the path taken by food as it is pushed through the system. [Adapted from Hall, 1992.]

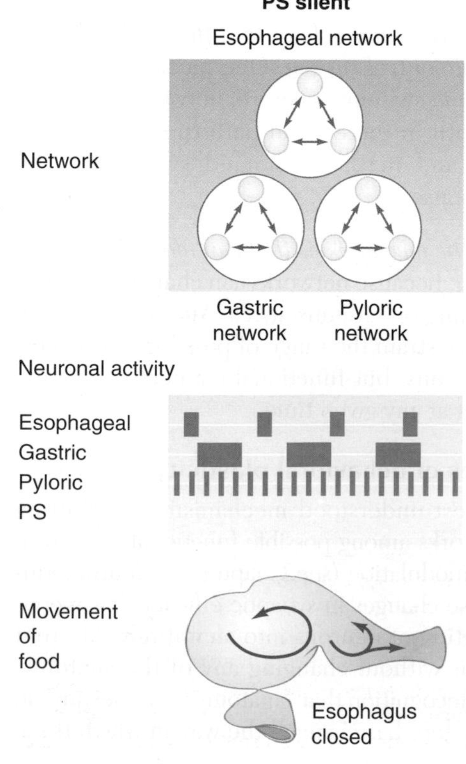

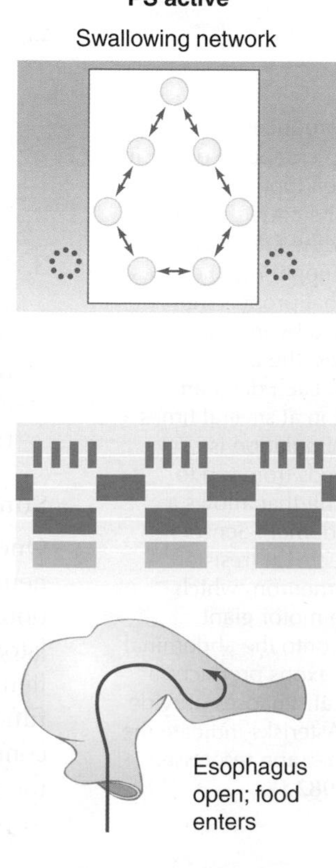

Figure 11-38 Modulatory inputs to the stomatogastric ganglion change the neuronal outputs dramatically, reconfiguring subnetworks in the ganglion. When the modulatory PS neurons are silent, neurons in the STG are organized into three separate subnetworks, the esophageal, gastric, and pyloric networks (top of figure). Each of these subnetworks produces a rhythmic output, but the outputs are not temporally coordinated with one another (middle panel). In this state, food is chewed and moved about within the cardiac sac and the pyloric cavity (indicated by blue arrows in the drawing at bottom left); no food enters or leaves this part of the digestive tract. When the PS neurons are active, neurons of all three subnetworks are recruited into a new network, in which their activity is coordinated to produce "swallowing" (indicated by the blue arrow in the drawing at bottom right). [Adapted from Meyrand, 1994.]

changes its response to stimulation. Injecting serotonin into a submissive crayfish makes it much less likely to back down from a fight; blocking serotonergic synaptic receptors causes an aggressive crayfish to back down from a challenge. This change in behavior depending on the presence of a neuromodulator must be caused by a resulting change in the functional properties of synapses linking sensory and motor neurons and suggests that neuromodulators can powerfully affect activity within command systems.

The crayfish escape response is a typical fixed action pattern, and the neurons that control it illustrate several of the features of command systems outlined earlier. Perhaps the most important feature is the existence of multiple control points within the network, which offer several ways to initiate or to slightly alter the performance of the behavior. This flexibility even within the constraints of a fixed action pattern has been a source of insight into the organization of behavior.

Neuromodulation of motor circuits

The recognition that neuromodulators can change the properties of a neuronal network has opened new avenues of thought. Central command systems, each of which was once believed to drive a single stereotypic behavioral pattern to completion, are now regarded as plastic, with the component neurons taking on different synaptic relations depending on circumstances. Neuromodulation has been found to change the nature of circuits in both vertebrates and invertebrates. A particularly dramatic example of how modification of connections among neurons can reshape output is found in the 30 large neurons that make up the stomatogastric ganglion (STG) of crustaceans. The esophagus and stomach of lobsters and crabs make up a complex structure that is responsible for swallowing, storing, chewing, grinding, and filtering food. There are four functional regions of the stomatogastric system: the esophagus, the cardiac sac, the gastric mill, and the pylorus (Figure 11-37). The neurons of the STG control all of the muscular chambers responsible for ingestion and peristaltic movement of food. They also control the position of the bony teeth in the gastric mill that are responsible for chewing and grinding food. Most neurons in the STG are motor neurons that innervate muscles in the stomatogastric system, but synaptic connections among them also contribute to pattern generation, so the neurons play both motor control and central pattern–generating roles. By studying this small set of neurons, physiologists hope to discover how the functional architecture of each potential subnetwork can be formed by changing the physiological properties of only a few neurons.

The stomatogastric ganglion can be divided into three subnetworks of neurons, which control muscles in the esophageal, gastric mill, and pyloric regions of the stomatogastric system. The esophageal, gastric, and pyloric networks can each generate patterns of rhythmic output that are independent of the other two (Figure 11-38, left). The frequency of output and the pattern of activity in individual neurons within the network are identifying characteristics of each subnetwork.

The presence of a number of neuromodulatory peptides greatly changes the behavior of many neurons within the STG. The ganglion has been found to receive inputs from neurons that release at least 19 different neurotransmitters and neuromodulators, and in addition, neurons within the STG are exposed to at least 18 circulating hormones. Several of these substances change the behavior of the neurons when they are applied experimentally. Figure 11-39 shows a more detailed version of the pyloric network, although it is still somewhat simplified. All of the connections illustrated have been described in an STG bathed in saline with no neuromodulators present. The figure illustrates

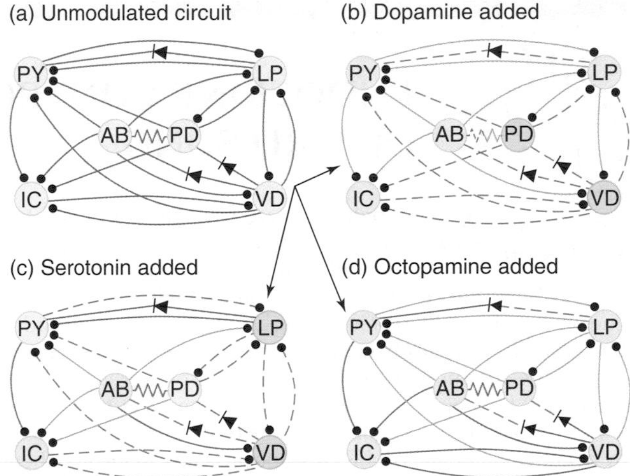

Figure 11-39 Peptide neuromodulators change connectivity within the crustacean stomatogastric ganglion. **(a)** A simplified diagram of the neurons in the pyloric network of the STG. Every connection illustrated in the diagram has been verified electrophysiologically and anatomically. This diagram shows the anatomic connections and the connections that can be recorded when the ganglion is bathed in physiological saline. Each neuron is labeled with the letters by which it is known. Lines represent connections; solid circles represent inhibitory chemical synapses. A resistor symbol represents symmetric electrical synapses, and a triangle plus a line represents a rectifying electrical synapse. **(b)** If dopamine is added to the saline bath, some neurons become more active (shown in green) and others become less active (shown in light orange). In addition, some synapses increase in efficacy (shown in solid green lines), while others become weaker (shown in dashed light orange lines). **(c)** If serotonin is added to the saline bath, a different set of neurons become excited and inhibited. Some neurons fail to respond to the presence of serotonin (shown in blue), but most neurons change their behavior. Notice that serotonin weakens more connections than it strengthens. **(d)** Adding octopamine to the saline bath excites all of the neurons in the pyloric network and strengthens only some of the synaptic connections. Although the neurons are more excited, the strength of several synapses remains unchanged, demonstrating that change in synaptic efficacy depends on more than just a shift in excitability of the neurons. [Adapted from Harris-Warrick et al., 1997.]

the changes in synaptic strengths produced by three common peptides, each of which is released by neurons innervating the STG. Notice that some neurons are excited by the modulator; others are inhibited. In addition, some synaptic connections are strengthened by the modulator and others are weakened. Notice, too, that excitation of a neuron can be accompanied by reduction in its synaptic efficacy, suggesting that the modulator changes membrane properties of the neurons as well as their activity. These shifts in the synaptic connectivity among the neurons of the pyloric network change the timing of activity in the neurons, which in turn changes the pattern of muscle contractions in the pylorus. The same types of modulations have been seen in circuits that control motor output from the vertebrate spinal cord, but these cells are technically much more difficult to study.

In some instances, particular neurons have been identified whose activity changes the behavior of cells in the STG. For example, two electrically coupled neurons, called PS neurons, whose somata lie outside the STG have been found to reconfigure subnetworks in the STG in much the same way that exogenous peptides do. When the PS neurons fire, a valve between the esophagus and the stomach opens and swallowing behavior is initiated. Then an entirely new rhythm begins in the neurons of the STG, coordinating all three parts of the stomatogastric system to produce a set of peristaltic waves that travel from the esophagus to the pylorus (Figure 11-38, right). All other rhythms are inhibited during this behavior; in fact, the three separate pyloric, gastric mill, and esophageal networks cease to exist in functional terms. When activity in the PS neurons ceases, yet another rhythm transiently appears,

ENDOCRINE CONTROL OF BIRD SONG

Many behaviors are strongly controlled by hormones, which act by modifying neuronal function. Courtship and parental behaviors in many species fall into this category, and one well-studied behavior is singing among the passerine birds. Bird song is a complex behavior. In most species, only the male sings a species-appropriate song. Although birds hatch with a propensity for singing, just as humans are thought to be born with a propensity for language, a male bird must learn his song by hearing adult males of his own species early in life and then practicing his song and comparing it with the remembered tutor song when he nears sexual maturity. Deafening the bird so that he cannot hear himself sing at this later stage interferes with song learning. Some male birds change their song each year, learning modifications during the early phases of the breeding season. Females of most species do not sing, although there are exceptions in which a male and female pair sing a duet as a part of their courtship and defense of territory.

Several regions in the avian brain participate in the learning and production of song. In some species, a subset of these regions increase in size before the breeding season and then decrease in size when breeding is over (see part a of the accompanying figure). The increase in size may depend on an increased number of neurons (in the case of the high vocal center, or HVC) or on an increase in the size of individual neurons and the complexity of their dendritic arbors (in the case of the robust nucleus of the archistriatum, or RA). These changes are produced primarily in response to the increased level of circulating testosterone that precedes and lasts through each breeding season *and* to lengthening days in the spring. Interestingly, the responding brain regions grow in a particular sequence once the level of testosterone has risen, suggesting that the growth of some regions depends not only on the hormone but also on input from other regions that have already enlarged (see part b of the figure).

The pattern of hormonal control of bird song is complex. One approach to studying it has been to treat female birds with the male hormone testosterone in an attempt to produce singing. Normal adult female birds cannot be induced to sing in this way. Experimenters reasoned that perhaps the brain of the adult female is already fixed, so instead they treated the brains of developing female birds with testosterone at a very early age. When these birds were treated with testosterone again as adults, they failed to sing. Only when female birds were treated with *estrogen* as young chicks and then treated with testosterone as adults did they

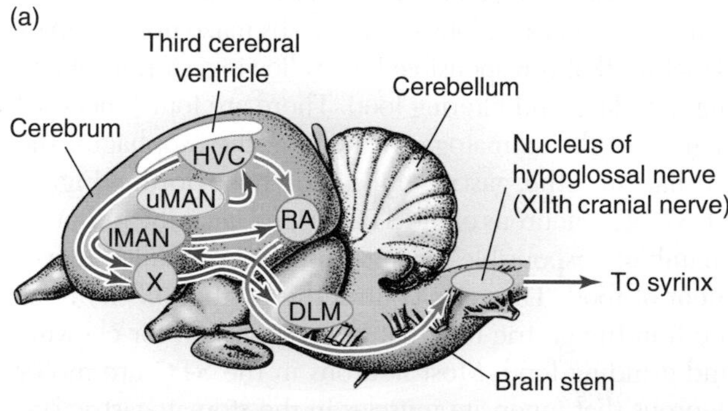

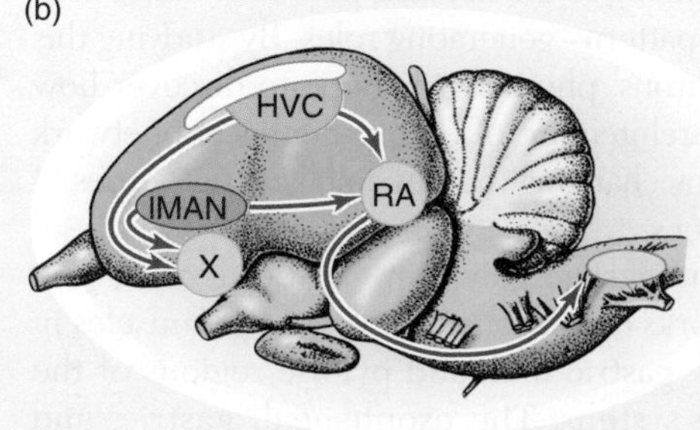

but eventually all of the neurons in the swallowing network return to their original activity patterns, and the three original networks are reconstituted.

The reconfiguration of this collection of 30 neurons into several distinct functional networks suggests a new view of neuronal circuits. Previous work showed that a single anatomically defined network can produce different forms of output in response to neuromodulatory agents, but the stomatogastric system suggests that the functional connections within the network can themselves be plastic. The ability of neuromodulators to functionally "rewire" a set of neurons without changing the anatomic synaptic connections provides a huge increase in the number of possible ways that information can be processed and motor output controlled. Clearly, one challenge will be to discover where the rewiring takes place and how it is regulated.

Many conclusions about signal processing in the mammalian brain are based primarily on observing anatomic connections. How might results showing neuromodulation of circuitry affect these ideas?

Neuroendocrine Control of Behavior

Although the stomatogastric ganglion offers a remarkable opportunity to study the cell-by-cell effects of neuromodulators, some might argue that the activity of a lobster's stomach is not the most exciting of behaviors. In fact, many more complicated behaviors are initiated or controlled by hormones or neuropeptides (Spotlight 11-3). Let's consider one of these as an example. Individuals from different species of American voles,

sing like males (see part c of the figure). Indeed, it is thought that estrogen—produced from testosterone—is the predisposing factor that permits the brains of male birds to respond later in life to testosterone.

A number of issues are being investigated in studies of bird song. Researchers are seeking the mechanisms by which steroid hormones modify the behavior of neuronal circuits. In addition, the annual increase in the number of neurons in the HVC is a striking example of new neurons arising in the adult brain, something that was until recently thought never to occur and is currently an exciting area of research.

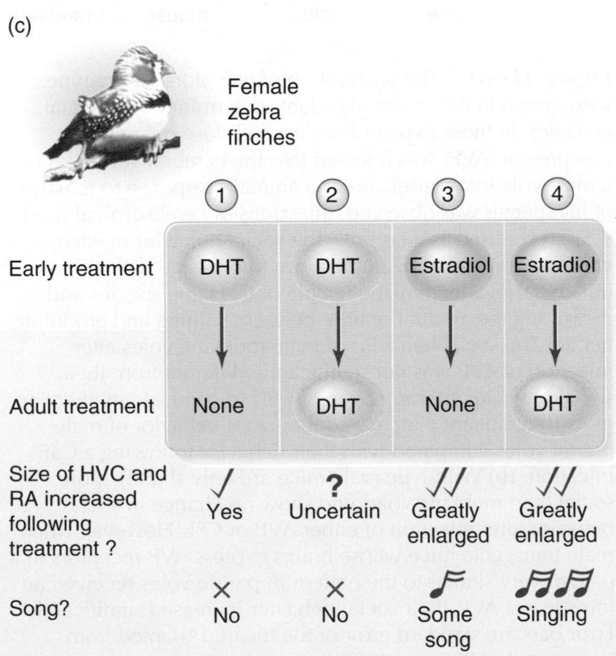

Several centers in the avian brain contribute to the learning and production of song. ***(a)*** *Sagittal section lateral to the midline, showing song-related centers. (Anterior is to the left.) The centers are grouped into three sets, connected by neuronal pathways drawn as arrows. The descending motor pathway to the syrinx (the song-producing organ) is shown in purple and includes the high vocal center (HVC), the robust nucleus of the archistriatum (RA), and the motor nucleus of the XIIth cranial nerve. The pathway shown in red, sometimes called the recursive pathway, plays an important role in the learning of song and includes the magnocellular nucleus of the anterior neostriatum (IMAN), a region called simply "area X" (X), the dorsolateral nucleus of the medial thalamus (DLM), and the RA. Finally, the pathway drawn in blue connects regions of the hypothalamus with the song pathways. It includes the dorsomedial nucleus of the posterior thalamus, the medial magnocellular nucleus of the anterior neostriatum (uMAN), and the HVC.* ***(b)*** *Some, but not all, vocal centers enlarge prior to the breeding season in response to increased circulating testosterone levels and lengthening days. The HVC enlarges earliest (shown in green), and its increased size results from the addition of many new neurons. Days later, area X and the RA (shown in blue) begin to enlarge as neuronal somata increase in size and dendritic complexity increases (causing the cell bodies to become less densely packed). Even later, the nucleus of the XIIth cranial nerve (shown in purple) enlarges. In contrast, the size of the IMAN (shown in gray) remains unchanged.* ***(c)*** *Treatment of very young female chicks with estrogen, but not testosterone, primes them to respond to testosterone treatment in adulthood by singing. In this experiment, young female zebra finches were treated with dihydrotestosterone (DHT), estradiol, or neither. When they became adults, some were treated with testosterone. Only the female birds that had been treated with estradiol as chicks sang well as adults following treatment with testosterone. (Zebra finches do not sing seasonally, so the relationship between this result and the patterns seen in seasonally breeding birds is uncertain.) [Parts a and b adapted from Tramontin and Brenowitz, 2000; part c adapted from Gurney and Konishi, 1980.]*

small mouselike rodents of the genus *Microtus*, display strikingly different levels of social behavior. The highly social prairie vole is monogamous; males help females raise young and defend a territory. In contrast, the related, but more isolationist, montane vole is promiscuous; males play no role in raising offspring, and they do not defend territory. Several lines of evidence suggest that the neuroendocrine peptide arginine vasopressin (AVP) plays a major role in shaping the behavior of these male voles. (Recall from Chapter 9 that vasopressin is also known as antidiuretic hormone (ADH). It was discovered independently by researchers in cardiovascular physiology and renal physiology. Each group named the molecule for its function in their system, and physiologists have continued to use both names interchangeably.) Vasopressin is one of two hormones synthesized by neurons in the hypothalamus of the brain and released in the posterior pituitary; the other peptide released in the posterior pituitary is oxytocin (see Chapter 9). Injecting AVP into the cerebral ventricles of a male prairie vole, a method that delivers the peptide widely throughout the brain, increases his social behavior with respect to female voles, causing him to sniff and to groom females more actively (Figure 11-40a). In contrast, injecting AVP into the cerebral ventricles of a male montane vole has no effect on his social behavior; instead, it increases the time he spends in grooming himself. Injecting oxytocin, which is very similar to AVP, has no effect on males of either species, but increases the probability of pair bonding when it is injected into the cerebral ventricles of female prairie voles.

At least part of the effect of AVP on male voles seems to depend on the location and distribution of receptor molecules for AVP. AVP receptors are much more broadly distributed in the brains of prairie voles than in the brains of montane voles, and in particular, they are found on neurons in parts of the brain that are associated with rewards and pleasure: the nucleus accumbens and amygdala. The importance of the presence and specific location of these receptors was tested by creating transgenic mice that received not only the gene for the prairie vole receptor, but also flanking sequences that regulate where the gene is expressed. One line of these transgenic mice reliably expresses the prairie vole receptor sequence in neurons of the brain, and the receptors are distributed in a pattern very similar to the pattern seen in prairie voles. This pattern is passed on to offspring of the mice. Mice are not very social rodents, and intraventricular injections of AVP have little effect on the social behavior of wild-type male mice. However, injecting AVP into the cerebral ventricles of the transgenic male mice increased their social behavior toward females, as measured by the time they spent sniffing an unfamiliar ovariectomized female and grooming her (Figure 11-40b). Thus, changing the pattern of neuronal expression of just one type of hormone receptor molecule in the brain appears to be capable of changing an entire set of behaviors, turning a nonsocial species into a social one.

Recent discoveries in the field of neuroethology have expanded our view of how behavior is controlled. Anatomically fixed networks of neurons may be functionally changed by the presence of modulatory peptides. In addition, changing the expression pattern of receptor molecules that already are expressed in a species, thus modifying which neurons are affected by a hormone or transmitter that binds to the receptors, can profoundly affect behavior. Our new understanding of how malleable neuronal circuits can be should facilitate progress toward determining how activity in the nervous system produces an animal's many and varied behaviors.

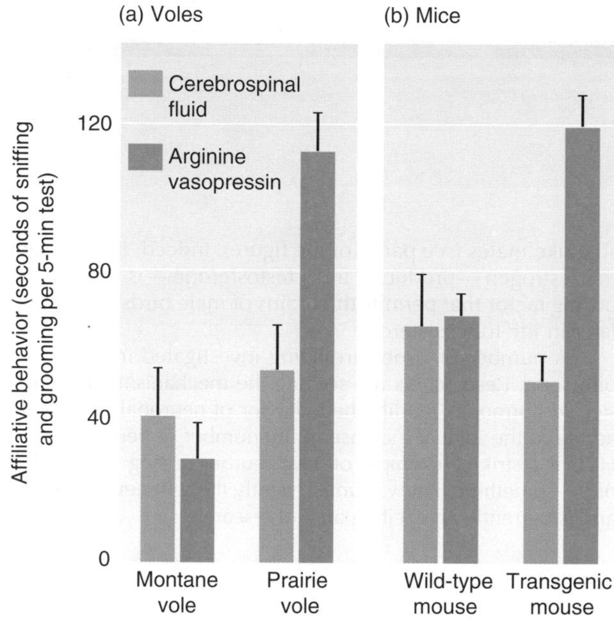

Figure 11-40 The distribution of receptors for arginine vasopressin in the brains of rodents determines their social behavior. In these experiments, a small dose of arginine vasopressin (AVP) was injected into the cerebral ventricles of a male vole (or mouse), and the animal's response to females of his species was observed. Injections of cerebrospinal fluid (CSF) were used as a control. The social behavior of each male was evaluated by putting him into a cage with an unfamiliar ovariectomized female of the same species and measuring the amount of time he spent sniffing and grooming her. **(a)** The social behavior of male montane voles after injection of AVP was not significantly different from their social behavior after injection of CSF. In contrast, an injection of AVP significantly increased the social behavior of male prairie voles compared with their behavior following a CSF injection. **(b)** Wild-type male mice are only slightly more social than montane voles and show no change in social behavior after injection of either AVP or CSF. However, when male transgenic mice whose brains express AVP receptors in a pattern very similar to the pattern in prairie voles received an injection of AVP, their social behavior increased significantly. Error bars are standard error of the mean. [Adapted from Young et al., 1998 and 1999.]

SUMMARY

The neuronal basis of a simple behavior

■ In a simple reflex arc, sensory information is transmitted to the central nervous system by neurons that synapse with motor neurons either directly (forming a monosynaptic reflex arc) or by way of some number of interneurons connected in series, the last of which synapses onto motor neurons. The vertebrate muscle stretch reflex is an example of a monosynaptic reflex arc.

The study of behavior

■ Ethologists typically study the behavior of animals in a natural setting. Neuroethologists study how activity in the nervous system underlies and generates behavior. Neuroethologists typically work in a laboratory and use a variety of experimental methods, including those that allow them to record activity in the nervous system.

■ Fixed action patterns are relatively complex behaviors that are produced repeatedly by an animal in response to particular key stimuli. The stereotypic nature of fixed action patterns has made them a good subject for the study of how complex behavior is produced by the nervous system.

■ Orientation and navigation are examples of complex behaviors that are observed even in relatively simple animals.

■ Echolocation allows bats and some marine mammals to navigate and to locate prey in situations in which vision fails.

Sensory circuits

■ The nervous system processes information and extracts and analyzes features of the environment as a signal is transmitted from one neuron to another.

■ A relatively simple example of feature extraction is the enhancement of contrast at the borders between objects, which takes place by way of lateral inhibition.

■ In vertebrates, visual information is acquired by photoreceptors in the retina. Initial information processing takes place as the information travels from the photoreceptors to the retinal ganglion cells. In birds and mammals, further processing takes place in the lateral geniculate (a nucleus of the thalamus) and continues as the signals are transmitted from one cortical center to another.

■ Most information received by the nervous system is sorted into separate parallel pathways, each of which extracts particular aspects of the signal. Eventually, the results of this parallel sensory processing are reunited at higher centers of the brain.

■ Within the cerebral cortex, and in other parts of the brain as well, sensory information is arranged into topologic maps.

Motor circuits

■ Neuronal centers of motor control are arranged in a hierarchy from higher brain centers to motor neurons. Feedback connections impinge at every level in the hierarchy.

■ Many rhythmic, repetitive behaviors are organized by networks of neurons called central pattern generators.

■ Some behaviors can be elicited by stimulating a single neuron. This powerful class of cells has been called command neurons. To confirm that a neuron is a command neuron, or that a small set of neurons is a command system, neuroethologists must demonstrate that activity in the neuron is both necessary and sufficient for producing the behavior.

■ Neuromodulatory transmitters or hormones can physiologically modify function in anatomically connected circuits and radically change the output of the circuit. This principle applies broadly from very simple systems such as the crustacean stomatogastric ganglion to very complex systems such as the brain circuitry that controls social behavior in rodents.

REVIEW QUESTIONS

1. Describe in detail the components of the vertebrate muscle stretch reflex. Compare it with the stretch reflex in a crayfish.
2. What would happen to your posture if all of your muscle spindles suddenly ceased to function?
3. How do γ-efferent fibers change the sensitivity of muscle spindles?
4. What kinds of behavior are called "fixed action patterns"? What is a "key stimulus"? Give at least one example of each.
5. Describe one mechanism for negative phototaxis.
6. How can a bat capture a moth in total darkness? How does an owl capture a mouse in total darkness?
7. What different kinds of information allow eels, pigeons, and bees to find their way around their environment?
8. Why does the evening sky appear to have a lighter band outlining the silhouette of a mountain range?
9. Describe the anatomic and functional organization of the vertebrate retina.
10. Because the eyes of primates are located on the front of the face, each eye sees approximately the same field. In contrast, the right hemisphere of the brain "sees" the left half of the visual field, whereas the left hemisphere "sees" the right half of the field. What is the circuitry behind this organization?

11. What is meant by the "receptive field" of a cortical neuron?
12. How can the receptive field of a simple cell in the visual cortex be a bar or a straight border, when cells of the lateral geniculate nucleus have circular receptive fields?
13. Discuss two of the general insights into neuronal organization that have resulted from studies of the retina and visual cortex.
14. The nervous system is sometimes compared to a telephone system or a computer. Discuss some properties of the nervous system that make this a good analogy and others that make it a poor analogy.
15. What is a central pattern generator? What are some of the properties of central pattern generators, and what roles do they play in the control of behavior? Describe some examples of central pattern generators.
16. What is a command neuron? What is a command system? Describe some properties and examples of command systems.
17. How can one neuron play different roles in several central pattern generators?
18. What allows neuromodulatory peptides to control behavior?
19. Propose one or more mechanisms by which peptide and steroid hormones might affect an animal's behavior.

SUGGESTED READINGS

Bentley, D., and R. R. Hoy. 1974. The neurobiology of cricket song. *Sci. Am.* 231:34–50. (A classic paper describing this complicated and interesting behavior.) Classic Work

Camhi, J. 1984. *Neuroethology.* Sunderland, MA: Sinauer Associates. (An excellent textbook, which is aging gracefully, summarizing many subfields of this rapidly growing discipline.)

Carew, T. J. 2000. *Behavioral Neurobiology.* Sunderland, MA: Sinauer Associates. (A recent textbook describing many of the principal animals and systems that have been studied by neuroethologists.)

Clarac, F., D. Cattaert, and D. Le Ray. 2000. Central control components of a "simple" stretch reflex. *Trends Neurosci.* 23:199–208. (Puts the stretch reflex into a broader context.)

Dickenson, M. H., C. T. Farley, J. R. Full, M. A. R. Koehl, R. Kram, and S. Lehman. 2000. How animals move: An integrative view. *Science* 288:100–106. (A comprehensive review of locomotion at all levels, from biomechanics to neuronal control. Excellent.)

Dowling, J. 1987. *The Retina: An Approachable Part of the Brain.* Cambridge, MA: Belknap Press. (A description of the structural and functional organization of the vertebrate retina, written by a major contributor to our current knowledge of this remarkable organ.)

Edwards, D. H., W. J. Heitler, and F. B. Krasne. 1999. Fifty years of a command neuron: The neurobiology of escape behavior in the crayfish. *Trends Neurosci.* 22:153–161. (A summary of work that introduced and has refined the concept of command neurons and systems.)

Grillner, S., and P. Wallen. 1985. Central pattern generators for locomotion, with special reference to vertebrates. *Annu. Rev. Neurosci.* 8:233–261. (A review of the properties of central pattern generators, with emphasis on the CPG for swimming in lampreys.)

Gwinner, E. 1986. Internal rhythms in bird migration. *Sci. Am.* 254:84–92. (A biological approach to this otherwise apparently mysterious navigational ability.)

Hubel, D. 1995. *Eye, Brain, and Vision.* New York: Scientific American Library Paperbacks. (An exceedingly enjoyable review of information processing in the visual system, written by one of the most prolific and creative researchers in the field.)

Katz, P. S., and R. M. Harris-Warrick. 1999. The evolution of neuronal circuits underlying species-specific behavior. *Curr. Opin. Neurobiol.* 9:628–633. (A brief paper that takes a broadly comparative approach to the neuronal basis of behavior.)

Knudsen, E. I. 1981. The hearing of the barn owl. *Sci. Am.* 245:113–125. (A discussion of the remarkable auditory nervous system of this bird, including a description of some very creative physiological experimentation.) Classic Work

Konishi, M. 1985. Birdsong: From behavior to neuron. *Annu. Rev. Neurosci.* 8:125–170. (A review of the neuronal basis of the production of bird songs, written by one of the most eminent experts on the physiology of the avian brain.)

McFarland, D. 1993. *Animal Behaviour: Psychobiology, Ethology, and Evolution.* New York: Wiley. (Superb introduction to the study of animal behavior.)

Nelson, R. J. 2000 *An Introduction to Behavioral Endocrinology.* 2d ed. Sunderland, MA: Sinauer Associates. (A textbook that discusses many aspects of this subject with a strong emphasis on vertebrates.)

Tramontin, A. D., and E. A. Brenowitz. 2000. Seasonal plasticity in the adult brain. *Trends Neurosci.* 23:251–258. (A concise review of the changes that take place in the song control centers of seasonally breeding birds.)

Van Essen, D. C., C. H. Anderson, and D. J. Felleman. 1992. Information processing in the primate visual system: An integrated systems perspective. *Science* 255:419–123. (A synthesis of an enormous amount of information regarding visual processing in the primate brain, giving a good sense of how information processing by the brain works.)

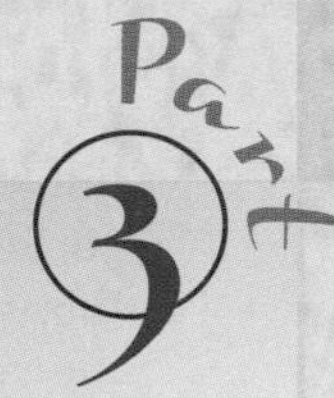

Integration of Physiological Systems

We have discussed the basic principles of animal physiology in Part 1 of this book (Chapters 1–4), followed by a discussion of the nervous, muscular, and endocrine systems and the processes by which they regulate physiological functions in Part 2. (Chapters 5–11). To be discussed in Part 3 (Chapters 12–17) are the various regulated physiological systems that are involved in the day-to-day efforts of animals to acquire and store nutrients and energy, to expel wastes, to respond to changing environments, and to reproduce.

Textbooks in animal physiology historically have treated each of these regulated physiological systems more or less separately, with relatively little focus on their mutual functional and structural interdependencies. This approach persists both for convenience and because, to some extent, it reflects the interests of biologists. Physiologists usually identify themselves as, for example, "cardiovascular physiologists"; few would stress the more integrated aspects of their field by identifying themselves, for example, as "energy transfer physiologists" who study the coordinated transport of nutrients, wastes, and heat between the environment and an animal's interior. Further, because there are similarities between, for example, the circulatory systems of all animals, it is convenient to discuss circulatory systems in a single chapter.

This division of physiological systems into units, so useful in organizing a course or a book, has, however, yielded generations of students with the mistaken impression that animals function as a series of loosely linked physiological systems that happen to be enclosed in a single organism. For that reason, we want to stress that animals operate as networks of integrated systems that are responsive to, and constrained by, their surrounding environment. These integrated systems act in a highly coordinated fashion when faced with environmental stresses that are either physical (such as temperature or pressure) or biological (such as predation or disease).

The actual design and function of an individual physiological system is modified by the constraints placed on it because it is part of a larger physiological network. Because the systems in the network are highly dependent upon one another, environmental stresses may make conflicting demands upon individual systems. It is important to think about these interactions, of which examples abound, in terms of both space and time. Lung vital capacity in some snakes is reduced following ingestion of a large prey item because of space limitations in the visceral cavity. Full lung capacity slowly returns as the meal is digested. A similar interaction between respiration and other body functions in space and time exists in humans after a large meal or during pregnancy. As another example, muscle power responds to physical training over time, but the result is not simply increased muscle mass. There must also be increased blood flow to the muscle, which may require changes in the heart and respiration (interactions between locomotor, cardiovascular, and respiratory systems over time). In addition, the skeletal frame must be strengthened to withstand the increased stress placed on the bones by training routines and stronger muscles.

Although we wish to emphasize the importance of taking an integrated view of the physiology of an animal, we realize that, at the same time, it is not practical to ask a student to learn simultaneously everything about all regulated physiological systems. Thus, we have divided the regulated systems into several different chapters. Although each chapter focuses on a particular system and its functions, examples are used throughout that will emphasize the interactions between the systems and the ways in which they respond in a coordinated manner to environmental change.

Chapters 12 through 14 of Part 3 discuss truly multifunctional systems. The circulation (Chapter 12) is a means of distributing material among tissues, in particular oxygen, carbon dioxide, and various nutrients and waste products. Acquisition of oxygen and elimination of carbon dioxide are the subjects of Chapter 13. The circulatory and respiratory systems of animals function together in maintaining homeostasis by regulating acid-base status and, in some cases, ionic and osmotic conditions within the animal (Chapter 14). Animals use a variety of mechanisms to acquire energy, as described in Chapter 15, which discusses the mechanics, control, and chemistry of food acquisition, digestion, and assimilation. The concluding chapters (Chapters 16 and 17) are, in many ways, a summary of all the themes in the book, delving into the energetics of animals. Energy use in movement, reproduction, growth, and maintenance of homeostasis is explored and placed in the overall evolutionary context of surviving to reproduce.

Circulation

In animals 1 mm or less in diameter, materials are transported within the body by diffusion. In larger animals, adequate rates of transport within the body can no longer be achieved by diffusion alone. In these animals, circulatory systems have evolved to transport respiratory gases, nutrients, waste products, hormones, antibodies, salts, and other materials among various regions of the body. Blood, the medium for transport of such materials, is a complex tissue containing many special cell types. It acts as a vehicle for most homeostatic processes and plays some role in nearly all physiological functions.

This chapter reviews the circulation of blood and how it is controlled to meet the requirements of the tissues. We focus on the mammalian circulatory system because it is the best known. Mammals are very active, predominantly aerobic, predominantly terrestrial animals, and their circulatory system has evolved to meet their particular requirements.

GENERAL PLAN OF THE CIRCULATORY SYSTEM

All circulatory systems comprise the following basic parts, which have similar functions in different animals:

- a main *propulsive organ,* usually a heart, that forces blood through the body
- an *arterial system* that distributes blood and acts as a pressure reservoir
- *capillaries,* in which transfer of materials between blood and tissues occurs
- a *venous system* that acts as a blood storage reservoir and as a system for returning blood to the heart

The heart and the major blood vessels leaving and entering it constitute the **central circulation.** The arterial system, capillaries, and venous system constitute the **peripheral circulation.**

The movement of blood through the body results from any or all of the following mechanisms:

- forces imparted by rhythmic contractions of the heart
- elastic recoil of arteries following filling by the action of the heart
- squeezing of blood vessels during body movements
- peristaltic contractions of smooth muscle surrounding blood vessels

The relative importance of each of these mechanisms in generating blood flow varies among animals. In vertebrates, the heart plays the major role in blood circulation; in arthropods, movements of the limbs and contractions of the dorsally located heart are equally important in generating blood flow. In the giant earthworm *(Megascolides australis),* peristaltic contractions of the dorsal vessel move blood in an anterior direction and fill a series of lateral hearts, which pump blood into the ventral vessel for distribution to the body (Figure 12-1a on the next page). This worm, which can be up to 6 m in length, is divided into segments separated by membranous structures called septa. Tracer studies have shown that the anterior 13 segments, each of which contains two lateral hearts, have a rapid circulation, but the remaining segments, which lack lateral hearts, have a very sluggish circulation. Because of the peristaltic contractions of the dorsal vessel, blood pressure is considerably higher there than in the ventral vessel (Figure 12-1b).

In all animals, valves or septa determine the direction of blood flow, and smooth muscle surrounding blood vessels alters vessel diameter, thereby regulating the amount of blood that flows through a particular pathway and controlling the distribution of blood within the body.

Open Circulations

Many invertebrates have an **open circulation,** in which blood pumped by the heart empties via an artery into an open, fluid-filled space, the **hemocoel,** which lies between the ectoderm and endoderm. The fluid

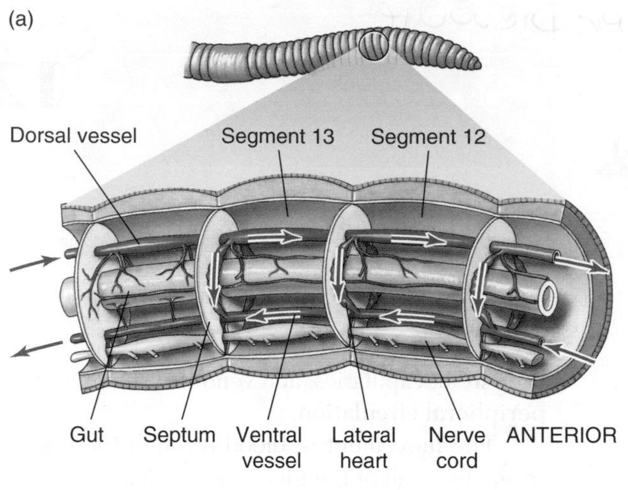

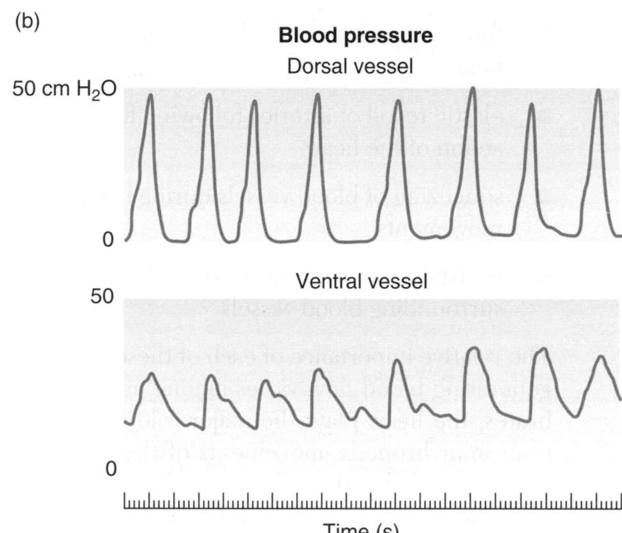

Figure 12-1 In the giant earthworm *(Megascolides australis),* peristaltic contractions of the dorsal vessel and pumping by the lateral hearts are both important in moving blood. **(a)** Blood flows from the dorsal vessel into the lateral hearts, present in the 13 anterior segments, and then is pumped into the ventral vessel. **(b)** Peak blood pressure in the dorsal vessel is about twice as high as in the ventral vessel owing to peristaltic contractions. [Adapted from Jones et al., 1994.]

contained within the hemocoel, referred to as **hemolymph**, or blood, is not circulated through capillaries, but bathes the tissues directly. Figures 12-2a and b illustrate the organization of the main vessels in the open circulatory systems of two groups of invertebrates. The hemocoel is often large and may constitute 20%–40% of body volume. The pressures in open circulatory systems are low, with arterial pressures seldom exceeding 0.6–1.3 kilopascals (kPa), or 4.5–9.7 mm Hg (1 kPa = 7.5 mm Hg). (See Spotlight 12-1 for an explanation of pressure measurement terms.) Higher pressures have been recorded in portions of the open circulation of the terrestrial snail *Helix* and in some bivalve mollusks, but these are exceptional. In snails, high pressures are generated by contractions of the heart; in some bivalve mollusks, high pressures are generated in the foot by contractions of surrounding muscles.

Animals with an open circulation generally have a limited ability to alter the velocity and distribution of blood flow. As a result, changes in oxygen uptake are

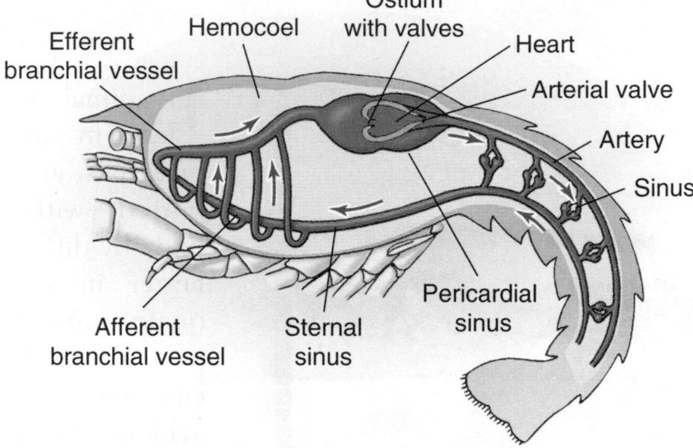

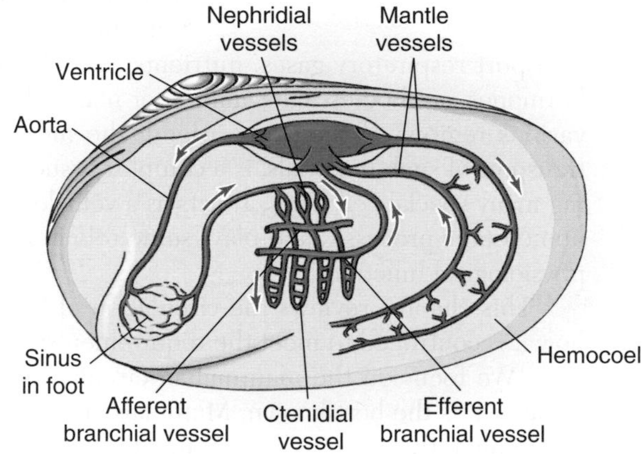

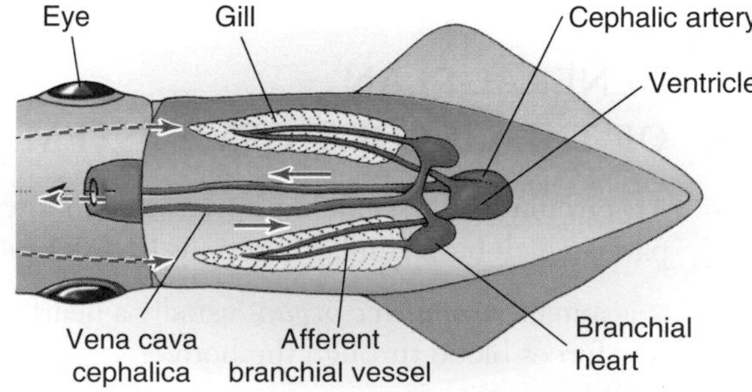

Figure 12-2 Most, though not all, invertebrates have an open circulation. The main blood vessels in the open circulation of **(a)** crayfishes and **(b)** bivalve mollusks empty into a large surrounding space, the hemocoel, which makes up about 30% of total body volume. **(c)** The closed circulation of cephalopods is characterized by a higher blood pressure and more efficient delivery of oxygen compared with open circulatory systems. In all diagrams, only the main blood vessels are shown. The arrows indicate direction of blood flow.

Spotlight 12-1

PRESSURE MEASUREMENT

The unit of pressure, the pascal, is named for Blaise Pascal (1623–1662), the French mathematician, physicist, and philosopher. Pascal stated that pressure differences between two points in a fluid at equilibrium while in a gravitational field are dependent only on the density of the liquid and the difference in elevation. The basic unit of force, the newton, was named for Sir Isaac Newton (1642–1727), born when Pascal was 20 years old. A newton (N) is the force required to produce an acceleration of one meter per second in a mass of one kilogram and is equal to 10^5 dynes. Pressure is force per unit area, and a pascal (Pa) is 1 N/m^2, or 10 $dynes/cm^2$.

The atmospheric pressure, due to the weight of the column of air above us, is 1 atmosphere at sea level, or 101.325 kPa. This is equivalent to the pressure of a column of mercury (Hg) only 760 mm in height, because of the large difference in density between air and mercury. Thus pressure was, and in some cases still is, expressed as mm Hg. One Torr is equal to 1 mm Hg at 0°C and standard gravity, equal to 133.32 Pa. Pressure can be expressed in terms of the height of a column of any liquid. The size of the column, as Pascal pointed out, depends on the density of the liquid. Thus 1 atmosphere of pressure is equivalent to 9.8 m of blood, 10.32 m of water, and 10.06 m of seawater, the height of the column varying with the density of the fluid. Hydrostatic pressure refers to that pressure due solely to the weight of the column of water (or other fluid) above the point of measurement. An animal living in seawater at 100 meters below the surface will be exposed to a pressure of 1 + 10 = 11 atmospheres (the combined pressure of the terrestrial atmosphere and the column of water above the animal), or 1.1 MPa.

Transmural blood pressure is the pressure difference between the inside and outside of a blood vessel. Human arterial blood pressure is usually expressed as a combination of the maximum (systolic) over the minimum (diastolic) pressure associated with each heartbeat, in units of mm Hg. It is usually around 120 mm Hg systolic and 80 mm Hg diastolic; that is, 120/80. This is equivalent to 16/10.66 kPa (1 mm Hg = 0.1333 kPa), 1.63/1.09 m H_2O (1 mm Hg = 1.36 cm H_2O), and 1.55/1.03 m blood (1 mm Hg = 1.29 cm blood). Thus, if an artery were severed, blood might squirt over 1.5 m into the air.

usually slow, and maximal rates of oxygen transfer are low per unit weight. Nevertheless, such animals exert some control over both the flow and distribution of hemolymph; moreover, the blood is distributed throughout the tissues in many small channels. In the absence of such features, even moderate rates of oxygen consumption would be impossible because the distances oxygen would have to diffuse between the hemolymph and the active tissue would be too great. Crabs, lobsters, and crayfishes, for example, exert considerable control over cardiac output and blood distribution and are able to manage fairly high rates of oxygen transfer and levels of exercise.

Insects also have an open circulation, but they do not depend on it for oxygen transport. Instead, they have evolved a tracheal system, in which respiratory gases are transported directly to tissues through air-filled tubes. The tracheal system bypasses the blood, which plays a negligible role in oxygen transport. Consequently, although insects have an open circulation, they have a large capacity for aerobic metabolism. The insect tracheal system is described in Chapter 13.

Closed Circulations

In a **closed circulation**, blood flows in a continuous circuit of tubes from arteries to veins through capillaries. All vertebrates and some invertebrates, such as cephalopods (octopuses, squids), have this type of circulation (Figure 12-2c). In general, there is a more complete separation of functions in closed circulatory systems than in open ones. The blood volume in the closed circulation of vertebrates typically is about 5%–10% of body volume, much smaller than that of open-circulation invertebrates. However, the volume of the extracellular space in vertebrates, expressed as a percentage of body volume, is similar to the hemocoel volume in invertebrates; the closed circulatory system of vertebrates is a specialized portion of that extracellular space.

In a closed circulation, the heart is the main propulsive organ, pumping blood into the arterial system and maintaining a high blood pressure in the arteries. The arterial system, in turn, acts as a pressure reservoir, forcing blood through the capillaries. In the capillary system, or **microcirculation**, the vessel walls are thin, thus allowing high rates of transfer of material between blood and tissues by diffusion, transport, or filtration. Each tissue has many capillaries, so that each cell is no more than two or three cells away from a capillary. Capillary networks have many branches running in parallel, allowing fine control of blood distribution and, therefore, oxygen delivery to tissues. Animals with a closed circulation can increase oxygen delivery to a tissue very rapidly. For this reason, squids, unlike many other invertebrates, can swim rapidly and maintain high rates of oxygen uptake; that is, their closed circulation permits sufficient flow and efficient enough distribution of hemolymph to the muscles to support short bursts of intense activity.

The blood is under sufficiently high pressure in a closed circulation to permit the ultrafiltration of blood in the tissues, especially the kidneys. **Ultrafiltration** refers to the separation of an ultrafiltrate—a fluid devoid of colloidal protein particles—from blood plasma by filtration though a semipermeable membrane (the capillary wall) using pressure (blood pressure) to force the fluid through the membrane. Ultrafiltration occurs in most vertebrate kidneys, resulting in the net movement of a protein-free fluid from the blood into the kidney tubules. In general, capillary walls are permeable, so when blood pressures are high, fluid slowly filters across the walls and into the spaces between cells. In vertebrates, a **lymphatic system** has evolved to recover fluid lost to tissues from the blood in this manner and return it to the venous system. The extent of filtration depends largely on the blood pressure and the permeability of the capillary wall. The permeability of capillaries varies among tissues, and the blood pressure varies with circumstances and also with the organization of the vessels in the tissue. For example, in the liver, high permeability permits rapid transfer of substrates and products of metabolism, and pressures are lower than in the rest of the body. Low pressure in the lung capillaries reduces filtration into the gas space of the lungs, which would impair gas transfer.

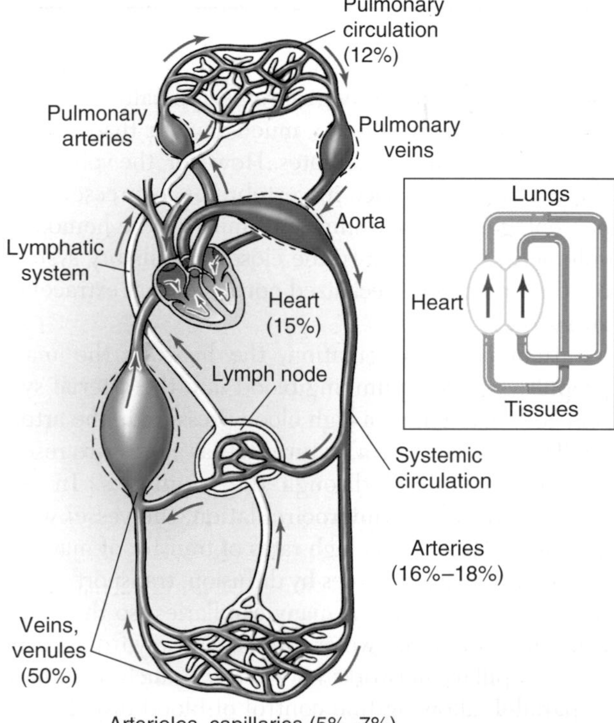

Figure 12-3 The closed circulation in mammals includes a fully divided heart, which permits different pressures to be maintained in the pulmonary and systemic circuits. The inset shows the general pattern of the mammalian circulation. The diagram illustrates the main components of the mammalian circulation, with oxygenated blood shown in red and deoxygenated blood shown in blue. The associated lymphatic system (yellow) returns fluid from the extracellular space to the bloodstream via the thoracic duct. The percentages indicate the relative proportion of blood in different parts of the circulation.

In vertebrates, the circulatory system is divided into a **systemic circuit** (which circulates blood throughout the body) and a **respiratory** or **pulmonary circuit** (which circulates blood to the organs of gas exchange). Mammals can maintain different pressures in these two circuits because they are equipped with a completely divided heart (Figure 12-3). The right side of the heart pumps blood through the pulmonary circuit, and the left side pumps blood through the systemic circuit. Note that the flows in the pulmonary and systemic circuits must be equal because blood returning from the lungs is pumped around the body. In other vertebrates, the heart is not completely divided, and flow to the lung can be varied independently of body blood flow.

The venous system collects blood from the capillaries and delivers it to the heart via the veins, which are typically low-pressure, flexible structures. Large changes in blood volume have little effect on venous pressure. Thus, the venous system contains most of the blood and acts as a large-volume reservoir. Blood donors give blood from this reservoir, and since there is little change in pressure as the venous volume decreases, the volumes and flows in other regions of the circulation are not markedly altered.

THE HEART

Hearts are valved, muscular pumps that propel blood around the body. Hearts consist of one or more muscular chambers connected in series and guarded by valves or, in a few cases, sphincters (e.g., in some molluscan hearts), which allow blood to flow in only one direction. The mammalian heart has four chambers: two **atria** and two **ventricles.** Contractions of the heart result in the ejection of blood into the circulatory system. Multiple heart chambers permit stepwise increases in pressure as blood passes from the venous to the arterial side of the circulation (Figure 12-4).

Vertebrate cardiac muscle fibers are similar to skeletal muscles in many respects (see Chapter 10). Their distinguishing feature is the presence of gap junctions, which means that they are electrically coupled to one another. Except for differences in the uptake and release of Ca^{2+}, the mechanisms of contraction of vertebrate skeletal and cardiac muscle are generally similar. The myocardium (i.e., heart muscle) consists of three types of muscle fibers, which differ in size and functional properties:

- The myocardial cells in the sinus node (or sinoatrial node) and in the atrioventricular node are often smaller than others. They are autorhythmic, only weakly contractile, and exhibit very slow electrical conduction between cells.
- The largest myocardial cells, found in the inner surface of the ventricular wall, are also weakly contractile but are specialized for fast electrical

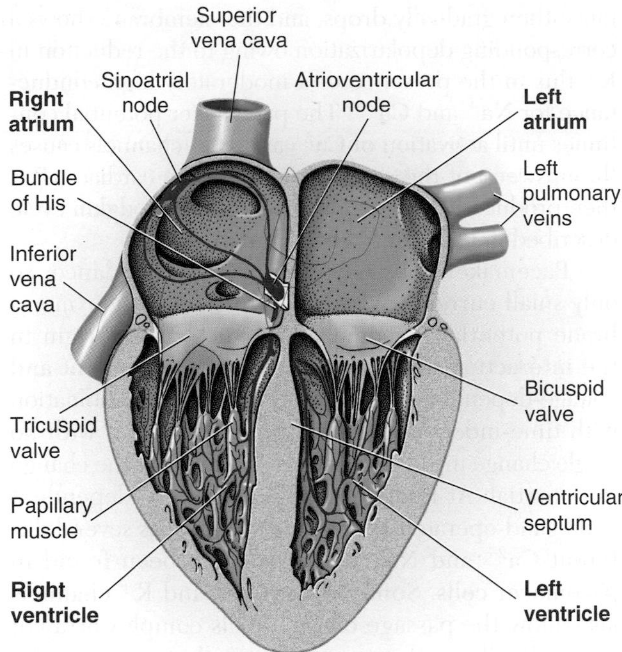

Figure 12-4 The multi-chambered mammalian heart permits the pressure to increase as blood moves from the venous to the arterial side. This cutaway view depicts the rear portion of the human heart, with the impulse pathways shown as blue lines. Impulses originate in the pacemaker region, located in the sinoatrial node, and spread to the atrioventricular node, from which they are transmitted to the ventricles. [Adapted from E. F. Adolph, 1967. Copyright © 1967 by Scientific American, Inc. All rights reserved.]

conduction and constitute the system for spreading excitation over the heart.

- The intermediate-sized myocardial cells are strongly contractile and constitute the bulk of the heart.

Why do you think so many species evolved a single large multi-chambered heart rather than a series of smaller hearts distributed throughout the circulation?

Electrical Properties of the Heart

A heartbeat consists of a rhythmic contraction (**systole**) and relaxation (**diastole**) of the whole cardiac muscle mass. The contraction of each cell is associated with an action potential (AP) in that cell. Electrical activity, initiated in the pacemaker region of the heart as described below, spreads over the heart from one cell to another because the cells are electrically coupled via gap junctions. The nature and extent of coupling determine the pattern by which the electrical wave of excitation spreads over the heart and also influence the rate of conduction.

Neurogenic and myogenic pacemakers

In fishes, the pacemaker is situated in the *sinus venosus* (see Figure 12-16); in other vertebrates, it is situated in a remnant of the sinus venosus called the **sinoatrial node** (see Figure 12-4). The **pacemaker** consists of a group of small, weakly contractile, specialized muscle cells that are capable of spontaneous activity. These cells may be either neurons, as in the neurogenic pacemaker in many invertebrate hearts, or muscle cells, as in the myogenic pacemaker in vertebrate and some invertebrate hearts. Hearts are often categorized by the type of pacemaker they have and hence are called either *neurogenic* or *myogenic* hearts.

In many invertebrates, it is not clear whether the pacemaker is neurogenic or myogenic. However, the hearts of decapod crustaceans (shrimps, lobsters, crabs) are plainly neurogenic. In these animals, the cardiac ganglion, situated on the heart, acts as a pacemaker. If the cardiac ganglion is removed, the heart stops beating, although the ganglion continues to be active and shows intrinsic rhythmicity. The cardiac ganglion consists of nine or more neurons (depending on the species), divided into small and large cells. The small cells act as pacemakers and are connected to large follower cells, which are all electrically coupled. Activity from the small pacemaker cells is fed into and integrated by the large follower cells and then distributed to the heart muscle. The crustacean cardiac ganglion is innervated by excitatory and inhibitory nerves originating in the central nervous system (CNS); these nerves can alter the rate of firing of the ganglion and therefore the rate at which the heart beats. Hormones released into the hemolymph can also modulate cardiac activity.

Vertebrate, molluscan, and many other invertebrate hearts are driven by myogenic pacemakers. These tissues have been studied extensively in a variety of species. A myogenic heart may contain many cells capable of pacemaker activity, but because all cardiac cells are electrically coupled, the cell (or group of cells) with the fastest intrinsic activity is the one that stimulates the whole heart to contract and determines the heart rate. These pacemaker cells normally overshadow those with slower pacemaker activity; however, if the normal pacemaker stops for some reason, the other pacemaker cells take over, producing a new, lower heart rate. Thus, cells with the capacity for spontaneous electrical activity may be categorized as pacemakers and latent pacemakers. In the event that a latent pacemaker becomes uncoupled electrically from the pacemaker, it may beat and control a portion of cardiac muscle—generally an entire chamber—at a rate different from that of the normal pacemaker. Such an **ectopic pacemaker** is dangerous because it desynchronizes the pumping action of the heart chambers.

Cardiac pacemaker potentials

An important characteristic of pacemaker cells is the absence of a stable resting potential. After each action

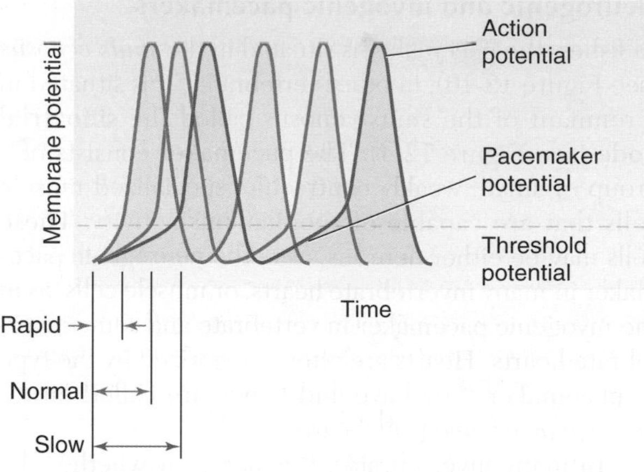

Figure 12-5 Pacemaker cells undergo a spontaneous depolarization of the membrane, referred to as the pacemaker potential, which triggers cardiac action potentials autorhythmically (red curve). A more rapid depolarization (green curve) increases the firing rate, and thus the heart rate, whereas a slower depolarization (blue curve) slows the firing rate and heart rate.

potential, the membrane of a pacemaker cell undergoes a steady depolarization, termed a *pacemaker potential.* The pacemaker potential brings the membrane to the threshold potential, usually in less than a second, giving rise to another all-or-none cardiac action potential. The interval between cardiac APs, which of course determines the heart rate, depends on the rate of the pacemaker potential, as well as on the extent of repolarization and the threshold potential for the cardiac AP. A faster pacemaker depolarization brings the membrane to a firing level sooner and thus increases the frequency of firing, leading to a faster heart rate, whereas a slower depolarization does the opposite (Figure 12-5).

In the frog sinus venosus, the pacemaker potential begins immediately after the previous AP, when the K^+ conductance of the membrane is high. The K^+ conductance then gradually drops, and the membrane shows a corresponding depolarization owing to the reduction in K^+ flux in the presence of a moderately high conductance for Na^+ and Ca^{2+}. The pacemaker potential continues until activation of Ca^{2+} and Na^+ channels causes the upsweep of the action potential. The cardiac AP is then produced by the completion of the Hodgkin cycle described in Chapter 5 (see Figure 5-24).

Pacemaker cells have high input impedance, so only small currents are needed to change their membrane potential. Pacemaker activity has its origin in the interaction between several time-dependent and voltage-dependent membrane currents in combination with time-independent background currents, with no single change in conductance accounting for the change in potential. At least six time- and voltage-dependent and ligand-operated K^+ channels, as well as several different Ca^{2+} and Na^+ channels, have been found in pacemaker cells. Some of the Ca^{2+} and K^+ channels also allow the passage of Na^+. This complex array of channels allows the pacemaker cells to initiate the heartbeat and incorporates multiple safety factors. As a result, the pacemaker cells can function for many years without interruption.

Acetylcholine, which is released from parasympathetic terminals of the vagus nerve (Xth cranial nerve) that innervate the heart, slows the heartbeat by increasing the K^+ conductance and reducing the Ca^{2+} conductance of pacemaker cells. The increase in K^+ conductance keeps the membrane potential near the K^+ equilibrium potential for a longer time, thereby slowing the pacemaker potential and delaying the onset of the next upstroke (Figure 12-6a). Adenosine also slows the heart via a mechanism similar to that of acetylcholine by modifying K^+ conductance. In contrast, norepinephrine released from sympathetic nerves accelerates the pacemaker potential, thus increasing the heart rate (Figure 12-6b). Catecholamines, such as

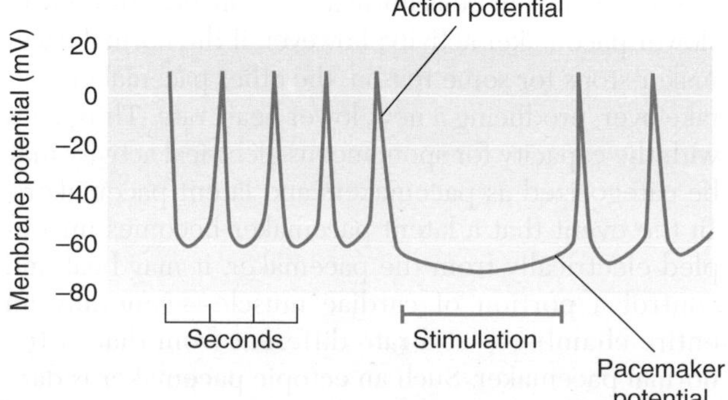

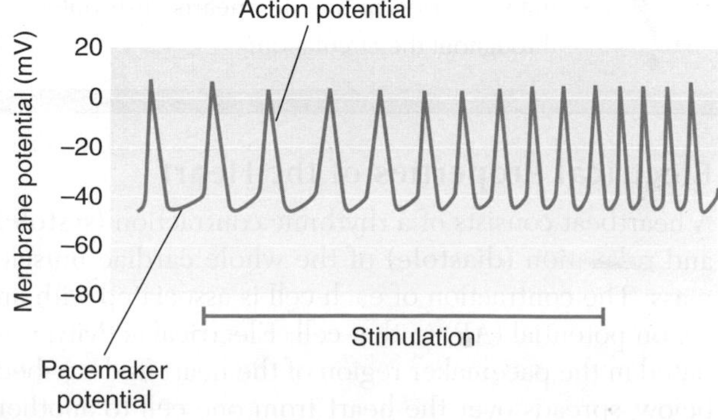

Classic Work *Figure 12-6* Parasympathetic stimulation via the vagus nerve and sympathetic stimulation have opposite effects on the pacemaker potential and heart rate. **(a)** Vagus stimulation produces a rise in diastolic (resting) transmembrane potential, a decrease in the rate of depolarization, and a decrease in the duration and frequency of the action potential. **(b)** Sympathetic stimulation produces an increase in the frequency of firing of the pacemaker cells. [From Hutter and Trautwein, 1956.]

norepinephrine, bind to β-adrenergic receptors on the cell surface, activating adenylate cyclase and increasing the levels of cyclic AMP (cAMP) in the pacemaker cell. The increase in cAMP activates cation channels triggered by Ca^{2+}, K^+, and hyperpolarization, as well as the Na^+/K^+ pump. The overall effect is a more rapid depolarization of the pacemaker cell and an increase in heart rate. Some of these same channels can be activated by an increase in the level of phosphorylation, which increases the intrinsic heart rate and results in a greater sensitivity to β-adrenergic agonists.

Cardiac action potentials

The action potentials that precede contraction in all vertebrate cardiac muscle fibers are of longer duration than those in skeletal muscle. In skeletal muscle, the AP is completed and the membrane is in a nonrefractory state before the onset of contraction; hence, repetitive stimulation and tetanic contraction are possible (Figure 12-7a). In cardiac muscle fibers, however, the AP reaches a plateau at which it remains for hundreds of milliseconds (Figure 12-7b), and the membrane remains in a refractory state until the heart has returned to a relaxed state. Thus, summation of contractions does not occur in cardiac muscle.

In most cardiac fibers, APs begin with a rapid depolarization that results from a large and rapid increase in Na^+ conductance. This depolarization differs from the slow depolarization of the pacemaker potential, which depends on a stable Na^+ conductance and decreasing K^+ conductance. Repolarization of the plasma membrane is delayed for hundreds of milliseconds during the so-called plateau phase (see Figure 12-7b). The long duration of the cardiac AP produces a prolonged contraction, so that an entire heart chamber can fully contract before any portion begins to relax—a process that is essential for efficient pumping of blood.

The prolonged plateau of the cardiac AP results from the maintenance of a high Ca^{2+} conductance and a delay in the subsequent increase in K^+ conductance (unlike the situation in skeletal muscle). The inflow of Ca^{2+} is especially important in lower vertebrates, in which a considerable proportion of the Ca^{2+} essential for activation of contraction enters through the plasma membrane. In birds and mammals, the surface-to-volume ratio of the larger cardiac muscle cells is too small to allow sufficient entry of Ca^{2+} to fully activate contraction. Most of the Ca^{2+} required to trigger contraction is released from the sarcoplasmic reticulum (extensive in cardiac muscle fibers of higher vertebrates) by calcium-mediated calcium release (see Chapter 10). A rapid repolarization, due to a fall in Ca^{2+} conductance and an increase in K^+ conductance, terminates the plateau phase.

The duration of the plateau phase and the rates of depolarization and repolarization vary among different cells of the same heart. The summation of these changes can be recorded by an **electrocardiogram** (ECG). Because of the large number of cells involved, the currents that flow during the synchronous activity of cardiac cells can be detected as small changes in potential from points all over the body, and can be easily recorded and then analyzed. Such a recording shows a characteristic pattern of electrical activity (Figure 12-8a on the next page). The initial P-wave is associated with depolarization of the atrium, the so called QRS complex with depolarization of the ventricle, and the T-wave with repolarization of the ventricle. The electrical activity associated with atrial repolarization is obscured by the much larger QRS complex. The exact form of the electrocardiogram varies by species and is affected by

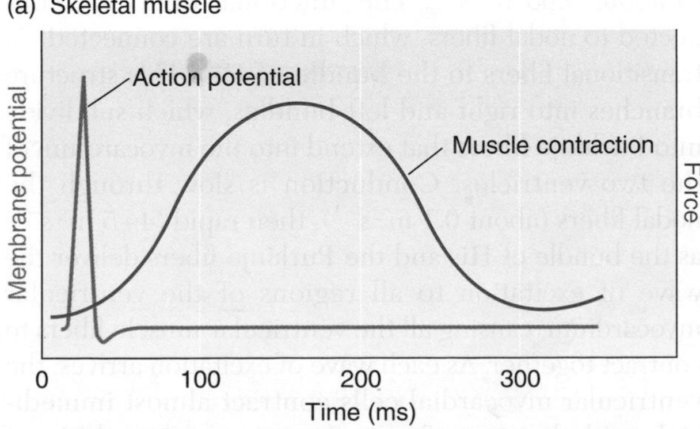

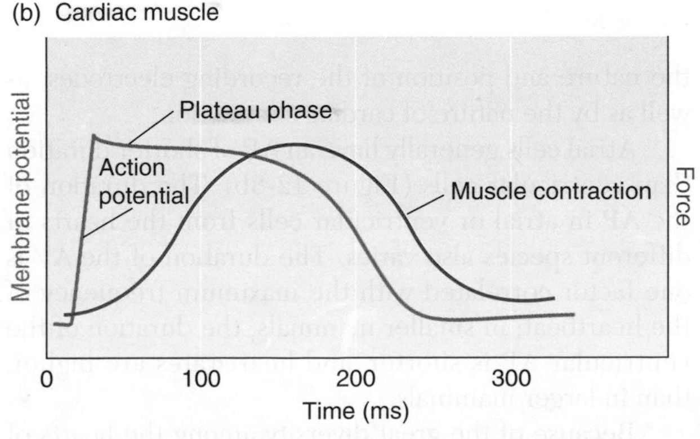

Figure 12-7 Action potentials differ in skeletal and cardiac muscle. **(a)** Action potentials in skeletal muscle are of very short duration. **(b)** In contrast, cardiac action potentials exhibit a prolonged repolarization, or plateau phase, during which the muscle fiber is refractory to stimulation. For this reason, repetitive stimulation during a contraction and summation of contractions can occur in the skeletal muscle, but not in cardiac muscle.

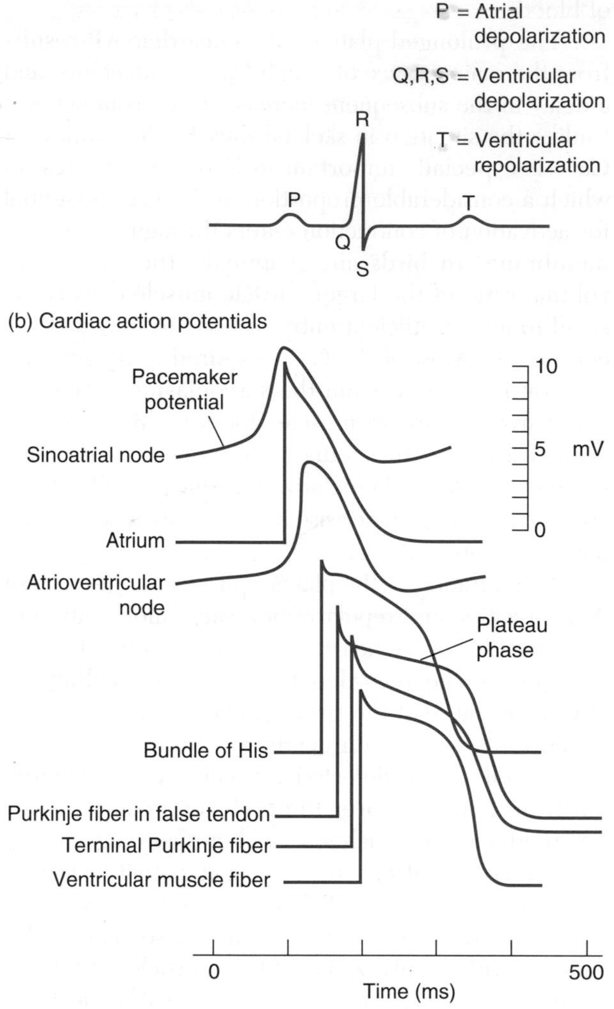

Classic Work **Figure 12-8** An electrocardiogram represents the summation of the electrical activity in various parts of the heart. **(a)** The major components of the electrocardiogram reflect atrial depolarization (P), ventricular depolarization (QRS), and ventricular repolarization (T). **(b)** The amplitude, configuration, and duration of cardiac action potentials differ at various sites in the heart. The sites shown fire in sequence from top to bottom. [Part b from Hoffman and Cranefield, 1960.]

the nature and position of the recording electrodes, as well as by the nature of cardiac contraction.

Atrial cells generally have an AP of shorter duration than ventricular cells (Figure 12-8b). The duration of the AP in atrial or ventricular cells from the hearts of different species also varies. The duration of the AP is one factor correlated with the maximum frequency of the heartbeat; in smaller mammals, the duration of the ventricular AP is shorter, and heart rates are higher, than in larger mammals.

Because of the great diversity among the hearts of different invertebrates, few generalizations can be made about the ionic mechanisms that generate the cardiac APs of these phyla. The one widespread characteristic is participation of Ca^{2+} in the initiation of the action potential.

Transmission of excitation over the heart

Electrical activity initiated in the pacemaker region is conducted over the entire heart as depolarization in one cell results in the depolarization of neighboring cells by virtue of current flow through gap junctions. These junctions are situated where neighboring myocardial cells are closely apposed, in a region termed the intercalated disk (see figure 10-48). The adhesion of cells at intercalated disks is strengthened by the presence of anchoring structures called desmosomes. The area of contact is increased by folding and interdigitation of the plasma membranes, the extent of which increases during development of the heart. Gap junctions allow current to flow with minimal resistance from one cell to the next across intercalated disks.

Although the junctions between myocardial cells can conduct electric current in both directions, transmission is generally unidirectional because the impulse spreads away from the pacemaker region. There are usually several pathways for the excitation of any single cardiac muscle fiber, since intercellular connections are numerous. If a portion of the heart becomes nonfunctional, the wave of excitation can easily flow around that portion, so that the remainder of the heart can still be excited. The prolonged nature of cardiac APs ensures that these multiple connections do not result in multiple stimulation and a reverberation of activity in cardiac muscle. An AP initiated in the pacemaker region results in a single AP being conducted through all the other myocardial cells, and another AP from the pacemaker region is required for the next wave of excitation.

In the mammalian heart, the wave of excitation begins in the sinoatrial node and spreads over both atria in a concentric fashion at a velocity of about $0.8\ m \cdot s^{-1}$. The atria are connected electrically to the ventricles only through the **atrioventricular node** on the right side of the heart; in other regions, the atria and ventricles are joined by connective tissue that does not conduct the wave of excitation (see Figure 12-4). Excitation spreads to the ventricle through small junctional fibers, in which the velocity of the wave of excitation is slowed to about $0.05\ m \cdot s^{-1}$. The junctional fibers are connected to nodal fibers, which in turn are connected via transitional fibers to the **bundle of His.** This structure branches into right and left bundles, which subdivide into **Purkinje fibers** that extend into the myocardium of the two ventricles. Conduction is slow through the nodal fibers (about $0.1\ m \cdot s^{-1}$), then rapid ($4–5\ m \cdot s^{-1}$) as the bundle of His and the Purkinje fibers deliver the wave of excitation to all regions of the ventricular myocardium, causing all the ventricular muscle fibers to contract together. As each wave of excitation arrives, the ventricular myocardial cells contract almost immediately, with the wave of excitation passing at a velocity of $0.5\ m \cdot s^{-1}$ from the internal lining of the heart wall (**endocardium**) to the external covering (**epicardium**). The functional significance of the electrical organization of the myocardium is its ability to generate separate, synchronous contractions of first the atria and then the ventricles. Thus, slow conduction through the atrioventricular node allows atrial contractions to precede ventricular contractions, providing time for blood to move from the atria into the ventricles.

As noted earlier, acetylcholine (ACh), released from parasympathetic nerve fibers, increases the interval between APs in pacemaker cells and thus slows the heart rate (see Figure 12-6a). This decrease in heart rate is sometimes referred to as a *negative chronotropic effect.* Parasympathetic cholinergic fibers in the vagus nerve innervate the sinoatrial node and atrioventricular node of the vertebrate heart. As the heart rate slows, acetylcholine also reduces the velocity of conduction from the atria to the ventricles through the atrioventricular node. High levels of acetylcholine block transmission through the atrioventricular node, so that only every second or third wave of excitation is transmitted to the ventricle. Under these unusual conditions, the atrial contraction rate will be two or three times that in the ventricle. Alternatively, high levels of acetylcholine may completely block conduction through the atrioventricular node (called *atrioventricular block*), allowing an ectopic pacemaker in the ventricle to take over. In this situation, the atria and ventricles are controlled by different pacemakers and contract at quite different rates, with the two beats uncoordinated. This event would be devastating for a fish, in which atrial contraction is very important for ventricular filling. It is not quite so devastating in mammals because atrial contraction only tops up the ventricles; they are filled mainly by the direct inflow of blood from the venous system through the relaxed atria.

The catecholamines epinephrine and norepinephrine have three distinct positive effects on heart function:

- They increase the rate of myocardial contractions, or heart rate (*positive chronotropic effect*).
- They increase the force of myocardial contraction (*positive inotropic effect*).
- They increase the speed of conduction of the wave of excitation over the heart (*positive dromotropic effect*).

The effect of epinephrine and norepinephrine on the rate of contraction is mediated via the pacemaker, whereas the increased strength of contraction is a general effect on all myocardial cells. Norepinephrine also increases conduction velocity through the atrioventricular node. It is released from adrenergic nerve fibers that innervate the sinoatrial node, atria, atrioventricular node, and ventricles, so that sympathetic adrenergic stimulation has a direct effect on all portions of the heart.

Mechanical Properties of the Heart

The vertebrate heart is a highly sophisticated pump that operates throughout the life of the animal, which in some instances exceeds 100 years. It must meet the requirements of the body for blood flow whether the animal is at rest, upside down, or exerting itself violently.

The chambers of the heart are not in a straight line, but are usually arranged in a looped pattern (Figure 12-9). The vessels delivering blood to and taking blood away from the heart are also usually curved, rather than

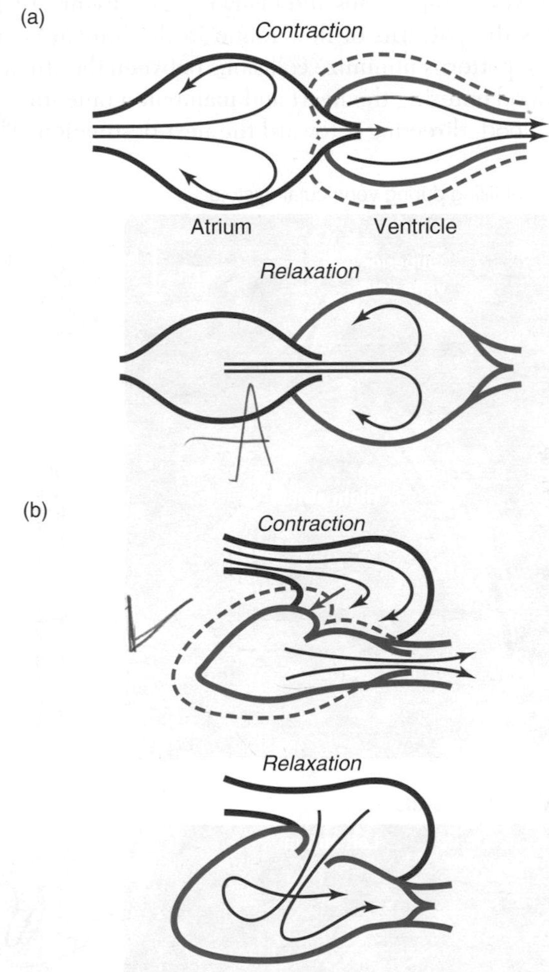

Figure 12-9 A looped atrioventricular arrangement enhances blood flow through the heart. **(a)** In heart chambers arranged linearly, contraction (systole) of the ventricle would pull on the atrium (blue arrow), contributing to atrial expansion. The flow of blood into the atrium would give rise to bilateral recirculation (curved arrows), a relatively unstable pattern of flow that would redirect blood inappropriately for subsequent ventricular filling. Any recoil of the ventricle away from the blood being expelled from it would push back against the atrium, counteracting atrial expansion. During relaxation (diastole), recirculation in the expanding ventricle would redirect fluid away from the outflow tract. **(b)** In a sinuously looped arrangement of heart chambers, ventricular contraction expands the adjacent atrium (blue arrow). Atrial filling is stable and streamlined, accommodated by wall curvatures that redirect momentum toward the atrioventricular valve. Recoil of the ventricle away from ejected blood now adds to the pull on the atrioventricular junction, enhancing rather than suppressing atrial expansion. Redirected blood flow in the atrium contributes to ventricular filling and, in turn, ventricular inflow is redirected toward the outflow tract. [Adapted from P. H. Kilner, G.-Z. Yang, J. A. Wilkes, R. H. Mohiaddin, D. N. Firmin, and M. H. Yacoub, 2000.]

straight. In addition, many hearts have grooves and ridges within the chambers and vessels. These features allow the contraction of one chamber to facilitate blood movement in other chambers. They also create complex flow patterns that direct and maintain momentum, reducing the energy spent by the heart in generating flow. They are found in all classes of vertebrates, including several amphibians and many reptiles. Figure 12-10 shows the patterns of flow found in the human heart. These patterns minimize collisions between the streams of blood entering the heart and maintain momentum in the blood, directing it toward the next destination. The blood vessels are curved to receive the swirling blood ejected from the ventricle without disrupting flow.

Heart rate, stroke volume, and cardiac output

Cardiac output is the volume of blood pumped per unit time from a ventricle. In mammals, it is defined as the volume ejected from the right or left ventricle, not the combined volume from both ventricles. The volume of blood ejected from a ventricle by each beat of the heart is termed the **stroke volume.** The mean stroke volume can be determined by dividing cardiac output by heart rate.

(a) Atrial filling during ventricular systole

Superior vena cava

cm

Right atrium

Inferior vena cava

0.7m s-1

0.1 m s-1

(b) Ventricular filling during diastole

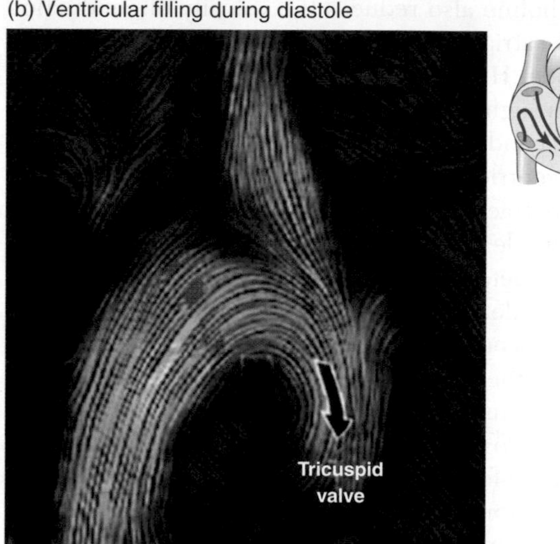

(c) Ventricular systole

Aortic valve

Left atrium

Left ventricle

(d) Ventricular diastole

Figure 12-10 Blood flow patterns through the human heart. **(a)** In ventricular systole, blood entering from superior and inferior venae cavae contributes to the forward rotation of blood in the expanding right atrium. Colored streamlines computed by magnetic resonance imaging show local speed, as indicated by the color scale. **(b)** In early ventricular diastole, further inflow of blood is again redirected forward and down the front of the right atrium, and away from the viewer (out of plane) through the open tricuspid valve. **(c)** In the left ventricle in systole, streamlines pass from the left ventricle through the aortic valve, viewed here in an oblique long-axis plane from above left, the front of the subject being located to the left of the image. **(d)** In early diastole, streamlines pass from the left atrium through the open mitral valve to the left ventricle, with asymmetric recirculation (curved arrow) round the anterior leaflet of the mitral valve. In c and d only, the color scale is modified to reach red at 1 $m \cdot s^{-1}$. [From Kilner, Yang, Wilkes, Mohiaddin, Firmin, and Yacoub, 2000.]

Stroke volume is the difference between the volume of blood in the ventricle just before contraction (end-diastolic volume) and the volume in the ventricle at the end of a contraction (end-systolic volume). Changes in stroke volume may therefore result from changes in either end-diastolic or end-systolic volume. The end-diastolic volume is determined by four parameters:

- venous filling pressure
- pressures generated during atrial contraction
- distensibility of the ventricular wall
- the time available for filling the ventricle

The end-systolic volume is determined by these two parameters:

- pressures generated during ventricular contraction
- the pressure in the outflow channels from the heart (the aortic and pulmonary arteries)

In an isolated mammalian heart, increasing the venous filling pressure causes an increase in end-diastolic volume and results in an increased stroke volume (Spotlight 12-2). End-systolic volume also increases, but not as much as end-diastolic volume. Thus, cardiac muscle behaves in a way similar to skeletal muscle in that stretch of the relaxed muscle within a certain range of length results in the development of increased tension during a contraction. Increases in arterial pressure also cause a rise in both end-diastolic and end-systolic volume, but with little change in stroke volume. In this instance, the increased mechanical work required to maintain stroke volume in the face of elevated arterial pressure results from the increased stretch of cardiac muscle during diastole.

Spotlight 12-2 THE FRANK-STARLING MECHANISM

Otto Frank observed that the more a frog heart was filled by venous return, the greater the stroke volume. Frank derived a length-tension relationship for the frog myocardium and demonstrated that its contractile tension increases with stretch up to a maximum, then decreases with further stretch. Ernest Starling, a dominant figure in many areas of physiology during the early 1900s, had come to similar conclusions. Although neither Starling nor Frank considered mechanical work in their calculations, the increase in mechanical work by the ventricle caused by an increase in end-diastolic volume (in this case, due to venous filling pressure) is termed the Frank-Starling mechanism (see part a of the accompanying figure). The curves derived from measuring work output from the ventricle at different venous filling pressures are known as Starling curves (part b of the figure).

No single Starling curve, however, describes the relationship between venous filling pressure and work output from the ventricle. The mechanical properties of the heart are affected by a number of factors, including the level of activity in the nerves innervating the heart and the composition of the blood perfusing the myocardium. For instance, the relationship between ventricular work output and venous filling pressure is markedly affected by stimulation of sympathetic nerves innervating the heart.

Starling was a versatile researcher who, along with William Bayliss, discovered the hormone secretin. He coined the term *hormones* and defined their basic properties (see Chapter 9). Starling also made many contributions to our understanding of the circulation. In addition to the observations described by the Frank-Starling mechanism, he proposed the Starling hypothesis, which states that the exchange of fluid between blood and tissues is due to differences in blood pressure and colloid osmotic pressures across the capillary wall (see Figure 12-38). This hypothesis was subsequently confirmed, largely by the work of E. Landis.

(a) Frank-Starling mechanism in frog heart

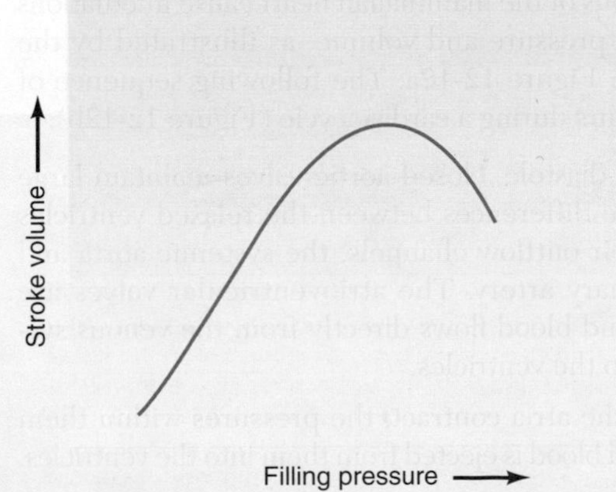

(b) Starling curves (measured in a mammalian heart)

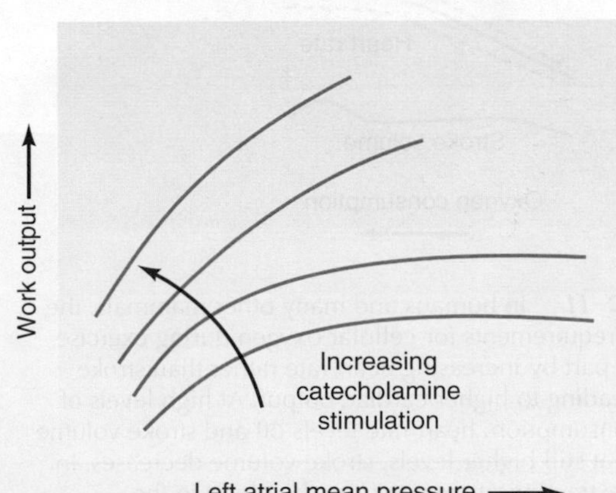

As noted above, epinephrine and norepinephrine increase the force of contraction of the ventricles; hence both the rate and the extent of ventricular emptying are increased by these catecholamines. The adrenergic sympathetic nerves have a markedly greater effect on the rate and force of ventricular output during each heartbeat than does the cholinergic contribution of the vagus nerve. This difference stems from the much more extensive innervation of the ventricles by adrenergic nerves than by cholinergic nerves.

The effects of sympathetic nerve stimulation or increased circulating levels of catecholamines represent a series of integrated actions. Stimulation of pacemaker cells leads to an increase in heart rate. The velocity at which the wave of excitation is conducted across the heart is increased throughout the heart to produce a more nearly synchronous beat of all parts of the ventricle. The production and consumption of ATP in ventricular cells increases, leading to an increase in ventricular work: the rate of ventricular emptying during systole increases, so that the same or a larger stroke volume is ejected in a much shorter time. This increased force of contraction is mediated by the action of catecholamines on both α- and β-adrenoreceptors (see Chapter 9 for more details). Thus, when adrenergic nerve stimulation increases the heart rate, the same stroke volume is ejected from the heart in a shorter time. Even though the time available for emptying and filling the heart is reduced as heart rate increases, stroke volume remains quite stable over a wide range of heart rates. For example, exercise in many mammals is associated with a large increase in heart rate with little change in stroke volume; only at the highest heart rates does the stroke volume fall (Figure 12-11). Stroke volume is relatively invariant because, over a wide range of heart rates,

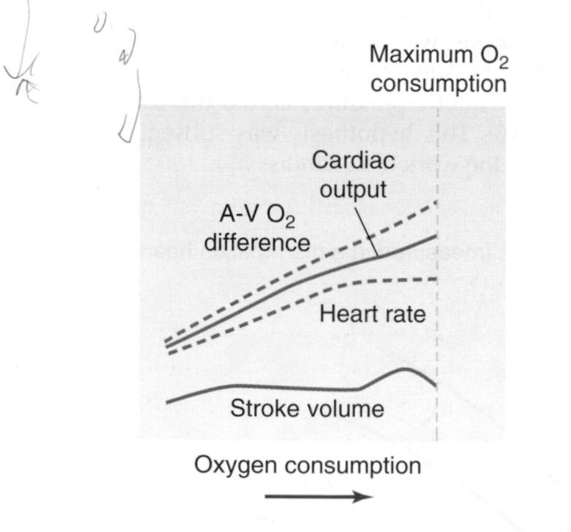

Figure 12-11 In humans and many other mammals, the increased requirements for cellular oxygen during exercise are met in part by increasing heart rate rather than stroke volume, leading to higher cardiac output. At high levels of oxygen consumption, heart rate levels off and stroke volume increases; at still higher levels, stroke volume decreases. In addition, extraction of oxygen from the blood in the capillaries increases during exercise, as indicated by the increase in the arterial-venous (A-V) O_2 difference.

increased sympathetic activity ensures more rapid ventricular emptying, and elevated venous pressures result in more rapid ventricular filling as heart rate increases.

There are limits to the amount by which diastole can be shortened, determined by the maximum possible rate at which the ventricles can be filled and emptied as well as by the nature of the coronary circulation (the circulation within the heart itself). Although it may seem paradoxical, coronary capillaries are occluded during contractions of the myocardium, so coronary blood flow is greatly reduced during systole. Coronary flow rises dramatically during diastole, but a decrease in the diastolic period reduces the period of flow.

Following experimental sympathetic denervation of the heart, exercise still results in increased cardiac output, but in this case there are large changes in stroke volume rather than in heart rate. The increased cardiac output is probably caused by an increase in venous return. It thus appears that sympathetic nerves are not involved in increasing cardiac output *per se,* but rather in raising heart rate and maintaining stroke volume, avoiding the large pressure oscillations associated with large stroke volumes and keeping the heart operating at or near its optimal stroke volume for efficiency of contraction. The sympathetic nerves thus play an important role in determining the relation between heart rate and stroke volume, but additional factors are involved in mediating the increase in cardiac output with exercise.

Cardiac regulation is typical of invertebrates too. For example, heart rate and stroke volume can be controlled independently in crabs and lobsters. In crustaceans, hemolymph leaves the heart via several arteries, and blood distribution is tightly regulated by cardio-arterial valves at the base of each artery. These muscular valves are innervated and are under both neuronal and neurohormonal control. Cardiac output in lobsters and crayfishes has been shown to increase with exercise, and adenosine, which stimulates heart rate and hemolymph flow in crayfishes, may be involved in the cardiovascular changes associated with exercise.

Changes in pressure and flow during a single heartbeat

Contractions of the mammalian heart cause fluctuations in cardiac pressure and volume, as illustrated by the tracings in Figure 12-12a. The following sequence of events occurs during a cardiac cycle (Figure 12-12b):

1. During diastole, closed aortic valves maintain large pressure differences between the relaxed ventricles and their outflow channels, the systemic aorta and pulmonary artery. The atrioventricular valves are open, and blood flows directly from the venous system into the ventricles.
2. When the atria contract, the pressures within them rise, and blood is ejected from them into the ventricles.
3. As the ventricles begin to contract, ventricular pressures rise and exceed those in the atria. At this point,

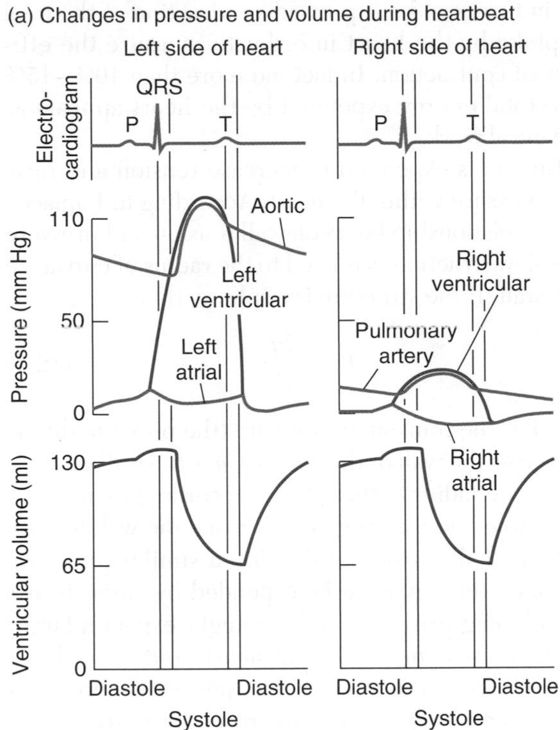

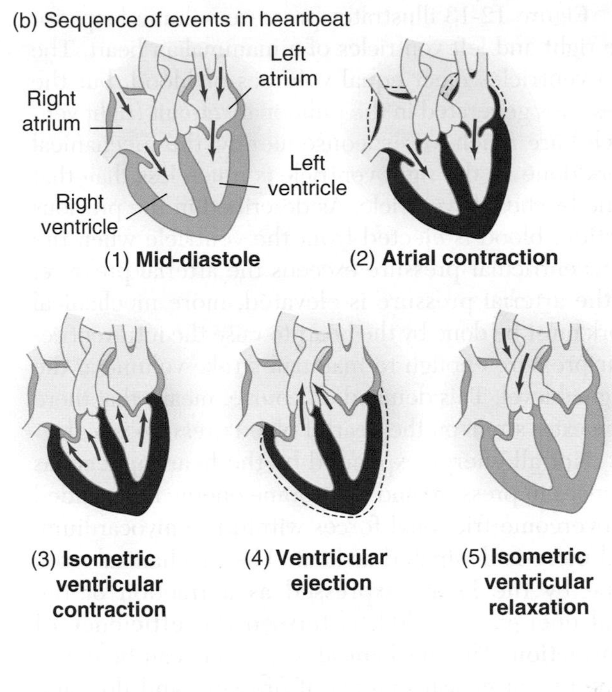

Figure 12-12 During a single cardiac cycle, sequential contraction of the atria and ventricles and the opening and closing of valves produce characteristic changes in pressure and volume. **(a)** Changes in pressure and volume in the ventricles and aorta (left) and pulmonary artery (right) during a single cardiac cycle. **(b)** Sequence of events in contraction of the mammalian heart. Dark red indicates contracted muscle.

the atrioventricular valves close, preventing backflow of blood into the atria, and ventricular contraction proceeds. During this phase, both the atrioventricular and the aortic valves are closed, so that the ventricles form sealed chambers, and there is no volume change. That is, the ventricular contraction is isometric.

4. Pressures within the ventricles increase rapidly and eventually exceed those in the systemic and pulmonary aortas. The aortic valves are then pushed open, and blood is ejected into the aortas, resulting in a decrease in ventricular volume.
5. As the ventricles begin to relax, intraventricular pressures fall below the pressures in the aortas, the aortic valves close, and there is an isometric relaxation of the ventricle.

Once the ventricular pressures fall below those in the atria, the atrioventricular valves are pushed open, ventricular filling starts again, and the cycle is repeated.

In the mammalian heart, the volume of blood forced into the ventricle by atrial contraction is about 30% of the volume of blood ejected into the aorta by ventricular contraction. Thus, ventricular filling is largely determined by the venous filling pressure, which forces blood from the venous system directly through the atria into the ventricles. Atrial contraction simply tops off the nearly full ventricles. Still, maximal cardiac output may be compromised if atrial contraction is impaired.

The contraction of cardiac muscle can be divided into two phases. The first is an isometric contraction, during which tension in the muscle and pressure in the ventricle increase rapidly. The second phase is essentially isotonic; tension does not change very much, for as soon as the aortic valves open, blood is ejected rapidly from the ventricles into the arterial system with little increase in ventricular pressure. Thus, tension is generated first with almost no change in muscle length; then the muscle shortens with little change in tension.

Work done by the heart

It is a simple principle of physics that mechanical work done is the product of mass times distance moved. In the context of the heart, work can be calculated as the change in pressure times flow. Flow is directly related to the change in volume with each contraction of the ventricle. With pressure given in grams per square centimeter and volume in cubic centimeters, pressure times volume equals grams times cubic centimeters divided by square centimeters, which equals grams times centimeters—the equivalent of mass times distance moved, or work. Thus, a plot of pressure times volume for a single contraction of a ventricle yields a pressure-volume loop whose area is proportional to the mechanical work done by that ventricle.

Figure 12-13 illustrates pressure-volume loops for the right and left ventricles of a mammalian heart. The two ventricles eject equal volumes of blood, but the pressures generated in the pulmonary circuit (right ventricle) are much lower; consequently, the mechanical work done by the right ventricle is much less than that done by the left ventricle. As described in the previous section, blood is ejected from the ventricle when the intraventricular pressure exceeds the arterial pressure. If the arterial pressure is elevated, more mechanical work must be done by the heart to raise the intraventricular pressure enough to maintain stroke volume at the original level. This demand, of course, means that there is an extra strain on the heart if blood pressure is high.

Not all energy expended by the heart appears as changes in pressure and flow. Some energy is expended to overcome frictional forces within the myocardium, and more is dissipated as heat. The mechanical work done by the heart, expressed as a fraction of the total energy expended, is termed the **efficiency of contraction.** The mechanical work done can be determined from measurements of pressure and flow and converted into milliliters of O_2 consumed. This measure, in turn, can be expressed as a fraction of the total O_2 uptake by the heart in order to measure the efficiency of contraction. In fact, no more than 10%–15% of the total energy expended by the heart appears as mechanical work.

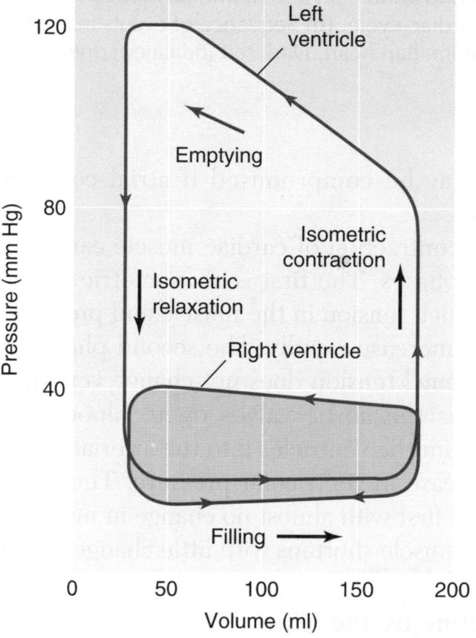

Figure 12-13 The area of a ventricular pressure-volume loop is proportional to the mechanical work done by a ventricle in one cardiac cycle. Shown here are loops for the right ventricle (red) and left ventricle (pink) of the mammalian heart. Once around a loop in a counterclockwise direction is equivalent to one heartbeat. Ventricular filling occurs at low pressure; pressure increases sharply only when the ventricles contract (the sharp upswing on the right-hand side of each loop). Ventricular volume decreases as blood flows into the arterial system, and ventricular pressure falls rapidly as the ventricle relaxes. Filling then begins again. Note that although the volume changes in both ventricles are similar, the pressure changes are much larger in the left ventricle than in the right one. Therefore, the left ventricle has a larger loop and hence does more mechanical work than the right ventricle.

Energy is expended to increase tension and raise blood pressure within the heart. According to Laplace's law, the relationship between wall tension and pressure in a hollow structure is related to the radius of curvature of the wall. If the structure is a sphere, then

$$P = \frac{2y}{R} \tag{12-1}$$

where P is the transmural pressure (the pressure difference across the wall of the sphere), y is the wall tension, and R is the radius of the sphere. According to this relation, a large heart must generate twice the wall tension of a heart half its size to develop a similar pressure. Thus, more energy must be expended by larger hearts in developing pressure, and we might expect a larger ratio of muscle mass to total heart volume in large hearts. Hearts are not, of course, perfect spheres, but have a complex gross and microscopic morphology; nevertheless, Laplace's law serves as a general guide. The energy expended in ejecting a given quantity of blood from the heart depends on the efficiency of contraction, the pressures developed, and the size and shape of the heart.

Coronary circulation

The coronary circulation supplies nutrients and oxygen to the heart. The blood supply to the heart is extensive—cardiac muscle has a much higher capillary density and more mitochondria than most skeletal muscles. It also has a high myoglobin content, which accounts for the typical red color of the heart. The blood flowing through the chambers of the heart supplies nutrients to the inner spongy layer of the heart in many fishes and amphibians, but even in these animals, an additional coronary supply is necessary to deliver oxygen and other substrates to the outer, denser regions of the heart wall. In general, hearts can use a wide variety of nutrients, including fatty acids, glucose, and lactate; the particular substrate used is determined largely by availability.

However, the heart relies primarily on aerobic pathways to generate energy, so it is very dependent on a continuous oxygen supply. An increase in cardiac activity depends on increased metabolism in the heart, which in turn requires increased coronary flow. Adenosine is a key metabolite in maintaining the relationship between coronary flow and cardiac activity. Adenosine and other local metabolic products cause dilation of coronary vessels and, therefore, increased coronary flow. The formation and release of adenosine increases with increased metabolism or during myocardial hypoxia (a drop in oxygen level). Increased adenosine levels also reduce heart rate and, therefore, energy expenditure. Thus the level of adenosine mediates the match between energy expenditure and supply. Sympathetic stimulation is a second, but less important, mechanism of increasing coronary flow. Circulating catecholamines increase cardiac contractility and cause coronary vasodilation mediated via β_1-adrenoreceptors.

Question Mammalian hearts have an extensive coronary circulation throughout the myocardium, but in some fish hearts, the coronary circulation is restricted to the epicardium, the external lining of the heart. What would be the consequences of this differing organization for the nutrition of the heart?

The pericardium

The heart is surrounded by a connective tissue membrane called the **pericardium.** The magnitude of the pressure changes within the pericardial cavity depend on the rigidity of the pericardium and on the magnitude and rate of change of the heart volume. In some animals, the pericardium is thin and flexible (compliant), in which case pressure changes within the pericardial cavity during each heartbeat are negligible. In others, such as sharks, the pericardium is quite rigid (noncompliant), in which case the intrapericardial pressure oscillates during each heartbeat.

The compliant pericardium enveloping the mammalian heart is formed of two layers, an outer fibrous layer and an inner serous layer. The serous layer is double, forming the inner lining of the pericardial cavity and the outer layer of the heart itself (the epicardium). In mammals, the serous layer secretes a fluid that acts as a lubricant, facilitating movement of the heart.

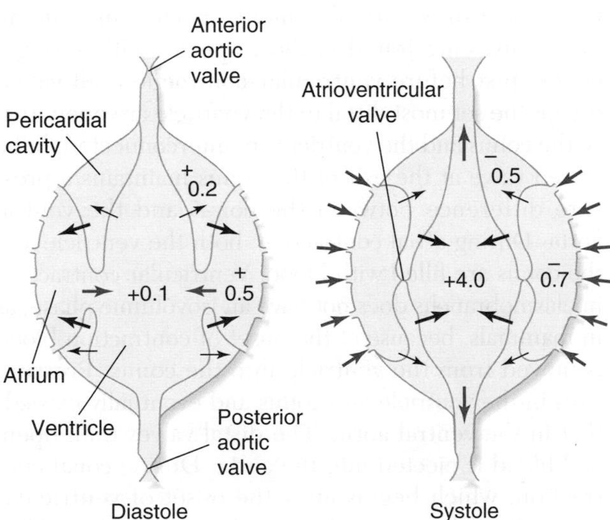

Figure 12-14 In the heart of the bivalve mollusk *Anodonta,* which has a noncompliant pericardium, ventricular contraction not only ejects blood, but also reduces pressure in the pericardial cavity, thus enhancing filling of the atria. Numbers are pressures in centimeters of seawater, which are expressed relative to ambient pressures. Large black arrows indicate movements of the walls of contracting chambers; small black arrows indicate movements of the walls of relaxing chambers. The red arrows indicate the direction of blood flow. [Adapted from Brand, 1972.]

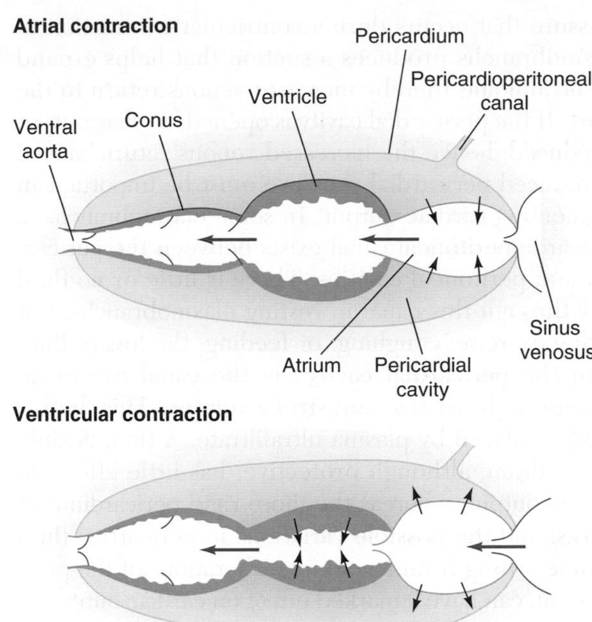

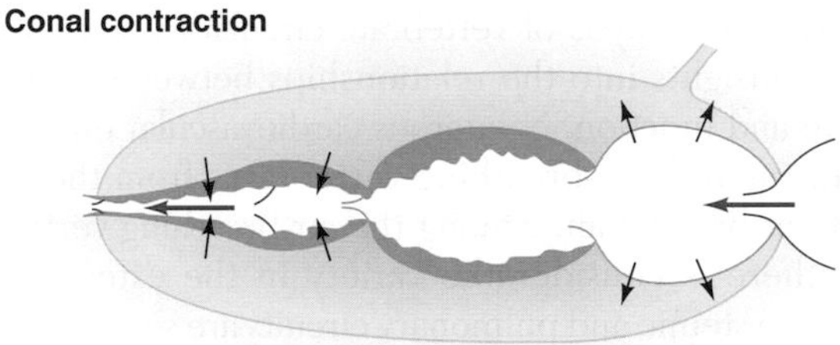

Figure 12-15 Because the elasmobranch heart is contained within a noncompliant pericardium, contraction of the ventricle reduces pressure in the pericardial cavity and assists atrial filling. In some elasmobranchs, fluid loss via the pericardioperitoneal canal during exercise, feeding, and coughing leads to an increase in heart size and stroke volume. Black arrows indicate the direction of wall movement during muscle contraction or relaxation; red arrows indicate the direction of blood flow.

Crustaceans and bivalve mollusks have a noncompliant pericardium. In these animals, contraction of the ventricle reduces pressure in the pericardial cavity and enhances flow into the atria from the venous system (Figure 12-14). Thus, tension generated in the ventricular wall is utilized both to eject blood into the arterial system and to draw blood into the atria from the venous system.

The pericardia of elasmobranchs (sharks) and lungfishes are also noncompliant, whereas those of bony fishes are compliant. The elasmobranch heart consists of four chambers—sinus venosus, atrium, ventricle, and **conus**—all contained within a rigid pericardium (Figure 12-15). The reduction in intrapericardial pressure that occurs during ventricular contraction in elasmobranchs produces a suction that helps expand the atrium and thereby increases venous return to the heart. If the pericardial cavity is opened, cardiac output is reduced; hence the increased venous return caused by reduced pericardial pressure must be important in augmenting cardiac output. In some elasmobranchs, a pericardioperitoneal canal exists between the pericardial and peritoneal cavities. There is little or no fluid flow through this canal in resting elasmobranchs, but during exercise, coughing, or feeding, the loss of fluid from the pericardial cavity via the canal causes an increase in heart size and stroke volume. This fluid is slowly replaced by plasma ultrafiltrate. A thin, flexible pericardium, although protective, has little effect on cardiac output, whereas the more rigid pericardium of sharks, and the possible variations in pericardial fluid volume arising from anatomic elaborations of the pericardium, can have a marked effect on cardiac output.

Vertebrate Hearts: Comparative Functional Morphology

A comparative analysis of vertebrate circulatory systems produces insights into the relationships between heart structure and function. Numerous cardiovascular differences distinguish air-breathing vertebrates from those that do not breathe air. Among the air-breathing vertebrates, there is considerable variety in the extent to which the systemic and pulmonary circuits are separated.

The pulmonary circuit of birds and mammals is maintained at much lower pressures than is the systemic circuit. This difference in pressure is possible because birds and mammals have two series of heart chambers in parallel. The left side of the heart ejects blood into the systemic circulation, and the right side ejects blood into the pulmonary circulation (see Figure 12-3). The advantage of maintaining a high blood pressure is that it allows rapid transit times and rapid changes in blood flow through small-diameter capillaries. However, when the difference in pressure across a vessel wall (i.e., the transmural pressure) is high, fluid filters across the capillary wall; as a result, extensive lymphatic drainage of the tissues is necessary. In the mammalian lung, capillary flow can be maintained by relatively low input pressures, reducing the requirements for lymphatic drainage and avoiding the formation of large extracellular fluid spaces that could increase diffusion distances between blood and air and impair the gas transfer capacity of the lung. One advantage of a divided heart is that blood flow to the body and the lungs can be maintained with different input pressures. A disadvantage of a completely divided heart is that in order to avoid shifts in blood volume from the systemic to the pulmonary circuit, or vice versa, cardiac output must be the same in both sides of the heart, independent of the requirements in the two circuits.

In contrast, lungfishes, amphibians, reptiles, bird embryos, and fetal mammals have either an undivided ventricle or some other mechanism that allows the shunting of blood from one circuit to the other. These shunts usually result in the movement of blood from the right (respiratory or pulmonary) to the left (systemic) side of the heart during periods of reduced gas transfer in the lung. At such times, blood returning from the body is shunted away from the lung to the left side of the heart and once again ejected into the systemic circuit, bypassing the lungs. In lungfishes, amphibians, and reptiles, flow to the lungs is commonly reduced during prolonged dives, during which gas transfer occurs across the skin, or oxygen stores in the body are used. Blood flow to the lungs is also reduced during development within the mother (mammals) or egg (birds), before the lungs become fully functional in gas exchange. Although a single undivided ventricle permits variations in the ratio of flows to the respiratory and systemic circuits, the same pressure must be maintained on both sides of the heart.

Water-breathing fishes

The hearts of water-breathing fishes, including elasmobranchs and most bony fishes (teleosts), consist of four chambers in series. All the chambers are contractile except the elastic **bulbus** of teleosts. Unidirectional flow of blood through the heart is maintained by valves at the sinoatrial and atrioventricular junctions and at the exit of the ventricle.

In elasmobranchs, the exit from the ventricle to the conus is guarded by a pair of flap valves, and there are from two to seven pairs of valves along the length of the conus, depending on the species (see Figure 12-15). Conus length is variable among species; in general, more valves are found in those species with a longer conus. Just before ventricular contraction, all valves except the set most distal to the ventricle are open; that is, the conus and the ventricle are interconnected, but a closed valve at the exit of the conus maintains a pressure difference between the conus and the ventral aorta. During atrial contraction, both the ventricle and the conus are filled with blood. Ventricular contraction in elasmobranchs does not have an isovolumic phase, as in mammals, because at the onset of contraction blood is moved from the ventricle into the conus. Pressure rises in the ventricle and conus and eventually exceeds that in the ventral aorta. The distal valves then open, and blood is ejected into the aorta. During conal contraction, which begins after the onset of ventricular contraction, the proximal valves close, preventing reflux of blood into the ventricle as it relaxes. Conal contraction proceeds relatively slowly away from the heart toward the aorta; each set of valves closes in turn to prevent backflow of blood.

As illustrated in Figure 12-16a, blood pumped by the heart in typical water-breathing fishes passes first through the gill (respiratory) circulation and then into a dorsal aorta that supplies the rest of the body (systemic circulation). Thus, the respiratory and systemic circuits

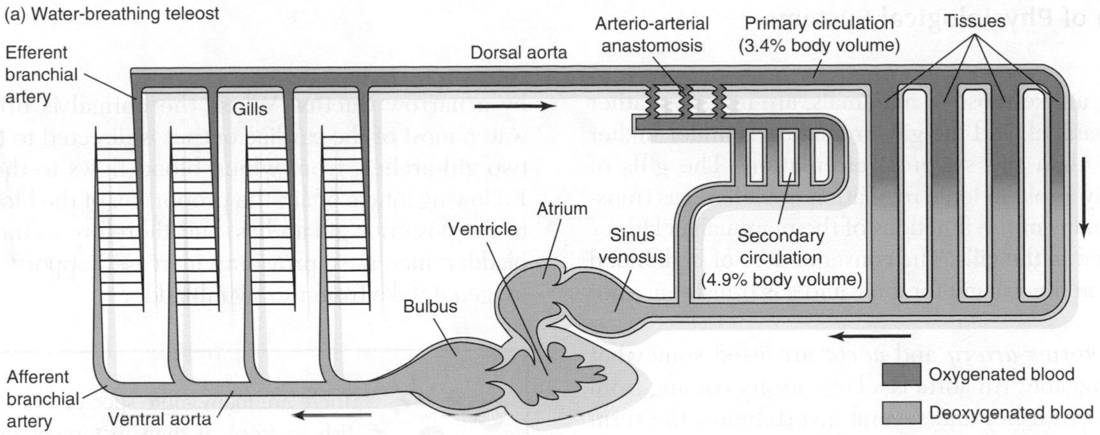

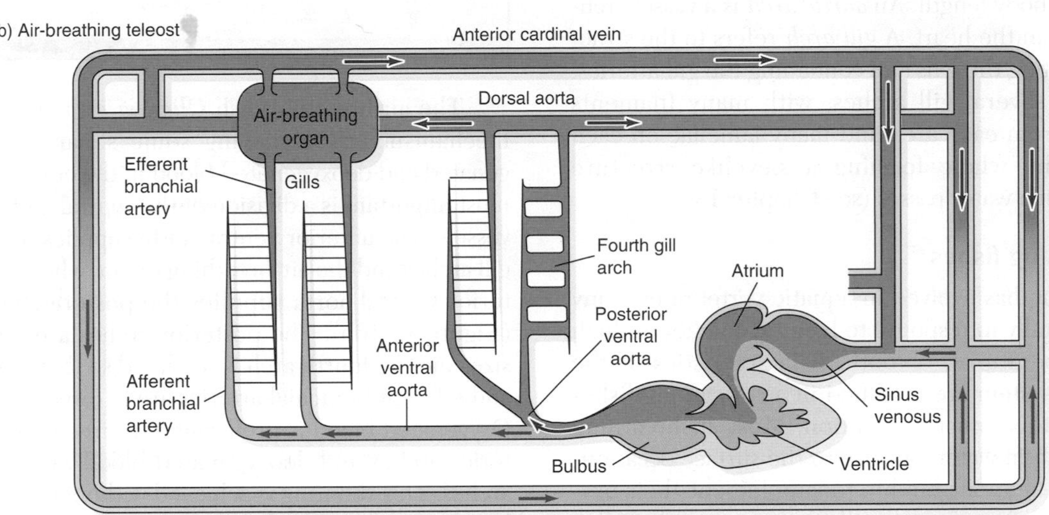

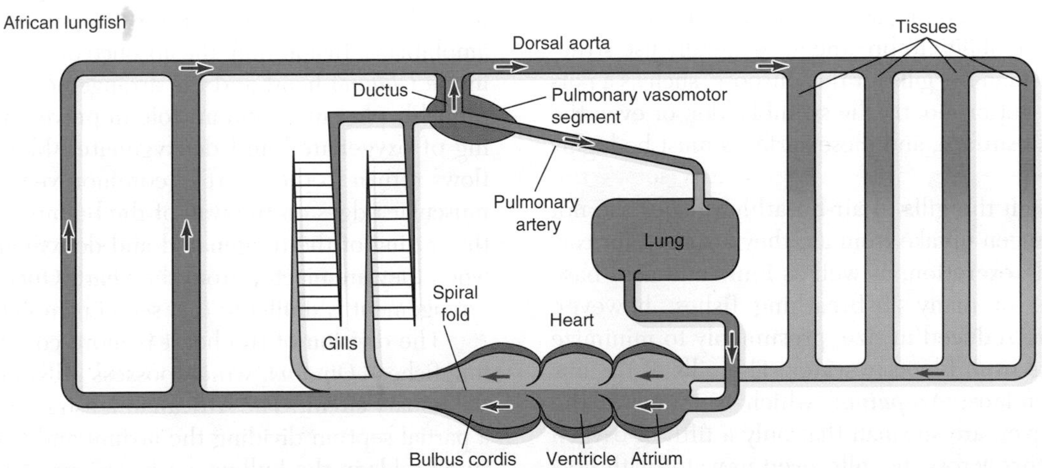

Figure 12-16 Comparison of the cardiovascular system in water-breathing and air-breathing fishes. **(a)** In a "typical" water-breathing teleost such as the trout, the respiratory circulation through the gills and the systemic circulation are in series. In the four-chambered, undivided heart, the pacemaker is in the sinus venosus. The ventricle ejects blood into the compliant bulbus and short ventral aorta. Blood flows through the gills into a long, stiff dorsal aorta. Most teleosts also have a low-hematocrit secondary circulation, which supplies nutrients, but not much oxygen, to the skin and gut. **(b)** Even though the heart of the air-breathing teleost *Channa argus* is undivided, the flows of oxygenated and deoxygenated blood are partially separated. Deoxygenated blood preferentially flows through the first two gill arches and air-breathing organ, whereas oxygenated blood flows through posterior arches into the dorsal aorta. The fourth gill arch is modified so that the afferent and efferent branchial arteries are connected. **(c)** The circulation of the African lungfish, *Protopterus,* is marked by nearly complete separation of oxygenated and deoxygenated blood. This separation is achieved by a septum partially dividing the atrial and ventricular chambers and a long spiral fold in the bulbus cordis. This fish possesses a lung and a distinct pulmonary circuit. The absence of lamellae in the anterior gill arches permits blood to flow directly to the systemic circuit via the dorsal aorta. The ductus and the pulmonary vasomotor segment act reciprocally to direct blood to the dorsal aorta or lungs, depending on whether the fish is breathing water or air. [Part b adapted from Ishimatzu and Itazawa, 1993; part c adapted from Randall, 1994.]

of fishes, unlike those of mammals, are in series rather than in parallel, and the gill circulation is under higher pressure than the systemic circulation. The gills of fishes play a role in ionic regulation as well as gas transfer, and many of the functions of the mammalian kidney are located in the gills. The consequences of high blood pressure in the fish gill for ionic and gas transfer are not clear.

The terms *artery* and *aorta* are used somewhat interchangeably. An aorta is a large artery coming from the heart. In fishes and several invertebrates the term also refers to the dorsal aorta (artery) extending along most of the body length. An *aortic arch* is a vessel arching away from the heart. A *gill arch* refers to the structure supporting the gills and containing the gill arteries. A fish has several gill arches, with many filaments extending from each arch and many lamellae on each filament, the whole forming a sievelike structure through which water passes (see Chapter 13).

Air-breathing fishes

Air breathing has evolved in aquatic vertebrates many times, generally in response to hypoxic conditions, high water temperatures, or both. Characteristics of the circulatory system are modified in air-breathing fishes to support this unusual adaption. Most air-breathing fishes remain in water, but rise to the surface occasionally to take in an air bubble to supplement their oxygen supplies. Because the gill filaments and lamellae usually collapse and stick together when exposed to air, they cannot be used for gas transfer in air. Hence, fishes that have the ability to breathe air generally use structures other than the gills for this purpose, such as a portion of the gut or mouth, the swimbladder, or even the general skin surface, and these surfaces must be highly vascularized.

Although the gills of air-breathing fishes are not used for oxygen uptake from air, they are used for carbon dioxide excretion, as well as ionic and acid-base regulation. In many air-breathing fishes, however, the gills are reduced in size, presumably to minimize oxygen loss from blood to water. The gills of the air-breathing teleost *Arapaima,* which is found in the Amazon River, are so small that only a fifth of oxygen uptake occurs across the gills, even in water with normal oxygen levels. The bulk of oxygen uptake by this fish occurs via the swimbladder, which is highly vascularized and has many septa that increase the surface area for gas exchange. These fish die if denied access to air; in other words, *Arapaima* is an obligate air-breathing fish.

Air-breathing fishes have evolved a variety of shunting mechanisms to permit changes in the distribution of blood to the gills and the air-breathing organ. In the tropical freshwater teleost *Hoplerythrinus,* the posterior gill arches give rise to the coeliac artery, which perfuses the swimbladder and connects to the dorsal aorta by a narrow ductus. When the animal is breathing water, most of the cardiac output is directed to the first two gill arches, from which blood flows to the body. Following intake of air, the proportion of the blood flow to the posterior gill arches, and therefore to the swimbladder, increases, providing increased opportunity for oxygen uptake from the swimbladder.

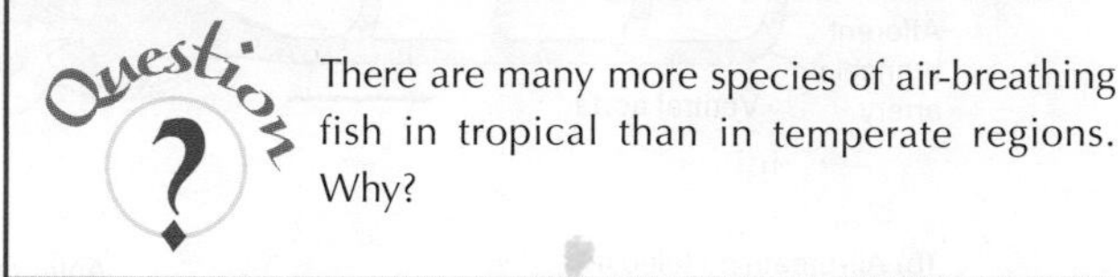

There are many more species of air-breathing fish in tropical than in temperate regions. Why?

The air-breathing fish *Channa argus* uses several mechanisms for achieving some separation of oxygenated and deoxygenated blood in the circulation. The most important is a division of the ventral aorta into two vessels. The anterior ventral aorta supplies the first two gill arches and the air-breathing organ, whereas the posterior ventral aorta supplies the posterior gill arches (Figure 12-16b). The posterior arches are reduced in size, and the fourth arch is modified so that the afferent and efferent branchial arteries are in direct connection. Oxygenated blood is preferentially directed to the posterior arches and deoxygenated blood to the first two arches. This shunting is achieved without division of the heart. The ventricle, however, is spongy (trabeculate), which may serve to prevent the mixing of blood within it, as has been suggested for the spongy heart of amphibians. In addition, the absence of sinoatrial valves in the *Channa* heart and the arrangement of the veins probably play an important role in preventing the mixing of oxygenated and deoxygenated blood as these flows return to the heart in common vessels. Finally, muscular ridges on the wall of the bulbus may prevent the mixing of the oxygenated and deoxygenated flows when they are ejected from the heart. Once again, this arrangement is similar to that seen in amphibians.

The division of the heart is more complete in the lungfishes (*Dipnoi*), which possess gills, lungs, and a pulmonary circuit. The African lungfish *Protopterus* has a partial septum dividing the atrium and ventricle and spiral folds in the bulbus cordis (Figure 12-16c). This arrangement maintains the separation of oxygenated and deoxygenated blood in the heart. The anterior gill arches lack lamellae, and oxygenated blood can flow from the left side of the heart directly through them to the tissues. Within the lamellae of the posterior gill arches is a basal connection that allows blood to bypass the lamellae when only the lung is in operation (e.g., during estivation, a state of torpor occurring in the summer). Blood from the posterior gill arches flows to the lungs or enters the dorsal aorta via a ductus. The ductus is richly innervated and is undoubtedly involved in controlling blood flow between the pulmonary artery and the systemic circulation. The initial segment of the pulmonary artery is muscular and is referred to as the pulmonary vasomotor segment. The pulmonary vasomotor segment and the ductus probably act in a reciprocal fashion: when one constricts, the other dilates. The ductus in lungfishes is analogous to the ductus arteriosus of fetal mammals, acting as a lung bypass when the lung is not functioning.

When air-breathing vertebrates evolved, were the initial species moving out of water onto land likely to have been large or small? Explain the reasons for your answer.

Amphibians

Amphibians have two completely separated atria, but a single ventricle. In the frog heart, the oxygenated and deoxygenated blood remain separated even though the ventricle is undivided. Oxygenated blood from the lungs and skin is preferentially directed toward the body via the systemic arch (the outflow tract leading to the systemic circuit), whereas deoxygenated blood from the body is directed toward the pulmocutaneous arch (the outflow tract leading to the pulmonary circuit through the pulmocutaneous arteries). This separation of oxygenated and deoxygenated blood is aided by a spiral fold within the conus arteriosus of the heart (Figure 12-17). Deoxygenated blood leaves the ventricle first during systole and enters the pulmonary circuit. Pressure rises in the pulmocutaneous arch and becomes similar to that in the systemic arch. Blood then begins flowing into both arches, with the spiral fold partially dividing the systemic and pulmocutaneous flows within the conus arteriosus.

The volume of blood going to the lungs or the body is inversely related to the amount of resistance that the flow encounters in each circuit. Immediately following a breath, the resistance to blood flow through the lungs is low and flow is high; between breaths, resistance gradually increases and is associated with a fall in blood flow. These oscillations in pulmonary blood flow are possible because of the partial division of the amphibian heart. Although deoxygenated blood is directed toward the pulmocutaneous arch, the ratio of pulmonary to systemic blood flow can be adjusted. That is, when the animal is not breathing, blood flow to the lungs can be reduced, so that most of the blood pumped by the ventricle is directed toward the body. When the animal is breathing, a more even distribution of flow to the lungs and body can be maintained. This distribution is possible only if the ventricle is not completely divided into right and left chambers as it is in mammals.

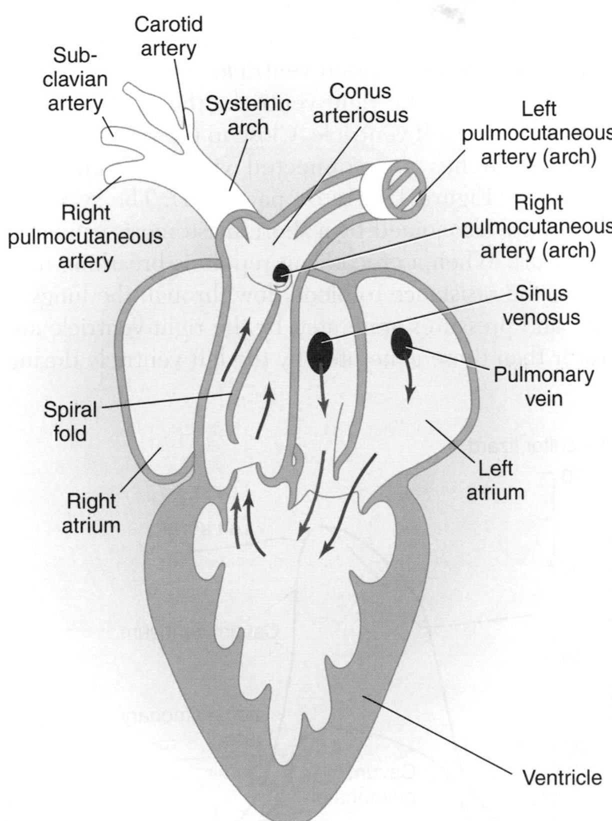

Figure 12-17 Even though the frog heart has a single ventricle, deoxygenated blood is directed to the lungs via the pulmocutaneous arches and oxygenated blood to the tissues via the systemic arch. This ventral view of the internal structure of the frog heart shows the position of the spiral fold, which aids in separating the two blood flows. [Adapted from Goodrich, 1958.]

Noncrocodilian reptiles

Most noncrocodilian reptiles, including turtles, snakes, and some lizards, have a partially divided ventricle and right and left systemic arches. In these animals, the ventricle is partially subdivided by an incomplete muscular septum referred to as the horizontal septum, Muskelleiste, or muscular ridge. This horizontal septum separates the cavum pulmonale from the cavum venosum and cavum arteriosum; the latter two are partially separated by the vertical septum (Figure 12-18 on the next page). The right atrium contracts slightly earlier than the left atrium and ejects deoxygenated blood into the cavum pulmonale across the free edge of the horizontal septum; ventricular contraction ejects this blood into the pulmonary artery. Oxygenated blood from the left atrium fills the cavum venosum and cavum arteriosum; from here the blood empties into the systemic arteries.

Measurements in turtles support the view that oxygenated blood from the left atrium passes into the systemic circuit, whereas deoxygenated blood from the right atrium passes into the pulmonary circuit. Pulmonary artery diastolic pressure is often lower than systemic

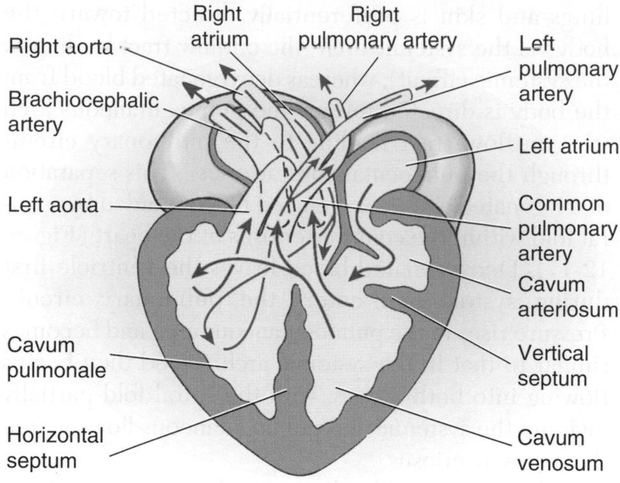

Figure 12-18 In noncrocodilian (chelonian) reptile hearts, the ventricle is partially divided by the horizontal septum into the cavum venosum and ventral cavum pulmonale. The common pulmonary artery arises from the cavum pulmonale, whereas all of the systemic arteries arise from the cavum venosum. In this ventral view of the turtle heart, the arrows schematically indicate movement of oxygenated blood (red) and deoxygenated blood (blue), but are not intended to represent the flow of separate bloodstreams through the heart. [Adapted from Shelton and Burggren, 1976.]

diastolic pressure; as a result, the pulmonary valves open first when the ventricle contracts. Thus, flow occurs earlier in the pulmonary artery than in the systemic arches during each cardiac cycle. In turtles, there may be some recirculation of arterial blood in the lung circuit; that is, there is a left-to-right shunt within the heart. The ventricle remains functionally undivided throughout the cardiac cycle, and the relative flow to the lungs and systemic circuits is determined by the resistance to flow in each part of the circulatory system. When the turtle breathes, resistance to flow through the pulmonary circuit is low and flow is high. When the turtle does not breathe, as during a dive, pulmonary vascular resistance increases, but systemic vascular resistance decreases, resulting in a right-to-left shunting of blood and a decrease in pulmonary blood flow. As in many other animals, there is a reduction in cardiac output associated with a marked slowing of the heart (**bradycardia**) during a dive.

The similarity of the pressures in the pulmonary and systemic outflow tracts in turtles, snakes, and some lizards indicates that their hearts have a single ventricular chamber partially divided into subchambers even during systole (Figure 12-19a). In monitor lizards and related varanid lizards, however, the pulmonary outflows are at much lower pressures than the systemic outflows during systole (Figure 12-19b). The pressure in the cavum pulmonale, for instance, may be only a third of that in the cavum venosum during systole in these animals. The pressure differential in varanid lizards is achieved by a pressure-tight contact between the muscular ridge (horizontal septum) and the wall of the heart during systole (Figure 12-20).

Crocodilian reptiles

Unlike other reptiles, crocodilian reptiles have a heart with a completely divided ventricle. The left systemic arch arises from the right ventricle; the right systemic arch, from the left ventricle. Close to the ventricles, the systemic arches are connected via the foramen of Panizzae (Figure 12-21a on page 494). The systemic arches are also joined by a short anastomosis caudal to the heart. When a crocodilian reptile is breathing normally, the resistance to blood flow through the lungs is low, and pressures generated by the right ventricle are lower than those generated by the left ventricle during

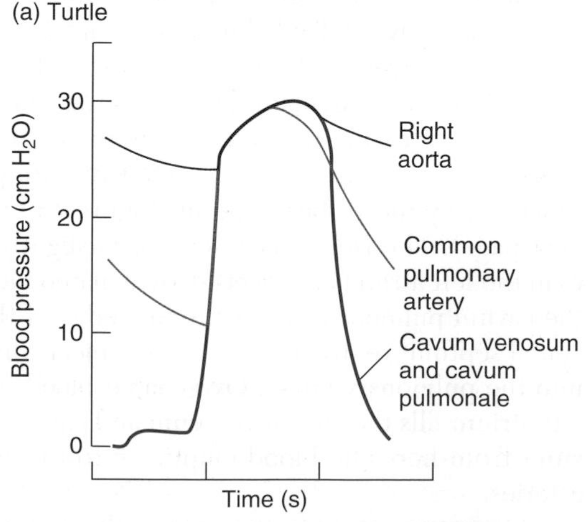

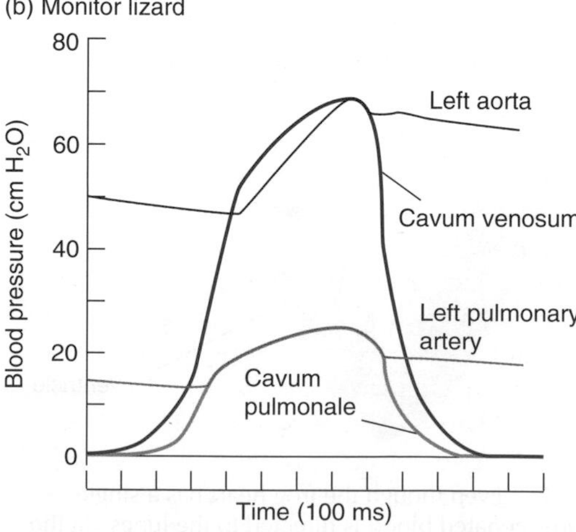

Figure 12-19 In turtles, the pressures in the systemic and pulmonary outflow tracts are nearly identical during systole, whereas in varanid lizards they differ considerably. Shown are blood pressures measured simultaneously at the indicated sites during a single heartbeat in **(a)** a turtle, *Chrysemys scripta,* and **(b)** a monitor lizard, *Varanus exanthematicus.* [Part a from Shelton and Burggren, 1976; part b from Burggren and Johansen, 1982.]

(a) Diastole

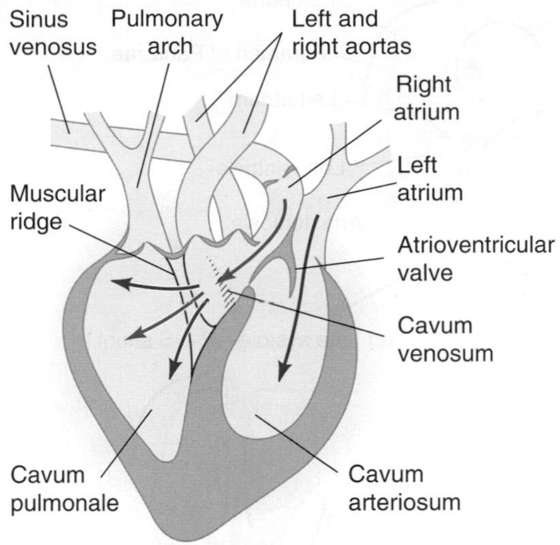

(b) Systole

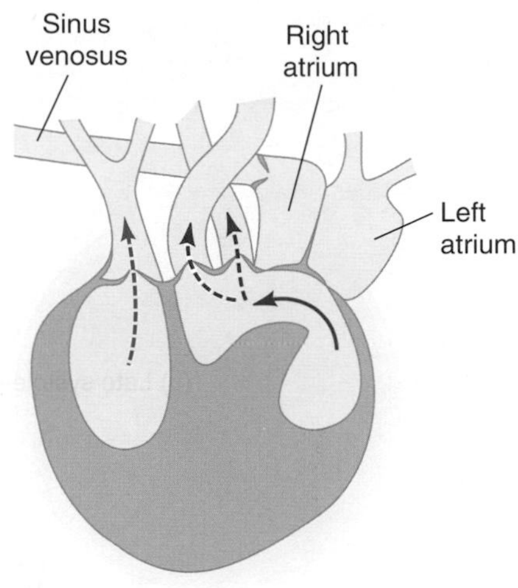

Figure 12-20 In varanid lizards, a pressure-tight separation between the cavum pulmonale and cavum venosum occurs during systole. **(a)** During diastole, the muscular ridge only partially separates the cavum venosum and cavum pulmonale. Thus, oxygenated blood (red arrows) remaining in the cavum venosum from the preceding systole is washed into the cavum pulmonale by deoxygenated blood (blue arrows). The cavum arteriosum is filled with oxygenated blood. Separation between the cavum arteriosum and the cavum venosum is provided by at least one atrioventricular valve. **(b)** During systole, the muscular ridge is pressed tight against the outer heart wall, forming a pressure-tight barrier. Deoxygenated blood remaining in the cavum venosum from the preceding diastole is mixed with oxygenated blood from the cavum arteriosum and flushed into the aortic arches. Deoxygenated blood with an admixture of oxygenated blood is expelled from the cavum pulmonale into the pulmonary arch. With no connection between the cavum venosum and cavum pulmonale, different pressures can develop in the outflow tracts. [Adapted from Heisler et al., 1983.]

all phases of the cardiac cycle. In this case, blood is pumped by the left ventricle into the right systemic arch during systole, with the open aortic valve closing off the foramen of Panizzae (Figure 12-21b). There is a small reflux of blood into the left aorta from the right aorta via the anastomosis during systole. Because of this connection, the pressure in the left systemic arch remains higher than the pressure in the right ventricle (Figure 12-21b, bottom); consequently, the valves at the base of the left systemic arch remain closed throughout the cardiac cycle. All blood ejected from the right ventricle passes into the pulmonary artery and flows to the lungs. Thus, the crocodilian circulation is functionally the same as the mammalian circulation in that there is complete separation of systemic and pulmonary blood flow.

Crocodilians, however, have the added capacity to shunt blood from the pulmonary to the systemic circuit. This P→S shunt is achieved by active closure of a valve at the base of the pulmonary outflow tract toward the end of systole. Under some experimental circumstances, peak right ventricular pressure becomes equal to left ventricular pressure and exceeds left systemic pressure. As a result, the valves at the base of the left systemic arch open, and blood from the right ventricle is ejected into the systemic circulation during late systole (Figure 12-21c). In this case, a portion of the deoxygenated blood returning to the heart from the body via the right atrium is recirculated in the systemic circuit. Exactly when the P→S shunt operates normally in the animal is not clear. The role of the foramen of Panizzae also remains enigmatic; it is open only during diastole, allowing flow between the aortic arches as the heart relaxes.

Mammals and birds

The four-chambered hearts of both mammals and birds are in fact two hearts beating as one. The heart originates as two separate tubes that join together during development to form the multi-chambered heart of the postnatal animal. The right side pumps blood to the lungs; the left side pumps blood around the body. Blood returning from the lungs enters the left atrium, passes into the left ventricle, and is ejected into the systemic circulation. Blood from the body collects in the right atrium, passes into the right ventricle, and is pumped to the lungs (see Figure 12-3).

Valves prevent backflow of blood from the aorta to the ventricle, the atrium, and the veins. These valves are passive and are opened and closed by pressure differences between the heart chambers. The atrioventricular valves (bicuspid and tricuspid valves) are connected to the ventricular wall by fibrous strands (see Figure 12-4), which prevent the valves from being everted into the atria when the ventricles contract and intraventricular pressures are much higher than those in the atria. The

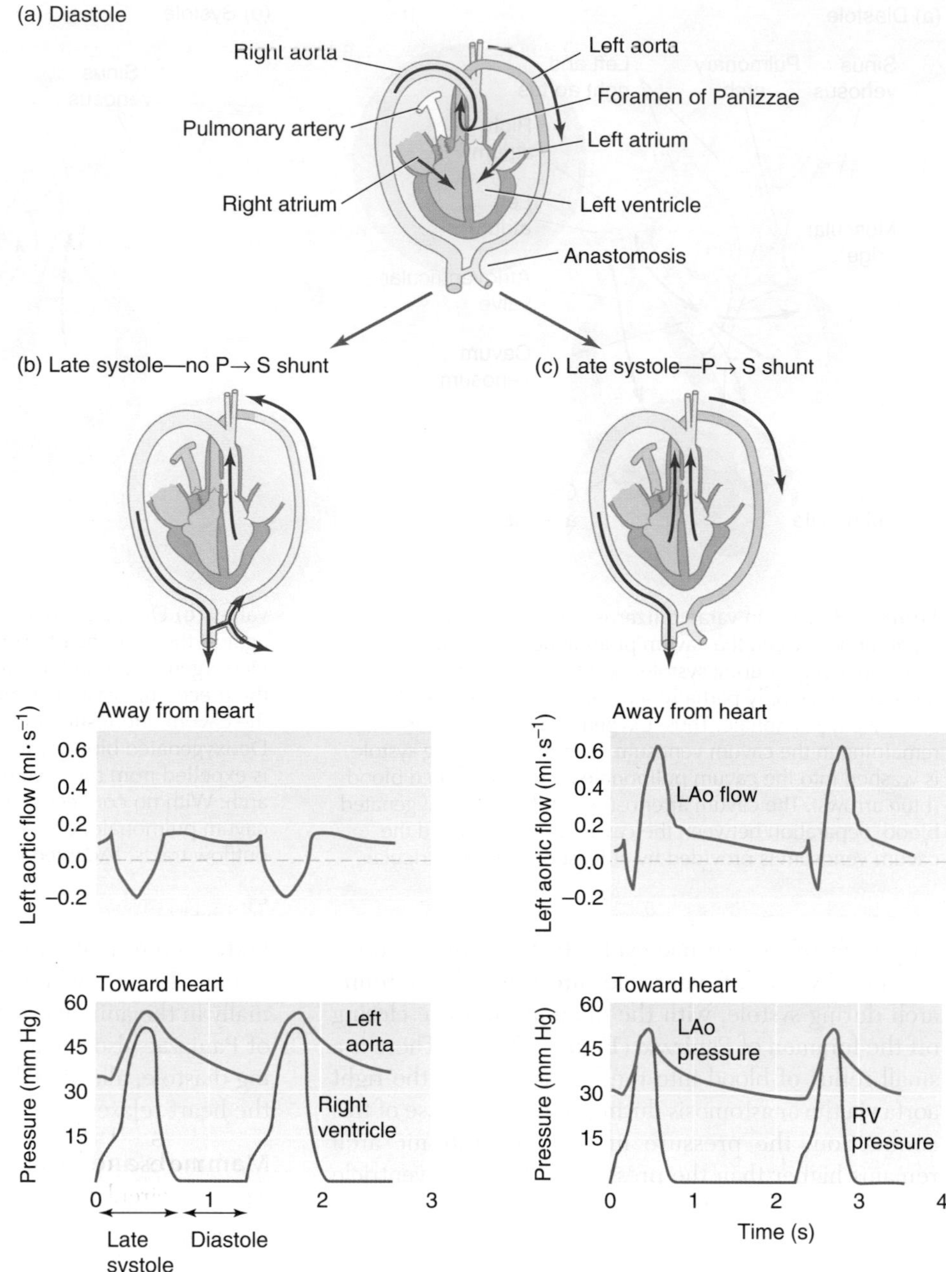

Figure 12-21 Under some conditions, a P ⟶ S shunt operates during late systole in crocodiles. These schematic diagrams and pressure and flow tracings illustrate what happens during the cardiac cycle with and without the shunt. [Adapted from Jones, 1995.]

walls of the ventricle, especially the left chamber, are thick and muscular. The inner surface of the ventricular muscle, or myocardium, is lined by an endothelial membrane, the endocardium. The outer surface of the ventricular myocardium is covered by the epicardium.

Mammalian fetus At birth, mammals must discard the placental circulation that has provided for all their needs and start breathing, a process that involves several central cardiovascular adjustments. In the mammalian fetus, the lungs are collapsed, presenting a high resistance to blood flow. The pulmonary artery is joined to the systemic arch via a short, but large-diameter, blood vessel, the **ductus arteriosus** (Figure 12-22). Heart function in the mammalian fetus exhibits several important features:

- A marked right-to-left (P⟶ S) shunt operates; that is, blood flows from the pulmonary to the systemic circuit.
- Most of the blood ejected by the right ventricle is returned to the systemic circuit via the ductus arteriosus.
- Blood flow through the pulmonary circuit is greatly reduced.

At birth, the lungs are inflated, reducing the resistance to flow in the pulmonary circuit. Blood ejected from the right ventricle now passes into the pulmonary arteries, resulting in increased venous return to the left side of the heart. At the same time, the placental circulation

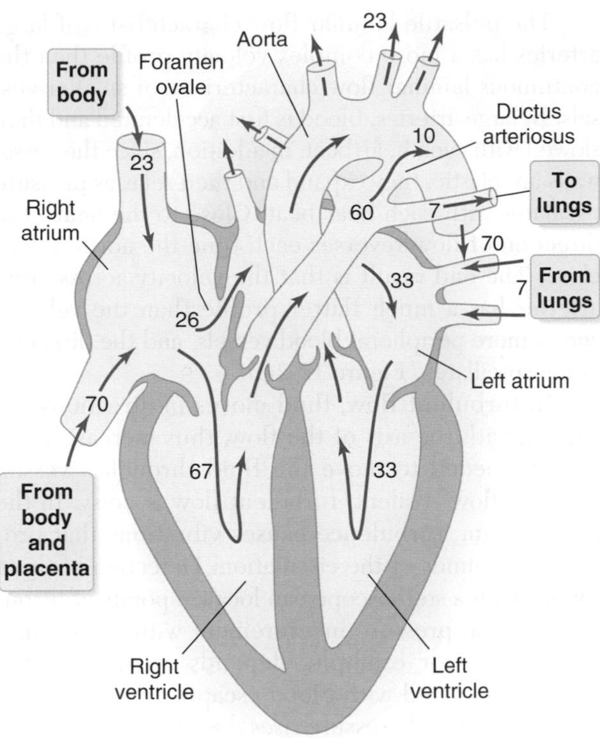

Figure 12-22 In the mammalian fetal heart, most of the blood ejected from the right ventricle returns to the systemic circulation via the ductus arteriosus. Oxygenated blood returning from the placenta is shunted from the right to the left atrium through the foramen ovale and then pumped into the aorta. After birth, the ductus arteriosus and the foramen ovale normally close, so that the systemic and pulmonary circulations are separated. The numbers refer to the percentage of the combined cardiac output from the right and left ventricles that flows to and from different regions of the body.

disappears, and the resistance to flow increases markedly in the systemic circuit. Pressures in the systemic circuit rise above those in the pulmonary circuit. If the ductus arteriosus fails to close after birth, this pressure difference results in a left-to-right (S→ P) shunt, with blood flowing from the systemic to the pulmonary circuit. Generally, however, the ductus arteriosus becomes occluded, and blood flow through the ductus does not persist.

If the ductus arteriosus remains open after birth, blood flow to the lungs exceeds systemic flow because a portion of the left ventricular output passes via the ductus into the pulmonary arteries. In these circumstances, systemic flow is often normal, but pulmonary flow may be twice the systemic flow, and cardiac output from the left ventricle may be twice that from the right ventricle. The result is a marked hypertrophy of the left ventricle. The work done by the left ventricle during exercise is also much greater than normal, and the capacity to increase output is limited. As a result, the maximum level of exercise is much reduced. Furthermore, this condition increases the blood pressure in the lungs, leading to a greater fluid loss across lung capillary walls and to possible pulmonary congestion. These problems become detrimental only when the left ventricle has become enlarged. An open ductus arteriosus is readily and easily correctable by surgery.

During gestation, fetal blood is oxygenated in the placenta and mixed with the blood returning to the heart from the lower body via the inferior vena cava, a vein that in turn empties into the right atrium. In the fetus, there is an opening in the interatrial septum, called the **foramen ovale**, that is covered by a flap valve. Oxygenated blood returning via the inferior vena cava is directed into the left atrium through the foramen ovale. It is then pumped from the left atrium into the left ventricle and ejected into the aorta, whence it flows to the head and upper limbs. Deoxygenated blood returning to the right atrium via the superior vena cava is preferentially directed toward the right ventricle, whence it flows into the systemic circuit via the ductus arteriosus. At birth, the pressure in the left atrium exceeds the pressure in the right atrium; as a result, the foramen ovale closes, but its position can be discerned later by the presence of a permanent depression.

Bird embryo In a bird egg, a network of blood vessels called the **chorioallantois** lies just under the shell. Oxygen diffusing across the shell is taken up by blood passing through the chorioallantois. Oxygenated blood leaving the chorioallantois and deoxygenated blood from the head and body enter the right atrium of the bird embryo's heart. Oxygenated blood from the chorioallantoic circulation passes from the right into the left atrium through several large and numerous small holes in the interatrial septum. The oxygenated blood is then pumped into the left ventricle and ejected into the aorta, whence it flows to the head and body. After the young bird hatches, the holes in the interatrial septum close, completely separating the pulmonary and systemic circulations.

HEMODYNAMICS

As we have noted, contractions of the heart generate blood flow through the vessels—arteries, capillaries, and veins—that form the circulatory system. Before examining the properties of these vessels in detail, it is necessary to discuss the general patterns of blood flow in these vessels and the relationship between pressure and flow in the circulatory system. The laws describing the relationships between pressure and flow apply to both open and closed circulatory systems.

In vertebrates and other animals with a closed circulation, the blood flows in a continuous circuit. Since water is incompressible, blood pumped by the heart causes the flow of an equivalent volume in every other part of the circulation. That is, at any one time, the same number of liters per minute flows through the arteries, capillaries, and veins. Furthermore, unless there is a change in total blood volume, a reduction in

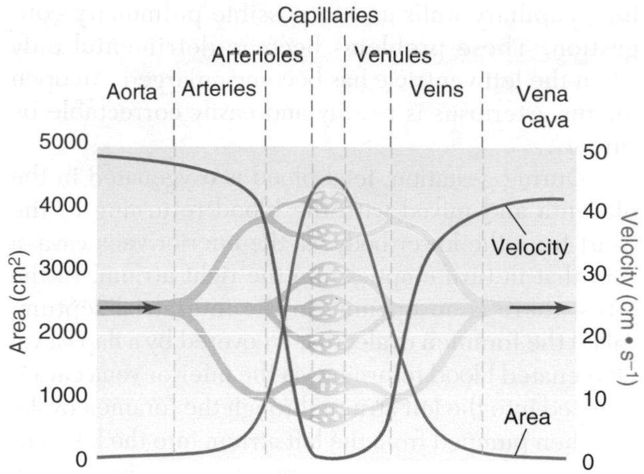

Classic Work **Figure 12-23** Blood velocity is inversely proportional to the cross-sectional area of the circulation at any given point. Blood velocity is highest in the arteries and veins and lowest in the capillaries; the converse is true for the cross-sectional area. [Adapted from Feigl, 1974.]

volume in one part of the circulation must lead to an increase in volume in another part.

The velocity of flow at any point in the circulation is related not to the proximity of the heart, but to the total cross-sectional area of that part of the circulation—that is, to the sum of the cross-sections of all vessels at that point in the circulation. The highest velocities of blood flow occur where the total cross-sectional area is smallest (and the lowest velocities where the cross-sectional area is largest). The arteries have the smallest total cross-sectional area, whereas the capillaries have by far the largest. Thus, the highest velocities of flow occur in the aorta and pulmonary artery in mammals; then velocity falls markedly as blood flows through the capillaries, but it rises again as blood flows through the veins (Figure 12-23). The slow flow of blood in capillaries has functional significance, because it is in capillaries that the time-consuming exchange of substances between blood and tissues takes place.

Laminar and Turbulent Flow

In many smaller vessels of the circulation, blood flow is streamlined. Such continuous **laminar flow** is characterized by a parabolic velocity profile across the vessel (Figure 12-24a). Flow is zero at the wall and maximal at the center along the axis of the vessel. A thin layer of blood adjacent to the vessel wall, called the boundary layer, does not move, but the next layer of fluid slides over this layer, and so on, with each successive layer moving at an increasingly higher velocity. A pressure difference supplies the force required to slide adjacent layers past each other, and *viscosity* is a measure of the resistance to sliding between adjacent layers of fluid. An increase in the viscosity of a fluid means that a larger pressure difference is required to maintain the same rate of flow, as we will see.

The pulsatile laminar flow characteristic of large arteries has a more complex velocity profile than the continuous laminar flow characteristic of smaller vessels. In large arteries, blood is first accelerated and then slowed with each heartbeat; in addition, since the vessel walls are elastic, they expand and then relax as pressure oscillates with each heartbeat. Close to the heart, the direction of flow reverses each time the aortic valves close. The end result is that the velocity across large arteries has a much flatter profile than the velocity across more peripheral blood vessels, and the direction of flow oscillates (Figure 12-24b).

In **turbulent flow**, fluid moves in directions not aligned with the axis of the flow, thus increasing the energy needed to move the fluid through a vessel. Laminar flow is silent; turbulent flow is noisy. In the bloodstream, turbulence causes vibrations that produce the sounds of the circulation. Detection of these sounds with a stethoscope can localize points of turbulence. Blood pressure measurement with a sphygmomanometer, for example, depends on hearing the sounds associated with blood escaping past the pressure cuff as blood pressure rises during systole. Sounds can be heard in vessels when blood velocity exceeds a certain critical value and in heart valves when they open and close.

Although turbulent flow is uncommon in the peripheral circulation, it does occur in some situations.

(a) Continous laminar flow

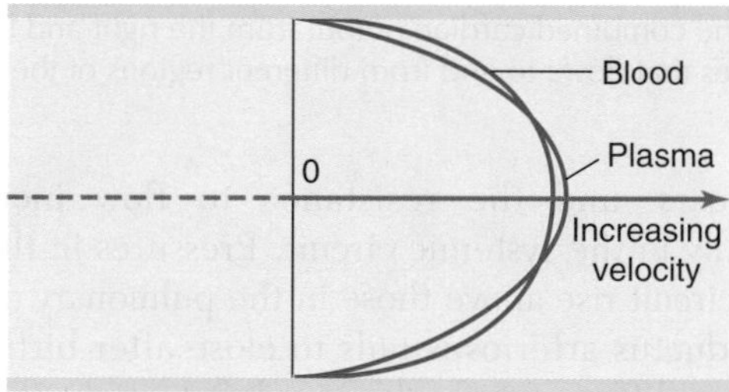

(b) Pulsatile laminar flow

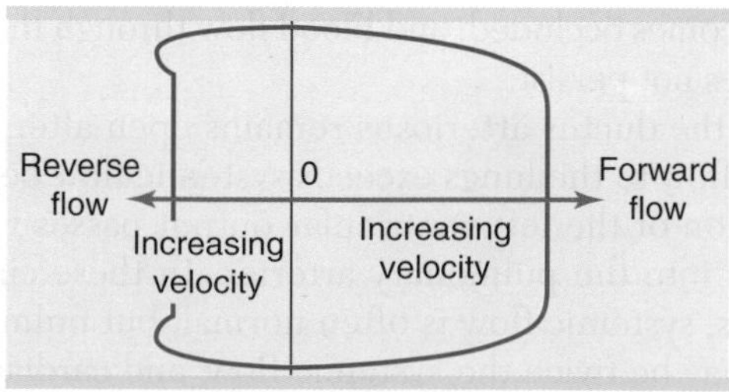

Figure 12-24 Blood flow through smaller vessels approximates continuous laminar flow, but in large, elastic arteries, pulsatile laminar flow is observed. As shown in these velocity profiles, the flow rate is higher toward the center of the vessel. **(a)** The presence of red blood cells flattens the velocity profile of blood compared with that of plasma. **(b)** Pulsatile flow is marked by a flat profile and reversal of flow during each heartbeat.

The **Reynolds number** *(Re)* is an empirically derived value that indicates whether flow will be laminar or turbulent under a particular set of conditions. A high Reynolds number indicates that flow will be turbulent; a low number indicates that flow will be laminar. The *Re* is directly proportional to the flow rate, $\dot{Q}$ (in milliliters per second), and density, ρ, of the blood, and inversely proportional to the inside radius of the vessel, r (in centimeters), and viscosity, η, of the blood:

$$Re = \frac{2\dot{Q}\rho}{\pi r \eta} \qquad (12\text{-}2)$$

The ratio of viscosity to density (η/ρ) is the *kinematic viscosity*. The greater the kinematic viscosity, the less the likelihood that turbulence will occur. The relative viscosity of blood (that is, its viscosity relative to water), and therefore its kinematic viscosity, increases with **hematocrit** (volume of red blood cells per unit volume of blood), so the presence of red blood cells decreases the occurrence of turbulence in the bloodstream.

In general, blood velocity is seldom high enough to create turbulence in undivided vessels with smooth walls, except during the very high blood flows associated with strenuous exercise. The highest flow rates in the mammalian circulation are in the proximal portions of the aorta and pulmonary artery, and turbulence may occur distal to the aortic and pulmonary valves at the peak of ventricular ejection or during backflow of blood as these valves close. Generally, in portions of the circulation where vessel walls are smooth and the vessels are undivided, flow will be turbulent only if *Re* is greater than about 1000, a value seldom observed. Small eddies may form at arterial branches and, like the eddies in rivers, may become detached from the main flow regime and be carried downstream as small discrete regions of turbulence. These eddies can form in the circulation when the *Re* is as low as 200.

Relationship Between Pressure and Flow

Flow can occur between two sites where there is a difference between them in potential energy, which can be measured as a difference in pressure. Thus differences in pressure between two points in a flow path establish a pressure gradient, and therefore the direction of flow—from high to low pressure. (An exception is a fluid at rest under gravity, in which pressure increases uniformly with depth, but flow does not occur.) When the heart contracts, the potential energy (pressure) in the ventricle increases. The pressure generated by the heart contraction is dissipated by the flow of blood because energy is used to overcome the resistance to flow through the vessels. For this reason, blood pressure falls as the blood passes from the arterial to the venous side of the circulation (Figure 12-25).

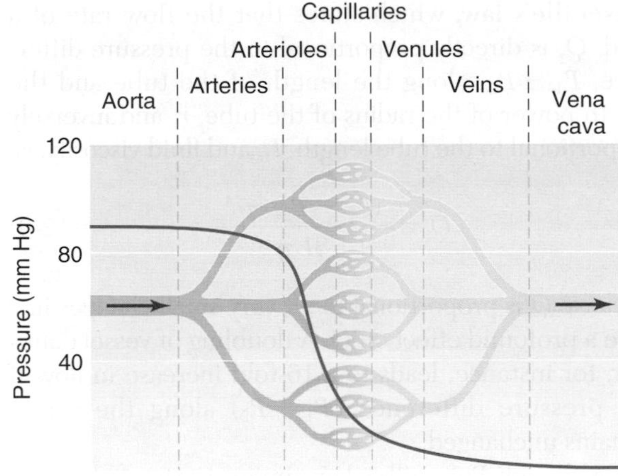

Classic Work **Figure 12-25** The pressure (potential energy) generated during each cardiac contraction is dissipated in overcoming the resistance to flow provided by the blood vessels. Because resistance is highest in the arterioles, the major pressure decrease occurs in this region of the circulation. [Adapted from Feigl, 1974.]

Role of kinetic energy

Energy is expended in setting the blood in motion, but once in motion, flowing blood has inertia; that is, fluids in motion possess kinetic energy. In static fluids, potential energy is measured in terms of pressure; in fluids in motion, potential energy is measured in terms of both pressure and kinetic energy. As we'll see, however, kinetic energy generally makes a negligible contribution to the flow rate of blood. The kinetic energy per milliliter of fluid is given by $\frac{1}{2}(\rho v^2)$, where ρ is the density of the fluid and v is the velocity of flow. If the velocity is given in centimeters per second and density in grams per millimeter, then kinetic energy can be measured in dynes per square centimeter, the same units as pressure.

The maximum velocity of blood flow occurs at the base of the aorta in mammals and is about 50 $\text{cm} \cdot \text{s}^{-1}$ at the peak of ventricular ejection; the density of the blood ejected is about 1.055 $\text{g} \cdot \text{ml}^{-1}$. Thus the kinetic energy of the blood in the aorta during peak ejection is calculated to be $\frac{1}{2} \times 1.055 \times 50^2 = 1318.75 \text{ dynes/cm}^2 = 131.875$ Pa, which is approximately 1 mm Hg (see Spotlight 12-1). This number is small compared with peak systolic transmural pressures of around 120 mm Hg. Blood velocity is low in the ventricle but accelerates as blood is ejected into the aorta; that is, blood gains kinetic energy as it leaves the ventricle. Pressure is converted into kinetic energy as blood is ejected from the heart, and this conversion accounts for most of the small drop in pressure that occurs between the ventricle and the aorta. Kinetic energy is highest in the aorta. In fact, the heart structure is adapted to maintain momentum in the blood flow (see Figure 12-10), reducing the cost of pumping blood. In the capillaries, the velocity of flow is only about 1 $\text{mm} \cdot \text{s}^{-1}$, and kinetic energy is therefore negligible.

Poiseuille's law

The relationship between pressure and continuous laminar flow of a fluid in a rigid tube is described by

Poiseuille's law, which states that the flow rate of a fluid, $\dot{Q}$, is directly proportional to the pressure difference, $P_1 - P_2$, along the length of the tube and the fourth power of the radius of the tube, r, and inversely proportional to the tube length, L, and fluid viscosity, η:

$$\dot{Q} = \frac{(P_1 - P_2)\,\pi r^4}{8L\eta} \tag{12-3}$$

Because $\dot{Q}$ is proportional to r^4, very small changes in r have a profound effect on $\dot{Q}$. A doubling of vessel diameter, for instance, leads to a 16-fold increase in flow if the pressure difference $(P_1 - P_2)$ along the vessel remains unchanged.

Although Poiseuille's law applies to steady flows in straight, rigid tubes, it can be used, with some limitations, to analyze the relationship between pressure and flow in small arteries (arterioles), capillaries, and veins, even though these are not rigid tubes. In arteries, blood pressure and flow are pulsatile, and blood is a complex fluid consisting of plasma and cells. Because the arterial walls are not rigid, the oscillations in the pressure and flow of blood are not in phase; consequently, the relationship between the two is no longer accurately described by Poiseuille's law.

How useful, then, is Poiseuille's law? The extent of the deviation of the relationship between pressure and flow from that predicted by Poiseuille's law is indicated by the value of a nondimensional constant α:

$$\alpha = r\frac{\sqrt{2\pi nf\rho}}{\eta} \tag{12-4}$$

where r and η are the density and viscosity of the fluid, respectively; f is the frequency of oscillation; n is the order of the harmonic component, and r is the radius of the vessel. If α is 0.5 or less, the relationship between pressure and flow is approximated by Poiseuille's law. Because the value for α in the small terminal arteries and veins is about 0.5, this equation can be used to analyze the relationship between pressure and flow in this portion of the circulation. In contrast, the values of α for the arterial systems of mammals and birds range from 1.3 to 16.7, depending on the species and the physiological state of the animal. Thus, Poiseuille's law is not applicable to this portion of the circulation.

There have been only a few studies *in vivo* of microcirculation due to the difficulty of measuring blood flow and pressure in capillaries. In those tissues for which the relationship between pressure and flow in the microcirculation has been measured, it has been found to be nonlinear, indicating that Poiseuille's law does not accurately describe the microcirculation. There are two reasons for this: the capillaries are branched, with collateral pathways that may open and close, and they are so small that red blood cells are deformed as they squeeze through the capillaries, complicating the mathematical description of the flow.

Resistance to flow

Because it is often difficult or impossible to measure the radii of all vessels in a vascular bed, we designate $8L\eta/\pi r^4$, the inverse of the term in Poiseuille's law (Equation 12-3), as the resistance to flow, R, which is equal to the pressure difference $(P_1 - P_2)$ across a vascular bed divided by the flow rate, $\dot{Q}$:

$$R = \frac{P_1 - P_2}{\dot{Q}} = \frac{8L\eta}{\pi r^4} \tag{12-5}$$

The resistance to flow in the peripheral circulation is sometimes expressed in peripheral resistance units (PRUs), with 1 PRU being equal to the resistance in a vascular bed when a pressure difference of 1 mm Hg results in a flow of 1 $ml \cdot s^{-1}$.

Blood flow through a vessel increases with an increased pressure differential along the vessel and decreased resistance to flow, which is inversely proportional to the fourth power of the radius of the vessel. As pressure increases in an elastic vessel, so does the radius; as a result, flow increases as well. Let us consider a blood vessel with a constant pressure differential along its length but operating at two pressure levels:

Example 1: Input pressure 100 mm Hg and outflow pressure 90 mm Hg; Δ10 mm Hg

Example 2: Input pressure 20 mm Hg and outflow pressure 10 mm Hg; Δ10 mm Hg

The flow rate in this vessel will be much greater at the higher pressures (example 1) if the vessel is distensible, simply because the radius of the vessel will be increased and the resistance to flow thus reduced.

Viscosity of blood

According to Poiseuille's law, the flow of blood is inversely related to its viscosity. Plasma has a viscosity relative to water of about 1.8. The addition of red blood cells further increases the relative viscosity of blood, so that mammalian and bird blood at 37°C have a relative viscosity of between 3 and 4. Thus, owing largely to the presence of red blood cells, blood behaves as though it were three or four times more viscous than water. This characteristic means that larger pressure gradients are required to maintain the flow of blood through a vascular bed than would be needed if the vascular bed were perfused by plasma alone. However, blood flowing through small vessels behaves as if its relative viscosity were much reduced. In fact, in vessels less than 0.3 mm in diameter, the relative viscosity of blood decreases with vessel diameter and approaches the viscosity of plasma. This phenomenon, called the **Fahraeus-Lindqvist effect**, will be explained momentarily.

As we saw earlier, the velocity profile across a vessel with continuous laminar fluid flow is parabolic, as is seen with plasma (see Figure 12-24a). The maximum velocity is twice the mean velocity, which can be determined by dividing the flow rate by the cross-sectional area of the tube. The rate of change in velocity is maximal near the walls and decreases toward the center of the vessel. In flowing blood, red cells tend to accumulate in the center of the vessel, where the velocity is highest but the rate of change in velocity between adjacent layers smallest. This accumulation leaves the walls relatively free of cells, so that fluid flowing from this area into small side vessels has a low concentration of red blood cells and consists almost entirely of plasma. This process is referred to as **plasma skimming.**

The accumulation of red blood cells in the center of the bloodstream means that blood viscosity is highest in the center and decreases toward the walls. This difference in viscosity between the center and the walls of the bloodstream alters the velocity profile of blood compared with that of plasma, resulting in a slight increase in blood flow at the walls and a slight reduction in flow at the center; that is, the parabolic shape of the velocity profile is flattened somewhat.

The hematocrit in small vessels is lower than that in larger ones. In small vessels, the boundary layer of plasma occupies a larger portion of the vessel lumen at a given flow velocity than in larger vessels. This axial flow of red blood cells in small vessels means that the greatest change in velocity occurs in the plasma layers close to the walls and explains why the apparent viscosity of blood flowing in these small vessels approaches that of plasma. Thus the Fahraeus-Lindqvist effect can be explained in terms of the reduced hematocrit seen in small vessels. This decrease in the apparent viscosity of blood, which occurs in arterioles, reduces the energy required to drive blood through the microcirculation.

In very small vessels—those with a diameter of approximately 5 to 7 μm—further decreases in diameter lead to an inversion of the Fahraeus-Lindqvist effect; namely, an increase in the apparent viscosity of blood. In such small vessels, a red blood cell completely fills the lumen and is distorted as it passes through. Because the plasma membrane of a mammalian red blood cell is not firmly anchored to the underlying structures, it can move over its own cell contents, acting somewhat like a tank tread as it moves along the walls of the vessel. Deformation of red blood cells in small vessels leads to a complex flow of plasma membrane and surrounding fluid as the cells squeeze through the narrow lumen. This tank-treading effect is probably limited to those red blood cells without a nucleus. Non-mammalian vertebrates have nucleated red blood cells with plasma membranes more firmly tied to the underlying structures.

Antarctic teleost fish operate at temperatures close to or even below 0°C. What effects might this have on pressure and flow in the circulation of these fish? What evolutionary modifications might be expected to compensate for these low temperatures?

Compliance in the circulatory system

A further consideration in analyzing the relationship between pressure and flow in the circulation is that blood vessels contain elastic fibers that enable them to be distended. Vessels are not, in fact, the straight, rigid tubes to which Poiseuille's law applies. Rather, as pressure in a vessel increases, its walls are stretched and its volume increases. The ratio of change in volume to change in pressure is termed the **compliance** of a system. The compliance of a system is related to its size and the elasticity of its walls. The greater the initial volume and the elasticity of the walls, the greater will be the compliance of the system.

The venous system is very compliant; that is, small changes in pressure produce large changes in volume. The venous system can act as a volume reservoir because large changes in volume have little effect on venous pressure (and therefore on the filling of the heart during diastole or on capillary blood flow). The arterial system, which overall is less compliant than the venous system, acts as a pressure reservoir in order to maintain capillary blood flow. Nevertheless, the portions of the arterial system near the heart are elastic enough both to damp the oscillations in pressure generated by contractions of the heart and to maintain flow in distal arteries during diastole.

THE PERIPHERAL CIRCULATION

Blood pumped from the left ventricle of the mammalian heart carries oxygenated blood via the arterial system to capillary beds in the tissues, where oxygen and nutrients are exchanged for carbon dioxide and waste products. The systemic venous system returns the deoxygenated blood to the right atrium (see Figure 12-3). Although all blood vessels share some structural features, the vessels in various parts of the peripheral circulation are adapted for the functions they serve. Figure 12-26 on the next page illustrates the structure of various-sized arteries and veins. A layer of endothelial cells, called the **endothelium,** lines the lumen of all blood vessels. In larger vessels, the endothelium is surrounded by a layer of elastic and collagenous fibers, but the walls of capillaries consist of a single layer of endothelial cells. Circular and longitudinal smooth muscle fibers may intermingle with or surround the elastic and collagenous fibers. The walls of larger blood vessels comprise three layers:

- **tunica adventitia,** the limiting fibrous outer coat
- **tunica media,** the middle layer, consisting of circular and longitudinal smooth muscle
- **tunica intima,** the inner layer, closest to the lumen, composed of endothelial cells and elastic fibers

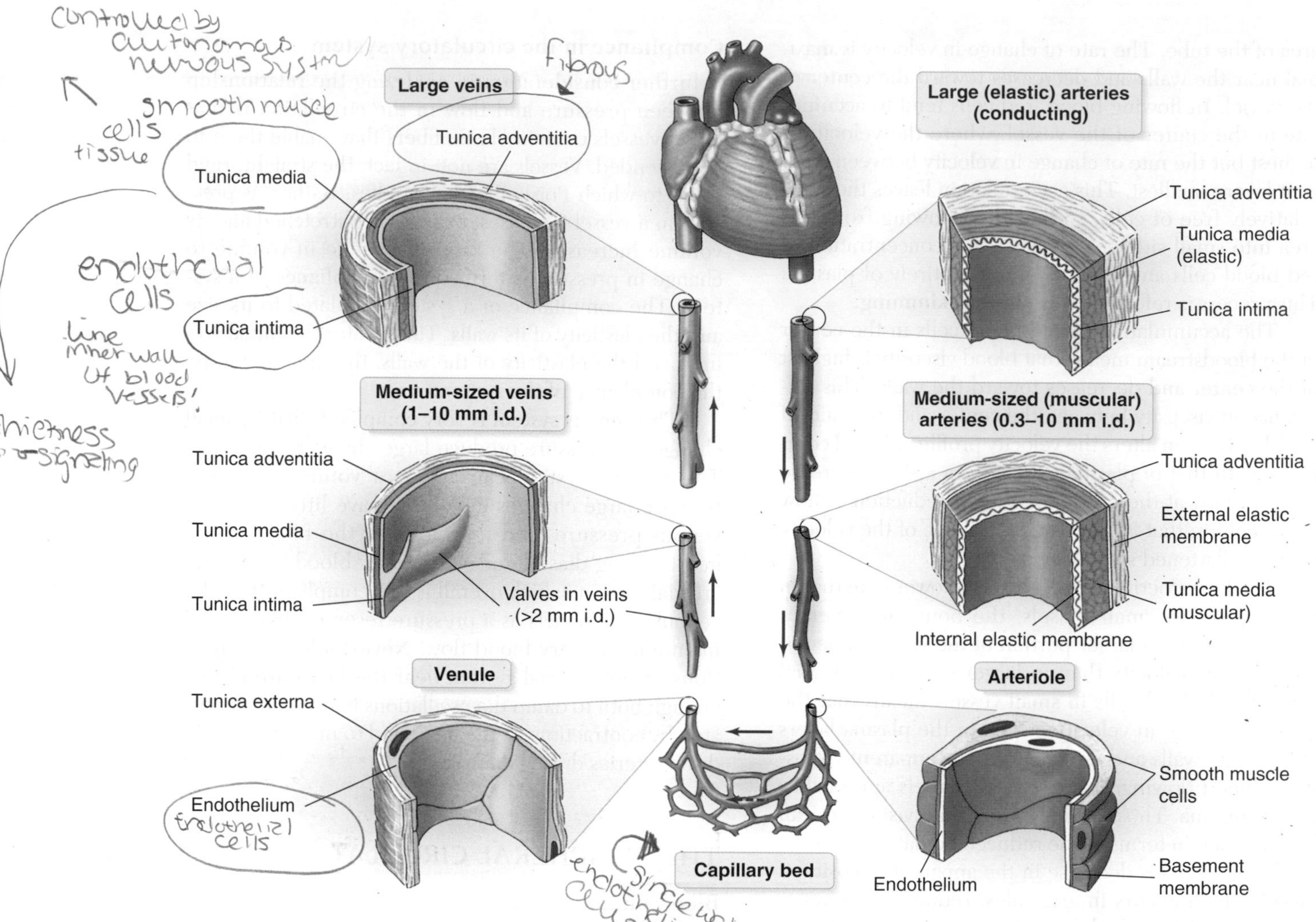

Figure 12-26 In the mammalian peripheral circulation, blood flows from the heart via progressively smaller arteries, then through the microcirculation, and finally back to the heart via progressively larger veins. A layer of endothelial cells, the endothelium, lines the lumen of all vessels. In larger vessels, the endothelium is surrounded by a muscle layer (tunica media) and outer fibrous layer (tunica adventitia). [Adapted from Martini and Timmons, 1995.]

The boundary between the tunica intima and the tunica media is not well defined; the tissues blend into one another. Because arteries are more muscular than veins, they have a thicker tunica media, and the larger arteries close to the heart are more elastic, with a wide tunica intima. The thick walls of large blood vessels require their own capillary circulation, termed the **vasa vasorum.** In general, arteries have thicker walls and much more smooth muscle than veins of similar outside diameter. In some veins, muscular tissue is absent.

Arterial System

The arterial system consists of a series of branching vessels with walls that are thick, elastic, and muscular—well suited to deliver blood from the heart to the fine capillaries that carry blood through the tissues.

Arteries serve four main functions, as illustrated in Figure 12-27:

1. They act as a conduit for blood between the heart and capillaries.
2. They act as a pressure reservoir for forcing blood into the small-diameter arterioles.
3. They damp the oscillations in pressure and flow generated by the heartbeat and produce a more even flow of blood into the capillaries.
4. They control the distribution of blood to different capillary networks via selective constriction of the terminal branches of the arterial tree.

Arterial blood pressure, which is finely controlled, is determined by the volume of blood the arterial system

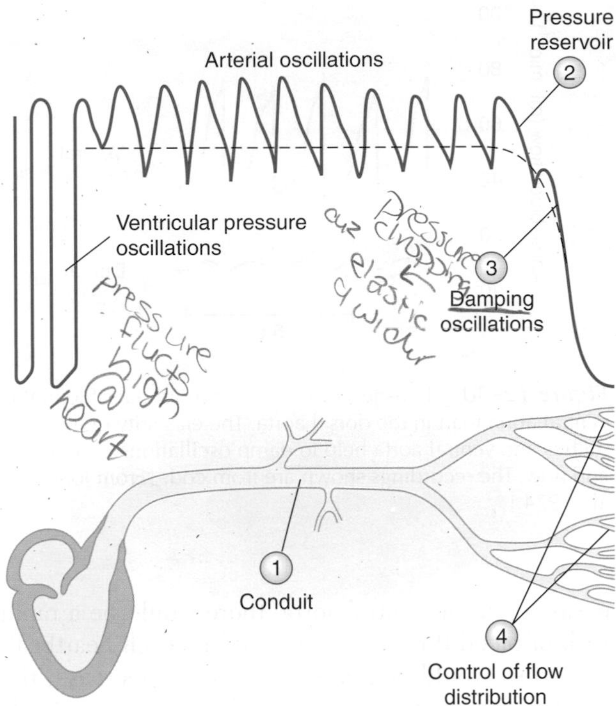

Classic Work ***Figure 12-27*** The systemic arterial system functions as a conduit and pressure reservoir; it also smoothes out pressure oscillations and controls the distribution of flow to capillaries. The conduit function (1) is served by the vascular channels along which blood flows toward the periphery with minimal frictional loss of pressure. The distensible walls and high outflow resistance of arteries account for the pressure reservoir function (2) and damping of oscillations in pressure and flow (3). Selective constriction in the peripheral arterial beds controls the distribution of blood to the various tissues (4). [Adapted from Rushmer, 1965a.]

contains and the properties of its walls. If either is altered, the pressure will change. The volume of blood in the arteries is determined by the rate of filling via cardiac contractions and of emptying via arterioles into capillaries. If cardiac output increases, arterial blood pressure will rise; if capillary flow increases, arterial blood pressure will fall. Normally, however, arterial blood pressure varies little because the rates of filling and emptying (i.e., cardiac output and capillary flow) are evenly matched.

Blood flow through the capillaries is proportional to the pressure difference between the arterial and venous systems. Because venous pressure is low and changes little, arterial pressure exerts primary control over the rate of capillary blood flow and is responsible for maintaining adequate perfusion of the tissues. Arterial pressure varies among species, generally ranging from 50 to 150 mm Hg. Pressure differences are small along large arteries (less than 1 mm Hg), but pressure drops considerably along small arteries and arterioles because of increasing resistance to flow with decreasing vessel diameter.

The oscillations in blood pressure and flow generated by contractions of the heart are damped in the arterial system by the elasticity of the arterial walls. As blood is ejected into the arterial system, pressure rises and the vessels expand. As the heart relaxes, blood flow to the periphery is maintained by the elastic recoil of the vessel walls, resulting in a reduction in arterial volume. If arteries were simply rigid tubes, pressures and flow in the periphery would exhibit the same stops and starts that occur at the exit of the ventricle during each heartbeat. Although elastic, arteries become progressively stiffer with increasing distension. As a result, they are easily distended at low pressures, but resist further expansion at high pressures. The response of arterial walls to distension is similar in a wide variety of animals, reflecting similar structural and functional characteristics (Figure 12-28).

According to Laplace's law, the wall tension required to maintain a given transmural pressure within a hollow structure increases with increasing radius (see Equation 12-1). Elastic vessels thus are unstable and tend to balloon; that is, since they cannot develop high wall tension as pressure increases, they tend to bulge. In blood vessels, this instability is prevented by a collagen sheath that limits their expansion. Ballooning of a blood vessel (aneurism) can occur, however, if the collagen sheath breaks down.

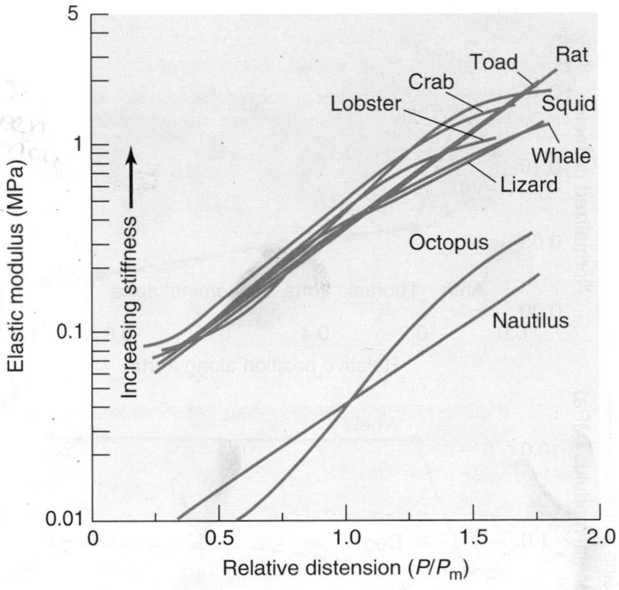

Figure 12-28 The elastic properties of arteries are surprisingly similar in a wide variety of animals, with nautilus and octopus being notable exceptions. This similarity is reflected in plots of the elastic modulus (the ratio of the applied load per unit area and the length change per unit length) versus relative distension, expressed as pressure (P) divided by the resting blood pressure (P_m) of the species. [Adapted from Shadwick, 1992.]

In general, the elasticity of the arterial wall, as well as the thickness of the muscular layer, decreases with increasing distance from the heart. Away from the heart, the arteries become more rigid and serve primarily as blood conduits. In a whale, the aortic arch at the exit from the heart is very elastic and has a large diameter, but the arterial system beyond the aortic arch narrows rapidly and becomes much more rigid than that of smaller animal, such as a dog (Figure 12-29). The elastic whale aortic arch expands with each heartbeat, accommodating about 50%–75% of the stroke volume; the remainder flows into the portion of the arterial system downstream of the aortic arch. The change in ventricular volume with each heartbeat can be as much as 35 liters in a large whale, which has a heart rate of around 12–18 beats per minute.

The extent of elastic tissue in arteries varies depending on the particular function of each vessel. In fishes, for example, blood pumped by the heart is forced into an elastic bulbus and ventral aorta (see Figure 12-16a). The blood then flows through the gills and passes into a dorsal aorta, the main conduit for the distribution of blood to the rest of the body. A smooth, continuous flow of blood through the gill capillaries is required for efficient gas transfer. The bulbus, the ventral aorta, and the afferent branchial arteries leading to the gills are very compliant and act to smooth and maintain flow in the gills in the face of the large oscillations produced by contractions of the heart. The dorsal aorta, which receives blood from the gills, is much less elastic than the ventral aorta. If the dorsal aorta were more elastic than the ventral aorta, there would be a rapid rush of blood through the gills during each heartbeat. This rush would increase, rather than decrease, the oscillations in flow through the gills. To ensure a steady blood flow through the gill capillaries, the major compliance must be placed before, not after, the gills (Figure 12-30).

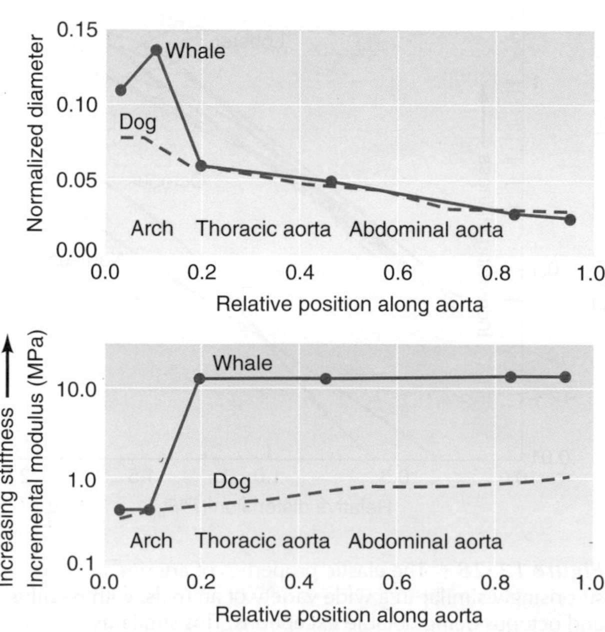

Figure 12-29 The arterial system of dogs and whales becomes stiffer and smaller in diameter with distance from the heart. In whales, there is an abrupt decrease in diameter and increase in stiffness between the aortic arch and the thoracic aorta. [Adapted from Gosline and Shadwick, 1996.]

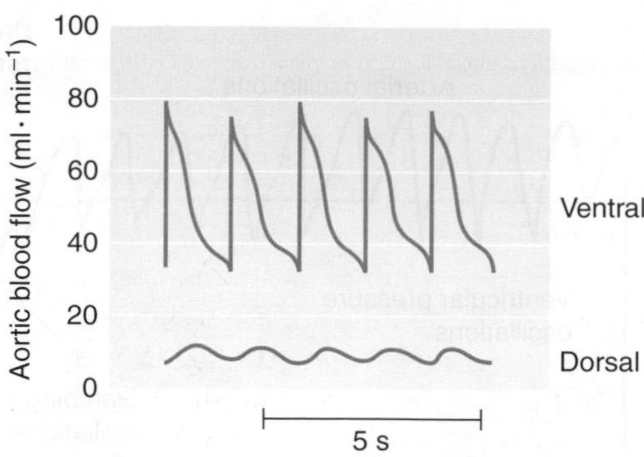

Figure 12-30 In fishes, blood flow is more pulsatile in the ventral aorta than in the dorsal aorta. The elasticity of the bulbus and ventral aorta help to damp oscillations in pressure and flow. The recordings shown are from cod. [From Jones et al., 1974.]

How might the arterial system of an invertebrate with an open circulation vary in structure and function from the vertebrate arterial system?

Blood pressure

Blood pressures in the arterial system are usually reported as transmural pressures (i.e., the difference in pressure between the inside and outside, across the wall of the blood vessel; see Spotlight 12-1). The pressure outside vessels is usually close to the ambient pressure, but changes in the extracellular pressure of tissues can have a marked effect on transmural pressure and therefore on vessel diameter and blood flow. For example, contractions of the heart raise pressure around coronary vessels and result in a marked reduction in coronary flow during systole. Inhaling is associated with a reduction in thoracic pressure and thus raises transmural pressure in veins leading back to the heart, increasing venous return to the heart.

The maximum arterial pressure during a cardiac cycle is referred to as systolic pressure and the minimum as diastolic pressure; the difference is the **pressure pulse.** The oscillations in pressure produced by the contraction and relaxation of the ventricle are reduced at the entrances to capillary beds and nonexistent in the venous system. Heart contractions cause small oscillations in pressure within capillaries. The velocity of the pressure pulse increases as the diameter of the artery decreases and the stiffness of the arterial wall increases. In the mammalian aorta, the pressure pulse travels at 3–5 $m \cdot s^{-1}$, increasing to 15–35 $m \cdot s^{-1}$ in small arteries.

Peak blood pressure and the size of the pressure pulse within the mammalian and avian aorta both increase with distance from the heart (Figure 12-31). This pulse amplification can be large during exercise. There are three possible explanations for this rather odd phenomenon. First, pressure waves are reflected from peripheral branches of the arterial tree. The initial and reflected waves sum, and where peaks coincide, the pressure pulse and peak pressure are greater than where they are out of phase. Where the initial and reflected waves are 180° out of phase, the oscillations in pressure will be reduced. It has been suggested that the heart is situated at a point where initial and reflected waves are out of phase, thus reducing peak arterial pressure in the aorta close to the ventricle. As distance from the heart increases, the initial and reflected pressure waves move into phase, and pressure peaks are observed in the aorta. Second, the decrease in the elasticity and diameter of arteries with distance from the heart may cause an increase in the magnitude of the pressure pulse. Third, the pressure pulse is a complex waveform, consisting of several harmonics. Higher frequencies travel at higher velocities, and it has been suggested that the change in waveform of the pressure pulse with distance is due to summation of different harmonics. This third explanation is open to question, as the distances appear to be too small to allow summation of harmonics.

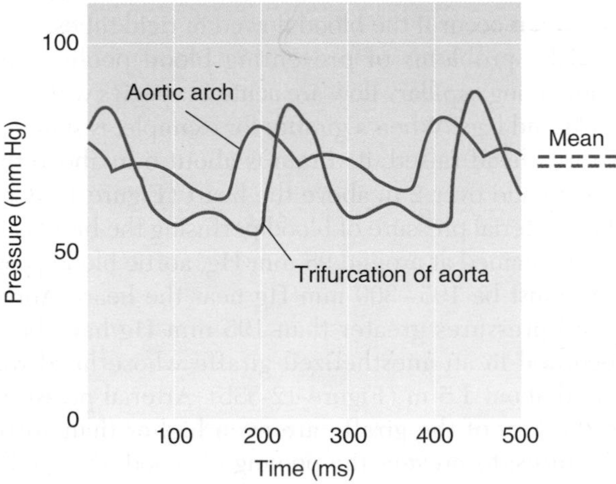

Figure 12-31 In the aorta of mammals and birds, the peak blood pressure and pressure pulse both increase with distance from the heart. Shown are simultaneous recordings of a rabbit's blood pressure in the aortic arch (2 cm from the heart) and at the location where the aorta divides into three branches (24 cm from the heart). Note that the mean pressure is slightly less at the trifurcation of the aorta than in the aortic arch, closer to the heart. [Adapted from Langille, 1975.]

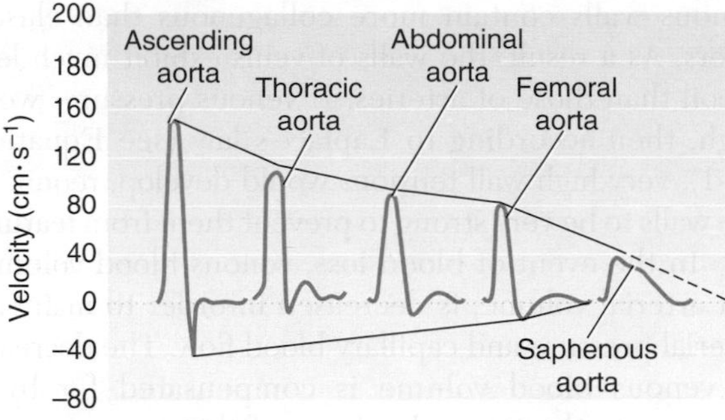

Classic Work *Figure 12-32* The maximal velocity of arterial blood flow and oscillations in flow decrease progressively with distance from the heart. A backflow phase is observed in the large arteries; in the ascending aorta, it is probably related to a brief reflux of blood through the aortic valves. These tracings were obtained from a dog's arteries. Oscillatory flow is damped out entirely in the capillaries. [Adapted from McDonald, 1960.]

Velocity of arterial blood flow

Blood flow and the oscillations in flow with each heartbeat are greatest at the exit to the ventricle, decreasing with increasing distance from the heart (Figure 12-32). At the base of the aorta, as noted earlier, flow is turbulent and reverses during diastole as closure of the aortic valves creates vortices in the blood ejected into the aorta during systole. In most other parts of the circulation, flow is laminar, and oscillations in velocity are damped by the compliance of the aorta and proximal arteries.

Mean velocity in the aorta—the point of maximal blood velocity—is calculated as about 33 $cm \cdot s^{-1}$ in humans, based on a cross-sectional area of about 2.5 cm^2 and cardiac output of about 5 $L \cdot min^{-1}$. If we assume that maximal velocity in a vessel is twice the mean velocity (valid only if the velocity profile is a parabola), then the maximal velocity of blood flow in the human aorta would be 66 $cm \cdot s^{-1}$. If cardiac output is increased by a factor of 6 during heavy exercise, maximal velocity is raised to 3.96 $m \cdot s^{-1}$. In contrast, the pressure pulse associated with each heartbeat travels through the circulation at 3–35 $m \cdot s^{-1}$; thus, the pressure pulse travels faster than the flow pulse.

Venous System

The venous system acts as a conduit for the return of blood from the capillaries to the heart. It is a large-volume, low-pressure system consisting of vessels with a larger inside diameter than the corresponding arteries (see Figure 12-26). In mammals, about 50% of the total blood volume is contained in the venous system (see Figure 12-3). Venous pressures seldom exceed 11 mm Hg (1.5 kPa, roughly 10% of arterial pressures). The large diameter and low pressure of the veins permits the venous system to function as a storage reservoir for blood. The walls of veins are much thinner, contain less smooth muscle, and are less elastic than arterial walls;

venous walls contain more collagenous than elastic fibers. As a result, the walls of veins exhibit much less recoil than those of arteries. If venous pressures were high, then according to Laplace's law (see Equation 12-1), very high wall tensions would develop, requiring the walls to be very strong to prevent them from tearing.

In the event of blood loss, venous blood volume, not arterial volume, is decreased in order to maintain arterial pressure and capillary blood flow. The decrease in venous blood volume is compensated for by a decrease in the internal volume of the venous system itself. The walls of many veins are covered by smooth muscle innervated by sympathetic adrenergic fibers. Stimulation of these nerves causes vasoconstriction and a reduction in the size of the venous reservoir. This reflex allows some bleeding to take place without a drop in venous pressure. Blood donors actually lose part of their venous reservoir; the loss is temporary, however, and the venous system gradually expands as blood is replaced by fluid retention. Volume can be replaced quickly through drinking; replacement of the other blood constituents, such as red blood cells, can take up to a week.

Venous blood flow

Blood flow in veins is affected by a number of factors other than contractions of the heart. Contractions of limb muscles and pressure exerted by the diaphragm on the gut both result in the squeezing of veins in those parts of the body. Because veins contain pocket valves that allow flow only toward the heart, this squeezing augments the return of blood to the heart. The resulting increase in venous return increases cardiac output. Activation of this skeletal muscle venous pump is associated with increased, localized activity in the sympathetic fibers innervating the venous smooth muscle, which increases smooth muscle tone. This sympathetic response ensures that the skeletal muscle pump specifically increases return to the heart, rather than simply distending another part of the venous system. In the absence of skeletal muscle contraction, there may be considerable pooling of blood in the venous system of the limbs.

Breathing in mammals also contributes to the return of venous blood to the heart. Expansion of the thoracic cage reduces pressure within the chest and draws air into the lungs; this pressure reduction sucks blood from the veins of the head and abdominal cavity into the heart and the large veins situated within the thoracic cavity. Peristaltic contractions of the smooth muscle of venules, the small vessels joining capillaries to veins, can also promote venous flow toward the heart. Such peristaltic activity has been observed in the venules of the bat wing.

Blood distribution in veins

When a person is lying down, the heart is at the same level as the feet and head, and pressures are similar in arteries in the head, chest, and limbs. When the person stands up, the relationship between the head, heart, and limbs changes with respect to gravity. The heart is now a meter above the lower limbs and lower than the head. Arterial pressure increases in the lower limbs and decreases in the head due to the different pressures associated with the height of a column of blood under gravity. In response to these changes, the sympathetic adrenergic fibers that innervate the limb veins are activated, causing contraction of venous smooth muscle and thereby promoting the redistribution of pooled blood. Such venoconstriction is inadequate, however, to maintain good circulation if the standing position is held for long periods in the absence of limb movements, as when soldiers stand immobile during a review. Under such circumstances, venous return to the heart, cardiac output, arterial pressure, and flow of blood to the brain are all reduced, which can result in fainting. Similar problems affect bedridden patients who attempt to stand after several days of inactivity and astronauts returning to Earth after a long period of weightlessness. In these instances, other control systems involving baroreceptors (pressure receptors) and arterioles may be disrupted as well. The reflex control of venous volume is normally reestablished with use.

Gravity has little effect on capillary flow, which is determined by the arterial-venous pressure difference. Gravity raises arterial and venous pressure by the same amount and therefore does not greatly affect the pressure gradient across a capillary bed. Because the vascular system is elastic, however, an increase in absolute pressure expands blood vessels, particularly the compliant veins. Thus, pooling of blood tends to occur, particularly in veins, in different regions of the body as an animal changes position with respect to gravity. This effect is related solely to the elasticity of blood vessels and would not occur if the blood flowed in rigid tubes.

The problems of preventing blood pooling and maintaining capillary flow are acute in species with long necks and legs. When a giraffe, for example, is standing with its head raised, its brain is about 6 m above the ground and over 2 m above the heart (Figure 12-33a). If the arterial pressure of blood perfusing the brain is to be maintained at around 98 mm Hg, aortic blood pressure must be 195–300 mm Hg near the heart. Aortic blood pressures greater than 195 mm Hg have been recorded in an anesthetized giraffe whose head was raised about 1.5 m (Figure 12-33b). Arterial pressures in the legs of the giraffe are even higher than aortic pressures; to prevent the pooling of blood, the giraffe has large quantities of connective tissue surrounding the leg vessels. As the giraffe lowers its head to the ground, arterial blood pressure at the level of the heart is reduced considerably, thus maintaining a relatively constant blood flow to the brain. The wide variation in aortic pressure as the giraffe moves its head could lead to extensive pooling of blood (head raised) or decreased flow (head lowered) in arterioles other than those of the

(a)

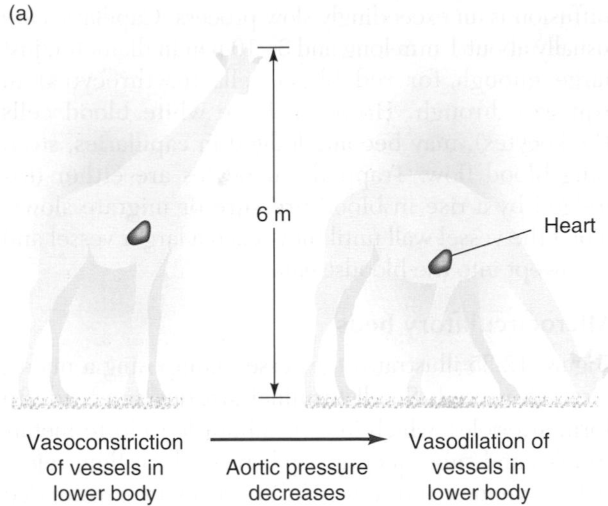

(b)

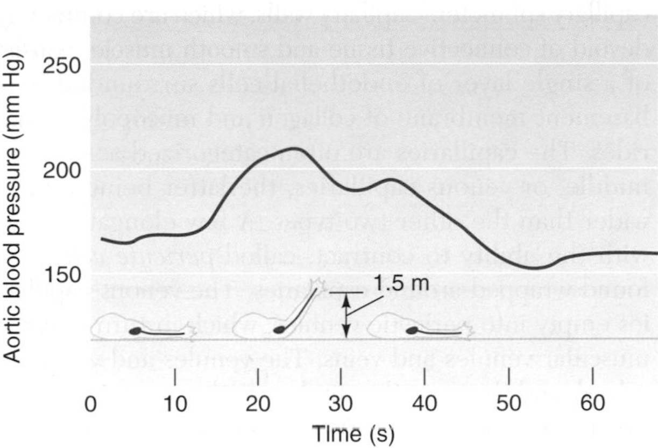

Classic Work ***Figure 12-33*** As animals with long necks, such as the giraffe, raise and lower their heads, the cardiovascular system must adjust to maintain blood flow to the brain and avoid pooling of blood in the lower part of the body. **(a)** A steady flow of blood is maintained by vasoconstriction in the limbs when the head is raised and vasodilation when it is lowered. **(b)** Aortic pressure rises dramatically when the head is raised or lowered. [Adapted from White, 1972.]

head. Pooling most likely is prevented by vasoconstriction of these peripheral vessels when the head is raised. Conversely, when the head is lowered, extensive vasodilation of arterioles leading to capillary beds other than those in the head probably maintains flow despite the lower aortic pressure.

The ability of the giraffe to regulate pressure and flow in peripheral vessels other than those of the head is particularly crucial for its kidney function. If the kidney tubules were subjected to the enormous changes in blood pressure associated with the raising and lowering of the giraffe's head, the rate of glomerular filtration would be chaotic. Each time the animal lifted its head, the large increase in arterial blood pressure would lead to a very high rate of ultrafiltrate formation in the kidney; this large volume of ultrafiltrate would in turn require that fluid be reabsorbed at an equally high rate. In the absence of any appropriate controls, the giraffe could lower its head to drink and then lose any fluid gained as it was filtered through the kidney when the head was raised. Thus, the giraffe must have mechanisms for adjusting peripheral resistance to flow in various capillary beds as it swings its head from ground level to a height of 6 m. Similar problems face camels and other animals with long necks.

The organization of the venous system is influenced by the degree of support offered by the medium in which the animal lives. There was an extensive reorganization of the venous system as vertebrates moved onto land and lost the support of water. Pooling of blood is not a problem for animals in water because the density of water is only slightly less than that of blood. In water, hydrostatic pressure increases with depth and effectively matches the increase in blood pressure due to gravity; thus transmural pressure does not change much, so the blood does not pool. Pooling became an immediate problem for terrestrial animals because air is much less dense than blood. The required changes in the venous system were in addition to those required to maintain separation of oxygenated and deoxygenated blood moving through the heart.

Although the effects of gravity are minimal in aquatic animals, swimming fish face a related problem affecting venous return to the heart. As the fish moves forward, blood collects in the tail due both to inertia and to compression waves associated with the swimming movements that pass down the body of the fish. To diminish these problems, most veins returning to the heart pass down the center of the fish's body. Some fishes also have an accessory caudal heart in the tail, which propels blood toward the central heart (Figure 12-34 on the next page). In addition, the flow of water over the pectoral region of some fishes may reduce hydrostatic pressure in that region, so that venous return to the heart is promoted with increased swimming speed.

Many dinosaurs had *very* long necks. How should long-necked dinosaurs have carried their heads when running, and why? How would problems of venous pooling differ for terrestrial and aquatic dinosaurs?

Countercurrent exchangers

Countercurrent exchangers are a common feature in the design of animals (see Spotlight 14-2). In many animals, arteries and veins run next to each other, with their blood flows moving in opposite directions (i.e., countercurrent blood flow). In many such instances, especially if the vessels are small, there is an exchange of heat between the countercurrent blood flows. Because heat is transferred much more readily than

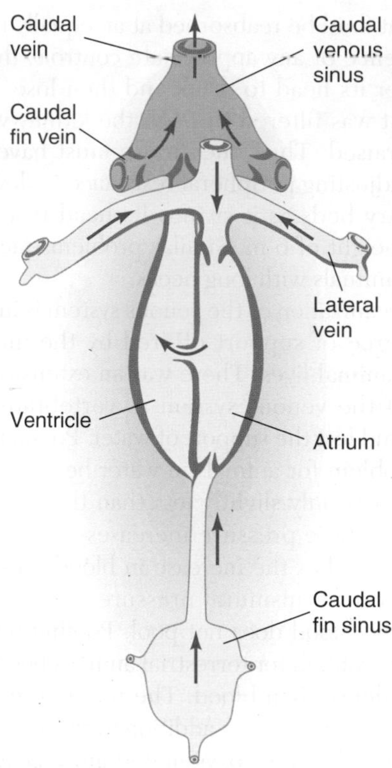

Figure 12-34 Some fishes have a caudal heart located in the tail, which aids in returning deoxygenated blood to the central heart. The walls of the heart contain skeletal muscle and beat rhythmically. [Adapted from Kampmeier, 1969.]

gases, it is possible to have heat exchange with little gas transfer. Countercurrent heat exchangers are common in the limbs of birds and mammals and are used to reduce the rate of heat transfer between the limb and the body and to regulate limb temperature.

A countercurrent arrangement of small arterioles and venules is referred to as a *rete mirabile*. Before entering a tissue, an artery divides into a large number of small capillaries that parallel a series of venous capillaries leaving the tissue. The "arterial" capillaries are surrounded by "venous" capillaries, and vice versa, forming an extensive exchange surface between inflowing and outflowing blood. These retial capillaries serve to transfer heat or gases between arterial blood entering a tissue and venous blood leaving it. Tuna have a large number of retia mirabile, which are used to regulate the temperature of the brain, muscles, and eyes (see Figures 16-22 and 16-23). The rete mirabile leading to the swimbladder of other fishes, such as the eel, also functions as a carbon dioxide countercurrent exchanger (see Figure 13-55).

Capillaries and the Microcirculation

Most tissues have such an extensive network of capillaries that no cell is more than three or four cells away from a capillary. This proximity is important for the transfer of gases, nutrients, and waste products because diffusion is an exceedingly slow process. Capillaries are usually about 1 mm long and 3–10 μm in diameter, just large enough for red blood cells (erythrocytes) to squeeze through. However, large white blood cells (leukocytes), may become lodged in capillaries, stopping blood flow. Trapped leukocytes are either dislodged by a rise in blood pressure or migrate slowly along the vessel wall until they reach a larger vessel and are swept into the bloodstream.

Microcirculatory beds

Figure 12-35 illustrates the vessels composing a microcirculatory bed. Small terminal arteries subdivide to form arterioles, which in turn subdivide to form metarterioles and subsequently capillaries, which then rejoin to form venules and veins. The arterioles are surrounded by smooth muscle, which becomes discontinuous in the metarterioles and ends in a smooth muscle ring, the precapillary sphincter. Capillary walls, which are completely devoid of connective tissue and smooth muscle, consist of a single layer of endothelial cells surrounded by a basement membrane of collagen and mucopolysaccharides. The capillaries are often categorized as arterial, middle, or venous capillaries, the latter being a little wider than the other two types. A few elongated cells with the ability to contract, called *pericyte cells*, are found wrapped around capillaries. The venous capillaries empty into pericytic venules, which in turn join the muscular venules and veins. The venules and veins are valved, and the smooth muscle sheath appears after the first postcapillary valve. Even though the walls of capillaries are thin and fragile, they require only a small wall tension to resist stretch in response to pressure because of their small diameter (see Equation 12-1).

The innervated smooth muscle of the arterioles and, in particular, the smooth muscle sphincter at the junction of arteries and arterioles control blood distribution to each capillary bed. Most arterioles are innervated by the sympathetic nervous system, a few (e.g., those in the lungs) by the parasympathetic nervous system. Different tissues have different numbers of capillaries open to flow at any one time and different degrees of control over blood flow through the capillary bed. In some tissues, the opening and closing of precapillary sphincters, which are not innervated and appear to be under local control, alters blood distribution within the capillary bed. In other tissues, however, most, if not all, of the capillaries tend to be either open (e.g., in the brain) or closed (e.g., in the skin) for considerable periods. All capillaries combined have a potential volume of about 14% of an animal's total blood volume. At any one moment, however, only 30%–50% of all capillaries are open, and thus only 5%–7% of the total blood volume is contained in the capillaries.

Transfer of substances across capillary walls

The transfer of substances between blood and tissues occurs across the walls of capillaries, pericytic venules,

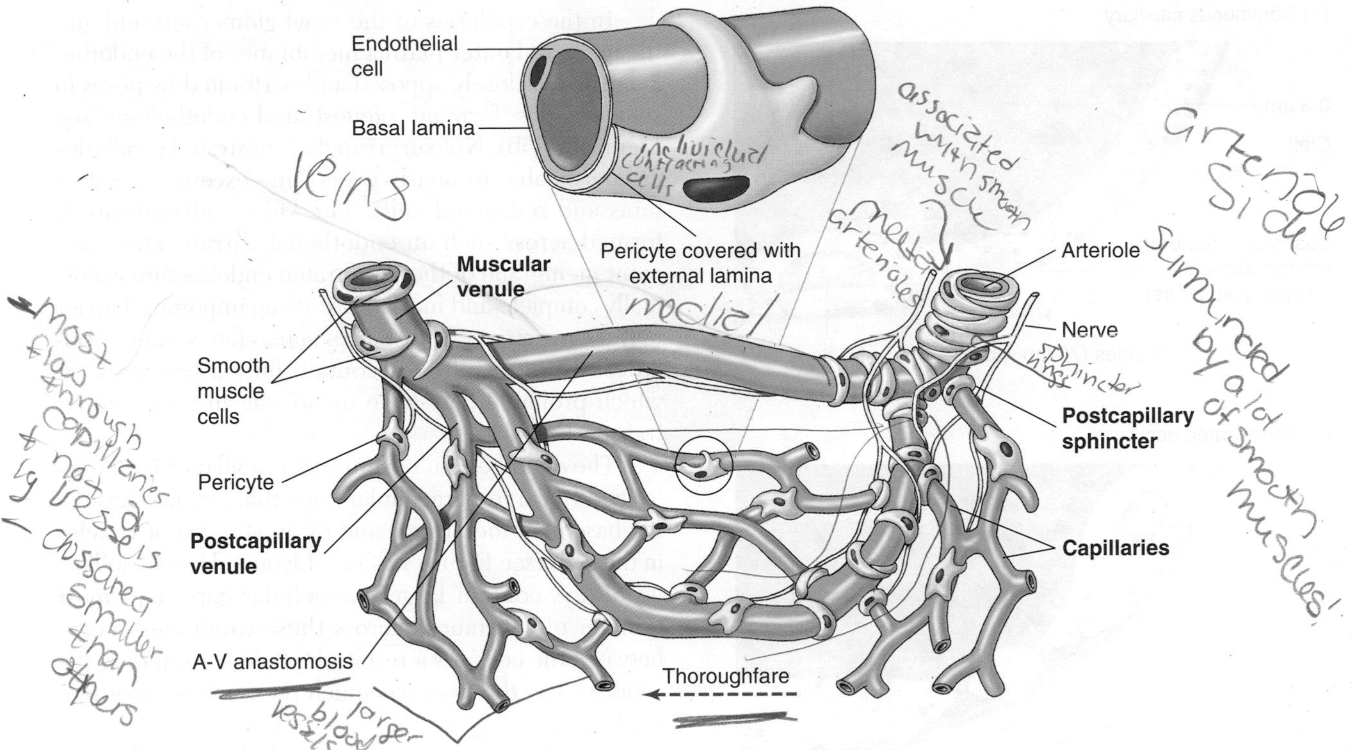

Figure 12-35 A microcirculatory bed consists of small arteries (arterioles), capillaries, and small veins (venules). Capillaries consist of a single layer of endothelial cells surrounded by a basement membrane and have occasional contractile cells called pericytes wrapped around them. Direct flow from the arterial to the venous system can occur via the thoroughfare channel or the A-V anastomosis, but most blood flows through the network of capillaries. The precapillary sphincter helps regulate flow into the capillary bed.

and to a lesser extent, metarterioles. The endothelium composing the capillary wall is several orders of magnitude more permeable than epithelial cell layers, allowing substances to move with relative ease in and out of capillaries. However, the capillaries in various tissues differ considerably in permeability, and these differences are associated with marked differences in the structure of the endothelium. Based on their wall structure, capillaries are classified into three types (Figure 12-36 on the next page):

- *Continuous capillaries,* which are the least permeable, are located in muscle, neuronal tissue, the lungs, connective tissue, and exocrine glands.
- *Fenestrated capillaries,* which exhibit intermediate permeability, are found in the renal glomerulus, intestines, and endocrine glands.
- *Sinusoidal capillaries,* which are the most permeable, are present in the liver, bone marrow, spleen, lymph nodes, and adrenal cortex.

In the continuous capillaries of skeletal muscle, which have been studied extensively, the endothelium is about 0.2–0.4 μm thick and is underlain by a continuous basement membrane (see Figure 12-36a). The individual endothelial cells are separated by clefts, which are about 4 nm wide at the narrowest point. Most of the cells contain large numbers of pinocytotic vesicles about 70 nm in diameter. Most of these vesicles are associated with the inner and outer plasma membranes of the endothelial cells; the rest are located in the cell matrix.

Substances can move across the walls of continuous capillaries either through or between the endothelial cells. Lipid-soluble substances diffuse through the cell membrane, whereas water and ions diffuse through the water-filled clefts between cells. In addition, at least in brain capillaries, there are carrier mechanisms for the transport of glucose and some amino acids. Evidence indicates that the numerous vesicles in the endothelial cells play a role in transferring substances across the capillary wall. For example, electron-microscopic studies have shown that when horseradish peroxidase is placed in the lumen of a muscle capillary, it first appears in vesicles near the lumen and then in vesicles close to the plasma membrane, but never in the cytoplasm. This finding suggests that material is packaged in vesicles and shuttled through the endothelial cells. Supporting this concept of vesicle-mediated transport is the observation that the endothelial cells of brain capillaries contain fewer vesicles and are less permeable than endothelial cells from other capillary beds. The lower permeability of brain capillaries, however, is also considered to result from the tight junctions between endothelial cells. Another possibility is suggested by microscopic observations of capillaries in the rat

(a) Continuous capillary

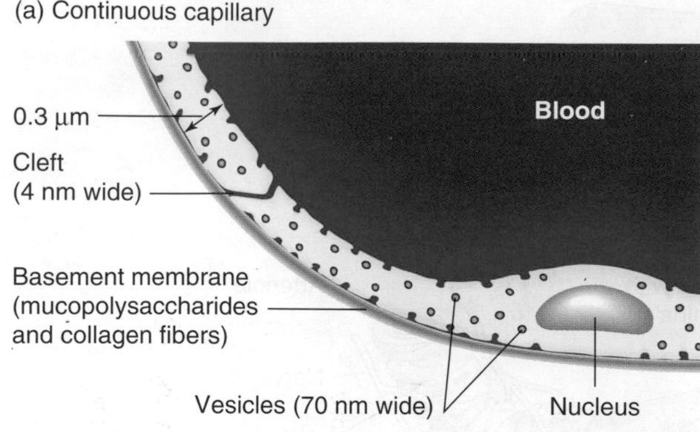

(b) Fenestrated capillary

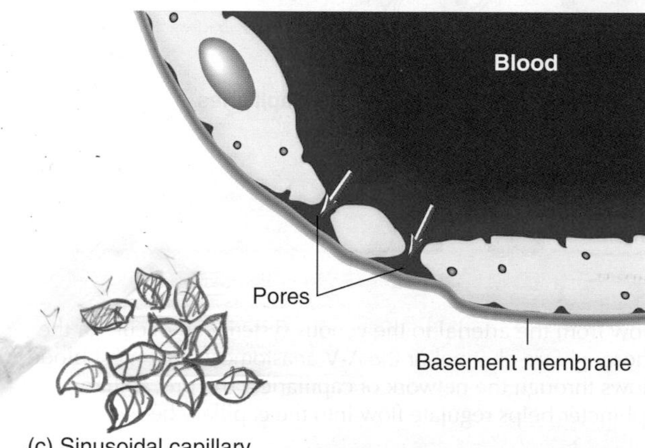

(c) Sinusoidal capillary

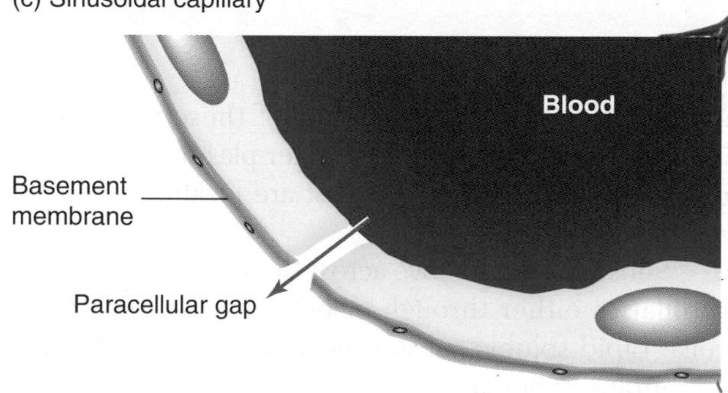

Figure 12-36 Differences in the structure of the capillary endothelium define three types of capillaries, which are found in characteristic tissues. Shown here are portions of the endothelial wall. **(a)** Continuous capillary, with 4 nm paracellular clefts, a complete basement membrane, and numerous vesicles. **(b)** Fenestrated capillary, with pores through a thin portion of the wall, few vesicles, and a complete basement membrane. **(c)** Sinusoidal capillary, with large paracellular gaps extending through the discontinuous basement membrane. In general, continuous capillaries are the least permeable and sinusoidal capillaries are the most permeable.

diaphragm, in which vesicles have been observed to coalesce, forming pores through the endothelial cells. Conceivably, then, substances, including large macromolecules, can diffuse through pores created by coalescence of nonmobile vesicles, rather than being packaged in vesicles that then move across the cell.

In the capillaries of the renal glomerulus and gut, the inner and outer plasma membranes of the endothelial cells are closely apposed and perforated by pores in some regions, forming a fenestrated endothelium (see Figure 12-36b). Not surprisingly, fenestrated capillaries are permeable to nearly everything except large proteins and red blood cells. The kidney ultrafiltrate is formed across such an endothelial barrier. The basement membrane of the fenestrated endothelium is normally complete and may constitute an important barrier to the movement of substances across fenestrated capillaries. These capillaries contain only a few vesicles, which probably play little or no role in transcellular transport.

The endothelium of sinusoidal capillaries is characterized by large paracellular gaps that extend through the basement membrane and by an absence of vesicles in the cells (see Figure 12-36c). Liver and bone capillaries always contain large paracellular gaps, and most transfer of substances across these capillaries occurs between the cells. As a result, the fluid surrounding the capillaries in the liver has much the same composition as plasma.

The clefts, pores, and paracellular gaps through which substances can freely diffuse across capillary walls are about 4 nm wide on average, but only molecules much smaller than 4 nm can get through, indicating the presence of some further sieving mechanism. The diameter of these openings varies within a single capillary network and usually is larger in the pericytic venules than in the arterial capillaries. This difference has functional significance because blood pressure, which is the filtration force for moving fluid across the capillary walls, decreases from the arterial to the venous end of the capillary network. Inflammation or treatment with a variety of substances (e.g., histamine, bradykinin, and prostaglandins) increases the size of the openings at the venous end of the capillary network, making it very permeable.

Capillary pressure and flow

The arrangement of arterioles and venules is such that all capillaries are only a short distance from an arteriole, so that pressure and flow are fairly uniform throughout the capillary bed. Transmural pressures of about 10 mm Hg have been recorded in capillaries (Figure 12-37). High pressures inside a capillary result in the filtration of fluid from the blood plasma into the interstitial space. This filtration pressure is opposed by the *colloid osmotic pressure* of the plasma, which results largely from the higher concentration of proteins in the blood plasma than in the interstitial fluid. Because these plasma proteins are large, they are retained in the blood and not transported across the capillary wall.

To visualize the relationship of these two pressures, consider the schematic situation depicted in Figure 12-38. Generally, blood pressure is higher than the colloid osmotic pressure of the plasma at the arterial end

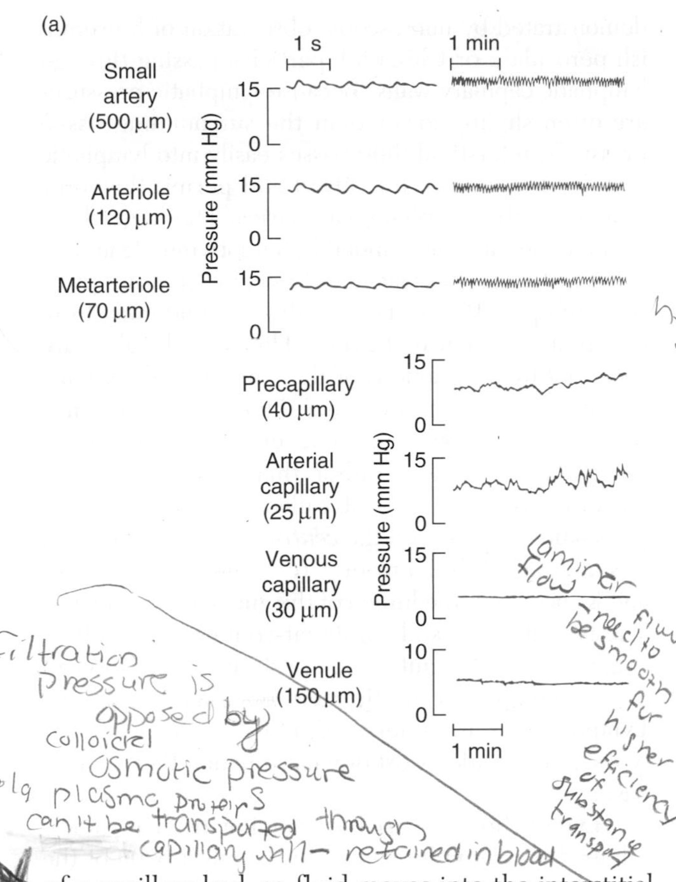

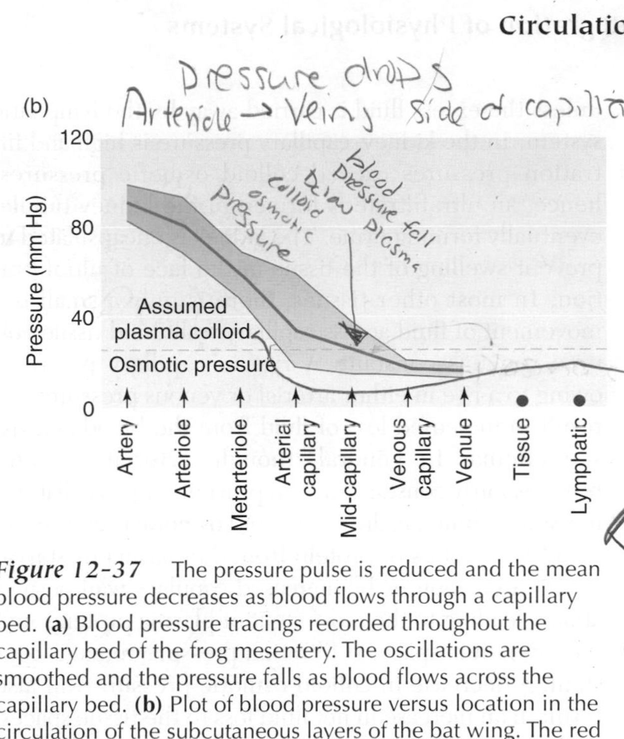

Figure 12-37 The pressure pulse is reduced and the mean blood pressure decreases as blood flows through a capillary bed. **(a)** Blood pressure tracings recorded throughout the capillary bed of the frog mesentery. The oscillations are smoothed and the pressure falls as blood flows across the capillary bed. **(b)** Plot of blood pressure versus location in the circulation of the subcutaneous layers of the bat wing. The red shaded area represents ±1 SE (standard error) from the average values indicated by the thick center line. Also plotted are typical tissue and lymphatic pressures for comparison (red dots indicate values). [Part a from Weiderhielm et al., 1964; part b from Weiderhielm and Weston, 1973.]

of a capillary bed, so fluid moves into the interstitial space (area 1 in Figure 12-38). The blood pressure steadily decreases along the length of the capillary, while the colloid osmotic pressure remains constant. Once the blood pressure falls below the colloid osmotic pressure, fluid in the interstitial space is drawn back into the blood by osmosis (area 2). Thus the net movement of fluid at any point along the capillary is determined by two factors: (a) the difference between blood pressure and colloid osmotic pressure and (b) the permeability of the capillary wall, which tends to increase toward the venous end. This concept is sometimes referred to as the *Starling hypothesis*, after its initial proponent, Ernest Starling (1866–1927), whose prolific research also included studies on the relationship between ventricular work output and venous filling pressure (see Spotlight 12-2).

In most capillary beds, the net loss of fluid at the arterial end is somewhat greater than the net uptake at the venous end of the capillary. The fluid does not accumulate in the tissues, however, but is drained by the lymphatic system and returned to the circulation. Thus, there typically is a circulation of fluid from the arterial end of the capillary bed into the interstitial spaces and back into the blood across the venous end of the capillary bed or via the lymphatic system. Because of this bulk flow of fluid, the exchange of gases, nutrients, and wastes between blood and tissues exceeds that expected by diffusion alone.

Net filtration of fluid across capillary walls will result in an increase in tissue volume, termed **edema**,

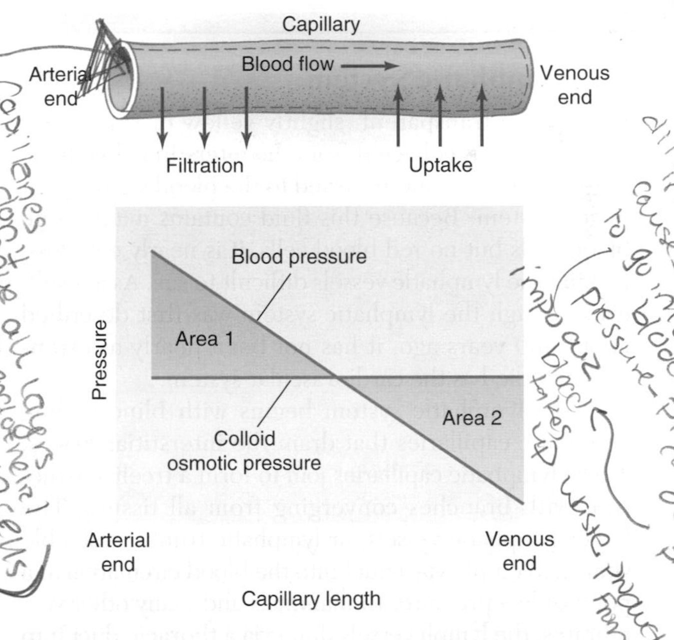

Figure 12-38 Net fluid flow across the capillary wall depends on the difference in blood pressure and colloid osmotic pressure between the extracellular fluid and the blood plasma. At the arterial end of the capillary, blood pressure exceeds the difference in colloid osmotic pressure, and fluid is filtered from the plasma into the extracellular space (area 1 in the graph). At the venous end, the reverse is true, and fluid is drawn back into the plasma from the extracellular space (area 2). Area 1 is somewhat larger than area 2 in most capillary beds; that is, there is a small net loss of fluid from the circulation to the extracellular space. In general, this tissue fluid is drained and returned to the bloodstream via the lymphatic system.

unless the excess fluid is carried away by the lymphatic system. In the kidney, capillary pressure is high and filtration pressures exceed colloid osmotic pressures; hence, an ultrafiltrate is formed in the kidney tubule, eventually forming urine. The kidney is encapsulated to prevent swelling of the tissue in the face of ultrafiltration. In most other tissues, there is only a small net movement of fluid across capillary walls, and tissue volume remains constant. A rise in capillary pressure, owing to a rise in either arterial or venous pressure, will result in increased loss of fluid from the blood and tissue edema. In general, though, arterial pressure remains fairly constant so as to prevent large oscillations in tissue volume. A drop in colloid osmotic pressure can result from a loss of protein from the plasma by starvation or excretion or by increased capillary wall permeability, leading to the movement of plasma proteins into the interstitial space. If filtration pressure remains constant, a decrease in colloid osmotic pressure will also result in an increase in net fluid loss to the tissue spaces.

Why does lying down and raising the legs reduce ankle swelling in humans? Swollen ankles are not a common malady in giraffes. Why not?

The Lymphatic System

Lymph—a transparent, slightly yellow or sometimes milky fluid—is collected from the interstitial fluid in all parts of the body and returned to the blood via the lymphatic system. Because this fluid contains many white blood cells but no red blood cells, it is nearly colorless, making the lymphatic vessels difficult to see. As a result, even though the lymphatic system was first described about 400 years ago, it has not been nearly as extensively studied as the cardiovascular system.

The lymphatic system begins with blind-ending lymphatic capillaries that drain the interstitial spaces. These lymphatic capillaries join to form a treelike structure with branches converging from all tissues. The larger lymphatic vessels, or lymphatic trunks, resemble veins and empty via a duct into the blood circulation at a point of low pressure. In mammals and many other vertebrates, the lymph vessels drain via a **thoracic duct** into a very low pressure region of the venous system, usually close to the heart (see Figure 12-3). The lymphatic system serves to return to the blood the excess fluid and proteins that filter across capillary walls into the interstitial spaces. Large molecules, particularly fats absorbed from the gut and probably high-molecular-weight hormones, reach the blood via the lymphatic system.

The walls of lymphatic capillaries consist of a single layer of endothelial cells. The basement membrane is absent or discontinuous, and there are large paracellular gaps between adjoining cells. This feature has been demonstrated by microscopic observation of horseradish peroxidase or China ink particles passing through lymphatic capillary walls. Because lymphatic pressures are often slightly lower than the surrounding tissue pressures, interstitial fluid passes easily into lymphatic vessels. The vessels are valved and permit flow only away from the lymphatic capillaries. The larger lymphatic vessels are surrounded by smooth muscle and, in some instances, contract rhythmically, creating pressures of up to 10 mm Hg and driving fluid away from the tissues (Figure 12-39). The vessels also are squeezed by contractions of the gut and skeletal muscles and by general movements of the body, all of which promote lymph flow. Fats are taken up from the gut by the lymphatic system rather than directly into the blood. Folds in the gut wall, called *villi*, each contain a lymphatic vessel (called the *central lacteal*), into which fats and fat-soluble nutrients (e.g., vitamins A, D, E, and K) pass from the lumen of the gut (see Chapter 15). The lacteal is "milked" of its fat-containing lymph by contractions of the gut, which push the lymph forward and eventually, via the thoracic duct, into the blood. Lymph vessels are innervated, but it is not clear what types of innervation exist or what function those nerves have.

Lymph flow is variable, ranging from 4 to 150 $ml \cdot hr^{-1}$ in resting humans. This is only 1/6000 of the cardiac output during the same time period. Nevertheless, lymphatic flow is important in draining tissues of excess interstitial fluid. If lymph production exceeds lymph flow, severe edema can result. In the tropical disease *filariasis*, larval nematodes, transmitted by mosquitoes to humans, invade the lymphatic system,

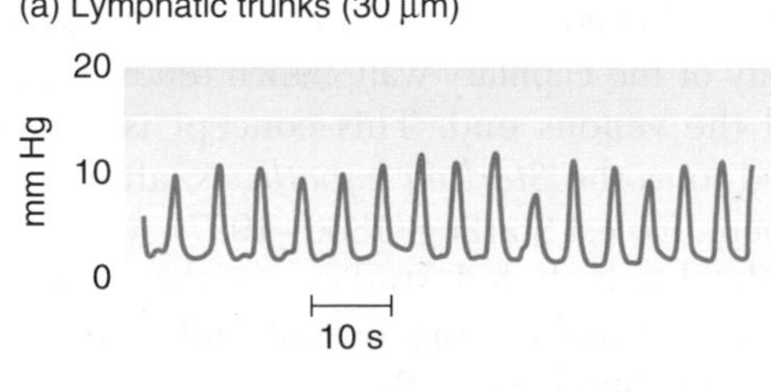

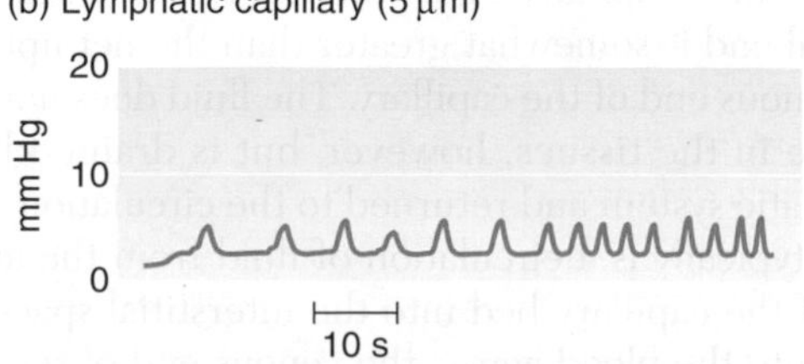

Classic Work **Figure 12-39** Pressures in the lymphatic system are similar to those in the venous system. These recordings are from **(a)** lymphatic trunks and **(b)** lymphatic capillaries in the wing of an unanesthetized bat. They were obtained by micropuncture without prior surgical intervention. [From Weiderhielm and Weston, 1973.]

causing blockage of lymph channels; in some cases, lymphatic drainage from certain parts of the body is blocked totally. The consequent edema can cause parts of the body to become so severely swollen that the condition has come to be called elephantiasis because of the resemblance of the swollen, hardened tissues to the hide of an elephant.

Reptiles and many amphibians have lymph hearts, which aid in the movement of fluid within the lymphatic system. Bird embryos have a pair of lymph hearts located in the region of the pelvis; these hearts persist in adult birds of a few species. Mammals lack these structures for moving lymph. Frogs have not only multiple lymph hearts but also a very large-volume lymph space, which serves as a reservoir for water and ions and as a fluid buffer between the skin and underlying tissues. The large lymph volume in amphibians is derived from both plasma filtration across capillaries and diffusion of water across the skin. The ratio of lymph flow to cardiac output is much higher in toads (approximately 1:60) than in mammals (approximately 1:6000). The toad lymph hearts, although having a much smaller stroke volume, can beat at rates higher than the blood heart.

Fishes appear to either lack a lymphatic system or have only a very rudimentary one, although they have a secondary circulation that in the past was described as a lymphatic system. This secondary circulation, which has a low hematocrit, is connected to the primary circulation via arterio-arterial anastomoses and drains into the primary venous system near the heart (see Figure 12-16a). The secondary circulation supplies nutrients, but not much oxygen, to the skin and gut, but it is not generally distributed to other parts of the body. The skin exchanges gases directly with the surrounding water. Because of its narrow distribution, it is unlikely that the secondary circulation performs the lymphatic function of maintaining tissue fluid balance. It is not clear how fish do perform this function, but the absence of a lymphatic system seems to be related to the fact that fish live in a medium with a density similar to that of their own bodies.

CIRCULATION AND THE IMMUNE RESPONSE

Both the circulatory and lymphatic systems contribute to the body's defense against infection. The crucial players in the immune response are **lymphocytes**, a type of leukocyte (white blood cell). The unique characteristic of lymphocytes is their ability to "recognize" foreign substances (antigens), including those on the surface of invading pathogens, virus-infected cells, and tumor cells. There are two main types of lymphocytes: B lymphocytes (B cells) and T lymphocytes (T cells). The latter are subdivided into helper T (T_H) cells and cytotoxic T (T_C) cells. Lymphocytes are aided by other types of leukocytes, particularly neutrophils and macrophages. Under certain conditions, both neutrophils and macrophages can engulf microorganisms and foreign particulate matter by phagocytosis. These phagocytic cells also produce and release various cytotoxic factors and antibacterial substances.

The immune response consists of recognizing the invader, then marking and destroying it. Recognition is carried out exclusively by lymphocytes, whereas destruction can be effected by both lymphocytes and phagocytes. The lymphocyte recognition system must be able to discriminate between natural constituents of the body and foreign invaders; that is, to distinguish between self and nonself. Failure to recognize self leads to autoimmune diseases, some of which can be fatal.

Lymphocytes respond in three ways to invasion (Figure 12-40):

1. B cells develop into plasma cells, which secrete antibodies that bind to pathogens, marking them for degradation by phagocytes.

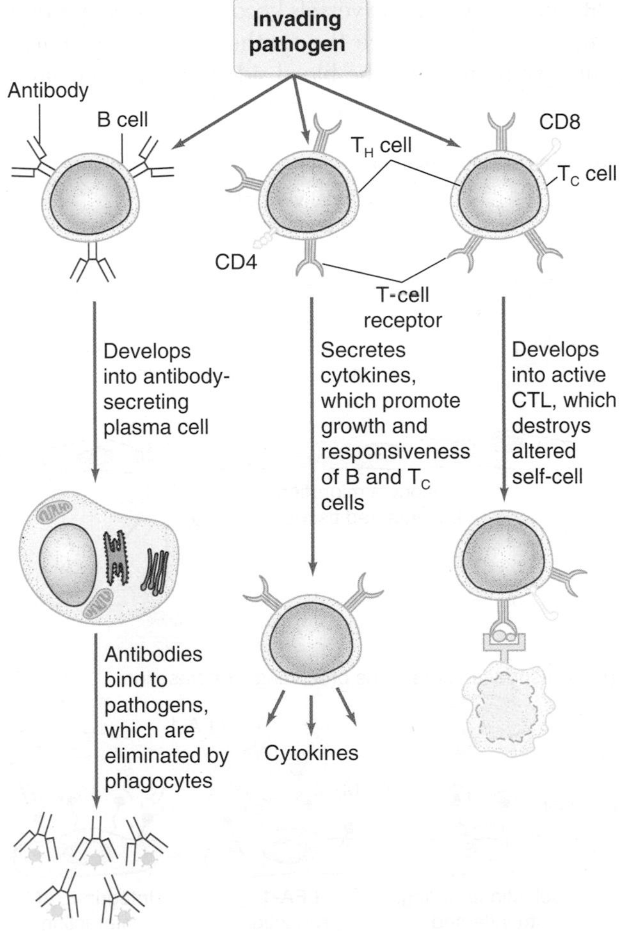

Figure 12-40 Three types of lymphocytes—B cells, helper T (T_H) cells, and cytotoxic T (T_C) cells—respond in different ways to antigens. Membrane-bound antibodies on B cells and T-cell receptors on T cells recognize and bind antigens specifically. T_H and T_C cells can be distinguished by the presence of membrane molecules called CD4 and CD8. [Adapted from Kuby, 1997.]

2. T_C cells recognize tumor cells, including those infected by pathogens; the recognition event stimulates T_C cells to mature into active cytotoxic T lymphocytes (CTLs), which destroy the altered self-cells.
3. Recognition of an antigen by T_H cells stimulates them to secrete cytokines, which in turn promote the growth and responsiveness of B cells, T_C cells, and macrophages, thereby increasing the strength of the immune response to a pathogen.

Leukocytes circulate in both the blood and the lymph. Large numbers of lymphocytes are present in lymph nodes, which are located along the lymphatic vessels (see Figure 12-3). These nodes filter the lymph and help bring antigens into contact with lymphocytes.

To get to tissues that have been invaded by pathogens, leukocytes must be able to leave the lymphatic and circulatory systems, a process termed **extravasation.** Normally, of course, leukocytes are swept along in the bloodstream and do not pass across vessel walls. At sites of infection, however, the vessels become inflamed, producing signals that induce the synthesis and activation of adhesive proteins on the blood side of the endothelium.

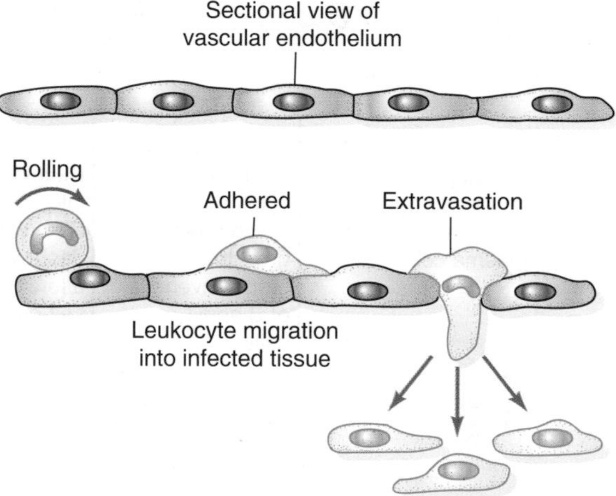

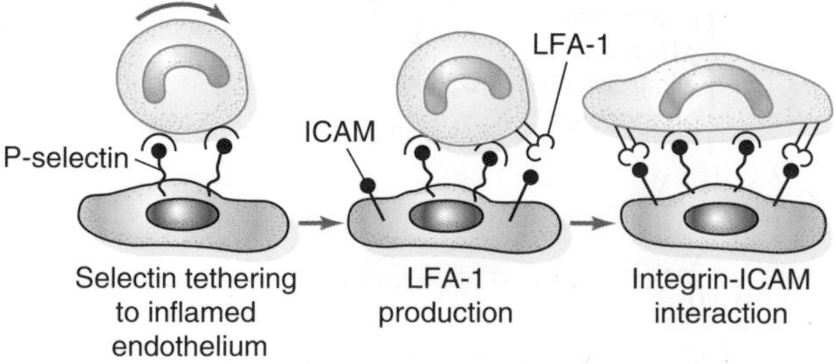

Figure 12-41 Leukocytes migrate from the circulation into tissues at sites of inflammation. **(a)** Overview of the process of leukocyte adhesion to and extravasation across inflamed vascular endothelium. **(b)** Some of the interactions between cell-surface molecules that cause leukocytes to adhere to an inflamed endothelium. [Adapted from Kuby, 1997.]

As leukocytes roll past an inflamed vascular endothelium, a molecule called P-selectin on the blood-facing surface binds to and slows a passing leukocyte (Figure 12-41). This interaction stimulates the leukocyte to produce integrin receptors (e.g., LFA-1), which then bind with intracellular adhesion molecules (ICAMs) on the surface of the endothelium. As a result of these and other interactions, the leukocyte adheres to the endothelium. Once firmly adhered, the leukocyte can move between the endothelial cells and migrate into the infected tissue.

REGULATION OF CIRCULATION

Regulation of circulation hinges on controlling arterial blood pressure so that three central priorities can be fulfilled:

1. delivering an adequate supply of blood to the brain and heart
2. supplying blood to other organs of the body, once the supply to the brain and heart is assured
3. controlling capillary pressure so as to maintain tissue volume and the composition of the interstitial fluid within reasonable ranges

The body employs a variety of receptors for monitoring the status of the cardiovascular system. In response to sensory inputs from these receptors, both neuronal and chemical signals induce appropriate adjustments to maintain an adequate arterial pressure. In this section, we first discuss regulatory features affecting the heart and main vessels, and then focus on the microcirculation.

Control of the Central Cardiovascular System

A number of sensory inputs contribute to the regulation of blood circulation. **Baroreceptors** monitor blood pressure at various sites in the cardiovascular system. Information from these baroreceptors, along with information from chemoreceptors monitoring the CO_2 and O_2 concentrations and the pH of the blood, is transmitted to the brain. Muscle contractions, as well as changes in the composition of the extracellular fluid of muscles, activate afferent nerve fibers embedded in muscle tissue, and this neuronal activity in turn causes changes in the cardiovascular system. In addition, inputs from cardiac mechanoreceptors and from a variety of thermoreceptors lead to reflex effects on the cardiovascular system.

In mammals, the integration of these sensory inputs occurs in a collection of brain neurons referred to as the **medullary cardiovascular center,** located at the level of the medulla oblongata and pons. The medullary cardiovascular center also receives inputs from other regions of the brain, including the medullary respiratory center, hypothalamus, amygdala nucleus, and cortex. The output from the medullary cardiovascular center is fed into sympathetic and parasympathetic autonomic motor neu-

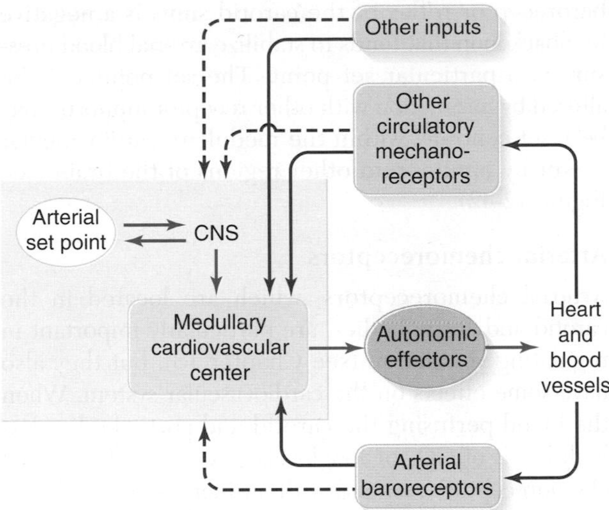

Figure 12-42 The circulatory control system in mammals involves a number of negative-feedback loops. Various sensory receptors monitor changes in the state of the cardiovascular system, sending inputs to the medullary cardiovascular center. After integrating these inputs and comparing them with the arterial set point, this center sends signals via autonomic nerves to maintain an appropriate arterial blood pressure. The arterial set point is altered by inputs from other areas of the brain, which are in turn influenced by a variety of peripheral inputs (dashed lines). [Adapted from Korner, 1971.]

rons that innervate the heart and the smooth muscle of arterioles and veins, as well as other areas of the brain.

Stimulation of sympathetic nerves increases the rate and force of heart contraction and causes vasoconstriction; the result is a marked increase in arterial blood pressure and cardiac output. In general, the reverse effects follow stimulation of parasympathetic nerves, the end result being a drop in arterial blood pressure and cardiac output. The medullary cardiovascular center can be divided into two functional regions, which have opposing effects on blood pressure:

- Stimulation of the *pressor center* results in sympathetic activation and a rise in blood pressure.
- Stimulation of the *depressor center* results in parasympathetic activation and a drop in blood pressure.

A variety of sensory inputs affect the balance between pressor and depressor activity: some activate the pressor center and inhibit the depressor center; others have the reverse effect. In this way, the various inputs that converge on the medullary cardiovascular center are modified and integrated. The result is an output that activates the pressor or the depressor center and produces cardiovascular changes in response to changing requirements of the body or disturbances to the cardiovascular system. Figure 12-42 presents an overview of this central circulatory control in mammals.

Arterial baroreceptors

Baroreceptors, which are widely distributed in the arterial system of vertebrates, show increased rates of firing with increases in blood pressure. Unmyelinated baroreceptors, which have been found in the central cardiovascular systems of amphibians, reptiles, and mammals, respond only to pressures above normal, initiating reflexes that reduce arterial blood pressure and thus protect the animal from damaging increases in blood pressure. Myelinated baroreceptors, which have been found only in mammals, respond to blood pressures below normal, thus protecting the animal from prolonged periods of reduced blood pressure. The baroreceptors located in the mammalian carotid sinus have been studied much more extensively than those in the aortic arch or the subclavian, common carotid, and pulmonary arteries. In mammals, there appear to be only minor quantitative differences between the baroreceptors of the carotid sinus and the aortic arch. Birds have aortic arch baroreceptors.

The carotid sinus in mammals is a dilation of the internal carotid artery at its origin, where the walls are somewhat thinner than in other portions of the artery. Buried in the thin walls are finely branched nerve endings that function as baroreceptors. Under normal physiological conditions, there is a resting discharge from these baroreceptors. An increase in blood pressure stretches the wall of the carotid sinus, causing an increase in discharge frequency. The relationship between blood pressure and baroreceptor discharge frequency is sigmoid, as the system is most sensitive over the physiological range of blood pressures (Figure 12-43). As in many sensory receptors, the baroreceptor

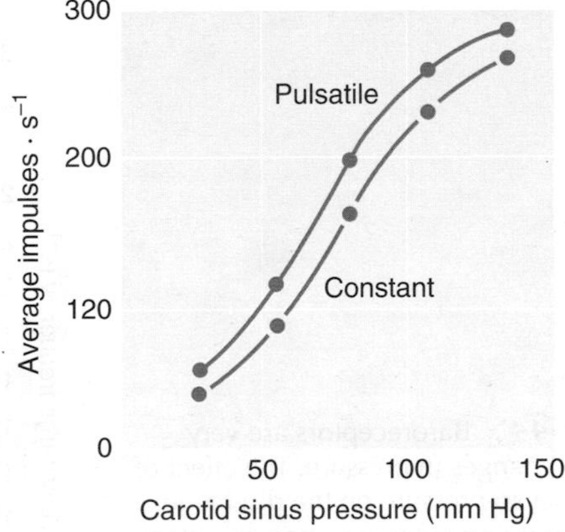

Classic Work ***Figure 12-43*** The discharge frequency of the baroreceptors of the carotid sinus increases with pressure in a sigmoid fashion. These receptors are most sensitive within the physiological range of pressures and when blood flow is pulsatile. These values were recorded from a multifiber preparation of the carotid sinus nerve and plotted against the mean pressure in the carotid sinus during pulsatile or constant flow. [Adapted from Korner, 1971.]

discharge frequency is higher when the stimulus is constantly changing, in this case, when the pressure is pulsatile rather than constant. The carotid sinus baroreceptors are most sensitive to frequencies of pressure oscillation between 1 and 10 Hz. Since arterial pressure increases and decreases with each heartbeat, this frequency range is within the normal physiological range of arterial pressure oscillations. Similar observations have been made on the relationship between discharge frequency and pressure for pulmocutaneous arterial baroreceptors in the toad (Figure 12-44). Sympathetic efferent fibers terminate in the arterial wall near the mammalian carotid sinus baroreceptors; stimulation of these sympathetic fibers increases the discharge frequency of these baroreceptors. Under normal physiological conditions, these efferent neurons may be utilized by the central nervous system to control the sensitivity of the receptors.

Signals from baroreceptors that arise in response to increased blood pressure are relayed through the medullary cardiovascular center to autonomic motor neurons, leading to a reduction in sympathetic outflow and resulting in a decrease in both cardiac output and peripheral vascular resistance. The reduction in cardiac output results from a drop in both the heart rate and the force of cardiac contraction. The culmination of these various autonomic effects is a decrease in arterial blood pressure. But as the arterial pressure decreases, so does the baroreceptor discharge frequency, which can cause a reflex increase in both cardiac output and peripheral resistance, which tends to increase arterial pressure. A fall in blood pressure also increases circulating levels of antidiuretic hormone and the activity of the renin-angiotensin-aldosterone system, both of which reduce urine production and enlarge blood volume. Thus, the baroreceptor reflex of the carotid sinus is a negative feedback loop that tends to stabilize arterial blood pressure at a particular set point. The set point may be altered by interaction with other receptor inputs or may be reset centrally within the medullary cardiovascular center by inputs from other regions of the brain (see Figure 12-42).

Arterial chemoreceptors

Arterial chemoreceptors, which are located in the carotid and aortic bodies, are particularly important in regulating ventilation (see Chapter 13), but they also have some effects on the cardiovascular system. When the blood perfusing the carotid and aortic bodies has high levels of CO_2 or low levels of O_2 and pH, arterial chemoreceptors respond with an increase in discharge frequency, which results in peripheral vasoconstriction and a slowing of the heart rate if the animal is not breathing (e.g., during submersion). Cardiac output is reduced while birds and mammals are diving; in these circumstances, peripheral vasoconstriction ensures the maintenance of arterial blood pressure and therefore blood flow to the brain.

Peripheral vasoconstriction can cause a rise in arterial pressure, which then evokes reflex slowing of the heart through stimulation of the arterial baroreceptors. Nevertheless, stimulation of arterial chemoreceptors results in a slowing of the heart even when arterial pressure remains at a constant level. Chemoreceptor stimulation thus has a direct effect on heart rate, as well as an indirect effect via changes in arterial pressure resulting from peripheral vasoconstriction.

Not surprisingly, there are many interactions between the control systems associated with the respiratory and the cardiovascular systems. For example, the

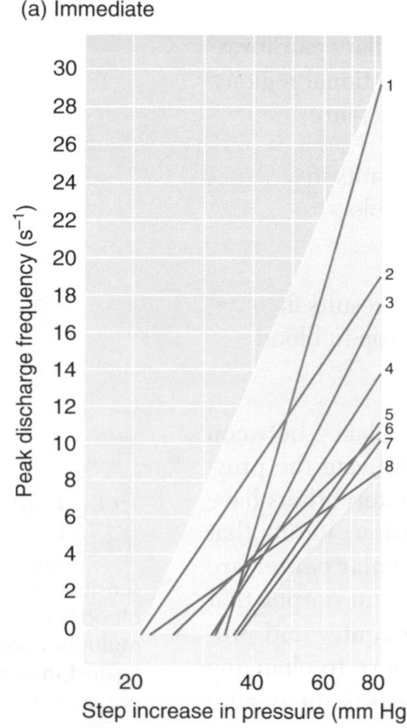

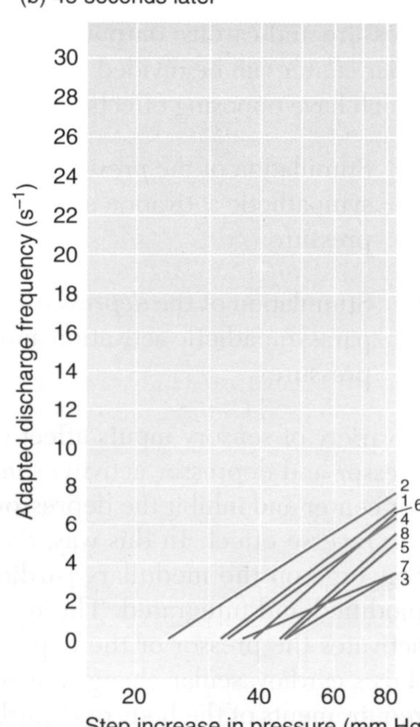

Figure 12-44 Baroreceptors are very sensitive to changes in pressure. The effect of step increases in pressure on the discharge frequency of pulmocutaneous baroreceptors in the toad was plotted **(a)** immediately after the pressure increase and **(b)** 45 seconds later. Each numbered red line represents one observation corresponding to the pressure increase shown on the horizontal axis. The rapid initial peak response is greater than the response 45 seconds after the pressure increase. [Adapted from Van Vliet and West, 1994.]

discharge pattern of stretch receptors in the lungs has a marked effect on the nature of the cardiovascular changes caused by chemoreceptor stimulation. If the animal is breathing normally, changes in gas levels in the blood will cause one set of reflex changes; if the animal is not breathing, chemoreceptor stimulation results in quite a different series of cardiovascular changes, as we will see below in our discussion on diving.

Cardiac sensory receptors

Both mechanoreceptive and chemoreceptive afferent nerve endings are located in various regions of the heart. Information on the state of the heart collected by these receptors is transmitted via the spinal cord to the medullary cardiovascular center and other regions of the brain. In addition, stimulation of some cardiac receptors causes hormone release either directly from the atria or from other endocrine tissues within the body. The stimulation of cardiac receptors evokes a series of reflex responses, including changes in heart rate and cardiac contractility and, under extreme conditions, the pain that can be associated with a heart attack.

Atrial receptors The atrial walls contain many mechanoreceptive afferent fibers, which are classified into three types. Myelinated *A-fiber* afferents respond to changes in heart rate and appear to relay information on heart rate to cardiovascular control centers located in the medulla oblongata. Myelinated *B-fiber* afferents respond to increases in the rate of filling and volume of the atria. Increases in venous volume result in an increase in venous pressure, which in turn increases atrial filling and stimulates the B-fibers, raising their discharge frequency. This increased activity is processed by the central cardiovascular centers, leading to two major effects, one on the heart and one on the kidney. Stimulation of atrial B-fibers leads to a faster heart rate mediated via increased activity in the sympathetic efferent fibers that innervate the sinoatrial node of the heart. Stimulation of the B-fibers also causes a marked increase in urine production (*diuresis*), probably mediated by a decrease in antidiuretic hormone (ADH) levels in the blood. Thus, there is a negative-feedback loop for regulating blood volume. An increase in blood volume raises venous pressure and atrial filling, which stimulates atrial B-fibers, leading to inhibition of ADH release from the pituitary. The resulting fall in blood ADH levels leads to diuresis and therefore a reduction in blood volume.

The third type of atrial mechanoreceptor comprises unmyelinated *C-fiber* afferents innervating the junction of the veins and atria. Stimulation of these afferent receptors affects both heart rate and blood pressure. If heart rate is low, stretching of this region results in an increase in heart rate, whereas if heart rate is high, stimulation results in a fall in heart rate. Stimulation of C-fibers also causes a fall in blood pressure.

The atrial wall also contains stretch-sensitive secretory cells that produce **atrial natriuretic peptide (ANP)**. This hormone, which is released into the blood when these cells are stretched, has several endocrine effects. As its name indicates, ANP causes an increase in Na^+ excretion (*natriuresis*) and in urine production, thereby effectively reducing blood volume and therefore blood pressure. ANP inhibits the release of renin by the kidney and the production of aldosterone by the adrenal cortex. It thus down-regulates the renin-angiotensin-aldosterone system, which normally stimulates Na^+ resorption and increases blood volume (see Chapter 14). In addition to these actions, ANP inhibits the release of ADH and acts directly on the kidney to increase water and Na^+ excretion. ANP has been demonstrated to have a depressor effect, reducing both cardiac output and blood pressure. In addition, ANP antagonizes the pressor effect of angiotensin.

Atrial natriuretic peptide belongs to a family of natriuretic peptides (A-, B-, C-, and V-type natriuretic peptide) sharing a common 17-amino-acid ring structure linked by a disulfide bridge. Since the initial investigations of ANP in the early 1980s, natriuretic peptides have been found in a wide variety of tissues, including the central nervous system. In many instances they may have an autocrine or paracrine function. For example, receptors for natriuretic peptides have been located in both the atria and ventricles of the hearts of several vertebrates. Binding of locally released natriuretic peptides to these receptors may reduce contractility, indicating a paracrine function within the heart.

Ventricular receptors The endings of both myelinated and unmyelinated sensory afferent fibers are embedded in the ventricle. The myelinated fibers are mechanoreceptive and chemoreceptive, with separate endings for each modality. Their mechanoreceptive endings are stimulated by the interruption of coronary blood flow. Their chemoreceptive endings are stimulated by substances such as plasma kinins, autocoids that are mediators of the inflammatory response. At low levels of stimulation, these fibers cause increased sympathetic outflow and decreased outflow from the vagus nerve to the heart, raising cardiac contractility as well as blood pressure. At higher levels of stimulation, these fibers are necessary for the perception of pain in the heart. These myelinated afferent fibers are much less numerous than unmyelinated C-fiber afferent endings in the left ventricle. Stimulation of the C-fiber afferents at low levels causes peripheral vasodilation and a reduction in heart rate. Increased stimulation of these fibers causes stomach relaxation and, at even higher frequencies, produces vomiting.

Skeletal muscle afferent fibers

Somewhat surprisingly, most nerves innervating skeletal muscles contain more afferent fibers than efferent fibers. The afferent fibers can be subdivided into four broad groups. Groups I and II are sensory fibers from muscle spindles and Golgi tendon organs, which record muscle tension and reflexively modulate the contraction of muscles, but seem to play little or no role in the control of the cardiovascular system. In contrast, stimulation of group III fibers, which are myelinated "free nerve endings," or group IV fibers, which are unmyelinated sensory endings, appears to have cardiovascular effects. These fibers are activated by either mechanical or chemical stimulation, with most fibers responding to only one or the other modality. Mechanical stimulation may be due to contraction, squeezing, or stretching of the muscle. Changes in the composition of the extracellular fluid around the muscle associated with contraction are thought to stimulate the chemoreceptors. Large changes in pH and osmotic pressure raise the activity of group IV fibers, but it is not clear whether the pH or osmotic changes occurring *in vivo* are adequate to mediate cardiovascular effects.

Electrical stimulation of muscle afferents can result in either an increase or a decrease in arterial blood pressure, depending on the fibers being stimulated or the frequency of stimulation of a particular group of afferent nerves. At low frequencies, stimulation of some afferent fibers results in a fall in arterial blood pressure, whereas stimulation of the same fibers at high frequencies results in a rise in blood pressure. Electrical stimulation of afferent nerves from muscles usually causes a change in heart rate in the same direction as the change in blood pressure; that is, if blood pressure is elevated, so is heart rate, and vice versa. In those instances in which electrical stimulation of muscle afferents causes an increase in heart rate and cardiac output, there is also a change in the distribution of blood in the body. Blood flow to the skin, kidney, gut, and inactive muscle is reduced, thus augmenting flow to the active muscles.

The cardiovascular response evoked by muscle contraction has been shown to disappear following dorsal root section, so the response is presumably reflex in origin, resulting from stimulation of afferent fibers in the muscle. The response varies depending on whether muscle contraction is isometric (static exercise) or isotonic (dynamic exercise). Static exercise is associated with an increase in arterial blood pressure with little change in cardiac output, whereas dynamic exercise results in a large increase in cardiac output with little change in arterial blood pressure. The sensory inputs from muscle afferent fibers are processed in the medullary cardiovascular center, leading to stimulation of the autonomic nerves innervating the heart and vessels, the efferent arm of these reflex arcs.

Control of the Microcirculation

Capillary blood flow can be adjusted to meet the demands of the tissues. If the requirements of a tissue change suddenly, as in skeletal muscle during exercise, then capillary flow to that tissue also changes. If requirements for nutrients and gas exchange in a tissue vary little with time, as in the brain, then capillary flow also varies little. The regulation of capillary flow can be divided into two main types, neuronal control and local control.

Neuronal control of capillary blood flow

Neuronal control serves to maintain arterial pressure by adjusting resistance to blood flow in the peripheral circulation. The vertebrate brain and heart must be perfused with blood at all times. An interruption in the perfusion of the human brain rapidly results in damage. Neuronal control of arterioles ensures that only a limited number of capillaries are open at any one moment, for if all capillaries were open, there would be a rapid drop in arterial pressure and blood flow to the brain would be reduced. The neuronal control of capillary flow operates under a priority system: If arterial pressure falls, blood flow to the gut, liver, and muscles is reduced to maintain flow to the brain and heart.

Most arterioles are innervated by sympathetic nerves, whose terminals release the catecholamine norepinephrine. Some arterioles, however, are innervated by parasympathetic nerves, whose terminals release acetylcholine. Binding of norepinephrine to α-adrenoreceptors in the smooth muscle of arterioles usually causes a decrease in the diameter of arterioles (vasoconstriction). This decrease in diameter causes an increase in resistance to flow, thus reducing blood flow through that capillary bed. The generalized effect of sympathetic nerve stimulation is thus peripheral vasoconstriction and a subsequent rise in arterial blood pressure.

Stimulation of β-adrenoreceptors in arterial smooth muscle, on the other hand, often results in relaxation of the muscle and an increase in the diameter of arterioles (i.e., vasodilation), thereby decreasing the resistance to flow and increasing the blood flow through that capillary bed. Because β-adrenoreceptors are rarely located near nerve terminals, they usually are stimulated by circulating catecholamines. These catecholamines are released into the bloodstream from adrenergic neurons of the autonomic nervous system and from chromaffin cells in the adrenal medulla. Circulating catecholamines are dominated by epinephrine released from the adrenal medulla (see Chapter 9). Epinephrine reacts with both α- and β-adrenoreceptors, causing vasoconstriction and vasodilation, respectively. Although α-adrenoreceptors are less sensitive to epinephrine, when activated, they override the vasodilation mediated by β-adrenoreceptors. The result is that high levels of circulating epinephrine cause vasoconstriction and thus an increase in peripheral resistance via α-adrenoreceptor stimulation. At lower levels of circulating epinephrine, however, β-adrenoreceptor stimulation dominates, producing an overall vasodilation and a decrease in peripheral resistance. Even at levels that produce vasodilation, epinephrine causes a rise in arterial blood pressure by stimulating β-adrenoreceptors in the heart, causing a marked increase in cardiac output.

The β-adrenoreceptors can be divided into two subgroups: β_1-adrenoreceptors, which are stimulated by both circulating catecholamines (mostly epinephrine) and adrenergic nerves (which release norepinephrine), and β_2-adrenoreceptors, which respond only to circulating catecholamines. In the peripheral circulation, only β_2-adrenoreceptors are present. In the heart and coronary circulation, however, both kinds of β-adrenoreceptors are present, so both circulating catecholamines and adrenergic neuronal stimulation can have marked effects.

The response of catecholamine receptors in any vascular bed depends on several things: the type of catecholamine, the nature of the receptors involved, and the relationship between stimulation of the receptors and the change in smooth muscle tone. Although stimulation of α-adrenoreceptors usually is associated with vasoconstriction and that of β-adrenoreceptors with vasodilation, it is not invariably the case. An additional complicating factor is that not all sympathetic fibers are adrenergic. Some are cholinergic, releasing acetylcholine from their terminals. Stimulation of sympathetic cholinergic nerves causes vasodilation in the vasculature of skeletal muscle.

The action of catecholamines is extensively modulated by a variety of substances, including neuropeptide Y and adenosine. **Neuropeptide Y,** first isolated from the porcine brain in 1982, is structurally related to mammalian pancreatic polypeptide and peptide YY. Neuropeptide Y is widespread throughout the animal kingdom and, so far, has been identified in many vertebrates and in insects. Neuropeptide Y is co-localized with norepinephrine in sympathetic ganglia and adrenergic nerves; it also is found in many non-adrenergic fibers. The atrial and ventricular myocardium and the coronary arteries are surrounded by nerve fibers that store and release neuropeptide Y. In addition, it appears that myocardial cells themselves can synthesize and secrete neuropeptide Y. In general, neuropeptide Y decreases coronary blood flow and the contraction of cardiac muscle. It appears to do so by reducing the level of inositol trisphosphate (IP_3), thereby ameliorating the effects of catecholamines on the heart and coronary circulation that are mediated via IP_3 (see Chapter 9). The role of neuropeptide Y in the peripheral circulation is less well understood, but it appears to ameliorate the increase in blood pressure resulting from norepinephrine-induced peripheral vasoconstriction mediated by α-adrenoreceptors.

ATP, as well as neuropeptide Y, is stored and co-released with catecholamines. ATP and its breakdown product, adenosine, act to inhibit the release of catecholamines. Adenosine is released by many tissues during hypoxia (low oxygen levels), but has only a paracrine or autocrine action because it is rapidly inactivated. Hypoxia tends to promote catecholamine release into the blood by the adrenal medulla, but this effect is modulated by the local release of adenosine.

Arterioles in the circulation to the brain and the lungs are innervated by parasympathetic nerves. These nerves contain cholinergic fibers, which release acetylcholine from their terminals when stimulated. In mammals, parasympathetic nerve stimulation causes vasodilation in arterioles. Some parasympathetic neurons release ATP and other purines from their terminals. Some of these purinergic neurons may participate in the control of capillary blood flow. ATP, for instance, causes vasodilation.

Local control of capillary blood flow

In addition to neuronal control of capillary blood flow, various mechanisms control the microcirculation at the local level. For example, if a vessel is stretched by an increase in input pressure, the vascular smooth muscle responds by contracting, thereby opposing any increase in vessel diameter. This tendency to maintain vessel diameter within narrow limits prevents large changes in resistance to flow and therefore maintains a relatively constant basal flow through the capillary bed. Local heating of a tissue, which may accompany inflammation, is associated with marked vasodilation, whereas a reduction in temperature causes vasoconstriction. Thus an ice pack can reduce the blood flow and, therefore, the swelling associated with damage to a tissue.

When the metabolic rate in a tissue increases, there must be a concomitant increase in blood flow. Local control of capillary flow ensures that the most active tissue has the most dilated vessels and, therefore, receives the most blood. Active tissues are marked by a decrease in O_2 and an increase in CO_2, H^+, various other metabolites (e.g., adenosine), and heat. Extracellular K^+ also rises in skeletal muscle during and following exercise. All of these activity-related metabolic changes, as well as certain compounds released by the vascular endothelium, have been shown to cause vasodilation and a local increase in capillary blood flow.

The term **hyperemia** means increased blood flow to a tissue; **ischemia** means the cessation of flow. **Active hyperemia** refers to the increase in blood flow that follows increased activity in a tissue, particularly skeletal muscle. If blood flow to an organ is stopped by clamping the artery or by a powerful vasoconstriction, there will be a much higher blood flow to that organ when the occlusion is removed than there was before the occlusion. This phenomenon is termed **reactive hyperemia.** Presumably, during the ischemic period (the period of no blood flow), O_2 levels are reduced, and CO_2, H^+, and other metabolites build up and cause local vasodilation. As a result, when the occlusion is removed, blood flow is much higher than normal.

Compounds produced in the endothelium Numerous compounds affecting capillary blood flow are produced by the vascular endothelium and other cells associated with the circulation. The endothelium is not merely a barrier between blood and the underlying tissues, but an active tissue, producing compounds such as nitric oxide, endothelin, and prostacyclin that affect vascular smooth muscle and, therefore, capillary blood flow.

Nitric oxide (NO) is produced and released continuously by the vascular endothelium. It regulates blood

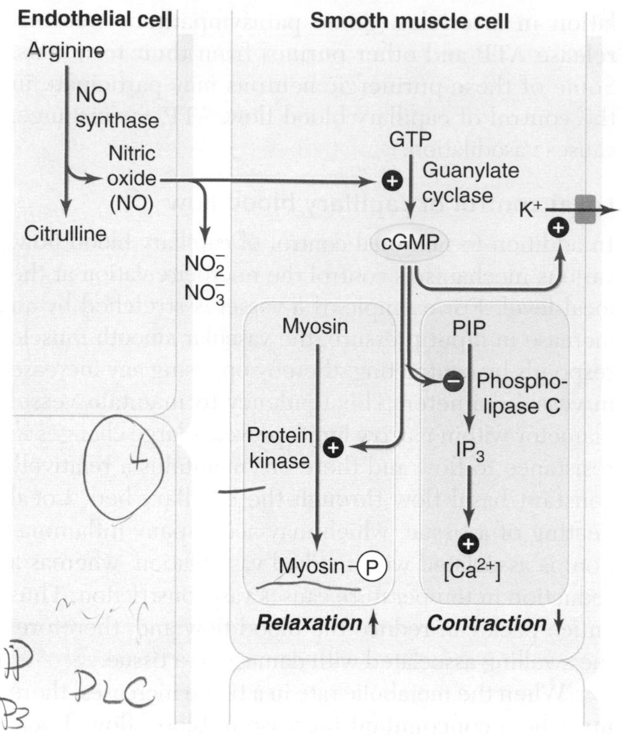

Figure 12-45 Nitric oxide (NO) is produced in vascular endothelial cells and diffuses into surrounding smooth muscle, where it causes relaxation of the muscle and dilation of the blood vessel and contributes to inhibition of the contraction pathway.

flow and pressure in mammals and other vertebrates by causing vascular smooth muscle to relax (Figure 12-45). Nitric oxide diffuses into smooth muscle cells and activates guanylate cyclase, leading to the production of the intracellular second messenger cGMP (cyclic 3′,5′ guanosine monophosphate; see Chapter 9). The resulting rise in cGMP in turn promotes muscle relaxation by stimulating the phosphorylation of myosin light chains and inhibiting the synthesis of inositol trisphosphate (IP_3), an important promoter of Ca^{2+} release. Nitric oxide also stimulates the activity of K^+ channels, causing muscle cell hyperpolarization and relaxation.

Nitric oxide is synthesized in the endothelium by the oxidation of arginine to citrulline and NO. Several NO synthases are calcium-dependent, and Ca^{2+} entry into endothelial cells has been shown to cause the production and release of NO and relaxation of surrounding smooth muscle. Some Ca^{2+} channels in the endothelium are stretch-gated, suggesting that NO may be produced in response to Ca^{2+} that enters the endothelial cell when the blood vessel is stretched. A variety of chemicals (e.g., acetylcholine, ATP, and bradykinin) stimulate release of NO, as does hypoxia, pH change, and mechanical stress on the blood vessel. There is evidence that NO is produced in response to the increased pressure associated with every heartbeat. Knockout mice lacking the gene for endothelial NO synthase have high blood pressure, indicating a general role for NO in blood pressure regulation. Nitric oxide can initiate erection of the penis by dilating the blood vessels leading to the erectile bodies. This knowledge has already led to the development of new drugs for treating impotence, such as Viagra, which prolongs the action of NO by inhibiting the breakdown of cGMP. Ingesting the explosive nitroglycerin has long been known to reduce chest pains associated with heart attacks; only recently have we learned that nitroglycerin works by releasing NO into the circulation.

Nitric oxide synthases have been found in a wide variety of animals, including horseshoe crabs, the blood-sucking bug *Rhodnius,* lampreys, and humans. Nitric oxide has been shown to have many functions other than vasodilation, so its presence in nonvascular tissues is not surprising. For example, NO is involved in modulation of synaptic activity and may also have a role in nonspecific defense reactions, the relaxation of nonvascular smooth muscle in the gastrointestinal and genito-urinary tracts, and the regulation of the release of some hormones. In addition, NO released by endothelial cells, blood platelets, and leukocytes modulates both cell adhesion and aggregation and inhibits blood clotting.

The vascular endothelium also releases endothelins and prostacyclin. *Endothelins* are small vasoconstrictive proteins that are released in response to stretch. *Prostacyclin* causes vasodilation and inhibits blood clotting. It is an antagonist of the prostaglandin *thromboxane* A_2, which promotes blood clotting and causes vasoconstriction.

Adenosine is produced in cells in response to hypoxia or anoxia, especially in the heart and kidneys. Formed from AMP and S-adenosylhomocysteine (Figure 12-46), adenosine is transported into the extracellular space, reaching concentrations of 0.1–0.3 μM. In a matter of seconds it is either deaminated to inosine and subsequently excreted, taken up by cells, or removed by the circulation. Thus adenosine acts locally, having only an autocrine and paracrine action. Adenosine acts on at least four cell-surface receptor subtypes, with the overall cellular response depending on the type and number of receptors present. During periods of reduced oxygen supply, adenosine coordinates energy expenditure with supply by decreasing the heart rate, which reduces the work done by the heart per unit time, and by causing coronary vasodilation, which increases the supply of oxygen and nutrients to the heart.

Hypoxic tissues also show an increase in the activity of hypoxia-inducing factor I (HIF-I), which slows gene transcription overall while stimulating the specific transcription of genes associated with the production of erythropoietin, vascular endothelial growth factors, endothelin-1, nitric oxide synthase, and glycolytic enzymes. HIF-I also plays a central role during development as a master regulator of cellular and developmental oxygen homeostasis; in the absence of HIF-I, the circulation does not develop. HIF-I is a heterodimer consisting of α- and β-subunits. The

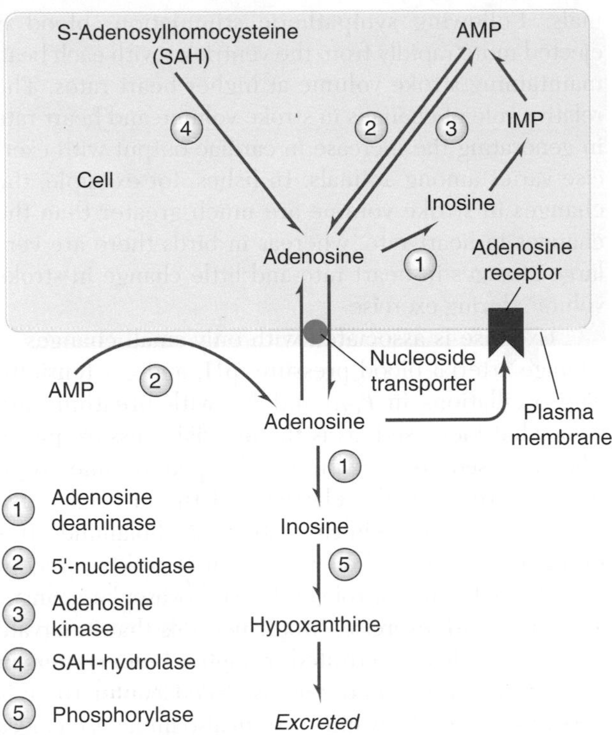

Figure 12-46 Adenosine is produced by cells in response to hypoxia or anoxia. It is transported rapidly out of the cell and binds with a variety of adenosine receptors on the plasma membrane, initiating several responses. Adenosine is removed from the extracellular space within seconds and has only a transient paracrine action.

α-subunit is normally degraded in the presence of oxygen, but in hypoxic conditions, the α-subunit is stabilized, accounting for the increase in activity of HIF-I when oxygen concentrations are low.

During hypoxia in vertebrates, production of erythropoietin by the kidney increases. Erythropoietin stimulates the production of red blood cells (erythropoiesis) in the bone marrow in mammals and in the kidney in fishes, leading to an increase in the number of circulating erythrocytes to compensate for low oxygen availability.

Inflammatory and other mediators Local injury in mammals is accompanied by a marked vasodilation of vessels in the region of the damage, due largely to the local release of histamine. **Histamine** is released not from endothelial cells, but from connective tissue and leukocytes in injured tissues. Antihistamines ameliorate, but do not completely remove, the inflammatory response, because another group of potent vasodilators, **plasma kinins**, are also activated in damaged tissues. Kinins, including bradykinin, are formed when proteolytic enzymes released from damaged tissue cleave kininogen to form active kinins. Hypoxia also stimulates formation of kinins.

Among the vasoconstrictors that act on arterioles are norepinephrine, released from sympathetic nerves, and angiotensin II. Angiotensin is formed, primarily in the lungs, from angiotensin I, which circulates in the blood (see Chapter 14). Finally, serotonin may act as a vasoconstrictor or vasodilator, depending on the vascular bed and the dose level. Serotonin is found in high concentration in the gut and blood platelets.

Histamine, bradykinin, and serotonin cause an increase in capillary permeability. As a result, large proteins and other macromolecules tend to distribute themselves more evenly between plasma and interstitial spaces, reducing the colloid osmotic pressure difference across the capillary wall. Filtration thus increases, and tissue edema occurs. On the other hand, norepinephrine, angiotensin II, and vasopressin tend to promote absorption of fluid from the interstitial spaces into the blood. This absorption can be achieved by reducing filtration pressure or the permeability of the capillaries.

Although low O_2 levels, indicative of tissue activity, cause vasodilation and increased blood flow in systemic capillaries, the lung capillary bed exhibits the opposite behavior. That is, low O_2 levels in the lung cause local vasoconstriction rather than vasodilation. The functional significance of this difference relates to the direction of gas transfer. In the lung capillaries, O_2 is taken up by the blood, and thus blood flow should be greatest in regions of high O_2. In systemic capillaries, however, O_2 leaves the blood for delivery to the tissues, and the highest blood flow should be to the area of greatest need, which is indicated by regions of low O_2.

The pulmonary circulation has high levels of an enzyme that converts angiotensin I to angiotensin II, and it is also involved in catecholamine metabolism. Why are these functions located in the pulmonary circulation?

CARDIOVASCULAR RESPONSE TO EXTREME CONDITIONS

In the previous sections, we've described the general organization of the circulation and its regulation under usual conditions. The cardiovascular system also responds in characteristic ways during exercise, diving, and hemorrhage to meet the physiological challenge of these extreme conditions.

Exercise

Regulation of the cardiovascular system during exercise is a complex process involving central neuronal control mechanisms, peripheral neuronal reflex mechanisms (especially those based on skeletal muscle afferent fibers), and local control. The importance of local control is evident in the finding that many cardiovascular changes associated with exercise occur even when neuronal mechanisms are absent. Nevertheless, neuronal control and reflexes triggered by mechanoreceptive and chemoreceptive inputs contribute importantly in ways that vary with the type of exercise.

During exercise, blood flow to skeletal muscle may increase by as much as 20 times, and the transfer of oxygen from blood to muscle may increase threefold, resulting in a 60-fold increase in oxygen utilization by the muscle. Active hyperemia is primarily responsible for increasing blood flow to muscle; the resulting decrease in peripheral resistance leads to an increase in cardiac output mediated by sympathetic nerves. At the same time, there is a reduction in flow to the gut, kidney, and at high levels of exercise, the skin (Figure 12-47). Cardiac output can increase up to 10 times above the resting level owing to large increases in heart rate and small changes in stroke volume. Much of the increase in cardiac output can be accounted for by a decrease in peripheral resistance to about 50% of the resting value and by an increase in venous return to the heart, due to both the pumping action of skeletal muscles on veins and the increase in breathing rate associated with exercise.

Increased sympathetic and decreased parasympathetic activity in nerves innervating the heart during exercise has the effect of increasing both heart rate and the force of contraction, so as to maintain stroke volume at a relatively constant level. In fact, despite the large increase in heart rate and the associated reduced time available for filling and emptying, stroke volume still increases by about 1.5 times during exercise in mammals. Following sympathetic stimulation, blood is ejected more rapidly from the ventricles with each beat, maintaining stroke volume at higher heart rates. The relative role of changes in stroke volume and heart rate in generating the increase in cardiac output with exercise varies among animals. In fishes, for example, the changes in stroke volume are much greater than the changes in heart rate, whereas in birds there are very large changes in heart rate and little change in stroke volume during exercise.

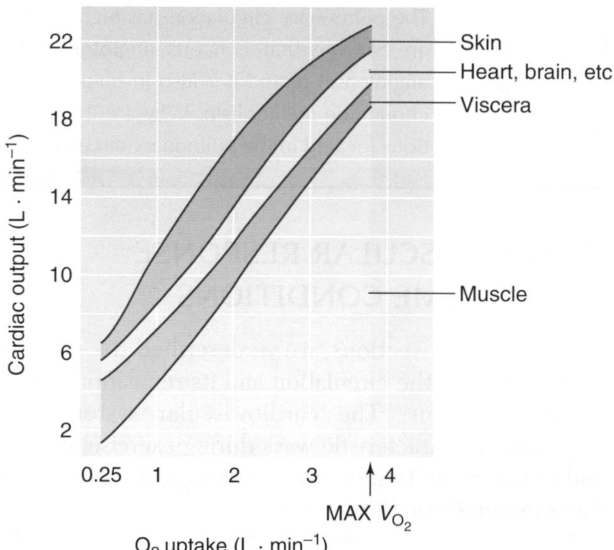

Figure 12-47 During exercise, total cardiac output increases and blood flow shifts to the active muscles. Shown is the approximate distribution of cardiac output at rest and at different levels of exercise up to the maximal oxygen consumption (Max V_{O_2}) in a normal young man. The progressive reduction in the percentage of cardiac output and absolute blood flow distributed to the viscera (splanchnic region and kidneys) augments muscle blood flow. Even skin is constricted during brief periods of exercise at high oxygen consumption.

Exercise is associated with only small changes in average arterial blood pressure, pH, and gas tensions. The oscillations in P_{CO_2} and P_{O_2} with breathing are somewhat increased, as is the arterial pressure pulse. The increased pressure pulse is damped to some extent by an increase in the elasticity of the arterial walls, which is due to a rise in circulating catecholamines. It is probable that arterial chemoreceptors and baroreceptors play only a minor role in the cardiovascular changes associated with exercise. Motor neurons that innervate skeletal muscle are activated by higher brain centers in the cortex at the onset of exercise (see Chapter 10); it is possible that this activating system also initiates changes in lung ventilation and blood flow. Proprioreceptive feedback from muscles may also play a role in increasing lung ventilation and cardiac output (see Chapter 13). A number of other changes augment gas transfer during exercise; for example, erythrocytes are released from the spleen in many animals, increasing the oxygen-carrying capacity of the blood. Thus, exercise is responsible for a complex series of integrated changes that lead to delivery of adequate oxygen and nutrients to the exercising muscle.

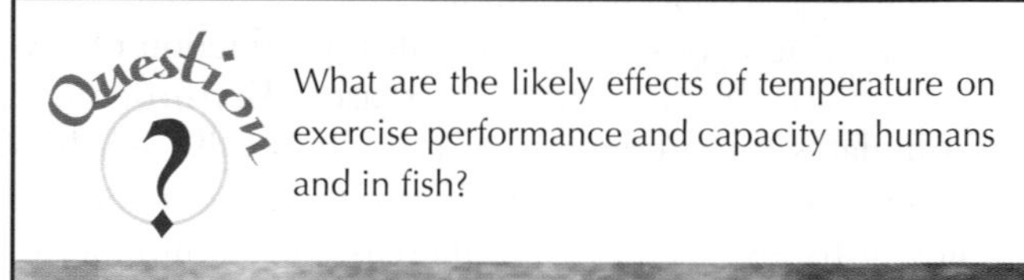

What are the likely effects of temperature on exercise performance and capacity in humans and in fish?

Diving

Many air-breathing vertebrates can remain submerged for prolonged periods. During submersion, the animal must rely on available oxygen stores in the blood and tissues (see Chapter 13). The cardiovascular system is adjusted to meter out the limited oxygen store to those organs—brain, heart, and some endocrine structures—that can least withstand anoxia.

Much of the information on responses to submersion has been collected from studies of animals forced to dive. Because naturally occurring dives vary considerably in depth, duration, and exercise level, information obtained on forced dives is not always directly applicable to natural dives. Whales and dolphins spend their lives in the water, going to the surface only to breathe, whereas seals may spend considerable time on land. Other animals may spend most of their time on land and dive only occasionally. Oxygen stores vary in animals, so metabolism may be completely aerobic during some dives but largely anaerobic during others.

Figure 12-48 illustrates the typical cardiovascular changes that occur when a seal dives and remains submerged. In mammals, but not in other vertebrates, stimulation of facial receptors inhibits breathing and causes a marked bradycardia. Although the initial pressurization of the lung as the animal descends in the water column can lead to a transient increase in blood O_2 and CO_2 levels, the continued utilization of O_2 during the dive results in a gradual fall in blood O_2 and a rise in blood CO_2 levels. This fall in blood O_2 stimulates the arterial chemoreceptors and, in the absence of lung stretch-receptor activity, causes peripheral vasoconstriction and a reduction in heart rate and cardiac output; thus blood flow to many tissues is reduced in order to maintain flow to the brain, heart, and some endocrine organs. The absence of lung stretch-receptor activity is due to the absence of breathing and the compression of the lung as the animal descends. The increase in peripheral resistance results from a marked rise in sympathetic output, which causes constriction of fairly large arteries. Reductions in blood flow to the kidney have been recorded in Weddell seals during a dive. In some instances, blood flow to muscle decreases, but this depends on the level of exercise associated with the dive and the species. Sometimes arterial pressure rises during a dive, causing stimulation of arterial baroreceptors; in such dives bradycardia is maintained by a rise in both chemoreceptor and baroreceptor discharge frequency. The bradycardia is caused by an increase in parasympathetic and, to some extent, a decrease in sympathetic activity in fibers innervating the heart.

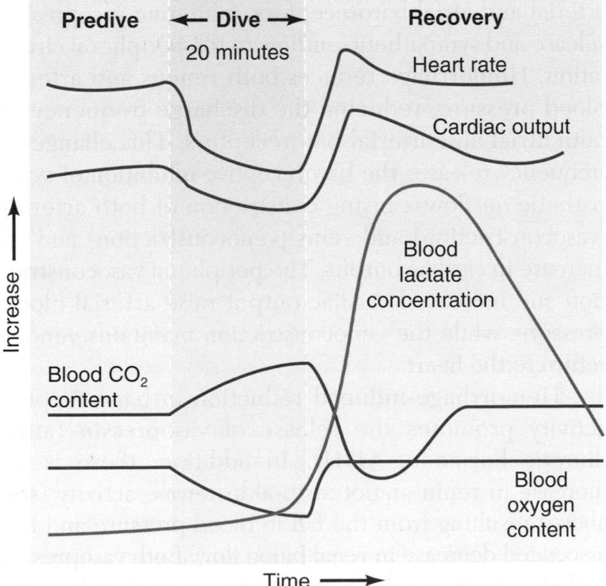

Figure 12-48 The cardiovascular system undergoes numerous adjustments when a seal dives. Heart rate, cardiac output, and blood O_2 content decrease during a dive, but blood CO_2 content increases. During the recovery period after a dive, blood lactate increases greatly; the other parameters first overshoot and then gradually return to predive values.

It has been shown in the seal that the generation of diving bradycardia can include some form of associative learning. In some trained seals, bradycardia occurs before the onset of the dive and, therefore, before the stimulation of any peripheral receptors. This psychogenic influence can have a marked effect on the change in heart rate during a dive in many animals. In general, if heart rate is low before a dive, there may be little or no change in heart rate during the dive. If heart rate is high, then there may be a marked bradycardia in response to wetting of the face and a decrease in lung stretch-receptor activity.

The "water" receptors present in birds are not directly involved in the cardiovascular changes associated with submersion. A decrease in heart rate is not observed either in submerged ducks breathing air through a tracheal cannula or in submerged ducks following carotid body denervation. Thus, activation of the "water" receptors causes suspension of breathing (apnea); the subsequent drop in blood P_{O_2} and pH and the rise in P_{CO_2} stimulates chemoreceptors, which then reflexively cause the cardiovascular changes.

Stimulation of lung stretch receptors in mammals modifies the reflex response initiated by chemoreceptor stimulation. In the absence of breathing, lung stretch receptors are not stimulated, and different reflex responses are elicited by chemoreceptor stimulation than when the animal is breathing. As a submerged animal rises in the water column, the lung becomes inflated, possibly activating stretch receptors in the lung and causing cardiac acceleration. Lung inflation tends to suppress the reflex cardiac inhibition and peripheral vasoconstriction caused by stimulation of arterial chemoreceptors. When the animal is breathing, stimulation of arterial chemoreceptors results in a marked increase in lung ventilation. In this case, low blood O_2 or high blood CO_2 levels cause peripheral vasodilation. This vasodilation leads to an increase in cardiac output to maintain arterial pressure in the face of increased peripheral blood flow. Thus, the hypoxia caused by cessation of breathing during a dive is associated with bradycardia and a reduction in cardiac output. In contrast, hypoxia that occurs when the animal is breathing (e.g., at high altitude) is associated with an increase in heart rate and cardiac output.

When mammals dive under water, the thorax is compressed, resulting in an increase in venous pressure. How might kidney function be affected during dives?

Hemorrhage

When blood vessels are damaged, blood clotting prevents uncontrolled loss of blood. Clotting occurs only where there is damaged tissue or exposed collagen. Blood is normally kept away from collagen by the intact endothelial wall. When damage occurs, a plug of blood platelets forms, followed by a blood clot. When the wound has healed, the clot is removed.

Platelets do not normally adhere to the endothelial wall unless damage is present or collagen is exposed. Under these conditions, platelets release ADP and become sticky. They also release arachidonic acid, which is converted to thromboxane A_2, which in turn attracts other platelets to the forming plug while also causing local vasoconstriction. The platelet plug does not spread to regions beyond the damaged area because the intact endothelium releases prostacyclin and nitric oxide, which both inhibit platelet aggregation. Prostacyclin opposes the action of thromboxane A_2.

Coagulation of the blood involves a cascade of events in which many factors normally found in the blood are activated in sequence, leading eventually to the formation of thrombin and activated factor XIII. Thrombin converts fibrinogen to fibrin, which then polymerizes into an insoluble fibrin clot (Figure 12-49). Sticky platelets have exposed fibrin receptors to which fibrin adheres, and the resulting fibrin-platelet mesh traps red blood cells and plasma. The sticky platelets contract, pull on the fibrin, and squeeze out serum (plasma minus fibrinogen), and the clot shrinks. The platelets also release phospholipid PF3, which is involved in activating one of the factors in the clotting cascade.

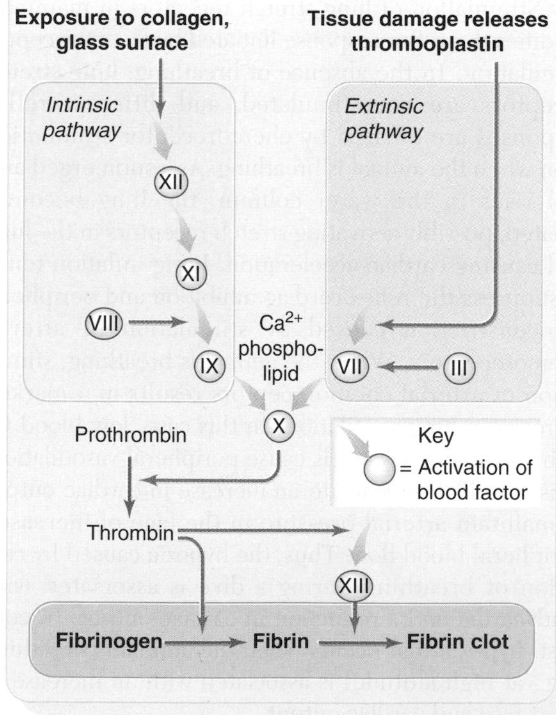

Figure 12-49 Blood coagulation involves a cascade of events in which many blood factors normally present in blood are activated in sequence. Tissue damage activates the extrinsic cascade, whereas exposure to collagen or glass surfaces activates the intrinsic cascade. Both pathways converge at the activation of factor X, leading eventually to the conversion of fibrinogen to the soluble monomer fibrin, which polymerizes into a fibrin clot.

There are two different pathways that may lead to this reaction. Tissue damage, which results in the release of thromboplastin, activates the so-called *extrinsic cascade*. The exposure of collagen, or exposure to glass or other surfaces, activates the *intrinsic pathway*. The extrinsic and intrinsic pathways converge on the activation of factor X, which along with a number of other cofactors catalyzes the cleavage of prothrombin to thrombin. Calcium is required for the activation of many factors in both the extrinsic and intrinsic pathways. Removal of Ca^{2+} prevents clotting.

An inappropriate clot that forms in the circulation, called a **thrombus**, can block blood flow. Animals produce a number of anticoagulants to prevent clotting and remove blood clots once they form. The anticoagulant heparin is found on the surfaces of endothelial cells, especially in the lungs, and inhibits platelet adhesion. Cell surfaces also have a protein, thrombomodulin, that binds thrombin. The complex thus formed activates protein C, which inhibits clotting by degrading factor V and catalyzing the production of plasmin from plasminogen. Plasmin dissolves fibrin and, therefore, blood clots. Thus thrombin has two major actions: first to initiate clot formation and then to promote clot dissolution.

If significant blood loss (hemorrhage) occurs, blood pressure falls. High blood pressure normally stimulates arterial and atrial baroreceptors, inhibiting vasopressin release and sympathetic outflow to the peripheral circulation. Hemorrhage reduces both venous and arterial blood pressure, reducing the discharge frequency of both atrial and arterial baroreceptors. This change in frequency releases the baroreceptive inhibition of sympathetic outflow, causing constriction of both arteries (vasoconstriction) and veins (venoconstriction) and an increase in cardiac output. The peripheral vasoconstriction and increased cardiac output raise arterial blood pressure, while the venoconstriction maintains venous return to the heart.

Hemorrhage-induced reduction in baroreceptor activity promotes the release of vasopressin (antidiuretic hormone, ADH). In addition, there is an increase in renin-angiotensin-aldosterone activity (see above) resulting from the fall in blood pressure and the associated decrease in renal blood flow. Both vasopressin and aldosterone reduce urine formation, thereby conserving plasma volume. There is also a marked stimulation of thirst, which restores plasma volume through drinking. The reduced renal blood flow promotes kidney production of erythropoietin, which stimulates erythrocyte production by the bone marrow. Thus, lost red blood cells are replaced by increased production in the days following the hemorrhage. The liver is also stimulated to increase the production of plasma proteins. The increase in production of erythrocytes and plasma proteins, along with the reduction in urine production and increased drinking rate, restores the blood to its original state.

SUMMARY

General plan of the circulatory system

■ Circulatory systems can be divided into two broad categories, open and closed.

■ In open circulatory systems, pressures are low, and blood pumped by the heart empties into a space called the hemocoel, in which blood bathes the cells directly.

■ In closed circulatory systems, blood passes via capillaries from the arterial to the venous circulation. Pressures are high, allowing ultrafiltration of the blood. Fluid that has leaked across capillary walls into the extracellular spaces is subsequently returned to the circulation via a lymphatic system.

The heart

■ The heart is a muscular pump that ejects blood into the arterial system. Excitation of the heart is initiated in a pacemaker and is conducted to the rest of the heart through gap junctions that connect the myocardial cells.

■ The initial phase of each heart contraction is isometric, followed by an isotonic phase in which blood is ejected into the arterial system.

■ Cardiac output is dependent on venous inflow. In mammals, changes in cardiac output are associated with changes in heart rate rather than in stroke volume.

Blood flow

■ Blood flow is generally streamlined (continuous laminar flow), but because the relationship between pressure and flow is complex, Poiseuille's law applies only to flow in smaller arteries and arterioles.

■ The arterial system acts as a pressure reservoir and a conduit for blood between the heart and capillaries. The elastic arteries damp oscillations in pressure and flow caused by contractions of the heart, and the muscular arterioles control the distribution of blood to the capillaries.

■ The venous system acts both as a conduit for blood between capillaries and the heart and as a blood reservoir. In mammals, 50% of the total blood volume is contained in veins.

■ Capillaries are the site of transfer of materials between the blood and tissues.

■ The walls of capillaries are generally much more permeable than other cell layers. Material is transferred between blood and tissues by passing either through or between the endothelial cells that form the capillary wall.

■ Endothelial cells contain large numbers of vesicles that may coalesce to form channels for the movement of material through the cell. Some endothelial cells have specific carrier mechanisms for transferring glucose and amino acids.

■ The size of the gaps between cells varies between capillary beds. Brain capillaries have tight junctions, whereas liver capillaries have large gaps between cells.

■ The lymphatic system drains excess fluid in tissues, resulting from the filtration of plasma across capillary walls, and returns it to the circulation at a point of low pressure (the thoracic duct in mammals).

Regulation of circulation

■ Arterial pressure is regulated via neuronal control mechanisms to maintain capillary blood flow, which can be further adjusted locally to meet the requirements of particular tissues.

■ Arterial baroreceptors monitor blood pressure and reflexly alter cardiac output and peripheral resistance to maintain arterial pressure.

■ Atrial and ventricular mechanoreceptors monitor venous pressure and heart rate to ensure that the activity of the heart is correlated with blood inflow from the venous system and blood outflow into the arterial system.

■ Arterial chemoreceptors respond to changes in the pH and gas levels of the blood.

■ All these sensory receptors feed information into the medullary cardiovascular center, where their inputs are integrated to ensure an appropriate response of the circulatory system to changing requirements of the animal, as during exercise.

■ Natriuretic peptides, vasopressin (ADH), and the renin-angiotensin-aldosterone system operate in conjunction with neuronal reflexes to maintain blood volume.

■ In general, stimulation of sympathetic nerves innervating vascular smooth muscle causes peripheral vasoconstriction and a rise in arterial blood pressure, whereas an increase in circulating catecholamines (especially epinephrine) causes a decrease in peripheral resistance accompanied by a rise in arterial pressure due to a concomitant rise in cardiac output.

■ As aerobic metabolism in a tissue increases, there is a local increase in capillary blood flow, termed active hyperemia. This assures that the most active tissues normally have the highest capillary blood flow.

■ The vascular endothelium releases various compounds (e.g., nitric oxide, endothelin, and prostacyclin) that cause localized vasoconstriction or vasodilation, thereby adjusting blood flow to tissue needs.

- Inflammatory mediators, including histamine and kinins, act to increase blood flow to sites of tissue injury.
- Blood clotting, which involves a cascade of events, prevents blood loss from damaged vessels. Clot formation begins with platelet aggregation, which is promoted by thromboxane A_2 but inhibited by nitric oxide and prostacyclin.

REVIEW QUESTIONS

1. Describe the properties of myogenic pacemakers.
2. Describe the transmission of excitation over the mammalian heart.
3. Describe the changes in pressure and flow during a single beat of the mammalian heart.
4. Discuss the factors that influence stroke volume of the heart.
5. What is the nature and function of the innervation of the mammalian heart?
6. What is the effect on cardiac function of a rigid versus a compliant pericardium?
7. What is the functional significance of the partially divided ventricle found in some reptiles?
8. Discuss the changes in circulation that occur at birth in the mammalian fetus.
9. Discuss the applicability of Poiseuille's law to the relationship between pressure and flow in the circulation.
10. What functions are served by the arterial system?
11. Describe the factors that determine capillary blood flow.
12. Describe the locations of the various baroreceptors and mechanoreceptors in the mammalian circulatory system and their role in cardiovascular regulation.
13. Compare and contrast cardiovascular responses to diving and to breathing air that is low in oxygen.
14. Describe the cardiovascular changes associated with exercise in mammals.
15. What are the consequences of raising or lowering arterial blood pressure for cardiac function and for exchange across capillary walls?
16. Discuss the relationship between capillary structure and organ function, comparing the capillary structures found in different organs of the body.
17. Describe the several ways in which substances are transferred between blood and tissues across capillary walls.
18. What are the functions of the venous system?
19. Describe the effects of gravity on blood circulation in a terrestrial mammal. How are these effects altered if the animal is in water?
20. Define Laplace's law. Discuss the law in the context of the structure of the cardiovascular system.
21. Discuss the role of the lymphatic system in fluid circulation. How does this role vary in different parts of the body?
22. What is the role of the endothelium in the regulation of the cardiovascular system?
23. Why does adenosine have only a paracrine action?
24. Discuss the roles of nitric oxide and adenosine in the circulation.

SUGGESTED READINGS

Crone, C. 1980. Ariadne's thread: An autobiographical essay on capillary permeability. *Microvasc. Res.* 20:133–149.

Goldsby, R. A., T. J. Kindt, and B. A. Osborne. 2000. Leukocyte migration and inflammation. In Goldsby et al., *Kuby Immunology,* 4th ed., 371–394. New York: W. H. Freeman.

Heisler, N., ed. 1995. *Mechanisms of Systemic Regulation: Respiration and Circulation.* Advances in Comparative and Environmental Physiology. Vol. 21. Berlin: Springer-Verlag.

Huang, P. L., Z. Huang, H. Mashimoto, K. D. Bloch, M. A. Moskowitz, J. A. Bevin, and M. C. Fishman. 1995. Hypertension in mice lacking the gene for endothelial nitric oxide synthase. *Nature* 377:239–242.

Irasawa, H., H. F. Brown, and W. Giles. 1993. Cardiac pacemaking in the sinoatrial node. *Physiol. Rev.* 78:197–225.

Kooyman, G. L. 1989. *Diverse Divers: Physiology and Behavior.* Zoophysiology. Vol. 23. New York: Springer-Verlag.

Murad, F. 1998. Nitric oxide signalling: Would you believe that a simple free radical could be a second messenger, autocoid, paracrine substance, neurotransmitter, and hormone? *Recent Prog. Horm. Res.* 53:43–60.

Perry, S., and B. Tufts, eds. 1998. Fish respiration. *Fish Physiology.* Vol 17. New York: Academic Press.

Poulsen, S. A., and R. J. Quinn. 1998. Adenosine receptors: New opportunities for future drugs. *Bioorg. Med. Chem.* 6:619–641.

Schmidt-Nielsen, K. 1972. *How Animals Work.* New York: Cambridge University Press.

Van Vilet, B. N., and N. H. West. 1994. Phylogenetic trends in the baroreceptor control of arterial blood pressure. *Physiol. Zool.* 67(6):1284–1304.

Weibel, E. R., C. R. Taylor, and L. Bolis, eds. 1998. *Principles of Animal Design: The Optimization and Symmorphosis Debate.* Cambridge: Cambridge University Press.

Gas Exchange and Acid-Base Balance

Just 200 years ago, Antoine Lavoisier showed that animals utilize oxygen and produce carbon dioxide and heat (Spotlight 13-1 on the next page). This process was later shown to take place at the level of the mitochondria (see Chapter 3). Animals obtain oxygen from the environment and use it for cellular respiration, producing carbon dioxide and water that are eventually liberated into the environment. For cellular respiration to proceed, there must be a steady supply of oxygen, and the waste product carbon dioxide must be continually removed.

In this chapter we review the transport of oxygen and carbon dioxide in the blood. We look at the systems that have evolved in animals to facilitate the movement of these gases between the environment and the blood and between the blood and the tissues. Our main focus is on systems found in vertebrates, particularly mammals, because these have been investigated the most thoroughly. A number of systems that transport oxygen between the environment and tissues are of particular interest, including the one that moves oxygen into the swimbladder of fishes against gradients of pressure that can be as much as several atmospheres. This system is described at the end of this chapter as an example of one the many intriguing problems of gas transfer in animals.

OXYGEN AND CARBON DIOXIDE IN THE PHYSICAL ENVIRONMENT

Oxygen is the most abundant element in the Earth's crust (49.2% by mass) and constitutes 20.93% of its atmosphere. The mass of oxygen in the atmosphere, which is mostly found as molecular oxygen (O_2), is around 1.2×10^{21} g, much higher than the carbon dioxide (CO_2) mass of 2.6×10^{18} g. In contrast, the total mass of free oxygen dissolved in water is only a small fraction of the mass in the atmosphere, whereas much more CO_2 is dissolved in water than is present in the atmosphere. Oxygen is added to the atmosphere by photosynthesis and by photodissociation of water vapor, with photosynthesis by far the dominant contributor. Oxygen is removed from the atmosphere principally by animal respiration, but it is also utilized in oxidizing organic matter, rocks, and gases and in burning various carbon fuels.

Oxygen is transferred between the atmosphere and the aquatic environment by turbulence and molecular diffusion. The solubility of oxygen in water decreases with increasing water temperature and salinity. The solubility of oxygen in water at 30°C and at a salinity of 35‰ (parts per thousand) is 1.2733 μmol $O_2 \cdot L^{-1} \cdot mm\ Hg^{-1}$. Seawater at this temperature equilibrated with air (termed *normoxic water*) contains 193.5 μmol $O_2 \cdot L^{-1}$ or 6.19 mg $O_2 \cdot L^{-1}$ or 4.34 ml $O_2 \cdot L^{-1}$. At 15°C, the amount of oxygen dissolved in seawater equilibrated with air increases to 5.75 ml $O_2 \cdot L^{-1}$ because the solubility of oxygen increases with decreasing temperature. Normoxic water is said to be 100% saturated with oxygen. *Hypoxic water* contains less oxygen that it would at normoxic levels, with the amount often expressed as a percentage of the normoxic value. *Anoxic water* contains no dissolved oxygen.

Oxygen is added to water as photosynthesis is carried out by aquatic plants and algae. Because photosynthesis occurs only when light is available, there are diurnal oscillations in oxygen levels in many aquatic environments. Marine lagoons, for example, often show marked circadian oscillations in oxygen levels, being *hyperoxic* during the day and hypoxic during the night. The magnitude of the oscillations depends on the relative levels of photosynthetic and respiratory activities and the degree of flushing of the lagoon. Tropical freshwater lakes show similar diurnal oscillations in oxygen levels. A layer of ice significantly affects the level of oxygen in the water below. The ice blocks some of the transfer of oxygen between the atmosphere and the

EARLY EXPERIMENTS ON GAS EXCHANGE IN ANIMALS

Poul Astrup and John Severinghaus, two prominent scientists in the field of gas exchange, describe many of the most significant experiments leading to our present understanding of gas transfer in animals in their book *The History of Blood Gases, Acids and Bases,* published in 1986. Studies of gas exchange in animals began as an extension of the work of Robert Boyle (1627–1691) on the properties of air. Boyle showed that both animals and flames die in a vacuum, suggesting that something in air is required both to maintain life and to keep a candle burning.

Joseph Priestley (1733–1804), who lived near a brewery, was fascinated by the large volumes of gas produced during the brewing process. Continuing Boyle's experiments in modified form, Priestley heated various chemicals, collected the gases produced, and then determined whether mice could live in those gases. He noticed that a mouse lived longer, and a flame burned brighter, in the gas produced by heating mercuric oxide than in the gases produced from other chemicals. He also observed that mice lived longer if plant material was present in their containers. Priestley's observations caused Benjamin Franklin to remark that the practice of cutting down trees near houses should cease since plants are able to restore air, which is spoiled by animals.

Priestley had demonstrated that plants, as well as certain chemicals when heated, could produce something that keeps animals and flames alive. Reasoning from a contemporary theory that held that a mysterious substance, *phlogiston,* was released when material was burned, Boyle had concluded that a substance was produced by plants and heated chemicals that absorbed phlogiston. Antoine Lavoisier (1743–1794), however, found that phosphorus gained weight when burned in air. He found that some other substances gained weight when heated in air, but not when heated in a vacuum. In other words, something in air was consumed, rather than given off, when some substances were heated. That was the end of the phlogiston theory. Lavoisier called the substance that was consumed during burning, and that was required to keep animals alive, *oxygen,* from Greek words meaning "to form acid."

Lavoisier repeated some experiments of Henry Cavendish (1731–1810), who found that the flammable gas released when metals are added to acid can combine with oxygen to form water. Lavoisier named this gas *hydrogen,* from Greek words meaning "to form water." By repeating and extending some of Priestley's experiments, he also found that if mercuric oxide was heated with coal, "fixed air" (carbon dioxide) was formed. Fixed air had been described earlier by Joseph Black (1728–1799), who produced it by adding acid to chalk.

Exhaled air was known to contain some fixed air, so Lavoisier took the next big step. He realized that burning coal and breathing animals both consume oxygen and produce heat and carbon dioxide. He then measured oxygen uptake and heat production in animals and found that the amount of heat produced relative to oxygen uptake was about the same for animals and burning coal, although the rates of these processes were much slower in animals.

Besides being a scientist, Lavoisier was also a tax collector. Such people are generally not held in high repute, and this brilliant man was no exception: he was sent to the guillotine during the French Revolution in 1794.

water, and the oxygen produced by photosynthesis is decreased due to low light conditions under the ice. Thus water under ice is often hypoxic.

The slow process of diffusion contributes little to the distribution of oxygen in water. In the absence of light, mixing of water is the only source of oxygen at depth. Oxygen is removed from water by respiration and by oxidation of chemicals and organic materials. Thus, if mixing does not occur, deep water becomes hypoxic or even anoxic. When layers of different temperature, salinity, or density (thermoclines, haloclines, or pycnoclines, respectively) form in a water column, vertical mixing is reduced or prevented, resulting in hypoxic or anoxic regions in the deep layers. Records of oxygen concentration in the oceans show an initial decrease from high levels at the surface and then an increase at greater depths. A layer called the *oxygen minimum zone* (OMZ) usually occurs below the thermocline at a depth of between 400 and 800 m. Photosynthetic activity in this region is low, respiratory activity is high, and mixing is limited. Zooplankton and bacteria often congregate in this region to feed on accumulated particulate matter.

Sediments receive oxygen by molecular diffusion from the adjacent water. Mud and silt are typically hypoxic or anoxic, with oxygen penetrating only a few millimeters into the top layer, unless water can filter through the sediment.

As a consequence of all these processes, the aquatic environment shows marked local variation in oxygen content, and aquatic hypoxia and anoxia are natural phenomena, both in time and space. Human activities also have a large effect on the levels of oxygen and carbon dioxide in aquatic environments as well as in the atmosphere (see Spotlight 13-2).

OXYGEN AND CARBON DIOXIDE IN LIVING SYSTEMS

Oxygen and carbon dioxide are transported in opposite directions in living systems, yet both processes have many elements in common. Both gases are transferred passively across the body surface by diffusion. The physical laws relevant to the behavior of gases, along with some of the terminology used in respiratory physiology,

EFFECTS OF HUMAN ACTIVITY ON ENVIRONMENTAL OXYGEN AND CARBON DIOXIDE LEVELS

Two billion years ago, the amount of oxygen in the atmosphere was negligible. The oxygen produced by photodissociation of water was consumed in the oxidation of rocks and gases. Early life evolved in the absence of oxygen. The appearance of photosynthetic organisms launched a great increase in oxygen production, and the levels of oxygen in the atmosphere gradually rose, with oscillations over time due to changes in rates of oxygen production and consumption. Oxygen levels today are higher than they were in the late Permian period, but lower than they were in the Carboniferous and Cretaceous periods. Aquatic vertebrates evolved and radiated when oxygen levels in the atmosphere were about half their present-day values. About 300 million years ago, however, oxygen levels in the atmosphere may have been as high as 35% (they are about 21% today). During this period, vertebrates first invaded the terrestrial environment and large insects first appeared in the fossil record, including giant dragonflies with wingspans of 70 cm. Oscillations in atmospheric oxygen content were a feature of the environment during much of vertebrate evolution, but recent times have been much more stable.

At the beginning of the twentieth century, a Swedish scientist, Svante Arrhenius (1859–1927), who received the Nobel Prize in chemistry in 1903, pointed out that industrial activity would result in increased levels of carbon dioxide in the atmosphere and that this increase would in turn cause global warming. Since that time, the human population has risen sharply, energy utilization has risen even faster, and human activity has become the single most important factor determining environmental change. The burning of fossil fuels in particular has resulted in a striking rise in carbon dioxide production and oxygen utilization. The change in atmospheric levels of carbon dioxide is especially noteworthy. There is 500 times more oxygen than carbon dioxide in the atmosphere, and burning all of the available fossil fuels would reduce atmospheric oxygen levels by only a fraction of one percent. Carbon dioxide levels, however, have already changed by much more than that due to human activity. Early in 2001, several international organizations released scientific reports documenting that global warming due to human activity must be regarded as a fact. We are watching Arrhenius's predictions about the atmosphere come true.

In the aquatic environment, human activity has resulted in increasing incidences of hypoxia and anoxia. Two-thirds of the world's human population lives close to the ocean, into which human sewage is discharged. Runoff of nutrients, such as animal waste and fertilizers from agricultural activities, contributes to the massive increase in pollution of the aquatic environment. Intensive fish farming in coastal waters, particularly in Southeast Asia, has significantly increased the total nutrient load. Sewage and other organic materials discharged into the oceans consume oxygen, and this effect, along with the increased productivity associated with nutrient loading, or *eutrophication,* causes aquatic hypoxia.

Dissolved oxygen levels have decreased in many water bodies around the world; in almost every region where there is a large human population close to the ocean, there is some degree of aquatic hypoxia. Eutrophication of the coastal regions of the Black Sea, largely due to materials carried into it by the Danube River, has resulted in growing periods of hypoxia and anoxia, and by 1980 had reduced the number of benthic species to a third of the 1960 estimate. Only 6 of the 26 commercial species harvested in 1960 supported a Black Sea fishery in 1980. Similar changes have been observed in other regions of the world. Hypoxic events in the Kattegat at the entrance of the Baltic Sea have resulted in poor catches of Norwegian lobster and shrinking schools of herring, with the fish more dispersed within the school.

Increasing periods of hypoxia have been recorded in the Pearl River, the Baltic Sea, the Gulf of Mexico, and many other rivers and marine areas. Summer hypoxia and anoxia in the Mississippi River delta and Chesapeake Bay, as well as many other regions, has reduced both community biomass and species richness but has increased the biomass of opportunistic species. In general, periodic hypoxia reduces total biomass and species numbers and alters species composition toward younger, smaller, shorter-lived fauna. The trophic structure of benthic communities also changes with decreasing oxygen levels. Some organisms move out of hypoxic or anoxic burrows and become more vulnerable to predation. Predators change their diets as some prey organisms disappear and others become more available.

All these changes in aquatic ecosystems have occurred in the last 30 to 50 years. Diaz and Rosenberg stated in 1995 that "there is no other environmental parameter of such ecological importance to coastal marine ecosystems that has changed so drastically in such a short period as dissolved oxygen. . . . If we do not move quickly to reduce or stop the primary cause of low oxygen . . . then the productivity structure of our major estuarine and coastal areas will be permanently altered."

are reviewed in Spotlight 13-3 on the next page. To maximize the rate of gas transfer for a given concentration difference, the respiratory surface area should be as large as possible and diffusion distances as small as possible.

The oxygen required and the carbon dioxide produced by an animal increase as a function of its mass, but the rate of gas transfer across the body surface is related primarily to surface area. The surface area of a sphere increases as the square of its diameter, whereas the volume increases as the cube. In very small animals, such as rotifers and protozoans, which are less than 0.5 mm in diameter, the distances for diffusion are small, and the ratio of surface area to volume is large. Diffusion alone is sufficient for the transfer of gases in these cases. When animal size increases, however, diffusion distances increase and the ratio of surface area to volume drops. Large surface-area-to-volume ratios are maintained in larger animals by the elaboration of special tissues for the exchange of gases. In some animals the whole body surface participates in gas transfer, but large, active animals have a specialized respiratory surface, the **respiratory epithelium,** made up of a thin

THE GAS LAWS

Over 300 years ago, Robert Boyle determined that at a given temperature, the product of pressure and volume is constant for a given number of molecules of gas. **Gay-Lussac's law** states that either the pressure or the volume of a gas is directly proportional to absolute temperature if the other is held constant. Combined, these laws are expressed in the equation of state for a gas:

$$PV = nRT$$

where P is pressure, V is volume, n is number of molecules of a gas, R is the universal gas constant (0.08205 $L \cdot atm \cdot K^{-1} \cdot mol^{-1}$, or 8.314×10^7 $ergs \cdot K^{-1} \cdot mol^{-1}$, or 1.987 $cal \cdot K^{-1} \cdot mol^{-1}$), and T is the absolute temperature in degrees kelvin.

The equation of state for a gas indicates that equal volumes of different gases at the same temperature and pressure contain equal numbers of molecules **(Avogadro's law).** One mole of gas occupies approximately 22.414 liters at 0°C and 760 mm Hg. Because the number of molecules per unit volume is dependent on pressure and temperature, these conditions should always be stated along with the volume of gas. Gas volumes in physiology are usually reported as being at body temperature, atmospheric pressure, and saturated with water vapor (BTPS); at ambient temperature and pressure and saturated with water vapor (ATPS); or at standard temperature and pressure (0°C, 760 mm Hg) and dry, or zero water vapor pressure (STPD).

Gas volumes measured under one set of conditions (e.g., ATPS) can be converted to another (e.g., BTPS) by using the equation of state for a gas. For example, the volume of air expired from a mammalian lung at a body temperature of 37°C (273 + 37 = 310 K) is often measured at ambient room temperature—say, 20°C (273 + 20 = 293 K). The drop in temperature will reduce the volume of the expired gas. A gas that is in contact with water will be saturated with water vapor. The water vapor pressure at 100% saturation varies with temperature. Expired air is saturated with water vapor, but as temperature decreases, the water vapor will condense, and this condensation will also reduce the expired gas volume. If the barometric pressure is 760 mm Hg, and the water vapor pressure at 37° and 20°C is 47 mm Hg and 17 mm Hg respectively, then a measured gas volume at 20°C of 500 ml is converted to BTPS expired volume as follows:

$$500 \text{ ml} \times \frac{(760 - 17)}{(760 - 47)} \times \frac{(273 + 37)}{(273 + 20)} = 551 \text{ ml}$$

Thus, under the conditions stated above, a gas volume of 551 ml within the lung is reduced to 500 ml following exhalation because of the drop in gas temperature and the condensation of water vapor.

Dalton's law of partial pressure states that the partial pressure of each gas in a mixture is independent of other gases present, so that the total pressure equals the sum of the partial pressures of all gases present. The **partial pressure** of a gas in a mixture depends on the number of molecules present in a given volume at a given temperature. Usually, oxygen accounts for 20.94% of all gas molecules present in dry air; thus, if the total pressure is 760 mm Hg, the partial pressure of oxygen, P_{O_2}, will be $760 \times 0.2094 = 159$ mm Hg. But air usually contains water vapor, which contributes to the total pressure. If the air is 50% saturated with water vapor at 22°C, the water vapor pressure is 18 mm Hg. If the total pressure is 760, the partial pressure of oxygen will be $(760 - 18) \times 0.2094 = 155$ mm Hg. If the partial pressure of CO_2 in a gas mixture is 7.6 mm Hg and the total pressure is 760 mm Hg, then 1% of the molecules in the air are CO_2.

Gases are soluble in liquids. The quantity of a gas that dissolves at a given temperature is proportional to the partial pressure of that gas in the gas phase **(Henry's law).** The quantity of gas in solution equals αP, where P is the partial pressure of the gas and α is the **Bunsen solubility coefficient,** which is independent of P. The Bunsen solubility coefficient varies with the type of gas, the temperature, and the liquid in question, but is constant for any one gas in a given liquid at constant temperature. The Bunsen solubility coefficient for oxygen decreases as the ionic strength and temperature of water increases.

layer of cells 0.5 to 15 μm thick. The respiratory epithelium constitutes the major portion of the total body surface area. In humans, for instance, the respiratory surface area of the lungs is between 50 and 100 m^2, varying with age and lung inflation.

Gas transfer between the environment and eggs, embryos, many larvae, and even some adult amphibians occurs by simple diffusion. Boundary layers of stagnant fluid low in oxygen and high in carbon dioxide are found whenever gas transfer occurs by diffusion alone. The thickness of this hypoxic layer increases with animal size, oxygen uptake, and decreasing temperature. Stagnation of the medium close to the gas-exchange surface is avoided in most animals by *ventilation,* which propels air or water over the respiratory surface. In larger animals, circulatory systems have evolved to transfer oxygen and carbon dioxide via blood flowing between the respiratory epithelium and the tissues. Blood passes through an extensive capillary network in both regions and is spread in a thin film just beneath the gas-exchange surface, an arrangement that minimizes the distance across which gases must diffuse and increases the area for diffusion. In the tissues, the diffusion distance between any cell and the nearest capillary is small.

Graham's law states that the rate of diffusion of a substance down a given gradient is inversely proportional to the square root of its molecular weight (or density). Because oxygen and carbon dioxide molecules are of similar size, they diffuse at similar rates in air; they are also utilized (O_2) and produced (CO_2) at approximately the same rates by animals. It can therefore be expected that a transfer system that meets the oxygen requirements of an animal will also ensure adequate rates of carbon dioxide removal.

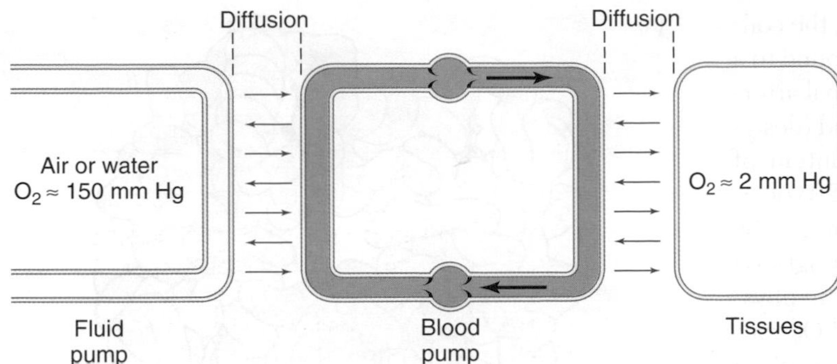

Figure 13-1 The gas-transport system of a vertebrate consists of two pumps and two diffusion barriers alternating in series between the external environment and the tissues. [Adapted from Rahn, 1967.]

Figure 13-1 illustrates the basic components of the gas-transfer system in many animals. The process involves four basic steps:

1. *Breathing movements,* which assure a continual supply of fluid (air or water) to the respiratory surface (e.g., lungs or gills)
2. *Diffusion* of O_2 and CO_2 across the respiratory epithelium
3. *Bulk transport* of gases by the blood
4. *Diffusion* of O_2 and CO_2 across capillary walls between the blood and mitochondria in tissue cells

The capacity of each of these steps is matched to the capacity of the other steps because natural selection tends to eliminate metabolically costly unutilized capacities. This matching of capacities in a chain of linked events has been referred to as *symmorphosis.* However, symmorphosis is not always evident in chains of biological events. One explanation for apparent over- or undercapacity is that a single element can be a link in several chains; thus, its capacity may be appropriate for one chain of events, but be in excess for another.

The rate of flux of respiratory gases varies enormously among animals, from 0.08 $ml \cdot g^{-1} \cdot h^{-1}$ in an earthworm to 40 $ml \cdot g^{-1} \cdot h^{-1}$ in a hovering hummingbird. The concentrations of the enzymes in the respiratory chain and the overall area of mitochondrial inner membrane in aerobic cells both increase with metabolic rate. Limits are established by physical constraints and physiological function. Clearly mitochondrial volume and density in muscles cannot be increased indefinitely without compromising the capacity of the muscles to contract; that is, there must be some relationship between the structures that supply energy (mitochondria) and the structures that use it (myofilaments). The space occupied by mitochondria never exceeds 45% of the total volume in the muscle of mammals, birds, and insects, the animals with the highest oxygen uptakes. There must also be limits to mitochondrial design in terms of the number of cristae (folds in the mitochondrial inner membrane) per unit mitochondrial volume, the ultimate miniaturization being determined by the minimum volume required by the respiratory enzymes. Hummingbirds, and perhaps some small mammals and a few insects, may have approached the limits of design in maximizing the rate of oxygen uptake.

Most insects are much smaller than the smallest birds and mammals. Some large extinct insects seem to have been displaced from their ecological niche by small birds, but smaller insects continue to thrive. We might surmise from these patterns that miniaturization in birds and mammals may be limited by the nature of their gas-transfer systems. In contrast, insects have a tracheal system that exchanges gases directly between the medium and the tissues, permitting high rates of oxygen uptake in these small animals and contributing to their hold on their ecological niche.

What are some advantages and disadvantages of a tracheal system, compared with a blood circulation system, in the transfer of gases between the environment and the tissues?

OXYGEN AND CARBON DIOXIDE IN BLOOD

In considering the movement of oxygen and carbon dioxide between the environment and the cells, we first discuss how these gases are transported in the blood, and then consider gas exchange with the environment and with the cell. We take this approach because the mechanisms by which oxygen and carbon dioxide are carried in the blood affect their transfer between the environment and the blood and between the blood and the tissues.

Respiratory Pigments

Once oxygen diffuses across the respiratory epithelium into the blood, it is bound by a **respiratory pigment.** The best-known respiratory pigment, **hemoglobin,** is what gives human blood its characteristic red color. The ability of respiratory pigments to bind oxygen greatly increases the carrying capacity of blood for molecular oxygen—in the absence of a respiratory pigment, the O_2 content of blood would necessarily be low. The Bunsen solubility coefficient (see Spotlight 13-3) of oxygen in blood at 37°C is 2.4 ml O_2 per 100 ml of blood per atmosphere of oxygen pressure. Therefore, the concentration of O_2 in physical solution (i.e., not bound to a respiratory pigment) in human blood at a normal arterial P_{O_2} would be only 0.3 ml O_2 per 100 ml blood (designated 0.3 vol % O_2). In fact, the total O_2 content of human arterial blood at a normal arterial P_{O_2} is 20 vol %. The 70-fold increase is due to the binding of oxygen by hemoglobin. In most animals for which hemoglobin serves as a respiratory pigment, the O_2 content in physical solution is only a small fraction of the total O_2 content of the blood. The Antarctic icefish is an exception among the vertebrates: the blood of this fish lacks a respiratory pigment and consequently has a low O_2 content. It compensates for the absence of hemoglobin by increased blood volume and cardiac output, but its rate of O_2 uptake is still low compared with that of species from the same habitat that have hemoglobin. Low temperatures have probably been a factor in the evolution of this fish. Oxygen, like all gases, has a higher solubility at low temperatures, and metabolic rates are low at low temperatures in poikilotherms (animals in which body temperature fluctuates more or less with ambient temperature).

Respiratory pigments are complexes of proteins and metal ions, and each one has a characteristic color that changes when it binds O_2. Thus, hemoglobin is bright red when it is loaded with O_2 and a dark maroon-red when deoxygenated. The hemoglobin of most animals is contained within red blood cells (erythrocytes). Vertebrate hemoglobin, except that of cyclostomes (jawless fishes such as lampreys and hagfishes), has a molecular weight of 68,000. It contains four iron-containing porphyrin prosthetic groups, called **heme**, associated with globin, a tetrameric protein (Figure 13-2a). The globin molecule consists of two dimers, $\alpha_1\beta_1$ and $\alpha_2\beta_2$, each of which is a tightly cohering unit; the two dimers associate with each other more loosely. These subunit interactions are altered when hemoglobin binds oxygen, leading to conformational changes in the hemoglobin molecule. Hemoglobin can be dissociated into two α and two β subunits of approximately equal weight, each containing one polypeptide chain and one heme group. Myoglobin, a respiratory pigment that stores O_2 in vertebrate muscles, is similar to a single hemoglobin subunit.

In a hemoglobin molecule, iron in the ferrous state (Fe^{2+}) is bound by the porphyrin ring of the heme, forming coordinate links with four pyrrole nitrogens (Figure 13-2b). The two remaining coordinate linkages are used to bind the heme group to an O_2 molecule and to the imidazole ring of a histidine residue in the globin (Figure 13-2c). Hemoglobin with O_2 bound is called *oxyhemoglobin;* when O_2 is absent, it is called *deoxyhemoglobin.* The binding of O_2 to the iron in heme does not normally oxidize the iron from the ferrous to the ferric (Fe^{3+}) state, as oxygen would when binding free iron, because of the coordination of iron to the nitrogen atoms in the heme molecule. Some oxidation of the ferrous iron in hemoglobin does occur, however, produc-

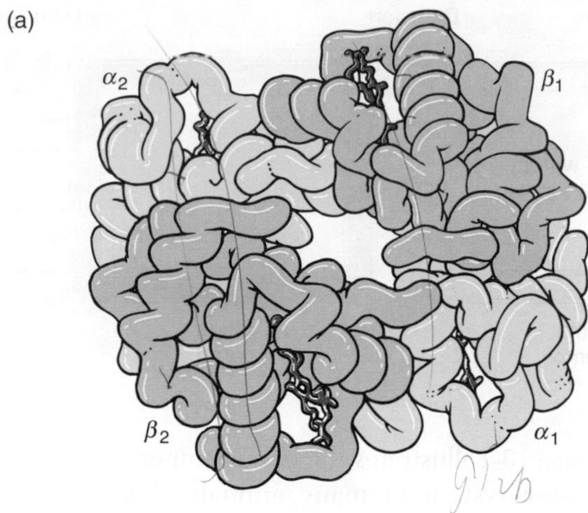

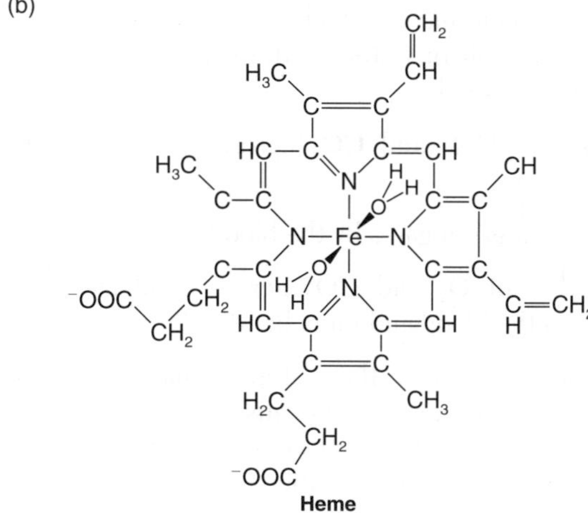

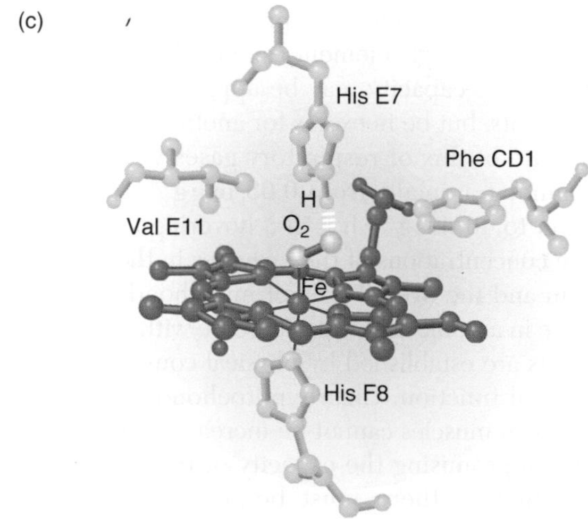

Figure 13-2 Hemoglobin, the main respiratory pigment in vertebrates, consists of four globin protein subunits, each containing one heme molecule. **(a)** Schematic diagram of a hemoglobin molecule, showing the relationship of the α and β chains. The four heme units (red) are visible in the folds formed by the polypeptide chains. **(b)** The structure of heme, formed by the combination of ferrous ion (Fe^{2+}) and protoporphyrin IX. **(c)** Heme in the pocket formed by the globin molecule. The side chain of a histidine (His) residue in globin acts as an additional ligand for the iron atom in heme. [Part c adapted from Nelson and Cox, 2000.]

ing *methemoglobin,* which does not bind O_2 and therefore is nonfunctional. Red blood cells contain the enzyme methemoglobin reductase, which reduces methemoglobin to the functional ferrous form. Certain compounds (e.g., nitrites and chlorates) act either to oxidize hemoglobin or to inactivate methemoglobin reductase, thereby increasing the level of methemoglobin and impairing oxygen transport.

The affinity of hemoglobin for carbon monoxide (CO) is about 200 times greater than its affinity for O_2. As a result, carbon monoxide will displace oxygen and saturate hemoglobin, even at very low partial pressures, causing a marked reduction in oxygen transport to the tissues. Hemoglobin saturated with carbon monoxide is called *carboxyhemoglobin.* The effect of CO saturation on oxidative metabolism is similar to that of oxygen deprivation, which is why the carbon monoxide produced by cars or improperly stoked coal or wood stoves is toxic. Even the levels found in city traffic can impair brain function by producing partial anoxia.

Hemoglobin is found in many invertebrate groups, but others possess different respiratory pigments, including hemerythrin (Priapulida, Brachiopoda, Annelida), chlorocruorin (Annelida), and hemocyanin (Mollusca, Arthropoda), and many invertebrates have no respiratory pigment at all. Hemocyanin, which is a large, copper-containing respiratory pigment, has many properties similar to those of hemoglobin. Hemocyanin binds oxygen in the ratio of 1 mol of O_2 to approximately 75,000 g of the respiratory pigment. In comparison, 4 mol of O_2 can bind to just 68,000 g of hemoglobin when it is completely saturated. Unlike hemoglobin in most species, hemocyanin is not packaged in cells. Moreover, it is light blue when oxygenated, colorless when deoxygenated.

Oxygen Transport in Blood

Each hemoglobin molecule can combine with four oxygen molecules, one per heme. The extent to which O_2 is bound to hemoglobin varies with the partial pressure of oxygen, P_{O_2}. When all four sites on the hemoglobin molecule are occupied by O_2, the blood is 100% saturated, and the oxygen content of the blood is equal to its oxygen capacity. A millimole of heme can bind a millimole of O_2, which represents a volume of 22.4 ml of O_2. Human blood contains about 0.9 mmol of heme per 100 ml of blood. Its oxygen capacity is therefore $0.9 \times 22.4 = 20.2$ vol %. The oxygen content of a unit volume of blood includes the O_2 in physical solution as well as that combined with hemoglobin, but in most cases the O_2 in physical solution is only a small fraction of the total O_2 content. Because the oxygen capacity of blood increases in proportion to its hemoglobin concentration, the oxygen content commonly is expressed as a percentage of the oxygen capacity, that is, the *percent saturation.* This makes it possible to compare the oxygen content of blood samples that contain different amounts of hemoglobin.

Oxygen dissociation curves describe the relationship between percent saturation and the partial pressure of oxygen. The oxygen dissociation curve of myoglobin is hyperbolic, whereas the oxygen dissociation curves of vertebrate hemoglobins, except those of cyclostomes, are sigmoid (Figure 13-3). The sigmoid shape of the hemoglobin dissociation curve results from *subunit cooperativity,* which means that oxygenation of the first heme group facilitates oxygenation of the others. The steep portion of the curve corresponds to oxygen levels at which at least one heme group is already occupied by an oxygen molecule, increasing the affinity of the remaining three heme groups for oxygen.

As a hemoglobin molecule is oxygenated, it goes through a conformational change from a tense (T) state to a relaxed (R) state. The molecule has a much higher affinity for ligands when it is in the T, or deoxygenated, state. Oxygenation is associated with changes in the tertiary structure near the hemes that weaken or break connections between the $\alpha_1\beta_1$ and $\alpha_2\beta_2$ dimers, leading to a large change in the quaternary structure from the T to the R state. The hemoglobins of cyclostomes appear not to bind ligands cooperatively because the subunits under physiological conditions are in monomer/dimer/tetramer equilibria rather than the dimer-dimer arrangement of other vertebrate hemoglobins.

An important property of respiratory pigments is their ability to combine reversibly with O_2 over the range of partial pressures normally encountered in the animal. At low P_{O_2}, only a small amount of O_2 is bound to the respiratory pigment; at high P_{O_2}, a large amount is bound. It is this property that allows the respiratory pigment to act as an oxygen carrier, loading at the

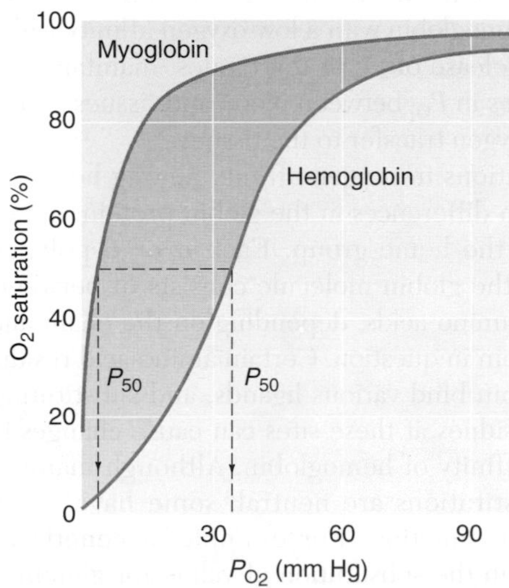

Figure 13-3 Hemoglobins with multiple heme groups have sigmoid oxygen dissociation curves, whereas myoglobin, with only a single heme group, has a hyperbolic dissociation curve. Lamprey hemoglobin, with a single heme group, has a dissociation curve similar to that of myoglobin. P_{50}, the partial pressure at which a respiratory pigment is 50% saturated with oxygen, is a measure of the pigment's oxygen affinity.

respiratory surface, where P_{O_2} is high, and unloading at tissues, where P_{O_2} is low. In some animals, the predominant role of respiratory pigments may be to serve as oxygen reservoirs, releasing O_2 to the tissues only when O_2 is relatively unavailable. In many animals at rest, the venous blood entering the lung or gills is about 70% saturated with oxygen; that is, most of the oxygen bound to hemoglobin is not removed during transit through the tissues. During exercise, when the oxygen demand by the tissues is increased, this venous reservoir of oxygen is tapped, and venous saturation may drop to 30% or less.

Respiratory pigments that have high oxygen affinities are saturated at low partial pressures of oxygen, whereas pigments with low oxygen affinities are completely saturated only at relatively high partial pressures of oxygen. The affinity of a respiratory pigment is expressed in terms of the $\boldsymbol{P_{50}}$, the partial pressure of oxygen at which the pigment is 50% saturated with oxygen; the lower the P_{50}, the higher the oxygen affinity. As the curves in Figure 13-3 demonstrate, myoglobin has a much higher oxygen affinity than hemoglobin.

The rate of oxygen transfer to and from blood increases in proportion to the difference in P_{O_2} across an epithelium. A hemoglobin with a high oxygen affinity facilitates the movement of O_2 into the blood from the environment because O_2 is bound to that hemoglobin at low P_{O_2}; that is, O_2 entering the blood is immediately bound to the hemoglobin, so O_2 is removed from solution and P_{O_2} is kept low. The difference in P_{O_2} across the respiratory epithelium remains large—and the rate of oxygen transfer into the blood remains high—until the hemoglobin is fully saturated. Only then does blood P_{O_2} rise. Hemoglobin with a high oxygen affinity will not release O_2 to the tissues until the P_{O_2} is very low. In contrast, a hemoglobin with a low oxygen affinity will facilitate the release of O_2 to the tissues, maintaining large differences in P_{O_2} between blood and tissues and a high rate of oxygen transfer to the tissues.

Variations in oxygen affinity among hemoglobins arise from differences in the globin protein, not differences in the heme group. Each α or β polypeptide chain of the globin molecule consists of between 141 and 147 amino acids, depending on the chain and the hemoglobin in question. Certain amino acid residues in hemoglobin bind various ligands, and substituting different residues at these sites can cause changes in the oxygen affinity of hemoglobin. Although many amino acid substitutions are neutral, some have a marked impact on function. For example, a genetic defect resulting in the substitution of valine for glutamic acid in position 6 of the β chain causes human hemoglobins to form large polymers that distort the erythrocyte into a sickle shape, giving rise to sickle cell anemia. Because sickle cells cannot pass through small blood vessels, oxygen delivery to tissues is impaired. Individuals with both normal and sickle cell hemoglobins suffer only mild debilitation but have greater resistance to malaria, which accounts for the continuation of this seemingly maladaptive gene in the population.

Changes in certain chemical and physical factors in the blood cause hemoglobin to favor oxygen binding at the respiratory epithelium and oxygen release in the tissues. Hemoglobin-oxygen affinity is reduced by the following conditions:

- elevated temperature
- binding of organic phosphate ligands, including 2,3-diphosphoglycerate (DPG), ATP, or GTP, by hemoglobin
- decrease in pH (increase in H^+ concentration)
- increase in CO_2

The reduction in the oxygen affinity of hemoglobin caused by a decrease in pH is termed the **Bohr effect**, or *Bohr shift* (Figure 13-4). As we will see below, carbon dioxide reacts with water to form carbonic acid and reacts with $—NH_2$ groups on plasma proteins and hemoglobin to form carbamino compounds. An increase in the partial pressure of CO_2 (P_{CO_2}) thus causes a reduction in the oxygen affinity of hemoglobin in two ways: by decreasing blood pH (the Bohr effect) and by promoting the direct combination of CO_2 with hemoglobin to form carbamino compounds. Therefore, when CO_2 enters the blood at the tissues, it facilitates the unloading of O_2 from hemoglobin; when CO_2 leaves the blood at the respiratory surface, it facilitates the uptake of O_2 by the blood.

Other respiratory pigments show a variety of responses to changes in pH. The oxygen dissociation

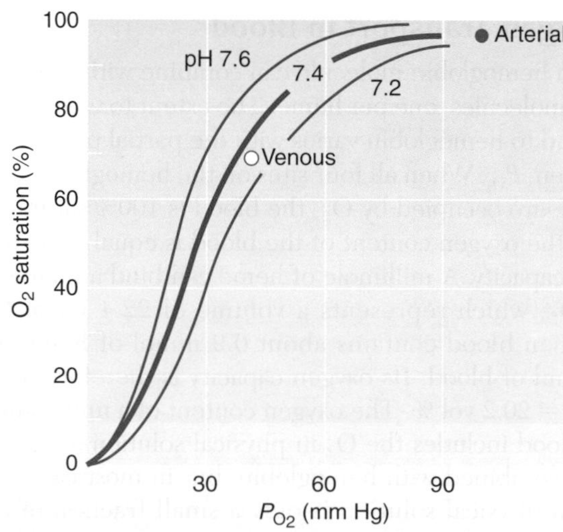

Figure 13-4 The oxygen affinity of hemoglobin decreases with decreasing pH, a phenomenon called the Bohr effect. As a result, changes in blood P_{CO_2}, which influence blood pH, indirectly affect hemoglobin-oxygen affinity. Shown are experimental oxygen dissociation curves in human blood at three pH values. The P_{O_2} values of mixed venous and arterial blood are indicated. [Adapted from Bartels, 1971.]

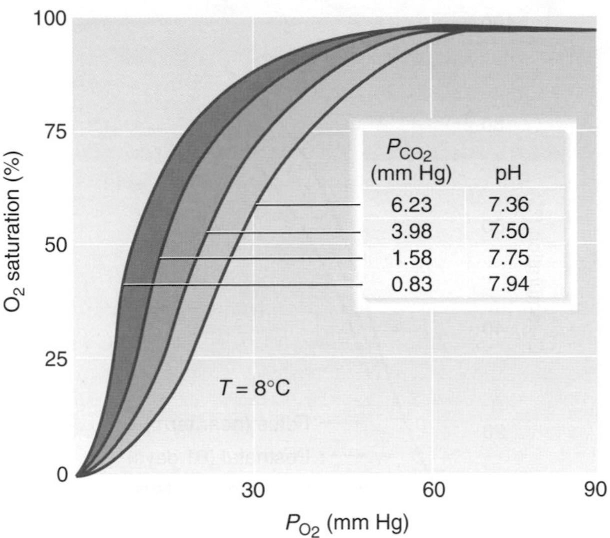

Figure 13-5 Some hemocyanins exhibit a Bohr shift like that of hemoglobin. Blood oxygen dissociation curves for *Cancer magister* reveal a marked Bohr shift in the hemocyanin from this crab. [Unpublished data supplied by D. G. McDonald.]

curve for myoglobin, for example, is relatively insensitive to changes in pH. Hemocyanins from the Dungeness crab *(Cancer magister)* and some other invertebrates exhibit a Bohr shift similar to that of hemoglobin (Figure 13-5), but hemocyanins from several gastropods and from the horseshoe crab *(Limulus)* show a *greater* oxygen affinity with a decrease in pH. This phenomenon, called a *reverse Bohr effect*, may facilitate oxygen uptake during periods of low oxygen availability, during which prolonged reductions in blood pH occur in these animals.

The binding of organic phosphate compounds reduces the oxygen affinity of most vertebrate hemoglobins, except those from cyclostomes, crocodiles, and ruminants. The dominant organophosphate found in erythrocytes differs among species. For instance, mammalian erythrocytes contain high levels of 2,3-diphosphoglycerate (DPG); indeed, hemoglobin and DPG are nearly equimolar in human erythrocytes. DPG binds to specific amino acid residues in the β chains of deoxyhemoglobin, but its binding decreases with increasing pH. Increases in DPG levels accompany reductions in blood O_2 or hemoglobin concentrations, increases in pH, or both. A climb to a higher altitude may result in low blood O_2 levels, as both barometric pressure and the partial pressure of O_2 in air decrease with altitude. In response to these conditions, the level of DPG rises; in humans, this response to high altitude is completed in 24 hours, with a half-time of about 6 hours. At elevations of 3000 m, the DPG concentration in erythrocytes is 10% greater than it is at sea level. Low blood O_2 levels also stimulate an increase in breathing. The resulting increase in gas exchange reduces CO_2 levels in the blood and raises blood pH, which in turn increases hemoglobin-oxygen affinity. The elevation in DPG at altitude offsets the effects of reduced CO_2 levels and maintains hemoglobin-oxygen affinity close to that at sea level.

In the erythrocytes of some vertebrates, other phosphorylated compounds are present in higher concentration than DPG and play a more important role in modulating the oxygen affinity of hemoglobin. In most fishes, ATP or GTP has this function, whereas inositol pentaphosphate (IP_5) is the dominant erythrocytic organophosphate in birds. In the Amazonian fish *Arapaima gigas,* ATP is the dominant erythrocytic organophosphate in the young aquatic form, but IP_5 is dominant in the obligate air-breathing adult.

Phosphorylated compounds in erythrocytes not only affect the oxygen affinity of hemoglobin, but also increase the magnitude of the Bohr effect and may affect subunit interaction. It appears that in mammals, the function of increased DPG levels is to maintain hemoglobin-oxygen affinity under hypoxic (low-oxygen) conditions, as at high altitude. In contrast, hypoxia reduces erythrocytic organophosphate levels in fishes. In these animals, however, hypoxia is often associated with a decrease in blood pH (acidosis), rather than the increase in pH (alkalosis) seen in mammals at altitude. The effect of organophosphate reduction in fishes is to offset the effects of this hypoxia-associated acidosis, thereby maintaining blood-oxygen affinity. Thus, in a functional sense, the effects of changing erythrocytic organophosphate levels are similar in both fishes and mammals; in both instances, the result is to maintain hemoglobin-oxygen affinity.

The binding of oxygen to hemoglobin is a rapid reaction and usually does not limit the rate of oxygen transfer. The rate at which oxygen is taken up by the blood depends in part, however, on its hemoglobin concentration. When more hemoglobin is present, a greater amount of oxygen is bound per unit time, allowing the large diffusion gradient for oxygen across the respiratory epithelium to persist for a longer period, which increases the rate of oxygen transfer.

The presence of a respiratory pigment also increases the transfer of oxygen through blood cells, because the oxygenated pigment co-diffuses with oxygen. A concentration gradient exists for both oxygen and the oxygenated pigment in the same direction through the solution; the gradient for the deoxygenated pigment is in the reverse direction. Hence, the oxygenated pigment diffuses in the same direction as oxygen, whereas the deoxygenated pigment diffuses in the reverse direction. Thus, a pigment such as hemoglobin may facilitate the mixing of gases in the blood, and myoglobin may play a similar role within tissues.

In some fishes, cephalopods, and crustaceans, an increase in CO_2 or a decrease in pH causes not only a reduction in the oxygen affinity of hemoglobin but also a reduction in oxygen binding capacity, which is termed the **Root effect**, or *Root shift*. In those hemoglobins showing a Root shift, low pH reduces oxygen binding to hemoglobin, so that even at high P_{O_2}, only some of the binding sites are oxygenated; that is, 100% saturation is never achieved.

An increase in temperature exacerbates problems of oxygen delivery in poikilothermic aquatic animals such as fishes. A rise in temperature not only reduces oxygen solubility in water, but also decreases the oxygen affinity of hemoglobin, making oxygen transfer between water and blood less efficient. At the same time, rising temperature causes an increase in tissue oxygen requirements.

It is generally assumed that a particular hemoglobin has evolved to meet the special gas-transfer and pH-regulation requirements of the animal. Not only do hemoglobins vary among species, but they also may change during development. In humans, for example, several genes encode β-like globin chains, and the relative expression of these chains differs during prenatal and postnatal life (Figure 13-6). Human fetal hemoglobin, which contains γ chains, rather than adult β chains, has a higher O_2 affinity than adult hemoglobin, which enhances oxygen transfer from mother to fetus. As the proportion of fetal hemoglobin decreases and adult hemoglobin increases following birth, the oxygen affinity of the blood decreases (Figure 13-7). Other mammals exhibit similar differences between fetal and adult hemoglobins.

It is important to remember that, although the hemoglobin of most animals is contained within red blood cells, the values of blood parameters usually refer to conditions in the plasma, not in the red blood cell. Differences in these parameters exist between the inside and outside of cells, including red blood cells. For example, the normal pH of mammalian arterial blood plasma at 37°C is 7.4; the pH inside the red blood cell is lower, about 7.2 at 37°C.

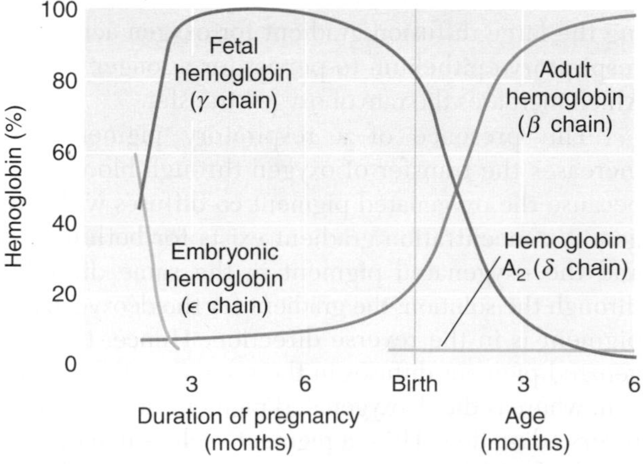

Figure 13-6 Different globin subunits constitute hemoglobin during development in humans. The relative amounts of the various β-like globin chains synthesized in the fetus change during the course of pregnancy. Fetal hemoglobin, which contains two α and two γ chains, has a higher oxygen affinity than adult hemoglobin ($\alpha_2\beta_2$). [Adapted from Young, 1971.]

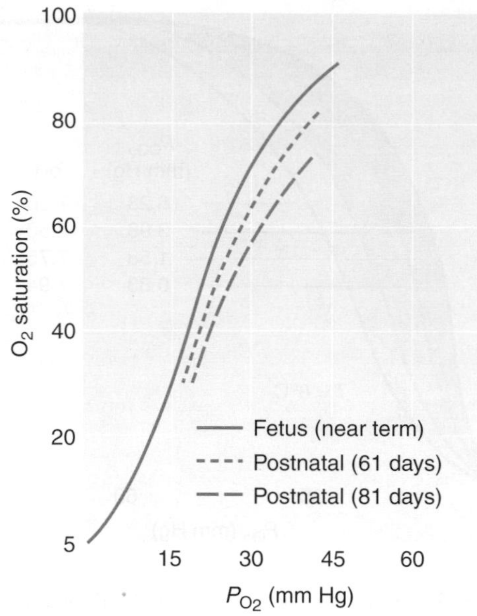

Figure 13-7 In humans, the oxygen affinity of blood decreases for about three months after birth as the fetal hemoglobin is replaced by adult hemoglobin (see Figure 13-6). These blood oxygen dissociation curves were determined at a pH of 7.40. [Adapted from Bartels, 1971.]

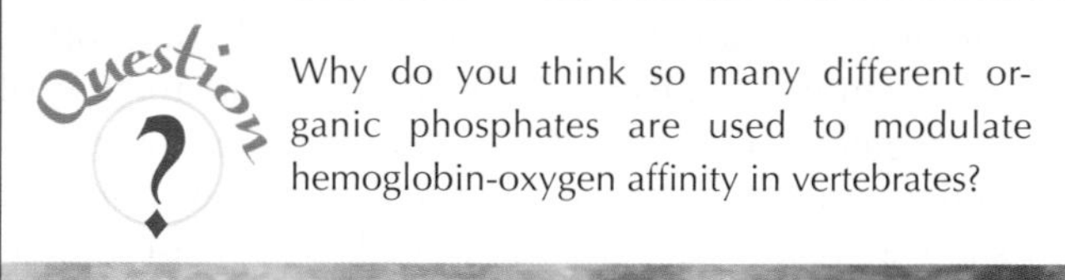

Why do you think so many different organic phosphates are used to modulate hemoglobin-oxygen affinity in vertebrates?

Carbon Dioxide Transport in Blood

Carbon dioxide diffuses into the blood from the tissues, is transported in the blood, and diffuses across the respiratory surface into the environment. Carbon dioxide reacts with water to form carbonic acid (H_2CO_3), a weak acid that dissociates into bicarbonate (HCO_3^-) and carbonate ions (CO_3^{2-}):

$$CO_2 + H_2O \rightleftharpoons H_2CO_3 \rightleftharpoons H^+ + HCO_3^-$$

$$HCO_3^- \rightleftharpoons H^+ + CO_3^{2-}$$

Carbon dioxide also reacts with hydroxyl ions (OH^-) to form bicarbonate:

$$H_2O \rightleftharpoons H^+ + OH^-$$

$$CO_2 + OH^- \rightleftharpoons HCO_3^-$$

The proportion of CO_2, HCO_3^-, and CO_3^{2-} in solution depends on pH, temperature, and the ionic strength of the solution. In mammalian blood at pH 7.4, the ratio of CO_2 to H_2CO_3 is approximately 1000 : 1, and the ratio of CO_2 to bicarbonate ions is about 1 : 20. Bicarbonate is, therefore, the predominant form of CO_2 in the blood at normal pH. The carbonate content of blood is usually negligible in birds and mammals. In poikilotherms, however, with their low temperatures and high blood pH, the carbonate content may approach 5% of the total CO_2 content of the blood, but bicarbonate is still the predominant form of CO_2.

Carbon dioxide also reacts with —NH_2 groups on proteins—particularly hemoglobin—to form carbamino compounds (—$NHCOO^-$):

$$\text{Protein—NH}_2 + CO_2 \rightleftharpoons H^+ + \text{protein—NHCOO}^-$$

In hemoglobin, carbamino groups are more likely to form when the molecule is oxygenated. The extent of carbamino formation increases with the concentration of CO_2 and the number of terminal —NH_2 groups available. The terminal —NH_2 groups of both the α and β chains of mammalian, avian, and reptilian hemoglobins are available for carbamino formation. The terminal —NH_2 group of the α chain of fish and amphibian hemoglobins, however, is acetylated, blocking the formation of carbamino groups. Organophosphates bind at the same terminal —NH_2 group of the β chain that forms carbamino groups, so organophosphate binding reduces carbamino formation. High pH reduces organophosphate binding and so augments carbamino formation by making more —NH_2 groups available. Because fish erythrocytes often have high organophosphate levels as well as acetylated α chains, fish rely less on carbamino formation for CO_2 transport than mammals.

The sum of all forms of CO_2 in the blood—that is, molecular CO_2, H_2CO_3, HCO_3^-, CO_3^{2-}, and carbamino compounds—is referred to as the *total CO_2 content* of the blood. The total CO_2 content varies with P_{CO_2}, and the relationship can be described graphically in the form of a CO_2 dissociation curve (Figure 13-8). As P_{CO_2} increases, the major change is in the bicarbonate content of the blood. The formation of bicarbonate is, of course, pH-

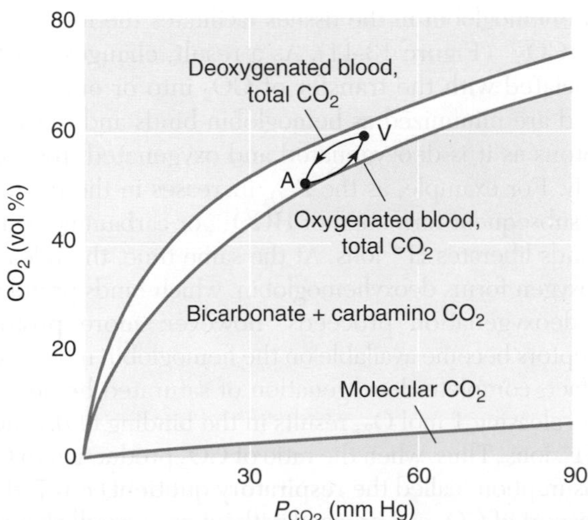

Figure 13-8 The total CO_2 content of blood increases with P_{CO_2}, but only the volume of molecular CO_2 increases linearly. Note that at a given P_{CO_2} oxygenated blood contains less CO_2 than deoxygenated blood (the Haldane effect). A and V refer to arterial and venous blood levels, respectively.

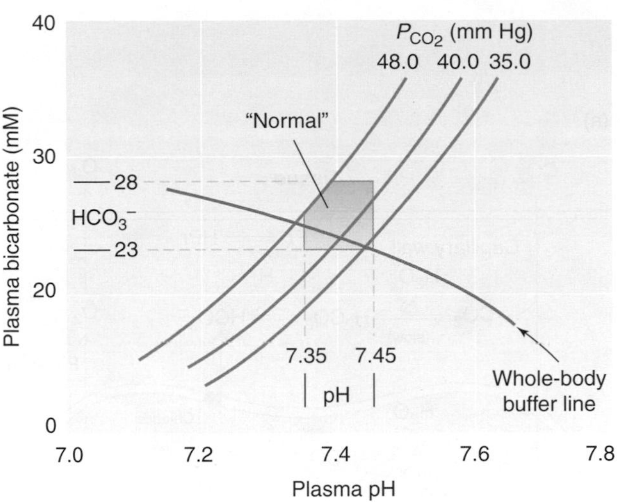

Figure 13-9 The pH, bicarbonate concentration, and P_{CO_2} in human plasma are interrelated and normally fall within quite narrow limits (indicated by the red box). However, when blood P_{CO_2} is altered *in vivo* by hyper- or hypoventilation, then plasma pH and bicarbonate are altered beyond the normal range, as indicated by the whole-body buffer line. [Adapted from Davenport, 1974.]

dependent. The relationships between plasma HCO_3^- concentration and plasma pH at three values of P_{CO_2} are shown graphically in Figure 13-9. A decrease in pH at a constant P_{CO_2} is associated with a fall in bicarbonate. The pH of red blood cells is less than that of plasma, but P_{CO_2} is in equilibrium across the cell membrane. Therefore, bicarbonate levels are lower in erythrocytes than in plasma. Erythrocytes usually constitute less than 50% of the blood volume (i.e., plasma volume is greater than erythrocyte volume), and the bicarbonate concentration is higher in plasma than in erythrocytes; it thus follows that most of the bicarbonate in the blood is in plasma.

Transfer of Gases to and from the Blood

As CO_2 is added to the blood in the tissues and removed from the blood at the respiratory surface, the levels of CO_2, HCO_3^-, and carbamino compounds all change. Carbon dioxide both enters and leaves the blood as molecular CO_2 rather than as bicarbonate because CO_2 molecules diffuse through membranes much more rapidly than HCO_3^- ions. In the tissues, CO_2 enters the blood and either is hydrated to form HCO_3^- or reacts with —NH_2 groups of hemoglobin and other proteins to form carbamino compounds. The reverse process occurs when CO_2 is unloaded from the blood at the respiratory surface. The greatest change is in the HCO_3^- concentration; changes in the levels of CO_2 and carbamino compounds usually represent less than 20% of total carbon dioxide excretion.

The uncatalyzed reaction of CO_2 with OH^- to form HCO_3^- is slow, requiring several seconds. However, in the presence of the enzyme *carbonic anhydrase*, this reaction approaches equilibrium in much less than a second. Although plasma has a higher total CO_2 content

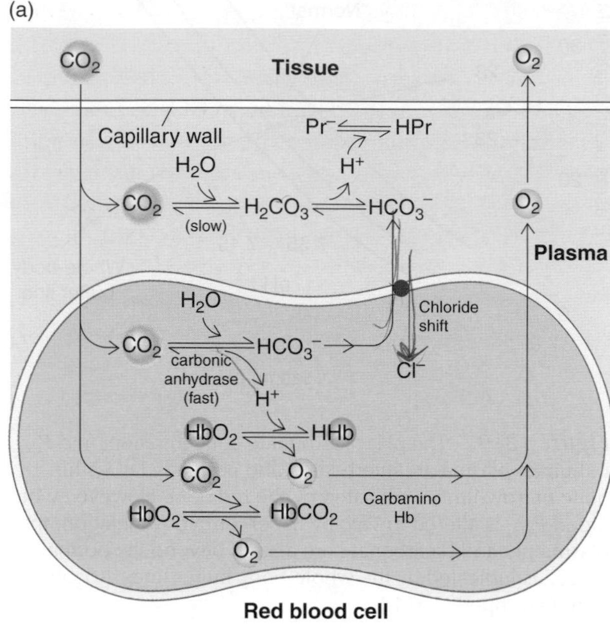

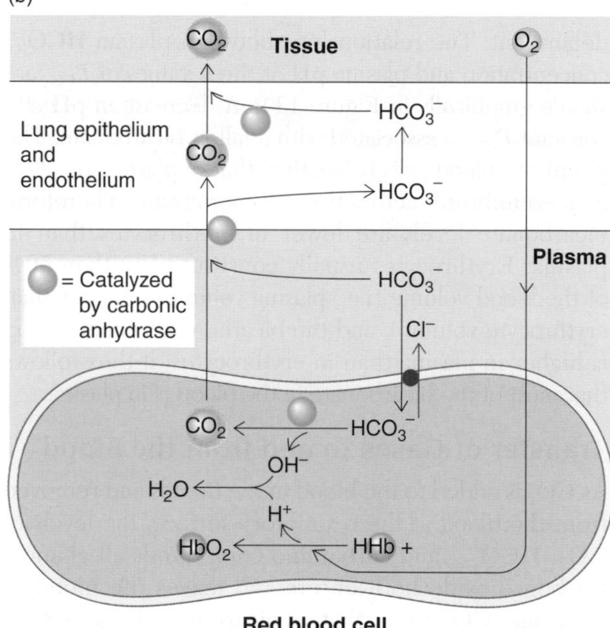

Figure 13-10 Most of the carbon dioxide entering the blood in the tissues and leaving the blood in the lungs passes through red blood cells. **(a)** Carbon dioxide produced in the tissues rapidly forms bicarbonate (HCO_3^-) in the red blood cell in a hydration reaction catalyzed by carbonic anhydrase. Bicarbonate leaves the erythrocyte in exchange for chloride, and excess protons are bound by deoxygenated hemoglobin (Hb). **(b)** These reactions are reversed in the lungs. Oxygen entering the red blood cell displaces protons from Hb, and carbon dioxide enters the plasma. Carbonic anhydrase (indicated by solid circles) in the membrane of the lung endothelial cells converts some of the plasma bicarbonate to carbon dioxide. Movement of carbon dioxide across the respiratory surface is augmented by the diffusion of bicarbonate and its conversion back to carbon dioxide at the outer surface, a process termed facilitated diffusion of CO_2.

than red blood cells, most of the CO_2 entering and leaving the plasma does so via erythrocytes, because carbonic anhydrase is present in red blood cells but not in the plasma. Therefore, formation of HCO_3^- ions in the tissues and CO_2 in the lungs occurs predominantly in red blood cells; once formed, HCO_3^- ions and CO_2 are transferred to or from the plasma.

On entering the blood from the tissues, CO_2 diffuses into red blood cells, where HCO_3^- is formed rapidly in the presence of carbonic anhydrase (Figure 13-10a). As the HCO_3^- level within erythrocytes rises, HCO_3^- ions move from the cells into the plasma. Charge balance within the cell is maintained by anion exchange; as HCO_3^- ions leave the red blood cells, there is a net influx of Cl^- ions from the plasma into the cells, a process called the **chloride shift.** Red blood cells, unlike many other cells, are very permeable to both Cl^- and HCO_3^- because their membranes have a high concentration of a special anion carrier, *band III protein*, which binds Cl^- and HCO_3^- and transfers them in opposite directions across the membrane. This anion exchange is passive and depends on concentration gradients to drive the process, which can occur in either direction, so bicarbonate can flow out of the erythrocyte in the tissues and into the erythrocyte at the respiratory surface (Figure 13-10b). Band III protein is present in all vertebrate erythrocytes except those of cyclostomes. In these animals, bicarbonate stays within the red blood cells, and there is no anion transfer between the erythrocyte and the plasma.

A second reason why most of the CO_2 entering or leaving the blood passes through the red blood cells is that the oxygenation of hemoglobin causes H^+ release, thereby acidifying the cell interior; conversely, deoxygenation results in the binding of H^+ to hemoglobin. Thus O_2 binding to hemoglobin at the respiratory surface facilitates the formation of CO_2, whereas release of O_2 from hemoglobin in the tissues facilitates the formation of HCO_3^- (Figure 13-11). As a result, changes in pH associated with the transfer of CO_2 into or out of the blood are minimized as hemoglobin binds and releases protons as it is deoxygenated and oxygenated, respectively. For example, as the P_{CO_2} increases in the tissues, the subsequent formation of HCO_3^- or carbamino compounds liberates H^+ ions. At the same time, the release of oxygen forms deoxyhemoglobin, which binds protons. As deoxygenation proceeds, however, more proton acceptors become available on the hemoglobin molecule. In fact, complete deoxygenation of saturated hemoglobin, releasing 1 mol O_2, results in the binding of 0.7 mol of H^+ ions. Thus, when the ratio of CO_2 production to O_2 consumption (called the **respiratory quotient**) is 0.7, the transport of CO_2 can proceed without any overall change in blood pH. (As discussed in Chapter 16, the respiratory quotient depends on the animal's diet.) Even when the respiratory quotient is 1, the additional 0.3 mol H^+ is buffered by blood proteins, including hemoglobin, and blood undergoes only a small change in pH. At a given

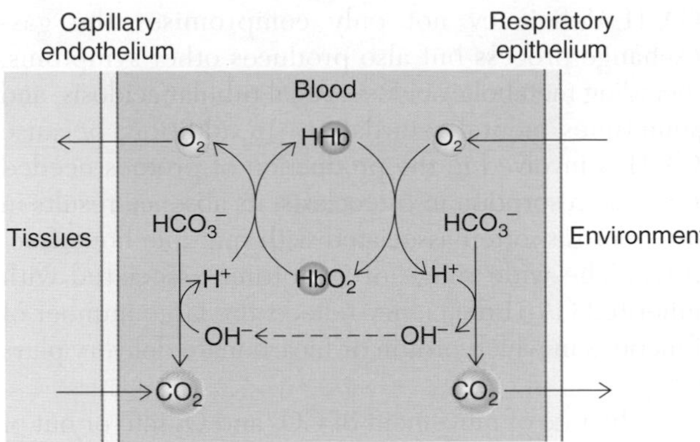

Figure 13-11 The pH changes associated with the changes in blood P_{CO_2} in the tissues and respiratory surface are offset by the binding and release of H^+ ions by deoxygenated and oxygenated blood. Transfer of CO_2 into the blood in the tissues causes a decrease in pH due to formation of bicarbonate; concomitant deoxygenation of hemoglobin frees proton acceptors, which bind the excess H^+ ions. The opposite reactions occur at the respiratory epithelium.

P_{CO_2}, deoxyhemoglobin binds more protons, thereby facilitating HCO_3^- formation, and reacts with CO_2 to form carbamino hemoglobin more easily than does oxyhemoglobin. As a result, the total CO_2 content of deoxygenated blood at a given P_{CO_2} is higher than that of oxygenated blood (see Figure 13-8). Thus, deoxygenation of hemoglobin in the tissues reduces the change in P_{CO_2} and pH as CO_2 enters the blood; this phenomenon is termed the **Haldane effect.**

As noted already, carbonic anhydrase is absent from the blood plasma of most vertebrates, and the interconversion of CO_2 and HCO_3^- occurs at a slow, uncatalyzed rate there. In the lungs, however, carbonic anhydrase is embedded in the endothelial cell membranes with its active site accessible to plasma, so HCO_3^- can be converted rapidly to CO_2 as blood perfuses the lung capillaries (see Figure 13-10b). In addition, oxygenation of hemoglobin acidifies erythrocytes in the lung capillaries, facilitating the conversion of HCO_3^- to CO_2, which then diffuses into the plasma and across the lung epithelium. The resulting decrease in the level of bicarbonate in erythrocytes results in an influx of HCO_3^- ions from the plasma, accompanied by the outward movement of Cl^- ions.

The relative quantities of HCO_3^- converted to CO_2 in the erythrocytes and in the plasma of the blood perfusing the respiratory epithelium are influenced by the extent of proton production associated with hemoglobin oxygenation and the amount of carbonic anhydrase activity in the membranes of the respiratory epithelium. In teleost fishes, such as the trout, the plasma perfusing the gills is not exposed to carbonic anhydrase. In these animals, almost all excretion of CO_2 occurs through the red blood cells and is tightly coupled to O_2 uptake through proton production by oxygenation of hemoglobin. Interestingly, the hemoglobins of several teleosts have a relatively low buffering capacity because of a reduced number of the histidine side chains that normally are involved in buffering H^+ (Figure 13-12). This

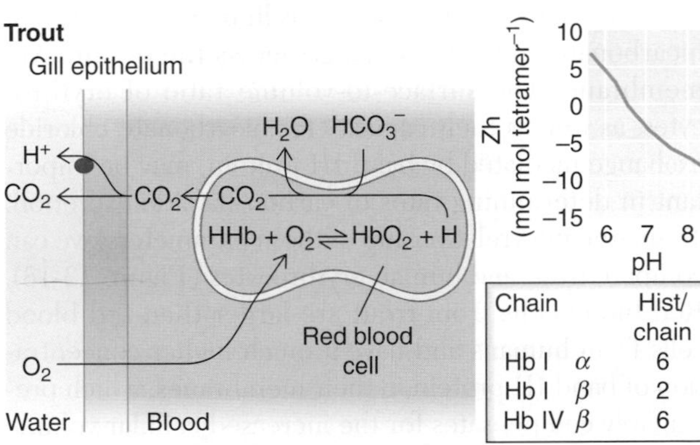

Chain		Hist/chain
Hb I	α	6
Hb I	β	2
Hb IV	β	6

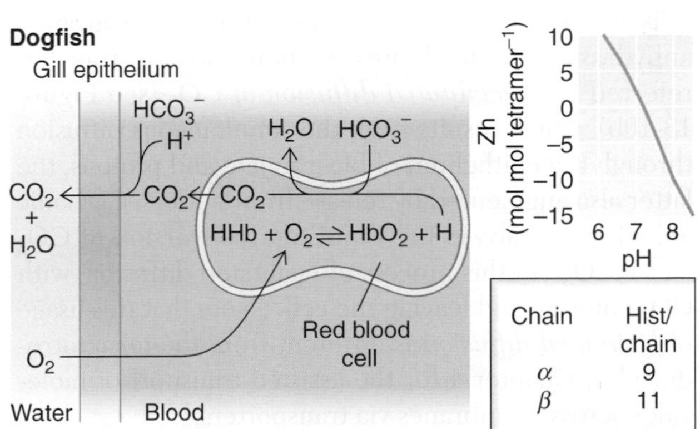

Chain	Hist/chain
α	9
β	11

Figure 13-12 The structures of trout and dogfish hemoglobins are different and are related to the functional role of the gill epithelium in each case. The gill epithelium of the trout, a freshwater teleost, is specialized for sodium uptake coupled to proton excretion, and carbonic anhydrase activity is restricted to the apical membrane. Carbonic anhydrase activity is absent from the basolateral membrane of the gills, and none is available to the plasma. This arrangement reduces the effect of oscillations in plasma pH on epithelial pH and proton excretion, and, therefore, on sodium uptake. Because carbonic anhydrase is absent from the plasma in trout, all excreted CO_2 passes through the red blood cells, and negligible amounts are excreted directly from the plasma. The dogfish gill, on the other hand, is not involved in sodium uptake, and carbonic anhydrase activity is found everywhere in the gill epithelium, plasma, and red blood cells. As a result a significant proportion of CO_2 is excreted directly from the plasma without passing through the red blood cells. Dogfish hemoglobin has a much higher buffering capacity than trout hemoglobin. The dogfish hemoglobin thus buffers the changes in pH associated with oxygenation, whereas in the trout, most of the protons are used in CO_2 excretion. The structure of hemoglobin is therefore influenced by the requirements for transfer of carbon dioxide as well as oxygen. The graphs on the right are the H^+ titration curves, Zh (net H^+ charge, mol $H^+ \cdot$ mol tetramer^{-1}) as a function of pH of hemoglobin from trout and dogfish. The number of histidine side chains for several α and β chains of hemoglobin from trout and dogfish is indicated to the right of each graph. [Adapted from Randall, 1998.]

means that protons released during oxygenation are available for bicarbonate dehydration, facilitating the tight coupling of oxygen and carbon dioxide transfer within the red blood cells seen in these animals. The absence of carbonic anhydrase from the plasma and the basolateral regions of the gill epithelium protects proton pumps in the gills from oscillations in plasma pH. The proton pump in the apical membrane of the gills of freshwater fishes excretes acid and develops an electrochemical gradient that brings sodium into the gill epithelium. The activity of the proton pump is pH-dependent, and oscillations in plasma pH could have a marked effect on the pump if they were rapidly transferred into the gill epithelium. Thus the structure of the hemoglobin molecule is finely tuned not only to the animal's requirements for oxygen and carbon dioxide transfer, but also for ion transfer across the respiratory epithelium. On the other hand, elasmobranchs, such as the dogfish, have lower levels of erythrocytic carbonic anhydrase than the trout and only a loose coupling between oxygen and carbon dioxide flux in the red blood cells. These fishes have significant levels of carbonic anhydrase activity in the blood plasma as well as a broad distribution of this enzyme in the gill epithelium, facilitating carbon dioxide excretion from plasma directly across the gill epithelium.

Carbonic anhydrase activity is also found on the endothelial surfaces of a number of systemic capillary beds, including those in skeletal muscle. In these capillaries, formation of HCO_3^- catalyzed by carbonic anhydrase can occur in the absence of red blood cells. Thus some of the CO_2 transferred into the blood in skeletal muscle does not pass through erythrocytes. Carbonic anhydrase also facilitates carbon dioxide transfer, referred to as *facilitated diffusion of* CO_2 (see Figure 13-10b), which results from the simultaneous diffusion through the epithelium of bicarbonate and protons, the latter also augmented by release from buffers. Carbonic anhydrase catalyzes the rapid interconversion of CO_2 and HCO_3^- in this process of facilitated diffusion, with CO_2 entering and leaving the cell. (Note that this usage of *facilitated diffusion* is different from the term introduced in Chapter 4 for the assisted transport of molecules across membranes via transporters.)

There are at least seven forms of carbonic anhydrase, designated CA-I through CA-VII. All are similar in structure and catalyze the interconversion of carbon dioxide and bicarbonate; they differ in location and catalytic potency. CA-II, an extremely efficient catalyst of the carbon dioxide–bicarbonate hydration-dehydration reactions, is found in a wide variety of tissues, including the brain, eye, kidney, cartilage, liver, lung, pancreas, gastric mucosa, skeletal muscle, and anterior pituitary, as well as red blood cells. CA-II is involved in the supply of bicarbonate and protons for a large number of cellular and metabolic processes. A few humans with an inherited CA-II deficiency have no detectable CA-II but have normal levels of CA-I in their red blood cells. CA-II deficiency not only compromises the gas-exchange process but also produces other symptoms, including metabolic acidosis, renal tubular acidosis, and sometimes mental retardation. In addition, because CA-II is involved in the production of protons needed for bone resorption in osteoclasts, its absence results in osteoporosis, often associated with multiple bone fractures. The wide range of symptoms associated with inherited CA-II deficiency reflects the large number of functions in which proton or bicarbonate delivery plays a role.

The rate of movement of CO_2 and O_2 into or out of the red blood cell is determined by the diffusion distance and the diffusion coefficient of these substances through the cell. Differences in diffusion distance, and hence the rate of erythrocyte oxygenation, might be expected to be related to red blood cell size, which varies considerably among vertebrates. The amphibian *Necturus,* for instance, has erythrocytes that are 600 times the volume of erythrocytes from a goat. Early studies demonstrated that small erythrocytes are oxygenated faster than larger cells *in vitro,* but this finding may have little relevance *in vivo.* Recent experiments using a whole-blood thin-film technique, which is analogous to the situation at the respiratory surface, have shown that oxygen uptake rates are independent of erythrocyte size. The explanation probably lies in the flattened shape of erythrocytes. If the large flat surface of the cells faces the respiratory medium as they pass single file through the respiratory capillaries, then their diffusion distances may be quite similar even though the volumes of the cells are very different.

Normally, excretion of CO_2 is limited by the rate of bicarbonate-chloride exchange across the erythrocyte membrane. The surface-to-volume ratio of erythrocytes, as well as their capacity for bicarbonate-chloride exchange mediated by band III protein, may be important in determining rates of carbon dioxide excretion. To see the interrelationship of these parameters, we can compare trout and human erythrocytes (Figure 13-13). Red blood cells from trout are larger than red blood cells from humans and have a much higher concentration of band III protein in their membranes, which presumably compensates for the increased cellular volume and offsets, at least to some degree, the effects of a lower body temperature on anion-exchange rates. Even so, anion exchange is slower across trout red blood cells at 15°C than across human red blood cells at 38°C. However, transit times for erythrocytes through the gills are longer than those in the lungs, allowing more time for anion exchange across the red blood cell.

Despite these considerations, it is still not clear why different species have evolved red blood cells of such different sizes. Those animals with large red blood cells also have large cells generally. Thus cell size may have been selected for reasons other than gas transfer and may be largely unrelated to gas-transfer rates. For instance, triploid salmon have red blood cells 1.5 times

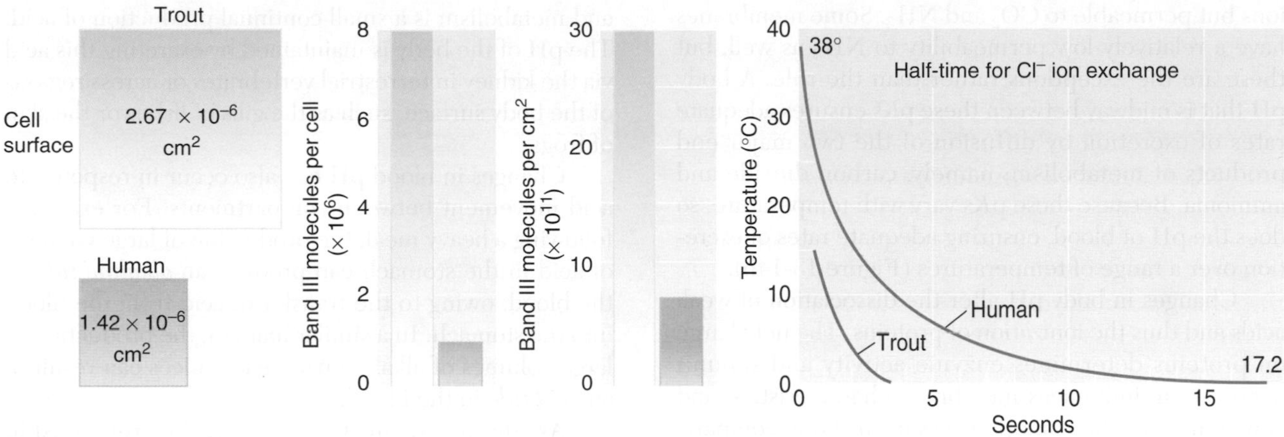

Figure 13-13 Comparison of surface area, band III protein, and half-time for exchange of Cl^- in trout and humans. [Data from Romano and Passow, 1984.]

the size of those of their diploid cousins but have the same hemoglobin concentration and swim just as fast, indicating that their efficiency of gas transfer is comparable.

It is important to remember that *in vivo* gas transfer is a dynamic process that takes place as blood moves rapidly through the capillaries. Rates of diffusion, reaction velocities, and steady-state conditions for gases in blood must all be taken into account in analyzing the process. For instance, a Bohr shift (e.g., a decrease in hemoglobin-oxygen affinity with decrease in pH) would have little importance if it occurred so slowly that the blood had left the capillaries supplying an active tissue before it took effect. The Bohr shift, in fact, occurs very rapidly, having a half-time at 37°C of 0.12 seconds in human red blood cells.

REGULATION OF BODY pH

Body pH in animals is normally slightly alkaline; that is, there are fewer hydrogen than hydroxyl ions in the body. The concentrations of both ions are very low in aqueous solutions because water is only weakly dissociated. Human blood plasma at 37°C has a pH of 7.4, or a hydrogen ion activity of 40 nanomoles per liter (1 nM = 10^{-9} M). Normal function can be maintained in mammals at 37°C over a blood plasma pH range of 7.0–7.8; that is, between 100 and 16 nM H^+. The tolerable percentage of deviation from the normal H^+ concentration of 40 nM is quite large compared with the tolerable variations in Na^+ or K^+ levels in the body. It is important to bear in mind, however, that the absolute changes in H^+ concentration are small, as are the actual concentrations of H^+ ions in the body.

Blood pH in vertebrates is midway between the negative logarithm of the dissociation constants (pK) of the carbon dioxide–bicarbonate and ammonia-ammonium reactions (Figure 13-14a). Most cell membranes are relatively impermeable to HCO_3^- and NH_4^+

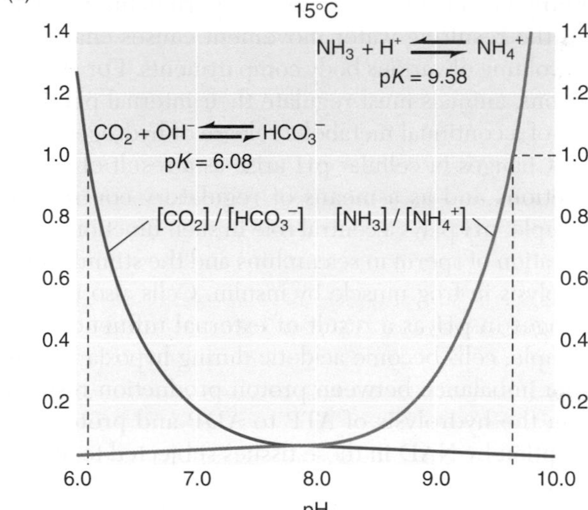

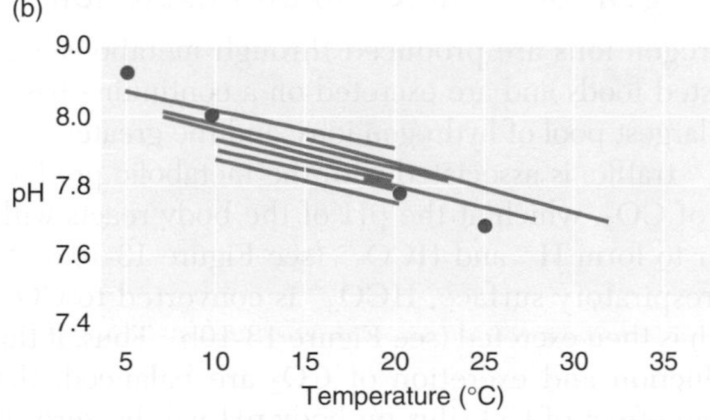

Figure 13-14 In vertebrates, the plasma pH is midway between the pKs of the ammonia-ammonium and carbon dioxide–bicarbonate reactions. **(a)** The effect of varying pH on the $[CO_2]/[HCO_3^-]$ and $[NH_3]/[NH_4^+]$ ratios in trout plasma at 15°C. The dashed lines mark the pH values at which the ratios equal 1 (i.e., the pK values). **(b)** Effect of temperature on plasma pH for several fishes. Red dots are the calculated pH values at which the CO_2/HCO_3^- and NH_3/NH_4^+ ratios are equal at various temperatures. Thus plasma pH is maintained at levels that ensure both NH_3 and CO_2 excretion. [Adapted from Randall and Wright, 1989.]

ions but permeable to CO_2 and NH_3. Some membranes have a relatively low permeability to NH_3 as well, but these are the exceptions rather than the rule. A body pH that is midway between these p*K*s ensures adequate rates of excretion by diffusion of the two major end products of metabolism; namely, carbon dioxide and ammonia. Because these p*K*s vary with temperature, so does the pH of blood, ensuring adequate rates of excretion over a range of temperatures (Figure 13-14b).

Changes in body pH alter the dissociation of weak acids and thus the ionization of proteins. The net charge on proteins determines enzyme activity and subunit aggregation, influences membrane characteristics, and contributes to the osmotic pressure of body compartments. Osmotic pressure is affected because the charge on proteins is a major contributor to the total fixed charge within cells. A change in the fixed charge alters the Donnan equilibrium of ions and therefore could affect the osmotic pressure. Differences in osmotic pressure between body compartments disappear rapidly because membranes are permeable to water, and the resulting water movement causes changes in the volume of various body compartments. For all these reasons, animals must regulate their internal pH in the face of a continual metabolic release of hydrogen ions.

Changes in cellular pH arise as a result of cellular functions and as a means of regulatory control. For example, pH plays a central role in such functions as the activation of sperm in sea urchins and the stimulation of glycolysis in frog muscle by insulin. Cells also undergo changes in pH as a result of external influences. For example, cells become acidotic during hypoxia because of an imbalance between proton production resulting from the hydrolysis of ATP to ADP and proton consumption by NAD in those tissues subjected to anaerobic metabolism.

Hydrogen Ion Production and Excretion

Hydrogen ions are produced through metabolism of ingested foods and are excreted on a continuing basis. The largest pool of hydrogen ions, and the greatest flux in H^+ traffic, is associated with the metabolic production of CO_2, which at the pH of the body reacts with water to form H^+ and HCO_3^- (see Figure 13-10a). At the respiratory surface, HCO_3^- is converted to CO_2, which is then excreted (see Figure 13-10b). Thus, if the production and excretion of CO_2 are balanced, the overall effect of CO_2 flux on body pH will be zero. If CO_2 excretion is less than production, however, so that CO_2 accumulates, the body will be acidified; if the reverse occurs, the body pH will rise. Terrestrial vertebrates can vary the rate of CO_2 excretion to maintain body pH.

Ingestion of meat usually results in a net intake of acid, whereas ingestion of plant food often results in a net intake of base. Generally, there is a small net production of hydrogen ions as a result of diet and metabolic activity. Thus the overall effect of food ingestion and metabolism is a small continual production of acid. The pH of the body is maintained by excreting this acid via the kidney in terrestrial vertebrates or across regions of the body surface, such as the gills of fishes or the skin of frogs.

Changes in blood pH can also occur in response to acid movement between compartments. For example, following a heavy meal, the production of large volumes of acid in the stomach can produce an *alkaline tide* in the blood, owing to the transfer of acid from the blood into the stomach. In a similar manner, the production of large volumes of alkaline pancreatic juices can result in an *acid tide* in the blood.

As discussed in Chapter 3, the relationship between pH and the extent of dissociation of a weak acid, HA, is described by the Henderson-Hasselbalch equation:

$$\mathrm{pH} = \mathrm{p}K' + \log\frac{[\mathrm{A}^-]}{[\mathrm{HA}]}$$

When the pH of a solution of a weak acid is equal to the p*K* of the acid, then 50% of the acid is in the undissociated form, HA, and 50% is in the dissociated form, $H^+ + A^-$. At 1 pH unit above the p*K*, the ratio of the undissociated to dissociated form is 10:90, and at 2 pH units above the p*K*, the ratio becomes 1:99. The Henderson-Hasselbalch equation can be rewritten for the CO_2/HCO_3^- acid-base pair as

$$\mathrm{pH} = \mathrm{p}K' + \log\frac{[\mathrm{HCO_3}^-]}{\alpha P_{\mathrm{CO_2}}}$$

where P_{CO_2} is the partial pressure of CO_2 in blood, α is the Bunsen solubility coefficient for CO_2, and pK' is the apparent dissociation constant. The term "apparent" is used because this pK' is a lumped value for the combined reactions of CO_2 with water and the subsequent formation of bicarbonate, and is not a true p*K*. We can see from this equation that changes in pH will affect the ratio of HCO_3^- to P_{CO_2}, and vice versa. The pK' of the CO_2/HCO_3^- reaction is about 6.1, and the pK' of the HCO_3^-/CO_3^{2-} reaction is about 9.4. At the pH of the body, about 95% of the CO_2 is in the form of HCO_3^-, the remainder being carbon dioxide and carbonic acid; the amount of CO_3^{2-} is negligible.

Weak acids have their greatest buffering action when pH = p*K*. Because the p*K* of plasma proteins and hemoglobin is close to the pH of blood, these compounds are important physical buffers in the blood. The CO_2/HCO_3^- pair, with an apparent pK' below the pH of the blood, is of less importance than either hemoglobin or plasma proteins in providing a physical buffer system. The importance of the carbon dioxide–bicarbonate system is that an increase in breathing can rapidly increase pH by lowering CO_2 levels in the blood and that the kidney can decrease blood pH by excreting HCO_3^-. Although bicarbonate is not an important chemical buffer in living systems, it is often referred to as a buffer because the ratio of CO_2 to bicarbonate can be adjusted by excretion in order to regulate pH. The most important true buffers in the blood are proteins, especially hemoglobin. Phosphates are also significant buffers in many cells.

The importance of buffers in ameliorating pH changes can be seen by considering the effects of acid infusion on mammalian blood: about 28 mmol of hydrogen ions must be added to the blood to reduce pH from 7.4 to 7.0. Only 60 nmol (about 0.2%) are required to change the pH of an aqueous solution to this extent; in blood, however, the bulk of the added 28 mmol of H^+ is buffered by conversion of HCO_3^- to CO_2 (18 mmol), by hemoglobin (8.0 mmol), by plasma proteins (1.7 mmol), and by phosphates (0.3 mmol). Thus, nearly 500,000 times as many hydrogen ions are buffered as are required to cause the pH to change from 7.4 to 7.0.

Clearly, if lung ventilation is reduced so that CO_2 excretion drops below CO_2 production, body CO_2 levels will rise and pH will fall. This decrease in body pH is referred to as **respiratory acidosis.** The reverse effect—a rise in pH due to increased lung ventilation—is termed **respiratory alkalosis.** The word "respiratory" is used to differentiate these pH changes from those caused by changes in metabolism or kidney function. Anaerobic metabolism, for example, results in net acid production, which reduces body pH; such changes are referred to as **metabolic acidosis.**

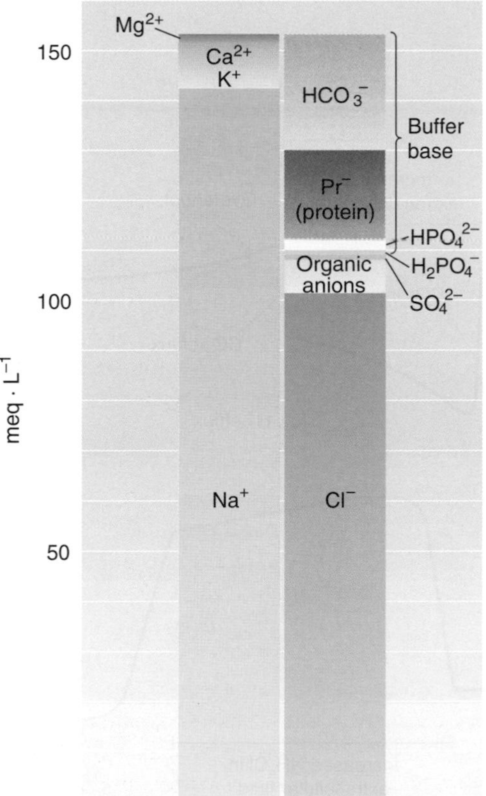

Figure 13-15 All body fluids are electroneutral, containing equal numbers of positive and negative charges. This diagram shows the equivalent concentrations ($meq \cdot L^{-1}$) of the major electrolytes in human plasma at normal pH. The concentration of the buffer base (the nonrespiratory acid-base displacement) depends on pH. Thus, a pH increase or decrease that changes the buffer base concentration must be accompanied by a corresponding change in the concentration of one or more strong ions, usually sodium or chloride. [Adapted from Siggaard-Andersen, 1963.]

Body fluids, like other solutions, are electroneutral; that is, the sum of the anions equals the sum of the cations. The normal electrolyte status of human plasma is illustrated in Figure 13-15. The sum of bicarbonate, phosphates, and protein anions is referred to as the **buffer base.** The remaining cations and anions are *strong ions* (i.e., they are completely dissociated in physiological solutions). The difference between the sum of strong cations and the sum of strong anions is referred to as the **strong ion difference (SID)** and is a reflection of the magnitude of the buffer base. Because a change in blood pH usually results in a change in the buffer base, the SID also must change to maintain electroneutrality. In this situation, the change in the SID usually involves either sodium or chloride, since these are the major ions in the blood. For example, a reduction in bicarbonate must be associated with an increase in chloride or a reduction in sodium. Conversely, a change in the ratio of sodium to chloride will be associated with a change in the buffer base, and therefore in blood pH. Vomiting the stomach contents results in chloride loss and a reduction in blood chloride levels; as a consequence, bicarbonate levels are increased, along with blood pH, without any change in P_{CO_2}; this effect is referred to as **metabolic alkalosis.** Vomit originating from the duodenum, rather than the stomach, results in the loss of more bicarbonate than chloride, causing metabolic acidosis.

A crocodile may eat a whole deer in a single meal! What changes would you expect in the pH of the various body compartments as digestion proceeds?

Hydrogen Ion Distribution Between Compartments

Cell membranes separating the intracellular and extracellular compartments and layers of cells between two body compartments are much more permeable to carbon dioxide than to either hydrogen or bicarbonate ions. The permeability of most cell membranes to H^+ ions, although usually low, is often greater than their permeability to K^+, Cl^-, and HCO_3^- ions; a notable exception is the erythrocyte membrane, which is very permeable to HCO_3^- and Cl^- ions, but not very permeable to H^+ ions. Red blood cells and cells in the collecting duct of the mammalian kidney have high levels of band III protein in their plasma membranes, but other cells do not. As discussed previously, band III protein mediates the rapid exchange of HCO_3^- for Cl^- ions. Thus, although all cell membranes are permeable to CO_2, only a few can transfer HCO_3^- at high rates via the band III anion-exchange mechanism.

An increase in extracellular P_{CO_2} causes an increase in both bicarbonate and hydrogen ion concentration, thereby creating gradients for CO_2, HCO_3^-, and H^+ across the cell membrane. In cells that are very permeable to CO_2 but not very permeable to H^+ or bicarbonate, such a situation leads to rapid movement of CO_2 into the cell; as the CO_2 is converted to HCO_3^-, the intracellular pH falls sharply. Acidification associated with increased P_{CO_2} often occurs much more rapidly in the intracellular compartment than in the extracellular compartment because carbonic anhydrase, which catalyzes the conversion of CO_2 to HCO_3^-, is present inside cells but not always in the extracellular fluid. Even when P_{CO_2} remains elevated, however, intracellular pH slowly returns to the initial level due to the slow extrusion of acid (or uptake of base) across the plasma membrane (Figure 13-16a). The rise in intracellular pH is such that if the P_{CO_2} is returned to the original value, the intracellular pH will be higher than its initial value; that is, there is a small overshoot in pH.

As noted earlier, most cell membranes are much more permeable to molecular ammonia (NH_3) than to ammonium ions (NH_4^+). If ammonium chloride (NH_4Cl) levels in the extracellular fluid increase, ammonia enters the cell, and the ratio of ammonia to ammonium ions within the cell increases. Ammonia combines with hydrogen ions to form ammonium ions within the cell, thus raising intracellular pH (Figure 13-16b). During prolonged NH_4Cl exposure, intracellular pH reaches a maximum and then starts to fall because of a slow influx of NH_4^+ along with other acid-base regulating mechanisms in the membrane. A return of the external NH_4Cl level to the original value results in a sharp fall in intracellular pH as NH_3 diffuses out of the cell. Because of the accumulation of intracellular NH_4^+, cell pH falls below the initial level, then slowly returns to the initial level as NH_4^+ diffuses out of the cell. NH_4^+ can be transported across plasma membranes by a variety of passive and active mechanisms that vary from cell to cell, including exchange for sodium, substitution for potassium in the transport

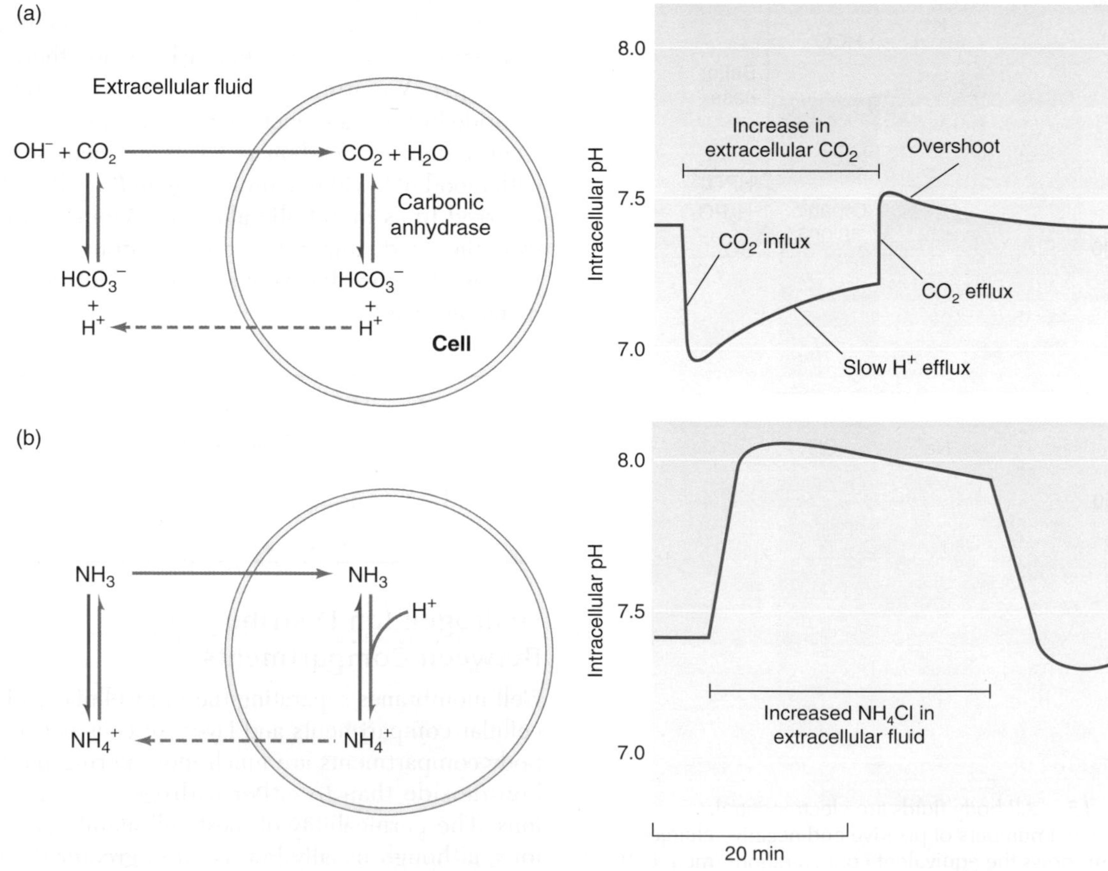

Figure 13-16 Changes in extracellular carbon dioxide or ammonium chloride levels cause changes in intracellular pH. **(a)** If CO_2 levels in the extracellular fluid are suddenly increased, CO_2 diffuses rapidly into the cell, forming bicarbonate and causing a sharp fall in intracellular pH. A subsequent slow efflux of H^+ ions (dashed line) leads to a gradual rise in intracellular pH. **(b)** If extracellular NH_4Cl levels rise sharply, NH_3 diffuses rapidly into the cell and combines with hydrogen ions to form ammonium ions, which diffuse slowly across the cell membrane (dashed line). As a result, the intracellular pH increases.

reaction carried out by the Na^+/K^+ pump, and transport through potassium channels.

Both of these mechanisms for pH adjustment are activated by a reduction in intracellular pH or an increase in extracellular pH. In mammalian cells, acid extrusion is reduced to low levels if extracellular pH falls below 7.0 or intracellular pH rises above 7.4. If an acid is injected into a cell, it is extruded from the cell at rates that increase in proportion to the decrease in cell pH. Although a portion of the H^+ efflux may be related to H^+ diffusion out of the cell, some of the efflux is coupled to sodium influx. This coupling of sodium and proton transport may be due to either a cation-exchange mechanism in the membrane or an electrogenic proton pump that increases membrane potential, thereby providing an electrochemical gradient that drives diffusion of Na^+ ions through Na^+-selective channels. For example, some cells can actively pump protons out via a proton ATPase in the membrane; this electrogenic proton efflux can result in a sodium influx. Often acid extrusion is accompanied by chloride efflux, presumably in exchange for extracellular HCO_3^-. For instance, the drug SITS (4-acetamido-4′-isothiocyanostilbene-2,2′-disulfonic acid), which blocks chloride-bicarbonate exchange in erythrocytes, also inhibits pH regulation in other cells.

Thus both proton-exchange and anion-exchange mechanisms in the plasma membrane play an important role in adjusting intracellular pH. An acid load in the cell is accompanied by H^+ efflux coupled to Na^+ influx and by HCO_3^- influx coupled to Cl^- efflux. The movement of HCO_3^- into the cell is equivalent to the movement of H^+ out of the cell because HCO_3^- ions that enter the cell are converted to CO_2, releasing hydroxyl ions and increasing pH. The CO_2 so formed leaves the cell and is converted to HCO_3^-, releasing protons. This cycling of CO_2 and HCO_3^-, referred to as the **Jacobs-Stewart cycle,** functions to remove H^+ ions from the cell interior in the face of an intracellular acid load, such as that generated by anaerobic metabolism (Figure 13-17).

In most vertebrate red blood cells, unlike most other cells, hydrogen ions are passively distributed across the membrane, and the membrane potential maintains a lower pH inside the red blood cell than in the blood plasma. A sudden addition of acid to the plasma (e.g., following anaerobic production of H^+) results in a fall in erythrocyte pH. The acid is transferred from the plasma to the interior of the erythrocyte not by diffusion of H^+ ions, but by bicarbonate-chloride exchange (see Figure 13-17). The addition of H^+ to the plasma causes the P_{CO_2} to increase as HCO_3^- is converted to CO_2, which then diffuses into the red blood cell and is converted to HCO_3^-, thereby reducing intracellular pH. Bicarbonate then diffuses out of the cell via the chloride-bicarbonate exchange mechanism. Thus, in erythrocytes, the Jacobs-Stewart cycle operates to transfer acid from the plasma to the cell interior.

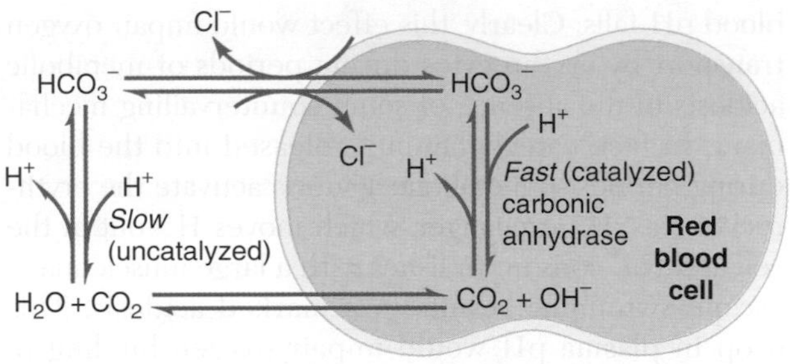

Figure 13-17 The Jacob-Stewart cycle is the cycling of carbon dioxide and bicarbonate that transfers acid between the extracellular and intracellular compartments. In a red blood cell, depicted here, the cycle generally operates to transfer acid from the blood plasma to the cell interior. Because carbonic anhydrase is present only inside the cells, the slow, uncatalyzed interconversion of CO_2 and HCO_3^- in the extracellular fluid determines the rate of acid transfer.

Factors Influencing Intracellular pH

Intracellular pH remains stable if the rate of acid loading, from metabolism or from influx into the cell, is equal to the rate of acid removal. Any sudden increase in cell acidity will be countered by the various mechanisms discussed above:

- buffering by physical buffers (e.g., proteins and phosphates) located within the cell
- reaction of HCO_3^- with H^+ ions, forming CO_2, which then diffuses out of the cell
- passive diffusion or active transport of H^+ ions from the cell
- cation-exchange mechanisms (Na^+/H^+ and Na^+/NH_4^+), anion-exchange mechanisms (HCO_3^-/Cl^-), or both in the plasma membrane

In addition, the generation of protons from metabolic activity may be modulated by pH. Many enzymes are inhibited by low pH, including enzymes of glycolysis and possibly some other metabolic pathways; inhibition of these pathways at low pH may serve to regulate intracellular pH by reducing the net production of protons during periods of increased acidity in cells.

In some instances, intracellular pH may be modulated to control or limit other cellular functions. It is not always clear whether these pH changes cause, or are caused by, the associated cellular activity. For example, when frog eggs are fertilized, intracellular calcium levels increase transiently, followed by a sustained increase in pH, and there is some evidence to indicate that the rise in pH may prolong the action of elevated calcium.

In a few cases, the regulation of intracellular pH has a clear effect on cellular function. For example, the erythrocytes of many teleosts have a Na^+/H^+ exchanger and a HCO_3^-/Cl^- exchanger in the plasma membrane. The hemoglobin of these animals exhibits a Root shift—that is, a decrease in blood oxygen capacity as blood pH falls. Clearly, this effect would impair oxygen transport by erythrocytes during periods of metabolic acidosis in the absence of some countervailing mechanism. In fact, catecholamines released into the blood during periods of metabolic acidosis activate the erythrocyte Na^+/H^+ exchanger, which moves H^+ out of the cell and Na^+ ions in. In fishes with a large muscle mass, escape swimming results in a marked acidosis. This drop in plasma pH would impair oxygen binding to hemoglobin and reduce the ability of these fishes to swim aerobically if erythrocytic pH were not regulated by this mechanism.

Factors Influencing Body pH

A stable body pH requires that acid production be matched to acid excretion. In mammals, this symmetry is achieved by adjusting the excretion of CO_2 via the lungs and the excretion of acid or bicarbonate via the kidneys. The collecting duct of the mammalian kidney has both A-type (acid-excreting) and B-type (base-excreting) cells, the activity of which can be altered to increase acid or base excretion. In aquatic animals, the external body surfaces have the capacity to extrude acid in ways similar to that seen in the collecting duct of the mammalian kidney (see Chapter 14). For example, the skin of frogs and the gills of freshwater fishes have an ATPase on the apical surface of the epithelium that excretes protons. Fish gills also have an apical HCO_3^-/Cl^- exchanger. If these mechanisms are inhibited by drugs, body pH is affected.

Temperature can have a marked effect on body pH. The dissociation of water varies with temperature, and the pH of neutrality (i.e., $[H^+] = [OH^-]$) is 7.00 only at 25°C. The dissociation of water decreases, and the pH of neutrality (pN) therefore increases, with a decrease in temperature. At 37°C, pN is 6.8, whereas at 0°C, it is 7.46. Human plasma at 37°C has a pH of 7.4, so it is slightly alkaline. The ratio of OH^- to H^+ ions increases with increasing alkalinity; at pH 7.4 at 37°C it is about 20. Most animals maintain almost the same alkalinity in many of their tissues relative to pN independent of the temperature of their bodies (Figure 13-18). Fishes at 5°C have a plasma pH of 7.9–8.0; turtles at 20°C, a plasma pH of about 7.6; and mammals at 37°C, a plasma pH of 7.4. Thus, all have a similar relative alkalinity and the same ratio of OH^- to H^+ ions (about 20) in plasma. Tissues are generally less alkaline than plasma; for example, the intracellular pH of erythrocytes is about 0.2 pH units less than the pH of plasma, and the intracellular pH of muscle cells is about 7.0.

Temperature also has a marked effect on the p*K* values of plasma proteins and the CO_2/HCO_3^- system, with the p*K* values increasing as temperature decreases. According to the Henderson-Hasselbalch equation, changes in p*K* will cause changes in pH or in the dissociation of weak acids. However, the temperature-induced changes in plasma pH will offset the temperature-dependent changes in the p*K* of plasma proteins so that

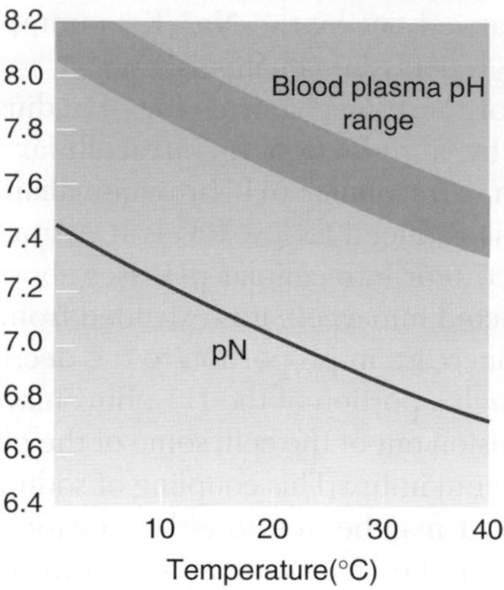

Classic Work **Figure 13-18** The pH at neutrality (pN) and plasma pH decrease with increasing temperature, but the relationship between the two is constant in most animals. In this graph, the effect of temperature on plasma pH in various turtles, frogs, and fishes is compared with the change in pN. [Adapted from Rahn, 1967.]

the extent to which the plasma proteins dissociate will remain constant. Because the p*K* of the CO_2 hydration-dehydration reaction changes less with temperature than does blood pH, animals must adjust the ratio of CO_2 to HCO_3^- in the blood. In general, it appears that as temperature falls, air-breathing, poikilothermic vertebrates keep their bicarbonate levels constant but decrease their molecular CO_2 levels. In aquatic animals, on the other hand, CO_2 levels remain the same and bicarbonate levels increase as temperature drops. This process results in the same adjustment of the carbon dioxide–bicarbonate ratio, and hence pH, in both aquatic and air-breathing vertebrates. The important point is that if body pH changes with temperature in the same way as the p*K* of proteins, then the Henderson-Hasselbalch equation predicts that the charge on proteins should remain unchanged. If there is little or no change in the net charge on proteins, their function will be retained over a wide range of temperatures.

The ability of the body to redistribute acid between body compartments has functional significance because some tissues are more adversely affected by changes in pH than others. The brain is particularly sensitive, whereas muscles tolerate much larger oscillations in pH. As a result, the brain has extensive, if poorly understood, mechanisms for regulating the pH of the cerebrospinal fluid. In the face of a sudden acid load in the blood, hydrogen ions are taken up by muscles, reducing oscillations in the blood and thus protecting the brain and other more sensitive tissues. Hydrogen ions are then slowly released into the blood by the muscles and excreted either via the lungs as CO_2 or via the kidney in acid urine. Thus, when there is a sudden acid load in the body, the muscles can act as a temporary H^+ reservoir, reducing the magnitude of the pH oscillations in other regions of the body.

GAS TRANSFER IN AIR: LUNGS AND OTHER SYSTEMS

In the previous sections, we considered the properties of oxygen and carbon dioxide and described how these gases are carried in the blood and the effects they have on body pH. In this section, we examine the ways in which O_2 and CO_2 are transferred between air and blood. We focus here on the vertebrate lung, but other gas-transfer systems are also considered.

The structure of a gas-transfer system is influenced by the properties of the respiratory medium as well as the requirements of the animal. Lungs and gills are quite different and are ventilated in different ways. These dissimilarities exist because the density and viscosity of water are both approximately 1000 times greater than those of air, and water contains only one-thirtieth as much molecular oxygen. Moreover, gas molecules diffuse 10,000 times more rapidly in air than in water. Thus, in general, air breathing consists of the reciprocal movement of air into and out of the lungs, whereas water breathing consists of a unidirectional flow of water over the gills (Figure 13-19a). The design objectives of fish gills are to minimize diffusion distances in water, creating a thin layer of water over the respiratory surface. The lung's location within the body protects the thin, delicate respiratory surfaces and facilitates the control of water loss from those surfaces. These variations in the environment, the nature of ventilation, and the structure of the respiratory apparatus result in differences in the partial pressures of gases in the blood and the tissues of air-breathing and water-breathing animals, particularly in P_{CO_2} (Figure 13-19b).

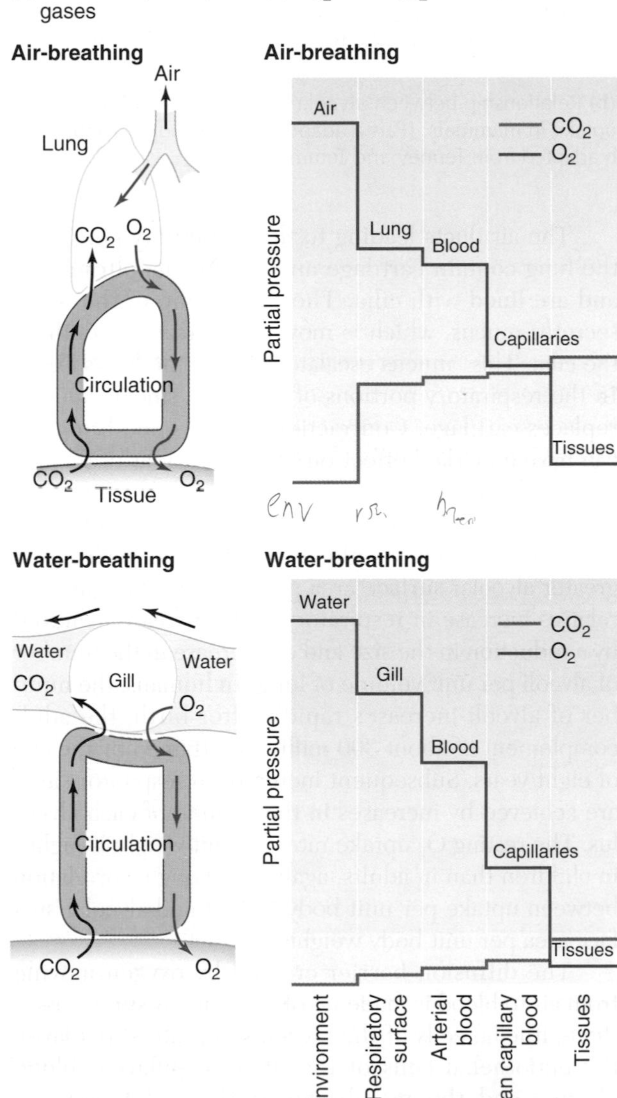

Figure 13-19 The different gas-transfer systems in air-breathing and water-breathing animals are associated with characteristic distributions of respiratory gases in the blood and tissues. **(a)** Schematic diagrams of O_2 and CO_2 flows in air-breathing and water-breathing animals. **(b)** Relative values of P_{O_2} and P_{CO_2} in the respiratory medium, blood, and tissues in air-breathing and water-breathing animals.

Functional Anatomy of the Lung

The vertebrate lung, which develops as a diverticulum of the gut, consists of a complex network of tubes and sacs, with the actual structure varying considerably among species. The sizes of the terminal air spaces in lungs become progressively smaller from amphibians to reptiles to mammals, while the total number of air spaces per unit volume of lung becomes greater. The structure of the amphibian lung is variable, ranging from a smooth-walled pouch in some salamanders to a lung subdivided by septa and folds into numerous interconnected air sacs in frogs and toads. The degree of subdivision increases in reptiles and increases even more in mammals, the total effect being an increase in respiratory surface area per unit volume of lung. In general, the area of the respiratory surface in mammals increases with body weight and the rate of oxygen uptake (Figure 13-20 on the next page). Teleost fishes typically have a smaller respiratory surface area than mammals of equivalent body weight.

The mammalian lung consists of millions of blind-ended, interconnected spaces, termed **alveoli.** The main airway, the **trachea**, subdivides to form **bronchi** and **bronchioles**, which branch repeatedly, leading eventually to *terminal bronchioles* and finally *respiratory bronchioles*, each of which is connected to terminal alveolar ducts and several alveoli (Figure 13-21 on the next page). The total cross-sectional area of the airways increases rapidly as a result of extensive branching, although the diameter of individual air ducts decreases from the trachea to the terminal bronchioles. The terminal bronchioles, the respiratory bronchioles, the alveolar ducts, and the alveoli constitute the respiratory portion of the lung. Gases are transferred across the thin-walled alveoli found in the regions distal to the terminal bronchioles. The airways leading to the terminal bronchioles constitute the nonrespiratory portion of the

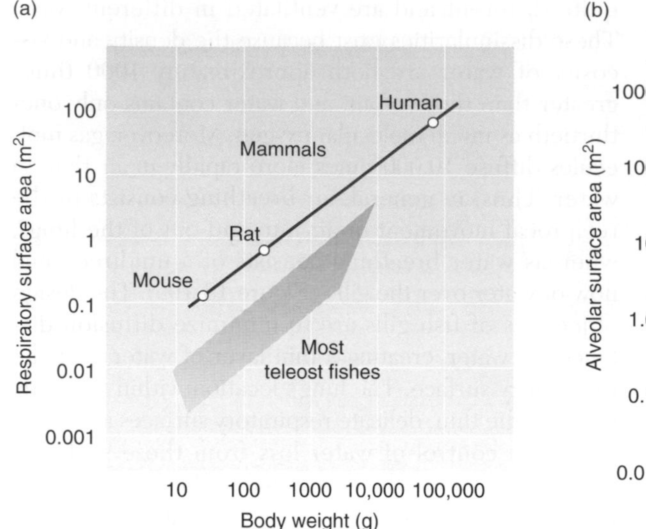

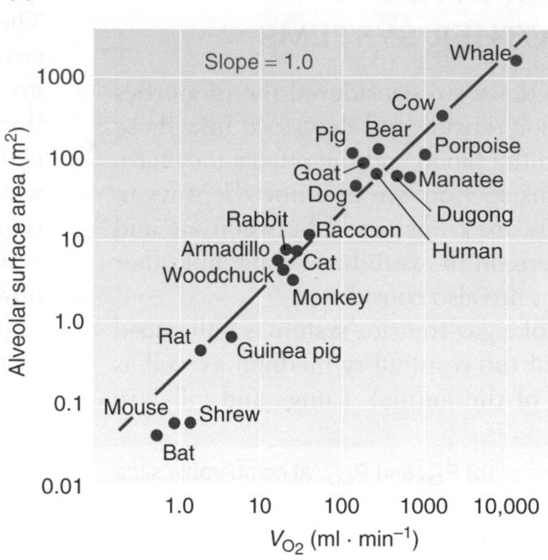

Figure 13-20 Respiratory surface area increases with body size. **(a)** Relationship between respiratory surface area and body weight in selected mammals and teleost fishes. **(b)** Relationship between alveolar surface area and oxygen uptake in mammals. [Part a adapted from Randall, 1970; part b adapted from Tenney and Temmers, 1963.]

lung. The alveoli are interconnected by a series of holes, the **pores of Kohn.** These pores allow collateral movement of air, which may be a significant factor in gas distribution during lung ventilation (Figure 13-22a).

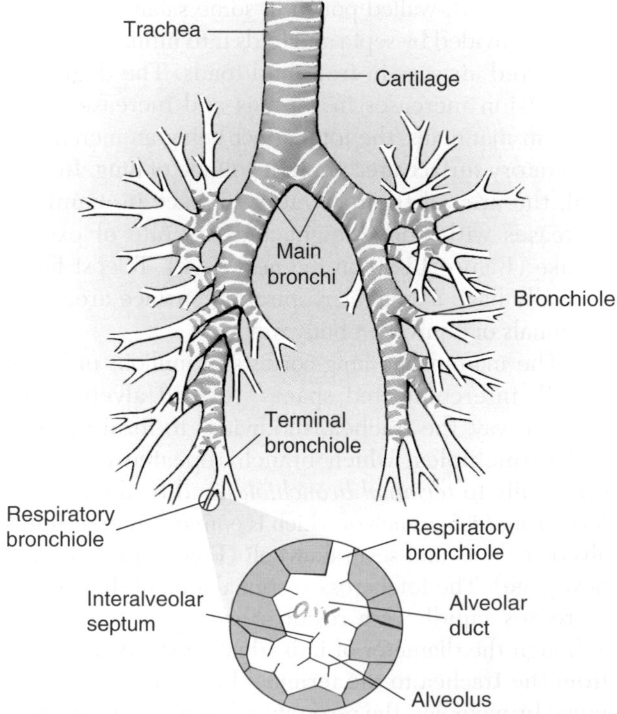

Figure 13-21 In the mammalian lung, a series of branching, progressively smaller ducts delivers air to the respiratory portion, consisting of terminal and respiratory bronchioles and alveolar ducts and sacs. Gas transfer occurs across the respiratory epithelium (shown in light red).

The air ducts leading to the respiratory portion of the lung contain cartilage and a little smooth muscle and are lined with cilia. The epithelium of the ducts secretes mucus, which is moved toward the mouth by the cilia. This "mucus escalator" keeps the lungs clean. In the respiratory portions of the lung, smooth muscle replaces cartilage. Contraction of this smooth muscle can have a marked effect on the dimensions of the airways in the lungs.

Small mammals have a higher resting O_2 uptake rate per unit body weight than large mammals and a greater alveolar surface area per unit body weight. The relative increase in respiratory surface area is achieved by a reduction in the size and an increase in the number of alveoli per unit volume of lung. In humans, the number of alveoli increases rapidly after birth; the adult complement of about 300 million is attained by the age of eight years. Subsequent increases in respiratory area are achieved by increases in the volume of each alveolus. The resting O_2 uptake rate per unit weight is higher in children than in adults, again reflecting a correlation between uptake per unit body weight and alveolar surface area per unit body weight.

The diffusion barrier crossed by oxygen moving from air to blood is made up of an aqueous surface film, the epithelial cells of the alveolus, an interstitial layer, the endothelial cells of the blood capillaries, blood plasma, and the membrane of the red blood cell (Figure 13-22b). Several cell types compose the lung epithelium. Type I cells, the most abundant, constitute the major part of the lung epithelium. They are squamous epithelial cells that have a thin platelike structure. A single type I cell extends between two adjacent alveoli, with its nucleus tucked away in a corner. Type II cells are characterized by a laminated body within the cell and have surface villi. These cells produce surfac-

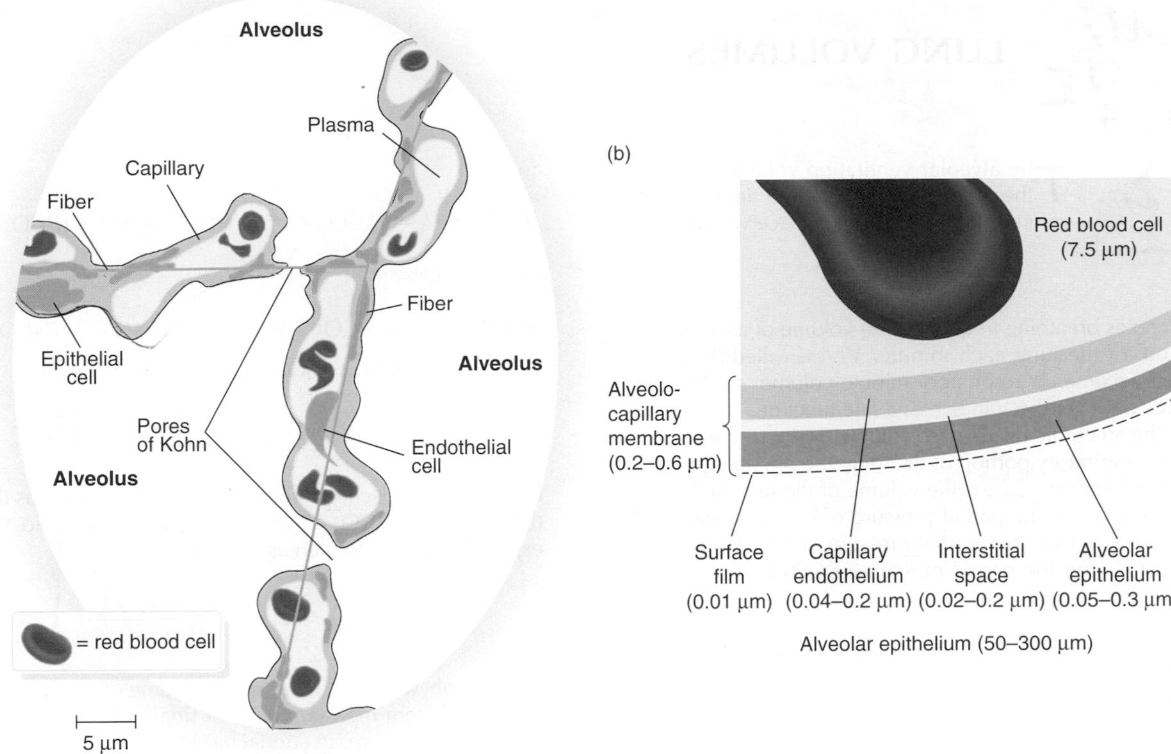

Figure 13-22 During ventilation of the mammalian lung, the respiratory gases move between the alveolar space and the blood in the pulmonary capillaries. **(a)** Three interalveolar septa in a dog lung meet at a junction line. Connective tissue fibers lie in the central plane, forming a continuous tensile network with which the capillary network is interwoven. Endothelial cells and type I epithelial cells form the lining of the thin air-blood barrier. Pores of Kohn connect alveoli. **(b)** Dimensions and structure of the diffusion barrier between air and blood. [Part a adapted from Weibel, 1973; part b adapted from Hildebrandt and Young, 1965.]

tants, which we will discuss below. Type III cells are rich in mitochondria and have numerous microvilli. These rare cells appear to be involved in NaCl uptake from lung fluid. A number of *alveolar macrophages* wander over the surface of the respiratory epithelium.

Lung Ventilation

The following terms are used to describe different types of breathing and lung ventilation:

- **Eupnea:** Normal, quiet breathing typical of an animal at rest
- **Hyperventilation** or **hypoventilation:** An increase or decrease, respectively, in the amount of air moved into or out of the lungs by changes in the rate or depth of breathing such that ventilation no longer matches CO_2 production and blood CO_2 levels change
- **Hyperpnea:** Increased lung ventilation due to increased breathing in response to elevated CO_2 production (e.g., during exercise)
- **Apnea:** Absence of breathing
- **Dyspnea:** Labored breathing associated with an unpleasant sensation of breathlessness
- **Polypnea:** An increase in breathing rate without an increase in the depth of breathing

The amount of air moved into or out of the lungs with each breath is referred to as the **tidal volume.** Air exchanged between the alveoli and the environment must pass through a series of tubes (trachea, bronchi, nonrespiratory bronchioles) not directly involved in gas transfer. At the end of an exhalation (expiration), the air contained in these tubes will have come from the alveoli and will be low in oxygen and high in carbon dioxide. This air will be the first to move back into the alveoli at the next breath. At the end of an inhalation (inspiration), the nonrespiratory tubes will be filled with fresh air, and this volume will be the first to be exhaled with the next breath. This volume of air is not involved in gas transfer and is referred to as the *anatomic dead-space volume.* Some air may be supplied to nonfunctional alveoli, or certain alveoli may be ventilated at too high a rate, increasing the volume of air not directly involved in gas exchange. This volume of air, the *physiological*

LUNG VOLUMES

The alveolar ventilation volume, V_A, equals the difference between the tidal ventilation volume, V_T, and the dead-space volume, V_D:

$$V_A = V_T - V_D$$

If f denotes breathing frequency, the volume of air moved into and out of the lung each minute, $V_A f$, is called the *alveolar minute volume,* or respiratory minute volume, symbolized as $\dot{V}_A$. The dot over the V indicates a rate function.

The anatomic dead-space volume, V_{Danat}, is the volume of the nonrespiratory portion of the lung; the physiological dead-space volume, $V_{Dphysiol}$, is the volume of the lung not involved in gas transfer. If the partial pressure of CO_2 in exhaled air is denoted by P_ECO_2, the partial pressure of CO_2 in alveolar air by P_ACO_2, and the partial pressure of CO_2 in inhaled air by P_ICO_2, then

$$P_ECO_2 \times V_T = (P_ACO_2 \times V_A) + (P_ICO_2 \times V_D)$$

But $V_A = V_T - V_D$, so substituting into this equation, we obtain

$$P_ECO_2 \times V_T = P_ACO_2(V_T - V_D) + (P_ICO_2 \times V_D)$$

and

$$P_ECO_2 \times V_T = (P_ACO_2 \times V_T) - (P_ACO_2 \times V_D) + (P_ICO_2 \times V_D)$$

By rearrangement,

$$(P_ACO_2 \times V_D) - (P_ICO_2 \times V_D) = (P_ACO_2 \times V_T) - (P_ECO_2 \times V_T)$$

$$V_D(P_ACO_2 - P_ICO_2) = V_T(P_ACO_2 - P_ECO_2)$$

$$V_{Dphysiol} = V_T \frac{(P_ACO_2 - P_ECO_2)}{(P_ACO_2 - P_ICO_2)}$$

But P_ICO_2 approaches zero, and P_ACO_2 is the same as the partial pressure of CO_2 in arterial blood, P_aCO_2. So the last expression can be written as follows:

$$V_{Dphysiol} = V_T \frac{(P_aCO_2 - P_ECO_2)}{P_aCO_2}$$

Thus the physiological dead-space volume of the lungs can be calculated from measurements of tidal volume, V_T, and the CO_2 partial pressures in arterial blood, P_aCO_2, and exhaled air, P_ECO_2.

dead-space volume, includes the anatomic dead-space volume (Spotlight 13-4).

The amount of fresh air moving into and out of the alveolar air sacs equals the tidal volume minus the anatomic dead-space volume and is referred to as the **alveolar ventilation volume.** Only this gas volume is directly involved in gas exchange. The lungs are not completely emptied even at maximal exhalation, leaving a residual volume of air in the lungs. The maximum volume of air that can be moved into or out of the lungs is referred to as the **vital capacity** of the lungs. These and other terms used to describe various volumes and capacities associated with lung function are illustrated in Figure 13-23.

The O_2 content is lower and the CO_2 content is higher in alveolar gas than in ambient air because only a portion of the lungs' gas volume is changed with each breath. Alveolar ventilation volume in humans is about 350 ml, whereas the functional residual volume of the lungs (see Figure 13-23) exceeds 2000 ml. During inhalation, the alveolar ducts elongate and widen, causing an increase in alveolar volume. During breathing, air moves into and out of the alveoli and may also move between adjacent alveoli through the pores of Kohn. Mixing of gases in the ducts and alveoli occurs by diffusion and by convection currents caused by breathing. In the alveolar ducts, O_2 diffuses toward the alveoli and CO_2 away from them. Partial pressures of O_2 and CO_2 are probably fairly uniform across the alveoli because diffusion is rapid in air and the distances involved are small. The partial pressures of gases within the alveoli

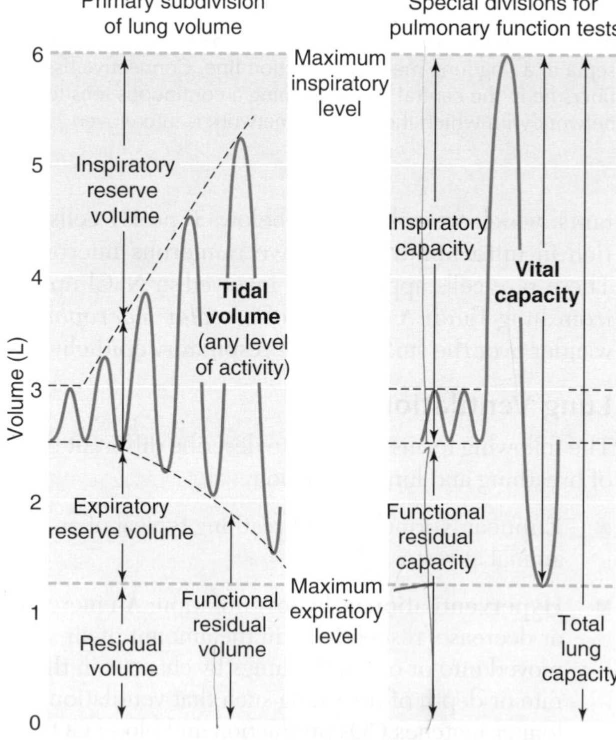

Figure 13-23 Numerous terms are used to describe various volumes and capacities associated with lung function. The tidal volume is the volume of air typically moved into and out of the lung, whereas the vital capacity is the maximum volume that can be moved into and out of the lung.

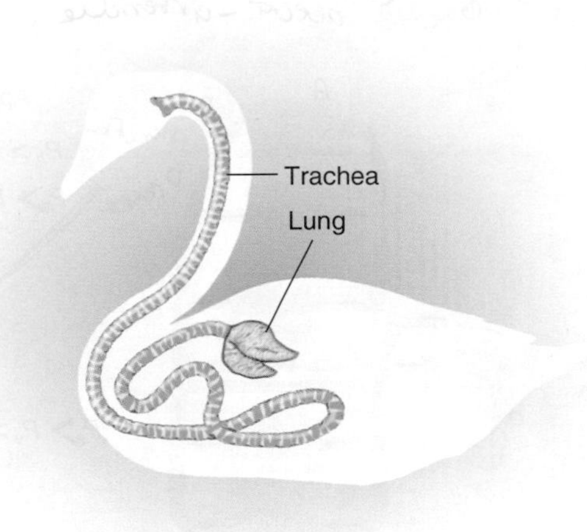

Figure 13-24 The extremely long trachea of the trumpeter swan results in a large anatomic dead-space volume. (For comparison, see Figure 13-28, illustrating the length of the human trachea.) [From Banko, 1960.]

oscillate in phase with the breathing movements, the magnitude of the partial pressures depending on the extent of tidal ventilation.

The O_2 and CO_2 levels in alveolar gas are determined by both the rate of gas transfer across the respiratory epithelium and the rate of alveolar ventilation. Alveolar ventilation depends on breathing rate, tidal volume, and anatomic dead-space volume. Variations in the magnitude of the anatomic dead space will alter partial pressures of gases in the alveolus in the absence of changes in tidal volume. Thus, artificial increases in anatomic dead space, such as those produced in human subjects breathing through a length of hose, result in a rise in CO_2 and a fall in O_2 in the lungs. As we will see below, these changes activate chemoreceptors, leading to an increase in tidal volume. In animals with long necks (e.g., the giraffe and trumpeter swan), the tracheal length, and therefore the anatomic dead-space volume, is greater than in those with short necks (Figure 13-24). In order to maintain adequate gas partial pressures in the lungs, long-necked animals have high tidal volumes.

Breathing rate and tidal volume vary considerably among animals. Humans breathe about 12 times per minute and have a tidal volume at rest of about 10% of total lung volume. Such relatively rapid, shallow breathing produces small oscillations in P_{O_2} in the lungs and blood. In contrast, the exclusively aquatic but air-breathing amphibian *Amphiuma,* which lives in swamp water, rises to the surface of the water about once each hour to breathe; its tidal volume, however, is more than 50% of its total lung capacity. This large tidal volume, coupled with infrequent breathing, produces large, slow oscillations in P_{O_2} in the lungs and blood, which are more or less in phase with the breathing movements (Figure 13-25). *Amphiuma* is preyed on by snakes and is most vulnerable when it rises to breathe. Because the water *Amphiuma* lives in has a low oxygen content, aquatic respiration is not a suitable alternative. The hazard of being eaten while surfacing to breathe may have influenced the evolution of its very low breathing rate, its large tidal and lung volume, and its ability to make cardiovascular adjustments that help maintain O_2 delivery to the tissues in the face of widely oscillating blood gas levels. Carbon dioxide levels in *Amphiuma* do not oscillate in the same way as oxygen levels because carbon dioxide is lost across the skin and is less dependent on lung ventilation.

In summary, O_2 and CO_2 levels in alveolar gas are determined by ventilation and the rate of gas transfer. Ventilation of the respiratory epithelium is determined by breathing rate, tidal volume, and anatomic dead-space volume. The nature and extent of ventilation also influence the magnitude of oscillations in O_2 and CO_2 in the blood during a breathing cycle.

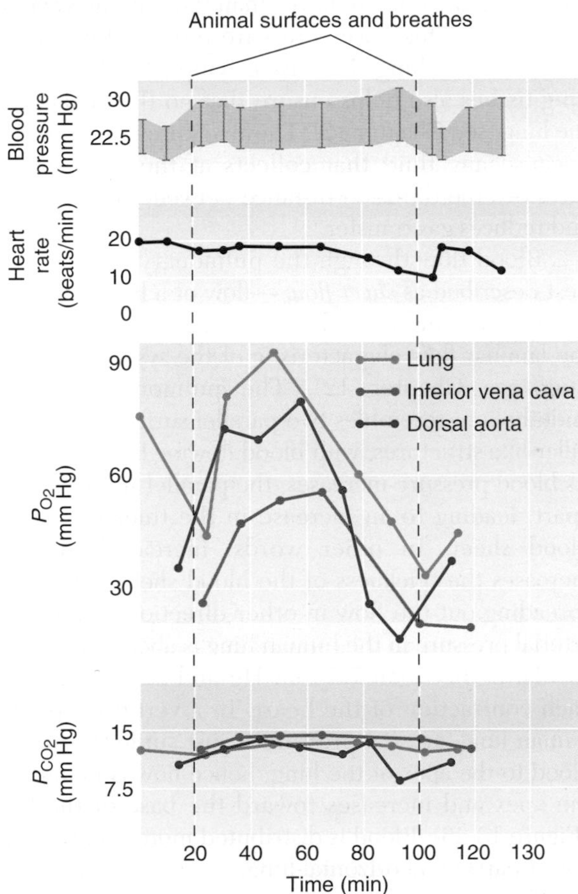

Figure 13-25 Breathing frequency tends to vary inversely with tidal volume and magnitude of oscillations in P_{O_2}. In *Amphiuma,* an aquatic, air-breathing amphibian that breathes infrequently, tidal volume and changes in P_{O_2} are large. Shown here are plots of blood pressure, heart rate, P_{O_2}, and P_{CO_2} in a 515 g *Amphiuma* during two breathing-diving cycles. Note that blood pressure, heart rate, and P_{CO_2} are nearly constant between breaths, whereas P_{O_2} shows large, slow oscillations in the lung and blood. [Adapted from Toews et al., 1971.]

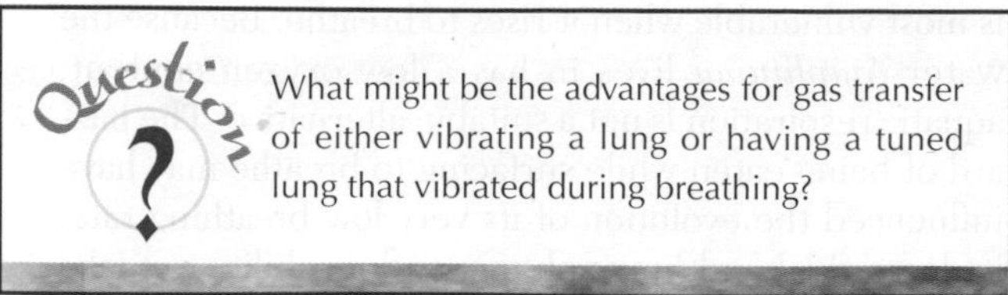

Pulmonary Circulation

The lung, like the heart, receives blood from two sources. The major flow is that of deoxygenated blood from the pulmonary artery that perfuses the lung, taking up O_2 and giving up CO_2; this flow is termed the *pulmonary circulation*. A second, smaller supply, the *bronchial circulation*, comes from the systemic (body) circulation and supplies the lung tissues themselves with O_2 and other substrates for growth and maintenance. Our discussion here is confined to the pulmonary circulation.

In birds and mammals, blood pressures in the pulmonary circulation are lower than those in the systemic circulation. This lower pressure reduces filtration of fluid into the lung. Extensive lymphatic drainage of lung tissues also helps ensure that no fluid collects in the lung (see Chapter 12). These features are important because any fluid that collects at the lung surface increases the diffusion distance between blood and air and reduces gas transfer.

Blood flow through the pulmonary circulation is best described as *sheet flow*—flow of a liquid between two parallel surfaces. This flow pattern contrasts with the laminar flow characteristic of the systemic circulation (see Chapter 12). The pulmonary capillary endothelium resembles two parallel surfaces, joined by pillar-like structures, with blood flowing between them. As blood pressure increases, the parallel surfaces move apart, leading to an increase in the thickness of the blood sheet. In other words, increased pressure increases the thickness of the blood sheet rather than spreading out the flow in other directions. The mean arterial pressure in the human lung is about 12 mm Hg, oscillating between 7.5 mm Hg and 22 mm Hg with each contraction of the heart. In a vertical (upright) human lung, arterial pressure is just sufficient to raise blood to the apex of the lung; hence flow is minimal at the apex and increases toward the base of the lung (Figure 13-26). Blood is distributed more evenly to different parts of a horizontal lung.

The pulmonary vessels are very distensible and subject to distortion by breathing movements. Small vessels within the interalveolar septa are particularly sensitive to changes in alveolar pressure. The diameter of these thin-walled, collapsible capillaries is determined by the transmural pressure (arterial blood pressure within capillaries, P_a, minus alveolar pressure, P_A). If the transmural pressure is negative (i.e., $P_A > P_a$), the capillaries collapse and blood flow ceases. This collapse may occur at the apex of the vertical human lung, where

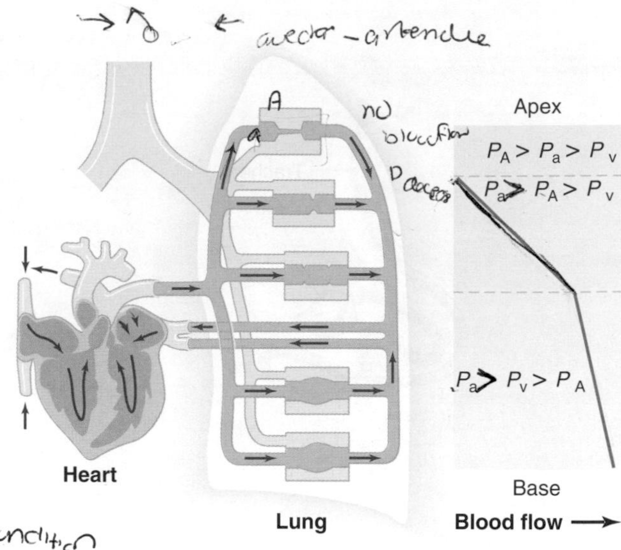

Figure 13-26 In the upper portion of the vertical lung, the diameter of alveolar capillaries, and hence blood flow through them, depends on the difference between the arterial pressure, P_a, and the alveolar pressure, P_A. In this schematic diagram of blood flow in the vertical human lung, the boxes represent the condition of vessels in the interalveolar septum in different portions of the lung. At the apex of the lung, P_A often exceeds P_a; as a result, the capillaries collapse and blood flow ceases. P_v is the venous pressure. [Adapted from West, 1970.]

P_a is low (see Figure 13-26). If pulmonary arterial pressure is greater than alveolar pressure, which in turn is greater than pulmonary venous pressure, then the difference between arterial and alveolar pressure will determine the diameter of capillaries in the interalveolar septa and, in the manner of a sluice gate, control blood flow through the capillaries. Venous pressure does not affect flow into the venous reservoir as long as alveolar pressure exceeds venous pressure. Flow in the upper portion of the vertical lung is probably determined in this way.

Arterial blood pressure (and therefore blood flow) increases with distance from the apex of the lung. In the bottom half of the vertical lung, where venous pressure exceeds alveolar pressure, blood flow is determined by the difference between arterial and venous blood pressures. This pressure difference does not vary with location in the lung, although both the arterial and venous pressures increase toward the base of the lung. This increase in absolute pressure results in an expansion of vessels and, therefore, a decrease in resistance to flow. Thus, flow increases toward the base of the lung, even though the arterial-venous pressure difference does not change (see Figure 13-26). The position of the lungs with respect to the heart is therefore an important determinant of pulmonary blood flow. The lungs surround the heart, thus minimizing the effect of gravity on pulmonary blood flow as an animal changes from a horizontal to vertical position. This close proximity of lungs and heart within the thorax also has significance for cardiac function: During inhalation, the pressures within the thorax are reduced, which aids venous return to the heart. This phenomenon is often referred to as the *thoraco-abdominal pump*.

Even though the mammalian pulmonary circulation lacks well-defined arterioles, both sympathetic adrenergic and parasympathetic cholinergic fibers innervate the smooth muscle around the pulmonary blood vessels and bronchioles. The pulmonary circulation, however, has much less innervation than does the systemic circulation and is relatively unresponsive to nerve stimulation or injected drugs. Sympathetic nerve stimulation or the injection of norepinephrine causes a slight increase in resistance to blood flow, whereas parasympathetic nerve stimulation or acetylcholine has the opposite effect.

Reductions in either oxygen levels or pH cause local vasoconstriction of pulmonary blood vessels. The vasoconstriction response to low oxygen levels, which is the opposite of that observed in systemic capillary networks, ensures that blood flows to the well-ventilated regions of the lung. Poorly ventilated regions of the lung will have low alveolar oxygen levels, causing local vasoconstriction and therefore a reduction in blood flow

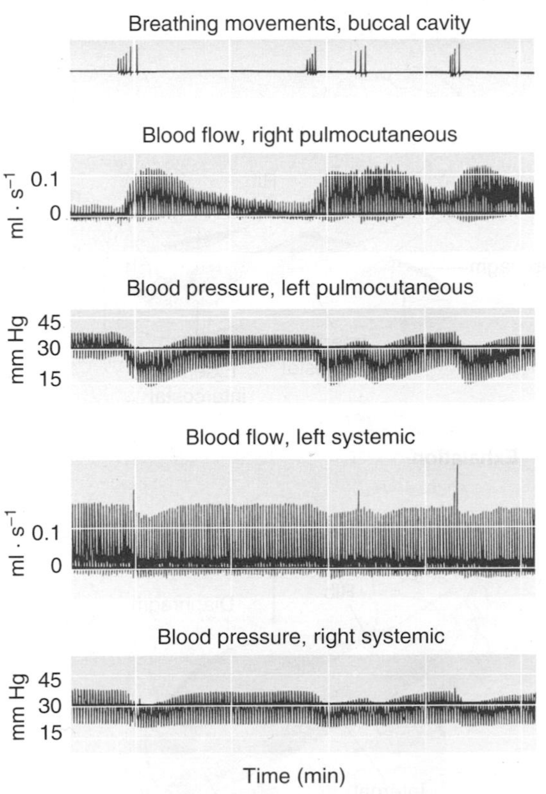

Figure 13-27 Pulmonary blood flow typically increases, whereas systemic flow remains constant, following breathing in turtles and frogs. These traces from the frog *Xenopus* record pressure changes in the buccal cavity (see Figure 13-33) produced by lung-ventilating movements of the buccal floor, as well as the corresponding flow and pressure in the pulmonary and systemic arterial arches. [From Shelton, 1970.]

to those areas. Alternatively, a well-ventilated area of the lung will have high alveolar oxygen levels, so that local blood vessels will be dilated and blood flow to that area will be high.

There appear to be several origins of hypoxic pulmonary vasoconstriction. Its initial rapid onset is related to hypoxic inhibition of potassium channels, smooth muscle depolarization, and calcium release followed by smooth muscle contraction. This initial phase is followed by a slowly developing, but sustained, phase that is dependent on factors released from the endothelium, including nitric oxide. Although pulmonary hypoxic vasoconstriction is important in directing blood flow to well-ventilated regions of the lung, it leads to problems when animals are exposed to general hypoxia, as may occur at high altitudes (see below).

Cardiac output to the pulmonary circuit is equal to cardiac output to the systemic circuit in mammals and birds. In amphibians and reptiles, which have a single or partially divided ventricle that ejects blood into both the pulmonary and the systemic circulation, the ratio of pulmonary to systemic blood flow can be altered. In turtles and frogs, there is a marked increase in blood flow to the lung following a breath due to pulmonary vasodilation. During periods between breaths in the frog *Xenopus*, pulmonary blood flow decreases, but systemic blood flow hardly changes (Figure 13-27). These animals breathe intermittently, and variable blood flow to the gas exchanger, independent of blood flow to the rest of the body, permits some control of the rate of oxygen use from the lung store and rapid renewal of blood oxygen stores during ventilation. In addition, cardiac work is reduced during apnea.

Mechanisms for Ventilation of the Lung

The mechanisms of lung ventilation vary considerably among animals, reflecting differences in the functional anatomy of the lungs and associated structures. First we look at ventilation of the mammalian lung, then we consider ventilation in birds, reptiles, frogs, and invertebrates.

Mammals

The lungs of mammals are elastic, multi-chambered bags, which are suspended within the **pleural cavity** and open to the exterior via a single tube, the trachea (Figure 13-28 on the next page). The walls of the pleural cavity, often referred to as the **thoracic cage**, are formed by the ribs and the **diaphragm**. The lungs fill most of the thoracic cage, leaving a low-volume pleural space between the lungs and thoracic wall; this space is sealed and filled with fluid. Because of their elasticity, isolated lungs are somewhat smaller than they are *in situ*. Their elasticity creates a pressure below atmospheric pressure in the fluid-filled pleural space. The fluid provides a flexible, lubricated connection between the outer lung surface and the thoracic wall. Body fluids

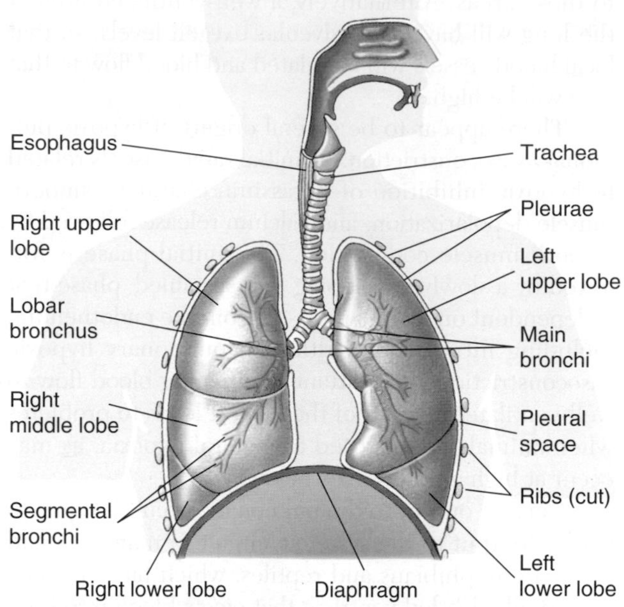

Figure 13-28 In mammals, the lungs fill most of the thoracic cavity, formed by the ribs and diaphragm. In humans, the right lung has three lobes and the left lung, two lobes. The low-volume pleural space between the lungs and thoracic wall is filled with fluid and sealed.

are essentially incompressible, so when the thoracic cage changes volume, the gas-filled lungs do too. If the thoracic cage is punctured, air is drawn into the pleural cavity and the lungs collapse—a condition known as **pneumothorax.**

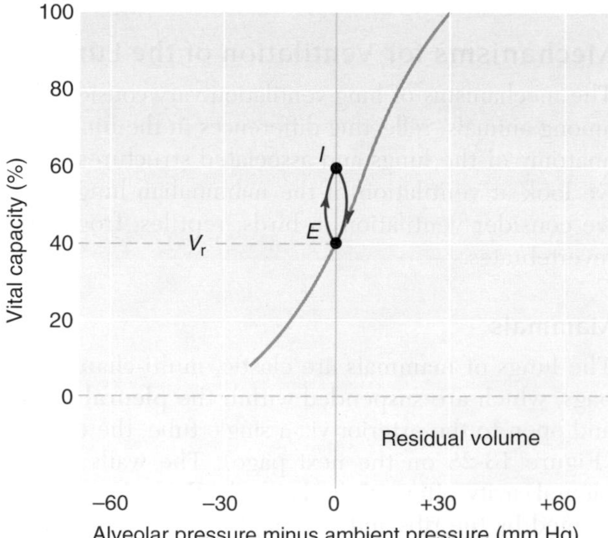

Figure 13-29 During quiet breathing with the thoracic muscles relaxed, the alveolar and ambient pressures are equal between breaths. This plot shows the relationship between lung volume and pressure within the thorax when muscles are relaxed but the glottis is closed. V_r is the lung volume at which alveolar pressure is the same as ambient pressure. The points *I* and *E* represent the pressure and volume of the system following inhalation and exhalation during quiet breathing.

When lungs *in situ* are filled to various volumes and the airway is closed with the muscles relaxed, alveolar pressure varies directly with lung volume. At low pulmonary volumes, alveolar pressure is less than ambient pressure owing to the resistance of the thorax to collapse, whereas at high pulmonary volumes, alveolar pressure exceeds ambient pressure because of the forces required to expand the thoracic cage. If lung volume is large, then once the mouth and glottis are opened, air will flow out of the lungs because the weight of the ribs will reduce pulmonary volume. At some intermediate volume, V_r, alveolar pressure in the relaxed thorax is equal to ambient pressure (Figure 13-29).

During normal breathing, the thoracic cage is expanded and contracted by a series of skeletal muscles, the diaphragm, and the external and internal intercostal muscles. Contractions of these muscles are determined by the activity of motor neurons controlled by the respiratory center within the medulla oblongata, as we will see below. The volume of the thorax increases as the ribs are raised and moved outward by contraction of the external intercostals and by contraction (and lowering) of the diaphragm (Figure 13-30a). Contractions of the

(a) **Inhalation**

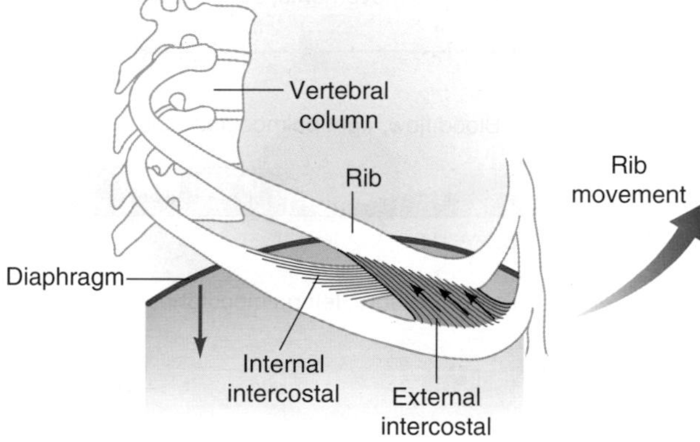

(b) **Exhalation**

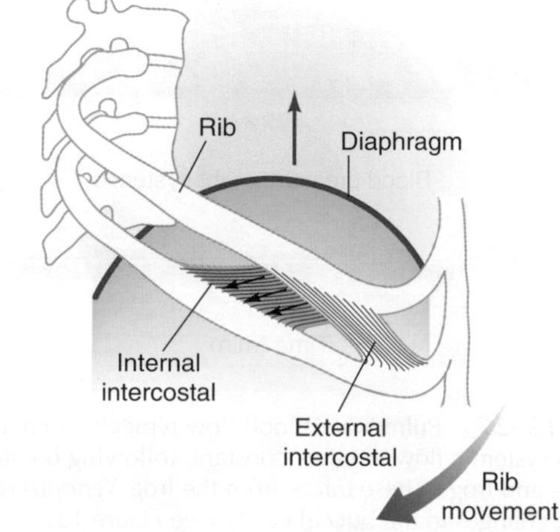

Figure 13-30 In mammals, the volume of the thorax increases during inhalation **(a)** and decreases during exhalation **(b)** due to movement of the ribs and diaphragm.

diaphragm account for up to two-thirds of the increase in pulmonary volume. The increase in thoracic volume reduces alveolar pressure, and air is drawn into the lungs. Relaxation of the diaphragm and external intercostal muscles reduces thoracic volume, thereby raising alveolar pressure and forcing air out of the lungs (Figure 13-30b). During quiet breathing, pulmonary volume between breaths is at an intermediate value, V_r, at which the alveolar and ambient pressures are equal (see Figure 13-29). Under these conditions, exhalation is often passive, due simply to relaxation of the diaphragm and external intercostals. With increased tidal volume, exhalation becomes active, owing to contraction of the internal intercostal muscles, which further reduces thoracic volume until it drops below V_r at the end of exhalation.

Birds

In birds, gas transfer takes place in small air capillaries (10 μm in diameter) that branch from tubes called **parabronchi** (Figure 13-31). The functional equivalent of mammalian alveoli, the small parabronchi extend between larger dorsobronchi and ventrobronchi, both of which are connected to an even larger tube, the mesobronchus, which joins the trachea anteriorly (Figure 13-32a). The parabronchi and their connecting tubes form the lung, which is contained within a thoracic cavity. A tight horizontal septum closes the caudal end of the thoracic cage. The ribs, which are curved to prevent lateral compression, move forward only slightly during breathing; as a result, the volume of the thoracic cage and lung changes little during breathing. The large flight muscles of birds are attached to the sternum and have little influence on breathing. Although there is no mechanical relation between flight and respiratory movements in birds, "in phase" flight and breathing movements may result from synchronous neuronal activation of the two groups of muscles involved.

(a)

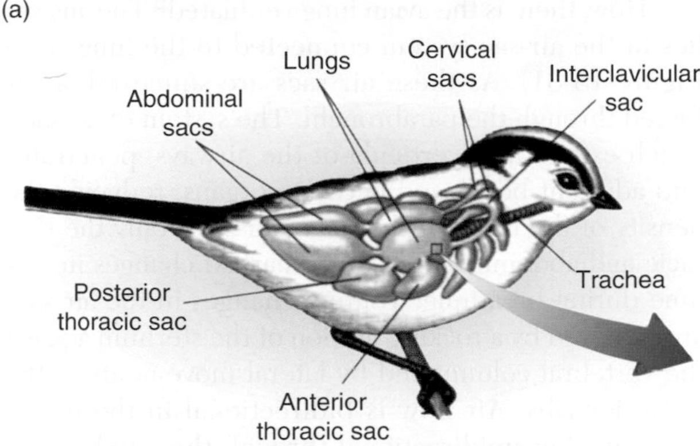

(b)

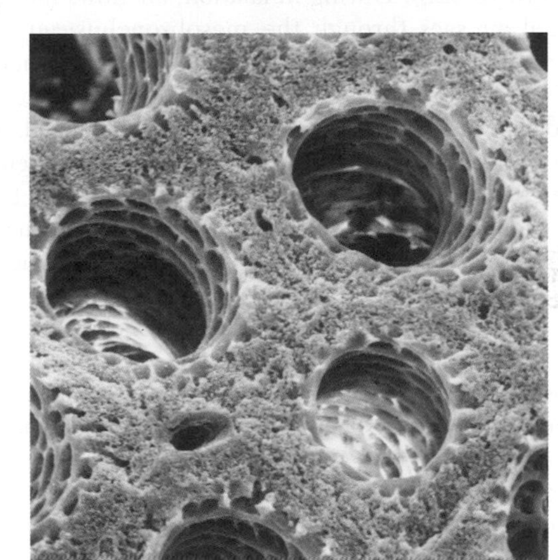

Figure 13-31 In bird lungs, gas exchange occurs in air capillaries extending from parabronchi, small tubelike structures that are the functional equivalent of alveoli in mammals. **(a)** During breathing, volume changes occur in the associated air sacs, not in the thoracic cage and lungs. **(b)** The parabronchi and connecting tubes form the lung. [Photograph courtesy of H. R. Duncker.]

(a)

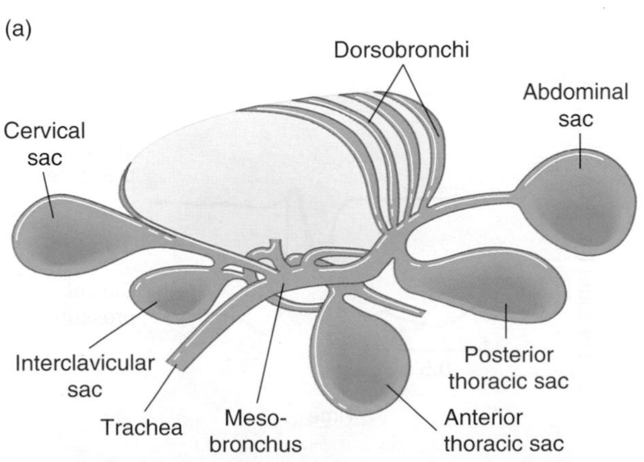

(b)

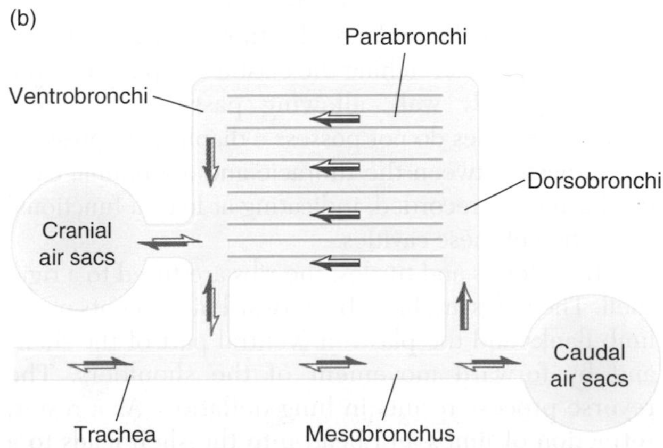

Figure 13-32 Squeezing of the airs sacs forces air through the parabronchi in bird lungs. **(a)** The avian bronchial tree and associated air sacs. The air sacs of the cranial group (cervical, interclavicular, and prethoracic sacs) depart from the three cranial ventrobronchi, whereas the air sacs of the caudal group (postthoracic and abdominal sacs) are connected directly to the mesobronchus. **(b)** Schematic diagram of air flow through the bird lung. Flow in the parabronchi is unidirectional. Solid arrows represent flow during inhalation; open arrows, flow during exhalation. [Adapted from Scheid et al., 1972.]

How, then, is the avian lung ventilated? The answer lies in the air-sac system connected to the lungs (see Figure 13-31). As these air sacs are squeezed, air is forced through the parabronchi. The system of air sacs, which extend as diverticula of the airways, penetrates into adjacent bones and between organs, reducing the density of the bird. Of the many air sacs, only the thoracic and abdominal sacs show marked changes in volume during breathing. Volume changes in the air sacs are achieved by a rocking motion of the sternum against the vertebral column and by lateral movements of the posterior ribs. Air flow is bidirectional in the mesobronchus but unidirectional through the parabronchi (Figure 13-32b). During inhalation, air flows into the caudal air sacs through the mesobronchus; air also moves into the cranial air sacs via the dorsobronchus and the parabronchi. During exhalation, air leaving the caudal air sacs passes through the parabronchi and, to a lesser extent, through the mesobronchus to the trachea. The cranial air sacs, whose volume changes less than that of the caudal air sacs, are reduced somewhat in volume by air moving via the ventrobronchi to the trachea during exhalation.

In this manner, the air in the parabronchi is changed during both inhalation and exhalation, enhancing gas transfer in the bird lung. The unidirectional flow is achieved not by mechanical valves, but by aerodynamic valving. The openings of the ventrobronchi and dorsobronchi into the mesobronchus show a variable, direction-dependent resistance to air flow. The structure of the openings is such that eddy formation, and therefore resistance to flow, varies with the direction of airflow.

Reptiles

The ribs of reptiles, like those of mammals, form a thoracic cage around the lungs. During inhalation, the ribs are moved cranially and ventrally, enlarging the thoracic cage. This expansion reduces the pressure within the cage below atmospheric pressure, and when the nares and glottis are open, air flows into the lungs. Relaxation of the muscles that enlarge the thoracic cage releases energy stored in stretching the elastic component of the lung and body wall, allowing passive exhalation. Although reptiles do not possess a diaphragm, pressure differences between the thoracic and abdominal cavities have been recorded, indicating at least a functional separation of these cavities.

In tortoises and turtles, the ribs are fused to a rigid shell. The lungs are filled by outward movements of the limb flanks and the plastron (ventral part of the shell) and by forward movement of the shoulders. The reverse process results in lung deflation. As a result, retraction of limbs and head into the shell leads to a decrease in pulmonary volume.

Frogs

In frogs, the nose opens into a buccal cavity, which is connected via the glottis to paired lungs. The frog can open and close its nares and glottis independently. Air is drawn into the buccal cavity with the nares open and glottis closed; then the nares are closed, the glottis is opened, and the buccal floor raised, forcing air from the buccal cavity into the lungs (Figure 13-33). This lung-filling process may be repeated several times in sequence. Exhalation may also occur in steps, with the lungs releasing air in portions to the buccal cavity. Exhalation may not be complete, so that some of the air from the lung is mixed with ambient air in the buccal cavity and then pumped back into the lungs. Thus, a mixture of pulmonary air, presumably low in O_2 and high in CO_2, is mixed with fresh air in the buccal cavity and returned to the lungs. The reason for this complex method of lung ventilation is not clear, but it may be directed toward reducing oscillations in CO_2 levels in

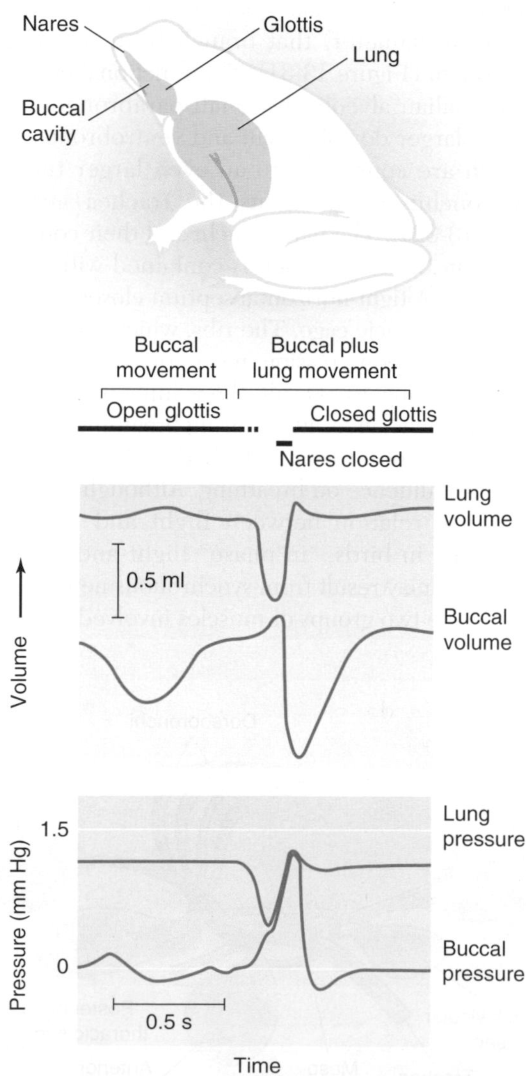

Figure 13-33 Ventilation in the frog is a stepwise process. Shown here are pressure and volume changes in the buccal cavity and lung of a frog during buccal movements alone with the glottis closed and during buccal and lung movements with the glottis open and the nares closed (i.e., lung filling). [Adapted from West and Jones, 1975.]

the lungs in order to stabilize and regulate blood P_{CO_2} and control blood pH.

Invertebrates

Invertebrates exhibit a variety of gas-transfer mechanisms. Ventilation does not occur in some invertebrates, which rely only on diffusion of gases between the lung and the environment. In spiders, which have paired ventilated lungs on the abdomen, the respiratory surface consists of a series of thin, blood-filled plates that extend like the leaves of a book into a cavity guarded by an opening (spiracle). The spiracle can be opened or closed to regulate the rate of water loss from these "book lungs." Snails and slugs have ventilated lungs that are well-vascularized invaginations of the body surface, the mantle cavity. Changes in the volume of the snail lung enable the animal to emerge from and withdraw into its rigid shell. When the snail retracts into its shell, the lungs empty, a situation similar to that seen in tortoises. In aquatic snails, the lungs serve to reduce the animal's density.

Pulmonary Surfactants

The lung wall tension depends on the properties of the wall and the surface tension at the liquid–air interface. Surface tension is a force that tends to minimize the area of a liquid surface, causing liquid droplets to form a sphere. It also makes a surface film resistant to stretch. Much of the resistance to stretch in the lung wall—about 70%—is supplied by the surface tension of the fluid lining the otherwise very compliant alveoli. This fluid lining is not simply water. If it were, the alveolar wall tension would be much higher than it is, and large forces would be required to inflate the lung. This awkward situation is avoided by the presence of **surfactants,** lipoprotein complexes that bestow a very low surface tension on the liquid-air interface.

Surfactants are produced by type II cells within the alveolar lining and have a half-life of about 12 hours in mammals. The lipid component in these lipoprotein complexes, predominantly dipalmitoyl phosphotidylcholine, forms a stable outer monolayer that is firmly associated with an underlying protein layer in contact with water. Surfactants are found in the lungs of amphibians, reptiles, birds, and mammals, and they may be present in some fishes that build bubble nests. In humans, surfactant release is stimulated by sighing.

Surfactants have a number of important roles in lung function:

1. As noted above, surfactants are responsible for the low surface tension of the fluid lining the alveoli, which allows the alveoli to expand easily during breathing and reduces the effort of inflating the lung.
2. Alveoli fold as their volume decreases and would become glued together by surface tension were it not for surfactant, which reduces the surface tension enough to allow easy inflation of the collapsed alveoli. When lung volume is reduced extremely, the lung will collapse, a condition called *atelectasis.* Due to the presence of surfactants, an atelectatic lung can be re-inflated relatively easily.
3. Surfactants allow newborn babies to inflate their lungs. In mammals, surfactants appear in the fetal lung prior to birth. Newborns who produce no lung surfactants cannot inflate their lungs, a condition called *neonatal respiratory distress syndrome.* Seen primarily in premature babies, the condition is addressed by forcing air into the lungs using positive pressure ventilation, and by surfactant replacement. Pregnant women likely to give birth prematurely are sometimes given an injection of cortisol during gestation, which stimulates production of type II alveolar cells in the fetus.
4. Surfactants reduce resistance to blood flow by increasing the compliance of the capillary-alveolar sheet. When surfactants are present, capillaries are less likely to collapse and, once collapsed, are easier to reopen.
5. Surfactants increase the osmotic pressure of the lung fluid, thereby reducing water flux across the lung epithelium.

Heat and Water Loss Across the Lung

Increases in lung ventilation not only increase gas transfer but also result in increased losses of heat and water. Thus, the evolution of lungs has involved some compromises. Cool, dry air entering the lungs of mammals is humidified and heated as air in contact with the respiratory surface becomes saturated with water vapor and comes into thermal equilibrium with the blood. Exhalation of this hot, humid air results in considerable loss of heat and water; the amount lost is proportional to the rate of ventilation. Many air-breathing animals live in very dry environments, where water conservation is of paramount importance. It is therefore not surprising that these animals in particular have evolved means of minimizing the loss of water from the lungs.

The rates of heat and water loss from the lung are intimately related. As air is inhaled, it is warmed and humidified by evaporation of water from the nasal mucosa. Because the evaporation of water cools the nasal mucosa, a temperature gradient exists along the nasal passages, which are cool at the tip of the nose and increase in temperature toward the glottis. Because the water vapor pressure for 100% saturation decreases with temperature, water from the humid air leaving the lung condenses on the nasal mucosa as it cools, and remains there ready to wet the next inflow of air. Thus the cooling of exhalant air in the nasal passages results in the conservation of both heat and water. There is an extensive blood supply to the nose to supply water to wet the incoming air. The circulation is arranged in a countercurrent fashion (see Spotlight 14-2) to prevent this high blood flow from warming the nose and destroying the temperature gradient. To increase heat loss, but at the expense of water loss, animals exhale through the mouth.

There is considerable structural variety in the nasal passages among vertebrates, and to some extent their structure can be correlated with the ability of animals to regulate heat and water loss. Humans have only a limited ability to cool exhaled air, which is saturated with water vapor and is at a temperature only a few degrees below core body temperature. Other animals have longer and narrower nasal passages for more effective water conservation, as we will see in Chapter 14.

Poikilotherms such as reptiles and amphibians, whose body temperatures adjust to the ambient temperature, exhale air saturated with water at temperatures about 0.5–1.0°C below body temperature. Pulmonary air temperatures and body surface temperatures are often slightly below ambient because of the continual evaporation of water. In some reptiles, however, body temperature is maintained above ambient. In the iguana, heat and water loss is controlled in a manner similar to that observed in mammals. In addition, this lizard conserves water by humidifying inhaled air with water evaporated from the excretory fluid of the nasal salt glands. The rate of water loss is closely correlated with lung ventilation and, therefore, oxygen uptake. Reptiles generally have much lower oxygen requirements than mammals or birds, and so their rate of water loss is much less.

Gas Transfer in Bird Eggs

The shells of bird eggs have fixed dimensions but contain an embryo whose gas-transfer requirements increase by a factor of 10^3 between laying and hatching. Thus, the transfer of O_2 and CO_2 must take place across the shell at ever-increasing rates while the dimensions of the transfer surface (the eggshell) do not change. Gases diffuse through small air-filled pores in the eggshell and then through the underlying membranes, including the chorioallantoic membrane (Figure 13-34a). The chorioallantoic circulation is in close apposition to the eggshell and increases with the development of the embryo. Several factors contribute to the increase in gas-transfer rates during development: development of the chorioallantoic circulation, an increase in blood flow and volume, an increase in hematocrit and blood oxygen affinity, and an increase in the P_{O_2} difference across the eggshell (Figure 13-34b). The eggshell itself, once produced, does not change during the development of the embryo.

Water is lost from the egg during development, causing the gradual enlargement of an air space within the egg. The volume of this air cell in the chicken egg is as much as 12 ml at hatching. Just before birds hatch, they ventilate their lungs by poking their beaks into the air cell. Blood P_{CO_2} is initially low in the embryo, but it gradually rises to about 45 mm Hg just before hatching (see Figure

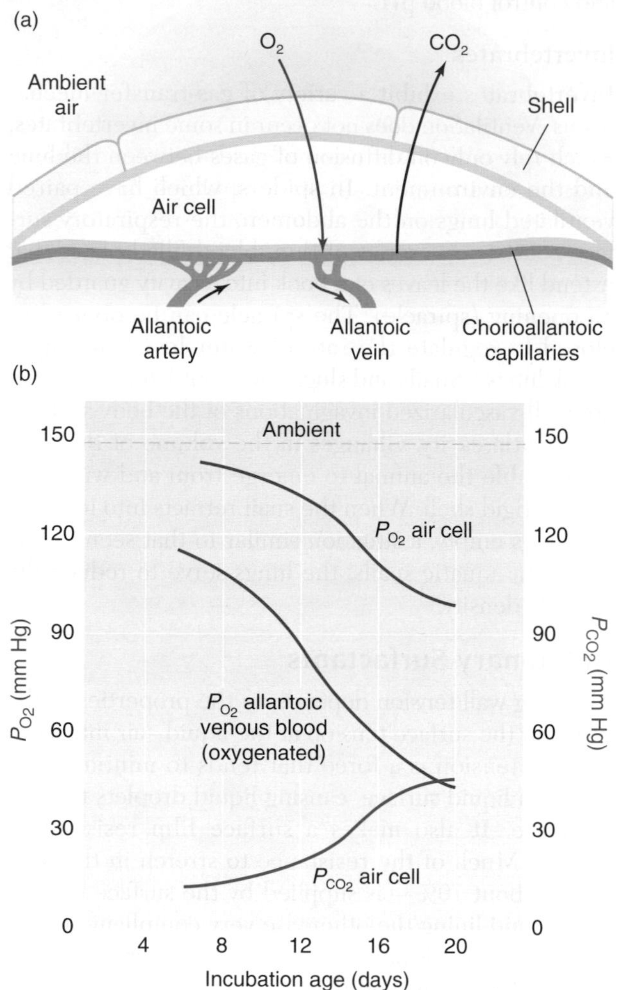

Figure 13-34 During the development of a bird embryo, gas transfer across the eggshell increases even though the shell structure does not change. **(a)** Diagram of the diffusion pathway between air and chicken embryo blood across the eggshell in the region of the air cell.
Classic Work **(b)** The P_{O_2} in the air cell (upper red trace) and in the allantoic venous blood (lower red trace) drop with incubation age. The P_{CO_2} in allantoic venous blood is identical to that in the air cell (blue trace); both increase with incubation age. [Adapted from Wangensteen, 1972.]

13-34b). This pressure is maintained after hatching, thus avoiding any marked acid-base changes when the bird switches from the shell to its lungs for gas exchange.

The shell and the underlying membranes, representing the barrier between ambient air and embryonic blood, can be divided into an outer gas phase (the air cell) and an inner liquid phase. At sea level, the outer gas phase represents about 30%–40% of the total diffusive resistance to O_2 transfer, 85% of that to CO_2, and 100% of that to water vapor. Eggs at altitude are exposed to a reduction in both P_{O_2} and total gas pressure. The rate of diffusion of gases increases as total pressure falls, so the reduced P_{O_2} at altitude is partially offset by increased rates of O_2 diffusion in the gas phase. Nevertheless, eggs become hypoxic at altitude. If eggs are kept in a hypoxic environment for a period of time, more capillaries develop in the chorioallantois, increasing oxygen-diffusing capacity and offsetting the effects of altitude on O_2 transfer across the eggshell. Because CO_2 and water vapor also diffuse more rapidly at the reduced pressures associated with altitude, eggs at altitude also have a reduced blood P_{CO_2} and lose water more rapidly than those at sea level. The properties of the shell are determined when the egg is laid, but it appears that some birds acclimated to altitude can reduce the effective pore area of the eggs they produce, compensating somewhat for the effects of reduced pressure.

Insect Tracheal Systems

The system that insects have evolved for transferring gases between their tissues and the environment differs fundamentally from that found in air-breathing vertebrates. The insect **tracheal system** takes advantage of the fact that oxygen and carbon dioxide diffuse 10,000 times more rapidly in air than in water, blood, or tissues. Tracheal systems consist of a series of air-filled tubes that penetrate from the body surface to the cells and act as a pathway for the rapid movement of O_2 and CO_2, thereby avoiding the need to transport gases through the circulatory system. These tubes, or tracheas, are invaginations of the body surface; thus, their wall structure is similar to that of the cuticle. Except in a few primitive forms, the tracheal entrances, called **spiracles,** can be adjusted to control air flow into the tracheas, regulate water loss, and keep out dust. The bug *Rhodnius* dies in three days if its spiracles are kept open in a dry environment. The tracheas branch everywhere in the tissues; the smallest, terminal branches, called **tracheoles,** are blind-ended and poke between and into individual cells (without disrupting the cell membrane), delivering O_2 to regions very close to the mitochondria. Air sacs are commonly located at intervals throughout the tracheal system; these sacs enlarge tracheal volume, and therefore oxygen stores, and sometimes reduce the specific gravity of organs, which may contribute to buoyancy or balance.

Tracheal ventilation

Diffusion of gases is a slow process. Much more rapid transfer of oxygen and carbon dioxide can be achieved by the mass movement of gases (**convection**). Large insects usually have some mechanism for generating air flow in the larger tubes of their tracheal system. The air sacs and tracheas are often compressible, allowing changes in tracheal volume. Some large insects ventilate the larger tubes and air sacs of the tracheal system by alternate compression and expansion of the body wall, particularly the abdomen. Different spiracles may open and close during different phases of the breathing cycle, controlling the direction of air flow. In the locust, for example, air enters through the thoracic spiracles but leaves through more posterior openings. Tracheal volume in insects is highly variable—40% of body volume in the beetle *Melolontha,* but only 6%–10% of body volume in the larva of the diving beetle *Dytiscus.* Each ventilation results in a maximum of 30% of tracheal volume being exchanged in *Melolontha* and 60% in *Dytiscus.*

Not all insects ventilate their tracheal systems; in fact, many calculations have shown that diffusion of gases in air is rapid enough to supply the tissue demands of many species. Ventilation of the tracheas occurs when simple diffusion must be augmented to meet the insect's needs, as in larger insects and during high levels of activity in some smaller insects.

In many insects the spiracles open and close in a sequence called the *discontinuous ventilation cycle.* This cycle can be divided into three phases: an open phase, a closed phase, and an intermediate flutter phase, during which the spiracle oscillates rapidly between the open and closed states. Oxygen utilization and carbon dioxide production by the tissues occur during all phases. When the spiracles are closed, oxygen is supplied from stores in the tracheal system. Pressure in the tracheal system falls during the closed phase because oxygen levels decrease more rapidly than carbon dioxide levels increase. Carbon dioxide levels rise slowly during the closed phase because most of the carbon dioxide produced by metabolism is stored in the tissues. Thus, during the flutter phase and at the onset of the open phase, gases move into the tracheas both by bulk flow down a pressure gradient and by diffusion. Carbon dioxide and water diffuse from the tracheal air space during the open phase and even during the flutter phase, but not during the closed phase (Figure 13-35 on the next page).

In theory, discontinuous ventilation could reduce water loss associated with respiration. The depression of oxygen levels in the tracheal air space during the closed phase ensures high rates of oxygen diffusion into the tracheal air space during the open phase compared with rates of water loss. The functional significance of the flutter phase in determining the rates of gas and water transfer is not clear, but it may enhance gas mixing in the tracheal space. In some instances, however, the role of discontinuous ventilation in water conservation appears to be of little significance. Many xeric species, which require little water, do not show discontinuous ventilation. The lubber grasshopper, for example, does not display discontinuous ventilation during desiccation even though it is capable of doing so. In this case, only about 5% of total water loss is via the tracheal system, so perhaps it is not surprising that the pattern of ventilation is not changed during desiccation. Water-stressed cockroaches lose water across the cuticle at a rate more than twice that of loss through the spiracles and can conserve water during periods of desiccation by closure of pore structures in the cuticle. Thus, it is not clear why many insects have adopted a pattern of discontinuous ventilation. Although opening and closing

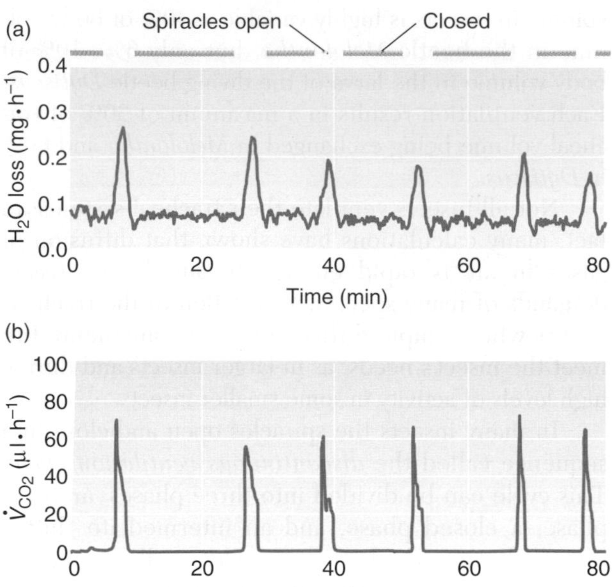

Figure 13-35 Some insects exhibit a discontinuous ventilation cycle. The traces of water loss **(a)** and carbon dioxide loss **(b)** are plotted against spiracle open and closed phases for an alate (would-be queen) harvester ant. Respiratory water loss is concentrated in the spiracle open phase associated with carbon dioxide excretion. The spiracles are closed between pulses of CO_2 excretion. The background cuticular water loss rates can be seen between the open phases. [From Lighton, 1994.]

the spiracles can reduce water loss in relation to oxygen uptake, the savings may not always be of much significance to the animal. The functional significance of the flutter phase is an even greater enigma.

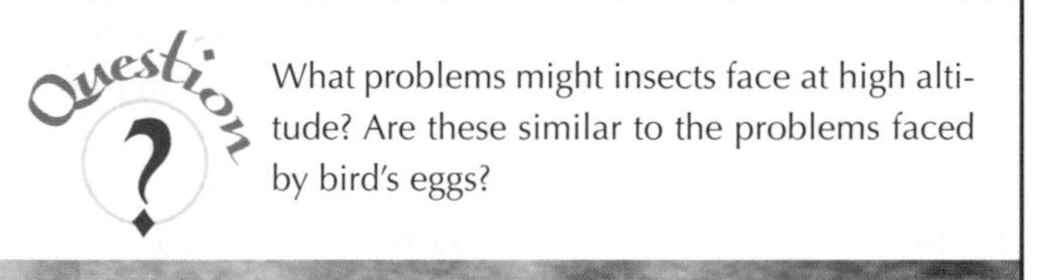

What problems might insects face at high altitude? Are these similar to the problems faced by bird's eggs?

Gas exchange across tracheolar walls

Gases are transferred between air and tissues across the walls of the tracheoles. These walls are very thin, approximately 40 to 70 nm across. The tracheolar surface area is very large, and only rarely is an insect cell more than three cells away from a tracheole. In all but a few species, the tips of the tracheoles are filled with fluid, so that oxygen diffusing from the tracheoles to the tissues moves through the fluid in the tracheoles, the tracheolar wall, the extracellular space (often negligible), and the cell membrane to the mitochondria. This diffusion distance can be altered in active tissues either by an increase in tissue osmolarity, which causes water to move out of the tracheoles and into the tissues, or by changes in the activity of an ion pump, which results in a net flow of ions and water out of the tracheoles. As fluid is lost from the tracheoles, it is replaced by air, so that oxygen can more rapidly diffuse into the tissues (Figure 13-36). Insect flight muscle has the highest recorded O_2 uptake rate of any tissue, with O_2 uptake increasing 10- to 100-fold above the resting value during flight. In general, more active tissues have more tracheoles, and in larger insects the tracheal system is more actively ventilated.

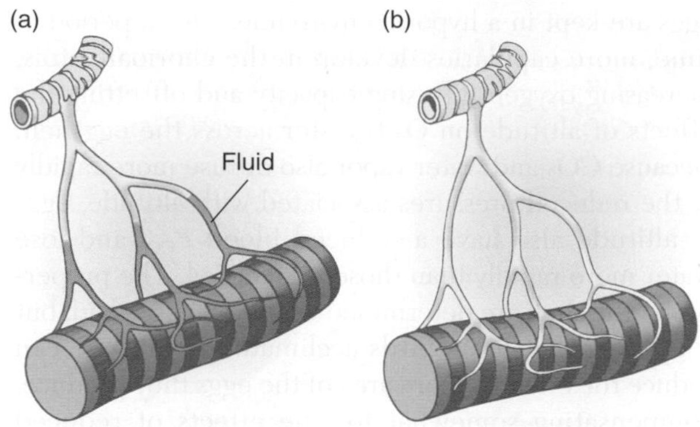

Figure 13-36 Insects can increase the rate of gas exchange in active muscles. **(a)** In resting insect muscle fibers, the terminal tracheoles contain fluid. **(b)** In active fibers, air may replace this fluid, thereby increasing diffusion of oxygen into the muscle. [Adapted from Wigglesworth, 1965.]

Modified tracheal systems

There are many modifications of the generalized tracheal system just described. Some larval insects, for example, rely on cutaneous respiration, the tracheal system being closed off and filled with fluid. Some aquatic insects have a closed, air-filled tracheal system in which gases are transferred between water and air across *tracheal gills.* The gills are evaginations of the body that are filled with tracheas. The air inside and the water outside are separated by a 1 μm membrane. Since this tracheal system is not readily compressible, the insect can change depth under water without the transfer of gases being impaired by changes in pressure.

Many aquatic insects, such as mosquito larvae, breathe through a hydrofuge (water-repellent) siphon that protrudes above the surface of the water. Others take bubbles of air beneath the surface with them. The water bug *Notonecta* carries air bubbles that cling to velvetlike hydrofuge hairs on its ventral surface when it is submerged. The water beetle *Dytiscus* dives with air bubbles beneath its wings or attached to its rear end. When such insects dive, gases are transferred between the bubble and the tissues via the tracheal system; gases can also diffuse, however, between the bubble and the water.

The rate of O_2 transfer between water and the interior of a bubble carried by an insect in a pond depends on the oxygen gradient established and the area of the air-water interface. The P_{O_2} in surface water is in equilibrium with that in the air above the surface; if the pond is well mixed and no oxygen is removed by aquatic animals or added through photosynthesis, the surface water mixes with deeper water and the P_{O_2} in the pond water does not vary with depth. As an air bubble is transported to depth by a water bug or beetle, it is compressed by hydrostatic pressure, causing the gas pressure within the bubble to rise and exceed that in the water. For every meter of depth, pressure in a bubble increases by approximately 0.1 atm. In a bubble just below the surface, the oxygen content decreases owing to uptake by the animal. This decrease establishes an O_2 gradient between the bubble and the water (assuming the water is in gaseous equilibrium with air), so that oxygen diffuses into the bubble from the water. The total pressure in a bubble just below the surface is maintained at approximately atmospheric pressure, but as P_{O_2} in the bubble is reduced, the nitrogen partial pressure, P_{N_2}, increases. Nitrogen therefore diffuses slowly from the bubble into the water (Figure 13-37). (Because of the high solubility of CO_2 in water, CO_2 levels in the bubble are always negligible.) If the bubble is taken deeper, however, the pressure increases by 0.1 atm for every meter of depth, increasing both P_{O_2} and P_{N_2} and speeding the diffusion of both N_2 and, initially, O_2 from the bubble into the water. Only when the P_{O_2} in the bubble drops below that in the water will oxygen move from the water into the bubble. The bubble will gradually get smaller and eventually disappear as nitrogen leaves it. Thus, the life of the bubble depends on the insect's metabolic rate, the initial size of the bubble, and the depth to which it is taken. It has been calculated that up to seven times the initial bubble O_2 content diffuses into the bubble from the water and is therefore available to the insect before the bubble disappears. It is possible that aquatic air-breathing vertebrates such as the beaver may also take advantage of the diffusion of oxygen from water into gas bubbles. These animals exhale under water, producing air bubbles that get trapped under the ice. These bubbles gain oxygen from the water, and later the animals can inhale the rejuvenated air.

If air bubbles were noncollapsible, the insects using them would not need to surface, because oxygen would continue to diffuse from the water via the bubble into the tracheal system and thence to the tissues. In some insects (e.g., *Aphelocheirus*), a thin film of air trapped by hydrofuge hairs, called a **plastron**, in effect provides a noncollapsible bubble (Figure 13-38a on the next page). The plastron can withstand pressures of several atmospheres before collapsing. In the small air space, N_2 is presumably in equilibrium with the water, P_{O_2} is low, and oxygen therefore diffuses from water into the plastron, which is continuous with the tracheal system (Figure 13-38b).

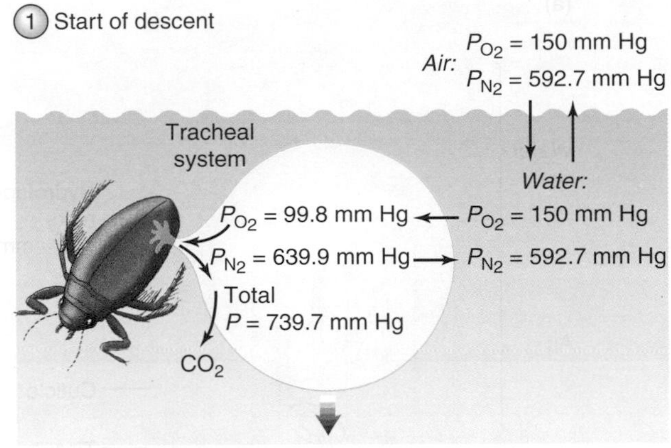

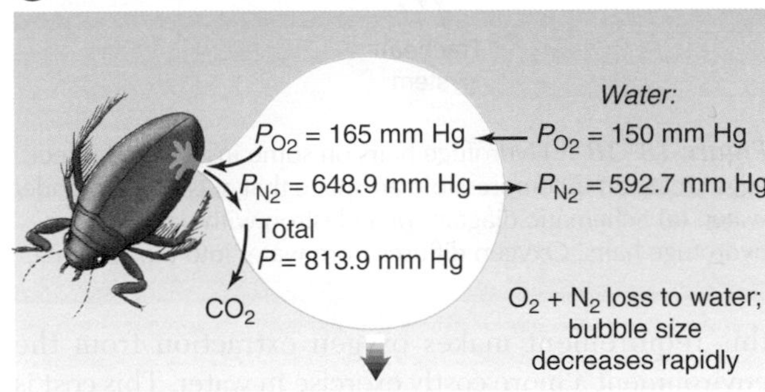

Figure 13-37 Some aquatic insects carry air bubbles when they dive. Under water, gas exchange occurs between the bubble and the insect's tracheal system and between the water and the bubble. The direction of gas flow depends on the partial pressures of O_2, CO_2, and N_2 and the total pressure (P) in the bubble. Arrows indicate diffusion of gas molecules. Note that the sum of the gas partial pressures in the water (and in the atmosphere) equals 742.7 mm Hg at the surface but increases when the bubble is taken to depth.

GAS TRANSFER IN WATER: GILLS

The gills of fishes, crabs, and other aquatic organisms are usually ventilated with a unidirectional flow of water (see Figure 13-19b). Because water has a much lower oxygen content than air, water-breathing animals require a much higher ventilation rate to achieve a given oxygen uptake than do air-breathing animals. Combined with the high density and viscosity of water,

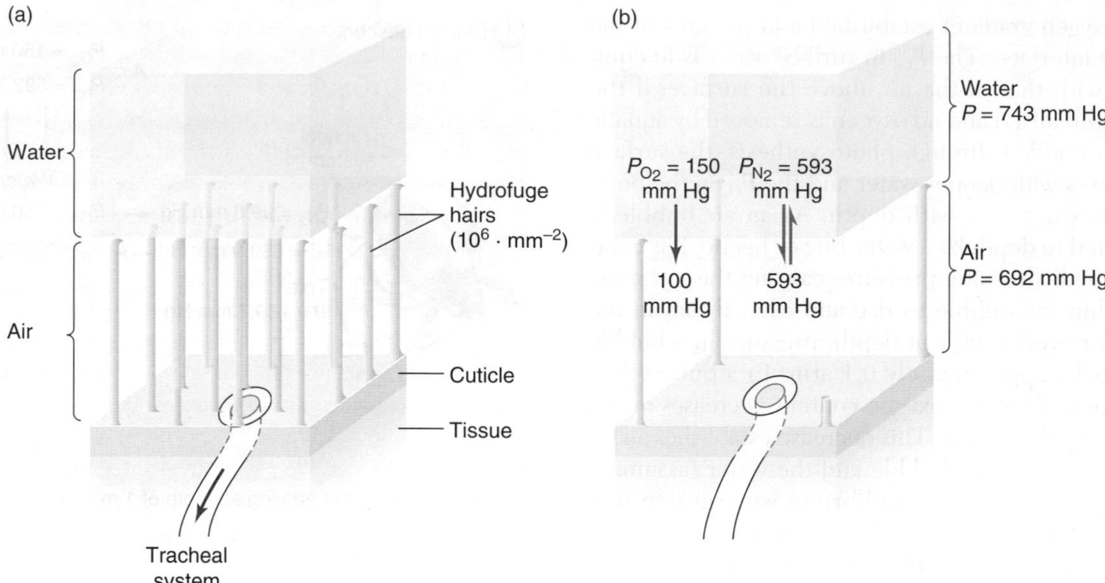

Figure 13-38 Hydrofuge hairs on some insects and insect eggs create an incompressible air space that acts as a gill under water. **(a)** Schematic diagram of a plastron with protruding hydrofuge hairs. Oxygen diffuses from water into the air space contained within the plastron and then into the animal via the tracheal system. Typically, there are about 10^6 hairs per mm^2; only a few are depicted here. **(b)** Partial pressures of oxygen and nitrogen in the air and water phases.

this requirement makes oxygen extraction from the environment a more costly exercise in water. This cost is offset somewhat by gills having a unidirectional, rather than tidal, flow of water. Tidal flow of water over gills, similar to that of air in the lung, would not be efficient because the energetic cost of reversing the direction of flow is simply too high.

The lamprey and sturgeon, however. are exceptions to the rule of unidirectional water flow. The mouth of the parasitic lamprey is often blocked by attachment to a host. Its gill pouches, although connected internally to the pharyngeal and mouth cavities, are ventilated by tidal movements of water through a single external opening to each pouch (Figure 13-39). This unusual method of gill ventilation is clearly associated with the animal's parasitic mode of life. The ammocoete larvae of lampreys are not parasitic and maintain a unidirectional flow of water over their gills. Water flow through the mouth and gills of sturgeon is normally unidirectional, but if the animal has its mouth in mud while searching for food, it can generate a tidal flow of water through slits in its gill coverings.

Flow of water over the gills of teleost fishes is maintained by the action of skeletal muscle pumps in the buccal and opercular cavities. Water is drawn into the mouth, passes over the gills, and exits through a cleft in the operculum (gill cover) (Figure 13-40; see also Figure 14-41). Valves guard the entrance to the buccal cavity and the opercular clefts, maintaining a unidirectional flow of water over the gills. The volume of the buccal cavity is changed by raising and lowering the floor of the mouth. The operculum swings in and out, enlarging and reducing the size of the opercular cavity. Changes in volume in these two cavities are nearly in

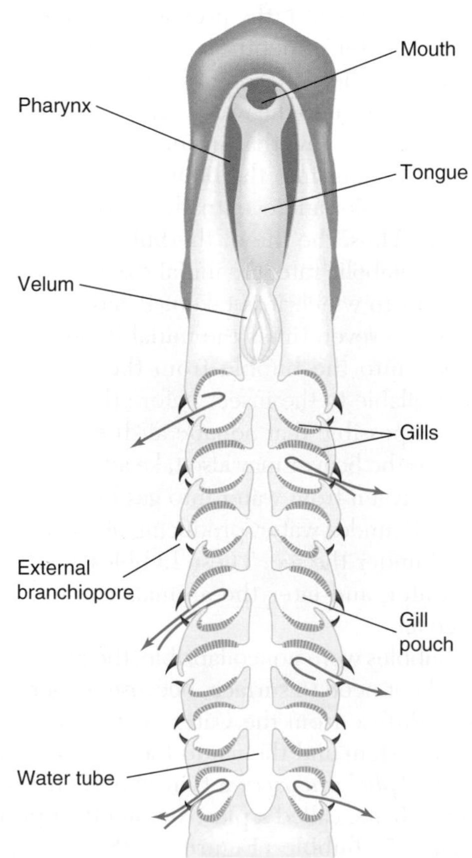

Figure 13-39 The adult lamprey is an exception to the rule that water flow through gills is unidirectional. Shown here is a longitudinal transverse section through the head of an adult lamprey. Arrows mark the direction of water flow. Water moves into and out of each gill pouch via an external branchiopore. The valves of the external branchiopores move in and out with the oscillating water flow.

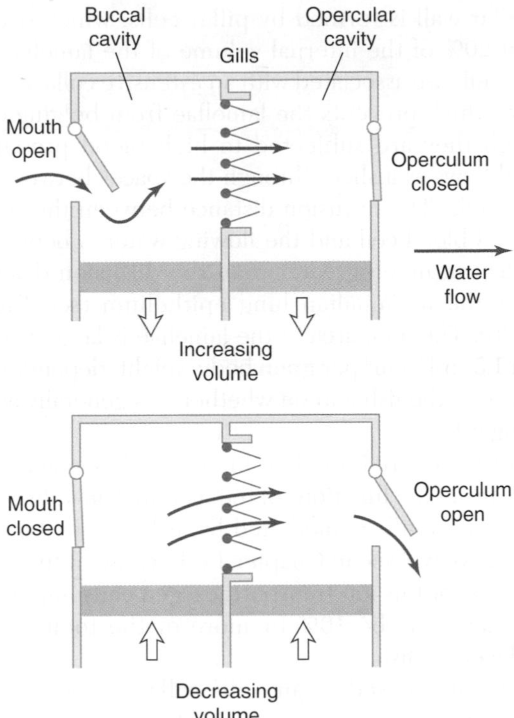

Figure 13-40 Unidirectional flow of water through the gills in teleost fishes is achieved by sequential opening and closing of the mouth and operculum and by a small pressure differential between the buccal and opercular cavities. Large hollow arrows indicate movement of the floor of the mouth.

phase, but a pressure differential is maintained across the gills throughout most of each breathing cycle. Because the pressure in the opercular cavity is slightly below that in the buccal cavity, a unidirectional flow of water is maintained throughout most, if not all, of the breathing cycle.

Many active fishes, such as tuna, "ram-ventilate" their gills, keeping their mouths open while swimming so as to ventilate the gills by the forward motion of the body. The remora, a fish that attaches itself to the body of a shark, relies on the forward motion of its host to ventilate its gills; only when the shark stops swimming does it ventilate its own gills.

Blood flow through fish gills can be described as sheet flow; that is, as pressure increases, the thickness, but not the other dimensions, of the blood sheet increases (Figure 13-41). In this respect, circulation through the gills is similar to the pulmonary circulation. The flow of blood relative to the flow of water in aquatic animals can be concurrent, countercurrent, or some combination of these two arrangements (Figure 13-42).

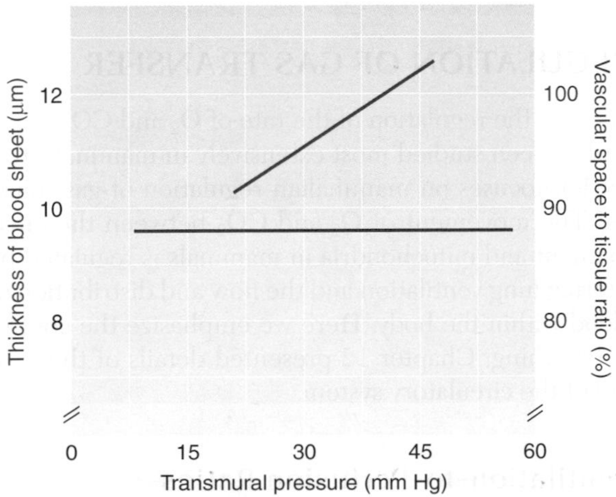

Figure 13-41 In gills, a rise in blood pressure increases the thickness of the blood sheet, but not its height or length. In this plot, based on measurements in gill lamellae of the lingcod *(Ophiodon elongatus)*, the blue line shows the thickness of the blood sheet and the red line, the vascular space to tissue ratio, a measure of the height and length of the blood sheet. [From Farrell et al., 1980.]

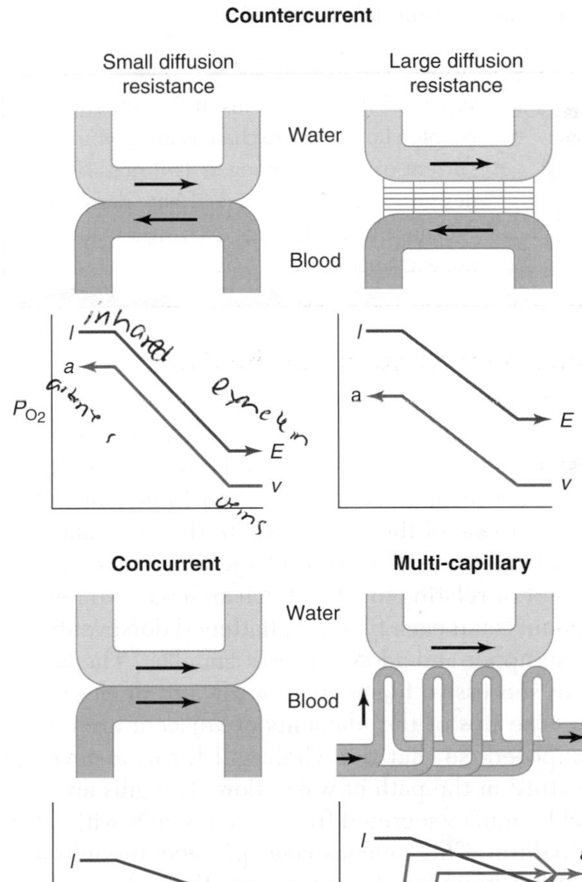

Figure 13-42 Various arrangements of the flows of water and blood at the respiratory surface are found in aquatic animals. Relative changes in P_{O_2} in water and blood are indicated below each diagram.

The advantage of a countercurrent flow of blood and water is that a larger difference in P_{O_2} can be maintained across the exchange surface, allowing more transfer of gases. A countercurrent flow is most advantageous if the values for O_2 content × flow (capacity-rate values) are similar in blood perfusing and water flowing over the gills. If the capacity-rate values for blood and water differ considerably, then countercurrent flow has little advantage over concurrent flow. For example, if water flow were very high in relation to blood flow, there would be little change in P_{O_2} in the water as it flowed over the gills, and the mean P_{O_2} difference across the gills would be similar in concurrent and countercurrent arrangements of flow. Although the flow rate of water across gills is much higher than the flow rate of blood, the O_2 content of fish blood is generally much higher than that of water. Thus the capacity-rate values are similar in blood and water in most fishes, and countercurrent flow is typical.

What are the differences in the design of a countercurrent heat exchanger and a countercurrent oxygen exchanger? Is it possible to design a gas exchanger that does not exchange heat, or a heat exchanger that does not exchange gases?

Functional Anatomy of the Gill

The details of gill structure vary among species, but the general plan is similar (see Figure 14-41). The gills of teleost fishes are taken to be representative of an aquatic respiratory surface. Four branchial arches on either side of the head separate the opercular and buccal cavities. (The term **branchial** refers to anything of or relating to gills.) Each arch has two rows of filaments, and each filament, flattened dorsoventrally, has an upper and a lower row of lamellae. The lamellae of successive filaments in a row are in close contact. The tips of the filaments of adjacent arches are juxtaposed, so that the whole gill forms a sievelike structure in the path of water flow. The gills are covered by mucus secreted from mucous cells within the epithelium. This mucous layer protects the gills and creates a boundary layer between the water and the epithelium.

Water flows through slitlike channels between neighboring lamellae. These channels are about 0.02–0.05 mm wide and about 0.2–1.6 mm long; the lamellae are about 0.1–0.5 mm high. As a result, the water flows in thin sheets between the lamellae, which represent the respiratory portion of the gill, and diffusion distances in water are reduced to a maximum of 0.01–0.025 mm (half the distance between adjacent lamellae on the same filament).

Gill lamellae are covered by thin sheets of epithelial cells, which are joined by tight junctions. The inner lamellar wall is formed by pillar cells, which occupy about 20% of the internal volume of the lamella. The pillar cells are associated with an extensive collagen network, which prevents the lamellae from bulging even though they are subjected to high blood pressures. Blood flows as a sheet through the spaces between the pillar cells. The diffusion distance between the center of a red blood cell and the flowing water is between 3 and 8 μm, much greater than the diffusion distance across the mammalian lung epithelium (see Figure 13-22b). The total area of the lamellae is large, varying from 1.5 to 15 cm^2 per gram body weight, depending on the size of the fish and on whether it is generally active or sluggish.

Fish gills are important in ion regulation and carry out many of the functions of the mammalian kidney. Ion exchange in gills is mediated by at least two types of cells, as we will see in Chapter 14. Because of the metabolic cost of this ion transport, oxygen consumption by gill tissue may be 10% or more of the total oxygen uptake of the fish.

When exposed to air, gills collapse and become nonfunctional, so a fish out of water usually becomes hypoxic, hypercapnic (excessive CO_2), and acidotic. A few fishes and crabs can breathe air, generally using a modified swimbladder, mouth, gut, or branchial cavity for this purpose (see Chapter 12). Air-breathing crabs usually show a decrease in oxygen consumption as well as a decrease in body carbon dioxide levels when they move from air to water breathing. The purple shore crab (*Leptograspus variegatus*), however, shows no change in body oxygen content as it moves between air and water. This crab may be able to regulate body carbon dioxide, and therefore pH levels, by adjusting the ratio of air to water breathing; thus, it is truly amphibious.

REGULATION OF GAS TRANSFER

Because the regulation of the rate of O_2 and CO_2 transfer has been studied most extensively in mammals, this section focuses on mammalian regulation of gas transfer. The movement of O_2 and CO_2 between the environment and mitochondria in mammals is regulated by altering lung ventilation and the flow and distribution of blood within the body. Here we emphasize the control of breathing; Chapter 12 presented details of the control of the circulatory system.

Ventilation-to-Perfusion Ratios

Energy is expended in ventilating the respiratory surface with air or water and in perfusing the respiratory epithelium with blood. The total cost of these two processes is difficult to assess but probably amounts to 4%–10% of the total aerobic energy output of an animal, depending on the species and the physiological state of the animal. Thus, there is significant selective

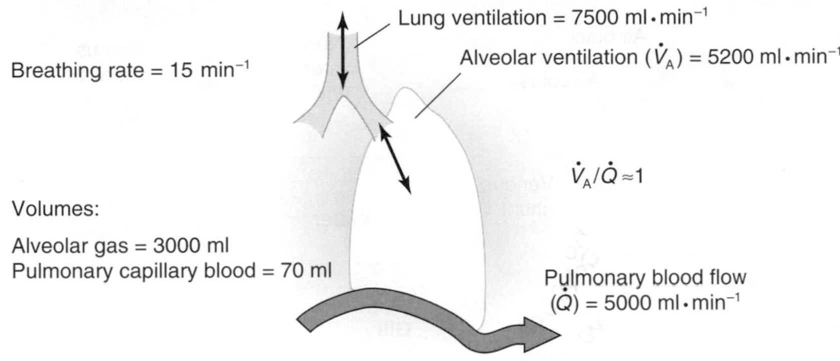

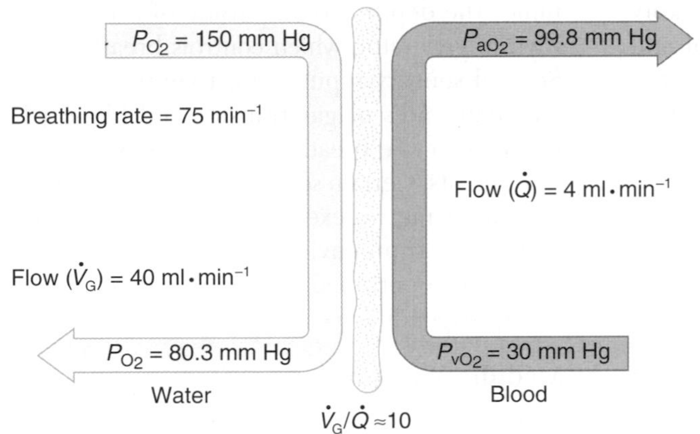

Figure 13-43 The ventilation-to-perfusion ratio in fish gills is much higher than that in the human lung. Approximations of volumes and flows in the human lung and trout gill are shown; actual values may vary considerably.

pressure in favor of the evolution of mechanisms for the close regulation of ventilation and perfusion in order to conserve energy.

The rate of blood perfusion of the respiratory surface is related to the requirements of the tissues for gas transfer and to the gas-transport capacity of the blood. To ensure that sufficient oxygen is delivered to the respiratory surface to saturate the blood with oxygen, the rate of ventilation, $\dot{V}_A$, must be adjusted in accord with the rate of perfusion, $\dot{Q}$, and the gas content of the two media so that the amount of oxygen delivered to the respiratory surface equals that taken away in the blood. The oxygen content of arterial blood in humans is normally similar to that of air. The $\dot{V}_A/\dot{Q}$ ratio is therefore about 1 in humans (Figure 13-43a). Water, however, contains only about one-thirtieth as much dissolved oxygen as an equivalent volume of air at the same P_{O_2} and temperature. Thus, in fishes, the ratio of water flow over the gills, $\dot{V}_G$, to blood flow through the gills, $\dot{Q}$, is between 10:1 and 20:1 (Figure 13-43b), much higher than the $\dot{V}_A/\dot{Q}$ ratio in air-breathing mammals. Based on the difference in the oxygen content of water and air, the $\dot{V}_G/\dot{Q}$ ratio in fishes might be expected to be 30:1; it is lower because the oxygen capacity of the blood of lower vertebrates is often only half that of mammalian blood.

Any changes in the oxygen content of the inhalant medium (air or water) will affect the ventilation-to-perfusion ratio. In order to maintain a given rate of oxygen uptake, a decrease in P_{O_2} of the inhalant medium must be compensated for by an increase in ventilation and hence an increase in the ventilation-to-perfusion ratio. Conversely, an increase in the inhalant P_{O_2} is accompanied by a decrease in ventilation.

The ventilation-to-perfusion ratio must be maintained over each portion of the respiratory surface as well as over the whole surface. The pattern of capillary blood flow can change in both gills and lungs, changing the distribution of blood over the respiratory surface. The distribution of air or water must reflect the blood distribution. Perfusion of an alveolus without ventilation is as pointless as ventilating an alveolus without blood perfusion of that alveolus. Although such extreme situations are unlikely to occur, the maintenance of too high or too low a blood flow or ventilation rate will result in energetically inefficient gas transfer per unit of energy expended. For efficient transfer, the optimal ventilation-to-perfusion ratio should be maintained over the whole respiratory surface. This optimal maintenance does not preclude differential rates of blood perfusion over the respiratory surface, but requires only that the flows of blood and inhalant medium be matched.

The efficiency of gas exchange is diminished if some of the blood entering the lungs or gills either bypasses the respiratory surface or perfuses a portion of

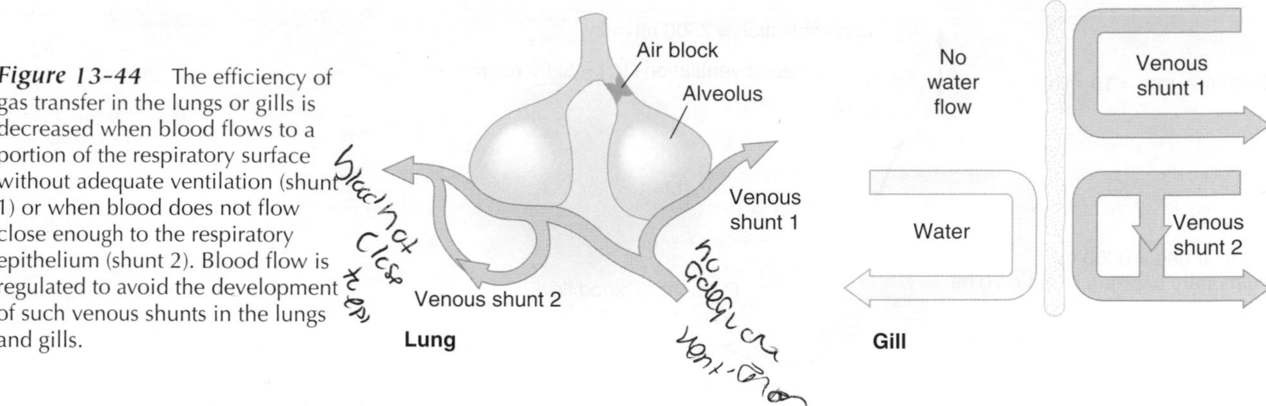

Figure 13-44 The efficiency of gas transfer in the lungs or gills is decreased when blood flows to a portion of the respiratory surface without adequate ventilation (shunt 1) or when blood does not flow close enough to the respiratory epithelium (shunt 2). Blood flow is regulated to avoid the development of such venous shunts in the lungs and gills.

the respiratory surface that is inadequately ventilated (Figure 13-44). The magnitude of such *venous shunts*, expressed as a percentage of total flow to the respiratory epithelium, can be calculated from the arterial and venous O_2 content, assuming an ideal arterial O_2 content. In the mammalian lung, for instance, the partial pressures of gases in blood are almost in equilibrium with alveolar gas pressures. If these pressures and the blood oxygen dissociation curves are known, the expected ideal O_2 content of arterial blood can be determined. Let us assume that this ideal content is 20 ml of O_2 per 100 ml of blood (20 vol %), and that the measured values for arterial and venous blood are 17 and 5 vol %, respectively. This reduction in measured arterial O_2 content from the ideal can be explained in terms of venous shunts in which oxygenated arterial blood (20 vol %) is being mixed with venous blood (5 vol %) in the ratio of 4 : 1, giving a final arterial O_2 content of 17 vol %; that is, 20% of the blood perfusing the lung is passing through one or more venous shunts. This is an extreme example to illustrate a point; in most cases, venous shunts are very small.

Flows of blood and inhalant medium are regulated to maintain a near-optimal ventilation-to-perfusion ratio over the surface of the respiratory epithelium under a variety of conditions. In general terms, $\dot{Q}$ is regulated to meet the requirements of the tissues, and $\dot{V}_A$ and $\dot{V}_G$ are matched to $\dot{Q}$. Mechanisms such as hypoxic vasoconstriction of blood vessels help to maintain optimal ventilation-to-perfusion ratios in different parts of the respiratory surface. As discussed earlier, low alveolar oxygen levels cause vasoconstriction in the lung capillaries, thereby reducing blood flow to poorly ventilated, and therefore hypoxic, regions and increasing blood flow to well-ventilated regions. Blood perfusion of the respiratory surface tends to be less well distributed in resting animals than in active ones. With exercise, blood pressure rises and blood is distributed more evenly, resulting in a more even ventilation-to-perfusion ratio over the respiratory surface.

Neuronal Regulation of Breathing

The integration of breathing movements in all vertebrates results from the processing of many sensory inputs by the central nervous system. The central processor consists of a pattern generator, which determines the depth and amplitude of each breath, and a rhythm generator, which controls breathing frequency. Several sensory inputs adjust ventilation to maintain adequate rates of gas transfer and blood pH. Other inputs integrate breathing movements with other body movements. Certain sensory inputs may cause coughing or swallowing reflexes, which protect the respiratory epithelium from environmental hazards. Other inputs function to optimize breathing patterns to minimize energy expenditure.

Medullary respiratory centers

As noted earlier, the mammalian lung is ventilated by the action of the diaphragm and muscles between the ribs (see Figures 13-28 and 13-30). These muscles are activated by spinal motor neurons, which receive inputs from groups of neurons that constitute the medullary respiratory center. The control of respiratory muscles can be very precise, allowing extremely fine control of air flow, as is required for such complex actions in humans as singing, whistling, and talking, as well as simply breathing.

Microsections of the neonatal rat brain stem indicate that the pre-Botzinger complex, an area in the ventral medulla, is capable of generating the respiratory rhythm and may represent the central rhythm generator that maintains breathing rhythm in the adult. Rhythmic activity is enhanced by neurons in the pons and medulla, and some neurons just anterior to the medulla cause prolonged inspiration in the absence of rhythmic drive from the pons. In 1868, Ewald Hering and Josef Breuer observed that inflation of the lungs decreases the frequency of breathing. This Hering-Breuer reflex can be abolished by cutting the vagus nerve. Inflation of the lungs stimulates pulmonary stretch receptors in the bronchi and bronchioles, which have a reflex inhibitory effect, via the vagus nerve, on the medullary inspiratory center (nucleus tractus solitarius) and therefore on inspiration. Thus the medulla contains a central rhythm generator that drives the pattern generator within the medullary respiratory center to cause breathing movements. This system is modified by inputs from other areas of the brain and from various peripheral receptors.

The medullary respiratory center contains inspiratory neurons, whose activity coincides with inspiration, and expiratory neurons, whose activity coincides with expiration. The respiratory rhythm was once considered to arise from reciprocal inhibition between inspiratory and expiratory neurons, with reexcitation and accommodation occurring within each set of neurons. But several lines of evidence indicate that this model of the central rhythm generator is not tenable, and more recent studies suggest that the respiratory rhythm depends primarily on the activity of inspiratory neurons driven by the pre-Botzinger complex.

Inspiratory neuronal activity is fed into the respiratory motor neurons, including the phrenic nerve, which projects to the diaphragm. Thus phrenic nerve activity is representative of activity in the inspiratory neurons. Inspiratory neuronal activity, recorded from either the phrenic nerve or from individual neurons in the medulla, shows a rapid onset, a gradual rise, and then a sharp cutoff with each burst of activity associated with inspiration. This neuronal activity results in a contraction of the inspiratory muscles and the diaphragm and results in a decrease in intrapulmonary pressure (Figure 13-45a). Increased blood CO_2 levels, acting through chemoreceptors, cause the inspiratory activity to rise more rapidly (Figure 13-45b), resulting in a more powerful inspiratory phase. The cutoff of inspiratory neuronal activity occurs once activity in the neuron has reached a threshold level. Expansion of the lung stimulates pulmonary stretch receptors, whose activity reduces the threshold for the cutoff (Figure 13-45c). Thus the pulmonary stretch receptors, through their action on inspiratory neurons, prevent overexpansion of the lungs.

The interval between breaths is determined by the interval between bursts of inspiratory neuronal activity, which is related to the level of activity in the previous burst and in afferent nerves from pulmonary stretch receptors. In general, the greater the level of inspiratory activity (i.e., the deeper the breath), the longer the pause between inspirations. The result is that the ratio of inspiratory to expiratory duration remains constant in spite of changes in the length of the breathing cycle. This ratio is affected by the level of activity in the pulmonary stretch receptors. If, for example, the lung empties only slowly during expiration, the pulmonary stretch receptors will remain active as long as the lung remains inflated; the continued activity of the stretch receptors will prolong the duration of expiration and increase the time available for expiration.

Expiration is often a largely passive process that does not depend on activity in expiratory neurons. This is especially true during quiet, normal breathing. Expiratory neurons are active only when inspiratory neurons are quiescent, in which case they show a burst pattern somewhat similar to, but out of phase with, that of inspiratory neurons. Inspiratory neurons' activity inhibits expiratory activity, showing the dominance of inspiratory neurons in the generation of rhythmic breathing. In the absence of inspiratory activity, expiratory neurons are continually active. However, inspiratory neurons impose a rhythm on expiratory neurons via periodic inhibition.

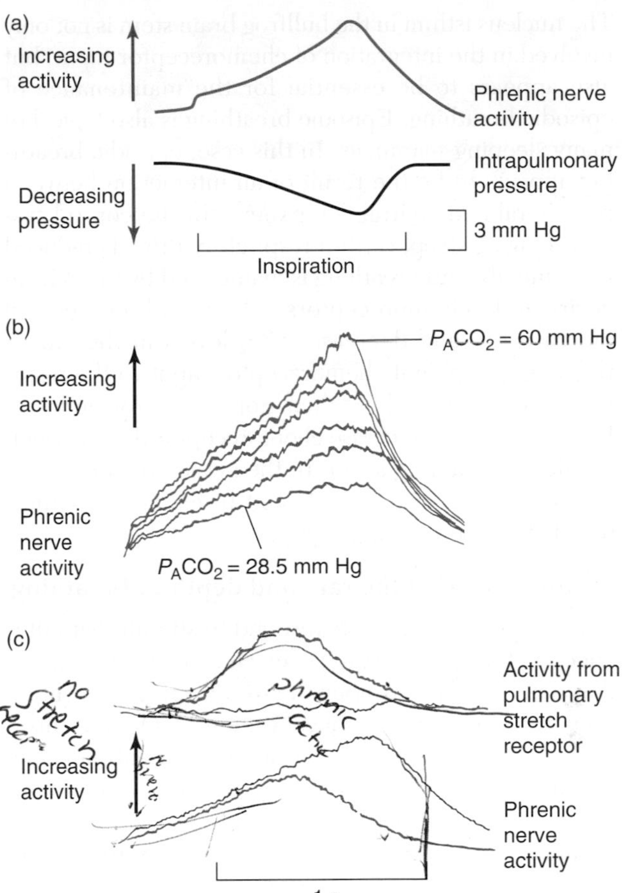

Figure 13-45 Phrenic nerve activity induces inspiration. **(a)** Relationship between phrenic nerve activity and intrapulmonary pressure during inspiration. Note the sudden onset, gradual rise, and rapid cutoff of inspiratory neuronal activity. **(b)** Effect of increasing alveolar P_{CO_2} (P_ACO_2) on activity in the phrenic nerve. The higher the P_ACO_2, the more rapid the rise of phrenic nerve activity during inspiration. **(c)** Effect of increasing activity in pulmonary stretch receptors on activity in the phrenic nerve. In the absence of stretch-receptor activity, the cutoff in phrenic nerve activity is delayed (red traces). An increase in receptor activity results in an earlier cutoff of activity in the phrenic nerve, but does not affect the rate of rise in phrenic activity before the cutoff (blue traces).

Fish, birds, and awake mammals usually breath rhythmically and continuously, whereas amphibians and reptiles often show episodic breathing, with pauses between episodes of rhythmic breathing. Recent studies of the bullfrog brain stem have shown that these episodic patterns of breathing are an intrinsic property of the brain stem, not dependent on sensory feedback.

The nucleus isthmi in the bullfrog brain stem is not only involved in the integration of chemoreceptor input, but also appears to be essential for the maintenance of episodic breathing. Episodic breathing is also typical of many sleeping mammals. In this case, episodic breathing appears to be the result of an interaction between peripheral and central components of the control system. During sleep, central respiratory drive is reduced in mammals, and breathing is maintained by input from peripheral chemoreceptors. A breathing period increases O_2 and decreases CO_2 levels in the blood, reducing peripheral chemoreceptor input to the respiratory center. Breathing then stops until oxygen levels fall sufficiently to increase chemoreceptor drive enough to initiate breathing again. In the awake mammal, central respiratory drive is sufficient to maintain continuous rhythmic breathing.

Factors affecting the rate and depth of breathing

Several types of receptors respond to stimuli that influence ventilation, causing reflex changes in the rate or depth of breathing. Among the stimuli affecting ventilation are changes in O_2 concentrations, CO_2 concentrations, and pH; lung inflation and deflation; lung irritation; emotions; sleep; variations in light and temperature; and the requirements for speech. These influences are integrated by the medullary respiratory centers. Breathing can also, of course, be controlled by conscious volition.

In most, if not all, animals, changes in O_2 and CO_2 levels lead to reflex changes in ventilation. The chemoreceptors involved have been localized in only a few groups of animals. Chemoreceptors in the carotid bodies and aortic bodies of mammals, in the carotid bodies of birds, and in the carotid labyrinth of amphibians monitor changes in O_2 and CO_2 in arterial blood. In teleost fishes, chemoreceptors located in the gills respond to reductions in O_2 levels in the water and the blood. In all cases, the chemoreceptors are innervated by branches of the IXth (glossopharyngeal) or Xth (vagus) cranial nerve. Mammals also have chemoreceptors located in the medulla that respond to decreases in the pH of the cerebrospinal fluid (CSF), usually caused by elevations in P_{CO_2}. Stimulation of this system is required to maintain normal breathing; if body P_{CO_2} falls, or is held at a low level experimentally, breathing ceases. These central chemoreceptors have little ability to respond to falling O_2 levels; the peripheral chemoreceptors have this role and are important in increasing ventilation during periods of hypoxia.

The carotid and aortic bodies of mammals receive a generous blood supply and have a high oxygen uptake per unit weight (Figure 13-46a). These arterial chemoreceptors consist of a number of lobules that surround a very convoluted network of blood vessels. These blood vessels can be divided into small and large capillaries and arteriovenous shunts. The arterioles leading to them are innervated by both sympathetic and

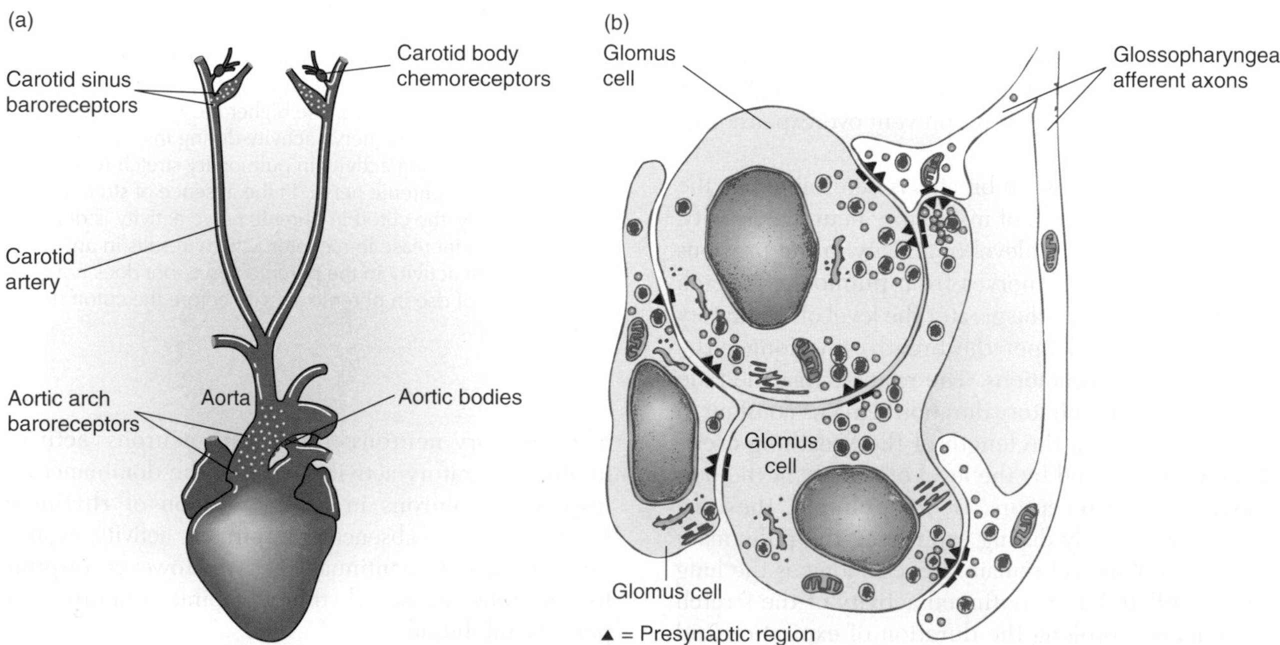

Figure 13-46 In mammals, chemoreceptors in the carotid and aortic bodies monitor blood gas levels and pH. **(a)** Diagram showing the location of carotid and aortic body chemoreceptors (dark areas) and carotid sinus and aortic arch baroreceptors (small yellow dots) in the dog. The baroreceptors help regulate arterial blood pressure (see Chapter 12). **(b)** A small portion of the rat carotid body, which consists of several lobules containing glomus cells. These cells are connected by synapses and innervated by glossopharyngeal afferent fibers. Some regions of afferent nerve endings are presynaptic to the glomus cell, some are postsynaptic, and some form reciprocal synapses. [Part a adapted from Comroe, 1962; part b adapted from McDonald and Mitchell, 1975.]

parasympathetic postganglionic efferents. Each lobule consists of several glomus (type I) cells covered by sustentacular (type II) cells. The glomus cells, thought to be the actual chemoreceptors, are small ovoid cells with a large nucleus and many dense-core vesicles, or granules (Figure 13-46b). These cells are interconnected by synapses and often possess cytoplasmic processes of different lengths. They are innervated by afferent fibers of the glossopharyngeal nerve and possibly by preganglionic sympathetic efferents. A single nerve fiber may innervate 10 or 20 glomus cells. A glomus cell may be either presynaptic or postsynaptic, or both (reciprocal), with respect to a nerve fiber. A single nerve fiber may be postsynaptic (afferent) to one glomus cell and presynaptic (efferent) to a neighboring glomus cell, or even another region of the same glomus cell. Many glomus cells lack innervation but are synaptically connected to other glomus cells in the lobule.

The chemoreceptors in the carotid and aortic bodies are stimulated by decreases in blood O_2 and pH and increases in blood CO_2. It is possible that the observed response to increasing CO_2 is due to changes in pH within these receptors, rather than to changes in CO_2 levels per se. The result of chemoreceptor stimulation is the recruitment of new fibers and an increase the firing rate in the afferent nerves innervating the glomus cells. The chemoreceptors adapt to changing arterial CO_2 levels. The carotid body chemoreceptors show a much greater response to pH or CO_2 changes than the aortic body chemoreceptors. Stimulation of the carotid body chemoreceptors leads to an increase in lung ventilation, mediated via the medullary respiratory center. The actual increase, in response to a given decrease in arterial P_{O_2}, depends on the blood CO_2 level, and vice versa (Figure 13-47). Efferent output to the carotid body modulates the response. Increased sympathetic efferent activity constricts arterioles in the carotid body via an α-adrenergic mechanism, thereby reducing blood flow, which in turn increases the chemoreceptor discharge and lung ventilation. Nonsympathetic efferent activity in the carotid nerve reduces the response of the carotid body. Increases in temperature and osmolarity also stimulate the arterial chemoreceptors, and stimulation of the carotid nerve causes an increase in the release of antidiuretic hormone (ADH). Thus the carotid body chemoreceptors may play a role in osmoregulation as well as in the control of breathing and circulation.

As mentioned previously, mammals, and possibly other air-breathing vertebrates, have central chemoreceptors that are necessary for normal breathing. These H^+-sensitive receptors are located in the region of the medullary respiratory center and are stimulated by a decrease in the pH of the cerebrospinal fluid. The CSF of mammals, and possibly of other vertebrates, is very low in protein and is essentially a solution of NaCl and $NaHCO_3$, with low but closely regulated levels of K^+, Mg^{2+}, and Ca^{2+}. The CSF is also poorly buffered;

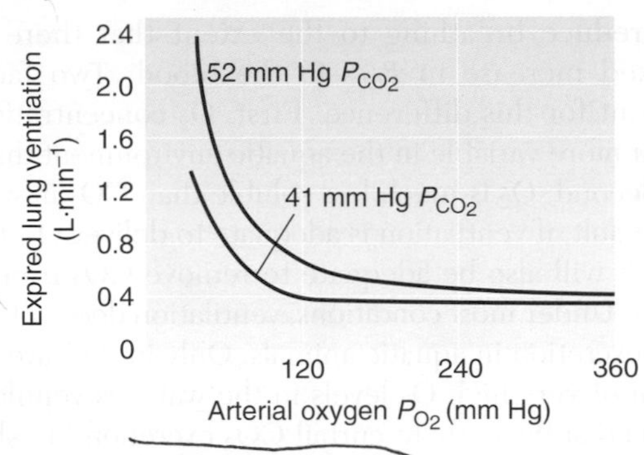

Figure 13-47 Lung ventilation rates increase with a decrease in arterial P_{O_2} and an increase in arterial P_{CO_2}. The relationships shown here are from measurements in the duck. [From Jones and Purves, 1970.]

therefore, small changes in P_{CO_2} have a marked effect on CSF pH. Because the blood-brain barrier is relatively impermeable to H^+, the central H^+-sensitive chemoreceptors are insensitive to changes in blood pH. However, changes in blood P_{CO_2} cause corresponding changes in the P_{CO_2} of the CSF, and these in turn result in changes in the pH of the CSF (Figure 13-48). An increase in blood P_{CO_2} leads to a decrease in the pH of the CSF; the resulting stimulation of the H^+-sensitive receptors causes reflex increases in breathing rate. Prolonged changes in P_{CO_2} are associated with an adjustment of pH by changes in HCO_3^- levels of the CSF.

In mammals and other air-breathing vertebrates, CO_2 levels, rather than O_2 levels, dominate in the control of breathing; in aquatic vertebrates, however, O_2 is the major factor. In fact, fishes exposed to high O_2 levels

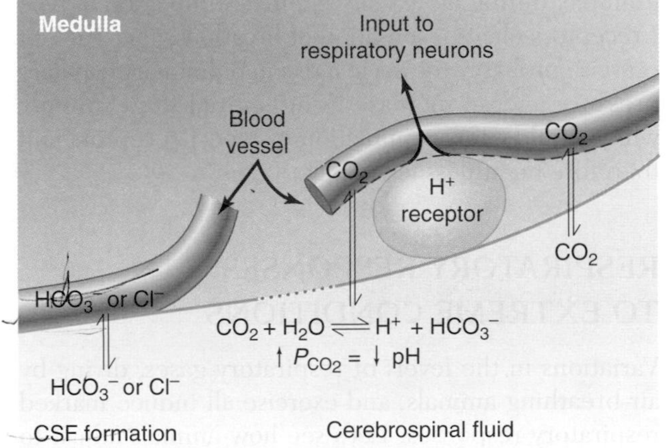

Figure 13-48 Central H^+-sensitive receptors are influenced by the pH of the cerebrospinal fluid (CSF) and by arterial P_{CO_2}. Carbon dioxide molecules diffuse readily across the walls of the brain capillaries and alter CSF pH, but the blood-brain barrier blocks other molecules. An increase in P_{CO_2} causes a decrease in the CSF pH; the resulting stimulation of H^+ receptors reflexly increases breathing. Across some capillary walls, exchange of HCO_3^- and Cl^- helps to maintain the pH of the CSF at a constant level in the face of a prolonged change in P_{CO_2}.

will reduce breathing to the extent that there is a marked increase in P_{CO_2} in the blood. Two factors account for this difference. First, O_2 concentration is much more variable in the aquatic environment than in air. Second, O_2 is much less soluble than CO_2 in water; as a result, if ventilation is adequate to deliver O_2 to the gills, it will also be adequate to remove CO_2 from the blood. Under most conditions, ventilation does not limit CO_2 excretion in aquatic animals. Only in the rare condition of very high O_2 levels in the water is ventilation reduced sufficiently to curtail CO_2 excretion in fish. In general, CO_2 levels in the blood are low in animals breathing water, and ventilation is not used to adjust blood CO_2 or pH.

The lungs contain several types of receptors that help regulate inflation and prevent irritation of the respiratory surface. We saw earlier that stimulation of pulmonary stretch receptors prevents overinflation of the lung (the Hering-Breuer reflex). In mammals, increased CO_2 levels act on these stretch receptors to reduce their inhibitory effects on the medullary respiratory center, thereby increasing the depth of breathing and lung ventilation.

In addition to pulmonary stretch receptors, a variety of irritant receptors are present in the lung. Stimulation of these receptors by mucus and dust or other irritant particles causes reflex bronchioconstriction and coughing. A third group of receptors in the lung are positioned close to the pulmonary capillaries in interstitial spaces; these are called *juxtapulmonary capillary receptors*, or *type J receptors*. These receptors were previously termed "deflation receptors," but their natural stimulus appears to be not lung deflation but an increase in the volume of interstitial fluid, as seen, for example, during pulmonary edema. Stimulation of type J receptors elicits a sensation of breathlessness. Violent exercise probably results in a rise in pulmonary capillary pressure and an increase in interstitial fluid volume, which could cause stimulation of type J receptors and therefore breathlessness.

RESPIRATORY RESPONSES TO EXTREME CONDITIONS

Variations in the levels of respiratory gases, diving by air-breathing animals, and exercise all induce marked respiratory responses. Let's see how animals adjust to these extreme conditions.

Reduced Oxygen Levels (Hypoxia)

Aquatic animals are subjected to more frequent and rapid changes in environmental oxygen levels than are air-breathing animals. Mixing and diffusion are both more rapid in air than in water, so regions of local hypoxia develop more often in aquatic environments. Although photosynthesis can result in very high oxygen levels during the day in some aquatic environments, oxygen consumption by both biological and chemical processes can produce localized hypoxic regions, especially at night. The changes in oxygen levels in water may or may not be accompanied by changes in carbon dioxide levels.

Many aquatic animals can withstand very long periods of hypoxia. Some fishes (e.g., carp) overwinter in the bottom mud of lakes, where the P_{O_2} is very low. Many invertebrates bury themselves in mud with low P_{O_2} but high nutrient content. Some parasites live in hypoxic regions, such as the gut of their hosts, during one or more phases of their life cycle. Limpets and bivalve mollusks close their shells during exposure at low tide to avoid desiccation, and are subject as a consequence to a period of hypoxia. Fish exposed to hypoxia first stop feeding and inhibit protein metabolism; as a result, their growth is impaired. As the oxygen level continues to fall, fish stop breeding. Thus, prolonged, severe hypoxia can decimate populations. Some fish swim less to reduce energy expenditure. During mild hypoxia, however, some aquatic animals become more active as they attempt to escape the hypoxic region. Many fish, when faced with hypoxia, move to cooler water; the resulting reduction in body temperature reduces energy metabolism. (Rats also reduce body temperature when exposed to hypoxia in air, presumably for the same reason.)

In general, animals reduce energy expenditures and utilize a variety of anaerobic metabolic pathways to survive periods of reduced oxygen availability. Animals also adjust their respiratory and cardiovascular systems to maintain oxygen delivery in the face of reduced oxygen availability. For instance, aquatic hypoxia causes an increase in gill ventilation in many fishes, as a result of stimulation of chemoreceptors on the gills. The increase in water flow offsets the reduction in oxygen content of the gill water and maintains delivery of oxygen to the fish. In fishes that ram-ventilate their gills by swimming forward with the mouth open, the size of the mouth gap increases with hypoxia to increase water flow over the gills.

Compared with aquatic environments, oxygen and carbon dioxide levels are relatively stable in air, and local regions of low oxygen or high carbon dioxide are rare and easily avoided. There is, of course, a gradual reduction in P_{O_2} with altitude, and animals vary in their capacity to climb to high altitudes and withstand the accompanying reduction in ambient oxygen levels. High altitudes are associated with low temperatures, as well as low pressures, and this also has a marked effect on animal distribution. The highest permanent human habitation is at about 5800 m, where the P_{O_2} is 80 mm Hg (compared with about 155 mm Hg at sea level). Many birds migrate over long distances at altitudes above 6000 m, where atmospheric pressures would cause severe respiratory distress in many mammals. It is not clear what confers this capacity on birds, although it is probably related to the differences in the mechanism of lung gas transfer between birds and mammals.

Balloonists in the nineteenth century, ascending to 9000 meters, reported that they had no sense of danger despite a numbness of mind and body and an inability to move. Some eventually went blind before passing out, and several died in the process. Many humans experience mountain sickness at high altitudes. Acute exposure is associated with fatigue, headaches, irritability, dizziness, nausea, loss of appetite, and insomnia. Chronic exposure can result in loss of memory and confusion as well as insomnia. It has been suggested that some of these problems may be related to the increased hematocrit associated with prolonged exposure to hypoxia, which may reduce cerebral blood flow, exacerbate cerebral hypoxia, and cause memory loss. Anemic individuals are reported to be less likely to suffer from mountain sickness. Interestingly, mountain sickness is common in Andean populations but rare among Tibetans. Thus Tibetans appear more adapted to high altitude than Andean peoples. It is not clear whether this pattern is due to lifestyle or genetic differences.

In mammals, a reduction in the P_{O_2} of ambient air results in a decrease in blood P_{O_2}, which in turn stimulates the carotid and aortic body chemoreceptors, causing an increase in lung ventilation. The rise in lung ventilation then leads to an increase in CO_2 elimination and a decrease in blood P_{CO_2}, which causes a reduction in the P_{CO_2} of the CSF and therefore an increase in its pH. Decreases in blood P_{CO_2} and increases in CSF pH tend to *reduce* ventilation, thereby attenuating the hypoxia-induced *increase* in lung ventilation. If hypoxic conditions are maintained, however, as occurs when animals move to high altitudes, both blood and CSF pH are returned to normal levels by the excretion of bicarbonate. This process takes a few days to about a week in humans. As CSF pH returns to normal, the reflex effects of hypoxia on ventilation predominate, resulting in a gradual increase in ventilation as the animal acclimatizes to altitude. The response to prolonged hypoxia may also involve modulation of the effects of CO_2 on the carotid and aortic bodies to reset these chemoreceptors to the new lower CO_2 level at high altitude. The hyperventilation associated with movement to altitude lowers blood CO_2 levels and raises blood pH. These changes increase the oxygen affinity of hemoglobin, which in turn enhances oxygen uptake in the lungs but decreases oxygen delivery to the tissues. In humans, movement to high altitude results in an increase in erythrocytic DPG levels (see page 533) after a day or so, which tends to reduce hemoglobin oxygen affinity, offsetting the effects of high blood pH.

Hypoxia resulting from travel to high altitude also results in systemic vasodilation and an increase in cardiac output. The higher cardiac output lasts only a few days, and cardiac output returns to normal or drops below normal as O_2 supplies to tissues are restored by the compensatory increases in ventilation and blood hemoglobin levels. During chronic exposure to hypoxia, blood volume increases by about one-third, so although cardiac output is normal, circulation time is reduced because of the increase in volume of the system.

As mentioned earlier, low oxygen levels cause local vasoconstriction in the pulmonary capillaries in mammals, producing a rise in pulmonary arterial blood pressure. This response normally has some importance in redistributing blood away from poorly ventilated, and therefore hypoxic, portions of the lung. When animals are subjected to a general hypoxic environment, however, an increase in the resistance to flow through the whole lung can have detrimental effects. Some mammals that live at high altitudes exhibit a reduced local pulmonary vasoconstriction in response to hypoxia; this response is probably a genetically determined acclimation. Humans residing at high altitudes are usually small and barrel-chested and have large lung volumes per unit weight. Lung development is oxygen-insensitive, but the growth of limbs is reduced under hypoxic conditions—humans living at altitude tend to have small bodies but normal-sized lungs. The high lung-to-body ratio enables these people to maintain oxygen uptake under hypoxic conditions. Their pulmonary blood pressures are high, and there is often hypertrophy of the right ventricle of the heart. High pulmonary pressures produce more even distribution of blood in the lung, and so augment the diffusion capacity for oxygen.

Hypoxic tissues produce adenosine and hypoxia-inducing factor 1 (HIF-1). As discussed in Chapter 12, HIF-1 causes a general reduction in gene transcription as well as the specific up-regulation of the production of erythropoietin, vascular endothelial growth factors, endothelin-1, nitric oxide synthase, and glycolytic enzymes. HIF-1 also plays a role in development and can be regarded as a master regulator of cellular and developmental oxygen homeostasis.

Under hypoxic conditions, increased intracellular adenosine levels result from increased metabolism of ATP/ADP/AMP and S-adenosylhomocysteine to adenosine. Adenosine is transported into the extracellular space, where it can reach concentrations of 0.1–0.3 μM, but persists for only a few seconds. Extracellular adenosine is rapidly taken up by cells or converted to inosine. Thus adenosine can act only locally. In mammals, and probably other vertebrates, adenosine plays a role of matching energy supply to energy demand (see Chapter 12).

Long-term adaptations occur during prolonged exposure to hypoxia. Most vertebrates respond by increasing the number of red blood cells and the blood hemoglobin content—and therefore the oxygen capacity of the blood. A reduction in blood oxygen levels stimulates production of the hormone erythropoietin in the kidney and liver. Erythropoietin acts on the bone marrow to increase production of red blood cells (erythropoiesis). Exposure to hypoxia also increases the production of vascular endothelial growth factors, which cause a proliferation of capillaries, ensuring a more adequate oxygen delivery to the tissues. The gills of fishes and amphibians are larger in species exposed to prolonged periods of hypoxia. A similar enlargement of the respiratory surface in response to hypoxia does not occur in mammals.

What effect would blocking the carbonic anhydrase activity in erythrocytes have on the ventilatory responses of humans observed at altitude?

Increased Carbon Dioxide Levels (Hypercapnia)

In many animals, an increase in blood P_{CO_2} results in an increase in ventilation. In mammals, the increase is proportional to the rise in the CO_2 level in the blood. The effect is mediated by modulation of the activity of several receptors that send messages to the medullary respiratory center. These receptors include the chemoreceptors of the aortic and carotid bodies and the mechanoreceptors in the lungs, but the response is dominated by the central H^+ receptors (see Figure 13-48). Correction of CSF pH, in the face of altered P_{CO_2}, is very important in the return of ventilation to normal.

A marked increase in ventilation occurs almost immediately in response to a rise in CO_2. The increase is maintained for long periods in the presence of increased CO_2, but ventilation eventually returns to a level slightly above the volume that prevailed before hypercapnia. This adjustment is related to increases in levels of plasma bicarbonate and CSF bicarbonate, with the result that pH returns to normal even though the raised CO_2 levels are maintained.

Diving by Air-Breathing Animals

Many air-breathing vertebrates live in water and dive for varying periods of time. Dolphins and whales rise to the surface to breathe but spend most of their lives submerged. The time between breaths varies with the diver but is around 10–20 minutes for many diving vertebrates (Table 13-1). The elephant seal dives regularly to depths of 400 m, subjecting itself to a pressure of over 40 atm at the bottom of the dive. The maximum depth recorded, achieved by a male northern elephant seal, was 1581 meters. The pressures at these depths would crush the thoracic cage of a human. The maximum dive duration recorded for elephant seals is about two hours. There are reports of sperm whales diving to nearly 2000 m and staying submerged for over an hour. Most dives are much shorter and shallower than these maximums.

Diving mammals and birds are, of course, subjected to periods of hypoxia during submergence. The mammalian central nervous system cannot withstand anoxia and must be supplied with oxygen throughout the dive. Diving animals solve this problem by utilizing oxygen stores in the lungs, blood, and tissues (Figure 13-49). Many diving animals have high hemoglobin and myoglobin levels, and their total oxygen stores generally are larger than those in nondiving animals. To minimize depletion of available stores, oxygen is preferentially delivered to the brain and the heart during a dive; blood flow to other organs may be reduced, and these tissues may adopt anaerobic metabolic pathways. There is a marked slowing of heart rate (bradycardia) and a reduction of cardiac output during a prolonged dive or if an animal is forcibly submerged in an experimental setting

Table 13-1 Total oxygen stores, mean dive time, and mean dive depth in diving vertebrates

Species	O_2 stores ($ml \cdot kg^{-1}$)	Mean dive time (minutes)	Mean depth of dive (meters)
Leatherback turtle	20	11	60
Emperor penguin	62	6	100
Weddell seal	87	15	100
Northern elephant seal	97	20	400
Human*	20	2	Shallow

*Leatherback turtles have oxygen stores similar to those of humans, but can dive for much longer times because they use oxygen at a lower rate.
Source: Kooyman et al., 1999.

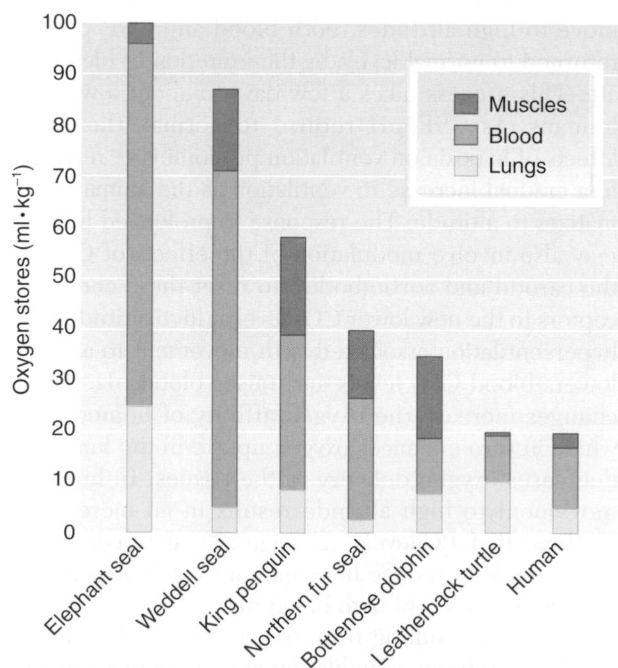

Figure 13-49 Air-breathing animals draw on oxygen stores in the lungs, blood, and tissues (especially muscles) when submerged. This graph compares the generalized total oxygen stores (expressed in ml $O_2 \cdot kg^{-1}$ body weight) of the major taxa of marine divers with those of humans. [Adapted from Kooyman, 1989 and Kooyman et al., 1999.]

(see Figure 12-48). Air-breathing animals that spend prolonged periods submerged must have sufficient oxygen stores to sustain aerobic metabolism, because they cannot tolerate the large amount of lactic acid that accumulates during anaerobic metabolism. During prolonged dives by these animals, metabolic rates, and thus oxygen needs, are often reduced.

Some diving animals, such as the Weddell seal, exhale before diving, thus reducing the oxygen store in their lungs. The increase in hydrostatic pressure during a deep dive results in lung compression. In those animals that reduce lung volume before a dive, air is forced out of the alveoli as the lungs collapse and is contained within the trachea and bronchi, which are more rigid but less permeable to gases. If gases remained in the alveoli, they would diffuse into the blood as pressure increased. At the end of the dive, the partial pressure of nitrogen in the blood would be high, and a rapid ascent could result in the formation of bubbles in the blood, the equivalent of decompression sickness in humans, also known as "the bends." Exhalation before diving reduces the likelihood of the bends. Since only about 7% of a Weddell seal's total oxygen stores are in the lungs, pre-dive exhalation appears to be a reasonable trade-off. Exhalation also reduces the buoyancy of the animal and decreases the energy required to swim to depth.

Receptors that detect the presence of water and that inhibit inspiration during a dive are situated near the glottis and near the mouth or nose (depending on the species). Thus, the decrease in blood O_2 levels and increase in CO_2 levels that occur during a dive do not stimulate ventilation because inputs from the chemoreceptors of the carotid and aortic bodies are ignored by the respiratory neurons while the animal is submerged. Increased chemoreceptor activity during a dive, however, generates bradycardia (see Chapter 12).

During birth, a mammal emerges from an aqueous environment into air and survives a short period of anoxia between the time the placental circulation stops and the time air is first inhaled. The respiratory and circulatory responses of the fetus during this period are similar in several respects to those of a diving mammal.

Exercise

Exercise increases O_2 utilization, CO_2 production, and metabolic acid production. Cardiac output increases to meet the higher demands of the tissues. Even though this increase reduces the transit time for blood through the lung capillaries, oxygen levels in blood leaving the lung remain in equilibrium with those in alveolar gas (Figure 13-50) because ventilation volume increases. The increase in ventilation in mammals is rapid, coinciding with the onset of exercise. This initial sudden increase in ventilation volume is followed by a more gradual rise until a steady state is obtained both for ventilation volume and oxygen uptake (Figure 13-51). When exercise is terminated, there is a sudden decrease

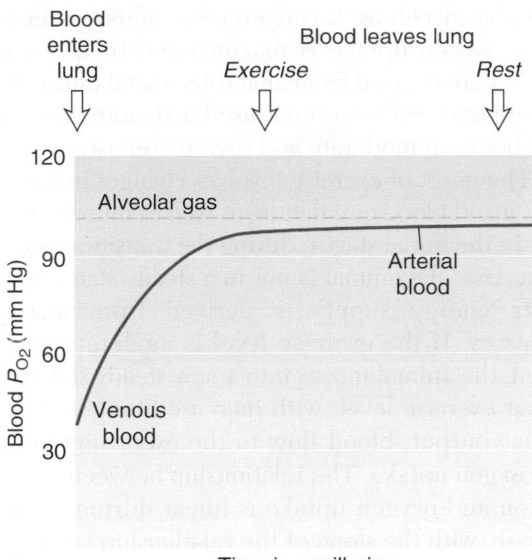

Figure 13-50 Blood P_{O_2} rapidly approaches equilibration with alveolar P_{O_2} even during vigorous exercise. Although blood flow increases during exercise, and blood therefore spends less time in the lung capillaries, the large diffusing capacity of the lung is sufficient to allow near-equilibration to occur. [Adapted from West, 1970.]

in ventilation rate, followed by a gradual decline in ventilation volume. During exercise, O_2 levels are reduced and CO_2 and H^+ levels are raised in venous blood, but the mean P_{O_2} and P_{CO_2} in arterial blood do not vary markedly, except during severe exercise. The oscillations in arterial blood P_{O_2} and P_{CO_2} associated with each breath increase in magnitude, although the mean level is unaltered.

Exercise covers a range from slow movements up to maximum exercise capacity. The phrase *moderate exercise* refers to exercise above resting levels that is aerobic, with only minor energy supplies derived from

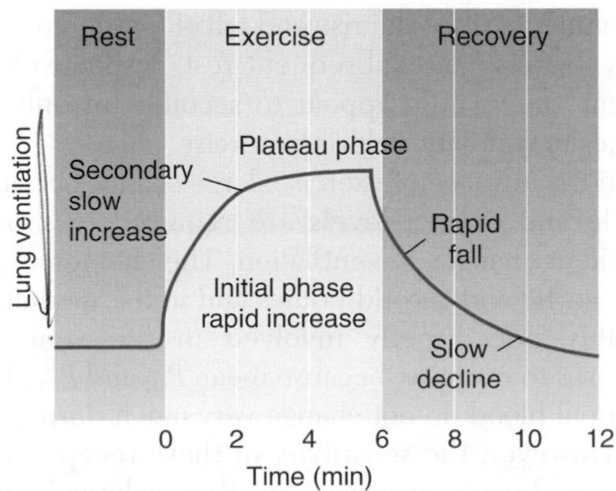

Figure 13-51 An increase in lung ventilation is one of several adjustments that occur in order to meet increased oxygen demand during exercise. Typical changes in lung ventilation during exercise and recovery in humans are depicted in this graph.

anaerobic glycolysis. *Severe exercise* refers to exercise in which oxygen uptake is maximal and further energy supplies are derived from anaerobic metabolism. *Heavy exercise* is a term sometimes used to denote the exercise level between moderate and severe exercise.

The onset of exercise involves changes in lung ventilation and blood circulation, as well as muscle contraction. In the initial stages, during the transition from rest to exercise, the animal is not in a steady state, and part of its energy supply is derived from anaerobic processes. If the exercise level is moderate and sustained, the animal moves into a new steady state typical of that exercise level, with increased lung ventilation, cardiac output, blood flow to the exercising muscles, and oxygen uptake. The relationship between lung ventilation and oxygen uptake is linear during moderate exercise, with the slope of the relationship varying with the type of exercise.

Several receptor systems appear to be involved in the respiratory responses to exercise. Muscle contractions stimulate stretch, acceleration, and position mechanoreceptors in muscles, joints, and tendons. Activity in these receptors reflexly stimulates ventilation, and this system probably causes the sudden changes in ventilation that occur at the beginning and end of a period of exercise. The increase in ventilation varies with the group of muscles being stimulated. Leg exercise, for example, results in a larger increase in ventilation than arm exercise; the same is true for bicycle exercise versus exercise on a treadmill. It has also been suggested that changes in neuronal activity in the brain and spinal cord leading to muscle contraction affect the medullary respiratory center, causing an increase in ventilation.

Muscle contraction generates heat and raises body temperature, thereby increasing ventilation via action on temperature receptors in the hypothalamus. The exact ventilatory response elicited by stimulation of the hypothalamus depends on the ambient temperature; the increase in ventilation is more pronounced in a hot environment. Since the rise and fall of temperature that follow exercise and subsequent rest, respectively, are gradual, they would appear to account for only slow changes in ventilation during exercise.

In the absence of exercise, large changes in carbon dioxide and oxygen levels are required to produce equivalent changes in ventilation. The chemoreceptors in the aortic and carotid bodies and in the medulla are probably not directly involved in the ventilatory responses to exercise, because mean P_{O_2} and P_{CO_2} levels in arterial blood do not change very much during exercise. However, the sensitivity of these receptors may increase during exercise, so that relatively small changes in gas partial pressures can cause an increase in ventilation. In this regard, it is significant that catecholamines, which are released in increased quantities during exercise, increase the sensitivity of medullary receptors to changes in carbon dioxide.

Threshold levels of carbon dioxide are required to drive ventilation during exercise, as under resting conditions. Exercising sheep connected to an external artificial lung that maintains low P_{CO_2} and high P_{O_2} in their blood do not breathe. Ventilation in the intact mammal increases in proportion to the CO_2 delivery to the lung, but the location of any receptors involved is not known. Chemical changes in exercising muscle may also play a role in reflexly stimulating ventilation via muscle afferent fibers.

Ventilation increases more during severe exercise than during moderate exercise, and the relationship between ventilation and oxygen uptake during severe exercise is no longer linear, but becomes exponential. This large increase in ventilation is probably driven by the same mechanisms as in moderate exercise, with the added stimulation of a marked metabolic acidosis and high circulating catecholamine levels.

SYMMORPHOSIS

Charles Richard (Dick) Taylor and Ewald R. Weibel (1981) defined **symmorphosis** as "the state of structural design commensurate to the functional needs resulting from regulated morphogenesis, whereby the formation of structural elements is regulated to satisfy but not to exceed the requirements of the functional system." In other words, there is an economy of structural design such that functional capacity does not exceed maximum expected loads (requirements) on the system. This idea has been applied to gas transfer, which involves a series of linked structures. It was found that structural parameters in mammals are adjusted to functional capacity for oxygen transfer, except for the lung, in a wide variety of animals of different sizes and exercise capacities.

In thoroughbred racehorses bred specifically for speed and high aerobic capacity, it was found that maximum oxygen transfer is limited by pulmonary diffusion capacity and lung ventilation, as well as circulatory convection and peripheral diffusion. In other words, all elements of the oxygen-transport system are limiting in racehorses, an observation consistent with symmorphosis. Racehorses, however, have been bred solely for running speed. More complex selective pressures operate on most animals, and design constraints are likely to be much more complex. Factors other than maximum gas-transfer rate would be expected to come into play, such as size of oxygen stores and rate of change in transfer rates. In addition, the body parts involved in gas transfer often participate in several functional systems. Blood flow, for example, must meet the requirements not only for oxygen delivery, but also for carbon dioxide removal, urine production, and many other functions. Thus, physiological systems are constrained in both design and capacity by the requirements of other linked systems in time and space, and cannot be viewed in isolation. Given that the level of integration varies, the extent and nature of such constraints need to be assessed when analyzing the optimization of animal design.

Alternatively, symmorphosis can be used as a means to estimate the degree of linkage between systems. If the capacity of a component part is matched perfectly to the requirements of one specific system, then there is probably little linkage between that component and other functional systems. Conversely, if the component part is not adjusted to the overall requirements of a particular system, then there is probably extensive linkage of that component to other systems. Thus, the concept of symmorphosis may be used as a tool in analyzing both the rate-limiting steps in a system and the extent of linkage between several systems.

SWIMBLADDERS: OXYGEN ACCUMULATION AGAINST LARGE GRADIENTS

Fish are denser than the surrounding water and must generate upward hydrodynamic forces if they are to maintain their position in the water column and not sink to the bottom. They can generate lift by swimming and using their fins and body as hydrofoils. For skipjack tuna, the minimum speed below which sufficient lift cannot be generated is about $0.6\ m \cdot s^{-1}$; thus, these fish must swim continually to maintain their position in the water column. Other fish hover like helicopters, using their pectoral fins to maintain their position. In both cases there is an energetic cost to maintaining position—an expenditure that can be reduced by incorporation of a buoyancy device.

Many aquatic animals reduce the expense of generating lift by maintaining neutral buoyancy, compensating for a dense skeletal structure by incorporating lighter materials into specialized organs. These "buoyancy tanks" may be ammonium chloride solutions (in squids), lipid layers (in many animals, including sharks), or air-filled swimbladders (in many fishes). Ammonium chloride and lipid floats have the advantage of being essentially incompressible in the face of the changing hydrostatic pressures accompanying vertical movement in water, but they are not much lighter than the other body tissues and so must be large if the animal is to achieve neutral buoyancy. Swimbladders are less dense and can be much smaller than NH_4Cl and lipid floats, but they are compressible—their volume, and thus the buoyancy of the animal, changes with depth.

Hydrostatic pressure increases by approximately 1 atm for every 10 m of depth in water. If a fish is swimming just below the surface and suddenly dives to a depth of 10 m, the total pressure in its swimbladder doubles from 1 to 2 atm, and the swimbladder volume is reduced by one-half, increasing the density of the fish. The fish should now continue to sink because it is more dense than water. Similarly, if the fish rises to a shallower depth, its swimbladder expands, decreasing the fish's density, and the fish should continue to rise. The low density of swimbladders is an advantageous feature,

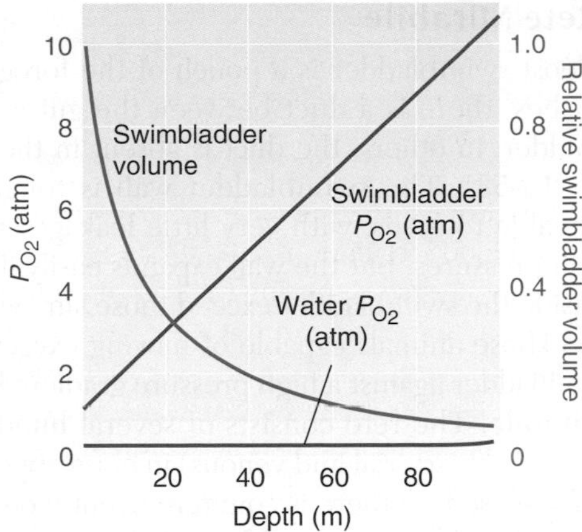

Figure 13-52 Swimbladder volume decreases and swimbladder P_{O_2} increases as a fish descends. Hydrostatic pressure increases by approximately 1 atm for every 10 m of depth. In this example, oxygen is assumed to be the only gas present and is neither added to nor removed from the swimbladder. Fishes can maintain constant density only by maintaining constant swimbladder volume, which is achieved by addition of oxygen to the swimbladder with increasing depth. Note the increasing P_{O_2} difference between water and swimbladder with depth. Oxygen must be moved from water into the swimbladder against this increasing P_{O_2} gradient.

but they are essentially unstable because of the volume changes they undergo with changes in depth. One means of preventing these volume changes is for gas to be added or removed as the fish descends or ascends, respectively. Many fishes have mechanisms for increasing or decreasing the amount of gas in the swimbladder in order to maintain a constant volume over a wide range of pressures.

Fishes with swimbladders spend most of their time in the upper 200 m of lakes, seas, and oceans. The pressure in the swimbladder ranges from 1 atm at the surface to about 21 atm at 200 m. Gases dissolved in water are generally in equilibrium with air, and neither the gas partial pressure nor the gas content in water varies with depth, because water is virtually incompressible (Figure 13-52). The swimbladder gas in most fishes consists of O_2, but in some species the swimbladder is filled with CO_2 or N_2. If the fish dives to a depth of 100 m, O_2 must be added to the swimbladder to maintain buoyancy. The aquatic environment is the source of this O_2, which is moved from the surrounding water into the swimbladder against a pressure difference—in this example, a difference of nearly 11 atm (water $P_{O_2} = 0.228$ atm; swimbladder $P_{O_2} = 11$ atm). To understand how this transfer occurs, let us look at the structure of the swimbladder.

What problems arise from using a swimbladder for buoyancy at great depths? Would you expect to find swimbladders in deep-sea fishes?

The Rete Mirabile

The teleost swimbladder is a pouch of the foregut. In some fishes, there is a duct between the gut and the swimbladder; in others, the duct is absent in the adult (Figure 13-53). The swimbladder wall is tough and impermeable to gases, with very little leakage even at very high pressures, but the wall expands easily if pressures inside the swimbladder exceed those surrounding the fish. Those animals capable of moving oxygen into the swimbladder against a high pressure gradient have a **rete mirabile.** The rete consists of several bundles of capillaries (both arterial and venous) in close apposition and arranged so that there is countercurrent blood flow between arterial and venous blood. It has been calculated that eel retia have 88,000 venous capillaries and 116,000 arterial capillaries containing about 0.4 ml of blood. The surface area of contact between the venous and arterial capillaries is about 100 cm^2. Blood passes first through the arterial capillaries of the rete, then through a secretory epithelium (gas gland) in the swimbladder wall, and finally back through the venous capillaries in the rete. The arterial blood and the venous blood in the rete are separated by a distance of about 1.5 μm.

The rete structure allows blood to flow into the swimbladder wall without a concomitant large loss of gas from the swimbladder. Blood leaves the gas gland at high P_{O_2} and passes into the venous capillaries. The P_{O_2} in both arterial and venous capillaries decreases with distance from the gas gland. The P_{O_2} difference between arterial and venous blood at the end of the rete distal to the swimbladder is small compared with the P_{O_2} difference between the environment and the swimbladder, reducing the loss of oxygen from the swimbladder. It was once thought that the reason oxygen levels dropped in the rete was because of the diffusion of oxygen from venous to arterial capillaries, with the rete acting as a countercurrent exchanger (see Spotlight 14-2). However, H. Kobayashi, B. Pelster, and P. Scheid (1993) were unable to detect any significant transfer of oxygen across the rete. The P_{O_2} does fall in the blood flowing away from the gas gland, but this happens because oxygen binds to hemoglobin, not because of any loss of oxygen to arterial blood entering the rete. We will see exactly how and why it happens below.

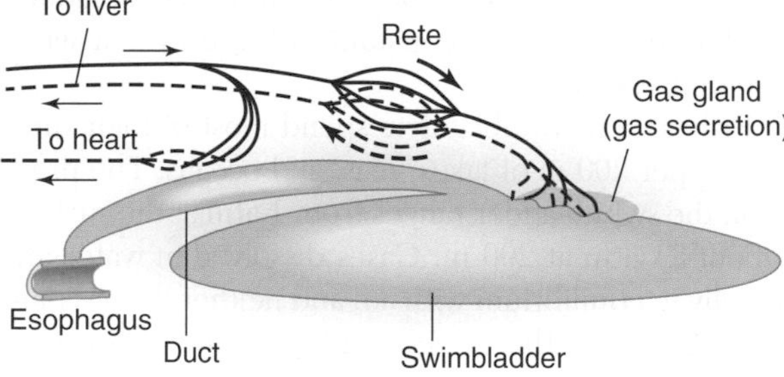

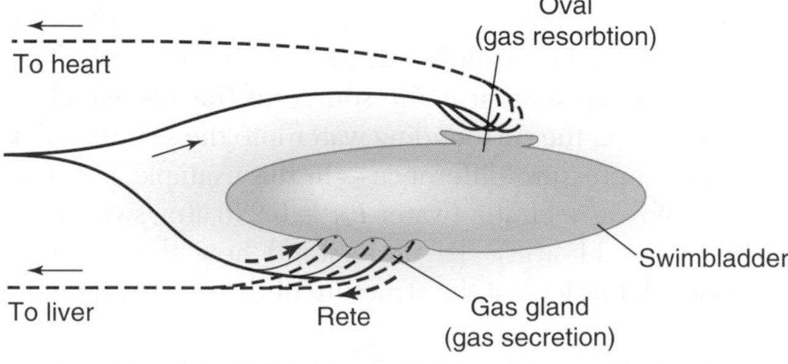

Figure 13-53 Two main types of swimbladders are found in fishes. **(a)** A physostome swimbladder (such as that in the eel *Anguilla vulgaris*) is connected to the outside via a duct to the esophagus. **(b)** A physoclist swimbladder (such as that in the perch, *Perca fluviatilis*) lacks a duct; gas enters and leaves the swimbladder via the blood. [Adapted from Denton, 1961.]

Oxygen Secretion

The rete structure reduces gas loss from the swimbladder, but how is oxygen secreted into the swimbladder? First, consider the relationship between P_{O_2}, oxygen solubility, and oxygen content. Oxygen is carried in blood bound to hemoglobin and in physical solution. If oxygen is released from hemoglobin into physical solution, P_{O_2} will increase. The release of oxygen from hemoglobin can be caused by a reduction in pH via the Root-off shift (Figure 13-54). An increase in the ionic concentration of a solution reduces oxygen solubility and also results in an increase in P_{O_2}, as long as the oxygen content in physical solution remains unchanged. Thus, an increase in blood P_{O_2} can be achieved by releasing oxygen from hemoglobin or by increasing the ionic concentration of the blood.

The cells of the gas gland have few mitochondria and negligible citric acid cycle activity. For this reason, even in an oxygenated atmosphere, glycolysis in the gas gland yields two lactate molecules and two protons for each glucose molecule. The pentose phosphate shunt, however, is active in the gas gland, producing carbon dioxide via decarboxylation of glucose without oxygen consumption. The production of carbon dioxide, lactate, and protons by gas-gland cells results in both a decrease in pH, which causes the release of oxygen from hemoglobin (Root-off shift), and an increase in ionic concentration, which in turn causes a reduction in oxygen solubility (sometimes termed the "salting-out effect"). Both changes cause the P_{O_2} in the gas gland to increase more than that in the swimbladder, so that oxygen diffuses from the blood into the gas space of the swimbladder (see Figure 13-54). The salting-out effect also reduces the solubility of other gases, such as nitrogen and carbon dioxide, and may explain the high levels of these gases sometimes observed in swimbladders.

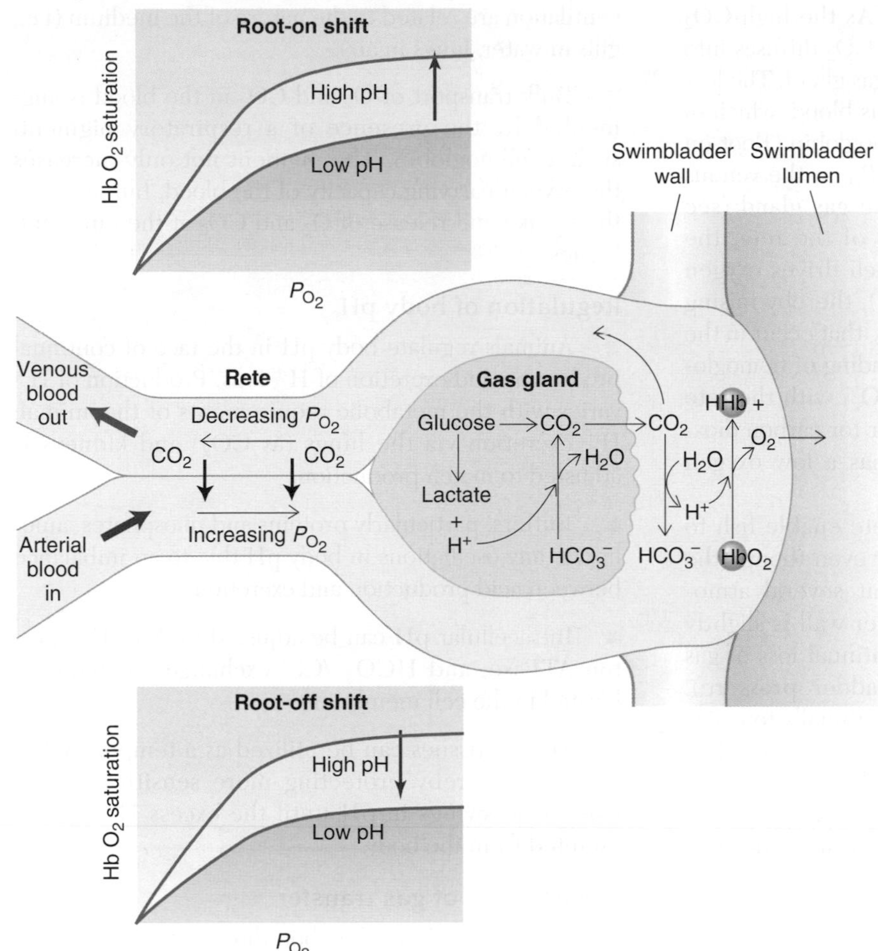

Figure 13-54 Anaerobic metabolism of glucose to lactate and CO_2 in the gas gland, located in the wall of the fish swimbladder, leads to a decrease in erythrocyte pH and the release of oxygen from hemoglobin. As a result, the P_{O_2} in the blood flowing through the gas gland becomes greater than the P_{O_2} in the lumen of the swimbladder, so oxygen diffuses into the lumen. A Root-on shift, leading to a decrease in P_{O_2}, occurs on the venous side of the rete, whereas a reverse Root-off shift, leading to an increase in P_{O_2}, occurs on the arterial side.

Let's return now to the situation in the rete. As discussed earlier, erythrocytes are not very permeable to H^+ ions, so the drop in pH in the gas gland is transferred to the red blood cells by CO_2, which crosses their plasma membranes with ease (Figure 13-55). Acid produced in the gas gland reacts with HCO_3^-, probably taken up from the plasma, producing CO_2. Thus, blood leaving the gas gland and entering the venous capillaries

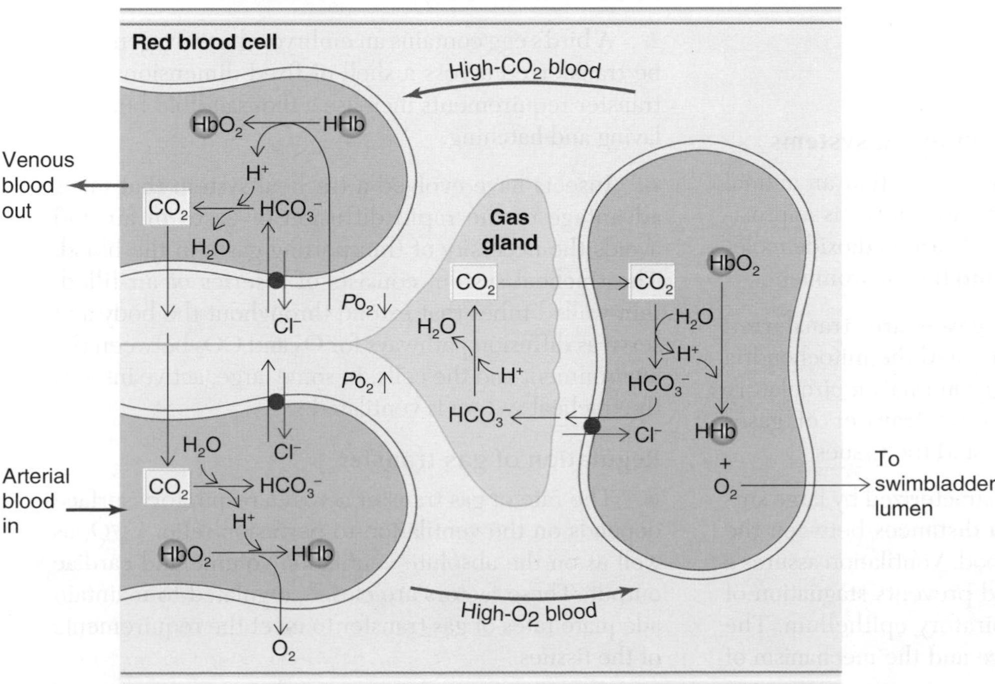

Figure 13-55 The rete, in association with the gas gland, acts as a countercurrent exchanger for carbon dioxide. Venous blood coming from the gas gland is high in CO_2, which diffuses into the arterial side of the rete, lowering pH and causing a Root-off shift (see Figure 13-54) and an increase in P_{O_2} in the arterial blood entering the gland. The CO_2 is recycled through the rete, further increasing P_{O_2} in the arterial blood and decreasing it in the venous blood.

of the rete has a high CO_2 content. As the high-CO_2 venous blood flows through the rete, CO_2 diffuses into the arterial blood flowing toward the gas gland. The loss of the CO_2 raises the pH of the venous blood, which in turn increases oxygen binding by hemoglobin (Root-on shift); as more oxygen is bound, the P_{O_2} in the venous blood falls as it flows away from the gas gland (see Figure 13-54). On the arterial side of the rete, the entering CO_2 lowers blood pH, which drives oxygen from the hemoglobin (Root-off shift), thereby raising the blood P_{O_2}. Thus, the changes in P_{O_2} that occur in the rete result from the loading and unloading of hemoglobin, caused by changing levels of CO_2, with the rete serving as a countercurrent exchanger for carbon dioxide, not oxygen. In fact, the rete has a low oxygen permeability.

The gas gland and associated rete enable fish to transfer oxygen into the swimbladder even though the swimbladder may contain oxygen at several atmospheres of pressure. The swimbladder wall is slightly permeable to gases, so there is a continual loss of gas that increases with depth (swimbladder pressure). Gases, therefore, must be secreted continually to maintain swimbladder volume in the face of this loss. The permeability of the swimbladder and, therefore, the rate of gas loss, along with the rate of gas secretion, determines the depth to which the swimbladder can function. Eels migrating at depth across the oceans enlarge their rete and gas gland and decrease the permeability of the swimbladder wall, which enables them to maintain swimbladder volume at higher pressures. The permeability of the swimbladder wall is decreased by an increase in its thickness due to increased deposition of guanine. Eels turn from yellow to silver as a result of these guanine deposits when they leave rivers and begin their migration across the oceans.

SUMMARY

Oxygen and carbon dioxide in living systems

■ The number of oxygen molecules that an animal extracts from the environment and utilizes is approximately the same as the number of carbon dioxide molecules it produces and releases into the environment.

■ In very small animals, gases are transferred between the respiratory medium and the mitochondria by diffusion alone, but in larger animals a circulatory system has evolved for the bulk transfer of gases between the respiratory surface and the tissues.

■ Respiratory surfaces are characterized by large surface areas and small diffusion distances between the respiratory medium and the blood. Ventilation assures a continual supply of oxygen and prevents stagnation of the medium close to the respiratory epithelium. The design of the respiratory surface and the mechanism of ventilation are related to the nature of the medium (i.e., gills in water, lungs in air).

■ Bulk transport of O_2 and CO_2 in the blood is augmented by the presence of a respiratory pigment, such as hemoglobin. The pigment not only increases the oxygen-carrying capacity of the blood, but also aids the uptake and release of O_2 and CO_2 at the lungs and tissues.

Regulation of body pH

■ Animals regulate body pH in the face of continual production and excretion of H^+ ions. Production of H^+ varies with the metabolic requirements of the animal; H^+ excretion via the lungs (as CO_2) and kidneys is adjusted to match production.

■ Buffers, particularly proteins and phosphates, ameliorate any oscillations in body pH due to an imbalance between acid production and excretion.

■ Intracellular pH can be adjusted by Na^+/H^+, proton ATPase, and HCO_3^-/Cl^- exchange mechanisms located in the cell membrane.

■ Muscle tissues can be utilized as a temporary H^+ reservoir, thereby protecting more sensitive tissues from wide swings in pH until the excess H^+ can be excreted from the body.

Mechanisms of gas transfer

■ Oxygen uptake and carbon dioxide excretion in air is achieved by tidal ventilation of a lung. The volume of each breath is usually much smaller than the total lung volume. Increased ventilation is achieved by increases in both rate and amplitude of breathing.

■ Gas transfer in water is achieved by a unidirectional flow of water over a gill.

■ A bird's egg contains an embryo whose oxygen must be transferred across a shell of fixed dimensions. The transfer requirements increase a thousandfold between laying and hatching.

■ Insects have evolved a tracheal system that takes advantage of the rapid diffusion of gases in air and avoids the necessity of transporting gases in the blood. The tracheal system consists of a series of air-filled, thin-walled tubes that extend throughout the body and serve as diffusion pathways for O_2 and CO_2 between the environment and the cells. In some large, active insects, the tracheal system is ventilated.

Regulation of gas transfer

■ The rate of gas transfer across a respiratory surface depends on the ventilation-to-perfusion ratio, $\dot{V}_A/\dot{Q}$, as well as on the absolute ventilation volume and cardiac output. These factors are closely regulated to maintain adequate rates of gas transfer to meet the requirements of the tissues.

■ The control system for ventilation, which has been studied extensively in mammals, consists of a number of mechanoreceptors and chemoreceptors that feed information into the medullary respiratory center. This center, through a variety of effectors, causes appropriate changes in breathing and blood flow to maintain rates of O_2 and CO_2 transfer at a level sufficient to meet the requirements of metabolism.

■ Animals adjust gas transfer during hypoxia, hypercapnia, exercise, and diving to maintain oxygen delivery to meet the requirements of the tissues. These adjustments require the control and coordination of a cascade of systems including ventilation, gas transfer across the respiratory surface, transport in the blood, and diffusion of oxygen into, and carbon dioxide out of, the tissues. The structural elements in the cascade appear to have similar functional capacities, supporting the concept of symmorphosis.

■ Fish swimbladders present interesting problems in gas transfer. Gas pressures in the fish swimbladder often exceed those in the blood by several orders of magnitude, but the design of the blood supply and gas gland is such that gases move from the blood into the swimbladder.

REVIEW QUESTIONS

1. Calculate the percentage change in volume when dry air at 20°C is inhaled into the human lung (temperature = 37°C).
2. Define the following terms: (a) oxygen capacity, (b) oxygen content, (c) percent saturation, (d) methemoglobin, (e) Bohr effect, and (f) Haldane effect.
3. Describe the role of hemoglobin in the transfer of oxygen and carbon dioxide.
4. Describe the effects of gravity on the distribution of blood in the human lung. What effect does alveolar pressure have on lung blood flow?
5. Compare and contrast ventilation of the mammalian lung and the bird lung.
6. What is the functional significance of the presence of surfactants in the lung?
7. How have insects avoided the necessity of transporting gases in the blood?
8. The number and dimensions of air pores in eggshells are constant for a given species. What effect would doubling the number of pores have on the transfer of oxygen, carbon dioxide, and water across the eggshell?
9. Discuss the role of the rete mirabile in the maintenance of high gas pressures in the fish swimbladder.
10. How is oxygen moved into the swimbladder of teleost fishes?
11. Describe the structural and functional differences between gills and lungs.
12. Why is the ventilation-to-perfusion ratio much higher in water-breathing than in air-breathing animals?
13. Describe the role of central chemoreceptors in the control of carbon dioxide excretion.
14. What is the importance of the Hering-Breuer reflex in the control of breathing?
15. Describe the processes involved in the acclimation of mammals to high altitudes.
16. What is the effect on intracellular pH of elevating extracellular NH_4Cl levels at either low or high extracellular pH?
17. Describe the role of the carbon dioxide–bicarbonate systems in pH regulation in mammals.
18. Explain the consequences of the localization of the enzyme carbonic anhydrase within the red blood cell and its absence in the plasma.
19. Describe the possible mode of operation of the medullary respiratory center.
20. Discuss the interaction between gas transfer and heat and water loss in air-breathing vertebrates.

SUGGESTED READINGS

Astrup, P., and J. W. Severinghaus. 1986. *The History of Blood Gases, Acids and Bases.* Copenhagen: Munksgaard.

Berner, R. A., S. T. Petsch, J. A. Lake, D. J. Beerling, B. N. Popp, R. S. Lane, E. A. Laws, M. B. Westley, N. Cassar, F. I. Woodward, and W. P. Quick. 2000. Isotopic fractionation and atmospheric oxygen: Implications for Phanerozoic O_2 evolution. *Nature* 287:1630–1633.

Budyko, M. I., A. B. Ronov, and A. L. Yanshin. 1985. *History of the Earth's Atmosphere.* Berlin: Springer-Verlag.

Dejours, P. 1988. *Respiration in Water and Air.* Amsterdam: Elsevier.

Diamond, J. 1982. How eggs breathe while avoiding desiccation and drowning. *Nature* 295:10–11.

Diaz, R. J., and R. Rosenberg. 1995. Marine benthic hypoxia: A review of its ecological effects and the behavioural responses of benthic macrofauna. *Oceanogr. Mar. Biol. Annu. Rev.* 33:245–303.

Heisler, N., ed. 1995. *Mechanisms of Systemic Regulation: Respiration and Circulation.* Advances in Comparative and Environmental Physiology. Vol. 21. Berlin: Springer-Verlag.

Hochachka, P. W., and G. N. Somero. 1983. *Strategies of Biochemical Adaptation.* 2d ed. Princeton, NJ: Princeton University Press.

Jensen, F. B. 1991. Multiple strategies in oxygen and carbon dioxide transport by haemoglobin. In A. J. Woakes, M. K. Grieshaber, and C. Bridges, eds. *Physiological Strategies for Gas Exchange and Metabolism,* 55–78. Society of Experimental

Biology Seminar Series. Vol. 41. Cambridge: Cambridge University Press.

Kooyman, G. L., P. J. Ponganis, and R. S. Howard. 1999. Diving animals. In C. E. G. Lungren and J. N. Miller, eds. *The Lung at Depth,* 587–620. New York: Marcel Dekker.

Krogh, A. 1968. *The Comparative Physiology of Respiratory Mechanisms.* New York: Dover.

Milvaganam, S. E. 1996. Structural basis for the Root effect in haemoglobin. *Nature Struct. Biol.* 3:275–283.

Nikinmaa, M. 1990. *Vertebrate Red Blood Cells: Adaptations of Function to Respiratory Requirements.* Zoophysiology. Vol. 28. Berlin: Springer-Verlag.

Perry, S. F., and B. Tufts, eds. 1998. *Fish Respiration.* Fish Physiology. Vol. 17. New York: Academic Press.

Perutz, M. F. 1996. Cause of the Root effect in fish haemoglobins. *Nature Struct. Biol.* 3:211–212.

Rahn, H. 1966. Aquatic gas exchange theory. *Respir. Physiol.* 1:1–12.

Richter, D. W., K. Ballanyi, and S. Schwarzacher. 1992. Mechanism of respiratory rhythm generation. *Curr. Opin. Neurobiol.* 2:788–793.

Roos, A., and W. F. Boron. 1981. Intracellular pH. *Physiol. Rev.* 61:296–434.

Schmidt-Nielsen, K. 1979. *How Animals Work.* Cambridge: Cambridge University Press.

Weber, R. E. 1992. Molecular strategies in the adaptation of vertebrate hemoglobin function. In S. C. Wood, R. E. Weber, A. R. Hargens, and R. W. Millard, eds. *Physiological Adaptations in Vertebrates: Respiration, Circulation and Metabolism,* 257–277. New York: Marcel Dekker.

Weibel, E. R., C. R. Taylor, and L. Bolis, eds. 1998. *Principles of Animal Design: The Optimization and Symmorphosis Debate.* Cambridge: Cambridge University Press.

West, J. B. 1974. *Respiratory Physiology: The Essentials.* Baltimore: Williams and Wilkins.

Zhu, X. L., and W. S. Sly. 1990. Carbonic anhydrase IV from human lung. *J. Biol. Chem.* 15:8795–8801.

Ionic and Osmotic Balance

The unique physical and chemical properties of water undoubtedly played a major role in the origin of life, and all life processes take place in a watery milieu (see Chapter 3). Indeed, the physicochemical nature of life as we know it is largely a reflection of the special properties of water. The presence of water here on Earth made it possible for life to arise several billion years ago in a shallow, salty sea. Extracellular fluids surrounding living cells to this day reflect the composition of the primeval sea in which life evolved (see Table 14-2).

The ability to survive in a variety of osmotic environments was achieved in the more advanced animal groups by the evolution of a stable internal environment, which protects the internal tissues from the vagaries and extremes of the external environment. Animals are restricted in their geographic distribution largely by two environmental factors, temperature and osmotic pressure. The evolution of mechanisms of osmoregulation has allowed organisms to penetrate into new and different osmotic environments. Such geographic dispersal is an important mechanism for the divergence of species in the process of evolution. If, for example, the arthropods and the vertebrates had not evolved mechanisms for regulating the osmolarity of their extracellular compartments, they would have been unsuccessful in their invasion of the osmotically hostile freshwater and terrestrial environments, where genetic isolation and new selective pressures spurred divergence and speciation. In the absence of terrestrial arthropods and vertebrates, other groups, such as plants, would have evolved differently, and terrestrial life would be quite different from what we now know.

In this chapter, we consider the osmotic environment, the exchange of water and salts between the animal and its environment, and the mechanisms used by various animals to cope with environmental osmotic extremes. The movement of water and solutes across cell membranes and multicellular epithelial layers was covered in Chapter 4. That discussion forms an essential background for understanding the osmoregulatory processes in organs such as the kidney, gill, and salt gland that we examine in this chapter. Toward the chapter's end we discuss the closely related problem of elimination of toxic nitrogenous wastes produced during the metabolism of amino acids and proteins.

PROBLEMS OF OSMOREGULATION

A major requirement for cell survival is the retention of appropriate quantities of water within the cell, along with appropriate concentrations of various solutes (e.g., salts and nutrient molecules) in the extracellular and intracellular compartments. Tissues can tolerate some dehydration, in some instances as much as 70% water loss by volume, but extensive dehydration is accompanied by a marked impairment of metabolism. Some tissues require an extracellular environment that is an approximation of seawater—namely, a fluid high in sodium and chloride and relatively low in the other major ions, such as potassium and the divalent cations. For many marine invertebrates, the surrounding seawater itself can act as the extracellular medium; for most of the more complex invertebrates, the internal fluids are in near ionic equilibrium with seawater. In contrast, the extracellular fluids of vertebrates, with the exception of the hagfishes, have an ionic concentration that is about one-third that of seawater, with much of the magnesium sulfate removed and some of the chloride replaced by bicarbonate anions (Table 14-1 on the next page). This un-oceanlike internal environment presumably reflects the freshwater origin of most vertebrates, including marine teleost fishes. The extracellular fluids of marine teleosts are much more dilute than seawater, and these fishes maintain both an ionic and an osmotic difference between their body fluids and seawater. Elasmobranchs, on the other hand, maintain an ionic difference, but only a minor osmotic difference,

Table 14-1 Composition of extracellular fluids of representative animals*

	Habitat*	Osmolarity (mosM)	Ionic concentrations (mM) Na^+	K^+	Ca^{2+}	Mg^{2+}	Cl^-	SO_4^{2-}	HPO_4^{2-}	Urea
Seawater†		1000	460	10	10	53	540	27		
Coelenterata										
Aurelia (jellyfish)	SW		454	10.2	9.7	51.0	554	14.6		
Echinodermata										
Asterias (starfish)	SW		428	9.5	11.7	49.2	487	26.7		
Annelida										
Arenicola (lugworm)	SW		459	10.1	10.0	52.4	537	24.4		
Lumbricus (earthworm)	Ter.		76	4.0	2.9		43			
Mollusca										
Aplysia (sea slug)	SW		492	9.7	13.3	49	543	28.2		
Liligo (squid)	SW		419	20.6	11.3	51.6	522	6.9		
Anodonta (clam)	FW		15.6	0.49	8.4	0.19	11.7	0.73		
Crustacea										
Cambarus (crayfish)	FW		146	3.9	8.1	4.3	139			
Homarus (lobster)	SW		472	10.0	15.6	6.7	470			
Insecta										
Locusta	Ter.		60	12	17	25				
Periplanta (cockroach)	Ter.		161	7.9	4.0	5.6	144			
Cyclostomata										
Eptatretus (hagfish)	SW	1002	554	6.8	8.8	23.4	532	1.7	2.1	3
Lampetra (lamprey)	FW	248	120	3.2	1.9	2.1	96	2.7		0.4
Chondrichthyes										
Dogfish shark	SW	1075	269	4.3	3.2	1.1	258	1	1.1	376
Carcharhinus	FW		200	8	3	2	180	0.5	4.0	132
Coelacantha										
Latimeria	SW		181	51.3	6.9	28.7	199			355
Teleostei										
Paralichthys (flounder)	SW	337	180	4	3	1	160	0.2		
Carassius (goldfish)	FW	293	142	2	6	3	107			
Amphibia										
Rana esculenta (frog)	FW	210	92	3	2.3	1.6	70			2
Rana cancrivora	FW	290	125	9			98			40
	80% SW	830	252	14			227			350
Reptilia										
Alligator	FW	278	140	3.6	5.1	3.0	111			
Aves										
Anas (duck)	FW	294	138	3.1	2.4		103		1.6	
Mammalia										
Homo sapiens	Ter.		142	4.0	5.0	2.0	104	1	2	
Lab rat	Ter.		145	6.2	3.1	1.6	116			

* The osmolarity and composition of seawater vary, and the values given here are not intended to be absolute.
The composition of body fluids of osmoconformers will also vary, depending on the composition of the seawater in which they are tested.
† SW = seawater; FW = freshwater; Ter. = terrestrial.
Sources: Schmidt-Nielsen and Mackay, 1972; Prosser, 1973.

between themselves and seawater; the high concentrations of urea in the body fluids of these fishes bring their osmolarity to slightly above that of seawater.

The intracellular environment of most animals is low in sodium but high in potassium, phosphate, and proteins (Table 14-2). There are only minor and transient osmotic differences between the intracellular and extracellular fluids of animals. Thus the plasma membrane maintains ionic, but not osmotic, differences between the intracellular and extracellular fluids, whereas the epithelium surrounding the body often maintains both ionic and osmotic differences between animals and their environments. In most multicellular animals, the entire body surface is not usually involved in ionic and osmotic regulation; rather, this regulation is effected by a specialized portion of the body surface,

Table 14-2 Electrolyte composition of the human body fluids

Electrolytes	Serum ($meq \cdot kg^{-1}\ H_2O$)	Interstitial fluid ($meq \cdot kg^{-1}\ H_2O$)	Intracellular fluid (muscle) ($meq \cdot kg^{-1}\ H_2O$)
Cations			
Na^+	142	145	10
K^+	4	4	156
Ca^{2+}	5		3
Mg^{2+}	2		26
Totals	153	149	195
Anions			
Cl^-	104	114	2
HCO_3^-	27	31	8
HPO_4^{2-}	2		95
SO_4^{2-}	1		20
Organic acids	6		
Proteins	13		55
Totals	153	145	180

Note: Some of the ions contained within cells are not completely dissolved within the cytosol, but may be partially sequestered within cytoplasmic organelles. Thus, the true free Ca^{2+} concentration in the cytosol is typically below the overall value given in the table for intracellular Ca^{2+}. Failure of anion and cation totals to agree reflects incomplete tabulation.

such as the gills of fishes, or an internalized epithelial structure, such as the salt gland of elasmobranchs or the kidney of mammals. The rest of the body surface, with the exception of the lining of the gut, is relatively impermeable to ions and water.

Cell membranes that are permeable to oxygen are also permeable to water, so energy must be expended to maintain the ionic and osmotic balance of the animal—an animal cannot reduce osmotic and ionic problems by sealing itself off from the environment because oxygen and nutrients must be acquired and waste products excreted. Some animals do encyst themselves, but this is feasible only if their metabolic rate is very much reduced. Brine shrimp larvae, for example, can survive in a state of suspended animation for many years. During encystment, energy turnover is extremely low, and there is little or no growth. Most animals, however, must ingest nutrients at a high rate and cope with the resulting osmotic and ionic problems.

Many waste products generated during metabolism are toxic and cannot accumulate to high levels in the body without serious consequences. Thus the cellular environment must be freed of these toxic by-products of metabolism. In the smallest aquatic organisms, this purification happens simply by diffusion of the wastes into the surrounding water. In animals that have circulatory systems, the blood typically passes through excretory organs, generally termed *kidneys.* In terrestrial animals, the kidneys not only play an important role in the removal of organic wastes, but are also the primary organs of osmoregulation.

A number of mechanisms are employed to handle osmotic problems and to regulate the differences between intracellular and extracellular compartments and between the extracellular compartment and the external environment. These are collectively termed *osmoregulatory mechanisms,* a term coined in 1902 by Rudolf Hober to refer to the regulation of osmotic pressure and ionic concentrations in the extracellular compartment of the animal body. The evolution of efficient osmoregulatory mechanisms had extraordinarily far-reaching effects on other aspects of animal speciation and diversification. The various adaptations and physiological mechanisms evolved by animals to cope with the rigors of the osmotic environment offer especially fascinating examples of the resourcefulness of evolutionary adaptation.

Although there may be hourly and daily variations in osmotic balance, an animal is generally in an osmotic steady state over the long term. That is, the input and output of water and of salts over an extended period are, on average, equal. Water enters terrestrial animals with their food and drink. Metabolic processes also produce water. For animals living in a freshwater environment, water enters the body primarily through the respiratory epithelium—the gill surfaces of fishes and invertebrates and the integument, or outer covering, of amphibians and many invertebrates. Water leaves the body in the urine, in the feces, and by evaporation through the lungs and integument.

The problem of osmotic regulation does not end with the intake and output of water. If that were so, osmoregulation would be a relatively simple matter: A frog sitting in freshwater far more dilute than its body fluids (i.e., its interstitial fluids and blood) would merely have to eliminate as much water as it took in through its skin, and a camel would simply stop producing urine between oases. Osmoregulation necessarily also involves the maintenance of favorable solute concentrations in the extracellular compartment. The frog immersed in hypo-osmotic pond water must not only eliminate excess water, but also retain salts, which tend to leak out through the skin. As we will see, diverse animals in a wide range of environments have evolved numerous mechanisms to solve a problem common to all: conserving a proper osmotic balance between cells, extracellular compartments, and the environment.

The rate of water turnover is different in a whale, a human, a shrimp, and a snake. What differences would you expect and why?

OSMOREGULATORS AND OSMOCONFORMERS

Animals that maintain an internal osmolarity different from the medium in which they are immersed have been termed **osmoregulators.** An animal that does not actively control the osmotic condition of its body fluids and instead conforms to the osmolarity of the surrounding medium is termed an **osmoconformer.** Most vertebrates, with the notable exception of the hagfishes, are strict osmoregulators, maintaining the composition of their body fluids within a small osmotic range. This constancy is also true of fishes such as salmon and eels that migrate between freshwater and saltwater environments. Some osmotic differences do exist among vertebrate species; elasmobranchs, for example, are slightly hyperosmotic to seawater. The blood and tissues of most vertebrates, however, are hypo-osmotic to seawater and significantly hyperosmotic to freshwater.

Marine invertebrates, as a rule, are in osmotic balance with seawater, and the ionic concentrations of their body fluids generally parallel those in seawater. This similarity has allowed the use of seawater as a physiological saline in studies of the tissues of marine species. For example, some large neurons removed from marine invertebrates continue to function for many hours when placed in seawater. In freshwater and terrestrial invertebrates, the body fluids are invariably more dilute than seawater but considerably more concentrated than freshwater.

Aquatic, brackish-water, and marine invertebrates are, of course, exposed to varying environmental osmolarities. Some aquatic invertebrates are strict osmoregulators like the vertebrates, some are limited osmoregulators, and some are strict osmoconformers (see Figure 1-4).

In osmoregulating animals, the internal tissues are generally not able to cope with more than minor changes in extracellular osmolarity and must depend entirely on osmotic regulation of the extracellular fluid to maintain cell volume. The cells of osmoconformers, on the other hand, are able to cope with high extracellular osmolarities by increasing their intracellular osmolarities, thereby maintaining cell volume. Osmolar homeostasis is achieved by increasing the concentration of intracellular organic **osmolytes,** substances that, by their presence in high concentrations, act to increase intracellular osmolarity. The use of such substances reduces the need to maintain osmotic pressure with inorganic ions, which could give rise to other problems, such as interference with enzyme efficiency. In some marine vertebrates and invertebrates, organic osmolytes are present in the blood and interstitial fluids as well as inside cells, so that both extracellular and intracellular osmolarity are brought close to that of seawater. The best-known examples of such organic osmolytes are **urea** and **trimethylamine oxide (TMAO),** both utilized by various marine elasmobranchs, the primitive coelacanth fish *Latimeria,* and the crab-eating, brackish-water frog *Rana cancrivora* of Southeast Asia (see Table 14-1).

The osmotic exchanges that take place between an animal and its environment can be divided into two classes (Figure 14-1):

- *Obligatory osmotic exchanges* occur mainly in response to physical factors over which the animal has little or no physiological control.
- *Regulated osmotic exchanges* are physiologically controlled and serve to aid in maintaining internal homeostasis.

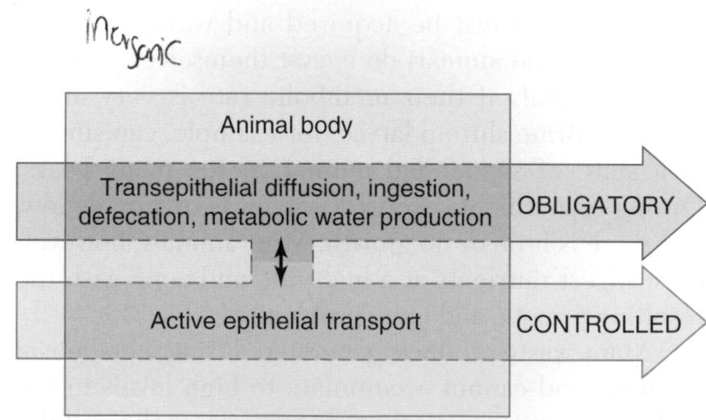

Figure 14-1 Two major classes of osmotic exchange—obligatory and controlled—occur between an animal and its environment. Obligatory exchanges occur in response to physical factors over which the animal has little short-term physiological control. Controlled exchanges are those that the animal can vary physiologically to maintain internal homeostasis. Substances entering the animal by either path can leave by the other.

Regulated exchanges generally serve to compensate for obligatory exchanges. In the next section, we consider obligatory exchanges and then, in the following sections, various mechanisms of regulated exchange.

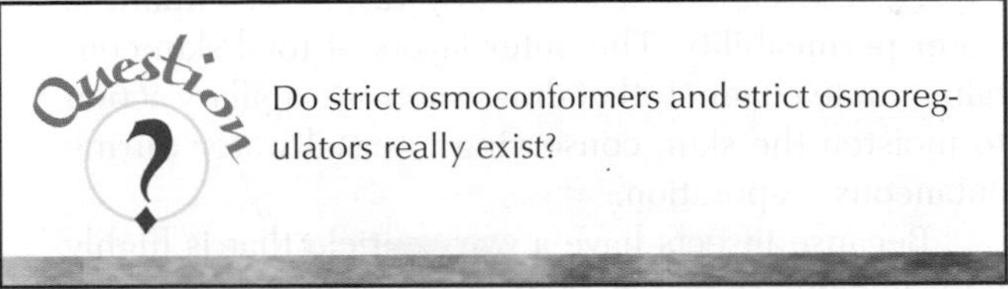

Do strict osmoconformers and strict osmoregulators really exist?

OBLIGATORY EXCHANGES OF IONS AND WATER

The integument, respiratory surfaces, and other epithelia in contact with the surrounding environment act as barriers to obligatory exchanges between an organism and its environment. We outline here the various factors that contribute to obligatory exchanges.

Gradients Between the Animal and the Environment

A frog immersed in a pond tends to take up water from its hypo-osmotic environment, and a bony fish in seawater is faced with the problem of losing water to its hypertonic environment. Similarly, a marine fish with a lower NaCl content than that of seawater faces a continual diffusion of salt into the body, whereas a freshwater fish faces a continual loss of salt. The rate of transfer depends on the surface area of the animal, the size of the gradient, and the permeability of the animal's surface.

Surface-to-Volume Ratio

The volume of an animal varies with the cube of its linear dimensions, but its surface area varies with the square of those dimensions. Consequently, the surface-to-volume ratio is greater for small animals than for large animals. It follows that the surface area of the integument, through which water or a solute can be exchanged with the environment, is greater relative to the water content of a small animal than it is for a large animal. This geometric relation means that for a given net rate of exchange across the integument (in moles per second per square centimeter), a small animal will dehydrate or hydrate more rapidly than a larger animal of the same shape (Figure 14-2). The above statement is based on the assumption that animals are round. In fact, in larger animals, surface area is largely determined by the size of the respiratory surface elaborated for oxygen uptake. Larger animals have a smaller oxygen requirement per unit volume than smaller animals and, therefore, have a smaller respiratory surface per unit volume. This difference contributes further to the lower rates of dehydration and hydration seen in larger animals.

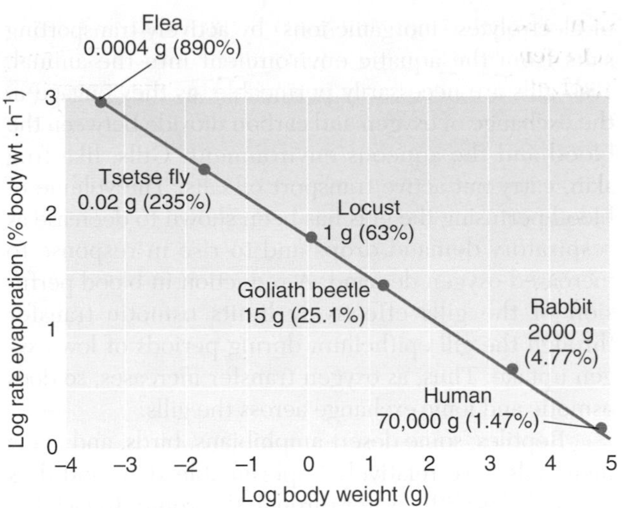

Figure 14-2 Small animals dehydrate more rapidly than large animals because of their high surface-to-mass (and thus surface-to-volume) ratios. This log-log plot shows the amount of water, as a percentage of body weight, that is lost per hour under hot desert conditions versus body weight. [Adapted from Edney and Nagy, 1976.]

Permeability of the Integument

The integument acts as a barrier between the extracellular compartment and the environment. Movement of water across the integument occurs through cells (transcellular movement) and between cells (paracellular movement). Tight junctions between cells reduce the permeability of the paracellular pathway to water. Pure phospholipid bilayers are not very permeable to water, so transcellular movement of water across biological membranes depends on the presence of water channels. For example, erythrocytes swell or shrink rapidly in response to changes in the osmotic strength of the extracellular fluid because of the presence in their plasma membrane of a 28 kDa protein called **aquaporin.** It appears that water channels in membranes are formed of a tetramer of identical aquaporin molecules. The role of aquaporin as a water channel was demonstrated in experiments with frogs eggs (oocytes), which are not very permeable to water and thus normally do not swell much when placed in pond water. When mRNA encoding the aquaporin protein was injected into frog oocytes, they became very permeable to water and swelled when placed in water. Membrane permeability to water is presumably related to the concentration of aquaporin within the phospholipid bilayer. The absence of aquaporin water channels in a membrane reduces transcellular permeability to water.

The permeability of the integument to water and solutes varies among animal groups. Amphibians generally have moist, highly permeable skins, through which they exchange oxygen and carbon dioxide with the environment and through which water and ions move by passive diffusion. Amphibian skin compensates for loss of electrolytes (inorganic ions) by actively transporting salts from the aquatic environment into the animal. Fish gills are necessarily permeable, as they engage in the exchange of oxygen and carbon dioxide between the blood and the aqueous environment. Gills, like frog skin, carry out active transport of salts. The volume of blood perfusing the gills has been shown to decrease as respiratory demand drops and to rise in response to increased oxygen demand. A reduction in blood perfusion of the gills effectively limits osmotic transfer through the gill epithelium during periods of low oxygen uptake. Thus, as oxygen transfer increases, so does osmotic and ionic exchange across the gills.

Reptiles, some desert amphibians, birds, and many mammals have relatively impermeable skins and thus generally lose little water through this route. In fact, the skin of some mammals (e.g., cowhide) is so impermeable that it can be used to carry water. The low permeability of the integument is maintained in those terrestrial species that have secondarily become aquatic, such as pond insects and marine mammals.

As mentioned above, not all vertebrates have an impermeable integument. Many amphibians, as well as mammals that perspire, can become dehydrated at low humidity because of water loss through the integument. These animals minimize water loss by behavioral strategies, avoiding desiccation by staying in cool, damp microenvironments during hot, dry times of day. Animals with highly permeable skin are simply not able to tolerate very hot, dry environments. Most frogs stay close to water. Toads and salamanders can venture a bit farther, but they are limited to moist woods or meadows not far from puddles, streams, or bodies of water in which they can replenish their supply of body water. The desert toads *Chiromantis xerampelina* and *Phyllomedusa sauvagii* have extremely low evaporative loss of water from their skin because it is covered by a secreted waxy coating. These toads also excrete uric acid rather than ammonia or urea (a water-saving strategy to which we will return below).

Frogs and toads are endowed with a large-volume lymphatic system and an oversized urinary bladder in which they can store water. When these animals wander from a body of water, or during periods of low rainfall, water moves osmotically from the lumen of the bladder into the partially dehydrated interstitial fluid and blood. The epithelium of the bladder, like the amphibian skin, is capable of actively transporting sodium and chloride from the bladder lumen into the body to compensate for the loss of salts that accompanies excessive hydration during times of plentiful water. Thus, the anuran bladder serves a dual function as a water reservoir in times of dehydration and as a source of salts during times of excessive hydration.

The high water permeability of amphibian skin is used to advantage to take up water from hypo-osmotic sources such as puddles. Many anurans have a specialized region of skin on the abdomen and thighs, termed the *pelvic patch*, that when immersed can take up water at a rate of three times the body weight per day. The permeability of amphibian skin is controlled by a hormone called *arginine vasotocin* (AVT) or, more simply, *vasotocin;* like the mammalian antidiuretic hormone (ADH; also called vasopressin), vasotocin enhances water permeability. The outer layers of toad skin contain minute channels that draw water by capillary action to moisten the skin, conserving internal water during cutaneous evaporation.

Because insects have a waxy cuticle that is highly impermeable to water, their evaporative water loss is much lower than that of many other animal groups (Table 14-3). The waxy layer is deposited on the surface of the exoskeleton through fine canals that penetrate the cuticle (Figure 14-3). The importance of the waxy layer for water retention has been demonstrated by measuring the rate of water loss in insects at different temperatures. In Figure 14-4, we see that there is a sudden jump in the rate of water loss coincident with

Table 14-3 Evaporative water loss of representative animals under desert conditions

Species	Water loss ($mg \cdot cm^{-2} \cdot h^{-1}$)	Remarks*
Arthropods		
Eleodes armata (beetle)	0.20	30°C; 0% r.h.
Hadrurus arizonensis (scorpion)	0.02	30°C; 0% r.h.
Locusta migratoria (locust)	0.70	30°C; 0% r.h.
Amphibians		
Cyclorana alboguttatus (frog)	4.90	25°C; 100% r.h.
Reptiles		
Gehrydra variegata (gecko)	0.22	30°C; dry air
Uta stansburiana (lizard)	0.10	0°C
Birds		
Amphispiza belli (sparrow)	1.48	30°C
Phalaenpitus nutalllii (poorwill)	0.86	30°C
Mammals†		
Peromyscus eremicus (cactus mouse)	0.66	30°C
Oryx beisa (African oryx)	3.24	22°C
Homo sapiens	22.32	70 kg; nude, sitting in sun; 35°C

* r.h. stands for relative humidity. Where not indicated, relative humidity is not available.

† The cactus mouse and African oryx are desert animals and employ various water-conservation measures. Thus their evaporative water loss is much less than that of humans.

Source: Hadley, 1972.

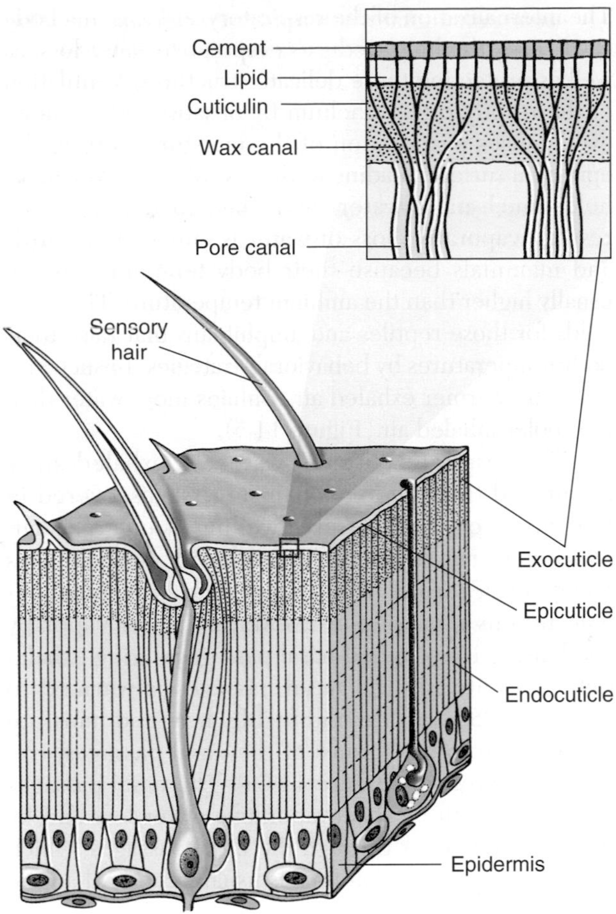

Figure 14-3 The waxy lipid layer on the outside of the insect cuticle serves as the major water barrier, reducing evaporative water loss. The waxy layer is deposited through minute canals in the integument. [Adapted from Edney, 1974.]

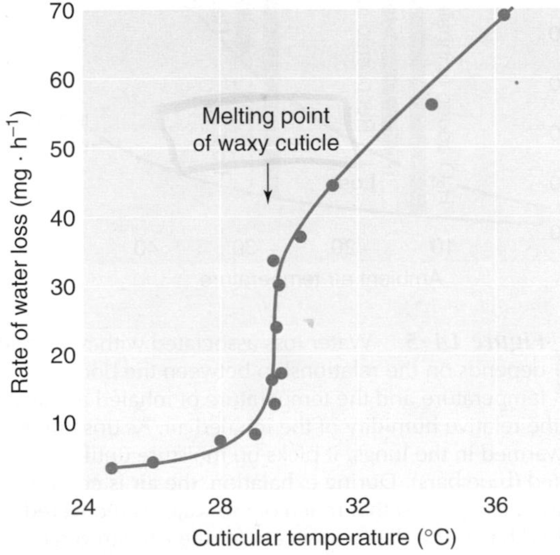

Figure 14-4 The rate of water loss from insects is much higher at temperatures above the melting point of the waxy layer covering the cuticle. The sharp break in this plot of water loss versus cuticular temperature in a cockroach corresponds to the melting point of the waxy cuticle. [From Beament, 1958.]

the melting point of the waxy coating. The major route of water loss in terrestrial insects is via the tracheal system, which consists of air-filled tracheoles that penetrate the tissues. As long as the tracheoles are open to the air, water vapor can diffuse out while oxygen and carbon dioxide diffuse down their respective gradients. The entrances to the tracheoles are guarded by valvelike spiracles that are closed periodically by the spiracular muscles, reducing water loss. The importance of this mechanism in water conservation in insects, however, has been questioned (see Chapter 13).

Feeding, Metabolic Factors, and Excretion

Water and solutes are taken in during feeding. End products of digestion and metabolism that cannot be used by the organism must be eliminated. Although water is an end product of cellular metabolism, it is produced in small enough quantities that its elimination is not problematic (Table 14-4). In fact, this so-called *metabolic water* is the major source of water for many desert dwellers. Osmotic problems are posed by the inevitable production of nitrogenous end products of metabolism (e.g., ammonia and urea) and by the ingestion of salts, because water is required for their elimination from the body.

An animal's diet may include excess water or excess salts. A seal feeding on marine invertebrates with an osmolarity similar to seawater ingests a relatively high quantity of salt relative to water and requires water to excrete the salt load. If the seal feeds on marine teleost fish, which are more dilute than seawater, the ingested salt load is much less. The seal metabolizes fat to produce both energy and water when eating marine invertebrates, but stores fat when eating fish. The metabolism of fat produces the water required to excrete the salt load associated with eating marine invertebrates (see Table 14-4). Seals become fat when eating fish but get thin eating marine invertebrates. The assumption that this pattern is related to water balance, however, should be treated with caution. Although burning fat produces metabolic water, the uptake of oxygen results

Table 14-4 **Production of metabolic water during oxidation of foods**

	Food		
	Carbohydrates	Fats	Proteins
Grams of metabolic water per gram of food	0.56	1.07	0.40
Kilojoules expended per gram of food	17.58	39.94	17.54
Grams of metabolic water per kilojoule expended	0.032	0.027	0.023

Source: Edney and Nagy, 1976.

in water loss across the respiratory surface. It is not known whether burning fat results in a net gain of water in seals.

In terrestrial animals, the regulation of ionic concentrations and the excretion of nitrogenous wastes are accompanied by unavoidable losses of body water. A number of physiological adaptations tend to minimize the loss of water associated with these important functions. Among terrestrial invertebrates, insects are highly effective in conserving water in the course of eliminating nitrogenous and inorganic wastes. The extent to which ions are reabsorbed in the insect rectum or eliminated with the feces is regulated according to the osmotic condition of the insect, as demonstrated by an experiment in which locusts were allowed to drink either pure water or a concentrated saline solution containing NaCl and KCl (450 mosm $\cdot$ L^{-1}). The salt concentration of the rectal fluid after the insect drank saline was several hundred times higher than after it drank pure water, whereas the salt concentration of the hemolymph increased by only about 50% after drinking saline.

The kidney is the chief organ of osmoregulation and nitrogen excretion in most terrestrial vertebrates, especially birds and mammals. The kidneys of birds and mammals utilize *countercurrent multiplication* to produce hyperosmotic urine, which is more concentrated than the blood plasma. This specialization, centered on a hairpin-like bend in the kidney tubule called the **loop of Henle**, has undoubtedly been of major importance in allowing birds and mammals to exploit dry terrestrial environments. The loop of Henle reaches its highest degree of specialization in desert animals such as the kangaroo rat, which can produce urine of up to 9000 milliosmoles per liter. In birds, the countercurrent organization of the loop of Henle is less efficient, perhaps because the avian kidney contains a mixture of tubules—"reptilian-type," which lack the loop of Henle, and "mammalian-type," which contain this specialized structure. The highest osmolarities found in avian urine (in the salt-marsh Savannah sparrow) have been around 2000 mosm $\cdot$ L^{-1}. Reptiles and amphibians, whose kidneys are not organized for countercurrent multiplication, are unable to produce hyperosmotic urine. As an adaptive consequence, some amphibians, when faced with dehydration, are able to cease urine production entirely during the period of osmotic stress.

Water Loss via Respiration

Because of its high heat of vaporization, water is ideally suited for the elimination of body heat by evaporation from epithelial surfaces (see Chapter 16). During evaporation, those water molecules with the highest energy content enter the gaseous phase, taking their thermal energy with them. As a result, the water left behind becomes cooler.

Respiratory surfaces are, by their very nature, a major avenue for water loss in all air-breathing animals. The internalization of the respiratory surfaces in a body cavity (i.e., the lung) reduces evaporative water loss, as well as protecting these delicate structures. Ventilation of the respiratory epithelium by unsaturated air, however, causes evaporation of the moisture wetting the epithelial surface, leading to the loss of water. Warm air holds much more water vapor than cool air, and as a result, evaporative loss of water is enhanced in birds and mammals because their body temperatures are usually higher than the ambient temperature. The same holds for those reptiles and amphibians that raise their body temperatures by behavioral strategies. In such animals, the warmer exhaled air contains more water than the cooler inhaled air (Figure 14-5).

The respiratory loss of water via exhaled air is minimized through a mechanism first discovered in the nose of the desert-dwelling kangaroo rat, *Dipodomys merriami,* by Knut Schmidt-Nielsen. This mechanism, termed a *temporal countercurrent system,* retains most of the respiratory water vapor by condensing it on cooled nasal passages during exhalation. Air entering the nasal passages is warmed to about 37–38°C and humidified by heat and moisture absorbed from the tissues of the nasal passages, trachea, and bronchi (Figure 14-6a). The nasal passages are cooled both by this evaporative water loss and by the flow of cool air through the nasal passages, which is why the nose of a mammal is usually cooler than the

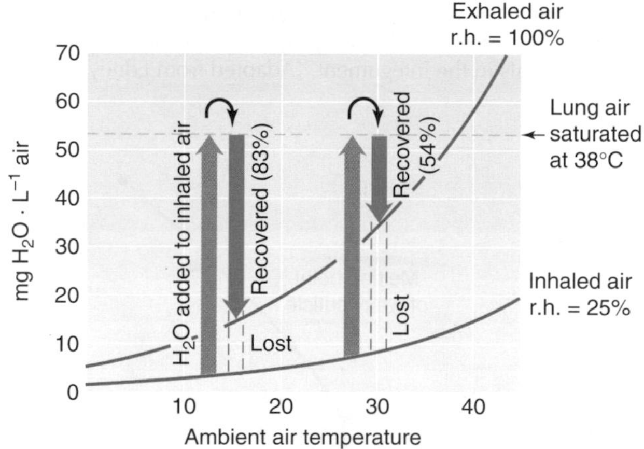

Classic Work **Figure 14-5** Water loss associated with respiration depends on the relationship between the body temperature and the temperature of inhaled air, as well as on the relative humidity of the inhaled air. As unsaturated air is warmed in the lungs, it picks up moisture until it is saturated (blue bars). During exhalation, the air is cooled in the nasal passages, so that much of the water is recovered there (red bars). The data shown are for the kangaroo rat when the inhaled air is at 25% relative humidity (r.h.) and 15°C (left bars) or 30°C (right bars). The amount of water recovered is obviously greater when the inhaled air is at the lower temperature. Indeed, under these climatic conditions, the kangaroo rat can exhale air at 13°C (lower than the ambient temperature). [Adapted from Schmidt-Nielsen et al., 1970.]

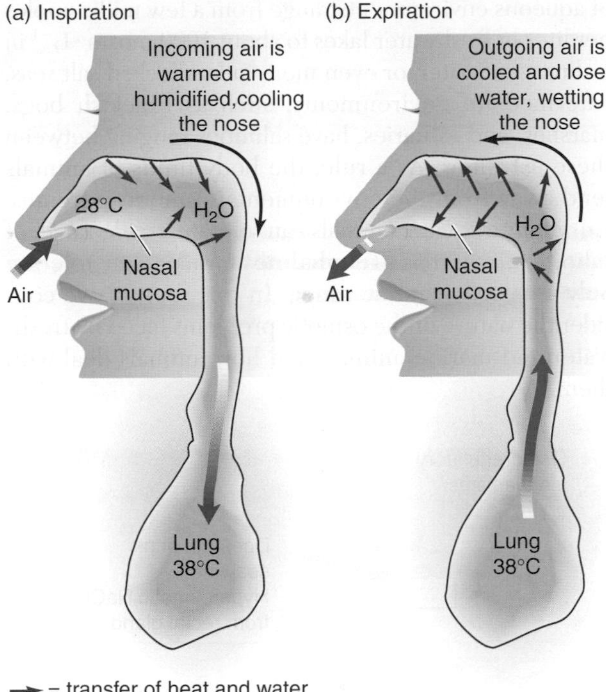

Figure 14-6 Temporal countercurrent exchange in the respiratory systems of many vertebrates acts to conserve body heat and body water. **(a)** During inhalation, cool air (e.g., at 28°C) is warmed and humidified as it flows toward the lungs, removing heat and water from the nasal passages. **(b)** During exhalation, the same air loses most of the heat and water it gained earlier as it warms and deposits water on the cooled nasal passages on its way out.

rest of the body. The tissue temperature is lowest at the tip of the nose and increases along the nasal passages toward the lung. The nose has a large blood supply that maintains the delivery of water to humidify the incoming air. The blood supply does not warm the nose because it is arranged in a countercurrent fashion, so that warm blood entering the nasal region is cooled by cold blood leaving the nose.

During exhalation, the process of heat exchange between air and nasal tissues is reversed. The warm expired air is cooled to somewhat above ambient temperature as it passes back out through the nasal passages, which were previously cooled by the same air during inhalation. As the exhaled air gives up some of its heat to the tissues of the nasal passages, most of the acquired moisture condenses back onto the cool nasal epithelium (Figure 14-6b). Mammals that employ this mechanism to humidify inhalant air (including humans) have "cold" noses, which are sometimes wet. With the next inhalation, this condensed moisture again contributes to the humidification of the inhaled air, and the cycle is repeated, with most of the vapor being recycled within the respiratory tract.

The nose, therefore, plays an important role in reducing the loss of water and heat from the body. The importance of the nose in cooling exhaled air can be detected easily by placing your hand in front of your nose and mouth and breathing out via your mouth and via your nose; the temperature difference is usually obvious. Because there is little cooling of air exhaled through the mouth, the loss of both water and heat is greater when breathing through the mouth (e.g., when the nose is clogged due to a cold) than through the nose. If air flow through the nasal passages is bypassed by placing a tube in the trachea, as during human or animal surgical operations, heat and water loss may increase; surgical patients must be given water to compensate for the loss. A common complaint after surgery is a sore throat, due to water loss from the trachea.

Strenuous exercise generates heat arising from muscle metabolism, which must be compensated for by a high rate of heat dissipation. The most effective routes for heat dissipation are evaporative cooling over respiratory surfaces (e.g., the lungs, air passages, and tongue) or evaporative water loss through the skin. If an animal exhales through its mouth to enhance heat loss, but breathes in through its nose, the nose will be cooled. In some very active mammals, body temperature rises during exercise but brain temperature remains normal, due to a countercurrent heat exchanger in the nasal region that cools the brain's blood supply.

Water loss via the lungs is small for mammals living in hot, humid climates and large for those living in cold, dry climates. The rate of ventilation and the ventilation pattern (e.g., breathing through either the mouth or nose) also affect the rate of water loss via the lungs. The problem of respiratory water loss is much less in animals with body temperatures similar to the ambient temperature; in this case, the air has only to be humidified at the ambient temperature. A reptile with a low metabolic rate and, therefore, a low ventilation rate and with a body temperature equal to the ambient temperature will have only minimal rates of water loss via the lung. This property confers on reptiles a selective advantage over mammals in regions where water is scarce. In marine iguanas, salt glands drain into the nasal passages, and the water in the excreted salt solution is used to wet the incoming air during inspiration. Occasionally all the water from the salt solution is evaporated, leaving salt deposits around the nose of these seawater-drinking lizards. The deposits are more obvious when the animal's temperature is well above that of the environment.

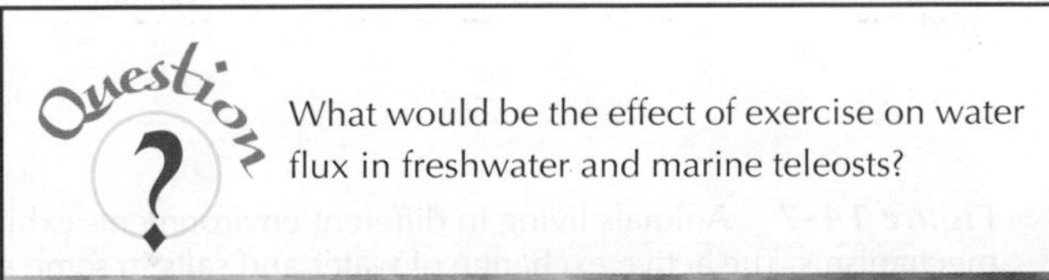

What would be the effect of exercise on water flux in freshwater and marine teleosts?

OSMOREGULATION IN AQUEOUS AND TERRESTRIAL ENVIRONMENTS

Animals face quite distinct osmotic problems in aqueous and terrestrial environments. In this section, we first discuss osmoregulation by water-breathing animals and then consider air-breathing animals. Figure 14-7 presents an overview of water and salt exchange in various osmoregulators.

Water-Breathing Animals

Many aquatic animals find themselves and all their respiratory surfaces immersed in water. The osmolarities of aqueous environments range from a few milliosmoles per liter in freshwater lakes to about 1000 $mosm \cdot L^{-1}$ in ordinary seawater, or even more in landlocked salt seas. Intermediate environments, such as brackish bogs, marshes, and estuaries, have salinities ranging between these extremes. As a rule, the body fluids of animals tend away from the environmental osmotic extremes. **Euryhaline** aquatic animals can tolerate a wide range of salinities, whereas **stenohaline** animals can tolerate only a narrow osmotic range. In this section, we consider the nature of the osmotic problems faced by freshwater and marine animals and how animals deal with them.

Type of animal	Blood concentration relative to environment	Urine concentration relative to blood	Osmoregulatory mechanisms
Marine elasmobranch	Slightly hyperosmotic	Iso-osmotic	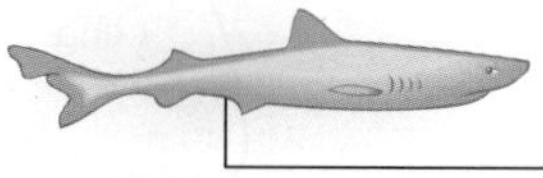 Does not drink seawater; Hyperosmotic NaCl from rectal gland
Marine teleost	Hypo-osmotic	Iso-osmotic	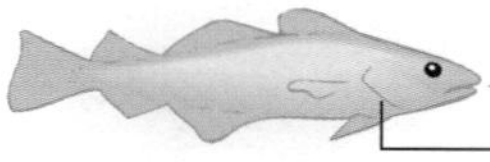 Drinks seawater; Secretes salt from gills
Freshwater teleost	Hyperosmotic	Hypo-osmotic	 Drinks no water; Absorbs salt with gills
Amphibian	Hyperosmotic	Hypo-osmotic	 Absorbs salt through skin
Marine reptile	Hypo-osmotic	Iso-osmotic	 Drinks seawater; Hyperosmotic salt-gland secretion
Desert mammal	–	Hyperosmotic	 Drinks no water; Depends on metabolic water
Marine mammal	Hypo-osmotic	Hyperosmotic	 Does not drink seawater
Marine bird	–	Hyperosmotic	 Drinks seawater; Hyperosmotic salt-gland secretion
Terrestrial bird	–	Hyperosmotic	 Drinks freshwater

Figure 14-7 Animals living in different environments exhibit different osmoregulatory mechanisms. The active exchange of water and salts in some vertebrates is illustrated here. Passive loss of water through the skin, lungs, and alimentary tract is not indicated.

Freshwater animals

The body fluids of freshwater animals, including invertebrates, fishes, amphibians, reptiles, and mammals, are generally hyperosmotic to their aqueous surroundings (see Table 14-1). Freshwater vertebrates have blood osmolarities in the range of 200–300 mosm · L^{-1}, while the osmolarity of freshwater is generally much less than 50 mosm · L^{-1}. Because they are hyperosmotic to their aqueous surroundings, freshwater animals face two kinds of osmoregulatory problems:

- They are subject to swelling owing to the movement of water into their bodies down the osmotic gradient.
- They are subject to the continual loss of body salts to the surrounding water.

Thus, freshwater animals must prevent the net gain of water and net loss of salts, which they accomplish by several means. Many freshwater animals possess an integument with a low permeability to both salts and water. As a general rule, animals living in freshwater refrain from drinking water, reducing the need to expel excess water. They get all the water they need from diffusion across the body surface.

One way to avoid a net gain of water is production of a dilute urine. Among closely related fishes, those that live in freshwater produce a more copious (i.e., plentiful and hence dilute) urine than their saltwater relatives (see Figure 14-7). The useful salts are largely retained by reabsorption into the blood from the ultrafiltrate in the tubules of the kidney. Nonetheless, some salts do pass out in the urine, so there is a potential problem of gradually washing out biologically important salts such as KCl, NaCl, and $CaCl_2$. Lost salts are replaced, in part, from ingested food. An important specialization for salt replacement in freshwater animals is the active transport of salts from the external dilute medium across the epithelium into the interstitial fluid and blood. This activity is accomplished by transporting epithelia such as those in the skin of amphibians and in the gills of fishes. In fishes and many aquatic invertebrates, the gills act as the major osmoregulatory organs, having many of the functions located in the kidneys of mammals.

Freshwater animals have remarkable abilities to take up salts from their dilute environment. The gills of freshwater fishes, for example, are able to extract Na^+ and Cl^- ions from water containing less than 1 mM NaCl, even though the plasma concentration of NaCl exceeds 100 mM. Thus, the active transport of NaCl in the gills takes place against a concentration gradient in excess of a hundredfold. The mechanisms of sodium reabsorption appear to have many similarities in the gills of freshwater fishes, frog skin, the turtle bladder, and the mammalian kidney. In all cases, the cells of these epithelia are joined together by tight junctions. The mechanisms of sodium reabsorption will be discussed in detail below.

Marine animals

In general, the intracellular and extracellular body fluids of marine invertebrates and ascidians (primitive chordates) are close to seawater both in osmolarity (iso-osmotic) and in the concentrations of the major inorganic salts (see Table 14-1). Such animals, therefore, need not expend much energy in regulating the osmolarity of their body fluids. A rare example of a vertebrate whose plasma is iso-osmotic to the marine environment is the hagfish. It differs from most marine invertebrates, however, in that it does regulate the concentrations of individual ions. In particular, blood Ca^{2+}, Mg^{2+}, and SO_4^{2-} are maintained at significantly lower concentrations than those in seawater, whereas the concentrations of Na^+ and Cl^- are higher. Since various functions of excitable tissues such as nerve and muscle in vertebrates are especially sensitive to concentrations of Ca^{2+} and Mg^{2+}, the regulation of these divalent cations may have evolved to accommodate the requirements of neuromuscular function.

Like the hagfishes, the elasmobranchs, such as sharks, rays, and skates, as well as the primitive coelacanth *Latimeria,* have plasma that is approximately iso-osmotic to seawater. They differ from the hagfishes, however, in that they maintain far lower concentrations of electrolytes, making up the difference with organic osmolytes such as urea and trimethylamine oxide. High urea concentrations tend to cause the breakup of proteins into their constituent subunits, whereas TMAO has the opposite effect, canceling the effect of urea and stabilizing protein structure in the face of high urea levels. In the elasmobranchs and coelacanths, excess inorganic electrolytes such as NaCl are excreted via the kidneys and also by means of a special excretory organ, the rectal gland, located at the end of the alimentary canal.

The body fluids of marine teleosts (modern bony fishes), like those of most higher vertebrates, are hypotonic to seawater, so there is a tendency for these fishes to lose water to the environment, especially across the gill epithelium. To replace the lost water, they drink saltwater. Most of their net salt uptake comes from drinking seawater rather than from uptake across the body surface or gills. Seventy to eighty percent of the ingested water is absorbed across the intestinal epithelium and enters the bloodstream, along with most of the NaCl and KCl it contains. Initially, the ingested seawater is diluted by about 50% by diffusion of salts across the esophagus. Active uptake of salts occurs in the small intestine. Left behind in the gut and expelled through the anus are most of the divalent cations such as Ca^{2+}, Mg^{2+}, and SO_4^{2-}. The excess salt absorbed along with the water is subsequently eliminated from the blood by active transport of Na^+, Cl^-, and some K^+ across the gill epithelium into the seawater and by secretion of divalent salts by the kidney. The urine is isotonic to the blood, but rich in those salts (especially Ca^{2+}, Mg^{2+}, and SO_4^{2-}) that are not secreted by the gills.

The gills of marine teleosts, as might be expected, are organized differently from those of freshwater fishes. The gill epithelium contains specialized cells, called **chloride cells**, that mediate transport of NaCl from the blood into the surrounding water. The mechanism of this transport, which makes it possible for these fishes to live in saltwater, is described below.

Some species—for example, the Pacific salmon and the eels in eastern North America and Europe—are able to maintain a more or less constant plasma osmolarity even though they migrate between marine and freshwater environments. Such fishes undergo physiological changes that enable them to maintain a more or less constant ionic composition in both environments. Some of these changes begin before the animal enters seawater. Eels, for instance, reduce the permeability of the integument, changing from yellow to silver in the process (see Chapter 13). We'll discuss the physiological changes found in migrating teleosts in detail later in this chapter.

How could you estimate the energetic cost of osmoregulation? Would you expect the cost to be less if freshwater and marine fish were placed in an iso-osmotic solution?

Air-Breathing Animals

Animals in a terrestrial environment can be thought of as being submerged in an ocean of air rather than water. Unless the humidity of the air is high, animals with a water-permeable epithelium are subject to dehydration very much as if they were submerged in a hypertonic medium such as seawater. Dehydration could be avoided if all epithelial surfaces exposed to air were totally impermeable to water. However, an epithelium that was impermeable to water (and thus dry) would have limited permeability to oxygen and carbon dioxide, and would thus be unsuited for the respiratory needs of a terrestrial animal. As a consequence, air-breathing animals are subject to dehydration through their respiratory epithelia, as we saw above. Air-breathing animals utilize various means to minimize water loss to the air by this route and others (see Figure 14-7).

Marine reptiles (e.g., marine iguanas, sea turtles, estuarine crocodiles, sea snakes) and marine birds drink seawater to obtain a supply of water, but, like marine teleosts, they are unable to produce a concentrated urine that is significantly hyperosmotic to their body fluids. Instead, they are endowed with glands specialized for the secretion of salts in a strong hyperosmotic fluid. These **salt glands** are generally located above the orbit of the eye in birds and near the nose or eyes in lizards. Brackish-water crocodiles were long suspected of using extrarenal means of excreting salts, and eventually salt glands were discovered in the tongue of this reptile. The salt glands of marine reptiles and birds secrete a sufficiently concentrated salt solution to compensate for the inability of their kidneys to excrete the salts they take in by drinking seawater (Figure 14-8a).

Human beings, like other mammals, are not equipped to drink seawater. The human kidney can remove up to about 6 g of Na^+ from the bloodstream per liter of urine produced. Because seawater contains about 12 $g \cdot L^{-1}$ of Na^+, imbibing seawater would cause a human to accumulate salt without adding a physiologically equivalent amount of water (Figure 14-8b). Stated differently, to excrete the salt ingested with a given volume of seawater, the human kidney would

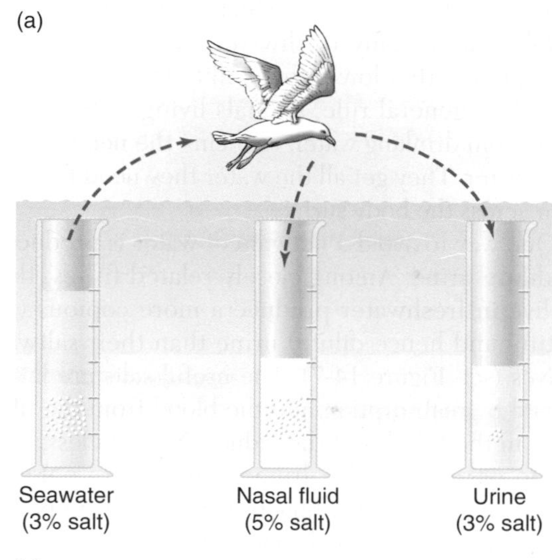

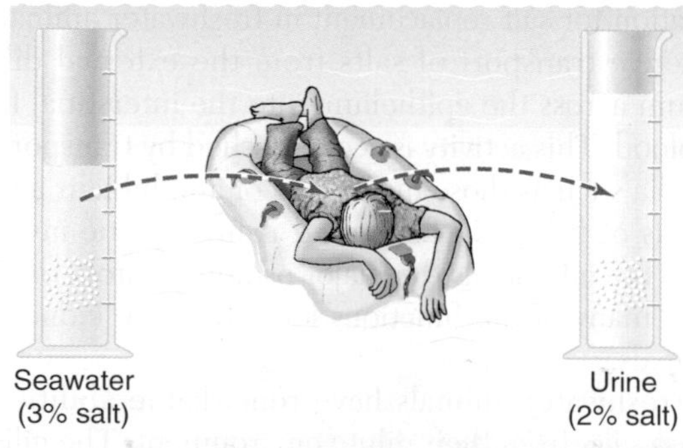

Classic Work ***Figure 14-8*** Marine reptiles and birds drink seawater to obtain water, whereas most mammals become dehydrated if they drink seawater. **(a)** When marine birds drink seawater, they secrete NaCl via the salt glands, thereby eliminating 80% of the ingested salt along with only 50% of the ingested water. As a result, these birds can produce a hypotonic urine without becoming dehydrated. **(b)** When humans and other mammals, which lack salt glands, drink seawater, they cannot concentrate their urine sufficiently to conserve water while eliminating the ingested salts. [From "Salt Glands" by K. Schmidt-Nielsen. Copyright © 1959 by Scientific American, Inc. All rights reserved.]

have to pass more water than was contained in that volume. Of course this net loss of water would lead rapidly to dehydration. The water produced by metabolism is not sufficient to offset losses through respiration and other routes, so humans stranded at sea will die unless they have access to drinking water.

Many other mammals, however, can survive without drinking water. Joseph Priestley (1733–1804), who was the first scientist to isolate many gases, including oxygen, observed that he could keep mice alive without water for three or four months in a cage on a shelf above the kitchen fireplace of his home in Yorkshire, England. Another small mammal, the kangaroo rat *(Dipodomys merriami)*, a native of the American Southwest, has become a classic example of a small mammal adapted to survive in the arid conditions of the desert without drinking water. Let's see how these mammals, as well as certain terrestrial arthropods, survive in the absence of freshwater.

Desert-dwelling mammals

The survival strategies of the kangaroo rat exemplify a variety of osmoregulatory adaptations characteristic of many small desert mammals (Figure 14-9). The kangaroo rat and other desert mammals are faced with a physiological double jeopardy—excessive heat and a near absence of free freshwater. Osmoregulation and temperature regulation are, of course, closely related, since one important means of channeling excess heat out of the body into the environment is by evaporative cooling. Since evaporative cooling is at odds with water conservation, most desert animals cannot afford to use this method and have devised means of circumventing it. The kangaroo rat, like many desert mammals, avoids much of the daytime heat by remaining in a burrow during the daylight hours and coming out only at night. This nocturnal lifestyle is an important and widespread behavioral adaptation to desert life. The cool burrow not only reduces the animal's heat load, but also reduces

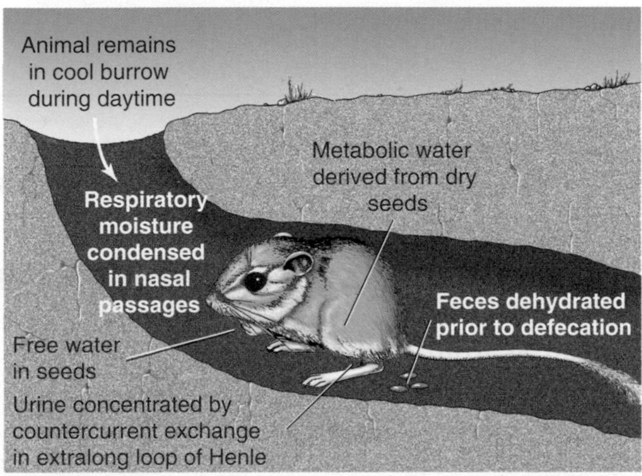

Figure 14-9 The water-conserving strategies of the kangaroo rat are characteristic of many small desert dwellers.

Table 14-5 **Sources of water gain and loss by the kangaroo rat**

Gains		Losses	
Metabolic water	90%	Evaporation and perspiration	70%
Free water in "dry" food	10%	Urine	25%
Drinking	0%	Feces	5%
	100%		100%

Source: Schmidt-Nielsen, 1972.

its respiratory water loss. The nasal countercurrent mechanism for conserving respiratory moisture depends, of course, on the ambient temperature in the burrow being significantly lower than the 37–40°C characteristic of the core body temperatures of birds and mammals. If the rodent ventures out of its cool burrow into air close to its own temperature, its respiratory water loss rises abruptly, since the cooling properties of the nasal epithelium are reduced. Desert mammals also generally avoid heat-generating exercise during the day. Because of its efficient kidneys, the kangaroo rat excretes a highly concentrated urine, and rectal absorption of water from the feces results in essentially dry fecal pellets. By using all these adaptations, the kangaroo rat greatly reduces its potential water loss. In spite of this extreme osmotic economy, the small amount of lost water must, of course, be replaced, or the animal will eventually dry up. Since the kangaroo rat eats dry seeds that contain only a trace of free water, is not known to drink, and in fact survives quite well in the near absence of free water, it must have a cryptic source of water. This source turns out to be the metabolic water noted earlier, produced as a by-product of the oxidation of foods (see Table 14-4). Its exquisite adaptations for water conservation allow the kangaroo rat to survive primarily on metabolic water, with water gains equal to water losses over the long term (Table 14-5).

Unlike the kangaroo rat, the camel is too large to hide from the hot desert sun in a burrow. When deprived of drinking water, camels do not sweat, but allow their body temperature to rise rather than losing water by evaporative cooling during the heat of the day. During the cooler night, the camel's body temperature drops. It increases only slowly the next day because of the animal's large body mass and thick fur, which acts as a heat shield. Nevertheless, the body temperature of a dehydrated camel may vary from 35°C at night to 41°C during the day (Figure 14-10a on the next page). This strategy is impossible in small rodents, in which body temperatures oscillate much more rapidly than in the larger camel (Figure 14-10b). Because of their small size, desert rodents heat up rapidly in the sun, so they must return to their burrows to cool down. The camel also reduces heating by orienting itself to give minimal

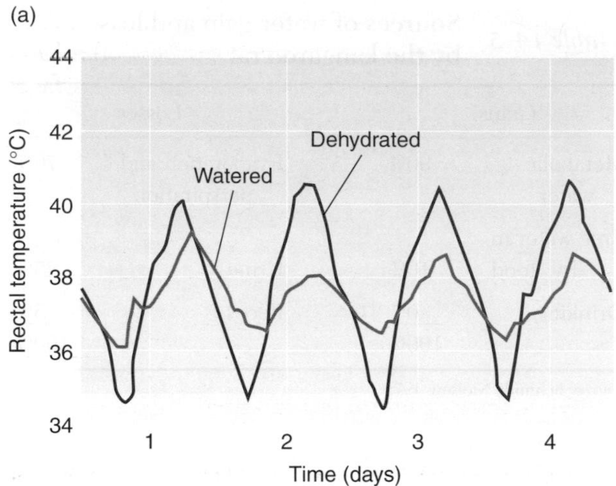

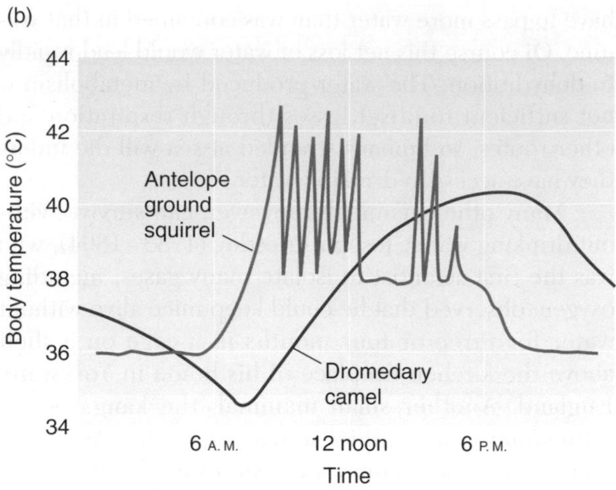

Classic Work **Figure 14-10** When water is scarce, large desert animals such as the camel exhibit a large, but slow, increase in body temperature during the day, whereas smaller animals heat up rapidly when exposed to the sun. **(a)** Daily temperature fluctuations in a well-watered and a dehydrated camel. When a camel is deprived of drinking water, its daily temperature fluctuation may increase to as much as 7°C. **(b)** Diagrammatic representation of the daily patterns of body temperature in a large and a small mammal subjected to heat stress under desert conditions. Small animals must enter burrows periodically to avoid overheating. [Part a from Schmidt-Nielsen, 1964; part b from Bartholomew, 1964.]

surface exposure to direct sunlight. The camel, like other desert animals, produces dry feces and concentrated urine. When water is not available, the camel does not produce urine, but stores urea in the tissues. The camel can tolerate not only dehydration but also high urea levels in the body. When water becomes available, these ships of the desert can rehydrate by drinking 80 liters in 10 minutes.

Marine mammals

Marine mammals face problems similar to those of desert animals because they live in an environment without available drinking water. The physiological responses of marine mammals, although different in detail, are generally similar to those of desert mammals. The emphasis is on water conservation. Marine mammals are endowed, as are other mammals, with highly efficient kidneys capable of producing a hypertonic urine. Seals have a characteristic labyrinth-like proliferation of epithelial surfaces in the nasal passages to reduce water loss by respiration. Whales and dolphins have a blowhole rather than the typical mammalian nose. These animals have large lung tidal volumes. The velocity of air flow through the blowhole is high because both inhalation and exhalation are rapid and large volumes of gas are moved with each breath. It is possible that the expansion of air passing through the blowhole also cools the air, resulting in water condensation in the region of the blowhole that can be used to wet inspired air.

A remarkable example of water retention in a marine mammal occurs in recently weaned baby elephant seals. After being abandoned by its mother, a baby seal goes for 8–10 weeks without food or water. During this time, the baby seal's only source of water is that derived from the oxidation of its body fat. It weighs about 140 kg at the time of weaning and loses only about 800 g of water per day, of which less than 500 g is lost through respiration. This economy is ascribed both to its nasal countercurrent heat exchanger and to a low metabolic rate, which allows it to stop breathing for 8 to 10 minutes. The ability to suspend breathing is, of course, not uncommon for marine mammals such as the elephant seal, which can dive for prolonged periods. The ability to conserve water is also seen in adult elephant seals. Adult females suckle their young for about four weeks on the beach, then abandon their pups and go to sea for four months. They return to land for about a month to molt, then go to sea again for another six months. While at sea, elephant seals do not drink, but rely on metabolic water and the water in the fish they consume. The large males spend up to three months on the beach during the breeding season, during which time they neither drink nor eat.

What are the osmoregulatory problems faced by young camels and whales? Can you suggest solutions to these problems?

Terrestrial arthropods

Certain terrestrial arthropods have the ability to extract water vapor directly from the air, even, in some species, when the relative humidity is as low as 50%. This ability has been demonstrated in certain arachnids (ticks, mites) and in a number of wingless forms of insects, primarily larvae. Many species that exhibit this ability live in desert habitats. The ability to remove water from the air is all the more remarkable in these arthropods because it normally occurs even when the vapor pressure of the hemolymph exceeds that of the air, which it does at all values of relative humidity below 99%. The water vapor pressure associated with a solution decreases with increasing ionic content, so highly concentrated salt solutions can absorb water from air. Insects take advantage of this effect by creating very concentrated solutions in their bodies. The site of water entry is often the rectum, which reduces the water content of fecal matter to remarkably low levels. As water is removed from the feces, the feces can take up new water from the air, if the water vapor pressure is high enough and if the air has access to the rectal lumen. In ticks, tissues in the mouth have been implicated in the uptake of water vapor. Here it appears that the salivary glands secrete a highly concentrated solution of KCl that in turn absorbs water from the air. Water vapor pressure associated with solutions in capillaries is also low (which is why velvet fabric is difficult to dry). Water capillaries formed between hairs can absorb water from air, a mechanism used by some insects to take up water.

OSMOREGULATORY ORGANS

The osmoregulatory capabilities of metazoans depend to a great extent on the properties of the transport epithelia located in the gills, skin, kidneys, and gut. The *apical* surface (sometimes referred to as the *mucosal* or *luminal* surface) of an epithelial cell faces a space that is continuous with the external world (such as the sea, the pond, the lumen of the gut, or the lumen of a kidney tubule). The other side of the epithelial cell, the *basal* surface (sometimes referred to as the *serosal* surface) generally bears deep basal clefts and faces an internal compartment containing extracellular fluid (see Chapter 4). This internal compartment is the one that contains all the other cells of the remaining body tissues. The cells of the body exist, so to speak, in their own private "pond," composed of the extracellular fluid in which they are bathed. The proper composition of this internal fluid depends on the osmoregulatory work and barrier functions performed by the epithelial cells.

The mechanisms for transporting substances across epithelia were discussed in Chapter 4. This same basic cellular machinery is used in all excretory and osmoregulatory organs. For example, similar salt-excreting cells are found in the nasal glands of birds and reptiles, the mammalian kidney, the rectal gland of elasmobranchs, and the gills of marine teleost fishes. Not only are the cells similar, but their activities are regulated by similar arrays of hormones. The cells themselves, however, are not identical. The basic plan is modified in each case to comply with the specific requirements of the osmoregulatory organ. We will consider several of these modifications here.

Energy, in the form of ATP, is expended directly or indirectly in transferring ions across epithelia against a concentration gradient. The role of ATP in the development of electrochemical gradients across cell membranes is dependent on three classes of ion-motive ATPases, or pumps (Figure 14-11; Table 14-6 on the next page). **F-ATP synthases,** found in mitochondria

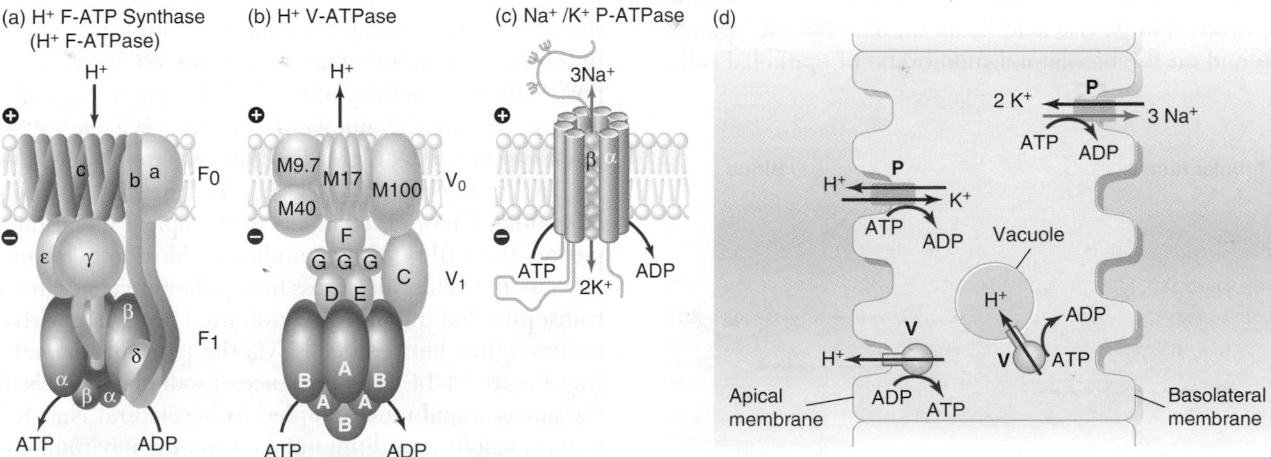

Figure 14-11 Several types of ATPases, each consisting of a large number of subunits, are involved in setting up electrochemical gradients across transporting epithelia. The catalytic/regulatory subunits are shown in red and the membrane domain subunits in purple, with connecting subunits in orange. **(a)** Protons from electron transport in the mitochondrion pass through the proton F-ATP synthase, driving ATP synthesis. **(b)** Proton V-ATPase pumps protons, utilizing energy from ATP. **(c)** Na^+/K^+ P-ATPase simultaneously pumps sodium out of cells and potassium into cells. **(d)** A model epithelial cell, showing the normal locations of various ion-motive ATPases. Proton V-ATPase pumps protons into vacuoles or out of the cell across the apical membrane. K^+/H^+ P-ATPases are also located in the apical membrane and pump protons out of the cell. Na^+/K^+ P-ATPase is found in the basolateral membrane. [Parts a–c from Nelson and Cox, 2000; part d from Wieczorek et al., 1999.]

Table 14-6 **Comparison of the three major classes of ion-motive ATPases**

Characteristic	H^+ F-ATP synthase	H^+ V-ATPase	Na^+/K^+ P-ATPase
Distribution	Mitochondria, chloroplast, bacterial plasma membrane	Eukaryote endomembranes, animal plasma membranes (apical)	Animal plasma membranes (basolateral)
Function	ATP synthesis, prokaryote plasma membrane energization	Vacuolar acidification, animal plasma membrane energization	Animal plasma membrane energization
Activity	Uses H^+ electrochemical gradient to make ATP; uses ATP to make H^+ electrochemical gradient	Uses ATP to make H^+ electrochemical gradient	Uses ATP to make electrochemical gradient for H^+, Na^+, K^+, Ca^{2+}
Close relatives	Na^+ F-ATP synthase, H^+ A-ATPase, H^+ V-ATPase	H^+ F-ATP synthase, H^+ A-ATPase	H^+/K^+ ATPase, Ca^{2+}P-ATPase
Inhibitor (concentration)	Azide (10^{-4} M)	Bafilomycin (10^{-9} M)	Ouabain (10^{-6} M)
Molecular mass	~380 kDa	~750 kDa	~100 kDa

Source: Adapted from Wieczorek et al, 1999.

and chloroplasts, use a proton electrochemical gradient to make ATP. **V-ATPases**, or vacuolar-type ATPases, hydrolyze ATP to generate electrochemical gradients, as do **P-ATPases**. These gradients serve many functions, including directing the movement of ions through associated channels, symporters, and antiporters (see Chapter 4).

Unlike those of V-ATPase, the reactions of the P-ATPases have a phosphorylated intermediate. P-ATPases include not only Na^+/K^+ P-ATPase (the Na^+/K^+ pump we have discussed in several earlier chapters), but also Ca^{2+} P-ATPase (the calcium pump involved in muscle contraction) and H^+/K^+ P-ATPase (involved in gastric acidification). The Na^+/K^+ pump, found on the basolateral membrane of epithelial cells, regulates intracellular sodium levels and cell volume by moving sodium out of the cell into the extracellular fluid. If the cell has K^+ channels in the apical membrane, then K^+ will be excreted into the extracellular space, as long as the electrochemical gradient is maintained in the appropriate direction (Figure 14-12). If there are K^+ channels in the basolateral membrane, there will be K^+ cycling between the cell and the extracellular fluid, driven by the activity of the Na^+/K^+ pump (Figure 14-13). These mechanisms are involved in the secretion of K^+ and the retention of Na^+ by the mammalian kidney, as we will see below.

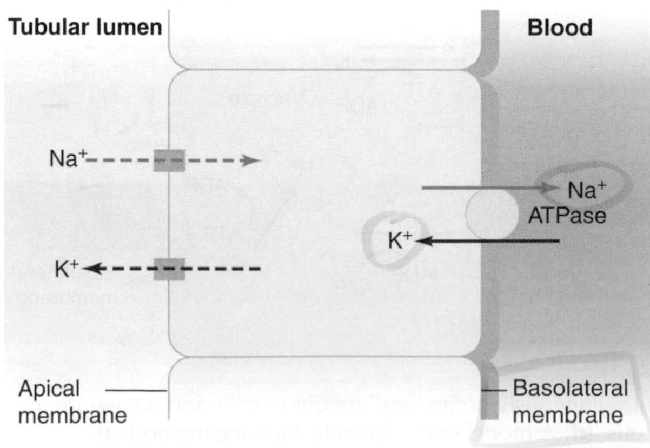

Figure 14-12 In the distal tubule and collecting duct of the mammalian kidney, epithelial cells secrete K^+ into the tubular filtrate. A Na^+/K^+ pump in the basolateral membrane actively transports K^+ into the cell; it then passively moves down its electrochemical gradient via K^+ channels in the apical membrane into the lumen.

The Na^+/K^+ ATPase directly or indirectly supports the movement of many substances. If the apical membrane contains a Na^+/glucose or a $Na^+/2Cl^-/K^+$ symporter, then the activity of the Na^+/K^+ pump can drive glucose, K^+, or Cl^- uptake. If the $Na^+/2Cl^-/K^+$ symporter is located on the basolateral membrane, the Na^+/K^+ pump can drive Cl^- uptake from the extracellular fluid. Chloride channels in the apical membrane permit the efflux of accumulated chloride. The net movement of chloride across the epithelium generates a transepithelial membrane potential that can drive sodium efflux between cells (via the paracellular pathway, Figure 14-14). The presence of sodium channels in the apical membrane, coupled to basolateral Na^+/K^+ pumps, results in sodium uptake from the environment as long as the sodium electrochemical gradient is favorable. If the apical membrane contains a Na^+/H^+ antiporter, then sodium uptake can be coupled to acid excretion. If environmental sodium levels are very low, however, the electrochemical gradient may not be sufficient to cause sodium uptake.

The addition of a proton pump (H^+ V-ATPase) to the apical membrane increases the electrochemical gradient and promotes sodium uptake from dilute solu-

(a)

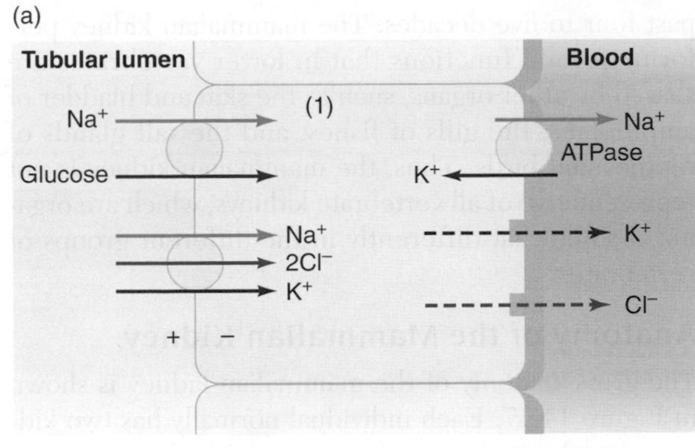

(b)

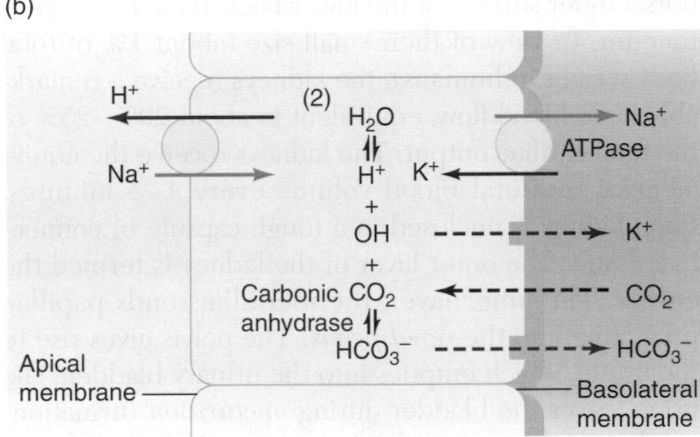

Figure 14-13 Several transport systems are involved in reabsorption of Na^+ in the proximal tubule and ascending limb of the loop of Henle in the mammalian kidney. **(a)** Sodium passively crosses the apical membrane via $Na^+/2Cl^-/K^+$ and glucose/Na^+ cotransporters. A Na^+/K^+ pump in the basolateral membrane actively removes Na^+ from the cell into the blood; K^+ and Cl^- exit via ion channels down their concentration gradients. **(b)** The movement of Na^+ down its electrochemical gradient into the cell also energizes the outward movement of protons via an electrically neutral Na^+/H^+ antiporter. Carbon dioxide in the blood diffuses into the cell, where carbonic anhydrase ensures a high rate of proton delivery to the exchanger. A basolateral sodium pump transports Na^+ from the cell into the blood. K^+ and HCO_3^- exit via ion channels down their electrochemical gradients.

tions, such as across frog skin or the freshwater fish gill (Figure 14-15). This mechanism also couples sodium uptake to acid excretion. The activity of the proton pump, however, could deplete the immediate intracellular site of protons, and this loss would slow the activity of the pump. Carbonic anhydrase, which catalyzes the interconversion of carbon dioxide and bicarbonate, is usually co-localized with the proton pump and maintains the supply of protons. The coupling of carbonic anhydrase with a proton pump results in bicarbonate accumulation in the cell; thus, an apical proton pump may be coupled to a chloride bicarbonate antiporter on the basolateral membrane. Thus carbon dioxide diffuses into a cell, but leaves as acid across the apical membrane along with bicarbonate in exchange for chloride (see Figure 14-15). The presence of a proton pump (H^+ V-ATPase) associated with a K^+/H^+ antiporter

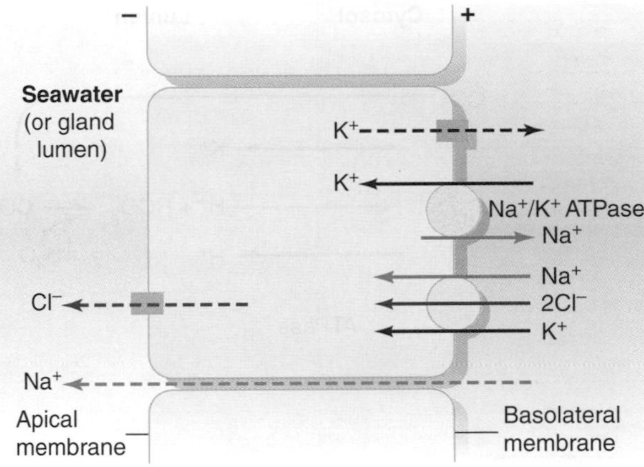

Figure 14-14 Salt-secreting cells present in the rectal gland of sharks, the avian and reptilian nasal gland, and the gills of marine teleosts all use the same basic mechanism for transporting salt from the blood. Operation of the Na^+/K^+ ATPase and the $Na^+/2Cl^-/K^+$ cotransporter in the basolateral membrane results in net movement of Cl^- from the blood into the cell. Chloride diffuses from the cell into the external environment via Cl^- channels in the apical membrane. The transmembrane potential created by this movement increases the Na^+ electrochemical gradient such that Na^+ can diffuse via paracellular channels even against a high sodium concentration gradient.

results in the secretion of potassium and the uptake of acid. The proton gradient drives the antiporter, but because two protons are transferred for each K^+ ion, the effect is to alkalinize the secretion (Figure 14-16 on the next page).

Numerous other factors also influence the movement of ions through epithelial cells. The cells forming

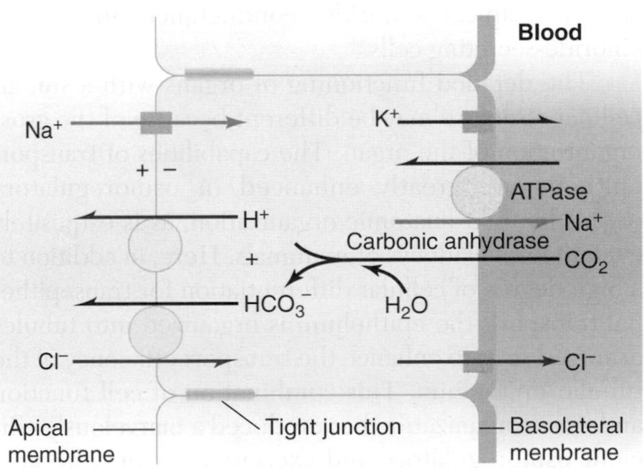

Figure 14-15 Epithelia can take up Na^+ from dilute media using a proton pump in the apical membrane to generate a gradient to move Na^+ into the cell. Sodium levels in the cell are kept low by the Na^+/K^+ pump in the basolateral membrane. Carbonic anhydrase catalyzes the hydration of CO_2, which supplies protons for the proton pump. The bicarbonate concentration gradient drives the anion antiporter in the apical membrane, taking up Cl^- into the cell.

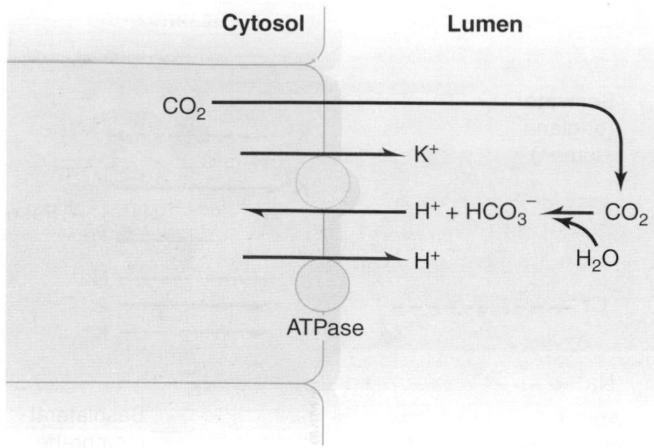

Figure 14-16 In the epithelia of the lepidopteran midgut, transport is energized by proton V-ATPase in the apical membrane. Because it is coupled to a $K^+/2H^+$ antiporter, the net effect is an alkalinization of the external medium. [Adapted from Wieczorek et al., 1999.]

the epithelium may be joined together by tight junctions, resulting in very low transepithelial permeability. Alternatively, there may be leaky tight junctions between cells, creating paracellular pathways that allow the passage of ions, as, for example, in salt-secreting epithelia (see Figure 14-14). There may be gap junctions between cells in an epithelium such that potentials generated in one region spread over the epithelium. The activity of Na^+/K^+ and proton ATPases can be modified by hormones; for example, aldosterone stimulates the activity of both ATPases. Cytosolic pH-sensitive proteins have been identified that either activate or inhibit V-ATPase activity. Hormones also influence channel activity; for example, epinephrine, prostaglandin E_2, or vasoactive intestinal peptide stimulate a cAMP-dependent protein kinase that in turn activates apical chloride conductance in several chloride-secreting cells.

The detailed functioning of organs with a similar cellular structure may be different because of the gross organization of the organ. The capabilities of transport epithelia are greatly enhanced in osmoregulatory organs by their anatomic organization, as is exquisitely evident in the kidneys of mammals. Here, in addition to a high degree of cellular differentiation for transepithelial transport, the epithelium is organized into tubules arranged so as to enhance the transport efficiency of the tubular epithelium. This combination of cell function and tissue organization has produced a marvelously efficient osmoregulatory and excretory organ. The next several sections describe and compare the operation of various types of osmoregulatory organs found in different animals.

THE MAMMALIAN KIDNEY

The mammalian kidney is the osmoregulatory organ we understand best, thanks to intensive research over the past four to five decades. The mammalian kidney performs several functions that in lower vertebrates are shared by other organs, such as the skin and bladder of amphibians, the gills of fishes, and the salt glands of reptiles and birds. Thus, the mammalian kidney is not representative of all vertebrate kidneys, which are organized somewhat differently in the different groups of vertebrates.

Anatomy of the Mammalian Kidney

The gross anatomy of the mammalian kidney is shown in Figure 14-17. Each individual normally has two kidneys, one located on each side of the body against the dorsal inner surface of the lower back, outside the peritoneum. In view of their small size (about 1% of total body weight in humans), the kidneys receive a remarkably large blood flow, equivalent to about 20%–25% of the total cardiac output. The kidneys receive the equivalent of the total blood volume every 4–5 minutes. Each kidney is enclosed in a tough capsule of connective tissue. The outer layer of the kidney is termed the **cortex.** The inner layer, the **medulla,** sends papillae projecting into the *renal pelvis*. The pelvis gives rise to the ureter, which empties into the urinary bladder. The urine leaves the bladder during *micturition* (urination) via the **urethra,** which leads to the end of the penis in males and into the vulva in females.

Human adults normally produce about a liter of slightly acidic (pH approximately 6.0) urine each day. The rate of urine production varies diurnally, being high during the day and low at night, reflecting the time course of water intake and metabolic water production.

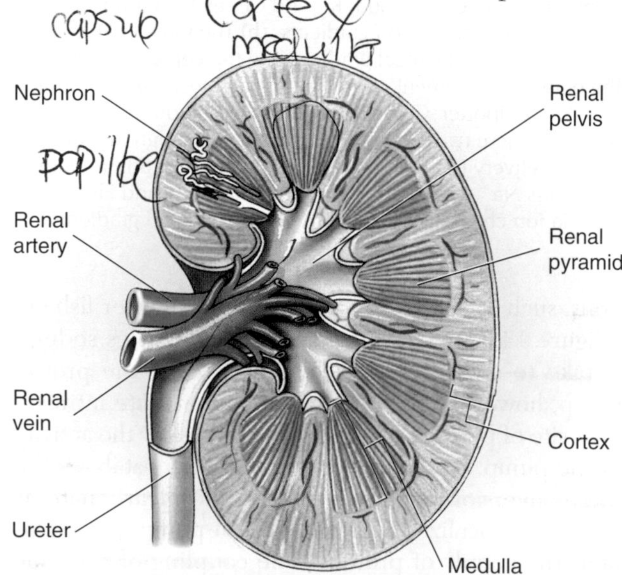

Figure 14-17 The functional units of the mammalian kidney, called nephrons, are arranged in a radiating fashion within the renal pyramids. The distal end of each nephron empties into a collecting duct, which drains into a central cavity termed the renal pelvis. The urine passes from the pelvis into the ureter, which takes it to the urinary bladder. In this cross-sectional drawing, only one nephron is depicted, although each pyramid contains many nephrons.

Urine contains water and other by-products of metabolism, such as urea, as well as NaCl, KCl, phosphates, and other substances that are present at concentrations in excess of the body's requirements. The function of the kidneys is to maintain a more or less constant body composition; hence, the volume and composition of urine reflects the volume of fluid taken in and the amount and composition of ingested food. The actual volume of urine produced is determined by the volume of water ingested plus the water produced during metabolism minus evaporative water loss via the lungs and sweating and, to a lesser extent, via the feces. When voided, urine is normally clear and transparent, but after a rich meal it may become alkaline and slightly turbid. The smell and color of urine are determined by the diet. For example, consumption of methylene blue will give urine, which typically is yellow, a distinctive blue color, and consumption of asparagus will completely change the more usual, slightly aromatic odor of urine.

The release of urine is accomplished by the simultaneous contraction of the smooth muscle of the bladder wall and relaxation of the skeletal muscle sphincter around the opening of the bladder. As the bladder is stretched by gradual filling, stretch receptors in the wall of the bladder generate nerve impulses that are carried by sensory neurons to the spinal cord and brain, producing the associated sensation of fullness. The sphincter can then be relaxed by inhibition of motor impulses, allowing the smooth muscle of the bladder wall to contract under autonomic control and empty the contents. The presence of a bladder allows the controlled release of stored urine rather than a continual dribble paralleling the flow of urine from the kidney into the bladder. Such controlled release is used by some animals to mark out their territory.

The functional unit of the mammalian kidney is the **nephron** (Figure 14-18), an intricate epithelial tube that is closed at its beginning but open at its distal end. Each kidney contains numerous nephrons, which empty into **collecting ducts.** These ducts combine to form papillary ducts, which eventually empty into the renal pelvis. At its closed end, the nephron is expanded—somewhat like a balloon that has been pushed in from one end toward its neck—to form the cup-shaped **Bowman's capsule.** The lumen of the capsule is continuous with the narrow lumen that extends through the renal tubule. A tuft of capillaries forms the renal **glomerulus** inside Bowman's capsule. This remarkable structure is responsible for the first step in urine formation. An ultrafiltrate of the blood moves from the capillaries into the lumen of Bowman's capsule, where it begins its trip through the various segments of the renal tubule, finally arriving at the collecting duct and eventually passing into the renal pelvis.

The wall of the renal tubule is one cell layer thick. This epithelium separates the lumen of the tubule, which contains the ultrafiltrate, from the interstitial fluid. In some portions of the nephron, the epithelial

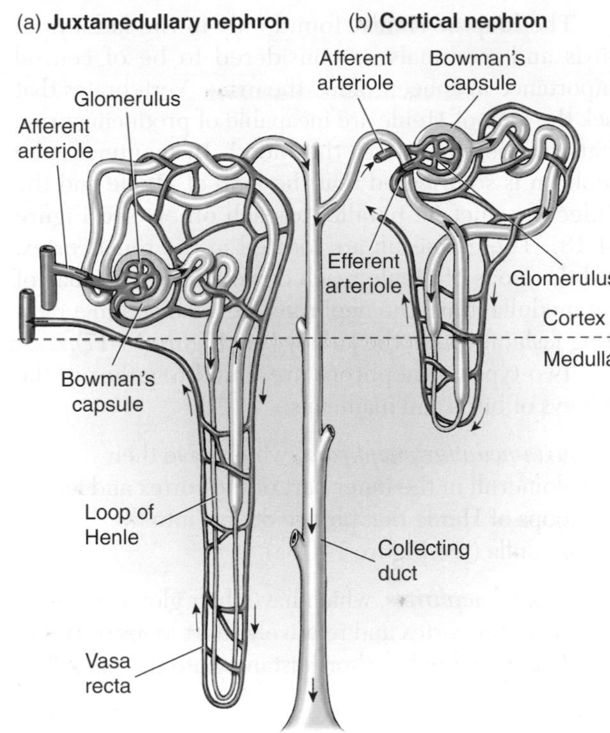

Figure 14-18 The mammalian nephron is a long tubular structure, which is closed at its beginning in Bowman's capsule but open at its terminus, where it empties into a collecting duct. The renal tubule and collecting duct are shown in yellow, the vascular elements in red and blue. **(a)** Juxtamedullary nephrons have a long loop of Henle, which passes deep into the renal medulla and is associated with a vasa recta. Blood first passes through the capillaries of the glomerulus and then flows through the hairpin loops of the vasa recta, which plunges into the medulla of the kidney along with the loop of Henle. **(b)** The more common cortical nephrons have a short loop of Henle, only a small portion of which enters the medulla, and lack a vasa recta. In these nephrons, blood passes from the afferent arteriole to the glomerular capillaries and then leaves the nephron via the efferent arteriole.

cells are morphologically specialized for transport, bearing a dense pile of microvilli on their apical surfaces and deep infoldings of their basolateral membranes (Figure 14-19 on the next page). The epithelial cells are tied together by leaky tight junctions, which permit limited paracellular diffusion between the lumen and the interstitial space surrounding the renal tubule.

The nephron can be divided into three main regions: the proximal nephron, the loop of Henle, and the distal nephron. The proximal nephron consists of Bowman's capsule and the **proximal tubule.** The hairpin loop of Henle comprises a descending limb and an ascending limb. The latter merges into the **distal tubule,** which joins a collecting duct serving several nephrons. The number of nephrons per kidney varies from several hundred in lower vertebrates to many thousands in small mammals to a million or more in humans and other large species.

The loop of Henle, found only in the kidneys of birds and mammals, is considered to be of central importance in concentrating the urine. Vertebrates that lack the loop of Henle are incapable of producing urine that is hyperosmotic to the blood. In mammals, the nephron is so oriented that the loop of Henle and the collecting duct lie parallel to each other (see Figure 14-18). The glomeruli are located in the renal cortex, and the loops of Henle reach down into the papillae of the medulla; thus, the nephrons are arranged in a radiating fashion within the kidney (see Figure 14-17).

Two types of nephrons are found together in the kidneys of birds and mammals:

- *juxtamedullary nephrons,* which have their glomeruli in the inner part of the cortex and long loops of Henle that plunge deeply into the medulla (see Figure 14-18a)
- *cortical nephrons,* which have their glomeruli in the outer cortex and relatively short loops of Henle that extend only a short distance into the medulla (see Figure 14-18b)

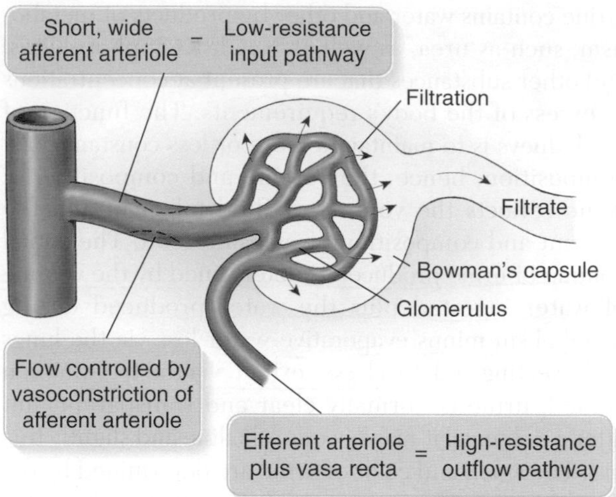

Figure 14-20 Blood pressure in the renal glomerulus is high because of the low-resistance input pathway (afferent arteriole) and the high-resistance output pathway. Regulation of glomerular blood pressure, which influences the filtration rate, is carried out largely through modulation of the diameter of the afferent arteriole.

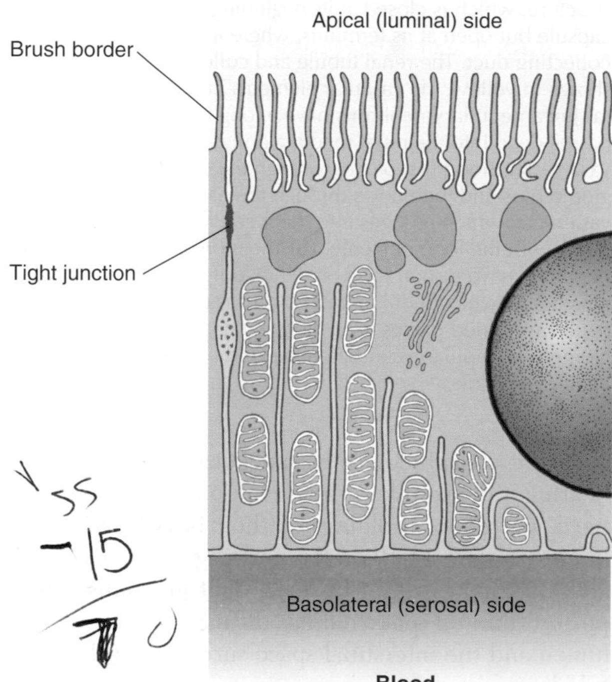

Figure 14-19 Some epithelial cells of the nephron are specialized for transport. The cells shown here, from the proximal tubule, are specialized for the transport of salts and other substances from the luminal (apical) side to the serosal (blood) side. The apical membrane facing the lumen is thrown into fingerlike projections (microvilli), greatly increasing its surface area. This kind of surface is referred to as a brush border. Mitochondria are concentrated near the basolateral (serosal) surface, which is thrown into deep basal clefts. These features allow the concentration of salts in the renal interstitium by active transport of salts across the basolateral membrane.

The anatomy of the renal circulatory system is important in the function of the nephron. The renal artery subdivides to form a series of short *afferent arterioles,* one of which supplies each nephron (see Figure 14-18). The glomerular capillaries within Bowman's capsule are subjected to somewhat higher pressures than other capillaries (Figure 14-20). Unlike most other capillary beds, which would join to form veins, the capillaries of the glomerulus come together to form an *efferent arteriole.* In juxtamedullary nephrons, the efferent arteriole then subdivides again to form a second series of capillaries surrounding the loop of Henle. Thus the blood, on leaving the glomerulus, enters the efferent arteriole and is carried into the medulla in a descending and subsequently ascending loop of anastomosing (interconnecting) capillaries before leaving the kidney via a vein. These hairpin loops, which parallel the loops of Henle of juxtamedullary nephrons, are referred to as the **vasa recta** (see Figure 14-18a). Flow in the efferent arteriole is less than that in the afferent arteriole because about 10% of the fluid in the blood is filtered across Bowman's capsule. For humans, this amounts to about a liter of filtrate formed every 10 minutes. Clearly, the flow rate of urine is much less, so much of the initial filtrate formed in Bowman's capsule must be reabsorbed into the blood.

Urine Production

Rather than simply secreting unwanted material across the body wall into the environment, the kidneys filter the blood plasma and then reabsorb needed substances into the blood from the ultrafiltrate; the substances left behind are excreted. In other words, the kidneys take back what the body needs and excrete everything else,

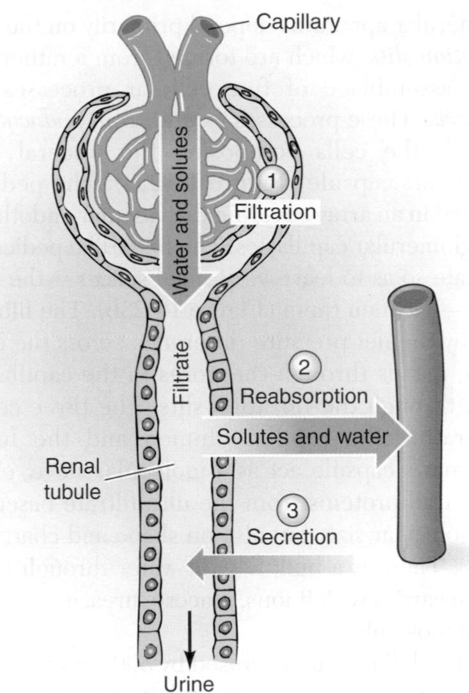

Figure 14-21 Urine formation in the mammalian nephron involves three main processes. Filtration, the initial step, takes place in Bowman's capsule, followed by reabsorption and secretion, which occur along the renal tubule. The final product of these processes is a hypertonic urine, whose composition differs from that of blood.

rather than defining what needs to be excreted. Some unwanted substances, however, are secreted across the tubular wall into the ultrafiltrate. Thus, three main processes contribute to the ultimate composition of the urine (Figure 14-21):

- filtration of blood plasma to form an ultrafiltrate in the lumen of Bowman's capsule
- tubular reabsorption of approximately 99% of the water and most of the salts from the ultrafiltrate, leaving behind and concentrating waste products such as urea
- tubular secretion of a number of substances, in nearly all instances via active transport

Formation of the ultrafiltrate in the glomerulus is the initial step in urine production; reabsorption and secretion occur along the length of the renal tubule. In addition to these processes, the excretion of nitrogenous wastes involves the synthesis of certain products in the tubular cells and lumen, as we will see at the end of the chapter.

Glomerular filtration

The glomerular ultrafiltrate contains essentially all the constituents of the blood, except blood cells and nearly all blood proteins. Filtration in the glomerulus is so extensive that 15%–25% of the water and solutes are removed from the plasma that flows through it. The glomerular ultrafiltrate is produced at the rate of about 125 ml · min^{-1}, or about 180 L · day^{-1}, in human kidneys. When this number is compared with the normal intake of water, it is evident that much of the ultrafiltrate is reabsorbed.

The process of ultrafiltration in the glomerulus (Figure 14-22) depends on three factors:

- the net hydrostatic pressure difference between the lumen of the glomerular capillaries and the lumen of Bowman's capsule (which favors filtration)
- the colloid osmotic pressure of the blood plasma (which opposes filtration)
- the hydraulic permeability (sievelike properties) of the three-layered tissue separating the two compartments

The net pressure gradient results from the sum of the hydrostatic pressure difference between the two compartments and the colloid osmotic pressure. The latter arises because of the separation of proteins during the filtration process. In humans, the proteins remaining in the capillary plasma produce an osmotic pressure of about −30 mm Hg, and the hydrostatic pressure difference (capillary blood pressure minus the back-pressure in the lumen of Bowman's capsule) is about +40 mm Hg

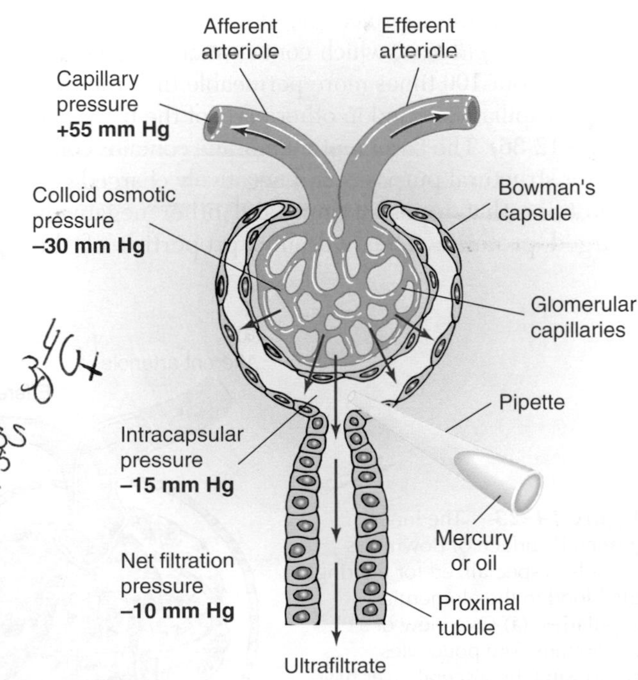

Figure 14-22 The net hydrostatic pressure affecting glomerular filtration is determined by the sum of the various forces indicated at the left. Samples of the glomerular filtrate can be obtained by insertion of a micropipette, as shown on the right. The mercury in the pipette is pushed to the tip by pressure before penetration of the capsule. A sample is then sucked into the calibrated tip for subsequent microanalysis. [Adapted from Hoar, 1975.]

Table 14-7 Balance sheet of pressures (in mm Hg) involved in glomerular ultrafiltration

	Salamander	Human
Glomerular capillary pressure	17.7	55
Intracapsular pressure	− 1.5	− 15
Net hydrostatic pressure	16.2	40
Colloid osmotic pressure	− 10.4	− 30
Net filtration pressure	5.8	10

Source: Pitts, 1968; Brenner et al., 1971.

(Table 14-7). The result is a net filtration pressure of only about +10 mm Hg. This small pressure differential acting on the high permeability of the glomerular sieve produces a phenomenal rate of ultrafiltrate formation by the millions of glomeruli in each human kidney. It is important to note that the filtration process in the kidney is entirely passive, depending on hydrostatic pressure that derives its energy from the contractions of the heart. In lower vertebrates such as the salamander, blood pressure in the glomerular capillaries is much lower than in humans, but the net filtration pressure is not much less than in the human kidney because of the lower intracapsular and osmotic pressures in these animals.

Fluid filtered from the blood into Bowman's capsule must cross the capillary wall, then the basement membrane of the capillary, and finally the inner (visceral) layer of the capsule. The glomerulus consists of *fenestrated capillaries,* which contain many large pores and are about 100 times more permeable than the continuous capillaries found in other parts of the body (see Figure 12-36). The basement membrane contains collagen for structural purposes and negatively charged glycoproteins that repel albumin and other negatively charged proteins. The hydraulic properties of the glomerular apparatus depend primarily on the sievelike *filtration slits,* which are formed from a rather remarkable assemblage of fine cellular processes termed *pedicels.* These processes extend from *podocytes* ("foot cells"), the cells composing the visceral layer of Bowman's capsule (Figure 14-23a). The pedicels are aligned in an array covering the vascular endothelium of the glomerular capillaries. The fingerlike pedicels interdigitate so as to leave very small spaces—the filtration slits—between them (Figure 14-23b). The filtrate, driven by the net pressure difference across the endothelium, passes through the pores in the capillaries and then through the filtration slits. The three cell layers separating the capillary lumen and the lumen of Bowman's capsule act as a molecular sieve, excluding almost all proteins from the ultrafiltrate based mainly on molecular size, but also on shape and charge (Table 14-8). There is a bulk flow of water through the sieve, which carries with it ions, glucose, urea, and many other small molecules.

The kidneys are perfused by 500–600 ml of plasma per minute. This perfusion takes place in a relatively low-resistance vascular bed within the kidney (see Figure 14-20). A high renal blood pressure is the result of the relatively direct arterial supply. Because these arteries and afferent arterioles are large in diameter and short in length, the loss of pressure due to friction is minimized. The efferent arterioles are of smaller diameter and, along with the capillaries of the vasa recta, constitute the major resistance of the renal vascular bed, ensuring high pressures within the glomeruli.

As noted already, the glomerular filtration rate depends largely on the net filtration pressure and the permeability of Bowman's capsule. Under normal conditions, the colloid osmotic pressure of plasma and the intracapsular pressure do not vary. However, the colloid osmotic pressure can be elevated during dehydration,

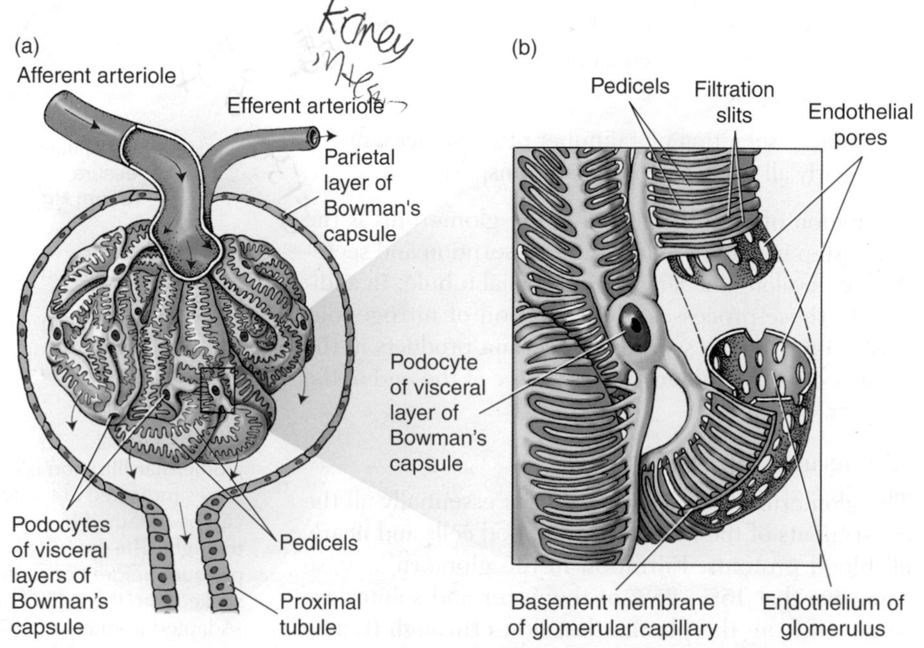

Figure 14-23 The inner (visceral) surface of Bowman's capsule is specialized for filtering the blood in the glomerular capillaries. **(a)** Overview of a glomerulus. The podocytes composing the visceral layer have long processes, termed pedicels, which cover the vascular epithelium. **(b)** Enlargement of the portion of part a enclosed in a box. Substances pass from the blood through the endothelial pores, across the basement membrane, and then through the filtration slits between pedicels.

Table 14-8 Relation between the molecular size of a substance and the ratio of its concentration in the filtrate appearing in Bowman's capsule to its concentration in the plasma [filtrate]/[filtrand]

Substance	Mol. wt.	Radius from diffusion coefficient (nm)	Dimensions from X-ray diffraction (nm)	[filtrate]/[filtrand]
Water	18	0.11		1.0
Urea	62	0.16		1.0
Glucose	180	0.36		1.0
Sucrose	342	0.44		1.0
Insulin	5500	1.48	54, 8	0.98
Myoglobin	17,000	1.95	88, 22	0.75
Egg albumin	43,500	2.85	54, 32	0.22
Hemoglobin	68,000	3.25		0.03
Serum albumin	69,000	3.55	150, 36	<0.01

Source: Pitts, 1968.

and the intracapsular pressure can be increased by the presence of kidney stones obstructing the renal tubules; in both cases, the rate of glomerular filtration will be reduced. In contrast, the seepage of plasma through burned skin can lower the colloid osmotic pressure of plasma, which in turn can increase the glomerular filtration rate. These examples, however, are exceptions rather than the rule.

In order to ensure that changes in blood pressure and cardiac output have little effect on the glomerular filtration rate under normal circumstances, a number of regulatory processes exist that control blood flow to the mammalian kidney. This regulation is achieved by modulating the resistance to flow in the afferent arteriole leading to each nephron and depends on a number of interrelated mechanisms involving both paracrine and endocrine secretions as well as neuronal control.

Several intrinsic mechanisms provide autoregulation of the glomerular filtration rate. First, an increase in blood pressure tends to stretch the afferent arteriole, which would be expected to increase the flow to the glomerulus. The wall of the afferent arteriole, however, responds to stretch by contracting, thus reducing the diameter of the arteriole and increasing the resistance to flow. This myogenic mechanism reduces variations in flow to the glomerulus in the face of oscillations in blood pressure.

Second, cells in the **juxtaglomerular apparatus,** which is located where the distal tubule passes close to Bowman's capsule between the afferent and efferent arterioles, secrete substances that modulate renal blood flow. The juxtaglomerular apparatus includes two types of specialized cells (Figure 14-24):

- modified distal-tubule cells, which form the *macula densa* and monitor the osmolarity and flow of fluid in the distal tubule
- modified smooth-muscle cells called *granular* or *juxtaglomerular cells,* which are located primarily in the wall of the afferent arteriole

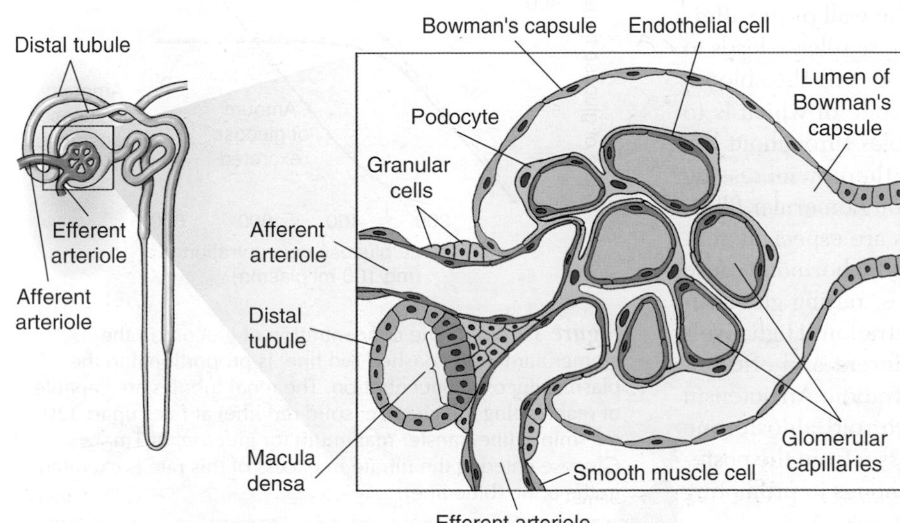

Figure 14-24 The juxtaglomerular apparatus plays a key role in controlling blood flow through the glomerulus. This structure contains two specialized cell types: modified distal-tubule cells, which constitute the macula densa, and secretory juxtaglomerular (granular) cells in the walls of the arterioles.

Under certain conditions, the granular cells release the enzyme **renin**, which indirectly affects blood pressure and, therefore, renal blood flow, as we will see below. The macula densa releases various substances that act in a paracrine fashion to cause vasoconstriction or vasodilation of the afferent arteriole in response to increased or decreased flow, respectively, through the distal tubule. Thus the myogenic and juxtaglomerular feedback-control mechanisms work together to autoregulate the glomerular filtration rate over a wide range of blood pressures.

Third, in addition to these autoregulatory mechanisms, the glomerular filtration rate is subject to extrinsic neuronal control. The afferent arterioles are innervated by the sympathetic nervous system. Sympathetic activation causes vasoconstriction of the afferent arterioles and a reduction in glomerular filtration. This response, which overrides any autoregulation, occurs when there is a sharp drop in blood pressure—for example, as a result of extensive blood loss. The reduction in filtration rate helps to restore blood volume and pressure to normal. Conversely, an elevation in blood pressure reduces sympathetic vasoconstriction and enhances glomerular filtration, decreasing blood pressure and volume.

Sympathetic activation can also cause contractions of cells within the glomerulus, closing off portions of the filtering capillaries and effectively reducing the area available for filtration. The podocytes are also contractile, and when they contract, the number of filtration slits decreases. Thus, contraction of either or both of these elements can effectively reduce the hydraulic permeability of Bowman's capsule. In the past, the hydraulic permeability of the glomerular membrane was thought to change only in disease states that caused the membrane to become leaky. It is now apparent that these changes are part of the normal regulation of the glomerular filtration rate.

A reduction in renal blood pressure, a fall in solute delivery to the distal tubule, or activation of the sympathetic innervation induces the release of renin from the secretory granular cells located in the wall of the afferent arteriole. Renin is an enzyme whose release leads to increased levels of angiotensin II in the blood. Angiotensin II has several actions, one of which is to cause general constriction of arterioles throughout the body, which raises blood pressure, thereby increasing both renal blood flow and the rate of glomerular filtration. Because the efferent arterioles are especially sensitive to angiotensin II, low levels of the hormone cause constriction of the efferent arterioles, raising glomerular blood pressure and increasing filtration. High levels of angiotensin II constrict both afferent and efferent arterioles and reduce glomerular filtration. Angiotensin II also stimulates release of the steroid aldosterone from the adrenal cortex and vasopressin from the posterior pituitary. The role of these hormones in promoting the tubular reabsorption of salts and water will be discussed below.

What advantages arise from filtering the equivalent of the blood volume every 4 or 5 minutes and then reabsorbing most of the salts and water? Why not simply excrete toxic waste by a process of secretion?

Tubular reabsorption

As the glomerular filtrate makes its way through the nephron, its original composition is quickly modified by reabsorption of various metabolites, ions, and water. The human kidneys produce about 180 liters of filtrate per day, but the final volume of urine is only about 1 liter. Thus, over 99% of the filtered water is reabsorbed. Of the 1800 g of NaCl typically contained in the original filtrate, only 10 g (or less than 1%) appear in the urine of a person consuming 10 g of NaCl per day. Varying amounts of many other filtered solutes are also reabsorbed from the tubular lumen. In addition, some substances are actively secreted into the tubular fluid. The **renal clearance** of a substance is a measure of the extent to which it is reabsorbed or secreted in the kidneys, as explained in Spotlight 14-1.

To understand the relationship between clearance and reabsorption, let's consider glucose. A healthy mammal exhibits a plasma glucose clearance of 0 ml $\cdot$ min^{-1}. That is, even though the glucose molecule is small and is freely filtered by the glomerulus, normally it is completely reabsorbed by the epithelium of the renal tubule (Figure 14-25). The loss of glucose in the urine would

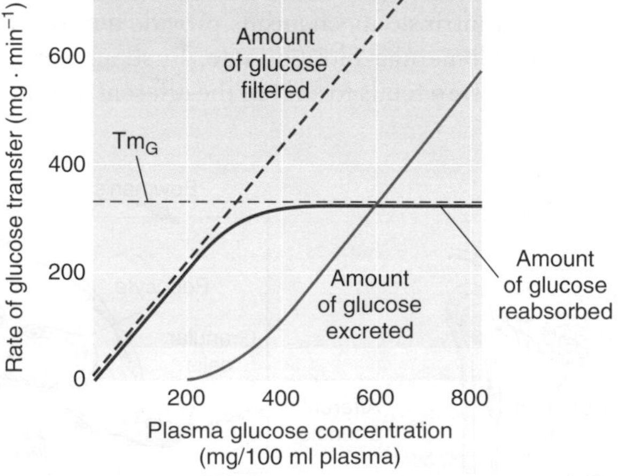

Figure 14-25 The concentration of glucose in the glomerular filtrate (dashed red line) is proportional to the plasma glucose concentration. The renal tubules are capable of reabsorbing the glucose (solid red line) at rates up to 320 mg $\cdot$ min^{-1} (the transfer maximum for glucose, or Tm_G). Glucose entering the filtrate in excess of this rate is excreted in the urine (blue line).

RENAL CLEARANCE

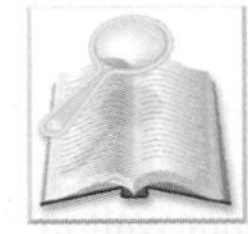

The renal clearance of a plasma-borne substance is the volume of blood plasma from which that substance is "cleared" (i.e., completely removed) per unit time by the kidneys. A substance that is freely filtered into the nephron along with water, but that is neither reabsorbed nor secreted in the kidney, permits the calculation of the *glomerular filtration rate* (GFR) merely by dividing the amount of the substance appearing in the urine by its concentration in the plasma. One such molecule is *inulin* (not insulin), a small starchlike carbohydrate (molecular weight 5000). Since the inulin molecule is neither reabsorbed nor secreted by the renal tubule, the inulin clearance is identical with the rate at which the glomerular filtrate is produced—that is, the GFR, generally given in milliliters per minute.

If we know the GFR and the concentration of a freely filtered substance in the plasma (and therefore its concentration in the ultrafiltrate), we can easily calculate whether the substance undergoes a net reabsorption or net secretion during the passage of the ultrafiltrate along the renal tubule. If less of the substance appears in the urine than was filtered in the glomerulus, some reabsorption of the substance must have occurred in the tubule. This is the case for water, NaCl, glucose, and many other essential constituents of the blood. If, however, the quantity of a substance appearing in the urine over a period of time is greater than the amount that passed into the nephron by glomerular filtration, it can be concluded that the substance is actively secreted into the lumen of the tubule. Unfortunately, the clearance technique is of limited usefulness in studies of renal function, because it indicates only the net output of the kidney relative to input and fails to provide insight into the physiological mechanisms involved.

In renal clearance studies, a test substance such as inulin is first injected into the subject's circulation and allowed to mix to uniform concentration in the bloodstream. A sample of blood is removed, and the plasma concentration of inulin, *P*, is determined from the sample. The rate of appearance of inulin in the urine is determined by multiplying the concentration of inulin in the urine, *U*, by the volume of urine produced per minute, *V*. The amount of inulin appearing in the urine per minute (*VU*) must equal the rate of plasma filtration (GFR) multiplied by the plasma concentration of inulin:

$$\frac{VU}{(\text{GFR})P} = \frac{\text{amount of inulin appearing in urine} \cdot \text{min}^{-1}}{\text{amount of inulin removed from blood} \cdot \text{min}^{-1}} = 1$$

In this special case, the substance used, inulin, is freely filtered and unchanged by tubular absorption or secretion. Therefore, the GFR and the clearance, *C*, of the substance are equal. Substituting *C* for GFR gives, for inulin,

$$\frac{VU}{CP} = 1$$

so the renal clearance is given by

$$\frac{VU}{P} = C = \text{renal clearance (ml} \cdot \text{min}^{-1})$$

If the amount of a substance, X, appearing in the urine per minute deviates from the amount of X present in the volume of plasma that is filtered per minute, this will be reflected in a value of C_X that differs from the inulin renal clearance, *C*. For example, if the inulin clearance of a subject, and hence the GFR, is 125 ml · min^{-1} and substance X exhibits a clearance of 62.5 m · min^{-1}, then

$$\frac{VU_X}{P_X} = C_X = 62.5 \text{ ml} \cdot \text{min}^{-1} = 0.5 \text{ GFR}$$

In this case, a volume of plasma equivalent to half that filtered each minute is cleared of substance X. Stated differently, only half the amount of substance X present in a volume of blood plasma equal to the volume filtered each minute actually appears in the urine per minute.

There are two possible reasons why the renal clearance for a substance would be less than the GFR. First, it may not be freely filterable. Filtration of a substance may be hindered if the substance is large, if it binds to serum proteins, or for some other reason. Second, a substance may be freely filtered, but then reabsorbed in the kidney tubules, thus reducing the amount that appears in the urine. As a matter of fact, most molecules below a molecular weight of about 5000 are freely filtered, but many of these are either partially reabsorbed or partially secreted (see Table 14-8). The extent of reabsorption or secretion of a substance can be gauged by its renal clearance. Reabsorption reduces the renal clearance to below the GFR. Tubular secretion, however, causes more of a substance to appear in the urine than is carried into the tubule by glomerular filtration.

mean a loss of chemical energy to the organism. Usually glucose appears in the urine only when the glucose concentration in the blood plasma, and hence in the glomerular filtrate, is very high. Figure 14-25 reveals that there is a maximum rate at which glucose can be removed from the tubular urine by reabsorption. This transfer maximum, or Tm, is about 320 mg · min^{-1} in humans. When plasma glucose levels are lower than about 1.8 mg · ml^{-1}, all the glucose appearing in the glomerular filtrate is reabsorbed. At about 3.0 mg · ml^{-1}, the glucose transport mechanism is fully saturated, and any additional amount of glucose appearing in the filtrate is passed out in the urine. The arterial plasma glucose concentration in humans is normally held at about 1 mg · ml^{-1} by an endocrine feedback loop involving insulin. Since this level is well below the Tm for glucose, normal urine contains essentially no glucose. Because the high plasma glucose levels typical of diabetes mellitus exceed the reabsorption ability of the renal tubule, diabetics commonly have glucose in their urine.

The details of tubular function vary from species to species. Our knowledge of the changes in the composition of urine as it moves through the nephron is based to a large extent on the technique of micropuncture, first developed by Alfred Richards and his coworkers in the 1920s. A glass capillary micropipette is used to remove a minute sample of the tubular fluid from the lumen of the nephron. The osmolarity of the sample (expressed as milliosmoles per liter) is then determined by measuring its freezing point: the lower the freezing point, the higher its osmolarity. The stopped-flow perfusion technique (Figure 14-26), a modification of Richards's original technique, can be used to isolate a portion of the lumen and analyze its action on injected samples of a defined solution *in vitro.*

Microchemical methods are now used to determine the concentrations of individual ionic species in a urine sample. In a technique developed recently, a given segment of renal tubule is dissected from the kidney and perfused *in vitro* with a defined test solution; analysis of the perfusate provides insight into the movement of substances across the isolated tubule segment (Figure 14-27). The results of numerous studies using these techniques have detailed the roles of various portions of the nephron in the reabsorption of salts and water, which are summarized in Figure 14-28.

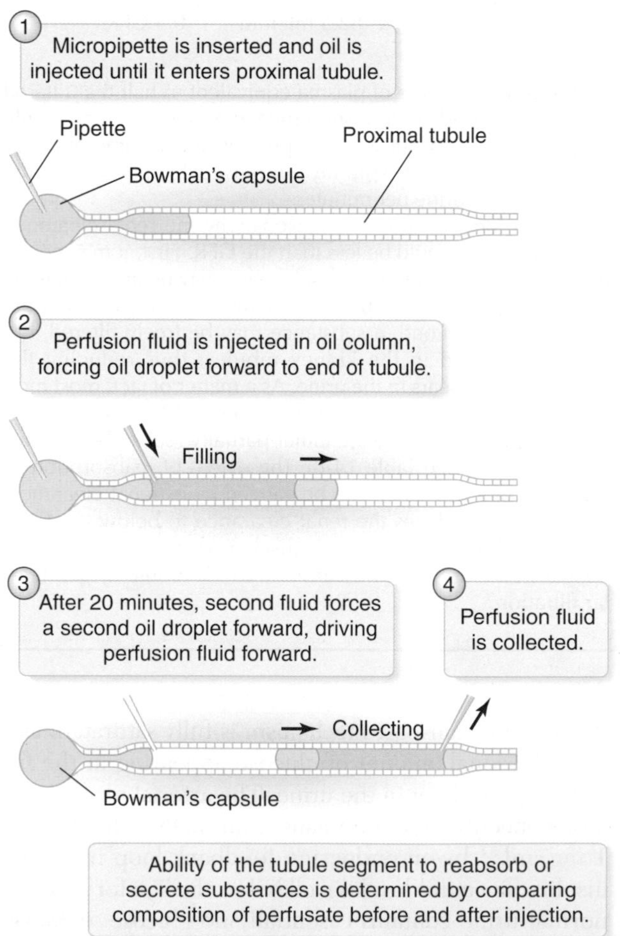

Figure 14-26 The stopped-flow perfusion technique is used to study the function of various portions of the renal tubule *in vitro.*

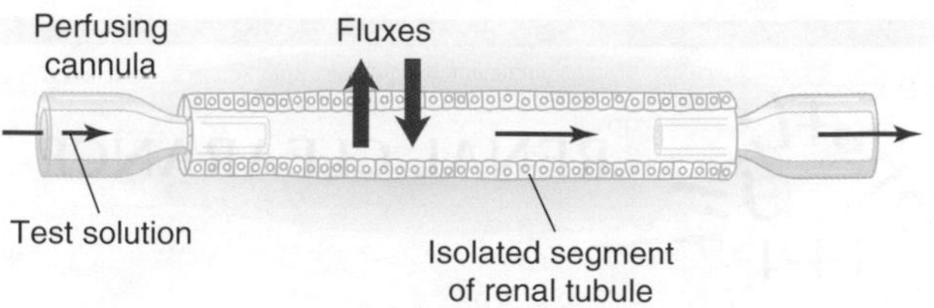

Figure 14-27 Perfusion of a dissected segment of a renal tubule and chemical analysis of the perfusate permit determination of the fluxes of ions across the tubular wall *in vitro.*

The proximal tubule, which begins the process of concentrating the glomerular filtrate, is the most important segment of the nephron in active reabsorption of salts. In this segment, about 70% of the Na^+ is removed from the lumen by active transport, and a nearly proportional amount of water and certain other solutes, such as Cl^-, follow passively. About 75% of the filtrate is reabsorbed before it reaches the loop of Henle. The result is a fluid that is iso-osmotic with respect to the plasma and interstitial fluids. Stopped-flow perfusion experiments have revealed that when the NaCl concentration inside the tubule is decreased, the movement of water also decreases. This result is just the opposite of what would be expected if the outward movement of reabsorbed water occurred by simple osmotic diffusion, and it indicates that water transport is coupled to active sodium transport (see Chapter 4). The actual pumping of Na^+ takes place at the basolateral surface of the epithelial cells of the proximal tubule, just as it does in frog skin and mammalian gallbladder epithelia. In the proximal tubule, $NaHCO_3$ is the major solute reabsorbed proximally (see Figure 14-13b), and NaCl is the major solute reabsorbed distally (see Figure 14-13a).

By the time it reaches the most distal portion of the proximal tubule (where it joins the thin descending limb of the loop of Henle), the glomerular filtrate has already been reduced to one-fourth of its original volume. As a result, substances that are not actively transported across the tubule or that do not passively diffuse across it are four times more concentrated than they were in the original filtrate. In spite of this great reduction in the volume of the tubular fluid, the fluid at this point is iso-osmotic relative to the fluid outside the nephron, having an osmolarity of about 300 mosm $\cdot$ L^{-1}. It is interesting to note that the active transport of NaCl alone can account for the changes in fluid volume and for the increasing concentrations of urea and many other filtered substances along the proximal tubule.

The proximal tubule is ideally structured for the massive reabsorption of salt and water. Numerous microvilli at the luminal border of the tubular epithelial cells form a so-called **brush border** (see Figure 14-19).

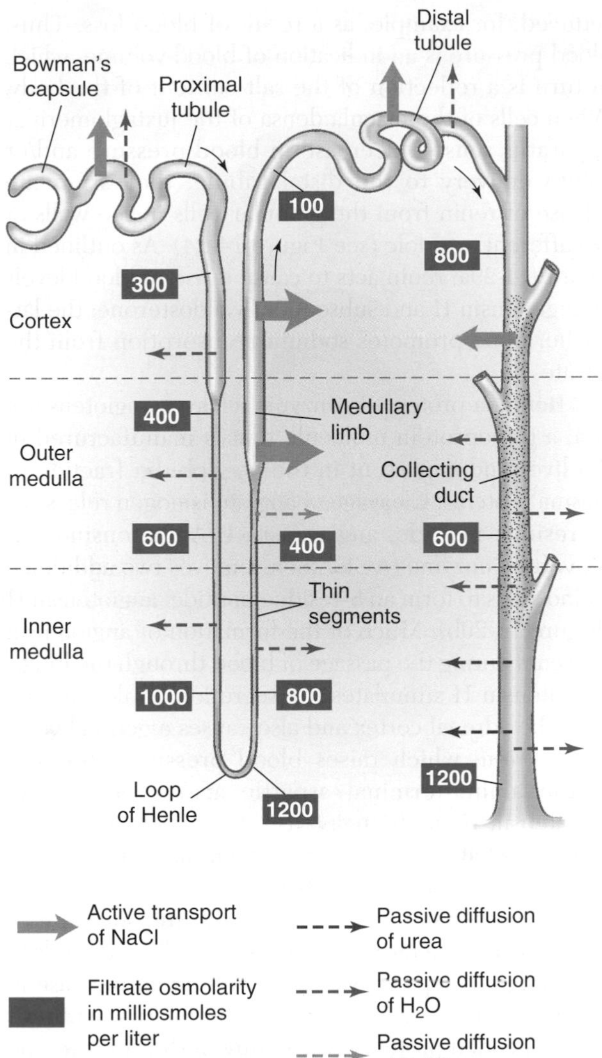

Figure 14-28 The movement of ions, water, and other substances into and out of the filtrate along the renal tubule determines the composition of the urine. In this schematic diagram, the fluxes of NaCl, water, and urea are shown in different portions of the mammalian renal tubule. The numbers indicate the filtrate tonicity in milliosmoles per liter. The relative rates of active transport of NaCl are indicated by the sizes of the arrows. The permeability of the stippled portion of the collecting duct is regulated by antidiuretic hormone (ADH). [Adapted from Pitts, 1959.]

These projections greatly increase the absorptive surface area of the apical membrane, thereby promoting diffusion of salt and water from the tubular lumen into the epithelial cell.

Glucose and amino acids are also reabsorbed in the proximal tubule by a sodium-dependent mechanism and are not normally present in the ultrafiltrate beyond the proximal tubule. Carrier proteins on the apical membrane cotransport sodium and glucose or amino acids from the ultrafiltrate into the cell (see Figure 14-13a). The uptake process—uphill for glucose and amino acids—depends on the sodium electrochemical gradient created by the Na^+/K^+ pump in the basolateral membrane of the tubular cell. Once in the tubular cell, glucose and amino acids diffuse into the blood.

Phosphates, calcium ions, and other electrolytes normally found in the blood are reabsorbed up to the amount required by the body, and any excess is excreted. Parathyroid hormone modulates the reabsorption of phosphates and calcium in the kidney by stimulating kidney 1α,25-hydroxylase activity, which in turn stimulates the production of *calcitriol*, the active form of vitamin D. Calcitriol released into the blood stimulates calcium reabsorption and phosphate excretion from the kidney, as well as calcium absorption from the gut and its release from bone (see Figure 9-45).

The *descending limb* and *thin segment of the ascending limb* of the loop of Henle are made up of very thin cells containing few mitochondria and no brush border. *In vitro* perfusion studies have demonstrated that there is no active salt transport in the descending limb. Moreover, this segment exhibits very low permeability to NaCl and low permeability to urea, but is permeable to water. The thin segment of the ascending limb has also been shown by perfusion experiments to be inactive in salt transport, although it is highly permeable to NaCl. Its permeability to urea is low and to water, very low. As we will see, the differing permeabilities of these segments play key roles in the urine-concentrating mechanism of the nephron.

The medullary *thick ascending limb* differs from the rest of the loop of Henle in that it actively transports NaCl outward from the lumen to the interstitial space (see Figure 14-28). This segment, along with the rest of the ascending limb, has a very low permeability to water. As a result of NaCl reabsorption, the fluid reaching the distal tubule is somewhat hypo-osmotic relative to the interstitial fluid. Salt reabsorption by the thick ascending limb is also important in the urine-concentrating mechanism, as we will see.

The movement of salt and water across the distal tubule and collecting duct is complex. This region of the tubule is important in the transport of K^+, H^+, and NH_3 into the lumen and of Na^+, Cl^-, and HCO_3^- out of the lumen and back into the interstitial fluid. As salts are pumped out of the tubule, water follows passively. The transport of salts in the distal tubule is under endocrine control and is adjusted in response to osmotic conditions.

Because the collecting duct is permeable to water, water flows from the dilute urine in the duct into the more concentrated interstitial fluid of the renal medulla (see Figure 14-28). This movement of water is the final step in the production of a hyperosmotic urine. The water permeability of the duct is controlled by vasopressin, also called antidiuretic hormone (ADH). Thus the rate at which water is absorbed is under delicate feedback control. The collecting duct reabsorbs NaCl by active transport of sodium. The inner medullary segment of the collecting duct, toward its distal end, is highly permeable to urea. The significance of these characteristics will become clear when we look at the countercurrent mechanism that concentrates the urine in the collecting duct.

Sodium chloride represents more than 90% of the osmotic activity of the extracellular fluid. Because reabsorption of salt results in the reabsorption of water, the amount of salt in the body is an important determinant of the volume of the extracellular fluid. If the fluid volume is large, then blood pressure tends to rise. Conversely, blood pressure drops when fluid volume is reduced, for example, as a result of blood loss. Thus, blood pressure is an indication of blood volume, which in turn is a reflection of the salt content of the body. When cells of the macula densa of the juxtaglomerular apparatus sense a decrease in blood pressure and/or solute delivery to the distal tubule, they stimulate release of renin from the granular cells in the walls of the afferent arteriole (see Figure 14-24). As outlined in Figure 14-29a, renin acts to cause a rise in blood levels of angiotensin II and subsequently aldosterone; the latter hormone promotes sodium reabsorption from the filtrate.

(a)

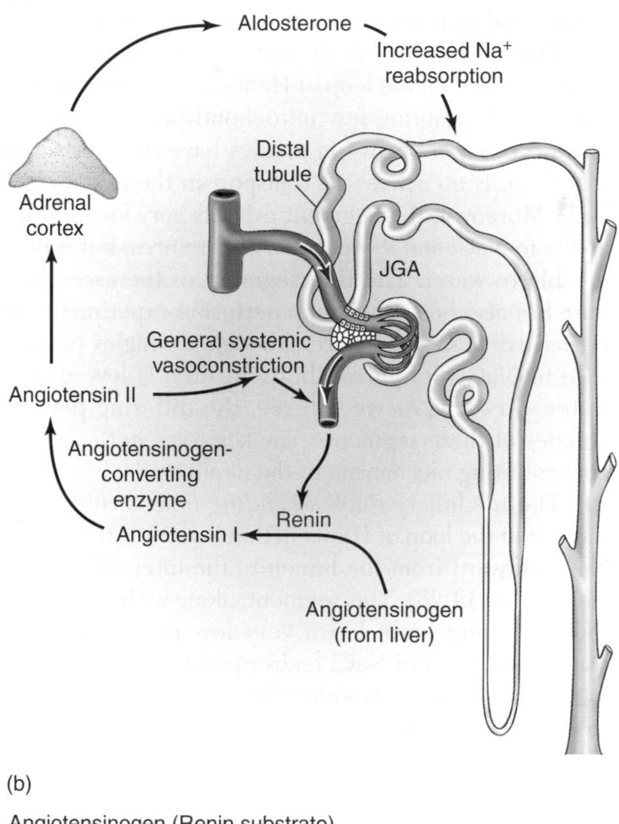

(b)

Angiotensinogen (Renin substrate)

Asp-Arg-Val-Tyr-Ile-His-Pro-Phe-His-Leu-Leu-Val-Tyr-Ser-Protein

↓ Renin

Angiotensin I

Asp-Arg-Val-Tyr-Ile-His-Pro-Phe-His-Leu

↓ Angiotensinogen-converting enzyme

Angiotensin II

Asp-Arg-Val-Tyr-Ile-His-Pro-Phe

Figure 14-29 The renin-angiotensin system plays an important role in controlling sodium reabsorption in the mammalian kidney. **(a)** Renin is released by secretory cells in the juxtaglomerular apparatus (JGA) in response to decreased pressure in the afferent arteriole and to low Na^+ concentration in the distal tubule. Circulating renin leads to an increase in the concentration of angiotensin II and aldosterone. Aldosterone stimulates Na^+ reabsorption from the filtrate in the renal tubule. **(b)** Renin is a proteolytic enzyme that cleaves angiotensinogen, an α_2-globulin, yielding angiotensin I. Angiotensinogen-converting enzyme then removes the two carboxyl-terminal residues to give angiotensin II.

Renin, a proteolytic enzyme, cleaves angiotensinogen, a glycoprotein molecule that is manufactured in the liver and is present in the α_2-globulin fraction of plasma proteins. Cleavage of angiotensinogen releases a 10-residue peptide, angiotensin I. Angiotensinogen-converting enzyme (ACE) then removes two additional amino acids to form an 8-residue peptide, angiotensin II (Figure 14-29b). Much of the formation of angiotensin II occurs during the passage of blood through the lungs. Angiotensin II stimulates the secretion of aldosterone from the adrenal cortex and also causes a general vasoconstriction, which raises blood pressure. Removal of the amino-terminal aspartic acid residue from angiotensin II yields angiotensin III, which also causes secretion of aldosterone from the adrenal cortex, but to a lesser extent than angiotensin II.

Like other steroid hormones, **aldosterone** diffuses across the plasma membrane of target cells and binds to cytoplasmic receptors within, leading to an increase in transcription of specific genes and ultimately synthesis of the encoded proteins (see Figure 9-18). Aldosterone acts on the cells of the tubular epithelium to increase their sodium reabsorption, but does so without affecting their water permeability. Three mechanisms have been proposed to account for the aldosterone-induced increase in sodium reabsorption across tubular epithelial cells:

1. *Sodium pump hypothesis:* Activity of the Na^+/K^+ pump in the basolateral membrane increases, perhaps due to changes in membrane structure that enhance ATPase activity as well as increased synthesis of the pump protein.
2. *Metabolic hypothesis:* The production of ATP increases, providing more ATP to power the Na^+/K^+ pump, perhaps due to an aldosterone-stimulated increase in fatty acid metabolism.
3. *Permease hypothesis:* The permeability of the apical membrane to Na^+ ions increases, presumably due to an increase in the number of sodium channels in the membrane.

It is quite possible that all three mechanisms operate in tubular cells stimulated by aldosterone.

A rise in the circulating levels of angiotensin II also increases the synthesis of antidiuretic hormone in the hypothalamus and its release from the posterior pituitary (see Figures 9-15 and 9-17). ADH acts to increase the water permeability of the distal tubule and collecting duct, promoting water reabsorption. Aldosterone then acts with ADH to enhance both sodium and water reabsorption by the kidney.

Atrial natriuretic peptide (ANP), released by cells in the atrium of the heart into the blood in response to an increase in venous pressure, causes an increase in urine production and sodium excretion. It thus has the opposite effect of the renin-angiotensin system on the kidney. ANP inhibits the release of ADH and renin and the production of aldosterone by the adrenal gland. ANP acts directly on the kidney to reduce sodium and, therefore, water reabsorption (see Chapter 12).

Diving to depth raises venous pressure in humans. What effect might this have on kidney function? What differences might be expected in the regulation of kidney function between humans and whales?

Tubular secretion

The nephron has several distinct systems that secrete substances by transporting them from the plasma into the tubular lumen. The substances transported include K^+, H^+, NH_3, organic acids, and organic bases. Many organic ions are reversibly bound to plasma proteins and are not readily filtered. There is, however, a small unbound fraction, and these ions are secreted across the wall of the proximal tubule. Organic anion secretion is driven by the sodium gradient established by the Na^+/K^+ pump (Figure 14-30). These secretory mechanisms are nonspecific; as a result, the nephron is capable of secreting innumerable "new" substances, including drugs and toxins, as well as endogenous, naturally occurring molecules. How is the nephron able to recognize and transport all these diverse substances? The answer resides in the role of the vertebrate liver, where many of these substances, along with normal metabolites, are conjugated with glucuronic acid or its sulfate, which allows them to react with the organic anionic and cationic transport systems. Since they are highly polar, these conjugated molecules cannot readily diffuse back across the wall of the nephron into the blood, so they are excreted in the urine. These secretory mechanisms remove many potentially dangerous substances from the blood.

A number of endogenous, as well as foreign, compounds are excreted via the organic cationic and anionic secretory mechanisms (Table 14-9). Because these secretory mechanisms are nonspecific, various organic ions compete for the same secretory pathway, and elevated levels of one organic ion can reduce the secretion of another organic ion. Drug activity can be prolonged by co-administering any of several cationic (or anionic)

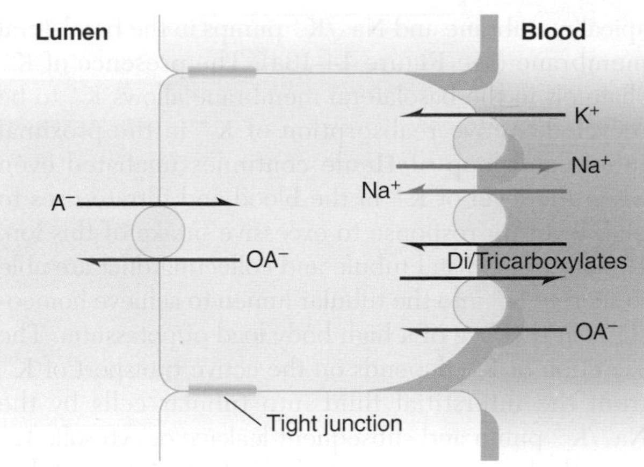

Figure 14-30 The transport of organic anions (OA^-) is driven by the sodium gradient created by the activity of the Na^+/K^+ pump. OA^- is taken from the blood into the cell in exchange for di- and tricarboxylates, and is then removed from the cell across the apical membrane via an anion antiporter.

drugs that compete for entry into the secretory pathway, thereby reducing the secretion of the therapeutic drug. Some substances, such as para-aminohippuric acid (PAH), are both filtered and secreted. PAH is completely removed from all plasma flowing to the kidney and is not reabsorbed. Thus, the plasma clearance for PAH is a measure of the rate of plasma flow through the kidney. The plant polysaccharide inulin is filtered, but not reabsorbed or secreted, and so can be used to estimate glomerular filtration rate (see Spotlight 14-1). Thus the fraction of plasma filtered can be determined from the ratio of inulin to PAH clearance. This ratio is usually around 20%.

Normally, most of the K^+ ions that are filtered at the glomerulus are reabsorbed from the filtrate in the proximal tubule and the loop of Henle, owing to the presence of a $Na^+/2Cl^-/K^+$ cotransport system in the

Table 14-9 **Some organic ions secreted by the proximal tubule**

Anions *Endogenous*	**Cations** *Endogenous*
Urates	Dopamine
Hippurates	Epinephrine
Oxalate	Norepinephrine
Prostaglandins	Creatinine
cAMP	
Exogenous	*Exogenous*
Furosemide	Morphine
Bumetanide	Amiloride
Penicillin	Quinine
Aspirin	Atropine
Chlorothiazides	Isoproterenol

apical membrane and Na^+/K^+ pumps in the basolateral membrane (see Figure 14-13a). The presence of K^+ channels in the basolateral membrane allows K^+ to be recycled. Active reabsorption of K^+ in the proximal tubule and loop of Henle continues unabated even when the level of K^+ in the blood and filtrate rises to high levels in response to excessive intake of this ion. However, the distal tubule and collecting duct are able to secrete K^+ into the tubular lumen to achieve homeostasis in the face of a high body load of potassium. The secretion of K^+ depends on the active transport of K^+ from the interstitial fluid into tubular cells by the Na^+/K^+ pump and subsequent leakage of cytosolic K^+ through channels in the apical membrane into the tubular fluid (see Figure 14-12). The tubular fluid is electronegative with respect to the cytosol, so K^+ can simply diffuse down its electrochemical gradient from inside the renal tubular cells into the lumen.

The rate of K^+ secretion (and Na^+ reabsorption) by these mechanisms is regulated by aldosterone, which is released in response to elevated plasma levels of K^+ as well as reduced Na^+ levels. Elevated K^+ levels directly stimulate the adrenal glands, whereas reduced blood Na^+ levels stimulate the adrenals via activation of the renin-angiotensin system. Stimulation of Na^+ reabsorption is therefore coupled to K^+ secretion through the action of aldosterone; one cannot be corrected without affecting the other. Release of aldosterone in response to low blood Na^+ levels enhances Na^+ reabsorption, but could also lead to abnormally low blood K^+ levels due to enhanced K^+ secretion and excretion.

Because high extracellular K^+ levels may cause cardiac arrest and convulsions, excess K^+ ions must be quickly removed from the plasma. Insulin released in response to high K^+ levels stimulates K^+ uptake by cells, especially fat cells. Potassium is then slowly released from these cells and removed by the somewhat slower renal mechanisms. Thus, like aldosterone, insulin release can lead to low plasma K^+ levels.

A few marine teleost fish species have aglomerular kidneys. What might have been the selective forces leading to the evolution of such structures?

Regulation of pH by the Kidney

As we saw in Chapter 13, the carbon dioxide–bicarbonate buffering system is primarily responsible for determining the pH of the extracellular space in mammals. Two factors combine to control pH in mammals: excretion of CO_2 via the lungs and excretion of acid via the kidneys. The ratio of lung ventilation to CO_2 production largely determines the CO_2 concentration of the body. Changes in ventilation can adjust carbon dioxide excretion, and therefore have a role in modulating body pH in the short term. The excretion of acid (H^+ ions) in the urine is ultimately responsible for maintaining the plasma bicarbonate (HCO_3^-) concentration in mammals. Acid excretion across the skin of amphibians or the gills of fishes supplements or takes over the role of acid excretion by the kidney in those animals.

The concentration of HCO_3^- in mammalian plasma is about 25×10^{-3} mol·L^{-1}, whereas the H^+ concentration is about 40×10^{-9} mol·L^{-1}. The concentrations of bicarbonate and protons in the glomerular ultrafiltrate are similar to those in the plasma; that is, the filtrate contains large quantities of bicarbonate but a very low concentration of protons. Yet urine has a pH of about 6.0 and contains little or no bicarbonate. Thus, acid must be added to the filtrate and most, if not all, of the bicarbonate removed in the process of urine formation. At pH 6, the urine still has a very low concentration of protons, and a change in H^+ concentration alone, as the filtrate flows down the tubule, would not be sufficient to maintain body pH in the face of continual metabolic production of acid. In fact, most of the acid added to the urine is buffered by either phosphate or ammonia.

Because protons are added to the tubular filtrate along the entire length of the tubule, the filtrate becomes progressively more acidic. In the proximal tubule and loop of Henle, protons are secreted via the Na^+/H^+ antiporter discussed earlier (see Figure 14-13b). The distal tubule and the collecting duct contain specialized cells, referred to as A-type cells, that have a proton pump in the apical membrane and a chloride-bicarbonate exchange system in the basolateral membrane. (This anion exchanger is similar to the band III protein in the red blood cell membrane.) These cells also contain high levels of carbonic anhydrase, which catalyzes the hydration of intracellular carbon dioxide, so that bicarbonate ions and protons are formed rapidly. The protons are transported across the apical membrane into the tubular fluid, and the bicarbonate ions move across the basolateral membrane into the interstitial fluid. The secreted protons can react with bicarbonate in the ultrafiltrate to form carbon dioxide and water, which can diffuse back into the cell. Thus, the secretion of protons from the A-type cell can result in a net uptake of bicarbonate into the blood through the cycling of carbon dioxide (Figure 14-31a). Clearly the A-type cell is an acid-secreting cell.

The removal of protons from an A-type cell makes the intracellular potential more negative, thereby enhancing sodium reabsorption from the filtrate. The intracellular sodium level is kept low by the activity of a Na^+/K^+ pump in the basolateral membrane, which transports Na^+ from the cell into the extracellular fluid. The basolateral membrane of the A-type cell also contains K^+ channels, and K^+ is cycled through this membrane by the Na^+/K^+ pump. Thus, acidification of the filtrate by A-type cells is coupled to sodium reabsorption.

The distal tubule and collecting duct also contain specialized base-secreting cells, called B-type cells.

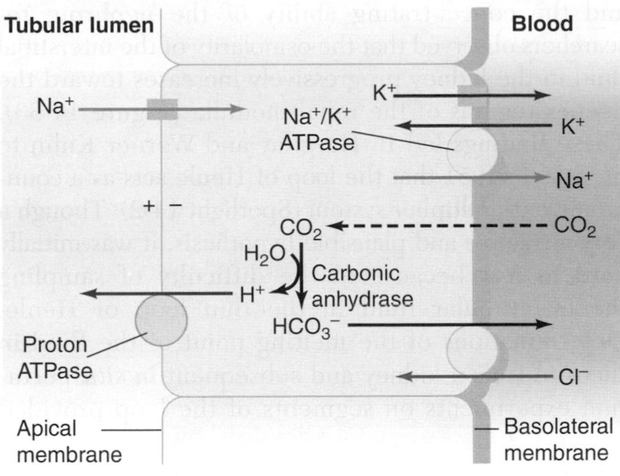

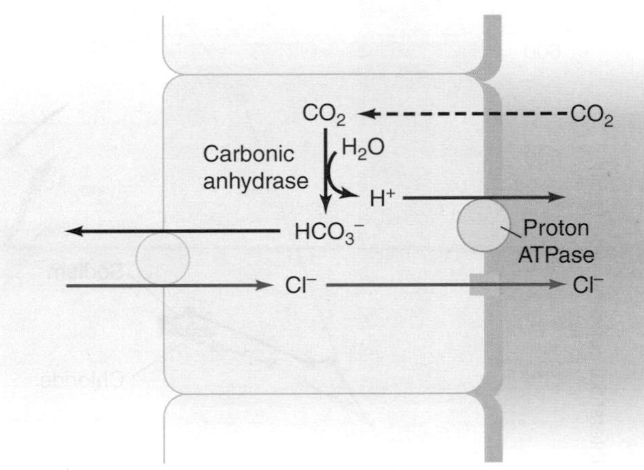

Figure 14-31 Body pH in mammals can be modulated by regulating the relative activity of acid-secreting (A-type) cells and base-secreting (B-type) cells in the distal tubule and collecting duct of the kidney. **(a)** A-type cells pump protons into the lumen via an apical H^+ ATPase, acidifying the filtrate; the resulting increase in the potential across the apical membrane favors reabsorption of Na^+. **(b)** B-type cells use the H^+ ATPase in the basolateral membrane to pump protons into the blood, accompanied by the reabsorption of Cl^-. Both cell types contain carbonic anhydrase, which rapidly forms H^+ and HCO_3^- ions from CO_2 diffusing into the cell from the blood.

These cells also have a chloride-bicarbonate exchanger in the apical membrane, but this exchanger differs from the band III–type protein found in the basolateral membrane of the A-type cell. As illustrated in Figure 14-31b, B-type cells contain carbonic anhydrase and secrete bicarbonate into the lumen of the tubule in exchange for chloride. Protons and chloride ions move across the basolateral membrane via a proton pump and chloride channels.

A mammal can regulate its body pH by altering the activity of the A-type and B-type cells. The activity of A-type cells, and therefore acid secretion, increases during acidosis, whereas increased B-type cell activity and bicarbonate secretion are associated with alkalosis. Changes in the activity of the A-type cells involve alterations in both the proton ATPase activity in the apical membrane and the number of bicarbonate-chloride exchangers present in the basolateral membrane. The cells contain pH-dependent cytosolic activators and inhibitors of V-ATPases, and aldosterone is known to stimulate proton ATPase activity.

Proton secretion by renal tubular cells reduces the pH of the ultrafiltrate, thereby increasing the gradient against which protons are transported. Thus the ability to secrete protons decreases with filtrate pH; when the pH of the filtrate drops below 4.5, acid secretion stops. If the ultrafiltrate is buffered, however, more protons can be secreted across the tubular epithelium without producing a drop in pH sufficient to inhibit the proton pump. The ultrafiltrate is buffered by bicarbonate, phosphates, and ammonia. Acid secreted into the ultrafiltrate reacts with bicarbonate to form carbon dioxide and water, with HPO_4^{2-} to form $H_2PO_4^-$, or with NH_3 (ammonia) to form NH_4^+ (ammonium) ions (Figure 14-32). The bicarbonate, phosphate, and ammonia buffer systems compete for protons secreted into the filtrate. The tubular membrane is essentially impermeable to both phosphates and ammonium ions; thus, these substances are trapped in the filtrate and then excreted from the body.

The phosphates found in the ultrafiltrate are filtered from the blood in the glomerulus, whereas ammonia diffuses from the blood across the tubular cells into the lumen, where it is converted to ammonium

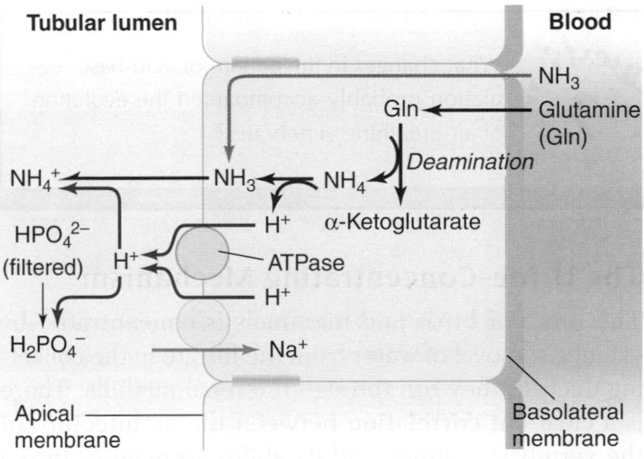

Figure 14-32 Buffering of the renal filtrate by $H_2PO_4^-$ and NH_4^+ permits greater secretion of protons. The phosphate ions reach the lumen by filtration, whereas the ammonium ions arrive there by passive diffusion of NH_3 from the blood across the tubular cells or by intracellular breakdown of glutamine. Glutamine (and other amino acids) enters the tubular cells via basolateral transporters and is deaminated, yielding NH_3, which diffuses across the apical membrane into the lumen. Because the membrane is largely impermeable to both $H_2PO_4^-$ and NH_4^+, both ions are trapped in the urine and excreted.

ions. Phosphate levels depend on the supply in the diet, with excess phosphate being filtered into the ultrafiltrate. Thus the capacity of the phosphate buffer system (i.e., the number of protons that it can bind) depends on what the animal eats and is independent of the acid-base requirements of the animal. Body pH is not generally regulated by selection of appropriate foods.

Under acidotic conditions, plasma bicarbonate levels often fall; as a result, bicarbonate levels in the filtrate are reduced, and less is available to act as a buffer. Under such conditions, ammonia is a major vehicle for the elimination of excess acid. Ammonia is produced within the renal tubular cells by enzymatic deamination of amino acids, especially glutamine (see Figure 14-32). Ammonia is formed continually in the liver, but because it is toxic, it must be converted to less toxic urea and glutamine. Ammonia is transported to the kidney as glutamine, which is taken up by the tubular cells and deaminated to release ammonia. In its nonpolar, unionized form, ammonia freely diffuses across the cell membrane into the lumen, where it reacts with protons to form NH_4^+ ions. Because the highly polar NH_4^+ is impermeant, it traps both nitrogen atoms and protons in the urine, thus serving as a vehicle for their excretion. If acidotic conditions continue in the body for a few days, ammonia production by the tubular epithelium increases, NH_4^+ concentration in the filtrate rises, and acid excretion by the kidney is increased. Mammals that have entered a state of metabolic acidosis (excess acid production) show dramatic increases in ammonia production and secretion, as this is the body's major long-term mechanism for correcting an acid load.

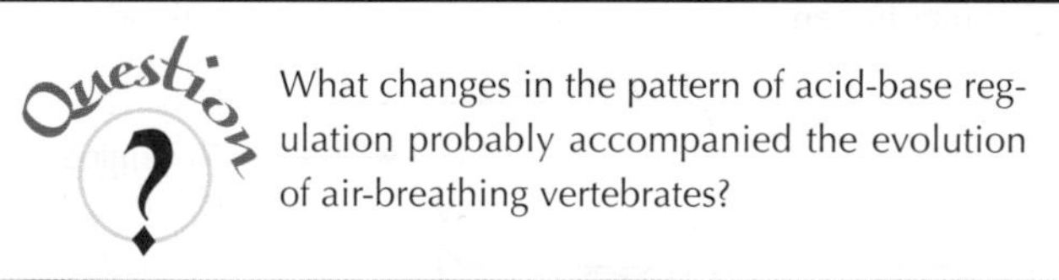

What changes in the pattern of acid-base regulation probably accompanied the evolution of air-breathing vertebrates?

The Urine-Concentrating Mechanism

The urine of birds and mammals is concentrated by osmotic removal of water from the filtrate in the collecting ducts as they run through the renal medulla. There is a clear-cut correlation between the architecture of the vertebrate kidney and its ability to manufacture a urine that is hypertonic relative to the body fluids. Kidneys capable of producing a hypertonic urine (i.e., those of mammals and birds) all have nephrons featuring the loop of Henle. Moreover, the ability of a mammal to concentrate urine is directly related to the length of the loops of Henle in its kidneys. The longer the loop, and the deeper it extends into the renal medulla, the greater the concentrating power of the nephron. The loops of Henle are longest, and the urine is most hypertonic, in desert dwellers, such as the kangaroo rat.

In addition to this correlation between anatomy and the concentrating ability of the nephron, researchers observed that the osmolarity of the interstitial fluid in the kidney progressively increases toward the deeper regions of the renal medulla (Figure 14-33). These findings led B. Hargitay and Werner Kuhn to propose in 1951 that the loop of Henle acts as a countercurrent multiplier system (Spotlight 14-2). Though a very attractive and plausible hypothesis, it was initially hard to test because of the difficulty of sampling the intratubular fluid in the thin loop of Henle. Determinations of the melting point of the fluid in slices of frozen kidney and subsequent *in situ* perfusion experiments on segments of the loop provided

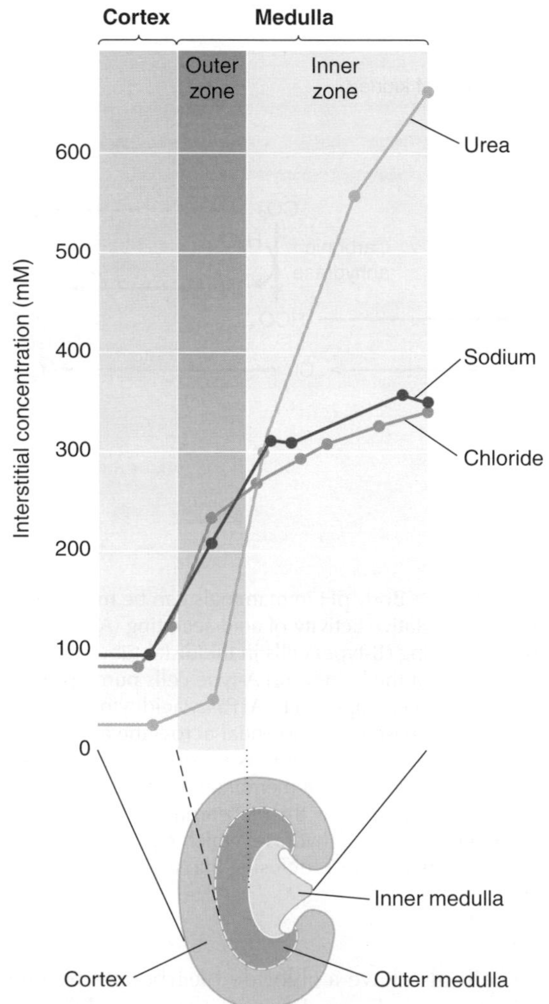

Figure 14-33 Solute concentrations in the interstitial fluid of the mammalian kidney progressively increase from the cortex to the depths of the medulla. Shown here are the interstitial concentrations (in millimoles per liter) of urea, sodium, and chloride at different depths. Note that most of the increase in urea concentration occurs across the inner medulla, whereas most of the increase in NaCl concentration occurs across the outer medulla. Since the osmotic contributions of Na^+ and Cl^- sum, the total osmotic contributions of NaCl and urea are about equal deep within the medulla. [Adapted from Ulrich et al., 1961.]

Spotlight 14-2 COUNTERCURRENT MULTIPLIER SYSTEMS

In 1944, Lyman C. Craig developed a method of concentrating chemical compounds based on the countercurrent multiplier principle. This method has proved useful in many industrial and laboratory applications. As often happens, human ingenuity turns out to be a reflection of nature's inventiveness. Countercurrent multiplier mechanisms have been found to operate in a variety of biological systems, including the vertebrate kidney, the gas-secreting organ of swimbladders and the gills of fishes, and the limbs of various birds and mammals that live in cold climates.

The countercurrent principle can be illustrated with a hypothetical countercurrent multiplier that employs an active-transport mechanism much like the one that operates in the mammalian kidney. The model shown in part a of the accompanying figure consists of a tube bent into a loop with a common dividing wall between the two limbs of the loop. A NaCl solution flows into one limb and then out the other. Let us assume that, within the common wall separating the two limbs, there is a mechanism that actively transports NaCl from the outflow limb to the inflow limb, without any accompanying movement of water. As bulk flow carries the fluid along the inflow limb, the effect of NaCl transport is cumulative, and the salt concentration becomes progressively higher. As the fluid rounds the bend and begins flowing out the other limb, its salt concentration progressively falls as a result of the cumulative effect of outward NaCl transport along the length of the outflow limb. By the time it reaches the end of that limb, its osmolarity is slightly lower than that of the fresh fluid beginning its inward flow in the other limb. This pattern establishes a salt gradient along the tube.

This example resembles the loop of Henle in principle, but not in detail. The loop of Henle has no common wall dividing the two limbs; nevertheless, the limbs are coupled functionally through the interstitial fluid, so that the NaCl pumped out of the ascending limb can diffuse the short distance toward the descending limb and cause osmotic reabsorption of water from that limb.

Several important points should be noted about countercurrent multiplier systems such as the loop of Henle and the simple model illustrated here. First, the standing concentration gradient set up in both limbs of the tube is due to both the continual movement of fluid through the system and the cumulative effect of transfer from the outflow limb to the inflow limb. The gradient would disappear if either fluid movement or transport across the membrane were to cease.

Second, the difference in concentration from one end of each limb to the other is far greater than the difference across the partition separating the limbs at any one point (part b of the figure). As a consequence, the countercurrent multiplier can produce greater concentration changes than would be attained by a simple transport epithelium without the configuration of a countercurrent system. The longer the multiplier, the greater the concentration differences that can be attained. Thus, in the kidney, longer loops of Henle produce larger overall gradients in osmolarity from renal cortex to medulla, thus permitting more efficient osmotic extraction of water from the collecting duct.

Third, the multiplier system can work only if it contains an asymmetry. In the model in part a of the figure, there is an active, energy-requiring net transport of NaCl in one direction across the wall. A passive countercurrent multiplier system, such as one used to conserve heat, does not require the expenditure of energy (part c of the figure). In the extremities of birds and mammals that inhabit cold climates, for example, a temperature differential exists between the arterial and venous flow of blood because the blood is cooled as it descends into the leg. As a result of this asymmetry and the countercurrent arrangement of the blood vessels, the arterial blood gives up some of its heat to the venous blood leaving the leg, thereby reducing the amount of heat lost to the environment.

(a)

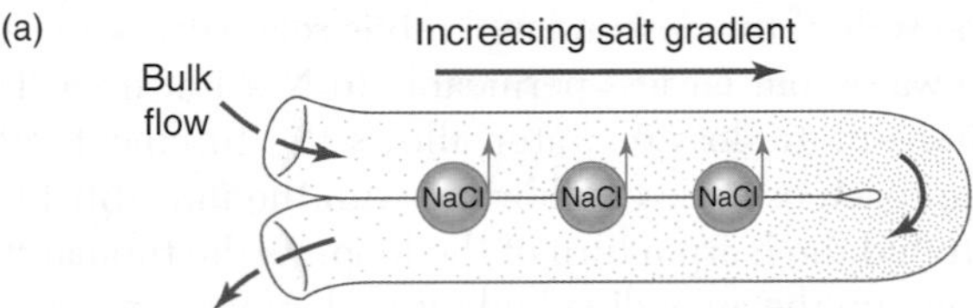

(b)

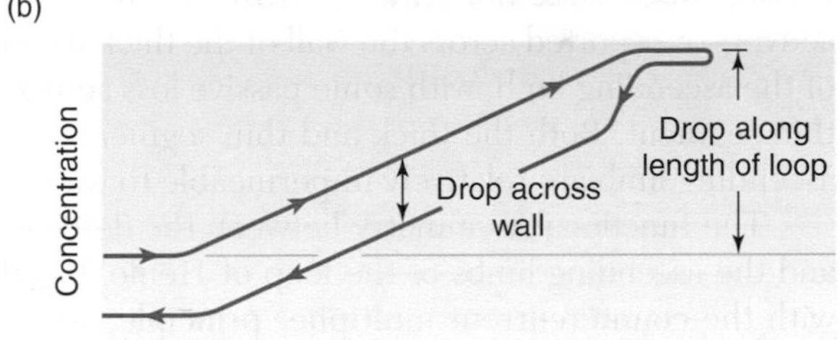

(c)

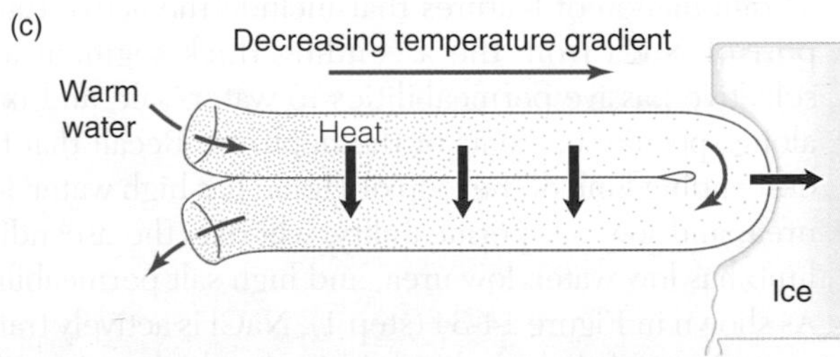

Active countercurrent multiplier systems require the expenditure of energy, whereas passive ones do not. ***(a)*** *Model of an active countercurrent multiplier system, in which a salt solution flows through a U-shaped tube with a common dividing wall. The active transport of NaCl from the outflow to the inflow limb constitutes an asymmetry necessary for the multiplier system to work.* ***(b)*** *A plot of salt concentration along the two limbs. Note that the concentration difference across the wall at any point is small relative to the total concentration difference along the length of the loop. The length of the loop and the efficiency of transport across the wall determine the overall concentration gradient along the entire length of the loop.* ***(c)*** *Model of a passive countercurrent multiplier system in which warm water in the inflow limb gives up part of its heat to cooler water flowing in the opposite direction in the outflow limb. Some heat is lost to the heat sink represented by ice in the environment, but much more of the heat is conserved by passive transfer from the inflow to the outflow limb.*

experimental support for the countercurrent multiplier hypothesis. These studies showed that filtrate entering the descending limb of the loop of Henle from the proximal tubule is iso-osmotic with respect to the extracellular fluid at that point (i.e., the outer portion of the renal medulla), having a concentration of about 300 mosm $\cdot$ L^{-1} (see Figure 14-28). The concentration of the fluid gradually increases as it makes its way down the descending limb toward the hairpin turn in the loop, where its concentration reaches 1000–3000 mosm $\cdot$ L^{-1} in most mammals. At this point, too, it is nearly iso-osmotic relative to the surrounding extracellular fluid in the deep portion of the renal medulla.

The increase in the osmolarity of the tubular fluid as it flows down the descending limb occurs because the wall of the descending limb is relatively permeable to water, but far less permeable to NaCl or urea. Thus the osmotic loss of water allows the tubular fluid to approach osmotic equilibrium with the interstitial fluid around the hairpin turn of the loop. As the tubular fluid flows up the ascending limb, it undergoes a progressive loss of NaCl (but not water). Most of this NaCl is actively transported across the wall of the thick segment of the ascending limb, with some passive loss across the thin segment. Both the thick and thin segments of the ascending limb are relatively impermeable to water.

The functional asymmetry between the descending and the ascending limbs of the loop of Henle, together with the countercurrent multiplier principle, accounts for the observed interstitial corticomedullary osmotic gradient represented by the blue wedge in Figure 14-34. The interstitial osmotic gradient is established by a combination of features that include the active transport of NaCl from the ascending thick segment and selective passive permeabilities to water, salt, and urea along specific segments of the nephron. Recall that the descending limb of the loop of Henle has high water, low urea, and low salt permeability, whereas the ascending limb has low water, low urea, and high salt permeability. As shown in Figure 14-34 (step 1), NaCl is actively transported out of the tubular fluid in the thick segment of the ascending limb and in the distal tubule. The loss of NaCl from these segments and its addition to the surrounding interstitium leads to the osmotic loss of water (step 2) from the salt-impermeant descending limb and from the distal tubule in the cortex and outer medulla.

Because of the net loss of water and salt from the filtrate in the loop of Henle and the distal tubule, the filtrate entering the collecting duct has a high urea concentration. The renal tubule up to this point is largely impermeable to urea, but as the collecting duct passes into the depths of the medulla, it becomes highly permeable to urea. As a result, urea leaks out down its concentration gradient (step 3), raising the interstitial osmolarity of the inner medulla. The resulting high interstitial osmolarity draws water from the descending limb of the loop of Henle (step 4), producing a very high intratubular solute concentration at the bottom of the loop. As the

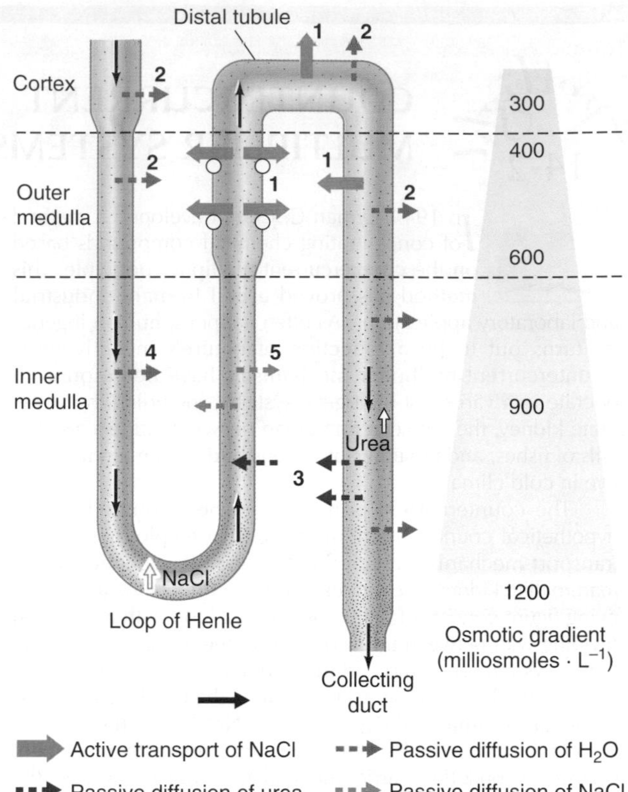

Figure 14-34 The steady-state corticomedullary osmotic gradient in the renal interstitium depends on differences in permeabilities and in active salt transport among different segments of juxtaglomerular nephrons, as well as on the anatomic layout of the nephrons and their circulatory supply (the vasa recta, not shown). The blue wedge depicts the osmotic gradient in the extracellular fluid; the small numbers indicate the total osmolarity. The bold numbers on the diagram indicate the various steps in the process of developing the osmotic gradient (see text for details). Active transport of NaCl from the ascending thick limb of the loop of Henle and distal tubule into the interstitial fluid (step 1) is largely responsible for the interstitial osmolarity of the cortex and outer medulla. The high osmolarity of the inner medulla depends largely on the passive diffusion of urea from the lower collecting duct (step 3), the only portion of the nephron that is highly permeable to urea. Some urea reenters the filtrate in the thin limb of the loop of Henle, where the urea level is relatively low, leading to recycling of urea (dashed red arrows). [Adapted from Jamison and Maffly, 1976.]

highly concentrated tubular fluid then flows up the highly salt-permeable thin segment of the ascending limb, NaCl leaks out (step 5) down its concentration gradient. The high osmolarity of the inner medullary interstitium thus depends largely on the passive accumulation of urea, which would not occur if the ascending limb were as permeable to urea as the collecting duct. Furthermore, if NaCl were not actively removed (with water following passively), urea would not become concentrated in the collecting duct, and the high medullary accumulation of urea would not take place.

It is interesting that the interstitial medullary urea gradient is established largely by passive means, although the active transport of NaCl is an essential component of the system and accounts for most of the metabolic energy expenditure necessary to set up the NaCl and urea gradients. The result of this combination of cellular specialization and tissue organization is a standing corticomedullary gradient of urea and NaCl in which the osmolarity becomes progressively higher with depth in the renal medulla, both inside the tubule and in the peritubular interstitium. This gradient is responsible for the final osmotic loss of water from the collecting ducts into the interstitium and the consequent production of a hyperosmotic urine.

A countercurrent feature in the organization of the vasa recta, the blood vessels around the nephron, is essential in maintaining the standing concentration gradient in the interstitium. Blood descends from the cortex into the deeper portions of the medulla in capillaries that form looplike networks around each juxtamedullary nephron and then ascends toward the cortex (see Figure 14-18a). As it moves through this circuit, the blood takes up salt and urea and gives up water osmotically as the surrounding interstitial fluid becomes increasingly hyperosmotic. Thus the osmolarity of the blood increases as it descends via the vasa recta into the medullary depths. The reverse occurs as the blood returns to the cortex and encounters an interstitium of progressively lower osmolarity. As a result, there is little net change in blood osmolarity during the circuit through the vasa recta. This countercurrent organization of the vasa recta allows a high rate of renal blood flow (essential for effective glomerular filtration) without disrupting the corticomedullary standing gradient of NaCl and urea concentration.

Control of Water Reabsorption

The concentration of the tubular fluid by the osmotic removal of water as it passes down the collecting duct provides a means of regulating the amount of water excreted in the urine. The rate at which water is osmotically drawn out across the wall of the collecting duct into the interstitial fluid depends on the water permeability of the wall of the collecting duct. Antidiuretic hormone regulates the water permeability of the collecting duct and thereby controls the amount of water leaving the animal via the urine. ADH acts, through the cAMP signaling system, to increase the number of water channels (aquaporins; see Chapter 4) in the apical membrane of the collecting duct and thereby promotes water reabsorption. The water channel in the collecting duct (WCH-CD) of the rat kidney has been cloned and expressed in other tissues. ADH up-regulates WCH-CD, moving water channel proteins from vesicular stores into the apical membrane. The higher the level of ADH in the blood, the more water channels there are in the apical membrane, and the more permeable the epithelial wall of the collecting duct becomes; hence, more water is drawn out of the urine as it passes down the duct toward the renal pelvis. ADH also causes a urea transporter protein (UT2) to be moved from vesicular stores into the apical membrane, increasing urea flux from the urine into the renal medulla. The transport of urea is coupled to sodium in an antiport fashion. Thus ADH stimulates both water and urea reabsorption across the collecting duct from tubular fluid and into the renal medulla.

The blood ADH level is a function of the osmotic pressure of the plasma and the blood pressure. The neurosecretory cells that produce ADH have their cell bodies in the hypothalamus and their axon terminals in the posterior pituitary gland (see Chapter 9). These osmotically sensitive cells respond to increased plasma osmolarity by increasing the rate at which ADH is released into the bloodstream from their axon terminals, thereby increasing the blood level of ADH and reabsorption of water from the collecting duct (Figure 14-35). If, for example, the osmolarity of the blood is increased as a result of dehydration, the activity of the neurosecretory cells increases, more ADH is released, the collecting ducts become more permeable, and water is osmotically drawn from the urine at a higher rate. This process results in the excretion of a more concentrated urine and the conservation of body water.

The hypothalamic cells that produce and release ADH receive inhibitory input from the arterial and atrial baroreceptors that respond to changes in blood

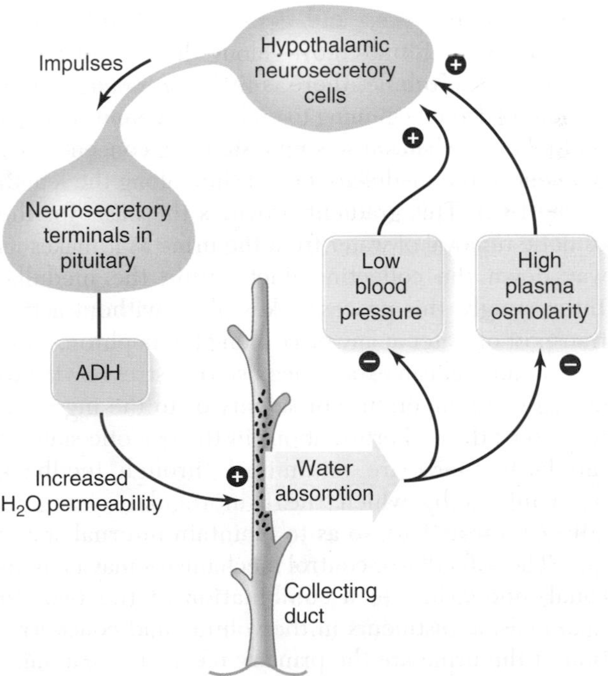

Figure 14-35 The osmolarity of the blood is under feedback regulation by the action of antidiuretic hormone on the collecting duct. Antidiuretic hormone (ADH) increases the water permeability of the stippled region of the duct, enhancing the rate of osmotic removal of water from the urine. The increased recovery of water counteracts low blood pressure and high plasma osmolarity, conditions that stimulate ADH secretion.

pressure. Hemorrhage, for example, results in a fall in blood pressure, reducing the activity of these baroreceptors (see Chapter 12); the resulting decreased inhibitory input to the ADH-producing cells in the hypothalamus leads to increased release of ADH and reduced loss of water in the urine, thus helping to maintain blood volume. Conversely, any factor that raises the venous blood pressure (e.g., an increase in blood volume due to dilution by ingested water) inhibits the ADH-producing hypothalamic cells, causing an increased loss of body water via the urine. The ingestion of drinks containing ethyl alcohol inhibits the release of ADH and therefore leads to copious urination and an increase of plasma osmolarity beyond the normal set point. Some degree of dehydration results, which contributes to the uncomfortable feeling of a hangover.

The action of mammalian ADH, and of the related peptide arginine vasotocin of non-mammalian vertebrate species, is not limited to the kidney. If these antidiuretic hormones are applied to frog skin and toad bladder, they increase the water permeability of those epithelia.

To review the mechanisms we've discussed, let's follow the path of the filtrate through the mammalian nephron. The formation of urine begins with the concentration of the glomerular filtrate into a hyperosmotic fluid in the proximal tubule. About 75% of the salt and water are removed from the filtrate in osmotically equivalent amounts as it passes through the proximal tubule, leaving urea and certain other substances behind. As the filtrate moves along the loop of Henle and the distal tubule, there is little net change in its osmolarity from beginning to end, but a countercurrent multiplier mechanism sets up a standing concentration gradient in the medullary interstitium along the length of the loop. This gradient provides the basis for the osmotic removal of water from the urine as it makes its way down the collecting duct within the medulla. Interestingly, this process takes place without active transport of water at any place along the nephron.

An animal can experience osmotic stress owing to changes in temperature or salinity or to the ingestion of food or drink. Perturbations in the osmotic state of the body fluids are minimized through feedback mechanisms by which the osmoregulatory organs adjust their activity so as to maintain internal status quo. These feedback-control mechanisms may be neuronal, endocrine, or a combination of the two. In mammals, adjustments in the volume and concentration of the urine are the primary means for maintaining osmotic homeostasis. In response to osmotic stress and other signals, mammals can regulate several aspects of urine formation, including the rate of glomerular filtration, the rate at which salts and water are absorbed from the lumen of the renal tubule, the secretion of unwanted substances, and the rate at which water is osmotically drawn out of the tubular fluid in the collecting duct.

NON-MAMMALIAN VERTEBRATE KIDNEYS

In the kidneys of the marine hagfishes (class Cyclostomata), the nephrons possess glomeruli, but no tubules, so the Bowman's capsules empty directly into collecting ducts. The kidneys are used largely to excrete divalent ions (e.g., Ca^{2+}, Mg^{2+}, and SO_4^{2-}) and carry out little or no osmoregulation. In fact, the extracellular fluids of the most primitive living vertebrates, the hagfishes, are relatively similar to seawater in salt concentration, and their plasma is essentially isotonic to seawater (see Table 14-1).

As a general rule, the kidneys of freshwater teleosts have more and larger glomeruli than the kidneys of their marine relatives. Because their bodies are hyperosmotic to the environment and water diffuses into their bodies, freshwater teleosts maintain osmotic balance by producing large volumes of dilute urine. In certain marine teleosts, the kidney nephrons have neither glomeruli nor Bowman's capsules. In such aglomerular kidneys, the urine is formed entirely by secretion because there is no specialized mechanism for the production of a filtrate. These fishes are hypo-osmotic to their environment and so lose water continually across the skin and gills. Their problem is water conservation, and they produce only small volumes of urine. Little urea is formed, and ammonia is excreted across the gills.

Amphibians and reptiles appear incapable of producing a hyperosmotic urine because they lack the countercurrent multiplier system of the loop of Henle that is necessary to produce urine of significantly greater osmolarity than the blood plasma. Only mammals and birds are known to have a loop of Henle, and thus only these animals, apparently, have their renal plumbing so organized as to allow osmotic countercurrent multiplication. The avian kidney contains a mixture of reptilian-type and mammalian-type nephrons. That is, some avian nephrons lack a loop of Henle, and in some birds, the loop is oriented perpendicular to the collecting duct, producing a less efficient concentrating mechanism. Many birds, however, have salt glands that excrete a concentrated salt solution, allowing them to drink seawater.

The elasmobranch *Raja erinacea* (a skate) has a complex renal tubule organization that has the anatomic requisites for countercurrent multiplication. However, the skate nephron is functionally quite different from the mammalian nephron. As we have seen, the mammalian kidney excretes urea and retains water to produce hyperosmotic urine. The elasmobranch kidney, in contrast, retains urea (which is used as an osmolyte) and does not produce concentrated urine. The countercurrent system in the kidneys of marine elasmobranchs is made up of tubular bundles. These fishes have high levels of urea in their tissues and reabsorb urea from the kidney ultrafiltrate. Freshwater stingrays, on the other hand, do not reabsorb filtered urea, and their kidneys lack tubular bundles, indicating that the bundles are the site of urea reabsorption. Thus, the function of the countercurrent organization of the elasmobranch nephron may be to conserve urea.

EXTRARENAL OSMOREGULATORY ORGANS IN VERTEBRATES

As we have already seen, many vertebrates rely on osmoregulatory organs other than kidneys to maintain osmotic homeostasis. First, we consider the specialized glands for excreting salt found in some animals, then we see how fish gills are used for osmoregulation.

Salt Glands

Elasmobranchs, marine birds, and some reptiles possess glands that secrete salt by cellular mechanisms similar to those used for sodium reabsorption in the mammalian kidney.

Elasmobranch rectal gland

Marine elasmobranchs, although slightly hypertonic to seawater, have a much lower NaCl content than seawater. As a result, there is a continuous influx of NaCl into the body of the animal. The excess salt is removed largely by the rectal gland, which produces and excretes a concentrated salt solution. The rectal gland is the major (perhaps the only) extrarenal site for secretion of excess NaCl by marine elasmobranchs. It regulates extracellular fluid volume by controlling the amount of NaCl in the body.

The rectal gland consists of a large number of blind-ending tubules surrounded by blood capillaries. The tubules drain into a duct, which opens into the intestine near the rectum. The fluid produced by the gland can have a slightly higher salt concentration than seawater, but is iso-osmotic to the blood plasma of the fish. Although the blood of elasmobranchs has a much lower salt concentration then seawater, it is slightly hyperosmotic to seawater, the osmolarity of the blood being made up by high concentrations of urea and trimethylamine oxide. Yet urea and TMAO do not appear in the rectal gland fluid; only NaCl does, so an iso-osmotic NaCl solution removes large amounts of excess sodium chloride.

The formation of the fluid secreted from the rectal gland does not involve filtration of the blood; rather, NaCl is actively secreted into the lumen of the tubules, and water follows. The tubule walls consist of a single type of salt-secreting cell similar to the chloride cells found in the gills of marine teleosts. The basolateral membrane of these cells is extensively folded and has a much larger surface area than the apical (mucosal) membrane. A Na^+/K^+ pump that is present in high concentrations in the basolateral membrane pumps Na^+ out of the cell and K^+ in; the K^+, however, cycles back out through the many K^+ channels that are also present in the basolateral membrane (see Figure 14-14). The activity of the Na^+/K^+ pump generates a large Na^+ gradient across the basolateral membrane, and that gradient drives NaCl uptake via a $Na^+/2Cl^-/K^+$ cotransport system also present in the basolateral membrane. Thus, as Na^+ and K^+ cycle across the basolateral membrane, the Cl^- level inside the cell rises above that in the tubular lumen. The Cl^- eventually exits via channels in the apical membrane, moving down its concentration gradient. The overall effect is the movement of Cl^- from the serosal (blood) side of the tubular wall into the lumen. This movement creates an electric potential across the tubule wall, with the serosal side positive and the lumen negative; the resulting electrochemical gradient for sodium permits diffusion of Na^+ from the serosal side through paracellular pathways into the lumen. Water follows NaCl and diffuses passively across the tubular wall, but the wall is impermeable to urea and TMAO. Thus the rectal gland produces a solution that is iso-osmotic with blood yet has a much higher NaCl concentration.

The hearts of dogfish sharks contain a natriuretic peptide hormone that stimulates chloride secretion in perfused rectal glands. Although circulating natriuretic peptide levels have not yet been measured in elasmobranchs, it is possible that natriuretic peptides released from the heart into the circulation stimulate secretion by the rectal gland, reducing extracellular fluid volume. The appropriate stimulus for the release of the natriuretic peptide would seem to be a rise in venous pressure—that is, the filling pressure of the heart. Supporting this conjecture is the finding that the heart of a teleost fish, the rainbow trout, releases natriuretic peptide into the circulation in response to increased venous pressure.

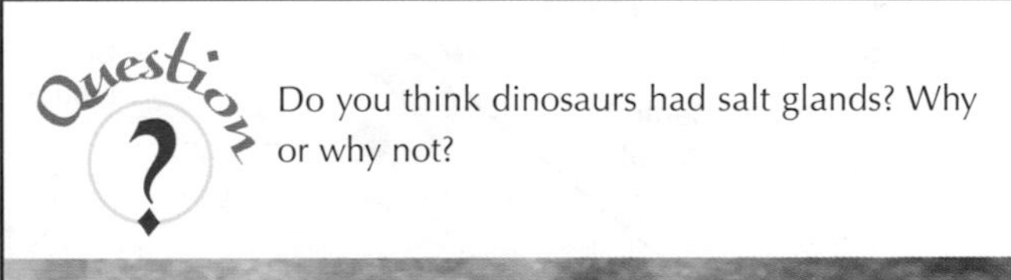

Do you think dinosaurs had salt glands? Why or why not?

Salt glands in birds and reptiles

In 1957, Knut Schmidt-Nielsen and his coworkers, while investigating the means by which marine birds maintain their osmotic balance without access to freshwater, discovered that the nasal salt glands secrete a hypertonic solution of NaCl. It was found in those early studies that if seawater is administered to cormorants or gulls by intravenous injection or by stomach tube, the resulting increase in plasma salt concentration leads to a prolonged nasal secretion of fluid with an osmolarity two to three times that of the plasma. Salt glands have subsequently been described in many species of birds and reptiles, especially those subjected to the osmotic stress of a marine or desert environment. These species include nearly all marine birds, ostriches, the marine iguana, sea snakes, and marine turtles, as well as many terrestrial reptiles. Crocodilians have a similar salt-secreting gland in the tongue.

The salt glands of birds and some reptiles occupy shallow depressions in the skull above the eyes. In birds, the salt gland consists of many lobes about 1 mm in diameter, each of which drains via branching secretory tubules and a central canal into a duct that, in turn, runs through the beak and empties into the nostrils (Figure 14-36a,b). Active secretion of salt takes place across the epithelium of the secretory tubules, which consist of characteristic salt-secreting cells. These cells have a profusion of deep infoldings in the basolateral membrane and are heavily laden with mitochondria, characteristics that are often associated with a highly active secretory epithelium. As in many other transport epithelia, adjacent cells are joined together by tight junctions that preclude massive leakage of water between the cells. The cell junctions in the salt gland are not as tight as those binding the cells of frog skin together, but allow some paracellular leakage of ions, as in the elasmobranch rectal gland.

The formation of fluid in the nasal gland, as in the rectal gland, does not involve filtration of the blood. The absence of filtration can be deduced from the nonappearance in the fluid of small filterable molecules (e.g., inulin or sucrose) that are injected into the bloodstream. High concentrations of a Na^+/K^+ pump have been found in the basolateral membrane of the tubular cells. Application of the ATPase inhibitor ouabain to the basal surface of the tubular epithelium blocks salt transport. Since ouabain does not cross epithelia and can block the pump only by direct contact with the ATPase, it appears that the sodium-transport mechanism operates in the basolateral membrane of the epithelial cells, as it does in the rectal gland. Increased salt secretion is associated with increased Na^+/K^+ pump activity in the salt gland. Some Na^+/K^+ pump molecules are found in the apical membrane as well. The basolateral membrane of the salt-gland epithelium also contains a $Na^+/2Cl^-/K^+$ cotransporter and K^+ channels, and the apical membrane contains Cl^- channels. The net result is movement of NaCl from the blood across the epithelium into the lumen of the salt gland (Figure 14-36c).

As we saw earlier, the salt solution produced by the elasmobranch rectal gland is iso-osmotic to plasma; in contrast, the fluid produced by the nasal gland is hyperosmotic to plasma. In both cases, the gland fluid has a high salt concentration, but the osmolarity of the blood of elasmobranchs is much higher than that of birds and reptiles. It is not clear how the solution produced by the

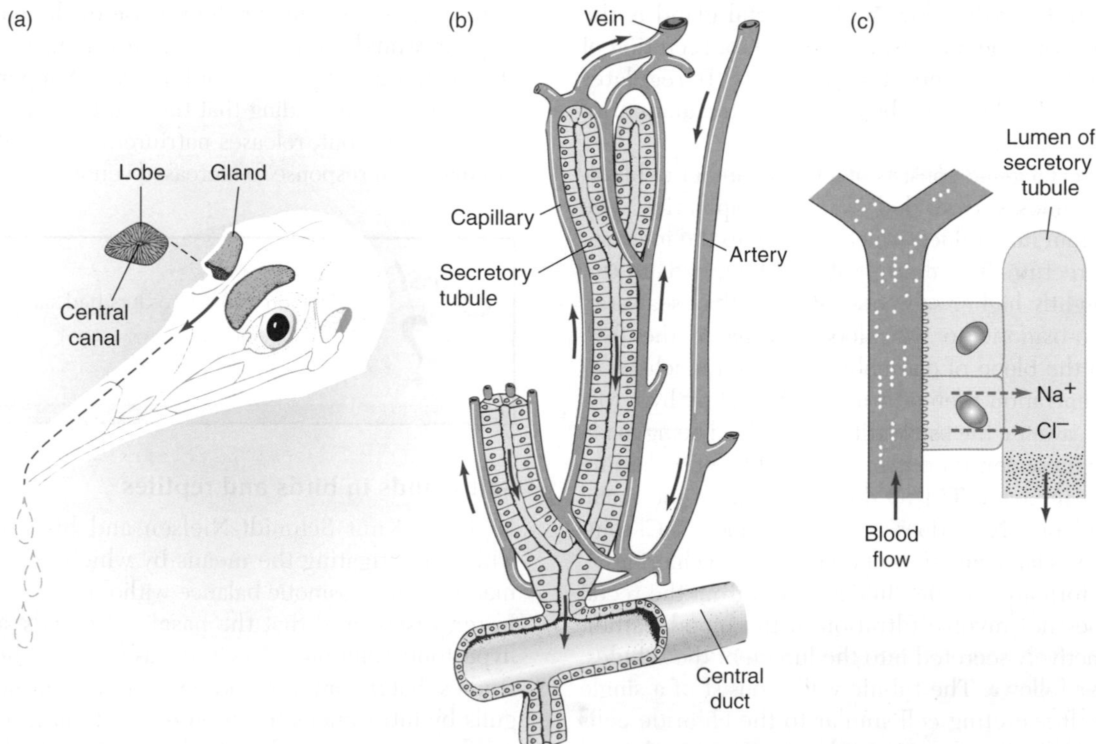

Figure 14-36 Marine birds maintain osmotic balance by excretion of a concentrated salt solution from glands located above the orbit of the eye. **(a)** The avian salt gland consists of a longitudinal arrangement of many lobes, which drain via a central canal into a duct that carries secretions to the nasal passages. **(b)** Each lobe consists of tubules and capillaries arranged radially around a central canal. Single tubules are surrounded by capillaries in which blood flows counter to the flow of secretory fluid in the tubule. This countercurrent flow facilitates the transfer of salt from the blood to the tubule, since the uphill gradient of salt concentration between capillary and tubule lumen is thereby minimized at each point along the length of the tubule. **(c)** The secretory cells constituting the tubular wall transport NaCl from the blood into the lumen via the mechanism depicted in Figure 14-14. These cells have a brush border and contain many mitochondria. [Part a adapted from Schmidt-Nielsen, 1960; part b from "Salt Glands," by K. Schmidt-Nielsen. Copyright © 1959 by Scientific American, Inc. All rights reserved.]

nasal glands of birds and reptiles is concentrated. It is possible that the initial solution in the apex of the tubule is iso-osmotic to plasma and becomes more concentrated as it passes down the tubule. The cells of the secretory epithelium of a single tubule become larger, and are surrounded by deeper paracellular channels, toward the base of the tubule, indicating that the fluid may become more concentrated toward the base of the tubule. Those birds that produce the most concentrated salt solutions have the largest secretory cells, with long paracellular channels between cells. In addition, the avian salt gland and the surrounding capillaries are organized as a countercurrent system, which may contribute to the concentration of the salt solution. The flow of blood is parallel to the secretory tubules and opposite in direction to the flow of secretory fluid (see Figure 14-36b). This arrangement maintains a minimal concentration gradient from blood to tubular lumen along the entire length of a tubule, thereby minimizing the concentration gradient for uphill transport from the plasma to the secretory fluid.

The salt gland becomes active in response to a salt load or expansion of the extracellular space. When a bird drinks seawater, water diffuses from the body into the gut because seawater has a higher osmolarity than the body fluids. At the same time, NaCl diffuses in the opposite direction, from seawater in the gut into the body. Thus, the initial effect of drinking seawater is a reduction in extracellular fluid volume and an increase in NaCl levels in the extracellular fluid and blood (Figure 14-37a). The salt level in the gut eventually drops because of salt loss to the body and diffusion of water from the body into the gut. Once the osmolarity of the gut fluids falls below that of the body, the movement of water from the body into the gut is reversed, and extracellular volume expands. The production of nasal fluid is initially inhibited by the reduction in extracellular volume, then strongly stimulated by the subsequent elevation of both extracellular volume and salt content. Consequently, there is often a short delay between drinking saltwater and secretion from the nasal gland. The solution secreted from the salt gland is more concentrated than the seawater taken in, so the bird ends up gaining osmotically free water, as illustrated in Figure 14-37b

Mammals have salt-secreting cells in the thick ascending limb of the loop of Henle that are similar to those found in the nasal gland of birds and the rectal gland of elasmobranchs, and which are even controlled, apparently, by the same array of hormones (natriuretic peptides and the renin-angiotensin system), but mammals cannot survive by drinking seawater. The arrangement of salt-secreting cells in mammals simply does not allow the production of a hypertonic salt solution that

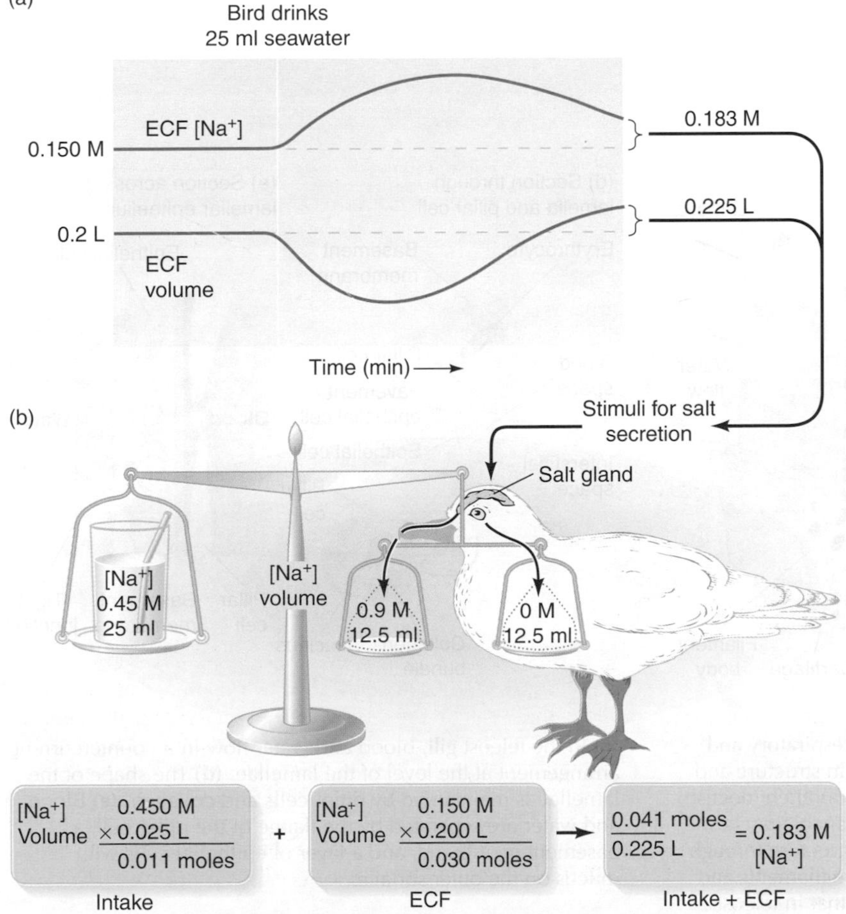

Figure 14-37 Because the salt-gland secretion is more concentrated than seawater, birds that drink seawater gain osmotically free water. Before drinking seawater, the gull in this example has an extracellular fluid (ECF) volume of 0.2 liters and an ECF sodium concentration of 0.15 M; thus the ECF contains 0.03 moles of Na^+. The bird then drinks 0.025 liters of seawater with a $[Na^+]$ of 0.45 M, so it ingests 0.011 moles of Na^+. **(a)** Initially, the ECF volume decreases and ECF $[Na^+]$ increases because Na^+ moves from the seawater in the gut into the ECF (down its concentration gradient), while water moves into the gut until osmotic equilibrium is established between the ECF and gut. The initial decrease in ECF volume inhibits salt-gland secretion. As the ECF $[Na^+]$ rises, water from the gut moves back into the ECF. When both the ECF volume and $[Na^+]$ are above their base levels, the salt gland is stimulated. **(b)** If the $[Na^+]$ of the secreted fluid is 0.9 M (twice the concentration of the ingested seawater), then the gull can secrete all the ingested salt in half the volume of water. In this example, the gull has a net profit of 12.5 ml of osmotically free water, which can be used to excrete other ions and molecules via the kidneys. [Adapted from unpublished material courtesy of Dr. Maryanne Hughes.]

can be excreted from the body. We see here a clear example of how the anatomic, cellular, and molecular organization of an animal determines its ability to survive in a particular environment.

Fish Gills

The surface area of a gill epithelium must be large if the gill is to function efficiently as a gas-exchange organ. Although a generous surface area makes gills an osmotic liability for animals such as teleost fishes, which are out of osmotic equilibrium with their aqueous environment, it also contributes to the suitability of gills as osmoregulatory organs. The gills of numerous aquatic species, vertebrate and invertebrate, are active not only in gas exchange, but also in such diverse functions as ion transport, excretion of nitrogenous wastes, and maintenance of the acid-base balance. In teleost fishes, gills play the central role in coping with osmotic stress.

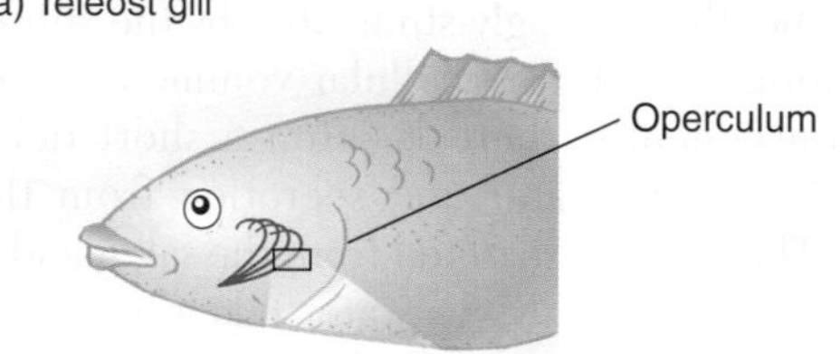

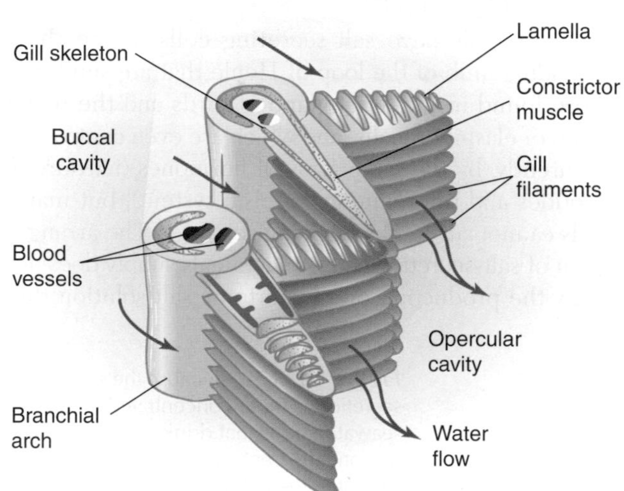

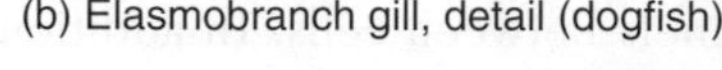

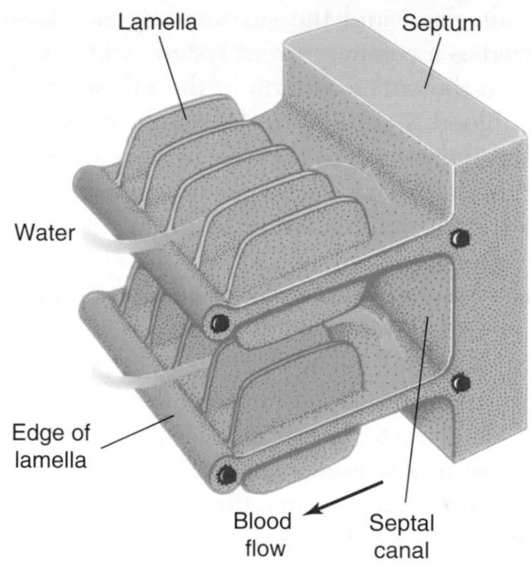

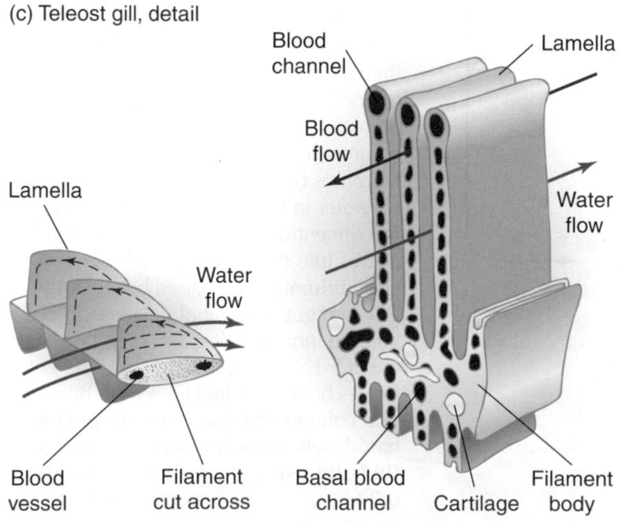

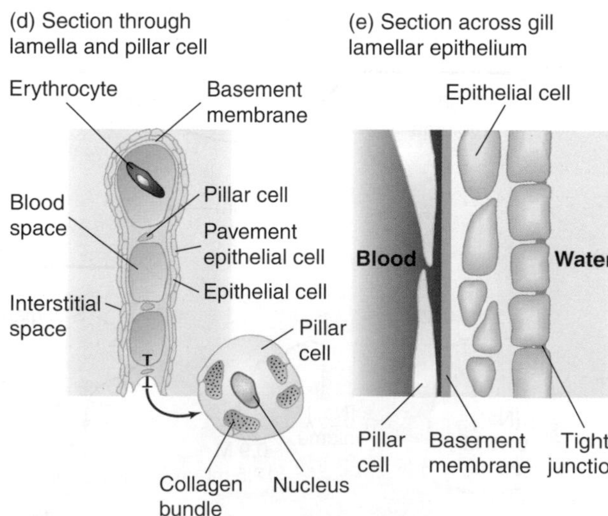

Figure 14-38 Fish gills function as both respiratory and osmoregulatory organs. Note the differences in structure and flow direction between the teleost and elasmobranch (dogfish) gill. **(a)** Teleost gills consist of a number of branchial arches, filaments, and lamellae, forming a sievelike structure through which water flows. **(b)** The arrangement of the filaments and flow direction in elasmobranchs differs from that in teleosts. **(c)** In the teleost gill, blood and water flow in a countercurrent arrangement at the level of the lamellae. **(d)** The shape of the lamellae is maintained by pillar cells and collagen. **(e)** Blood and water are separated by the flange of the pillar cell, a basement membrane, and a layer of epithelial cells with mucus on the outer surface.

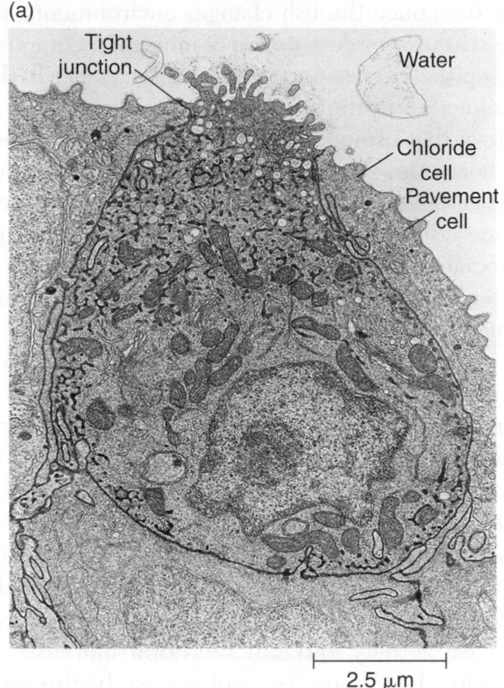

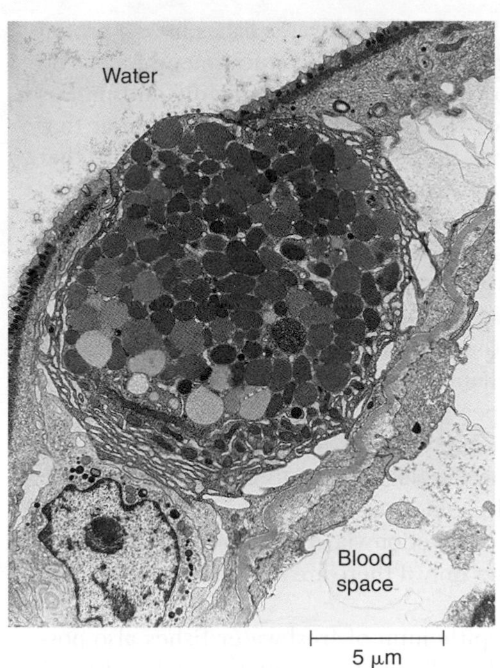

Figure 14-39 The epithelium of the gill is composed largely of pavement cells interspersed with a few mucous and chloride cells. The chloride cells tend to be at the base of the lamellae. **(a)** Electron micrograph of a freshwater teleost chloride cell with adjacent pavement cells. **(b)** Electron micrograph of a dogfish gill mucous cell containing many large mucous granules. [Electron micrographs courtesy of Jonathan Wilson.]

The structure of a teleost gill is illustrated in Figure 14-38. The epithelium separating the blood from the external water consists of several cell types, including mucous cells, chloride cells, and pavement cells (Figure 14-39). The epithelium of the lamellae consists mostly of flat pavement cells no more than 3–5 μm thick, containing some mitochondria. These cells are clearly best suited for respiratory exchange, acting as minimal barriers for diffusion of gases. The epithelium covering the gill filaments, and less frequently that of the gill lamellae, contains chloride cells, which are more columnar in shape and several times thicker from base to apex than the pavement cells. Chloride cells are deeply invaginated by infoldings of the basolateral membrane and are heavily endowed with mitochondria and enzymes related to active salt transport. Pavement cells and chloride cells are joined by tight junctions that limit the paracellular movement of water and ions. Mucous cells distributed unevenly over the gill epithelium release mucus that forms a boundary between the epithelium and the water.

Secretion of salt in seawater

Chloride cells were first described in 1932 by Ancel Keys and Edward Willmer. Their role in the transport of chloride was suggested by histochemical similarities to cells that secrete hydrochloric acid in the amphibian stomach and by the previous demonstration that the gill of marine teleosts is the site of extrarenal excretion of chloride (and sodium). Subsequent histochemical studies confirmed the presence of high levels of chloride in these cells, especially near the pit that develops on the apical border in fish that have become adapted to high salinities.

The mechanism of salt transport by chloride cells is similar to that of the salt-secreting cells illustrated in Figure 14-14. Thus, chloride cells have high levels of a Na^+/K^+ pump associated with a $Na^+/2Cl^-/K^+$ cotransporter in the basolateral membrane and a chloride channel in the apical membrane. Each chloride cell is associated with an accessory cell (distinct from a pavement cell). Sodium diffuses from blood to seawater through the paracellular channel between the chloride cell and accessory cell. In marine teleosts, the secretion of salt occurs against an osmotic gradient, so no movement of water follows the movement of salt. The shark rectal gland, the avian nasal salt gland, the marine teleost gill, and the thick ascending limb of the loop of Henle in the mammalian kidney tubule all appear to contain salt-secreting cells that transport NaCl by the same basic mechanism (see Figures 14-14 and 14-36). In the mammalian kidney, however, the direction of salt transport is into the blood rather than into the environment.

Chloride cells have been found to mediate the exchange of ions other than chloride, including Ca^{2+}. Calcium in the aqueous environment is taken up via Ca^{2+} channels in the apical membrane of the chloride cells and then actively transported into the blood via a calcium pump, which is present at high levels in the basolateral membrane.

Uptake of salt in freshwater

The cells in the gills of freshwater fishes have a proton pump and Na^+ channels in the apical membrane. The proton pump is presumably electrogenic and pumps protons out of the gills, generating a potential that draws Na^+ into the cell, a mechanism similar to that demonstrated in frog skin and the mammalian kidney (see Figure 14-31a). A clear relationship between the activity of the proton pump and the apical membrane potential has yet to be demonstrated in fish gills. A Na^+/K^+ pump in the basolateral membrane pumps Na^+ out of the cell into the blood, and K^+ cycles through K^+ channels in the membrane. Thus, the proton pump appears to energize Na^+ uptake across the apical membrane, whereas the Na^+/K^+ pump moves Na^+ across the basolateral membrane of the gills into body fluids. In some freshwater fishes, Na^+ uptake is coupled to H^+ excretion via an antiporter. This passive system operates only if there are adequate levels of Na^+ in the water.

The gill epithelium of freshwater fishes also possesses chloride cells, which mediate uptake of Ca^{2+} from the water. These cells, which differ from the chloride cells in marine teleost fishes, have an anion transporter on the apical membrane and high levels of proton pumps within the cell. They may be involved in the uptake of Na^+ and Cl^-, as well as Ca^{2+}, by freshwater fishes.

Physiological acclimatization in migrating fish

In fishes that regularly migrate between seawater and freshwater, such as salmon and eels, the gill epithelium adapts to changes in environmental salinity. These fishes actively take up NaCl in freshwater and actively excrete it in saltwater by the mechanisms described above. The physiological acclimatization of the gills, which takes a few days once the fish changes environments, involves the synthesis or destruction of molecular components of the epithelial transport systems and changes in the morphology and number of the chloride cells. It is now known that osmoregulatory acclimatization is mediated by hormones, which influence epithelial differentiation and metabolism. Growth hormone and the steroid hormone cortisol stimulate the changes in gill structure associated with the transition from freshwater to seawater, whereas prolactin stimulates the changes in gill structure that accompany the reverse transition.

Let's first consider what happens when a fish migrates from freshwater to seawater (Table 14-10, part A). The proton pump is quickly down-regulated, reducing Na^+ uptake. The influx of sodium from seawater causes plasma Na^+ levels to rise, which in turn stimulates secretion of cortisol (Figure 14-40a). Cortisol, along with growth hormone, induces an increase in the number of typical seawater chloride cells. As a result of these changes, the gill Na^+/K^+ ATPase activity and salt secretion increase (Figure 14-40b). In salmon, cortisol release begins while the fish is moving downriver, thus pre-acclimatizing the fish for a seawater existence. This process is termed *smolting*, and the end product is a *smolt*, a fish ready for the transfer to seawater. The increase in plasma Na^+ that occurs when the smolt enters seawater causes additional release of cortisol and completes the changes that will allow the fish to live in seawater. Normally about a week passes before the level of sodium in the plasma falls back to levels similar to those of freshwater fish (see Figure 14-40a).

When a marine teleost moves from seawater to freshwater, an opposite strategy for acclimatization is seen (Table 14-10, part B). Initially, the paracellular gaps between chloride and accessory cells in the gill epithe-

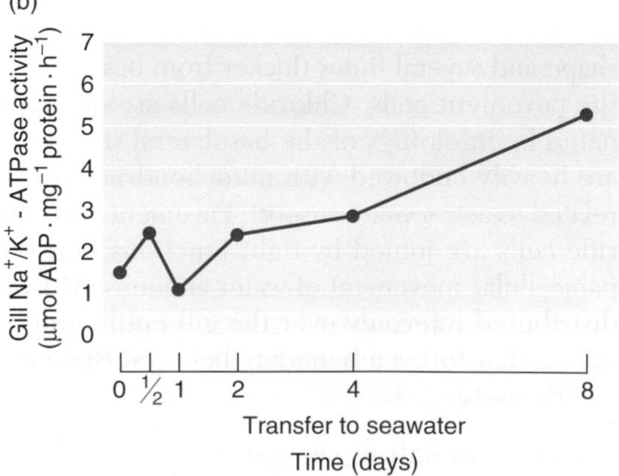

Figure 14-40 Cortisol plays a major role in inducing the physiological acclimatization that occurs when coho salmon are transferred from freshwater to seawater. **(a)** When a fish first moves from freshwater to seawater, the plasma Na^+ level begins to rise, stimulating secretion of cortisol. **(b)** The spike in plasma cortisol levels mediates a number of changes in the gills, including an increase in the gill Na^+/K^+ ATPase activity. As this activity increases, Na^+ secretion from the gills rises; thus, after several days in seawater, plasma Na^+ returns to values close to those observed in freshwater fish. [Unpublished data supplied by N. M. Whiteley and J. M. Wilson.]

Table 14-10 **Physiological acclimatizations that accompany the movement of fish to water of differing salinity**

(A) From freshwater to seawater

1. The proton pump that powers active uptake of NaCl is down-regulated.
2. The rise in the flux of Na^+ into the body raises plasma Na^+, stimulating an increase in plasma cortisol and growth hormone levels.
3. Hormones induce the proliferation of chloride cells and an increase in the infolding of their basolateral membranes.
4. The changes above cause an increase in the activity of the Na^+/K^+ pump and the secretion of NaCl.
5. Plasma Na^+ levels return to normal.

(B) From seawater to freshwater

1. The paracellular gaps between chloride and accessory cells close in response to low external Na^+ levels, causing NaCl efflux to fall rapidly.
2. Plasma prolactin levels increase.
3. Prolactin causes the number of chloride cells to decrease and the apical pits to disappear.
4. As a result, the activity of the Na^+/K^+ pump falls.
5. Up-regulation of the proton pump returns the fish to the freshwater condition.

lium close, reducing salt loss. A rise in the plasma prolactin level stimulates changes in the chloride cells, with an accompanying decrease in the activity of the Na^+/K^+ pump. Finally, up-regulation of the proton pump allows the uptake of salt necessary for survival in freshwater.

What design conflicts arise from the use of gills for both gas exchange and ionic regulation?

INVERTEBRATE OSMOREGULATORY ORGANS

In general, invertebrate osmoregulatory organs employ mechanisms of filtration, reabsorption, and secretion similar in principle to those of the vertebrate kidney, producing urine that is significantly different in osmolarity and composition from the body fluids. Insects and possibly some spiders are the only invertebrates known to produce a concentrated urine. Osmoregulatory mechanisms are used to differing extents in various organs and in different groups of invertebrate animals. The utility of these mechanisms is underscored by the widespread convergent evolution of osmoregulatory mechanisms in nonhomologous organs.

Filtration-Reabsorption Systems

Several lines of evidence indicate that the primary urine in both mollusks and crustaceans is formed by a filtration process that is similar in principle to that which occurs in the Bowman's capsule of vertebrates. When the undigestible polysaccharide inulin is injected into the bloodstream or coelomic fluid of a mollusk or crustacean, it appears in high concentrations in the urine, as it does in mammals. Since it is unlikely that a substance such as inulin is actively secreted, it must enter the urine during a filtration process in which all molecules below a certain size pass through a sievelike structure. When water and essential solutes are absorbed, inulin remains behind in the urine.

As in vertebrates, the normal urine of some invertebrates contains little or no glucose, even when the level of glucose in the blood is substantial. However, studies with several mollusks have shown that when the blood glucose is elevated by artificial means (such as direct injection), glucose appears in the urine at a characteristic threshold concentration of blood glucose. Beyond that threshold level, the urine glucose concentration rises linearly with blood glucose concentration. This behavior, which parallels that in the mammalian kidney (see Figure 14-25), probably results from saturation of the transport system by which glucose filtered into the tubular fluid is reabsorbed from the filtrate into the blood. Once the transport system is saturated, the "spillover" of glucose in the urine would be expected to be proportional to its concentration in the blood. More conclusive evidence has been obtained with the drug phlorizin, which is known to block active glucose transport. When phlorizin is administered to mollusks and crustaceans, glucose appears in the urine even at normal blood glucose levels. A reasonable conclusion is that glucose enters the urine as part of a filtrate and remains in the urine when the reabsorption mechanism is blocked by phlorizin.

Further support for a filtration-reabsorption mechanism comes from analyses of tubular fluids near suspected sites of filtration, which indicate that their composition is similar to that of the plasma. Finally, the rate of urine formation in some invertebrates has been found to depend on blood pressure. This relationship is consistent with a filtration mechanism, although a change in blood pressure may also produce a change in the circulation to the osmoregulatory organ.

The site of primary urine formation by filtration is known for only a few invertebrates. In a number of marine and freshwater mollusks, filtration takes place across the wall of the heart into the pericardial cavity, and the filtrate is conducted to the "kidney" through a special canal. Glucose, amino acids, and essential electrolytes are reabsorbed in the kidney. In the crayfish, the major organ of osmoregulation is the so-called antennal gland (Figure 14-41 on the next page). Part of that organ, the coelomosac, resembles the vertebrate

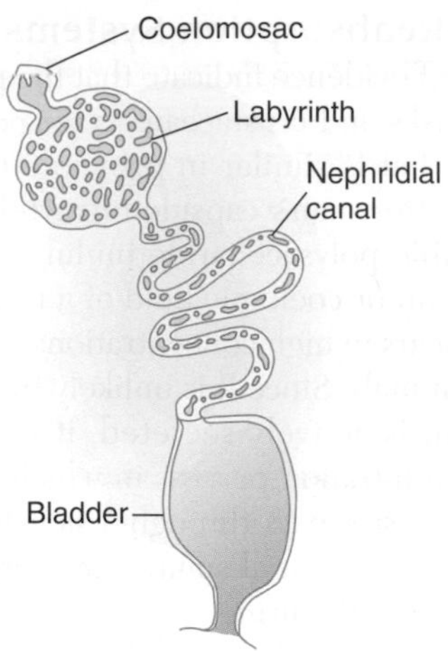

Figure 14-41 Osmoregulation in some invertebrates depends on filtration-reabsorption organs that differ structurally from the mammalian kidney but are functionally analogous to it. Shown is a schematic diagram of the crayfish antennal gland. Filtration of the blood produces the initial excretory fluid, which then is modified by selective reabsorption of various substances. [Adapted from Phillips, 1975.]

glomerulus in its ultrastructure. Micropuncture measurements have shown that the excretory fluid that collects in the coelomosac is produced by ultrafiltration of the blood. The antennal gland clearly plays an important role in regulating the concentration of ions (e.g., Mg^{2+}) in the hemolymph.

Since the final urine in mollusks and crustaceans differs in composition from the initial filtrate, the osmoregulatory organs of these animals must either secrete substances into the filtrate or reabsorb substances from the filtrate. The reabsorption of electrolytes is well established in freshwater species, since the final urine has a lower salt concentration than either the plasma or the filtrate. Glucose must also be reabsorbed, since it is present in the plasma and in the filtrate but is either absent or present at very low concentrations in the final urine.

The filtration-reabsorption type of osmoregulatory system has evolved separately in at least three phyla (Mollusca, Arthropoda, Chordata) and perhaps more. This kind of system has the important advantage that all the low-molecular-weight constituents of the plasma are filtered into the ultrafiltrate in proportion to their concentration in the plasma. Such physiologically important molecules as glucose and, in freshwater animals, such ions as Na^+, K^+, Cl^-, and Ca^{2+} are subsequently removed from the filtrate by tubular reabsorption, leaving toxic substances or unimportant molecules behind to be excreted in the urine. This process avoids the need for active transport into the urine of toxic metabolites or, for that matter, artificial substances of a neutral or toxic nature encountered in the environment. Thus, an advantage of the filtration-reabsorption system is that it permits the excretion of unknown and unwanted chemicals taken in from the environment without the necessity for a large number of distinct transport systems.

A disadvantage of the filtration-reabsorption osmoregulatory system is its high energy cost. The filtering of large quantities of plasma requires the active uptake of large quantities of salts, either by the excretory organ itself or by other organs, such as gills or skin. In frog skin, for example, it has been shown that 1 mol of oxygen must be reduced in the synthesis of ATP for every 16–18 mol of Na^+ ions transported. In freshwater clams, the cost of maintaining sodium balance amounts to about 20% of the total energy metabolism. The process is much less expensive for marine invertebrates, since salt conservation in a saltwater environment is much less of a problem.

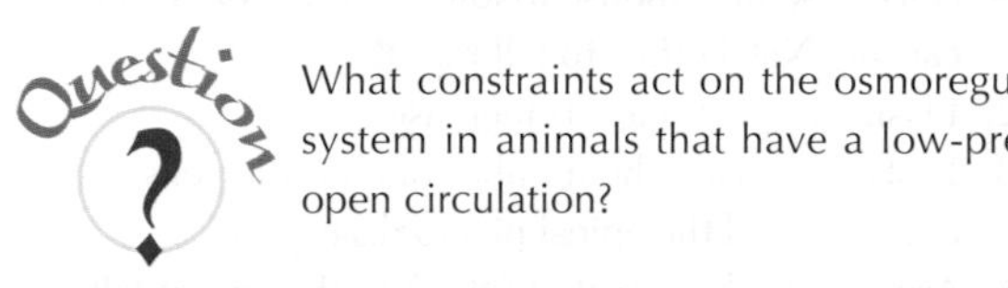

What constraints act on the osmoregulatory system in animals that have a low-pressure open circulation?

Secretion-Reabsorption Systems

Insects can survive in both freshwater and arid terrestrial environments. Given their often large surface-to-volume ratios, the osmotic demands placed on these animals can be extreme. The locust, for example, has a large capacity to regulate the ionic strength of the hemolymph. During dehydration, the hemolymph volume may decrease by up to 90%, yet the ionic composition of the hemolymph is maintained. When locusts drink solutions ranging in osmotic strength from seawater to tapwater, hemolymph osmotic pressure changes by only 30%. This capacity to regulate hemolymph composition depends on a secretion-reabsorption osmoregulatory system.

In broad outline, the osmoregulatory system of locusts and other insects consists of the **Malpighian tubules** and the hindgut (ileum, colon, and rectum). The closed ends of the long, thin Malpighian tubules lie in the hemocoel (the hemolymph-containing body cavity); the tubules empty into the alimentary canal at the junction of the midgut and hindgut (Figure 14-42). The secretion formed in the tubules passes into the hindgut, where it is dehydrated, passed into the rectum, and voided as concentrated urine through the anus. The presence of a tracheal system for respiration in insects (described in Chapter 13) diminishes the importance of an efficient circulatory system. As a consequence, the Malpighian tubules do not receive a direct, pressurized arterial blood supply, as the mammalian nephron does. Instead, they are surrounded with hemolymph at a pressure essentially no greater than the pressure within the tubules. Since there is no significant pressure differ-

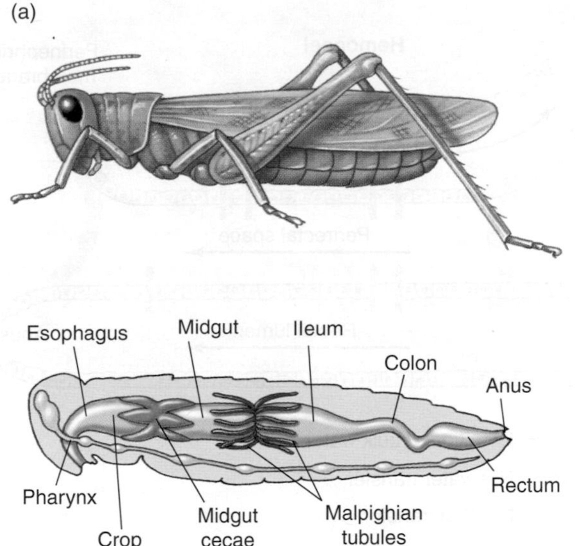

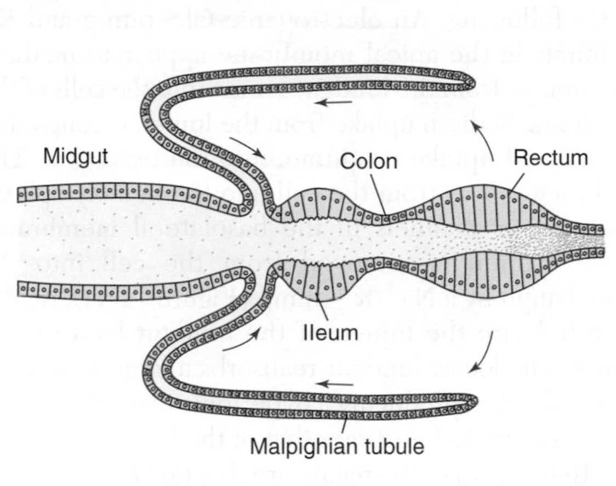

Figure 14-42 Osmoregulation in insects involves a secretion-reabsorption mechanism. **(a)** External side view and cross-sectional view of a locust. **(b)** Simplified diagram showing the relation of the Malpighian tubules to the gut of the locust. The primary urine is produced by secretion into the lumen of the numerous Malpighian tubules, which lie in the hemolymph-containing hemocoel. The primary urine flows into the rectum, where it is concentrated by the reabsorption of water. Although ions are also reabsorbed, the final urine is hypertonic to the hemolymph. The arrows indicate the circular pathway of water and ion movement.

ential across the walls of the Malpighian tubules, filtration cannot play a role in urine formation in insects. Instead, the urine must be formed entirely by secretion, with the subsequent reabsorption of some constituents of the secreted fluid. This process is analogous to the formation of urine by secretion in the aglomerular kidneys of marine teleosts. The serosal surface of a Malpighian tubule exhibits a profusion of microvilli and mitochondria.

The details of urine formation by tubular secretion differ among insects, but some major features seem to be common throughout this group. KCl and, to a lesser extent, NaCl are transported from the hemocoel into the lumen of the Malpighian tubules, along with such waste products as uric acid and allantoin. Most of the NaCl and KCl is returned to the hemolymph across the rectal wall (see Figure 14-42). The transport of K^+ is the major driving force for the formation of the primary urine in the Malpighian tubules, with most of the other substances following passively. Although K^+ is osmotically the most important substance actively transported, there is evidence that active transport plays an important role in the secretion of uric acid and some other nitrogenous wastes. The primary urine formed in the Malpighian tubules is relatively uniform in composition from one species to another, and in each species it remains iso-osmotic to the hemolymph under different osmoregulatory demands. The fluid formed in the Malpighian tubules passes into the hindgut, where water and ions are removed in amounts that maintain the proper composition of the hemolymph. Thus, it is in the hindgut that the composition of the final urine is determined.

The most complete study of the osmoregulatory function of the hindgut has been done with the desert locust *Schistocerca.* The serosal surface of the ileum and rectum is a highly specialized secretory epithelium (Figure 14-43). When a solution similar to hemolymph is injected into the hindgut of this insect, water, K^+, Na^+, and Cl^- are absorbed into the surrounding

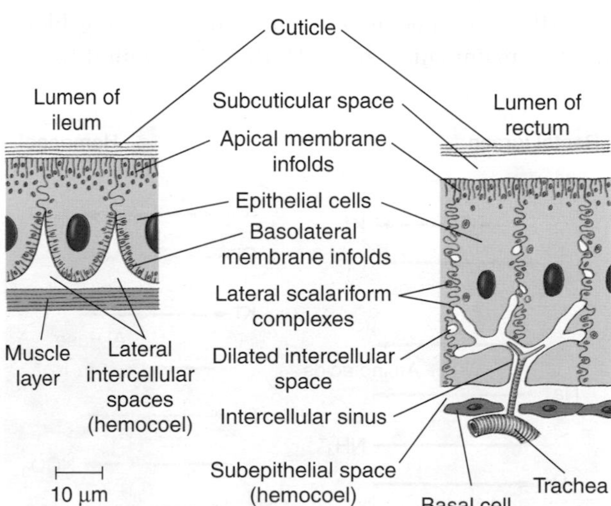

Figure 14-43 The hindgut of insects is specialized for transport of water and ions from the lumen to the surrounding hemocoel. Shown here are the ultrastructural organization and gross dimensions of the epithelium of the ileum and rectum in the locust, both of which are involved in reabsorption. Note the extensive infolding of the apical membrane and extensive lateral intercellular spaces. [Adapted from Irvine et al., 1988.]

hemolymph. Evidence from electrical measurements suggests that these ions are transported actively, with water following. An electrogenic Cl^- pump and K^+ channels in the apical membrane appear to mediate KCl uptake from the hindgut lumen into the cells of the gut lining. Sodium uptake from the lumen is coupled to amino acid uptake or ammonium ion excretion. The KCl then moves from the cell into the hemolymph via appropriate channels in the basolateral membrane, while sodium is removed from the cell into the hemolymph by a Na^+/K^+ pump (Figure 14-44). Acid is excreted into the lumen of the hindgut by a proton pump. The locust hindgut reabsorbs a large amount of ions and water, producing a hypertonic urine that has an osmolarity up to four times that of the blood.

In the larva of the mealworm beetle *(Tenebrio),* the urine-to-blood osmolarity ratio can be as high as 10, suggesting that the osmoregulatory system of this insect has a concentrating ability comparable to that of the most efficient mammalian kidneys. It has been suggested that uphill transport of water in *Tenebrio* and some other species results from a countercurrent arrangement of the Malpighian tubules, the perirectal space, and the rectum (Figure 14-45). Water is drawn osmotically from the rectum into the Malpighian tubules by the KCl gradient produced by active transport. The direction of bulk flow in these compartments is such that the osmotic gradient along the length of the complex is maximized, with the absolute osmolarities highest toward the anal end of the rectum. This gradient may allow the concentrations near the anal end to exceed those of the hemolymph by several times.

There is evidence for feedback regulation of osmolarity among the invertebrates, especially in insects. The bug *Rhodnius* becomes bloated after sucking blood from a mammalian host. Within 2–3 minutes, the

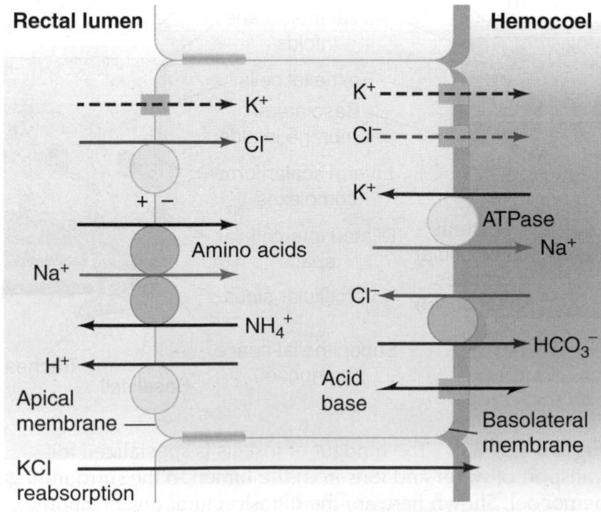

Figure 14-44 Ions are transported in and out of locust rectal cells by numerous mechanisms. Their primary effect is the reabsorption of KCl and water and the excretion of ammonia and acid.

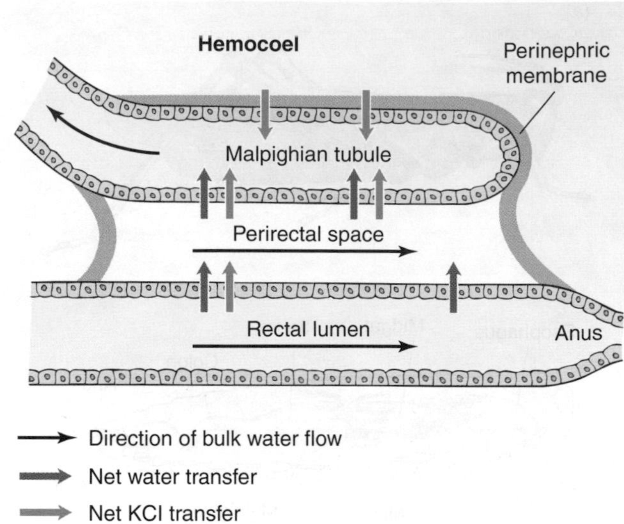

Figure 14-45 The countercurrent arrangement of the water-extraction apparatus of the rectum of the mealworm beetle, *Tenebrio,* probably accounts for the high urine-concentrating ability of this organism. Most of the water and KCl entering the rectal lumen is recycled into the Malpighian tubules. [Adapted from Phillips, 1970.]

Malpighian tubules increase their secretion of fluid by more than a thousand times, producing a copious urine. Artificially bloating the insect with a saline solution does not produce such diuresis in an unfed *Rhodnius.* It has also been found that isolated Malpighian tubules immersed in the hemolymph of unfed individuals remain quiescent, but if immersed in the hemolymph of a recently fed *Rhodnius,* they produce a copious secretion. A factor that stimulates secretion by these tubules can be extracted from neuronal tissue containing the cell bodies or axons of neurosecretory cells, primarily those of the metathoracic ganglion. Thus, it appears that these cells release a diuretic hormone in response to something present in the ingested blood. The only neurohormone known to stimulate the diuretic action of the neurosecretory cells is serotonin. Similar findings in other insect species suggest that diuretic and antidiuretic hormones produced in the nervous system regulate the secretory activity of the Malpighian tubules or the reabsorptive activity of the rectum. In earthworms, removal of the anterior ganglion results in the retention of water and a concomitant decrease in plasma osmolarity. Injection of homogenized brain tissue reverses these effects, suggesting a hormonal mechanism.

EXCRETION OF NITROGENOUS WASTES

When amino acids are catabolized, the amino group ($—NH_2$) is released (deamination) or transferred to another molecule for removal or reuse. Amino groups not salvaged for resynthesis of amino acids must be dis-

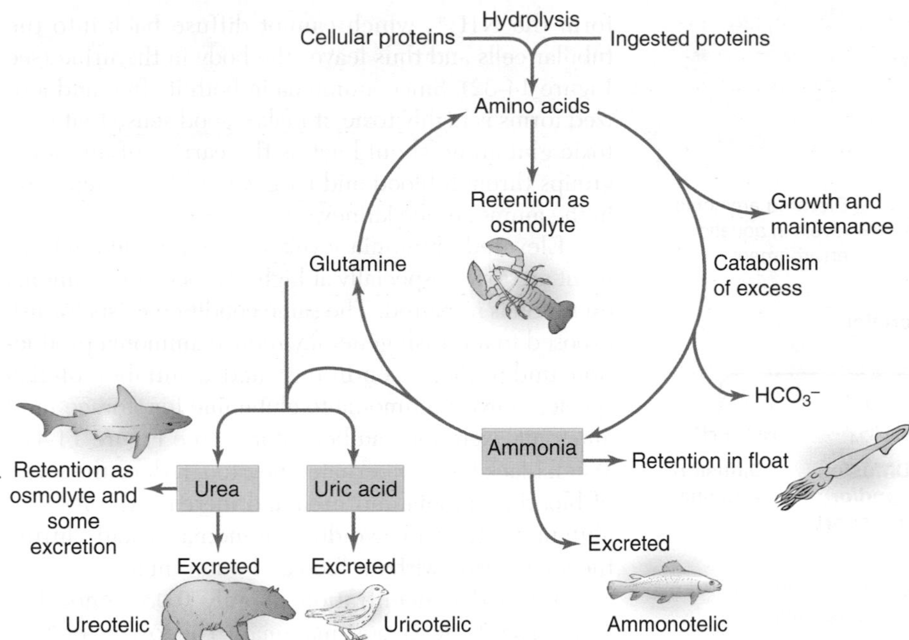

Figure 14-46 Water availability generally correlates with the predominant form of nitrogen (purple boxes) excreted by different animals. This general overview of nitrogen metabolism and excretion in animals shows the points at which they differ. An animal can be classified as ammonotelic, ureotelic, or uricotelic based on the main nitrogenous excretory compound it uses. Some animals use certain nitrogenous compounds as osmolytes to adjust body osmolarity. [Adapted from Wright, 1995.]

solved in water and excreted to avoid a toxic rise in the plasma concentration of nitrogenous wastes. Elevated levels of these wastes can cause convulsions, coma, and eventually death. Most excess nitrogen is excreted by animals as ammonia, urea, or uric acid. Lesser quantities of nitrogen are excreted in the form of other compounds, such as creatinine, creatine, or trimethylamine oxide, and, in very small quantities, amino acids, purines, and pyrimidines. The different properties of the three primary nitrogenous excretory compounds affect how animals have evolved to produce one or the other of these forms during all or part of their life cycles (Figure 14-46).

Ammonia is more toxic than urea or uric acid and must be kept at low levels in the body. Because excretion of ammonia occurs by diffusion, maintaining the concentration of ammonia in the excretory fluid below that in the body requires a large volume of water—about 0.5 liter of water is needed to excrete 1 g of nitrogen in the form of ammonia. Urea is less toxic than ammonia, and only 0.05 liter of water is required to excrete 1 g of nitrogen as urea, just 10% of that needed to excrete the equivalent amount of nitrogen as ammonia. However, urea synthesis consumes ATP, so energy can be saved if enough water is available to support the excretion of nitrogenous waste as dilute ammonia. Even less water is required to excrete uric acid—only 0.001 liter per gram of nitrogen, 1% of that needed for ammonia excretion. Uric acid is only slightly soluble in water and is excreted as the white, pasty precipitate characteristic of bird feces. The low solubility of uric acid has adaptive significance, since precipitated uric acid contributes nothing to the tonicity of the "urine" or feces.

In general, it is the availability of water that determines the nature and pattern of nitrogen excretion. Aquatic animals normally excrete ammonia across their gills, whereas terrestrial animals normally excrete urea or uric acid via their kidneys. Terrestrial birds excrete about 90% of their nitrogenous wastes as uric acid and only 3%–4% as ammonia, but semiaquatic birds, such as ducks, excrete only 50% of their nitrogenous wastes as uric acid and excrete 30% as ammonia. Mammals excrete most of their wastes as urea. Frog tadpoles are aquatic and excrete ammonia; after they metamorphose into the terrestrial adult form, they excrete urea. Avian embryos produce ammonia for the first day or so, and then switch to uric acid, which is deposited within the egg as an insoluble solid and thus has no effect on the osmolarity of the small amount of precious fluid contained in the egg. Lizards and snakes have various developmental schedules for switching from the production of ammonia and urea to the production of uric acid. In species that lay their eggs in moist sand, the switch to uric acid production occurs late in development, but before hatching. The switch to uric acid production is a kind of biochemical metamorphosis that prepares the organism for a dry, terrestrial habitat. It is evident, however, that there are many overlaps of different excretory products in animals in similar environments.

Ammonia-Excreting (Ammonotelic) Animals

Most teleosts and aquatic invertebrates are **ammonotelic**; that is, they excrete their nitrogenous wastes primarily as ammonia, producing little or no urea (Figure 14-47 on the next page). The terrestrial isopod (sowbug), as well as some terrestrial snails, crabs, and an air-exposed fish called the Chinese weather loach *(Misgurnus anguillicaudatus)*, also excrete a significant portion of their nitrogenous wastes by ammonia volatilization, a process in which ammonia gas is released into the air.

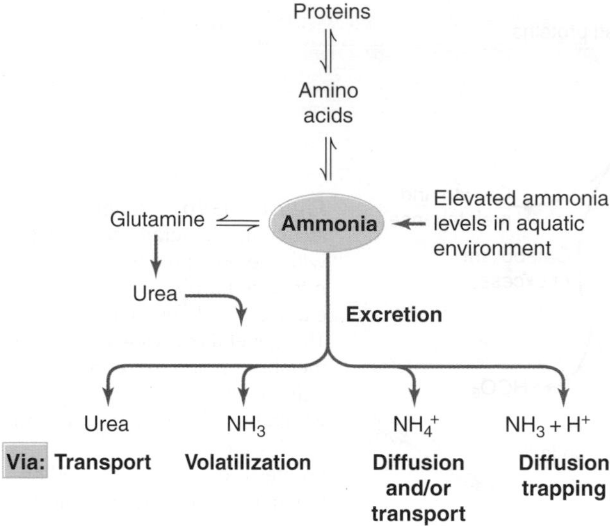

Figure 14-47 Ammonia accumulates in the body due to proteolysis and deamination of amino acids. Ammonia toxicity is avoided by either excreting ammonia or converting it to less toxic compounds, which are in turn either stored or excreted. [Adapted from Ip et al., 2000.]

Cell membranes are generally permeable to un-ionized ammonia (NH_3), but not very permeable to ammonium ions (NH_4^+). Most ammonia excretion occurs via passive diffusion of un-ionized ammonia. In most teleosts, nearly all ammonia is excreted as NH_3, along with H^+ and CO_2. The latter two substances acidify the water next to the gill surface, trapping NH_3 as the largely impermeant NH_4^+ and enhancing ammonia excretion. The mudskipper (*Periophthalmodon schlosseri*) excretes ammonium ions actively, which contributes to its ability to survive concentrations of ambient ammonia that would be toxic to other fishes. Active excretion of NH_4^+ would be costly if associated with high membrane NH_3 permeability and would result in ammonia cycling with the transfer of protons across the membrane. Thus, some membranes have a low permeability to NH_3 as well as NH_4^+. Membranes of *Xenopus* eggs and those of the cells of the thick ascending limb of the loop of Henle in the mammalian kidney are examples of structures having a low NH_3 permeability.

The amino groups of many amino acids are transferred to α-ketoglutarate, forming glutamate, which is then deaminated back to α-ketoglutarate and ammonium ions in the liver. Glutamate is also converted in the liver to glutamine, which is much less toxic than ammonia but is not normally excreted in significant quantities. Mammals excrete most nitrogenous wastes as urea, but they also excrete small amounts of ammonia in the urine, as we saw earlier. The less toxic glutamine, rather than ammonia, is released from the mammalian liver into the blood and taken up by the kidney. The glutamine is then deaminated in the cells of the kidney tubules, and ammonia is released into the tubular fluid. The excreted ammonia can take up a proton to form the NH_4^+, which cannot diffuse back into the tubular cells and thus leaves the body in the urine (see Figure 14-32). Since ammonia in both its free and ionized forms is highly toxic, it makes good sense that nontoxic glutamine should act as the carrier of ammonia groups through blood and tissues until its deamination in the ammonotelic kidney.

Elevated ammonia levels in the aquatic environment are toxic, especially at high pH, because ammonia excretion is impaired. The same condition exists for fish exposed to air. Fish generally reduce ammonia production under these conditions, and a number of fish species convert ammonia to glutamine for storage until ammonia excretion can be restored (see Figure 14-47). When black bears hibernate, they, too, reduce the levels of blood ammonia and urea and increase the level of glutamine, thereby avoiding ammonia toxicity during the long period without food and water intake.

A blood concentration of only 0.05 $mmol \cdot L^{-1}$ ammonia is toxic to most mammals. The toxicity of NH_3 is due in part to the elevation of pH it produces, which causes changes in the tertiary structure of proteins. Ammonia also interferes with some ion-transport mechanisms by substituting for K^+. Ammonia can also affect blood flow in the brain and some aspects of synaptic transmission, particularly glutamate metabolism. Acute toxic effects have been observed in many other animals as well, including birds, reptiles, and fish. Mexican guano bats are unusual among mammals in that they can withstand the very high levels of ammonia (1800 ppm) in the atmosphere of the caves in which they live. This level is sufficient to kill humans—enter guano bat caves with care!

Some squids, shrimps, and tunicates sequester NH_4^+ at high concentrations in specialized acidified chambers that act as floats, substituting NH_4^+ for heavier Ca^{2+} and Mg^{2+} ions (see Figure 14-49). Ammonium levels in the floats are very high, and the tissues that make up the float must be resistant to the toxic effects of ammonia. Ammonia levels in other regions of the body are relatively low.

How do variations in the pH of body compartments affect the transfer and distribution of ammonia and ammonium ions within the body?

Urea-Excreting (Ureotelic) Animals

Ureotelic animals excrete most nitrogenous wastes as urea, which is quite soluble in water, is far less toxic than ammonia, requires much less water for excretion than ammonia, and contains two nitrogen atoms per molecule. Ureotelic animals utilize one of two pathways for urea formation. Vertebrates, with the exception of most teleost fishes, synthesize urea primarily in the liver

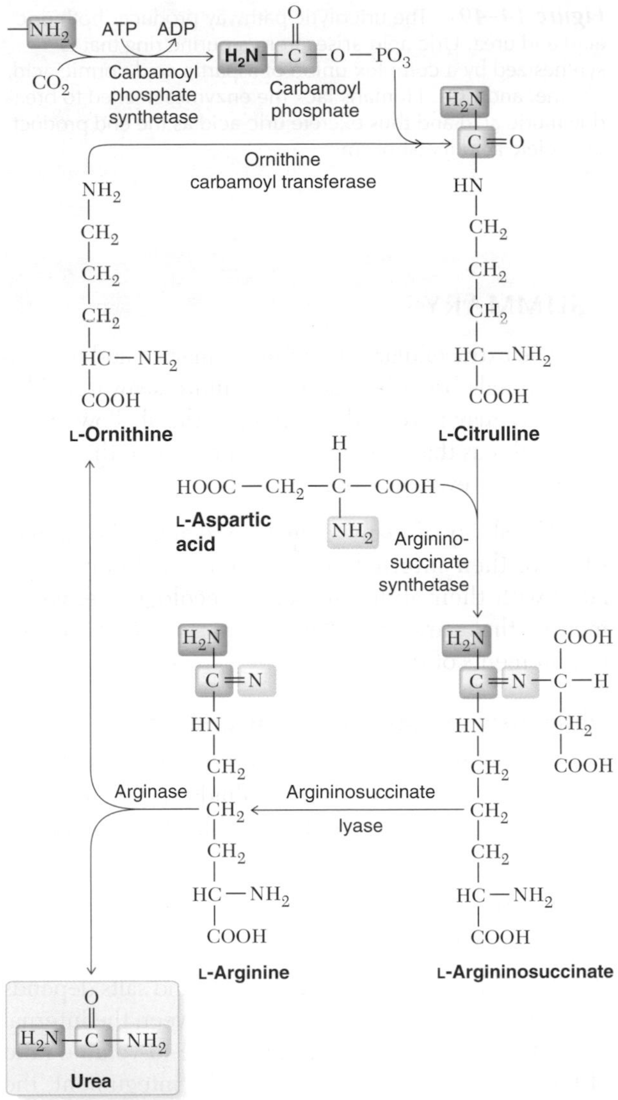

Figure 14-48 Urea is formed by the ornithine-urea cycle in most vertebrates. Because ATP is required for the first step, nitrogen excretion in ureotelic animals is more energetically expensive than in other animals.

via the **ornithine-urea cycle** (Figure 14-48). In this pathway, two —NH_2 groups and a molecule of CO_2 are added to ornithine to form arginine. The enzyme arginase, present in relatively large quantities in these animals, then catalyzes removal of the urea molecule from arginine, regenerating ornithine.

Most aquatic animals excrete ammonia, but not all. The Lake Magadi tilapia *(Oreochromis alcalicus grahami)* differs from most teleosts in excreting all nitrogenous wastes as urea. The high pH of Lake Magadi (about 10) impairs ammonia excretion, leading to ammonia accumulation and death in other fish. This tilapia can live in Lake Magadi because it converts ammonia to urea via the ornithine-urea cycle, thereby avoiding ammonia toxicity. Elasmobranch fishes use urea produced from ammonia via the ornithine-urea cycle as an osmolyte to increase their body osmolarity; they also excrete most of their nitrogenous wastes as urea across their gills. Elasmobranchs have high concentrations of urea in their bodies, which is prevented from leaking into the environment by the very low permeability to urea of their gill epithelia. This low permeability is due to the high cholesterol composition of the gill membrane, coupled to a urea/sodium antiporter in the basolateral membrane that transports urea back into the blood. The elasmobranch kidney also functions to conserve urea.

Most teleosts and many invertebrates utilize the so-called **uricolytic pathway**, in which urea is produced from uric acid generated by transamination via aspartate or the metabolism of nucleic acids. In this pathway, uric acid is converted first to allantoin and allantoic acid in reactions catalyzed by the enzymes uricase and allantoinase, respectively, and then to urea in a reaction catalyzed by allantoicase (Figure 14-49 on the next page). During evolution, most mammals have lost the ability to produce allantoicase and allantoinase; hominoid primates also cannot synthesize uricase, and thus excrete uric acid as the end product of nucleic acid metabolism. Uric acid excretion normally constitutes only about 1% by weight of urea excretion in humans. Should uric acid production or intake increase, however, uric acid levels in the blood may rise because excretion may be compromised by the low solubility of uric acid, especially if urine volume is small. The low solubility of uric acid may also result in the precipitation of uric acid crystals, which causes the painful condition called gout.

Lipid membranes are rather impermeable to urea, so its transport across membranes requires specialized membrane transporters. Specific urea transporters are thought to be present in elasmobranchs and in several other vertebrates, including mammals. Urea transporters are widely distributed in tissues, and in some instances they may be involved in rapid urea transport to stabilize cell volume in the face of osmotic shock.

Uric Acid–Excreting (Uricotelic) Animals

Uricotelic animals—birds, reptiles, and most terrestrial arthropods—excrete nitrogenous wastes chiefly in the form of uric acid or the closely related compound guanine. Uric acid and guanine have the advantage of carrying away four nitrogen atoms per molecule. The nitrogen atoms in uric acid arise from the breakdown of the amino acids glycine, aspartate, and glutamine (see Figure 14-49). Because these animals lack uricase, they cannot break down uric acid. Thus the catalysis of nitrogenous molecules is terminated at uric acid. This poorly soluble compound largely precipitates and is excreted as the end product, with very little urinary water required. In general, uricotelic animals are adapted to conditions of limited water availability.

Uric acid is secreted by the proximal tubule of the nephron via the organic anion secretory mechanism. In

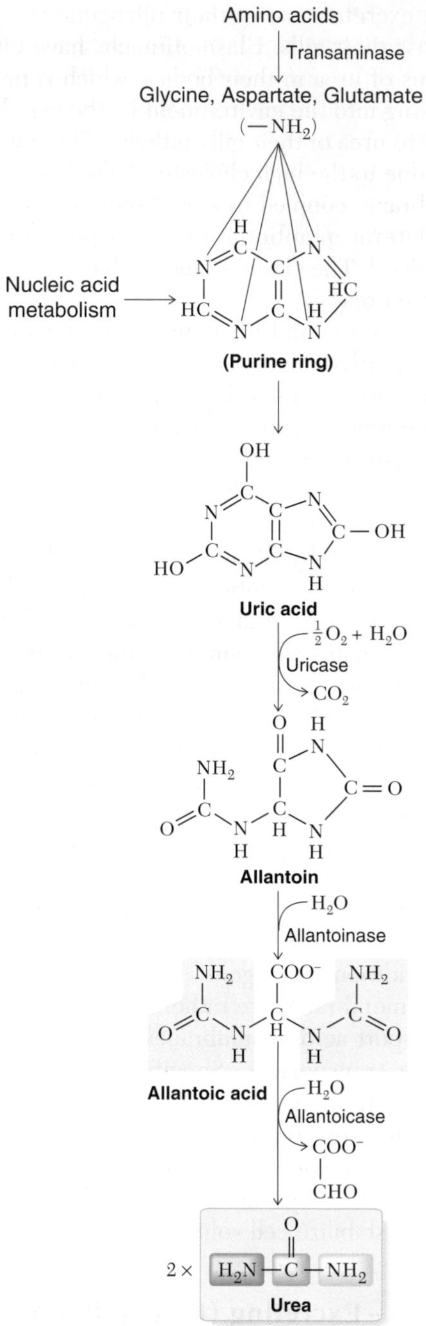

Amino acids → Transaminase → Glycine, Aspartate, Glutamate ($-NH_2$) → (Purine ring) ← Nucleic acid metabolism → Uric acid → ($\frac{1}{2}O_2 + H_2O$; Uricase; CO_2) → Allantoin → (H_2O; Allantoinase) → Allantoic acid → (H_2O; Allantoicase; COO^- CHO) → $2\times$ $H_2N-C(=O)-NH_2$ Urea

Figure 14-49 The uricolytic pathway produces both uric acid and urea. Uric acid arises from a purine ring that is synthesized by a complex union of aspartic acid, formic acid, glycine, and CO_2. Humans lack the enzymes needed to break down uric acid and thus excrete uric acid as the end product of nucleic acid metabolism.

birds, uric acid competes for transport in the kidney tubule with para-aminohippuric acid (PAH), but it does not do so in reptiles, suggesting a difference in the secretory mechanism. Two unusual arid-land toads, *Chiromantis xerampelina* and *Phyllomedusa sauvagii,* not only have an extremely low evaporative water loss from their skin, but, like reptiles and unlike most amphibians, they excrete nitrogen as uric acid rather than ammonia or urea. The low solubility of uric acid causes it to precipitate readily in the cloaca, allowing these toads to minimize the volume of urine necessary to eliminate their excess nitrogen.

SUMMARY

- The extracellular fluid of many marine and nonmarine animals broadly resembles dilute seawater. This similarity may have had its origin in the shallow, dilute primeval seas that are believed to have been the setting for the early evolution of animal life.

- The ability of many animals to regulate the composition of their internal environments is closely correlated with their ability to occupy ecological environments that are osmotically at odds with the requirements of their tissues.

Obligatory exchanges of ions and water

- Osmoregulation requires the exchange of salts and water between the extracellular fluids and the external environment to compensate for obligatory, or uncontrolled, losses and gains.

- The transport of solutes and water across epithelial layers is fundamental to all osmoregulatory activity.

- The obligatory exchange of water and salts depends on the osmotic gradient that exists between the internal and external environments, the surface-to-volume ratio of the animal, the permeability of the integument, the intake of food and water, evaporative losses required for thermoregulation, and the disposal of digestive and metabolic wastes in urine and feces.

Osmoregulation in aqueous and terrestrial environments

- Marine and terrestrial animals are faced with dehydration, whereas freshwater animals must prevent hydration by uncontrolled osmotic uptake of water.

- Freshwater animals take in water passively and remove it actively through the osmotic work of kidneys (vertebrates) or kidneylike organs (invertebrates). They lose salts to the dilute environment and replace them by actively absorbing ions from the surrounding fluids into their bodies through skin, gills, or other actively transporting epithelia.

- Marine birds, reptiles, and teleosts replace lost water by drinking seawater and actively secreting salt through secretory epithelia.

- Marine fishes lose water osmotically, especially through the gills. To replace lost water, they drink seawater and secrete the excess salt ingested with it back into the environment. This process takes place through active transport by extrarenal osmoregulatory organs such as the gills and the rectal gland.

■ Birds and mammals are the only vertebrates that secrete hyperosmotic urine. In addition, many desert species utilize mechanisms for minimizing respiratory water loss.

Osmoregulatory organs

■ Energy is expended in transferring ions across epithelia. H^+ V-ATPases and Na^+/K^+ P-ATPase hydrolyze ATP to generate electrochemical gradients, which serve to move ions through associated channels, symporters, and antiporters.

■ The Na^+/K^+ pump is found on the basolateral membrane of epithelial cells and moves Na^+ out of the cell into the extracellular fluid. Intracellular sodium levels are kept low and cell volume is regulated by the activity of this ATPase. The presence of various ion channels in the basolateral and apical cell membranes results in ion movement through the epithelium driven by the activity of the Na^+/K^+ pump.

■ The addition of a proton pump (H^+ V-ATPase) to the apical membrane increases the electrochemical gradient and promotes sodium uptake from dilute solutions. This mechanism also couples sodium uptake to acid excretion. The presence of a proton pump associated with a K^+/H^+ antiporter results in the secretion of potassium and the uptake of acid. The proton gradient drives the antiporter, but because two protons are transferred for each K^+ ion, the effect is to alkalinize the secretion.

■ Cells forming the transporting epithelium may be joined together by tight junctions, resulting in very low transepithelial permeability. Alternatively, there may be leaky tight junctions between cells, creating paracellular pathways that allow the passage of ions.

■ The activities of Na^+/K^+ and proton pumps, as well as that of channels, can be modified by hormones.

The mammalian kidney

■ Most vertebrate kidneys utilize filtration, reabsorption, and secretion to form urine. These processes allow the urinary composition to depart strongly from the proportions of substances occurring in the blood.

■ The filtration of the plasma in the glomerulus is dependent on arterial pressure. Most substances present in the blood plasma are filtered, leaving only blood cells and large proteins behind.

■ Salts and organic molecules such as sugars are reabsorbed from the glomerular filtrate in the renal tubules.

■ Certain substances are actively secreted into the tubular fluid.

■ A countercurrent multiplier system allows the production of hyperosmotic urine. The collecting duct and the loop of Henle set up a steep extracellular concentration gradient of salt and urea that extends deep into the medulla of the mammalian kidney. Water is drawn osmotically out of the collecting duct as it passes through high medullary concentrations of salt and urea toward the renal pelvis.

■ Endocrine control of the water permeability of the collecting duct determines the volume of water reabsorbed and retained in the circulation.

Extrarenal osmoregulatory organs

■ Birds and reptiles can drink seawater, excreting the salt load through a nasal salt gland. Elasmobranchs excrete salt via a rectal gland, which is made up of salt-secreting cells similar to those found in the thick ascending limb of the loop of Henle in the mammalian kidney, the avian and reptilian salt glands, and the chloride cells in the gills of marine teleosts. The hormonal regulation of the activity of these cells is also similar in sharks, birds, reptiles, and mammals.

■ The gills of teleost fishes and many invertebrates perform osmoregulation by active transport of salts, the direction of transport being inward in freshwater fishes and outward in marine fishes.

Invertebrate osmoregulatory organs

■ The formation of urine follows the same major outline in most vertebrates and invertebrates. A primary urine is formed that contains essentially all the small molecules and ions found in the blood. In most vertebrates and in the crustaceans and mollusks, this formation is accomplished by ultrafiltration; in insects, by the secretion through the epithelium of the Malpighian tubules of KCl, NaCl, and phosphate, with water and other small molecules, such as amino acids and sugars, following passively by osmosis and diffusion down their concentration gradients.

■ The primary urine is subsequently modified by the selective reabsorption of ions and water and, in some animals, by secretion of waste substances into the lumen of the nephron by the tubular epithelium.

Excretion of nitrogenous wastes

■ The nitrogen produced in the catabolism of amino acids and proteins is concentrated into one of three forms of nitrogenous waste, depending on the osmotic environment of different animal groups.

■ Ammonia, highly toxic and soluble, requires large volumes of water for dilution and subsequent excretion across the gills of teleosts.

■ Uric acid is less toxic and poorly soluble; it is excreted as a semisolid suspension via the kidneys of birds and reptiles.

■ Urea is the least toxic nitrogenous compound, and its excretion requires a moderately small amount of water. Mammals convert most of their nitrogenous wastes into urea, which is excreted in the urine. Elasmobranchs use urea as an osmotic agent in their blood and excrete most excess nitrogen as urea across their gills.

REVIEW QUESTIONS

1. How has the development of osmoregulatory mechanisms affected animal evolution?
2. What factors influence obligatory osmotic exchange with the environment?
3. Explain why respiration, temperature regulation, and water balance in terrestrial animals are closely interrelated. Give examples.
4. Describe three anatomic or physiological mechanisms used by insects to minimize dehydration in dry environments.
5. How do marine and freshwater fishes maintain osmotic homeostasis?
6. Name and describe the three major processes used by the vertebrate kidney to achieve the final composition of the urine.
7. What factors determine the rate of ultrafiltration in the glomerulus?
8. What is meant by the renal clearance of a substance?
9. If the tubular fluid in the loop of Henle remains nearly iso-osmotic relative to the extracellular fluid along its path, and is even slightly hypotonic on leaving the loop, how is the final urine made hypertonic in the mammalian kidney?
10. Explain why the consumption of 1 liter of beer will lead to a greater urine production than the consumption of an equal volume of water.
11. What role does the kidney play in the regulation of blood pressure?
12. Discuss the role of the kidney in the control of plasma pH.
13. How do insects produce concentrated, hypertonic urine and feces?
14. In the course of evolution, terrestrial organisms have come to excrete mainly uric acid and urea rather than ammonia. What are the adaptive reasons for such a change?
15. Explain why gulls can drink seawater and survive but humans cannot.
16. After the injection of inulin into a small mammal, the plasma inulin concentration was found to be 1 $mg \cdot ml^{-1}$, the concentration in the urine 10 $mg \cdot ml^{-1}$, and the urine flow rate through the ureter 10 $ml \cdot h^{-1}$. What was the rate of plasma filtration and the clearance in milliliters per minute? How much water was reabsorbed in the tubules per hour?
17. What evidence is there that the mammalian nephron employs tubular secretion as one means of eliminating substances into the urine?
18. Why is a countercurrent system more efficient in physical transport and transfer than a system in which fluids in opposed vessels flow in the same direction?
19. What are the similarities and differences between the elasmobranch rectal gland and the avian salt gland?

SUGGESTED READINGS

Fushimi, K., S. Uchida, Y. Hara, Y. Hirata, F. Marumo, and S. Sasaki. 1993. Cloning and expression of apical membrane water channel of rat kidney collecting tubule. *Nature* 361:549–552.

Krogh, A. 1939. *Osmotic Regulation in Aquatic Animals.* Cambridge: Cambridge University Press. Classic Work

Phillips, J. E., et al. 1994. Mechanisms of acid-base transport and control in locust excretory system. *Physiol. Zool.* 67:95–119.

Pitts, R. F. 1974. *Physiology of the Kidney and Body Fluids.* 3d ed. Chicago: Year Book Medical Publishers. Classic Work

Riordan, J. R., B. Forbush III, and J. W. Hanrahan. 1994. The molecular basis of chloride transport in shark rectal gland. *J. Exp. Biol.* 196:405–418.

Rodriguez-Boulan, E., and W. J. Nelson, eds. 1993. *Epithelial and Neuronal Cell Polarity and Differentiation.* Journal of Cell Science, Supplement 17. Cambridge: Company of Biologists.

Smith, H. W. 1953. *From Fish to Philosopher.* Boston: Little, Brown. Classic Work

Wieczorek, H., D. Brown, S. Grinstein, J. Ehrenfeld, and W. R. Harvey. 1999. Animal plasma membrane energization by proton-motive V-ATPases. *BioEssays* 21:637–648.

Wood, C. M., and T. J. Shuttleworth, eds. 1995. *Cellular and Molecular Approaches to Fish Ionic Regulation.* San Diego: Academic Press.

Wright, P. A. 1995. Nitrogen excretion: Three end products, many physiological roles. *J. Exp. Biol.* 198:273–281.

You, G., C. P. Smith, Y. Kanai, W.-S. Lee, M. Stelzner, and M. Hediger. 1993. Cloning and characterization of the vasopressin-regulated urea transporter. *Nature* 365:844–847.

Acquiring Energy: Feeding, Digestion, and Metabolism

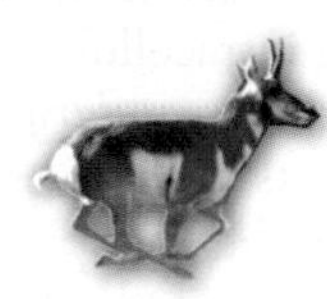

Every animal requires both raw materials and energy to grow, maintain itself, and reproduce. The materials and energy that are used in metabolism come from food, but what actually constitutes food varies greatly among animals, ranging from individual molecules absorbed across the body surface to living prey swallowed whole. Regardless of its origin, which can be plant, animal, or inorganic, food is used as material for production of new tissue, for the repair of existing tissue, and for reproduction. Food also serves as an energy source for ongoing processes, such as movement and metabolism. This chapter focuses on the acquisition, digestion, and metabolism of the wide variety of substances used as food by animals.

FEEDING METHODS

The effort to obtain food that is adequate in both quantity and quality accounts for much of the routine behavior of most animals. Certainly an animal's physiology and morphology are the result of natural selection that favors effective acquisition of energy from food while avoiding becoming someone else's food. The complexity and sophistication of the nervous and muscular systems, for example, attest to the power of the selective forces acting on the organism. As these systems vary, so do the variety of methods by which animals feed. Sessile (nonmobile) bottom-dwelling species commonly resort to surface absorption, filter feeding, or trapping. Mobile animals follow a more active sequence, which in the extreme case of many carnivores (meat eaters) includes searching, stalking, pouncing, capturing, and killing.

Food Absorption Through Exterior Body Surfaces

The feeding method that is least dependent on specialized capture and digestive mechanisms is the absorption of nutrients directly across the body wall. Many protozoans, endoparasites (animals that live within other animals), and aquatic invertebrates are able to take up nutrient molecules from the surrounding medium directly through the soft body wall. Endoparasites such as parasitic protozoans, tapeworms, flukes, and certain mollusks and crustaceans are surrounded by the tissues or the alimentary canal fluids of their hosts, both of which are high in nutrients. Tapeworms, which in humans may be many meters long, lack even a rudimentary digestive system. Tapeworms evolved from a primitive flatworm that lacked a body cavity. Some other endoparasites, however, appear to have secondarily lost the digestive apparatus that was present in their ancestors. Parasitic crustaceans, for example, which belong to the Cirripedia (barnacle group), lack an alimentary canal, but they appear to have evolved from nonparasitic ancestors possessing a gut.

Some free-living protozoans and invertebrates derive some of their nutrients by direct surface uptake from the surrounding medium. These animals take up small molecules such as amino acids from a dilute solution by transport mechanisms (described in Chapter 4) against what can often be a huge concentration gradient. Some of these organisms take up larger molecules or particles by a bulk process such as endocytosis, which is described next.

Endocytosis

Endocytosis represents a more active form of "feeding" than passive absorption directly across the body wall. Like direct nutrient absorption, however, it occurs at the local cellular level, rather than at the tissue or organismal level. Endocytosis includes two processes, as we saw in Chapter 4. In phagocytosis ("cell eating"), pseudopod-like protuberances extend out and envelop relatively large nutrient particles. Pinocytosis ("cell drinking") occurs when a smaller particle binds to the cell surface and the plasma membrane invaginates (folds inward) under it, forming an endocytotic cavity. Whether captured by phagocytosis or pinocytosis, the morsel is then engulfed in a membrane-enclosed vesicle that pinches off from the bottom of the cavity. The vesicle (called a *food vacuole* in protozoans) then fuses with lysosomes, organelles containing intracellular digestive enzymes, after which it is called a *secondary vacuole*. After digestion, the usable contents of the vacuole pass through the vacuole wall into the cytoplasm. The remaining undigested material is excreted externally by exocytosis. You are probably familiar with feeding by endocytosis in protozoans such as *Paramecium*, but these processes also occur in the lining of the alimentary canals and other tissues of many multicellular animals.

Filter Feeding

Many aquatic animals use **filter feeding** (also called *suspension feeding*) to capture food. These animals capture food items (usually phytoplankton or zooplankton) suspended in the water using specialized entrapment devices either on the body surface or within it.

Most marine filter feeders are small, sessile animals, such as sponges, brachiopods, bivalve mollusks, and tunicates. Food items are carried along on water

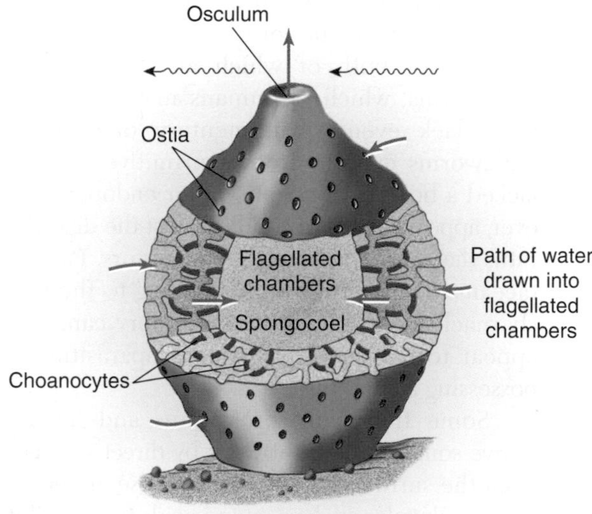

Classic Work **Figure 15-1** Water flows in an organized fashion through syconoid sponges. A significant proportion of this flow (blue arrows) results from the reduction in hydrostatic pressure at the osculum due to the Bernoulli effect, produced by transverse water currents (black arrows) flowing over the osculum at relatively high velocity. Water flow is also generated by the activity of flagellated choanocytes that line the flagellated chambers (lined in red). Water enters the sponge through the ostia, passes through the flagellated chambers, and ends up in the inner cavity, the spongocoel. Nutrients are then taken up into individual cells by endocytosis. [Adapted from Hyman, 1940; Vogel, 1978.]

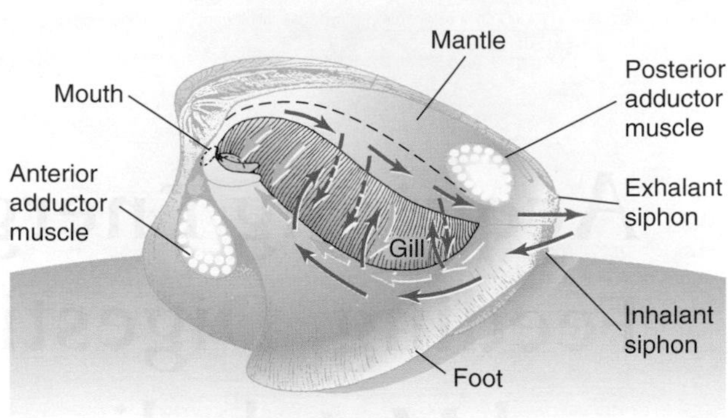

Figure 15-2 Bivalve mollusks employ ciliary feeding. This diagram shows a side view of a generalized bivalve with right valve removed. Blue arrows show the path of water into the inhalant siphon, over the surfaces of the gills, and out the exhalant siphon. The green arrows show the pathway of food particles borne on cilia and passing down the gills. Food particles are passed by cilia to the mouth, while sand and other indigestible materials are eliminated out the exhalant siphon.

currents, which may occur naturally or may be generated by movements of the filter-feeding animal or its internal or external cilia and flagella. A number of other sessile animals located in moving water make use of the Bernoulli effect (i.e., a drop in fluid pressure as fluid velocity increases) to increase the rate of water flowing through their entrapment sites, at no energy cost to themselves. An example of such passively assisted filter feeding is seen in sponges (Figure 15-1). The flow of water across the large terminal opening, the osculum, causes a drop in pressure (the Bernoulli effect) outside the osculum. As a result, water is drawn out of the sponge through the osculum, and is drawn in through the numerous ostia (mouthlike openings) in the body wall. The drop in pressure is facilitated by the shape of the sponge's exterior, which causes the water moving over the osculum to flow with greater velocity than the water moving past the ostia. Food particles, swept into the ostia along with the water, are engulfed by choanocytes, the flagellated cells lining the hollow, water-filled body cavity, or spongocoel. The flagella of the choanocytes also create internal water currents within the spongocoel.

Cilia propel water through many sessile animals, not only to capture suspended food, but also to aid in respiration. This process is of greatest importance in still water. In mollusks such as the mussel *(Mytilus)*, beating cilia on the surface of the gills draw a stream of water through the inhalant siphon, passing it between the gill filaments (Figure 15-2). Waterborne microorganisms and food particles are trapped in a layer of sticky mucus that covers the gill epithelium. The cilia keep a constant flow of mucus traveling down along the filaments (i.e., 90° to the water flow) to the tip of the gill, where it travels in a special groove under ciliary

(a)

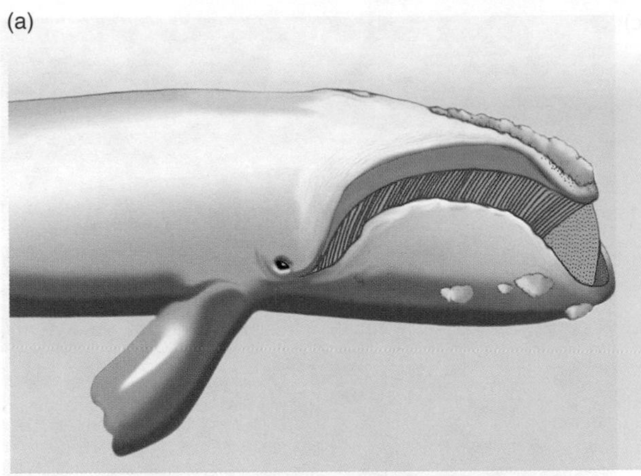

(b)

Figure 15-3 Convergent evolution has led to filter-feeding mechanisms in both **(a)** the black right whale *(Eubalaena glacialis)* and **(b)** the lesser flamingo *(Phoeniconaias minor).* Both the baleen in the whale's mouth and the fringe along the edge of the flamingo's bill act as strainers.

power toward the mouth in a ropelike string. Sand and other indigestible particles are sorted out and rejected (presumably on the basis of texture), passing out with the water leaving the exhalant siphon.

Mobile animals filter-feed by various mechanisms. A number of fishes are planktivorous, using modified gill rakers to strain plankton out of the flow of water passing through the mouth and over the gills. Filter feeding is also very common in amphibian larvae. In *Xenopus laevis,* the South African clawed frog, the gills bear mucus-covered filter plates that entrap suspended organic material. The mucus is then swept by cilia into the esophagus to be swallowed.

The largest filter feeders are the baleen whales, such as the right whale. Horny baleen plates in the whale's mouth bear a fringe of parallel filaments of hair-like keratin that hang down from the upper and lower jaws and act as strainers analogous to the gill rakers of fishes or larval amphibians (Figure 15-3a). These whales swim with jaws open into schools of pelagic crustaceans such as krill, engulfing vast numbers suspended in tons of water. As the jaws close, the water is squeezed back out through the baleen strainers with the help of the large tongue, and the crustaceans, left behind inside the mouth, are swallowed. Clearly, filter feeding can be a very effective form of food capture, if it can support an animal of such huge dimensions. Birds such as flamingos also use filter feeding to capture small animals and other morsels they find in the muddy bottoms of their freshwater habitat (Figure 15-3b).

Fluid Feeding

Fluid feeders use a variety of mechanisms, including piercing and sucking (the pesky mosquito), and cutting and licking (the vampire bat).

Piercing and sucking

Feeding by piercing a food item and sucking fluids from it occurs among the platyhelminths, nematodes, annelids, and arthropods. Leeches, among the annelids, are true bloodsuckers, using an anticoagulant in their saliva to prevent their prey's blood from clotting. In fact, the anticoagulants in leeches have been chemically isolated and are being used clinically. Some free-living flatworms seize their invertebrate prey by wrapping themselves around it. They then penetrate its body wall with a protrusible pharynx, through which they suck out the victim's body fluids and viscera. Penetration and liquefaction of the victim's tissues are facilitated by proteolytic enzymes secreted by the muscular pharynx.

Large numbers of arthropods feed by piercing and sucking. Most familiar and irksome of these to humans are mosquitoes, fleas, bedbugs, and lice, which can be vectors of disease. The majority of sucking arthropods victimize animal hosts. However, a number of species, especially among the Hemiptera (true bugs), pierce and suck plants, from which they draw sap. Sucking insects generally possess fine piercing mouthparts in the form of a proboscis (Figure 15-4a on the next page). Often, the two maxillae are shaped so that they make up two channels that run to the tip of the proboscis (Figure 15-4b). One of these channels is the passage for blood or sap sucked from the host. The other carries saliva, containing anticoagulants or enzymes, from the salivary glands into the host. Suction is created by the action of a muscular pharynx. After feeding, most insects are able to fold the proboscis back out of the way (Figure 15-4c).

Fluid feeders like mosquitoes or ticks can ingest a huge amount of fluid relative to their body mass in a very short period of time. While this practice allows the animal to acquire an abundance of food, what type of short-term physiological imbalances could the ingestion of blood, sap, or other fluids pose for the fluid feeder?

Cutting and licking

Numerous invertebrates and a few vertebrates feed by cutting the body wall of their prey and then licking or sponging the body fluids that leak from the cut. The blackfly and related biting flies have mouthparts with a sharpened mandible for cutting and a large, spongelike

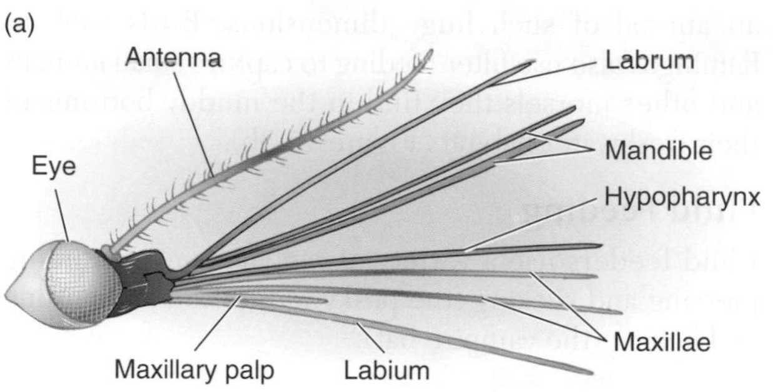

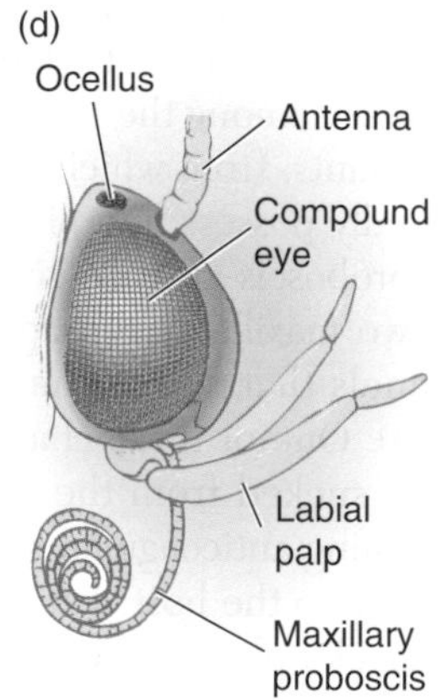

Figure 15-4 Sucking insects use tubular mouthparts for feeding. **(a)** Lateral view of the proboscis of a mosquito, with the mouthparts separated for identification. **(b)** Cross-section of the assembled mouthparts of the mosquito, showing separate channels for moving blood to the mouth and saliva to the wound. **(c)** Mosquito. **(d)** Lateral view of the head of a moth, showing the sucking mouthparts curled up between feeding bouts. **(e)** Moth. [Parts a, b, and d adapted from Rupert and Barnes, 1994; part c courtesy of Dwight R. Kuhn; part e courtesy of Holt Studios/Peter Wilson/ Photo Researchers.]

labium for transferring the body fluid (usually blood) to the esophagus. Among the chordates, the cyclostomes (phyletically ancient fishes such as lampreys and hagfishes) use rasplike mouths to make large, circular flesh wounds on their hosts. They feed on the blood that flows from these wounds. Vampire bats use their teeth to make puncture wounds in cattle, from which they lick oozing blood. The saliva of these bats contains an anticoagulant, as well as an analgesic to prevent the host from feeling the effects of the bite, at least until the bat has finished feeding.

Seizing Prey

Predators use various types of mouthparts and other appendages to capture and process food. Some predators use toxins to immobilize prey.

Jaws, teeth, and beaks

Although the invertebrates lack true teeth, various invertebrates have beaklike or toothlike chitinous structures for biting or feeding. Invertebrates such as the preying mantis and the lobster also have anterior limbs modified for prey capture (Figure 15-5). Spiders and their relatives have needlelike mouthparts for injection of venom, while cephalopods, such as the octopus, have a sharp, tearing beak. Among the vertebrates, hagfishes, sharks, bony fishes, amphibians, and reptiles have pointed teeth, mounted on the jaws or palate, that aid in holding, tearing, and swallowing prey.

The teeth of non-mammalian vertebrates are usually undifferentiated—a single tooth type is found throughout the mouth. One notable exception is found among the poisonous snakes, such as vipers, cobras, and

(a)

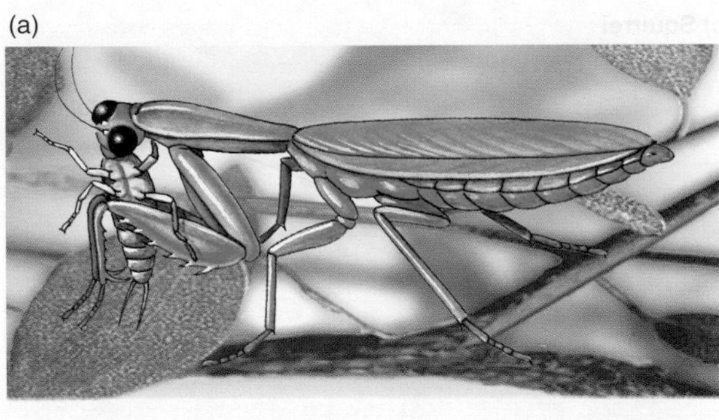

(b)

Figure 15-5 The anterior limbs of many arthropods are modified for seizing prey and holding it while the mouthparts tear off small pieces to be swallowed. **(a)** A mantispid insect (order Neuroptera) uses its anterior set of modified walking legs to capture and grasp its prey. **(b)** Lobsters (order Decapoda) have one claw modified for tearing and one modified for crushing.

rattlesnakes, which have modified teeth, called fangs, that they use to inject venom (Figure 15-6). These fangs either are equipped with a groove that guides the venom or are hollow, very much like a syringe. In rattlesnakes, the fangs fold back against the roof of the mouth, but extend perpendicularly when the mouth is opened to strike at prey. A snake's jaws are held together by an elastic ligament that allows them to spread apart during swallowing. This mechanism enables the snake to swallow animals larger than the diameter of its head. Swallowing prey whole is relatively common among snakes.

Mammals use their teeth for seizing and masticating their prey. Considerable variety has arisen among mammalian teeth during the course of evolution (Figure 15-7 on the next page). Chisel-like incisors are used for gnawing, especially by rodents and rabbits. In the elephants (and before them, mammoths), the incisors are modified into a pair of tusks. Pointed, daggerlike canines are used by the carnivores, insectivores, and primates for piercing and tearing food. In some groups, such as the wild pigs and walruses, the canines are elongated as tusks, which are used for prying and fighting. Most complex and interesting in their form are the molars of some herbivorous groups, including pigs and horses. These teeth, which are used in a side-to-side grinding motion, are composed of folded layers of enamel, cement, and dentine, all of which differ in hardness and in rate of wearing. Because the softer dentine wears rather quickly, the harder enamel and cement layers form ridges that enhance the effectiveness of the molars for chewing grass and other tough vegetation. Many mammals, such as the cats (the domestic cat and the great cats, such as the lion), use limbs equipped with sharp claws to supplement the teeth as food-capturing structures.

Instead of teeth, birds have horny beaks, in a multitude of shapes and sizes, adapted to each species' unique food sources and methods of obtaining them. For instance, beaks may have finely serrated edges, sharp, hooklike upper bills, or sharp points for drilling into wood (Figure 15-8 on the next page). Charles Darwin's observations of variation in beak structure among the finches of the Galápagos Islands helped give form to his emerging ideas of how natural selection operates. Seed-eating birds eat their food whole (perhaps after removing the outer hull), but may grind the

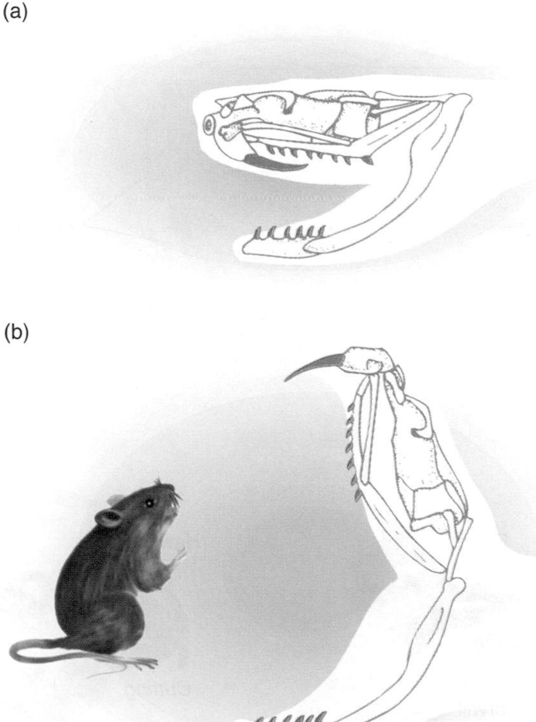

Figure 15-6 Rattlesnakes have modified teeth, known as fangs, which they use to inject venom into their prey. These side views of a rattlesnake skull show **(a)** a non-striking position, with the jaws only partially open and the hinged fangs folded into the roof of the mouth, and **(b)** a striking position, in which the jaws are open wide and the fangs extended. The extraordinary flexibility of the lower jaws allows the snake to swallow prey whole after injecting it with deadly venom.

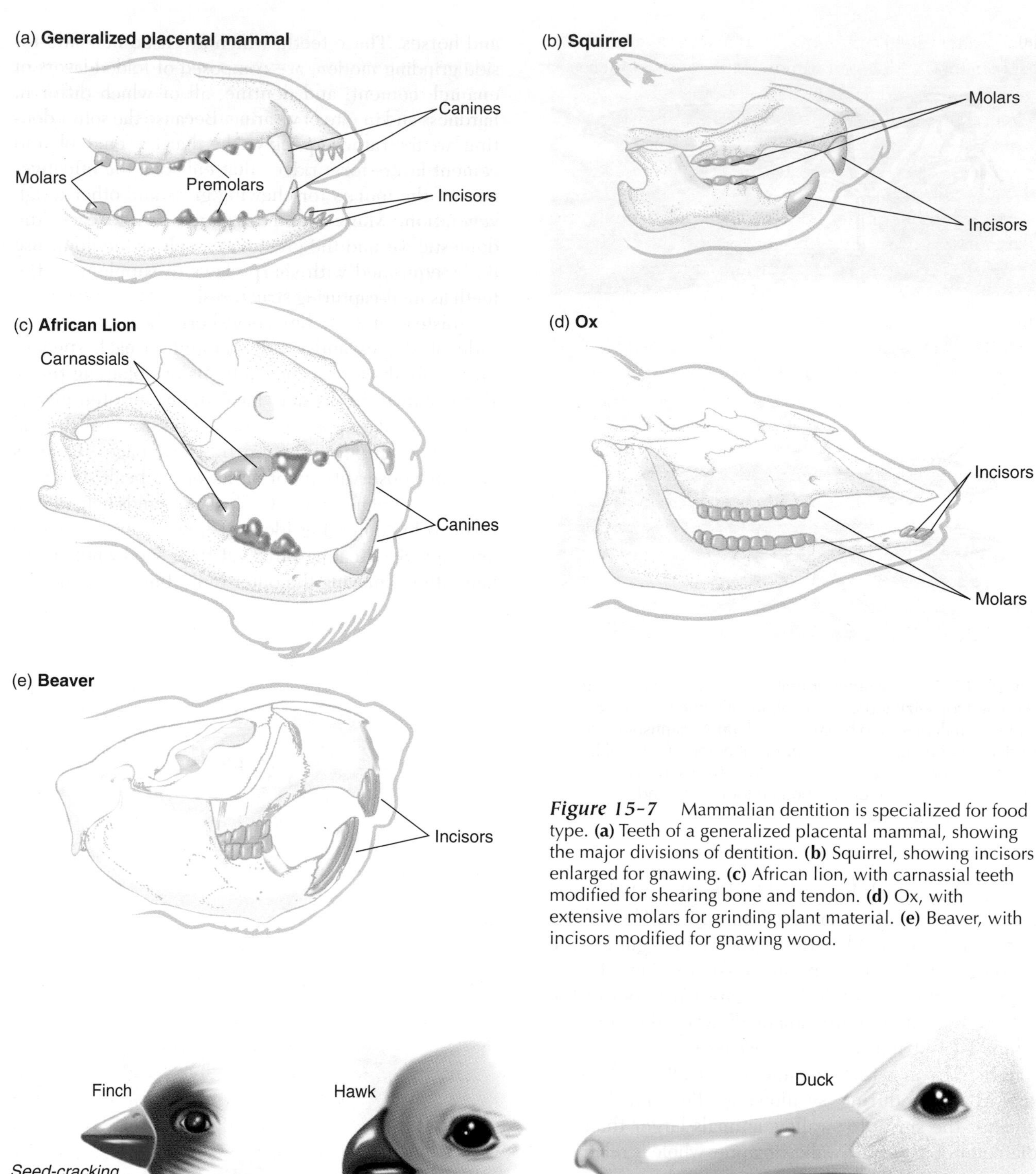

Figure 15-7 Mammalian dentition is specialized for food type. **(a)** Teeth of a generalized placental mammal, showing the major divisions of dentition. **(b)** Squirrel, showing incisors enlarged for gnawing. **(c)** African lion, with carnassial teeth modified for shearing bone and tendon. **(d)** Ox, with extensive molars for grinding plant material. **(e)** Beaver, with incisors modified for gnawing wood.

Figure 15-8 Bird beaks are adapted to particular modes of feeding. Beaks for seed cracking (finch) are typically short and powerful. Birds that catch insects on the fly (poorwill) typically have beaks with a potentially wide gape, often assisted in insect capture by hairlike feathers surrounding the mouth. Birds of prey (hawk) have sharp, pointed beaks whose overlap allows for a shearing action to cut the flesh of prey. Birds that probe holes in search of earthworms and other invertebrates (snipe) have a long, pointed beak. Birds that forage underwater (duck) often have broad beaks that allow water to be scooped and sieved for its contents.

swallowed seed in a muscular crop or gizzard containing pebbles that act like millstones. Raptorial birds (hawks, eagles), endowed with excellent vision and mobility, capture prey with their talons as well as with their beaks.

Toxins

A large number of animals from different phyla use toxins either to subdue prey or to fend off predators. Most of these toxins act at synapses in the nervous system. The coelenterates (hydras, jellyfishes, anemones, corals), for example, employ nematocysts. Concentrated in large numbers on the tentacles, these stinging cells inject paralytic toxins into prey and immobilize it while the tentacles carry it to the mouth (Figure 15-9). Many nemertine worms paralyze their prey by injecting venom through a stiletto-like proboscis. Venoms are also used by annelids, gastropod mollusks, and cephalopods, including the blue-ringed octopus of the Indo-Pacific region, a tiny creature (just 20 cm across its extended tentacles) that kills a number of humans every year.

Among the arthropods, scorpions and spiders are most notorious for their toxins, which are usually compounds that bind with great specificity to certain receptor types. After grabbing its prey with its large chelae (pincerlike organs), a scorpion arches its tail and plunges its stinger into the prey. The scorpion then injects the victim with a poison containing a neurotoxin that interferes with the proper firing of nerve impulses. Spider poisons also contain neurotoxins. The black widow spider's venom induces massive release of neurotransmitter at the motor endplate of the victim's muscles. A neurotoxin, α-Bungarotoxin (see Spotlight 6-4), found in the venom of the cobralike krait, binds to nicotinic acetylcholine (ACh) receptors, thereby blocking neuromuscular transmission in vertebrates. The venoms of various species of rattlesnakes contain hemolytic (blood cell–destroying) substances.

Toxins are usually delivered by bites or stings in carefully measured doses. They must also be carefully stored before administration to avoid self-poisoning. Toxins are generally proteins and, as such, are rendered harmless by the proteolytic enzymes of the predator's digestive system when it ingests its poisoned prey.

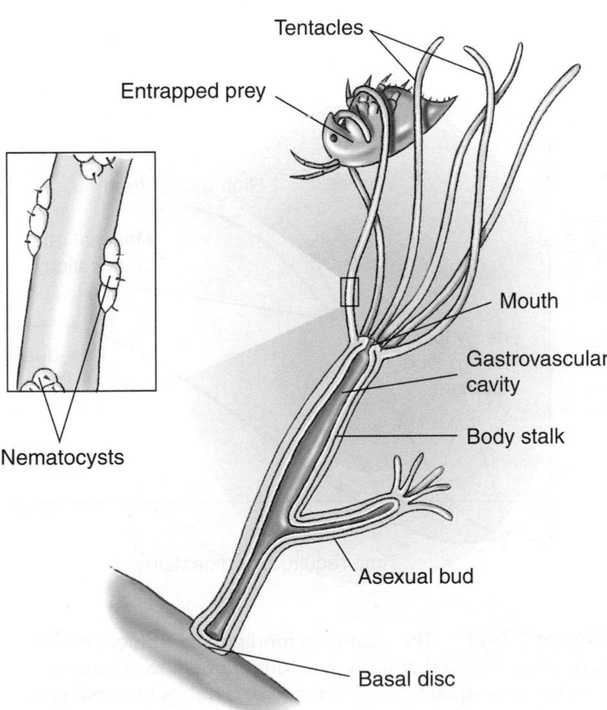

Figure 15-9 Tentacles bearing stinging nematocysts dangle from around the mouth of the hydra. Small prey items (generally zooplankton) are stung, paralyzed, and then transferred to the mouth for ingestion. [Adapted from Rupert and Barnes, 1994.]

Herbivory and Grazing

Many herbivores have mouthparts specialized for feeding on plant material. The mouthparts of many gastropods include a rasplike structure called a *radula* that they use to scrape algae from rock surfaces or to rasp through vegetation. Vertebrate herbivores have bony plates (some fishes and reptiles) or teeth (mammals) that are specialized for grinding plant material. Plants (especially some grasses) contain relatively large amounts of silicates and can be tremendously abrasive. Consequently, the molars of herbivores often are coated in especially tough enamel to resist wear, as we saw above. Alternatively, some herbivores such as small rodents (microtines) have continuously growing, rootless teeth.

OVERVIEW OF ALIMENTARY SYSTEMS

Alimentary systems play an essential role in nutrition by digesting and absorbing food and removing indigestible materials and toxic by-products of digestion from the body. The most primitive "alimentary system" is the plasma membrane of unicellular organisms, which engulfs microscopic food particles, undigested, by endocytosis. Once in the cell, food particles undergo intracellular digestion by acids and enzymes in food vacuoles before being absorbed into the cytoplasm. More complex multicellular animals rely primarily on extracellular digestion carried out by true alimentary systems.

From an anatomic perspective, there are myriad designs of alimentary systems. From a physiological perspective, however, alimentary systems fall into one of three categories based on how they process food. So-called **batch reactors** are blind tubes or cavities that receive food and eliminate wastes in a pulsed fashion; that is, one batch is processed and eliminated

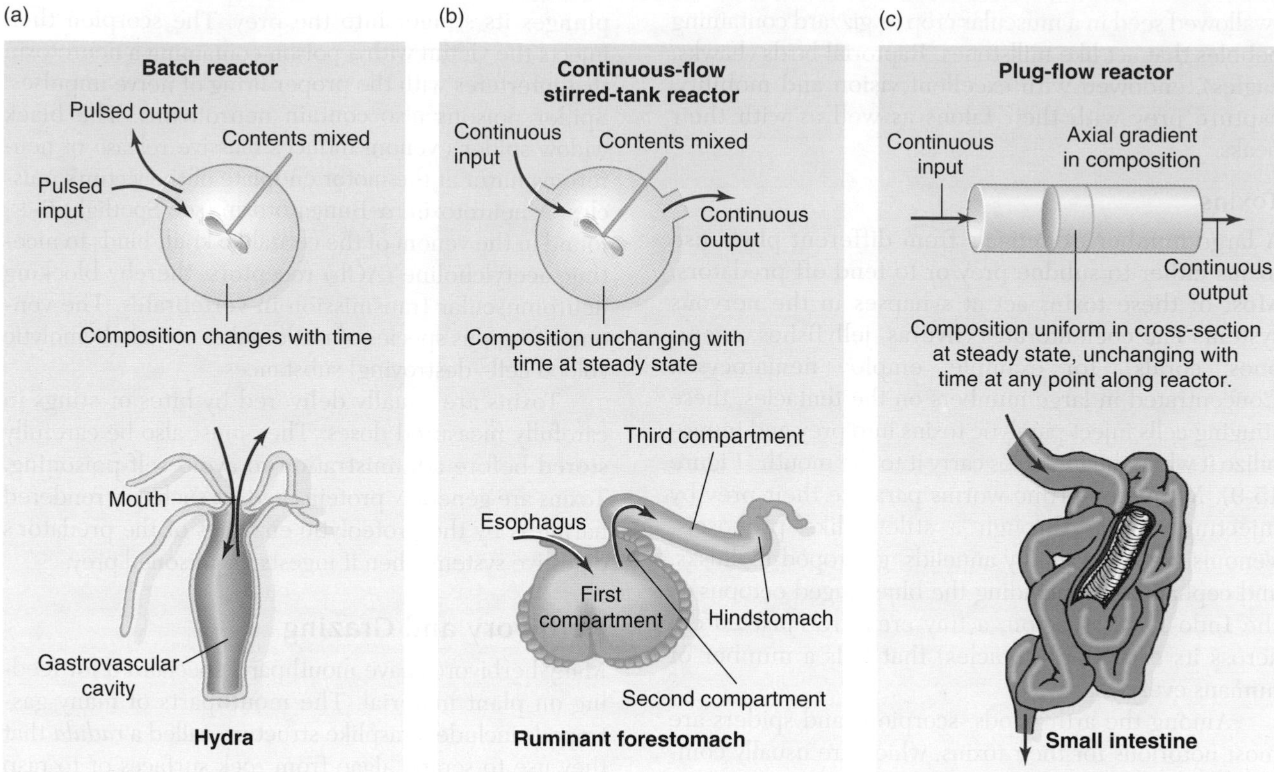

Figure 15-10 Digestive systems are classified according to the type of chemical reactor they form. **(a)** Batch reactors are found in simple organisms such as *Hydra*. **(b)** Ruminants have a continuous-flow, stirred-tank reactor in the form of a forestomach. **(c)** The small intestine of many vertebrates acts as a plug-flow reactor. The digestive systems of many animals combine features of both continuous-flow and plug-flow reactors. [Adapted from Stevens and Hume, 1995.]

before the next one is brought in (Figure 15-10a). Coelenterates, for example, have a blind tube or cavity, called the coelenteron, which opens only at a "mouth" that serves double duty as an "anus" for the expulsion of undigested remains. In all phyla higher than the flatworms, ingested material passes through a hollow, tubular cavity—the **alimentary canal**, also called the *gastrointestinal tract* or *gut*—extending through the organism and open at both ends. Processing goes on continuously, rather than in pulses, with new food ingested while older food is still being processed. Some alimentary canals can be modeled as **continuous-flow, stirred-tank reactors,** in which food is continuously added and mixed into a homogeneous mass, and the products of digestion are continuously eliminated, overflowing from the reactor (Figure 15-10b). An example is the forestomach of ruminants. The third category is the **plug-flow reactor,** in which a bolus (a discrete plug or collection) of food is progressively digested as it winds its way through a long, tubelike digestive reactor (Figure 15-10c). In contrast to the stirred-tank reactor, the composition of the food varies according to its position along the reactor tube. Typically, the vertebrate midgut functions as a plug-flow reactor. It is important to recognize that the alimentary canals of many animals

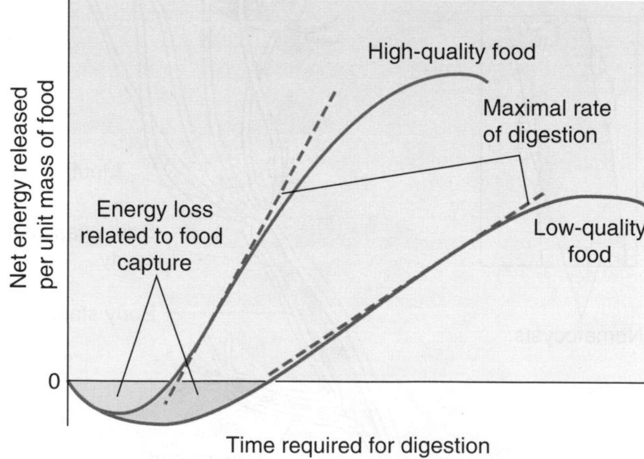

Figure 15-11 The quality of food greatly influences the time required for digestion in a continuous-flow digestive reactor. High-quality food (red curve) requires minimal energy to capture and eat and, once eaten, is quickly digested to release large amounts of energy. The maximum rate of digestion occurs at the point of the curve with the steepest slope, as indicated by the dashed line. Low-quality food (blue curve) requires considerable energy to capture and eat, has a low rate of digestion, and yields only low amounts of energy. [Adapted from Hume, 1989; Sibley, 1981.]

combine features of both continuous-flow and plug-flow reactors. In many animals, chemical digestion begins in the stomach, configured as a continuous-flow, stirred-tank reactor, and then continues in the small intestine, configured as a plug-flow reactor.

It is critically important that the design of the alimentary canal and the reactors it contains match the quality of the food that the animal routinely eats. Maximum energy can be extracted from high-quality food with minimal time spent in the digestive reactor (Figure 15-11). Lower-quality food requires a longer period of digestion to release its energy, which in turn requires longer periods spent in the reactor and longer transit times through the alimentary canal. As also indicated in Figure 15-11, the amount of energy spent in capturing and eating a particular food must also be factored into consideration of food quality. Transit time through the alimentary canal (also called mean retention time) varies not only with anatomic design—which

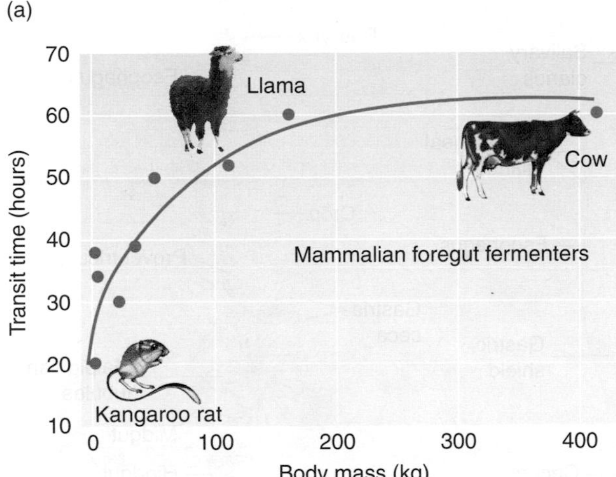

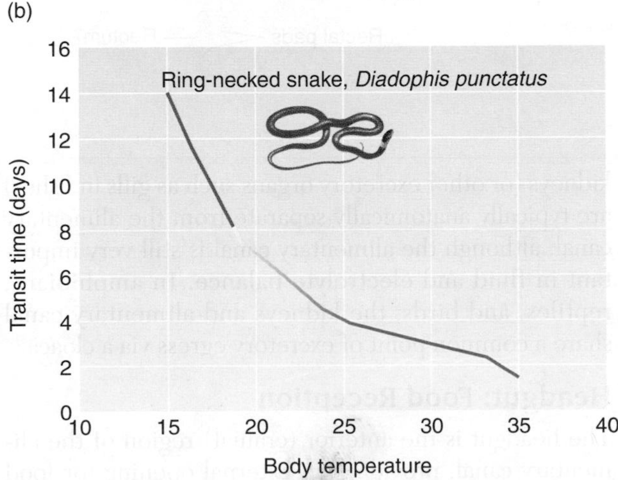

Figure 15-12 The transit time of ingested material through the vertebrate alimentary canal varies with **(a)** body mass and **(b)** body temperature. Transit times for both data sets were measured by observing the first appearance in fecal material of particular markers added to the ingested food. [Adapted from Stevens and Hume, 1995.]

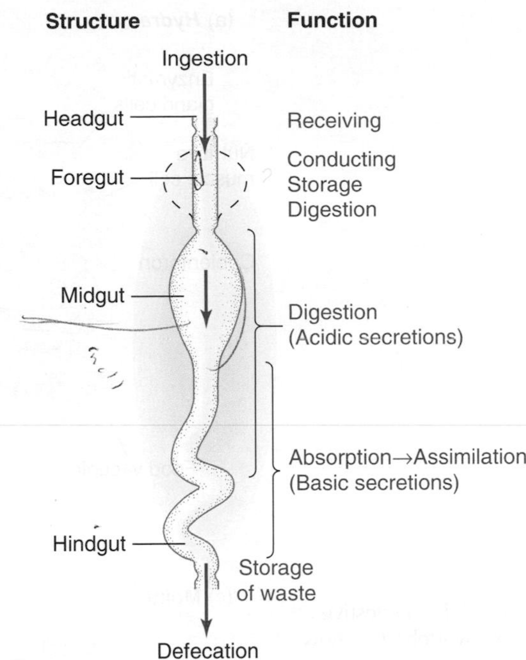

Figure 15-13 A digestive tract with one-way passage of food allows simultaneous operation of the sequential stages in the processing of food and reduces mixing of digested and undigested matter. The dashed outline represents a "crop," a storage region found in some animals.

is related to the quality of the typical food—but also with body mass, and with body temperature in the case of ectotherms (Figure 15-12).

A generalized alimentary canal, or digestive tract, is illustrated in Figure 15-13. The lumen of this alimentary canal is topologically external to the body. Sphincters and other devices guard the entrance and exit of the canal, preventing uncontrolled exchange between the lumen and the external environment. Ingested material is subjected to various mechanical, chemical, and bacterial treatments as it passes through this canal, and digestive juices (primarily enzymes and acids) are mixed with the ingested material at appropriate regions in the alimentary canal. As the ingested material is first mechanically broken down and then chemically digested, nutrients undergo absorption and are then transported into the circulatory system. Undigested, unabsorbed material is stored briefly until it, along with the remains of symbiotic bacteria (which make up a surprisingly large proportion of most animal waste), is expelled as feces by the process of defecation.

The overall tubular organization of the alimentary canal is efficient because it allows ingested material to travel in one direction, passing through different regions that can then be specialized for particular digestive tasks. For example, the alimentary canal near the point of ingestion is often specialized for the secretion of acidic compounds, while the secretions of more distant regions are alkaline. This regional specialization

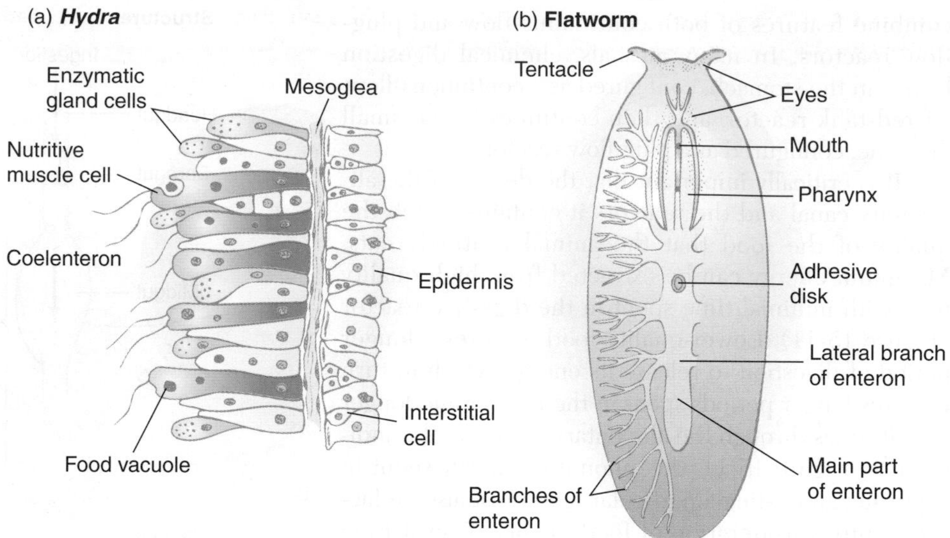

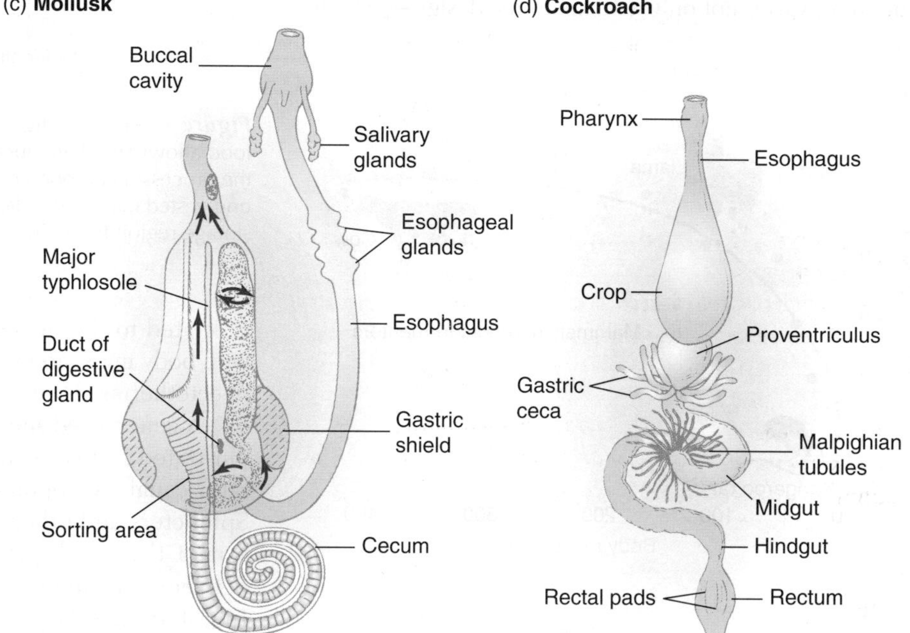

Figure 15-14 Digestive systems of invertebrates show great variation, ranging from simple to highly complex. **(a)** Section through the body wall of *Hydra,* a coelenterate. The epithelial lining of the coelenteron includes phagocytosing cells (called nutritive muscle cells) and gland cells that secrete digestive enzymes. **(b)** Digestive system of a flatworm. **(c)** Digestive system of a prosobranch gastropod mollusk. Arrows show ciliary currents and the rotation of the mucus mass. **(d)** Digestive system of the cockroach *Periplaneta.* The proventriculus (or gizzard) contains chitinous teeth for grinding food. [Part c from Rupert and Barnes, 1994; part d from Imms, 1949.]

allows both acid and base secretion to occur at the same time and to permit different types of digestive action.

In general, alimentary canals can be readily subdivided on a structural and functional basis into four major divisions (see Figure 15-13). The **headgut** receives ingested material; the **foregut** conducts, stores, and digests ingested material; the **midgut** digests and absorbs nutrients; and the **hindgut** absorbs water before the digestive materials are expelled during defecation. Within this general classification, considerable diversity exists (Figures 15-14 and 15-15).

Many invertebrate circulatory systems reflect a close integration of digestive functions with fluid and ionic balance, and the delineation between digestive and excretory organs is rather blurred. Vertebrates tend to have a more discrete separation of renal (urinary) and digestive organs and functions (Figure 15-16). The kidneys (or other excretory organs such as gills in fishes) are typically anatomically separate from the alimentary canal, although the alimentary canal is still very important in fluid and electrolyte balance. In amphibians, reptiles, and birds, the kidneys and alimentary canal share a common point of excretory egress via a cloaca.

Headgut: Food Reception

The headgut is the anterior (cranial) region of the alimentary canal, providing an external opening for food entry (see Figure 15-13). It consists of organs and structures for feeding and swallowing, including the mouthparts, buccal (oral) cavity, pharynx (throat), and associated structures such as bills, teeth, tongue, and salivary glands. In many species, a common pathway leads to both the alimentary canal and the passageway ending in the organ of internal gas exchange (e.g., the trachea).

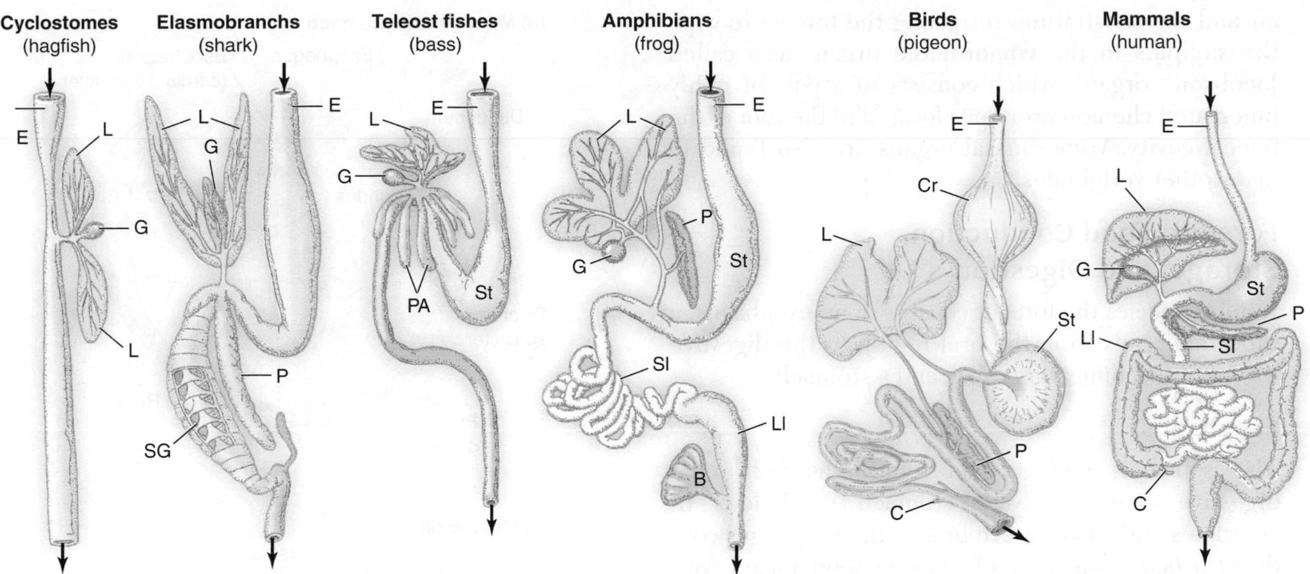

Figure 15-15 The tubelike digestive system of vertebrates has one basic organizational plan, with common elements of esophagus, stomach, intestine, and colon. B, bladder; C, cecum; Cr, crop (gizzard); E, esophagus; G, gallbladder; L, liver; LI, large intestine; P, pancreas; PA, pyloric appendices; SG, spiral gut; SI, small intestine; St, stomach. [Adapted from Florey, 1966; Stempell, 1926.]

Thus, there may be additional sphincterlike or valvelike structures that direct the flow of ingested material to the digestive system and inspired water or air to gas-exchange organs.

In small-particle feeders such as coelenterates, flatworms, and sponges, the headgut is devoid of glandular secretions. Most other metazoans have **salivary glands,** whose secretions aid ingestion and the mechanical (and often chemical) digestion of food. The primary function of the salivary secretion, **saliva,** is lubrication of the headgut to assist swallowing. The lubrication is provided in many cases by a slippery mucus of which the chief constituent is the protein *mucin.* The saliva often contains additional agents, such as digestive enzymes, toxins, or anticoagulants (in blood-lapping or blood-sucking animals such as vampire bats and leeches).

Tongues, an innovation of the chordates, assist in the mechanical digestion and swallowing of food. Some animals use their tongues to grasp food. Tongues are also used in chemoreception, bearing gustatory receptors called *taste buds* (see Figure 7-18). Snakes use their forked tongues to take olfactory samples from the

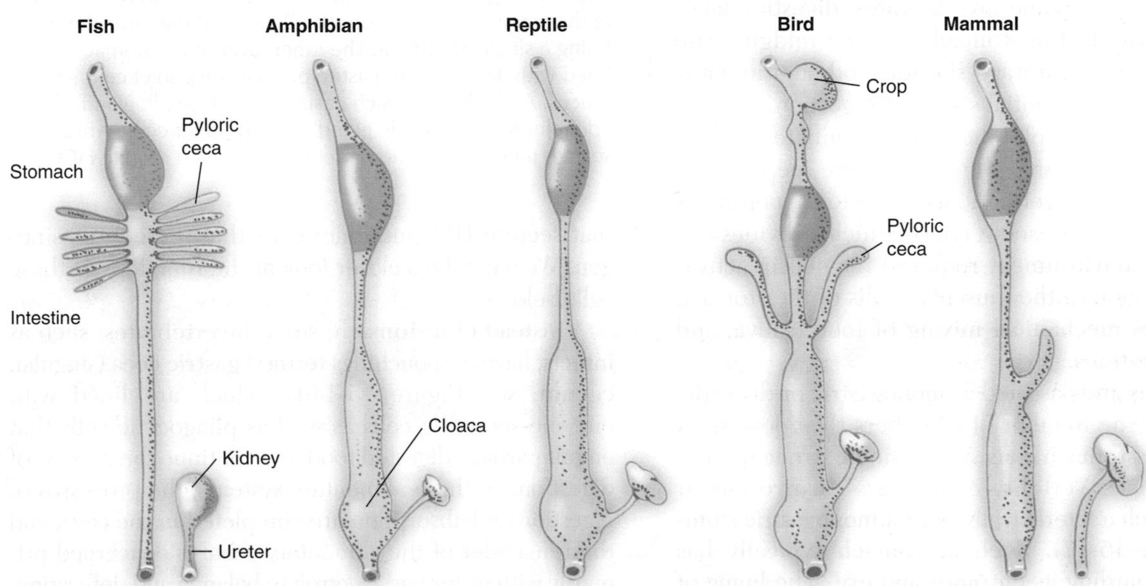

Figure 15-16 The alimentary canal of vertebrates typically is anatomically distinct from the gill-based or kidney-based excretory system. An exception is the emptying of the ureters into the cloaca in amphibians, reptiles, and birds. The alimentary canal of many fishes has, in addition to a stomach and intestine, a large number of pyloric ceca. In birds there is typically a crop and often ceca, sometimes paired.

air and the substratum, retracting the tongue to wipe the samples in the vomeronasal organ (also called Jacobson's organ), which consists of a pair of richly innervated chemosensory pits located in the roof of the buccal cavity. Vomeronasal organs are also found in many other vertebrates.

Foregut: Food Conduction, Storage, and Digestion

In most species the foregut consists of an **esophagus**, a tube that leads from the oral region to the digestive region of the alimentary canal, and a **stomach.**

Esophagus

The esophagus conducts food from the headgut to the digestive areas, usually the stomach (see below). In chordates and some invertebrates, the esophagus conducts a *bolus*—a mass of chewed food mixed with saliva—by peristaltic movement from the buccal cavity or pharynx through the esophagus. In some animals, this conducting region contains a saclike expanded section, the **crop,** which is used to store food before digestion. A crop, which is generally found in animals that feed infrequently, allows quantities of food to be stored for digestion at a later time. Leeches, for example, feed very infrequently, going for weeks or months between feeding periods. They ingest large quantities of blood at a meal, storing it for many weeks and digesting it in small amounts between their rare feedings. The crops of some animals are also used to ferment or digest foods for purposes other than their immediate assimilation. Some parent birds prepare food in the crop to be regurgitated for their nestlings.

Stomach

In vertebrates and some invertebrates, digestion takes place primarily in the stomach and the midgut. The stomach serves as a storage site for food, and in many species it begins the initial stages of digestion. In most vertebrates, for example, the stomach initiates protein digestion by secreting the proenzyme pepsinogen (which is quickly converted to its active form, pepsin) as well as hydrochloric acid (HCl), which provides the highly acidic environment required for pepsin activation. Contraction of the muscular walls of the stomach also provides mechanical mixing of food, saliva, and stomach secretions.

Stomachs are classified as monogastric or digastric, according to the number of chambers they possess. A **monogastric stomach** consists of a single strong muscular tube or sac. Vertebrates that are carnivorous or omnivorous characteristically have a monogastric stomach (Figure 15-17). Such a stomach typically has sphincters guarding its entrance and exit. The lining of the stomach bears many small invaginations of the gastric mucosa that form **gastric pits,** lined with goblet cells that produce mucus. The bottom of the gastric pit contains the gastric gland, which includes parietal cells that secrete HCl and chief cells that secrete pepsinogen. We will take a closer look at the functions of these cells below.

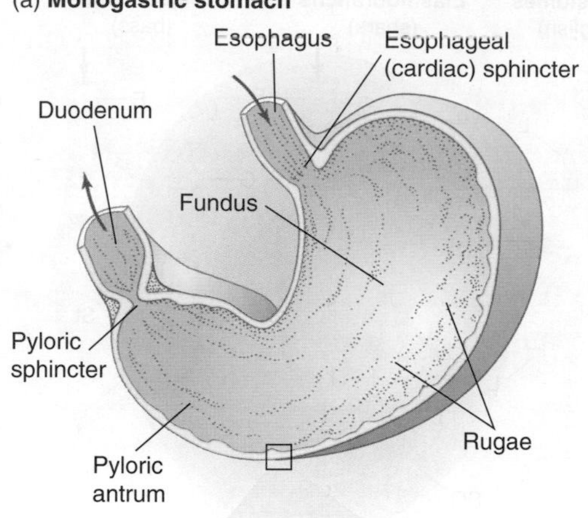

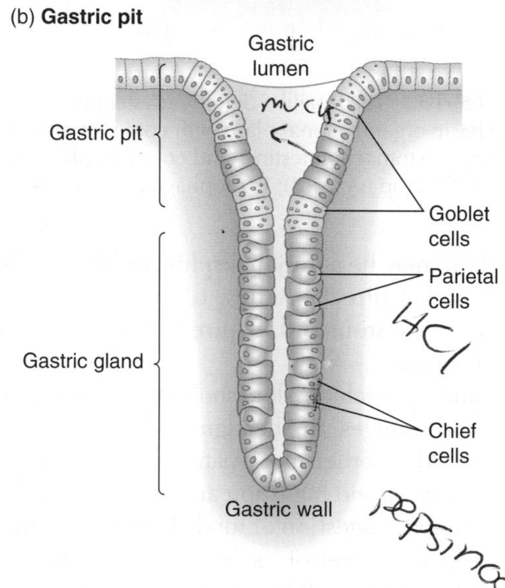

Figure 15-17 The monogastric stomach is a single chamber lined with a specialized epithelium. **(a)** Major parts of the mammalian stomach. **(b)** Detail of the gastric glands lining a single gastric pit. The inner layer of the stomach is lined with thousands of gastric pits, whose goblet cells secrete mucus. At the base of each gastric pit is a gastric gland. The epithelium of the gastric gland contains chief cells, which secrete pepsinogen, and parietal cells, which secrete HCl.

Instead of a stomach, some invertebrates, such as insects, have outpouchings termed **gastric ceca** (singular, **cecum;** see Figure 15-14d), which are lined with enzyme-secreting cells as well as phagocytic cells that engulf partially digested food and continue the process of digestion. In these alimentary systems, the processes of digestion and absorption are completed in the ceca, and the remainder of the alimentary canal is concerned primarily with water and electrolyte balance and defecation.

Some birds have a tough, muscular gizzard, or a crop, or both (see Figure 15-15). These birds swallow sand, pebbles, or stones, which lodge in the gizzard, where they aid in the grinding of seeds and grains. The proventriculus of insects and the stomach of decapod crustaceans, comparable to the bird's gizzard, contain grinding structures for chewing swallowed food. Some fishes also have gizzards. On the other hand, some fishes and larval toads lack stomachs altogether, with material from the esophagus entering what is functionally the midgut.

Multi-chambered **digastric stomachs** (Figure 15-18) are found in the mammalian suborder Ruminantia (herbivores such as deer, elk, giraffes, bison, sheep, and cattle). Somewhat similar digastric stomachs occur in other herbivores outside this suborder, in particular in the suborder Tylopoda (camels, llamas, alpacas, vicuñas). Symbiotic microorganisms in the first division of the stomach carry out fermentation, the anaerobic conversion of organic compounds to simpler compounds, yielding energy as ATP. All of the above-named groups carry out rumination, in which partially digested food is regurgitated (transported back to the mouth) for remastication (additional chewing). This process allows the ruminant (a gazelle grazing on the open savanna, for example) to swallow food hastily and then chew it more thoroughly later when at rest in a place of relative safety from predators. After the regurgitated food is chewed, it is swallowed again. This time it passes into the second division of the digastric stomach and begins the second stage of digestion. In this stage, hydrolysis takes place with the assistance of digestive enzymes secreted by the stomach lining.

The digastric stomach of the Ruminantia has four chambers that are separated into two divisions. The first division consists of the **rumen** and **reticulum**; the second division comprises the **omasum** and the **abomasum** (true stomach). The rumen and reticulum act as a fermentation vat that receives grazed vegetation. Bacteria and protozoans in these chambers thrive on this vegetation, fermenting its carbohydrates into products such as butyrate, lactate, acetate, and propionate, as well as the gases carbon dioxide and methane. The nongaseous products of fermentation, along with some peptides, amino acids, and short-chain fatty acids, are absorbed into the bloodstream from the rumen fluid. Microorganisms from the rumen, along with undigested particles, are passed into the omasum (absent in the Tylopoda) and then into the abomasum. Only the abomasum secretes digestive enzymes; this chamber is homologous to the monogastric stomach of nonruminants. The gases produced by fermentation can be substantial, amounting to as much as a liter per minute in domestic cattle. While the CO_2 can be absorbed, methane (CH_4) is much less soluble, and at the height of fermentation both CO_2 and CH_4 must be expelled by **eructation**, the release of gas from the stomach via the esophagus ("burping"). An animal's inability to rid itself of the gaseous products of fermentation results in "bloat," a potentially lethal situation that is often associated with the rapid consumption of large amounts of leguminous vegetation.

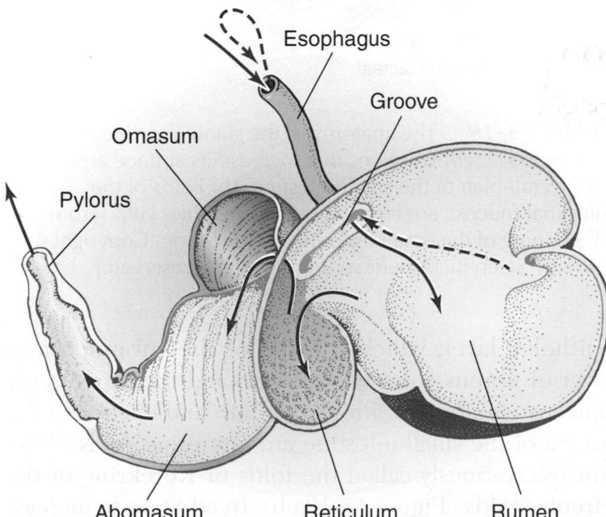

Figure 15-18 The digastric stomach of ruminants has multiple chambers for storing and digesting food materials. This sheep stomach, characteristic of the ruminants, has two divisions made up of four chambers. The rumen and reticulum make up the fermentative division; the omasum and abomasum (true stomach) make up the digestive division.

Fermentation in the stomach is not limited to ruminant animals. It is found in some other species in which the passage of food to the stomach is delayed, allowing the growth of symbiotic microorganisms in a zone anterior to the digestive stomach, as in the kangaroo and in the crops of galliform (chickenlike) birds.

Midgut: Chemical Digestion and Absorption

In vertebrates, the midgut is the major site for the chemical digestion of proteins, fats, and carbohydrates. Once digested to their component molecules, these materials are then absorbed in the midgut and transported away from the alimentary canal in the blood. As food is ready to pass on from the stomach, it is released into the midgut through the **pyloric sphincter**, which relaxes as the peristaltic movements of the stomach squeeze the acidic contents into the **duodenum** (see Figure 15-17a), the initial segment of the vertebrate midgut, or small intestine. Digestion continues in the small intestine, generally in an alkaline environment.

General structure and function of the midgut

Among the vertebrates, carnivores have shorter and simpler intestines than do herbivores, reflecting the shorter time required to digest meat than vegetation. Amphibian tadpoles, for example, which are almost always herbivorous, have a much longer intestine relative to body size than adult frogs, which are carnivorous. A particularly striking exception is seen in the normally planktivorous larvae of the salamander *Ambystoma tigrinum*, which have the long midgut typical of herbivores. When crowded and starved, a small proportion of these larvae become cannibalistic, and this change in diet is accompanied by a rapid conversion of the midgut from a long, winding canal into a much shorter passageway.

The vertebrate small intestine is typically divided into three distinct portions. The first, rather short, section is the duodenum, the lining of which secretes mucus and fluids and receives secretions carried by ducts from the liver and pancreas. Next is the **jejunum**, which also secretes fluids and is involved in digestion and absorption. The most posterior section, the **ileum**, primarily absorbs nutrients digested in the duodenum and jejunum, although some secretion also occurs in this section.

The cells of the liver produce bile salts, which are carried in the bile fluid to the duodenum through the bile duct. Bile fluid has two important functions: it emulsifies fats, and it helps neutralize acid introduced into the duodenum from the stomach. The pancreas, an important exocrine organ discussed in Chapter 9, produces pancreatic juice, which is released through the pancreatic duct into the duodenum. Pancreatic juice contains many of the proteases, lipases, and carbohydrases essential for intestinal digestion in vertebrates. Like bile, it is also important in neutralizing gastric acid in the intestine.

The intestine of most animals is a complex ecosystem, containing large numbers of bacteria, protozoans, and fungi (and usually some metazoan parasites). These single-celled community members multiply, contribute enzymatically to digestion, and are usually, in turn, digested themselves. An important function of some intestinal symbionts is the synthesis of essential vitamins.

There is great functional and structural variety in the midgut region across the various animal groups. In many invertebrates, especially those with extensive ceca and diverticula (blind outpouchings of the alimentary canal), the intestine serves no digestive function. In some air-breathing fishes, such as the weather loach, the midgut is modified into a gas-exchange organ, where O_2 from gulped air is exchanged with CO_2 from the cells, and residual gas is then expelled out the anus.

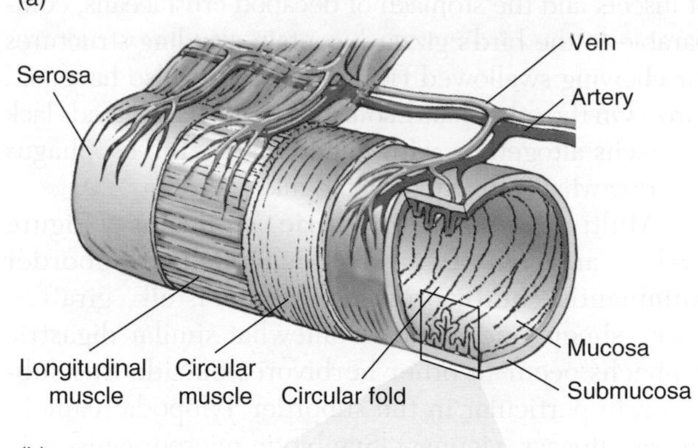

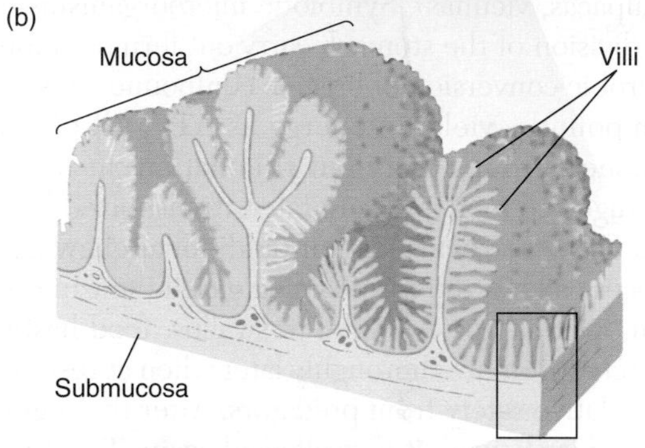

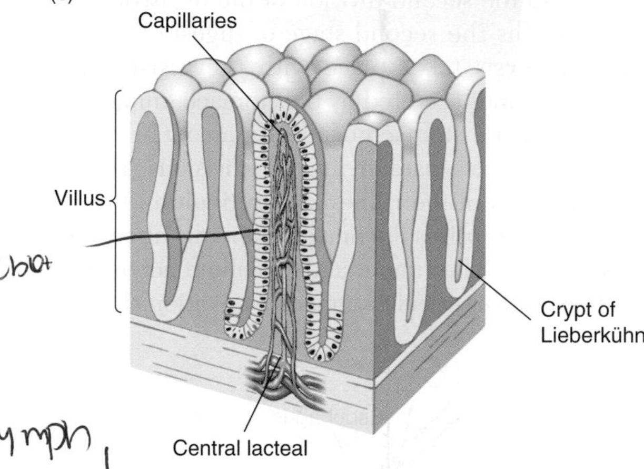

Figure 15-19 The anatomy of the small intestine is dominated by convolutions that increase its surface area. **(a)** Overall plan of the small intestine. **(b)** Folds of the intestinal mucosa are covered by **(c)** fingerlike villi. [From "The Lining of the Small Intestine," by F. Moog. Copyright © 1981 by Scientific American, Inc. All rights reserved.]

Intestinal epithelium

The vertebrate small intestine has anatomic adaptations at every organizational level, from its gross anatomy to the organelles of its individual cells. All of these adaptations are designed to amplify the surface area available for the absorption of nutrients. Since the rate of absorption is generally proportional to the area of the apical surface membrane of the cells lining the epithelium, the huge increase in surface area provided by intestinal convolutions greatly aids the absorption of digested substances from the fluid within the intestine. We will now examine this remarkable system of valleys and peaks, peninsulas and inlets.

The general organization of the vertebrate small intestine is shown in Figure 15-19a. The outermost layer is the **serosa**, which is the same tissue that covers the other visceral organs of the abdomen. The serosa overlies an outer layer of longitudinal smooth muscle. An inner layer of circular smooth muscle surrounds the epithelial layer, which consists of the **submucosa** (a layer of fibrous connective tissue) and the **mucosa** (or mucous membrane). Projecting into and encircling the lumen of the small intestine are numerous folds of the mucosa, variously called the **folds of Kerckring** or the **circular folds** (Figure 15-19a,b). In addition to increasing surface area, these folds slow the progress of food through the intestine, allowing more time for digestion.

At the next anatomic level are the fingerlike **villi** (Figure 15-19b,c), which line the folds and stand about 1 mm tall. Each villus sits in a circular depression known as the **crypt of Lieberkühn** (see Figure 15-19c).

Within each villus is a network of blood vessels formed by arterioles, capillaries, and venules. The villus also contains a network of lymph vessels, the largest of which is the **central lacteal.** Nutrients taken up from the intestine are transferred into these blood and lymph vessels for transport to other tissues; the central lacteal can, in addition, take up larger particles. The villi are covered by the actual absorptive surface of the small intestine, the cells of the digestive epithelium. The epithelium consists of goblet cells interspersed among columnar absorptive cells (Figure 15-20a). The absorptive cells proliferate at the base of the villus and steadily

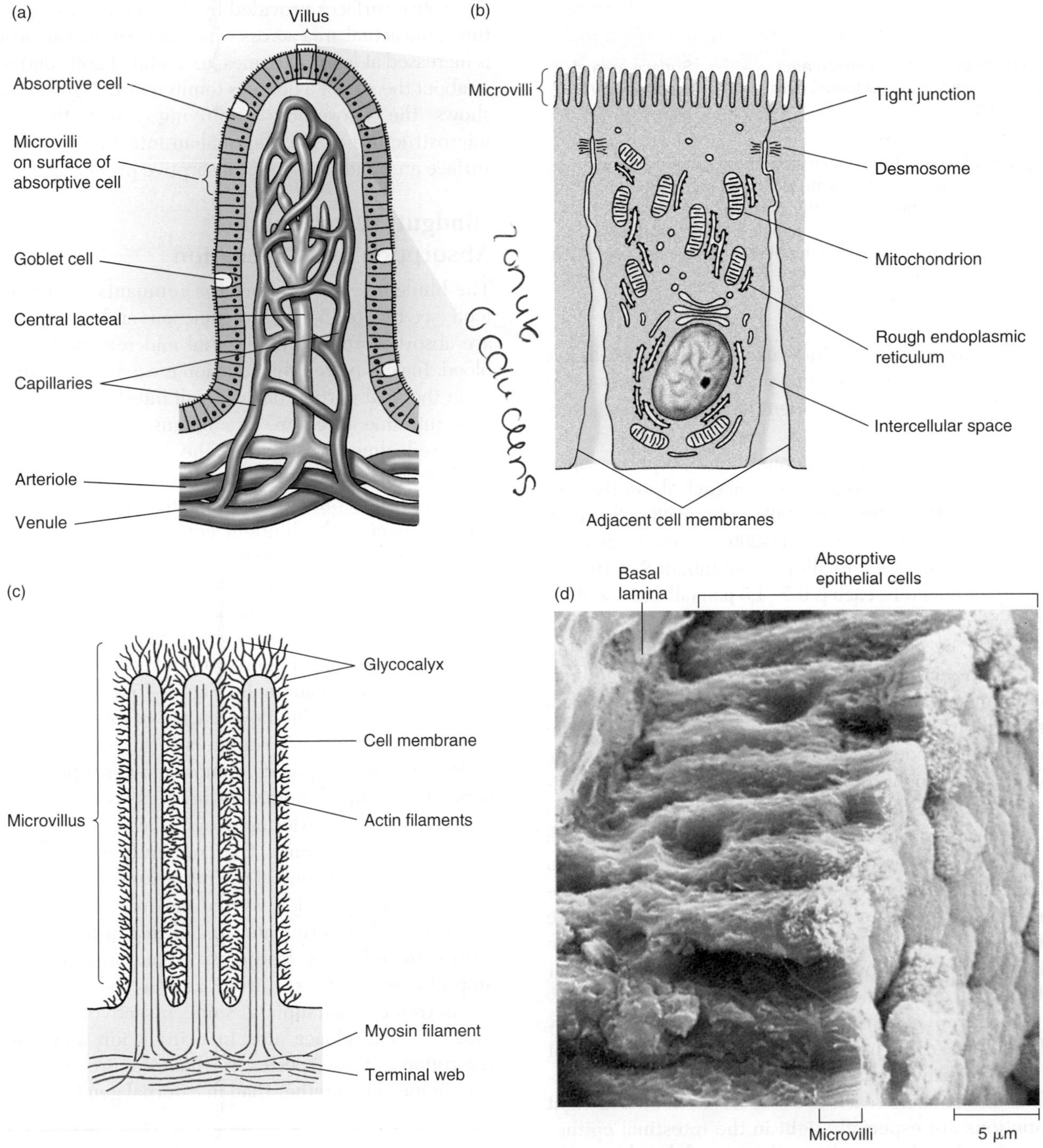

Figure 15-20 The lining of the mammalian small intestine has a complex microanatomy specialized for absorption and secretion. **(a)** A villus covered with the mucosal epithelium, which consists primarily of absorptive cells and occasional goblet cells. **(b)** An absorptive cell. The luminal, or apical, surface of the absorptive cell bears a brush border of microvilli. **(c)** The microvilli consist of evaginations of the plasma membrane, enclosing bundles of actin filaments. **(d)** Scanning electron micrograph of a group of absorptive cells from the human small intestine, showing the brush border. [Parts a–c from "The Lining of the Small Intestine," by F. Moog. Copyright © 1981 by Scientific American, Inc. All rights reserved; part d from R. Kessel and R. Kardon, 1979, *Tissues and Organs: A Text-Atlas of Scanning Electron Microscopy,* W. H. Freeman and Company, p. 176.]

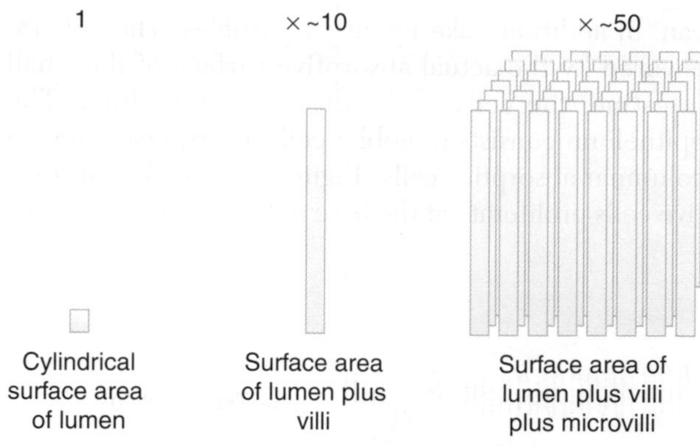

Figure 15-21 The surface area of the small intestine of mammals is increased enormously by its microstructure. This diagram shows the scale of this increase by comparing the cylindrical surface area of the lumen, the combined surface area of lumen plus villi, and the overall surface area of lumen plus villi plus microvilli.

migrate toward its tip. In humans absorptive cells are sloughed off at the rate of about 2×10^{10} cells per day, meaning that the entire midgut lining is replaced every few days.

The next level in the hierarchy of absorptive adaptations is found at the apical surface of each absorptive cell, where fingerlike structures called microvilli collectively form a brush border (Figure 15-20b–d). There are up to several thousand microvilli per cell (about 2×10^5 per square millimeter); each is 0.5–1.5 μm tall and about 0.1 μm wide. The membrane of the microvillus is continuous with the plasma membrane of the epithelium and surrounds actin filaments that form cross-bridge links with myosin filaments present at the base of each microvillus (Figure 15-20c). Intermittent actin-myosin interaction produces rhythmic motions of the microvilli, which are thought to help mix and exchange the intestinal chyme (the semifluid mass of partially digested food) near the absorptive surface. The surfaces of the microvilli are covered by the glycocalyx, a meshwork up to 0.3 μm thick made up of acidic mucopolysaccharides and glycoproteins (see Figure 15-20c). Water and mucus are trapped within the interstices of the glycocalyx. The mucus is secreted by the goblet cells, named for their shape, that are found among the absorptive cells.

Adjacent absorptive cells are held together by desmosomes (Figure 4-34). Near the apex, the tight junctions with neighboring cells form the *zonula occludens* that encircles each cell (Figure 15-20b). The tight junctions are especially tight in the intestinal epithelium, so that the apical membranes of the absorptive cells effectively form a continuous sheet of membrane, without breaks between cells. Because of the virtual impermeability of the tight junctions, all nutrients must pass across the apical membrane and through the absorptive cell cytoplasm to get from the lumen to the blood and lymph vessels within the villi. Little, if any, paracellular passage occurs.

The architecture of the intestine provides for an enormously increased absorptive surface area and thus a rapid rate of absorption of intestinal contents. In the adult human, for example, the lumen of the small intestine has a gross cylindrical surface area of only about 0.4 m^2, the equivalent of about 7 to 8 pages of this book. However, because of the enormous elaboration of absorptive surfaces provided by this hierarchy of structures, the actual area across which absorption can occur is increased at least 500 times, to a total of 200–300 m^2, or about the size of a doubles tennis court. Figure 15-21 shows the enormous amplifying effect that the microstructure of the mammalian intestine has on its surface area and thus on its absorptive properties.

Hindgut: Water and Ion Absorption and Defecation

The hindgut serves to store the remnants of digested food (see Figure 15-13). Inorganic ions and excess water are absorbed from this material and returned to the blood. In vertebrates, this function is carried out primarily in the final portion of the small intestine and in the large intestine, or **colon.** In some insects, the feces are rendered almost dry within the last portion of the hindgut by a specialized mechanism for removing water from the rectal contents (see Chapter 14). In many vertebrate species, the hindgut consolidates undigested material and bacteria growing in the hindgut into feces. The feces pass into the cloaca or rectum and are then expelled through the anus in the process of defecation.

The hindgut is the major site for bacterial digestion of intestinal contents in herbivorous reptiles, birds, and most herbivorous mammals (Figure 15-22). The colon acts as a modified plug-flow reactor in most large animals that are hindgut fermenters (e.g., horses, zebras, tapirs, sirenians, elephants, rhinos, and marsupial wombats). In smaller hindgut fermenters (rabbits, many rodents, hyraxes, howler monkeys, koalas, and brushtail and ringtail opossums), the tremendously enlarged cecum, an outpocketing of the posterior small intestine, acts as a continuous-flow, stirred-tank reactor.

The hindgut terminates in a **cloaca** in many vertebrates, including cyclostomes, elasmobranchs, adult amphibians, reptiles, birds, and a few mammals (monotremes, marsupials, some insectivores, a few rodents). The cloaca aids in urinary ion and water resorption in those species in which the ureters terminate in the cloaca rather than in external genitalia.

Several species of air-breathing catfishes swallow air and extract oxygen from the gas bubble across the wall of a specially modified intestine. What anatomic and physiological features of an alimentary canal could serve the concurrent needs of digestion, absorption, and gas exchange?

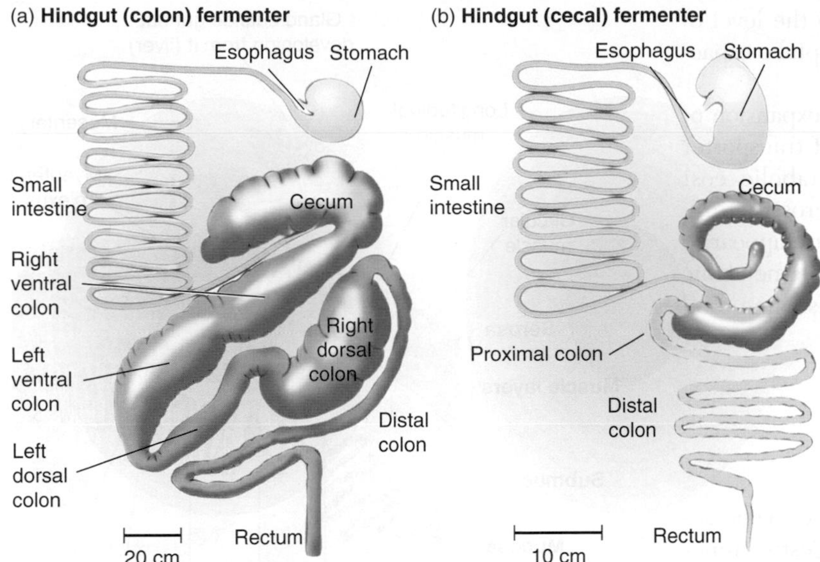

Figure 15-22 The digestive tract of a colon fermenter has an enlarged colon compared with that of a cecal fermenter, which has an enlarged cecum. **(a)** Digestive tract of the horse *(Equus caballus)*, a colon fermenter. Site of fermentation is shown in red. **(b)** Digestive tract of the rabbit *(Oryctolagus cuniculus)*, a cecal fermenter. [Adapted from Stevens and Hume, 1995.]

Dynamics of Gut Structure and the Influence of Diet

Research over the past few decades has changed the traditional view of the gut as a relatively static set of organs and tissues. In fact, we now know that the gut's size, structure, enzymatic activity, and absorptive capabilities are quite dynamic, responding to changes in both energy demand and food quality in most animals, be they carnivores or herbivores. Most responsive is overall gut size. House wrens that were induced to increase their food intake through exposure to combinations of lowered ambient temperature and enforced exercise over several months responded by increasing the overall length of the small intestine by about one-fifth; efficacy of nutrient uptake increased as a result. The mass of the empty stomach of the thirteen-lined ground squirrel *(Spermophilus tridecemlineatus)* increases threefold to fourfold within a few months after the animal rouses from hibernation. Although reptiles have a much lower rate of metabolism than birds and mammals (see Chapter 16), some reptiles appear to remodel the gut in response to food intake much more rapidly than birds and mammals do, sometimes within a few hours or days. In the Burmese python *(Python molurus)*, anterior small intestine mass increases by over 40% over fasting levels within six hours of a large meal (defined as 25% of body mass), and reaches double the fasting mass two days after a meal. These changes are due largely to proliferation of the mucosal layer, rather than the serosal layer. Associated with these morphologic changes were increases in the capacity for amino acid uptake that ranged from 10 to 24 times the fasting values.

Even when the overall length and diameter of the gut are not affected by dietary changes, the "microstructure" formed by the microvilli may change, resulting in increases in absorptive surface area. These changes can lead to an overall increase in nutrient absorption when an animal's energy demands are great and may also enhance the extraction of nutrients by slowing the passage of food through the gut. This latter situation is particularly evident in cecal fermenters.

Dietary changes can similarly alter the cellular and macromolecular makeup of the gut. Research by Jared Diamond, Steve Secor, William Karasov, and others has shown that most intestinal membrane transport proteins are regulated by dietary levels of their substrates. Increasing substrate levels stimulate an increase in the concentration or activity of transporters for glucose, fructose, some nonessential amino acids, and peptides (Figure 15-23). The proliferation

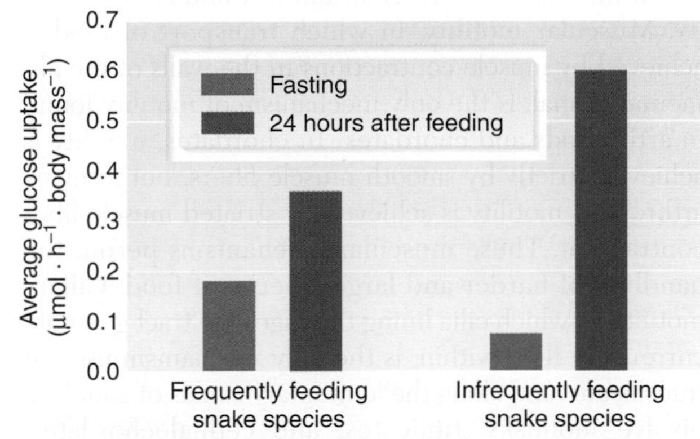

Figure 15-23 The nutrient uptake capacity of the gut of snakes is dependent upon both the natural history of the snake and whether it is fasting or digesting a meal. Shown is the average glucose uptake by the alimentary canal of a group of four frequently feeding snake species and four infrequently feeding snake species. Note that the glucose uptake capacity is moderate and similar during both fasting and digesting in frequently feeding snakes, while in infrequently feeding snakes the very low level of glucose uptake during fasting increases manyfold in the 24-hour period following feeding. These changes in glucose uptake capacity in both groups of snakes are mirrored by changes in the uptake capacity for several amino acids. [Data from Secor and Diamond, 2000]

of transporters appears to be matched to the level of nutrient intake, providing only as much uptake capacity as diet requires.

It is important to emphasize that an expansion of gut surface area or an increase in nutrient transporter proteins carries with it significant metabolic cost in support of this new macro- or microstructure. Consequently, most changes in gut structure appear to be completely reversible, which reduces the metabolic cost of maintaining the gut during periods when food resources are scarce.

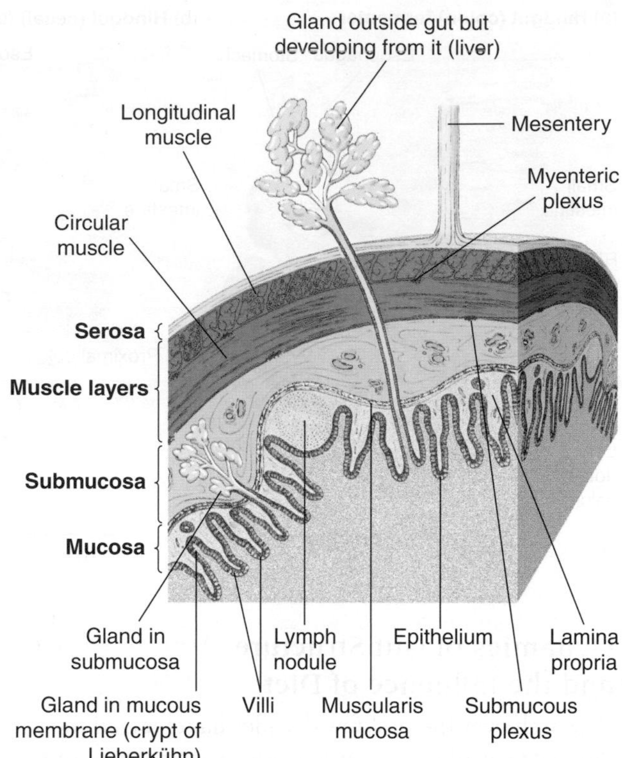

Figure 15-24 A generalized cross-section of the vertebrate intestine reveals a many-layered wall with substantial smooth muscle tissue. The wall consists of four layers: the serosa, made of connective tissue; a layer of longitudinal and circular muscle; the submucosa; and the innermost mucosa.

MOTILITY OF THE ALIMENTARY CANAL

The ability of the alimentary canal to contract, a characteristic called **motility**, is important to digestive function in several ways:

1. Propulsion of food along the entire length of the alimentary canal and the final expulsion of fecal material
2. Mechanical treatment of food by grinding and kneading to mix it with digestive juices and convert it to a soluble form
3. Stirring of the gut contents so that there is continual renewal of material in contact with the absorbing and secreting surfaces of the epithelial lining

Muscular and Ciliary Motility

Motility in the alimentary canal is achieved by two different mechanisms: muscular motility and ciliary motility. Muscular motility, in which transport of food is achieved by muscle contractions in the walls of the alimentary canal, is the only mechanism of motility found in arthropods and chordates. In chordates, motility is achieved strictly by smooth muscle fibers, but in many arthropods motility is achieved by striated muscle fiber contraction. These muscular mechanisms permit the handling of harder and larger pieces of food. Ciliary motility, in which cilia lining the digestive tract generate currents of fluid within, is the only mechanism used to translocate food along the alimentary canals of annelids, bivalve mollusks, tunicates, and cephalochordates. However, ciliary motility is used in conjunction with muscular mechanisms in echinoderms and most mollusks.

Peristalsis

The arrangement of the alimentary musculature in vertebrates consists of an inner circular layer and an outer longitudinal layer of smooth muscle (Figure 15-24; see also Figure 15-19). Contraction of the circular layer coordinated with relaxation of the longitudinal layer produces an active constriction with an elongation of the gut. Active shortening of the longitudinal layer coordinated with relaxation of the circular layer produces distension.

Peristalsis consists of a traveling wave of constriction produced by contraction of the circular muscle that is preceded along its length by a simultaneous contraction of the longitudinal muscle and relaxation of the circular muscle (Figure 15-25). This pattern of contraction "pushes" the luminal contents in the direction of the peristaltic wave. Mixing and then packing of the luminal contents is achieved primarily by a process called **segmentation**, which consists of rhythmic contractions of the circular muscle layer that occur asynchronously along the intestine at various points without participation of the longitudinal muscle.

Swallowing propels a bolus from the buccal cavity to the stomach. In vertebrates, swallowing involves the integrated movements of muscles in the tongue and pharynx as well as peristaltic movements of the esophagus. Both actions are under direct neuronal control by control centers in the medulla oblongata of the brain. Regurgitation occurs when peristalsis takes place in the reverse direction, moving the luminal contents back into the buccal cavity.

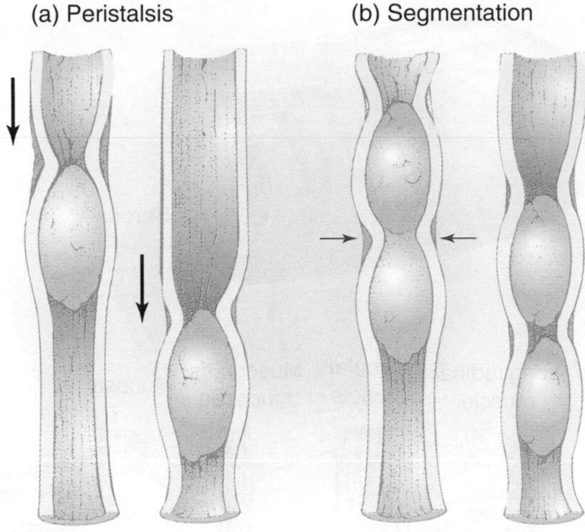

Figure 15-25 Coordinated contraction of the alimentary canal propels material through its lumen. **(a)** Peristalsis occurs as a traveling wave of contraction of circular muscle preceded by relaxation, which produces longitudinal movement of the bolus through the alimentary canal. **(b)** Segmentation occurs as alternating relaxations and contractions, primarily of circular muscle. The result is a kneading and mixing of the intestinal contents.

Control of Motility

A combination of distinct mechanisms controls the coordinated contractions of circular and longitudinal smooth muscle that provide alimentary canal motility in vertebrates.

Intrinsic control

Contraction of the smooth muscle tissue in the wall of the alimentary canal is myogenic—that is, the muscles are capable of producing an intrinsic cycle of electrical activity that leads to muscle contraction without external neuronal stimulation. In mammals, pacemaker cells of the alimentary canal generate rhythmic depolarizations and repolarizations called the **basic electric rhythm (BER)**. Voltage-clamp studies have revealed that, at least in mice, the spontaneous activity of the ICC results from nonselective cation conductance; research based on cloned ion channel genes is closing in on the roles that each cation plays under specific physiological conditions. Since the ICC are electrically coupled to the smooth muscle cells of the alimentary canal walls, spontaneous waves of depolarization progress slowly along the muscle layers (Figure 15-26). Some of these slow waves give rise to action potentials (APs) produced by an inward current carried by calcium ions. These calcium "spikes" lead to contractions of the smooth muscle cells in which they occur. The amplitude of the slow-wave BER is modulated by local influences, such as stretching of the muscle tissue, which occurs when a chamber of the alimentary canal is stretched by food in its lumen. Another influence on contraction is chemical stimulation of the mucosa by substances in the chyme.

Extrinsic control

The intrinsic patterns of the BER are modulated by locally released gastrointestinal peptide hormones. Thus, a chemical stimulant in the chyme can cause the release of a paracrine hormone, and this hormone, in turn, can modulate the motility of the muscle tissue.

In addition to local stimuli, intestinal motility is influenced by diffuse innervation from the autonomic nervous system (see Chapter 8). Sympathetic and parasympathetic postganglionic neurons form networks

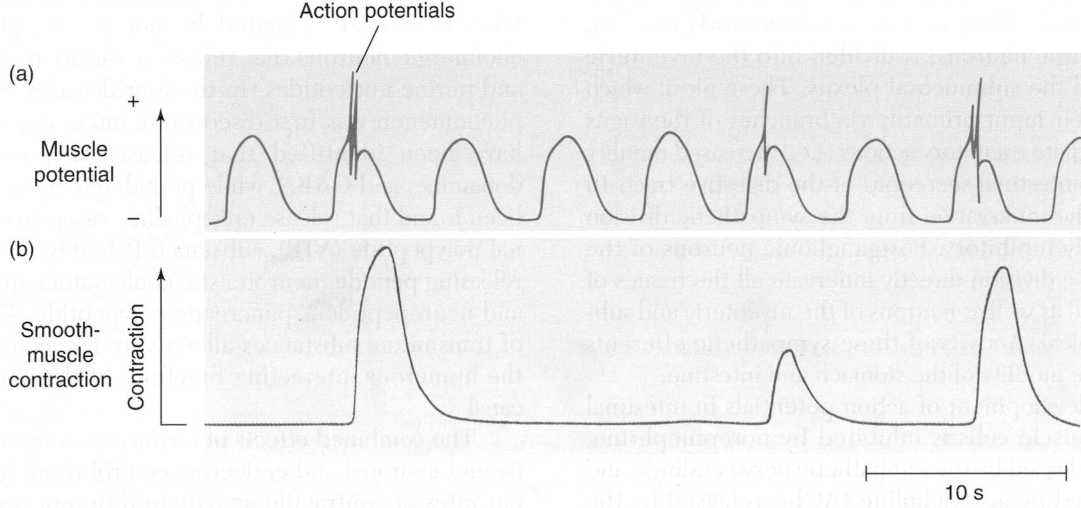

Figure 15-26 Electrical and mechanical (contraction) activity are coordinated in the cat jejunum. **(a)** The slow basic electric rhythm (BER), evident as oscillations in the membrane potential of the muscle fiber, occasionally gives rise to action potentials derived from Ca^{2+} movements at the peaks. **(b)** Calcium action potentials elicit contractions of the smooth muscle in which they occur. [Adapted from Bortoff, 1985.]

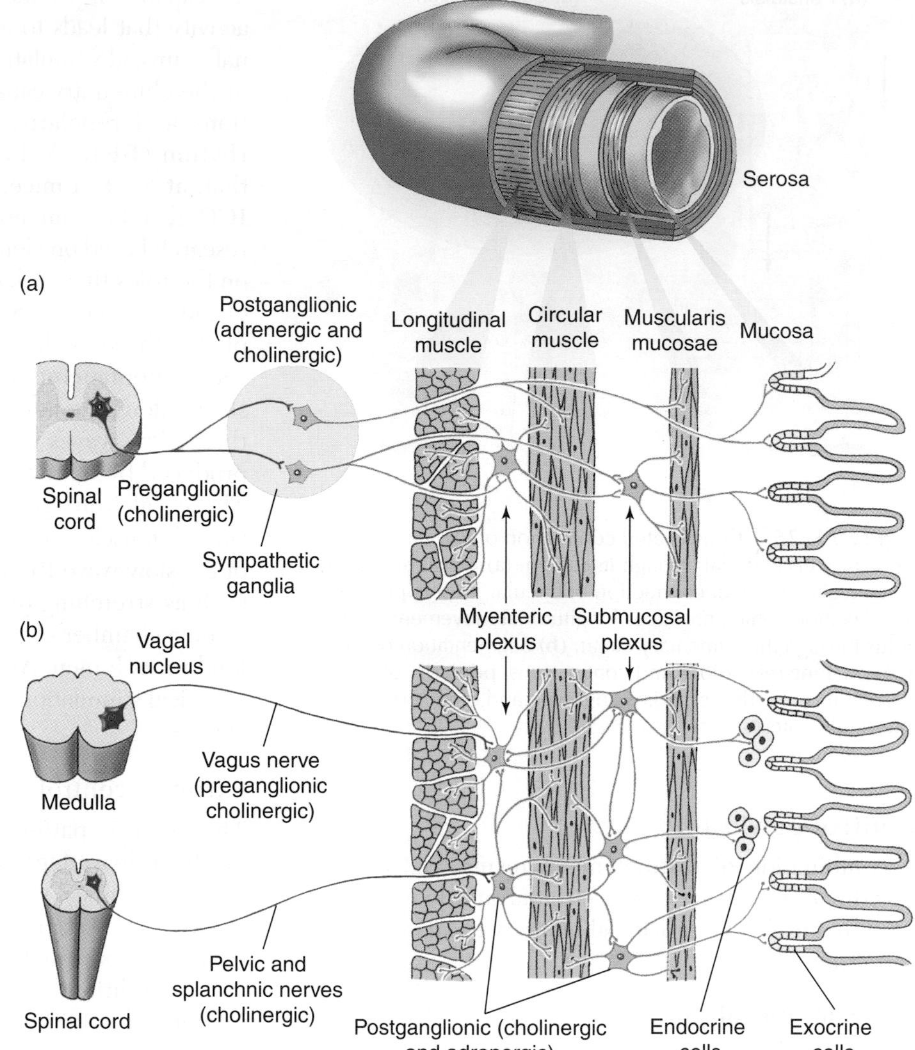

Figure 15-27 The alimentary canal has rich sympathetic and parasympathetic innervation. **(a)** Efferent sympathetic innervation. **(b)** Parasympathetic innervation. All neurons synapsing on the gastrointestinal target tissues (muscle, glands) are postganglionic. [Adapted from Davenport, 1977.]

that are dispersed throughout the smooth muscle layers (Figure 15-27). The parasympathetic network, made up of cholinergic neurons, is divided into the **myenteric plexus** and the **submucosal plexus.** These plexi, which receive their input primarily via branches of the vagus nerve, mediate excitatory actions (i.e., increased motility and gastrointestinal secretion) of the digestive tract. In contrast, the innervation from the sympathetic division is primarily inhibitory. Postganglionic neurons of the sympathetic division directly innervate all the tissues of the gut wall as well as neurons of the myenteric and submucosal plexi. Activity of these sympathetic efferents inhibits the motility of the stomach and intestine.

The development of action potentials in intestinal smooth muscle cells is inhibited by norepinephrine, which is released by the sympathetic nerve endings, and is promoted by acetylcholine (ACh), released by the parasympathetic nerve endings (Figure 15-28a). Each impulse associated with excitation produces an increment of tension, which subsides with cessation of impulses (Figure 15-28b).

Smooth muscle in the alimentary canal of vertebrates is also regulated by non-adrenergic, noncholinergic neurons that release a variety of peptides and purine nucleotides. In the four decades since this phenomenon was first discovered, aminergic neurons have been identified that release ATP, serotonin, dopamine, and GABA, while peptidergic neurons have been found that release enkephalins, vasoactive intestinal polypeptide (VIP), substance P, bombesin/gastrin-releasing peptide, neurotensin, cholecystokinin (CCK), and neuropeptide Y/pancreatic polypeptide. This host of transmitter substances allows very fine control over the numerous interacting functions of the alimentary canal.

The combined effects of intrinsic pacemaker activity and neuronal and endocrine control result in different rates of contractile activity in different regions of the alimentary canal (Figure 15-29). In mammals, stomach contractions tend to be slow and powerful to provide a churning action on the as yet undigested food material, while those of the small intestine are more fre-

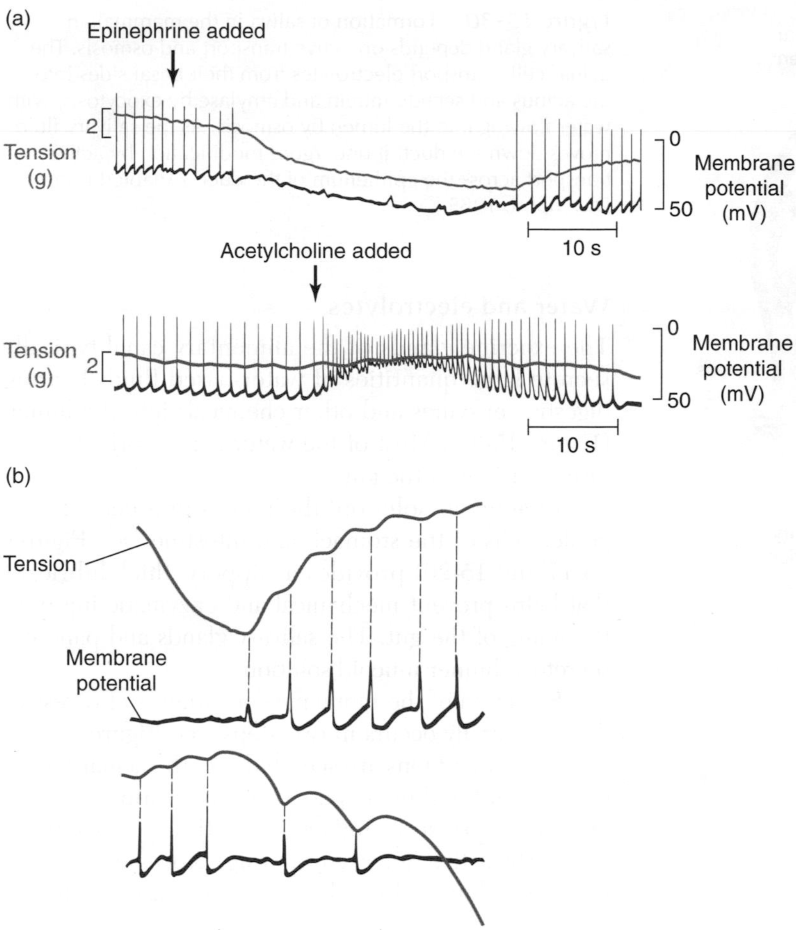

Figure 15-28 Membrane potentials (red spikes) dictate tension (blue lines) in the tenia coli, a longitudinal muscle along the colon. **(a)** Effects of topically applied epinephrine and acetylcholine. **(b)** Time correlation between action potentials (red) and tension (blue). [Part a adapted from Bülbring and Kuriyama, 1963; part b adapted from Bülbring, 1959.]

quent and of shorter duration, appropriate for kneading and transporting the semi-digested bolus within. The colon provides stronger contractions of greater duration to ensure that the forming fecal mater is compacted and ejected.

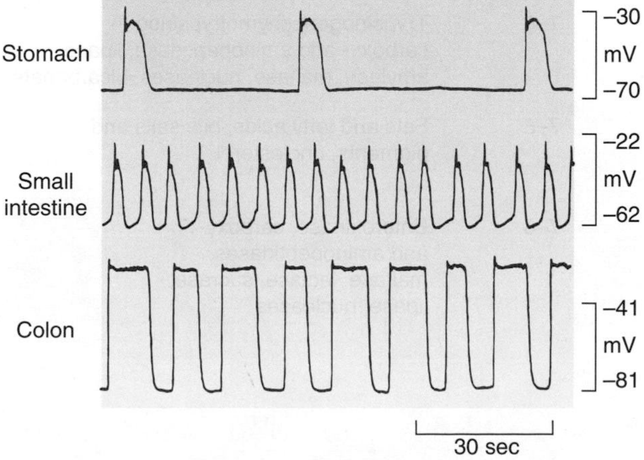

Figure 15-29 Electrical activity recorded in the smooth muscle of the alimentary canal of the dog shows regional variation in both frequency and duration. [From Sanders, 1992]

GASTROINTESTINAL SECRETIONS

The alimentary canal has been described as the largest endocrine and exocrine gland of the body. Exocrine gastrointestinal secretions usually consist of aqueous mixtures of substances rather than a single species of molecule. Exocrine tissues of the alimentary canal include the salivary glands, the secretory cells in the stomach and intestinal epithelium, and the secretory cells of the liver and pancreas. The primary secretions of these exocrine glands enter the acinar lumen of the gland, and then generally become secondarily modified in the gland's secretory duct. This secondary modification can involve further transport of water and electrolytes into or out of the duct to produce the final secretory juice, as illustrated for the salivary gland (Figure 15-30 on the next page) and described in detail in Chapter 9.

Exocrine Secretions of the Alimentary Canal

The secretions from different regions of the alimentary canal vary greatly in composition. However, these mixtures usually consist of some combination of water, ions, mucus, and enzymes.

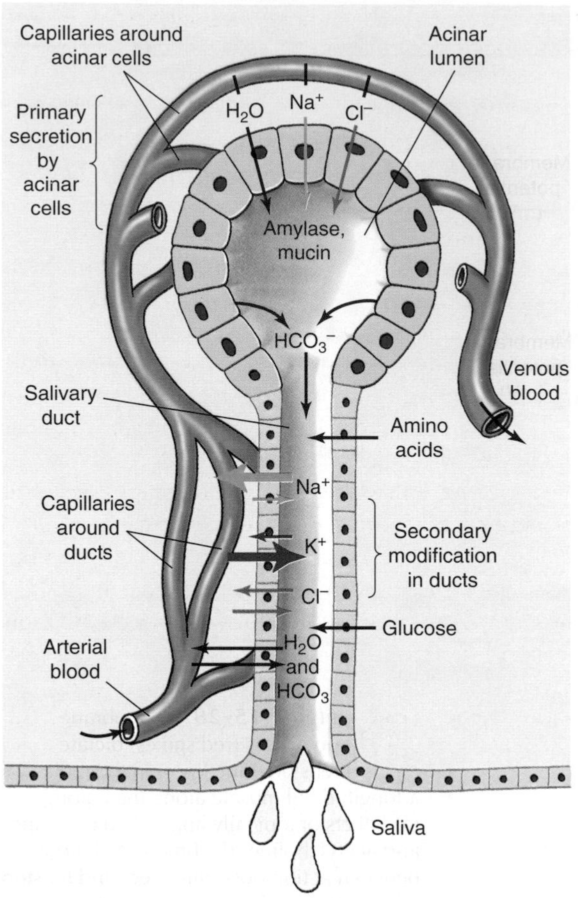

Figure 15-30 Formation of saliva in the mammalian salivary gland depends on active transport and osmosis. The acinar cells transport electrolytes from their basal sides into the acinus and secrete mucin and amylase by exocytosis, with water flowing into the lumen by osmosis. As the salivary fluid moves down the duct, it undergoes modification by active transport across the epithelium of the duct. [Adapted from Davenport, 1985.]

Water and electrolytes

The exocrine glands of the alimentary canal typically secrete large quantities of water-based fluids bearing digestive enzymes and other chemicals into the lumen (Figure 15-31). Most of the water is reabsorbed in the distal portions of the gut.

In aqueous solution, the mucus produced by the goblet cells of the stomach and intestine (see Figures 15-17 and 15-20) provides a slippery, thick lubricant that helps prevent mechanical and enzymatic injury to the lining of the gut. The salivary glands and pancreas secrete a thinner mucoid solution.

Secretion of the inorganic constituents of digestive fluids generally occurs in two steps (see Figure 15-30). First, water and ions are secreted into the acinar lumen of the gland, either by passive ultrafiltration due to a hydrostatic pressure gradient across the luminal epithelium or by active (energy-requiring) processes from the interstitial fluid bathing the basal portions of the acinar

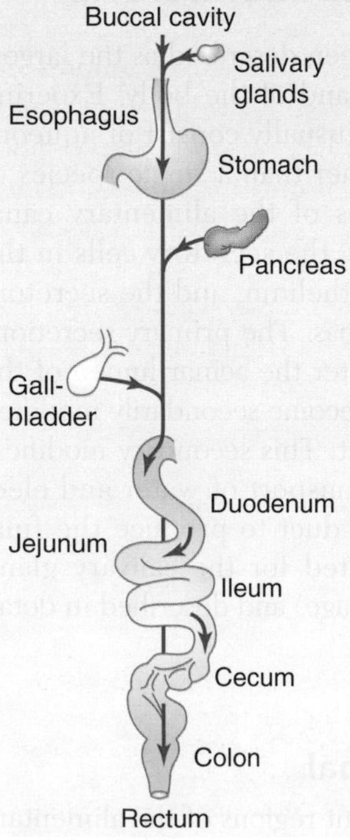

Region	Secretion	Daily amount (L)	pH	Composition*
Buccal cavity				
Salivary glands	Saliva	1+	6.5	Amylase, bicarbonate
Esophagus				
Stomach	Gastric juice	1–3	1.5	Pepsinogen, HCl, rennin in infants, intrinsic factor
Pancreas	Pancreatic juice	1	7–8	Trypsinogen, chymotrypsinogen, carboxy- and aminopeptidase, lipase, amylase, maltase, nucleases, bicarbonate
Gall-bladder	Bile	1	7–8	Fats and fatty acids, bile salts and pigments, cholesterol
Duodenum	Succus entericus	1	7–8	Enterokinase, carboxy- and aminopeptidases, maltase, lactase, sucrase, lipase, nucleases
Jejunum				
Ileum				
Cecum				
Colon				
Rectum				

*Excluding mucus and water, which together make up some 95% of the actual secretion.

Figure 15-31 Important digestive secretions occur at all points along the human alimentary canal. The approximate volume and pH of each secretion is shown on the right.

cells. The latter is believed to entail active transport of ions by these cells, which is followed by the osmotic flow of water into the acinus. This fluid undergoes secondary modification by active or passive transport across the epithelium lining the exocrine ducts as it passes along the ducts toward the alimentary canal.

Bile and bile salts

The vertebrate liver does not produce digestive enzymes. However, it does secrete **bile**, a fluid essential for digestion of fats. Bile consists of water and a weakly basic mixture of cholesterol, lecithin, inorganic salts, bile salts, and bile pigments. The bile salts are organic salts composed of cholic acids manufactured by the liver from cholesterol and conjugated with amino acids complexed with sodium. The bile pigments derive from biliverdin and bilirubin, which are products of the breakdown of hemoglobin that has spilled into the blood plasma from old, ruptured red blood cells. Bile produced in the liver is transported via the hepatic duct to the gallbladder, where it is concentrated and stored. Water is removed osmotically, following active transport of Na^+ and Cl^- from the bile across the gallbladder epithelium.

Bile serves numerous functions important to digestion. First, its high alkalinity is important in the terminal stages of digestion because it buffers the high acidity provided by the gastric juice secreted earlier in the digestive process. Second, bile salts facilitate the digestion of fats by breaking them down into multiple microscopic droplets and dispersing them in aqueous solution for more effective attack by digestive enzymes. Bile salts act as a detergent for the emulsification of fat droplets, dispersing them in aqueous solution for more effective attack by digestive enzymes. Ultimately, the bile salts are removed from the large intestine by highly efficient active transport and returned to the bloodstream, where they become bound to a plasma carrier protein and are returned to the liver to be recycled. Bile salts also disperse lipid-soluble vitamins for transport in the blood. Third, bile contains waste substances removed from the blood by the liver, such as hemoglobin pigments, cholesterol, steroids, and hydrophobic drugs. These substances are either digested or excreted in the feces.

Digestive enzymes

An animal must break food into its constituent parts before it can be used for tissue maintenance and growth or as a source of chemical energy. **Digestion** is a complex chemical process in which special digestive enzymes catalyze the hydrolysis of large food molecules into simpler compounds that are small enough to cross the cell membranes of the intestinal barrier. Starch, for example, a long-chain polysaccharide, is degraded to much smaller disaccharides and monosaccharides; proteins are hydrolyzed into polypeptides and then into tripeptides, dipeptides, and amino acids.

Digestive enzymes carry out hydrolysis by breaking the peptide bonds that link amino acids in proteins or the glycosidic bonds that link sugar monomers in polysaccharides. Addition of H^+ to one residue and OH^- to the other frees the constituent monomers (e.g., monosaccharides, amino acids, monoglycerides) from which the polymer is formed (Figure 15-32), making them small enough for absorption from the alimentary canal into the circulating body fluids and for subsequent entry into cells to be metabolized.

Digestive enzymes, like all enzymes, exhibit substrate specificity and are sensitive to temperature, pH, and certain ions (Table 15-1 on the next page). Corresponding to the three major types of foods are three major groups of digestive enzymes: proteases, carbohydrases, and lipases.

Proteases Proteases are proteolytic enzymes, categorized as either endopeptidases or exopeptidases. Both types of enzymes attack the peptide bonds of proteins and polypeptides. They differ in that endopeptidases confine their attacks to bonds well within (*endo-*, "within") the protein molecule, breaking large peptide chains into shorter polypeptide segments. These shorter segments provide a much greater number of sites of action for the exopeptidases. These enzymes attack only peptide bonds near the end (*exo-*, "outside") of a peptide chain, producing free amino acids, plus dipeptides and tripeptides. Some proteases exhibit

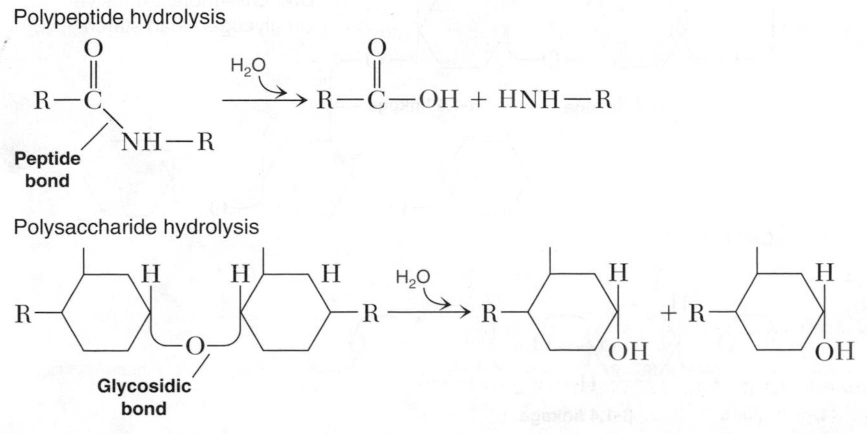

Polypeptide hydrolysis

$$R{-}C(=O){-}NH{-}R \xrightarrow{H_2O} R{-}C(=O){-}OH + HNH{-}R$$

Peptide bond

Polysaccharide hydrolysis

H_2O ; H ; OH ; R

Glycosidic bond

Figure 15-32 During digestion, proteins and polysaccharides are broken down by hydrolysis. In these reactions, catalyzed by several enzymes with different specificities, a molecule of water is added to the bond linking the monomers, cleaving the bond and splitting the polymer into pieces that undergo further degradation.

Table 15-1 **Action of the major enzymes secreted in the mouth, stomach, pancreas, and small intestine**

Enzyme	Site of action	Substrate	Products of action
Mouth			
Salivary α-amylase	Mouth	Starch	Disaccharides (few)
Stomach			
Pepsinogen:pepsin	Stomach	Proteins	Large peptides
Pancreas			
Pancreatic α-amylase	Small intestine	Starch	Disaccharides
Trypsinogen:trypsin	Small intestine	Proteins	Large peptides
Chymotrypsin	Small intestine	Proteins	Large peptides
Elastase	Small intestine	Elastin	Large peptides
Carboxypeptidases	Small intestine	Large peptides	Small peptides (oligopeptides)
Aminopeptidases	Small intestine	Large peptides	Oligopeptides
Lipase	Small intestine	Triglycerides	Monoglycerides, fatty acids, glycerol
Nucleases	Small intestine	Nucleic acids	Nucleotides
Small intestine			
Enterokinase	Small intestine	Trypsinogen	Trypsin
Disaccharidases	Small intestine*	Disaccharides	Monosaccharides
Peptidases	Small intestine*	Oligopeptides	Amino acids
Nucleotidases	Small intestine*	Nucleotides	Nucleosidases, phosphoric acid
Nucleosidases	Small intestine*	Nucleosides	Sugars, purines, pyrimidines

*Intracellular

marked specificity for particular amino acid residues located on either side of the bonds they attack. For example, the endopeptidase trypsin attacks only those peptide bonds in which the carboxyl group is provided by arginine or lysine, regardless of where they occur within the peptide chain.

In mammals, protein digestion usually begins in the stomach with the action of the gastric protease pepsin. The most powerful form of pepsin functions best in an extremely acidic environment, with a pH of only 2. The action of pepsin is aided by secretion of HCl in the stomach and results in the hydrolysis of proteins into polypeptides and some free amino acids. In the mammalian intestine, several proteases produced by the pancreas continue the proteolytic process, yielding a mixture of free amino acids and small peptide chains. Finally, proteolytic enzymes intimately associated with the epithelium of the intestinal wall hydrolyze the polypeptides into oligopeptides, which consist of residues of two or three amino acids, and then break these down further into individual amino acids.

Carbohydrases Carbohydrases can be divided functionally into polysaccharidases and glycosidases. Polysaccharidases hydrolyze the glycosidic bonds of long-chain carbohydrates such as cellulose, glycogen, and starch (Figure 15-33). The most common polysac-

starch, glucus

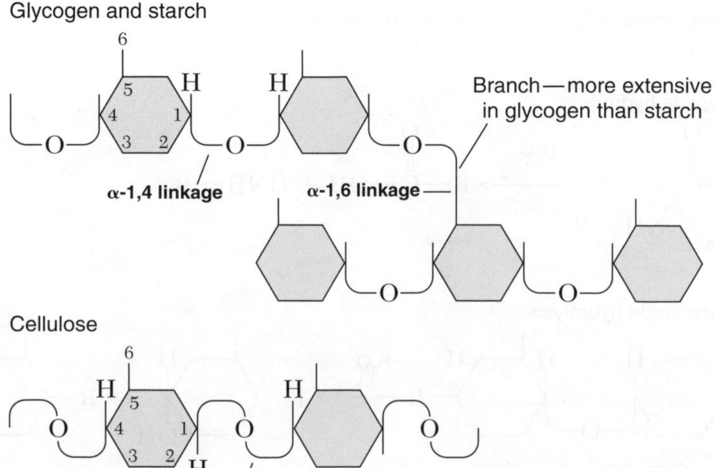

Figure 15-33 Starch and glycogen are polymers of glucose units connected by α-1,4 linkages with periodic branches connected by α-1,6 linkages. Cellulose consists of glucose subunits connected by β-1,4 linkages. Hydrolysis of the labeled bonds yields monosaccharides for absorption.

charidases are the amylases, which hydrolyze all but the terminal glycosidic bonds within starch and glycogen, producing disaccharides and oligosaccharides. Amylases are secreted in vertebrates by the salivary glands and pancreas and in small amounts by the stomach, and in most invertebrates by salivary glands and intestinal epithelium. The glycosidases, which occur in the glycocalyx attached to the surface of the absorptive cells, act on disaccharides such as sucrose and fructose by hydrolyzing the remaining α-1,6 and α-1,4 glycosidic bonds. Their action breaks these sugars down into their constituent monosaccharides for absorption.

Many herbivores consume large amounts of plant cell walls, containing cellulose, hemicellulose, and lignin. Cellulose, which is in greatest abundance, consists of glucose molecules polymerized via β-1,4 bonds. Cellulase, a polysaccharidase that digests cellulose and hemicellulose, is produced by symbiotic microorganisms in the guts of host animals as diverse as termites and cattle, which themselves are incapable of producing this enzyme. In termites, cellulase is released into the intestinal lumen by the symbiont and functions extracellularly to digest the ingested wood. In cattle, the symbiotic microbes take up cellulose molecules (from ingested grass, etc.), digest them intracellularly, and pass some digested fragments into the surrounding fluid. These symbiotic gut bacteria, in turn, multiply and are themselves subsequently digested. Were it not for these symbiotic microorganisms, cellulose (the major nutritional constituent of grass, hay, and leaves) would be unavailable as food for grazing and browsing animals. Only a few animals, such as the shipworm *Toredo* (a wood-boring clam), *Limnoria* (an isopod), and the silverfish (an insect common in houses), can secrete cellulase without the help of symbionts.

Lipases Fats are water-insoluble, which presents a special problem for their digestion. First, fats must be **emulsified**—that is, they are rendered water-soluble by dispersing them into small droplets through the mechanical churning of the intestinal contents and by the chemical action of detergents such as bile salts and the phospholipid lecithin. Bile salt molecules have a hydrophobic, fat-soluble end and a hydrophilic, water-soluble end. A lipid molecule attaches to the hydrophobic end, while water attaches to the hydrophilic end. This reaction disperses the fat in the water-based fluid of the digestive tract. The overall effect is comparable to making mayonnaise, in which salad oil is dispersed in vinegar with egg yolk acting as the emulsifier.

The second step in vertebrates is the formation of **micelles** (see Figure 3-9), aided by bile salts. Micelles are exceedingly small spherical structures formed from molecules that have polar hydrophilic groups at one end and nonpolar hydrophobic groups at the other. These molecules assemble so that their polar ends face outward into the aqueous solution. The lipid core of each micelle is about 10^{-6} times the size of the original emulsified fat droplets. Since the ratio of surface area to volume increases as spheres get smaller, this decrease in size greatly increases the surface area available for digestion. Lipids are degraded by lipases into fatty acids plus monoglycerides and diglycerides. In the absence of sufficient bile salts, fat digestion by lipases is incomplete, and undigested fat is allowed to enter the colon and be eliminated without being absorbed.

Proenzymes Certain digestive enzymes—proteolytic enzymes in particular—are synthesized, stored, and released in an inactive molecular form known as a **proenzyme**, or **zymogen**. Proenzymes require activation, usually by hydrochloric acid in the lumen of the gastric glands, before they can carry out their degradative functions. Initial packaging of the enzyme in an inactive form prevents self-digestion of the enzyme and the tissue container in which it is stored. The proenzyme is activated by the removal of a portion of the molecule, either by the action of another enzyme specific for this purpose or through a rise in ambient acidity.

Other digestive enzymes In addition to the major classes of digestive enzymes just described, there are others that play a less important role in digestion. Nucleases, nucleotidases, and nucleosidases, as their names imply, hydrolyze nucleic acids and their residues. Esterases hydrolyze esters, which include the fruity-smelling compounds characteristic of ripe fruit. These and other minor digestive enzymes are not essential for nutrition, but they enhance the efficient use of ingested food.

Control of Gastrointestinal Secretions

Among vertebrates, the primary stimulus for secretion of digestive juices is the presence of food in a given part of the digestive tract. The presence of food molecules stimulates chemoreceptors in the alimentary canal, leading to the reflex activation of autonomic efferent neurons that activate or inhibit motility and exocrine secretion. Appropriate food molecules also directly stimulate epithelial endocrine cells by contact with their membrane receptors, causing reflex secretion of gastrointestinal hormones into the local circulation. These reflexes permit secretory organs outside the alimentary canal proper (the liver and pancreas, for example) to be properly coordinated with the need for digestion of food passing along the digestive tract. While the presence of food in the digestive tract is correctly viewed as the primary digestive stimulus, the role of cognition or thought processes in the control of digestive secretion is often overlooked. Cephalic influences, such as mental images of food as well as learned behaviors, also stimulate digestive secretion, at least in mammals (Spotlight 15-1 on the next page).

Gastrointestinal secretion is largely under the control of gastrointestinal peptide hormones secreted by endocrine cells of the gastric and intestinal submucosa (Table 15-2 on the next page). Several of these hormones turn out to be identical to neuropeptides that act

BEHAVIORAL CONDITIONING IN FEEDING AND DIGESTION

The experiments of the Russian physiologist Ivan Pavlov (1849–1936) figure prominently in the histories of both psychology and physiology. Pavlov, who in 1904 won the Nobel Prize for his research, demonstrated reflexive secretion of saliva in dogs. In his experiments, a dog was given food following the sounding of a bell. Normally, a dog will salivate in response to the sight or taste of food, but not in response to a bell. However, after several presentations of the bell (conditioned stimulus) together with food (unconditioned stimulus), the bell alone elicited salivation. Pavlov's discovery was the first recognition of a conditioned reflex. These experiments were important in the development of theories of animal behavior and psychology. In the context of this chapter, they demonstrated that some secretions of the digestive tract are under cephalic control (i.e., controlled by the brain). Thus, in vertebrates, neuronal control of digestive secretions operates by two mechanisms. In the first, neuronal output to gland tissues is generated by an unconditioned reflex elicited directly by food in contact with chemoreceptors in the alimentary canal. In the second, neuronal output is evoked indirectly by association of a conditioned stimulus with an unconditioned stimulus.

Another example of cephalic control of digestive secretions is the reflexive secretion of salivary and gastric fluids evoked by the sight, smell, or anticipation of food. (Try closing your eyes and concentrating intently on the taste and texture of your favorite snack food. Has your salivary flow increased?) This reflex is based on past experience (i.e., associative learning). Closely related to this phenomenon is the discovery that some animals exhibit one-trial avoidance learning in response to noxious foods. In this case, an animal will reject a meal even before tasting it if it looks or smells like something previously sampled that proved to be noxious. Insect-eating birds have been found to avoid a particular species of bad-tasting insect prey on the basis of a single experience with that prey, particularly when the prey is strikingly colored and easily recognized. Examples of avoidance of noxious foods by one-trial learning have also been described in several mammals. This phenomenon is called Garcia conditioning.

Table 15-2 The major gastrointestinal peptide hormones

Hormone	Tissues of origin	Target tissue	Primary action	Stimulus to secretion
Gastrin	Stomach and duodenum	Secretory cells and muscles of stomach	HCl production and secretion; stimulation of gastric motility	Vagus nerve activity; peptides and proteins in stomach
Cholecystokinin (CCK)*	Upper small intestine	Gallbladder	Contraction of gallbladder	Fatty acids and amino acids in duodenum
		Pancreas	Pancreatic juice secretion	
Secretin*	Duodenum	Pancreas, secretory cells, and muscles of stomach	Water and $NaHCO_3$ secretion; inhibition of gastric motility	Food and strong acid in stomach and small intestine
Gastric inhibitory peptide (GIP)	Upper small intestine	Gastric mucosa and musculature	Inhibition of gastric secretion and motility	Monosaccharides and fats in duodenum
Bulbogastrone	Upper small intestine	Stomach	Inhibition of gastric secretion and motility	Acid in duodenum
Vasoactive intestinal peptide (VIP)*	Duodenum	Stomach, intestine	Increase of blood flow; secretion of thin pancreatic fluid; inhibition of gastric secretion	Fats in duodenum
Enteroglucagon	Duodenum	Jejunum, pancreas	Inhibition of motility and secretion	Carbohydrates in duodenum
Enkephalin*	Small intestine	Stomach, pancreas, intestine	Stimulation of HCl secretion; inhibition of pancreatic enzyme secretion and intestinal motility	Basic conditions in stomach and intestine
Somatostatin*	Small intestine	Stomach, pancreas, intestine, splanchnic arterioles	Inhibition of HCl secretion, pancreatic secretion, intestinal motility, and visceral blood flow	Acid in lumen of stomach

*These peptides are also found in central nervous tissue as neuropeptides. Additional unlisted neuropeptides identified in both brain and gut tissue include substance P, neurotensin, bombesin, insulin, pancreatic polypeptide, and ACTH.

as transmitters in the central nervous system. This finding indicates that the genetic machinery for producing these biologically active peptides has been put to use by cells of both the central nervous system and the alimentary canal.

The characteristics of digestive secretion (rate of secretion, quantity of secretion) depend on several interacting features, including whether secretion is neuronally or hormonally controlled, where in the alimentary canal the secretion occurs, and how long food is normally present in the region being stimulated. Salivary secretion, for example, is very rapid and is entirely under involuntary neuronal control. Gastric secretions are under hormonal as well as neuronal control, and intestinal secretions are slower and are primarily under hormonal control. As in other systems, neuronal control predominates in rapid reflexes, whereas endocrine mechanisms produce reflexes that develop over minutes or hours.

Compared with vertebrates, very little is known about the control of digestive secretions in invertebrates. Filter feeders evidently maintain a steady secretion of digestive fluids as they continuously feed. Other invertebrates secrete enzymes in response to the presence of food in the alimentary canal, but the precise control mechanisms involved have yet to be intensively studied. The formidable variety of invertebrates further precludes generalizing about their digestive systems.

Salivary secretions

Mammalian saliva contains water, electrolytes (primarily Na^+, K^+, Cl^-, HCO_3^-, and in ruminants, PO_4^{3-}), mucin, amylase, and antimicrobial agents such as lysozyme and thiocyanate (see Figure 15-33). The water and electrolytes are derived from blood plasma, while the HCO_3^- is provided by secretory cells. The tonicity (see Chapter 4) and ionic content of saliva vary with species and with rate of flow. In the absence of food, the salivary glands produce a slow flow of watery saliva. Secretion of saliva is stimulated by the presence of food in the mouth, or by any mechanical stimulation of tissues within the mouth, via cholinergic parasympathetic nerves projecting to the salivary glands. Cognitive awareness of food has an identical effect (see Spotlight 15-1). Salivary secretion is decreased by release of norepinephrine from sympathetic innervation of the salivary glands. The amylase in saliva mixes with food during chewing and digests starches. The mucin and watery fluid condition the food bolus to help it slide smoothly toward the stomach with the peristaltic movements of the esophagus.

Gastric secretions

A major secretion of the stomach lining is hydrochloric acid (HCl). This secretion is produced by the **parietal cells**, or *oxyntic cells,* located in the stomach's gastric pits within the gastric mucosa (see Figure 15-17b). The secretion of HCl is stimulated by parasympathetic activity in the vagus nerve, by the action of the hormone gastrin in conjunction with histamine, and by secretagogues (substances that stimulate secretion) in food, such as caffeine, alcohol, and the active ingredients of spices. The secreted HCl helps break the peptide bonds of proteins, activates some gastric enzymes, and kills microorganisms that enter with the food. In some animals, the amount of H^+ used to produce secreted HCl is so great that blood and other extracellular fluids may actually become alkalotic for hours or days after ingestion of a large meal. This so-called *alkaline tide* can result in a rise in blood pH of 0.5 or even 1.0 pH unit in crocodiles, snakes, and other predators that have large, infrequent meals.

The parietal cells produce a concentration of hydrogen ions in the gastric juice that is 10^6 times greater than that in plasma. They do this with the aid of the enzyme carbonic anhydrase, which catalyzes the reaction of water with carbon dioxide:

$$CO_2 + H_2O \underset{\text{carbonic anhydrase}}{\rightleftharpoons} H_2CO_3$$

The H_2CO_3 then dissociates into HCO_3^- and H^+ in the parietal cell. The resulting HCO_3^- is exported from the parietal cell into the plasma in exchange for Cl^- via a HCO_3^-/Cl^- antiporter in the basolateral membrane (Figure 15-34). The imported Cl^- diffuses along an electrochemical gradient to the apical membrane, where it exits via a Cl^- channel and enters the lumen of the gastric gland. The H^+ is actively transported by the apical cell membrane into the lumen of the gastric gland in exchange for K^+ (which tends to leak out of the cell through both the apical and basolateral membranes). These processes of importing and exporting ions allow the parietal cell to maintain a constant pH while providing the stomach with a highly acidic solution.

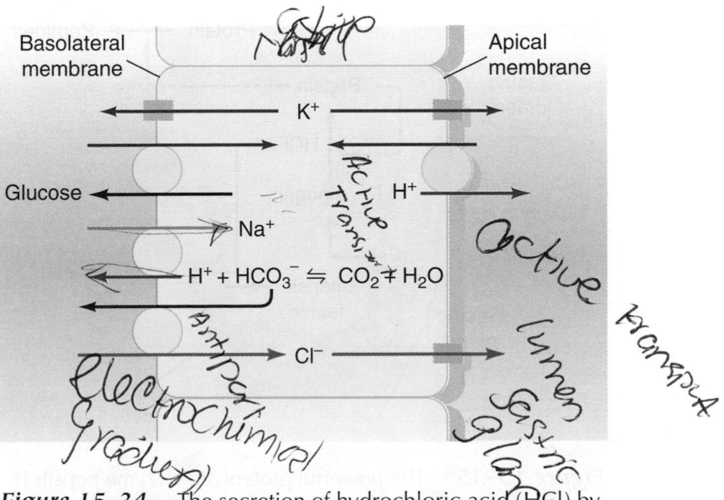

Figure 15-34 The secretion of hydrochloric acid (HCl) by gastric parietal cells employs primary active transport of H^+ produced from the breakdown of H_2CO_3 and electrochemical transport of Cl^- acquired in exchange for HCO_3^-.

Pepsin is the major enzyme secreted by the stomach. This proteolytic enzyme is secreted in the form of a proenzyme, pepsinogen, by exocrine cells called **chief cells,** or *zygomatic cells* (Figure 15-35). Chief cell secretion is under the control of the gastric branch of the vagus nerve and is also stimulated by the hormone gastrin, which is secreted by endocrine cells in the pyloric region of the gastric mucosa (Figure 15-36). Pepsinogen, of which there are several variants, is converted to active pepsin by a low-pH-dependent cleavage of a part of the peptide chain. Pepsin, an endopeptidase, then selectively cleaves inner peptide bonds that occur adjacent to carboxylic side groups of large protein molecules.

Goblet cells in the lining of the stomach secrete a gastric mucus containing various mucopolysaccharides. The mucus coats the gastric epithelium, protecting it from digestion by pepsin and HCl. Although HCl can penetrate the layer of mucus, alkaline electrolytes trapped within the mucus neutralize the acid.

Gastric secretion in mammals occurs in three distinct phases: the cephalic, the gastric, and the intestinal phase. In the cephalic phase, gastric secretion occurs in response to the sight, smell, or taste of food, or in response to conditioned reflexes (see Spotlight 15-1). This phase is mediated by the brain (hence the term *cephalic*) and is abolished when the vagus nerve leading to the stomach is cut.

In the gastric phase, the secretion of HCl and pepsin is stimulated directly by the presence of food in the stomach, which stimulates both chemoreceptors and mechanoreceptors. The gastric phase is mediated by the polypeptide hormone gastrin in conjunction with histamine, a local hormone with paracrine actions that is synthesized in the mast cells of the gastric mucosa. (Both substances are required for HCl secretion because they bind to different receptors on the parietal cell membrane, both of which must be activated for HCl secretion to occur.)

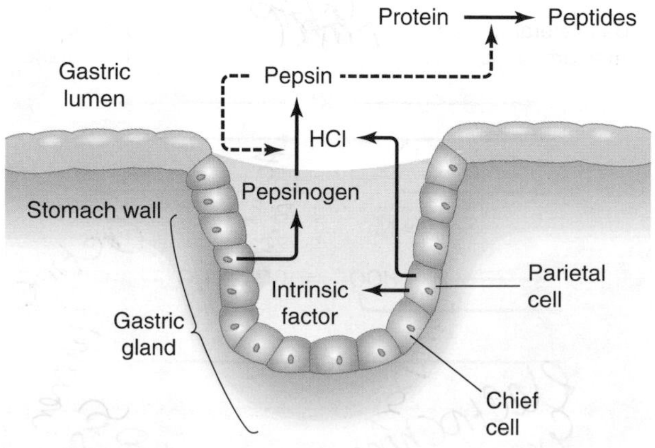

Figure 15-35 The powerful proteolytic enzyme pepsin is secreted in an inactive form (pepsinogen), which is then activated by HCl. The chief cells in the gastric glands secrete pepsinogen, while the parietal cells secrete HCl as well as intrinsic factor.

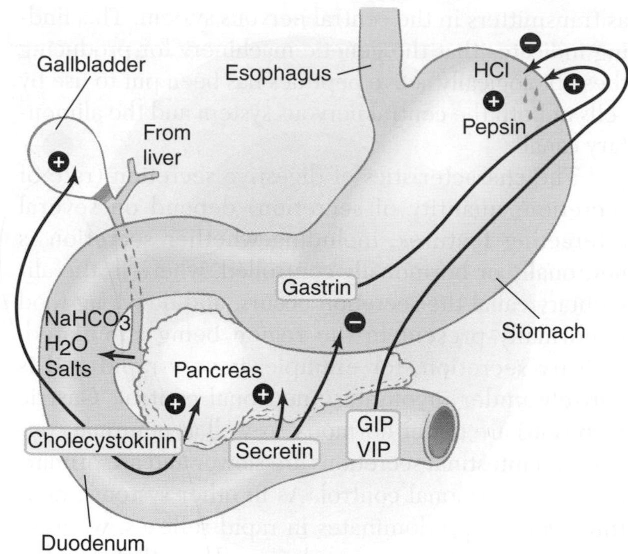

Figure 15-36 Vertebrate gastrointestinal hormones influence the secretory and mechanical activity of the digestive tract. Gastrin from the lower stomach stimulates the flow of both HCl and pepsin from secretory cells in the gastric glands as well as the churning action of the muscular walls. Gastrin is secreted in response to the presence of protein in the stomach, stomach distension, and input from the vagus nerve. Gastric inhibitory peptide, released from the small intestine in response to high levels of fatty acids, inhibits these activities. Neutralization and digestion of the chyme is accomplished by pancreatic secretions stimulated by cholecystokinin, which also induces contraction of the gallbladder, releasing the fat-emulsifying bile into the small intestine. Cholecystokinin is secreted in response to the presence of amino acids and fatty acids in the duodenum. Secretin stimulates pancreatic secretion, but inhibits gastric activity. The plus and minus signs indicate stimulation and inhibition, respectively.

Gastrin is secreted from endocrine cells of the pyloric mucosa of the stomach in response to the presence of gastric chyme containing protein and to distension of the stomach. It stimulates stomach motility by binding to smooth muscle, and it induces a strong secretion of HCl and moderate secretion of pepsin by binding to secretory cells in the stomach lining. When the pH of the gastric chyme drops to 3.5 or below, gastrin secretion slows, and at pH 1.5 it stops. As already noted, secretion of histamine by the gastric mucosa also stimulates secretion of HCl, as does mechanical distension of the stomach.

The intestinal phase is more complex than the other phases, being controlled not only by gastrin, but also by the hormones secretin, vasoactive intestinal peptide (VIP), and gastric inhibitory peptide (GIP) (see Figure 15-36, Table 15-2). GIP, for instance, is released by endocrine cells in the mucosa of the upper small intestine in response to the entry of fats and sugars into the duodenum. As food enters the duodenum, partially digested proteins in acidic chyme directly stimulate the duodenal mucosa to secrete enteric gastrin (also called intestinal gastrin). Enteric gastrin has the same action as stomach gastrin, stimulating the gastric glands to increase their rate of secretion. In humans, at least, the intestinal phase is thought to play a relatively small role in overall regulation of gastric secretion.

The secretion of gastric juice can be reduced both by the absence of stimulating factors and by reflex inhibition. The enterogastric reflex, which inhibits gastric secretion, is triggered when the duodenum is stretched by chyme pumped from the stomach and when this chyme contains partially digested proteins or has particularly low pH. Gastric secretion can also be inhibited by strong activation of the sympathetic nervous system. Action potentials in the sympathetic nerves terminating in the stomach release norepinephrine, which inhibits both gastric secretion and gastric emptying. Thus, the fight-or-flight reaction, characterized by strong sympathetic neuronal stimulation and release of circulating catecholamines (see Chapter 9), involves the temporary suspension of gastric function.

Intestinal and pancreatic secretions

The epithelium of the mammalian small intestine secretes a mixture of two fluids called *intestinal juice,* or **succus entericus. Brunner's glands,** located in the first part of the duodenum between the pyloric sphincter and the pancreatic duct, secrete a viscous, enzyme-free, alkaline mucoid fluid. This secretion enables the duodenum to withstand the acidic chyme coming from the stomach until it can be neutralized by the alkaline pancreatic and biliary secretions coming from the pancreatic duct. A thinner, enzyme-rich alkaline fluid arises in the crypts of Lieberkühn (see Figure 15-19) and mixes with duodenal secretions. The secretion of intestinal juice is regulated by several hormones, including secretin, gastric inhibitory peptide (GIP), and gastrin, and additionally is under neuronal control. Distension of the wall of the small intestine elicits a local secretory reflex. Vagal innervation also stimulates secretion.

The large intestine secretes no enzymes. However, it does secrete a thin alkaline fluid containing HCO_3^- and K^+ ions, plus some mucus that binds the fecal matter together.

The pancreas, in addition to its endocrine secretion of insulin from the islets of Langerhans (see Chapter 9), contains exocrine tissue that produces several digestive enzymes that enter the small intestine through the pancreatic duct. The pancreatic enzymes, including α-amylase, trypsin, chymotrypsin, elastase, carboxypeptidases, aminopeptidases, lipases, and nucleases, are delivered in an alkaline, bicarbonate-rich fluid that helps neutralize the acidic chyme formed in the stomach. This buffering is essential, since the pancreatic enzymes require a neutral or slightly alkaline pH for optimum activity.

Exocrine secretion by the pancreas is controlled by the peptide hormones produced in the upper small intestine. Acidic chyme reaching the small intestine from the stomach stimulates the release of secretin and VIP, both produced by endocrine cells in the upper small intestine (see Table 15-2). These peptides are transported in the blood, reaching the duct cells of the pancreas and stimulating them to produce its thin bicarbonate fluid. Gastrin secreted from the stomach lining also elicits a small flow of pancreatic juice in anticipation of the food that will enter the duodenum. However, peptide hormones only weakly stimulate the secretion of pancreatic enzymes.

Secretion of pancreatic enzymes is elicited by another upper intestinal hormone, the peptide cholecystokinin (CCK; see Table 15-2), which is secreted from epithelial endocrine cells in response to fatty acids and amino acids in the intestinal chyme. CCK stimulates pancreatic secretion of enzymes as well as contraction of the smooth muscle wall of the gallbladder, forcing bile into the duodenum (see Figure 15-36).

The neuropeptides somatostatin and enkephalin have also been identified in endocrine cells of the upper intestinal mucosa in vertebrate guts. Both hormones have a variety of actions on gastrointestinal function. Somatostatin, which normally acts through paracrine effects, inhibits gastric acid secretion, pancreatic secretion, and intestinal motility, as well as blood flow. The enkephalins inhibit gastric acid secretion, stimulate pancreatic enzyme secretion, and inhibit intestinal motility.

Reinforcing the notion that the alimentary canal is the largest endocrine and exocrine gland of the body, endocrinologists have recently discovered in rats and pigeons that the esophagus, stomach, duodenum, and colon all synthesize melatonin. Removal of the pineal gland—long considered the source of melatonin—has no effect on alimentary melatonin levels, proving that this substance is synthesized within the alimentary canal. Melatonin has several effects in the alimentary canal, including inhibition of epithelial growth in the jejunum, reduced Na^+ transport in the colon, and inhibition of serotonin-induced smooth-muscle contraction. Because melatonin is often implicated in the seasonal changes in secretory activity of the pineal gland (see Chapter 8), it is tempting to speculate that melatonin could be related in some way to the seasonal changes in gut structure that are observed in a variety of vertebrates. This possibility is ripe for further experimentation.

What effect would a diminished flow of secretions from the pancreatic duct have on digestion? Why does blockage of the pancreatic duct of mammals sometimes lead to rapid death?

ABSORPTION

The breakdown products of digestion are transported from the gut to the animal's tissues and cells. In a unicellular organism, the products of digestion leave the

food vacuole to enter the surrounding cytoplasm. In a multicellular animal, these products must be transported across the absorptive epithelium into the circulation, then move from the blood into the tissues.

Digestion products are absorbed mainly through the microvilli of the apical membrane of the absorptive cells (see Figure 15-23). The digestive and absorptive mechanisms of the microvilli include the glycocalyx, digestive enzymes intimately associated with the membrane, and specific intramembrane transporter proteins. In the basolateral membrane, other mechanisms transfer these substances out of the absorptive cells into the interstitial fluid and eventually into the general circulation.

Nutrient Uptake in the Intestine

The carbohydrate-rich filaments composing the glycocalyx arise from, and are continuous with, the plasma membrane of the microvillus itself. The filaments of the glycocalyx appear to be the carbohydrate side chains of glycoproteins embedded in the membrane. Further, the brush border (microvilli plus glycocalyx) has been found to contain digestive enzymes for the final breakdown of various small food molecules. These enzymes are also membrane-associated glycoproteins having carbohydrate side chains protruding into the lumen. The enzymes associated with the brush border include disaccharidases, aminopeptidases, and phosphatases. Thus, some of the terminal stages of digestion are carried out at the brush border, close to the sites of uptake from the lumen into the absorptive cells.

Absorption involves several transfer processes, including simple (passive) diffusion, facilitated diffusion, cotransport, active transport, and endocytosis (see Chapter 4). The type of transfer mechanism used depends on the type of molecule being absorbed.

Simple diffusion

Simple (passive) diffusion can take place across the lipid bilayer (provided the diffusing substance has a high lipid solubility) or through water-filled pores. Substances that diffuse across the apical membrane of the brush border include fatty acids, monoglycerides, cholesterol, and other fat-soluble substances. Substances that pass through water-filled pores include water, certain sugars, alcohols, and other small, water-soluble molecules. For nonelectrolytes, the net diffusion rate is proportional to their chemical concentration gradient; for electrolytes, it is proportional to their electrochemical gradient. In passive diffusion, net transfer is always "downhill," using the energy of the concentration gradient. Channels expressly for water movement—aquaporins—are under hormonal regulation, even though passage through them is passive (see Chapter 4).

Facilitated diffusion

The absorption of monosaccharides and amino acids presents two problems. First, these molecules are hydrophilic because of their —OH groups, because of charges they may bear, or both. Second, they are too large to be transported by simple diffusion through water-filled pores. These problems are overcome by facilitated diffusion across the absorptive cell membrane via transporter proteins (Figure 15-37). Hydrophilic, lipid-insoluble sugars such as fructose are carried down their concentration gradient by facilitated diffusion via specific transporter proteins located in the absorptive cell membrane, in a process powered by coupling sugar transport to the electrochemical gradient for Na^+ across the plasma membrane. The Na^+/glucose cotransporter, SGLT1, is the integral membrane protein that couples the transport of Na^+ with glucose across the brush border. According to this model, GLUT5 is the brush border fructose transporter, and GLUT2 is the basolateral membrane transporter for fructose as well as glucose and galactose (not shown in Figure 15-37).

Some monosaccharides are taken up into the absorptive cells by a related mechanism, hydrolase transport, in which a glycosidase attached to the apical membrane hydrolyzes the parent disaccharide (e.g., sucrose, maltose) and also acts as, or is coupled to, the

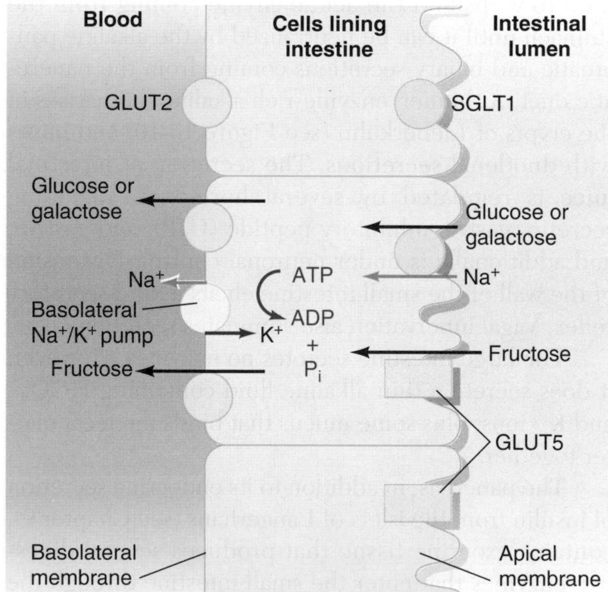

Figure 15-37 Cotransport proteins are central to this model of sugar transport across the absorptive cells lining the small intestine. SGLT1 is the Na^+/glucose cotransporter, an integral membrane protein that couples the transport of Na^+ to that of glucose or galactose. Fructose passes through the brush border by way of the GLUT5 transport protein. Sugar uptake is powered by the electrochemical gradient for Na^+ across the apical membrane. Sugars are then transported through the basolateral membrane down their concentration gradients via the GLUT2 transport protein. A basolateral Na^+/K^+ pump pumps out Na^+, creating the gradient that powers the whole process. [Adapted from Wright, 1993.]

mechanism that transfers the resulting monosaccharide into the absorptive cell.

Active transport

In the mammalian intestine, sodium-driven transport of amino acids into absorptive cells takes place via four separate and noncompeting cotransport systems, each transporting just one of four groups of amino acids:

1. The three dibasic amino acids having two basic amino groups each (lysine, arginine, and histidine)
2. The two diacidic amino acids having two carboxyl groups each (glutamate and aspartate)
3. A special amino acid class consisting of glycine, proline, and hydroxyproline
4. The remaining neutral amino acids

Dipeptides and tripeptides are transported into absorptive cells by yet another separate transport system. Once inside the cell, dipeptides and tripeptides are cleaved into their constituent amino acids by intracellular peptidases. This system has the advantage of preventing a buildup of oligopeptides within the cell, so that there is always a large inwardly directed concentration gradient promoting their inward transport.

Once through the epithelium, sugars and amino acid molecules enter the blood by diffusion into the capillaries within the villi. Upon reaching other tissues of the body, sugars and amino acids are transferred into cells by the same types of active transport and facilitated diffusion mechanisms.

Special handling of lipids

The digestion products of fats—monoglycerides, fatty acids, and glycerol—diffuse through the brush border membrane and are reconstructed within the absorptive cell into triglycerides. They are then collected, together with phospholipids and cholesterol, into tiny droplets termed **chylomicrons,** which are about 150 μm in diameter (Figure 15-38). Chylomicrons are coated with a layer of protein, and are loosely contained in vesicles formed by the Golgi complex. They are subsequently expelled by exocytosis through the fusion of these vesicles with the basolateral membrane of the absorptive cell. The chylomicrons then enter the central lacteal, from which they are conveyed in lymph on to the blood for broad distribution throughout the body.

Endocytosis

Transport of sugars and amino acids across the basolateral membrane of the absorptive cells occurs by facilitated diffusion, as noted earlier. Some oligopeptides, however, are taken up by absorptive cells through endocytosis. In newborn mammals, this process is responsible for the uptake in the intestine of immunoglobulin molecules derived from the mother's milk that escape digestion.

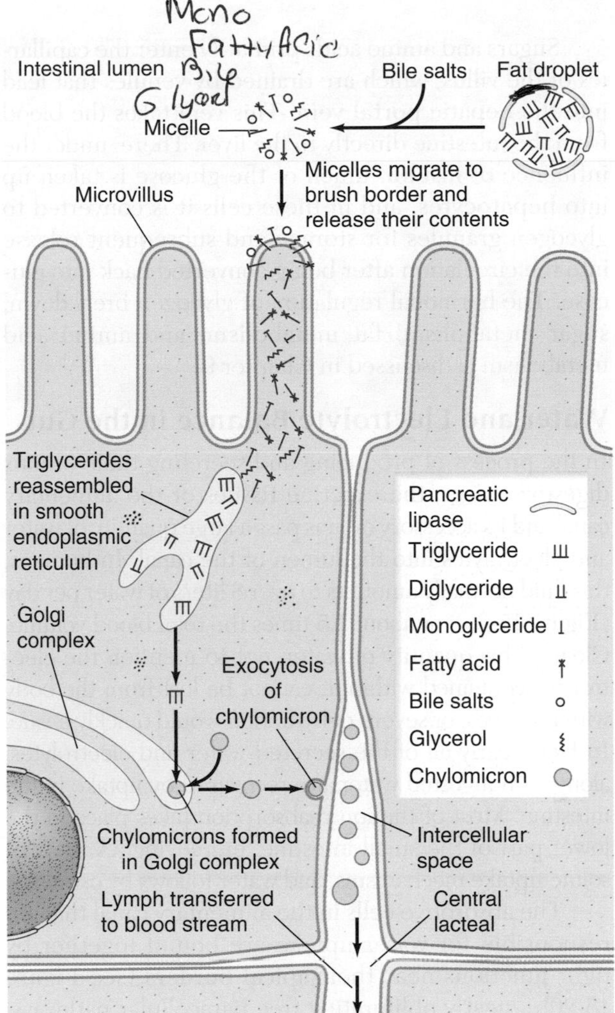

Figure 15-38 Lipids are transported from the intestinal lumen through absorptive cells and into the interstitial space. Hydrolytic products of triglyceride digestion—monoglycerides, fatty acids, and glycerol—form micelles with the bile salts in solution. The micelles transport these materials to the brush border, where their contents enter the absorptive cell by passive diffusion across the lipid bilayer. Within the cell they are resynthesized into triglycerides in the smooth endoplasmic reticulum and, together with a smaller amount of phospholipids and cholesterol, are stored in the Golgi complex as chylomicrons—droplets about 150 μm in diameter. The chylomicrons then leave the basolateral portions of the cell by exocytosis and enter the central lacteal, which conveys them to the circulating blood.

Transport of Nutrients in the Blood

Once inside the absorptive cell, digestion products pass through the basolateral membrane into the interior of the villus. From the interstitial fluid of the villus, they enter the blood or the lymphatic circulation (see Figure 15-38). In humans, about 80% of chylomicrons enter the bloodstream via the lymphatic system, while the rest enter the blood directly. The pathway into the lymphatic system begins with the blind central lacteal of the villus (see Figure 15-20a). Lymph is returned to the blood circulation via the thoracic duct.

Sugars and amino acids primarily enter the capillaries of the villus, which are drained by venules that lead into the **hepatic portal vein.** This vein takes the blood from the intestine directly to the liver. There, under the influence of insulin, much of the glucose is taken up into **hepatocytes,** and in these cells it is converted to glycogen granules for storage and subsequent release into the circulation after being converted back into glucose. The hormonal regulation of glycogen breakdown, sugar metabolism, fat metabolism, and amino acid metabolism is discussed in Chapter 9.

Water and Electrolyte Balance in the Gut

In the process of producing and secreting their various digestive juices, the exocrine tissues of the alimentary canal and its accessory organs pass a large quantity of water and electrolytes into the lumen of the canal. In humans, this fluid normally amounts to over 8 liters of water per day (Figure 15-39), or about 1.5 times the total blood volume. Clearly, this quantity of water, not to mention the electrolytes contained within it, cannot be lost from the body with the feces, or severe dehydration would quickly ensue. In fact, nearly all of the secreted water and electrolytes, along with ingested water, are recovered by uptake in the intestine. Most of the ion reabsorption takes place in the lower part of the small intestine, employing a variety of solute uptake mechanisms, and water follows by osmosis.

The absorptive cells in the alimentary canal that are responsible for water uptake are bound together by tight junctions near their apical borders (see Figure 15-20b), nearly obliterating free paracellular pathways. Tracer studies using deuterium oxide (D_2O) indicate that water leaves the intestinal lumen through aqueous channels that occupy only 0.1% of the absorptive epithelial cell surface. Flux studies employing isotopically labeled solutes indicate that these channels exclude water-soluble molecules with molecular weights exceeding 200 $g \cdot mol^{-1}$. Smaller solute molecules are carried passively along with the water as it flows down its osmotic gradient through the hydrated channels. The osmotic gradient driving this water movement is set up primarily by the active transport of substances from the lumen into the villus, in particular the transport of salts, sugars, and amino acids. The elevated osmotic pressure within the villus that results from this active transport, especially in the intercellular clefts of the epithelium (see Figures 4-38 and 4-39), draws water osmotically from the absorptive cells. This water is then replenished by water entering osmotically across the apical membrane from the lumen.

Most of the absorption of water and electrolytes occurs at or near the tips of the villi. The greater proportion of water absorption at the villus tip results from an elevated concentration of Na^+ near the upper end of the villus lumen, which decreases with increasing distance from the villus tip. There are two reasons for this concentration gradient. First, most of the active absorption of Na^+ (in exchange for H^+) takes place across absorptive cells located at the tip of each villus. Chloride follows through exchange for HCO_3^-, and NaCl accumulation is therefore greatest at the blind upper end of the villus lumen. Second, the organization of the capillaries within the villus leads to a further concentration of NaCl at the upper end of the villus lumen through a countercurrent mechanism (see Spotlight 14-2). Arterial blood flowing toward the tip of the villus picks up Na^+ and Cl^- from NaCl-enriched blood leaving the villus in a descending venule. The "short-circuiting" of NaCl in this manner recirculates and concentrates it in the villus tip, promoting osmotic flow of water from the intestinal lumen into the villus. The absorption of Na^+ and Cl^- into the villus is enhanced by high concentrations of glucose and certain other hexose sugars in the intestinal lumen, which stimulate Na^+/sugar cotransport.

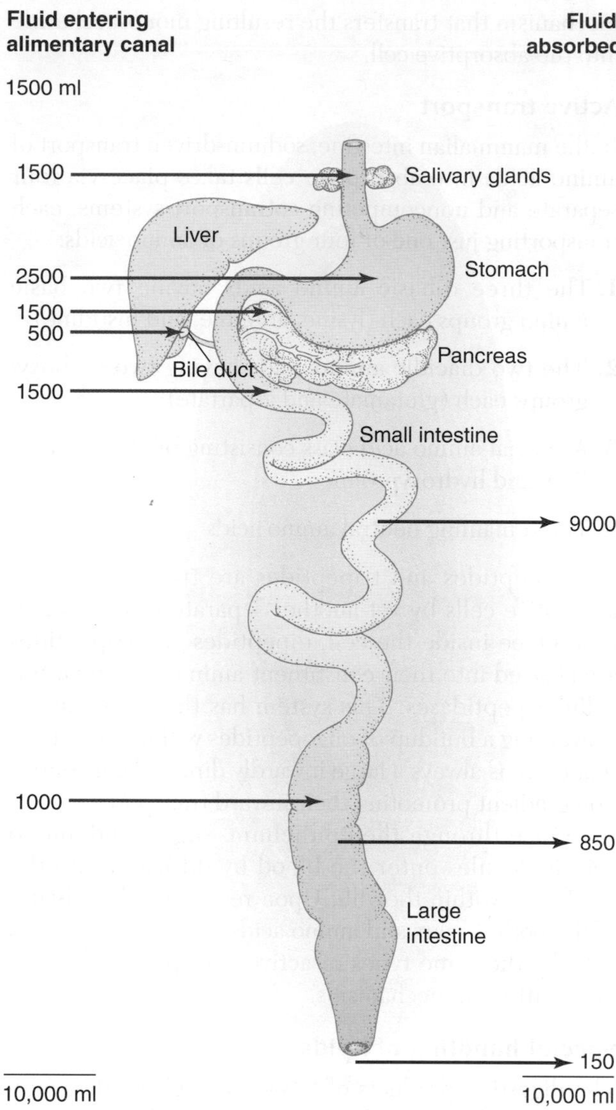

Figure 15-39 Fluid fluxes occur all along the length of the human alimentary canal. Flux volumes vary with the condition and body mass of the individual. Values in black are the amounts of fluid entering the alimentary canal, and those in red are the amounts reabsorbed from the lumen. [Adapted from Madge, 1975.]

Excessive uptake of water from the lumen across the intestinal wall results in abnormally dry lumen contents (and hence constipation). This situation is normally prevented by an inhibitory action on electrolyte and water uptake by some of the gastrointestinal hormones. Gastrin acts indirectly to inhibit water absorption from the small intestine, while secretin and CCK reduce the uptake of Na^+, K^+, and Cl^- in the upper jejunum. Cholic acids and fatty acids also inhibit the absorption of water and electrolytes.

Unlike water, Ca^{2+} requires a special active transport mechanism for absorption from the gut. The calcium ion is first bound to a calcium-binding protein found in the microvillus membrane and then transported as a complex into the absorptive cell by an energy-consuming process. From the absorptive cell, the Ca^{2+} then passes into the blood. The presence of calcium-binding protein is regulated by the hormone *calcitriol* (formerly known as 1,25-dihydroxy-vitamin D_3). The release of Ca^{2+} from the absorptive cell into the blood is accelerated by parathyroid hormone.

Vitamin B_{12}, which has a molecular weight of 1357, is the largest water-soluble essential nutrient taken up intact across the intestinal lumen. Its transport occurs in the region of the distal ileum. This highly charged cobalt-containing compound is associated with food protein, to which it is bound as a coenzyme. In the process of absorption, B_{12} is transferred from the dietary protein to a mucoprotein known as **intrinsic factor** (or hemopoietic factor), which is produced by the H^+-secreting parietal cells of the stomach. B_{12} is essential for the synthesis and maturation of red blood cells, and pernicious anemia occurs when B_{12} absorption is prevented by interference with its binding to intrinsic factor. Some tapeworms "steal" B_{12} in the intestine of the host by producing a compound that removes it from intrinsic factor, making it unavailable to the host but available to the tapeworm.

NUTRITIONAL REQUIREMENTS

All animals must acquire an appropriate variety and amount of nutritive substances. **Nutrients** are substances that serve as sources of metabolic energy and as raw materials for growth, repair of tissues, and production of gametes. Nutrients also include essential trace elements such as iodine, zinc, and other metals that may be required in extremely small quantities. There is enormous variation among species in nutritional needs. Within a species, nutritional needs vary according to phenotypic differences in body size, composition, and activity, and also with age, sex, and reproductive state. A gravid (egg-bearing) or pregnant female may require more nutrients than a male, while a male that is producing sperm may have greater nutritive needs than one that is not, particularly in those species that produce large sperm packets. Regardless of reproductive state, a small animal requires more food per unit body weight than does a larger animal because its metabolic rate per unit body weight is higher (see Chapter 16). Similarly, an animal with a high body temperature requires more food, to satisfy greater energy needs, than does an animal with a lower body temperature (see Chapter 17).

Nutrient Molecules

A wide variety of molecules serve as nutrients for animals, including water, proteins and amino acids, carbohydrates, fats and lipids, nucleic acids, inorganic salts, and vitamins.

Animals differ in their abilities to synthesize the substances fundamental to their maintenance and growth. Thus, for a given animal species, certain metabolic cofactors (such as vitamin B_{12}, mentioned earlier) or building blocks (such as amino acids) essential for important biochemical reactions or for the production of tissue molecules must be obtained from food sources because they cannot be produced by the animal itself. Such substances are known as **essential nutrients.**

Water

Of all the constituents of animal tissues, none is more pervasively important to living tissue than water. This unique and marvelous substance can constitute 95% or more of the weight of some animal tissues. It is replenished in most animals by drinking and by ingestion as a constituent of food. Some marine and desert animals depend almost entirely on metabolic water—water produced during the oxidation of fats and carbohydrates—to replace water lost by evaporation, defecation, and urination (see Chapter 14).

Proteins and amino acids

Proteins are used as structural components of tissues, as channels, transporters, and regulatory molecules, and as enzymes. They can also be utilized as energy sources if first broken down to amino acids (see Chapter 3). The proteins of animal tissues are composed of about 20 different amino acids. The ability to synthesize these amino acids differs among species. **Essential amino acids** are those that cannot be synthesized by an animal, but are required for synthesis of essential proteins. The recognition of essential amino acids has been of enormous economic significance in the poultry industry. The growth rate of chickens at one time was limited by too small a proportion of a few essential amino acids in the grain diet they were provided. Supplementing the diet with these amino acids allowed full utilization of the other amino acids present in the feed, greatly increasing the rate of protein synthesis and hence the rate of poultry growth and egg laying. Microbiologists sometimes artificially induce the limiting condition of a specific amino acid requirement by genetically engineering microbes that lack the ability to synthesize some amino acid (e.g., lysine) not normally found in their environment. Thus, the microbes will grow only in an environment enriched with that amino acid, a characteristic used to prevent the spread of these modified bacteria into normal populations in our environment.

Carbohydrates

Carbohydrates are used primarily as sources of chemical energy, to be either metabolized immediately as glucose or stored as glycogen. However, they may also be converted to metabolic intermediates for the formation of other biomolecules, including amino acids and fats. Conversely, proteins can be converted by most animals into carbohydrates. The major food sources of carbohydrates are the sugars, starches, and cellulose found in plants and the glycogen stored in animal tissues.

Lipids

Lipid (fat) molecules are especially suitable as concentrated energy reserves. Each gram of fat provides over 20% more caloric energy than a gram of carbohydrate, and the nonpolar nature of fat molecules allows them to be stored in large quantities without binding water, whereas stored carbohydrates are richly hydrated, binding about twice their dry weight in water molecules. Far more energy can therefore be stored in the form of fat per unit volume of tissue. Fat is commonly stored by animals to be used during periods of **caloric deficit**, such as hibernation, when energy expenditure exceeds energy intake. Lipids are also important in certain tissue components, such as the plasma membranes and other membrane-based organelles of the cell (see Chapter 4).

Nucleic acids

Although nucleic acids are essential for the genetic machinery of the cell, all animal cells appear to be capable of synthesizing them from simple precursors. Thus, the intake of intact nucleic acids is not necessary from a nutritional perspective.

Inorganic salts

Some chloride, sulfate, phosphate, and carbonate salts of the metals calcium, potassium, sodium, and magne-

Table 15-3 Some mammalian vitamins

Vitamin	Major dietary sources; solubility*	Uptake; storage	Function in mammals†	Deficiency symptoms
Ascorbic acid (C)	Citrus fruits; WS	Absorbed from gut; little storage	Vital element for collagen; antioxidant	Scurvy (failure to form connective tissue)
Biotin	Egg yolk, tomatoes, liver, synthesis by intestinal flora; WS	Absorbed from gut	Protein and fatty acid synthesis; CO_2 fixation; transamination	Scaly dermatitis, muscle pains, weakness
Cyanocobalamin (B_{12})	Liver, kidney, brain, fish, eggs, synthesis by intestinal flora; WS	Absorbed from gut; stored in liver, kidney, brain	Nucleoprotein synthesis; formation of erythrocytes	Pernicious anemia, malformed erythrocytes
Folic acid (folacin, pteroylglutamic acid)	Meats; WS	Absorbed from gut; utilized as acquired	Nucleoprotein synthesis; formation of erythrocytes	Failure of erythrocytes to mature, anemia
Niacin	Lean meat, liver, whole grains; WS	Absorbed from gut; distributed to all tissues	Coenzyme in hydrogen transport (NAD, NADP)	Pellagra, skin lesions, digestive disturbances, dementia
Pantothenic acid	Many foods; WS	Absorbed from gut; stored in all tissues	Constituent of coenzyme A (CoA)	Neuromotor, cardiovascular disorders
Pyridoxine (B_6)	Whole grains, traces in many foods; WS	Absorbed from gut; half appears in urine	Coenzyme for amino and fatty acid metabolism	Dermatitis, nervous disorders
Riboflavin (B_2)	Milk, eggs, lean meat, liver, whole grains; WS	Absorbed from gut; stored in kidney, liver, heart	Flavoproteins in oxidative phosphorylation	Photophobia, fissuring of the skin
Thiamine (B_1)	Brain, liver, kidney, heart, whole grains, nuts, beans, potatoes	Absorbed from gut; stored in liver, brain, kidney	Formation of cocarboxylase enzyme involved in decarboxylation (citric acid cycle)	Stoppage of CH_2O metabolism at pyruvate, beriberi, neuritis, heart failure

*FS = fat-soluble; WS = water-soluble.
†Most vitamins have numerous functions; the functions listed are a mere sampling.

sium are important constituents of intracellular and extracellular fluids. Calcium phosphate occurs as hydroxyapatite $[Ca_{10}(PO_4)_6(OH)_2]$, a crystalline material that lends hardness and rigidity to the bones of vertebrates and the shells of mollusks. Iron, copper, and other metals are required for redox reactions (as cofactors) and for oxygen transport and binding (in hemoglobin and myoglobin). Many enzymes require specific metal atoms to complete their catalytic functions. Animal tissues need moderate quantities of some elements (Ca, P, K, Na, Mg, S, and Cl) and trace amounts of others (Mn, Fe, I, Co, Cr, Cu, Zn, and Se).

Vitamins

Vitamins are a diverse and chemically unrelated group of organic substances that generally are required in small quantities, primarily to act as cofactors for enzymes. Some vitamins important in mammalian nutrition are listed in Table 15-3, along with some of their diverse functions. Detailed vitamin requirements are known primarily for humans and for domesticated animals grown for their meat, eggs, or other products. Very little is known about the vitamins involved in the metabolism of lower vertebrates and especially invertebrates.

The ability to synthesize different vitamins differs among species, and those essential vitamins that an animal cannot produce itself must be obtained from other sources—primarily from plants, but also from dietary animal flesh or from intestinal microbes. Ascorbic acid (vitamin C), for example, is synthesized by many animals, but not by humans, who acquire it mainly from citrus fruits. Scurvy, a condition of ascorbic acid deficiency in humans, was common on board ships before the British admiralty instituted the use of citrus fruits—especially limes—to supplement the diet of the crews. Their use of limes led to the use of the term "limey" to describe the English. Humans also are unable to produce vitamins K and B_{12}, which are produced by intestinal bacteria and then absorbed for distribution to the tissues. Fat-soluble vitamins such as A, D_3, E, and K are stored in body fat deposits. Water-soluble vitamins such as ascorbic acid are not stored in the body, however, and so must be ingested or produced continually to maintain adequate levels.

SUMMARY

■ Animals obtain food in many different ways, including absorption through the body surface, endocytosis, filter feeding, trapping with mucus, sucking, biting, and chewing.

■ Digestion consists of the enzymatic hydrolysis of large molecules into their monomeric building blocks. In multicellular animals, this process takes place extracellularly in an alimentary canal.

Motility of the alimentary canal

■ The motility of the vertebrate alimentary canal depends on the coordinated activity of longitudinal and circular layers of smooth muscle. Peristalsis occurs when a ring of circular contraction proceeds along the gut, preceded by a region in which the circular muscles are relaxed. Motility is stimulated by parasympathetic innervation and inhibited by sympathetic innervation.

■ The motility of smooth muscle, as well as the secretion of digestive juices, is under precise neuronal and endocrine control.

Gastrointestinal secretions

■ Digestion in vertebrates begins in a region of low pH, the stomach, and proceeds to a region of higher pH, the small intestine.

■ Proteolytic enzymes are released as proenzymes, or zymogens, which are inactive until a portion of the peptide chain is removed by digestion. This avoids the problem of proteolytic destruction of the enzyme-producing tissues themselves.

■ Other exocrine cells secrete digestive enzymes (e.g., carbohydrases and lipases), mucin, or electrolytes such as HCl or $NaHCO_3$.

■ All gastrointestinal hormones are peptides; many also function as neuropeptides in the central nervous system, where they act as transmitters or short-range neurohormones.

■ Both direct activation by food in the gut and neuronal activation stimulate the endocrine cells of the gastrointestinal mucosa that secrete peptide hormones.

■ Peptide hormones either stimulate or inhibit the activity of the various kinds of exocrine cells in the gut that produce digestive enzymes and juices.

Absorption

■ Products of digestion are taken up by the absorptive cells of the intestinal mucosa and transferred to the lymphatic and circulatory systems. The area of the absorptive surface is hugely amplified by the presence of microvilli. The absorptive cells cover larger fingerlike villi that reside on convoluted folds and ridges in the wall of the intestine, further increasing its surface area.

■ The terminal processes of digestion take place in the brush border on the apical surface of the absorptive cells.

■ Sugars and peptides are hydrolyzed into monomeric residues before transport through the intestinal epithelium takes place.

■ Some sugars are transported by facilitated diffusion, which requires a membrane transport protein, but no metabolic energy. Most sugars and amino acids require energy expenditure for adequate rates of absorption.

- In many cases, sugars and amino acids are cotransported with Na^+, utilizing a common membrane protein and the potential energy of the electrochemical gradient driving Na^+ from the lumen into the cytoplasm of the absorptive cell.
- Endocytosis plays a role in the uptake of small polypeptides and, rarely, of larger proteins, such as immunoglobulin in newborn animals.
- Fatty substances enter the absorptive cell by simple diffusion across the plasma membrane.
- Water and electrolytes enter the alimentary canal as constituents of digestive juices, but these quantities are nearly all recovered. Active transport of solutes from the intestinal lumen results in the passive osmotic movement of water from the lumen into the cells and eventually back into the bloodstream. Without such recycling of electrolytes and water, the digestive system would impose a lethal osmotic load on the animal.

REVIEW QUESTIONS

1. Define the terms *digestion, absorption,* and *nutrition.*
2. Give three examples of proteins produced specifically for the purpose of obtaining or utilizing food.
3. What is an essential amino acid?
4. Why would it be inadvisable for the digestive system to fragment amino acids, hexose sugars, and fatty acids into still smaller molecular fragments, even though doing so might facilitate absorption?
5. What prevents proteolytic enzymes from digesting the exocrine cells in which they are produced and stored before release?
6. State two adaptive advantages of the digastric stomach. Since it has three or four chambers, why is it referred to as digastric?
7. How does bile aid the digestive process, even though it contains few or no enzymes?
8. Outline the autonomic innervation of the intestinal wall, explaining the organization and functions of its sympathetic and parasympathetic innervation.
9. How is HCl produced and secreted into the stomach by parietal cells?
10. Describe the roles of gastrin, secretin, and cholecystokinin in mammalian digestion.
11. Why are some gastrointestinal hormones also classified as neuropeptides? Give examples.
12. What is meant by the cephalic, the gastric, and the intestinal phases of gastric secretion, and how are they regulated?
13. How are amino acids and some sugars transported against a concentration gradient from the intestinal lumen into epithelial cells?
14. Why is the countercurrent principle important in the removal of water from the intestinal lumen?
15. How are vitamin B_{12} deficiency and pernicious anemia related to intestinal function?

SUGGESTED READINGS

Chivers, D. J., and P. Langer. 1994. *The Digestive System in Mammals: Food, Form and Function.* New York: Cambridge University Press. (An integrative view.)

Diamond, J. M. 1991. Evolutionary design of intestinal nutrient absorption: Enough but not too much. *News Physiol. Sci.* 6:92–96. (Digestion as an example of how evolutionary design matches a system's needs and capabilities.)

Hofmann, A. F. 1999. Bile acids: The good, the bad and the ugly. *News Physiol. Sci.* 14:24–29. (Discusses the complex functions of cholic acids as surfactants as well as their role in stimulating biliary secretion and influencing cholesterol metabolism.)

Horowitz, B., S. M. Ward, and K. M. Sanders. 1999. Cellular and molecular basis for electrical rhythmicity in gastrointestinal muscles. *Annu. Rev. Physiol.* 61:19–43. (Describes the electrical pacemaker of the gastrointestinal tract, and shows how ongoing molecular studies are revealing the ion channels that determine its pacemaker properties.)

Karasov, W. H., and I. D. Hume. 1997. Vertebrate gastrointestinal system. In *Handbook of Physiology,* Section 13, *Comparative Physiology,* Vol. 1, 409–480. New York: Oxford University Press. (The complex structure and function of the vertebrate digestive system are clearly dissected and reviewed in this extensively referenced chapter.)

Mawe, G. M. 1998. Nerves and hormones interact to control gallbladder function. *News Physiol. Sci.* 13: 84–90. (Discusses how gallbladder function is regulated, presenting a wonderful example of the nearly seamless interactions of the nervous and endocrine systems in regulating a physiological process.)

Stevens, C. E., and I. D. Hume. 1996. *Comparative Physiology of the Vertebrate Digestive System.* 2d ed. Cambridge: Cambridge University Press. (This updated compendium describes the basic structural and functional characteristics as well as taxonomic variation in the vertebrates.)

Energy Expenditure: Body Size, Locomotion, and Reproduction

Animals derive chemical energy from food, which they use to perform work, maintain their structural integrity, and, ultimately, reproduce. In Chapters 3 and 15, we saw how animals degrade large organic compounds to transfer some of their chemical energy to special "high-energy" molecules, such as ATP. These molecules are subsequently used to drive endergonic (energy-requiring) reactions. Thus, animals eventually use the chemical energy of food to produce electrical, ionic, and osmotic gradients, synthesize metabolites, and carry out work via muscle contraction. Animals that are more effective at capturing and using the energy resources available in their environment are better able to compete against members of their own and other species, and are therefore more fit in an evolutionary sense.

This chapter explores the various factors that affect energy expenditure by animals. It considers in particular the relations between energy costs and body size, mode of locomotion, and reproduction. In other words, we consider here the effects of the intrinsic characteristics of animals on their energy expenditures—that is, their energy costs based on "who they are." The next chapter examines how environment and temperature affect the metabolism of animals—that is, their energy costs based on "where they live."

THE CONCEPT OF ENERGY METABOLISM

Metabolism, in its broadest sense, is the sum total of all the chemical reactions occurring in an organism (see Chapter 3). Because the rate of a chemical reaction increases with temperature, the metabolic activity of an animal is closely linked to its body temperature (a topic we explore in depth in Chapter 17). Animals, like machines, are less than 100% efficient in their energy conversions; a large fraction of metabolic energy is lost in the form of heat produced as a by-product of the release of free energy during exergonic reactions, such as those occurring in muscle contraction. This metabolic heat is comparable to the waste heat produced by a gasoline engine as it converts chemical energy into mechanical work. Yet, in many animals, this heat is not "wasted" because it raises the temperature of the animal's tissues to levels that significantly enhance the rates of biochemical reactions.

Body mass strongly affects an animal's energy metabolism. Smaller animals tend to have higher metabolic rates per unit body mass than do larger animals. Thus, body mass affects many different physiological processes and influences the performance of most physiological systems.

Muscular activity also affects the rate of energy metabolism. A hummingbird hovering over a nectar-laden flower consumes far more metabolic fuel than the same bird roosting at night. Animal behaviors and lifestyles are associated with different degrees of energetic efficiency. Flying is metabolically more expensive than running, and running requires more energy than swimming. All require more energy than patiently stalking.

Finally, the process of reproduction may require a major commitment of acquired and stored energy. In some animals, the release of gametes results in relatively little energy loss; in others, a huge percentage of the annual energy intake is used in producing eggs or sperm and in caring for young.

DESCRIBING AND MEASURING METABOLIC RATE

Metabolic pathways fall into two major categories:

- **Anabolic pathways** assemble simple substances into more complex molecules required by the organism, a process that requires energy. Anabolism is associated with repair, regeneration, and growth. Although it is difficult to measure anabolic metabolism quantitatively, one index of anabolic activity is a positive nitrogen balance. That is, anabolic activity leads to a net incorporation of nitrogen-containing molecules through protein synthesis, rather than a net loss due to protein breakdown.

- **Catabolic pathways** break down complex, energy- or material-rich molecules into simpler molecules, releasing chemical energy. Some of the energy derived from catabolism is stored as high-energy phosphate compounds, such as ATP, which are subsequently used to power cellular activities. Metabolic intermediates, such as glucose or lactate, can serve as energy-storage compounds until they are mobilized as substrates for additional exergonic reactions.

All energy released during metabolic processes is either used to perform work, stored as chemical energy, or eventually lost as heat. This simple fact makes it possible to use heat production as an index of energy metabolism, provided the organism is in a thermal steady state with its environment. The conversion of chemical energy into heat is expressed as the **metabolic rate**—measured as heat energy released per unit time. Although heat production is a useful measure of metabolic rate, other common and traditional measures are also used, such as oxygen consumption and, less often, carbon dioxide production. Localized metabolism within tissues can also be measured by a variety of noninvasive techniques (see Chapter 2). Currently, nuclear magnetic resonance (NMR) imaging is being used to characterize directly the metabolism of high-energy phosphate groups taking place within muscle and other tissues.

Measurements of metabolic rates are useful not only to physiologists, but also to ecologists, animal behaviorists, evolutionary biologists, and many other scientists because they can be used to calculate an animal's energy requirements. To survive over the long term, an animal must take in as much energy, in the form of energy-yielding food molecules, as it releases. Measurements of metabolic rate at different ambient (environmental) temperatures may provide information about the heat-conserving or heat-dissipating mechanisms of an animal. Measurements of metabolic rate during different types of exercise help us to understand the energy costs of such activities and allow us to calculate the metabolic costs of being large or small, flying, swimming, running, or surviving.

The metabolic rate of an animal varies with the type and intensity of the processes taking place. These processes include tissue growth and repair; chemical, osmotic, electrical, and mechanical internal work; and external work for locomotion and communication (Figure 16-1).

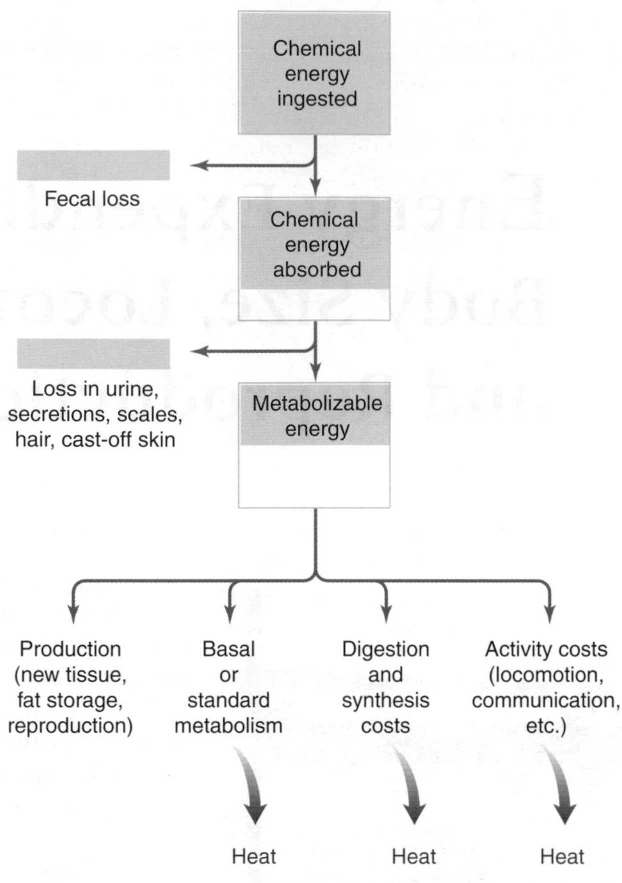

Figure 16-1 Animals must take in and utilize chemical energy from food. Part of the potential chemical energy ingested is unabsorbed and is degraded by intestinal flora or passed out in fecal matter. Of the chemical energy absorbed, some is lost in urine and other secretions and cast-off structures, and some appears directly as heat resulting from exergonic metabolic reactions (chemical, electrical, or mechanical work). Any energy left over is conserved in the anabolic buildup of tissues.

In addition to body and environmental temperatures, body mass, reproductive status, and activity, other factors that affect metabolic rate are time of day, season, age, sex, shape, stress, and the type of food being metabolized. Consequently, the metabolic rates of different animals can be meaningfully compared only under carefully chosen and closely controlled conditions.

Basal and Standard Metabolic Rates

A bird can be sleeping quietly with its head tucked under its wing one moment and in the next moment be launched into full flight—two normal and radically different states. Animal physiologists who measure metabolic rates recognize several normal metabolic levels, or states, and control for these states or weave them into their experimental design.

The **basal metabolic rate (BMR)** is the stable rate of energy metabolism measured in mammals and birds under conditions of minimum environmental and physiological stress (namely, at rest with no temperature stress) and after fasting has temporarily halted digestive and absorptive processes. Ambient temperature affects body temperature in almost all animals other than birds and mammals. Because the minimum metabolic rate varies with body temperature, it is necessary to measure the equivalent of the basal metabolic rate in these animals at a controlled, specified body temperature at which the animal is not expending additional metabolic energy just to warm or cool itself. For that reason, the **standard metabolic rate (SMR)** is defined as an animal's resting and fasting metabolism at a given body temperature. Interestingly, the SMR of some ectotherms depends on their *previous* temperature history, owing to metabolic compensation or thermal acclimation, which is described in Chapter 17.

Basal and standard metabolic rates are useful measurements for comparing baseline metabolic rates both within and between species. However, they give little information about the metabolic costs of normal activities carried out by the animals because the conditions under which they are measured differ greatly from natural conditions—typically, the animal is in an unnaturally controlled and quiet state for these measurements. The measure that best describes the metabolic rate of an animal in its natural state is its **field metabolic rate (FMR)**, which is the average rate of energy utilization as the animal goes about its normal activities, which may range from complete inactivity during rest to maximum exertion when chasing prey (or escaping a predator). The field metabolic rate is the hardest form of metabolic rate to measure accurately.

The Cellular and Biochemical Components of Basal Metabolic Rate

The BMR of an animal reflects the sum of all of the cellular and subcellular processes occurring in its tissues. Examination of both interspecific and intraspecific variation in BMR suggests that maintaining membrane form and function is a major component of an animal's energy expenditure. In particular, the maintenance of transmembrane electrochemical gradients is a major consumer of energy. For example, the proton pumps of the mitochondrial membranes and the Na^+/K^+ pumps of the plasma membranes both consume large amounts of energy to maintain cellular homeostasis. Other cellular activities making major contributions to BMR include protein synthesis and, of course, the formation of ATP.

Cellular contributions to BMR vary throughout an animal's life cycle. During early development, for example, rates of protein synthesis are very high, and this is reflected in a typically high BMR. Na^+/K^+ ATPase activity also increases during the early stages of development as nerve, muscle, and osmoregulatory tissues proliferate. The liver is a highly active metabolic organ in many animals, especially mammals, and developmental increases in hepatocyte metabolism have major influences on BMR. In newborn rat pups, for example, the metabolic rate of the liver per unit tissue mass nearly doubles within a few hours and is soon nearly four times higher than that in an adult rat. This increase in metabolic rate can be blocked with ouabain, which inhibits Na^+/K^+ pumps.

Metabolic Scope

The range of metabolic rates of which an animal is capable is called its **aerobic metabolic scope.** It is defined as the ratio of the maximum sustainable metabolic rate to the BMR (or the SMR) determined under controlled, resting conditions. This dimensionless number (e.g., 5, 7, or 14) indicates the increase in an animal's maximal energy expenditure over and above the amount of energy that it expends under resting conditions. Maximal energy expenditure, usually measured as oxygen consumption, is typically evoked by stimulating maximal rates of locomotor activity. Metabolic rate increases by as much as 10 to 15 times with activity in many animals. Note that, because sustained activity is normally powered by aerobic metabolism, this type of measurement does not take into account the contribution of anaerobic processes to activity.

The concept of metabolic scope applies to all animals, regardless of their mode of locomotion. In fish swimming in flow tanks, for example, faster swimming can be stimulated by increasing the rate of water flow. Such experiments have shown that metabolic scope varies with body size. The ratio of active to standard metabolism in salmon, for example, increases from less than 5 in young specimens weighing 5 g to more than 16 in those weighing 2.5 kg. This general relation is compounded when comparisons are made between species with different modes of activity. Although smaller than a 5 g larval salmon, flying insects, especially those that sustain high body temperatures during flight, can exhibit an aerobic metabolic scope of as much as 100—probably the highest in the animal kingdom.

Studies on metabolic scope have inherent complexities and pitfalls. As mentioned above, anaerobic metabolism may make a significant contribution to activity, leading to a state of **oxygen debt**, especially during short periods of intense exertion (Figure 16-2 on the next page). Oxygen debt is "repaid" through the delayed oxidation of anaerobic breakdown products; for short periods these molecules may be allowed to build up in the tissues, but eventually they must be oxidized and removed, consuming O_2 in the process. White muscles in some vertebrates are specially adapted to operate under anaerobic conditions and are therefore particularly suitable for short-term spurts of intense activity that result in oxygen debt. This component of total metabolism may not be detected in short-term metabolic measurements because the aerobic breakdown of anaerobic products is delayed. For this reason, it is best

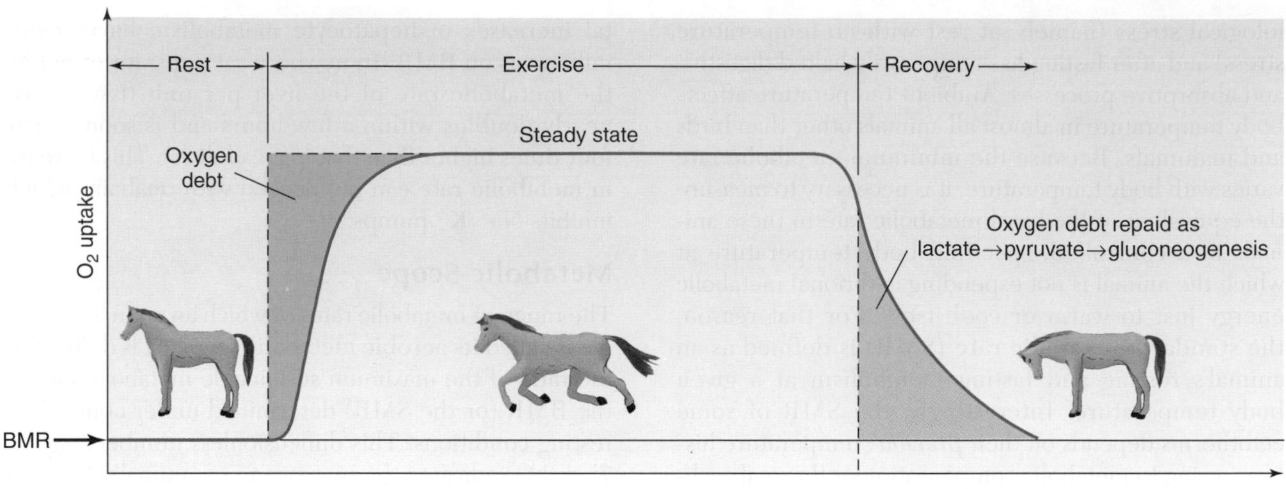

Figure 16-2 An oxygen debt typically develops during a period of sustained, intense activity. Active muscle tissue with anaerobic capabilities can acquire an oxygen debt, which is subsequently paid off in the form of delayed oxidation of an anaerobic product such as lactic acid. As a result, an elevated metabolic rate continues after cessation of activity but gradually subsides with time. The initial oxygen debt is incurred by the use of preexisting stores of high-energy phosphate compounds stored during rest. Replenishment of these stores is included in the repayment of the oxygen debt.

to make measurements of metabolic scope only during sustained activity at a constant level of exertion.

An additional practical problem in determining metabolic scope is that the maximum exertion evoked in an apparatus designed by the experimenter may not be the animal's actual maximum possible exertion, because the animal's cooperation and motivation may not be maximized.

Finally, measurements of metabolic scope can be skewed by variations in SMR. Very low SMRs measured in animals that are asleep or in torpor can lead to very high estimates of metabolic scope.

Would a similarly sized factorial increase in aerobic metabolism during activity have the same energetic implications in an animal with a very high BMR (e.g., a bird) and one with a very low BMR (e.g., a cockroach)?

Direct Calorimetry: Measuring Metabolism from Heat Production

If no physical work were being performed and no chemical synthesis were occurring, all of the chemical energy released by an animal in carrying out its metabolic functions would ultimately flow from the animal as heat. This principle is formalized in **Hess's law** (1840), which states that the total amount of energy released in the breakdown of a fuel to a given set of end products is always the same, irrespective of the intermediate chemical steps or pathways used. The metabolic rate of an organism can therefore be determined by measuring the amount of energy released as heat over a given period. This method of measurement is called **direct calorimetry.** An animal, usually unrestrained and minimally disturbed, is placed in a well-insulated chamber, called a calorimeter. The amount of heat lost from the animal is determined from the rise in temperature of a known mass of water used to trap that heat. The earliest and simplest calorimeter was that devised in the 1780s by Antoine Lavoisier and Pierre de Laplace, in which the heat given off by an animal in a chamber melted ice packed around that chamber. The heat loss was calculated from the mass of the collected water and the latent heat of melting ice. In one type of modern calorimeter, water flows through coiled copper pipes in the measuring chamber. The total heat lost by the animal is the sum of the heat gained by the water plus the latent heat present in the water vapor of expired air and evaporated skin moisture. To measure this latent heat, the mass of the water vapor is determined by passing the air through sulfuric acid, which absorbs the water. The energy content of each gram of water absorbed is 2.45 kJ (0.585 kcal), the latent heat of vaporization of water at 20°C. The results are generally reported in calories or kilocalories per hour (see Spotlight 16-1 for a discussion of the numerous, and sometimes confusing, ways in which energy units are expressed).

Although simple in principle, direct calorimetry can be rather cumbersome in practice. The technique may be too imprecise for animals having very low metabolic rates, and very large animals require a calorimetry chamber of impractical dimensions. Consequently, direct calorimetry has been used most often for birds and small mammals with high metabolic rates. Another disadvantage of direct calorimetry is that an animal's behavior (and therefore its metabolism) is unavoidably altered by the restrictions imposed by the measurement conditions, such as confinement in a calorimetry chamber.

16-1

ENERGY UNITS (OR WHEN IS A CALORIE NOT A CALORIE?)

The most commonly used unit of measurement for heat is the calorie, abbreviated "cal," which is defined as the quantity of heat required to raise the temperature of 1 g of water 1°C. This heat quantity varies slightly with temperature; the calorie is, more precisely, the amount of heat required to raise the temperature of 1 g of water from 14.5°C to 15.5°C.

Because a calorie is a very small quantity of heat relative to many biological processes, a more practical unit of heat energy is the kilocalorie (1 kcal = 1000 cal). Unfortunately, confusion has arisen because of the use of the popular term Calorie (note the capital C), which designates 1000 calories (cal). When the label of a soft drink can states that it contains "125 Calories," the drink actually contains 125 kilocalories. These uses of "calorie" and "kilocalorie" have persisted largely because they are familiar to most people.

According to the International System of Units (Système International d'Unités, SI), heat is defined in terms of energy, and the unit of measurement is the joule (J). Again, the more useful version is the kilojoule (1 kJ = 1000 J). Calories and joules are easily interconverted: 1 cal = 4.184 J, and 1 kcal = 4.184 kJ. If we assume a respiratory quotient (R_Q) of 0.79, which is a typical value, 1 liter of oxygen used in the oxidation of a substrate will release 4.8 kcal, or 20.1 kJ, of heat energy.

Power is the amount of energy expended per unit time, and is measured in SI units of watts (W), with $1\ W = 1\ J \cdot s^{-1}$. Conversion tables for units of energy are given in Appendix 3.

Indirect Calorimetry: Measuring Metabolism from Food Intake and Waste Excretion

Metabolic rate can be estimated from a "balance sheet" that sums energy gain and compares it with energy loss. Living organisms obey the laws of energy conservation and transformations, which were initially derived for nonliving chemical and physical systems (see Chapter 3). In principle, we can therefore determine the metabolic rate of an animal in an energy steady state (i.e., with a general constant metabolic rate) by using the following formulation:

rate of chemical energy intake − rate of chemical energy loss = metabolic rate (heat production) (16-1)

Total energy intake over a given period equals the chemical energy content of ingested food over that same period. Energy loss is the unabsorbed chemical energy that is eliminated in the feces and urine produced by the animal over the same period. The energy content of both food and wastes can be measured from the heat of combustion of these materials in a **bomb calorimeter.** In this method, the material to be tested is first dried and then placed inside an ignition chamber enveloped in a jacket containing a known amount of water. The material is completely burned to ash (using no additional fuel) with the aid of oxygen gas. The resulting heat is captured in the surrounding water jacket. The amount of energy released from the burning of the test substance is then determined from the increase in temperature of the surrounding water. The energy released from the burned material is equivalent to that which would be released if all of this material were to be passed through aerobic metabolic pathways. By obtaining a quantity for the energy in the original food substances and subtracting the energy that remains in waste products, one obtains the amount of energy acquired from the materials that entered and passed through metabolic pathways.

When using the balance-sheet approach to energy metabolism, one must contend with variables that are difficult to control. For example, not all the energy extracted from food is actually available to the animal. Depending on the food type, only a variable fraction may actually be digested and absorbed in the digestive tract (see Chapter 15). One must correct for this fraction when calculating total energy intake. Another complicating factor is that energy can be obtained during the period of measurement from an animal's tissue reserves (e.g., stored fat). Because these energy reserves can be exhausted, the animal will eventually lose weight, signaling a nonsteady state (a violation of one of the assumptions of this technique).

The balance-sheet method does not measure BMR, SMR, or **resting metabolic rate** (**RMR**, measured in a fed, thermoregulating, inactive animal), which are best measured by more direct methods.

Radioisotopes: Measuring Metabolism by Tracking Atom Movements

Another important method for measuring metabolic rate employs radioisotopes (see Chapter 2). These techniques first came to prominence in the measurement of water fluxes in animals. Deuterium- or tritium-labeled water is injected into an animal, and the radioactivity of serum, blood, or other body fluid samples is determined. The decline in radioactivity with time indicates the loss of labeled water, which correlates with outward water flux. The use of isotopes was then extended to the measurement of CO_2 production as an index of metabolic rate. In essence, radioisotopes of oxygen and hydrogen are injected into an animal. The subsequent decline in the amount of isotopic oxygen (^{18}O) in body water is related to the rates of CO_2 loss through exhalation and through water loss (the latter being measured by the disappearance of deuterium- or tritium-labeled water). Although numerous assumptions require validation for each experimental setting, the great advantage of this technique is that it can be employed on intact, unrestrained animals behaving normally in the field. The many studies by Ken Nagy and his colleagues have shown the usefulness of this technique in measuring field metabolic rate.

Respirometry: Measuring Metabolism from Gas Exchange

Other indirect measures of metabolic rate depend on the measurement of some variable other than heat production that is related to energy utilization. The energy contained in food molecules becomes available for use by an animal when those molecules or their products are subjected to oxidation (see Chapter 3). In aerobic oxidation, the quantity of heat produced is directly related to the quantity of oxygen consumed. Thus, measurements of oxygen uptake ($\dot{M}_{O_2}$) and carbon dioxide production ($\dot{M}_{CO_2}$), expressed, for example, as micromoles of gas per hour (or per minute), can be used to calculate metabolic rate.*

Respirometry is the measurement of an animal's respiratory gas exchange—that is, its $\dot{M}_{O_2}$ and $\dot{M}_{CO_2}$. In **closed-system respirometry**, an animal is confined to a closed, water- or air-filled chamber in which the amounts of oxygen consumed and/or carbon dioxide produced are monitored for a given time period. Oxygen consumption is measured by successive determinations of the decreasing amount of oxygen dissolved in the water or present in the air contained in the chamber. These measurements can be obtained very conveniently with the aid of an oxygen electrode and the appropriate electronic circuitry. The partial pressure of oxygen of the water or air directly determines the signal produced in the electrode. Oxygen in the gaseous phase (that is, in an air-filled respirometer) can also be measured by a mass spectrometer or an electrochemical cell. In water or air, CO_2 can be measured with a CO_2 electrode, but the complex chemistry of CO_2 dissolved in water (see Chapter 13) makes the interpretation of these measurements more difficult. In gases, CO_2 can be accurately measured with an infrared device, a gas chromatograph, or a mass spectrometer. Generally, O_2 is easier to measure than CO_2, especially in water, so $\dot{M}_{O_2}$ is more commonly reported than $\dot{M}_{CO_2}$ as a measure of metabolic rate.

All of these methods of gas analysis also allow mass-flow analytic techniques in which the flow of gas or water into and out of a chamber is monitored, and the difference in gas concentrations or partial pressures between inflow and outflow is used to calculate respiratory exchange. Such systems employ **open-system** or **flow-through respirometry**. Importantly, the chamber in which the animal resides must be well stirred to ensure that the medium exiting the chamber is in equilibrium with that throughout the chamber. Open-system respirometry can also be carried out on animals fitted with breathing masks, a method that is especially useful in wind-tunnel tests of flying animals or in tests of animals running on treadmills.

Closed and open respirometry systems can be combined in a single apparatus, as illustrated in Figure 16-3. Such combined systems can be used in partitioning total gas exchange into pulmonary, branchial, and cutaneous exchange in amphibian larvae and air-

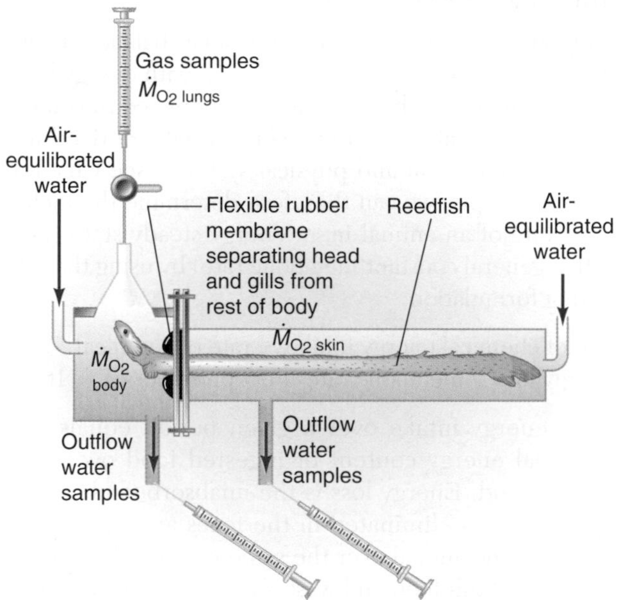

Figure 16-3 Open and closed respirometry can be used independently to measure whole-animal gas exchange or can be combined in various ways in a single experiment to measure an animal's gas exchange partitioning between various sites. In this relatively complex experiment, oxygen consumption is measured from three sites in the air-breathing, eel-like reedfish, *Calamoicthys calabaricus*. Two independent open, flow-through systems are used to determine the separate branchial oxygen uptake ($\dot{M}_{O_2\,gills}$) and cutaneous aquatic oxygen uptake ($\dot{M}_{O_2\,skin}$). A third, closed respirometry system ($\dot{M}_{O_2\,lungs}$) measures the gas in the funnel above the fish's head, from which the fish intermittently breathes. Gas samples taken after an air breath are used to calculate aerial oxygen consumption. The total oxygen consumption of the animal is derived from the sum of oxygen consumption measured by each of the three systems. [Adapted from Sacca and Burggren, 1982.]

*Oxygen uptake and carbon dioxide production are also commonly expressed as volume of gas, $\dot{V}_{O_2}$ and $\dot{V}_{CO_2}$, respectively. This is less desirable than expressing these values as molar amounts, which by definition are completely independent of measurement temperature and atmospheric pressure; $\dot{V}_{O_2}$ and $\dot{V}_{CO_2}$ reported in a paper can accurately be converted into other units such as $\dot{M}_{O_2}$ or $\dot{M}_{CO_2}$ only if the author has reported temperature and pressure, which is not always the case.

breathing fishes that simultaneously employ water and air breathing.

The determination of metabolic rate from O_2 consumption rests on several important assumptions:

1. *The relevant chemical reactions are assumed to be aerobic.* This assumption holds for most animals at rest because the energy acquired from anaerobic reactions is relatively minor except during vigorous activity. However, anaerobiosis is important in animals that live in environments that are oxygen-poor or lack oxygen altogether, such as gut parasites and invertebrates that dwell in mud at the bottom of deep lakes. Oxygen consumption in such animals would be an unreliable index of metabolic rate and would greatly underestimate the true metabolic rate.

2. *The amount of heat produced (i.e., energy released) when a given volume of oxygen is consumed is assumed to be constant irrespective of the metabolic substrate.* This assumption is necessary but is not precisely true: more heat is produced when a liter of O_2 is used in the breakdown of carbohydrates than when fat or protein is the substrate. However, the error resulting from this assumption is no greater than about 10%. Unfortunately, it is generally difficult to identify the substrate(s) being oxidized so as to correct for differences in caloric yield.

3. *The oxygen stores in the bodies of most animals are small, so the minute-to-minute oxygen consumption from air or water flowing over the gas-exchange organs is assumed to accurately represent the metabolic rate.* (Note that the ability to store CO_2 in body tissues is much greater than the ability to store O_2, so the minute-to-minute elimination of CO_2 is a much less reliable indicator of metabolic rate.)

Respiratory Quotient

Because each type of food molecule has a different characteristic energy yield and waste production per liter of O_2 consumed, we can use the ratio of CO_2 production to O_2 consumption to determine the proportions of carbohydrates, proteins, and fats being metabolized. To convert the amount of oxygen consumed into the equivalent in heat production, we must know the relative amounts of carbon and hydrogen oxidized. The oxidation of hydrogen atoms is hard to determine, however, because metabolic water (i.e., that produced by the oxidation of hydrogen atoms available in food), together with other water, is lost in the urine and from a variety of body surfaces at a rate that is irregular and determined by unrelated factors (e.g., osmotic stress and ambient relative humidity). It is more practical to measure, along with the O_2 consumed, the amount of CO_2 produced, as explained above. As noted in Chapter 13, the ratio of the volume of CO_2 produced to the volume of O_2 consumed within a given time is called the **respiratory quotient** (R_Q):

Table 16-1 **Heat production and respiratory quotient for the three major food types**

	Heat Production (kJ)			
	Per gram of food	Per liter of CO_2 produced	Per liter of O_2 consumed	R_Q
Carbohydrates	17.1	21.1	21.1	1.00
Fats	38.9	19.8	27.9	0.71
Proteins (to urea)	17.6	18.6	23.3	0.80

$$R_Q = \frac{\text{rate of } CO_2 \text{ production}}{\text{rate of } O_2 \text{ consumption}} \qquad (16\text{-}2)$$

Under resting, steady-state conditions, the R_Q is characteristic of the type of molecule being catabolized (carbohydrate, fat, or protein: Table 16-1). Thus, the R_Q reflects the proportions of carbon and hydrogen in food molecules.

The following examples illustrate how the R_Q of the major food types may be calculated from a formulation of their oxidation reactions.

Carbohydrates

The general formula of carbohydrates is $(CH_2O)_n$. In the complete oxidation of a carbohydrate, oxygen is used, in effect, only to oxidize the carbon to form CO_2. Upon complete oxidation, each mole of a carbohydrate produces n moles of both H_2O and CO_2 and consumes n moles of O_2. The R_Q for carbohydrate oxidation is thus 1. The overall catabolism of glucose, for example, may be formulated as

$$C_6H_{12}O_6 + 6\ O_2 \longrightarrow 6\ CO_2 + 6\ H_2O$$

$$R_Q = \frac{6 \text{ volumes of } CO_2}{6 \text{ volumes of } O_2} = 1$$

Fats

The R_Q characteristic of the oxidation of a fat, such as tripalmitin, may be calculated as follows:

$$2\ C_{51}H_{98}O_6 + 145\ O_2 \longrightarrow 102\ CO_2 + 98\ H_2O$$

$$R_Q = \frac{102 \text{ volumes of } CO_2}{145 \text{ volumes of } O_2} = 0.70$$

Different fats contain different ratios of carbon, hydrogen, and oxygen, so they have slightly different R_Qs when oxidized.

Proteins

The R_Q of protein catabolism presents a special problem because proteins are not completely broken down in oxidative metabolism. Some of the oxygen and carbon of the constituent amino acid residues remains combined with nitrogen and is excreted as nitrogenous wastes in urine and feces. In mammals, the excreted end product is urea, $(NH_2)_2CO$. In birds, it is primarily uric acid, $C_5H_4N_4O_2$. To obtain the R_Q for protein catabolism, it is therefore necessary to know the amounts and kinds of nitrogenous wastes excreted as well as the amount of ingested protein. The oxidation of carbon and hydrogen in the catabolism of protein typically produces the following R_Q:

$$R_Q = \frac{77.5 \text{ volumes of } CO_2}{96.7 \text{ volumes of } O_2} = 0.80$$

Several assumptions are routinely made when making deductions from R_Q:

1. The only substances metabolized are carbohydrates, fats, and proteins.
2. No synthesis takes place alongside breakdown.
3. The amount of CO_2 exhaled in a given time equals the CO_2 produced by the tissues in that interval.

These assumptions are not invariably true, so caution must be exercised in using R_Q values for animals at rest and in postabsorptive (fasting) states. Under such conditions, protein utilization is negligible and carbohydrate utilization is minor, so the animal is considered to be metabolizing primarily fat. The oxidation of 1 gram of mixed carbohydrate releases about 17.1 kJ (4.1 kcal) as heat (see Table 16-1). When 1 liter of O_2 is used to oxidize carbohydrate, 21.1 kJ (5.05 kcal) of heat is obtained. The value for fats is 19.87 kJ (4.7 kcal) and for protein (metabolized to urea), 18.6 kJ (4.46 kcal). A fasting aerobic animal presumed to be metabolizing mainly fats produces about 20.1 kJ (4.80 kcal) of heat for every liter of oxygen consumed.

Another term often used to describe the ratio of $\dot{M}_{O_2}$ to $\dot{M}_{CO_2}$ is the **respiratory exchange ratio** (R_E), which is a measure of the instantaneous relation between $\dot{M}_{O_2}$ and $\dot{M}_{CO_2}$. Whereas R_Q reflects the overall metabolism of an animal, R_E reflects the minute-to-minute exchange of O_2 and CO_2 between the animal and its environment. In practice, measures of R_E are taken from gas or water exiting a respirometer, expired gas or water collected from a face mask or similar device, or even from samples of lung or bladder gas itself.

The measured R_E is often quite different from R_Q in an animal that is breathing intermittently, in part because of the very different solubilities for O_2 and CO_2 in gas and tissues (see Chapter 13). In diving freshwater turtles, for example, CO_2 is temporarily stored in body tissues during long periods of breath-holding, rather than being eliminated into lung gas. Thus, during breath-holding, the apparent $\dot{M}_{CO_2}$ is far lower than the actual $\dot{M}_{CO_2}$ at the tissue level, where R_Q may still be 0.7–0.8. This temporary sequestering of CO_2 in the body tissues is reflected in an R_E that might approach 0.3, showing that far more O_2 is being removed from lung gas than CO_2 is being added (Figure 16-4). As dive time becomes longer, the R_E becomes lower. When the turtle eventually surfaces and begins to breathe, however, the stored CO_2 floods out of the tissues, and the R_E actually rises to 1.5 or higher, indicating that CO_2 is being eliminated much faster than O_2 is being taken up. Thus, in intermittently breathing animals, R_E can show enormous variation even as R_Q is quite steady.

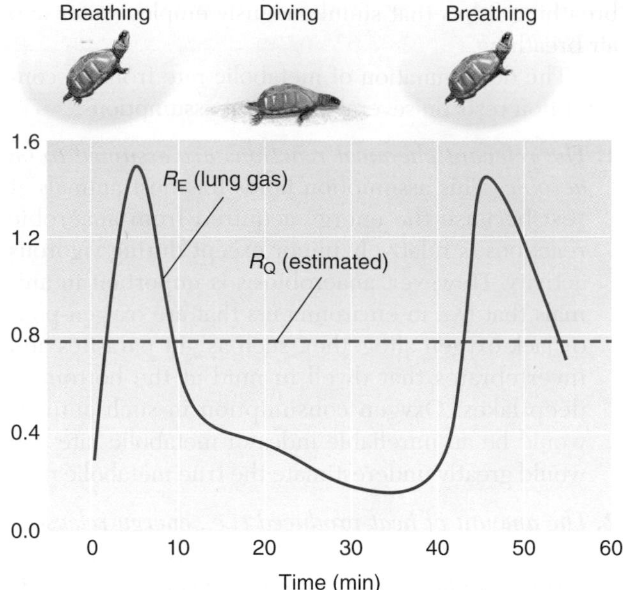

Figure 16-4 The respiratory exchange ratio, R_E, reflects the minute-to-minute gas exchange between tissues and the respiratory medium. The R_E measured from pulmonary gas samples during intermittent diving and breathing in the freshwater turtle *Chrysemys picta* changes drastically as CO_2 is alternately sequestered in the tissues during long bouts of diving and then quickly released during brief periods of lung ventilation. Note that the changes in R_E occur despite a relatively stable R_Q of just under 0.8. [Unpublished data from W. Burggren.]

Energy Storage

Although animals expend metabolic energy continually, most ingest food intermittently. Consequently, they do not strike a moment-to-moment balance between food intake and energy expenditure. As food is taken in in bursts (i.e., in discrete meals), the animal's immediate energy requirements are exceeded. The excess is stored for later use, primarily as fats and carbohydrates.

Protein is not an ideal storage material for energy reserves because nitrogen is a relatively scarce commodity and is often the major limiting factor in growth and reproduction; it would be wasteful to tie up valuable nitrogen in energy reserves. Fat is the most effective form of energy storage because its oxidation yields 38.9 $kJ \cdot g^{-1}$ ($9.3\ kcal \cdot g^{-1}$), nearly twice the yield per gram for carbohydrate or protein (see Table 16-1). This efficiency is of great importance in animals such as migrating birds or insects, in which economy of weight and volume is essential. Not only is the energy yield per gram of carbohydrate lower than that of fat, but carbohydrates must be stored in a bulky hydrated form, carrying along more than two grams of water per gram of carbohydrate. Fats, in contrast, are stored in an anhydrous state.

Nonetheless, carbohydrates are important in energy storage. Glycogen, a branched, starchlike carbohydrate polymer, is stored as granules in the skeletal muscle fibers and liver cells of vertebrates. Muscle glycogen is rapidly converted into glucose for oxidation within the muscle cells during intense activity, and liver glycogen is used to maintain blood glucose levels. Glycogen can be broken down directly into glucose 6-phosphate, providing fuel for metabolism more directly and rapidly than the mobilization of fat. Thus, carbohydrates tend to be used to power short-term increases in metabolism—during sudden activity, for example. Fats, which cannot be directly metabolized anaerobically, are metabolized aerobically in response to longer-term demands for energy and during fasting, when carbohydrate stores have been depleted.

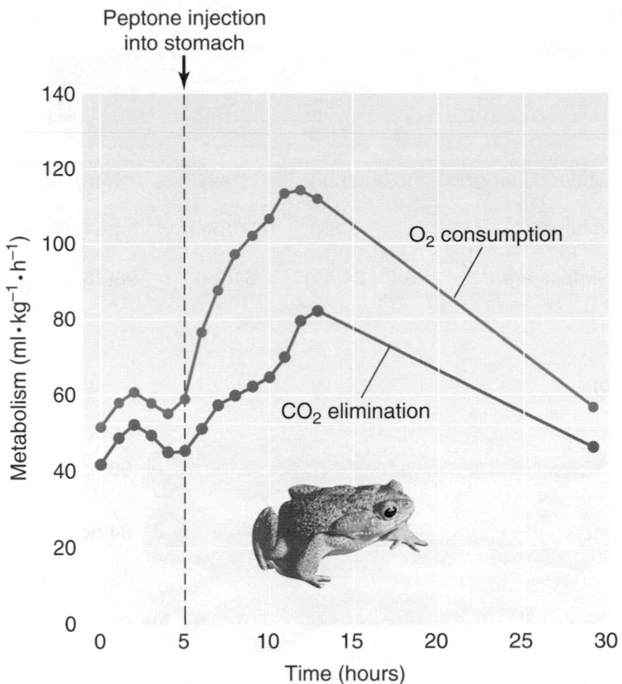

Figure 16-5 Specific dynamic action—a sharp rise in metabolic rate that occurs shortly after feeding—can be stimulated in the toad *Bufo marinus* by injecting peptone (a mixture of amino acids produced from chemically digested meat protein) into the animal's stomach. Both O_2 consumption (red) and CO_2 production (blue) are shown. [Adapted from Wang et al., 1995.]

Specific Dynamic Action

Max Rubner reported in 1885 that a marked increase in metabolism accompanies the processes of digestion and assimilation of food, independent of other activities. He gave this phenomenon the rather awkward name **specific dynamic action (SDA).** Since then, SDA has been documented in all five vertebrate classes, as well as in many invertebrates, including crustaceans, insects, and mollusks. Generally, an animal's oxygen consumption and heat production increase within about an hour after a meal is eaten, reaching a peak some 3–6 hours later and remaining elevated above the basal value for several hours (Figure 16-5). In fish, amphibians, and reptiles, which have an SDA equivalent to a doubling or tripling of metabolic rate, there are also attendant large increases in heart rate and cardiac output and a temporary redistribution of blood toward the gut. Similar cardiovascular changes of lesser magnitude occur in animals with less prominent SDA responses (such as humans).

The mechanism of SDA is not clearly understood, but apparently the work of digestion (and the concomitant increase in metabolism in the tissues of the gastrointestinal tract) is responsible for only a small part of the elevated metabolic rate. A more likely explanation may be that certain organs, such as the liver, expend extra energy processing recently absorbed nutrients for entry into metabolic pathways. The extra energy consumed by such processes is lost as heat. The increase in heat production varies depending on the ingested food materials. The magnitude of the increase ranges from 5% to 10% of the total energy of ingested carbohydrates and fats and from 25% to 30% of that of proteins.

Specific dynamic action probably accounts for some of the variation in metabolic rates reported by different researchers for a single species. Very different metabolic rates could be obtained, depending on whether the animals measured were in a postabsorptive state or were in some part of the SDA response. Consequently, basal metabolism must be measured only during the postabsorptive state so as to minimize any contribution of SDA.

BODY SIZE AND METABOLIC RATE

Body size is one of the more important physical characteristics affecting an animal's physiology. How both anatomical and physiological characteristics change with body mass is determined by considering **scaling effects**—that is, how any given parameter changes with body mass.

Isometric and Allometric Scaling of Metabolism

The simplest form of scaling is **isometry,** in which dimensions remain proportional independent of size. When an object increases in size isometrically, its

(a)

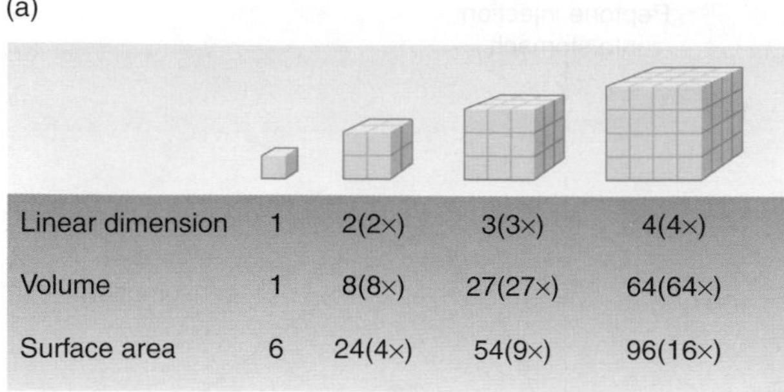

(b)

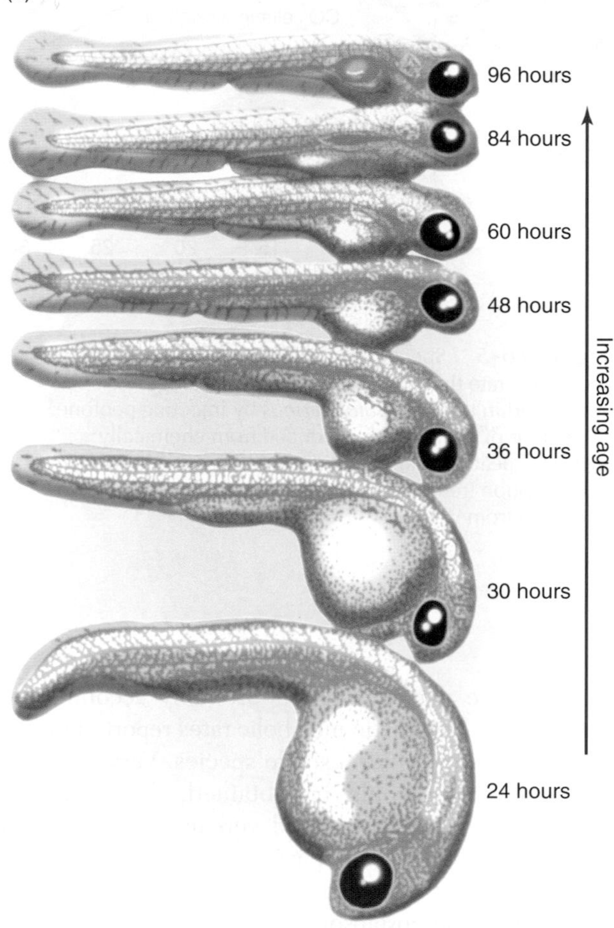

Figure 16-6 The dimensions and properties of objects, both inanimate and living, show scaling as their size changes. **(a)** In isometric (geometric) scaling, the linear dimensions of an object change in proportion to one another, and its volume and surface area change in predictable ways. **(b)** In allometric scaling, the body proportions (and subsequently volume and surface area) change in nongeometric fashion. This composite of a growing zebrafish shows that the size of the head, and especially the eyes, changes radically in proportion to the size of the body as body length increases. Similar changes occur during growth in most animals.

relative height, width, and depth stay constant, as Figure 16-6a shows. As the linear dimensions of the depicted block double, its volume increases by the cube, whereas its surface area increases by the square.

Isometry is mathematically straightforward but is only infrequently found in animals and the environments they inhabit. Changes in animal body size, in particular, introduce changes that frequently are not simple and proportional. The necessity of nongeometric scaling for the functional anatomy and physiology of the animal is immediately evident. You can carry out a thought experiment by imagining a spherical animal that absorbs nutrients uniformly across its surface. If its linear dimensions double, its volume will increase by eight times, whereas its surface area will increase by only three times. Its surface area will not rise fast enough to keep up with the heightened requirements of its increased volume. Similarly, imagine a mouse scaled up to the size of an elephant while retaining its mouse-like body proportions. Clearly, the body proportions of this imaginary mouse would be different from those of an elephant. Its relatively slender legs would collapse under the weight of its newly massive body. After all, for each doubling of the height of the mouse, its mass would increase by a factor of eight (height cubed) while the cross-sectional area of its leg bones would increase by a factor of only four (height squared). These same scaling factors make a real mouse capable of jumping to, and landing from, a height many times its own body length without harm, whereas an elephant is essentially "earthbound."

Most groups of animals show nonisometric changes in body proportion as well as size as they change in body mass (Figure 16-6b). Systematic change in body proportions with increasing body size is called **allometry.** Allometric comparisons have usually been applied to different species, considering changes that have occurred during the evolution of species of notably different size. Recently, however, researchers have been paying attention to intraspecific allometric comparisons, considering changes within a species as animals grow and develop.

Changes in body mass have great effects on an animal's metabolic rate. Consider, for example, the respiratory and metabolic requirements of a tiny diving water shrew compared with those of a submerged whale. Although both whales and water shrews normally dive, a whale can hold its breath and remain under water far longer than a shrew. The reason stems from the general principle that small animals must respire at higher rates per unit body mass than large animals. In fact, there is an inverse relation between the rate of O_2 consumption per gram of body mass and the total mass of the animal. Thus, a 100 g mammal consumes far more energy per unit mass per unit time than does a 1000 g mammal. The nonproportionality of the basal metabolic rates of mammals ranging from very small to very large is illustrated by the long-recognized "mouse-to-elephant" curve (Figure 16-7a). A similar relation holds not only for other vertebrate groups, but throughout the animal and plant kingdoms. Few biological principles are so widely applicable.

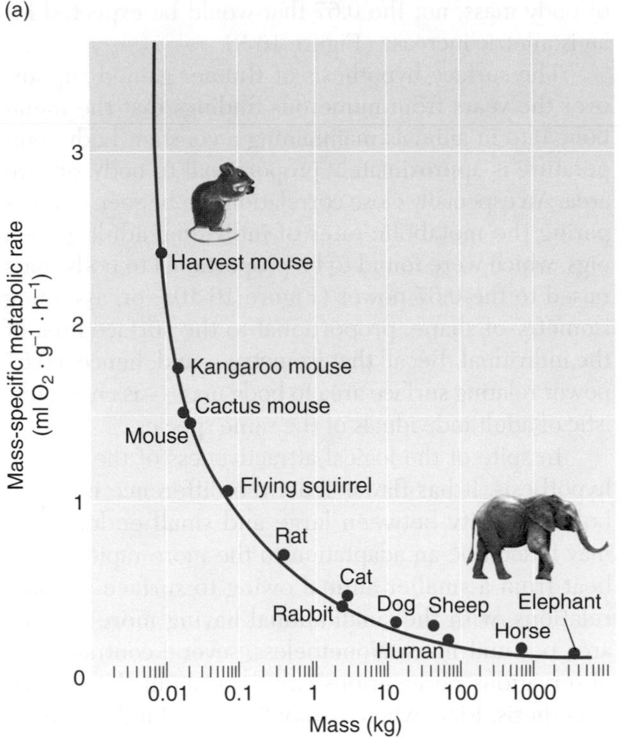

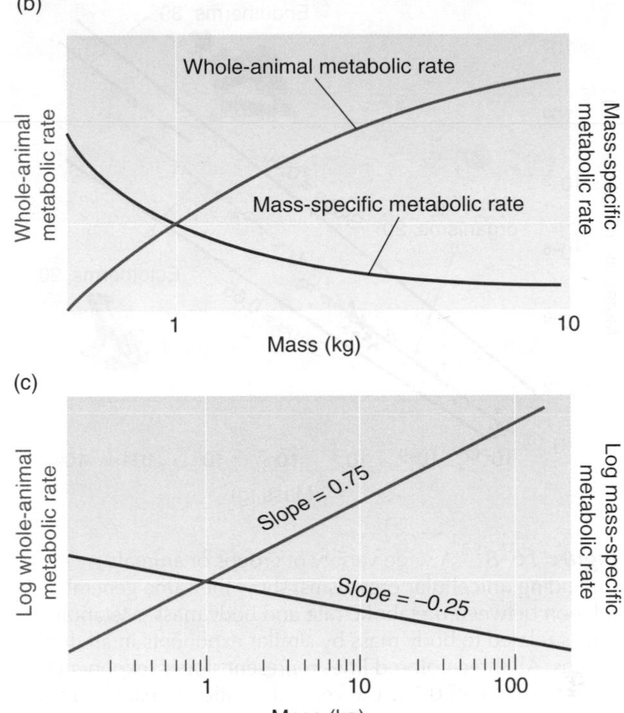

Classic Work ***Figure 16-7*** Mass-specific metabolic rate in mammals declines as body mass increases. **(a)** The mouse-to-elephant curve, with metabolic intensity (mass-specific metabolic rate, given as O_2 consumption per unit mass per unit time) plotted against body mass. Note the logarithmic scale of body mass. **(b)** Generalized relations between overall metabolic rate and body mass (blue curve) and between metabolic intensity and body mass (red curve). **(c)** Logarithmic plots of the plots in part b. The plots for the log of whole-animal metabolic rate and mass-specific metabolic rate cross at mass = 1 kg.

The inverse relation between metabolic rate and body mass applies within species as well as between species. Thus, a small child, an early-instar cockroach, and fish fry all tend to have a higher metabolic rate per unit mass per unit time than larger, adult members of the same species. However, this relation is often difficult to demonstrate within a single species, in which the overall range of body mass may be quite small compared with that between species and in which other factors, such as sex, nutrition, and season, may exert confounding effects.

Metabolic rate is a power function of body mass, as described by the simple relation

$$\mathrm{MR} = aM^b \tag{16-3}$$

where MR is the basal or standard metabolic rate, M is the body mass, a is the intercept of the log-log regression line (which differs among species), and b is an empirically determined exponent that expresses the rate of change of MR with change in body mass.

Mass-specific metabolic rate, also termed **metabolic intensity,** is the metabolic rate of a unit mass of tissue (i.e., amount of O_2 consumed per unit mass per unit time). It is determined by dividing both sides of Equation 16-3 by M:

$$\frac{\mathrm{MR}}{M} = \frac{aM^b}{M} = aM^{(b-1)} \tag{16-4}$$

The relation described by Equation 16-3 is shown by the red curve in Figure 16-7b. Because it is often more convenient to work with straight-line rather than curved plots, Equations 16-3 and 16-4 are often put into their logarithmic forms. Thus, Equation 16-3 becomes

$$\log \mathrm{MR} = \log a + b(\log M) \tag{16-5}$$

and Equation 16-4 becomes

$$\frac{\log \mathrm{MR}}{M} = \log a + (b-1)\log M \tag{16-6}$$

The logarithmic equations above are plotted in Figure 16-7c. (See Appendix 2 for a discussion of logarithmic equations.)

Notice the difference in how whole-animal metabolic rate (the blue plots in Figures 16-7b and c) and mass-specific metabolic rate (the red plots) change with body mass. As body mass increases, overall metabolic rate rises, whereas the mass-specific metabolic rate falls.

The value of exponent b lies close to 0.75 for many different taxonomic groups of vertebrates and

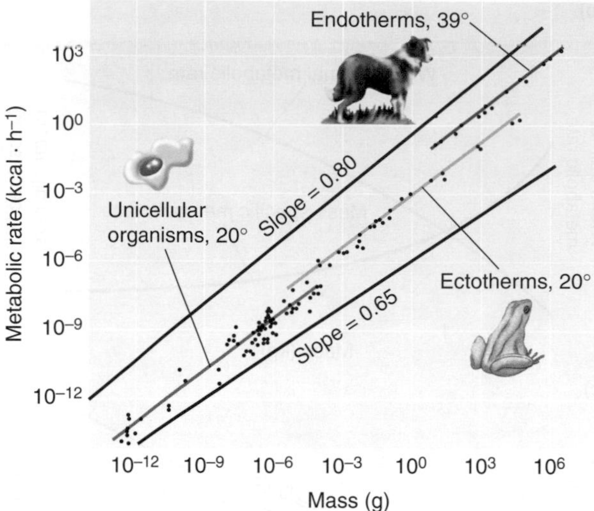

Figure 16-8 A wide variety of groups of animals (including unicellular organisms) show the same general relation between metabolic rate and body mass. Metabolic rate is related to body mass by similar exponents in all three groups. All three colored lines represent slopes (exponent *b* in Equation 16-3) of 0.75. The vertical position of each group on the graph is related to coefficient *a* in Equation 16-3. [Adapted from Hemmingsen, 1969.]

invertebrates and even for various unicellular taxa (Figure 16-8). This exponential relation between body size and metabolic rate has been attracting the attention of physiologists since it was first recognized more than a century ago. Many attempts have been made to supply a rational explanation for this nearly universal logarithmic relation. In 1883, Max Rubner proposed an attractive theory known as the *surface hypothesis.* Rubner reasoned that the metabolic rate of birds and mammals that maintain a more or less constant body temperature should be proportional to body surface area because the rate of heat transfer between two compartments (i.e., warm animal body and cool environment) is proportional, all else being equal, to their area of mutual contact (see Chapter 17). The surface area of an object of isometric shape (i.e., of nonvarying proportions) and uniform density varies as the 0.67 (or 2/3) power of its mass as the object increases in size because, as we saw above, mass increases as the cube of linear dimension, whereas surface area increases only as the square. As noted, this relation would hold true for a series of animals of different mass only if their body proportions remained constant. This provision is generally satisfied only by adult individuals of different size within a species, which do tend to obey the principle of isometry. The principle of isometry is not followed, however, in individuals of different sizes belonging to related but different species (such as the mouse and the elephant). Instead, they tend to follow the principle of allometry. In a comparison of surface-to-mass relations among mammals of different species ranging from mice to whales, surface area was found to vary as the 0.63 power of body mass, not the 0.67 that would be expected for an isometric increase (Figure 16-9).

The surface hypothesis of Rubner gained support over the years from numerous findings that the metabolic rate in animals maintaining a constant body temperature is approximately proportional to body surface area. An especially close correlation can be seen in comparing the metabolic rates of individual adult guinea pigs, which were found to be proportional to body mass raised to the 0.67 power (Figure 16-10a) or, assuming isometry of shape, proportional to the surface area of the individual. Recall that isometry—and, hence a 0.67 power relating surface area to body mass—is characteristic of adult individuals of the same species.

In spite of the logical attractiveness of the surface hypothesis, it has flaws. True, the difference in metabolic intensity between large and small endotherms may indeed be an adaptation to the more rapid loss of heat from a smaller animal owing to surface-to-mass relations, with the small animal having more surface area per unit mass. Nonetheless, several contradictory observations raise serious concerns about the surface hypothesis. First, when metabolic rates of individuals of different species of mammals are plotted against body mass, the exponent relating metabolic rate to body mass is found to be approximately 0.75 (Figure 16-10b), not 0.63 or 0.67. The 0.75 exponent relating metabolic rate to body mass was first discovered by Max Kleiber (1932) and is often referred to as **Kleiber's law.** It is significantly higher than the 0.63 predicted by the surface

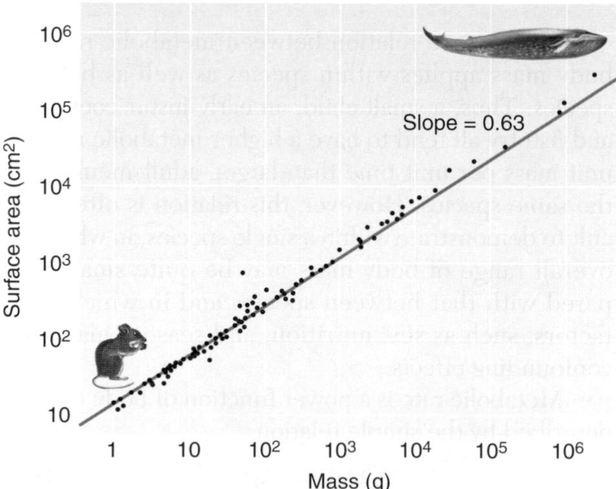

Figure 16-9 The body surface area of mammals ranging from mice to whales is very closely correlated with their body mass. The slope of the line is 0.63, rather than the 0.67 predicted for isometric (i.e., proportional) scaling. This allometric (i.e., nonproportional) scaling arises from the fact that, with increasing body size, there is a progressive relative thickening of body structures (i.e., bones, muscles), so that a large species has relatively less surface area than would be predicted from isometric scaling (recall the relative proportions of mouse and elephant). [Adapted from McMahon and Bonner, 1983.]

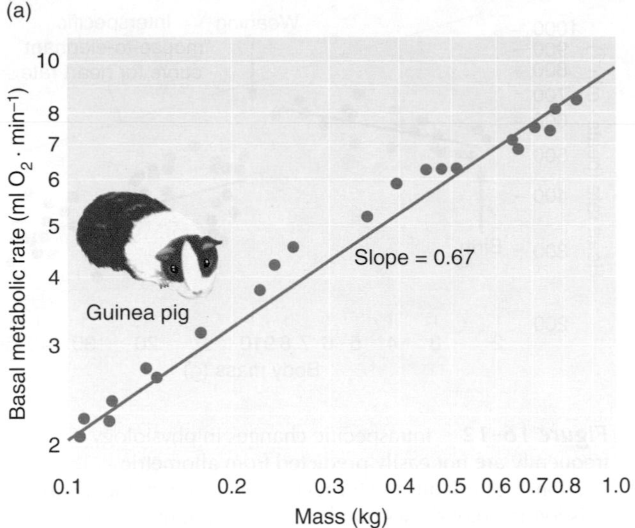

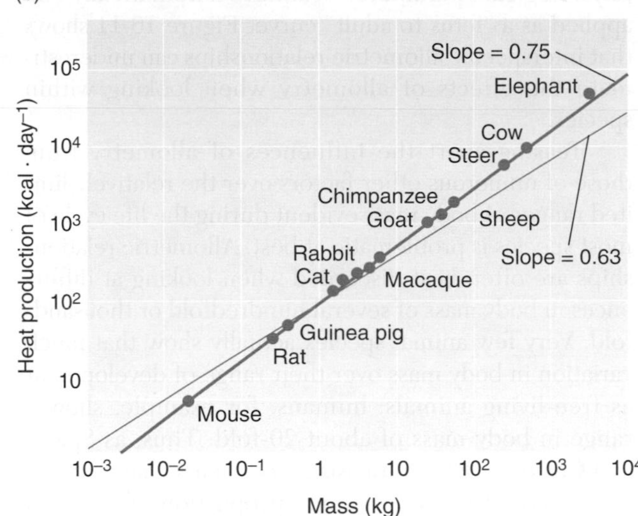

Figure 16-10 The basal metabolic rate of animals is closely correlated with body mass. **(a)** Basal metabolic rate of individuals of the same species (adult guinea pigs) plotted against individual body mass. The slope of the line indicates that the basal metabolic rate is proportional to the body mass raised to the 0.67 power. **(b)** Basal metabolic rate of animals of different species plotted against body mass. Although basal metabolic rate is closely correlated over a tenfold variation in body mass, note that the thick line through the points has a slope of 0.75. The thin line has the slope of 0.63 predicted by the surface hypothesis. This discrepancy is statistically significant, indicating that metabolic rate is not strictly related to surface area in mammals. [Part a adapted from Wilkie, 1977; part b adapted from Kleiber, 1932.]

hypothesis (see Figure 16-9). While this difference may seem trivial, it can account for major differences in predicted and measured metabolic rates as body mass changes. Thus, in comparing different species, differences in metabolic rate clearly cannot be predicted simply on the basis of differences in body surface area.

A second flaw of the surface hypothesis arises from the simple observation that the metabolic rates of animal groups whose body temperatures vary with that of their surroundings (ectotherms, such as fishes, amphibians, reptiles, and most invertebrates) exhibit nearly the same relation to body mass as the metabolic rates of animals that actively maintain a constant, high body temperature (endotherms, such as mammals; see Figure 16-8). The surface hypothesis, which is based on the need to compensate for heat dissipation across the body surface, should not so readily predict the metabolic rate of animals with variable body temperatures. Thus, the surface hypothesis appears to be insufficient to explain the quantitative aspects of scaling of body mass, surface area, and metabolism.

Scaling effects also occur at the cellular level in a wide variety of animals. There is a correlation between metabolic intensity in animals of differing sizes and the number of mitochondria per unit volume of tissue. The cells of a small mammal contain more mitochondria and mitochondrial enzymes in a given volume of tissue than do the cells of a large mammal. Because mitochondria are sites of oxidative respiration, this correlation comes as no surprise. However, we are still left wondering how metabolic intensity is functionally related to body size.

Why do large animals have lower metabolic rates per volume of tissue than small animals? Numerous investigators have tried to explain this phenomenon since the time of Rubner and Kleiber. McMahon and Bonner (1983) pointed out that the scaling of the cross-sectional area of the body (or rather of its parts) to body mass more closely resembles the scaling of metabolic rate to body mass than does that of body surface area to mass. Indeed, the cross-sectional area of any body part in a series of animals of increasing size should be proportional to the 0.75 power of body mass, owing to the allometric principles that require an elephant's leg to be proportionately thicker than a mouse's leg. Remember that metabolic rate bears the same (0.75) power relation to body mass in a wide range of animals (see Figures 16-8 and 16-10b).

Although the allometry of metabolic rate (and numerous other physiological variables) is well documented, comparative physiologists have yet to "prove" definitively why this relation exists, and experiments and speculations on the subject continue. However, there is no doubt of the physiological implications of allometry for animals. Small animals with proportionately higher metabolic rates must spend more of their time looking for food and may also be more susceptible to temporary shortages of metabolic substrates or oxygen.

Interspecific versus Intraspecific Allometry

Intraspecific allometry is very clear in a host of developing animals (see Figure 16-6b). However, intraspecific allometry is not necessarily predicted by interspecific allometry. That is, the "mouse-to-elephant" curve established for metabolism, cardiac output, or any other physiological parameter cannot automatically be applied as a "fetus to adult" curve. Figure 16-11 shows that interspecific allometric relationships can underestimate the effects of allometry when looking within species.

Teasing apart the influences of allometry from those of numerous other factors over the relatively limited ranges of body mass evident during the life cycle of most species is problematic at best. Allometric relationships are often best discerned when looking at differences in body mass of several hundredfold or thousandfold. Very few animal species actually show that much variation in body mass over their range of development as free-living animals: humans, for example, show a range in body mass of about 20-fold. Thus, as Spicer and Gaston (1999) emphasize, body mass changes typically account for only a small proportion of the total variation in metabolic rate and other physiological variables. In the tree swallow *(Tachycineta bicolor)*, for example, only one-third of the variation in metabolic rate over its life cycle can be explained by body mass changes, whereas for a grouping of 22 bird species,

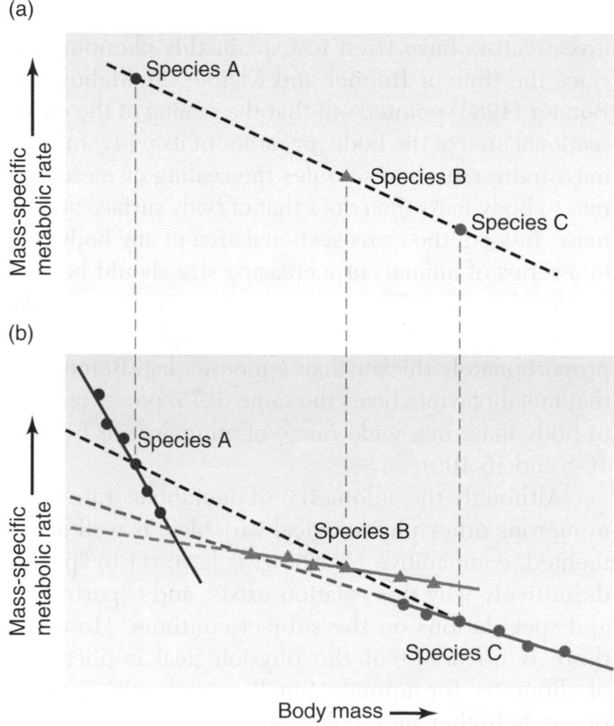

Figure 16-11 The intraspecific relationship between body mass and metabolism (or any other physiological variable), described by the slope and intercept, can be quite different from the relationship derived from limited data from multiple species. **(a)** Interspecific mass-specific metabolic rate for a broad taxonomic group (e.g., birds, mammals, or crustaceans) calculated from mean values for individual species A, B and C. **(b)** The slope and intercept of the line describing intraspecific mass-specific metabolic rate for each of the three species do not necessarily reflect the more general interspecific relationship for that broad taxonomic group and can be higher, lower, or the same.

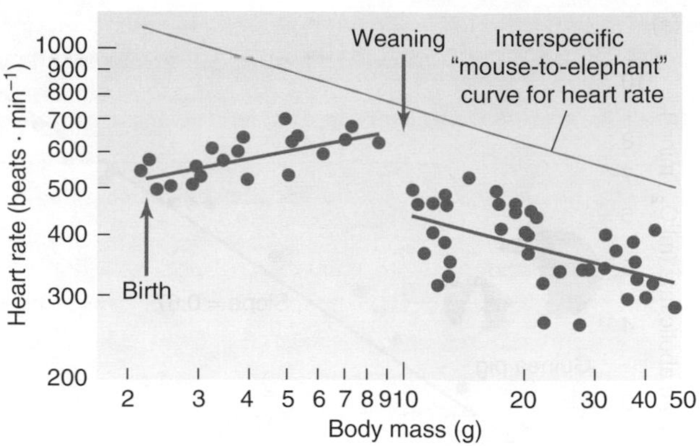

Figure 16-12 Intraspecific changes in physiology frequently are not easily predicted from allometric relationships established through interspecific comparisons. This log-log plot shows that heart rate in the mouse *(Mus musculus)* increases from birth to weaning (contrary to predictions based on increasing body mass), then decreases with further growth, during which it rather closely follows the slope and direction of the "mouse-to-elephant" curve for heart rate (gray line). Note, however, that whereas the dark orange lines predicting heart rate change before and after weaning have statistically significant slopes, there is still considerable individual variation about the lines, typical of intraspecific allometric analyses over small body mass ranges. [Adapted from Hou and Burggren, 1989.]

body mass accounts for a full 95% of the variation in metabolic rate among them (Daan et al., 1991).

Another important, but complicating, factor is development *per se.* In most species, body mass increases as animals develop and mature. Thus, how do we differentiate between trends due to allometry and trends due to maturation that are only tangentially related to body mass? Many species show size-related trends in metabolism and other physiological changes early in development that are eliminated or even reversed as development proceeds. In the mouse *(Mus musculus)*, for example, heart rate (and presumably cardiac output) increases during early development in a pattern counter to that predicted by allometric relationships based on the mouse-to-elephant curve; after weaning, however, heart rate begins to decline at a rate predicted by allometry (Figure 16-12). Thus, allometric relationships based on interspecific comparisons are completely invalid in early development, then become extremely accurate later in development.

Life cycles that include complex changes in body mass may provide an opportunity to separate the influences of body mass from other factors. In the paradoxical frog *(Psuedis paradoxus)*, for example, body mass rises as the fertilized egg grows to a tadpole of over 150 grams, only to decrease rapidly to 4–5 grams at metamorphosis (suggesting a potentially wonderful study in apoptosis, or programmed cell death). The froglet then grows back to a maximum of about 50 grams as an adult. While this is an extreme example, it illustrates the point that not all animals grow larger as they mature, and such animals may prove useful in determining the complex interactions between development and growth within an individual.

> **Question ?**
> In *Gulliver's Travels* (written by Jonathan Swift, 1667–1745), Gulliver travels to a land peopled with giants (the Brobdingnagians) and to a land peopled with tiny persons (Lilliputians). Knowing what you do now about allometry, what physiological and structural problems might be faced by each population that Gulliver visited?

ENERGETICS OF LOCOMOTION

To this point, we have been concerned with the rates of metabolism in resting animals. The energy expended by an animal exceeds the basal rate by a considerable margin when the animal is active. The most readily quantified type of muscle activity in most animals is *locomotion.* Because locomotion is required for finding food and mates and for escaping predators, it is one of the more important types of routine activity. We now turn to an examination of the metabolic cost of animal locomotion.

Animal Size, Velocity, and Cost of Locomotion

The metabolic cost of locomotion is the amount of energy required, beyond the basal requirements of the animal at rest, to move a unit mass of animal a unit distance, usually expressed in units of kilocalories per kilogram per kilometer. Measurements of O_2 consumption or CO_2 production associated with locomotion are generally made while an animal is running on a treadmill, swimming in a tank of flowing water, or flying in a wind tunnel. The measured rate of gas exchange is then translated into a rate of energy conversion.

It is a complex task to relate the net work done in locomotion to the gross energy conversion powering the underlying muscle activity because a significant percentage of muscular effort during locomotion does not contribute directly to the production of forward motion. Some muscle contraction holds limb joints in their proper articulating positions. Another large percentage of muscle work is performed as muscles counteract gravity, absorb shocks, and finely tune the movements of the limbs during contraction of antagonist muscles (see Chapter 10). The comparative energetics of animal locomotion are further complicated by the well-documented inverse relation between the force produced by a contracting muscle and its rate of shortening (i.e., muscle length or sarcomere length per second; see Figure 10-13). The higher the rate of cross-bridge cycling, the higher the metabolic cost of shortening the muscle by a given distance. Small animals show higher limb stride, tailbeat, or wingbeat rates, and thus employ higher rates of muscle shortening (and hence cross-bridge cycling) to achieve a given velocity of locomotion, than do larger animals. For that reason, smaller animals must convert correspondingly larger amounts of metabolic energy to produce a given amount of force per unit cross-section of muscle tissue in moving their limbs.

We can make several generalizations that relate the overall energy cost of locomotion to the size and velocity of an animal. Locomotion is a metabolically expensive process. The rate of oxygen consumption in excess of the basal metabolic rate increases linearly with the velocity of forward motion (Figure 16-13a). It is noteworthy, however, that the increase in energy utilization per unit weight for a given increment in speed is less for larger animals than for smaller ones. This can be seen in the different slopes of the plots in Figure 16-13a. When the cost of locomotion is plotted as energy utilization

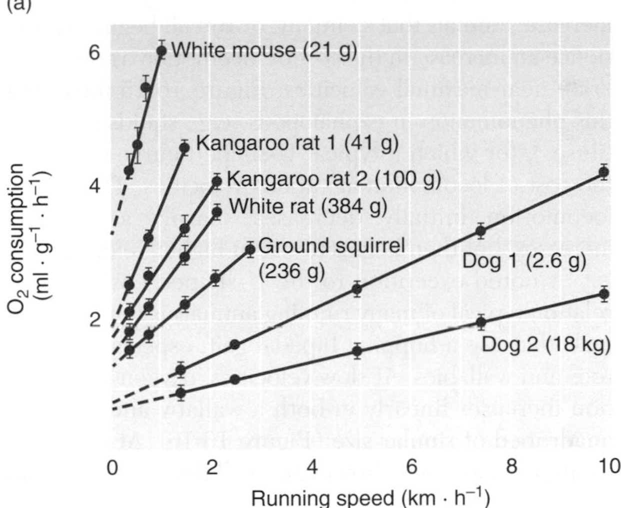

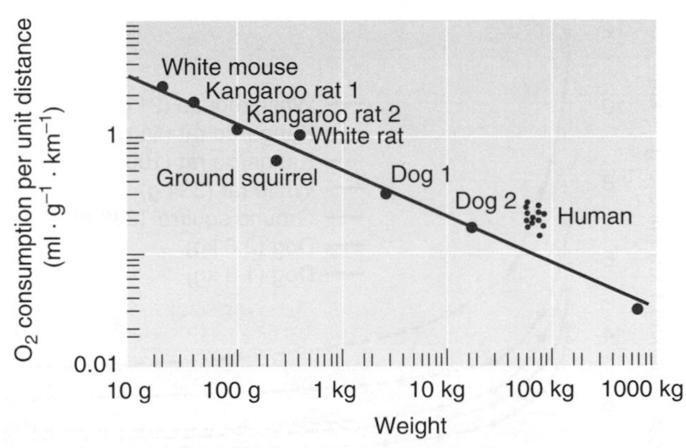

Figure 16-13 Metabolic rate during locomotion depends on both body size and velocity. **(a)** Relation between rate of oxygen consumption and running speed in mammals of different sizes. The slope of each plot represents the cost of transporting a unit mass over a unit distance. **(b)** Log-log plot of the metabolic cost of transporting 1 g a distance of 1 km in running mammals of different sizes. The cost of basal metabolism was subtracted before plotting the values. Data are from slopes of plots in part a. Values for tetrapods lie close to a straight line. [Adapted from Taylor et al., 1970.]

per gram of tissue per kilometer against body mass, it is again apparent that larger animals expend less energy to move a given mass a given distance (Figure 16-13b). The lower energy efficiency of small animals during locomotion may, to a limited degree, be due to the greater drag that they experience in moving through air or water (to be discussed shortly), but this explanation certainly does not suffice for terrestrial animals moving at low and moderate speeds through air, where drag is negligible. More likely, their lower energy efficiency is related to the lower efficiency of rapidly contracting muscle.

The relation between the velocity of an animal and the cost of achieving that velocity is complex. As running velocity increases in quadrupedal mammals, for example, the metabolic cost of traveling a given distance initially decreases (Figure 16-14) because nonlocomotory costs account for a progressively smaller fraction of the total energy expended. However, as velocity continues to increase, animals that swim, fly, or run all begin to experience an increase in the cost of locomotion as they generate near-maximal velocities. Figure 16-15 illustrates this phenomenon in cephalopods (e.g., squids and nautiluses), for which a typical U-shaped curve describes the cost of locomotion at various velocities. The cost of locomotion initially decreases sharply as velocity increases, but then begins to rise at higher velocities.

A noted exception to the U-shaped cost-velocity relation typical of many running animals is found in animals that use a hopping bipedal gait, especially kangaroos and wallabies. At slow velocities, oxygen consumption increases linearly in both a wallaby and a typical quadruped of similar size (Figure 16-16). At moderate to high velocities, however, wallabies are able to

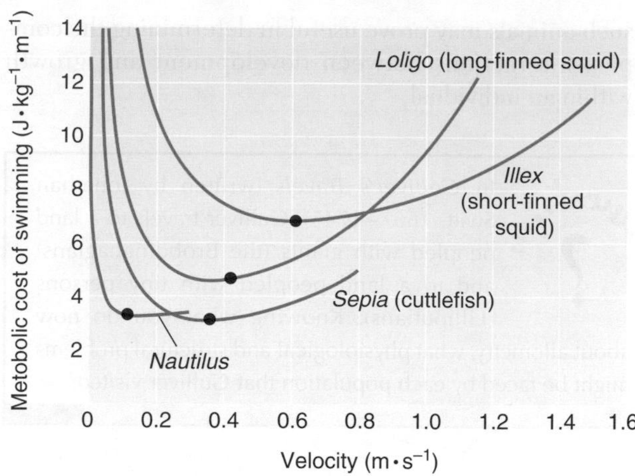

Figure 16-15 The metabolic cost of locomotion (swimming) plotted against velocity in cephalopods shows a U-shaped curve typical of many flying, swimming, and running animals. Both very slow and very rapid locomotion are relatively expensive. All cephalopods plotted weigh approximately 0.6 kg. Black dots indicate the point at which the cost of transport is lowest for each cephalopod. [Adapted from O'Dor and Webber, 1991.]

increase velocity steadily without increasing oxygen consumption—a seemingly impossible achievement. They do so by switching from a running to a hopping gait, which uses their powerful hind legs as springs; the hind legs store much of the kinetic energy expended in elevating the animal's body mass during leg extension. Thus, the type of gait employed becomes an additional factor in analyzing the cost of locomotion.

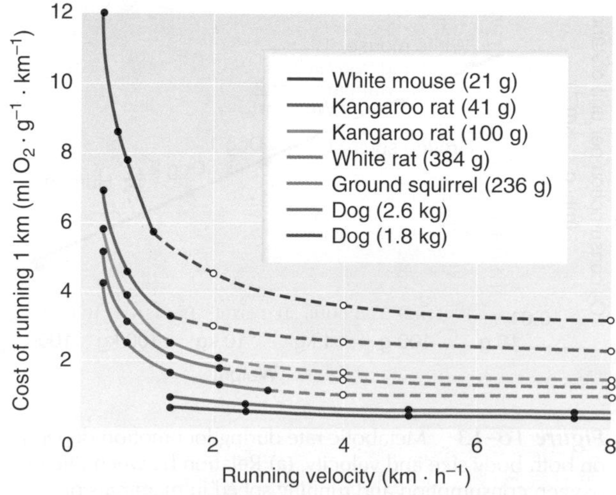

Figure 16-14 Among quadrupedal mammals, the energetic cost of transporting a unit of body mass by running decreases as body size increases. The cost of running 1 km drops and levels off with increasing velocity. Dashed portions of curves and open circles are extrapolated. [Adapted from Taylor et al., 1970.]

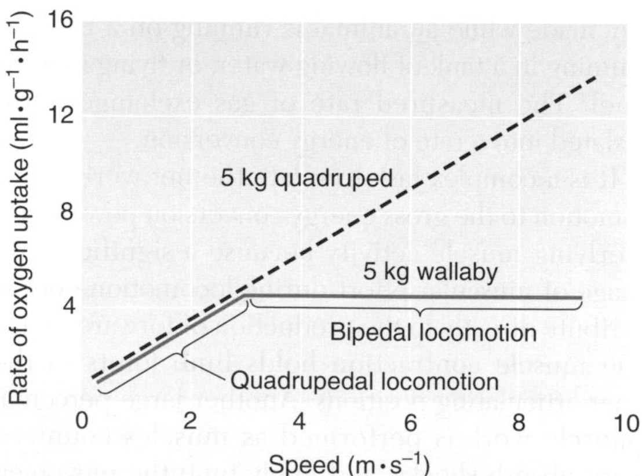

Figure 16-16 Hopping bipedal animals such as wallabies and kangaroos can increase forward velocity with no increase in oxygen consumption. Similar-sized quadrupeds and wallabies initially show a linear increase in oxygen consumption as velocity increases. As wallabies switch to bipedal locomotion, however, their rate of oxygen consumption does not increase with further increases in velocity. Blood lactate does not rise following the switch to bipedal locomotion even as velocity continues to increase, indicating that the higher speeds are not achieved by an increase in anaerobic metabolism. [Adapted from Baudinette, 1991.]

Physical Factors Affecting Locomotion

The metabolic cost of moving a given mass of animal tissue over a given distance also depends on the purely physical factors of inertia and drag.

Inertia is the tendency of a mass to resist acceleration, whereas *momentum* refers to the tendency of a moving mass to sustain its velocity. These concepts are closely related and are collectively referred to as **inertial effects.**

Every animate and inanimate object possesses both inertia and momentum proportional to its mass. The larger the object, the greater its inertia and the greater its momentum when it is in motion. The high inertial forces that must be overcome during the acceleration of a large animal like a moose account for a significant utilization of energy during the period of acceleration (Figure 16-17a). Small animals require less energy to accelerate to a given velocity and to decelerate to a complete stop. Therefore, a small animal starts and stops abruptly at the beginning and end of a locomotory effort (imagine a scurrying ghost crab on the beach), whereas a large animal accelerates more slowly after locomotion begins and slows down more gradually as locomotion ends (Figure 16-17b). Similarly, in terrestrial quadrupeds, the limbs are engaged in back-and-forth movements during running. The limbs are subject to inertial forces related to their mass as they accelerate and decelerate during locomotion. The limbs of a large animal exhibit greater inertia and momentum than do those of a small animal (consider the stride of a cat and a giraffe).

(a)

Rate of energy expenditure per unit weight

Large animal

Small animal

Period of locomotion

Time

(b)

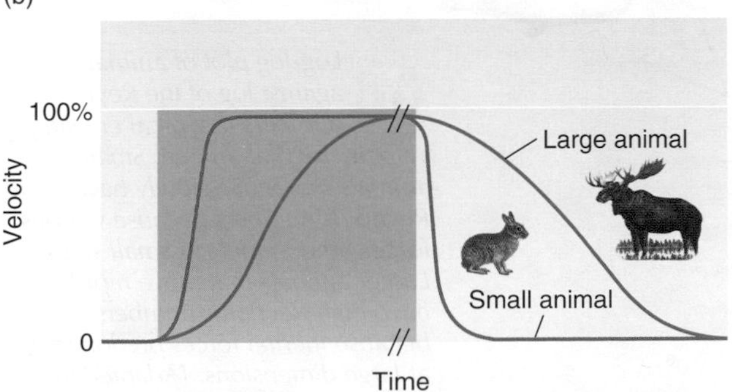

Figure 16-17 Body mass affects both rate of energy expenditure and acceleration during locomotion. **(a)** Rate at which energy is used per unit mass during onset and maintenance of locomotion (shaded region) in a large and a small animal of similar shape. **(b)** Velocity of a small and a large animal during acceleration and deceleration at the beginning and end of a period of locomotion (shaded region).

The energetics of sustained locomotion are also affected by the physical properties of the gas or liquid through which the animal moves. **Drag** is the force exerted in the opposite direction of an animal's movement by the viscosity and density of the medium. The drag produced in a given medium depends on the velocity, surface area, and shape of a moving object. For an object of a given shape, drag is proportional to surface area. Because larger animals have lower surface-to-mass ratios, they experience less drag per unit mass than do smaller animals, for which overcoming drag is energetically more costly. Once it is under way, a larger animal expends less energy per unit mass to propel itself at a given velocity than does a smaller animal of similar shape (see Figure 16-17a). Drag is also proportional to the square of an animal's velocity, meaning that the energy required to overcome drag and propel the animal at faster speeds increases sharply with velocity.

These effects are far more pronounced in water than in air because water has a higher viscosity and density, and produces far more drag on a moving object, than air does. Drag is a major factor in the energetics of swimming animals, but of little importance to running animals, because the viscosity of air is low. These relations are quantified in the **Reynolds number** (see Spotlight 16-2 on the next page). While air is much less viscous in absolute terms than water, its viscosity relative to its density—termed its **kinematic viscosity**—is surprisingly high. That is, air puts up surprising resistance given its low mass. This characteristic of air, combined with the often high velocities of flying animals, makes drag a significant factor for them but also permits high-speed maneuvering and the other aerial acrobatics that one witnesses when, for example, a blackbird harasses a small hawk that has strayed too near its nest.

Aquatic, Aerial, and Terrestrial Locomotion

Animals have evolved highly diverse mechanisms for moving in water, on land, and in air. Yet each mode of locomotion is similarly constrained by the medium in which the animal moves and by the laws of physics.

Swimming

Animals that swim in water need to expend little or no energy to support their own weight. Many aquatic animals have evolved flotation bladders or large amounts of body fat that suspend them in the water column with little expenditure of energy. Although the high density of water allows them to be neutrally buoyant, it also produces high drag, and thus a potentially high cost for locomotion. This hindrance to objects moving through a fluid has led to a convergence of body forms among

THE REYNOLDS NUMBER: IMPLICATIONS FOR LARGE AND SMALL ANIMALS

The energy expended in propelling an animal through a fluid medium (water or air) depends in part on the flow pattern set up in the medium. The flow pattern is determined not only by the density and viscosity of the medium, but also by the dimensions and velocity of the animal. Osborn Reynolds (1842–1912), a British engineer with an intense interest in fluid dynamics, combined these four factors in a dimensionless ratio that relates inertial forces (proportional to density, size, and velocity) to viscous forces. This ratio is the so-called Reynolds number (*Re*), which is calculated as

$$Re = \frac{\rho VL}{\mu}$$

where ρ is the density of the medium, V is the velocity of the object, L is an appropriate linear dimension of the object, and μ is the viscosity of the medium. Thus, when an object moves through a fluid such as water or air, the flow pattern depends on its *Re*. The larger the object, or the higher its speed in water, the higher is its *Re*. The same object moving at the same speed in air as in water would be characterized by a lower *Re* (about 15 times lower) because of the much lower density of air.

A Reynolds number below 1.0 characterizes movement in which the object produces a purely laminar pattern of flow in the fluid passing over its surface. If *Re* is above about 40, turbulence begins to appear in the wake of the object. As *Re* exceeds about 10^6, the fluid in contact with the object becomes turbulent. At this point, the energy needed to increase the velocity further rises steeply. The velocity at which turbulence appears is higher for a streamlined object such as a dolphin than for a nonstreamlined object, such as a human scuba diver with tanks and gear strapped on. Because the benefit of a streamlined form is a reduction in turbulence, streamlining offers no advantage for small organisms operating at very low Reynolds numbers—these organisms do not experience turbulence.

To a small organism such as a bacterium, spermatozoan, or ciliate, the watery medium appears far more viscous than it does to a human. The viscosity encountered by *Paramecium* swimming through water has been compared to the viscosity that would be experienced by a person swimming through a pool of honey. This is another example of an allometric scaling effect. Viscous effects are proportional to surface area, which rises with the square of animal length. Inertial effects, due to momentum of the moving animal, are proportional to mass, which rises with the cube of length. Thus, the movements of small organisms are dominated by viscous effects, whereas those of large animals are dominated by inertial effects.

The relative importance of these two factors for animals of different sizes coasting through a medium can be illustrated by forcing a tiny toothpick (low *Re*) floating on water to a given velocity, say 0.1 $m \cdot s^{-1}$, and then doing the same to a large floating log (high *Re*) of similar physical proportions but larger size. When let go, the little toothpick abruptly comes to rest due to the drag exerted on it by the viscosity and cohesion of the water. In contrast, the massive log coasts for many seconds after its release because its far greater momentum (deriving from its much greater mass) overcomes the drag (based on its surface area). Similarly, *Paramecium* comes to an abrupt halt when its cilia stop their rapid beating, whereas a whale coasts with little loss of velocity between the slow thrusts of its flukes.

The role of the Reynolds number and the effects of turbulence have become apparent in the recent spate of new human swimming speed records, due in part to new materials developed for swimming suits. Typically, these materials reduce turbulence, and thus drag on swimmers, shaving time off their performance.

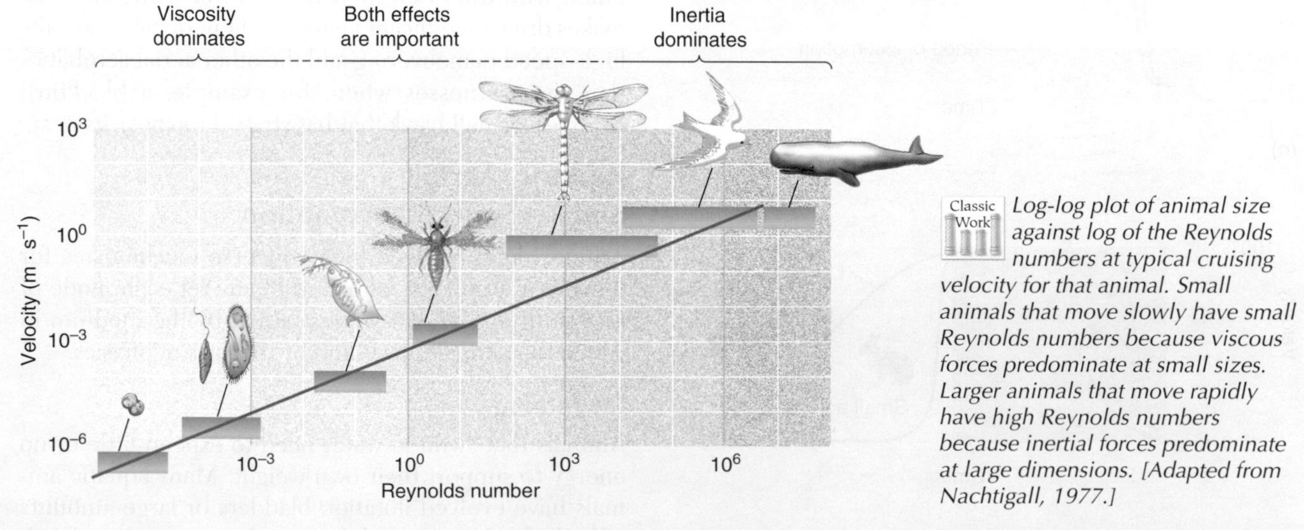

Classic Work *Log-log plot of animal size against log of the Reynolds numbers at typical cruising velocity for that animal. Small animals that move slowly have small Reynolds numbers because viscous forces predominate at small sizes. Larger animals that move rapidly have high Reynolds numbers because inertial forces predominate at large dimensions. [Adapted from Nachtigall, 1977.]*

marine mammals and fishes. The streamlined, fusiform (torpedo-like) body shape is wonderfully developed in most sharks, teleost fishes, and dolphins. Why is this shape so effective for movement through water?

The ease with which an object moves through water depends in part on how water flows over its surface. The fluid at the immediate surface of the object moves at the same velocity as the object; fluid at a distance from the object is undisturbed. If the transition in fluid velocity is smoothly continuous at distances progressively farther from the object's surface, then flow at the boundary layer—the layer of unstirred fluid immediately adjacent to the object's surface—is *laminar* (Figure 16-18). In contrast, *turbulent* flow results when there are sharp gradients and inconsistencies in the velocity of flow. Because of conservation of energy, pressure and velocity are reciprocally related in a given fluid system, and the higher the velocity of fluid at a given site, the lower the pressure. Thus, strongly differing flow rates around an object cause eddy currents owing to secondary flow patterns set up between regions of high and low pressure. Moreover, the higher the viscosity of the medium, or the higher the relative motion between the object and the surrounding fluid, the greater the shear forces produced and, hence, the greater the tendency toward turbulence. Because turbulence dissipates energy as heat, it retards the efficient conversion of metabolic energy into forward propulsive movement.

Long, streamlined shapes promote laminar flow with minimal eddy current formation. Fishes and marine mammals such as seals, porpoises, and whales are admirably streamlined, exhibiting nearly turbulence-free passage through water, even at high speeds. Birds are similarly streamlined in flight. An additional factor reducing turbulence in these animals is the compliance (deformability) of the body surface. A high body compliance damps small perturbations in the pressure of the water flowing over the body surface and thereby lessens the local variations in water pressure that give rise to energy-dissipating turbulence. Thus, the thick, compliant blubber layer of whales, for example, serves

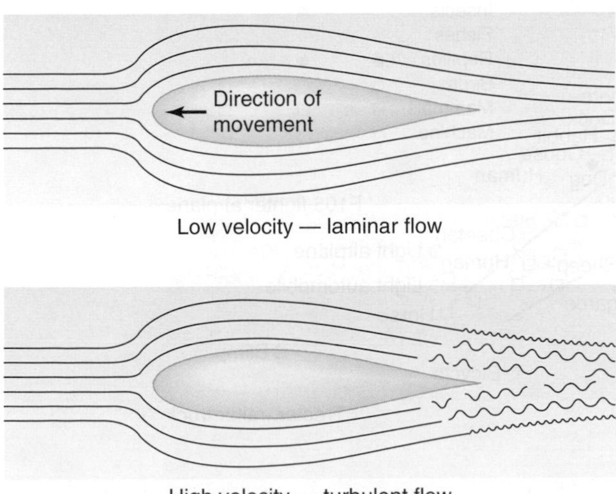

Figure 16-18 The pattern of fluid flow around a body moving through water depends on the body's forward velocity. Rapid movement through a fluid creates turbulence owing to uneven fluid pressures. Laminar flow occurs where pressure gradients are minimal. The larger the body and the less viscous the fluid, the higher the velocity before turbulent flow is created.

the dual purpose of insulating the animal and reducing the cost of locomotion.

The speed of a swimming animal is proportional to the power-to-drag ratio. The power developed by contracting muscle is directly proportional to muscle mass, and, if we assume that muscle mass increases in proportion to total body mass, we can assume that power (thrust) rises in proportion to body mass. On the other hand, for a large swimming animal at a given velocity, total drag increases, but drag per unit mass decreases, as body mass increases. If the shape remains constant, the surface and cross-sectional areas (which determine the drag) increase as a function of some linear dimension squared, whereas body mass (which determines the power available) increases as a function of that linear dimension cubed. Thus, a large aquatic animal can develop power out of proportion to its drag forces, which is why large fishes and marine mammals can swim faster than their smaller counterparts. However, because of the high drag forces developed in water and because drag increases as the square of velocity, aquatic animals can reach the speed of a bird in flight only if they are much larger and more powerful than the bird.

Flying

Unlike water, air offers little buoyant support, so all fliers must overcome gravity by utilizing the principles of aerodynamic lift. Although the effects of drag increase with speed, there is still less need for streamlining among birds than among fish because of the very low density of air. Thanks to the relatively low drag forces developed, birds can therefore achieve much higher speeds than fish. The production of propulsive force, which drives a bird forward, and lift, which keeps it aloft, is accomplished simultaneously during the downstroke of the bird wing (Figure 16-19 on the next page). The wing is driven downward and forward with an angle of attack that pushes the air both downward and backward, creating an upward and forward thrust. The components of lift and forward propulsion overcome the bird's weight and drag, respectively.

The body shapes of fish and birds demonstrate the great differences that exist between the physical properties of water and air, as well as the biological divergence that results from adaptation to these two dissimilar media. When a bird is gliding, its elongated, extended wings take the form of an airfoil and produce excellent lift; if the bird were in water, extended wings would obviously generate far too much drag. Thus, the wings of penguins are modified to serve as short paddles and are folded against the body while the birds are coasting under water. Because drag forces are much higher in water than in air, only medium-sized to large animals can coast in water. In contrast, only very small flying animals, smaller than a dragonfly, are unable to coast (glide) in air. Small insects such as flies and mosquitoes must continually beat their wings to maintain headway because they have very little momentum.

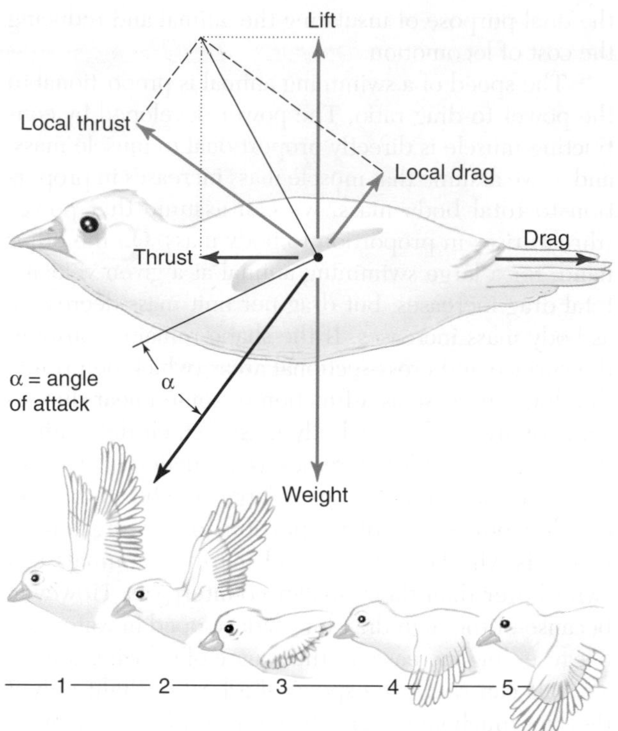

Figure 16-19 Wing movements during bird flight generate and experience numerous forces in several directions. The downstroke of a bird wing (stage 3) creates thrust and lift that oppose drag and weight. All the forces shown relate to wingbeat except for weight, which relates to the body itself. Local drag equals the drag produced as a consequence of lift production, while local thrust is complementary to local drag.

Running

When swimming, flying, and running are compared with respect to the energy cost of moving a given body mass a given distance (Figure 16-20), terrestrial locomotion (i.e., running) is clearly the most costly, whereas swimming is the least expensive. A swimming fish expends less energy in locomotion than a flying bird because, as already noted, the fish is close to neutral buoyancy, whereas a bird must expend energy to stay aloft. But why is running less efficient than either flying or swimming?

Running differs from swimming and flying in the way the limb muscles are used, and this difference accounts for the low work efficiency of running. When a bipedal or quadrupedal animal runs, its center of mass (CM) rises and falls cyclically with the gait. A rise in the CM occurs when the foot and leg extensors push the body up and forward, and a fall occurs as gravity inexorably tugs at the body, bringing it back to earth between extensions. Locomotor efficiency is reduced because the antigravity extensor muscles that contract to propel the CM upward and forward must also break the fall of the CM that occurs before the next stride. To control the fall, the extensor muscles must expend energy to resist lengthening as they slow the rate of descent of the body in preparation for the next cycle. Yet all this work expended in raising and lowering the CM does not directly contribute to the forward velocity of the animal. Such mechanically unproductive use of muscle energy to counteract the pull of gravity pro-

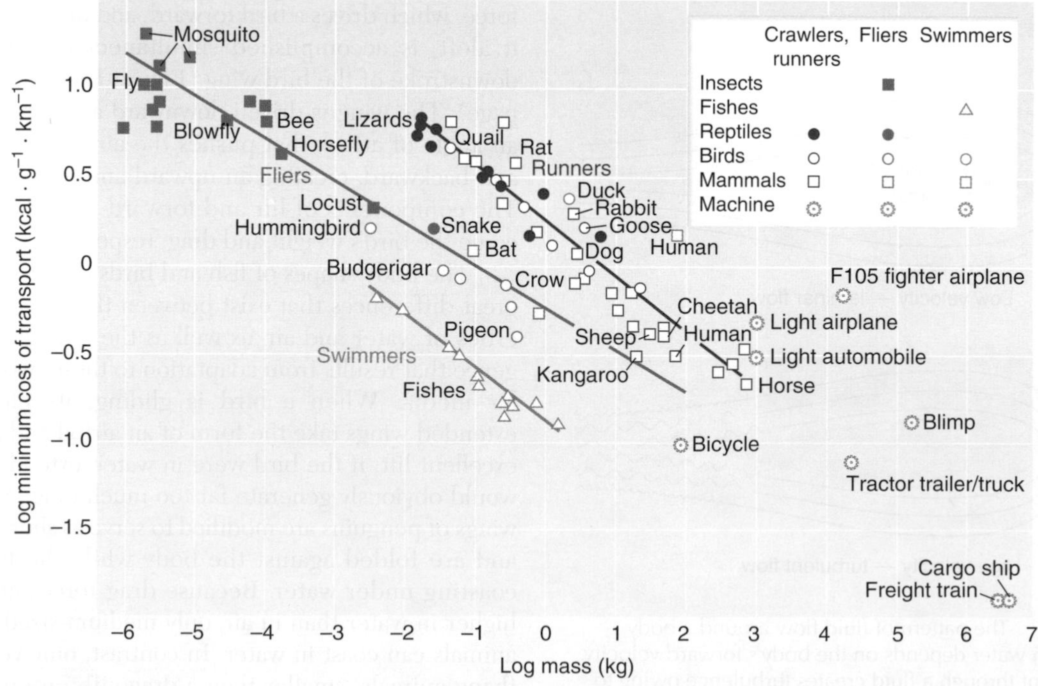

Figure 16-20 An animal's metabolic cost of transport is more closely related to its mode of locomotion (e.g., swimming, running, or flying) than to its taxonomy (e.g., bird, insect, or mammal). For comparison, cost of transport is also given for several artificial mechanical forms of transportation. [Data from Tucker, 1975.]

duces negative work. You experience negative work of this sort in the extensor muscles of your legs when you hike down a steep trail.

In short, running or walking is less efficient than flying, swimming, or bicycling because the muscles must be used for deceleration (negative work) as well as acceleration (positive work). One reason that a person riding a bicycle is so efficient (and thus why people can cycle many times faster and farther than they can run) is that the CM does not rise and fall, which means that more muscular energy can be transferred into forward velocity.

Elastic energy storage in elastic components of the limbs appears to be especially important in running and hopping animals. Consider the bounding of a kangaroo. The greater the height achieved during the hop, the greater the speed of descent; when the legs strike the ground, energy is transferred to the elastic components in the limbs, and the faster the descent, the greater the force of elastic recoil of the limbs when they subsequently extend to begin the next hop. Not many terrestrial animals actually hop, but the concept of elastic energy storage is important when considering changes in gait (e.g., walking to trotting to galloping in a horse). By changing gait at appropriate speeds, land animals enhance their locomotory efficiency and avoid potentially injurious forces on their legs. Consider a pony trained to trot on a treadmill at a speed at which it would normally gallop, or to gallop when it would normally trot, or to trot when it would normally walk. In all cases, it expends more energy than it would if allowed to change its gait naturally. Optimum gaits result from the relative amounts of energy stored in the elastic components of the body, such as tendons, when performing the different gaits. Little energy is stored when an animal walks; more is stored when it trots. When the animal is galloping, its entire trunk is involved in elastic storage. At least half the negative work done in absorbing kinetic energy during the stretching of an active muscle is released as heat; the remainder is stored in stretched elastic components such as the muscle crossbridges, sarcoplasmic reticulum, and Z lines of muscle cells and tendons. Only the elastically stored energy is available for recovery on the rebound, and only from 60% to 80% of that energy is recovered on its release. The energy converted into heat is not available for conversion into mechanical work in living tissue, and in fact accounts for some of the typically large heat load that an animal encounters during sustained running.

Intermittent Locomotion

Irrespective of whether an animal runs, flies, or swims, moving through the environment can be metabolically costly. One way to reduce the cost of locomotion (and, potentially, to reduce the attention of predators) is to exhibit **intermittent locomotion**, which in its most simple form means that an animal inserts periods of muscular inactivity between bursts of locomotion. For an animal that walks, runs, crawls, or climbs, this means simply moving in short, discrete bursts of activity. In swimming or flying animals, it means gliding between active power strokes. The cost savings of gliding can be substantial. Diving mammals such as seals and whales use intermittent swimming throughout a dive, and in doing so expend less energy, perform less work, and consume less oxygen (Figure 16-21a). For example, a Weddell seal that consumes about 100 ml $O_2 \cdot kg \cdot min^{-1}$ while swimming constantly during a dive uses only about 80 ml $O_2 \cdot kg \cdot min^{-1}$ when it alternates between active swimming and gliding (Figure 16-21b). Even taking into account its slower overall velocity, intermittent

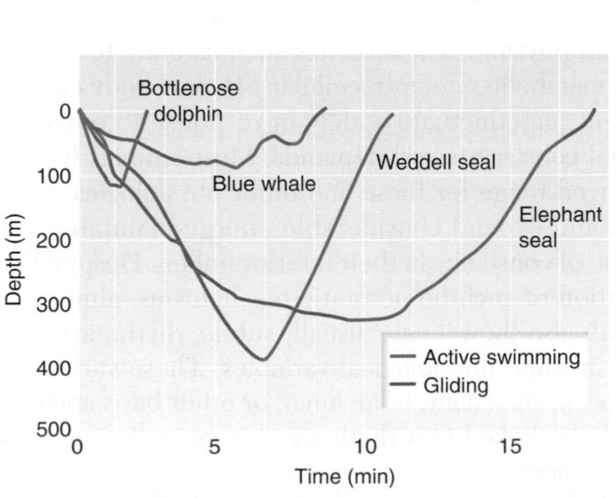

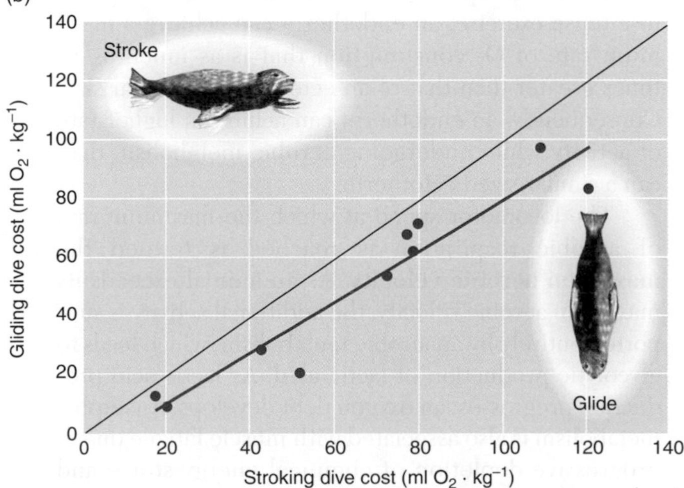

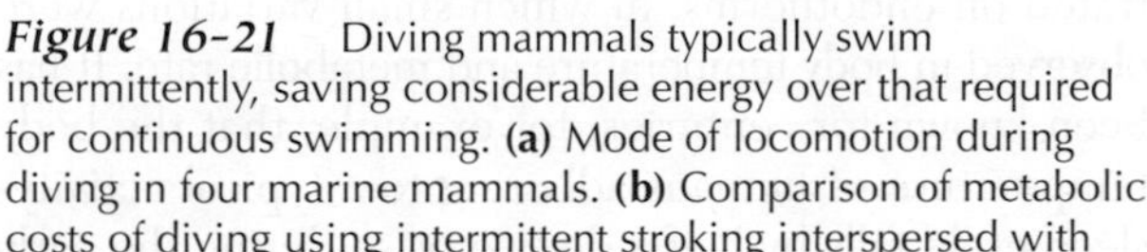

Figure 16-21 Diving mammals typically swim intermittently, saving considerable energy over that required for continuous swimming. **(a)** Mode of locomotion during diving in four marine mammals. **(b)** Comparison of metabolic costs of diving using intermittent stroking interspersed with periods of gliding, versus continuous swimming. At any given cost of locomotion, the intermittent gliding mode is about 20% less expensive than continuous stroking. [Adapted from Williams et. al., 2000.]

locomotion allows diving mammals to dive deeper and longer than if they swam continuously.

Locomotory Energetics of Ectotherms versus Endotherms

Endothermic animals generate heat as a by-product of metabolism, raising their body temperature above ambient temperature; ectotherms, with low metabolic rate and low heat production, rely on environmental heat for the degree of temperature regulation that they are able to achieve. (See Chapter 17 for detailed discussion of endothermy versus ectothermy.) Terrestrial endotherms and ectotherms of the same size typically expend a similar amount of metabolic energy to run at a given velocity. When O_2 consumption is plotted against running velocity for a lizard and a mammal of similar size, the aerobic regions of the two plots exhibit similar slopes. That is, when movement begins, a similar increment in metabolic energy expenditure is required for a similar incremental increase in velocity for both the mammal and the lizard. The difference between the two animals lies in the lower y-intercept of the lizard's plot relative to its standard metabolic rate while at rest. The reason for the differences between the resting metabolic rates and the y-intercepts is not completely understood, but these differences may represent the "postural cost" of locomotion—this cost being higher for a mammal than for a lizard.

As noted above, O_2 consumption rises linearly with increasing velocity of locomotion in ectotherms as well as endotherms. An endotherm of a given mass typically has a basal metabolic rate about 6 to 10 times that of an ectotherm of similar mass. In both groups, a similar relation exists between the basal rate and the maximum metabolic rate that can be achieved with intense exercise. That is, the factorial scope for locomotion exhibited by both groups is about the same. Thus, in response to intense exercise, an endotherm can achieve a maximum rate of O_2 consumption that is as much as 10 times greater than that of an ectotherm of similar size. Consequently, an endotherm can achieve a higher rate of activity while undergoing aerobic metabolism than can a similar-sized ectotherm.

The locomotor speed at which the maximum rate of aerobic respiration is reached is termed the **maximum aerobic velocity.** As an animal exceeds its maximum aerobic velocity, the additional activity is supported entirely by anaerobic metabolism, which leads to glycolytic production of lactic acid. As lactic acid production progresses, an oxygen debt develops. Anaerobic metabolism is also associated with muscle fatigue due to progressive depletion of chemical energy stores and metabolic acidosis, which if extreme can disrupt tissue metabolism. Consequently, anaerobic metabolism is unsuitable for sustained activity. Thus, only locomotion below the maximum aerobic speed can be sustained by either endotherms or ectotherms. Because endotherms are capable of far higher rates of aerobic metabolism than are ectotherms, they are generally capable of higher rates of sustained locomotor activity.

Clearly, the implications of ectothermy and endothermy are not limited to mechanisms of temperature control, but are also of great importance to the kinds of activity that an animal can undertake. The metabolic differences between ectotherms and endotherms determine, for example, how far and how fast they can travel. This is not to say that ectotherms cannot achieve rates of activity and speeds of locomotion as high as those of endotherms (consider a barracuda knifing through the ocean after its prey). However, because locomotor activity in excess of the maximum aerobic velocity relies on prodigious rates of anaerobic metabolism, high rates of locomotion can be sustained by ectotherms only for brief periods. This limitation can be observed in ectothermic vertebrates such as some lizards and frogs, in which a rapid burst of activity, seldom lasting more than a few seconds, takes the animal quickly to a new resting or hiding place. In some fishes, more than 50% of body mass is glycolytic white muscle fibers specialized for very short bursts of locomotor activity. The disadvantages of ectothermy, such as the inability to sustain high rates of activity, are offset by its more modest energy requirements, which enable ectotherms to spend more time hiding and less time exposing themselves to predators as they look for a meal.

Why can fleas leap hundreds of times their own body length, whereas most medium-sized mammals can jump only a few body lengths, and large mammals do not even attempt to leap? Consider both the physical surroundings of the animals and their structural makeup.

BODY RHYTHMS AND ENERGETICS

Certain physiological variables such as body temperature, metabolic rate, intracellular pH, and body energy content may fluctuate widely in response to environmental constraints and demands. Most animals have a preferred range for these and other physiological variables and expend considerable energy maintaining a degree of constancy in their interior milieu. Despite the evolution of such homeostatic mechanisms, almost all animals also show innate, usually subtle, rhythmic variations in some physiological variables. These variations may occur on a daily, tidal, lunar, or other basis and can usually be linked to a rhythmic change in the animal's environment.

Early experiments on biological rhythms concentrated on endotherms, in which small variations were observed in body temperature and metabolic rate. It has been known for centuries, for example, that the body temperature of humans adhering to a typical activity-sleep cycle falls by half a degree or so during the early morning hours (about 3:00–5:00 A.M.), only to rise again at about normal waking time. In fact, virtually all animals and plants show some type of rhythmic variation in metabolism or some other physiological variable. Biological rhythms are so innate to animals that even individual cells in a cell culture show rhythms in the rate of cell division. The cell does not have to be particularly complex—a daily rhythm occurs, for example, even in prokaryotic cells such as nitrogen-fixing cyanobacteria.

Circadian Rhythms

Biological rhythms lasting from milliseconds (at the cellular level) to years (at the whole-animal level) have been identified in a wide variety of animals. Most of these rhythms (and certainly many of the most prominent and well-studied ones) relate to daily cycles and are called **circadian rhythms.** A true circadian rhythm, which is endogenously generated, can be distinguished from a physiological or other variable that just happens to track daily changes in the environment by the use of four criteria: persistence, temperature independence, susceptibility to disruption, and entrainment. Let's examine each criterion in turn.

Persistence

The first criterion for a circadian rhythm is that it must persist for at least several days or weeks in an animal that has been removed from its natural environment and placed in a laboratory setting with constant environmental conditions (constant temperature, constant light or dark, etc.). A true circadian rhythm will persist in the animal, manifested most often as a continuation of that animal's normal daily cycles, with a period of approximately 24 hours. One of the most frequently measured indices of circadian rhythm is the general activity level. Figure 16-22 shows a typical apparatus for measuring the locomotor activity of a rodent. When the animal runs in the exercise wheel, its activity is recorded using a computer as a data-logging device. Appropriate activity-measuring devices can be substituted to record activity in birds, fish, or virtually any other animal.

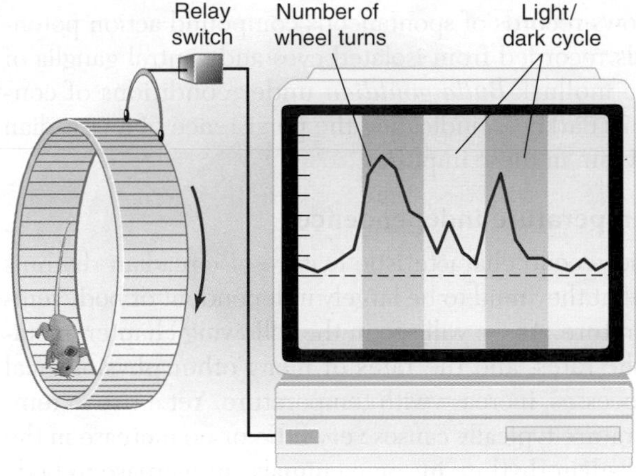

Figure 16-22 Circadian rhythms can be recorded by monitoring spontaneous activity. In this example, a rat exposed to a day of 12 hours of light and 12 hours of darkness is allowed free access to an exercise wheel. Every turn of the wheel activates a relay switch that temporarily closes an electric circuit. The resulting electrical record of activity is then logged in a computer for later analysis of the rat's activity patterns, in this case over a two-day period.

While general activity is a useful indicator of circadian rhythms, many different physiological or behavioral characteristics can be measured. Figure 16-23a

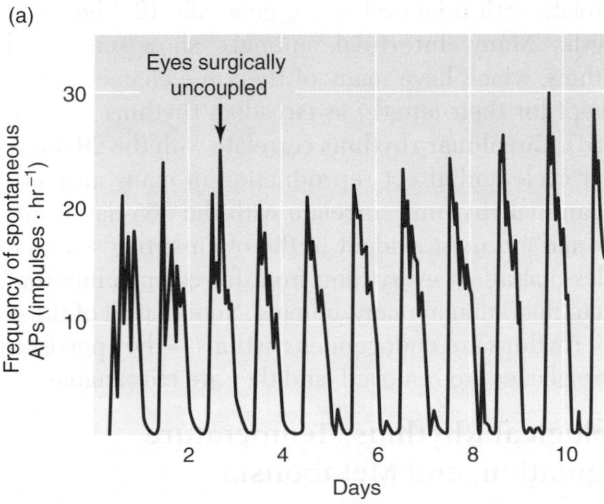

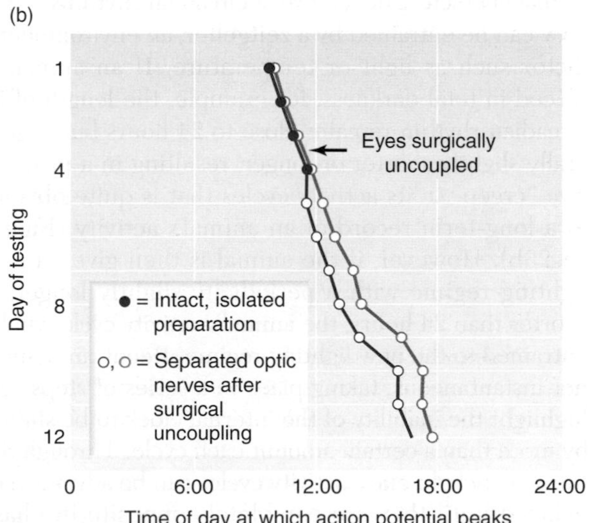

Figure 16-23 Circadian rhythms persist after removal of environmental cues such as light. **(a)** This recording shows spontaneous action potentials detected in the optic nerves of the mollusk *Bulla gouldiana*. An isolated preparation consisting of the two eyes and connected central ganglia was bathed in artificial seawater and kept in total darkness. A strong circadian rhythm persisted in the isolated, darkened preparation and continued unabated in each of the two eyes even after they were surgically uncoupled from each other (arrow). **(b)** Time of day at which the frequency of action potentials is highest in the preparation kept in total darkness. Note that the circadian rhythm persists in darkness, but like many such rhythms, begins to shift slowly without daily entrainment by light. [Adapted from Page and Nalovic, 1992.]

shows records of spontaneous compound action potentials recorded from isolated eyes and central ganglia of the mollusk *Bulla gouldian* under conditions of constant darkness, indicating the persistence of a circadian rhythm in these impulses.

Temperature independence

The second characteristic feature of circadian rhythms is that they tend to be largely independent of body temperature. As we will see in the following chapter, metabolic rates, and the rates of many other physiological processes, increase with temperature. Yet a rise in temperature typically causes very little or no increase in the circadian rhythm; in some animals, an increase in body temperature may actually slow down the circadian rhythm.

Susceptibility to disruption

Circadian rhythms are also characterized by the fact that they are conditionally arrhythmic—that is, a certain regime of modified temperature, lighting, oxygen level, or some other environmental parameter can disrupt the normal circadian rhythm. Often there is a threshold temperature above or below which the circadian rhythm is interrupted due to other physiological responses initiated to counter the effects of cold or heat. The effects of light are more graded and sometimes less intuitive. In the mosquito, for example, a circadian activity rhythm is evident when the light phase of an experimental light-dark cycle consists of only low light, but this rhythm gradually diminishes as light intensity increases.

Entrainment

A final characteristic feature of circadian rhythms is that they can be entrained by a **zeitgeber**, an environmental factor such as light or temperature. If an animal is placed in total darkness, for example, the length of its circadian rhythm remains close to 24 hours but is generally slightly shorter or longer, resulting in a progressive "creep" in its activity cycles that is quite obvious in a long-term record of an animal's activity (Figure 16-23b). However, if the animal is then given a new lighting regime with a periodicity slightly longer or shorter than 24 hours, the animal's activity cycle will be entrained to the new lighting regime. Reentrainment is not instantaneous, taking place in a series of steps that highlight the inability of the internal clock to be shifted by more than a certain amount each cycle. Through the use of new light cues, activity cycles can be advanced or delayed until they are completely opposite in phase with the original circadian rhythm. (If you've been lucky enough to travel internationally across multiple time zones, consider the striking changes in your activity cycle.)

Light is the most effective zeitgeber, but environmental temperature, food availability, interactions with other animals of the same or different species, and even scaling effects that operate on the rate and timing of physiological processes all may influence rhythms of metabolism, activity, and many other basic functions of animal life.

If the most basic biochemical reactions of cells are temperature sensitive, how can an intrinsic "biological clock" controlling circadian rhythms itself be temperature independent?

Noncircadian Endogenous Rhythms

With the circadian rhythm as the baseline standard, endogenous biological rhythms can be further classified as infradian rhythms (less than a day in length) and ultradian rhythms (greater than a day in length).

Infradian rhythms are usually related to aspects of cell function. In fact, more than 400 distinct infradian rhythms in cell function have been identified to date. These cycles greatly affect animal energetics, but the effects are sometimes subtler and more difficult to measure than changes occurring on a daily or longer basis. Many infradian rhythms, such as those related to certain aspects of cell division, have not yet been correlated with any type of rhythmic environmental change. In some cases, the external environment may have little or no role; in other cases, we have probably just failed to identify the environmental factor that entrains the rhythm.

Ultradian rhythms are very common in animals. The influence of the moon, through both its light and its production of ocean tides, greatly affects the physiology of many intertidal animals. **Circatidal rhythms**, which correlate with tidal cycles, are generally 12.4 hours in length. Many intertidal animals show circatidal rhythms, which have many of the same characteristics (except for their length) as circadian rhythms (Figure 16-24). **Circalunar rhythms** correlate with the 29.5-day lunar cycle and affect reproduction in many animals. **Circannual rhythms** correlate with the 365-day Earth year and are most evident in the often strong seasonal cycles that affect everything from fur color to hibernation to migration in many animals. Note that all of these *circ-* rhythms are endogenous rhythms—they persist if external cues are removed, and they are entrainable.

Biological Rhythms, Temperature Regulation, and Metabolism

Many animals show circadian or other rhythms in body temperature. In endotherms, considerable amounts of energy are used in maintaining a constant body temperature, either directly through thermogenesis or indirectly by powering physiological mechanisms that regulate heat loss or gain. In ectotherms, body temperature directly affects the animal's metabolism. Thus, for both

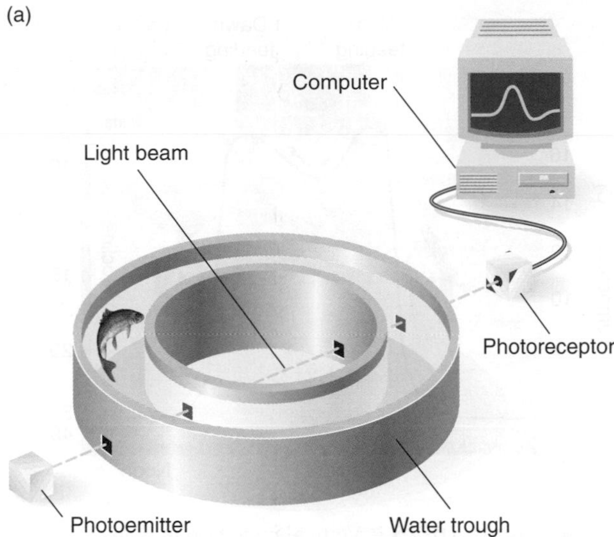

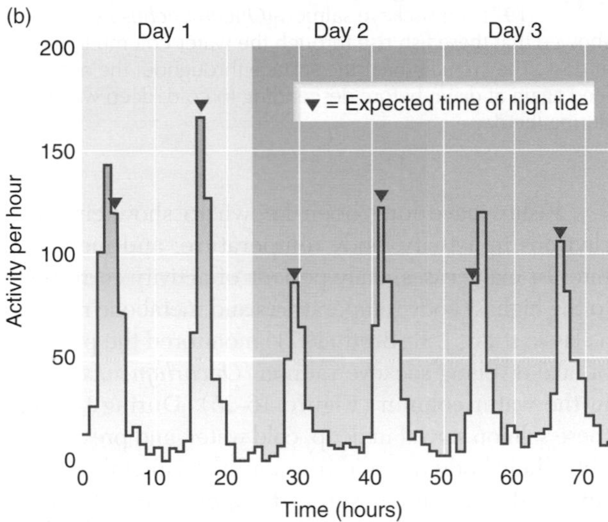

Figure 16-24 A shanny *(Lipophrys pholis)*, a small intertidal fish, displays activity rhythms related to tidal cycles. **(a)** Aquatic treadmill used to monitor swimming activity. Interruption of a light beam when the fish passes through it is recorded on a computer. **(b)** Activity records acquired in the laboratory under constant conditions indicate the persistence of a strong tidal activity rhythm. There are two high tides each day, which occur, on average, 50 minutes later each day. [Adapted from Horn and Gibson, 1989.]

endotherms and ectotherms, circadian and other rhythms affecting body temperature also affect an animal's energy metabolism. Because the consequences of biological rhythms for temperature regulation and for metabolism are inseparable, we will consider them together.

Vertebrate endotherms

Circadian rhythms in body temperature have been identified in most birds and mammals. There is a relatively strong scaling effect, with smaller animals showing larger circadian variation in body temperature. Thus, in very small endotherms such as shrews, deer mice, and hummingbirds, each weighing only a few grams, daily temperature fluctuations are often as much as 20°C. In humans weighing from 50 to 80 kg, the daily variation in body temperature is only about 0.6°C. The large daily temperature variation in small endotherms is probably due to the fact that many of them enter into a nightly (or, in some cases, daily) torpor, coupled with the greater metabolic cost of maintaining body temperature in very small animals. The smaller the animal, the greater the energy savings from allowing body temperature to fall by several degrees. (Chapter 17 deals more extensively with energetics related to torpor and hibernation.)

Why do circadian rhythms in body temperature occur in endotherms? Body temperature in an endotherm is a function of its heat production and heat loss and gain from the environment, so it follows that one or more of these factors must show rhythmic change to account for daily or other variations in body temperature. Relatively few studies have examined the total heat production and heat conduction budgets of an endotherm in the context of circadian or other rhythms. However, experiments in human subjects have simultaneously measured the circadian rhythms in body temperature, heat conductance, and heat production. Changes in heat production (i.e., changes in metabolic rate) account for about one-quarter of the daily swing of 0.6°C in core temperature, with the remainder resulting from changes in heat conductance between the core and the environment.

Some endotherms modify the amplitude of circadian body temperature rhythms (but not their periodicity) when exposed to stresses ranging from temperature extremes to inadequate food or water. In a classic study in the late 1950s on endotherm thermoregulation, K. Schmidt-Nielson and his colleagues studied African camels *(Camelus dromedarius)* in the Algerian Sahara desert. Well-hydrated and well-fed camels showed a circadian body temperature rhythm with an amplitude of about 2°C, but this increased to about 6°C when the camels were unable to drink water. The maximum core temperature (which occurred in late afternoon) increased and the minimum core temperature (in early morning) decreased. These changes reduced water loss from evaporative cooling during the day and decreased heat loss in the cooler nighttime by reducing the thermal gradient between the animal and its surroundings (see Chapter 14). Both reduced the need for metabolic heat production. Birds such as kestrels and pigeons, which also exhibit circadian body temperature rhythms, similarly show a large daily core temperature range, mainly owing to a low nighttime core temperature.

Field observations alone may be insufficient to identify circadian rhythms in body temperature because many endotherms also show strong daily rhythms in activity levels. Depending on the animal's ability to dissipate metabolically produced heat, a rhythmic rise in body temperature could result solely from increased locomotor activity that is itself a manifestation of a circadian rhythm. However, two lines of evidence indicate that there is usually a distinct intrinsic rhythm in body temperature independent of activity level: first, temperature rhythms persist in animals in the laboratory in which activity levels have been corrected for or controlled, and second, temperature rhythms persist in humans over several days of complete bed rest. Thus we find that circadian rhythms in body temperature are often imposed on activity rhythms with a similar time component, amplifying the daily range of body temperature.

Ultradian rhythms in body temperature are best exemplified by hibernators, which may lower their core temperature by 20–35°C for periods of weeks or months, punctuated by brief periods of arousal. In hibernating golden-mantled ground squirrels (genus *Citellus*), these rhythms are apparently entrained *in utero*, since they can persist for at least four years in squirrels isolated at birth from any light or temperature zeitgebers. In hibernating bats, circadian rhythms in body temperature and metabolism can be detected not only during active periods, but also at the much lower mean core temperature typical of hibernation. This observation emphasizes the generally temperature-independent nature of the biological clock that is responsible for circadian rhythms. However, the circadian rhythm eventually disappears as hibernation continues. Rodents such as the thirteen-lined ground squirrel (*Spermophilus tridecemlineatus*) show no evidence of a continuing circadian rhythm in oxygen consumption with the onset of hibernation.

Mammals living in environments with very stable conditions of temperature, light, and food availability often show little or no evidence of a circadian rhythm in either body temperature or metabolic rate. Fossorial (burrow-dwelling) pocket gophers and moles, for example, live in constant darkness with little temperature variation, and they exhibit no metabolic circadian rhythms. It is unclear what advantage there would be to a significant rhythm of body temperature and metabolism in such animals. Some small mammals, such as voles, that have a herbivorous diet high in bulk must feed nearly constantly to derive sufficient energy. These animals similarly show little or no metabolic rhythmicity.

Vertebrate ectotherms

All ectotherms, by definition, rely on external heat sources or heat sinks to modify their body temperature. The behavioral and physiological mechanisms used by ectotherms to regulate body temperature reflect circadian rhythms in the temperatures they prefer during different parts of the day. Because metabolic rate is closely related to body temperature, circadian rhythms in O_2 consumption and CO_2 production are closely correlated with body temperature changes.

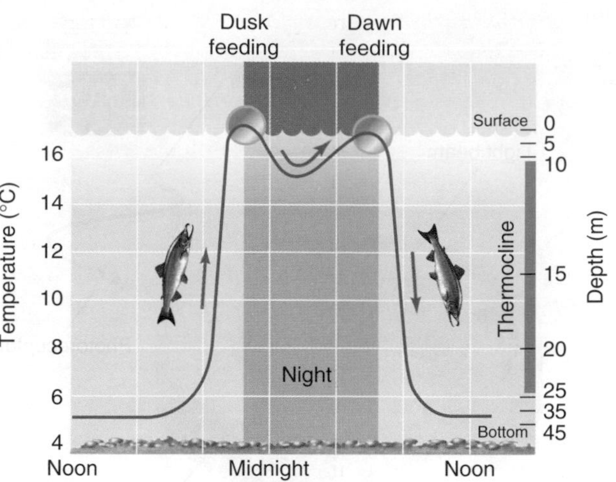

Classic Work **Figure 16-25** Vertical migration in many fishes shows a strong diurnal rhythm. A classic study by Brett (1971) on sockeye salmon *(Oncorhynchus nerka)* showed that these fish rise through the water column to feed at dusk. They remain near the surface throughout the night to feed again at dawn before descending to cold, deep water during the day.

Fishes have long been known to show circadian rhythms in activity, body temperature, and metabolic rate. In many cases, daily periods of activity correspond to the highest body temperatures and metabolic rates. In a classic study, J. R. Brett (1971) monitored the positions of lake-dwelling sockeye salmon *(Oncorhynchus nerka)* in the water column (Figure 16-25). During the day, these salmon stayed in deep, cold water, and presumably both their body temperature and their metabolic rate reflected the low water temperature. As dusk approached, the fish rose to the surface to feed, which brought them through the thermocline into water of about 17°C, where they remained for a bout of dawn feeding before descending to cold water for the day. This overall rhythm allows the sockeye salmon to conserve energy by maintaining a cold-induced low metabolic rate during periods of inactivity. Such circadian rhythms of vertical migration related to feeding are very common in pelagic fishes. When migrations take ectotherms through temperature gradients, their metabolic rate similarly shows daily variations. But are these changes in metabolic rate true circadian rhythms, or are they merely reflections of changes in body temperature? In fact, daily rhythms in oxygen consumption persist under conditions of constant temperature and light in many fish species.

Studies on circadian rhythms in temperature regulation and metabolism in amphibians are few. Circadian rhythms in preferred body temperature and activity level have been found in the aquatic salamander *Necturus maculosus,* but not in the toad *Bufo boreas,* the larvae of the frog *Rana cascadae,* or the salamander *Plethodon cinereus.* Many species of toads and frogs show pronounced activity patterns related to feeding, predation, and so forth, but we currently know very little about which of these patterns are cued by the external environment and which are due to innate circadian rhythms—that is, which are controlled by "biological clocks."

Many reptiles show a circadian rhythm in preferred body temperature. Resting metabolism tracks body temperature in these animals; consequently, the metabolic rate ranges up and down in the course of a 24-hour period. Again, establishing whether these thermoregulatory and metabolic rhythms are true innate rhythms requires that we monitor animals under constant environmental conditions. In the lizard *Sceloporus occidentalis* and other species, a circadian rhythm in preferred body temperature persists for several days under constant light conditions in the laboratory. Similarly, circadian rhythms in oxygen consumption are found in some species of the lizard *Lacerta*.

Invertebrates

Some degree of thermoregulatory capacity has been identified in invertebrates from many different phyla. Control is exerted primarily through behavioral means, but also by physiological means (such as metabolic heat production in flying insects). Daily or other rhythms in preferred body temperature have been observed in numerous arthropods, including crayfishes, shrimps, and a variety of insects such as silkmoths, bees, and wasps. Oxygen consumption tracks body temperature in these animals, so daily cycles in metabolic rate are seen as the animal's body temperature changes. Daily rhythms in oxygen consumption have also been observed in earthworms, amphipods, sea pens, and mollusks. Intertidal animals may show combinations of circadian, tidal, and lunar metabolic cycles.

With few exceptions, we don't know whether these rhythms are endogenous or are exogenously triggered. Male American silkmoths *(Hyalophora cecropia)* show endogenous rhythms of endothermic warming in preparation for flight, and these rhythms are unaffected by ambient temperature (recall that temperature independence is one of the criteria for a circadian rhythm). Honey bees *(Apis mellifera)* show very prominent circadian rhythms in oxygen consumption when kept in constant darkness, with oxygen consumption increasing from 20 to 30 times over resting levels during the periods that correspond to daylight activity and foraging. The fruit fly *Drosophila*, when kept in constant darkness, similarly shows an increase in metabolic rate during the period corresponding to daylight.

Unicellular organisms

The presence of true circadian or other rhythms in unicellular organisms is of particular interest to chronobiologists. Although the study of more complex animals has implicated the brain, pineal gland, and other tissues as sites of a biological clock, the presence of true biological rhythms in a unicellular organism indicates that all the necessary components for a clock can be found among its cellular organelles.

Unicellular organisms exhibit rhythms in rates of photosynthesis, oxidative metabolism, bioluminescence, cell division, growth, phototaxis, and vertical migration, to name but a few variables. The first definitive demonstration of an innate circadian rhythm in a unicellular organism was made in 1948 in the alga *Euglena gracilis*, which shows phototactic rhythms (migration toward light). Since that time, a variety of other eukaryotic cells, including *Paramecium*, have been shown to exhibit true circadian rhythms that persist under constant conditions and can be entrained. More recently, circadian rhythms have been found in the much simpler prokaryotic cells of cyanobacteria.

The knowledge that individual cells show circadian cycles in cell division has been considered in designing more effective chemotherapy for human cancer patients. Different chemotherapeutic agents act on different phases of the cell division cycle, which in humans has a circadian rhythm. By timing the administration of the drug to the vulnerable period of the cell cycle (typically, two o'clock in the morning rather than during the usual clinic hours), effectiveness can be increased by as much as tenfold. Moreover, undesirable and deleterious side effects of the treatment are also greatly reduced.

ENERGETICS OF REPRODUCTION

Reproduction, simply put, is the ultimate goal of all organisms, and the evolution of virtually all specializations can be linked directly or indirectly to improvement in an animal's reproductive fitness. Not surprisingly, reproduction often accounts for a considerable proportion of an animal's energy budget. Exactly what proportion depends on many factors, including mode of reproduction, body size, and whether an animal is ectothermic or endothermic. We begin by considering the different general patterns that have evolved for the investment of energy in reproduction.

Patterns of Energetic Investment in Reproduction

Natural selection on reproductive structures and processes has resulted in a wide variety of reproductive patterns. The most favorable mode of reproduction for a species is that which results in the largest possible number of offspring reaching sexual maturity. By the mid-1960s, several groups of ecologists and evolutionary biologists had recognized that, collectively, animals have evolved toward either of just two general patterns of energy investment in reproduction. These two patterns were called ***r*-selection** and ***K*-selection**, with the letters r and K coming from the logistics equation that models the growth rate of continually reproducing animal populations. (Consult a text in ecology or evolution for further details on the logistics equation and the growth of animal populations.)

r-Selection: "Smaller, more, on their own"

In the first pattern of energetic investment in reproduction, an animal produces offspring that are extremely small at the start of their development. By virtue of each offspring having a very small energy content (and thus a small energy cost to the parents), the parent(s) can produce very large numbers of offspring. Females of some species of sea urchins, for example, release as many as *100,000,000* eggs in a single spawning! Among vertebrates, some pelagic fishes similarly release huge numbers of eggs. The mackerel *(Scomber scombrus)*, for example, releases tens of thousands of eggs in one spawning. These are extreme examples, but most invertebrates and many ectothermic vertebrates produce dozens or more offspring in a single breeding episode.

The trade-off in producing large numbers of small offspring is that the parent is less able to afford parental care for each of so many offspring. Consequently, most *r*-selected animals simply release their offspring into the environment to fend for themselves. Because they are so small, vulnerable, and without parental protection, relatively few survive to reproduce. Only six out of a million mackerel spawn are estimated to survive to reproductive age.

K-Selection: "Larger, fewer, protected"

The second pattern of energetic investment in reproduction is exhibited by *K*-selected animals. These animals produce relatively large offspring—offspring that initially have a high energy content and represent a large energetic investment by the parents. As a consequence, the total number of offspring produced is far smaller than that produced by *r*-selected animals. Mammals and birds, for example, tend to produce offspring that, at birth or hatching, are at least a few percent of the body mass of the mother, and may be much larger. Litters or egg clutches rarely consist of more than 8 to 10 offspring because of the huge energy cost of producing each of them. However, because the number of offspring is small, they can be offered parental protection and nurturing, so *K*-selected animals usually invest additional energy in parental care. Because the offspring of *K*-selected animals tend to be larger at the outset and cared for during their early development, their chances of successfully reaching reproductive age are far greater than those of *r*-selected animals. Contrast the six-in-a-million chance of a mackerel surviving to reproductive age with that of a bird or mammal, whose chances may approach 50% or higher.

Like many classification systems, the classification of animals as *r*-selected or *K*-selected species is not absolute, and many species show characteristics of both patterns. Many cichlid fishes, for example, produce hundreds of small larvae (an *r*-selected characteristic). However, they then invest huge amounts of time and energy in parental care (a *K*-selected characteristic)—females may even stop their own feeding for weeks to protect their young.

Which kind of animal population, *r* selected or *K*-selected, would experience a greater proportion of juvenile mortality if the availability of energy in the form of food were to become severely limited?

Allometry and the Energetic Costs of Reproduction

Like virtually all other aspects of an animal's physiology, the energy cost of reproduction is affected by allometric scaling. Generally, larger animals invest relatively less of their energy in producing and nurturing (if *K*-selected) their offspring than do smaller animals. The value of the exponent in the allometric equation relating total energetic cost of reproduction to body mass ranges from a low of 0.52 in ducks and geese to 0.95 in hoverflies, with mammals showing a range of about 0.69–0.83.

If we consider both invertebrates and vertebrates, larger females within a species devote relatively more energy to reproduction than do smaller females. One can see this clearly in mammals: a large, fat rodent, feline, or canine produces a larger litter than does a thin female with few energy reserves. Yet in species with a sexual dimorphism in which males are larger than females, the data suggest that larger males expend relatively less of their energy on reproduction than do smaller males.

The Cost of Egg and Sperm Production

Reproduction begins with the production of gametes (eggs and sperm). The energy cost of gamete production varies greatly. Gamete production is a costly business in most invertebrates, with half or even more of the total energy assimilated being diverted into the gametes. Usually, the female makes the largest energy expenditure in producing yolky eggs that will nurture the growing embryos. Sperm production usually requires a smaller expenditure of energy, but there are some dramatic exceptions. The testes and associated accessory organs in males of the cricket *Acheta domesticus* account for 25% of the animal's body mass. The sperm packet that is passed to the female during copulation—the spermatophore—is about 2.5% of the male's body mass, and he produces two or more spermatophores per day.

Although the energy cost of gamete production may be high in invertebrates, very few of them expend energy on protecting and nurturing their offspring after they are produced. Some notable exceptions are found among the arthropods. Female scorpions and some spiders carry their eggs, and then their newly hatched offspring, on their backs. The lives of social insects such as ants and bees are entirely organized around the care and nurturing of offspring, as we will see.

Ectothermic vertebrates, like invertebrates, spend as much as half of their total energy budget on gamete production and reproduction. Sperm production is relatively inexpensive for vertebrate males, but egg production can be energetically costly for vertebrate females. As already mentioned, many fishes produce large clutches of eggs. Different species of oviparous (egg-laying) amphibians produce egg clutches ranging from 4 to 15,000 eggs. Estimates of the energy costs of these eggs are few, but in the salamander *Desmognathus ochrophaeus*, about 48% of the female's energy is used in a combination of egg production and parental care. In the lizard *Uta stansburiana*, reproductive behavior during spring consumes 32% of the energy expended by males and as much as 83% of the energy expended by females (26% for increased metabolic costs and 57% for producing eggs). Female crocodiles, alligators, and, especially, brooding snakes such as pythons similarly invest large amounts of energy in combined egg production and parental care.

The energy cost of gamete production is highly variable among endothermic animals. In birds, some of which produce large numbers of relatively large, yolky eggs, estimates of the cost of egg laying range from less than 10% to more than 30% of the animal's total energy budget. Domestic hens (white leghorns), which have evolved in response to human selection to maximize efficiency of egg production, expend about 15% to 20% of their energy on egg production. The energy cost of producing sperm in roosters, however, is negligible. Similarly, the cost of sperm production in mammals is of little import. Almost all female mammals produce very small numbers of very small eggs and, again, their energy cost is negligible. However, as we will see next, almost all mammals invest a considerable amount of energy in protecting their developing offspring after fertilization.

Parental Care as an Energy Cost of Reproduction

The energetic costs of parental care fall into two categories. First is the cost of the actual transfer of nutrients from a parent to the developing offspring. An excellent example is lactation in mammals, in which large amounts of milk rich in lipids and carbohydrates are secreted to nourish the newborns. Typically, lactating mammals expend as much as 40% of their energy on producing milk. In dairy cattle, which have been artificially selected for copious milk production, as much as 50% of total energy may be used for milk production.

Animals other than mammals also produce secretions for their developing offspring. One or both parents of many bird species regurgitate semi-digested food into the mouths of offspring. Although this practice is not as energetically costly as producing the same weight of milk, it has a real cost in that the regurgitated food might otherwise be digested and assimilated into the parent's own body. Doves produce "crop milk," a viscous fluid formed from the initial digestion of material temporarily stored in the crop and then given to offspring. Certain species of viviparous and ovoviviparous amphibians and reptiles produce uterine secretions ("uterine milk") that nourish the embryos until birth. Embryos of viviparous caecilian (apodan) amphibians use specialized dentition to scrape away and ingest the lining of their mother's oviduct. The female poison-dart frog (*Dendrobates pumilio*) returns to the small ponds that hold her larval offspring and deposits unfertilized eggs for her larvae to eat. Among the invertebrates, ants, bees, and wasps expend large amounts of their total energy gathering raw materials and producing honey or equivalent substances for the nourishment of the developing animals in the colony.

The second category includes the indirect costs of parental care–those costs associated with the behaviors and physiological responses specifically associated with parenting, including egg incubation. Complex parental care is evident in the numerous invertebrate taxa, including mollusks (octopus), polychaete worms, and the social insects (ants, bees, and wasps), that exhibit elaborate nest-building behaviors, brooding, and other forms of parental care. Among the vertebrates, parental care is found in all classes and is widespread in birds and mammals. In birds, which incubate their eggs, the metabolic cost of the heat produced for incubation can be sizeable. Incubating blue tits (*Parus caerulescens*), for example, expend up to 1/3 more energy than non-incubating birds. The Indian python (*Python molurus*) also incubates its eggs, expending energy by shivering to produce heat to warm them (see Chapter 17). Another indirect cost of reproduction associated with parenting may arise from the inability or unwillingness of parents to feed themselves during incubation of their eggs or protection of their newborns. This is most evident in some birds that forgo foraging for food to incubate their eggs. The male emperor penguin (*Aptenodytes forsteri*) continuously incubates his mate's eggs for 105–115 days without feeding. The food he failed to acquire—akin to "lost income"—must be part of the overall consideration of the energy budget for reproduction for such species. It has also been argued, however, that the calculation of energy cost should include energy *savings* that may accrue to some incubating birds because while they are nesting, they are not engaging in the energetically expensive behavior of flight to forage for food.

Ultimately, quantification of the energy cost of parental care is a difficult enterprise because parenting behaviors often include such activities as brooding, foraging, grooming, and so forth—complex behaviors, not easily duplicated in the laboratory, and not easily separated from ongoing energy costs not related to parenting. But one might equally make the argument that there can be no meaningful separation between the costs of reproduction and the cost of survival for an organism or a species.

SUMMARY

Describing and measuring metabolic rate

■ The total energy released in the conversion of a higher-energy compound into a lower-energy end product is independent of the chemical route taken. The utilization of chemical energy in tissue metabolism is accompanied by the inevitable release of heat as a by-product.

■ Food substances consistently release the same amount of heat and require the same amount of O_2 when oxidized to H_2O and CO_2. The rate of heat production or the rate of O_2 consumption (and CO_2 production) can therefore be used as a measure of metabolic rate.

■ The respiratory quotient—the ratio of CO_2 production to O_2 consumption—is useful in determining the proportions of carbohydrates, proteins, and fats metabolized, each of which has a different characteristic energy yield per liter of O_2 consumed.

Body size and metabolic rate

■ The basal metabolic rate and the standard metabolic rate are related to body size—the smaller the animal, the higher the metabolic rate per unit mass of tissue (termed metabolic intensity).

■ Although there is a fairly good correlation between metabolic intensity and body surface-to-mass ratio in endotherms, which suggests that metabolic rate is determined by heat balance mechanisms, this correlation may be incidental. Similar correlations are seen in ectotherms that are in temperature equilibrium with their environment.

Energetics of locomotion

■ The energetics of locomotion are also related to body size. The smaller the animal, the higher the metabolic cost of transporting a unit mass of body tissue over a given distance.

■ The Reynolds number (Re) of a body moving through a liquid or gaseous medium is the ratio of the relative importance of inertial and viscous forces in the medium.

■ The rate of energy utilization during different kinds of locomotion typically increases with velocity. By changing gait from walking to running, hopping, or trotting, terrestrial animals increase their efficiency. Increased efficiency is achieved when the energy of falling at the end of the stride is elastically stored for release during the next stride.

Body rhythms and energetics

■ Metabolic rate in many animals shows distinct endogenous rhythms that may be manifested in locomotor activity or changes in body temperature (in ectotherms). These rhythms may be circadian (daily), infradian (shorter than a day), or ultradian (longer than a day). Circadian rhythms are characterized by their persistence in the absence of environmental cues, are temperature-independent, can be disrupted, and are entrainable by zeitgebers such as light.

Energetics of reproduction

■ *r*-Selected animals produce large numbers of very small offspring and offer them no parental care. The low rate of offspring survival is offset by the large number of offspring produced. *K*-Selected animals produce small numbers of large offspring. Owing in part to parental care, the survival rate of offspring is high.

■ Energy costs of reproduction to parents include the costs of producing gametes, of providing nutrition, as in mammalian lactation, and of behaviors constituting parental care.

REVIEW QUESTIONS

1. Define *basal metabolic rate, standard metabolic rate,* and *respiratory quotient.*
2. Explain why the rate of heat production can be used to accurately measure metabolic rate.
3. Explain why respiratory gas exchange is a workable measure of metabolic rate.
4. Why is the surface hypothesis inadequate as an explanation for the high metabolic intensity of small animals?
5. Why does the viscosity of the medium in which it travels affect the locomotion of a small aquatic animal more than that of a large animal?
6. Summarize the factors that affect the flow pattern of fluid around a swimming animal. What factors minimize turbulence?
7. Why does riding a bicycle for 10 km require less energy than running the same distance at the same speed?
8. Why are the 100-foot ants and other scaled-up insect monsters of old grade-B science fiction movies anatomically untenable?
9. Give examples of the effect of body size on the metabolism and locomotion of animals.
10. The potency of some medications depends on metabolic factors. Explain why it would be risky to give a 100 kg person a drug dose 100 times greater than the dose proved to be effective in a 1 kg guinea pig.
11. How would you distinguish between (1) a true circadian rhythm, (2) a metabolic rhythm, and (3) a rhythm that is triggered by rhythmic changes in the environment?
12. Compare and contrast reproduction in *r*-selected and *K*-selected animals. What are the advantages of each pattern?

SUGGESTED READINGS

Blake, R., ed. 1991. *Efficiency and Economy in Animal Physiology.* Cambridge: Cambridge University Press. (Considers, in an evolutionary framework, the efficiency and metabolic costs of various types of animal locomotion.)

Dickson, M. H., C. T. Farley, R. J. Full, M. A. R. Koehl, R. Kram, and S. Lehman. 2000. How animals move: An integrative view. *Science* 288:100–106. (Examines the general principles of energy storage and exchange in running, flying, and swimming animals.)

Harrison, J. F., and S. P. Roberts. 2000. Flight respiration and energetics. *Annu. Rev. Physiol.* 62:179–206. (A broadly comparative review of the literature on flight energetics, dealing with insects, birds, and flying mammals.)

Hulbert, A. J., and P. L. Else. 2000. Mechanisms underlying the cost of living in animals. *Annu. Rev. Physiol.* 62:207–235. (A review article describing the cellular energetics that contribute to whole-body BMR.)

McMahon, T. A., and J. T. Bonner. 1983. *On Size and Life.* New York: Scientific American Books. (A delightfully illustrated book that examines how scaling and allometry pervade the world around us.)

McNeil, R. A. 1992. *Exploring Biomechanics: Animals in Motion.* New York: Scientific American Library. (A comprehensive treatment of the anatomy and biomechanics of animal locomotion.)

Schmidt-Nielsen, K. 1983. *Scaling: Why Is Animal Size So Important?* New York: Cambridge University Press. (A now-classic book that provides a wealth of information on how anatomy and physiology are affected by body size, and why physiologists should care.) Classic Work

Spicer, J. I., and K. J. Gaston. 1999. *Physiological Diversity and Its Ecological Implications.* Oxford: Blackwell. (A highly insightful book that discusses many aspects of allometry, metabolism, growth, and other physiological variables in the infrequently considered but highly relevant context of ecology and evolution.)

Underwood, H. A., G. T. Wassmer, and T. L. Page. 1997. Daily and seasonal rhythms. In *Handbook of Physiology,* Section 13, *Comparative Physiology,* Vol. 2, 1653–1763. New York: Oxford University Press. (Presents a thorough definition and discussion of the concepts surrounding biological rhythms.)

Wood, C. M., and McDonald, G., eds. 1997. *Global Warming: Implications for Freshwater and Marine Fish.* Cambridge: Cambridge University Press. (Describes in detail the effects of temperature on animals from molecular to population levels, focusing on examples from fishes.)

Energetic Costs of Meeting Environmental Challenges

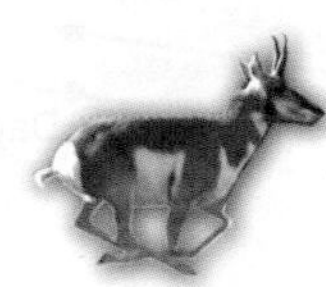

An animal's patterns of energy expenditure closely reflect the *who* and *how* of that animal. That is, we can tell much about an animal's energy budget by determining its size, the mechanism by which it moves, and how it reproduces, as explored in the previous chapter. An equally important set of energetic characteristics revolve around *where* an animal lives. In this chapter we consider how an animal's environment—and in particular, the temperature of that environment—affects that animal's metabolism and subsequent energy costs.

TEMPERATURE DEPENDENCE OF METABOLIC RATE

Few environmental factors have a larger influence on animal energetics and metabolism than temperature. Body temperature is a critical variable that touches on all aspects of animal function. Low body temperatures preclude high metabolic rates because of the temperature dependence of enzymatic reactions. High metabolic rates, on the other hand, with their high rates of heat production, may lead to overheating and attendant deleterious effects on tissue function, especially in hot climates. In cold climates, excessive heat loss can depress body temperature to dangerously low levels, slowing the metabolic rate and body heat production and launching a vicious cycle of declining temperature that can end in death. Those animals whose body temperature fluctuates with that of the environment experience corresponding temperature-induced changes in metabolic rate, whereas those that maintain a constant body temperature in fluctuating environmental temperatures have to expend metabolic energy to do so.

The Biochemical and Molecular Basis for Thermal Influences on Metabolism

The rates of chemical reactions, including those catalyzed by enzymes, are highly temperature-dependent. Tissue metabolism, as well as whole-body oxygen consumption and carbon dioxide production, depend on maintenance of the internal environment at temperatures compatible with metabolic reactions facilitated by enzymes. In those many animals that do not maintain a constant body temperature, even seemingly small changes in temperature can have major influences on metabolic rate through their effect on enzymes that regulate metabolism. In the zebrafish *(Danio rerio)*, for example, a relatively modest increase in temperature from 25°C through 28°C (the fish's preferred temperature) to 31°C increases the oxygen consumption of 10- to 50-day-old larvae by approximately twofold (Figure 17-1a on the next page). Put differently, an increase in body temperature of just 6°C causes a doubling of metabolic rate, and thus a doubling of the fish's energy requirements.

When we consider the effect of temperature on the rate of a biochemical reaction, it is useful to obtain a temperature quotient by comparing the rate of that reaction at two different temperatures. A temperature difference of 10°C has become a standard (if arbitrary) span over which to determine the temperature sensitivity of a biological function. The so-called Q_{10} is calculated by using the **van't Hoff equation:**

$$Q_{10} = (k_2/k_1)^{10/(t_2-t_1)} \quad (17\text{-}1)$$

where k_1 and k_2 are rates of the reaction (rate constants) at temperatures t_1 and t_2, respectively.

The beauty of the Q_{10} concept is that it can be applied both to simple processes such as single enzymatic reactions and to complex processes such as resting metabolic rate or even locomotion and growth. To

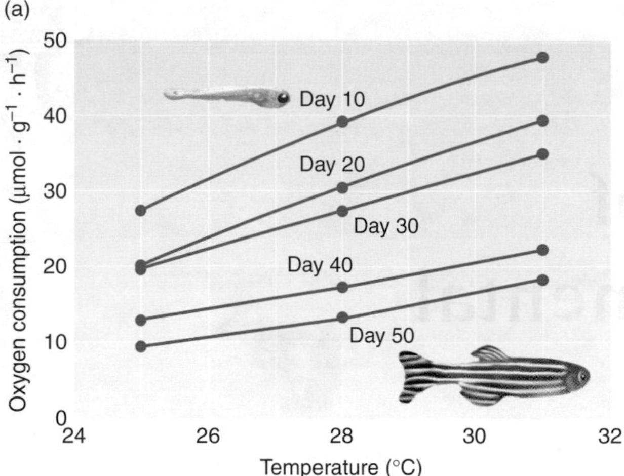

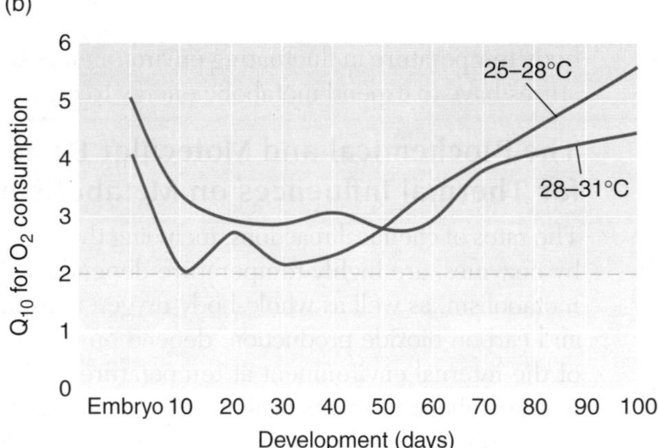

Figure 17-1 The oxygen consumption of larvae and juveniles of the zebrafish *(Danio rerio)* increases sharply as body temperature increases. **(a)** Metabolic rate approximately doubles over a 6°C increase in body temperature in zebrafish ranging from 10 to 50 days of age. **(b)** The Q_{10} for oxygen consumption in developing zebrafish depends greatly on the temperature range over which it is calculated. The dashed line indicates a Q_{10} of 2. [Adapted from Barrionuevo and Burggren, 1999.]

relate the van't Hoff equation to metabolic rate, consider the following form of the van't Hoff equation:

$$Q_{10} = (\mathrm{MR}_2/\mathrm{MR}_1)^{10/(t_2-t_1)} \tag{17-2}$$

in which MR_1 and MR_2 are the metabolic rates at temperatures t_1 and t_2, respectively. For temperature intervals of precisely 10 degrees, the following simpler form of Equation 17-2 can be used:

$$Q_{10} = \frac{\mathrm{MR}_{(t+10)}}{\mathrm{MR}_t} \tag{17-3}$$

where MR_t is the metabolic rate at the lower temperature and $MR_{(t+10)}$ is the metabolic rate at the higher temperature. However, it is often not convenient, or necessary, to calculate Q_{10} for MR (or any other physiological variable) by measuring at exactly 10-degree intervals.

As a rule of thumb, chemical reactions (and such physiological processes as metabolism, growth, and locomotion) have Q_{10} values of about 2 to 3, whereas purely physical processes (such as diffusion) have lower temperature sensitivities (i.e., Q_{10} values closer to 1.0). The Q_{10} of a given enzymatic reaction depends on the particular temperature range being considered, so it is important, when citing a Q_{10} value, to clearly indicate the range of temperatures (i.e., t_1 to t_2) for which it was determined. For example, the Q_{10} for oxygen consumption measured between 25°C and 28°C ranges from 3 to 5 during zebrafish development, while the Q_{10} for oxygen consumption measured between 28°C and 31°C ranges from 2 to 4 (Figure 17-1b).

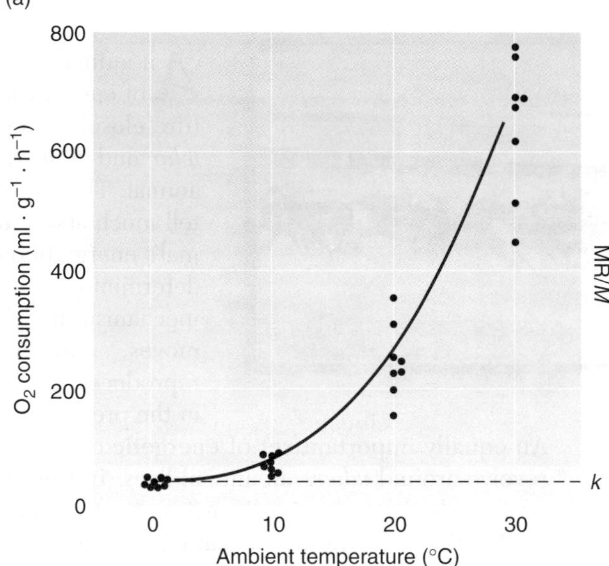

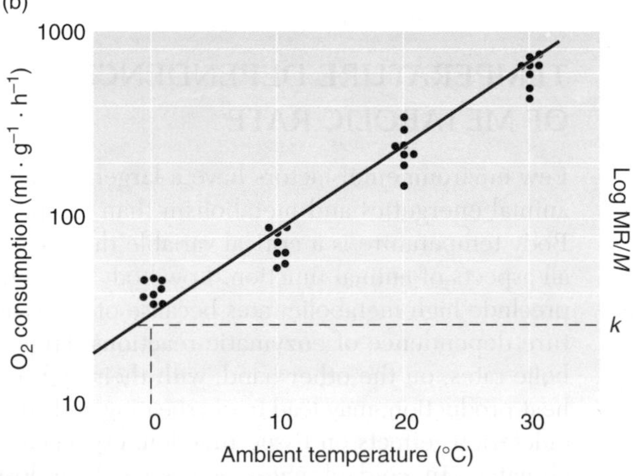

Classic Work *Figure 17-2* The oxygen consumption of the tiger moth caterpillar increases sharply as its body temperature increases. **(a)** Geometric plot. **(b)** Semilog plot. The generalized ordinates are shown in red at the right in reference to Equations 17-4 and 17-5. The constant k is obtained by extrapolating the metabolic rate to a body temperature of 0°C and is the proportionality factor in Equations 17-4 and 17-5. [Adapted from Scholander et al., 1953.]

Over relatively small ranges in body temperature, changes in enzyme function, expressed as metabolic rates, may appear to be linear (Figure 17-2a). However, over larger ranges in body temperature, the effects of temperature on enzymes cause the metabolic rate of an animal to increase exponentially with body temperature, as described by the equation

$$\frac{MR}{M} = k10^{b_1 t} \qquad (17\text{-}4)$$

where MR is metabolic rate and M is body mass (making MR/M the metabolic intensity in kilocalories per kilogram per hour), k and b_1 are constants, and t is temperature in degrees Celsius. It is useful to transform this relation into a logarithmic one to produce a linear plot. Thus, Equation 17-4 becomes

$$\frac{\log MR}{M} = \log k + b_1 t \qquad (17\text{-}5)$$

Now the coefficient b_1 gives the slope of the line—that is, the rate of increase in log MR/M per degree Celsius (Figure 17-2b).

Enzymatic basis of thermal acclimation

In many species, environmental heat or cold elicits compensatory changes in physiology and, in some cases, morphology. These changes help an individual organism to cope with temperature stress. An animal that cannot escape the winter cold (e.g., a pond-dwelling fish in a temperate climate) gradually undergoes, over the course of several weeks, a suite of compensatory biochemical adaptations to winter's low temperatures. Recall from Chapter 1 that the overall change that an animal undergoes in its natural setting is termed *acclimatization*. We will confine ourselves here to a more restricted concept, *acclimation*, the specific physiological change(s) that occur over time in the laboratory in response to variation in a single environmental condition, such as temperature. (The term *adaptation*, often loosely used as a synonym for these terms, should be reserved for genetically based evolutionary changes over many generations.)

Acclimation in whole animals occurs through the acclimation of individual cells and tissues. At a given experimental temperature, for instance, the actin and myosin in the skeletal muscles of winter-acclimated frogs and summer-acclimated frogs have different contractile properties, and the frogs exhibit different heart rates due to changes in cardiac pacemaker activity. Similarly, nerve conduction persists at low temperatures in cold-acclimated fish, but it is blocked at these same temperatures in warm-acclimated ones. How can we explain these observations? Given that enzymatic reactions underlie almost all key metabolic reactions, and that enzymes are temperature-sensitive, it is reasonable to suppose that the kinetics of enzymatic reactions are involved in temperature acclimation. In Figure 17-3a,

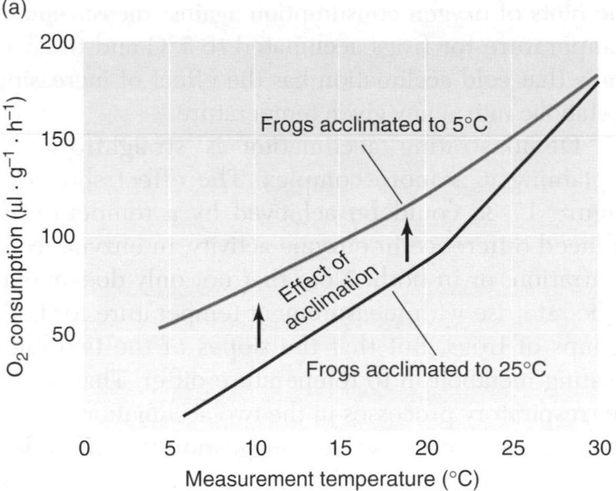

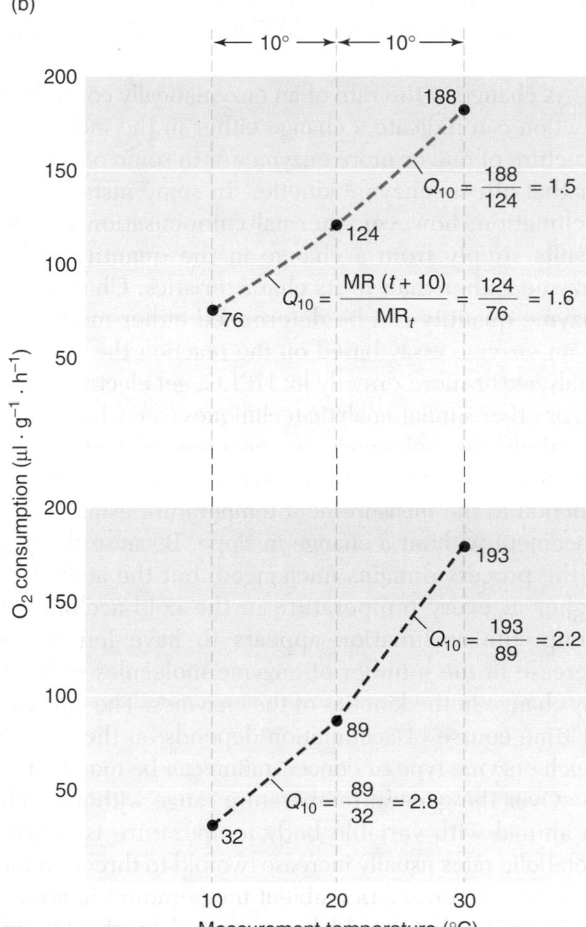

Figure 17-3 Temperature acclimation greatly influences temperature effects on metabolic rate. **(a)** At any given measurement temperature, oxygen consumption in frogs acclimated to 5°C is greater than oxygen consumption in frogs acclimated to 25°C. This phenomenon, the result of changes in the performance of metabolic enzymes, minimizes the disruptive effects of temperature change in these and other ectotherms. **(b)** The calculated Q_{10} values for each group of frogs in (a) over subsets of their total temperature range show additional effects of acclimation, as well as emphasizing that Q_{10} values often vary over an animal's entire body temperature range.

the plots of oxygen consumption against measurement temperature for frogs acclimated to 5°C and to 25°C show that cold acclimation has the effect of increasing metabolic rate at any given temperature.

Demonstrating acclimation is straightforward; explaining it is more complex. The effect shown in Figure 17-3a could be achieved by a temperature-induced difference in enzyme activity, in enzyme concentration, or in both. Note that not only does metabolic rate rise with measurement temperature for both groups of frogs, but that the slopes of the two lines relating metabolism to temperature differ. That is, the net respiratory processes in the two acclimation groups exhibit different Q_{10} values, suggesting that there has been a modification in the temperature sensitivity of enzyme activity. In this particular case, the effects of acclimation revealed by Q_{10} are quite different between the temperature ranges 5–20°C and 20–25°C (Figure 17-3b).

A change in the rate of an enzymatically controlled reaction can indicate a change either in the molecular structure of one or more enzymes or in some other factor that affects enzyme kinetics. In some instances of acclimation, however, thermal compensation clearly results simply from a change in the quantity of an enzyme rather than in its characteristics. Changes in enzyme quantity can be determined either indirectly, by an enzyme assay based on the reaction the enzyme catalyzes, or more directly by HPLC, gel electrophoresis, or other similar analytic techniques (see Chapter 2). Metabolically, changes in enzyme concentration become apparent when the plot relating a metabolic function to the measurement temperature exhibits displacement without a change in slope. Because the Q_{10} of the process remains unchanged, but the activity is higher at every temperature in the cold-acclimated group, the acclimation appears to have led to an increase in the number of enzyme molecules without any change in the kinetics of the enzymes. The particular time course of acclimation depends on the rate at which enzyme type or concentration can be modified.

Over the general temperature range within which an animal with variable body temperature is active, metabolic rates usually increase twofold to threefold for every 10°C increase in ambient temperature, in accordance with what would be predicted by the Q_{10} of enzymes. Yet there are important exceptions. The metabolic rates of some ectotherms exhibit a remarkable temperature independence. Some intertidal invertebrates, for example, experience large swings in ambient temperature as the tide alternately covers them with cold seawater and exposes them to the sun. Some of these animals have metabolic rates with a Q_{10} very close to 1.0; their rate of metabolism changes very little with temperature changes as large as 20 degrees. These animals appear to possess enzyme systems with extremely broad temperature optima, which prevent their inactivation during environmental temperature swings. Such broad temperature optima may be due to a staggering of the temperature optima of sequential enzymes in a reaction, such that a drop in the rate of one step in a sequence of reactions "compensates" for an increased rate of other steps in the sequence. Another exception is found toward either end of many animals' normally encountered temperature ranges. Thus, while Q_{10} may be 2–3 over the range of frequently encountered temperatures, it may show much higher or much lower values at temperature extremes (see Figure 17-3b). These effects are best shown by calculating Q_{10} values for small temperature ranges of 5°C or less. A third, important exception to the temperature dependency of enzymatic reactions occurs in cells constituting the "biological clocks" of animals, which have a temperature-insensitive Q_{10} of 1. Otherwise, the "time" kept by these "clocks" would be utterly dependent on an animal's body temperature, and even a fever in a mammal or bird would throw off its body rhythms (see Chapter 16).

Homeoviscous membrane adaptation

Biological membranes are very sensitive to temperature change. Low temperatures can cause a membrane to enter a gel-like phase with very high membrane lipid viscosity (imagine a stick of margarine in the refrigerator), whereas high temperatures can cause a membrane to become "hyperfluid," with very little viscosity (imagine the same stick of margarine melting on the counter in a warm kitchen at 30°C). Either change is increasingly disruptive as temperatures move away from the optimal values for a particular animal. The many functions of cell membranes—compartmentation, selective transport, maintenance of ion gradients, and so on—can be jeopardized if membrane lipid viscosity becomes too high or too low.

The temperature dependence of biological membranes is also affected by the lipid composition of the membrane, a relationship we can visualize with another kitchen example: At room temperature, cooking grease is a solid and cooking oil is a liquid. Both are composed of structurally similar lipid molecules that differ in the degree of hydrogenation of the carbon backbone. The greater the proportion of unsaturated carbon-carbon bonds (double bonds) in a lipid's fatty acid molecules, the lower its melting point. In mammals, different tissues contain lipids with different properties, including different melting points. In the limbs, which may be subjected to near-freezing temperatures, the lipids in the tissues are less saturated, and have a lower melting point, than those in the body core. At 37°C, the lipids in the limbs are much "oilier" than the waxier lipids of the warmer body regions.

Part of the acclimatization of ectothermic animals to cold or hot environments involves the membrane lipids becoming more saturated during acclimatization to warmth and less saturated during acclimatization to cold. These changes help stabilize the form of the membrane and the cellular functions that spring from it. This

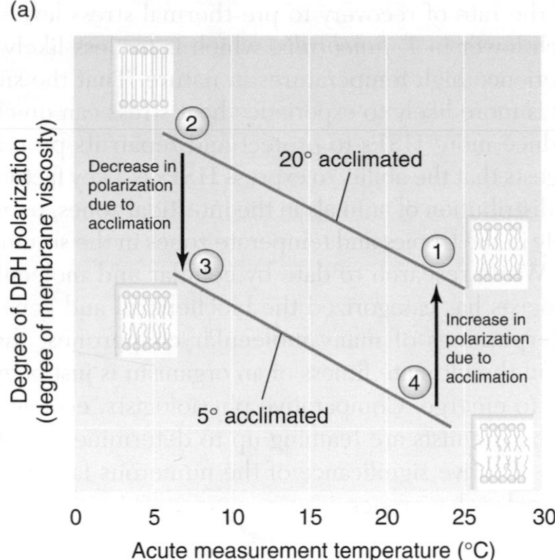

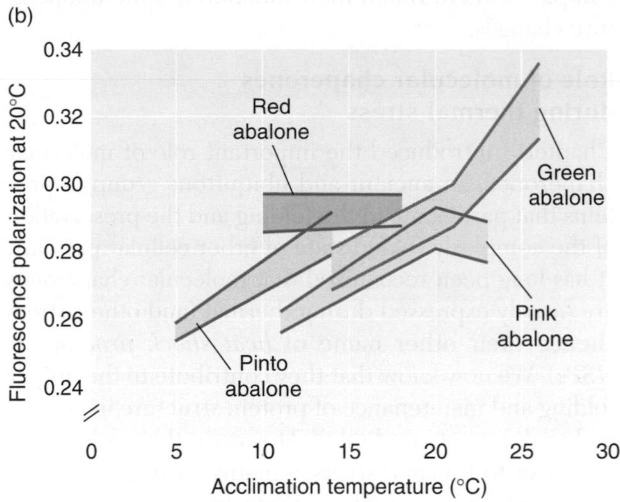

Figure 17-4 Homeoviscous adaptation maintains relatively constant lipid properties in animal cell membranes. **(a)** After an initial measurement at 25°C (point 1), intestinal cells from a warm-acclimated (20°C) rainbow trout are rapidly cooled to 5°C. Initially, the membranes of the cells become more polarized and more viscous (point 2), but as homeoviscous adaptation occurs, they become less polarized and regain their fluidity (point 3). Similarly, if intestinal cells from a trout acclimated to 5°C are rapidly warmed to 25°C, the lipid membranes are initially highly depolarized (point 4) but become more polarized with acclimation (point 1). **(b)** Fluorescence polarization increases (indicating decreasing membrane viscosity) with increasing acclimation temperature in the mitochondrial membranes of three of four different species of abalone. [Part a adapted from Hazel, 1995; part b adapted from Dalhoff and Somero, 1993.]

phenomenon is called **homeoviscous adaptation** (the term refers to evolutionary adaptations at the molecular level that help minimize temperature-induced differences in membrane viscosity).

Unfortunately, there is no simple measure of membrane viscosity. The index most often used is the *steady-state fluorescence anisotropy,* a measure of the lack of symmetry of a molecule or structure. A commonly used probe of membrane viscosity is 1,6-diphenyl-1,3,5-hexatriene (DPH). A high fluorescence anisotropy indicates a high degree of lipid polarization and a membrane order consistent with a low membrane viscosity. Figure 17-4a shows changes in the DPH polarization of the basolateral membranes of intestinal cells isolated from rainbow trout *(Oncorhynchus mykiss).* Initially, acute temperature changes are accompanied by changes in membrane polarization and viscosity. However, with acclimation to the new temperature, homeoviscous adaptation results in a lipid polarization and membrane viscosity that are similar to those at the initial temperature.

Cholesterol is another important membrane constituent, adding rigidity to fluid-phase cell membranes. Higher concentrations of cholesterol in the cell membranes of warm-acclimated animals could be expected to contribute to homeoviscous adaptation, and indeed, cholesterol concentrations tend to be higher in endotherms than in ectotherms (the latter typically having lower body temperatures than the former). However, studies on acclimated populations of crabs, trout, and other animals have been inconclusive, with cholesterol increasing, decreasing, or staying the same following acclimation. The reason for this variation may lie in membrane-specific responses to cholesterol and in the degree of lipid saturation. Increasingly, animal physiologists are appreciating the heterogeneous nature of the multitude of membrane types in various organelles and cell types. For example, although the basolateral membranes of the intestinal cells of trout acclimated to 20°C had a lower degree of saturation than those in trout acclimated to 5°C, the brush border membranes showed no such changes.

Homeoviscous adaptation appears to be a very general phenomenon, evident in invertebrates as well as vertebrates (Figure 17-4b). Typically, homeoviscous adaptation (as well as changes in key critical enzymes and the attendant rates of metabolic processes they support) is most evident over the actual temperature range of a species' or population's habitat, indicating a functional linkage between molecular and biochemical acclimation to temperature and a species' or population's biogeographic distribution.

Homeoviscous adaptation is a powerful explanation for acclimation and adaptation in animals with variable body temperatures, but researchers in this area readily admit that it cannot be the only explanation. Some animals become fully acclimated to new temperatures with moderate or even no change in lipid membrane properties. In any case, the altered expression of membrane proteins and the proliferation of mitochondrial and sarcoplasmic reticular membranes, along with homeoviscous adaptation of membrane lipids, present a picture of cell membranes as dynamic structures that change in complex ways to retain their function despite temperature changes.

Role of molecular chaperones during thermal stress

Chapter 3 introduced the important role of molecular chaperones, an ancient and ubiquitous group of proteins that participate in the folding and the preservation of the complexly folded state of other cellular proteins. It has long been recognized that molecular chaperones are heavily expressed during thermal (and other) stress (hence their other name of *heat shock proteins* or *HSPs*). We now know that they contribute to the proper folding and maintenance of protein structure in general and under most conditions. However, their role in the response to thermal stress remains an important and closely studied aspect of their biological function. Tomanek and Feder (1996) genetically engineered fruit flies *(Drosophila melanogaster)* to hyperexpress the protein hsp70 when thermally stressed by inserting 12 extra copies of the *hsp70* gene at a single site. This fruit fly strain was far more thermotolerant over ecologically relevant temperature ranges than other strains that lacked the additional *hsp* genes.

Molecular chaperones may also play a role in the geographic distribution of species. Two species of marine snails, *Tegula funebralis* (a low- to mid-tidal species that spends most of its time submerged) and *Tegula brunnea* (an intertidal species that is regularly exposed to the air), both express chaperone proteins when exposed to high temperatures mimicking summer exposure at low tide (Figure 17-5). However, both the rate of expression of HSPs of 38, 70, 77, and 90 kDA

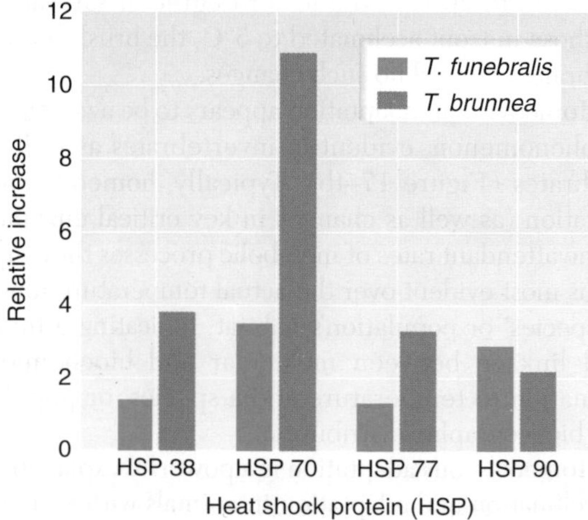

Figure 17-5 Heat shock proteins are expressed in the marine snails *Tegula funebralis* and *Tegula brunnea* when exposed to thermal stress. Shown are the relative increases in newly synthesized HSPs following a 2.5-hour aerial exposure at 30°C, followed by return to seawater at 13°C. Both species express HSPs, but the intertidal species, *T. brunnea,* expresses greater amounts of them. [Adapted from Tomanek and Somero, 2000.]

and the rate of recovery to pre-thermal stress levels is much lower in *T. funebralis,* which is far less likely to experience high temperatures in nature. That the snail that is more likely to experience heat stress can quickly produce more HSPs to protect and repair its proteins suggests that the ability to express HSPs is a key factor in the distribution of animals in the intertidal zones, particularly in the tropics and temperate zones in the summer.

While research to date by cellular and molecular biologists has categorized the biochemical and genetic underpinnings of many molecular chaperones, their role in the ultimate fitness of an organism is just beginning to emerge. Comparative physiologists, ecologists, and biochemists are teaming up to determine the ultimate adaptive significance of the numerous families of molecular chaperones.

Intertidal animals in temperate climates can experience changes in ambient temperature of as much as 50°C in the summer as they are alternately covered by cool water and then exposed to the hot sun. How would the concept of thermal acclimation apply to these animals, and what types of physiological or biochemical adaptations—or both—would you expect them to have to help them cope?

DETERMINANTS OF BODY HEAT AND TEMPERATURE

An animal's body temperature depends on the amount of heat (calories) contained per unit mass of body tissue. Because tissues consist primarily of water, the heat capacity of tissues between 0°C and 40°C approximates 1.0 cal $\cdot$ °C$^{-1} \cdot$ g^{-1}, the heat capacity of pure water. It follows that the larger the animal, the greater its body heat content at a given temperature. A key factor in every animal's thermal biology is the rate of change of body heat—that is, the relationship between heat loss and heat gain.

The Physics of Heat Gain, Loss, and Storage

The rate of change of body heat depends on the rate of heat production through metabolic means, the rate of external heat gain through radiation, conductance, and convection, and the rate of heat loss to the environment, again through radiation, conductance, and convection. We can summarize these relationships as

body heat = heat produced + (heat gained − heat lost) = heat produced + heat transferred

Thus, body heat, and hence the body temperature of an animal, can be regulated by changes in the rate of heat production and heat transfer or exchange (i.e., heat gained minus heat lost).

Numerous factors affect the rate of body heat production. Locomotor activity causes an increase in heat production by elevating metabolism. The activation of autonomic mechanisms leading to the release of hormones can produce accelerated metabolism of energy reserves. Acclimatization mechanisms, which are slower than behavioral and hormonal processes, often lead to an elevation in basal metabolism and its associated heat production.

The total heat content of an animal is determined by the metabolic production of heat and the thermal flux between the animal and its surroundings, as shown in Figure 17-6. The relation between these factors can be represented as

$$H_{tot} = H_v + H_c + H_r + H_e + H_s$$

in which H_{tot} is the total heat content, H_v is the heat produced metabolically, H_c is the heat lost or gained by conduction and convection, H_r is the net heat transfer by radiation, H_e is the heat lost by evaporation, and H_s is the heat stored in the body. Heat leaving the animal has a negative value, whereas heat entering the body from the environment has a positive value.

Heat can be transferred to and from an animal by conduction, convection, radiation, and evaporation. Let's consider each of these key physical processes that are so important to an animal's thermal biology.

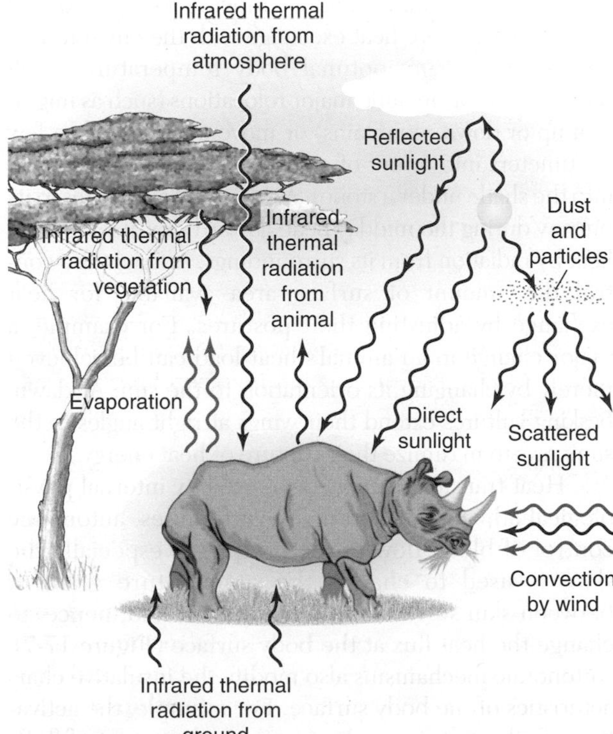

Figure 17-6 Heat is transferred between an animal and its environment in numerous ways, all based on convection, conduction, radiation, and evaporation. Infrared thermal radiation and direct and reflected sunlight transfer heat into the animal, whereas radiation and evaporation transfer heat out of the body and into the environment.

Conduction

The transfer of heat between objects in contact with each other is called *conduction.* It results from the direct transfer of the kinetic energy of molecular motion from molecule to molecule, with the net flow of energy being from the warmer to the cooler region. The rate of heat transfer through a solid conductor of uniform properties can be expressed as

$$Q = kA\frac{(t_2 - t_1)}{\ell} \qquad (17\text{-}6)$$

where Q is the rate of heat transfer (in joules per centimeter per second) by conduction; k is the *thermal conductivity* of the conductor; A is the cross-sectional area (in square centimeters); and l is the distance (in centimeters) between points 1 and 2, which are at temperatures t_1 and t_2, respectively. Conduction is not limited to heat flow within a given substance; it may also occur between two phases, as when heat flows from skin into the air or water in contact with the body surface.

Convection

The transfer of heat contained in a mass of gas or liquid by the movement of that mass is called *convection.* Convection may result from an externally imposed flow (e.g., wind) or from changes in the density of the mass produced by heating or cooling of the gas or fluid. Convection can accelerate heat transfer by conduction between a solid and a fluid because continuous replacement of the fluid (e.g., air, water, or blood) in contact with a solid of a different temperature maximizes the temperature difference between the two phases and thus facilitates the conductive transfer of heat between the solid and the fluid.

Radiation

The transfer of heat by electromagnetic radiation takes place without direct contact between objects. All physical bodies at a temperature above absolute zero emit electromagnetic radiation in proportion to the fourth power of the absolute temperature of the surface. As an example of how radiation works, the sun's rays may warm a black stone (or a dark reptile) to a temperature well above the temperature of the air surrounding the body. A dark body both radiates and absorbs radiation more strongly than does a more reflective body. As a general rule of thumb, when the temperature difference between the surfaces of two objects is about 20°C or less, the net radiant heat exchange is approximately proportional to the temperature difference.

Evaporation

Every liquid has its own latent heat of vaporization, the amount of energy required to change the liquid to a gas of the same temperature—that is, to evaporate it. The energy required to evaporate 1 g of water into water vapor is relatively high, approximately 585 cal. Many animals use this basic physical property of water to dissipate heat by allowing water to evaporate from moist body surfaces.

Heat storage

An object's temperature reflects its capacity for heat storage. The larger the object's mass, or the higher its specific heat, the smaller its rise in temperature for a given quantity of heat absorbed from the environment and stored. Therefore, a large animal, which has a small surface-to-mass ratio, tends to heat up more slowly in response to a given environmental heat load than does a small animal, which has a relatively high surface-to-mass ratio. This follows from the simple fact that heat exchange with the environment must take place through the body surface.

The rate of heat transfer (kilocalories per hour) into or out of an animal depends on several key factors. Changing the value of any one of them alters heat flow across the body surface down the temperature gradient and influences the overall heat stores of the animal:

- *Surface area.* As we saw in Chapter 16, surface area per gram of tissue decreases with increasing body mass, providing small animals with a high heat flux per unit of body weight. Animals can sometimes control their apparent surface area by changing posture (e.g., by extending limbs or drawing them close to the body).
- *Temperature gradient.* A temperature difference between the environment and an animal's body has a large effect by altering the temperature gradient (i.e., change in temperature per unit distance) for heat transfer. The closer the temperature of the animal to the ambient temperature, the less heat will flow into or out of its body.
- *Specific heat conductance.* The specific heat conductance of the animal's body surface varies with the nature of that surface. Animals whose surface tissues have a high heat conductance typically remain close to the temperature of their surroundings (with some exceptions, such as when an animal basks in sunlight). Animals that actively maintain a constant body temperature (birds, mammals) typically have insulation in the form of feathers, fur, or blubber that decreases the heat conductance of their body surfaces. An important feature of fur and feathers is that they trap and hold air, which has a very low thermal conductivity and therefore further retards the transfer of heat. Such insulation spreads out the temperature difference between the body core and the animal's surroundings over a distance of several millimeters or centimeters so that the temperature gradient is less steep and thus the rate of heat flow is reduced.

The body temperatures of most animals are similar to the environmental temperature. The maintenance of a sharp temperature differential between body tissues and their surroundings is restricted to a relatively small number of species, as we will soon see. A key factor in whether an animal can maintain an elevated body temperature is how the animal breathes. Animals breathing water can maintain only very limited regions of the body above ambient temperature because oxygen transfer is slower than heat transfer. Water contains little oxygen but has a high specific heat, so delivery of oxygen to the respiratory surface inevitably removes all the heat produced by metabolism. Air-breathing animals, on the other hand, can obtain sufficient oxygen from a small volume of air and can easily heat inspired air to a higher temperature. Air, unlike water, has a high O_2 content and a low specific heat. Thus, air-breathing animals can raise their body temperatures above the ambient temperature, whereas water-breathing animals cannot. Large animals in particular have heat to spare; the real challenge for them may be getting rid of excess heat in warm environments.

General Biological Mechanisms for Regulating Heat Transfer

Animals use several different mechanisms to regulate the exchange of heat between themselves and their environments. These mechanisms include behavioral modifications, autonomic changes in blood flow and sweating, and long-term morphologic changes.

Behavioral control of heat transfer between the body and the environment includes moving to a part of the environment where heat exchange with the environment favors attaining an optimal body temperature. Such movements may include major relocations (such as migration up or down mountains) or movements of only a few centimeters into or out of a "microhabitat" (for example, into the shade under a stone). A desert rodent retires to its burrow during the midday heat; a lizard suns itself to gain heat by radiation from its surroundings. Animals also control the amount of surface area available for heat exchange by adjusting their postures. For example, a major change in an animal's heat load can be achieved merely by changing its orientation to the sun. At dawn, basking vultures extend their wings at right angles to the sun's rays to maximize their capture of heat energy.

Heat transfer can also be altered by internal physiological adjustments. In many vertebrates, autonomic control of blood flow to the periphery, especially the skin, is used to change the temperature gradient between skin surface and environment and, hence, to change the heat flux at the body surface (Figure 17-7). Autonomic mechanisms also modify the insulative characteristics of the body surface. For example, the activation of piloerector muscles increases the extent of fluffing of pelage and plumage, which increases the effectiveness of insulation by increasing the amount of trapped, unstirred air next to the skin (Figure 17-8). Sweating and salivation are also under autonomic neuronal control and influence body temperature through evaporative cooling.

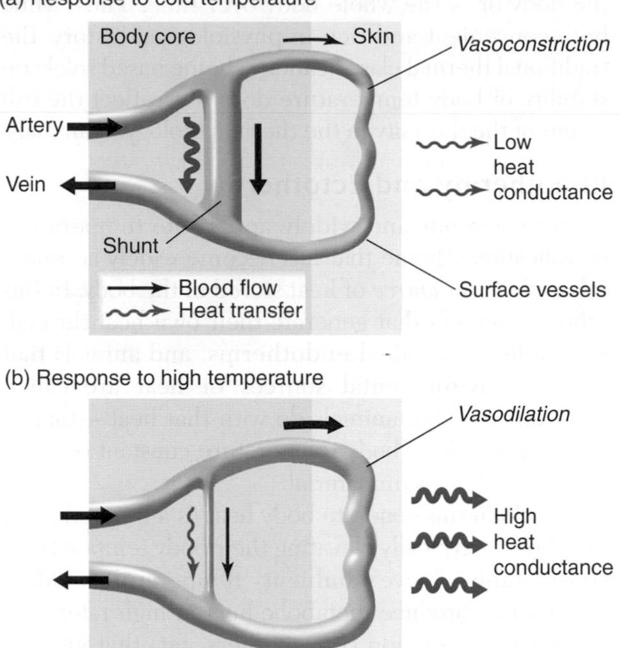

Figure 17-7 Changes in blood flow to the skin help regulate the heat conductance of the body surface. Vasomotor control of peripheral arterioles shunts arterial blood either to the skin or away from it. **(a)** In endotherms, peripheral blood vessels constrict in response to environmental cold, shunting blood away from the body surface. **(b)** In response to high environmental temperatures, blood flow is directed to the skin, where it approaches temperature equilibrium with the environment. Ectotherms often use this same mechanism of peripheral vasodilation to absorb heat from the environment.

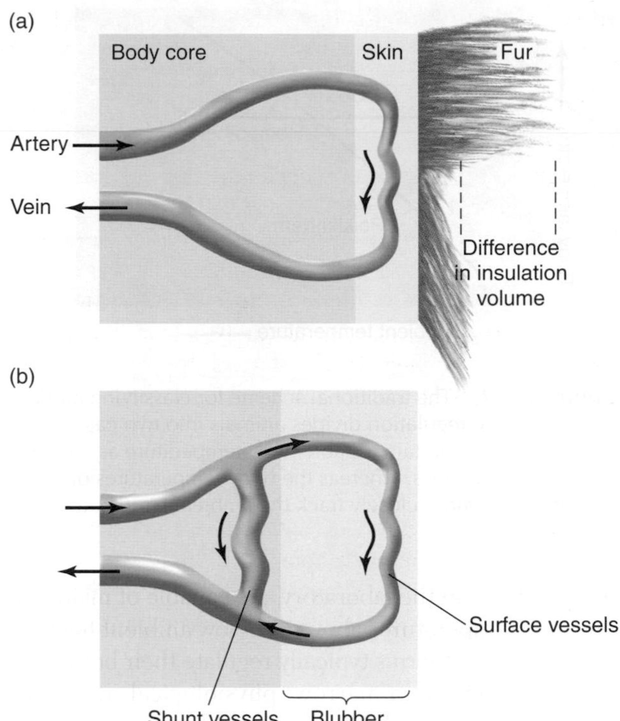

Figure 17-8 Fur and blubber provide thermal insulation. **(a)** Fur is located outside the skin and the circulation, and its insulating properties can be changed rapidly only by flattening or fluffing through pilomotor control. **(b)** Blubber is located under the skin and is supplied with blood vessels, so its insulating value can be regulated by shunting blood to or away from the surface below the blubber.

Long-term morphologic changes through the process of acclimatization also regulate heat transfer between an animal and its environment. Examples include long-term changes in pelage (think of a long-haired dog in the spring, shedding its winter coat by the handful) or subdermal fatty-layer insulation, which alter the conductance of the animal's body surface. Acclimatization also influences physiological processes—for example, by changing the capacity for autonomic control of evaporative heat loss through sweating.

Let's now consider in detail the specific mechanisms used by animals as they meet the thermal challenges of their environments.

PHYSIOLOGICAL CLASSIFICATION USING THERMAL BIOLOGY

Almost all habitable environments on our planet are characterized by fluctuations in temperature. Temperature variation is maximized in terrestrial temperate environments, some of which may have daytime summer surface temperatures of nearly 40°C (or higher) and nighttime winter temperatures of −40°C. Most animals—particularly small ones—occupy microhabitats that offer less extreme temperature fluctuations than their general surroundings. A few very specialized environments have highly stable temperatures, varying no more than a degree or two throughout the year. Examples are the shallow marine waters under the Arctic and Antarctic ice, the deep regions of the seas, the deep interiors of many caves, and the microenvironments within deep groundwater. With these few exceptions, almost all environments show significant long-term or short-term variations in temperature, and a wide variety of mechanisms have evolved through natural selection to sustain animal life through these variations. The diversity of temperature-control mechanisms is reflected in the history of classification schemes used by comparative physiologists over the past century.

Homeothermy and Poikilothermy

The traditional scheme used by comparative physiologists to classify the thermoregulatory modes of animals was based on the stability of body temperature. Animals were thus classified as either homeotherms or poikilotherms.

Homeotherms (also spelled "homoiotherms") are animals that, when exposed to changing environmental

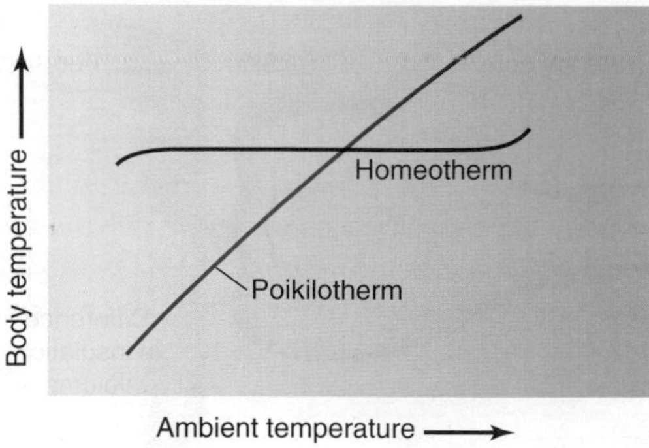

Figure 17-9 The traditional scheme for classifying modes of temperature regulation divides animals into two categories. Homeotherms maintain a stable body temperature as ambient temperature changes, whereas the body temperatures of poikilotherms more closely track the ambient temperature.

temperatures in the laboratory, are capable of maintaining body temperatures above or below ambient temperatures. Homeotherms typically regulate their body temperatures within a narrow physiological range by controlling heat production and heat loss (Figure 17-9). Some vertebrates other than birds and mammals and some invertebrates also can control their body temperatures in this manner, although such control is often limited to periods of activity or rapid growth.

Poikilotherms are those animals in which body temperature tends to fluctuate, more or less tracking ambient temperatures when they are varied experimentally. The colloquial term "cold-blooded" used to describe poikilotherms is inappropriate because many poikilotherms can actually become quite warm. A locust flying in the equatorial sun or a lizard running across the sand at midday in a hot desert may have blood temperatures considerably in excess of "warm-blooded" homeothermic mammals.

Early comparative physiologists painted with a broad brush, considering all fishes, amphibians, reptiles, and invertebrates to be poikilotherms because all of these animals were thought to lack the high rates of heat production that are found in birds and mammals. Several difficulties with the traditional homeotherm-poikilotherm classification scheme became apparent with the completion of more field studies (especially with the use of radiotelemetry to record body temperature). For example, some deep-sea fishes have more stable body temperatures than do many higher vertebrates simply because these fishes live in specialized environments that are thermally very stable (less than one degree of variation annually). Moreover, many so-called poikilotherms (e.g., lizards) are able to regulate their body temperatures quite well in their natural surroundings by controlling heat exchange with their environment or by moving back and forth between microenvironments. On the other hand, numerous birds and mammals have been known for some time to allow their body temperatures to vary widely, either in some part of the body or in the whole body over time. Thus, while both convenient and rich in physiological history, the traditional thermal classification scheme based solely on stability of body temperature does not reflect the full extent of the diversity in the thermal biology of animals.

Endothermy and Ectothermy

A more rigorous and widely applicable temperature classification scheme that has become widely accepted is based on the *source* of heat stored in the body. In this scheme, animals that generate their own heat through metabolism are called **endotherms,** and animals that rely on environmental sources of heat are called **ectotherms.** What animals do with that heat—that is, whether they hold body temperature constant or allow it to fluctuate—is immaterial.

Endotherms generate body heat as a by-product of metabolism, typically elevating their body temperatures considerably above ambient temperatures. Most endotherms produce metabolic heat at high rates, and many have insulation (fur, feathers, fat) that enables them to conserve heat despite a high temperature gradient between the body and the environment. Because endotherms (all birds and mammals plus many terrestrial reptiles and a number of insects) maintain their body temperatures well above ambient temperatures in cold climates, they have been able to invade habitats that are too cold for most ectotherms. Endotherms keep warm at considerable metabolic cost: the metabolic rate of an endotherm at rest is usually at least five times that of an ectotherm of equal size and body temperature.

Ectotherms produce metabolic heat at comparatively low rates—rates that are normally too low to allow for endothermy—and many of them are poorly insulated. As a result, heat derived from metabolic processes is quickly lost to cooler surroundings. On the other hand, their high thermal conductance allows ectotherms to absorb heat readily from their surroundings. Accordingly, heat exchange with the environment is much more important than metabolic heat production in determining an ectotherm's body temperature. The principal means by which ectotherms regulate their body temperatures is through behavioral mechanisms. This can be demonstrated in the laboratory by placing animals in an artificial environment with a thermal gradient and monitoring their preferred body temperatures. Alternatively, animals can be placed in a "shuttle box," which consists of one chamber that is well below the preferred body temperature and a connected chamber that is well above it. The animal will "shuttle" back and forth, keeping its body temperature at a level that is between those of the two available thermal environments.

Field observations combined with radiotelemetry have revealed the considerable thermoregulatory ability of reptiles through behavioral as well as physiological means. Many ectotherms needing to change their body

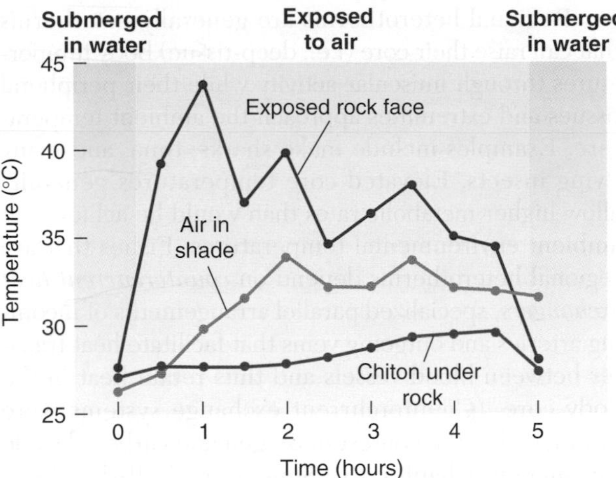

Figure 17-10 Microclimates afford ectotherms protection from harsh thermal environments. The tropical chiton *Chiton stokesii* can live in an intertidal zone whose surface reaches lethally high temperatures at low tide during the day by seeking cool microclimates under rocks. The temperatures in this graph were recorded from the exposed face of a chiton-bearing rock, in shaded air, and in the space beneath the attached foot of a chiton hidden under the rock. [Adapted from McMahon et al., 1991.]

temperature behave in a way that facilitates heat absorption from the environment or helps the animal unload heat to the environment (or minimizes heat uptake from the environment). A lizard or a snake may bask in the sun with its body oriented for maximal warming until it achieves a temperature suitable for efficient muscular function. Small ectotherms in hot environments (lizards, ants) often elevate their bodies to the extent that their legs allow to avoid the hottest temperatures immediately adjacent to the sand or rock over which they are moving.

In general, the most effective thermoregulatory action taken by ectotherms is movement into a suitable microclimate or microhabitat in the environment. A burrow under a rock, for example, is often much more moderate in temperature than the surface (Figure 17-10). Intertidal zones in the tropics often appear to be devoid of invertebrate life during the heat of the day, but that same habitat may be teeming with life at night as animals emerge from their microclimates underneath rocks and in burrows.

Thermal Classification of Real versus Ideal Animals

It is important to emphasize that there is no absolutely "correct" thermal classification scheme. The concepts of homeothermy, poikilothermy, endothermy, and ectothermy all represent idealized extremes. Most animals are not at these extremes, but as illustrated in Figure 17-11, operate closer to the "edges" of their somewhat simplified classifications. Moreover, classification schemes may be combined: a cat or human, for example, can be referred to as a "homeothermic endotherm." Such combinations of the two prevalent classification schemes—"homeothermic endotherm," "poikilothermic ectotherm," and so forth—may well generate the most accurate classifications.

Recognition of the fuzzy nature of the boundaries between classifications is no more apparent than when considering **heterotherms**, animals that are capable of varying degrees of endothermic heat production but that generally do not regulate body temperature within a narrow range. Heterotherms can be divided into two groups, *regional heterotherms* and *temporal heterotherms.* A few large fishes and some flying insects are termed *regional heterothermic endotherms* because they maintain regions of their bodies above ambient temperatures. Temporal heterotherms constitute a broad category of animals whose temperatures vary widely over time. Monotremes (egg-laying mammals) such as the echidna are temporal heterotherms (Figure 17-12 on the next page), as are other mammals and birds that undergo torpor and hibernation. Temporal heterothermy is also shown by many flying insects, pythons, and some fishes, which can raise the temperature of their bodies (or regions of their bodies, as in the fishes) well above ambient temperature by virtue of heat generated as a by-product of intense muscular activity. Some insects prepare for flight, for example, by

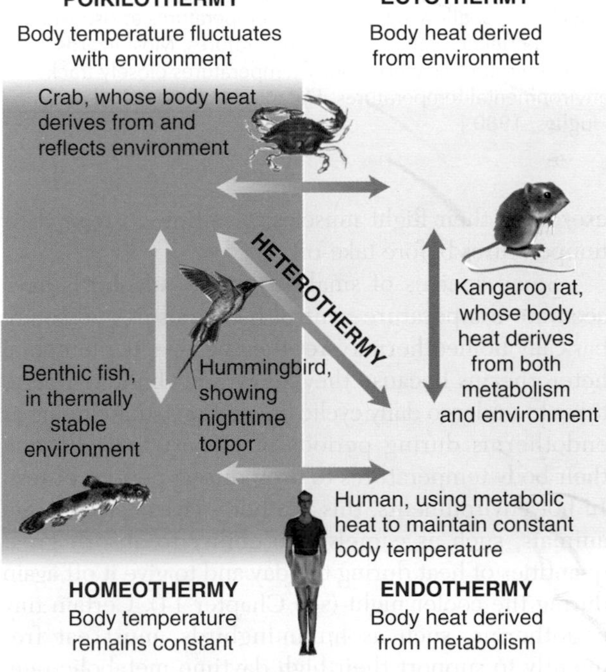

Figure 17-11 The thermal biology of animals can be classified in terms of the stability of body temperature (left side of diagram) or in terms of the source of body heat (right side of diagram). Many animals do not easily fit in either classification scheme, both of which refer to "ideal" animals. Heterotherms, for example, transcend these two classification schemes by using metabolic heat to regulate body temperature, but also allowing body temperature to vary widely depending on circumstances.

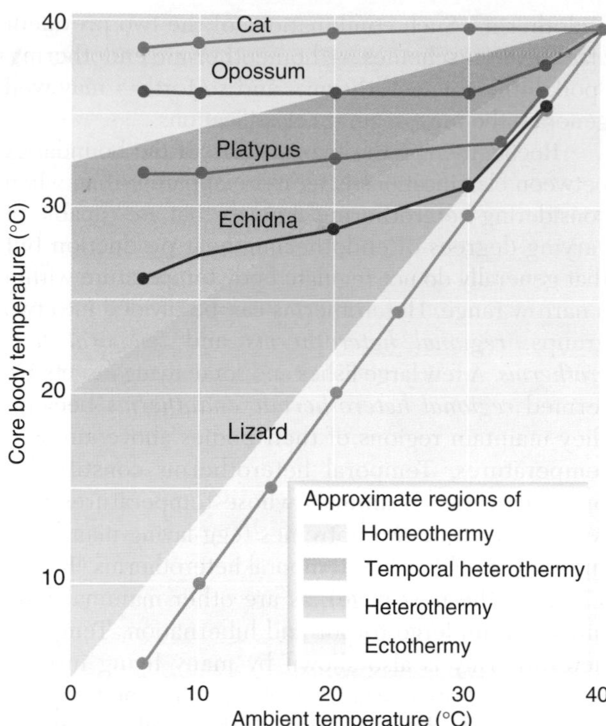

Figure 17-12 The relation between core body temperature (measured rectally) and ambient temperature differs among homeotherms, temporal heterotherms, and heterotherms. The cat is a strict homeotherm, maintaining a body temperature independent of ambient temperature. Monotremes (platypus and echidna) are temporal heterotherms, allowing their body temperatures to rise when they experience high ambient temperatures. Most lizards are strict heterotherms whose body temperatures closely track environmental temperatures. [Adapted from Marshall and Hughes, 1980.]

exercising their flight muscles for a time to raise their temperatures before take-off.

Some species of small mammals and birds have accurate temperature-control mechanisms and so are basically homeothermic. Yet they behave like temporal heterotherms because they allow their body temperatures to undergo daily cyclic fluctuations, functioning as endotherms during periods of activity and allowing their body temperatures to drop during periods of rest. In hot environments, this flexibility gives certain large animals, such as camels, the ability to absorb great quantities of heat during the day and to give it off again during the cooler night (see Chapter 14). Certain tiny endotherms, such as hummingbirds, must eat frequently to support their high daytime metabolic rate. To avoid running out of energy stores at night when they cannot feed, they enter a state of *torpor*, during which they allow their metabolic rates to decline and their body temperatures to drop toward the ambient temperature. Even some large endotherms resort to a long winter torpor, reducing their body temperature to save energy. We will revisit torpor and hibernation later in this chapter.

Regional heterotherms are generally ectotherms that can raise their core (i.e., deep-tissue) body temperatures through muscular activity while their peripheral tissues and extremities approach the ambient temperature. Examples include mako sharks, tuna, and many flying insects. Elevated core temperatures generally allow higher metabolic rates than would be achieved at ambient environmental temperatures. Fishes that are regional heterotherms depend on *countercurrent heat exchangers,* specialized parallel arrangements of incoming arteries and outgoing veins that facilitate heat transfer between blood vessels and thus retain heat in the body core. (Countercurrent exchange systems were introduced in the context of oxygen and carbon dioxide exchange in Chapter 13.) Some large billfishes (e.g., marlin) use specialized ocular muscle called "heater tissue" to elevate their brain temperature, as we will see below. Another example of regional heterothermy is seen in the scrotum of some mammals, including canines, cattle, and humans, which hold the testes outside the body core to keep them at a slightly lower temperature. The muscles suspending the testicles in the scrotum shorten in cool air, drawing the testes against the warmer body, and lengthen as body temperature rises. These actions regulate testicular temperature and, in particular, prevent overheating of the testes, which can interfere with sperm production.

While recognizing that thermal biological classification schemes are imperfect, let's focus on the concepts of ectothermy and endothermy to frame our in-depth exploration of thermal physiology of animals.

THERMAL BIOLOGY OF ECTOTHERMS

Despite the inability of ectotherms to use their metabolic furnace to maintain a stable body temperature, they should not be viewed as physiologically disadvantaged, for they have occupied almost every habitable environment on the planet—from hot to cold.

Ectotherms in Cold and Freezing Environments

Many animals living in cool (but not freezing) environments survive through their ability to generate an adequate level of metabolism at the very low levels of enzyme activity characteristic of low temperatures. Enzymes in these animals show maximal activity at temperatures many degrees below those of homologous enzymes in animals living in warmer environments. Figure 17-13 documents thermal adaptation in the Michaelis-Menten constant (K_m; see Chapter 3) of A_4-lactate dehydrogenase for its substrate, pyruvate. The K_m value for this enzyme in *Trematomus centronotus*, a fish that lives in Antarctic water that is almost always at −1.9°C, is far higher than that in fishes and other vertebrates that inhabit warmer and more thermally varied environments. Even within a single species of bar-

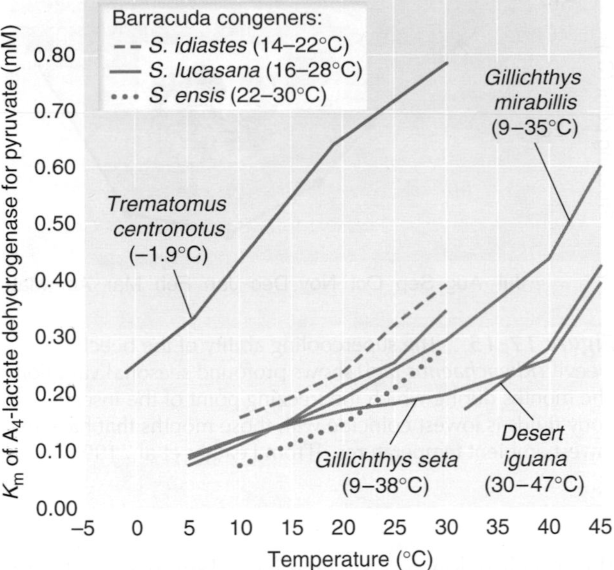

Figure 17-13 In some animals living in cold environments, natural selection has led to higher K_m values of A_4-lactate dehydrogenase for its substrate, pyruvate. This relation holds both between species and within species. Shown are comparisons between *Trematomus,* a fish that lives at very cold temperatures, and *Gillichthys,* which lives at warmer temperatures, and comparisons among several species of barracuda that live at different water temperatures. [Adapted from Somero, 1995.]

racuda, individual fish that inhabit cooler waters possess a lactate dehydrogenase having a K_m for pyruvate that is higher than that in individuals that inhabit warmer waters. This biochemical adaptation allows the cooler-water fish to maintain a higher metabolism in relation to their body temperature than the warmer-water fish. Doubtless numerous other such adaptations in key metabolic enzymes remain to be identified.

Because the body temperatures of many ectotherms closely track ambient temperatures, freezing is a threat to those species living in environments where ambient temperatures can extend below 0°C. The formation of ice crystals within cells is usually lethal because as the crystals grow in size, they rupture and destroy the cells. In contrast, ice crystals that form outside the cells in extracellular compartments do little damage. Thus, those ectotherms that overwinter in freezing, even subzero, conditions typically have adaptations that prevent intracellular ice crystal formation. These adaptations include preemptive ice crystal formation, antifreezes, and supercooling.

Preemptive ice crystal formation

It may seem counterintuitive, but one adaptation that helps prevent the formation of intracellular ice crystals is the rapid formation of ice crystals in extracellular tissues. Certain beetles, for example, can withstand freezing temperatures because their extracellular fluids (but not their intracellular fluids) contain a substance that accelerates *nucleation* (the process that initiates crystal formation). As the temperature falls below the freezing point of water, the extracellular fluids, with their nucleating agent, freeze more readily than the intracellular fluids (Figure 17-14a). Ice crystals form in the extracellular fluid, excluding solutes and concentrating them in the remaining unfrozen fluid. The elevated extracellular solute concentration draws water out of surrounding cells by osmosis, which in turn raises the solute concentration of the aqueous solution in the cell, lowering its freezing point. As the temperature drops further, the process continues, producing further depression of the freezing point of the remaining intracellular water. This process preserves intracellular integrity up to the point at which the elevated intracellular solute concentration proves damaging to intracellular organelles (Fig 17-14b). A similar protective effect by preemptive ice crystal formation stimulated by a nucleating agent occurs in the freshwater larvae of the midge *Chironomus,* which can survive repeated freezing as the water at the bottom of the shallow ponds in

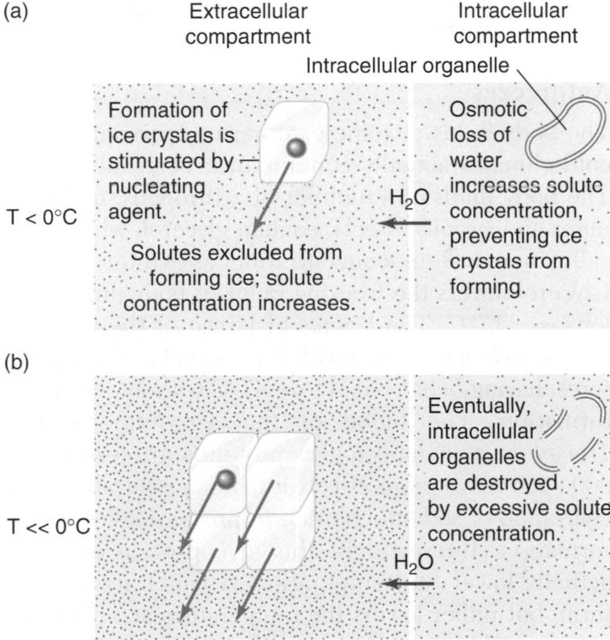

Figure 17-14 Destruction of cells by freezing can be averted or delayed by nucleating agents that differentially promote extracellular rather than intracellular ice crystal formation. **(a)** In some animals resistant to freezing, nucleating agents are present in extracellular, but not intracellular, fluids. When body temperature falls below freezing, ice crystals form around the nucleating agent. Solutes are excluded from the ice crystals, raising the extracellular solute concentration. Water flows down its concentration gradient from intracellular to extracellular fluid. The consequent increased intracellular solute concentration lowers the freezing point of the intracellular fluid, preventing ice crystal formation. **(b)** Eventually, so much intracellular water is drawn from the cell that the elevated intracellular solute concentration proves disruptive to cell organelles, leading to cell death.

which they live freezes and thaws. In the laboratory, *Chironomus* larvae yield some unfrozen liquid when their body temperatures are as low as −32°C.

Some cell types are more resistant to damage by freezing than others. Red blood cells, yeast, sperm, and other cell types can withstand freezing damage, provided intracellular ion concentrations are not driven above levels that cause damage to cell organelles. Experiments on storage of blood and sperm cells have shown that the *rate* of freezing is also critical, with rates of cooling of several hundred degrees per second (as would occur when a small number of cells are flash-frozen by plunging them directly into liquid nitrogen) providing optimal protection against cell damage.

A few vertebrates, primarily anuran amphibians, also can withstand freezing. Paradoxically, given the higher survivorship of individual cells that are flash-frozen, as described above, frogs survive freezing only if ice crystal formation occurs over 24 hours or longer. The wood frog (*Rana sylvatica*) can survive the freezing of up to 65% of its total body water. Moreover, more than half of the water in the heart, liver, intestine, and peripheral nerves translocates to the lymph sacs and coelom, where it freezes. In frogs that survive freezing, both nucleating agents and cryoprotectants ("antifreezes") are employed.

Antifreezes

The body fluids of some ectotherms overwintering in cold climates actually contain antifreeze substances. The body fluids of a number of arthropods, including mites and various insects, contain glycerol, which typically increases in concentration as winter sets in. Glycerol lowers the freezing point of body fluids to as low as −17°C. The tissues of larvae of the parasitic wasp *Brachon cephi* can withstand even lower temperatures, having been cooled to −47°C without ice crystal formation. The blood of the Antarctic ice fish *Trematomus* contains a glycoprotein antifreeze that is 200–500 times more effective than an equivalent concentration of sodium chloride in preventing ice crystal formation. Interestingly, this glycoprotein does not actually lower the temperature at which ice crystals form, but rather lowers the temperature at which they enlarge and begin to destroy cell organelles.

Supercooling

Some animals can support supercooling, a state in which body fluids are cooled to below their freezing temperature yet remain unfrozen because ice crystals fail to form. Ice crystals cannot form and spread if they have no nuclei (mechanical "seeds," so to speak) to initiate crystal formation. Certain fishes dwelling at the bottom of Arctic fjords live in a continually supercooled state and normally do not freeze. However, they must not brush against frozen ice on the water surface—to do so causes the extremely rapid formation of ice crystals throughout the body and instant death. Survival for these fish depends on behavioral adaptations that keep them well below any surface ice. Many overwintering insects also show supercooling. In fact, the majority of cold-tolerant insects are actually freezing-intolerant and will die if their tissues actually freeze. The ability of insects to supercool, which is aided by antifreeze agents such as glycerol, typically shows profound seasonal changes (Figure 17-15).

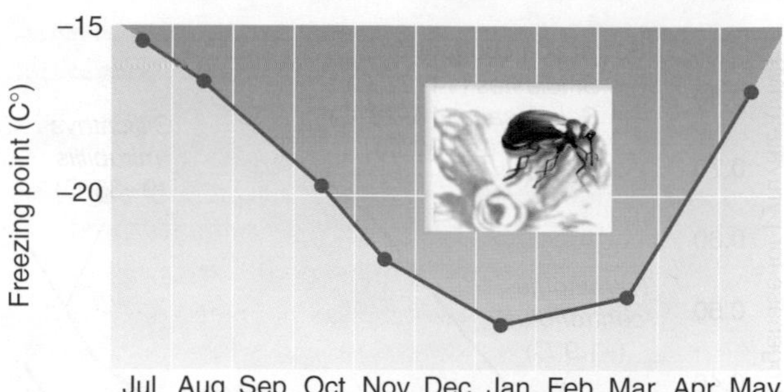

Figure 17-15 The supercooling ability of the beech weevil *(Rhynchaenus fagi)* shows profound seasonal variation. The months during which the freezing point of the insect's body fluids is lowest coincide with those months that have the lowest ambient temperatures. [From Leather et al., 1993.]

Ectotherms in Warm and Hot Environments

All ectotherms have a *critical thermal maximum* (CTM), a temperature threshold above which long-term exposure proves lethal. Generally, the CTM is determined by measuring the temperature at which 50% mortality occurs in a population. The critical thermal maximum varies enormously among ectothermic organisms. Some thermophilic bacteria and hydrothermal vent invertebrates can thrive at temperatures above 90°C, although almost all metazoans have CTMs below 45°C. There are a variety of physiological reasons for a CTM. One ultimate upper limit is the temperature at which proteins are denatured, although enzymes usually fail to function at levels well below the temperature of denaturation. Heat-shock proteins may provide some moderation of enzymatic degradation at high temperatures, and their presence has been correlated with increased CTMs in insects and gastropods. Often, an animal's CTM relates to a breakdown in some critical physiological process. In many ectotherms, for example, tissue functions are handicapped by a decreased affinity of the respiratory pigment for oxygen in the upper ranges of tolerated body temperature. At 50°C, the arterial blood of a chuckwalla *(Sauromalus obesus)* cannot achieve more than 50% O_2 saturation, which prevents vigorous aerobic activity by this lizard.

Most ectotherms never experience sustained temperatures approaching the critical thermal maximum. Yet even in temperate climates, many experience general environmental temperatures that are high enough to require an active response to avoid unacceptably elevated body temperatures. Many ectotherms seek shade to reduce heat gain by radiation from the environment. Once the body temperature drops, they may return to full sun exposure. The effectiveness of such behavioral thermoregulation is affected by the typically high thermal conductance and low thermal capacity resulting from the low body mass of many ectotherms.

In addition to behavioral mechanisms, certain reptiles use physiological means to control the cooling and heating rates of their bodies. An inanimate object generally cools and heats at the same rate because there are no active, internal processes that affect heat conduction through the object. In contrast, many ectotherms—including most reptiles, as well as some amphibians and arthropods—heat and cool at quite different rates, indicating the involvement of an active physiological mechanism. In the Galápagos marine iguana *(Amblyrhynchus cristatus)*, for example, body temperature can rise at about twice the rate at which it drops. The iguana regulates the rate of change by moving between land and water, as well as by adjusting its heart rate and the flow of blood to its surface tissues. To maximize warming rates, the iguana basks in the sun and simultaneously increases its heart rate and peripheral blood flow, diverting cooler core blood to the body surface (see Figure 17-16a). The increased cutaneous blood flow increases the skin's heat conductance and speeds absorption of environmental heat into the animal. Increased blood flow also accelerates the transfer of absorbed heat from surface tissues to deeper tissues at the animal's core. Transport of heat by convection, a very rapid process, thus circumvents the much slower conductive pathway for heat from peripheral to core tissues. During the marine iguana's prolonged feeding dives in the cool ocean, a diving bradycardia develops (see Chapter 12), probably accompanied by a major redistribution of systemic blood flow leading to reduced cutaneous perfusion (Figure 17-16b). The combined result of these circulatory changes during diving is a reduced rate of heat loss to the surrounding cool water.

The involvement of an integrated physiological response in differential rates of heat loss and gain is clearly evident when we plot physiological factors against body temperature as temperature changes in both directions. Figure 17-17 on the next page shows the relationship between heart rate and body temperature during heating and cooling in air in the marine iguana (652–1360 g), the green iguana (*Iguana iguana*, about 1000 g), and the eastern fence lizard (*Sceloporus undulatus*, about 10 g). The plots reveal a mild yet unmistakable hysteresis (an asymmetric response). Interestingly, the marine iguana shows a much more pronounced hysteresis when it cools in water. The heat capacity of water is much greater than that of air, so

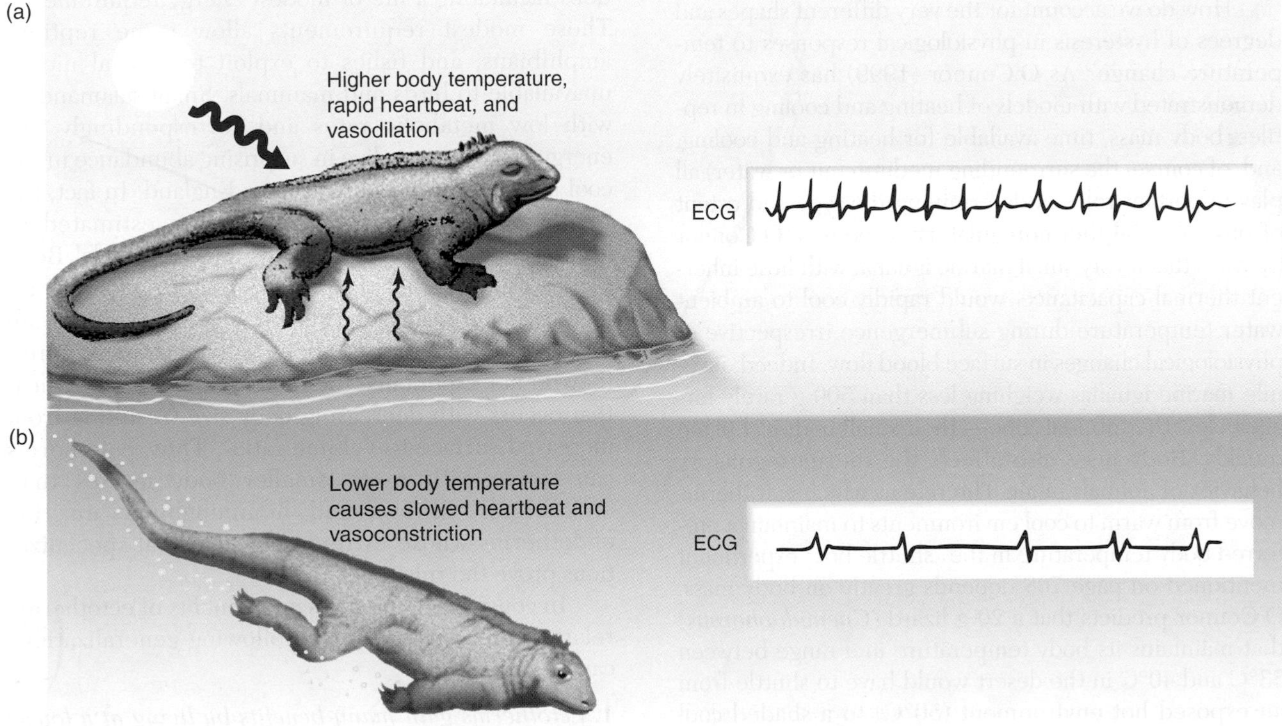

Figure 17-16 The Galápagos marine iguana heats and cools at different rates, indicating an active regulation of heat exchange with its environment. **(a)** On land, the basking marine iguana absorbs heat from the sun's rays. Cutaneous vasodilation and a rapid heartbeat (as recorded in the electrocardiogram) assure heating of the blood and efficient circulation, which quickly distributes the heat throughout the body. **(b)** When the iguana is under water, heat loss is retarded by a slowed heartbeat and cutaneous vasoconstriction, both of which minimize the flow of blood to the skin.

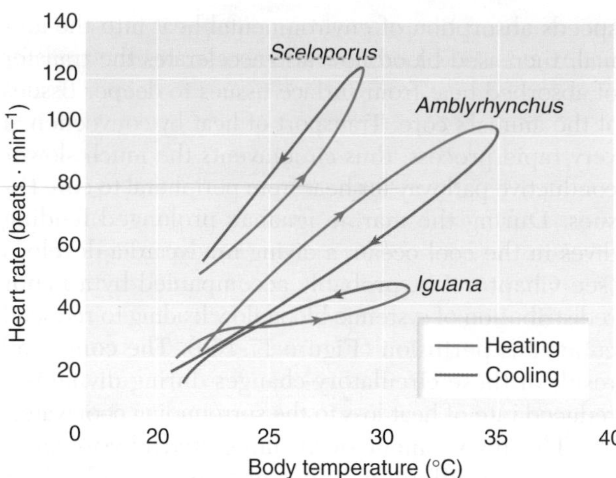

Figure 17-17 A hysteresis is seen in the relation between heart rate and body temperature during heating (red) followed by cooling (blue) in many ectotherms, including these three species of lizards as they undergo temperature changes in air. [Data on marine iguana, *Amblyrhynchus,* from Bartholomew and Lasiewski, 1965; data on common iguana, *Iguana,* and fence lizard, *Sceloporus,* from Dzialowski and O'Connor, 2001.]

water conducts heat from the surface of the marine iguana much more rapidly than air does; thus, it is especially important for heat conservation that circulation to the skin be slowed during diving. The pronounced diving bradycardia associated with the overall diving reflex is an integral part of this thermoregulatory response.

How do we account for the very different shapes and degrees of hysteresis in physiological responses to temperature change? As O'Connor (1999) has exquisitely demonstrated with models of heating and cooling in reptiles, body mass, time available for heating and cooling, and, of course, the surrounding medium (air or water) all play important roles in determining the type and extent of physiological thermoregulatory responses. O'Connor predicts that a very small marine iguana, with little inherent thermal capacitance, would rapidly cool to ambient water temperature during submergence irrespective of physiological changes in surface blood flow. Indeed, juvenile marine iguanas weighing less than 500 g rarely forage below the subtidal zone—their small bodies chill too quickly. Body mass also affects the thermoregulatory behavior of animals in air. The rate at which ectotherms move from warm to cool environments to maintain a preferred body temperature in the "shuttle box" experiment mentioned on page 708 depends greatly on body mass. O'Connor predicts that a 20 g lizard *(Cnemidophorous)* that maintains its body temperature in a range between 33°C and 40°C in the desert would have to shuttle from an exposed hot environment (50°C) to a shaded cool environment (20°C) every 3 minutes. Yet a 3 kg tortoise *(Gopherus)* under the same conditions would have to shuttle back and forth only every 90 minutes.

In summary, we know that major cardiovascular adjustments are coupled to thermoregulatory responses to actively regulate the heating and cooling rates of many ectotherms. However, as we have seen in earlier chapters, the same cardiovascular responses (changes in heart rate, systemic blood flow patterns) have also been invoked to explain optimization of gas exchange, ionic balance, and so on. Assembling a complete understanding of the integrated response to simultaneous physiological challenges suggests a program of experiments upon which an inspired student could build a career.

Costs and Benefits of Ectothermy Relative to Endothermy

Early comparative physiologists assumed that ectothermy was somehow inferior to endothermy as a way of life (which required ignoring the evidence of the relative success of each lifestyle, based on species numbers). The endothermic vertebrates (primarily birds and mammals) were viewed as more complex and more recent evolutionary arrivals than the primarily ectothermic "lower vertebrates" (fishes, amphibians, and lizards). The term "lower vertebrates" is rapidly losing favor as we realize that these animals are as highly adapted for their way of life as birds and mammals.

Endotherms and ectotherms have undeniably different lifestyles, the former representing a fast, high-energy way of life and the latter representing a slower, low-energy approach. Many of the anatomic and functional properties of ectothermic vertebrates are adaptations facilitating a life of modest energy requirements. Those modest requirements allow some reptiles, amphibians, and fishes to exploit terrestrial niches unavailable to birds and mammals. Small salamanders with low metabolic rates and correspondingly low energy requirements live in surprising abundance in the cool litter on forest floors of New England. In fact, the total biomass of such salamanders is estimated to exceed that of the forest's birds and mammals! Body size is often a critical factor in considering the advantages of ectothermy. Because few ectotherms elevate their body temperatures above ambient temperatures, they do not experience the increased loss of body heat that occurs with decreasing body size (resulting from increased surface-to-volume ratio). Thus, ectotherms can thrive with much smaller body masses than endotherms. Shrews and hummingbirds are tiny endotherms whose extreme physiological specializations prove the rule.

In considering the costs and benefits of ectothermy relative to endothermy, the following generalizations can be made:

1. *Ectotherms gain many benefits by living at a lower metabolic rate.* Because their body temperatures generally remain close to ambient temperatures, they expend relatively little energy on thermoregulation. As a consequence, ectotherms can invest a larger proportion of their energy budget in growth

and reproduction. Ectotherms require less food, so they can spend less time foraging and more time quietly avoiding predators. They also need less water because they lose less by evaporation from their typically cooler surfaces, and they need not be massive in order to achieve an effective surface-to-volume ratio.

2. *The benefits to ectotherms of a low metabolic rate are balanced by the costs,* including the inability of ectotherms to regulate their body temperatures (unless their environments permit behavioral thermoregulation). A lizard can elevate its temperature by basking only if there is sufficient solar radiation, which limits the times of day and seasons of the year when such activity is possible. Furthermore, a low rate of aerobic metabolism, coupled with the need to accommodate oxygen debt during anaerobic respiration, limits the duration of bursts of high activity. Such factors have been evoked to argue that large dinosaurs must have been endotherms.

3. *The respective costs and benefits of ectothermy are the benefits and costs of endothermy.* Because of their high rates of aerobic respiration and their elevated temperatures, endotherms can generally sustain longer periods of intense activity. Thus, the endotherms can be thought of as energetic high rollers compared with the more thrifty ectotherms, which are characterized by lower energy intake and lower energy expenditure. Another advantage of endothermy is that the constancy of body temperature allows enzymes to function more efficiently over a relatively narrow range.

Endothermic animals can do certain things on a bigger, faster scale, but they do so at a price. The field metabolic rates of endotherms are up to 20 times higher than the field metabolic rates of ectotherms. The price paid by endotherms for their high metabolic rate includes the requirement that they take in correspondingly larger amounts of food and water daily. Thus, a 300 g rodent needs 17 times as much food per day as a 300 g lizard living in the same habitat on the same diet of insects. Their high rate of respiratory gas exchange also makes endotherms susceptible to dehydration in hot, dry climates. High body temperature relative to ambient temperature makes a very small body mass problematic for endotherms because of surface-to-volume considerations that cause a small animal to lose heat faster than a larger one. Finally, because such a large quantity of energy is consumed by an endotherm to elevate and maintain its body temperature, only a relatively small proportion of its energy can be budgeted for growth and reproduction.

This list of costs and benefits shows that ectothermy and endothermy constitute a metabolic dichotomy affecting far more than just body temperature. Indeed, the implications of these two types of energy economies also extend to such areas as morphology, physiology, behavior, and the evolution of these attributes. Moreover, the advantages of the two lifestyles are not necessarily exclusive. Some thermoregulating terrestrial ectotherms are capable of regulating their body temperatures with precision and at levels as much as 30°C higher than air temperature. Among endotherms, which typically maintain a relatively constant temperature set point, some can vary that set point quite dramatically depending on activity requirements, achieving notable economy in the use of metabolic fuels. Endotherms that allow the body temperature to rise and fall during activity and rest, respectively, echo the basking behavior of some reptiles.

Endothermy and ectothermy also offer animals different advantages in different climates. In the tropics, ectotherms such as reptiles compete successfully with, or even outcompete, mammals both in the abundance of species and the number of individuals. This competitive success is thought to be due in part to the warm, relatively stable tropical climate, which allows reptiles to expand into nocturnal activity rhythms, whereas tropical mammals tend to be diurnal in habit; and enhances the energy economy enjoyed by ectotherms because they need not expend energy to elevate body temperature. The metabolic energy thus saved by tropical ectotherms can be diverted to reproduction and to other uses that promote species survival. In moderate and cold climates, ectotherms are necessarily more sluggish, are thus less competent as predators, and are generally less abundant (in terms of numbers and species) than mammals in the same climates. Endotherms have a significant competitive edge over ectotherms in the cold because they keep their tissues warm. Generally speaking, the farther from the equator, the higher the prevalence of terrestrial endothermy. Only a few genera of amphibians and insects occupy subpolar environments, and in the Arctic, where the polar bear reigns supreme, there are no ectothermic vertebrates and almost no insects.

THERMAL BIOLOGY OF HETEROTHERMS

Heterotherms show characteristics of both ectotherms and endotherms. Some flying insects, including locusts, beetles, cicadas, and arctic flies, can be considered both temporal and regional heterotherms. When inactive, these insects behave strictly as ectotherms. At moderate ambient temperatures, they are unable to take off and fly spontaneously without warming up, because their flight muscles contract too slowly to produce sufficient power for flight at muscle temperatures much below 30–40°C. To prepare for flight, they raise the core temperature in the thorax to a regulated level that is

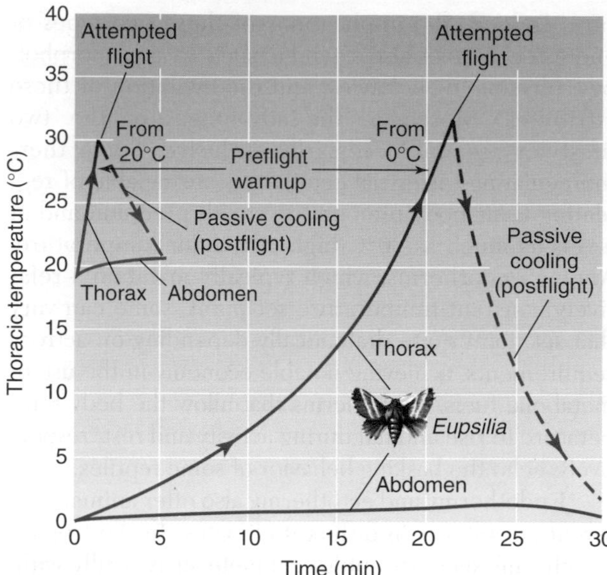

Figure 17-18 The noctuid winter moth *Eupsilia* uses preflight thermogenesis to remain active even in the cold winter conditions of its temperate habitat. In studies of tethered moths, voluntary shivering of the thoracic flight muscles caused a steep increase in thoracic temperature prior to attempted flight, which was followed by a post-flight cooling period. At an ambient temperature of 20°C, the warmup takes only a few minutes, compared with nearly 20 minutes at an ambient temperature near 0°C. Note that in both situations the abdomen remains very close to ambient temperature. [Adapted from Heinrich, 1987.]

species-specific, but is generally in the range of 30–40°C (Figure 17-18). Warmup is achieved by activating the large thoracic flight muscles, which are among the most metabolically active tissues known. Antagonistic muscles work against each other, producing heat without much wing movement other than small, rapid vibrations akin to shivering. Countercurrent heat exchangers in some flying insects help retain heat within the thorax, further increasing the attainable temperature for flight muscle. Flight is finally initiated when the thoracic temperature has reached the temperature that is maintained during flight. During this time, the abdomen typically remains close to ambient air temperature. Not surprisingly, the warmup period is greatly reduced at higher ambient air temperatures. Once a heterothermic insect is aloft, its flight muscles produce enough heat to maintain elevated muscle temperatures, and the insect may even need to deploy heat-dissipating mechanisms to prevent overheating.

Like "ideal" endotherms, heterothermic flying insects face the problem of regulating their body temperature in environments that have large temperature gradients. At ambient temperatures approaching 0°C, convective heat loss is generally so rapid that the high temperatures necessary for flight cannot be maintained. Typically, these flying insects have a large mass, and some, such as bumblebees, butterflies, and moths, are covered with insulating "hairs" or scales that help them retain body heat during their pre-flight warmup. High ambient temperatures, on the other hand, place the insect in danger of overheating. At ambient temperatures above 20°C, the hovering sphinx moth *(Manduca sexta)* prevents thoracic overheating by regulating the flow of warm hemolymph to the abdomen. The flow of heat from the active flight muscles in the thorax to the relatively inactive and poorly insulated abdomen increases the convective loss of heat to the environment through the body surface and especially through the tracheal system.

An interesting and somewhat unusual example of true shivering thermogenesis in an insect is found in honey bee swarms, in which individual honey bees contribute to the regulation of the swarm's core temperature by a combination of shivering movements and alterations in the overall swarm structure. At low ambient temperatures (e.g., 5°C), the bees in the swarm pack together tightly, restricting the free flow of air into and out of the swarm to the minimum required for respiration. Through shivering activity, the core of the swarm can be maintained as high as 35°C. In warm weather, the swarm loosens, providing ventilatory passages for air flow so that the core temperature exceeds the outside temperature by only a few degrees.

Another example of muscle-generated heat in a heterothermic species is found in the brooding female Indian python *(Python molurus)*. The snake coils herself around her eggs and elevates her body temperature—and that of her eggs—with shivering thermogenesis. Laboratory observations revealed that the rate of muscle contractions increased as ambient temperature decreased, and that the increase in contractions was accompanied by an increased difference between the ambient and body temperatures.

Unlike terrestrial ectothermic vertebrates, which can bask in the sun to warm up, aquatic ectotherms cannot obtain radiant energy as a source of heat, due to the high infrared absorption of water. As a result, fishes and other aquatic ectotherms can rise above ambient temperature only through very intense metabolic activity. Many teleost fishes are strictly ectothermic, operating at core temperatures close to ambient temperatures. However, as already mentioned, some fishes, such as tunas and lamnid sharks (e.g., the mako shark), are classified as regional heterotherms because they have specializations for generating and retaining sufficient heat to raise the temperature of body muscle, brain, or eyes as much as 10–15°C above that of the surrounding water. Typically, these fishes have a large body mass, and at first glance one might presume that their smaller surface-to-volume ratio relative to smaller fishes would help them attain a relatively constant muscle temperature. However, the relatively enormous gill surface area of fishes (see Chapter 13) largely negates this advantage.

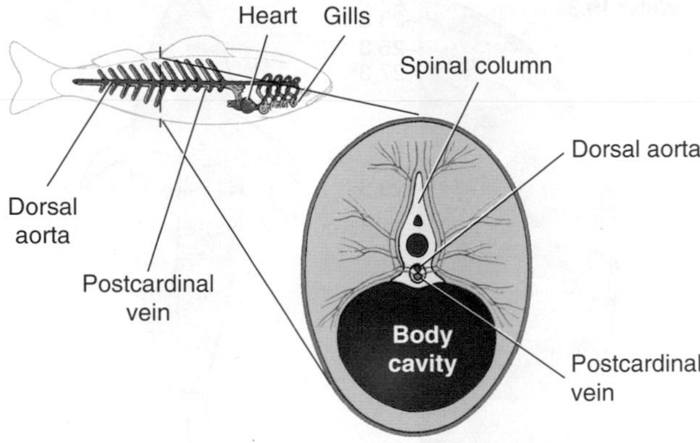

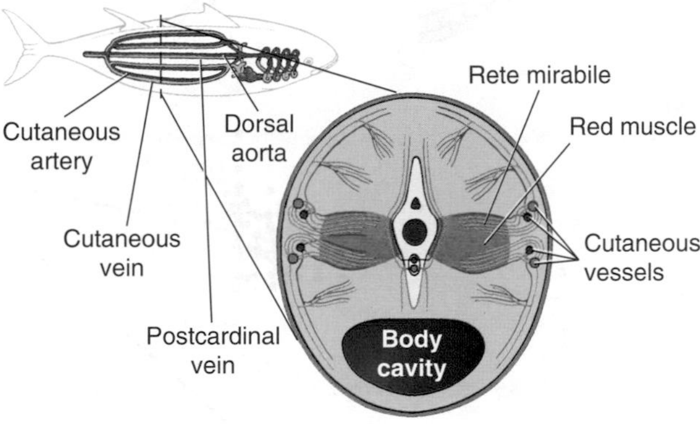

Figure 17-19 Differences in the vascular anatomy of a typical ectothermic fish, such as a trout, and a heterothermic fish, the bluefin tuna *(Tunnus thynnus),* account for the differential ability of the tuna to raise the temperature of its muscles. **(a)** The trout has its major blood vessels located centrally. **(b)** The tuna has its major blood vessels located under the skin and uses a rete mechanism to conserve deep body heat by countercurrent exchange. The advantage of this arrangement is that no body heat is lost in warming arterial blood that is unavoidably cooled while passing through the gills.

Instead, the retention of heat in the body core depends on the organization of the vascular system. Unlike ectothermic fishes, which have a centrally located aorta and postcardinal vein (Figure 17-19a), heterothermic fishes have major blood vessels (lateral cutaneous arteries and veins) located under the skin (Figure 17-19b). The red (dark) aerobic swimming muscles are located deep in the core of the fish's body. Blood is delivered to these muscles through a rete mirabile, a uniquely structured network of fine arteries that intermingle with small veins carrying warm blood away from the muscles. The classic work of Frank Carey and his colleagues during the 1960s and 1970s revealed that the rete acts as a countercurrent heat-exchange system (Figure 17-20 on the next page). Arterial blood, which is rapidly and unavoidably cooled during passage through surface vessels and the extensive and extensively perfused respiratory tissues of the gills passes from the cool periphery into the warm deeper muscle tissue through the rete. The cool arterial blood passing from the surface toward the core picks up heat from the warm venous blood leaving the muscle tissue and flowing toward the periphery. This process retains metabolically produced heat within the deep red muscle tissue and minimizes heat loss to the surroundings. Thus, the fish can maintain its swimming muscles at a temperature suitable for vigorous muscular activity even when the temperature of the surface tissues approaches that of typically cooler surrounding water. An important behavioral factor is that these regional heterotherms typically swim continuously, so their red muscle never cools down to the ambient temperature.

One of the implications of regional heterothermy, in other regional heterotherms as well as fishes, is that energy savings accrue in their cool tissues because the temperatures of only certain tissues, such as the swimming muscles, are elevated (expending large amounts of energy in the process).

THERMAL BIOLOGY OF ENDOTHERMS

In homeothermic endotherms (almost all mammals and birds), body temperature is closely regulated by homeostatic mechanisms that regulate the rates of heat production and loss to maintain a relatively constant body temperature independent of environmental temperatures. These animals maintain their core temperatures within a narrow range, between 37°C and 38°C in mammals and at about 40°C in birds. The temperatures of peripheral tissues and extremities are held less constant and are sometimes allowed to approach environmental temperatures. Basal metabolic heat production for a homeothermic endotherm of a given size is about the same irrespective of taxonomy. Typically, the basal metabolic rate of endotherms can be 7–20 times as high as the standard metabolic rate of comparably sized ectotherms measured at the same body temperature. This relatively high basal metabolism, in conjunction with heat-conserving and heat-dissipating mechanisms, allows homeothermic endotherms to maintain constant body temperatures many degrees above ambient temperatures. In polar bears actively ranging over polar ice, body temperature can be >70°C above that of the environment!

Question

Almost all mammals are endothermic, with the exception of hibernators, marsupials (e.g., opossums), and the monotremes (e.g., the echidna), which are heterothermic. Without concern for whether these heterothermic mammals evolved from endothermic ancestors or have retained the original ancestral condition, what environmental selection pressures led to the persistence of heterothermy in marsupials and monotremes?

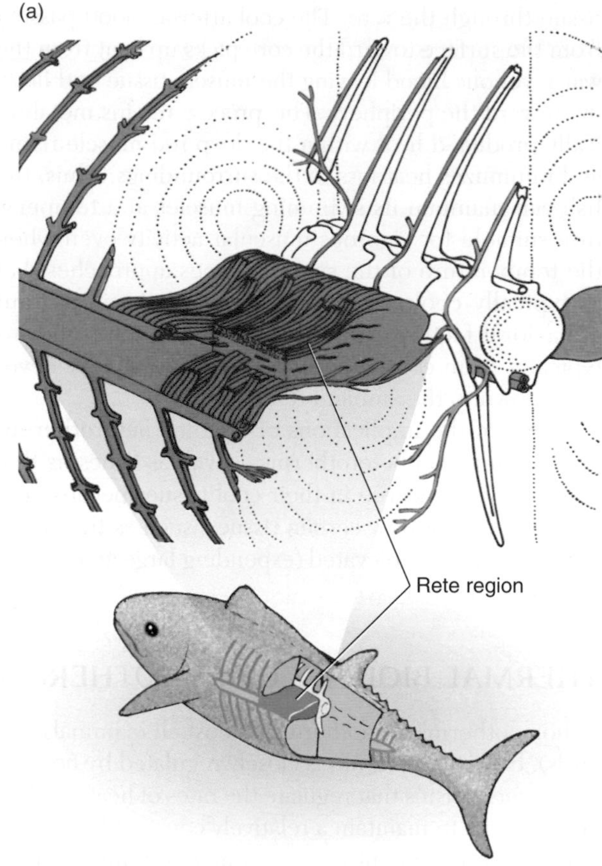

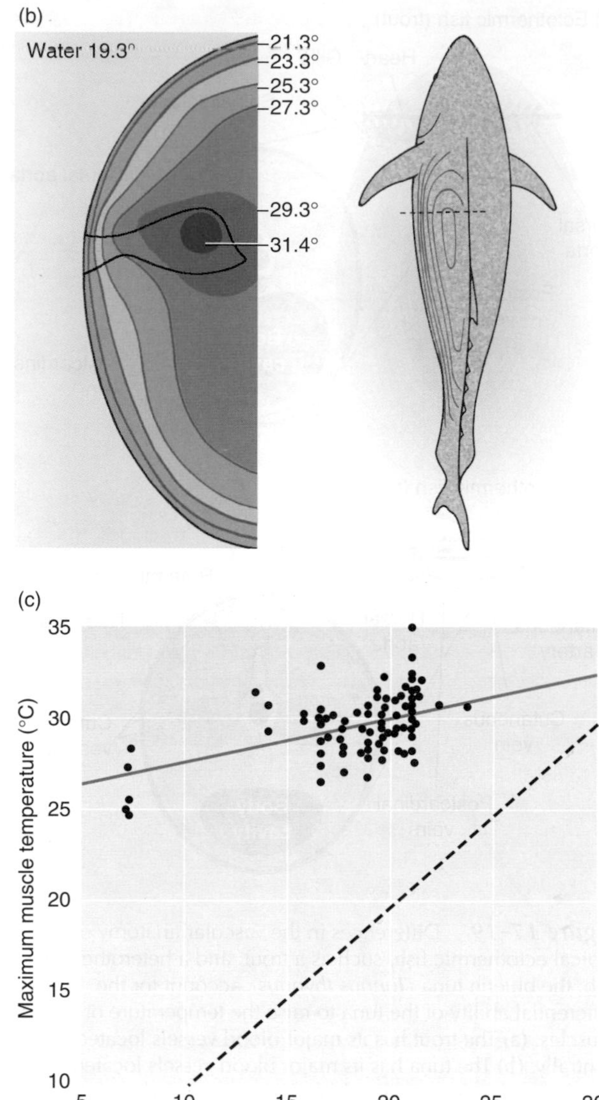

Classic Work **Figure 17-20** The bluefin tuna controls regional body temperature with a countercurrent arterial-venous heat-exchange rete. The rete (shown in red) helps the tuna retain heat produced in its active deep muscles. **(a)** Enlargement of the rete area. **(b)** Isotherms (left), plotted at intervals of 2°C, show temperature distribution in cross-section (right). **(c)** Maximum muscle temperatures of bluefins caught in waters of different temperatures. The dashed line indicates equality between body temperature and water temperature. [From Carey and Teal, 1966.]

The Concept of the Thermal Neutral Zone

Before we consider how polar bears and other less environmentally challenged endotherms regulate their body temperature, we must first introduce the concept of the thermal neutral zone. The degree of thermoregulatory activity that homeothermic endotherms require to maintain a constant core temperature increases with increasing extremes of environmental temperature. At moderate temperatures, the basal rate of heat production nicely balances heat loss to the environment. Within this temperature range, called the **thermal neutral zone** (Figure 17-21), an endotherm need not expend large amounts of energy to maintain its body temperature. Instead, it can regulate body temperature by adjusting the rate of heat loss through metabolically inexpensive adjustments in the thermal conductance of the body surface, including

- *Vasomotor responses:* Blood flow between the periphery and the core can be regulated to control heat loss or gain across the body surface (see Figures 17-7 and 17-8).
- *Postural changes:* Body shape or body orientation relative to sun or shade can be changed to alter the exposed surface area and the associated heat exchange by radiation, conduction, and convection.
- *Insulation adjustments:* The insulating effectiveness of the pelage or plumage can be adjusted by using pilomotor muscles to raise ("fluff") or lower hairs or feathers, thus altering their thickness and insulative qualities. (The "goose bumps" of humans are a vestige of the pilomotor control of body fur long lost through evolution.)

As the ambient temperature (T_a) decreases, an endotherm eventually reaches a *lower critical temperature* (LCT), the T_a below which the basal metabolic rate is insufficient to balance heat loss despite all the

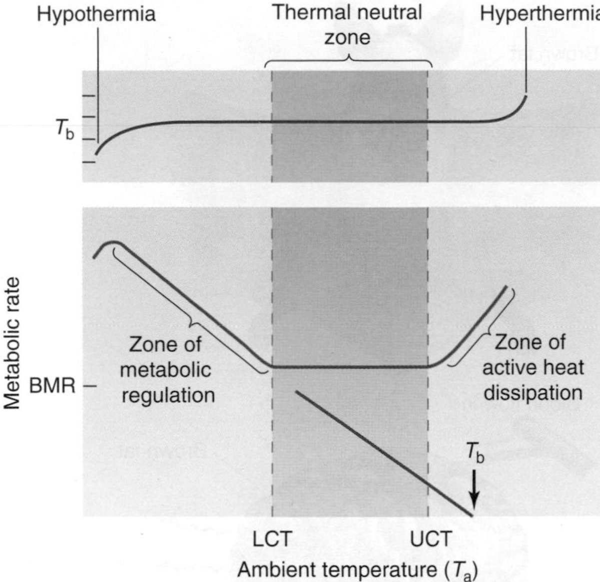

Figure 17-21 The resting metabolic rate of an endothermic homeotherm (blue curve) is higher at extremes of ambient temperature. Within the thermal neutral zone, body temperature is regulated entirely by changing the heat conductance of the body surface, which requires essentially no change in metabolic effort. The thermal neutral zone extends from the lower critical temperature (LCT) to the upper critical temperature (UCT). Above and below this zone, the metabolic rate must rise, either to increase thermogenesis in the zone of metabolic regulation or to increase active dissipation of heat by evaporative cooling, if body temperature, T_b (red curve), is to remain essentially constant. At ambient temperatures far below the LCT, thermogenesis is unable to replace body heat at the rate at which it is lost to the environment, and hypothermia sets in. At ambient temperatures far above the UCT, heat production and gain exceed the rate of heat loss, and hyperthermia occurs.

adjustments in thermal conductance listed above. Below its LCT, which is unique to each species, an endotherm must increase its heat production above basal levels to offset heat loss. Heat production then rises linearly with decreasing temperature below the LCT in what is called the *zone of metabolic regulation* (see Figure 17-21, left). If the environmental temperature drops below the zone of metabolic regulation, compensating mechanisms fail, the body temperature (T_b) falls, and the metabolic rate drops. Many endothermic animals can tolerate small decreases in T_b during their normal rest period (including humans during sleep). However, if an endotherm's body temperature falls much below its normal range, it enters a state of *hypothermia.* If this condition persists, the endotherm grows progressively cooler and, because cooling lowers the metabolic rate even further, it spirals into a positive feedback cycle that proves fatal.

The thermal neutral zone typically lies entirely below an endotherm's normal body temperature. Why? To answer this question, consider that heat loss by mechanisms that boost thermal conductance cannot be increased beyond a certain point. As the environmental temperature rises, it eventually reaches an *upper critical temperature* (UCT) because the surface insulation has already been minimized and peripheral vasodilation maximized. Any further increase in T_a above that temperature will therefore cause a rise in T_b, unless active heat-dissipating mechanisms such as sweating or panting are brought into play (see Figure 17-21, right). Without evaporative heat loss, the UCT will be exceeded, leading to *hyperthermia,* because the heat produced by basal metabolism cannot escape passively from the body as fast as it is being produced. (Sauna and hot-tub enthusiasts should bear this in mind.)

Metabolic rate rises linearly with temperature below the lower critical temperature, along a line that extrapolates to zero at an ambient temperature equal to body temperature (see Figure 17-21). To understand why, consider Fourier's law of heat flow:

$$Q = C\,(T_b - T_a) \tag{17-7}$$

where Q is the rate of heat loss from the body (in calories per minute) and C is the animal's thermal conductance. Because T_b is constant, Q varies linearly with the ambient temperature. The thermal conductance determines the slope of the plot below the thermal neutral zone; the better the insulation (i.e., the lower the value for C), the shallower the slope and the less heat must be produced metabolically at low temperatures.

The extrapolated intercept with zero is at T_b because, if $T_a = T_b$, then $C\,(T_b - T_a) = 0$. With $Q = 0$, there is no net heat loss. We know that the metabolic rate does not normally drop below the basal metabolic rate. When $T_a = T_b$, body temperature must be above the thermal neutral zone because there is no gradient for heat loss, so the animal will tend to warm up. The animal must cool itself by some means other than heat conduction. As mentioned above, the only means of cooling when T_a lies above the upper critical temperature is by evaporation.

Endothermy in Cold Environments: Producing and Retaining Body Heat

Endotherms use a wide variety of physiological and behavioral mechanisms to maintain body temperature within a narrow range in both cool and warm environments. The major thermal adaptations of endotherms living in cool or cold environments revolve around producing sufficient body heat and then retaining it.

Thermogenesis

When the ambient temperature drops below the lower critical temperature, an endothermic animal responds by generating large amounts of additional heat from its energy stores, thereby preventing a decrease in its core temperature. There are two primary means of heat production, or **thermogenesis,** other than locomotor activity: shivering and nonshivering thermogenesis. Both processes convert chemical energy into heat by a normal energy-converting metabolic mechanism that has been adapted to produce primarily heat. Essentially all the chemical-bond energy released in these processes is fully degraded to heat rather than being converted to chemical or mechanical work.

Shivering is a means of using muscle contraction to produce heat. Shivering thermogenesis is employed by most endothermic vertebrates as well as by some insects. The nervous system activates groups of antagonistic skeletal muscles so that there is little net muscle movement other than the shivering action itself. The activation of the muscles causes ATP to be hydrolyzed to provide energy for their contraction. Because the muscle contractions are inefficiently timed and mutually opposed, however, they produce no useful physical work, and the chemical energy released during contraction appears as heat.

In *nonshivering thermogenesis*, enzyme systems for the metabolism of fats are activated throughout the body so that fats are broken down and oxidized to produce heat. Very little of the energy released is conserved in the form of newly synthesized ATP. A few mammals have evolved an adipose tissue called **brown fat** that is specialized for fat-fueled thermogenesis. Brown fat contains such extensive vascularization and so many mitochondria that it is brown (owing largely to mitochondrial cytochrome oxidase) rather than white. Generally found as small deposits in the neck and between the shoulders (Figure 17-22), brown fat is an adaptation for rapid, massive heat production. In the routine metabolism of ordinary body fat, deposits of fat are first reduced to fatty acids, then exported through the circulation to be taken up by other tissues, where they are oxidized. In brown fat, oxidation takes place within the fat cells themselves, which are richly endowed with fat-metabolizing enzyme systems.

Nonshivering thermogenesis in both regular and brown fat is activated by the sympathetic nervous system through the release of norepinephrine, which binds to receptors on the adipose cells. Through a second-messenger mechanism (see Chapter 9), this signal stimulates thermogenesis by two routes:

1. The signal from the sympathetic nervous system stimulates an increase in the hydrolysis of ATP through processes such as nonproductive, cyclic ion pumping across the plasma membrane, with the energy of ATP released as heat.

2. Electron transport through the respiratory chain is uncoupled from the synthesis of ATP. Electron transport is normally coupled to the extrusion of protons (H^+) out of the mitochondrial matrix. Reentry of protons into the matrix is coupled to the synthesis of ATP from ADP and P_i. Thermogenesis in brown fat is characterized by the appearance in the inner mitochondrial membrane of an uncoupling protein, *thermogenin,* that provides a pathway for protons to leak back into the mitochondrial matrix without driving the phosphorylation of ADP to ATP. The energy of oxidation usually captured in ATP is instead dissipated as heat.

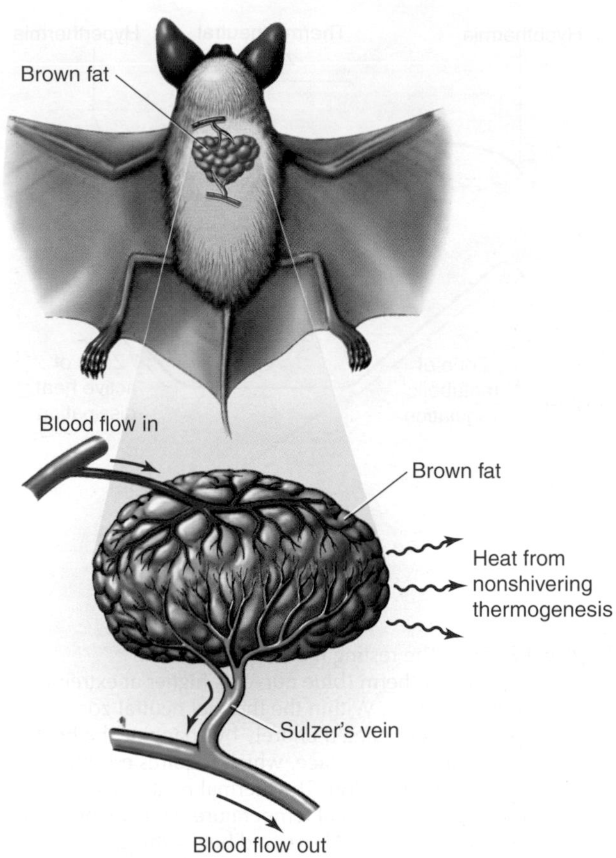

Figure 17-22 Brown-fat deposits are found between the scapulae in bats and several other mammals. The detail shows the extensive vascularization of this tissue. During brown-fat oxidation, this tissue is detectable as a warm region by its infrared emission.

Fat heats up significantly during thermogenesis. The heat is rapidly dispersed to other parts of the body by blood flowing through the fat's vasculature, particularly in the very heavily perfused brown-fat deposits.

Nonshivering thermogenesis is especially pronounced during arousal of hibernating or torpid mammals, when it supplements shivering to facilitate rapid warming. One consequence of acclimation to cold by mammals is an increase in brown-fat deposits, which allows for a gradual changeover from shivering to nonshivering thermogenesis at low ambient temperatures. The acclimatory increase in brown-fat thermogenesis is mediated by the thyroid hormones. Brown fat is also present in some mammalian infants, including human infants, where it is generally located in the region of the neck and shoulders, the spine, and the chest. Because neonatal mammals are relatively small and inactive at birth, deposits of brown fat provide an

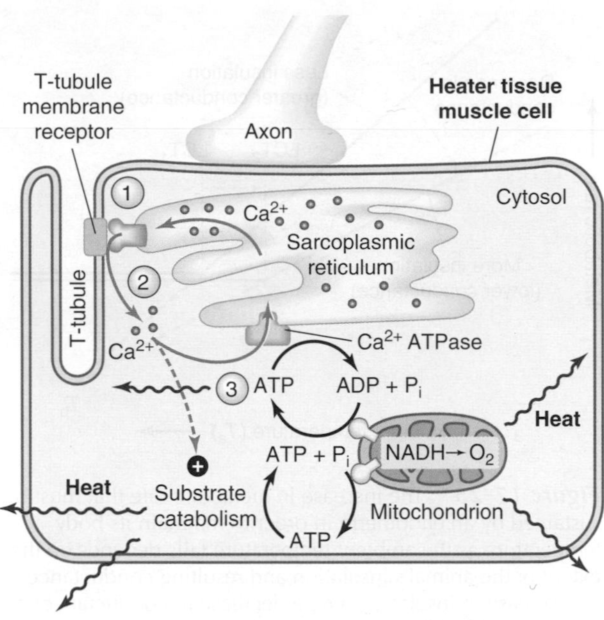

Figure 17-23 The metabolism of heater tissues in billfishes is specialized for heat production. Upon exogenous stimulation, a T-tubule receptor activates a calcium channel in the sarcoplasmic reticulum (1), causing the release of stored Ca^{2+} into the cytosol (2). The increased concentration of Ca^{2+} ions in the cytosol stimulates ATP-consuming catabolic processes (3) that release energy in the form of heat. [Adapted from Block, 1994.]

important and rapid means of warming should hypothermia threaten.

Nonshivering thermogenesis has also evolved in billfishes (e.g. marlin, sailfish), in which heater tissues formed from modified superior rectus eye muscles are specialized for heat production rather than force generation. Barbara Block and her colleagues have used molecular and immunological techniques to investigate the physiology and biochemistry of these tissues, which have an enormous capacity to generate heat (as high as 250 $W \cdot kg^{-1}$). Heater cells lack myofibrils and sarcomeres, but retain a sarcoplasmic reticulum. They produce heat through the release of Ca^{2+} from internal cytoplasmic stores, which then stimulates catabolic processes and mitochondrial respiration (Figure 17-23).

Countercurrent heat exchange

If endotherms are to move effectively, their limbs cannot be hindered by a massive layer of heat-retaining insulation. Yet limbs are well-vascularized, thin, and have large surface areas—characteristics that combine to offer great potential for the dissipation of body heat. Heat loss from the periphery can be reduced drastically by countercurrent heat exchange, as already mentioned in the context of the rete mirabile of heterothermic fishes (see Figures 17-19 and 17-20). Arterial blood is warm as it emerges from the animal's core. Conversely, the venous blood returning from peripheral tissues may be cold. Warm arterial blood passing into a limb or flipper passes through arteries that lie next to veins carrying blood back from the extremity. As the arteries and veins pass each other, the warm arterial blood gives up heat to the returning venous blood, becoming successively cooler as it traverses the extremity. By the time it reaches the periphery, the arterial blood has been cooled to within a few degrees of the ambient temperature, and has little heat left to lose. Conversely, the returning venous blood is warmed by the arterial blood and is nearly at core temperature as it flows into the core.

A highly evolved example of countercurrent heat exchange is found in the rete mirabile of the flipper of the porpoise. Here, the artery carrying warm blood flowing toward the extremity is completely encased in a circlet of veins carrying cold blood back from the extremity. Birds and Arctic land mammals also use countercurrent heat exchange to minimize heat loss from their extremities in cold climates, and to some extent this mechanism operates in the extremities of humans. As a result, the extremities of cold-climate endotherms are maintained at temperatures that are far below T_b, the core temperature, and often approach T_a (Figure 17-24).

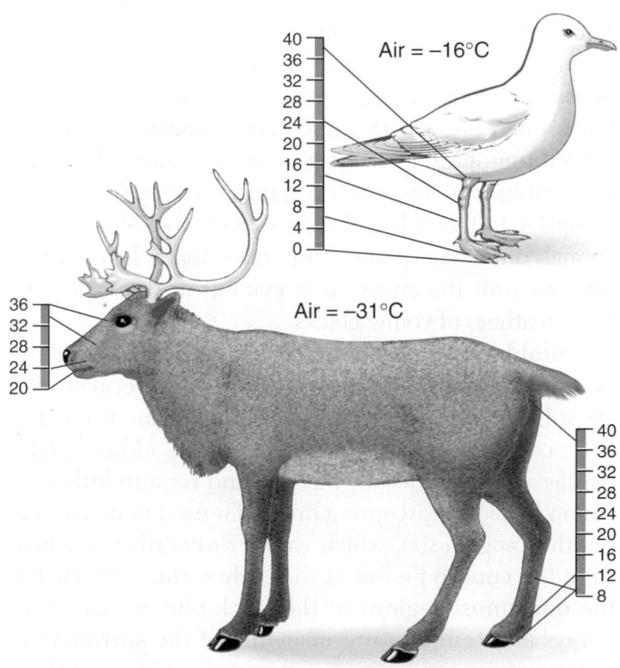

Figure 17-24 Endotherms can be regionally heterothermic. Temperatures in the extremities of Arctic birds and mammals are much lower than their core temperature of about 38°C.

The degree of efficiency of countercurrent heat exchangers is generally under vasomotor control. Major changes in the extent of blood flow to the limb would compromise gas exchange with muscle tissues. However, sympathetic and parasympathetic innervation can regulate the amount of blood that is shunted past the heat exchanger network via parallel vessels, thereby fine-tuning heat loss and retention in the extremity.

Adaptation and acclimatization of body insulation

An endotherm sensing heat loss in a windy place will fluff its fur or feathers and move to a more sheltered area, reducing convection and the dissipation of body heat by the wind. More enduring adaptations to cold include the increased insulation found in many Arctic animals in the form of subcutaneous fat or thick pelage or plumage. The insulating effectiveness of these layers in Arctic and sub-Arctic animals shows acclimatization, changing with both season and latitude to match insulation qualities with insulation needs. Animals living in more temperate zones exhibit seasonal variations by shedding old fur or feathers and growing a new body covering, thereby providing thick insulation during the winter yet preventing overheating during the summer.

The specific heat conductances of homeothermic endotherms vary over a large range and decrease with body size. Larger animals have lower specific heat conductances owing to their generally thicker coats of fur or feathers. In addition, they face smaller heat losses in cold climates because of their relatively smaller surface areas. Thus, one adaptation of endotherms to cold latitudes is an increase in body size. As surface-to-volume ratio becomes smaller, pelage becomes thicker and conductance decreases. With increased insulation, the lower critical temperature decreases and the thermal neutral zone extends to lower temperatures (Figure 17-25). An exception to this rule is that small animals and immature animals often have plumage or pelage that is less conductive per unit thickness, as is evident in the especially fluffy feathers of young chicks.

Blubber, a fatty tissue typically found under the skin in cetaceans, is an effective insulator because, like air, it has a lower thermal conductivity than water (the main constituent of nonfatty tissues). In addition, fatty tissues are metabolically inactive and require little perfusion by blood (excepting brown fat used in nonshivering thermogenesis), which would ordinarily carry heat from the core to be lost at the body surface. In whales, the outermost regions of the thick blubber layer are always at a temperature near that of the surrounding water.

An important means of controlling heat loss from the body surface is the diversion of blood flow to or away from the skin (see Figure 17-7). Vasoconstriction of arterioles leading to the skin keeps warm blood from perfusing cold skin and conserves the heat of the body core. An interesting advantage of blubber over pelage in the control of heat loss is illustrated in Figure 17-8b, which highlights that fur is located outside the body proper, whereas blubber is contained within the body and is supplied with blood vessels. Thus, whereas the insulating properties of fur remain unaffected by circulatory adjustments, the insulating properties of blubber depend on whether blood flow to the surface is restricted or not. Diverting more blood away from the vessels within the blubber increases the effective thickness of the insulating layer. This ability to regulate heat transfer through blubber also allows a marine mammal to get rid of excess body heat by shunting its surface blood to the outer regions of the insulating layer during periods of intense activity in warmer waters or when lying on the shore in warm air or exposed to the sun.

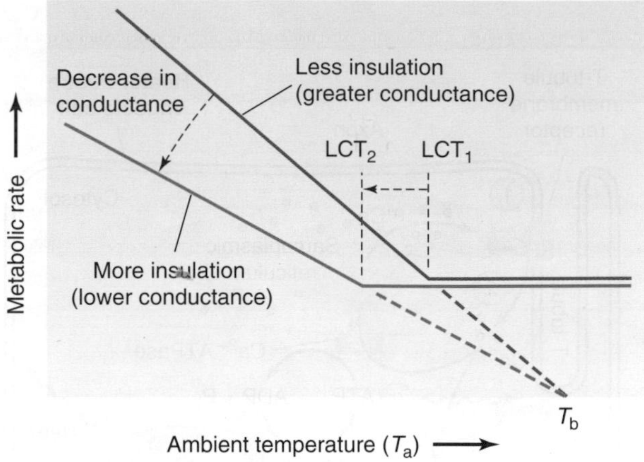

Figure 17-25 The increase in metabolic rate that must be sustained by an endotherm in order to maintain its body temperature as the ambient temperature falls depends on the extent of the animal's insulation and resulting conductance. An increase in insulation (i.e., a decrease in conductance), caused by either fluffing existing hair or feathers or growing thicker ones, depresses the lower critical temperature (LCT_1 to LCT_2) and reduces the slope of increasing metabolism. Both slopes, however, still extrapolate to body temperature (T_b) at zero metabolic rate.

Endothermy in Hot Environments: Dissipating Body Heat

Endotherms living in hot, dry environments face the challenge of getting rid of excess heat—and of doing so while conserving body water.

Limited heterothermy

Large endotherms in very hot, dry climates have the dual advantages of relatively low surface-to-mass ratios and large heat capacities. Camels, legendary for their ability to tolerate heat, have not only a large body mass, but also a thick pelage that helps insulate them from external heat. These characteristics retard the camel's absorption of heat from its surroundings by radiation and convection. Furthermore, because of their large mass and the high specific heat of tissue water, the camel and other large mammals can absorb relatively large quantities of heat for a given rise in body temperature. These features also result in a slow loss of heat during the cool hours of the night. Thus, large mass acts as a heat buffer that, by reducing rates of both absorption and loss of heat, minimizes temperature fluctuations.

A dehydrated camel can further increase its heat-absorbing capacity by tolerating an elevation of its core temperature by several degrees. The large amounts of heat that accumulate gradually during daytime hours are then dissipated efficiently in the cool of the night when the temperature gradient from body to environment is greatest. In preparation for the next onslaught of daytime heat, the dehydrated camel allows its core temperature to drop during the night to several degrees below typical daytime levels. As a consequence, the camel starts the day with a heat deficit, which allows it to absorb an equivalent amount of additional heat during the hot part of the day without reaching a harmful body temperature. This practice, called **limited heterothermy**, allows the camel to tolerate the extreme daytime desert heat without using much water for evaporative cooling. It also calls into question the concept of "normal body temperature" as it applies to camels and other heterotherms.

Limited temporal heterothermy is also practiced by the antelope ground squirrel *(Ammospermophilus leucurus)*, a small diurnal desert mammal. The antelope ground squirrel cannot emulate a camel and continuously gain heat for several hours in the hot sun, as the body temperature of an animal with such a small heat capacity would rise much too rapidly. Instead, the squirrel exposes itself to high temperatures at the desert surface for a maximum of about 8 minutes per excursion from its burrow. It then returns to its cooler burrow and allows the heat it has gained to escape into the cool underground air. By allowing its temperature to drop a bit below "normal" before returning to the hot desert surface, the antelope ground squirrel is able to extend its stay a few minutes without lethal overheating.

Regulating temperature with "heat windows"

Body surface temperature is an important factor affecting heat loss to the environment because it determines the temperature gradient, $T_b - T_a$, along which heat will flow. Heat can be lost by conduction, convection, and radiation as long as the ambient temperature is below the body surface temperature. The closer the body surface temperature is to the core temperature, the higher the rate of heat loss through the surface to cooler surroundings. Heat is transferred from the core to the surface primarily by the circulation; the rate of heat loss to the environment is regulated by the flow of blood to surface vessels (see Figures 17-7 and 17-8).

Endotherms thus can be thought of as using a variety of "heat windows" to regulate the loss of body heat, opening or shutting these windows by regulating blood flow to them. Heat windows permit the loss of heat from the body surface by radiation, conduction, and in some cases, evaporative cooling. An example of a heat window can be seen in the thin, lightly furred, very large ears of rabbits, with their extensively interconnected arterioles and venules. Another example is seen in the horns of various mammals. In goats and cattle, the horns are highly vascularized by a network of blood vessels that, under conditions of heat load, dilate and act as radiators of heat. Similarly, the legs and snouts of many mammals, which have large surface-to-volume ratios, are used as thermal windows for the dissipation of heat by regulation of blood flow through the arterioles serving the skin of these areas. Some mammals living under conditions of intense solar radiation or high temperatures have certain areas of the body surface that are exceptionally lightly furred, or even naked, to facilitate heat loss by radiant, evaporative, or conductive means. Such areas generally include the axilla (armpit), groin, scrotum, and parts of the ventral surface. Some of these areas, such as the udder and scrotum, carry additional temperature sensors that are used to detect changes in air temperature with minimal interference from the core temperature. By this means, the animal can anticipate changing heat loads and make the corresponding adjustments in advance.

Variations in posture or body orientation also can affect rates of heat absorption or loss, making the entire body a "heat window." For example, the guanaco, a medium-sized camel-like inhabitant of the Andes, has very densely matted hair on its back and a lighter covering of fur on its head and neck and the outer sides of its legs. The inner sides of the upper thighs and the underside are nearly naked, acting as thermal windows covering nearly 20% of the body surface. By adjusting the posture and orientation of its body with respect to solar radiation and cooling breezes, the guanaco can adjust the degree to which its thermal windows are open or shut, permitting a fivefold change in thermal conductance. While the guanaco is an extreme example, posturally controlled flexibility in surface insulation allows many endotherms to regulate heat transfer across their surface.

Evaporative cooling

The evaporation of 1 g of water requires a great deal of energy—2448 J (585 cal), to be specific. Consequently, evaporative cooling is a highly effective means of removing excess body heat, provided that there is enough body water available to expend in this fashion. Certain reptiles and birds and some mammals use available body water (saliva and urine) or standing water from the environment to spread on various body surfaces, allowing it to evaporate at the expense of body heat. Animals with naturally moist skin, such as amphibians, may have a body temperature slightly lower than ambient temperature because of evaporative cooling, though this is not an effect that has been selected for.

Some vertebrates sweat or pant to produce evaporative cooling. In sweating, exhibited by some (but not all) mammals, sweat glands in the skin actively extrude water through pores onto the surface of the skin (see Chapter 9). Sweating is under autonomic control. Although it is a mechanism for evaporative cooling, sweating can persist in the absence of evaporation when the relative humidity of air is very high. Water may continue to be secreted from sweat glands even when the relative humidity of the air is too high for evaporation to keep up with the rate of sweating, leading not only to elevated body temperature, but also to elevated water (and salt) loss.

In panting, mammals and birds use the respiratory system to lose heat by evaporative cooling (see also Chapter 14 on osmoregulation in desert environments). As noted earlier, the highly vascularized nasal passages play an important role in many mammalian species in retaining both water and body heat. Thus, panting mammals breathe rapidly and shallowly through the mouth instead of through the nose. Heat is carried away in warmed exhalant air because the relatively low surface area of the oral cavity absorbs less heat from the exhalant air than the nasal cavity does. How does the hyperventilation of panting mammals affect gas exchange? An increase in alveolar ventilation would result in a fall of blood P_{CO_2} and a rise in blood pH. Respiratory alkalosis is avoided during panting by a disproportionate increase in dead-space ventilation (i.e., flow through the mouth and trachea) without an increase in ventilation of the alveolar respiratory surface (Figure 17-26). Breathing rate is increased, but tidal volume is reduced. Overheated canines and birds pant by inhaling through the nose and exhaling through the mouth, exposing the tongue and other mouth structures to encourage further water evaporation and therefore heat loss. This behavior is readily evident in the gaping beaks exhibited by birds on very hot days. Panting produces a one-way air flow over the nonrespiratory surfaces of the nose, trachea, bronchi, and mouth, increasing evaporation without causing stagnation of saturated air in these passages. The amount of respiratory work required is minimized because the panting animal causes its respiratory system to oscillate at the frequency that requires the least muscular effort to sustain. Panting is accompanied by an autonomically controlled increase in secretions from the salivary glands of the nose. Most of the water that is not evaporated by panting is swallowed and conserved.

Because evaporation from the skin or respiratory epithelium is the most effective means of ridding the body of excess heat, there is a close link between water balance and temperature control in hot environments (see Chapter 14). In hot, arid environments, animals can be faced with the unpleasant choice of either overheating or desiccating. Dehydrated mammals conserve water by reducing evaporation caused by panting or sweating, instead allowing the body temperature to rise. When a small mammal, with its small heat capacity, is exposed to the desert heat in the absence of thermoregulatory water, it will undergo a rise in temperature that is far more rapid and more threatening than a larger animal would. To survive, small mammals must either drink water or stay out of the heat.

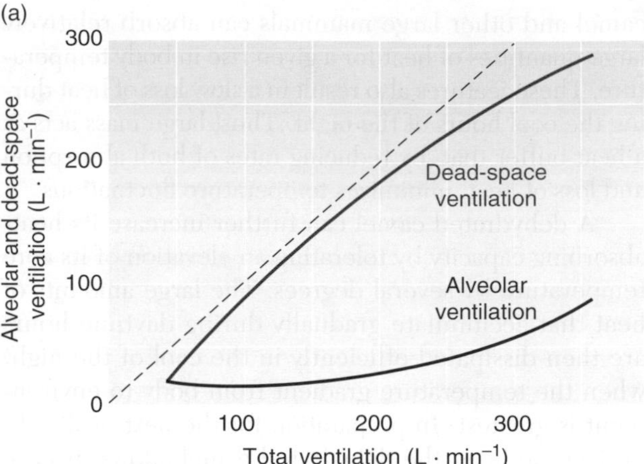

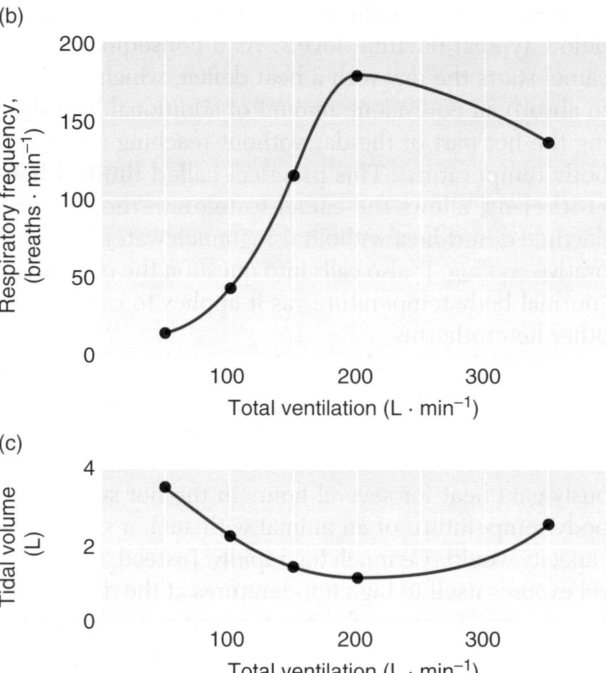

Classic Work **Figure 17-26** Panting induces a shift from primarily alveolar ventilation to mixed alveolar and dead-space ventilation. **(a)** In a panting ox, the dead-space ventilation (flow through the mouth and trachea) increases steadily as the total respiratory ventilation (*x* axis) increases, increasing heat loss without increasing alveolar ventilation. Panting is produced by increasing breathing frequency with little or no change in tidal volume. When total ventilation exceeds about 200 $L \cdot min^{-1}$, alveolar ventilation also starts to rise. In extreme panting, the respiratory frequency decreases **(b)** as tidal volume increases **(c)**. [After Hales, 1966.]

The reciprocity of water conservation and heat dissipation in a small desert animal can be illustrated by considering water balance and temperature control in the kangaroo rat *(Dipodomys)*. To conserve water, this animal uses a countercurrent heat-exchange system located in the nasal passages. The nasal epithelium is cooled during inhalation by the incoming air. During exhalation, most of the moisture picked up by the air in the warm, humid respiratory passages is conserved by its condensation on the cool nasal epithelium. Of course, this mechanism also retains body heat and requires that the inhaled air be cooler than the body core. As a consequence, the kangaroo rat is confined to its cool burrow during the hot times of day. If inhaled air were at or above body temperature, the kangaroo rat's loss of respiratory moisture would increase. Although the evaporative loss of water would help cool the animal, it would also seriously disturb its water balance.

NEURONAL MECHANISMS OF TEMPERATURE CONTROL

Along with the myriad physiological, behavioral, morphologic, and biochemical mechanisms for producing, retaining, or dissipating body heat, most animals—both ectotherms and endotherms—also control body temperature via the central nervous system by a mechanism that operates surprisingly like a common household thermostat.

Thermostatic Regulation of Body Temperature

Consider a laboratory water bath in which a temperature comparator compares the water temperature, T_w, detected by a temperature sensor with a set-point temperature, T_{set}. If T_w is below T_{set}, the thermostat closes a circuit that activates the water bath heater, producing extra heat and raising T_w until $T_w = T_{set}$, after which the thermostat contacts open and heat production ceases. The cycle is repeated each time T_w drops. By analogy, T_w is equivalent to T_b in an animal's zone of metabolic regulation (see Figure 17-21), in which heat production increases with decreasing ambient temperature. Of course, both homeothermic endotherms and homeothermic ectotherms also use nonmetabolic means to regulate body temperature; those mechanisms would be analogous to adding or removing an insulating cover from the water bath.

While simple in principle, the thermostatic regulation of body temperature is only now coming to be well understood after many decades of research on the topic. Most animals have not one but many temperature sensors, called thermoreceptors, located in various regions of the body, which participate in a negative-feedback system (see Spotlight 1-1). Furthermore, to maintain T_b near T_{set}, homeothermic animals can call on several heat-producing and heat-exchanging mechanisms, so the thermostat controls heat-conserving and heat-dissipating mechanisms as well as heat production. Body temperature control is analogous to the microprocessor-controlled heating and cooling system of the "smart house" of the future, in which the thermostat, in addition to turning the furnace and air conditioner on and off, controls the position of window shades, the opening and closing of windows, the conductance of the wall and roof insulation, and so forth. Furthermore, control of thermogenesis in a homeotherm is not an all-or-none phenomenon, like the turning on and off of a furnace. Instead, the rate of metabolic heat production is graded according to need. The colder the temperature sensors become (within limits), the higher the rate of thermogenesis. This phenomenon is referred to as *proportional control* because heat production and conservation are more or less proportional to the difference between T_b and T_{set}.

The hypothalamus: The mammal's "thermostat"

Mammalian body temperature can vary widely (as much as 30°C) between the periphery and the body core, with the extremities undergoing far more variation than the core. Temperature-sensitive neurons and nerve endings exist in the mammalian brain, spinal cord, skin, and at sites in the body core, providing input to thermostatic centers in the brain. Although a mammal may have several thermoregulatory centers, the most important one is located in the hypothalamus (see Figure 9-15), considered to be the body's "thermostat." The thermoregulatory role of the hypothalamus was discovered by Henry G. Barbour in 1912 during a series of experiments in which a small temperature-controlled probe was implanted in different parts of the rabbit brain. The probe evoked strong thermoregulatory responses only when it heated or cooled the hypothalamus. Cooling the hypothalamus produced an increase in metabolic rate and a rise in T_b, whereas heating it evoked panting and a drop in T_b. Barbour's manipulations with the probe were analogous to aiming a hot blow dryer at a house thermostat. As the thermostat is warmed above its set-point temperature, it shuts down the furnace, causing the room temperature to drop below the set point.

Experimental procedures following in Barbour's footsteps have shown that the mammalian hypothalamic thermostat is highly sensitive to temperature. Variations in mammalian brain temperature of only a few degrees Celsius seriously affect brain function, so it is not surprising to find that the brain is the location of the major thermoregulatory center, complete with "thermostat." At least three different subsets of neurons in the anterior hypothalamus are very sensitive to temperature change, with different sensitivities and different responses:

- One subset shows a sharply defined increase in firing frequency with increased hypothalamic temperature (red plot in Figure 17-27 on the next page). These neurons are believed to activate heat-dissipating responses such as vasodilation and sweating.
- Another subset fires less frequently as hypothalamic temperature rises (purple plot in Figure 17-27).

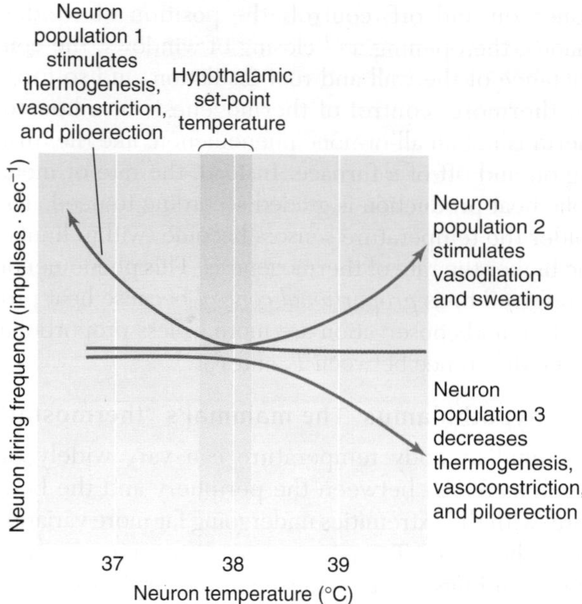

Figure 17-27 Different hypothalamic neurons show distinctly different temperature-activity patterns in mammals. Neurons that fire at higher frequencies when the hypothalamic temperature rises above 38–39°C (red curve) stimulate heat-dissipating mechanisms and inhibit heat-producing mechanisms, whereas neurons that increase their firing rate as temperature decreases (green curve) stimulate heat-producing mechanisms. In either case, the response ultimately returns T_b toward the set point, and firing frequency decreases. A third set of neurons whose firing frequency drops as hypothalamic temperature rises (purple curve) stimulates heat-dissipating mechanisms.

These neurons are thought to down-regulate heat-producing mechanisms (metabolism, shivering, nonshivering thermogenesis).

- A third subset increases its firing frequency when the brain temperature drops below the set-point temperature (Figure 17-27, green plot). This group of neurons appears to control the activation of heat-producing responses (e.g., shivering, nonshivering thermogenesis, brown-fat metabolism) and heat-conserving responses, such as pilomotor activation.

In addition to information about its own temperature, the hypothalamus also receives input from thermoreceptors in other parts of the body. All this thermal information is integrated and used to control the hypothalamic motor output. Neuronal pathways leaving the hypothalamus make connections with peripheral neuronal pathways that affect heat production and heat loss. Some of these pathways are reflexly activated both by signals from peripheral and spinal thermoreceptors and by the hypothalamic temperature-sensitive neurons. When temperatures are high, the efferent pathways activate increased sweating and panting, as well as a lowered peripheral vasomotor tone that produces increased blood flow to the skin. Conversely, detection by the thermoreceptors of body cooling triggers reflexes leading to thermogenesis and increased peripheral vasomotor tone, reducing heat loss from the periphery. These same physiological responses can be elicited without cooling the whole body by simply cooling the neurons of the hypothalamus. Experimentally lowering hypothalamic temperature in a dog leads to elevated metabolic heat production by shivering. Warming the dog's hypothalamus elicits the heat-dissipating response of panting.

In most mammals, a rise in core temperature of only 0.5°C causes such extreme peripheral vasodilation that the blood flow to the skin can increase several times above normal. In humans, this response produces the flushed appearance that follows sustained exercise. The effect of elevated core temperature on peripheral vasodilation and, hence, skin temperature is illustrated in Figure 17-28, which shows that the skin temperature of a rabbit's ear rose very sharply from less than 15°C to about 35°C at the point at which the rabbit's core temperature exceeded 39.4°C. Because the temperature of the rabbit's ear reached a maximum, it can be assumed that the vessels of the ear dilated fully as soon as the core temperature exceeded this limit.

There appears to be a strong bias in the hypothalamus toward certain sources of neuronal input on temperature variation. The response of mechanisms controlled by the hypothalamus is about 20 times greater when the sensory information relayed to the hypothalamus originates from thermoreceptors that are centrally located (including anterior hypothalamic thermoreceptors) than when it comes from receptors in the periphery. This bias is significant for the precise regulation of brain temperature. Without dominance of the hypothalamic thermostat, an internally overheated animal exercising in a cold environment would fail to activate heat-dissipating blood flow to the surface capillaries, and its core temperature would continue to rise to dangerous levels.

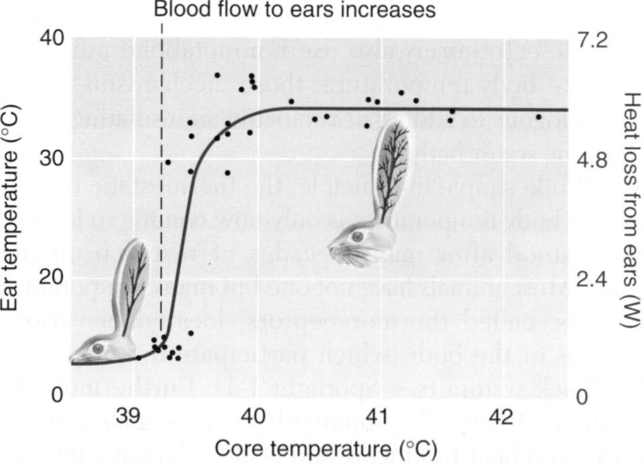

Figure 17-28 Heat loss through the ear of a rabbit running on a treadmill at an ambient air temperature of 10°C rises sharply as blood flow to the ear increases. This circulatory response was stimulated when core temperature increased to above 39.5°C. Heat loss from the ears is calculated from the thermal gradient between the ears and ambient air. [From Kluger, 1979.]

In some homeothermic endotherms, especially small mammals subject to rapid cooling at low ambient temperatures, the set-point temperature of the hypothalamic thermostat changes with ambient temperature, presumably because ambient deviations are sensed by peripheral receptors. In the kangaroo rat, for example, a sudden drop in ambient temperature is quickly followed by a rise in set-point temperature. This rise causes an increase in metabolic heat production in anticipation of increased heat loss to the environment.

The relationships between core temperature and the various thermoregulatory responses reflexly stimulated by peripheral thermoreceptors and hypothalamic centers are summarized in Figure 17-29. Small deviations in core temperature from the set point produce primarily changes in thermal conductance through peripheral vasomotor and pilomotor responses (red plot). These small deviations usually result from moderate variations within a range of ambient temperatures corresponding to the thermal neutral zone. When the core temperature is forced out of this range by more extreme deviations in T_a or by locomotor activity that generates excess heat, thermoregulatory responses that merely alter conductance no longer suffice, and the hypothalamic centers stimulate additional mechanisms, including thermogenesis or evaporative heat loss (light blue and dark blue plots in Figure 17-29).

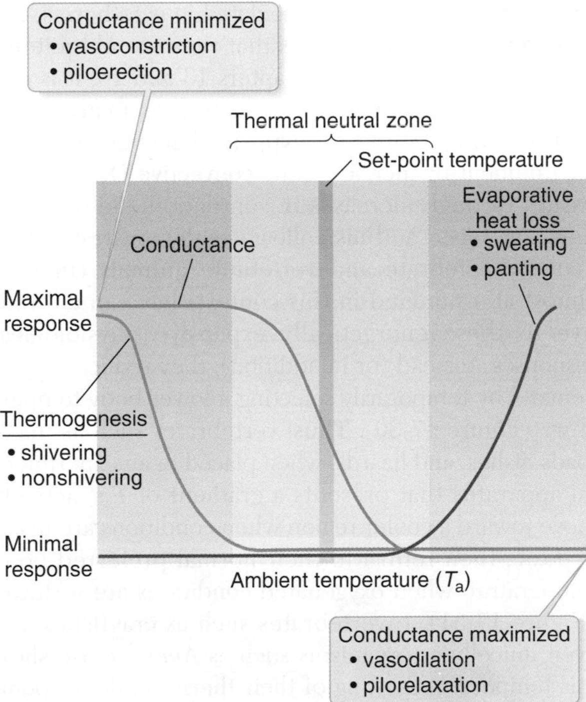

Figure 17-29 The degree of an animal's thermoregulatory response is greatest at body temperatures above or below the normal core temperature. Within a range (light blue) on either side of the set-point temperature (dark blue), body temperature is regulated primarily by control of heat conductance to the environment, achieved by varying the peripheral blood flow or the insulating effectiveness of fur or feathers. Above and below this range, the potential of these measures to change thermal conductance is exhausted, and active thermogenesis at low temperature or evaporative heat loss at high temperature develops.

While the role of the temperature-sensitive anterior hypothalamus of mammals is relatively well understood with respect to short-term physiological changes that promote thermoregulation, less is known about the regulation of T_b through behavior, which must also involve higher brain centers. (Consider the conscious decision of a human sunbather on the beach to move from the hot sand to the shade under an umbrella.) Much research will also be required to understand more fully the complex interactions of nervous and endocrine systems that stimulate the morphologic changes associated with acclimatization to changing thermal environments.

Thermoregulatory centers in non-mammals

Thermostatic control of body temperature has received far less attention in birds than in mammals, in part because the mechanism of control seems to be more complex in birds, but also because of entirely practical issues such as the greater difficulty of maintaining appropriate levels of anesthesia in birds during laboratory experiments than in mammals. The region of the hypothalamus that serves as the thermoregulatory center in mammals is largely insensitive to temperature changes in birds, whose responses to heating or cooling of that region are muted or nonexistent. Instead, the spinal cord appears to be the site of central temperature sensing in pigeons, penguins, and ducks. Yet another difference between birds and mammals is that body core thermoreceptors lying outside the central nervous system dominate reflex thermoregulatory activity in birds. As in mammals, the temperature sensors in the core presumably signal the avian thermostat, which integrates their input and activates the peripheral thermoregulatory effectors. The intrinsic thermal responses of the avian hypothalamus appear to be primarily involved in behavioral, rather than physiological, thermoregulation.

Fishes and reptiles, like mammals, have a center in the hypothalamus that is temperature-sensitive. Heating the hypothalamus with an implanted, electronically activated probe produces hyperventilation in the scorpion fish; cooling leads to slower ventilatory movements. Peripheral cooling produces similar ventilatory responses. At first glance, these responses seem paradoxical—as we have already discussed, changes in ventilation of the gills have virtually no effect on body temperature in fishes since the water flowing over the gills is such an enormous heat sink. Recall, however, that a fish's rate of metabolism varies with body temperature, so a rise in temperature leads to an increase in oxygen demand. The temperature-determined adjustment in the rate of respiration stimulated by hypothalamic heating in fishes is thus acclimative in that it anticipates changes in respiratory need and minimizes fluctuations in blood oxygen levels.

The reptilian response to cooling of the hypothalamus is to engage in thermophilic (i.e., heat-seeking) behavior, such as moving toward a warmer microhabitat (e.g., from the shade under a rock onto the sun-warmed upper surface of the rock). Correspondingly, heating of the hypothalamus elicits thermophobic (i.e., heat-avoiding) behavior, including changes in posture that facilitate heat dissipation. Similar findings have been described for both amphibians and fishes, which migrate to cooler environments when the rostral region of the brain stem is heated.

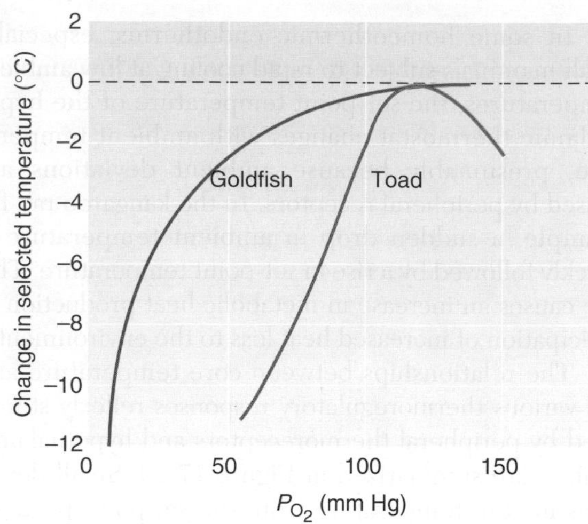

Figure 17-31 Hypoxia induces changes in selected body temperature in the goldfish *(Carassius auratus)* and the marine toad *(Bufo marinus)*. [Adapted from Wood, 1991.]

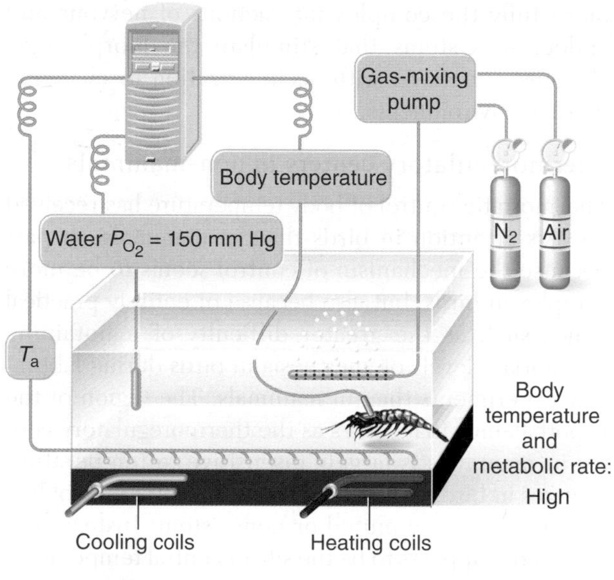

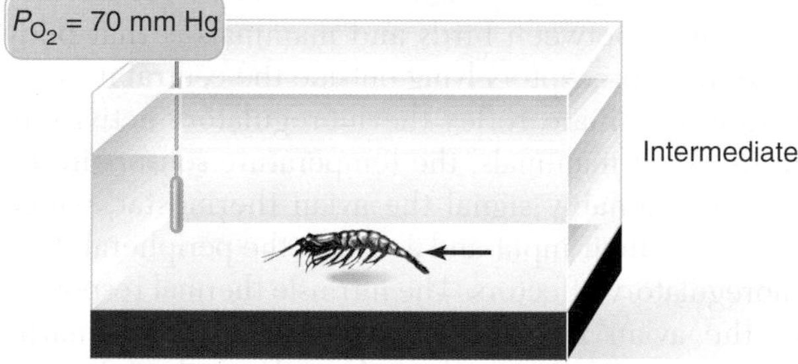

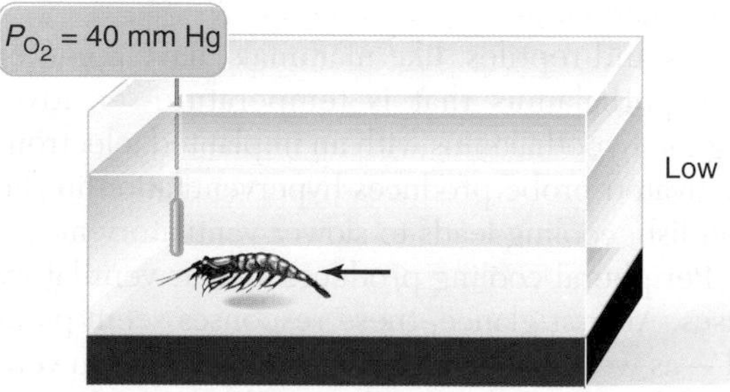

Figure 17-30 Voluntary selection of a lower body temperature driven by environmental hypoxia can be demonstrated in an apparatus that provides a distinct thermal gradient. In this apparatus designed for an aquatic ectotherm, a range of T_as can be freely selected by the animal in response to changes in ambient oxygen levels. In practice, the creation in the laboratory of a smooth gradient of environmental temperatures is difficult.

Experiments by S. C. Wood and his colleagues (1991) have drawn some intriguing links between behavioral thermoregulation and hypoxic exposure. These findings were first elucidated in ectotherms, but subsequent research reveals that they are ubiquitous. From our discussions in Chapters 13 and 16, it is evident that most invertebrates, as well as many vertebrates, can respond to a hypoxic challenge from the environment by increasing the convective O_2 supply to tissues through increases in ventilation and/or blood flow. Yet Wood and his colleagues demonstrated that many invertebrate and vertebrate animals (in fact, almost all examined in this context) have an alternative to these energetically expensive physiological responses. Instead (or in addition), they reduce oxygen demand by temporarily selecting a lower body temperature (Figure 17-30). Thus, vertebrates such as mice, toads, fishes, and lizards, when placed in an experimental apparatus that presents a gradient of T_as, actively move toward a cooler region when conditions are made hypoxic, then return to their normal preferred body temperature when oxygenated conditions are restored (Figure 17-31). Invertebrates such as crayfishes and even unicellular organisms such as *Amoeba* also show this temporary resetting of their thermostatic set point during hypoxic exposure.

Thermoregulation During Activity

The energy efficiency of muscle contraction is only about 25%. For every joule of chemical energy converted into mechanical work, 3 joules of energy are released as heat. During locomotor activity, the extra heat produced is added to the heat generated by basal metabolism and will cause a rise in T_b above the set-point temperature unless it can be dissipated to the environment at the same rate at which it is produced.

Most of the excess heat can be transferred to the environment, but the core temperatures of endotherms still rise during locomotor activity, indicating incomplete removal of the excess heat. Elevations of 2–4°C in core temperature are commonly observed in humans after strenuous, sustained running—the temperature of blood in the femoral arteries draining the legs can reach temperatures of 42°C or higher. Similar increases in core temperature occur during prolonged running in race horses, greyhounds, and sled dogs.

The rise in an endotherm's core temperature during activity is moderately useful in several respects:

- It increases the difference between T_b and T_a and thereby increases the effectiveness of heat-dissipating processes by increasing the gradient for heat loss.
- It leads to an increased rate of metabolic reactions, including those that support physical activity.
- It facilitates unloading of oxygen in metabolically active, warmer tissues (see Chapter 13).

However, the core temperature can rise to dangerously high levels during heavy locomotor activity in warm environments, and this excess heat has to be dealt with.

The level to which the core temperature rises in homeothermic endotherms is proportional to the rate of muscular work. During light or moderate locomotor activity in cool environments, T_b rises to a new level and is regulated at that level as long as the locomotor activity continues. Thus, T_b appears to remain under the control of the body's thermostat. The rise in T_b appears to be a consequence primarily of an increase in the error signal, $T_b - T_{set}$, of the hypothalamic thermostat. The error signal is the difference between the thermostat's set point and the actual core temperature. The greater this difference (i.e., the greater the error signal), the greater the activation of heat-loss mechanisms. Thus, the rate of heat dissipation increases as the core temperature rises above the set point, and a new equilibrium becomes established between heat production and heat loss. During heavy locomotor activity, especially in warm environments, the heat-dissipating mechanisms are not able to balance heat production until body temperature rises several degrees, increasing the $T_b - T_a$ difference.

The rise in T_b (and in the error signal as T_b rises above T_{set}) is kept small by the high sensitivity of the feedback-control mechanism. For example, a small increase in T_b above the set-point temperature produces a strong and steep increase in the rate of sweating (Figure 17-32). Heat-loss mechanisms are initiated by vigorous locomotor activity even before peripheral body temperature has undergone any significant increase. For example, in humans, sweating begins within 2 seconds after onset of heavy physical work, even though there is no detectable increase in skin temperature in that time. However, core blood temperature shows a

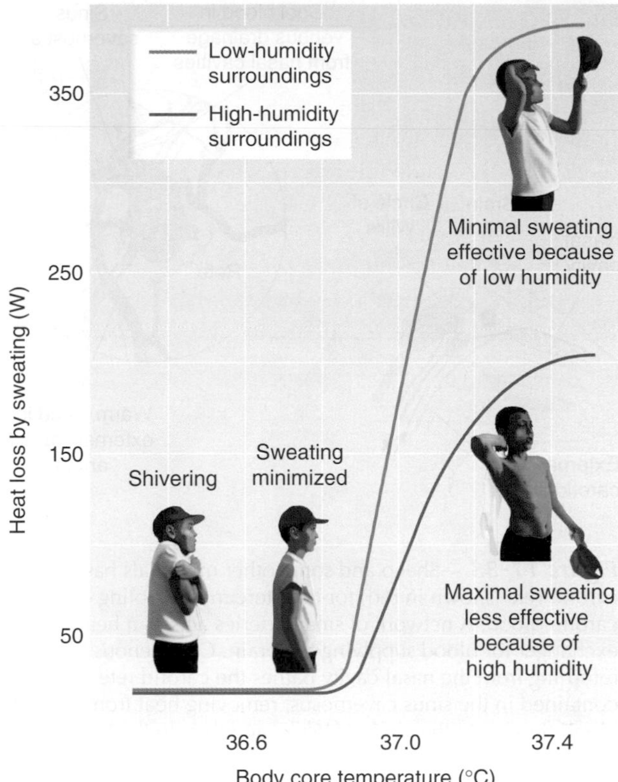

Figure 17-32 The rate of sweating in humans increases sharply as body temperature approaches, and then exceeds, 37°C due to an increase in T_a or the onset of exercise. Note that, in humans, the addition or removal of clothing and hats constitutes behavioral changes in thermal conductance.

detectable rise within 1 second after locomotor activity has begun. Apparently, the onset of sweating, nearly concurrent with the onset of neuronal activity underlying locomotor activity, results from the reflex activation of the sweating response by central temperature receptors. The set points for heat loss are lower in well-trained athletes, especially in warm weather. The effectiveness of sweating is affected by humidity of the surrounding air—the higher the humidity, the less effective it is as a heat-dissipating mechanism (as any person familiar with either desert or high-humidity environments can testify).

Some hoofed mammals (e.g., sheep, goats, and gazelles) and carnivores (e.g., cats and dogs) have evolved a special countercurrent heat exchanger to prevent overheating of the brain during strenuous, prolonged activity. This system, the carotid rete, uses cool venous blood returning from the respiratory passages to remove heat from hot arterial blood traveling toward the brain (Figure 17-33 on the next page). In these animals, most of the blood to the brain flows through the external carotid artery. At the base of the skull, the carotid subdivides into hundreds of small arteries that form a vascular rete, the vessels of which rejoin just before passage into the brain. These arteries pass through the *sinus cavernosus*, a large sinus of venous

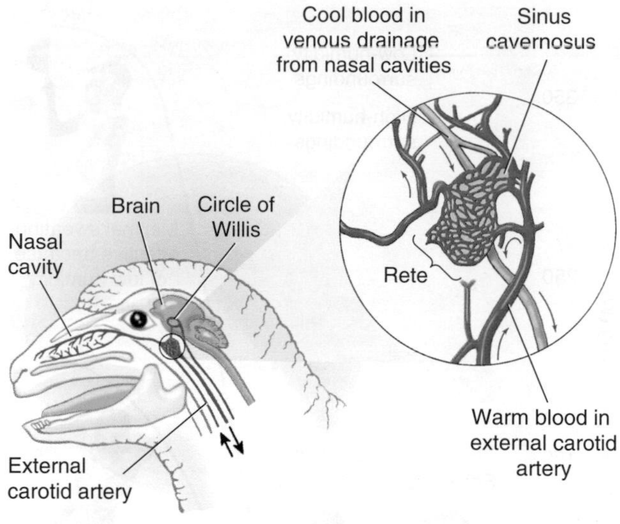

Figure 17-33 Sheep and some other mammals have a carotid rete (shown in red) for countercurrent cooling of carotid blood. A network of small arteries acts as a heat exchanger for blood supplying the brain. Cool venous blood returning from the nasal cavity bathes the carotid rete contained in the sinus cavernosus, removing heat from arterial blood flowing to the circle of Willis and then to the brain.

blood. This venous blood is significantly cooler than the arterial blood because it has come from the walls of the nasal passages, where it is cooled by respiratory air flow. Thus, the hot arterial blood flowing through the rete gives up some of its heat to the cooler venous blood before it enters the skull. As a result, brain temperature may be 2–3°C lower than the core body temperature. Although sustained running in hot surroundings inevitably places a heat load on these animals, the most serious and acute consequence of overheating—spastic brain function—is thereby prevented. This system of cooling is most effective when the animal is breathing hard during strenuous locomotor activity.

THERMOREGULATION AND SPECIALIZED METABOLIC STATES

For many endotherms with intrinsically high metabolic rates, metabolic substrates can quickly become a limiting factor unless more or less unlimited food supplies are available. Since few environments offer this luxury, many animals have evolved various forms of **dormancy** (a general term for states of lowered metabolism). Dormancy lowers animals' food requirements, allowing them to manage their energy stores more efficiently. Dormant states are characterized not only by significant changes in metabolism, but also by quantitative and sometimes qualitative changes in thermoregulation. As we will see, the pathological condition of fever also introduces interesting changes in thermoregulation. It is important to emphasize that all of these phenomena—characterized in part by changes in body temperature—do not represent a loss of thermoregulatory control, but rather reflect an active resetting of thermoregulatory set points.

Dormancy can be classified according to its severity or depth (indicated both by decrease in T_b and ability for arousal) or its duration. Dormant states include sleep, torpor, hibernation, winter sleep, and estivation. Sleep has been the most thoroughly investigated of these states (probably because it is the only dormant state experienced by humans). The remaining four categories are less well understood; however, in homeothermic endotherms, all appear to be manifestations of physiologically related processes.

Sleep

Sleep entails extensive adjustments in brain function. In mammals, slow-wave sleep is associated with a drop in both hypothalamic temperature sensitivity and body temperature, as well as with changes in respiratory and cardiovascular reflexes. During rapid-eye-movement (REM) sleep, hypothalamic temperature control is suspended. Although there may be a variety of triggers of sleep, in mammals there is evidence of sleep-inducing substances that build up during wakefulness, accumulating in extracellular fluids of the central nervous system.

The time course and extent of sleep vary greatly among animals. Seals resting on pack ice sleep for only a few minutes at a time before rousing to scan the ice for approaching polar bears. Humans and many other mammals sleep for hours at a time. Many of the big carnivores (e.g., lions and tigers) sleep for as long as 20 hours a day, especially after a meal, while domestic cats sleep as much overall but in numerous, shorter periods ("catnaps"). Evidence for "sleep" has also been accumulating in arthropods (insects, scorpions), in which denial of a period of inactivity leads to reduced efficiency at negotiating mazes and other tests of motor function.

Torpor

As T_b falls, so does basal metabolism and the rate of conversion of energy stores, such as fatty tissues, into body heat. Thus, it is generally advantageous to allow T_b to decrease during periods of nonfeeding and inactivity. Small endotherms, because of their high rates of metabolism, are subject to starvation during periods of inactivity when they are not feeding. During those periods, some animals enter a state of **torpor**, in which temperature and metabolic rate drop. Thus they can be classified as heterothermic endotherms (see Figure 17-11). Before the animal becomes active again, its T_b rises as a result of a burst of metabolic activity, often through shivering or oxidation of brown-fat stores or, in mammals only, both.

Daily torpor is practiced by many terrestrial birds. The rufous hummingbird (*Selasphorus rufus*) is a classic example, allowing its body temperature to fall from a daytime level of about 40°C to as low as 13°C when a low T_a permits. Several species of small mammals also undergo daily torpor (e.g., shrews), but large mammals have too much thermal mass to cool down quickly enough.

Hibernation and Winter Sleep

Many characteristics of torpor are shared by **hibernation**, or winter dormancy, a period of deep torpor that lasts for weeks or even several months in cold climates. Hibernation is entered through slow-wave sleep and is devoid of rapid-eye-movement sleep. Hibernation is common in mammals of the orders Rodentia, Insectivora, and Chiroptera, and allows them to survive on their energy reserves during periods when their normal food is limited or nonexistent. Many hibernators arouse periodically (as often as once a week or as infrequently as every 4–6 weeks) to urinate and defecate. During hibernation, the set point of the hypothalamic thermostat is lowered by as much as 20°C or more. At a T_a of 5–15°C, many hibernators keep their temperatures as little as 1°C above the ambient temperature. If the air temperature falls to dangerously low levels, the animal increases its metabolic rate to maintain a constant low T_b or becomes aroused.

Thermoregulatory control during hibernation, as during other forms of dormancy, is not suspended—it simply continues with a lowered set point and reduced sensitivity (amplification). In a hibernating marmot (*Marmota* sp.), for example, experimental cooling of the anterior hypothalamus with an electronically controlled, implanted probe increases metabolic production of body heat. The increase in heat production is proportional to the difference between the set-point temperature and the actual hypothalamic temperature, just as in the non-hibernating marmot. The set-point temperature drops about 2.5°C within a day or two as the animal enters a deeper state of hibernation.

As might be expected, body functions are greatly slowed at the lowered body temperatures characteristic of torpor and hibernation. The effect of reduced T_b on

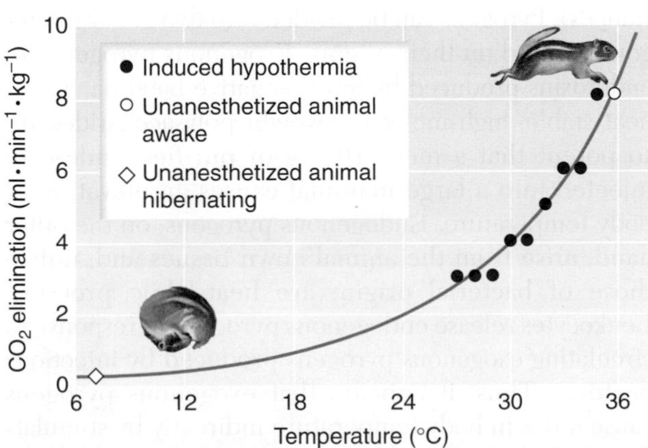

Figure 17-34 Both experimentally induced hypothermia and natural hibernation reduce the metabolic rate of the golden-mantled squirrel *(Spermophilus saturatus)*. [From Milsom, 1992.]

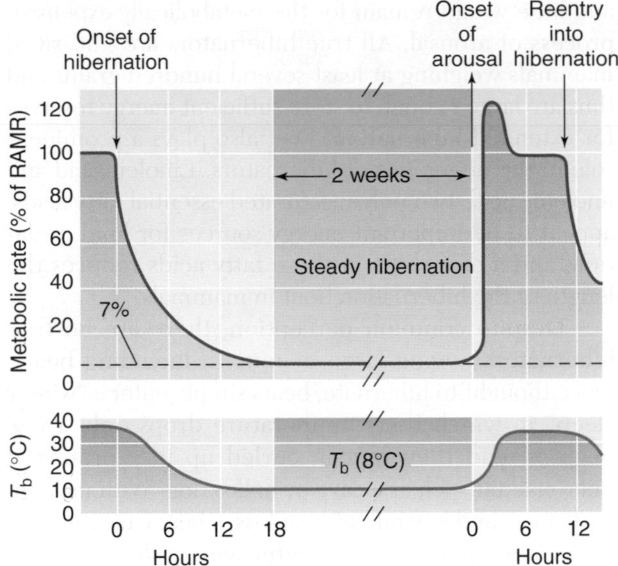

Classic Work ***Figure 17-35*** Metabolism increases briefly during an episode of arousal from hibernation in a ground squirrel. The squirrel was kept in a chamber with a temperature of 4°C. The period of steady-state hibernation is shaded red. At the onset of hibernation, the set point for body temperature is depressed. Metabolism decreases, allowing T_b to drop to 1–3°C above T_a throughout hibernation. Arousal occurs when the set-point temperature climbs to 38°C, and a strong surge of metabolic heat production raises T_b to the new set-point level. RAMR, resting average metabolic rate. [Adapted from Swan, 1974.]

the metabolic rate (expressed as $\dot{V}_{CO_2}$) of golden-mantled ground squirrels *(Spermophilus saturatus)* is shown in Figure 17-34. In conjunction with a decrease in metabolism, cardiac output in hibernating mammals is typically reduced to about 10% of prehibernation values, although the head and brown-fat deposits receive a proportionately much higher blood flow than other tissues. The reduction in cardiac output is accomplished by a drastic slowing of the heartbeat, with stroke volume remaining essentially unchanged. As a result of reduced respiratory exchange, the blood of many hibernators becomes more acidic. This acidosis may further lower enzyme activity because of the departure from the pH optimum of metabolic enzymes.

The rate of arousal from hibernation is often much higher than the rate of entry into hibernation. In the ground squirrel, for example, the transition into the dormant state requires 12–18 hours (Figure 17-35), whereas arousal typically occurs in less than 3 hours. This rapid emergence from hibernation depends on rapid heating initiated by intense oxidation of brown fat, accompanied by shivering, both reflected in a large surge in metabolic rate.

Many small heterothermic endotherms undergo daily torpor, but their high rates of metabolism preclude extended periods of torpor in the form of hibernation because, even in the hibernating state, they would quickly consume their stored energy reserves, and little would remain for the metabolically expensive process of arousal. All true hibernators are mid-sized mammals weighing at least several hundred grams and thus are large enough to store sufficient energy reserves for extended hibernation. Diet also plays a prominent role in the energetics of hibernators. Linoleic acid and linolenic acid, two polyunsaturated essential fatty acids, appear to be important energy sources for lipid oxidation, and a diet short in these fatty acids reduces the length of the hibernation bout in mammals.

Despite common perception, there are no true hibernators among large mammals, including bears. Once thought to hibernate, bears simply enter a "winter sleep" in which body temperature drops only a few degrees, and they remain curled up in a protected microhabitat such as a cave or hollow log. With its large body mass and low rate of heat loss, a bear can store sufficient energy reserves to enter winter sleep without dropping its body temperature dramatically. Overwintering bears remain dangerous to encounter, since they are able to wake up and become active quickly. Typically, however, bears stay in winter sleep for weeks or months at a time, retaining metabolic wastes in their bodies. Winter sleep, with its relatively high body temperature, does not offer the same degree of energy savings as deep hibernation, but a fall in body temperature of even a few degrees saves energy.

Why are there no large hibernators? First, they have less need to save metabolic fuels, because their normal basal metabolic rates are low relative to their energy stores based on the allometry of metabolism and energy storage. Second, a large mass and a correspondingly low rate of metabolism would require a prolonged and energetically very expensive metabolic effort to raise body temperature from a low level near ambient temperature to normal. It has been calculated, for example, that a 200–300 kg bear would require at least 24–48 hours and untold calories to warm up from a hibernating temperature of 5°C to a normal T_b of 37°C.

A final example of torpor occurs in the male emperor penguin, which lives in the Antarctic. In this species, the male incubates the egg, staying on the nest until the egg successfully hatches. To conserve body heat and energy, groups of incubating males huddle together with their backs to the wind, conserving heat. They also enter into a modest torpor with a 15%–20% reduction in metabolic rate. The occasional male that does not enter torpor in such a protective huddle cannot afford to go without food throughout the required incubation period, and eventually abandons the egg to forage.

Estivation

Estivation, also called "summer sleep," is a loosely defined term for a dormancy exhibited by some vertebrates and invertebrates in response to high ambient temperatures, danger of dehydration, or both. Faced with long periods of low humidity, the land snails *Helix* and *Otala* seal the entrance to their shell by secreting a diaphragm-like *operculum*, which retards loss of water by evaporation, and enter dormancy. Many land crabs (e.g., the blue land crab, *Cardisoma*) similarly spend dry seasons in an inactive state at the bottom of their burrows. The African lungfishes (*Protopterus*) are well known as estivators. These air-breathing fish survive periods of drought during which their ponds dry up by estivating in the semidry bottom until the next rainy season floods the area. The lungfish seals itself inside a secreted "cocoon"; a small tube leads from the fish's mouth to the exterior to allow ventilation of the lungs. Such cocoons have kept estivating lungfish alive for seven years—their viability can be demonstrated in the laboratory by showing that the dried dirt ball containing the cocoon still has a miniscule O_2 consumption. Interestingly, the chemical factors that induce estivation, isolated by high-performance liquid chromatography from the plasma of estivating lungfish, produce a torporlike state when injected into mammals.

Estivation *per se* is rare in mammals. Some small mammals, such as the Columbian ground squirrel (*Spermophilus columbianus*), remain inactive in their burrows during the late days of their hot summer, allowing their T_b to approach T_a. This poorly understood state is probably similar physiologically to hibernation but differs in seasonal timing.

What challenges would you anticipate to the renal and digestive systems of a lungfish arousing from years of estivation?

Pyrogens and Fever

An interesting feature of the thermoregulatory center of vertebrates is its sensitivity to certain chemicals collectively termed **pyrogens** (fever-producing substances). Pyrogens can be divided into two general categories, based on their origins. Exogenous pyrogens are endotoxins produced by gram-negative bacteria. These heat-stable, high-molecular-weight polysaccharides are so potent that a mere 10^{-9} g of purified endotoxin injected into a large mammal causes an elevation of body temperature. Endogenous pyrogens, on the other hand, arise from the animal's own tissues and, unlike those of bacterial origin, are heat-labile proteins. Leukocytes release endogenous pyrogens in response to circulating exogenous pyrogens produced by infectious bacteria. Thus, it appears that exogenous pyrogens cause a rise in body temperature indirectly by stimulating the release of endogenous pyrogens, which act directly on the hypothalamic center. This idea is supported by evidence that the hypothalamus is more sensitive to direct application of endogenous pyrogens than to exogenous ones.

The sensitivity of the hypothalamic temperature-sensing neurons to these pyrogenic molecules leads to an elevation in the set point to a higher temperature than normal. The result is that the body temperature rises several degrees, and the animal experiences a fever. Anesthetics and opiates such as morphine, in contrast to pyrogens, cause a lowering of the set-point temperature and hence a drop in body temperature.

Pyrogenic bacteria elevate body temperature in some poikilothermic ectotherms as well as in homeothermic endotherms. In a now classic experiment, Bernheim and Kluger (1976) monitored the body temperature of a desert iguana *(Dipsosaurus dorsalis)* under laboratory conditions simulating a desert environment, and then injected it with pyrogenic bacteria known to produce fever in mammals (Figure 17-36). In response to the fever-producing bacteria, the lizard positioned itself more frequently in radiantly heated zones of its artificial environment, effectively raising its T_b to abnormally high levels (i.e., giving itself a fever).

Fevers, whether physiologically or behaviorally induced, confer protection against bacterial infection on the afflicted animal. This protection is thought to take two forms: first, antiviral and antitumor agents such as interferon are more effective at higher temperatures, and second, elevated temperatures diminish the growth rates of some microbes.

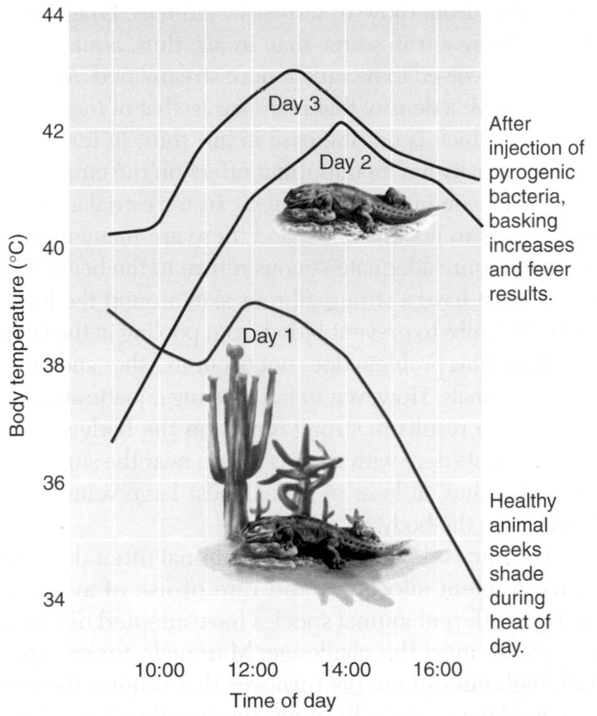

Classic Work **Figure 17-36** An ectotherm responds to injections of pyrogenic bacteria with a fever. Like other lizards, the desert iguana *(Dipsosaurus dorsalis)* regulates its body temperature behaviorally by adjusting its location and posture with respect to radiant heat from the sun or hot objects such as dark rocks. After infection by injected pyrogenic bacteria, lizards in an artificial laboratory environment raised their body temperatures above normal levels for two successive days by increased basking behavior. [Adapted from Bernheim and Kluger, 1976.]

ENERGY, ENVIRONMENT, AND EVOLUTION

In the introduction to this book, we described the interdependency of numerous physiological systems. As Donald Jackson (1987) commented so succinctly in considering the problems faced by interacting physiological systems, "A disturbance to one part reverberates throughout the organism, and produces responses, compromises and adaptations of various functions." Any animal must resolve potential conflicts between the differing demands of networked physiological systems in the context of both space and time. A conflict between the demands of two physiological systems with different needs—for example, the need of the thermoregulatory system to expend water to prevent overheating through evaporative cooling and the need of the osmoregulatory system to preserve water to prevent desiccation—can be tolerated for short periods of time. In fact, a simple snapshot of an animal's complete physiology would reveal a myriad of such conflicts. However, sooner or later (usually sooner), such conflicts must be resolved by an appropriate physiological adjustment.

Moreover, different environmental stresses carry with them very different senses of urgency. Figure 17-37a on the next page indicates the vastly different amounts of time during which an absence of oxygen, water, and food and an excess of body heat can be tolerated by a typical large mammal. In most animals, a physiological conflict that denies an animal food or water, for example, can be tolerated far longer than a physiological conflict that denies an animal oxygen. Additionally, different animal species have very different tolerances for a single category of stress, such as oxygen deprivation (Figure 17-37b). The brain cells of humans, for example, begin to die within a few minutes of oxygen deprivation, but those of diving birds and mammals can last for many minutes, while the neurons of some species of turtles can survive for days or even months in the complete absence of oxygen. Some simple metazoans (e.g., many intestinal parasites) can switch over to anaerobic metabolism and survive indefinitely without oxygen. Adding another level of complexity, there may be intraspecific differences. A desert tortoise with copious fat and water stores can last for months without additional food or water, while another tortoise lacking these reserves may succumb within days. Consequently, physiological conflicts are typically resolved on the basis of which of two resulting conditions is the least threat to homeostasis, and there are great interspecific and intraspecific differences in which two (or more) physiological systems may be in conflict under a given set of stressors.

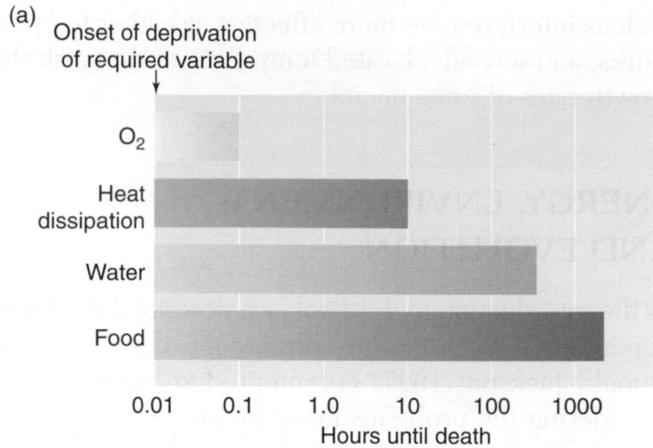

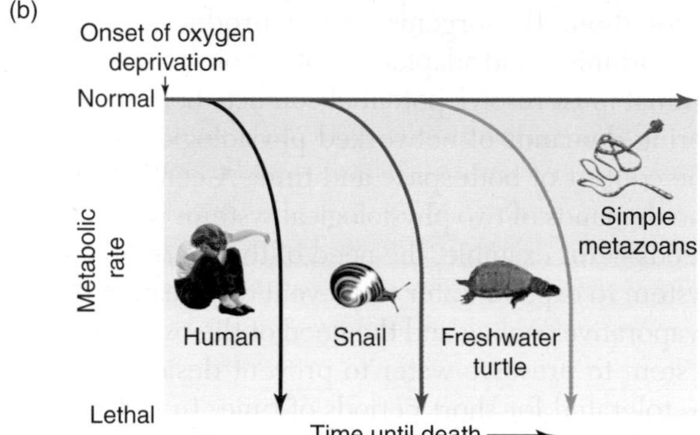

Figure 17-37 Different environmental stresses convey different levels of threat to animal species. **(a)** Oxygen deprivation is typically most rapidly detrimental, while starvation is typically the least immediately threatening. This graph shows approximate times leading to death following deprivation of oxygen, water, and food and the failure to dissipate heat in a typical large mammal. Note that the time scale is logarithmic (1000 hours is about 42 days). **(b)** The rate at which cellular metabolic rate declines toward lethally low levels differs greatly among species after the abrupt removal of oxygen, food, or water or due to failure to eliminate metabolically produced heat.

Traditionally, animal physiologists have focused on the characteristics of individual physiological systems as if they operated in isolation from one another. This historical approach is quite understandable, for until recently, the problem of unraveling the secrets of any one physiological system seemed almost intractable even without the introduction of additional complexity of input from, interactions with, and impositions by other physiological systems. It was not that comparative physiologists lacked an integrated view of an animal's physiology, but that the reductionist approach was a practical response to what at times seemed to be an overwhelming amount of unexplored territory. Perhaps reflecting an emerging confidence in our understanding of the basic physiology of these systems, many animal physiologists have begun to focus on *interactions* between different physiological systems, rather than on the isolated characters of the individual systems. This integrated approach will by necessity draw on systems-level physiology, ecological physiology, behavioral physiology, environmental physiology, and evolutionary and developmental physiology.

A salmon swimming up a stream and traversing a series of waterfalls sends large quantities of blood through almost all of the respiratory surfaces of its gills in an attempt to acquire oxygen and eliminate carbon dioxide. At the same time, it produces huge quantities of urine. Why is urine production so elevated, and what physiological systems have been thrown into conflict?

A Final Word . . .

This concluding chapter has shown how a consideration of metabolism—so fundamental to all physiological systems—integrates the animal and its physiology into its environment. Of course, the integration between physiology and environment affects virtually all physiological systems, all physiological process, and the morphology that supports them.

Environments impose constant constraints on animals' activities, placing limitations and demands on their design and function and shaping their evolution through natural selection. Simply put, there is no benign, nurturing environment. Consequently, environments, through natural selection, have shaped the animal species that live in them. Consider how the shapes of animals living in water differ from those of terrestrial animals. Drag forces are much greater in water than in air; thus, aquatic animals have evolved to be much more streamlined. Animals in water have a density that is similar to that of their environment, which is not the case in air; thus, in terrestrial animals, gravity has an important effect on the circulation that is not seen in aquatic animals. In terrestrial animals, blood tends to pool in veins, and there are many mechanisms to ensure adequate venous return to the heart. The giraffe must have a strong, fibrous skin around the lower part of its limbs to prevent blood from pooling in the veins of its legs. This problem does not occur in fishes and other aquatic animals. However, in fast-moving aquatic animals, acceleration results in strong forces on the body surface that could interfere with venous return near the surface of the body. Thus, at least in fishes, most large veins travel deep within the body.

The survival of an individual animal often depends on its efficient allocation and rate of use of available energy. Different animal species have adopted different lifestyles to meet this challenge. Mammals, for example, have high rates of energy turnover that require them to seek food continually. Reptiles, on the other hand, have much lower energy turnover rates and can survive on much less energy. Different environments favor different lifestyles at different times. For example, reptiles seem to have the advantage in water and in food-scarce desert environments during the day, but mammals seem to have the edge during the cooler desert nights. Mammals spend energy on maintaining a high body temperature and therefore need more food, but they can retain high levels of activity during cold nights.

The success of an individual animal is measured by the genetic legacy left by that animal—that is, by whether it survives long enough to reproduce. Reproduction occurs when the animal is mature and conditions are favorable for survival of the young. During periods of decreased energy availability unfavorable for reproduction, the animal may enter a dormant state, such as hibernation. Essentially, the animal is cutting its energetic costs during the "bad times" to balance energy input and output. Many animals migrate to avoid certain environments where, for example, food is short and temperatures are low. Migration has significant energetic costs, with flying being much less costly per unit distance than walking and running on land or swimming in water.

The results of the process of animal evolution have given us numerous examples of physiological adaptations for survival in a multitude of different habitats. Each example is a variation in the organization of a series of basic component parts that make up the vast panoply of life.

SUMMARY

■ The dependence of enzymatic reactions and metabolic rate on tissue temperature is described by the Q_{10}, the ratio of metabolic rate at a given temperature to the metabolic rate at a temperature 10°C lower. This ratio typically lies between 2 and 3.

Physiological classification using thermal biology

■ Endotherms are animals that generate most of their own body heat, which allows them to elevate their core temperature above that of their environment. Ectotherms obtain most of their body heat from their surroundings. Some elevate their temperatures by various behavioral means, such as basking.

■ Poikilothermy, homeothermy, and heterothermy refer to varying degrees of control over body temperature. The most descriptive terms combine classifications, as in "homeothermic endotherm."

Thermal biology of ectotherms

■ Ectotherms use a variety of mechanisms for survival in temperature extremes. Some species cope with subzero temperatures by using "antifreeze" substances or by supercooling without ice crystal formation, but no animals have been shown to survive freezing of water within cells. Other ectotherms elevate body temperature by shivering or nonshivering muscle contraction at certain times or in certain parts of the body. Such heat production is used by some insects and large fishes to warm locomotor muscles to optimal operating temperatures.

■ Heat absorption or heat loss to the environment is regulated in some ectothermic species by changes in blood flow to the skin. In this way, heat absorbed from the sun's rays can be quickly transferred by the blood from the body surface to the body core. Conversely, core heat can be conserved in a cold environment by restricting blood circulation to the skin.

Thermal biology of endotherms

■ Within the thermal neutral zone, endotherms use changes in surface conductance to compensate for changes in ambient temperature. Below this temperature zone, thermogenesis compensates for increased heat loss to the environment. Heat may be generated by shivering, oxidation of fats, locomotor activity, and other mechanisms.

■ At ambient temperatures above the thermal neutral zone, homeothermic endotherms actively dissipate heat by means of evaporative cooling, either by sweating or by panting.

■ Endotherms in cold environments conserve body heat by increasing the effectiveness of their surface insulation. They do this by decreasing peripheral circulation, increasing the fluffiness or thickness of pelage or plumage, or adding fatty insulating tissue. Endotherms also conserve heat by means of countercurrent heat-exchange mechanisms in the circulation to the limbs.

■ The use of water for evaporative cooling places an osmotic burden on desert dwellers. Most small desert inhabitants, which are subject to rapid changes in body temperature, minimize such changes by remaining in cool microenvironments during the day to avoid heat gain. Large desert mammals, buffered against rapid temperature changes by more favorable surface-to-volume ratios and large heat capacities, can conserve water by slowly absorbing heat during the day and dissipating it during the cool night.

Neuronal mechanisms of temperature control

■ Body temperature in endotherms and some ectotherms is regulated by a neuronal thermostat sensitive to differences between the actual temperature of neuronal sensors and the thermostatic set-point temperature. Differences result in neuronal output to thermoregulatory effectors for corrective heat loss or heat gain.

■ The brain is specially protected from overheating in some mammals by a highly developed carotid rete in which cool venous blood from the nasal epithelium removes heat from arterial blood heading toward the brain.

Thermoregulation and specialized metabolic states

■ Sleep, torpor, hibernation, winter sleep, and estivation are all neurophysiologically and metabolically related forms of dormancy. During periods when food supply is absent or restricted, small and medium-sized endotherms allow their body temperatures to drop in accord with a lowered thermostatic set-point temperature. By lowering body temperature to within a few degrees of the ambient temperature, the animal conserves energy stores. Oxidation of brown fat and shivering thermogenesis are used to produce rapid warming at the termination of torpor or hibernation.

■ Fever develops when the set-point temperature is raised by the action of endogenous pyrogens, which are protein molecules released by leukocytes in response to exogenous pyrogens produced by infectious bacteria.

REVIEW QUESTIONS

1. Compare and contrast the two major classification systems of animals based on thermal biology. Do any animals fit neatly in either system?
2. Give examples of low-temperature adaptations of some ectotherms and some endotherms.
3. What are some of the factors that determine the limits of the thermal neutral zone of a homeotherm?
4. What thermoregulatory mechanisms are available to a homeotherm at temperatures below and above the thermal neutral zone?
5. What do the terms *temporal* and *regional heterothermy* mean, and why do these phenomena occur?
6. Explain and give examples of the relations that exist between water balance and temperature regulation in a desert animal.
7. Describe the integration of peripheral and core temperatures in the thermostatic control of temperature in a mammal.
8. Describe two naturally occurring situations in which the set-point temperature of the hypothalamic thermostat is changed and the body temperature correspondingly changes.
9. Explain the mechanism of heat production in two different kinds of thermogenesis.
10. What is the role of countercurrent heat exchange in porpoises, Arctic mammals, tunas, and sheep?
11. What are the sphinx moth's two major means of regulating thoracic temperature?
12. By what means does the marine iguana speed the elevation of its body temperature and then retard cooling during diving?

SUGGESTED READINGS

Block, B. A. 1994. Thermogenesis in muscle. *Annu. Rev. Physiol.* 56:535–577. (A comprehensive review describing the biochemical and cellular mechanisms behind specialized heater tissues in endothermic animals.)

Carrey, C., ed. 1993. *Life in the Cold: Ecological, Physiological and Molecular Mechanisms.* Boulder: Westview Press. (A collection of multilevel reviews on hibernation, torpor, and other mechanisms by which animals survive life in the cold.)

Feder, M. E., and G. E. Hofmann. 1999. Heat-shock proteins, molecular chaperones, and the stress response: Evolutionary and ecological physiology. *Annu. Rev. Physiol.* 61:243–282. (Integrates the molecular and cellular biology of HSPs with the ecophysiology of the animals that express them.)

Fletcher, G. L., L. H. Choy, and P. L. Davies. 2001. Antifreeze proteins of teleost fishes. *Annu. Rev. Physiol.* 63:359–390. (A comprehensive review of the remarkable diversity of antifreezes in fishes.)

Heinrich, B. 1993. *The Hot-Blooded Insects: Strategies and Mechanisms of Thermoregulation.* Cambridge, MA: Harvard University Press. (Explores the physiological and biochemical mechanisms of endothermy among the insects.)

Leather, S. R., K. F. A. Walters, and J. S. Bale. 1993. *The Ecology of Insect Overwintering.* Cambridge: Cambridge University Press. (Describes the many behavioral, physiological, and biochemical mechanisms for insect freezing avoidance and torpor.)

Lee, R. E., Jr., and J. P. Costanzo. 1998. Biological ice nucleation and ice distribution in cold-hardy ectothermic animals. *Annu. Rev. Physiol.* 60:55–72. (A thorough treatment of the biochemistry and physiology underlying ice crystal formation in animal tissues.)

Mortola, J. P., and P. B. Frappell. 2000. Ventilatory responses to changes in temperature in mammals and other vertebrates. *Annu. Rev. Physiol.* 62:847–874. (Describes changes in ventilation, oxygen extraction, and oxygen consumption, as well as the role of the ventilatory system in thermoregulation during normal activity, sleep, torpor and hibernation in a wide variety of endotherms and ectotherms.)

Reppert, S. M., and D. R. Weaver. 2001. Molecular analysis of mammalian circadian rythms. *Annu. Rev. Physiol.* 63:647–676. (Describes the cellular and molecular mechanisms of the biological clock in mammalian brains.)

Ruben, J. 1995. The evolution of endothermy in mammals and birds: From physiology to fossils. *Annu. Rev. Physiol.* 995:69–95. (A comprehensive review that speculates on the evolutionary processes leading to endothermy in vertebrates.)

Somero, G. N. 1997. Temperature relationships: From molecules to biogeography. *Handbook of Physiology,* Section 13, *Comparative Physiology,* Vol. 2, 1391–1444. New York: Oxford University Press. (Presents a very thorough definition and discussion of the concepts surrounding biological rhythms.)

Trayhurn, P., and D. G. Nicholls, eds. 1986. *Brown Adipose Tissue.* London: E. Arnold. (Covers neuronal control mechanisms, biochemistry, metabolism, physiology, and anatomy of brown fat.)

Wood, C. M., and G. McDonald, eds. 1997. *Global Warming: Implications for Freshwater and Marine Fish.* Cambridge: Cambridge University Press. (Focusing on examples from fish, describes in detail the effects of temperature on animals from molecular to population levels.)

Glossary

A band The region of a muscle sarcomere that corresponds to the myosin thick filaments.
abomasum The true digestive stomach of the ruminant digastric stomach.
absolute refractory period The period during or after an action potential when a second stimulus produces no action potential. *See also* relative refractory period
absolute temperature Temperature measured from absolute zero, the state of no atomic or molecular thermal agitation. The absolute scale is divided into kelvins (K), with 1 K having the same size as 1 Celsius degree. Thus, 0 K is equal to –273.15°C or –459.67°F.
acclimation A persisting change in a specific function due to prolonged exposure to an environmental condition such as high or low temperature.
acclimatization A persisting spectrum of changes due to prolonged exposure to environmental conditions such as high or low temperature.
accommodation The temporary increase in threshold that develops during the course of a stimulus.
acetylcholine (ACh) An acetic acid ester of choline, important as a synaptic transmitter in most animal species and in many different types of neurons.
acetylcholinesterase An enzyme that hydrolyzes acetylcholine into acetate and choline.
acid A proton donor.
acidosis Excessive body acidity.
acid tide Acidification of the blood following pancreatic secretion.
acinus (singular; plural, acini) A small sac or alveolus, sometimes lined with exocrine cells.
acromegaly Enlargement of the skeletal extremities and facial structures, caused by hypersecretion of growth hormone in adulthood.
ACTH *See* adrenocorticotropic hormone
actin A ubiquitous protein that participates in muscle contraction and other forms of cellular motility. G-actin is the globular monomer that polymerizes to form F-actin, the backbone of the thin filaments of muscle.
action potential (AP; nerve impulse, spike) Transient all-or-none reversal of a membrane potential produced by regenerative inward current in excitable membranes.
action spectrum The pattern of response to each of many different wavelengths of incident light.
activation energy The energy required to bring reactant molecules to velocities sufficiently high to break or make chemical bonds.
active electrical properties The cellular properties of neurons and muscle fibers that allow them to generate and transmit action potentials.
active hyperemia The increase in blood flow during and after increased activity in a tissue, particularly skeletal muscle.
active site The catalytic region of an enzyme molecule.
active state In muscle fibers, the condition in which myosin crossbridges are attached to actin, causing the fibers to resist a force that would pull them apart.
active transport Energy-requiring translocation of a substance across a membrane, usually against its concentration or electrochemical gradient. *See also* primary active transport, secondary active transport
active zone The local region, within a presynaptic terminal, at which synaptic vesicles dock and are prepared for release by exocytosis.
activity The capacity of a substance to react with another substance; the effective concentration of an ionic species in the free state.
activity coefficient A proportionality factor obtained by dividing the effective reactive concentration of an ion (as indicated by its properties in a solution) by its molar concentration.
actomyosin A complex of muscle proteins formed when myosin cross-bridges bind to actin.
acuity Resolving power.
adaptation Evolution through natural selection leading to an organism whose physiology, anatomy, and behavior are matched to the demands of its environment. *See also* sensory adaptation
adductor muscle A muscle that brings a limb toward the median plane of the body.
adenine 6-amino-purine, $C_5H_5N_5$; a base component of nucleic acids.
adenohypophysis *See* anterior pituitary
adenosine diphosphate (ADP) A nucleotide formed by hydrolysis of ATP, with the release of one high-energy bond.
adenosine triphosphatases *See* ATPases
adenosine triphosphate (ATP) An energy-rich nucleotide used as a common energy currency by all cells.
adenylate cyclase (adenyl cyclase) A membrane-bound enzyme that catalyzes the conversion of ATP to cAMP.
ADH *See* antidiuretic hormone
adipose Fatty.
ADP *See* adenosine diphosphate
adrenaline *See* epinephrine
adrenal medulla The central portion of the adrenal gland.
adrenergic Pertaining to epinephrine and norepinephrine.
adrenergic receptors (adrenoreceptors) Receptors on cell surfaces that bind norepinephrine and epinephrine; binding activates a G protein in the cells.
adrenocorticotropic hormone (ACTH; adrenocorticotropin; corticotropin) A hormone released by cells in the anterior pituitary that acts mainly on the adrenal cortex, stimulating growth and corticosteroid production and secretion in that organ.
adrenoreceptors *See* adrenergic receptors
aequorin A protein extracted from the jellyfish *Aequorea;* on combining with Ca^{2+}, it emits a photon of blue-green light.
aerobic Utilizing molecular oxygen.
aerobic metabolic scope The ratio of the maximum sustainable metabolic rate to the basal metabolic rate (or the standard metabolic rate).
aerobic metabolism Metabolism utilizing molecular oxygen.
afferent Transporting or conducting toward a central region; centripetal.
afferent fiber (afferent neuron) An axon that carries sensory information to the central nervous system or within the CNS toward the highest centers.
affinity sequence (selectivity sequence) The order of preference with which an electrostatic site will bind different species of counterions.
after-hyperpolarization The transient period at the end of an action potential when V_m is more negative than V_{rest}; also called undershoot.
aglomerular Lacking glomerulae in the kidney.
agonist A substance that can interact with receptor molecules and mimic the action of an endogenous signaling molecule.
aldehydes A large class of substances derived from the primary alcohols by oxidation and containing the —CHO group.
aldosterone A mineralocorticoid secreted by the adrenal cortex; the most important electrolyte-controlling steroid, which acts on the renal tubules to increase the reabsorption of sodium.
alimentary canal A hollow, tubular cavity extending through an animal and open at both ends; functions in ingestion, digestion, and absorption of food materials.
alkaline tide A period of increased body and urinary alkalinity associated with increased gastric HCl secretion during digestion.
alkaloids A large group of organic nitrogenous bases found in plant tissues, many of which are pharmacologically active (e.g., codeine, morphine).
alkalosis Excessive body alkalinity.
allantoic membrane One of the membranes within a bird eggshell; important in the respiration of the unhatched chick.
allantoin Waste product of purine metabolism.
allometry Systematic change in body proportions with increasing body size.
all-or-none A condition in which the magnitude of a cell's response is independent of the strength of a stimulus above some threshold value. If an input signal brings the cell to its threshold, the amplitude of the response is maximal; if the stimulus fails to bring the cell to threshold, there is no response.
allosteric site An area of an enzyme that binds a substance other than the substrate, changing the conformation of the protein so as to alter the catalytic effectiveness of the active site.
α-adrenergic receptors Receptors on cell surfaces that bind norepinephrine and, less effectively, epinephrine; the binding activates a G protein, causing slow synaptic responses in the cells.
α-Bungarotoxin *See* Bungarotoxin

alpha (α) helix Helical secondary structure of many proteins in which each NH group is hydrogen-bonded to a CO group at a distance equivalent to three amino acid residues; the helix makes a complete turn for each 3.6 residues.
α-motor neurons Large spinal neurons that innervate extrafusal skeletal muscle fibers of vertebrates.
alveolar ventilation volume The volume of fresh atmospheric air entering the alveoli during each inhalation.
alveoli Small cavities, especially those microscopic cavities that are the functional units of the lung.
amacrine cells Neurons without axons that are located in the inner plexiform layer of the vertebrate retina.
amide An organic derivative of ammonia in which a hydrogen atom is replaced by an acyl group.
amine A derivative of ammonia in which at least one hydrogen atom is replaced by an organic group.
amino acids A class of organic compounds containing at least one carboxyl group and one amino group. The alpha-amino acids, $RCH(NH_2)COOH$, make up proteins.
amino group $—NH_2$.
ammonia NH_3; a toxic, water-soluble, alkaline waste product of deamination of amino acids and uric acid.
ammonotelic Pertaining to the excretion of nitrogen in the form of ammonia.
ampere (A) A unit of electric current equal to the current produced through a 1 ohm (V) resistance by a potential difference of 1 volt (V); the movement of 1 coulomb (C) of charge per second.
amphipathic Bearing both hydrophilic and hydrophobic groups.
amphoteric Having opposite characteristics; behaving as either an acid or a base.
amygdala The region of the brain that processes information and organizes output related to the emotions.
amylases Carbohydrases that hydrolyze all but the terminal glycosidic bonds within starch and glycogen, producing disaccharides and oligosaccharides.
anabolism Synthesis by living cells of complex substances from simple substances.
anaerobic Oxygen-free.
anaerobic metabolism Metabolism not utilizing molecular oxygen.
anatomic dead space The nonrespiratory conducting pathways in the lung.
androgens Hormones having masculinizing activity.
aneurism A localized dilation of an artery wall.
angiotensin A protein in the blood that functions as a potent vasopressor and stimulator of aldosterone secretion. It is converted from angiotensinogen by the action of renin; it first exists as a decapeptide (angiotensin I) that is acted upon by a peptidase, which cleaves it into an octapeptide (angiotensin II).
animal rights The concept that animals have intrinsic and inalienable rights, just as humans have rights.
animal welfare The principle that animals are entitled to humane treatment in terms of comfort and well-being.
anion A negatively charged ion; an ion attracted to the anode or positive pole.
anode A positive electrode or pole to which negatively charged ions are attracted.
anoxemia A lack of oxygen in the blood.
anoxia A lack of oxygen.
antagonist An agent that inhibits, blocks, or counteracts an effect; for example, antagonists of synaptic transmitters typically block the postsynaptic receptor molecules that bind the neurotransmitters.
antagonist muscle A muscle acting in opposition to the movement of another muscle.
antennal gland A crustacean osmoregulatory organ.
anterior pituitary (adenohypophysis, anterior lobe) The glandular anterior lobe of the pituitary gland, consisting of the pars tuberalis, pars intermedia, and pars distalis.
antibody An immunoglobulin; a four-chain protein molecule of a specific amino acid sequence. An antibody will interact only with the antigen that brought about its production, or one very similar to it.
antidiuretic hormone (ADH, vasopressin) A hormone made in the hypothalamus and released from storage in the posterior pituitary; acts on the epithelium of the renal collecting duct by stimulating osmotic reabsorption of water, thereby producing a more concentrated urine; also acts as a vasopressor.
antigen A substance capable of bringing about the production of antibodies and of then reacting with them specifically.
antimycin An antibiotic that is isolated from a *Streptomyces* strain; acts to block electron transport from cytochrome *b* to cytochrome *c* in the electron-transport chain.
antiporter A carrier protein that transfers two solutes in opposite directions across a biological membrane.
anus The opening of the alimentary canal through which feces are expelled.
aorta The main artery leaving the heart.
aortic body A nodule on the aortic arch containing chemoreceptors that sense the chemical composition of the blood.
apical Pertaining to the apex; opposite the base.
apnea The suspension or absence of breathing.
apocrine secretion Secretion by sloughing off of the apical portion of secretory cell.
apoenzyme The protein portion of an enzyme, which combines with a coenzyme to form a functioning enzyme.
apolysis Release; loosening from.
aquaporin A 28 kDa protein, tetramers of which form water channels in membranes.
area centralis *See* fovea
arginine phosphate A phosphorylated nitrogenous compound found primarily in muscle; acts as a storage form of high-energy phosphate for the rapid phosphorylation of ADP to ATP.
arteriole A tiny branch of an artery; in particular, one nearest a capillary.
association cortex Areas of the cerebral cortex that neither directly receive sensory information nor directly contribute to motor output; instead, these areas typically receive input from many sensory modalities and are broadly connected to other areas in the cortex and other brain centers.
asynchronous muscle (fibrillar muscle) A type of flight muscle found in the thorax of some insects in which contractions are neither individually controlled by nor synchronized with motor impulses.
ATP *See* adenosine triphosphate
ATPases (adenosine triphosphatases) A class of enzymes that catalyze the hydrolysis of ATP.
ATPS Ambient temperature and pressure, saturated with water vapor; referring to gas volume measurements.
atrial natriuretic peptide (ANP) One of a family of peptide hormones cleaved from a single precursor peptide and produced in the cardiac atria; its physiological effects include increased urine output, increased sodium excretion, and receptor-mediated vasodilation, the net result of which is lowered blood pressure.
atrioventricular node Specialized conduction tissue in the heart that, along with Purkinje fibers, forms a bridge for electrical conduction of excitation waves from the atria to the ventricles.
atrium (singular; plural, atria) A chamber that gives entrance to another structure or organ; usually used to refer to an atrium of the heart.
auditory cortex Regions of the cerebral cortex that are associated with hearing.
auditory ossicles The bones of the middle ear (the malleus, the incus, and the stapes) that connect the tympanic membrane and the oval window.
autacoid hormones Endogenous substances with hormonal properties that are not amines, eicosanoids, steroid hormones, or peptide hormones.
autocrine Pertaining to a hormone that binds to receptors on, and initiates a cellular response in, the same cell that produced it.
autoinhibition Self-inhibition.
autonomic nervous system The efferent pathways of the nervous system that control involuntary visceral functions; classically subdivided into the sympathetic and parasympathetic divisions.
autoradiography The process of making a photographic record of the internal structures of a tissue by utilizing radiation emitted from incorporated radioactive material.
autorhythmicity The generation of rhythmic activity without extrinsic control.
autotroph An organism that synthesizes food from inorganic substances by utilizing the energy of the sun or of inorganic compounds.
Avogadro's law The principle that equal volumes of different gases at the same temperature and pressure contain equal numbers of

molecules. One mole (mol) of an ideal gas at 0°C and 1 standard atmosphere (atm) occupies 22.414 liters (L). Avogadro's number equals 6.02252×10^{23} molecules/mol.

axon The elongated cylindrical process of a neuron along which action potentials are conducted; a nerve fiber.

axoneme A complex of microtubules and associated structures within the flagellar or ciliary shaft.

axon hillock The transitional region between the soma and the axon of a neuron.

axon terminal The end of an axon; typically, the site where signals are passed to another cell.

axoplasm The cytoplasm within an axon.

azide Any compound bearing the N_3^- group.

baroreceptor A sensory receptor that is stimulated by changes in pressure, as in the walls of blood vessels.

basal cell A cell in a chemoreceptive organ that regularly gives rise to new chemoreceptor cells throughout the animal's life.

basal metabolic rate (BMR) The rate of energy conversion in an animal while it is resting quietly within the thermal neutral zone without food in the intestine.

base A proton acceptor.

basic electric rhythm (BER) Rhythmic depolarizations and repolarizations generated by pacemaker cells (interstitial cells of Cajal in mammals) that control electrical activity in the alimentary canal.

basilar membrane The delicate ribbon of tissue bearing the auditory hair cells in the cochlea of the vertebrate ear.

batch reactor An alimentary system consisting of a blind tube or cavity that receives food and eliminates wastes in a pulsed fashion; each batch is processed and eliminated before the next is brought in.

Bell-Magendie rule The dorsal root of the spinal cord contains only sensory axons, whereas the ventral root contains only motor axons.

Bernoulli effect Fluid pressure drops as fluid velocity increases.

***β*-adrenergic receptors** Receptors on cell surfaces that bind epinephrine and norepinephrine equally well; the binding normally leads to the activation of adenylate cyclase.

beta (*β*) pleated sheet A protein secondary structure in which two or more distinct amino acid chains lie side by side, held together by hydrogen bonds.

bile A viscous yellow or greenish alkaline fluid produced by the liver and stored in the gallbladder; it contains bile salts, bile pigments, and certain lipids, and is essential for digestion of fats.

bile duct The duct carrying bile fluid from the liver to the duodenum.

bile pigments Pigments in bile fluid derived from breakdown products of hemoglobin.

bile salts Bile acids such as cholic acid conjugated with glycine or taurine, which promote emulsification and solubilization of intestinal fats.

binocular convergence Positioning of the eyes so that the images formed fall on analogous portions of the two retinas.

biogenic amine Any of a number of signaling molecules that are synthesized in the body from single amino acid molecules.

bipolar cell A neuron with two major processes that emerge from opposite sides of the soma; one class of these neurons is found in the vertebrate retina, where they transmit signals from the photoreceptor cells to the retinal ganglion cells.

birefringence Double refraction; the ability to pass preferentially light that is polarized in one plane.

bleaching The fading of photopigment upon absorption of light.

blubber A fatty, insulating tissue typically found under the skin in cetaceans.

Bohr effect (Bohr shift) A change in hemoglobin-oxygen affinity due to a change in pH.

bolus A discrete plug or collection of food material moving through the alimentary canal.

bombykol The sex attractant pheromone of the female silkworm moth *(Bombyx mori)*.

book lungs The respiratory surface in spiders.

Bowman's capsule (glomerular capsule) A globular expansion at the beginning of a renal tubule that surrounds the glomerulus.

Boyle's law The principle that, at a given temperature, the product of the pressure and volume of a given mass of gas is constant.

bradycardia A reduction in heart rate from the normal level.

bradykinin A hormone formed from a precursor normally circulating in the blood; a potent cutaneous vasodilator.

brain The major neuronal center within the body; typically located at the anterior of the body.

brain hormone *See* prothoracicotropic hormone

branchial Pertaining to the gills.

bronchi Conducting airways in the lung; branches of the tracheae.

bronchioles Small conducting airways in the lung; branches of the bronchi.

brood spot A prolactin-induced bald area on the ventral surface of some brooding birds that receives a rich supply of blood for the incubation of eggs.

brown fat Adipose tissue with extensive vascularization, numerous mitochondria, and enzyme systems for oxidation. Found in small, specific deposits in a few mammals and used for nonshivering thermogenesis.

Brunner's glands Exocrine glands that are located in the intestinal mucosa and secrete an alkaline mucoid fluid.

brush border A free epithelial cell surface bearing numerous microvilli.

BTPS Body temperature, atmospheric pressure, saturated with water vapor; referring to gas volume measurements.

buccal Pertaining to the mouth cavity.

buffer A chemical system that stabilizes the concentration of a substance; acid-base systems serve as pH buffers, preventing large changes in hydrogen ion concentration.

bulbus A heart chamber.

bundle of His The conducting tissue within the interventricular septum of the mammalian heart.

Bungarotoxin (BuTX) A blocking agent composed of a group of neurotoxins isolated from the venom of members of the snake genus *Bungarus* (the krait) of the cobra family; one component, α-BuTX, binds selectively and irreversibly to nicotinic acetylcholine receptors.

Bunsen solubility coefficient The quantity of gas at STPD that will dissolve in a given volume of liquid per unit partial pressure of the gas in the gas phase. This coefficient is used only for gases that do not react chemically with the solvent.

bursicon A hormone secreted by neurosecretory cells of the insect central nervous system; tans and hardens the cuticle of freshly molted insects.

cable properties The passive electrical properties (resistance and capacitance) of a cell; the physics of transmission through long, narrow cylinders was first worked out for submarine cables, hence the name.

calcitonin (thyrocalcitonin) A protein hormone secreted by the parafollicular cells of the mammalian thyroid in response to elevated plasma calcium levels.

calcitriol A steroidlike compound produced from vitamin D or D_3, whose physiological actions are similar to those of parathyroid hormone.

calcium response A graded depolarization due to a weakly regenerative inward calcium current.

caldesmon A calcium-binding regulatory protein that plays a role in the "latch" mechanism of some smooth muscles.

calmodulin A troponin-like calcium-binding regulatory protein found in essentially all tissues.

caloric deficit A state during which energy expenditure exceeds energy intake.

calorie (cal) The quantity of heat required to raise the temperature of 1 g of water from 14.5° to 15.5°C; the unit most commonly used is the kilocalorie (kcal = 1000 cal).

calorimetry The measurement of heat production in an animal.

calsequestrin A calcium-binding protein that contributes to the regulation of $[Ca^{2+}]$ in muscles.

cAMP *See* cyclic AMP

capacitance The ability to store electric charge by electrostatic means. The unit of measure for capacitance is the farad (F), which describes the proportionality between charge stored and potential for a given voltage: $C = q/V$ = coulombs per volt.

capacitative current Current entering and leaving a capacitor.

carbohydrases Enzymes that break down carbohydrates.

carbohydrates Aldehyde or ketone derivatives of alcohol; utilized by animal cells primarily for the storage and supply of chemical energy; most important are the sugars and starches.

carbonic anhydrase An enzyme that catalyzes the reversible interconversion of carbonic acid to carbon dioxide and water.
carbonyl The organic radical —C═O, which occurs in such compounds as aldehydes, ketones, carboxylic acids, and esters.
carboxyhemoglobin The compound formed when carbon monoxide combines with hemoglobin; carbon monoxide competes successfully with oxygen for combination with hemoglobin, producing tissue anoxia.
carboxylates R—COO—, salts or esters of carboxylic acids.
carboxyl group The radical —COOH, which occurs in the carboxylic acids.
cardiac muscle Heart muscle; individual fibers differ from skeletal muscle fibers in being small, elongated, tapered, and mononucleated. Includes noncontractile conducting fibers and contractile fibers.
cardiac output The total volume of blood pumped by the heart per unit time; cardiac output equals heart rate times stroke volume.
carotid body A nodule on the occipital artery just above the carotid sinus, containing chemoreceptors that sense the chemical composition of arterial blood.
carotid rete Countercurrent heat exchanger at the base of the skull that helps prevent overheating of the brain of certain hoofed mammals and carnivores.
carotid sinus A dilation of the internal carotid artery at its origin, containing many baroreceptors.
carrier-mediated transport Transmembrane transport of solutes achieved by membrane-embedded carrier proteins (e.g., facilitated diffusion).
catabolism Disassembly of complex molecules into simpler ones.
catalysis An increase in the rate of a chemical reaction promoted by a substance—the catalyst—not consumed by the reaction.
catalyst A substance that increases the rate of a reaction without being used up in the reaction.
cataract A condition in which proteins of the lens of the eye become cross-linked, causing the lens to become cloudy and reducing visual acuity.
catecholamines A group of related compounds that act as both neurotransmitters and hormones; examples are epinephrine, norepinephrine, and dopamine.
cathode Negative electrode or pole to which positively charged ions are attracted.
cation A positively charged ion; attracted to the cathode or negative electrode in a solution connected to a power source.
caudal Pertaining to or located toward the tail end.
C cells *See* parafollicular cells
CCK *See* cholecystokinin
cecum A blind pouch in the alimentary canal.
cellulase An enzyme that digests cellulose and hemicellulose, produced by symbiotic microorganisms in the gut.
cellulose An unbranched plant polymer consisting of glucose molecules polymerized via β-1,4 bonds.
central chemoreceptors Sensory structures in the brain that monitor pH and initiate appropriate changes in breathing.
central circulation The circulatory system that comprises the heart and the major blood vessels entering and leaving it.
central lacteal A small, blind-ended lymph vessel in the center of an intestinal villus that functions in the uptake of fats and some vitamins.
central nervous system The collection of neurons and parts of neurons that are contained within the brain and spinal cord of vertebrates or within the brain, main nerve cord, and major ganglia of invertebrates.
central pattern generator A group of neurons that produces and maintains a rhythmic pattern of action, such as breathing, walking, chewing, or swimming.
central sulcus A deep, almost vertical furrow on the cerebrum, dividing the frontal and parietal lobes.
cephalic Pertaining to the head.
cephalic phase The initial phase of digestion, in which gastric secretion occurs in response to the sight, smell, or taste of food, or in response to conditioned reflexes.
cephalization The evolutionary tendency for the neurons of higher organisms to be concentrated in a brain located at the anterior end of the animal.
cerebellum A part of the hindbrain that contributes to the coordination of motor output and participates in the learning of motor skills.
cerebral cortex The thin layer of gray matter that covers the cerebrum of mammals and birds.
cerebral hemispheres The large paired structures of the cerebrum, connected by the corpus callosum.
cerebral ventricles A series of interconnected cavities within the brain of vertebrates containing cerebrospinal fluid.
cerebrospinal fluid A clear fluid that fills the cerebral ventricles of the brain and the central canal of the spinal cord in vertebrates.
cerebrum The largest part and highest center of the mammalian brain; evolved from the olfactory centers of lower vertebrates.
cGMP *See* cyclic GMP
charge, electric (q) Measured in units of coulombs (C). To convert 1 g · equivalent weight of a monovalent ion to its elemental form (or vice versa) requires a charge of 96,500 C (1 faraday, F). Thus, in loose terms, a coulomb is equivalent to 1/96,500 g · equiv of electrons. The charge on one electron is -1.6×10^{-19} C. If this is multiplied by Avogadro's number, the total charge is 1 F, or $-96{,}487\ C \cdot mol^{-1}$.
chelating agent A chemical that binds calcium or other ions and removes them from solution.
chemical energy Energy contained in the chemical bonds holding molecules together.
chemical synapse A junction between a neuron and another cell in which the signal from the presynaptic neuron is carried across the synaptic cleft by neurotransmitter molecules.
chemoreceptor A sensory receptor that is specifically sensitive to certain molecules.
chief cells (zygomatic cells) Epithelial cells of the gastric epithelium that release pepsin.
chitin A structural polymer of D-glucosamine that serves as the primary constituent of arthropod exoskeletons.
chloride cells Epithelial cells of fish gills that engage in active transport of salts.
chloride shift The movement of chloride ions across the red blood cell membrane to compensate for the movement of bicarbonate ions.
choanocytes Flagellated cells lining the body cavity of sponges.
cholecystokinin (CCK; pancreozymin) A hormone released by the upper intestinal mucosa that induces gallbladder contraction and release of pancreatic enzymes.
cholesterol A natural sterol; precursor to the steroid hormones.
cholinergic Pertaining to acetylcholine or substances with actions similar to that of ACh.
choroid plexus A series of highly vascularized, furrowed projections into the brain ventricles that secrete cerebrospinal fluid.
choroid rete A countercurrent arrangement of arterioles and venules behind the retina in the eyes of teleost fishes.
chromaffin cells Epinephrine-secreting cells of the adrenal medulla; named for their high affinity for chromium salt stains.
chromatography A general technique that exploits the fact that different components in a sample will move at different rates through a substrate such as chromatography paper or another porous solid matrix.
chromophore A chemical group that lends a distinct color to a compound containing it.
chronobiology The study of biological rhythms.
chronotropic Pertaining to rate or frequency, especially in reference to the heartbeat.
chylomicrons Tiny protein-coated droplets of triglycerides, phospholipids, and cholesterol formed within vesicles of absorptive cells from the digestion produces of fats, monoglycerides, fatty acids, and glycerol.
chyme The mixture of partially digested food and digestive juices found in the stomach and the intestine.
chymotrypsin A proteolytic enzyme that breaks peptide bonds in which the carboxyl group is provided by tyrosine, phenylalanine, tryptophan, leucine, or methionine.
chymotrypsinogen The proenzyme of chymotrypsin.
ciliary body A thick region of the anterior vascular tunic of the eye; joins the choroid and the iris.
ciliary muscle A muscle of the ciliary body of the vertebrate eye; influences the shape of the lens in visual accommodation.
cilium (singular; plural, cilia) A motile organelle with a "9 + 2" microtubular substructure; generally found in large numbers; a small flagellum.
circadian rhythm A biological rhythm with a daily cycle.
circalunar rhythm A biological rhythm related to the lunar cycle.

circannual cycle A biological rhythm with a yearly cycle.
circatidal cycle A biological rhythm related to the tidal cycle.
circular folds *See* folds of Kerckring
circular smooth muscle The inner layer of smooth muscle that encircles the small intestine.
cis A molecular configuration with similar atoms or groups on the same side of the molecular backbone.
***cis-trans* isomerization** Conversion of a *cis* isomer into a *trans* isomer.
citric acid cycle (Krebs cycle, tricarboxylic acid cycle) A series of eight major reactions following glycolysis, in which acetate residues are degraded to CO_2 and H_2O.
cladogram A form of diagrammatic analysis of taxonomic relationships among organisms that relates animals according to suites of common characters.
clathrin In endocytosis, the protein surrounding the cytoplasmic surface of a coated vesicle membrane.
cloaca The terminal area of the hindgut in some fishes, amphibians, reptiles, birds, and a few mammals; aids in urinary ion and water reabsorption.
clone A population of genetically identical cells all derived from a single original cell.
closed-system respirometry An experimental technique in which an animal is confined to a closed, water- or air-filled chamber, allowing direct measurement of oxygen consumed and carbon dioxide produced.
coated pit A receptor-lined membrane depression that eventually forms a coated vesicle during the process of receptor-mediated endocytosis.
coated vesicle A vesicle whose cytoplasmic surface is covered with the protein clathrin, formed in the process of receptor-mediated endocytosis.
cochlea A portion of the inner ear of many vertebrates; a tapered tube wound in mammals into a spiral like the shell of a snail, containing hair-cell receptors for detecting sound.
coelenteron A blind-ended tube or cavity in coelenterates that serves as a "batch reactor" site for chemical digestion.
coelom The body cavity of higher metazoans, situated between the gut and the body wall and lined with mesodermal epithelium.
coenzyme An organic molecule that combines with an apoenzyme to form a functioning enzyme.
coenzyme A (CoA) A derivative of pantothenic acid to which acetate becomes attached to form acetyl CoA.
cofactor An atom, ion, or molecule that combines with an enzyme to activate it.
coitus Sexual intercourse.
colchicine An antimitotic agent that disrupts microtubules by interfering with the polymerization of tubulin monomers.
collagen The major protein of connective tissue.
collateral processes Branches of an axon that terminate in locations other than the major target location.
collaterals Side branches of a nerve or blood vessel.
collecting duct The portion of the mammalian renal tubule in which the final concentration of urine occurs.
colligative properties Characteristics of a solution that depend on the number of molecules in a given volume.
colloid A system in which fine solid particles are suspended in a liquid.
colon The part of the large intestine the extends from the cecum to the rectum.
command neuron A neuron that when activated has the ability to organize and produce an entire behavior by controlling the activity of many other neurons.
command system A set of neurons that, when stimulated, elicit a set pattern of coordinated movements.
comparative physiology A subdiscipline in which diverse species are compared to discern physiological principles and evolutionary patterns.
competitive inhibition Reversible inhibition of enzyme activity caused by competition between a substrate and an inhibitor for the active site of the enzyme.
compliance The change in length or volume per unit change in the applied force.
compound eye The multifaceted arthropod eye; its functional unit is the ommatidium.
concentration gradient The difference in solute concentration across a membrane or between two different regions of a solution.
condensation A reaction between two or more organic molecules leading to the formation of a larger molecule and the elimination of a simple molecule, such as water or alcohol.
conditioned reflex A reflex that is learned or modified through behavioral repetition.
conductance, electrical (G) A measure of the ease with which a conductor carries an electric current; the unit of measure is the siemen (S), reciprocal of the ohm (Ω).
conductance, thermal *See* thermal conductance
conductivity The intrinsic ability of a substance to conduct electric current; the reciprocal of resistivity.
conductor A material that carries electric current.
cone A vertebrate photoreceptor cell that has a tapered outer segment in which the lamellar photosensitive membranes remain continuous with the plasma membrane. *See also* rod
confocal scanning microscope A microscope using a focused laser beam to rapidly scan different areas of the specimen in a single plane. Light reflected or emitted from that plane is assembled by a computer into a composite image of the specimen.
conformer An animal whose internal conditions tend to parallel those of the external environment.
conjugate acid-base pair Two substances related by gain or loss of an H^+ ion (proton).
connective A collection of axons that carry information between neuronal centers, such as ganglia, in many invertebrate nervous systems.
continuous-flow stirred-tank reactor An alimentary system in which food is continually added and mixed into a homogeneous mass, and the products of digestion are continuously eliminated.
contracture A sustained contraction in response to an abnormal stimulus.
contralateral Pertaining to the opposite side.
controlled variable The variable that is regulated in a feedback system.
controller An element in a feedback system that regulates a controlled variable.
conus A chamber invested with cardiac muscle that is found in series with and downstream from the ventricle in elasmobranchs.
convection The mass transfer of heat due to mass movement of a gas or liquid.
convergence A pattern in which inputs from many different neurons impinge upon a single neuron.
cornea The clear surface of the eye through which light passes as it enters the eye.
corneal lens The clear structure at the outside surface of an ommatidium; it admits and focuses light entering the ommatidium.
corpora allata Nonneuronal insect glands that exist as paired organs or groups of cells dorsal and posterior to the corpora cardiaca and that secrete juvenile hormone.
corpora cardiaca Major insect neurohemal organs that exist as paired structures immediately posterior to the brain and that release prothoracicotropic hormone.
corpus luteum In mammals, a yellow ovarian glandular body that arises from a mature follicle that has released its ovum. If the ovum released is fertilized, the corpus luteum grows and secretes progesterone during gestation; if not, it atrophies and disappears.
cortex The external or surface layer of an organ.
corticospinal tract A group of axons projecting from neurons whose somata and dendrites are located in the motor cortex of the brain and whose axon terminals synapse on motor neurons in the spinal cord.
corticotropin *See* adrenocorticotropic hormone
cortisol A steroid hormone secreted by the adrenal cortex.
cotransmitter A second neurotransmitter molecule synthesized in and released from an axon terminal along with a small transmitter molecule such as acetylcholine or GABA.
cotransport Carrier-mediated transport across a membrane in which two dissimilar molecules bind to two specific sites on the carrier protein for transport in the same direction.
coulomb (C) A unit of electric charge; equal to the amount of charge transferred in 1 second (s) by 1 ampere (A) of current. *See also* charge, electric
countercurrent heat exchanger A specialized parallel arrangement of incoming arteries and outgoing veins forming a heat exchanger that conserves heat in the body core.
countercurrent multiplier A pair of opposed channels containing

fluids flowing in opposite directions and having an energetic gradient directed transversely from one of the channels into the other. Since exchange due to the gradient is cumulative with distance, the exchange per unit distance will be multiplied, so to speak, as a function of the total distance over which exchange takes place.

counterion An ion associated with, and having a charge opposite to, another ion or an ionized group of a molecule.

countertransport The uphill transport of one substance across a membrane driven by the downhill diffusion of another substance.

coupled transport Simultaneous transport via a membrane protein of two substances in the same or opposite directions.

covalent bond A bond formed by the sharing of electrons between two atoms.

creatine phosphate (phosphocreatine) A phosphorylated nitrogenous compound, found primarily in muscle, that acts as a storage form of high-energy phosphate for the rapid phosphorylation of ADP to ATP.

cretinism A chronic condition caused by hypothyroidism in childhood; characterized by arrested physical and mental development.

cristae The folds of the inner mitochondrial membrane.

critical fusion frequency The minimum number of light flashes per second at which a light is perceived to be continuous.

critical thermal maximum The temperature above which long-term survival is not possible.

crop (gizzard) A muscular organ in the foregut, found in birds, fishes, and some invertebrates. The crop receives swallowed food items and churns the mixture together to break down the food into more digestible particles.

crop milk A nutrient-rich substance fed by regurgitation to pigeon chicks by both parents.

cross-bridges Spirally arranged projections from myosin thick filaments that bind to sites on actin thin filaments during muscle contraction.

crypt of Lieberkühn A circular depression around the base of each villus in the intestine.

cupula A small upside-down cup or domelike cap; in the lateral line and the organs of equilibrium, the cupula contacts the cilia of hair cells by way of a gelatinous matrix.

curare (D-tubocurarine) A South American arrow poison; blocks synaptic transmission at the neuromuscular junction by competitive inhibition of nicotinic acetylcholine receptors.

current, electric The flow of electric charge. A current of 1 coulomb (C) per second is called an ampere (A). By convention, the direction of current flow is the direction in which a positive charge moves (i.e., from the anode to the cathode).

cuticle The hard outer coat of insects and crustaceans.

cyanide CN^-; blocks the transfer of electrons in the respiratory chain.

cyclic AMP (cAMP) A ubiquitous cyclic nucleotide (adenosine 3′5′-cyclic monophosphate) produced from ATP by the enzymatic action of adenylate cyclase; an important cellular regulatory agent that acts as the second messenger for many hormones and transmitters.

cyclic GMP (cGMP) A cyclic nucleotide (guanosine 3′5′-cyclic monophosphate) analogous to cAMP but present in cells at a far lower concentration and producing target cell responses that are frequently opposite to those of cAMP.

cyclostomes A group of jawless vertebrates, including lampreys and hagfishes.

cytochalasin A drug that disrupts cytoplasmic microfilaments.

cytochromes A group of iron-containing proteins that function in the electron-transport chain in aerobic cells by accepting and passing on electrons.

cytoplasm The semifluid substance within a cell, exclusive of the nucleus but including other organelles.

cytosine Oxyamino-pyrimidine, $C_4H_5N_3O$; a base component of nucleic acids.

cytosol The unstructured aqueous phase of the cytoplasm between the structured organelles.

Dalton's law The principle that the partial pressure of a gas in a mixture is independent of other gases present. The total pressure is the sum of the partial pressures of all gases present.

dark current A steady sodium current that leaks into a vertebrate photoreceptor cell at the outer segment. The sodium is actively pumped out of the inner segment, completing the circuit. The dark current is reduced by photoexcitation.

decerebration Experimental elimination of cerebral activity by section of the brain stem or by interruption of the blood supply to the brain.

decremental transmission (passive electrotonic transmission) Electrical signal transmission in which signals are conducted, but not regeneratively, so that their amplitude drops with distance traveled.

decussation Crossing over from one side to the other.

defecation The process of expelling feces.

dehydrogenase An enzyme that "loosens" the hydrogen of a substrate in preparation for its passage to a hydrogen receptor.

dehydroretinol (retinol 2; vitamin A_2) The form of vitamin A found in the liver and the retina of freshwater fishes, some invertebrates, and some amphibians. *See also* 3-dehydroretinal

delayed outward current (late outward current) A current carried by K^+ through channels that open with a time lag after the onset of a depolarization; responsible for repolarization during the action potential.

denaturation Alteration or destruction of the normal nature of a substance by chemical or physical means.

dendrites Fine processes of a neuron that typically provide the main receptive area for synaptic inputs from other neurons.

densitometer An instrument that measures the amount of exposure of the film emulsion during autoradiography.

deoxyhemoglobin Hemoglobin in which no oxygen is bound to the Fe atom of the heme group.

deoxyribonucleic acid *See* DNA

depolarization The reduction or reversal of the potential difference that exists across a membrane compared with V_m at rest.

desmosome A type of cell junction that serves primarily to aid the structural bonding of neighboring cells.

developmental physiology A subdiscipline that concentrates on how physiological processes unfold during the course of animal development from embryo through larva or fetus to adulthood.

diabetes mellitus A metabolic malady in which derangements in the insulin signaling pathway lead to defective control of blood glucose levels and loss of glucose in the urine.

diacylglycerol (DAG) A diglyceride that is part of the inositol phospholipid signaling cascade. It is present as a constituent of cellular phospholipids; when released from these phospholipids by an agonist-activated phospholipase, it serves as the endogenous activator of the calcium- and calmodulin-dependent protein kinase (protein kinase C).

diaphragm The dome-shaped muscle that separates the thoracic and abdominal cavities in mammals and serves as the chief muscle of respiration.

diastole The phase in the heartbeat during which the myocardium is relaxed and the chambers are filling with blood.

dielectric constant A measure of the degree to which a substance is able to store electric charge under an applied voltage.

diffusion Dispersion of atoms, molecules, or ions as a result of random thermal motion.

diffusion coefficient A coefficient relating the rate of diffusional flux to concentration gradient, path length, and the area across which diffusion occurs.

digastric stomach The multichambered stomach of ruminants.

digestive enzymes Enzymes secreted by the alimentary canal to aid in chemical digestion.

digestive epithelium Epithelium lining the small intestine.

dimer A molecule made by the joining of two simple molecules (monomers) of the same kind.

dipole A molecule having separate regions of net negative and net positive charge, so that one end acts as a positive pole and the other as a negative pole.

direct chemical synaptic transmission *See* fast (direct) chemical synaptic transmission

disaccharide sugar A double sugar formed when two monosaccharides (single sugars) are joined together by dehydration synthesis.

disinhibition Release of a neuron from inhibitory input when the inhibitory neuron is itself synaptically inhibited.

dissociation Separation; resolution by thermal agitation or solvation of a substance into simpler constituents.

dissociation constant (K′) The empirical measure of the degree of dissociation of a conjugate acid-base pair in solution.

distal More distant from a point of reference in the centrifugal direction (i.e., away from the center).
distal tubule The portion of a renal tubule located in the renal cortex, leading from (and continuous with) the ascending limb of the loop of Henle to the collecting duct.
disulfide linkage A bond between sulfide groups that determines protein tertiary structure by linking together portions of polypeptide chains.
diuresis An increase in urine excretion.
diuretic An agent that increases urine secretion.
divalent Carrying an electric charge of two units; a valence of 2.
divergence A pattern in which the axon of a single neuron branches, allowing it to synapse on more than one target cell.
DNA (deoxyribonucleic acid) The class of nucleic acids responsible for hereditary transmission and for the coding of amino acid sequences of proteins.
Donnan equilibrium The electrochemical equilibrium that develops when two solutions are separated by a membrane permeable to only some of the ions in the solutions.
dormancy The general term for states of reduced body activity, including sleep, torpor, hibernation, "winter sleep," and estivation.
dorsal horn The dorsal part of the gray matter in the vertebrate spinal cord, which contains the somata of neurons that receive, process, and transmit sensory information.
dorsal root A nerve trunk that enters the spinal cord near the dorsal surface; contains only sensory axons.
dorsal root ganglion (DRG) A collection of somata of sensory neurons located on the surface of a dorsal root, which send processes into the region of the body that is innervated by that spinal segment; each spinal segment contains bilaterally paired DRGs.
down-regulation Control of physiological activity by a decrease in receptor density in a target cell membrane.
drag The resistance to movement of an object through a medium, which increases with the viscosity and density of the medium and varies with the surface area and shape of the object.
D-tubocurarine *See* curare
ductus arteriosus The vessel connecting the pulmonary artery and the aorta in fetal mammals.
duodenum The initial section of the small intestine, situated between the pylorus of the stomach and the jejunum.
dwarfism An abnormally small size in humans; a result of insufficient growth hormone secretion during childhood and adolescence.
dynamic equilibrium *See* steady state
dynamic range The range of energy over which a sensory system is responsive and can encode information about stimulus intensity.
dynein A ciliary protein with magnesium-activated ATPase activity.
dynein arms Projections from tubule A of one microtubule doublet toward tubule B of the next.
dyspnea Labored, difficult breathing.

early inward current A depolarizing current of excitable tissues, carried by Na^+ or Ca^{2+}; responsible for the rising phase of the action potential.
eccentric cell In the horseshoe crab, the afferent neuron of each ommatidium; it is surrounded by and receives information from photoreceptive retinular cells.
eccrine gland An exocrine gland with a coiled, unbranched duct leading from the secretory region.
ecdysis Shedding of the outer shell; molting in an arthropod.
ecdysone A steroid hormone secreted by the thoracic gland of arthropods that induces molting.
echolocation The ability of some species to recognize and locate objects in their environment by emitting sounds and perceiving and interpreting the echoes reflected back.
eclosion hormone An insect hormone that induces the emergence of the adult from the puparium.
ectopic pacemaker A pacemaker situated outside the area where it is normally found.
ectotherm An animal that cannot generate enough heat through metabolism to elevate its body temperature above the ambient temperature, and whose body temperature is dependent on heat flow to or from the environment.
edema Retention of interstitial fluid in organs or tissues.
EDTA Ethylenediaminetetraacetic acid; a calcium- and magnesium-chelating agent.
effector A cell, tissue, or organ that acts to change the condition of an organism (e.g., by contracting muscles or secreting a hormone) in response to neuronal or hormonal signals.
efferent Transporting or conducting from a central region toward the periphery.
efflux The movement of solute or solvent out of a cell across the plasma membrane.
EGTA Ethyleneglycol-bis(β-aminoethylether)-N,N′-tetraacetic acid; a calcium-chelating agent.
eicosanoids Paracrine hormones derived from fatty acids; known to be involved in reproduction, inflammation, fever, pain, and a variety of other processes.
elastic Capable of being distorted, stretched, or compressed with subsequent spontaneous return to original shape; resilient.
electrical mobility The rate at which an ionic species migrates in solution.
electrical synapse A junction between two cells at which a signal is carried from one cell to the other by the passage of ions through gap junctions.
electric potential Electrostatic force; an electric potential difference (i.e., voltage) is required to propel the flow of electric current across a resistance.
electrocardiogram (ECG) A record of electrical events associated with contractions of the heart; typically obtained with electrodes placed on the surface of the body.
electrochemical equilibrium The state in which the concentration gradient of an ion across a membrane is precisely balanced by the electric potential across the membrane.
electrochemical gradient The sum of the combined forces of the concentration gradients and electrical gradients acting on an ion.
electrochemical potential The electric potential developed across a membrane by a chemical concentration gradient of an ion that can diffuse across the membrane.
electrode An electrical circuit element used to make contact with a solution, a tissue, or a cell interior; used to measure potential or to carry current.
electrogenic Giving rise to an electric current or voltage.
electrolyte A compound that dissociates into ions when dissolved in water.
electromotive force (emf) The potential difference across the terminals of a battery or any other source of electric energy.
electronegativity Affinity for electrons.
electroneutrality rule The principle that in any macroscopic region the number of positive and negative charges must be equal. For example, an ionic solution must contain as many anions as cations; on a charged capacitor, excess positive charge stored on one plate must equal excess negative charge stored on the other.
electron pressure A measure of the tendency to donate electrons.
electron-transport chain (respiratory chain) A series of enzymes that transfer electrons from substrate molecules to molecular oxygen.
electro-olfactogram An extracellular electrical recording of the summed activity in many olfactory receptors.
electrophoresis A technique for separating proteins using the net positive or negative charge on their surface amino acids.
electroplax organ An organ in electric fishes such as *Torpedo* that builds up and delivers a powerful electric charge sufficient to stun or even to kill prey. The electroplax organ is derived from embryonic muscle and uses synapses similar to neuromuscular junctions to build up the charge.
electroreceptors Sensory receptors that detect electric fields.
electroretinogram (ERG) An extracellular electrical recording, made at the surface of the eye, of activity in many photoreceptors and other retinal neurons.
electrotonic potential A graded potential generated locally by currents flowing across a membrane; not actively propagated and not all-or-none.
ELISA *See* enzyme-linked immunosorbent assay
emulsification A process by which water-insoluble substances such as fats are rendered water-soluble by dispersion into small droplets and combination with detergents, such as bile salts.
endergonic Characterized by a concomitant absorption of energy.
endocardium The internal lining of the heart chamber.
endocrine Characterized by the production of a biologically active substance by a ductless gland; the substance is carried through the bloodstream to initiate a cellular response in a distant target cell or tissue.

endocrine glands Ductless organs or tissues that secrete a hormone into the bloodstream.
endocytosis Bulk uptake of material into a cell by membrane inpocketing to form an internal vesicle.
endogenous Arising within the body or within an organ.
endogenous opioids Neurotransmitter or neuromodulator molecules (e.g., endorphins and enkephalins) whose receptors also bind opioid drugs, such as opium and heroin.
endolymph The aqueous solution surrounding many hair-cell sensory receptors, including cochlear hair cells; it has a high K^+ concentration and a low Na^+ concentration.
endometrium An epithelium that lines the uterus.
endopeptidases Proteolytic enzymes that break up large peptide chains into shorter polypeptide segments.
endorphins Endogenous neuropeptides that exhibit morphinelike actions, found in the central nervous system of vertebrates; different types consist of 16, 17, or 31 amino acid residues.
endothelium A single cell layer forming the internal lining of blood vessels.
endotherm An animal that regulates its body temperature using heat produced by its own metabolic activity.
endplate potential (epp) A postsynaptic potential in a muscle fiber at the neuromuscular junction (or motor endplate).
end-product inhibition Inhibition of a biosynthetic pathway by the end product of the pathway.
energy Capacity to perform work.
energy metabolism The complex collection of biochemical reactions within cells that generate ATP and other high-energy compounds, which serve as the immediate source of energy for all biological events.
enkephalins Endogenous neuropeptides exhibiting morphinelike actions, found in the central nervous system of vertebrates; these peptides consist of five amino acid residues.
enterogastric reflex A reflex that inhibits gastric secretion, triggered when the duodenum is stretched by chyme pumped from the stomach.
enterokinase An intestinal proteolytic enzyme.
enthalpy The heat produced or taken up by a chemical reaction.
entropy A measure of that portion of energy not available for work in a closed system; a measure of molecular randomness.
environmental physiology A subdiscipline that examines animals in the context of their environment, with a particular focus on evolutionary adaptations to environmental conditions.
enzyme A protein with catalytic properties.
enzyme activity A measure of the catalytic potency of an enzyme: the number of substrate molecules that react per minute per enzyme molecule.
enzyme induction Enzyme production stimulated by the specific substrate (inducer) of that enzyme or by a molecule structurally similar to the substrate.
enzyme-linked immunosorbent assay (ELISA) A technique for identifying and quantifying antigens. The sample is fixed, treated with primary antibody, washed, then incubated with antibody-enzyme complex that binds the primary antibody. Subsequent detection of the reaction product of the complexed enzyme reveals the presence of the original antigen.
epicardium The external covering of the heart wall.
epididymis A long, stringlike duct along the dorsal edge of the testis; its function is to store sperm.
epinephrine (adrenaline) A catecholamine released from the adrenal cortex.
equilibrium The lowest energy state of a system; may result from equal action by opposing forces arising from within the system.
equilibrium potential The voltage difference across a semipermeable membrane at which an ionic species that can diffuse across the membrane is in electrochemical equilibrium; it is dependent on the concentration gradient of the ion and is described by the Nernst equation.
equimolar Having the same molarity.
equivalent pore size The pore diameter that accounts for the rate of diffusion of polar substances across a plasma membrane.
eructation Release of gas from the stomach via the esophagus ("burping").
erythrocytes Red blood cells.
eserine (physostigmine) An alkaloid ($C_{15}H_{21}N_3O_2$) of plant origin that blocks activity of the enzyme cholinesterase.
esophagus The region of the alimentary canal that conducts food from the headgut to the digestive areas.
essential amino acids Amino acids that cannot be synthesized by an animal, but are required for synthesis of essential proteins and must therefore be obtained in the diet.
esterases Enzymes that hydrolyze esters.
estivation Dormancy in response to high ambient temperatures or danger of dehydration.
estradiol-17β The most active natural estrogen.
estrogens A family of sex steroids responsible for producing female reproductive cycles and secondary sex characteristics, and which also prepare the reproductive system for fertilization and implantation of the ovum; synthesized primarily in the ovary, although some are made in the adrenal cortex and male testis.
ethers A class of compounds in which two organic groups are joined by an oxygen atom, $R_1—O—R_2$.
ethology The study of animal behavior under natural conditions.
eupnea Normal breathing.
euryhaline Able to tolerate wide variations in salinity.
evolutionary physiology A subdiscipline that uses the techniques of evolutionary biology and systematics to understand the evolution of animals from a physiological viewpoint using physiological markers (e.g., maintenance of constant body temperature) rather than anatomic markers (e.g., feathers).
exchange diffusion A process by which the movement of one molecule across a membrane enhances the movement of another molecule in the opposite direction; most likely involves a common carrier molecule.
excitability The capacity for altered membrane conductance (and often membrane potential) in response to stimulation.
excitation-contraction coupling In muscle fibers, the process by which electrical excitation of the plasma membrane leads to activation of the contractile process.
excitatory In neurophysiology, pertaining to an increase in the probability of producing an action potential.
excitatory postsynaptic potential (epsp) A change in the membrane potential of a postsynaptic cell that increases the probability of an action potential in that cell.
exergonic Characterized by a concomitant release of energy, often accompanied by a release of heat.
exocrine Pertaining to organs or structures that secrete substances via a duct.
exocrine gland A gland that secretes a fluid via a duct.
exocytosis Fusion of the membrane of a vesicle to the plasma membrane and subsequent expulsion of the vesicle contents to the cell exterior.
exopeptidases Proteolytic enzymes that hydrolyze peptide bonds only near the end of a peptide chain, resulting in free amino acids, plus dipeptides and tripeptides.
exothermic Characterized by a release of heat.
expiratory neurons Neurons in the medulla of the brain that control the motor neurons innervating muscles involved in exhalation.
explants Small pieces of tissue removed from a donor animal and kept alive in a container filled with the appropriate mixture of nutrients.
extensor A muscle that extends or straightens a limb or other extremity.
exteroceptors Sense organs that detect stimuli arriving at the surface of the body from a distance.
extracellular digestion Digestion occurring outside of the cell in an alimentary system.
extrafusal fibers Contractile muscle fibers that make up the bulk of skeletal muscle. *See also* intrafusal fibers
extravasation The forcing of fluid—usually blood, serum, or lymph—out of blood vessels.
eye An organ of visual reception that includes parts specialized for optical processing of light as well as photoreceptive neurons.

facilitated diffusion *See* passive transport
facilitation An increase in the efficacy of a synapse as the result of a preceding activation of that synapse.
factorial scope for locomotion The ratio between the basal metabolic rate and the maximum metabolic rate that can be achieved with intense exercise.
FAD *See* flavin adenine dinucleotide
Fahraeus-Lindqvist effect The reduction in the apparent viscosity of blood as it flows into small arterioles.

farad (F) The unit of electrical capacitance.
faraday A measure of electric charge, $-96{,}487\ C \cdot mol^{-1}$.
Faraday's constant (F) The equivalent charge of a mole of electrons, equal to 9.649×10^4 coulombs (C) per mole of electrons.
fast (direct) chemical synaptic transmission Synaptic transmission at a chemical synapse mediated by neurotransmitters that bind to receptor protein complexes in the postsynaptic membrane, each of which includes an ion channel. Binding of the transmitter to the receptor complex is sufficient to open (or to close) the channel.
feces Undigested material and bacteria eliminated from the hindgut.
feedback The return of output to the input part of a system. In negative feedback, the sign of the output is inverted before it is fed back to the input so as to stabilize the output. In positive feedback, the output is unstable because it is returned to the input without a sign inversion, and thus becomes self-reinforcing, or regenerative.
fermentation Enzymatic decomposition; anaerobic transformation of nutrients without net oxidation or electron transfer.
ferritin A large protein molecule that is opaque to electrons, used as a marker in electron microscopy; normally present in the spleen as a storage protein for iron.
fever A disease-induced increase in body temperature above normal levels.
fibrillar muscle *See* asynchronous muscle
fibroblasts Connective tissue cells that can differentiate into chondroblasts, collagenoblasts, or osteoblasts.
Fick diffusion equation An equation defining the rate of solute diffusion through a solvent.
field metabolic rate (FMR) The average rate of energy utilization as an animal goes about its normal activities, which may range from complete inactivity during resting to maximum exertion.
filter feeding (suspension feeding) Food capture by straining waterborne particles through specialized entrapment devices on or within the body surface.
final common pathway The motor neurons through which the sum total of all in motor output is channeled to the muscles.
first law of thermodynamics The principle that energy is neither created nor lost in any process.
first-order enzyme kinetics A kinetic pattern exhibited by enzymatic reactions whose rates are directly proportional to one reactant's concentration (either substrate or product).
fixation In microscopy, the addition of a specialized chemical, such as formalin, that kills cells and immobilizes their constituents, typically by cross-linking amino groups of proteins with covalent bonds.
fixed action pattern A behavior that is performed in a stereotyped fashion in response to specific stimuli.
flagellum A motile, whiplike organelle similar in organization to a cilium, but longer and generally present on a cell in small numbers.
flavin adenine dinucleotide (FAD) A coenzyme formed by the condensation of riboflavin phosphate and adenylic acid, which performs an important function in electron transport and is a prosthetic group for some enzymes.
flexor A muscle that flexes or bends a limb or other extremity.
fluid mosaic model The accepted model for biological membranes, in which globular proteins are integrated into the lipid bilayer.
fluorescence The property of emitting light following molecular excitation by incident light; the emitted light is always less energetic (has a longer wavelength) than the light producing the excitation.
flux The rate of flow of matter or energy across a unit area.
folds of Kerckring (circular folds, plica circularis) Extensive folds of the intestinal mucosa.
follicle-stimulating hormone (FSH) An anterior pituitary gonadotropin that stimulates the development of ovarian follicles in the female and spermatogenesis in the male.
follicular phase The phase of the female reproductive cycle that is characterized by the formation of, and secretion by, ovarian follicles.
food vacuole An enzyme-filled vesicle in protozoans that functions in digestion.
foramen An orifice or opening.
foramen ovale A hole in the interatrial septum covered by a flap valve, which directs oxygenated blood returning to the heart via the inferior vena cava into the left atrium.
foregut The upper region of the alimentary canal, which is involved in food conduction, storage, and digestion.
Fourier's law The principle that the rate of heat flow in a conducting body is proportional to its thermal conductance and to the temperature gradient.
fovea (area centralis) The portion of the mammalian retina that has the highest visual resolution due to the small divergence and convergence in the pathways linking photoreceptors to retinal ganglion cells; in primates, contains closely packed cone cells.
Frank-Starling mechanism An increase in mechanical work by the ventricle that is caused by an increase in end-diastolic volume (or venous filling pressure).
free energy The energy available to do work at a given temperature and pressure.
fructose A ketohexose, $C_6H_{12}O_6$, found in honey and many fruits.
FSH *See* follicle-stimulating hormone

GABA (γ-aminobutyric acid) A ubiquitous inhibitory transmitter.
gain The increase in a signal produced by amplification.
gallbladder An organ associated with the liver that concentrates and stores bile for eventual discharge into the intestine.
γ-efferents (gamma motor neurons) The motor axons innervating the intrafusal muscle fibers of spindle organs.
gamma rays Electromagnetic radiation of very short wavelength (10^{-12} cm) and very high energy.
ganglion (singular; plural, ganglia) An anatomically distinct collection of neuronal somata.
ganglion cells A nonspecific term applied to some neuronal somata, especially those located in ganglia of invertebrates or outside the vertebrate central nervous system proper. *See also* retinal ganglion cells
gap junctions Specialized cell junctions that allow for electrical coupling between cells, in which plasma membranes about 2 nm apart are linked by tubular assemblies of proteins that are called connexons in vertebrates.
gastric Pertaining to the stomach.
gastric cecum An outpouching of the insect alimentary canal, lined with enzyme-secreting and phagocytic cells, that serves as a stomach.
gastric inhibitory peptide (GIP) A gastrointestinal hormone released into the bloodstream from the duodenal mucosa that inhibits gastric secretion and motility.
gastric juice Fluid secreted by the cells of the gastric epithelium.
gastric phase The secretion phase of digestion, stimulated by the presence of food in the stomach.
gastrin A protein hormone that is released by endocrine cells in the pyloric region of the gastric mucosa and induces gastric secretion and motility.
gastrointestinal peptide hormones Hormones that regulate the basic electric rhythm of smooth muscle in the alimentary canal.
Gay-Lussac's law The principle that either the pressure or the volume of a gas is directly proportional to absolute temperature if the other is held constant.
Geiger counter An instrument that detects the presence of ionizing radiation.
gene A region of encoded information in DNA.
generator potential A change in transmembrane potential within the receptive portion of a sensory neuron whose amplitude is graded with stimulus intensity and is sufficiently large at the spike-initiating zone to produce action potentials. *See also* receptor potential.
gestation Pregnancy.
gigantism Excessive growth due to hypersecretion of pituitary growth hormone from birth.
GIP *See* gastric inhibitory peptide
gizzard *See* crop
gland An aggregation of specialized cells that secrete or excrete substances, such as the pituitary gland, which produces hormones, and the spleen, which takes part in blood production.
glial cells (neuroglia) Nonexcitable supportive cells associated with neurons in nervous tissue.
globin The protein constituent of hemoglobin, which is composed of two pairs of polypeptide chains.
glomerular filtrate rate (GFR) The total amount of glomerular

filtrate produced per minute by all nephrons of both kidneys; equal to the clearance of a freely filtered and nonreabsorbed substance such as insulin.

glomerulus A coiled mass of capillaries.

glucagon A protein hormone released by the alpha cells of the pancreatic islets of Langerhans whose secretion is induced by low blood sugar or growth hormone and which stimulates glycogenolysis in the liver.

glucocorticoids Steroids synthesized in the adrenal cortex that have wide-ranging metabolic activity; they include cortisone, cortisol, corticosterone, and 11-deoxycorticosterone.

gluconeogenesis The synthesis of carbohydrates from noncarbohydrate sources, such as amino acids.

glucose A six-carbon sugar constituting the cell's primary metabolic fuel; blood sugar.

glutamate The most ubiquitous excitatory synaptic transmitter in the vertebrate central nervous system and the excitatory transmitter at arthropod neuromuscular junctions.

glutamine An amino acid; the form in which ammonia is transported between the liver and kidneys in mammals.

glycocalyx A meshwork of acid mucopolysaccharide and glycoprotein filaments that arises from the membrane covering the microvilli of the intestinal brush border.

glycogen A highly branched D-glucose polymer found in animals.

glycogenesis The synthesis of glycogen.

glycogenolysis The breakdown of glycogen to glucose 6-phosphate.

glycogen synthetase An enzyme that catalyzes the polymerization of glucose to glycogen.

glycolipid A lipid containing carbohydrate groups, in most cases galactose.

glycolysis The metabolic pathway by which hexose and triose sugars are broken down to simpler substances, especially pyruvate or lactate.

glycosidases Carbohydrases that break down disaccharides (sucrose, fructose, maltose, lactose) by hydrolyzing α-1,6 and α-1,4 glucosidic bonds into their constituent monosaccharides for absorption.

glycosuria The excretion of excessive amounts of glucose in the urine.

goblet cells Mucus-secreting cells found in most epithelia.

goiter An abnormal increase in the size of the thyroid gland, usually resulting from a lack of dietary iodine.

Goldman equation The equation describing the steady-state potential for a system in which two or more species of diffusible ions are separated by a semipermeable membrane; if only one species can diffuse across the membrane, the equation reduces to the Nernst equation.

Golgi tendon organs Tension-sensing nerve endings of the Ib-afferent fibers found in muscle tendons.

gonadotropins (gonadotropic hormones) Hormones that influence the activity of the gonads; in particular, those secreted by the anterior pituitary.

G protein A GTP-binding protein that plays a critical role in signal transduction pathways across membranes.

Graafian follicle A mature ovarian follicle in which fluid is accumulating.

graded response A response that increases as a function of the energy of the stimulus; a membrane response that is not all-or-none.

Graham's law The principle that the rate of diffusion of a gas is proportional to the square root of the density of that gas.

granular cells Specialized secretory cells situated in the afferent glomerular arterioles that respond to low blood pressure by secreting renin.

gray matter Tissue of the vertebrate central nervous system consisting of neuronal somata, unmyelinated fibers, and glial cells.

growth hormone (GH, somatotropin) A protein hormone that is secreted by the anterior pituitary and stimulates growth; directly influences protein, fat, and carbohydrate metabolism and regulates growth rate.

GTP *See* guanosine triphosphate

guanine 2-amino-6-oxypurine ($C_5H_5N_5O$); a base component of nucleic acids.

guano The white, pasty waste product of birds and reptiles; high in uric acid content.

guanosine triphosphate (GTP) A high-energy molecule similar to ATP that participates in several energy-requiring processes, such as peptide bond formation.

guanylate cyclase An enzyme that converts GTP to cyclic GMP.

gustation The sense of taste; chemoreception of ions and molecules in solution by specialized epithelial sensory receptors.

gustatory receptors Receptor molecules involved in taste.

habituation A progressive reduction in the probability of a behavioral response with repeated stimulation.

hair cell A mechanosensory epithelial cell bearing stereocilia and in some cases a kinocilium.

Haldane effect A reduction in the total CO_2 content of the blood at constant P_{CO_2} when hemoglobin is oxygenated.

half-width The length of time during which a transient physiological variable is at half of its maximum value or greater.

halide A binary compound of a halogen and another element.

halogens A family of related elements that form similar saltlike compounds with most metals, comprising fluorine, chlorine, bromine, and iodine.

headgut The anterior (cranial) region of the alimentary canal, which provides an external opening for food entry.

heat Energy in the form of molecular or atomic vibration that is transferred by conduction, convection, or radiation down a thermal gradient.

heat capacity The amount of heat required to raise 1 g of a substance 1°C.

heater tissues Tissues specialized for heat production; an example is the modified eye muscles in billfishes.

heat of vaporization The heat necessary per mass unit of a given liquid to convert all of the liquid to gas at its boiling point.

heat-shock proteins Molecules involved in protein folding and cellular response to stresses such as heat. *See also* molecular chaperones.

heavy meromyosin The "head" and "neck" of the myosin molecule; the portion of the myosin molecule that has ATPase activity.

helicotrema The opening that connects the scala tympani and the scala vestibuli at the cochlear apex.

hematocrit The percentage of total blood volume occupied by red blood cells; in humans, normally 40%–50%.

heme An iron protoporphyrin portion of many respiratory pigments.

hemerythrin An invertebrate respiratory pigment protein that lacks heme.

hemimetabolous Showing incomplete metamorphosis. *See also* holometabolous

hemocoel A space between ectoderm and endoderm that contains blood (hemolymph) in many invertebrates.

hemocyanin An invertebrate respiratory pigment protein that contains copper; found in mollusks and crustaceans.

hemoglobin The oxygen-carrying pigment of red blood cells; a complex protein composed of four heme groups and four globin polypeptide chains. Also found in invertebrates.

hemolymph The blood of invertebrates with open circulatory systems.

hemopoietic factor *See* intrinsic factor

Henderson-Hasselbalch equation $pH = pK' + \log ([H^+ \text{ acceptor}]/[H^+ \text{ donor}])$. The formula for calculation of the pH of a buffer solution.

Henry's law The principle that the quantity of gas that dissolves in a liquid is nearly proportional to the partial pressure of that gas in the gas phase.

hepatic portal vein A large vein that delivers blood directly from the intestines to the liver.

hepatocyte A liver cell.

Hering-Breuer reflex A reflex in which lung inflation activates pulmonary stretch receptors that inhibit further inhalation during that cycle.

hertz (Hz) Cycles per second.

Hess's law The principle that the total energy released in the breakdown of a fuel to a given set of end products is always the same, irrespective of the intermediate chemical steps or pathways used.

heterodimer A dimer consisting of two nonidentical subunits.

heterosynaptic facilitation Increased efficacy of synaptic transmission between two neurons as a result of activity in a third neuron.

heterosynaptic modulation A change in the efficacy of synaptic

transmission at one synapse that is due to activity at another, separate synapse.
heterotherm An animal that derives essentially all of its body heat from the environment.
heterotroph An organism that depends for its nourishment on energy-yielding carbon compounds derived from the ingestion of other organisms.
hexose A six-carbon monosaccharide.
hibernation A period of deep torpor, or winter dormancy, in animals living in cold climates, lasting weeks or months.
high-performance liquid chromatography (HPLC) A chromatographic procedure that permits rapid and precise separation of molecular mixtures based on size, charge, and solubility.
hindgut The terminal region of the alimentary canal, where the remnants of digested food are stored and eventually eliminated.
hindgut fermentation Fermentative digestion occurring in the terminal portion of the alimentary canal.
histamine The base formed from histidine by decarboxylation; responsible for dilation of blood vessels.
Hodgkin cycle The regenerative, or positive-feedback, loop responsible for the upstroke of the action potential, in which depolarization causes an increase in sodium permeability, permitting an increased influx of Na^+, which further depolarizes the membrane.
holocrine secretion A secretory mechanism in which entire secretory cells are cast off and break up to release their contents.
holometabolous Showing complete metamorphosis. *See also* hemimetabolous
homeostasis The condition of relative internal stability maintained by physiological control systems.
homeotherm An animal that regulates its own internal temperature within a narrow range, regardless of the ambient temperature, by controlling metabolic heat production and heat loss.
homeoviscous adaptation Molecular-level adaptations, especially of biological membranes, that help minimize temperature-induced differences in viscosity.
homonymous Having the same origin.
homosynaptic modulation A change in the efficacy of a synapse that results from activity at that synapse.
horizontal cell A neuron whose fibers extend laterally in the outer plexiform layer of the vertebrate retina, and which carries photoreceptor signals in the plane of the retina to more distant bipolar cells.
hormone A chemical compound that is synthesized and secreted by an endocrine tissue into the bloodstream and that influences the activity of a target tissue.
horseradish peroxidase A large protein molecule used to trace neuronal pathways in the central nervous system.
HPLC *See* high-performance liquid chromatography
hybridoma A hybrid cell formed by the fusion of two different cell types, used in the production of monoclonal antibodies and other cellular products.
hydration Solvation when the solvent is water.
hydraulic permeability The sievelike properties of the Bowman's capsule in the kidney.
hydride A compound consisting of an element or a radical combined with hydrogen.
hydrofuge Water-repellent.
hydrogen bond A weak electrostatic attraction between a hydrogen atom bound to a highly electronegative element in a molecule and another highly electronegative atom in the same or a different molecule.
hydrolase transport The mechanism by which monosaccharides are taken up into absorptive cells, using a membrane-bound glycosidase to break down and transport the parent disaccharide across the absorptive cell's membrane.
hydrolysis Fragmentation or splitting of a compound by the addition of water, whereupon the hydroxyl group joins one fragment and the hydrogen atom the other.
hydronium ion (H_3O^+) A hydrogen ion (H^+) combined with a water molecule (H_2O).
hydrophilic Having an affinity for water.
hydrophobic Lacking an affinity for water.
hydrostatic pressure Force exerted over an area due to pressure in a fluid.
hydroxyapatite $Ca_{10}(PO_4)_6(OH)_2$, a crystalline material lending hardness and rigidity to the bones of vertebrates and the shells of mollusks.
hydroxyl group (radical) The $-OH^-$ group.
hydroxyl ion OH^-.
hypercalcemia Excessive plasma calcium levels.
hypercapnia Excessive levels of carbon dioxide.
hyperemia Increased blood flow to a tissue or an organ.
hyperglycemia Excessive blood glucose levels.
hyperosmotic Containing a greater concentration of osmotically active constituents than the solution of reference.
hyperpnea (hyperventilation) Increased lung ventilation.
hyperpolarization An increase in the potential difference across a membrane, making the cell interior more negative than it is at rest.
hyperthermia A state of abnormally high body temperature.
hypertonic Having a higher tonicity than a reference solution.
hypertrophy Excessive growth or development of an organ or a tissue.
hyperventilation *See* hyperpnea
hypoglycemia Low blood glucose levels.
hypo-osmotic Containing a lower concentration of osmotically active constituents than the solution of reference.
hypophysis *See* pituitary gland
hypopnea (hypoventilation) Decreased lung ventilation.
hypothalamo-hypophyseal portal system A system of portal veins linking the capillaries of the hypothalamic median eminence with those of the anterior pituitary, which transport hypothalamic neurosecretions directly to the anterior pituitary.
hypothalamus The part of the diencephalon that forms the floor of the median ventricle of the brain; includes many subregions that contribute to control the autonomic nervous system and of endocrine function.
hypothermia A state of abnormally low body temperature.
hypothesis A specific prediction that can be tested by performing experiments.
hypothyroidism Reduced thyroid activity.
hypotonic Having a lower tonicity than a reference solution.
hypoventilation *See* hypopnea
hypoxia Reduced oxygen levels.
hysteresis A nonlinear change in the physical state of a system, such that the state depends in part on the previous history of the system.
H zone (H band) The light zone in the center of a resting muscle sarcomere where myosin filaments do not overlap with actin filaments; the region between actin filaments.

I band The light region between the A band and the Z disk of a resting muscle sarcomere, which contains the part of the actin thin filaments that do not overlap with myosin filaments.
ileum The posterior section of the small intestine.
immunoblot assay A technique in which bands of protein separated by gel electrophoresis are subjected in sequence to primary antibody, secondary antibody linked to an enzyme, and then substrate for a colorimetric reaction, as in ELISA.
impedance The dynamic resistance to flow met by fluids or electric currents moving in a pulsatile manner.
incisors Chisel-like teeth used for gnawing.
incus The middle auditory ossicle, which connects the malleus and the stapes.
indirect chemical synaptic transmission *See* slow (indirect) chemical synaptic transmission
influx The movement of solute or solvent into a cell across the plasma membrane.
infradian rhythm A biological rhythm with a periodicity of less than a day.
infrared radiation Thermal radiation; electromagnetic radiation of wavelengths greater than 7.7×10^{-5} cm and less than 12×10^{-4} cm; the region of the electromagnetic spectrum between red light and radio waves.
inhibitory In neurophysiology, tending to reduce the probability of an action potential.
inhibitory postsynaptic potential (ipsp) A change in the transmembrane potential of a postsynaptic cell that reduces the probability of an action potential in that cell.
inner plexiform layer In the vertebrate retina, the layer of connecting processes that lies between the bipolar cells and the ganglion cells; the location of amacrine cells.

inner segment The portion of a vertebrate photoreceptor cell that contains the organelles and synaptic contacts.
inositol trisphosphate (IP_3) The intracellular second messenger produced by the action of phospholipase C on membrane phosphatidylinositolphosphate in response to stimulation of cell-surface receptors by growth factors, hormones, or neurotransmitters.
inotropic Pertaining to the strength of contraction of the heart.
input resistance The total resistance encountered by electric current flowing into or out of a cell.
inspiratory neurons Neurons in the medulla of the brain that control the motor neurons innervating muscles associated with inhalation.
instars The stages between molts in insect development.
insulin A protein hormone synthesized and secreted by the beta cells of the pancreatic islets of Langerhans; controls cellular uptake of carbohydrates and influences lipid and amino acid metabolism.
integral proteins Proteins spanning the plasma membrane that form selective filters and active transport devices that get nutrients into and cellular products and waste out of the cell.
intercalated disk The junctional region between two connected cardiac muscle cells.
intercellular clefts Lateral intercellular spaces between adjacent cells in epithelia, restricted at the luminal ends by tight junctions and open at the basal ends.
interferon An antiviral and antitumor agent produced by animal cells.
intermittent locomotion Bursts of locomotion interspersed with periods of muscular inactivity.
interneuron A neuron that is entirely contained within the central nervous system; typically connects two or more other neurons.
internode The space along a myelinated axon that is covered by a myelinating glial cell (i.e., a Schwann cell or oligodendrocyte).
interoceptive receptors Sensory receptors that respond to changes inside the body.
interstitial Between cells or tissues.
interstitium The tissue space between cells.
intestinal juice *See* succus entericus
intestinal phase The phase of digestion controlled by gastrin and other hormones.
intracellular digestion Nutrient breakdown that occurs within cells.
intracellular milieu The general physiochemical characteristics within the cell.
intrafusal fibers The muscle fibers within a muscle spindle organ. *See also* extrafusal fibers
intrinsic factor (hemopoietic factor) A mucoprotein produced by the H^+-secreting parietal cells of the stomach; involved in vitamin B_{12} absorption.
inulin An indigestible vegetable starch; used in studies of kidney function because it is freely filtered and not actively transported.
in vitro "In a glass"; in an artificial environment outside the body.
in vivo Within a living organism or tissue.
ion An atom or molecule bearing a net charge due to loss or gain of electrons.
ion-binding site A partially ionized region of a protein or other molecule that interacts electrostatically with ions in the surrounding solution.
ionic bond An electrostatic bond.
ionization The dissociation into ions of a compound in solution.
ionophore A molecule that facilitates the diffusion of ions across membranes.
ionotropic glutamate receptor (iGluR) A glutamate receptor that includes an ion channel in the protein complex and produces excitatory neurotransmission. *See also* metabotropic glutamate receptor
ipsilateral Pertaining to the same side.
iris The pigmented circular diaphragm located behind the cornea of the vertebrate eye.
ischemia The absence of blood flow to an organ or a tissue.
islets of Langerhans Microscopic endocrine structures dispersed throughout the pancreas, which consist of three cell types: alpha cells, which secrete glucagon; beta cells, which secrete insulin; and delta cells, which secrete gastrin.
isoelectric point The pH of a solution at which an amphoteric molecule has a net charge of zero.
isomer A compound having the same chemical formula as another, but with a different arrangement of its atoms.
isometric contraction Contraction during which a muscle does not shorten significantly.
isometry Proportionality of shape regardless of size.
iso-osmotic Having the same osmotic pressure.
isotonic Having a tonicity equivalent to that of a reference solution.
isotonic contraction Contraction in which the force generated remains constant while the muscle shortens.
isotope Any of two or more forms of an element with the same number of protons (atomic number), but a different number of neutrons (atomic weight).
isovolumic Having the same volume.
isozymes Multiple forms of an enzyme found in the same animal species or even in the same cell.

Jacobs-Stewart cycle The cycling of CO_2 and HCO_3 between intracellular and extracellular compartments, which functions to transfer H^+ ions between the cell interior and the extracellular fluid.
jejunum The portion of small intestine between the duodenum and the ileum.
joule (J) A unit of work equivalent to 0.239 calories (cal).
junctional fold A fold in the plasma membrane of a postsynaptic cell that lies under an active zone in the axon terminal of a presynaptic cell; typical of skeletal muscle fibers at neuromuscular junctions.
juvenile hormone (JH) A class of insect hormones that are secreted by the corpora allata and which promote retention of juvenile characteristics.
juxtaglomerular apparatus A group of specialized cells situated between the distal renal tubule and the afferent glomerular arterioles that modulate renal blood flow.
juxtaglomerular cells *See* granular cells
juxtapulmonary capillary receptors (type J receptors) Sensory receptors found in the lung that, when stimulated, elicit the sensation of breathlessness.

kelvin (K) *See* absolute temperature
keratin A structural protein found in skin, feathers, nails, and hooves.
ketone A compound having a carbonyl group (CO) attached (by the carbon) to hydrocarbon groups.
ketone bodies Acetone, acetoacetic acid, and β-hydroxybutyric acid; products of fat and pyruvate metabolism formed from acetyl CoA in the liver; oxidized in muscle and by the central nervous system during starvation.
key stimulus (sign stimulus, releaser) A stimulus that is effective in producing a fixed action pattern.
kinematic viscosity Viscosity divided by density; gases of equal kinematic viscosity will become turbulent at equal flow rates in identical airways.
kinetic energy Energy inherent in the motion of a mass.
kininogen Precursor of bradykinin.
kinocilium A true "9 + 2" or "9 + 0" cilium present in sensory hair cells.
Kirchhoff's laws First law: The sum of the currents entering a junction in a circuit equals the sum of the currents leaving the junction. Second law: The sum of the potential changes encountered in any closed loop in a circuit is equal to zero.
Kleiber's law The finding that metabolic rate is related to body mass across mammalian species by the exponent 0.75.
knockout An animal in which a functional gene has been removed or replaced, and which therefore cannot express the protein originally coded for by that gene.
Krebs cycle *See* citric acid cycle
***K*-selection** A pattern of energetic investment in reproduction in which small numbers of large offspring are produced and extensive parental care is provided.

labeled line coding A pattern of information processing in the nervous system in which each neuron encodes only one particular type of information (e.g., sour stimuli in the taste system), and all of the axons that carry that type of information project to the same location or locations.
lactation The production of milk by the mammary glands (breasts).
lamella A thin sheet or leaf.

laminar flow A pattern of turbulence-free flow of fluid in a vessel or past a moving object; a gradient of relative velocity exists in which the fluid layers closest to the wall or object have the lowest relative velocity.

Laplace's law The principle that transmural pressure in a thin-walled tube is proportional to the wall tension divided by the inner radius of the tube.

larva The immature, active feeding stage characteristic of an animal that must metamorphose before achieving its adult form.

latent heat of vaporization The amount of energy required to change a liquid to a gas of the same temperature.

latent period The interval between an action potential in a muscle fiber and the initiation of contraction.

lateral geniculate nuclei The bilaterally paired relay nuclei between the retina and the visual cortex in the mammalian visual system, located in the thalamus.

lateral inhibition Reciprocal suppression of excitation by neighboring neurons in a sensory network; produces enhanced contrast at boundaries and an increase in dynamic range.

lateral-line system A series of hair cells in canals running the length of the head and body in fishes and many amphibians; these canals have openings to the outside, and the system is sensitive to water movement.

lateral plexus In the compound eye of the horseshoe crab, the collection of neurons that interconnect eccentric cells of the ommatidia, producing lateral inhibition.

lecithin Any of a group of phospholipids found in animal and plant tissues; composed of choline, phosphoric acid, fatty acids, and glycerol.

length constant (λ) The distance along a cell over which a potential change decays in amplitude by $(1 - 1/e)$, or 63%.

length-tension relation Relation between the amount of overlap between actin and myosin filaments and the tension developed by an active sarcomere; also, the relation between the initial length of a whole muscle and the maximum tension it can produce during isometric contraction.

lens The major light-focusing structure in the vertebrate eye.

leukocytes White blood cells.

Leydig cells (interstitial cells) Cells of the testes that are stimulated by luteinizing hormone to secrete testosterone.

LH *See* luteinizing hormone

ligand-gated ion channel A channel that opens when a specific molecule or molecules bind to the extracellular domain of the channel protein.

light Electromagnetic radiation with wavelengths between those of X rays and those of heat (infrared radiation).

light meromyosin The rodlike portion of the myosin molecule that constitutes most of the molecule's backbone.

limited heterothermy A survival mechanism used by camels and other normally homeothermic animals in which body temperature is allowed to rise and fall somewhat as the ambient temperature goes to extremes.

Lineweaver-Burk equation Straight-line transformation of the Michaelis-Menten equation.

lipase An enzyme that specifically breaks down lipids.

lipid Any of the fatty acids, neutral fats, waxes, steroids, and phosphatides; lipids are hydrophobic and feel greasy.

lipid bilayer The continuous double layer of lipid molecules that forms the basic structure of a biological membrane.

lipogenesis The formation of fat from nonlipid sources.

lipophilic Having an affinity for lipids.

lipoprotein A protein-lipid complex in the plasma membrane.

local circuit current The current that spreads electrotonically from the excited portion of an axon during conduction of an action potential, flowing longitudinally along the axon, across the membrane, and back to the excited portion.

longitudinal smooth muscle The outer layer of smooth muscle running along the long axis of the small intestine.

long-term potentiation (LTP) An increase in synaptic efficacy that develops as a result of sustained synaptic input and that lasts for a relatively long time—even days, weeks, or months.

long-term depression (LTD) A decrease in synaptic efficacy that develops as a result of sustained synaptic input and that lasts for a relatively long time—even days, weeks, or months.

loop of Henle The U-shaped portion of the renal tubule that lies in the renal medulla.

lower critical temperature (LCT) The ambient temperature below which the BMR becomes insufficient to balance heat loss, resulting in falling body temperature.

lumen The interior of a cavity or duct.

luminosity Brightness; the relative quantity of light reflected or emitted.

luteal phase The phase of the female reproductive cycle characterized by the formation of, and secretion by, the corpus luteum.

luteinizing hormone (LH) A gonadotropin that is secreted by the anterior pituitary and that acts with follicle-stimulating hormone (FSH) to induce ovulation and release of estrogen from the ovary; also influences the formation of the corpus luteum and stimulates growth in and secretion from the testicular Leydig cells.

lymph The plasmalike fluid collected from interstitial fluid and returned to the bloodstream via the thoracic duct; contains white, but not red, blood cells.

lymphatic system A collection of blind-ended tubes that drain filtered extracellular fluid from tissues and return it to the blood circulation.

lymph heart A muscular pump found in fishes and amphibians that causes movement of lymph.

lymph nodes Aggregations of lymphoid tissue in the lymphatic system that produce lymphocytes and filter the lymphatic fluid.

lymphocytes White blood cells produced in lymphoid tissue that lack cytoplasmic granules and have a large, round nucleus.

lysolecithin A lecithin without the terminal acid group.

lysosomes Minute organelles that occur in many cell types, contain hydrolytic enzymes, and are normally involved in localized intracellular digestion.

macula densa A cluster of cells in the juxtaglomerular apparatus; they sense the NaCl concentration of tubular fluid.

maculae Organs of equilibrium in the vertebrate inner ear.

magnetite A magnetic mineral composed of Fe_3O_4 and found in some animals; believed to play a role in geomagnetic orientation.

malleus The outermost auditory ossicle, which connects the tympanic membrane with the incus.

Malpighian tubules Insect excretory organs responsible for the active secretion of waste products and the formation of urine.

mass action, law of The principle that the velocity of a chemical reaction is proportional to the active masses of the reactants.

mass-specific metabolic rate The metabolic rate of a unit mass of tissue.

mass spectrometry A technique that can identify the gases in a gaseous mixture based on their mass and charge.

mastication The chewing or grinding of food with the teeth.

mastoid bone The posterior process of the temporal bone, situated behind the ear and in front of the occipital bone.

maximum aerobic velocity The locomotor speed at which the maximum rate of aerobic respiration is reached.

mechanoreceptor A sensory receptor tuned to respond to mechanical distortion or pressure.

median eminence A structure at the base of the hypothalamus that is continuous with the hypophyseal stalk; contains the primary capillary plexus of the hypothalamo-hypophyseal portal system.

medulla oblongata In vertebrates, a cone-shaped neuronal mass that lies between the pons and the spinal cord.

medullary cardiovascular center A group of neurons in the medulla involved in the integration of information used in the control and regulation of circulation.

medullary respiratory centers Groups of neurons in the medulla that control the activity of motor neurons associated with breathing.

melanocyte-stimulating hormone (MSH) A peptide hormone released by the anterior pituitary that affects melanin distribution in mammals and creates skin color changes in fishes, amphibians, and reptiles.

melting point The lowest temperature at which a solid will begin to liquefy.

membrane potential The electric potential measured from within a cell relative to the potential of the extracellular fluid, which is by convention considered to be zero; the potential difference across a membrane.

membrane recycling The recovery of membrane through endocytosis and its re-formation into new secretory vesicles.

membrane transport proteins Integral proteins that transport particular classes of molecules across membranes. *See also* active transport

menarche The onset of menstruation during puberty.
menopause The cessation of the menstrual cycle in the mature female human.
menses The shedding of the endometrium during a menstrual cycle.
menstrual cycle A reproductive cycle that includes menstruation.
menstruation The shedding of the endometrium, an event that usually occurs in the absence of conception throughout the reproductive life of the female in certain primate species, including humans.
mesencephalicus lateralis dorsalis (MLD) The nucleus to which auditory information projects in the owl, contributing to the bird's ability to locate objects in its environment based entirely on audition.
messenger RNA (mRNA) A type of RNA that is responsible for transmission of the informational base sequence of DNA to the ribosomes.
metabolic acidosis A decrease in blood pH at constant P_{CO_2}, usually as a result of metabolism or kidney function.
metabolic alkalosis An increase in blood pH at constant P_{CO_2}, usually as a result of metabolism or kidney function.
metabolic intensity *See* mass-specific metabolic rate
metabolic pathway A sequence of enzymatic reactions that changes one substance into another.
metabolic rate The rate of conversion of chemical energy into heat, commonly measured as heat energy released per unit time.
metabolic water Water derived from cellular oxidation.
metabolism The totality of physical and chemical processes in anabolism, catabolism, and cell energetics.
metabotropic glutamate receptor (mGluR) A G protein–linked glutamate receptor that lacks an ion channel in the receptor protein complex. *See also* ionotropic glutamate receptor
metamorphosis A change in morphology—in particular, from one stage of development to another, such as juvenile to adult.
metarhodopsin The product of the absorption of light by rhodopsin; decomposes to opsin and *trans*-retinal.
metarteriole An arterial capillary.
Metazoa Multicellular organisms.
methemoglobin Hemoglobin in which the Fe^{3+} of heme has been oxidized to Fe^{2+}.
micelle A microscopic particle made from an aggregation of amphipathic molecules in solution.
Michaelis-Menten equation The equation describing the dependence of initial reaction velocity on substrate concentration for catalyzed reactions.
microcirculation Circulation through capillaries.
microclimate A small refugium (e.g., a burrow, a crack in bark) that provides protection from general climatic conditions.
microelectrode A tiny hollow glass needle that can be inserted into tissues or cells to record physiological data.
microfilaments Actin filaments within the cytoplasm, which have a diameter of less than 10 nm.
micromanipulator A mechanical device that holds and moves microelectrodes incrementally in three dimensions.
microtome A device used in microscopy to cut ultrathin sections from small blocks of tissue.
microtubule A cylindrical cytoplasmic structure made of polymerized tubulin that is found in many cells, especially motile cells, as a constituent of mitotic spindles, cilia, and flagella.
microvilli Tiny cylindrical projections on a cell surface that greatly increase the cell-surface area; frequently found on absorptive epithelia, but also found in photoreceptors.
micturition Urination.
midgut The region of the alimentary canal where chemical digestion of proteins, fats, and carbohydrates takes place.
mineralocorticoids Steroid hormones that are synthesized and secreted by the adrenal cortex and that influence body fluid volume—in particular, by causing sodium and chloride reabsorption in the kidney tubules.
miniature endplate potential (mepp) A tiny depolarization (generally 1 mV or less) of the postsynaptic membrane at a neuromuscular junction (motor endplate); produced by presynaptic release of a single packet of transmitter.
miniature postsynaptic potential (mpsp) A potential produced in a postsynaptic neuron by presynaptic release of a single packet of transmitter.
mitochondria Membrane-enclosed organelles where ATP is produced during aerobic metabolism.
mixed nerve A nerve that contains axons of both sensory and motor neurons.
M line In a muscle sarcomere, the darkly staining structure in the middle of the H zone.
modulatory agent A substance that either increases or decreases the response of a tissue to a physical or chemical signal.
molality The number of moles of solute in a kilogram of a pure solvent.
molarity The number of moles of solute in a liter of solution.
molars Teeth used in a side-by-side grinding motion to break down food.
mole Avogadro's number (6.023×10^{23}) of molecules of an element or compound; equal to the molecular weight in grams.
molecular chaperones A family of proteins that features prominently in protein folding and the preservation of the complexly folded state of proteins.
molecular phylogeny A system of phylogenetic relations inferred from similarities and differences in the nucleic acid sequences coding for identified proteins.
monoclonal antibody A homogeneous antibody that is produced by a clone of antibody-forming cells and that binds a single type of antigen.
monocyte A white blood cell that lacks cytoplasmic granules and has an indented or horseshoe-shaped nucleus.
monogastric stomach A stomach consisting of a single strong muscular tube or sac.
monomer A molecule capable of combining in repeating units to form a dimer, trimer, or polymer.
monosaccharide sugar An unhydrolyzable carbohydrate or simple sugar; a sweet-tasting, colorless crystalline compound with the formula $C_n(H_2O)_n$.
monosynaptic Requiring or transmitted through only one synapse.
monosynaptic reflex arc A simple reflex in which a sensory neuron (the receptor) synapses in the central nervous system onto a motor neuron that innervates an effector.
monozygotic Arising from one ovum or zygote.
motility In digestion, the ability of the alimentary canal to contract and transport ingested material along its length.
motor cortex The part of the mammalian cerebral cortex that controls motor function; situated anterior to the central sulcus, which separates the frontal and parietal lobes.
motor endplate *See* neuromuscular junction
motor neuron (motoneuron) A neuron that innervates muscle fibers.
motor pool The collection of motor neurons that innervate a particular muscle.
motor unit The unit of motor activity, consisting of a motor neuron and all of the muscle fibers it innervates.
mRNA *See* messenger RNA
mucin The mucopolysaccharide that forms the chief lubricant of mucus.
mucosa A mucous membrane facing a body cavity or the exterior of the body; in particular, the layer of the alimentary canal facing the lumen.
mucosal Pertaining to the side of an epithelial tissue facing the lumen of a body cavity or the exterior of the body. *See also* serosal
mucous cells Mucus-secreting cells of the intestine.
mucus A viscous, protein-containing mixture of mucopolysaccharides secreted from specialized mucous membranes; often plays an important role in filter feeding (in invertebrates) or in lubricating or protecting body surfaces.
Müllerian ducts Paired embryonic ducts originating from the peritoneum that connect with the urogenital sinus to develop into the uterus and fallopian tubes.
multineuronal innervation Innervation of a muscle fiber by several motor neurons; found in many invertebrates, especially arthropods.
multiterminal innervation Numerous synapses made by a single motor neuron along the length of a muscle fiber.
multi-unit smooth muscle A smooth muscle in which individual muscle fibers contract only when they receive excitatory input from neurons.
muscarinic Pertaining to muscarine, a toxin derived from mushrooms, or to acetylcholine receptors that respond to muscarine, but not to nicotine.

muscle fiber A skeletal muscle cell.
muscle spindle A length-sensitive receptor organ located between and in parallel with extrafusal muscle fibers; gives rise to the myotatic, or stretch, reflex of vertebrates.
muscle stretch reflex *See* myotatic reflex
mutagens Compounds that produce mutations in the germ cell line.
mutation A heritable alteration in genetic material.
myelination The formation of a myelin sheath.
myelin sheath A membranous sheath formed by Schwann cells or oligodendrocytes that are wrapped tightly around segments of a vertebrate axon; serves as electrical insulation in saltatory conduction.
myenteric plexus A parasympathetic network of the autonomic nervous system that mediates excitatory actions of the digestive tract.
myoblast An embryonic precursor of skeletal muscle fibers.
myocardium Heart muscle.
myofibril A longitudinal unit of a muscle fiber, made up of sarcomeres and surrounded by sarcoplasmic reticulum.
myogenic Initiated by rhythmic electrical activity within a muscle cell or cells in the absence of neuronal input. *See also* neurogenic
myoglobin An iron-containing protoporphyrin-globin complex found in muscle; serves as a reservoir for oxygen and gives some muscles their red or pink color.
myoplasm (sarcoplasm) The cytoplasm in a muscle cell.
myosin A protein that makes up the thick filaments and cross-bridges in muscle fibers; it is also found in many other cell types and is associated with cellular motility.
myotatic reflex (muscle stretch reflex) The reflex contraction of a muscle in response to stretch that activates stretch receptors.
myotube A developing muscle fiber.

NAD *See* nicotinamide adenine dinucleotide
Na^+/K^+ pump (sodium-potassium pump) A membrane protein responsible for maintaining asymmetric concentrations of Na^+ and K^+ ions across the plasma membrane; actively extrudes Na^+ from the cell and takes up K^+ from extracellular fluids at the expense of metabolic energy.
naloxone An analog of morphine that acts as an opioid antagonist.
nares Nostrils.
negative feedback *See* feedback
nematocysts Stinging cells of hydras, jellyfishes, and anemones.
nephron The morphologic and functional unit of the vertebrate kidney; composed of the glomerulus and Bowman's capsule, the proximal and distal tubules, the loop of Henle (in birds and mammals), and the collecting duct.
Nernst equation An equation for calculating electrochemical equilibrium conditions; defines the electric potential difference across a membrane that will just balance the concentration gradient of a single permeant ionic species.
nerve (n.) A bundle of axons held together as a unit by connective tissue.
nerve fiber *See* axon
nerve impulse *See* action potential
nerve net A collection of interconnected neurons that is distributed through the body, rather than concentrated in a central location; most typical of lower organisms, such as coelenterates.
nerve processes (neurites) Fibers that emanate from the soma of neurons; include dendrites and axons.
nerve-specific energy The term used by Johannes Müller in his hypothesis that the sensory modality of a stimulus is encoded in the projection pattern of sensory neurons and does not depend on particular features of the cellular response in the stimulated neurons.
nervous system The collection of neurons in an animal's body.
net flux The sum of influx and efflux through a membrane or other material.
neurites *See* nerve processes
neuroethology A subdiscipline in which neuronal activity during a behavior is studied to determine how activity in neuronal circuits generates and controls behavior.
neurogenic Initiated and maintained by activity in motor neurons. *See also* myogenic
neuroglia *See* glial cells
neurohemal organ An organ for storage and discharge into the blood of the products of neurosecretion.
neurohormone A substance that is released by neurons and exerts hormonal effects outside the nervous system.
neurohypophysis *See* posterior pituitary
neuromast A collection of hair cells embedded in a cupula in the lateral-line system of aquatic lower vertebrates.
neuromodulation A change in neuronal function caused by chemical messengers (neuromodulators) that are released from axon terminals but that diffuse more widely than do typical neurotransmitters; neuromodulatory effects can be relatively long-lasting.
neuromuscular junction (motor endplate) The synapse that connects a motor neuron with a skeletal muscle fiber.
neuron A nerve cell.
neuronal circuit A set of interconnected neurons.
neuronal integration The ongoing summation of all synaptic inputs to a postsynaptic cell, which determines whether or not that cell will produce an action potential.
neuronal network (neuronal circuit) A system of interacting neurons.
neuronal plasticity The capacity for modification of the activity in a neuronal circuit based on experience and changes in input.
neuropeptide A peptide molecule identified as a neurotransmitter.
neuropeptide Y A 36-amino-acid peptide, co-localized with norepinephrine in sympathetic ganglia and adrenergic nerves and localized in some nonadrenergic fibers, the physiological effects of which include reduction in the action of catecholamines on the mammalian heart and potentiation of the action of catecholamines on fish hearts.
neurophysins Proteins associated with neurohypophyseal hormones stored in granules in the neurosecretory terminals; cleaved from the hormones before secretion.
neuropil A dense mass of closely interwoven and synapsing neuronal processes (axon collaterals and dendrites) and glial cells.
neurosecretory cell A neuron that releases neurohormones.
neurotoxin A substance that interferes with normal neuronal function.
neurotransmitter A chemical messenger that is released by a presynaptic nerve ending and that interacts with receptor molecules in the postsynaptic membrane.
nicotinamide adenine dinucleotide (NAD) A coenzyme widely distributed in living organisms that participates in many enzymatic reactions; made up of adenine, nicotinamide, and two molecules each of D-ribose and phosphoric acid.
nicotinic Pertaining to nicotine, an alkaloid derived from tobacco, or to acetylcholine receptors that respond to nicotine but not to muscarine.
node of Ranvier One of the regularly spaced interruptions of the myelin sheath along an axon.
noncompetitive inhibition Enzyme inhibition due to alteration or destruction of the active site.
nonsaturation kinetics A kinetic pattern in which the reaction rate increases in proportion to the concentration of reactant.
nonshivering thermogenesis A thermogenic process in which enzyme systems for fat metabolism are activated, breaking down and oxidizing conventional fats to produce heat.
nonspiking neuron A neuron that receives and transmits information without action potentials.
nonspiking release The release of neurotransmitter from a presynaptic neuron without action potentials; typically, graded changes in membrane potential modulate the activation of voltage-gated Ca^{2+} channels, and changes in the intracellular concentration of free Ca^{2+} modulate the release of transmitter.
norepinephrine (noradrenaline) A neurotransmitter secreted by postganglionic sympathetic neurons, some cells of the central nervous system, and the adrenal medulla.
nuclease An enzyme that hydrolyzes nucleic acids and their residues.
nucleic acids Nucleotide polymers of high molecular weight. *See also* DNA, RNA
nucleosidase An enzyme that hydrolyzes nucleic acids.
nucleotide A component of nucleic acids, made up of a purine or pyrimidine base, a ribose or deoxyribose sugar, and a phosphate group.
nucleus (atomic) The central, positively charged mass surrounded by a cloud of electrons.
nucleus (cellular) The membrane-bounded body within a eukaryotic cell that houses the genetic material of the cell.

nucleus (neuronal) An anatomically and functionally distinct group of neuronal somata in the central nervous system.
nymph A juvenile developmental stage in some arthropods, whose morphology resembles that of the adult.
nystatin A rod-shaped antibiotic molecule that creates channels through membranes that allow the passage of molecules of a diameter less than 0.4 nm.

obligatory osmotic exchange An exchange between an animal and its environment that is determined by physical factors beyond the animal's control.
occipital lobe The most posterior region of the mammalian cerebral hemisphere, which receives visual information.
ocular dominance column A set of neurons arranged vertically through the mammalian visual cortex, all of which receive input from only one of the two eyes.
ohm (Ω) A unit of electrical resistance, equivalent to the resistance of a column of mercury 1 mm^2 in cross-section and 106 cm long.
Ohm's law $I = V / R$; the strength of an electric current, I, varies directly as the voltage, V, and inversely as the resistance, R.
olfaction The sense of smell; chemoreception of molecules released at some distance from the animal.
oligodendrocytes A class of glial cells with few processes that wrap axons in the central nervous system, forming myelin sheaths.
oligopeptide A peptide made up of several amino acid residues.
oligosaccharide A carbohydrate made up of a small number of monosaccharide residues.
omasum The chamber of the ruminant stomach lying between the rumen and the abomasum.
ommatidium (singular; plural, ommatidia) The functional unit of the invertebrate compound eye, consisting of an elongated structure with a corneal lens, a focusing cone, and several photoreceptor cells.
oncotic pressure Osmotic pressure plus hydrostatic pressure caused by distribution of ions according to the Donnan equilibrium.
1a-afferent fiber The axon of a sensory neuron whose receptor endings innervate a muscle spindle organ and respond to stretch of that organ; the neuron's central terminals synapse directly onto α-motor neurons of the homonymous muscle, as well as other neurons.
1b-afferent fiber The axon of a sensory neuron whose receptor endings innervate the tendons of skeletal muscle and respond to tension.
oocyte A developing ovum.
open-system respirometry An experimental technique in which the oxygen consumption and carbon dioxide production of an animal are determined by measuring the gas content of air before and after it flows through a chamber in which the animal is confined.
operator A gene that regulates the synthetic activity of closely linked structural genes via its association with a regulator gene.
operon A segment of DNA consisting of an operator and its associated structural genes.
opiates Opium-derived narcotic substances.
opioids Substances that exert opiatelike effects; some opioids are synthesized endogenously by neurons within the vertebrate central nervous system.
opsin The protein component of visual pigments.
optic axis An imaginary straight line passing through the center of curvature of a simple lens.
optic chiasm (optic chiasma) A swelling under the hypothalamus of the vertebrate brain where the two optic nerves meet; in some species, some axons cross the midline here and project to the contralateral side of the brain.
optic lobe *See* tectum
organ of Corti The part of the cochlea that contains the hair cells.
organs of equilibrium Regions of the inner ear that sense the position of the body relative to gravity or changes of the body's position with respect to gravity.
ornithine-urea cycle A cyclic succession of reactions that eliminate ammonia and produce urea in the liver of ureotelic organisms.
osmoconformer An organism that exhibits little or no osmoregulation, so that the osmolarity of its body fluids changes with the osmolarity of the environment.
osmolarity The effective osmotic pressure of a solution.
osmole The standard unit of osmotic pressure.
osmolyte A substance that serves the purpose of raising the osmotic pressure or lowering the freezing point of a body fluid.
osmometer An instrument for the measurement of the osmotic pressure of a solution.
osmoregulator An organism that maintains an internal osmolarity different from that of the environment.
osmosis The movement of pure solvent from an area of low solute concentration to an area of high solute concentration through a semipermeable membrane.
osmotic flow The solvent flux due to osmotic pressure.
osmotic pressure The pressure that can potentially be created by osmosis between two solutions separated by a semipermeable membrane; the amount of pressure necessary to prevent osmotic flow between the two solutions.
ostia The small, mouthlike openings in the body wall of sponges.
otolith A calcareous particle that lies on hair cells in the vertebrate organs of equilibrium.
ouabain A cardiac glycoside that is capable of blocking some sodium pumps.
outer plexiform layer In the vertebrate retina, the layer of connecting processes that lies between the photoreceptor cells and the bipolar cells; the location of horizontal cells.
outer segment The portion of a vertebrate photoreceptor that contains the photoreceptive membranes; it is attached to the inner segment by a thin bridge containing microfilaments arranged as in a cilium.
oval window The input connection between the middle ear and the cochlea; covered by the base of the stapes.
overshoot The reversal of membrane potential during an action potential; the period of time during which the cell becomes inside-positive.
ovulation The release of an ovum from the ovarian follicle.
ovum An egg cell; the reproductive cell (gamete) of the female.
oxidant The electron acceptor in a redox reaction.
oxidation A loss of electrons or an increase in the net positivity of an atom or molecule. Biological oxidations are usually achieved by removal of hydrogen from a molecule.
oxidative phosphorylation (respiratory-chain phosphorylation) The formation of high-energy phosphate bonds via phosphorylation of ADP to ATP, accompanied by the transport of electrons to oxygen from the substrate.
oxyconformer An animal that allows its oxygen consumption to fall as ambient oxygen falls.
oxygen debt The extra oxygen necessary to oxidize the products of anaerobic metabolism that accumulate in muscle tissue during intense physical activity.
oxygen dissociation curve A curve that describes the relationship between the extent of combination of oxygen with a respiratory pigment and the partial pressure of oxygen in the gas phase.
oxyhemoglobin Hemoglobin with oxygen bound to the Fe atom of the heme group.
oxyntic cells *See* parietal cells
oxyregulator An animal that maintains oxygen consumption as ambient oxygen falls.
oxytocin An octapeptide hormone secreted by the posterior pituitary; stimulates contractions of the uterus at parturition and the release of milk from mammary glands.

pacemaker An excitable cell or tissue that fires spontaneously and rhythmically.
pacemaker potentials The spontaneous and rhythmic depolarizations produced by pacemaker tissue.
Pacinian corpuscles Pressure receptors found in the skin, muscles, joints, and connective tissues of vertebrates, consisting of a nerve ending surrounded by a laminated capsule of connective tissue.
pancreas An organ that produces exocrine secretions, such as digestive enzymes, as well as endocrine secretions, including the hormones insulin and glucagon.
pancreatic duct A duct carrying secretions from the pancreas to the small intestine.
pancreatic juice Pancreatic secretion containing proteases, lipases, and carbohydrases essential for intestinal digestion.
pancreozymin *See* cholecystokinin
parabiosis An experimental connection of two individuals to allow mixing of their body fluids.
parabronchi Air-conduction pathways in the bird lung.

paracellular pathway A route through an epithelium that passes between, rather than through, cells.
paracrine Pertaining to a biologically active substance that passes by diffusion within the extracellular space from the source to a nearby cell, where it initiates a response.
parafollicular cells (C cells) Cells in the mammalian thyroid that secrete calcitonin.
parallel processing A pattern of information processing in the nervous system in which multiple pathways simultaneously carry information about a particular input or output; the information is resynthesized where the pathways converge.
parasympathetic nervous system The division of the autonomic nervous system in which increased activity generally supports functions such as digestion and sexual activity.
parathyroid glands Small masses of endocrine tissue (usually two pairs) close to the thyroid gland that secrete parathyroid hormone.
parathyroid hormone (PTH; parathormone) A polypeptide hormone secreted by the parathyroid glands in response to a low plasma calcium level; stimulates calcium release from bone and calcium absorption by the intestines while reducing calcium excretion by the kidneys.
paraventricular nucleus A group of neurosecretory neurons in the supraoptic hypothalamus whose axons project to the posterior pituitary.
parietal cells (oxyntic cells) Cells of the stomach lining that secrete hydrochloric acid.
pars intercerebralis The dorsal part of the insect brain; contains the somata of neurosecretory cells that secrete prothoracicotropic hormone from axon terminals in the corpora cardiaca.
pars nervosa *See* posterior pituitary
partition coefficient Ratio of the distribution of a substance between two different liquid phases (e.g., oil and water).
parturition The process of giving birth.
passive electrical properties The physical properties of cells that control the movement of an electrical signal through a nerve in the same way that electrons move along a wire or ions move through an aqueous solution in a beaker.
passive electrotonic transmission *See* decremental transmission
passive transport (facilitated diffusion) Transport across a membrane that does not require the direct input of energy.
patch clamping A recording technique in which a microelectrode is sealed tightly against a cell membrane and the transmembrane potential is held constant; can be used to measure ionic currents through single ion channels or across the membrane of an entire cell.
patella The bone of the kneecap.
pentose A five-carbon monosaccharide sugar.
pepsin A proteolytic enzyme secreted by the stomach lining.
pepsinogen The proenzyme of pepsin.
peptide A molecule consisting of a linear array of amino acid residues. Protein molecules are made of one or more peptides; short peptide chains are called oligopeptides and long chains are called polypeptides.
peptide bond The center bond of the —C(O)—NH— group, created by the condensation of amino acids into peptides.
perfusion The passage of fluid over or through an organ, tissue, or cell.
pericardium The sac of connective tissue that encloses the heart.
perilymph The aqueous solution contained within the scala tympani and scala vestibuli of the cochlea, which is similar in composition to other body fluids. *See also* endolymph
peripheral circulation Circulation through the arteries, capillaries, and veins.
peripheral nervous system The entire collection of neurons and parts of neurons that lie outside of the central nervous system.
peripheral proteins Membrane-linked proteins that do not extend through the membrane.
peripheral resistance unit (PRU) A measure of the drop in pressure (in millimeters of mercury, mm Hg) along a vascular bed divided by mean flow in milliliters per second.
peristalsis A traveling wave of constriction in tubular tissue produced by the contraction of circular muscle.
peritoneum The membrane that lines the abdominal and pelvic cavities.
permeability The ease with which substances can pass through a membrane.
P_{50} The partial pressure of oxygen at which hemoglobin is 50% saturated with oxygen.
phagocyte A cell that engulfs other cells, microorganisms, or foreign particulate matter.
phagocytosis The uptake of particles, cells, or microorganisms by a cell via invaginations in the plasma membrane that seal off to become cytoplasmic vacuoles.
phase contrast microscopy A microscopic technique using differential light refraction by different components of the specimen to enhance viewed images.
phasic Transient.
phasic receptor A sensory receptor that produces action potentials during only part of a sustained stimulus, typically at the onset or offset of the stimulus, or in some cases both.
phasic response The response of a neuron that, when stimulated continuously by a current of constant intensity, accommodates rapidly and generates action potentials only during the beginning of the stimulus period.
pheromone A species-specific substance released into the environment for the purpose of signaling between individuals.
phlorizin A glycoside that inhibits active transport of glucose.
phonon A quantum of sound energy.
phonotaxis Movement of an animal relative to a sound source.
phosphagens High-energy phosphate compounds (e.g., phosphocreatine) that serve as phosphate group donors for rapid rephosphorylation of ADP to ATP.
phosphocreatine *See* creatine phosphate
phosphodiesterase A hydrolytic cytoplasmic enzyme that degrades cAMP to AMP.
phosphodiester bonds Bonds that link the individual nucleotides in nucleic acids.
phosphodiester group —O—P—O—.
phosphoglycerides Glycerol-based lipids found in biological membranes.
phospholipid A phosphorus-containing lipid that hydrolyzes to fatty acids, glycerol, and a nitrogenous compound.
phosphorylase *a* The activated (phosphorylated) form of phosphorylase that catalyzes the cleavage of glycogen to glucose 1-phophate.
phosphorylase kinase An enzyme that, when phosphorylated by a protein kinase, converts phosphorylase *b* to the more active phosphorylase *a*.
phosphorylation The incorporation of a phosphate group into an organic molecule.
photon A quantum of light energy (the smallest amount of light energy that can exist at each wavelength).
photopigment A pigment molecule that changes its energy state when it absorbs one or more photons of light.
photoreceptor A sensory neuron that is tuned to receive light energy.
phototaxis Movement of an animal relative to light.
pH scale Negative log scale (base 10) of hydrogen ion concentration of a solution. $pH = -\log[H^+]$.
physiological dead space That portion of inhaled air not involved in gas transfer in the lung.
pilomotor Pertaining to the autonomic control of smooth muscle for the erection of body hair.
pinna (singular; plural, pinnae) The outer structure of the mammalian ear, which can be more or less elaborate and which captures and funnels sound into the ear.
pinocytosis Fluid intake by a cell via invaginations of the plasma membrane that seal off to become vacuoles filled with liquid.
pituitary gland (hypophysis) A complex endocrine organ situated at the base of the brain and connected to the hypothalamus by a stalk. It is of dual origin: the anterior pituitary is derived from embryonic buccal epithelium, whereas the posterior pituitary is derived from neuronal tissue of the diencephalon.
p*K*′ The negative log (base 10) of an ionization constant, $K \cdot pK' = -\log_{10} K'$.
placebo A physiologically neutral substance that elicits curative or analgesic effects, apparently through psychological means.
placental lactogen A hormone from the placenta that prepares the breasts for milk production.
plane-polarized light Light vibrating in only one plane.
plasma kinins Peptide hormones formed in the blood after injury—for example, bradykinin.
plasma membrane Cell membrane; surface membrane.

plasma skimming The separation of plasma from blood within the circulation.
plasticity The capacity for change in response to an external influence.
plastron The ventral shell of a tortoise or turtle; also, a gas film held in place under water by hydrofuge hairs, creating a large air-water interface.
pleural cavity The cavity between the lungs and the wall of the thorax.
plicae circularis *See* folds of Kerckring
plug-flow reactor An alimentary system in which a bolus of food is progressively digested as it wends its way through a long, tubelike digestive reactor.
pN The pH of a solution at neutrality.
pneumothorax Collapse of the lung due to a puncture that penetrates the pleural cavity.
poikilotherm An animal whose body temperature tends to fluctuate with the ambient temperature.
Poiseuille's law The principle that, in laminar flow, the flow is directly proportional to the driving pressure, and resistance is independent of flow.
Poisson distribution A theoretical description of the probability that random, independent events, based on a unitary event of a particular size, will occur.
polyclonal Derived from different cell lines or clones.
polymer A compound composed of a linear sequence of simple molecules or residues.
polypeptide A long chain of amino acids linked by peptide bonds.
polypnea Rapid breathing.
polysaccharidases Carbohydrases that hydrolyze the glycosidic bonds of long-chain carbohydrates (cellulose, glycogen, and starch).
polysynaptic Characterized by transmission through multiple synapses in series.
polytene Having many duplicate chromatin strands.
pons The region of the vertebrate brain that lies just rostral to the medulla oblongata.
pores of Kohn Small holes between adjacent regions of the lung, permitting collateral air flow.
porphyrins A group of cyclic tetrapyrrole derivatives.
porphyropsin A purple photopigment, based on 3-dehydroretinal, that is present in the retinal rods of some freshwater fishes.
portal vessels Blood vessels that carry blood directly from one capillary bed to another.
positive feedback *See* feedback
posterior pituitary (neurohypophysis, pars nervosa) A neuronally derived reservoir that releases neurohormones synthesized in the hypothalamus.
postganglionic neuron An autonomic neuron that has its soma located in a peripheral ganglion, receives synaptic input from preganglionic neurons, and synapses onto target organs.
postsynaptic Located on the receiving side of a synaptic connection.
postsynaptic current (psc) A change in the rate of ion flow across a postsynaptic membrane in response to a synaptic signal.
postsynaptic potential (psp) A change in V_m in a postsynaptic neuron in response to a synaptic signal.
posttetanic depression Reduced postsynaptic response following prolonged presynaptic stimulation at a high frequency; believed to be due to presynaptic depletion of transmitter.
posttetanic potentiation Increased efficacy of synaptic transmission following presynaptic stimulation at a high frequency; often follows posttetanic depression.
potential energy Stored energy that can be released to do work.
preganglionic neuron An autonomic neuron that has its soma located in the central nervous system, sends an axon into the periphery, and synapses onto postganglionic cells.
presbyopia The tendency for human eyes to become less able to focus on close objects ("far-sighted") with age; occurs as the lens becomes less compliant.
pressure pulse The difference between the systolic and diastolic pressures.
presynaptic Located on the sending side of a synaptic connection.
presynaptic inhibition Neuronal inhibition resulting from the action of an inhibitory terminal that synapses on the presynaptic terminal of an excitatory synapse, which reduces the amount of transmitter released.
primary active transport Transport across a membrane driven by hydrolysis of ATP.
primary fluid A fluid secreted by exocrine glands to which other substances (ions, hormones, etc.) are added.
primary follicle An immature ovarian follicle.
primary projection cortex A region of the cerebral cortex that directly receives sensory signals from lower centers; the first cortical cells to receive sensory information projected to the brain.
primary sensory neurons Neurons that directly receive sensory stimulation.
primary structure The sequence of amino acid residues of a polypeptide chain.
primary transport Transport of a substance directly coupled to the hydrolysis of ATP.
proboscis An elongated, protruding mouthpart, typical of sucking insects.
proenzyme (zymogen) The inactive form of an enzyme before it is activated by removal of a terminal segment.
progesterone A hormone secreted by the corpus luteum, adrenal cortex, and placenta that promotes the growth of a uterine lining suitable for implantation and development of the fertilized ovum.
prolactin A hormone secreted by the anterior pituitary that stimulates milk production and lactation after parturition in mammals.
propeptide A large peptide that contains the amino acid sequences of several smaller peptides, which are released when the large peptide is enzymatically cleaved.
proprioceptors Sensory receptors situated primarily in muscles and tendons that relay information about the position and relative motion of parts of the body.
prostaglandins A family of natural fatty acids that arise in a variety of tissues and are able to induce contraction in uterine and other smooth muscle, lower blood pressure, and modify the actions of some hormones.
prostate gland A gland located around the neck of the bladder and urethra in males that contributes to the seminal fluid.
prosthetic group An organic compound essential to the function of an enzyme. Prosthetic groups differ from coenzymes in that they are more firmly attached to the enzyme protein.
protease An enzyme that breaks the peptide bonds of proteins and polypeptides.
protein A large molecule composed of one or more polypeptide chains.
protein kinase An enzyme that catalyzes the transfer of a phosphate group from ATP to a protein, creating a phosphoprotein.
proteolysis The splitting of proteins by hydrolysis of peptide bonds.
proteolytic Protein-hydrolyzing.
prothoracic glands Ecdysone-secreting tissues situated in the anterior thorax of insects.
prothoracicotropic hormone (PTTH; brain hormone) A neurohormone produced by neurosecretory cells in the pars intercerebralis of the insect brain and released by the corpora cardiaca; activates the prothoracic gland to synthesize and secrete ecdysone.
proximal tubule The coiled portion of the renal tubule located in the renal cortex, beginning at the glomerulus and leading to (and continuous with) the descending limb of the loop of Henle.
pseudopodium Literally, false foot; a temporary projection of an amoeboid cell for engulfment of food or for locomotion.
pseudopregnancy False pregnancy.
psychophysics The branch of psychology concerned with relationships between physical stimuli and perception.
pulmonary Pertaining to the lungs.
pulmonary circuit *See* respiratory circuit
pupa A developmental stage of some insects between the larva and the adult.
pupil The opening at the center of the iris through which light passes into the eye.
purinergic Pertaining to purines or their derivatives in their role as transmitter substances.
purines A class of nitrogenous heterocyclic compounds, $C_5H_4N_4$, derivatives of which (purine bases) are found in nucleotides; they are colorless and crystalline.
Purkinje fibers Junctional fibers that convey the wave of excitation from the bundle of His after it branches and that extend into the myocardium of the two ventricles.

P-wave That portion of the electrocardiogram associated with depolarization of the atria.
pyloric sphincter A sphincter guarding the opening of the stomach into the duodenum.
pylorus The distal stomach opening that releases the stomach contents into the duodenum.
pyrimidines A class of nitrogenous heterocyclic compounds, $C_4H_4N_2$, derivatives of which (pyrimidine bases) are found in nucleotides.
pyrogen A substance that leads to the resetting of a homeotherm's body thermostat to a higher set point, thereby producing fever.

QRS complex That portion of the electrocardiogram associated with depolarization of the ventricle.
Q_{10} The ratio of the rate of a reaction at a given temperature to its rate at a temperature 10°C lower.
quality A property that distinguishes sensory stimuli within a sensory modality (e.g., color is a quality of visual stimuli).
quantal release The release of neurotransmitter in discrete packets that correspond to vesicles containing transmitter molecules.
quaternary structure The characteristic way in which the subunits of a protein containing more than one polypeptide chain are combined.

radiation The transfer of energy by electromagnetic radiation without direct contact between objects.
radioimmunoassay (RIA) An immunological technique for the measurement of minute quantities of antigen or antibody, hormones, certain drugs, and other substances with the use of radioactively labeled reagents.
radioisotope A radioactive isotope.
radula A rasplike structure in the mouth of many gastropods.
range fractionation The arrangement by which receptors within a sensory modality are tuned to receive information within relatively narrow, but not identical, intensity ranges, so the entire dynamic range of the modality is divided among different classes of receptors.
rate constant (specific reaction rate) The proportionality factor by which the concentration of a reactant in an enzymatic reaction is related to the reaction rate.
reactive hyperemia Higher than normal blood flow that occurs following a brief period of ischemia.
receptive field That area of an organism's body (e.g., on the skin or in the retina) that, when stimulated, influences the activity of a given neuron.
receptor cell A neuronal cell that is specialized to respond to some particular sensory stimulus.
receptor current A stimulus-induced change in the movement of ions across a receptor cell plasma membrane.
receptor-mediated endocytosis A specialized process of endocytosis in which solutes temporarily bind to receptor molecules embedded in the plasma membrane prior to transport across the membrane.
receptor potential A change in membrane potential elicited in sensory receptor cells by sensory stimulation, which changes the flow of ionic current across the plasma membrane.
receptor protein A molecule that is situated on the plasma membrane of a cell and that interacts specifically with messenger molecules, such as hormones or transmitters.
receptor tyrosine kinase (RTK) A receptor protein with intrinsic tyrosine kinase activity, e.g., the receptors for insulin and a number of growth factors. When activated by external signal binding, the RTK transfers the phosphate group from ATP to the hydroxyl group on a tyrosine residue of selected proteins in the cytosol. RTKs also phosphorylate themselves when activated; this autophosphorylation enhances the activity of the kinase.
reciprocal inhibition Inhibition of the motor neurons innervating one set of muscles during excitation of their antagonists.
recombinant DNA An engineered DNA molecule.
recruitment A pattern in which neurons with increasingly high thresholds become active as stimulus or output intensity increases; seen in sensory neurons (range fractionation) and in motor neurons.
rectal gland An organ near the rectum of elasmobranchs that excretes a highly concentrated NaCl solution.
redox pair Two compounds, molecules, or atoms involved in mutual reduction and oxidation.
reductant The electron donor in a redox reaction.
reduction The addition of electrons to a substance.
reductionism The study of complex systems by reducing them to their constituent parts, followed by detailed analysis of the components rather than the integrated whole.
reduction potential A measurement of the tendency of a reductant to yield electrons in a redox reaction, expressed in volts.
reflex An action that is generated without the participation of the highest neuronal centers and thus is not entirely voluntary; an involuntary motor response mediated by a neuronal arc in response to sensory input.
reflex arc A neuronal pathway that connects sensory input and motor output; consists of afferent nerve input to a nerve center that produces activity in efferent nerves connected to an effector organ, such as a muscle.
refraction The bending of light rays as they pass from a medium of one density into a medium of another density.
refractive index The refractive power of a medium compared with that of air, which is designated 1.
refractory period The period of increased membrane threshold during and immediately following an action potential.
regenerative Self-reinforcing; utilizing positive feedback; autocatalytic.
regulator An animal that uses biochemical, physiological, behavioral, or other mechanisms to maintain internal homeostasis.
regulator gene A gene that codes for a protein that affects the expression of structural genes.
regurgitation Reverse movement of intestinal luminal contents produced by reverse peristalsis.
Reissner's membrane A membrane within the mammalian cochlea.
relative refractory period The period following an action potential during which another action potential, possibly partial, may be triggered, but only if the stimulus is more intense than usual. *See also* absolute refractory period
release-inhibiting hormones (RIH; release-inhibiting factors, RIF) Hypothalamic neurohormones that are carried by portal vessels to the anterior pituitary, where they restrain the release of specific hormones.
releaser *See* key stimulus
releasing hormones (RH, releasing factors) Hypothalamic neurohormones that are carried by portal vessels to the anterior pituitary, where they stimulate the release of specific hormones.
renal clearance That volume of plasma containing the quantity of a freely filtered substance that appears in the glomerular filtrate per unit time. Total renal clearance is the amount of ultrafiltrate produced by the kidney per unit time.
renin A proteolytic enzyme produced by specialized cells in renal arterioles, which converts angiotensinogen to angiotensin.
rennin An endopeptidase that coagulates milk by promoting the formation of calcium caseinate from the milk protein casein; found especially in the gastric juice of young mammals.
Renshaw cells Small inhibitory interneurons in the ventral horn of the spinal cord that are excited by terminals of α-motor neuron axons and that feed back to the motor pool.
repolarization The return of a membrane that has been depolarized to its resting potential.
repressor gene (regulator gene) A gene that produces a substance (repressor protein) that shuts off the structural-gene activity of an operon by an interaction with its operator gene.
reserpine A botanically derived tranquilizing agent that interferes with the uptake of catecholamines from the cytosol by secretory vesicles; its effect is to deplete the catecholamine content of adrenergic cells.
residual volume The volume of air left in the lungs after maximal expiratory effort.
resistance (R) The property that hinders the flow of electric current. The unit of measure is the ohm (Ω), defined as the resistance that allows 1 ampere (A) of current to flow when a potential drop of 1 volt (V) exists across the resistance.
resistivity (ρ) The resistance of a conductor 1 cm in length and 1 cm^2 in cross-sectional area.

respiratory acidosis A decrease in blood pH associated with a fall in blood P_{CO_2} as a result of lung hypoventilation.
respiratory alkalosis An increase in blood pH associated with a rise in blood P_{CO_2} as a result of lung hyperventilation.
respiratory chain *See* electron-transport chain; oxidative phosphorylation
respiratory circuit The circuit that carries blood between the heart and lungs. *See also* systemic circuit
respiratory pigment A substance that reversibly binds oxygen (e.g., hemoglobin).
respiratory quotient (RQ) The ratio of CO_2 production to O_2 consumption; depends on the type of food oxidized by the animal.
respirometry Measurement of an animal's respiratory gas exchange.
resting potential The normal membrane potential of a cell at rest.
rete mirabile An extensive countercurrent arrangement of arterial and venous capillaries.
reticulum A small network.
retina The photosensitive inner surface of the vertebrate eye.
retinal The aldehyde of retinol obtained from the enzymatic oxidative cleavage of carotene; in the 11-*cis* form, it unites with opsins in the retina to form the visual pigments.
retinal ganglion cells Third-order neurons in the visual pathway that receive information from photoreceptors by way of bipolar cells and whose axons make up the optic nerve.
retinol Vitamin A ($C_{20}H_{30}O$), an alcohol of 20 carbons; converted reversibly to retinal by enzymatic dehydrogenation.
retinular cell A photoreceptor cell of the arthropod compound eye.
reversal potential The membrane potential at which no current flows through membrane ion channels, even though the channels are open; equal to the equilibrium or steady-state potential for the ion or ions that are conducted through the open channels.
Reynolds number (Re) A unitless number; the tendency of a flowing gas or liquid to become turbulent in proportion to its velocity and density and in inverse proportion to its viscosity.
rhabdome The aggregate structure consisting of a longitudinal rosette of rhabdomeres arranged axially in the ommatidium.
rhabdomere The light-absorbing part of a retinular cell that faces the central axis of an ommatidium; the photopigment-bearing plasma membrane is expanded into closely packed microvilli, increasing the amount of photosensitive membrane.
rheogenic Producing electric current.
rhodopsin (visual purple) A purplish-red, light-sensitive pigment with 11-*cis*-retinal as its prosthetic group; found in the rods and cones of the vertebrate retina and in the eyes of many other phyla; bleaches to "visual yellow" (all-*trans*-retinal) when it absorbs incident light.
ribonucleic acid *See* RNA
ribose A pentose monosaccharide with the chemical formula $HOCH_2(CHOH)_3CHO$; a constituent of RNA.
ribosome A ribonucleoprotein particle found within the cytoplasm; the site of intersection of mRNA, tRNA, and amino acids during the synthesis of polypeptide chains.
rigor mortis The rigidity that develops in dying muscle as ATP becomes depleted and cross-bridges remain attached.
Ringer solution Physiological saline solution.
RNA (ribonucleic acid) A nucleic acid made up of adenine, guanine, cytosine, uracil, ribose, and phosphoric acid; responsible for the transcription of DNA and its translation into protein.
rod A vertebrate photoreceptor cell that is very sensitive to light, owing to its cellular properties and a high degree of convergence onto second-order cells. *See also* cone
Root effect (Root shift) A change in blood oxygen capacity as a result of a pH change.
round window A membrane-covered opening, separating the middle ear and the cochlea, through which pressure waves leave after traveling through the cochlea.
***r*-selection** A pattern of energetic investment in reproduction in which large numbers of very small offspring are produced, generally with no parental care provided.
rumen The storage and fermentation chamber in the digastric stomach of ruminants.
rumination The chewing of partially digested food brought up by reverse peristalsis from the rumen in ungulates and in other ruminants.
saccharides A family of carbohydrates that includes the sugars and that are grouped according to the number of saccharide ($C_nH_{2n}O_{n-1}$) groups comprising them into the mono-, di-, tri-, oligo-, and polysaccharides.
sacculus One of the vertebrate organs of equilibrium.
safety factor A factor relating input and output in a system, describing the likelihood that transmission through the system will fail.
saliva A watery fluid secreted in the headgut that aids in mechanical and chemical digestion.
salivary glands Glands that secrete saliva into the headgut.
saltatory conduction Discontinuous conduction of action potentials that takes place at the nodes of Ranvier in myelinated axons.
salt glands Osmoregulatory organs that form a hypertonic aqueous exudate by means of active salt secretion into small tubules situated above the eyes, which is excreted via the nostrils; found in many birds and reptiles that live in desert or marine environments.
sarcolemma The plasma membrane of muscle fibers.
sarcomere The contractile unit of a myofibril, bounded at each end by a Z disk.
sarcoplasm *See* myoplasm
sarcoplasmic reticulum (SR) A smooth, membrane-limited network surrounding each myofibril. Calcium is stored in the SR and released as free Ca^{2+} during excitation-contraction coupling.
saturated In reference to fatty acid chains, indicates absence of double bonds.
saturation kinetics A kinetic pattern in which the concentration of a substrate or carrier limits the rate of reaction.
scala media One of the cochlear ducts; a membranous labyrinth containing the organ of Corti and the tectorial membrane and filled with endolymph.
scala tympani A cochlear chamber connected with the scala vestibuli through the helicotrema and filled with perilymph.
scala vestibuli A cochlear chamber beginning at the oval window, connected with the scala tympani through the helicotrema and filled with perilymph.
scaling The study of how both anatomic and physiological characteristics change with body mass.
Schwann cell A glial cell that wraps around peripheral axons to produce an insulating myelin sheath.
scintillation counter An instrument that detects and counts tiny flashes of light produced in scintillation fluid by particles emitted from radioisotopes.
SDA *See* specific dynamic action
secondary active transport Transport across membranes driven by a preexisting electrochemical gradient.
secondary structure The repeating conformation adopted by a polypeptide segment (e.g., alpha helix).
secondary vacuole A vacuole formed when a food vacuole merges with enzyme-containing lysosomes.
second law of thermodynamics The principle that all natural or spontaneous processes are accompanied by an increase in entropy.
second messenger An intracellular regulatory agent such as cAMP, cGMP, or Ca^{2+} that is itself under the control of an extracellular first messenger, such as a hormone.
second-order enzyme kinetics A kinetic pattern reflecting the concentrations of two reactants multiplied together or of one reactant squared.
second-order neuron A neuron that receives input from primary sensory neurons.
secretagogue A substance that stimulates or promotes secretion.
secretin A polypeptide hormone secreted by the duodenal and jejunal mucosa in response to the presence of acidic chyme in the intestine; induces pancreatic secretion into the intestine.
secretory granules (secretory vesicles) Membrane-bounded cytoplasmic vesicles containing secretory products of a cell.
segmentation Rhythmic contractions of the circular muscle layer of the intestine that mix the intestinal contents.
selectivity sequence *See* affinity sequence
semicircular canals The vertebrate organs of equilibrium, which sense acceleration of the head with respect to the gravitational field.
seminal vesicles Paired sperm storage sacs attached to the posterior urinary bladder and connected to the vas deferens.
semipermeable membrane A membrane that allows certain molecules, but not others, to pass through it.

sensation The perception of a sensory stimulus (as opposed to the reception of stimulus energy by a primary sensory receptor).

sensillum (singular; plural, sensilla) A chitinous, hollow, hairlike projection of the arthropod exoskeleton that serves as an auxiliary structure for sensory neurons.

sensor A mechanical, electrical, or biological device that detects changes in its immediate environment.

sensory adaptation The process by which a sensory system becomes less sensitive to stimuli during prolonged or repeated stimulation.

sensory fiber An axon that carries sensory information to the central nervous system.

sensory filter A neuronal circuit that selectively transmits some features of a sensory input and ignores other features.

sensory modality A category of sensory receptors that are tuned to receive a particular class of energy (e.g., vision, hearing, and olfaction are three different sensory modalities).

sensory reception Absorption of stimulus energy by a neuron that produces a receptor potential, leading to action potentials that travel to the central nervous system.

series elastic components (SEC) Elastic structures (such as tendons and other connective tissues) that are arranged in series with the contractile elements in muscle.

serosa The outermost layer of the alimentary canal.

serosal Pertaining to the side of an epithelial tissue facing the blood. *See also* mucosal

serotonin 5-hydroxytryptamine (5-HT); a neurotransmitter, $C_{10}H_{12}N_2O$.

serum The clear component of blood plasma.

servomotor system A control system that utilizes negative feedback to correct deviations from a selected level, the set point.

set point In a negative-feedback system, the state to which feedback tends to return the system.

SID *See* strong ion difference

siemen (S) The unit of electrical conductance; reciprocal of the ohm.

signal inversion In feedback systems, a response to a signal in which the sign or direction of the output is opposite to that of its input.

signal-to-noise ratio The relation between a signal and the random background activity that arises as the result of kinetic energy or other irrelevant events

sign stimulus *See* key stimulus

single-unit smooth muscle A smooth muscle in which individual fibers are coupled through gap junctions, allowing excitation to spread through the muscle independent of neuronal activity; contraction in these muscles is often myogenic, driven by internal pacemaker cells.

sinoatrial node A mass of specialized cardiac tissue that lies at the junction of the superior vena cava with the right atrium and acts as the pacemaker of the heart.

sinus A cavity or sac; a dilated part of a blood vessel.

sinus venosus A membrane chamber attached to the heart that receives venous blood and transmits it to the atrium.

skeletal muscle The striated muscle whose neurogenic contraction is responsible for moving the bodies of animals.

sliding-filament theory The theory that muscle sarcomeres shorten when actin thin filaments are actively pulled toward the middle of myosin thick filaments by the action of myosin cross-bridges.

slow (indirect) chemical synaptic transmission Synaptic transmission at a chemical synapse mediated by neurotransmitters that bind to receptor molecules in the postsynaptic membrane and activate intracellular second-messenger systems, typically through G proteins.

smooth muscle Muscle without sarcomeres and hence without striations, whose myofilaments are nonuniformly distributed within small, mononucleated, spindle-shaped cells.

sodium activation An increased conductance for sodium ions across excitable membranes in response to membrane depolarization.

sodium-potassium pump *See* Na^+/K^+ pump

solvation The clustering of solvent molecules around a solute.

soma (singular; plural, somata) The cell body of a neuron; in general, the body.

somatic Referring to the body tissues, as distinct from the germ cells.

somatic nervous system (voluntary nervous system) The part of the nervous system that mediates conscious perception and controls voluntary activity in the skeletal muscles.

somatosensory cortex The region of the cerebral cortex that receives sensory input from the body surface.

somatostatin (GH-inhibiting hormone, GIH) A hypothalamic neurohormone that inhibits growth hormone release from the pituitary.

spatial summation The integration by a postsynaptic neuron of simultaneous synaptic potentials that arise from the terminals of different presynaptic neurons.

specific dynamic action (SDA) The increment in metabolic energy cost that can be ascribed to the digestion and assimilation of food; it is highest for proteins.

specific resistivity (R_m) Resistance per unit area of a membrane in ohms per square centimeter.

spectrophotometer A device that passes a beam of visible or ultraviolet light through a fluid-filled vial and measures the intensity of emerging wavelengths.

spectrum The collection of individual wavelengths of electromagnetic radiation produced by refraction or diffraction.

sphincter A ring-shaped band of muscle capable of constricting an opening or a passageway.

sphingolipid A lipid formed by a fatty acid attached to the nitrogen atom of sphingosine, a long-chain, oily amino alcohol ($C_{18}H_{37}O_2N$); occurs primarily in the membranes of neurons.

spike-initiating zone The region of an axon where an action potential is initiated; in many, but not all, neurons, located at the axon hillock.

spinal canal The fluid-filled cavity that runs longitudinally through the vertebrate spinal cord; it is confluent with the cerebral ventricles and contains cerebrospinal fluid.

spinal cord The portion of the vertebrate central nervous system that is encased in the vertebral column, extending from the caudal end of the medulla; consists of a core of gray matter and an outer layer of white matter.

spinal root A large bundle of axons that enters or leaves the spinal cord at each spinal segment.

spiracle A surface opening of the tracheal system in insects.

SR *See* sarcoplasmic reticulum

standard metabolic rate (SMR) A measure that is similar to basal metabolic rate but used for a heterotherm maintained at a selected body temperature.

standard temperature and pressure (STP; dry, STPD) 25°C, 1 atmosphere (atm).

standing-gradient hypothesis A hypothesis describing the process of solute-coupled water transport, based on the active transport of salt across the portions of the epithelial cell membranes facing intercellular clefts.

standing wave A resonating wave with fixed nodes. *See also* traveling wave

stapes The innermost auditory ossicle, which articulates at its apex with the incus and whose base is connected to the oval window.

starch A polysaccharide of plant origin $(C_6H_{10}O_5)_n$.

Starling curve A curve that describes the relationship between work done by the heart and filling pressure.

statocyst A gravity-sensing organ made up of mechanoreceptive hair cells and associated particles called statoliths.

statolith A small, dense, solid granule found in a statocyst.

steady state (dynamic equilibrium) A condition in which the value of a variable does not change, but continuous expenditure of energy is required to maintain that constancy. *See also* equilibrium

stenohaline Able to tolerate only a narrow range of salinities.

stereocilia Nonmotile filament-filled projections of hair cells, lacking the internal structure of motile "9 + 2" cilia.

steric Pertaining to the spatial arrangement of atoms.

steroid hormones Cyclic hydrocarbon derivatives synthesized from cholesterol that function as chemical messengers.

sterols A group of solid, primarily unsaturated polycyclic alcohols.

stimulus A substance, action, or other influence that when applied with sufficient intensity to a tissue causes a response.

stomach The major digestive region of the alimentary canal.

stomatogastric ganglion (STG) A ganglion in crustaceans that

contains the central pattern generator neurons controlling the rhythmic activity of the stomach and associated organs.
STP *See* standard temperature and pressure
STPD *See* standard temperature and pressure
stretch receptor A sensory receptor that responds to stretch, typically associated with lungs, blood vessels, or muscle tissue.
striated muscle Muscle characterized by sarcomeres aligned in register; includes skeletal and cardiac muscle.
stria vascularis A vascular tissue layer over the external wall of the scala media; it secretes endolymph.
stroke volume The volume of blood pumped by one ventricle during a single heartbeat.
strong ion difference (SID) In body fluids, the difference between the sum of strong cations and the sum of strong anions.
strychnine A poisonous alkaloid ($C_{21}H_{22}N_2O_2$) that blocks inhibitory synaptic transmission in the vertebrate central nervous system.
submucosa The secondmost inner layer of the alimentary canal, underlying the mucosa.
submucosal plexus A neuronal network that acts to stimulate gut motility and secretion.
substrate A substance that is acted on by an enzyme.
substrate-level phosphorylation The mechanism by which chemical energy of oxidation is stored in the form of ATP.
succus entericus (intestinal juice) A digestive juice secreted by Brunner's glands and the glands of Lieberkühn in the small intestine.
sulfhydryl group The radical —SH.
summation of contraction The addition of muscle tension due to repeated rapid stimulation of muscle.
supercooling Cooling of a fluid below its freezing temperature without actual freezing because ice crystals fail to form.
supraoptic nucleus A group of neurosecretory neurons in the hypothalamus, just above the optic chiasm, whose axons project to the posterior pituitary.
surface charge Electric charge at the membrane surface, arising from fixed charged groups associated with the membrane surface.
surface hypothesis The hypothesis proposed by Rubner that the metabolic rate of birds and mammals should be proportional to body surface area.
surface tension The elasticity of the surface of a substance (particularly a fluid), which tends to minimize the surface area at each interface.
surfactant A surface-active substance that tends to reduce surface tension, for example, in the lung.
suspension feeding *See* filter feeding
swimbladder A gas-filled bladder used for flotation; found in many teleost fishes.
symmorphosis The concept that the functional capacity of each of a series of linked physiological components of a system is generally well matched to typical demands on the system.
sympathetic nervous system The division of the autonomic nervous system in which increased activity typically provides metabolic support for vigorous physical activity; sometimes called "the fight-or-flight system."
symporter A carrier protein that transfers two solutes in the same direction across a biological membrane.
synapse A connection between a neuron and a target cell at which activity in the presynaptic (transmitting) cell influences the activity of the postsynaptic (receiving) cell.
synaptic cleft The thin space separating the cells at a synapse.
synaptic current The ionic current that flows across a postsynaptic membrane when ion channels open following the binding of neurotransmitter molecules to membrane receptors.
synaptic delay The time separating the arrival of an action potential at a presynaptic nerve terminal and a change in the membrane potential of the postsynaptic cell.
synaptic depression A decrease in synaptic efficacy following prolonged, high-frequency stimulation.
synaptic desensitization The failure of a postsynaptic membrane to respond to neurotransmitter due to the inactivation or loss of postsynaptic receptor molecules.
synaptic efficacy The effectiveness of a presynaptic impulse in producing a postsynaptic potential change.
synaptic facilitation An increase in synaptic efficacy.
synaptic inhibition A change in a postsynaptic cell that reduces the probability of its generating an action potential; produced by a transmitter substance that elicits a postsynaptic current having a reversal potential more negative than the threshold for the action potential.
synaptic noise Irregular changes in the transmembrane potential of a postsynaptic cell, produced by random subthreshold synaptic input.
synaptic plasticity The capacity for long-lasting or permanent changes in synaptic efficacy as a result of experience.
synaptic summation The integration of multiple postsynaptic potentials, resulting in a change in V_m in a postsynaptic neuron.
synaptic transmission The transfer of a signal between a neuron and a target cell at a synapse.
synaptic vesicles Membrane-bounded vesicles containing neurotransmitter molecules; located within axon terminals and typically clustered at active zones.
systemic Pertaining to or affecting the whole body.
systemic circuit The circuit that carries blood throughout the body except to the respiratory surfaces. *See also* respiratory circuit
systole The portion of the hearbeat when the heart muscle is contracting; it takes place between the first and second heart sounds as the blood flows through the aorta and pulmonary artery.

tachycardia An increase in heart rate above the normal level.
taxis A locomotory response that is oriented with respect to a stimulus direction or gradient.
tectorial membrane A fine gelatinous sheet overlying the hair cells of the organ of Corti.
tectum (optic lobe) The highest center for processing visual information in fishes and amphibians.
teleost A bony fish of the infraclass Teleostei.
temporal lobe A lobe of the cerebral hemisphere, situated in the lower lateral area of the brain, at the temples.
temporal summation The summation of postsynaptic membrane potentials that are elicited close to one another in time.
tendon A band of tough, fibrous connective tissue that anchors a skeletal muscle to the skeleton, allowing contraction of the muscle to move the body of an animal.
terminal cisternae The closed spaces that make up part of the sarcoplasmic reticulum on both sides of the Z line, making close contact with the T tubules.
tertiary structure The way in which a polypeptide chain is folded or bent to produce the overall conformation of the molecule.
testosterone A steroid androgen synthesized by the Leydig cells of the testes; responsible for the production and maintenance of male secondary sex characteristics.
tetanus An uninterrupted muscular contraction caused by high-frequency motor impulses. Also the name of a disease caused by a bacterial neurotoxin that is transported in retrograde fashion (toward the soma) in axons and that causes prolonged excitation of muscle fibers, causing tetanic contraction.
tetraethylammonium (TEA) A quarternary ammonium agent, $(C_2H_5)_4N$, that blocks many voltage-gated potassium channels in membranes.
tetrodotoxin (TTX) The puffer fish poison, which selectively blocks voltage-gated sodium channels in the membranes of excitable cells.
thalamus A major center in the midbrain of birds and mammals that receives and transmits both sensory and motor information.
theca interna The internal vascular layer encasing an ovarian follicle; responsible for the biosynthesis and secretion of estrogen.
thermal conductance A quantity describing the ease with which heat flows by conduction down a temperature gradient across a substance or an object.
thermal neutral zone (thermoneutral zone) That range of ambient temperatures within which a homeotherm can control its temperature by passive measures and without elevating its metabolic rate to maintain thermal homeostasis.
thermogenesis The production of body heat by metabolic means such as brown-fat metabolism or muscle contraction during shivering.
thermophilic behavior Heat-seeking behavior.
thermophobic behavior Heat-avoiding behavior.
thermoreceptor A sensory nerve ending specifically responsive to temperature changes.
thick filament A myofilament made primarily of myosin.
thin filament A myofilament that contains actin and regulatory proteins such as troponin and tropomyosin.

thoracic cage The chest compartment formed by the ribs and diaphragm, containing the lungs and heart.
thoracic duct A duct through which fluid from the lymphatic system returns to the bloodstream.
3-dehydroretinal An aldehyde of vitamin A_2 that is found in the visual pigments of freshwater fishes and amphibians.
3,5,3-triiodothyronine (T_3) An iodine-bearing tyrosine derivative synthesized in and secreted by the thyroid gland; raises cellular metabolic rate, as does thyroxine.
threshold of detection In sensory transduction, the minimum stimulus energy that will produce a response in a receptor 50% of the time.
threshold potential The potential at which a response (e.g., an action potential or a muscle twitch) is produced.
threshold stimulus The minimum stimulus energy necessary to produce a detectable response or an all-or-none response 50% of the time.
thrombus A clot in the circulation that blocks blood flow.
thymine A pyrimidine base, 5-methyluracil ($C_5H_6N_2O$); a component of DNA.
thyroid-stimulating hormone (TSH) A hormone secreted by the anterior pituitary that stimulates the secretory activity of the thyroid gland.
thyroxine (T_4) An iodine-bearing, tyrosine-derived hormone that is synthesized and secreted by the thyroid gland and raises cellular metabolic rate.
tidal volume The volume of air moved in or out of the lungs with each breath.
tight junction An area of membrane fusion between adjoining cells that prevents the passage of extracellular material between adjoining cells.
time constant (τ) A measure of the rate of accumulation or decay in an exponential process; the time required for an exponential process to reach 63% completion. In electricity, it is proportional to the product of resistance and capacitance and determines how rapidly a membrane charges or discharges.
tonic Steady; slowly adapting.
tonic fibers Muscle fibers that contract very slowly and do not produce twitches. Found in the postural muscles of amphibians, reptiles, and birds, as well as in the muscle spindles and extraocular muscles of mammals. Tonic fibers normally produce no action potentials, and action potentials are not required to spread excitation, because the innervating motor neuron runs the length of the muscle fiber and makes repeated synapses all along it.
tonicity The relative osmotic pressure of a solution under given conditions. *See also* hypertonic, hypotonic, isotonic
tonic receptors Sensory receptors that continue to fire action potentials throughout the duration of a stimulus and can thus directly convey information about how long the stimulus lasts.
tonic response The response pattern of neurons that, when stimulated continuously by a current of constant intensity, accommodate slowly and fire repetitively with gradually decreasing frequency.
tonotopic map A pattern of auditory projection to the brain in which neurons are arranged based on the frequency of the sound to which they respond.
tonus Sustained resting contraction of muscle, produced by basal neuromotor activity.
torpor A state of inactivity, often with lowered body temperature and reduced metabolism, that some endotherms enter to conserve energy stores.
trachea The large respiratory passageway that connects the pharynx with the bronchi of the vertebrate lung.
tracheal system A system of air-filled tubules that carry respiratory gases between the tissues and the exterior in insects.
tracheoles Minute subdivisions of the tracheal system in insects.
tract Within the central nervous system, a collection of axons that have related functions.
tragus The tab that extends from the ventral (anterior) edge of the outer ear and partially covers the opening of the ear.
train of impulses A number of action potentials propagated in rapid succession down a nerve fiber.
trans A molecular configuration with particular atoms or groups on opposite sides.
transcellular pathway A route through an epithelium in which substances cross through cells.
transcription The formation of an RNA chain of a complementary base sequence from the informational base sequence of DNA.
transducin The G protein that links the capture of light by rhodopsin molecules with a change in the current flowing across the membrane of photoreceptors.
transduction The transformation of one kind of energy or signals into another kind of energy or signals (e.g., sense organs transduce sensory stimuli into action potentials).
transfer RNA (tRNA) A small RNA molecule that is responsible for the transfer of amino acids from their activating enzymes to the ribosomes; there are 20 tRNAs, one for each amino acid.
transgenic animal An animal whose genetic constitution has been experimentally altered by the addition or substitution of genes from other animals of its own or another species.
translation Utilization of the mRNA base sequence for linear organization of amino acid residues on a polypeptide.
transmitter *See* neurotransmitter
transmural pressure The difference in pressure across the wall of a structure, such as a blood vessel.
transverse tubules (T tubules) Tubules found in striated muscle fibers that are continuous with the plasma membrane, extend deep into the fiber, and are closely apposed to the terminal cisternae of the sarcoplasmic reticulum.
traveling wave A wave that moves through the propagating medium, as opposed to a standing wave, which remains stationary.
tricarboxylic acid cycle *See* citric acid cycle
trichromacy theory The theory that three kinds of photoreceptor cone cells exist in the human retina, each with a characteristic maximal sensitivity to a different part of the color spectrum.
triglyceride A neutral molecule composed of three fatty acid residues esterified to glycerol; formed in animals from carbohydrates.
trimer A molecule made up of three simple molecules (monomers) of the same kind.
trimethylamine oxide (TMAO) A nitrogenous waste product, probably from choline decomposition.
tritium A radioactive isotope of hydrogen with an atomic mass of three (H_3).
Triton X-100 A nonionic detergent used in cell biology to solubilize lipids and certain cell proteins.
tRNA *See* transfer RNA
trophic level The position of an organism within a food chain.
tropic Pertaining to hormones that act on other endocrine tissues
tropomyosin A long protein molecule located in the grooves between actin filaments of muscle; inhibits muscle contraction by blocking the binding of myosin cross-bridges to actin filaments.
troponin A calcium-binding complex of globular proteins associated with actin and tropomyosin in the thin filaments of muscle. When troponin binds Ca^{2+}, it undergoes a conformational change, causing tropomyosin to reveal myosin-binding sites on the actin filament.
trypsin A proteolytic enzyme that breaks peptide bonds in which the carboxyl group is provided by arginine or lysine.
trypsinogen The proenzyme of trypsin.
tubulin A 4 nm globular protein molecule that is the building block of microtubules.
tunica adventitia The fibrous outer layer of blood vessel walls.
tunica intima The inner lining of blood vessel walls.
tunica media The middle layer of blood vessel walls, consisting of smooth muscle and elastic tissue.
turbinates Chambers of the nasal passages with olfactory receptors in the surface epithelium.
turbulent flow A flow pattern in which sharp gradients and inconsistencies in velocity and direction of fluid flow exist.
turgor Distension; swollenness.
turnover number A measure of the catalytic potency of an enzyme, usually given as reactions catalyzed per second.
T-wave That portion of the electrocardiogram associated with repolarization (and usually relaxation) of the ventricle.
twitch fibers (phasic fibers) The most common striated vertebrate skeletal muscle; produces an all-or-none twitch in response to an all-or-none action potential.
2-deoxyribose A five-carbon sugar that forms a major structural component of DNA.
tympanic membrane The eardrum.
tympanum The middle ear cavity; houses the auditory ossicles.
type J receptors *See* juxtapulmonary capillary receptors

ultradian rhythm A biological rhythm with a periodicity greater than a day.
ultrafiltrate The product of ultrafiltration.
ultrafiltration The process of separating colloidal or molecular particles from a solute by filtration, using suction or pressure, through a colloidal filter or semipermeable membrane.
ultraviolet light Light of wavelengths between 180 and 390 nm.
undershoot *See* after-hyperpolarizaton
uniporter A carrier protein that transports a single solute from one side of a biological membrane to the other.
unitary current An electric current that results from the sudden opening of individual channels in the plasma membrane.
unit membrane The sandwichlike profile of a biological membrane seen in electron micrographs and believed to represent the lipid bilayer with a hydrophobic center region between hydrophilic surfaces.
unsaturated In reference to fatty acid molecules, having some carbon-carbon double bonds.
upper critical temperature (UCT) The ambient temperature above which normal heat loss mechanisms cannot prevent an increase in body temperature.
up-regulation Control of physiological activity by an increase in receptor density in a target cell membrane.
uracil A pyrimidine base, $C_4H_4O_2N_2$; a component of RNA.
urea $(NH_2)_2CO$, the primary nitrogenous waste product in the urine of mammals.
ureotelic Excreting nitrogen in the form of urea.
ureter A muscular tube that passes urine to the bladder from the kidney.
urethra The channel that passes urine from the bladder out of the body.
uric acid A crystalline waste product of nitrogen metabolism found in the feces and urine of birds and reptiles; poorly soluble in water.
uricolytic pathway The pathway through which uric acid or urates are broken down.
uricotelic Excreting nitrogen in the form of uric acid.
Ussing chamber A chamber used to suspend tissue in order to measure its epithelial transport properties.
utriculus One of the vertebrate organs of equilibrium.

vacuole A membrane-limited cavity in the cytoplasm of a cell.
vagus nerve (Xth cranial nerve) A major cranial nerve that carries sensory fibers from the tongue, pharynx, larynx, and ear; motor fibers to the esophagus, larynx, and pharynx; and parasympathetic and afferent fibers to and from the viscera of the thoracic and abdominal regions.
van der Waals forces The close-range, relatively weak attraction exhibited between atoms and molecules with hydrophobic properties.
van't Hoff equation An equation used to calculate the Q_{10} of a biological function.
varicosities Swellings along the length of a vessel or fiber.
vasa recta The capillary network that surrounds the loop of Henle in the tubules of the mammalian kidney.
vasa vasorum The tiny arteries and veins that supply nutrients to and remove waste products from the tissues in the walls of large blood vessels.
vas deferens A testicular duct that joins the excretory duct of the seminal vesicle to form the ejaculatory duct.
vasoactive intestinal peptide (VIP) A peptide hormone that regulates the intestinal phase of gastric secretion.
vasoconstriction Contraction of the circular muscle of blood vessels, decreasing their volume and increasing the vascular resistance.
vasodilation A widening of the lumen or interior space of blood vessels, increasing blood flow.
vasomotor Pertaining to the autonomic control of vasoconstriction or vasodilation by contraction or relaxation of circular muscle.
vasopressin *See* antidiuretic hormone
vector A carrier; an animal that transfers a parasite or pathogen from host to host. Also, a mathematical term for a quantity with direction, magnitude, and sign.
venous shunt A direct connection between arterioles and venules, bypassing the capillary network.
ventilation In respiratory physiology, the process of air exchange between the lungs and the ambient air.
ventral Toward the belly surface.
ventral horn The ventral part of the gray matter in the vertebrate spinal cord, in which the somata of motor neurons are situated.
ventral root A nerve trunk that leaves the spinal cord near its ventral surface; contains only motor axons.
ventricle A small cavity; also, a chamber of the vertebrate heart.
ventricular zone The region of the brain that surrounds the cerebral ventricles; in embryonic vertebrates, cells of the ventricular zone remain mitotic and generate the neurons of the brain and spinal cord.
venule A small vessel that connects a capillary bed with a vein.
vestibular apparatus The collection of vertebrate organs of equilibrium in the inner ear.
villi Small, fingerlike projections (e.g., on the intestinal epithelium).
viscosity A physical property of fluids that determines the ease with which layers of a fluid move past each other.
visible light Light of wavelengths between 390 and 740 nm.
visual cortex The cerebral cortex in the occipital region of the mammalian cerebrum; devoted to processing visual information.
visual streak A horizontal region of the retina with a high concentration of photoreceptors, which provides high resolution along the visual horizon; found in the eyes of some species that inhabit plains. *See also* fovea
vital capacity The maximum volume of air that can be inhaled into or exhaled from the lungs.
volt (V) A unit of electromotive force; the force required to induce a 1 ampere (A) current to flow through a 1 ohm (Ω) resistance.
voltage (*E* or *V*) The electromotive force, or electric potential, expressed in volts. When the work required to move 1 coulomb (C) of charge from one point to a point of higher potential is 1 joule (J), or 1/4.184 calories (cal), the potential difference between these points is said to be 1 volt (V).
voltage clamping An experimental method of imposing a selected membrane potential across a membrane by means of feedback control.
voltage-gated ion channel A channel whose conductance depends upon the transmembrane electric potential difference.
voluntary nervous system *See* somatic nervous system

watt (W) A unit of electrical power; the work performed at 1 joule (J) per second.
Weber-Fechner law The principle that sensation increases arithmetically as a stimulus increases geometrically; the least perceptible change in stimulus intensity above any background bears a constant proportion to the intensity of the background.
white matter Tissue of the central nervous system that consists mainly of myelinated nerve fibers.
Wolffian ducts The embryonic ducts that are associated with the primordial kidney and that become the excretory and reproductive ducts in the male.
work Force exerted upon an object over a distance; force times distance.

X-ray diffraction A method of examining the crystalline structure of a substance using the pattern of scattered X rays.

Z disk (Z line, Z band) A narrow zone at either end of a muscle sarcomere, consisting of a latticework to which the actin thin filaments are anchored.
zeitgeber An environmental factor that entrains biological rhythms.
zero-order kinetics A kinetic pattern in which the rate of the reaction is independent of the concentration of any of the reactants, which would occur if the enzyme concentration were the limiting factor.
zonula Zone.
zonula adherens A form of desmosome in epithelial cells that forms a belt of cell-to-cell adhesion under tight junctions.
zonula occludens A series of tight junctions between epithelial cells, usually having a ring-shaped configuration, which occlude transepithelial extracellular pathways.
zwitterion A molecule carrying both negatively and positively ionized or ionizable sites.
zygomatic cells *See* chief cells
zygote A fertilized ovum before first cleavage.
zymogen *See* proenzyme

References

Adams, P. R., S. W. Jones, et al. 1986. Slow synaptic transmission in frog sympathetic ganglia. *J. Exp. Biol.* 124:259–285.

Adolph, E. F. 1967. The heart's pacemaker. *Sci. Am.* 216(3):32–37.

Ahlquist, R. P. 1948. A study of the adrenotropin receptors. *Am. J. Physiol.* 153:586–600.

Ashley, C. C. 1971. Calcium and the activation of skeletal muscle. *Endeavor* 30:18–25.

Astrup, P., and J. Severinghaus. 1986. *The History of Blood Gases, Acids, and Bases.* Copenhagen: Munksgaard.

Audesirk, T., and G. Audesirk. 1996. *Biology: Life on Earth.* 4th ed. Upper Saddle River, N.J.: Prentice-Hall.

Avenet, P., S. C. Kinnamon, and S. D. Roper. 1993. Peripheral transduction mechanisms. In S. A. Simon and S. D. Roper, eds., *Mechanisms of Taste Transduction.* Boca Raton, Fla.: CRC Press.

Bajjalieh, S. M. 1999. Synaptic vesicle docking and fusion. *Curr. Opin. Neurobiol.* 9:321–328.

Baker, J. J. W., and G. E. Allen. 1965. *Matter, Energy, and Life.* Reading, Mass.: Addison-Wesley.

Banko, W. E. 1960. *The Trumpeter Swan.* North American Fauna, no. 63. Washington, DC: U.S. Department of the Interior, Fish and Wildlife Service.

Barrionuevo, W. R., and W. W. Burggren. 1999. O_2 consumption and heart rate in developing zebrafish *(Danio rerio):* Influence of temperature and ambient O_2. *Am. J. Physiol.* 276:R505–R513.

Bartels, H. 1971. Blood oxygen dissociation curves: Mammals. In P. L. Altman and S. W. Dittmer, eds., *Respiration and Circulation.* Bethesda, Md.: Federation of American Societies for Experimental Biology.

Bartholomew, G. A. 1964. In *Homeostasis and Feedback Mechanisms,* Symposia of the Society for Experimental Biology, no. 18, 7–29. New York: Academic Press.

Bartholomew, G. A., and R. C. Lasiewski. 1965. Heating and cooling rates, heart rates and simulated diving in the Galápagos marine iguana. *Comp. Biochem. Physiol.* 16:573–582.

Baudinette, R. V. 1991. The energetics and cardiorespiratory correlates of mammalian terrestrial locomotion. *J. Exp. Biol.* 160: 209–231.

Baylor, D. 1996. How photons start vision. *Proc. Natl. Acad. Sci. USA* 93:560–565.

Baylor, D., T. D. Lamb, and K.-W. Yau. 1979. Responses of retinal rods to single photons. *J. Physiol.* 288:613–634.

Beament, J. W. L. 1958. The effect of temperature on the waterproofing mechanism of an insect. *J. Exp. Biol.* 35:494–519.

Bear, M. F., B. W. Connors, and M. A. Paradiso. 1996. *Neuroscience: Exploring the Brain.* Baltimore: Williams and Wilkins.

Beck, W. S. 1971. *Human Design.* New York: Harcourt, Brace, Jovanovich.

Bell, G. H., J. N. Davidson, and H. Scarborough. 1972. *Textbook of Physiology and Biochemistry.* 8th ed. Edinburgh: Churchill Livingstone.

Bendall, J. R. 1969. *Muscles, Molecules, and Movement.* New York: Elsevier.

Bennett, M. V. L. 1968. Similarities between chemical and electrical mediated transmission. In F. D. Carlson, ed., *Physiological and Biochemical Aspects of Nervous Integration.* Englewood Cliffs, N.J.: Prentice-Hall.

Bentley, D. R., and R. R. Hoy. 1972. Genetic control of the neuronal network generating cricket *(Teleogryllus gryllus)* song patterns. *Anim. Behav.* 20:478–492.

Benzinger, T. H. 1961. The diminution of thermoregulatory sweating during cold reception at the skin. *Proc. Nat. Acad. Sci. USA* 47:1683–1688.

Berg, H. C., and E. M. Purcell. 1977. Physics of chemoreception. *Biophys. J.* 20:193–219.

Bernard, C. 1872. *Physiologie Generale.* Paris: Hachette.

Berne, R. M., and M. V. Levy. 1998. *Physiology.* 4th ed. St. Louis: Mosby.

Bernheim, H. A., and M. G. Kluger. 1976. Fever and antipyresis in the lizard *Dipsosaurus dorsalis. Am. J. Physiol.* 231:198–203.

Berridge, M. J. 1985. The molecular basis of communication within the cell. *Sci. Am.* 253:124–125.

Berridge, M. J. 1993. Inositol trisphosphate and calcium signalling. *Nature* 361:315–325.

Berthold, A. A. 1849. Transplantation der hoden. *Arch. Anat. Physiol. Wiss. Med.* 16:42–46.

Bigge, C. F. 1999. Ionotropic glutamate receptors. *Curr. Opin. Chem. Biol.* 3:441–447.

Block, B. A. 1994. Thermogenesis in muscle. *Annu. Rev. Physiol.* 56:535–577.

Block, B., T. Imagawa, K. P. Campbell, and C. Franzini-Armstrong. 1988. Structural evidence for direct interaction between the molecular components of the transverse tubule/sarcoplasmic reticulum junction in skeletal muscle. *J. Cell Biol.* 107:2587–2600.

Bortoff, A. 1976. Myogenic control of intestinal motility. *Physiol. Rev.* 56:416–434.

Bourne, H. R., and E. C. Meng. 2000. Rhodopsin sees the light. *Science* 289:733–734.

Brand, A. R. 1972. The mechanisms of blood circulation in *Anodonta anatina* L. (Bivalvia: Unionidae). *J. Exp. Biol.* 56:362–379.

Brenner, B. M., J. L. Troy, and T. M. Daugharty. 1971. The dynamics of glomerular ultrafiltration in the rat. *J. Clin. Invest.* 50:1776–1780.

Bretscher, M. S. 1985. The molecules of the cell membrane. *Sci. Am.* 253:100–108.

Brett, J. R. 1971. Role of thermoregulation in salmon physiology and behavior. *Am. Zool.* 11:99–113.

Brown, K. T. 1974. Physiology of the retina. In V. B. Mountcastle, ed., *Medical Physiology.* 13th ed. St. Louis: Mosby.

Brownell, P., and R. D. Farley. 1979a. Detection of vibrations in sand by tarsal sense organs of the nocturnal scorpion *Paruroctonus mesaensis. J. Comp. Physiol.* 131:23–30.

Brownell, P., and R. D. Farley. 1979b. Orientation to vibrations in sand by the nocturnal scorpion *Paruroctonus mesaensis:* Mechanism to target location. *J. Comp. Physiol.* 131:31–38.

Bruns, D., and R. Jahn. 1995. Real-time measurement of transmitter release from single synaptic vesicles. *Nature* 377:62–65.

Buddenbrock, W. von. 1956. *The Love of Animals.* London: Muller.

Bülbring, E. 1959. *Lectures on the Scientific Basis of Medicine.* Vol. 7. London: Athlone.

Bülbring, E., and H. Kuriyama. 1963. Effects of changes in ionic environment on the action of acetylcholine and adrenaline on smooth muscle cells of guinea pig. *J. Physiol.* 166:59–74.

Bullock, T. H., and F. P. J. Diecke. 1956. Properties of an infrared receptor. *J. Physiol.* 134:47–87.

Bullock, T. H., and G. A. Horridge. 1965. *Structure and Function in the Nervous Systems of Invertebrates.* New York: W. H. Freeman.

Burggren, W. W. 2000. Developmental physiology, animal models, and the August Krogh principle. *Zoology* 102:148–156.

Burggren, W. W., and K. Johansen. 1982. Ventricular hemodynamics in the monitor lizard, *Varanus exanthematicus:* Pulmonary and systemic pressure separation. *J. Exp. Biol.* 96:343–354.

Cahalan, M., and E. Neher. 1992. Patch clamp techniques: An overview. *Methods Enzymol.* 207:3–14.

Camhi, J. M. 1984. *Neuroethology.* Sunderland, Mass.: Sinauer Associates.

Cannon, W. 1929. Organization for physiological homeostatics. *Physiol. Rev.* 9:399–431.

Capecchi, M. R. 1994. Targeted gene replacement. *Sci. Am.* 270: 52–59.

Carey, F. G. 1973. Fishes with warm bodies. *Sci. Am.* 228:36–44.

Carey, F. G., and J. M. Teal. 1966. Heat conservation in tuna fish muscle. *Proc. Nat. Acad. Sci. USA* 56:1464–1469.

Catania, K. C., and J. H. Kaas. 1996. The unusual nose and brain of the star-nosed mole. *BioScience* 46(8):578–586.

Cattaert, D., A. el Manira, and F. Clarac. 1992. Direct evidence for presynaptic inhibitory mechanisms in crayfish sensory afferents. *J. Neurophysiol.* 67:610–624.

Chen, J.-N., and M. Fishman. 1996. Genetic dissection of heart development. In W. W. Burggren and B. Keller, eds., *Development of Cardiovascular Systems: Molecules to Organisms.* New York: Cambridge University Press.

Chess, A., L. Buck, et al. 1992. Molecular biology of smell: Expression of the multigene family encoding putative odorant receptors. *Cold Spring Harb. Symp. Quant. Biol.* 57:505–516.

Cheung, W. Y. 1979. Calmodulin plays a pivotal role in cellular regulation. *Science* 207:17–27.

Clarac, F., D. Cattaert, and D. Le Ray. 2000. Central control components of a "simple" stretch reflex. *Trends Neurosci.* 23:199–208.

Cole, K. S., and H. J. Curtis. 1939. Electric impedance of the squid giant axon during activity. *J. Gen. Physiol.* 22:640–670.

Comroe, J. H. 1962. *Physiology of Respiration.* Chicago: Year Book Medical Publishers.
Codina, J., A. Yatani, et al. 1987. The alpha subunit of the GTP binding protein Gk opens atrial potassium channels. *Science* 236: 442–445.
Cornwall, I. W. 1956. *Bones for the Archaeologist.* London: Phoenix House.
Curran, P. F. 1965. Ion transport in intestine and its coupling to other transport processes. *Fed. Proc.* 24:993–999.
Dalhoff, E., and G. N. Somero. 1993. Effects of temperature on mitochondria from abalone (genus *Haliotis*): Adaptive plasticity and its limits. *J. Exp. Biol.* 185:151–168.
Darnell, J., H. Lodish, and D. Baltimore. 1990. *Molecular Cell Biology.* 2d ed. New York: Scientific American Books.
Davenport, H. W. 1974. *The A.B.C. of Acid-Base Chemistry.* 6th rev. ed. Chicago: University of Chicago Press.
Davenport, H. W. 1985. *Physiology of the Digestive Tract.* 5th ed. Chicago: Year Book Medical Publishers.
Davis, H. 1968. Mechanisms of the inner ear. *Ann. Otol. Rhinol. Laryngol.* 77:644–655.
Del Castillo, J., and B. Katz. 1954. Quantal components of the end-plate potential. *J. Physiol.* 124:560–573.
Denton, E. J. 1961. The buoyancy of fish and cephalopods. *Prog. Biophys. Mol. Biol.* 11:178–234.
Diamond, J. M, and K. Hammond. 1992. The matches, achieved by natural selection, between biological capacities and their natural loads. *Experientia* 48:551–557.
Diamond, J. M., and J. M. Tormey. 1966. Studies on the structural basis of water transport across epithelial membranes. *Fed. Proc.* 25:1458–1463.
Douglas, W. W. 1974. Mechanism of release of neurohypophyseal hormones: Stimulus-secretion coupling. In R. O. Greep, ed., *Handbook of Physiology,* Section 7, *Endocrinology* (Vol. 4, Part 1, Pituitary Gland). Washington, DC: American Physiological Society.
Douglas, W. W., J. Nagasawa, and R. Schulz. 1971. Electron microscopic studies on the mechanism of secretion of posterior pituitary hormones and significance of microvesicles ("synaptic vesicles"): Evidence of secretion by exocytosis and formation of microvesicles as a by-product of this process. In H. Heller and K. Lederis, eds., *Subcellular Organization and Function in Endocrine Tissues.* Memoirs of the Society for Endocrinology, no. 19. New York: Cambridge University Press.
Doyle, D. A., J. M. Cabral, et al. 1998. The structure of the potassium channel: Molecular basis of K^+ conduction and selectivity. *Science* 280:69–77.
Dudel, J., and S. W. Kuffler. 1961. Presynaptic inhibition at the crayfish neuromuscular junction. *J. Physiol.* 155:543–562.
Dzialowski, E. M., and M. P. O'Connor. 2001. Physiological control of warming and cooling during shuttling and basking in three lizards. *Physiol. Biochem. Zool.* 74(5):679–693.
Eakin, R. 1965. Evolution of photoreceptors. *Cold Spring Harb. Symp. Quant. Biol.* 30:363–370.
Ebashi, S., M. Endo, and I. Ohtsuki. 1969. Control of muscle contraction. *Q. Rev. Biophys.* 2:351–384.
Ebashi, S., K. Maruyama, and M. Endo, eds. 1980. *Muscle Contraction: Its Regulatory Mechanisms.* New York: Springer-Verlag.
Eccles, J. C. 1969. Historical introduction to central cholinergic transmission and its behavioral aspects. *Fed. Proc.* 28:90–94.
Eckert, R. O. 1961. Reflex relationships of the abdominal stretch receptors of the crayfish. *J. Cell. Comp. Physiol.* 57:149–162.
Eckert, R. O. 1972. Bioelectric control of ciliary activity. *Science* 176:473–481.
Edgar, W. M. 1992. Saliva: Its secretions, compositions and functions. *Br. Entomol. J.* 172:305–312.
Edney, E. B. 1974. Desert arthropods. In G. W. Brown, ed., *Desert Biology.* Vol. 2. New York: Academic Press.
Edney, E. B., and K. A. Nagy. 1976. Water balance and excretion. In J. Bligh, J. L. Cloudsley-Thompson, and A. G. MacDonald, eds., *Environmental Physiology of Animals.* Oxford: Blackwell Scientific Publications.
Eiduson, S. 1967. The biochemistry of behavior. *Science J.* 3:113–117.
Erlanger, J., and H. S. Gasser. 1937. *Electrical Signs of Nervous Activity.* Philadelphia: University of Pennsylvania Press.
Eyzaguirre, C., and S. W. Kuffler. 1955. Processes of excitation in the dendrites and in the soma of single isolated sensory nerve cells of the lobster and crayfish. *J. Gen. Physiol.* 39:87–119.
Farrell, A. P., S. S. Sobin, D. J. Randall, and S. Crosby. 1980. Intralamellar blood flow patterns in fish gills. *Am. J. Physiol.* 239:R429–R436.
Fatt, P., and B. Katz. 1951. An analysis of the endplate potential recorded with an intracellular electrode. *J. Physiol.* 115: 320–370.
Fatt, P., and B. Katz. 1952. Spontaneous subthreshold activity at motor nerve endings. *J. Physiol.* 117:109–128.
Fawcett, D. W. 1986. *Bloom and Fawcett: A Textbook of Histology.* 11th ed. Philadelphia: Saunders.
Feigl, E. O. 1974. Physics of the cardiovascular system. In T. C. Ruch and H. D. Patton, eds., *Physiology and Biophysics.* 20th ed. Vol. 2. Philadelphia: Saunders.
Felleman, D. J., and D. C. Van Essen. 1991. Distributed hierarchical processing in the primate cerebral cortex. *Cerebral Cortex* 1:1–47.
Fessenden, R. J., and J. S. Fessenden. 1982. *Organic Chemistry.* 2d ed. Boston: Willard Grant Press.
Firestein, S., G. M. Shepherd, and F. S. Werblin. 1990. Time course of the membrane current underlying sensory transduction in salamander olfactory receptor neurones. *J. Physiol.* 430:135–158.
Flock, A. 1967. Ultrastructure and function in the lateral line organs. In P. H. Cahn, ed., *Lateral Line Detectors.* Bloomington: Indiana University Press.
Florey, E. 1966. *General and Comparative Animal Physiology.* Philadelphia: Saunders.
Franzini-Armstrong, C., F. Protasi, and V. Ramesh. 1999. Shape, size, and distribution of Ca^{2+} release units and couplons in skeletal and cardiac muscles. *Biophys. J.* 77:1528–1539.
Frieden, E. H., and H. Lipner. 1971. *Biochemical Endocrinology of the Vertebrates.* Englewood Cliffs, N.J.: Prentice-Hall.
Furshpan, E. J., and D. D. Potter. 1959. Transmission at the giant motor synapses of the crayfish. *J. Physiol.* 145:289–325.
Gesteland, R. C. 1966. The mechanics of smell. *Discovery* 27(2).
Gilman, A. G. 1987. G proteins: Transducers of receptor-generated signals. *Annu. Rev. Biochem.* 56:615–649.
Goldberg, N. D. 1975. Cyclic nucleotides and cell function. In G. Weissman and R. Claiborne, eds., *Cell Membranes: Biochemistry, Cell Biology, and Pathology.* New York: Hospital Practice Publishing.
Goldman, D. E. 1943. Potential, impedance, and rectification in membranes. *J. Gen. Physiol.* 27:37–60.
Goldsby, R. A. 1967. *Cells and Energy.* New York: Macmillan.
Goodrich, E. S. 1958. *Studies on the Structure and Development of Vertebrates.* Vol. 2. New York: Dover.
Gordon, A. M., E. Homsher, and M. Regnier. 2000. Regulation of contraction in striated muscle. *Physiol. Rev.* 80:853–924.
Gordon, A. M., A. F. Huxley, and F. J. Julian. 1966. The variation in isometric tension with sarcomere length in vertebrate muscle fibres. *J. Physiol.* 184:170–192.
Gorski, R. A. 1979. Long-term hormonal modulation of neuronal structure and function. In F. O. Schmitt and F. G. Worden, eds., *The Neurosciences: Fourth Study Program.* Cambridge, Mass.: MIT Press.
Gosline, J., and R. E. Shadwick. 1996. The mechanical properties of fin whale arteries are explained by novel connective tissue. *J. Exp. Biol.* 199:985–995.
Grantham, J. J. 1971. Mode of water transport in mammalian renal collecting tubules. *Fed. Proc.* 30:14–21.
Grell, K. G. 1973. *Protozoology.* New York: Springer-Verlag.
Grigg, G. C. 1970. Water flow through the gills of Port Jackson sharks. *J. Exp. Biol.* 52:565–568.
Gurney, M. E., and M. Konishi. 1980. Hormone induced sexual differentiation of brain and behavior in zebra finches. *Science* 208:1380–1383.
Hadley, M. E. 1992. *Endocrinology.* 3d ed. Englewood Cliffs, N.J.: Prentice-Hall.
Hadley, N. 1972. Desert species and adaptation. *Am. Sci.* 60:338–347.
Haggis, G. H., D. Michie, et al. 1964. *Introduction to Molecular Biology.* London: Longmans.
Hagins, W. A. 1972. The visual process: Excitatory mechanisms in the primary receptor cells. *Annu. Rev. Biophys. Bioeng.* 1: 131–158.
Hales, J. R. S. 1966. The partition of respiratory ventilation of the panting ox. *J. Physiol.* 188:45–68.

Hall, Z. 1992. *An Introduction to Molecular Neurobiology.* Sunderland, Mass.: Sinauer Associates.

Ham, A. W. 1957. *Histology.* Philadelphia: Lippincott.

Hanamori, T., I. J. Miller, Jr., and D. V. Smith. 1988. Gustatory responsiveness of fibers in the hamster glossopharyngeal nerve. *J. Neurophysiol.* 60:478–498.

Hanaway, J., T. A. Woolsey, M. H. Gado, and M. P. Roberts, Jr. 1998. *The Brain Atlas: A Visual Guide to the Human Central Nervous System.* Bethesda, Md.: Fitzgerald Science Press.

Hargitay, B., and W. Kuhn. 1951. Das Multiplikationspringzip als Grundlage der Harnkonzentrierung in der Niere. *Z. Electrochem.* 55:539–558.

Harris, G. G., and A. Flock. 1967. Spontaneous and evoked activity from *Xenopus laevis* lateral line. In P. H. Cahn, ed., *Lateral Line Detectors.* Bloomington: Indiana University Press.

Harris-Warrick, R. M., D. J. Baro, L. S. Coniglio, B. R. Johnson, R. M. Levini, J. H. Peck, and B. Zhang. 1997. Chemical modulation of crustacean stomatogastric pattern generator networks. In P. S. G. Stein, S. Grillner, A. I. Selverston, and D. G. Stuart, eds., *Neurons, Networks, and Motor Behavior.* Cambridge, Mass.: MIT Press.

Harteneck, C., T. D. Plant, and G. Schultz. 2000. From worm to man: Three subfamilies of TRP channels. *Trends Neurosci.* 23: 159–166.

Hartline, H. K. 1934. Intensity and duration in the excitation of single photoreceptor units. *J Cell. Comp. Physiol.* 5:229–274.

Hartline, H. K., H. G. Wanter, and F. Ratliff. 1956. Inhibition in the eye of *Limulus. J. Gen. Physiol.* 39:651–673.

Hayward, J. N., and M. A. Baker. 1969. A comparative study of the role of the cerebral arterial blood in the regulation of brain temperature in five mammals. *Brain Res.* 16:417–440.

Hazel, J. R. 1995. Thermal adaptation in biological membranes: Is homeoviscous adaptation the explanation? *Annu. Rev. Physiol.* 57:19–42.

Hebb, D. O. 1949. *The Organization of Behaviour.* New York: Wiley.

Heinrich, B. 1974. Thermoregulation in endothermic insects. *Science* 185:747–756.

Heinrich, B. 1987. Thermoregulation in winter moths. *Sci. Am.*, March, 1987.

Heisler, N., P. Neuman, and G. M. O. Maloiy. 1983. The mechanism of intracardiac shunting in the lizard *Varanus exanthematicus. J. Exp. Biol.* 105:15–31.

Hellam, D. C., and R. J. Podolsky. 1967. Force measurements in skinned muscle fibres. *J. Physiol.* 200:807–819.

Heller, H. C., L. I. Crawshaw, and H. T. Hammel. 1978. The thermostat of vertebrate animals. *Sci. Am.* 239:102–113.

Hellon, R. F. 1967. Thermal stimulation of hypothalamic neurones in unanaesthetized rabbits. *J. Physiol.* 193:381–395.

Hemmingsen, A. M. 1969. Energy metabolism as related to body size and respiratory surfaces, and its evolution. *Rep. Steno. Mem. Hosp. Nordisk Insulinlaboratorium* 9:1–110.

Herkenham, M., et al. 1991. Characterization and localization of cannabinoid receptors in rat brain: A quantitative in vitro autoradiographic study. *J. Neurosci.* 11(2):563–583.

Herness, M. S., and T. A. Gilbertson. 1999. Cellular mechanisms of taste transduction. *Annu. Rev. Physiol.* 61:873–900.

Hildebrandt, J., and A. C. Young. 1965. Anatomy and physiology of respiration. In T. C. Ruch and H. D. Patton, eds., *Physiology and Biophysics.* 19th ed. Philadelphia: Saunders.

Hill, A. V. 1938. The heat of shortening and the dynamic constants of muscle. *Proc. R. Soc. Lond.* B 126:136–195.

Hill, A. V. 1964. The efficiency of mechanical power development during muscular shortening and its relation to load. *Proc. R. Soc. Lond.* B 159:319–324.

Hille, B. 1992. *Ionic Channels of Excitable Membranes.* 2d ed. Sunderland, Mass.: Sinauer Associates.

Hirakow, R. 1970. Ultrastructural characteristics of the mammalian and sauropsidan heart. *Am. J. Cardiol.* 25:195–203.

Hoar, W. S. 1975. *General and Comparative Physiology.* 2d ed. Englewood Cliffs, N.J.: Prentice-Hall.

Hodgkin, A. L. 1937. Evidence for electrical transmission in nerve. *J. Physiol.* 90:183–232.

Hodgkin, A. L., and P. Horowicz. 1959. The influence of potassium and chloride ions on the membrane potential of single muscle fibres. *J. Physiol.* 148:127–160.

Hodgkin, A. L., and P. Horowicz. 1960. Potassium contractures in single muscle fibres. *J. Physiol.* 153:386–403.

Hodgkin, A. L., and A. F. Huxley. 1939. Action potentials recorded from inside a nerve fibre. *Nature* 144:710–711.

Hodgkin, A. L., and A. F. Huxley. 1952a. Currents carried by sodium and potassium ions through the membrane of the giant axon of *Loligo. J. Physiol.* 116:449–472.

Hodgkin, A. L., and A. F. Huxley. 1952b. A quantitative description of membrane current and its application to conduction and excitation in nerve. *J. Physiol.* 117:500–544.

Hodgkin, A. L., and A. F. Huxley. 1952c. Properties of nerve exons. I. Movement of sodium and potassium ions during nervous activity. *Cold Spring Harb. Symp. Quant. Biol.* 17:43–52.

Hodgkin, A. L., A. F. Huxley, and B. Katz. 1952. Measurement of current-voltage relations in the membrane of the giant axon of *Loligo. J. Physiol.* 116:424–448.

Hodgkin, A. L., and B. Katz. 1949. The effect of sodium ions on the electrical activity of the giant axon of the squid. *J. Physiol.* 108:37.

Hoffman, B. F., and P. F. Cranefield. 1960. *Electrophysiolgy of the Heart.* New York: McGraw-Hill.

Hokin, M. R., and L. E. Hokin. 1953. Enzyme secretion and the incorporation of ^{32}P into phospholipids of pancreas slices. *J. Biol. Chem.* 203:967–977.

Holland, R. A. B., and R. E. Forster. 1966. The effect of size of red cells on the kinetics of their oxygen uptake. *J. Gen. Physiol.* 49:727–742.

Horridge, G. A. 1968. *Interneurons.* New York: W. H. Freeman.

Hou, P.-C. L., and W. W. Burggren. 1989. Interaction of allometry and development in the mouse *Mus musculus:* Heart rate and hematology. *Respir. Physiol.* 78:265–280.

Hoy, R. R., J. Hahn, and R. C. Paul. 1977. Hybrid cricket auditory behavior: Evidence for genetic coupling in animal communication. *Science* 195:82–84.

Hoy, R. R., and D. Robert. 1996. Tympanal hearing in insects. *Annu. Rev. Entomol.* 41:433–450.

Hoyle, G. 1967. Specificity of muscle. In C. A. G. Wiersma, ed., *Invertebrate Nervous Systems.* Chicago: University of Chicago Press.

Hubbard, R., and A. Kropf. 1967. Molecular isomers in vision. *Sci. Am.* 216(6):64–76. (Offprint 1075.)

Hubel, D. H. 1963. The visual cortex of the brain. *Sci. Am.* 209:54–62.

Hubel, D. H. 1995. *Eye, Brain, and Vision.* New York: Scientific American Library.

Hughes, C. M. 1964. How a fish extracts oxygen from water. *New Scientist* 11:346–348.

Hume, I. D. 1989. Optimal digestive strategies in mammalian herbivores. *Physiol. Zool.* 62(6):1145–1163.

Hutter, O. F., and W. Trautwein. 1956. Vagal and sympathetic effects on the pacemaker fibres in the sinus venosus of the heart. *J. Gen. Physiol.* 39:715–733.

Huxley, A. F., and R. Niedergerke. 1954. Structural changes in muscle during contraction: Interference microscopy of living muscle fibres. *Nature* 173:971–973.

Huxley, A. F., and R. M. Simmons. 1971. Proposed mechanism of force generation in striated muscle. *Nature* 233:533–538.

Huxley, A. F., and R. E. Taylor. 1958. Local activation of striated muscle fibers. *J. Physiol. (Lond.)* 144:426–441.

Huxley, H. E. 1963. Electron microscope studies on the structure of material and synthetic protein filaments from striated muscle. *J. Mol. Biol.* 7:281–308.

Huxley, H. E. 1969. The mechanism of muscular contraction. *Science* 164:1356–1365.

Hyman, L. H. 1940. *The Invertebrates: Protozoa through Ctenophora.* New York: McGraw-Hill.

Imms, A. D. 1949. *Outlines of Entomology.* London: Methuen.

Ip, Y. K., S. F. Chew, and D. J. Randall. 2001. Ammonia toxicity, tolerance and excretion. In P. Wright and P. Anderson, eds., *Nitrogen Excretion: Metabolic and Environmental Perspectives.* New York: Academic Press. In press.

Irvine, B., N. Audsley, et al. 1988. Transport properties of locust ileum in vitro: Effects of cAMP. *J. Exp. Biol.* 137:361–385.

Irving, L. 1966. Adaptations to cold. *Sci. Am.* 214:94–101.

Ishimatzu, A., and Y. Itazawa. 1993. Difference in blood oxygen levels in the outflow vessels of the heart of an air-breathing fish, *Channa argus:* Do separate bloodstreams exist in teleostean heart? *J. Comp. Physiol.* 149:435.

Jackson, D. C. 1987. Assigning priorities among interacting physiological systems. In M. E. Feder, A. F. Bennett, W. W. Burggren, and R. B.

Huey, eds., *New Directions in Ecological Physiology.* New York: Cambridge University Press.

Jamison, R. L., and R. H. Maffly. 1976. The urinary concentrating mechanism. *N. Engl. J. Med.* 295:1059–1067.

Jan, Y. N., and L. Jan. 1983. Coexistence and corelease of cholinergic and peptidergic transmitters in frog sympathetic ganglia. *Fed. Proc.* 42:2929–2933.

Jennings, J. B. 1972. *Feeding, Digestion and Assimilation in Animals.* New York: St. Martin's Press.

Jewell, R. R., and J. C. Ruegg. 1966. Oscillatory contraction of insect fibrillar muscle after glycerol extraction. *Proc. R. Soc. Lond.* B 164:428–459.

Jones, D. R. 1995. Crocodilian cardiac dynamics: A half-hearted attempt. *Physiol. Zool.* 68(4):9–15.

Jones, D. R., P. G. Bushnell, B. K. Evans, and J. Baldwin. 1994. Circulation in the Gippsland giant earthworm *Megascolides australis. Physiol. Zool.* 67(6):1383–1401.

Jones, D. R., B. L. Langille, D. J. Randall, and G. Shelton. 1974. Blood flow in dorsal and ventral aortas of the cod *Gadus morhua. Am. J. Physiol.* 226:90–95.

Jones, D. R., and M. J. Purves. 1970. The effect of carotid body denervation upon the respiratory response to hypoxia and hypercapnia in the duck. *J. Physiol.* 211:295–309.

Josephson, R. K. 1985. The mechanical power output of a tettigoniid wing muscle during singing and flight. *J. Exp. Biol.* 117:357–368.

Kampmeier, O. F. 1969. *Evolution and Comparative Morphology of the Lymphatic System.* Springfield, Ill.: Thomas.

Kandel, E. R. 1976. *Cellular Basis of Behavior.* New York: W. H. Freeman.

Kandel, E. R., T. Abrams, et al. 1983. Classical conditioning and sensitization share aspects of the same molecular cascade in *Aplysia. Cold Spring Harb. Symp. Quant. Biol.* 48:821–830.

Katz, B., and R. Miledi. 1966. Input-output relation of a single synapse. *Nature* 212:1242–1245.

Katz, B., and R. Miledi. 1967. Tetrodotoxin and neuromuscular transmission. *Proc. R. Soc. Lond.* B 167:8–22.

Katz, B., and R. Miledi. 1968. The role of calcium in neuromuscular facilitation. *J. Physiol.* 195:481–492.

Katz, B., and R. Miledi. 1970. Further study of the role of calcium in synaptic transmission. *J. Physiol.* 207:789–801.

Katz, P. S., P. A. Getting, and W. N. Frost. 1994. Dynamic neuromodulation of synaptic strength intrinsic to a central pattern generator circuit. *Nature* 367:729–731.

Kauer, J. S. 1987. Coding in the olfactory system. In T. E. Finger and W. L. Silver, eds., *Neurobiology of Taste and Smell.* New York: Wiley.

Kemali, M., and V. Braitenberg. 1969. *Atlas of the Frog Brain.* New York: Springer-Verlag.

Kenagy, G. J., R. D. Stevenson, and D. Masman. 1989. Energy requirements for lactation and postnatal growth in captive golden-mantled ground squirrels. *Physiol. Zool.* 62(2):470–487.

Kerkut, G. A., and R. C. Thomas. 1964. The effect of anion injection and changes in the external potassium and chloride concentration on the reversal potentials of the IPSP and acetylcholine. *Comp. Physiol. Biochem.* 11:199–213.

Kessel, R., and R. Kardon. 1979. *Tissues and Organs: A Text-Atlas of Scanning Electron Microscopy.* New York: W. H. Freeman.

Keynes, R. D. 1958. The nerve impulse and the squid. *Sci. Am.* 199(6):83–90.

Keynes, R. D. 1979. Ion channels in the nerve-cell membrane. *Sci. Am.*, 240:126–135.

Keynes, R. D., and K. J. Aidley. 1981. *Nerve and Muscle.* Cambridge: Cambridge University Press.

Keys, A., and E. N. Wilmer. 1932. "Chloride secreting cells" in the gills of fishes with special reference to the common eel. *J. Physiol.* 76: 368–378.

Kirschfeld, K. 1971. *Verhandlungen der Gesellschaft Deutscher Naturforscher und Ärtze.* Berlin: Springer-Verlag.

Kleiber, M. 1932. Body size and metabolism. *Hilgardia* 6:315–353.

Kluger, M. J. 1979. *Fever: Its Biology, Evolution, Function.* Princeton, N.J.: Princeton University Press.

Knudsen, E. I. 1981. The hearing of the barn owl. *Sci. Am.* 245: 113–125.

Knudsen, E. I., and M. Konishi. 1978. A neural map of auditory space in the owl. *Science* 200:795–797.

Kobayashi, H., B. Pelster, and P. Scheid. 1993. Gas exchange in fish swimbladders. In P. Scheid, ed., *Respiration in Health and Desease.* Funktionsanalyse biologischer Systeme 23:113–120.

Koefoed-Johnsen, V., and H. H. Ussing. 1958. The nature of frog skin. *Acta Physiol. Scand.* 42:298–308.

König, J. F. R., and R. A. Klippel. 1967. *The Rat Brain.* Huntington, N.Y.: Robert E. Krieger Publishing.

Konishi, M. 1993. Listening with two ears. *Sci. Am.* 268(4):66.

Kooyman, G. L. 1989. Diverse divers: Physiology and behaviour. In W. W. Burggren, D. S. Farner, et al., eds., *Zoophysiology.* Vol. 23. New York: Springer-Verlag.

Korner, P. I. 1971. Integrative neural cardiovascular control. *Physiol. Revs.* 51(2):312–367.

Kotyk, A., and K. Janáĉek. 1970. *Cell Membrane Transport.* New York: Plenum.

Krebs, H. A. 1975. The August Krogh principle: "For many problems there is an animal on which it can be most conveniently studied." *J. Exp. Zool.* 194:309–344.

Kuby, J. 1997. *Immunology.* 3d ed. New York: W. H. Freeman.

Kuby, J. 2000. *Immunology.* 4th ed. New York: W. H. Freeman.

Kuffler, S. W. 1942. Further study on transmission in an isolated nerve-muscle fibre preparation. *J. Neurophysiol.* 6:99–110.

Lamb, G. D. 2000. Excitation-contraction coupling in skeletal muscle: Comparisons with cardiac muscle. *Clin. Exp. Pharmacol. Physiol.* 27:216–224.

Land, M., and R. Fernald. 1992. The evolution of eyes. *Annu. Rev. Neurosci.* 15:1–29.

Langille, B. J. 1975. A comparative study of central cardiovascular dynamics in vertebrates. Ph.D. dissertation. University of British Columbia, Vancouver, Canada.

Leather, S. R., K. F. A. Walters, and J. S. Bale. 1993. *The Ecology of Insect Overwintering.* Cambridge: Cambridge University Press.

Lehninger, A. L. 1975. *Biochemistry.* 2d ed. New York: Worth.

Lehrer, S. S., and E. P. Morris. 1982. Dual effects of tropomyosin and troponin-tropomyosin on actomyosin subfragment 1 ATPase. *J. Biol. Chem.* 257:8073–8080.

Lighton, J. R. B. 1994. Discontinuous ventilation in terrestrial insects. *Physiol. Zool.* 67:142–162.

Lindemann, W. 1955. Über die Jugendentwicklung beim Luchs (*Lynx l. lynx* Kerr.) und bei der Waldkatze (*Felis sylvestris* Schreb). *Behavior* 8:1–45.

Lissman, H. W. 1963. Electric location of fishes. *Sci. Am.* 208(3): 50–59. (Offprint 152.)

Llinás, R., and C. Nicholson. 1975. Calcium role in depolarization-secretion coupling: An aequorin study in squid giant synapse. *Proc. Nat. Acad. Sci. USA* 72:187–190.

Lodish, H., D. Baltimore, et al. 1995. *Molecular Cell Biology.* 3d ed. New York: Scientific American Books.

Lodish, H., A. Berk, S. L. Zipursky, P. Matsudaira, D. Baltimore, and J. Darnell 2000. *Molecular Cell Biology.* 4th ed. New York: W. H. Freeman.

Loeb, J. 1918. *Forced Movements, Tropisms, and Animal Conduct.* Philadelphia: J. B. Lippincott.

Loewenstein, W. R. 1960. Biological transducers. *Sci. Am.* 203: 98–108.

Loewenstein, W. R. 1971. *Handbook of Sensory Physiology: Principles of Receptor Physiology.* New York: Springer-Verlag.

Loewi, O. 1921. Uber humoral Ubertragbarkeit der Herznervenwirkung. *Pflugers Arch. Ges. Physiol.* 189:239–242.

Lohmann, K. J., and K. Johnsen. 2000. The neurobiology of magnetoreception in vertebrate animals. *Trends Neurosci.* 23:153–159.

Lorenz, K. Z. 1954. *Man Meets Dog.* Translated by M. K. Wilson. London: Methuen.

Lorenz, K., and N. Tinbergen. 1938. Taxis und Instinkhandlung in der Eirolbewegung der Graugans. *Z. Tierpsychol.* 2:1–29.

Loumaye, E., J. Thorner, and K. J. Catt. 1982. Yeast mating pheromone activates mammalian gonadotrophs: Evolutionary conservation of a reproductive hormone? *Science* 218:1323–1325.

Lutz, G. J., and L. C. Rome. 1994. Built for jumping: The design of the frog muscular system. *Science* 263:370–372.

Lutz, G. J., and L. C. Rome. 1996a. Muscle function during jumping in frogs. I. Sarcomere length change, EMG pattern, and jumping performance. *Am. J. Physiol.* 271:C563–C570.

Lutz, G. J., and L. C. Rome. 1996b. Muscle function during jumping in frogs. II. Mechanical properties of muscle: Implications for system design. *Am. J. Physiol.* 271:C571–C578.

Madge, D. S. 1975. *The Mammalian Alimentary System.* London: Arnold.

Malenka, R. C., and R. A. Nicoll. 1999. Long-term potentiation: A decade of progress? *Science* 285:1870–1874.

Marks, W. B. 1965. Visual pigments of single goldfish cones. *J. Physiol.* 178:14–32.

Marshall, P. T., and G. M. Hughes. 1980. *Physiology of Mammals and Other Vertebrates.* 2d ed. Cambridge: Cambridge University Press.

Martini, F., and M. J. Timmons. 1995. *Human Anatomy.* Englewood Cliffs, N.J.: Prentice-Hall.

Matsumoto, A., and S. Ischii, eds. 1992. *Atlas of Endocrine Organs.* Heidelberg: Springer-Verlag.

Matsunami, H., and L. B. Buck. 1997. A multigene family encoding a diverse array of putative pheromone receptors in mammals. *Cell* 90:775–784.

Mazokhin-Porshnyakov, G. A. 1969. *Insect Vision.* New York: Plenum.

McDonald, D. A. 1960. *Blood Flow in Arteries.* Baltimore: Williams and Wilkins.

McDonald, D. M., and R. A. Mitchell. 1975. The innervation of the glomus cells, ganglion cells, and blood vessels in the rat carotid body: A quantitative ultrastructural analysis. *J. Neurocytol.* 4:177–230.

McGilvery, R. W. 1970. *Biochemistry: A Functional Approach.* Philadelphia: Saunders.

McMahan, U. J., N. C. Spitzer, and K. Peper. 1972. Visual identification of nerve terminals in living isolated skeletal muscle. *Proc. R. Soc. Lond.* B 181:421–430.

McMahon, B. R., W. W. Burggren, A. W. Pinder, and M. G. Wheatly. 1991. Air exposure and physiological compensation in a tropical intertidal chiton, *Chiton stokesii* (Mollusca: Polyplacophora). *Physiol. Zool.* 64(3):728–747.

McMahon, T. A. 1983. *Muscles, Reflexes and Locomotion.* Princeton, N.J.: Princeton University Press.

McMahon, T. A., and J. T. Bonner. 1983. *On Size and Life.* New York: Scientific American Books.

McNaught, A. B., and R. Callander. 1975. *Illustrated Physiology.* New York: Churchill Livingstone.

Meyrand, P., J. Simmers, and M. Moulins. 1994. Dynamic construction of a neural network from multiple pattern generators in the lobster stomatogastric nervous system. *J. Neurosci.* 14:630–644.

Michael, C. R. 1969. Retinal processing of visual images. *Sci. Am.* 205:104–114.

Mikiten, T. M. 1967. Electrically stimulated release of vasopressin from rat neurohypophyses *in vitro.* Ph.D. dissertation, Yeshiva University, New York.

Miller, C. 1983. Integral membrane channels: studies in model membranes. *Physiol. Rev.* 63(4):1209–1242.

Miller, W. H., F. Ratliff, and H. K. Hartline. 1961. How cells receive stimuli. *Sci. Am.* 205:222–238.

Milner, A. 1981. Flamingos, stilts and whales. *Nature* 289:347.

Milsom, W. K. 1992. Control of breathing in hibernating animals. In S. C. Wood, R. E. Weber, A. R. Hargens, and R. W. Millard, eds., *Physiological Adaptations in Vertebrates,* 119–148. New York: Marcel Dekker.

Moffett, D., S. Moffett, and C. L. Schauf. 1993. *Human Physiology: Foundations and Frontiers.* St. Louis: Mosby.

Montagna, W. 1959. *Comparative Anatomy.* New York: Wiley.

Moog, F. 1981. The lining of the small intestine. *Sci. Am.* 245:154–176.

Morad, M., and R. Orkand. 1971. Excitation-contraction coupling in frog ventricle: Evidence from voltage clamp studies. *J. Physiol.* 219:167–189.

Morris, J. F., and D. V. Pow. 1988. Capturing and quantifying the exocytotic event. *J. Exp. Biol.* 139:81–103.

Mountcastle, V. B., and R. J. Baldessarini. 1968. Synaptic transmission. In V. B. Mountcastle, ed., *Medical Physiology.* 13th ed. St. Louis: Mosby.

Muller, K. J. 1979. Synapses between neurones in the central nervous system of the leech. *Biol. Rev.* 54:99–134.

Murrary, J. M., and A. Weber. 1974. The cooperative action of muscle proteins. *Sci. Am.* 230(2):58–71.

Murray, R. G. 1973. The ultrastructure of taste buds. In I. Friedmann, ed., *The Ultrastructure of Sensory Organs.* New York: Elsevier.

Murray, R., and A. Murray. 1970. *Taste and Smell in Vertebrates.* London: Churchill.

Nachtigall, W. 1977. On the significance of Reynolds number and the fluid mechanical phenomena connected to it in swimming physiology and flight biophysics. In W. Nachtigall, ed., *Physiology of Movement—Biomechanics.* Stuttgart: Fischer Verlag.

Nagy, K. A. 1989. Field bioenergetics: Accuracy of models and methods. *Physiol. Zool.* 62:237–252.

Nakajima, S., and K. Onodera. 1969. Membrane properties of the stretch receptor neurones of crayfish with particular reference to mechanisms of sensory adaptations. *Am. J. Physiol.* 200:161–185.

Nathans, J., and D. S. Hogness. 1984. Isolation and nucleotide sequence of the gene encoding human rhodopsin. *Proc. Nat. Acad. Sci. USA* 81:4851–4855.

Nathans, J., D. Thomas, and D. S. Hogness. 1986. Molecular genetics of human color vision: The genes encoding blue, green, and red pigments. *Science* 232:193–202.

Neal, H. V., and H. W. Rand. 1936. *Comparative Anatomy.* Philadelphia: Blakiston.

Neher, E., and B. Sakmann. 1976. Single channel currents recorded from membrane of denervated frog muscle fibres. *Nature* 260:799–802.

Nelson, D. L., and M. M. Cox. 2000. *Lehninger Principles of Biochemistry.* 3d ed. New York: W. H. Freeman.

Nickel, E., and L. Potter. 1970. Synaptic vesicles in freeze-etched electric tissue of *Torpedo. Brain Res.* 23:95–100.

Noback, C. R., and R. J. Demarest. 1972. *The Nervous System: Introduction and Reviews.* New York: McGraw-Hill.

Nobili, R., F. Mammano, and J. Ashmore. 1998. How well do we understand the cochlea? *Trends Neurosci.* 21:159–167.

O'Connor, M. P. 1999. Physiological and ecological implications of a simple model of heating and cooling in reptiles. *J . Thermal. Biol.* 24:113–136.

O'Dor, R. K., and D. M. Webber. 1991. Invertebrate athletes: Trade-offs between transport efficiency and power density in cephalopod evolution. *J. Exp. Biol.* 160:93–112.

O'Mally, B. W., and W. T. Schrader. 1976. The receptors of steroid hormones. *Sci. Am.* 234(2):32–43.

Palmer, J. 1973. Tidal rhythms: The clock control of the rhythmic physiology of marine organisms. *Biol. Rev. Camb. Philos. Soc.* 48:377–418.

Parker, H. W. 1963. *Snakes.* London: Hale.

Patlack, J., and R. Horn. 1982. Effect of *N*-bromoacetamide on single sodium channel currents in excised membrane patches. *J. Gen. Physiol.* 79:333–351.

Peachey, L. D. 1965. Transverse tubules in excitation-contraction coupling. *Fed. Proc.* 24:1124–1134.

Pearse, B. 1980. Coated vesicles. *Trends Biochem. Sci.* 5:131–134.

Pearson, K. G., and J.-M. Ramirez. 1997. Sensory modulation of pattern-generating circuits. In P. S. G. Stein, S. Grillner, A. I. Selverston, and D. G. Stuart, eds., *Neurons, Networks, and Motor Behavior.* Cambridge, Mass.: MIT Press.

Penfield, W., and T. Rasmussen. 1950. *The Cerebral Cortex of Man.* New York: Macmillan.

Penry, D. L., and P. A. Jumars. 1986. Chemical reactor analysis and optimal digestion. *BioScience* 36:310–315.

Phillips, G. N., Jr., J. P. Filliers, and C. Cohen. 1986. Tropomyosin crystal structure and regulation. *J. Mol. Biol.* 192:111–131.

Phillips, J. E. 1970. Apparent transport of water in insect excretory systems. *Am. Zool.* 10:416–436.

Phillips, J. G. 1975. *Environmental Physiology.* New York: Wiley.

Pitts, R. F. 1959. *The Physiological Basis of Diuretic Therapy.* Springfield, Ill.: Thomas.

Pitts, R. F. 1968. *Physiology of the Kidney and Body Fluids.* 2d ed. Chicago: Year Book Medical Publishers.

Pitts, R. F. 1974. *Physiology of the Kidney and Body Fluids.* 3d ed. Chicago: Year Book Medical Publishers.

Porter, W. P., and D. M. Gates. 1969. Thermodynamic equilibria of animals with environment. *Ecol. Monogr.* 39:227–244.

Pough, F. H., J. B. Heiser, and W. N. McFarland. 1996. *Vertebrate Life.* 4th ed. Upper Saddle River, N.J.: Prentice Hall.

Prosser, C. L. 1973. *Comparative Animal Physiology.* Vol. 1. Philadelphia: Saunders.

Rahn, H. 1967. Gas transport from the external environment to the cell. In A. V. S. de Reuck and R. Porter, eds., *Development of the Lung.* London: Churchill.

Randall, D. J. 1968. Functional morphology of the heart in fishes. *Am. Zool.* 8:179–189.

Randall, D. J. 1970. Gas exchange in fish. In W. S. Hoar and D. J. Randall, eds., *Fish Physiology.* Vol. 4. New York: Academic Press.

Randall, D. J. 1994. Cardiorespiratory modeling in fishes and the consequences of the evolution of airbreathing. *CardioScience* 5:167–171.

Randall, D. J., and P. A. Wright. 1989. The interaction between carbon dioxide and ammonia excretion and water pH in fish. *Can. J. Zool.* 67:2936–2942.

Reggiani, C., R. Bottinelli, and G. J. M. Stienen. 2000. Sarcomeric myosin isoforms: Fine tuning of a molecular motor. *News Physiol. Sci.* 15:26–33.

Rhoades, R., and R. Pflanzer. 1996. *Human Physiology.* 3d ed. Fort Worth: Saunders College Publishing.

Riddiford, L. M., and J. W. Truman. 1978. Biochemistry of insect hormone and insect growth regulators. In M. Rockstein, ed., *Biochemistry of Insects.* New York: Academic Press.

Romano, L., and H. Passow. 1984. Characterization of anion transport system in trout red blood cell. *Am. J. Physiol.* 62A:257–271.

Rome, L. C., R. P. Funke, R. M. Alexander, et al. 1988. Why animals have different muscle fibre types. *Nature* 355:824–827.

Rome, L. C., R. P. Funke, and R. M. Alexander. 1990. The influence of temperature on muscle velocity and sustained performance in swimming carp. *J. Exp. Biol.* 154:163–178.

Rome, L. C., and M. J. Kushmerick. 1983. The energetic cost of generating isometric force as a function of temperature in isolated frog muscle. *Am. J. Physiol.* 244:C100–C109.

Rome, L. C., and A. A. Sosnicki. 1991. Myofilament overlap in swimming carp. II. Sarcomere length changes during swimming. *Am. J. Physiol.* 260:C289–C296.

Rome, L. C., D. Swank, and D. Corda. 1993. How fish power swimming. *Science* 261:340–343.

Rome, L. C., D. A. Syme, S. Hollingworth, et al. 1996. The whistle and the rattle: The design of sound producing muscles. *Proc. Nat. Acad. Sci. USA.* 93:8095–8100.

Romer, A. S. 1955. *The Vertebrate Body.* Philadelphia: Saunders.

Romer, A. S. 1962. *The Vertebrate Body.* 3d ed. Philadelphia: Saunders.

Rosenthal, J. 1969. Post-tetanic potentiation at the neuromuscular junction of the frog. *J. Physiol.* 203:121–133.

Rowell, L. B. 1974. Circulation to skeletal muscle. In T. C. Ruch and H. D. Patton, eds., *Physiology and Biophysics.* 20th ed., Vol. 2. Philadelphia: Saunders.

Rupert, E. W., and R. D. Barnes. 1994. *Invertebrate Zoology.* 6th ed. Philadelphia: Saunders.

Rushmer, R. F. 1965a. The arterial system: Arteries and arterioles. In T. C. Ruch and H. D. Patton, eds., *Physiology and Biophysics.* 19th ed. Philadelphia: Saunders.

Rushmer, R. F. 1965b. Control of cardiac output. In T. C. Ruch and H. D. Patton, eds., *Physiology and Biophysics.* 19th ed. Philadelphia: Saunders.

Russell, I. J. 1980. The responses of vertebrate hair cells to mechanical stimulation. In A. Roberts and B. M. Bush, eds., *Neurones without Impulses.* Cambridge: Cambridge University Press.

Ryan, A., and P. Dallos. 1996. The physiology of the cochlea. In *Hearing Disorders,* 15–31. Boston: Allyn and Bacon.

Sacca, R., and W. W. Burggren. 1982. Oxygen partitioning between the skin, gills and lungs of the air-breathing reedfish, *Calamoicthys calabaricus. J. Exp. Biol.* 97:179–186.

Sakmann, B. 1992. The patch clamp technique. *Sci. Am.* 266(3): 44–51.

Samsó, M., and T. Wagenknecht. 1998. Contributions of electron microscopy and single-particle techniques to the determination of the ryanodine receptor three-dimensional structure. *J. Struct. Biol.* 121:172–180.

Saudou, F., and R. Hen. 1994. 5-Hydroxytryptamine receptor subtypes in vertebrates and invertebrates. *Neurochem. Int.* 25(6): 503–532.

Scheid, P., H. Slama, and J. Piiper. 1972. Mechanisms of unidirectional flow in parabronchi of avian lungs: Measurements in duck lung preparations. *Respir. Physiol.* 14:83–95.

Schmidt, R. F. 1971. Möglichkeiten und Grenzen der Hautsinne. *Klin. Wochenschr.* 49:530–540.

Schmidt-Nielsen, B. M., and W. C. Mackay. 1972. Comparative physiology of electrolyte and water regulation, with emphasis on sodium, potassium, chloride, urea, and osmotic pressure. In M. H. Maxwell and C. R. Kleeman, eds., *Clinical Disorders of Fluid and Electrolyte Metabolism.* New York: McGraw-Hill.

Schmidt-Nielsen, K. 1959. Salt glands. *Sci. Am.* 200:109–116.

Schmidt-Nielsen, K. 1960. The salt-secreting gland of marine birds. *Circulation* 21:955–967.

Schmidt-Nielsen, K. 1964. *Desert Animals: Physiological Problems of Heat and Water.* London: Oxford University Press.

Schmidt-Nielsen, K. 1972. *How Animals Work.* Cambridge: Cambridge University Press.

Schmidt-Nielsen, K. B. 1975. *Animal Physiology, Adaptation and Environment.* New York: Cambridge University Press.

Schmidt-Nielsen, K., W. L. Bretz, and C. R. Taylor. 1970. Panting in dogs: Unidirectional air flow over evaporative surfaces. *Science* 169:1102–1104.

Scholander, P. F., W. Flagg, V. Walters, and L. Irving. 1953. Climatic adaptation in arctic and tropical poikilotherms. *Physiol. Zool.* 26:67–92.

Schultz, S. G., and P. F. Curran. 1969. The role of sodium in nonelectrolyte transport across animal cell membranes. *Physiologist* 12:437–452.

Secor, S. M., and J. M. Diamond. 2000. Evolution of regulatory responses to feeding in snakes. *Physiol. Biochem. Zool.* 73(2): 123–141.

Shadwick, R. E. 1992. Circulatory structure and mechanics. In A. A. Biewener, ed., *Biomechanics, Structures and Systems: A Practical Approach,* 233–261. Oxford: IRL Press.

Shaw, E. A. T. 1974. Transformation of sound pressure level from the free field to the eardrum in the horizontal plane. *J. Acoust. Soc. Am.* 56:1848–1871.

Shelton, G. 1970. The effect of lung ventilation on blood flow to the lungs and body of the amphibian *Xenopus laevis. Respir. Physiol.* 9:183–196.

Shelton, G., and W. W. Burggren. 1976. Cardiovascular dynamics of the Chelonia during apnoea and lung ventilation. *J. Exp. Biol.* 64(2):323–343.

Shepherd, G. M. 1994. *Neurobiology.* 3d ed. New York: Oxford University Press.

Sherrington, C. S. 1906. *The Integrative Activity of the Nervous System.* New Haven, CT: Yale University Press.

Sherwood, L. 1993. *Human Physiology: From Cells to Systems.* 2d ed. New York: West Publishing.

Sherwood, L. L. 2001. *Human Physiology: From Cells to Systems.* 4th ed. Pacific Grove, Calif.: Brooks/Cole.

Sibley, A. P. 1981. Strategies of digestion and defecation. In C. R. Townsend and P. Calow, eds., *Physiological Ecology: An Evolutionary Approach to Resource Use.* Sunderland, Mass.: Sinauer Associates.

Siegelbaum, S. A., J. S. Camardo, and E. R. Kandel. 1982. Serotonin and cyclic AMP close single K^+ channels in *Aplysia* sensory neurones. *Nature* 299:413–417.

Siggaard-Andersen, O. 1963. *The Acid-Base Status of the Blood.* Copenhagen: Munksgaard.

Simmons, J. A., B. M. Fenton, and M. J. O'Farrell. 1979. Echolocation and pursuit of prey by bats. *Science* 203:16–21.

Singer, S. J., and G. L. Nicolson. 1972. The fluid mosaic model of the structure of cell membranes. *Science* 175:720–731.

Smith, D. S. 1965. The flight muscle of insects. *Sci. Am.* 212(6): 76–88.

Smith, E. L., et al. 1983. *Principles of Biochemistry: Mammalian Biochemistry.* 6th ed. New York: McGraw-Hill.

Solomon, A. K. 1962. Pumps in the living cell. *Sci. Am.* 207(2): 100–108.

Somero, G. N. 1995. Proteins and temperature. *Annu. Rev. Physiol.* 57:43–68.

Sperry, R. W. 1959. The growth of nerve circuits. *Sci. Am.* 201: 100–108.

Spicer, J. I., and K. J. Gaston. 1999. *Physiological Diversity and Its Ecological Implications.* Oxford: Blackwell.

Spratt, N. T., Jr. 1971. *Developmental Biology.* Belmont, Calif.: Wadsworth.

Squire, J. M., and E. P. Morris. 1998. A new look at thin filament regulation in vertebrate skeletal muscle. *FASEB J.* 12:761–771.

Staehelin, L. A. 1974. Structure and function of intercellular junctions. *Int. Rev. Cytol.* 39:191–283.

Starling, E. H. 1908. The chemical control of the body. *Harvey Lect.* 3:115–131.

Steen, J. B. 1963. The physiology of the swimbladder of the eel *Anguilla vulgaris*. I. The solubility of gases and the buffer capacity of the blood. *Acta Physiol. Scand.* 58:124–137.

Steinbrecht, R. A. 1969. Comparative morphology of olfactory receptors. In C. Pfaffman, ed., *Olfaction and Taste*. Vol. 3. New York: Rockefeller University Press.

Stempell, W. 1926. *Zoologie im Grundriss*. Berlin: G. Borntraeger.

Stent, G. S. 1972. Cellular communication. *Sci. Am.* 227:42–51.

Stevens, C. E. 1988. *Comparative Physiology of the Vertebrate Digestive System.* Cambridge: Cambridge University Press.

Stevens, C. E., and I. D. Hume. 1995. *Comparative Physiology of the Vertebrate Digestive System.* 2d ed. Cambridge: Cambridge University Press.

Storey, K. B., and J. M. Storey. 1992. Natural freeze tolerance in ectothermic vertebrates. *Annu. Rev. Physiol.* 54:619–637.

Stryer, L. 1988. *Biochemistry*. 3d ed. New York: W. H. Freeman.

Swan, H. 1974. *Thermoregulation and Bioenergetics.* New York: Elsevier.

Taylor, C. R., K. Schmidt-Nielsen, and J. L. Raab. 1970. Scaling of energy costs of running to body size in mammals. *Am. J. Physiol.* 219:1104–1107.

Taylor, C. R., and E. R. Weibel. 1981. Design of the mammalian respiratory system. *Respir. Physiol.* 44:1–164.

Tenney, S. M., and J. E. Temmers. 1963. Comparative quantitative morphology of the mammalian lung: Diffusing area. *Nature* 197:54–57.

Thomas, D. H., and J. G. Phillips. 1975. Studies in avian and adrenal steroid function. Parts 4–5. *Gen. Comp. Endocrinol.* 26:427–450.

Threadgold, L. J. 1967. *Ultra-structure of the Animal Cell.* New York: Academic Press.

Thurm, U. 1965. An insect mechanoreceptor. *Cold Spring Harb. Symp. Quant. Biol.* 30:75–82.

Tinbergen, N. 1951. *The Study of Instinct.* Oxford: Clarendon.

Tirindelli, R., C. Mucignat-Caretta, and N. J. P. Ryba. 1998. Molecular aspects of pheromonal communication via the vomeronasal organ of mammals. *Trends Neurosci.* 21:482–486.

Toews, D. P., G. Shelton, and D. J. Randall. 1971. Gas tensions in the lungs and major blood vessels of the urodele amphibian, *Amphiuma tridactylum. J. Exp. Biol.* 55:47–61.

Tomanek, L., and G. N. Somero. 2000. Time course and magnitude of synthesis of heat-shock proteins in congeneric marine snails (genus *Tegula*) from different tidal heights. *Physiol. Biochem. Zool.* 73(2):249–256.

Tomita, T., A. Kaneko, M. Murakami, and E. L. Pautler. 1967. Spectral response curves of single cones in the carp. *Vision Res.* 7:519–531.

Tootell, R. B., M. S. Silverman, E. Switkes, and R. L. DeValois. 1982. Deoxyglucose analysis of retinotopic organization in primate striate cortex. *Science* 218:902–904.

Torre, V., J. F. Ashmore, T. D. Lamb, and A. Menine. 1995. Transduction and adaptation in sensory receptor cells. *J. Neurosci.* 15:7757–7768.

Tramontin, A. D., and E. A. Brenowitz. 2000. Seasonal plasticity in the adult brain. *Trends Neurosci.* 23:251–258.

Tucker, V. A. 1975. The energy cost of moving about. *Am. Sci.* 63:413–419.

Ulrich, K. J., K. Kramer, and J. W. Boylan. 1961. Present knowledge of the countercurrent system in the mammalian kidney. *Prog. Cardiovasc. Dis.* 3:395–431.

Unwin, N. 1993. Nicotinic acetylcholine receptor at 9 Å resolution. *J. Mol. Biol.* 229:1101–1124.

Vale, R. D., and R. A. Milligan. 2000. The way things move: Looking under the hood of molecular motor proteins. *Science* 288:88–95.

Vander, A. J., J. H. Sherman, and D. S. Luciano. 1975. *Human Physiology: The Mechanisms of Body Function.* 2d ed. New York: McGraw-Hill.

Van Essen, D. C., C. H. Anderson, and D. J. Felleman. 1992. Information processing in the primate visual system: An integration systems perspective. *Science* 255:419–423.

Van Vliet, B. N., and N. H. West. 1994. Functional characteristics of arterial chemoreceptors in an amphibian *Bufo marinus. Respir. Physiol.* 88:113–127.

Verdugo, P. 1990. Goblet cell secretion and mucogenesis. *Annu. Rev. Physiol.* 52:157–176.

Verdugo, P., M. Aitken, L. Langley, and M. J. Villalon. 1987. Molecular mechanisms of product storage and release in mucin secretion. II. The role of extracellular Ca^{++}. *Biorheology* 24: 625–633.

Vogel, S. 1978. Organisms that capture currents. *Sci. Am.* 239: 128–139.

Vollrath, F. 1992. Spider webs and silks. *Sci. Am.* 266(3):70–76.

von Békésy, G. 1960. *Experiments in Hearing.* New York: McGraw-Hill.

von Euler, U. S., and J. H. Gaddum. 1931. An unidentified depressor substance in certain tissue extracts. *J. Physiol.* 72:74–87.

Walker, R. G., A. T. Willingham, and C. S. Zuker. 2000. A *Drosophila* mechanosensory transduction channel. *Science* 287:2229–2234.

Wang, T., W. W. Burggren, and E. Nobrega. 1995. Metabolic, ventilatory and acid-base responses associated with specific dynamic action in the toad, *Bufo marinus. Physiol. Zool.* 68(2):192–205.

Wangensteen, O. D. 1972. Gas exchange by a bird's embryo. *Respir. Physiol.* 14:64–74.

Waterman, T. H., and H. R. Fernández. 1970. E-vector and wavelength discrimination by retinular cells of the crayfish *Procamberus. Z. Vergl. Physiol.* 68:157–174.

Weibel, E. R. 1973. Morphological basis of alveolar-capillary gas exchange. *Physiol. Rev.* 53:419–495.

Weiderhielm, C. A., and B. U. Weston. 1973. Microvascular lymphatic and tissue pressures in the unanesthetized mammal. *Am. J. Physiol.* 225:992–996.

Weiderhielm, C. A., J. W. Woodbury, S. Kirk, and R. F. Rushmer. 1964. Pulsatile pressures in the microcirculation of frog's mesentery. *Am. J. Physiol.* 207:173–176.

Wenning, A. 1999. Sensing effectors make sense. *Trends Neurosci.* 22:550–555.

Werblin, F. S., and J. E. Dowling. 1969. Organization of the retina of the mudpuppy, *Necturus maculosus:* II. Intracellular recording. *J. Neurophys.* 32:339–355.

West, E. S. 1964. *Textbook of Biophysical Chemistry.* New York: Macmillan.

West, J. B. 1970. *Ventilation/Blood Flow and Gas Exchange.* 2d ed. Oxford: Blackwell Scientific Publications.

West, N. H., and D. R. Jones. 1975. Breathing movements in the frog *Rana pipiens*. I. The mechanical events associated with lung and buccal ventilation. *Can. J. Zool.* 52:332–334.

White, F. N. 1972. Circulation: Environmental correlation. In M. S. Gordon, ed., *Animal Physiology: Principles and Adaptation.* 2d ed. New York: Macmillan.

White, J. G., W. B. Amos, and M. Fordham. 1987. An evaluation of confocal versus conventional imaging of biological structures by fluorescence light microscopy. *J. Cell. Biol.* 105:41–48.

Wiedersheim, R. E. 1907. *Comparative Anatomy of Vertebrates.* London: Macmillan.

Wigglesworth, V. B. 1965. *The Principles of Insect Physiology.* 6th ed. London: Methuen.

Wilkie, D. R. 1977. Metabolism and body size. In T. J. Pedley, ed., *Scale Effects in Animal Locomotion.* New York: Academic Press.

Williams, P. L., ed. 1995. *Gray's Anatomy.* 38th ed. New York: Churchill Livingstone.

Williams, T. M., R. W. Davis, L. A. Fuiman, J. Francis, B. J. Le Boeuf, M. Horning, and D. A. Croll. 2000. Sink or swim: Strategies for cost-efficient diving by marine mammals. *Science* 288 (5463):133–136.

Wilson, D. M. 1964. The origin of the flight-motor command in grasshoppers. In R. F. Reiss, ed., *Neural Theory and Modeling: Proceedings of the 1962 Ojai Symposium.* Stanford, Calif.: Stanford University Press.

Wilson, D. M. 1971. Neural operations in arthropod ganglia. In F. O. Schmitt, ed., *The Neurosciences: Second Study Program.* New York: Rockefeller University Press.

Wilson, W. J., T. Friedmann, et al. 1972. *Biology: An Appreciation of Life.* Del Mar, Calif.: CRM Books.

Wine, J. J., and F. B. Krasne. 1972. The organization of the escape behavior in the crayfish. *J. Exp. Biol.* 56:1–18.

Wine, J. J., and F. B. Krasne. 1982. The cellular organization of crayfish escape behavior. In D. E. Bliss, H. Atwood, and D. Sandeman, eds., *The Biology of Crustacea.* Vol. IV, *Neural Integration.* New York: Academic Press.

Winlow, W., and E. Kandel. 1976. The morphology of identified neurons in the abdominal ganglion of *Aplysia californica. Brain Res.* 112:221–249.

Withers, P. D. 1992. *Comparative Animal Physiology.* Fort Worth: Saunders College Publishing.

Wood, S. C. 1991. Interactions between hypoxia and hypothermia. *Annu. Rev. Physiol.* 53:71–85.

Wright, E. M. 1993. The intestinal Na^+/glucose cotransporter. *Annu. Rev. Physiol.* 55:575–589.

Wright, P. A. 1995. Nitrogen excretion: Three end products, many physiological roles. *J. Exp. Biol.* 198:273–281.

Yau, K.-W. 1976. Receptive fields, geometry and conduction block of sensory neurones in the central nervous system of the leech. *J. Physiol.* 263:513–538.

Yau, K.-W., and K. Nakatani. 1985. Light-suppressible, cyclic GMP-sensitive conductance in the plasma membrane of a truncated rod outer segment. *Nature* 317:252–255.

Young, L. J., Nilsen, K., G. Waymire, G. R. MacGregor, and T. R. Insel. 1999. Increased affiliative response to vasopressin in mice expressing the V1a receptor from a monogamous vole. *Nature* 400:766–768.

Young, L. J., Z. Wang, and T. R. Insel. 1998. Neuroendocrine bases of monogamy. *Trends Neurosci.* 21:71–75.

Young, M. 1971. Changes in human hemoglobins with development. In P. L. Altman and D. W. Dittmer, eds., *Respiration and Circulation.* Bethesda, Md.: Federation of American Societies for Experimental Biology.

Young, R. W. 1970. Visual cells. *Sci. Am.* 223:80–91.

Zotterman, Y. 1959. Thermal sensations. In H. W. Magoun, ed., *Handbook of Physiology,* Section 1, *Neurophysiology.* Vol. I. Baltimore: Williams and Wilkins.

Zuker, C. S. 1996. The biology of vision in *Drosophila. Proc. Natl. Acad. Sci. USA* 93:571–576.

Appendixes

Appendix 1: Units, Conversions, Constants, and Definitions

Basic SI units

Physical quantity	Name of unit	Symbol for unit
Length	meter	m
Mass	kilogram	kg
Time	second	s
Electric current	ampere	A
Temperature	kelvin	K
Luminous intensity	candela	cd

SI multipliers

Multiplier	Prefix	Symbol	Multiplier	Prefix	Symbol
10^{12}	tera	T	10^{-1}	deci	d
10^{9}	giga	G	10^{-2}	centi	c
10^{6}	mega	M	10^{-3}	milli	m
10^{3}	kilo	k	10^{-6}	micro	μ
10^{2}	hecto	h	10^{-9}	nano	n
10	deka	da	10^{-12}	pico	p

Derived SI units

Physical quantity	Name of unit	Symbol for unit	Definition of unit
Acceleration	meter per second squared	$m \cdot s^2$	
Activity	1 per second	s^{-1}	
Electric capacitance	farad	F	$A \cdot s \cdot V^{-1}$
Electric charge	coulomb	C	$A \cdot s$
Electric field strength	volt per meter	$V \cdot m^{-1}$	
Electrical resistance	ohm	Ω	$V \cdot A^{-1}$
Entropy	joule per kelvin	$J \cdot K^{-1}$	
Force	newton	N	$kg \cdot m \cdot s^2$
Frequency	hertz	Hz	s^{-1}
Illumination	lux	lx	$lm \cdot m^2$
Luminance	candela per square meter	$cd \cdot m^{-2}$	
Power	watt	W	$J \cdot s^{-1}$
Pressure	newton per square meter	$N \cdot m^2$	
Voltage, potential difference	volt	V	$W \cdot A^{-1}$
Work, energy, heat	joule	J	$N \cdot m$

Physical and chemical constants

Avogadro's number	$N_A = 6.022 \times 10^{23}$
Faraday constant	$F = 96{,}487\ C \cdot mol^{-1}$
Gas constant	$R = 8.314\ J \cdot K^{-1} \cdot mol$
	$= 1.98\ cal \cdot K^{-1} \cdot mol$
	$= 0.082\ L \cdot atm \cdot K^{-1} \cdot mol$
Planck's constant	$h = 6.626 \times 10^{-34}\ J \cdot s^{-1}$
	$= 1.58 \times 10^{-34}\ cal \cdot s^{-1}$
Speed of light in a vacuum	$c = 2.997 \times 10^{8}\ m \cdot s^{-1}$
	$= 186{,}000\ mi \cdot s^{-1}$

Dimensions of plane and solid figures

Area of a square $= l^2$
Surface area of a cube $= 6l^2$
Volume of a cube $= l^3$
Circumference of a circle $= 2\pi r$
Area of a circle $= \pi r^2$
Surface area of a sphere $= 4\pi r^2$
Volume of a sphere $= 4/3\pi r^3$
Surface area of a cylinder $= 2\pi rh$
Volume of a cylinder $= \pi r^2 h$

Conversion factors

To convert from	to	multiply by
angstroms	inches	3.937×10^{-9}
	meters	1×10^{-10}
	micrometers (μm)	1×10^{-4}
atmospheres	bars	1.01325
	dynes per square centimeter	1.01325×10^{6}
	grams per square centimeter	1033.23
	torr (= mm Hg; 0°C)	760
	pounds per square inch	14.696
	pascals	1.013×10^{5}
bars	atmospheres	0.9869
	dynes per square centimeter	1×10^{6}
	grams per square centimeter	1019.716
	pounds per square inch	14.5038
	torr (= mm Hg; 0°C)	750.062
	pascals	10^{5}
calories	British thermal units	3.968×10^{-3}
	ergs	4.184×10^{7}
	foot-pounds	3.08596
	kilocalories	10^{-3}
	horsepower-hours	1.55857×10^{-6}
	joules	40184
	watt-hours	1.1622×10^{-3}
	watt-seconds	4.184
ergs	British thermal units	9.48451×10^{-11}
	calories	2.39×10^{-8}
	dynes per centimeter	1
	foot-pounds	7.37562×10^{-8}
	gram-centimeters	1.0197×10^{-3}
	joules	1×10^{-7}
	watt-seconds	1×10^{-7}
grams	daltons	6.024×10^{23}
	grains	15.432358
	ounces (avdp)	3.52739×10^{-2}
	pounds (avdp)	2.2046×10^{-3}

To convert from	to	multiply by
inches	angstroms	2.54×10^{8}
	centimeters	2.54
	feet	8.333×10^{-2}
	meters	2.54×10^{-2}
joules	calories	0.239
	ergs	1×10^{7}
	foot-pounds	0.73756
	watt-hours	2.777×10^{-4}
	watt-seconds	1
liters	cubic centimeters	10^{3}
	gallons (US, liq)	0.2641794
	pints (US, liq)	2.113436
	quarts (US, liq)	1.056718
lumens	candle power	7.9577×10^{-2}
lux	lumens per square meter	1
meters	angstroms	1×10^{10}
	micrometers (μm)	1×10^{6}
	centimeters	100
	feet	3.2808
	inches	39.37
	kilometers	1×10^{-3}
	miles (statute)	6.2137×10^{-4}
	millimeters	1000
	yards	1.0936
newtons	dynes	10^{5}
pascals	bars	10^{-5}
	atmospheres	9.87×10^{-6}
	dynes per square centimeter	10
	grams per square centimeter	1.0197×10^{-2}
	torr (= mm Hg; 0°C)	7.52×10^{-3}
	pounds per square inch	1.450×10^{-4}
watts	British thermal units per second	9.485×10^{-4}
	calories per minute	14.3197
	ergs per second	1×10^{7}
	foot-pounds per minute	44.2537
	horsepower	1.341×10^{-3}
	joules per second	1

Temperature conversions

$°C = 5/9\,(°F - 32)$

$°F = 9/5\,(°C) + 32$

$0°K = -273.15°C = -459.67°F + 32$

$0°C = 273.15\ K = 32°F$

Useful formulas

Electric potential	$E = IR = q/C$ E = electric potential (voltage) I = current R = resistance q = charge C = capacitance
Power	$p = w/t$ w = work t = time
Electric power	$p = RI^2 = EI$ E = electric potential
Work	$W = RI^2t = EIt = Pt$
Pressure	P = force (f)/unit area
Weight	$W = mg$ m = mass g = acceleration of gravity
Force	$f = ma$ m = mass a = acceleration
Dalton's law of partial pressures	$PV = V(p_1 + p_2 + p_3 + \cdots + p_n)$ P = pressure of gas mixture V = volume p = pressure of each gas alone
Electrostatic force of attraction	$F = \frac{q_1 q_2}{\varepsilon r^2}$ r = distance separating q_1 and q_2 ε = dielectric constant
Potential energy	$E = mgh$ h = height of mass above surface of Earth
Kinetic energy	$E = 1/2mv^2$ v = velocity of mass
Energy of a charge	$E = 1/2qV$ q = charge V = electric potential
Perfect gas law	$PV = nRT$ P = pressure V = volume n = number of moles R = gas constant T = absolute temperature
Hooke's law of elasticity	$F = kT$ k = spring constant F = force T = tension
Energy of a photon	$E = h\nu$ h = Planck's constant ν = frequency

Chemical definitions

1 mol = the mass in grams of a substance equal to its molecular or atomic weight: this mass contains Avogadro's number (N_A) of molecules or atoms

Molar volume = the volume occupied by a mole of gas at standard temperature and pressure (25°C, 1 atm) = 22.414 L

1 molal solution = 1 mol per 1000 g of solvent

1 molar solution = 1 mol of solute in 1 L of solution

1 equivalent = 1 mol of 1 unit charge

1 einstein = 1 mol of photons

Appendix 2: Logs and Exponentials

Straight-Line Equations

If a straight line describes a plot of x against y, then b is the value of the intercept of the line on the ordinate and a is the slope of the line. The relationship beteween x and y is

$$y = ax + b$$

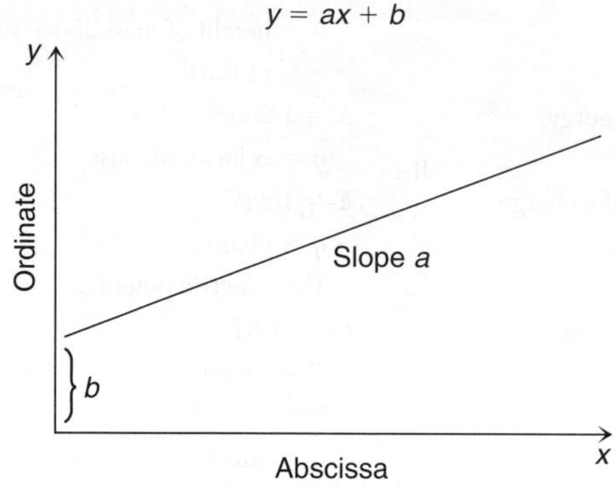

Exponential Equations

In biological systems there is often an exponential relationship between values, described by the equation

$$y = b \cdot a^x$$

The logarithmic form of this equation is

$$\log y = \log b + x \log a$$

Using semi-log graph paper (linear abscissa), log y can be plotted against x, giving a straight line to determine the values of slope a and intercept b.

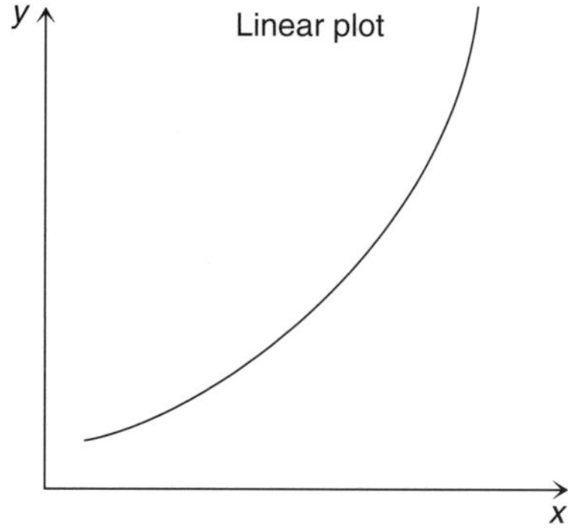

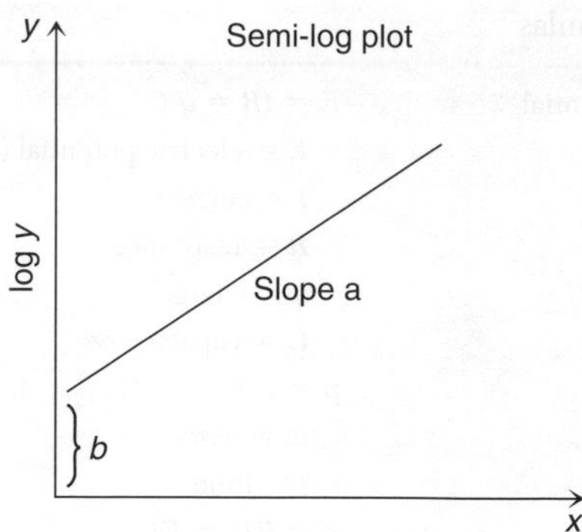

(Examples of the use of straight-line and exponential equations may be found throughout the book, especially in Chapters 3, 16, and 17.)

Use of Logarithm Terms

In the equation $y = 10^x$, x is the logarithm of y. That is, x is the power to which 10 must be raised in order to yield y. For example, the logarithm of 10 is 1, and the logarithm of 100 is 2. The equation

$$y = \frac{a}{b}$$

can be transformed into a logarithmic equation:

$$\log y = \log \frac{a}{b} = \log a - \log b$$

just as multiplication can be transformed into the addition of logarithms. A convenient identity that is useful in calculating equilibrium potentials using the Goldman equation is

$$\log \frac{a}{b} = -\log \frac{b}{a}$$

This identity is true because $\log \frac{a}{b} = \log a - \log b$ and $\log \frac{b}{a} = \log b - \log a$. Notice that $\log a - \log b = -(\log b - \log a)$, which proves the identity.

Index

Page numbers in *italics* indicate illustrations.